4th International School and Conference on Optoelectronics, Photonics, Engineering and Nanostructures (Saint Petersburg OPEN 2017)

Journal of Physics: Conference Series Volume 917

Saint-Petersburg, Russia
3 - 6 April 2017

Part 1 of 2

Editors:

Alexey Toropov
Leonid Karachinsky
Sergei Alexandrov
Sergey Morozov
Alexander Marmaluk
Nika Akopian
Anatoly Evstrapov
Seppo Honkanen

ISBN: 978-1-5108-5340-9
ISSN: 1742-6588

Printed from e-media with permission by:

Curran Associates, Inc.
57 Morehouse Lane
Red Hook, NY 12571

proceedings.com

Some format issues inherent in the e-media version may also appear in this print version.

Printed by Curran Associates, Inc. (2018)

Institute of Physics
Dirac House, Temple Back
Bristol BS1 6BE UK

Phone: 44 1 17 929 7481
Fax: 44 1 17 920 0979

techtracking@iop.org

Additional copies of this publication are available from:

Curran Associates, Inc.
57 Morehouse Lane
Red Hook, NY 12571 USA
Phone: 845-758-0400
Fax: 845-758-2633
Email: curran@proceedings.com
Web: www.proceedings.com

TABLE OF CONTENTS

PART 1

NANOBIOTECHNOLOGY, BIOPHYSICS AND BIOPHOTONICS

LASERS, SOLAR CELLS AND OTHER OPTOELECTRONIC DEVICES

PART 2

NANOPHOTONICS, SPECTROSCOPY, MICROCAVITIES, OPTICS, PLASMONICS

SPINTRONICS, ELECTRO- AND MAGNETOOPTICS

ELECTRIC, MAGNETIC AND MICROWAVE DEVICES

OTHER ASPECTS OF NANOTECHNOLOGY

4th International School and Conference
Saint Petersburg OPEN 2017

Optoelectronics, Photonics, Engineering and Nanostructures

Saint Petersburg, April 3–6, 2017

Acknowledgements

RFBR
Russian Foundation for Basic Research

RSF
Russian Science Foundation

SPIE. STUDENT CHAPTER
SAINT PETERSBURG ACADEMIC UNIVERSITY RUSSIAN ACADEMY OF SCIENCES

OSA®
The Optical Society

IOP Institute of Physics

IOP Conf. Series: Journal of Physics: Conf. Series **917** (2017) 012001 doi:10.1088/1742-6596/917/1/012001

Editorial Preface

Dear Colleagues,

4th International School and Conference "Saint Petersburg OPEN 2017" on Optoelectronics, Photonics, Engineering and Nanostructures was held on April 3 - 6, 2017 at St. Petersburg Academic University. The School and Conference included a series of invited talks given by leading professors with the aim to introduce young scientists with actual problems and major advances in physics and technology. The keynote speakers were

Alexey Toropov *(Ioffe Physical Technical Institute RAS, Russia)*
Leonid Karachinsky *(Connector Optics LLC, Russia)*
Sergei Alexandrov *(Peter the Great St Petersburg Polytechnic University, Russia)*
Sergey Morozov *(Institute for Physics of Microstructures RAS, Russia)*
Alexander Marmaluk *(JSC M. F. Stelmakh Research Institute "Polyus", Russia)*
Nika Akopian *(Technical University of Denmark, Denmark)*
Anatoly Evstrapov *(Institute for Analytical Instrumentation RAS, Russia)*
Seppo Honkanen *(University of Eastern Finland, Finland)*

During the poster session all undergraduate and graduate students attending the conference presented their works. Sufficiently large number of participants with 286 student attendees from all over the world allowed the Conference to provide a fertile ground for the fruitful discussions between the young scientists as well as to become a perfect platform for the valuable discussions between student authors and highly experienced scientists. The best student papers, which were selected by the Program Committee and by the invited speakers basing on the theses and their poster presentation, were awarded with diplomas of the conference and prizes – see the photos.

This year "**Saint Petersburg OPEN 2017**" is organized by **St Petersburg Academic University** in cooperation with **Peter the Great St. Petersburg Polytechnic University**. School and Conference is supported by **Russian Foundation for Basic Research** (*Project #17-32-10035*), **Russian Science Foundation**, **SPIE** and **OSA**. It is a continuation of the annual schools and seminars for youth on topical problems of physics and technology that is organized by the Academic University since 2009.

We invite all the students and young scientists to attend "Saint Petersburg OPEN" in 2018!

More details at http://spbopen.spbau.com/

IOP Conf. Series: Journal of Physics: Conf. Series **917** (2017) 012001 doi:10.1088/1742-6596/917/1/012001

The head of Program Committee,
rector of St Petersburg Academic University
academic *Zhores Alferov* is making opening remarks

The head of Organizing Committee Professor *Alexey Zhukov*

IOP Conf. Series: Journal of Physics: Conf. Series **917** (2017) 012001 doi:10.1088/1742-6596/917/1/012001

The invited speakers

Alexey Toropov

Leonid Karachinsky

Sergei Alexandrov

Sergey Morozov

Alexander Marmaluk

Nika Akopian

IOP Conf. Series: Journal of Physics: Conf. Series **917** (2017) 012001 doi:10.1088/1742-6596/917/1/012001

Seppo Honkanen

Anatoly Evstrapov

Poster Session

IOP Conf. Series: Journal of Physics: Conf. Series **917** (2017) 012001 doi:10.1088/1742-6596/917/1/012001

Elevator speech session

Award ceremony
Prof. *Alexey Zhukov* and *Andrey Lipovskii* with students

Peer review statement

All papers published in this volume of *Journal of Physics: Conference Series* have been peer reviewed through processes administered by the proceedings Editors. Reviews were conducted by expert referees to the professional and scientific standards expected of a proceedings journal published by IOP Publishing.

Single-photon sources of visible light

A A Toropov[1], M V Rakhlin[1], K G Belyaev[1], S V Sorokin[1], G V Klimko[1], S V Gronin[1], I V Sedova[1], ISMukhin[2,3], T V Shubina[1], S V Ivanov[1]
[1]Ioffe Institute, Saint Petersburg 194021, Russia

[2]St. Petersburg Academic University, St. Petersburg 194021, Russia

[3]ITMO University, St. Petersburg 197101, Russia

toropov@beam.ioffe.ru

Abstract. We report on emission properties of single-photon sources of visible light, based on InAs/AlGaAs, CdTe/ZnTe, and CdSe/ZnSe quantum dot (QD) heterostructures grown by molecular beam epitaxy. For all types of QDs, the studies performed by micro-photoluminescence and photon correlation spectroscopies demonstrated the single-photon nature of emission with typical values of the second-order correlation function $g^{(2)}(0)$ less than 0.2 and allowed evaluation of sub-microsecond kinetics of blinking governed by local charging effects.

1. Introduction

The sources of single-photon emission (SPE) and entangled photon pairs, based on epitaxial semiconductor quantum dots (QDs) emitting in the visible spectral range, are promising for applications in the systems of quantum cryptography, intended for secure communications through free-space atmospheric optical connection lines or plastic fibers. Commercial devices of this type are currently absent due to both technological complexity of reproducible fabrication of the samples with the isolated single QDs and material constraints limiting possibility to distinguish emission lines of single excitons, charged excitons (trions), and biexcitons at elevated device operation temperatures. In this work, we present comparative optical studies of the SPE sources based on different QD heterostructures: $InAs/Al_xGa_{1-x}As$ (x>0.35), CdTe/ZnTe, and CdSe/ZnSe, whose emission spectrum jointly covers extremely wide spectral range between 500 and 1000 nm.

2. Samples and experimental techniques

2.1. InAs/AlGaAs quantum dots

Advantages of the $InAs/Al_xGa_{1-x}As$ QDs consist in possibility to achieve SPE in the red spectral range using well-developed epitaxial techniques for the growth of the required QD heterostructures [1-5]. A potential disadvantage of previously studied single $InAs/Al_xGa_{1-x}As$ QDs is enhancement of the spectral width of the single-QD emission lines with an increase of the Al content in the $Al_xGa_{1-x}As$ barriers [4] that is a signature of the emission blinking and spectral diffusion. Besides, the enlarged content of Al results in a significant (up to 1 meV) enhancement of the anisotropic exchange splitting of a single-QD exciton [5], that hampers generation of polarization entangled photons during a

IOP Conf. Series: Journal of Physics: Conf. Series **917** (2017) 022001 doi:10.1088/1742-6596/917/2/022001

cascade recombination of biexcitons. To the best of our knowledge, single-photon characteristics of the QDs of this type have never been studied.

To elucidate these issues we have investigated systematically emission properties of single self-organized InAs/$Al_xGa_{1-x}As$ (0.35<x<0.5) QDs formed by molecular beam epitaxy (MBE) on (001)GaAs substrates, using the Stranski-Krastanow growth mechanism. Two different designs were considered. In the first one (type A), the layer of InAs QDs was directly embedded in an $Al_xGa_{1-x}As$ matrix. In the second design (type B) a two-monolayer-thick interlayer of GaAs was deposited prior to the QDs growth. The developed MBE growth regimes allowed achievement of the QDs surface density below10^{10} cm^{-2}.

2.2. CdTe/ZnTe and CdSe/ZnSe quantum dots
Single QDs based on wide bandgap II-VI compounds CdTe and CdSe are of particular interest due to the possibility to obtain SPE in green, yellow, and orange spectral ranges [6-8]. Yet unsolved problem (especially for CdSe QDs) has been the reproducible growth of the structures with small enough QD surface density. For both types of heterostructures, we have developed MBE growth regimes resulting in the formation of a single sheet of QDs with the QD surface density in the range of 10^{10}- 10^{11} cm^{-2}. The achieved decrease in the surface density of CdTe/ZnTe QDs relied on further development of so called thermal activation technique, where the QD formation is induced by the reduction of the surface energy of a strained 2D CdTe layer by means of covering it with an amorphous Te layer, followed by its fast desorption [9,10]. The formation of a low-density array of CdSe/ZnSe QDs was realized using a reorganization process occurring during growth interruption and thermal activation [11].

2.3. Micro-photoluminescence and photon correlation spectroscopy
For single-QD micro-photoluminescence (μ-PL) measurements, cylindrical mesa-structures containing only a few emitting QDs were fabricated by electron-beam lithography and Ar^+-ion beam etching. The samples with mesa-structures were attached to a 3D Attocube piezo-driver located within a Janis continuous flow He micro-cryostat. The QD PL was excited by a continuous wave semiconductor laser diode at 405 nm and collectedby using a confocal microscope with a 50x Mitutoyo micro-objective. The emission was dispersed by a 50 cm grating monochromator and registered by a liquid-nitrogen-cooled CCD camera. Statistical characteristics of the single-QD emission were evaluated by means of photon correlation measurements performed under the continuous wave excitation with a Hanbury Brown – Twiss intensity interferometer equipped with two single-photon avalanched Si photodiodes providing time resolution better than 40 ps. To obtain the second-order correlation function $g^{(2)}(\tau)$, the measured histogram of coincidences was normalized by count rates of the detectors, width of the time bins, and the integration time [12].

3. Results and discussion
The type-A heterostructures with InAs QDs embedded immediately in $Al_xGa_{1-x}As$ (x>0.35) demonstrate a wide PL spectrum extending over the spectral range between ~1000 and ~620 nm (see Fig. 1). From the blue side, the spectrum is limited by the wavelength corresponding to the direct-indirect bandgap transition in the $Al_xGa_{1-x}As$ ternary alloy.No distinct PL line related to a wetting layer can typically be observed in the type-A structures. In contrast to that the type-B structures demonstrate a narrower spectrum whose maximum is located between 900 and 1000 nm, while the blue boundary of the QD emission spectrum corresponds to a distinct narrow PL peak near 700 nm, attributed to the emission of the wetting layer. The emission spectrum of CdTe QDs is adjacent to that of InAs QDs and spreads approximately between 650 and 550 nm (see Fig. 2). From the blue side, it overlaps with the PL band of CdSe QDs, extending towards 500 nm.

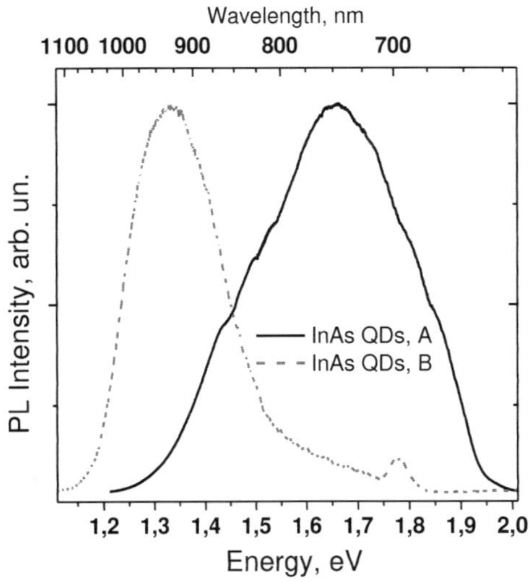

Figure 1. PL spectra measured at 77 K in $InAs/Al_xGa_{1-x}As$ QD samples with a 0.3-mm-size excitation spot. Solid and dashed lines correspond to a type-A (x=0.4) and type-B (x=0.35) samples, correspondingly.

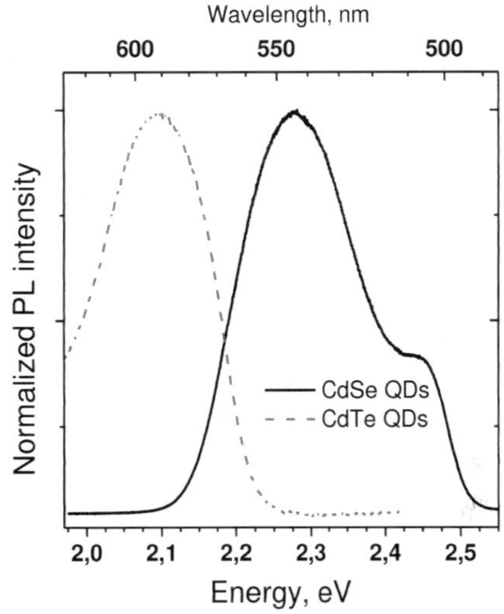

Figure 2. Typical PL spectra measured at 77 K in CdSe/ZnSe (solid line) and CdTe/ZnTe (dashed line) QD samples with a 0.3-mm-size excitation spot.

For both types of $InAs/Al_xGa_{1-x}As$ QD structures, the μ-PL spectra measured at 8 K in small enough mesa-structures demonstrate a set of ultra-narrow lines attributed to the emission of single individual QDs (see Fig. 3 (a)). Many of them possess the spectral width that is below resolution of the used monochromator (~50 μeV). In accordance with Ref. [5], some of these lines demonstrate a fine structure splitting of linearly polarized exciton lines, which can be as large as ~800 μeV. It was found that an increase in the Al content leads to enhancement of the average density of the QDs with the largest splitting at the expense of the QDs with the smallest one. Nevertheless, for both kinds of structures and any Al mole fraction in the range of 35-50% there always exists a pronounced number of QDs demonstrating emission lines with the anisotropic exchange splitting smaller than theset-up spectral resolution.These lines presumably correspond to the emission of either single excitons confined in relatively isotropic QDs or charged excitons (trions) and are promising for obtaining efficient SPE. The respective emission lines of individual CdTe QDs (Fig. 3 (b)) and CdSe QDs (Fig. 3 (c)) are noticeably wider: 0.2 – 0.8 meV for the former structures and 0.1 – 0.2 meV for the latter.

Figure 4 represents the second-order correlation functions obtained in representative mesa-structures containing $InAs/Al_xGa_{1-x}As$ QDs (Fig. 4 (a)), CdTe/ZnTe QDs (Fig. 4 (b)), and CdSe/ZnSe QDs (Fig. 4 (c)). The strongest and narrowest emission lines were selected for these measurements. The functions are fitted with the expression

$$g^{(2)}(\tau) = 1 + c_1 e^{-\tau/\tau_1} + c_2 e^{-\tau/\tau_2}. \qquad (1)$$

For all types of QDs, the data exhibit the antibunching dip at zero time delay with the average $g^{(2)}(0)$ as low as 0.1 – 0.2 that is a clear signature of the single-photon nature of the emission. The time width of the dip, described by the factor τ_1, corresponds to the PL lifetime being in the range 1-1.5 ns. The second exponential term is responsible for the appearance of two symmetric bunching peaks, which reflect the kinetics of blinking induced by local charging effects [13]. One can see in Fig. 4 that the emission of single CdSe/ZnSe QDs is practically free of these parasitic effects, while both InAs/$Al_xGa_{1-x}As$ and CdTe/ZnTe QDs suffer from certain blinking caused most probably by the QDs dynamical charging taking place on the time scale of ~10 ns.

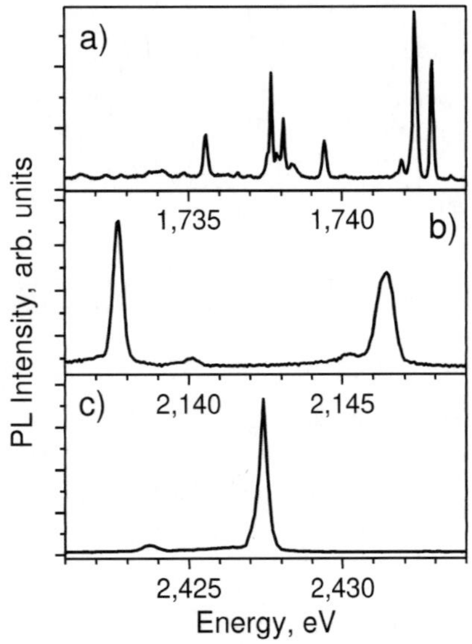

Figure 3. μ-PL spectra measured at 8 K in representative InAs/AlGaAs (a), CdTe/ZnTe (b), and CdSe/ZnSe (c) mesa-structures.

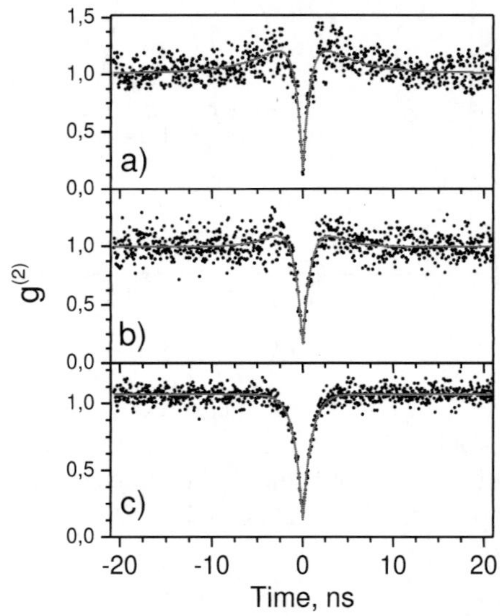

Figure 4. Typical second-order correlation $g^{(2)}$ functionsmeasured at 8 K in representative InAs/AlGaAs (a), CdTe/ZnTe (b), and CdSe/ZnSe (c) mesa-structures. Solid lines represent fits of the experimental data, obtained with Eq. (1).

4. Conclusions

We have demonstrated possibility to fabricate sources of single photons, based on low-density arrays of InAs/$Al_xGa_{1-x}As$ (x>0.35), CdTe/ZnTe, and CdSe/ZnSe self-organized QDs, whose emission spectra progressively cover all visible spectral range above~500 nm. While both CdTe/ZnTe and InAs/$Al_xGa_{1-x}As$ QDs suffer from certain blinking taking place on the time scale of ~10 ns, all the types of QDs show the second-order correlation function at zero delay $g^{(2)}(0)$ as low as 0.1 – 0.2, that is promising for applications in effective systems of quantum cryptography.

Acknowledgments

Authors gratefully acknowledge the financial support by the Russian Science Foundation (project #14-22-00107) of the studies of InAs/$Al_xGa_{1-x}As$ (x>0.35) and CdTe/ZnTe QD heterostructures.

References

[1] Leon R, Fafard S, Leonard D, Merz J L, and Petroff P M 1995 *Appl. Phys. Lett.* **67** 521

[2] Ballet P, Smathers J B, Yang H, Workman C L, and Salamo G J 2001 *J. Appl. Phys.***90** 481
[3] Polimeni A, Patanć A, Henini M, Eaves L, and Main P C 1999 *Phys. Rev. B***59** 5064
[4] Grijseels S C M, van Bree J, Koenraad P M, Toropov A A, Klimko G V, Ivanov S V, Pryor C E, Silov A Yu 2016 *J. of Lum.***176** 95
[5] Finley J J, Mowbray D J, Skolnick M S, Ashmore A D, Baker C, Monte A F G, Hopkinson M 2002 *Phys. Rev. B* **66** 153316
[6] Couteau C, Moehl S, Tinjod F, Gćrard J M, Kheng K, Mariette H, Gaj J A, Romestain R, Poizat J P 2004 *Appl. Phys. Lett.***85** 6251
[7] Wojnar P, Janik E, Baczewski L T, Kret S, Karczewski G, Wojtowicz T, Goryca M, Kazimierczuk T, and Kossacki P 2011 *Appl. Phys. Lett.***99** 113109
[8] Fedorych O,Kruse C, RubanA, HommelD, BacherG,andKummel T 2012*Appl.Phys.Lett.***100**061114
[9] Tinjod F, Gilles B, Moehl S, Kheng K, Mariette H 2003 *Appl. Phys. Lett.***82** 4340
[10] Sorokin S V, Sedova I V, Gronin S V, Klimko G V, Belyaev K G, Rakhlin M V, Mukhin I S, Toropov A A, and Ivanov S V2017*J. Cryst. Growth*(to be published)
[11] Rabe M, Lowisch M, Henneberger F 1998 *J. Cryst. Growth***184/185** 248
[12] Brouri R, Beveratos A, Poizat J P, and Grangier P 2000 *Opt. Lett.***25** 1294
[13] Santori C, Fattal D, Vučković J, Solomon G S, Waks E, and Yamamoto Y 2004 *Phys. Rev. B***69** 205324

IOP Conf. Series: Journal of Physics: Conf. Series **917** (2017) 022002 doi:10.1088/1742-6596/917/2/022002

Micro- and nanofluidic systems in devices for biological, medical and environmental research

A A Evstrapov [1,2,3]
[1] Laboratory of information and measurement biosensor and chemosensor microsystems, Institute for Analytical Instrumentation RAS, St. Petersburg 198095, Russia
[2]Nanobiotech Lab, St. Petersburg Academic University, St. Petersburg 194021, Russia
[3] Department of nanophotonics and metamaterials, ITMO University, St. Petersburg 197101, Russia

Abstract. The use of micro- and nanofluidic systems in modern analytical instruments allow you to implement a number of unique opportunities and achieve ultra-high measurement sensitivity. The possibility of manipulation of the individual biological objects (cells, bacteria, viruses, proteins, nucleic acids) in a liquid medium caused the development of devices on microchip platform for methods: chromatographic and electrophoretic analyzes; polymerase chain reaction; sequencing of nucleic acids; immunoassay; cytometric studies. Development of micro and nano fabrication technologies, materials science, surface chemistry, analytical chemistry, cell engineering have led to the creation of a unique systems such as "lab-on-a-chip", "human-on-a-chip" and other. This article discusses common in microfluidics materials and methods of making functional structures. Examples of integration of nanoscale structures in microfluidic devices for the implementation of new features and improve the technical characteristics of devices and systems are shown.

1. Introduction

Modern technologies allow to form repeatable micro- and nanoscale structures with desired characteristics in various materials. Proximity of the size of the nanostructures formed to the size of biological molecules (nucleic acids, amino acids, proteins) allows to make precision operation with them. The integration of nanostructures into compact microfluidic devices (chips) promotes creation of new devices and systems for biological and medical research, which are implemented the unique properties of nanomaterials, especially nanofluidics and the broadest opportunities of microfluidics.

The global market for microfluidic devices in 2015 was about $ 2.56 billion (according to Yole Development, Emerging Markets for Microfluidic Application). Experts estimate the growth of the market by 2020 to about $ 5.95 billion. The largest growth in the market is predicted for the near future for systems aimed at pharmaceutical research and on-site diagnostic systems (Point Of Care - POC). At the current time, more than 50% of microfluidic devices are used in pharmaceutical research (search for medicines) and in Life Sciences, about 14% in clinical and veterinary diagnostics.

The advantages of systems on microchip platform include are: a) the total control of all (or basic) stages of analysis or synthesis, b) a low reagent consumption, and c) the possibility of operations with a very small sample volume or the amount of analyte, g) the high sensitivity of detection analytes, d) the high speed of analysis or high throughput synthesis of materials. Manipulations with individual biological objects (cells, bacteria, viruses, proteins, nucleic acids) in a liquid medium may be performed on microchip devices. These advantages led to development of the devises on microchip platform for: electrophoretic and chromatographic methods of analysis; methods based on polymerase chain reaction; sequencing of nucleic acids; immunoassay methods; cytometric studies [1,2], and others. Although studies have been lead in microfluidics for more than 25 years, only a few companies have been able to fabricate commercial instruments [3]. For example, based on the methods of "droplet" microfluidics are produced commercial systems for digital PCR: QX200™ Droplet Digital™ PCR System (Bio-Rad) and RainDrop™ Digital PCR System (Raindance Tech., Inc.). An emulsion PCR is used in the Genome Sequencer FLX (Roche).

It is possible to identify the main trends in the development of micro- and nanofluid systems in the analytical instrumentation, aimed at the creation of: a) automated systems for "full" analysis or the of certain key stages («micro Total Analysis System») [4]; b) relatively cheap microchips and specialized devices for rapid diagnosis (including «Point Of Care» system [5]); c) devices that mimic the basic physiological functions of the "living" systems for biological, pharmacological, toxicological studies («Organ-on-a-chip», «tissue-on-a-chip», etc.). [6]; g) systems, embedded in the laboratory devices (eg, mass spectrometers, optical microscopes), to solve research problems («chip-in-a-lab») [7].

One of the fundamental advantages of microfluidic devices is the ability to integrate multiple chemical processing and fluidic manipulation operations into a small sizes. In many cases an automated sample-in/answer-out system is the ultimate goal. Such systems should provide accurate results in real time. The integration of automated sample collection, processing, and real time analysis on microfluidic platforms with mass spectrometry (MS) continued to show progress in analytical research. [8] For example, a microextraction system was designed to automatically extract samples from biological object without any pretreatments. This mechanical extraction system was integrated with both MS and a fluorescent microscope in order to monitor the time resolved extraction progress.

The POC applications continue to be popular. A number of new portable microfluidic devices were reported [9] with many built-in functions for creating simpler ones [10], faster and cheaper devices [11] with better detection limits. Microfluidic device consisting of 500 channels is intended for the rapid analysis (within 20 minutes) of the integrity of the sperm DNA using fluorescent imaging [11]. In experiments a significant increase in the viability of the bull and human sperm was demonstrated in comparison with existing methods.

The cost of developing medicines is growing every year. Preparations considered as drug candidates often fail in clinical trials. Today, the cost of developing medicines in the US has reached the level of $ 2.6 billion. The limited predictability of *in-vitro* results relative to the results *in vivo* and the small number of patients participating in phase I and phase II clinical trials do not allow an adequate estimate of the likely clinical effectiveness [12]. This situation led to the creation of systems “organ-on-a-chip” for the detail study of physical and biochemical processes at the cellular level using biomimetic research platforms [6]. Separate systems of "organs" can be combined into a "human-on-a-chip." Systemic interactions between of various organs can significantly affect the health of the organism as the functions of the organs are highly dependent on their interaction [13]. With regard to testing preparations, individual drugs may well influence the target organ, but his the metabolites can lead to toxic effects on another organ [14]. An important task has been formed now, which is to develop new high-performance toxicity tests for a wide range of substances for prediction of toxic effects in the human body. Biochemical processes in the human body under the influence of toxic substances are complex and not thoroughly studied. The efforts of many researchers [15, 16] focus on the development of "organ-on-a-chip" devices that mimic the cellular environment in the organ or several interrelated organs. Thus, devices have been created to simulate the functioning of organs under the influence of drugs or changing metabolism to determine the toxicity of the drug. Many devices try to reproduce the processes occurring in the liver, since many pharmaceutical problems are caused by interactions occurring in the liver [for example, 14]. Systems that simulate several interconnected "human" "organs-on-a-chip" can help in solving problems of identifying new biomarkers, determining the toxicity of substances, studying the body's response to viral or bacterial infections, which may be important for clinical tests.

IOP Conf. Series: Journal of Physics: Conf. Series **917** (2017) 022002 doi:10.1088/1742-6596/917/2/022002

2. Materials and methods for manufacturing of micro- and nanofluidic systems

Development and manufacturing of the prototype device is an important stage in the creation of micro and nanofluidic systems for analytical instruments. Polydimethylsiloxane (PDMS) is the most popular material for such prototypes. In 1997 group prof. G. M. Whitesides (Harvard) demonstrated the possibilities of: production of reproducible fingerprint patterns from the master-form in PDMS with nanometer resolution, the usage of PDMS replicas for microcontact printing and micro-transfer molding elastomer [17]. The method was called Soft lithography. PDMS chips have high optical transparency, elasticity, gas and water-permeability and good biocompatibility. However, because of its porosity and permeability of PDMS is unacceptable for some applications.

Low molecular weight oligomer chains in PDMS can leach out into solution, negatively impacting cellular studies [18]. PDMS is hydrophobic material and susceptible to nonspecific adsorption and permeation by hydrophobic molecules [19]. Chemical modification of PDMS can address these issues. Plasma exposure will hydrophilize the PDMS surface but only a short time [20]. Reaction of silanes with silanols formed by plasma activation or other methods slows this change in surface properties [21]. PDMS is a valuable material for rapid prototyping and is commonly used in microfluidics studies, but it is not suitable for commercial use. The efforts of researchers are aimed at finding other polymers (for example, variants of polyurethanes) because it has been found that they have the same benefits as PDMS but do not have permeability for small molecules [22]. Thermoplastic polymers: Poly (methyl methacrylate), Polycarbonate, Polystyrene, Cyclic-olefin copolymer are cheaper, and are considered as candidates for mass production of microchips. [23]. Technology micro / nano hot embossing and injection molding are promising for mass production of microchips. But to get appropriate the good reproducibility of the sizes and shape of elements necessary avoid structural damage due to thermal stress, adhesion and friction at the interface between the workpiece and the shape is necessary to improve the surface properties of the master-form [24].

For the direct formation of micro-dimensional structures in polymeric materials, laser ablation is used often. The ablation mechanisms are different for nano-, pico- and femtosecond laser pulses [25]. With femtosecond pulses, it is possible to achieve more accurate and clean laser processing of metals and other solid materials than in pico and nano-second generation regimes. This technology is convenient for the production of a small series of microfluidic devices because it allows flexible enough to change the topology and design of microchips, but has limited spatial resolution capabilities of structures. In addition, laser ablation creates a rough surface, which is unacceptable for some applications.

Micro- and nano-fabrication technology is beautifully designed for silicon and glass, also used in the creation of microfluidic devices. Silicon surface chemistry based on the silanol group (−Si−OH) is well developed, so modification is easily accomplished via silanes. For example, nonspecific adsorption can be reduced or cellular growth improved through chemical modification of the surface [26]. Silicon is transparent to infrared but not visible light and not used for optical detection. This issue can be overcome by having a transparent material (polymer or glass) bound to silicon in a hybrid system. Microstructures in glass are created by etching into the glass through wet or dry methods [27]. Formation of fluidic communication requires bonding or protection layer attachment to enclose. Glass has low background fluorescence, and as with silicon, modification chemistries are silanol based. Glass is compatible with biological samples, has relatively low nonspecific adsorption, and is not gas permeable. Glass microfluidic systems are currently available to us as commercial products of Dolomite, Micralyne, Agilent and Caliper. The glass microfluidic chip with a flow-through reaction chambers containing an array nanocells for clusters with high density, is used in devices for nucleic acid sequencing HiSeq (Illumina) (Chromatin Immunoprecipitation Sequencing technology - ChIP-Seq), another microfluidic chip with a system of interlocking microchannels cameras is used in Agilent 2100 Bioanalyzer system (Agilent Technologies) for the electrophoretic separation of the sample and flow cytometry.

3. Nanoscale structures in microfluidic devices

The integration of nano-sized structures in microfluidic devices can improve the specifications and realize new possibilities for detecting an analyte. Among the requested features of nanoscale structures are the following: a) highly sensitive detection of the desired object / analyte (electric detection, plasmon resonance methods, detection with use nanopores, etc.) [28, 29]. b) transport (control the motion of molecules, particles - such as electroosmotic system) [30]; c) filtering and separation [31]; g) control (valves, gates) [32], and others. (heat exchangers, mixers, reactors). There are difficulties in detecting for nanofluidic devices due to: 1) a small number of analyte molecules at low signal / noise ratio; 2) small size detection region; 3) considerable time scanning study area for high spatial resolution, if necessary. Since the optical detection methods are highly compatible with microfluidic devices, they often try to adapt to nanofluidics, such as thermal lens spectrometry method [33]. Other techniques such as surface plasmon resonance spectroscopy, Raman spectroscopy, may be useful for the detection of non-fluorescent molecules [34, 35]. Methods of electrical detection of biomolecules are promising. When the size of the holding space is close to the size of the molecule, its intrusion into that space causes a change in electrical properties that can be registered.

In systems for nanopore sequences that are widely applied at present [36], two types of nanopores are used: biological and solid state. Biological nanopores created through the introduction channel protein (alpha-hemolysin (α-HL) and MSPA) into the lipid bilayer [37]. Solid-state nanopores are usually formed by a focused ion or electron beam in the dielectric membrane [38] There are problems using nanopore in particular: a) necessity to ensure extremely fast rate of DNA translocation through a nanopore thread [39]; b) nanopores must have reduced thickness to achieve single nucleotide resolution (application of modified graphene membranes [40].

Other promising nanoscale structures are "nanowhiskers» (nanowires - NW), which may be made of different materials (eg, semiconductors or metals). Semiconductor NW have narrow fluorescence band and higher photostability than organic fluorophores, and other useful properties. The area of possible application of NW is very wide. For example, a group of prof. CM Lieber (Department of Chemistry and Chemical Biology Harvard University) is developing technologies and methods for design and synthesis of nanoscale components, which can be used for creating any functional structures or nanosystems: from ultrasensitive sensors to nano interface devices for the study of living cells and tissues [41, 42].

The research at creating devices on a microfluidic platform with highly sensitive optical detection systems for express analysis of biological samples is conduct in the Institute for Analytical Instrumentation RAS. One of the research direction is the creation of nanoscale functional structures integrated into a microfluidic chip. Together with the University ITMO, lithography technologies are developed, which oriented to the creation of nanoscale structures directly in a microfluidic chip for mechanical fixation of cells [43]. Integration of porous meso- and nanostructures (for example, porous glasses) into a microfluidic chip allows us to realize new possibilities in the study of biological samples. Porous structures can be used as: filtration and separation units, elements of electroosmotic pump, base for sensitive sensor, etc. The possibility of using porous sodium borosilicate glass of SBS-MAP as a sensory element when insulin detection in the sample at competitive immunoassay was shown [44, 45]. This approach can be applied when creating microfluidic devices with biosensor elements for the detection of other proteins.

4. Conclusion

Micro- and nanofluid technologies are widely used in the modern developments of systems for sequencing nucleid acids, devices for digital amplification of nucleid acids, systems for testing diagnostic and therapeutic approaches in medicine. Microfluidic chips are considered an ideal platform for biochemical assays, tissue engineering, drug screening, detection of biomolecules, research and analysis of cells, toxicology studies [46, 47]. The development of microfluidic devices with integrate functional nanostructures will create fundamentally new devices with unique capabilities for biological and medical research.

References

[1] Tetala K K R and Vijayalakshmi M A 2016 *Analytica Chimica Acta* **906** 7e21
[2] Yousuff C M, Ho E T W, Hussain K I and Hamid N H B 2017 *Micromachines* **8** 15
[3] Chin C D, Linder V and Sia S K 2012 *Lab Chip* **12** 21
[4] Kim J, Jensen E, Stockton A and Mathies R A 2013 *Anal. Chem.* **85** 16
[5] Kulinsky L, Noroozi Z and Madou M 2013 *Methods Mol Biol.* **949** 3
[6] Bhatia S N and Ingber D E. 2014 *Nature Biotechnology* **32** 760
[7] Streets A M and Huang Y. 2013 *Biomicrofluidics* **7** 011302
[8] Hu J-B; Chen S-Y; Wu J-T; Chen Y-C and Urban P L 2014 *RSC Adv.* **4** 10693
[9] Zhu Z, Guan Z, Jia S, Lei Z, Lin S, Zhang H, Ma Y, Tian Z-Q and Yang C J 2014 *Angew. Chem. Int.* **53**, 12503
[10] Trantum J R, Baglia M L, Eagleton Z E, Mernaugh R L and Haselton F R 2014 *Lab Chip* **14**, 315
[11] Nosrati R, Vollmer M, Eamer L, San Gabriel M.C, Zeidan K, Zini A and Sinton D 2014 *Lab Chip* **14** 1142
[12] Kaushik G, Leijten J, Khademhosseini A 2017 *Stem Cells* **35** 51
[13] Ulgherait M, Rana A, Rera M, Graniel J and Walker D W 2014 *Cell Rep.* **8** (6) 1767
[14] Polini A, Prodanov L, Bhise NS, Manoharan V, Dokmeci M R and Khademhosseini A 2014 *Expert Opin Drug Discov* **9** 335
[15] Roth A, Singer T 2014 *Adv. Drug Delivery Rev.* **69-70** 179
[16] Esch EW, Bahinski A and Huh D 2015 *Nat. Rev. Drug Discovery* **14** 248
[17] Zhao X-M, Xia Y, Whitesides G M 1997 *J. Mater. Chem.* **7** 1069
[18] Berthier E, Young E W K and Beebe D 2012 *Lab Chip* **12** 1224
[19] Roman G T, Hlaus T, Bass K J, Seelhammer T G and Culbertson C T 2005 *Anal. Chem.* **77** 1414
[20] Hillborg H, Tomczak N, Olah A, Schonherr H and Vancso G J 2004 *Langmuir* **20** 785
[21] Sui G, Wang J, Lee C-C, Lu W, Lee S P, Leyton J V, Wu A M and Tseng H-R 2006 *Anal. Chem.* **78**, 5543
[22] Domansky K, Leslie D C, McKinney J, Fraser J P, Sliz J D, Hamkins-Indik T, Hamilton G A, Bahinski A and Ingber D E 2013 *Lab Chip* **13** 3956
[23] Nge P N., Rogers C I and Woolley AT 2013 *Chem. Rev.* **113** 2550
[24] Saha B, Toh W Q, Liu E, Tor S B, Hardt D E and Lee J 2016 *J. Micromech. Microeng.* **26** 013002
[25] Kuznetsov A I, Koch J and Chichkov B N 1996 *Appl. Phys.* A **63** 109
[26] Li L, Marchant R E, Dubnisheva A, Roy S and Fissell W H J 2011 *Biomater. Sci.* 22 91
[27] Iliescu C, Taylor H, Avram M, Miao J and Franssila S 2012 *Biomicrofluidics* 6 016505
[28] Choi S, Goryll M, Yi L, Sin M, Wong P K, and Chae J 2011 *Microfluid Nanofluid* **10** 231
[29] Lopez G A, Estevez M-C, Solera M and Lechuga L M 2017 *Nanophotonics* **6** 1
[30] Wang X, Cheng C, Wang S and Liu S. 2009 *Microfluidics and Nanofluidics* **6** 2
[31] Levy S L and Craighead H G 2010 *Chem. Soc. Rev.* **39** 1133
[32] Siwy Z S and Howorka S 2010 *Chem. Soc. Rev.* **39** 1115
[33] Shimizu H, Mawatari K and Kitamori T 2010 *Anal. Chem.* **82** 7479
[34] Lopez G A, Estevez M-C, Solera M and Lechuga L M 2017 *Nanophotonics* **6** 1
[35] Oh Y, Park S, Kang M, Choi J, Nam Y and Jeong K 2011 *Small* **7** 184
[36] Bayley H 2015 *Clinical Chemistry* **61** 1
[37] Rincon-Restrepo M, Mikhailova E, Bayley H and Maglia G 2011 *Nano Lett.* **11** 2
[38] Storm A, Chen J, Ling X, Zandbergen H and Dekker C 2003 *Nat. Mater.* **2** 8
[39] Ying Y L, Zhang J, Gao R and Long Y T 2013 *Angew. Chem. Int. Ed. Engl.* **52**, 50
[40] Merchant C A, Healy K, Wanunu M, Ray V, Peterman N, Bartel J, Fischbein M D, Venta K, Luo Z, Johnson A T C and Drndic M 2010 *Nano Lett.* **10** 8
[41] Guihua Yu and Lieber C M 2010 *Pure Appl. Chem.* **82** 12

[42] Zheng G, Gao X and Lieber CM 2010 *Nano Lett.* **10** 3179
[43] Evstrapov A A, Mukhin I S, Bukatin A S and Kuhtevich I V 2012 *Nuclear Instr. and Meth. in Phys. Res. B* **282** 145
[44] Evstrapov A, Esikova N, Rudnitskaya G and Antropova T V 2010 *Optica Applicata* **XL** (2) 333
[45] Esikova N A, Evstrapov A A, Bulyanitsa A L and Antropova T V 2015 *Glass Physics and Chemistry* **41** (1) 89
[46] Wu J, He Z, Chen Q and Lin J-M 2016 *Trends in Analytical Chemistry* **80** 213
[47] Kolahchi A R, Mohtaram N K, Modarres H P, Mohammadi M H, Geraili A, Jafari P, Akbari M and Sanati-Nezhad 2016 *Micromachines* **7** 162

IOP Conf. Series: Journal of Physics: Conf. Series **917** (2017) 022003 doi:10.1088/1742-6596/917/2/022003

Semiconductor light sources for near- and mid-infrared spectral ranges

L Ya Karachinsky [1,2,3*], **A V Babichev** [1], **A G Gladyshev** [1], **D V Denisov** [1], **A V Filimonov** [1], **I I Novikov** [1,3], **A Yu Egorov** [1,3]

[1]Connector Optics LLC, St. Petersburg194292, Russia
[2]Ioffe Institute, St. Petersburg 194021, Russia
[3]ITMO University, St. Petersburg 197101, Russia

Abstract. 1550 nm band wafer-fused vertical-cavity surface-emitting lasers (VCSELs) and 5-10 μm band multi-stages quantum-cascade lasers (QCL) grown by molecular beam epitaxy (MBE) were fabricated and studied. VCSELs show high output optical power up to 6 mW in single-mode regime (SMSR > 40 dB) and open-eye diagrams at 30 Gbps of standard NRZ at 20°C. QCL heterostructures show high structural quality (fluctuations of composition and thickness < 1%). 20-μm-stripe width QCLs mounted on copper heatsinks show lasing at ~ 6, 7.5 and 9 μm.

E-mail: leonid.karachinsky@connector-optics.com

1. Introduction

Researchers all over the world focus their research on semiconductor lasers of various types due to the wide range of potential applications. For example, semiconductor lasers of the near infrared spectral range are widely used for datacom and telecom applications, optical pumping of solid-state and fiber lasers, material processing etc. Vertical-cavity surface-emitting lasers (VCSELs) have several advantages in comparison with conventional Fabry–Perot lasers - low angular divergence, symmetric far field diagram, low threshold currents, low power consumption, increased temperature stability and radiation hardness, possibility of on-wafer testing of devices.

GaAs-based VCSELs [1] emitting at 850 nm are widely used in optical data communications at continuously shrinking distances and increasing bit data rates and steadily replace copper links [2]. Long-wavelength (LW) vertical-cavity surface-emitting lasers (VCSELs) emitting at 1300 – 1550 nm looks promising for optical interconnects, fiber to the home applications and other Si-photonics related technologies [3]. Compared to GaAs-based short-wavelength VCSELs, LW devices have lower power consumption and optical loss in silica optical fibers [4-6].

There are two main techniques which are used for 1550 nm band VCSEL manufacturing. The first one is based on direct growth of a VCSEL structure on InP substrate followed by a buried tunnel junction (BTJ) formation [47] and a second one that is based on wafer fusion (WF) of InP-based active area with GaAs-based Distributed Bragg Reflectors (DBRs) [8]. At present time commercially available 1550 nm band VCSELs demonstrate data transfer rate of about 3-10 Gbps [9]. The highest data transfer rate of 56 Gbps was reached on 1530 nm band VCSELs with BTJ for 4-5 μm aperture size devices [10].

Quantum cascade lasers (QCL) are lasers which operate based on quantum transitions of one type charge carriers between subbands. Such lasers were proposed theoretically in 1971 [11] but manufactured only in the mid-1990s due to the progress of molecular beam epitaxy (MBE) [12]. QCLs provide the possibility to get lasing over a broad wavelength range from several micrometers to ~100 μm range [12]. Another advantage of QCLs is relatively small dependence of the lasing wavelength shift versus temperature.

Despite alternative approaches to the creation of quantum-cascade lasers based on Si/Ge heterostructures [13], the main research has focused on lasers based on AIIIBV compounds [12]. To date, lasing of QCL in continuous wavelength (cw) mode based on ternary solid alloys InGaAs/InAlAs, on an InP substrates is demonstrated [14]. In spite of some success obtained in the fabrication of QCL [15] there is an evident needs for further research on increasing of wall-plug efficiency (WPE) of such lasers [16]. One approach to raise the laser efficiency is to use multi-stages QCL [17, 18]. With increase in the number of stages, the threshold power of the QCL weakly changes. (Threshold current density of QCL is inversely proportional to the number of stages N, while its threshold voltage is directly proportional to N) In turn, the use of multi-stage QCL reduces the intersubband light absorption and the heating of the structure (due to decreasing of the threshold current density) and improves the whole laser performance.

This paper is focused on 1550 nm band wafer-fused vertical-cavity surface-emitting lasers (VCSELs) and 5-10 μm band multi-stages quantum-cascade lasers grown by MBE.

2. 1550 nm band wafer-fused VCSELs

Wafer-fused 1550 nm VCSELs comprising active region with seven thin (2.8 nm), compressively strained InAlGaAs multiple quantum wells (MQWs) [19] and AlGaAs/GaAs DBRs that are fully-grown by MBE have been manufactured and studied. Devices with the TJ aperture size of 8 μm demonstrate lasing threshold ~ 1.5 mA and output optical power ~ 6 mW at 20°C (Figure 1). VCSELs show single mode CW operation with a longitudinal side mode suppression ration (SMSR) in the range of 40 - 45 dB with emission close to 1566 nm (see inset in Figure 1).

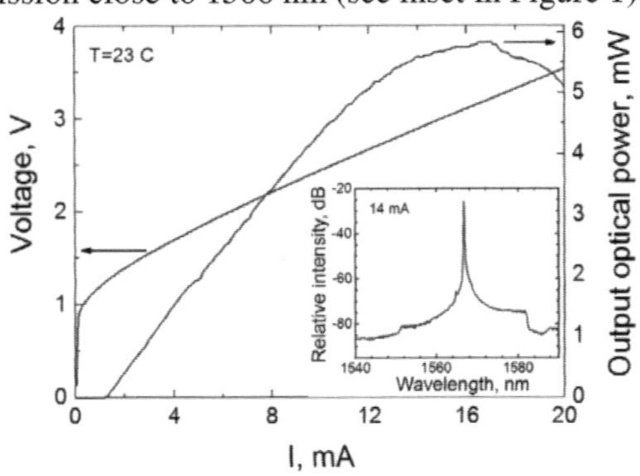

Figure 1. LIV characteristics. Inset shows a spectrum of VCSEL measured at bias current I=14 mA and 23 °C.

An open-eye diagram at 30 Gbps of standard NRZ is demonstrated at 20°C (Figure 2). With the temperature increasing up to 85°C threshold current increases but is still < 2.5 mA, devices show single mode operation with > 40 dB SMSR and demonstrate open-eye diagram at 20 Gbps of standard NRZ.

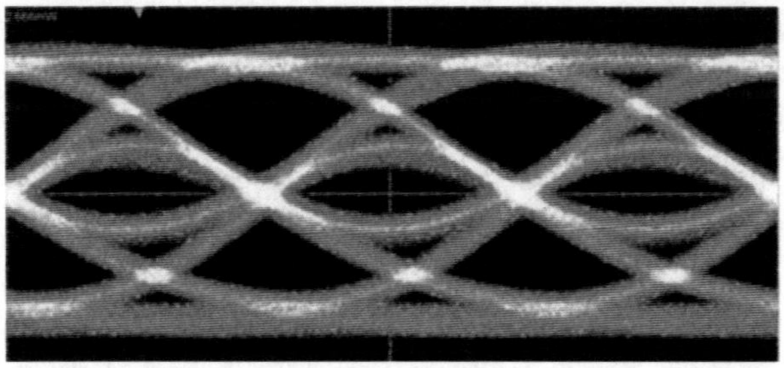

Figure 2. Optical eye diagram at bias current of 14 mA, modulation voltage of 0.5 V peak-to-peak, 30 Gbps and 20°C.

3. 5-10 μm band multi-stage quantum-cascade lasers

Three different designs of multi-stages quantum-cascade lasers were grown by molecular-beam epitaxy on InP (100) substrates using the Riber MBE49 machine. Grown QCL heterostructures were studied by XRD and TEM. Obtained results indicate high quality of the structures and high precision of the growth technique. Fluctuations of composition and thickness was less than 1% for all 3 structures. To study electroluminescent characteristics of the QCL heterostructures were processed into 20-μm-stripe-width lasers. Cavity length of ~ 1.5 mm was chosen. Lasers were mounted on copper heatsinks and studied in pulsed regime.

The first design with the active region based on bound-to-continuum design [20] was focused on lasing at ~ 6 μm. Before forming the stages of the laser on an InP substrate, an 100 nm InGaAs layer doped by silicon (Si) lattice matched to InP was grown. The heterostructure contains 60 stages formed by the strain-compensated $In_{0.44}Al_{0.56}As/In_{0.60}Ga_{0.40}As$ heteropair provided the necessary elastic balance in the QCL stages and a band offset ΔE_C of no less than 630 meV at the heterointerface [21-23]. The InAlAs cap layer and InGaAs contact layer doped by Si are of thickness 1500 and 50 nm, respectively. The structure demonstrates a lasing at room temperature in pulse mode. LI characteristic is depicted at Figure 3. The threshold current density is about 4.3 kA/cm^2. It is observed the multi-mode lasing with emission maximum close to 5.77 μm (see inset in Figure 3). The intermode distance $\Delta\lambda$ between longitudinal modes is ~ 3.7 nm. The effective refractive index for the fundamental mode is $n_{eff} = \lambda^2/(2L\Delta\lambda) = 3.26$, where L is the cavity length. The calculated reflectance of the mirrors, $R_{eff} = (n_{eff} - 1)^2/(n_{eff} + 1)^2$ is 0.28. The loss factor of the mirrors $\alpha_m = \ln(R_{eff})/L$ is 9.1 cm^{-1}.

IOP Conf. Series: Journal of Physics: Conf. Series **917** (2017) 022003 doi:10.1088/1742-6596/917/2/022003

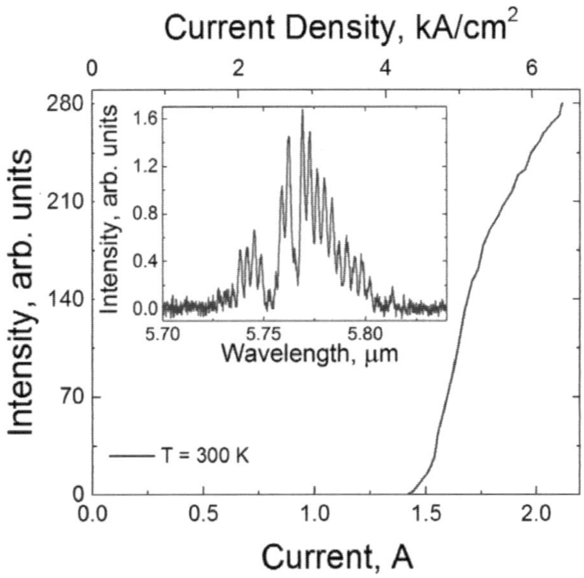

Figure 3. LI characteristic measured at T=300 K. Inset shows a spectrum of QCL with lasing about 5.77 μm at T = 300 K.

The second design with the active region based on two-phonon scheme [24] was focused on lasing at ~ 7.5 μm. Design of the active region is based on $In_{0.53}Ga_{0.47}As/Al_{0.48}In_{0.52}As$ and consists of 50 stages. The bottom cladding is formed by consistently grown Si-doped InP (thickness ~ 2000 nm) and InGaAs (thickness ~ 500 nm) layers. After growth of stages Si-doped InGaAs contact layer is grown (thickness ~ 120 nm). Edge-emitting lasers based on the grown heterostructures demonstrate lasing with threshold current density below 1.6 kA/cm^2 at a temperature of 78 K.

The third design with the active region based on three-phonon resonance scheme [25] was used for 9 μm range QCL. It contains the 40 stages active region based on $In_{0.53}Ga_{0.47}As/Al_{0.48}In_{0.52}As$. The bottom cladding is formed by Si-doped InGaAs layer (thickness ~ 2000 nm). The top cladding is based on InGaAs and InAlAs with thickness of 1830 and 4000 nm, respectively. n++ InGaAs contact layer is fabricated on heterostructure surface. The spectra of spontaneous emission demonstrate a single emission peak with a width at half maximum (FWHM) of about 50 cm^{-1} with a center at 1030-1040 cm^{-1} with slight blue shift with the increase of current. A typical lasing spectrum demonstrates a single line at the frequency of 1043 cm^{-1} (wavelength 9.6 μm) measured at 78 K.

4. Conclusions

1550 nm band wafer-fused VCSELs and 5-10 μm band multi-stages QCL grown by MBE were fabricated and studied. VCSELs show high output optical power up to 6 mW in single-mode regime (SMSR > 40 dB) and open-eye diagrams at 30 Gbps of standard NRZ at 20°C. Different scheme of QCL active region was applied to realize lasing at different wavelength. Using X-ray diffraction and transmission electron microscopy analysis, QCL heterostructures high structural quality is demonstrated (fluctuations of composition and thickness < 1%). The heterostructures were used for manufacturing of stripe lasers with 20 μm stripe width. QCLs mounted on copper heatsinks show lasing at ~ 6, 7.5 and 9 μm.

Acknowledgments
This study was supported by the Russian Ministry of Education and Science, project ID: RFMEFI61617X0074. Authors are grateful to Iakovlev V, Sirbu A, Mikhailov S, Ledentsov N N, Agustin M, Kurochkin A,Voropaev K, Nevedomsky V N, Il'inskaya N D, Sofronov A, Sokolovskii G S and Bousseksou A for their help in the experiments and interesting discussions.

References

[1] Iga K, Koyama F and Kinoshita S 1988 *IEEE J. Quant. Electron.* **24** 1845
[2] Ji C et al 2009 *Proc. SPIE* **7229** 722904
[3] Bimberg D, Larsson A and Joel A 2014 *Compd. Semicond.* **1** 22-26
[4] Babichev A V et al 2017 *Proc. of SPIE (San-Francisco)* vol 10122 ed K D Choquette and C Lei (Bellingham: SPIE) p 1012208-1
[5] Kryzhanovskaya N V et al 2017 *Proc. of SPIE (San Francisco)* vol 10098 ed B Witzigmann et al (Bellingham: SPIE), p. 1009811-1
[6] Egorov A Y, Karachinsky L Y, Novikov I I, Babichev A V, Nevedomskiy V N and Bugrov V E 2016 *Semiconductors 50* 612-15
[7] Gründl T, Debernardi P, Mueller M, Grasse C, Ebert P, Geiger K, Ortsiefer M, Bohm G, Meyer R and Amann M C 2013 *IEEE J. Sel. Top. Quantum Electron.* **19** 1700913
[8] Kapon E and Sirbu A 2009 *Nat. Photonics* **3** 27-29
[9] VCSEL product guide http://www.beamexpress.com/vcsel2.php (accessed Dec 15, 2016)
[10] Spiga S, Andrejew A and Amann M C 2016 *Proc. 18th ICTON2016* (Trento), (Piscataway: IEEE) pp 1-4
[11] Kazarinov R F and Suris R A 1971 *Sov. Phys. Semicond.* **5** 707
[12] Faist J, Capasso F, Sivco D L, Sirtori C, Hutchinson A L and Cho A Y 1994 *Science* **264** 553-56
[13] Diehl L et al 2002 *Appl. Phys. Lett.* **81** 4700-02
[14] Beck M, Hofstetter D, Aellen T, Faist J, Oesterle U, Ilegems M, Gini E and Melchior H 2002 *Science* **295** 301-05
[15] Owschimikow N, Gmachl C, Belyanin A, Kocharovsky V, Sivco D L, Colombelli R, Capasso F and Cho A Y 2003 *Phys. Rev. Lett.* **90** 043902
[16] Yao Y, Hoffman A J and Gmachl C F 2012 *Nature Photon.* **6** 432-39
[17] Razeghi M and Slivken S 2003 *Opto-Electron. Rev.* **11** 85-92
[18] Lyakh A et al 2008 *Appl. Phys. Lett.* **92** 111110
[19] Novikov I I et al 2016 *Semiconductors* **50** 1412-15
[20] Faist J, Beck M, Aellen T and Gini E 2001 *Appl. Phys. Lett.* **78** 147-49
[21] Egorov A Y, Brunkov P N, Nikitina E V, Pirogov E V, Sobolev M S, Lazarenko A A, Baidakova M V, Kirilenko D A and Konnikov S G 2014 *Semiconductors* **48** 1600-04
[22] Babichev A V et al 2016 *Semiconductors* **50** 1299-1303
[23] Egorov A Y et al 2015 *Semiconductors* **49** 1527-30
[24] Faist K, Hofstetter D, Beck M, Aellen T, Rochat M and Blaser S 2002 *IEEE J. Quantum Electron.* **38** 533-46
[25] Wang Q J, Pflügl C, Diehl L, Capasso F, Edamura T, Furuta S and Kan H 2009 *Appl. Phys. Lett.* **94** 011103

IOP Conf. Series: Journal of Physics: Conf. Series **917** (2017) 032001 doi:10.1088/1742-6596/917/3/032001

InGaAs/GaAs hybrid quantum well-dot nanostructures: Impact of substrate orientation and recombination mechanisms

A M Nadtochiy[1,2,3], M V Maximov[1,2], S A Mintairov[2,1,3], N A Kalyuzhnyy[2,1], S Rouvimov[4] and A E Zhukov[1]

[1]St. Petersburg Academic University, 8/3 Khlopina, St Petersburg 194021, Russia
[2]Ioffe Institute, 26 Polytechnicheskaya, St Petersburg 194021, Russia
[3]Solar Dots Ltd, 26 Polytechnicheskaya, St Petersburg 194021, Russia
[4]University of Notre Dame, Notre Dame, IN 46556, USA

e-mail: al.nadtochy@mail.ioffe.ru

Abstract. We study the impact of substrate orientation on growth of quantum well-dots (QWD) hybrid low dimensional structures that are intermediate in properties between quantum wells and quantum dots. QWD are formed by thickness and composition modulations of InGaAs quantum well, which localize electrons and holes. We show that in case of growth on 6 degree misoriented substrates such modulations are much more pronounced than in case of growth on exact oriented substrate. The modulations result in efficient strain relaxation, which allows vertical stacking of at least fifteen 8 ML-thick $In_{0.4}Ga_{0.6}As$ QWD layers without dislocation formation. Fitting of experimental dependencies of integrated PL intensity on excitation power by using ABC model allowed us determining relative contribution of different recombination processes.

1. Introduction

Semiconductor quantum wells (QWs) and quantum dots (QDs) have been intensively studied all over the world owing to their unique fundamental properties as well as advantages for device applications [1-2]. In an ideal QW, the carriers are confined in the direction perpendicular to the QW but can freely move in the QW plane. In an ideal QD, electrons and holes are strongly confined in all three spatial directions and there is no carrier exchange between individual QDs within the ensemble. However, even using modern epitaxial technologies such as MBE or MOCVD it is difficult to fabricate ideal semiconductor QWs and QDs. For instance GaAs/AlGaAs QWs usually contain monolayer thickness fluctuations that confine carriers in lateral direction and play the role of shallow QDs at helium temperatures. In case of InGaAs/GaAs QWs there are fluctuations of In composition that act as localization potential for electrons and holes. In a typical array of self-organized In(Ga)As/Al(Ga)As QDs there is temperature escape of electrons and holes from QDs to wetting layer or GaAs matrix and recapture by other QDs. This carrier exchange is more pronounced for QD array with small localization energy and is getting more intensive as temperature increases. In a very dense QD array, carriers may tunnel in lateral direction between neighboring QDs. These processes result in carrier lateral transport in realistic QD array making it "non-ideal". Technologists usually try to fabricate as "ideal" as possible QWs or QDs. However, nanostructures that are intermediate in properties between QWs and QDs may attract special fundamental interest and be beneficial for certain practical applications. Recently we have developed novel quantum well-dots hybrid nanostructures (hereinafter QWD structures) that show some advantages over both QWs and QDs [3-4]. This hybrid type of the active region represents a QW, containing ultra-dense array of narrow-gap In-rich regions with increased thickness, which are capable of localizing electrons and holes. Using "QD language" the QWD structures can be considered as a high-density array of shallow $In_xGa_{1-x}As$ QDs inside thin $In_yGa_{1-y}As$ layer (x>y). In this paper, we focus on the influence of substrate orientation on structural and optical properties of QWDs and consider mechanisms of carrier recombination in QWD structures.

2. Experimental

Epitaxial structures were grown on either exact oriented or vicinal (100) GaAs substrates using a MOCVD installation with horizontal reactor. Metal alkyls (threemethylgallium, threemethylaluminum, threemethylindium and diethylzinc) as well as hydrides (arsine and silane) were used as precursors. The QWDs were formed using 3D growth mode at low growth temperature by the deposition of $In_xGa_{1-x}As$ with In concentration x from 30% to 50% and average thickness D (3-27 monolayers) on exact oriented as well as on 6° and 10° misoriented GaAs substrates.

Structures for studying photoluminescence (PL) and structural properties by transmission electron microscopy (TEM) contained one or several InGaAs QWD layers in a GaAs matrix surrounded with 50 nm-thick $Al_{0.3}Ga_{0.7}As$ barrier layers. PL spectra were recorded with the use of a 2nd harmonics of YAG:Nd laser (532 nm) and cooled Ge photodiode employing a conventional lock-in technique.

Photovoltaic structures represented GaAs *p-i-n*-diodes with a 3 μm-thick base, a 0.5 μm-thick emitter and contained 15 InGaAs QWD layers in the *i*-region. $Al_{0.8}Ga_{0.2}As$ and $Al_{0.3}Ga_{0.7}As$ layers were utilized for window and back surface field layer, respectively. Photovoltaic characteristics were studied under illumination with terrestrial and space spectra.

A Titan 80–300 (FEI, USA) transmission electron microscope (TEM) with resolution of 0.136 nm and with 0.1 nm information limit in scanning TEM (STEM) and high resolution TEM mode, respectively, was used for structural and compositional characterization at nanometer and atomic levels. The Titan TEM was equipped with an energy dispersive x-ray spectrometer (EDS, Oxford Inca) with a spectral energy resolution of 130 eV. Cross-sectional TEM samples were prepared with a focused ion beam technique using a FEI Helios SEM/FIB Nanofactory system.

3. Results and discussion

Cross-sectional TEM images of QWD InGaAs layers grown on both exact (001) oriented (a, c) and vicinal (b, d) GaAs substrates are shown in Figure 1. The images were taken along <110> zone at medium (a, b) and high (c, d) magnifications to demonstrate the structure difference associated with substrate orientation. QD-like features are seen in both QWD InGaAs layers as thickness and compositional fluctuation, but they are much more pronounced in case of vicinal substrates due to presence of surface steps on the substrate. Compositional fluctuation result in contrast variations inside the QWD layers. Thicker areas have darker contrast indicating the presence of composition modulations along the QWD layers.

The thickness modulations may result from 3D growth mode employed for the growth of QWD layers being further enhanced by surface migration of In atoms due to lateral strain. High-resolution TEM images (figure 1, c) reveal that the thickness modulations often follow

IOP Conf. Series: Journal of Physics: Conf. Series **917** (2017) 032001 doi:10.1088/1742-6596/917/3/032001

the atomic steps on the vicinal surfaces. This explains stronger modulations of the thickness as well as lager lateral sizes (~ 20 nm) of QD-like features grown on vicinal surface (figure 1, c). Large lens-shaped islands on top of QW-like InGaAs layers are also seen in case of the sample grown on vicinal surface (figure 1, a) resembling QDs formed in Stranski-Krastanow growth mode. Such QDs cover more than one atomic step and may appear due to coalescence of individual QWDs. In case of exact (001) oriented surface, the thickness/composition modulations have smaller lateral size ~ 10 nm being more periodic (figure 1, d).

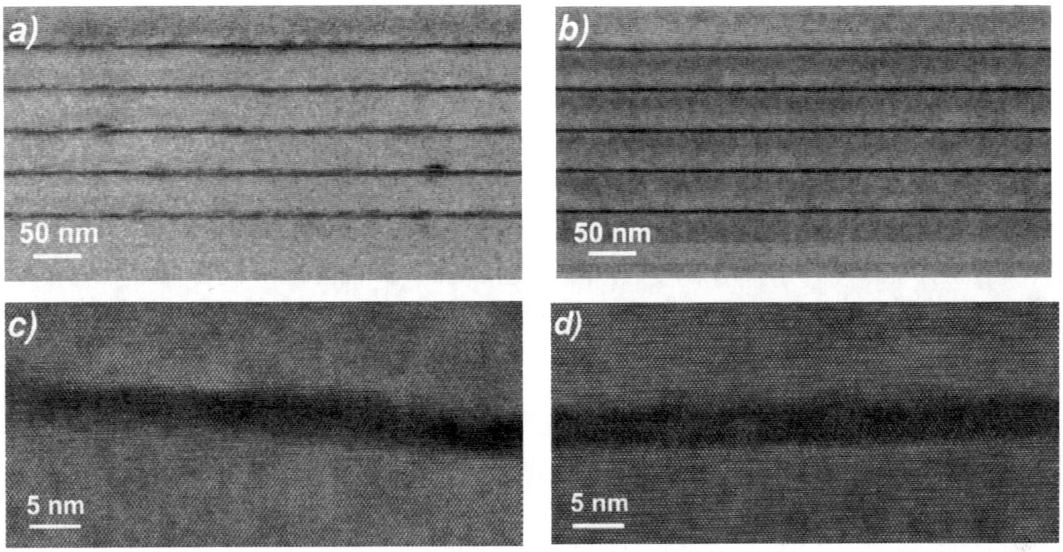

Figure 1. Cross-sectional TEM images of QWD InGaAs layers taken along <110> zone at medium (a,b) and high (c,d) magnifications. InGaAs layers were grown on both exact <001> oriented (a,c) and vicinal (b,d) GaAs substrates. One can see much higher variations in both thickness and composition of InGaAs layers in case of vicinal substrates due to presence of surface steps on the substrate. The compositional fluctuation are visible as changes in the gray scale variations.

The thickness and composition modulations in QWD structures that result in partial strain relaxation in the layers provide an opportunity to deposit more layers without formation of misfit dislocations in comparison with uniform QW of the same thickness and composition.

Figure 2 demonstrates the impact of substrate orientation on PL spectra. The QWDs were formed by the deposition of 8.1 ML of $In_{0.4}Ga_{0.6}As$ on exact oriented as well as on 6° and 10° misoriented GaAs substrates. Samples grown on misoriented substrates show significantly broader PL spectra. This fact is in agreement with stronger modulation of QWD thickness and composition revealed by TEM studies. The inset of figure 2 shows the dependence of integrated PL intensity on substrate orientation for QWDs formed by the deposition of 7, 8, 9 and 10 ML of $In_{0.4}Ga_{0.6}As$. The highest PL intensity is achieved in case of 6° misoriented GaAs substrates and $In_{0.4}Ga_{0.6}As$ effective thickness of 8-9 ML.

IOP Conf. Series: Journal of Physics: Conf. Series **917** (2017) 032001 doi:10.1088/1742-6596/917/3/032001

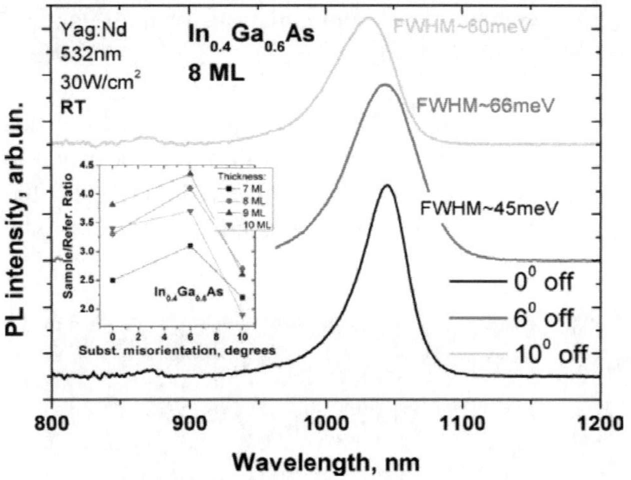

Figure 2. PL spectra of QWD structures formed by the deposition of 8.1ML of $In_{0.4}Ga_{0.6}As$ on exact oriented as well as on 6° and 10° misoriented GaAs substrates. Inset: integrated PL intensity of QWDs formed by the deposition of 7, 8, 9 and 10 ML of $In_{0.4}Ga_{0.6}As$ vs substrate orientation.

Substrate orientation significantly influences properties of photovoltaic QWD structures. GaAs solar cells (SCs) were based on 15 QWD layers grown under optimal conditions (8.1 ML-thick $In_{0.4}Ga_{0.6}As$ QWDs separated by 40 nm thick GaAs spacers) on exact oriented and 6° or 10° misoriented GaAs substrates. Insertion of QWD layers in the SC grown on vicinal surface results in a broadening of SC sensitivity range from 900 nm to 1050 nm (figure 3) without any change (for 6° off) or with minimal changes (for 10° off) of the spectrum in the GaAs photosensitivity range (less than 900 nm). Thus, incorporation of 15 strained QWD layers (in case of 6° misoriented substrate) does not deteriorate crystal quality of GaAs *p-n*-junction. For the SC grown on exact GaAs substrate the form of QWD related part of QE spectrum looks similar to that for SC on vicinal surface (figure 3) but the QE in entire spectral range is about 30% lower. This fact can be explained by non-elastic strain relaxation accompanied by formation of misfit dislocations in case of QWDs grown on exact GaAs substrate. In contrast, more pronounced thickness and composition modulations in QWD structures on vicinal surface (figure 1, a) provide better elastic strain redistribution and correspondingly, high crystalline quality.

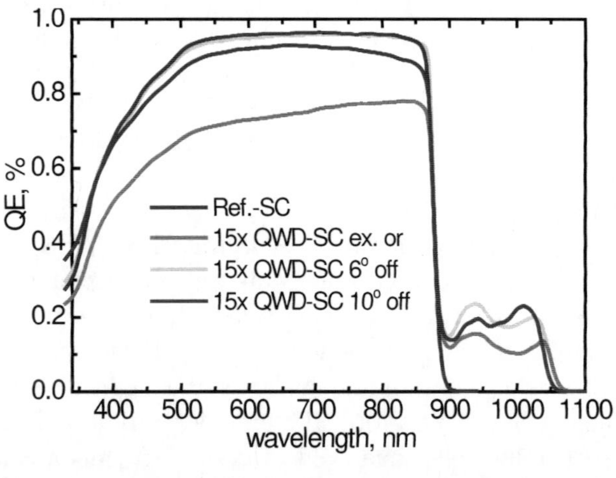

Figure 3. Internal quantum efficiency for reference GaAs SC (green) as well as GaAs SCs with 15 QWD layers grown on vicinal (6° off – green, 10° off – blue) and exact oriented (red) substrates.

4. Recombination mechanisms in QWD structures

Analysis of recombination process was done by fitting logarithm of excitation level as a function of logarithm of square root of integrated PL. The recombination rate of carriers is given by the so-called ABC model [5].

$$P \sim An+Bn^2+Cn^3 \quad (1)$$

where P is carrier generation rate (excitation power density), n is carrier concentration (we assume that electron and hole concentrations are equal), An is the Shockley-Read-Hall (SRH) recombination rate (non-radiative recombination on defects), Bn^2 is the rate of radiative recombination, Cn^3 is the Auger recombination rate. It is obvious that at very low excitation conditions SRH mechanism dominates the others. As the excitation increases, an increasing number of high-order terms come into action. At high excitation densities the Auger mechanism becomes the dominant recombination process. Depending on relations of recombination coefficients in the intermediate excitation range one can expect radiative recombination process to be the most intensive one.

Fitting of dependencies of integrated PL intensity on excitation power by equation (1) allowed us determining relative contribution of different recombination processes (figure 4). Experimental dependencies are well fitted without consideration of Auger recombination ($P \sim An+Bn^2$). Obviously, investigated range of excitation densities is insufficient for arising intensive Auger recombination.

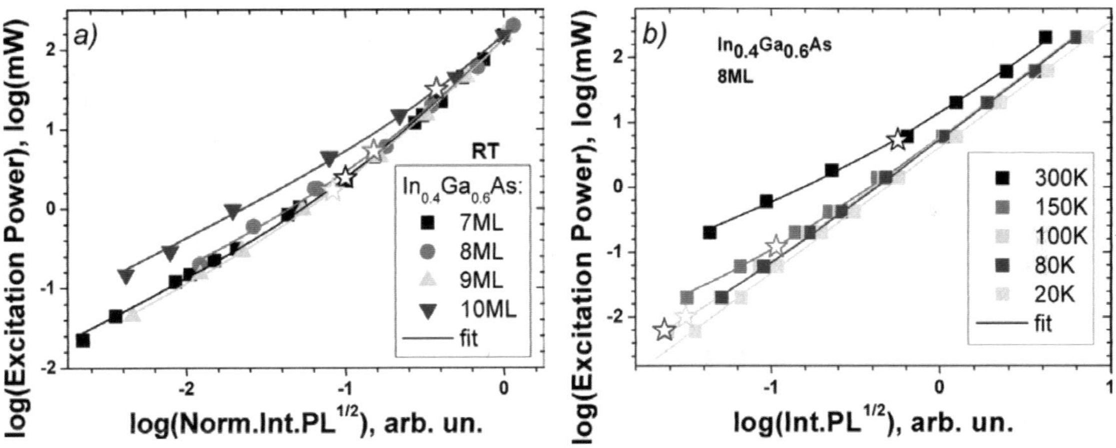

Figure 4. Experimental data and fits of logarithm of excitation level as a function of logarithm of square root of integrated PL for QWD structures with In content of 40% and thicknesses 7, 8, 9 and 10 ML (a), temperature dependent experimental data and fits for QWD structures with In content of 40% and thicknesses of 8 ML (b). The points corresponding to equal SRH and radiative recombination rates are marked by stars.

Since we measured PL in arbitrary units we can estimate only interrelation of recombination constants or, that is more informative, estimate excitation level at which rates of non-radiative and radiative recombination become equal. Such inflection point on the dependency of integrated PL intensity on excitation power is marked by star in figure 4, a. The higher is the excitation density corresponding to inflection point the stronger is contribution of SRH (non-radiative) recombination. For the 10 ML-thick QWDs the inflection point is located at the highest excitation density as compared to the other samples and thus, the contribution of SRH (non-radiative) recombination is the most intensive. We attribute this to the fact that deposition of too thick QWD layer results in the formation of misfit dislocation. 7 ML-thick and 9 ML-thick QWDs show small and nearly the same excitation levels corresponding to the inflection point that implies the weakest non-radiative recombination. In case of 7 ML-thick QWDs this fact can be explained by the low total strain. In case of 9 ML-thick QWDs we assume effective strain relaxation without creation of dislocations via formation of QWDs. Slightly higher excitation density corresponding to the inflection point for the 8 ML-thick QWDs may be due to some fluctuation of technological parameters during the MOCVD deposition.

As temperature decreases, a dramatic increase of radiative recombination rate in relation to SHR one (figure 4, b) occurs. Such behaviour is typical of different semiconductor systems [6-8]. Temperature lowering from 300 K to 150 K results in decrease of excitation density corresponding to the inflection point by more than an order of magnitude. Dependence taken at 80K shows no SHR as well as no Auger non-radiative recombination in the entire investigated range of excitation densities (4 orders of magnitude).

5. Conclusions

We described structural and optical properties of novel quantum well-dots hybrid nanostructures grown on exact oriented and vicinal (100) GaAs substrates using MOCVD. Thickness and composition modulations along the QWD layers are resolved by TEM for the structures grown on both exact oriented and vicinal substrates but for the latter ones they are much more pronounced. The thickness modulations appear due to 3D growth mode used to form the QWD structures. TEM images at high magnification reveal that the thickness modulations often follow the atomic steps existing on the vicinal surface. Effective strain relaxation in QWD provides an opportunity to deposit more layers without formation of misfit dislocations in comparison with uniform QW of the same thickness and composition. Structures grown on 6° misoriented GaAs substrates show higher PL intensity and quantum efficiency as compared to the ones grown on exact oriented substrate. Relative contribution of different recombination processes is evaluated by fitting logarithm of excitation level as a function of logarithm of square root of integrated PL with ABC model.

Acknowledgments

The work was supported by the Russian Science Foundation (project 16-12-10269).

References

[1] Nag B R 2002 *Physics of Quantum Well*, Springer
[2] Bimberg D, Grundman M and Ledentsov N N 1999 *Quantum Dot Heterostructures*, Wiley
[3] Mintairov S A, Kalyuzhnyy N A, Lantratov V M, Maximov M V, Nadtochiy A M, Rouvimov S and Zhukov A E 2015 *Nanotechnology*, **26** 385202
[4] Mintairov S A, Kalyuzhnyy N A, Maximov M V, Nadtochiy A M, Rouvimov S and Zhukov A 2015 *IEEE Electron. Lett.*, **51** 1602
[5] Agrawal G P and Dutta N K 1986 Long Wavelength Semiconductor lasers, New York: V.N. Reinhold
[6] Gurioli M, Vinattieri A, Colocci M, Deparis C, Massies J, Neu G, Bosacchi A and Franchi S 1991 *Phys. Rev. B*, **44** 3115
[7] Scott R *et al* 2016, *Phys. Chem. Chem. Phys.*, **18** 3197
[8] Nguen H P T, Djavid M, Cui K and Mi Z 2012, *Nanotechnology*, **23** 194012

Separation of Ga-polar GaN layer from Si substrate by wet chemical etching

K Yu Shubina, T N Berezovskaya, D V Mokhov, A M Mizerov and E V Nikitina

Nanoelectronics Lab, St. Petersburg Academic University, Khlopina 8/3, 194021 St. Petersburg, Russia

Abstract. In this work the effects of H_3PO_4:CH_3COOH:HNO_3:HF etching solution taken with different concentrations to the Ga-polar GaN/Si_xN_y/Si(111) epitaxial structures is investigated. Possibility of at least partial separation of Ga-polar GaN film from the silicon substrate without any GaN surface morphology changes by the selected etchants is demonstrated. The etching process mechanism is shown. Resistivity of the photoresist mask to etchants used in the experiments is found.

1. Introduction

Wide bandgap semiconductor materials A^3N and their alloys have a set of unique electrical, optical and mechanical properties, which are important in modern optoelecronics, high power electronics and microelectromechanical systems (MEMS) industry. Since natural substrates for A^3N materials are very expensive, silicon carbide (SiC), sapphire (*c*-Al_2O_3) and silicon (Si(111)) substrates are used for epitaxial growth of III-nitrides.

One of the key technological problems of GaN synthesis on silicon substrates is dramatic lattice mismatch (16.9%) between GaN and Si(111). Moreover, there is a large difference in their thermal expansion coefficients. These two factors lead to relaxation of elastic strain through the formation of misfit dislocations at the GaN/Si interface. Not the least important problem is interdiffusion of Si and Ga atoms under growth conditions, which are typical for molecular-beam epitaxy (MBE). Therefore, the local etching of Si(111) substrate is possible with macroscopic voids formation during the growth process [1]. One way to overcome these problems is forming a thin Si_xN_y layer on the Si(111) substrate surface before GaN epitaxial growth [2].

Crystallographic polarity is an essential feature of nitride semiconductor materials with wurtzite crystal structure. Polarity affects the growth kinetics, chemical, structural, optical and electrical properties of heterostructures [3]. N-polar A^3N films are chemically active, while metal-polar layers, which are used more often, are known to be almost chemically inert. Thus the post-growth treatment of metal-polar nitrides by wet chemical etching is challenging. But it can be an advantage for MEMS production. Sacrificial layers are always used to from MEMS. This step can be skipped when etching of the substrate by solution to which the nitride layer is resistant.

2. Experimental details

In this work the experiments for investigation the effects of H_3PO_4:CH_3COOH:HNO_3:HF etching solution, taken with different concentrations, to the GaN/Si_xN_y/Si(111) epitaxial structures were carried out. The aim of the experiments was at least partial separation of GaN epitaxial layer from the Si(111) substrate. The Ga-polar GaN films of 0.6-0.8 mkm thickness were obtained by plasma assisted molecular beam epitaxy (PA-MBE) on Veeco Gen 200 MBE system equipped with RF plasma source. The layers were grown on the semi-insulating (R > 10 kΩcm) Si(111) substrates, which were cleaned according to the Shiraki method [4]. Before the growth, the Si(111) substrates were annealed at T_{sub} = 850 °C and then nitridated in the growth chamber using different nitridation temperatures and times. Thin Si_xN_y layer, formed during the nitridation process, was used as a buffer layer.

To identify crystallographic polarity of the samples obtained the express-technique was developed (patent application №2016144845 15.11.2016). The meaning of developed technique is the observation of the surface morphology modifications as a result of anisotropic wet chemical etching by hot aqueous KOH solution in case of N-face structure, while surface and thickness of Ga-face material remain unchanged. Checking for surface morphology changes was carried out using scanning electron microscope 25 Supra Zeiss.

To investigate the effects of selected H_3PO_4:CH_3COOH:HNO_3:HF (5:1:1:1) and H_3PO_4:CH_3COOH:HNO_3:HF (5:1:3:0.33) etchants, Ga-polar sample,s coated with photoresist mask, were immersed in these solutions at room temperature without heating and auxiliary illumination. The duration of etching process was 40-150 s.

3. Results and discussion

3.1. Crystallographic polarity

As a result of the polarity identification experiments, Ga polarity of the samples was confirmed. SEM images of Ga-polar sample surface before and after polarity identification etching are demonstrated in figure 1.

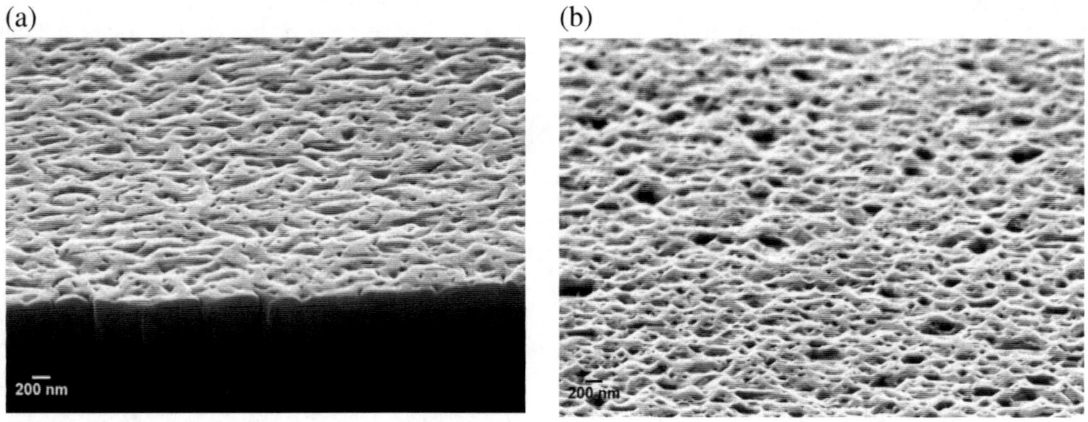

Figure 1. SEM image of Ga-polar GaN/Si_xN_y/Si(111) sample before (a) and after etching in hot aqueous KOH solution (b). Formation of hexagonal etch pits is a result of defect-selective etching of GaN layer.

Despite the Ga-polarity, some changes in the surface morphology of the sample (hexagonal etch pits) are clearly visible (see figure 1, b). The primary cause of their appearance can be selective etching of various structural defects present in the GaN epitaxial layer: dislocations [5], nanotubes [6] or inversion domains [7]. However, the morphology of the defect-free areas and the thickness of the epitaxial layer remained unchanged, that indicates the Ga-polarity of the sample.

3.2. GaN epitaxial film separation from the substrate

As a result of the etching in H_3PO_4:CH_3COOH:HNO_3:HF (5:1:1:1) and H_3PO_4:CH_3COOH:HNO_3:HF (5:1:3:0.33) solutions, samples with GaN film, partially separated from the substrate, were obtained. The thickness of the GaN epitaxial layer has not changed. SEM image of the sample after etching is demonstrated in figure 2. Slight upward curvature of the GaN layer appeared after its separation from the substrate by the etching process. It could be due to gradient in the strain of the material in growth direction [8].

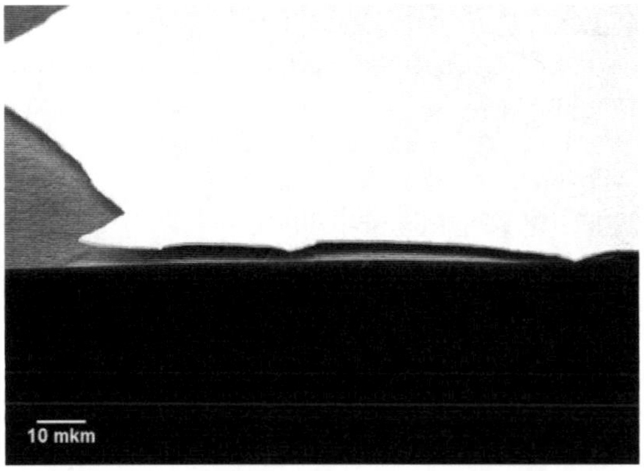

Figure 2. SEM image of the sample after etching in H_3PO_4:CH_3COOH:HNO_3:HF (5:1:1:1) solution. GaN film is partially separated from the Si substrate.

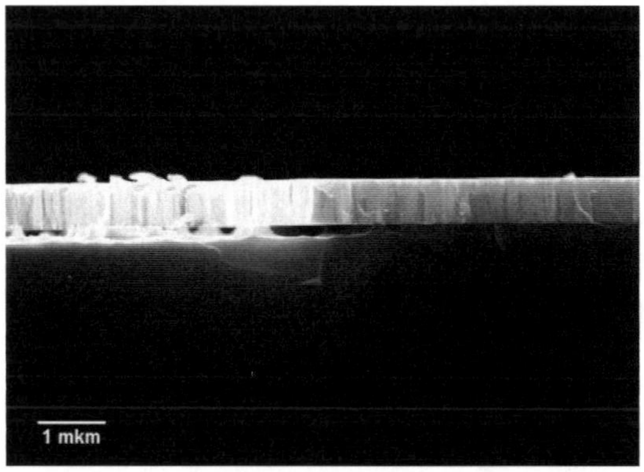

Figure 3. SEM cross-sectional image of the sample after etching. The mechanism of the separation process is clearly visible.

The H_3PO_4:CH_3COOH:HNO_3:HF etching solution includes a mixture of nitric and hydrofluoric acids, which is commonly used for silicon etching. Therefore the separation mechanism of the GaN film from the Si_xN_y/Si(111) substrate is based on the silicon etching. However it should be noted that the etching occurs on the GaN/Si interface (see figure 3). Wherein, thin Si_xN_y layer dissolves completely, as it was confirmed by additional experiments.

The etchant is assumed to react with the interface not only from the side faces of the sample, but also penetrating through the GaN epitaxial layer structural defects. What is more, the etching is not defect-selective and the surface morphology of the GaN epitaxial film remains unchanged (figure 4, a).

(a) (b)

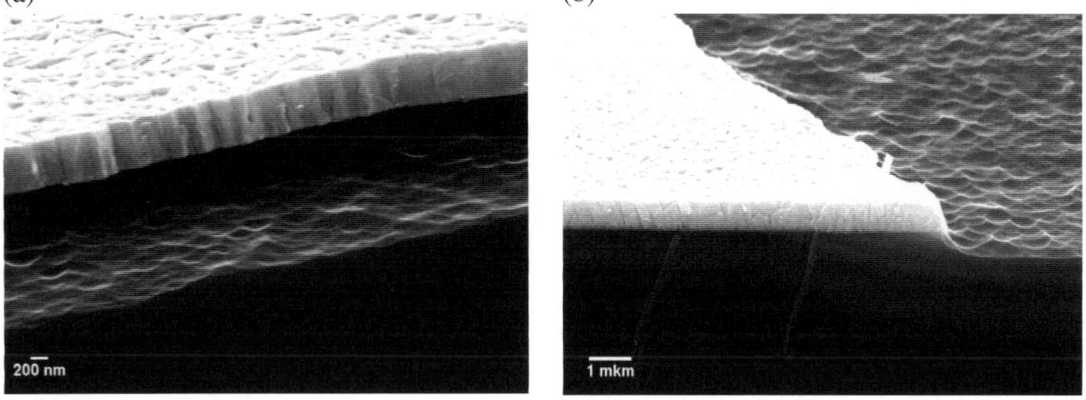

Figure 4. SEM image of Ga-polar GaN/Si_xN_y/Si(111) sample after etching in H_3PO_4:CH_3COOH:HNO_3:HF (5:1:1:1) etching solution. The morphology of GaN epitaxial layer remain unchanged (a). There is no evidence of etching in the area protected by photoresist mask (b).

In the sample areas coated with photoresist, the evidence of etching is not found (figure 4, b) and the photoresist mask is not dissolved. Variations in the composition of the etching solution did not lead to significant changes in the etching process and its results.

4. Conclusion

In this work the effects of etchants H_3PO_4:CH_3COOH:HNO_3:HF (5:1:1:1) and H_3PO_4:CH_3COOH:HNO_3:HF (5:1:3:0.33) to GaN/Si_xN_y/Si(111) Ga-polar epitaxial heterostructures was investigated. As a result, the possibility of at least partial separation of the Ga-polar GaN epitaxial film from the Si(111) substrate without additional lighting and heating was demonstrated. The mechanism of the process based on penetration of the etchant through structural defects of the epitaxial layer to the interface layer/substrate and etching Si_xN_y/Si(111) substrate was shown. It was experimentally established that these etchants do not damage the Ga-polar epitaxial layer and do not change its morphology. Resistance of the photoresist mask to H_3PO_4:CH_3COOH:HNO_3:HF (5:1:1:1) and H_3PO_4:CH_3COOH:HNO_3:HF (5:1:3:0.33) solutions was determined. It was found that results of etching in the etchants mentioned above are not significantly different.

Acknowledgments

The work was done as a part of the state assignment (№ 16.9789.2017/BCh) of the Ministry of Education and Science of the Russian Federation

References

[1] Sobanska M, Klosek K, Zytkiewicz Z R, Borysiuk J, Witkowski B S, Lusakowska E, Reszka A and Jakiela R 2012 *Cryst. Res. Technol.* **47** 307-12
[2] Kumar M, Rajpalke M K, Roul B, Bhat T N, Sinha N, Kalghatgi A T and Krupanidhi S B 2011 *Appl. Surf. Sci.* **257** 2107-10
[3] Hellman E S 1998 *MRS Internet J. Nitride Semicond. Res.* **3** e11
[4] Ishizaka A and Shiraki Y 1986 *J. Eltrochem. Soc.* **133** 666-71
[5] Visconti P, Huang D, Reshchikov M A, Yun F, Cingolani R, Smith D J, Jasinski J, Swider W, Liliental-Weber Z and Morkoc H 2002 *Mat. Sci. Eng.* B **93** 229-33
[6] Hong S K, Yao T, Kim B J, Yoon S Y and Kim T I 2000 *Appl. Phys. Lett.* **77** 82-84
[7] Ng H M, Weimann N G and Chowdhury A 2003 *J. Appl. Phys.* **94** 650-53
[8] Stonas A R, MacDonald N C, Turner K L, DenBaars S P and Hu E L 2001 *J. Vac. Sci. Technol.* B **19** 2838-41

Preparation of a silicon surface for subsequent growth of dilute nitride alloys by molecular-beam epitaxy

A A Lazarenko, T N Berezovskaya, D V Denisov, M S Sobolev, E V Pirogov and E V Nikitina
Nanoelectronics Lab, St. Petersburg Academic University, St. Petersburg 194021, Russia

Abstract. This article discusses the process of preparation of a silicon surface for subsequent growth of dilute nitride alloys by molecular-beam epitaxy. The method of preparation of Si (100) and Si (111) substrates was developed. This method provides reproducible high-quality silicon surface for molecular-beam epitaxy of Si-GaP heterostructures. As a result, it managed to reduce the eviction oxide temperature below 800 °C, which is an important parameter for the MBE technology.

1. Introduction

The pre-growth preparation of substrates is one of the most important step in the formation of structures of molecular-beam epitaxy (MBE) technology for opto- and microelectronics, since the presence of chemically pure, structurally perfect, atomically smooth surface is a prerequisite for producing epitaxial layers of device quality. Nowadays the most common methods of surface cleaning are the high-temperature annealing and the formation of a protective layer on the surface, subsequently removed in ultra-high vacuum thermally.

High-temperature annealing (850-1200 °C) found wide application in silicon MBE, as it allows to obtain a surface of the necessary growth quality. However, high-temperature heating causes unwanted diffusion of impurities and the appearance of a near-surface p-type layer, thereby changing the doping profile of the structures. In addition, the growth of such crystal defects as dislocations and packing defects, the appearance of extended microsteps and pits of thermal etching is observed [1]. Thus, the use of a relatively low-temperature (not higher than 900 °C) pre-epitaxial preparation method, using the passivation effect of the Si surface with nonstoichiometric SiO_2 or hydrogen, is more preferable and, accordingly, more common.

The main impurity remaining on the surface of the semiconductor substrates after degassing in the ultra-high vacuum (UHV) is carbon, which moreover, at temperatures above 700 °C, reacts chemically with silicon and form a hard-to-remove SiC, so the passivation layer must have carbon-photophobic properties. Also, it must be sufficiently resistant to the action of chemically active contaminants that are always present in the environment, easily formes during the chemical preparation of the substrates and completely removes from the surface with moderate warming along with traces of residual contaminants.

The objective was to prepare the silicon surface for subsequent growth of dilute nitride alloys by MBE.

2. Experiment

Based on the methods proposed in the works [2,3] the method of preparation of Si (100) and Si (111) substrates was developed.

The procedure of the chemical treatment of substrates consists of the following operations (HNO_3 ~ 70%; HF ~ 46%; H_2O_2 ~ 30%; NH_4OH ~ 28%; HCl ~ 36%; distilled deionized water with a resistance of at least 15 MOhm·cm):

1. Washing off traces of organic contaminants by boiling first in CCl_4, then in acetone.
2. Oxidation of the upper layer of silicon in boiling nitric acid and bleeding of the formed SiO_2 in a solution of hydrofluoric acid ($HF:H_2O$ = 1:3) for 20-30 seconds (this operation is repeated several times).

IOP Conf. Series: Journal of Physics: Conf. Series **917** (2017) 032003 doi:10.1088/1742-6596/917/3/032003

3. Boiling in a peroxide-ammonia solution ($NH_4OH:H_2O_2:H_2O$ = 1:1:3) for 10 minutes, followed by removal of contaminants together with silicon oxide in a solution of hydrofluoric acid.
4. Formation on the surface of a thin protective oxide by boiling in peroxide-acid solution ($HCl:H_2O_2:H_2O$ = 3:1:1) for 10 minutes.
5. Final washing with deionized water and drying with ethanol in centrifuge.

Samples prepared in this way were immediately loaded into the MBE chamber, where they were before the final stage of pre-epitaxial preparation in the conditions of UHV. The pre-epitaxial preparation of the silicon surface was tested on a SUPRA-32 (Riber) whose growth module contains an electron-beam evaporator to obtain the required Si flux.

Directly before the growth experiments, the substrate was sequentially degassed at the temperatures of 400 and 750 °C for 60 and 40 minutes, respectively, after which, at a temperature of 750 °C in a weak Si stream ($\sim 1 \cdot 10^{13}$ atoms·cm^{-2}·s^{-1}) or a layer of silicon oxide was removed from the surface only thermally at a temperature of 800 °C for 15-20 minutes.

In order to test the quality of the prepared surface the control layers of Si were grown by MBE at a temperature of (600 ÷ 650) °C (optimal for silicon MBE) and thickness from several hundred to several thousand angstroms. Processes of the pre-growth preparation and growth under vacuum were monitored in-situ by reflection high-energy electron diffraction (RHEED) and ex-situ by atomic-force microscopy (AFM).

3. Results and discussion

According to AFM, silicon plates initially had an atomically smooth surface (figure 1). This is indicated by the image of the surface relief in the AFM pictures and the profile parameters characterizing the surface rms = 3 Å, peak-peak = 20 Å.

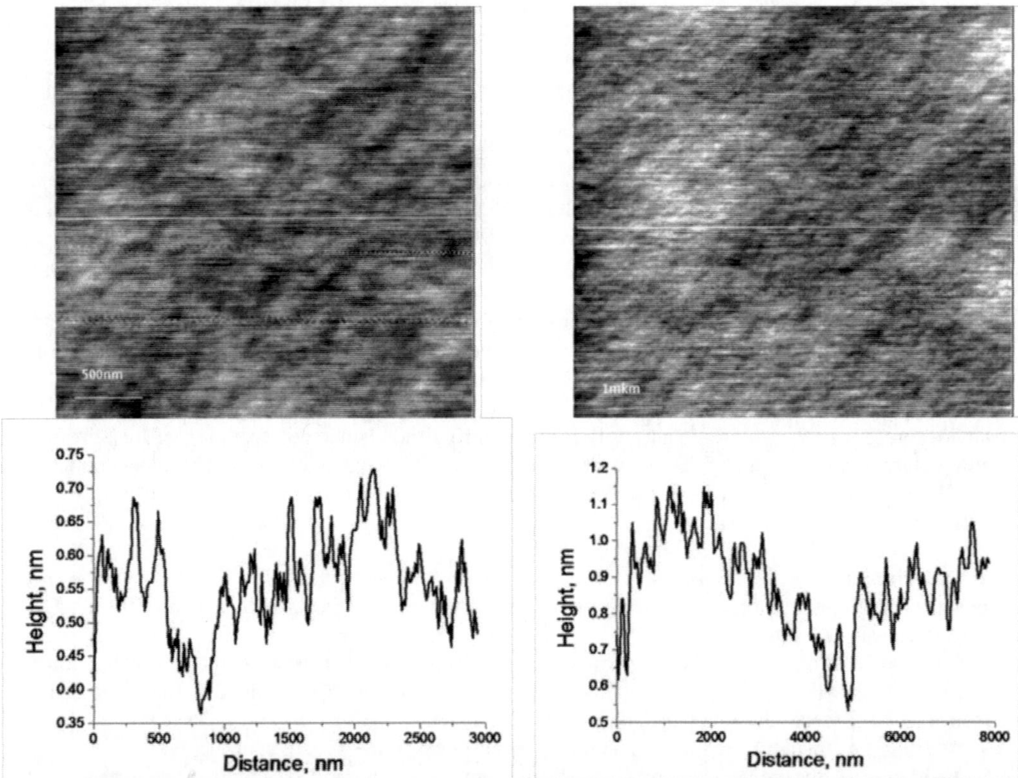

Figure 1. AFM image of the original surface Si (100) rms = 3 Å, peak-peak = 20 Å.

After chemical treatment, the morphology of the substrate surface did not change. According to the AFM, chemical preparation also does not lead to coarsening of the surface relief in the nanoscale range (rms = 3 Å, peak-peak = 20 Å). The presence of a pattern of elongated diffraction reflections on this surface (figure 2), corresponding to the Si lattice, indicates that the surface of the substrates is atomically smooth and that the thickness of the formed SiO_2 layer does not exceed several monolayers (for Si (100) surface 1ML ~ $6.8 \cdot 10^{14}$ atoms/cm^2).

Figure 2. The RHEED image from the Si (100) surface after chemical preparation.

In UHV conditions, chemically prepared substrates were sequentially degassed in a growth module at 450 °C for 60 minutes and 750 °C for 40 minutes until the pressure in the growth chamber stabilized at a minimum level (3÷9 10^{-10}) Torr). Removal of the protective oxide thermally at 800 °C (t ~ 10-15 minutes), or by heating at 750 °C in a stream of Si atoms (~ 1 10^{13} atoms · cm^{-2}·s^{-1}, t~15-20 minutes), according to RHEED, it is possible to obtain an atomically-smooth structurally-ordered silicon surface (figure 3).

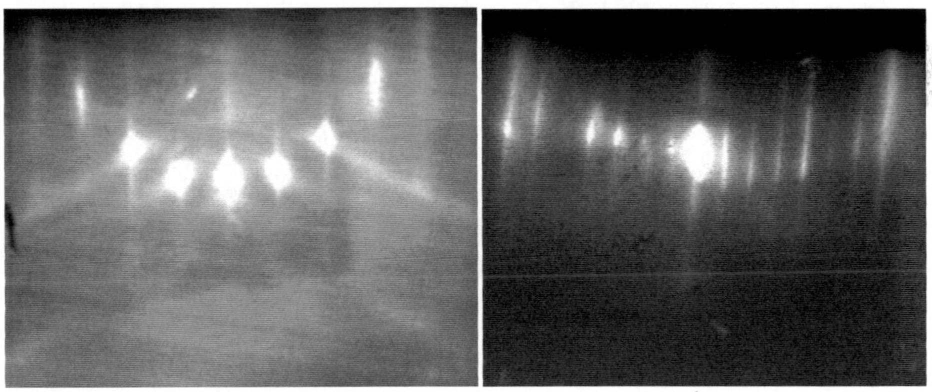

a b

Figure 3. The RHEED image from the Si (100) (a) and Si (111) (b) substrates after removal of the oxide layer. The electron energy is 10 keV. The azimuth of the incidence of the electron beam [011] and [112] for (100) and (111) surfaces, respectively.

Control Si layers with a thickness of 100 to 5000 Å were grown on the prepared substrates at temperatures from 350 °C to 750 °C and deposition rates of 1.0 Å/s. All the samples showed linear (2x2) and (7x7) diffraction for Si (100) and Si (111) substrates, respectively, with well-defined long narrow strands with presence of a superstructure and Kikuchi lines, which indicates a two-dimensional nature of the surface, its structural order and absence of microroughnesses (figure 4).

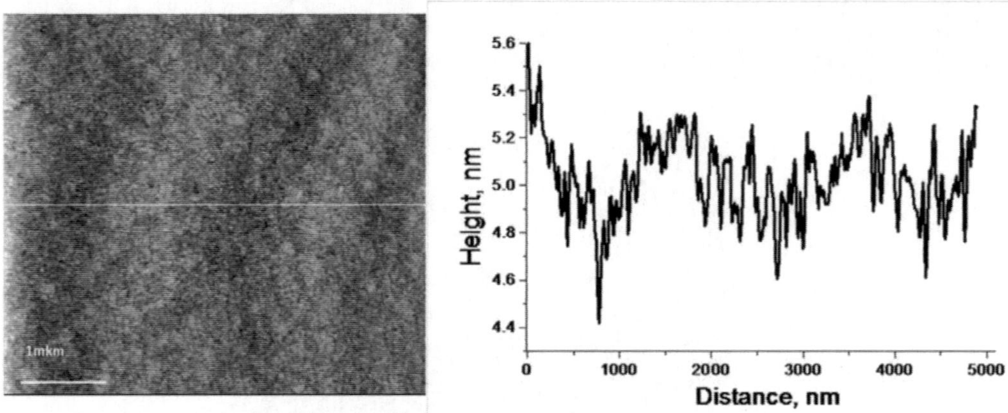

Figure 4. AFM image of the Si[500Å]\Si(100) structure. rms = 3 Å, peak-peak = 10 Å.

Grown structures demonstrate good morphological surface quality, without sharp jumps of relief, significant changes in the level of the average relief and areas of the heterogeneous roughness. The surface roughness is less than 1 nm. As a result, it managed to reduce the eviction oxide temperature below 800 °C, which is an important parameter for the MBE technology.

Acknowledgments
This work is supported by RFBR Grant № 16-32-00322 and № 16-58-150006.

References

[1] Liehr M, Renier M, Wachnik R A and Scilla G J 1987 *J. Appl. Phys.* **61** 4619
[2] Ishizaka A and Shiraki Y 1986 *J. Electrochem. Soc.* **133** 666-71
[3] Ioshkin V A, Orlikovskii A A, Oktyabr'skii S R, Kvit A V, and Dovydenko E Yu 1994 *Proceedings FTIAN* (Nauka, Moscow) **8** 58

IOP Conf. Series: Journal of Physics: Conf. Series **917** (2017) 032004 doi:10.1088/1742-6596/917/3/032004

Simulation of the formation of polymorphic varieties of nanodiamonds

V A Greshnyakov[1] and E A Belenkov[1]

[1] Physics Department, Chelyabinsk State University, Chelyabinsk 454001, Russia

Abstract. Structural stability and the possible pathways to experimental formation of lonsdaleite and other polymorphic modifications of diamond have been studied in the framework of the density functional theory. It was established that the structural transformation of orthorhombic *Cmmm* graphite to lonsdaleite must take place at a pressure of 61 GPa, while the formation of lonsdaleite from hexagonal *P6/mmm* graphite must take place at 56 GPa.

1. Introduction

Cubic diamond is a reference material setting the upper limit of hardness on the Mohs scale. However, according to the theoretical research, performing in [1], hardness of hexagonal polytype of diamond (as called as lonsdaleite) exceeds hardness of cubic diamond twice. How lonsdaleite can be synthesized till now remains insufficiently clear. Therefore in this work, formation conditions of lonsdaleite have been theoretically studied at the strong compression of graphite.

2. Methods

The model investigation of the structural stability and possible pathways of formation of lonsdaleite and other polymorphs of diamond was based on calculations of the «graphite ↔ lonsdaleite (hexagonal diamond)» and «graphite ↔ diamond-like phases» phase transitions using the method described in [2, 3]. Calculations of the crystalline structures and energy characteristics of carbon phases were performed by methods of the density functional theory (DFT) [4]. These calculations employed the Perdew-Zunger density functional for the exchange-correlation energy in the local density approximation (LDA) [5]. Figures 1 and 2 show unit cells of structural modifications of graphite and diamond used for the DFT-LDA calculations.

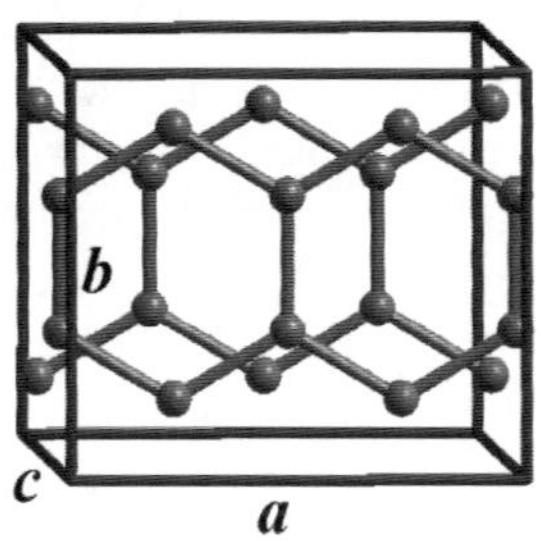

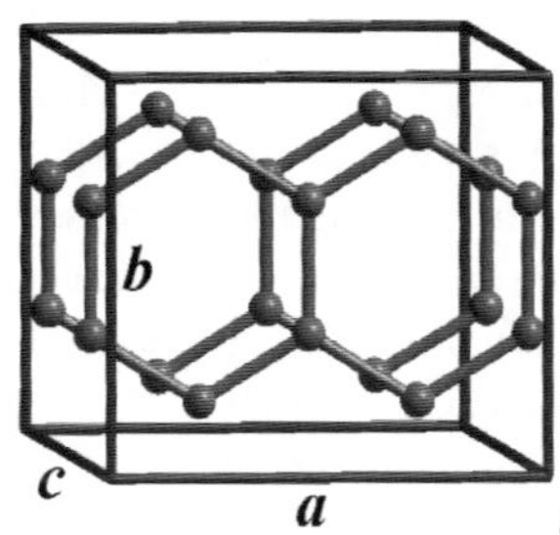

Figure 1. Unit cells of orthorhombic (a) and hexagonal (b) graphites used for DFT-LDA calculations.

IOP Conf. Series: Journal of Physics: Conf. Series **917** (2017) 032004 doi:10.1088/1742-6596/917/3/032004

The heats of structural transitions of graphite to diamond polymorphs under isothermal compression conditions were determined by calculating the difference of enthalpies at the point of transition. The enthalpy was calculated using the standard relation $H = E_{total} + P \cdot V_{at}$, where E_{total} is the total energy, P is the pressure, and V_{at} is the volume per atom.

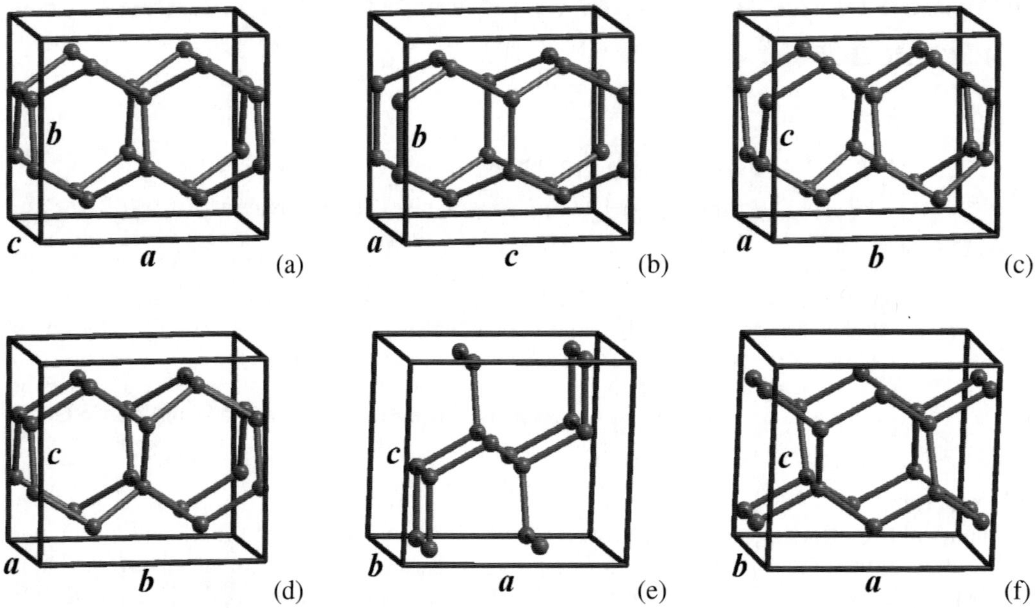

Figure 2. Orthorhombic unit cells of hexagonal diamond (a), LA3 (b), LA5 (c), *Pmma* (d), $P2_1/m$ (e), and *C2/m* (f) diamond-like phases, respectively.

3. Results

As a result of the first stage of DFT calculations, we determined the values of equilibrium unit cell parameters for the initial and final structures. At the second stage of calculations, we simulated the direct and reverse phase transitions from three- to four-coordinated carbon compounds under uniaxial deformation conditions.

Figures 3 and 4 show dependences of the total energy difference (ΔE_{total}) on the atomic volume (V_{at}) characterizing direct and reverse phase transitions between polymorphic modifications of carbon. Investigations of the possible synthesis pathways to lonsdaleite (hexagonal diamond) showed that this carbon structure can be formed from two metastable graphite phases, *Cmmm* and *P6/mmm*. The structural transition from *Cmmm* graphite to hexagonal diamond (with the [100] strain direction) requires surmounting an energy barrier of 0.241 eV per atom (Figure 3) at an atomic volume of 6.332 A^3 per atom, when the pressure reaches 61.0 GPa. In the case of initial *P6/mmm* graphite, the potential barrier height slightly increases to 0.275 eV per atom (Figure 4), whereas the phase transition pressure decreases to 56.3 GPa. Since the pressure required for the structural transition from graphite to hexagonal diamond is lower that the values calculated for the formation of diamond-like LA3, LA5, *Pmma*, $P2_1/m$, and *C2/m* phases, the formation of lonsdaleite from *P6/mmm* graphite must take place before the formation of other polymorphic modifications of diamond.

For estimating the stability of hexagonal diamond in comparison to other diamond polymorphs, we have studied the reverse phase transitions caused by the uniaxial tensile deformation of this diamond until complete transformation to graphite. The calculations were performed for the stress applied along axes parallel to the [001], [100], and [110] crystallographic directions. The values of potential barriers for the destruction of hexagonal diamond by tension along the [001] and [110] axes amounted to 0.361 and 0.383 eV per atom, respectively (Figure 3). The value of energy barrier to be surmounted for the transition from hexagonal diamond to orthorhombic graphite under applied stress of 112 GPa was 0.310 eV per atom (Figure 4). The minimum barrier value for hexagonal diamond significantly exceeds the corresponding values of energy barriers for diamond-like LA3, LA5, *Pmma*, $P2_1/m$, and *C2/m* phases (Figures 3 and 4), which is indicative of a higher thermodynamic stability of lonsdaleite in comparison to all diamond-like phases with equivalent and non-equivalent atomic positions. It should be noted that a potential barrier for the transition from hexagonal diamond to graphite is only 0.003 eV per atom smaller than the corresponding value for cubic diamond [3]. Since the phase transition from diamond to graphite under normal pressure takes place at temperatures above 1500 Kelvin and taking into account the theoretically calculated value of «cubic diamond→graphite» barrier, the temperature of phase transition from hexagonal diamond to graphite can be estimated at approximately 1470 Kelvin.

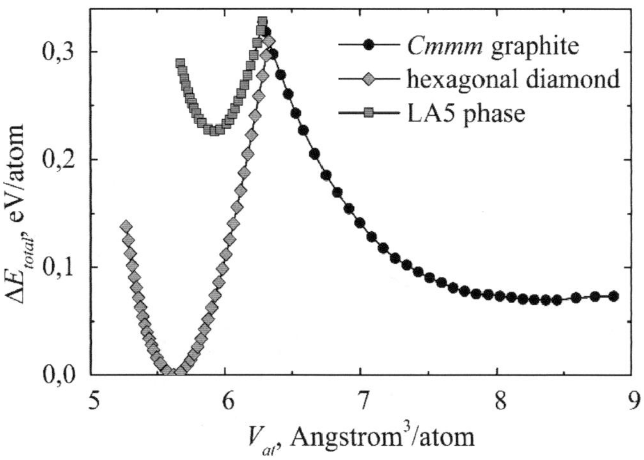

Figure 3. The relationship between total energy difference (ΔE_{total}) and atomic volume (V_{at}) for orthorhombic graphite, hexagonal diamond, and LA5.

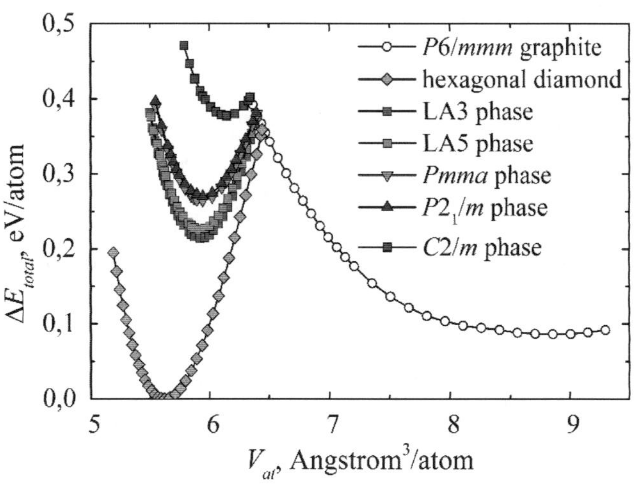

Figure 4. The relationship between total energy difference and atomic volume for *P6/mmm* graphite, hexagonal diamond, and diamond-like LA3, LA5, *Pmma*, $P2_1/m$, and *C2/m* phases.

The calculated enthalpy of the carbon phases is shown in Figures 5 and 6. The structural transition of *P6/mmm* graphite into hexagonal diamond is accompanied by a heat evolution of 0.43 eV per atom (Figure 5). The difference of enthalpies of the initial and final carbon phases for the formation of hexagonal diamond from *Cmmm* graphite amounts to $\Delta H = -0.69$ eV per atom (Figure 6).

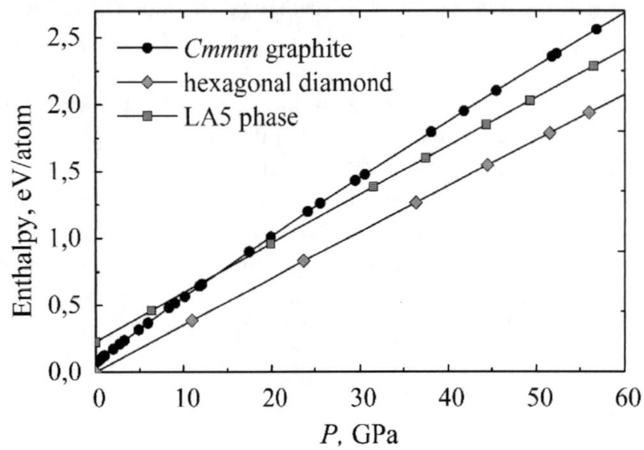

Figure 5. Dependence of the enthalpy (*H*) on the pressure (*P*) for orthorhombic graphite, hexagonal diamond, and LA5.

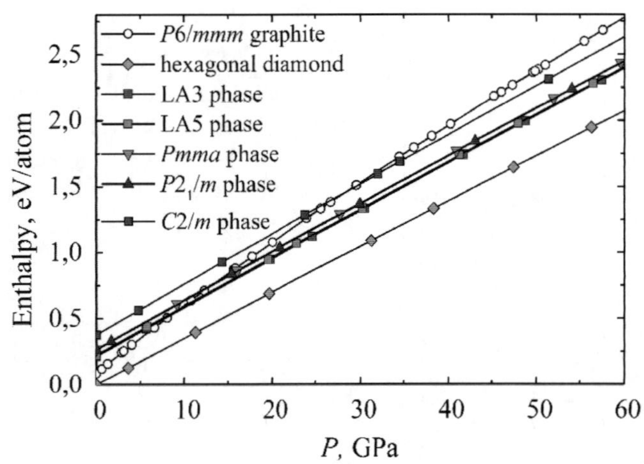

Figure 6. Dependence of the enthalpy on the pressure for *P6/mmm* graphite, hexagonal diamond, and diamond-like LA3, LA5, *Pmma*, *P*2$_1$/*m*, and *C*2/*m* phases.

4. Conclusions

It was established that lonsdaleite (hexagonal diamond) is stable at temperatures up to about 1470 Kelvin and can be synthesized from graphite as a result of compression at pressures of 56–61 GPa, which are close to those necessary for the formation of cubic diamond. It has been also found that the lonsdaleite can be obtained from other structural varieties of graphite at lower pressures necessary for the formation of most stable diamond-like phases [6].

Acknowledgments

V A Greshnyakov would like to thank the Russian Foundation for Basic Research for the financial support (project No. 16-33-00030 mol_a). The authors also thank Foundation of Perspective Research of Chelyabinsk State University for the financial support.

References

[1] Pan Z, Sun H, Zhang Yi, Chen C 2009 *Phys. Rev. Lett.* **102** 055503
[2] Greshnyakov V A, Belenkov E A 2016 *Tech. Phys.* **61** 1462
[3] Greshnyakov V A, Belenkov E A 2017 *J. Exp. Theor. Phys.* **124** 265
[4] Hohenberg P, Kohn W 1964 *Phys. Rev.* **136** 864
[5] Perdew J P, Zunger A 1981 *Phys. Rev. B* **23** 5048
[6] Belenkov E A, Greshnyakov V A 2015 *Phys. Solid State* **57** 205

IOP Conf. Series: Journal of Physics: Conf. Series **917** (2017) 032005 doi:10.1088/1742-6596/917/3/032005

Conductive indium nanowires deposited on silicon surface by dip-pen nanolithography

A S Kozhukhov[1, 2], D V Shcheglov[1], A V Latyshev[1]
[1]Rzhanov Institute of Semiconductor Physics, Siberian Branch of Russian Academy of Sciences, Novosibirsk 630090, Russia
[2] Novosibirsk State University, Novosibirsk 630090, Russia

Abstract. Conductive indium nanowires up to 50 nm in width and up to 10 μm in length are fabricated on the surface of silicon by local resputtering from the probe of an atomic-force microscope. To accomplish this, indium was transferred from an atomic force microscopy tip to the surface by applying a potential difference between the tip and substrate. The fabricated indium nanowires were several micrometers in length. Unlike thermal DPN, our DPN method hardly oxidized the indium, producing nanowires with conductivities from $5.7 \cdot 10^{-3}$ to $4 \cdot 10^{-2}$ Ω·cm.

1. Introduction

Dip-pen nanolithography (DPN)[1,2] is a method by which metals or organic compounds can be directly deposited on a substrate from an atomic force microscopy (AFM) tip. This technique is useful in nanoelectronics as a way to fabricate conducting nanowires[3,4] and in studying biomolecules and polymers in controlled locations[5-7]. The DPN method makes possible the formation of conductive structuresat semiconductor surfaces with a resolution of no worse than 50 nm. In thermal DPN, metal is transferred from an AFM probe onto the surface by heating the probe to the melting point of the loaded metal. Metals with low melting points such as indium, tin, and lead are required. When the metal on the probe tip is heated to its melting point, it transfers to the surface of the semiconductor by a metal–surface adhesion force. However, heating the metal oxidizes it, significantly reducing the conductivity of the fabricated nanostructures.[1] To reduce oxidation, in this paper, we deposited metal from an AFM probe by applying an electric potential between the tip and substrate.

2. Materials and methods

Indium (99.99%) was loaded onto the AFM probe tip by thermal vacuum spraying. To control the indium thickness, thermal spraying was performed through a mask with a hole of 6 μm diameter located 2 μm from the probe tip. To improve adhesion of indium to the probe, a layer of Ni and 30 nm of Cr were first deposited on the probe. Then, indium was deposited with a thickness from 80 to 200 nm (Figure 1a). Indium was deposited from the AFM tip onto the stepped surface of Si (111) with an interaction force of 10^{-7} N between the probe and the surface at room temperature. We fabricated structures with a vertical resolution of 0.2 nm on the smooth surface of Si (111)[8] covered with 15 nm of anodic oxide as an isolation layer, matching our earlier study of local anodic oxidation[9] caused by AFM. A positive potential of 10 V was applied to the AFM cantilever, during which the indium-coated probe was moved parallel to the surface in lines for 1–2 μm at 0.1 μm/s.

IOP Conf. Series: Journal of Physics: Conf. Series **917** (2017) 032005 doi:10.1088/1742-6596/917/3/032005

3. Results and discussion

3.1. Nanowire formation

Using the indium-coated AFM tip, we fabricated several indium nanowires on the stepped surface of Si(111). After fabricating the nanowires, we investigated a section of each sample using AFM, which had a sharp tip with the radius of curvature of less than 10 nm. Figure 1b shows an AFM image of indium nanowires on the Si(111) surface. Analyzing the relief profile of the fabricated nanowires, we found their width to be 50–70 nm and their height to be 0.8–1 nm.

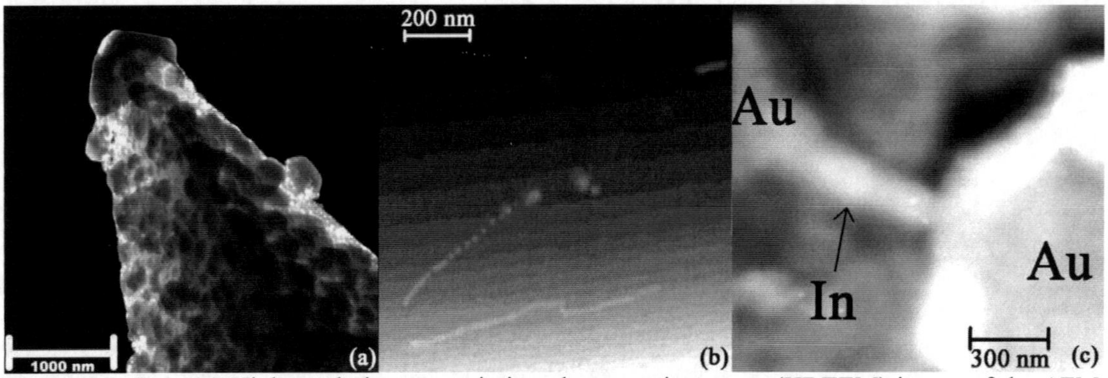

Figure 1 (a,b,c). (a) High-resolution transmission electron microscopy (HRTEM) image of the AFM probe coated with indium. The total volume of the coating is about 0.1 μm^3. **(a)**, **(b)** AFM topographic image of the surface of silicon with formed indium nanowires, **(c)** AFM image of area with connected contacts.

3.2. The mechanisms of nanowire formation

To establish the factors that influence the geometric dimensions of the resultant indium nanowires, during the transfer of indium, we varied the potential between the probe and the surface from 3 to 10 V and the force of interaction between the probe and the surface from 10^{-7} to $5 \cdot 10^{-6}$ N. It is found that the dependences of the height of the resultant structures on the applied potential between the probe and the surface and on the force of interaction between the probe and the surface are linear. This means that the nanowire height depends on both the applied potential and the force of interaction between the probe and the surface. It is conceived that the transfer of indium proceeds, on the one hand, due to a high-strength electric field (up to 10^8 V cm^{-1}) formed in the region of the tip edge and, on the other hand, due to current flow between the substrate and the tip, with the local heating and melting of indium in the current channel. In this case, the application of the voltage results in the reproducible resputtering of about 1% indium from the tip to the surface. At the same time, if the potential is not applied, the transfer of indium to the surface stops immediately after lubrication of the edge region (0.01%); in this case, the process sometimes does not occur, i.e., is irreproducible. The amount of transferred indium (~1% of the total coverage) can be attributed to localization of the high field at the very edge of the tip. 1% portion of the material is located within a region ~0.3 μm from the center. On the assumption that resputtering is initiated by an electric field, the critical-field strength at such a distance from the tip is $\sim 3 \cdot 10^5$ V cm^{-1}. Such a field is not high enough for molecules to dissociate. However, if the tip is pressed to the substrate, this field is high enough to activate the surface drift of indium from the tip to the substrate surface. In general, the physical nature of the transfer of indium from the tip to the surface in a high field with a current invites further investigation.

3.3. Nanowire electrical resistivity

To study the electrical properties of the resultant indium films, we fabricated two separated 10 nm thick gold contacts on an insulating substrate. The spacing between the contacts was no larger than 1–

2 μm. Then, using the method described above, we deposited indium onto the substrate region between the gold contacts (Figure 1c). As a result, the contacts were connected by an indium strip with a width of 0.3–1.3 μm and a height of 10–260 nm. Measurements show that the resistivity of the deposited indium corresponds to $5.7\cdot10^{-3}$–$4\cdot10^{-2}$ Ω·cm.

4. Conclusion

In summary, we developed a new DPN technique that allows one to fabricate indium nanowires with diameters of 30–50 nm on semiconductors and insulators. Unlike thermal DPN, our DPN technique does not require the AFM tip to be heated to indium's melting point. Indium transport was possible because of local heating generated by electrical current from the AFM probe tip to the surface under positive potential. The main advantage of this technique is the high conductivity of the produced indium nanowires because of decreased indium oxidation. The measured indium electrical resistivity varied from $5.7\cdot10^{-3}$ to $4\cdot10^{-2}$ Ω·cm.

Acknowledgments

This work was supported by the Russian Science Foundation, project no. 14-22-00143.

References

[1] B A Nelson, W P King, A R Laracuente, P E Sheehan, and L J Whitman, Direct deposition of continuous metal nanostructures by thermal dip-pen Nanolithography, Applied Physics Letters **88**, 033104, 2006.

[2] Wechung Maria Wang, Melburne C LeMieux, Selvapraba Selvarasah, Mehmet R Dokmeci and Zhenan Bao, Dip-Pen Nanolithography of Electrical Contacts to Single-Walled Carbon Nanotubes, ACS Nano, 2009, **3** (11), pp 3543–3551.

[3] Woo-Kyung Lee, Minchul Yang, Arnaldo R Laracuente, William P King, Lloyd J Whitman and Paul E Sheehan, Direct-write polymer nanolithography in ultra-high vacuum, , Beilstein J. Nanotechnol, 2012, **3**, 52–56.

[4] Jiang H Z, Stupp S I, Dip-pen patterning and surface assembly of peptide amphiphiles. Langmuir **21**, 5242–5246 (2005).

[5] Lim, J H et al. Direct-write dip-pen nanolithography of proteins on modified silicon oxide surfaces. Angew. Chem. Int. Edn **42**, 2309–2312 (2003).

[6] Weeks B L, Noy A, Miller A E, De Yoreo J J, Effect of dissolution kinetics on feature size in dip-pen nanolithography, Phys. Rev. Lett. **88**, 255505 (2002).

[7] David Bullen, Chang Liu, Electrostatically actuated dip pen nanolithography probe arrays, Sensors and Actuators A, **125** (2006) 504–511.

[8] A V Latyshev, L V Litvin, A L Aseev, Peculiarities of step bunching on Si(001) surface induced by DC heating, Applied Surface Science, Volumes **130–132**, 1998, Pages 139–145.

[9] D V Sheglov, A V Prozorov and et al, Physics of Low-Dimensional Structures, 2002, **5/6**, 239-246.

IOP Conf. Series: Journal of Physics: Conf. Series **917** (2017) 032006 doi:10.1088/1742-6596/917/3/032006

Influence of annealing temperature on the upconversion luminescence properties of $NaYF_4$:Er,Yb@SiO_2 particles

E A Sagaydachnaya[1], **V I Kochubey**[1, 2] and **Ju G Konyukhova**[1]
[1]Saratov National Research State University, Russian Federation
[2]National Research Tomsk State University, Russian Federation

Abstract. Upconversion phosphors are promising materials for a medical application. Investigation of influence of annealing temperature of particles on the upconversion luminescence properties is relevant. The synthesized $NaYF_4$:Er,Yb@SiO_2 particles were annealed at different temperatures in the range of 300 °C to 550 °C. Dependence of the luminescence intensity on the annealing temperature takes the maximum value at 550 °C and has local maximum at 400 °C.

1. Introduction

Innovative luminescence materials arouse substantial interest due to the extensive perspective, which they open in different areas of science and technology. Particularly, upconversion $NaYF_4$:Er,Yb phosphors are promising materials for the photodynamic therapy, creation of temperature micro-nanosensors and bioimaging of cells and tissue [1]. The effects of the reagent concentration molar ratio, temperature and time of hydrothermal synthesis on the morphology, size and luminescence properties of the particles is actively investigated [2, 3, 4, 5]. Point out that creation of the heterogeneous core/shell structure promotes the enhancing upconversion luminescence of the particles [6]. In addition, important factor influenced to the effective of upconversion is the annealing temperature of obtained particles. The aim of this work is investigation of influence of annealing temperature of $NaYF_4$:Er,Yb@SiO_2 particles on the upconversion luminescence properties.

2. Experimental Section

In a typical hydrothermal synthesis, 2 mmol trisodium citrate (Na_3Cit) and 114 mmol sodium chloride (NaCl) were firstly dissolved in 16,7 mL H_2O with magnetic stirring. Solution of $YCl_3{\cdot}6H_2O$ (0,5 M), $Yb(CH_3COO)_3{\cdot}4H_2O$ (0,5 M) and $ErCl_3{\cdot}5H_2O$ (1 M) (molar ratio of rare earth Y/Yb/Er = 0,5/0,33/0,05) was added slowly into the above solution to form the metal-citrate complex. After vigorous stirring for 30 min, 20 mL aqueous solution of NaF (1 M) was introduced. After additional stirring for 60 min, the as-obtained mixing solution was transferred into a teflon bottle held in a stainless steel autoclave and maintained at 200°C for 20 h. After the autoclave was cooled to room temperature naturally, the resulting precipitates were separated by centrifugation and washed with weak sodium citrate solution (5g/L) three times. Obtained NaYF4:Yb^{3+}/Er^{3+} particles were dispersed in 36 mL of isopropanol. Then 1,0 mL of ammonia aqueous solution (12 wt %) and 80 μL of TEOS (99 wt %) were added to the solution. After stirring at room temperature for 1 h, the obtained particles were separated, washed with distilled water three times and dried in air at 75°C for 12 h. The as-prepared $NaYF_4$:Yb,Er@SiO_2 particles were annealed at different temperatures in the range of 300 °C to 550 °C.

The structure, morphology and size of obtained $NaYF_4$:Er,Yb@SiO_2 particles were characterized by XRD, thermogravimetric analysis, scanning electron microscopy (SEM) and method of dynamic light scattering. Upconversion luminescence of Er^{3+} ions under irradiation of the particles by NIR-laser (980 nm, 2,8 W/cm^2) was observed using a spectrometer (Ocean Optics QE6500 FL, USA).

3. Results and discussion

The XRD-patterns indicate that the lattice of the synthesized particles is hexagonal with the residual cubic phase. In addition, the half-width of reflexes decreases and the intensity increases with annealing. It is worth noting that the XRD-patterns of the annealed particles demonstrate larger crystallite sizes what causes increasing the luminescence intensity. From the data of dynamic light scattering, the average diameter of the not annealed and annealed particles is about 0,3 μm and 1 μm, respectively. Upon that two phase of the not annealed particles with diameters about 5 μm and 300 nm are observed in the SEM image.

The luminescence spectra of the particles annealed at the different temperature are presented in Figure 1. It should be note that luminescence spectra of $NaYF_4$:Yb,Er@SiO_2 particles are sensitive to their temperature. Exciting IR-radiation heats the particles in the absence of an external temperature support system. For example, the particles are heated to 80°C at a power density of 2,8 W/cm^2. Figure 1 demonstrates spectra of heated particles by the laser to the equilibrium temperature.

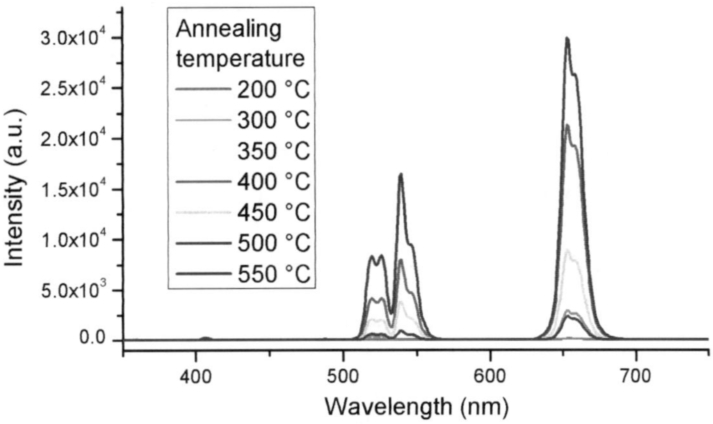

Figure 1. Luminescence spectra of the particles.

In order to easily compare upconversion between samples, the area under the peaks was calculated over the wavelengths within each of the bands. Integral values of luminescence were obtained for the three emission bands corresponding to the energy transition: $^4H_{11/2}\rightarrow{}^4I_{15/2}$ (525 nm), $^4S_{3/2}\rightarrow{}^4I_{15/2}$ (543 nm) and $^4F_{9/2}\rightarrow{}^4I_{15/2}$ (655 nm). The obtained dependences of luminescence intensity on the annealing temperature are presented in Figure 2 (right ordinate axis). The ratio of the intensities of 'red' and 'green' emission bands ('red/green' ratio) as a function of annealing temperature is also presented in Figure 2 (left ordinate axis). The intensities at 200 °C (the temperature of hydrothermal synthesis) correspond with not annealed particles.

The luminescence intensity is observed to increase as the annealing temperature is raised up to 400 °C. This is because impurities desorb from the particle surface. According to data of thermogravimetric analysis, the bound water is removed by heating. When the temperature is raised above 400 °C, luminescence intensity decreases. This can be caused by the decomposition of the stabilizing shell of sodium citrate which melts at a temperature above 300 °C and decomposes without evaporation. Recrystallization takes place at a temperature above 500 °C what provides an increase in the luminescence intensity. It should be noted that the sintering of the particles annealed at 550 °C is observed.

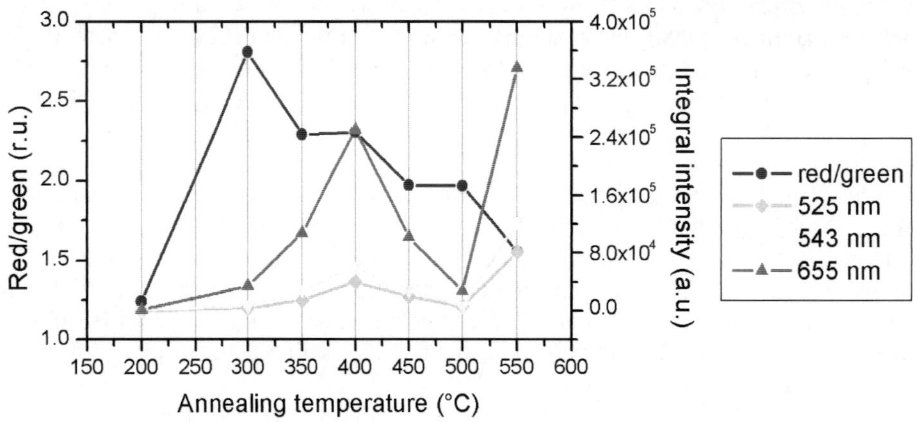

Figure 2. Luminescence intensity (right ordinate axis) and 'red/green' ratio (left ordinate axis) versus annealing temperature.

According to the obtained data, the 'red/green' ratio has peak value at the annealing temperature 300°C. It decreases with increasing the annealing temperature to 550°C. The change of the 'red/green' ratio is consistent with the data [7, 8].This is the results of the rate reducing of thermal relaxation which in turn can be the consequence of reducing the number of defects and increasing the crystallites size.

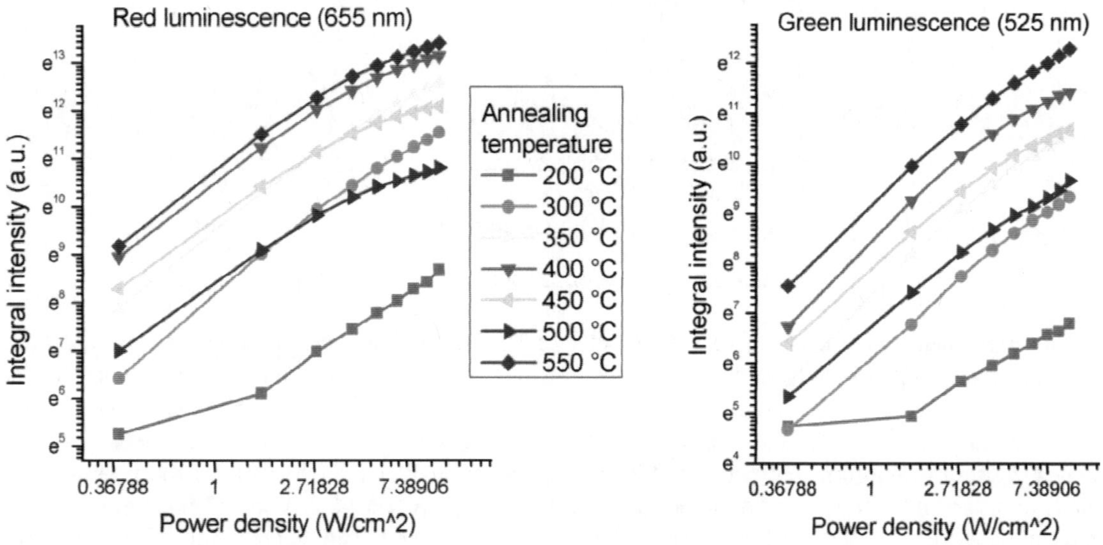

Figure 3. Log–log plot of upconversion emissions of green and red luminescence versus excitation power density of annealed at different temperatures $NaYF_4$:Er,Yb@SiO_2 particles.

Upconversion emission of green and red luminescence versus excitation power density is presented in Figure 3. The experimentally observed saturation of the upconversion-induced luminescence with increasing excitation is consistent with the existing model [9]. In the low-power limit an n-photon process will characterized by the slope of n when the luminescence intensity is plotted in a double-logarithmic representation versus absorbed pump power. In the high-power limit slope of all processes is less than or equal to 1 depending on the prevailing mechanism of upconversion (energy transfer upconversion or exited-state absorption). It is worth mentioning that the measurements carried out in this research correspond to the low and medium pump power. The slopes of the approximating lines in log-log scale at this pump powers are given in the Table 1.

Table 1. The slope at low and medium pump powers.

Annealing temperature	Red luminescence (655 nm)		Green luminescence (543 nm)		Green luminescence (525 nm)	
	low pump power	medium pump power	low pump power	medium pump power	low pump power	medium pump power
200°C	0.4	0.9	0.3	0.2	0.8	1.5
300°C	1.5	1.1	1.4	0.5	1.8	1.2
350°C	1.8	1.1	1.6	0.5	1.7	1.0
400°C	1.7	0.8	1.4	0.2	1.6	0.7
450°C	1.6	0.8	1.2	0.1	1.5	0.5
500°C	1.5	1.2	1.1	0.4	1.4	0.6
550°C	1.7	1.1	1.3	0.5	1.6	0.7

Red and green luminescence as known is a result of two-photon excitation. Deviation of the slope from two with increasing the pump power is conditioned by the complex interplay between the relevant energy transfer processes, exited-state absorption, upconversion from the intermediate state, cross-relaxation, radiative and nonradiative decay. The obtained values of slope for the green luminescence at the different pump powers are consistent with simplify model (when only energy transfer upconversion on the acceptor ion is included), whilst slope for the red luminescence correspond to the more complex model described in [9]. However, keeping in mind that the particles were investigated in the range of low and medium pump power, obtained results may indicate different values of the high-power limit for different luminescence bands.

4. Conclusion

In results it can be concluded that complicated changes of luminescence intensity occur due to the existence of several processes such as removing bound water, decomposition of the sodium citrate sell and recrystallization. More effective conversion of NIR excitation to visible light energy occurs at annealing temperatures at 400 °C and 550 °C. However, after annealing at 550 °C in contrast to the annealing at 400 °C the particles are sintered. At the same time, the conversion efficiency of the exciting radiation to green luminescence increases with the increasing of the annealing temperature.

Acknowledgments

This work was supported by Russian Presidential under Grant No. NSh-7898.2016.2

References

[1] Chatterjee D K, Fong L S and Zhang Y 2008 *Advanced Drug Delivery* **60** 1627–37
[2] Zhao J, Sun Y, Kong X, Tian L, Wang Y, Tu L, Zhao J and Zhang H 2008 *J. Phys. Chem. B* **112** 15666–72
[3] Sun Y, Chen Y, Tian L, Yu Y, Kong X, Zhao J and Zhang H 2007 *Nanotechnology* **18** 275609 (9pp)
[4] Li C, Quan Z, Yang J, Yang P and Lin J 2007 *Inorg. Chem.* **46** 6329-37
[5] Shan J and Ju Y 2007 *Applied Physics Letters* **91** 123103
[6] Chen G, Qiu H, Prasad P N and Chen X 2014 *Chem. Rev.* **114** 5161–214
[7] Dyck N C, van Veggel F C J M and Demopoulos G P 2013 *ACS Appl. Mater. Interfaces* **5** 11661–67
[8] Assaaoudi H, Shan G, Nathan Dyck N and Demopoulos G P 2013 *Cryst. Eng. Comm.* 2013 **15** 4739–46

IOP Conf. Series: Journal of Physics: Conf. Series **917** (2017) 032006 doi:10.1088/1742-6596/917/3/032006

[9] Suyver J F, Aebischer A, García-Revilla S, Gerner P and Güdel H U 2005 *Physical review B* **71** 125123

IOP Conf. Series: Journal of Physics: Conf. Series **917** (2017) 032007 doi:10.1088/1742-6596/917/3/032007

Novel approach for III-N on Si (111) templates fabrication by low-temperature PA MBE using porous Si layer

Zolotukhin D*[1], Seredin P[1], Lenshin A[1], Goloshchapov D[1,3], Mizerov A[2]

[1]Voronezh State University, Universitetskaya pl. 1, Voronezh, 394018, Russia
[2] St. Petersburg Academic University, Khlopina 8/3, 194021 St. Petersburg, Russia
[3]Voronezh State Technical University, Voronezh, Russia

Abstract. We report on successful growth of GaN nanorods by low-temperature plasma-assisted molecular beam epitaxy on a Si(111) substrate with and without preformed thin porous Si layer (*por-Si*). The deposited GaN initially forms islands which act as a seed for the wires. Porous structure of the por-Si layer helps to control nucleation islands sizes and achieve homogeneous distribution of the nanorods diameters. In addition 850 nm-thick crack-free GaN layer was formed on Si(111) substrate with preformed por-Si layer.

1. Introduction

For decades, there have been numerous attempts to solve a well-known problem of III-nitride material family, i.e., defects and dislocations due to a lack of proper substrates. Nanorods have attracted much interest in recent time because of their potential for the fabrication of electronic and optoelectronic devices [1-3]. Moreover, nanorods properties and presence of large free surface helps to suppress propagation of the threading dislocations and stress generation [4] arising from lattice and thermal expansion coefficient mismatches (17% and 54% for GaN on Si(111), respectively [5]). Conventionally nanorods are formed by using complex approaches such as selective growth, Vapor-Liquid-Solid (VLS) technique, etc. [6]. However, these approaches involve the usage of complex technological processes, which significantly increased the cost of the resulting device. The solutions for this problem can be the use of cheap compliant Si substrates with a por-Si layer. Self-organized array of nanopores in a por-Si layer leads to the formation of GaN nucleation islands with the same sizes, which in turn leads to the formation of nanorods with the same diameters

In this paper we report on fabrication of self-organized nanorods with the same diameters on a compliant Si-substrate by a low temperature plasma-assisted MBE (LT PA MBE). In addition 850 nm thick crack-free GaN/por-Si/Si(111) template was formed to demonstrate a possibility of tensile stress generation suppression using a por-Si layer.

2. Experimental

All Si substrates were pre-treated using the Shiraki method [7]. Then a 20 nm-thick por-Si layer was formed on several substrates using the original method of selective etching. After that all the substrates was annealed at 820°C for an hour until the 7×7 surface reconstruction was observed by RHEED. All growth processes were performed by using Veeco Gen 200 PA MBE setup. For comparison nanorods were grown at the one growth process on conventional Si(111) substrate and substrate with a por-Si layer (sample **a** and **b**, respectively) without nitridation step. LT GaN nanorods were formed at

F_{Ga}=F_N=0.1µm/h flux ratio and the substrate temperature of T_s=400°C. Prior to sample **c** growth, the Si(111) substrate with a por-Si layer underwent a high-temperature (T_s=850°C) nitridation step for one hour at nitrogen plasma source forward power of 500 W and reflected power 0 Wr, with a nitrogen flow rate of 3.7 sccm. After that the 10-nm-thick GaN buffer layer was grown at T_s=650°C and F_{Ga}/F_N=1. Then 840 nm of HT GaN was formed at higher values of T_s=700°C and F_{Ga}=0.6µm/h (at fixed F_N=0.1µm/h). RHEED was used for *in situ* control of the surface morphology of the layers. The surface morphology also was investigated *ex situ* by an atomic force microscope (AFM) scanning electron microscopy (SEM) and optical microscope (OM).

3. Results

Cross section SEM images of all heterostructures (Figure 1) illustrate rather planar morphology for all the interfaces. One can also observe a more pronounced nanocolumnar structure in case of sample **b**.

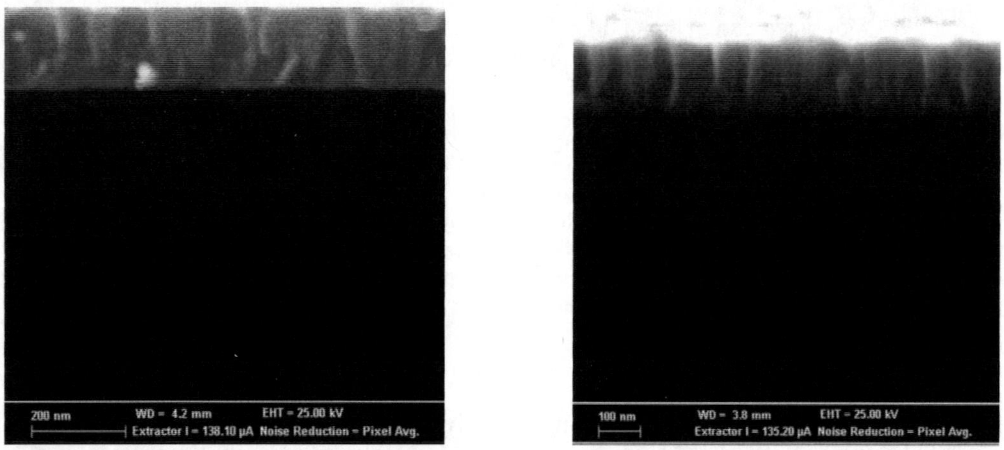

Figure 1 Cross-section SEM images of the GaN/Si(111) (left) and GaN/por-Si/Si (right) nanorods.

In addition, SEM image of GaN/por-Si/Si heterostructure revealed an increase in the distances between nanorods as well as in homogeneity of the distribution of their diameters.

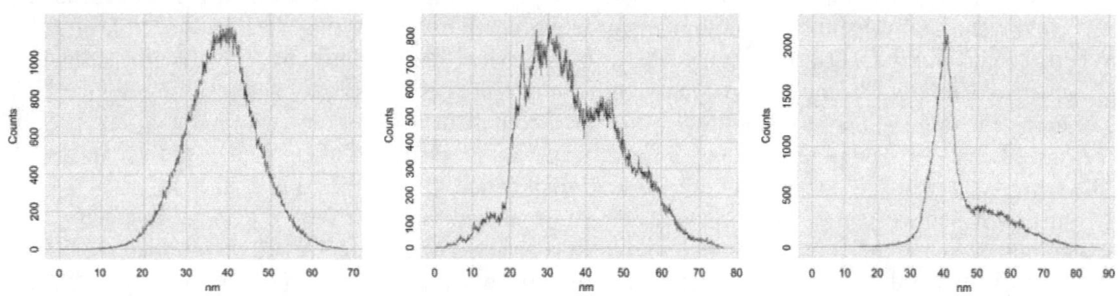

Figure 2. The lateral sizes distribution histogram obtained from AFM study of por-Si layer (left), GaN/Si(111) layer (centre) and GaN/por-Si/Si(111) layer (right).

The lateral sizes of the distribution histogram obtained from the AFM study of the substrate and samples surfaces (Figure 2) shows that the use of a por-Si layer has a crucial influence on the distribution of the diameters of the nanorods.

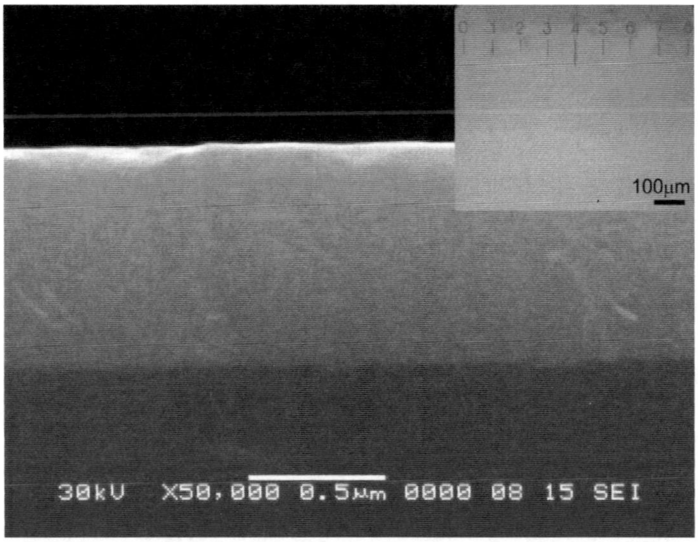

Figure 3. Cross-section SEM and OM image of GaN/por-Si/Si(111) heterostructure.

As can be seen from Figure 3, we achieved a flat GaN/Si interface in GaN(850 nm)/por-Si/Si(111) heterostructure without pronounced interaction. Moreover, OM revealed a crack-free surface, which corresponds to a low level of tensile stress at room temperature.

4. Discussion

As shown in Nechayev *et. al.* paper [8], growth temperature significantly affects on nucleation islands sizes due to increase of surface adatom mobility. The higher surface mobility led to greater average nucleation islands sizes. However, it is necessary to take into account that we are talking only about the average size, while the real distribution of nucleation islands sizes remains fairly chaotic. Lee S *et. al.* [9] reported that the most energetically favorable crystallographic plane for GaN growth on silicon is the {111} plane. We believe that etching of Si(111) substrate led to similar as described in [10] surface morphology with isotropically distributed etching hillocks. In the case of a thin (20 nm) porous layer, the etching hillocks becomes truncated with {111} plane at the top and roughly same diameter. Thus, GaN was nucleated on the tops of isotropically distributed hillocks with a further increase in grain sizes until nucleation island completely covers the entire surface of the truncated hillock. This fact can account for better uniformity of the distribution of the diameters of the nanorods.

The cross section SEM image of sample **c** revealed flat GaN-Si interface without well known Ga-Si interaction which related with Ga-Si phase diagram [11]. A long nitridation step led to Si_xN_y layer formation which acts as a mask and prevent Ga-Si interaction was formed on Si(111) substrate surface GaN buffer layer with the thickness of 10 nm in sample **c**. The stress arising from thermal expansion coefficient difference of $\Delta\alpha_{Si-GaN}$=2.9×10^{-6} K^{-1} [12] between GaN and Si can be estimated using formula (1)

$$\sigma_{th} = M_{GaN} \int_{RT}^{T_s} \Delta\alpha_{Si-GaN} dT \qquad (1)$$

where M_{GaN} is the GaN biaxial modulus. Using M_{GaN}=478 GPa [13] and T_s=700° the tensile stress can be evaluated as σ_{th}=970 MPa. As described in [14], for films on substrates with different elastic properties, this critical thickness for channeling steady-state propagation parallel to the interface by cracks that extend to the film-substrate interface is given by the relation (2)

$$h_{critical} = \Gamma \overline{E} / Z\sigma^2 \qquad (2)$$

where Γ is the fracture resistance of the material, $\overline{E}$ is the plane strain elastic modulus $\overline{E} = E/(1-\nu)$ (using elastic constants from [15] can be evaluated as 379 GPa for GaN), $Z = g(\alpha,\beta) * \pi / 2$. If we assume that GaN layer grown is fully unstrained, then using the values listed in [14] and σ_{th}=970 MPa, the critical thickness can be evaluated as $h_{critical}$=804 nm. Nevertheless, the OM study of GaN (850nm)/por-Si/Si(111) heterostructure (Figure 3) revealed crack-free surface, despite a lower calculated critical thickness value. We believe that this excess of the critical thickness is related with stresses relaxation at the por-Si / GaN interface.

Summarizing, by using compliant substrates with a porous Si layer and different PA MBE regimes, we have obtained self-organized array of nanorods with approximately the same diameters as well as a crack-free thick GaN layer on Si(111) substrate. The use of low growth temperature for nanorods growth and Si_xN_y mask layer for GaN layer growth helps to suppress Ga-Si interaction and prevent deterioration of the Si-GaN interface. In addition, we confirmed a positive role of a porous Si layer for relaxation of the tensile stress. Thus, complaint Si(111) substrates with a porous Si layer seems to be suitable for GaN based device fabrication by PA MBE.

Acknowledgments
The research as part of the development of nanoheterostructures as well as monitoring and investigation of the fundamental properties of samples is conducted with the support of grant of President of Russian Federation MD-188.2017.2 and grant of the Russian Ministry of Education as part of the state's task for higher-education institutions for scientific research 2017-2019 11.4718.2017/BCH. This growth experiments were supported financially by the Ministry of Education and Science of the Russian Federation № 16.9789.2017/BCH.

References
[1] Cui Y, Lieber C 2001 Science **291**, 851.
[2] Qian F *et al.* 2005 Nano Lett. **5**, 2287.
[3] Greytak A *et al.* 2005 Appl. Phys. Lett. **87**, 151103.
[4] Kishino K, Ishizawa S 2015 Nanotechnology **26,** 225602.
[5] Rahman M *et al.* 2016 Appl. Surf. Sci. **362** 572.
[6] Dubrovskii V, Cirlin G, Ustinov V 2009 Semiconductors **43**, 1539.
[7] Lenshin A *et al*. 2014 Technical Physics **59**, No. 2, 224.
[8] Nechaev D *et. al.* 2013 J. Crys. Growth **378** 319.
[9] Lee S *et. al.* 2004 Appl. Phys. Lett., **84**, 2079
[10] Houbertz K *et. al.* 1994 J. Vac. Sci. Technol. **12**, 3145
[11] Olesinski R, Kanani N, Abbaschian 1985 G Bull. Alloy Phase Diagr. **6**, 362
[12] Y.Cordier *et.al.* 2009 J. Crys. Growth **311**, 2002
[13] A. Ubukata, K. Ikenaga, N. Akutsu, A. Yamaguchi, K. Matsumoto, T. Yamazaki, T. Egawa 2007 J. Cryst. Growth **298**, 198.
[14] Raghavan S, Redwing J 2005 J. Appl. Phys. **98**, 023514
[15] Vurgaftman I, Meyer J 2003 J. Appl. Phys. **94**, 3675

IOP Conf. Series: Journal of Physics: Conf. Series **917** (2017) 032008 doi:10.1088/1742-6596/917/3/032008

Germanium layers grown by zone thermal crystallization from a discrete liquid source

A N Yatsenko[1], S N Chebotarev[1,2], V N Lozovskii[1], A A A Mohamed[1], G A Erimeev[1], L M Goncharova[1], A A Varnavskaya[1]

[1]Department of Physics and Electronics, Platov South-Russian State Polytechnic University (NPI), Novocherkassk 346421, Russia

[2]Laboratory of Crystals and Structures for Solid State Electronics, Southern Scientific Center RAS, Rostov-on-Don, 344006, Russia

Abstract. It is proposed and investigated a method for growing thin uniform germanium layers onto large silicon substrates. The technique uses the hexagonally arranged local sources filled with liquid germanium. Germanium evaporates on very close substrate and in these conditions the residual gases vapor pressure highly reduces. It is shown that to achieve uniformity of the deposited layer better than 97% the critical thickness of the vacuum zone must be equal to l_{cr} = 1.2 mm for a hexagonal arranged system of round local sources with the radius of r = 0.75 mm and the distance between the sources of h = 0.5 mm.

1. Introduction

One of the main tendencies in modern semiconductor technology is development of growth methods to produce thin layers and nanoheterostructures [1]. Molecular-beam epitaxy is the most suitable technique to fabricate such structures [2]. Despite of the high quality of the grown structures, there are some problems which make this method not suitable for mass production [3]. These lacks are small transfer coefficients of main substances and impurities, obtaining difficulties of uniform thickness layers on large substrates. Therefore, it is important to development another alternative growth methods such as ion-beam crystallization [4-6], ion-beam sputtering [7-9], laser beam crystallization [10] and zone sublimation regrowth [11]. The goal of this work is to development of zone thermal recrystallization method using a germanium discrete source for producing uniform thickness layers on large substrates.

2. Calculation

The main idea of our experimental set up is shown in figure 1. The local sources are arranged hexagonally, where $2r$ is the diameter of the local sources, h is the minimal distance between the neighboring local sources, T_1 and T_2 are temperature of the evaporator and the substrate respectively, d is the thickness of the deposited layer. The filling factor γ is one of the important characteristics. The filling factor is the ratio of the area of all local sources to the area of the wafer where the evaporators are located. The value of γ indicates how many times the flow of the evaporating material is weakened if the discrete source is used instead of a continuous one. We used a discrete source with hexagonal arranged local evaporators and fill factor for this case is defined as $\gamma = \dfrac{2\pi r^2}{\sqrt{3}(2r+h)^2}$.

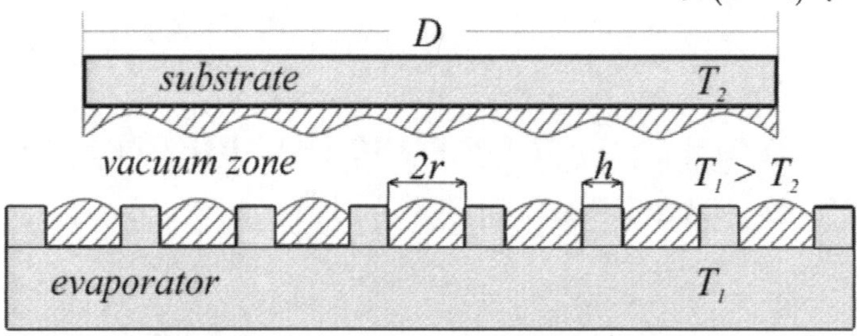

Figure 1. Discrete source-substrate system

Deposition probability ω for a molecule evaporated from a circular local source to a point of the substrate, is defined as

$$\omega(\rho) = \frac{1}{2}\left(1 - \frac{l^2 - r^2 + \rho^2}{\sqrt{(l^2 - r^2 + \rho^2)^2 + 4r^2 l^2}}\right), \qquad (1)$$

where ρ is the distance from the center of the projection of the local source on the substrate to the point on the substrate. The layer thickness depends on the technological and geometrical parameters and is described by the equation:

$$d(x,\varphi) = v_{ucn} \cdot t \left\{ \omega(\rho_0) + \sum_{n=1.}^{k} \sum_{m=0}^{n-1} \sum_{z=0}^{5} \omega(\rho_i) \right\} \qquad (2)$$

,

where $\rho_0 = x$, $\rho_i^2 = x^2 + L^2(n^2 - nm + m^2) - 2xL\sqrt{n^2 - nm + m^2} \cdot \cos(arctg(\frac{\sqrt{3}m}{2n-m}) + z\frac{\pi}{3} - \varphi)$.

The position of i-th local source in the hexagonal configuration is characterized by three coordinates n, m and z. Here n is the serial number of the hexagon, z - serial number of the side of a hexagon, m is the serial number of the source located on the side of the hexagon. The value k is determined by the diameter D of the discrete source. For the hexagonal discrete source k is equal to 3. The central local source described by the first separate term in equation (2). The dependence of the layer thickness on temperature-time mode determined by the factor $v_{ev} \cdot t$, where v_{ev} - the rate of evaporation from the surface of each local sources, t is the time of evaporation. In the derivation of equation (2) it is assumed the additivity of molecular flows from the local sources. It does not take into account that the surface diffusion and the atoms multipath from the source and substrate. These factors can lead to a weakening of the modulation of thickness deposited layer $d(x,\varphi)$. The position of the point, in which is calculated d, is given by polar coordinates x and φ. To determine the non-uniformity of the grown layer is sufficient to consider the direction, which is given by formula $\varphi = \pi/6 + q\pi/3$, where q is an integer.

3. Experiment and discussion

Experimental verification of the possibility of obtaining uniform thickness layers on large substrates was carried out using germanium as a model material. The evaporator is made of chemical pure finely porous graphite. The system of the local sources with hexagonal geometry was drilled in the evaporator. Graphite vapor pressure at the operating temperature of the deposition process is not more than 10^{-9} Pa. Graphite is essentially insoluble in germanium and quite easily handled mechanically. After machining the graphite wafer was hardened by the pyrolytic carbon layer. Then this hardened wafer was thermal cleaned in a vacuum. The 100 mm standard polished silicon wafers with crystallographic orientation (100) is used as substrates. Further purification was achieved by sublimation etching of the substrate at the temperature of 1150 °C in a vacuum. The drilled local sources were loaded with germanium powder. Then the loaded germanium powder was melted. Note that the vapor pressure of liquid germanium during the deposition of the layer in the temperature range 650 to 920 °C is less than 10^{-2} Pa. In this experimental conditions $l/R < 10^{-3}$, $l/\lambda \approx 10^{-4}$. Scanning electron microscopy were used to investigate the thickness of the grown layers using position mark technique [12].

The non-uniformity of the deposited layer is calculated by the formula:

$$\delta = 1 - d_{\min} / d_{\max} , \qquad (3)$$

The calculated (curves) and experimental (crosses) results of δ for the different parameters of r and l are shown in Figure 2. The curves 1-3 is for the center of the substrate and the curves 4-6 is for the edge of the substrate.

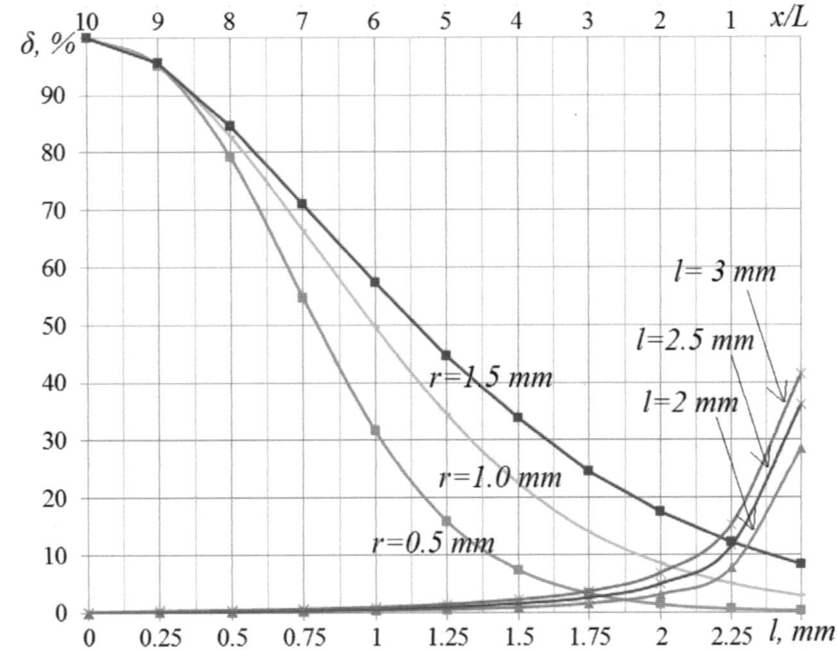

Figure 2. The dependence of non-uniformity thickness on the distance between the source and the substrate l at different radius of the local sources r

The ordinate axis is expressed in term of the relative layer thickness $d(l)$. Increasing the parameter l at $x = 0$ leads to a decrease in the layer thickness $d_1(l)$. Otherwise, increasing the value of l at $x = L/2$ leads to a increase in the layer thickness $d_2(l)$. At the point $l = 2.5$ mm, the values of both functions $d_1(l)$ and $d_2(l)$ are equal. Under these conditions the non-uniformity for local sources with radius $r = 1$ mm is less than 0.03. In other words, the deposited layers becomes uniform.

Thereby a model of zone thermal recrystallization of germanium on silicon from a discrete liquid-phase source has been developed and an experimental verification of theoretical calculations has been carried out. It is established that an increase in the thickness of the growth zone l most significantly affects the inhomogeneity of the grown layer. It is shown that at thicknesses less than $l < 1$ mm the inhomogeneity of the layer is more than 30% for all technologically realizable radiuses of local evaporators and the distances between them. In the general case an increase in the distance between the local evaporators h leads to a deterioration in the homogeneity of the layer, but essentially depends on the distance between the local evaporators r. An increase in the radius r also increases the heterogeneity of the growing layers.

4. Conclusions

The method of zone thermal recrystallization from a discrete source to obtain uniform thin semiconductor layers was proposed. The method uses the system of geometrical arranged local sources filled with liquid germanium. It is shown that to achieve uniformity of better than 97% the critical thickness must be equal to $l_{cr} = 1.2$ mm for a hexagonal arranged system of round local sources with the radius of $r = 0.75$ mm and the distance between sources of $h = 0.5$ mm.

Acknowledgements

The authors would like to thank for financially supporting this work the Russian Foundation for Basic Research [grant number 15-08-08263_a]. The reported study was also funded by the Grant of the President of Russian Federation for young PhD [project MK-5115.2016.8].

References

[1] Zhukov A E, Maximov M V, Savelyev A V, Shernyakov Y M, Zubov F I, Korenev V V, Martinez A, Ramdane A, Provost J-G, Livshits D A 2013 *J. Appl. Phys.* **113** 233103
[2] Nikiforov A I, Timofeev V F, Pchelyakov O P 2015 *Appl. Surf. Sci.* **354** 450
[3] Shengurov V G, Denisov S A, Chalkov V Y, Buzynin Y N, Drozdov M N, Buzynin A N, Yunin P A 2015 *Tech. Phys. Lett.* **41** 36
[4] Chebotarev S N, Pashchenko A S, Williamson A, Lunin L S, Irkha V A, Gamidov V A 2015 *Tech. Phys. Lett.* **41** 661
[5] Lunin L S, Chebotarev S N, Pashchenko A S, Dudnikov S A 2013 *Journal of Surface Investigation* **7** 36
[6] Lunin L S, Sysoev I A, Alfimova D L, Chebotarev S N, Pashchenko A S 2011 *Inorganic Materials.* **47** 816
[7] Lunin L S., Chebotarev S N, Pashchenko A S, Bolobanova L N 2012 *Inorganic Materials* **48** 439
[8] Chebotarev S N, Pashchenko A S, Lunin L S, Zhivotova E N, Erimeev G A, Lunina M L 2017 *Beilstein Journal of Nanotechnology* **8** 12
[9] Chebotarev S N, Pashchenko A S, Lunin L S, Irkha V A 2016 *Nanotechnologies in Russia* **11** 435
[10] Balandin V Yu, Aleksandrov L N, Dvurechenskii A V, Kulyasova O A 1994 *Phys. Stat. Sol. (a)* **142** 99
[11] Lozovskii V N, Lozovskii S V, Valov G V 2013 *Tech. Phys. Lett.* **39** 175
[12] Lozovskii V N, Chebotarev S N, Irkha V A, Valov G V 2010 *Tech. Phys. Lett.* **36** 737

IOP Conf. Series: Journal of Physics: Conf. Series **917** (2017) 032009 doi:10.1088/1742-6596/917/3/032009

Structural and phase transformations on zirconium surface after heat treatment with high-frequency currents

V Koshuro[1], M Fomina[1], A Voyko[1], A Shelkunov[1], I Egorov[1], A Zakharevich[2], A Skaptsov[2], V Atkin[2], V Galushka[2], I Rodionov[1]
[1]Yuri Gagarin State Technical University of Saratov, Saratov 410054, Russia
[2]Saratov State University named after N.G. Chernyshevsky, Saratov 410012, Russia

Abstract. Research results on the surface morphological characteristics of zirconium samples after heat treatment with high-frequency currents have been described. It has been established that at temperature 600–1200 °C and duration of heat treatment 30–300 s oxide coatings consisting of nano-grains are formed. The resulting oxide ZrO_2 coatings are a combination of two phases – monoclinic (baddeleyite) and cubic.

1. Introduction

In order to protect metal surfaces from the environmental factors different methods of oxidation of metals are actively used, the basic of them comprise gas-thermal and electrochemical methods [1-5]. ZrO_2 coatings can also be produced by deposition metods [6]. In this regard, to create protective film systems and special functional coatings on zirconium surface a highly effective method of induction-thermal oxidation can be used, which has been tested successfully for commercially pure titanium, its constructional alloys, and chromium-nickel stainless steel [1,7]. As a result of this treatment on the surface of zirconium the processes of diffusion saturation with oxygen and formation of oxide compounds in the form of films and coatings are possible [8].

2. Methodology

The samples were cylinders with a diameter of 5.8–6 mm and a length of 2–2.5 mm fabricated from commercially pure zirconium grade E110. Their surface was exposed to the microtexturing air-abrasive (corundum with an average fineness of 200–300 μm) treatment. The resulting metal substrates of samples were also subjected to ultrasonic cleaning in aqueous-alcoholic solution. Further, the surfaces of the prepared samples were oxidized in the course of treatment with HFC [1]. Maximum power consumption of the laboratory device for the treatment with HFC during heating of the products with the geometry of this type was less than 550 W, the frequency of the current in the inductor – 89.5±0.5 kHz when the treatment temperature reached 600, 800, 1000 and 1200 °C. In the course of the studies the effect of treatment with HFC was established in the temperature range of 600–1200 °C on the morphological changes occurring in the surface during the formation of oxide coatings. The modes of coating production were given double numbers: the first number designated the heat treatment temperature of zirconium substrate measured in °C; the second – treatment duration counted in seconds upon reaching the necessary temperature of treatment. For example, mode 800-030 corresponds to the temperature of 800±10 °C and treatment duration of 30±1 s.

The surface morphology of the samples after the selected technological treatment was examined using scanning electron microscopy (SEM) in the micro- and nanoscale in order to identify the patterns of structure formation of the surface layer and coatings. SEM combined with energy-dispersive X-ray spectroscopy (EDX) of the chemical composition of samples was performed on MIRA II LMU with INCA PentaFETx3 detector. Changes in the concentrations of zirconium and oxygen as well as other impurities measured in atomic percent (at.%) were analyzed in the areas of samples with different morphology types. Phase-structural surface state of the samples after IHT was studied by X-ray diffraction (XRD) on Gemini/Xcalibur (CuKα-radiation, λ = 1.541874 Å, 20° < 2Θ < 110°).

3. Results

High oxygen content in the surface layer of zirconium samples showed its high adsorption capacity. Surface of zirconium was most active with respect to oxygen that can be traced throughout the range of heat treatment. The content of oxygen reached 49–57 at.%, which affected both the stabilization of initial phase α-Zr and formation of zirconium dioxide.

Treatment of zirconium with HFC lead to the activation of diffusion saturation of the near-surface layer with oxygen, oxidation, and formation of films of ZrO_{2-x} and thicker coatings of ZrO_2 in different allotropic modifications: stable baddeleyite m-ZrO_2 (monoclinic system) and unstable c-ZrO_2 (cubic system) [8,9].

XRD data proved that untreated zirconium is characterized by the presence of α-Zr phase (Figure 1a). When the treatment temperature reached 600 ºC about 20–40% of the high-temperature phase of β-Zr appeared. In the near-surface layer of zirconium samples at a continuous treatment of about 300 s no more than 38% of baddeleyite m-ZrO_2 and about 11% of c-ZrO_2 were formed. At a short exposure of about 30 s no oxide phase was detected. Increase of the treatment temperature to 800 °C lead to more intense oxidation of zirconium. At a short exposure the amount of zirconia (m-ZrO_2 + c-ZrO_2) equaled 18–20%. At a maximum exposure time of about 300 s the amount of zirconia reached a maximum of 75–80%. At the same time the number of the most stable baddeleyite was 55 % (Figure 1b).

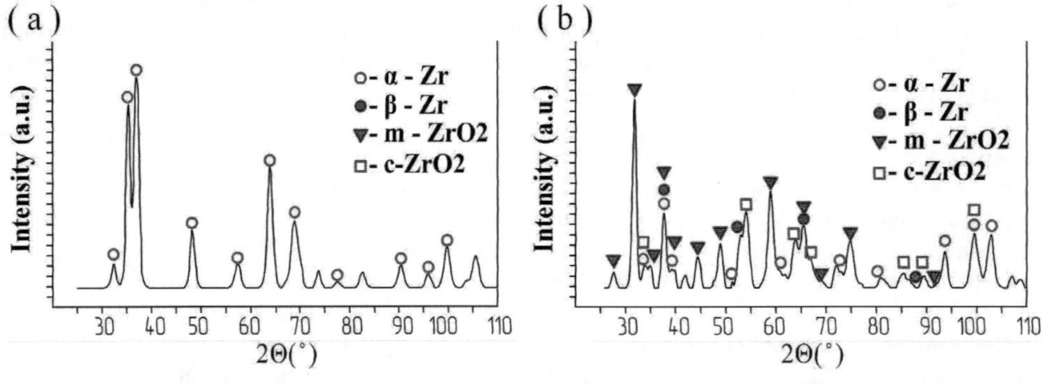

Figure 1(a, b). Results of XRD analysis of the zirconium surface: the untreated sample **(a)**; the sample oxidized by HFC at 800 °C (800-300 mode) **(b)**.

In the high-temperature range of treatment at about 1000 °C the amount of zirconium dioxide reached 45–47% at the treatment time of 30 s. However, the maximum content of zirconia did not exceed 70% at the treatment time of about 300 s. This was probably related to the increase in defects in the form of pores and cracks.

With the temperature growth to 1200 °C and at duration of HFC treatment of about 30 s there was an increase in the oxide phases to 59–60%. At the treatment duration of about 300 s the total amount of ZrO_2 did not exceed 65%.

Morphology of the initial microstructure of zirconium samples remained almost unchanged to the temperature of 1000 ºC and maximum duration of the oxidation (Figure 2a). When the temperature

IOP Conf. Series: Journal of Physics: Conf. Series **917** (2017) 032009 doi:10.1088/1742-6596/917/3/032009

reached about 1200 °C and the treatment duraion equalled about 300 s the crystals of the formed high-temperature phase of zirconia had time to grow to submicron values that significantly altered the surface relief after turning and air-abrasive treatment (Figure 2b).

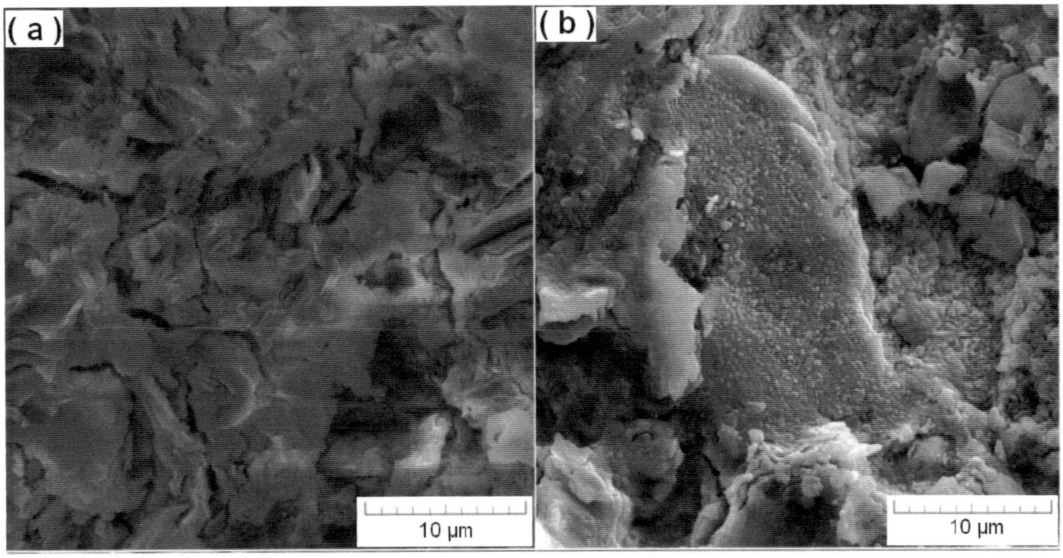

Figure 2(a, b). Microstructure of the surface after oxidation of zirconium samples at HFC mode 800-300 at different magnifications: ×10,000 **(a)**; ×200,000 **(b)**.

The formed nanocrystals of the oxide film rapidly started to increase in size, which lead to their further growth to submicron size and cracking (Figure 3d).

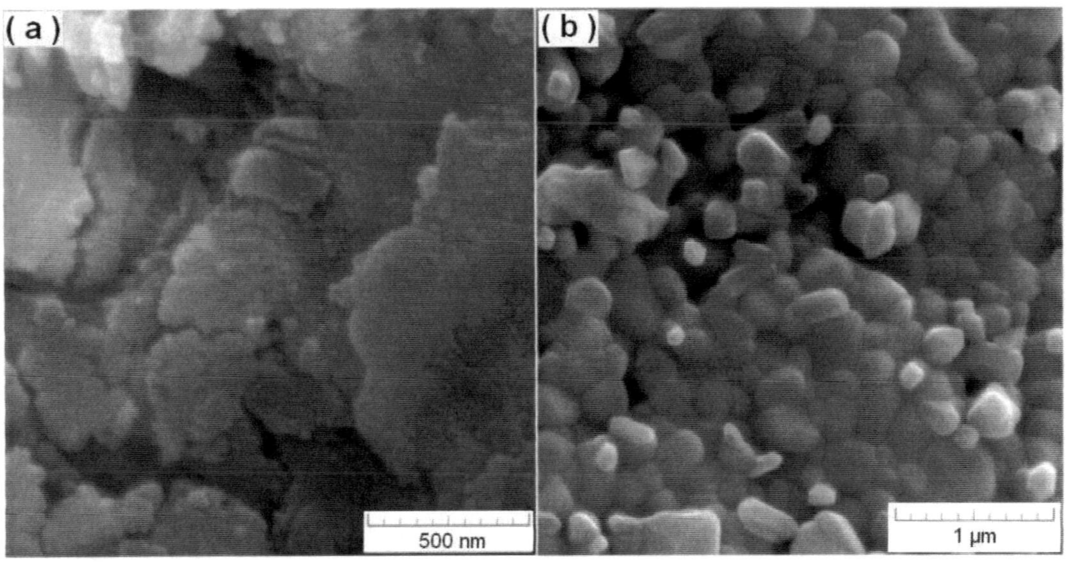

Figure 3(a, b). Nanostructure of the surface after oxidation of zirconium samples at various modes of treatment with HFC: mode 800-300 (×200,000) **(a)**; mode 1200-300 (×100,000) **(b)**.

The resulting nanostructure of the oxide coating had grains with an average size within 40±10 nm. Intensive growth was observed with the increase in heat treatment temperature to 1200±30 °C and when the duration exceeded 30±1 s. Maximum values of zirconia grains after high temperature treatment (mode 1200-120 and 1200-300) were 100–250 nm. However, there was no spontaneous separation of the outer thick-layer oxide, which is typical for, e.g. commercially pure titanium and its alloys in this range of treatment with HFC [1].

4. Conclusions

Thus, the surface structure of zirconium samples treated with HFC was characterized by the formation of zirconia coatings, which are a combination of two phases: stable baddeleyite m-ZrO_2 (monoclinic system) and unstable c-ZrO_2 (cubic system). Functional coatings were formed mostly by nano-sized and submicron crystals of round and prismatic shapes. The samples of rough coatings with the highest parameters of morphological heterogeneity in the nanoscale corresponded to the following treatment modes: 800-120 (at $P = 150$ W, $t = 800$ °C, $\tau = 120$ s), 800-300 (at $P = 150$ W, $t = 800$ °C, $\tau = 300$ s), 1000-030 (at $P = 260$ W, $t = 1000$ °C, $\tau = 30$ s), 1000-120 (at $P = 260$ W, $t = 1000$ °C, $\tau = 120$ s), 1200-001 (at $P = 520$ W, $t = 1200$ °C, $\tau \approx 1$ s), and 1200-030 (at $P = 520$ W, $t = 1200$ °C, $\tau = 30$ s).

It should also be noted that zirconium products manufactured by conventional machining can be subjected to nanostructuring treatment with HFC in order to obtain the required surface morphology heterogeneity. Improving the quality of zirconium products is also be related to the control of the tribological parameters, particularly hardness, wear resistance, and coefficient of friction.

Acknowledgments

The research was supported by the Ministry of Education and Science of the Russian Federation in the framework of the Program of Scientific Research in Universities (project No. 11.1943.2017/PP).

References

[1] Fomin A, Dorozhkin S, Fomina M, Koshuro V, Rodionov I, Zakharevich A, Petrova N and Skaptsov A 2016 *Ceram. int.* **42** 10838
[2] Fomin A A, Steinhauer A B, Rodionov I V, Fomina M A and Zakharevich A M 2013 *Tech. Phys. Lett.* **39** (11) 969
[3] Fomin A A, Steinhauer A B, Rodionov I V, Petrova N V, Zakharevich A M, Skaptsov A A and Gribov A N 2013 *Biomed. Eng.* **47** (3) 138
[4] Fomin A A, Fomina M A, Koshuro V A, Rodionov I V, Voiko A V, Zakharevich A M, Aman A, Oseev A, Hirsch S and Majcherek S 2016 *Tech. Phys. Lett.* **42** (9) 932
[5] Fomin A, Fomina M, Koshuro V, Rodionov I, Zakharevich A and Skaptsov A 2017 *Ceram. int.* **43** 11189
[6] Ni J, Noh K, Frandsen C J, Kong S D, He G, Tang T and Jin S 2013 *Mat. Sci. Eng.* **C33** 259
[7] Wang Q M, Gong J, Sun C, Lee H W and Kim K H 2011 *J. Ceram. Proc. Res.* **12** (3) 259
[8] Abriata J P, Garcés J and Versac R 1986 *Bulletin of Alloy Phase Diagrams* **7** (2) 116
[9] Yashima M and Tsunekawa S 2006 *Acta Cryst.* **B62** 161

IOP Conf. Series: Journal of Physics: Conf. Series **917** (2017) 032010 doi:10.1088/1742-6596/917/3/032010

Structural and optical properties of epitaxial $Ga_xIn_{1-x}P$ alloys with atomic ordering

P V Seredin[1*], D L Goloshchapov[1], Yu Yu Khudyakov[1], A S Lenshin[1], I N Arsentyev[2] and Tatiana Prutskij[3]
[1] Voronezh State University, Universitetskaya pl., 1, 394006 Voronezh, Russia
[2] Ioffe Physical and Technical Institute, Polytekhnicheskaya, 26, 194021 St-Petersburg, Russia
[3] Instituto de Ciencias, Benemérita Universidad Autónoma de Puebla, Privada 17 Norte, No 3417, Col San Miguel Hueyotlipan , 72050, Puebla, Pue., Mexico

Abstract. A range of structural and spectroscopic techniques were used for the study of the properties of epitaxial $Ga_xIn_{1-x}P$ alloys with an ordered arrangement of atoms in a crystal lattice grown by MOCVD on GaAs (100). The appearance of atomic ordering in the coherent growth conditions of the ordered $Ga_xIn_{1-x}P$ alloy resulted in cardinal changes of the structural and optical properties of semiconductor in comparison to disordered alloys, including the change of the crystal lattice parameter and, consequently, reduced crystal symmetry, decreased band gap and formation of two different types of surface nanorelief. Based on the variance analysis data with regard to the IR-reflection spectra as well as the UV-spectroscopy data obtained in the transmission-reflection mode, the main optical characteristics of the ordered $Ga_xIn_{1-x}P$ alloys were determined for the first time, namely, refractive index dispersion and high-frequency dielectric constant.

1. Introduction

Alloys based on A_3B_5 semiconductors have attracted the interest of researchers of semiconductor optoelectronics cause they have a positive formation enthalpy that facilitates superstructure ordered phases on the basis of these alloys [1–3]. This phenomena of the ordering appearance in A_3B_5 allows the inclusion of narrow-band semiconductors in a wide-band gap matrix and vice versa, thus creating a localizing potential for the current carriers. Furthermore, the spontaneous appearance of the nanostructures provides a basis for a new technology for the production of ordered arrays of irregularities, moreover, the basis for the next generation opto- and microelectronics [1,4]. The urgent problem of atomic ordering is firstly connected with the modification of the fundamental properties of semiconductor systems caused by the change of symmetry of the crystalline structure [4,5]. However, to date the findings only confirm the theoretical data regarding the change in band gap for alloys with an ordered arrangement of the atoms dependent on the composition and degree of ordering. There is a lack of information on the optical properties of $Ga_xIn_{1-x}P$ alloys with ordering in IR and UV spectral ranges for the epitaxial alloys, as well as the direct data and the values of parameters for the crystal lattice in the ordered $Ga_xIn_{1-x}P$ alloys dependent on the degree of ordering and conditions of epitaxial growth. For this reason, the aim of this study was to investigate the structural and optical properties of epitaxial $Ga_xIn_{1-x}P$ alloys with atomic ordering grown by MOCVD in a coherent way on GaAs (100) substrates.

2. Objects and Methods of Investigations

Epitaxial layers of $Ga_xIn_{1-x}P$ with ordering matched by the parameter of the crystal lattice with single-crystalline GaAs (100) substrates were grown by MOCVD in a horizontal reactor under a pressure of 60 mbar. The reactants for III group elements were tri-methyl gallium (TMG) and tri-methyl indium (TMI), and the reactant for the V group element was phosphine in a 10% mixture with hydrogen. $Ga_xIn_{1-x}P$ layers with a composition of $x \sim 0.50$ were grown on semi-insulating substrates of GaAs (100), misoriented by 2° relative to the direction of [110], with epitaxial films ~ 1 μm thick. The technological growth parameters are presented in Table 1.

Table 1. Parameters of growth and thickness of the epitaxial layers in the samples of heterostructures with ordering in the alloy.

Sample	Growth temperature, ^0C	V/III ratio	Thickness, μm
#S1	700	160	1
#S2	650	100	1
#S3	600	135	0.85

The structural quality of the heterostructures were performed by X-ray diffraction. The IR reflection spectra of the samples were obtained with the use of spectrometer Vertex-70 Bruker. The study of the surface morphology was performed by AFM. PL spectra were obtained from the samples according to the standard technique, with the use of a monochromator TRIAX550 and CCD detector and 514.5 nm excitation wavelength. The optical properties in the range of 190–900 nm were studied using a LAMBDA 650.

3. Experimental Results and Discussion

3.1. Structure investigations

Experimental profiles of X-ray diffraction from 400 planes for the investigated heterostructures are presented in Figures 1.*a* – *d* by the solid lines. As seen from the figures, reflections from the single-crystalline substrate and the alloy can be clearly observed in all the diffraction patterns; clearly, the diffraction from the ordered alloy is characterized by the different angle position. Taking into account that all of the investigated $Ga_xIn_{1-x}P$ alloys were actually of the same composition, it was assumed that the essential differences in the interplanar distances are probably due to a different degree of ordering in the epitaxial film.

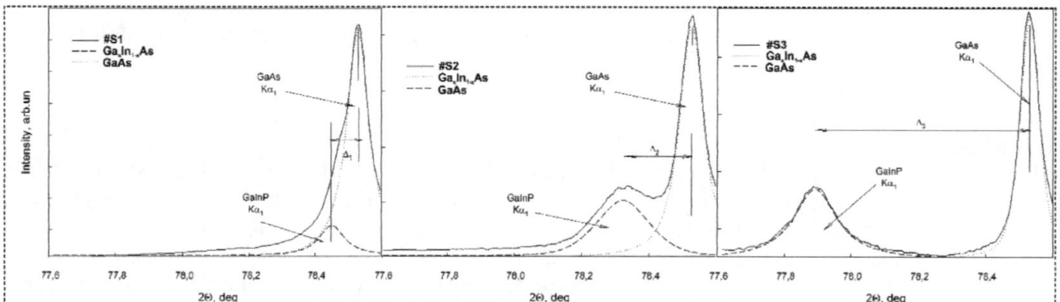

Figure 1. Experimental profiles of X-ray diffraction obtained from (400) plane of the ordered $Ga_xIn_{1-x}P$/GaAs(100) heterostructures as well as the results of profiles decomposition over components.

Table 2. Experimental and calculated data obtained on the basis of results of X-ray diffraction and microanalysis.

Sample	Perpendicular deformation $\varepsilon^{\perp}$	The ratio of the half-widths $FWHM_{GaInP}$/ $FWHM_{GaAs}$	Long range order parameter η calculated from [2]	Atomic content calculated from XRD and Zunger data [2]
#S1	0.0003	1.11	0.50	~0.51
#S2	0.001	1.61	0.50	~0.50
#S3	0.003	2.01	0.60	~0.50

The calculated lattice parameters in accordance with the linear elasticity theory [6,7] taking into account the internal stresses in heteropeitaxial layer a^v and half widths of the diffraction lines are presented in Table 2. It is quite clear that these parameters do not coincide, which is due to the change in the structure of the alloy under atomic ordering in the metallic sublattice. Calculated perpendicular deformations in $Ga_xIn_{1-x}P$ films for the samples investigated in this work are presented in Table 2.

3.2. Atomic force microscopy

It is well known that the epitaxial growth of $Ga_xIn_{1-x}P$ alloys with an ordering in the crystal lattice is often accompanied by formation of close-packed clusters on the surface of the epitaxial film [5,8]. Therefore, the morphology and phase distribution was investigated on the surface of the samples of $Ga_xIn_{1-x}P$/GaAs (100) heterostructures with ordering in the alloy using atomic-force microscopy (AFM). Figure 2 represents 3D images of the characteristic areas in the heterostructures.

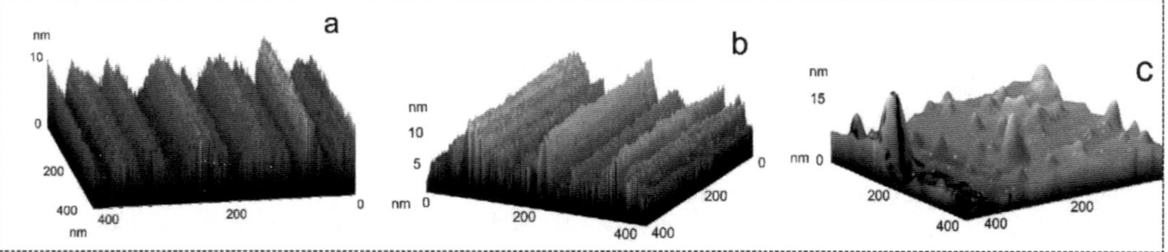

Figure 2. Results of atomic-force microscopy for the samples of the ordered $Ga_xIn_{1-x}P$/GaAs(100) heterostructures. (a,b,c) – images of characteristic area of the sample (#S1, #S2, #S3).

The results of AFM demonstrated that the areas of the ordered ridge-shaped nanorelief oriented along a single direction are present on the surface of the epitaxial film for the samples #S1 and #S2. The ordering period of nanorelief for the samples of #S1 and #S2 is ~50 – 75 nm for the mean relief height of 5 – 10 nm. The phase contrast data showed that only one phase was present on the surface of these samples, $Ga_xIn_{1-x}P$ alloy with an ordering. Unlike the first two samples on the surface of heterostructure #S3 the areas with the ordered nanorelief are not observed. However, one can observe the appearance of an array of nano-sized non-uniformities in the form of islets on the surface of the epitaxial film. The composition of the nanoislets, according to the phase contrast data, differs from the matrix composition of the $Ga_xIn_{1-x}P$ matrix. The characteristic size of the non-uniformities is 20 – 50 nm in diameter and 10 – 15 nm in height.

3.3. IR-spectroscopy

IR-spectroscopy technique is widely used for studies of fine lattice properties and estimation of the structural quality of epitaxial films. It allows the estimation of internal stresses in the crystal lattice, since this method is very sensitive to the stresses. The IR reflection spectra for all of the heterostructures demonstrate longitudinal LO and transverse TO phonon modes of Ga – P and In – P vibrations characteristic for $Ga_xIn_{1-x}P$ alloy, in agreement with data previously reported [9]. However, additional phonon vibrations were observed in the range of 200 – 220 cm^{-1} in the spectrum of sample #S3, which according to the data of theoretical calculations taken from [9], belong to the vibrations of an ordered alloy with a large order factor η. Moreover, the frequencies of the basic phonon modes for the samples #S1 and #S2 are quite close, while for sample #S3, the shift of the phonon vibration frequencies to the low-frequency range is observed (see Figure 3).

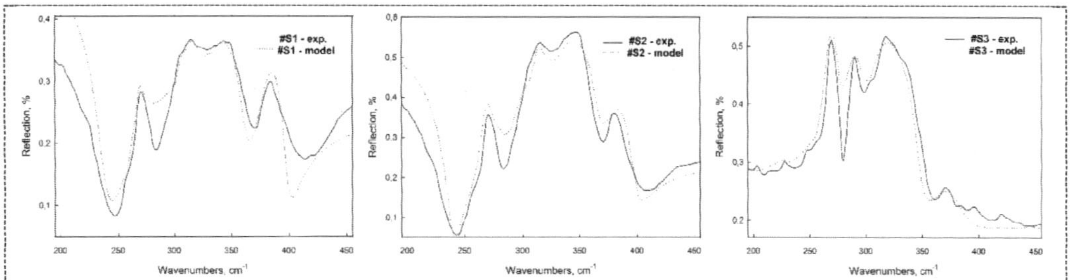

Figure 3. Experimental and simulated on the basis of dispersion analysis IR reflection spectra from single-crystalline GaAs substrate and epitaxial heterostructures $Ga_xIn_{1-x}P$/GaAs(100) with ordering.

In addition, a noticeable splitting of the frequency bands can be clearly seen. During modelling of IR-reflection spectra for the lattice vibrations in heterostructures, these spectra can be calculated in the model "film-substrate", as described in our previous studies [10–12]. This model makes it possible to perform dispersion analysis of the IR reflection spectra taking into account the phonon modes appearing in the film, as in the substrate. The value of high-frequency dielectric permeability ε_∞ for epitaxial $Ga_xIn_{1-x}P$ alloy with the ordering determined in the dispersion analysis as well as with the use of Kramers-Kronig relation is also presented in Table 3.

Table 3. Results of dispersion analysis for IR reflection spectra and photoluminescence spectroscopy for the investigated samples.

Sample	High-frequency dielectric permeability	Experimental PL energy at 300 K, eV	Calculated for x_{EDX} PL energy at 300 K, eV	Long range order parameter h, from [2]
#S1	16.2	1.841	1.893	0.50
#S2	17.3	1.845	1.893	0.50
#S3	12.5	1.831	1.885	0.60

3.4. PL spectroscopy

As noted above, the appearance of atomic ordering in the $Ga_xIn_{1-x}P$ alloy inevitably affects its energy characteristics by changing the band gap as compared with the disordered alloy of similar composition [5,8,13]. So, the application of photoluminescence spectroscopy as a tool for the determination of band gap changes and the influence of the ordering degree in the alloy on the band gap value is appropriate. Experimental photoluminescence spectra for the samples are presented in Figure 4.

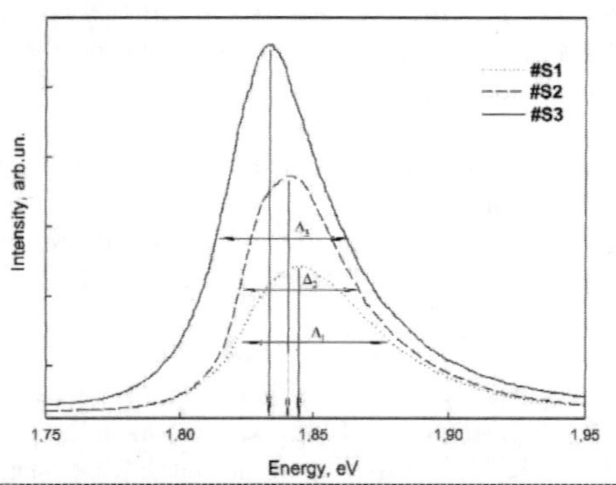

Figure 4. Experimental photoluminescence spectra of epitaxial ordered $Ga_xIn_{1-x}P/GaAs(100)$ heterostructures.

Emission from the ordered $Ga_xIn_{1-x}P$ alloy for all samples appears as a peak with a half-width of $\Delta \sim 0.05$eV. It should be noted that the intensity of photoluminescence for sample #S3 exceeds the intensity of the #S1 and #S2 samples by more than 1.5 – 2 times. The experimental values of E_g are presented in Table 3, as well as the values of the band gap for disordered $Ga_xIn_{1-x}P$ alloys of similar composition determined in [14]. From the data in Table 4, it is evident that the value of the band gap energy for the epitaxial film of each sample is less than it should be in case of a disordered alloy of similar composition. Based on the theoretical calculations reported by Zunger *et al.* [2] in combination with the value of band gap energy determined in this experimental study, the values of the coefficients of the perpendicular distortion (the value of elastic stresses) and parameter ΔEg (decrease of the band gap value), it was possible to specify the degree of ordering in the $Ga_xIn_{1-x}P$ alloy grown on GaAs (100) (see Table 3). This data is in good agreement with the similar data from [13].

3.5. UV spectroscopy

In our previous studies [15,16], the use of surveying spectra technique in reflection mode within ultraviolet and visible ranges of electromagnetic emission was shown to be a convenient approach for the study of the optical properties of thin semiconductor and dielectric films. Using the experimental data obtained from the analysis of maximums and minimums in the transmission-reflection spectra for two different incidence angles, it is possible not only to determine refractive index of the film in the sample but also to calculate dispersion of the refractive coefficient in rather wide interval of the angles. The calculated values (points) of the refractive coefficient for the investigated $Ga_xIn_{1-x}P/GaAs$ structures and the dependences approximated by splines are presented in Figure 5.

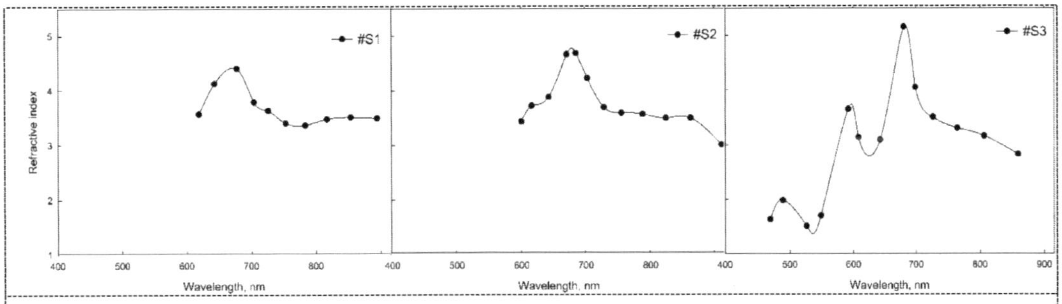

Figure 5. Reflection-transmission spectra of epitaxial ordered $Ga_xIn_{1-x}P/GaAs(100)$ heterostructures in the range of 400 – 900 nm. a – sample #S1, b – sample #S2, c – sample #S3.

4. Discussion of the obtained results

Based on the XRD data (Figure 1, Tables 1 & 2), one can conclude that growth of the $Ga_xIn_{1-x}P$ alloy with ordering within the range of compositions (x ~ 0.50) that provides matching of the parameters of the epitaxial film and a single-crystalline GaAs (100) substrate results in the appearance of elastic stresses in the alloy. This is due to changes in the crystal lattice parameter in the ordered alloy in comparison with that of the disordered alloy with a similar composition. Furthermore, the effects of atomic ordering and distortion (elastic stresses) result in the appearance of the areas with the gradient of composition in the epitaxial film, as well as the formation on the surface of the film either of areas with the ordered nanorelief, or an array of nano-size irregularities with a composition different from the composition of the matrix ($Ga_xIn_{1-x}P$ alloy). With an increase in the ordering factor (long-range order) at the invariable composition of the alloy, the tetragonal distortion also increased, since it is due to the growth of a crystal lattice parameter in the ordered alloy. An increase in the elastic stresses results in the modification of the shape of the epitaxial film surface. Consequently, at low values of the order/distortion factor, the areas of ordered nanorelief predominate the surface of the film (see Figure 2). The maximum values of the order/distortion factor lead to the formation of an array of nano-size irregularities on the surface. According to the IR-spectroscopy data, it is not only the changes in the frequencies for the main phonon modes that are observed in the spectrum of the ordered alloy with a high degree of order in comparison with disordered alloy of similar composition, but also some additional phonon modes appear shifted towards the low-frequency range. This is in good agreement with the theoretical representations on the vibrational spectra in $Ga_xIn_{1-x}P$ alloys with ordering and they also coincide with similar data for the ordered $Al_{0.50}Ga_{0.50}As$ alloys [12,17]. Comparing the value of high-frequency dielectric permeability ε_∞ for the epitaxial ordered $Ga_xIn_{1-x}P$ alloys with that for the disordered alloys of the same composition x ~ 0.50 [18] determined from the analysis of IR reflection spectra, it was for the first time shown that this parameter for the ordered alloy was 1.5 – 2 times higher (see Table 3). The study of the optical and photoluminescence properties of the ordered $Ga_xIn_{1-x}P$ alloys in the UV spectral range confirms previous data that the MOCVD not only gives rise to a strong CuPt-B type ordering but also ensures a good uniformity of the film and its transmission capacity. A decrease in the band gap energy on the ordered $Ga_xIn_{1-x}P$ alloy determined in our experiments at the specified level of distortion and factor of the order is in good agreement with the theoretical data [2]. A decrease in the epitaxy temperature from 700 to 600°C increased the degree of order in $Ga_xIn_{1-x}P$ alloys, enhanced intensity of photoluminescence and also reduced the band gap value. However, none of the previously performed investigations provided any information on the dispersion of the refractive coefficient in the ordered $Ga_xIn_{1-x}P$ alloys. Based on the results of our calculations obtained from the direct experimental data, it was shown for the first time (see Figure 5) that the refractive index for the epitaxial films with ordering is 1.5 – 2.5 times higher than that for disordered values of the same composition x ~ 0.50. Moreover, the maximum value of refractive index for the sample with a high degree of order was n = 5.45 at the wavelength of ~ 680 nm, while the close value for the disordered alloy with a similar composition of x ~ 0.50 was attained in the wavelengths in the range of ~ 330 – 340 nm [18].
In summary, the findings from this study confirm the perspectives and possibilities for engineering of the band gap in heterostructures by combining the effects of tetragonal distortion under epitaxial growth with the effects of atomic ordering.

Acknowledgments
The research is conducted with the support of grant of President of Russian Federation MD-188.2017.2.

References

[1] Zunger A 1997 Spontaneous Atomic Ordering in Semiconductor Alloys: Causes, Carriers, and Consequences *MRS Bull.* **22** 20–6

[2] Wei S-H and Zunger A 1994 Optical properties of zinc-blende semiconductor alloys: Effects of epitaxial strain and atomic ordering *Phys. Rev. B* **49** 14337–51

[3] Lee H S and Lee J Y 2003 CuAu-I-type ordered structures in InxAl1−xAs epilayers grown on (001) InP substrates *Appl. Phys. Lett.* **82** 2999–3001

[4] Seredin P V, Domashevskaya P, Arsentyev I N, Vinokurov D A, Stankevich A L and Prutskij T 2013 Superstructured ordering in AlxGa1−xAs and GaxIn1−xP alloys *Semiconductors* **47** 1–6

[5] Ahrenkiel S P, Jones K M, Matson R J, Al-Jassim M M, Zhang Y, Mascarenhas A, Friedman D J, Arent D J, Olson J M and Hanna M C 1999 CuPt-B Ordered Microstructures in GaInP and GaInAs Films *MRS Proc.* **583**

[6] Zhou D and Usher B F 2001 Deviation of the AlGaAs lattice constant from Vegard's law *J. Phys. Appl. Phys.* **34** 1461–5

[7] Seredin P V, Glotov A V, Ternovaya V E, Domashevskaya E P, Arsentyev I N, Vavilova L S and Tarasov I S 2011 Spinodal Decomposition of GaxIn1−xAsyP1−y Quaternary Alloys *Semiconductors* **45** 1433–40

[8] Gomyo A, Suzuki T and Iijima S 1988 Observation of Strong Ordering in GaxIn1−x P alloy semiconductors *Phys. Rev. Lett.* **60** 2645–8

[9] Pagès O, Chafi A, Fristot D and Postnikov A V 2006 (Ga,In)P: A standard alloy in the classification of phonon mode behavior *Phys. Rev. B* **73** 165206

[10] Verleur H W 1968 Determination of Optical Constants from Reflectance or Transmittance Measurements on Bulk Crystals or Thin Films *J. Opt. Soc. Am.* **58** 1356

[11] Seredin P V, Glotov A V, Domashevskaya E P, Arsentyev I N, Vinokurov D A, Tarasov I S and Zhurbina I A 2010 The substructure and luminescence of low-temperature AlGaAs/GaAs(100) heterostructures *Semiconductors* **44** 184–8

[12] Domashevskaya É P, Seredin P V, Lukin A N, Bityutskaya L A, Grechkina M V, Arsent'ev I N, Vinokurov D A and Tarasov I S 2006 Infrared reflectance spectra and morphologic features of the surface of epitaxial AlxGa1−xAs/GaAs(100) heterostructures with the ordered AlGaAs2 phase *Semiconductors* **40** 406–13

[13] Ernst P, Geng C, Scholz F and Schweizer H 1996 Ordering in GaInP2 studied by optical spectroscopy *Phys. Status Solidi B* **193** 213–29

[14] Adachi S 2009 *Properties of semiconductor alloys: group-IV, III-V and II-VI semiconductors* (Chichester, U.K: Wiley)

[15] Seredin P V, Lenshin A S, Kashkarov V M, Lukin A N, Arsentiev I N, Bondarev A D and Tarasov I S 2015 Ultrathin nano-sized Al2O3 strips on the surface of por-Si *Mater. Sci. Semicond. Process.* **39** 551–8

[16] Seredin P V, Kashkarov V M, Arsentyev I N, Bondarev A D and Tarasov I S 2016 Distinctions of the growth and structural-spectroscopic investigations of thin AlN films grown on the GaAs substrates *Phys. B Condens. Matter* **495** 54–63

[17] Domashevskaya E P, Seredin P V, Lukin A N, Bityutskaya L A, Grechkina M V, Arsentyev I N, Vinokurov D A and Tarasov I S 2006 XRD, AFM and IR investigations of ordered AlGaAs2 phase in epitaxial AlxGa1–xAs/GaAs (100) heterostructures *Surf. Interface Anal.* **38** 828–32

[18] Schubert M, Gottschalch V, Herzinger C M, Yao H, Snyder P G and Woollam J A 1995 Optical constants of GaxIn1−xP lattice matched to GaAs *J. Appl. Phys.* **77** 3416

IOP Conf. Series: Journal of Physics: Conf. Series **917** (2017) 032011 doi:10.1088/1742-6596/917/3/032011

Novel approach to investigation of semiconductor MOCVD by microreactor technology

S A Konakov[1], **V V Krzhizhanovskaya**[2,3]

[1] Peter the Great St. Petersburg Polytechnic University, 195251, St. Petersburg, Russia
[2] University of Amsterdam, 1098 XH Amsterdam, The Netherlands
[3] ITMO University, 197101 St. Petersburg, Russia

Abstract. Metal-Organic Chemical Vapour Deposition is a very complex technology that requires further investigation and optimization. We propose to apply microreactors to (1) replace multiple expensive time-consuming macroscale experiments by just one microreactor deposition with many points on one substrate; (2) to derive chemical reaction rates from individual deposition profiles using theoretical analytical solution. In this paper we also present the analytical solution of a simplified equation describing the deposition rate dependency on temperature. It allows to solve an inverse problem and to obtain detailed information about chemical reaction mechanism of MOCVD process.

1. Introduction and methodological issues of MOCVD technology

Metal-Organic Chemical Vapour Deposition (MOCVD) is the one of the most popular processes for deposition of semiconductor thin films [1,2]. Single-specie (Si, Ge) or compound semiconductors (GaAs, CdTe and others) are deposited as monocrystalline, polycrystalline or amorphous films [3,4]. Materials produced by MOCVD process are widely used in microelectronics and photonics devices [5]. The MOCVD technology has been actively investigated in the past few decades, however there are still open questions and technological issues to be tackled for new materials and reactors [6-8].

MOCVD is a complex process with a large number of technological parameters, which strongly influence film properties. The main process parameters are temperature, pressure, flow rate, chemical composition of gas phase, reactor geometry, residence time and surface properties. This is a big advantage of the MOCVD technology, because we can accurately tune the deposition process to produce a film with the optimal properties for a specific application. At the same time it is a disadvantage, because it is very difficult to find these optimal conditions from the thousands of possible parameter sets. To solve the problem of multi-parameter optimization, researchers and process engineers carry out many laboratory and numerical experiments. To cover all possible sets of process parameters, ideally more than 10^5 experiments shall be carried out – infeasible in current experimental or computational research. In this paper we propose a new approach to investigation of semiconductor MOCVD by microreactor technology.

Microreactor Chemical Vapor Deposition (μR CVD) is a variation of the CVD method, where the main distinguishing feature is a localized material deposition achieved by a special configuration of gas streams in the microreactor. This gas stream configuration is provided by the microsystem devices located near the substrate, called microreactors

IOP Conf. Series: Journal of Physics: Conf. Series **917** (2017) 032011 doi:10.1088/1742-6596/917/3/032011

There are many types of microreactors. The simplest configuration has just one microchannel, as shown in Figure 1. Advanced modifications have additional channels for precise deposition control, investigated in our previous work [9,10].

Typical size of deposition area is around 100 μm. It allows using this approach in two different ways: (1) replacing many time-consuming macroscale experiments with one microreactor deposition with many points on one substrate; (2) deriving chemical reaction rates from individual deposition profiles using theoretical analytical solution. In the following two sections we describe these methods in more detail.

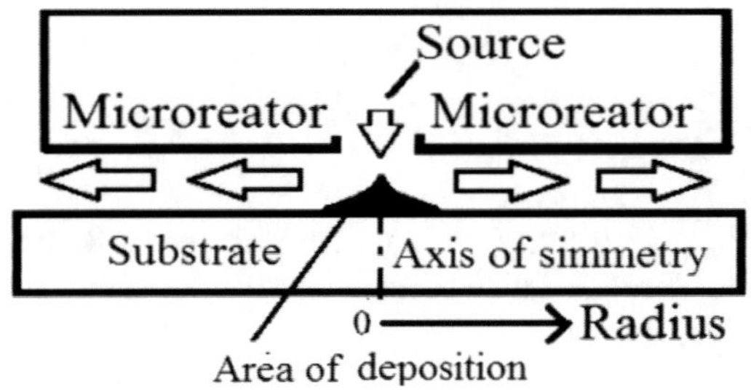

Figure 1. Scheme of a one-jet axisymmetric microreactor for deposition of one point-sample.

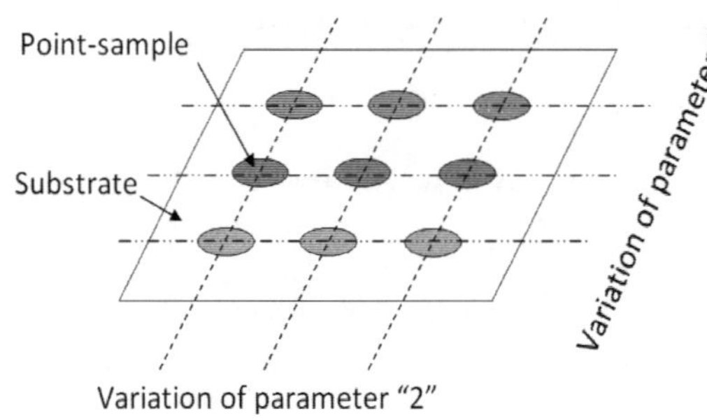

Figure 2. Scheme of point-samples array deposited by μR CVD.

2. Experimental method for deposition of thousands of point-samples by μR CVD process

Equipment for implementation of μR CVD technology can be built from a standard CVD tool, which includes a process chamber, substrate holder with a heater, gas supply and exhaust or vacuum system. We only need to add a microreactor and a system (robot "arm") for its movement above the substrate.

During the deposition process gases (precursors and carriers) are supplied to the microreactor, which than generates specially configured gas streams. Because the microreactor is located very close to the substrate (at a distance of up to hundreds micrometers), deposition is localized on a small area (less than 100 μm diameter [10]). This way, one point-sample is produced. An array of these point-samples is created by step-by-step movement of the microreactor over the substrate, like in a standard ink printing technology (as shown in Figure 2). For each point we can modify process parameters in an automated programmed regime.

On one 4-inch substrate, we can deposit an array of around 7000 point-samples, so we need only 15 wafers to cover 10^5 experimental points. Then we use standard automatic inspection equipment to investigate the deposition results. They provide chip inspection (optic, electron microscopy, atomic force microscopy, electric testing, spectrum analysis, etc.) on wafers with a rate over 1 wafer per hour. These tools are widely spread on any semiconductor or MEMS factory.

This way, application of μR CVD technology provides abundant experimental information about the CVD process from a single wafer. The number of experimental samples in μR CVD is several orders of magnitude higher than in the traditional CVD experimental technique. Our approach promises a faster progress in CVD research. It is a powerful instrument for development and optimisation of deposition technology.

3. Mathematical model of chemical kinetics mechanism in μR CVD process

The second application of μR CVD for investigation of CVD technology is deriving chemical reaction rates from individual deposition profiles using theoretical analytical solution. We developed a mathematical model of one-jet axisymmetric microreactor to show a dependency between the profile and kinetic parameters of chemical processes during the deposition. In the model we make several assumptions: constant gas density, one-dimentional "ideal plug flow" reactor, the microchannel is replaced by a point gas source of zero width (see Figure 3).

We consider two ring cross-sections (L) and (N) as shown in Figure 3. In cross-section (L) at a distance r [m] from the point gas source, volume concentration of gas precursor "A" is $C_A(r)$ [mol/m^3] and velocity is $V(r)$ [m/s]. In cross-section (N) at the distance δr [m] from cross-section (L), concentration and velocity are $C_A(r+\delta r)$ and $V(r+\delta r)$ respectively. Point gas source provides gas flow G [kg/s] (for gas with constant density ρ [kg/m^3], volumetric flow rate is Q [m^3/s]). Gas precursor "A" is chemically reacting and leading to material deposition on the substrate. This deposition flux on the surface has rate χ [mol/(s·m^2)], from which we can calculate deposition rate X [μm/s] of CVD process using density of material ρ_m [kg/m^3].

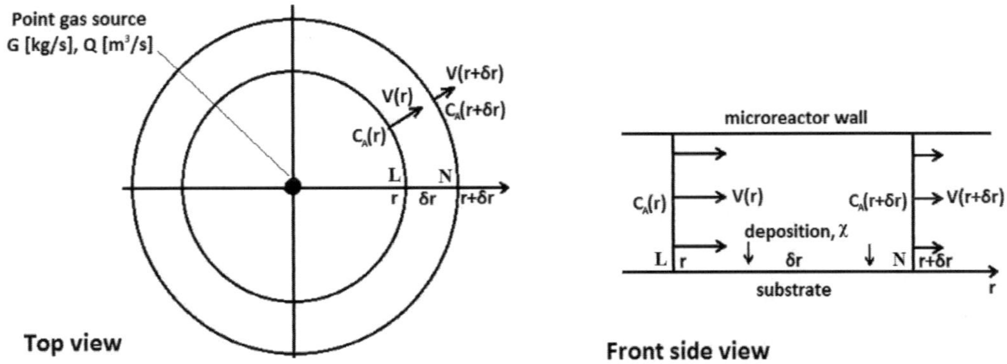

Figure 3. Explanatory scheme for a mathematical model of one-jet axisymmetric microreactor.

As the first approximation, we neglect mass transfer by diffusion along the radius, which is justified a small radius and high gas velocities. In this case the mass balance equation for volume between the cross-sections (L) and (N) can we written as

$$-\frac{G}{\rho}C_A(r)+\frac{G}{\rho}C_A(r+\delta r)+\pi(-\chi)((r+\delta r)^2-r^2)=0, \tag{1}$$

where the first term is molar flux of precursor "A" through cross-section (L), the second term is molar flux of precursor "A" through cross-section (N) and the third term stands for consumption of precursor "A" due to the surface chemical reaction. Using a Tailor series and omitting the terms of $O(\delta r^2)$, we transform (1) in (2):

$$-\frac{G}{\rho}C_A(r)+\frac{G}{\rho}[C_A(r)+\frac{\mathrm{d}C_A(r)\delta r}{\mathrm{d}r}]+2\pi(-\chi)r\delta r=0,$$

$$Q\frac{\mathrm{d}C_A(r)\delta r}{\mathrm{d}r}+2\pi(-\chi)r\delta r=0,$$

$$-\frac{\mathrm{d}C_A(r)}{(-\chi)}=\frac{2\pi}{Q}r\mathrm{d}r,$$

$$-\int_{C_{A0}}^{C_A}\frac{\mathrm{d}C_A(r)}{(-\chi)}=\frac{2\pi}{Q}\int_0^r r\mathrm{d}r,$$

$$-\int_{C_{A0}}^{C_A}\frac{\mathrm{d}C_A(r)}{(-\chi)}=\frac{\pi}{Q}r^2 \tag{2}$$

Equation 2 is the basic equation for calculation of precursor concentration in a one-jet axisymmetric microreactor. We consider two cases. The first case is a zero-order surface chemical reaction, meaning

$$(-\chi)=k_0,$$

where k_0 [mol/s·m^2] is a reaction rate coefficient. In this case equation (2) can be re-written as

$$-\int_{C_{A0}}^{C_A} \frac{dC_A(r)}{k_0} = \frac{\pi}{Q} r^2 \quad (3)$$

After integration we obtain

$$C_A(r) = C_{A0} - \frac{\pi k_0}{Q} r^2, \quad (4)$$

where C_{AO} is the concentration of precursor "A" at the point source. The zero-order reaction is independent from the concentration of precursor "A", thus the deposition profile is uniform ($(-\chi) = k_0$) from the center till r_{crit}, which is calculated from conditions

$$C_A(r_{crit}) = 0,$$

$$r_{crit} = \left(\frac{QC_{A0}}{\pi k_0}\right)^{1/2} \quad (5)$$

During experimental investigation we set C_{AO} and Q and measure the deposition profiles. In this way we obtain value of r_{crit} and from (5) we calculate k_0 for the surface reaction. Carrying out two experiments for different Q we find C_{AO} if its value is not known with sufficient accuracy. In addition, based on equation (5) we find gas flow Q and precursor concentration C_{AO} to obtain a deposition area of a specified size.

The second case is the first order of surface chemical reaction. It means

$$(-\chi) = k_1 C_A(r),$$

where k_1 [m/s] is a reaction rate coefficient. In this case equation (2) can be written as

$$-\int_{C_{A0}}^{C_A} \frac{dC_A(r)}{k_1 C_A(r)} = \frac{\pi}{Q} r^2 \quad (6)$$

After integration we obtain

$$C_A(r) = C_{A0} \exp\left(-\frac{\pi k_1}{Q} r^2\right) \quad (7)$$

Expression for a deposition rate distribution X [m/s] is

$$X(r) = \frac{M\, k_1 C_A(r)}{\rho_m}, \quad (8)$$

where ρ_m [kg/m^3] is density of deposited material, M is molar mass of deposited material. Expressing $C_A(r)$ from (8) and substituting it in (7), we derive an expression for the deposition rate distribution:

$$X(r) = X_0 \exp\left(-\frac{\pi k_1}{Q} r^2\right) \quad (9)$$

From (9) we derive an equation for the reaction rate coefficient (10). With a known flow rate Q we measure the deposition rate in two points $X(0)$ and $X(r)$ to find this coefficient:

$$k_1 = \frac{Q}{\pi r^2} \ln \frac{X(0)}{X(r)} \quad (10)$$

Reaction rate coefficients usually depend on temperature, as described by equation (11)

$$k(T) = A_p \exp\left(-\frac{E_a}{RT}\right), \quad (11)$$

where A_p is pre-exponent factor and E_a/R is activation temperature. It is easy to obtain a temperature dependence of surface reaction rate coefficient providing process of μR CVD at different temperature. It is possible to calculate pre-exponent factor and activation temperature from the experimental data.

4. Results of mathematical modelling of chemical kinetics mechanism in μR CVD process

To illustrate a microreactor deposition process described in our model, we consider an example of the microreactor with the following parameters: source concentration C_{AO} is 0.01 mol/m^3, density ρ_m and molar mass M of deposited material are 2000 kg/m^3 and 0.046 kg/mol respectively. For these conditions, we consider the zero-order and first-order surface reaction. We then investigate the distribution of precursor "A" and deposition profiles in the microreactor at different temperatures. The results are presented in Figures 4 and 5.

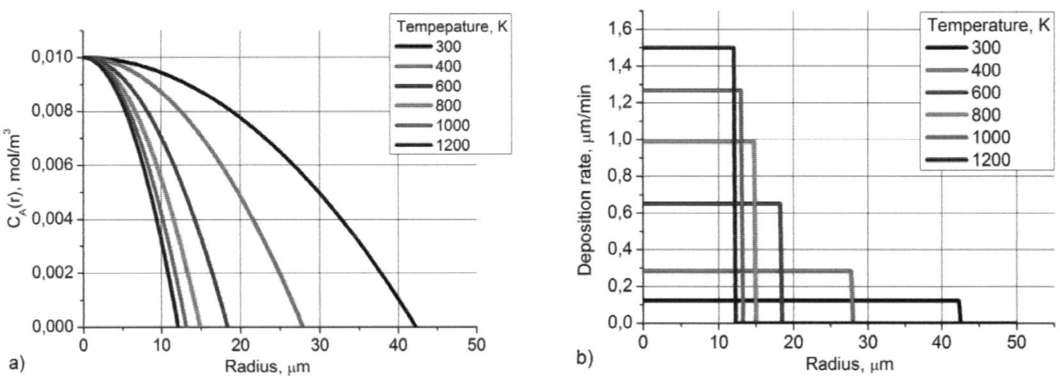

Figure 4. Distribution of precursor concentration (a) and deposition rate (b) in microreactor for the zero-order reaction and flow rate Q=5·10^{-11} m^3/s. Surface chemical reaction has pre-exponent factor A_p=0.0025 and activation temperature E_a/R=1000 K.

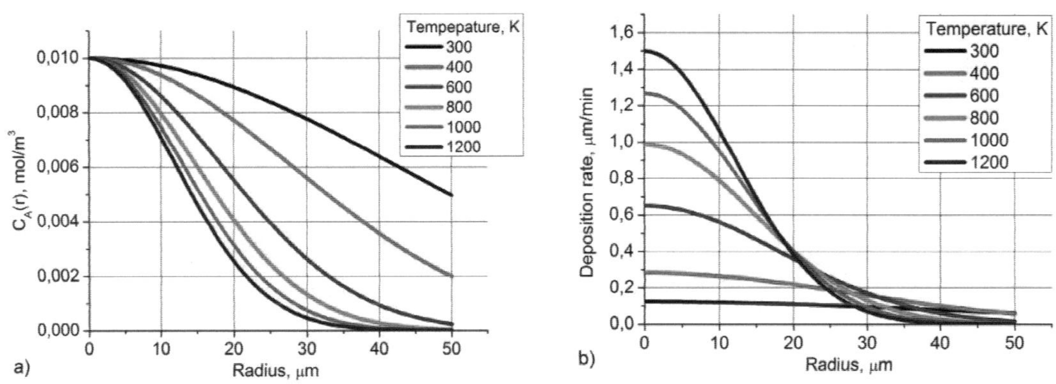

Figure 5. Distribution of precursor concentration (a) and deposition rate (b) in microreactor for the first-order reaction and flow rate Q=10^{-10} m^3/s. Surface chemical reaction has pre-exponent factor A_p=0.25 and activation temperature E_a/R=1000 K.

Figures 4 and 5 illustrate the property of microreactors to provide localized deposition at different temperatures. It is possible to design reactors for μR CVD process taking into account these graphs and equations (4)-(11). In addition these equations give a way to analyze the deposition profiles and to calculate rate coefficient for surface chemical reaction.

5. Conclusions and future work

We propose microreactors for replacing time-consuming macroscale experiments with one microreactor deposition and for extracting chemical reaction rates from individual deposition profiles. This way, a very large number of experiments (around 7000) can be carried out on one wafer. This abundant amount of experimental results is a great value in itself. The second way to use microreactor deposition is deriving chemical reaction rates from individual deposition profiles. In this paper we developed a simplified mathematical model of a one-jet microreactor and provided an analytical solution to calculate chemical kinetics parameters. Therefore we propose to apply the method of μR CVD for detailed investigation of chemical processes during MOCVD.

Our model makes some assumptions. We plan to develop more detailed analytical models with multistep reactions and diffusion to study the influence of these assumptions. A complete mathematical model cannot be solved analytically, thus we will use numerical approaches and computer simulations based on our previous research [9,10]. In addition the next step of our research will include experimental application of microreactors in investigation of semiconductor MOCVD.

Acknowledgements
This work was supported by the Ministry of Education and Science of the Russian Federation (grant №14.578.21.0123, RFMEFI57815X0123)

References

[1] Kern W 2012 *Thin Film Processes II* (San Diego: Academic press) 888 p.
[2] Morosanu C E 2016 *Thin Films by Chemical Vapour Deposition Vol. 7* (Amsterdam: Elsevier)
[3] Jones A C, O'Brien P 2008 *CVD of Compound Semiconductors: Precursor Synthesis, Development and Applications* (Weinheim:John Wiley & Sons).
[4] Kartopu G, et al. 2016 *Progress in Photovoltaics: Research and Applications* **24**(3) 283-291.
[5] Braniste T, et al. 2017 *Superlattices and Microstructures* **102** 221-234.
[6] Keller S, et al. 2014 *Semiconductor Science and Technology* **29**(11) 113001.
[7] Ortner H M et al.2014 *International Journal of Refractory Metals and Hard Materials* **44** 148-159.
[8] Kurek A, et al. 2014 *Australian Journal of Chemistry* **67**(7) 989-996.
[9] Konakov S A, Krzhizhanovskaya V V 2015 *Journal of Physics: Conference Series* **574**(1) 012145
[10] Konakov S A, Krzhizhanovskaya V V 2016 *Journal of Physics:Conference Series* **741**(1) 012018

IOP Conf. Series: Journal of Physics: Conf. Series **917** (2017) 032012 doi:10.1088/1742-6596/917/3/032012

Estimation of critical thickness of Stranski–Krastanow transition in GeSi/Sn/Si system

K A Lozovoy, A A Pishchagin, A P Kokhanenko, A V Voitsekhovskii

Tomsk State University, Faculty of Radiophysics, 36 Lenin Av., Tomsk, 634050, Russia

Abstract. In this paper Stranski–Krastanow growth of Ge_xSi_{1-x} epitaxial layers on the Si(001) surface with pre-deposited tin layer with the thickness less than 1 ML is considered. For the calculations of critical thickness of transition from 2D to 3D growth in this paper a theoretical model based on general nucleation theory is used. This model is specified by taking into account dependencies of elastic modulus, lattices mismatch and surface energy of side facet on the composition x, as well as change in the adatoms diffusion coefficient and surface energy of the substrate in the presence of tin. As a result, dependencies of critical thickness of Stranski–Krastanow transition on compositon x and temperature are obtained. The simulated results are in a good agreement with experimentally observed results.

1. Introduction

Since 1990s $Si_{1-x}Ge_x$/Si material system with nano-dimensional inclusions of germanium has acquired a reputation of very promising system for various applications in electronics and optoelectronics. Great attention to the $Si_{1-x}Ge_x$/Sn/Si and $Ge_{1-y}Sn_y$/Si systems was also attracted recently because they allow one to effectively manage the band-gap energy and to obtain direct band semiconductors. These materials may be used for creation of photodetectors, solar cells and light-emitting devices and compete with traditional materials based on AIII-BV compounds [1, 2].

One of the most promising methods of creation of $Si_{1-x}Ge_x$/Si heterostructures is molecular beam epitaxy. During the growth of quantum dots in this method one can manage the morphology of islands' ensemble by changing growth temperature, deposition rate, and volume of deposited material.

One of the most important values, characterizing the process of molecular beam epitaxy is critical thickness for Stranski–Krastanow transition from 2D to 3D growth. But to date there is no theoretical models capable to adequately describe initial stages of growth in GeSi/Sn/Si system and to predict values of critical thickness of 2D-3D transition and its dependencies on growth temperature and composition of solid solution.

It is well known that thin tin layer act as a surfactant during deposition of germanium on silicon [3–6]. Meanwhile, tin segregates to the surface of both silicon and germanium and increases the mobility of adatoms on the surface of substrate. Therefore, in case of GeSi layers deposition on the silicon surface with pre-deposited 0.5–1.0 ML thick tin layer it necessary to additionally take into account the increase in diffusion coefficient of atoms of growing layer on the surface of substrate covered by tin. Besides that, the presence of surfactant on the substrate's surface alters its surface energy. In the case of tin layer on the silicon surface it is known that this energy slightly decreases [3, 4]. So, in our estimations we should also take this fact into account by changing the value of surface energy.

2. Theoretical model, results and discussion

For description of the Ge_xSi_{1-x} layers deposition on the silicon surface it is essential to take into account in the formulae above the dependence of thermodynamic parameters on the composition x. For this purpose the values of physical constants for pure materials (Si, Ge) and the Vegard's law are used:

$$\varepsilon(x) = \varepsilon_0 x, \quad (1)$$

$$\lambda(x) = x\lambda(Ge) + (1-x)\lambda(Si), \quad (2)$$

$$\gamma(x) = x\gamma(Ge) + (1-x)\gamma(Si), \quad (3)$$

where $\varepsilon(x)$ is the lattice mismatch of Ge_xSi_{1-x} layer and silicon substrate, ε_0 is the lattice mismatch in the case of pure Ge, $\lambda(x)$ is the elastic modulus, $\gamma(x)$ is the specific surface energy [10].

If the modelling of quantum dots formation kinetics in Ge_xSi_{1-x}/Sn/Si system is required for estimation of surface density and size distribution function of quantum dots, it is carried out similarly to Ge/Si system [7–10] but with the parameters recalculated by the formulae described above.

For calculations of the characteristics of quantum dots growth in Ge_xSi_{1-x}/Sn/Si and $Ge_{1-y}Sn_y$/Si systems for small tin contents we used the following values for the parameters of the model [8–10]: $l_0 = 0.395$ nm, $d_0 = 0.145$ nm, $\gamma(Ge) = 800$ erg/cm^2, $\gamma(Si) = 1400$ erg/cm^2, $\lambda(Ge) = 1.27\cdot10^{12}$ dyn/cm^2, $\lambda(Si) = 1.8\cdot10^{12}$ dyn/cm^2, $\varepsilon_0 = 0.042$, $\Psi_0 = 450$ erg/cm^2, $\varphi = 20°$, $\nu = 10$, $D_0(T) = 10^{-4}\cdot\exp(-1.21 / k_BT)$ cm^2/s.

For estimation of critical thickness of transition in the case of Ge_xSi_{1-x} layers deposition on silicon surface with pre-deposited tin layer with the thickness of 0.5–1.0 ML it is necessary to additionally take into account the increase in diffusion coefficient of atoms of growing layer on the surface of substrate with tin. This question is poorly studied in literature but it was shown [11] that the diffusion coefficient of Ge on Si in the presence of 1 ML thick Sn layer increases by four orders of magnitude. So, for the calculations of critical thickness of transition in Ge_xSi_{1-x}/Sn/Si system we used two values of the adatoms diffusion coefficient: $D = D_0\cdot10^2$ and $D = D_0\cdot10^4$, where D_0 is the diffusion coefficient of atoms on pure Si surface. It is known that presence of Sn reduces surface energy insignificantly, but numerical values of this reduction are not given in the literature. We will assume in our calculations that surface energy of silicon decreases by 100 erg/cm^2 and 200 erg/cm^2.

It is known, that in case of Sn-mediated epitaxy of Ge on Si with the ~ 0.5–1 ML thick tin layer critical thickness of Stranski–Krastanow transition decreases from ≈ 5 ML, typical for clean Si surface, to ≈ 4 ML in the presence of tin [3–6]. Calculations with the formulae described in this article give for the critical thickness values of $h_c = 4.3$ ML and $h_c = 4.0$ ML for two stated above diffusion coefficients correspondingly, that is in very good agreement with the experimental data.

Dependencies of critical thickness on composition x in $Si_{1-x}Ge_x$/Sn/Si system for the mentioned diffusion coefficients and surface energies at $T = 300$, 500, 700 °C were calculated (Figure 1). It is obvious, that the curve behaviors qualitatively resemble the case of deposition on clean silicon surface. This means that with increasing of Ge content x critical thickness of transition decreases in the presence of Sn. But for any composition x critical thickness of transition decreases in the presence of tin compared to the case of deposition in GeSi/Si system.

There are few data on the values of critical thickness of transition in $Si_{1-x}Ge_x$/Sn/Si system for various compositions. Figure 2 shows calculated and experimental [3] values of critical thickness of transition to 3D growth in $Si_{1-x}Ge_x$/Sn/Si system for different diffusion coefficients and substrate's surface energies. Comparison with experimental results (Figure 2) indicates that theoretical predictions are in rather good agreement with experiments and describes the observed behavior of critical thickness with variation in composition correctly, especially with the assumption of the surface energy reducing by 200 erg/cm^2. Some discrepancy in obtained values may be caused by variations in growth rate and temperature, as well as difference in techniques of critical thickness estimation. So, it is quite probable that energetic parameters of the system slightly alter in the presence of tin and the critical thickness increases significantly, but this question demands further experimental and theoretical investigations.

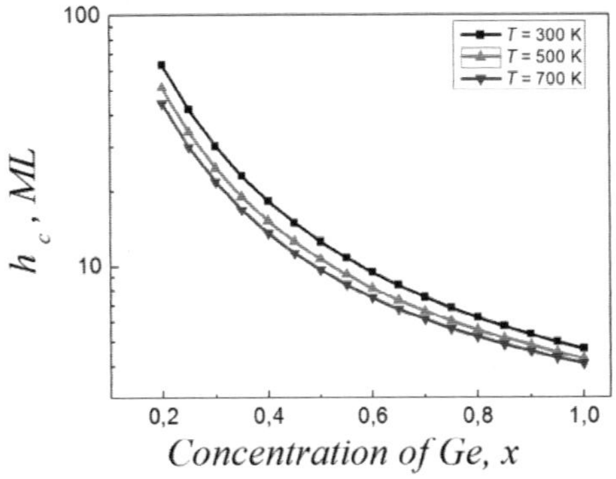

Figure 1. Critical thickness of transition to 3D growth in the presence of tin with assumption of diffusion coefficient increasing by four orders of magnitude.

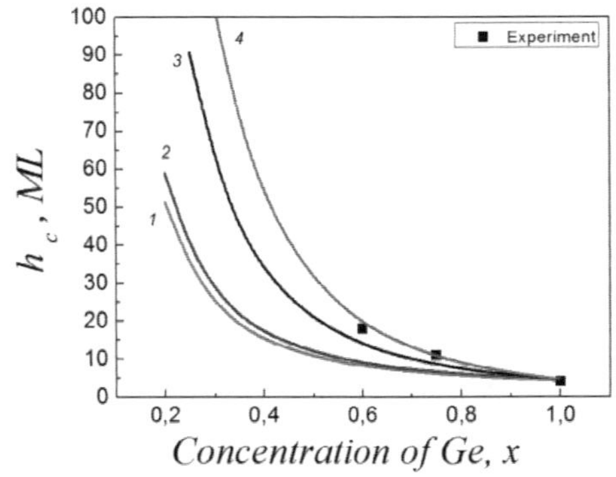

Figure 2. Comparison of calculated critical thicknesses with experimental data [3] with assumption of diffusion coefficient increasing by two (*2*) and four (*1*) orders of magnitude and with assumption of the substrate surface energy decreasing by 100 erg/cm^2 (*3*) and 200 erg/cm^2 (*4*).

3. Conclusions

Thus, in this paper theoretical model for calculation of dependencies of critical thickness of Stranski–Krastanow transition on composition x for $Si_{1-x}Ge_x$/Sn/Si system at various temperatures is proposed. Moreover, the described model allows one to estimate surface density and size distribution function of quantum dots in these systems. Although there are a very limited number of experimental data on the critical thickness of transition from 2D to 3D growth in $Si_{1-x}Ge_x$/Sn/Si systems, the results of calculations demonstrate good agreement with the data available. In addition, derived expressions explain the experimentally observed temperature dependencies of critical thickness at various germanium contents.

Acknowledgements

This study was supported by the Ministry of education and science of the Russian Federation, State task № 3.8177.2017/ITR.

References

[1] Wu J, Chen S, Seeds A and Liu H 2015 *J. Phys. D: Appl. Phys.* **48** 363001
[2] Aqua J, Berbezier I and Favre L 2013 *Phys. Rep.* **522** 59
[3] Wakahara A, Vong K K, Hasegawa T, Fujihara A and Sasaki A 1995 *J. Cryst. Growth* **151** 52
[4] Lin X W, Liliental-Weber Z, Washburn J, Weber E R, Sasaki A, Wakahara A and Hasegawa T 1995 *Phys. Rev. B* **52** 16581
[5] Kandel D and Kaxiras E 1999 *Solid State Physics* **54** 219
[6] Bratland K A, Foo Y L, Desjardins P and Greene J E 2003 *Appl. Phys. Lett.* **82** 4247
[7] Pchelyakov O P, Dvurechenskii A V, Nikiforov A I, Voitsekhovskii A V, Grigor'ev D V and Kokhanenko A P 2011 *Russian Physics Journal* **53** 943
[8] Dubrovskii V G 2006 *Semiconductors* **40** 1123
[9] Lozovoy K A, Kokhanenko A P and Voitsekhovskii A V 2015 *Phys. Chem. Chem. Phys.* **17** 30052
[10] Lozovoy K A, Kokhanenko A P and Voitsekhovskii A V 2016 *Appl. Phys. Lett.* **109** 021604
[11] Dolbak A E and Olshanetsky B Z 2008 *Cent. Eur. J. Phys.* **6** 634

Hybrid sp^2+sp^3 carbon phases created from carbon nanotubes

M I Tingaev[1] **and E A Belenkov**[1]
[1]Department of Solid State Physics, Chelyabinsk State University, Chelyabinsk 454001, Russia

Abstract. Using the density functional theory in the gradient approximation (DFT–GGA) methods was calculated the geometrically optimized structure and electronic properties for six new hybrid carbon phases. These hybrid phases consists of atoms in three - and four-coordinated (sp^2+sp^3-hybridized) states. The initial structure of the carbon phases was constructed by partial cross-linking of (8,0) carbon nanotube bundles. Sublimation energies calculated for hybrid phases above the sublimation energy of cubic diamond, however, fall into the range of values typical for carbon materials, which are stable under normal conditions. The density of electronic states at the Fermi energy for the two phases is non-zero and these phases should have metallic properties. The other hybrid phases should be semiconductors with a band gap from 0.5 to 1.1 eV.

1. Introduction

Hybrid carbon phases are crystalline compounds consisting of atoms in states with different coordination and with the hybridization of the electronic orbitals [1,2]. Hybrid carbon phases can be assigned to one of four classes – sp+sp^2 sp+sp^3 sp^2+sp^3 and sp+sp^2+sp^3 [1,2]. Graphite-diamond carbon phases consist of atoms in three- and four-coordinated (sp^2+sp^3-hybridized) atoms and have a solid three-dimensional structure. Such compounds are very interesting because of their properties that can be changed by varying the ratio of carbon atoms in the states with different hybridization. A variety of hybrid carbon materials with sp^2+sp^3-hybridized atoms have been synthesized experimentally, when exposed to fullerene condensates high pressures [3,4]. The possibility of even more variety of graphite-diamond carbon phases and nanostructures have been established theoretically [5-7]. Model build the structure of various compounds of this type are possible from the structures of the precursor (graphene layers, carbon nanotubes (CNT), or fulleren-like clusters) as a result of partial "cross-linking" or "combining" [1,2,8]. However, some of the constructed models of the hybrid phases are unstable – for example, all hybrid sp^2+sp^3 phases from graphene layers are unstable. In this study, we investigated the structure and properties of a range of hybrid phases based on carbon nanotubes bundles.

2. Methods

The initial structures of the hybrid phases were built as a result of a partial "cross-linking" or "combining" bundles of single-layer zig-zag (8,0) carbon nanotubes (Figure 1a). From built phases were selected only phases, in which the atoms with sp^2 and sp^3 hybridizations were in crystallographically equivalent positions. Nanotubes in bundles were in four-coordinated states (each nanotube have four neighbors). When "cross-linking" or "combining" between the carbon atoms of neighboring nanotubes are formed bonds (Figure 1b). The atoms forming additional covalent bonds with atoms of the neighboring nanotubes, passed from three-coordinated states in four-coordinated states (from sp^2 hybridized to sp^3-hybridized states). As a result of this process the initial structure of the hybrid sp^2+sp^3 phases were formed (Figure 1b). These structures were geometrically optimized by methods of the density functional theory in the gradient approximation [9] (Figure 1c). The calculations used a set of k-points: 12×12×12. To limit the dimensionality of a set of basis functions, the value of E_{cutoff} was set at 950 eV. Structural parameters characterizing the hybrid phases were L_i interatomic bonds and angles between them φ_{ij} (Figure 2).

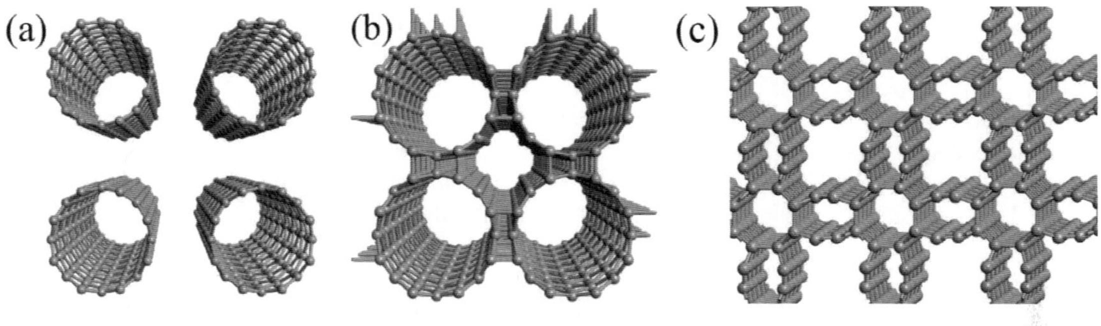

Figure 1. Bundle of carbon nanotubes (a), partially cross-linked carbon nanotubes (b), and phase structure after geometrical optimization (c).

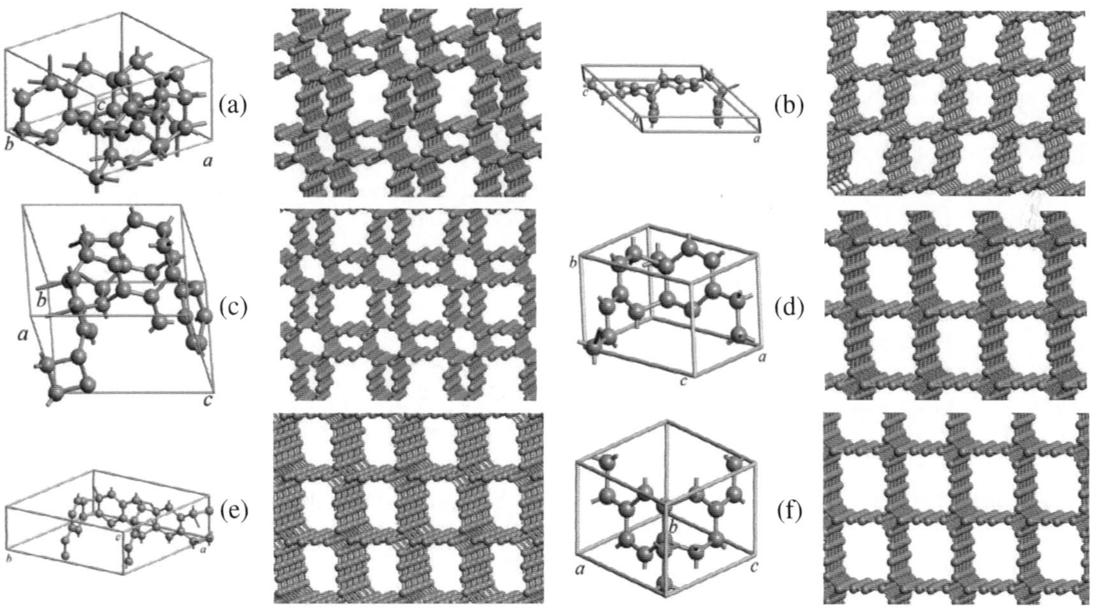

Figure 2. Geometrically optimized structure of the unit cells and the crystal lattice of sp^2+sp^3 carbon phases, obtained by cross-linking and combining nanotubes (8,0) bundles: (a) $T_{(8,0)}a_{A4a}$, (b) $T_{(8,0)}b_{A4a}$; (c) $T_{(8,0)}a_{A4b}$, (d) $T_{(8,0)}b_{A4b}$, (e) $T_{(8,0)}a_{A4c}$, (f) $T_{(8,0)}a_{A4d}$.

3. Results and Discusion

From different variants of cross-linking and combining of single-walled nanotubes (8,0) bundles were constructed six hybrid sp^2+sp^3 phases with various structures. Four phases ($T_{(8,0)}a_{A4a}$, $T_{(8,0)}a_{A4b}$,$T_{(8,0)}a_{A4c}$, $T_{(8,0)}a_{A4d}$) were obtained as a result of "cross-linking" nanotube bundles, and two phases ($T_{(8,0)}b_{A4a}$, $T_{(8,0)}b_{A4b}$) as a result of "combining". Image of crystal structure phases and their unit cells after geometrical optimization are shown in Figure 3. All six phases constructed after the geometrical optimization were sp^2+sp^3-hybridized and their structure were not transformed into the structure of the compounds consisting of only sp^3 or only sp^2 atoms, just as observed previously for hybrid phases based on the graphene layers [7].

The estimated values of the lengths of interatomic bonds and angles between them are given in Table 1. Lengths of bonds vary in the range from 1.387 to 1.622 Å (see Table 1). Different lengths of the covalent bonds due to the fact that they are formed of different number of electron pairs. The maximum length of the bonds formed by one pair of electrons. These are bonds with lengths L_1, L_2, L_3 between – four-coordinated (sp^3-hybridized) atoms. The minimum length of bonds L_5, L_6 is observed between three-coordinated (sp^2-hybridized) atoms. The length of the bond L_4 between three-coordinated and four-coordinated atoms has an intermediate length. The short lengths of the bonds, seems to indicate that the order of bond is in the range from 1.2 to 1.5. The lengths of the bonds in the hybrid phases differ from the values characteristic of graphite and diamond. The degree of these deviations is characterized by the strength parameters Str_1 and Str_2, which are calculated as the sum of modules of the difference of bond lengths in the hybrid phase, and in the ideal structures of the diamond (1) graphite (2) (Table 2). The highest values for the strength is for the phase of $T_{(8,0)}b_{A4b}$ ($Str_1 = 0.335$, $Str_2 = 0.234$).

Table 1. Values of the lengths of interatomic bonds and angles between them for sp^2+sp^3 carbon phases, obtained by cross-linking and combining bundles of single-layer CNT (8,0).

	$T_{(8,0)}a_{A4a}$	$T_{(8,0)}b_{A4a}$	$T_{(8,0)}a_{A4b}$	$T_{(8,0)}b_{A4b}$	$T_{(8,0)}a_{A4c}$	$T_{(8,0)}a_{A4d}$
	1.563	1.535	1.557	1.621	1.599	1.481
L_2 (Å)	1.598	1.603	1.539	1.562	1.569	1.621
L_3 (Å)	1.581	1.574	1.606	1.622	1.618	1.563
L_4 (Å)	1.552	1.536	1.532	1.482	1.504	1.621
L_5 (Å)	1.387	1.404	1.402	1.424	1.446	1.563
L_6 (Å)	1.433	1.407	1.424	1.393	1.411	1.481
φ_{12} (°)	115.91	107.52	112.70	110.36	112.73	116.38
φ_{13} (°)	111.83	114.89	89.51	83.01	90.03	115.96
φ_{14} (°)	104.69	100.73	115.83	116.39	116.24	116.35
φ_{23} (°)	109.91	110.49	129.11	110.38	90.09	110.33
φ_{24} (°)	99.76	107.54	102.83	115.97	117.82	83.09
φ_{34} (°)	114.20	114.94	107.38	116.29	124.77	110.31
φ_{45} (°)	121.64	114.91	115.97	123.06	119.57	110.33
φ_{46} (°)	114.20	122.26	119.62	115.97	117.77	116.36
φ_{56} (°)	120.54	121.37	121.32	120.97	122.63	115.96

The angles between the bonds in the hybrid phases have values that differ from the typical values for ideal structures of diamond and graphite. So, four-coordinated atoms, the angles between bonds vary in the range from 83.01° to 129.11°, which differs significantly from the angle of 109.47°, which is characteristic for diamond. The angles between bonds for three-coordinated atoms change from 110.33° to 123.06°, which is also different from the angle of 120°, typical for graphite. Such deviations of the angles of atomic bonds indicate the presence of significant stresses in the structure of the hybrid phase. Quantitative characteristics of these stresses are deformation parameters Def_1 and Def_2, which is calculated as the sum of modules of differences of the angles in a hybrid phase in the ideal structures of the diamond (1) graphite (2) (Table 2). In the most stressed condition of sp^3 atoms in the phase $T_{(8,0)}a_{A4c}$ ($Def_1 = 72.50°$) and sp^2 atoms in phase $T_{(8,0)}a_{A4d}$ ($Def_2 = 17.35°$). However, the values of deformation parameters for the other phases are of the same order as for the most stressed. The density of the hybrid phases is less than the density of cubic diamond. Two phases have a density of ~ 2.7 g/cm^3, and the density of the other phases is less than the density of graphite (Table 2).

The unit cell of the hybrid phases containing from 16 to 32 atoms. The number of atoms in sp^2-hybridized state to the number of atoms in the sp^3-hybridized state for the five phases ratio of 1:1. In the phase of $T_{(8,0)}b_{A4a}$ number of sp^2-hybridized atoms is twice the number of sp^3-hybridized atoms. The ring parameters of atoms in four-coordinated conditions in the different hybrid phases differ significantly (Table 2). In addition to hexagonal rings, there are also four-, eight - and twelve-coordinated rings, which, apparently, are the reason for the presence of stresses in the structure phases.

Table 2. Structural parameters, unit cell parameters and values for the complete and total energy (E_{tot}), differential energy (ΔE_{tot}), sublimation energy (E_{sub}), density (ρ) and band gap (ΔE_g) of carbon sp^2+sp^3 phases obtained by cross-linking and combining the bundles of single layer (8,0) CNT (for comparison, G – graphite, D – diamond).

Phase	$T_{(8,0)}a_{A4a}$	$T_{(8,0)}b_{A4a}$	$T_{(8,0)}a_{A4b}$	$T_{(8,0)}b_{A4b}$	$T_{(8,0)}a_{A4c}$	$T_{(8,0)}a_{A4d}$	G	D
a (Å)	7.451	9.958	7.686	6.399	8.950	6.399	2.487	3.596
b (Å)	7.451	4.271	4.293	4.253	4.380	4.253	2.487	3.596
c (Å)	4.229	7.361	7.816	6.399	12.114	6.399	7.333	3.596
α (°)	90	73.14	106.04	109.41	90.05	70.60	90	90
β (°)	90	47.32	93.39	96.36	46.19	83.72	90	90
γ (°)	90	90	73.30	70.59	104.36	70.59	120	90
N (at)	32	24	32	16	32	16	4	2
Rng_l	$6^5 8^1$	$6^5 12^1$	$4^1 6^3 8^2$	$4^1 6^5$	$4^2 6^6 8^1$	$4^1 6^5$	-	6^6
Str_1 (Å)	0.276	0.287	0.276	0.335	0.273	0.159	0	-
Str_2 (Å)	0.114	0.106	0.092	0.234	0.193	0.235	-	0
Def_1 (°)	28.47	24.53	57.92	48.50	72.50	48.36	-	0
Def_2 (°)	7.970	8.720	5.730	8.060	5.290	17.35	0	-
sp^2:sp^3	1:1	2:1	1:1	1:1	1:1	1:1	∞	0
E_{tot} (eV)	-5021.2	-3768.1	-5018.2	-2511.9	-5021.7	-2511.9	-629.35	-314.55
E_{tot} (eV/at)	-156.91	-157.01	-156.82	-156.99	-156.93	-156.99	-157.34	-157.28
ΔE_{tot} (eV)	0.43	0.33	0.52	0.35	0.41	0.35	0	0.06
E_{sub} (eV/at)	7.35	7.45	7.26	7.43	7.37	7.43	7.78	7.72
ΔE_g (eV)	0.5	0	0	0.196	0.126	0.418	0	5.6
ρ (g/cm³)	2.718	1.529	2.774	1.832	1.982	2.059	2.2	3.5

Total energy E_{tot} (per atom) for all phases is less than for graphite and diamond (Table 2). The difference of total energy of the hybrid phases relative to the energy of graphite ΔE_{tot} is changed from 0.33 to 0.52 eV/at. Sublimation energies for hybrid phases lies in the range of 7.26 to 7.45 eV/at. Most sublimation energy is observed for the phase of $T_{(8,0)}b_{A4a}$ (E_{sub} = 7.45 eV/at). Apparently, this phase should be the most stable. However, the sublimation energy of all investigated hybrid phases are in the range of values typical for carbon materials, available stable under normal conditions, and all these phases can be synthesized.

The results of calculations of the density of electronic states of the hybrid phases is shown in Figure 4. For the two phases, $T_{(8,0)}b_{A4a}$ and $T_{(8,0)}a_{A4b}$ the density of electronic states at the Fermi level is nonzero, the bandgap is zero and the phase should have metallic properties. For the remaining four phases of the bandgap ranges from 0.126 to 0.5 eV. These structures must be narrow-gap semiconductors.

4. Conclusions

Thus, the calculations of the structure and properties of six new hybrid sp^2+sp^3 carbon phases based on bundles of carbon nanotubes (8,0). These phases can stably exist under normal conditions. It is possible to synthesize them as a result of polymerization of carbon nanotubes or molecular compounds having the carbon structure of the frame close to the structure of fragment phases. The conductive properties of the hybrid phases can be metal or semiconductor, so these phases may find applications as materials for electronic equipment.

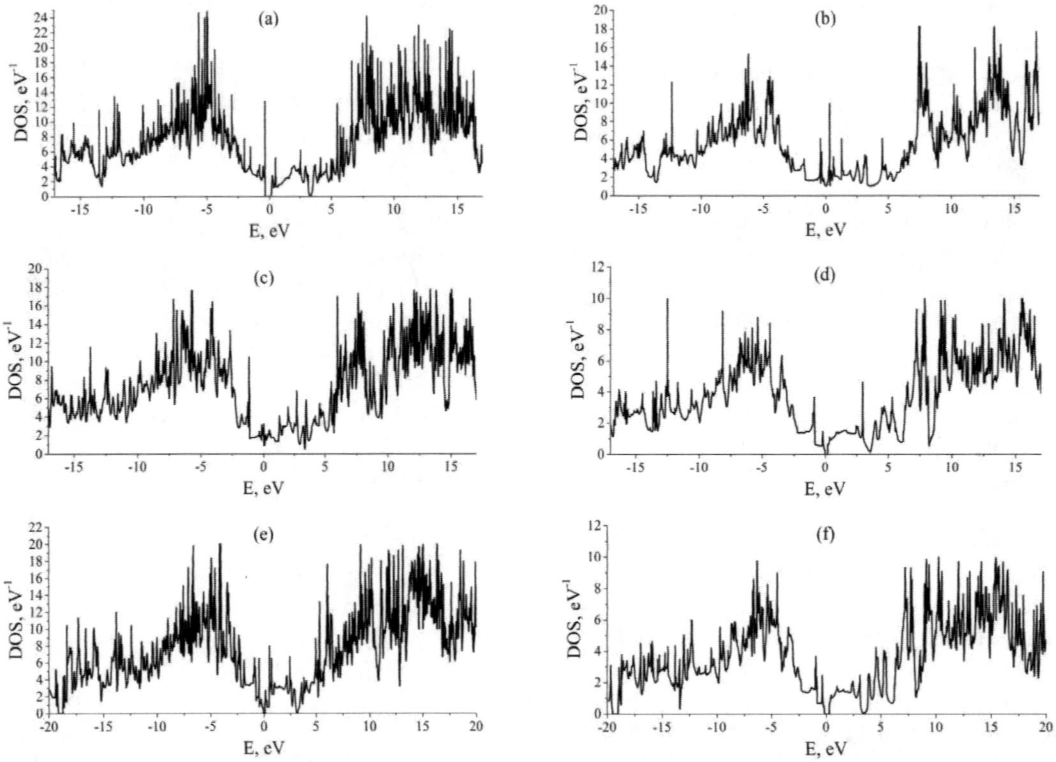

Figure 3. Bandstructures of sp^2+sp^3 carbon phases, obtained by merging and combining bundles of single-layer CNT (8,0): (a) $T_{(8,0)}a_{A4a}$; (b) $T_{(8,0)}b_{A4a}$; (c) $T_{(8,0)}a_{A4b}$; (d) $T_{(8,0)}b_{A4b}$; (e) $T_{(8,0)}a_{A4c}$; (f) $T_{(8,0)}a_{A4d}$.

Acknowledgments
The authors thank Foundation of Perspective Research of Chelyabinsk State University for the financial support.

References
[1] Belenkov E A, Greshnyakov V A 2013 Phys. Solid State 55(8) 1754
[2] Belenkov E A, Greshnyakov V A 2013 Carb. Mat. 28(4) 273
[3] Nunez-Regueiro M, Monceau P, Rassat A, Bernier P, Zahab A 1991 Nature **354** 289
[4] Brazhkin V V, Lyapin A G 1995 JETP Let. 42 350
[5] Bucknum M J, Hoffmann R 1994 J. Am. Chem. Soc. 116 (25) 11456
[6] Kuc A, Seifert G 2006 Phys. Rev. B. 74 214104
[7] Belenkov E A, Ivanovskii A L, Ul'yanov S. N, Shabiev F K 2005 J. Str. Chem. 46(6) 961
[8] Belenkov E A, Tingaev M I 2015 Let. on mat. 5(1) 15
[9] Koch W A, Holthausen M C 2001 *Chemist's Guide to Density Functional Theory. 2nd edition* (Weinheim: Wiley-VCH Verlag GmbH) 293 p

MBE growth and optical properties of GaN layers on SiC/Si(111) hybrid substrate

R R Reznik[1, 2, 4, 7], K P Kotlyar[1], I P Soshnikov[1, 3, 4], S A Kukushkin[6], A V Osipov[6], E V Nikitina[1] and G E Cirlin[1-4, 7]

[1]St Petersburg Academic University, Khlopina 8/3, 194021, St-Petersburg, Russia
[2]Peter the Great St.Petersburg Polytechnic University, Polytechnicheskaya 29, 195251, St-Petersburg, Russia
[3]Ioffe Physical Technical Institute RAS, Politekhnicheskaya 26, 194021, St-Petersburg, Russia
[4]Institute for Analytical Instrumentation RAS, Rizhsky 26, 190103, St-Petersburg, Russia
[5]Fock Institute of Physics, St-Petersburg state University, 198504, Petrodvorets, Russia
[6]Institute of Problems of Mechanical Engineering Russian Academy of Science, Bolshoj 6, 199178, St. Petersburg, Russia
[7]ITMO University, Kronverkskiy pr. 49, 197101, St. Petersburg, Russia

E-mail: moment92@mail.ru

Abstract. The fundamental possibility of the growth of GaN layers by molecular-beam epitaxy on a silicon substrate with nanoscale buffer layer of silicon carbide without any AlN layers has been demonstrated for the first time. Morphological properties of the resulting system have been studied.

1. Introduction

Wide-gap nanoheterostructures based on GaN are advantageous for the fabrication of electronic [1] and optoelectronic [2] devices. The high mechanical strength and melting temperature (2500°C) of GaN combined with its chemical stability ensure stable operation of such devices in rough conditions. The lack of gallium nitride substrates motivates the search for substrates suitable for growing such structures. Sapphire substrates yield high dislocations densities due to a considerable (13%) mismatch between interatomic distances in the (0001) interface plane. The use of silicon carbide substrates, which offer the closest match between lattice parameters at the interface, is limited by their small area and high cost. The experiments on growing GaN layers on silicon are very promising, since this material is relatively cheap and has an acceptable thermal conductivity; it is also possible to fabricate large-area silicon substrates. This growth technique provides an opportunity to integrate optoelectronic devices based on gallium nitride into silicon micro- and nanoelectronics. However, the mismatch of lattice parameters at the interface between Si(111) and GaN(0001) planes is 17%, while the coefficients of thermal expansion differ by 33%. This results in a high density of defects of various types, which impair the performance of devices to be fabricated, in the epitaxial layer. For example, threading dislocations are the centers of nonradiative recombination (Coulomb scattering) and thus reduce the electron mobility. Dislocations enhance the reverse current in p–n junctions and the dark current in photodetectors. It is known that optoelectronic devices based on GaN may be operated for a long time with no degradation in spite of a high density of linear defects. The density of dislocations in GaN-based light-emitting diodes may exceed the maximum permissible density of dislocations in GaAs-based devices by five orders of magnitude. The reason for this is as follows: although dislocations serve as recombination centers in GaN, the diffusion length of minority carriers is shorter than the distance between dislocations (shorter than 250 nm). The mobility of linear defects in gallium nitride is much lower than that in other A3B5 compounds. However, the quality of GaN structures should be perfected in order to extend the service life of optoelectronic devices.

In this work, in order to increase the optical quality of GaN layers nanometer (about 50 nm) buffer layer of SiC was used. Because of small difference in the lattice parameters between SiC and GaN (only 3%), we were able to grow thick GaN layer (about 20 micrometers) on SiC/Si(111) hybrid substrate by molecular-beam epitaxy without any AlN layers, which adversely affect the structures properties for the first time.

2. Experiments and results.

The method developed and described in [3, 4, 5] was used to form a SiC buffer layer on a Si(111) substrate. This method for synthesizing silicon carbide films on silicon differs fundamentally from all the currently known techniques for single crystal, film, and nanostructure growth. It is based on the concept of substituting a fraction of silicon atoms with carbon atoms within the silicon substrate. In order to do this, a carbon atom is introduced in advance into an interstitial position of the silicon lattice, while a neighboring silicon atom is removed. As a result, a silicon vacancy is produced, and an ensemble of dilatation dipoles, which are stable complexes formed by dilatation centers (a silicon atom in the interstitial position and a silicon vacancy), is formed in the near-surface region of the silicon lattice. In a crystal with cubic symmetry, these two dilatation centers interact elastically with each other. Specifically, if dilatation dipoles are arranged perpendicularly to the (111) silicon plane, they are attracted to each other, and almost the entire dilatation elastic energy, which was produced by the introduction of a carbon atom and the formation of a vacancy, relaxes. Such elastic dipoles are synthesized via a chemical reaction between silicon and carbon monoxide. The rate of this reaction is the highest in the direction along which dilatation dipoles attract each other ([111] of the Si substrate). After the process of chemical transformation is complete, mechanical dipoles, which have fulfilled their function, decay into a silicon carbide film and pores under its surface. The film orientation is set not just by the substrate surface (as is often the case in traditional film growth techniques), but by the "old" crystal structure of the initial Si matrix. The gas temperature and pressure are chosen so that the processes of formation of SiC and pore seeds run alongside each other with equal rates. The formation of elastic carbon atom–Si vacancy dipoles provides an opportunity to synthesize high-quality silicon carbide films. The potential to grow not only cubic, but also several hexagonal polytypes of SiC is one of the most important properties of nano-SiC synthesized by the atomic substitution method [4, 5].

Growth experiments are carried out using Riber Compact12 MBE setup equipped with the effusion Ga cell and the nitrogen source. Firstly, the substrate was transferred into the growth chamber, and the substrate temperature was set at 950°C for further purification. Then, the substrate temperature was lowered to 730°C, after that, using nitrogen source plasma was ignited and nitrogen flow on the sample was formed, and the gallium source was opened too, the temperature of which was chosen 1000°C, which corresponds to growth rate 1 ML/s. Growth time of GaN NWs was 24 hours. The layer growth was monitored in situ using (RHEED) showing GaN layer wurtzite phase after an incubation period (typically, 10 min) (Figure 1).

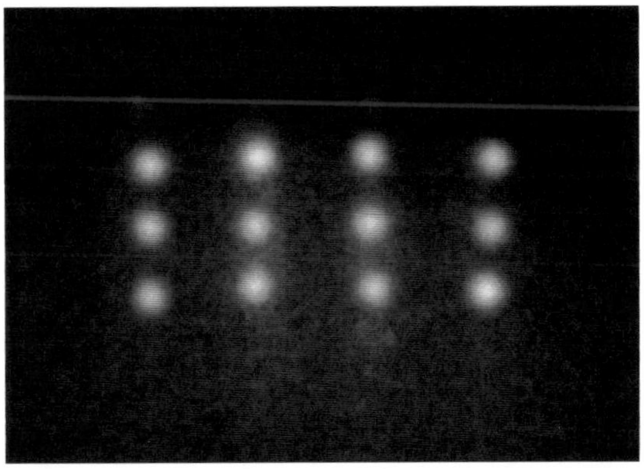

Figure 1. RHEED pattern of the GaN NWs on the SiC buffer layer after 120 minutes of growth.

After the growth, the samples were studied using the scanning electron microscopy (SEM). Figure 2 shows typical SEM images of GaN layers grown on SiC/Si combined substrate. It is seen that GaN layers on the substrate Si/SiC (111) average height is about 24 μm. It should be noted that layers samples were formed by a columnar growth mechanism.

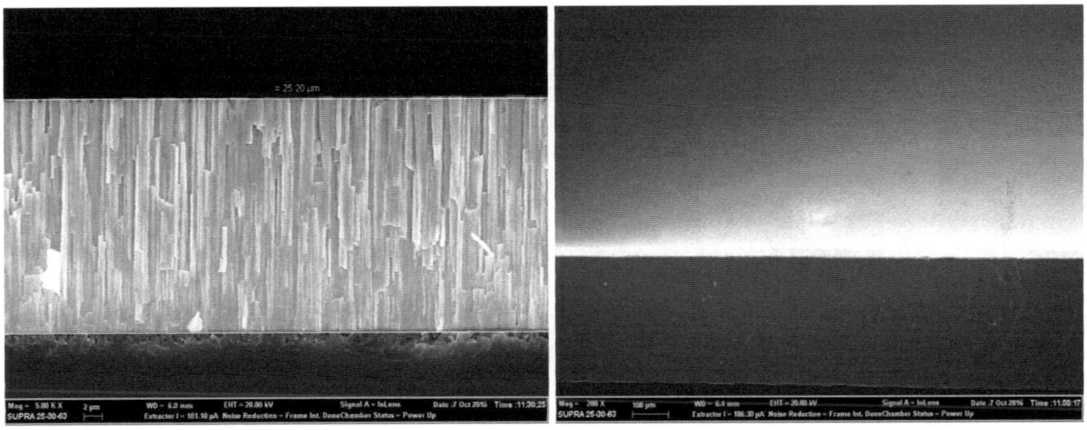

Figure 2. Typical SEM images of GaN layers grown on SiC/Si combined substrate.

In conclusion, we demonstrate the possibility of the growth of GaN layers by molecular-beam epitaxy on a silicon substrate with nanoscale buffer layer of silicon carbide without any AlN layers has been demonstrated for the first time. Optimization of growth parameters and the investigation of optical properties will be studied in the following works.

Acknowledgments
We grateful the support of the Ministry of education and science of Russian Federation (state task, project No. 16.2483.2017/4.6). The nanowire samples were grown under the support of Russian Science Foundation (Project No 14-12-00393).

References

[1] Pearton S J, Ren F 2000 *Adv. Mater.* **12** 1571
[2] Nakamura S, Fasol G 1997 *New York: Springer-Verlag* 368

[3] Kukushkin S A, Osipov A V 2008 *Phys. of the Solid State* **50** 1238
[4] Kukushkin S A, Osipov A V, Feoktistov N A 2014 *Phys. Solid State* **56** 1507
[5] Kukushkin S A, Osipov A V 2014 *J. Phys. D: Appl. Phys.* **47** 313001

IOP Conf. Series: Journal of Physics: Conf. Series **917** (2017) 032015 doi:10.1088/1742-6596/917/3/032015

Structure of graphane polymorphs

T E Belenkova[1], V A Greshyakov[1], V M Chernov[1] and E A Belenkov[1]

[1] Physics Department, Chelyabinsk State University, Chelyabinsk 454001, Russia

Abstract. Calculations of the structure and electronic properties for five structural variations of graphane were performed within the framework of density functional theory (DFT) with generalized gradient approximations (GGA). The electron densities of states and band structure of graphene crystals have been calculated. It has been established that the band gap for graphane polymorphs varies from 5.50 eV to 5.65 eV. Sublimation energy of graphane layers with different structure was varying from 11.33 to 11.48 eV per C-H molecular group.

1. Introduction

Graphane layers obtain by joining one hydrogen atom to each of the carbon atoms of the graphene layer [1]. Hydrogenographane layers were experimentally synthesized in 2009 [2]. Graphane (unlike graphene) is a dielectric. Hydrogenographane layers consist of carbon atoms in the sp^3 hybridization [3]. Perhaps, structural types of hydrogenographane layers with different properties can exist that may be of interest in terms of using them as a basis for electronic devices. In this study, we investigated the structure and electronic properties of the five main polymorphic types of hydrogenographane layers.

2. Methods

For practical use of graphane layers it is desirable obtaining layers of graphane with non-carbon atoms attached to only one side of the layer. However, studies carried out earlier in [5] are showed, that these layers are strongly deformed and tend to form structure like a scroll, which makes them unsuitable for use as a basis for nanoelectronic devices. Therefore, in this study we analyzed only hydrogenographane layers with hydrogen atoms attached to both sides of graphane layer.

As shown by theoretical analysis, there are only five different ways in connection with non-carbon atoms of graphane layer on both sides, so that the positions of all the carbon atoms were crystallographically equivalent. Various structural types of graphane layers with non-equivalent positions of carbon atoms may be more, but layers with equivalent positions of the atoms should be the most thermodynamically stable.

Figure 1 shows the initial graphene layers to which the non-carbon atoms can join with the upper side of layer (atoms highlighted in gray), or with the lower side of layer (atoms marked black color). The first of these types of connection, T1, corresponds hydrogenographane layer described in [2,3]. Three ways to join T2, T3, and T4 were considered earlier in [6,7]. The possibility of existence the fifth main structural type, T5, of hydrogenographane layers studied for the first time in this paper.

Theoretically built hydrogenographane layers of the five main types were subjected to geometrical optimization by the methods of density functional theory (DFT) [8] within the generalized gradient approximations (GGA) [9]. Calculations were carried out for the layers arranged in a three-dimensional structure, but the distance between the layers (at the same time it was a unit cell period along the crystallographic c-axis) was chosen sufficiently large (10 Å), so that layers do not affect the structure of the adjacent layers and their properties. Densities of electron states calculated using k-point set of 12 × 12 × 12. The wave functions are displayed on a truncated basis set of plane waves. The dimension of the set of basis functions has been limited by the value E_{cutoff} = 1 keV.

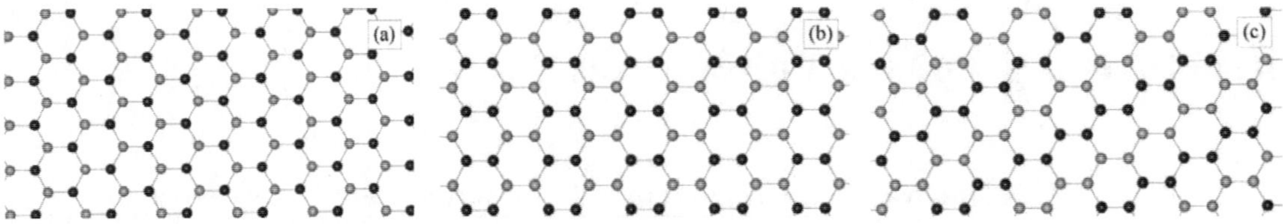

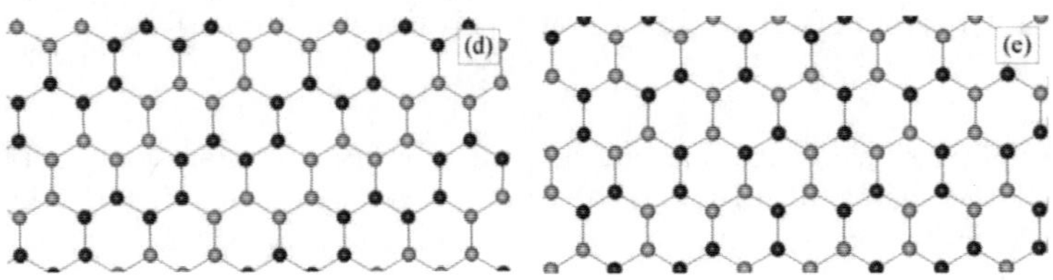

Figure 1. Possible ways of attachment of hydrogen atoms on two sides of the graphene layer: (a) T1; (b) T2; (c) T3; (d) T4; (e) T5.

3. Results and Discussion

Images of five structural types of graphane layers, geometrically optimized by the DFT-GGA, are shown in Figure 2. The structural characteristics of the layers are shown in Table 1. Hydrogenographane layers of various types have very different structures. The layers also have various unit cells which differ by parameters and number of contained atoms (Figure 2 and Table 1). The unit cells of graphane layers contain from four to sixteen atoms and characterized by hexagonal (Hex), triclinic (Tr), and orthorhombic (Ort) crystal systems. The density of hydrogenographane layers ranges from 0.77 to 0.88 mg per m^2 that exceeds the density of graphene layers of 0.74 mg per m^2.

Table 1. The structural parameters and properties the main types of graphane layers (T1–T5), as well as hexagonal graphene (L_6).

Type of layer	T1	T2	T3	T4	T5	L_6
Crystal system	Hex	Tr	Tr	Ort	Ort	Hex
a, Å	2.5532	2.542	4.616	4.571	5.006	2.491
b, Å		5.021	2.561	4.350	4.440	
γ, °	120	59.56	56.26	90	90	120
E_{total}, eV/u.c.	-346.74	-693.08	-693.29	-1385.99	-1385.74	-314.64
E_{total}, eV/(CH)	-173.37	-173.27	-173.32	-173.25	-173.22	-
E_{sub}, eV	11.48	11.38	11.43	11.36	11.33	7.76
ΔE_{sub}, eV	3.72	3.63	3.67	3.60	3.57	0
Δ, eV	5.55	5.50	5.54	5.51	5.65	0.00
ρ, mg/m^2	0.77	0.79	0.88	0.87	0.78	0.74
N, at.	4	8	8	16	16	2

The total energy per one molecular C-H group (E_{total}) ranges from -173.37 to -173.22 eV (Table 1). Sublimation energy (E_{sub}) was made as a result of the calculation of the difference between the total energy of a single C-H molecular group in hydrogenographane layer and the energy of isolated carbon and hydrogen atoms (the estimated value of these energies equal to -149.56 and -12.33 eV/atom, correspondently).

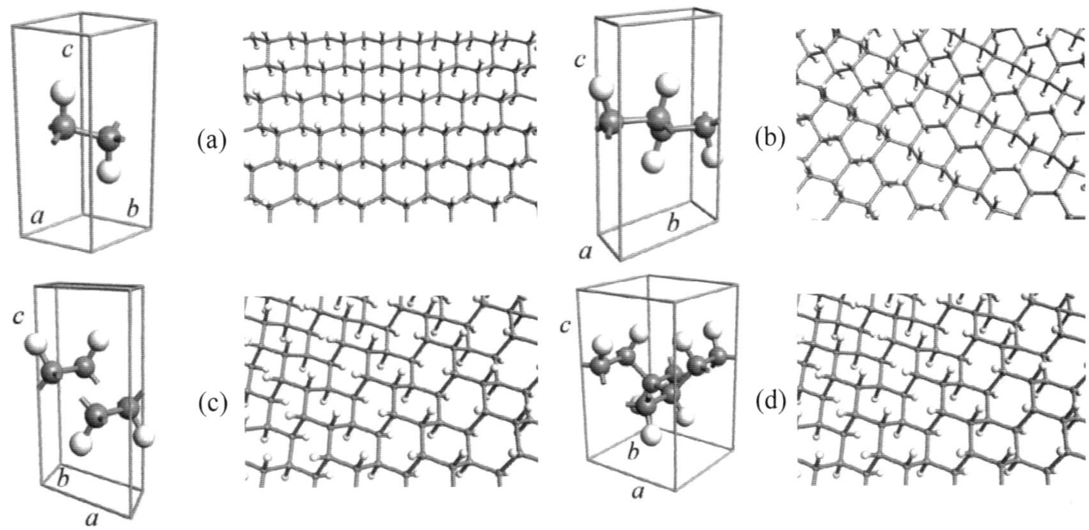

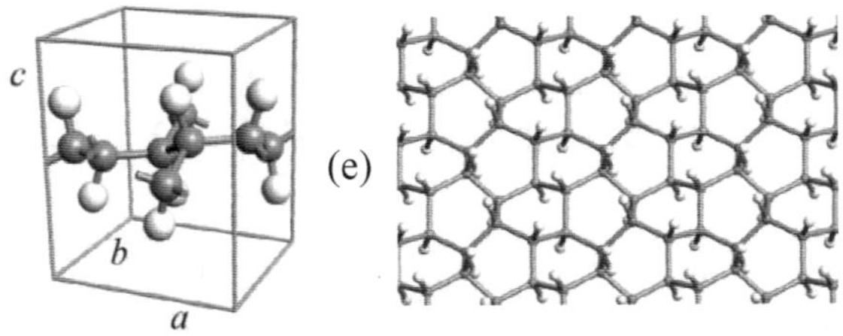

Figure 2. Unit cells and structures of hydrogenographane layers of various types: (a) T1; (b) T2; (c) T3; (d) T4; (e) T5.

In result of calculations (In the calculation process,) the specific sublimation energy per one carbon atom of graphane layer was found. The numerical values of sublimation energy hydrogenographane layers are in the range of 11.33 to 11.48 eV/atom which are 3.57–3.7 eV greater than the value of the specific sublimation energy of hexagonal graphene layer (Table 1). Apparently this is due to higher strength carbon-hydrogen bond as compared with the carbon-carbon bond between the π electrons of graphene layer. The maximum sublimation energy is observed for the T1-type layer and the minimum is observed for T5, but the relative energy difference is only 1.3 %, which indicates the possibility of stable existence of all main polymorphic types of graphane.

The results of measuring the lengths of interatomic bonds and the angles between them are shown in Table 2. The carbon-carbon bond lengths vary in the range of 1.5459 to 1.5830 Å, whereas the carbon-hydrogen bond lengths are in the interval of 1.1094 to 1.1183 Å. The bond angles in graphane layers differ from the angle of 109.47 ° which is characteristic of the atomic positions in the diamond structure. The total deviation of bond angles from the diamond angle is characterized by a deformation parameter Def, which quantifies the degree of tension of the structure in comparison with the diamond structure. The minimum value of Def (10.87 °) was observed for T1-type graphane layer, the maximum value (22.33 °) was observed for T5-layer. This is a good explanation of the sublimation energy difference of structural types, because the sublimation energy minimum is observed for strained layers, while the maximum of E_{sub} is typical for the layer with the smallest deformation parameter (Tables 1 and 2).

The calculated electron densities of states for five graphane layers are shown in Figure 3. The numerical values of band gaps are shown in Table 1. The band gap (Δ) of graphene layers were in the range of 5.50 to 5.65 eV. This indicates that all hydrogenographane layers, regardless of the type, should exhibit structural dielectric properties.

IOP Conf. Series: Journal of Physics: Conf. Series **917** (2017) 032015 doi:10.1088/1742-6596/917/3/032015

Table 2. The bond lengths and bond angles in graphane layers.

Type of layer	T_1	T_2	T_3	T_4	T_5
$L_{1(C\text{-}C)}$, Å	1.5467	1.5796	1.5530	1.5506	1.5701
$L_{2(C\text{-}C)}$, Å	1.5467	1.5459	1.5530	1.5830	1.5476
$L_{3(C\text{-}C)}$, Å	1.5467	1.5459	1.5530	1.5830	1.5559
$L_{4(C\text{-}H)}$, Å	1.1183	1.1124	1.1128	1.1094	1.1139
φ_{12}, °	111.25	112.27	112.28	112.44	112.00
φ_{13}, °	111.25	112.27	112.28	112.44	117.77
φ_{14}, °	107.63	106.61	107.85	107.42	105.94
φ_{23}, °	111.25	110.60	111.04	112.39	107.44
φ_{24}, °	107.63	107.41	106.50	106.47	106.55
φ_{34}, °	107.63	107.41	106.50	106.47	106.45
Def, °	10.87	13.71	14.74	16.91	22.33

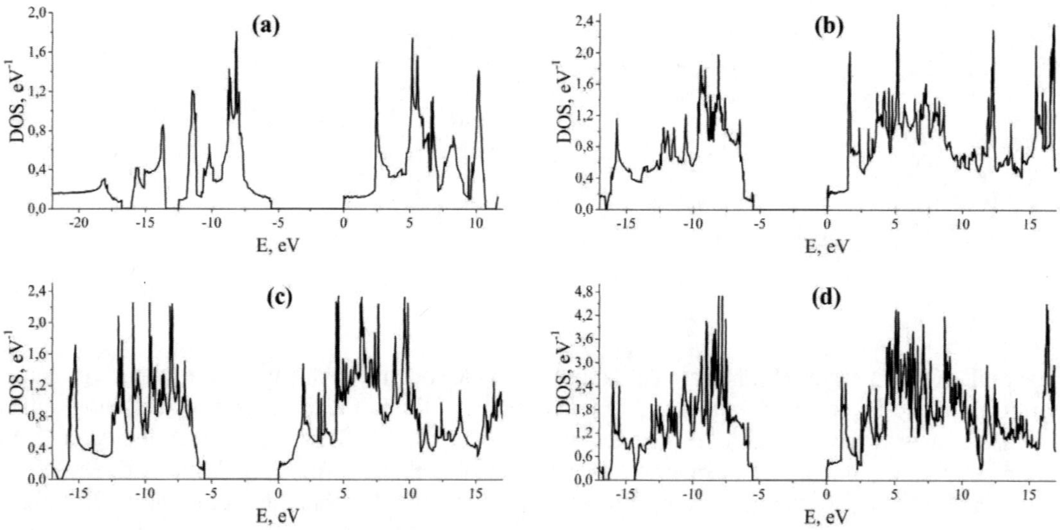

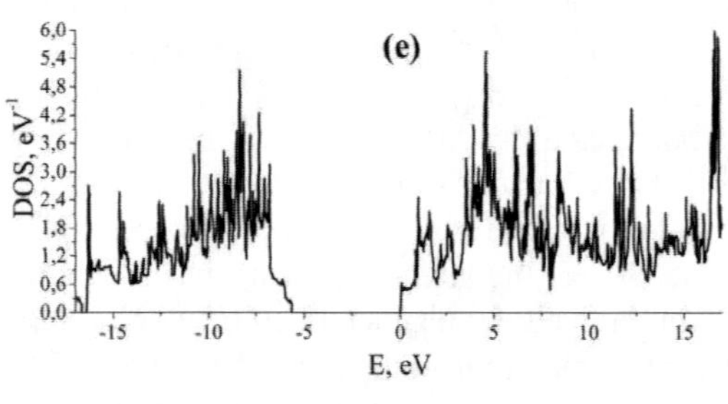

Figure 3. The density of electron states hydrogenographane layers of different types: (a) T1; (b) T2; (c) T3; (d) T4; (e) T5 (zero energy was taken by the Fermi energy).

4. Conclusions

Thus, the crystal structure and electronic properties of five main structural types of hydrogenographane layers were calculated by the DFT-GGA method. The layers were obtained by attachment of hydrogen atoms to the graphene layer L_6. Calculations of the band structure and density of electronic states graphane compounds indicate that their electronic properties to be different from each other slightly and all hydrogenographane layers should bad conduct electricity. Local carbonization and making of graphene ribbons on dielectric layer of graphane are possible for hydrogenographane layers of any of the five main structural types. Different nanoelectronic devices can be obtained as a result of this process. Perhaps some of the polymorphic types of graphane may be preferable to deal with these problems; therefore this question remains open and should be studied in future research.

Acknowledgments
The authors thank Foundation of Perspective Research of Chelyabinsk State University for the financial support.

References

[1] Novoselov K S, Geim A K, Morozov S V, Jiang D, Zhang Y, Dubonos S V, Grigorieva I V, Firsov A A 2004 *Science* **306** 666
[2] Sofo J O, Chaudhari A S, Barber G D 2007 *Phys. Rev.* B **75** 153401
[3] Elias D C, Nair R R, Mohiuddin T M, Morozov S V, Blake P, Halsall M P, Ferrari A C, Boukhvalov D W, Katsnelson M I, Geim A K, Novoselov K S 2009 *Science* **323** 610
[4] Openov L A, Podlivaev A I 2010 *Tech. Phys. Lett.* **36** 31
[5] Belenkov E A, Shabiev F K 2015 *Letters on Materials* **5** 459
[6] Wen X D, Hand L, Labet V, Yang T, Hoffmann R, Ashcroft N W, Oganov A R, Lyakhov A O 2011 *PNAS* **108** 6833
[7] Sorokin P B, Chernozatonskii L A 2013 *Phys. Usp.* **56** 105
[8] Koch W A, Holthausen M C 2001 *Chemist's Guide to Density Functional Theory. 2nd edition* (Weinheim: Wiley-VCH Verlag GmbH) 293 p
[9] Perdew J P, Chevary J A, Vosko S H, Jackson K A, Pederson M R, Singh D J, Fiolhais C 1992 *Phys. Rev.* B **46** 6671

IOP Conf. Series: Journal of Physics: Conf. Series **917** (2017) 032016 doi:10.1088/1742-6596/917/3/032016

Quinary InAlGaPAs/GaAs solid solutions grown by temperature gradient zone melting

D A Arustamyan[1], M L Lunina[2], L S Lunin[1,2] and A E Kazakova[1]

[1] Platov South-Russian State Polytechnic University (NPI), Novocherkassk, 346428, Russia
[2] Southern Scientific Center Russian Academy of Sciences, Rostov-on-Don, 3464006, Russia

Abstract. In this paper we present the results of calculating lattice constant and bandgap $In_xAl_yGa_{1-x-y}P_zAs_{1-z}$/GaAs heterostructures at room temperature. The features of growing InAlGaPAs/GaAs quinary solid solution by the temperature gradient zone melting are discussed.

1. Introduction

Using a quinary solid solutions of $A^{III}B^{V}$ compounds enable to independently modify the lattice constant, the bandgap and the coefficient of thermal expansion (CTE) [1]. This create opportunities for obtaining isoparametric (izoperiodic and CTE matched) heterostructures based on the binary compounds and working in a wider spectral range than quaternary solid solution [2]. These opportunities open up fundamentally new perspectives for the creation of optoelectronic devices and instruments[3-5]. For use in optoelectronics, such heterostructures must meet high requirements for their crystal structure [6, 7].

One of the available methods for obtaining heterostructures continues to be epitaxy from the liquid phase (LPE) [8]. Despite a number of advantages of the LPE method [9], growing by this method of multicomponent heterostructures of $A^{III}B^{V}$ compounds with a controlled composition is difficult due to a temperature change in the growth process. The method of temperature gradient zone melting (TGZM) is economical, since it does not require a large number of initial materials due to small thicknesses of the zones and repeated use of the same melt. In the TGZM method, by programmatically changing the process parameters is possible to control the properties and structural perfection of multicomponent heterostructures of $A^{III}B^{V}$ compound.

2. Theoretical analysis

For calculate the lattice constant a(x, y, z), and the bandgap (E_g) using equation from [10]. The $In_xAl_yGa_{1-x-y}P_zAs_{1-z}$/GaAs heterostructures even with a slight increase concentration of indium (x) isoperiodic line shifted to the region of high phosphorus concentrations (figure 1). This is important, as in a quaternary AlGaPAs solid solution for matching lattice constant of the epitaxial layer and the GaAs substrate required the small amount of phosphorus and high partition ratio, therefore, growing by TGZM method without feeding is difficult. Thus, to varying the composition of the solid phase is needed feeding of the liquid phase with strongly segregate components, for example, P and Al.

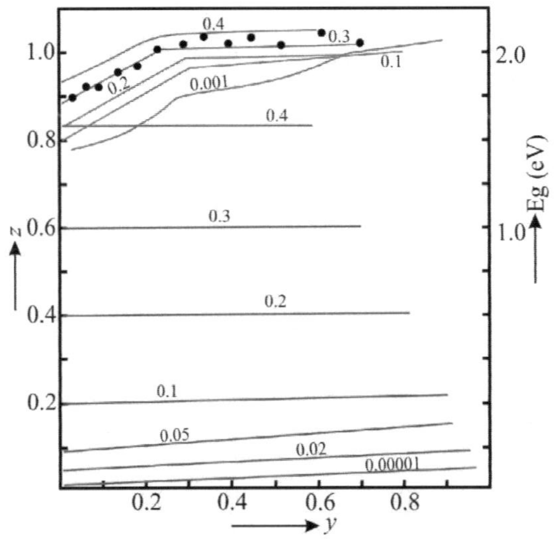

Figure 1. Concentration dependences: isoperiodic lines (red lines), the bandgap (green lines), the value of x given on the curves, dots indicate the experimental data.

At tightening the requirements for the matching ratio of the CTE of the layer and the substrate, area of the izoperiodic compositions on y is narrowed. Calculations showed that usually CTE of quinary solid solutions less than that of GaAs, and the difference of CTE increases with increasing distance of the composition of the solid solution from the composition of a binary compound. When $\delta a = 0.2$, then cooling heterostructure from T_{epit} (1100 K) to room temperature leads to rise the difference between the lattices constants of the substrate and epitaxial layer less than 0.1%.

Figure 1 shows the functions of the bandgap E_g of the compositions (green lines). It is seen that E_g rise with increasing *y* for certain *x*.

3. Experimental part

Growing InAlGaPAs/GaAs heterostructure performed by TGZM method. As a substrate used monocrystalline gallium arsenide wafers without doping, Si-doped with concentration $1\text{-}5\cdot10^{18}cm^{-3}$ (n-type) and Zn-doped with concentration $1\text{-}8\cdot10^{18}cm^{-3}$. Diameter wafers was 50 nm with planes orientation (100) and (111). Plates polished and subjected to chemical etching to remove the outer surface layer. Processing performed in etchants: $HCl:CrO_3 = 3:1$ or $H_2SO_4:H_2O_2:H_2O = 3:1:1$.

The thickness of the substrate after the treatment was 400-450 µm. The thickness of the gallium zone varied from 10 to 200 µm. At the growth of thin (less than 10 µm) layers, InP, GaP, and GaAs plates were used as additional feeding. In order to obtain thick (more than 10 µm) layers of InAlGaPAs solid solutions, were used previously synthesized InAlPAs polycrystals containing components with a large distribution coefficient (Al, P). The resulting ingot of the polycrystal was cut into plates with a thickness of 500-600 µm.

«GaAs-Ga - polycrystal InAlGaPAs» sandwiches placed in a chamber which was vacuated to 10^{-3} Pa, and then inflated hydrogen to 0.5 Pa, TGZM process was carried out in a hydrogen stream. Heterostructure InAlGaPAs/GaAs grown at temperatures $837 \leq T \leq 937$ K and a temperature gradient $10 \leq G \leq 30$ K. Choice of temperature is carried out by examining the liquidus temperature for each composition by using visually-thermal analysis «in situ» by the method of [11].

The determination of the mismatch between the lattice parameters of the substrate and the layer and also the estimation of the crystal perfection of the heterostructures were carried out by the method of recording X-ray diffraction.

The compositions of the resulting solid solutions were determined on an X-ray microprobe analyzer "Camebax" with an accelerating voltage of 20 kV and a primary beam current of 0.1-1.0 µA. Photoluminescence measurements were carried out in the spectral range from 950 to 1500 nm at a temperature of 300 K and in liquid nitrogen at 77 K in a cryostat with quartz windows. The source of exciting optical radiation was an injection laser with wavelength of 402 nm and radiation power of 8.5 mW. The photoluminescence was excited from the side of the epitaxial layers.

4. Results and discussion

Investigation and analysis of luminescent properties of quinary solid solutions is conveniently carried out in comparison, starting from binary compounds, increasing number of components to five (figure 2).

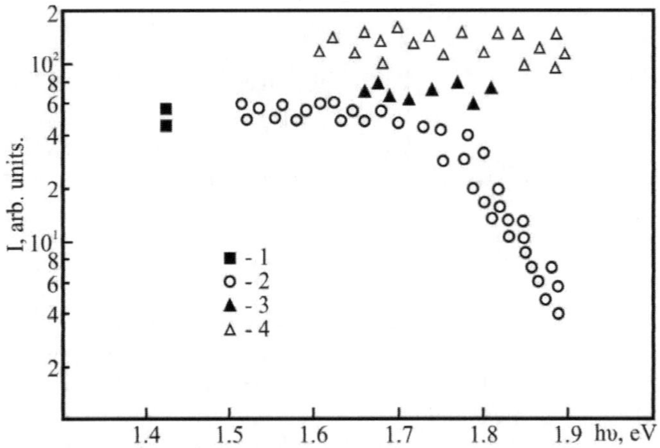

Figure 2. Dependence of the intensity of photoluminescence of $A^{III}B^{V}$ solid solutions on the composition (bandgap): 1 -p-GaAs, 2-p-AlGaAs, 3-p-AlGaPAs, 4-p-InAlGaPAs/GaAs.

For AlGaAs solid solutions in the region of "direct" transitions, the radiation intensity is almost constant, and a further increase in the content of AlAs in the solid solution causes a rapid decrease in the intensity of radiative recombination.

The physical reason for the sharp decrease in the intensity of radiation in the region of transition to "indirect" compositions is the transition of electrons from the "direct" minimum to "indirect" and the increase, as a consequence, of the fraction of nonradiative recombination.

The use of quaternary solid solutions isoperiodic for binary is reduces the effect of structure defect associated with the mismatch between the lattice parameters of materials mating at the heterointerface and, thereby, increases the efficiency of radiative recombination. As shown in figure 2, the photoluminescence efficiency of epitaxial AlGaPAs layers is practically independent of the composition of the solid solution. In addition, the intensity of radiative recombination increases with the introduction of the fourth component into the ternary solid solution.

Analysis showed that the introduction of the fifth component into the quaternary solid solution makes it possible to obtain, on the basis of binary substrates, strictly isoperiodic and isoexpandic heterostructures and minimizes defectiveness at the heterointerface, which entails a significant increase in the efficiency of radiative recombination and an improvement in the luminescence characteristics of such heterostructures.

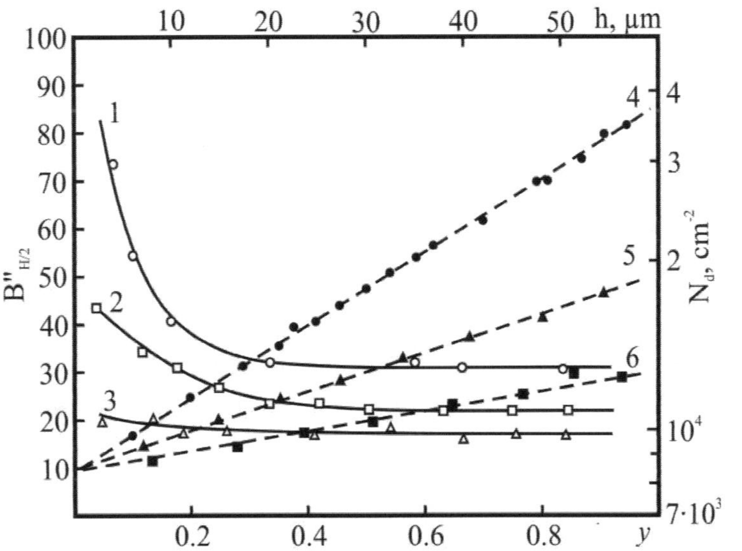

Figure 3. The distribution of the density of dislocations along the thickness of the $GaAs/In_xAl_yGa_{1-x-y}P_zAs_{1-z}$ layers: 1 - x = 0, y = 0.35, z = 0; 2 - x = 0, y = 0.35, z = 0.05; 3 - x = 0.08, y = 0.35, z = 0.05; (T = 1133 K, l = 300 μm, G = 30 °C/sec); Changes in the structural perfection of the epitaxial layer and the bordering regions of the substrate, depending on the composition of the film: 4 - reflection from AlGaAs; 5 - reflection from AlGaPAs; 6 - reflection from InAlGaPAs.

Experimental studies (figure 2) showed that for the isoperiodic with the substrate quinary $In_xAl_yGa_{1-x-y}P_zAs_{1-z}/GaAs$ solid solutions not observed a significant change in the intensity of the photoluminescence when their composition varies, if the band structure of the solid solution does not change.

In figure 3 shows the results of studies of the dislocation density distribution over the thickness of various heterostructures. It is seen that the density of dislocations is the same at all points of a homogeneous composition of the layer (figure 3, curves 1-3), except for regions near the heterointerface.

The dislocation densities in the layer and the substrate usually do not coincide. Moreover, in the layer on the boundary with the substrate, the dislocation density exceeds that in the layer and the substrate. This character of the distribution of the dislocation density along the thickness indicates that, for at large values of the jump in the composition at the ternary layer-binary substrate boundary, the heterojunction is tense.

Measurements of the width of the rocking curve ($B_{H/2}$''), carried out on the angle lap of $Al_xGa_{1-x}As/GaAs$ heterostructures, confirm this assumption. The results of the ($B_{H/2}$'') measurement for $Al_xGa_{1-x}As$ solid solutions and the adjacent regions of GaAs substrates are respectively shown in figure 3, curves 4, 5. It can be seen that structural perfection decreases with increasing concentration of aluminum at the heterointerface. Due to the difference in the CTE of GaAs and AlGaAs, on the heterointerface the value of Δa at the crystallization temperature is lower than at room temperature. Therefore, after cooling to room temperature, the heterojunction is in a stressed state.

Studies have shown that elastic stresses are concentrated in the region of ~ 10 μm on both sides of the heterointerface and are absent in the volume of the layer, and their value in the substrate and layer is practically the same and depends on the composition of the solid solution. Evaluation of the elastic stress level near the $Al_xGa_{1-x}As/GaAs$ heterojunction give values not exceeding $1.8\cdot10^7$ Pa. The stressed state of the heterojunction can be reduced by decreasing Δa due to the addition of the fourth component (P) in $Al_xGa_{1-x}As$ (figure 3, curve 5) and InP (figure 3, curve 6). The results of measurements of the dislocations density along the thickness of the $Al_xGa_{1-x}As_{1-y}P_y$ layers (figure 3, curve 2) showed that the defect area was significantly reduced. The addition of the fourth component reduces the level of residual stresses to $0.5\cdot10^7$ Pa in $Al_{0.35}Ga_{0.65}As_{0.95}P_{0.05}$ heterostructures. Consequently, the addition of the fourth component significantly reduces the stresses by providing a lattice matching of the layer and substrate at epitaxial temperatures. However, the defectiveness of the heterointerface, even for isoperiodic compositions of quaternary solid solutions remains quite large, due to the difference in the CTE of the interface layer and substrate, and can be reduced by going to the quinary solution (figure 3, curve 6). In heterostructures based on quaternary solid solutions in which the lattice parameters of the mating materials under epitaxy temperatures coincide, the main source of defects heterojunctions are the thermal stresses caused by the difference in CTE. Thermal stresses are removed by introducing the fifth component. In figure 3 shows the distribution of the dislocation density over the thickness of the $In_xAl_yGa_{1-x-y}P_zAs_{1-z}/GaAs$ layers (curve 3). It can be seen that the introduction of a small addition of In ($x < 0.1$) almost completely nullifies the defect transition range.

Consequently, in quinary solid solutions, due to the matches of lattice parameters and CTE, it is possible to obtain an "ideal" heterocomposition (in which both misfit dislocations and stresses are absent). However, in such cases, the quality of the substrate material plays an important role, since the main contribution to defect formation in quinary heterostructures will be the inheritance of defects from the substrate. Thus, in the epitaxy of quinary solid solutions, the requirements on the quality of the substrate material significantly increase.

5. Conclusions

InAlGaPAs/GaAs quinary solid solutions, with controlled thickness, composition, and structural perfection can be grown in a temperature gradient field from the liquid phase. A homogeneous distribution of the components across the thickness was achieved using recrystallising source InAlPAs. The main parameters determining the quality of the surface, structural perfection and luminescent properties of heterostructures are the thickness and composition of the liquid zone, temperature and temperature gradient of the TGZM process.

Quinary InAlGaPAs/GaAs heterostructures have high crystalline perfection and better luminescence characteristics in comparison with the corresponding ternary and quaternary heterostructures, and therefore are more preferable for highly efficient devices based on radiative recombination.

Acknowledgments

This work was financial supported by the Russian Foundation for Basic Research [grant No. 17-08-01206-a] and State Assignment from the Ministry of Education and Science of the Russian Federation [grant No. 16.4757.2017/8.9].

References

[1] Lozovskii V N, Lunin L S 1992 *Quinary solid solutions of compounds (new optoelectronics materials)* (Rostov-on-Don: NCSC of Higher School,) p 193

[2] Kuznetsov V V, Lunin L S, Ratushnyi V I 2003 *Heterostructures based on quaternary and quinary solid solutions of compounds* (Rostov-on-Don: NCSC of Higher School,) p 376

[3] Lunin L S, Sysoev I A, Alfimova D L, Chebotarev S N, Pashchenko A S 2011 *Inorganic Materials* **47** 816
[4] Lunin L S, Chebotarev S N, Pashchenko A S, Dudnikov S A 2013 *Journal of Surface Investigation* **7** 36
[5] Chebotarev S N, Pashchenko A S, Williamson A, Lunin L S, Irkha V A, Gamidov V A 2015 *Tech. Phys. Lett.* **41** 661
[6] Alfimova D L, Lunin L S, Lunina M L, Pashchenko A S, Chebotarev S N 2016 *Physics of the Solid State* **58**(9) 1751-1757
[7] Ma C, Dech M, Tarof L, Yn I 1995 *IEEE. Trans. Electron Devices* **42**(5) 810
[8] Xu Z L, Xu W J, Li L, Yang C Q, Liu H D 1998 *Applied Physics A* **66**(5) 565-567
[9] Mukai S, Yajima H, Mitsuhashi Y, Yanagisawa S, Kutsuwada N 1984 *Appl. Phys. Lett.* **44** 904
[10] Lozovskii V N, Lunin L S, Askaryan T A 1989 *Soviet Physics Journal* **32**(7) 537-542
[11] Kuznetsov V V, Kognovitskaya E A, Lunina M L, Rubtsov E R 2011 *Russian Journal of Physical Chemistry A* **85**(12) 2062-2067

The structure of carbon nanotubes formed of graphene layers $L_{4\text{-}8}$, $L_{5\text{-}7}$, $L_{3\text{-}12}$, $L_{4\text{-}6\text{-}12}$

K E Shapovalova[1] **and E A Belenkov**[1]
[1] Physics Department, Chelyabinsk State University, Chelyabinsk 454001, Russia

Abstract. We geometrically calculate the optimized structure of nanotubes based on the graphene layers, using the method of molecular mechanics MM+. It was found that only the nanotubes, based on the graphene layers $L_{4\text{-}8}$, $L_{5\text{-}7}$, $L_{3\text{-}12}$, $L_{4\text{-}6\text{-}12}$, have a cylindrical form. Calculations of the sublimation energy, carried out using the semi-empirical quantum-mechanic method PM3, show that energy increases with the increase of nanotube diameters.

1. Introduction

From model of carbon nanotubes (CNT) can be obtained by folding narrow strips, cut from graphene layers [1]. The real mechanism of carbon nanotubes structure formation is other; however, the model approach allows to classify carbon nanotubes and to explain the possibility of varying from semiconductor to metal properties. The surface of the CNTs, synthesized to the current time, consists of carbon atoms located on the nanotube surface at the vertices of hexagons, i.e. these nanotubes are formed on the basis of hexagonal graphene. The possible graphene layers can be of different types [2-5], therefore, an existence of different structural varieties of carbon nanotubes on the basis of these layers is possible. In this study, we investigate the carbon nanotubes-based graphene layers $L_{4\text{-}8}$, $L_{5\text{-}7}$, $L_{3\text{-}12}$, $L_{4\text{-}6\text{-}12}$.

2. Methods

The calculations of the geometrically optimized structures of carbon nanotubes were performed by using molecular mechanics method MM+. The ribbons of a various width and of the various carbon-carbon bonds orientation, relatively to their axis, are cut out from the $L_{4\text{-}8}$, $L_{5\text{-}7a}$, $L_{5\text{-}7b}$, $L_{3\text{-}12}$, $L_{4\text{-}6\text{-}12}$ graphene layers (Figure 1), then, in the result of folding these layers and stitching them around the edges, the original CNT structure was obtained for a further geometric optimization. After the geometric optimization by the semi-empirical quantum-mechanical method PM3, we have calculated the sublimation energy of all CNTs.

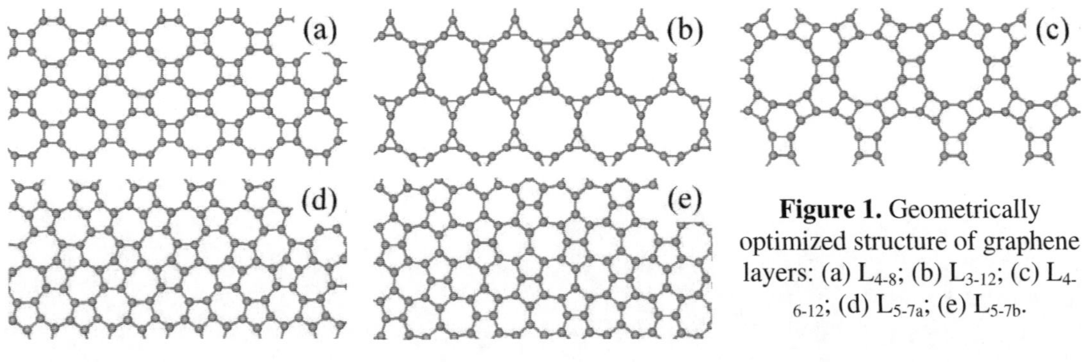

Figure 1. Geometrically optimized structure of graphene layers: (a) $L_{4\text{-}8}$; (b) $L_{3\text{-}12}$; (c) $L_{4\text{-}6\text{-}12}$; (d) $L_{5\text{-}7a}$; (e) $L_{5\text{-}7b}$.

3. Results and Discusion

In the result of modelling researches, the structure of 23 different carbon nanotubes were calculated on the basis of graphene layers $L_{4\text{-}8}$, $L_{5\text{-}7a}$, $L_{5\text{-}7b}$, $L_{3\text{-}12}$, $L_{4\text{-}6\text{-}12}$. The zigzag, armchair and chiral carbon nanotubes were calculated. The structure of the CNTs formed from $L_{4\text{-}8}$, $L_{5\text{-}7a}$, $L_{5\text{-}7b}$, $L_{4\text{-}6\text{-}12}$ layers was cylindrical – all atoms, of which these CNTs are composed, were located on smooth cylindrical surfaces, constituting the walls of nanotubes (Figures 2-5). However, for nanotubes based on the graphene layer $L_{3\text{-}12}$, the cylindrical structure violates, so that the walls of the nanotube become wrinkled and in a cross-section, the nanotube becomes more oval than round. Structural and energy characteristics of carbon nanotubes are given in Table 1. Calculations were carried out for fragments of nanotubes. The number of carbon atoms (N_C) in the fragments was from 48 to 240 atoms. At the ends of fragments of nanotubes, broken bonds were compensated by hydrogen atoms. The diameters of nanotubes varied from 3.33 to 11.6 Å. The minimum energy of sublimation (E_{sub}=88.34 kcal/mol) was in the nanotube (3.0) formed from the $L_{4\text{-}8}$ layer. The maximum energy of sublimation (E_{sub}=225.97 kcal/mol) was observed for the tube (7.0) which was obtained from layer $L_{4\text{-}8}$. Sublimation energy of carbon nanotubes depends on their diameter – the larger is the diameter, the greater is the sublimation energy value.

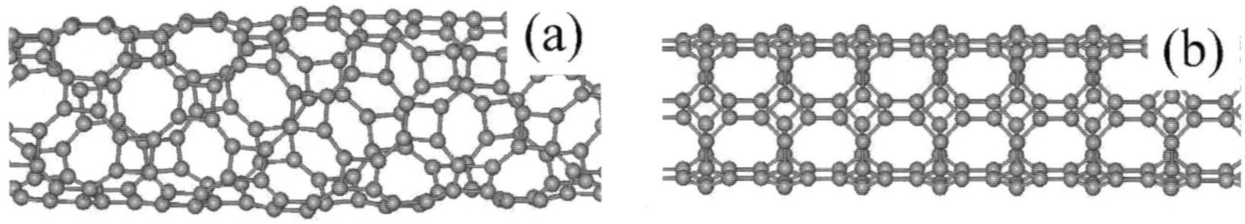

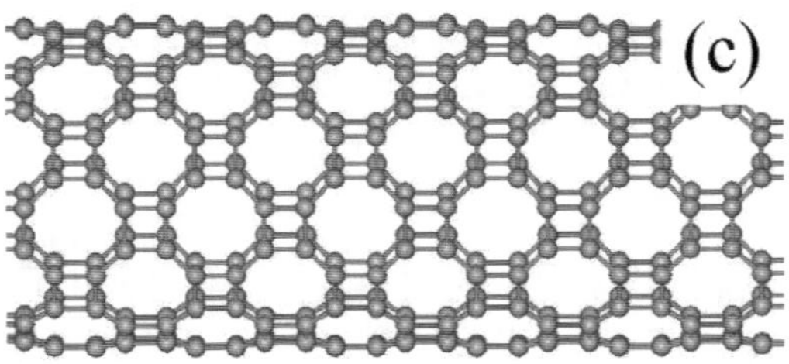

Figure 2. Geometrically optimized CNTs structure, obtained by the folding of ribbons of cut from graphene layers $L_{4\text{-}8}$: (a) (5,1); (b) (6,0); (c) (7,0).

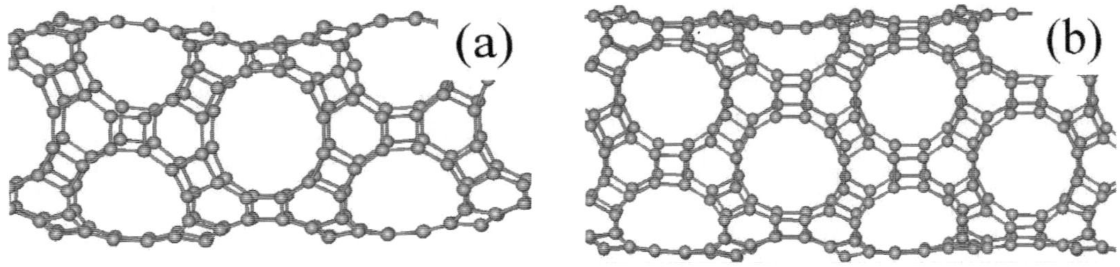

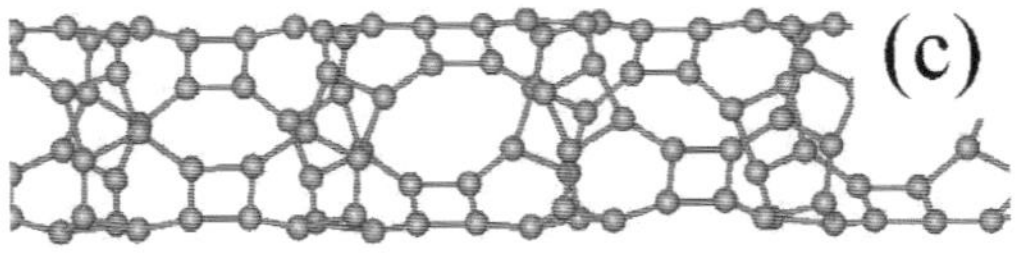

Figure 3. Geometrically optimized CNTs structure, obtained by the folding of ribbons of cut from graphene layers $L_{4\text{-}6\text{-}12}$: (a) (4,0); (b) (5,0); (c) (2,0).

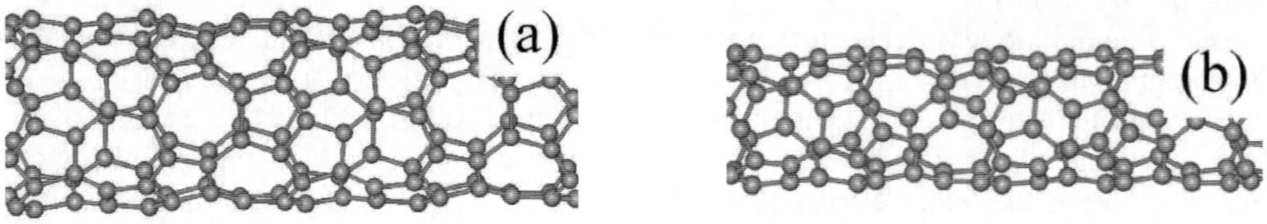

Figure 4. Geometrically optimized CNTs structure, obtained by the folding of ribbons of cut from graphene layers $L_{5\text{-}7a}$: (a) (3,0); (b) (2,1).

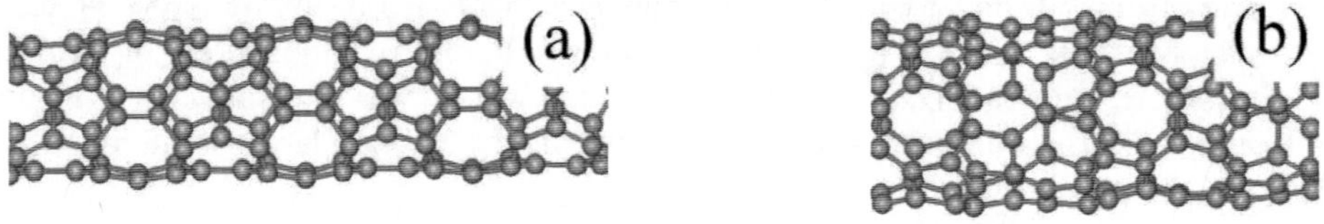

Figure 5. Geometrically optimized CNTs structure, obtained by the folding of ribbons of cut from graphene layers $L_{5\text{-}7b}$: (a) (4,0); (b) (2,2).

Table 1. Structural and energy characteristics of carbon nanotubes formed of $L_{4\text{-}8}$, $L_{5\text{-}7}$, $L_{3\text{-}12}$, $L_{4\text{-}6\text{-}12}$ graphene layers.

Layer	CNT	N_C (at)	N_H (at)	D (Å)	E_{totol} (kcal/mol)	$E_{C\text{-}C}$ (kcal/mol)	E_{sub} (kcal/mol)
$L_{4\text{-}8}$	(3,0)	60	6	3.39	-6297.1	-58.89	88.34
	(4,0)	56	16	4.47	-9070.1	-103.9	155.85
	(6,0)	120	12	6.58	-18102	-97.29	145.94
	(2,0)	64	8	3.33	-8954	-84.57	126.86
	(5,0)	180	20	7.45	-40342	-143.83	215.75
	(7,0)	224	28	10.56	-47488	-150.65	225.97
	(3,2)	52	8	4.32	-6799.9	-76.03	114.05
	(4,1)	76	14	5.71	-11647	-97.88	146.82
	(5,1)	128	18	7.06	-19885	-102.25	153.37
$L_{4\text{-}6\text{-}12}$	(2,0)	72	16	5.02	-10686	-95.12	142.68
	(4,0)	192	16	9.55	-28225	-96.61	144.92
	(5,0)	240	20	10.58	-36696	-101.49	152.24
	(3,2)	106	12	5.7	-15804	-95.62	143.43
	(4,1)	200	12	6.93	-28853	-94.15	141.22
	(5,1)	158	20	9.69	-24150	-101.22	151.83
$L_{5\text{-}7a}$	(2,0)	56	8	3.6	-8444.3	-100.78	151.18
	(3,0)	84	12	4.97	-13552	-108.55	162.83
	(4,0)	154	16	6.7	-23483	-101.92	152.89
	(2,2)	112	16	5.78	-18473	-111.21	166.81
$L_{5\text{-}7b}$	(2,0)	60	12	4.6	-9862.7	-111.35	167.03
	(3,0)	90	18	7.34	-15051	-113.55	170.32
	(5,0)	100	20	11.6	-16964	-115.40	173.10
	(2,1)	48	8	4,9	-7364,8	-102,81	154,22

IOP Conf. Series: Journal of Physics: Conf. Series **917** (2017) 032017 doi:10.1088/1742-6596/917/3/032017

Acknowledgments
The authors thank Foundation of Perspective Research of Chelyabinsk State University for the financial support.

References

[1] Dresselhaus M S, Dresselhaus G and Avouris Ph 2001 *Carbon nanotubes: synthesis, structure, properties, and applications* (Berlin: Springer-Verlag) p 80
[2] Belenkov E A, Kochengin A E 2015 *Physics of the Solid State* **57** 2126
[3] Kochergin A E, Belenkov E A 2017 Perspektivnye materialy **2** 5
[4] Belenkov E A, Greshnyakov V A 2013 *Phys. Solid State* **55** 1754
[5] Belenkov E A, Greshnyakov V A 2015 *New Carbon Materials* **28** 273

IOP Conf. Series: Journal of Physics: Conf. Series **917** (2017) 032018 doi:10.1088/1742-6596/917/3/032018

Stress in (Al, Ga)N heterostructures grown on 6H-SiC and Si substrates byplasma-assisted molecular beam epitaxy

O A Koshelev[1,2], D V Nechaev[1], A A Sitnikova[1], V V Ratnikov[1], S V Ivanov[1], and V N Jmerik[1]

[1]Ioffe Institute, Polytekhnicheskaya 26, St. Petersburg 194021, Russia
[2]Micro- and NanoelectronicsDepartment, St. PetersburgElectrotechnical University "LETI", Prof.Popova 5, St. Petersburg 197376, Russia

Abstract. The paper describes experimental results on low temperature plasma-assisted molecular beam epitaxy of GaN/AlN heterostructures on both 6H-SiC and Si(111) substrates. We demonstrate that application of migration enhanced epitaxy and metal-modulated epitaxy for growth of AlN nucleation and buffer layers lowers the screw and edge(total)threading dislocation (TD) densities down to $1.7 \cdot 10^8$ and $2 \cdot 10^9$ cm^{-2}, respectively, in a 2.8-μm-thick GaN buffer layer grown atop of AlN/6H-SiC. The screw and total TD densities of $1.2 \cdot 10^9$ and $7.4 \cdot 10^9$ cm^{-2}, respectively, were achieved in a 1-μm-thickGaN/AlNheterostructure on Si(111).Stress generation and relaxation in GaN/AlN heterostructures were investigated by using multi-beam optical stress sensor (MOSS) to achieve zero substrate curvature at room temperature. It isdemonstrated that a 1-μm-thick GaN/AlN buffer layer grown by PA MBE provides planar substrate morphology in the case of growth on Si substrates whereas 5-μm-thick GaN buffer layers have to be used to achieve the same when growing on 6H-SiC substrates.

1. Introduction

Development of high energy/frequency electronicsis based largely on progress in the epitaxial technologies of wide bandgap (Al,Ga)N-based heterostructures.Amongdifferent technologies used to fabricate these heterostructuresa plasma-assisted molecular beam epitaxy (PA MBE) hasa set of unique features, namely, low-temperature epitaxial growth (T_S<800°C), precise growth control of quantum-sized heterostructures with monolayer accuracy*etc.*

Until now, the absence of commercially available homoepitaxial substrates causes employingdifferent lattice-mismatched substrates. Among them, two play a crucial role in manufacturing (Al,Ga)N HEMT transistors. First,6H-SiC substrates have arelatively small lattice mismatch (~1%)and the same coefficient of thermal expansion (TEC)as compared to AlN buffer layers. The highest thermal conductivity of SiC provides the most efficient heat dissipation in high-power devices. Despite the small crystallographic mismatch,of emerging of threading dislocations (TD) is an importantissue for these substrates. In addition, GaN layers usually involved in device heterostructures have a largerTEC mismatch with the 6H-SiC substrates, which can generate a tensile residual stress resulting in cracking the device heterostructures.Finally, the high cost of these substrates still limits their wide applications. Second type of substrates isSi(111) which is attractive for mass-production of (Al,Ga)N devices due to their low cost,large size, and well-developed technology. However, these substrates haverather high crystallographic and TECmismatches with AlN epitaxial layers (~+19%,~-13.5% respectively), that inevitably leads to the generation of strains in the heterostructures, which relax through the formation of TDswith a density up to ~10^{10}cm^{-2}. These TDs lead to degradation of carrier transport characteristics in the device structures, whereas the high level of residual stress in the heterostructures followed by their significant bending makesit difficult to perform the post-growth devicefabricationand even can result in device cracking in the case of positive (tensile) strain.Thus, the growth of thick buffer layers for reduction of TDdensity through their interactions and re-orientation in the basal plane as well as strain engineering to provide a small residual stress (preferably compressive) arethe principal issues for the epitaxial technology of high-performance device heterostructures.

This paper describes an experimental study of the stress generation and relaxation in GaN/AlNheterostructures grown on both Si and 6H-SiC substrates by using low-temperature (<800°C) PA MBE. Special attention is paid to the development of migration enhanced epitaxy (MEE) for growth of nucleation layers with a minimum TD density and metal-modulated epitaxy (MME) for the growth of subsequentAlN and GaNbuffer layersto achieve high compressive incremental stresses inthe heterostructures, compensating the tensile stress arising during heterostructurecooling, and in this way to reduce the residual stress therein.

2. Experiment and estimations

Two structures A and B were grown by using a PA MBE set up Compact21T on commercial 2 inch6H-SiC and Si(111) substrates, respectively.These structures are schematically shown in Figure 1.The same nitrogen fluxof ~0.5ML/s was used for growth of all layers in both structures. The AlNnucleation layers (NLs) inboth structures were grown by MEE with alternating Al and N fluxes having an intensity ratio of 1.2,as described in [1].The NLs with thickness of 65(30) nm were grown at substrate temperatures of T_S=700(740)°Cin samples A(B), respectively. Then, thick AlN buffer layers of 455 and 170 nmin thickness were grown by MMEwith the same flux ratio III/N ~1.7 in samples A and B, respectively, as described in [2]. Finally, top GaN layers with thickness of 2.8 and 1 μm in structures A and B were grown by MME and standard PA MBE at T_S=700°C, respectively.

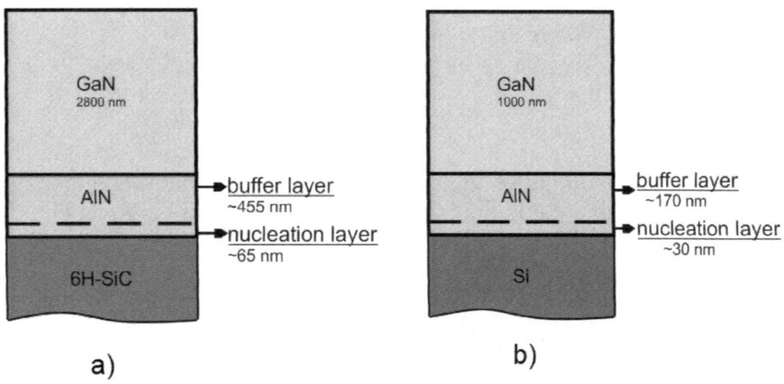

Figure 1. Sketchesof structures A (a) and B (b) under study.

A home-made multi-beam optical stress sensor (MOSS) was used for *in situ* evaluation of incremental stresses $\sigma(h_l)$in the heterostructures from the experimental data on the substrate curvature change through Stoney's differential equation

$$\sigma(h_l) = \frac{M_s h_s^2}{6} \frac{\partial k}{\partial h_l},$$

whereM_S isthebiaxial module of substrate, k isthe substrate curvature; h_l and h_s are the thicknesses of substrate and epilayer, respectively. The values of biaxial modules,a-lattice constants and TECs are represented in Table 1. Residualstressin the heterostructures ($\langle\sigma_l\rangle$) was calculated through an integration of incremental stresses

$$\langle\sigma_l\rangle = \frac{1}{h_l}\int_0^{h_l} \sigma(z)\, dz\,.$$

Table 1. The reference data of III/N compounds and heteroepitaxial substrates.

Compound	*a*-latticeconstant (Å)	Thermalexpansion coefficient ($10^{-6}K^{-1}$)	Biaxialmodulus (GPa)	Latticemismatch[6]	
				toAlN (%)	toGaN (%)
AlN	3.111[3]	4.20[3]	470[4]	0	-2.5
GaN	3.189[3]	5.59[3]	449[4]	+2.4	0
6H-SiC	3.081[3]	4.20[3]	602[4]	-1	-3.5
Si(111)	5.431[5]	3.7[5]	229[5]	+19	+17

The crystallographic quality of the heterostructures was evaluated by high angle annular dark field scanning Transmission Electron Microscopy (HAADF STEM) and conventional TEM as well as X-ray diffraction (XRD) analysis of symmetric (0002) and antisymmetric ($10\bar{1}5$) reflexes.

3. Resultsand discussion

Figure 1 demonstrates different behaviors of the substrate curvature during PA MBE growth of heterostructures A and B. For the former, the constant compressive incremental stresses of -3(-2.2)GPa were measured during growth of AlN nucleation (buffer) layer, respectively, that corresponds to the stress relaxation degree as large as 36(53)% in these layers (Figure 2a). The low degree of stress relaxationin these layers is related to a two-dimensional (2D) growth mechanism due to MEE and MME techniques used for their growth. Moreover, XRD data presented in Table 2 and TEM images (see below) indicate that the MEE techniqueeffectively suppresses the TDoccurrence in this layer.It is likely due to largesize of 2D grains formed during MEE, which results insignificant reduction of total length of the grain boundaries serving as preferential sites for the TD generation.

In contrast, in the case of growth of structure Bthe constant tensile stress of 0.2GPa was measured in AlN nucleation layer,whereas the stresses with the same sign but gradually raised from 0.2 up to 0.8 GPawere foundin the AlN buffer layer (Figure 2b).

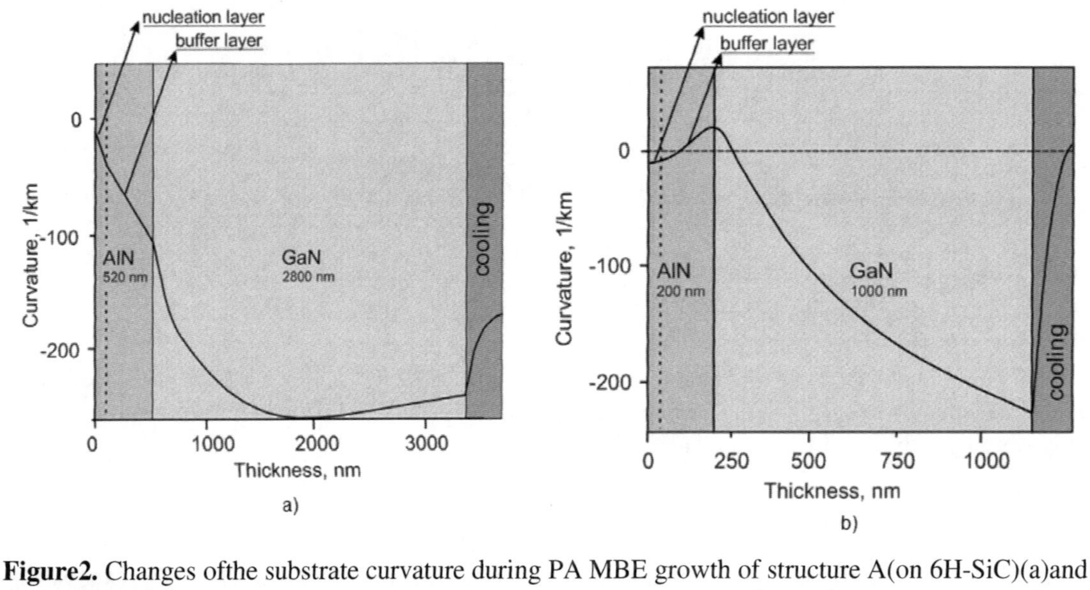

Figure2. Changes ofthe substrate curvature during PA MBE growth of structure A(on 6H-SiC)(a)and B (onSi (111))(b).

These results appear to be related to different lattice mismatchbetweenAlNlayers and substrates in structures A and B,as listed in Table 1. Indeed, the relatively small lattice mismatchin structure Aleads to a gradual relaxation of the initial moderate compressive stress (~-1%)during the entire AlN growth run,while a huge mismatch in structureB (~19%) results inthe practically instantstress relaxation through generation of TDsin AlNlayers. This is confirmed by XRD data summarized in Table 2, which indicatea much higher TD density in AlN layers grown on Si as compared to the layer grown at the same modes on SiC.

Table 2. The resultsof X-ray diffraction analysis.

Reflex	Sample 1		Sample 2	
	FWHM, arcsec	TD density, cm $^{-2}$	FWHM, arcsec	TD density, cm $^{-2}$
(0002)	290	$1.7\cdot10^{8}$	780	$1.2\cdot10^{9}$
$(10\bar{1}5)$	620	$2\cdot10^{9}$	1180	$7.4\cdot10^{9}$

During the next stageof growingthick GaN layers on AlNbuffersthe compressive stresseswith maximum values of -6.1 and -3.8 GPawere observed in structures A and B, respectively. Figure 2 shows that these stresses are almost completely relaxed at the GaN thickness of about 1μm in the both structures. One should note also the generation of constant tensile stress of +0.4 GPain sample A after the complete stress relaxation at around 2μm. This stress can be induced bydifferent processes occurring in the growing GaNlayer with the large TD density,namelyTD inclination and annihilation [7].

Figure 1 indicates also a tensile stress generation during post-growth cooling in both structures,that results in moderate residual compressive stresses in sample Awith a convex bending, and practically zero curvature (planar substrate) for the sample B.In addition, using the data from Table 1 and assuming the constant incremental stress,we estimatea GaNthickness as large as 5 μm to achievethe planar substrate morphology in GaN/AlN heterostructures on 6H-SiC substrate typically used forHEMTsmanufacturing.

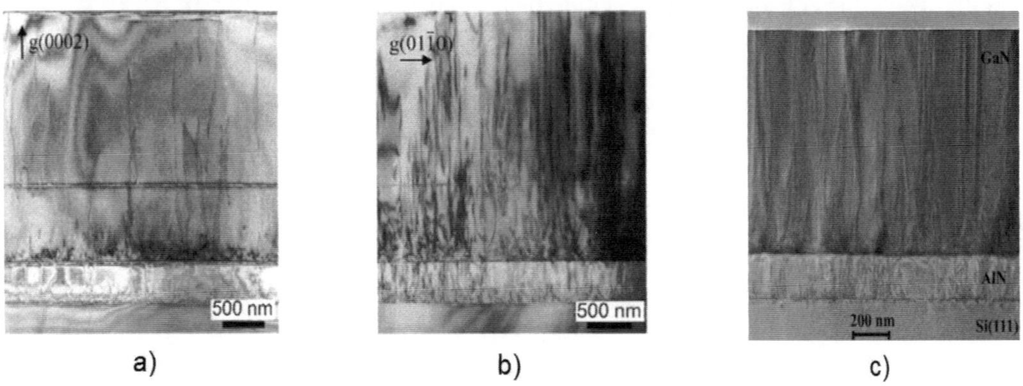

Figure 3. TEM imagesof sample A with reflexes g(0002) (a) and g(01$\bar{1}$0) (b) as well as HAADF STEM imageof sample B(c).

The observed thickness of 520 nm for highly stressed AlN layer on 6H-SiC substrate exceeds the critical thickness calculated from the Matthews and Blakeslee model [8] and corresponds to the data obtained by Okumura et al. [9] who used extra-low PA MBE growth temperatures (~650°C). Thus, our results obtained at higher temperatures of 740°C indicate a high potential of MEE technique for growinghighly stressed III-N films at the moderate substrate temperatures.It should be noted also, that in the case of MOCVD of analogous heterostructures at the much high substrate temperature (~1100°C) the instant stress relaxation has been observed [10].Furthermore, in accordance with our data the compressive stress of 6.1GPa induced during PA MBE growth of subsequent GaN/AlNheterostructure are much higher in comparison with the stress of 1.04 GPa measured for such heterostructures grown by MOCVD [10].

4. Conclusion

The efficiency of a MEE mode for PA MBE growth of AlN nucleation layers on both 6H-SiC and Si substrates was confirmed by the complex study of GaN/AlNheterostructures by using MOSS (in situ) as well as XRD and TEM (ex situ) techniques.The TD densities of $1.7\cdot10^{8}(2\cdot10^{9})cm^{-2}$ were measured for screw and edge dislocations in the 3µm-thick GaN/AlNheterostructure grown on 6H-SiC substrate, whereas analogous densities of $1.2\cdot10^{9}$ $(7.4\cdot10^{9})cm^{-2}$ were revealed in the 1.2-µm-thick GaN/AlNheterostructure grown on Si(111) substrate at close growth parameters.Also, we demonstrate that PA MBE can provide the relatively high compressive stress in GaN/AlN heterostructures grown on both SiC and Si substrates. As a result, these stresses can completely compensate the harmful tensile stress inevitably generating during post-growth cooling of HEMT device heterostructures.

Acknowledgement

This work has been supported by Russian Science Foundation, grant № 14-22-00107.

References

[1] Nechaev D V, Aseev P A, Jmerik V N, Brunkov P N, Kuznetsova Y V, Sitnikova A A, Ratnikov V V, and Ivanov S V 2013 J. CrystalGrowth **378**, 319.
[2] Jmerik V N, Mizerov A M, Nechaev D V, Aseev P A, Sitnikova A A, Troshkov S I, Kop'ev P S, and Ivanov S V 2012 J. CrystalGrowth**354**, 188.
[3] Ponce F A, Krusor B S, Major J S.Plano W E and Welch D F 1995 Appl. Phys. Lett. **67**410.
[4] Wright A F 1997 J. Appl. Phys. **82** 2833.
[5] Ayers J E 2007Heteroepitaxy of semiconductors: theory, growth, and characterization (Storrs-CRC Press).
[6] Zhu D, Wallis D J, and Humphreys C J 2013 J. Rep. Prog. Phys.**76** 106501.
[7] Romanov A E and Speck J S 2003 J. Appl. Phys. **83**2569.
[8] Matthews J W and Blakeslee A E1974J. Crystal Growth **27** 118.
[9] Okumura H, Kimoto T andSudaJ2012 Appl. Phys. Express **5** 105502.
[10] Redwing J M and Al Balushi Z Y 2015 J. Mater. Res.**30** 2900.

Investigation of milling processes of semiconductor zinc oxide nanostructured powders by X-ray phase analysis

I A Pronin[1,2], I A Averin[1], N D Yakushova[1], G V Vishnevskaya[1], M M Sychov[3], V A Moshnikov[1,2], E I Terukov[4]

[1] Department of Nano- and Microelectronics, Penza State University, Penza 440026, Russia

[2] Department of Micro-and Nanoelectronics, Saint-Petersburg Electrotechnical University "LETI", St. Petersburg 197376, Russia

[3] Department of Theoretical Foundations of Materials, Saint-Petersburg State Technological Institute, St. Petersburg 190013, Russia

[4] Ioffe Institute RAS, Saint Petersburg 194021, Russia

Abstract. The processes of mechanical activation of nanostructured zinc oxide powders are investigated by X-ray phase analysis. It was determined, that samples remain in a single phase state during the milling process. The particle size decreases according to the linear time law, and microstrains grow parabolically.

1. Introduction

Semiconductor zinc oxide (ZnO) has unique optical and electrical properties. It is used when creating nanolasers, piezoelectric nanogenerators, solar cells, and gas sensors [1-4]. The aim of the paper is to study the characteristics of milling processes of nanostructured zinc oxide powders by X-ray phase analysis.

2. Experiment

Samples of zinc oxide nanopowders were produced by high-energy mechanical milling out of ZnO commercial powder (pure for analysis, CJSC Vekton) in the attritor (Type: BATCH-LAB, Model: HD/01, Union Process, Inc., USA) under the following conditions: the rotational speed of the drum axis is 450 rpm; milling time is 1, 3 and 5 hours; the ratio of milling bodies mass (ZrO_2 balls of 3 mm diameter) and the powder is 10:1; milling medium is isopropyl alcohol absolutized. All samples have been dried for 24 hours at a temperature of 70 °C after the milling process.

X-ray diffraction spectra of the samples were obtained by means of DRON-3 diffractometer.

3. Results and discussion

Figure 1 shows the diffractograms of all the samples obtained. In all the samples, only the wurtzite phase of ZnO with reflexes corresponding to the families of the (100), (002), (101), (102), (110), (103), (112) planes is present. No traces of other phases are detected.

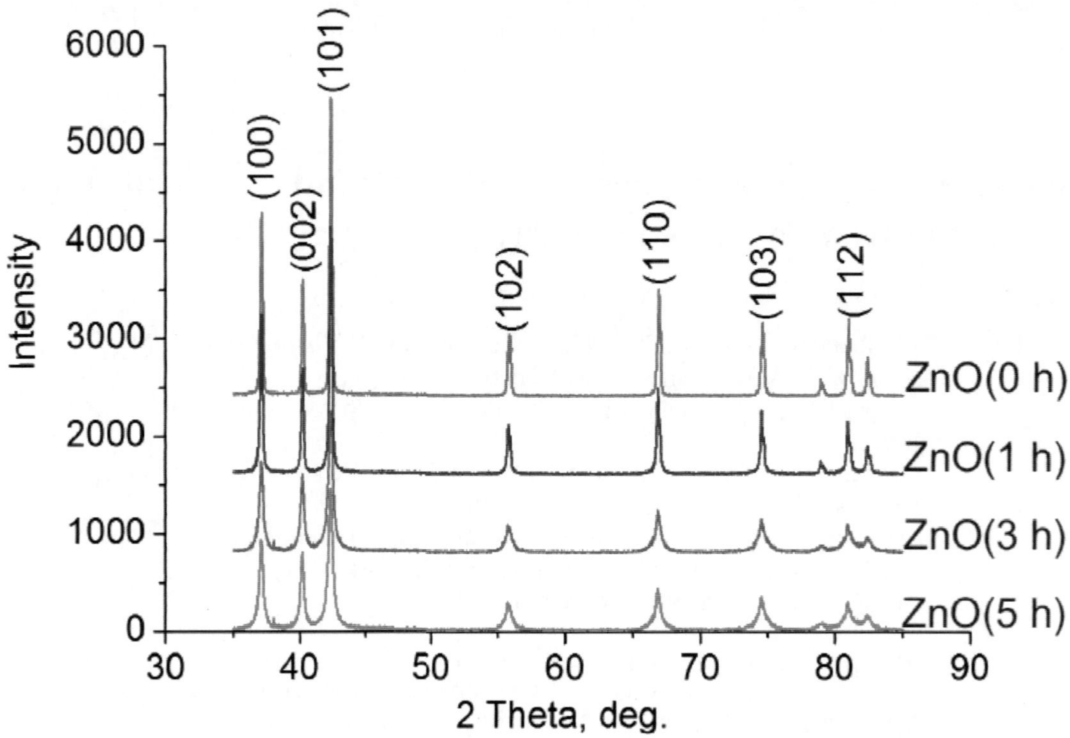

Figure 1. Diffractograms of the obtained samples.

A qualitative analysis of the obtained diffractograms shows that as the grinding time increases, the intensity of all reflexes decreases, and their broadening increases. This indicates a possible appearance of amorphization in the material, as well as a decrease in the average size of the coherent scattering regions, and also the accumulation of microdeformations.

The width of the lines in the diffraction pattern is due to the two contributions: a tool component and a physical component. The physical component is determined by the two factors: line broadening, depending on the size of coherent scattering regions (CSR), and broadening, depending on the lattice distortion.

Broadening of B_D line, associated with the CSR size, can be described by the following relation:

$$B_D = \frac{K\lambda}{D \cdot \cos\Theta}$$

where K is form factor; λ is the X-ray wavelength; D is the CSR size; Θ is the diffraction angle.

Broadening of D_S line, associated with the lattice microstrains, is described by Stoke's formula:

$$B_S = 4\eta \cdot \mathrm{tg}\Theta$$

where η is the relative lattice strain, equal to $\eta = \frac{\Delta a}{a}$; a is identity period of the lattice.

Since the physical broadening is a mathematical convolution of both factors, then describing the real profile of reflexes by Cauchy's functional equation, we can assume that $B = B_D + B_S$, whence it follows:

$$B \cdot \cos\Theta = \frac{K\lambda}{D} + 4\eta \cdot \sin\Theta \quad (1)$$

From (1) it is possible to find the desired parameters D and η graphically, by plotting sinΘ on the horizontal axis, and by plotting BcosΘ on the vertical axis.
Fig. 2 shows the obtained dependence of the crystallite sizes on microstrain values for the samples, activated mechanically for 1, 3 and 5 hours.

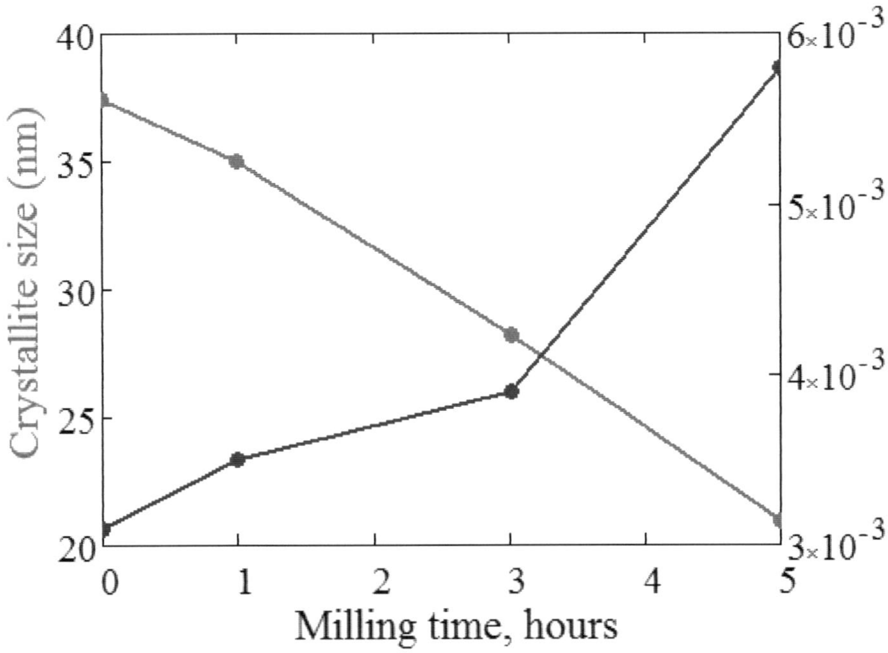

Figure 2. Crystallite sizes and microstrain values for the samples, activated mechanically for 1, 3 and 5 hours.

The Figure shows that while milling the crystallite size decreases almost linearly, and the microstrains in the samples increase according to the law close to the parabolic one.

4. Conclusions

Thus, the processes of mechanical activation of nanostructured zinc oxide powders have been investigated by the method of X-ray phase analysis. It was determined, that samples remain in a single phase state during the milling process. The particle size decreases according to the linear time law, and microstrains grow parabolically.

Acknowledgments
The results were obtained within the framework of the government task of the Ministry of Education and Science of the Russian Federation (project No. 16.897.2017/4.6).

References

[1] Dimitrov D Tz, Nikolaev N K, Papazova K I, Krasteva L K, Pronin I A, Averin I A, Bojinova A S, Georgieva A Ts, Yakushova N D, Peshkova T V, Karmanov A A, Kaneva N V, Moshnikov V A 2017 *Appl. Surf. Sci.* **392** 95
[2] Kononova I E, Moshnikov V A, Krishtab M B, Pronin I A 2014 *Glass Phys. Chem.* **40** 190

[3] Pronin I A, Donkova B V, Dimitrov D Tz, Averin I A, Pencheva J A, Moshnikov V A 2014 *Semiconductors* **48** 842
[4] Korotcenkov G 2007 *Mater. Sci. Eng. B* **139** 1

Copper (I) oxide rf-magnetron sputtering at elevated substrate temperatures

D Kudryashov[1], A Monastyrenko[1], A Mozharov[1,2], A Bukatin[1], E Nikitina[1], E Pirogov[1] and A Gudovskikh[1,3]

[1] St. Petersburg Academic University, St. Petersburg 194021, Russia
[2] ITMO University, 197101, St. Petersburg, Russia
[3] St. Petersburg Electrotechnical University "LETI", St. Petersburg 197376, Russia

Abstract. The structural and optical properties of Cu_2O thin films grown by magnetron sputtering under elevated substrate temperatures were investigated. All samples show clear XRD patterns and photoconductivity response. An influence of Ar/O_2 ratio and sputtering power on Cu_2O structural properties was demonstrated. A ZnO/Cu_2O-based solar cell was grown by magnetron sputtering at a temperature of 450 °C. An open circuit voltage of the solar cell is 80 mV.

1. Introduction

Copper (I) oxide is a promising candidate for use in oxide photovoltaics. It is a semiconductor with a band gap of 2.2 eV and a high optical absorption coefficient [1]. A growth of low defect Cu_2O films is a main problem due to high formation temperatures (up to 900 °C). Magnetron sputtering is a simple and not expensive method of thin films growth. Due to high kinetic energy of sputtering atoms it is possible to provide a films growth with good crystallinity at much lower temperatures. For example as previously reported [2] it is possible to deposit ZnO films at room temperature with a clearly observed band-edge photoluminescence. But Cu_2O films deposited at low temperature unfortunately consist of amorphous phase and thus an increasing of substrate temperature is needed. The aim of this work is to investigate the structural and optical properties of Cu_2O thin films grown by magnetron sputtering under elevated substrate temperatures.

2. Experimental

All films were grown by rf-magnetron sputtering using Boc Edwards Auto 500 technique with a modified substrate holder. Due to this upgrade it is possible to provide a high temperature growth up to 500 °C. Cu_2O films with thickness of 3 microns were deposited on Si(100) substrates at different oxygen (99.999 %) and argon (99.999 %) partial pressures, rf-power of 150-300 W and substrate temperature of 450 °C. Copper (99.99%) from LTS Chemicals is used as magnetron target. Every time before the deposition a Cu target was presputtered for 5 minutes to remove contaminations. SEM images were analyzed using a scanning electron microscope SEM (Zeiss SUPRA 25). XRD spectra were measured using DRON-8 equipment. Raman measurements were carried out using a Horiba LabRam HR spectrometer. The optical transmittance and external quantum efficiency were measured using equipment based on Solar Laser Systems M266 monochromator.

IOP Conf. Series: Journal of Physics: Conf. Series **917** (2017) 032020 doi:10.1088/1742-6596/917/3/032020

3. Results and discussion

SEM images of copper oxide films grown by magnetron sputtering at a temperature of 450 °C and different rf-power are shown in figure 1. It can be seen that increasing the sputtering power at fixed substrate temperature leads to a growth of crystallites with distinctly pronounced facets. All subsequent experiments were performed using rf-power value of 300 W.

Figure 1. SEM images of copper oxide (plan-view) deposited at rf-power of 150 W - (a) and 300 W - (b). Argon pressure was 0.17 Pa. Oxygen pressure was 0.033 Pa.

XRD patterns and raman spectra of copper oxides deposited on Si(100) at fixed Ar pressure (0.17 Pa) and different O_2 pressure are shown in figure 2. Increasing oxygen concentration results in changes in the films composition – from Cu_2O to CuO.

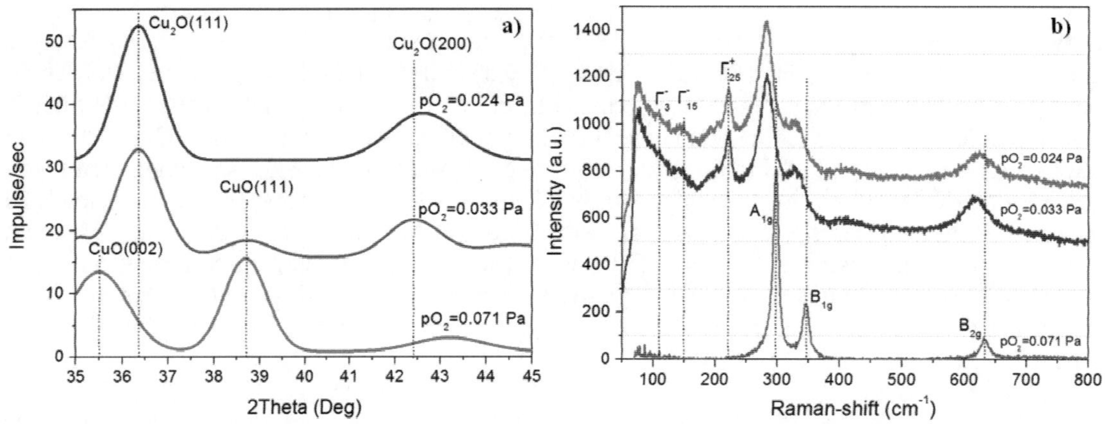

Figure 2. XRD patterns (a) and raman spectra (b) of copper oxide films deposited on silicon as a function of the oxygen partial pressure.

SEM images of Cu_2O films deposited on Si-wafer at fixed O_2 pressure (0.024 Pa) and different Ar pressure values are shown in figure 3. Increasing argon partial pressure in a range of 0.079-0.17 Pa leads to surface roughening due to decrease of the deposited atoms kinetic energy. Decreasing Ar concentration leads to surface smoothing. Corresponding XRD spectra are shown in figure 4. High argon pressure leads to growth of copper oxide film with low XRD peak intensity. On the contrary, low Ar concentration promotes crystallites growth with dominated crystal plane orientation.

Figure 3. Plan-view (top) and cross-section (bottom) SEM images of Cu_2O films deposited on silicon as a function of Ar partial pressure: 0.17 Pa (a,e), 0.12 Pa (b,f), 0.096 Pa (c,g), 0.079 Pa (d,h), fixed oxygen pressure (0.024 Pa) and sputtering power (300 W).

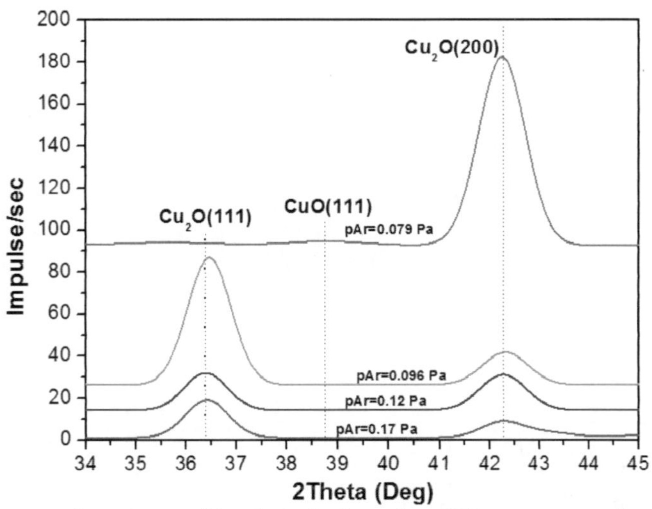

Figure 4. XRD spectra of copper oxide deposited under different argon flow rate, fixed oxygen pressure (0.024 Pa) and sputtering power (300 W).

Copper (I) oxide films deposited by magnetron sputtering at 450 °C and low Ar partial pressure (0.096 Pa) show an interference in transmittance spectra (figure 5a) at a wavelength range of 600-1000 nm with a band-edge absorption at 520 nm (2.3 eV). After applying external bias of 5V to Al contact pads deposited on Cu_2O films it possible to observe a photocurrent response in a wavelength region of 350-650 nm (figure 5b). Oxide films grown at high oxygen concentration (0.071Pa) show very low transmittance and photoresponse. Based on demonstrated experimental data a copper (I) oxide sputtered at rf-power of 300 W, temperature of 450 °C argon partial pressure of 0.096 Pa and O_2 partial pressure of 0.023 Pa was chosen as an active (base) layer for ZnO/Cu_2O-based solar cell growth.

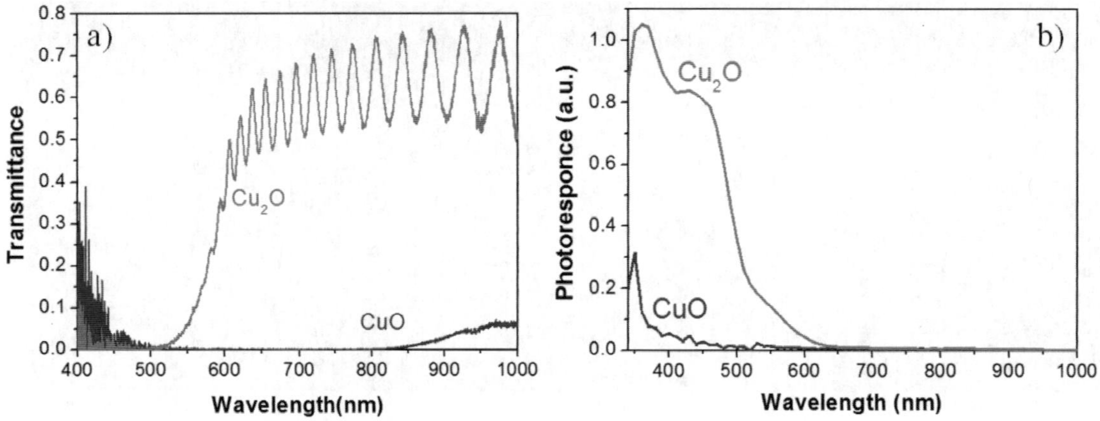

Figure 5. Transmittance (a) and photoconductivity response spectra (b) of copper oxides deposited under: for Cu_2O - pAr=0.096 Pa and pO_2=0.023 Pa, for CuO - pAr=0.17 Pa and pO_2=0.071 Pa.

To avoid problems with shunts the solar cell was grown on polished Si(100) wafer covered with magnetron deposited 60 nm silicon oxide and 400 nm of copper film (figure 6). Silicon oxide acts as a diffusion barrier for copper migration in silicon. 3000 nm of copper (I) oxide (at 450 °C) and 60 nm of zinc oxide (at 25 °C) were deposited in sequence by magnetron sputtering. Top transparent indium tin oxide (ITO) contacts (4x4 mm^2) also were deposited by magnetron sputtering at room temperature in oxygen-free environment [3]. On the top of ITO layer a small (1 mm^2) silver pads were formed. Then solar cells were annealed for 30 minutes at 100 °C in an air atmosphere.

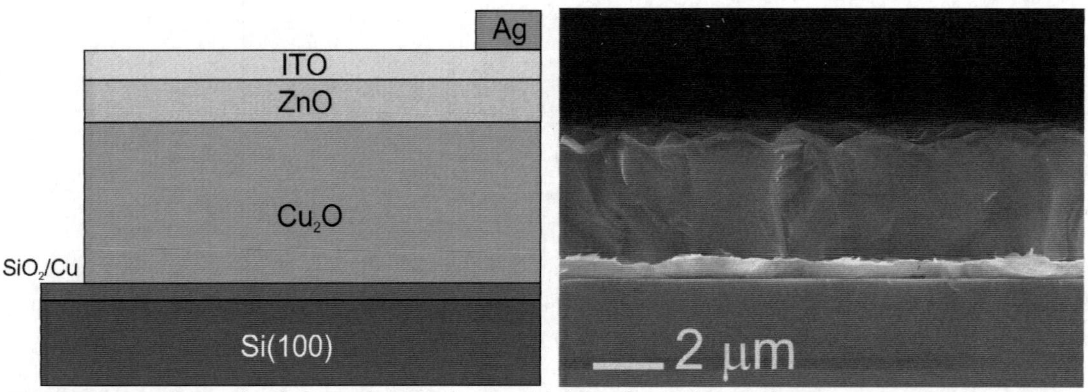

Figure 6. A schematic (left) and SEM image (right) of ZnO/Cu_2O based solar cell deposited by magnetron sputtering.

I-V curve measured under AM1.5 solar illumination for the ZnO/Cu_2O solar cell structure is shown in figure 7a. An open circuit voltage of ZnO/Cu_2O-based solar cell is 80 mV and a short-circuit current is 0.017 mA/cm^2. Figure 7b shows the external quantum efficiency curve of ZnO/Cu_2O solar cell deposited by magnetron sputtering. The clear sharp peak at 365 nm may be due to photogeneration in copper (I) oxide or ZnO. In this region a strong absorption in zinc oxide is possible due to its bandgap of 3.4 eV [2]. A photoresponse at higher wavelengths up to 700 nm is attributed to copper (I) oxide.

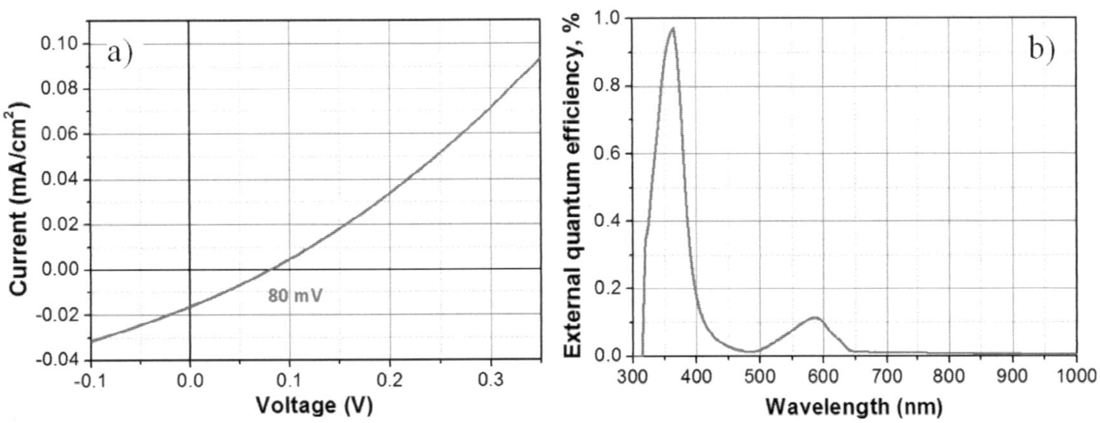

Figure 7. Current-Voltage characteristics under illumination for ZnO/Cu_2O based solar cell deposited by magnetron sputtering (a) and its external quantum efficiency (b).

4. Conclusions

The structural and optical properties of Cu_2O thin films grown by magnetron sputtering under elevated substrate temperatures were investigated. All samples show clear XRD patterns and photoconductivity response. An influence of Ar/O_2 ratio and sputtering power on Cu_2O structural properties was demonstrated. A ZnO/Cu_2O-based solar cell was grown by magnetron sputtering at a temperature of 450 °C. An open circuit voltage of the solar cell is 80 mV.

Acknowledgment

The reported study was partially supported by RFBR research project № 15-08-06645A

References

[1] Ruhle Sven et al. 2012 *J. Phys. Chem. Lett.* **3** 3755
[2] D Kudryashov et al. 2015 *J. Phys.: Conf. Ser.* **643** 01201
[3] D Kudryashov et al. 2013 *Solid State Phenomena.* **200** 10

The concept of the phases ratio control during the formation of composite filamentary nanocrystals xInSe-(1–x)In_2O_3 on glass substrates

S A Gasanly[2], V V Tomaev[1,2], T V Stoyanova[2]
[1] Institute of Chemistry, Saint-Petersburg State University Saint-Petersburg, 198504, Russia
[2] Department of General and Applied Physics, Saint-Petersburg Mining University, Saint-Petersburg, 199106, Russia

Abstract. The method of vacuum deposition on substrates of glass marks the C-29 series PbSe deposited film and the film In the area 3x3 mm^2 and a thickness of ~1 μm. Films are oxidized in dry air at a temperature of 550 °C. Based on studies by X-ray microanalysis and scanning electron microscopy shows the principal possibility of formation of nanowires xInSe-(1–x)In_2O_3 on the PbSe/In structure. The results allowed to formulate the concept of the control of phases ratio in the forming nanowires xInSe-(1–x)In_2O_3 on glass substrates.

1. Introduction

Metal oxide nanostructures with a one-dimensional morphology surface are of great importance both in fundamental and applied fields of use [1]. In comparison with the other forms of nanostructures the dimensional ones have several advantages such as extremely high ratio of surface to the largest volume of nanowhiskers, excellent stability due to its high degree of crystallinity of the individual elements, multifunctional surface, the possibility of the obtaining of transistor structure in the presence of the field effect at the surface, and etc. [2].

The fundamental interest in InSe is due to the display of the diverse unique photoelectric properties [3- 6]. The practical use of layered InSe as a material for various areas of electronics is also very promising [7-12]. Another semiconductor wide-gap indium oxide (In_2O_3) due to a combination of high electrical conductivity and optical transparency is of great scientific interest and widely applied as a transparent electrode, antireflection coating, passivation layer in optoelectronic devices, and also as a sensitive layer in gas sensors [13-15]] etc. A biphase composite based on these indium compounds (InSe and In_2O_3) is in demand in fundamental and applied research as well [16, 17]. The purpose of this paper was the synthesis of composite material by an original method based on the In-Se-O system on glass substrates with the ability to control the ratio of the components (InSe and In_2O_3).

2. Materials and methods

PbSe film with area 3×3 mm^2 and thickness of 1.0 - 1.5 μm was applied on the type C-29 glass substrate. Indium layer with thickness of ~1.6 μm (Figure 1), was deposited on the PbSe film. The films were oxidized in dry air at a temperature of 550 °C. Using the SEM method the surface and chip both the origin films and the oxidized ones have been studied. With X-ray microanalysis method the qualitative and quantitative assessment of the elemental composition in the subsurface area of the investigated films has been carried out.

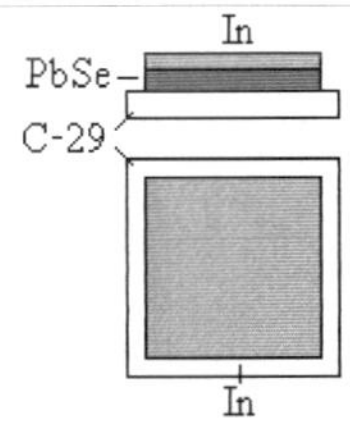

Figure 1. Scheme of the deposited on the glass substrate films

3. Thermodynamic analysis.

When lead selenide is heated at 363 ° C above the solid phase of PbSe, the vapor pressure of selenium (Se) ~ 1 mm Hg is reached, while the vapor pressure of lead (Pb) and lead selenide (PbSe) is less a lot than of the magnitude compared with the pressure of selenium vapor [18, 19].

When the double-coated film consisting of PbSe and In films is heated up to 550 °C, the indium layer transforms into liquid state (the melting point of indium is 156.4 °C [20]). While abundant selenium from the solid lead selenide changes into a liquid film, Indium reacts with the chemical reaction and diffuses to the external surface of indium, interacting further with oxygen. Figure 2 shows the In–Se diagram constructed from the literature review [21-23]. The investigations have been carried out by differential thermal, microstructural and X-ray analysis methods and a microhardness measurement method. Alloys were synthesized in evacuated quartz ampules using In and the simple Se with the purity above 99.9% (by the weight).

The compound In_2Se is metastable, melts incongruently, and has a high volatility; and the InSe and In_2Se_3 melt congruently at the temperatures of 660 and 900 °C, respectively. The system has established the existence of four compounds: In_2Se, InSe, In_5Se_6, and In_2Se_3. The In_2Se and In_5Se_6 compounds have been developed with the help of peritectic reactions at 540 and 660 °C, respectively, and are highly volatile. Solid solutions based on In and Se are practically absent. Both these components form degenerate eutectics: (In) + In_2Se at the temperature of 156 °C and (Se) + β-In_2Se_3 at the temperature of 219 °C. The value of the latter temperature is indicated to be lower than in the original works (225 °C), in accordance with the adopted lower (at 6 °C) melting point of Se equal to 221 °C.

In the In-rich alloys, incompatibility in the liquid state takes place. The polymorphic transformations of compounds In_5Se_6 (β ↔ α at 550 °C) and In_2Se_3 (β ↔ α at 200° C, γ ↔ β at 650 °C, δ ↔ γ at 750 °C) have been discovered in the papers [21-23]. The presence of a polymorphic transformation in the compound In_5Se_6 of the paper [22] has not confirmed. The course of the liquidus line from the Se side at high temperatures, namely the practically horizontal portion at 770 °C, suggests that in this concentration range, a second exfoliation in the liquid state in the system can take place. However, according to [21-23], in these alloys the formation of two layers in the liquid state has not been observed.

The analysis of the phase diagram T-x, for the In-Se system [21-24] presented in the Figure 2. The technological conditions of heat treatment at 550 ° C in an dry air atmosphere of liquid indium with gaseous selenium and then with the subsequent cooling to room temperature indicate the formation of InSe and In_2Se compounds inside the indium liquid film with limited access to oxygen. In the region of rich in In alloys (the left side of the diagram, see Figure 2), incompatibility in the liquid state takes place. The course of the liquidus line on the Se side (the right-hand side of the diagram, see Figure 2) has a practically horizontal section at 770 °C at high temperatures, which suggests that in this concentration range the second layering in the liquid state in the system may take place [21 - 24].

IOP Conf. Series: Journal of Physics: Conf. Series **917** (2017) 032021 doi:10.1088/1742-6596/917/3/032021

To complete the picture associated with the oxidation of the two-layer PbSe and In film at 550 °C in the dry air atmosphere, it is also necessary to take into account the possibility of chemical interaction of indium at the interface with air oxygen under these conditions.

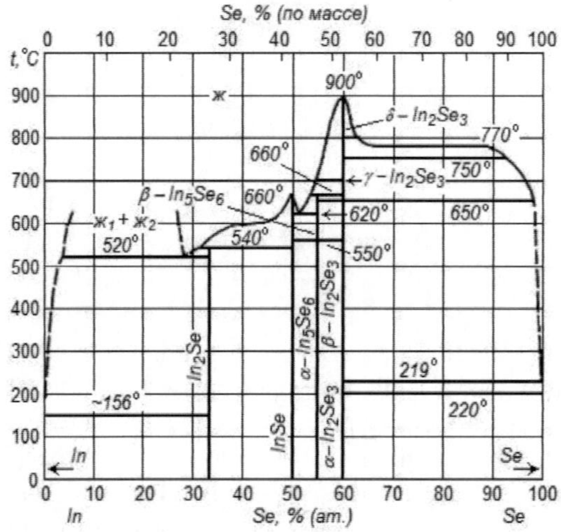

Figure 2. The state diagram of the In-Se system [21].

In the interaction of In with O, the formation of In_2O, InO, and In_2O_3 compounds with an O O – 6.52, 12.24, and 17.30% (by mass) was established [21]. The most stable oxide in this series is In_2O_3, which has two modifications, with a body-centered cubic and trigonal structure.
In_2O_3 with the trigonal structure of corundum type was obtained by the hydrothermal method and, it appears, to be a metastable phase formed at high pressures (lattice parameters in hexagonal system: $a = 0.549$ nm, c = 1.452 nm, Pearson symbol hP10, $R\overline{3}c$. The cubic modification of In_2O_3 is isostructural one with Mn_2O_3 (Pearson symbol cI8, space group Ia3) a = 1.012 nm. An increase in the parameter of the body-centered cubic lattice In_2O_3 from 1.0119 nm at 30 °C up to 1.0199 nm at 968° C was distinguished.

4. Results and discussion.

The sequence of changes on the surface of the two-layer film is shown using a Zeiss Supra 40VP scanning electron microscope in the Figure 3. The comparison was made with the original film.

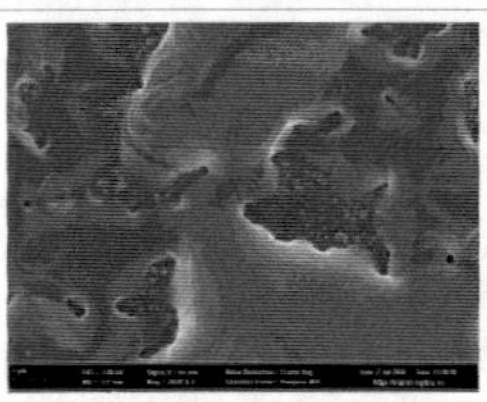	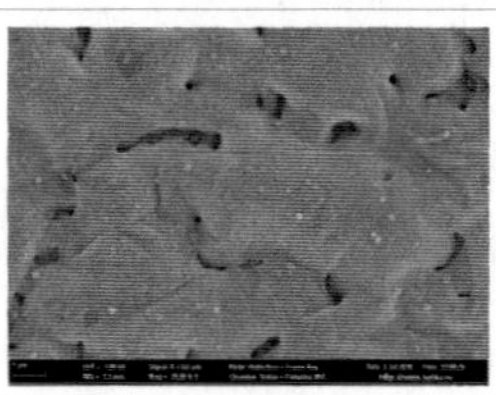	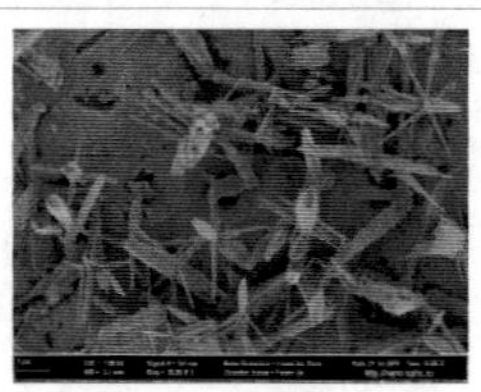
(a)	**(b)**	**(c)**

Figure 3(a, b, c). Surface of the initial sample PbSe/In **(a)** and oxidized at 550 °C samples PbSe/In for t=30 min **(b)** and t=60 min **(c).**

Fragments of the initial surfaces in the (Figure 3(a)) and oxidized at 550 °C for 60 (Figure 3(b)) and 120 min (Figure 3(c)) of the two-layer films of lead and indium selenide have been presented in photomicrographs. For all three cases, the formed islets of indium on the surface of both the initial and oxidized films are visible. Compared with the initial surface of the film (Figure 3(a)) on the film surface, an oxidized at 550 °C C for 60 min (Figure 3(b)), the precipitates can be observed, similar droplet size ranges 150÷250 nm.

The X-ray spectral analysis data for the film oxidized at 550 °C for 120 min, with the choice of the analyzed region at the point on the surface of the island In was shown in the Figure 4 as well.

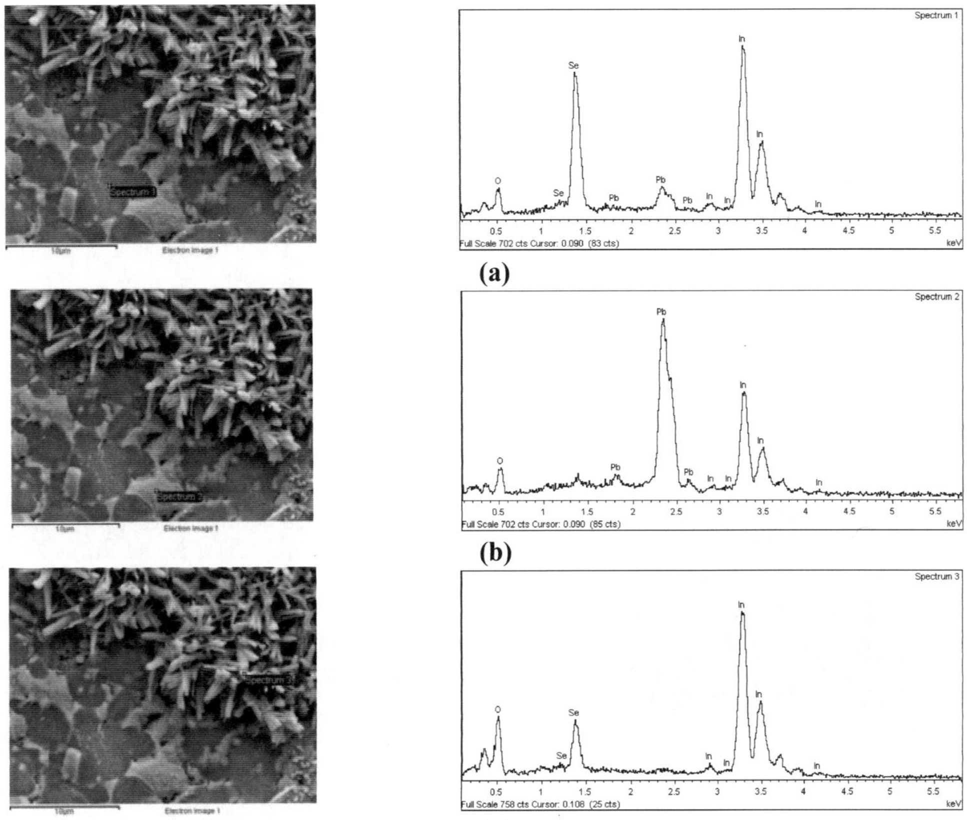

(a) **(b)** **(c)**

Figure 4(a, b, c). Express analysis of the elemental composition of two-layer PbSe/In films oxidized at 550° C for 120 min with the choice of the analyzed points: on the surface of the "island" - (a); In the troughs (b); and at the point of the nanowires - (c).

The elemental composition in the selected region consists of oxygen atoms, selenium, lead and indium. In the temperature range 470÷610 °C, lead selenite decomposes to form gaseous selenium dioxide and $2PbO{\cdot}PbSeO_3$ and $4PbO{\cdot}PbSeO_3$ compounds. With a long exposure (120 min), there is interaction of selenium dioxide with indium and indium oxide. Figure 4(b) shows X-ray spectral analysis data for the film oxidized at 550 °C for 120 min, with the choice of the region to be analyzed on the surface of the In film at the troughs point. The phase analysis in the investigated point showed the presence of lead oxide and indium oxide.

Figure 4(c) presents X-ray spectral analysis data for the two-layer film oxidized at 550 °C for 120 min with a choice of the point where the nanowires are located. It can be seen that the elemental composition at the selected point includes the oxygen atoms, selenium and indium atoms. By the value of the elements mass values, it can be assumed that the composition of the nanowires includes indium indium selenide and indium oxide, the content of indium selenide being much lower than that of indium oxide.

According to the results of X-ray microanalysis, there are atoms of indium, selenium and oxygen in the spectrum of whiskers. The investigation of individual nanowires was carried out. After mechanical removal from the sample surface and ultrasonic dispersion in an organic solvent, the whiskers precipitated on a graphite substrate were carried out.

X-ray microanalysis spectra taken from the surfaces of individual nanowires were recorded at accelerating voltages of 10, 15 and 20 kV (Figure 5(a, b, c)).

The length of the nanowire is ~ 1.6 μm, and the diameter is ~ 90 nm. It can be seen that the elemental composition of the nanowire surface for three different points at the beginning, middle and end of the filament is the same and consists of the atoms of oxygen, selenium and indium. A large background signal coming from the substrate and bonded to the carbon atom was not taken into account.

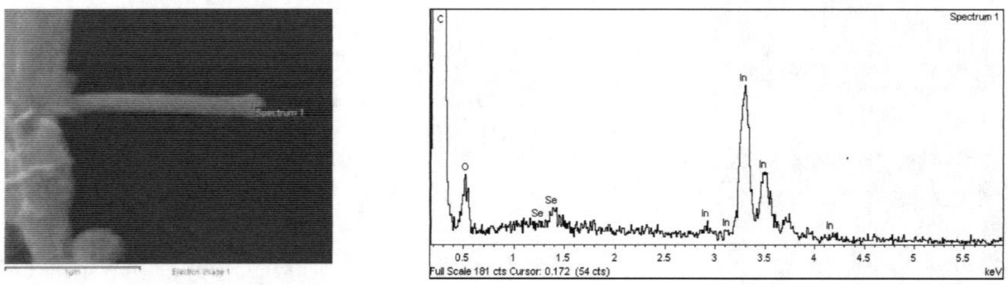

(a)

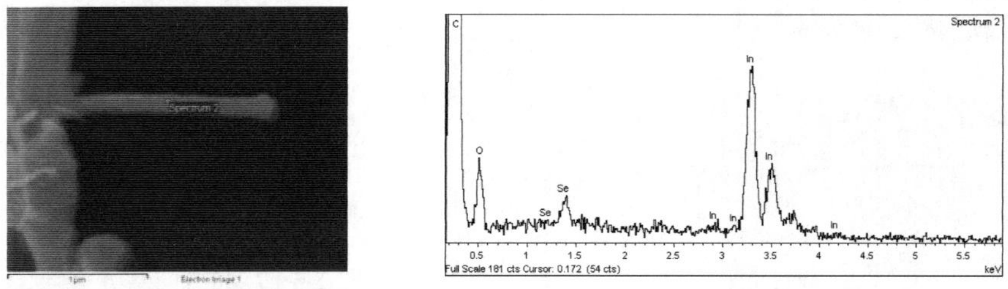

(b)

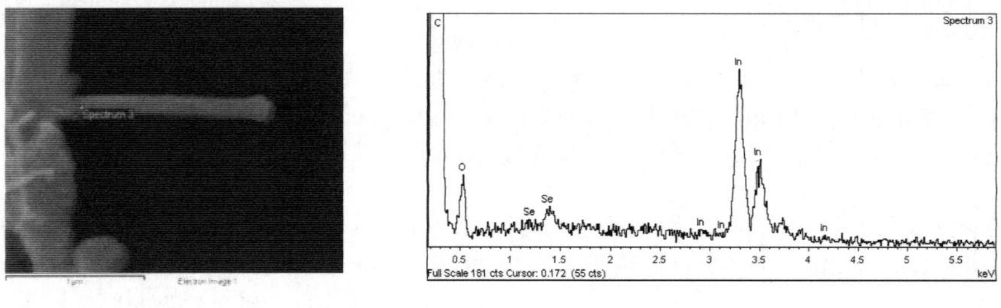

(c)

Figure 5(a, b, c). The analysis of the elemental composition of a whisker crystal deposited on a graphite substrate, with the choice of the analyzed region at the point.

Examinations showed that the mass quantity of each element for each point has the same value within the error of measurement. From the X-ray spectral analysis data, the mass values of the elements were determined, which turned out to be equal to: In – 36±13%; Se – 4±2%; O – 60±15.

In thermal oxidation result of indium and thermal oxidation of lead selenide to lead selenite and its further decomposition, the upper layer of indium in the initial film was transformed into filamentary crystals of composition (1-x)In_2O_3 – xInSe with the range of x variation from 0.10 to 0.15, and the lower layer of lead selenide was transformed into a friable and thicker layer consisting of lead selenite and lead monoxide. The material consists of a number of disoriented filamentary nanocrystals (1-x)In_2O_3·xInSe with a range of x variation from 0.10 to 0.15, which are mounted in the matrix of the composite consisting of indium oxide, selenite and lead oxide.

5. Conclusions

It was shown that nanowhiskers xInSe-(1–x)In_2O_3 were formed on the surface of the films by oxidation at a temperature of 550 °C. Sensor material that consists of a mixture of selenium and indium oxides possesses high gas sensitivity to different liquids and gases.

References

[1] Comini E, Baratto C, Faglia G, Ferroni M, Vomiero A, and Sberveglieri G. 2009 *Progress in Materials Science* 54 1

[2] Stoyanova T V, Tomaev V V, Stoyanov N D, Andreev S K 2014 *Journal of Physics: Conference Series* 558 012039

[3] Shi W & Ding Y J A monochromatic and high-power terahertz source tunable in the ranges of 2.7–38.4 and 58.2–3540 μm for variety of potential applications 2004 *Appl. Phys. Lett.* **84** 1635

[4] Bringuier E, Bourdon A, Piccioli N. & Chevy A Optical second-harmonic generation in lossy media: Application to GaSe and InSe 1994 *Phys. Rev. B* **49** 16971

[5] Manjo F J, Mun V, Segura A, Bouvier J & Andre M V Strong optical nonlinearities in gallium and indium selenides related to inter-valence-band transitions induced by light pulses 1997 **56**

[6] Sánchez-Royo J F, Segura A & Muñoz V Anisotropy of the refractive index and absorption coefficient in the layer plane of gallium telluride single crystals 1995 *Phys. Status Solidi* **151** 257

[7] Segura A, Guesdon J P, Besson J M & Chevy A Photoconductivity and photovoltaic effect in indium selenide 1983 *J. Appl. Phys.* **54** 876

[8] Martínez-Pastor J, Segura A, Valdés J L & Chevy A Electrical and photovoltaic properties of indium-tin-oxide/p-InSe/Au solar cells 1987 *J. Appl. Phys.* **62** 1477

[9] Segura A, Besson J M, Chevy A & Martin M S Photovoltaic properties of GaSe and InSe junctions 2007 *Nuovo Cim.* B Ser. 11 **38** 345

[10] Sánchez-Royo J F *et al.* Optical and photovoltaic properties of indium selenide thin films prepared by van der Waals epitaxy 2001 *J. Appl. Phys.* **90** 2818

[11] Geim A K & Grigorieva I V Van der Waals heterostructures 2013 *Nature* **499** 419

[12] Ferrari A C et al. Science and technology roadmap for graphene, related two-dimensional crystals, and hybrid systems 2015 *Nanoscale* **7** 4598

[13] Zhang D, Liu Z, Li C, Tang T, Liu X, Han S, Lei B, Zhou C Detection of NO_2 down to ppb levels using individual and multiple In_2O_3 nanowire devices 2004 *Nano Lett.* **4** 1919

[14] Francioso L, Forleo A, Capone S, Epifani M, Taurino A M, Siciliano P Nanostructured In_2O_3–SnO_2 sol-gel thin film as material for NO_2 detection 2006 *Sensors and Actuators* B **114** 646

[15] Comini E, Faglia G, Sberveglieri G Solid state gas sensing. 2009 New York: Springer

[16] Katerynchuk V N, Kovalyuk Z D Production of oxide-p-InSe heterostructures with improved photoelectric characteristics 2004 *Physics and Technology of Semiconductors* vol 38 **4** 417

[17] Lang O, Pettenkofer C, Sanchez-Royo J F, Segura A, Klein A, Jaegermann W. Thin film growth and band lineup of In_2O_3 on the layered semiconductor InSe 1999 *J.Appl. Phys.* vol 86 **10** 5667

[18] Zlomanov V P, Novoselova A V P-T-x-diagrams of the state of metal-chalcogen systems 1987 (Moscow: Science) p 208

[19] Maier H, Hesse J Growth, Properties and application of narrow-gap semiconductors 1980 Organic crystals germinates semiconductors Berlin p 145

[20] Rabinovich V A, Havin Z Ya Brief 1978 Chemical Handbook (Leningrad: Chemistry) p 392.

[21] Diagrams of the state of double metal systems 2001 Handbook (Moscow: Mechanical Engineering) **1** p 872

[22] Slavnova G K, Luzhnaya N P, Medvedeva Z S New data on the state diagram of the In-Se system 1963 *J.Organ Chemistry* vol 8 **5** p 1199

[23] Semiletov SbA Electron-diffraction determination of the structure of InSe 1958 *Crystallography* vol **3** issue 3 p 288

[24] Physico-chemical properties of semiconductor substances 1979 Handbook (Moscow: Science) p 340

IOP Conf. Series: Journal of Physics: Conf. Series **917** (2017) 032022 doi:10.1088/1742-6596/917/3/032022

Formation of ZnO nanorods on seed layers for piezoelectric nanogenerators

A A Semenova [1], N A Lashkova [1], A I Maximov [1], V A Moshnikov [1], D A Kudryashov [2], A M Mozharov [2], V N Verbitsky [3], P A Somov [4]

[1] Department of Micro- and Nanoelectronics, St Petersburg Electrotechnical University "LETI", Saint-Petersburg 197376, Russia

[2] Renewable Energy Lab, St Petersburg Academic University, Saint-Petersburg 194021, Russia

[3] Lab of Physico-Chemical Properties of Semiconductors, Ioffe Physical-Technical Institute of the Russian Academy of Sciences, Saint-Petersburg 194021, Russia

[4] CJSC «Svetlana-Rost», Saint-Petersburg 194156, Russia

E-mail: anastassiyan@mail.ru

Abstract. The influence of zinc oxide seed layers on morphological and electrophysical properties of ZnO nanorods was studied. Sublayers were obtained via spin-coating and magnetron sputtering methods. Nanorods were produces via low-temperature hydrothermal synthesis in the presence of anionic and zwitterionic surface-active substances. Characterization of the obtained coatings was carried out using scanning electron microscopy and atomic-force microscopy methods.

1. Introduction

In the 21st century, the development of nanoelectronics sets special terms for the elemental base of energy sources. Piezoelectric nanogenerators (PENG), which are devices that convert scattered mechanical energy in the pulses of electric current [1], are promising for powering medical devices inside a human body (e.g. pacemakers) and for a vehicles pressure conversion on the roads. Such devices are sources of renewable energy, furthermore, PENGs open-circuit voltage can reach 30 V [2].

Piezoelectric crystals – dielectric and semiconductor materials with a lack of a symmetry center in a crystal structure - are used as an active part of the PENG. Inorganic materials are mainly used because of rapid thermal and mechanical degradation of the organic one.

Among the wide variety of materials (PZT, GaN, CdS etc.) for formation of the active part of the PENG one of the most promising is zinc oxide ZnO – wide bandgap semiconductor (E_g = 3.36 eV), which has a *n*-type conductivity [3]. The advantages of this material are high values of piezocoefficients in the direction of the 6th order axis, high chemical resistance, technological ease of preparation, possibility to use this material at the high and ultra-high frequencies. Materials based on low-dimension ZnO structures are being applied in gas sensors [4, 5], photocatalytic systems and new generation solar cells.

ZnO one-dimensional (1*D*) nanostructures (nanorods) form an active part of PENGs [6]. In comparison with bulk materials using 1*D*-nanostructures allows to increase the output signal due to their parallel connection between the electrodes, to improve the mass-dimensional characteristics and effectively change the properties of an active layer. To obtain highly oriented nanorods it is necessary to ensure their growth from the seed layer. In this paper, the formation of sublayers was carried out by spin-coating and magnetron sputtering methods because of their efficiency and the possibility to obtain thin films with a small size of crystallites.

2. Experiment

Spin-coating is the method of applying thin films based on the coating a substrate with an agent solution with the further rapid rotation leading to the removal of a solvent and formation of a thin film. 10 mM ethanol solution of chemically pure zinc acetate $Zn(CH_3COO)_2$ was used as the source of the zinc oxide in this method. Spin-coating was carried out for 60 s at the rate of 3000 revs per min. Annealing of samples on Si-(100) substrates was held at 350°C, and for samples on PET-substrates (polyethylene terephthalate) with ITO (60 Ohm/□) layer (Sigma Aldrich) – at 120°C.

Also sublayers were made by magnetron sputtering. Deposition of ZnO and ITO thin films was carried out using RF magnetron sputtering setup (BOC EDWARDS Auto 500RF) from ZnO and In_2O_3 (90%) SnO_2 (10%) sintered targets, respectively. Deposition of thin films perfomed at room temperature. Pure argon (99.999%) and oxigen (99.999%) were used as working gases.

Finally, ZnO nanorods were produced on the obtained thin films via low-temperature ($T = 85$°C) hydrothermal synthesis. The solution of $Zn(NO_3)_2$ 25 mM was used as a Zn^{2+} precursor. As the result of two-stage hydrolysis reaction unstable zinc hydroxide is formed:

$$Zn(NO_3)_2 + H_2O \leftrightarrow Zn(OH)NO_3 + HNO_3 \quad (1)$$
$$Zn(OH)NO_3 + H_2O \leftrightarrow Zn(OH)_2 + HNO_3 \quad (2),$$

which converts by the reaction $Zn(OH)_2 \rightarrow ZnO_{solid} + H_2O$ into zinc oxide.

Hexamethylenetetramine was used (HMTA) $C_6H_{12}N_4$ for the creation of a buffer solution and consequently providing the possibility of carrying out synthesis at the low (less than 100°C) temperature [7] due to its slow two-step OH^- groups release:

$$C_6H_{12}N_4 + 6H_2O \leftrightarrow 6HCHO + 4NH_3\uparrow \quad (3)$$
$$NH_3 + H_2O \leftrightarrow NH^{4+} + OH^- \quad (4)$$

Formed in the process of the hydrolysis zinc cations Zn^{2+} react with anions of alkali OH^- with various hydroxides formation: $ZnOH^+$ (aqueous), $Zn(OH)_2$ (aqueous), $[Zn(OH)_3]^-$ (aqueous) and $[Zn(OH)_4]^{2-}$:

$$Zn^{2+} + OH^- \leftrightarrow [Zn(OH)]^+ \quad (5)$$
$$Zn^{2+} + 2OH^- \leftrightarrow Zn(OH)_2 \quad (6)$$
$$Zn^{2+} + 3OH^- \leftrightarrow [Zn(OH)_3]^- \quad (7)$$
$$Zn^{2+} + 4OH^- \leftrightarrow [Zn(OH)_4]^{2-} \quad (8)$$

Charged hydroxocomplexes of zinc are adsorbed on the respective surfaces of oxygen O^{2-} and the zinc Zn^{2+} due to the presence of the pyroelectric effect in seed ZnO crystallites with a further condensation in the oxide form. Zinc hydroxide $Zn(OH)_2$ (6) is not involved in the process of growth 1*D*-nanostructures ZnO because of an absence of charge. One of the possible ways of hydroxide – oxide conversion is the process (9):

$$[Zn(OH)_4]^{2-} \leftrightarrow ZnO + H_2O + 2OH^- \quad (9)$$

Overall, the chemical reaction can be written as:

$$Zn^{2+} + 2OH^- \leftrightarrow ZnO + H_2O \quad (10)$$

Reactions take place in the solution lead to minimization of the free energy of the entire system. As the polar planes (0001) have the highest energy, an adsorption of occurred during reactions (1)-(4) molecules happens on these polar planes with the further nanostructures growth in the *c* axis direction. In particular, the adsorption of molecules occurs due to formation of a monolayer which has an opposite polarity to the existing (0001) plane. Thus, on the surface of Zn^{2+} atoms, the precipitate hydroxides are formed in (7) and (8), then, according to reaction (9), ZnO is formed. At the end, the plane (0001) ends again with Zn^{2+} ions.

Dielectric insulation of the nanorods, as well as control of the morphology of the obtained structures was achieved by addition to stock solution anionic (sodium lauryl sulfate $C_{12}H_{25}SO_4Na$, SLS) and zwitterionic (cysteine $C_3H_7NO_2S$, Cys) surface-active substances (surfactants) with concentrations 1 surfactant molecule per 10 Zn^{2+} cations. Using SLS reduces the rate of growth of zinc faces due to the adsorption of $C_{12}H_{25}SO_4^-$. The use of Cys moreover leads to the diminution of O^{2-} edges growth rate, in addition to zinc one, due to the partial hydrogen sulphide release and $C_3H_6NO_2^+$ formation.

3. Results and discussion

Obtained structures were investigated using scanning electron microscope (SEM Seron Technology AIS 2300C, TESCAN MIRA LMU with an energy dispersion spectrometer EDX). Morphological and electrophysical characteristics of ZnO layers were studied using atomic force microscopy (AFM Therma NTEGRA NT-MDT).

In the case of the seed layers obtained via spin-coating method and annealed at temperatures less than 330°C the formation of zinc oxoacetate crystallites takes place:

$$4Zn(CH_3COO)_2 \rightarrow Zn_4O(CH_3COO)_6 + (CH_3)_2CO + CO_2$$

For a sample obtained on Si-substrate and annealed at T_{an}=120°C elemental composition analysis was carried out by electron probe micro analysis method (EPMA) (figure1).

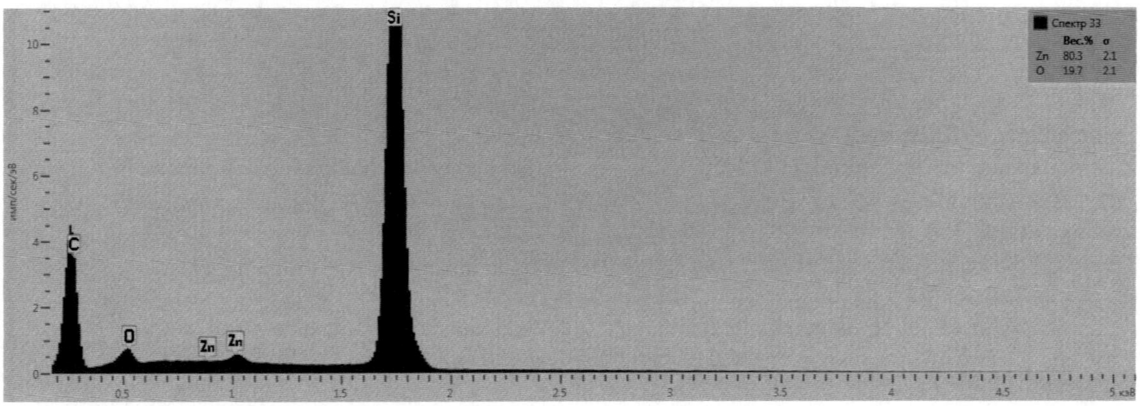

Figure 1. EPMA-data of ZnO seed layers obtained via spin-coating method on Si-substrate at an annealing temperature T_{an}=120°C.

Peak corresponding to carbon indicates the zinc oxoacetate crystallites formation at low temperatures of the synthesis. These coatings cannot be used as the seed layers for the formation of ZnO nanorods due to the absence of a pyroelectric effect in the material.

The optimal samples annealing temperature for the ZnO solid phase formation is 350°C, herewith the reaction proceeds on a substrate surface with the formation of homogeneous layers:

$$Zn(CH_3COO)_2 + 4O_2 \rightarrow ZnO + 4CO_2\uparrow + 3H_2O$$

The average size of obtained on Si-substrates at T_{an}=350°C zinc oxide crystallites is 50 nm (figure 2).

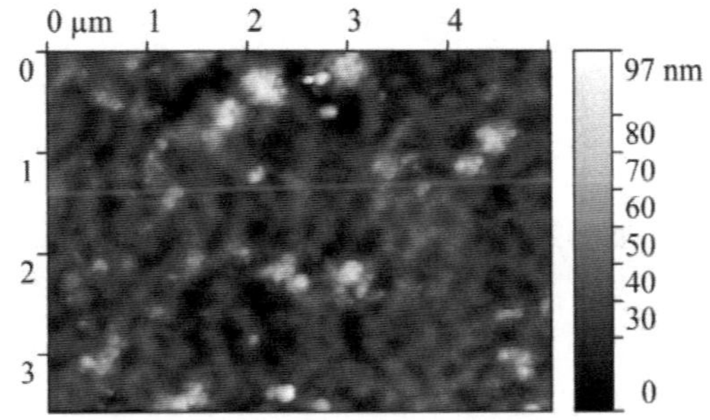

(a)

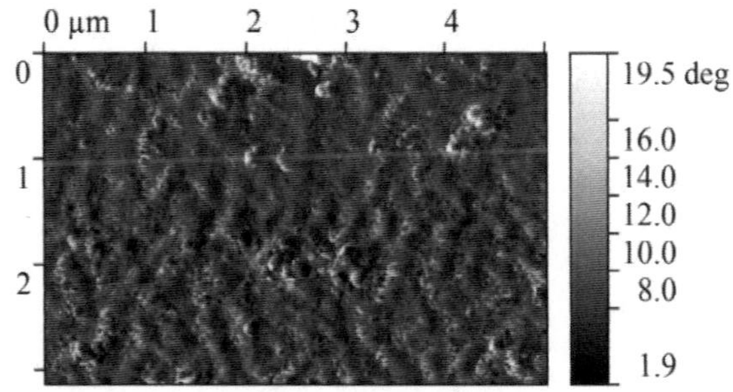

(b)

Figure 2. AFM-data of ZnO thin film (5 layers): *a* - topography, *b* – phase contrast.

With spin rate increasing a porosity of the layers decreases, however, at speeds above 5000 rpm is the uneven distribution profile of the film along the radius of the sample is observed.

ZnO 1*D*-nanostructures were produced on the obtained sublayers via low-temperature hydrothermal synthesis. The best ratio of a diameter to length *X* was achieved using cysteine as the surfactant (table 1). This effect may be associated with the possibility of zwitterionic surfactants adsorption on both sublattices, while anionic surfactants can be adsorbed only on the zinc one. As the growth of ZnO nanorods in the *c* axis direction ends by ions Zn^{2+}, the presence of anionic surfactants in solution reduces the rate of growth in this direction.

Table 1. The dependence of the ZnO nanorods aspect ratio obtained on spin-coating seed layers (5 layers) on the nature of the surfactant.

Surfactant	The dimensions of nanorods
Cysteine	*D*= 30 нм, *L*=600 нм, *X*=0.05
Sodium lauryl sulfate	*D*= 60 нм, *L*= 630 нм, *X*=0.09

Changing pH by adding an aqueous solution of ammonia leads to a uniform one-dimensional arrays of ZnO nanorods formation, moreover, their average diameter increases (figure 3).

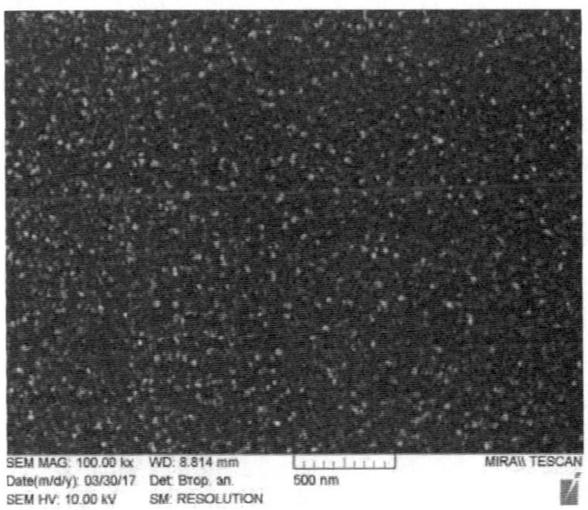

(a)

(b)

Figure 3. SEM-data of ZnO nanorods (5 seed layers obtained with the spin rate 3000 revs per min on Si-substrates): *a* – pH=10, *b* – pH=13.5.

In the absence of NH_4OH addition solution become depleted during the process of growth, whereby the geometrical parameters of the nanorods depend on the size of the sublayers crystallites, moreover, increase in time of synthesis does not change the sample structure. With increasing in the concentration of OH^- groups the solution is less depleted in the synthesis process, therefore, the size of the nanorods is mainly determined by the time of synthesis.

IOP Conf. Series: Journal of Physics: Conf. Series **917** (2017) 032022 doi:10.1088/1742-6596/917/3/032022

For the creation of a seed layers on flexible PET-substrates magnetron sputtering method was used. The deposition of thin films was carried out at room temperature. Thickness of obtained layers reached 10, 20, 30 nm.

The dependence of the nanorods sizes obtained on 20 nm seed layers (via magnetron sputtering method) on the nature of the surfactant is presented in the table 2. The best aspect ratio (as in the case of layers obtained by spin-coating method) refers to the 1*D*-nanostructures formed in the presence of Cys.

Table 2. The dependence of the ZnO nanorods aspect ratio obtained on magnetron sputtering seed layers (20 nm) on the nature of the surfactant.

Surfactant	The dimensions of nanorods
Cysteine	D=25 нм, L=390 нм, X=0.06
Sodium lauryl sulfate	D=40 нм, L= 350 нм, X=0.11

According to SEM-data disordered growth of nanorods with increasing thickness of seed layers was found (figure 4).

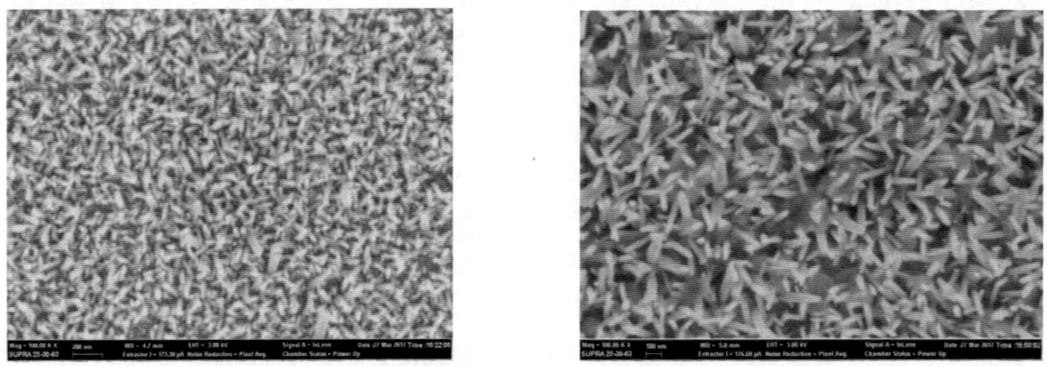

(a) (b)

Figure 4. SEM-data of ZnO nanorods, obtained on PET-substrates in the presence of SLS with a seed layers thickness: *a* – 10 nm, *b* – 30 nm.

This effect occurs due to increased probability of seed crystallites formation in different crystallographic directions, beside this increases thickness of nanorods and their length.

Electrophysical parameters of the structures obtained in the initial stages of growth with using SLS as surfactant were investigated using piezo-response force microscopy PFM (figure 5). Probes NSG10 were used in this paper (tip curvature radius 10 nm, resonant frequency 240 kHz, forse constant 11.8 N/m, cantilever length 95 mkm, cantilever width 30 mkm, cantilever thickness 2 mkm):

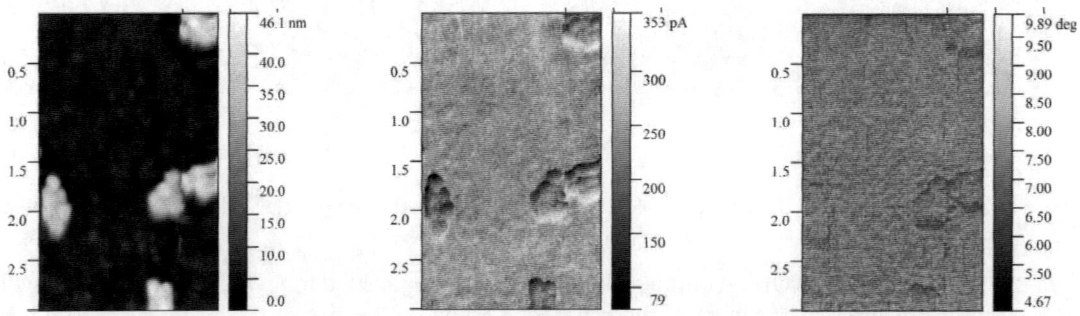

(a) (b) (c)

Figure 5. PFM-data of ZnO nanorods, obtained on PET-substrates in the presence of SLS: *a* – topography, *b* – amplitude, *c* – phase.

The piezoelectric response was registered on nanostructured coatings; moreover, its maximum value reached 350 pA. Detailed piezo-response force microscopy data analysis allows concluding that the main directions of the piezoelectric response in the initial stages of nanorods formation are directions along the surface of the substrate. The emerging crystallites are polarized in lateral directions. This conclusion is confirmed by increase in the number of disordered ZnO nanorods formed in these conditions in the later stages of hydrothermal synthesis (figure 4).

4. Conclusion

In this paper, arrays of ZnO nanorods were obtained via soft hydrothermal synthesis in the presence of different surfactants (cysteine, sodium lauryl sulfate) on the seed layers obtained by spin-coating and magnetron sputtering methods. These methods allow forming thin films with high homogeneity of the crystallites.

The using of zinc acetate as the source of zinc cations for the seed layers formation via spin-coating method is not possible on PET substrates due to the low melting temperature of the polymer.

The optimal thickness of the sublayer, is obtained by magnetron sputtering is 10 nm. With appropriate thickness of ZnO sublayer increases the probability of tunneling of charge carriers in the lower electrode increases, growth oriented nanorods perpendicular to the substrate.

Thus, the most optimal method of the seed layers formation is the method of magnetron sputtering, which provides the possibility of ZnO films deposition at low temperatures.

The best aspect ratio X=0.05-0.06 independent of the thickness of the nucleation layer was achieved using cysteine as the surfactant. The increase in the concentration of OH^- groups leads to the formation of nanorods with an equal length. These nanostructured coatings can be used as an active part of piezoelectric nanogenerators.

References

[1] Wang Z L, Song J 2006 Science **312** (5771) 242
[2] Hu Y, Wang Z L 2015 *Nano Energy* **14** 3
[3] Klingshirn C F, Meyer B K, Waag A, Hoffmann A, Geurts J 2010 *Zinc Oxide. From Fundamental Properties Towards Novel Applications* (Berlin: Springer) p. 374
[4] Karpova S S, Moshnikov V A, Maksimov A I et al. 2013 *Semiconductors* **47** (8) 1026
[5] Karpova S S, Moshnikov V A, Mjakin S V et al. 2013 *Semiconductors* **47** (3) 392
[6] Gavrilov S A, Gromov D G, Kozmin A M et al. 2013 *Physics of the solid state* **55** (7) 1376
[7] McPeak K M, Le T P, Britton N G, Nickolov Zh S, Elabd Y A, Baxter J B 2011 *Langmuir* **27** 3672

Investigation of resistive switching of $Zn_xTi_yHf_zO_i$ nanocomposite for RRAM elements manufacturing

R V Tominov[1], E G Zamburg[1], D A Khakhulin[1], V S Klimin[1], V A Smirnov[1], Y H Chu[2,3,4,5] and O A Ageev[1]

[1] Department of Nanotechnology and Microsystems, Southern Federal University, Taganrog, 347928, Russia
[2] Department of Materials Science and Engineering, National Chiao Tung University, Hsinchu, 1001, Taiwan
[3] Department of Electrophysics, National Chiao Tung University, Hsinchu, 1001, Taiwan
[4] Institute of Physics, Academia Sinica, Taipei, 11529, Taiwan
[5] Material and Chemical Research Laboratories, Industrial Technology Research Institute, Hsinchu, 31040, Taiwan

Abstract. The resistive switching effect in $Zn_xTi_yHf_zO_i$ nanocomposite film, grown by pulsed laser deposition technique, was investigated. It was shown that $Zn_xTi_yHf_zO_i$ film surface had a granular structure with 0.8±0.4 μm^2 grain size and 7.3±5.1 nm grain height. Resistive switching from high resistance state (HRS) to low resistance state (LRS) was occurred at 0.9±0.4 V, and from LRS to HRS at 1.5±0.2 V. HRS/LRS ratio was 2.6. The results can be used for nanocomposite-metal-oxide-film RRAM fabrication.

1. Introduction

Development of the electronic devices technology in respect to a size reduction of integrated circuit elements and chip density is increasing using nanotechnology that allows fabrication of nanoscale structures with dimensions less than 50 nm. There are many types of electronic devices, including, in particular, high-speed memory devices with high-density integration and low power consumption [1, 2].

Pulsed laser deposition (PLD) is a promising technique as it allows growing films with control of crystalline quality and surface adhesion to obtain materials with complex stoichiometry [3-5]. There are a lot of structures that can be obtained in wide range of controlled properties by PLD: nanostructured films, single crystalline films, nanowhiskers, nanotetrapods, etc.

Resistive switching effect in oxide films is attractive for resistive random access memory (RRAM) development [6-10]. The main advantages of RRAM are non-volatility, high performance, small cell size, and low power consumption [6-10]. Resistive switching effect was observed in many oxide film (ZnO, TiO_2, HfO_2, NiO, Al_2O_3, SnO_2, WO_3, SiO_2, etc.) and it is expressed in the switching of resistance of oxide film between high and low resistance states after threshold electric field applying [6-10].

One of the main problems of RRAM is a large variation of switching parameters, both in write/reset of the same cell and in write/reset of the different cells. Using of nanocomposite metal oxides allows to solve this problem [11-14].

The aim of this work is the investigation of resistive switching effect in $Zn_xTi_yHf_zO_i$ nanocomposite film, grown by pulsed laser deposition.

2. Experiment details

Nanocomposite $Zn_xTi_yHf_zO_i$ thin film was grown by pulsed laser deposition. Al_2O_3 was used as a substrate. To deposite TiN bottom electrode the following conditions were used: substrate temperature 600°C, target–substrate distance 50 mm, laser pulse energy 300 mJ, number of pulses 7,500. After deposition, sample was taken out from a chamber in order to put a mask for forming clear boundary for measurements of nanocomposite film thickness. To form nanocomposite film, ZnO, HfO_2, TiO_2 targets were used for 51 cycles (2/3 of layers = 17 pulses (2 Hz), 1/6 of layers = 10 pulses (2 Hz), 1/6 of layers = 4 pulses (2 Hz), respectively). The deposition was performed under the following conditions: substrate temperature 400°C, target–substrate distance 50 mm, O_2 pressure 1 mTorr, laser pulse energy 300 mJ.

An AFM-image of the $Zn_xTi_yHf_zO_i$ film surface was obtained using atomic-force microscope Solver P47 Pro (NT-MDT, Russia) in semi-contact mode. The AFM-image processing was performed using Image Analysis software. The $Zn_xTi_yHf_zO_i$ film thickness was also measured using Solver P47 Pro by TiN/ $Zn_xTi_yHf_zO_i$ boundary scanning, and was equaled 12.3±0.7 nm. Investigation of structural parameters of the nanocomposite film was carried out using X-ray Powder Diffraction (XRD) and Energy-dispersive X-ray spectroscopy (EDX). Electrical measurements were carried out using semiconductor characterization system Keithley 4200-SCS (Keithley, USA) with W probes. During experiment TiN layer was grounded. Current-voltage curves were obtained from –2 to +2 V sweep for 50 cycles at the same point. To prevent thermal breakdown of the nanocomposite film, 1 mA compliance current was set during electric measurements. Read voltage was at 0.4 V.

3. Results and discussion

Figure 1 shows experimental investigation of $Zn_xTi_yHf_zO_i$ film morphology. It is shown that $Zn_xTi_yHf_zO_i$ film surface has a granular structure (figure 1 a) with 0.8±0.4 μm^2 grain size (figure 1 c) and 7.3±5.1 nm grain height (figure 1 b). The nanocomposite film thickness was investigated using AFM by scanning bottom electrode/nanocomposite film boundary and was equaled 12.3±3.7 nm.

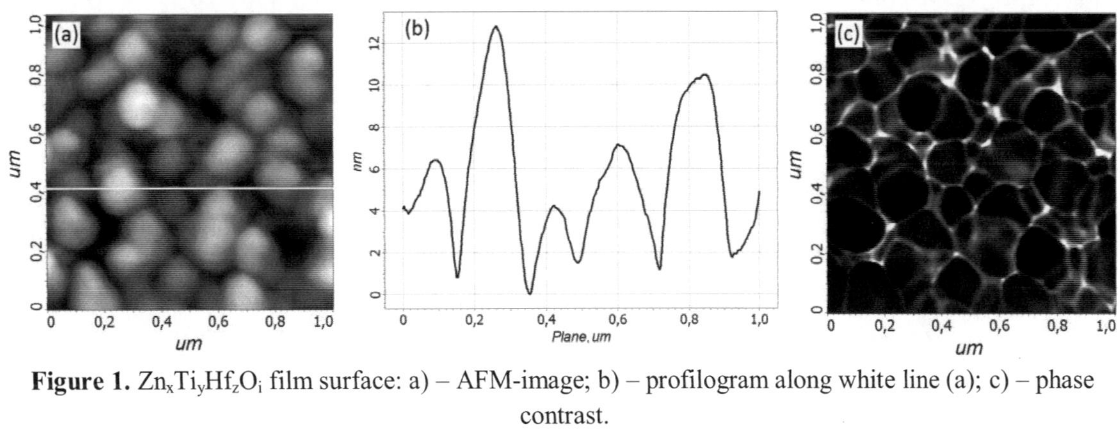

Figure 1. $Zn_xTi_yHf_zO_i$ film surface: a) – AFM-image; b) – profilogram along white line (a); c) – phase contrast.

Figure 2 shows investigation of structural parameters of $Zn_xTi_yHf_zO_i$ film. It follows from diffractogram (figure 2 a), (100) and (111) crystallites prevails for ZnO and HfO_2 phases, respectively. Figure 2 b and table 1 show results of energy-dispersive X-ray spectroscopy. It's shown that there is 7.93% of Ti in the nanocomposite film, 3.93% of Zn, 4.42% of Hf.

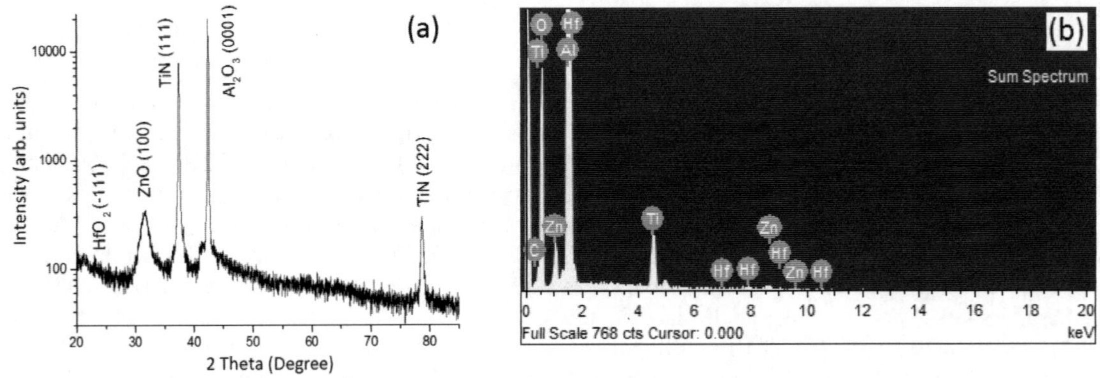

Figure 2. Investigation of structural parameters of the $Zn_xTi_yHf_zO_i$ film: a) – X-ray Powder Diffraction (XRD); b) – Energy-dispersive X-ray spectroscopy (EDX).

Table 1. Energy-dispersive X-ray spectroscopy of the $Zn_xTi_yHf_zO_i$ nanocomposite film.

Element	C	O	Al	Ti	Zn	Hf	Totals
Weight(%)	2.46	31.68	49.58	7.93	3.93	4.42	100.00
Atomic(%)	4.79	46.34	43.01	3.88	1.41	0.58	

Average current-voltage curve of structure TiN/$Zn_xTi_yHf_zO_i$/W was performed from 50 current-voltage curves (figure 3 a) measured at the same point on the nanocomposite surface. It was shown, resistive switching from high resistance state (HRS) to low resistance state (LRS) was occurred at 0.9±0.4 V, and from LRS to HRS at -1.5±0.2 V. Resistance-cycle dependence was built based on obtained curves. It was shown that HRS was 6.5±1.8 kOhm, LRS was 2.5±1.1 kOhm. HRS/LRS ratio was 2.6 (figure 3 c).

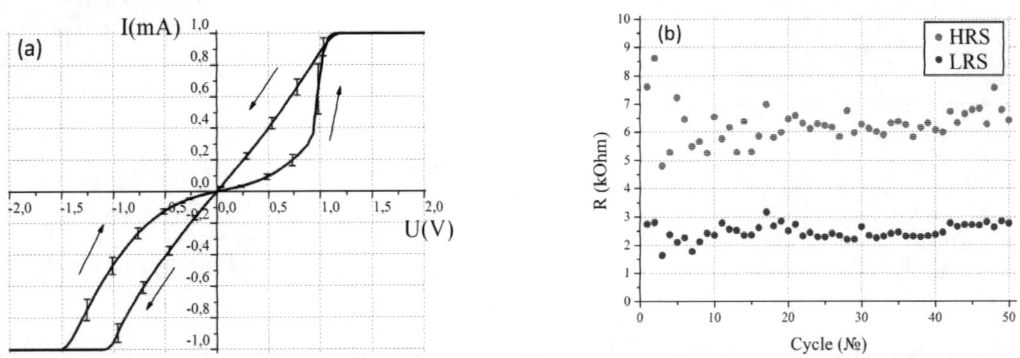

Figure 3. Experimental studies of resistive switching effect in $Zn_xTi_yHf_zO_i$ film: a) – averaged current-voltage characteristic of 50 sweep cycles at the same point on in $Zn_xTi_yHf_zO_i$ film surface; b) – endurance test of $Zn_xTi_yHf_zO_i$ film for 50 sweep cycles.

4. Conclusion

In summary, we have investigated resistive switching effect in nanocomposite $Zn_xTi_yHf_zO_i$ thin film, grown by pulsed laser deposition. We have shown that HRS was 6.5±1.8 kOhm, LRS was 2.5±1.1 kOhm. HRS/LRS ratio was 2.6. The results can be used for nanocomposite-metal-oxide-film RRAM fabrication.

Acknowledgments

The work was supported by and RFBR (№ 16-32-00069 mol_a project).

References

[1] Misty Blowers 2015 *Evolution of Cyber Technologies and Operations to 2035* (Switzerland: Springer International Publishing)

[2] Klimin V S, Solodovnik M S, Smirnov V A, Eskov, A.V., Tominov R V, Ageev O A 2016 *Proceedings of SPIE* **10224** 102241Z

[3] Vakulov Z E, Zamburg E G, Khakhulin D A, Ageev O A 2017 *Materials Science in Semiconductor Processing* **66** 21

[4] Ageev O A, Dostanko A P, Zamburg E G, Konoplev B G, Polyakov V V, Cherednichenko D I 2015 *Physics of the Solid State* **57(10)** 2093

[5] Dostanko A P, Ageev O A, Golosov D A, Zavadski S M, Zamburg E G, Vakulov D E, Vakulov Z E 2014 *Semiconductors* **48(9)** 1242

[6] Avilov V I, Ageev O A, Kolomiitsev A S, Konoplev B G, Smirnov V A 2014 *Semiconductors* **48** 1757

[7] Ageev O A, Blinov Y F, Il'in O I, Konoplev B G, Rubashkina M V, Smirnov V A, Fedotov A A 2015 *Physics of the Solid State* **57** 825

[8] Ageev O A, Alyab'eva N I, Konoplev B G, Polyakov V V, Smirnov V A 2010 *Semiconductors* **44** 1703

[9] Avilov V I, Ageev O A, Jityaev I L, Kolomiytsev A S, Smirnov V A 2016 *Proceedings of SPIE* **10224** 102240T

[10] Ageev O A, Gusev E Y, Zamburg E G, Vakulov D E, Vakulov Z E, Shumov A V, Ivonin M N 2014 *Applied Mechanics and Materials* **475** 446

[11] Perkins J D, Cueto J A, Alleman J L, Warmsingh C, Keyes B M, Gedvilas L M, Parilla P A, To B, Readey D W, Ginley D S 2002 *J.Thin Solid Films* **411** 152

[12] Ohkubo I, Christen H M, Khalifah P, Sathyamurthy S, Zhai H Y, Rouleau C M, Mandrus D G, Lowndes D H 2004 *J. Appl. Surf. Sci* **223** 35

[13] Au K, Gao X S, Wang J, Bao Z Y, Liu J M, Dai J Y 2016. *J. Appl. Phys.* **114** 027019

[14] Liu Q, Long S, Wang W, Zuo Q, Zhang S, Chen J, and Liu M 2009 *J.Elect. dev. lett.* **30** 12

IOP Conf. Series: Journal of Physics: Conf. Series **917** (2017) 032024 doi:10.1088/1742-6596/917/3/032024

Influence of target-substrate distance during pulsed laser deposition on properties of $LiNbO_3$ thin films

Z Vakulov[1], E Zamburg[2], D A Golosov[3], S M Zavadskiy[3], A P Dostanko[3], A V Miakonkikh[4], I E Klemente[4], K V Rudenko[4], O A Ageev[1]

[1]Research and Education Centre «Nanotechnology», Southern Federal University, Taganrog 347922, Russian Federation
[2]Institute of Microengineering, École polytechnique fédérale de Lausanne, Neuchâtel CH-2002, Switzerland
[3]Department of Research and Development, Belarusian State University of Informatics and Radioelectronics, Minsk 220013, Belarus
[4]Institute of Physics and Technology RAS, Moscow 117218, Russian Federation

Abstract. This paper shows the results of studying the influence of target-substrate distance during pulsed laser deposition on electrical properties of $LiNbO_3$ thin films. It has been shown that changing the target-substrate distance we can obtain thin films with a different composition. EDX spectra indicate that the $LiNbO_3$ thin films fabricated under 120 mm target-substrate distance depleted Nb (0.7 atm. %) compare to $LiNbO_3$ films fabricated under 20 mm target-substrate distance (11.0 atm. %). Varying target-substrate distance in the range from 20 mm to 120 mm charge carrier mobility increasing from 24 $cm^2/V{\cdot}s$ to 395 $cm^2/V{\cdot}s$ and concentration of charge carrier decreasing from $3{\cdot}10^{13}$ cm^{-3} to $1{\cdot}10^{12}$ cm^{-3}. Obtained results can be used under development and fabrication of integral acousto-optic and surface acoustic wave (SAW) devices.

1. Introduction

Lithium niobate ($LiNbO_3$) is the material which is widely used for fabrication of SAW devices due to its piezo-, pyroelectrics, acousto-optics, and electro-optics properties [1]. Development of integral acousto-optic cell for radio engineering processors of ultra-wideband microwave signals is one of the prospective areas of acousto-optic. Currently, the task of obtaining of $LiNbO_3$ thin films with low optic losses by traditional microelectronic techniques is most important. $LiNbO_3$ thin films could be fabricated by different techniques: molecular beam epitaxy (MBE), RF-sputtering, sol-gel process, pulsed laser deposition (PLD). PLD is a prospective technique for fabrication $LiNbO_3$ thin films due to a numerous of technological parameters and the ability to keep stoichiometric composition of a target [2, 3].

Besides, $LiNbO_3$ films are multicomponent oxides and its properties (crystallographic orientation, resistance, concentration and mobility of charge carriers, optical and piezoelectric properties, and surface roughness) depend on the exact observance of the stoichiometric composition and structure, which in turn depends on the method and technological modes. In that light research influence of technological modes on properties of $LiNbO_3$ thin films in order to fabricate the films with target properties is important. Since target-substrate distance is one of the key parameters of the PLD [4], the task of this work is researching the influence of target-substrate distance during PLD on $LiNbO_3$ thin films electrical properties.

2. Experiment

In order to fabricate of $LiNbO_3$ thin films, we used nanocluster NANOFAB NTK-9 (NT-MDT, Russia), comprising PLD equipment Pioneer 180 (Neocera Co., USA). The $LiNbO_3$ target was ablated by excimer KrF laser (λ=248 nm) (Coherent Inc., USA). Energy density on target surface was maintained at 2.0 J/cm^2. The quantity and frequency of laser pulses were 50.000 and 10 Hz, respectively. Target-substrate distance had been varied from 20 mm to 120 mm. $LiNbO_3$ films were fabricated on Al_2O_3 substrates (1 cm x 1 cm). All substrates were subjected to preliminary purification in inorganic solvents to remove surface contamination.

The morphology of obtained film was studied by scanning electron microscopy (SEM, Nova Nanolab 600 with EDX system (FEI.Co, Netherlands)) and atomic-force microscopy in semi-contact mode (AFM, Ntegra probe nanolab (NT - MDT, Russia)). Electrical properties of the $LiNbO_3$ films were measured by Ecopia HMS-3000 equipment (Ecopia Co., Republic of Korea). Optical properties of obtained films (refractive index n, absorption coefficient k) were studied by spectral ellipsometry technique on spectral ellipsometer M-2000X (Woollam J.A. Co, USA) in the range of wavelength from 240 nm to 1000 nm.

3. Results and discussion

Figure 1 shows AFM images and dependencies of $LiNbO_3$ thin films morphological properties on target-substrate distance.

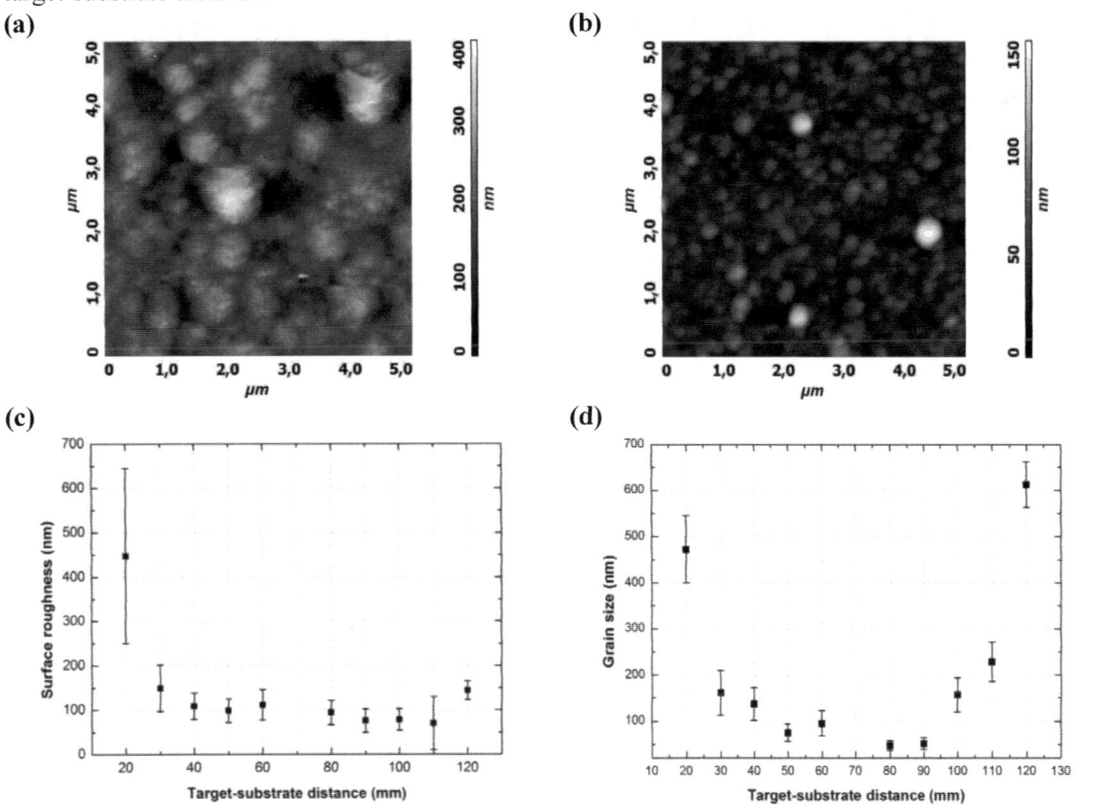

Figure 1 (a, b, c, d). AFM images of $LiNbO_3$ thin films obtained under 30 mm (a) and 60 mm (b) target-substrate distance, dependences of surface roughness (c) and grain size (d) of $LiNbO_3$ films on target-substrate distance.

IOP Conf. Series: Journal of Physics: Conf. Series **917** (2017) 032024 doi:10.1088/1742-6596/917/3/032024

Increase target-substrate distance from 20 mm to 110 mm surface roughness of obtained films decrease from 447.5 nm to 69.8 nm. Further increase target-substrate distance from 110 mm to 120 mm increase surface roughness to 144,6 nm. Wherein, increase target-substrate distance from 20 mm to 80 mm results in reducing of grain size from 471 nm to 27.4 nm. Further increase of target-substrate distance from 80 mm to 120 mm results in increasing grain size up to 612 nm. Such behaviour of the dependences may be linked to additional heating from plasma plume (in case of target-substrate distance from 20 mm to 60 mm). The subsequent increase of grain size (target-substrate distance > 80 mm) may be related to deficiency of Li in the films and Nb deposition in the form of a solid phase.

Figure 2 shows dependencies of $LiNbO_3$ thin films electrical properties on target-substrate distance.

(a)

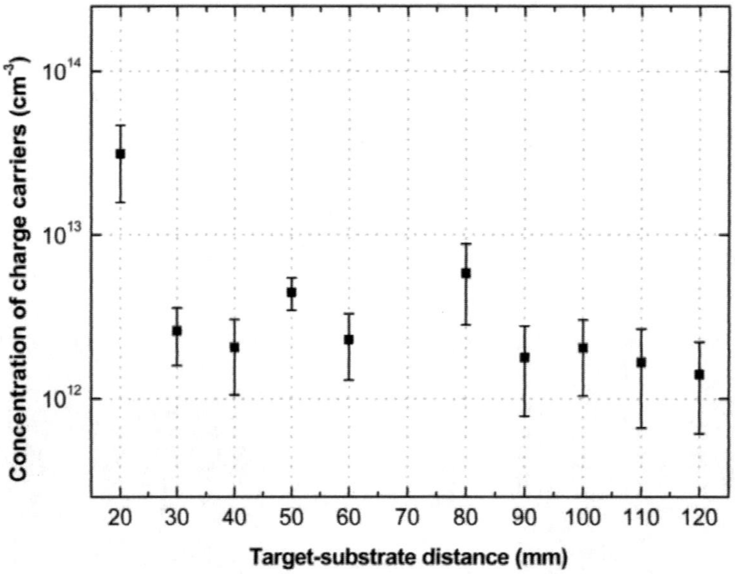

(b)

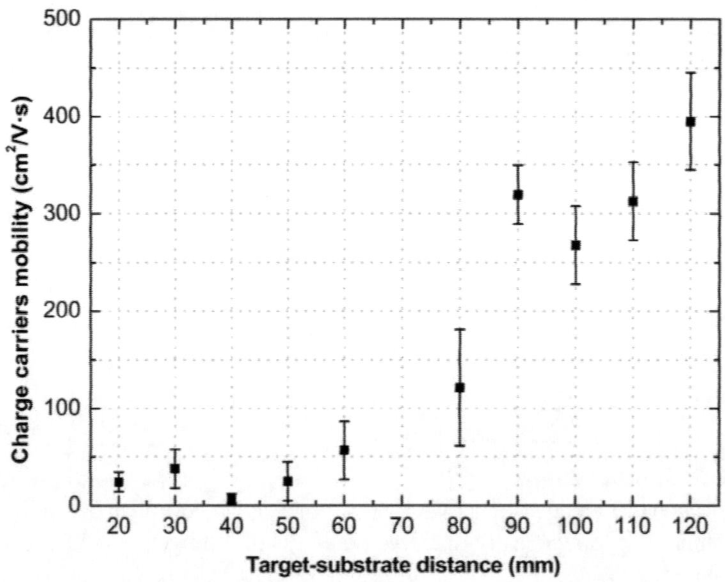

(c)

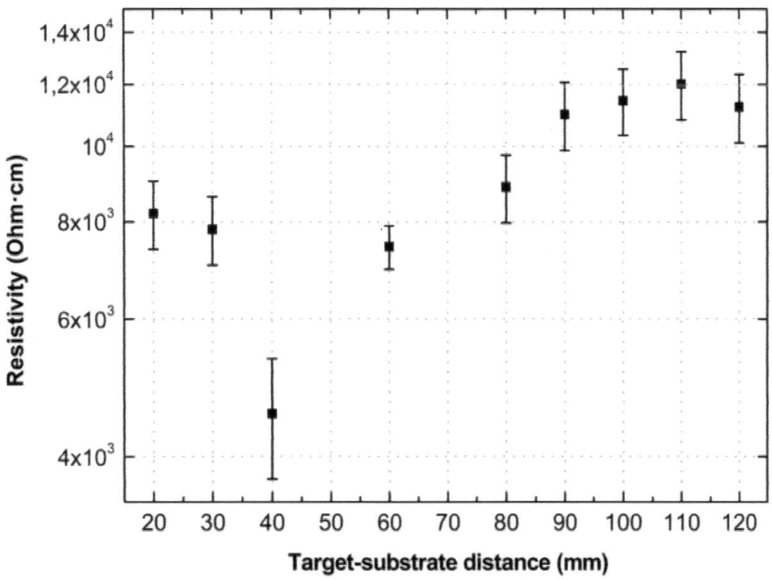

Figure 2 (a, b, c). Dependencies of concentration (a), mobility (b), and resistivity (c) of charge carrier on target-substrate distance.

Varying target-substrate distance in the range from 20 mm to 120 mm charge carrier mobility increasing from 24 $cm^2/V \cdot s$ to 395 $cm^2/V \cdot s$ and concentration of charge carrier decreasing from $3 \cdot 10^{13}$ cm^{-3} to $1 \cdot 10^{12}$ cm^{-3}. According to the results of morphological study this fact indicates enhancing of crystalline quality of the films. The resistivity of the films was changed insignificantly (from $8.3 \cdot 10^3$ Ohm·cm to $1.27 \cdot 10^4$ Ohm·cm).

Figure 3 shows the results of EDX analysis of $LiNbO_3$ films fabricated under different target-substrate distance. However, the EDX of the matrices could not show the presence of lithium due to its low atomic number. EDX of the matrices showing the presence of the niobium (Nb) and oxygen (O) in obtained $LiNbO_3$ films.

(a) **(b)**

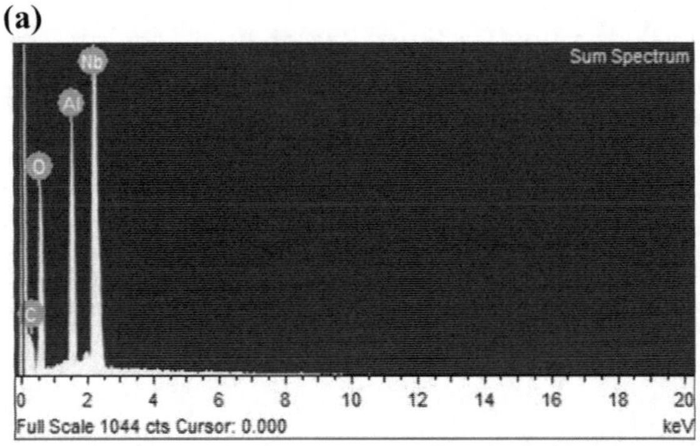

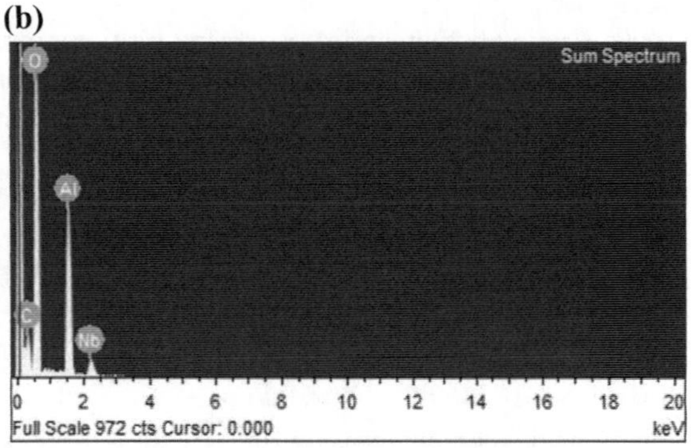

Figure 3 (a, b). The results of EDX analysis of $LiNbO_3$ films fabricated under different target-substrate distance: 20 mm (a), and 120 mm (b).

The results of EDX analyses shows that $LiNbO_3$ thin films obtained under target-substrate distance 120 mm contain less Nb (0.7 atm. %) compare to the same films obtained under target-substrate distance 20 mm (11.0 atm. %). Presence of Al peak is related to Al_2O_3 substrate.

Figure 4 shows spectral dependence of refractive index *n* and absorption coefficient *k* of the $LiNbO_3$ film fabricated under target-substrate distance 80 mm.

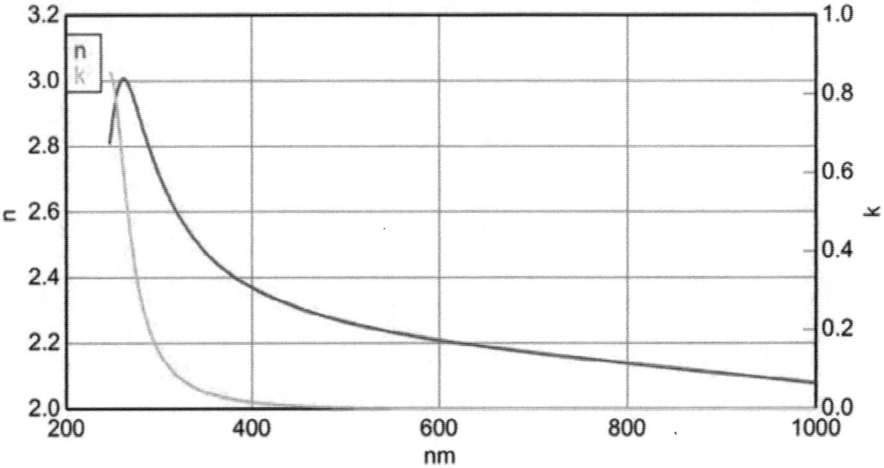

Figure 4. Spectral dependence of refractive index *n* and absorption coefficient *k* of the $LiNbO_3$ film fabricated under target-substrate distance 80 mm.

It was established, that the results of optical study didn't depend on orientation of samples. This fact shows isotropic character of the optical properties. In the visible wavelength range the refractive index decreased from 3.03 (350 nm) to 1.95 (800 nm).

4. Conclusion

Obtained results show that morphological, electrical, and optical properties of $LiNbO_3$ strongly depend on the target-substrate distance. EDX analysis indicate that $LiNbO_3$ thin films fabricated under 120 mm target-substrate distance depleted Nb (0.7 atm. %) compare to $LiNbO_3$ films fabricated under 20 mm target-substrate distance (11.0 atm. %). The ability of formation $LiNbO_3$ thin films with target properties and different composition by PLD has been shown. The results can be used under development and fabrication of integral acousto-optic and SAW devices.

Acknowledgments
This work was financially supported by the Russian Foundation for Basic Research № 16-57-00028-Bel_a, and № T16P-094. The work was done on the equipment of Research and Education Centre «Nanotechnology» and Collective Use Centre «Nanotechnology», Southern Federal University.

References

[1] Wong K K 2002 *Properties of lithium niobate*. (London: INSPEC)
[2] Lanzhong Hao, Jun Zhu, Yunjie Liu, Shuili Wang, Huizhong Zeng, Xiuwei Liao, Yingying Liu, Huawei Lei, Ying Zhang, Wanli Zhang and Yanrong Li 2012 *Thin Solid Films* **520** 3035–38
[3] Xinchang Wang, Zhizhen Ye and Binghui Zhao 2007 *Solid State Commun.* **142** 694–97
[4] Ageev O A, Dostanko A P, Zamburg E G, Konoplev B G, Polyakov V V and Cherednichenko D I 2015 *Phys. Solid State* **57** 2093–98

IOP Conf. Series: Journal of Physics: Conf. Series **917** (2017) 032025 doi:10.1088/1742-6596/917/3/032025

Doping profiles during nanowire growth via the vapor-liquid-solid mechanism

E D Leshchenko[1,2], **V G Dubrovskii**[1,2,3]

[1] St. Petersburg Academic University, Khlopina 8/3, 194021 St. Petersburg, Russia
[2] ITMO University, Kronverkskiy pr. 49, 197101 St. Petersburg, Russia
[3] Ioffe Physical Technical Institute of the Russian Academy of Sciences, Politekhnicheskaya 26, 194021 St. Petersburg, Russia

Abstract. We report on a dopant incorporation mechanism in core-shell nanowires fabricated via the vapor-liquid-solid growth method. In particular, a solution to the diffusion equation for the dopant profile in the core in the general case of spatial anisotropy is obtained. This allows us to describe the evolution of the dopant concentration during the nanowire growth. Within the model, we show that the dopant inhomogeneity arises from the NW core morphology. We also describe qualitatively the entire diffusion process. It is shown how to achieve the homogeneous dopant distribution.

1. Introduction

In spite of the fact that the p-n junction was first formed in free-standing nanowires (NWs) over a quarter of a century ago [1], controlled doping with atomic accuracy required for the majority of nanoscale devices remains challenging [2,3]. These difficulties include impossibility to achieve an ideal abruptness, interface chemistry control and the inhomogeneous dopant distribution and result in the significant NW conductivity decrease. In order to overcome these obstacles and improve efficiency of nanodevices based on NWs, comprehensive research efforts in recent years have been directed towards the study of NW doping including such methods as standard Hall measurements [4], atom probe tomography [5] and photoluminescence emission [6]. Nowadays, semiconductor NW investigations greatly outnumber investigations of any other nanostructures. Such tremendous interest is determined by NW material properties and potential prospects for future use in nanotechnology [7-9]. Particularly, NWs are basic building blocks for a broad spectrum of electronic and photovoltaic devices including gas sensors [10], light-emitting diodes [11], low cost solar cells at high enough efficiency [12] and flexible panel displays [13]. The solid source diffusion process is widely applied as a method for industrial applications allowing one to avoid the non-uniform dopant distribution in semiconductor NWs. Ion-implantation is another technique for NW material modification which opens the possibility of the steerable position of doping ions [14]. However, despite of advanced NW growth and doping technologies, the dopant incorporation mechanism has essential influence on the resulting doping profile. In this contribution, we demonstrate that radial dopant inhomogeneity arises from the NW core morphology. Additionally, we describe the evolution of the doping distribution in the general case of anisotropic diffusion coefficient.

IOP Conf. Series: Journal of Physics: Conf. Series **917** (2017) 032025 doi:10.1088/1742-6596/917/3/032025

2. Theoretical model

A critical step in fabrication of semiconductor NWs with the p-n junction is the synthesis of the homogeneously doped NW core. The vapor-liquid-solid mechanism [15-17] looks very promising and allows one to grow high quality NWs with simultaneous doping. Such method involves liquid catalyst droplet feeding from the supply phase and NW growth as the result of crystallization under the droplet. But the NW morphology can be different and depends on growth conditions. Under widely used conditions for III-V NW growth, one-dimensional nanostructures have the hexagonal cross-sections with truncated corners. According to the work [18], the dopant distribution symmetry is broken in such geometry and segregation behavior when dopants arriving to the catalyst droplet concentrate near truncated facets is observed. It is noted, that such behavior is reproducible for different semiconductors and doping agents. The growth scheme of the NW with hexagonal cross-section and truncated facets {112} is presented in Fig. 1. The sizes of the {112} facets can be ignored in view of its small value in comparison with the {111} top facet size. Therefore we believe that the dopant incorporation takes place only at the regular hexagon corners, after that, dopants diffuse into the two-dimensional lateral layer. Its diffusion along the growth axis is neglected because of the fact that dopants do not spread far and wide during new monolayer formation due to the small value of the diffusion coefficient in solid semiconductor (which is of the order of 0.05 nm^2/s). However, in the end of growth process difference between the top and base dopant profiles will be notable. Moreover, dopant diffusion along the NW core during shell growth does not change the situation because of the small diffusion coefficient at low temperatures corresponding to NW shell growth conditions.

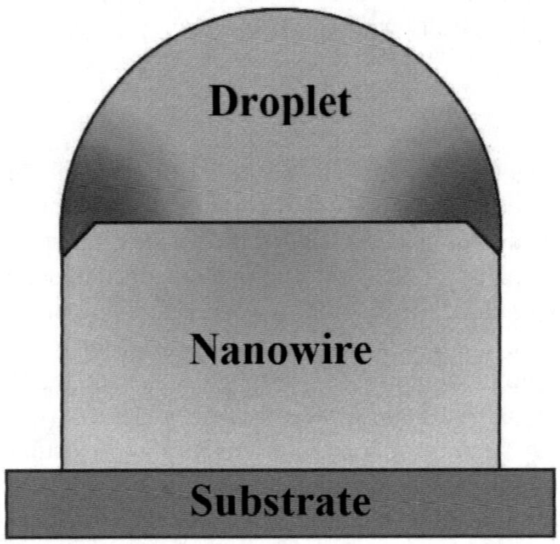

Figure 1. Scheme of NW growth. The red gradient corresponds to the dopant distribution in the catalyst droplet with maximum concentration near the truncated facets.

In the general case of spatial anisotropy, the two-dimensional diffusion equation is

$$\frac{\partial C(x,y,t)}{\partial t} = D_x \cdot \frac{\partial^2 C(x,y,t)}{\partial x^2} + D_y \cdot \frac{\partial^2 C(x,y,t)}{\partial y^2} \tag{1}$$

where $C(x,y,t)$ is the dopant concentration at the time moment t, D_x and D_y are the diffusion coefficients in the different directions.

As the initial condition, we use the sum of delta functions with coordinates (x_n, y_n) corresponding to the dopant sources:

$$C(x, y, 0) = \frac{A}{N}\sum_{n=1}^{N}\delta(x - x_n, y - y_n) \tag{2}$$

Here A is a normalization constant, which allows one to get the real concentration value, N is sources number. It is well understood that $1/N$ coefficient provide the normalization of the integral concentration for 2-dimensional layer to unity in the isotropic case and if $A = 1$.
We ignore the vapor-solid influx though the NW sidewalls, whereas it should be taken into account for NW shell doping modeling. Therefore the boundary condition involves zero vapor-solid influx though the NW core boundary S

$$\left.\frac{\partial C(x, y, t)}{\partial \vec{n}}\right|_S = 0 \tag{3}$$

with $\vec{n}$ being the normal to the NW boundary.
The solution of the non-stationary diffusion equation is the sum contributions from the real and fictitious sources:

$$C(x, y, t) = \frac{3A}{4\pi\sqrt{D_x D_y}tN}\sum_m \exp\left(-\frac{1}{4t}\left(\frac{(x - x_m)^2}{D_x} + \frac{(y - y_m)^2}{D_y}\right)\right) \tag{4}$$

Here, summation is performed by all real and fictitious sources which are obtained by the method of mirror images where the NW core boundary plays role a reflecting boundary. Thereby infinite number of fictitious sources is obtained as the result of reflection of real sources from the regular hexagon sides. However, as distance to the fictitious sources raises, their contributions in overall concentration inside the NW core decrease and far sources can be ignored in computations.

3. Results and discussion

Let us first consider an example of anisotropic dopant diffusion from a corner. In this case Eq. (4) takes form

$$C(x, y, t) = \frac{3A}{4\pi\sqrt{D_x D_y}t}\sum_{k,l=-\infty}^{+\infty} \exp\left(-\frac{\left(x - \sqrt{3}ak/2\right)^2}{4D_x t} - \frac{(y - al/2)^2}{4D_y t}\right) \tag{5}$$

for even $k + l$. Here a is the distance between the nearest NW crystal planes.
Fig. 2 shows the evolution of the dopant distribution migrated from the corner for $D_x = 0.2$ nm^2/s and $D_y = 0.04$ nm^2/s. These values inside the obtained range [19,20]. Obviously, under the conditions of presence of at least 3 initial incorporation positions, we will have the three-fold symmetric doping profile. Thus the truncated facets are attractive positions for dopant incorporation and are responsible for doping inhomogeneity during NW growth via the vapor-liquid-solid mechanism.

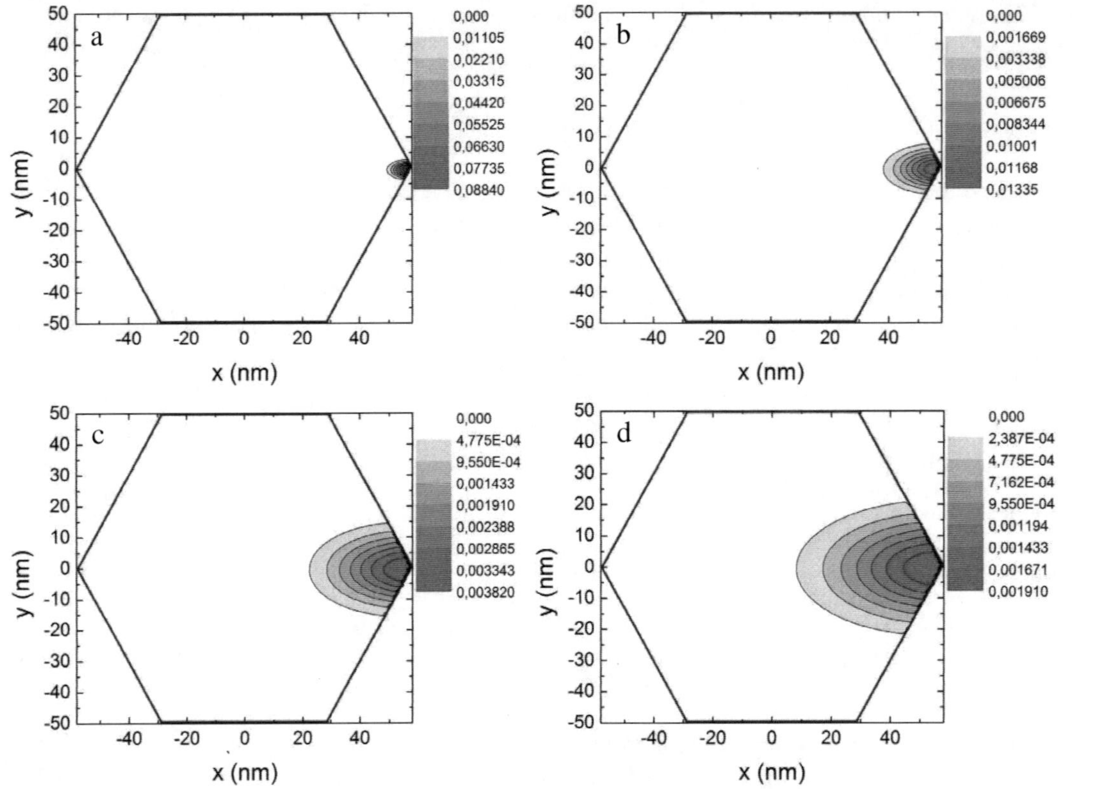

Figure 2. Dopant concentration profiles in case of spatial anisotropy at time moments $t = 30$ s (a), $t = 200$ s (b), $t = 700$ s (c) and $t = 1400$ s (d).

Now let us consider the case of isotropic diffusion from six corners of the NW core. The dependences of dopant concentration on NW radius along the line connected farthest corners for $D = 0.1$ nm^2/s at different times are presented in Fig. 3. It is seen that the equation of the dopant concentration appears at long times – of the order of 100 minutes, while an average time of NW growth core is twice smaller - of the order of 40-50 minutes.

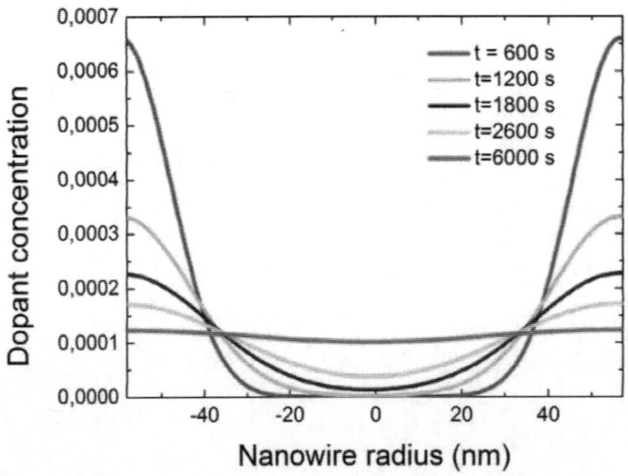

Figure 3. Radius dependences of dopant concentrations in NW core.

4. Conclusion

The evolution of the dopant distribution function in the general case of the anisotropic diffusion coefficient.is obtained during semiconductor NW core growth via the vapor-liquid-solid mechanism. It is shown that radial dopant inhomogeneity arises from the NW core morphology. The possibility of fabrication NWs with homogeneous doped core using the solid-source diffusion method is presented. These theoretical results can be used for understanding and optimization of the vapor-liquid-solid III-V NWs [21] and NWs of elemental semiconductors [22]. It will also be interesting to consider the strain-induced effects [23] and composition of ternary III-V NWs [24] and across axial or radial NW heterostructures [25,26] influencing the doping profiles. These studies will be presented elsewhere.

Acknowledgement

Authors are grateful for the financial support from Russian Ministry of Education and Science Project RFMEFI61316X0055 “Advanced heterostructures based on III-V nanostructures for photonics”

References

[1] Haraguchi K., Katsuyama T., Hiruma K., Ogawa K. 1992 *Appl. Phys. Lett.* **60** 745
[2] Gudiksen M.S. *et al*, 2002 *Nature* **415** 617
[3] Stamplecoskie K.G. *et al*, 2008 *Nano Lett.* **8** 2674
[4] Blömers Ch., Grap T., Lepsa M.I., Moers J., *et al* 2012 *Appl. Phys. Lett.* **101** 152106
[5] Perea D.E., Hemesath E.R., Schwalbach E.J., *et al* 2009 *Nature Nanotech.* **4** 315
[6] Arab S., Yao M., Zhou C., Dapkus P.D., *et al* 2016 *Appl. Phys. Lett.* **108** 182106
[7] Jiang N., Wong-Leung, J., Joyce H. J., Gao Q., *et al* 2014 *Nano Lett.* **14** 5865
[8] Darbandi A., Kavanagh K. L., Watkins S. P., 2015 *Nano Lett.* **15** 5408
[9] Qian F., Gradecak S., Li Y., Wen C., Lieber C. M., 2005 *Nano Lett.* **5** 2287
[10] Chen X., Wonga C., Yuanc C.A., *et al*, 2013 *Sensors and Actuators B: Chemical* **177** 178
[11] Hua B., Motohisa J., Kobayashi S., Hara S., Fukui T., 2009 *Nano Lett.* **9** 112
[12] Law M., Greene L.E., Radenovic A., Kuykendall T. *et al*, 2006 *J. Phys. Chem. B* **110** 22652
[13] Lee H., Kim M., Kim I., *et al* 2016 *Advanced Materials* **28** 4541
[14] Cohen G.M, Rooks M.J., Chu J.O., Laux S.E., *et al* 2007 *Appl. Phys. Lett.* **90** 233110
[15] Dubrovskii V.G., Cirlin G.E., Sibirev N.V. *et al.* 2011 *Nano Lett.* **11** 1247
[16] Dubrovskii V. G., Xu T., Díaz Álvarez A. *et al.* 2015 *Nano Lett.* **15** 5580
[17] Tersoff J. 2015 *Nano Lett.* **15** 6609
[18] Connell J.G., Yoon K., Perea D.E., E.J. Schwalbach, *et al.* 2013 *Nano Lett.* **13** 199
[19] Chen W., Dubrovskii V.G., Liu X., Xu T., *et al.* 2012 *J. Appl. Phys.* **111** 094909
[20] Koren E., Berkovitch N., Rosenwaks Y. 2010 *Nano Lett.* **10** 1163
[21] Yee R J, Gibson S J, Dubrovskii V G, LaPierre R R 2012 *Appl. Phys. Lett.* **101** 263106
[22] Chen W, Dubrovskii V G, Liu X, Xu T, Larde R, Nys J P, Grandidier B, Stievenard D, Patriarche G, Pareige P 2012 *J. Appl. Phys.* **111** 094909
[23] Zhang X, Dubrovskii V G, Sibirev N V, Ren X 2011 *Cryst. Growth Des.* **11**, 5441
[24] Cirlin G E, Dubrovskii V G, Sibirev N V, Soshnikov I P, Samsonenko Yu B, Tonkikh A A, Ustinov V M 2005 *Semiconductors* **39** 557
[25] Ng K W, Ko W S, Tran T T D *et al.* 2013 *ACS Nano* **7** 100
[26] Dubrovskii V. G., Sibirev N. V. 2016 *Cryst. Growth Des.* **16** 2019

IOP Conf. Series: Journal of Physics: Conf. Series **917** (2017) 032026 doi:10.1088/1742-6596/917/3/032026

Ligands Exchange, Studying the Stability and Optical Properties of CdSe/CdS/ZnS Quantum Dots with Liquid Crystal

A J Al-Alwani[1,2], A S Chumakov[1], M S Albermani[3], O A Shinkarenko[1], N N Begletsova[1], A M Vostrikova[1], I A Gorbachev[1], S B Venig[1], E G Glukhovskoy[1]

[1]Saratov State University named after N.G. Chernyshevsky, Astrakhanskay st. 83, 410012, Saratov, Russia
[2]Babylon University, Babylon, Iraq
[3]Al-Qasim Green University, Babylon, Iraq

Abstract. Liquid crystal (LC) ligands have the ability to control the dispersion and the homogeneity of quantum dots (QDs) in solution. LC phase transition can be used as a tool to control the assembly of dispersed QDs. Comparing the results before and after the ligands exchange (oleic acid to liquid crystal) highlights some important differences in optical and morphology properties of QDs. We have studied changes in optical properties and a shift in the spectral wavelength band of QDs with LC ligands. Application of LC ligands leads to quenching of the emission and excitation spectra intensity of CdSe/CdS/ZnS QDs. Such hybrid materials have interesting potential for usage in a variety of applications such as photonic materials or (bio) chemical sensors.

1. Introduction

Quantum dots (QDs) have a significant advantage because of their unique tunable properties due to quantum confinement effects. Also, the emitted wavelength range depends on the size of the quantum dot. The properties of desired thin film structure can be controlled by using various types of QDs or using different stabilizer molecules, etc. [1]. Liquid crystals (LC) can also be used as stabilizers and have the potential to control the dispersion of quantum dots (QDs) in liquid crystal phase. The challenge in creating hybrid material based on QDs incorporated to LC (QD:LC) is their stabilization as a composite material over time. QDs without stabilizer (not surface-functionalized) are strongly attracted to each other by electromagnetic forces lead to aggregates and separation from an organic solvent. Recently, there are some studies have been carried out on developing LC /nanoparticle materials hybrid [2, 3]. But these studies didn't cover the properties of nanoparticles in 2 dimensions (2D) form. The 2D monolayer study can provide information about the interaction between molecules and homogeneity of monolayer surface. The dispersed QDs in the LC material will respond to the local order phase and have been found that QDs preferentially located in the low molecular order such as isotropic phase [4, 5]. LC was used to organize the QDs in the form of their self-assembly [6]. We have studied the designing system of the QDs:LC mixture, the ligands of LC can act as a surfactant between QDs and solvent. We used core/shell/shell system CdSe/CdS/ZnS QDs stabilized by oleic acid (OA).Some studies are reported that various ligands of fatty acid could effects on the optical properties of QDs [7]. Previously we have studied the influence of separation excess ligands of oleic acid from QDs solution on the morphology and optoelectronic properties of QDs thin film [8]. QDs were separated from ligands of OA and investigated their dispersion in the LC solution after ligand exchange. The LC phase, structure and intermolecular interactions depends on different parameters such as temperature, electric field, concentration, etc., these could be tuning properties of hybrid materials. The using of stabilizer can enhance the dispersion and control the assembly of QDs [9]. Moreover, the existence of LC in the QDs solutions can modify their optical properties (emission and excitation). LC mixture can be tuning QDs properties for a particular application such as chemical and biochemical optical sensing, etc. [10].

Therefore, in this work we have studied the formation of the 2D monolayer at air-water interface by Langmuir-Blodgett (LB) technique and investigated an influence of LC on the properties of QDs solution and structure of their monolayers.

2. Materials and methods

4′-octyl-4-biphenylcarbonitrile (8CB) was purchased from (Sigma-Aldrich) and dissolved in chloroform at a concentration of 10^{-3} M. CdSe/CdS/ZnS QDs were prepared by hot-injection method [11]. We have replaced the ligands (OA) with mesogenic ligands of LC to produce QDs stabilized by LC ligands (see figure 1). QDs were washed by using a precipitation and re-dispersion. 2 mL of QD solution was precipitated with 2 mL of toluene and centrifuged. The OA ligands were removed by two times rinsing with centrifugation at 6000 rpm for 8 mins. Afterward, the precipitate was washed by adding ethanol solution and finally resuspended in 2 ml chloroform with concentration about 10^{-7} M – such low concentration was used to ensure a uniform dispersion of CdSe/CdS/ZnS QDs in the LC phase. A mixture of QD:LC was prepared at a different ratio 1:1 and 1:2. The QD:LC mixture in bath sonicated at temperature 45°C for 4 hours to obtain QDs dispersion in the isotropic LC medium. The mixture of QD:LC is then taken away from the heating and was left to cool at room temperature. The formation of the monolayer at air-water interface was performed by LB bath (KSV Nima LB trough KN 1003, Finland). A 50 µl of QD solution was spread on the sub-phase surface of the deionized water (18.2 Ω × cm). Surface pressure isotherms of the monolayers were recorded at room temperature. The movable barriers compressed monolayer at a specific area with constant rate 12 mm/min. The same LB experiments were done for LC and a mixture of QDs:LC at ratio (1:1 and 1:2). The monolayers were transferred on a glass substrate and their morphology was studied by atomic force microscope (AFM) (Nanoeducator II AFM microscope).The emission and excitation spectra of solutions were obtained using Cary Eclipse Fluorescence Spectrophotometer (Agilent Technologies).

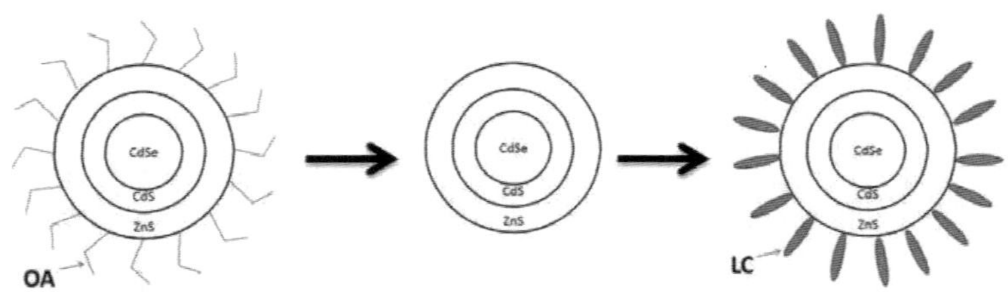

Figure 1. Schematic representation of ligands exchange of QDs that was performed by oleic acid (OA) substitution with liquid crystal (LC) molecules.

3. Results and Discussion

The replacement of ligands capped QDs by chemical treatments was investigated. The estimated areas of monolayers (when monolayer in solid state) for LC, QDs and QDs: LC at ratio (1:1 and 1:2) solutions are equal to 164.1, 41.41, 108.91 and 141.5 cm^2, respectively (figure 2). LC isotherm (curve 1) shows higher occupied area and lowest surface pressure 7.5 mN/m. The isotherm of QDs (curve 2) shows low occupied area with the high surface pressure of 29 mN/m. We assume that the

IOP Conf. Series: Journal of Physics: Conf. Series **917** (2017) 032026 doi:10.1088/1742-6596/917/3/032026

changes in surface pressure (π) and occupied area of QDs:LC monolayer (curve 3 and 4) are influenced by LC ligands, and then isotherms proportionally shift towards the larger area with increasing the concentration of LC (figure 2). The isotherms areas (curve 2, 3 and 4) before collapse point region II indicate formation homogeneous monolayer solid-like state.

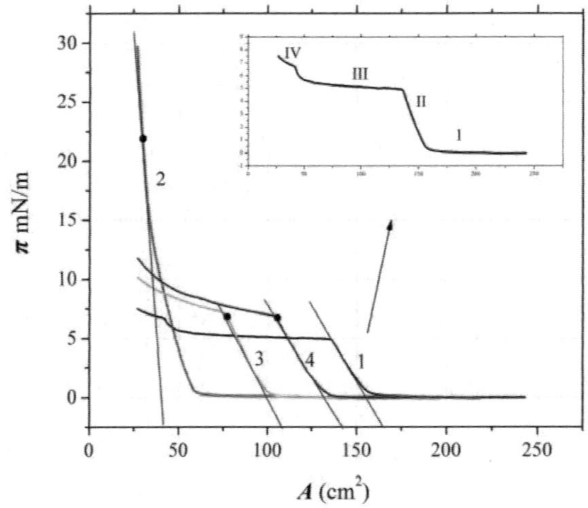

Figure 2. LB isotherms of LC (1), QDs (2), QDs:LC 1:1 (3) and QDs:LC 1:2 (4).

After the collapse point region III, the surface pressure of LC raised slightly (the plateau region), we postulate the existence of a bilayer on top of a monolayer (trilayer) [12]. The isotherm (curve 4) shows increased in occupied area and surface pressure due to presence higher concentration of LC. The shape of the isotherms (curve 3 and 4) of the QDs:LC monolayer reveals that the surface pressure rises monotonically instead of creating the plateau region and didn't show second collapse similar to pure LC (region IV) (see figure 2). This result indicates the enhancement of the mixture molecular compression and stability, the mixture highly compressed and formed a densely packed multilayer.

The structure of the hybrid material depends on the surface properties of QDs and the liquid-crystalline properties of the host phase. The thin films of QDs and QDs:LC (1:1 and 1:2) transferred on glass substrates at surface pressure 22, 6.9 and 6.9 mN/m, respectively (see figure 3). AFM images in figure 3(b) and 3(c) show the more homogenous morphology of QDs:LC monolayer compared with the morphology of only QDs monolayer and the occupied area of QDs decrease with increase the concentration of LC.

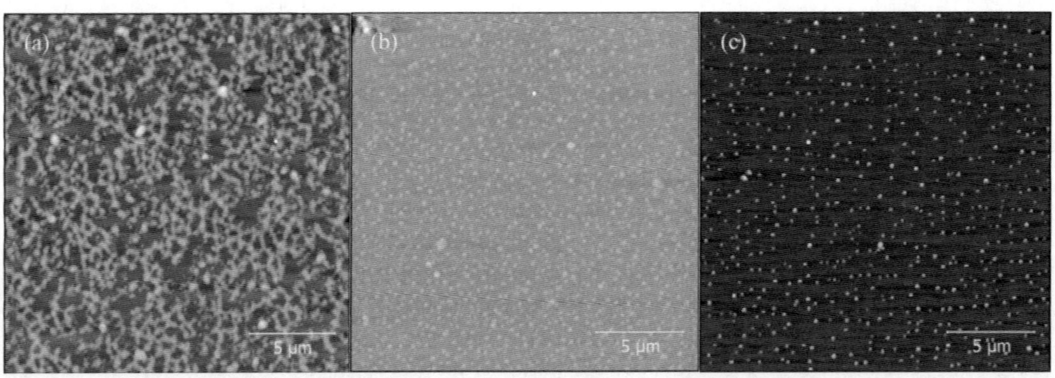

Figure 3. AFM images of QDs at 22 mN/m (a), QDs:LC 1:1 at 6.9 mN/m (b) and QDs:LC 1:2 at 6.9 mN/m (c).

IOP Conf. Series: Journal of Physics: Conf. Series **917** (2017) 032026 doi:10.1088/1742-6596/917/3/032026

Emission and excitation spectra of the solutions before and after the ligand exchange are shown in (figure 4). The QDs were capped with OA ligands have an optical emission band at 618 nm and excitation band at 597 nm. When surface capping ligands of oleic acid are separated, the emission and excitation spectra of QDs solution have shown red shifting to 623.5 and 640 nm, respectively. The emission and excitation spectra band are shifted to longer wavelength thus can emphasize that QDs became un-stabilized and aggregated after separating OA thus lead to increasing the physical size of QDs. However according to the previous study reported that the possibility of induced red shift in absorption and emission spectra band due to aggregation of QDs [13].

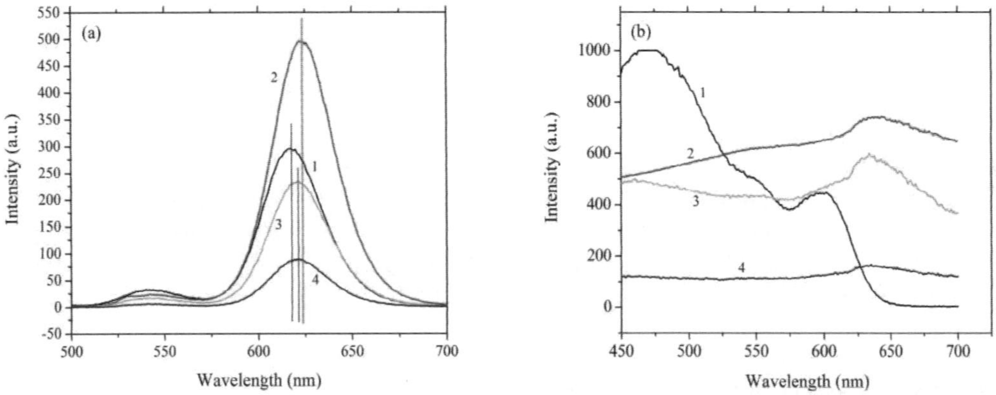

Figure 4. Emission spectra (a) and excitation spectra (b) for solutions of QDs with OA stabilizer (1), QDs solution without OA stabilizer (2), QDs:LC 1:1 (3) and QDs:LC 1:2 (4).

Figure 4a and 4b depict that the emission and excitation intensity of QDs:LC at a ratio (1:1 and 1:2) (curve 3 and 4) shifted to 621.2 and 634 nm, respectively. This result due to the influence LC phase on the surface properties of QDs, and according to the previous study the changing in the "refractive index" can cause a red or blue shift in spectra of QDs [14,15].

The intensity of emission and excitation spectra of QDs after ligand removal integrated intensity increased. The peak excitation spectra (figure 4b) were expanded could be explained by such ligands have an influence on the surface charges, interparticle separation and carrier mobility [16]. The understanding and designing surface-modified QDs in LC phases is an important role in the photonic applications that require soft-switchable materials.

4. Conclusion

This study provides a procedure of ligand exchange that plays an important role in control the stability and optical properties of CdSe/CdS/ZnS QDs. We compared the effects of removing and exchanging the ligands on the stability, morphology, emission and excitation spectra of QDs. We have studied the formation of QDs and LC mixture monolayer in 2D by LB method which helps our understanding of QDs organization within the host material of LC phase. CdSe/CdS/ZnS QDs solution incorporating the LC ligand show good dispersion in the isotropic phase and give enhancement in optical properties of QDs. Increases the concentration of LC lead to decrease the intensity of emission spectra of QDs solution and reduce the occupied area of the gas phase in QDs: LC monolayer. CdSe/CdS/ZnS QDs exhibit a significantly red shift in both the emission and excitation spectra after removing the ligand of OA. The results could be useful in the creation of optical, photonic device or molecular encapsulation applications.

Acknowledgments
This work was supported by grants from the Russian Foundation for Basic Research project № 16-07-00093/16 and № 16-07-00185/16.

References

[1] Gorbachev I A, Goryacheva I Y, Brezesinski G, Gluhovskoy E G. 2016 *Proc. of SPIE* 9917 99171J-1
[2] Rodarte A L, Nuno Z S, Cao B H, Pandolfi R J, Quint M T, Ghosh S, Hein J E, Hirst L S 2014 *ChemPhysChem* **15(7)** 1413–1421
[3] Chen L-J, Lin J-D, Rong C 2014 *J. Mater. Chem. C* **2** 4388-4394
[4] Rodarte A L, Pandolfi R, Ghosh S, Hirst L S 2013*J. Mater. Chem.* **1** 5527-5532
[5] Muševič I, Škarabot M, Tkalec U, Ravnik M, Žumer S 2006 *Science* **313** 954-958
[6] Kumar A, Prakash J, Deshmukh A D, Haranath D, Silotia P, Biradar A M 2012 *Applied Physics Letters***100** 134101
[7] Kalyuzhny G, Murray R W 2005 *J. Phys. Chem. B* **109 (15)** 7012–7021
[8] Al-Alwani A J, Chumakov A S, Gorbachev I A, Kuznetsov N, Kletsov A A, Glukhovskoy E G 2017 *BioNanoScience* 1-5
[9] RodarteA L, Ferri C G L, Grey C, Hirst L S, Ghosh S2012 *Proc.of SPIE* **8279** 82790H
[10] Costa-Fernández J M, Pereiro R, Sanz-Medel A 2006 *TrAC Trends Anal. Chem* **25** 207– 218
[11] Speranskayaa E S, Beloglazovab N V, Lenainb P, Saegerb S D, Wangc Z, Zhangc S, Hensd Z, Knoppe D, Niessnere R, Potapkina D V, Goryachevaa I Y 2014 *Biosensors and Bioelectronics* **53** 225–231
[12] Xue J, Jung C S, Kim M W 1992 *Physical Review Letters* **69 (3)** 474
[13] Sarkar S K, Hodes G 2005 *J. Phys. Chem. B* **109** 7214
[14] Singha G, Fischb M R, Kumar S 2016 *Liquid Crystals* 1-9
[15] Chen C-T, Liu C-C, Wang C-H, Chen C-W, Chen Y-F 2011 *Appl. Phys. Lett.* **98** 261918
[16] Porter V J, Geyer S, Halpert J E, Kastner M A, Bawendi M G 200*8 J. Phys.Chem. C* **112** 2308-2316

Evolution of size distribution and structure of Si and SiO_2 nanoparticles: laser-assisted formation and fragmentation

K. Eidelman[1], D. Gudkov[1], O. Segbefia[1], E. Ageev[2], A. Krivonosov[2], A. Matuhina[2]
[1]National University of Science and Technology "MISiS", Moscow, Russia.
[2]ITMO University, Saint Petersburg, Russia

Abstract. In this work, Si and SiO_2, nanoparticles (NPs) was prepared by pulsed laser ablation (PLA) in distilled water. The radiation of a ytterbium fiber laser (repetition rate f = 50 kHz, wavelength λ = 1064 nm and pulse duration τ = 8 ns and 100 ns) at different laser intensities was utilized to ablate the Si target (99.999%, cubic, 7x7 mm^2) under liquid layer to synthesize and to fragment the silicon colloidal NPs. Studies of morphology and size distribution of silica NPs were conducted using Transmission Electron Microscopy (TEM). The NPs of crystalline and amorphous phases were founded. Most of the NPs in the nano colloids were found to have dimensions less than 100 nm, and a few of them were between 100 nm and 700 nm. Dependence of average NP size on the number of laser passes was revealed. The average size of the nanoparticles obtained by TEM was confirmed by dynamic light scattering (DLS) measurements.

1. Introduction

Last 50 years the great interest in nanoscale objects was growing. Technologies allow obtaining nanoparticles (NPs) with size from one to several hundred nanometers. The special feature of these particles is that their spectral characteristics can be tuned in a wide range of wavelengths. With a variation of the composition and size of nanoparticles (NPs), it is possible to cover the spectral range from 400 nm to 3 microns [1].

Silicon and silica (SiO_x) are an extensively used photovoltaic material as a fabrication component of different photovoltaic products, because of their superior biological, magnetic, thermal, electronic and electrical properties [2]. In this regard, the Si-based NPs have been attracted great attention and are widely used in various scientific and technical fields [3,4], among which photovoltaics [5–7] can be called as one of the most important.

The synthesis of nanocolloids (NCs) is the vital step in the application of NPs. Recently laser-assisted methods of NPs formation attracted great attention due to provided chemical purity and the possibility to control such NPs features as size, morphology, and aggregation of the NPs which in turns govern their electrical and chemical properties [2]. The final NPs characteristics are determined not only by the properties of the environment but also by the properties of the irradiation process (wavelength, pulse duration, frequency, polarization, fluence), which can be adjusted via changing the parameters of the system. Among the variety of laser-assisted NPs synthesis methods, pulsed laser ablation (PLA) is considered as an effective and "pure" technique of creating semiconductor, oxide or even organic nanoparticles. During NPs synthesis by PLA in the gas or liquid environments, NPs formation occurs because of target material molecules colliding with one another in the process of adiabatic expansion and plasma plume recombination.

Numerous works on PLA of silicon target under the action of femtosecond pulses can be found, it was reported in vacuum [8], in various gaseous atmospheres [9] and liquid mediums [10,11]. Recently, PLA of bulk targets in water with pulses of nanosecond duration was used to produce NPs with the ability to control sizes, morphology, and aggregation [12,13]. Nonetheless, these detailed investigations utilized various physical and chemical techniques that have been conducted have failed to address the challenges in synthesizing the desired NPs for photovoltaic applications.

IOP Conf. Series: Journal of Physics: Conf. Series **917** (2017) 032027 doi:10.1088/1742-6596/917/3/032027

The goal of this study is to investigate the structure properties of silicon NPs obtained by laser ablation with nanosecond pulses under a liquid layer for further application in photovoltaics.

2. Methods and techniques

The laser system used in this study is comprised of a ytterbium fiber laser operated at 1064 nm wavelength, 8 ns pulse duration, a repetition rate of 50 kHz. The silicon wafer was used as the target plate; the Si target surface is cleaned ultrasonically in acetone before ablation. The laser beam was focused on the Si target immersed in distilled water with a spot size about 50 microns, and scanned in a path of parallel lines with speed up to 8 m/s. First, ablation was carried out for 1000 number of passes (at a laser intensity of $7.32 \cdot 10^7$ W/cm^2, $\tau = 8$ ns) to produce the silicon colloidal NPs (solution 1). Afterward, the bulk target is removed from the solution and a laser beam with a laser intensity of $4.05 \cdot 10^7$ W/cm^2, $\tau = 100$ ns, was applied into the solution for additional 250 (solution 2) or 500 (solution 3) passes to provide a fragmentation of obtained NPs. To characterize structural properties of the nanoparticles, transmission electron microscopy (TEM, JEOL-2100) at 200 kV was performed for Si-NPs dispersed on a holey carbon coated copper grid. Selected area electron diffraction (SAED) patterns of nanoparticles are also collected to understand crystallinity of the prepared NPs. Further quantitative analysis of the NPs sizes was performed by dynamic light scattering analysis (DLS, SZ100, Horiba Jobin Yvon).

3. Discussion

A series of TEM-images was obtained for each solution, as shown in Figure 1. SAED patterns (fig.1) demonstrated that both a crystalline and an amorphous phase was presented in each solution. The crystalline phase corresponded to silicon (in the right corner fig.1a,b,c), while an amorphous phase was not recognized.

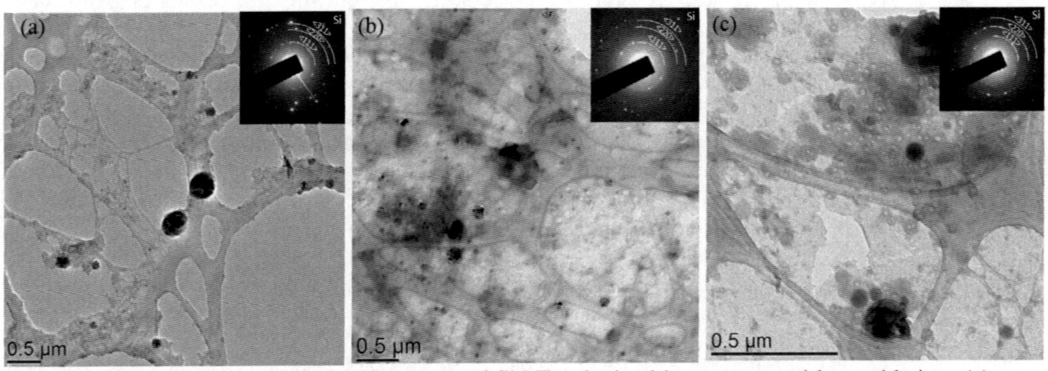

Figure 1. TEM-image and their SAED pattern of Si-NPs obtained by nanosecond laser ablation: (a) as-prepared (solution 1): (b) after an additional 250 passages (solution 2) and (c) after an additional 500 passes.

The size of NPs was measured by TEM (fig.2). Histograms of the particle size distribution were obtained by statistical calculation as a result of measuring of 200 particles by transmission electron microscopy. The general size of NPs was ranged from 10 to 200 nm. Samples included clusters ranged in size from 300 to 900 nm. Probably, the reason of such large scale is an agglomerate of small NPs. For the first solution, the average size of NPs was 150 nm. For the second solution (250 extra passage) the average NPs size was decreased to 120 nm. TEM-image from the third solution demonstrated a presence of great amount of amorphous silicon (fig.1c) and silicon oxide. The average size of NPs is 120 nm.

IOP Conf. Series: Journal of Physics: Conf. Series **917** (2017) 032027 doi:10.1088/1742-6596/917/3/032027

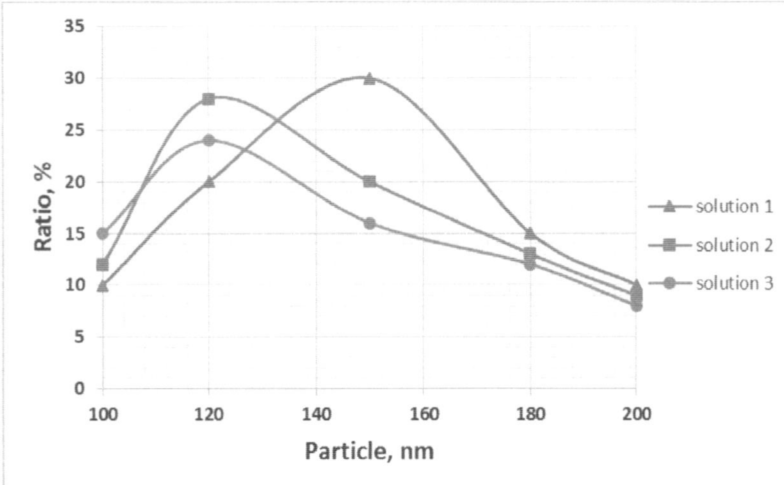

Figure 2. NPs size distribution obtained by TEM.

Determination of the average NPs size was difficult due to the large number of small NPs with a size of 20-30 nm. To determine the NPs size DLS method was used. The results of this experiment are shown in Figure 3.

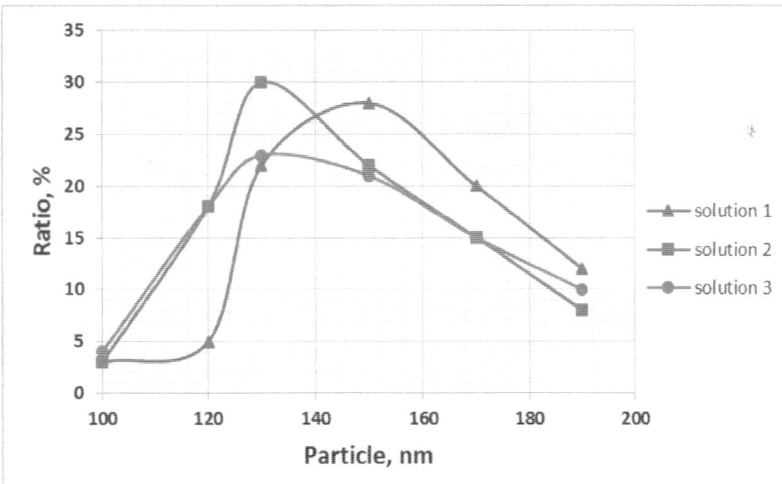

Figure 3. NPs size distribution obtained by DLS.

The results show that in as-prepared solution NPs with 150 nm size were dominated in a quantitative ratio of 30%. After additional 250 passes the average size of NPs logically decreased to 130 nm (30% of the total). The last sample, after 500 passages had the average size about 130 nm, but the total amount was reduced from 30 to 25%.

4. Conclusion

In this work, Si-NPs obtained by laser ablation in liquids and with a different number of laser passes was investigated. Each solution presented a crystalline and amorphous phase of silicon and silicon oxide. Repeated ablation (additional fragmentation) of colloidal solutions leads to decrease in the average size of NPs and growth of the percentage of amorphous silicon. TEM and DLS studies revealed the shape, average size and ratio of NPs in the samples. NPs had a spherical shape, NPs agglomerates up to 900 nm were also found. Increasing the number of ablation passes leads to decreasing the size of NPs. These results can be a good base for using NPs like QDs, for example in solar cell structure.

IOP Conf. Series: Journal of Physics: Conf. Series **917** (2017) 032027 doi:10.1088/1742-6596/917/3/032027

References

[1] Otanicar T P, DeJarnette D, Hewakuruppu Y, Adv. Opt. Photonics 8 (2016) 541.
[2] Kasaeian A, Eshghi S, Sameti M, Renew. Sustain. Energy Rev. 43 (2015) 584–598.
[3] Pavesi L, Turan R, Silicon Nanocrystals: Fundamentals, Synthesis and Applications, Wiley, 2010.
[4] A Ischenko, G Fetisov, L Aslalnov, Nanosilicon: Properties, Synthesis, Applications, Methods of Analysis and Control, CRC Press, 2015.
[5] K Nakayama, K Tanabe, H Atwater, Appl. Phys. Lett. 93 (2008) 121904.
[6] K. Reddy, N Kamnapure, S Srivastava, Int. J. Low-Carbon Technol. (2016) ctw007.
[7] V Milichko, A Shalin, I Mukhin, C Simovski. Physics-Uspekhi 59 (2016) 727–772.
[8] S Amoruso, R Bruzzese, N.Spinelli, Appl. Phys. Lett. 84 (2004) 4502–4504.
[9] B Tull, J Carey, M Sheehy. Appl. Phys. A Mater. Sci. Process. 83 (2006) 341–346.
[10] N Semaltianos, S Logothetidis, W Perrie. Res. 12 (2010) 573–580.
[11] P Kuzmin, G Shafeev, V Bukin, J. Phys. Chem. C 114 (2010) 15266–15273.
[12] L Vaccaro, R Popescu, F Messina, J Appl. Phys. 120 (2016) 24303.
[13] E Solati, L Dejam, D Dorranian, Opt. Laser Technol. 58 (2014) 26–32.

IOP Conf. Series: Journal of Physics: Conf. Series **917** (2017) 032028 doi:10.1088/1742-6596/917/3/032028

GaN growth via HVPE on SiC/Si substrates: growth mechanisms

Sh Sh Sharofidinov[1,4], **A V Redkov**[2,3,5], **A V Osipov**[2-4] **and S A Kukushkin**[2-4]

[1]Ioffe Physical-Technical Institute RAS, St. Petersburg 194021, Russia
[2]Institute of Problems of Mechanical Engineering RAS, St. Petersburg 199178, Russia
[3]St. Petersburg Academic University, St. Petersburg 194021, Russia
[4]ITMO University, St. Petersburg 197101, Russia
[5]E-mail: avredkov@gmail.com

Abstract. The article focuses on the study of GaN thin film growth via chloride epitaxy on SiC/Si hybrid substrate. SiC buffer layer was grown by a method of substitution of atoms, which allows one to reduce impact of mechanical stress therein on subsequent growth of III-nitride films. It is shown, that change in GaN growth conditions leads to change in its growth mechanism. Three mechanisms: epitaxial, spiral and stepwise growth are considered and mechanical stresses are estimated via Raman spectroscopy.

1. Introduction

Nowadays wide-bandgap semiconductors like GaN, AlN, SiC and others are becoming more widespread in view of huge prospects of their use in power electronics and LED technology, HEMT-transistors and much more [1-3]. However, large-scale production of such materials is still difficult, since certain problems arise during the growth of thin films of these semiconductors. In particular, today sapphire [4] is used mainly as substrates for the growth of GaN thin films and heterostructures, which has a number of drawbacks, due to which it is impossible to grow a defect-free film: sapphire has a large lattice mismatch and different coefficient of thermal expansion comparing to GaN [5]. Therefore, the search for new types of substrates, allowing one to obtain more structurally perfect GaN films, heterostructures and devices continues. In a series of papers [6,7] a method for the synthesis of SiC/Si substrates has been developed theoretically and put into practice, based on the chemical substitution of Si atoms for C atoms directly in the silicon substrate. It was demonstrated that SiC/Si substrates obtained by this method can be used for the growth of high-quality GaN layers. Moreover, use of silicon as a starting material will allow integrating wide-bandgap semiconductors into traditional silicon electronics. Therefore, it seems expedient to further study and optimize the growth of III-nitrides on SiC/Si substrates.

In addition to choosing a suitable substrate for the growth of GaN layers, a well-developed technique for their deposition is also required. One of such techniques, along with molecular-beam (MBE) [8] and metal-organic (MOCVD) [9] epitaxy, is chloride epitaxy (HVPE) [10],

$$GaCl + NH_3 = GaN + HCl + H_2 \quad (1)$$

which allows one to ensure a high growth rate of thin films and to simplify the mass production of GaN-based devices. In this paper, we investigate mechanisms of GaN growth via HVPE on SiC/Si substrates formed by the method of substitution of atoms [6] and influence of the growth conditions on growth mechanisms.

2. Experiments

The first stage of the experiment was the deposition of the SiC buffer layer on silicon substrates with an orientation of <111>. In the growth process via method [7] the following parameters were used: growth temperature 1250°C, total pressure of the gas mixture 2 Torr, synthesis time 15 min, gas mixture flow 100 ml/min, silane fraction in the mixture 16%. We have grown three similar substrates. Then, buffer layers of AlN were deposited on the obtained SiC/Si samples via chloride epitaxy, and then GaN layers were grown in a single process. For the growth we used HVPE reactor of horizontal type, having symmetric parallelepiped shape, the pressure in which was close to atmospheric. It should be noted, that three samples were placed in different places of the reactor. The first one SH295-1 was the closest to the reactor inlet nozzle, the second SH295-2 was at the center of the reactor, and the third SH295-3 was near the outlet. In this connection we suppose, that growth conditions were the same for all samples except total supersaturation, since when the gas mixture arrive to the third sample, it has already reacted with the first sample and the second one, etc. Therefore, we consider, that supersaturation gradually decreased from the first sample to the third one. Deposition of AlN films occurred as a result of a gas-phase reaction between aluminum trichloride ($AlCl_3$) and ammonia, GaN deposition was carried out as a result of reaction (1). Growth temperature for AlN and GaN was 1050°C. NH_3 flow was 1.5 l/min, flow of HCl through gallium was 200 ml/min during GaN growth, and the same through aluminum during AlN growth. After the growth process the samples were studied by optical microscopy, Raman spectroscopy and optical profilometry.

3. Results and discussion

The study showed that the growth mechanism of GaN depends essentially on the supersaturation. Figure 1 shows optical images of the samples surface demonstrating various growth mechanisms of GaN film: epitaxial growth, stepwise growth, and spiral growth.

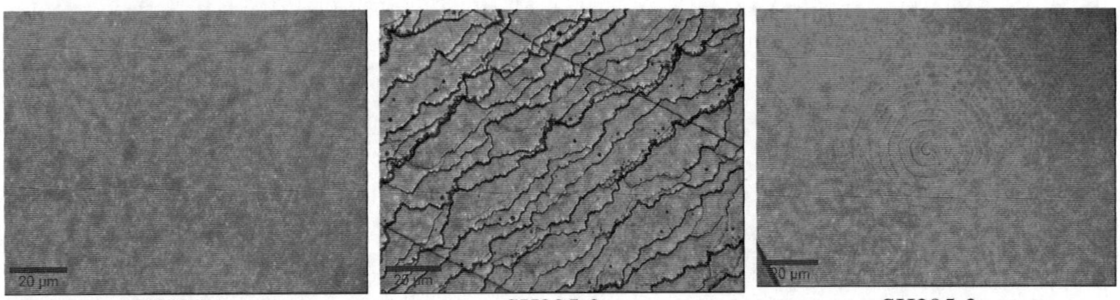

SH295-1 SH295-2 SH295-3

Figure 1. Optical image of GaN/AlN/ SiC/Si samples growing by different mechanisms. As the total supersaturation decreases, the growth mechanism changes from epitaxial (a) and stepwise (b), to spiral one (c).

In epitaxial mode (SH295-1) film grows uniform in thickness and does not have expressed defects on the surface. Its average roughness (rms) is 15 nm according to optical profilometry. On the GaN sample growing by a stepwise mechanism (SH295-2) average distance between the steps is of the order of 10-20 µm, and their height is 200-300 nm according to optical profilometry. GaN sample growing in conditions of lower supersaturation (SH295-3) demonstrate growth due to screw dislocations and growth spirals are clearly visible on the surface. Characteristic sizes of the spirals are

40-70 microns in diameter, and 250-400 nm in height according to optical profilometry (Fig. 2). The height of the individual steps of the spiral is much smaller than in the case of stepwise growth and is of the order of 25-40 nm (Fig. 2b).

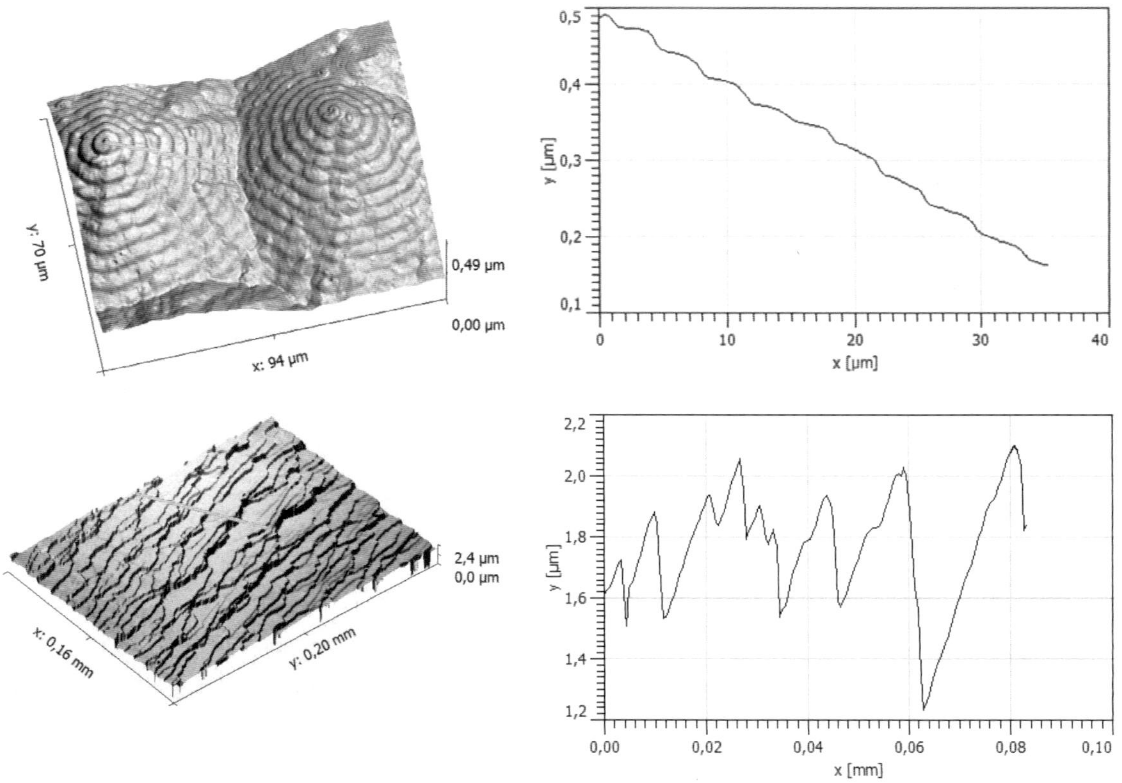

Figure 2. 3D surface maps of the samples SH295-3, SH295-2 (the cases of spiral and stepwise growth correspondingly) obtained by optical profilometry and their profiles along the green lines.

The change in growth mechanism seems to be related to change in supersaturation and agrees with previous results on crystal growth. In [11] it was shown that at low supersaturations growth occurs mainly due to screw dislocations, whereas with increasing supersaturation, concentration of adatoms on the surface becomes sufficient for 3D and 2D nucleation. The question arises about the samples SH293-1, and SH293-2. It seems that stepwise growth should occur at higher supersaturations than flat epitaxial growth, and this is not fully understood. One of possible reasons is a small inhomogeneity in the temperature at the center (SH295-2) and along the edges of the reactor (SH295-1, SH295-3), which affects reaction rate near the center.

The Raman spectra of the samples are shown in Figure 3. The spectra contain GaN bands (567.5 cm^{-1}, 734 cm^{-1}), AlN (658 cm^{-1}), Si (521 cm^{-1}), and SiC band (796 cm^{-1}) of low intensity due to the small thickness of the film. Using the Raman spectra mechanical stresses arised in GaN films grown via different mechanisms were estimated. For evaluation we used spectral position of 567.5 cm^{-1} gallium nitride band, which is depicted in the inset in Fig. 3.

The evaluation was carried out using the formula [12]:

$$\Delta\omega_{GaN} = 4.3\sigma\ \text{cm}^{-1}/\text{GPa} \tag{2}$$

where $\Delta\omega_{GaN}$ is the deviation from the band position at unstressed state (567.5 cm^{-1})[12], and σ is the mechanical stress. Obtained stress values for the sample SH295-1 is 200-250 MPa, for SH295-2 is 100-150 MPa, for SH295-3 is 250-300 MPa. It should be noted that on the sample SH295-3 demonstrating spiral growth, fewer cracks were visually observed than on the other two samples. We suppose increased elastic stresses in this sample are associated with this fact, since elastic energy remains stored instead of being used for the formation of cracks.

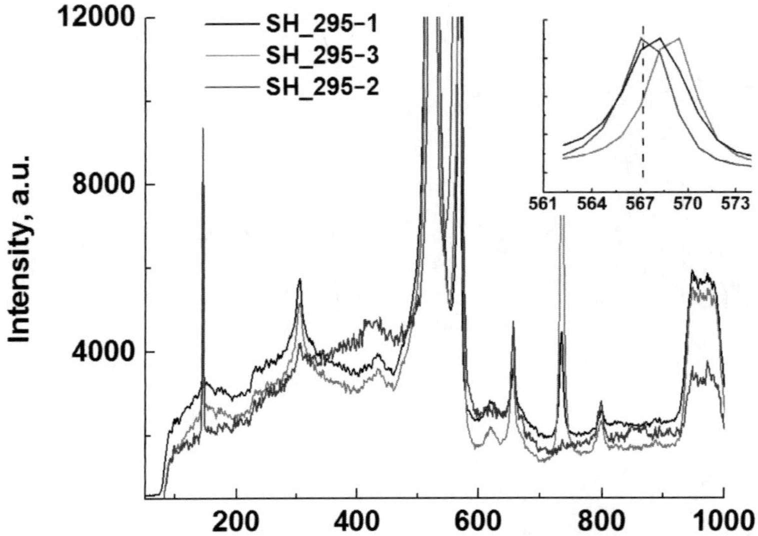

Figure 3. Raman spectra of the samples. Inset demonstrates position of GaN band (567.5 cm^{-1}) on a larger scale.

4. Conclusions

In this paper the growth of GaN via the method of chloride epitaxy is considered. A series of samples of thin GaN films were grown in a single process on SiC/Si substrates obtained by the method of substitution of atoms. It is found that while supersaturation increases, the growth mechanism of GaN changes from growth via screw dislocations (growth spirals are observed on the surface), to epitaxial growth (the surface remains relatively smooth), and stepwise growth (on the surface there are clearly discernible parallel steps). Using optical profilometry characteristic sizes of the spirals were estimated to be ~ 40-70 μm in diameter and 250-400 nm in height. The height of the steps of the growth spirals is about 25-40 nm. The characteristic height of the steps during stepwise growth is about 200-300 nanometers and the distance between them is 10-20 microns. The mechanical stresses in the samples were estimated and it is shown that in the sample on which the spiral growth was observed mechanical stresses are maximal and are equal to about 300 MPa, whereas this sample is the least cracked. The results can be used in further findings of optimal growth conditions of GaN on SiC/Si substrates via HVPE.

Acknowledgments

This study was supported by Russian Science Foundation (RNF grant N 14-12-01102). Part of the results were obtained unique scientific facility "Physics, chemistry, and mechanics of crystals and thin films" (IPME RAS, St. Petersburg, Russia).

References

[1] Casady J B and Johnson R W 1996 *Solid-State Electronics* **39** 1409
[2] Pearton S J, Abernathy C R, Norton D P, Hebard A F et al 2003 *Materials Science and Engineering: R: Reports* **40** 137
[3] Tran N H, Le B H, Zhao S and Mi Z 2017 *Applied Physics Letters* **110** 032102
[4] Amano H, Akasaki I, Hiramatsu K, Koide N et al 1988 *Thin Solid Films* **163** 415
[5] Kukushkin S, Osipov A, Bessolov V, Medvedev B, Nevolin V and Tcarik K 2008 Rev. Adv.

Mater. Sci **17** 1-32
[6] Kukushkin S A and Osipov A V 2014 *J. Phys. D: Appl. Phys.* **47** 313001
[7] Kukushkin S A and Osipov A V 2013 *Journal of Applied Physics* **113** 024909
[8] Nikishin S A, Faleev N N, Antipov V G, Francoeur S et al 1999 *Applied Physics Letters* **75** 2073
[9] Kukushkin S A, Osipov A V, Rozhavskaya M M, Myasoedov A V et al 2015 *Physics of the Solid State* **57** 1899
[10] Bessolov V, Konenkova E, Shcheglov M, Sharofidinov S et al 2013 *Physica status solidi (c)* **10** 433
[11] Tsukamoto K (2016) *Progress in Crystal Growth and Characterization of Materials* **62** 111
[12] Tripathy S, Chua S J, Chen P and Miao Z L 2002 *Journal of Applied Physics* **92** 3503

Effects of PECVD temperature and RF power on surface structure and refractive index of amorphous and polycrystalline silicon films

E Yu Gusev[1], J Y Jityaeva[1], A A Geldash[1], O A Ageev[1]
[1] Research and Education Centre "Nanotechnologies", Institute of Nanotechnology, Electronics and Electronic Equipment Engineering, Southern Federal University, Taganrog 347900, Russia

Abstract. The amorphous and polycrystalline silicon films were obtained by plasma enhanced chemical vapor deposition. The effect of power and temperature over a wide range on the surface morphology and structure, deposition rate films and refractive index have been investigated by atomic force microscopy, profilometry and ellipsometry. The deposition rate increased from 19.1 to 58.2 nm/min with RF power from 10 to 40 W. The refractive index of the films obtained at a temperature of 550 to 700 °C was in the range from 3.5 to 5.5.

1. Introduction

Silicon in all phase states is extremely demanded material in electronics, e.g. solar cells [1], thin films transistors (TFTs), micro- and nanoelectromechanical systems (MEMS/NEMS) [2,3], and others. Solar cells based on monocrystalline silicon are most effective maximum up to 21.3%, however their spread is limited by high cost [4]. To reduce the cost, the multi-layer structure of a thin film solar cell is formed on inexpensive glass and plastic substrates, which greatly limits the maximum process temperature. The elements based on amorphous hydrogenated silicon (a-Si:H, nc-Si:H) are actively used in mass solar cell production [1,4]. This is a consequence of a combination of low deposition temperature, high productivity, ease of manufacture, a sufficiently high optical absorption coefficient, and relatively low cost compared with the elements based on GaAs, InP, ZnP. It has been possible to increase the efficiency of solar cells based on a-Si: H to a maximum of 15% in experimental studies, by improving the properties of materials and developing new multilayer structures [5]. However, the main problem of thin-film silicon solar cells, which limits their effectiveness, is the light-induced degradation effect (Staebler-Wronski Effect) [6]. To reduce this effect, nano- and microcrystalline hydrogenated silicon films (nc-Si:H, μc-Si:H) with grain sizes from several units to tens of nanometers placed in an amorphous matrix were used [7]. They demonstrate greater stability in lighting, better electrophysical properties and lower defect density than a-Si:H. To increase the efficiency of thin-film solar cells, a combination of carbon nanotubes and silicon was used [8]. CNT are not only transport charge carriers, but also act as some photoactive thin films use in elements of solar cells, due of the large number of physical effects that occur in the bulk and at grain boundaries [9]. The high temperature and low deposition rate is limiting factor.

Amorphous, nano- and microcrystalline hydrogenated silicon and crystalline silicon films depending on the application can be prepared by various methods of chemical vapor deposition (CVD). This method allows deposition amorphous hydrogenated, nano- and polycrystalline silicon films in a wide temperature range from 100 to 1300°C [3]. However, the possibility of reducing temperature to 700°C and less is interesting. To date, the most flexible method in terms of parameters process is plasma enhanced chemical vapor deposition (PECVD) [10]. Obtaining silicon films with reproducible properties requires the establishment of key parameters of process and fine adjustment in the deposition process.

The purpose of this study is to investigate the grain size and shape, deposition rate and refractive index of amorphous and polycrystalline silicon films depending on the temperature and RF power of plasma enhanced chemical vapor deposition.

2. Experiments

Polycrystalline silicon and silicon dioxide films were prepared by PECVD used Oxford Instruments PlasmaLab 100 [10, 11]. The substrate was n-type silicon wafer (100). Previously, the substrates were covered with a layer of silicon oxide in plasma of $Ar:SiH_4:N_2O$ (161.5:8.5:710 sccm). Pressure, RF power and temperature were 1 Torr, 20 W and 350 C, respectively. The film thickness of 0.45 μm, RMS roughness not more than 0.6 nm and compressive stresses of 380 MPa observed by profilometry. Further, plasma deposition of amorphous and polycrystalline silicon films was carried according to [11]. The surface morphology and grain size of the films were studied by atomic force microscopy (NTEGRA Probe Nanolaboratory) [12]. The thickness of the films was determined by profilometry (KLA Tencor D-100) of patterned steps formed by photolithography (MJB4). The refractive index was determined by ellipsometry (LEF-3M).

3. Results and discussion

Silicon films on the SiO_2/Si (100) structure were obtained at temperature from 550 to 700°C (power was fixed to 40 W). The deposition rate was 150 nm/min. Investigations of the films by AFM and profilometry showed an increase in the grain size with increasing temperature and inflection of RMS curve (Fig. 1). The grain size of the films was in the range from 40 to 140 nm, and RMS roughness – 1.79-3.54 nm. Analysis of the AFM data (Figure 1,2) and previous RHEED findings of the same films [10] allowed to compare the flex point to the temperature of the phase transition from amorphous to polycrystalline at 550-600°C, which corresponds to the transition temperature of LPCVD films [9].

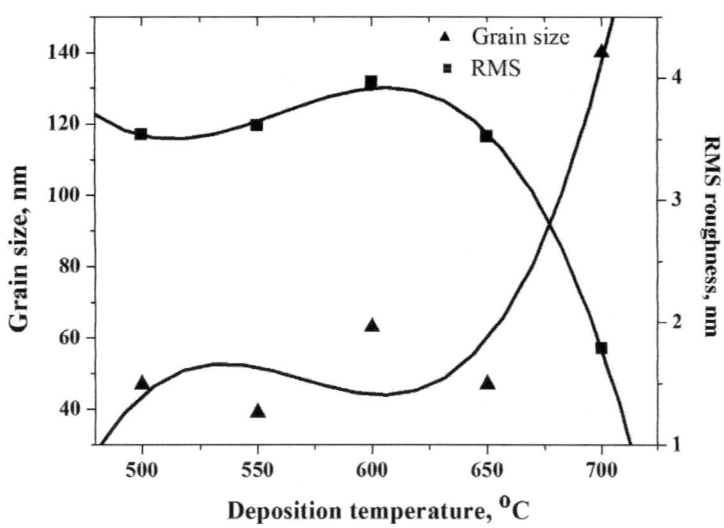

Figure 1. Grain size and RMS roughness vs deposition temperature curves.

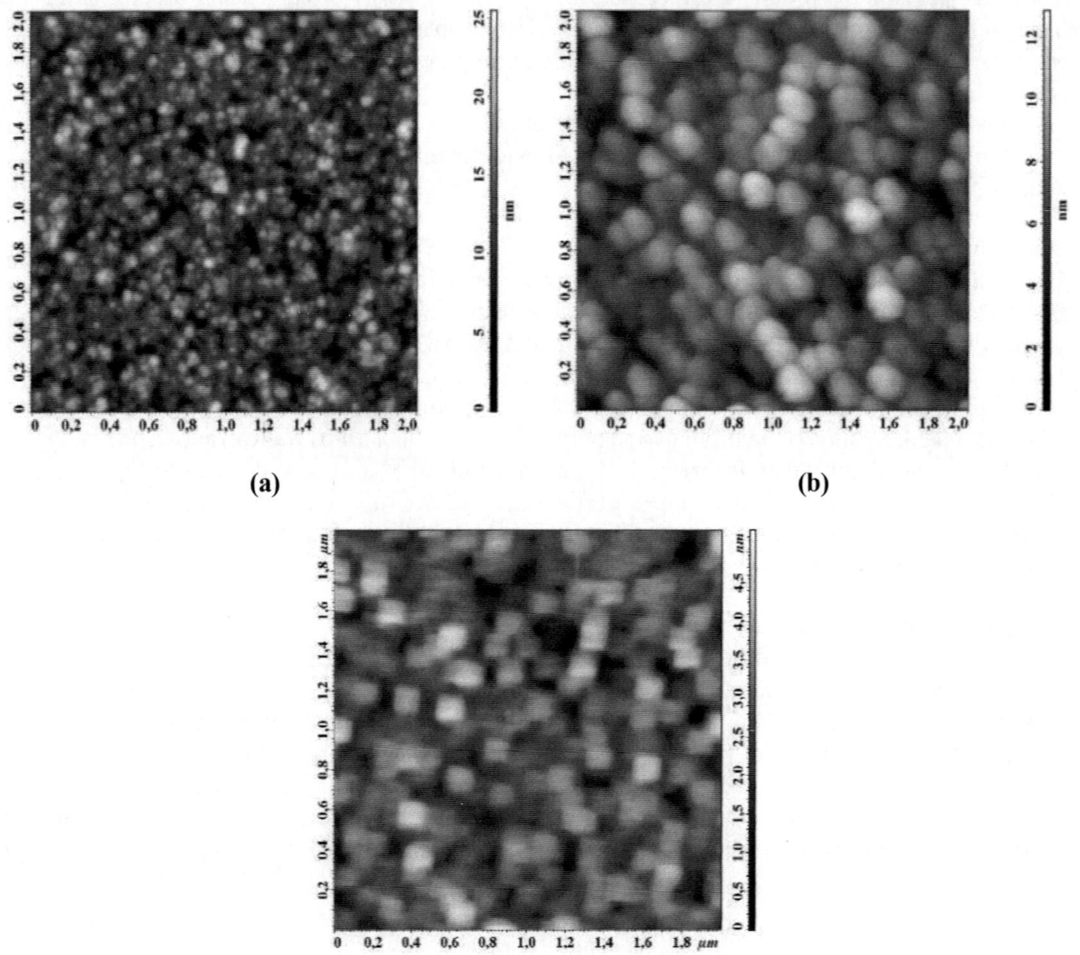

Figure 2 (a-c). AFM images of silicon films : (a) amorphous and (b,c) textured polycrystalline.

Moreover, this process is characterized by a transition from tensile stresses, prevailing in the amorphous phase, to compressive stresses (- 300 MPa). Further increasing in temperature above 600 C, contributes to the gradual decrease of internal stresses.

It is established that with increasing RF power from 10 to 40 W the deposition rate substantially increases almost linearly from 19.1 to 58.2 nm/min. Thus, the use the RF plasma allows to increase the deposition rate film by 6 times, compared to LPCVD. The shape of grains also changed (Figure 2). At a power of 40 W, the films have spherical and elliptical textures. With power decreasing below 20 W, it turns to columnar texture with a square cross-section. This behavior can be explained by the different kinetics of nuclear centers growth and deposition mechanism. The columnar structure is obtained under more equilibrium conditions (normal growth), and more dense. Whereas the elliptic structure is related to the inclusion of silicon particles formed in plasma. The density of such films is lower.

Values of the refractive index obtained at temperatures of 550-700°C ranged from 3.5 to 5.5 with the change from amorphous phase to polycrystalline.

4. Conclusion

In order to achieve efficiency enhancement of thin film-devices, it is desirable to increase structural and optical properties of the film by optimization deposition conditions. A study has been conducted on estimating effects of deposition parameters on surface morphology and optical parameter of silicon films by using plasma-enhanced chemical vapor deposition.

Films of amorphous and polycrystalline silicon on the PECVD-SiO_2/Si(100) structure were obtained. Deposition conditions providing grain size control have been identified in the range of 40-140 nm and ensure the control of the shape from nanoscale spherical crystallites to columnar elliptical and squared. The deposition rate of PECVD films was up to 60-90 nm/min, which is several times higher typical rates of low pressure chemical vapor deposition. Results of the investigations have shown that the fine adjustment of the deposition conditions allows to obtain a silicon films with the desired properties (grain size and shape, roughness and refractive index), and increase their stabilized performance and simultaneously increase their deposition rates. It can be used in the design and development of multilayer solar cells and microelectromechanical systems and devices fabrication.

Acknowledgment

This work was financially supported Southern Federal University (grant VnGr-07/2017-02). The equipment of the Research and Educational Centre "Nanotechnologies" of Southern Federal University was used for the study.

References

[1] Kondo M, Matsuda A 2004 *Thin solid films* **457** 97-102
[2] Petersen Kurt E 1982 *Proceeding of the IEEE* **20.5** 420-457
[3] Adams T, Layton R 2010 *Introduction MEMS: fabrication and applications* 447
[4] McEvoy A, Markvart T, Castaner L 2012 *Practical Handbook of Photovoltaics*
[5] Kabir M I, Shahahmadi S A, Lim V, Zaidi S, Sopian K, Amin N 2012 *International Journal Photoenergy* 460919
[6] Staebler D L, Wronski C R, 1977 *Appl. Phys. Lett* . **31** 292–294
[7] Hamui L, Monroy B M, Kim K H, Lopez-Suarez A 2016 *Material Science in Semiconductor Processing* **41** 390-397
[8] Tune D D, Flavel B S, Krupke R, Shapter J G 2012 *Advanced Energy Materials* **13**
[9] French P J 2002 *Sensors and actuators A Physical* **99** 3
[10] Ageev O A, Gusev E Yu, Jityaeva J Y, Bykov Al V, Bespoludin V V 2016 *Advanced Materials, Mechanical and Structural Engineering* 13-16
[11] Ageev O A, Gusev E Yu, Jityaeva J Y, Bykov Al V 2016 *Journal of Physics: Conference Series* **741** 012001
[12] Avilov V I, Ageev O A, Jityaev I L, Kolomiytsev A S, Smirnov V A 2016 *Proceedings of SPIE* 102240T
[13] Ageev O A, Balakirev S V, Bykov A V, Gusev E Yu, Fedotov A A, Jityaeva J Y, Il'in O I, Il'ina M V, Kolomiytsev A S, Konoplev B G, Krasnoborodko S U, Polyakov V V, Smirnov V A, Solodovnik M S, Zamburg E G 2016 *Advanced Materials. Manufacturing, physics, mechanics and applications* **175** 707

IOP Conf. Series: Journal of Physics: Conf. Series **917** (2017) 032030 doi:10.1088/1742-6596/917/3/032030

The study of effect of solid electrolyte on charge-discharge characteristics of thin-film lithium-ion batteries

L A Mazaletskiy[1,2], M E Lebedev[2,1], A A Mironenko[2], V V Naumov[2], A V Novozhilova[1,2], I S Fedorov[1], A S Rudy[2,1]

[1] Department of Nanotechnology in Electronics, Yaroslavl State University, Sovetskaya Street 14, Yaroslavl 150003, Russia
[2] Yaroslavl Branch of the Institute of Physics and Technology, Russian Academy of Sciences, Universitetskaya Street 21, Yaroslavl 150007, Russia

E-mail: l.mazaletskiy@uniyar.ac.ru

Abstract. Results of studies of the solid electrolyte effect on capacitance of thin-film electrodes on the basis of Si-O-Al and V_xO_y nanocomposites are presented. The studies were carried out by comparing the charge-discharge characteristics of two pairs of the identical electrodes, one of which was covered by LiPON film, within prototypes with two lithium electrodes - the counter and the reference electrode.

1. Introduction
Experimental studies in the field of all solid-state lithium-ion battery (LIB) and micro accumulators have started in the beginning of 90s of the last century. In that time have appeared the first versions of thin-film LIB fabricated on integral technology with lithium anode and cathode based on V_2O_5 [1], TiS_2 (S.D. Jones) and Li_xCoO_2 (R.A. Goldner). In all of these LIBs LiPON was used as solid electrolyte. Later it has become widely used in the production of all solid-state LIB.

Materials based on silicon and vanadium have the highest capacity towards reversible lithium intercalation. At the same time these materials are traditional materials of microelectronics, what makes it possible to create all solid-state thin-film LIB using microelectronics technologies, for example, vacuum magnetron sputtering. As one of the variants of solid-state LIB, which can be fabricated by the method of magnetron deposition, the integral structure Ti / $Li_xV_yO_z$ / LiPON / Si-O-Al / Ti is considered. Here, as a negative electrode, a nanocomposite of silicon, aluminium and their oxides Si-O-Al is used. However, its fabrication raises a number of problems such as the development of a technological routine for the structure formation, lithiation of one of the electrodes and improvement of the adhesive strength of the layers, forming the battery.

2. Creation of prototypes
Among these problems arises the problem of the resistance of LiPON-electrode junction. To estimate the value of this resistance within this study were made two prototypes consisting of two half-cells. Each of prototypes was composed of two identical electrodes deposited by magnetron sputtering in one process. One prototype was assembled of electrodes based on V_xO_y, the other one - of electrodes made of Si-O-Al nanocomposite. One of electrodes in each prototype was covered by LiPON layer.

Thus, every prototype consisted of two half-cells with a common lithium counter electrode and lithium reference electrode:

- Ti / Si-O-Al / LiPON - (LP-71) - Li - (LP-71) - (Si-O-Al) / Ti - (LP-71) - Li,
- Ti / V_xO_y / LiPON - (LP-71) - Li – (LP-71) - V_xO_y / Ti – (LP-71) - Li.

The structure of these prototypes is presented in figure 1.

Given the fact that positive electrodes (in that case Si-O-Al and V_xO_y) were manufactured in one technological process, such structure provides the most reliable data on the influence of solid-state electrolyte on capacitance of LIB cell.

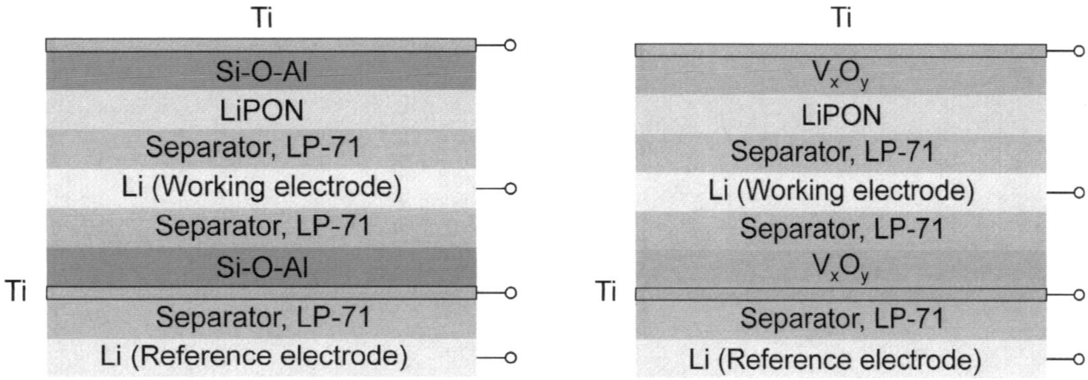

Figure 1. The structure of the half-cells for the comparative tests of the electrodes with solid electrolyte LiPON and without the solid electrolyte

3. Cycling tests

Both prototypes were tested in galvanostatic regime at the measuring complex "АЗРВК-0,05". More than 20 charge-discharge cycles were performed at different currents. Charge-discharge curves of a half-cell Ti / Si-O-Al / LiPON - (LP-71) - Li are shown in figure 2. Charge-discharge curves of a half-cell Ti / V_xO_y / LiPON - (LP-71) - Li are displayed in figure 3. The results of the prototypes cycling are shown in figure 4.

The half-cell Ti / V_xO_y / LiPON - (LP-71) - Li has demonstrated low capacitance (figure 2). At the operating current of 50 μA its capacitance was approximately 2.5% of the capacitance of the half-cell Ti / V_xO_y - (LP-71) - Li and when operating current was risen up to 100 μA, capacitance has dropped down to the noise values.

The half-cell Ti / Si-O-Al / LiPON - (LP-71) - Li has demonstrated good efficiency, although it had a lower specific capacitance as compared with Ti / (Si-O-Al) - (LP-71) - Li half-cell. When operating at various current values its capacitance varied from 50% to 65% of specific capacitance of Ti / (Si-O-Al) - (LP-71) - Li half-cell. Given that the LiPON resistivity is higher by two orders than that of LP-71 and its thickness is 15 times less, the resistance of the whole structure should rise 7 times approximately. Nevertheless the capacitance has reduced only by 50%, that is obviously a good result.

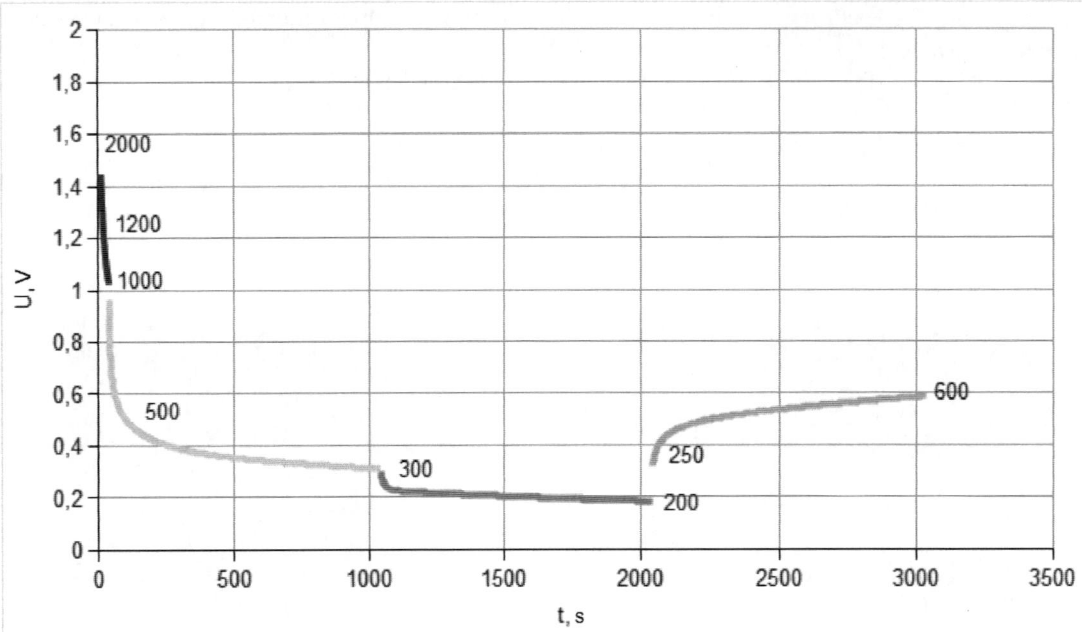

Figure 2. Charge-discharge curves of a half-cell Ti / Si-O-Al / LiPON - (LP-71) - Li. Discontinuities show the potentials at which the electrode impedance measurements were made

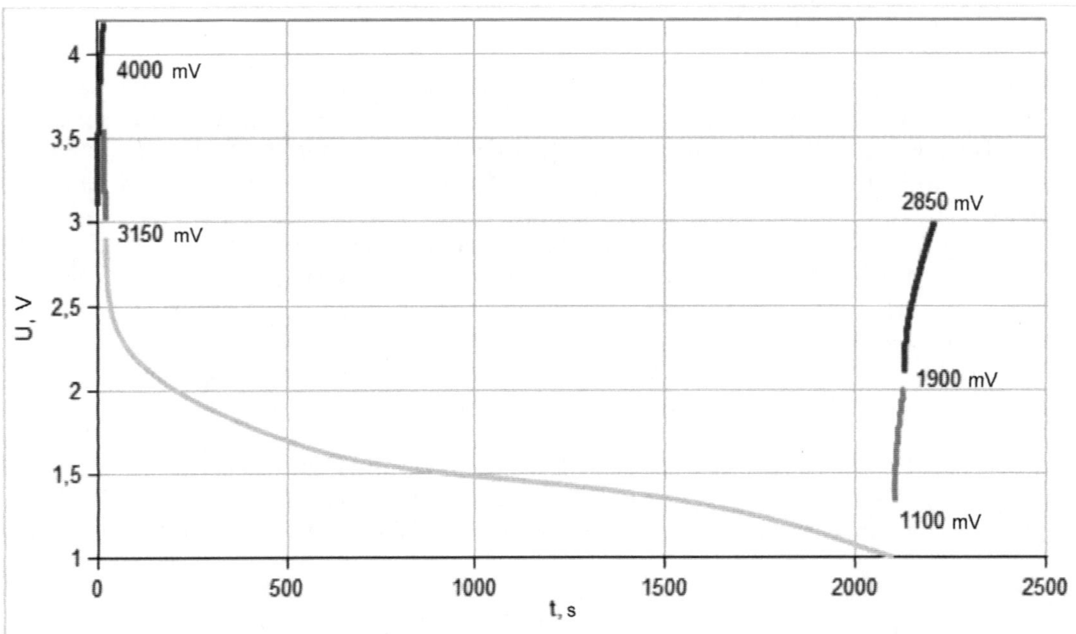

Figure 3. Charge-discharge curves of a half-cell Ti / V_xO_y / LiPON - (LP-71) - Li. Discontinuities show the potentials at which the electrode impedance measurements were made

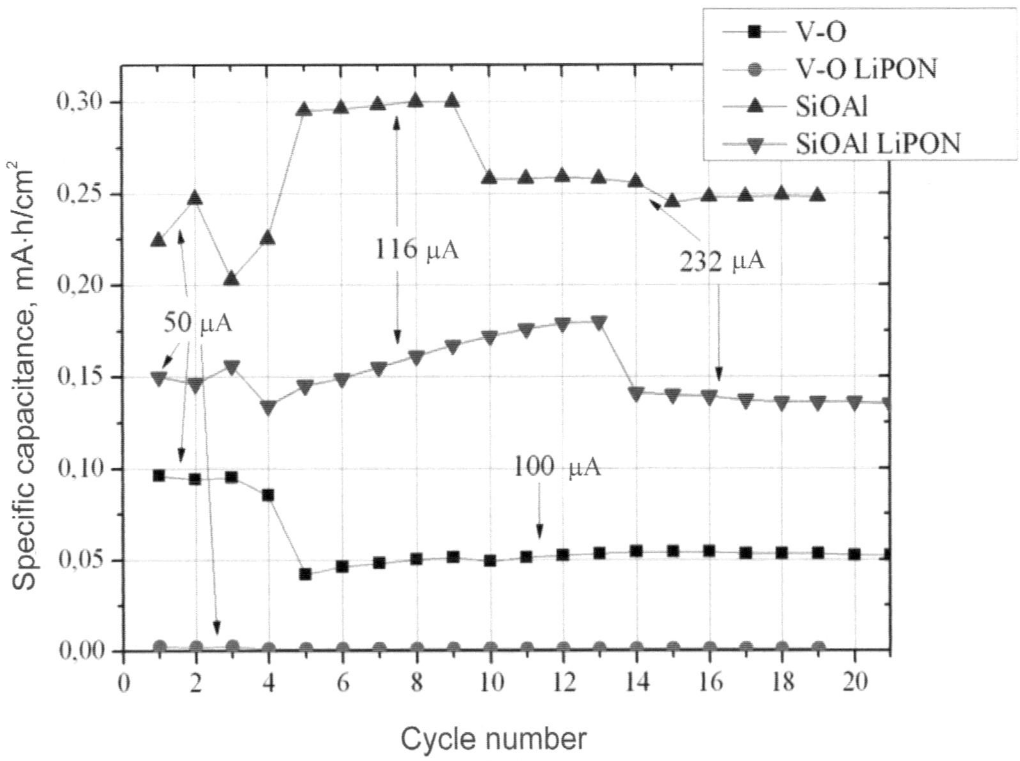

Figure 4. Comparative results of half-cells testing by means of galvanostatic cycling: Ti / Si-O-Al / LiPON - (LP-71) - Li versus Ti / (Si-O-Al) - (LP-71) - Li and Ti / V_xO_y / LiPON - (LP-71) - Li versus Ti / V_xO_y - (LP-71) - Li. Operating currents are specified in figure

4. Impedance measurements

During prototypes testing the impedance of half-cell Si-O-Al / LiPON / LP-71 / Li was measured. The measurements were carried out in the frequency range from 10^{-2} Hz to 500 kHz with excitation signal amplitude of 10^{-2} V. The impedance was measured within galvanostatic regime at potential range from 1 to 4.2 V relatively to Li/Li^+. Basing on experimental impedance spectra of prototype and with the help of EIS Specrum Analyser program the equivalent scheme (figure 5) was created and its parameters were defined. The example of approximation of experimental impedance hodographs by curves, calculated on the basis of equivalent scheme is shown in figure 6.

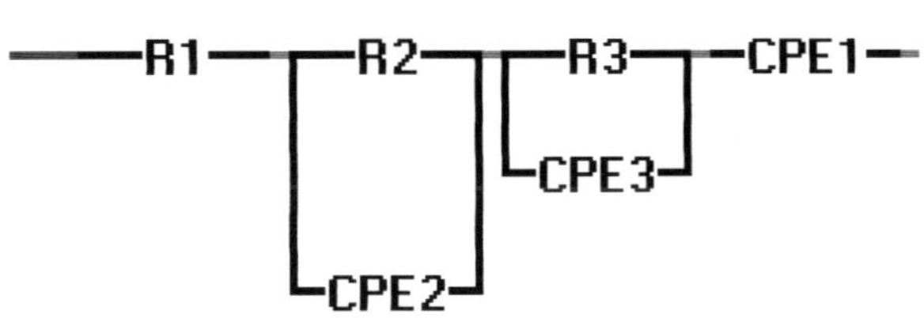

Figure 5. Equivalent scheme of the half-cell Ti / Si-O-Al / LiPON - (LP-71) – Li. R is an active resistance, CPE is a constant phase element

IOP Conf. Series: Journal of Physics: Conf. Series **917** (2017) 032030 doi:10.1088/1742-6596/917/3/032030

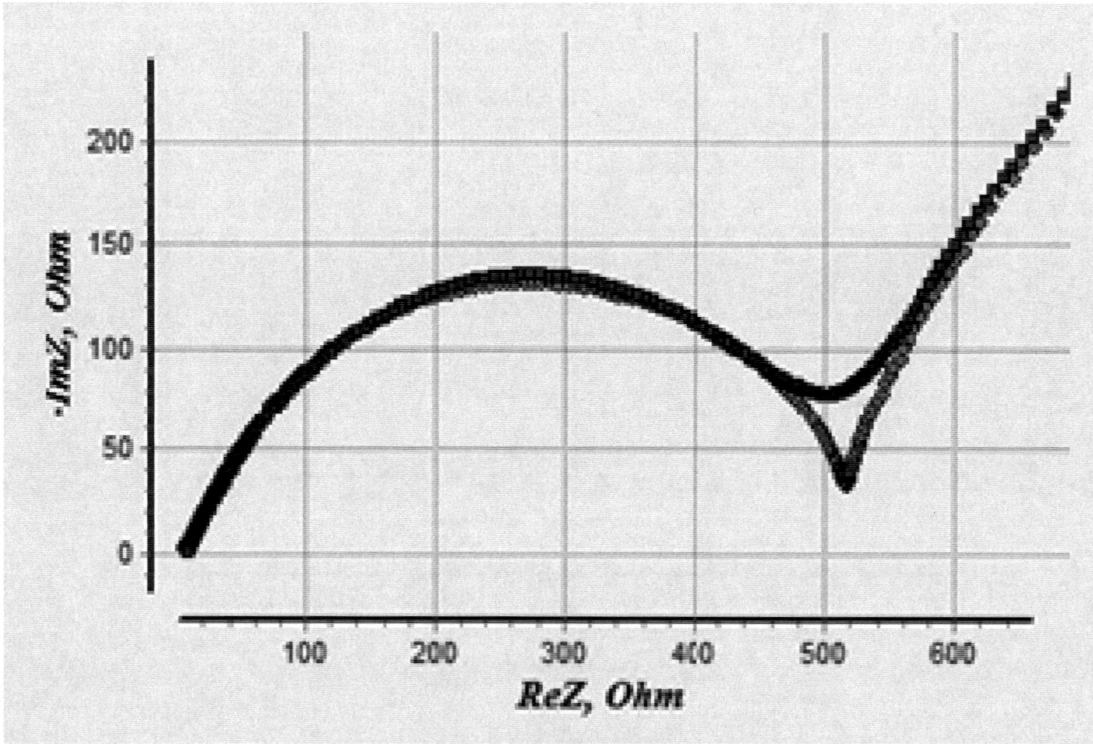

Figure 6. Experimental and simulated hodographs of an impedance of half-cell Ti / Si-O-Al / LiPON - (LP-71) – Li at a potential of 200 mV relative to Li/Li^+, corresponding to straight and reverse sweeps: red dots - simulated data, blue dots – experimental data

Acknowledgments
This work has been done in Facilities sharing centre "Diagnostics of Micro- and Nanostructures" and has been supported by Foundation for Assistance to Small Innovative Enterprises "UMNIK" (Agreement № 8036ГУ2015 from 23.11.2015).

References
[1] Bates J B, Dudney N J, Gruzalski G R, Zuhr R A, Choudhury A, Luck C F, Robertson J D 1993 *J. of Power Sources* **43** 103-110.

Synthesis of Ge and Si nanoparticles by spark discharge

A A Lizunova[1], D A Mylnikov[1], A A Efimov[1], V V Ivanov[1]

[1]Research Centre for Functional Materials, Moscow Institute of Physics and Technology, Dolgoprudny 141700, Russia

Abstract. The production of airborne nanoparticles by multi gap spark discharge generators is shown to be feasible for semiconductors on the example of erosion germanium and silicon electrodes in inert atmosphere. Pure silicon and germanium crystalline nanoparticles with the mean diameter less than 15 nm combined in submicron agglomerates were obtained. Since the presence of residual oxygen in the setup partially oxidized silicon nanoparticles and core-shell germanium structures with pure Ge core and oxidized shell were also observed in the samples. This continuous nanoparticles production technique enables direct impaction of prepared nanoparticles onto a liquid or substrate for applications in the fields of electronics, biomedicine and energetics.

1. Introduction

In recent years semiconductor nanoparticles are widely studied and used in various applications in the fields of science and technology due to their unique optical and electronic properties [1,2]. For example, silicon and germanium nanoparticles can be injected into gate oxide of MOS memory devices to prevent the leakage current [3], they can be exploited as anode materials for Li-ion batteries, fluorescent bioprobes [2], functionalized Si and Ge nanoparticles deposited on wool improve photostability toward UV radiation of the fabric [4].

There are several methods for creating semiconductor nanoparticles: inorganic solution phase chemical approaches and vapor-based techniques [1, 2]. The spark discharge generation of nanoparticles permits to generate aerosol nanoparticles with a diameter of primary particles less than 10 nm from metals and oxides materials showing high chemical purity of the synthesized nanoparticles and low agglomeration properties. Due to problems with removing residual oxygen from camera setup to establish inert atmosphere it is known only one study of silicon nanoparticles generation by spark discharge [5].

The present paper discusses the investigation of nanoparticle properties produced by spark discharge method by electrical erosion of silicon and germanium electrodes in inert atmosphere.

2. Experimental

The multi gap spark discharge generator described in the paper [6] was used in the experiments with following conditions: 6 pair of single-crystal germanium (n-type) or silicon (p-type) electrodes 8 mm in diameter and 30 mm in length with resistivity of 0.005 Ohm·cm, 2 mm gap spacing, total capacitance 21 nF, frequency of the discharges was 310 Hz. Argon gas with purity of 99.9999 % from a bottle was used as a carrier gas at flow rate of 5 litres/min. The amount of residual oxygen in the discharge chamber was less than 20 ppm. The pressure of the argon gas in camera setup was held up by approximately 1.5 bar.

The particle size distribution, morphology, crystal structure and element composition of nanoparticles were investigated using transmission electron microscope (TEM) JEOL JEM2100 with energy dispersive X-ray (EDX) spectrometer X-MAX^N OXFORD Instruments. For microscopic studies, the particles were deposited directly to a TEM grid with carbon film in the setup, and transferred to a microscope in hexane to prevent extra oxidation. The real-time measurements of size and concentration of the nanoparticles in the aerosol flow were determined by an aerosol spectrometer SMPS 3936 (TSI Inc.).

3. Results and discussions

TEM analysis of nanoparticles produced by electrical erosion of germanium electrodes reveals the typical fractal-like structure of large agglomerates (Figure 1). According to aerosol spectrometer measurements the mean size of agglomerates is equal to 370 nm (Figure 2). The primary particles are spherical with quite uniform diameters, as typical for spark-generated particles. The average diameter of particles is 13 ± 2 nm. A typical high resolution electron microscopic (HRTEM) image of the germanium nanoparticles is shown in Figure 3. The visible lattice lines confirm the presence of crystalline material; the spacing distance is 3.22 Å, which is in agreement with theoretical value between the (111) Ge crystalline planes. The specific EDX line profiles of 35 nm nanoparticle show the increase in the germanium signal with the constant oxygen concentration (Figure 4). It was also observed several nanoparticles with core-shell structure, where crystalline germanium form the core of nanoparticles and germanium oxide built the amorphous shell.

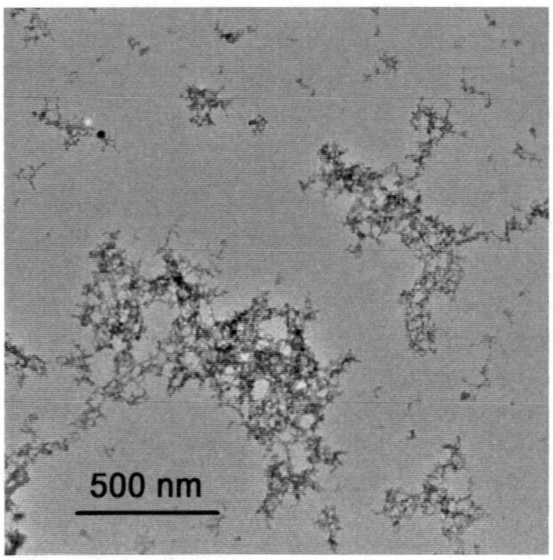

Figure 1. TEM image of Ge primary particles in large agglomerates.

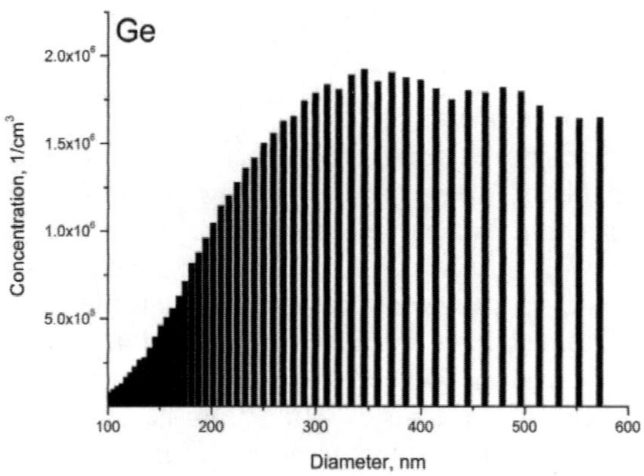

Figure 2. Size distribution of number concentration of airborne Ge nanoparticles in agglomerates.

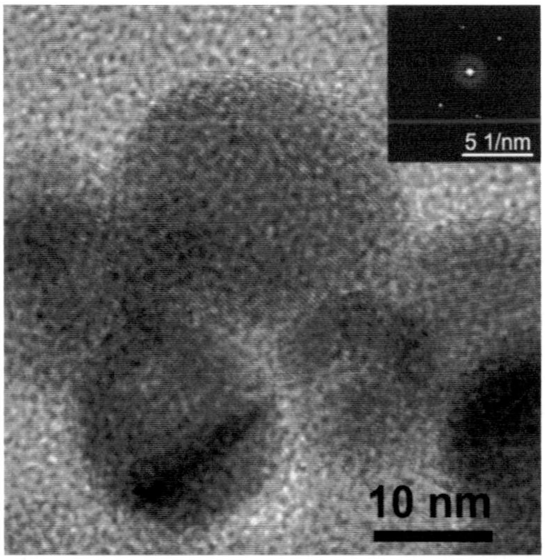

Figure 3. HRTEM image and the corresponding FFT pattern of germanium nanoparticles.

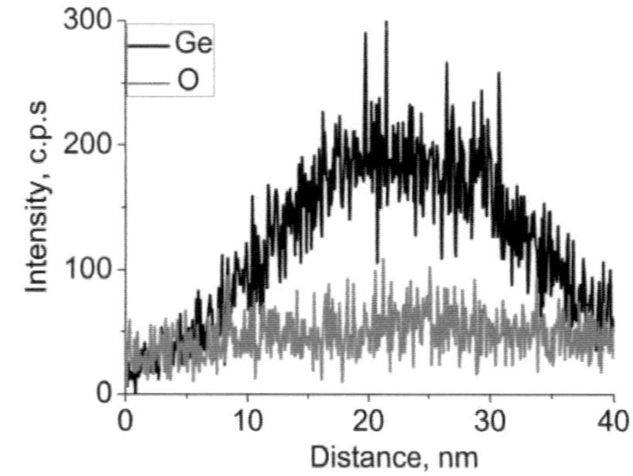

Figure 4. Representative EDX line scan with Ge Kα and O Kα signals obtained on germanium nanoparticle with the size of 35 nm.

The HRTEM images and electron diffraction patterns of obtained silicon nanoparticles show that the sample consist of agglomerates of silicon nanocrystals encapsulated within an amorphous matrix of silicon oxide (Figure 5). The size of the primary semiconductor crystalline particles is in the range of 3 to 30 nm with the mean size of 10 ± 2 nm (Figure 6). The average size of the agglomerates is about 350 nm by the aerosol spectrometer which is consistent with the TEM-data.

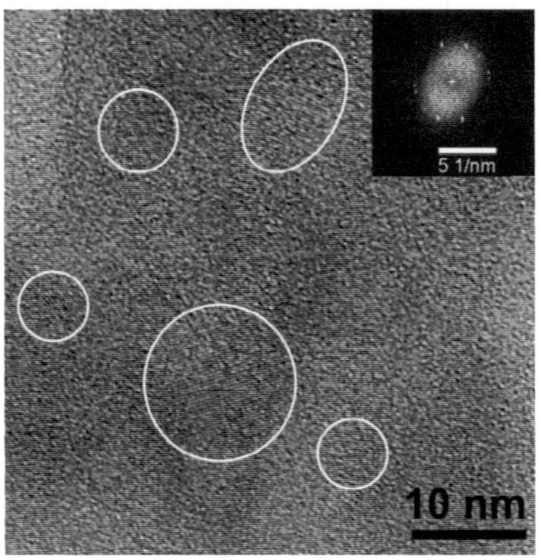

Figure 5. HRTEM image and the corresponding FFT pattern of silicon nanoparticles in amorphous matrix. The lattice fringes corresponding to the (111) of individual Si nanocrystalls are shown by white rings.

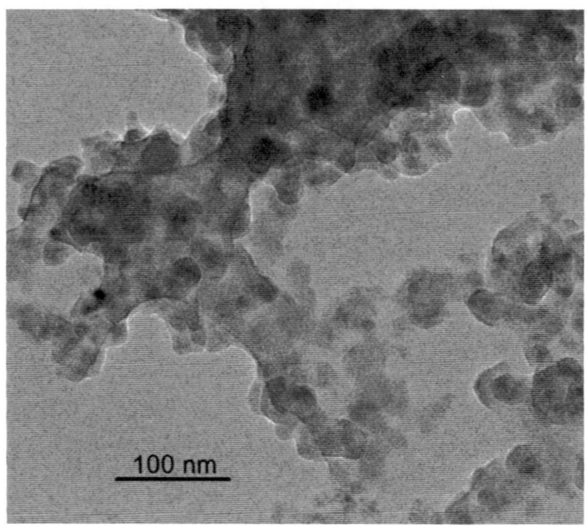

Figure 6. Typical TEM image of silicon nanoparticles in agglomerate.

Difference in morphology structure of silicon and germanium nanoparticles can be explained in terms of energy formation of semiconductor oxides. Since the energy of formation of silicon oxide (-904 kJ/mole) is two times higher than for germanium oxide (-515 kJ/mole) germanium oxidation reaction is slower. So we can synthesis only partially crystalline silicon nanoparticles in amorphous silicon oxide matrix.

4. Results

We have shown application of multi-spark discharge nanoparticle generator for the production of spherical semiconductor nanoparticles, including agglomerates of silicon nanocrystals encapsulated within an amorphous matrix of silicon oxide and core-shell germanium nanoparticles. The primary particle size was between 3 to 30 nm. In further experiments by controlling the residual oxygen concentration in the discharge chamber it is possible to achieve only pure semiconductor, oxides or core-shell nanoparticle structures of silicon and germanium nanoparticles.

Acknowledgments

This work was supported by the Russian Science Foundation (project # 15-19-00190).

References

[1] Huan C and Shu-Qing S 2014 *Chin. Phys. B* **23** 8
[2] Vaughn II D D and Schaak R E 2013 *Chem. Soc. Rev.* **42** 2861
[3] Chakraborty G, Sengupta A, Requejo F G, Sarkar C K 2011 *J. Appl. Phys.* **109** 64504
[4] Deb-Choudhury S, Prabakar S, Krsinic G, Dyer J M, Tilley R D 2013 *J. Agric Food Chem.* **61** 7188
[5] Vons V A, de Smet L C P M, Munao D, Evirgen A, Kelder E M, Schmidt-Ott A 2011 *J.Nanopart. Res.* **13** 4867
[6] Ivanov V V, Efimov A A, Mylnikov D A, Lizunova A A, Bagazeev A V, Beketov I V and Shcherbinin S V 2016 *Tech. Phys. Lett.* **42** 876

Site-controlled GaN nanocolumns with InGaN insertions grown by MBE

D V Nechaev[1], A N Semenov[1], O A Koshelev[1] V N Jmerik[1], V Yu Davydov[1], A N Smirnov[1], G Pozina[2], T V Shubina[1], and S V Ivanov[1]

[1]Ioffe Institute, 26 Politekhnicheskaya, St Petersburg 194021, Russia
[2]Department of Physics, Chemistry and Biology (IFM), Linköping University, S-581 83 Linköping, Sweden

Abstract. The site-controlled plasma-assisted molecular beam epitaxy (PA MBE) has been developed to fabricate the regular array of GaN nanocolumns (NCs) with InGaN insertions on micro-cone patterned sapphire substrates (μ-CPSSs). Two-stage growth of GaN NCs, including a nucleation layer grown at metal-rich conditions and high temperature GaN growth in strong N-rich condition, has been developed to achieve the selective growth of the NCs. Micro-cathodoluminescence measurements have demonstrated pronounced emission from the InGaN insertions in 450-600 nm spectral range. The optically isolated NCs can be used as effective nano-emitters operating in the visible range.

1. Introduction

Regular arrays of III-Nitride nanocolumns (NCs) with controlled geometrical sizes, uniformity, composition and selectivity can be useful for a lot of applications in nanophotonics, including room-temperature quantum information processing and single photon emitters [1]. Typically, the selective area growth (SAG) technique has been exploited to grow site-controlled III-Nitride NCs with small dispersions in size, position, and alloy composition by using different metal or dielectric nanomasks and lithography techniques [2]. These techniques allow to fabricate preferential nucleation sites for group III adatoms followed by the growth of vertical NCs. Unfortunately, the insufficient selectivity of the NCs growth and low-yield in case of large scale process are still restrict such techniques [3]. Recently, we have demonstrated the benefits of plasma-assisted molecular beam epitaxy (PA MBE) to grow site-controlled III-nitride monocrystals and NCs on micro-cone-shaped patterned c-sapphire substrates (μ-CPSS) without exploiting any nanomasks-based technology [4,5]. The CPSSs have attracted significant attention due to their ability to realized SAG without any complex nanomasks. In this study, we report on the successful growth of GaN NCs with optically-active InGaN insertions at their tops using the same approach.

2. Growth and characterization

The GaN NCs were grown by PA MBE on commercial μ-CPSS substrates by using Compact 21T. The cones have a diameter of 3.5 μm and a height of 1.7 μm. The distance between them is 2.4 μm. Prior to the growth, the μ-CPSSs were annealed and nitridated by a plasma-activated N_2-flux at 800°C. The GaN NCs were grown using a two-stage process comprising the nucleation and growth stages. Standard PA MBE as well as migration enhanced epitaxy MEE were used for growth of nucleation layers (NL) with a nominal thickness of ~65 nm. Different nucleation layers, such as AlN and GaN, and different growth conditions were examined to achieve the selective growth on the cone apexes

IOP Conf. Series: Journal of Physics: Conf. Series **917** (2017) 032032 doi:10.1088/1742-6596/917/3/032032

only with the suppression of material deposition on the cone planes. All NLs were grown under metal-rich conditions. The GaN NCs were grown with typical III/N flux ratio ~0.2 at substrate temperature 760°C. The InGaN insertions were deposited at the top part of GaN NC at temperature 620°C with nominal thickness of 3 nm and then covered by a thin layer of GaN.

These InGaN/GaN NCs were investigated by scanning electron microscopy (SEM) and micro-cathodoluminescence (μ-CL) at 15 K using the LEO 1550 Gemini analytical setup with 10 kV beam voltage.

3. Basic results and discussion

It was found that NLs are strongly effect the further GaN NCs growth and result in different degree of selectivity. In the case of AlN NLs the aluminum covers whole μ-CPSS surface including cone slops (figure 1, left) that result in uniform growth of AlN layers. The further growth of GaN NCs on AlN underlayer leads to the NCs growth in different areas and directions without any selectivity (figure 1, right). The same situation was observed in case of the GaN NLs grown at substrate temperature around 700°C.

Figure 1. The SEM image of AlN NL (left) and GaN NCs grown on AlN NL (right).

The best selective growth was achieved by using the GaN nucleation layer deposited at the growth temperature around 760-765°C. Figure 2 manifests the selective growth of GaN NCs on such GaN NL. We suppose that the high Ga-adatoms mobility at these conditions, along with the lowest surface energy at (000-1) GaN planes on the cone's apexes, similar to that in planar layer [6], provides the selective growth in these areas, while the semi-polar planes of the cones remain almost bare. The redistribution of Ga adatoms occurs due to the surface diffusion enhanced by high growth temperature.

IOP Conf. Series: Journal of Physics: Conf. Series **917** (2017) 032032 doi:10.1088/1742-6596/917/3/032032

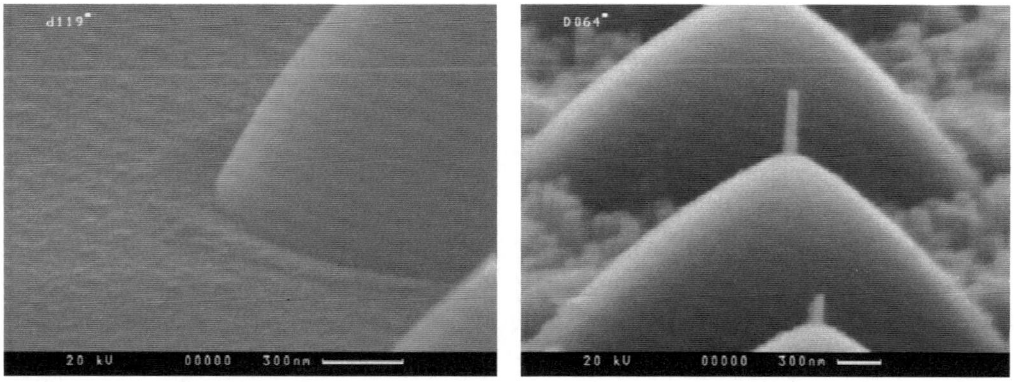

Figure 2. The SEM image of GaN NL grown at 760°C (left) and GaN NCs grown on GaN NL (right).

The grown InGaN/GaN NCs demonstrate bright CL emission in the 450-550 nm spectral range. Essentially, all NCs are surrounded by a bare area with a diameter of about 2 μm, where emission is suppressed as shown in pan-chromatic μ-CL images (figure 3).

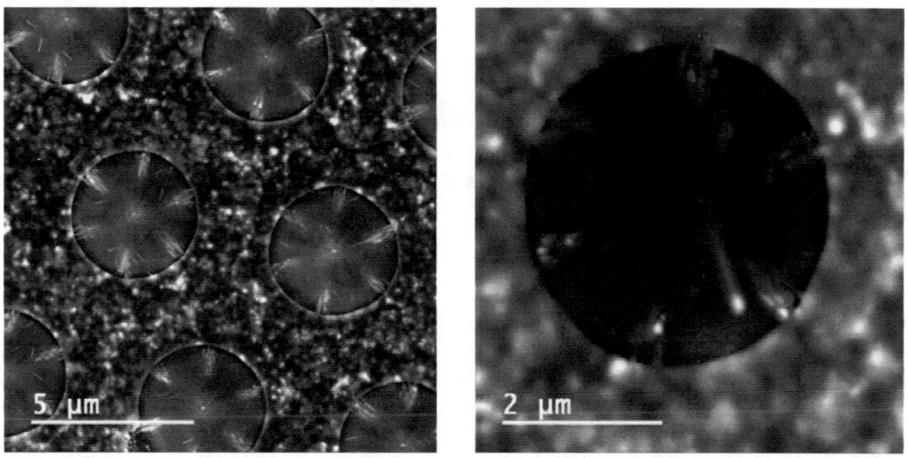

Figure 3. Top-view pan-chromatic SEM image of the cones with GaN/InGaN/GaN NCs on their apexes, visible as small bright sports in the cone centers (left) and he magnified CL image of a cone with an inclined NC (right).

In general the deposited InGaN demonstrates emission in 450-600 nm spectral range which can originate from different places – the InGaN/GaN NCs on the cone apexes, planar areas between cones, and short inclined NCs formed on the cone facets. The μ-CL spectrum measured in the spot regime in case of beam focused in area between cones is shown at figure 4. In this case, the single broad peak around 500 nm wavelengths was observed. Figure 5 presents the μ-CL spectra of broken NC (see inset on figure 5) measured in different parts of NC along of NC axis: located at 100, 350 and 800 nm. The bright emission at slightly higher energy (~450 nm) and with a narrower spectral width than the rest areas was observed in case of measurement near to the top of NC (marked as 800 nm on NC axis) where InGaN insertion was deposited (figure 5 a). We believe that this difference could be explained by the different growth conditions in the vertically oriented and inclined NCs. In case of pumping beam was focused far from the tip of NC (at 100 and 350 nm along NC axis) the emission at the week emission at the wavelength 550 nm was dominated.

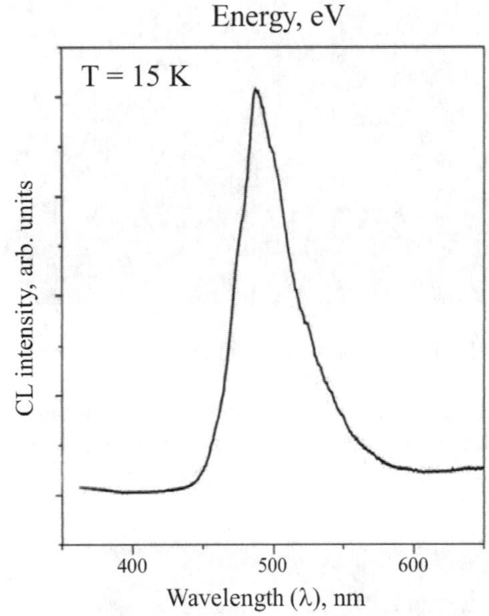

Figure 4. μ-CL spectrum recorded in the spot regime between cones.

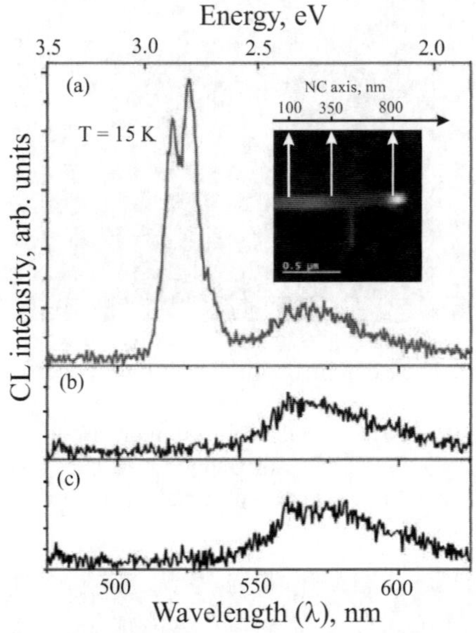

Figure 5 μ-CL spectra recorded in the spot regime along broken InGaN/GaN NC: a – pumping near to the tip of NC where InGaN insertion was deposited (800 nm), b and c – far away from the InGaN insertion (100 and 350 nm, respectively).

4. Conclusions

In conclusion, the selective area growth of GaN NCs with InGaN insertions on μ-CPSS by PA MBE has been successfully developed. Two stage process was used including high temperature NLs deposition and NCs growth. The highest selectivity was achieved in case of GaN NL grown at 760°C. The InGaN/GaN insertions demonstrate strong emission in 450-600 nm spectral range. We believe that created arrays of optically and spectrally isolated NCs can be used as effective site-controlled nano-emitters for the visible spectral range.

Acknowledgement
This work has been supported by Russian Science Foundation, grant № 14-22-00107.

References

[1] Holmes M *et al.*, Nano Lett. **14**, 982 (2014).
[2] Li S and Waag A, J. Appl. Phys. **111** 071101 (2012).
[3] T. Yamamoto et al., Jpn. J. Appl. Phys. 55, 04EK03 (2016).
[4] Shubina T V *et al.*, Scientific Reports **5**, 17970 (2015).
[5] Shubina T V *et al.*, Phys. Status Solidi B **253**, 845 (2016).
[6] H.Li *et al.*, PRL **115**, 085503 (2015)

IOP Conf. Series: Journal of Physics: Conf. Series **917** (2017) 032033 doi:10.1088/1742-6596/917/3/032033

Kinetic Monte Carlo simulation of the indium droplet epitaxy on the Ga-terminated GaAs(001) surface

S V Balakirev, M S Solodovnik, O A Ageev

Southern Federal University, Research and Educational Center "Nanotechnologies", 2 Shevchenko Street, Taganrog 347922, Russia

Abstract. Model of In/GaAs(001) molecular beam epitaxial growth is developed to study the processes of droplet epitaxy on the Ga-terminated surface. The model allows computation of significant geometrical characteristics of nanostructures on a large scale and estimating a phase of material by the crystal density. The nucleation and growth mechanisms in a range of indium near-melting temperatures are investigated. Average size of islands is shown to increase with rising substrate temperature whereas the island density demonstrates an inverse relationship between these parameters. An increase of the growth rate intensifies the nucleation processes and leads to an increase in the droplet surface density and decrease in the average droplet diameter. The deposition thickness is observed to have an influence on the island size, but not impact on the surface density of islands.

1. Introduction

Ample opportunities for application of $A^{III}B^{V}$ nanostructures in nanoelectronic and photonic devices lead to the necessity of better controlling their parameters. Therefore, methods of the formation of nanostructures with specified characteristics have been actively developed. One of these methods is droplet epitaxy which allows the formation of nanostructure arrays with either ultralarge or ultrasmall surface density, a wide range of sizes and in a wide range of technological parameters.

Although AlGaAs droplet epitaxial system has been successfully investigated for a long time [1–3], there is a lack of studies on InAs/GaAs nanostructures formed by this method. Because of difficulties in simulation of InAs/GaAs growth processes, theoretical works on the droplet epitaxy of InAs on the GaAs substrate are rare or absent at all. Experimental studies made it possible to examine basic nucleation, growth and crystallization processes [4–6] and determine the dependences of diffusion parameters and geometrical characteristics on the growth conditions [4,5,7,8]. However, in all these works the growth was carried out on As-terminated surfaces because obtaining Ga-stabilized conditions (especially under low growth temperatures) is a challenge. At the same time, droplet nucleation on the metal-terminated surfaces is expected to occur without wetting layer formation which could be useful for many applications.

In the present work, kinetic Monte Carlo method is used to simulate the MBE growth of indium on the Ga-terminated GaAs(001) surface. This technique permits to reproduce the growth process on a large time and dimensional scale and consider many relevant microscopic processes and features of the material system [9,10].

2. Description of the model

2.1. Crystal

The simulation area represents a two-layered square lattice with a lateral period of 0.2 nm and 0.14 nm in vertical direction. The nearest sites are located diagonally to each other so that the distance between two layers of the simulation lattice is equal to 0.2 nm. Thus, the (1+1)-dimensional approach of the crystal simulation is used [1] and the zinc blende structure of GaAs is taken as a basis.

Each atom in the lattice is bound to atoms belonging to seven coordination spheres characterizing distances between sites. The distance r between an atom and its neighbour enters into the expression of the Lennard–Jones potential [11]:

$$U(r) = \varepsilon\left[\left(\frac{r_m}{r}\right)^{12} - 2\left(\frac{r_m}{r}\right)^{6}\right]$$

where r_m is the distance at which the potential reaches its minimum, ε is the interaction energy of a pair of atoms at the distance r_m. Values for r_m are taken to be average inter-atomic distances in the substance to which a pair of atoms belongs. The binding energy of all pairs "atom–neighbour" amounts to the total binding energy of an atom in a certain environment.

Before the growth starts, the simulation lattice is filled with atoms of the GaAs substrate. The upper layer is set to be gallium monolayer implying that Ga-terminated conditions are created on the surface. This situation can be achieved by appropriate simultaneous control of the substrate temperature and arsenic flux before cooling down the sample. Low temperatures (100–200°C), at which the simulations are carried out in this work, allow maintaining these conditions.

2.2. Microscopic processes on the surface

Atoms on the substrate can occupy each site of the lattice, even in case they do not fit in the starting zinc blende structure. The deposition is implemented as an appearance of an atom in a random site next layer above the substrate or the growing layer. Deposited atoms are considered to be adsorbed immediately. The probability of depositing a new indium atom is predefined by the growth rate.

A deposited atom, then, can diffuse in one of empty site of the 1st or 2nd coordination sphere. Therefore, six directions for the atom migration are available. In order to estimate the activation barrier for the diffusion, the difference between binding energies with initial and final neighbourhoods is calculated. The probabilities obtained for each possible direction are summed up and normalized to unity. The direction is selected by a random number distributed uniformly over the interval (0–1). If a site where an atom can move to is occupied by a neighbouring atom, the probability of the migration in this direction goes to zero because atom exchange is impossible.

The desorption of In atoms is neglected since the growth temperature under consideration is low.

2.3. Kinetic Monte Carlo procedure

The algorithm is based on the traditional kinetic Monte Carlo technique [12]. The activation energy E_i for an event i defines its probability expressed by the Arrhenius equation [13,14]:

$$p_i = \nu_0 \exp(E_i / k_B T)$$

where ν_0 is the attempt frequency, k_B is the Boltzmann constant, T is the growth temperature.

When all events are processed and unnecessary ones are rejected, the expectation time for each event is calculated as follows:

$$\tau_i = -\frac{1}{p_i}\ln\delta$$

where δ is a random number distributed uniformly over the range (0–1).

After an event is executed, its expectation time is added to the total growth time (settled-life time) An event with the minimal settled-life time is executed during each iteration.

3. Results and discussion

Using the model, the MBE growth of indium on the GaAs(001) substrate in a range of the indium near-melting temperatures (100–200 °C) is investigated. The simulation results show that all indium atoms conglomerate into islands of near-segmental shape (figure 1). A wetting layer does not form on the surface. However, because of the attractive force of the substrate the island shape is far from semi-spherical and the contact angle is small.

In figure 1, the difference between the crystal densities in different regions of the island can be discerned. In some regions atoms are located closer to each other and some regions are more rarefied. In the framework of the zinc blende-based lattice, it is impossible to simulate the ideal structure of the solid indium. However, the density of atoms in a part of an island allows approximate estimating the substance phase. At $T = 150$ °C (which is 6.6 °C below the indium melting temperature) the aggregate state of islands is expected to be solid. However, a quantity of experimental works is evidence of the liquid droplet formation in a range of temperatures from 100 to 200 °C and more which can be attributed to the dimensional effect. As figure 1 demonstrates, the crystal in larger islands is denser as well as in the depth of an island. This observation confirms the fact that nanometer islands can be liquid under the melting temperature due to the dimensional effect.

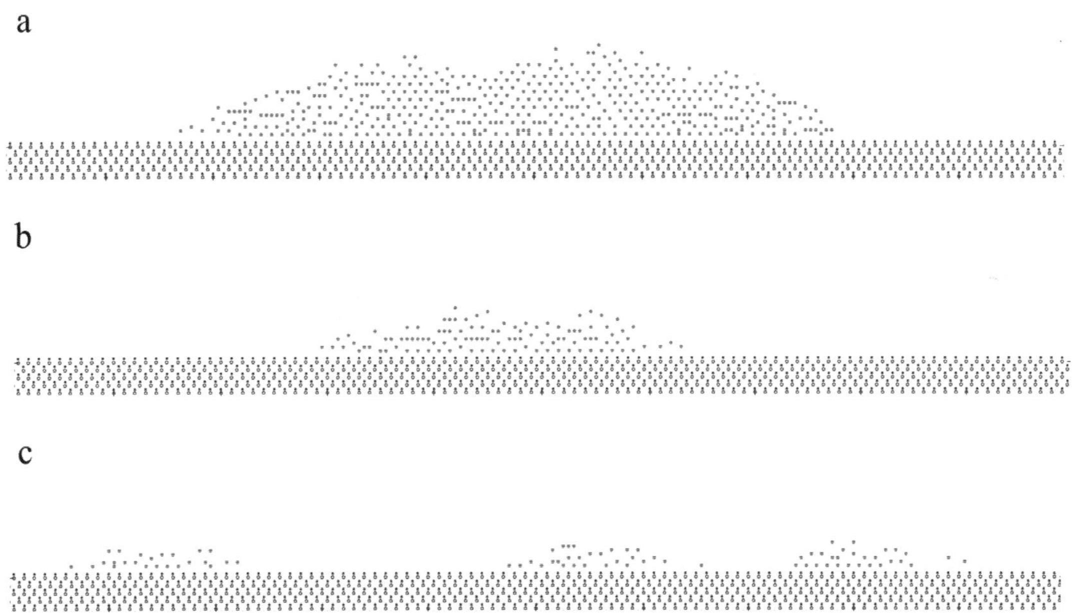

Figure 1. Morphology of In droplets after deposition of 3 monolayers of indium on the Ga-terminated GaAs(001) surface at $T = 150$ °C: a) $v = 0.5$ ML/s, B) $v = 1$ ML/s, C) $v = 2$ ML/s.

Figure 2 shows the dependences of the droplet array characteristics on the growth rate v and temperature T. An increase of the droplet density is observed both with decreasing substrate temperature (figure 2a) and increasing growth rate (figure 2c). This behavior is attributed to the enhancement and suppression of the surface diffusion of In adatoms, respectively. It is known [15,16] that geometry of MBE grown islands strongly depends on the migration of metal atoms on the surface. Since the critical island size falls with increasing supersaturation, the average island size is observed to decrease at the same time (figures 2b and 2d). At $T = 100$ °C and $v = 1$ ML/s, the droplet density is equal to $6 \cdot 10^{10}$ cm^{-2} and average size is about 8 nm, whereas these values are $8 \cdot 10^{8}$ cm^{-2} and 80 nm, respectively, at $T = 200$ °C and $v = 0.25$ ML/s. The simulation values are in good agreement with experimental data [8,17].

An increase of the growth rate leads to the gradual saturation of the temperature dependence of the droplet density. This phenomenon can be explained by the fact that the concentration of indium nuclei on the GaAs(001) surface cannot reach the adatom concentration or even the doublet concentration according to Frenkel's theory [18]. There is a critical size under which an indium cluster is unstable and exposed to the decay.

The growth rate dependence of the droplet diameter (figure 2d) is observed to become stronger when the temperature exceeds the indium melting point (156.6 °C). The most probable reason of this behaviour is the exponential growth of the island diameter with increasing temperature. However, the island crystal phase or shape dynamics could have a significant influence on this dependence too.

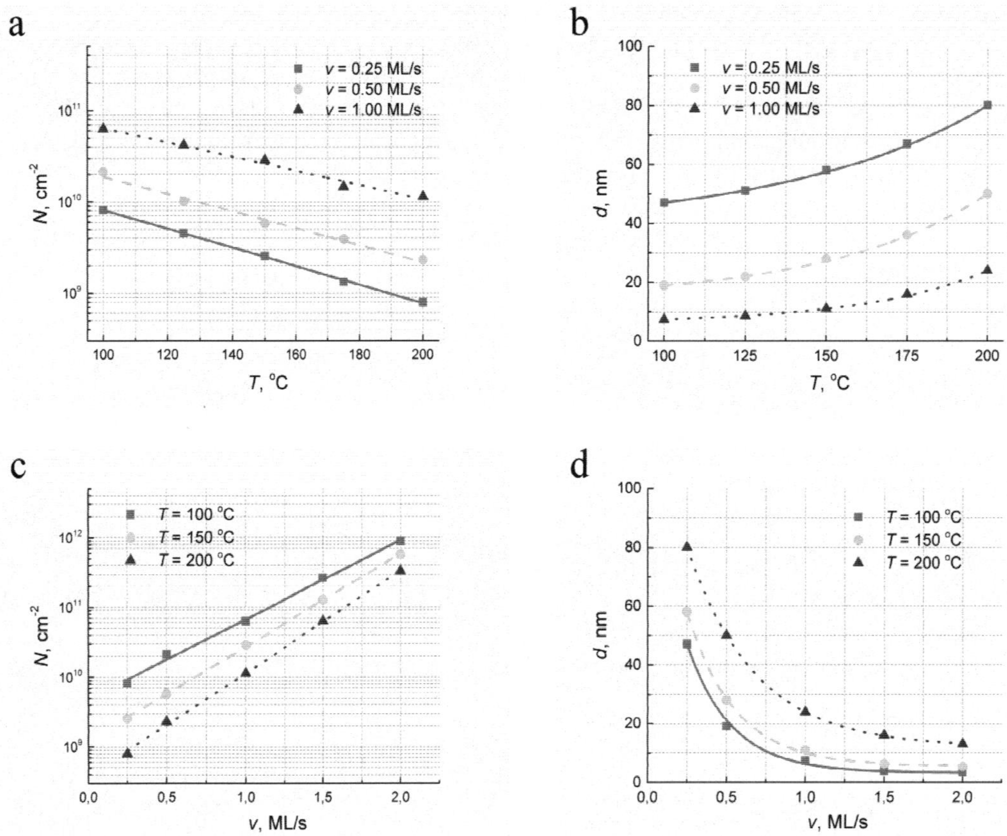

Figure 2. Temperature (a, b) and growth rate (c, d) dependences of the In/GaAs(001) nanostructure surface density (a, c) and average diameter (b, d).

The possibility to control island characteristics by altering the deposition thickness is also found. As figure 3 shows, the average size of droplets steadily increases with increasing amount of the deposited material whereas the island density remains almost steady. This means that the deposition thickness does not have an impact on the nucleation processes and only leads to the island growth and shape transformation. Therefore, the average size of In/GaAs nanostructures can be adjusted while maintaining their density with a value defined by the other technological parameters.

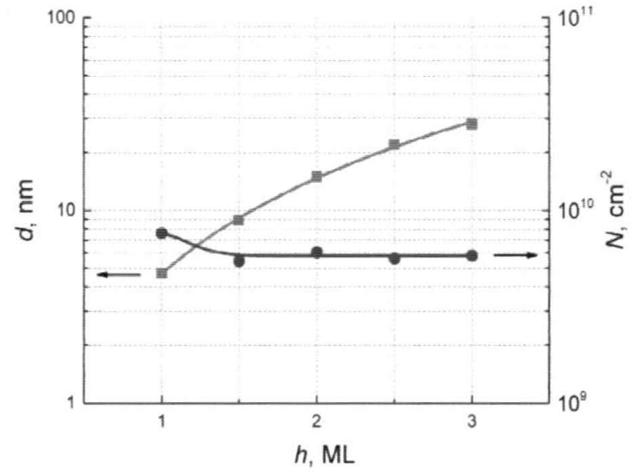

Figure 3. Dependence of the In/GaAs(001) nanostructure density and average diameter on the indium deposition thickness.

4. Conclusion

In summary, the low-temperature droplet epitaxy of indium on the Ga-terminated GaAs(001) surface was investigated using the kinetic Monte Carlo model. Using the simulations, we considered the significant microscopic processes on the surface and the crystal phase of growing material and calculated the geometrical characteristics of In/GaAs(001) nanostructures. The average island size was found to increase with rising substrate temperature and decreasing growth rate whereas the island surface density is exposed to the opposite influence. It was shown that the size of nanostructures can be adjusted by varying the deposition thickness without changing the island density. As the simulations results are in good agreement with experimental data, the model is applicable to the description of the initial stage of the $A^{III}B^{V}$ nanostructure formation and is intended to be developed to consider the processes of the droplet crystallization and formation of nanostructures on modified surfaces with complicated morphology.

Acknowledgements

This work was supported by the Russian Science Foundation Grant No. 15-19-10006. The results were obtained using the equipment of the Center for Collective Use and Research and Education Center "Nanotechnologies" of Southern Federal University.

References

[1] Reyes K, Smereka P, Nothern D, Millunchick J M, Bietti S, Somaschini C, Sanguinetti S and Frigeri C 2013 *Phys. Rev. B* **87(16)** 165406
[2] Somaschini C, Bietti S, Koguchi N and Sanguinetti S 2010 *Appl. Phys. Lett.* **97(20)** 203109
[3] Somaschini C, Bietti S, Koguchi N and Sanguinetti S 2009 *Nano Lett.* **9** 3419–24
[4] Mano T, Watanabe K, Tsukamoto S, Fujioka H, Oshima M and Koguchi N 1999 *Jpn. J. Appl. Phys.* **38** L1009–11
[5] Noda T, Mano T and Sakaki H 2011 *Crystal Growth & Design* **11(3)** 726–8
[6] Lee J H, Wang Z M and Salamo G J 2007 *J. Phys. Condens. Matt.* **19(17)** 176223
[7] Lee J, Wang Z and Salamo G 2009 *IEEE Transactions on Nanotechnology* **8** 431–6
[8] Lee J H, Wang Z M, Kim E S, Kim N Y, Park S H and Salamo G J 2009 Nanoscale Research Letters **5(2)** 308–14
[9] Ageev O A, Solodovnik M S, Balakirev S V and Mikhaylin I A 2016 *Tech. Phys.* **61(7)** 971
[10] Ageev O A, Solodovnik M S and Balakirev S V 2016 Advanced Materials, Mechanical and Structural Engineering (Proceedings of AMMSE 2015) eds S Hong *et al* (London: CRC Press) chapter 12 pp 57–60

[11] Lennard-Jones J E 1924 *Proc. R. Soc. Lond. A* **106(738)** 463
[12] Vasilenko M A, Neizvestny I G, Shwartz N L 2015 *Comp. Mat. Sci.* **102** 286–92
[13] Amrani A, Djafari Rouhani M and Mraoufel A 2011 *Appl. Nanosci.* **1** 59
[14] Ageev O A, Solodovnik M S, Balakirev S V and Mikhaylin I A 2016 *J. Phys. Conf. Ser.* **681(1)** 012036
[15] Ageev O A, Solodovnik M S, Balakirev S V, Mikhaylin I A and Eremenko M M 2017 *J. Cryst. Growth* **457** 46
[16] Ageev O A, Solodovnik M S, Balakirev S V and Eremenko M M 2016 *J. Vac. Sci. Technol. B* **34(4)** 041804
[17] Mano T, Watanabe K, Tsukamoto S, Imanaka Y, Takamasu T, Fujioka H, Kido G, Oshima M and Koguchi N 2000 *Jpn. J. Appl. Phys.* **39** 4580–3
[18] Frenkel J 1924 *Zeitschrift für Physik* **26** 117

IOP Conf. Series: Journal of Physics: Conf. Series **917** (2017) 032034 doi:10.1088/1742-6596/917/3/032034

Monte Carlo investigation of the MBE growth of GaAs on the surfaces with different crystallographic orientations

S V Balakirev, M S Solodovnik, I A Mikhaylin, M M Eremenko, O A Ageev

Southern Federal University, Research and Educational Center "Nanotechnologies", 2 Shevchenko Street, Taganrog 347922, Russia

Abstract. MBE growth of GaAs on the surfaces with different orientations is studied by kinetic Monte Carlo simulations. The model allows taking into account the most relevant processes and reconstructions on the surfaces. The simulations show that islands grown on the (111)A surface have a chaotic shape whereas those on the (001) and (110) surfaces have more ordered structure due to the features of the surface structure. The diffusion length of Ga adatoms is observed to be dependent on the growth rate and V/III flux ratio. This influence is stronger for the (111) and (110) surfaces demonstrating that diffusion is weaker on the (001) surface. However, the diffusion length reduces with increasing growth rate and As_2/Ga flux ratio tending to become equal for all surfaces under consideration.

1. Introduction

The formation of $A^{III}B^{V}$ nanostructures is realized by a number of techniques. Many advantages make molecular beam epitaxy (MBE) one of the most promising methods for the fabrication of nanostructure devices for nanophotonics, opto- and nanoelectronics [1].

In the growth process of $A^{III}B^{V}$ materials, the most attention is paid to the characteristics of grown structures which are controlled by the substrate temperature. However, most studies do not consider the V/III flux ratio and growth rate dependences, although these parameters have a huge influence on the characteristics of growing films [2–6].

Despite the fact that studies of the epitaxial growth of semiconductor materials based on GaAs have been carried out for a long time, the growth mechanisms have been mainly studied on a substrate with the (001) orientation. The substrates with other orientations, such as (110) and (111) A, which have other properties and characteristics in comparison with (001), have become no less significant in recent times.

The study of the (111)A orientation of GaAs is important due to the possibility to realize laterally symmetric quantum dots belonging to the C_{3v}point symmetry group of the (111) surfaces. At such symmetric points, effective spin orientation through optical pumping and minimal multiplet splitting is observed. This makes it possible to efficiently generate pairs of entangled photons [7,8]. In addition, the surface of GaAs (111) A is subject to increased attention due to its unique technological properties: silicon-doped GaAs layers grown on such a surface by the MBE method exhibit either n- or p-type conductivity depending on the growth conditions [9]. $A^{III}B^{V}$ semiconductors with zinc blende structure give almost atomically clean cleavage surfaces (110) in view of their charge neutrality, which are used as mirror planes of optical cavities.

The relevance of studying epitaxial growth on surfaces with different orientations is also due to the possibility of forming low-dimensional structures on pre-structured substrates [10–13]. In such cases, the initial surface is a system, in most cases, of low-index crystallographic planes. For the controlled formation of arrays of nanostructures and heterostructures based on them, precise control of the growth process and a deep understanding of elementary growth processes are needed, including adsorption-desorption, surface diffusion and two-dimensional nucleation. Growth by this method is very sensitive to technological parameters and can be accompanied by the formation of undesirable defects. In connection with this, it is necessary to study the mechanisms of the formation of epitaxial layers at an atomic scale on surfaces with different crystallography.

In this paper, we investigate the influence of the growth rate and the V/III flux ratio on the diffusion characteristics of Ga adatoms with the growth of GaAs on surfaces with different crystallographic orientations. To implement this study, we used the kinetic Monte Carlo method which takes into account such important factors as multicomponent growth, the surface structure, molecular form of species, modelling in a wide range of growth temperatures and growth rates, taking into account the growth time corresponding to the experimental growth, and also features of the growth system, which is not feasible with the help of analytical models.

2. Description of the model

2.1. Growth system

Within the framework of the model under consideration, it is assumed that growth occurs by the "vapor-solid" mechanism. The initial simulated GaAs (001) surface has the crystalline structure of zinc blende and the reconstruction of $\beta2(2\times4)$, and the GaAs (111) A and GaAs (110) have a reconstruction of (2×2) and (1×1), respectively.

2.2. Algorithm

The algorithm for the growth simulation is based on the kinetic Monte Carlo method. The minimal step of the model is a simple event specified randomly in the timeline, the probability of which is determined by the activation barrier for an event, each of which is related to the type of microscopic process, the location of the particle on the reconstructed surface and its closest surroundings in the interaction radius of two interatomic distances for the surface with orientation (001), and for (111)A and (110) the interaction radius is one. The values of the activation energies for the main events were chosen on the basis of both experimental and earlier theoretical studies [2,14–18]. For intermediate values, the activation energy was modified by adding binding energies of the conglomerates obtained as a result of calculations in the density functional theory [18] to the activation barrier for the corresponding jump of a free atom. The frequency of an event is determined by the Arrhenius equation [19–23]:

$$p_i = \nu_0 \exp\left(E_i / kT\right)$$

where ν_0 is the vibration frequency of the lattice atoms estimated at 10^{13} s^{-1} [24], E_i is the activation energy of a microscopic process, and k is the Boltzmann constant.

2.3. Microscopic processes

The following microscopic processes are taken into account in the present simulation.

2.3.1. Adsorption and desorption of Ga. Ga atoms are deposited at random lattice sites from a flux determined by the growth rate and equal to:

$$J_{Ga} = 0.5\frac{\nu}{\sigma}$$

where σ is the area occupied by one lattice atom. Ga atoms are adsorbed with a single sticking coefficient, desorption is neglected [25,26].

2.3.2. Surface diffusion of Ga. Migration of a Ga adatom is implemented as successive diffusion jumps to neighboring lattice sites. The surface diffusion of Ga is anisotropic for GaAs (001), and for

(110) and (111)Ait is isotropic. Ga adatom can diffuse from any state, but if it is in a strongly bound state [22], then the probability of its jump becomes close to zero, and the adatom can be considered as incorporated into the crystal.

2.3.3. Physisorption and surface diffusion As_2. Molecules of As_2 can be adsorbed without dissociation in the form of a dimer [14].

Therefore, modelling in this paper assumes that arsenic dimers precipitate into random lattice sites. As_2 molecules are chemically adsorbed (chemisorbed) only on adatoms of Ga, whereas the interaction of As_2 with an uncovered surface occurs through weak molecular adsorption (physisorption) in several places [14]. Diffusion of As takes place in the form of dimer molecules in adjacent rows. However, this process competes with the desorption and chemisorption processes of As_2, which determines the necessity of introducing an activation barrier for the diffusion [16]. The mobility of arsenic on the surface is mainly due to the presence of the physisorped state of As_2. Any contribution of atomic As on the surface is neglected in connection with the absence of dissociative chemisorption As_2 [22].

2.3.4. Chemisorption of As_2. Chemosorption of As_2 on one or more gallium atoms is possible. The activation energy of this process also depends on the location and environment of the arsenic dimer, depending on the binding energy of the complex of atoms [14].

2.3.5. Desorption of As_2. In the model, it is accepted that desorption of arsenic is possible only from the physisorbed state [17,23].

3. Results and discussion

The morphology of the surface at various technological parameters is shown in Figure1. The essential difference between the morphology on various surfaces is obvious.This is primarily due to the difference in the surface structure. While the surface with orientations (110) and (111)A have isotropic reconstructions(1×1) and (2×2), respectively, the surface (001) is (2×4)-reconstructed. This is appeared in the presence of alternating arsenic dimericrows and missing trenches. As a result, the migration of adatoms is limited in the direction transverse to the dimer rows which leads to anisotropy in the shape of islands [2]. According to the results of experimental studies [27,28], the diffusion on the (111)A surface is isotropic, while on the (001) surface the diffusion intensity essentially depends on its direction, which is confirmed by the results of the simulation.

Also in Figure 1, there is a significant difference in the geometric characteristics of islands on the (001) surface and on the (110) and (111)A surfaces. In addition to the fact that the dimer rows limits the surface diffusion of adatoms, the density of packing of the crystal atoms differs on the surfaces under consideration. In the case of (001), the interatomic distance is larger, as a result of which the energy barrier for diffusion increases. In connection with this, nucleation processes prevail over the growth processes of two-dimensional islands and the island density increases with a corresponding decrease in their average size [29].

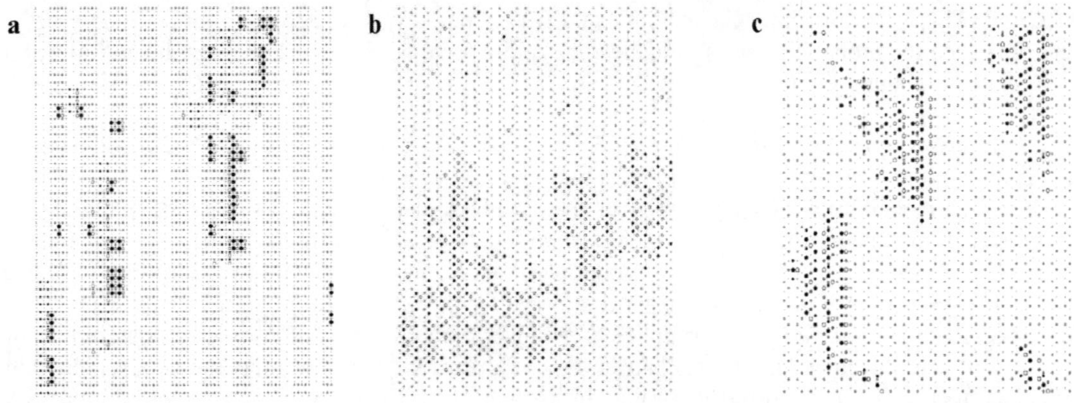

Figure 1. Morphology of the GaAs surfaces in a simulation area of 160 Å × 200 Å after deposition of 0.1 ML GaAs (○ – Ga atoms, ● – As atoms): a) GaAs(001): $T = 580°C$, $R_{As/Ga} = 6$, $v = 0.1$ ML/s, b) GaAs(111): $T = 500°C$, $R_{As/Ga} = 6$, $v = 0.1$ ML/s, c) GaAs(110): $T = 500°C$, $R_{As/Ga} = 6$, $v = 0.1$ ML/s.

The surfaces (110) and (111) A demonstrate a similar morphology and close in size and surface density of islands. However, the islands on the (110) surface have sharper facets, while on the (111) they are chaotic in shape, as evidenced by the results of STM studies of two-dimensional islands[15]. This phenomenon can be explained by the features of the crystal structure of the surface.

Using the developed model, the diffusion characteristics of Ga adatoms were also calculated as a function of the kinetic parameters of the MBE method (Figures 2 and 3). It is shown that on the (001) surface the diffusion length is really much smaller than on the other surfaces under consideration (Figures 2c and 3c) which is confirmed in [27,30].

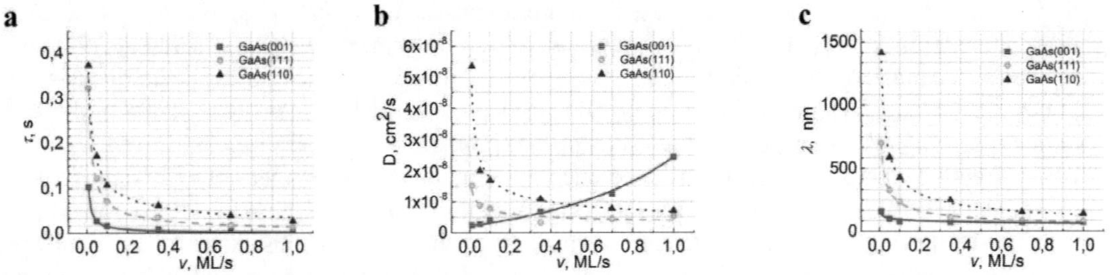

Figure 2. Growth rate dependences of the a) lifetime, b) diffusion coefficient, c) diffusion length of Ga adatoms on the GaAs(001), GaAs(110) and GaAs(111) surfaces at $R_{As/Ga} = 6$.

The simulation also shows that the diffusion of Ga adatoms exhibits a different behavior when the V/III flux ratio changes. Both the growth rate and the V/III flux ratio have a significant effect on the diffusion parameters of adatoms, but adatoms on the (110) and (111)A surfaces are most influenced, in good agreement with the works [2,15]. Although the diffusion length on the GaAs(001) is less dependent on the technological parameters, both the growth rate and V/III flux ratio have an influence on the diffusion characteristics of Ga adatoms which was shown in previous works [2,31]. It can be seen from the results that the diffusion length and lifetime (Figures 2a and 3a) of Ga adatoms on the (110) surface are significantly larger than those of (001) and (111)A in connection with the lower binding energy of adatoms with the surface. The diffusion characteristics on the (110) and (111)A surfaces are similar, but diffusion is more intense at the (110) surface. Despite the weak dependence of the diffusion length and lifetime of Ga adatoms on the (001) surface, the behaviour of these characteristics with changing technological parameters are the same.

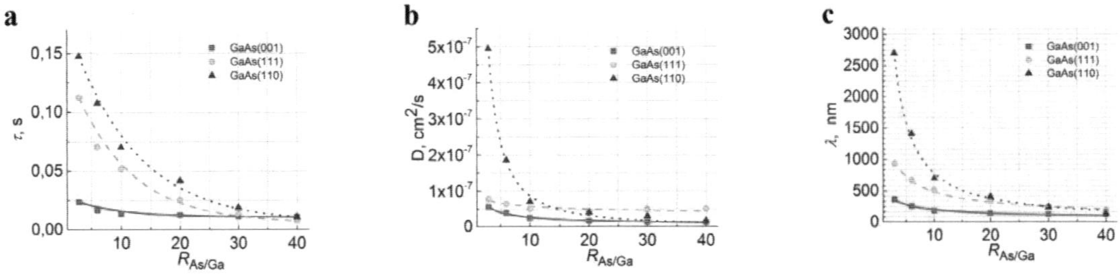

Figure 3. V/III flux ratio dependences of the a) lifetime, b) diffusion coefficient, c) diffusion length of Ga adatoms on the GaAs(001), GaAs(110) and GaAs(111) surfaces at $v = 0.1$ ML/s.

With an increase in the growth rate from 0.01 to 1 ML/s, the diffusion length and lifetime of Ga adatoms decrease for all surfaces, and it is most significant when changing at low growth rates. A similar dependence is observed when the V/III flux ratio is increased from 3 to 40, but the effect of this quantity is greater than the growth rate. This regularity is also observed in [11,30].

An important feature of the diffusion on the (001) surface is the reverse dependence of the diffusion coefficient on the V/III flux ratio compared to other surfaces (Figure 2b). With an increase in the flux ratio, the diffusion coefficient on the (001) surface increases, while the diffusion length and lifetime decrease. This indicates that a change in the lifetime of adatoms before incorporation does not lead to a sufficiently significant change in the diffusion length. A similar result was obtained in [20] for a temperature dependence in which the lifetime of Ga adatoms before incorporation increased with temperature, while the diffusion coefficient exhibited the opposite dependence.

4. Conclusion

In summary, the influence of the crystallographic orientation of GaAs on the MBE growth kinetics was investigated using the kinetic Monte Carlo method. The island morphology was shown to be strongly dependent on the orientation due to the difference between the surface reconstructions. The diffusion of Ga adatoms on the GaAs(001) surface was observed to be anisotropic and more suppressed in comparison with the (110) and (111)A surfaces. It was found that the diffusion length is the largest on the GaAs(110) surface, and it is the most strongly influenced by the growth rate and As_2/Ga flux ratio. The surface lifetime and diffusion length are weakly dependent on these parameters but decrease with increasing growth rate and V/III flux ratio as well.

Acknowledgements

The reported study was funded by Grant of the President of the Russian Federation, project No. MK-2629.2017.8. The results were obtained using the equipment of the Center for Collective Use and Research and Educational Center "Nanotechnologies" of Southern Federal University.

References

[1] Bhattacharya A and Bansal B 2015 *Handbook of Crystal Growth* vol 26, ed T Kuech (Amsterdam: Elsevier) pp 1057–99
[2] LaBella V P, Bullock D W, Ding Z, Emery C, Harter W G and Thibado P M 2000 *J. Vac. Sci. Technol.* A. **18** 1526–31
[3] Ledentsov N N *et al.* 1996 *Sol. St. Electron.* **40** 785–98
[4] Riel B J, Hinzer K, Moisa S, Fraser J, Finnie P, Piercy P, Fafard S, Wasilewski Z R 2002 *J. Cryst. Growth* **236** 145–54
[5] Ageev O A, Solodovnik M S and Balakirev S V 2015 *Advanced Materials, Mechanical and Structural Engineering* (South Korea: CRC Press) pp 57–60
[6] Ageev O A, Solodovnik M S, Balakirev S V and Mikhaylin I A 2016 *Tech. Phys.* **61** 971–7
[7] Bietti S, Esposito L, Fedorov A, Ballabio A, Martinelli A and Stefano S 2015 *Nanoscale Res.*

Lett. **10(247)** 1–7.

[8] Ha N, Mano T, Kuroda T, Mitsuishi K, Ohtake A, Castellano A, Sanguinetti S, Noda T, Sakuma Y and Sakoda K 2015 *Jpn. J. Appl. Phys.* **54(4S)** 04DH07

[9] Fujita K, Ohnishi H, Hirai M, Shimada K and Watanabe T 1996 *Solid-State Electron.* **40** 633–36

[10] Kishimoto D, Nishinaga T, Naritsuka S, Noda T, Nakamura Y and Sakaki H 2000 *J Cryst. Growth* **212(3)** 373–78

[11] Shen X-Q, Nishinaga T 1993 *Jpn. J. Appl. Phys.* **32(8B)**

[12] Hata M, Isu T, Watanabe A and Katayama Y 1990 *Appl. Phys. Lett.* **56** 2542–4

[13] Nishinaga T. 2004 *Prog. Cryst. Growth Charact. Mater.* **48** 104–22

[14] Morgan C G, Kratzer P and Scheffler M. 1999 *Phys. Rev. Lett.* **82** 4886–9

[15] Avery A R, Dobbs H T, Holmes D M, Joyce B A and Vvedensky D D 1997 *Phys. Rev. Lett.* **79** 3938–41

[16] Tok E S, Neave J H, Allegretti F E, Zhang J, Jones T S and Joyce B A 1997 *Surf. Sci.* **371** 277–88

[17] Garcia J C, Neri C and Massies J 1989 *J. Cryst. Growth* **98** 511–8

[18] Foxon C T and Joyce B A 1975 *Surf. Sci.* **50** 434–50

[19] Amrani A, DjafariRouhani M and Mraoufel A. 2011 *Appl. Nanosci.* **1** 59–65

[20] Kangawa Y, Ito T, Taguchi A, Shiraishi K, Irisawa T and Ohachi T 2002 *Appl. Surf. Sci.* **190** 517–20

[21] Ageev O A, Solodovnik M S, Balakirev S V and Mikhaylin I A 2016 *J. Phys.: Conf. Ser.* **681** 012036

[22] Kratzer P and Scheffler M 2002 *Phys. Rev. Lett.* **88(3)** 036102

[23] Banse B A and Creighton J R 1992 *Appl. Phys. Lett.* **60** 856

[24] Kley A, Ruggerone P and Scheffler M 1997 *Phys. Rev. Lett.* **79** 5278

[25] Foxon C T and Joyce B A 1977 *Surf. Sci.* **64** 293

[26] Tok E S, Neave J H, Zhang J, Joyce B A and Jones T S 1997 *Surf. Sci.* **374** 374

[27] Takebe T, Fujii M, Yamamoto T, Fujita K and Watanabe T 1997 *J. Appl. Phys.* **81** 7273–81

[28] Hata M, Isu T, Watanabe A and Katayama Y 1990 *J. Vac. Sci. Technol., B* **8** 692–6

[29] Ageev O A, Solodovnik M S, Balakirev S V and Eremenko M M 2016 *J. Vac. Sci. Technol., B* **34(4)** 041804

[30] Lopez M and Nomura Y 1995 *J. Cryst. Growth* **150** 68–72

[31] Ageev O A, Solodovnik M S, Balakirev S V, Mikhaylin I A and Eremenko M M 2017 *J. Cryst. Growth* **457** 46–51

IOP Conf. Series: Journal of Physics: Conf. Series **917** (2017) 032035 doi:10.1088/1742-6596/917/3/032035

MBE growth of GaAs and InAs nanowires using colloidal Ag nanoparticles

I V Ilkiv[1], R R Reznik[1], K P Kotlyar[1], A D Bouravleuv[1] and G E Cirlin[1]

[1]Laboratory of epitaxial nanotechnologies, St. Petersburg Academic University, St. Petersburg 194021, Russia

Abstract. Ag colloidal nanoparticles were used as a catalyst for molecular beam epitaxy of GaAs and InAs nanowires on the Si(111) substrates. The scanning electron microscopy measurements revealed that nanowires obtained are uniform and have small size distribution.

1. Introduction

During recent years, semiconductor nanowires (NWs) attract great attention due to their distinctive optical and electronic properties. On the practical side, semiconductor NWs are very attractive for direct integration of *III-V* materials on Si, which provide new opportunities for the design and fabrication of numerous applications, including solar cells [1], light emitting diodes [2], field-effect transistors [3]. For a large number of devices, it is essential to fabricate NWs with well-controlled dimension, orientation, structure, phase purity and density on the surface. Nanowire size and shape are important parameters that have to be considered. At these scales quantum mechanical effects play a key role; even small variations in size of NWs can influence on overall device performance. Up to now, the most common technique for nanowire synthesis is vapor-liquid-solid growth using Au metal seed particle, which acts as a catalyst to initiate the growth. However, the use of Au can result in unwanted effects, such as formation of deep-level defect states and Au migration on the semiconductor surface. This leads to deterioration of own NW optical and electrical properties [4]. To solve this problem, emerging methods based on self-catalyzed [5] or template-assisted growth of NWs [6] can be used. In addition, the use of foreign metals as a catalyst (e.g. Ag [7]) is also of great interest.

The position and size control of seed can be achieved by pattering the substrate using lithographic technique, e.g. electron beam lithography [8], nanoimprint lithography [9] and subsequent Au deposition using physical vapor deposition. However, lithographic processes are intrinsically wasteful, not suitable for large scale-production and challenging for the deposition of small-diameter catalyst particles. Other methods based on nanosphere lithography [10] and laser interference lithography [11] remain quite complex to deal with.

Ag colloidal nanoparticles can potentially overcome the above-mentioned limitations. Nowadays colloidal nanoparticles have gained widespread acceptance in different areas (markers and biosensors [12], optoelectronic applications [13], etc.) due to their properties, withal they are commercially available in a wide range of different sizes and shapes. Therefore, they are an attractive alternative to use as a catalyst of NW growth.

In this study we present MBE growth of GaAs and InAs nanowires arrays using Ag colloidal nanoparticles.

IOP Conf. Series: Journal of Physics: Conf. Series **917** (2017) 032035 doi:10.1088/1742-6596/917/3/032035

2. Experiment

The investigations were carried out using epi-ready silicon (111) oriented p-type substrates. The substrate surface was covered with a native oxide layer. We used Ag colloidal nanoparticles with 20 nm diameter, which were obtained from BBI Solutions. Deposition of Ag nanoparticles was performed by placing a drop of colloidal solution directly on the silicon surface. The typical deposition time was about 30s. Then the solution remnants were blown away and the substrates were dried with N_2 gas.

For the growth experiments Compact 21 MBE system was used. MBE system consist of the main chamber equipped with standard Ga, In effusion cells and As cracking cell, and additional high vacuum connected chamber, which allows to perform the annealing of the samples at the temperatures up to 950°C. The process of NWs growth was monitored in situ by the reflection high energy electron diffraction (RHEED). RHEED was also used to calibrate material growth rate. The deposition rates of GaAs and InAs were set at 0.7 and 0.3 ML/s, respectively.

A Zeiss Supra 24 field emission scanning electron microscope (SEM) was used to investigate the morphology and dimensions of the Ag seeds and NW arrays.

3. Results and discussion

Pregrowth Ag nanoparticles distribution on the Si (111) substrate is shown in Figure 1a. As it was expected, Ag nanoparticles are randomly distributed, but they homogeneously cover the whole silicon surface.

The substrates with deposited Ag nanoparticles were loaded into additional chamber of MBE system. An annealing step at a set temperature of 850°C was carried out for 5 min to remove surface contaminants and native oxide (see Figure 1 b). Then the temperature was decreased, the samples were transferred to the growth chamber. After reaching the growth temperature, group III metal flux and As shutters were simultaneously opened. The GaAs and InAs NWs growth temperatures were set at 580°C and 400°C, respectively. In-situ RHEED observation showed that pattern changes appeared after 10-20 seconds of the growth. According to the RHEED patterns NWs have mixed wurzite/zinc blende phases. This effect has been previously found experimentally for most III–V NWs [14]. The growth was terminated by switching off the group III metal flux while maintaining the As flux until the substrate temperature dropped to below 300°C. Total growth time for GaAs and InAs NWs was equal to 15 and 20 min, respectively.

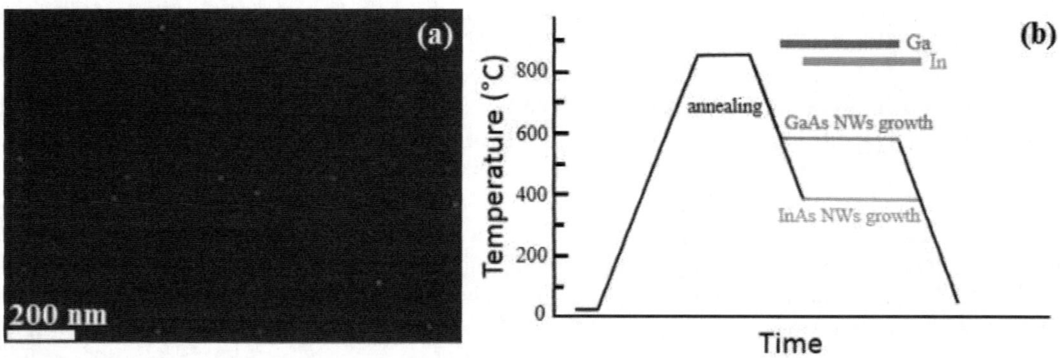

Figure 1. (a) - SEM image of the silicon surface with deposited Ag nanoparticles, (b) - schematic illustrations of the temperature sequence for annealing process and the subsequent InAs, GaAs nanowire growth.

Typical arrays of NWs formed under conditions described above are shown in Figure 2. The result obtained demonstrate than both GaAs and InAs NWs can be grown using colloidal Ag nanoparticles by MBE. InAs and GaAs nanowires exhibit prevalent vertical directionality, indicating a direct relationship to the underlying Si(111) substrate. Moreover, no droplets related to the group III element (i.e. Ga or In) rich phase could be resolved at the individual nanowire tips by SEM. As can be seen also, the temperature where InAs NW formation was observed strongly differ from that for GaAs NWs (see Figure 1 b). This is in good agreement with other reports [15, 16] and connected with different solubility of group III elements in the Ag catalyst. It is noteworthy, that temperatures ranges are much narrower; no NW formation was observed at temperature changes greater than 30°C.

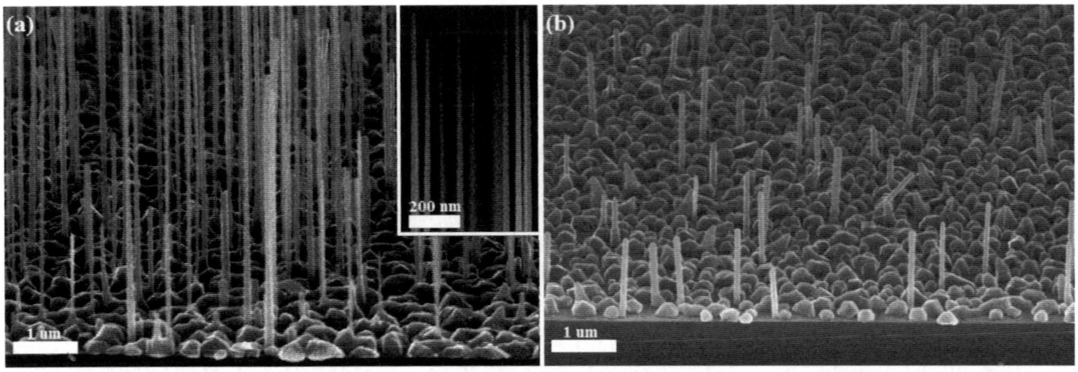

Figure 2. SEM images of (a) – GaAs, insert shows the shape changes of NWs along the growth direction; (b) – InAs NWs grown on Si(111) using Ag nanoparticles.

The cross-sectional SEM images were used to determine the mean NW height and diameter by making a statistical average over about 10 wires. As it was expected, GaAs NWs have small size distributions. They have typical 6.8 μm length. However, the diameters of these NWs are irregular, NWs smoothly tapered from 80 nm at the base to sub-30 nm at the top (see insert in Figure 2 a). It could be caused by the limitations of Ga adatom diffusion length. Besides, there are significant amount of low-tier NWs with length of about 3 um and 40 nm diameter. The size distribution of the InAs NWs also have no discernible differences, they have 1.2 μm length and 100 nm diameter, which is constant along the growth direction. It can be seen that average GaAs NWs growth rate is about 7 times higher than InAs growth rate (7.5 nm/s versus 1 nm/s). This could be explained by the different temperatures of the substrates during the growth experiments (Figure 1 b).

According to the top-view SEM images (see Figure 3), NWs in both arrays have hexagon-shaped geometries. GaAs NWs grown on Si using Ag nanoparticles have higher surface density that InAs NWs. It is likely that low-tier GaAs NWs are obtained by the self-catalyzed method, i.e. Ga droplets were formed and acted as additional catalyst for the NW growth [5]. But this question requires an additional clarification.

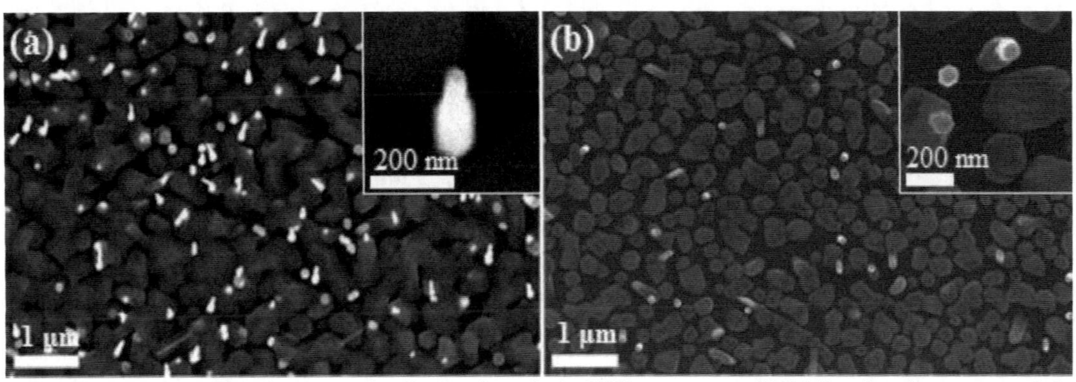

Figure 3. Top-view SEM images of – GaAs, (b) – InAs NWs grown on Si(111) using Ag nanoparticles. Inserts show the shape of NWs.

4. Conclusion

In summary, the GaAs and InAs NWs have been obtained by molecular beam epitaxy using Ag colloidal nanoparticles as a growth catalyst on the Si(111).The majority of the NWs have preferential growth direction along <111>. The NWs obtained are uniform and have small size distribution. The ability to prepare uniform nanowires on Si substrates may provide new opportunities for studying their size-dependent physical properties and fabrication promising nano- and optoelectronic devices.

Acknowledges

This work was carried out with the support of the Ministry of education and science of Russian Federation (state task, project No. 16.2483.2017/PCh). The nanowire samples were grown under the support of Russian Science Foundation (Project No 14-12-00393).

References

[1] Åberg I, Vescovi G, Asoli D, Naseem U, Gilboy J P, Sundvall C and Samuelson L 2016 *IEEE J. Photovolt.* **1** 185
[2] Mi Z and Chang Y L 2009 *J. Nanophotonics* **1** 031602
[3] Bryllert T, Wernersson L E, Froberg L E and Samuelson L 2006 *IEEE Electron Device Lett.* **5** 323-325
[4] Breuer S, Pfüller C, Flissikowski T, Brandt O, Grahn H T, Geelhaar L and Riechert H 2011 *Nano Lett.* **3** 1276
[5] Samsonenko Y, Cirlin G, Khrebtov A, Bouravleuv A, Polyakov N, Ulin V and Werner P 2011 *Semiconductors* **4** 431-435
[6] Soo M T, Zheng K, Gao Q, Tan H H, Jagadish C and Zou J 2016 *Nano Lett.* **7** 4189
[7] Dick K A and Caroff P 2014 *Nanoscale* **6** 3006
[8] Law M, Greene L E, Johnson J C, Saykally R and Yang P 2005 *Nat. Mater.* **6** 455-459
[9] Mårtensson T, Carlberg P, Borgström M, Montelius L, Seifert W and Samuelson L 2004 *Nano Lett.* **4** 699
[10] Fuhrmann B, Leipner H S, Höche H R, Schubert L, Werner P and Gösele U 2005 *Nano Lett.* **12** 2524-2527
[11] Kauppinen C, Haggren T, Kravchenko A, Jiang H, Huhtio T, Kauppinen E and Sopanen M 2016 *Nanotechnology* **13** 135601
[12] Howes P D, Chandrawati R and Stevens M M 2014 *Science* **6205** 1247390
[13] Hsieh H H, Sung Y M, Hsu F C, Hsiao K J, Lee Y J and Chen Y F 2015 *RSC Adv.* **2** 1549
[14] Caroff P, Dick K A, Johansson J, Messing M E, Deppert K and Samuelson L 2009 *Nat. Nanotechnol.* **4** 50-55
[15] Lindberg C., Whiticar A, Dick K A, Sköld N, Nygård J and Bolinsson J 2016 *Nano Lett.* **4** 2181
[16] Pan D, Fu M, Yu X, Wang X, Zhu L, Nie S and Zhao J 2014 *Nano Lett.* **3**, 1214-1220

IOP Conf. Series: Journal of Physics: Conf. Series **917** (2017) 032036 doi:10.1088/1742-6596/917/3/032036

Modeling of semiconductor nanowire selective-area MOCVD growth

A A Koriakin[1,2,3], **M Reiter**[1], **Zh V Sokolova**[2], **N V Sibirev**[1,2]
[1] St. Petersburg Academic University, St. Petersburg 194021, Russia
[2] ITMO University, St. Petersburg 197101, Russia
[3] Saint-Petersburg Branch of the Joint Supercomputer Centre RAS, Saint Petersburg 194021, Russia

Abstract. A numerical approach to the determination of gas kinetics in the case of non-planar nanostructure growth via the selective-area metal-organic chemical vapor deposition is developed. The direct simulation Monte-Carlo method is utilized to model the rarefied gas flow of precursor particles nearby the substrate. The computation is performed for the GaAs nanowire growth via the selective-area metal-organic chemical vapor deposition. The model allows the quantitative description of the decrease of nanowire length with the increase of distance between nanowires (the so-called synergetic effect). The optimal pitch of the mask that corresponds to the maximal nanowire length is found for typical growth conditions. In particular, our calculation shows that the optimal pitch increases with the increase of the nanowire diameter.

1. Introduction

During the last decade, much interest has been focused on the selective-area metal–organic chemical vapor deposition (SA-MOCVD) growth of semiconductor III-V nanowires (NW) [1-3]. This method facilitates the formation of highly uniform NW array and has high productivity compared to molecular-beam epitaxy technics. As a result of the intensive studies, the growth conditions of III-V NW have been found. Also, the following facts regarding the impact of the mask size and pitch on the NW morphology have been estimated: (i) at the mask edges, the NW growth rate is usually several times higher than that in the middle part of the mask [3, 4]; (ii) the NW growth rate decreases with the increase of the mask window size [1, 5]; (iii) the increase of the mask pitch p leads to the increase [6] or decrease [1] of the NW length. The first dependence is explained by the additional gas-phase and surface fluxes of species from the area without the mask windows to the NW array [3, 4]. The second dependence is explained within the models [4, 5] from which it follows that the NW growth rate is inversely proportional to the NW diameter or squared diameter. The models also claim that the NW length is proportional to the area per one NW, i.e. squared pitch, at any value of p. Consequently, the third dependence can not be completely described by these models. The dependence of the NW length on the pitch can be explained quantitatively within the model [7] constructed for the catalytic MOCVD growth of NW. The key assumption of the model is that two material fluxes should be considered: the initial precursor flux coming to the substrate and reflected flux produced by the precursor decomposition on the surface of catalyst droplet seated at the NW top facet. At large values of p, the decrease of the NW length is associated with the decrease of the reflected flux from the neighboring NW (the so-called synergetic effect).

In this paper, a quantitative model of SA-MOCVD NW growth is developed. The model enables the description of the non-monotonic dependence of the NW length on pitch. At large distances from the substrate, the precursor flux in the gas-phase is found by solving the diffusion equation assuming that the macroscopic velocity of gas flow in the boundary layer equals to zero (the boundary layer model [8]). Nearby the substrate, the Knudsen number for the precursor molecules can be large $Kn \sim 1$. In this case, the rarefied gas fluxes are determined by the direct simulation Monte-Carlo method [9]. The computation is performed for the GaAs NW growth on GaAs(111)B substrate coated by SiO_2 mask [1]. Within the model, the optimal pitch of NW growth is found to minimize the NW growth time.

2. Model

Consider a semiconductor substrate coated by a dielectric mask as shown in figure 1. The hexagonal-

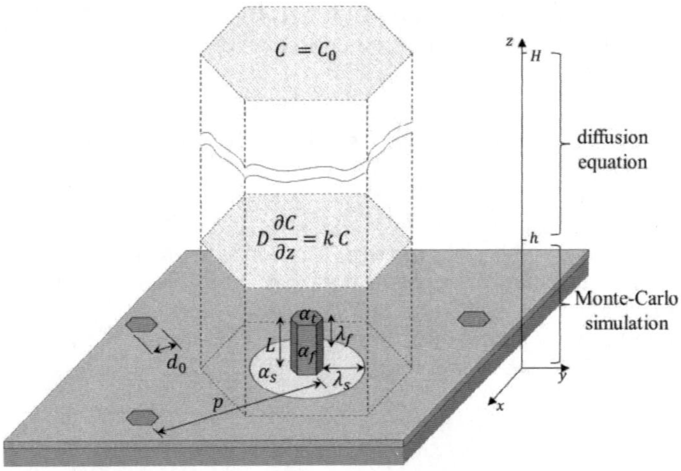

Figure 1. Schematic illustration of the substrate coated by the mask. The prism of the height H is the region of modeling. The NW is shown as the prism of the height L. The neighboring NWs are not shown.

shaped mask windows of the size d_0 form the rectangular lattice of the pitch p. For simplicity, we consider a region of the substrate located far from the mask edges to neglect their influence. For all NWs within this region, the growth conditions are identical. Then, we divide the mask area into equal hexagons and construct regular prisms with the height H. It is sufficient to consider one of the equivalent prisms due to the translational symmetry of the system. Orient the z axis perpendicular to the substrate in the direction from the lower base of the prism to the upper base. The height H equals to the thickness of the boundary layer in which the material transport over the distances much larger than the diffusion length λ in the gas-phase can be described by the diffusion equation. The ratio of the diffusion coefficient of the precursor to the boundary layer height characterizes the material transport to the substrate. Following the work [10], the value of H is estimated by using the equation $H = (1/0.89)(D s_z s_y / V)^{1/3}$, where D is the diffusion coefficient of precursor, s_z, s_y are the height and length of the reactor chamber, respectively, V is the velocity of the gas flow.

The group V precursor pressure is much higher than the group III precursor pressure [1], therefore, the crystal growth is limited by the group III flux. Following the work [7], we consider the initial flux of group III precursor coming to the substrate and reflected flux of group III produced by collisions of precursor molecules with the NW facets or mask openings. In the experiment [1], the GaAs NWs were synthesized at the growth temperature of $750\,^{o}C$ using trimethylgallium (TMGa) as a group III precursor. It is known that at such a temperature and the pressure of the carrier gas (H_2 in our case) of 0.1 atm, TMGa molecules partially decompose in the gas-phase [8]. In the general case, the decomposition of TMGa molecules proceeds through a series of reactions that is often unknown [8]. For the sake of concreteness, the flux of monomethylgallium (MMGa) is considered as the initial flux in the further discussion [11]. As a result of single or multiple collisions with the GaAs crystals, MMGa molecules decompose producing gallium atoms that recoil back into gas phase. We neglect the possibility of the Ga atom to incorporate into a growing crystal without re-evaporation. Also, we suppose that MMGa decomposition on the mask is negligible [11]. To consider the surface diffusion of Ga atoms, the diffusion length on NW side facets and mask are introduced, λ_{f} and λ_{s}, respectively [12]. Their values determine the collection areas of the growth material (figure 1). In the model, the surface diffusion lengths of MMGa molecules are put to be zero. Estimates made for the standard GaAs NW growth conditions [1] show that the gas-phase diffusion lengths λ for the MMGa molecules and Ga atoms equal approximately 0.4 μm and 0.9 μm, respectively. The values of λ are of the same order of magnitude as the sizes of system (L, d_0, p, where L is the NW length) or larger. Therefore, the material transport is governed by diffusion only in the bounded region $h<z<H$, where $h>>\lambda$ (figure 1). In the region $0<z<h$, the direct simulation Monte-Carlo method is used for the determination of the rarefied gas fluxes.

For the MMGa flux, the stationary diffusion equation

$$D\Delta C=0, \tag{1}$$

where D and C are the diffusion coefficient and concentration of MMGa molecules, respectively, is solved with the boundary conditions

$$C(H)=C_0 \text{ and } j(h)=-k\,C(h), \tag{2}$$

where j is the MMGa flux density, k is the effective reaction rate constant of the surface $z=h$. The solution of equation (1) is obtained in the form of $C=A\,z+B$, where A, B are constant, neglecting the concentration dependence on x and y. Then, the solution for the flux density j is

$$j(h)=-\frac{k\,C_0}{1+\dfrac{k\,(H-h)}{D}}. \tag{3}$$

Following the work [11], the effective reaction rate constant k can be expressed in terms of the effective sticking coefficient α

$$k=\frac{\alpha/4}{1-\alpha/2}v. \tag{4}$$

Here $v=(8k_{B}T/\pi m)^{1/2}$ is the mean velocity of the MMGa molecules, k_B is the Boltzmann constant, T is the growth temperature, m is the MMGa molecule mass. The coefficient α equals to N_{d}/N, where N_d is the number of MMGa molecules decomposed on the GaAs crystals per second, N is the total number of MMGa molecules crossed the surface $z=h$ in the direction to the substrate per second. Introduce the sticking coefficient α' that equals N_{s}/N_{d}, where N_s is the number of Ga atoms stuck to the GaAs crystals per second. The total Ga flux feeding the NW growth can be found by the formula $J=|\,j(h)\,|\,\alpha' S$ (in s^{-1}), where $S=3^{3/2}p^{2}/6$ is the area of substrate per one NW, or

$$J=\frac{3^{3/2}\alpha\,\alpha' v\,C_0 p^2}{24(1-\alpha/2)\left(1+\dfrac{\alpha v(H-h)}{4(1-\alpha/2)D}\right)}. \tag{5}$$

Note, we neglect the term to the total flux J from the Ga atoms reflected from the region $h<z<H$. Strictly speaking, the diffusion equation for the Ga atoms should be solved with the boundary

conditions similar to (2) in the region $h' < z < H'$, where h' and H' are the boundary layer heights for Ga atoms. Expand (5) in powers of α (our calculations show that α can be treated as a small parameter, $\alpha < 0.5$) and, for a simple analysis, retain only the first term

$$J \approx (3^{3/2} / 24)\, v\, C_0 \alpha_{\text{tot}} p^2 . \quad (6)$$

Here $\alpha_{\text{tot}} = \alpha \alpha'$ is the total sticking coefficient, that equals N_s / N. In the general case, this coefficient depends on the geometry of the system, sticking coefficients and diffusion lengths of MMGa and Ga on the GaAs crystals and mask surfaces.

To find the NW length at the growth time *t*, we integrate the material balance equation

$$\frac{dL}{dt} = \frac{\Omega}{S_t} J \quad (7)$$

with the initial condition $L(t_0) = 0$, where dL / dt is the NW growth rate, Ω is the volume of the GaAs pair in solid, $S_t = 3^{3/2} / 6\, d^2$ is the area of the top NW facet, $d \approx d_0$ [1] is the NW diameter, t_0 is the time between the switching on the material sources and complete filling of the mask openings (the so-called incubation time). By substitution (6) into (7), the following formula for the NW growth rate is obtained

$$\frac{dL}{dt} \approx \Omega v\, C_0 \alpha_{\text{tot}} p^2 / 4d^2 . \quad (8)$$

As the total flux, the incubation time is a function of the geometry of the system and material constants. If the reverse gallium flux to the substrate is neglected, the incubation time equals to infinity. This work is focused on the NW growth stage, therefore, a model scheme is used to account for the incubation time. For this purpose, the following initial condition is used $L(0) = L_0$, where L_0 is the effective NW length such that the NW growth rate equals to the mask filling rate during the incubation time. This length can be estimated by means of equating the area of the mask opening $3^{3/2} / 6\, d_0^{\,2}$ to the area of the NW side facets $\pi d L_0$, $L_0 \approx d_0 / 4$.

3. Results and discussion

Consider the case when the sticking coefficients of MMGa on the NW side facets, mask openings and mask surface [11] are negligibly small. Introduce the sticking coefficients of Ga on the top NW facets, side NW facets and mask, α_t, α_f and α_s, respectively (figure 1). Also, we take into consideration the Ga diffusion length on the NW side facets and mask, λ_f and λ_s, respectively. The total sticking coefficient α_{tot} and, consequently, the NW growth rate are determined by the Ga atoms sticking to the part of the NW side facets of length λ_f after single or multiple collisions with the NW side facets, mask surface or facets of the prism of height *H*. Note, that the reflection of species from the NW facets and mask surface is assumed to be diffuse, but the reflection from the facets of the system is specular. If $L \le \lambda_f$ then the total sticking coefficient α_{tot} has a term of the diffusion flux from the substrate region of the radius $\lambda_s + d/2$. For the determination of the dependence of α_{tot} on the parameters of the system in the first approximation, we neglect the collisions between species in the region $0 < z < h$ considering the free molecular flows. Thus, the coefficient α_{tot} is a function of the following parameters of the model: p, d, L, α_f, α_s, λ_f, λ_s. The parameters p and d are determined by the mask shape. The NW length L is found during modelling. The material constant α_f, α_s, λ_f and λ_s strongly depend on the surface reconstructions that are often unknown. For the particular experiment, it is easy to estimate only the orders of these constants. Therefore, the parameters α_f, α_s, λ_f and λ_s should be treated as fitting parameters of the model.

The results of the calculation of α_{tot} by the direct simulation Monte-Carlo method are shown in figure 2. The dependences of α_{tot} on the NW length have two linear regions: with a high slope at

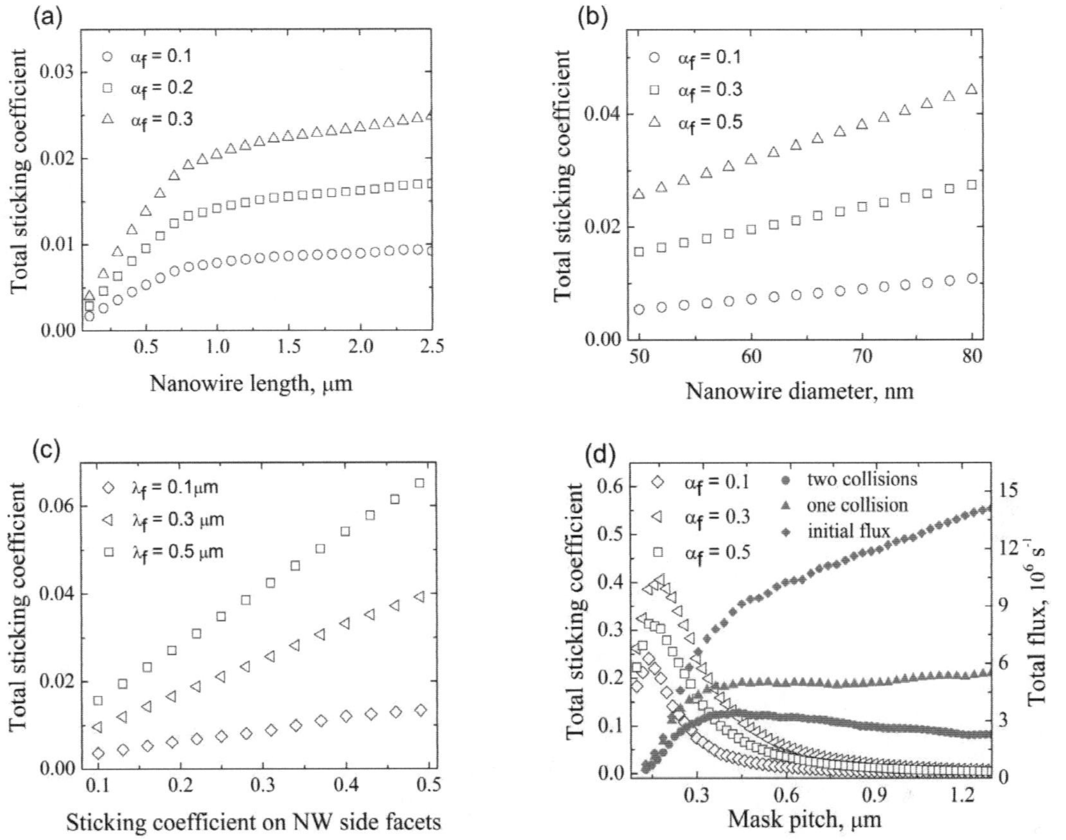

Figure 2. The dependence of the total sticking coefficient α_{tot} on the NW length L at d = 80 nm, λ_f = 0.3 μm, p = 0.6 μm (a), NW diameter d at L = 3 μm, λ_f = 0.3 μm, p = 0.6 μm (b), sticking coefficient α_f at d = 80 nm, L = 3 μm, p = 0.6 μm (c) and pitch p at L = 3 μm, d = 80 nm, λ_f = 0.5 μm (d) (shown by the white dots). The dependences of the total flux on the pitch at L = 3 μm, d = 80 nm, α_f = 0.3, λ_f = 0.5 μm, $C_0 = 0.646\ \mu m^{-3}$ are shown by the gray dots (d). The diffusion length on the mask λ_s equals to 0 in the computation.

$L < 2.5\lambda_f$ and low slope at $L > 2.5\lambda_f$. The dependences of α_{tot} on the NW radius and sticking coefficient α_f are close to linear. The dependences of α_{tot} on the pitch have a maximum. At small values of p, α_{tot} increases as $1 - d^2/p^2$, because the MMGa molecules adsorbed on the top NW facets recoil and escape the system. If a MMGa molecule enter the space between the NWs it sticks to the NW side facets with a high probability after several collisions. At large values of p, the decrease of α_{tot} is caused by the decrease of the probability that the MMGa molecules adsorb on the NW side facets and, consequently, decompose and, at the same time, by the probability that the Ga atoms escape the system increases. Here, it is necessary to emphasize the importance of the introduction of the reflected flux for the explanation of the experimental data. When we only consider the initial flux, the MMGa molecules decompose on the NW side facets as a result of single collisions and produce the Ga atoms that instantly incorporate with the probability α_f. In such a case, the coefficient α_{tot} decreases slower than $1/p^2$. Then, the total flux $J \sim \alpha_{tot} p^2$ is the increasing function at the large p (figure 2, d). Let us assume that, as a result of single collisions, the MMGa molecules decompose

IOP Conf. Series: Journal of Physics: Conf. Series **917** (2017) 032036 doi:10.1088/1742-6596/917/3/032036

producing the Ga atoms, but the direct incorporation is negligible. Then, the coefficient α_{tot} is approximately a constant at large p (figure 2, d). Note, if $\alpha_f < 0.1$ the dependence $J(p)$ is a weak decreasing function in the range of p about 0.5- 0.9 μm and this function becomes a constant in the range of p larger than 1 μm. In this case, the calculations are not in quantitatively agreement with the experiment [1]. To obtain a satisfactory agreement with the data [1], we should assume that, after two collisions with the NW side facets, the MMGa molecules decompose producing the Ga atoms. Also, we neglect the direct incorporation of Ga atoms. Then, the coefficient α_{tot} decreases faster than $1/p^2$ and the total flux is the decreasing function at large p (figure 2, d). This result can be interpreted as follows. The probability of decomposition of the MMGa molecule is small during the first collision. When a MMGa molecule collides with the GaAs crystals, the molecule becomes activated and, after the second collision, decomposes with a high probability.

It should be also noted that the NW growth rate dL/dt decreases with an increase of *d*, because α_{tot} increases linearly with an increase of d and $dL/dt \sim \alpha_{tot}/d^2$. We compare the results of modeling with the experimental data. The dependence of the NW length on the mask pitch is shown in figure 3. The parameters used in the computation are: $\alpha_f = 0.18$ (at the AsH_3 pressure 1×10^{-3} atm),

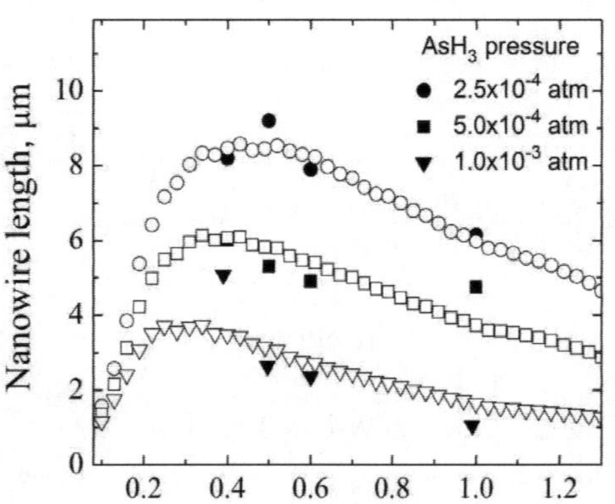

Figure 3. The dependence of the NW length on the mask pitch. The black dots are the experimental data [1], the white dots are the results of the calculation.

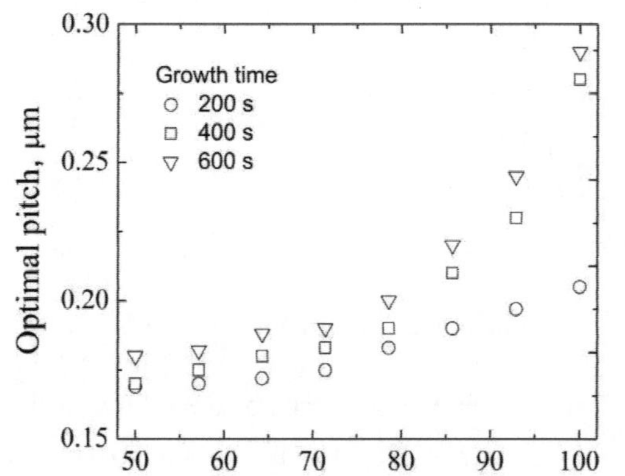

Figure 4. The dependence of the optimal pitch on the NW diameter at $\alpha_f = 0.2, \lambda_f = 0.3\ \mu m$, $C_0 = 0.646\ \mu m^{-3}$.

$\alpha_f = 0.32$ (at the AsH_3 pressure 5×10^{-4} atm), $\alpha_f = 0.45$ (at the AsH_3 pressure 2.5×10^{-4} atm), $\lambda_f = 0.3\ \mu m$, $\lambda_s = 0$, $C_0 = 0.646\ \mu m^{-3}$, $t = 1200$ s, $T = 750^oC$, $d = 80$ nm. At the large values of the mask pitch, the model predicts the decrease of the NW length that is in agreement with the experiment [1]. Thus, the synergetic effect can be described by the consideration of the gas fluxes with neglecting the surface diffusion on the substrate, i.e. at $\lambda_s = 0$. In the model, the decrease of the NW length with the increase of the AsH_3 pressure (figure 3) is described by the decrease of the sticking coefficient on the NW side facets α_f. In figure 3, it is seen that there is a maximum of the NW length at $p_0 \approx$ 0.4-0.5 μm. This maximum was not observed in the work [1], because the measurements were not performed at the small values of the mask pitch. By growing NWs at $p_0 \approx$ 0.4-0.5 μm, it is possible to minimize the growth time. Therefore, the value of pitch $p = p_0$ can be called the optimal pitch for the particular growth condition. The figure 4 shows the optimal pitch dependence on the NW diameter.

The optimal pitch increases with the increase of d. The best-fit analysis shows that this dependence is close to a parabolic dependence. Note that the dependence of p_0 on the NW length (not shown) is increasing at $L < 0.3\text{-}0.5\ \mu\text{m}$ and very weak at $L > 0.5\ \mu\text{m}$ for $\lambda_\text{f} = 0.3\ \mu\text{m}$.

In summary, the model of the SA-MOCVD growth of III-V NWs was developed. The model allows the quantitative description of the decreasing dependence of the NW length on the mask pitch. The role of the reflected flux of precursor in elucidating of this dependence was shown. The results of simulation are in agreement with the experimental data on the GaAs NW growth. The dependence of the optimal pitch that corresponds to the maximal NW length was found as a function of the diameter and length of NW.

Acknowledgments

The authors gratefully acknowledge financial support received from the Ministry of Education and Science of the Russian Federation under grant 14.587.21.0040 (project ID RFMEFI58717X0040).

References

[1] Noborisaka J, Motohisa J, Fukui T 2005 *Appl. Phys. Lett.* **86** 213102
[2] Bouravleuv A D, Sibirev N V, Statkute G, Cirlin G E, Lipsanen H, Dubrovskii V G 2010 *J. Cryst. Growth* **312** 1676
[3] Chu H J, Yeh T W, Stewart L, Dapkus P D 2010 *Phys. Status Solidi C* **7** 2494
[4] Xu L, Huang Q 2014 *IEEE Trans. Nanotechnol.* **13** 1093
[5] Ikejiri K, Sato T, Yoshida H, Hiruma K, Motohisa J, Hara S, Fukui T 2008 *Nanotechnology* **19** 265604
[6] Wang X *et al.* 2014 *J. Appl. Phys.* **115** 163104
[7] Borgstrom M T, Immink G, Ketelaars B, Algra R, Bakkers E 2007 *Nature Nanotechnol.* **2** 541
[8] Stringfellow G B 1999 *Organometallic vapor-phase epitaxy: theory and practice* (London: Academic Press)
[9] Coronell D G, Jensen K F 1992 *J. Electrochem. Soc.* **139** 2264-2273
[10] Ven J, Rutten G, Raaijmakers M, Giling L 1986 *J. Cryst. Growth* **76** 352
[11] Oh H, Sugiyama M, Nakano Y, Shimogaki Y 2003 *Jpn. J. Appl. Phys.* **42** 6284
[12] Cirlin G E, Dubrovskii V G, Sibirev N V *et al.* 2005 *Semiconductors* **39** 557

IOP Conf. Series: Journal of Physics: Conf. Series **917** (2017) 032037 doi:10.1088/1742-6596/917/3/032037

Droplet epitaxy of GaAs nanostructures on the As-stabilized GaAs(001) surface

M S Solodovnik, S V Balakirev, M M Eremenko, I A Mikhaylin, V I Avilov, S A Lisitsyn, O A Ageev

Research and Educational Center "Nanotechnologies", Southern Federal University, 2 Shevchenko Street, Taganrog 347922, Russia

Abstract. The paper presents the results of experimental studies of the formation of self-assembled Ga and GaAs nanostructures on the As-stabilized GaAs(001) surface by droplet epitaxy. The results of investigations of the influence of the effective deposition thickness and surface temperature on the characteristics of arrays of formed structures are presented. The possibility of forming superdense arrays of GaAs nanostructures with a density of up to $4 \cdot 10^{11}$ cm^{-2} is shown. Using a combination of *in-situ* removal of GaAs oxide and deposition with interruption, regular arrays of GaAs quantum dots of 78 ± 6 nm in size and a density of $4.6 \cdot 10^{8}$ cm^{-2} are obtained at temperatures above 500°C on the GaAs(001) surface structured by local anodic oxidation technique.

1. Introduction

The analysis of instrumental applications of arrays of self-organizing nanostructures (quantum dots, nanowires, nanoclusters, etc.) shows the need for precise control of the resulting structures characteristics: size, density, mutual arrangement, etc [1–8].The existing methods for the formation of quantum dots using the technology of molecular beam epitaxy are based on the Stransky-Krastanov mechanism – the elastic relaxation of mechanical stresses in lattice mismatched systems. In this technique the growth is achieved by the simultaneous deposition of components of III and V groups. The disadvantages of this approach are the need for the mismatch of crystal lattices and the interdependence between the density and the size of quantum dots formed by such a mechanism [9, 10], which sharply narrows the range of combinations of materials and limits the scope of their application. The weakening of the negative correlation can be achieved through the use of various technological methods (structured surfaces, submonolayer epitaxy, etc.), but obtaining arrays of quantum dots with ultra-low density is still very difficult [11–13].

At the same time, droplet epitaxy, based on separate and sequential deposition of the components of groups III and V, allows not only independently to control the density and size of quantum dots, but also to form quantum dots in any A^3B^5 systems [14–16]. Moreover, the method is promising for the integration of low-dimensional A^3B^5 systems with silicon technology, as well as for creating hybrid metal/semiconductor systems of great interest for optoelectronics and THz devices due to the appearance of plasmon effects. The use of multi-stage droplet epitaxy techniques makes it possible to effectively minimize the influence of negative factors associated with pre-surface treatment of the surface in the case of using modified substrates, which also advantageously distinguishes this method from other A^3B^5 quantum dot synthesis techniques.

The purpose of this work is to experimentally study the formation of various types of GaAs nanostructures on an As-stabilized GaAs(001) surface by droplet epitaxy.

2. Description of the model

Experimental studies were carried out on a molecular-beam epitaxy system SemiTEq STE 35, equipped with solid-state sources of group III components and a valve arsenic source. A GaAs(001) epi-ready wafers were used as growth substrates. *In-situ* growth control was carried out by observation of RHEED pattern. The temperature T in the 450-600°C range was monitored by a pyrometer, and below 450°C it was controlled by a thermocouple, previously calibrated according to the procedure [17]. After standard procedure of native oxide thermal desorption at 580°C with subsequent surface smoothing by annealing in As_4 flux, GaAs buffer layer 500 nm thick was grown at 590°C and growth rate 1 ML/s. Then the source Ga was closed and the substrate cooled to a growth temperature. The source of As_4 was closed at 550°C to provide As-surface stabilized conditions and surface reconstruction (2×4).

In the first part of study, the effect of droplet epitaxy parameters – effective deposition thickness and substrate temperature – on the sizes and density of Ga droplets was studied.

The substrate temperature was varied from 150°C to 450°C. The thickness of Ga deposition H was calibrated by the oscillations in the intensity of the RHEED pattern and was varied in the 3-4 MLs range. The effective growth rate V for all samples was 0.5 ML/s. The reconstruction (2×4) was maintained during the cooling, which indicates that there is no accumulation of additional excess arsenic on the surface. After the substrate was cooled to a predetermined temperature, a Ga source was opened and Ga droplets were formed. At the same time, the surface reconstruction in the low-temperature region (up to 300°C) changed from (2×4) to 1×, and at temperatures above 300°C it was changed by 2×.

In the second part of the experiment, after the formation of the GaAs buffer layer, the substrates were removed from the growth chamber, subjected to structuring by local anodic oxidation (LAO), and then used to study growth on modified surfaces. To increase the growth selectivity deposition of Ga was carried out at 500°C with interruptions of 60 s after each 1 ML, which on the one hand made it possible to provide relatively high surface mobility of Ga adatoms, and on the other, to eliminate desorption and rearrangement of the modified regions of the substrate. With the same purpose, the Ga deposition rate was reduced to 0.25 ML/s. After Ga deposition and droplet formation, the samples were exposed in an As_4 flux during 15 min at the same temperature, after which the substrate was cooled to room temperature. For comparison, the growth was carried out on flat and modified substrates. After growth, all samples were investigated by AFM and SEM.

3. Results and discussion

As was mentioned in section 2, in the first part of the experiment, the influence of the growth temperature and deposition thickness in the process of droplet MBE on the density and dimensions of the Ga nanostructures formed, without the crystallization stage, was studied. Figure 1 shows SEM images of Ga droplets obtained at an effective deposition thickness of 4 ML and a growth rate of 4 ML/s and different substrate temperatures. It is seen that as the growth temperature increases, the average droplet size increases, and their density, on the contrary, is expected to decrease, which is caused by an increase in the surface mobility of Ga adatoms. Thus, at $T = 150$°C, the density of the structures is $3.6\cdot10^{11}$ cm^{-2} with an average diameter of .4.54 ± 3.53 nm (according to AFM measurements). With an increase in temperature to 450°C, the droplet density decreases by an order of magnitude to $3.36\cdot10^{10}$ cm^{-2}, and the diameter increases to 17.65 ± 9.8 nm. The size dispersion thus decreases from 78% to 55%, respectively, which is also related to the growth of the diffusion length of adatoms Ga and the intensification of the exchange processes between droplets of various sizes.

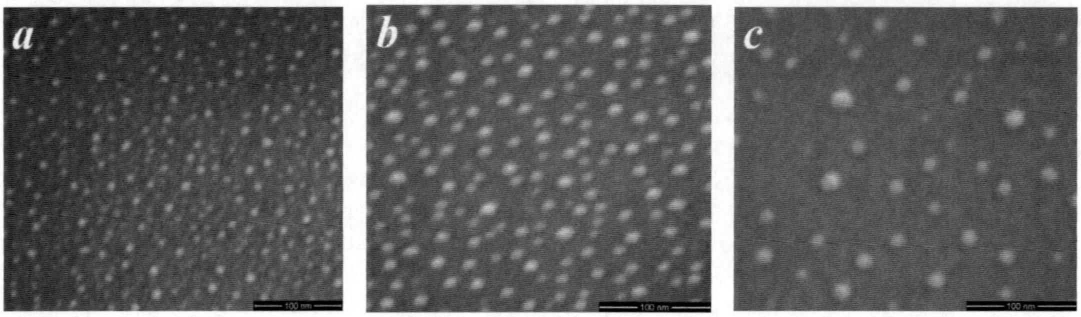

Figure 1. SEM images of GaAs(001) surface after Ga deposition at $H = 4$ ML and different growth temperature: *a*) $T = 200°C$, *b*) $T = 300°C$, *c*) $T = 400°C$. The scale bar is 100 nm.

Figure 2 shows the approximation of the AFM data for the dependence of the density and diameters of Ga droplets on the growth temperature at different effective thicknesses of deposition. It can be seen that a decrease in the thickness of deposition from 4 to 3 ML practically does not affect the density of the formed droplets. The discrepancy between the values is in the range of 4-6%, which is significantly lower than the measurement error. At the same time, the diameter of the Ga droplets decreases, on average, by 24% over the entire temperature range: from 4.54 ± 3.53 nm to 3.67 ± 2.54 nm at $T = 150°C$ and from 17.65 ± 9.8 nm to 14.12 ± 9.4 nm at $T = 450°C$.

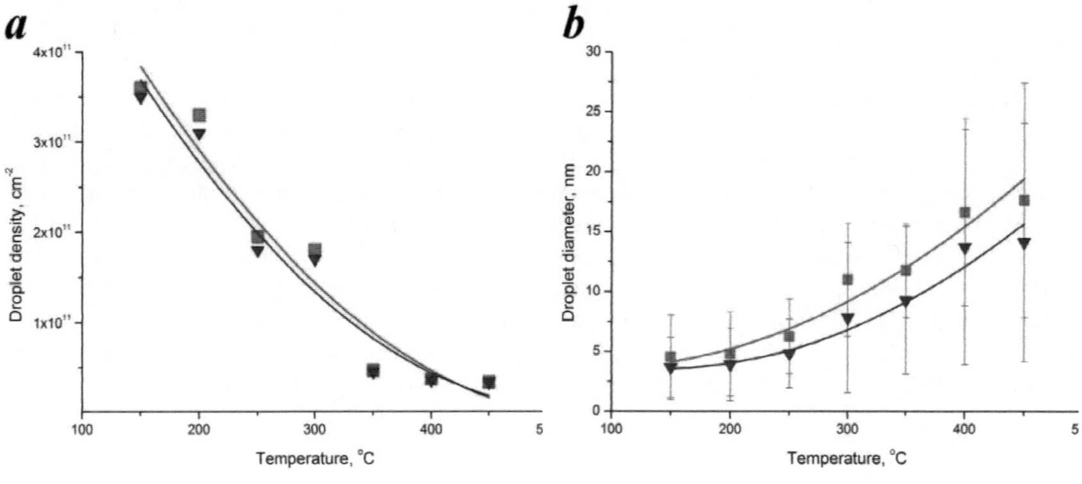

Figure 2. Temperature dependencies of Ga droplet density (*a*) and diameter (*b*) at different effective thicknesses: $H = 4$ ML (red) and $H = 3$ ML (blue). Dots – experimental data, line – approximation.

In the second part of the work, we investigated the processes of droplet epitaxy on LAO-modified substrates. After the epitaxial layer surface was structured by the LAO method, a GaAs oxide film was formed on the samples. Therefore, an *in-situ* removal of the oxide was first carried out according to the procedure [17–19].We have experimentally determined that to completely remove the oxide layer at a temperature of 500°C, it is necessary to deposit about 6 ML of Ga. Then part of the structured samples was extracted, and the relief studies were performed by the AFM technique. It was found that with selected LAO regimes, pits with a depth of 6 ± 1 nm and a diameter of 182 ± 29 nm were formed on the surface with a rms surface roughness of 0.1 nm.

Thus, taking into account the sacrificial Ga layer with a thickness of 6 ML and an effective deposition thickness of 4 ML, the total thickness of the Ga deposition on the structured substrates was 10 ML.

The density of Ga droplets on the flat areas of all the samples was $2\cdot10^8$ cm^{-2} with an average droplet diameter of 255 ± 78 nm and a height of 31 ± 3 nm (Fig. 3*a*). Such a rapid increase in droplet sizes and a decrease in density with a slight (at 50°C) increase in temperature in comparison with the results obtained in the first stage of the study is primarily due to a decrease in the growth rate to 0.25 ML/s.

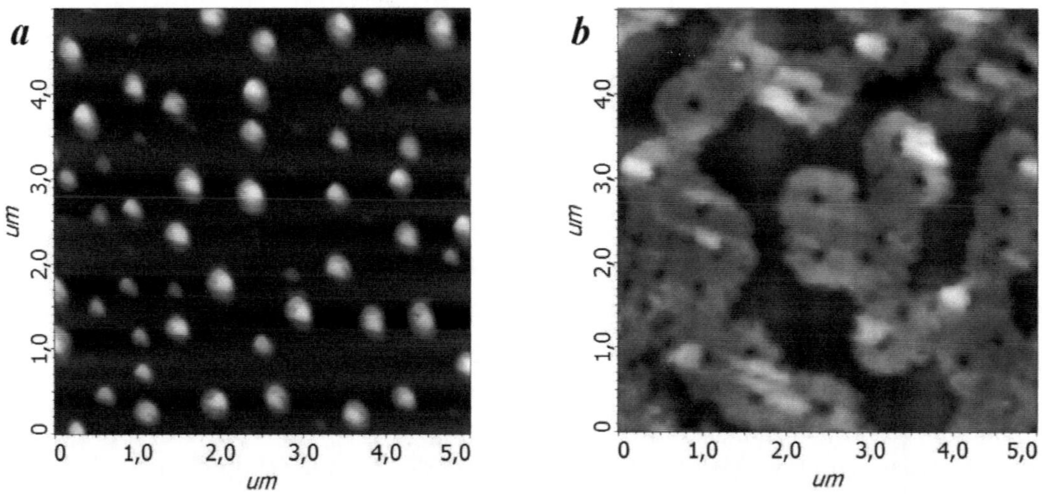

Figure 3. AFM images of GaAs(001) surface after Ga deposition by droplet MBE: *a*) Ga droplets before annealing under As_4 flux, *b*) GaAs nanostructures after annealing under As_4 flux.

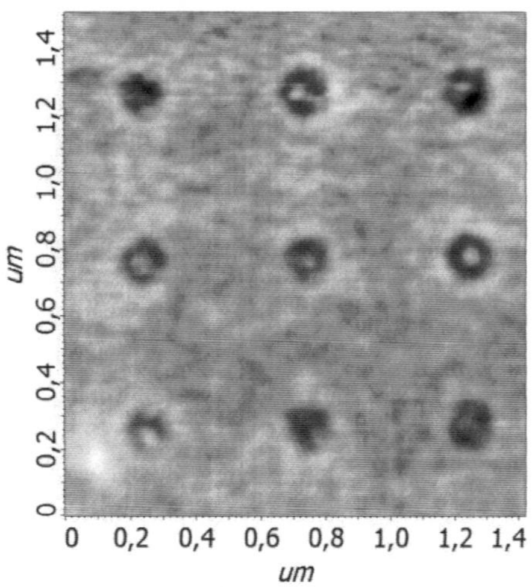

Figure 4. AFM image of GaAs quantum dot regular array obtained on LAO modified GaAs(001) surface by droplet MBE.

After the formation of Ga droplets, their crystallization in the As_4 flux was carried out at a constant surface temperature. On unmodified areas of samples, Ga droplets were transformed into disc-shaped structures with pits in the place of droplets (Fig. 3*b*). The height of the disk was 20 nm above the level of the initial surface of epitaxial layer. The average radius of the formed disks was 262 ± 37 nm and is limited, apparently, by the diffusion length of Ga adatoms under the given conditions. It should be noted that the set value correlates well with the work data [20, 21].

On modified LAO sites, after annealing in the As_4 flux, regular arrays of GaAs quantum dots were formed, whose density was equal to the density of the formed pits and amounted to $4.6 \cdot 10^8$ cm^{-2} (Fig. 4). The size of the formed quantum dots was 78 ± 6 nm, and the degree of filling of the pits was 89%. As can be seen from the presented data, the density of Ga droplets (and GaAs quantum dots) is 2.5 times higher in comparison with unmodified areas, and the size of the structures is 3 times smaller. The obtained ratios of the values of the densities and the diameters of the structures for different sample regions are in good agreement with the redistribution of the material during the Ga deposition due to the stimulated by high temperature and low growth rate in the absence of arsenic and desorption. Since the distances between LAO-formed pits (500 nm) are less than twice the value of the diffusion length of Ga adatoms, determined earlier by us, the growth of structures is suppressed by diffusion gathering into the pits.

4. Conclusion

Experimental studies of the droplet epitaxy of arrays of self-organizing GaAs nanostructures (droplets and quantum dots) on an As-stabilized GaAs(001) surface are carried out. The possibility of obtaining ordered arrays of Ga droplets with a density up to $4 \cdot 10^{11}$ cm^{-2} is shown. The possibility of independent control of the density and size of the structures formed by droplet epitaxy is experimentally confirmed. The regimes for the formation of regular arrays of Ga droplets and GaAs quantum dots with reduced dispersion on the LAO-modified surfaces at relatively high growth temperatures are determined. Moreover, it was shown, in this case the density of the structures is set, in the first place, by the parameters of the modified substrate.

Acknowledgements

This work was supported by the Russian Science Foundation Grant No. 15-19-10006. The results were obtained using the equipment of the Center for Collective Use and Research and Education Center "Nanotechnologies" of Southern Federal University.

References

[1] Kianpoura M, Sabbaghi-Nadooshanb R and Navic K 2014 *J. Computer and System Sc.* **80** 1404–14
[2] Karimia M J, Rezaei G and Nazaria M 2014 *J. of Luminescence* **145** 55–60
[3] Ageev O A, Solodovnik M S, Balakirev S V, Eremenko M M and Mikhaylin I A 2016 *J. of Phys.: Conf. Series* **741(1)** 012012
[4] Shahazadeh M and Sabaeian M 2014 *Superlattices and Microstructures* **75** 514–522
[5] Ageev O A, Konoplev B G, Rubashkina M V, Rukomoikin A V, Smirnov V A and Solodovnik M S 2014 *Nanotech. in Rus.* **8** 23–28
[6] Ramos E, R. Franco R, Silva-Valencia J and Figueira M S 2014 *Physica E* **64** 39–44.
[7] Ageev O A, Solodovnik M S, Balakirev S V and Eremenko M M 2016 *Phys. of the Solid State* **58(5)** 1045–1052
[8] Sayahian J A and Rezarei G 2015 *Physica B* **456** 103–107
[9] Wang Z M 2008 *Self-assembled quantum dots* (Springer) 468 p
[10] Atkinson P, Schmidt O G, Bremner S P and Ritchie D A 2008 *C.R. Physique* **9** 88–803
[11] Lee B C, Lin S D, Lee C P, Lee H M, Wu J C and Sun K W 2002 *Appl. Phys. Lett.* 2. 326–328
[12] Ageev O A, Smirnov V A, Solodovnik M S, Rukomoikin A V and Avilov V I 2012 *Semiconductors* **46(13)** 1616–1621
[13] M. Zander M, Nishinaga J, Iga K and Horikoshi Y 2011 *J. Cryst. Growth* **323** 9–12
[14] Panyakeow S 2009 *Engineering J.* **13(1)** 51–56

[15] Mano T, Watanabe K, Tsukamoto S, Fujioka H, Oshima M and Koguchi N 1999 *Jpn. J. Appl. Phys*. **38** 1009–11
[16] Lee J H, Wang Z M and Salamo G J 2009 *IEEE Transactions on Nanotech.* **8(4)** 431–36
[17] Avilov V I, Ageev O A, Smirnov V A, Solodovnik M S and Tsukanova O G 2015 *Nanotech. in Rus.* **10(3-4)** 214–219
[18] Atkinson P, Schmidt O G 2009 *J. Cryst. Growth* **311** 1815–18
[19] Asaoka Y 2003 *J. Cryst. Growth* **251** 40–45
[20] Ageev O A, Solodovnik M S, Balakirev S V, Eremenko M M and Mikhaylin I A 2017 *J. Cryst. Growth* **457** 46–51
[21] Ageev O A, Solodovnik M S, Balakirev S V and Eremenko M M 2016 *J. Vac. Sci. Technol. B* **34(4)** 041804

IOP Conf. Series: Journal of Physics: Conf. Series **917** (2017) 032038 doi:10.1088/1742-6596/917/3/032038

Effect of SiC buffer layer on GaN growth on Si via PA-MBE

S A Kukushkin[1-3], A M Mizerov[1], A V Osipov[1-3], A V Redkov[1,2,4], R S Telyatnik[2], S N Timoshnev[1]

[1]St. Petersburg Academic University, St. Petersburg 194021, Russia
[2]Institute of Problems of Mechanical Engineering RAS, St. Petersburg 199178, Russia
[3]ITMO University, St. Petersburg 197101, Russia
[4]E-mail: avredkov@gmail.com

Abstract. The study is devoted to comparison of GaN thin films grown on SiC/Si substrates made by the method of atoms substitution with the films grown directly on Si substrates. The growth was performed in a single process via plasma assisted molecular beam epitaxy. The samples were studied via optical microscopy, Raman spectroscopy, ellipsometry, and a comparison of their characteristics was made. Using chemical etching in KOH, the polarity of GaN films grown on SiC/Si and Si substrates was determined.

1. Introduction

Today, the development of electronics is skyrocketing, since it is increasingly being implemented in our life. This is the reason why one of the main goals of thin film technology is to develop new and prospective methods to grow cheaper semiconductor thin films of higher quality. One of the most promising material is gallium nitride (GaN) [1-3], which can be successfully used in power electronics, light emitting diodes and many other devices. However, there are still some limitations on the quality of growing GaN films caused by the absence of suitable substrates [4], since the common substrates, such as sapphire, silicon, etc., are not well-suited to GaN lattice and its thermal expansion coefficient. More suitable materials like silicon carbide [5] are still too expensive. Thereby, researchers are trying to find new approaches, for instance, deposition of different buffer layers [6] or even multi-layered buffers in efforts to grow high-quality GaN. To solve this problem and to produce cheap suitable substrates for GaN, we suggest to use the method of atoms substitution proposed in [7,8], which allows one to grow a thin SiC buffer layer of high quality directly on Si substrate. In the current work, we compared the crystalline properties of GaN films grown on SiC/Si substrate and directly on Si substrate via plasma assisted molecular beam epitaxy (PA MBE) [9]. This method has a series of advantages. Firstly, it allows one to grow films without using hydrogen and at low substrate temperatures. Secondly, depending on growth conditions and substrate type it allows producing GaN films of different polarity [10] which is important in view of further GaN use in various devices. The grown samples were studied via optical microscopy, ellipsometry, Raman spectroscopy and SEM. After growth, the samples were etched in KOH to determine their polarity: Ga-polar or N-polar.

2. Experiments

We used semi-insulating Si(111) substrates and SiC/Si(111) substrates made by the method of atoms substitution [7]. For the growth of the latter, we used p-type silicon. The growth parameters were as follows: total pressure ($CO+SiH_4$) 1.8 Torr, temperature 1250°C, content of SiH_4 in a gas mixture 16%.

The growth time was 15 min. Thickness of the resulting SiC buffer layer was 72 nm, according to ellipsometry. Epitaxial GaN layers were grown using Veeco Gen 200 facility, which allows one to use multiple substrates in one PA MBE growth process. Before the growth, silicon samples were cleaned via Shiraki [11] method. For nitrogen activation, a high-frequency plasma source RFN 50/63 (Riber) was used. The growth was performed in two stages: firstly, 200 nm thick LT-GaN layer has been grown at low temperature (T=650°C) and after that HT-GaN layer has been grown at higher temperature (T=730°C).

3. Results

The ellipsometric data of the samples is presented in Figure 1. According to ellipsometry, the thickness of the GaN film on Si(111) was ~800 nm, whereas the GaN film on SiC/Si(111) was ~780 nm in thickness. Ellipsometry also demonstrates that the GaN/Si(111) sample is slightly more transparent than the GaN/SiC/Si(111) sample. We also measured Raman spectra of the samples. Absolute position of the band of GaN E_2(high) 568 cm^{-1}[12], which is dependent on the mechanical stress in the film, was different for both samples. The estimation of residual mechanical stresses, based on the position of the band, gave 150 MPa tensile stress for GaN/Si sample and 190 MPa compressive stress for GaN/SiC/Si sample.

The samples were also subjected to X-ray diffraction, which has shown that FWHM of the rocking curve for GaN/SiC/Si sample was 36 arcmin, whereas for the sample GaN/Si it was 53 arcmin, and thus GaN grown on SiC/Si substrate has a higher crystalline quality.

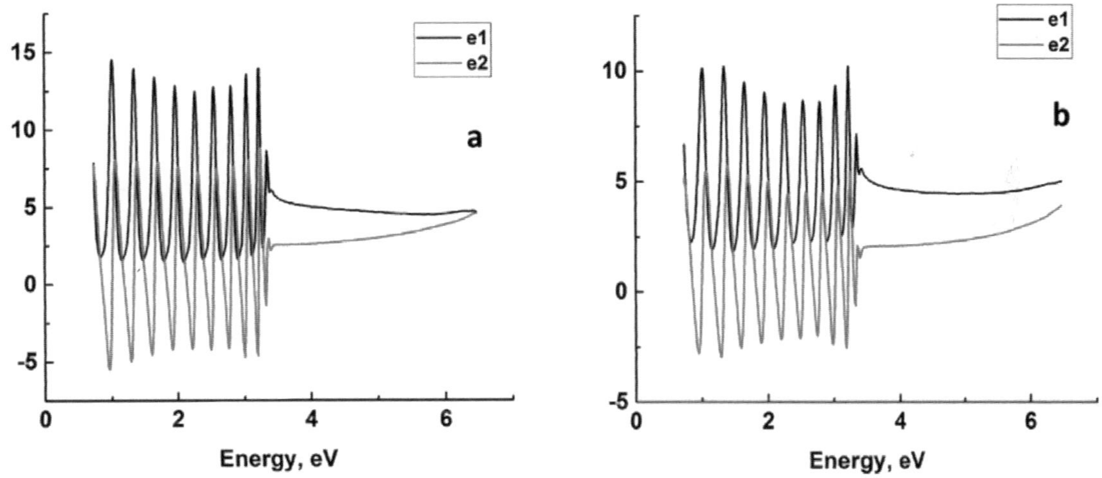

Figure 1. Ellipsometric data of GaN/Si(111) sample (a) and GaN/SiC/Si sample (b).

During our studies, we have noticed formation of untypical islets of triangular shape with a characteristic size of the order of a few tens of microns on the surface of both samples in the regions far away from the center. Since the islets were found on both substrates and were of a similar nature but had some differences in the formation (on GaN/Si substrate they were more clearly visible and of higher concentration than on GaN/SiC/Si substrate), we studied them via Raman spectroscopy to understand their origin. Optical image of the islets on GaN/Si sample is presented in Figure 2a.

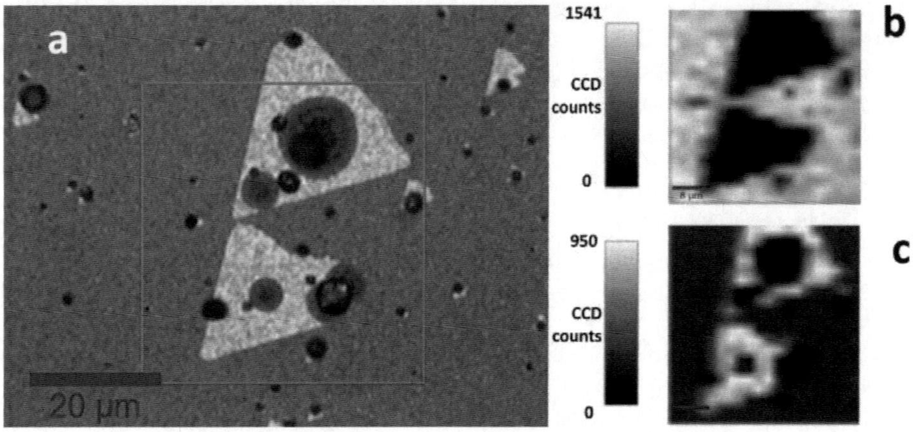

Figure 2. Optical image of the GaN/Si surface demonstrating triangular islets (a) and Raman maps constructed using intensity of silicon band (521 cm^{-1}) (b) and GaN band (568 cm^{-1}) (c).

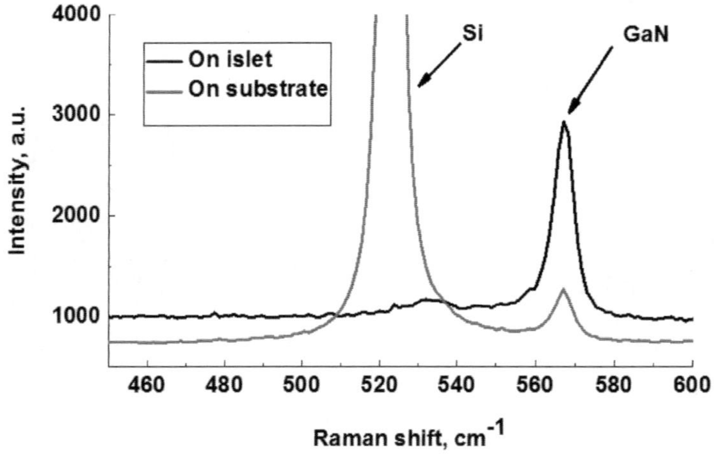

Figure 3. Raman spectra of the sample on the substrate far from islet (red line) and on the islet (black line).

Using a confocal micro-Raman setup Witec Alpha 300R equipped with a piezo-table, the area containing triangular islets was studied. Figures 2b and 2c show the Raman maps (20 * 20 points) constructed using the intensity of silicon band (521 cm^{-1}) (Fig. 2b) and GaN band (568 cm^{-1}) (Fig. 2c). The maps correspond to the rectangular region marked in Figure 2a. It is clearly seen that triangular islet has composition different from surrounding substrate. The substrate dominates on the Raman map of silicon band (see figure 3b), whereas the islets dominate on the Raman map of GaN. Consequently, it can be concluded that these triangular islets consist of gallium nitride. One should note that Raman spectra of the substrate far from the islets also contain low-intensity band of GaN (Fig. 3). Thus, it can be supposed that the substrate is covered by thin GaN film and by triangular GaN islets on it, like in Stranski-Krastanow growth mode, which takes place when mechanical stresses in the film are quite high. It should be noted that difference in GaN band intensity on such islets and on substrate is much greater on the GaN/Si sample than on GaN/SiC/Si sample, which confirms conclusions about elastic stresses in the GaN film grown on SiC/Si. The origin of dark round defects (Fig.2a) is not fully understood.

Since many optical and electrical properties of GaN thin films and devices are determined by the polarity [13] of these films, we conducted a series of experiments to determine the influence of the substrate on the polarity of the grown structures. For this, we etched samples in KOH via standard procedure [14] and after that studied surface morphology via scanning electron microscopy (see Figure 4). One of the samples (GaN/SiC/Si) demonstrated pyramidal structures on the surface, which are typical for N-polar GaN. The other sample (GaN/Si) surface was not affected by etching procedure, and therefore it is Ga-polar.

4. Conclusions

Figure 4. SEM image of cross-sections of GaN/Si(111) sample (a) and GaN/SiC/Si sample (b) etched in KOH for 5 minutes.

In this work, we considered the growth of GaN films on SiC/Si(111) substrate made by the method of atoms substitution and directly on Si(111) substrate in one process of plasma assisted molecular beam epitaxy. Resulting samples were studied via different diagnostic methods: optical microscopy, ellipsometry, Raman spectroscopy, SEM. The total thickness of the films was estimated, and it was shown that GaN/Si sample is slightly more transparent than GaN/SiC/Si sample. Also, by means of X-ray diffraction, it was demonstrated that crystalline quality of GaN film grown on SiC/Si substrate is higher than on Si(111) substrate. The estimation of mechanical stress gave the value of 150 MPa tensile stress in GaN/Si(111) sample and 190 MPa compressive stress in GaN/SiC/Si(111) sample. Formation of triangular-shaped islets on the surface of the samples has been revealed, corresponding Raman maps of the islets were constructed, and it was shown that the islets consist of GaN. The islets are forming more actively on GaN/Si substrate than on GaN/SiC/Si. The obtained results suggest that SiC buffer plays a positive role during GaN growth.

Acknowledgments
This work was supported by the Russian Science Foundation (№ 14-12-01102). Part of experimental results was obtained using unique scientific facility "Physics, chemistry, and mechanics of crystals and thin films" (IPME RAS, St. Petersburg, Russia).

References

[1] Nakamura S and Krames M R (2013) *Proceedings of the IEEE* **101** 2211
[2] Den Baars S P, Feezell D, Kelchner K, Pimputkar S, Pan C C, Yen C C et al (2013) *Acta Materialia* **61** 945
[3] Baliga B J (2013). *Semiconductor Science and Technology* **28** 074011
[4] Kukushkin S A, Osipov A V, Bessolov V N, Medvedev B K, Nevolin V K and Tcarik K A (2008) *Rev. Adv. Mater. Sci.* **17** 1
[5] Zakharov D N, Liliental-Weber Z, Wagner B, Reitmeier Z J, Preble E A and Davis R F (2005)

Physical Review B **71** 235334.

[6] Amano H, Sawaki N, Akasaki I and Toyoda Y (1986) *Appl. Phys. Lett.* **48** 353

[7] Kukushkin S A and Osipov A V (2014) *Journal of Physics D: Applied Physics* **47** 313001

[8] Kukushkin S A and Osipov A V (2013) *Journal of Applied Physics* **113** 024909.

[9] Jmerik V N, Mizerov A M, Shubina T V, Listoshin S B, Ivanov S V (2007) *Technical Physics Letters* **33** 333

[10] Mizerov A M, Jmerik V N, Kaibyshev V K, Komissarova T A, Masalov S A, Ivanov S V (2009) *Semiconductors* **43** 1058

[11] Ishizaka A, Shiraki Y (1986) *J. Eltrochem. Soc.* **133** 666

[12] Kisielowski C, Krüger J, Ruvimov S, Suski T, Ager III J W, Jones E, Liliental-Weber Z, Rubin M, Weber E R, Bremser M D, Davis R F (1996) *Physical Review B* **54** 17745

[13] Hellman E S (1998). *MRS Internet Journal of Nitride Semiconductor Research*, **3** 1

[14] Li D, Sumiya M, Fuke S, Yang D, Que D, Suzuki Y, Fukuda Y (2001) *J. Appl. Phys.* **90** 4219

IOP Conf. Series: Journal of Physics: Conf. Series **917** (2017) 032039 doi:10.1088/1742-6596/917/3/032039

Comparative study of thermal and plasma enhanced atomic layer deposition of aluminum oxide on graphene

I Clemente[1,2], A Miakonkikh[1,2], O Kononenko[3,4], V Matveev[3], K Rudenko[1,2]
[1]Institute of Physics and Technology, Moscow 117218, Russia
[2]Moscow Institute of Physics and Technology, Dolgoprudny 147001, Russia
[3]Institute of Microelectronics Technology and High Purity Materials, Russian Academy of Sciences, Chernogolovka, Moscow Region 142432, Russia
[4]National University of Science and Technology "MISiS", Moscow 119049, Russia

Abstract. Atomic layer deposition of alumina on graphene was studied in thermal and plasma enhanced process. Deposition was controlled by in situ ellipsometry every half cycle, which allows measuring of Al_2O_3 thickness and graphene equivalent thickness during the process. Properties of graphene were measured by Raman spectroscopy prior and after deposition of dielectric layer. It was shown that plasma enhanced deposition leads to decrease of effective graphene thickness, while thermal deposition does not affect graphene layer. No substantial nucleation lag was observed in both types of deposition.

1.Introduction

Graphene became one of promising materials that could improve the performance of high-frequency and low-power information processing, and THz detectors. Developing of graphene based nanoelectronics devices requires deposition of ultrathin gate dielectric on graphene in a process that do not lead to structural changes in graphene. Alumina has high dielectric constant (9.0), excellent barrier properties for moisture and for diffusion of impurities, and high breakdown field which make it candidate for gate dielectric and passivation layer for advanced electronic devices [1]. Although plasma enhanced ALD produce alumina with higher purity and excellent dielectric properties oxygen plasma could damage graphene layer, because carbon is subject to oxidation reaction with oxygen plasma. Thermally activated ALD growth uses water vapor as a source of oxygen. Another effect which could present in the ALD process is so called nucleation lag. This effect consists in the fact that the growth of the film does not begin immediately, but only after several cycles. As a rule, this is due to the peculiarities of the adsorption of a precursor by an untreated surface. The effect of ALD growth of dielectric on graphene layer is studied by Raman spectroscopy and spectral ellipsometry.

2.Experimental

In present work 5 nm thick Al_2O_3 films were deposited in FlexAl ALD tool (Oxford instruments plasma technology, UK) accompanied by *in situ* ellipsometry measurements and *ex situ* Raman spectroscopy. We used CVD deposited one-layer and multilayer graphene on Si substrate with 300-500 nm of SiO_2 insulating layer. Samples 2x2 cm^2 in size were placed on the heated table of ALD reactor with controllable temperature (300°C).
A typical ALD cycle consists of four steps. The cycle starts by introducing the first precursor to the reactor chamber which is purged then by flow of inert gas. Then the second precursor is introduces and purged. As a result of surface reaction one complete monolayer of deposited layer is added to the surface.
For plasma enhanced ALD (PEALD) process TMA ($Al(CH_3)_3$) as metal precursors and remote ICP plasma of O_2 as non-metal precursor were used. ICP plasma was excited at pressure of oxygen of 15 mTorr with applied power of 250W. This step lasts for 2 seconds.
Thermal ALD (TALD) process was performed using water vapor as oxidizing precursor. Growth rate was 0.1 nm per cycle for both processes.
Samples of graphene were placed in the vacuum chamber from atmosphere through the loadlock. No additional annealing was done prior the process. Thermostabilisation of sample on the heated table takes

about 3 minutes at the pressure of 200 Torr. Wafer temperature was also controlled by spectral ellipsometry measurements and did not differ from table temperature by more than 25°C [2].
Spectral ellipsometry (SE) was performed *in situ* by J.A. Woollam Co M2000X 65 degrees ellipsometer in wide optical range of wavelength (245-1000 nm). SE measurements were done on every half of ALD cycle (during purge by argon after metalorganic precursor dosage and after oxidizing step as well). Special corrections were done to take into account sample tilt which cannot be adjusted in reactor chamber. In order to prevent optical windows contaminations automated shutters were used to close windows during precursor dosage to the chamber.
Ellipsometry measurement takes about 3 seconds and doesn't substantially affect whole duration of ALD cycle which lasts for about 10 seconds. The method is sensitive enough to measure one monolayer adsorption of precursor and deposition of dielectric [2]. Temperature effects could be corrected. Size of light spot was 2×5 mm^2. That means that effective thickness is averaged on that area.
We used four layer optical models. The first layer is a bulk of silicon of infinite depth, second is the predetermined thickness of silicon oxide, third is graphene, and final layer is alumina. We did measurements on pristine samples to determine thickness of three initial layers. Thickness of silicon oxide was used in subsequent calculations of apparent thickness of alumina and graphene. Due to possible removal of existing on the surface polar adsorbed molecules like water or –OH groups apparent thickness could become negative during in the beginning of deposition. Optical properties of graphene were taken from [3], assuming that one monolayer graphene should have 0.34 nm effective thickness.
Since ellipsometry is able to measure single monolayer of deposited or adsorbed molecules it could provide information not only referred to growth rate, but to adsorption of first precursor. Apparent thickness of adsorbed first precursor ($Al(CH_3)_3$) is higher than apparent thickness of Al_2O_3. That is why after first half cycle increase of apparent thickness can be seen (by more than thickness of one monolayer), and after second half cycle, when first precursor is oxidized, apparent thickness is slightly reduced. Total increase in whole cycle is the one monolayer thickness.
Micro Raman spectra of graphene films before and after Al_2O_3 film deposition were obtained with a SENTERRA BRUKER Raman microscope using 532 nm excitation wavelengths. A diameter of the laser spot was about 1 µm.

3.Results and discussion

Decrease in effective thickness of graphene exposed to oxygen plasma was measured. In conditions same to deposition, but without dosing precursor, mean rate of decreasing in effective graphene thickness was 0.014 nm/sec.
In case of plasma enhanced ALD (Figure 1) we observe linear growth of graphene since second cycle. Decrease of graphene thickness lasts for about 10 cycles of deposition. It could be concluded that 2 nm layer of alumina could prevent further degradation of graphene.
In thermal process decrease of effective graphene thickness does not exceed 0.02 nm over 20 cycles (Figure 2). Growth of Al_2O_3 is completely linear with small nucleation delay of no more than 1 cycle. There is not substantially degradation of effective graphene thickness in thermal ALD on multilayer graphene (Figure 3).

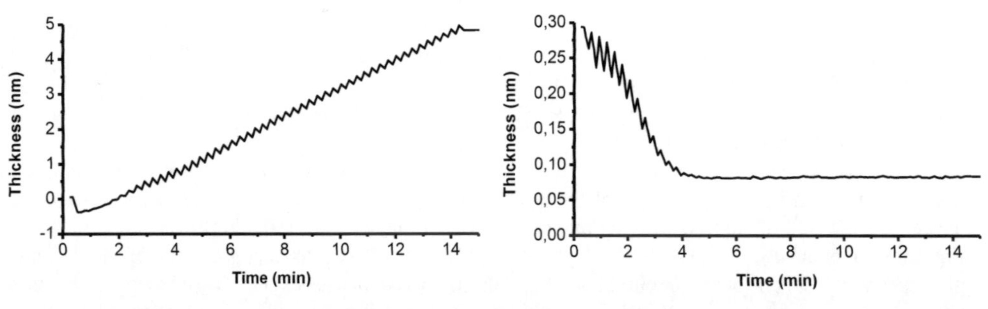

Figure 1. Apparent thickness of deposited alumina (left) and effective thickness of graphene (right) during 50 cycles of plasma enhanced ALD by in situ ellipsometry.

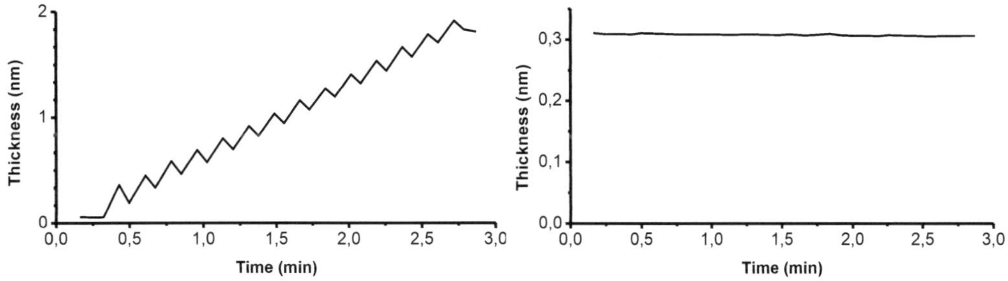

Figure 2. Apparent thickness of deposited alumina (left) and effective thickness of graphene (right) during 50 cycles of thermal ALD by in situ ellipsometry.

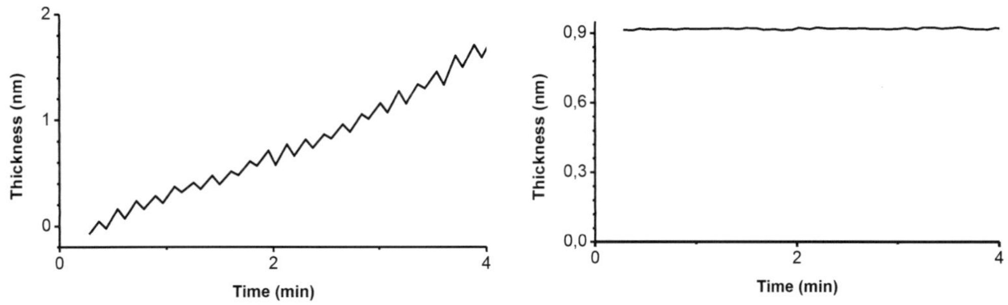

Figure 3. Thickness of deposited alumina (left) and effective thickness of multilayer graphene (right) during 50 cycles of thermal ALD by in situ ellipsometry.

Figure 4 shows Raman spectra collected from pristine graphene and from graphene with deposited PEALD and TALD Al_2O_3. The defect-activated peaks (D, D' and D+D') are seen in spectrum obtained from graphene coated by PEALD Al_2O_3. G- and 2D- peaks have low intensities and are broaden. D'- peak almost merges with the G- peak. From the ratio of the D and G peak intensities, we can estimate the concentration of defects formed as a result of exposure to oxygen plasma. In [4] it was shown that defect density in graphene after plasma etching can be determined using the formula:

$$n_d(cm^{-2}) = (7.3 \pm 2.2) \times 10^9 E_L^4 \left(\frac{I_D}{I_G}\right),$$

E_L – laser excitation energy.

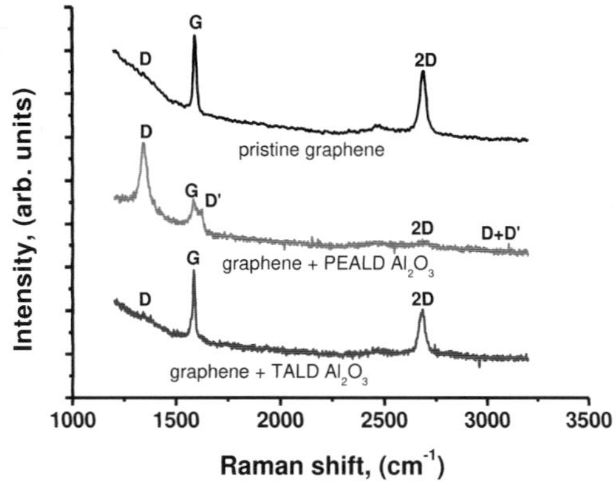

Figure 4. Raman spectra collected from pristine graphene and from graphene coated by alumina deposited by plasma enhanced ALD (PEALD) and and thermal ALD (TALD).

In graphene used in this work the defect density n_D is not more than 10^9 cm^{-1}, I_D/I_G is less then 0.01. After exposure to oxygen plasma I_D/I_G is around 2.2, the defect density is around 5×10^{11} cm^{-1}. Raman spectrum obtained from graphene coated by TALD Al_2O_3, practically does not differ from that of the pristine graphene. Thus we can conclude that the number of defects in graphene after deposition of TALD alumina is almost not increased.

In multilayer graphene there are areas with different number of layers, so Raman spectra (Figure 5) for one-layer areas of graphene sample exhibits plasma damage. Areas with multilayer graphene is less damaged. Ellipsometric observations during PEALD on multilayer graphene is similar to data of PEALD on graphene. This is due to the fact that damage appears only in the upper layers of graphene, and the lower layers remain intact.

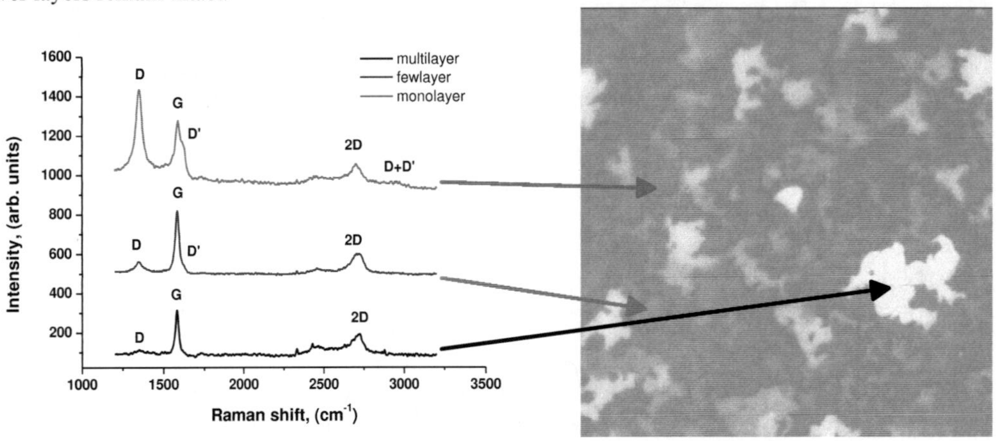

Figure 5. Micro Raman spectrum of multilayer graphene after exposure to oxygen plasma.

4.Conclusion

Atomic layer deposition of alumina on graphene was studied in thermal and plasma enhanced ALD process. Deposition was controlled by in situ ellipsometry every half cycle, which allows measuring of Al_2O_3 thickness and graphene equivalent thickness during the process. Properties of graphene were measured by Raman spectroscopy prior and after deposition of dielectric layer. It was shown that plasma enhanced deposition leads to decrease of effective graphene thickness, while thermal deposition does not affect graphene layer. In multilayer graphene there are areas with different number of layers, so Raman spectra for one-layer areas of graphene sample exhibits plasma damage. Areas with multilayer graphene is less damaged.

References

[1] V. P. Popov, M. A. Ilnitskii, E. D. Zhanaev, A. V. Myakon'kich, K. V. Rudenko, and A. V. Glukhov. Biosensor properties of SOI nanowire transistors with a PEALD Al_2O_3 dielectric protective layer. Semiconductors, 50(5), 632–638 (2016).

[2] I. E. Clemente and A. V. Miakonkikh. Application of spectral ellipsometry to in situ diagnostics of atomic layer deposition of dielectrics on silicon and AlGaN, Proc. SPIE 10224, International Conference on Micro- and Nano-Electronics 1022425 (2016).

[3] F. J. Nelson, V. K. Kamineni, T. Zhang, E. S. Comfort, J. U. Lee, and A. C. Diebold. . Appl. Phys. Lett. 97, 253110 (2016).

[4] L. G. Cancado, A. Jorio, E. H. Martins, Ferreira, F. Stavale, C. A. Achete, R. B. Capaz, M. V. O. Moutinho, A. Lombardo, T. S. Kulmala, and A. C. Ferrari. Quantifying Defects in Graphene via Raman Spectroscopy at Different Excitation Energies, Nano Lett. 11, 3190–3196 (2011).

IOP Conf. Series: Journal of Physics: Conf. Series **917** (2017) 032040 doi:10.1088/1742-6596/917/3/032040

Effect of Ga seeding layer on formation of epitaxial Y-shaped GaN nanoparticles on silicon

V V Fedorov[1,2,3], A D Bolshakov[1,2], A M Mozharov[1,2], G A Sapunov[1], I V Shtrom[3,4], D A Kirilenko[3], A A Sitnikova[3] and I S Mukhin[1,2]

[1] St.Petersburg Academic University, 194021, St. Petersburg, Russia
[2] ITMO University, 197101, St. Petersburg, Russia
[3] Ioffe Physical-Technical Institute of the Russian Academy of Sciences, 194021 St. Petersburg, Russia
[4] St. Petersburg State University, 7/9 Universitetskaya emb., St. Petersburg, 199034, Russia
E-mail: vfedorov@fl.ioffe.ru

Abstract. Silicon and aluminium nitrides, commonly used as buffer layers for GaN growth on Si are wide gap insulators, preventing barrier free charge-carrier transport across the heterojunction and limiting the functionality of GaN-on-silicon technology. In this work we explore possibility of direct growth of GaN on Si nano-heterostructures by PA-MBE with use of Ga-nanodroplets as seeds. It is demonstrated that use of seeding layer can result in formation of Y-shaped planar GaN nanoparticles (GaN tripods) along with commonly observed GaN nanowires. Growth mechanism, morphology and structural characterization of GaN/Si nano-heterostructures is discussed.

1. Introduction

Synthesis of group-III nitride nanostructures attracted a lot of attention due to their importance both in fundamental research and potential application in future opto-electronic nano-scaled devices [1]. Geometric confinement and reduced dimension effects, can significantly affect physical properties of nanostructures, as well as promote formation of defect-free hetero-interfaces and provide efficient strain relaxation even on highly mismatched substrates, as it has been demonstrated for epitaxial growth of GaN nanowires on Si [2]. The latter nanostructures are nowadays object of growing interest not only due to high potential of GaN-on-Si technology, but also because of functional GaN/Si heterojunction realization possibility, as it was shown in our previous work [3]. However, majority of today's GaN-on-Si technologies involves formation of silicon nitride or aluminum nitride buffer layer [4], which are wide-gap insulators, prohibiting barrier free charge-carrier transport.

Thus, in this study we investigate possibility of direct growth of GaN/Si (111) nano-heterostructures by plasma-assisted molecular beam epitaxy (PA-MBE) with use of monolayer thick Ga-seeding layer yielding the formation of liquid Ga nano-droplets. Effect of the Ga seeding layer on GaN structures growth and morphology is studied by use of reflection high-energy electron diffraction (RHEED), atomic force (AFM), scanning (SEM) and high-resolution transmission electron microscopy (HRTEM).

2. Experimental

GaN nanostructures were synthesized by PA-MBE in Veeco GEN-III MBE machine with use of Riber valved RF-plasma (13.56 Mhz) source of activated nitrogen. To avoid presence of charged species and to increase angular uniformity of nitrogen flux, additional molybdenum aperture plate was installed in front of the source. Silicon (111) *p*-type wafers with a 4° miscut oriented towards <$1\bar{1}0$>, cleaned with a Shiraki method were used as substrates. After thermal annealing (10 min @ 950° C) an atomically clean Si surface was obtained as confirmed by observation of a 7x7 RHEED reconstruction pattern and smooth surface morphology on AFM images. Total nitrogen BEP was in the range of 2..3x10^{-7} Torr. To maintain nitrogen-rich (N-rich) conditions Ga beam equivalent pressure (BEP) was kept sufficiently small - 1..2x10^{-8} Torr. Ga beam equivalent pressure (BEP) was kept constant during deposition of the seeding layer and growth process. In-growth n-type doping of GaN was done with Si effusion cell.

Structure and morphology of nanoheterostructures were studied with scanning electron microscopy (Zeiss SUPRA 25-30-63) and atomic-force microscopy (Bruker Bioscope Catalyst SPM). High-resolution transmission electron microscopy (HR-TEM) studies were carried out using a JEOL 2100F microscope (200kV) The samples for HRTEM studies were prepared by standard methods involving ion sputtering at the last stage.

3. Results and Discussion

3.1. Seeding layer formation

In order to obtain direct nucleation of GaN on Si, deposition of the Ga droplet seeding layer was performed prior to ignition of the RF-plasma. In the following droplet layer was annealed under the nitrogen flow exposure up to growth temperature (790-810 °C) and further deposition of GaN was carried out. Ga droplets were formed by exposing the reconstructed Si (111)-(7x7) substrate surface at 200 °C to a Ga molecular beam. Such relatively low substrate temperature was chosen to obtain dense arrays of nano-droplets [5]. Equivalent Ga thickness was estimated by monitoring the Ga-induced surface reconstruction with RHEED. It is known that deposition of 0.3 ML of Ga leads to formation of $\sqrt{3}\times\sqrt{3}$ -R30° Ga-reconstructed surface [6] and deposition of 0.6 ML leads to formation of 6.3x6.3-Ga induced superstructural phase [7] (1 ML here means 6.81 atoms/cm^2, the surface density of the truncated (111) plane of bulk Si) [8]. Seeding layers with equivalent thickness in range of 0.6-2.5 ML were deposited while monitoring the resulting surface reconstruction in-situ with RHEED. Diffuse semi-circular halos of liquid Ga was observed for the exposition with equivalent thickness more than 1.5 ML. Use of thicker seeding layers was avoided due to risk of eutectic Ga-Si solution formation, unwanted Si substrate doping with Ga atoms and melt-back etching of Si [4].

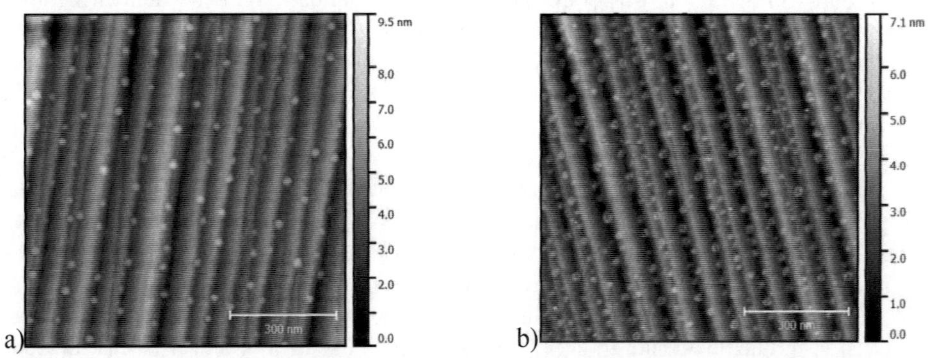

Figure 1. AFM images (900x900 nm) of the Ga nanodroplet layer with equivalent thickness of 1.6 ML deposited at 200 °C (a) before and (b) after nitridation.

Surface morphology of the Ga droplet layer with 1.2 ML equivalent thickness deposited on vicinal Si (111) is presented on AFM image in Figure 1 (a). As it can be seen from AFM images high temperature annealing of vicinal silicon surface leads to coalescence of surface steps, i.e. step bunching, and ordered step-terrace arrays with average terrace width of ~80 nm appear on the substrate surface.

Ga droplets are located at the surface steps edges and have a height about 2-3 nm, average distance between droplets is 100 nm, lateral resolution is limited by AFM tip size.

Surface morphology obtained after nitridation of Ga droplet layer with equivalent thickness of 1.2 ML is presented on AFM image in figure 1 (b). Comparing figure 1 (a) with figure 1 (b) we deduce that nitridated GaN nanoislands are also located at the surface step edges, but do not longer take spherical shape and present ring-like morphology with a 0.5-1 nm thick pit in the ring center. After nitridation islands possess decreased height (1-2 nm) and increased surface density comparing with Ga nano-droplets. Similar nanoring morphology can be obtained during droplet-epitaxy of GaAs [9] and related to localized III-V growth at the Ga-droplet edge. Surface morphology of the nitridated layer did not change after etching with the hydrochloric acid (HCl) aqueous solution, which confirms that there was no metallic Ga or eutectic Si-Ga solution on the surface after sample nitridation in our experiments.

3.2. Structure and morphology of Y-shaped GaN nanoparticles (GaN tripods)

Surface morphology of GaN/Si nanoheterostructures grown with use of seeding layer are presented on the SEM images shown in figure 2. It was found that use of nano-droplet seeding layer can result in formation of oriented Y-shaped (tripod) GaN nanoparticle arrays. As can be seen from top- and side- view SEM images (figure 2 (a) and (b)) Y-shaped nanoparticles have planar structure and are formed with three equally sized elongated branches lying at an angle of 120° to each other. Single and double branched V-shaped GaN nanoparticles can be found along with the Y-shaped ones. Unfortunately, nanoparticle faceting cannot be resolved by means of AFM or SEM. In our case growth of vertical GaN nanowires are also observed, and as will be shown further, surface densities for both NWs and Y-shaped nanoparticles strongly depend on equivalent thickness of the seeding layer and growth parameters and as a consequence can be controlled. According to analysis of SEM images Y-shaped nanoparticles demonstrate several preferential in-plane orientations corresponding to alignment of the nanoparticle branches along the $<1\bar{1}0>$ and $<11\bar{2}>$ directions in Si. Nanoparticles orientations are indicated in figure 2 (a) with colored solid Y-marks. Crystal rotational twinning corresponding to 180° rotations around normal to the Si substrate are indicated with dashed Y-marks. Mosaic spread of in-plane orientation in the range of ±5° is observed. According to the [10] similar tripod shaped nanoparticles were observed during hydride vapour phase epitaxy (HVPE) of GaN on sapphire (0001) substrates. The authors suggest that GaN tripod growth is related with the formation of nanoparticle in the zinc blende (ZB) phase and further nucleation and growth of the wurtzite (WZ) nanorods on the {111} facets of the ZB core [10].

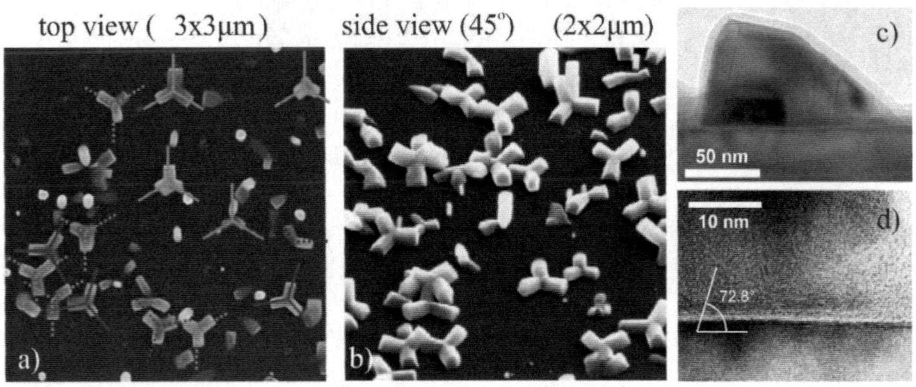

Figure 2. (a) Top and (b) side view (45°) SEM images of GaN/Si nanoheterostructure surface; (c) and (d) HRTEM images of a single branched GaN nanoparticle.

More detailed analysis of crystal structure and morphology of the Y-shaped nanoparticles was performed with HRTEM. Cross-section images of the single branched GaN nanoparticle are presented in figure 2 (c) and (d). Presence of thin (~0.6 nm) amorphous layer at the heterointerface between GaN nanoparticle and silicon substrate can be clearly seen on HRTEM images. According to the transmission electron microscopy microdiffraction (not presented here) nanoparticles have a hexagonal wurtzite GaN crystal structure. GaN c-planes (0001) can be seen on close-up image in figure 2 (d). Estimated angle between silicon (111) and GaN (0001) planes is around 72.8° which is close to value of the angle between (0001) and ($11\bar{2}1$) planes of GaN (72.92°) [10]. There is no any visible presence of the ZB-phase acting as a seed on the obtained HRTEM images. This can be related with small size of the core and sample preparation procedure. Linear contrast features at the HRTEM image (figure 2 (c)) are related to planar defects perpendicular to the (0001) direction, in particular, stacking faults in the hexagonal lattice caused by disorder in stacking sequence and monolayer-thick inclusions of zinc-blende GaN polytype. On the other hand, visible planar defects can be caused by polarity inversion at domain boundaries.

3.3. The role of growth parameters on surface density and morphology of GaN/Si nanoheterostructures

3.3.1. Role of the seeding layer thickness. As was reported previously [12], [13] and according to our studies there is no direct correlation between surface density of the nanoislands of the seeding layer and density of the synthesized GaN nanoparticles. But despite that fact, it has been found that the effective thickness of the seeding layer significantly affects further GaN growth. Thus nitridated Ga-droplets are not acting as nucleation centres, but affect surface diffusion processes. For instance, at specific growth conditions (substrate temperature of 800° C and Ga BEP of 1×10^{-8} Torr) use of a seeding layer with an equivalent thickness less than 0.6 ML results in formation of vertical arrays of GaN NWs with low surface density (~1 μm^{-2}) – nucleation and growth of Y-shaped nanoparticles are absent. For the seeding layers thicker than 0.8 ML the growth of Y-shaped nanoparticles appears. With an increase of the seeding layer equivalent thickness the surface density of the Y-shaped particles and NWs increases. It has been found that use of the seeding layer with an equivalent thickness of more than 2.4 ML, leads to formation of two-dimensional compact layer which significantly alters further growth of GaN.

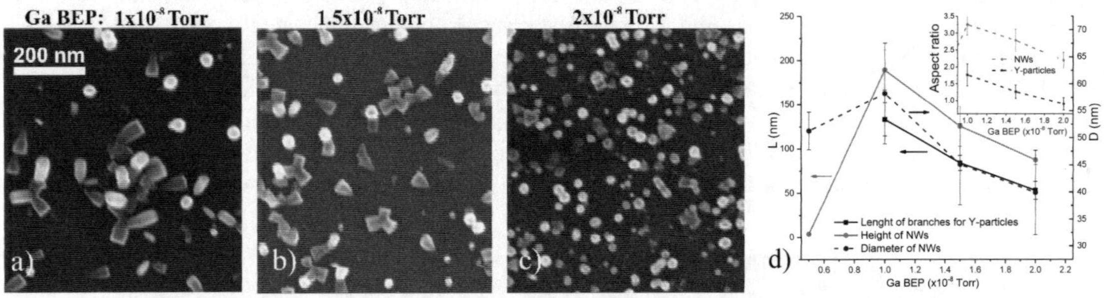

Figure 3. (a),(b), (c) Top view SEM images (WxH: 660x740 nm) of GaN/Si nanoheterostructures grown at different Ga fluxes, (d) size and aspect ratio dependence on the Ga flux.

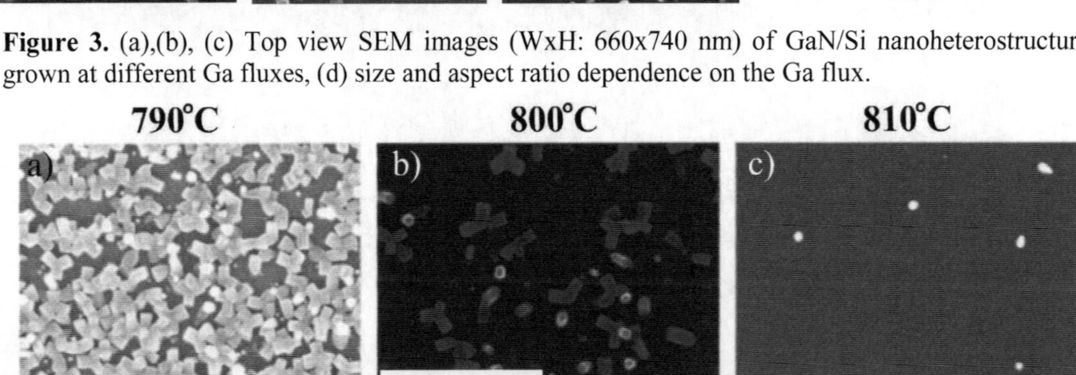

Figure 4. Top view SEM images (WxH: 2x1.5 μm) of GaN/Si nanoheterostructure surface grown at different substrate temperatures (a) 790° C, (b) 800° C, (c) 810° C.

3.3.2. Role of III- V fluxes ratio. As will be shown further N- rich growth conditions lead to Y-shaped particle branches elongation in the similar way as it affects growth of GaN nanowires [2]. Typical surface morphology of GaN/Si nanoheterostructures grown at different Ga BEP (1, 1.5, $2x10^{-8}$ Torr) is presented in figures 3 (a), (b) and (c). In general, increase of the Ga flux and excess of Ga leads to increase of the nucleation centers density, stimulates radial growth rate – and subsequently increases surface density of GaN nanoparticles with decrease in their dimensions (see graph in figure 3 d)). It was found that aspect ratio (relation between the length along the *c*-axis and width – lateral size of the *c*-plane) for the Y-shaped nanoparticles and NWs decreases with increase of Ga flow and can vary in the range of 2 to 1 and 3.5 to 2 respectively - see inset in figure 3 (d). Formation of coalescent two-dimensional compact layer occurs in case of Ga BEP larger than $2x10^{-8}$ Torr.

3.3.3. Influence of growth temperature. It is found that due to small values of the group III- and V- elements fluxes, growth of GaN nanostructures occurs only in a narrow temperature window. The SEM images of the GaN/Si heterostructures surface morphology grown at 790, 800 and 810° C are shown in figures 4 (a), (b), (c) respectively. It can be seen that at temperatures above 805° C, nucleation of Y-shaped nanoparticles can be suppressed almost completely, leaving only NWs on the surface. On the other hand, at a substrate temperature of 790° C and below, the surface density of the Y-shaped nanoparticles becomes so high that they begin to coalesce and form a continuous layer.

4. Conclusions

Formation of epitaxial arrays of Y-shaped GaN nanoparticles on Si (111) by PA-MBE is observed for the first time. Surface density, size and aspect ratio of the structures can be controlled by choosing the growth temperature and group III-, V- fluxes and the equivalent thickness of the seeding layer. It is shown that nucleation can be completely suppressed in favor of the NWs or Y-shaped GaN nanoparticles by choosing the appropriate growth conditions. Epitaxial relations for the Y-shaped nanoparticles are found - the $(11\bar{2}1)$ plane of GaN is parallel to the (111) plane of the silicon substrate, $<11\bar{2}\bar{6}>$ direction of GaN can be aligned along the $<1\bar{1}0>$ and $<11\bar{2}>$ silicon directions with a ±5° mosaic spread.

Acknowledgements

This work was carried out with the support of President grant (14 Z56.17.3632-MK), the Russian Foundation for Basic Research (15-02-06839 A, 16-32-60094 mol_a_dk and 16-32-00560 mol_a), the leading universities of the Russian Federation (grant 074-U01), grant of government of the Russian Federation (16.2593.2017/8.9, 16.2593.2017/4.6).

References

[1] Mandl M et al 2013 *Phys. Status Solidi RRL* **7** 800–814
[2] Consonni V 2013 *Phys. Status Solidi RRL* **7** 699–712
[3] Mozharov A, Bolshakov A, Cirlin G, and Mukhin I 2015 *Phys. Status Solidi RRL* **9** 507–510
[4] Dadgar A 2015 *Phys. Status Solidi B* **252** 1063–68
[5] Ristić J, Calleja E, Fernández-Garrido S, Cerutti L, Trampert A, Jahn U and Ploog K H 2008 *J. Cryst. Growth* **310** 4035–45
[6] Kawazu A and Sakama H 1988 *Phys. Rev. B* **37** pp. 2704–2706
[7] Čechal J, Kolíbal M, Kostelník P and Šikola T 2007 *J. Phys.: Condens. Matter* **19** 16011
[8] Kumar P, Kuyyalil J and Shivaprasad S M 2010 *Appl. Phys. Lett.* **97** p. 221913
[9] Zhou Z Y, C. X. Zheng C X, Tang W X, Tersoff J, and Jesson D E 2013, *Phys. Rev. Let.* **111** 36102
[10] Lee S, Sohn Y, Kim C, Lee D R and Lee H -H 2010 *J. Appl. Cryst.* **43** 1300-04
[11] Wang T 2016 *Semicond. Sci. Technol.* **31** 93003
[12] Yu I S, Chang C P, Yang C P, Lin C T, Ma Y R, and Chen C C 2014 *Nanoscale Res Lett.* **9** 682
[13] Debnath R K, Stoica T, Besmehn A, Jaganathan K, Meijers R, Lüth H and Calarco R 2009 *J. Cryst. Growth* **311** 3389–94

IOP Conf. Series: Journal of Physics: Conf. Series **917** (2017) 032041 doi:10.1088/1742-6596/917/3/032041

Electronic and optical properties of HEMT heterostructures with δ-Si doped GaAs/AlGaAs quantum rings — quantum well system

Y D Sibirmovsky[1], I S Vasil'evskii[1], A N Vinichenko[1], D M Zhigunov[2], I S Eremin[1], O S Kolentsova[1], D A Safonov[1] and N I Kargin[1]

[1] National Research Nuclear University "MEPhI", 115409, Moscow, Russia
[2] Lomonosov Moscow State University, 119992, Moscow, Russia

E-mail: `YDSibirmovsky@mephi.ru`

Abstract. Samples of δ-Si doped AlGaAs/GaAs/AlGaAs HEMT heterostructures with GaAs quantum rings (QRs) on top of the quantum well (QW) were grown by molecular beam epitaxy and their properties were compared to the reference samples without QRs. The thickness of the QW was 6 – 10 nm for the samples with QRs and 20 nm for the reference samples. Photoluminescence measurements at low temperatures for all samples show at least two distinct lines in addition to the bulk GaAs line. The Hall effect and low temperature magnetotransport measurements at 4 – 320 K show that conductivity with and without illumination decreases significantly with QRs introduction, however the relative photoconductivity increases. Samples with 6 nm QW are insulating, which could be caused by the strong localization of the charge carriers in the QRs.

1. Introduction

Semiconductor quantum rings promise a wide range of applications, especially for optoelectronics, such as lasers, solar cells and photodetectors [1]. They could also be used as a low mobility layer in velocity-modulation transistors [2]. Unlike quantum dots, their conductivity can be strongly affected by magnetic field, which makes it possible to create spintronics devices without the need for magnetic materials. These cases make it necessary to study the lateral transport in the arrays of doped QRs. There are few investigations into the properties of the doped QRs. In such cases, the dopant (usually Si) is introduced directly into the QRs [1]. However, the fact that Ga and As are deposited separately during QRs growth [3] influences Si incorporation, and could even change the doping type from n to p [4]. Thus, it is prudent to investigate δ-Si doped QR layers.

2. Experimental samples

In this work, electron Hall concentration and mobility, as well as PL spectra are compared for the samples with GaAs QRs-QW system surrounded by $Al_{0.3}Ga_{0.7}As$ barrier (samples 339, 343, 344) and the reference samples without QRs (samples 320, 321 and 342) for $T = 4 - 320$ K. Special care was taken to ensure the high quality of the quantum well, which was achieved by

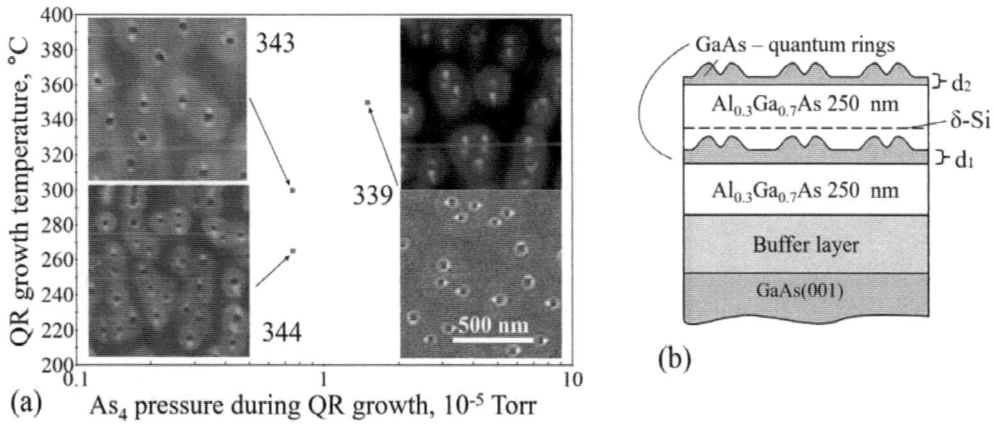

Figure 1. (a) QRs growth conditions and 1x1 μ AFM scans (for sample 339 a SEM scan as well); (b) layer structure of the samples, details are provided in the text.

fully reproducing a reference HEMT heterostructure until the QRs growth step. For example, electron mobility in sample 342 is $4.2 \cdot 10^5$ cm^2/V·s ($1.65 \cdot 10^5$ cm^2/V·s) at T= 4.2 K (77 K). The dopant sheet concentration in the δ-Si layer is of the order of 10^{12} cm^{-2} for all the samples. It is important to note that the buffer layer contains two 10 period GaAs/$Al_{0.3}Ga_{0.7}As$ superlattices with layer thickness values of 2.4/4.7 nm and 3.5/14 nm respectively.

The QRs growth conditions, the corresponding surface scans and the layer structure of the samples are shown in Figure 1. d_1 = 10 nm for sample 339 and 6 nm for samples 343, 344. The top QR layer was grown for AFM measurements and was not doped. d_2 = 3 nm for all samples. More information is provided in Table 1. For details on the growth procedure see [5]. The spacer layer thickness is 15 nm.

Table 1. The growth conditions and resulting parameters of the QRs. n is the surface density, λ is the mean distance between the centers, d is the inner ring diameter, D is the outer ring diameter, h is the height.

No.	P_{As4}, Torr	T, °C	n, μm^{-2}	λ, nm	d, nm	D, nm	h, nm
339	$1.5 \cdot 10^{-5}$	350	15	250	86	233	14
343	$0.75 \cdot 10^{-5}$	300	13	267	91	242	12
344	$0.75 \cdot 10^{-5}$	265	28	192	66	132	16

3. Measurement results and discussion

Electron transport was studied by Hall effect (Van der Pauw geometry) at T = 77-320 K and low temperature magnetotransport measurements (LTM) at T = 2-300 K and B = 0-6.5 T. The measurements were conducted for the dark and illuminated samples. Illumination for constant field Hall measurements was provided by a white LED kept at room temperature, while for LTM we used IR LED kept at the same temperature as the sample. The current was 4 μA for Hall and 6 μA for LTM.

IOP Conf. Series: Journal of Physics: Conf. Series **917** (2017) 032041 doi:10.1088/1742-6596/917/3/032041

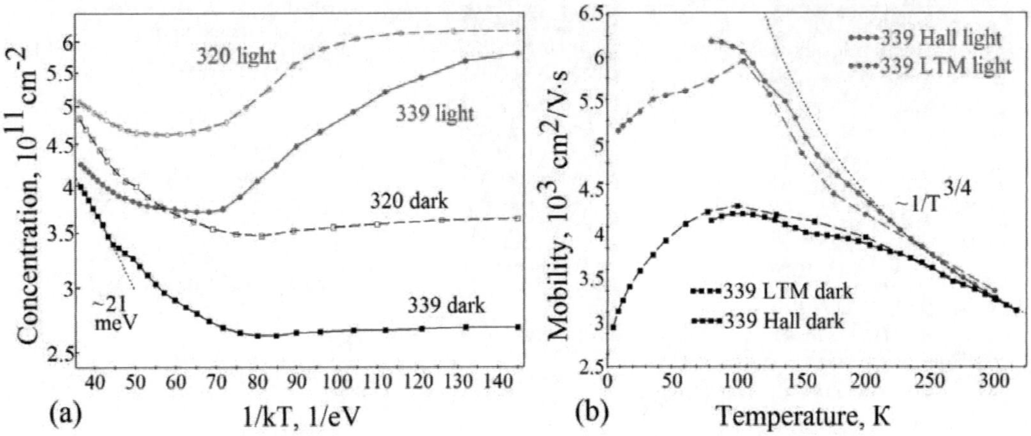

Figure 2. (a) Hall concentration (logarithm) for samples 339 and reference sample 320; (b) mobility for sample 339 from constant field Hall measurements ($B = 0.585$ T) and low temperature magnetotransport measurements (LTM).

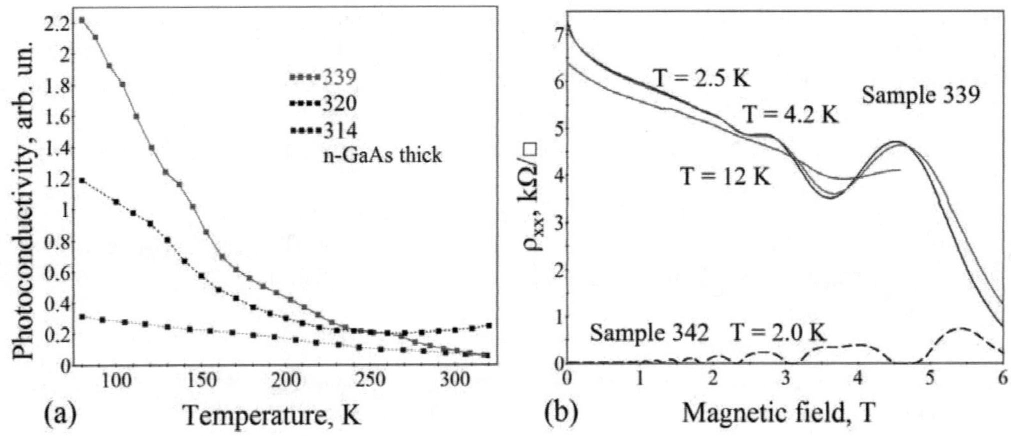

Figure 3. (a) Relative photoconductivity for sample 339 and reference samples; (b) $\rho_{xx}(B)$ for sample 339 and reference sample 342.

Samples 343 and 344 are insulating which can be attributed to the decreased QW thickness and stronger localization. However, it's possible that sample 343 was not doped due to the Si source shutter error. As for sample 344, comparing it to conducting sample 339, we can conclude that 6 nm thick QW is not enough to prevent localization of electrons inside QRs, or at least strong interface roughness scattering which starts to play a role for thin QWs [6].

Since the dopant concentrations in sample 339 and reference sample 320 are the same, we can directly compare the measurement results. Hall concentration in sample 339 has decreased by approximately $1 \cdot 10^{11}$ cm^{-2} (Figure 2a), while mobility decreased more than 2 times at room temperature and around 100 times at low temperatures compared to sample 320. The main reason for both of these results is electron scattering and localization in the disordered QR layer on top of the QW. Due to low temperature of the droplet epitaxy process it is also possible that the spacer layer contains additional defects, which could trap electrons.

Temperature dependence of the Hall concentration for dark and illuminated sample has the same character for samples 339 and 320. Without illumination the concentration is almost constant at low temperatures and slowly grows for $T > 150$ K, which is caused by thermal ionization of deep donors [7, 8]. For high temperatures the estimated energy is around 21 meV (Figure 2a), however using the low compensation model $n \sim \exp(-E/2kT)$ we obtain the activation energy $E = 42$ meV. Illumination at low temperatures leads to ionization of deep donors and persistent photoconductivity, the mechanism of which is described in [8, 9].

Unlike the reference samples, the mobility for sample 339 decreases as the temperature falls below 100 K, which is more common for bulk semiconductors. At low temperatures the main contributors to the scattering are the ionized shallow donors as well as extreme interface roughness caused by quantum rings. Which is why illumination of this sample leads to a significant increase in mobility, related to the screening of the potential roughness by generated electrons.

For higher temperatures, the mobility of the illuminated sample has the temperature dependence $\sim T^{-3/4}$ typical for phonon scattering. However, in the dark the dependence is weaker, which means that other significant scattering mechanisms are sill involved.

Comparing concentration of photogenerated electrons (the difference between measurements with and without illumination), we can see that it's approximately the same for samples 339 and 320. On the other hand, the relative photoconductivity (ratio of light to dark conductivity) is much higher for sample 339.

Weak Shubnikov-de-Haas oscillations were observed for sample 339 in high magnetic fields, while the reference samples show at least 4 full oscillations and quantum Hall plateaux (Figure 3b). This is expected due to low mobility (Figure 2b).

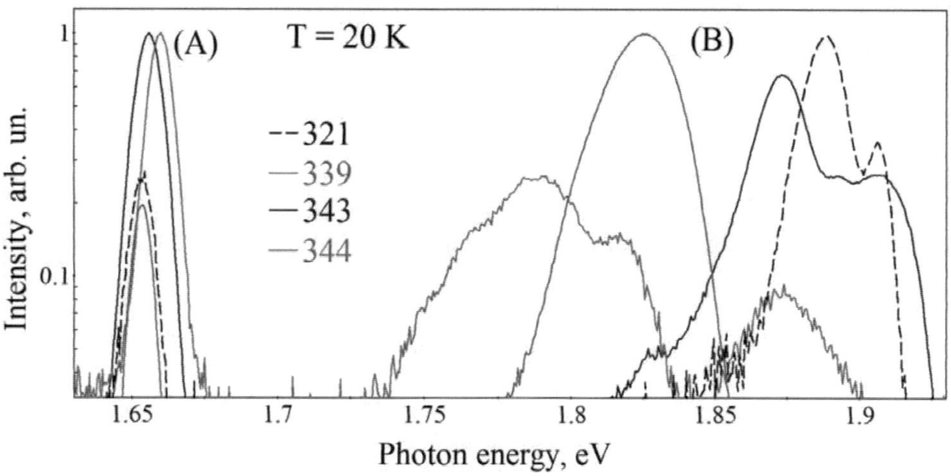

Figure 4. Normalized PL spectra for samples 339, 343, 344 and reference sample 321. Bulk GaAs line at ~ 1.52 eV is omitted for all samples. Labels (A) and (B) are introduced for convenience.

For PL measurements 488 nm Ar^+ laser was used. Attempt to identify PL lines was made by comparing their energy with GaAs band gap (1.52 eV at 20 K) and $Al_{0.3}Ga_{0.7}As$ band gap (1.90 eV at 20 K) [10]. Bulk GaAs line was observed for all samples, as well as the line labelled (A) in Figure 4. Since line (A) exists for reference sample 321 without QRs, we have initially assumed it to be caused by the QW or the buried QW-QRs system. However, calculations indicate that

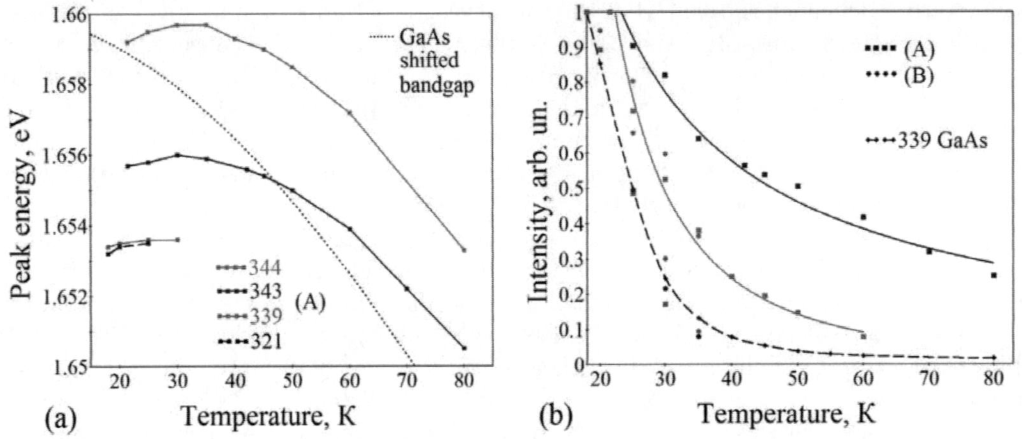

Figure 5. Temperature dependence of (a) peak energy of (A) lines and (b) normalized maximum intensity for (A) and (B) lines for all samples.

line (A) is most likely caused by the recombination inside the buffer layer, namely the second GaAs/$Al_{0.3}Ga_{0.7}As$ superlattice. Which is why the lines are so close for all the samples.

As illustrated by Figure 4, line (or lines) labelled (B) is different in both shape and position for all the samples. On the one hand, this line could be caused by the recombination inside the $Al_{0.3}Ga_{0.7}As$ barrier. For samples 321 and 343 the position of line (B) is in good agreement with the calculated band gap. Sample 344 has a weak line there as well. On the other hand, for sample 339 line (B) is centred around 1.83 eV, while sample 344 has a line at 1.79 eV. The exact source for these lines is still undefined.

For line (A) the peak energy initially increases with temperature growth, then starts to decrease (Figure 5a), in qualitative agreement with the temperature dependence of GaAs band gap [10] (as could be seen for samples 343 and 344, the line is quickly quenched for the other samples). All lines are eventually quenched with temperature, however the rate is different (Figure 5b). Intensity of line (A) for sample 343 decreases slower than all the other lines.

Acknowledgements
This work was supported by the Russian Foundation for Basic Research (RFBR): grant 16-32-00897.

References

[1] Wu, J et al. 2009 *Appl. Phys. Lett.* **94** 171102
[2] Sakaki, H. 1982 *Jpn. J. Appl. Phys.* **21** L381L383
[3] Lee, C-D et al. 1998 *Appl. Phys. Lett.* **73** 26152617
[4] Lamas, T E et al. 2005 *Thin Solid Films* **474** 2530
[5] Sibirmovskii Yu D, Vasil'evskii I S, Vinichenko A N et al. 2014 *Bull. Lebedev Phys. Inst.* **41** 243
[6] Sakaki, H, Noda, T, Hirakawa, K, Tanaka, M, Matsusue, T 1987 *Appl. Phys. Lett.* **51** 19341936
[7] Schubert, E F, Knecht, J, Ploog, K 1985 *Journal of Physics C* **18** L215
[8] McKinnon, W R, Hurd, C M 1987 *J. Appl. Phys.* **61** 22502256
[9] Hurd, C M et al. 1987 *J. Appl. Phys.* **61** 22442249
[10] Levinshtein M, Rumyantsev S, Shur M 1996 *Handbook Series on Semiconductor Parameters* (Singapore: World Scientific)

Researching the electrical properties of single A3B5 nanowires

A A Vasiliev[1], A M Mozharov[1,2], F E Komissarenko[2], G E Cirlin[3], D A Buravlev[3] and I S Muckin[1,2]

[1]Renewable Energy Lab, St. Petersburg Academic University, St. Petersburg 194021, Russia
[2]The metamaterial Lab, University ITMO, Saint-Petersburg 197101, Russia
[3]Epitaxial nanotechnology Lab, St. Petersburg Academic University, St. Petersburg 194021, Russia

E-mail: ftf.vasiliev@yandex.ru

Abstract. We investigate electrical characteristics of GaN, GaAs and GaP NWs which are grown with MOCVD and MBE. We developed measurement technique and it allows to determine the required properties of the structures.

1. Introduction

The A3B5 material systems can be applied to create LEDs, detectors, transistors, solar cells and others. However, there are several technologic difficulties such as obtaining high-quality and effective planar structures (in the case of GaN, etc.) and the high cost of production (GaAs, etc.). Therefore, it's required to find new approaches to create the semiconductor structures. Nowadays, researchers carry out many experimental and theoretical work to understand the process of formation NWs and how they can be doped. NWs are rods, which diameter is much smaller than their length. Since the geometrical parameters, composition and doping level can be controlled, NWs can be successfully used to develop a lot of semiconductor devices [1,2].

1.1.Synthesis Nws

There are two most widely use growth technology: metal-organic chemical vapour deposition (MOCVD) and molecular beam epitaxy (MBE). Both technologies have some advantages and limitations. For example, MBE gives us atomic layered control of growth. In the case of using MOCVD the average growth speed is higher than in MBE. Using these technologies and different methods of growth NWs can be synthesized. The most common method is based on the vapour-liquisolid mechanism with metal catalysts [4,5]. In this work we investigate GaN, GaAs NWs which were produced by means of MOCVD and MBE [6,7,8].

1.2. Measurements

There are different measurement techniques which are used to determine electrical properties of planar semiconductor heterostructures. However, not some of them are suitable for measuring properties of NWs. Therefore, we developed a methodology for measuring the properties of single NWs.

In order to determine the electrical properties we need to create the ohmic contacts to single NW. Measuring the VAC we can calculate conductivity $\sigma=1/R=1/V$. In order to independently determine the carrier concentration and their mobility in the NW, it is necessary to use a transistor structure. It's known, that conductivity is proportional to carrier concentration and mobility. However, field effects mainly depends on the concentration, therefore we can calculate carrier concentration using a transistor structure $n=1/e\rho\mu$.

2. Experiment

In order to form ohmic contact to a single nanowire the following procedure was applied. The synthesized NWs were separated from the growth substrate with ultrasonic method and transferred to the quartz or Si/SiO2. Contacts were created using methods of the laser photolithography or electron lithography and thermal evaporation in a vacuum. For some samples annealing is required (annealing temperature depends on the sample). For producing the transistor structure, it's necessary to form the third contact. It can be placed on the back side of the substrate, or on top of the structures.

2.1.Investigation of GaN NWs

We studied GaN NW's conductivity, carrier concentration and mobility. We used MESFET structure to independently determine the carrier concentration and mobility. The drain/source ohmic contacts Ti/Al/Ti/Ag (30/150/30/100 nm, they were annealed at 710 °C) were formed on a quartz substrate as shown in Figure 1 (a). We measured VAC and determined that conductivity was 16 S/cm. The gate contact was evaporated with Pd (50 nm). Then we measured VAC of the gate and I_{ds} from V_g as shown on Figure 1 (b, c).

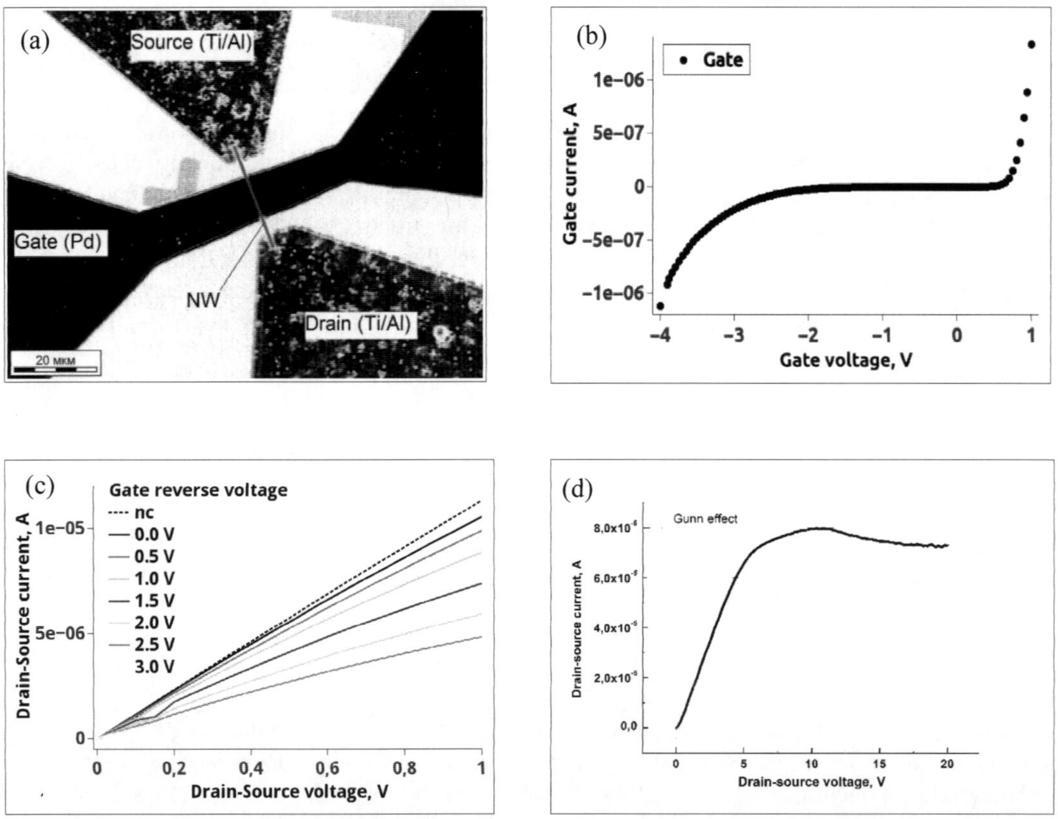

Figure 1(a,b,c). Scheme of the transistor **(a)**, VAC of the transistor **(b)**, VAC through the gate for the source and drain contacts **(c)**.

The contact of gate is made of aluminium (it is 300 nm thick and it was annealed at 690 °C). Contacts drain/source were formed from Ni/Pd/Zn/Pd (5/15/10/30 nm, without annealing). As a result of measurements, the following VAC characteristics of the NWs were obtained when different voltages were applied to the source/gate as shown in Figure 2 (b, c).
When the gate voltage increases, the NWs's conductivity increases too. This is due to growth of the carriers density in the NW-electrons. Therefore, it can be concluded that the NWs is n-type.

2.2. Investigation of the type of doping of GaAs NWs
We investigated a type of GaAs NWs and needed to produce a transistor structure on a NW (figure 2a). For this we used a Si/SiO2 substrate (substrate is p-type, thickness of Si02 layer is 224 nm).

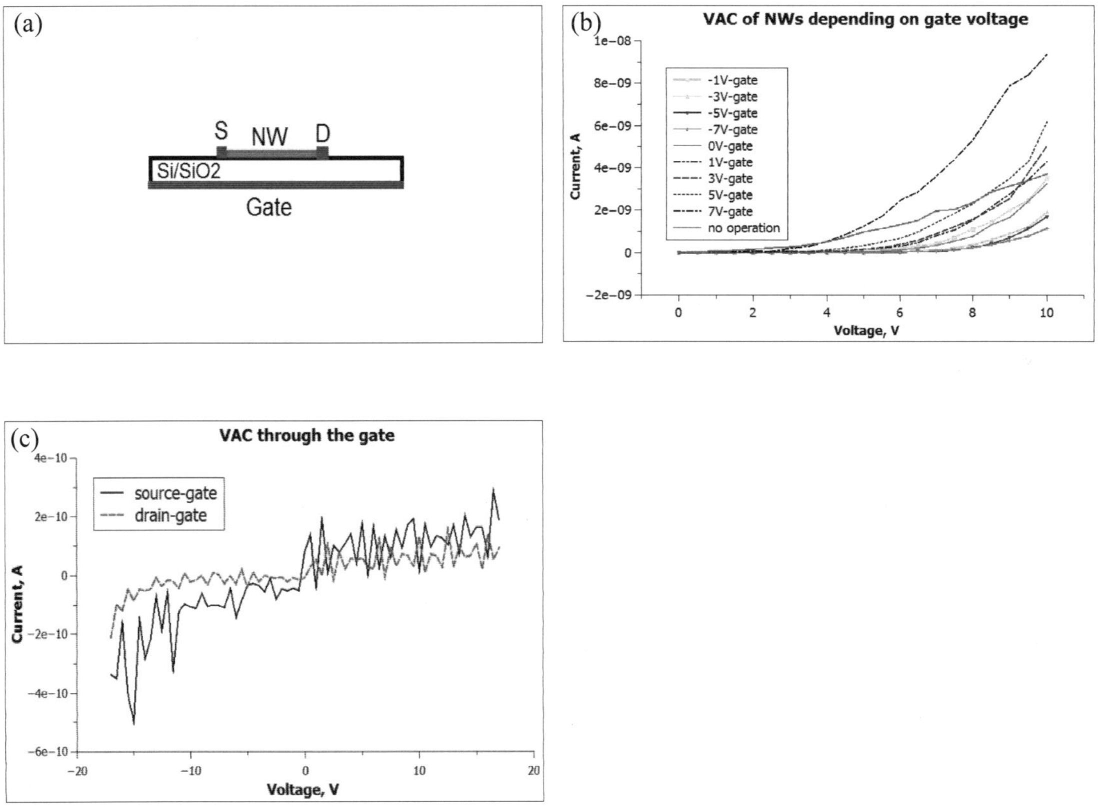

Figure 2(a,b,c). Scheme of the transistor **(a)**, VAC of the transistor **(b)**, VAC through the gate for the source and drain contacts **(c)**.

The contact of gate is made of aluminium (it is 300 nm thick and it was annealed at 690 °C). Contacts drain/source were formed from Ni/Pd/Zn/Pd (5/15/10/30 nm, without annealing). As a result of measurements, the following VAC characteristics of the NWs were obtained when different voltages were applied to the source/gate as shown in Figure 2 (b, c).

When the gate voltage increases, the NWs's conductivity increases too. This is due to growth of the carriers density in the NW-electrons. Therefore, it can be concluded that the NWs is n-type.

2.3. Investigation of the mobility and carriers concentration in GaAs-p type NWs
To estimate mobility and carrier concentration contacts were formed to a single NW on a quartz substrate. Contacts were created from Ni/Pd/Zn/Pd (5/15/10/70 nm). Because NWs were doped we got an ohmic contacts to some NWs without annealing as shown in Figure 3, annealing temperature is 405 °C).

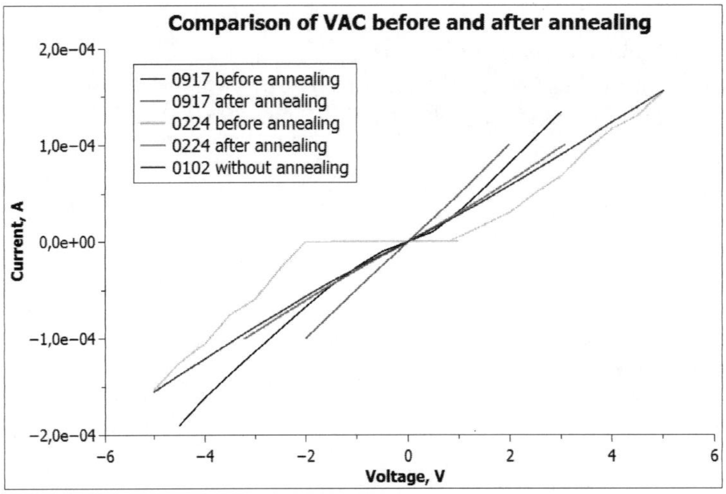

Figure 3. VAC of the contact.

From the obtained data, we estimated the value of $n\mu \simeq (0.5/5)10^{21}$ $(V \cdot s \cdot cm)^{-1}$. Using the dependence of n on μ in the GaAs-p type, we could established that $\mu \simeq 50$ $cm^2/(V \cdot s)$, $n \simeq 10^{19} \div 10^{20}$ cm^{-3}.

3. Summary
In this work we developed the technique of measurements the electrical characteristics of single NWs. In according to this technique the properties of the single GaN and GaAs NWs were carried out. We can conclude that method allows to investigate different structures.

4. Acknowledgments
This work was carried out with the support of President grant (14.Z56.17.3632-MK), the Russian Foundation for Basic Research (150206839 A, 16-32-60094 mol_a_dk and 163200560 mol_a), the leading universities of the Russian Federation (grant 074U01), grant of government of the Russian Federation (16.2593.2017/ПЧ).

References
[1] Li Y, Xiang J, Quang F, Gradecak S, Wu Y, Yan H, Blom D A and Lieber C M. 2006 *Nanoletters* **6** 1468
[2] Bjork M T, Ohlsson B J, Sass T, Persson A I, Thelander C, Magnusson M H, Deppert K,Wallenberg L R and Samuelson L 2002 *Appl. Phys. Lett.* **80** 1058
[3] Li S, Waag A 2012 *J. Appl. Phys.* **111** 071101
[4] Kihyun C, Munetaka A and Yasuhiko A 2012 *J. of Crystal Growth* **357** 58-61

[5] Rozhavskaya M M, Lundin V V, Lundina E, Davydov S, Troshkov S I, Vasilyev A A, BrunkovP, Baklanov A, Tsatsulnikov A F, Dubrovskii V G 2014 *J. Appl. Phys.* **14** 11399
[6] Stephen D H, Xinyu S and Xin W 2006 *Nanoletters* **6(8)** 1808-11
[7] Bouravleuv A D, Sibirev N V, Gilstein E P, Brunkov P N, Mukhin I S, Tchernysheva M,Khrebtov A I, Samsonenko Yu B and Cirlin G E 2014 *Semiconductors* **48(3)** 344-49
[8] Bolshakov A D, Sapunov G A, Mozharov A M, Cirlin G E, Shtrom I V and Mukhin I S 2016 *J. Phys.* **741** 012044

Length distributions of Au-catalyzed III-V nanowires in different regimes of the diffusion-induced growth

Y Berdnikov[1,2], A A Zhiglinsky[2], M V Rylkova[2] and V G Dubrovskii[1,2,3]

[1] St. Petersburg Academic University, Khlopina 8/3, 194021, St. Petersburg, Russia
[2] ITMO University, Kronverkskiy pr. 49, 197101 St. Petersburg, Russia
[3] Ioffe Physical Technical Institute RAS, Politekhnicheskaya 26, 194021, St. Petersburg, Russia

Abstract. We present a model for kinetic broadening effects on the length distributions of Au-catalyzed III-V nanowires obtained in the growth regime with adatom diffusion from the substrate and the nanowire sidewalls to the top. We observe three different regimes for the length distribution evolution with time. For short growth times, the length distribution is sub-Poissonian, converting to broader than Poissonian with increasing the mean length above a certain threshold value. After the diffusion flux from the nanowire sidewalls has stabilized, the length distribution variance increases linearly with the mean length, as in the Poissonian process.

1. Introduction

Semiconductor nanowires (NWs) and in particular III-V NWs are considered promising for fundamental nanoscience as well as building blocks of the next generation electronic and optoelectronic devices. NWs allow for almost unlimited bottom-up bandgap engineering and can be grown on dissimilar substrates without dislocations [1,2]. For example, it is very difficult to grow defect-free InAs Stranski-Krastanow islands on silicon substrates (lattice mismatch 11.6%) [3], while high quality epitaxial InAs NWs can easily be grown on Si(111) if their diameter is smaller than the critical dimension for plastic relaxation (~ 25 nm) [4]. However, size (length and diameter) uniformity within the NW ensembles remains an issue for various electronic and photonic applications [5-7]. NWs obtained by Au-catalyzed vapor-liquid-solid (VLS) growth technique were previously thought, and sometimes without grounds, to have high degree of the length uniformity [8,9]. Broadening of the length distributions (LDs) induced by kinetic fluctuations are expected to be Poissonian [10,11] in the case of length-independent axial growth rates. Recent studies of III-V NW length statistics show that slow NW nucleation from the catalyst droplets resting on the substrate surface and surface diffusion of group III adatoms from the NW sidewalls lead to a considerable broadening of the LDs [12,13]. So far, the NW LDs were analysed for systems with negligible adatom diffusion from the substrate. However, in some VLS systems this transport mechanism dominates in the initial growth stage [14-17]. Consequently, in this work we investigate the effect of different kinetic pathways of surface diffusion on the NW LDs.

2. Model

We consider diffusion-induced growth of Au-catalysed III-V NWs limited by transport of group III atoms, deposited from vapor with the influx I onto the droplet surface. We use the rate equations (REs) approach to describe the LDs over the ensemble of NWs [11,12]

$$\frac{dn_0}{dt} = -K_0 n_0, \tag{1}$$

$$\frac{dn_s}{dt} = K_{s-1} n_{s-1} - K_s n_s, \tag{2}$$

where n_s is the surface density of NWs having the length of $s = L/h$ monolayers, with L as the NW length and h as the height of a monolayer.

According to the model of Ref. [17], the diffusion-induced axial growth rate of NWs at a time-independent radius R depend on the NW length L and the diffusion lengths of surface (λ_s) and sidewall (λ_f) group III adatoms:

$$\left(\frac{dL}{dt}\right)_{diff} = V\frac{BU(L/\lambda_f)\ +C}{U\prime(L/\lambda_f)}. \tag{3}$$

Here, $V \equiv I\Omega_s \cos\alpha$ is the normalized influx from vapour, with Ω_s as the elementary volume in the solid phase, α is the beam angle with respect to the substrate normal (we write all the equations using the language of molecular beam epitaxy). The radius-dependent coefficients B and C describe the diffusion fluxes originating from the NW sidewalls and the substrate, respectively, which are determined by the adatom activities in the liquid droplet (θ_l), on the NW sidewalls (θ_f) and on the substrate (θ_s) according to [17]

$$B = \frac{2\lambda_f \tan\alpha}{2\pi R}\left(1 - \frac{\theta_l}{\theta_f}\right), \tag{4}$$

$$C = \frac{2\lambda_s F}{R}\left(1 - \frac{\theta_l}{\theta_s}\right). \tag{5}$$

For a dilute ensemble of NWs, the F function is given by $F = K_1(R/\lambda_s)/K_0(R/\lambda_s)$, where K_i are the modified Bessel function of second kind of the order I [17].

When the diffusion lengths and the NW radius are much larger than the lattice constant, the function U is reduced to [17]

$$U(L/\lambda_f) = \sinh\left(\frac{L}{\lambda_f}\right) + b\frac{\lambda_f}{\lambda_s}F\left(\cosh\left(\frac{L}{\lambda_f}\right) - 1\right), \tag{6}$$

with $U'\left(\frac{L}{\lambda_f}\right) = dU/d\left(\frac{L}{\lambda_f}\right)$ in Eq. (3). In Au-catalysed VLS process, NWs emerge from pre-existing gold droplets, which initially have the surface density n_{tot}. During growth, the sum of the surface densities of differently sized NWs and the remaining surface droplets stays constant, yielding the conservation of zero moment of the LD: $n_{tot} = \sum_{s\geq 0} n_s = const$. Using Eq. (3) and neglecting the direct impingement term with respect to diffusion, we can rewrite the REs given by Eqs. (1) and (2) in terms of the normalized LD $f_s = n_s/n_{tot}$ and the dimensionless time $\tau = \frac{V}{h}t$

$$\frac{df_0}{d\tau} = -v_0 f_0, \tag{7}$$

$$\frac{df_s}{d\tau} = v_{s-1} f_{s-1} - v_s f_s \,, \tag{8}$$

with

$$v_s \equiv \frac{ds}{d\tau} = \frac{BU\left(\frac{s}{\Lambda_f}\right)+C}{U'\left(\frac{s}{\Lambda_f}\right)} \,. \tag{9}$$

Here, $\Lambda_s = \lambda_s/h$ and $\Lambda_f = \lambda_f/h$ are the diffusion lengths measured in numbers of monolayers, $C = \frac{2\Lambda_s F}{r}\left(1 - \frac{\theta_l}{\theta_s}\right)$, and $F = K_1(r/\Lambda_s)/K_0(r/\Lambda_s)$, with $r = R/h$ as the dimensionless NW radius.

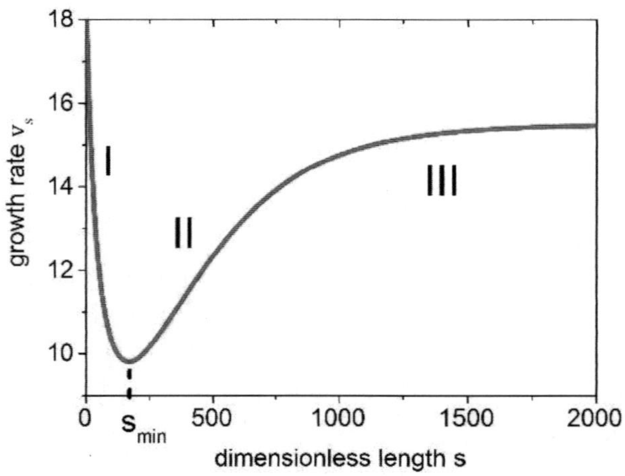

Figure 1. Three regimes of the axial growth rate versus the NW length: decrease of v_s due to diminishing diffusion flux from the substrate (I), its increase in the intermediate range where group III atoms are collected from the entire NW length (II), and stabilization after a certain length when the group III atoms are collected only from the topmost section of a NW (III).

3. Results and discussion

We can now numerically solve the REs given by Eqs. (7) to (9). We observe different growth regimes corresponding to different dependences of the v_s on the NW length. Figure 1 shows the growth rate versus length obtained from Eq. (9) at $r = 40, \Lambda_s = 150\,, \Lambda_f = 310,\ B = 15.5, C = 5$. The v_s first decreases due to diminishing diffusion flux from the substrate, then increases at intermediate lengths as long as the material is collected from the entire NW length [11] and finally saturates to a constant for long NW that collects the material only from its topmost section with a length on the order of Λ_f. Such a behavior has been observed experimentally and described theoretically [17]. In regime (I) with decreasing v_s, numerical solution of the REs gives rather uniform LDs which appear narrower than Poissonian as shown in Figure 2 (a).When the NW length exceeds the threshold value s_{min} corresponding to the minimum of the growth rate, increasing v_s (regime II) leads to rapid spreading of the LDs which become broader than Poissonian as shown in Figure 2 (b). This process is similar to the diffusion-induced broadening described in Ref. [11]. For longer NWs whose length exceeds the sidewall diffusion length, the growth rate stabilizes to a constant, which corresponds to regime III.

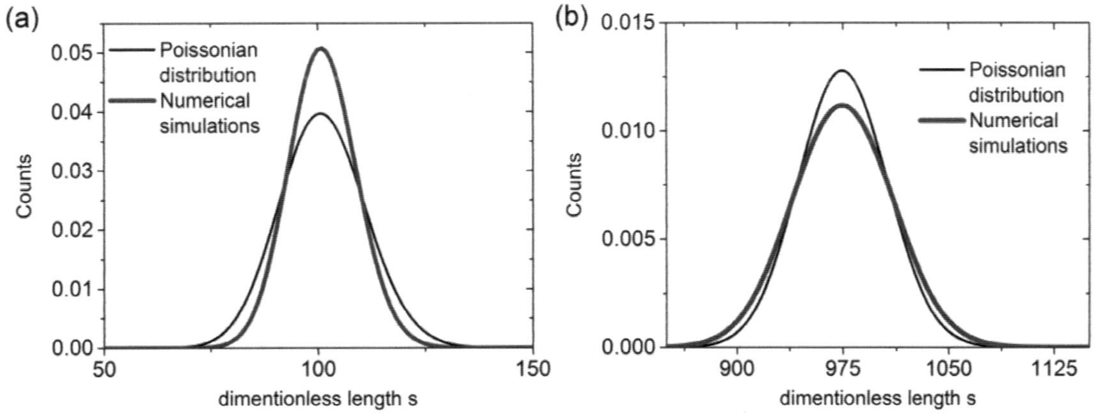

Figure 2. NW LDs obtained by numerical integration of Eqs. (7) and (8) with the growth rates given by Eq. (9) (thick lines) in regimes I (a) and II (b), compared to the Poissonian LDs (thin lines).

Analytical solution for the LDs obtained in Ref. [11] predicts that the LD variance scales as the mean length $\langle s \rangle$ in the case of s-independent growth rate and is proportional to $\langle s \rangle^2$ in case of s-linear growth rate. This property is confirmed by Figure 3 showing the variance versus the mean length, which becomes linear for NWs much longer than the diffusion length Λ_f. The variance increases faster than linear in the intermediate growth stage corresponding to regime II with an increasing growth rate. In contrast to the results of Ref. [11], a sharp change from superlinear to linear regime just above Λ_f is replaced by a smooth transition between them. This difference is related to a more precise expression for the growth rate used here compared to the one in Ref. [11].

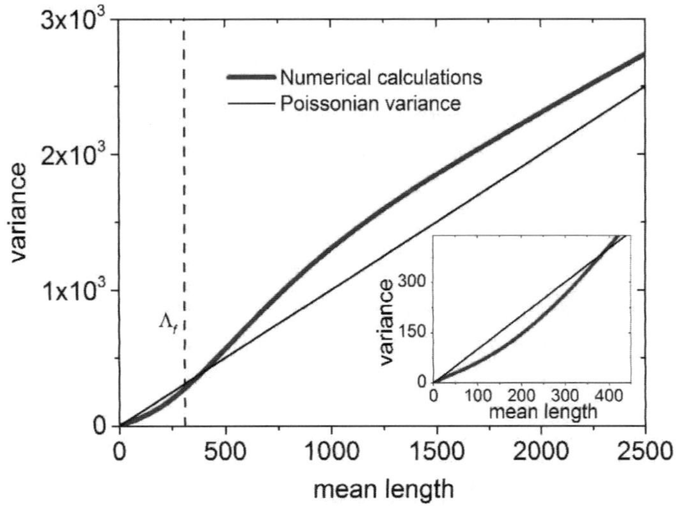

Figure 3. Variance of the NW LD versus the mean length obtained from Eqs. (7) to (9) (thick lines), compared to the linear Poissonian variance (thin lines). The inset shows the same dependence for short NWs. Dashed line shows the diffusion length of adatoms on the NW sidewalls Λ_f.

Quite interestingly, ensembles of short NWs feature sub-Poissonian variance that has not been reported so far, even in modelling. This is explained by the fact that a very limited number of works consider growth rates that decrease with the NW length. In real systems, this initial narrowness may be hidden by the nucleation randomness for the NW themselves [12], disregarded in this study. In our modelling, the LDs narrower than Poissonian can only be observed in regime I and strongly depend on the NW radius, assumed identical for all NWs in the ensemble. Figure 4 shows the variations of v_s and the LD shapes with the NW radius r. For large radius $r = 100$, the growth rate shows very slow length dependence that corresponds to almost Poissonian LD. For smaller radii, the length dependence of the growth rate becomes steeper and this effect narrows the LD.

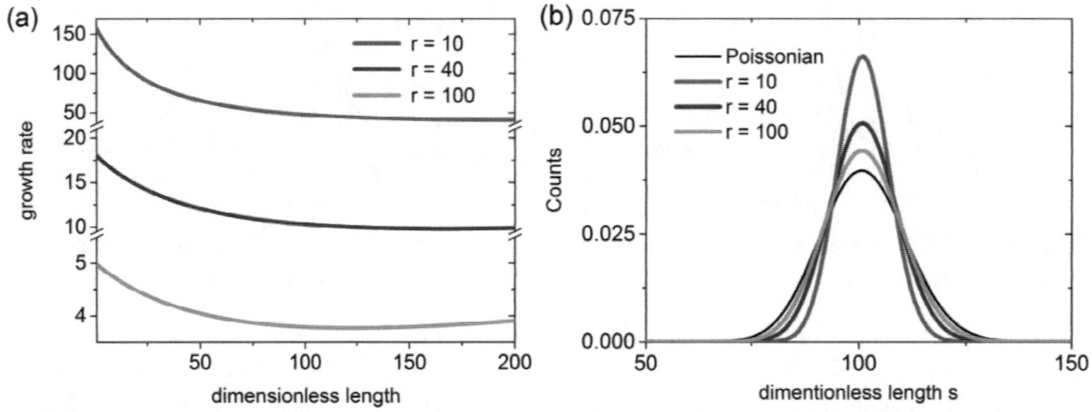

Figure 4. Growth rate dependences on the NW length **(a)** and the corresponding LDs **(b)** for different NW radii r.

In conclusion, we have developed a model for the LDs of Au-catalyzed NWs growing by surface diffusion from the substrate surface and the NW sidewalls. Non-linear length dependence of the axial NW growth rate yields an interesting time evolution of the LDs whose nucleation rate is fast. At the beginning, the growth rate decreases with length and the LDs of sufficiently short NWs is sub-Poissonian. After the substrate diffusion flux is disabled, the axial growth rate increases with length and the LD rapidly spreads. Finally, after the diffusion flux from the NW sidewalls has stabilized, the LD variance becomes linear in the mean length, as in the Poissonian growth. These predictions are better for highly desired length uniformity than the earlier results of Ref. [11]. Overall, this study identifies major factors influencing the LD shapes in different stages of the diffusion-induced growth of NWs.

Acknowledgement

The authors gratefully acknowledge financial support received from the Ministry of Education and Science of the Russian Federation under grant 14.587.21.0040 (project ID RFMEFI58717X0040).

References:

[1] Glas F 2015 *Strain in nanowires and nanowire heterosructures* Semiconductors and Semimetals 93 ed. Anna Fontcuberta I Morral, Shadi A Dayeh and Chennupati Jagadish (Burlington: Academic Press) pp.79-123
[2] Zhang X, Dubrovskii V G, Sibirev N V, Ren X 2011 *Cryst. Growth Des.* **11** 5441
[3] Cirlin G E, Dubrovskii V G, Petrov V N et al. 1998 *Semicond. Sci. Technol.* **13** 1262
[4] Chuang L C, Moewe M, Crankshaw S, Chase C, Kobayashi N P, Chang-Hasnain C 2007 *Appl. Phys. Lett.* **90** 043115
[5] Yang P, Yan R and Fardy M 2010 *Nano Lett.* **10** 1529.
[6] Dubrovskii V G 2014 *Nucleation Theory and Growth of Nanostructures* (Heidelberg, New York, Dordrecht, London: Springer).

[7] Johansson J and Dick K A 2011 *Cryst. Eng. Comm* **13** 7175.
[8] Ohlsson B J, Björk M T, Magnusson M H, Deppert K, Samuelson L and Wallenberg L R 2001 *Appl. Phys. Lett.* **79** 3335.
[9] Qian F, Li Y, Gradečak S, Park H-G, Dong Y, Ding Y, Wang Z, Lieber C M 2008 *Nature Materials* **7** 701.
[10] Dubrovskii V G 2013 *Phys. Rev. B.* **87** 195426.
[11] Dubrovskii V G, Berdnikov Y, Schmidtbauer J, Borg M, Storm K, Deppert K and Johansson J 2016 *Cryst. Growth Des.* **16** 2167.
[12] Dubrovskii V G, Sibirev N V, Berdnikov Y, Gomes U P, Ercolani D, Zannier V and Sorba L 2016 *Nanotechnology* **27** 375602.
[13] Dubrovskii V G, Xu T, Lambert Y, Nys J-P, Grandidier B, Stievenard D, Chen W and Pareige P 2012 *Phys. Rev. Lett.* **108** 105501.
[14] Cirlin G E, Dubrovskii V G, Sibirev N V, Soshnikov I P, Samsonenko Yu B, Tonkikh A A, Ustinov V M 2005 *Semiconductors* **39** 557
[15] Johansson J, Svensson C P T, Mårtensson T, Samuelson L and Seifert W 2005 *J. Phys. Chem. B* **109** 13567.
[16] Sibirev N V, Tchernycheva M, Timofeeva M A, Harmand J C, Cirlin G E, Dubrovskii V G 2012 *J. Appl. Phys.* **111** 104317.
[17] Dubrovskii V G, Hervieu Yu Yu 2014 *J. Cryst. Growth* **401** 431.

IOP Conf. Series: Journal of Physics: Conf. Series **917** (2017) 032044 doi:10.1088/1742-6596/917/3/032044

Epitaxial growth and investigation of GaP/GaP(As)N heterostructures on Si (100) 4^0 substrates

A A Lazarenko, M S Sobolev, E V Pirogov, N V Kryzhanovskaya, E V Nikitina
St. Petersburg Academic University, St. Petersburg 194021, Russia

Abstract This article demonstrates the phased creation of heterostructures containing GaPN and GaPAsN layers on Si substrates grown by molecular-beam epitaxy. A technique of preparing a silicon surface for epitaxial growth, the creation of a GaP initial layer, and optical studies of GaPN and GaPAsN layers are described. It is shown that high-quality GaP(As)N epitaxial layers can be created on Si substrates by molecular-beam epitaxy.

1. Introduction

Nowadays heterostructures based on dilute nitrides are of great interest in the world, since they open the possibility of integrating silicon technology and AIIIBV technology for creating light emitting devices, solar cells and integrated circuits [1-4]. Addition of nitrogen in GaP modifies conduction band and makes the $GaP_{1-x}N_x$ compound direct-gap, besides addition of nitrogen leads to lattice matching of the Si and $GaP_{1-x}N_x$. However, during the epitaxial growth of AIIIBV compounds on Si substrates, many defects and dislocations arise, such as anti-phase domains, misfit and threading dislocations etc. That is why the preparation of the silicon surface and the growth of the initial layer are important stages in the creation of heterostructures on Si substrates.

2. Pre-growth preparation of silicon substrates

The pre-growth preparation of substrates is one of the most important step in the formation of structures of molecular-beam epitaxy (MBE) technology for opto- and microelectronics, since the presence of chemically pure, structurally perfect, atomically smooth surface is a prerequisite for producing epitaxial layers of device quality.

Based on the methods proposed in the works [5,6] the method of preparation of Si (111) substrates was developed.

The procedure of the chemical treatment of substrates consists of the following operations (HNO_3 ~ 70%; HF ~ 46%; H_2O_2 ~ 30%; NH_4OH ~ 28%; HCl ~ 36%; distilled deionized water with a resistance of at least 15MOhm·cm):

1. Washing off traces of organic contaminants by boiling first in CCl_4, then in acetone.
2. Oxidation of the upper layer of silicon in boiling nitric acid and bleeding of the formed SiO2 in a solution of hydrofluoric acid (HF: H_2O = 1: 3) for 20-30 seconds (this operation is repeated several times).
3. Boiling in a peroxide-ammonia solution (NH_4OH: H_2O_2: H_2O = 1:1:3) for 10 minutes, followed by removal of contaminants together with silicon oxide in a solution of hydrofluoric acid.
4. Formation on the surface of a thin protective oxide by boiling in peroxide-acid solution (HCl: H_2O_2: H_2O = 3: 1: 1) for 10 minutes.
5. Final washing with deionized water and drying with ethanol in centrifuge.

Samples prepared in this way were immediately loaded into the MBE chamber, where they were before the final stage of pre-epitaxial preparation in the conditions of ultra-high vacuum (UHV). In UHV conditions, chemically prepared substrates were sequentially degassed in a growth module at 450°C for 60 minutes and 750°C for 40 minutes.

3. Epitaxial growth of GaP initial layer

The growth of the GaP initial layer was carried out on VEECO GEN III MBE on silicon substrates. To overcome the problem of the formation of antiphase regions during the growth of AIIIBV compounds on silicon, vicinal silicon substrates were used with misorientations at 4^0 relative to the (001) surface in the [110] direction.

The GaP semiconductor initial layer was formed by sequential deposition of Ga and P monolayers on a silicon substrate.

During the opening of the shutter of the Ga source, gallium atoms accumulate on the epitaxial surface. When the Ga shutter is closed and the P shutter opens, the accumulated gallium is consumed by the growth of the GaP film. In each cycle, a significant part of the time, the GaP film grows in metal-enriched conditions. The time to hold the growth surface under a stream of phosphorus atoms is selected in such a way that all the accumulated gallium reacts with phosphorus. After all the metal is consumed to form a GaP film, the surface goes into a phosphorus-stabilized state, and each new cycle begins with the deposition of gallium on the phosphorus-stabilized surface. The GaP initial layer consisted of 30 periods of Ga and P monolayers at characteristic thicknesses of monolayers of 0.13 nm and 0.19 nm, respectively. Further, a 200-nm-thick GaP buffer layer was grown on top of the initial layer (figure 1).

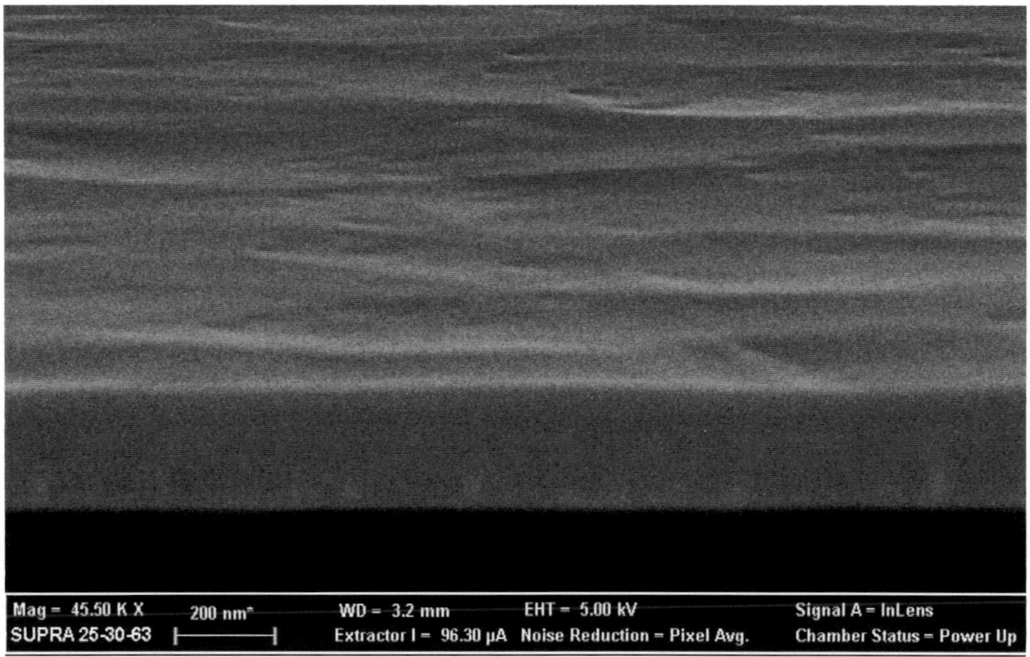

Figure 1. SEM image of GaP/Si heterostructure with GaP initial layer

It is seen that dislocation cones are formed in the initaial layer, but then annihilate and do not penetrate in surface of the buffer layer.

4. Low-temperature photoluminescence of GaP(As)N layers

GaP (As) N layers were grown on GaP and Si substrates by MBE (table 1).

Table 1. Experimental samples of heterostructures with epitaxial layers GaPN and GaPNAs GaP and Si.

№ sample	layer/substrate
1	$GaP_{0.918}As_{0.077}N_{0.005}$/GaP
2	$GaP_{0.993}N_{0.007}$/GaP
3	$GaP_{0.994}N_{0.006}$/Si
4	$GaP_{0.937}As_{0.047}N_{0.016}$/GaP

Photoluminescence (PL) spectra recorded at cryogenic temperature, 10 K, for the samples given in table 1 are shown on figure 2.

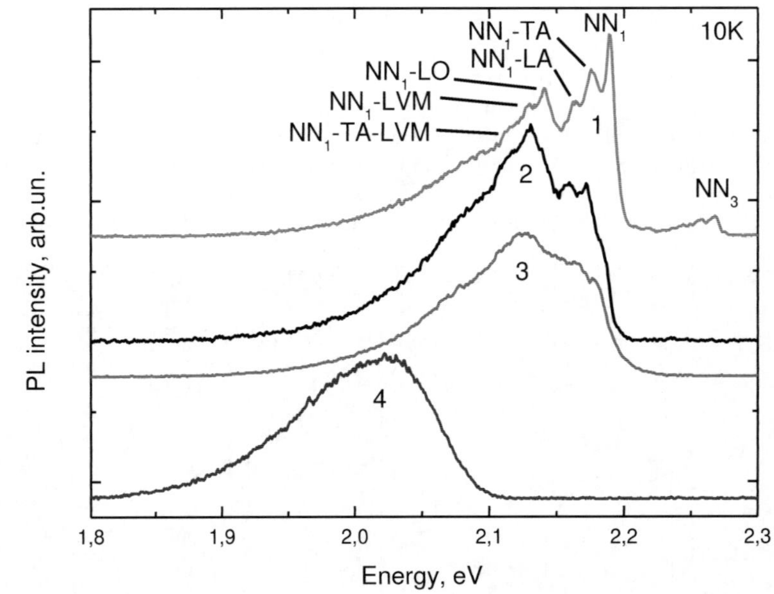

Figure 2. The PL spectra of GaPN / GaP, GaPAsN / GaP and GaPN / Si heterostructures at T = 10 K

Starting from temperatures close to the temperature of liquid nitrogen, 80 K, and at lower temperatures, PL spectra are split into a series of lines (samples 1,2 and 3). The effect was most distinctly in the study of a sample with the lowest molar fraction of nitrogen (sample 1). The most intense emission line, designated NN_1, is associated with the recombination of carriers through a localized state formed by the closest pair of nitrogen atoms. In a general form, the possible nitrogen pairs are designated NN_i, for i = 1 nitrogen atoms are located at the nearest anion positions [7]. The peaks with lower energy associated with the state of NN_1 are its phonon replicas, including optical longitudinal and transverse phonons (LO and TO, respectively), acoustic phonons (LA, TA), and local phonon vibrational modes (LVM). Replicas associated with different phonons are indicated in figure 2 by symbols corresponding to specific phonons. Additionally, at a temperature of 10 K to the right of the most intense series of phonon replicas associated with the pair NN_1, a pair of NN_3 lines and associated replicas arise. The energy position of the observed lines NN_1 and NN_3, as well as the energy distance between the phonon lines, are in good agreement with the known energy values of these lines in GaPN [7]. The presence of arsenic in moderate concentrations in sample 1 does not affect the energy position of the observed lines NN_1 and NN_3. In addition, on the spectra of samples 1, 2 and 3 to the left of the phonon replicas, a shoulder is observed associated with optical transitions between the E_- band and the top of the valence band. An example of the decomposition of the observed spectrum into possible component spectra for sample 2 is shown in figure 3.

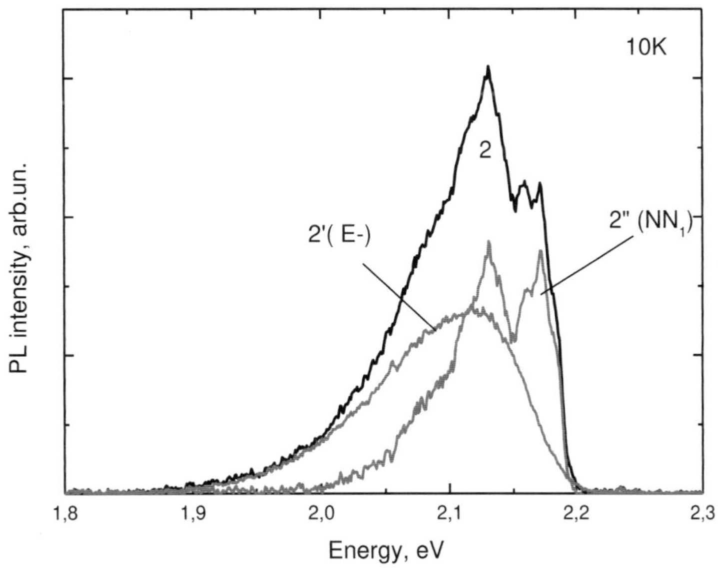

Figure 3. An example of the decomposition of the observed integral PL spectrum (2) into possible component spectra (2 'and 2 ") for sample 2 at T = 10 K.

The change in the mutual arrangement of the bottom of the E_- band relative to the level of the pair NN1, which is in a clearly fixed position, leads to an energy shift of the PL related to the transitions between the bottom of the E_- subband and the top of the valence band. As a result, the shape of the integrated PL spectrum, which we observe, is modified. For example, one can note the difference between the spectra of samples 1 and 2. The energy distribution of the levels of nitrogen pairs NN_i and isolated nitrogen atoms in GaP and the subband E_- of GaPN solid solution is shown schematically in figure 4.

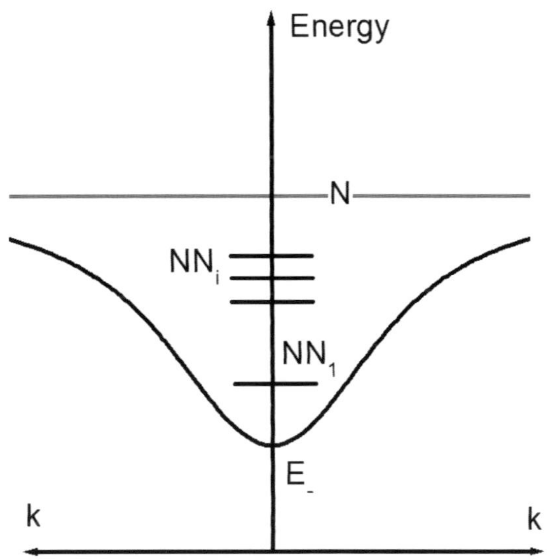

Figure 4. The energy distribution of the levels of nitrogen pairs NN_i and isolated nitrogen atoms in GaP and the subband E_- of GaPN solid solution

For samples with a small fraction of nitrogen (samples 1, 2 and 3), the energy gap between E_- bottom and the NN_1 level is small and radiative recombination is carried out on both channels. For sample 4, with a higher fraction of nitrogen, the energy gap increases and the PL spectrum (see figure 2) is determined only by optical transitions between the E_- band and the top of the valence band.

In general, the photoluminescence spectra allow us to conclude that the molecular beam epitaxy is suitable for the creation of GaPN and GaPAsN layers both on the surface of gallium phosphide and on the surface of silicon. Critical fall PL signal intensity when replacing a GaP substrate to the Si substrate is not observed.

5. Conclusion

This article demonstrates the phased creation of heterostructures containing GaPN and GaPAsN layers on Si substrates grown by molecular-beam epitaxy. A technique of preparing a silicon surface for epitaxial growth, the creation of a GaP initial layer, and optical studies of GaPN and GaPAsN layers are described. It is shown that high-quality GaP(As)N epitaxial layers can be created on Si substrates by molecular-beam epitaxy.

Acknowledgments

This work is supported by the Russian Foundation for Basic Research, projects 17-08-00840, 16-29-03037.

References

[1] A. Tukiainen, A. Aho, V. Polojärvi, R. Ahorinta and M. Guina 2016 *Journal of Green Engineering* Vol 5 pp 113–132
[2] B. Royall, N. Balkan 2011 *Physica status solidi (b)*Vol 248 Issue 5 pp 1203–1206
[3] Chee-Keong Tan, Damir Borovac, Wei Sun, Nelson Tansu 2016 *Scientific Reports* 6 Article number: 24412
[4] Tatjana Wegele, Andreas Beyer, Peter Ludewig, Phil Rosenow, Lennart Duschek, Kakhaber Jandieri, Ralf Tonner, Wolfgang Stolz and Kerstin Volz 2016 *Journal of Physics D: Applied Physics* Vol. 49 Number 7
[5] A.Ishizaka, Y.Shiraki *J. Electrochem. Soc.* 1986 vol.133 p. 666
[6] V. A. Ioshkin, A. A. Orlikovskii, S. R. Oktyabr'skii, A. V. Kvit, and E. Yu. Dovydenko 1994 *Proceedings FTIAN* (Nauka, Moscow) Vol. 8 p. 58.
[7] D.G. Thomas, J.J. Hopfield, C.J. Frosch 1965 *Phys. Rev. Lett.*15 p. 857

IOP Conf. Series: Journal of Physics: Conf. Series **917** (2017) 042001 doi:10.1088/1742-6596/917/4/042001

Comparative NEXAFS study of the selected icefish hard tissues and hydroxyapatite

O V Petrova[1], S V Nekipelov[1,2], D V Sivkov[1], A E Mingaleva[1], A Nikolaev[3], O V Frank-Kamenetskaya[3], V V Bazhenov[4,5], D V Vyalikh[6,7], S L Molodtsov[4,5], V N Sivkov[1], H Ehrlich[4]

[1]Institute of Physics and Mathematics, Komi Science Centre Ural Division, RAS, 167982, Syktyvkar, Russia
[2] Syktyvkar State University, Syktyvkar, Russia
[3] Institute of Earth Sciences, Department of Crystallography, St. Petersburg State University, St. Petersburg, Russia
[4]Institute of Experimental Physics, TU Bergakademie Freiberg, Freiberg, Germany
[5] European XFEL GmbH, Albert-Einstein-Ring 19, 22761 Hamburg, Germany
[6] IKERBASQUE, Basque Foundation for Science, 48011 Bilbao, Spain
[7] Departamento de Fisica de Materiales and CFM-MPC UPV/EHU, Donostia International Physics Center (DIPC), 20080 SanSebastian, Spain

Abstract. The structure of native *Champsocephalus gunnari* icefish otoliths, scales, teeth, bones and pristine hydroxyapatite (HA) were examined using Near Edge X-ray Absorption Fine Structure (NEXAFS) spectroscopy. NEXAFS C1s-absorption spectra of the selected icefish hard tissues indicate that otoliths contain anion $[CO_3]^{2-}$. NEXAFS P2p-spectra clearly indicate the absence of phosphorus atoms only within otoliths and scales samples. However, the icefish teeth and bones P2p-spectra demonstrate identical spectral feature typical for the HA. NEXAFS Ca2p-spectra of the icefish hard tissues studied also shows features, which are in good correspondence with HA spectra. Interestingly, there is a red shift ≈ 0.1 eV of the $2p_{1/2,3/2} \rightarrow 3d$ transition energies in NEXAFS Ca2p-spectra of teethes and bones of the *C. gunnari* in comparison to HA.

1. Introduction

Near Edge X-ray Absorption Fine Structure (NEXAFS) spectroscopy become to be an important and highly informative method in modern studies related to chemistry and structure of diverse biomacromolecules including biominerals and biocomposites. Recent records as discoveries of novel hydroxylated collagen in deep-sea glass sponges [1], or the oldest exceptionally preserved chitin in Cambrian fossilized sponges [2] have been made also using NEXAFS. Our preliminary studies also provide strong evidence that NEXAFS spectroscopy at the C1s, P2p and Ca2p-edge represents the most suitable method to examine structural changes and differences within organic and inorganic functional groups in both natural and biomimetically developed Ca-containing biomaterials [3-5]. Currently attractive studies on the biomineral-based composites are related to synthesis under ambient laboratory conditions. Surprisingly, extreme biomineralization, which seems to be one of the most ancient naturally occurring phenomenon, and which is to be found up today in diverse natural extremes (hot springs, Arctic and Antarctic seas, etc.) is less considered. Psychrophilic biomineralization is one direction of extreme biomineralization that includes chemical, biogeochemical and biochemical processes at very low temperatures. Typical examples of calcium-based biominerals within psychrophilic icefishes are their teeth, scales, otoliths and bones (see Figure 1.). However, information on characterization of their structure and chemical composition is still limited. Antarctic ice fishes habituate under extreme condition in deep, oxygen-reach, cold waters and due to specific adaptation can withstand temperature about -1,9°C that freeze the blood of all other types of fishes. To our best knowledge, there is still no any direct study with regard to investigation of chemical and mineral composition and nanostructure of ice- fish hard tissues. Our preliminary X-ray diffraction and NEXAFS studies showed that mineral component of the icefish bone is presented by the poorly crystallized hydroxyapatite (HA) [5].

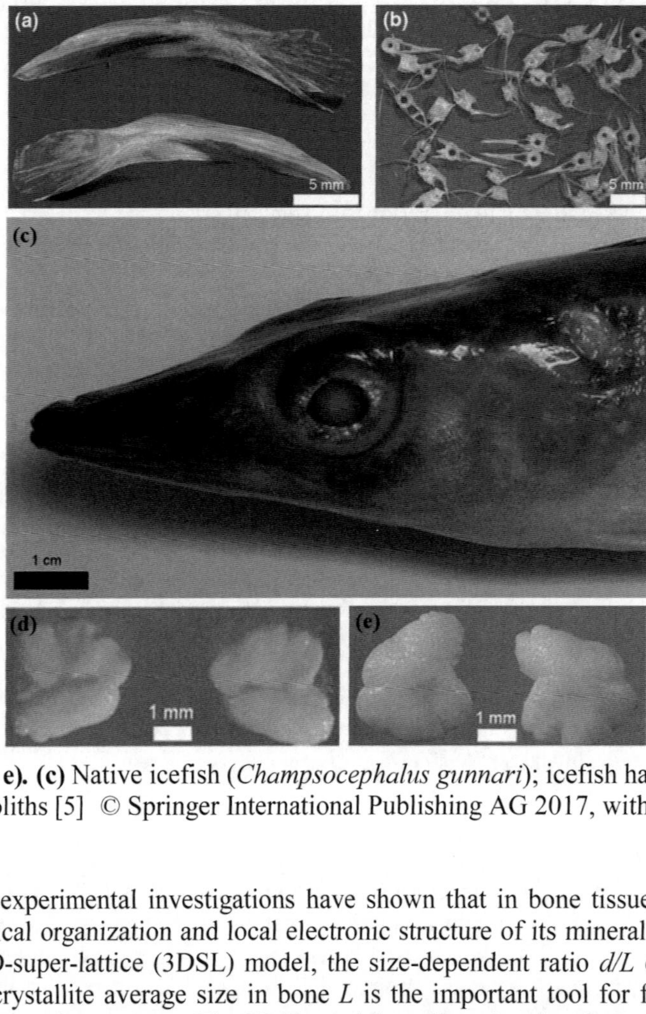

Figure 1(a, b, c, d, e). (c) Native icefish (*Champsocephalus gunnari*); icefish hard tissues: **(a, b)** bones and **(d, e)** otoliths [5] © Springer International Publishing AG 2017, with permission of Springer.

The theoretical and experimental investigations have shown that in bone tissue a close relationship between its hierarchical organization and local electronic structure of its mineralized phase exists [6]. According to the 3D-super-lattice (3DSL) model, the size-dependent ratio d/L of the hydrated layer thickness d to nanocrystallite average size in bone L is the important tool for functionalizing of the mineralized bone electronic structure. The 3DSL model predicts the size-dependent spectral changes in electronic structure of mineralized phase relative to the band structure of HA. In particular it predicts the red shift δE_{3d} of the band in the assembly of nanocrystallites and roughly estimate it as $\delta E_{3d}=E_{3d}\cdot d/L$, where E_{3d} is the energy above the bottom of the conduction band HA. According to the band structure calculations of HA [7, 8], the energy $E_{3d} \approx 2$ eV above the bottom of the conduction band. In addition to the red shift, the narrowing of the band in the assembly is also expected [6]. These effects can be investigated by studying the NEXAFS Ca2p-spectra of the pure HA deposited layer and bone nanocrystallites. NEXAFS allows probing the atomic composition and angular-momentum-character specific densities of unoccupied electronic states and may give information about chemical bonds in investigated samples [9]. This method is using to measure the photoabsorption cross-section for the excitation of primarily bound core electrons that makes NEXAFS spectra element specific since each atom has characteristic binding energies of core levels. Thus, we have a great opportunity to apply this method to obtain information about individual units and substitutional groups, and to explore their fingerprinting character using the building block concept. The purpose of this comparative work is to study the NEXAFS Ca2p-spectra of pure HA and selected *C. gunnari* icefish hard tissues and to determine the average sizes of nanocrystallites in them.

2. Experimental details

The samples for absorption measurements were prepared ex situ in air by pressing of powders into the clean copper plate surface. The *C. gunnari* icefish bones were put into test tubes with H_2O_2 solution for 24 hours and then mechanically cleaned from muscles. For finalize cleaning of attached tissues the separated bones were put into the test tubes with 50% ethanol solution or NaOH solution, exposured in ultrasonic bath for 2-4 hours, washed in distilled water and, finally, dried at 50^0C. For investigation of Ca2p-, P2p- and N1s-spectra the selected powdered icefish structures were mixed with pure fullerene C_{60} for better conductivity and NEXAFS P2p-spectra calibration (C1s in second order diffraction). For investigation of C1s-spectra the pure powdered icefish samples were used. Icefish otoliths and scales have been extracted mechanically, separated, washed in 50% ethanol solution and dried. Powered otoliths samples were mixed with C_{60} too, while scales were not powdered, but covered by the gold layer except thin central band, for better charge draining. The NEXAFS studies on the powdered samples were carried out using monochromatic synchrotron radiation (SR) at the RGBL at the BESSY-II electron storage ring (the Helmholtz-Zentrum Berlin, German) [9]. All measurements of the absorption spectra were performed under vacuum conditions $\sim 10^{-9}$ torr in the experimental chamber. The samples were placed at the angle of $\sim 45°$ with respect to the incident beam of SR and the size of the focal spot on the sample was around 100×100 mkm^2. The NEXAFS-spectra at the Ca2p-, P2p- and C1s-absorption edge of the bone samples were measured by the total electron yield (TEY), which is proportional to the absorption cross section [9, 10]. All the X-ray absorption spectra were recorded with an energy resolution about 0.01eV. The spectra were normalized to the incident photon flux, which was defined as a ratio between TEY-signal from a clean Au-plate and Au-atom cross section. The suppression and estimation of the nonmonochromatic background in TEY-signal was carried out by absorptionTi-filter method [3]. The energy over the range of the Ca2p-spectrum was calibrated by the first narrow peak in the F1s spectra of solid K_2TiF_6 (683.9 eV [13]) in the second order diffraction, C1s- and P2p- spectra by the peak in the C1s-absorption spectra of fullerite C_{60} (284.5 eV [14]) in the first and second order diffraction, respectively. All spectra were collected by TEY-signal monitoring by the drain current detection with Keithley 617 microampermeter.

3. Results and discussion

C1s-absorption spectra of the selected *C. gunnari* icefish biomineralized structures and calcite $CaCO_3$ as reference are shown in Figure. 2(a). Almost all icefish samples spectra, with exception of the scales, demonstrate feature at 290.2 eV similar to flat group $[CO_3]^{2-}$-anion that indicates the presence of carbonates in otoliths, bones and teeth samples and carbonates absence in scales. Concerning C1s-spectra of $CaCO_3$, the structure presented before strong resonance at 290.2 eV is connected with sample contamination. In the case of the icefish samples the same structure is determined by the presence of specific chemical compounds in them. P2p-absorption spectra of the icefish bones, teeth and hydroxyapatite (HA, $Ca_{10}(PO_4)_6OH_2$) spectrum [15] contain identical structure (see Figure. 2(b).). This structure include 3 groups of features: **a** (136.5 eV) and **a′** (137.5 eV), **b** (138.2 eV) and **b′** (139.1 eV), **c** (141.5 eV) and **d** (146.8 eV), which correspond with $P2p_{1/2,3/2}$ electron transition to molecular unoccupied states of $[PO_4]^{3-}$-anion with predominant contribution of P3s, P3p and P3d component, respectively.

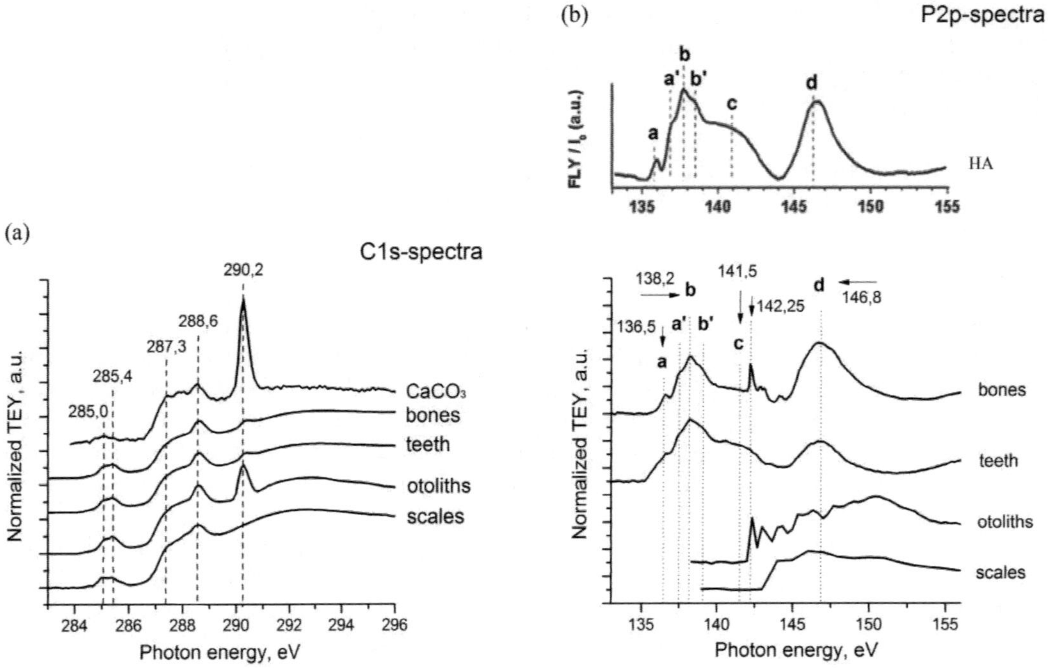

Figure 2(a, b). (a) NEXAFS C1s-absorption spectra of the *C. gunnari* icefish biomineralized structures and $CaCO_3$; **(b)** NEXAFS P2p-absorption spectra of the icefish biomineralized structures and the standard hydroxyapatite (HA) spectrum [15]. Reprinted from [15] Copyright © 2010, with permission from Elsevier.

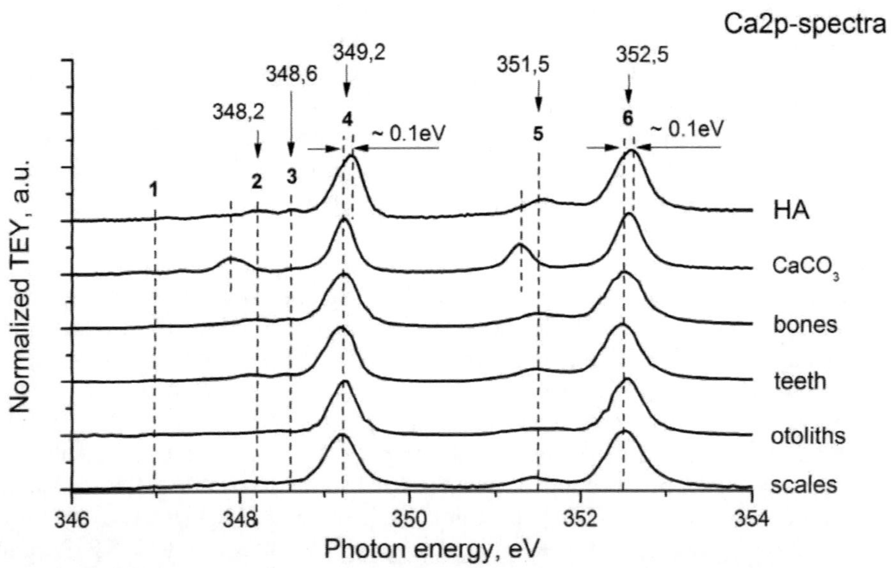

Figure 3. NEXAFS Ca2p-absorption spectra of the icefish biomineralized structures and HA.

Figure 3. shows the NEXAFS Ca2p spectra of the different icefish specimens (bones, teeth, otoliths and scales) as well as the HA reference. In all absorption spectra dominate two main spin–orbit-related The icefish otolithes and scales absorption spectra indicate the lack of phosphorus in them and HA presence in icefish bones and teeth. The absorption spectra of HA was taken from [15] and measured by Fluorescent Yield detection mode. The structure presented in the energy range of 140-160 eV is connected with the second order contribution of fullerene C1s-edge. peaks 4 and 6, along with a number of smaller peaks preceding the $2p_{3/2}$ and $2p_{1/2}$ main peaks. The origin of these multi-peak patterns is known to be the crystal field arising from the symmetry of the Ca^{2+} ion surrounded by oxygen atoms with close to octahedral configuration in the first coordination sphere [16]. Small shift $\approx$ 0.1 eV detected for the Ca $2p_{1/2,3/2} \rightarrow$ 3d transition energies in the icefish biomineralized structures relative to HA is a very interesting phenomena. It is clearly seen that the magnitude of the red shift differs for diverse icefish hard tissues: for otoliths, the shift is minimal and increases when passing from the scales and bones to the teeth. The red shift also were detected in Ca2p-spectra of the bones of white mongrel male rats [6] and has been interpreted by the influence of the super periodicity on the band structure of HA and are assigned with the hierarchy-induced effect on the ground level in the bone tissue. Taking into account the fact that E_{3d} = 2 eV, our experimental results allow us to determine the value of L/d $\approx$ 20.

4. Conclusion

Comparative NEXAFS C1s-absorption spectra study of the selected icefish hard tissues, hydroxyapatite and calcite show the presence of carbonates in otoliths, bones and teeth samples and absence in scales. The selected icefish hard tissues NEXAFS P2p-absorption spectra indicate the lack of phosphorus in otolithes and scales as well as HA presence in icefish bones and teeth. NEXAFS Ca2p-absorption spectra comparative analysis of the icefish bones and teeth and hydroxyapatite indicate the red shift for the Ca2p-spectra main features in icefishes bones and teeth relative to HA. This red shift is estimated $\approx$ 0.1 eV. According to 3DSL-model the presence of 0.1 eV red shift allow to define the nanocrystallite size as L=20d where d is hydrated layer thickness between neighboring HA nanocrystalites. The obtained experimental results demonstrate the prospects of using of NEXAFS-spectroscopy for studying the hierarchical organization structure of HA containing icefishes hard tissues and, thus, stimulate further research in this area.

Acknowledgments
This work was supported by State Contract П-337, Minfin KR, grants RFBR 16-32-00441мол-а, 16-42-110610 р-а, 16-43-110350 р-а, Basic Research Program of UrD RAS 15-9-1-13, GRISC-Programme, by the Bilateral Program of the RGBL at BESSY II and the DFG Grant EH 394/3-2.

References

[1] Ehrlich H, Deutzmann R, Capellini E, Koon H, Solazzo C, Yang Y, Ashford D, Thomas-Oates J, Lubeck M, Baessmann C, Langrock T, Hoffmann R, Wörheide G, Reitner J, Simon P, Ereskovsky A V, Mertig M, Vyalikh D V, Molodtsov S L, Worch H, Brunner E, Smetacek V, Collins M 2010 *Nature Chemistry* **2** 1084
[2] Ehrlich H, Rigby J K, Botting J, Tsurkan M, Werner C, Schwille P, Petrasek Z, Pisera A, Simon P, Sivkov V, Vyalikh D, Molodtsov S, Kurek D, Kammer M, Hunoldt S, Born R, Stawski D, Steinhof A, Bazhenov V, Geisler T 2013 Scientific *Reports* **3** 3497
[3] Kummer K, Sivkov V N, Vyalikh D V, Maslyuk V V, Blüher A, Nekipelov S V, Bredow T, Mertig I, Mertig M, Molodtsov S L 2009 *Phys. Rev. B* **80** 155433
[4] Ehrlich H, Elkin Yu N, Artyukov A A, Kozlovskaya E, Stonik V A, Safronov P P, Kurek D V, Bazhenov V V, Kummer K, Vyalikh D, Molodtsov S L, Sivkov V N, Born R, Ruhnow M, Meissner H, Richter G 2010 *J Adv Micro Res.* **5** 100
[5] Ehrich H 2016 *Extreme Biomimetics* (Springer) p. 276

[6] Pavlychev A A, Avrunin A S, Vinogradov A S, Filatova E O, Doctorov A A, Krivosenko Yu S, Samoilenko D O, Svirskiy G I, Konashuk A S and Rostov D A 2016 *Nanotechnology* **27** 504002
[7] Rulis P, Ouyang L, Ching W Y 2004 *Phys. Rev. B* **70** 155104
[8] Matsunaga K, Kuwabara A 2007 *Phys. Rev. B* **75** 014102
[9] Stöhr J 1992 *NEXAFS spectroscopy* (Springer: Berlin) p.403
[10] Fedoseenko S I, Vyalikh D V, Iossifov I E, Follath R, Gorovikov S A, Puttner R, Schmidt J-S, Molodtsov S L, Adamchuk V K, Gudat W, Kaindl G 2003 *Nucl. Instrum. Methods Phys. Res. A* **505** 718
[11] Lukirskii A P and Brytov I A 1964 *Sov. Phys-Solid State* **6** 43
[12] Gudat W and Kunz C 1972 *Phys. Rev. Lett.* **29** 169
[13] Vinogradov A S, Fedoseenko S I, Krasnikov S A, Preobrajenski A B, Sivkov V N, Vyalikh D V, Molodtsov S L, Adamchuk V K, Laubschat C, Kaindl G 2005 *Phys. Rev. B* **71** 045124
[14] Brühwiler P A, Maxwell A J, Nilsson A, Whetten R L, Martensson N 1992 *Chem. Phys. Lett.* **193** 311
[15] Demirkiran H, Hu Y, Zuin L, Appathurai N, Aswath P B 2011 *Mater. Sci. Eng. C* **31** 134
[16] de Groot F M F, Fuggle J C 1990 *Phys Rev B* **41** 928

Application of Marcus theory for modeling proton transfer in cytochrome *c* oxidase

A A Garbuz[1], S E Boronovskiy[1] and Ya R Nartsissov[1]

[1]Institute of cytochemistry and molecular pharmacology, Moscow 115404, Russia

Abstract. The process of proton transport in cytochrome *c* oxidase is studied in the framework of stochastic modeling. The activation energies are calculated using Marcus theory. This model allows to define the key amino acid residues and water molecules which form the main H^+ transduction pathway. According to the simulation results, Asn-207 and Asn-121 are not involved in direct proton translocation. The estimated rate of the proton transfer through the D-channel of cytochrome *c* oxidase is $(1.43\pm0.18)\cdot10^4\ s^{-1}$.

1. Introduction

Oxygen is necessary as an oxidizing agent for the process of cellular respiration of aerobic organisms. The reduction of molecular oxygen to water requiring electrons and protons to be taken up from the opposite sides of the inner mitochondrial or cytoplasmic membrane is carried out in cytochrome *c* oxidase, which is the terminal complex of the respiratory chain. The number of its subunits differs for mammalian and bacterial forms, but the structure and amino acid sequence of three core subunits (I-III) are highly conserved [1]. This enzyme is also known to be a generator of the electrochemical proton gradient due to the fact that electron transfer is coupled with proton pumping across the membrane. The process of proton transfer in cytochrome *c* oxidase is provided by the D-channel and K-channel, which names correspond to the entrance amino acid residues of aspartic acid (e.g. D) and lysine (e.g. K) respectively. Of particular interest is the D-channel, which is considered to transfer four pumped protons and at least two chemical protons required for the oxygen reduction [2,3]. Its orifice represented by Asp-132 (*Rhodobacter sphaeroides* nomenclature) can be probably protonated both directly from the matrix and by protein surface groups named proton-collecting antenna. They comprise of several carboxylates and histidine rings proceeding a rapid reprotonation of the entrance aspartate due to the fact that proton exchange between these residues is faster than proton diffusion in the bulk solution [4,5]. The final point of the D-channel is Glu-286, which is known to be a valve in the H^+ pumping. Since it is capable to switch conformations of the side chain from «up» to «down» and vice versa, Glu-286 provides proton transfer both to binuclear center and proton-loading site [6,7]. Different theoretical approaches, in particular, multistate empirical valence bond model, *ab initio* molecular dynamics and quantum mechanics/molecular mechanics hybrid method, are applied to studying the process of proton translocation in cytochrome *c* oxidase [8,9,10]. It is a question of current interest, because proton transport dysfunction in this enzyme caused, for example, by mutations in the mDNA leads to some mitochondrial diseases [11]. In spite of intensive research, the mechanism of H^+ pumping is one of the key biophysical issues, because its atomistic details remain unexplored yet. In this work Marcus theory is considered as a tool of revealing the main transduction pathway of proton transport. Constructing a model defining key amino acid residues and water molecules allows us to study pH and pK_a dependence of the proton transfer rate through the D-channel of cytochrome *c* oxidase.

2. Theoretical background

According to Marcus theory, the reaction of proton transport can be considered as a sequence of three explicit steps. The first one is mutual attraction of reagents due to change of free energy, which is followed by hydrogen bonding of acceptor and proton. Proton translocation from donor to acceptor occurring afterwards involves intramolecular reorganization and solvent reorganization required for formation of the transition state. The last step is mutual repulsion of products with breaking of a hydrogen bond between proton and donor [12].

In this theoretical approach proton is represented as a harmonic oscillator. The potential-energy profile of the proton transfer reaction consists of two parabolas. Their intersection corresponds to the transition state characterized by the activation energy $\Delta G^{\neq}$, which is connected with the Gibbs free energy ΔG_0 and the reorganization energy λ:

$$\Delta G^{\neq} = w_r + \frac{\lambda}{4}\left(1 + \frac{\Delta G_0 + w_p - w_r}{\lambda}\right)^2 \qquad (1)$$

where w_r and w_p are the electrical work of attraction of the reagents and repulsion of the products respectively. The reorganization energy consists of two terms: the intramolecular reorganization energy λ_i and the energy of solvent reorganization λ_o. The first one is determined by the bond-length change between donor or acceptor and proton, while the second one is defined by the donor-acceptor distance and properties of the medium:

$$\lambda = \lambda_i + \lambda_o = \frac{1}{2}\sum_k \left[\frac{2f_k^D f_k^A}{f_k^D + f_k^A}(\Delta d_k)^2\right] + (\Delta e)^2\left(\frac{1}{\varepsilon_{op}} - \frac{1}{\varepsilon_{st}}\right)\left(\frac{1}{2R_D} + \frac{1}{2R_A} - \frac{1}{r_{DA}}\right), \qquad (2)$$

where f_k^D and f_k^A are the force constants of donor and acceptor respectively, Δd_k is the bond-length change, Δe is a transferred charge, ε_{op} and ε_{st} are the optical and static permittivities respectively, R_D and R_A are the radii of donor and acceptor respectively, r_{DA} is the donor-acceptor distance [13]. Obtaining the activation energy of the reaction allows to calculate the proton transfer rate using the Eyring equation, which describes the dependence between the energy barrier and the reaction rate:

$$k = \kappa \frac{k_B T}{h} e^{-\frac{\Delta G^{\neq}}{RT}}, \qquad (3)$$

where κ is a transmission coefficient, k_B is the Boltzmann constant, T is a temperature, h is the Planck constant and R is the gas constant.

The structure of the D-channel involves several highly conserved polar amino acid residues and water molecules, which can play a direct role in the proton transfer. Since the proton pathway is represented as a set of donors and acceptors shown in Figure 1, it is reasonable to apply the stochastic approach based on the continuous-time Markov chain for the system with discrete states. Transition probability $P[i \to j \mid \Delta t]$ is supposed to be directly proportional to transition time Δt:

$$\begin{cases} P[i \to j \mid \Delta t] = q_{ij}\Delta t \\ P[i \to i \mid \Delta t] = 1 - \sum_{i \neq j} q_{ij}\Delta t \end{cases}, \qquad (4)$$

where q_{ij} is the transition intensity from state *i* to state *j* and $P[i \to i \mid \Delta t]$ is the probability to stay in state *i*. Simulation modeling of the process of proton transfer is performed by generation of continuous and discrete random variables. The former is used for obtaining sojourn time in state *i*, which is exponentially distributed, while the latter defines what state *j* is chosen for the transition from state *i*.

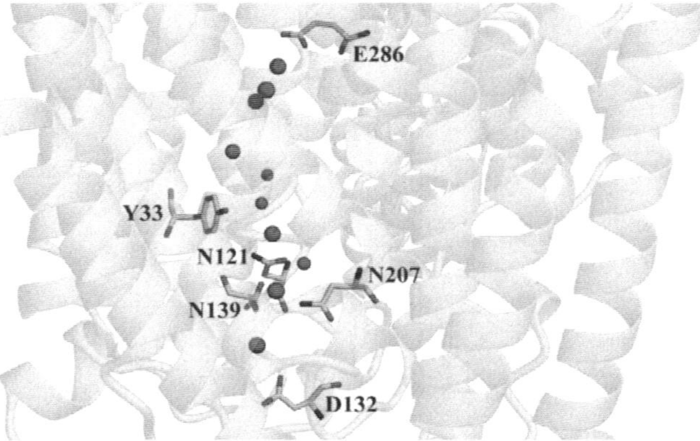

Figure 1. Key amino acid residues and water molecules of the D-channel (represented by labeled sticks and balls respectively; PDB code 1M56). The rest of the subunit I is shown in ribbons.

3. Results and discussion

The main advantage of our constructed model is that it requires only structural data and pK_a values to reveal important details concerning the process of proton transport. The scheme of the spatial arrangement of the D-channel was based on the X-ray crystal structure of cytochrome *c* oxidase from *Rhodobacter sphaeroides*, which was solved at 2.3/2.8 Å (PDB ID: 1M56). Markov-like stochastic model was used for analyzing of the D-channel structural elements, involved in the process of proton translocation in cytochrome *c* oxidase.

According to the simulation results, the states Asp-132, Asn-139, Tyr-33, Glu-286 and the water molecules sequence extending from Asn-139 to the terminal residue of the D-channel form the main transduction pathway (Figure 2). Thus, under normal physiological conditions Asn-207, Asn-121 and the water molecules 2052, 2083 and 2037 do not participate in direct proton transfer. It is consistent with the fact that the asparagine residues form a flexible gate, which influences a spatial organization of hydrogen network, instead of being directly protonated. MS-RMD results reported the asparagine side chains to undergo conformational changes, which control the hydration state of the central region of the D-channel [14]. The proton transport rate through the D-channel calculated during our simulation is $(1.43 \pm 0.18) \cdot 10^4\ s^{-1}$, which is confirmed by the experimental data [15,16].

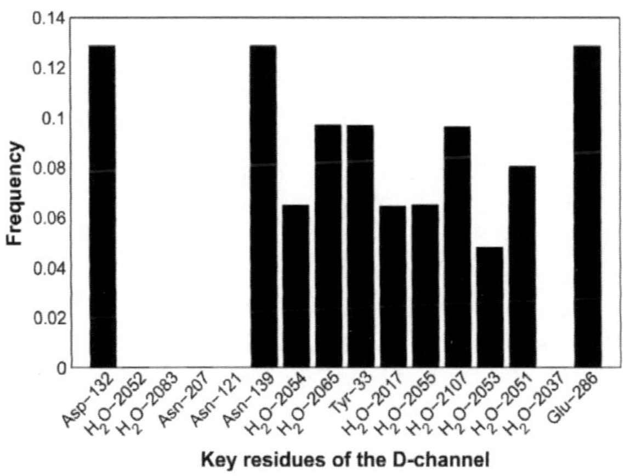

Figure 2. The essential D-channel residues involved in proton transfer. The results of 10^4 simulation steps of Markov chain are represented by frequency bars.

Since the key amino acid residues located in the core subunits of cytochrome *c* oxidase are highly conserved both for mammalian and bacterial forms of the enzyme, one can talk about the protonation of Asp-132 from the mitochondrial matrix. Its actual mechanism still remains unclear, therefore in our study we consider the entrance aspartate to be protonated only from the bulk solution directly. On this point, the dependence of the proton transfer rate on the matrix pH varying from 4 to 10 was obtained. Figure 3 shows that at physiological pH the rate values range from $1.5 \cdot 10^3\ s^{-1}$ to $7.7 \cdot 10^3\ s^{-1}$. The upper limit of $1.5 \cdot 10^4\ s^{-1}$ is observed at pH values below 5.

IOP Conf. Series: Journal of Physics: Conf. Series **917** (2017) 042002 doi:10.1088/1742-6596/917/4/042002

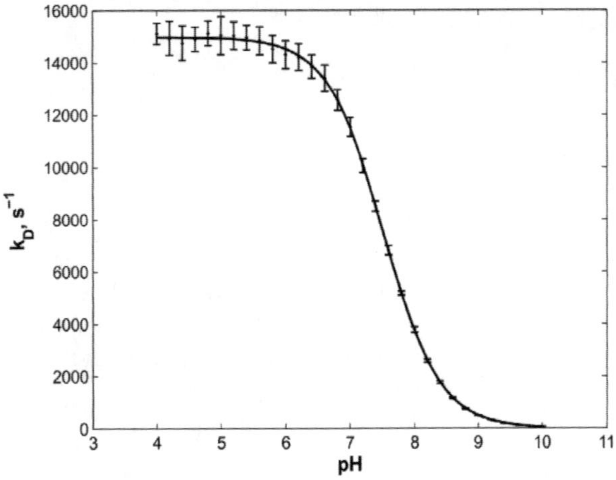

Figure 3. The dependence of the proton transfer rate on the mitochondrial matrix pH. The points are the means±SD of 10 independent computer simulations for each pH value.

Due to the fact that the task of the experimental estimation of pK_a corresponded to the particular amino acid residues buried in protein is not trivial, there is a wide range of their values estimated computationally. Experiments based on Fourier transform infrared (FTIR) spectroscopy showed that the apparent pK_a of Glu-286 is 9.4 [17], while its continuum electrostatic evaluation is from approximately 8.0 to 10.6 according to the catalytic state of cytochrome *c* oxidase [18]. Similar calculations done for Asp-132 produced a range from 3.2 to 6.2 [19,20], besides, its minimum value is close to pK_a of aspartate in solution. In case of Tyr-33, the pK_a range 12.5–21.7 was achieved [19]. Based on the above data, it is appropriate to speak about pK_a shift for the key amino acid residues of the D-channel. The dependence of the proton transfer rate on the pK_a shift obtained for Tyr-33 is represented in Figure 4. This residue is known to be hydrogen-bonded with one of the water molecules located within the pathway [21]. From this result, one can conclude that Tyr-33 is not robust and may play an important role in the process that triggers direction of proton transitions.

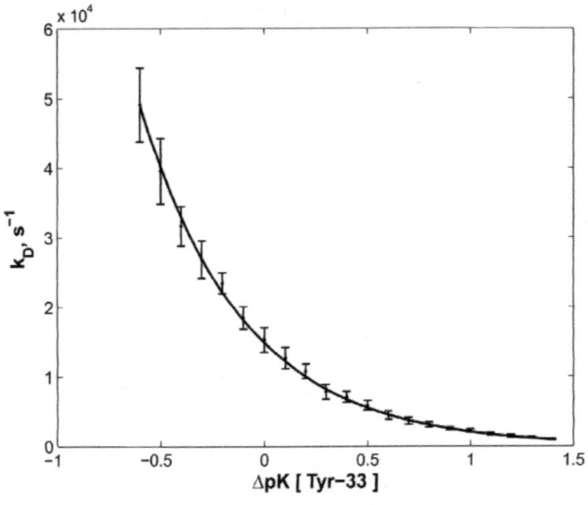

Figure 4. The dependence of the proton transfer rate on the pK_a shift for Tyr-33 with $pK_a = 9.8$. The points indicate the mean±SD of 10 independent computer simulations for each ΔpK value.

Similar plots relating to Asp-132 and Glu-286 show that these residues do not affect the total k_D value (Figure 5). Thus, the proton-pumping stoichiometry depends to a great extent on changes from "up" conformation of the side chain of Glu-286 to "down".

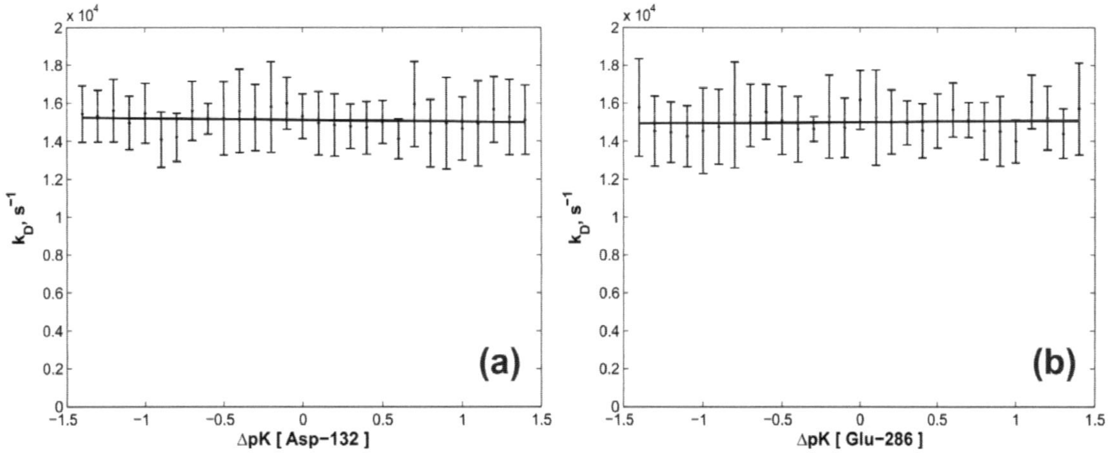

Figure 5. The dependence of the proton transfer rate on the pK_a shift for a) Asp-132 with $pK_a = 3.2$ and b) Glu-286 with $pK_a = 9.4$. The points are the means±SD of 10 independent computer simulations for each ΔpK value.

References

[1] Pierron D, Wildman D E, Hüttemann M, Markondapatnaikuni G C, Aras S and Grossman L I 2012 *Biochimica et Biophysica Acta* **1817(4)** 590-7
[2] Lepp H, Salomonsson L, Zhu J P, Gennis R B and Brzezinski P 2008 *Biochimica et Biophysica Acta* **1777** 897-903
[3] Konstantinov A A, Siletsky S, Mitchell D, Kaulen A and Gennis R B 1997 *Proc. Natl. Acad. Sci. USA* **94** 9085-90
[4] Marantz Y, Nachliel E, Aagaard A, Brzezinski P and Gutman M 1998 *Proc. Natl. Acad. Sci. USA* **95** 8590-5
[5] Marantz Y, Einarsdóttir Ó, Nachliel E and Gutman M 2001 *Biochemistry* **40(50)** 15086-97
[6] Samudio B M, Couch V and Stuchebrukhov A A 2016 *J. Phys. Chem. B.* **120(9)** 2095-105
[7] Kaila V R I, Verkhovsky M I, Hummer G and Wikström M 2009 *Biochimica et Biophysica Acta* **1787** 1205-14
[8] Peng Y and Voth G A 2012 *Biochimica et Biophysica Acta* **1817(4)** 518-25
[9] Kamiya K, Boero M, Tateno M and Oshiyama A 2007 *J. Phys.: Condens. Matter* **19(36)** 365220
[10] Goyal P, Yang S and Cui Q 2015 *Chem. Sci.* **6(1)** 826-41
[11] Chatterjee A, Mambo E and Sidransky D 2006 *Oncogene* **25** 4663-74
[12] Marcus R A 1975 *Faraday Symp. Chem. Soc.* **10** 60-8
[13] Marcus R A 1993 *Rev. Mod. Phys.* **65** 599
[14] Liang R, Swanson J M J, Peng Y, Wikström M and Voth G A 2016 *Proc. Natl. Acad. Sci. USA* **113(27)** 7420-5
[15] Brzezinski P and Johansson A L 2010 *Biochimica et Biophysica Acta* **1797** 710-23
[16] Wikström M and Verkhovsky M I 2011 *Biochimica et Biophysica Acta* **1807(10)** 1273-8
[17] Namslauer A, Aagaard A, Katsonouri A and Brzezinski P 2003 *Biochemistry* **42** 1488-98
[18] Fadda E, Yu C H and Pomès R 2008 *Biochimica et Biophysica Acta* **1777** 277-84
[19] Olkhova E, Helms V and Michel H 2005 *Biophysical Journal* **89** 2324-31
[20] Popovic D M, Leontyev I V, Beech D G and Stuchebrukhov A A 2010 *Proteins* **78(12)** 2691-8
[21] Cukier R I 2004 *Biochimica et Biophysica Acta* **1656** 189-202

Fabrication and characterization of biological tissue phantoms with embedded nanoparticles

A A Skaptsov[1], S O Ustalkov[1], A H M Mohammed[1], O A Savenko[2], A S Novikova[2], E A Kozlova[1] and V I Kochubey[1]
[1]Department of Physics, Saratov State University, Saratov, 410012, Russia
[2] Institute of Chemistry, Saratov State University, Saratov, 410012, Russia

Abstract. Phantoms are imitations of biological tissue, which are used for modelling of the light propagation in biological tissues. Carrying out any biophysical experiments requires an indispensable constancy of the initial experiment conditions. The use of solid undegradable phantoms is the basis to obtain reliable reproducible experimental results. The fabrication of biological tissues phantoms containing high absorbance or fluorescence nanoparticles and corresponding to specific mechanical, optical properties is an actual task. This work describes development, fabrication and characterization of such solid tissue phantoms with embedded CdSe/ZnS quantum dots, gold and upconversion nanoparticles. Luminescence of samples with CdSe/ZnS quantum dots and upconversion nanoparticles were recorded. A sample of gold nanorods was analyzed using thermal gravimetric analysis. It can be concluded that the samples are well suited for experiments on laser thermolysis.

1. Introduction

Development of new kinds of cancer diagnosis and treatment methods requires extensive testing before the introduction into clinical practice. Using natural biological tissues instead natural human or animal tissue is practical for several reasons. Optical properties of the natural biological tissues depend on temperature [1, 2] and denaturation degree [3, 4] and may vary from source to source, change rapidly in various atmospheric conditions and quickly degrade over time [5]. Thus, the topical task of this work is the developing method to fabricate biological tissue phantoms with the specified optical properties that can be used as standard samples.

Despite there are a variety of liquid and solid phantoms, in carrying out experimental work preference is given to solid phantoms [6]. Solid phantoms consist of solid transparent polymeric elastic material with scattering and absorbing particles uniformly distributed in the matrix material. Creating those phantoms requires management methods of controlling its optical properties during manufacture [7]. To set scattering and absorbance coefficients and scattering anisotropy factor of phantom the special concentration and size of the scattering particles, such as TiO2, ZnO is used [8]. Thus, it is possible to create a phantom with the process of light diffusion which is identical to the process occurring in tissues.

Research on the therapeutic thermal effects of laser radiation on biological tissue, as well as plasmonic photothermal therapy[9] require both the identity of the optical properties of phantom and biological tissues, and the equivalence of the thermal properties[10]. However, it must be mentioned that the thermal properties of phantoms are slightly different from those of real skin: the value of skin thermal conductivity–for human epidermis, 0.209 $Wm^{-1}K^{-1}$; dermis, 0.293–0.322 $Wm^{-1}K^{-1}$ [11-13].

The purpose of this article is to provide a method for creating a phantom of biological tissue containing various nanostructures such as gold nanorods (AuNRs), upconversion nanoparticles (UCNPs) based on NaYF4: Yb, Er, and CdSe/ZnS quantum dots (QDs). And also to study optical properties, characterization and well whether thermolysis samples are suitable.

2. Materials and methods

2.1. Materials and equipment

Trisodium citrate dihydrate (ACS reagent, ≥99.0%), $HAuCl_4$ (ACS reagent, ≥99.0%), hexadecyltrimethylammonium bromide (CTAB) solution (ACS reagent, ≥99.0%), $NaBH_4$ (ACS reagent, ≥99.0%), solution of polyvinylpyrrolidone (PVP) (10 kDa, ACS reagent, ≥99.0%), isopropyl alcohol, aqueous ammonia (25%), tetraethoxysilane (TEOS-SH) (ACS reagent, ≥99.0%) were applied for synthesis of nanoparticles.

Spectrometer Ocean QEPro(Ocean Optics,USA), laboratory centrifuge (Eppendorf 5804, Germany), thermal analyzer Q500 (TA instruments, USA), and scanning electron microscopy (Mira II LMU, Tescan, UK) were used for characterization of the nanoparticles and phantoms with them.

2.2. Synthesis of nanoparticles

2.2.1. Synthesis of $NaYF_4$:Yb,Er

Nanoparticles $NaYF_4$-Yb, Er were synthesized by hydrothermal method using trisodium citrate as a capping agent. In relation to Er the Yb concentration was 30 mol% Yb and Er one was 5 mol%. The synthesized particles were rinsed several times with a dilute solution of trisodium citrate. After rinsing, the particles were heated in air for 1 h at 500°C.

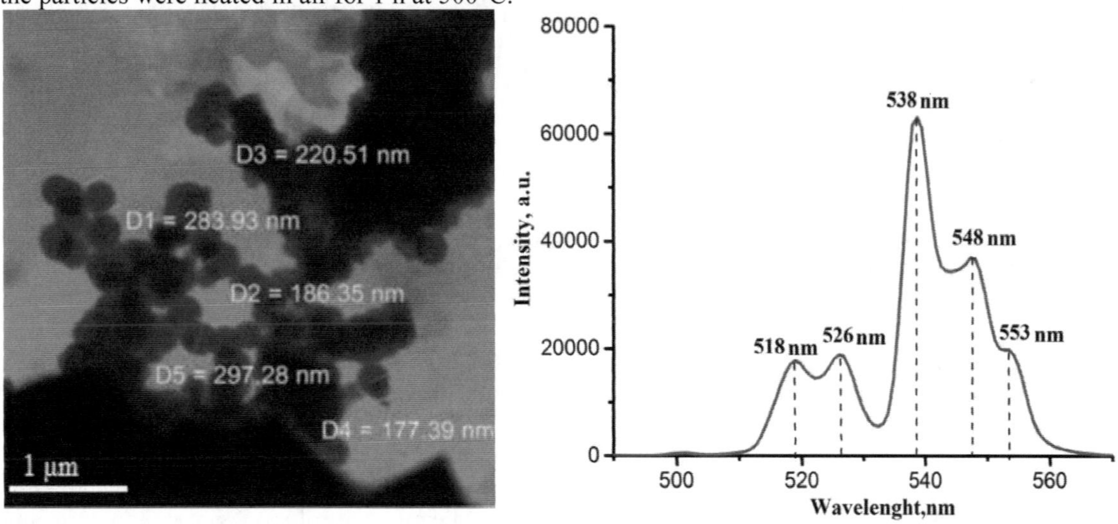

(a) (b)

Figure 1. (a) - SEM image of upconversion nanoparticles ($NaYF_4$:Yb^{3+}, Er^{3+}); (b) - their fluorescence spectrum (excitation wavelength – 980 nm, size of the laser beam 1x3 mm, temperature 25°C).

2.2.2. Synthesis of gold nanorods

Synthesis of AuNRs was done according to early described protocol [14]. Absorbance spectrum and TEM image of synthesized AuNRs solution are shown in Figure 2.

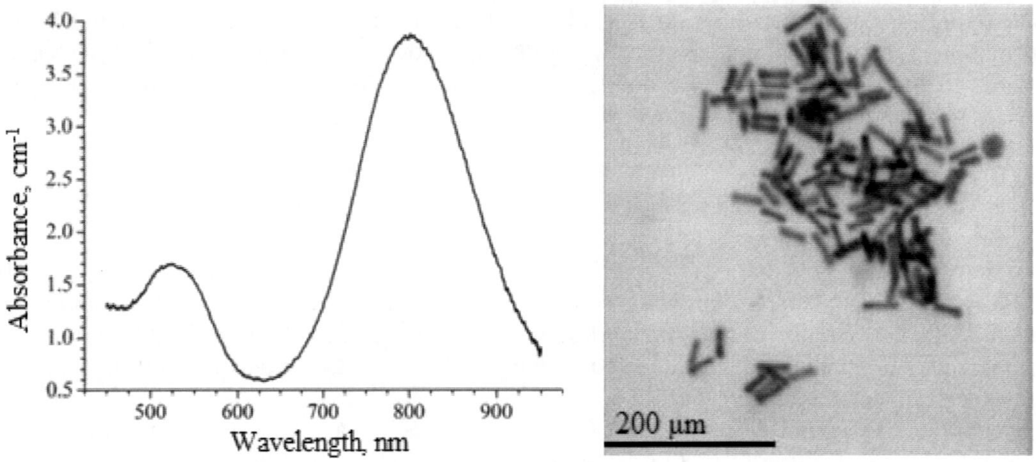

(a) (b)

Figure 2. (a) – absorbance spectrum of AuNRs solution, (b) – TEM image of AuNRs.

2.2.3. Coating of gold nanorods

Obtained phantoms have a uniform distribution of the nanoparticles only if the nanoparticles are stable in acetone solution, thus it is needed to coat AuNRs by silica shell. Initial nanorods solution (50 mL) was centrifuged for 60 min at RCF equals to 12000g. The upper phase was discarded and the residue was left (1.5 mL). A solution of PVP of 30 mg per 50 mL of AuNRs was prepared. Solution PVP added to nanorods precipitate, made up to 100 mL in volume, was left for 12 min to took place the reaction. Then solution is centrifuged for 60 min at RCF equals to 12000g. After centrifugation the residue was left (4.5 mL).

Five PVP solutions were prepared by serial dilution (concentration of 8, 4, 2, 0.5, and 0.25 mg to 3g of isopropyl alcohol). In each solution was added 0.75 mL of concentrated aqueous nanorods coated with PVP and 0.5 mL ammonia. In 5 min after the addition of ammonia, 0.25 mL of TEOS-SH was added and kept for 40 min. Over time, whitening solution was observed. Then solution is centrifuged during 5 min at RCF 5000g and dissolved in ethanol, then centrifuged for 2 min at RCF 2000g and dissolved in ethanol, then centrifuged during 1 min at RCF 1000g, two cycles 1 min at RCF 2000g and 2 min at RCF 2000g. Finally result was ethanol solution of coating gold nanorods.

2.3. Fabrication of tissue phantoms

The components are optical epoxy casting compounds (PEO-90, Laboratory of polymer's physics, Saint-Petersburg, Russia), acetone, isopropyl alcohol, nanoparticle aqueous solutions.

The method of fabrication consists in follows. Firstly, 0.2 mL of acetone is added to 1.2 mL of epoxy and heated to 55°C. Secondly, 0.2 mL of isopropyl alcohol is added. Thirdly, 0.3 mL of thickener is added. Fourthly, 0.2 mL of aqueous solution of nanoparticles is added. After adding of each new component mixture is mixed by magnetic mixer. Finally, the solution is heated to 75°C and keep day. If nanoparticles are stable in acetone they will be distributed regularly in the phantoms.

IOP Conf. Series: Journal of Physics: Conf. Series **917** (2017) 042003 doi:10.1088/1742-6596/917/4/042003

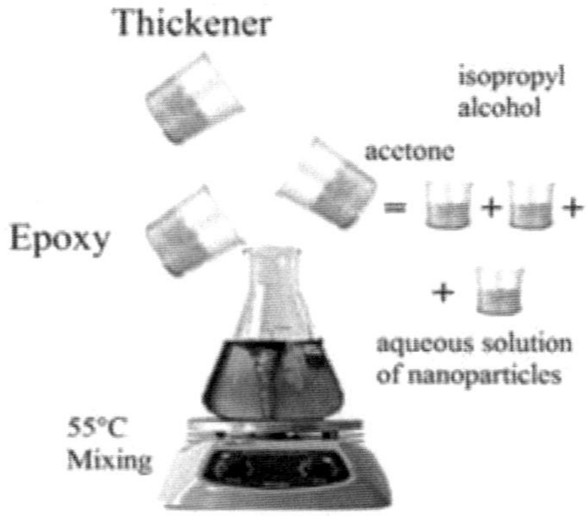

Figure 3. Scheme of preparation of tissue phantoms containing nanoparticles.

3. Results and discussions

Luminescent properties of obtained samples were investigated. The sample made without nanoparticles has own weak luminescence spectrum, where the wavelength of maximum luminescence is 450 nm. The CdSe/ZnS quantum dots have a maximum of luminescence at 590 nm. The epoxy sample with CdSe/ZnS Nps shown luminescence in green region and have two maximums of luminescence (at 470 nm and 585nm). Created epoxy sample with CdSe / ZnS QDs causes distortion of the luminescence spectra of quantum dots (Figure 4).

Absorbance spectrum of AuNRs and luminescence spectrum of UCNPs are shown in Figure 5. According to these results, we can conclude that AuNRs distort the true spectrum of luminescence upconversion nanoparticles.

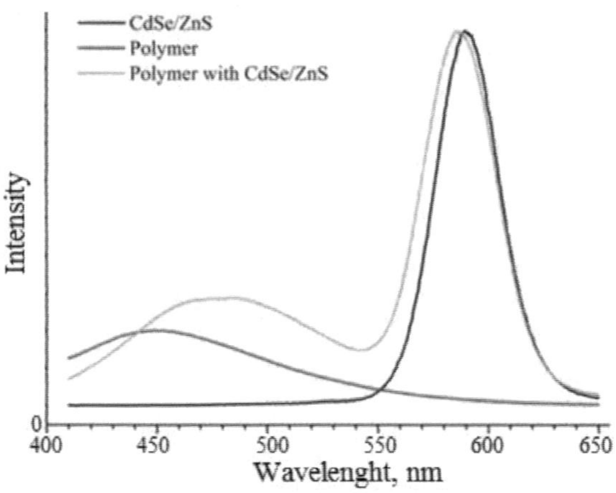

Figure 4. The experimental luminescence spectra of CdSe/ZnS, polymer, polymer with CdSe/ZnS.

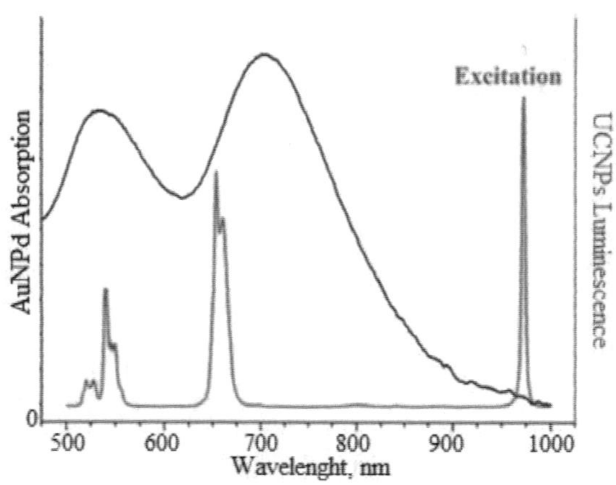

Figure 5. Absorbance spectrum of AuNRs and luminescence spectrum of UCNPs.

We have studied the thermal stability of the phantoms. A sample of gold nanorods was subjected to thermal gravimetric analysis (TGA). Operating method: the sample was heated to 450°C with 1°C per min step size. The temperature of phase transition to the gaseous state was determined.

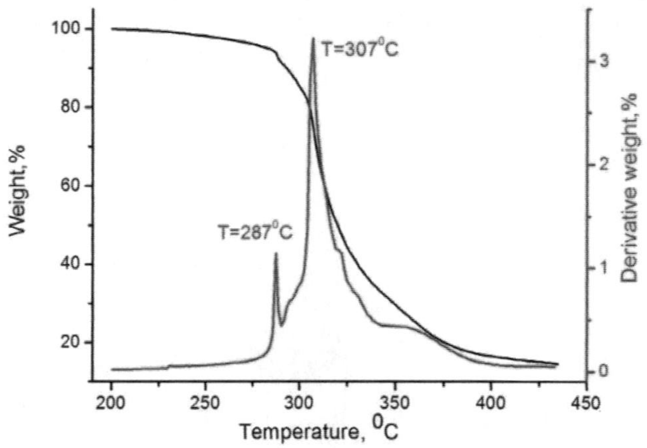

Figure 6. Results of TGA of phantom sample loaded with gold nanorods.

4. Conclusion

Using solid phantoms provide repeatability of the initial experimental conditions, the statistical validity of the results, and no degradation of the test system under the influence of external factors that allows carry out a series of experiments on the same sample. The optical properties of the phantoms can be easily customized for any type of biological tissue by varying the concentration of the scattering particles. Thermal properties of phantoms play a key role in carrying out the experiments on laser thermolysis.

According to our experiments, creating phantom distorts the luminescence of the quantum dots. Own weak luminescence of the sample causes small distortions in the measurement. Phantoms containing gold and upconversion nanoparticles well suited for experiments on laser thermolysis. Received phantoms have high temperature of degradation and resistant to thermal stresses. Thus, these samples can be used as biological tissue phantoms for experiments on laser thermolysis.

Acknowledgments

The work was supported by Russian Presidential grant NSh-7898.2016.2

References

[1] Bashkatov A N, Genina E A, Kochubey V I, Rubtsov V S, Kolesnikova E A and Tuchin V V 2014 *Quant. Electron.* **44** 779-784

[2] Bashkatov A N, Genina E A, Kozintseva M D, Kochubey V I, Gorodkov S Yu and Tuchin V V 2016 *Opt. Spectrosc.* **120** 1-8

[3] Konyukhova J, Skaptsov A, Volkova E, Galushka V and Kochubey V 2014 *Proc. SPIE* **9031** 90310L

[4] Lazareva E N and Tuchin V V 2016 *Proc. SPIE* **9917** 99170U

[5] Tuchina D K, Genin V D, Bashkatov A N, Genina E A and Tuchin V V 2016 *Opt. Spectrosc.* **120** 28-37

[6] Pogue B W, Patterson M S, 2006 *J. Biomed. Opt.* **11** 041102

[7] Wróbel M S, Popov A P, Bykov A V, Kinnunen M, Jędrzejewska-Szczerska M and Tuchin V V 2015 *J. Biomed. Opt.* **20** 045004

[8] Wróbel M S, Popov A P, Bykov A V, Kinnunen M, Jędrzejewska-Szczerska M and Tuchin V V 2015 *J. Innov. Opt. Health Sci.* **8** 1541005

[9] Huang X, Jain P K, El-Sayed I H and El-Sayed M A 2008 *Lasers Med. Sci.* **23** 217-228

[10] Jędrzejewska-Szczerska M, Wróbel M S, Galla S, Popov A P, Bykov A V, Tuchin V V and Cenian A 2015 *Proc. SPIE* **9447** 944715

[11] Cohen M L 1977 *J. Invest. Dermatol.* **69** 333-338

[12] Skaptsov A A, Novikova A S, Galushka V V, Markin A V, Kochubeya V I and Goryacheva I Yu 2016 *Proc. of SPIE* **9917** 991711

[13] Skaptsov A A, Goftman V V, Galushka V V, Markin A V, Kochubeya V I and Goryacheva I Yu 2016 *Proc. of SPIE* **9917**, 991712

[14] Nikoobakht B and El-Sayed M A 2013 *Chem. Mater.* **15** 1957-1962

Quantum-mechanical analysis of amino acid residues function in the proton transport during F_oF_1-ATP synthase catalytic cycle

L A Ivontsin, E V Mashkovtseva and Ya R Nartsissov

Institute of Cytochemistry and Molecular Pharmacology, 24/14 6th Radialnaya str., Moscow 115404, Russia

Abstract. Implications of quantum-mechanical approach to the description of proton transport in biological systems are a tempting subject for an overlapping of fundamental physics and biology. The model of proton transport through the integrated membrane enzyme F_oF_1-ATP synthase responsible for ATP synthesis was developed. The estimation of the mathematical expectation of the proton transfer time through the half-channel was performed. Observed set of proton pathways through the inlet half-channel showed the nanosecond timescale highly dependable of some amino acid residues. There were proposed two types of crucial amino acids: critically localized (His245) and being a part of energy conserving system (Asp119).

1. Introduction

In 1941 Fritz Lipmann discovered that the main storage of energy in the cell is adenosine triphosphate (ATP). These molecules belong to the so-called macroergic compounds containing bonds, hydrolyzed with a release of a significant amount of energy. In living organisms, ATP is synthesized by substrate-level, oxidative and photosynthetic phosphorylation of adenosine diphosphate (ADP) [1].

There are several types of ATPases, which can hydrolyze ATP – F-type, V-type, A-type. Nevertheless, the only enzyme capable of synthesizing ATP in the process of oxidative phosphorylation is F_oF_1-ATP synthase [2]. F_oF_1-ATPsynthase is a complicated molecular motor that is a unique energy transformer in a living cell. This highly conserved protein complex uses the energy of the electrochemical gradient of hydrogen ions created by the electron transport chain [3].

Despite the long-term biochemical and structural studies and theoretical modeling [4-7], many issues remain unanswered. Among them are the mechanisms of energy, time and proton transfer coupling. In addition, the question of the precise enzyme structure also remains open. It is assumed that the membrane part of the enzyme contains two half-channels for proton transport, however, direct measurement of any characteristics of this process is challenging [8]. The key unresolved issue is how the process of proton transfer through the membrane is coupled with the synthesis of ATP.

One of the most precise ways to describe deep molecular processes is the application of quantum-mechanical principles. It can be widely used for current calculation in nanoscale field-effect transistors, drug design, enzyme-ligand docking, protein folding and many other processes. Wave function of proton in gramicidin channel, which is a pore in a lipid membrane, was previously described elsewhere [9]. Meanwhile processes in more complicated biological objects consisting of a complex channels or half-channels can be also characterized using quantum-mechanical approach.

It is generally accepted that proton transport is carried out along the chain of amino acid residues of the enzyme. Therefore, the problem of identifying the essential amino acids for proton transport became very important in the last decade [5]. Studies of the protein structure were performed by X-ray diffraction analysis, cryo-EM, site-specific mutagenesis and cross-linking [10, 11]. Thus, a number of amino acid residues that are supposedly involved in proton transport has been identified. Being a part of a channel wall and taking part in the cumulative charge forming these residues can also serve as intermediate energy storages or sources.

New data delivered using quantum-mechanical approach can be very useful not only for description of proton movement in F-ATP synthase half-channels but also for the understanding of other transmembrane ion movement processes.

2. Methods

Currently, there is a list of different pdb-structures of F_o-ATP synthase, but many of them are incomplete or have low resolution (7-8 Å). In this paper we use one of the first pdb-files 1C17 for *E.Coli*, which contains the a_1c_{12} subunits [12].

Thorough study of the published data revealed the following amino acid residues important for proton transport in the inlet half-channel: Gln234, Asn116, Asp119, His132, Ser144, Asn148, Ser152, Asn214, Glu219, Ser233, Asn238, His245, Thr250, Gln252 [13, 14, 15].

There are different approaches to the description of proton transport in various environments: Grotthuss mechanism, the stochastic modeling, the molecular dynamics and so on. In this paper, in order to simulate proton transport through the input half-channel, a combined approach is used. It allows obtaining information that is more detailed. Probability of proton transfer from one charged center to another is determined in quantum-mechanical model of one-dimensional motion. This allows us to consider the purely quantum effects inherent in microparticles. The general problem of proton movement through the sequence of charged centers in the half-channel is solved using stochastic approach [16].

In order to describe the mechanism of proton transport it is necessary to define the potential of interaction between a proton and charged centers (amino acid residues and water molecules), which can bind protons on its way through the half-channel. Energy spectra of bound states in potential wells were determined by Bohr-Sommerfeld quantization. Occupation numbers of the quantum levels were given by Boltzmann distribution. Hopping motion including only amino acid residues leads to rather low probability values of a transition from one charged center to another due to their considerable distance. Therefore, the distribution of water molecules was obtained in the input half-channel. Possible pathways of proton movement through half-channel were revealed using Monte-Carlo simulation. An exact description of the model and the spatial arrangement of water molecules were previously described elsewhere [16].

To determine the direction of a proton motion when choosing from adjacent centers, it is necessary to calculate the characteristic times of elementary steps. Using Monte-Carlo simulation, the transfer from the current center to another is selected considering the probability of the transition, which is inversely proportional to the calculated transfer time.

The universality of this approach consists in the ease including and excluding any number of essential amino acid residues based on the analysis of published data and depending on the specific research tasks. In this paper, we consider the case of 15 amino acids in the input half-channel.

3. Results

The model described in this paper allows one to obtain sets of pathways for the transport of a single proton through an inlet half-channel. The motion of a proton is not limited in direction and is determined only by the probability of a transition. The existence of adjacent charged centers in some regions of the half-channel lead to the set of equiprobable transfers that may cause possible cyclic transitions, thus increasing the total number of steps. The calculation of the proton path was carried out 100,000 times in each set. The number of steps varied from 600 to 200,000 in one pathway, however, more than 60% of the pathways included no more than 5000 steps. The estimation of the mathematical expectation of the proton transfer time through the half-channel was $(1.66 \pm 0.06) \cdot 10^{-10}$ s. Errors were estimated from the obtained sample using the formula of an unbiased dispersion estimate.

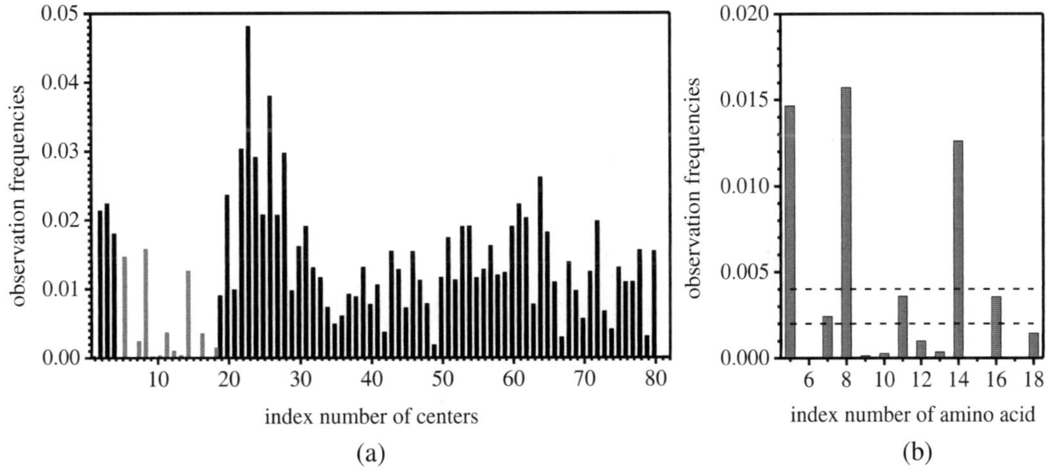

Figure 1. Histogram of the observation frequencies of individual centers in the resulting set of paths. Black bars represent water molecules and red bars represent amino acid residues. (a) For all centers; (b) For amino acids.

Figure 1 shows a histogram of the observation frequencies of individual centers in the resulting set of paths. Black bars represent water molecules, the first four of which are located at the beginning of the half-channel. Therefore, participation of one of them is inevitable in every pathway. Red bars represent amino acid residues, numbered in the order described above. It is easily seen that water molecules are encountered more frequently in the proton pathways. However, the effect of some amino acid residues on proton transport is also significant. All amino acid residues can be divided into three groups according to their observation frequencies. Often found (frequencies > 0.004) are 5(Gln234), 8(His132), 14(Ser233); rarely found (frequencies > 0.002) are 7(Asp119), 11(Ser152), 16(His245), 18(Gln252). All the other amino acids practically do not participate in proton transport. To analyze the contribution of the specific amino acid the presented data should be analyzed in details. An amino acid residue is assumed essential in the transfer process if its exclusion from the pathway will lead to the increase in overall proton transport time. However, it should be mentioned that in this model the proton transport could not be seized by exclusion of any charged center due to stochastic mode of motion and high density of centers in the half-channel.

In two cases of exclusion of an amino acid residue from the second group, the significant changes in the time of movement through the input half-channel were observed. When Asp119 was excluded the time of proton transport increased by (23±7)%, however, no visible redistribution of observation frequencies of the other centers occurred. Therefore, there is no transport stoppage, however, the increase in time may indicate the non-optimal character of the observed paths, which may be due to the need for energy accumulation at this step. Asp119 is known to be one of the important amino acid residues in the proton pathways. Substitutions of this residue leads to the reduce but do not the abolishing of proton conductivity [15].

Another key amino acid residue is His245. Due to its highly conservative localization, it is often used to reconstruct the structure of the *a*-subunit obtained by the cryo-EM method [10]. The exclusion of this residue in the model calculations also led to an increase in proton transport time by (28±8)%. Observed redistribution of the participation frequency for both water molecules and amino acid residues consisted of an increase in the frequency of the centers located in the first half of the channel (1-30) along with the decreased frequency of the centers at the end of the channel (35-80) (Figure 2). This caused to the elongation of the pathway set. His245 located in the center of the membrane with a rather dense protein structure seems to be an important link between the inlet and the end of the half-channel.

IOP Conf. Series: Journal of Physics: Conf. Series **917** (2017) 042004 doi:10.1088/1742-6596/917/4/042004

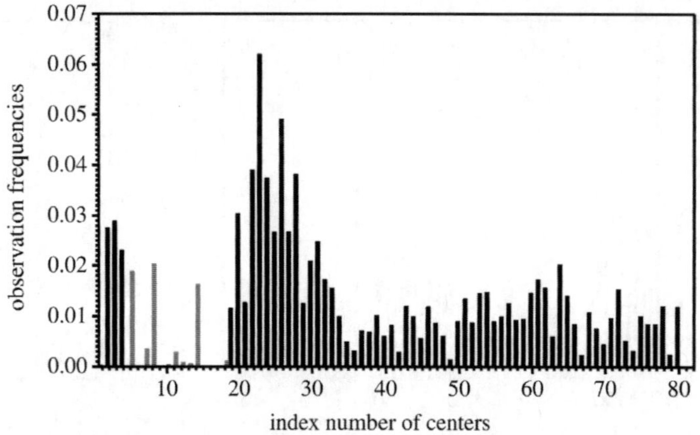

Figure 2. Histogram of the observation frequencies of individual centers in the resulting set of paths without His245. Black bars represent water molecules and red bars represent amino acid residues.

4. Discussion

The proton transport via F_o-ATP synthase half-channels is an important step of energy formation in the living cell. Meanwhile the energy storage and transformation mechanism is still under consideration. Quantum-mechanical approach appeared to be fully applicable to the description of ion transport in this system. Observed set of proton pathways through the inlet half-channel showed the nanosecond timescale highly dependable of the protein structure. Most of the considered essential amino acid residues revealed to be an important part of the channel wall. Nevertheless, the observation frequencies of Asn116, Ser144, Asn148, Asn214, Glu219, Asn238, Thr250 tended to the zero that can be caused by its localization and water environment.

The often and rarely found amino acid residues are considered to be significant in the proton transfer. They can be crucially localized (His245) or being a part of energy conserving system (Asp119). Each separate pathway may consist of a huge number of steps, but short pathways (up to 5000 steps) are more often. A prevalence of water molecules as participating charged centers leads to the appearance of possible cyclic transitions regions located in the inlet part of the half-channel and near Asp61 (Figure 3).

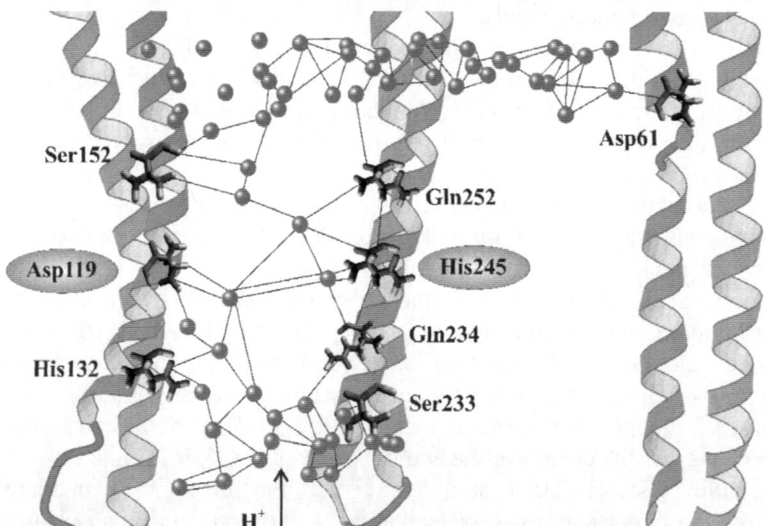

Figure 3. Illustration of a proton movement through the inlet half-channel. The ribbons are 2-5 TMC of *a*-subunit and the nearest *c*-subunit. Red circles are the water molecules, essential amino acids are pointed. Lines show the trajectory of a proton.

IOP Conf. Series: Journal of Physics: Conf. Series **917** (2017) 042004 doi:10.1088/1742-6596/917/4/042004

The data discussed in the paper provide a quantitative estimation of proton movement through the inlet half-channel of F_oF_1-ATP synthase. Its stochastic behavior leads to a huge variability of certain pathways meanwhile conserving a high sensitivity to the types and the number of charged centers on the proton way. Further investigations of the system and its parameters under the different conditions and modifications can be an important step in the comprehensive theory of energy storage and transformation.

References

[1] Boyer P 1999 *Nature* **402** 247-249
[2] Cortes-Hernandez P Vazquez-Memije M E and Garcia J J 2007 *J. Biol. Chem.* **282** 1051-1058
[3] Deckers-Hebestreit G and Altendorf K 1996 *Microbiol.* **50** 791–824
[4] Romanovsky Yu M and Tikhonov A N 2010 *Phys. Usp.* **53** 893–914
[5] Fillingame R H and Steed P 2014 *Biochim. Biophys. Acta* **1837** 1063–68
[6] Junge W *et al* 2004 *Biophysical Journal* **86** 4094–4109
[7] Mashkovtseva E Boronovsky S and Nartsissov Ya 2013 *Math. Biosci.* **243** 117–25
[8] Altendorf K Stalz W Greie J and Deckers-Hebestreit G 2000 *J. Exp. Biol.* **203** 19-28
[9] Chen D and Wei G W 2012 *J. Chem. Phys* **136** 134109
[10] Zhou A *et al* 2015 *eLife* **4** e10180
[11] Schwem B E and Fillingame R H 2006 *J. Biol. Chem.* **281** 37861–37867
[12] Rastogi V and Girvin M 1999 *Nature* **402** 263-68
[13] Allegretti M *et al* 2015 *Nature* **521** 237-40
[14] Steven B V Patterson A R and Antonio B J 1998 *J. Biol. Chem.* **273** 16229–16234
[15] Bartl F *et al* 1995 *Biophysical Journal* **68** 104-110
[16] Ivontsin L Mashkovtseva E and Nartsissov Ya 2017 *J. Phys.* **784** 012021

A stochastic modeling of isotope exchange reactions in glutamine synthetase

N V Kazmiruk[1], S E Boronovskiy[1] and Ya R Nartsissov[1]

[1]Institute of cytochemistry and molecular pharmacology, Moscow 115404, Russia

Abstract. The model presented in this work allows simulation of isotopic exchange reactions at chemical equilibrium catalyzed by a glutamine synthetase. To simulate the functioning of the enzyme the algorithm based on the stochastic approach was applied. The dependence of exchange rates for ^{14}C and ^{32}P on metabolite concentration was estimated. The simulation results confirmed the hypothesis of the ascertained validity for preferred order random binding mechanism. Corresponding values of $K_{0.5}$ were also obtained.

1. Introduction

Up to the present time, the isotope exchange kinetics at chemical equilibrium has been recognized as an important experimental approach in the study of chemical reactions mechanisms. In particular, it appears as a powerful tool for validation of their pathways and structure of transition states [1,2]. In biomedical research the method of equilibrium isotope exchange is often used for the investigation of the substrate binding in the active site of the enzyme. Variation of structurally related reactant-product pairs yields information about substrate binding order and the relative rates of association/dissociation, whereas the equilibrium exchange rates are used to make a distinction between the processes of fully random or preferred order random binding of substrates [3]. Glutamine synthetase (GS), which is known to be a key component of the metabolism of glutamic acid, is among the enzymes that are under active consideration with the help of this technique.

$$\text{L-glutamate} + \text{ATP} + \text{NH}_4^+ \xrightarrow{Mg^{2+}\ or\ Mn^{2+}} \text{ADP} + \text{P}_i + \text{L-glutamine}$$

It catalyzes the synthesis of glutamine and provides the regulation of ammonium level in almost all organisms, in particular in astrocytes of human brain, where it is a critical enzyme for metabolism of glutamate, GABA and ammonia. Since GS is an essential part of a complex astrocyte–neuron signaling process, changes in enzyme expression and activity will lead to neurological dysfunction. A number of different brain pathologies are associated with alterations in GS expression/activity [4]. The mechanisms by which the expression and activity of this enzyme are regulated under normal and pathological conditions (e.g. under hyperammonemia) still remain unclear [5]. It is important to note that there is the major evidence that the active site of the enzyme is rather conservative suggesting an identical catalytic mechanism of the various forms of the enzyme [6]. The biosynthesis reaction of GS is quite complex, and as a first approximation it can be described in two processes Figure 1. The first step consisting in the formation of the activated intermediate γ-glutamyl phosphate is followed by the second one, which is the ammonia attack on the intermediate and glutamine release [7-9].

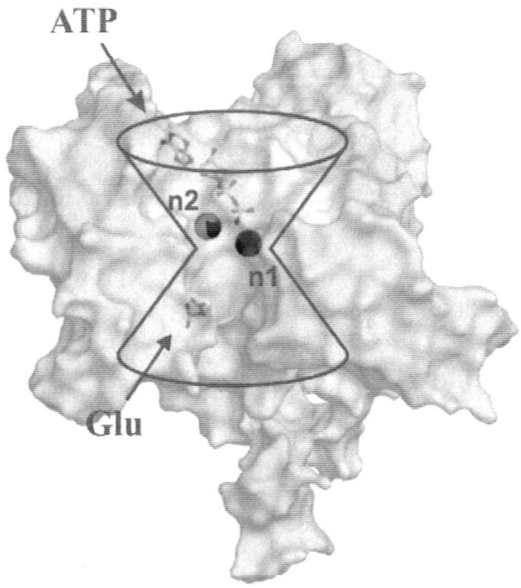

Figure 1. Structure of active center of bacterial glutamine synthetase (PDB code 1F1H); n1 and n2 are the high- and low-affinity binding sites for divalent cations [10].

Nevertheless, stochastic simulation requires the detailed mechanism, taking into account all the possible variations of binding metabolites with a certain probability. Such a mechanism is «Random Ter-Ter», which has been proposed in the earlier studies [11]. Thus, the study of bacterial glutamine synthetase using the computational model of such a method allows to obtain the dependence of exchange rate between the substrates and the reaction products on the concentrations of various metabolites and makes the conclusions concerning the substrates binding mechanism. Furthermore, it yields a great perspective of Gln-Glu cycle studies *in silico*.

2. Model

2.1. Equilibrium Isotope Exchange in Enzyme Catalysis

Generally, isotope exchange reactions can be defined as the spontaneous processes of redistribution of isotopes between the molecules, which are not accompanied by chemical changes of the molecules (i.e., changing only their isotopic composition). However, the exchange may occur between the molecules located in the same or different phases and within the molecules. In the study of GS exchange reactions $[^{14}C]Glu \leftrightarrow [^{14}C]Gln$, $[^{14}C]ADP \leftrightarrow [^{14}C]ATP$, $[^{32}P]P_i \leftrightarrow [^{32}P]ATP$ were simulated, which were used for carrying out classical earlier experimental studies [12,13].

The equation for calculating speeds of the equilibrium exchange reaction $X \leftrightarrow Y$ is as follows:

$$R = -\frac{[X][Y]}{[X]+[Y]}\ln(1-f) \quad \textbf{(1)}$$

where f is the fraction of isotopic equilibrium, defined as:

$$f = \frac{[X]+[Y]}{[Y]} \times \left(\frac{y}{x+y}\right) \quad \textbf{(2)}$$

where x and y are disintegrations per minute (dpm) of radiolabel in the X and Y pools. The dependence of the fraction of isotopic equilibrium from the half-time period of is determined by the formula:

$$-\ln(1-f) = \frac{\ln 2}{t_{0.5}} \quad \textbf{(3)}$$

This equation allows to determine f according to experimental data and to find the value of the rate of exchange R.

2.2. Stochastic description of the enzyme functioning
The stochastic algorithm based on probabilistic approach was applied for modeling of an isotopic exchange reaction. In contrast to the kinetic approach, which is employed to describe the behavior of the protein ensemble and has been previously used for exchange reactions modeling, a stochastic one allows us to use thermodynamic parameters (e.g. Gibbs energy) and to work directly with a single protein [14]. Therefore, in this work the equilibrium constants for the substrate binding and conformational changes acted as the basic parameters. The binding process can be represented using a simple scheme:

$$E + A \underset{k_{off}}{\overset{k_{on}}{\rightleftarrows}} EA \quad \Rightarrow \quad K_d = \frac{[A]\cdot[E]}{[EA]} \qquad \textbf{(4)}$$

where A is a substrate, E is a protein (GS), and the value of K_d can be determined according to changes in the thermodynamic functions or measured in the experiments. The algorithm is based on a step-by-step calculation of state probabilities with the subsequent estimation of both rate constant and exchange rates for each «substrate-product» pair. If it is assumed that there are a number of independent states of the protein, it is necessary to calculate the probability of finding the protein in the free state or in complex with the substrate:

$$\begin{cases} P_E = \dfrac{N_E}{N_{EA+E}} \\ P_{EA} = \dfrac{N_{EA}}{N_{EA+E}} \\ N_E + N_{EA} = N \end{cases} \Rightarrow \quad \frac{K_d}{[A]} = \frac{N_E}{(N - N_E)} \qquad \textbf{(5)}$$

where N_E is a number of protein molecules in the free state, and N_{EA} is a number of protein molecules in a complex with the substrate. Then an average probability vector can be written as the following:

$$\vec{P} = \left(\frac{1}{1+\sum_{j=1}^{N}\left(K_j^{d_j}\cdot[A_j]^{-s_j\cdot d_j}\right)}, \ldots, \frac{K_k^{d_k}\cdot[A_k]^{-s_k\cdot d_k}}{1+\sum_{j=1}^{N}\left(K_j^{d_j}\cdot[A_j]^{-s_j\cdot d_j}\right)}, \ldots, \frac{K_N^{d_N}\cdot[A_N]^{-s_N\cdot d_N}}{1+\sum_{j=1}^{N}\left(K_j^{d_j}\cdot[A_j]^{-s_j\cdot d_j}\right)} \right) \qquad \textbf{(6)}$$

where d and s are the direction of a single transition and metabolite consumption respectively.

3. Results and discussion
Generally, to obtain saturation curves for an isotopic exchange reaction while maintaining chemical equilibrium, it is necessary to vary substrate and product concentration in a constant ratio. The rates for isotope exchanges $[^{14}C]Glu \leftrightarrow [^{14}C]Gln$, $[^{14}C]ADP \leftrightarrow [^{14}C]ATP$ and $[^{32}P]P_i \leftrightarrow [^{32}P]ATP$ were obtained using the single protein stochastic model.

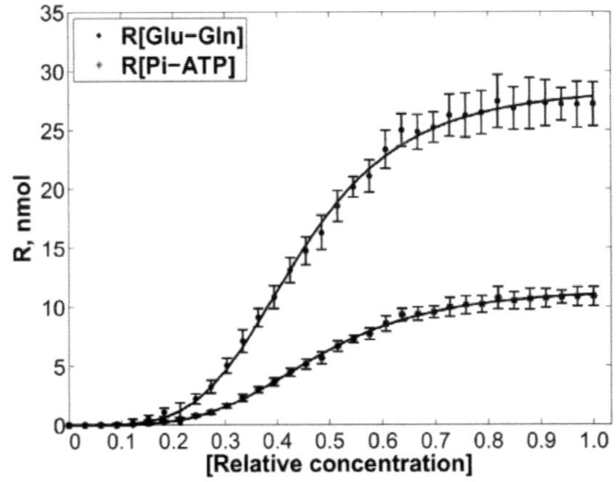

Figure 2. Equilibrium exchange rates. The effect of varying all metabolite concentrations. The points indicate the mean±SD of 20 independent computer simulations of GS activity.

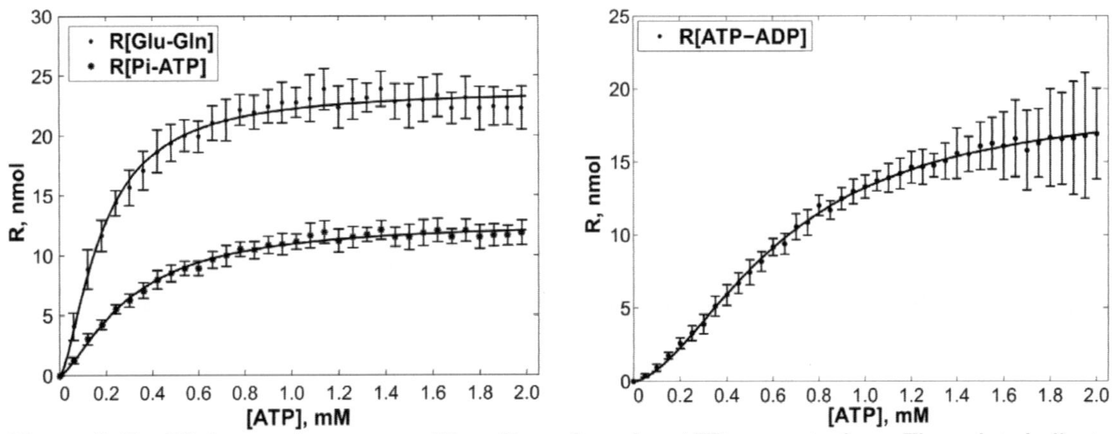

Figure 3. Equilibrium exchange rates. The effect of varying ATP concentrations. The points indicate the mean±SD of 20 independent computer simulations of GS activity. The value of equilibrium ratio is $[ATP]/[P_i] = 1/10$, which is similar to original experiments.

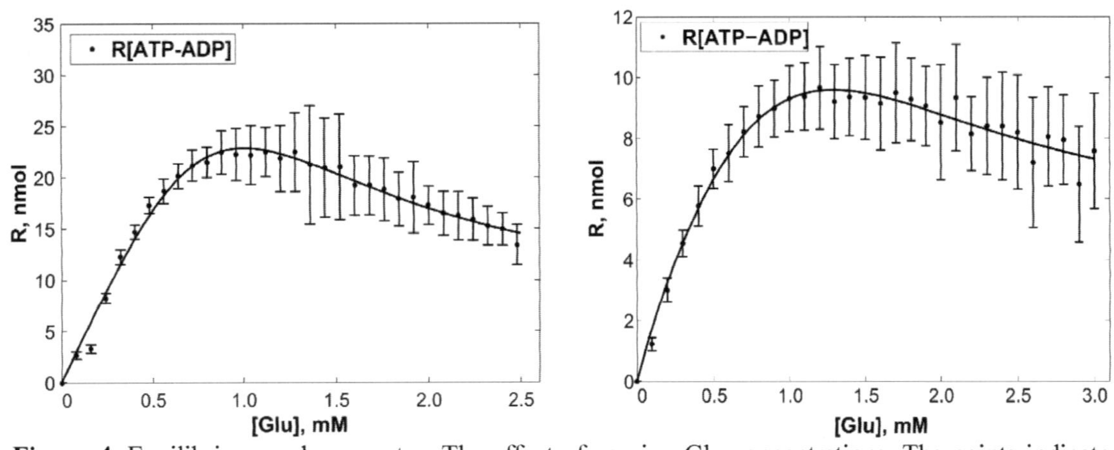

Figure 4. Equilibrium exchange rates. The effect of varying Glu concentrations. The points indicate the mean±SD of 20 independent computer simulations of GS activity. The values of equilibrium ratio are $[Glu]/[P_i] = 1/10$ and $[Glu]/[ADP] = 1/4$ respectively, which are similar to original experiments.

An information about the equilibrium level achieved in the experiments allowed to select the following values for the parameter *f*: 0.7 for the exchanges [^{14}C]Glu↔[^{14}C]Gln and [^{14}C]ADP↔[^{14}C]ATP, and 0.4 for the exchange [^{32}P]P_i↔[^{32}P]ATP. It is important to note that there is no decrease in the exchange rate in the case of relative concentration modeling Figure 2. This result is the major evidence of the hypothesis which excludes strictly ordered sequential binding. Thus, there is a tendency for preferred ordered binding of substrates (i.e. presence of dominant pathways) as a part of completely random mechanism. The fact that saturation curves increase in a sigmoidal manner is in consistence with the «Random Ter-Ter» mechanism of the reaction. Obtained values of $K_{0.5}$ are in a good agreement with the *in vitro* experimental data.

References

[1] Mauve C and Giraud N et al 2016 *Plant Physiology et Biochemistry* **108** 203
[2] Tyapochkin E, Cook P and Chen G 2008 *Biochemistry* **47** 11894
[3] Wedler F C 1995 *Methods in Enzymology* **249** 443
[4] Rose C F et al 2013 *Biochemical Society Transactions* **41** 1518
[5] Eid T et al 2013 *Neurochemistry International* **63** 670
[6] Eisenberg D et all 2000 *Biochimica et Biophysica Acta* **1477** 122
[7] Krishnaswamy P R et al 1962 *The Journal of Biological Chemistry* **237** 2932
[8] Meek T D and Villafranca J J 1980 *Biochemistry* **19** 5513
[9] Liaw S H and Eisenberg D 1994 *Biochemistry* **33** 675
[10] Hunt J B and Ginsburg A 1979 *The Journal of Biological Chemistry* **255** 590
[11] Wedler F C and Boyer P 1972 *The Journal of Biological Chemistry* **247** 984
[12] Wedler F C 1974 *The Journal of Biological Chemistry* **249** 5080
[13] Wedler F C and Barkley R W 1989 *Analytical Biochemistry* **177** 268
[14] Boronovsky S E and Nartsissov Ya R 2016 *Bulletin of Mathematical Biology* **78** 1380

IOP Conf. Series: Journal of Physics: Conf. Series **917** (2017) 042006 doi:10.1088/1742-6596/917/4/042006

A Single-Cell Electronic Sensor of Toxins

D D Stupin
St. Petersburg Academic University, Khlopina 8/3, 194021 St. Petersburg, Russia
E-mail: Stu87@ya.ru, Stupin@spbau.ru

Abstract. Here we propose a simple label-free bio-electronic toxin detector based on non-destructive impedance spectroscopy (IS) method with a single living cell as a sensing element. The toxins distort cell membrane, which significantly affects on the impedance level of an electrode, which covered by a cell. This effect could be used for toxin detection. We believe that our bio-sensor will open a new roadmap in water purity purposes and will save many a one lives.

1. Introduction

Research of living matter has become a hot topic in experimental physics in the resent years. Numerous efforts are being made to develop the new nondestructive and label-free methods for cell diagnostics *in vitro*. The impedance spectroscopy is one of the promising methods for this [1]. The developed by Ivar Giaever electrical cell-substrate impedance sensing technology have become practical in the areas of cancer research, wound healing research, pharmacology, *etc* [2, 3]. Here we show a possibility of the usage IS based bio-sensor as selective detector, which *a priory* reacts on any bio-active toxiferous substances.

2. Theory

According to the Giaever-Keese theory [2], the cell on the electrode acts as an added serial active resistance R_s in the impedance, because current flow not only through electrochemical impedance $Z_{e/c}$ (electrode located in phosphate buffered saline), but also through seal between cell and electrode (figure 1). If electrode is not covered by cell or if cell-membrane is distorted, the value of added resistance R_s decrease. The total impedance Z_{total} of the electrode/cell/electrolyte interface reads as

$$Z_{\text{total}} = Z_{e/c} + R_{\text{sol}} + R_s, \tag{1}$$

where R_{sol} is active bulk resistance of the electrolyte. The action of a toxin on a cell leads to the distortion of the cell membrane. Therefore, after the action of a toxin R_s value should become much lower than $Z_{e/c} + R_{\text{sol}}$. Accordinly, the impedance spectrum of a cell-contained electrode becomes similar to the impedance spectrum of the empty electrode. This phenomenon can be used for a selective toxin detection.

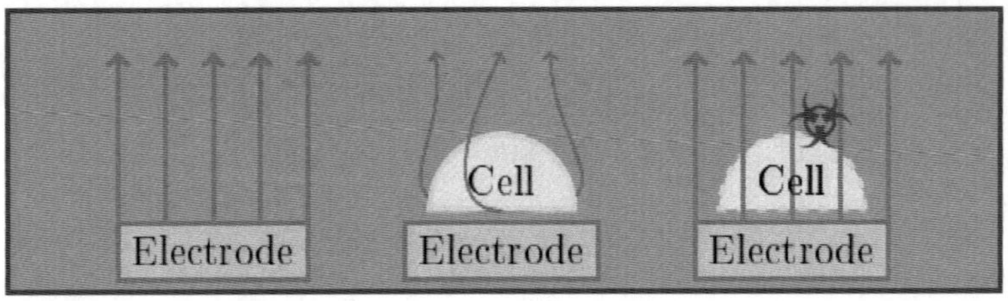

Figure 1. Principle of the operation of the proposed toxin detector (color on-line). When an electrode is covered by a cell (in the middle), a current flow (red arrows) is possible not only through the electrochemical impedance as for empty electrodes (on the left), but also through the contact between a cell membrane and the surface of an electrode. The presence of toxins destroying the cell membrane leads to an increase of the current to the value of that for an empty electrode (on the right).

3. Materials and methods

We have used multielectrode array MEA 200/30 (Multi Channel Systems GmbH, Germany, 30 μm electrode diameter) covered by HeLa cells in 0.5 ml of the phosphate buffered saline (Biolot, Russia) with 7.5 μl propidium iodide *in vitro* under microscope study and home-made advanced device and processing for impedance measurements [4, 5]. Triton X-100 (Union Carbide, USA) was used as model toxin. HeLa cells were obtained from the Bank of Cell Cultures of the Institute of Cytology of the Russian Academy of Sciences. Impedance spectra were measured between large rectangle reference electrode (50 μm $\times$ 250 μm, not presented on photographs) and empty (control) electrode (50 μm diameter, not presented on photographs), and between reference electrode and electrode covered by a single cell [see figures 1(b,e,h))]. Photographs were made by Leica DM4000 microscope (Leica, Germany).

During the IS measurement 2.5 μl of the toxin were added into matrix Petri dish. The final concentration of the toxin was 2.5 μl per 1 ml of the phosphate buffered saline.

4. Results and discussion

The results are shown on figure 2. This data indicate, that impedance of the covered by cell electrode is higher (by magnitude) than impedance of empty electrode before adding toxin and close to empty control electrode after adding toxin. This is result of the significant cells' membrane distortion, what additionally justified by the fluorescence of the propidium iodide.

So, criteria of the close equality between impedance spectra of the cell-contained and empty electrode could be used as danger alarm. We believe that our bio-sensor will open a new roadmap in water purity purposes and will save many a one lives.

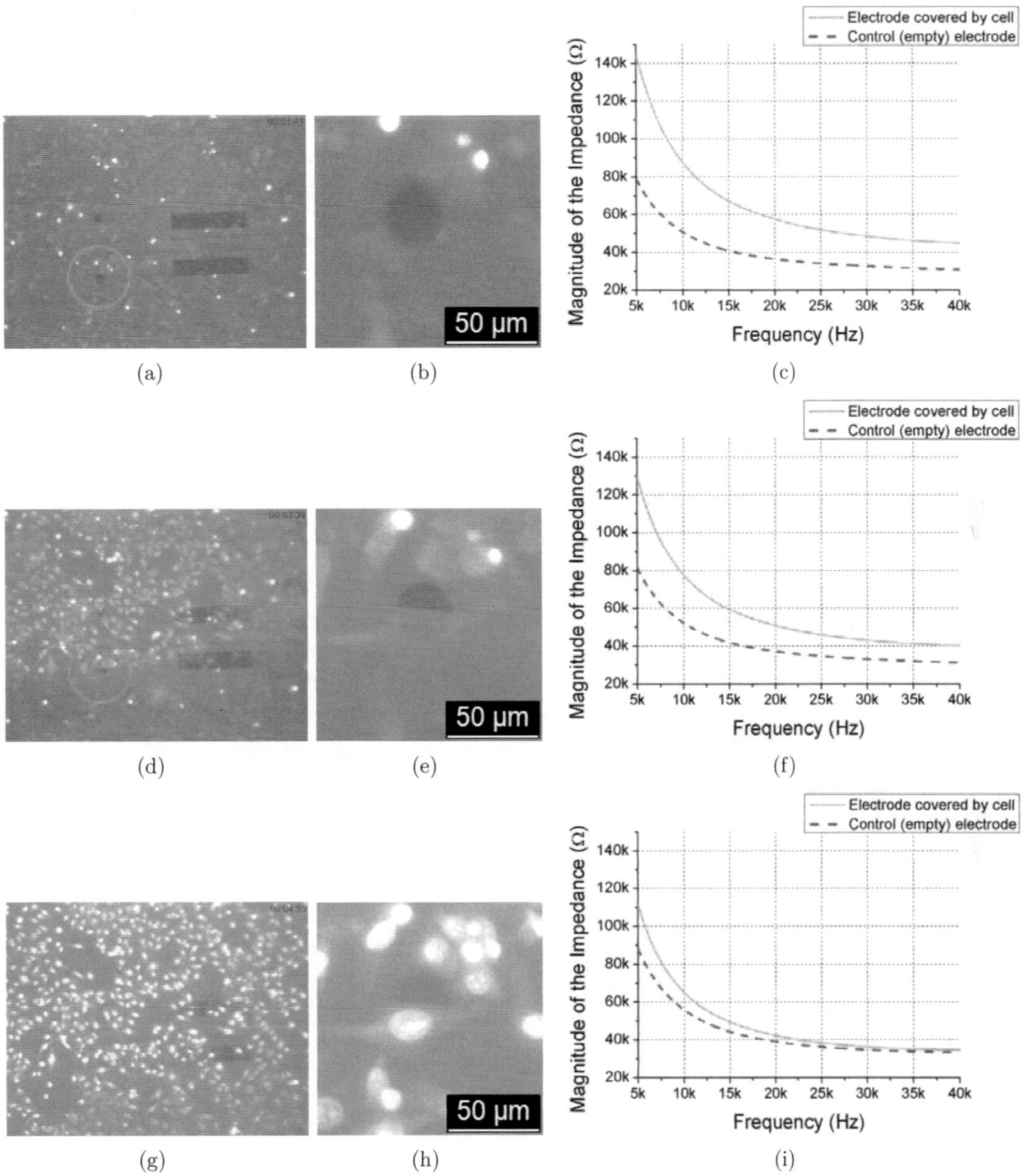

Figure 2. Toxin effect on the impedance (color on-line). Photographs of the electrode with cell (in green circle), magnified image of the electrode with cell, and impedance spectra: (a), (b), (c) – before adding toxin; (d), (e), (f) – 1 minute after adding toxin, arising propidium iodide fluoresce; (g),(h), (i) – 2.5 minutes after adding toxin, cell membrane distorted. Green solid lines corresponds to covered by cell electrode, red dashed line corresponds to control (empty) electrode (not presented on photographs).

Acknowledgments
The author acknowledges SPbOPEN Organizing Committee for holding wonderful conference, and Nikolay A. Verlov, Nikolay A. Knyazev, Sergei V. Koniakhin, Mikhail N. Ryazantsev and Michael V. Dubina for multifaceted assistance and support.

IOP Conf. Series: Journal of Physics: Conf. Series **917** (2017) 042006 doi:10.1088/1742-6596/917/4/042006

References

[1] Barsoukov E and Macdonald J R 2005 *Impedance Spectroscopy: Theory Experiment and Applications* 2nd ed (Wiley-Interscience)
[2] Wegener J, Keese C R and Giaever I 2000 *Experimental cell research* **259** 158–166
[3] Banerjee P and Bhunia A K 2009 *Trends in biotechnology* **27** 179–188
[4] Stupin D D, Koniakhin S V, Verlov N A and Dubina M V 2017 *arXiv preprint arXiv:1701.06503*
[5] Stupin D D, Koniakhin S V, Verlov N A and Dubina M V 2017 *Phys. Rev. Applied* **7**(5) 054024

The deconvolution of complex spectra by artificial immune system

D I Galiakhmetova[1], M E Sibgatullin[1], D Z Galimullin[2] and D I Kamalova[1]

[1]Institute of Physics, Kazan Federal University, Kazan 420008, Russia
[2]Kazan Innovation University, Kazan 420111, Russia

Abstract. An application of the artificial immune system method for decomposition of complex spectra is presented. The results of decomposition of the model contour consisting of three components, Gaussian contours, are demonstrated. The method of artificial immune system is an optimization method, which is based on the behaviour of the immune system and refers to modern methods of search for the engine optimization.

1. Introduction
The problem of separating complex spectral contours into components is important in the study of optical spectra in spectral analysis. The decomposition of a complex spectral contour into components can be reduced to finding the optimal values of the parameters of the elementary components, with which the reconstructed total contour will have minimum deviation from the experimentally obtained one.

Currently, one of the directions for the solution of the optimization problem is the so-called Natural Computing [1-3], which represents dynamic models containing the laws and processes of nature. Natural Computing is divided into several types, namely, Artificial Neural Network, Autonomic Computing, Membrane Computing, Artificial Life, Amorphous Computing. One of the classes of Autonomic Computing is Artificial Immune System. This approach is based on the processes and principles of the vertebrate immune system.

In this work, the optimization method was used, which is based on the behavior of an artificial immune system, in order to solve the problem of decomposition of a complex spectral contour into elementary components. The biological prototype of the artificial immune system is the human immune system, the information inside which is processed by protein molecules (peptides). This system is a complex adaptive structure using a combination of different mechanisms for protection against external pathogens, which can be any microorganisms (including viruses and bacteria) that can cause a human pathological condition (disease). The protective system has a hierarchical multilevel structure. Each successive level increases the specificity of protection. The congenital and acquired immune systems are the protection methods.

2. Results and Discussion
Biological immune system solves the problems of searching and classification. It is able to learn, and it possesses memory. The immune system processes huge amounts of information, producing complex parallel computations, so the algorithms using it as a basis have proved to be effective in mathematical problems of search, recognition and optimization.

An artificial immune network is a set of B-lymphocytes, which are interconnected and are subject to certain operations of cloning and mutation. When a foreign agent enters the body, molecules called antigens are formed, in response to their appearance, antibodies reacting with them are formed. Antibodies are the molecules of B-lymphocytes that circulate in the blood circulatory and lymphatic systems and wait for collision and neutralization of antigens. If, by the recognition of antigens, the degree of closeness (similarity) of genetic sets of antigen and antibodies (R^{gb} - affinity) is small, mutation of antibodies occurs. Then, for the protection of the body, the best antibodies are cloned with the highest possible affinity. If contact with a foreign agent occurs not for the first time, the immune memory of the system, that is, the best antibodies produced earlier by the first contact with these pathogens, is used.

In the artificial immune system, affinity measures are used for the description of the interaction of the components of the system (antibodies), and immune algorithms are used as adaptation mechanisms [4]. To build such a system, the knowledge of the application area or objective function is required [5]. The following biological concepts and corresponding mathematical definitions are used in this work:

- antigen – the initial spectral contour obtained experimentally;
- antibody – the decision x∈D (D is the tolerance range), which is a sum of spectral components, each of which depends on three parameters (frequency, intensity and half-width);
- affinity – the measure of the usefulness of antibodies, it determines the difference between the experimental contour and the solution x;
- population – the set of antibodies;
- clone – the antibody formed from another antibody by cloning (copying);
- mutation – the change in antibody corresponding to a random alteration in the components of the vector x.

The following notation is used:

- b – antibody;
- $f(b)$ – the value of the objective function corresponding to the antibody;
- $B=\{b_i, i\in[1:n_b]\}$ – population of antibodies;
- $b_{i,1}, b_{i,2}, \ldots, b_{i,n}$ – components of antibodies $b_i\in B$;
- $\varphi_i = -f(b)$ – the affinity of the antibody b_i;
- M – population of memory cells (antibodies that are not part of the population B);
- C_i – set of antibody clones b_i;
- $C=\{C_i, i\in[1:n_b]\}$ – set of all antibody clones;
- $b_j^{C_i}$ – j-th clone of antibody b_i;
- $b_{best}^{C_i}$ – the best clone antibodies b_i (the clone corresponds to the minimum value of the objective function);
- n_c – the number of clones generated by each antibody in the population B;
- α – step of mutation.

The main steps of the artificial immune system method are following:

1) Initializing the antibody population B, where the antibody $b_i = (b_{i,1}, b_{i,2}, \ldots, b_{i,n})$ is a vector in the space of variates.
2) Determining the usefulness φ_i of each antibody b_i in the population B.
3) Performing the cloning operation on the antibody with the greatest usefulness. A lot of antibodies C_i are created, which are identical to the original one $C_i=\{b_j^{C_i}: b_j^{C_i}=b_i, j\in[1:n_c]\}$.
4) Performing the mutation $b_j^{C_i}=b_j^{C_i}+\alpha$ on the set of clones C, where α is determined by the formula

$$\alpha = 10^{-3} b_j^{C_i}.$$

5) Calculating the usefulness of all the cloned antibodies from the set C and finding the antibody $b_{best}^{C_i}$ with the highest affinity.

IOP Conf. Series: Journal of Physics: Conf. Series **917** (2017) 042007 doi:10.1088/1742-6596/917/4/042007

6) Performing the clonal selection. The best of the antibodies' descendants replace the parents in accordance with the formula

$$b_i = \begin{cases} b_i, \; if \; f(b_{best}^{C_i}) > f(b_i), \\ b_{best}^{C_i}, \; if \; f(b_{best}^{C_i}) \leq f(b_i). \end{cases}$$

If the usefulness of the best antibody is higher than that of the corresponding memory cells, then the latter is replaced by the above mutated clone.

7) Replacing the worst solution b_j, b_k by the new antibodies provided that

$$\|b_j - b_k\| < \beta, j \neq k,$$

where $\|b_j - b_k\|$ is a measure of the closeness of the cells (root-mean-square deviation); β is the compression threshold.

8) The set of memory cells *M* stores the best solutions found during the work of the method.

To study the effectiveness of the optimization method based on the algorithm of the artificial immune system in solving problems of complex spectral contour decomposition, the model spectra obtained by summing elementary contours were used [6]. The solid curve in figure 1 shows the model spectrum, the dotted lines show the components of the total contour calculated with the artificial immune system method. For the reconstruction of the elementary components, the number of components in the model contour and the frequency values corresponding to them were set [7]. The values of intensity and half-width of the reconstructed components were obtained. For comparison, table 1 represents the results of the initial parameter of the elementary components of the model contour consisting of two Gaussian contours and the parameters of the reconstructed elementary components. The following designations for the contours' parameters are used in table 1: δ is the half-width, Δ is frequency, *I* is intensity.

Table 1. The parameters of the initial and reconstructed two-component contour.

	Initial contour			Reconstructed contour		
	δ	Δ	*I*	δ	Δ	*I*
1st Gaussian contour	70.0	20.0	5.0	71.6	20.0	5.2
Gaussian contour	120.0	200.0	3.0	115.3	200.0	2.9

The error of half-width reconstruction is 2.28% and 3.92% for the 1st and 2nd contours, respectively. The error of reconstructing the intensity of the first contour is 4.00%, and for the second one it is 3.33%.

In this work, the decomposition of a three-component contour was carried out. Table 2 shows the parameters of the initial and reconstructed elementary components used by the separation of a three-component spectral contour. When the three-component contour was reconstructed, the error of intensity reconstruction did not exceed 1.43%.

Table 2. Parameters of the initial and reconstructed three-component contour.

	Initial contour			Reconstructed contour		
	δ	Δ	*I*	δ	Δ	*I*
1st Gaussian contour	70.0	10.0	7.0	70.0	10.0	7.1
Gaussian contour	120.0	630.0	4.0	120.0	630.0	4.0
Gaussian contour	220.0	190.0	5.0	220.0	190.0	5.0

In addition, the decomposition of a three-component contour with closely spaced components was carried out. Figure 1 shows the result of this decomposition using the artificial immune system method. The values of the parameters of the model and reconstructed elementary components in the case of closely spaced components are indicated in table 3.

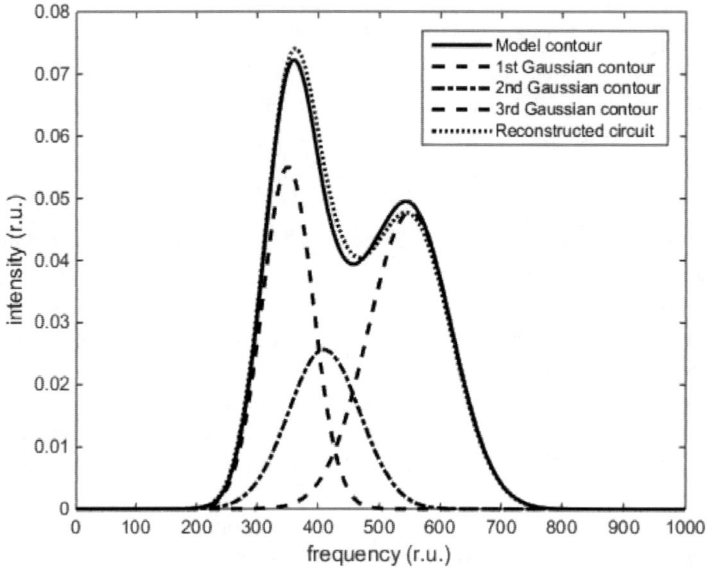

Figure 1. Deconvolution of a three-component model contour consisting of closely spaced Gaussian components.

Table 3. Parameters of the initial and reconstructed three-component contour in the case of closely spaced components.

	Initial contour			Reconstructed contour		
	δ	Δ	*I*	δ	Δ	*I*
1st Gaussian contour	59.0	350.0	5.8	60.9	350.0	6.0
Gaussian contour	83.0	410.0	3.8	79.1	410.0	4.1
Gaussian contour	96.0	550.0	8.2	95.6	550.0	7.9

When the parameters of three elementary components of the initial contour were reconstructed in the case of closely spaced components, the half-width reconstruction error did not exceed 4.70%, the intensity reconstruction error was 7.89% of the set values of the model loop parameters.

In this work, an optimization method based on the artificial immune system algorithm was adapted to decomposition of complex spectra into elementary components. The basic elements and rules of the model were compared to mathematical concepts. The algorithm was tested on model two- and three-component contours.

References

[1] Karpenko A P 2014 Modern algorithms of search engine optimization. *Nature-inspired optimization algorithms* (Moscow: Bauman MSTU Press) p 446 (in Russian)

[2] Yagfarov R R, Sibgatullin M E, Galimullin D Z, Kamalova D I and Salakhov M Kh 2016 Deconvolution of complex spectra into components by the bee swarm algorithm *J. Phys.: Conf. Series* **714**

[3] Kamalova D I, Galimullin D Z, Sibgatullin M E, Shaymuhametova M E and Salakhov M Kh 2013 Separation of closely located Fourier transform Fourier bands by means of a genetic algorithm *Optics and Spectroscopy* **114** (1) 44-50 (in Russian)

[4] Kushnir N V, Kushnir A V, Anatskaya Ye V, Katysheva P A and Ustinov K G 2015 Artificial immune systems: overview and current state *Scientific works of KubSTU* **12** 1-10

[5] Karpenko A P and Shurov D L 2012 Hybrid method of global optimization based on the artificial immune system *Science and Education* **8** (Moscow: Scientific Publication of BMSTU) pp 339-378 (in Russian)

[6] Geri M and Dzhonson D 1979 Computers and Intractability: A Guide to the Theory of NP Completeness. A Series of Books in the Mathematical Sciences (San Francisco, Calif.: W. H. Freeman and Co) p 338

[7] Attetkov A V, Galkin S V and Zarubin V S 2001 Optimization methods: Textbook for High Schools (Moscow: Bauman MSTU Press) p 439 (in Russian)

IOP Conf. Series: Journal of Physics: Conf. Series **917** (2017) 042008 doi:10.1088/1742-6596/917/4/042008

Fluorescence imaging of the nanoparticles modified with indocyanine green

K G Gareev[1], K Y Babikova[1], V N Postnov[2,3], E B Naumisheva[2,3], D V Korolev[3,4,5]

[1]SaintPetersburg Electrotechnical University "LETI", Saint Petersburg, 197376, Russia
[2]Saint Petersburg State University, Saint Petersburg, 199034, Russia
[3]Federal Almazov North-West Medical Research Centre, Saint-Petersburg, 194156, Russia
[4]First Pavlov State Medical University of Saint Petersburg, Saint-Petersburg, 197101, Russia
[5]Saint Petersburg National Research University of Information Technologies, Mechanics and Optics, Saint Petersburg, 197101, Russia

Abstract
The comparative research of silica, the magnetite and magnetite-silica nanoparticles modified with fluorescent dyes using gas-phase and liquid-phase methods was conducted. At the content of fluorescent dye comparable in size a particular spectrophotometric method, nanoparticles with fluorescein have up to 1000 times larger overall luminous efficiency. It is revealed that magnetic nanoparticles are characterized by a smaller light efficiency in comparison with silica particles, at the same time particles of a magnetite are most effective at modification with fluorescein, and magnetite-silica particles – at modification with indocyanine green.

1. Introduction
History of fluorescent visualization in medicine contains more than 50 years. At the same time only two fluorophores are allowed for use in many countries of the world today, it is indocyanine green (ICG) or as it is called still, cardiogreen and various modifications of fluorescein (FS) [1]. Today, these two fluorophores are used mainly for receiving angiograms of a retina of an eye which demand rather high dose of substance. At the same time there can be side effects, such as hypersensitivity reactions, however, any specific toxicity it was revealed not [2]. For molecular visualization much lower dosage of a fluorophore having minimum toxicity is used [1].

Now the fissile are used and the activated composite fluorescent constructions [3]. The fissile constructions represent the fluorescent dyes conjugated with nanoparticles, monoclonal antibodies, enzymes and so on [4]. The activated constructions contain a fluorophore the secondary radiation of which is extinguished in a routine state. To receive induced radiation it is necessary to suppress bloom clearing with chemical [5], specific to this case. A classic example of such extinguishing of a fluorescence is formation of dimers in aqueous solutions of ICG and solution activation by an albumin [6].

In this work the possibility of application for fluorescent visualization of various types of nanoparticles (NPs) modified via various methods with two fluorophores was investigated: fluorescein and indocyanine green.

2. Experimental methods
In present micrographs of NPs were obtained with a use of the transmission electron microscope (TEM) JEM-1400 STEM cathode (JEOL, Japan).

In this work the NPs of following types were used: industrially produced silica nanoparticles - Aerosil A-200 of the Evonik Degussa GmbH, NPs of Aerosil are spherical, of magnetite - a needle-form (MNP-

1) and NPs of magnetite-silica (MNP-2)were obtained according to [7]. All nanoparticles had the average size about 10 nanometers. NPs of A-200 represented particles in a form the close to spherical, MNP-1– particles of a needle form, MNP-2 – separate particles and their agglomerates. The average form factor of A-200 and MNP-2 was equal to 1, of MNP-1 – 3.5.

For an immobilization of fluorophores on a surface of NPs amination was carried out. Two methods of amination were used: gas-phase and liquid-phase. The gas-phase technique as these NPs initially were in a type of dry powder was applied to NPs of Aerosil, and there was an opportunity to carry out synthesis in the pseudo-boiling gas-phase mode. The NPs received in the form of suspensions — magnetite and magnetite-silica – amination in the fluid environment. In both cases 3-aminopropyltriethoxysilane (APTES) was used as the amination agent .

For carrying out gas-phase modification with NPs amino groups the technique including APTES chemisorption from a gas phase in the flowing reactor was applied. The design of the reactor provided evaporation of reagent at a temperature 220 °C, a chemisorption, removal of excess of reagent and side products of reaction. As gas carrier nitrogen was used. The technique included physical removal of the binded water from a NPs surface at a temperature of 250 °C within 2 hours, chemisorption of APTES at a temperature of 250 °C within 1 hour, removal of excess of reagents and side products of reaction at the same temperature within 2 hours [8].

Liquid-phase amination of magnetic NPs was carried out with two methods: from organic solvent and from an aqueous medium.

For amination from organic solvent the aqueous suspension containing 50 mg of NPs in 2 ml of water was centrifuged for 5 minutes at a speed of 3000 min^{-1}. Supernatant merged, and the insoluble precipitate was five times washed out with ethanol. Then the NPs dispersion was carried out with use of UZD-2 ultrasound disperser (Ultrasound Technics, Russia) within 5 minutes in ethanol.

Liquid-phase amination was carried out as follows. The suspension containing 50 mg of nanoparticles in 2 ml of ethanol or distilled water was placed in the polypropylene tube with a volume of 15 ml then 200 μl of APTES was added and mixed on LS-220 shaker (LOIP, Russia) within 30 minutes at a speed of 300 min^{-1} at room temperature. Then suspension was centrifuged within 5 minutes at a speed of 3000 min^{-1} and washed out 5 times with distilled water with centrifugation.

Chemisorption of fluorescent dyes on aminated NPs was carried out. For all NPs types operation was performed identically.

Solution of fluorescent dye with concentration of 1 mg/ml in a volume of 1 ml and distilled water in a volume of 1 ml were added to the suspension containing 50 mg aminated nanoparticles in 2 ml of water. The sorption was carried out in 15 ml polypropylene test-tubes on LS-220 shaker at a speed of 300 min^{-1} within 2 hours. Then solution was centrifuged within 5 minutes at a stirring speed of 3000 min^{-1} and washed out 5 times with distilled water and centrifuged. The presumable mechanism of chemisorption of FLN is shown in the figure 1a, ICG – in the figure 1b.

The amount of sorbed fluorescent dye was defined after sorption. 0.1 N Solution of sodium hydroxide in a volume of 10 ml was added for this purpose to a washed precipitate and then desorption was carried out within 15 minutes. Then solution was centrifuged and 1 ml of supernatant liquid was taken. The received solution was analyzed on the content of fluorescent dye with a spectrophotometric method.

The wavelength of absorption maximum was defined with a use of a spectrophotometer Unico 2802s and equaled 485 nanometers for FLN and 700 nanometers for ICG.

Calibration of dependence of attenuation range of aqueous solutions on the maintenance of both fluorophores was carried out at these wavelengths. Since the maximum of a wavelength of absorption of FLN is blocked by the line of absorption of cations of iron, and alkali in turn partially dissolves MNP for MNP-FLN samples the content of fluorescent dye with this method did not manage to be estimated. For MNP-ICG samples, on the contrary, cation of iron does not prevent determination of content of a fluorophore.

Table 1. Properties of the modified nanomaterials

Sample No.	Type of NPs	Fluorophore	Method of amination	Fluorophore content, mg/ml
1.0	—	FLN	—	0,005
1.1	A-200	FLN	Gas-phase	0,013
1.2	MNP1	FLN	Water	0,005
1.3	MNP1	FLN	Ethanol	0,020
1.4	MNP2	FLN	Water	0,005
1.5	MNP2	FLN	Ethanol	0,020
2.0	—	ICG	—	0,005
2.1	A-200	ICG	Gas-phase	0,007
2.2	MNP1	ICG	Water	—
2.3	MNP1	ICG	Ethanol	—
2.4	MNP2	ICG	Water	—
2.5	MNP2	ICG	Ethanol	—

Samples (Table 1) which then were used in an experiment on fluorescent imaging were synthesized. For comparison of samples FLN and ICG solution in distilled water were prepared (samples 1.0 and 2.0 respectively).

For a research of a possibility of the synthesized samples *in vitro* fluorescence the IVIS Lumina LT Series III (Perkin Elmer, the USA) device was used. Researches were carried out in the tablet with the small holes in which with a use of a batcher with removable tips was put on 100 μl of the studied solution. The colloidal solution of NPs in distilled water was prepared from beforehand the washed out samples by means of an ultrasonic disperser, concentration for NPs labeled FLN was 4 mg/ml, for NPs labeled ICG – 2 mg/ml. For induction of fluorescence embedded light filters for FLN, 465 nanometers, for ICG, 745 nanometers, were used. Registration was carried out with the embedded FITS and ICG filters respectively having the characteristics suitable for selection of induced radiation of the corresponding fluorescent dyes according to documentation on the device.

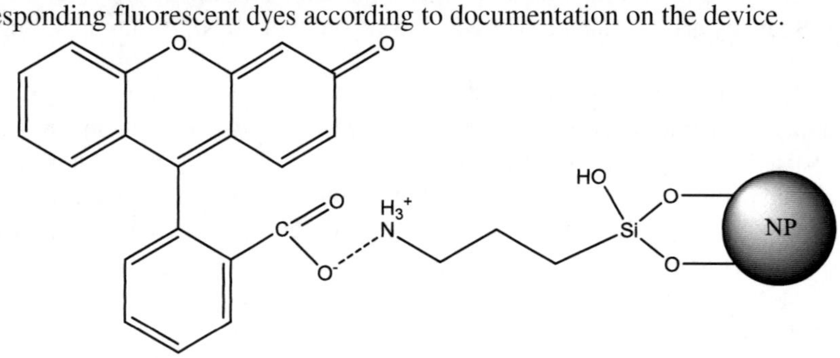

a

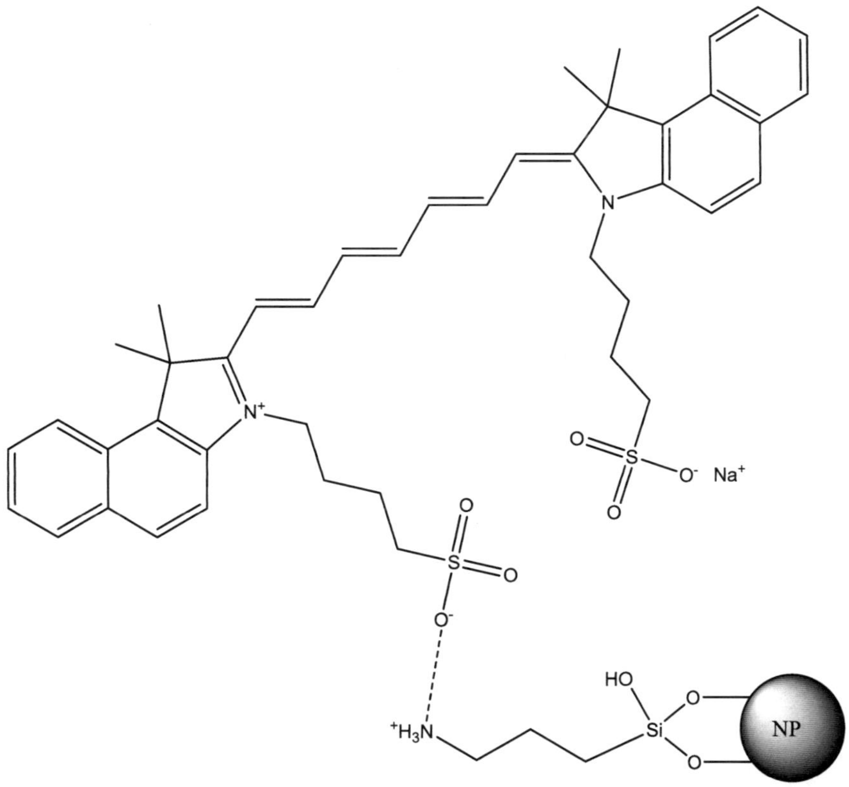

b

Figure 1. Scheme of synthesis: a — amination of nanoparticles; b — a fluoresceine immobilization; c — an immobilization of indocyanine green

3. Results and discussion

Aminated samples were investigated using TEM. The analysis did not allow to estimate APTES shell's thickness. It means that a layer of the amination agent rather thin to be identified by means of this method. Also lack of agglomeration in comparison with initial samples was revealed that in turn testifies to applicability of all three methods of amination for the studied NPs.

Data of two channels of visualization of samples on both fluorophores are shown on the Figure 2. Numerical characteristics of light return of samples fluorescent radiation are provided in Table 2.

The overall light return of samples with FLN nearly three orders higher than for samples with ICG. However at the same time the area of a fluorescence of this dye coincides with an autofluorescence of many materials and living tissues that is the known fact [9]. It limits use of dyes on the basis of FLN. At the same time the maximum level of a fluorescence in comparison with control (sample 1.0) is observed at an sample 1.1 in which dye is immobilized on A-200 surface. The under level of a fluorescence of samples based on the magnetic nanoparticles apparently is bound to an overlap of iron and FLN absorption ranges. Despite it samples 1.2 and 1.3 based on MNP also show good level of fluorescence suitable for practical application.

Absolutely other picture is observed when using ICG. Samples 2.1, 2.4 and 2.5 on the basis of A-200 and MNP2 were suitable for practical use. Apparently it is bound to the fact that when using MNP1 the effect of extinguishing of fluorescence on magnetite surface takes place. It is absolutely true that such fact is excluded at NPs based on A-200. MNP2 have an envelope from silica which isolates a fluorophore from magnetite. More high level of fluorescence of samples 2.4 and 2.5 in comparison with the sample 2.1 apparently is bound to method of amination of a surface. A thin covering is formed at gas-phase amination. NPs are covered with thicker APTES polymer containing respectively larger quantity of amino groups at liquid-phase amination. The experiment shows that the number of available groups $-NH_2$ for fluorescent dye is also higher in this case.

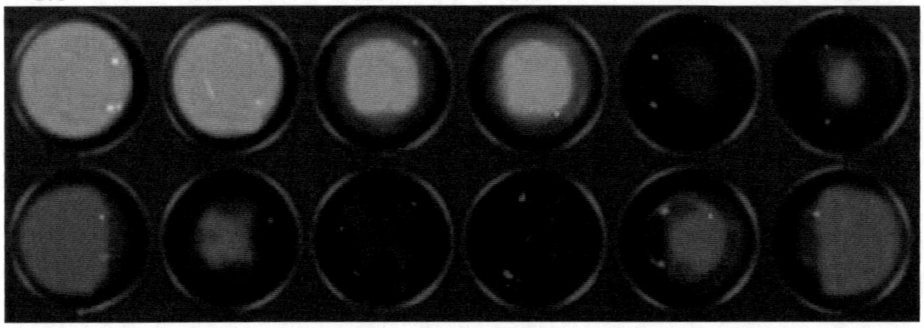

Figure 2. *In vitro* testing of the received exemplars by means of a fluorescent visualizer of IVIS Lumina III LT

Table 2. Measured values of output fluorescent radiation of samples

Sample No.	Overall light output [p/s] / [μW/cmI] ×10^{-8}	Average light output [p/s/cmI/sr] / [μW/cmI] ×10^{-8}	Standart deviation [p/s/cmI/sr] / [μW/cmI] ×10^{-8}	Min light output [p/s/cmI/sr] / [μW/cmI] ×10^{-8}	Max light output [p/s/cmI/sr] / [μW/cmI] ×10^{-8}
1.0	2583,000	950,000	180,100	605,600	1300,000
1.1	2418,000	644,400	109,300	408,700	934,700
1.2	564,900	153,300	25,490	97,890	224,700
1.3	596,500	137,100	22,220	80,980	200,800
2.0	26,550	9,256	1,733	6,008	13,060
2.1	4,763	1,904	0,415	1,169	3,197
2.4	7,445	2,694	0,568	1,412	4,540
2.5	14,840	6,473	1,384	2,934	8,923

4. Conclusions

Nanoparticles of three types: Aerosil, magnetite and magnetite-silica were investigated as possible agents for fluorescent imaging. For the purpose of modification of nanoparticles surface with molecules of fluorescein and indocyanine green fluorescent dyes amination of particles with use of liquid-phase and gas-phase methods was provided. Comparison of results of amination on the content of fluorescent dye showed that for modification of magnetic nanoparticles of both types application of a liquid-phase method of amination from ethanol solution is the most efficient. Nanoparticles modified with the molecules of fluorophores were compared for light output efficiency.

The overall light output of nanoparticles with fluorescein is up to three orders higher than a light output of nanoparticles with indocyanine green. Nevertheless, application of such particles is limited to a presence of lines of an autofluorescense of other substances and living tissues at a fluorescein fluorescence wavelength range. Magnetite particles showed the highest level of the overall light output (among magnetic nanoparticles) using modification with fluorescein, and magnetite-silica NPs showed the highest overall light output at modification with indocyanine green.

Acknowledgments
Work was performed on the basis of an inventory of the Resource center of St. Petersburg State University "Innovative technologies of composite nanomaterials".

The technique of magnetite-silica NPs synthesis was developed thanks to the support of the Russian Foundation for Basic Research (grant 16-32-60010).

References

[1] Kobayashi H, Ogawa M, Alford R, Choyke P L, Urano Y 2010 Chem. Rev. 110 2620–2640
[2] Butner R W, McPherson A R Ann. 1983 Ophthalmol. 15 1084–1089
[3] Ntziachristos V 2006 Annu. Rev. Biomed. Eng. 8 1–33
[4] Neri D, Carnemolla B, Nissim A, Balza E, Leprini A, Querze' G, Pini A, Tarli L, Halin C, Neri P, Zardi L, Winter G 1997 Nat. Biotechnol. 15 1271–1275
[5] Weissleder R, Tung C H, Mahmood U, Bogdanov A 1999 Nat. Biotechnol. 17 375–378
[6] Sonin D, Papayan G, Pochkaeva E 2016 Biomed. Opt. Express 8 151−161
[7] Toropova Y G, Golovkin A S, Gareev K G, Malashicheva A B, Gorshkov A N, Afonin M V, Korolev D V, Galagudza M M 2017 Int. J. Nanomedicine 12 593–603
[8] Korolev D V, Babikova K Y, Postnov V N, Naumysheva E B, Papayan G V, Pochkaeva E I, Sonin D L 2016 Biotechnosphere 47 42−47
[9] Galagudza M M, Korolev D V, Sonin D L, Aleksandrov I V, Postnov V N, Papayan G V, Shlyakhto E V 2010 Nanotech. in Russia 5 844−850

Immobilization of cardioprotective drug phosphocreatine on a surface of nanoparticles of silica

D V Korolev[1,5,6], N V Evreinova[1,2], E V Zakharova[4], K G Gareev[4], E B Naumysheva[1,3], V N Postnov[1,3], M M Galagudza[1,5,6]

[1]Almazov Federal North-West Medical Research Centre, Saint-Petersburg, 194156, Russia
[2]Saint Petersburg State Technological Institute (Technical University), Saint Petersburg 190013, Russia
[3]Saint Petersburg State University, Saint Petersburg, 199034, Russia
[4]Saint Petersburg Electrotechnical University "LETI", Saint Petersburg, 197376, Russia
[5]Pavlov First State Medical University of Saint Petersburg, Saint Petersburg, 197101, Russia
[6]Saint Petersburg National Research University of Information Technologies, Mechanics and Optics, Saint Petersburg, 197101, Russia

Abstract. In this work silica aminated nanoparticles were used to show capability for chemisorbing organic compound having a carboxyl group. Phosphocreatine (creatine phosphate) was used as an active ingredient. Since the method for determination of phosphocreatine with the sample analysis using Jaffe reaction didn't give a positive result, the definition of free phosphocreatine was carried out by the method of diacetyl in the presence of α-naphthol.

1. Introduction

The development of nanochemistry and nanomedicine allowed to look at the use of nanoparticles as a transporter for medicines. More recently, there have been works demonstrating the possibilities of prolongation of the action [1] and address delivery in a myocardium of cardioprotectors with a use of different carriers [2], the authors proposed to use nanoparticles of silica (NPS) for these purposes earlier [3].

Colloidal silicon dioxide which is a pyrogenic silica nanoparticle produced under the trade name of Aerosil is used in the pharmaceutical industry as an auxiliary. As a traditional sliding agent Aerosil provides the optimum flowability of powders necessary for modern high-speed tableting [4]. Unique properties of Aerosil allow it to be used in solid dosage forms, gels and ointments as a filler.

We used the silica nanoparticles (SN) Polysorb-300 pre-modified with the method [6].

SN aminated via specific way (Figure 1*a*) using 3-aminopropyltriethoxysilane can chemisorb organic compound having a carboxyl group [7]. We used the phosphocreatine (creatine phosphate) as this active substance has cardioprotective properties.

The phosphocreatine which is produced under a trademark Neotone [8] is the medicine improving metabolism in warm and other muscular tissue. On chemical structure it is similar to macroergic endogenous phosphocreatine. The drug slows down process of destruction of a sarcolemma of ischemic cardiomyocytes and other muscular elements, provides intracellular transport of energy.

Due to microcirculation improvement medicine reduces the size of a zone of a necrosis and ischemia. It causes an antiarrhythmic effect in case of ischemia and post-ischemic reperfusion which is associated with a decrease in the ectopic activity of the ventricles and the preservation of the physiological function of the Purkinje fibers.

2. Experimental methods

The scheme of immobilization of phosphocreatine on the surface of SN is shown in Figure 1b.

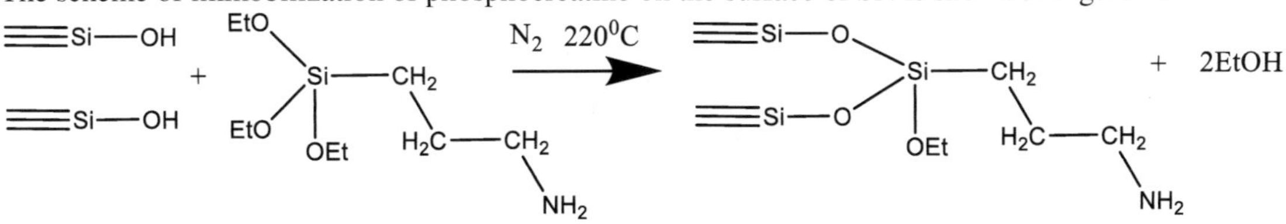

a

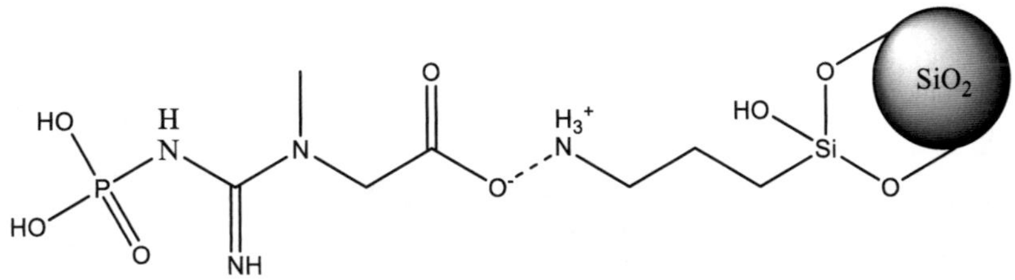

b

Figure 1. The scheme of synthesis: amination of the nanoparticles (a) and immobilization of Phosphocreatine (b)

Phosphocreatine in an aqueous-alkaline solution easily passes into creatinine by cleaving phosphoric acid (Figure 2). This transition is easily evaluated with the reaction of Jaffe [9] based on the reaction of creatinine with sodium picrate.

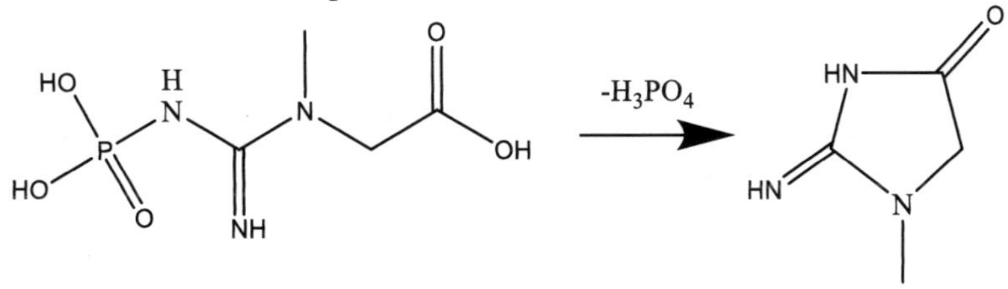

Figure 2. Diagram of the transition creatine phosphate-creatinine

Creatinine reacts with alkaline picrates forming a red complex. The change in optical density of the formed complex is proportional to the concentration of creatinine in the sample. A solution of phosphocreatine concentration 2 mg/ml was prepared as follows, a portion of the phosphocreatine weight of 105.5 mg dissolved in 50 ml of 0.1 N hydrochloric acid. A solution of phosphocreatine was prepared similarly in 0.3 N sulfuric acid.Immobilization of phosphocreatine on the surface of SN was carried out as follows. 1 ml of 0.1 N sodium hydroxide solution and 1 ml of an aqueous solution of phosphocreatine in 0.1 N hydrochloric acid of 2 mg/ml concentration were added to 50 mg of aminated nanoparticles. Sorption was carried out in polypropylene tubes with a capacity of 15 ml on a shaker LS-220 (LOIP, Russia) at a stirring rate of 300 min^{-1} for 2 hours. The solution was then centrifuged for 5 minutes at 3000 min^{-1} and washed 5 times with distilled water and centrifuged. The resulting precipitate was dispersed on ultrasonic disperser UZD-2 (Ultrasound Technics, Russia) for 5 minutes in distilled water in the required concentration.

IOP Conf. Series: Journal of Physics: Conf. Series **917** (2017) 042009 doi:10.1088/1742-6596/917/4/042009

3. Results and discussion

Micrographs of NPs were obtained with a use of the transmission electron microscope (TEM) JEM-1400 STEM cathode (JEOL, Japan). Micrograph of modified Aerosil is shown in Figure 3.

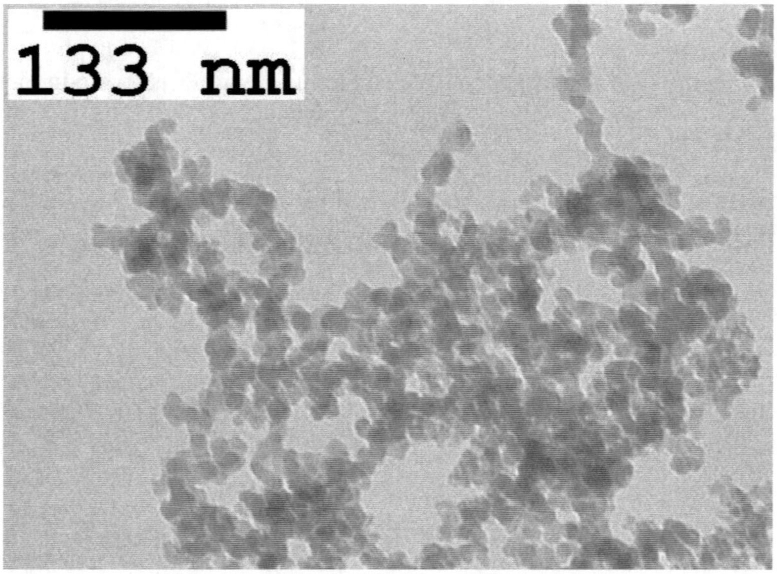

Figure 3. Micrograph of modified Aerosil

To determine the content of active substance in the sample calibration dependence was built. In samples containing phosphocreatine in quantities 4.216, 8.432, 12.648, 16.864, 21.080 μg 3 ml of aqueous solution of picric acid 17.5 mmol/l and 0.2 ml of an aqueous solution NaOH concentration of 2.5 mol/l was added. The dolution was left for 10 minutes, then the solution volume was adjusted to 100 ml and the optical density relative to the blank sample only containing picric acid and NaOH in specified quantities was measured at a wavelength $\lambda = 510$ nm. Experiments were repeated 3 times. 2802S Unico spectrophotometer (Unico Sys, USA) was used for determining the optical density of the absorption spectra.

The measured optical density of a solution containing phosphocreatine in different concentrations did not differ from the optical density of the blank sample. This indicates the absence of transition creatine phosphate-creatinine.

A blank sample solution of picric acid and a solution with a maximum content of phosphocreatine (21.080 μg) was prepared for a more detailed study of the discovered fact. With the help of spectrophotometer the absorption spectra of the solutions in the wavelength range 450−1100 nm was received. The obtained results (Figure 4) indicated the absence of absorption bands of creatinine (500−600 nm) with a maximum content of phosphocreatine in the sample.

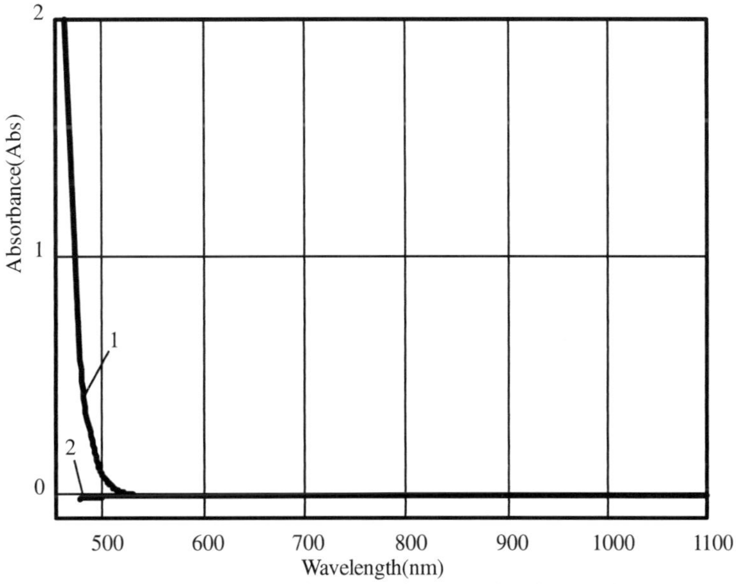

Figure 4. Absorption spectra of a blank sample (1) and a sample of Phosphocreatine relative to a blank sample (2)

Since the method for determination of Phosphocreatine with the sample analysis using Jaffe reaction didn't give a positive result the definition of free Phosphocreatine was carried out with the method [10] of diacetyl in the presence of α-naphthol.

We used 1% alkaline solution of α-naphthol. The alkaline solution contained 6 g of NaOH and 16 g of Na_2CO_3 in 100 ml of water.

A solution of diacetyl was prepared as follows. We took 1.6 g of dimethylglyoxime in Wurtz flask adding 5N H_2SO_4 with a volume of 200 ml. Flask was placed on a sand bath and distillate was distilled. The first 50 ml of distillate was collecting and the volume was adjusted with water to 100 ml. Obtained 1% solution of diacetyl was stored at +4 °C in the refrigerator. The solution was diluted to the concentration of diacetyl is 0.05% before use.

To study the applicability of the method for determination of the content of phosphocreatine, a sample was prepared with a zero (blank sample) and its maximum concentration. 1.0 ml of a 1% solution of α-naphthol and 0.5 ml of a 0.05% solution of diacetyl were added for this purpose containing no phosphocreatine and to a sample containing 21.080 μg of the latter. The volume of the solution was adjusted to 5 ml with distilled water, thoroughly mixed and left at room temperature for 30 minutes in the dark.

Samples were analyzed spectrophotometrically for absorption in the wavelength range 450–1100 nm. The obtained spectra show that at a wavelength of 540 nm there is an amplitude shift dependent on the concentration of the active substance (Figure 5).

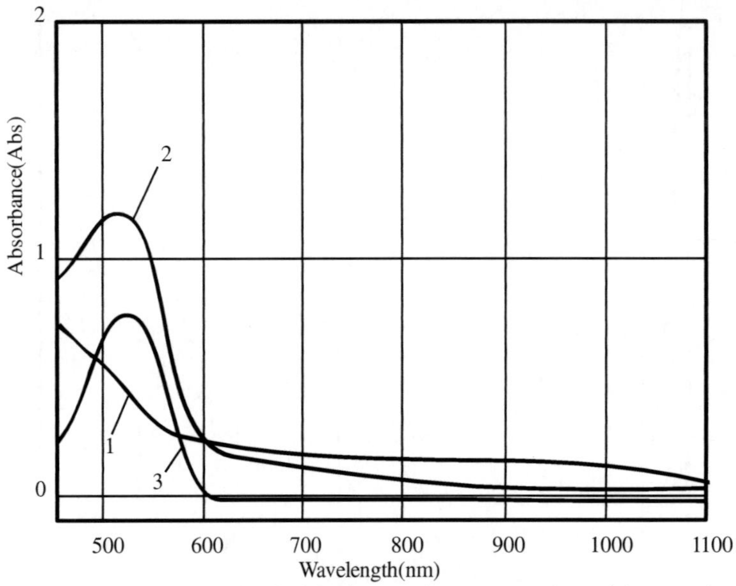

Figure 5. Absorption spectra of a blank sample (1), samples with a maximum content of phosphocreatine (2), samples with a maximum content of phosphocreatine relative to a blank sample (3)

The determination of the calibration curve was made for samples with a phosphocreatine content of 4.216, 8.432, 12.648, 16.864, 21.080 μg. Then the optical density of the solution relative to the distilled water was measured at a wavelength λ = 540 nm.

The results of determining the optical density of solutions depending on the content of the active substance are shown in Figure 6.

The procedure was then repeated in a more acidic medium replacing the acid from the monobasic to the dibasic acid (sulfuric) in order to separate the inorganic phosphate.

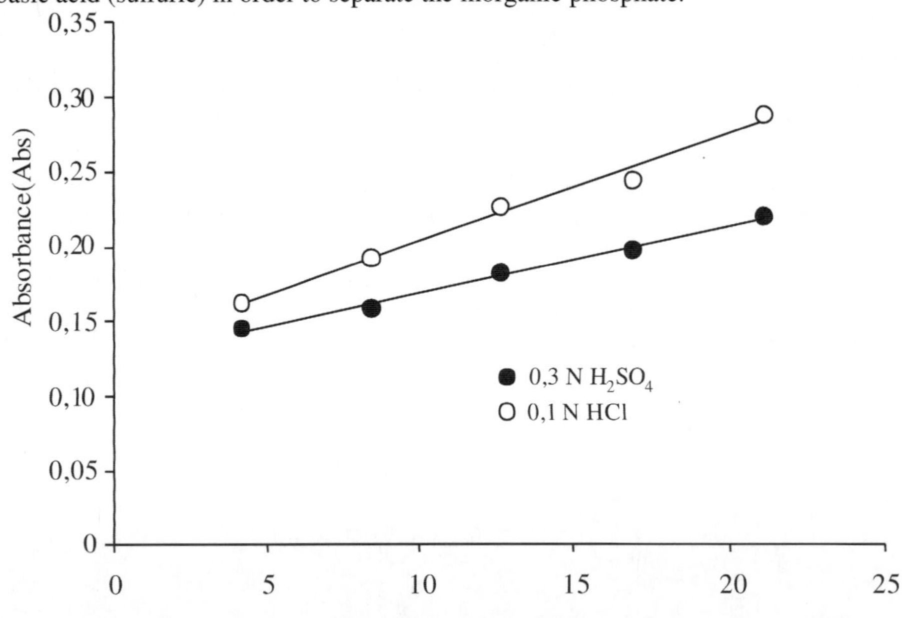

Figure 6. Calibration curves of the average optical density of the solution on the content of phosphocreatine

As it can be seen from the figure, increasing of the acidity of the solution affects the concentration dependence of the average optical density. In further work we used the solution of phosphocreatine prepared with 0.1 N HCl since the increased acidity affects the sorption capacity of aminated nanoparticles.

Sodium hydroxide in a quantity of 5 ml was added to determine the concentration of phosphocreatine chemisorbed on the surface of the SN and desorption was carried out for 15 minutes with manual shaking. The resulting solution was centrifuged and the above the sediment was taken in a volume of 1 ml and analyzed for the content of free phosphocreatine with the method described above at a wavelength $\lambda = 540$ nm.

The content of phosphocreatine determined using a pre-constructed calibration curve equaled to 0.04 mg which corresponds to a concentration of 0.8 mg or 4 μmol on 1 g of silica.

References

[1] Takahama H, Minamino T, Asanuma H Fujita M, Asai T, Wakeno M, Sasaki H, Kikuchi H, Hashimoto K, Oku N, Asakura M, Kim J, Takashima S, Komamura K, Sugimachi M, Mochizuki N, Kitakaze M 2009 J. Am. Coll. Card. **53** 709–717

[2] Luan J, Zhai G 2016 Curr. Drug Targets **17** 467–474

[3] Galagudza M M, Korolev D V, Sonin D L, Postnov V N, Papayan G V, Uskov I S, Belozertseva A V, Shlyakhto E V 2010 Int. J. Nanomed. **5** 231–237

[4] Maizels A 2008 Pharmaceutical technologies and packaging 5 20–23

[5] Markelov D A, Nitsak O V, Gerashchenko I I 2008 Pharmaceutical Chem. J. 7 30–33

[6] Korolev D V, Babikova K Yu, Postnov V N, Naumysheva E B, Papayan G V, Pochkaeva E I, Sonin D L 2016 Biotechnosphere 47 42–47

[7] Galagudza M, Korolev D, Sonin D et al. Targeted drug delivery to ischemic heart with use of nanoparticulate carriers: Concepts, pitfalls and perspectives // Journal of Manufacturing Technology Management, 2010, N 8. – P. 930–949

[8] Kovalev S A, Belov V N, Osipova O A 2015 Bulletin of Experimental and Clinical Surgery **29** 314–317

[9] Nessar A Clinical Biochemistry. New York: Oxford University Press. 2011 72.

[10] Severin S E Practical work on biochemistry: teaching. Moscow: Publishing House of Moscow State University. 1989. 189–190.

Registration of fluorescence in biomolecular solutions using dynamic pin photodiode

E A Savchenko[1], E K Nepomnyashchaya[1], D B Dyubo[1], E N Velichko[1], O Yu Tsybin[1]
[1]Institute of Physics, Nanotechnology and Telecommunications, Peter the Great Saint Petersburg Polytechnic University, Saint Petersburg 195251, Russia

Abstract. The application of a dynamic pin-photodiode as a recording device of the fluorescence radiation is considered. An experimental setup for registration of luminescence signals from biomolecules excited by laser radiation was developed. Rhodamine 6G and chlorophyll were investigated.

1. Introduction

Fluorescent diagnostic methods are becoming more and more popular in clinical practice [1]. For example, in medical research special fluorescent dyes introduced into the body and distributed to tissues according to their properties are used [2]. The prospects of development of such testing methods are in their high sensitivity and speed of analysis of signals. there are various new areas of fluorescence research and technology based on the migration of excitation energy in biological material and fluorescent probes, such as FRET and plasmon resonance sensors. In addition, by using a combination of different fluorescence methods it is possible to study the movement of cell structures, to measure temporal kinetics of fluorescence imaging with temporal selection and spectral resolution for detection of molecular interactions, and to study the fluorescence spectra, from which the information on fluorescent molecules, their concentration, conformation, and interaction within cells and tissues (aggregation) can be obtained [3].

To register fluorescent signals, sufficiently sensitive hardware systems should be used [4]. Registered signals typically consist of not only the fluorophore fluorescence, but also background fluorescence of solvents and scattering from turbid solutions: Rayleigh or Raman scattering. The measurement of the fluorescence intensity can be influenced by the polarization of emitted light [5]. Therefore, to obtain reliable spectral data, it is necessary to take into account all types of registered fluorescence and scattering signals and to modify some spectrofluorimeter units.

In our work we suggest a new modification of the spectrofluorimeter: a dynamic pin-photodiode (DPD) with an insulated gate is used as a registration element of the device (instead of a standard photomultiplier).

2. Techniques

Standard devices for registration of fluorescence typically have the following disadvantages. Ordinary vacuum photomultipliers have slow speeds, high dark currents and high noises; detectors of THz radiation are expensive, a thermistor type of detectors of microwave radiation has no high enough sensitivity, and conventional semiconductor photodetectors have a high noise level. Recently a new device based on a pin photodiode [6] with an insulated gate operating in the dynamic mode has appeared. This device is promising for application in systems of registration of weak radiation from activated biomolecular solutions. A pin-photodiode contains two strongly p- and n-doped regions separated by a weakly doped (or undoped) region (shown in Figure 1,a).

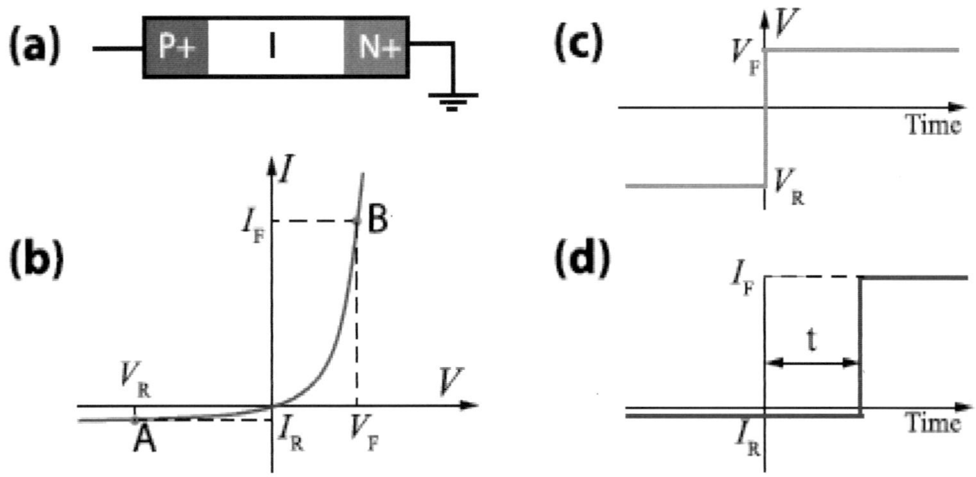

Figure. 1. PIN diode in static and dynamic modes: (a) a schematic structure of a PIN diode; (b) PIN diode dark I-V curve; (c) applied voltage; (d) current-time dependence for DPD. Triggering time t depends on the light intensity [6].

In the static regime, the diode is kept at a fixed reverse bias (point A in Figure 1,b), where the dark current is very low. Under illumination, the device current increases proportionally to the number of absorbed photons.

In our work we used the dynamic photodiode where the applied voltage was switched, as shown in Figure1,c, from the reverse to a forward regime (from point A to point B in Figure1,b). The applied forward bias induced a high forward current after a time delay (Figure 1,d). The measured photocurrent of the same device used in the static mode of the photodiode was up to 18 nA at maximum light intensity. At the same time, the output current of DPD was 0.8 mA, which was more than four orders of magnitude higher than the photocurrent. The magnitude of the dc current was controlled only by the value of the applied dc voltage and did not depend on the light intensity. In contrast, the delay time was a function of the absorbed light power.

The physics of DPD can be explained as follows. When the reverse bias is applied, the excess carriers are removed from the depleted region. After switching to the forward bias, the charge in the depleted regions adjacent to n+ and p+ creates potential barriers, which strongly suppress the dc current. In the end, hole and electron currents from the p+ and n+ regions reduce these potential barriers and the dc current starts to flow.

A sharp increase in the measured current is due to a positive feedback mechanism: holes coming from contact p+ reduce the potential barrier near the contact n+ and increase electronic injection, which, in turn, lowers the potential barrier near the contact p+ and increases the current holes. In the presence of incident light, barriers still exist after switching the voltage, but the induced photocarriers accelerate the reduction of the barrier and reduce the switching time. A simple device without any optimization has a delay time of 100 ps. A short light pulse considerably reduces the switching time (triggering time) of the device. In our work the modifications of the pin structure with the delay time of the nanosecond–millisecond range was used [6]. The dark triggering time of the device (self-triggering time) is about 0.6 μs.

The DPD has several important features: it has a high output signal, which is normally several orders of magnitude higher than that of a standard pin diode; the output signal can be digitalized without any analog-to-digital converter [6].

The model of the diode with the dimensions of the actual device was created in the program Comsol. The simulation provided an opportunity to obtain the total charge (Q_{tot}) needed to calculate the intensity of the radiation incident on the diode. Q_{tot} is the total charge of current carriers which are necessary for neutralization of volume charge under the gate in the depletion mode region.

To evaluate the value of Q_{tot}, we should calculate the charge in the depletion mode region. The charge may be calculated as an area integral of carrier concentration in the presence of a positive gate bias and without voltage. Figure 2 presents the model of the hole concentration distribution. The area integral was calculated in the area marked by the rectangle in the upper left corner. This area corresponds to the depletion mode region where probabilities of photon absorption and electron-hole generation are the highest.

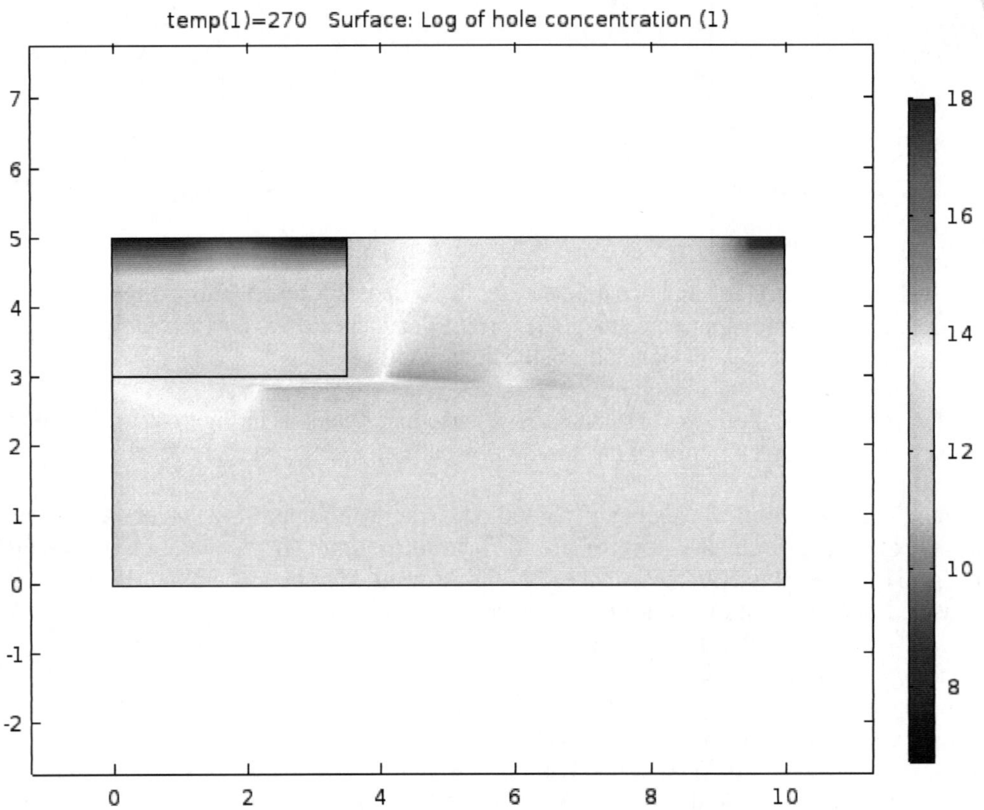

Figure. 2. Distribution of hole concentration.

After calculating the total charge Q_{tot} of the media, the intensity was calculated from the following formulas [7]:

$$\frac{1}{T_{trig}} = \frac{1}{N_{tot}}\left(n_{phot} + n_{dark}\right) = \frac{q}{Q_{tot}}\left(\frac{A*QE_{\lambda}*\lambda}{h*c} I_{light} + n_{dark}\right) \quad (1)$$

$$I_{light} = \frac{Q_{tot}*h*c}{A*QE_{\lambda}*\lambda}\left(\frac{1}{T_{trig}} - \frac{1}{T_{strig}}\right), \quad (2)$$

here n_{phot} is the number of integrated photogenerated charges per second [s^{-1}]; n_{dark} is the dark carrier generation [s^{-1}]; A is the active area [m^2] (for our device 9^{-12} m^2) ; T_{trig} is the triggering time [s]; T_{strig} is the self-triggering time [s]; QE_{λ} is the quantum efficiency for a different wavelength; N_{tot} is the number of carriers needed for triggering.

3. Experimental setup

A scheme of the experimental setup is shown in Figure 3. Light from the laser sources with wavelengths $\lambda_1 = 405$ nm and $\lambda_2 = 532$ nm passes through the prism and is reflected from the side with the sample in the total reflection regime.

The total internal reflection fluorescence (TIRF) is an optical technique that provides for the excitation of fluorophores in an extremely thin axial region ("optical section"). The method is based on the following principle: when excitation light is totally internally reflected in a transparent solid (e.g., coverglass) at its interface with a liquid, an electromagnetic field (the evanescent wave) is generated in the liquid at the solid-liquid interface and is the same frequency as the excitation light. Since the intensity of the evanescent wave exponentially decays with the distance from the surface of the solid, only fluorescent molecules in a layer a few hundred nanometers thick are efficiently excited.

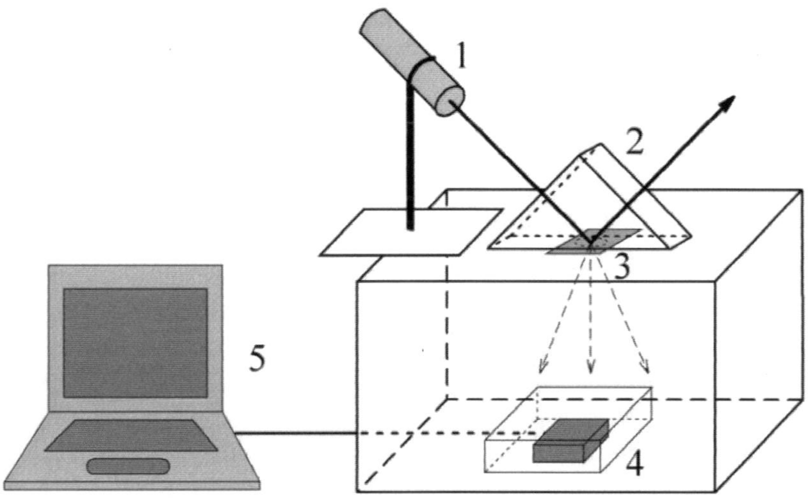

Figure.3. 1 — laser, 2 — prism TIR, 3 — sample, 4 — pin diode, 5 — computer.

The fluorescence intensity of biomolecules was recorded by a computer using a dynamic pin diode placed directly under the sample at a distance of 0.1 m. Due to this design using a prism-total internal reflection, the main beam was not registered by the device, only the fluorescence signal through the filter was received by the diode, then the output signal was entered to the computer to record the triggering time.

4. Samples

Rhodamine 6G and chlorophyll were used as test objects. It was necessary to conduct experiments on the registration of fluorescence in different wavelength ranges to test the sensitivity of the recording instrument.

(a) (b)

Figure.4. Fluorescence of chlorophyll (a) and rhodamine 6G (b).

Rhodamine 6G

Rhodamine 6G derivatives are molecules extensively used as a fluorescent labeling reagent due to their excellent photophysical properties, such as absorption and emission extended to the visible region, high fluorescence quantum yield, and high absorption coefficient [8]. Rhodamine 6G has the lasing range at 555 to 585 nm with a maximum at 566 nm [9]. It is often used as a tracer dye in water to determine the rate and direction of flow and transport. Rhodamine 6G dyes are extensively used in biotechnology applications, such as fluorescence microscopy, flow cytometry, fluorescence correlation spectroscopy and enzyme-linked immunosorbent assay. Rhodamine 6G is soluble in water, methanol and ethanol [8].

Chlorophyll

Chlorophyll is the major light-harvesting pigment in photosynthesis. As known, photosynthesis is a powerful indicator of the internal state of the green cells and the amount of chlorophyll shows the ecological status of water objects and natural reservoirs. The fluorescence spectrum of chlorophyll is in the range from 600 to 700 nm.

5. Results

Test measurements on the prism without a sample and on the prism with water as a sample were performed. An object in a volume of 0.1 ml was placed on the bottom face of the prism to form a thin layer. The type of the excitation laser, the concentration of the investigated objects and calculated fluorescence intensity are summarized in Table 1.

Table 1. Experimental results

Object	Concentration	Type of laser (λ, nm)	Intensity, mW/mm^2
Clean prism	—	405	0.01
		532	0.00
Distilled water	100 %	405	0.09
		532	0.06
Chlorophyll	25 % (dissolved in spirit)	405	11.98
Rhodamine 6G	2 % (dissolved in water)	405	0.08
		532	12.81

It may be noted that there is a low scattering signal from water (this effect may be caused by the presence of a small number of dust particles in the water). The highest luminescence intensity is exhibited by molecules of chlorophyll (in case of excitation by 405 nm laser) and rhodamine 6G (in case of excitation by 532 nm laser). The signal from rhodamine is not observed on wavelength 405 nm as it is out of lasing range of rhodamine 6G. Dynamic pin-photodiode demonstrated high sensitivity and measurements will be continued in our further works.

6. Conclusions

The preliminary results of the study of fluorescence by a dynamic pin photodiode were obtained. A high sensitivity of the suggested device is shown. In our further studies more measurements of fluorescence from different biological solutions will be conducted.

Acknowledgments

The authors are grateful to ActLight company [10] for providing the dynamic pin photodiode.

References

[1] Tuchin V V 2007 *Tissue Optics: Light Scattering Methods and Instruments for Medical Diagnosis* (SPIE Tutorial Texts in Optical Engineering TT38, Bellingham, WA, USA) PM 166 p 882

[2] Akbar N, Sokolovski S, Dunaev A, Belch J J F, Rafailov E, Khan F 2014 *J. of Microscopy* **255** 42
[3] Palmer S, Litvinova K, Dunaev A, Fleming S, McGloin D, Nabi G 2016 *J. Biomedical Optics Express* **7** 1193
[4] Dremin V V, Zherebtsov E A, Rafailov I E, Vinokurov A Y, Novikova I N 2016 *Proc. SPIE 9917, Saratov Fall Meeting 2015: Third International Symposium on Optics and Biophotonics and Seventh Finnish-Russian Photonics and Laser Symposium (PALS)* 99170Y-1
[5] Lakowicz J R 2006 *Principles of Fluorescence Spectroscopy* (N.Y.: Springer Science) p 960
[6] Okhonin S, Gureev M, Salin D, Appel J, Kvasov A, Pastre M, Polzik E S, Tagantsev A K, Uddeegard F, Kayal M 2015 *J.Applied Physics Letters*. **06** 3111
[7] Sallin D 2015 (PhD Thesis 6869, EPFL, Lausanne, Switzerland) 110
[8] Sameiro M, Gonçalves T 2009 *J. Chem. Rev*. **109** 190
[9] Zehentbauer F M et al. 2014 *J. Spectrochimica Acta Part A: Molecular and Biomolecular Spectroscopy* **121** 147
[10] http://act-light.com/technology/

IOP Conf. Series: Journal of Physics: Conf. Series **917** (2017) 042011 doi:10.1088/1742-6596/917/4/042011

Red blood cells aligning inside innovative liquid crystal cell

S V Likhomanova[1,*], A A Kamanin[2], N V Kamanina[1,3,]**

[1]Department for "Photophysics of media with nanoobject", Vavilov State Optical Institute, St. Petersburg, 199053, Russia
[2]Pavlov First Saint Petersburg State Medical University, Saint Petersburg, 197022, Russia
[3]Saint Petersburg Electrotechnical University "LETI", St. Petersburg, 197376, Russia

*e-mail: lsv-87@bk.ru
**e-mail: nvkamanina@mail.ru

Abstract. Investigation results of red blood cells (human erythrocytes) aligning and fixing inside the liquid crystal (LC) cell have been presented in the present paper. LC cells have been modified through the improved nanostructured relief and LC sensitized with intermolecular charge transfer complex COANP-C_{70}.

1. Introduction

Currently liquid crystal (LC) medium is used in broad areas of science and technology. It's well known applying system based on LC in display technology, filters, limiters, etc. LC mesophase can be efficiently applied to solve biomedical problems such as the study and analysis of biological objects [1, 2] as well due to their aligning ability and unique combination of physical and chemical properties.

The size and the form of biological objects such as red blood cell, namely, the erythrocytes are important parameters for human disease diagnosis. It is well known that the normal erythrocytes are biconcave disk with diameter approximately 6-10 μm and a thickness at the thickest point of 2–2.5 μm and a minimum thickness in the centre of 0.8–1 μm [3]. The changes of colour, form and size of erythrocytes occurs during blood diseases, inflammation, etc. The erythrocytes investigation is conducted in two ways: using an automated analyzer and a haemocytometer. The most part of blood samples are examined by automated devices. This method allows carrying out analysis fast and accurate, but it monitors only cells of an average size and form. In the case of abnormal cells the manual method a haemocytometer is used. Haemocytometer consists of 2 glass substrates, one of which has a special chamber with a grid of perpendicular lines (figure 1). Number of blood cells is directly counted using a microscope for each lattice region and total count is performed using special math. The difficulty of performing a manual method consists in uncertainty of determining the position of cells located at the lattice region boundary.

IOP Conf. Series: Journal of Physics: Conf. Series **917** (2017) 042011 doi:10.1088/1742-6596/917/4/042011

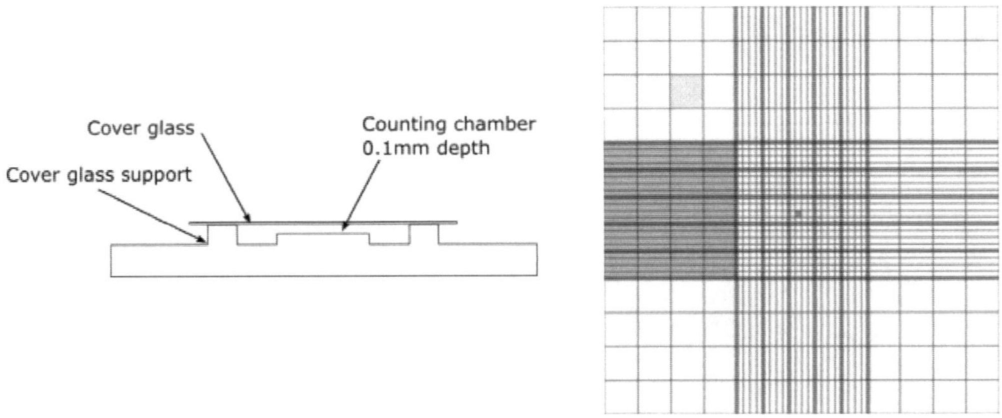

a)

b)

Figure 1. Haemocytometer: a) the parts [4] and b) grid [5].

The LC molecules ability to align red blood cells has been demonstrated in previous works [6, 7]. The experiments have been carried out in LC-cells with the thickness of 10 μm. Relief for LC molecule aligning has been made through polishing method of glass substrate by the polymer coating. LC cell content has been investigated via optical microscope. Photomicrography has demonstrated the areas of erythrocytes aligning and fixing. The interaction between a residual potential of erythrocytes membranes and uncompensated LC molecules dipoles has been considered as an aligning mechanism one. The feedback has been detected too: erythrocytes introduction resulted to LC selforganization and their division of the field domain. [8].

In the current paper both the sensitized LC medium with the increased polarization due to incorporation of the intermolecular charge transfer complex (CTC) and the nontoxic contactless laser oriented deposition (LOD) method to relief formation has been proposed. LOD method consists of deposition on the conductive ITO (indium tin oxide) layer covered with the carbon nanotubes (CNTs) additionally treated with the surface electromagnetic waves (SEW). Use efficiency of modified LC-cell to improve optical and photorefractive properties has been shown in work [9], the ability of LC polarization increase via intermolecular CTC introduction has been presented in papers [10, 11].

2. Experiments and Results

In the current work blood stable with sodium citrate has been added in concentration of 1:5 to mixture LC+COANP (2-cyclooctylamino-5-nitropyridine)+C_{70} in modified LC-cell. As a basic matrix LC from cyanobiphenyl class has been used. This matrix has been sensitized with the intermolecular complex based on COANP+fullerenes C_{70}. The influence of intermolecular CTC COANP+C_{70} on optical and nonlinear-optical medium properties has been shown in work [11, 12]. Intermolecular CTC COANP-C_{70} lead to LC selforganization and to transfer of the nematic LC phase to quasi-smectic one that influenced at transmittance, refraction and dynamics increase well. The transmittance growth and laser strength improvement of the LC cells with the modified relief via LOD applying have been described by the Fresnel losses decrease due to lower CNTs refractive index (n=1.05-1.1) and their high hardness [13]. Moreover, the relief with the CNTs modification can possibly reveal the disinfection of the biological mixture [14].

To confirm the erythrocytes orienting effect under the improved conditions explained above, the ocular micrometer has been used for the registration in this work. The obtained photomicrographs are presented at figure 2.

IOP Conf. Series: Journal of Physics: Conf. Series **917** (2017) 042011 doi:10.1088/1742-6596/917/4/042011

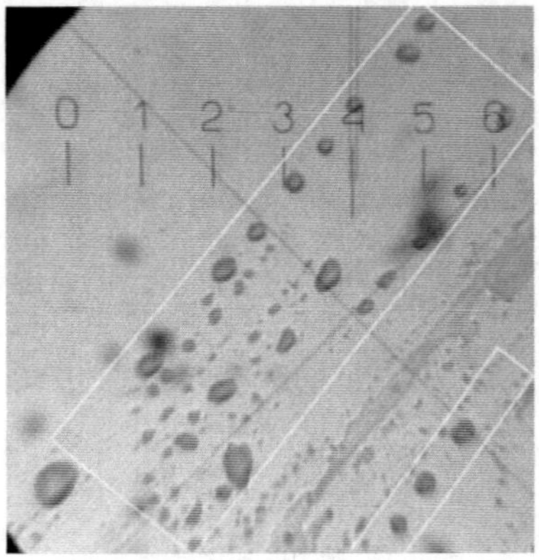

a

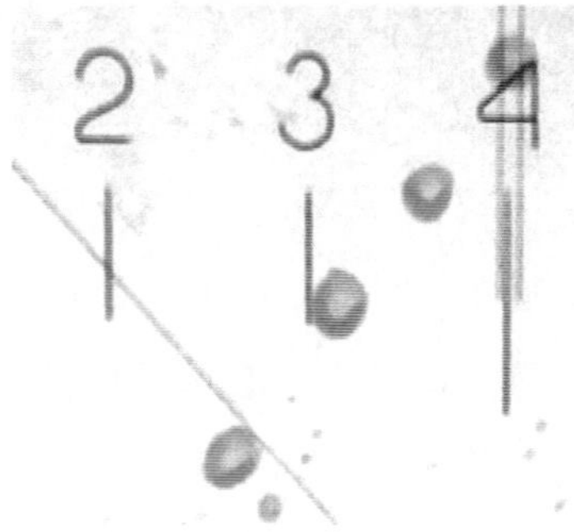

b

Figure 2. Photomicrographs of the erythrocytes orienting effect existing in the modified LC-cell with blood in mixture LC+COANP+C_{70}: a) selected area with aligning erythrocytes, b) increased erythrocytes image (the dimension is 31 μm).

One can see at the presented images the clear areas of aligning as blood cells and LC in a direction of created relief.

3. Discussion

To discuss erythrocytes aligning and fixing in LC-cell we should to observe the electrochemical property of their membranes which play important roles in interface. The membrane surface has a negative charge due to carboxyl group of sialic acid [15]. We consider the interaction between charged erythrocytes membrane and LC dipole with the increased polarizability. The increased polarizability has been obtained due to the intermolecular CTC COANP+C_{70} incorporation in the LC matrix. This interaction between two charged items can be considered as a main mechanism for aligning of the human blood cells (figure 3).

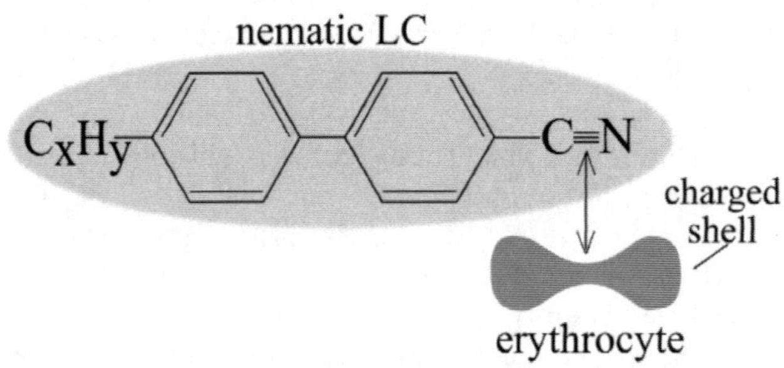

Figure 3. Model of the interaction the erythrocytes-nematic LC molecule.

4. Conclusion

To summarize the results it can be concluded, for the first, the most successfully use of LC-cells with LOD modified relief (without toxic polymer layer) and, for the second, the predominant applying of the polarized LC with intermolecular CTC as well. These facts are important for the work with the bioobjects. The obtained aligning and fixing of the blood cells affect can be used for manual analysis of size and shape of erythrocytes as a human disease indicator under the express diagnostic conditions. Moreover, the procedure explained can be used for the DNA visualization and fixing as well.

Acknowledgments
The current work has been supported by RFBR grant #13-03-00044 (2013-2015), project "Nanocoating-GOI" (2012-2015) and international FP7 Program, Marie Curie Action, Project "BIOMOLEC" (2011-2015). The authors appreciate their colleagues from the department for "Photophysics of media with nanoobject", Vavilov State Optical Institute and abroad too for the helpful discussion.

References

[1] Garner A.L., Deminsky M., Neculaes V.B., Chashihin V., et. al. 2013, *J. Appl. Phys.*, **113**, 214701
[2] Aguirre L.E., de Oliveira A., Sec D., Copar D., Almeida P.L., et. al., 2016, *PNAS*, **113**, №5, 1174
[3] Sackmann E 1995 *Handbook of Biological Physics* vol.1, ed R Lipowsky and E Sackmann (North Holland, Elsevier) p 1052
[4] https://commons.wikimedia.org/wiki/File:Haemocytometer_side.svg
[5] en.wikipedia to Commons by Sreejithk2000 using CommonsHelper
[6] Kamanina N.V., Kidalov V.N. 1996, *Tech. Phys. Lett.,* **22(7),** 571
[7] Kamanina N.V., 1997, *Tech. Phys. Lett.*, **23(12)**, 902
[8] Kamanin A., Kamanina N., 2008, *Mol. Cryst. Liq. Cryst.*, **486**, 50
[9] Kamanina N.V., Vasilyev P.Ya., 2007, Russian Patent № 2355001 (RU 2 355 001 C2)
[10] Kamanina N.V., 2005, *Physics-Uspekhi* **48 (4)**, 419-427
[11] Likhomanova S.V., Kamanina N.V. 2016, *Journal of Physics: Conference Series*, **741**, №1, 012146
[12] Kamanina N.V., 2002, *Journal of Optics A: Pure and Applied Optics*, **4(4)**, 571
[13] Kamanina N.V., Kukharchik A.A., Kuzhakov P.V., Zubtcova Yu.A., Stepanov R.O., Baryshnikov N.V. 2015, *Liq. Cryst. and their Appl*, **15(3)**, 109 (in Russian)
[14] http://perst.issp.ras.ru, April, 2017, Vol. 24, Iss.7-8, page 1.
[15] Tokumasu F, Ostera G R, Amaratunga C, Fairhurst R M 2012 *Exp Parasitol*, **131(2),** 245 doi: 10.1016/j.exppara.2012.03.005

IOP Conf. Series: Journal of Physics: Conf. Series **917** (2017) 042012 doi:10.1088/1742-6596/917/4/042012

Polymer-based microfluidic chips for isothermal amplification of nucleic acids

Y S Posmitnaya[1,2], G E Rudnitskaya[2], A N Tupik[2], T A Lukashenko[2], A C Bukatin[2,3] and A A Evstrapov[1,2,3]

[1]Department of Nanophotonics and Metamaterials, ITMO University, Saint Petersburg 197101, Russia

[2]Lab of information and measurement biosensor and chemosensor microsystems, Institute for Analytical Instrumentation RAS, Saint Petersburg 190103, Russia

[3]Nanobiotech Lab, St. Petersburg Academic University, Saint Petersburg 194021, Russia

Abstract. Creation of low-cost compact devices based on microfluidic platforms for biological and medical research depends on the degree of development and enhancement of prototyping technologies. Two designs of polymer and hybrid microfluidic devices fabricated by soft lithography and intended for isothermal amplification and polymerase chain reaction are presented in this paper. The digital helicase-dependent isothermal amplification was tested in the device containing a droplet generator. Polymerase chain reaction was carried out in the hybrid microfluidic device having ten reaction chambers. A synthesized cDNA fragment of GAPDH housekeeping gene was used as a target.

1. Introduction

Nucleic acid amplification methods serve as an important tool for biological and medical research, as they significantly increase the sensitivity of molecular genetic analysis. The implementation of these methods in microfluidic devices (MFD) is a promising approach to developing compact, relatively inexpensive, sensitive and effective diagnostic tools for clinical use, food safety control and environmental monitoring [1].

Technologies for isothermal amplification of nucleic acids operate at a single constant temperature, which allows the creation of simpler, cheaper and more reliable devices that are suitable for use in resource-limited conditions (point-of-care systems). For isothermal amplification, a water bath, resistive heaters or the heat of exothermic chemical reactions can be used as heating means. Isothermal amplification methods are becoming a promising alternative to polymerase chain reaction (PCR) and greatly simplify the implementation of amplification methods in medical diagnostic devices and analytical equipment. These methods are characterized by high productivity, specificity, promptness and simplified sample preparation. Peculiarities of the method include: requirement for specific pairs of primers, inability to use the amplification product in further studies, restriction on the length of fragments used, a probability of obtaining a false-positive result [2].

The development of microfluidic devices for PCR is associated with the need for complex thermal cycling equipment to amplify the target DNA. However, the enhancement of prototyping techniques for fabrication of microfluidic chips (MFC) allows for the development of new approaches, research strategies and improvement of methods for nucleic acids analysis, in particular, the digital amplification method (digital isothermal/polymerase chain reaction). Digital methods allow to achieve high sensitivity (at the level of single molecules, which is necessary, for example, when diagnosing diseases at early stages and identifying residual signs of the disease) and productivity. However, in this cases complex devices are required to create stable fluid flows (in the case of droplet microfluidics) and thermocycling of the sample. When developing methods for droplet microfluidics, it is also necessary to select surfactants for ensure thermal stability of the emulsion.

At a present, disposable MFCs for amplification, which contain all the necessary reagents, enzymes, primers in a liquid or dehydrated state perform a special interest. These devices reduce user's errors, exclude potential sources of pollution and facilitate the analysis in the field conditions or at the site of medical care. In addition, holding the multiplex reactions if different pairs of primers are loaded into different reaction chambers on a chip becomes possible. In this case, several pathogens or genetic modifications of the organisms in the sample are simultaneously detected.

The development of MFDs for isothermal amplification based on droplet microfluidics is a perspective approach.

Soft lithography is applied for rapid fabrication of polydimethylsiloxane-based (PDMS) MFCs [3]. There are reveals the disadvantage of this material: porosity and high gas permeability, which leads to the evaporation of the reaction mixture during the thermocycling (in case of PCR). In the case of droplet-based amplification, the porosity of PDMS does not affect the reaction process, but certain requirements are imposed on the properties of the chip surface which provide the required conditions for droplet generation. Studies aimed at finding materials for prototyping, free from shortcomings of PDMS, are being carried out. Particularly, UV-curing polymer compositions and epoxy resins are considered promising [4].

Possible design of MFC prototypes for detection of nucleic acids are considered in this paper: 1) based on droplet microfluidics for isothermal amplification, 2) using the hybrid constructions for PCR.

2. Experimental setup

2.1. Material inhibition testing

A high surface-to-volume ratio of microstructures leads to an increase in the probability of interaction between the inner surface of the microchip and the molecules of the reaction mixture, which affects the efficiency of the reaction down to its termination (inhibition). Presumably, the mechanism of inhibition is determined by two major factors: physical adsorption and chemical interaction between the mixture components and the microchip material.

During the process of selecting materials for fabrication of microchip devices for amplification, preference should be given to those that do not affect the effectiveness of the reaction. Therefore, it is necessary to confirm that the materials do not inhibit the reaction.

The following materials were studied: PDMS Sylgard®-184 (Dow Corning, USA), single-component UV-curing adhesive Permabond UV630 (Permabond Engineering Adhesives Ltd, USA), adhesive compound PEO-221K (Saint-Petersburg State Institute of Technology (Technical University), Russia), cyclo-olefin polymer film Zeonor Film® ZF14-188 (Zeon Europe GmbH, Germany), polyethylene terephthalate film (DuPont, USA).

The tested materials were added to the tube during a real-time PCR, while a tube without additives was placed for comparison. ANK-32 thermocyclers (IAI RAS, Russia), as well as specific primers and probes (DNA-Synthesis, Russia) and reaction mixture M-428 «PCR mix» (Syntol, Russia) were employed for the experiments. A synthesized DNA fragment of GAPDH housekeeping gene 226 b.p. was used as a target in a concentration of 10^5 copies/μl. The value of the threshold cycle (C_t) was determined using the ANK-32 software for three to five repetitions. If the difference in the average threshold cycle values in the test tubes with a sample and the tubes without additives exceeded 1.0 cycle of real-time PCR, it was assumed that the samples inhibit the reaction.

The experimental results of polymers testing show that the materials do not inhibit the real-time PCR.

2.2. Amplification of nucleic acids in MFC prototypes

For experimental studies, MFCs of two configurations were fabricated: 1 – with a droplet generator for digital isothermal amplification; 2 – with ten reaction chambers for PCR.

For the fabrication of the MFCs, PDMS Sylgard®-184 and adhesive compound PEO-221K were used; and microstructures were formed by soft lithography. PDMS replicas with microstructures were sealed with a PDMS-film or a glass plate by high-frequency (13.6 MHz) plasma treatment at low pressure (1 mbar) in oxygen for 1 min (Denier Zepto, Denier electronic GmbH, Germany). When using PEO-221K, the sealing was carried out by a capillary method with UV-adhesive Permabond UV630.

The configuration of the MFC (figure 1, a) based on droplet microfluidics contained a droplet generator, which ensures the sample encapsulation by a focused flow of fluid into separated droplets with diameter of 10-40 μm in mineral oil. The width of the channel for the continuous phase is ~28 μm, for discrete – ~32 μm; the width of the generator is ~13 μm; the width of the reaction chamber is ~195 μm, its depth is ~31 μm.

a b

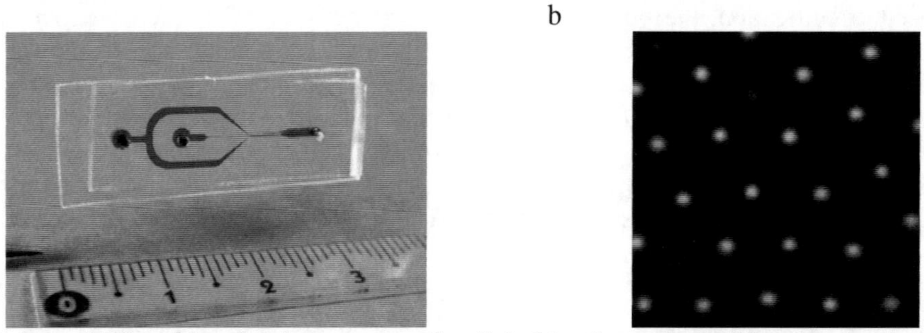

Figure 1. a – MFC with a droplet generator for digital isothermal amplification; b – image of emulsion flow after amplification, detected by CLSM in fluorescence mode. Size of the fragment is 100 x 100 μm^2.

BioXtra mineral oil (Sigma-Aldrich Corp., USA) with the addition of 5% Span® 80 surfactant and 0.5% Brij L4 (w/w) (Sigma-Aldrich Corp., USA) was used as the continuous phase (of transport flow); and, as the discrete phase, IsoAmp® II Universal tHDA Kit (BioHelix Corporation, USA) was used for helicase-dependent isothermal amplification. It was found that the combination of these surfactants ensures thermal stability of the droplets. The cDNA of GAPDH housekeeping gene was applied as a target, the amplification proceeded for 1 hour at 65 °C. The amplification results were registered by a fluorescence signal in the droplet flow (the volume of each droplet being ~29 nl); a confocal laser scanning microscope (CLSM) Leica TCS SL (Leica Microsystems, Germany) was used for this purpose. Detection conditions: excitation at 488 nm, emission at 510-530 nm (figure 1, b).

The obtained data: a) confirm the possibility of registration the results of amplification in the flow; b) indicate that the surfactant combination used does not affect the fluorescence of EvaGreen®.

The second configuration of the MFC had five pairs of reaction chambers (4 for reaction mixture, 1 for negative control) (figure 2, a), of ~160 μm depth and ~1.4 μl volume each. The design featured integration of the polyolefin film (Sarstedt AG & Co., Germany) into a PDMS-replica to prevent evaporation of fluid through the material when heated. Spectral measurements of MFC light transmittance showed that, in the spectral range of 500-850 nm, the transmission exceeds 94%, which ensures a reliable detection of the fluorescence signal of the most common fluorophores. Spectral characteristics of PDMS integrated with a cyclic olefin copolymer (COC) film (ZEONEX®, ZEON EUROPE GmbH, Germany) were also studied. The results of measuring the PDMS autofluorescence with polyolefin and COC films showed that the constructions do not have intrinsic fluorescence when excited at 650 nm and recorded at 655-750 nm, which is necessary when using a Cy-5 probe.

a b

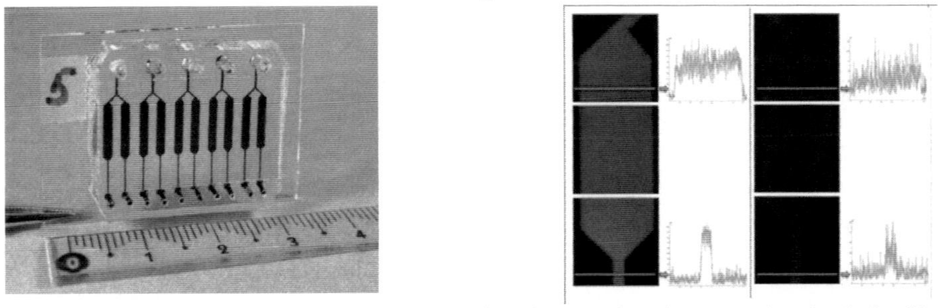

Figure 2. a – MFC for PCR with ten reaction chambers and an integrated polyolefin film; b – images of fragments of the reaction chambers with PCR-mixture and cDNA target after amplification (left); with negative control (right), detected with CLSM in the mode of measuring fluorescence. Intensities of fluorescence signal of reaction chamber and input channels are shown on the right side of the images. The size of each fragment is 1.5 x 1.5 mm^2.

After the chambers were filled with the reaction mixture, the input/output channels were sealed with silicone compound Pentelast®-712 A (Silikonovye Materialy, Russia), which cured during the first stage of MFC heating. PCR was carried out using TaqMan technology, specific primers and Cy-5 probe and reaction mixture M-428 «PCR-Mix». A synthesized DNA fragment of GAPDH housekeeping gene 226 b.p. was used as a target in a concentration of 10^5 copies/μl. The PCR results were registered by CLSM in the fluorescence detection mode at 650-670 nm with excitation at 633 nm (figure 2, b).

The obtained data show: a) the possibility of using Pentelast®-712 to seal the sample input/output channels; b) the effectiveness of the polyolefin film in preventing evaporation of the sample from the reaction chamber.

A hybrid construction was made of adhesive compound PEO-221K and a glass plate by soft lithography using the elastic PDMS-master mold. This combination of the materials also helps to prevent evaporation of the reaction mixture; however, it is necessary to select a compound for sealing of the sample reservoirs.

3. Conclusion

The results of the research showed that the selected materials for MFC prototyping: PDMS Sylgard®-184, single-component UV-curing adhesive Permabond UV630, adhesive compound PEO-221K, cyclo-olefin polymer film Zeonor Film® ZF14-188 and polyethylene terephthalate film have no significant effect on the amplification (do not inhibit the real-time PCR).

For the experimental studies, two microfluidic chip configurations were made: 1 – with a droplet generator for digital isothermal amplification; 2 – with ten reaction chambers for PCR. MFCs were made by soft lithography in the PDMS.

The results confirm the possibility of registration the results of droplet isothermal amplification in the transport flow, as well as in the reaction chambers of the MFC for PCR. The choice of the surfactants combination and their concentrations ensures thermal stability of the emulsion during isothermal amplification and does not affect the fluorescence of EvaGreen®.

A special feature of the MFC design for PCR was the integration of a polyolefin film to prevent evaporation of the sample, the effectiveness of which is shown by images of fragments of the reaction chambers. Pentelast®-712 A was used to seal the sample reservoirs.

Acknowledgments
This work was supported by the Program of the Presidium of RAS No. 8, the project Microdevices Based on the Principles of Droplet Microfluidics for Purposes of Chemical and Biological Analysis (Institute for Analytical Instrumentation, RAS).

References
[1] Ahrberg C D, Manz A and Chung B G 2016 *Lab Chip* **16** 3866
[2] Yan L, Zhou J, Zheng Y, Gamson A S, Roembke B T, Nakayama S and Sintim H O 2014 *Mol. BioSyst.* **10** 970
[3] Zhao X-M, Xia Y, Whitesides G M 1997 *J. Mater. Chem.* **7** 1069
[4] Alvankarian J and Majlis B Y 2012 *J. Micromech. and Microeng.* **22** (3) 035006

ssDNA damage dependence from singlet oxygen concentration at photodynamic interaction

V.V. Klimenko[1], N.E. Kaydanov[1], A.K. Emelyanov[1], A.A. Bogdanov[1]

[1] St. Petersburg Academic University, St. Petersburg, Russia

Abstract. Single stranded DNA damage at photodynamic treatment with Radachlorin photosensitizer was investigated. Chemical trap method was used to evaluate generation of singlet oxygen in water solution. Interaction of singlet oxygen with ssDNA resulted into decrease of the replication activity of ssDNA. DNA stopped replicating during PCR at irradiation doses greater than 15 J/cm^2 and concentration of photosensitizer [PS] = 3.8 μM. The dependence of replication activity of ssDNA on generated singlet oxygen concentration was identified.

1. Introduction

Photodynamic therapy (PDT) is an innovative method for the treatment of many tumors [1,2]. The main mechanism of PDT involves the interaction between the photoactive state of the photosensitizer (PS) and molecular oxygen ($^3O_2$) to produce singlet oxygen ($^1O_2$). Generally, it is believed that $^1O_2$ is the main cytotoxic species in PDT [3]. $^1O_2$ mainly damages lipids, proteins, RNA and DNA. There was a lot of interest in how singlet oxygen affects DNA chain in recent studies, because such interaction may lead to mutations and carcinogenesis [4,5,6]. But we have not found any data presenting dependence of DNA replication activity from concentrations of singlet oxygen generated during irradiation at PDT. The analysis of DNA molecular damages can be applied to investigate the efficiency of irradiation modes at photodynamic therapy. In present study we analyzed the dependence of ssDNA damage on generated singlet oxygen concentration.

2. Materials and methods

2.1. ssDNA damage by singlet oxygen

ssDNA (cDNA) was taken as a model system to investigate DNA oxidation at photodynamic therapy. RNA was isolated from k562 cells using the GeneJET RNA purification Kit (Thermo Scientific, USA) according to the manufacturer's instructions. cDNA was synthesized with an input of 2 μg of total RNA using Maxima First Strand cDNA Synthesis Kit with dsDNase (Thermo Scientific, USA), following the manufacturer's instructions. Concentration of cDNA was measured on NanoDrop 2000c (Thermo Scientific, USA). 20μl water solution of cDNA (9.65 ng/μl) and 3.8 μM photosensitizer Radachlorin (Rada-Pharma Co, Ltd., Moscow, Russia) in individual tubes was prepared. This solution was irradiated with 662 nm laser «LAHTA – MILON» (MILON Laser, St. Petersburg, Russia) at power density I = 50 mW/cm^2 with energy doses: 2.5 J/cm^2, 5 J/cm^2, 7.5 J/cm^2, 10 J/cm^2, 15 J/cm^2. cDNA damage after photodynamic treatment was analyzed using polymerase chain reaction - diagnosis method (PCR). The PCR was performed in a 40 μL volume using 20 μl water solution of cDNA (9.65 ng/μl) after irradiation, 0.25 μM forward (5'-TGGTTCATCATCATTCAACGGTGG-3') and reverse (5'-GTTCCCGTAGGTCATGAACTCAG-3') primers [7], Taq polymerase (1.25 units per reaction), dNTP (200 μM) and 10x Mg^{2+} Tris-HCl buffer («Sileks», Russia). Used primers were specific to the *BCR-ABL* gene fragment of cDNA. The PCR samples were initially denaturated for 1 minute at 95°C followed by 35 cycles of 95°C for 15 seconds, 60°C for 15 seconds, and 72°C for 15 seconds, with a final extension at 72°C for 5 minutes using a thermocycler (T100, Bio-Rad, USA). The PCR products (374 b.p.) were separated by electrophoresis in 8% polyacrylamide gel. Results were analyzed with Gel Doc XR+ (Bio-Rad, USA). The results were reported as the mean values with standard deviation from three experiments.

2.2. *Singlet oxygen generation in solution*
Chemical trap method was used to register singlet oxygen emerged in solution under photodynamic interaction [8]. Ethanol solution of chemical trap 1,3-diphenylisobenzofuran (DPBF) at concentration [DPBF] = 42 μM and Radachlorin at concentration [PS] = 3.5 μM (ε_{655} = 28300 $M^{-1}cm^{-1}$ in PBS) was prepared. Solution was irradiated with 662 nm laser «LAHTA – MILON» at power I = 10 mW/cm^2 during t = 100 sec. Absorption spectra were measured each 10 second on NanoDrop 2000c (Thermo Scientific, USA). The decrease of absorbance on 412 nm wavelength corresponds to the interaction of singlet oxygen with chemical traps DPBF.

3. Results

3.1. *ssDNA damage by singlet oxygen after PDT*
Fig. 1 shows UV picture of DNA polyacrylamide electrophoresis after photodynamic treatment. Fig. 2 shows dependence of PCR product from the irradiation dose. PCR product was calculated as I/I_0, where I is intensity volume of the analyzed band, I_0 is intensity volume of the control group band.

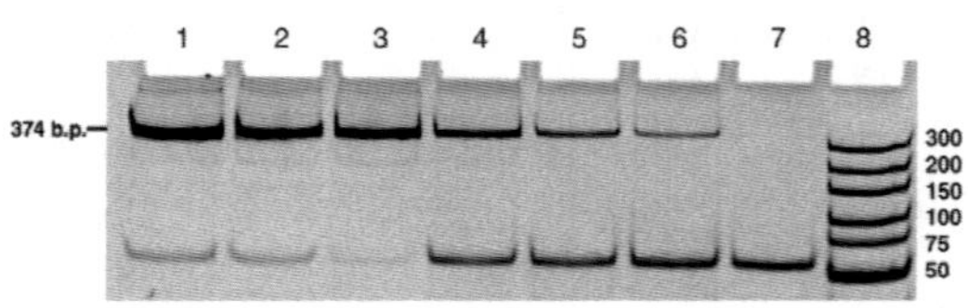

Figure 1. DNA polyacrylamide electrophoresis after photodynamic treatment. Lane 1 — control PS-; Lane 2 — control PS+; Lanes 3-7 — PDT with irradiation doses 2.5, 5, 7.5, 10, 15 J/cm^2; Lane 8 — 300-bp DNA ladder.

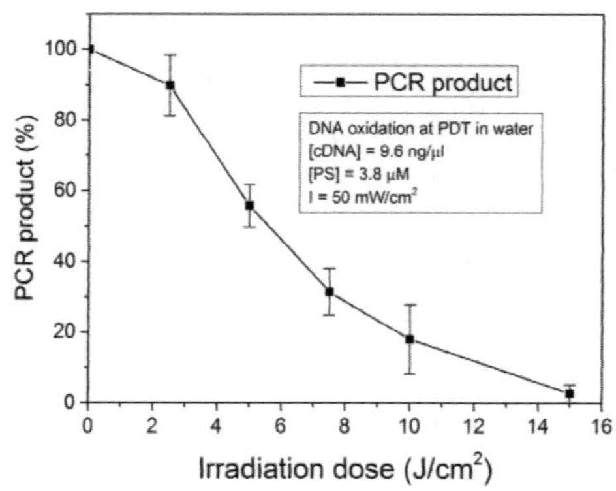

Figure 2. Irradiation dose dependent PCR product (DNA replication activity) after photodynamic treatment with 3.8 μM of Radachlorin PS and irradiation doses 2.5-15 J/cm^2.

3.2. *Detection of singlet oxygen generation in solution at PDT*
Fig. 3 shows photodegradation of DPBF by irradiation with red laser in the presence of Radachlorin in ethanol. Absorbance decrease can be observed on λ = 412nm. Fig. 4 shows concentration dependences of oxidized DPBF and calculated concentration of $^1O_2$ (Eq. (1)) from irradiation time.

$$\left|\frac{dA}{dt}\right| = \frac{d\left[{}^1O_2\right]}{dt}\frac{\varepsilon k_{ox}\left[DPBF\right]}{\tau_\Delta^{-1} + k_{ox}\left[DPBF\right]} \tag{1}$$

where A is absorption of DPBF at 412nm, ε is molar absorption coefficient equal to 2.35×10^4 M^{-1} cm^{-1} (412 nm) for DPBF, τ_Δ is life time of $^1O_2$ in ethanol equal to 13.5 μs, k_{ox} is a constant of oxidation of DPBF equal to 1.4×10^9 M^{-1} s^{-1} in ethanol, [DPBF] is concentration of active DPBF. The same experiment was conducted in H_2O solution. Fig. 5 shows absorbance decrease of DPBF from irradiation time at different power densities I = 10, 20, 40 mW/cm^2. Using Eq. 1 and parameters for water solution ($^1O_2$ life time τ_Δ = 4μs, $k_{ox} = 2.8 \times 10^9$ M^{-1} s^{-1} [8]) singlet oxygen concentration was calculated. Fig. 6 shows linear dependence of $^1O_2$ generation rate from laser power density.

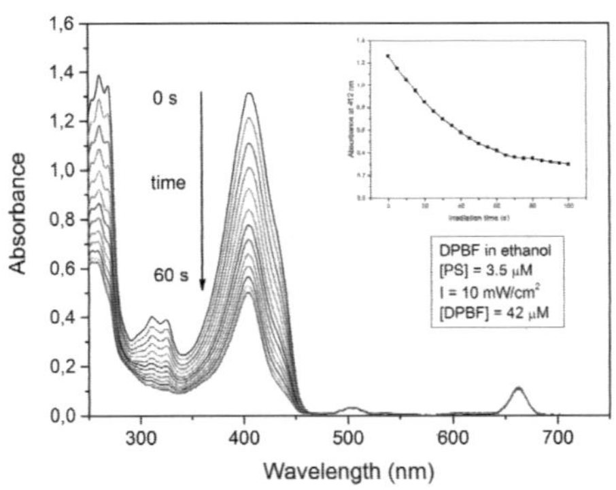

Figure 3. Photodegradation of DPBF by irradiation with red laser in the presence of Radachlorin in ethanol.

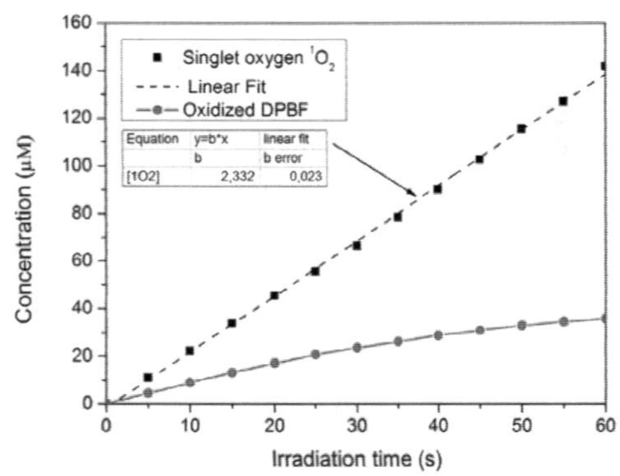

Figure 4. Time dependences of singlet oxygen and oxidized DPBF concentrations.

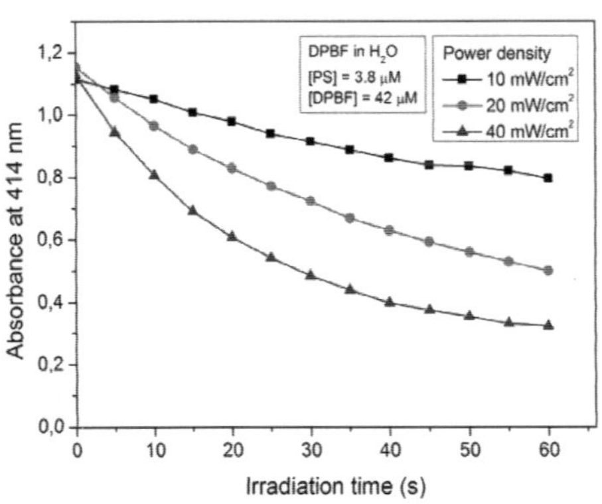

Figure 5. Photodegradation of DPBF by irradiation with red laser with different irradiation power in the presence of Radachlorin in H_2O + 0.2M SDS.

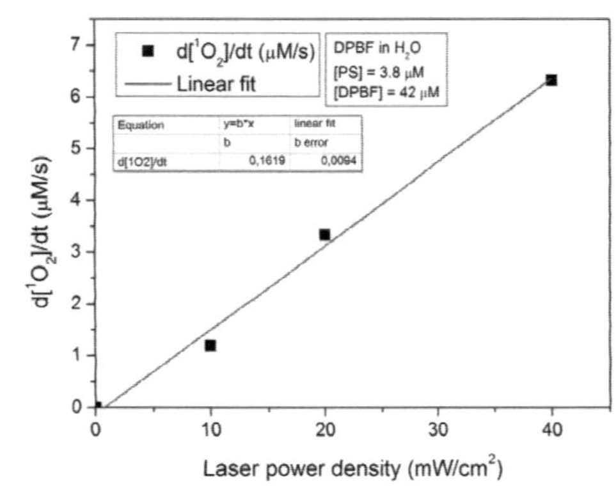

Figure 6. Singlet oxygen generation rate dependence of laser power density in H_2O + 0.2M SDS.

η parameter was calculated using Eq. 2 and linear approximations of data from fig. 4 and fig. 6. It stands for the [$^1O_2$] generation rate per I = 1 mW/cm^2 irradiation and 1 μM of photosensitizer per second.

$$\frac{d\left[{}^{1}O_2\right]}{dt} = \eta \cdot \left[PS\right] \cdot I \qquad (2)$$

Table 1. Life time of $^1O_2$ molecule and its generation rate in ethanol and water solutions upon irradiation of Radachlorin photosensitizer by 662 nm laser.

Solvent	**τ_Δ, μs**	**η, mW^{-1}cm^{2}s^{-1}**
Ethanol	13.5	(6.7±0.7)*10^{-2}
Water + 0.2M SDS	4	(4.3±0.3)*10^{-2}

This value can be used to recalculate generation of singlet oxygen in solutions with other concentrations of photosensitizer and intensities of irradiation. This constant was further used in our study to calculate cumulative singlet oxygen concentration at photodynamic therapy.

3.3. DNA damage dependence from cumulative singlet oxygen concentration

With η parameter we calculated cumulative singlet oxygen concentrations generated in DNA solution at PDT using macroscopic singlet oxygen model with photobleaching [9]. It allowed us to recalculate dependence of PCR product from the concentration of total singlet oxygen generated in water. Concentration of singlet oxygen that oxidized DNA $[^1O_2]_q$ was calculated using Eq. (3).

$$\left[{}^{1}O_2\right]_q = \left[{}^{1}O_2\right]\frac{k_{ox}\left[cDNA\right]}{\tau_\Delta^{-1} + k_{ox}\left[cDNA\right]} \qquad (3)$$

where τ_Δ is life time of $^1O_2$ in water equal to 4 μs, k_{ox} is a constant of oxidation of cDNA equal to $5.1*10^6$ M^{-1} s^{-1} in water [4], [cDNA] is DNA nucleotide concentration equal to $9.65*10^{-3}$ (g/L) / 300 (g/mol) = 32.2 μM. In our study $[^1O_2]_q/[^1O_2] = 6.6*10^{-5}$. Table 2 shows energy-dose dependent concentrations of cumulative singlet oxygen, singlet oxygen that oxidized cDNA and PCR product after PDT. At 15 J/cm^2 the replication of cDNA terminates. For this irradiation dose the ratio of oxidized guanines to total concentration of guanines was calculated and it equals 352 8-oxodG per 10^5 dG.

Table 2. Energy-dose dependent concentrations of cumulative singlet oxygen at [PS] = 3.8μM, singlet oxygen that oxidized cDNA and PCR product.

Dose, J/cm^2	$[^1O_2]$, μM (±7%)	$[^1O_2]_q$, μM	8-oxodG per 10^5 dG	PCR product, %
0 (control)	0	0.000	0	100±0
2.5	370	0.024	76	89.9±8.5
5	702	0.046	144	55.8±5.9
7.5	999	0.066	205	31.5±6.5
10	1266	0.084	259	18.0±9.8
15	1719	0.113	352	2.7±2.4

4. Discussion

We showed that PDT influence leads to dose-dependent decrease of ssDNA replication activity. For the first time we calculated the dependence of replication activity of ssDNA from cumulative concentration of singlet oxygen generated in solution during irradiation (Table 2). At irradiation doses greater than 15 J/cm^2 and concentration of photosensitizer [PS] = 3.8 μM PCR product was not observed (Fig. 1). It matches cumulative singlet oxygen concentration [$^1O_2$] = 1.7 mM. Decrease in DNA replication of target gene can be explained by oxidation of ssDNA by singlet oxygen. It is known that singlet oxygen reacts with DNA molecule and generally oxidizes guanine nucleotides forming 8-oxodG [5]. There are two general mechanisms of ssDNA damage by singlet oxygen: 8-oxodG generation and single-strand breakage of ssDNA [4]. According to our PCR results single-strand breakage of DNA was not observed. We suggest that first mechanism of damage is the main reason for decrease in replication activity of DNA. In our study, PCR started from binding of reverse primers on cDNA (5'-GTTCCCGTAGGTCATGAACTCAG-3'). Most primers are synthesized so that they have high GC-content for their strong binding. If guanines in cDNA are oxidized by singlet oxygen with generation of 8-oxodG then reverse primer cannot bind to complementary DNA fragment. This leads to inhibition of replication activity of target gene and decrease of PCR product. If 8-oxodG is "inside" target gene of cDNA, complementary DNA chain generates but it has G to T mutations [5]. However, these mutations have no effect on PCR product. Finally, this experiment shows that decrease in replication activity of DNA is dependent from cumulative concentration of singlet oxygen generated during irradiation. Efficiency of photodynamic treatment strongly depends on generated singlet oxygen concentration [10]. Therefore, this simple method of registration of DNA molecular damages can be applied to investigate the biotoxic efficiency of irradiation modes at photodynamic therapy.

Acknowledgments
The reported study was funded by RFBR according to the research project № 16-04-01858 a.

References
[1] Allison R R, Sibata C H, 2010 *Photodiagnosis Photodyn.* **7** 61–75.
[2] Agostinis P *et al* 2011 *CA Cancer J Clin.* **61**(4) 250-81.
[3] Soukos N S, Goodson J M, 2011 *Periodontol.* **55** 143–146.
[4] Sies H, Menck C F M, 1992 *Mutation Research/DNAging* **275** 367-375.
[5] Agnez-Limaa L F, Meloa J T A, Silvaa A E, Oliveiraa A H S, Timoteoa A R S, Lima-Bessaa K M, Martinezb G R, Medeirosc M H G, Mascioc P, Galhardod R S, Menckd C F M, 2012 *Reviews in Mutation Research* **751**(1) 15–28.
[6] Dizdaroglu M 2012 *Cancer Letters* **307** *26-47.*
[7] Wongboonma W, Thongnoppakhun W, Auewarakul C U 2011 *Journal of Hematology & Oncology* **4** 7.
[8] Krasnovsky A A Jr., Kozlova A S, Roumbala Ya V *Photochem. Photobiol. Sci.* 2012 **11** 988-997
[9] Klimenko V V, Knyazev N A, Moiseenko F V, Rusanov A A, Bogdanov A A, Dubina M V 2016 *Photodiagnosis Photodyn.* **13** 101–107.
[10] Kim M M, Penjweini R, Gemmell N R, Veilleux I, McCarthy A, Buller G S, Zhu T C, Hadfield R H, Wilson B C 2016 *Cancers* **8** (12) 1–15.

IOP Conf. Series: Journal of Physics: Conf. Series **917** (2017) 042014 doi:10.1088/1742-6596/917/4/042014

Super-resolution microscopy of living bacterial cells

E V Ponomareva[1], I E Vishnyakov[2], N E Morozova[1], V S Polinovskaya[1], M A Khodorkovskii[1] and A D Vedyaykin[1, a]

[1] Peter the Great St.Petersburg Polytechnic University, Saint-Petersburg, 195251, Russia

[2] Institute of Cytology, Russian Academy of Sciences, Tikhoretsky av. 4, Saint Petersburg 194064, Russia

Email: misterkotlin@gmail.com

Abstract. Currently several methods of super-resolution optical microscopy are known. One of them – super-resolution radial fluctuations (SRRF) microscopy – was successfully used in present work to study the structures, formed by FtsZ protein in living *E.coli* cells with a resolution well below the diffraction limit.

1. Introduction

The aim of this work was to visualize cell division process (in particular, structures formed by FtsZ protein) in living *Escherichia coli* cells using super-resolution microscopy. It is well-known that FtsZ protein plays one of the main roles in this process [1]. FtsZ forms so-called Z-ring at the site of future septum, which is considered to be a scaffold for several dozens of downstream proteins, thereby it directs the cytokinesis. FtsZ is a tubulin homologue and is able to polymerize *in vitro* into several high-order structures, including linear filaments, bundles and helices [2]. One of the most striking observations about FtsZ behavior *in vitro* is its ability to generate contractile force upon conditions close to *in vivo* situation [3], although the actual role and significance of this force *in vivo* seems to be doubtful [4]. To better understand the molecular mechanisms of FtsZ role in division, for instance, whether FtsZ actually generates the force, further study is needed.

In previous studies we successfully utilized the method of single-molecule localization microscopy (SMLM) in combination with immunofluorescence for visualization of FtsZ structures in fixed *E. coli* cells [5-7]. However for better understanding of mechanisms of bacterial division, it is essential to visualize the process of division in dynamics, i.e. in living cells. Thus, a significant interest for researchers is the dynamics of FtsZ structures in the course of fission, which is largely a result of the development of new methods of super-resolution microscopy. The study of dynamics testifies to the high dynamism of the processes of assembly and disassembly of the Z-ring. In particular, it has recently been possible to observe such a phenomenon as treadmilling [8]. Until recently the tools for FtsZ visualization in living cells were limited by the observation that all *E.coli* fluorescent fusions with FtsZ protein used for super-resolution visualization before were not fully-functional, thereby this fact could substantially distort the actual FtsZ action. Recently several fully-functional fluorescent FtsZ fusions were developed, including those incorporated into bacterial chromosome [9]. In current work we visualized FtsZ in living *E.coli* cells with super-resolution using aforementioned fully-

[a] To whom any correspondence should be addressed

functional proteins in combination with a novel technique of Super-resolution radial fluctuations (SRRF) [10].

2. Materials and methods

All experiments were conducted using *E. coli* strain harbouring fully-functional FtsZ fusion with mNeonGreen incorporated into genome [9]. Cells were grown overnight in LB medium at 37°C, then passaged to fresh medium and grown until $OD_{600}\approx 0.5$. Then 1 µl of cell suspension was placed into agarose pad slide, which preparation was described previously [11].

Visualization was carried out using custom-made set-up based on upright optical microscope which was previously described [12]. To excite fluorescence the set-up was equipped with a set of solid-state and diode lasers as well as Hg lamp. High-sensitive cooled EM-CCD camera (Andor Ixon 896) was used for registration of fluorescence. Series of frames (from 100 to 1000) were acquired with an exposition of 20 ms and maximum camera speed using MicroManager [13]. Consequently these series were processed using NanoJ-SRRF [10] software package a freely available open-source plugin for the popular ImageJ or Fiji image analysis software. Data processing was performed using ImageJ [14].

3. Results and discussion

SRRF method in combination with fully-functional fluorescent FtsZ fusion allowed us to successfully visualize the structures formed by FtsZ while bacteria were dividing (see Figure 1 and Figure 2). These structures are in strong agreement with data obtained previously using combination of SMLM and immunofluorescence [5, 12, 15]. It is important to note that dynamics of the process of division is extremely important because FtsZ forms various transient spatial structures, and their interpretation is indispensible for understanding of molecular mechanisms of bacterial division. Studying of FtsZ represents not only fundamental importance, since understanding of the mechanisms of bacterial cells division seems to be a crucial step for development of new antibiotics.

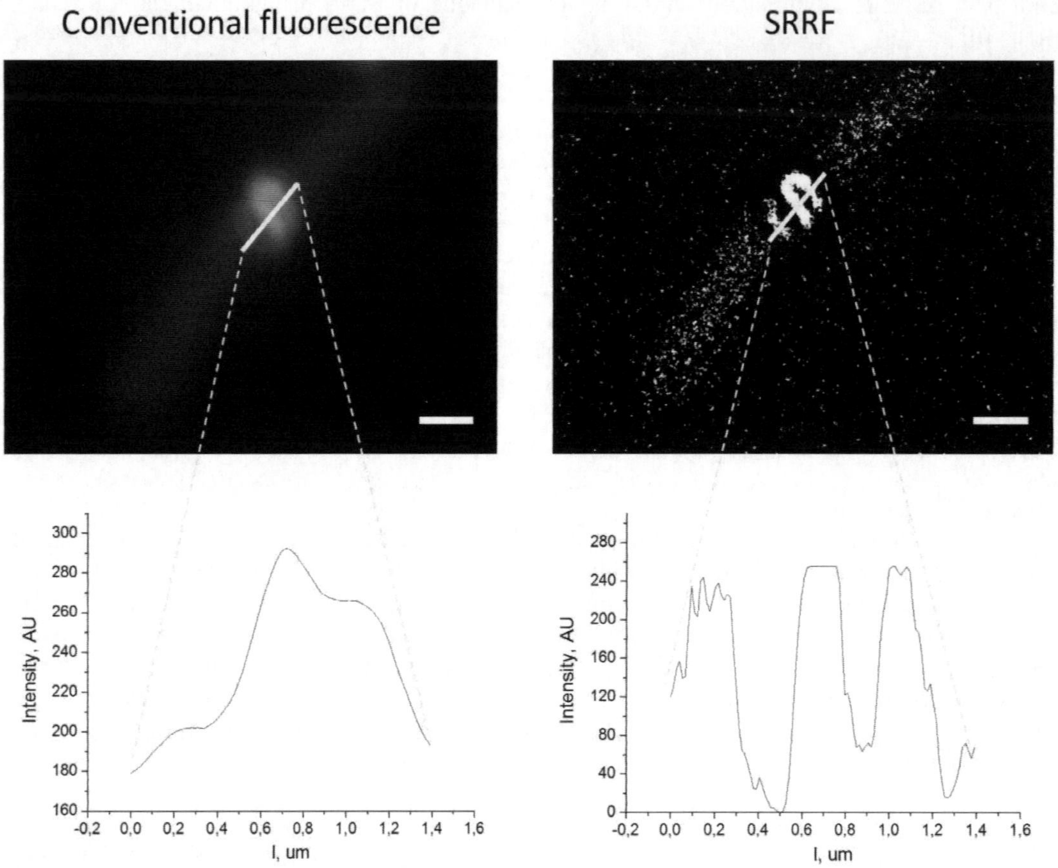

Figure 1. Comparison of image of FtsZ obtained using conventional fluorescence (left) and corresponding image obtained using SRRF technique. Charts correspond to intensity profiles along yellow lines on both images. Scale bar corresponds to 1 μm at all images.

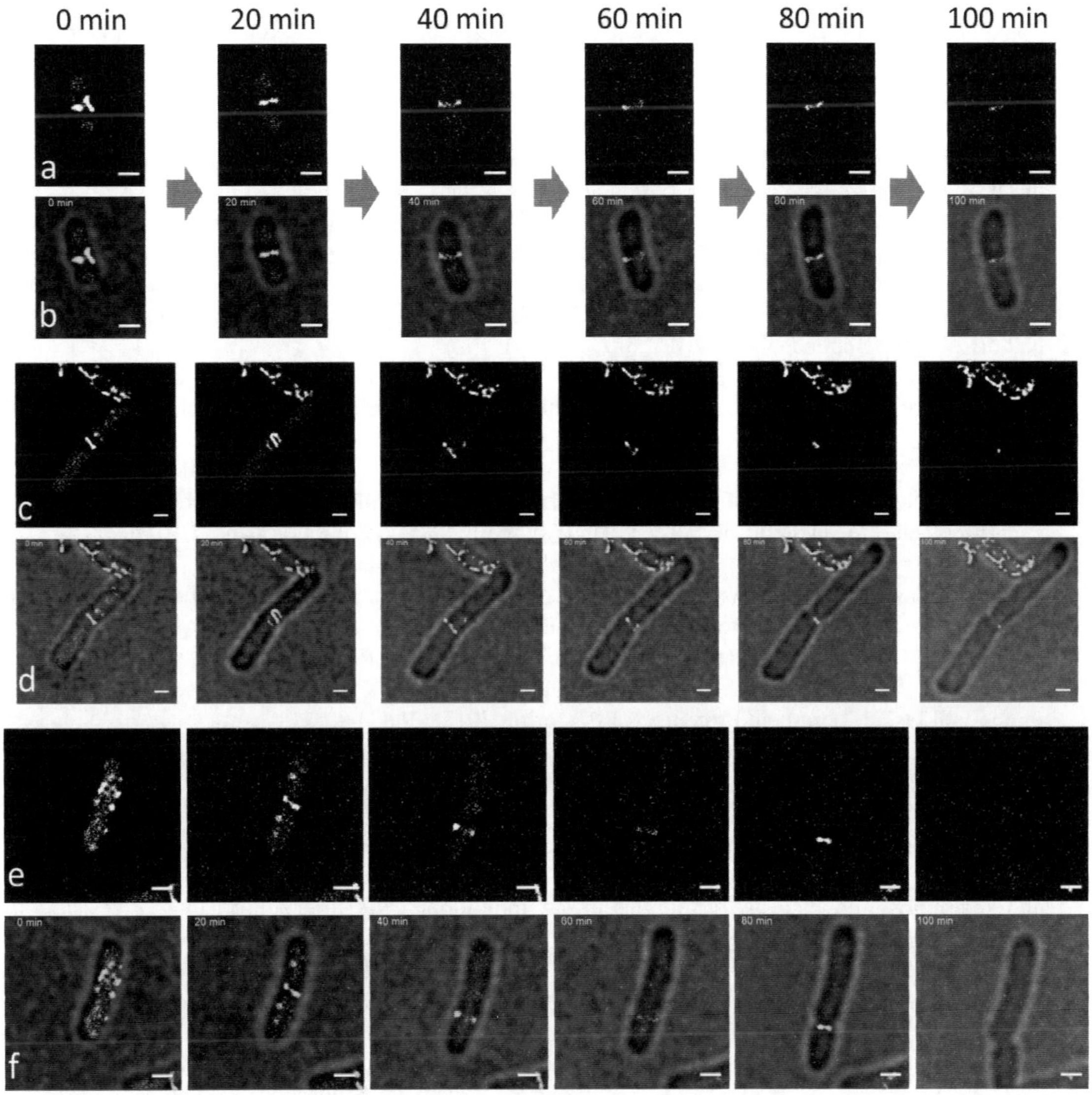

Figure 2. Time-lapse imaging of cell division. Images of three individual cells during 100 minutes are presented. a, c, e – SRRF images of FtsZ, b, d, f – composite images of super-resolution image of FtsZ (green) and conventional transmitted light image (gray) in living *E.coli* cells. Scale bar corresponds to 1 μm at all images.

4. Acknowledgements

Authors would like to thank Dr. Harold Erickson for providing strains with fully-functional fluorescent FtsZ fusions. The work was carried out using the unique installation «Laser tweezers. The work was supported by the Russian Foundation for Basic Research (project 15-04- 07472).

References

1. D.P. Haeusser, W. Margolin. Splitsville: structural and functional insights into the dynamic bacterial Z ring. // Nat Rev Microbiol, 2016. 14(5): p. 305-19.

2. H.P. Erickson, D.E. Anderson, M. Osawa. FtsZ in Bacterial Cytokinesis: Cytoskeleton and Force Generator All in One. // Microbiology and Molecular Biology Reviews, 2010. 74(4): p. 504-528.
3. M. Osawa, H.P. Erickson. Liposome division by a simple bacterial division machinery. // Proc Natl Acad Sci U S A, 2013. 110(27): p. 11000-4.
4. D.O. Daley, U. Skoglund, B. Soderstrom. FtsZ does not initiate membrane constriction at the onset of division. // Sci Rep, 2016. 6: p. 33138.
5. A.D. Vedyaykin, V.V. Gorbunov, A.V. Sabantsev, V.S. Polinovskaya, I.E. Vishnyakov, A.S. Melnikov, P.Y. Serdobintsev, M.A. Khodorkovskii. Multi-color localization microscopy of fixed cells as a promising tool to study organization of bacterial cytoskeleton. // 2nd International School and Conference Saint-Petersburg Open on Optoelectronics, Photonics, Engineering and Nanostructures (Spbopen2015), 2015. 643.
6. A.D. Vedyaykin, A.V. Sabantsev, I.E. Vishnyakov, N.E. Morozova, V.S. Polinovskaya, M.A. Khodorkovskii. 3D super-resolution microscopy of bacterial division machinery. // Journal of Physics: Conference Series, 2016. 741(1): p. 012066.
7. A.D. Vedyaykin, I.E. Vishnyakov, V.S. Polinovskaya, M.A. Khodorkovskii, A.V. Sabantsev. New insights into FtsZ rearrangements during the cell division of Escherichia coli from single-molecule localization microscopy of fixed cells. // Microbiologyopen, 2016. 5(3): p. 378-86.
8. M. Loose, T.J. Mitchison. The bacterial cell division proteins FtsA and FtsZ self-organize into dynamic cytoskeletal patterns. // Nat Cell Biol, 2014. 16(1): p. 38-46.
9. D.A. Moore, Z.N. Whatley, C.P. Joshi, M. Osawa, H.P. Erickson. Probing for Binding Regions of the FtsZ Protein Surface through Site-Directed Insertions: Discovery of Fully Functional FtsZ-Fluorescent Proteins. // J Bacteriol, 2017. 199(1).
10. N. Gustafsson, S. Culley, G. Ashdown, D.M. Owen, P.M. Pereira, R. Henriques. Fast live-cell conventional fluorophore nanoscopy with ImageJ through super-resolution radial fluctuations. // Nat Commun, 2016. 7: p. 12471.
11. N. Morozova, A. Sabantsev, E. Bogdanova, Y. Fedorova, A. Maikova, A. Vedyaykin, A. Rodic, M. Djordjevic, M. Khodorkovskii, K. Severinov. Temporal dynamics of methyltransferase and restriction endonuclease accumulation in individual cells after introducing a restriction-modification system. // Nucleic Acids Res, 2016. 44(2): p. 790-800.
12. A.D. Vedyaykin, A.V. Sabantsev, I.E. Vishnyakov, S.N. Borchsenius, Y.V. Fedorova, A.S. Melnikov, P.Y. Serdobintsev, M.A. Khodorkovskii. Localization microscopy study of FtsZ structures in E. coli cells during SOS-response. // 1st International School and Conference Saint-Petersburg Open 2014 on Optoelectronics, Photonics, Engineering and Nanostructures, 2014. 541.
13. A. Edelstein, N. Amodaj, K. Hoover, R. Vale, N. Stuurman. Computer control of microscopes using microManager. // Curr Protoc Mol Biol, 2010. Chapter 14: p. Unit14 20.
14. C.A. Schneider, W.S. Rasband, K.W. Eliceiri. NIH Image to ImageJ: 25 years of image analysis. // Nat Methods, 2012. 9(7): p. 671-5.
15. A.D. Vedyaykin, I.E. Vishnyakov, V.S. Polinovskaya, M.A. Khodorkovskii, A.V. Sabantsev. New insights into FtsZ rearrangements during the cell division of Escherichia coli from single-molecule localization microscopy of fixed cells. // Microbiologyopen, 2016. 5(3): p. 378-386.

IOP Conf. Series: Journal of Physics: Conf. Series **917** (2017) 042015 doi:10.1088/1742-6596/917/4/042015

The study of monodisperse water-in-oil macroemulsion dynamics in a microfluidic chip

D V Nozdriukhin[1,2], **K I Belousov**[3,4], **N A Filatov**[1] **and A S Bukatin**[1,3]
[1]Nanobiotech Lab, St. Petersburg Academic University, St. Petersburg 194021, Russia
[2]Peter the Great St. Petersburg Polytechnic University, St. Petersburg 195251, Russia
[3]Department of biophysics, Siberian Federal University, Krasnoyarsk 660041, Russia
[4]Department of Nanotechnologies and Materials Science ITMO University, St. Petersburg 197101, Russia

Abstract. Emulsion dynamics is little-studied, but important subject for biological, medical and physical applications. In this work microfluidic droplet generator was used to experimentally study two-dimensional flow of monodispersed macroemulsion with 25 – 50 μm droplet size. Quantity of droplets, their trajectories and velocity profiles were studied. Obtained results showed, that velocity profile of droplets flow does not coincide with the Poiseuille profile of laminar oil flow in the channel. Droplet velocity decrease with the increasing of droplets concentration, which can be explained by the hydrodynamic interactions between droplets. The law of motion can be well described by the Greenberg's traffic flow model.

1. Introduction

An emulsion is a mixture of two immiscible substances. It usually contains microdrops of one fluid, called dispersed phase, suspended in another fluid, called continuous phase. Nowadays, these systems are used in various fields such as food industry, car chemicals, beauty products etc.

Modern microfluidic technologies make it possible to produce monodisperse macroemulsions, which can be used as analytical framework for clinical diagnostics, pharmacological screening of drugs, biomarker analysis etc. Developing the theory of emulsion dynamics is important for understanding streams of blood and other body fluids. In addition, emulsions can be used for modeling some micro- and macroscopic interactions inside many-body systems. It is known from hydrodynamics that one moving liquid droplet creates 2D-dipole-like fluid flow field around it [1, 2]. Because of that, it is possible to use a system of droplets as an illustrative model of magnetic or electric dipole flows. Emulsions can also be used for self-assembling applications [3], which is another reason of researching their dynamics.

In this work, we studied dependence of droplet velocities from the emulsion concentration in 2D fluid flow inside a microfluidic channel.

2. Methods

The microfluidic chip with a flow-focusing droplet generator and elongated (10 mm) outlet channel (Figure 1a) was fabricated using soft lithography method from PDMS Sylgard 184 (Dow Corning) [4].

The droplet generator's operating principle is based on the fact that the continuous phase flowing from two side channels meets the dispersed phase at the channel's intersection, where the dispersed phase is squeezed by the continuous phase and breaks up into droplets (Figure 1b) [5]. Changing the flow rates of the liquids, it is possible to produce monodisperse macroemulsions with defined droplet size (Figure 1c).

In this work, we studied droplets 25 - 50 µm in diameter. Mineral light oil (cat. N. 330779 Sigma-Aldrich) with 3.5% ABIL EM 180 surfactant (Evonik Industries) was used as a continuous phase and deionized water as a dispersed phase. Chip's sidewalls were hydrophobized with ClearVue Rain Repellent (Turtle Wax Inc.). For introducing continuous and dispersed phases into the microfluidic chip under the constant pressure in the range 8 – 30 kPa, we used a microfluidic pressure controller based on ITV001 electro-pneumatic regulators (SMC, Japan). The chip was mounted on an optical microscope Leica DM4000 B LED (Leica Microsystems) and after the generator stabilization, the droplets motion was recorded using a camera Pike F100B (Allied Vision Technologies) at 60 fps in the middle of the outlet channel.

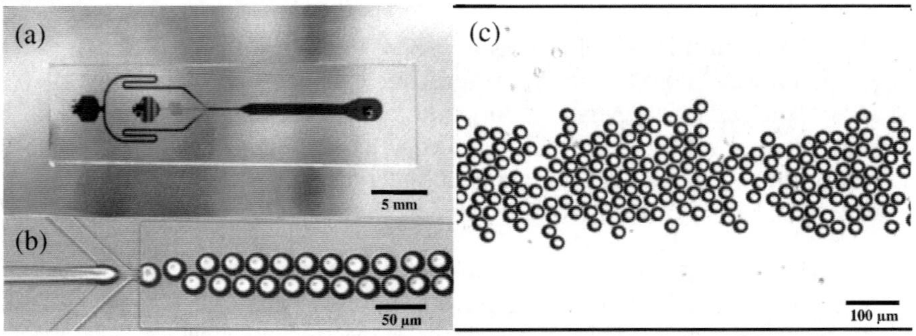

Figure 1. a) Chip with long outlet capillary; **b)** Flow-focusing generator; **c)** Droplets in the outlet channel at oil pressure 8 kPa, water pressure 6.5 kPa, droplet size 27 µm.

Data was detected far from the droplet generator, so perturbations, created by it, did not influence on the flow. Matlab R2014a (MathWorks) was used for data processing, which included recognition of droplets, definition of their trajectories and individual droplets velocities calculation.

3. Results and discussion

It is well known that laminar fluid flows occur at low Reynolds numbers (in our case Re ~ 10^{-4}) and have a Poiseuille parabolic velocity profile with the maximum in the center of the channel [6]. Droplets in such a flow interact with each other hydrodynamically. Dipole forces in such systems are opposite to the direction of the flow. In a 1D droplet chain, this interaction causes reduction of droplet's velocity, known as a collective drag reduction or peloton effect [1]. In a 2D droplet flow, there are possible options: droplets occupy the entire channel and droplets flow mostly in the center of the channel. In the first case, velocities of all droplets are close to each other and emulsion is moving as a uniform substance. In the second case despite the fact that the continuous phase has a Poiseuille velocity profile, the droplets velocity in the center of the channel is less than near the sidewalls (Figure 2a).

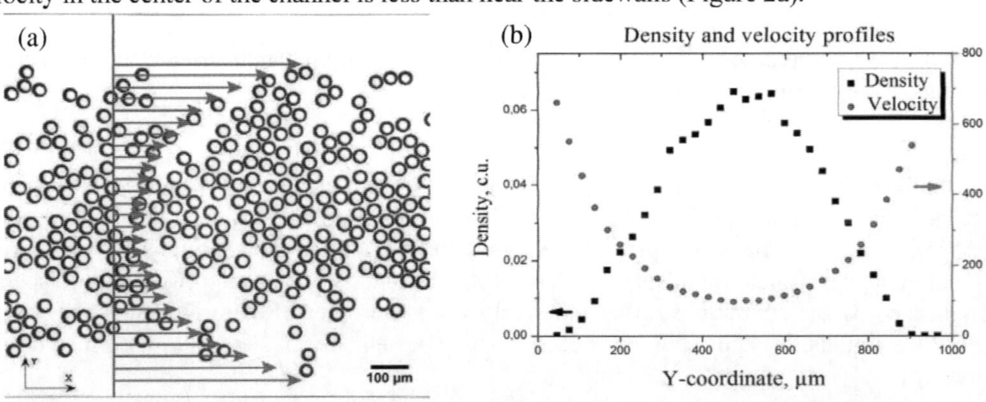

Figure 2. a) Droplet flow and velocity profile at oil pressure 11 kPa, water pressure 8.7 kPa, droplet size 35 µm; **b)** Density and velocity profiles at above-mentioned pressures.

After the analysis of about 10^4 droplets' trajectories, we have seen that near the center of the channel, where the droplet flow density is higher, droplet velocities are lower, than the velocities of drops, moving far from channel's center (Figure 2b). In the areas with the highest droplets' concentration, their velocity decreased thereby collective drag reduction – two dimensional peloton effect.

Theoretical analysis of droplet flows can be performed in terms of continuity models and models of automodel motion, which are based on the continuity equation. Due to low interaction between moving droplets in the orthogonal direction to the direction of motion friction forces can be neglected and the Euler equation can be used to describe the flow:

$$\frac{dv}{dt} = -\rho^{-1}\nabla p = -\rho^{-1}\frac{\partial p}{\partial \rho}\frac{\partial \rho}{\partial x}$$

Hence, equation of motion is:

$$\frac{dv}{dt} = -C^2\rho^{-1}\frac{\partial \rho}{\partial x}$$

where $C = \frac{\partial p}{\partial \rho} > 0$ – sound velocity in the emulsion (cm/s).

From this equation, dependence of the velocity on the density (Figure 3) can be approximated by Greenberg's model of traffic flow [7] with an addition, describing the floor value of velocity:

$$v = v_0 + C ln\frac{\rho_{max}}{\rho}$$

where ρ_{max} is the density of close-packing system of droplets (cm^{-1}), v_0 - velocity of close-packing system of droplets (cm/s), at above-mentioned pressures ρ_{max} = 137 cm^{-1}, v_0= 0.016 cm/s.

The difference between this model and Greenberg's model is in the velocity of the close-packing system: on the road this case means, that the traffic flow velocity is zero, but in our case, the oil flow pushes it with the constant velocity.

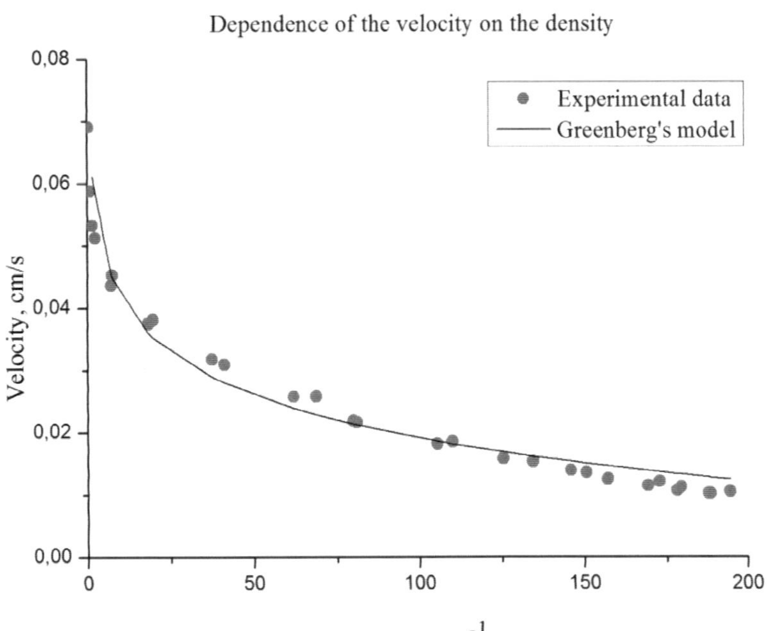

Figure 3. Example of dependence of droplet flow's velocity on density at oil pressure 11 kPa, water pressure 8.7 kPa.

4. Conclusion

The velocity and density profiles of monodisperse water-in-oil macroemulsion flow were obtained. Investigated the mismatch between velocity profiles of laminar flow of classical fluid and macroemulsion, caused by collective hydrodynamic interactions. The dependence of the velocity on the droplet concentration was studied. Proved, that macroemulsion's flow can be described in terms of motion of continuous compressible fluid. Based on the study of dependence of the velocity on the density of the droplet flow, it shown that such system motion could be approximated by modified Greenberg's model of traffic flow. That means, macroemulsion flows can be used for demonstrable simulation of the traffic on the road. In addition, at oil pressure 11 kPa and water pressure 8.7 kPa, the sound velocity in emulsion with respect to Greenberg's model is 0.01 cm/s, which correlates with results, obtained in [1].

Acknowledgments

The research was supported by the grant of the Russian Science Foundation (project no.15-19-10041).

References

[1] Beatus T, Tlusty T, Bar-Ziv R 2006 *Nature Physics* **2** 743-748
[2] Tlusty T 2006 *Macromolecules* **39** 3927-3930
[3] Meissner M, Dong J, Eggers J, Seddon A M, Royall C P 2017 *Soft Matter* **13** 788-794
[4] Bukatin A S, Mukhin I S, Malyshev E I, Kukhtevich I V, Evstrapov A A, Dubina M V 2016 *Technical Physics* Vol 61 No 10 pp 1566-1571
[5] Kukhtevich I V, Posmitnaya Y S, Belousov K I, Bukatin A S, Evstrapov A A 2015 *NAUCHNOE PRIBOROSTROENIE* Vol 25 No 3 pp 65–85
[6] Bruus H 2006 *Theoretical microfluidics* (MIC Lecture notes third edition) pp 28-30
[7] Greenberg H 1959 *Operations Research* Vol 7 pp 79–85

IOP Conf. Series: Journal of Physics: Conf. Series **917** (2017) 042016 doi:10.1088/1742-6596/917/4/042016

The sensitivity research of multiparameter biosensors based on HEMT by the mathematic modeling method

V G Tikhomirov[1,3], A G Gudkov[2], S V Agasieva[2], E N Gorlacheva[2], V D Shashurin[2], A A Zybin[3], A S Evseenkov[1,3], Y M Parnes[1,3]

[1]Saint-Petersburg State Electrotechnical University "LETI", St. Petersburg 197376, Russia
[2]Bauman Moscow State Technical University, Moscow 105005, Russia
[3] Closed Joint Stock Company "Svetlana-Electronpribor", St. Petersburg 194156, Russia

Abstract. The numerical impact modeling of some external effects on the CVC of biosensors based on AlGaN/GaN heterostructures (HEMT) was carried out. The mathematical model was created that allowed to predict the behavior of the drain current depending on condition changes on the heterostructure surface in the gate region and to start the process of directed construction optimization of the biosensors based on AlGaN/GaN HEMT with the aim of improving their performance. The calculation of the drain current of the biosensor construction was carried out to confirm the reliability of the developed mathematical model and obtained results.

1. Introduction

It was proposed to use heterostructures AlGaN/GaN (HEMTs) without the gate to create multiparameter biosensors by many authors [1-3]. The absorption of biological molecules on the gate region causes the new charge state of the surface that affects the flow of current in two-dimensional channel of the HEMT. Thus, the adsorption of molecules controls current drain-source similarly to classic gate.

2. Simulation results

Different coats (platinum, gold, etc.) in the form of droplet or solid films are suggested for the selective adsorption of biological molecules on the gate region. Nitrides of metals of the 3rd group – semiconductors allowed the creation of the new devices of opto- and microelectronics with unique characteristics, were considered as the most promising semiconductor material for these biosensors. An example of such biosensor is shown in Figure 1. It is important for invasive biosensors that devices based on GaN, AlGaN have exceptional chemical and temperature resistance. The temperature resistance is ensured not only by the properties of semiconductor material but also by contact resistance of the drain-source.

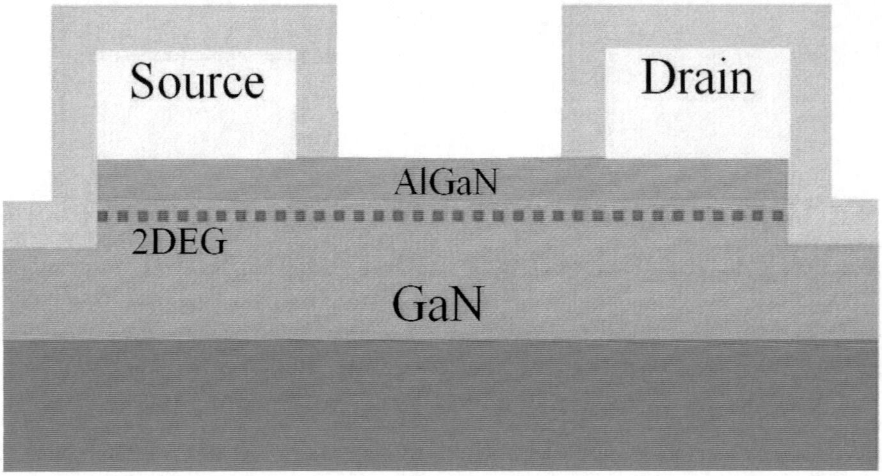

Figure 1. A schematic representation of a biosensor based on GaN HEMT without the gate.

Nowadays the reliable confirmation of the impact of heterostructure parameters and its surface state on the performance of the multibiosensors based on HEMTs without the gate continues to be complex, time-consuming and costly procedure.

The numerical impact modeling of some external effects on the CVC of biosensors based on AlGaN/GaN heterostructures (HEMT) was carried out in the framework of the present study. Well-known software packages for the numerical modeling of semiconductor devices were used as basic tools for the modeling and the optimization of HEMTs [4]. The simulation results were compared with the experimental results of electrophysical parameters measurement of heterostructures and HEMTs. Figure 2 shows a schematic cross section of the base of AlGaN/GaN HEMTs with a fragment of the estimated region of the transistor for the numerical modeling.

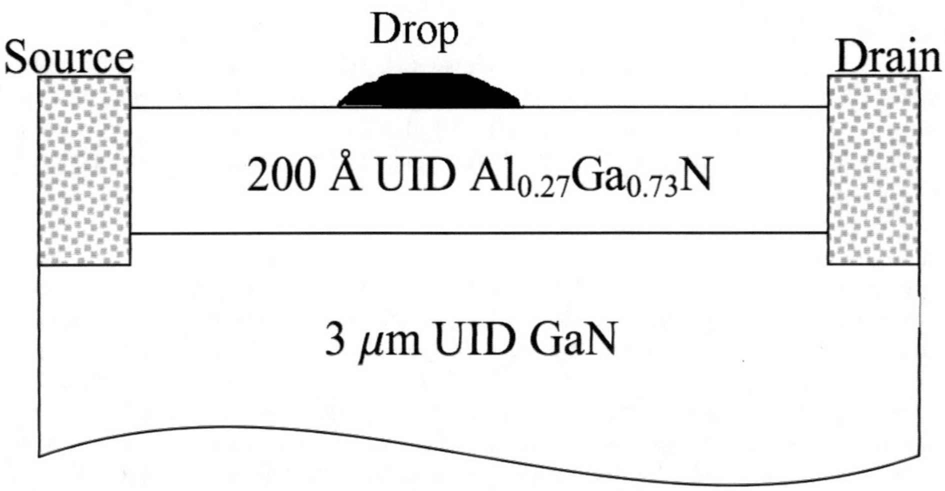

Figure 2. A schematic cross section of the base of AlGaN/GaN HEMTs with a fragment of the estimated region of the transistor for the numerical modeling.

Before considering of calculation results, it should be discussed about some important problems and ways to solve them for multiparameter biosensors based on GaN HEMT. The authors of modern scientific articles [5-8] propose several constructive solutions providing high sensitivity and easy operation of biosensor. For example, it is suggested using biosensor construction with the coat of the gold in the gate area and special acid (thioglycolic acid) to determine the presence of heavy metals ions in researched liquid.

Several authors bring the sequence of technological operations to get finished biosensor design on the silicon substrate, in particular, with plating of Ti and Ag/AgCl [6]. The researchers proposed a planar design providing both the comfortable deposition of researched liquid in the area between source and drain and getting small quantities of the metallization resistance of electrodes to maintain a high sensitivity predicted in laboratory experiments or using the mathematical modeling.

It is especially interesting for the possibility of the mathematical modeling of biosensors based on AlGaN/GaN HEMT that there is a change of GaN surface potential when the pH value of the sample at the gate changes.

Thus, authors of article [8] considered 3 designs of biosensors based on AlGaN/GaN HEMT and received the dependence of surface potential on the heterostructure border from the pH value of the liquid put in the gate area for all 3 variants.

It was possible due to the obtained results to create mathematical model that allows to predict the behavior of the drain current depending on condition changes on the heterostructure surface in the gate region, in particular, application of an electrolyte with known pH value, and in addition, allows to start the process of directed construction optimization of the biosensors based on AlGaN/GaN HEMT with the aim of improving their performance.

The calculation of drain current of the biosensor construction described and researched in [8] was carried out to confirm the reliability of the developed mathematical model and obtained results. Figure 3 shows the results of mathematical modeling with the potential on the drain 0.5 V and the experimental research data obtained in the aforementioned study. Mathematical modeling predicted the sensitivity of the biosensor at 80 μA/*p*H that was confirmed by experimental data. According to the results of the researched and the mathematical modeling the planar design of biosensors with different distance of the source-drain and metallization with gold, see Figure 4.

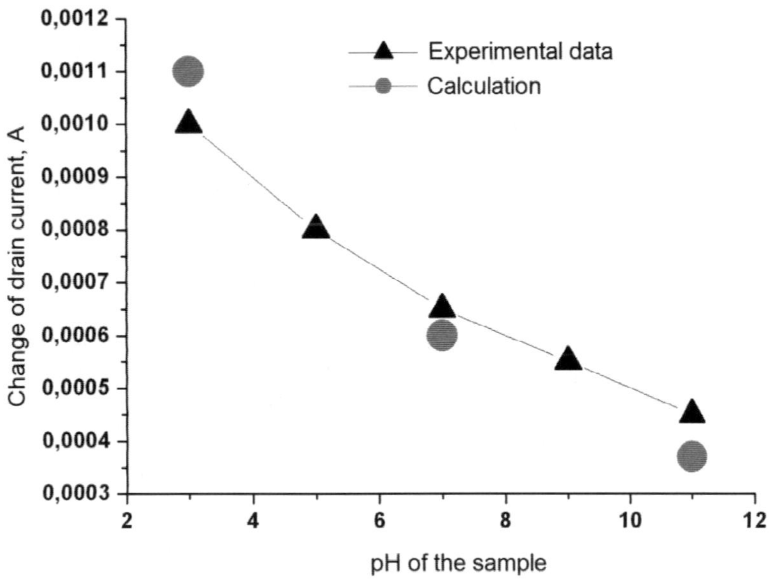

Figure 3. The dependence of the drain current change on the pH value of the sample at the gate. Circles show values predicted using of mathematical modeling.

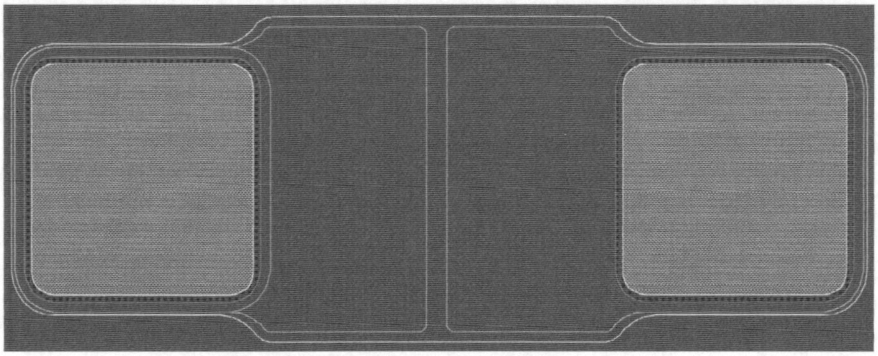

Figure 4. Plan view photomicrograph of a completed device with Au film on the gate region.

3. Conclusion

The results of accomplished studies state the following. Numerical calculations allowed to suggest that the impact of specified external factor (location of electrolyte with known pH at the gate) on the CVC of researched device exists and this dependence has a monotonous character. More researches are needed for directional changes of the heterostructure and of the overall design of the biosensor to increase the sensitivity in the considered cases and to expand the range of detectable substances.

Acknowledgments

This study was supported by the Ministry of Education and Science of the Russian Federation (agreement on grant no. 14.574.21.0116, the unique project identifier is RFMEFI57414XO116).

References

[1] Ambacher O, Eickhoff M, Steinhoff G, Hermann M, Gorgens L, and Werss V, *Proc. ECS 214*, 27 (2002).

[2] Stutzmann M, Steinhoff G, Eickhoff M, Ambacher O, Nobel C, Schalwig J, et al. *Diamond Rel Mater* 2002; **11**:886.

[3] Kang B S, Wang H T, Ren F, Pearton S J. Electrical detection of biomaterials using AlGaN/GaN HEMTs *J Appl Phys*. 2008;**104**(8):031101.

[4] Tikhomirov V, Zemlyakov V, Volkov V, Parnes Ya, Vyuginov V, Lundin W, Sakharov A, Zavarin E, Tsatsulnikov A, Cherkashin N, Mizerov M, and Ustinov V, *Semiconductors*, 2016, Vol. **50**, N. 2, pp. 244–248.

[5] Kang B, Wang H, Ren F, and Pearton S. Electrical detection of biomaterials using AlGaN/GaN high electron mobility transistors. *Journal of Applied Physics* **104**, 031101 (2008).

[6] Hung S, Wang Y, Hicks B, Pearton S, Ren F, Johnson J, Rajagopal P, Roberts J, Piner E, K Linthicum, and Chia G. Integration of Selective Area Anodized AgCl Thin Film with AlGaN/GaN HEMTs for Chloride Ion Detection. *Electrochemical and Solid-State Letters*, **11** (9) H241-H244 (2008).

[7] Wang H, Kang B, Chancellor T, Lele T, Tseng Y, Ren F, Pearton S, Dabiran A, Osinsky A, Chow P. Selective detection of Hg(II) ions from Cu(II) and Pb(II) using AlGaN/GaN high electron mobility transistors. *Electrochem Solid State Lett*. 2007;**10**:J150-3.

[8] Steinhoff G, Hermann M, Schaff W, Eastmann L, Stutzmann M, and Eickhoff M, *Appl. Phys. Lett*. **83**, 177 (2003).

New nondestructive method for determining the composition of components in biological objects in express mode

N S Myazin[1], V V Davydov[1, 2, 3], V V Yushkova[4], T I Davydova[1], V Yu Rud'[3]
[1]Peter the Great Saint Petersburg Polytechnic University, Saint Petersburg 195251, Russia
[2]The Bonch-Bruevich Saint - Petersburg State University of Telecommunications, Saint Petersburg 193232, Russia
[3]All-Russian Research Institute of Phytopathology, Moscow Region 143050, Russia
[4]Saint Petersburg University of Management Technologies and Economics, Saint Petersburg 190103, Russia

Abstract. The new method for determining the components composition and their concentration in the express mode in biological objects and liquid suspensions is proposed. This method is based on the Bloch equations solutions for the components of the magnetization vector in case of registration of the NMR signal using a modulation technique in a weak magnetic field. It has allowed to measure not only the relaxation constants T_1 and T_2, using which it is possible determine the deviation degree of the medium from the standard state, but also the concentrations of the components included in mediums composition, at the site of sampling by a compact NMR spectrometer.

1. Introduction

Currently, one of the urgent problems of applied physics is the development of fast and reliable methods for the express control of the condensed medium state [1–6]. Such studies are particularly in demand in determining the liquid state (various aqueous solutions, gasolines, oil, biological complexes or suspensions, etc.), as well as granular media (various salts, biological objects in the form of powders, etc.) in express mode [6–10]. To solve these problems, a large number of instruments have been developed, but most of them are mainly designed for research of a certain type of medium (for example, oil, motor oil, milk, etc.). It should also be noted that the sample with the researched medium after the measurements performed on some of such devices cannot be used for re-investigation on high-resolution spectrometers in stationary laboratories in order to confirm the result [1, 2, 4–7, 10–13]. This is because research using such devices can cause irreversible changes in the medium structure. In accordance with the latest requirements for rapid control methods, such changes are unacceptable.

One of the possible solutions to this problem is the small-sized nuclear magnetic spectrometer (relaxometer) considered in [1, 5, 6]. We can determine the degree of deviation from the standard state in the express mode using measured values of the relaxation constants T_1 and T_2. Experiments showed that the express method of control using a nuclear magnetic spectrometer (relaxometer) is very much in demand when carrying out various researches of condensed media, biological solutions and medical suspensions, environmental monitoring, and also in cases of product quality testing [1, 5, 6]. In addition, studies of various media conducted by the method of nuclear magnetic resonance (NMR) do not cause irreversible changes in their chemical composition and physical structure, unlike other express methods [1, 3, 5–9, 13–16]. Recently before conducting various researches, especially when working with biological solutions and suspensions, as well as acids, salts, it has increasingly become necessary to determine in express mode not only the medium state but also components concentration of which it is made (e.g., HCl, NaF, NaOH or FeCl3), acidity (pH), etc. Experiments have shown that this is especially important if the medium is used after prolonged storage, transport, container changes, etc.

In previously developed NMR spectrometer (relaxometer), registering of the NMR signal was possible only at the resonant frequency f_p of the protons contained in the researched medium. It does not allow solving new tasks of express control.

2. Compact NMR spectrometer and measurement methods

To solve the considered new problems, we developed a new small-size design of the magnetic system of the spectrometer and a scheme for registering the NMR signal (figure 1)

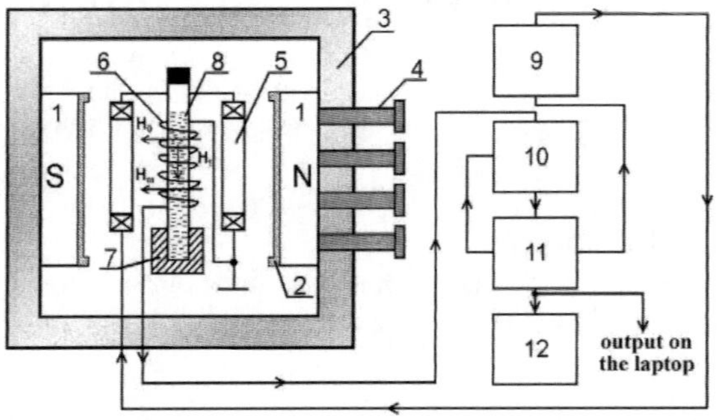

Figure 1 Structural scheme of compact NMR spectrometer: 1 — permanent magnet; 2 — inserts; 3 — neutral for the placement and alignment of the magnets; 4 — adjusting screws; 5 — modulation coil; 6 — NMR signal registration coil; 7 — locking device for the container with the researched medium; 8 — container with the researched medium; 9 — RF generator; 10 — registration scheme; 11 — processing and control unit; 12 — oscilloscope

In the new design of compact NMR spectrometer by reducing the degree of inhomogeneity of the magnetic field more than tenfold and increasing the induction B_0 by several times in the registration coil placement area 6 S/N of the detected NMR signal is increased by 13 times. This made it possible to register the NMR signal at the resonance frequencies of the other elements nuclei, for example, sodium or chlorine. For mediums consisting only of elements containing nuclei with magnetic moments (for example, HCl, NaF, or ZnCl2) we have created a method that allows determining concentration of elements in a medium using amplitudes of the registered NMR signals from different nuclei.

However, the most significant result obtained on newly developed design is NMR signal registration with amount of "wiggles" (oscillating peaks) not less than six. This allows us to carry out many measurements with higher precision and to develop a new method for determining components concentrations of a mixture, which is formed by substances that not enter into a chemical reaction (e.g., fats, multicomponent medical suspensions, oils, gasoline, etc.). NMR signal from such mixtures due to specific of its registration with use of modulation technique is the total signal from each of the components.

In our new method for describing the registered NMR signal, we use solutions of the Bloch equations in a rotating coordinate system [16, 17].

$$\begin{aligned} \frac{du}{dt} + \frac{u}{T_2^*} + \Delta\omega \cdot v = 0 \\ \frac{dv}{dt} + \frac{v}{T_2^*} - \Delta\omega \cdot u = -\gamma H_1 M_z \\ \frac{dM_Z}{dt} + \frac{M_Z}{T_1} - \gamma H_1 v = \frac{M}{T_1} \end{aligned} \quad (1)$$

where $\Delta\omega = \gamma H_0 - \omega$ — field frequency deviation from resonance, $M = \chi_0 H$— magnetization of the investigated medium in the magnetic field of the spectrometer, $v(t), u(t)$ — absorption and dispersion signals.

If case of using a modulation technique to detect the NMR signal in a weak magnetic field, the value of H varies as follows:

IOP Conf. Series: Journal of Physics: Conf. Series **917** (2017) 042017 doi:10.1088/1742-6596/917/4/042017

$$H = H_0 + H_m sin(\omega_m t), \tag{2}$$

where H_0 — constant magnetic field, H_m — modulation coil field, ω_m — modulation frequency (figure 1).

In this case, the change in the frequency deviation of the field from the resonance in the system of Bloch equations, with allowance for (2), will have the following dependence:

$$\Delta\omega = \gamma H_0 + \gamma H_m sin(\omega_m t) - \omega \tag{3}$$

One of the features of registering the NMR signal in a weak magnetic field using the modulation technique is that its registering should be performed only at the resonance frequency ($\omega = \omega_0 = \gamma H_0$). In this case (3) takes the following form:

$$\Delta\omega = \gamma H_m sin(\omega_m t) \tag{4}$$

At the resonant registering frequency $\omega = \omega_0$, taking into account (4), the system of Bloch equations (1) takes the following form:

$$\begin{aligned} u'(t) + \frac{u(t)}{T_2} + \gamma H_m sin(\omega_m t) v(t) = 0 \\ v'(t) + \frac{v(t)}{T_2} - \gamma H_m sin(\omega_m t) u(t) - \gamma H_1 M_z(t) = 0 \\ M_z'(t) + \frac{M_z(t)}{T_1} - \gamma\, v(t)\, H_1 - \frac{M}{T_1} = 0 \end{aligned} \tag{5}$$

where $M = \chi_0 H_0 + \gamma H_m sin(\omega_m t)$ — magnetization of the investigated medium in the magnetic field of the spectrometer, $\chi_0 = NI(I+1)\frac{\mu^2}{3kT}$ — statistical nuclear magnetic susceptibility, N — concentration of paramagnetic particles; μ — particle magnetic moment; k — Boltzmann's constant; T — absolute temperature.

This system of equations is solved for $v(t)$, $u(t)$ and $M_z(t)$ components, taking into account the initial conditions:

$$M_z(0) = \chi_0 H_0;\ u(0) =; v(0) = 0$$

The experimental researches carried out by us showed that the shape of the line $G(t)$ of the registered NMR signal from the substance in a weak magnetic field when tuning the autodyne detector circuit to the maximum of the S/N ratio is described by the following relation:

$$G(t) = \sqrt{Av^2(t) + BU^2(t)}, \tag{6}$$

where $v(t), U(t)$ — absorption and dispersion signals, A, B — coefficients determining the contribution to the NMR signal of the absorption and dispersion signals.

The obtained solution (6), which is in good agreement with the experimental results, made it possible to develop a technique for simulating the detected NMR signal from the mixture, by dividing the received signal into signals from its constituent components. Following relation can represent the shape of the registered NMR signal $G_m(t)$ from the mixture:

$$\begin{gathered} G_m(t) = \sqrt{A_m v_m^2(t) + B_m U_m^2(t)} = \sum_{i=1}^{k} V_i \cdot N_i \sqrt{A_i v_i^2(t) + B_i U_i^2(t)} \\ \sum_{i=1}^{k} V_i = V_r \end{gathered} \tag{7}$$

where $v(t), U(t)$ — absorption and dispersion signals, A, B — coefficients determining the contribution to the NMR signal of the absorption and dispersion signals (m — mixture, i — mixture components).

Since in the experiment all NMR signals, both from the mixture and its components, are formed in the same fields H_0, H_1 and H_m, the solutions obtained for their components $v(t)$ and $U(t)$ from the Bloch equations (5), differ only in the relaxation constants. For the mixture itself, T_1 and T_2 can be determined using NMR signal detected from it. In addition, we know relaxation constants for one of the mixture components, because initially this mixture should be this component in its pure form. The relaxation constants of the remaining components of the mixture, as well as the volumes in which they are contained in the researched medium, are chosen so that (7) is satisfied, the temperature of the researched mixture is known. When relation (7) is fulfilled, through relaxation constants we can determine mixture components, as well as their relative concentrations through their volumes.

3. Results of the research and their discussion

Figure 2 shows, as an example, the NMR signals from sodium hydroxide NaOH registered on sodium in the new design of the NMR spectrometer.

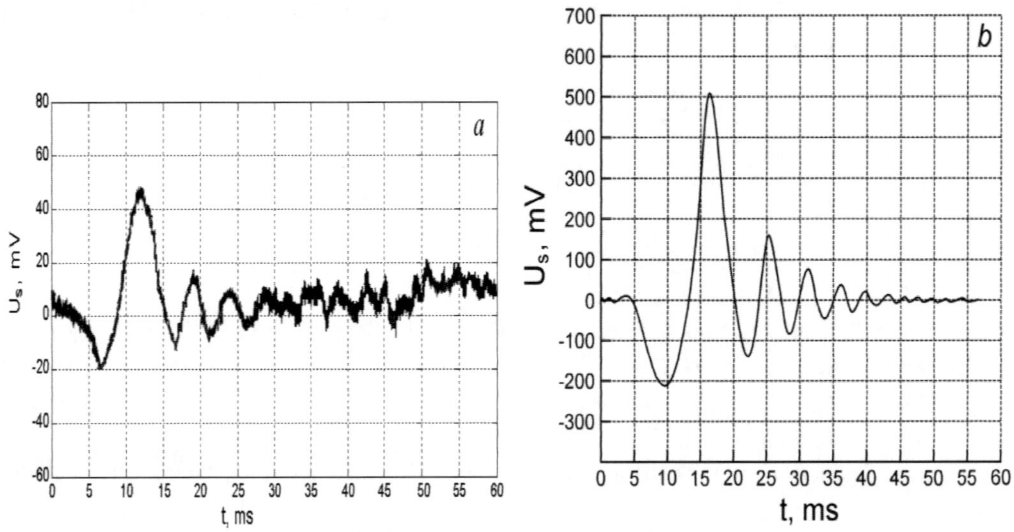

Figure 2 The registered NMR signal from sodium hydroxide at T = 291.3 K: a) without accumulation; b) output of accumulation scheme

The analysis of NMR signals presented in figure 2 shows that when they are registered at the resonance frequency of sodium nuclei ω_{Na}, signal-to-noise ratio more than 1.3. This makes it possible to perform the AFC on the resonance of sodium nuclei. However, it is impossible to carry out measurements of T_1 and T_2 with an error no more than 1.0% (which is necessary for unambiguously determining the medium state medium [1–4, 6, 8]) without usage of an accumulation scheme since the SNR < 3.0 (figure 2.a). In addition, there are noises at the peaks of the NMR signal. The subsequent accumulation of the NMR signal makes it possible to obtain a SNR > 10.0 (figure 2.b), which ensures measurement of T_1 and T_2 with the required accuracy. Moreover, the relative concentrations of protons and sodium in sodium hydroxide can be determined using ratio between their amplitudes U_s in registered NMR signal.

Figure 3 shows, as an example, the possibility of determining the components and their concentrations in a mixture of two gasolines AI-95 and A-76 using developed by us method applied to registered NMR signal.

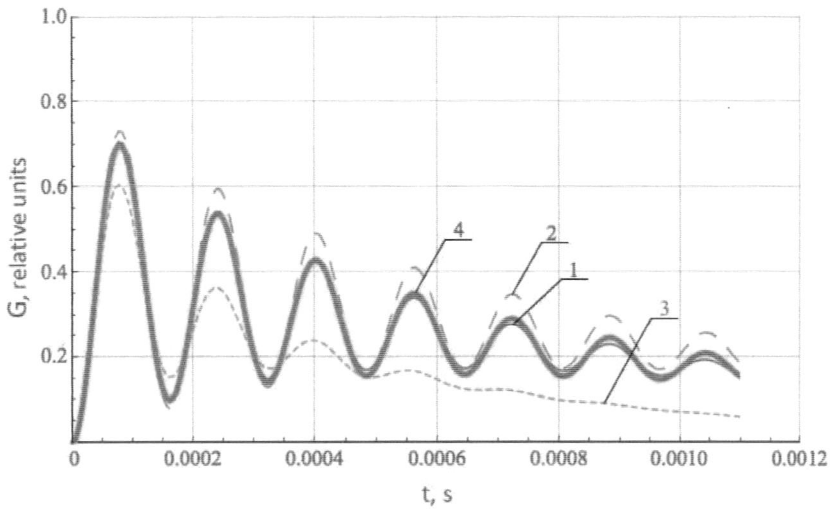

Figure 3 NMR signal line forms. Graph 1 corresponds to the experimental signal from a mixture of gasoline AI-95 and A-76 in the proportion of 75% to 25%. The NMR signal simulation: from pure gasoline AI-95 and A-76 - graphs 2 and 3, mixture of gasoline AI-95 and A-76 in a proportion of 75% to 25% - graph 4.

Comparison of the results of NMR signals simulation from gasoline AI-95 and A-76, as well as on their mixture, with experiment shows the reliability of proposed method. The calculated concentrations of gasolines AI-95 and A-76 in the researched medium are within the measurement error with the concentrations that we have used to prepare this mixture.

4. Conclusion

The obtained results shows that the usage of proposed by us method while carrying out researches in a new design of a small-size NMR spectrometer allows obtaining information on the composition of the medium at the site of sampling, and drawing a conclusion about its further use without additional measurements in a stationary laboratory. Earlier in [1, 5, 6], the measurement of T_1 and T_2 only provided information on the presence of deviation from the medium standard state and required its additional study in a stationary laboratory to make a reliable decision on the further use of this medium.

In addition, in the new design of the spectrometer, realization of the NMR signal registration from various nuclei with magnetic moments greatly expands its functional capabilities both in the number of possible researched mediums and in the information obtained about the state of the medium itself.

References

[1] Karseev A Yu, Cheremiskina A V, Davydov V V, Velichko E N 2014 J. of Physics: Conference Series **541** 012006

[2] Arkhipov V V 2012 Instruments and Experimental Techniques **55** 692

[3] Kashaev R S, Gazizov E G 2010 Journal of Applied Spectroscopy 77 321

[4] Mussil V V, Pilipenko V V, Lemeshevskaya E T, Keremzhanov K D 2011 Instruments and Experimental Techniques **54** 397

[5] Karseev A Yu, Vologdin V A, Davydov V V 2015 J. of Physics: Conference Series **643** 012108

[6] Davydov V V, Velichko E N, Dudkin V I, Karseev A Yu 2015 Instruments and Experimental Techniques **58** 234

[7] Alexandrov A S, Archipov R V, Ivanov A A, Gnezdilov O I, Gafurov M R, Skirda V D 2014 Applied Magnetic Resonance **45** 1275

[8] Zaporozhets O V, Shkurdoda V F, Peregudov O N, Zaporozhets V K 2010 Instruments and Experimental Techniques **53** 718

IOP Conf. Series: Journal of Physics: Conf. Series **917** (2017) 042017 doi:10.1088/1742-6596/917/4/042017

[9] Ryzhov V A, Pleshakov I V, Nechitailov A A, Glebova N V, E.N. Pyatyshev E N 2014, Applied Magnetic Resonance **45** 339

[10] Panchenko P A, Fedorov Y V, Perevalov V P, Jonusauskas G P, Fedorova O A 2010 Journal Phys. Chem. A. **114** 4118

[11] Nikolaeva M V, Puzyk M V 2014 Optics and Spectroscopy **116** 589

[12] Erdogru Y M, Manimaran D K, Gulluolu M T, Amalanathan M K, Hubert J I, Yurdakul N S 2013 Optics and Spectroscopy **114** 525

[13] Davydov V V, Dudkin V I, Karseev A Yu 2014 Measurement Techniques **57** 912

[14] Giulotto L, Lanzi G, Tosca L 1956 Journal of Chem. Physics **24** 632

[15] Chen B, Ivanov I V. Klein M L, Parrinello M S 2003 Phys. Rev. Lett. **91** 2155

[16] Leshe A Nuclear induction 1963 (Berlin: Veb Deustscher Verlag Der Wissenschaften)

[17] Abragam A Principles of nuclear magnetism 1961 (Oxford: Qxford at the Clarendon Press)

IOP Conf. Series: Journal of Physics: Conf. Series **917** (2017) 042018 doi:10.1088/1742-6596/917/4/042018

Hybrid mode piezoresponse force microscopy for compositional electromechanical study of biopiezoelectrics

A S Kalinin[1,2], **V V Polyakov**[1] **and V A Bykov**[1,2]

[1] NT-MDT Spectrum Instruments, Proezd 4922, 4/3 Zelenograd, Moscow 124460, Russian Federation

[2] Moscow Institute of Physics and Technology, 9 Institutskiy per., Dolgoprudny, Moscow Region, 141700, Russian Federation

Abstract. We present our recent development of new approach for compositional study of topography, nanomechanical and piezoelectric properties with nanometer spatial resolution and demonstrate it in application to two types of biopiezoelectrics: type I collagen and diphenylalanine peptide nanotubes.

1. Introduction

Atomic force microscopy (AFM) is a powerful tool for surface imaging and examination of a material's local properties with nanometer-level spatial resolution. Since an AFM's working principle is based on the direct interaction between sharp tip and sample, a variety of unique AFM measurement techniques have been developed: quantitative nanomechanical measurements, conductivity mapping, local electromagnetic studies etc. One of these AFM techniques is Piezoresponse Force Microscopy (PFM) – where we explore the electromechanical performance of ferroelectric and piezoelectric materials in terms of their domain morphology with nanometer spatial resolution in different environments and various temperatures. Since the development of this technique by Guthner and Dransfeld in 1991 [1], PFM has become a widely used technique for ferro- and piezoelectric crystals research.

Piezoelectricity was also discovered for biological objects. This effect, which arises from the electromechanical coupling, was named biopiezoelectricity. Biopiezoelectricity was observed in certain types of muscular movement, the nervous system, ion transportation, amino acids etc. [2–8] and as a result, its detection has become important for nanomedicine and biomedical applications. This requires a new method for investigating of electromechanical coupling in life systems. PFM is great candidate for this purpose since it allows piezoresponse measurement with nanometer-level resolution. However, PFM being a contact mode technique, it is unsuitable for studying biological samples: the lateral tip-sample interaction arising from the constant contact of AFM tip with the surface can be significant enough to destroy or deform softer and fragile materials. We introduce a new approach for PFM investigation of such soft and fragile objects by utilizing the reduced lateral tip-sample interaction in HybriD Piezoresponse Force Microscopy mode (HD PFM)

2. Instrumental part

HD PFM is extension of recently introduced HybriD mode (HD mode) – scanning technique based on fast force-distance curves measurements with real-time processing of tip response (Figure 1).

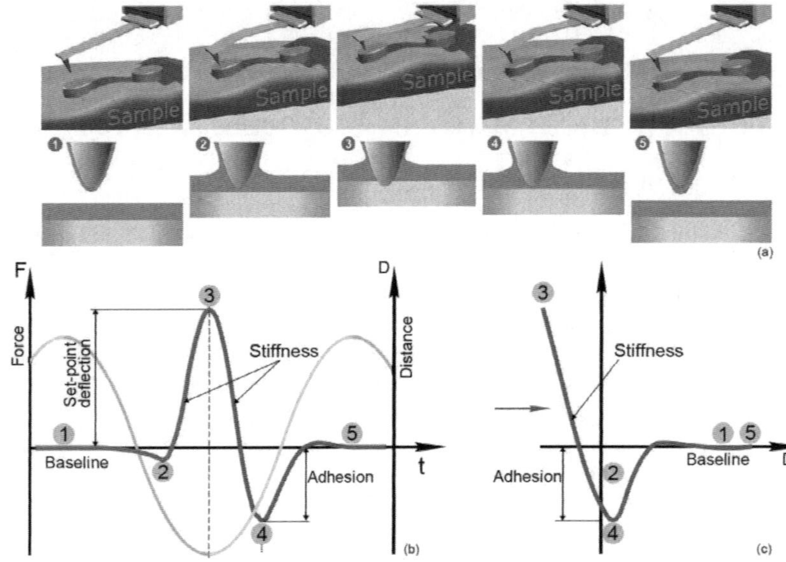

Figure 1. HD mode working principle: a) a model illustrating a performance of HD mode; b) an idealized temporal deflection curve during an oscillatory cycle; c) an idealized force-distance curve.

In HD mode, the sample or the tip is driven into a vertical oscillation by a Z piezoscanner with hundreds of Hertz frequency. As the result, each oscillatory cycle AFM tip goes through 5 main points: point 1 – away from the surface where only long-distance forces can be detected, point 2 – bends down in response to adhesive or capillary forces, point 3 – bending reverses upwards until it reaches the set-point level, point 4 – experiences adhesive interactions, point 5 – goes back to the baseline. The resultant vertical and lateral displacement curve of the tip is recorded and analysed to get: topography, Young's modulus, tip-surface adhesion value and long-distance electrostatic or magnetic forces map simultaneously.

One of the features of HD mode is ability of "time window" measurements: ability of switching on AC or DC voltage supply, signal recording and processing in the user-defined part of oscillatory cycle. This feature allowed for example to investigate conductivity of carbon nanotubes by allying DC voltage and detecting tip-sample current signal in the "time window" corresponded to the tip-sample contact (range 2-4 on the figure 1) [9].

HD PFM mode working principle also based on the "time window" approach (Figure 2). In the user-defined "time window" referred to the tip-sample contact the AC voltage applied between conductive coating of the tip and investigated object. AC voltage causes mechanical oscillations of the sample depending on its local polarization. Corresponding vertical (DFL siganl) and lateral (LF signal) motion of AFM tip is recorded in defined "time window" and processed to get amplitude and phase signals. Amplitude of DFL and LF signals characterize local piezoelectric coefficient of the material whereas the phase signals give information about local polarization direction. The full DFL(t) curve of each HybriD mode circle is also processed to calculate adhesion, E modulus, feedback input signals and long-distance electrostatic forces. Thus, HD PFM provides parallel study of different sample's electromechanical properties through a single measurement cycle, and at the same time, makes piezoresponse measurements nondestructive by retracting the tip from the sample at each HybriD oscillating cycle.

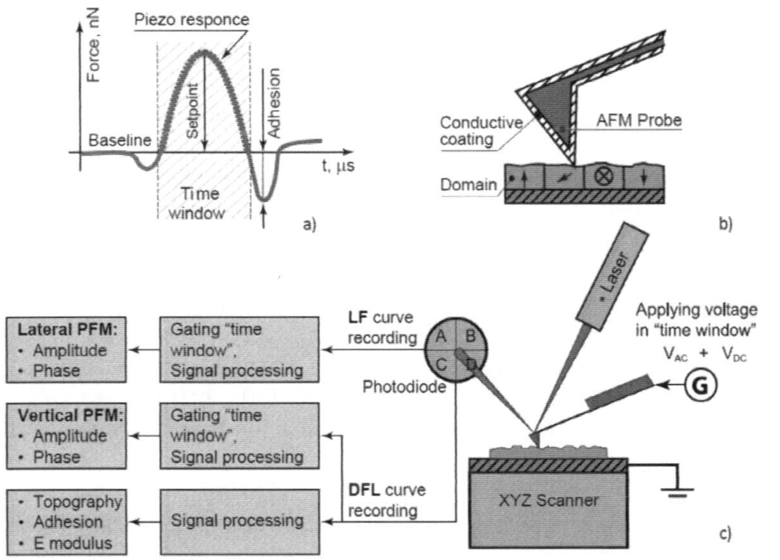

Figure 2. Working principle of HD PFM: a) an idealized temporal deflection curve during an oscillatory cycle; b) tip-sample interaction scheme; c) piezoresponse detection scheme.

3. Results and discussion

Above discussed HD PFM was implemented using Hybrid 2.0 control electronics (NT-MDT S.I., Moscow, Russia). Two types of biopiezoelectrics was studied: type I collagen matrix and diphenylalanine peptide nanotubes using VEGA atomic force microscope (NT-MDT S.I., Moscow, Russia)

3.1. Study of type I collagen matrix

One of the most interesting biomaterials demonstrating piezoelectricity are collagen fibrils – the main building component of bones, teeth, corneal stroma and blood vessels. Collagen is composed of aligned polar protein molecules (fibrils) [6] that form a strong organic crystalline matrix. Piezoelectric properties of single collagen fibrils were recently studied with nanometer-level resolution [10,11]. Despite this advance, the scientific challenge of any piezoresponse study of collagen matrix is still tricky. The main issue for traditional contact PFM measurements of these structures is complicated by rough surfaces where there is height variation of almost one micrometer. This makes AFM tip to cling to the single fibrils, and vice versa, highly distorting the topography and piezoresponse images.

Due to its working principle, HD PFM mode was found to be a superior technique for studying collagen matrix. The samples were provided by Fibralign Corporation, where a unique technique of depositing animal collagen on glass substrate was developed [12]. Figure 3 illustrates results of HD PFM study of type I collagen matrix isolated from bovine corneas.

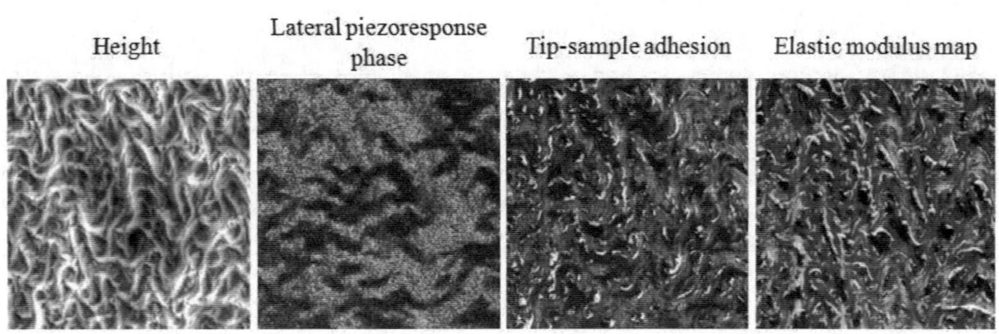

Figure 3. HD PFM images of animal collagen's matrix. Images size is 15×15 um. Obtained using NT-MDT S.I. Etalon HA_HR probes with force constant of 14N/m.

Measurements were performed with 280 kHz AC voltage of 8 V applied to conductive tip of Etalon HA_HR/W2C+ probe with force constant 14 N/m (NT-MDT S.I., Moscow, Russia). The contrast in height covers variation in 0-800 nm range. Lateral PFM phase demonstrates the domain distribution, darker and brighter areas correspond to the different direction of polarization. Adhesion demonstrates tip-sample adhesion force, brighter areas correspond to higher adhesion. E modulus in demonstrates collagen matrix E modulus distribution, brighter areas correspond to higher E modulus.

3.2. Study of peptide nanotubes

Peptide nanotubes (PNTs) self-assembled from diphenylalanine monomers were recently discovered to exhibit strong piezoelectric properties. Kholkin et al. demonstrated in-plane PFM contrast and high effective d_{15} piezoelectric coefficient of at least 60 pm/V [13] (for tubes 200 nm in diameter) which is the highest value for known biopiezoelectrics. Together with intrinsic biocompatibility this makes PNTs to be very promising materials for developing piezonanodevices that are potentially compatible with human tissue.

PNTs are challenging samples for traditional contact PFM investigation due to their fragility and loose contact with a substrate. However, HD mode was recently demonstrated to be nondestructive method for investigation of thin PNTs (around 30 nm in diameter) in terms of obtaining information on topography, adhesion and Young's modulus [9]. Therefore, it was logical to apply HD PFM to these fragile samples in order to measure piezoresponse of PNTs to demonstrate advantages of this mode. Figure 4 demonstrates topography, lateral PFM phase, tip-sample adhesion force and long-distance electrostatic forces map.

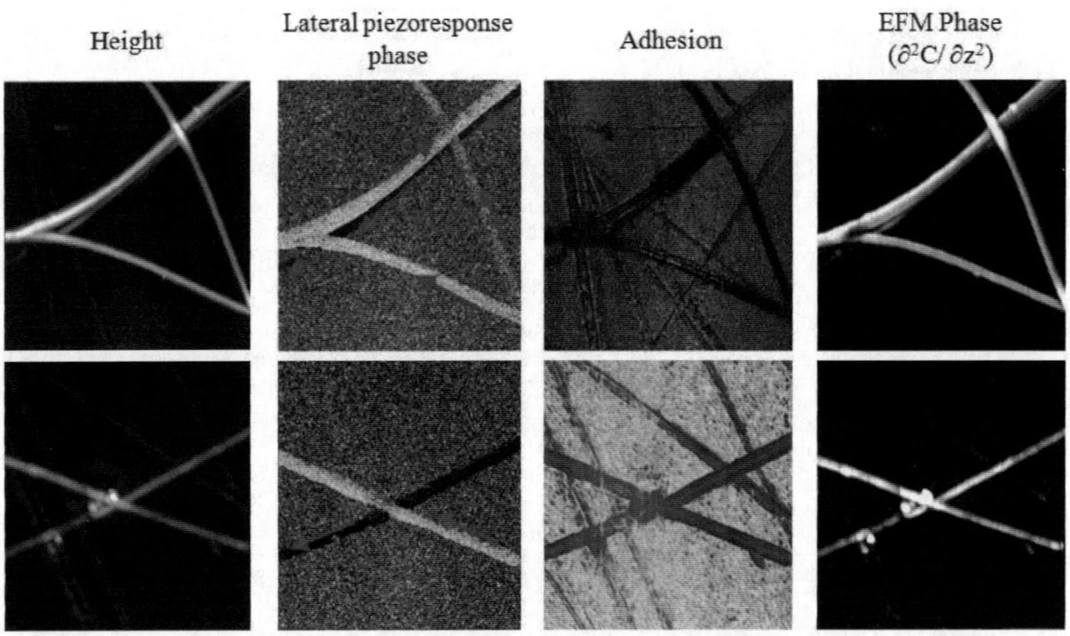

Figure 4. HD PFM images of peptide nanotubes. Arrows demonstrate the direction of polarization. Scan size 8×8 um (top row) and 7×7 um (bottom row). Sample courtesy: Prof. V.Y. Shur, S.G. Vasiliev UCSU "Modern Nanotechnologies" UFU; A.L. Kholkin, University of Aveiro.

Measurements were performed with 100 kHz AC voltage of 10 V applied on conductive tip coating of Etalon HA_FM/W2C+ probe with force constant of 3 N/m ((NT-MDT S.I., Moscow, Russia)). Lateral PFM phase also demonstrates PNTs with opposite polarization direction, that is marked by arrows.

4. Conclusion

The novel approach for non-destructive electromechanical studies of soft and fragile piezoelectrics is demonstrated. It has been successfully applied for simultaneous investigation of topography and local piezoresponsese, adhesion and E modulus of challenging biological samples – corneal stoma collage matrix and self-assembled diphenylalanine peptide nanotubes.

References

[1] Güthner P and Dransfeld K 1992 *Appl. Phys. Lett.* **61** 1137–9
[2] Leuchtag H R 2008 *Voltage-sensitive ion channels: Biophysics of molecular excitability* (Dordrecht: Springer Netherlands)
[3] Leuchtag H R and Bystrov V S 1999 *Ferroelectrics* **220** 157–204
[4] Amdursky N, Beker P, Schklovsky J, Gazit E and Rosenman G 2010 *Ferroelectrics* **399** 107–17
[5] Hladky S B and Haydon D A 1970 *Nature* **225** 451–3
[6] Athenstaedt H 1974 *Ann. N. Y. Acad. Sci.* **238** 68–94
[7] Lemanov V V, Popov S N and Pankova G 2002 *Phys. Solid State* **44** 1929–35
[8] Lemanov V V, Popov S N and Pankova G 2003 *Ferroelectrics* **285** 207–16
[9] Montenegro J, Vázquez-Vázquez C, Kalinin A, Geckeler K E and Granja J R 2014 *J. Am. Chem. Soc.* **136** 2484–91
[10] Minary-Jolandan M and Yu M-F 2009 *Nanotechnology* **20** 85706
[11] Minary-Jolandan M and Yu M F 2009 *ACS Nano* **3** 1859–63
[12] Muthusubramaniam L, Peng L, Zaitseva T, Paukshto M, Martin G R and Desai T A 2012 *J. Biomed. Mater. Res. - Part A* **100 A** 613–21
[13] Kholkin A, Amdursky N, Bdikin I, Gazit E and Rosenman G 2010 *ACS Nano* **4** 610–4

IOP Conf. Series: Journal of Physics: Conf. Series **917** (2017) 042019 doi:10.1088/1742-6596/917/4/042019

The importance of the biomimetic composites components for recreating the optical properties and molecular composition of intact dental tissues.

P V Seredin[1], D L Goloshchapov[1,3], M S Gushchin[1], Y A Ippolitov[2], T Prutskij[4]

[1]Department of Solid State Physics and Nanostructures, Voronezh State University, Voronezh, Russia

[2]Department of Pediatric Dentistry with Orthodontia, Voronezh State Medical University, Voronezh, Russia

[3]Department of Physics, Voronezh State Technical University, Voronezh, Russia

[4]Instituto de Ciencias, Benemérita Universidad Autónoma de Puebla, Puebla, Mexico

Abstract. The objective of this paper was to investigate whether it is possible to obtain biomimetic materials recreating the luminescent properties and molecular composition of intact dental tissues. Biomimetic materials were produced and their properties compared with native dental tissues. In addition, the overall contribution of the organic and non-organic components in the photoluminescence band was investigated. The results showed that it is possible to develop biomimetic materials with similar molecular composition and optical properties to native dental tissues for the early identification of dental caries.

1. Introduction

The discovery and detection of carious lesions is the starting point in dentistry. The benefits of the use of infrared, Raman and luminescent spectroscopy to differentiate intact and carious tooth enamel and dentin have been shown, but these methods do not provide information about the types of disturbance present in the tooth matrix. The present study investigates tooth and biomimetic composites (BC), demonstrating the importance of biomimetic composite components for recreating the optical properties and molecular composition of intact dental tissues.

First, several ionic components (L-arginine hydrochloride, L-histidine, L-lysine hydrochloride and hyaluronic acid) were introduced into the biocomposite via the synthesis of biomaterials using calcium hydroxyapatite. Based on the results of our previous studies in which other amino acids were used to study the molecular composition of the samples [1,2], it was necessary to use all of the selected ionic components together to recreate the molecular composition of intact dentin as in [3].

2. Materials and methods

2.1. Preparing dental tissue samples

Voronezh State University Ethics Committee approved of this study (approval number 001.027-2016).

In order to compare the characteristics of the native dental tissues and biomimetic materials modeling their composition, 10 samples of intact enamel and 10 samples of intact dentin were investigated. The samples (cuts) were extracted from clinically healthy teeth that were removed from patients at the age of 18-25 according to the orthodontic indications. The cuts were prepared in the following manner. Originally the teeth were washed in the running water, mechanically cleared of dental plaque, then their surface was dried using filter paper. After cleaning stage, the teeth were divided at a specialized setup with a diamond disk using water cooling into the plates with the thickness of up to 1 mm. Small rotation speed of the disk was used to make sure the samples were not overheated. The resulting microsections of the teeth were polished with the use of a diamond paste. Then the microsections of the teeth were once again washed and cleaned.

2.2. Preparing biomimetic samples

The inorganic component of the elaborated biocomposite is a nanocrystalline carbonate-substituted calcium hydroxyapatite (C-HAP). It was obtained with the use of an organic source of calcium (egg shell) according to the methods described previously [4]. The morphological arrangement of C-HAP synthesized using this technology is identical to the morphology of human hard dental tissue and is necessary to obtain biomimetic samples capable of supplying the natural biogenic C-HAP [5–7]. The organic component of the synthesized biocomposite is a mix of polar amino acids found in the enamel dental tubule, L-arginine hydrochloride, L-histidine, L-lysine hydrochloride and hyaluronic acid, with nanocrystalline C-HAP in an isotonic solution in a proportion typical of native human dental tissue. Note that the ratios of the organic and inorganic components were taken as 5/95, 25/75, 40/60, thus, approximately reproducing the composition of human tooth enamel and dentin.

2.3. Research methods

The diffractometer studies (X-ray diffraction using a Seifert XRD 3003 PTS-HR) and IR-spectroscopy data (diamond PLATINUM ATR, Vertex-70 (Bruker, Germany)), shows that the defined biocomposite composition used yielded results in good agreement with those in [1–3].

Luminescence spectra were obtained at the room temperature from the samples surface according to the technique described in [8] with the use of the unit supplied with TRIAX 550 monochromator and CCD detector. The laser output power was of 30 mW.

3. Results

3.1. IR-data

Figure 1 shows IR-spectra of synthetical materials and native dental tissue.

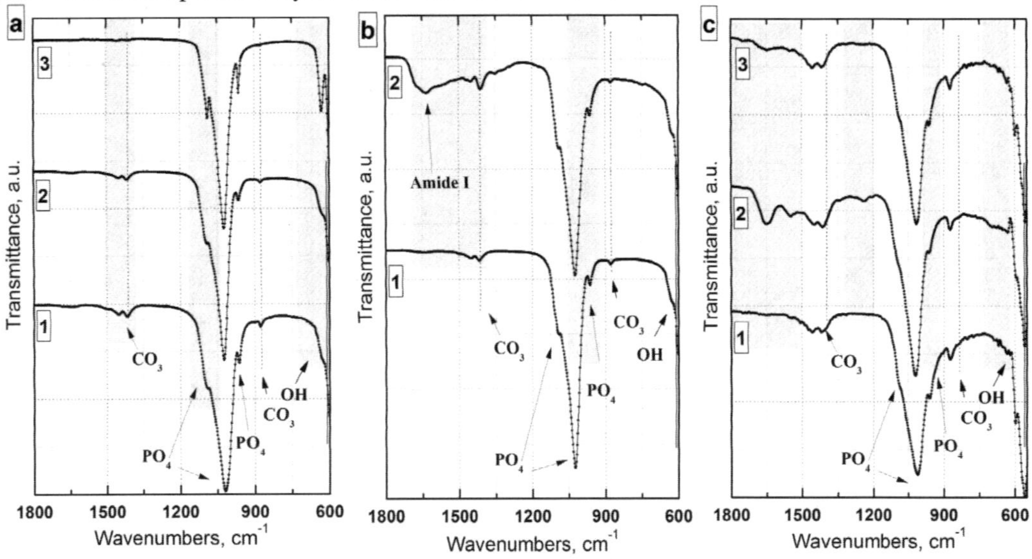

Figure 1. The FTIR spectra of: **(a)** 1,2 - HAP (nanocrystalline), 3 - HAP (microcrystalline); **(b)** 1 - nanocrystalline HAP, 2 – BC 25/75; **(c)** 1- sound enamel, 2 -sound dentin, 3 - carious enamel.

IOP Conf. Series: Journal of Physics: Conf. Series **917** (2017) 042019 doi:10.1088/1742-6596/917/4/042019

Figure 1a, shows the transmission IR-spectra of two nanocrystalline samples of C-HAP with different amounts of a carbonate anion (spectra 1, 2). The various CO_3 content can be seen from the comparison of vibration band intensities in the region 1400 – 1450cm^{-1}, which are related to a B-type carbonate substitution in C-HAP. Figure 1a also presents the IR-spectrum of a microcrystalline HAP (curve 3). Figure 1b shows the transmission IR-spectrum of the biocomposite with the ratio of organic/non-organic component 25/75 (curve 2) and an IR-spectrum of the HAP used to synthesize it (curve 1). Figure 1 shows the IR-spectra of native dental tissues: 1 sound enamel, 2 sound dentin, 3 carious enamel. Note that the vibration bands from the amide groups in the biocomposite spectra in the range of 1400 – 1800 cm^{-1} are related to vibration modes of the organic component. These vibration modes are identical to similar vibration modes of native dental tissues, see figure 1c, curves 1, 2, 3. Comparison of the data shown in figures 1a and b with the results of figure 1c makes it possible to evaluate the correspondence of the molecular composition of the biocomposite to the composition of native tissue as well as to identify its structural disorganization in carious processes. It should be noted that all IR-spectra of biogenic samples are characterized by the redistribution of the intensity of the vibration band intensity in the range 960 – 1100 cm^{-1}.

3.2. Data of laser induced luminescence

The laser-induced luminescence data confirms the differences in the composition between the samples of micro- and nanocrystalline HAP and native dental tissues that have been identified using IR-spectroscopy. The study of the contribution of the organic and non-organic components in the photoluminescence band showed that certain contributions of different organic components are found in a specific region of the photoluminescence band. Note that this type of research has already been described in detail in [9].

The comparative analysis of all spectra figure 2a, b, c shows that synthetic HAP, biocomposite and biological materials have a wide luminescence band in the range of 540 – 620 nm, which agrees with published data for dental enamel and dentin. Ideally, biomimetic materials should be structurally arranged like enamel and tooth dentin, so the luminescence spectra of nanocrystalline HAP (different CO_3 concentration) - figure 1a, curve 1, 2 and microcrystalline HAP - figure 1a, curve 3 samples were compared to sound enamel -figure 1c, curve 1, 2, sound dentin - figure 1c, curve 3 and carious enamel - figure 1c, curve 3.

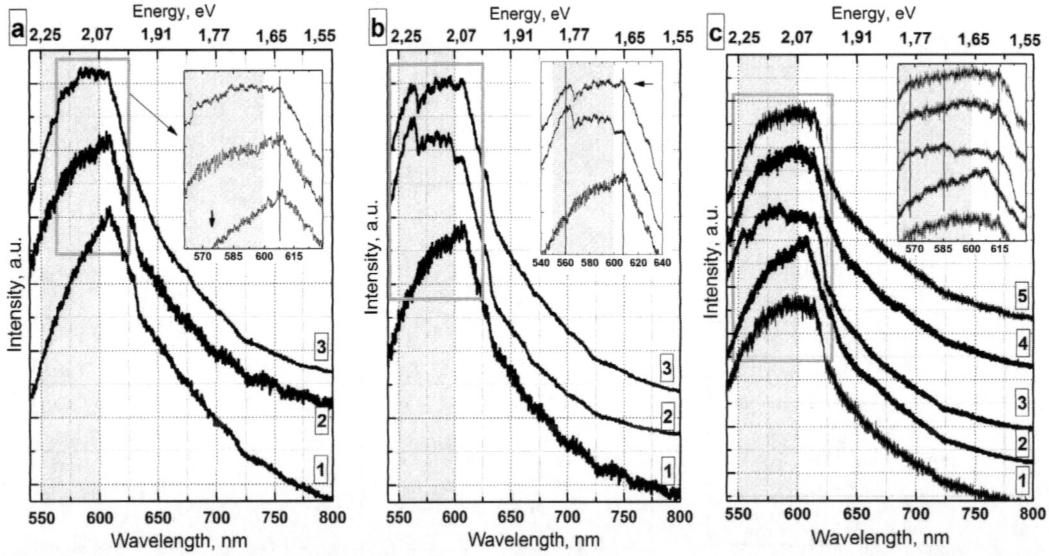

Figure 2. The luminescence spectra of: **(a)** 1,2 - HAP (nanocrystalline), 3 - HAP (microcrystalline); **(b)** 1 - nanocrystalline HAP, 2 - organic component of BC, 3 — BC 25/75; **(c)** 1,2 - sound enamel, 3 - sound dentin, 4,5 - carious enamel.

4. Discussion

4.1. Features of IR-spectra of the transmission of the samples

Comparison of the IR-spectra of the HAP samples presented in figure 1 allows one to identify special features of the enamel associated with stoichiometry and periodic structure of the enamel hydroxyapatite. The size of hydroxyapatite nanocrystals is known to vary from 70 to 500 nm in one direction of growth in native dental tissue [4,10] and was confirmed by the redistribution of the band intensity (maxima) in the IR-spectra in the range from 960 to 1100 cm^{-1}. Note that nanocrystalline HAP with a crystal length of 50 – 70 nm was used for the synthesis of the biocomposite sample. The IR-spectrum of the nanocrystalline HAP - curve 1, and the biocomposite with the ratio of organic/non-organic component of 25/75 are shown in figure 1b. Clearly, the IR-spectra of native tissues are in excellent agreement with those of the microcrystalline HAP sample - figure 1a, curve 3. At the same time in the IR-spectrum of dentin, the ratio of intensities in the range from 960 to 1100 cm^{-1} experiences changes and corresponds with the nanocrystalline HAP sample, with the entire spectrum being consistent with that of the biocomposite - figure 1b, curve 2. The same is observed in the IR-spectra of caries-damaged enamel - figure 1c, curve 3, but is caused by structural disorganization of the enamel tissue and a reduction in its crystallinity, which is consistent with the previous research [11].

Note that while comparing curve 1 in figure 1b (the biocomposite spectrum) and curves 1, 2, and 3 in figure 1c (native dental tissues), the position and intensity of vibratory modes from the organic component are found to correspond. This indicates that ionic components, L-arginine hydrochloride, L-histidine, L-lysine hydrochloride and hyaluronic acid, could be employed to recreate the molecular composition of enamel and dentin to develop biomimetic materials. In addition, as suggested by the IR-spectroscopy data, the effect of the organic component is less pronounced in biocomposites with a high level of HAP [3].

4.2. Features of laser-induced luminescence

The results of the analysis of the luminescence spectra - figures 2a, b, c show the importance of using C-HAP as the non-organic component as well as the organic component to develop materials for recreating the optical properties of human teeth.

Regarding hydroxyapatite, note that the change in the shape of the luminescence band from nano- to microcrystalline HAP - figure 2a, curves 1 & 3 appears to be caused by the structural perfection and dimensional weight of apatite. In the present report, a theoretical and experimental study was performed to compare the structure of tooth enamel and dentin microcrystals with the synthetic nanocrystalline HAP samples [11,12], indicating that the synthetic material and biogenic tissues are identical in size. Thus, note that luminescence of the healthy enamel behaves identically to HAP - figure 2c, curves 1, 2. The change in the shape of the maximum luminescence band for intact enamel is due to its complex hierarchical structure. The luminescence spectrum 1 in figure 2c obtained from intact enamel is more consistent with the spectrum of microcrystalline HAP in figure 2a, curve 3, while spectrum 2 in figure 2c also obtained from enamel is consistent with the luminescence spectrum of nanocrystalline HAP - figure 2a, curve 1. Enamel in differently-oriented crystals and sizes is in agreement with the available data. However, with regard to the material structure and its effect on the photoluminescence spectrum, it is important to note that the data from [13] indicated that the maximum luminescence band of a defect-free hydroxyapatite is within another spectrum. Changes in the local atomic structure of HAP have a considerable effect on the position of the luminescence band as the shape and size of the HAP crystals remain the same. The study of the luminescence of nanocrystals and microspheres of the hydroxyapatite [14,15] indicates that the change in the position of the maximum luminescence is due not only to the shape and size of the HAP, but is related to defects in the HAP crystals (i.e., with the local atomic structure of the HAP). Hence, considering the composition of the apatite of tooth dentin as well as HAP obtained using the liquid-phase method, defects in the HAP structure have more effect on the shape and position of the luminescence band

[13,16–18]. It should also be noted that the materials were synthesized with a microelement composition identical to that of enamel and dentin, having the essential structural B-defect (replacement of the PO_4 and CO_3 groups). It has been suggested that this defect is responsible for changes in the intensity of the luminescence band in such samples. In [19], there is much discussion of how many defects in the human dental enamel are associated with the replacement of HAP atoms by foreign ions, despite good crystallization of the HAP crystals of human dental enamel.

The analysis of the shape and features of the luminescence spectrum of the synthesized biocomposite - figure 2b, curve 3 allows one to identify the contribution of C-HAP - figure 2b, curve 1) as well as the introduced organic substances - figure 2b, curve 2. A detailed investigation of the resulting data indicates that the typical features in the luminescence spectra of native dentin - figure 2c, curve 3 are present in the spectra of native tissue and biomimetic materials, and that the highlighted areas - figure 2a, b, c can be used to identify similarities and differences between native tissues and synthesized biocomposite.

The comparative analysis of the spectra of native sound and caries tissue as well as that of the biocomposite - figure 2b, c shows that the luminescence largely depends on the type of structural defect and quantitative ratio of the organic and non-organic component in intact dental tissues. Changes in the regions of 550 – 575 and 600 – 625 nm for the biocomposite luminescence spectra suggest that the changes in luminescence of native tissue are due to structural disorganization and demineralization processes, including tooth organic transformation. Therefore, it can be assumed that the luminescence largely depends on the type of structural defect and quantitative ratio of the organic and non-organic component in intact dental tissues and less on the hierarchy.

5. Conclusion

The identified features in the IR and luminescence spectra in the HAP samples with varying stoichiometry and dimensionality, biocomposite and native human dental tissue can be employed to develop the technology for biomimetic materials with a similar molecular composition and optical properties to native dental tissues. These findings indicate that such research can assist in the development of a highly efficient method for the identification of dental caries at an early stage.

Acknowledgments
This work was supported by the grant of Russian Science Foundation, grant number 16-15-00003.

References

[1] Jahromi M T, Yao G and Cerruti M 2013 The importance of amino acid interactions in the crystallization of hydroxyapatite *J. R. Soc. Interface* **10** 1–14
[2] Lee W-H, Loo C-Y, Van K L, Zavgorodniy A V and Rohanizadeh R 2012 Modulating protein adsorption onto hydroxyapatite particles using different amino acid treatments *J. R. Soc. Interface* **9** 918–27
[3] Seredin P V, Goloshchapov D L, Kashkarov V M, Ippolitov Y A and Prutskij T 2016 Emission properties of biomimetic composites for dentistry *Results Phys.* **6** 447–8
[4] Goloshchapov D L, Kashkarov V M, Rumyantseva N A, Seredin P V, Lenshin A S, Agapov B L and Domashevskaya E P 2013 Synthesis of nanocrystalline hydroxyapatite by precipitation using hen's eggshell *Ceram. Int.* **39** 4539–49
[5] Liu Q, Huang S, Matinlinna J P, Chen Z and Pan H 2013 Insight into Biological Apatite: Physicochemical Properties and Preparation Approaches *BioMed Res. Int.* **2013** 1–13
[6] Palmer L C, Newcomb C J, Kaltz S R, Spoerke E D and Stupp S I 2008 Biomimetic Systems for Hydroxyapatite Mineralization Inspired By Bone and Enamel *Chem. Rev.* **108** 4754–83
[7] Frank R M, Klewansky P, Hemmerle J and Tenenbaum H 1991 Ultrastructural demonstration of the importance of crystal size of bioceramic powders implanted into human periodontal lesions *J. Clin. Periodontol.* **18** 669–80

[8] Seredin P, Goloshchapov D, Prutskij T and Ippolitov Y 2015 Phase Transformations in a Human Tooth Tissue at the Initial Stage of Caries *PLoS ONE* **10** 1–11

[9] Bachmann L, Zezell D M, Ribeiro A da C, Gomes L and Ito A S 2006 Fluorescence Spectroscopy of Biological Tissues—A Review *Appl. Spectrosc. Rev.* **41** 575–90

[10] Gadaleta S J, Mendelsohn R, Paschalis E L, Camacho N P, Betts F and Boskey A L 1995 Fourier Transform Infrared Spectroscopy of Synthetic and Biological Apatites *Mineral Scale Formation and Inhibition* ed D Z A M.Sc (Springer US) pp 283–94

[11] Seredin P, Kashkarov V, Lukin A, Ippolitov Y, Julian R and Doyle S 2013 Local study of fissure caries by Fourier transform infrared microscopy and X-ray diffraction using synchrotron radiation *J. Synchrotron Radiat.* **20** 705–10

[12] Xue J, Zhang L, Zou L, Liao Y, Li J, Xiao L and Li W 2008 High-resolution X-ray microdiffraction analysis of natural teeth *J. Synchrotron Radiat.* **15** 235–8

[13] Anon Annual Report Photon Science DESY - 20101246.pdf

[14] Zhang C, Yang J, Quan Z, Yang P, Li C, Hou Z and Lin J 2009 Hydroxyapatite Nano- and Microcrystals with Multiform Morphologies: Controllable Synthesis and Luminescence Properties *Cryst. Growth Des.* **9** 2725–33

[15] Dai C, Duan J, Zhang L, Jia G, Zhang C and Zhang J 2014 Biocompatibility of Defect-Related Luminescent Nanostructured and Microstructured Hydroxyapatite *Biol. Trace Elem. Res.* **162** 158–67

[16] Kataoka T, Shiba K and Tagaya M 2016 Preparation of europium(III)-doped hydroxyapatite nanocrystals in the presence of cationic surfactant *Colloid Interface Sci. Commun.* **13** 1–5

[17] Jiang F, Wang D-P, Ye S and Zhao X 2014 Strontium-substituted, luminescent and mesoporous hydroxyapatite microspheres for sustained drug release *J. Mater. Sci. Mater. Med.* **25** 391–400

[18] Nishikawa H 2001 Thermal behavior of hydroxyapatite in structural and spectrophotometric characteristics *Mater. Lett.* **50** 364–70

[19] Ghadimi E, Eimar H, Marelli B, Nazhat S N, Asgharian M, Vali H and Tamimi F 2013 Trace elements can influence the physical properties of tooth enamel *SpringerPlus* **2** 499–512

Application of the spectral-correlation method for diagnostics of cellulose paper

D Kiesewetter[1], V Malyugin[2], A Reznik[1], A Yudin[2], N Zhuravleva[1]
[1]Institute of Power Engineering and Transportation, Peter the Great Saint Petersburg Polytechnic University, Saint Petersburg 195251, Russia
[2]Institute of Physics, Nanotechnology and Telecommunications, Peter the Great Saint Petersburg Polytechnic University, Saint Petersburg 195251, Russia

Abstract. The spectral-correlation method was described for diagnostics of optically inhomogeneous biological objects and materials of natural origin. The interrelation between parameters of the studied objects and parameters of the cross correlation function of speckle patterns produced by scattering of coherent light at different wavelengths is shown for thickness, optical density and internal structure of the material. A detailed study was performed for cellulose electric insulating paper with different parameters.

1. Introduction

Determination of physical properties of biological objects by contact methods is difficult or impossible in many cases. Therefore, the application of non-destructive contactless methods, in particular, spectral-correlation method [1-2], for diagnostics of biological objects is very important. Of particular importance in high voltage technique has electrical insulation based on cellulose of plant origin, as well as prospective bacterial cellulose. The use of spectral-correlation method gives the opportunity to measure some of the parameters of different biological objects and materials in the process of exposure to different factors, in particular, in the artificial aging. We can assume that the use of this method will expand the possibilities of the study of biological objects and materials and will improve their performance.

2. Basic principles of the spectral-correlation method

Spectral correlation method of diagnosis is based on finding statistical regularities of speckles produced by scattering of coherent light at changing the wavelength (λ) of the probing radiation. Cross-correlation functions (CCF) of the intensity distributions of the scattered radiation at different wavelengths [1-2] were calculated for this purpose:

$$CCF(\Delta x, \Delta y) = \frac{1}{4x_s y_s} \int_{y_0-y_s}^{y_0+y_s} \int_{x_0-x_s}^{x_0+x_s} I_{\lambda 1}(x, y) I_{\lambda 2}(x-\Delta x, y-\Delta y) dxdy \quad (1)$$

where x_0, y_0, x_s, y_s – are the coordinates of the centre of integrating area and halfwidth of the area, $I_{\lambda 1}$, $I_{\lambda 2}$ are the intensity distributions at wavelengthes λ_1 and λ_2. At usage of the CCD camera with the size of the matrix 736x572 pixels, the values x_s and y_s were chosen 50 pixels. It has allowed us consider speckles within the limits of the area of integrating like isotropic, and too, quantity of speckles are sufficient for correct calculation of the CCF.

It is known that the spots of speckles of a scattered radiation when changing the wavelength of the radiation acquire directional movement and the transformation of the shape [2]. Directed movement depends on the angle of scattering and does not depend on the parameters of the scatterer. Therefore, it is of not interesting for this work.

Transformation of speckles ("boiling speckles") depends on the optical characteristics of the scattering object. In order to separate these two effects, the search of the maximum value of CCF (by two parameters – Δx, Δy) at each wavelength (λ_2) was used. The dependences CCF_{max} on the wavelength can be considered as independent on the scattering angle, respectively, depending only on the scattering properties of the investigating surface.

The CCF and CCF_{max} parameters (the width and form of dependence) are characteristics of optical inhomogeneities of the investigated objects. The more the longitudinal typical size of the optical inhomogeneities of the object, the more narrow characteristic range of variation of the CCF_{max}. There are other methods of processing of imaging of the speckles, such as [3-4].

3. Experimental setup

In this work, semiconductor laser ($\lambda \approx 650$ nm) was used as the radiation source. The wavelength tuning is carried out by changing the temperature of the heterojunction using the Peltier element [1, 6]. The CCD television camera was used to register distribution of intensity of radiation scattered by the investigated sample. The angle of incidence of the probing beam relative to the normal to the surface of the paper was 10 - 30 degrees. The intensity distribution of diffuse scattered radiation is produced approximately in the direction of specular reflection. Processing of the obtained images was performed with a personal computer connected to the television camera. The gauge dependence of λ on temperature T was measured experimentally using the spectrometer "Avantes-20148". The degree of polymerization of the cellulose was determined for aged samples of the paper, and images of the fiber structure (Fig. 1b) were obtained using a scanning electron microscope (SEM).

4. Data obtained

The CCF study were conducted on samples of insulating paper of various types (made from pulp of coniferous trees, as well as bacterial cellulose) and thickness of both new and artificially aged according to the method [5]. The paper samples with different density were investigated also. An example of the obtained dependences is shown in Figure 1a.

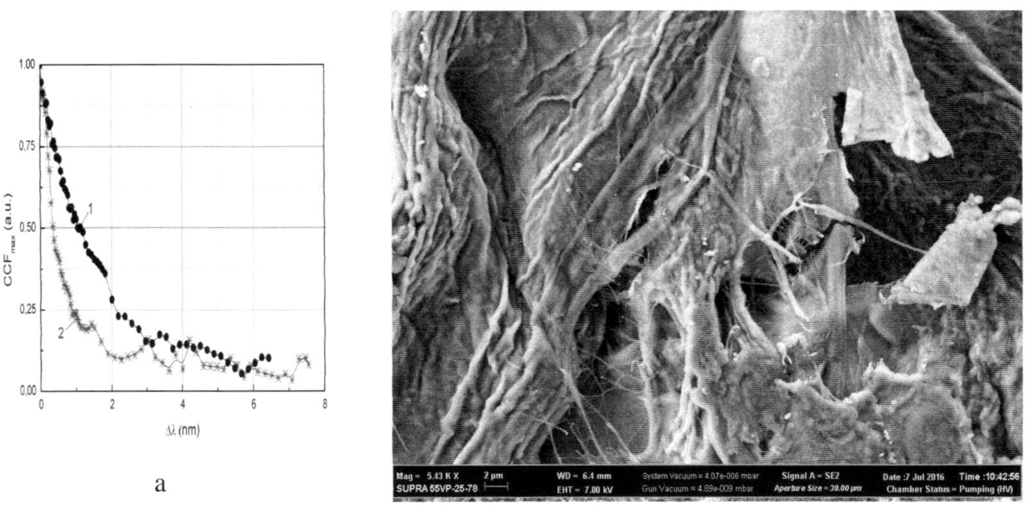

a b

Figure 1 (a, b). (a) CCF_{max} for insulating paper thickness of: 1 – 40 μm, 2 – 175 μm; **(b)** The structure of the paper after thermal aging for about 240 hours at 140^0 C. The destruction of the structure of the paper is visible.

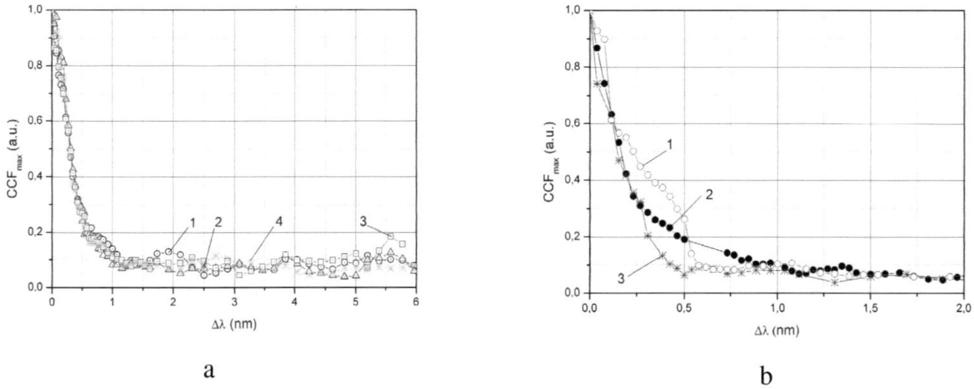

Figure 2 (a, b). (a) CCF$_{max}$ for insulating paper K140 after thermal aging for about 240 hours at 140^{0}C: 1–3 – three measurements in the same area, 4 – measurement at the different area of the sample; **(b)** The change of the CCF at the thermal aging process of insulation paper: 1 – new (befor thermal aging), 2 – after 170 h, 3 – after 530 h.

The half-width of the dependence CCF (Figure 1a) for the paper with thickness of 40 μm is substantially greater than that for the paper with thickness of 175 μm, which confirm the regularity mentioned above. However, the ratio of the obtained values of half-width of CCFs is significantly smaller than the ratio of the sample thickness of the papers.

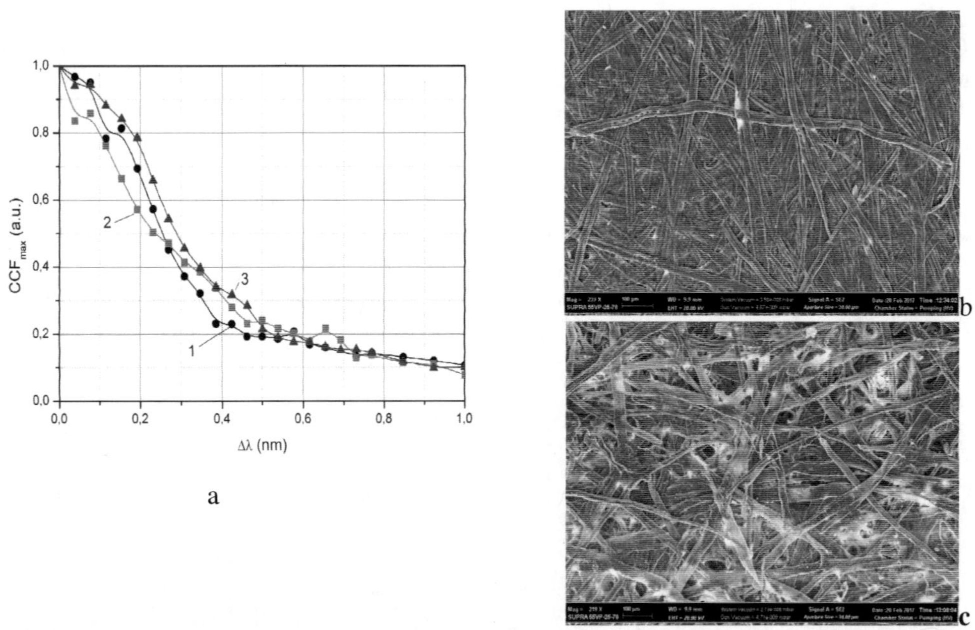

Figure 3 (a, b). (a) CCF$_{max}$ for new (befor thermal aging) insulating paper: 1 – K140 (Rus), 2 – DLZ-130 (produced by Henglong Plastics, China), 3 – experimental sample with a reduced density (ESRD) of the cellulose fibres; **(b, c)** The structure of the paper of the paper obtained using SEM: **(b)** – DLZ-130, **(c)** – ESRD.

IOP Conf. Series: Journal of Physics: Conf. Series **917** (2017) 042020 doi:10.1088/1742-6596/917/4/042020

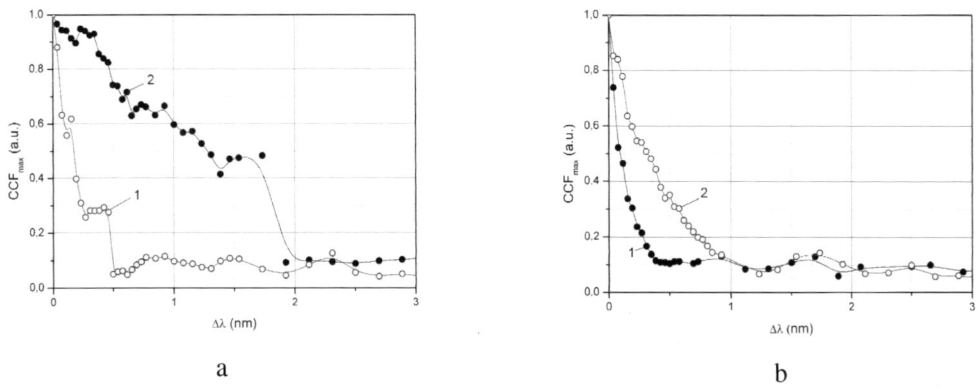

Figure 4 (a, b). (a) CCF_{max} of insulating (1) and electrically conductive (2) cable paper; **(b)** CCF_{max} of the dry (1) and oil-impregnated (2) insulation paper.

This may be explained by the penetration of light not the entire thickness of the paper. We have investigated the reproducibility of measurement of CCF_{max} to assess the reliability of the obtained results (Figure 2). For these purposes, distribution of scattered intensity on the same area of illuminated paper surface (type K140) was three times measured at 50 – 60 wavelengths. The CCFs were determined based on the distributions. The measurement of the CCF for the other area of the surface of the paper was carried out also. The calculated CCF_{max} are shown in the Figure 2. As follows from the graphs, there is good reproducibility of the results. Therefore, at case of a significant difference between the CCF parameters of different samples the parameters of the paper can be considered as different. Figure 3 shows the dependence of CCF (Figure 3a) on the structure of the paper. We also fulfilled a comparison of CCF for the scattered radiation of the cable paper K140 (light brown) and the electrically conductive cable paper with the carbon filler in the form of soot (black). A large half-width of the CCF_{max} of the black paper caused by the greater magnitude of the absorption of the radiation in the paper, respectively, the smaller the penetration depth of the radiation (Figure 4a). A significant difference of CCF_{max} has dry and oil-impregnated insulation paper (Figure 4b).

The obtained data confirm the basic principles mentioned above: the relationship of thickness, density and structure of the paper, as well as the parameters of the impregnation composition with the parameters of the CCF. The results showed the possibility of using spectral correlation method for the diagnosis of paper-impregnated high-voltage insulation, in particular insulation of power transformers, as well as other optically heterogeneous biological objects.

References

[1] Boldyreva E O, Kiesewetter D V, Malyugin V I, Modanov A V 2008 *Proc. SPIE* **7006** 700618

[2] Malyugin V I, Kiesewetter D V, Boldyreva E O, Modanov A V *Technical Physics Letters* 2008 **34** 754

[3] Francon M 1978 *La granulate laser (speckle) et ses applications en optique* Masson Paris NewYork Barcelone p 162

[4] Jacquot P. Blackwell Publishing Ltd *Strain* 2008 **44** 57

[5] Zhuravleva N, Reznik A, Kiesewetter D, Tukacheva A, Smirnova E *Proc. 57th International Scientific Conference on Power and Electrical Engineering of Riga Technical University* 2016 1

[6] Kiesewetter D V, Malyugin V I, Ilyin N V, Changsen Sun *Proc. Laser Optics-2016, St. Petersburg,* **2016 5**

IOP Conf. Series: Journal of Physics: Conf. Series **917** (2017) 042021 doi:10.1088/1742-6596/917/4/042021

Applicability of submerged jet model to describe the liquid sample load into measuring chamber of micron and submillimeter sizes

A L Bulyanitsa[1,2], K I Belousov[3,4], A A Evstrapov[1,3,4]
[1] Laboratory of information and measurement biosensor and chemosensor microsystems, Institute for Analytical Instrumentation RAS, St. Petersburg 198095, Russia
[2] Department of Higher Mathematics, Peter the Great St. Petersburg Polytechnic University, St. Petersburg 195251, Russia
[3] Department of nanophotonics and metamaterials, ITMO University, St. Petersburg 197101, Russia
[4] Department of biophysics, Siberian Federal University, Krasnoyarsk 660041, Russia

Abstract. The load of a liquid sample into a measuring chamber is one of the stages of substance analysis in modern devices. Fluid flow is effectively calculated by numerical simulation using application packages, for example, COMSOL MULTIPHYSICS. In the same time it is often desirable to have an approximate analytical solution. The applicability of a submerged jet model for simulation the liquid sample load is considered for the chamber with sizes from hundreds micrometers to several millimeters. The paper examines the extent to which the introduction of amendments to the jet cutting and its replacement with an energy equivalent jet provide acceptable accuracy for evaluation of the loading process dynamics.

1. Introduction
Mass transfer in laminar flow with transverse Reynolds numbers less than 1 is often used during the liquid injection from a channel with small cross-section to a large tank in devices for the analysis of liquid samples. Understanding the dynamics of the input is very important, as a chemical reaction starts at the moment of the sample load in a number of analytical schemes [1]. If measuring chamber has an aspect ratio (diameter/depth) of 10 or more, submerged jet scheme can be applied only at the initial stage, when the expanding jet is entirely placed within the chamber boundaries. Otherwise, the jet will be "cut" in the vertical plane. This effect can be taken into account. When the liquid front becomes close to the chamber wall, the fundamental limitation of the model - the principle of momentum flux conservation - breaks down by the addition of the momentum reflected from the wall. The problem of the jet dynamics into a submerged space strictly mathematically formulated and solved as a first approximation by L.D. Landau [2]. We have proposed a scheme of conversion of "vertically cut" jet into energy equivalent one with another geometrical parameters.

2. Methods

2.1 Analytical solution of submerged jet
According to [2] the radial flow velocity vector component is:

$$v_r = \frac{1}{r} F(\theta) \tag{1}$$

$$F(\theta) = 2\nu \left(\frac{A^2 - 1}{(A - \cos(\theta))^2} - 1 \right) \tag{2}$$

where θ is the angle from the jet axis, r is the radial displacement, ν is the kinematic viscosity. Parameter A is the constant of integration determined by the radial component of the momentum flux tensor in the jet:

$$\Pi_{rr} = p + \rho V_r V_r \tag{3}$$

where p is the pressure, ρ is the density of the liquid. For complete non-truncated jet we get the momentum flux W:

$$W = 2\pi \int_0^{\pi} r^2 \Pi_{rr} \cos(\theta)\sin(\theta) d\theta \tag{4}$$

$$\Pi_{rr} = \frac{4\nu^2 \rho}{r^2} \left(\frac{(A^2 - 1)^2}{(A - \cos(\theta))^4} - \frac{A}{A - \cos(\theta)} \right) \tag{5}$$

The equation (6) takes into account the truncate of the jet (up and down limitation by the depth of the reaction chamber H):

$$W = \int_{\pi/2-\theta^*}^{\pi/2+\theta^*} r^2 \Pi_{rr} \cos(\theta)\sin(\theta)(2\pi - 4\arcsin\sqrt{1 - (H / 2r\sin(\theta))^2}) d\theta + \int_0^{\pi/2-\theta^*} r^2 \Pi_{rr} \cos(\theta)\sin(\theta) 2\pi d\theta$$
$$+ \int_{\pi/2+\theta^*}^{\pi} r^2 \Pi_{rr} \cos(\theta)\sin(\theta) 2\pi d\theta \tag{6}$$

Here, the boundary angle θ^* is determined from the condition:

$$\sin(\theta^*) = H / 2r \tag{7}$$

where H is the depth of the reaction chamber, r is the displacement of the loaded liquid front.
The idea of an equivalent non-truncated jet is to find A_{corr} from:

$$W'(A_{beg}) = W(A_{corr}) \tag{8}$$

From (1) with the zero initial condition:

$$r = \sqrt{2F(\theta)t} \tag{9}$$

Therefore, if the jet parameter A and the physical characteristics (ρ, ν) of the medium are constant, the radial displacement of the loaded liquid front is described by a simple relationship:

$$r = const \cdot \sqrt{t} \tag{10}$$

2.2 Numerical study of liquid injection

The Navier-Stokes and the continuity equations were used for velocity profile simulation:

$$\rho\left(\frac{\partial \mathbf{u}}{\partial t}+\mathbf{u}\cdot\nabla\mathbf{u}\right)=-\nabla p+\mu\nabla^2\mathbf{u}$$
$$\nabla\cdot\mathbf{u}=0 \qquad (11)$$

Simulation was carried in 3D by finite element method using COMSOL Multiphysics package [3]. Stability of the solution was checked by comparing the results with different split rank of the space and time. The results had small variations that allows us presume the small error.

3. Results and discussion

Accounting vertical cutting (6) and clarifying the radial component of the momentum flux tensor, one specifies parameter A of the equivalent jet on which its shape depends (figure 1), calculates the flow velocity which differs from the original non-truncated jet velocity and the displacement.

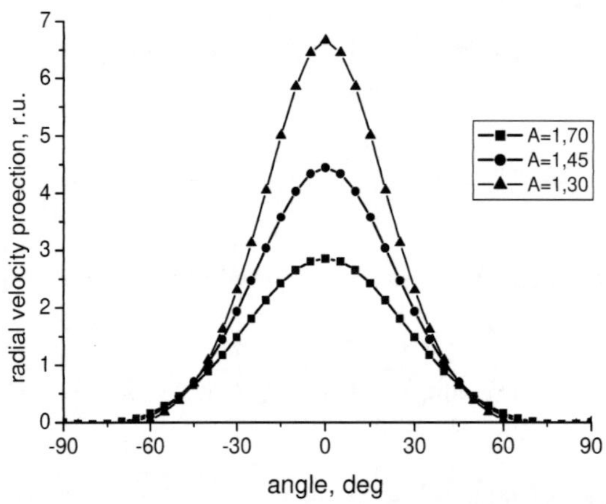

Figure 1. Dependence of the projection of the radial velocity on the axis of the jet for various values of the constant A.

The analytical expression for the radial velocity component was calculated for the jet axis. Jet parameters were defined by the displacement during the first 0.02 seconds and were used in the analytical solution. Figure 2 (a) shows the results of simulation of the liquid front displacement during its input through 0.4 mm wide channel into a circular chamber with a diameter of 12 mm (curve (1), equation (11)) and analytical expression (the curve (2), equations (9)-(10)):

$$2.105\sqrt{t} \qquad (12)$$

which corresponds to non-truncated jet parameters.

After the 12 seconds the curve (1) satisfies a power law with an exponent greater ½, which is associated with a significant effect of its cut and with a decrease in the parameter A of equivalent jet.

There is the comparison of the normalized flow velocity obtained by simulation and the normalized averaged along height radial velocity calculated analytically according to [2] in the figure 2 (b).

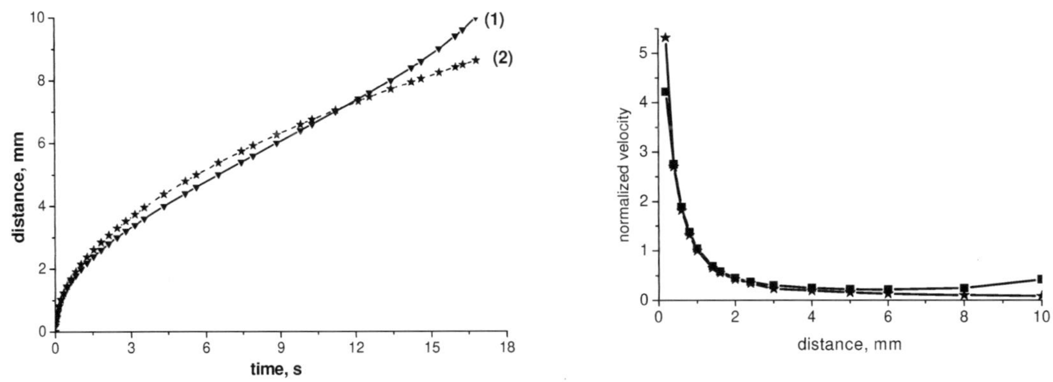

Figure 2(a,b). (a) Comparison of the simulation (1) of liquid sample front displacement and the analytical solutions (2); **(b)** Comparison of normalized velocities of liquid injection calculated numerically (■) and by submerged jet model (★) with H = 1 mm.

Approximating dependence was built for the averaged along vertical axis of jet radial velocity component. The simulation results of the liquid injection velocity into the reaction chamber with the depth of 1 mm have good qualitative and quantitative agreement with estimates obtained by the model of truncated jet. However, the averaged value of the radial velocity projection:

$$\int_0^{\pi} \left(\frac{A^2 - 1}{(A - \cos(\theta))^2} - 1 \right) \cos(\theta) d\theta / \pi \tag{13}$$

provides better quantitative agreement (figure 2 (b)).

Comparison of the data presented in figure 1 showed: i) displacement dependence on the time interval 0 - 9 seconds up to 6 mm has a good quality and satisfactory quantitative agreement (coefficient of determination R^2 = 0.992); ii) the changes of radial velocity over time obtained by simulation and submerged truncated jet model (5) and (6) have shown little difference for the chamber with 1 mm depth. Discrepancy of velocities (figure 1 (b)) after 6-8 mm can be attributed to the vicinity of the chamber outer wall. In this case: a) the boundary condition is the zero normal velocity, b) velocity momentum reflects from the walls is appeared and the jet model is not correct under these circumstances.

Thus, the load of a sample through a narrow channel into the chamber can be simulated with a model of a submerged Landau jet with corrections for its vertical cut (equations (1) - (8)).

During estimation the area in which the submerged jet approximation is applicable, the important geometric parameters are chamber's length and width.

Rectangular chambers (e. g. figure 3 (a, b)) with a widths of 12, 6, 2, 1, and 0.4 mm and depths of 200 and 1000 μm and rounded ones (e. g. figure 3 (c, d)) with the same widths (except 0.4 mm) and depths were studied by simulation (equation (11)). One of the variants of the latter is a round reaction chamber. The width of the inlet channel remained constant and equal to 0.4 mm. We considered only a part of a chamber near the sample inlet since at small Reynolds numbers (~ 1) the velocity magnitude is symmetrical with respect to an entrance and an exit of a chamber. The simulation area consisted of an input channel 0.4 mm long and a chamber with a length two times bigger than width. There was a zero pressure boundary condition at the outlet.

The dependences of the velocity along the chip axis on the distance from the input were obtained as a result of the simulation. The movement of the front was calculated from these velocities. The velocity stopped changing with increasing distance to the inlet at some distance from it as the chamber ends with a straight channel. Therefore, we were also interested in the velocity v_n normalized to its steady-state value:

$$v_n = \frac{v_r - v_0}{v_0} \tag{14}$$

where v_0 is steady-state value of velocity Also, the width of the chamber normalized to the width of the input channel and the distance from the inlet normalized to the width of the chamber were used:

$$D_n = \frac{D}{d} \tag{15}$$

$$r_n = \frac{r}{D} \tag{16}$$

where D is the width of the chamber, d is the width of the input channel equal to 0.4 mm, D_n is the normalized width of the chamber, r is the distance from the inlet to the chamber, r_n is the normalized distance from the inlet to the chamber.

The figure 3 shows velocity distribution in different types of chambers with the widths 1 and 6 mm. Here, x and y are the coordinates counted from the beginning of the input channel. The dependences of the normalized velocities v_n on the normalized distance r_n for different chambers widths shown in the figure 4 (a) coincide qualitatively and quantitatively for different chambers' depths and types.

It can be seen that the velocity almost does not change since r_n is equal to 0.5. The consequence is a transition to a uniform motion of the sample at late stages of loading. This transition tends to be linear and is more different from the submerged jet approach (10) with channel's width decrease. This is evidenced by the approximation of dependences of displacement on time by a power function at^b. Figure 4 (b) shows the dependence of b on the normalized width of the chamber D_n. It was found that the index b equal 0.52 and 0.57 for the chambers' width 12 and 6 mm, which is close to 0.5 that corresponds to a non-truncated jet according to equation (10). b increases to 0.7 when a width is 2 mm and becomes equal to 1 in the limiting case of the absence of the chamber. This is due to the fact that one of the conditions of the submerged jet model is zero velocity at an infinite distance from the inlet. A wide chamber can provide a sufficient difference between the velocity at the inlet and the steady-state velocity, while the value of v_0 begins to affect the transient process with decreasing the ratio D_n. This can be seen from the decrease in the difference between the initial and final velocities and changes in the shape of the curves presented in figure 4(a).

The values of the exponents b remain fairly constant when the depth or type of the chamber changes. The approximations differ from each other mainly by the value of the coefficient a.

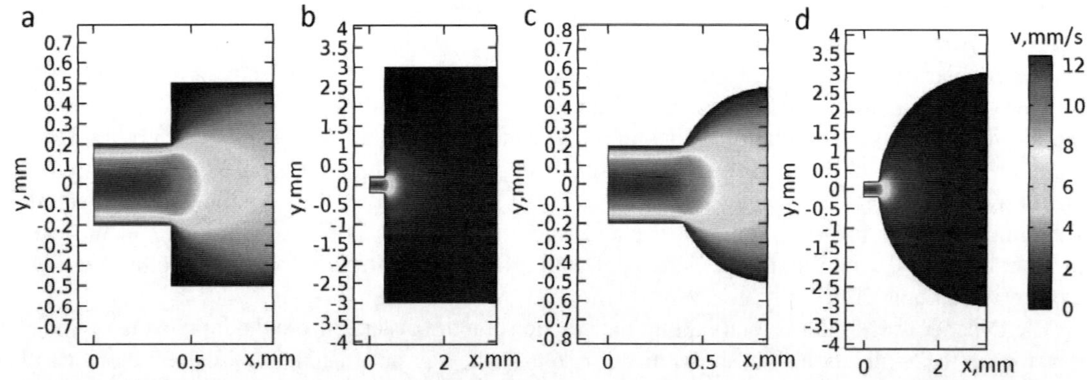

Figure 3(a,b,c,d). The distribution of liquid velocities in the horizontal cross-section along the center of the channels obtained during the simulation. **(a)** rectangular chamber 1 mm wide; **(b)** rectangular chamber 6 mm wide; **(c)** chamber with rounded edges 1 mm wide; **(d)** chamber with rounded edges 6 mm wide. The depth is 200 μm.

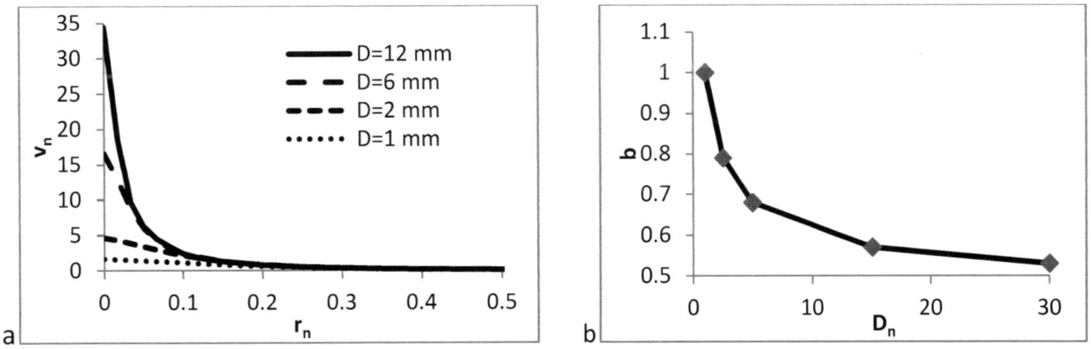

Figure 4(a,b). (a) The dependence of the normalized velocity v_n on the normalized distance r_n for different widths of a rectangular chamber 200 μm deep; **(b)** The change of a function at^b exponent obtained in approximation of the fluid front displacement with the increase of the normalized width D_n of a rectangular chamber 200 μm deep.

4. Conclusions

Thus, the analytical expression based on a submerged jet approach [2] accurately describes the load of liquid sample into the reaction chamber with a ratio of its width to the width of the input channel more than 15 at a distance of half-width of the chamber from the inlet. The amendment to the jet "cut" allows to specify the dynamics of the injection during the longer time.

Acknowledgments

The research was supported by the grant of the Russian Science Foundation (project no.15-19-10041).

References

[1] Lukyanenko K, Denisov I, Yakimov A, Esimbekova E, Belousov K, Bukatin A, Kukhtevich I, Sorokin V, Evstrapov A, Belobrov P 2016 *Biotechnology* **32** (5) 69–76

[2] Landau L D, Lifshitz E M 2006 *Theoretical physics. T.VI. Hydrodynamics* (Moscow: Physmatlit) p 118-121

[3] www.comsol.com

Scanning ion-conductance and atomic force microscope with specialized sphere-shaped nanopippettes

M V Zhukov[1,2], I D Sapozhnikov [2], A O Golubok[1,2], V I Chubinskiy-Nadezhdin[3], F E Komissarenko[1,4], S Y Lukashenko[1,2]

[1] Department of nanophotonics and metamaterials, ITMO University, St. Petersburg 197101, Russia
[2] Laboratory of scanning probe microscopy and spectroscopy, Institute for Analytical Instrumentation RAS, St. Petersburg 198095, Russia
[3] Laboratory of Ionic Mechanisms of Cell Signaling, Institute of Cytology RAS, St. Petersburg 194064, Russia
[4] Laboratory of renewable energy sources, St Petersburg Academic University RAS, 194021 Khlopina, 8/3, Saint-Petersburg, Russian Federation

Abstract. A scanning ion-conductance microscope was designed on the basis of scanning probe microscope NanoTutor. The optimal parameters of nanopipettes fabrication were found according to scanning electron microscopy diagnostics, current-distance I (Z) and current-voltage characteristics. A comparison of images of test objects, including biological samples, was carried out in the modes of optical microscopy, atomic force microscopy and scanning ion-conductance microscopy. Sphere-shaped nanopippettes probes were developed and tested to increase the stability of pipettes, reduce invasiveness and improve image quality of atomic force microscopy in tapping mode. The efficiency of sphere-shaped nanopippettes is shown.

1. Introduction

Atomic force microscopy (AFM) is one of the principal diagnostic methods to visualize the native biological objects (cells, bacteria, DNA, etc.) in a liquid with a high spatial resolution. However, AFM method leads to distortion of the soft biological objects while scanning process even in the less invasive tapping mode. Recently, in connection with the possibility of reducing the invasiveness of studying the native object in a fluid, a new method called scanning ion-conductance microscopy (SICM) was developed [1,2]. The interest to SICM is also explained by new mechanisms of image contrast that can give additional information of the object being studied, as well as the possibility of combining this method with "patch-clamp" technique [3,4].

Unlike tunneling and atomic-force microscopy, in SICM the ions transfer in electrolyte through narrow (D ~ 200 nm) and relatively long (L ~ 20 mm) pipette channel with charge exchange process at the electrodes. Emerging microfluidic flows are very sensitive to micro- and nanobubbles formed in the electrolyte, as well as to presence of microparticles and the quality of micropipette tip surface. Therefore, at the present there is an issue to find optimum schemes and methods for SICM [5,6]. A particular problem is the preparation of micropipette probe for SICM and fixation of biological objects while scanning. The ability to create a combined technique such as SICM, AFM and "patch-clamp" mode is of particular interest nowadays.

The aim of this work was to create a prototype of the scanning ion-conductance with integrated atomic force mode and develop technology for nanopipette probe fabrication for non-invasive studying objects of organic and inorganic nature.

2. Experimental setup

We have created a SICM based on scanning probe microscope NanoTutor (NT-SPb, Russia). Fabrication of nanopipettes probes (NP) was performed by means of PMP 107 Micropipette puller (WPI, USA). NP study was carried out by a scanning electron microscope SNE-4500M (SEC Co., Korea). Fabrication of NP was carried out on borate glass bar with a melting point of 820 ^{0}C. NaCl buffer was selected as electrolyte for SICM. A calibration grid TGZ (NT-MDT, Russia) was used as a test structure. Fabrication of the NP with spheres and control of their geometrical parameters were performed by means of scanning electron microscope (SEM) CrossBeam Neon 40 (Carl Zeiss, Germany).

3. Results and Discussions

SICM principle of operation is based on ionic conduction currents flowing between the two electrodes of Ag/AgCl, one of which is placed in bath with sample, while the second - in the probe itself (NP). Ag / AgCl electrode is characterized by a reversible redox reaction, which resulted that chloride atoms and electrons get into the solution, metalizing silver and vice versa. The current value depends on the diameter of the output MP and decreases when approaching MP tip to the object. At the same time the stability of MP, which depends on a number of factors, is effected on the stability of the servo system of SICM.

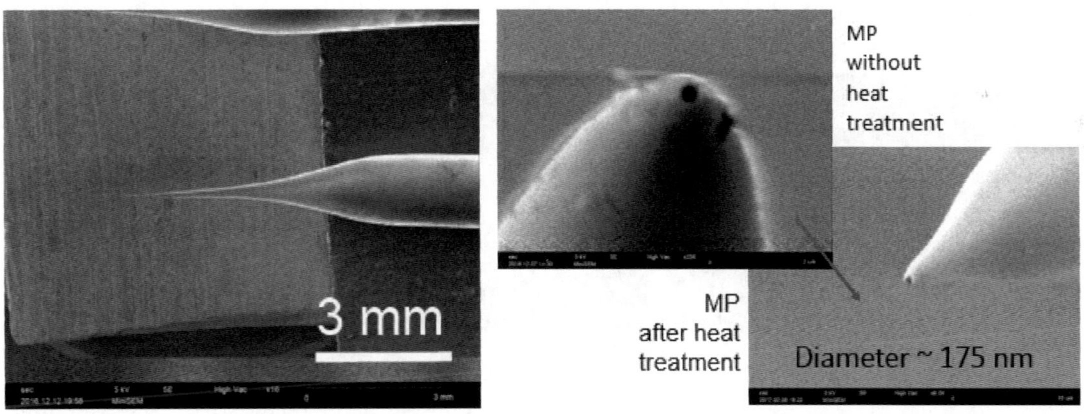

Figure 1. General view of the NP and magnified areas of pipettes tips with hollows about 175 nm.

During the study the loss of SICM image contrast was found with the small smoothness of MP tip front surface. Instability in the ion current also can be caused by small angles of verge convergence of the MP. To improve the stability of MP and contrast of SICM images the MP was fabricated by double exposure. In the process of heating the initial glass bar was gradually stretched and then abruptly torn. Moreover, in the stretch process MP twist along its axis was carried out. To smooth surface of the MP it was repeatedly short-term heat to reflow the tip (Figure 1).

Optimum contrast and image stability of SICM were found with MP convergence angle ~ (5-8)^o and a diameter ~ (175-200) nm.

The dependence of the ion current from the distance to sample surface I (Z) and current-voltage characteristics I (V) was studied, the steepness of which is possible to diagnose the quality of the MP. Revealed offset potential in the process of SICM is indicating the necessity for periodic recharging of the electrode potential (electrochemical formation of AgCl layer on Ag). Topology visualization of polyethylene substrate by SICM and AFM indicates good correlation of results, which was obtained by different methods. To determine the resolution of SICM on the surface of the polyethylene sample was created specifically periodic structure (replica from TGZ grid) with a period of about 3 μm and height of about 100 nm (Figure 2).

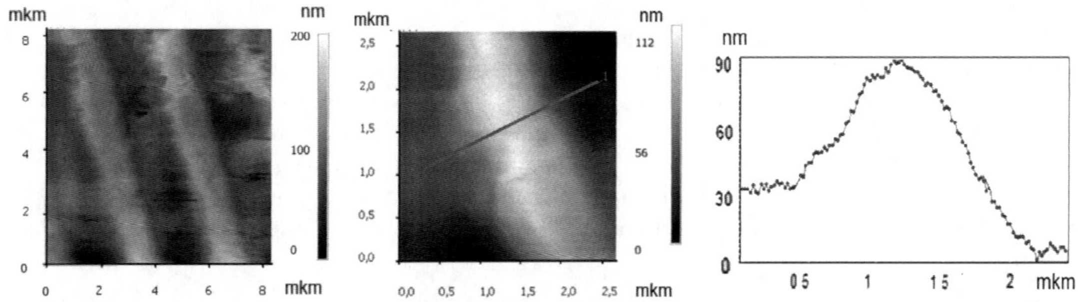

(a) (b) (c)

Figure 2. SICM image of the calibration grid TGZ (a), magnified image of a single strip with cross-section line (b) and cross-section area (c).

The small breakdowns on SICM was noticed on the border of steps compared to the semi-contact AFM mode, but in general the AFM and SICM images demonstrated a good correlation. In addition, visualization of CHO cell culture (Chinese Hamster Ovary) was carried out by means of optical microscopy, AFM and SICM modes (Figure 3). The effectiveness of non-invasive diagnostics by means of SICM method was demonstrated.

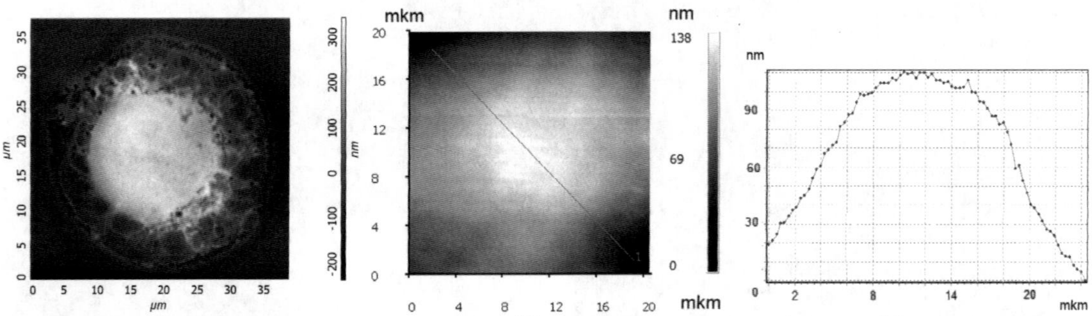

(a) (b) (c)

Figure 3. AFM image of the CHO cell in air (a), SICM image of native CHO cell in liquid with cross-section line (b) and cross-section area of the cell (c).

In SPM NanoTutor it is possible to perform a simultaneous study of objects both in SICM and AFM tapping mode using a pipette as a probe. However, when using a pipette in the AFM tapping mode leading to image noise and the gradual destruction of the pipette. To improve the stability of pipette, provide non-invasive interaction and improve the image quality, a calibrated Au sphere with a radius of about 200 nm was attached to the side of the NP (Figure 4). The attachment was carried out in two ways: with the help of Van-der-Waals forces between sphere and tip under the electron beam and by welding the sphere to the pipette by depositing the precursor gas (C). The second variant is preferable because of the greater stability of the NP and the dielectric characteristics of the C, which is important for the SICM studies.

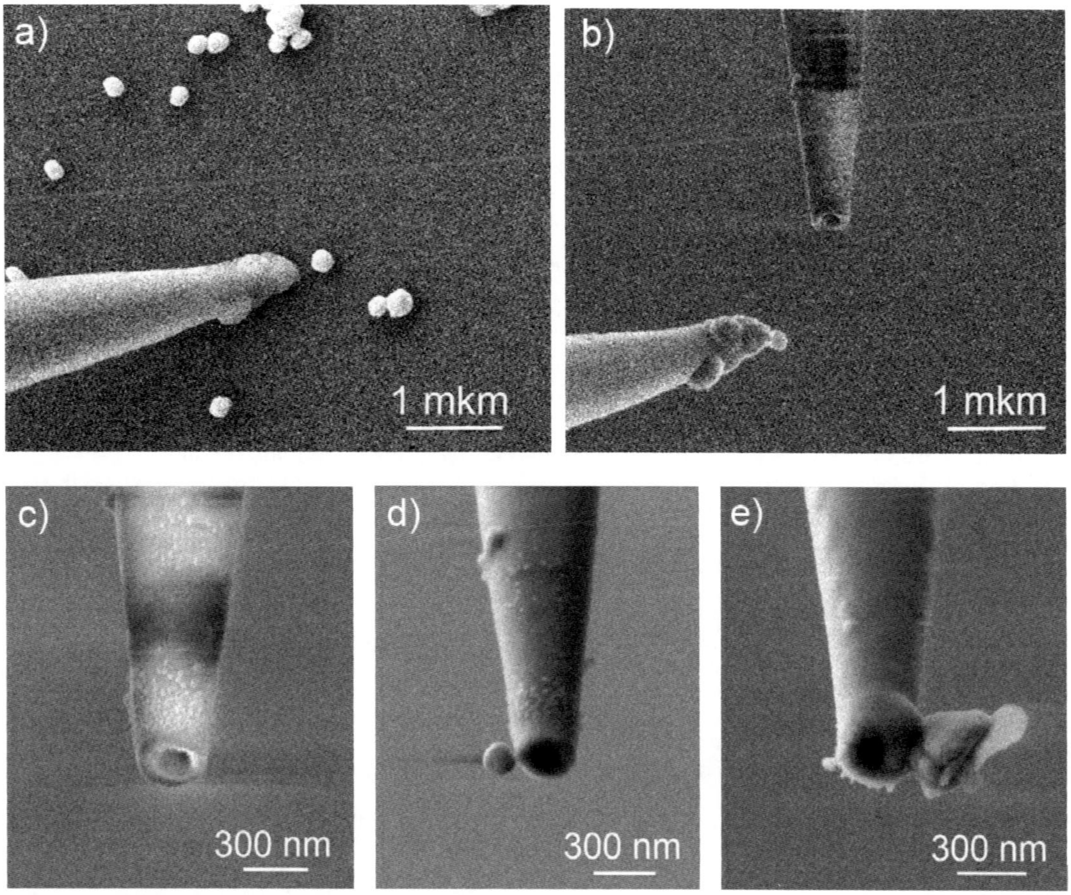

Figure 4. SEM images of sphere-shaped NP fabrication: search for a sphere on a substrate (a), grabbing a sphere on a manipulator (b), image of standard NP (c), attachment of sphere by Van-der-Waals forces (d) and welding of sphere by depositing the precursor gas C (e).

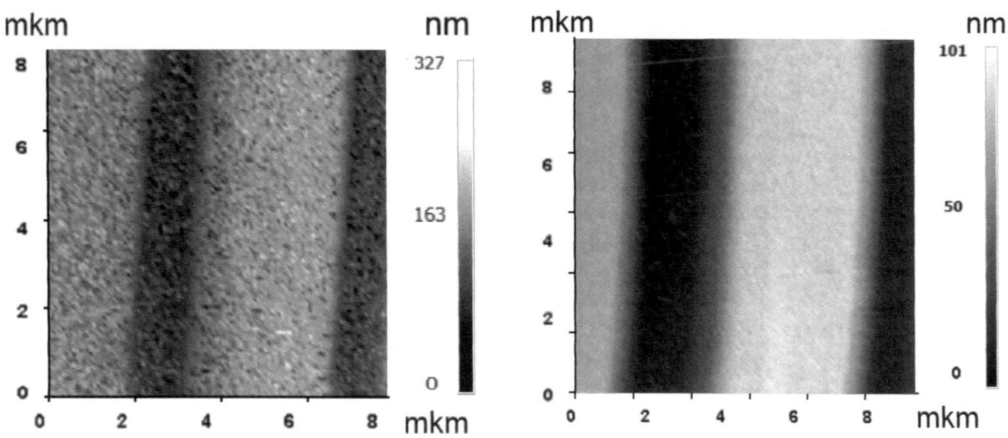

(a) (b)

Figure 5. AFM image of the calibration grid TGZ, measured by standard NP (a) and NP with sphere (b).

Figure 5 shows an improvement of the image quality using the NP with a sphere. The local soft contact of the sphere provides noise reduction and allows scanning of grid with greater accuracy (compliance in height and gap between grid lines). It should be noted that when using the sphere, the probe's resonance has decreased from about 9 kHz to about 5 kHz. The sphere does not affect the flow of ion current in the SICM mode, since it locates at a distance from the sample in SICM mode and does not block the hollow of the NP. In addition, the spheres reduce the invasiveness of the AFM method, which is important in the study of the biological objects.

4. Conclusions

Scanning ion-conductance microscope was presented based on «NanoTutor» microscope. It is shown that stable and contrasting SMIP-images obtained by nanopipette with the convergence angle ~ $(5\text{-}8)^0$, diameter ~ (175-200) nm and smoothed tip obtained by local heat treatment. The dependence of the ion current from the distance to sample surface I (Z) and current-voltage characteristics were studied for nanopipettes, which is useful to obtain optimum MT fabrication regimes.

Revealed offset potential in the process of SICM was indicating the necessity for periodic recharging of the electrodes. A good correlation of SMIP images of test surfaces, including the biological samples, with the images obtained by means of AFM and optical microscopy was demonstrated.

Sphere-shaped nanopippettes probes were developed and tested. It is noticed that spheres increase the stability of pipettes, reduce invasiveness and improve image quality of atomic force microscopy in tapping mode. The efficiency of sphere-shaped nanopippettes is shown.

Acknowledgments
Work carried out with the support of the Russian Foundation for Basic Research (16-32-00806), the leading universities of the Russian Federation (grant 074-U01).

References

[1] P. Happel, F. Wehner and I. D. Dietzel 2007 *Modern Research and Educational Topics in Microscopy* (Spain: Formatex Press) p 968
[2] Patrick Happel, Denis Thatenhorst and Irmgard D. Dietzel 2012 *Sensors* **12** 14983-15008
[3] Anamika Bhargava, Xianming Lin,Pavel Novak, Kinneri Mehta, Yuri Korchev, Mario Delmar, and Julia Gorelik 2013 *Circ. Res.* **112(8)** 1112–1120
[4] Pavel Novak, Julia Gorelik, Umesh Vivekananda, Andrew I. Shevchuk, Yaroslav S. Ermolyuk, Russell J. Bailey, Andrew J. Bushby, Guy W.J. Moss, Dmitri A. Rusakov, David Klenerman, Dimitri M. Kullmann, Kirill E. Volynski, and Yuri E. Korchev 2013 *Neuron.* **79(6)** 1067–1077.
[5] David Pastre, Hideki Iwamoto, Jie Liu, Gabor Szabo, Zhifeng Shao 2001 *Ultramicroscopy* **90** 13–19
[6] Alex Zhukov, Owen Richards, Victor Ostanin, Yuri Korchev, David Klenerman 2012 *Ultramicroscopy* **121** 1-7

Optogenetic light pulses generator

A I Erofeev[1], M V Matveev[1], O A Zakharova[1], S G Terekhin[1], V A Kilimnik[1], I B Bezprozvanny[1,2] and O L Vlasova[1]

[1]Molecular Neurodegeneration Lab, Peter the Great Polytechnic University, Saint-Petersburg, 195251, Russia
[2]Department of Physiology, University of Texas Southwestern Medical Center, Dallas, TX, USA

Abstract. To date, optogenetics is one of the most popular methods in the world in neuroscience. There are new equipment and devices created to keep the progress of this method. This article describes a light pulse generator developed at the Laboratory of Molecular Neurodegeneration, designed for optogenetic experiments.

1. Introduction

On the world market there are a large number of equipment able to manage specified wavelength LEDs, but most of them do not allow to set the necessary parameters of the light stimulation, such as duration, frequency, number of light pulses, which would be required for optogenetic experiments [1, 2, 3]. Therefore was developed light pulse generator, which allows to set required parameters of the light and which was tested on neuron cultures.

2. Materials and methods

2.1. Primary hippocampal neuron cultures

The hippocampal cultures of mice were established from postnatal day 0–1 pups and maintained in culture as we described previously [4, 5]. Briefly, after dissection and dissociation, neurons were plated on coverslips (pre-treated with poly-lysine) and cultured in neurobasal A medium with addition of 1% FBS and 2% B27. At third day in vitro (DIV3), Ara-C (4 _M) was added to prevent glial cell growth. At DIV7 and DIV14, 50% of medium was exchanged with fresh neurobasal A medium containing 2% B27 without FBS. In these culture conditions, the astrocytes constitute about 10–20% in total cells in our cultures at DIV15 as determined by GFAP staining (data not shown). For assessment of synapse morphology, hippocampal cultures were transfected with TD-tomato for easily detection of transfected cells and for optogenetics experiments -FCK-CHR$_2$ plasmids at DIV7 using the calcium phosphate method.

2.2. Whole cell patch recordings in hippocampal cultures

Whole cell recordings in ACSF external solution (124mM NaCl, 26mM $NaHCO_3$, 10mM glucose, 5mM KCl, 2.5mM $CaCl_2$, 1.3mM $MgCl_2$, 1mM NaH_2PO_4) were performed in a current-clamp mode (Axopatch-200B amplifier) using 5–10 MΩ pipettes filled with internal solution (K-Gluconate 140mM, MgCl2 2mM, NaCl 2mM, ATP-Na2 2mM, GTP Mg 0.3mM,HEPES 10mM). Following establishment of whole-cell configuration, the light with period 1 s in duration from 10 ms to 500 ms were illuminated and the corresponding potential changes were recorded.

3. Light pulses generator

The developed device (4) is a part of the optogenetic research system, which illustrated in Figure 1. It allows to set the specific light pulses settings. The duration, frequency and number of optogenetic impulses are controlled by the integrated microcontroller.

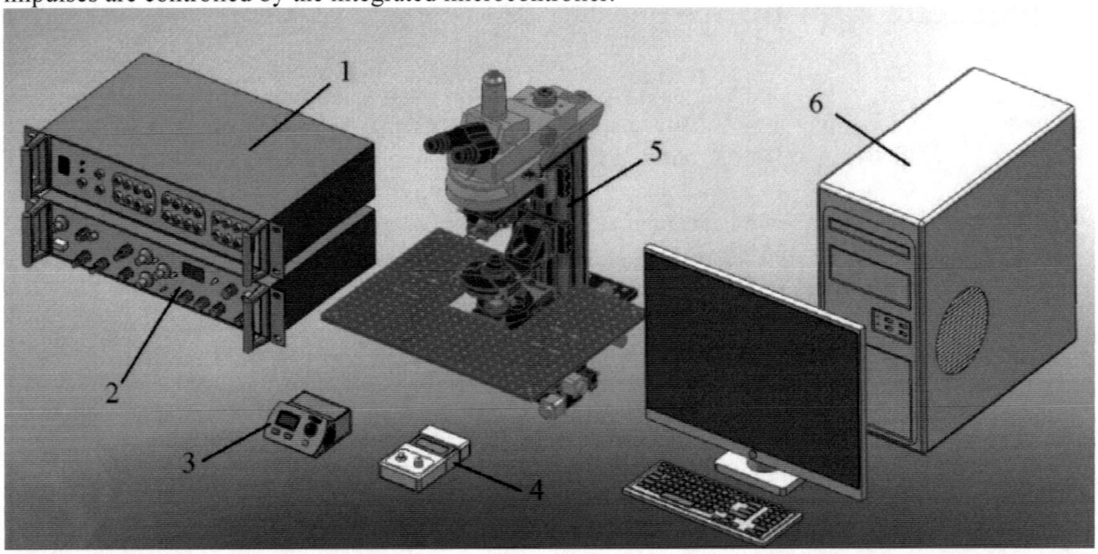

Figure 1. Optogenetic research complex 1-Digidata 1440A - Data Acquisition System; 2-Axopatch 200B - Patch-clamp amplifier; 3-Inverted microscope; 4-Optogenetic impulse generator; 5-DC4100 - 4-channel LED driver; 6-Personal Computer.

The developed device is a driver for a LED source controlling. The pulse generator is used in the complex of optogenetic studies. This unit is used both as an independent regulator of light pulses, and as a programmable unit. In the second case, the device is built into the chain of adaptive feedback and performs the function of signal tuning. The control signal is generated based on the data from the biopotential amplification block. Figure 2-3 shows a schematic diagram and 3D scheme of the developed device.

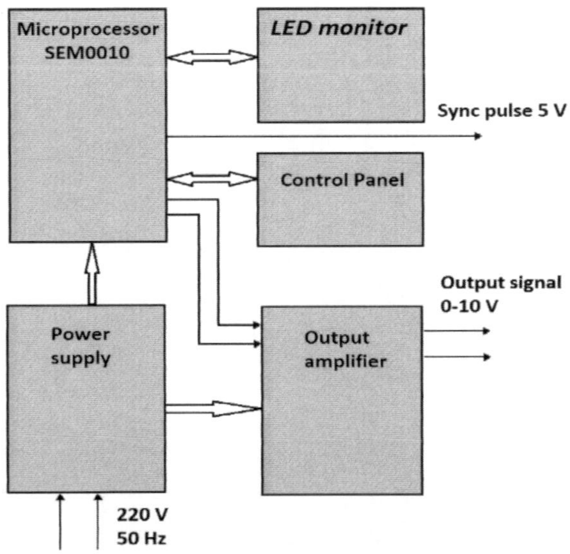

Figure 2. Optogenetic pulse generator (General scheme).

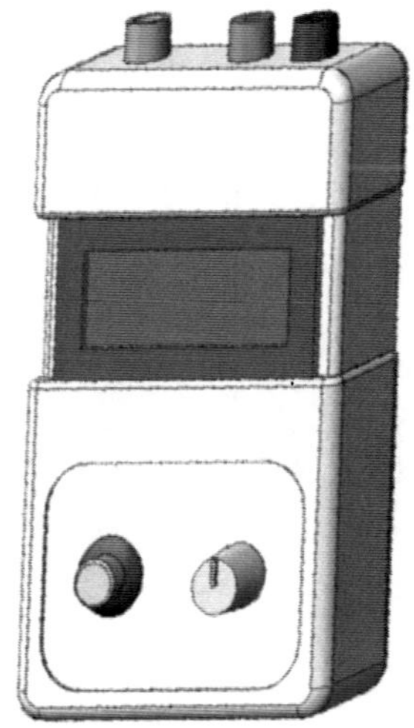

Figure 3. Optogenetic pulse generator (3D scheme).

4. Test light pulses generator on the hippocampal neurons culture

Checking the device was carried out in conjunction with thortlabs DC4100. Experiments were conducted in transfected with ChR2-GFP plasmid mouse primary hippocampal neurons. Traces of neurons activity were recorded in current-clamp mode (whole-cell configuration) in response of 470-nm light. The results are shown in Figure 4.

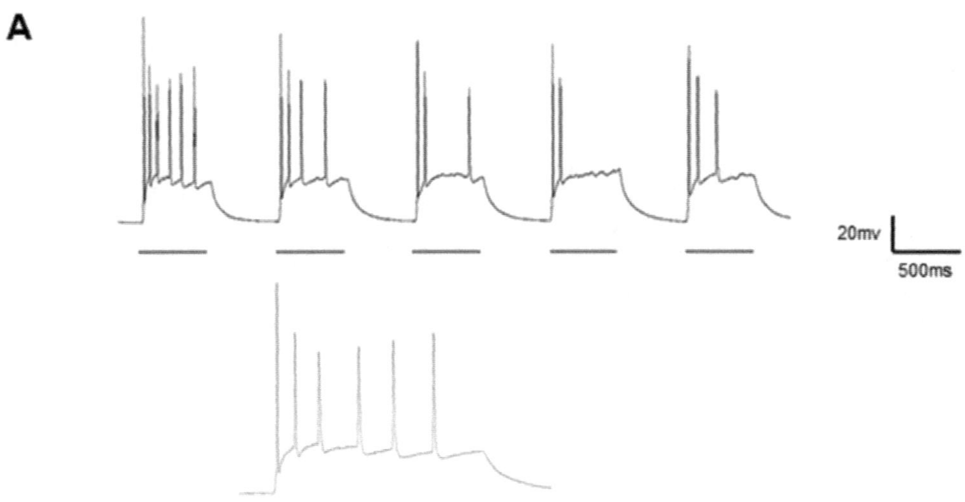

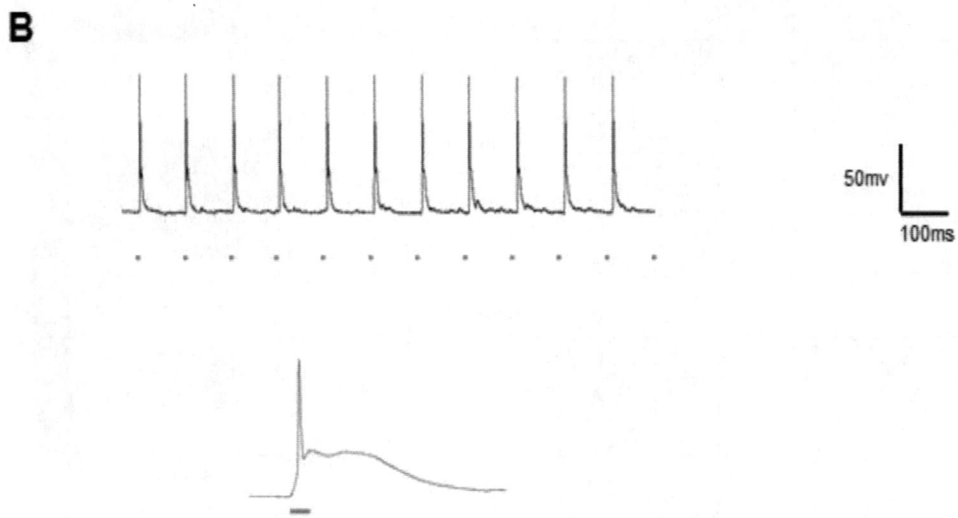

Figure 4. Action potentials evoked by 470 nm light (indicated by blue bar, T=1s) with duration A – 500 ms, B – 10 ms recorded in current-clamped neuron in whole-cell configuration.

5. Future plans and conclusions

The created devices could affect neurons by light stimulation with the specified parameters, particularly close to physiological conditions. Our future plans include the development of optogenetic brain implantable system for in vivo experiments.

Acknowledgments

This work was supported by the state grant 17.991.2017/ПЧ, by the Russian Science Foundation Grant 14-25-00024. The financial support was divided in the following way: experiments depicted on Figure 1-3 was supported by the state grant 17.991.2017/ПЧ, experiments depicted on Figure 4 was supported by the Russian Science Foundation Grant.

References

[1] Deisseroth K 2010 *Scientific American* **303** 48
[2] Boyden ES, Zhang F, Bamberg E, Nagel G and Deisseroth K 2005 *Nature Neuroscience* **8** 1263
[3] Windhorst U and Johansson H 1999 190
[4] He B and Soderlund D M 2010 *Neuroscience Letters* **469** 268
[5] Zhang H, Jie Liu J, Sun S, Pchitskaya E and Popugaeva E and Bezprozvanny I 2015 *Journal of Alzheimer's Disease* **45** 561

The kinetic study of solidification PEGDA microparticles in flow-focusing microfluidic chip

N A Filatov[1], D V Nozdriukhin[1,2] and A S Bukatin[1,3]

[1]Nanobiotech Lab, St. Petersburg Academic University, St. Petersburg 194021, Russia
[2]Peter the Great St. Petersburg Polytechnic University, St. Petersburg 195251, Russia
[3]Department of biophysics, Siberian Federal University, Krasnoyarsk 660041, Russia

Abstract. Polyethylene glycol diacrylate (PEGDA) microparticles (MPs) have a great potential in a wider range of applications including cell carriers, drug delivery agents, and multiplexing assays. Among existing MP fabricating methods, only the microfluidic droplet-based methods provide both morphology complexity and substance diversity. This work is concerned with the production PEGDA-based droplets with the photoinitiator (Irgacure 2959) in a flow-focusing microfluidic generator. The kinetics of MPs solidification in mineral oil with various common surfactants (Abil EM 180, Span 80) were investigated. Also, the Young's modulus of solid PEGDA were obtained.

1. Introduction

Typically, microparticles (MPs) have been prepared through traditional emulsion polymerization using rotors/stators or high-pressure valve homogenizers. Nevertheless, these techniques have a confinement in achieving unique characteristics such as well-controlled size, various morphologies and multi-compartments. The droplet microfluidics, electrohydrodynamics-based, centrifugation-based, and template-based methods have been recently proposed to overcome this limitation [1]. However, microfluidics has shown enormous potential both in providing control over chemical, physical, mechanical, and morphological characteristics and robust approaches for the production of monodisperse geometries [2]. Meanwhile, polyethylene glycol diacrylate (PEGDA) MPs have become widely used as cell carriers, drug delivery and tissue scaffolding materials [3]. Acrylate-based polymers are preferred due to their transparency, color variation, robust mechanical properties, elasticity, chemical versatility, bioinertia and biocompatibility. Moreover, acrylates can be readily photopolymerized on industrial scales [3, 4]. A little attention is observed to study the kinetics of droplet solidification that is very important for design and optimization of drug delivery systems.

Microfluidics provides control of fluids at the microscale and generate micrometer diameter droplets with a narrow size distribution and well-tailored structures [5]. In droplet-based methods, microdroplets are generated using shear forces induced by immiscible fluids inside a pinched region of a microfluidic device and then solidified to be MPs. In cell carrying and drug delivery agents, typically used MPs diameter ranges from about 90 to 200 um are used, in some special circumstances size was up to 600 um [1].

In this work, we studied conditions of PEGDA droplet formation and solidification: solidification time from photoinitiator concentration (Irgacure 2959) and mechanical properties of solid PEGDA (Young's modulus).

2. Methods

In this work, droplet microfluidic generator, based on flow focusing principle, was used (figure 1). Its operating principle is based on the fact that the continuous phase flowing through two side channels meets the dispersed phase at a channels' intersection, where the dispersed phase is squeezed by the continuous phase and breaks up into droplets. It allows avoiding a contact of forming droplets with channels' walls and helps to prevent potential negative effects on sample's components of the dispersed phase [6].

The PDMS microchips were fabricated by soft lithography method [7]. Mineral oil (330779 light, Sigma-Aldrich) with different widespread surfactants (Abil EM 180, Span 80) was used as a continuous medium. PEGDA mixed with photoinitiator (Irgacure 2959) was used as a disperse phase. The UV lamp 365 nm and 150 mW/cm^2 was used for droplets solidification.

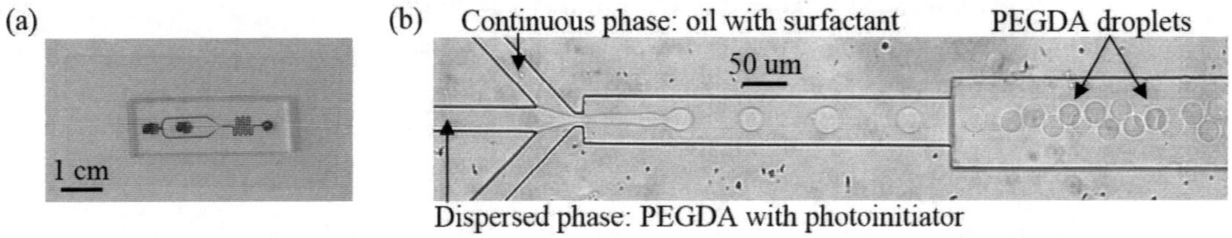

Figure 1. Flow focusing microfluidic droplet generator: (a) general view; (b) active area.

To obtain Young's modulus of solid PEGDA, the atomic force microscope (AFM) Bruker Bioscope Catalyst with RTESPA probe (stiffness 42 N/m) was used. Force curves analysis was proceeded by NanoScope Analysis software. To estimate Young's modulus, the Sneddon (Conical) fit model was used.

3. Results and discussion

During a UV solidification, a PEGDA+photoinitiator compound changes color (figure 2), which was used to determine the solidification time. The last phase occurs when the color stops changing. Furthermore, the color does not change, but the compound saturate until completely solidified. In addition, droplets with different diameters have the same solidification time.

It was found that the surfactant Span 80 influence on solidification because color changing does not appear in contradistinction to pure solution or with Abil EM 180.

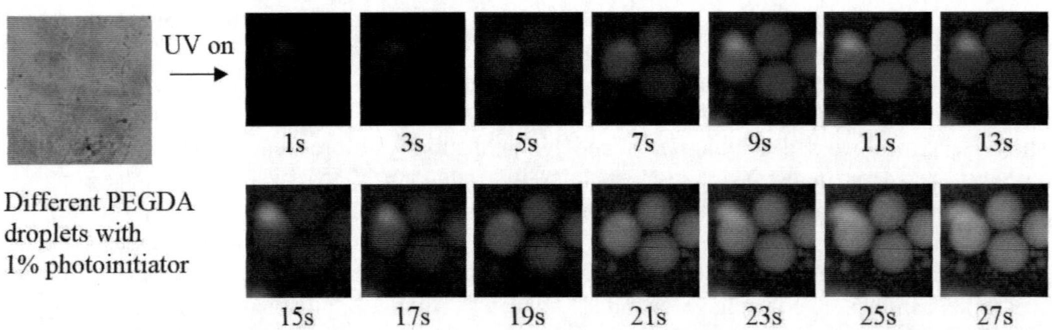

Figure 2. Solidification process of PEGDA droplets with different diameters at 1% w/w photoinitiator.

We suggest the next qualitative scheme of the PEGDA solidification (figure 3a). When the UV is switched on, the first instantaneous solidification takes place (flash point). The flash point value depends on photoinitiator concentrations. In some cases, the value can even tending to zero. Further, there is a gradual change of color PEGDA or maturation. Based on the color changes this process can be divided into three stages. At the first stage, the color changes smoothly from dark blue to light brown. Then, at the second stage, the color changes from light brown to blue. Finally, at the third stage, there is no color shift. Importantly, there are three dominant colors, so three stages. The colors themselves may depend on the experimental conditions. The beginning of the third stage is called t_s. The last stage is called the saturation stage. Since the color does not change at the third stage, it is impossible to estimate accurately the value $t_{100\%}$ (figure 3a).

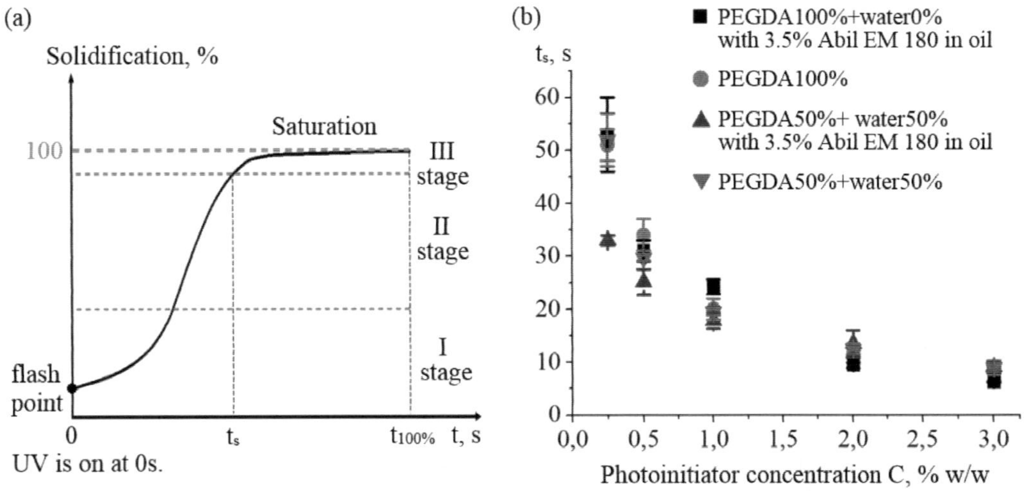

Figure 3. (a) Qualitative solidification scheme; (b) Dependences of solidification time t_s on photoinitiator concentration for the droplet diameter 200 μm.

The dependences of t_s on photoinitiator concentration is shown in figure 4a. Four cases were considered. First, it was the bulk solution of PEGDA with photoinitiator. Second, it was droplets of PEGDA with photoinitiator in oil phase with 3,5% w/w surfactant Abil EM 180. Third, it was the bulk solution of 50% w/w water and 50% w/w PEGDA with photoinitiator. And, fourth, it was droplets of 50% w/w water and 50% w/w PEGDA with photoinitiator in oil phase with 3,5% w/w surfactant Abil EM 180. In all cases, the photoinitiator concentration was 0.25, 0.5, 1, 2, 3 % w/w (to the whole volume). The points of time values on the graph (figure 3b) are averaged value of five experiments. To better capture the color changing during the UV solidification, large drops with the diameter 200 μm were used.

From experiments, it can be concluded that solidification time t_s does not depend on droplet size. The higher concentration, the faster solidification. The data was approximated by exponential fitting:

$$t_s = Ae^{-\frac{C}{C_0}} + t_0 \tag{1}$$

Values $C_0 = 1.3$, $A = 55$ s, $t_0 = 4$ s were obtain from approximation fitting. The optimal concentration is about 1%.

To estimate the Young's modulus behavior, PEGDA solutions in DI water with concentration of PEGDA 10-100% at 1% w/w photoinitiator were used. Young's modulus of PEGDA droplets is shown in figure 4. According to the shape of obtained force curves, it can be indicated if the droplet contains a solid core and a soft shell or only a solid core (figures 4b, 4c). It was found that ratio 50% w/w PEGDA and 50% w/w water is critical (figure 4b). If the concentration of PEGDA is less than 50% w/w, solidification occurs evenly without shell formation. Otherwise, shell appears. In addition, Young's modulus of shell is smaller than modulus of core.

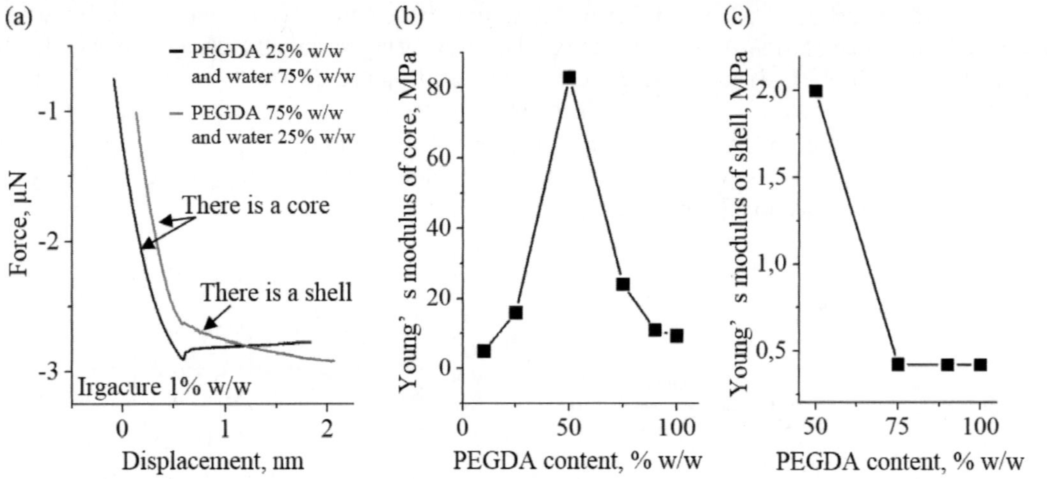

Figure 4. Dependences of force acting the sample on displacement (a); Dependences of Young's modulus on PEGDA content in solution: (b) core; (c) shell.

4. Conclusions

To sum up, the kinetic study of solidification PEGDA microparticles in flow-focusing microfluidic device was investigated. The data was collected by the optical microscopy and besides, kinetic data were processed statistically. The results show the behavior of droplets formation at different PEGDA concentration in solution. It was concluded that the solidification time t_S depends is exponentially on photoinitiator concentration. Moreover, optimal concentration is about 1%. The results indicate that t_S does not depend on droplet size.

Young's modulus was studied by the AFM force curve measurements. According to the shape of obtained force curves it can be indicated if the droplet contains a solid core and a soft shell or only a solid core. In addition, the Young's modulus of the droplet shell is smaller than the modulus of the core. We obtained that the maximum Young's modulus is 83 MPa on condition 50% w/w PEGDA and 50% w/w water. Young's modulus changes in range from 5 to 83 MPa for core, and from 0.4 to 2 MPa for shell.

Acknowledgments

This work was financially supported by the grant from the President of Russian Federation (project MK-2131.2017.4).

References

[1] Choi A, Seo K D, Kim D W, Kim B C and Kim D C 2017 *Lab Chip* **17** 591-613
[2] Martino C, Vigolo D, Solvas X C, Stavrakis S and deMello A J 2016 *Adv. Mater. Technol.* **1** 1600028
[3] Krutkramelis K, Xia B and Oakey J 2016 *Lab Chip* **16** 1457-65
[4] Xue P, Wu Y, Menon N V and Kang Y 2015 *Microfluid Nanofluid* **18** 333
[5] Wan J 2012 *Polymers* **4** 1084-1108
[6] Filatov N A, Belousov K I, Bukatin A S, Kukhtevich I V and Evstrapov A A 2016 *J. Phys.: Conf. Ser.* **741** 012052
[7] Bukatin A S, Mukhin I S, Malyshev E I, Kukhtevich I V, Evstrapov A A and Dubina M V 2016 *Tech. Phys.* **61** 1566-71

IOP Conf. Series: Journal of Physics: Conf. Series **917** (2017) 042025 doi:10.1088/1742-6596/917/4/042025

Physical Principles of Discrete Hierarchies Formation in Protein Macromolecules

E V Malyshko, V A Tverdislov

Faculty of Physics, Lomonosov Moscow State University, Moscow, 119991 Russia

E-mail: katyamalyshko@mail.ru, tverdislov@mail.ru

Abstract. A model for chiral periodicity with alternating chiral sense in hierarchies of protein and nucleic acid structures is proposed and substantiated. Regular alternation of the chirality sense is revealed in transitions from the lowest to higher levels of structural-functional organization in proteins where it is L-D-L-D. The stratification principle combines the ideas of biomacromolecules folding and molecular biological machines.

1. Chiral hierarchies of intra- and intermolecular structures of proteins and nucleic acids

Symmetries and symmetry breaking form a physical basis for the structure and functioning of living systems, their origin and evolution. We are talking about symmetry relations in the primary, secondary, etc. structural levels of nucleic acids and proteins. However, both systems must be constructed hierarchically to have executive and regulating subsystems. Chiral duality becomes a natural tool in combinatorics of structural correlations in these subsystems at all levels of structural organization. The purpose of the biological hierarchies is the ability to match the processes of different time and space scales.

New aspects of the theory of principles and mechanisms of sign-alternating chiral stratification in hierarchies of biological structures are proposed. It has previously been shown that the described discreteness principle combined ideas of biomacromolecules folding and molecular biological machines [1].

Earlier we detected the succession of the hierarchical levels with an alternating chirality sign in the structural hierarchy of DNA and proteins (Figure 1): from the "lower" asymmetric carbon atom to the helices, superhelices and supramolecular structures [1, 2]. The regular alternations of the chirality sign D-L-D-L for DNA and L-D-L-D for protein structures were observed during the transition to a higher level of the structural and functional organization [3]. There is a shift in chiral phase in proteins, because the protein primary structure is a sequence of L-amino acids residues, when the structure of the deoxyribose polymeric chain is taken to be the primary structure in DNA molecules.

Two basic primary polymer chains of DNA are composed of D-deoxyribose molecules. Nucleotides in the "left-handed" gosh-conformation together form "right-handed" DNA double helix (in the most common B-form). In bacteria, the circular "right-handed" helix is twisted into the "left-handed" superhelix. Negative supercoiling facilitates the melting of the double helix which is necessary for transcription and replication. In eukaryotes, the next level of organization is nucleosomal. The nucleosome is formed by a histone globule around which DNA wrapped in a superhelical manner. A histone octamer is the eight protein complex; it is a so-called "left-handed" superhelix. The right-handed DNA double helix wraps the histone octamer in the "left" way. It can be

IOP Conf. Series: Journal of Physics: Conf. Series **917** (2017) 042025 doi:10.1088/1742-6596/917/4/042025

assumed that such configuration creates the possibility for an easier way to open the helical structure of a lower scale. Structural hierarchy of proteins will be discussed more thoroughly.

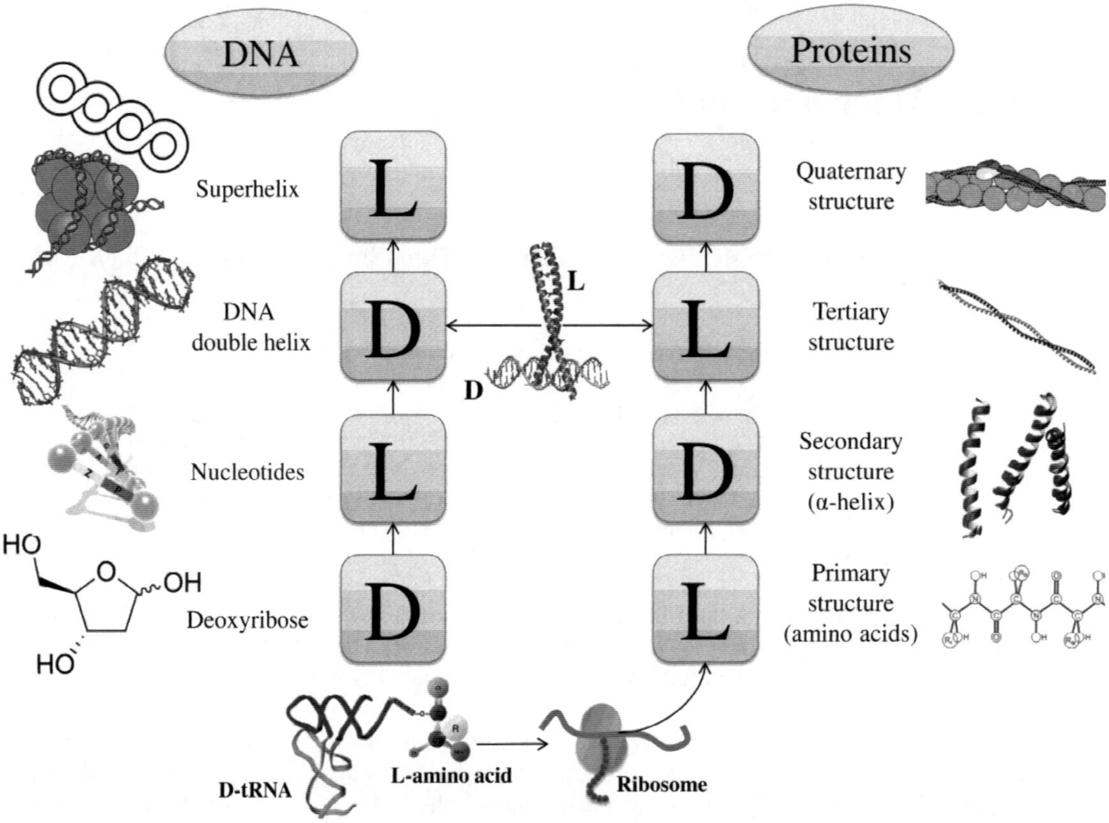

Figure 1. Periodic system of sign-alternating hierarchies from primary to quaternary chiral (helical) structures for DNA (left column) and proteins (right column). L denotes the left enantiomer or the helical configuration; D, the right enantiomer or the helical configuration.

The change of chirality sign provides for the evolutionary irreversibility of stratification (stratification is understood as the splitting of a hierarchical level of a system into two levels with different chirality signs, i.e. formation of chiral layers in a system). Indeed, to switch its chirality sign a chiral object, which does not possess a symmetry axis, must be disassembled and then reassembled from the same elements. The chirality of biological structures of different levels ensures that the process of L/D stratification is universal and hierarchical levels are deterministic and stable.

Our concept assumes that the same type of macromolecules tend to interact in the space of one type of chiral symmetry, L or D, and different types - in complementary pairs of different-valued chiral enantiomorphs.

One of the key properties of tRNA molecules is chiral selectivity for amino acids. Each amino acid has its own one or sometimes several different tRNAs. The addition of amino acids to tRNA is catalyzes by enzymes - aminoacyl-tRNA synthetases. Every such synthetase attaches a specific amino acid to the corresponding tRNA. The ATP molecule acts in this process as a source of energy. The process of aminoacylation of tRNA with aminoacyl-tRNA synthetase occurs in two stages: first, the enzyme activates the corresponding amino acid, and then a covalent bond occurs between the amino acid and tRNA. All amino acids (except glycine) are “left”, whereas tRNA is “right” at all structural levels. Thus, in the aminoacylation reaction, D- tRNA interacts with L-amino acids. Evolutionary such

IOP Conf. Series: Journal of Physics: Conf. Series **917** (2017) 042025 doi:10.1088/1742-6596/917/4/042025

chiral selectivity fixation could happen both from the excess of D-RNA and from the excess of L-amino acids in the RNA world.

A distributed resource of free energy appears due to the L-amino acids selection by tRNA from the racemate.

2. Structural and functional hierarchy of proteins

The protein primary structure is a sequence of L-amino acids residues. The polypeptide chain is laid in the helix or in the pleated sheet. Generally, α-helix is D-enantiomer, because the right-handed α-helix is more stable than the left-handed one [4]. The twist of β-sheet is right-handed. Interacting right-handed α-helices tend to form left superhelices. The quaternary structure of proteins is represented by the supramolecular structures, mainly formed with the left-handed superhelices interacting with the right-coiling structures. A phase shift is obvious because the protein hierarchy starts from L-amino acids, and the nucleotide hierarchy starts from D-carbohydrates deoxyriboses.

The "chiral stairs" in the intramolecular protein structure were thoroughly studied. To prove the dominance of the left-handed superhelices in all types of configurations, we analyzed all available protein structures from the "periodic table" [4, 5]. In almost all the cases where intermolecular overlap of α-helices was detected there was a tendency for the secondary left superhelix formation (Figure 2).

The structure analysis gives quantitative evaluations of the presence of lefthanded superhelical motif in each class selected from the table [4]. In the simplest case - for superhelix consisted of two α-helices, - left twisting tendency was observed almost in all structures in this class. Increasing a number of α-helices, number of structures with a similar trend is reduced, but in particular classes, which are presented by more complex superhelices, containing 5-9 α-helices, the percentage of structures with a tendency to left twisting increases sharply. In some cases it was difficult to classify the twisting due to accumulations of α-helices and β- sheets near the coiled coil. Sometimes it was difficult to observe the tendency because of too short fragments of α-helices in a superhelix. In addition in some cases (most often in superhelices based on 4 α-helices) there was almost a parallel layer of α-helices with a light left-handed twisting tendency, and we did not rank these structures as having left-handed motif. Today creating the program which can quantitative reveal the left-handed and right-handed motives in the local protein structures draws towards completion.

Figure 2. Right-handed α-helices are twisted around each other forming a left-handed supercoil (coiled-coil structure): 1AA0 (based on 3 α-helices), 1QU7 (based on 4 α-helices) and 1MZ9 (based on 5 α-helices) in [6].

Also we analyzed the functional class distribution of proteins containing certain types of superhelices from the database CC + [5]. Predominance of enzymes (25-45%) was observed in proteins with the most common types of superhelices - consisted of two, three and four α-helices. Furthermore, viral proteins (20%) and chaperones (9%) were frequently observed. This result essentially specifies the statement about "selected mechanical degrees of freedom" in the concept of "protein-enzyme-machine" [1, 2].

The evaluation of included in the α-helices and β-sheets in proteins amino acids percentage was made. For this purpose we used obtained from PDB [6] protein structure data for the samples from the table [4]. The calculations show that the average ratio of the quantity of making up α-helices amino acids to the total quantity of amino acids in protein is about 8% and about it is 2% for making up β-sheets amino acids to the total quantity of amino acids in protein.

The obtained results were used for quantitative thermodynamic analysis of hierarchical structure formation in typical protein molecules.

In our paper [3] there is a description of possible solution of Levinthal paradox: complementary sign-changing chiral sequences D–L–D–L and L–D–L–D for DNA and proteins form an Ariadne's thread that directs macromolecular folding along the necessary trajectory toward Levinthal's trap. It is known that a homochiral molecular substance (e.g., an amino acid or carbohydrate solution) undergoes racemization to equalize the concentration of enantiomers, raising the system's entropy to its maximum and lowering its free energy. Meanwhile, a linear homochiral polymer is able to reduce the free energy not only through monomer racemization ("horizontal racemization") but through "vertical racemization" as well (i.e., by creating higher level structures with the different chirality sign). In this case the "right-handed" and the "left-handed" structural units of different structural levels are the particles of the system. The system thus distributes its homochirality, lowering its free energy and leading to the development, among some (or all) macromolecules, of a more stable (longer lived, harder) shell, as opposed to the initial state.

3. Conclusions

A systemic regularity of molecular biology is considered: starting from the level of asymmetric carbon in the deoxyribose and amino acids there is a tendency toward alternating chirality sense of intramolecular structural levels of DNA and proteins, namely D-L-D-L for DNA and L-D-L-D for proteins. A special case of chirality is helicity.

In intermolecular interactions the sense of chirality of the highest intramolecular structural level directly involved in the interaction prevails in each of the participants. For molecules of the same nature (protein-protein, DNA-RNA, tRNA-mRNA, ribozymes) the interaction is realized mainly in the case of the same sense of chirality, either L-L, or D-D, and for molecules of different types (DNA-protein. tRNA-amino acids, enzyme-substrate) in the case of different senses, either D-L, or L-D.

Alternating sense of chiral hierarchy of conjugated levels of macromolecular structures in proteins and nucleic acids is of general biological importance: it determines discreteness of the levels, serves as an instrument of the folding, provides a structural basis for "preferred collective" (or "macroscopic mechanical") degrees of freedom in constructions of macromolecular machines, as well as one of the mechanisms of block/saltatory development of the evolutionary process.

A new principal concept is put forward: homochirality of primary structures of DNA and proteins determines the amount of entropic component of the free energy which is used in the processes of folding and molecular rearrangements.

The resulting hierarchy of conjugated sign-alternating chiral structures kinetically stabilizes them by inhibiting spontaneous racemization; forms a conjugated system with selected degrees of freedom, which makes the work of biological machines possible; in living systems macromolecular and autowave chiral structures are conjugated as a consequence of their channeling through the structures of chiral biological machines; defines the vector of the general development of a system in the direction of an upper, "open-ended" hierarchical level [10].

Acknowledgements

The authors would like to thank Yakovenko L V, Sidorova A E, Ilchenko S A, Skurida V D and Smirnov A V for their contribution to this research.

This work was supported by the grant of the Russian Scientific Fund (project №14-50-0029).

References

[1] Tverdislov V A, Malyshko E V and Ilchenko S A 2015 From autowave mechanisms of selfassembly to molecular machines *Bulletin of the Russian Academy of Sciences. Physics* **79(12)** 1516-20

[2] Tverdislov V A 2013 Chirality as a primary switch of hierarchical levels in molecular biological systems *Biophysics* **58(1)** 128–32

[3] Tverdislov V A, Sidorova A E and Yakovenko L V 2012 From symmetries to the laws of evolution. I. Chirality as a means of active media stratification *Biophysics* **57(1)** 120

[4] Moutevelis E and Woolfson D 2009 A Periodic Table of Coiled-Coil Protein Structures *J. Mol. Biol.* **385** 726–32

[5] Testa O D, Moutevelis E and Woolfson D N 2009 CC+: a relational database of coiled-coil structures *Nucleic Acids Res.* **37** D315-22

[6] http://www.rcsb.org/; Berman H M, Westbrook J, Feng Z, Gilliland G, Bhat T N, Weissig H, Shindyalov I N and Bourne P E 2000 The Protein Data Bank *Nucleic Acids Research* **28** 235-42

[7] Tverdislov V A Chirality as an Instrument of Stratification of Hierarchical Systems in Animate and Inanimate Nature *arXiv*:1212.1677

Visualization of morphological features of chitosan microtubes during biodegradation

T S Babicheva, N O Gegel and A B Shipovskaya
Educational and Research Institute of Nanostructures and Biosystems, Saratov State University 83 Astrakhanskaya St., Saratov 410012, Russian Federation

Abstract. Morphological features were visualized and the weight loss of chitosan microtubes during *in vitro* biodegradation was evaluated in two model media (an enzyme medium of lysozyme and the phosphate buffer) by means of scanning electron microscopy and gravimetry. It was observed that during incubation in the enzymatic medium, the relief and morphology of the microtube surface were changing, as expressed in the appearance of micropores, cracks and bulk defects. When keeping the samples in the phosphate buffer, no substantial defects were formed, only few micropores appeared. The weight loss of the samples after 180 days of storage in the lysozyme medium and phosphate buffer was 30 and 7%, respectively.

1. Introduction

Biodegradable polymeric materials in the form of films, pins, and complex 3D structures attract increasing interest in reconstructive medicine [1-4]. Such materials are used as drainage-separation systems and connecting elements. Topical is the design of bioengineering structures, wherein a polymer serves as a matrix for cell adhesion and proliferation. To obtain such materials, biodegradable synthetic [4, 5] or natural [6-8] polymers are used. For example, in the case of synthetic polymers, linear aliphatic polyesters (polyglycolic acid, polylactic acid) [5] are most often used. In the case of natural polymers, proteins, such as collagen [6] and fibrin [7], are used, as well as polysaccharides, among which the use of the aminopolysaccharide chitosan [8, 9] is most common. For example, the use of bioresorbable microtubes can avoid their further remodeling. In previous papers [10, 11] we described the preparation and formation mechanism of microtubes from chitosan solutions. The results of our study of the physicochemical, mechanical and biological properties of such microtubes have shown that they may be a promising material for designing biodegradable vascular prosthesis [12]. In this case, a microtube would act as a hollow frame, whose surface the endothelial cells would adhere to and proliferate on, thereby forming a new living organ. The polymer backbone would bioresorb in a natural way under the influence of host enzymes. In this context, further research should involve studying of the biodegradation process of chitosan microtubes. Scanning electron microscopy (SEM) is known to be an informative tool of research of synthetic and biological polymer objects. This method allows obtaining information of the surface morphology and topography of the material examined with sufficient accuracy. SEM has been successfully used to study structural features of chitosan-containing materials, such as composite films [13], nanoparticles [14], hydrogels [15], the transformation of films into hollow cylindrical structures [16], etc. It provides a basis for the possible use of SEM for visualizing structural changes occurring in biodegradable materials.

In this paper, SEM was used to visualize morphological changes on chitosan microtube walls during their *in vitro* biodegradation in two modeling media, namely: an enzymatic lysozyme medium and phosphate buffer. Chitosan-based materials are known to biodegrade in the presence of non-specific enzymes of hydrolase class, among which lysozyme is most common [17, 18]. Therefore, our choice of a hydrolytic enzyme was due to the ability of lysozyme to break the β–(1, 4)–glycosidic bonds between the glucopyranose cycles of macromolecular chains [19].

2. Methodology

Microtubes were obtained from a chitosan solution in glycolic acid by an interfacial reaction of salting out chitosan glycolate into the form of a polybase. Chitosan with a molecular mass of 700 kDa, a deacetylation degree of 80 mol. % (Bioprogress Ltd., Russian Federation) and 1.5% aqueous solution of glycolic acid (Sigma-Aldrich, England) were used. A 4 wt. % chitosan solution was prepared by dissolving an air-dry weighed sample of the polymer powder in the presence of an acid while stirring with a magnetic stirrer at room temperature for 7 h. The finished solution was left for 24 h to remove air bubbles and used to form microtubes. NaOH was used as the salting-out agent. To obtain microtubes, a rod with a layer of the chitosan solution was first held in the salting-out medium for 1 min, then dried at a temperature of 50°C until the solvent completely evaporated and re-immersed into the salting-out solution for 3 h. The process of obtaining microtubes with a detailed description of all the stages is described elsewhere [10, 11].

To simulate the biodegradation process, chitosan microtubes were placed into a solution of the immobilized enzyme lysozyme (Medigen Ltd., Russian Federation) with a concentration of 2 mg/ml in phosphate buffer (pH = 7.4). A phosphate buffer solution with no enzyme (pH = 7.4) was used as a control. All solutions used in our experiments were sterilized by autoclaving at 120°C under 0.11 MPa for 45 min. The enzymatic and control model systems were incubated in a thermostat at 37°C throughout the experiment.

The biodegradation process was studied as follows. Samples were weighed and placed into the model medium: either the lysozyme solution or phosphate buffer (control). The fermentation medium was freshened every 30 days to avoid loss of the lysozyme activity. The experiment lasted 6 months. The total number of samples taken for this experiment was 48. Every 30 days, 3 samples were withdrawn from the experiment to estimate the weight loss, and one served to examine changes in the surface morphology. Microtubes were dried by standard methods.

The weight loss was calculated by the formula:

$$\Delta m = ((m_0 - m) / m_0) \cdot 100\% ,$$

where m_0 is the mass of the original microtube sample, considering its humidity (the moisture content was determined in preliminary experiments), *g*; *m* the mass of the sample after exposure to the model medium and dried to constant weight, *g*.

In special experiments, the microtube was separated from the enzymatic and control systems with a cellulose dialysis membrane. Dialysis was performed against phosphate buffer. To analyze biodegradation products, the qualitative Rouemann reaction for free amino acid was used. Optical density was measured at λ = 545 nm on a semiautomatic flatbed photometer Stat Fax 4200 (USA). A calibration line for the glucosamine concentration evaluation was plotted using the amino acid glycine (chemically pure, Vekton, Russian federation) and ninhydrin (chemically pure, Diaem Lyd., Russian Federation).

The surface morphology of chitosan microtubes was examined by SEM. Photos were obtained on a MIRA \\ LMU microscope (Tescan, Czech Republic) at a voltage of 15 V and a conductive current of 400 pA. Sample preparation was performed as follows: a sample was removed from the model media and filled with chlorhexidine for 10 min. It was then washed with sterile water, cut along and kept in a 96% ethanol solution for 30 min. Then it was picked up from the alcoholic medium and dried in air at room temperature for 24 h. A gold layer was sputtered onto the resulting air-dried sample (a 5 nm thickness) on a K450X Carbon Coater device (Germany).

3. Results and discussion

Degradation is performed in microtubes two systems: in the enzyme lysozyme and the medium of phosphate buffer medium. It is known from the literature that lysozyme is contained in many human biological fluids and serves as a nonspecific antibacterial barrier [20]. In this connection, the enzyme solution was chosen as the model of wound, phosphate buffer - as a model of the blood plasma. By assessing the kinetics of biodegradation experiment results microtubes samples held for 6 months are shown in the table.

Table. The kinetics of microtubule mass loss of the sample of chitosan in phosphate buffer and lysozyme fermentation medium.

The holding time, day	Weight loss (Δm), %	
	Phosphate buffer (control)	Lysozyme solution
30	3	12
60	3	23
90	3	24
120	3	25
150	5	27
180	7	30

As seen from the table, the highest weight loss was realized in the fermentation medium, namely: 30% after 180 days of incubation. Just in the first 60 days of incubation the sample loses 23% of its weight. In the next 120 days of incubation, the overall weight loss is 7%. In the phosphate buffer medium (control) the weight loss was only 7% for the entire duration of the experiment, probably due to dissolution of the low-molecular-weight polymer fractions. The higher weight loss of the microtube in the lysozyme medium is logically explained by hydrolytic chitosan degradation, which proceeds by rupture of the β–(1, 4)–glycosidic bonds in the macromolecules by the enzyme.

Carrying out the Reuemann qualitative reaction confirmed the course of the biodegradation process of our chitosan microtubes by the mechanism of breakage of the β– (1,4) –glycosidic bond. In the enzymatic hydrolysis process, the samples are destroyed to chitosan *n*-mers and glucosamine. This is confirmed by the fact that a blue-violet complex was formed as a result of the interaction of the biodegradation products with ninhydrin, the color intensity being proportional to the free amino acid amount. It was established spectrophotometrically that the total concentration of the hydrolyzed chitosan products was 3 μg/ml for 10 days of the experiment.

At the next stage, the surface morphology of both source microtubes and those kept in the model media was evaluated by SEM. The outer and inner wall surfaces of the source samples taken for the experiment were visually smooth without any visible defects. In the process of keeping in the model media, the microtube wall surface morphology underwent changes. Typical SEM photos of the inner surface wall of the tested samples are shown in Figure. Let us analyze the morphological changes occurred in the samples after 6 months of storage in the lysozyme medium and phosphate buffer.

No occurrence of serious defects on the sample surface was observed when keeping a microtube in phosphate buffer during 180 days (Figure a–c). Only the formation of micropores was observed in the process of storage and even only under prolonged storage of the sample in the buffer medium (Figure c). For the sample kept in the fermentation lysozyme medium, the formation of micropores was also characteristic. However, the changes in the surface morphology of these samples were significantly different from the control ones. E.g., small bulk defects appeared along with micropores on the microtube surface after 120 days of storage (Figure d). At the 150th day of storage, both the pore size and their total number over the entire wall surface increased. The bulk defects acquired the view of extended cracks and depressions (Figure e). After 180 days of exposure in the lysozyme medium, the microtube surface micro-relief changed more and acquired a pronounced loose structure (Figure f).

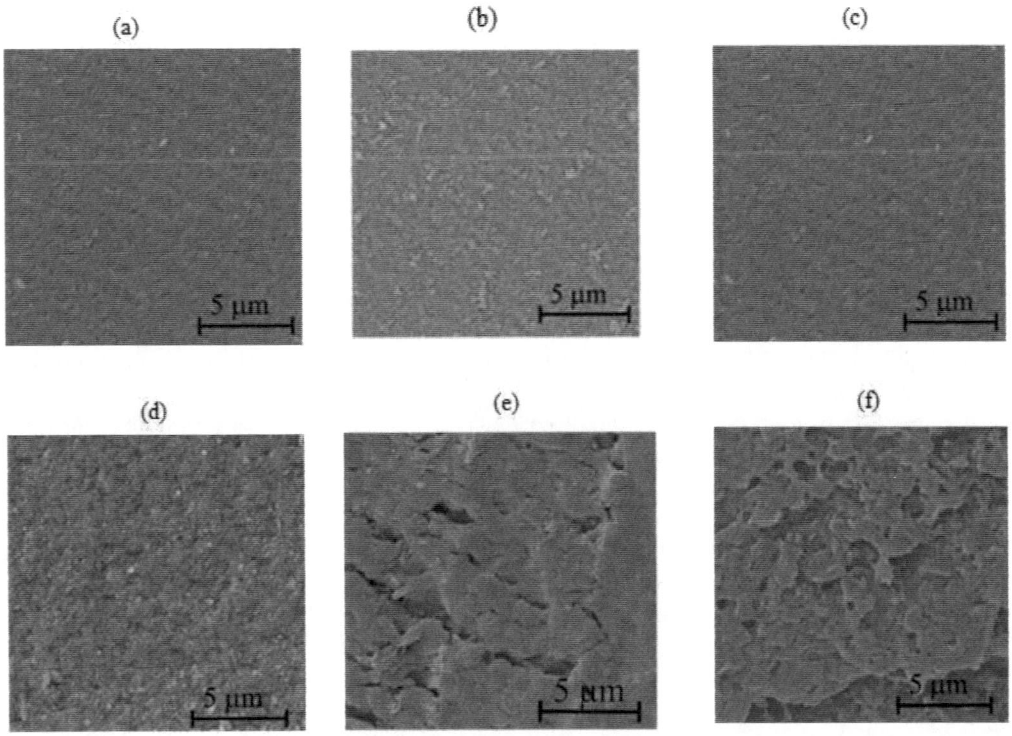

Figure. SEM photos of the inner chitosan microtube surface stored at 37°C during 120 (a, d), 150 (b, e), and 180 (c, f) days in phosphate buffer (a–c) and in the fermentative lysozyme medium (d–f).

Therefore, SEM has allowed visualization of the changes occurring in the surface topography and structure of our chitosan microtubes under the influence of the model media. The most significant changes in the sample surface were realized under the influence of the non-specific enzyme lysozyme (a hydrolytic enzyme).

4. Conclusions

The biodegradation of our chitosan microtubules was studied *in vitro* in the fermentation lysozyme medium and phosphate buffer. It was established that after 6 months of enzymatic hydrolysis and storage in the buffer the sample weight decreased by 30 and 7%, respectively. In the fermentation medium, the maximum mass loss was realized during the first 60 days of storage. Further, the rate of destructive processes reduced. During enzymatic degradation, micropores, cracks, and other bulk defects appeared on the walls of the microtubular substrates, up to the formation of a pronounced loose structure. It is suggested that the degradation process in the lysozyme medium goes through the mechanism of rupture of chitosan macrochains by the β–(1, 4)–glycosidic bonds. The rate and degree of biodegradation of the examined samples of chitosan microtubes are comparable with the corresponding values of these parameters demanded for prosthetic blood vessels made of bioresorbable materials [21]. Our experiments allow considering chitosan microtubes as promising bioartificial materials with controlled duration and degree of biodegradation *in vivo*.

Acknowledgments

This work was supported by the Russian Foundation for Basic Research, project No 16-33-00953 «Scientific fundamentals of the preparation of 3D biodegradable structures from natural polysaccharides to create blood vessel analogs».

References

[1] Teo A J, Mishra A, Park I, Kim Y J, Park W T and Yoon Y J 2016 *ACS Biomater. Sci. Eng.* **2** **4** 454-472
[2] Annabi N, Tamayol A, Uquillas J A, Akbari M, Bertassoni L E, Cha C and Khademhosseini A 2014 *Advanced Mater.* **26** **1** 85-124
[3] Moisenovich M M, Pustovalova O L, Arhipova A Yu, Vasiljeva T V, Sokolova O S and Bogush V G 2010 *Mater. Research.* **96A** **1** 125–131
[4] Pham Q P, Sharma U and Mikos A G 2006 *Biomacromol.* **7** 2796–2805
[5] Uematsu K 2005 *Biomaterial.* **26** **20** 4273-79
[6] Pabbruwe M B, Esfandiari E, Kafienah W, Tarlton J F and Hollander A P 2009 *Biomaterial.* **30** **26** 4277-86
[7] Altman G H, Diaz F, Jakuba C, Calabro T, Horan R L, Chen J, Lu H,Richmond J and Kaplan D L 2003 *Biomater.* **24** 401- 416
[8] Dmitriev Y A, Shipovskaya A B and Kossovich L Y 2011 *Izvestiya Vysshikh Uchebnykh Zavedenii. Seriya Khimiya I Khimicheskaya Tekhnologiya* **54** **11**
[9] Kashyap M, Archana D, Semwal A, Dutta J and Dutta P K 2016 *Chitin and Chitosan for Regenerative Medicine* 261-277
[10] Gegel N O, Shipovskaya A B, Vdovykh L S and Babicheva T S 2014 *J. Soft Mat.* Article ID 863096 p.9
[11] Babicheva T S, Gegel N O and Shipovskaya A B 2015 *J. Nat. Sci. Sustainable Techn.* **9** **2** 285-295
[12] Gegel N O, Babicheva T S and Shipovskaya A B 2015 *Butlerov Comm.* **41** **3** 44-53
[13] Cárdenas G, Anaya P, Del Rio R and Schrebler R 2010 *J. Chil. Chem. Soc.* **55** **3** 352-58
[14] Fazilova S A, Yugai S M and Rashidova S S 2010 *Chem.of plant raw mat.* **1** 13-19
[15] Malaise S, Rami L, Montembault A, Alcouffe P, Burdin B, Bordenave L, Delmond S and David L 2014 *Mater. Sci. Eng. Mater. Biol. Appl.* **42C** **5** 374-384
[16] Gulati K, Johnson L, Karunagaran R, Findlay D and Losic D 2016 *Biomacromolec.* **17** **4** 1261 71
[17] Kean T and Thanou M 2010 *Adv. Drug Delivery Rev.* **62** **1** 3-11
[18] Kulish E I, Chernov V V, Volodin V P and Kolesov S V 2008 *Bull. Bashkir University.* **13** **1** 23-26
[19] Zhang L, Dou S, Li Y, Yuan Y, Ji Y, Wang Y and Yang Y 2013 *Mater. Sci. Eng.* **C. 33** **5** 2626-31
[20] Cutting K F 2003 *British j. of community nursing.* **8** 4-9
[21] Cleary M A, Geiger E, Grady C, Best C, Naito Y and Breuer C 2012 *Trends in molecular medicine.* **18** **7** 394-404

Mass spectrometry for identification of proteins that specifically bind to a distal enhancer of the *Oct4* gene

E I Bakhmet[1], I B Nazarov[1], T O Artamonova[2], M A Khodorkovsky[2], A N Tomilin[1]
[1]Laboratory of Molecular Biology of Stem Cells, Institute of Cytology, RAS, St-Petersburg 194064, Russia
[2]Polytechnical State University, St-Petersburg 195251, Russia

Abstract. Transcription factor Oct4 is a marker of pluripotent stem cells and has a significant role in their self-renewal. *Oct4* gene is controlled by three cis-regulatory elements – proximal promoter, proximal enhancer and distal enhancer. All of these elements are targets for binding of regulatory proteins. Distal enhancer is in our research focus because of its activity in early stages of embryonic development. There are two main sequences called site 2A and site 2B that are presented in distal enhancer. For this moment proteins which bind to a site 2A (CCCCTCCCCCC) remain unknown. Using combination of in vitro method electrophoretic mobility shift assay (EMSA) and mass spectromery we identified several candidates that can regulate *Oct4* gene expression through site 2A.

1. Introduction

In the past decade, the topic of induced pluripotent stem cells (iPSCs) is developed rapidly [1]. Like embryonic stem cells (ESCs), these cells can differentiate into all three germ layers - endoderm, mesoderm and ectoderm. Besides, iPSCs may be derived from terminally differentiated adult cells, unlike ESCs that are presented only in the early stages of embryogenesis. Transcription factor Oct4 is a marker of iPSCs and ESCs and indispensable for the induction of pluripotency [2, 3]. *Oct4* gene is exposed for a very delicate regulation - as an increase or decrease of its expression in less than 50% leads to a differentiation of pluripotent stem cells [4]. Oct4 transcription is controlled through the proximal promoter (PP), a proximal enhancer (PE) and a distal enhancer (DE). These DNA sequences are heavily methylated in differentiated cells, where Oct4 is not expressed, but during the onset of *Oct4* transcription, these sequences become targets for binding of regulatory proteins. The DE, required for Oct4 expression in the inner cell mass and cultured ESCs, contains two key elements - sites 2A and 2B. Heterodimer Oct4-Sox2 is bound to a site 2B, but proteins that interact with the site 2A (CCCCTCCCCCC) remain unknown [5]. Our work is dedicated to the identification of 2A site-binding proteins and dissection of their roles in the regulation of expression of the *Oct4* gene.

2. Results

It has been shown that proteins that are specifically bound with site 2A are present not only in stem cells D3, but also in differentiated NIH 3T3 cells [5]. Using electrophoretic mobility shift assay (EMSA) and biotinilated DNA-probes with site 2A sequence we have shown that similar proteins are also presented in murine thymus and brain (Figure 1).

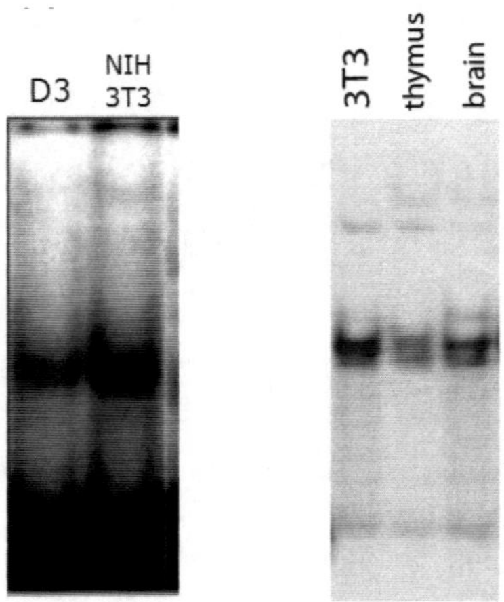

Figure 1. Proteins that interact with site 2A are presented not only in stem cells.

For derivation of these proteins we used mouse brain. Using affinity chromatography, the first purification was performed (Figure 2). Fractions were checked for proteins of interest through EMSA.

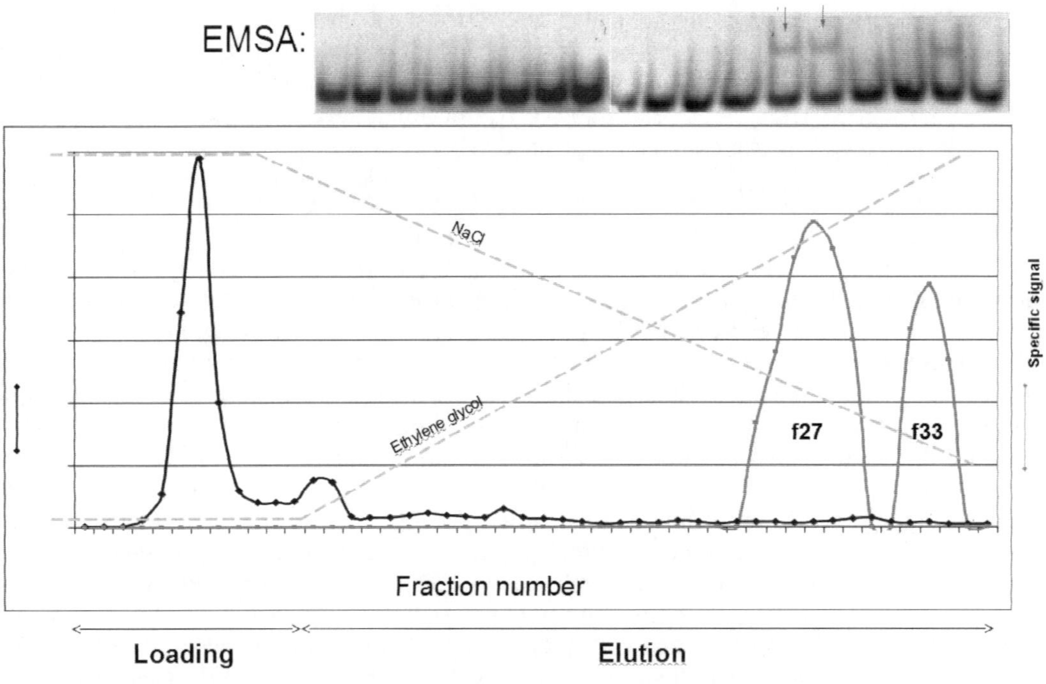

Figure 2. Chromatography of murine brain extract on phenyl-sepharose CL 4B. The elution was performed with linear gradient of NaCl (0.5 M – 0.1 M) and ethylene glycol (0 – 50%) concentrations.

Further, we used fractions that brought proteins of interest. Four different labeled probes were used to verify proteins that interact with site 2A. WT – wild type, and three mutant probes: M1, M2, M3 in order to get rid of nonspecific proteins (Figure 3).

WT GATCTAGCCCGACCCTGcccctcccccAGGATC
CTAGATCGGGCTGGGACggggagggggTCCTAG

M1 GATCTAGCCCGACCCTGcaaagaaaacCAGGATC
CTAGATCGGGCTGGGACgtttcttttggTCCTAG

M2 GATCTAGCCCGAAAAGAcccctcccccAGGATC
CTAGATCGGGCTTTTCTggggagggggTCCTAG

M3 GATCTAGCCCGACCCTGcccctaaaaaaAGGATC
CTAGATCGGGCTGGGACggggattttttTCCTAG

Figure 3. Sequences that were used for derivation of proteins that interact with site 2A.

For futher purification of the specifically interacting proteins with DE site 2A, we used the affinity chromatography and checked fractions from first cleaning step. Wild type (wt) and mutant (m1) site 2A sequences were immobilized on sepharose particles. These particles were used for derivation of target proteins from fractions. Elution was performed in gradient of salt. Received elutions were checked by electrophoresis in acrylamide gel (Figure 4).

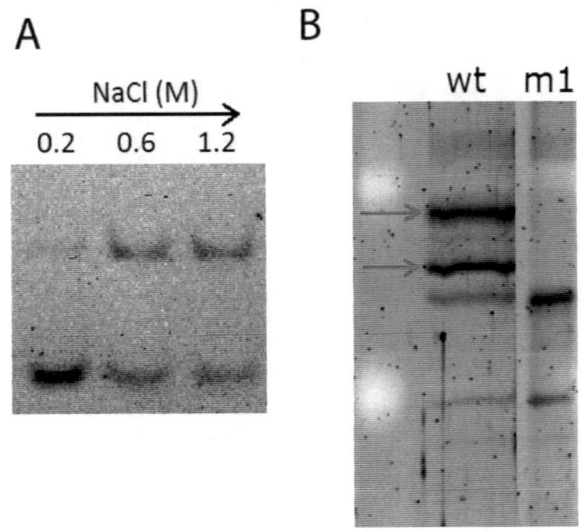

Figure 4. A - EMSA of eluted proteins with stepwise gradient of salt. B - SDS PAGE of proteins eluted from wt and m1 columns. Red arrows points the protein bands that were specific to eluates from wt column.

Specific bands were cut from gel and proteins were identified by Varian 902-MS MALDI Mass Spectrometer (ICR FTMS). A third harmonic of the Nd: YAG laser (355 nm) was used to irradiate the sample. Spectra were removed in the regime of positive ions. The accuracy of the measured monoisotopic masses after the calibration of the autolysis peaks of trypsin (842, 5094 Da) and keratin (2383.9525 Da) was less than 6 ppm. The spectra were obtained in the mass range 500-4500 m/z, choosing the laser power is optimal for achieving the best resolution.

We have found that in vitro, the 2A site specifically binds to poly-(rC) binding proteins, such as Pcbp1 and Pcbp2 (Table1 and Table2). These proteins are known to be involved in regulation of transcription of several genes such as BRCA1 and μ-opioid receptor [6].

Table 1. Data of mass spectrometry analysis for Pcbp1 (red peptides were identified in mass spectrometry analysis)

Name	Poly(rC)-binding protein 1
Amino acid sequence	1 MDAGVTESGL NVTLTIRLLM HGKEVGSIIG KKGESVKRIR EESGARINIS 51 EGNCPERIIT LTGPTNAIFK AFAMIIDKLE EDINSSMTNS TAASRPPVTL 101 RLVVPATQCG SLIGKGGCKI KEIRESTGAQ VQVAGDMLPN STERAITIAG 151 VPQSVTECVK QICLVMLETL SQSPQGRVMT IPYQPMPASS PVICAGGQDR 201 CSDAAGYPHA THDLEGPPLD AYSIQGQHTI SPLDLAKLNQ VARQQSHFAM 251 MHGGTGFAGI DSSSPEVKGY WASLDASTQT THELTIPNNL IGCIIGRQGA 301 NINEIRQMSG AQIKIANPVE GSSGRQVTIT GSAASISLAQ YLINARLSSE 351 KGMGCS
Score	64 (p<0.05)
Covering	48%

Table 2. Data of mass spectrometry analysis for Pcbp2 (red peptides were identified in mass spectrometry analysis)

Name	Poly(rC)-binding protein 2
Amino acid sequence	1 MDTGVIEGGL NVTLTIRLLM HGKEVGSIIG KKGESVKKMR EESGARINIS 51 EGNCPERIIT LAGPTNAIFK AFAMIIDKLE EDISSSMTNS TAASRPPVTL 101 RLVVPASQCG SLIGKGGCKI KEIRESTGAQ VQVAGDMLPN STERAITIAG 151 IPQSIIECVK QICVVMLESP PKGVTIPYRP KPSSSPVIFA GGQDRYSTGS 201 DSASFPHTTP SMCLNPDLEG PPLEAYTIQG QYAIPQPDLT KLHQLAMQQS 251 HFPMTHGNTG FSGIESSSPE VKGYWAGLDA SAQTTSHELT IPNDLIGCII 301 GRQGAKINEI RQMSGAQIKI ANPVEGSTDR QVTITGSAAS ISLAQYLINV 351 RLSSETGGMG SS
Score	57 (p<0.05)
Covering	46%

At the moment, we are focused on confirming our results in vivo using methods of chromatin immunoprecipitation (ChIP) and gene knockout based on CRISPR-Cas9 system.

Acknowledgments
The work was supported by the Russian Science Foundation (grant 17-14-01407).

References

[1] Takahashi K, Yamanaka S. Cell 2006;126:663-76.
[2] Wu G, Scholer HR. Cell Regeneration 2014;3:10.
[3] Kim JB, Sebastiano V, Wu G, Arauzo-Bravo MJ, Sasse P, Gentile L, et al. Cell 2009;136:411-9.
[4] Niwa H, Miyazaki J-i, Smith AG. Nat Genet 2000;24:372-6.
[5] Okumura-Nakanishi S, Saito M, Niwa H, Ishikawa F. The Journal of biological chemistry 2005;280:5307-17.

[6] Choi Hyun W, Joo Jin Y, Hong Yean J, Kim Jong S, Song H, Lee Jeong W, et al. Stem Cell Reports 2016;7:911-26.

Biocompatible Er:(Na,K)NbO3 nanofibers

N P Markova[1], A M Grishin[1,2,3]
[1]Petrozavodsk State University, Petrozavodsk 185000, Russia
[2] KTH Royal Institute of Technology, Stockholm-Kista SE-164 40, Sweden
[3] INMATECH Intelligent Materials Technology, Skärholmen SE-127 51, Sweden

Abstract. Dense homogeneous fabric composed from continuous bead-free erbium-doped sodium potassium niobate (Er:NKN) 100 μm long and 100-200 nm in diameter nanofibers was sintered by sol-gel calcination assisted electrospinning technique. Er doping with the concentration of 2 at.% provides readily detectable room-temperature broad-band photoluminescence (PL) centered at λ_{PL} = 0.55 and 0.98 μm being pumped, respectively, with 532 and 785 nm lasers. Electric field induced resistance switching and strong electric rectification effect were found in nanoporous sandwich Au/Er:NKN/Pt capacitive cell. Memristor-type current-voltage *I-V* characteristics originate from the electrochemical migration of oxygen vacancies at the *n*-type Er:NKN oxide/high work function Pt cathode junction interface.

1. Introduction

Ferroelectric sodium potassium niobate (hereinafter NKN) ceramics was patented and FDA-approved (U.S. Food and Drug Administration) as a biocompatible material for implants.[1] Thorough toxicology tests showed there were no any bacterial products (endotoxin) appear as well as viability of human monocytes was not negatively affected by the presence of NKN ceramics. In this paper, we report electrical and optical properties of Er-doped NKN nanofibers. Erbium doping promises new photonic applications to $(Na,K)NbO_3$ ceramics which is already abundantly supplied with piezoelectric, electrostrictive, low microwave loss voltage-tunable dielectric, and electro-optic properties.

2. Experimental details

In this work, fabrication of $Er{:}Na_xK_{1-x}NbO_3$ fibers was performed by the electrospinning method. Electrospinning is a simple and inexpensive method for producing nanofibers with submicron and nanometer diameters. At first, fibers are formed by injecting onto metallic collector a stream of polymer – precursors solution in electrostatic field. Then, fibers are subjected to drying and annealing at high temeparture to vuporize a polymer and to crystallize a rest of inorganic material. Actually, electrospinning is a version of the sol-gel technology.

To prepare NKN precursor solution by the sol-gel method, sodium $NaO_2C_2H_3{\times}H_2O$ and potassium $KO_2C_2H_3{\times}3H_2O$ acetates were mixed in 2-methoxyethanol $CH_3OCH_2CH_2OH$ at room temperature and stirred for 1 hour to achieve a clear and transparent solution. Niobium ethoxide $C_{10}H_{25}NbO_5$ and erbium nitrate pentahydrate $Er(NO3)_3{\times}5H_2O$ (chosen with the concentration of 2 at.%) were dissolved at room temperature in acetyl acetone $C_5H_8O_2$ (used as a chelating agent), stirred for 1 hour, then added to the Na/K precursor solution, and continued stirring for 24 hours in a closed cap glass ware. Finally, to prepare the solution for the electrospinning, high-molecular polyvinylpyrrolidone PVP (PVP, 0.05 g/ml) was added and completely dissolved in the precursor mixture.

IOP Conf. Series: Journal of Physics: Conf. Series **917** (2017) 042028 doi:10.1088/1742-6596/917/4/042028

Viscous polymer stream was ejected from a plastic syringe pump NE-300 that feeds PVP/Er:NKN solution at a constant rate of 0.5 mL/hour in electric field of 1.8 kV/cm applied between metallic needle and aluminum foil collector (Figure 1). Bead-free composite polymer-hydrolyzed Er:NKN precursor fiber mat was collected from the surface of the collector and dried at 100 °C in nitrogen atmosphere for 12 hours. After drying, as-spun mat consists of interlaced binder-contained jelly-like 350 nm thick threads with a smooth surface (Figure 2). After the process of electrospinning, to remove a polymer binder, the annealing of obtained fibers was carried out in a high-temperature vacuum furnace OTF-1200X in air at 800 °C. The rate of heating/cooling was 5 °C/min.[2]

Surface morphology of annealed nanofibers was characterized using scanning electron microscope (SEM) Hitachi SU1510. Phase content and crystalline structure were examined by X-ray diffraction (XRD) using Siemens D5000 diffractometer. Quantitative elemental analysis was performed by using SEM energy-dispersive X-ray spectroscopy (EDX) analyzer with Thermo Scientific UltraDryTM detector.

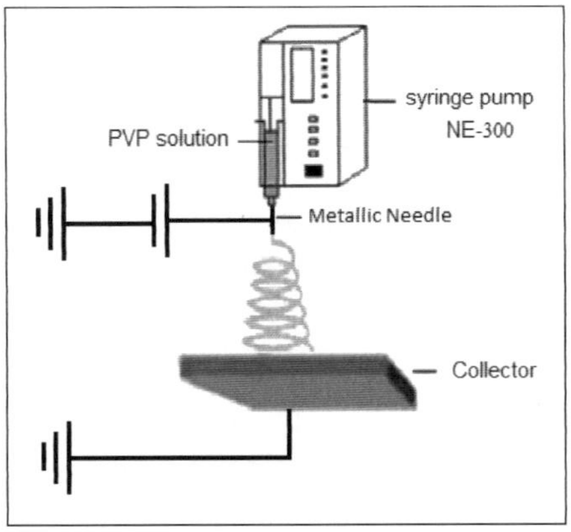

Figure 1. Schematic diagram for the electro-spinning setup.

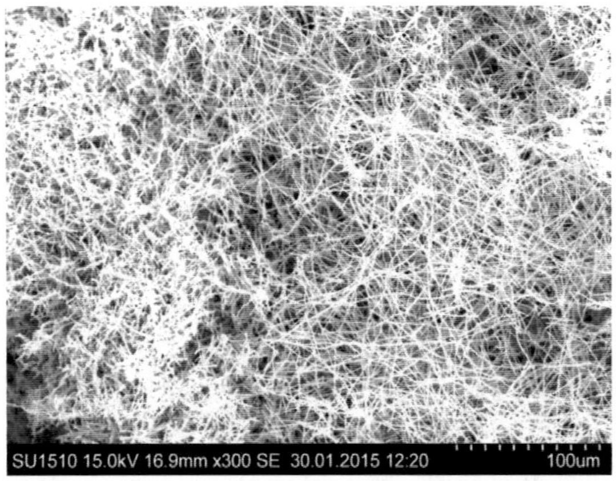

Figure 2. SEM images of as-spun and dried fibers.

3. Conclusion

Scanning electron microscope images of as-spun and dried NKN fibers captered at accelerating voltage of 20 kV are shown in Figure 2. As-spun nanofibers have diameter of about 350 nm. After annealing diameter of fibers significantly reduced to values of about 100 nm. Nanothreads experience strong shrinkage due to the vaporization of PVP and solvents during the annealing process. Because of nitrogen and carbon presence in NKN precursor solution, big amount of these elements was expected in fibers before their calcination. Based on our recent obervations of nanofibers fabrication by electrospinning using PVP, we concluded that complete polymer and solvent removal occurs at temperature around 550 °C. This was confirmed from quantitative elemental analysis of EDX spectra recorded at accelerating voltage of 10 kV. These results are collected in Table 1. After calcination, the nitrogen was not detected at all and carbon content was about 3–4 at.%. This indicates thermal decomposition of PVP into volatile substances that leaving nanofibers. In fact the concentration of the carbon content in calcined nanofibers can be even lower. Additional contamination might occurs as an unnecessary carbon deposition during EDX measurements. This is so called black box effect is commonly attributed to the oil carbon vapour from a fore vacuum pump of SEM.

Figure 3 depicts the trace of thermogravimetric analysis (TGA) during calcination of our as-spun Er-doped Er:(Na,K)NbO_3 fibers. The shape of the TGA curve is similar to that for (Na,K)NbO_3 nanofibers sintered earlier (see Ref. [4,5]). It suggests the presence of water and/or residual solvent. The main weight loss occurs below 550 °C indicating that the most of organics and other volatiles have been already removed at this temperature. Noticeable weight loss about 8 % is observed in TGA curve between 25 °C and 100 °C. This weight loss at relatively low temperatures we rely upon the water vaporization. Degradation of PVP exhibits itself at 150–250 °C and causes 15 % of weight loss.

Big and sharp step in the profile of the weight's loss change between 250 °C and 450 °C represents the pyrolysis and/or oxidation of PVP. The total loss of fibers in flowing dry air equals approximately to 63 wt% and signifies almost complete at 550 °C removal of PVP initially added to the precursor colloidal dispersion. Further annealing up to 600–800 °C leads to the crystallization of nanothreads and accomplishes synthesis of NKN fibers.

Table 1 Results of quantitative EDX elemental analysis of Er:(Na,K)NbO_3 nanofibers after calcination.

Element	Mass concentration [%]	Mass concentration error [%]	Atomic concentration [%]	Atomic concentration error [%]
C	1.83	± 0.22	4.1	± 0.35
O	36.58	± 0.35	65.32	± 0.58
Na	8.97	± 0.10	11.24	± 0.06
K	10.17	± 0.08	7.46	± 0.05
Nb	37.56	± 1.06	10.86	± 0.31
Er	4.89	± 0.26	1.02	± 0.03

XRD spectrum of crystallized Er-doped NKN fibers is presented in Θ-2Θ scan in Figure 4. $Na_{0.35}K_{0.65}NbO_3$ perovskite phase predominates there. The relative intensity ratios of NKN Bragg reflections I_{hkl} indicate noticeable preferential NKN(001) orientation: intensity reflection ratio $I_{001}/I_{011} = 0.90$ in Er:NKN fibers compared to 0.58 in "ideal" $Na_{0.35}K_{0.65}NbO_3$ powder.[3]

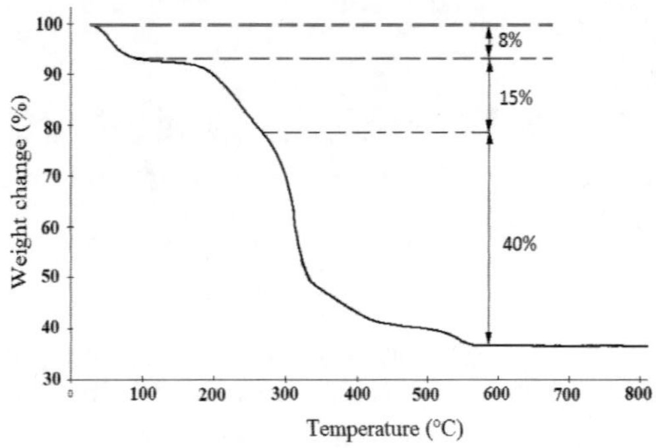

Figure 3. TGA trace during annealing of Er:(Na,K)NbO_3 nanofibers.

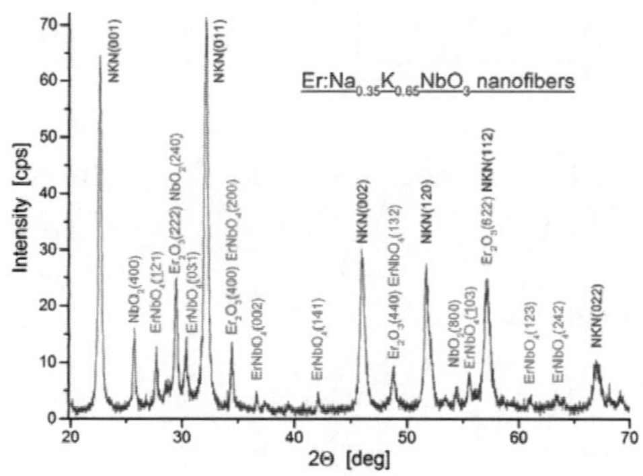

Figure 4. XRD patterns in CuK_α radiation of Er:NKN nanofibers annealed at 800 °C in air.

Electric properties of Er:NKN nanofibers were investigated by a measurement of current-voltage *I-V* characteristics. They were traced for the planar Au/Er:NKN/Au cell with 350 μm thick Er:NKN fiber fabric placed onto glass-ceramic Sitall substrate as well as for the vertical sandvich Pt/Er:NKN (260μm)/Au structure shown, respectively, in the left inset and in the main frame of Figure 5. Circular 0.95 mm in diameter ohmic Au electrodes were thermally evaporated through a shadow mask on the top of nanofiber fabric specimens.

Hysteretic character of I-V plots at weak electric field in the planar Au/Er:NKN/Au cell and positive bias voltage shift U_o when current goes to zero indicate ferroelectric properties of Er:NKN fibers. Pt/Er:NKN/Au sandwich structure exhibits a strong rectification of direct current as well as resistance switching from low-resistance to high-resistance state with the resistance ratio of 10^2. Quasi-static *I-V* characteristics correlate with the results obtained by impedance spectroscopy carried out at frequencies up to 1 MHz.

Intensive room temperature luminescence was observed by pumping fibers with 532 and 785 nm lasers light. Photoluminescence spectra are presented in Figure 6. Excitations at 532 nm and 785 nm laser light illumination are shown with green and red colors, correspondingly. They produce intensive luminescence at λ_{PL} = 0.55 μm (Raman shift 556 cm^{-1}, $^4S_{3/2} \rightarrow {}^4I_{15/2}$) and within the λ_{PL} band from 0.96 to 1.11 μm (Raman shifts from 2300 to 3750 cm^{-1}, $^4I_{11/2} \rightarrow {}^4I_{15/2}$). Radiation of 532 nm excites Raman-active internal vibrations of NbO_6 octahedra that correspond to stretching $1A_{1g}$ (ν_1 = 632 nm), $1E_g$ (ν_2 = 559 nm), and bending $1F_{1u}$ (ν_4 = 437 nm) modes.

Radiation of 785 nm laser excites strong fluorescence lines with the wavelengths from 0.96 to 1.11 μm (Raman shifts from 2300 to 3750 cm^{-1} in the right panel of Figure 6). They are attributed to radiative intra-4f Er^{3+} ion $^4I_{11/2} \rightarrow {}^4I_{15/2}$ transitions. Also, $^4I_{9/2} \rightarrow {}^4I_{15/2}$ transitions cause weaker PL at wavelengths from 0.83 to 0.87 μm (Raman shifts from 700 to 1200 cm^{-1} in the left panel of Figure 6). In addition, $^4I_{15/2} \rightarrow {}^4S_{3/2}$ and $^4I_{15/2} \rightarrow {}^4F_{9/2}$ transitions clearly manifest a resonant Er^{3+} ions absorption of the light reflected from Er:NKN fiber fabric. Corresponding ligh reflection spectrum contains two distinct pits at λ = 524 nm and 658 nm which are in respectable accord with an excitation of $^4S_{3/2}$ and $^4F_{9/2}$ energy states of Er^{3+} ion.

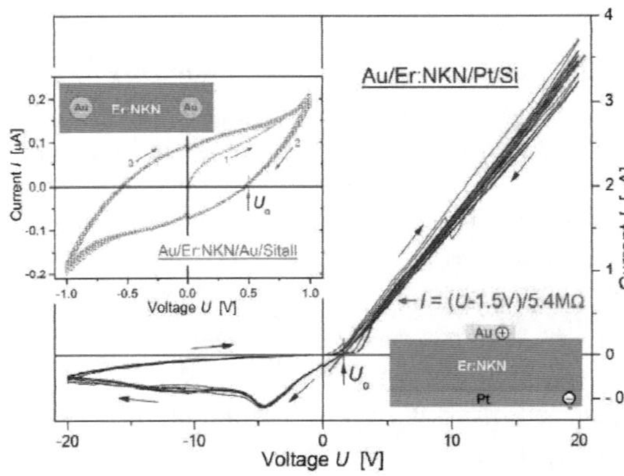

Figure 5. Current-voltage I-V characteristics traced with Keithley 2410 SourceMeter. Main frame - in the vertical Au/NKN (260μm)/Pt/Si cell. Inset - at low voltages in the planar Au/NKN(350μm)/Au cell onto Sitall substrate.

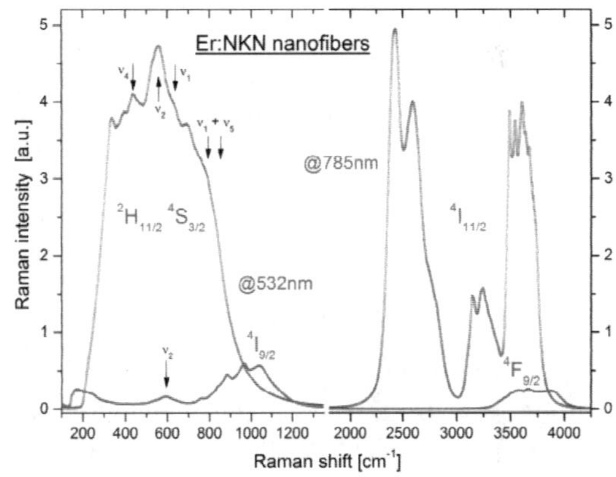

Figure 6. Unpolarized backscattered Raman spectra of Er:(Na,K)NbO_3 nanofibers calcined at 800 °C in air. Excitations at 532 nm and 785 nm laser light pumping are shown with green and red colors, respectively.

4. Conclusion

Erbium doping opens up serious prospects for new photonic applications of ferroelectric (Na,K)NbO_3 ceramics which has already demonstrated a wide range of piezoelectric, electrostrictive, low microwave loss voltage-tunable dielectric, and electro-optic properties.

Acknowledgments

This work was supported by the RF Ministry of Education and Science as a base part of state program no.16.5857.2017 in the scientific field.

References

[1] Nilsson K, Lidman J, Ljungstrom K and Kjellman C. U.S. patent 6,526,984 (4 March 2003).
[2] Grishin A and Markova N, 2017 *J. Am. Ceram. Soc.* **100** 1051.
[3] 2001 JCPDS-International Center for Diffraction Data, Card Nos.: $Na_{0.35}K_{0.65}NbO_3$ (77-0038); NbO_2 (85-0379); Er_2O_3 (76-0159); and $ErNbO_4$ (22-1095).
[4] Jalalian A and A.M. Grishin A 2012 *Appl. Phys. Lett.* **100** 012904.
[5] Jalalian A and Grishin A 2014 *Appl. Phys. Lett.* **104** 243701.

IOP Conf. Series: Journal of Physics: Conf. Series **917** (2017) 052001 doi:10.1088/1742-6596/917/5/052001

The effect of *p*-doping on multi-state lasing in InAs/InGaAs quantum dot lasers for different cavity lengths

V V Korenev[1,2], A V Savelyev[1], M V Maximov[1,2,3], F I Zubov[1], Yu M Shernyakov[1,3], A E Zhukov[1,2]
[1]St. Petersburg Academic University RAS, St. Petersburg 194021, Russia
[2]Peter the Great St. Petersburg Polytechnic University, St. Petersburg 195251, Russia
[3]Ioffe Physico–Technical Institute RAS, St. Petersburg 194021, Russia
E-mail: korenev@spbau.ru

Abstract. The effect of modulation *p*-doping on multi-state lasing in InAs/InGaAs quantum dot (QD) lasers is studied for different levels of acceptor concentration. It is shown that in case of the short laser cavities, *p*-doping results in higher output power of the ground-state optical transitions of InAs/InGaAs QDs whereas in longer samples *p*-doping may result in the decrease of this power component. On the basis of this observation, the optimal design of laser active region and optimal doping level are discussed in details.

1. Introduction

Long-wavelength InAs/InGaAs quantum dot (QD) lasers emitting via QD ground-state (GS) optical transitions of quantum dots near 1.3 μm are important light sources for many practical applications and, in particular, for ultrafast data transmission [1 – 2]. Such lasers are characterized with small injection currents and high temperature stability [3]. However, even if lasing starts at QD GS, further increase in injection may result in an additional short wavelength spectral line centered at 1.2 μm associated with first excited-state (ES) optical transition of QDs [4 – 8]. This, in its turn, results in GS-lasing quenching with the increase in injection current shortly after the onset of multi-state lasing [6, 8]. It is known that *p*-doping may be an effective way to increase the power of GS spectral component [9, 10]. However, such an important effect as an influence of *p*-doping level on multi-state lasing phenomenon was not studied yet. The key goal of this work is to shed light on this question both experimentally and theoretically.

2. Sample growth and experimental results

To study this question experimentally, we have fabricated two series of samples with short and long cavities. All InAs/InGaAs QD lasers studied in the paper were grown using molecular beam epitaxy (MBE) and had the same stripe width of 50 μm. Cavity length was either 0.5 mm (short samples) or 1.0 mm (long samples). The active region of each sample was comprised of 10 layers of InAs/InGaAs QDs separated by 35 nm-thick GaAs spacers, which were *p*-doped with different concentrations of carbon dopant atoms (0, 3, and 5×10^{17} cm^{-3}). The experimental light-current curves of these samples were measured for these two crucial cavity lengths as it is shown in Fig. 1.

As it can be seen from the Fig. 1a, in case of the long samples, the increase in *p*-doping level results in the decrease of W_{max}^{GS} – the maximal output power corresponding to the GS of QDs. At the same time, in case of the short samples the situation is opposite – see Fig. 1b. Such a behaviour may be somehow surprising, as previously it was typically assumed that *p*-doping always leads to the enhancement of the output power corresponding to the GS of QDs as well as improves the other characteristics of QD lasers [8 – 10].

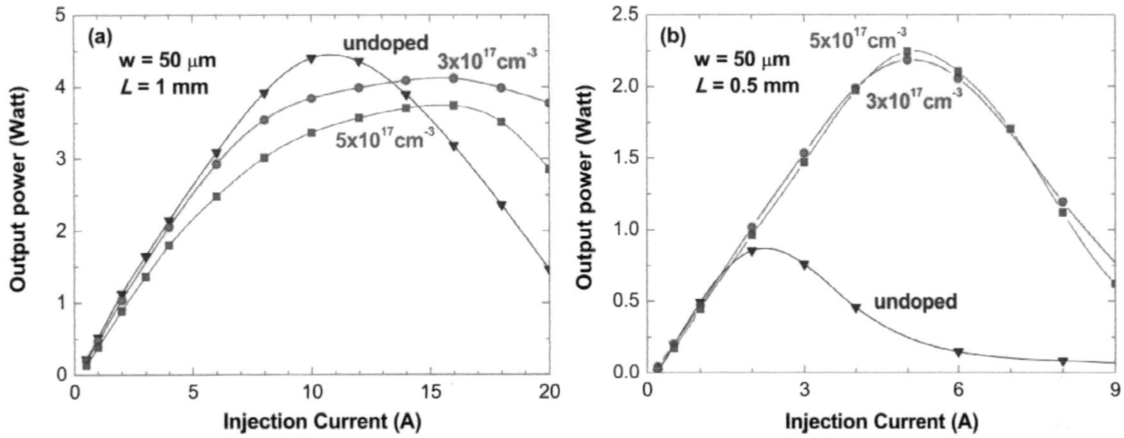

Figure 1(a, b). Experimental dependence of GS and ES components of the output optical power on injection current for *p*-doped InAs/InGaAs QD lasers having different concentrations of carbon dopant atoms at different cavity lengths of (a) 1.0 mm and (b) 0.5 mm. For both figures, the concentration *p*-type dopant varies from 0 to 5×10^{17} cm^{-3}.

3. Theoretical description

To describe theoretically an influence of *p*-doping level on the operation of QD lasers, we used a rate-equation model describing charge carrier dynamics in QDs and photon dynamics in laser cavity in accord with [8]. As it can be seen from the experimental data shown in the Fig. 1, there is a distinct trade-off between the influence of modulation *p*-doping on multi-state lasing in the case of short and long samples. Qualitatively, such a discrepancy can be interpreted as follows.

On the one hand, the increase in *p*-doping level leads to the increase in the maximal output power corresponding to the GS of QDs (W_{max}^{GS}) due to the increased hole capture rate because of the reasons discussed in [8, 9]. In particular, this is due to the higher occupancy of the hole energy levels in InAs/InGaAs QDs. This mitigates the competition for the common holes between GS and ES optical transitions of InAs/InGaAs QDs that is one of the key limiting factors for W_{max}^{GS}. On the other hand, the increase in *p*-doping level typically results in the increase in internal loss (α_{in}), e.g. due to the growth in free carrier absorption and, thus, in the decrease in laser efficiency as it can be seen from the *L-I* curve shown in the Fig. 1a, which becomes less steep as *p*-doping level increases.

In the case of shorter samples having sufficiently high output loss, GS gain is comparable with its saturated value. Therefore, an effect of *p*-doping on the suppression of ES-lasing is significant. At the same time, the increase in the internal loss (α_{in}) is insignificant as compared with radiation output loss (α_{out}). For instance, in the case of the sample with L=0.5 mm, the increase in *p*-doping level from 0 to 3×10^{17} cm^{-3} results in the increase in α_{in} by 0.2 cm^{-1}, while the output loss has a value of ~24 cm^{-1}. Therefore, the external differential efficiency of such a laser remains practically unchanged and W_{max}^{GS} increases from 0.8 to 2.2W as the level of modulation *p*-doping increases from 0 to 5×10^{17} cm^{-3} – see Fig. 1b.

In the case of long samples, the output loss (α_{out}) is much smaller as compared to the short samples and GS gain is far from its saturated value. Therefore, the usage of modulation *p*-doping does not contribute significantly to the ES-lasing suppression. At the same time, in the case of long samples, the increase in α_{in} plays a more substantial role and contributes to the decrease in the external differential efficiency of such lasers – compare the slopes of the red (3×10^{17} cm^{-3}) and the green (5×10^{17} cm^{-3}) in the Fig. 1a. As it can be seen from the comparison of the experimental results presented in the Fig. 1a and Fig. 1b, the effect of the decrease in W_{max}^{GS} due to the decreased efficiency in case of the *p*-doped samples is much more pronounced in long samples and tends to dominate over the positive effect of the *p*-doping. At the same time, in the case of short saples the situation is opposite and the increase in *p*-doping level has a positive influence on the GS-power. As a result for a given cavity length there is an optimal concentration of the *p*-dopant maximally enhancing the output power corresponding to the GS of QDs.

4. Conclusion

In conclusion, the effect of multi-state lasing was studied both theoretically and experimentally. It was shown that the higher *p*-doping level does not necessarily lead to the higher output power corresponding to the GS optical transitions of quantum dots. On the one hand, the increase in the *p*-dopant concentration leads to the higher hole capture rate, increases GS gain and mitigates the competition for the common holes between the GS and ES optical transitions. This makes GS-lasing more preferable as compared to ES-lasing. On the other hand, an introduction the *p*-dopant atoms increase the internal loss resulting in the reduction in the slope efficiency.

In case of the short samples having high loss, the introduction of modulation *p*-doping increases the maximal power of the GS spectral component. In long samples, the effect is opposite because of its influence on internal loss. Thus, the usage of *p*-doping occurs to be most beneficial in case of the short samples allowing to increase GS output power.

Acknowledgements

This work is supported by the Russian Scientific Foundation (project #14-42-00006).

References

[1] Zhukov A E, Kovsh A R 2008 *Quantum Electronics* **38** 409
[2] Lam C F 2007 Passive Optical Networks: Principles and Practice San–Diego (Elsevier) 158
[3] Savelyev A V, Novikov I I, Maximov M V, Shernyakov Yu M, Zhukov A E 2009 Semicond. **43** 1597
[4] Lee J, Lee D 2011 J. Korean Phys. Soc. **58** 239
[5] Markus A, Fiore A 2004 Phys. Stat. Sol. (a) **201** 338
[6] Kim Y J, Joshi Y K, Fedorov A G 2010, J. Appl. Phys. **107** 073104
[7] Sugawara M, Hatori N, Ebe H, Ishida M, Arakawa Y, Akiyama K, Otsubo T, Nakata Y 2005 J. Appl. Phys. **97** 043523
[8] Korenev V V, Savelyev A V, Zhukov A E, Omelchenko A V, Maximov M V 2013 Appl. Phys. Lett. **102** 112101
[9] Maximov M V, Shernyakov Yu M, Zubov F I, Zhukov A E, Gordeev N Yu, Korenev V V, Savelyev A V, Livshits D A 2013 Semicond. Sci. Technol. **28** 105016
[10] Smowton P M, Sandall I C, Mowbray D J, Liu H Y, Hopkinson M 2007 IEEE J. Sel. Top. Quantum Electron. **13** 1261

Investigation of the effect of surface passivation on microdisk lasers based on InGaAsN/GaAs quantum well active region

E I Moiseev[1], N V Kryzhanovskaya[1,2], M V Maximov[1,2,3], A M Mozharov[1,4], A S Gudovskikh[1,5], A S Polushkin[1], I S Mukhin[1,4], Yu A Guseva[3], M M Kulagina[3], S I Troshkov[2], T Niemi[6], R Isoaho[6], M Guina[6], A E Zhukov[1,2]

[1] St Petersburg Academic University, Khlopina 8(3), St Petersburg 194021, Russia
[2] Peter the Great St Petersburg Polytechnic University, St Petersburg 195251, Russia
[3] Ioffe Institute, St Petersburg, Russia
[4] ITMO University, 197101, St. Petersburg, Russia
[5] St. Petersburg Electrotechnical University "LETI", St. Petersburg 197376, Russia
[6] Optoelectronics Research Centre, Tampere University of Technology, 33720, Tampere, Finland

Abstract. Microdisk lasers based on three InGaAsN/GaAs quantum wells with different types of surface passivation are fabricated and studied under optical pumping. Room temperature lasing at 1.3 μm in 7 μm in diameter microdisks with InGaAsN/GaAs QW is demonstrated. We evaluated the thermal resistance as 1 °C/mW.

1. Introduction

Semiconductors microdisk and microring lasers have been widely investigated as possible building blocks for photonic integrated circuits [1]. The circular symmetry of the cavity results in unique advantages such as low threshold, small foot-print, small mode volume, in-plane emission, control of emission wavelength by the device size, etc. The use of quantum dots (QDs) as active region in such lasers have the advantages of low threshold, high quantum efficiency, and high thermal stability of characteristics [2]. However since the ground state QDs optical gain is limited due to the finite number of the QDs the decrease of the resonator diameter below a certain value (typically several microns) results in lasing via excited states of QDs. Quantum well (QW) active region provides higher gain as compared to QDs. Thus, the use of QW active region may help to overcome the problem of gain saturation when the laser`s size is scaled down. As device dimensions decrease, the recombination of carriers at the semiconductor surface becomes increasingly important [3]. GaAs-based materials are especially prone to nonradiative recombination at unpassivated surfaces. Sulfide treatment of the GaAs surface, as well as other III–V semiconductors, has been demonstrated as an effective method for reducing the surface recombination rate [4]. In this work we have studied microdisk lasers based on InGaAsN/GaAsN QW active region with different types of surface passivation.

2. Experiment and results

Epitaxial structures were grown with molecular beam epitaxy on GaAs(100) substrate. Three $Ga_{0.7}In_{0.3}N_{0.02}As_{0.98}$ QWs were separated with 10-nm-thick GaAs layers and have ground-state emission peak around 1.2 μm at room temperature. The active region was placed in the middle of the GaAs layer confined from both sides with 10 nm thick $Al_{0.3}Ga_{0.7}As$ barriers to prevent carrier leakage.

The total thickness of the waveguide layer was about 230 nm. This waveguide layer was grown on top of the 1 μm thick $Al_{0.8}Ga_{0.2}As$ layer that later formed a pedestal of the microdisks. Microdisks were fabricated using photolithography and two step wet etching. A circular mesa was formed by etching in $HBr:K_2Cr_2O_7:CH_3COOH$ solution. The pedestal was formed using diluted HF. The structures were investigated under different optical excitation powers using Nd:YLF laser (λ = 527 nm) . The samples were mounted in a flow cryostat Janis ST-500, the temperature was varied from 78 to 300K. Microphotoluminescence signal (μPL) was measured by an Optical Spectrum Analyzer AQ6370. The microdisks were passivated to suppress surface non-radiative recombination at microdisk sidewalls. The passivation was performed with aqueous or hydroalcoholic solution of Na_2S or benzene solution SeS_2 during different holding time (4 min, 10 min, 20 min) and then, the mesas were washed in water (in benzene for SeS_2 solution) and dried in nitrogen. To achieve a good stability of the operation after treatment in sulfur solution, the microdisks were encapsulated with a thin (the thickness was approximately 30 nm) SiN layer deposited by PECVD.

Optical properties of the GaInNAs/GaAs QW were investigated using an unprocessed sample in the temperature range of 78–300 K under excitation of 0.37 mW. We observe a moderate decrease (by 13.7 times) in GaInNAs/GaAs QWs integrated photoluminescence intensity as temperature increases in this temperature interval evidencing high optical quality of the GaInNAs/GaAs QWs (figure 1). Temperature dependence of the PL peak maximum and full width at half maximum (FWHM) are shown in figure 2. At low temperature (78K), FWHM of the spectrum is ~17 meV, revealing reasonable homogeneity of the GaInNAs/GaAs QW thickness and composition. FWHM monotonically increases up to 26 meV as temperature grows to 293K. The ground-state emission peak is around 1.024 eV at room temperature.

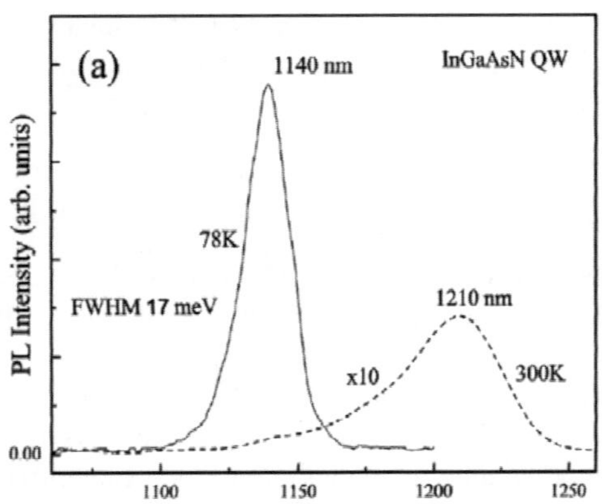

Figure 1. μPL spectra of the unprocessed sample with GaInNAs/GaAs QWs obtained at 78K and 300 K.

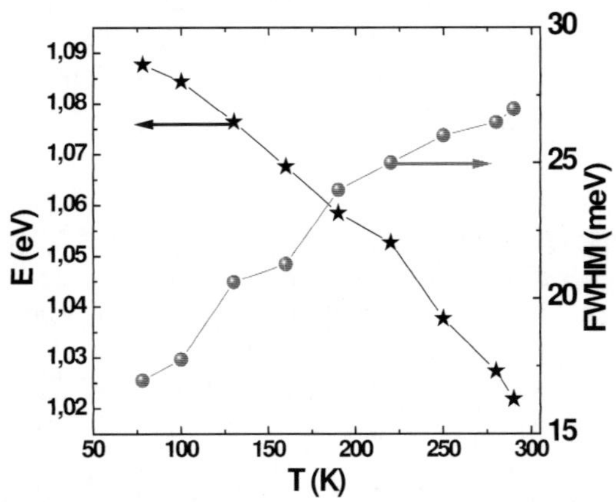

Figure 2. Temperature dependence of PL peak maximum and FWHM of GaInNAs/GaAs QW.

Microlasers with a diameter of 7 μm without passivation demonstrated lasing only up to 170 K. A scanning electron micrograph of a 7 μm in diameter microdisk laser is shown in the inset in figure 3. Passivation significantly improves maximal operation temperature and threshold. The highest operation temperature and the smallest threshold were obtained for passivation in hydroalcoholic Na_2S solution. The μPL spectra at RT of the 7 μm in diameter microdisk after passivation in this type of Na_2S solution is shown in figure 3. Sharp lines at the spectra correspond to high quality whispering gallery modes. Lasing threshold was determined from the dependence of the output intensity versus input optical pumping power for the lasing mode. The temperature dependence of the threshold pump power (figure 4) can be approximated by exponential relationship with characteristic temperature $T_0 = 60$ K. Passivation in SeS_2 solution showed significantly higher threshold as compared to passivation in Na_2S. We evaluated the thermal resistance for the passivated structures as 1 °C/mW. A similar value was obtained in [5]. A low thermal resistance can increase the current at which thermal rollover begins and thus, enhance maximal output power.

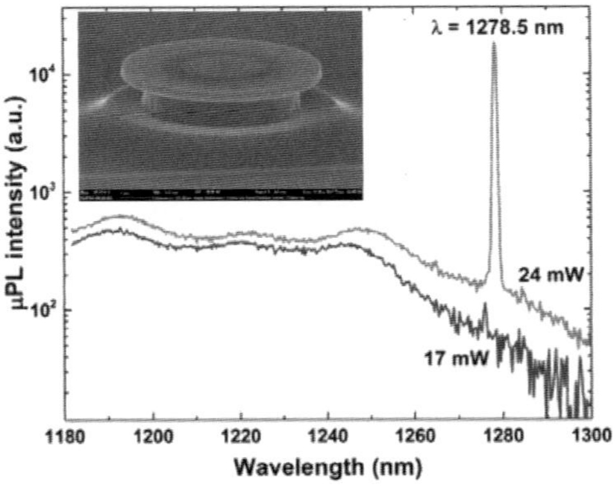

Figure 3. μPL spectra of a 7 μm in diameter microdisk laser at RT. Inset: SEM image of the microdisk laser.

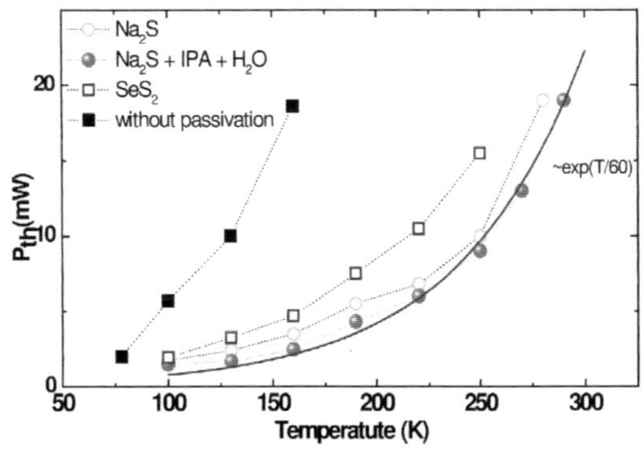

Figure 4. Temperature dependence of threshold pump power in case of different types of surface passivation.

Acknowledgments
The work is supported in different parts by RFBR projects 16-29-03127, 16-29-03111 and the Ministry of Higher Education and Science of the Russian Federation (project No 3.9787.2017/8.9).

References

[1] S. Koseki et al, Applied Physics Letters, 94, 051110 (2009)
[2] Y. Arakawa, H. Sakaki, Appl. Phys. Lett. 40 939 (1982)
[3] B. Corbett and W. M. Kelley, Appl. Phys. Lett. 62, 87 (1993)
[4] C. J. Sandroff, R. N. Nottenburg, J.-C. Bischoff, and R. Bhat, Appl. Phys.Lett. 51, 33 (1987)
[5] Demir, A., Zhao, G., & Deppe, D. G., Electronics letters, 46(16), 1147-1149 (2010)

IOP Conf. Series: Journal of Physics: Conf. Series **917** (2017) 052003 doi:10.1088/1742-6596/917/5/052003

Free-space subcarrier wave quantum communication

S M Kynev[1], V V Chistyakov[1], S V Smirnov[1], K P Volkova[2], V I Egorov[1], A V Gleim[1]

[1]Department of Photonics and Optical Information Technology, ITMO University, Saint Petersburg 197101, Russia

[2]Department of Quantum Electronics, Peter the Great Saint-Petersburg Polytechnic University, Saint Petersburg, 195251, Russia

Abstract. We experimentally demonstrate quantum communication in 10 dB loss outdoor atmospheric channel with 5 kbit/s bitrate using subcarrier wave coding method. Free-space link was organized by telescoping system with symmetric fiber-optic collimators.

1. Introduction

Quantum communication (QC) is a new technique of secure key exchanging between parties in communication network. Nowadays different groups demonstrating quantum communication in optical fibers up to 404 km [1] and up to 144 km in free space [2]. Free space experiments demonstrating the feasibility of QC for sattelite-to-sattelite and ground-to-ground application. For expample in ground-to-ground applications free space QC can solve last-mile problem in quantum networks. To date, mostly method based on polarization-coding has been implemented in free space[3- 5]. In this work we show that subcarrier wave coding technique [6-8] can be adapted to free space quantum communication.

2. Experimental setup

Experimental setup is shown in figure 1. Free-space subcarrier wave QC system consists of Alice and Bob modules connected by free-space atmospheric quantum channel.

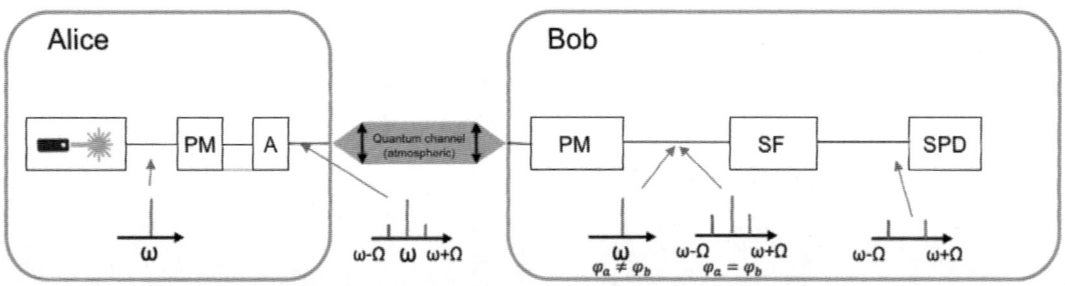

Figure 1. Experimental setup of free-space subcarrier wave quantum communication system. PM – phase modulator, A – attenuator, SF – spectral filter, SPD – single photon detector.

Semiconductor laser at the Alice module emits a signal at optical frequency ω which is directly modulated in $LiNbO_3$ phase modulator by an electrical signal with frequency Ω and modulation depth m<<1. The optical spectra are composed of central peak and two subcarriers at frequencies $\omega + \Omega$ and $\omega - \Omega$ with phase shift φ_A randomly chosen from four-state protocol. After modulation signal is attenuated and passed through beam expander and to receiver side by atmospheric quantum channel. The total optical power at the output of Alice module was equal to mean photon number 1 per pulse at subcarriers.

Free space atmospheric quantum channel was organized by telescoping system with symmetric fiber-optic collimators (Fig. 2.) An ideal atmospheric channel can be descripted by Gaussian beam solution of paraxial wave equation. Maximal length between Alice and Bob modules is limited by beam diffraction and depended on the collimator lens diameter.

$$L = \frac{2\pi\omega^2}{\lambda}\frac{\omega_0}{\omega}\sqrt{1-\left(\frac{\omega_0}{\omega}\right)^2},$$

where ω is the beam waist, and ω_0 – radius of the beam at the lens. The length is maximized when $\omega_0/\omega = \frac{1}{\sqrt{2}}$.

At the receiver side signal is collected by symmetric beam collimator and coupled in an optical fiber and passed to Bob module. Here signal passed through phase modulator where random phase shift φ_B is introduced according to four-state protocol. Then signal transmitted to an optical spectral filter that separates the carrier from subcarriers. After spectral filter two subcarriers at frequencies $\omega + \Omega$ and $\omega - \Omega$ passed to single photon detector. When Alice and Bob introduced equal phase shifts ($\varphi_A - \varphi_B = 0$), constructive interference was observed in the side frequency optical signal. When the difference in phase shifts was a multiple of π, destructive interference was observed. Key generation and sifting was performed experimentally using the BB84 protocol.

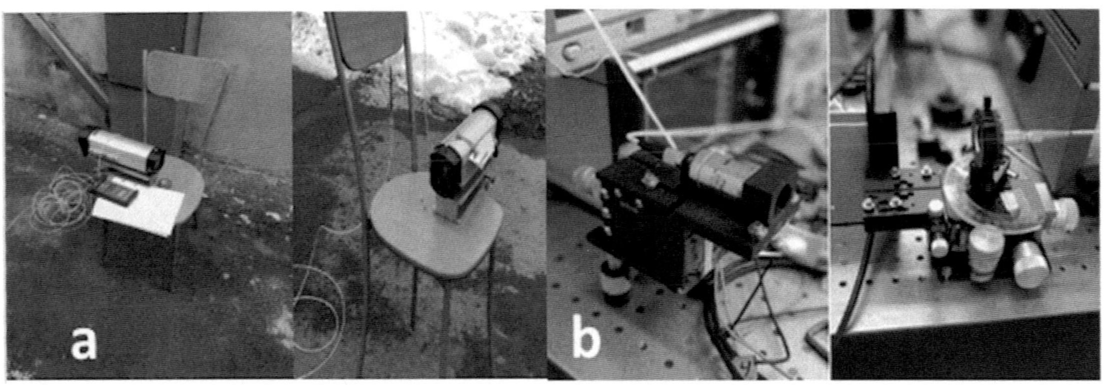

Figure 2. Free-space channels with beam collimators in outdoor (a) and indoor (b) experiments.

The experiment was perfomed two times: indoor in lab with 1 m on-table-link (Fig.2a.) and outdoor on a cloudy day (Fig. 2b.) with atmospheric channel between two beam collimators ~20 m. . QC system parameters are shown in Table 1.

Table 1. QC system parameters.

Parameter	Value
Central wavelength	1550 nm
Modulation clock frequency	100 MHz
Modulation frequency	4.8 GHz
Mean photon number	1
Measured losses in quantum channel	20 dB
Losses in Bob module	7 dB
Quantum efficiency of detector	20%
Spectral filter coefficient	99,99%
Spectral filter band	7.5 GHz

For measured losses in outdoor 20 m free-space link of 10 dB the sifted key rate was about 5kbit/s and QBER value was 6%.

For measured losses in indoor 1 m free-space link of 8 dB the sifted key rate was about 15 kbit/s and average QBER value was 3,28 %. At the reciever side beam collimator was rotated 45 degrees 8 times. Figure 3 shows that QBER pproximately the same for different positions of the telescope system.

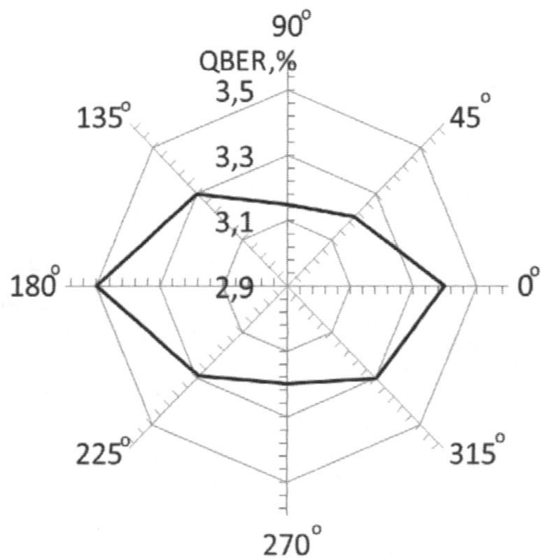

Figure 3. Dependence of QBER on collimator rotation angle

3. Conclusion

In this work we have implemented subcarrier wave technique for free-space quantum communication using standard fiber-optical components in Alice and Bob modules. In free-space

IOP Conf. Series: Journal of Physics: Conf. Series **917** (2017) 052003 doi:10.1088/1742-6596/917/5/052003

regime this technique offers invariance to rotation of the telescope system and good capabilities in multiplexing.

Acknowledgments

This work was financially supported by Government of Russian Federation, Grant 074-U01 and by the Ministry of Education and Science of Russian Federation (project № 14.578.21.0112 № 02.G25.31.0229).

References

[1] Yin H-L et al 2016 *Phys.Rev. Lett* **117(19)** 190501
[2] Ursin R et al 2007 *Nat. Phys.* **3** 481
[3] Elser D et al 2009 New J. Phys. **11** 045014
[4] Heim B et al 2014 New J. Phys. **16** 113018
[5] Xavier G B 2009 New J. Phys **11 045015**
[6] Mora J et al 2012 *Opt.Lett.* **37(11)** 2031-2033.
[7] Gleim A V et al 2014 *Bull. Russ.Acad. Sci., Phys.* **78(3)** 171-175.
[8] Gleim A V et al 2016 *Opt.Express* **24(3)** 2619-2633.

Lateral conductivity of n-GaP/p-Si heterojunction with an inversion layer

A V Uvarov[1], A S Gudovskikh[1,2], D A Kudryashov[1]

[1]St. Petersburg Academic University, St. Petersburg 194021, Russia
[2]St. Petersburg Electrotechnical University "LETI", St. Petersburg 197376, Russia

Abstract. This article is concerned with investigation of n-GaP/p-Si heterostructure for using in solar cells. We examine the contribution of the inversion layer to lateral transport for (n)GaP wide gap emitter. We evaluate the carrier density in the inversion layer by varying the temperature and the doping level of the emitter. Simulations predict a formation of the inversion layer in p-Si close to the interface with a thickness being in the range of 10 - 20 nm and electron concentration, which is strongly depends on doping level and temperature. It is shown that the temperature dependence of surface conductivity n-GaP/p-Si is linear and could be related to dependence of electron concentration in the inversion layer.

1. Introduction

Hydrogenated amorphous silicon (a-Si:H)/crystalline silicon (c-Si) heterojunction solar cells have achieved the record efficiency for Si based solar cells (26.3%). a-Si:H serves as a wide bandgap emitter in this type of solar cells. The bandgap of a-Si:H is about 1.7 eV, and consequently, it absorbs a significant portion of short-wave radiation. It is possible to reduce this parasitic absorption using emitter materials with higher bandgap. One of such materials is a gallium phosphide (Eg = 2.26 eV). When the band diagram for the heterojunction between highly doped n-GaP and p-Si is considered the region with n-type conductivity is formed in the silicon close to the interface due to conduction band discontinuity [1], which called the inversion layer. The inversion layer has a higher conductivity due to the high carrier mobility of silicon and can significantly influence the collection of electrical charge [2]. This paper presents the results of experimental study and numerical simulations of the inversion layer in the n-GaP/p-Si heterojunction. Temperature dependence and emitter doping level contributions to lateral conduction are discussed.

2. Experimental study

The of GaP/Si structures were fabricated by plasma-enhanced atomic layer deposition (PEALD) of GaP doped with silicon on c-Si(100) and quartz substrates using a modified Oxford Plasmalab 100 PECVD setup [3]. The concentration of the acceptor impurity in the p-Si substrate is 1×10^{16} cm^{-3}. As precursors of gallium and phosphorus, trimethylgallium and phosphine were used, respectively. The doping of gallium phosphide was carried out by the addition of SiH_4 diluted in hydrogen during the deposition process. The thickness of the GaP layer was determined by SEM and was 50-52 nm for 200 deposition cycles, which indicates a deposition rate of 0.93-0.96 monolayers per cycle. Two types of GaP layers with a different crystalline fraction were studied. According to Raman spectroscopy one GaP layer was amorphous (a-GaP) and other one crystalline or micro-crystalline (μc-GaP). Ohmic contacts of a given geometry were deposited to determine the surface conductivity and its temperature dependence.

The surface resistance of 50 nm thick gallium phosphide layer on the surface of the quartz substrate is more than 10^9 Ω/□ and cannot be directly measured. This indicates a low mobility of the main carriers, which is usual for amorphous layers deposited by the plasma-chemical method. However, when measuring the surface conductivity of GaP on a silicon surface, relatively low values of the surface resistance were obtained. Current-voltage characteristics between ohmic contacts have a linear shape which excludes current flow through the silicon substrate because of the presence of p-n junction. According to the results of the Hall measurement under temperature of 300K, it was found that the main

IOP Conf. Series: Journal of Physics: Conf. Series **917** (2017) 052004 doi:10.1088/1742-6596/917/5/052004

carriers are electrons with a concentration of 1.4×10^{19} cm^{-3} and a mobility of 15 cm^2/V/s with a nominal thickness of 50 nm. To exclude the effect of n-type doping of silicon surface by phosphorus due to diffusion during the process of plasma-chemical deposition, the following experiment was carried out. A layer of gallium phosphide between metal contacts was removed by selective chemical etching (figure 1). At the same time, the surface of silicon remained intact. Then a strong decrease of the lateral conductivity was observed (figure 2). Therefore, the surface conductivity is not a consequence of doping of silicon due to the diffusion of phosphorus under the influence of temperature or plasma.

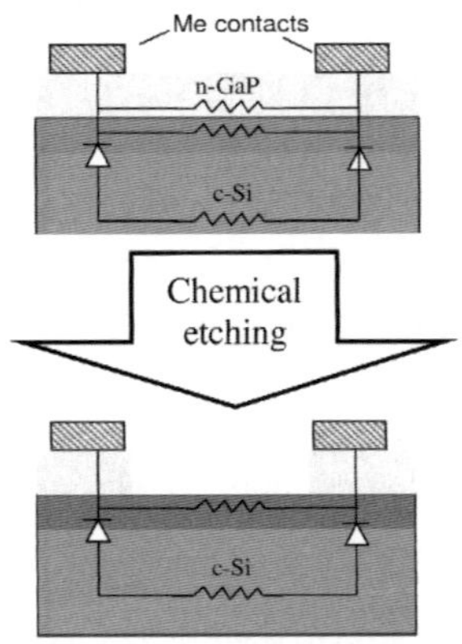

Figure 1. Experiment with etching of gallium phosphide.

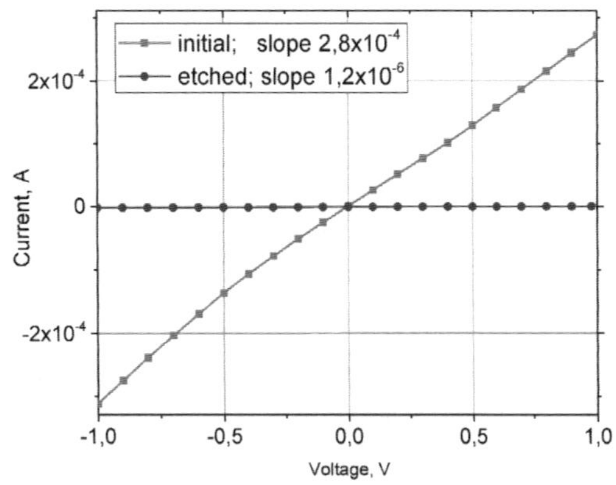

Figure 2. I-V curve before and after etching.

So, the lateral conductivity of the structure obtained by the plasma-chemical method is related to the high concentration of carriers at the GaP/Si heterointerface. On the one hand a high conductivity at the interface can be due to high mobility in the near-surface region of the heavily doped GaP layer. On the other hand (being the most likely case) this can be due to the injection of minority carriers into the p-type silicon layer and the formation of an inversion layer. In this case, a thin layer with a high electron concentration is formed at the near-surface region of the silicon substrate(figure 3). Together with the high mobility of electrons in crystalline silicon, a region of 10-20 nm thick with high conductivity values is formed.

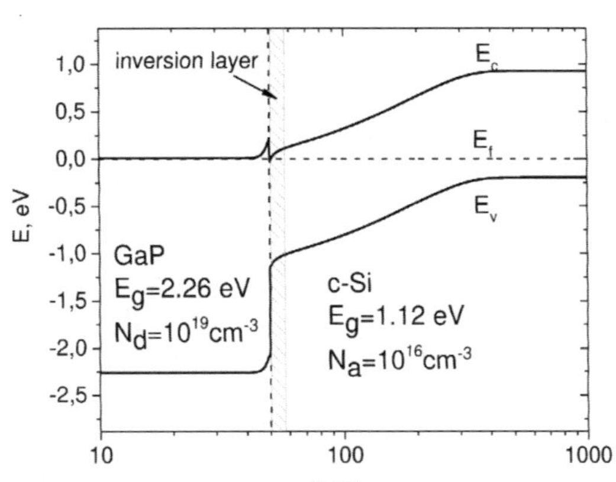

Figure 3. The band diagram and the position of the inversion layer.

The temperature dependence of the surface conductivity of the n-GaP/p-Si structures was measured. The measurements were carried out in a nitrogen cryostat in the temperature range 80-320 K and presented in 'figure 4'.

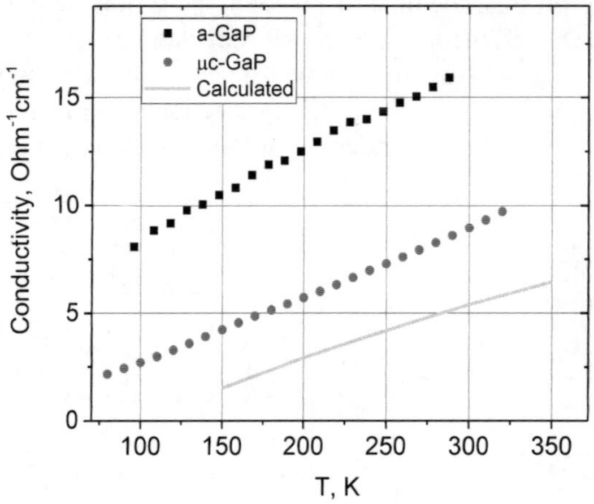

Figure 4. Experimental temperature dependence of the surface conductivity of GaP/Si (symbol) in comparison with the simulation results(line).

From the measurements it was found that the conductivity increases linearly with temperature over a wide range, which is not typical for semiconductors. However, this can be explained by the temperature dependence of the inversion layer, which requires a more detailed consideration.

3. Simulations

Modeling of a heterojunction was performed using COMSOL Multiphysics software package for one-dimensional and two-dimensional cases. The simulations were carried out by finite element method in the approximation of the drift-diffusion model. The temperature dependences of the effective density of states and band gap are taken into account [4]. Based on the results of the calculation of the structure of n-GaP/c-Si, it was shown that an inversion layer with an electronic type of conductivity could be formed in the near-surface region of silicon.

Table 1. Input parameters of modeling.

		n-GaP	p-Si
Band gap	E_g, eV	2.26	1.12
Electron affinity	X_s, eV	3.8	4.05
Doping level	$N_d(N_a)$, cm^{-3}	5×10^{16} - 10^{20}	1×10^{16}
Activation energy	$\Delta E_d(\Delta E_a)$, eV	0.085	0.045

Moreover, the values of electron concentration were calculated in the temperature range of 150-350 K. When performing the simulations, we used the Fermi-Dirac statistics with a degeneracy degree of 2 because of high electron concentration in the inversion layer of silicon. The effect of incomplete ionization of impurities with a decrease in temperature was taken into account in accordance with the activation energies. The surface concentration of electrons was determined by integrating the volume value according to the formula

$$n_{inv} = \int_{x_0}^{x_{inv}} n_{vol} dx,$$

where the distance between the interface (x_0) and the conditional point in Si (x_{inv}) at which the minority carrier concentration is $1x10^{16}$ cm^{-3} is equal to the width of the inversion layer.

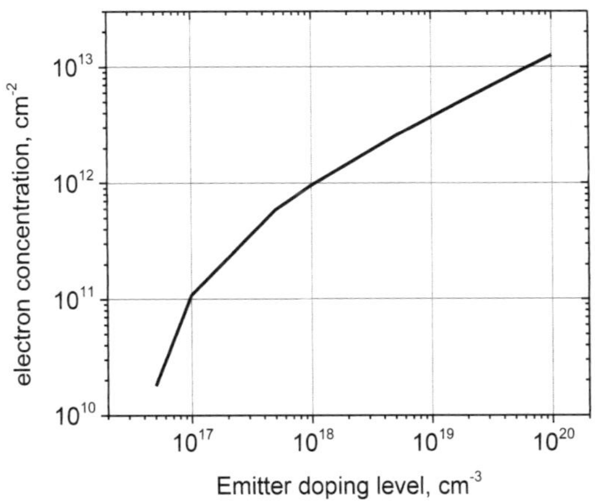

Figure 5. The electron concentration in the inversion layer as a function of the doping of the emitter for ΔE_c=0.25 eV.

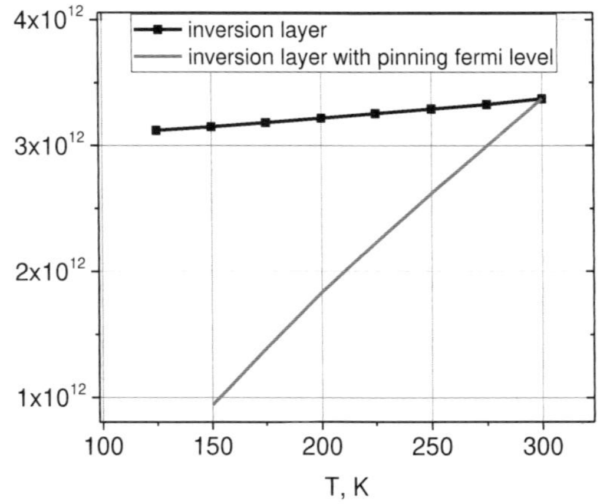

Figure 6. Calculated temperature dependence of electron concentration in the inversion layer.

Based on the simulation results, the electron concentration in the inversion layer was found to depend on the emitter doping level and on the temperature (figure 5,6). To form an inversion region with high concentration values (close to the measured), it is necessary to have an emitter with a doping level of more than 1×10^{18} cm^{-3}. With a decrease in the emitter doping level, the inversion layer electron concentration is sharply reduced. The dependence of the electron concentration on the temperature was calculated for two cases. In the first case, the position of the Fermi level in GaP varied with the temperature in accordance with the impurity (Si) activation energy (table 1). In the second case, the pinning of the Fermi level in GaP was considered and therefore the Fermi level position was fixed independent of temperature. In case of pinning of the Fermi level, a significant change in the surface concentration of electrons in the inversion layer with temperature is observed. Based on the electron concentration, one can obtain the conductivity values taking into account the mobility values determined by the Hall measurements (15 cm^2/V/s). The temperature dependence of the lateral conductivity calculated in approximation of temperature independent mobility is presented in 'figure 4'.

According to the simulation results, the obtained temperature dependence qualitatively coincides with the experimental data. Differences in absolute values be related to incertitude of the estimation for the values of conduction band offset, GaP doping level as well as a presents of interface states, which were no considered. The simulation results demonstrates that the linear dependence of the lateral conductivity of the GaP/Si heterojunction on the temperature can be explained in terms of the temperature dependence of the electron concentration in the inversion layer with approximation of the Fermi level pinning in GaP. However, the precise analysis requires a knowledge of the temperature dependence of the mobility which should be measured in the future.

4. Conclusions

The effect of the inversion layer on the surface conductivity of the GaP/Si heterojunction was considered. It was noted experimentally that the temperature dependence of the surface conductivity of inversion layer is linear in the temperature range of 100-350 K. Numerical simulations show that such temperature dependence could be related to that of electron concentration in the inversion layer which is formed in p-Si near to n-GaP/p-Si interface.

Acknowledgements

This work was carried out in the framework of the joint Russian-French project LAMSOL supported by Ministry of Science and Education of Russian Federation (RFMEFI61615X0040) and by PHC Kolmogorov Program (35522TL).

IOP Conf. Series: Journal of Physics: Conf. Series **917** (2017) 052004 doi:10.1088/1742-6596/917/5/052004

References

[1] J.-P. Kleider, Y. M. Soro, R. Chouffot, A. S. Gudovskikh, P. Roca i Cabarrocas, J. Damon-Lacoste, D. Eon, and P.-J. Ribeyron, J. Non-Cryst. Solids, 354, 2641, 2008.
[2] Miha Filipič, Zachary C. Holman, Franc Smole, Stefaan De Wolf, Christophe Ballif, Marko Topič, J. Appl. Phys, Vol. 114, No. 7, 074504, 2013.
[3] R.L. Puurunen, J. Appl. Phys. 97, 121301, 2005.
[4] M. Levinshtein, S. Rumyantsev, M. Shur. Handbook Series on Semiconductor Parameters, Vol 1, Chapter 5, 1996.
[5] F. Fang and S. Triebwasser, IBM J. Res. Dev. 8, 410, 1964.

IOP Conf. Series: Journal of Physics: Conf. Series **917** (2017) 052005 doi:10.1088/1742-6596/917/5/052005

Electroluminescence of InAsSb-based mid-infrared LEDs in 4.2–300 K temperature range

A A Semakova[1,2], K D Mynbaev[1,2], N L Bazhenov[2], A V Chernyaev[2,3], S S Kizhaev[3] and N D Stoyanov[3]

[1] ITMO University, Saint-Petersburg 197101, Russia
[2] Ioffe Institute, Saint-Petersburg 194021, Russia
[3] Microsensor Technology, LLC, Saint-Petersburg 194223, Russia

Abstract. Electroluminescence of mid-infrared LEDs with active layer made of InAs and InAsSb in a wide temperature range (4.2–300 K) was studied. At low temperatures (T=4.2–100 K), stimulated emission from the LEDs was observed. The emission became spontaneous at higher temperatures due to the effect of CHHS Auger recombination process, when the energy of a recombining electron-hole pair was transferred to another hole with the latter transitioning to the spin-orbit-splitted band. The spontaneous character of recombination held up to the room temperature due to the effect of other Auger processes. Still, the results obtained show that structures based on InAsSb are a promising material for fabrication of mid-infrared lasers.

1. Introduction

The mid–infrared wavelength range (MWIR, wavelengths λ 2–6 μm) represents great interest for the development of optoelectronic devices with applications in optical gas sensing, environmental control, biomedical imaging, thermal imaging, etc. From the viewpoint of sensor technology, light-emitting diodes (LEDs) represent an almost ideal source of MWIR radiation, as they demonstrate reasonably wide emission band, room temperature operation, very high modulation rate, small size and low power consumption [1]. Their problem is low emissivity, which is mostly due to the specifics of the electronic structure of narrow-bandgap AIIIBV semiconductors, which serve as a basis for MWIR LED heterostructures.

To enhance the efficiency of optoelectronic devices, it is important to understand and control the processes of carrier recombination, which govern emission of light (or the lack of thereof). For that task, it is worth studying the operation of the devices not only at working temperatures, but also at lower temperatures. Within the frames of such an approach, we have studied electroluminescence (EL) of InAsSb-based LEDs in a temperature range, 4.2–300 K.

2. Experimental details

The LED heterostructures were grown with the use of metal-organic chemical vapour deposition at Microsensor Technology, LLC, using synthesis method described elsewhere [2]. For most of the heterostructures, a strongly sulphur-doped InAs substrate (n-type conductivity, electron concentration at 77 K $n_{77}\approx(1\text{-}2)\times10^{18}$ cm^{-3}) was used. An active layer of the heterostructures was made of InAs or InAsSb. This layer was not intentionally doped and had electron concentration $n_{77}\approx1\times10^{16}$ cm^{-3}, presumably due to background donors. A typical thickness of the layer was 2 to 3 μm. On top of the active layer, a *p*-type InAsSb(P,Ga) barrier layer was grown, this layer was doped with zinc and had hole concentration at 77 K $p_{77}{\approx}2{\times}10^{18}$ cm^{-3}. Parameters of the heterostructures are given in Table 1. For the growth of structure of type *D*, a *p*-type InAs substrate was used. On this substrate, an InAsSbP barrier layer was grown first ($p_{77}{\approx}2{\times}10^{18}$ cm^{-3}) and only then, an *n*-type InAs active layer.

Table 1. Parameters of the studied heterostructures.

Structure type	Active layer	Barrier layer
A	InAs	$InAs_{0.15}Sb_{0.31}P_{0.54}$
B	InAs	$InAs_{0.25}Sb_{0.25}P_{0.50}$
C	InAs	$In_{0.85}Ga_{0.15}As_{0.72}Sb_{0.28}$
D	InAs	$InAs_{0.15}Sb_{0.31}P_{0.54}$
E	$InAs_{0.94}Sb_{0.06}$	$InAs_{0.40}Sb_{0.20}P_{0.40}$
F	$InAs_{0.93}Sb_{0.07}$	$InAs_{0.70}Sb_{0.10}P_{0.20}$
G	$InAs_{0.91}Sb_{0.09}$	$InAs_{0.48}Sb_{0.18}P_{0.34}$

LED chips with 380×380 μm size were fabricated with the use of standard photolithography and wet chemical etching. Electrical contacts were based on a multi-layer Cr–Au–Ni–Au composition. A non-transparent solid contact was placed on the top epitaxial layer, while a ring-type contact with 35 μm thickness and 200 μm internal diameter was placed on the InAs substrate. The emission was collected from the side of the substrate. For the measurements, the chips were placed on TO-18 holders. EL spectra were recorded under pulse excitation (frequency 1 kHz, pulse duration 1 μs) and InSb photodiode was used as a detector.

3. Experimental results and discussion

Figure 1 shows normalized EL spectra of four types of LEDs at *T*=300 K, which is a typical working temperature for the diodes employed in gas sensors (Fig. 1 (a)), and at *T*=4.2 K (Fig. 1 (b)). In Fig. 1(a) one can observe a red-shift of the maximum of the spectrum due to the introduction of InSb, which has the energy bandgap value E_g smaller than that of InAs, and some broadening of the EL line typical of alloys as compared to binary compounds. Figure 1(b), in its turn, shows normalized EL spectra for three types of studied heterostructures at *T*=4.2 K at various driving currents *I*. As can be seen, these spectra drastically differ both in shape and full-width at half-maximum (FWHM) values from those recorded at 300 K. For example, for the structure of type *A*, at the high-energy side of the barely visible wide emission band with FWHM 20 meV, there appeared a second narrow line with FWHM of 2 meV. For other structures at the temperatures close to 4.2 K we also observed quite narrow emission bands.

With temperature increasing, for all the structures we observed that the spectra with narrow EL peaks transformed back to broad bands. Temperature dependence of the calculated energy bandgap of some of the studied structures is presented in Fig. 2. This figure also shows experimental data. In the graph, for narrow emission lines (low temperature) we simply plotted out the position of the narrow peak of the EL spectrum. Though the position of this peak did not coincide with the maximum of the broad emission band (see Fig. 1(b)), it was not always possible to define the latter, as in most of the structures one could not distinguish the broad band with low intensity against the narrow peak with much higher intensity. For higher temperatures, when the EL spectra contained only broad bands, we derived the value of optical bandgap by calculating the EL spectrum and fitting its high-energy part to the experimental one with carrier concentration as a parameter. For the calculations, we took into account both electron–heavy hole and electron–light hole transitions and non-parabolic dependence of the conduction band and the valence band for light holes on energy, typical of narrow-bandgap semiconductors. The appropriate transition matrix element was calculated in Ref. [3]. The material parameters necessary for calculations were taken from Ref. [4]. The calculated spectra (not shown) at T=77 K typically gave the best fit to the experimental ones at the values of carrier concentration under injection $n \approx p \approx (3–5)\times10^{16}$ cm^{-3}. The values of 'optical' bandgap were derived from the intercept formed by the low-energy part of calculated spectra on the abscissa axis. These values are shown in in Fig. 2 with empty symbols.

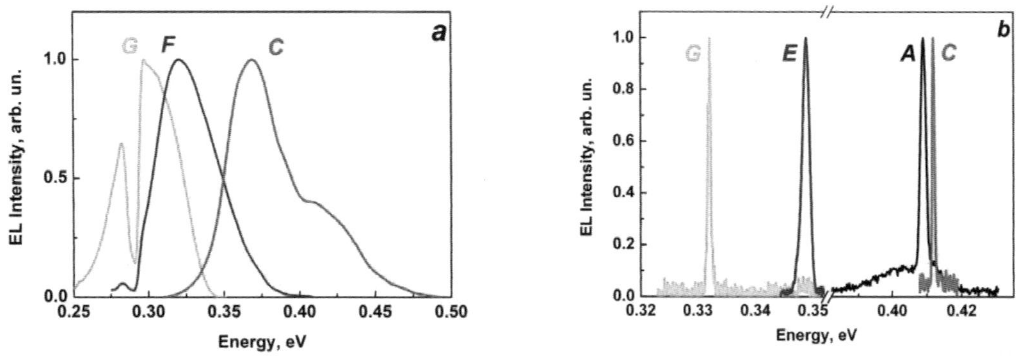

Figure 1. Normalized EL spectra of the heterostructures of type C, F and G at T=300 K at driving current I=1 A (a), and spectra of the heterostructures of type A (I=0.5 A), C (I=0.6 A), E (I=0.4 A) and G (I=1.2 A) at 4.2 K (b).

The appearance of the strong narrow peak in the EL spectra suggested that at low temperatures (from 4.2 up to 50–100 K, depending on the type of heterostructure), conditions for stimulated emission held true in the studied LEDs (see also Ref. [5]). For two structures of type F at temperatures, at which stimulated emission was still observed, we measured the light–current characteristics. With such measurements, if a pronounced "break" is observed in the recorded curve, a threshold current of stimulated emission I_{th} can be determined from the intercept formed by the characteristic on the abscissa axis. For the samples under study, it was possible to obtain a pronounced $I_{th}(T)$ dependence in the temperature range 25–130 K (not shown). The experimental dependences appeared to be exponential ($I_{th}\sim\exp(T/T_0)$, where T_0 is a characteristic temperature) in the portion ranging from 70 to 130 K. According to the known concepts about the temperature dependence of I_{th} of a semiconductor laser, this is indicative of the influence of Auger recombination [6]. The T_0 value was found to be 25±5 K for the 80–120 K range. No pronounced $I_{th}(T)$ dependence was observed at low temperatures (T<70 K), which could be indicative of the predominance of Shockley–Read recombination.

4. Discussion

The analysis of the obtained results indeed shows that at low temperatures (4.2–50 K), in the studied structures conditions for stimulated emission held true. Interestingly, this effect was observed not only for the LEDs with active layer sandwiched between the substrate and the barrier layer with wider bandgaps, but also for the structure of type D with active layer grown on top of the barrier layer. As it was shown that in LED structures of this type the optical resonator spontaneously forms normal to the growth plane, most probably, between the surface of the chip with a gold Ohmic contact and the substrate [5], the heterostructures look promising for fabrication of the vertical-emitting MWIR lasers, which are currently of great demand [7].

Figure 2 shows that transition from stimulated to spontaneous emission occurred at different temperatures (50 to 100 K) for different types of heterostructures. Our data showed that actually this transition occurred when the energy of the bandgap coincided with the energy of the spin-orbit splitting Δ_{SO} in the material of the active layer (Δ_{SO} was calculated on the basis of the data by Vurgaftman *et al.* [4]). With temperature increasing and E_g decreasing, we observed a resonant 'switch-on' of the CHHS Auger process, when the energy of recombining electron-hole pair was transferred to a hole transitioning to the spin-orbit-splitted band. With further temperature increasing and E_g continuing to decrease, the resonance condition $E_g=\Delta_{SO}$, that was responsible for quenching of the effect of stimulated emission, disappeared, but the emission remained spontaneous. This was obviously due to the influence of Auger processes of other types, which suppressed amplification. We calculated temperature dependence of carrier lifetime limited by radiative recombination and two most likely Auger processes: the one with the participation of two electrons and a heavy hole with excitation of electron in more energetic state (CHCC) and the one involving two heavy holes and an electron, with the conversion of a heavy hole into a light hole (CHHL). For these calculations, expressions derived in terms of the microscopic model for Auger recombination were used [8]; those describing radiative recombination were taken from Ref. [9]. The results of the calculation performed for structure of type *C* are presented in Fig. 3 . It is seen that under the considered injection level ($n \approx p \approx 4\times10^{16}$ cm^{-3}, derived from the data used to fit the calculated spectra to experimental ones as discussed above and applied to the whole temperature range), the dominating Auger process is CHHL. Still, at T<170 K, radiative lifetime remains smaller than the one limited by Auger recombination.

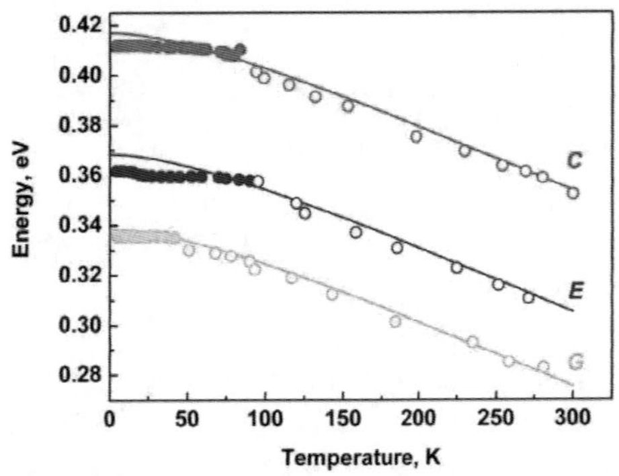

Figure 2. Calculated dependences E_g (T) in the active layer of LEDs (lines), E_{EL} of stimulated emission (filled symbols) and E_g values derived from the emission spectra (empty symbols)

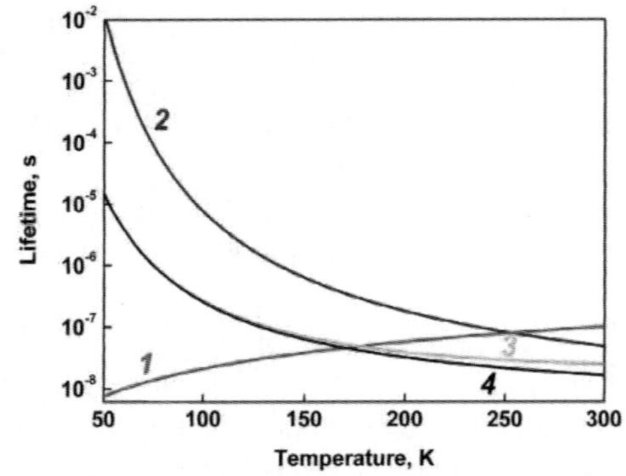

Figure 3. Calculated carrier lifetime limited by radiative recombination (*1*), CHCC (*2*), CHHL (*3*), and the sum of CHCC and CHHL (*4*)

5. Conclusion

When studying electroluminescence of InAsSb–based mid-infrared LEDs in the temperature range 4.2–300 K, we were able to follow the effect of three recombination mechanisms on the properties of the LEDs: Shockley–Read recombination at T<70 K, CHHS Auger process at 70<T<130 K and CHHL Auger process at T>130 K. Despite the influence of non-radiative recombination, however, stimulated emission from the structures was observed at T<100 K. This shows that InAsSb-based LED heterostructures are promising for fabrication of vertical-emitting MWIR lasers.

References

[1] Krier A, Ed. 2006 *Mid-Infrared Semiconductor Optoelectronics*. Springer Series in Optical Sciences, Vol. 118 (Springer: Berlin)
[2] Sopanen M, Koljonen T, Lipsanen H and Tuomi T 1994 *J. Cryst. Growth* **145** 492
[3] Bazhenov N L, Mynbaev K D and Zegrya G G 2015 *Semiconductors* **49** 1170
[4] Vurgaftman I, Meyer J R and Ram-Mohan L R 2001 *J. Appl. Phys.* **89** 5815
[5] Mynbaev K D, Bazhenov N L, Semakova A A, Mikhailova M P, Stoyanov N D, Kizhaev S S, Molchanov S S, Astakhova A P, Chernyaev A V *et al* 2017 *Semiconductors* **51** 239
[6] Gelmont B L and Zegrya G G 1991 *Sov. Phys. – Semicond.* **25** 1216
[7] Laurain A, Cerutti L, Myara M and Garnach A 2012 *IEEE Photon. Technol. Lett.* **24** 246
[8] Bazhenov N L, Mynbaev K D and Zegrya G G 2015 *Semiconductors* **49** 239
[9] Vorob'ev L E, Ivchenko E L, Firsov D A and Shalygin V A 2001 *Optical Properties of Nanostructures* (Nauka: Saint-Petersburg) (in Russian)

IOP Conf. Series: Journal of Physics: Conf. Series **917** (2017) 052006 doi:10.1088/1742-6596/917/5/052006

Structural and optical properties of pulsed laser deposited cadmium telluride thin films

H R Dashtoyan[1], K E Avjyan[2], L A Matevosyan[2] and S V Smbatyan[1]

[1]National Polytechnic University of Armenia, Yerevan, Republic of Armenia
[2]Institute of Radiophysics & Electronics, NAS of RA, Ashtarak, Republic of Armenia

Abstract. Cadmium telluride thin (< 1 μ) films were grown by pulsed laser deposition method on glass substrates from room temperature up to 300°C. The structural and optical properties were analyzed as a function of substrate temperature and laser radiation intensity.

1. Introduction

Thin films of $A^{II}B^{VI}$ semiconductors, in particular of cadmium telluride (CdTe), have wide application in electronics, integrated optics, optoelectronics, currently being used in many devices, such as field-effect transistors, photodiodes and photovoltaic solar energy converters. One of the most promising application of single and zinc doped crystals of CdTe is their use as a not requiring of additional cooling x-ray and radioactive radiation detectors. This paper presents the results of structural and optical properties of pulsed laser deposited CdTe thin films, obtained under high super-saturation conditions of growth.

Nowadays, PLD being the part of laser technologies and a kind of flexible, universal vacuum deposition, recognized as a simple and versatile method for deposition of thin films and structures based on them [1-3]. This method based on the using of physical phenomena occurring under the influence of laser radiation on solid targets, which reduced to ablation of a substance from the irradiation zone. The method holds of unique characteristics: possibility of evaporation practically all the materials including as well high-melting; short times of deposition (~10^{-6} sec.), which is equivalent to improve the "effective vacuum" on a several orders. Practically all materials needed for functional electronics are obtained by this technology. PLD is indispensable for synthesis of a number of complex composite materials with extraordinary properties.

2. Instruments and Experimental Techniques

CdTe films were obtained by the vacuum pulsed laser deposition (PLD) method from crystalline CdTe target on glass substrates from room temperature up to 300°C. The PLD unit consists of a Q-switched YAG:Nd^{3+} laser (1.064 μm wavelength, 30 ns pulse duration, laser energy – 0.35 J per pulse) and a vacuum chamber with residual gas pressure $2x10^{-5}$ mm Hg (figure 1). Laser intensity on the CdTe target was varied from 10^8 W/cm^2 up to 10^9 W/cm^2. The thickness of deposited layer from a single laser pulse was determined by dividing measured by optical interference method of a thick layer on a number of laser pulses (2.5 nm per pulse for $2x10^8$ W/cm^2 and 4.2 nm for $6x10^8$ W/cm^2 laser radiation intensities).

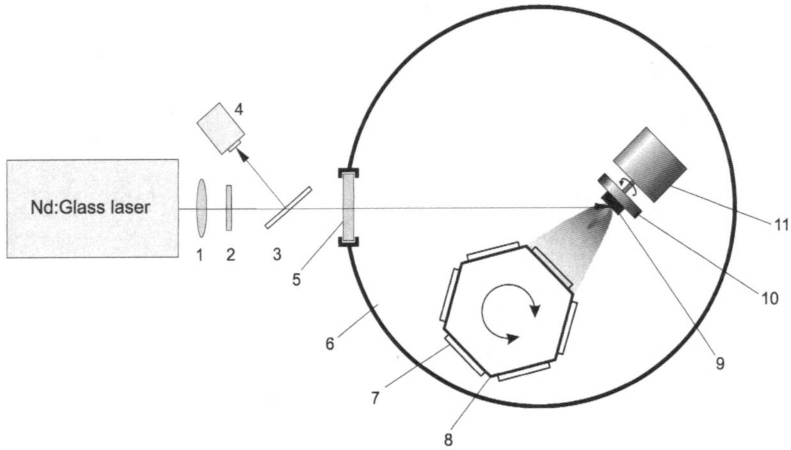

Figure 1. PLD system.
1. Quartz lens, 2. Filter, 3. Beam splitter, 4. Calorimeter, 5. Quartz window, 6. Vacuum chamber, 7. Substrate, 8. Substrate holder and heater, 9. Target, 10. Target holder, 11. Motor

The crystal structure of films was investigated by reflection high-energy electron diffraction method (accelerating voltage 75 kV). Independent measurements transmittance (T) and reflectance (R) of films were made at room temperature by means of spectrometer Filmetrics F20 (spectral range 400-1000 nm) at the normal incidence of light. Absorption coefficient (α) was calculated from measured data R and T by using of well known approximate relation in the interference free region:

$$T = (1 - R)^2 e^{-\alpha d} / (1 - R2 e^{-2\alpha d}),$$

where d is the film thickness [4]. In order to establish the nature of optical transitions, α was approximated with the empirical Urbach law $\alpha = \alpha_0 \exp(h\omega/E_u)$ (α_0 and E_u are the Urbach parameters) and power law $\alpha h\omega = B(h\omega - E_g)^m$, which under certain assumptions describes interband transitions.

3. Results

Beforehand we say that an explicit dependence of the physical properties of obtained films on their thickness is not observed. Electron diffraction studies showed that deposited at room temperature films has amorphous structure regardless laser intensity. CdTe films exhibit polycrystalline structure at 280°C and low (2x10^8 W/cm^2) laser intensities (figure 2).

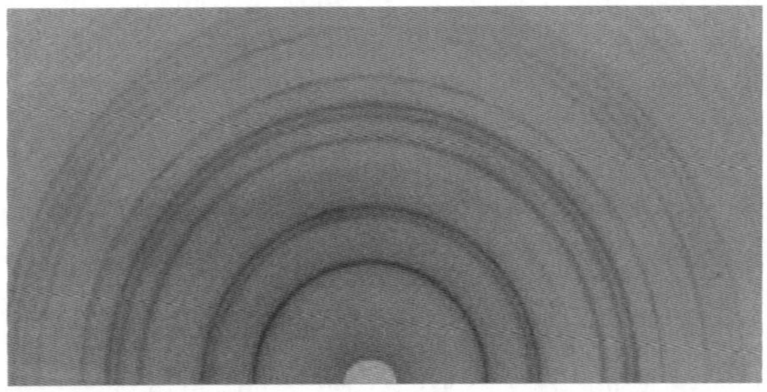

Figure 2. Electron diffraction pattern from CdTe film.

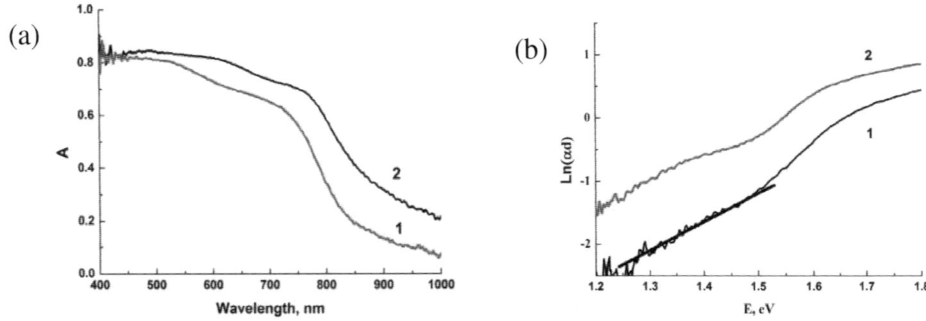

Figure 3. Absorption of films (a), Ln (ad) plot (b).

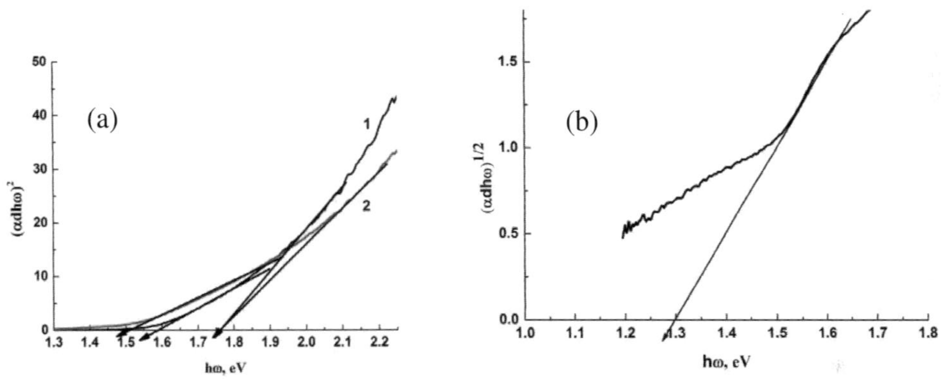

Figure 4. Absorption coefficient by power law of 2 (a) and power law of ½ (b).

Figure 3 (a) shows absorption ($A = 1 - R - T$) of films (thickness – 800 nm) deposited at room (1) and 280°C (low laser intensities). It can be seen that a significant absorption observes in films at photon wavelengths less than 850 nm, which is due to the existence of optical absorption edge corresponding to fundamental absorption edge in the CdTe single crystal. Figure 3 (b) shows a plot of Ln (ad) for the amorphous (1) and polycrystalline (2) CdTe film. We can see that these dependencies linearized in the energy range 1.3-1.5 eV for the amorphous film (calculated Urbach "tail" is $E_u = 0.2$ eV. The appearance of the "tails" Urbach in the gap points to an existing structural disorder [5].

Approximation of the absorption coefficient by power law (Fig. 4 (a)) gives the following results:

1. There are optical transitions (1.53 eV for a polycrystalline film and 1.6 eV for the amorphous film, Figure 4 (a)) with a minimum of photon energy (direct allowed transition, m = 1/2) as well as in a single crystal CdTe with the sphalerite structure (transition Γ8v → Γ6c at the high Γ point of Brillouin zone). Direct and allowed optical transition at 1.75 eV is associated with the spin-orbit splitting of the valence band.
2. There is also indirect-allowed optical transition (m = 2) in amorphous film at 1.3 eV (Figure 4 (b)), which is associated with the structural disorder.

4. Conclusion

CdTe thin films were grown on glass substrates by PLD method. The structural and optical properties were investigated as a function of substrate temperature and laser radiation intensity. Electron diffraction studies showed that deposited at room temperature films has amorphous structure regardless laser intensity. Polycrystalline structure is exhibit at low laser intensities and high substrate temperatures. Both amorphous and polycrystalline films have high absorption coefficient above the fundamental absorption edge (1.53 eV for a polycrystalline film and 1.6 eV for amorphous film).

IOP Conf. Series: Journal of Physics: Conf. Series **917** (2017) 052006 doi:10.1088/1742-6596/917/5/052006

References

[1] Miller J C 1994 *Laser Ablation - Principles and Applications* (Berlin: Springer-Verlag)
[2] Eason R 2006 *Pulsed Laser Deposition of Thin Films* (John Wiley & Sons Ltd)
[3] Alexanian A G, Aramyan N S, Avjyan K E, Khachatryan A M, Grigoryan R P and Yeremyan A S 2007 Technology of PLD for Photodetector Materials *Combinatorial and High-Throughput Discovery and Optimization of Catalysts and Materials* (CRC/Taylor & Francis) chapter 23 pp 447–468
[4] Willardson R K and Beer A C 1967 *Semiconductors and Semimetals* (New York: Academic Press)
[5] Brodsky M H 1979 *Amorphous Semiconductors* (Berlin: Springer-Verlag)

IOP Conf. Series: Journal of Physics: Conf. Series **917** (2017) 052007 doi:10.1088/1742-6596/917/5/052007

Laser marking of contrast images for optical read-out systems

O S Yulmetova [1,2] and M A Tumanova[1]

[1]Concern CSRI Elektropribor, JSC, Saint Petersburg 197046, Russian Federation
[2]ITMO University, 49 Kronverksky Pr., St. Petersburg, Russian Federation

Abstract. In the present study the formation of contrast images that provide functionality of optical read-out systems is considered.The image contrast is determined by the difference of reflection coefficients of the beryllium surface covered with titanium nitride film (*TiN)* formed by physical vapor deposition and the image created on it by laser oxidation. Two ways of contrast variation are studied: by regulating both *TiN* reflection coefficient during vapor deposition and the reflection coefficient of the image obtained with the laser. The test results show the efficiency of the proposed approach.

1. Introduction

Traditional techniques for marking metal surfaces include electrochemical, chemical, etching, and mechanical processes that are typically used for functional image formation. All these 'standard' marking techniques damage the surface or strongly change its surface properties. However, in some application areas, contrast image is required to be created with minimum changes of surface properties. A clear example is a rotor of an electrostatic gyroscope that contains a contrast image on its surface, providing functionality of an optoelectronic read-out system of the gyroscope [1]. Surface homogeneity determines dynamic characteristics of a spherical rotor during its high speed rotation. The quality of the pattern defines the accuracy of data acquisition from a spinning rotor. For this reason, the technical process of pattern formation is of great importance [2]. The work described in this paper is focused on the development of a process for contrast image formation on titanium nitride surface without surface roughness and shape deviation. This task is proposed to be solved using a localised laser oxidation process. A pulsed fiber laser was used to treat a titanium nitride film deposited on beryllium substrates in the air with intensities below an ablation threshold to provide oxide formation.

2. The object and purpose of investigation

The objects of the research were spherical beryllium samples with a diameter of 10 mm and surface roughness Ra=0.05 μm, with a 1 μm titanium nitride coating using physical vapor deposition (PVD) before laser treatment.

The main purpose of the work was to create a pattern consisting of 8 stripes on *TiN* surface with the required contrast ratio K=0.5±0.1 measured on a read-out system with a wave length of 860 nm without deviation of pattern roughness from the basic surface roughness.

3. Theoretical issues

One of the most important characteristics of the pattern is its optical contrast K, which is defined by the difference between the reflection coefficients of the basic surface R_{TiN} and those of the pattern surface R_r formed by the local laser modification of the basic surface, described by the following formula:

$$K = \frac{R_{TIN}(\lambda) - R_r(\lambda)}{R_{TIN}(\lambda) + R_r(\lambda)} \tag{1}$$

R_{TIN} and R_r – reflection coefficients of titanium nitride and pattern surfaces, respectively, λ – the wavelength of the optoelectronic read-out system equal to 860 nm. It is possible to plot lines of equal contrast (figure 1) for various combinations of R_{TiN} and R_r, using formula (1).

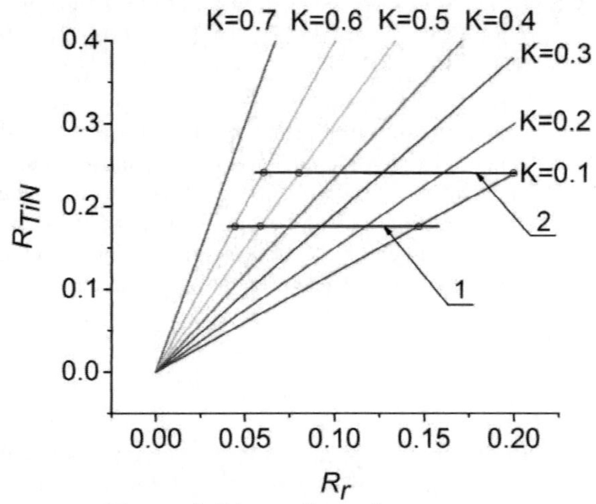

Figure 1. Lines of equal contrast.

Lines 1 and 2 from figure 1 show the possibility of getting the desired contrast of 0.5-0.6 for different combinations of R_{TiN} and R_r. Obviously, the contrast value K can be regulated by changing reflection coefficients R_{TiN} during vapor deposition of titanium nitride thin film and R_r during laser treatment. The laser oxidation technique was chosen because of its high performance and versatility in comparison with anodizing or photolithography [3, 8].

It is required to provide the contrast value K with the minimum level of the thickness of the pattern layer formed by a localised laser oxidation process. This condition can be met if the contrast value K is achieved with a minimum level of R_r. According to figure 1, for line 1, the minimum value of R_r is equal to 0.038 and the corresponding level of R_{TiN} is 0.15. Such calculations determine how to achieve the maximum contrast with the minimum impact of a laser treatment.

Regulation of R_{TiN} is possible by forming *TiN*-coating with non-stoichiometric composition described by formula TiN_x, where x is taken from the range of values from 0.58 to 1, corresponding to the coating color from light to dark golden. It is important to note that hardness of non-stoichiometric composition of titanium nitride films is much higher than that of stoichiometric one [4]. This phenomenon is very useful in our research because it is possible to create a strong thin film on the rotor surface that protects beryllium surface from damage in the case of the emergency shutdown.

The exact composition of titanium nitride coating is regulated by the partial pressure of nitrogen gas used during the vapor deposition process. For example, the nitrogen pressure of 0.013 Pa provides a coating having the light golden colour, the range of gas pressure from 0.1 to 0.8 Pa forms a coating having the golden colour and it's possible to get the dark golden colour using the pressure of 1.04 Pa.

Minimisation of R_r is correlated with decreasing laser power that defines the roughness and the depth of the modified layer. The reflection coefficient of the pattern R_r is regulated by laser modes. During the research the laser power was determined as the most significant parameter of laser treatment. Considering laser marking as a thermally activated process, described by the Arrhenius equation [5], assuming that R_r decreases exponentially under decreasing laser power, using nonparametric approximation, it is possible to obtain the following equation:

$$K \approx K_{\max} - Ae^{-P/P_0} \tag{2}$$

K_{max} is the maximum available value of the contrast, P_0 is the power corresponding to the level when R_r decreases by a factor of e, A is a constant that depends on the material properties. The approximation of equation (2) by the experimental data is shown in figure 2.

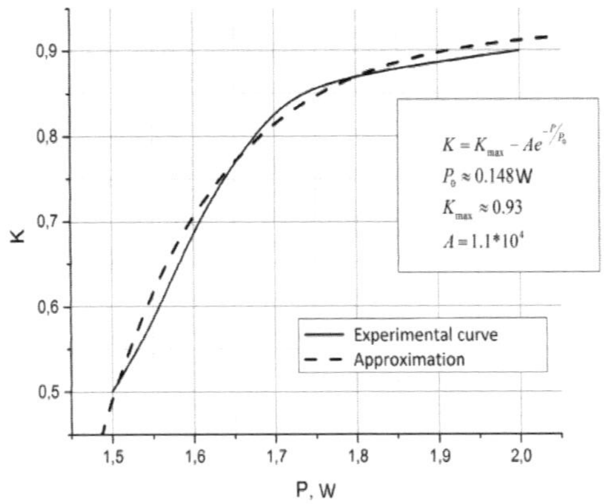

Figure 2. Experimental and approximation curves for *K=f(P)*.

The curves presented in figure 2 show that the experimental results are in a good agreement with the results of calculations, for more detail, see [6]. For laser marking of titanium nitride with non-stoichiometric composition described by formula $TiN_{0,9}$, such approximation allows one to determine the values of K_{max}, P_0 and constant A, as follows $K_{max} = 0.93$, $P_0 = 0.148$W and $A = 1.1\cdot10^4$.

4. Experimental details

A thin film of titanium nitride coating was formed by cathodic arc deposition (PVD method). Nitrogen pressure was selected in the range from 0.1 to 0.2 Pa to create golden colour. The choice of medium pressure is explained by the additional requirements for *TiN* coating to provide a high level of wear resistance and hardness with minimum defects. The dependence of *TiN* coating hardness from the level of nitrogen pressure was considered in [7]. It was shown that the maximum of hardness with minimum defects corresponds to the range of nitrogen pressure from 0.081 to 0.104 Pa.

TiN coating formed by cathodic arc deposition with partial pressure of nitrogen equal to 0.16 Pa has surface reflection coefficient R_{TiN} equal to 0.15. The dependence of nitrogen pressure on the reflection coefficient was determined empirically.

The following pattern was created in the air using a commercially available machine based on an ytterbium pulse fiber laser with λ = 1.06 μm wavelength. In the process, the pulse width τ was 4 ns, the average power P_{aver} and the pulse repetition frequency f were 5 W and 99 kHz, correspondingly. The laser spot with a focal diameter of d_0 = 50 μm moved over the sample surface with a velocity of 85 mm/s. This mode provides an optical contrast without changes of surface roughness R_a=0.05 μm and was obtained experimentally, based on the theoretical issues mentioned before.

Four colours (golden, brown, purple and blue) were obtained using the described mode by changing numbers of passes. Golden colour was obtained after the first pass, brown - after the second laser scanning of exactly the same place, purple - after the third pass, and blue - after the forth pass. Multipass laser treatment was used as the method to increase contrast gradually with the minimum thermal exposure at the maximum resolution. The following values of the contrast, calculated according to formula (1), were obtained from optical analysis of the treated areas: 0.14 for golden colour area, 0.39 – for brown, 0.53 – for purple, and 0.6 – for blue colour. The laser mode forming the blue colour pattern with contrast 0.6 was used in the further study.

The surface roughness of the treated layers was measured using the roughness measuring station Hommel Tester T8000. The surface roughness of the samples was evaluated by calculation of the standard roughness parameters presented in table 1.

Table 1. The roughness parameters before and after laser oxidation of *TiN* coating.

Roughness parameters, μm	R_a	R_q	R_z	R_{max}
Before laser oxidation	0.05	0.08	0.38	0.52
After laser oxidation	0.05	0.08	0.39	0.58

The experimental data from table 1 shows that the set of roughness parameters remains unchanged after laser treatment, which indicates that the formation of a color image is carried out by the laser local oxidation without initialization of evaporation process. Contrast image formation was realised by localised laser oxidation with generation of color thin films. Approximate thickness of color films is about 100 – 400 nm. The titanium oxide film described by formula TiO_2 and oxynitride with composition $TiO_{0.34}N_{0.74}$ were detected by XDR analysis after multiple laser treatment of the titanium nitride thin film.

5. Experimental results

The pattern consisting of 8 stripes (4 wide and 4 narrow) was marked by laser on the spherical sample covered with *TiN* coating. The optimal modes for vapor deposition and laser oxidation were obtained experimentally, based on the theoretical issues mentioned before. Then the sample included in read-out system was subjected to test, using the emitting-receiving optical sensor. Results are shown in figure 3.

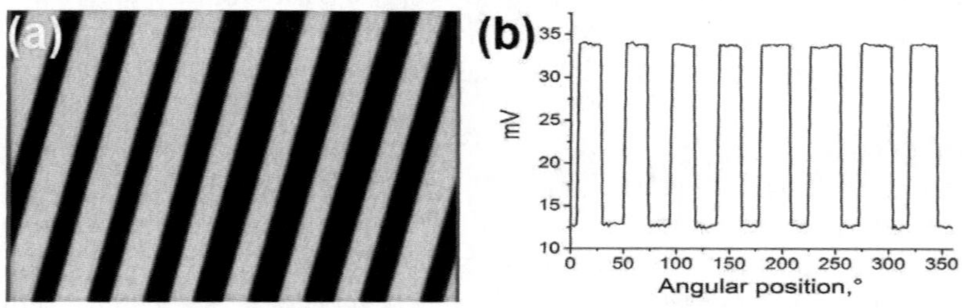

Figure 3(a,b). (a) View of the pattern obtained from the optical read-out system; **(b)** signal from the optical sensor in the equatorial zone of the spherical sample.

From figure 3(a) one can see that the read-out system scans eight stripes formed by a laser. These stripes have clear boundaries and the high level of the homogeneity. Figure 3(b) shows that the optical sensor detects 4 wide and 4 narrow stripes rather clearly. Detailed description of functionality of the optoelectronic read-out system with a pattern created by laser oxidation process can be found in [8].

References

[1] Egorov A V, Landau B E, Levin S L, Romanenko S G 2012 Rotor motion in a strapdown electrostatic gyro onboard an orbiting spacecraft *GyroscopyNavig. Springer* **3** 144–51

[2] Yulmetova O S, Yulmetova R R, Matyzhonok V N 2010 Research on the contrast of laser marks on precise surfaces of metal *Proc. SPIE* **7996**, *Fundam. Laser Assist. Micro-and Nanotechnologies* 799604.

[3] Yulmetova O S, Scherback A G, Landau B E, Butsyk A Ya, Veiko V P and Valetov V A 2013 Patent RU no. 2498224

[4] F Vaz, J Ferreira, E Ribeiro, L Rebouta, S Lanceros-Méndez, J A Mendes, E Alves, Ph Goudeau, J P Rivière, F Ribeiro, I Moutinho, K Pischow, J de Rijk 2005 Influence of

nitrogen content on the structural, mechanical and electrical properties of TiN thin films *Surface and Coating Technology* **191** 317–23

[5] Laidler K J 1984 The development of the Arrhenius equation *Journal of Chemical Education* **61** 494–8.

[6] Yulmetova O S, Scherback A G, Veiko V P, Yulmetova R F, Scherbak S A 2015 Control over raster image contrast in laser marking: technological process modeling. *Journal of instrument engineering* **6** DOI 10.17586/0021-3454-2015-58-6-485-491 (in Russian)

[7] Matlakhov V P 2006 Dependence of physico-mechanical properties of nitride-titanium coatings on nitrogen pressure *Bulletin of BSTU* **2** 93-6 (in Russian)

[8] Yulmetova O S, Shcherbak A G, Belash A A, Landau B E, Filippov A Yu 2016 Comparative evaluation of electrophysical and electrochemical methods of creating patterns on rotors of electrostatic gyroscopes *Metaloobrabotka* **4 (94)** 39-45 (in Russian)

Water Splitting Using Porous Silicon Photo-electrodes for Hydrogen Production

M Ali[1], V V Starkov[2], E A Gosteva[1], A V Druzhinin[1], S Sattar[1]

[1] National University of Science and Technology «MISiS», Moscow, Russia
[2] Institute of Microelectronics Technology and High Purity Materials, Russian Academy of Sciences, Chernogolovka, Russia

Abstract. This paper presents the efficiency study results of using gradient-porous silicon structures with different morphology, as photo-anodes for photo-electrochemical dissociation of water. The results of a study of the physicochemical properties of gradient-porous silicon structures show the relatively low cost and simplicity of the technological process, as well as the possibility of forming structures with predefined properties, allow the creation of effective devices for artificial photosynthesis based on porous silicon for subsequent use in hydrogen energy.

1. Introduction

Constantly increasing energy consumption throughout the world leads to the inevitable depletion of various fossil fuel sources. This encourages intensive research for practical application of so-called alternative or renewable energy sources. A special place is given to solar energy efficiency research. It is believed that the main supplier of solar energy will be batteries based on solar cells (SE). Direct conversion of solar radiation into the most convenient for human use electric energy with the help of modern solar cells is approaching its possible limit in efficiency. At the same time, the cost of a unit of generated electricity is already comparable to the same amount of electricity produced by burning coal, which is one of the cheapest fossil sources. In addition, the use of solar cell batteries does not require any special operating costs and, at the same time, does not lead to undesirable pollution of the environment. However, the receipt of environmentally friendly electricity with the help of solar cells is accompanied, by the problem associated with the inconstancy of solar radiation falling on the solar cell battery.

On the other hand, there is a volatility in the consumption of electricity over time. There is a need to accumulate surplus electricity generated at maximum intensity of solar radiation and their use under peak loads or minimum intensity of solar radiation. In this case, the preference should be given to the scheme of accumulation, which would not destroy the level of ecology that was achieved in the production of electricity using solar cells. For such purpose, the use of solar energy for the splitting water to produce and store hydrogen and oxygen is considered. The subsequent use of hydrogen as a fuel, and oxygen as an oxidant in hydrogen fuel cells (for example, PEMFC) will virtually balance the generation and consumption of electricity from solar power plants. At the same time as a result of the application of such a scheme there is no pollution of the environment, as the result of the operation of fuel cells in addition to electricity is water and heat. This work presents the results of development and analysis of the efficiency of gradient-porous silicon structures with variable pore morphology, designed to create photo-electrochemical dissociation devices for water based on photo-electrodes.

2. Artificial photosynthesis of hydrogen fuel

Preservation and accumulation of energy are the most important directions for the successful use of various renewable energy sources and, in particular, the sun in the future. An ideal system will simulate a natural cycle of photosynthesis, but with much greater efficiency. One of the most promising candidates in this regard is hydrogen, since it can be obtained from water by electrolysis. Direct production of hydrogen by splitting water with the help of sunlight has a theoretical efficiency (efficiency) of up to 20% using a unijunction semiconductor electrode. More than 30% can be achieved using multi-converter photoelectric converters, for example, a combination of two semiconductor materials with different band gap widths. In the laboratory conditions, the efficiency is more than 12% and 18% for unipolar or tandem semiconductor electrodes, respectively. Structurally - technological aspects of unijunction and multi-transition electrodes are considered in [1]. It should be noted, however, that all materials used are too expensive or do not provide long-term stability of electrochemical devices. For the first time, a device for the decomposition of water under the influence of sunlight in photo-cells was demonstrated in the works of Fujishima and Honda [2].

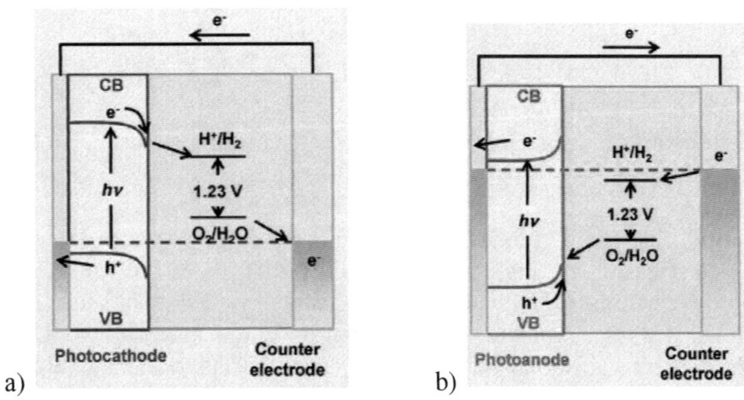

Figure 1. Schematics of water electrolysis with semiconductor and metal electrodes: a) p-type, b) n-type.

The photo-electrolysis schematics using semiconductor electrodes of p- or n-type conductivity are shown in figure 1 a, b, respectively. A quantum of light generates electron-hole pairs in a semiconductor electrode. The hole diffuses to the interface between the semiconductor and the aqueous electrolyte, and the water molecule is oxidized there. Electrons pass along the external circuit to a metal cathode, where hydrogen ions are reduced.

The main problem is the choice of material for photo-electrodes. It should be inexpensive and affordable, resistant to photo-corrosion. The most important characteristic for a semiconductor photo-electrode is the width of the forbidden band and its position relative to the electrochemical potentials of water oxidation and hydrogen reduction [2, 3]. The values of the band gap should be in the range from 1.7 to 3 eV (figure 2). These parameters are determined by the fact that the photo-electrode must create a potential sufficient for the electrolysis of water and at the same time efficiently absorb and transform light into electricity in the widest possible wavelength range. In comparison with the binding energy of water molecules (E c.b = 1.23 eV), mono-crystalline silicon is characterized by a narrower width of the forbidden band (E_g = 1.1 eV). For this reason, silicon has not been used for a long time as a material for photo-electrodes and its application in photolysis has not been practically considered [3]. Further development of the field of application of silicon is associated with the development of various composite materials based on it. One of the most known and easily obtainable composites is porous silicon (pSi).

3. Sicon photo-electrodes with gradient porosity

Porous silicon was discovered at the beginning of the second half of the last century. Since then, a large volume of scientific research has been carried out both in methods of obtaining and studying properties and in the variety of practical use of the unique properties of this promising nano-composite material [4-8]. The most significant characteristic of pSi for photo-electrolysis, as noted above, is the band gap (Eg pSi). According to the results of studies by different authors, Eg pSi is essentially dependent on porosity.

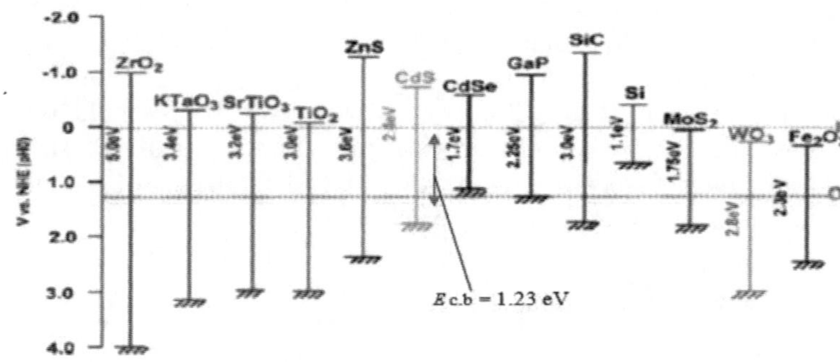

Figure 2. The diagram of forbidden bands of semiconductor materials most commonly used for photo-electrolysis [3].

When the porosity is increased from 5% to 95%, the width of the forbidden band varies from 1.1 eV, which in practice corresponds to single crystal silicon (c-Si), to Eg PC = 3.05 eV [9], which can correspond to a porous crystalline (PC) with characteristic pore wall sizes less than Units of nanometers. In addition, a material with such porosity has an extremely large surface area, which can reach 1000 m^2 / cm^3 [4]. Thus, nano-porous silicon is very attractive option for use as a material for photo-electrodes [9, 10].

An important limitation for the practical application of nano-porous structures is the extreme fragility of the material. If the Young's modulus is 160 GPa for c-Si, then in a PC with a porosity of 90% this value decreases to 0.87 GPa [4]. This does not allow the use of nano-porous silicon in the form of free membranes or electrodes. Therefore, to support nano-porous films it is necessary to use special support elements. As a carrier of the nano-porous layer, it is proposed to use gradient-porous silicon structures. The first gradient-porous structures (GPSi), in the form of free membranes, are described in [11]. The developed technology of deep anodic etching makes it possible to form macro-porous membranes with thickness from 150 μm to 550 μm. Such membranes were used as electrodes for micro-fuel elements [4, 12]. The strength of porous membranes was ensured by the successive formation of layers with different morphologies of macro-pores as shown in figure in 3a, b, and c. In contrast to the hetero porous crystalline (HPC) structures with several spatially separated layers, the GPSi-var structures are characterized by a smoother change in pore morphology in the depth of the porous layer (figure 3d). In the present experiment, the GPSi-var structures were fabricated by the method of deep anodic etching of p-type silicon wafers with a resistivity of 10 Ω cm.

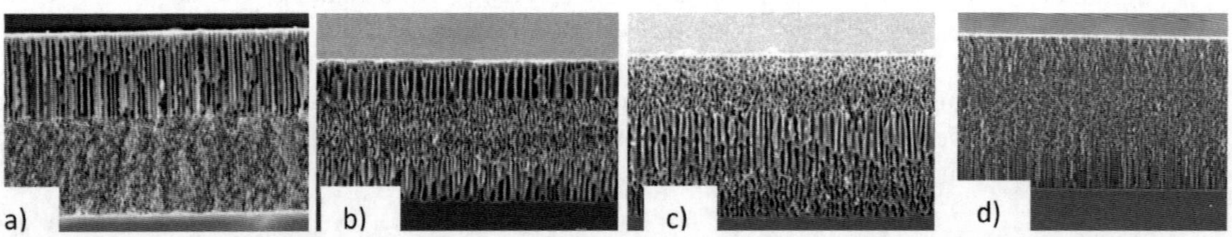

Figure 3. Gradient-porous silicon structures: a) two-layered (GPSi-2); b), c) three-layered (GPSi-3); d) with the smoothly variable morphology (GPSi-var).

The structures were formed on plates with the orientation of the surface of the plates (111) and (100). The thickness of the porous film (figure 4) was 320 μm. The etching was carried out at a constant current density of 2.5 mA / cm^2 in a solution of hydrofluoric acid in alcohol HF: C2H5OH: CTAC = 200: 50: 1 (CTAC-cetyltrimethylammonium chloride). The etching time was 480 minutes. In contrast to GPSi-2 and GPSi-3, the production mode of GPSi-var structures does not provide a change in the electrolyte during the etching process. Spontaneous formation of the morphology of the porous structure occurs at a constant density of the anodic etching current [13].

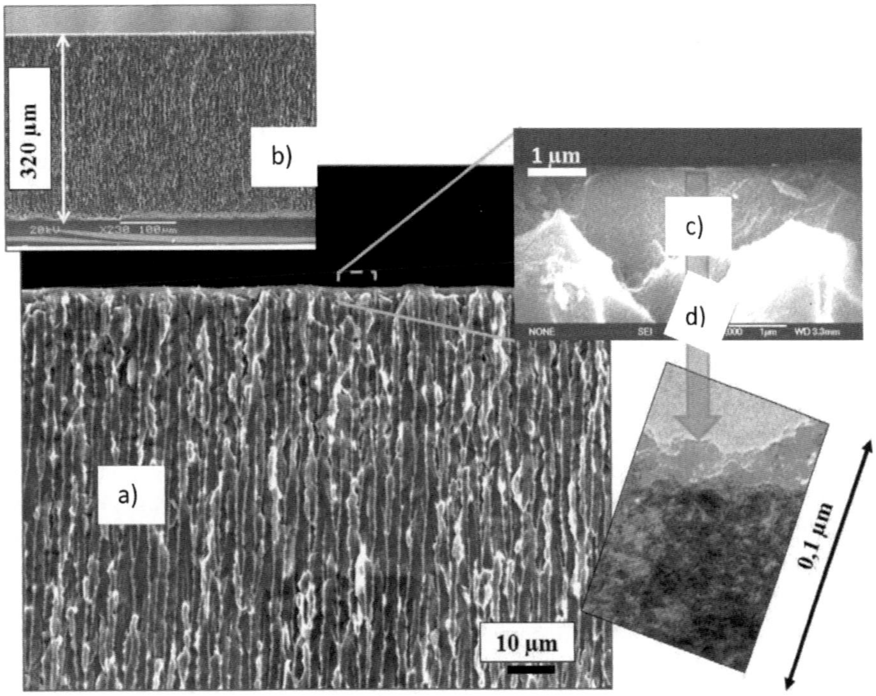

Figure 4. Images of the fragments of the end face of an arbitrary cleavage of the **GPSi-var** structure formed on p-type conductivity plates, ρ_V = 10 Ω cm, **orientation** (100).

The porous surface, in comparison with the surface of the original silicon wafer, acquires a brilliant black color (black Si). However, there is no significant increase in the roughness of the porous surface, which is typically characteristic of the porous surface of single-layered or multilayer macro-porous films. This is due to the fact that a nano-porous layer is located on the surface of the macro-porous layer (figure 4b, c, d). This layer is formed during the process of disproportionation occurring during the anode etching regimes used. The thickness of the nano-porous layer in the structures depicted in figure. 4 c, d is between 0.2 and 3 μm. A similar structure was observed in the etching of silicon wafers with a different orientation of the initial surface. As shown in [13], a nano-porous film is formed on the surface at the initial stages of anodic etching and is retained throughout the etching process. To obtain porous membranes after the anodic etching process was completed, the remaining un-etched portion of the silicon wafer on the bottom of the pores was removed by mechanical polishing.

An essential characteristic of the surface of photo-electrodes is the ability of absorption over a wide range of frequencies and angles of the incident light. Reflectance spectra obtained from the surfaces were studied in Ref. [14]. Figure 5 shows the comparative reflection spectra, the front side (marked in the figure as "shiny side") and the back side of the membrane (marked in the figure as "matte side").

IOP Conf. Series: Journal of Physics: Conf. Series **917** (2017) 052008 doi:10.1088/1742-6596/917/5/052008

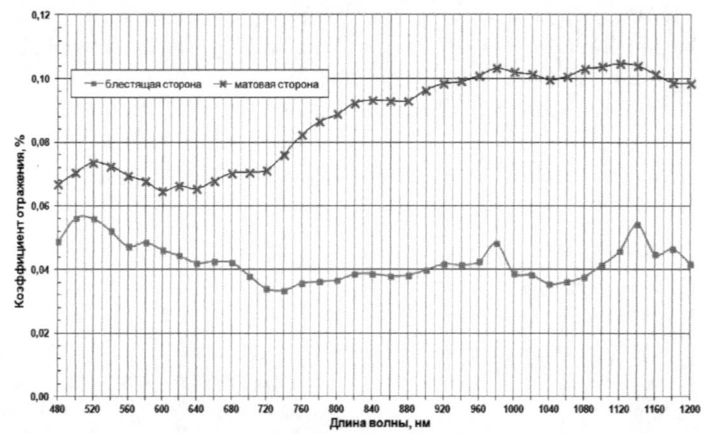

Figure 5. Spectra of reflection of GPSi-var membranes with the front (shiny) and back (matte) sides.

The analysis of the obtained spectra indicates, first of all, that in the considered wavelength range, the reflection of light is at least an order of magnitude lower than on the surface of the original silicon wafers. It should be noted that the roughness of the surface of the initial surface of the silicon wafers was Rz = 16.56 nm, and for the nano-porous region this value was Rz = 30.3 nm. In addition, we noted a weak dependence of the reflection coefficient on the angle of incidence of light on the nano-porous area of the electrode. At the same time, the reflection coefficient strongly depended on the angle of incidence for the original silicon. As follows from the dependencies shown in figure 5, the presence of the nano-porous layer on the surface of the macro-porous layer (front, shining side) compared to the mechanically polished macro-porous side provides higher antireflection characteristics.

4. Water splitting method

Figure 6 shows the experimental design, which will confirm the operation of a single-layer porous silicon electrode. The structure is a Pt-H_2O-por-Si / p-Si system. The counter electrode (4) is a platinum wire, and the sample itself is opposite (2). The contact is made to the back side by the liquid method. In the center of the bulb (1) is a gas outlet. The experiment was conducted under two conditions when illuminated with light and in the dark, and its reaction to light was also tested (figure. 6). The voltage in the absence of light was about 40 mV, with a light supply of -65 mV. The current was 1.2 μA and 1.8 μA, respectively. The working area in this experiment is about 0.5 cm^2. There is an interesting feature that with a completely darkened sample, a current of the order of 0.8 μA. The flowing current has sufficiently low values, which corresponds to a slow flow of the decomposition reaction of water. Therefore, it is necessary to reduce the resistance of the sample, because it is about 30 kOhm. The main contribution to this value is made by the porous film [10].

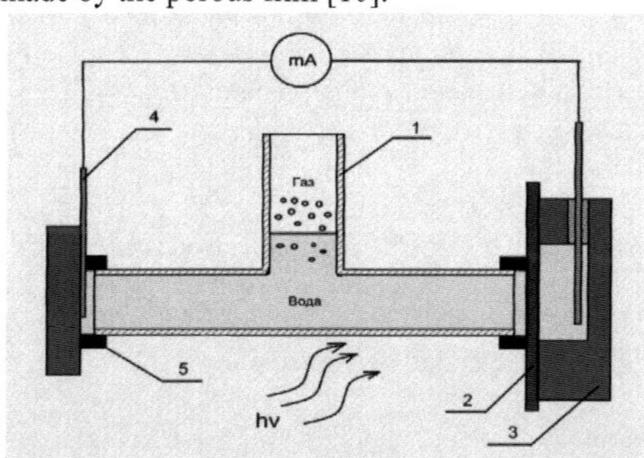

Figure 6. Scheme of the experiment for testing the performance of a porous silicon electrode.

5. Conclusions

In the work it is proposed to use the GPSi structures as photo-electrodes for devices of artificially photosynthetic hydrogen. The results of the investigation of the physicochemical properties of HPC-var silicon structures indicate the greatest attraction of the use of such structures based on silicon for the photoelectric decomposition of water. The use of GPSi structures allows to reduce the reflection coefficient below the level of 0.1%, and also extend the range of use of the light range of wavelengths. Photo-electrodes based on HPC structures significantly reduce the dependence of the magnitude of the generated photocurrent on the angle of incidence of light. Further optimization of the structural and technological design of photo-electrodes based on HPC-var silicon structures and, in particular, reduction of the series resistance of the electrode is necessary.

References

[1] Bogdanoff P, Stellmach D, Gabriel O, Stannowski B, Schlatmann R, Krol R, and Fiechter S 2016 *Energy Technol.* **4** 230
[2] Fujishima A, Honda K 1972 *Nature* **238** 37
[3] Kudo A, Miseki Y 2009 *Chem.Soc.Rev.* **38** 253
[4] Starkov V V 2009 *All materials. Encyclopedic reference book* **4** 13
[5] Granitzer P and Rumpf K 2010 *Materials* **3** 943
[6] Kotkovskii G E, Kuzishchin Yu A, Martynov I L, Nabiev I R, and Chistyako A A 2013 *Nuclear Physics and Engineering* **4** 174
[7] Yuzova V A, Levitsky A A and Harlashin P A 2011 *Journal of Siberian Federal University. Engineering & Technologies 1* 2011 **4** 92
[8] Lotuhina N, Chepurnov V, Pisarenko G 2013 *Elektronika* №4 (00126) s.106-109 (in Russ)
[9] Tynyshtykbayev K B, Borisovich G V, Muratov D, Rakhmetov B, Tokmoldin N S, Tokmoldin S Z 2016 *Russian Microelectronics* **45** 603
[10] Gosteva E A, 2015 *Mezhdunarodnyj Nauchnyj Zhurnal Al'ternativnaya Energetika i Ekologiya* –T. 183. – S. 14-19 (in Russ)
[11] Starkov V V and Gavrilin E 2007 *Phys. Stat. Sol. (c)* **4** 2026
[12] Starkov V V, Lyskov N V, Dobrovolsky Yu A 2009 *International Scientific Journal for Alternative Energy and Ecology* **76** 78
[13] Starkov V V, Irzhak D V, Roshchupkin D V. 2008 *Deformaciya i razrushenie materialov* №10 str.6-11 (in Russ)
[14] Gosteva E A, Starkov V V, Parhomenko YU N, Kah M, Iwe I A *Mezhdunarodnyj Nauchnyj Zhurnal Al'ternativnaya Energetika i Ekologiya* (in press., In Russ)

IOP Conf. Series: Journal of Physics: Conf. Series **917** (2017) 052009 doi:10.1088/1742-6596/917/5/052009

Laser induced nanostructures created from Au layer on polyhydroxybutyrate

I Michaljaničová[1], P Slepička[1], P Juřík[1] and V Švorčík[1]
[1]Department of Solid State Engineering, University of Chemistry and Technology, Prague 166 28, Czech Republic

Abstract. Nanostructures as well as composite materials expand the range of materials properties and allow use of these materials in new and highly specific applications. In this paper, we described laser modification of polyhydroxybutyrate films covered with thin gold layer, which led to the formation of various composite structures. The crucial for the composite structures creation was setting of appropriate laser parameters; 15 mJ cm^{-2} laser fluence and 6 000 pulses were recognized as the best. The morphology of structures was determined by the thickness of the Au layer. The most interesting formations, very porous with the biggest roughness, were observed after treatment of foils covered with 10 nm of Au. The morphology was observed by atomic force microscopy. The influence on roughness and the difference between projected area and surface area was also determined.

1. Introduction

Spontaneous formation of complex structures offers an interesting alternative for the preparation of surfaces with specific properties. The structures formed by self-organization currently have a wide variety of uses, e.g. optical devices [1], the flexible electronics [2] systems for separating particles [3].

One of the self-organizing mechanisms is wrinkling instability. By this bottom-up method is possible to prepare various surface structures with patterns in a wide range of length scales. Wrinkles are characterized as smooth and shallow surface undulations as a result of the uneven expansion of a surface consisting of materials with different mechanical properties [4]. The wrinkles occur, when the residual stresses exceed a critical value [4], which can be induced by heating [5], solvent swelling [6], mechanical stretching or compression [3] and others. The effect of wrinkling on a thin bi-layer film (polymeric film metalized with a noble metal) was described in the works [7, 8].

Polyhydroxybutyrate (PHB) is an interesting biocompatible and biodegradable polymer, which can be synthesized by bacterial fermentation. PHB is frequently used in medical and pharmaceutical applications as a material for implants, scaffolds or drug delivery systems [9]. To improve processability, PHB is frequently produced in copolymer form with polyhydroxyvalerate (PHV), because the homopolymers of PHB are very fragile and crystalline [10]. Further development of this promising substrate can broaden its applications and eliminate disadvantages of the material.

This paper deals with surface modification of biopolymer foil of polyhydroxybutyrate covered with thin Au layer and treated by excimer laser beam. Observation of such treated samples showed various patterns depending on the Au layer thickness. The surface morphology of these structures was investigated by AFM.

2. Material and Methods

2.1. Materials and modification

For experiments we used biopolymer foils of polyhydroxybutyrate (PHB, with 8% polyhydroxyvalerate) supplied by Goodfellow Ltd., Cambridge, Great Britain (density 1.25 g cm^{-3}, 50 µm thick foils).

Gold layers were deposited on the foils from Au target (99.999 %) by diode sputtering technique (BAL-TEC SCD 050 equipment, Switzerland) [11]. Typical sputtering conditions were: room temperature, sputtering times 20, 50, 100 and 280 s to achieve the layers of 5, 10, 20 and 50 nm, argon (purity 99.997 %) pressure of about 5 Pa, electrode distance of 50 mm, and electric current of 20 mA [11].

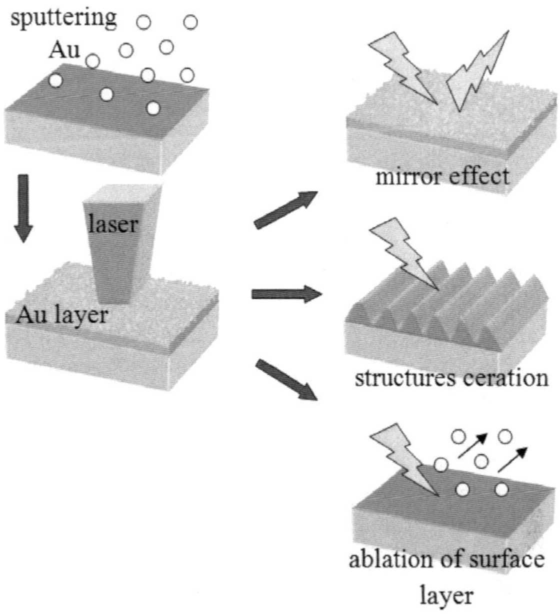

Figure 1. The scheme of modification: polymer substrate is sputtered with Au and then the bilayer substrate is modified by laser beam.

For modification of the foils covered with Au layer and also the pristine polymer foils we employed a KrF excimer laser (Coherent Compex Pro 50 wavelength of 248 nm, pulse duration of 20–40 ns, repetition rate 10 Hz). The beam of KrF laser was polarized linearly with cube of UV grade fused silica 25×25×25 mm^3 with active polarization layer. For homogeneous illumination of the samples we used only the central part of the beam by means of an aperture (0.5×1.0 cm^2). The samples were mounted onto a translation stage at perpendicular position of the sample and laser beam. [12-14] We used 6000 pulses with laser fluences in interval 5–30 mJ cm^{-2}, but the most interesting laser fluence for experiments was 15 mJ cm^{-2}. **Figure 1**. shows the scheme of modification, where the polymer substrate is first of all sputtered with gold and then the bilayer substrate is modified by excimer laser beam.

2.2. Characterization techniques

As the main method for surface morphology and roughness examination, of the pristine and modified PHB foils, we used atomic force microscopy (AFM) (instrument VEECO CP II in tapping mode and a Si probe RTESPA-CP with the spring constant 20–80 N m^{-1}) [12, 13].

The mean roughness values (R_a) and (RMS) represent the arithmetic and root mean square average of the deviations from the centre plane of the sample. The both roughness values were calculated by software SPMLab.

The mean roughness values (R_a) represents the average of the deviations from the centre plane of the sample. The R_a determination is constructed from the matrix 512 × 512 points, the equation is

$$R_a = \frac{\sum_{i=1}^{N} |Z_i - Z_{cp}|}{N} \quad (1)$$

where Z_{cp} is the value of the central plane, Z_i is the actual Z value, and N is the number of points, where the Z_i is evaluated. The room mean square (RMS) is calculated from the formula

$$RMS = \left(\frac{\sum_{i=1}^{N}(Z_i - Z_{ave})^2}{N}\right)^{\frac{1}{2}} \tag{2}$$

where Z_{ave} is the average value Z of measured area, Z_i is the actual Z value, and N is the number of points, where the Z_i is evaluated.

3. Results and discussion

The laser treatment of polymer foils covered by metal layer can have three types of effect: (1) mirror effect or reflection, (2) structures creation and (3) ablation effect. **Figure 1**. shows the scheme of modification, where the polymer substrate is first of all sputtered with gold and then the bilayer substrate is modified by excimer laser beam. Modification causes the effect mentioned above or their combination, because the combination of these effects is also possible.

The mirror effect occurs if the metal does not absorb laser radiation. In this case the incident beam is reflected from the surface without causing changes of the substrate. We observed the reflection when the applied laser fluence was too low. When the laser fluence in above the ablation threshold of used metal layer, the surface material is sputtered out and the ablation occurs. There is also the possibility that laser beam radiation induces a reaction of metal with polymer and this creates structured composite material. What effects, structures or results are obtained depends not only on the metal composition and the thickness of the layer, but also on the polymer substrate.

The PHB substrates were deposited with Au layer of thicknesses 5, 10, 20 and 50 nm. How the thickness of Au layer affected the composite structure is evident from **Figure 2**. The pristine PHB obtained from a supplier exhibits specific hilly structure with fibrous surface, which disappear after the Au overlay. Gold surfaces seemed flat and high resolution allowed observing Au clusters. The modification of pristine polymer (without metal layer) by laser treatment led to obliteration of fibrous surface and creation of cracks in the material.

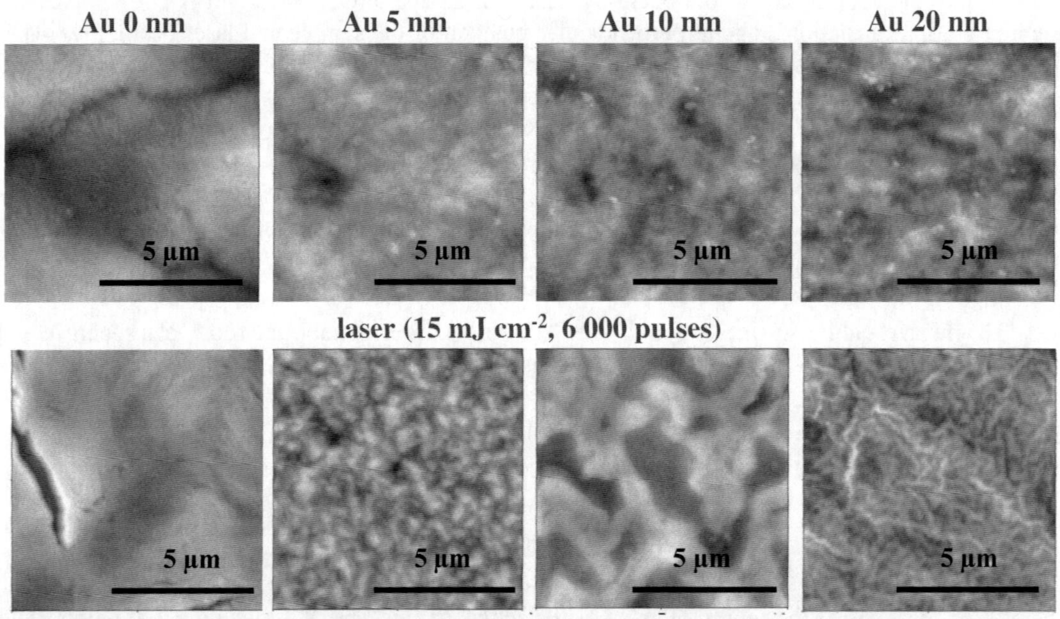

Figure 2. AFM scans of PHB covered with Au layer of thickness mentioned up in the figure. First line represents scans without the laser treatment and the second line of scans represents samples above modified by KrF laser (15 mJ cm^{-2} and 6 000 pulses).

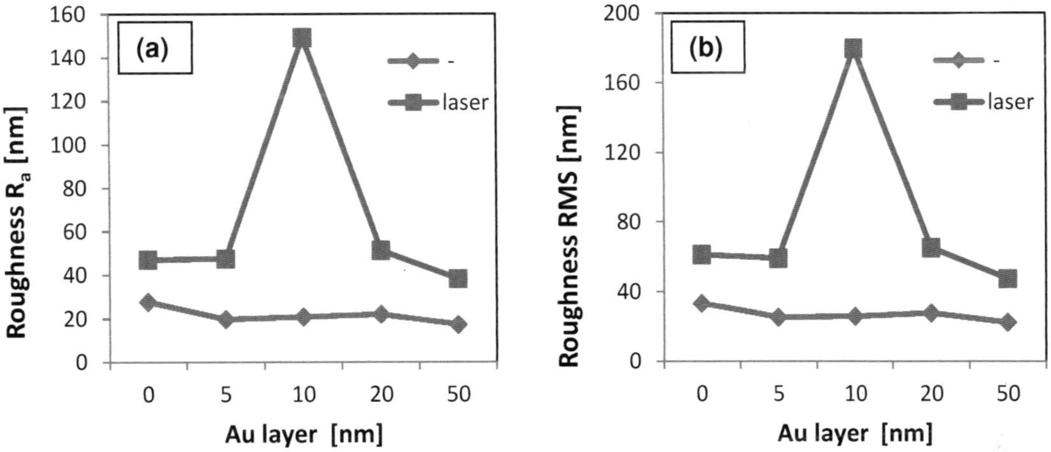

Figure 3 (a, b). The roughness **(a)** R_a and **(b)** *RMS* of PHB covered with different thickness of Au layer for non-treated samples and samples modified by excimer laser (15 mJ cm^{-2} and 6 000 pulses).

Conversely, the modification of bilayer substrate caused composite structures according to the thickness of the Au layer. The thinnest layer (5 nm) formed in combination with the polymer goblet structures on the surface. With increasing thickness of the Au layer (10 nm), the spongy or lamellar structure was created. Even higher thickening of Au layer (20 nm or more) lead to wrinkle-like structure.

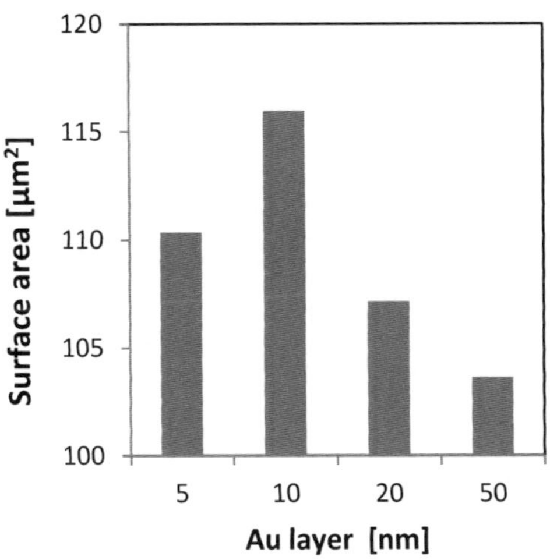

Figure 4. The surface area of modified PHB (laser treatment: 15 mJ cm^{-2} and 6 000 pulses) depending on the thickness of Au layer. Projected area of each sample was 100 μm^2.

At lower laser fluencies, no modification of the surface was observed and the morphology corresponded to the surface morphology before laser treatment. Conversely, at high laser fluencies, the gold layer was ablated without interconnection of materials.

From the **Figure 2.**, it is evident that the structures differs also in roughness, because of that the separate charts of R_a and *RMS* roughness are included in the **Figure 3**. As suspected, the trend *RMS* follows the trend of R_a. Without a doubt, the highest value of roughness (R_a =149 nm or *RMS* = 180 nm) belongs to lamellar structure created from Au layer of 10 nm thickness. Other laser modifications significantly differed in morphology, but the roughness remained similar for all other modifications. For laser non-modified samples was roughness quite identical and lower than in case of laser modified substrates. Only sample without gold layer was slightly rougher because sputtered Au caused slight surface leveling.

The surface area documented in **Figure 4**. naturally corresponded with the trend of roughness. Also in this case the biggest value of surface area was measured and calculated for the lamellar structure (116 μm^2). The surface area was calculated from the projected area 100 μm^2.

4. Conclusion

The laser treatment of polyhydroxybutyrate substrate covered with Au layer of different thickness was confirmed to be a successful tool for production variously structured composite material. The laser treatment of polymer foils covered by metal layer may have in addition to structuring effect also the ablation or reflection effect. The highlights in morphology was achieved by laser modification (by laser fluence 15 mJ cm^{-2} and 6 000 pulses) of PHB covered by 10 nm Au. The gained structure had laminar or spongy character with significant roughness. Also modification with other Au thickness were interesting and caused formation of goblet structure (for 5 nm Au) or wrinkle-like structure (20 nm or more).

Acknowledgments

This work was supported by the GACR under project 13-06609S.

References

[1] Lee S G, Lee D Y, Lim H S, Lee D H, Lee S and Cho K 2010 *Adv. Mater.* **22** 5013
[2] Kim D H, Ahn J H, Choi W M, Kim H S, Kim T H, Song J Z, Huang Y G Y, Liu Z J, Lu C and Rogers J A 2008 *Science*. **320** 053301
[3] Efimenko K, Rackaitis M, Manias E, Vaziri A, Mahadevan L and Genzer J 2005 *Nat. Mater.* **4** 293
[4] Chen C-M and Yang S 2012 *Polym. Int.* **61** 1041
[5] Yoo P J 2011 *Electron. Mater. Lett.* **7** 17
[6] Chung J Y, Nolte A J and Stafford C M 2009 *Adv. Mater.* **21** 1358
[7] Kim K J and Shahinpoor M 2003 *Smart Mater. Struct.* **12** 65
[8] Slepicka P, Jurik P, Kolska Z, Malinsky P, Mackova A, Michaljanicova I and Svorcik V 2012 *Nanoscale Res. Lett.* **7** 671
[9] Chen G Q and Wu Q 2005 *Biomaterials*. **26** 6565
[10] Pascu M, Vasile C and Gheorghiu M 2003 *Mater. Chem. Phys.* **80** 548
[11] Juřík P, Slepička P, Nagyová M and Švorčík V 2017 *Surf. Coat. Technol.* **311** 344
[12] Slepička P, Michaljaničová I, Rimpelová S and Švorčík V 2017 *Mater. Sci. Eng. C.* **76** 818
[13] Slepicka P, Michaljanicova I, Sajdl P, Fitl P and Svorcik V 2013 *Appl. Surf. Sci.* **283** 438
[14] Michaljanicova I, Slepicka P, Heitz J, Barb R A, Sajdl P and Svorcik V 2015 *Appl. Surf. Sci.* **339** 144

IOP Conf. Series: Journal of Physics: Conf. Series **917** (2017) 052010 doi:10.1088/1742-6596/917/5/052010

Numerical Modelling of a Bidirectional Long Ring Raman Fiber Laser Dynamics

S V Sukhanov[1], **L A Melnikov**[1], **Yu A Mazhirina**[1]
[1]Yuri Gagarin State Technical University of Saratov, Saratov, Russian Federation.

Abstract. The numerical model for the simulation of the dynamics of a bidirectional long ring Raman fiber laser is proposed. The model is based on the transport equations and Courant-Isaacson-Rees method. Different regimes of a bidirectional long ring Raman fiber laser and long time-domain realizations are investigated.

1. Introduction

Experimental realizations of ultralong laser systems [1-4] demonstrated that in these systems some optical effects appear in a special way. Cavity length is one of the most important laser parameters, which determines the properties of laser radiation. Particularly, the cavity length determines spectral interval between longitudinal modes, their interaction and output spectrum.

There are two important phenomena that put obstacles to increasing the cavity length – the existence of large amount of weakly interacting longitudinal modes, and Rayleigh backscattering in optical fiber. So numerical schemes for modeling of fiber lasers with long cavity must take into account these phenomena.

Reasonable representation of the first spatial derivative can be obtained by the Courant–Isaacson–Rees (CIR) numerical scheme [5]. It allows to develop a numerical model for calculating the dynamics of long ring fiber lasers on times corresponding to large amount of roundtrips and to study the space-temporal behavior of intensities of counterpropagating waves [6].

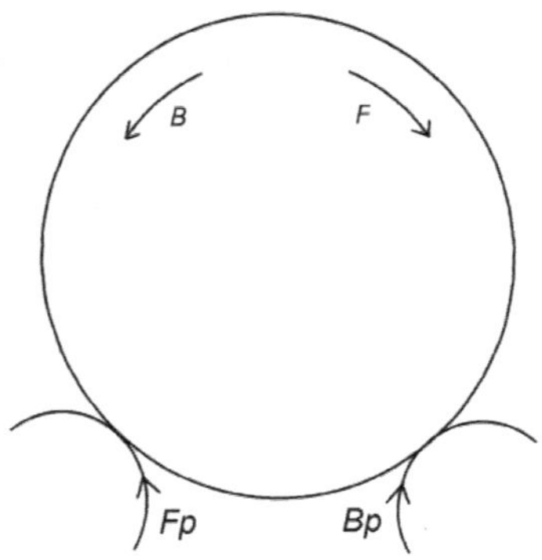

Figure 1. Scheme of a long ring Raman fiber laser.

On the Figure 1 the scheme of a long ring Raman fiber laser under consideration is represented. The length of the cavity is in the order of several tens of kilometers. The pump is induced from both directions, and two counterpropagating waves are generated in the cavity.

2. Numerical model

For the basis of our numerical model we use the following equations (1) and boundary conditions (2) [5]:

$$\begin{aligned}
&\frac{\partial F_p}{\partial t}+\frac{\partial F_p}{\partial z}+\bar{\gamma}_p F_p=-\bar{\alpha}F_p\left(|F|^2+|B|^2\right),\\
&\frac{\partial B_p}{\partial t}-\frac{\partial B_p}{\partial z}+\bar{\gamma}_p B_p=-\bar{\alpha}B_p\left(|F|^2+|B|^2\right),\\
&\frac{\partial F}{\partial t}+\frac{\partial F}{\partial z}+\bar{\gamma}F=\frac{\bar{\alpha}\omega}{\omega_p}F\left(|F_p|^2+|B_p|^2\right)+c(z)B+i\varphi F,\\
&\frac{\partial B}{\partial t}-\frac{\partial B}{\partial z}+\bar{\gamma}B=\frac{\bar{\alpha}\omega}{\omega_p}B\left(|F_p|^2+|B_p|^2\right)+c^*(z)F-i\varphi B,
\end{aligned} \tag{1}$$

$$\begin{aligned}
&F_p(0)=F_p(L)+f_p,\\
&B_p(L)=B_p(0)+b_p,\\
&F(0)=F(L),\\
&B(L)=B(0),
\end{aligned} \tag{2}$$

where F and B are the amplitudes of generated (Stokes) clockwise and counterclockwise waves respectively, index p means pump, γ_p, γ – losses for pump and generated waves respectively, α – unsaturated coefficient of Raman gain, L is the cavity length and $c(z)$ – the coupling coefficient of the oppositely running waves due to scattering, φ is a phase difference between counterpropagating waves. Laser configuration includes ring fiber loop and WDM directional coupler with 100% cross transmission at ω_p and 100% direct transmission at ω.

In numerical model the equations (1) were rewritten for the intensities, and relative phases of the scattered field were assumed as random number in the range $0\ldots2\pi$ with random amplitudes of coupling coefficient $0<|c(z)|<\kappa$, where κ – portion of scattered radiation, getting into the fiber mode relatively to the fiber losses.

3. Results

First, the dependences of Stokes wave intensities from input pump power were calculated: the example is shown on Figure 2. On the picture we can see the occurrence of instabilities at the pump power near 0.4 W.

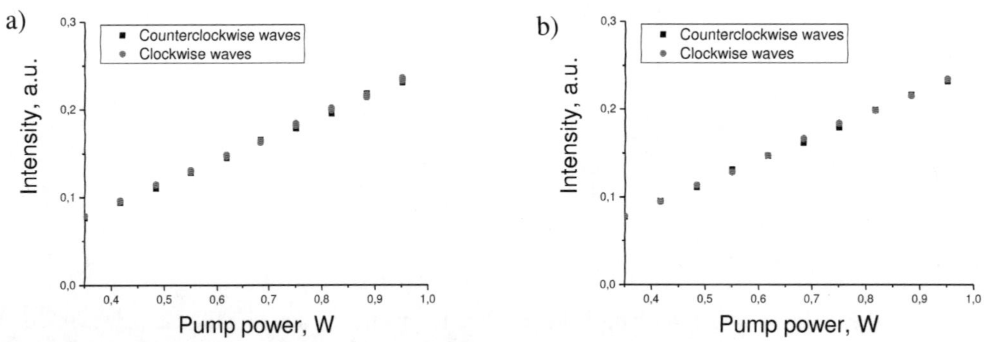

Figure 2. Dependence of the intensities of generated waves from input pump power. a) L=60 km, α=1.0 W^{-1}, dφ=0.1; b) L=80 km, α=1.0 W^{-1}, dφ=0.1.

Temporal dependence of Stokes wave intensities are shown in Figure 3, demonstrating anti-phase oscillations with period equal to roundtrip time (each 40 dots correspond to 1 roundtrip).

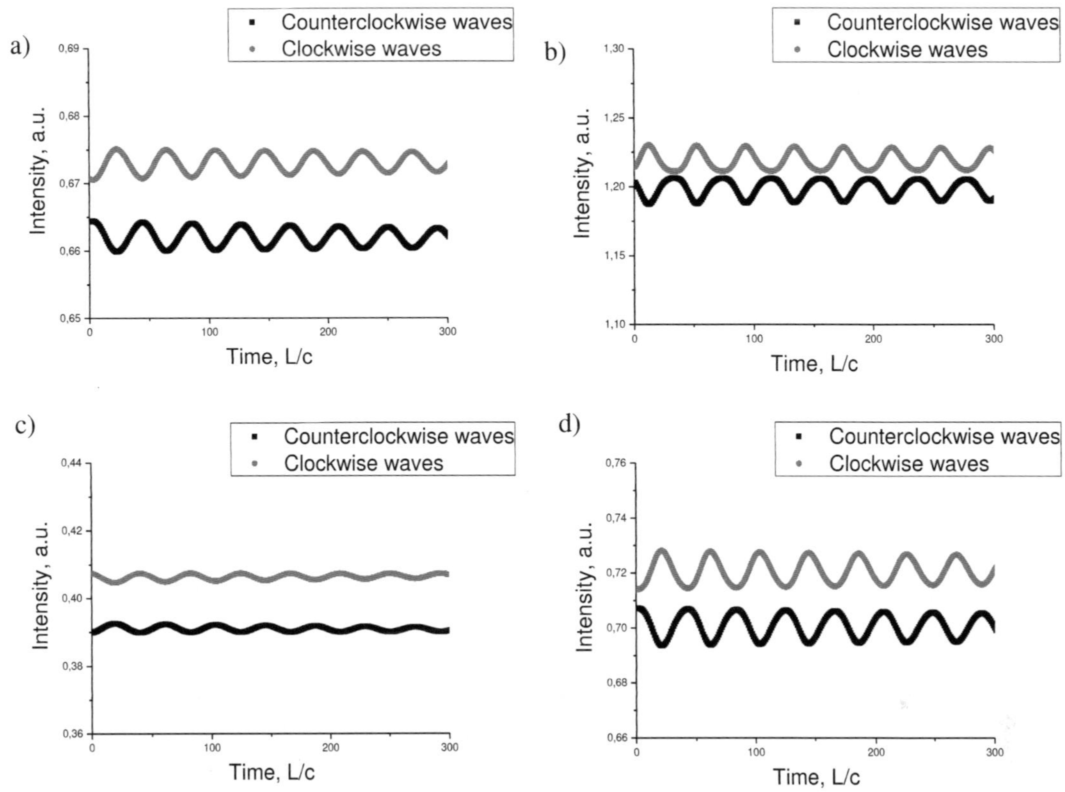

Figure 3. Dependence of the intensities of generated waves from time at particular value of pump power: a) L=80 km, α=1.0 W^{-1}, dφ=0.1, pump=2,55. b) L=80 km, α=1.0 W^{-1}, dφ=0.1, pump=4,35. c) L=60 km, α=1.0 W^{-1}, dφ=0.1, pump=2,55. d) L=60 km, α=1.0 W^{-1}, dφ=0.1, pump=4,35.

4. Conclusion

The presented numerical model, based on upwind (CIR) method, allows investigation in details the dynamics of long ring Raman fiber laser dynamics, to study different regimes and to find optimal parameters for long fiber laser cavity. This model also allows to analyze gyroscopic properties of such laser.

References

[1] Churkin, D., El-Taher, A., Vatnik, I., Ania-Castacyn, J., Harper, P., Podivilov, E., Babin, S., and Turitsyn, S., "Experimental and theoretical study of longitudinal power distribution in a random DFB fiber laser," Opt. Express, Vol. 20, 11178 (2012).

[2] Turitsyn, S.K., Ania-Castanon, J.D., Babin, S.A., Karalekas, V., Harper, P., Churkin, D., Kablukov, S.I., El-Taher, A.E., Podivilov, E.V., and Mezentsev, V.K., "270-km ultralong Raman fiber laser," Phys. Rev. Lett.. Vol. 103, 133901 (2009).

[3] Turitsyn, S.K., Babin, S.A., El-Taher, A.E., Harper, P., Churkin, D.V., Kablukov, S.I., Ania-Castanon, J.D., Karalekas, V., and Podivilov, E.V., "Random distributed feedback fibre laser," Nature Photonics, Vol. 4, 231 (2010).

[4] Turitsyn, S.K., Babin, S.A., Churkin, D.V., Vatnik, I.D., Nikulin, M., Podivilov, E.V., "Random distributed feedback fibre lasers," Physics Reports, Vol. 542, 133 (2014).
[5] R. Courant, E. Isaacson, and M. Rees, "On the solution of nonlinear hyperbolic differential equations by finite differences," Communications on Pure and Applied Mathematics, Vol. 5(3), p. 243, 1952.a S, Iwase N, Yamada T, Matsushita T, Kiyoku H and Sugimoto Y 1996 *Japan. J. Appl. Phys.* 35 L74
[6] Yu.A. Mazhirina, L.A. Melnikov, S.K. Turitsyn, D.V. Churkin, N.S. Tarasov, "Nonlinear dynamics of long mirrorless fiber Raman laser," «Izvestia VUZ. Applied Nonlinear Dynamics», Vol. 22, № 5, p. 73, 2014 (in Russ).

Features of the induced photopleochroism oscillations in a photosensitive structures based on $CuInSe_2$

I A Zharikov[1], R V Davydov[1], V A Lyapishev[1], V Yu Rud'[2,3], Yu V Rud'[3], A P Glynushkin[2]
[1]Peter the Great Saint Petersburg Polytechnic University, Saint Petersburg 195251, Russia
[2]Department of Ecology, All-Russian Research Institute of Phytopathology, 143050, Moscow Region, Odintsovo district, B.Vyazyomy, Russia
[3]Department of Solid State Physics, Ioffe Physico-Technical Institute, Russian Academy of Sciences, St Petersburg 194021, Russia

Abstract. The work is devoted to study of the phenomenon of oscillations induced photopleochroism depending on the different technologies for the creation of photosensitive structures CuInSe2. Induced photopleochroism was observed, and peculiarities of this phenomenon are discussed. A conclusion is made such structures beneficial to apply as sensors of polarization light for fiber optic.

1. Introduction

Improvement of technologies for producing CuInSe2 has allowed to increase the energy conversion efficiency of up to 18-20 %. The heat treatments have been found to be one important process step for improving the efficiency of solar cells on CIS. This contribution presents the first research of polarization photosensitivity of two different types structures which were obtained by vacuum and air het treatment of the CIS as grown bulk crystals.

However, these materials with sufficiently advanced technology can find its application as sensors linearly-polarized radiation in optical communication systems. The successful implementation of this direction depends on the polarization study of photosensitivity in these materials. It requires thorough research of photopleochroism induced in the different structures based on these materials.

2. The method

$CuInSe_2$ belongs to the group of ternary chalcopyrite compounds which derives from the group IV class of tetrahedral bonded semiconductors according to the Grimm-Sommerfeld rule, i.e. there must be an average of 4 valence atoms per atomic site. In these structures each atom has four neighbors arranged at the corners of a regular tetrahedron bonded by sp^3 bonds. The tetrahedral structure of the chalcopyrites can be considered as a superlattice structure of the sphalerite or zincblende structure which has a diamond like structure (such as Si) consisting of two inter-penetrating face centered cubic lattices, separated by a translation vector of (1/4 , 1/4 , 1/4). One sublattice is occupied by cations and the other by anions (II-VI and III-V compounds). In a ternary chalcopyrite compounds the cations are furthermore replaced by a one cation of higher valency and one cation of lower valency which occupy the cation sublattice in an ordered manner as shown in Figure 1. In that sense $CuInSe_2$ can be envisioned as the ternary analogue of the binary ZnS. The reduced symmetry, due to the two kinds of cations, leads to a primitive cell of eight atoms in the chalcopyrite structure compared to a primitive cell of two atoms in the zincblende structure. The Bravais lattice of the chalcopyrite is body centered tetragonal. Compared to the face centered cubic Bravais-cell of the zincblende the unit cell is doubled along the crystal c axis. If the different cations are distributed at random, the ternary compound has a sphalerite structure.

The electronic band structure of a chalcopyrite compound can in principle be derived from its II-VI analogue since in both structures (zincblende and chalcopyrite) the chemical binding is realized via hybrid sp3-orbitals. The resulting band structure at the Γ-point is determined by a s-type conduction band and a p-type valence band [1].

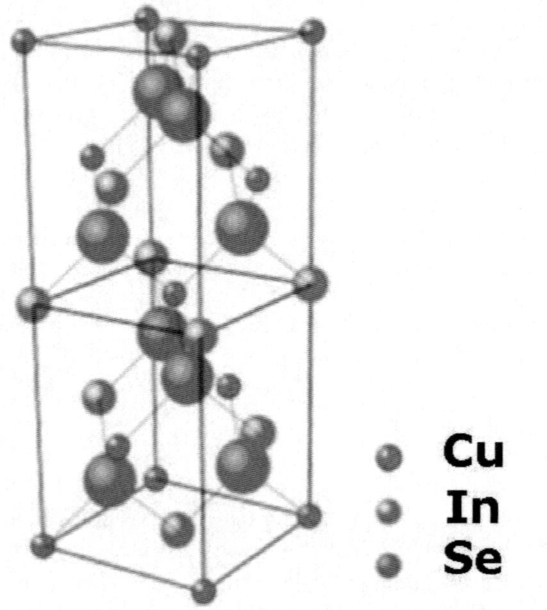

Figure 1. Crystal structure of chalcopyrite lattice .

However, it was shown that the uppermost valence bands of a I-III-VI2 compound are largely influenced by the d-levels of the group I metal in the valence band [1]. As a consequence the valence band is split into three single bands due to spin-orbit interaction and due to the crystalld in the tetragonal structure (Figure 2). The interaction between the Γ15 levels in figure 2 will raise the uppermost Γ15 to higher energies, so that the direct band gap at the Γ-point of a chalcopyrite is in general 1 eV lower than the value of the respective II-VI compound. Theoretical calculations by Jaffe and Zunger revealed that the reduced band gap in chalcopyrites can be explained by two effects – a chemical effect $\Delta E^{chem}{}_{g}$ which is due to the p-d hybrids and a structural $\Delta E^{struc}{}_{g}$ which is related to the anion displacement in the chalcopyrite lattice.

The CIS polycrystalline ingots were prepared by directional crystallization. As grown n- and p-type conductivity crystals were electrically homogeneous and had the free carrier concentrations about 10^{18} cm^{-3} at room temperatures. The CIS wafers were exposed to heat treatment at temperature 500 C in vacuum quartz ampoules (HV) at the constant volume and directly in the dry air (HA).

The HV is usually accompanied by conversion of conductivity of a p-type crystals. At heat treatment times about 1 min we prepare n-p-homotransitions (HT). With increase of time up to 10 min the conversion becomes bulky and from these wafers it is possible to make only the isotopic n'-n –HT.

It is important to emphasize, that the surface of plates after such a treatment does not visually differ from an initial state, and the conversion of the conductivity type is caused by formation of the V_{se} donors.

3. Results and discussion

At the illumination of such homotransitions the created films always charged negatively, that corresponds to a pass direction of these HTs. The photosensitivity of such HTs dominates at illumination on the formed films side and looks like narrow band spectrum with full width at half-height $\delta_{1/2}$ about 40 meV. The energy maximum hw^m in the photoconversion efficiency spectrum is always located slightly below the CIS energy gap E_G = 1.02 eV. Typical sharp short wavelength fall down of the photosensitivity for such HTs in conditions of illumination from the side of obtained HW films is caused by the increase of influence of the optical absorption as approaching EG.

The films where formed on the surface of wafers by the air treatment of CIS. The coloring of these films changed with the time of treatment and the conductivity type of initial wafers remained the same. The HA methods allow to prepare rectifying HT, the forward direction of which always corresponds to positive polarity of the external voltage displacement on a film. Under the illumination of the isotypic p'-p and n'-n HTs obtained by HA of the CIS crystals, the films always charged positively and corresponded to forward direction of such HTs. The photosensitivity maximum was achieved in n'-n-HTs, prepared by HA method at t = 5 min. The sign of photovoltaic effect was independent of location light probe and of photon energy and, hence, the CIS air treatment could be connected to active area formation in these structures.

Both types of obtained homojunctions for linearly-polarized light irradiation along a normal to the illumination plane (angle of incidence $\Theta=0^\circ$) did not show and dependence on spatial orientation of the electrical field vector of light wave [1]. The main laws for the polarization photosensitivity investigations of different two types of $CuInSe_2$ homotransitions consists in following. All this different types of vacuum treatments of HTs demonstrates the angular dependencies of short circuit

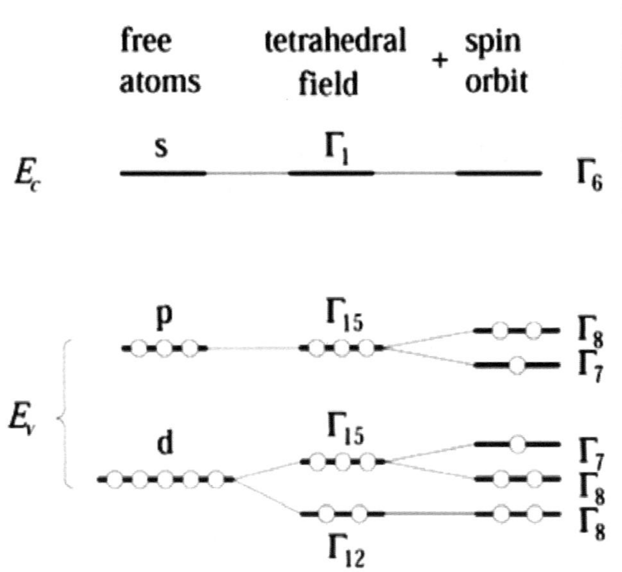

Figure 2. Splitting of energy bands for CIS.

photocurrent i^p (where the vector of electrical field parallel to the irradiation plane of incidence) and i^s (when the vector of electrical field perpendicular to the irradiation plane of incidence) are consists with the analysis based on Freshnel relations for passage process of light wave to the air/$CuInSe_2$.

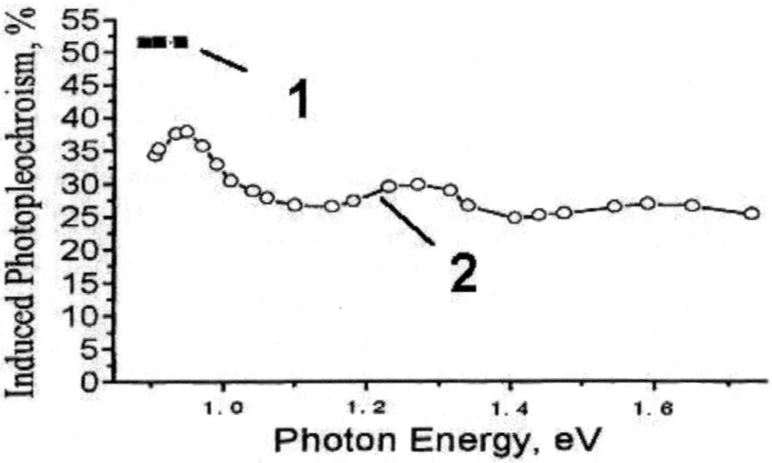

Figure. 3. Induced photopleochroism spectra for $CuInSe_2$ for both type treatment (1- in vacuum; 2- on air).

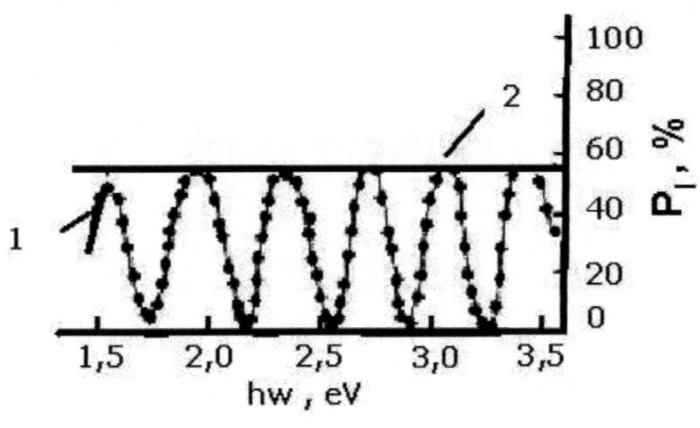

Figure. 4. Induced photopleochroism spectra for ZnO:Al/p-GaAs:Mn structures at Θ =85° (1- experiment; 2- calculation without interferention).

In the case of heterojunctions obtained at the air, photocurrent is, as well as for i^p, at $\Theta>0^\circ$ in the certain region of incidence angle shows an increase. This behavior may be connected with the decrease reflection losses also in the case of a s-wave [2]. This peculiarity can be explained [3] by interference of linearly polarized light in the entrance film of such structures. Which leads to oscillations in the spectral dependence of such homojunctions (Fig. 3).

The same dependences was obtained for example for ZnO:Al/p-GaAs:Mn structures(Fig. 4). The oscillatory character of the induced photopleochroism, which attributed to the interference of incident linearly polarized light [4] in the entrance window of such photosensitive structures, is clearly seen. Turning to the spectral dependences of the quantum efficiency of p- and s- wave phototransducers

IOP Conf. Series: Journal of Physics: Conf. Series **917** (2017) 052011 doi:10.1088/1742-6596/917/5/052011

it is evident that the oscillations of the induced photocurrents i^p and is oscillate in antiphase [5]. For example a minimum of i^s corresponds to a maximum of i^p.

4. Conclusions

Presented here results show that structures based on CIS and thin films of different natures cold find an application as narrow band photovoltaic devices for unpolarized and linearly polarized light. The main result of this research can be considered concluded that such structures advantageously be used as selective sensors of polarized light for optical fiber.

References

[1] Shay J L, Tell B, Kasper H M, and Schiavone L M. 1972 *Physics Review* B 3 5003.

[2] Bairamov B Ch, Rud' V Yu, Rud' Yu V 1998 MRS Bull 23 91.

[3] Shpunt V Ch, Rud' Yu V, Tsendin K D 1993 *Journal Technical Physics* **19** 13 41

[4] Kesamanly F P, Rud' V Yu, Rud' Yu V 1999 *Semiconductors* **5** 483

[5] Rud`V Yu, Rud` Yu V, Shpunt V Ch, Terukov E I 2016 *Optical Memory & Neural Networks (Information Optics)* **25** 1 40

IOP Conf. Series: Journal of Physics: Conf. Series **917** (2017) 052012 doi:10.1088/1742-6596/917/5/052012

MBE growth, structural and optical properties of multilayer heterostructures for quantum-cascade lasers

R R Reznik[1, 2, 4, 8], **N V Kryzhanovskaya**[1], **F I Zubov**[1], **A E Zhukov**[1, 3], **R A Khabibullin**[5], **S V Morozov**[6, 7] **and G E Cirlin**[1, 2, 4, 8]

[1]St Petersburg Academic University, Khlopina 8/3, 194021, St-Petersburg, Russia
[2]Institute for Analytical Instrumentation RAS, Rizhsky 26, 190103, St-Petersburg, Russia
[3]Saint-Petersburg Science Center, Russian Academy of Sciences, Universitetskaya nab. 5, 199034, St. Petersburg, Russia
[4]ITMO University, Kronverkskiy pr. 49, 197101, St. Petersburg, Russia
[5]Institute of Ultrahigh Frequency Semiconductor Electronics Russian Academy of Sciences, Nagornyi pr. 7, str.5, 117105, Moscow, Russia
[6]Institute for Physics of Microstructures of the Russian Academy of Sciences, Academicheskaya 7, 603087, Afonino, Russia
[7]Lobachevsky State University of Nizhni Novgorod, Pr. Gagarina 23/2, 603950, Nižnij Novgorod, Russia
[8]Peter the Great St. Petersburg Polytechnic University, St. Petersburg 195251, Russia

E-mail: moment92@mail.ru

Abstract. The results obtained in MBE growth and study of the structural and optical properties of GaAs/AlGaAs heterostructures with 228 quantum cascades are presented.

1. Introduction

Sources of the terahertz (THz) frequency range are in demand for fundamental science, commercial interest and special purposes, including spectroscopy, determination of trace amounts of various substances, development of systems forming images of objects concealed from ordinary optical systems, wide-bandwidth communication systems, etc. One of the main problems is the lack of compact coherent sources of radiation of sufficiently high-power for this spectral range. Among various techniques for THz generation [1-3], THz quantum cascade lasers (THz QCL) are compact, coherent, continuous wave (cw) solid-state source with electrical pumping considered to be the most promising THz source. The first QCLs for the THz range were fabricated in the early 2000s [4, 5]. To date THz QCLs of this kind have been developed, which operate up to 200°K [6] and have an emission power of several tens of mW at liquid-nitrogen temperature. Also, room temperature laser sources of terahertz radiation have been developed [7] based on the intraband amplification of the differential frequency of two QCLs of the mid-infrared (IR) spectral range [8].

In this study, we have been investigated the possibility of MBE synthesis of multiple-period multilayer heterostructures for use in the fabrication of QCLs for the frequency range around 3 THz.

We have perform the MBE growth of two multilayer GaAs/AlGaAs heterostructures with different GaAs active layer thicknesses, examine their structural and optical properties by the X-ray diffraction method and photoluminescence (PL) spectroscopy.

2. Experiments.

The epitaxial structures QCL-1 and QCL-2 were synthesized by molecular-beam epitaxy using a Riber 21 MBE machine. Growth was performed on semi-insulating GaAs (100) substrates under arsenic-stabilized conditions. Particular attention was given to accurately setting the growth rates and to maintaining their stability during whole the growth run. The growth rates were carefully calibrated on a separate sample directly prior to growth of the laser structures. The GaAs and AlAs growth rates were set at 0.85 and 0.15 monolayers per second (ML/s) respectively for QCL-1 and 0.7 and 0.105 ML/s for QCL-2. Special high-speed shutters were used to obtain high quality imterfaces. In our case, the actuation time of the shutters of the aluminium and gallium sources did not exceed 0.15 s. A 200-nm-thick $Al_{0.8}Ga_{0.2}As$ stop-layer was deposited onto a GaAs buffer layer. The active region contained 228 periods. Each cascade contains a GaAs/AlGaAs double quantum well (QW), between whose levels a laser transition occurs, and a wider QW serving as the injector/extractor of electrons. The active region was bound from above and below by GaAs:Si contact layers (5×10^{18} cm^{-3}) with thicknesses of 75 and 50 nm, respectively. The middle part of the injector/extractor layers also had n-type doping to a concentration of $\sim 5 \times 10^{16}$ cm^{-3}. When heterostructures intended for the generation of light in the THz range are synthesized by the MBE method with total thicknesses of ~10 μm, the GaAs growth rate may, nevertheless, decrease due to exhaustion of the gallium source in the course of prolonged deposition. Figure 1 shows the schematic diagram of the grown structure. The same circumstance may lead to irreproducibility of the layer thicknesses from one structure to another. In order to elucidate the influence exerted by accurate maintenance of the gallium-arsenide deposition rate, we synthesized and studied two structures, QCL-1 and QCL-2, designed in such a way that the thicknesses of all GaAs layers in QCL-1 are 15% smaller than those of the corresponding layers in QCL-2.

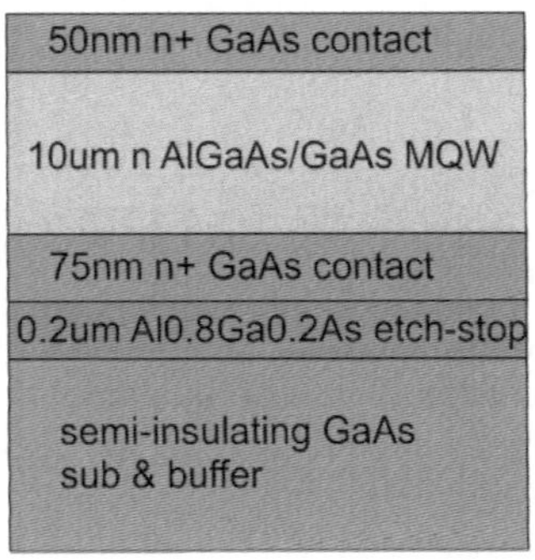

Figure 1. The schematic diagram of the QCL-1.

3. Results and Discussion.

The structural properties of the epitaxial samples were examined by the high-resolution X-ray diffraction (HR-XRD) method with a D8 DISCOVER Bruker AXS diffractometer (radiation wavelength λ =0.15406 nm) with a primary beam half-width of <12 arcsec in the Ω–2θ scanning mode. Figure 2 shows the rocking curve around the symmetric GaAs (004) reflection, measured for the QCL-1 structure. The full width at half-maximum (FWHM) of the satellite peaks due to the periodic repetition of QCL cascades is 15–19 arcsec. It is noteworthy that the full width of the superstructure peaks in the model spectrum, found with allowance for bending of the structure under elastic stresses, is 22.4 arcsec. This means that both the possible effect of inaccurately maintaining the cascade thickness within the whole structure and the roughness of the heterointerfaces can be neglected, which confirms that the technological parameters for the case of synthesis of the active region are chosen correctly. The calculated rocking curve for the model structure providing the best agreement with the experimental data is also shown in Figure. 1. The thickness of the cascade, determined by simulation of the X-ray rocking curves, was 38.3 and 43.6 nm for, respectively, the QCL-1 and QCL-2 structures, instead of the expected values of 38.95 and 43.91 nm. Thus, the discrepancy between the expected and experimentally measured values of the period was 1.7 and 0.7% for the QCL-1 and QCL-2 structures, respectively.

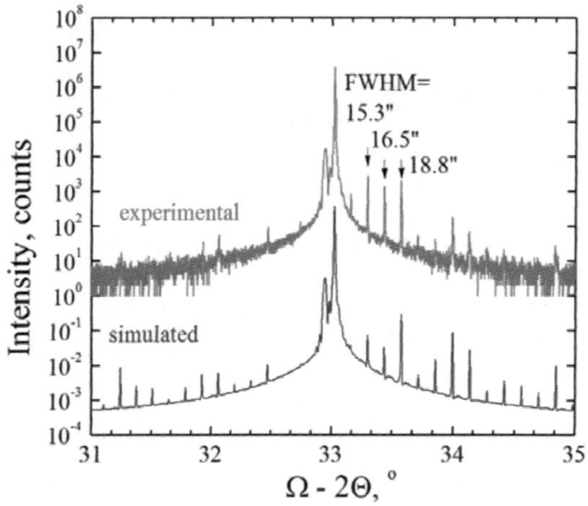

Figure 2. X-ray rocking curve of the QCL-1 structure near the GaAs (004) reflection and the simulated curve.

In near- and mid-IR lasers operating at interband optical transitions, the lasing wavelength is found to be close to the PL wavelength of the active region. The use of rapid luminescence diagnostics facilities that can be applied to epitaxial structures before performing processes associated with device fabrication strongly simplifies the development of a technology for the fabrication of lasers of this kind, with the design of the laser structure modified when necessary. In the case of THz QCLs, however, there are no direct methods of this kind for rapidly determining the expected lasing wavelengths. In this context, it seems to be effective to compare the results obtained in measuring the energies of electron–hole optical transitions and correlate these energies with calculation results. In the case of their good agreement, it can be assumed that the energies of the intraband transitions between electron levels of the cascades will also coincide with those predicted by simulation results.

The optical properties were studied by PL spectroscopy. Measurements were made in the temperature range T = 77–300°K. Optical pumping was provided by a YAG:Nd laser operating at the second harmonic in the continuous-wave mode (wavelength λ =527 nm), with the pump power density varied from 2.5 to 15 W/cm^2. The PL signal was detected with an FHR 1000 monochromator and a single-channel cooled Si photodiode.

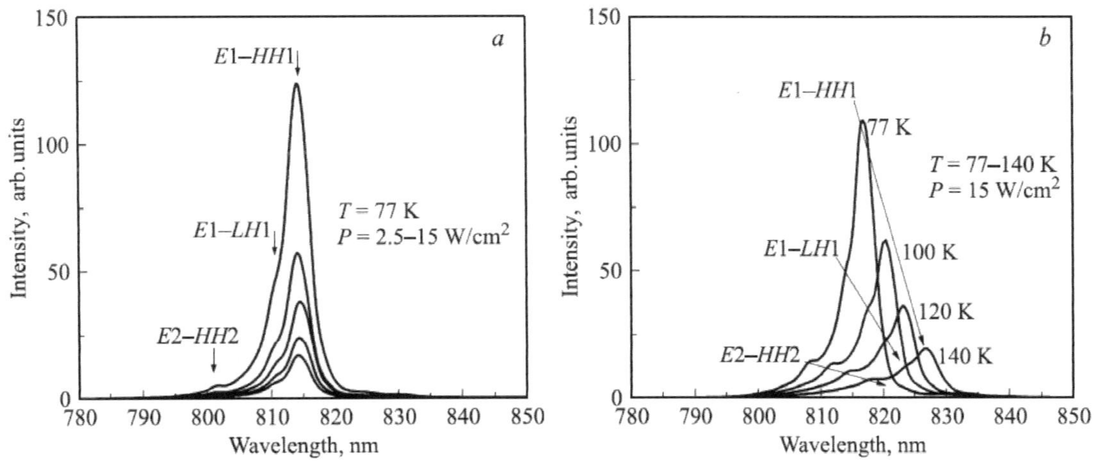

Figure 3. (a) Band diagram of a biased QCL-2 structure and (b) relationship between the oscillator strength and transition frequency for QCL-1 and QCL-2.

Figure 3a shows the PL spectra of the QCL-1 structure, measured at 77 K and various excitation densities in the range P = 2.5–15 W/cm^2, and Figure 3b, spectra of the QCL-2 structure, measured at a fixed power density (15 W/cm^2) and various temperatures in the range T = 77–140°K. Three emission peaks can be distinguished in the spectra of both structures, identified as being associated with transitions between the first electron level and the first level of heavy holes (E1–HH1), the first electron level and the first level of light holes (E1–LH1), and the second electron level and the second level of heavy holes (E2–HH2). This interpretation of the peaks agrees with the fact that the relative intensity of shorter wavelength peaks increases with increasing excitation level, and when the temperature changes, the spectral position of the peaks moves in accordance with the temperature behavior of the band gap of GaAs. It is noteworthy that highintensity luminescence at the wavelength of the ground optical transition E1–HH1 was observed in both structures up to room temperature.

Thus, the method of molecular-beam epitaxy was used to synthesize multiperiod (228 cascades, ~10 μm) GaAs/AlGaAs epitaxial structures intended for use in the fabrication of quantum-cascade lasers for the terahertz range. The experimental value of the cascade thickness corresponds to the expected value with an accuracy of better than 2%. The angular width of the superstructure peaks in the X-ray rocking curve does not exceed 20 arcsec. A study of the photoluminescence spectra demonstrated good agreement (within 1 meV) between the interband optical-transition energies and the calculated values. A 15% change in the thickness of the GaAs layers in two different structures leads to expected shift of the positions of the optical transitions. This means that the positions of energy levels in complex structures of this kind can be estimated by various express optical diagnostic procedures operating in the near-IR range.

The technology for the fabrication of waveguides of the metal–semiconductor–metal type is being developed in order to fabricate laser diodes from the structures has been investigated this report. Finally, the laser-related I-V curves were recorded at 10°K.

Acknowledgments
The authors gratefully acknowledge financial support received from the Ministry of Education and Science of the Russian Federation under grant 14.587.21.0040 (project ID RFMEFI58717X0040). The study was supported by the Program of Basic Research of the Presidium of the Russian Academy of Sciences "Nanostructures: physics, chemistry, biology, and fundamentals of technology." The samples were grown under support of the Ministry of education and science of Russian Federation (state task, project No. 16.9791.2017/8.9).

References

[1] Galiev R R, Yachmenev A E, Bugaev A S, Galiev G B, Fedorov Yu V, Klimov E A, Khabibullin R A, Ponomarev D S, Maltsev P P 2016 *Bul. Rus. Acad. Sci:. Phys.* **80** 476–478
[2] Lavrukhin D V, Yachmenev A E, Galiev R R, Bugaev A S, Fedorov Yu V, Khabibullin R A, Ponomarev D S, Maltsev P P 2015 *Int. J. High Speed Electron. Syst.* **24** 1520001
[3] Lavrukhin D V, Yachmenev A E, Galiev R R, Khabibullin R A, Ponomarev D S, Maltsev P P 2014 *Semiconductors* **48** 69–72
[4] Koehler R, Tredicucci A, Beltram F, Beere H E, Linfield E H, Davies A G, Ritchie D A, Iotti R C, Rossi F 2002 *Nature* **417** 156
[5] Rochat M, Ajili L, Willenberg H, Faist J, Beere H E, Davies A G, Linfield E, Ritchie D A 2002 *Appl. Phys.Lett.* **81** 1381
[6] Fathololoumi S, DupontS, Chan C W I, Wasilewski Z R, Laframboise S R, Ban D, Matyas A, Jirauschek C, Hu Q, Liu H C 2012 *Opt. Express* **20** 3866
[7] Lu Q Y, Bandyopadhyay N, Slivken S, Bai Y, Razeghi M 2014 *Appl. Phys. Lett.* **104** 221105
[8] Belkin M A, Wang Q J, Pflugl C, Belyanin A, Khanna S P, Davies A G, Linfield E H, Capasso F 2009 *IEEE J. Sel. Top. Quantum Electron.* **15** 952

IOP Conf. Series: Journal of Physics: Conf. Series **917** (2017) 052013 doi:10.1088/1742-6596/917/5/052013

Waveguide and Γ-factor optimization for low-divergence ridge lasers

I K Boikov[1], A V Savelyev[2] and A E Zhukov[2]

[1] Department of Applied Mathematics and Physics, St. Petersburg Academic University of the Russian Academy of Sciences, St. Petersburg 194021, Russia

[2] Nanophotonics Lab, St. Petersburg Academic University of the Russian Academy of Sciences, St. Petersburg 194021, Russia

E-mail: ik.boikov@gmail.com

Abstract. The work is aimed for analytical approach to maximize ridge laser modal gain at a given beam divergence in a vertical direction. When the gain/loss profile of planar waveguide is known the equation for optimum near-field profile was obtained. Surprisingly, it is appeared to be Schrödinger type equation where normalized gain/loss profile plays role of potential energy. The developed approach was applied to a few examples widely met in practice.

1. Introduction

Semiconductor ridge emitting lasers have a broad range of applications from medical applications to material processing, therefore, lasers parameters optimization is a matter of particular interest. One of goals is maximizing Γ-factor, which means increased field profile confinement. But it's shown that it results in rising far-field beam divergence, so a trade-off takes place. In this abstract we look for analytical approach to this problem.

2. Analytical approach

Let us consider planar waveguide with TE-polarized wave propagating along x-direction and let $F(z)$ be its near-field profile of y-component of electric field, where z-axis points growth direction in planar waveguide. As was shown in [1], measure of beam divergence can be introduced as squared average angle Θ:

$$\Theta^2 = -k_0^2 \int\limits_{-\infty}^{\infty} F(z)\frac{d^2F}{dz^2}dz \text{ , while } \int\limits_{-\infty}^{\infty} |F(z)|^2 dz = 1 \text{ ,} \tag{1}$$

where k_0 is the light wavenumber in vacuum for the lasing wavelength $\lambda_0 = 2\pi/k_0$. Let $g(z)$ be gain profile. Modal gain G can be found via:

$$G = \int\limits_{-\infty}^{\infty} |F(z)|^2 g(z) dz \text{ .} \tag{2}$$

In order to find best possible modal gain keeping the value of Θ fixed we use variational approach with the Lagrange multipliers, which is described in [2]. To do so, we introduce functional $S[F(z)]$ as well as Lagrange multipliers ρ and λ:

$$S[F(z)] = \int_{-\infty}^{\infty} |F(z)|^2 g(z)dz + \rho\left(\Theta^2 k_0^2 + \int_{-\infty}^{\infty} F(z)\frac{d^2F}{dz^2}(z)dz\right) + \lambda\left(\int_{-\infty}^{\infty} |F(z)|^2 dz - 1\right) . \quad (3)$$

Hence, we convert initial task of maximum value of G to the problem of finding extremum of $S[F(z)]$. The conventional variational procedure lead us to the following equation on $F(z)$:

$$-\frac{d^2F}{dz^2} - \frac{1}{\rho}F(z)g(z) = \frac{\lambda}{\rho}F(z) . \quad (4)$$

One can clearly see here, that this is Shr'odinger eigenvalue problem where $-\rho^{-1}g(z)$ plays role of potential energy. Hence we can use well developed methods to solve this problem for different $g(z)$ and to interpret physical meaning of its solution.

We can introduce $\alpha = \rho^{-1}$ as well as $\xi = \lambda\rho^{-1}$ as only their quotient plays role. From this point onwards α is considered a given constant.

$$\frac{d^2F}{dz^2} + F(z)\left(\alpha g(z) + \xi\right) = 0 \quad (5)$$

This is a Schrödinger's equation for particle in potential $-\alpha g(z)$ with energy ξ. The variable ξ can be found during solution of this eigenproblem. Hence, only α could be varied during optimization procedure. From point of view of Shrödinger equation varying α would lead to changing of depth of "potential well" that corresponds to the gain region. Higher values of the α corresponds to more tight localization of the field profile $F(z)$ around gain region, higher Γ-factor and higher beam divergence. Still the solution at any fixed α is optimal in the sense that Γ could not be better for a given $g(z)$ profile and beam divergence. While $g(z)$ is arbitrary, we cannot proceed any further analytically and bring here few examples.

3. Examples

3.1. Example 1: single gain layer

In that case gain be introduced as $g(z) = A\delta(z)$. Solution of (5) is well known to be

$$F(z) \sim \exp\left(-\sqrt{\alpha A}|z|\right) , \quad (6)$$

where α could be any positive value. Properties of that solution have been studied in details in [1]. We note here that actual value of A is of no importance, since α is arbitrary and should be "scanned" in order to find $F(z)$ with required divergence and, therefore, we ommit it in the following.

3.2. Example 2: two gain layers

Two gain layers in an active media can represent a couple of quantum wells as an example. With the gain profile written as

$$g(z) = \delta(z-a) + \delta(z+a) , \quad (7)$$

the solution of (5) is following:

$$F(z) = \begin{cases} N\exp\left(\sqrt{\xi}(z+a)\right), & z < -a \\ N\,\mathrm{sech}\left(\sqrt{\xi}a\right)\cosh\left(\sqrt{\xi}z\right), & z \in [-a, a], \\ N\exp\left(-\sqrt{\xi}(z-a)\right), & z > a \end{cases} \quad (8)$$

where N is normalization constant and ξ can be found as the solution of following transcendent equation:

$$\tanh\left(a\sqrt{\xi}\right) = 2\alpha\xi^{-1/2} - 1 \tag{9}$$

The profiles of $F(z)$ are shown in Figure 1 for different values of a. At the Figure 2 best achievable values of Θ are shown as a dependence of a modal gain (g_0 is unity for the choice $A = 1$, see above).

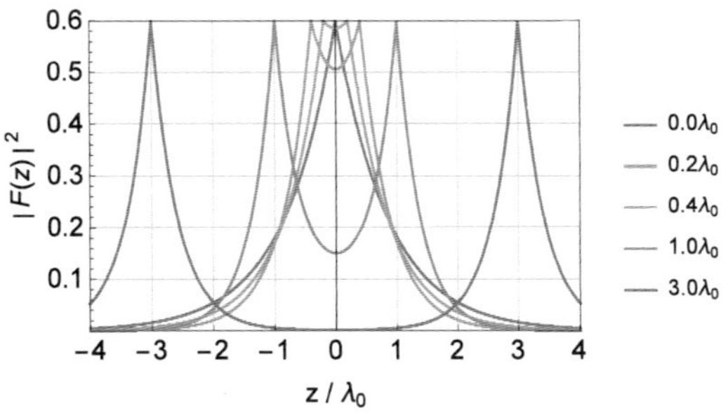

Figure 1: Optimal near-field intensity profiles for different values of a that are shown at the right side of graph.

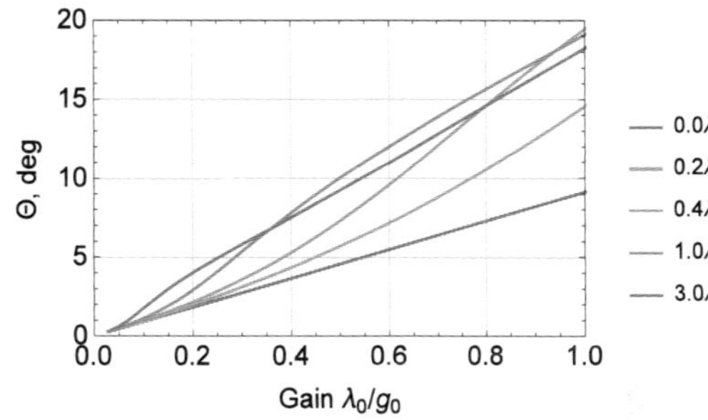

Figure 2: $\Theta(G)$ dependency on the gain for various a.

As we can see from the Figure 2, at low gain and $a < \lambda_0$ small increase in gain results in the proportional increase in divergence for all relevant a due to field distribution being "blob" of width much more than λ_0 so field will not respond to small a change as much as at high gain. Also it is worth noting that best divergence angle is reached at $a = 0$, thus gain profile confinement is preferred.

3.3. Example 3: symmetric slab

There are cases when gain of single layer is not sufficient for effective laser operation, namely, in lasers with quantum dot active media. This problem is typically solved by stacking of gain layers. As an example, in InGaAs/GaAs quantum dot lasers up to 10 layers may be stacked. The total width of the gain area in that case is up to few hundreds nanometres and is compared with the wavelength in the material. In that case gain profile could well approximated as a "gain slab" of thickness $2a$ (see Figure 3):

$$g(z) = g_0\theta(a+z)\theta(a-z)\ , \tag{10}$$

where $\theta(z)$ is Heaviside step function.
The optimum field profile is a symmetric one:

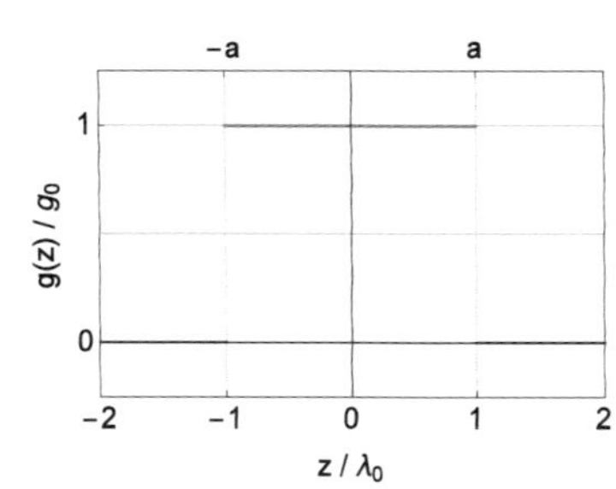

Figure 3: Gain profile with $a = \lambda_0$.

IOP Conf. Series: Journal of Physics: Conf. Series **917** (2017) 052013 doi:10.1088/1742-6596/917/5/052013

$$F(z) = \begin{cases} A\exp\left(k_1(z+a)\right), & z < -a \\ A\,\mathrm{sech}\left(k_2 a\right)\cosh\left(k_2 x\right), & z \in [-a, a] \\ A\exp\left(-k_1(z-a)\right), & z > a \end{cases} \quad \text{where} \begin{cases} k_1 = \sqrt{-\xi} \\ k_2 = \sqrt{-\xi - \alpha} \end{cases} \tag{11}$$

ξ can be found as solution of following equation:

$$\tanh(k_2(\xi)a) = -\frac{k_1(\xi)}{k_2(\xi)} \tag{12}$$

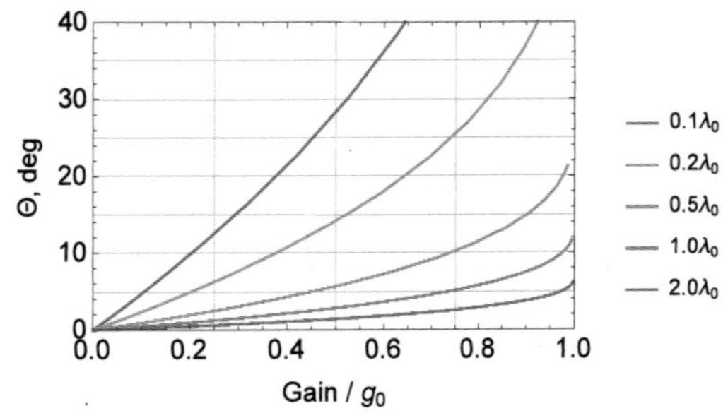

Figure 4: $\Theta(G, a)$ characteristics.

Note that there is modal gain limit g_0, which means field is being contained in $g(z)$ completely. Also, better result is reached for greater slab width.

4. Conclusion

We have developed variation approach that in analytical terms describes the near-field profiles that corresponds to better possible modal gain for a given divergence angle or vice versa. Though it is obvious that this approach could not be always used in real material systems due to limited refractive index variations, the approach still is useful to determine how far from the "ideal" any given structure is.

Acknowledgements

This work is supported by the Russian Science Foundation (Project No. 14-42-00006).

References

[1] Korenev V V, Savelyev A V, Zhukov A E, Maximov M V and Omelchenko A V *Waveguide and active region structure optimization for low-divergence InAs/InGaAs quantum dot comb lasers* Proc. SPIE 9503, 950305-1-7 (2015)

[2] Dimitri P Bertsekas 1982 *Constrained Optimization and Lagrange Multiplier Methods.* (Massachusetts: Academic Press) p 72

[3] Bruce van Brunt 2003 *The Calculus of Variations* (New York: Springer Science & Business Media) p 28

IOP Conf. Series: Journal of Physics: Conf. Series **917** (2017) 052014 doi:10.1088/1742-6596/917/5/052014

Optical Rectification of Phase Modulated Signal Based on Injection Locking

A Lukashchuk[1,2], V A Neskorniuk[1,2], V S Lyubopytov[3,4], I R Gabitov[1,5], M Malekizandi[6], A Chipouline[6] and F Küppers[6]

[1] Skolkovo Institute of Science and Technology, Skolkovo 143025, Russia
[2] Moscow Institute of Physics and Technology, Dolgoprudny141700, Russia
[3] Telecommunication Systems Dept., Ufa State Aviation Technical University, Ufa 450000, Russia
[4] Department of Photonics Engineering, Technical University of Denmark (DTU), Ørsteds Plads 343, 2800 Kgs. Lyngby, Denmark
[5] Department of Mathematics, University of Arizona, Tucson 85721-0089, USA
[6] Institute for Microwave Engineering and Photonics, TU Darmstadt, 64283 Darmstadt, Germany

Abstract. We experimentally demonstrate feasibility of simultaneous use of Differential Phase Shift Keying (DPSK) and Amplitude Shift Keying (ASK) formats (orthogonal modulation) using injection-locked semiconductor laser. Experimental study shows significant improvement of the bit-error-rate (BER) and doubling of the system capacity.

1. Introduction

Fiber-optic communication lines is a back bone for data transmission. Present and future demand on high bandwidth Internet services leads to the requirement of increase of channel's capacity at all the stages of network. Here we investigate an efficient and cheap technique to double the capacity of metro-level networks or inter data centre communications based on optical injection locking.

Phase locking of a laser with an external seeding signal has been investigated in details [1], and stable locking conditions have been determined [2]. It has been found, that the laser under certain parameters can follow the phase of the external seeding, while keeping its own output amplitude, almost independent from the amplitude of the input seeding signal. Semiconductor edge-emitting lasers locked by phase-modulated external signal have been also tested as phase preserving amplitude limiters [4].

In this work we show how a distributed feed-back (DFB) laser under special injection-locking conditions can be used in fiber optic communication systems for the suppressing of an amplitude modulation. To distinguish DPSK and ASK signals DFB laser is utilized to extract the phase information before receiver. This method is simple, requires only relatively small modification in the infrastructure of communication systems, and hence is supposed to be cost effective.

Fragkos et al. have already demonstrated an experimental possibility of successful orthogonal modulation. In contrast to the previously published data [5], we show that the relation between BER and amplitude modulation Extinction Ratio (ER) depends critically on the correlation between ASK and DPSK bit sequences.

IOP Conf. Series: Journal of Physics: Conf. Series **917** (2017) 052014 doi:10.1088/1742-6596/917/5/052014

2. Operation principle and experimental setup

CW light from a tunable external cavity laser (ECL), used as a master laser, is consecutively and independently modulated by amplitude and phase with 10 Gbit/s rate using two pseudorandom binary sequences of lengths 2^9–1 and 2^7–1 bits. Pulse pattern generators (PPG) are synchronized in such a way that the modulated amplitude and phase symbols in optical signal coincide in time. Afterwards signal is divided into two parts. One goes directly to the photodiode to decode ASK component. Second part is directed to the Slave laser to rectify phase. The DFB laser, used as slave one, is pumped up to 15.5mA, while its lasing threshold is near 8.7mA. The output signal of DFB laser after the circulator is detected by a differential phase-shift keying (DPSK) receiver. Analysis of the detected bit sequences is performed by bit error rate tester (BERT). For capturing eye diagrams and measuring extinction ratios (ER) before and after DFB laser, digitizing oscilloscope is used.

The desirable regime of injection locking can be obtained when the detuning between master and slave lasers approaches zero (from the shorter wavelengths). Depending on both DFB laser current and resulting detuning, an optimal injection power was empirically adjusted in order to minimize BER in the demodulated signal after DPSK receiver.

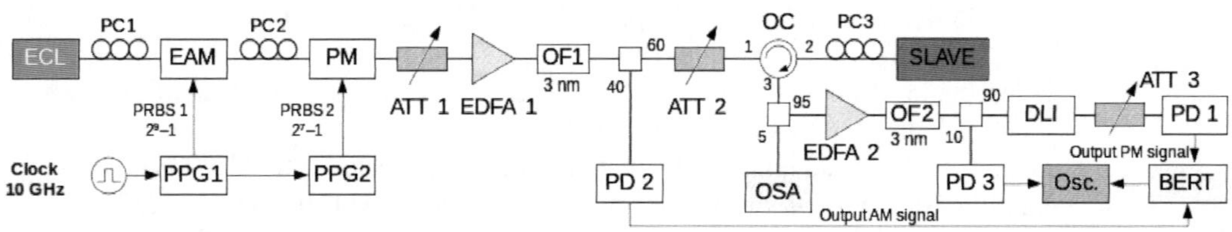

Figure 1. Scheme of the experimental setup (ECL - external cavity laser; PC – circular polarizer; EAM - Mach Zehnder amplitude modulator; PM - phase modulator; PPG - pulse pattern generator; ATT - variable attenuator; EDFA - erbium doped fiber amplifier; OF - optical filter; OC - optical circulator; PD - p-i-n photodiode; DLI – delay line interferometer; OSA – optical spectrum analyser; Osc. – digitizing oscilloscope; BERT – bit error tester; DFB – distributed feedback laser;).

3. Experiment results and discussion

Figure 2 shows BER vs. AM (amplitude modulation) ER dependences. As follows from these curves, injection-locked DFB laser provides significant improvement of the transmission reliability for the DPSK signal component, and great BER improvement can be achieved. Furthermore, amplitude and advanced phase modulations can be both recognized error free. This way we can double the capacity of transmission lines.

Figure 3 demonstrates dependence of correlation between ASK and DPSK signals. In non-correlated case Pseudo Random Bit Sequence of 2^7-1 bits to generate DPSK signal and 2^9-1 for ASK were used. In correlated – 2^{31}-1 for both. It is worth noting that using same data streams for ASK and DPSK modulators is not reliable. In this case, it is possible to vary the level of correlation between two bit sequences from non-correlated to fully correlated simply by changing the length of SMA cables connecting PPGs and modulators or by switching PPGs at different times.

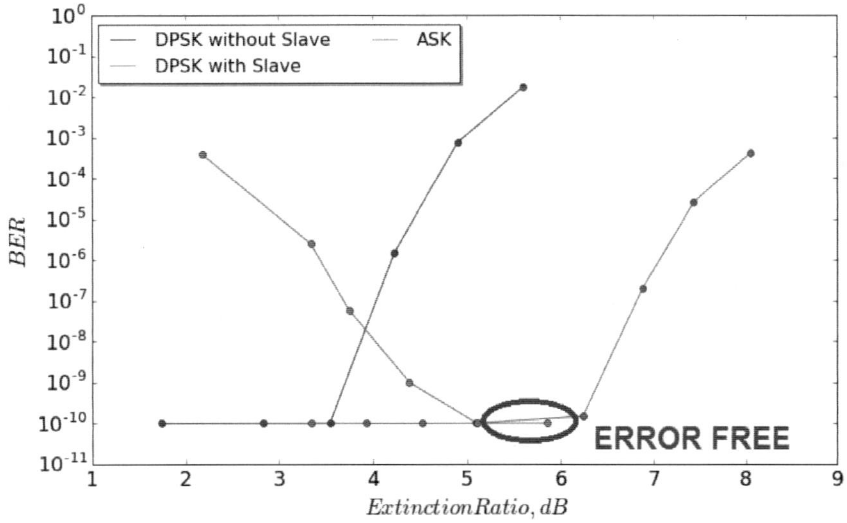

Figure 2. BER vs. Amplitude modulation ER for cases without Slave laser and with Slave laser.

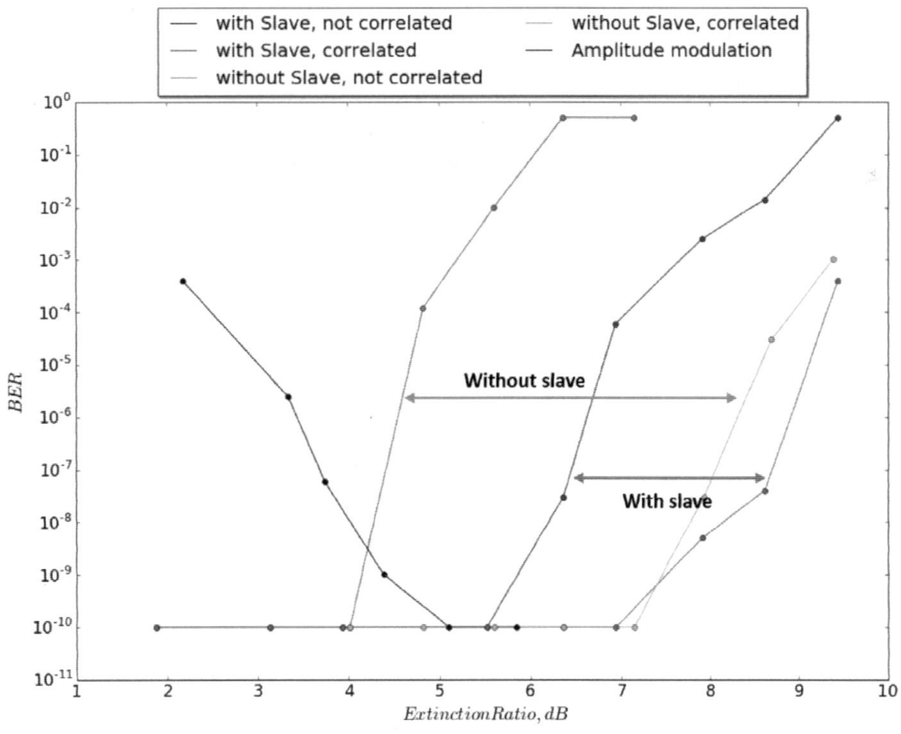

Figure 3. BER vs Amplitude modulation ER for correlated and non-correlated bit sequences.

4. Conclusion

We have found that the special operation mode of the externally seeded DFB laser can be used for effective suppression of the amplitude modulation in a phase modulated optical signals. Based on this effect, we have successfully demonstrated BER improvement of the phase modulated signal by several orders of magnitude. It leads to effective decoding of ASK/DPSK modulated channel and increasing its capacity. In contrast to the previous publications, it was found that the system performance depends critically on the bit correlation of both orthogonal channels.

References

[1] H. Haken, *Laser Light Dynamics*, Vol. 2 (North-Holland Physics Publishing, 1986).
[2] E.K. Lau, L.J. Wong, and M.C. Wu, “Enhanced Modulation Characteristics of Optical Injection Locked Lasers: A Tutorial,” IEEE J. Sel. Topics Quantum Electron. **15(3)**, 618-633 (2009).
[3] A. Pikovsky, M. Rosenblum and J. Kurth, *Synchronization. A Universal Concept in Nonlinear Science* (Cambridge University Press, 2001), p. 46.
[4] Fragkos et al. "Amplitude noise limiting amplifier for phase encoded signals using injection locking in semiconductor lasers." *Journal of Lightwave Technology* 30.5 (2012): 764-771.
[5] A. Fragkos, A. Bogris, D. Syvridis, “Efficient orthogonal modulation enabled by injection locked limiting amplifiers”, *IEEE Photonics Technology Letters*, Vol 25, Issue 7, no. 6461056, Pages 667-670, Apr. 1 2013.

Energy transfer in Gd_2O_3:Er nanoparticles applying as a down-conversion layer for solar cell

Yu A Kuznetsova, A F Zatsepin, V A Pustovarov, M A Mashkovtsev, V N Rychkov
Institute of Physics and Technology, Ural Federal University, Yekaterinburg, Russia

Abstract. The optical properties of Gd_2O_3:Er (0.25-8 mol%) nanoparticles have been investigated. The observed effect of energy transfer from lattice Gd^{3+} to activator Er^{3+} ions can be used for enhancement of solar cells efficiency. The general scheme showing the principle of operation of down-conversion layer based on Gd_2O_3:Er was presented. The concentration dependences of emission intensity and energy transfer efficiency were obtained.

1. Introduction

Gadolinium oxide is of interest as host material for functional devices of nanophotonics, optoelectronics, alternative energy and conversion systems because of its low phonon energy, good thermal stability and chemical durability [1]. Among the rare-earth ions, Er^{3+} is one of the most perspective activator. Optical transitions from the Er^{3+} long-lived excited states result in the intensive emission in green and red spectral regions. Moreover, specialties of Er^{3+} energy levels structure provide an opportunity to convert of UV and near infrared radiation to visible light [2].

The efficiency of energy conversion is determined by a number of factors, associated with both the host lattice material (crystal structure, defectiveness, size and surface effects) and properties of activator ions (qualitative and quantitative compositions of donor-acceptor pairs, absorption cross section, quantum yield of luminescence) [3]. Along with the above, the determination of optimal concentrations of dopants is one of the key tasks to achieve the highest energy transfer efficiency [4]. The aim of this work is to investigation the luminescent behavior of nanostructured Gd_2O_3:Er as a function of different Er^{3+} concentration.

2. Experimental

The nanoparticles of Gd_2O_3:Er with Er^{3+} concentration from 0.25 to 8 mol% were synthesized by chemical precipitation method. The precursors $Gd(NO_3)_2$ and $Er(NO_3)_2$ with appropriate molar concentrations were dissolved in aqueous ammonia solution (pH=8.5). After the reaction was completed, obtained $Gd(OH)_3$:Er precipitates were washed with distilled water several times and dried at 120°C. Subsequent calcination at 1000°C during for 2 hours had led to the formation of Gd_2O_3:Er nanoparticles.

The crystallinity, phase composition and particle size of obtained samples were investigated by X-ray diffraction patterns recorded by the XPertPro MPD diffractometer equipped by the solid state pixel detector for Cu-Kα radiation. Morphology study of prepared nanoparticles was performed by Carl Zeiss SIGMA VP scanning electron microscope in high-vacuum mode with the InLens detector. Photostimulated emission and excitation spectra were measured on a setup assembled with the 400 W deuterium lamp, R6358-10 Hamamatsu photomultiplier and two prismatic DMR-4 computer-controlled monochromators.

3. Results and discussion

3.1 Structure and morphology

Figure 1 shows the XRD patterns of undoped as well as Er^{3+}-doped Gd_2O_3 nanoparticles. All peaks are corresponding to the cubic single-phase structure of Gd_2O_3 with Ia-3 space group (JCPDS No. 65-3181). No impurity peaks and other possible phases were not observed due to the incorporation of Er^{3+} ions. The values of lattice parameter and average crystalline grain size for all samples are listed in table 1. It is seen that increasing of Er^{3+} concentration results in the decreasing of lattice constant and particle size. This fact probably results from smaller radius of Er^{3+} (0.881 Å) in comparison with the Gd^{3+} (0.938 Å) and can indicate the successful doping the activator ions into the host lattice. The SEM images (figure 2 shows the image for sample with 2 mol% of Er^{3+} concentration) reveal that nanoparticles are different in size and shape. In addition, the micrographs clearly indicate the porosity and agglomeration of obtained crystallites.

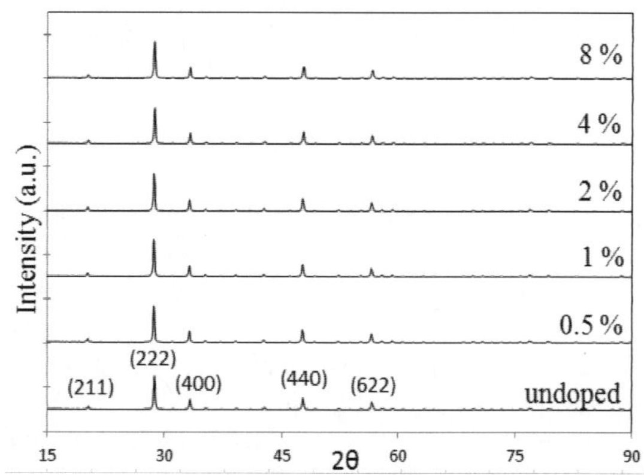

Figure 1. XRD patterns of undoped and Er^{3+}-doped (0.5-8 mol%) Gd_2O_3 nanoparticles.

Figure 2. SEM image of Gd_2O_3:Er nanoparticles with 2 mol% of Er^{3+} concentration.

Table 1. Average particle size and lattice parameter of Gd_2O_3:Er nanoparticles with different Er^{3+} concentrations.

Concentration of Er^{3+} ions (mol%)	0	0.5	1	2	4	8
Average particle size (nm)	57	56	54	53	50	48
Lattice parameter (Å)	10.813	10.812	10.810	10.808	10.804	10.793

3.2 Luminescent properties and energy conversion parameters

The excitation spectrum of Gd_2O_3:Er (2 mol%) shown at figure 3 was recorded under the monitoring of the most intensive emission of Er^{3+} at 540 nm. The three bands with maximums at 232, 275 and 310 nm correspond to the 4f-4f optical transitions of Gd^{3+} ion that indicates the realization of energy transfer between lattice cations and activator ions [5,6]. Optically-active Gd^{3+} ions provide the absorption of near UV radiation followed with the energy transfer to Er^{3+} ions that in turn emit in the visible spectral range, so we observe down-conversion process [7].

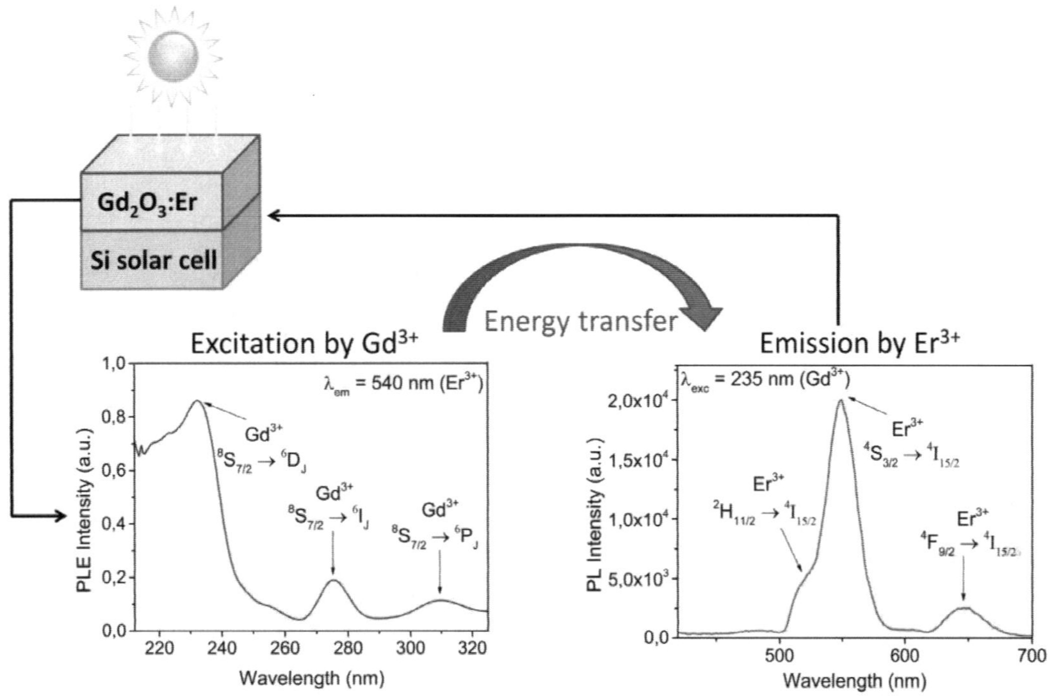

Figure 3. The explaining scheme of the operating principle of Gd_2O_3:Er additional layer on the top of silicon solar cell in accordance with the down-conversion mechanism that is realized due to the Gd^{3+}-Er^{3+} energy transfer.

This mechanism of radiation conversion is applicable for special additional layer that is placed in the front surface of traditional silicon solar cells in order to enhancement of their efficiency. Down-conversion layer is intended to reduce the thermalization losses that arise due to the absorption of photons with energy higher than the Si band-gap [8].

In order to determine the influence of Er^{3+} concentration on luminescent behavior of Gd_2O_3:Er nanoparticles as well as to obtain the information about energy transfer efficiency in Gd^{3+}-Er^{3+} ion pair we have performed the photoluminescence measurements for all samples from the Gd_2O_3:Er (0.25-8 mol%) concentration series (figure 4 and figure 5). It was observed that the intrinsic emission of Gd^{3+} ions acting as donors of excitation continuously decreases with Er^{3+} concentration increases. This fact points out the growth of Gd^{3+}-Er^{3+} energy transfer efficiency under the increasing of acceptor concentration (table 2). Herewith the intensity of Er^{3+} emission under the indirect excitation (energy transfer from Gd^{3+}) increases up to the 2 mol% of Er^{3+} concentration at which it reaches a maximum. A decrease in the luminescence intensity with a further increase of Er^{3+} concentration can be explained by the non-radiative losses related with the energy migration of excitation along a chain of Er^{3+} ions (so-called concentration quenching effect) [9,10]. In other words, for concentrations of acceptor ions greater than 2 mol% the efficiency of Er^{3+}-Er^{3+} energy migration process begins to prevail over the efficiency of Gd^{3+}-Er^{3+} energy transfer.

Table 2. Energy transfer efficiency in Gd^{3+}-Er^{3+} pair as function of Er^{3+} concentration.

Concentration of Er^{3+} ions (mol%)	0	0.5	1	2	4	8
Energy transfer efficiency (%)	25	41	50	59	50	76

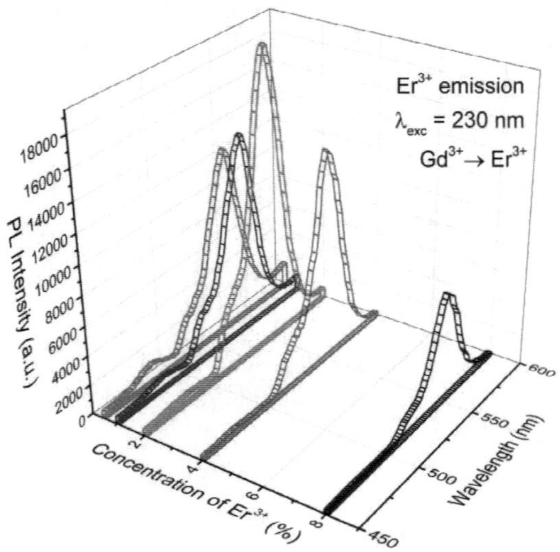

Figure 4. Emission spectra of Gd_2O_3:Er (0.5-8 mol%) in the region of 450-600 nm. Bands at 525 and 540 nm correspond to the optical transitions in Er^{3+} ions under the indirect excitation by the energy transfer from Gd^{3+} ions.

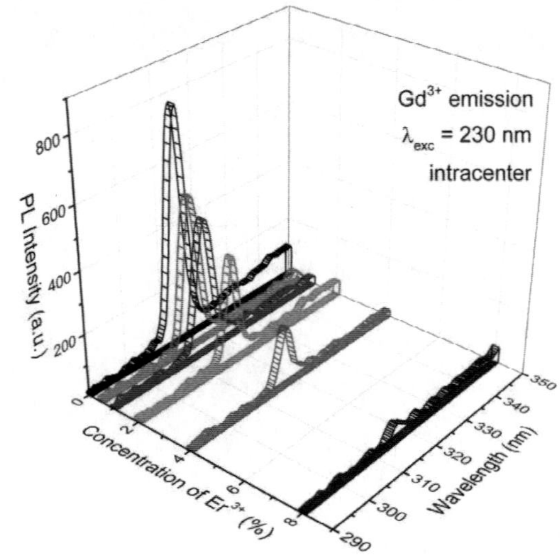

Figure 5. Emission spectra of Gd_2O_3:Er (0.5-8 mol%) in the region of 290-350 nm. Band at 315 nm corresponds to the optical transition in Gd^{3+} ions under the direct intracenter excitation.

4. Conclusion

The Gd_2O_3:Er (0.5-8 mol%) nanoparticles with cubic crystalline structure and average particle sizes of 48-57 nm were obtained by the chemical precipitation method and characterized by XRD and SEM techniques. The emission and excitation spectra reveal the energy transfer from Gd^{3+} to Er^{3+} ions that can be applicable for improvement of solar cells efficiency. The structural scheme demonstrating the applying Gd_2O_3:Er as a down-conversion layer is presented. It was established that the intensity of Er^{3+} emission achieves maximum at concentration of 2 mol% while the efficiency of Gd^{3+}-Er^{3+} energy transfer is 50%.

Acknowledgments

The work has been funded by the Ministry of Education and Science of the Russian Federation (Government task №3.1485.2017/PP).

References

[1] Guo H, Dong N, Yin, M et all 2004 *J. Phys. Chem B* **108** 19205-19209
[2] Chen X, Ma E, Liu G et all 2007 *J. Phys. Chem. C* **111** 9638-9643
[3] Shan J, Uddi M, Wei R et all 2010 *J. Phys. Chem. C* **114** 2452-2461
[4] Linganna K, Jayasankar C K 1981 *Luminescence spectroscopy of lanthanides: Optical properties of lanthanide-doped oxide materials* (Scholar's Press)
[5] Zatsepin A F, Kuznetsova Yu. A. 2015 *J. Phys. Conf. Ser.* **643** 012057
[6] Zatsepin A F, Kuznetsova Yu A, Spallino L et all 2016 *Energy Proc.* **102** 144-151
[7] Mora M B, Amelines-Sarria O, Monroy B M et all 2017 *Sol. Energ. Mat. Sol. C.* **165** 59-71
[8] Richards B S 2006 *Sol. Energ. Mat. Sol. C.* **90** 1189-1207
[9] Lakowicz J R 2006 *Principles of Fluorescence Spectroscopy* (Springer US)
[10] Meza O, Villabona-Leal E G, Diaz-Torres L A et all 2014 *J. Phys. Chem. A* **118** 1390-1396

IOP Conf. Series: Journal of Physics: Conf. Series **917** (2017) 052016 doi:10.1088/1742-6596/917/5/052016

MBE growth and characterization of InAlAs/InGaAs 9 μm range quantum cascade laser

A S Kurochkin[1,2], I I Novikov[2], L Ya Karachinsky[2,4], D V Denisov[2], A G Gladyshev[2], G A Gusev[3], A N Sofronov[3], A A Usikova[4], Yu M Zadiranov[4], G S Sokolovsky[4], V M Ustinov[4], A Yu Egorov[1,2]

[1] ITMO University, Saint Petersburg, Russia
[2] Connector Optics LLC, St. Petersburg, Russia
[3] Peter the Great St.Petersburg Polytechnic University, St. Petersburg, Russia
[4] Ioffe Institute, St. Petersburg, Russia

Abstract. Quantum cascade laser heterosructure consisting of $In_{0,52}Al_{0,48}As/In_{0,53}Ga_{0,47}As$ short-period superlattices have been grown by MBE and investigated by X-ray diffraction, electron microscopy and electroluminescent techniques.

1. Introduction

The idea of quantum cascade lasers (QCL) was proposed in 1971 [1]. The first experimental realization was done in 1994 by F.Capasso's group [2]. For a now QCLs are the most useful sources of mid-infrared laser radiation [3]. QCLs are used for air pollution detection, gas sensor systems, LIDARs, systems for explosives and chemical warfare agent's detection, clinical medicine etc. [4]
Widespread use of 9 μm wavelength range QCL is provided by several factors. This wavelength matches the atmospheric transparency window of 8-12 μm which allow using 9 μm range QCL for free-space data transfer systems. In addition, this radiation is not harmful to the human body which allow to use these lasers for various medical applications.
It is important to note that for the growth of QCL heterostructures very high precision of composition and thickness of layers is needed. As a result molecular beam epitaxy (MBE) looks like the most suitable approach for the growth of such lasers. The creation of 9 μm range QCLs by MBE is relevant because of the lack of this growth technology in Russia.

2. Experiment

QCL heterostructure under investigation was grown by Connector Optics LLC by using Riber 49 MBE machine. Heterostructure consisted of the following parts: $In_{0.53}Ga_{0.47}As$: Si (1e18 cm^{-3}) layer with a thickness of 50 nm was deposited on n^+ InP substrate, after that $In_{0.53}Ga_{0.47}As$: Si (5e16 sm^{-3}) layer with a thickness of 4000 nm was grown and 40 cascades of short-period $In_{0.52}Al_{0.48}As$ / $In_{0.53}Ga_{0.47}As$ superlattice were embedded in the middle of this layer. The last layers were 4000 nm thick step-doped $In_{0.52}Al_{0.48}As$:Si (1e17/1e18 cm^{-3}) and 200 nm thick contact layer of $In_{0.53}Ga_{0.47}As$:Si (5e18 $cм^{-3}$). The active region based on 3-phonon resonant design [5] was calculated to produce radiation at 9 μm at 77 K. A theoretical calculation of the energy levels of a single quantum cascade is indicated in Figure 1. The calculated energy of the intersubband transition is 128.6 meV. Each cascade contained the following sequence of layers $In_{0.52}Al_{0.48}As/In_{0.53}Ga_{0.47}As$ (in nanometers): 4.0/**2.0**/0.9/**5.8**/0.9/**5.5**/0.9/**5.1**/1.2/**4.4**/2.2/**4.2**/1.8/**3.8**/1.6/**3.3**/1.8/**3.1**/2.3/**3.1**/2.7/**2.9**, where bold and underlined layers correspond to $In_{0.53}Ga_{0.47}As$ and doped with silicon to concentration 5e16cm^{-3} layers, respectively. Table 1 shows the QCL heterostructure design.

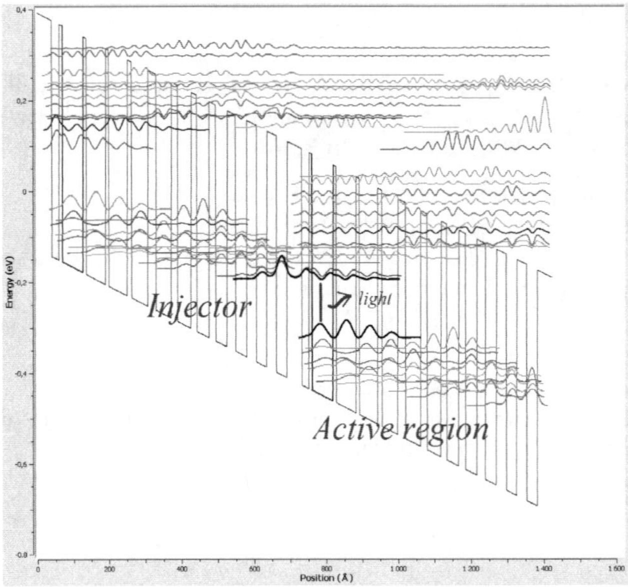

Figure 1. Conduction band profile and moduli squared wavefunctions of active/injector region of the QCL at 77K under the applied field of 61 kV/cm

Table 1. QCL heterostructure design.

Composition	Thickness, nm
InP substrate	-
GaInAs:Si	2000
InAlAs/InGaAs	40 times repeat
GaInAs:Si	2000
InAlAs:Si	4000
GaInAs:Si	200

Growth was carried out at As_4 beam equivalent pressure (BEP) of 5×10^{-5} Torr and substrate temperature of 480°C. InAs growth rate was about 1 Å/s. Prior to the final heterostructure growth the test sample with the superlattice containing 10 cascades of $In_{0.52}Al_{0.48}As/In_{0.53}Ga_{0.47}As$ 10 nm thick each was grown to calibrate compositions and thicknesses of the epitaxial layers.

Structural properties of the heterostructure were studied using PANalytical X'Pert Pro X-ray diffractometer. For simulation and fitting of structure parameters the software X'Pert Epitaxy and

Smoothfit of PANanalytical company has been used. According to x-ray diffraction measurements the growth rates of the binary compounds InAs, GaAs, AlAs were determined and adjusted.

Figure 2 shows the theoretical and experimental rocking curve of QCL heterostructure, produced by high resolution X-ray diffraction method (HRXRD) near the symmetrical reflection (004) of InP. There are diffraction peaks of InP substrate, cladding layers and periodic cascade structure. Zero peak of satellite structure (sl0) is close to InP substrate peak, which shows the good match of the epitaxial layers and substrate lattice constants. A satellite peak structure is observed in a wide angular range, indicating that there are continuity and planarity of the interfaces forming the periodic structure. Narrow satellite peaks (FWHM less than 20 arcsec) without any splitting indicate that there are no significant variations in composition of cascade layers. The total thickness of one cascade was determined by the difference between the angular positions of the satellites and was equal to 70.0 ± 0.2 nm, which with high precision corresponds to the nominal value of the cascade thickness 69.8 nm. In general, the observed diffraction pattern indicates a high structural quality of QCL heterosructure and good matching of the layer thicknesses for all 40 cascades. This in turn shows a high degree of the growth parameters stability.

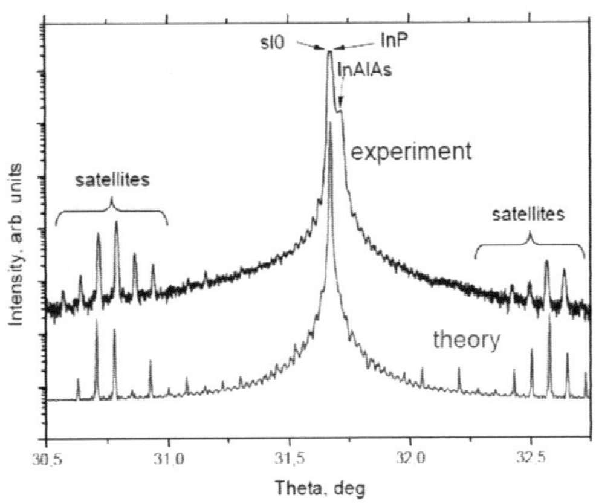

Figure 2. Theoretical and experimental rocking curve of QCL heterostructure, produced by high resolution X-ray diffraction method (HRXRD) near the symmetrical reflection (004) of InP.

Additionally the good quality of QCL heterostructure is confirmed by the results of transmission electron microscopy (TEM) investigation (see Figure 3). It can be seen that within the cascade all layers have planar boundaries and there are no changes in the thicknesses of the cascade layers.

For electroluminescent (EL) measurements QCL heterostructure was processed into 22-μm-stripe-width lasers with ~ 1 mm cavity length. Figure 4 and Figure 5 show scanning electron microscopy (SEM) images of the QCL stripe.

IOP Conf. Series: Journal of Physics: Conf. Series **917** (2017) 052016 doi:10.1088/1742-6596/917/5/052016

Figure 3. The dark-field TEM image of the one cascade of QCL heterostructure (cross section (1-10)).

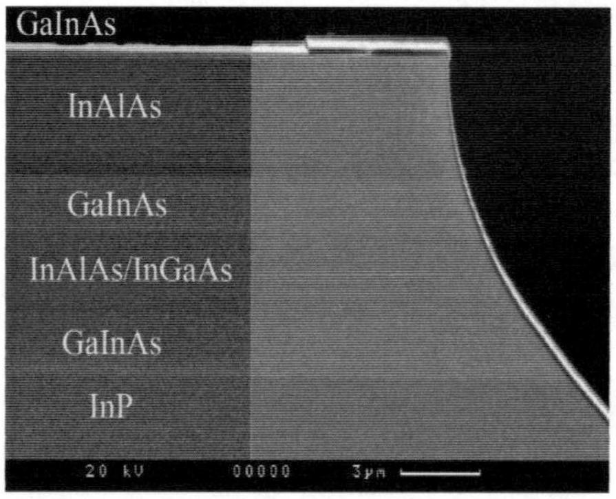

Figure 4. SEM image of the QCL stripe cross-section.

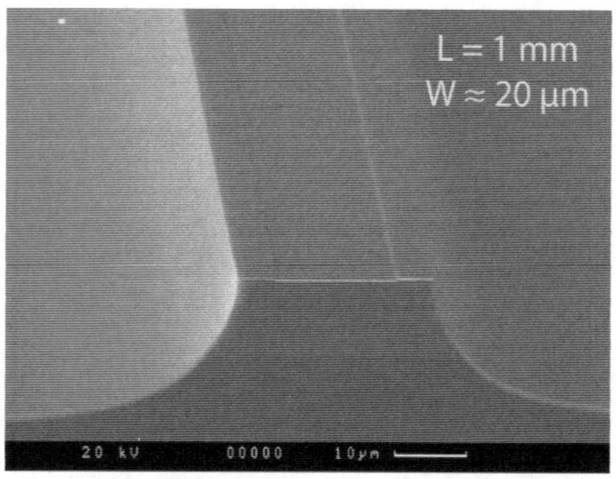

Figure 5. SEM image of the QCL stripe.

Figure 6 shows the results of the electroluminescence (EL) measurements of QCL. The peak position is in a good agreement with the calculated energy of the intersubband transition of 128.6 meV. EL peak at 9.6 μm has FWHM of about 8.1 meV [6], which is comparable with the thermal broadening energy kT = 7.5 meV.

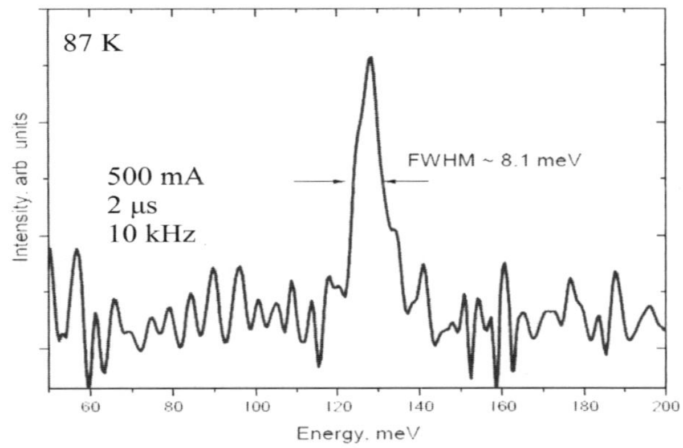

Figure 6. Electroluminescence spectrum of the QCL.

3. Conclusions

QCL heterostructure based on $In_{0,52}Al_{0,48}As/In_{0,53}Ga_{0,47}As$ short-period superlattices have been grown by MBE technique. The heterostructure shows a high accuracy of compositions and layer thicknesses and high structural quality. EL measurements show emission at ~ 9 μm.

Acknowledgments

This study was supported by the Russian Ministry of Education and Science, project ID: RFMEFI61617X0074.

References

[1] Kazarinov R F and Suris R A 1971 *Sov. Phys. Semicond.* **5** 207
[2] Faist J et al 1994 *Science* **264** 553
[3] Robert F C et al 2010 *Chem. Phys. Lett.* **487.1** 1-18.
[4] Y Yu, A J Hoffman and C F Gmachl. 2012 *Nature Photonics* **6.7** 432-439.
[5] Wang Q J et al 2009 *Appl. Phys. Lett.* **94** 011103.
[6] Sirtori C et al 1996 *Appl. Phys. Lett.* **68** 1745

IOP Conf. Series: Journal of Physics: Conf. Series **917** (2017) 052017 doi:10.1088/1742-6596/917/5/052017

Degradation of green InGaN-based LEDs during the tests at the constant current

I V Frolov[1], O A Radaev[1,2], V A Sergeev[1,2]

[1] Ulyanovsk Branch of Kotel'nikov Institute of Radio-Engineering and Electronics of Russian Academy of Sciences, Ulyanovsk, 432071, Russia
[2] Ulyanovsk State Technical University, Ulyanovsk, Russia

Abstract. The article presents the results of investigation of the degradation of electrical and optical characteristics of green InGaN-based LEDs under tests at constant direct current for 5000 hours. It is determined that LEDs which were subjected to a 4-hours exposure of high-density pulsed current degrade slowly than LEDs not exposed to this exposure. The external quantum efficiency (EQE) of the short-wave components of the radiation spectrum of the LEDs when testing reduced by a greater amount than the EQE of the long-wave components of the spectrum. For LEDs that have a current-voltage characteristic with a non-ideality factor of $m \approx 2$, the rate of decrease of optical power during testing is lower than for LEDs with $m > 2$.

1. Introduction

The effects of decreasing of the EQE of InGaN-based LEDs with multiple quantum wells in the process of exploitation have been actively investigated over the last two decades [1,2], however, the physical mechanisms of these effects have not been fully studied and the problem of degradation remains relevant at the present time. One of the tasks aimed to improving the stability of the characteristics of LED-based devices in the process of their operation is the development of methods for forecasting of the service life of LEDs according to their initial characteristics [3]. The article presents the results of investigation of degradation of commercial green InGaN-based LEDs of type ARL-5213 PGC during tests at the constant direct current and results of estimates of the relationship between the rate of decrease of optical power and LEDs initial characteristics.

2. Experimental conditions

We investigated the commercial green InGaN-based LEDs in an amount of N = 30 pieces with following parameters: central wavelength of the radiation spectrum λ_{max} = 525 nm, maximum forward current is 30 mA, chip sizes is 270x320 μm, maximum pulse forward current is 100 mA at the duty cycle ≤ 1/10 and the pulse width ≤ 10 ms, the maximum dissipated power is 120 mW.

Before the tests, the LEDs were separated into three groups of 10 pieces each. The first group of LEDs was tested for 4 hours under the pulses of direct current with an amplitude of 0.5 A, a duration of 100 μs and a duty cycle of 1/100 at an ambient temperature of 25 °C. The average value of the power dissipation in this mode was 42 mW. According to the estimates obtained by measuring of the thermal resistance and the thermal time constant of the LED crystal, the average temperature of the LEDs active region during the tests did not exceed 30 °C. The first group of LEDs, which passed 4-hour pulse tests, and the second group of LEDs which not subjected to pulse tests, were put on long-term tests under the direct current of 20 mA (j = 23 A/cm^2). The third group of LEDs was put on testing under the direct current of 25 mA (j = 29 A/cm^2). The LEDs were tested for 5000 hours. Measurement of the radiation power of the LEDs was performed on the measuring sphere TKA-KK1.

3. Decrease of optical power of LEDs during testing

Figure 1 presents the time dependencies of the optical power during the testing of LEDs subjected to and not subjected to 4-hour tests under a pulsed current. The radiation power was measured at a current of 10 mA. The graphs are plotted from the average values of the radiation power of the LEDs within each group. There are three regions on the graphs: a region of a slight change of optical power during the first 1000 hours of testing, a region of rapid decrease from 1000 to 2000 hours, and a region of slow decrease after 2000 hours. It can be seen from the graphs that the LEDs subjected to 4-hour pulse tests degrade slowly than the LEDs that did not pass the pulse tests: the average value of the radiation power of the LEDs of this group after 5000 hours of testing is 6% higher.

Figure 2 shows a optical power reduction of the third group of LEDs when tested at a direct current of 25 mA. There are three regions on the graphs also. The region of a slight change of optical power is much smaller and amounts to 150 hours. The region of fast decrease of optical power lies in the interval from 150 to 2000 hours, and the slow decrease also occurs after 2000 hours. Under the influence of a direct current of 25 mA, the radiation power of the LEDs decreases faster than at a current of 20 mA: in the time interval from 1000 to 2000 hours – by 1.2 times, in the interval from 2000 to 5000 hours – by 2.7 times.

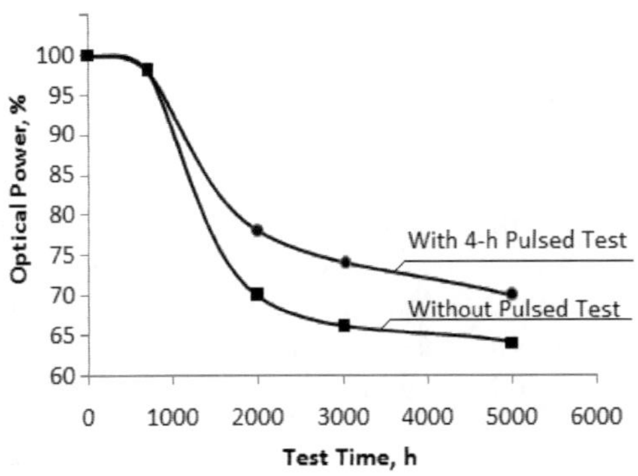

Figure 1. Optical power degradation measured during test at the constant current 20 mA (measuring current 10 mA).

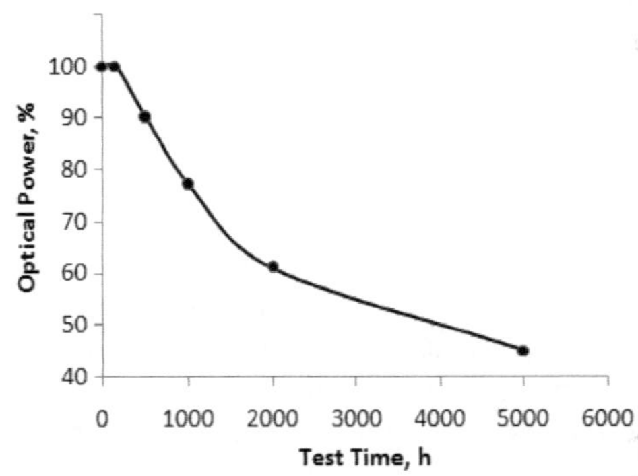

Figure 2. Optical power degradation measured during test at the constant current 25 mA (measuring current 10 mA).

During the tests, the EQE of the LEDs was measured at several current values in the range from 1 μA to 50 mA. Figure 3 shows the graphs of the current dependences of the EQE of one of the investigated LEDs before the tests and after 5000 hours of testing, measured in the full electroluminescence (EL) spectrum.

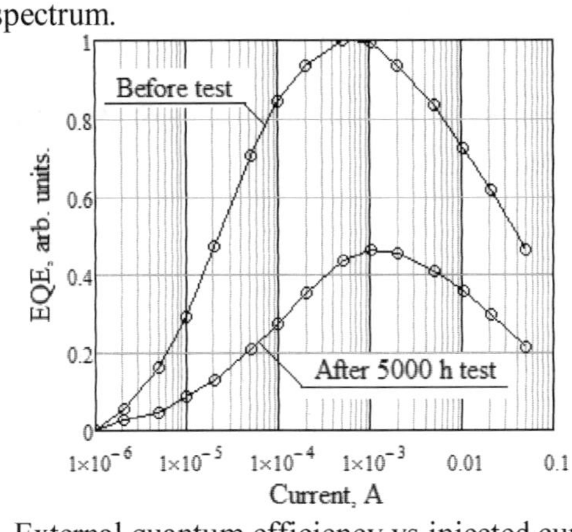

Figure 3. External quantum efficiency vs injected current curve measured on full electroluminescence spectrum.

IOP Conf. Series: Journal of Physics: Conf. Series **917** (2017) 052017 doi:10.1088/1742-6596/917/5/052017

It was found that the largest decrease of EQE occurs in the range of low currents corresponding to the growth area of the EQE (I) dependence. During the test, the maximum of the EQE shifts towards higher currents: after 5000 hours of testing, the current value has changed from 500 μA to 1 mA.

4. The decrease of EQE measured at different components of the EL spectrum during the test

It is known that the inhomogeneous distribution of indium over the area of the LEDs active region leads to an inhomogeneous distribution of the current density over the chip area [4]. Through the areas with low concentration of indium currents of higher density flow. As a consequence, the generation of defects is irregular in the area of the structure, which should lead to a change in the EL spectrum during the degradation process.

The results of measuring of the EQE current dependences on different spectral components of the full EL spectrum have shown that they reach a maximum at different current values (Figure 4). In the wavelength range corresponding to the short-wave region of the EL spectrum (λ = 500 nm), the saturation of the current dependence of the EQE is not observed in the range of operating currents. As the wavelength increases the maximum of the current dependence of the EQE shifts to the region of low currents and at a wavelength of λ = 520 нм the EQE reaches a maximum at a current of 1 mA. In the long-wave range of EL spectrum (λ = 540 nm), the maximum of EQE is achieved at a current of 200 μA.

Thus, it follows from the results obtained that the intensity of the radiative recombination of charge carriers located at different energy levels is saturated at different currents of the LED, and when the current changes, the energy levels participate in the formation of the full EL spectrum.

It is determined that the EQE of the short-wave components of the electroluminescence spectrum decreases more intensively than the EQE of the long-wavelength components of the spectrum (Figure 5) during the degradation process. In Fig. 5 the EL spectra before the tests and after 5000 hours of testing shown, as well as the EQE ratio after 5000 hours to the initial EQE values shown. It follows from the graph that the component of the spectrum at a wavelength of 500 nm drops by 7% more, than the spectral component at a wavelength of 550 nm. The obtained results agree with [5].

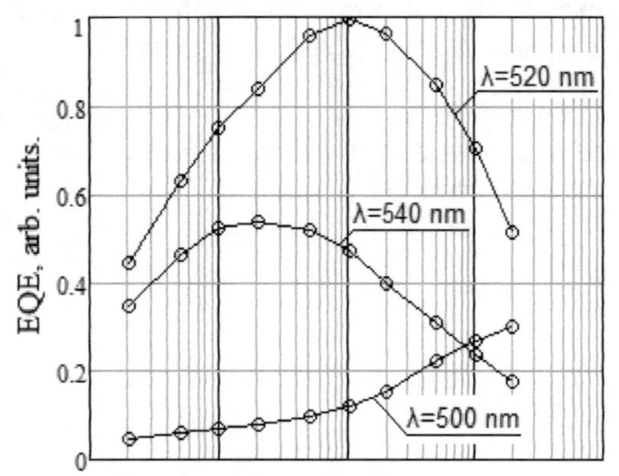

Figure 4. EQE vs injected current curve measured on the spectral components within a full LED EL spectrum.

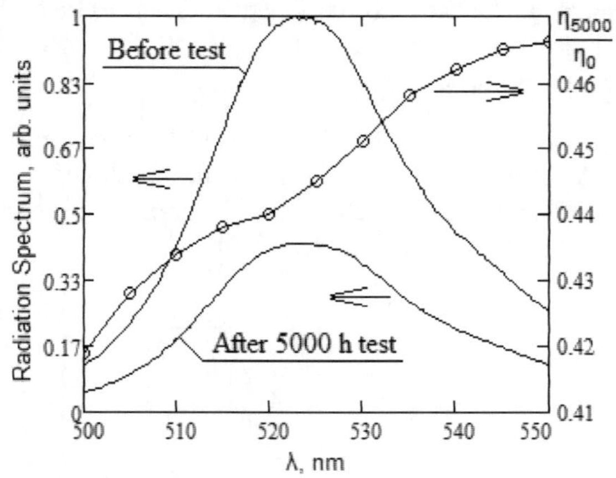

Figure 5. EL spectrums measured on one of the analyzed LEDs and relation between EQE measured on the spectral components within a full LED EL spectrum before and after test.

5. The changes of current-voltage (I-V) and capacitance-voltage (C-V) characteristics during test

The I-V characteristics of the studied LEDs differ quite strongly in form, especially in the range of low currents. Most of the investigated samples in the range of direct voltages of the order of 2 V have a inflection of I-V characteristics (Figure 6).

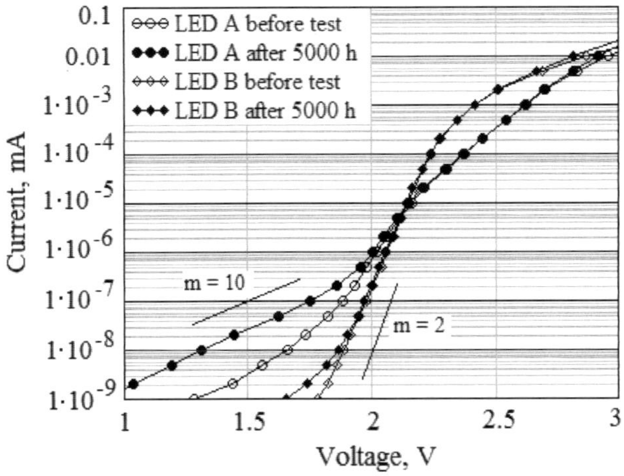

Figure 6. I-V characteristics of two LEDs measured before and after 5000 h test at the current 20 mA.

The inflection point of the I-V characteristic corresponds to the current at which the optical emission of the LED is registered. It is determined that by the value of the threshold current, investigated LEDs have an exponential distribution [6, 7]. The LEDs whose I-V characteristics in the voltage range 1 ... 2 V have a large value of the non-ideality factor m are characterized by larger threshold current values than the LEDs with $m = 2$. The large values of the coefficient m are due to the tunneling mechanism of current flow through the levels of defects, in which radiative recombination is absent. For most of the studied LED samples, the I-V characteristic changes during the test. In the range of low currents (up to 1 μA), the non-ideality factor m increases. It has been determined that the LEDs which have value of m close to 2 are characterized by lower values of the threshold current and have a smaller rate of decrease of the optical power than the LEDs having the value $m = 3 ... 8$. The optical power of the LED A after 5000 hours of testing at a current of 25 mA decreased by 67% (the largest value within the investigated sample), and the optical power of the LED B decreased by 48% (the smallest value within the sampled sample).

During the testing of LEDs, the change of their barrier capacity was controlled. The measurement of the C-V characteristics is performed in the range of bias voltages from +2 to -20 V in the installation [8]. The apparent profiles of charge carrier concentration distribution of LED A and LED B shown in Figure 7.

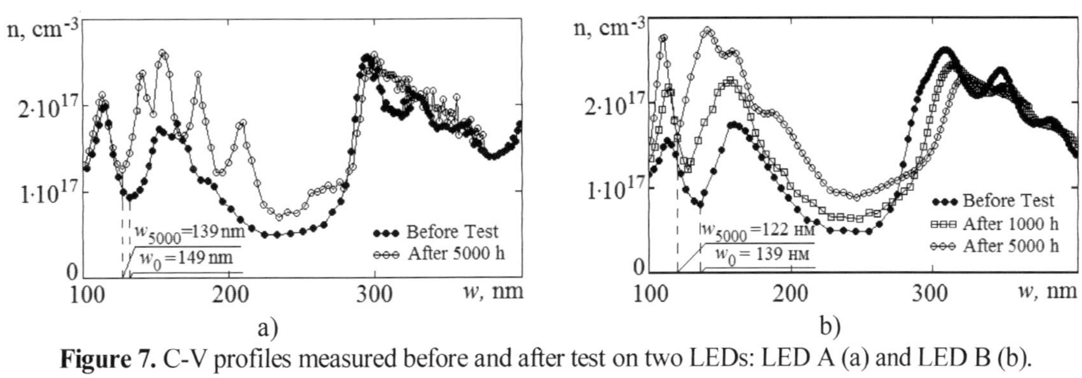

Figure 7. C-V profiles measured before and after test on two LEDs: LED A (a) and LED B (b).

It is determined that when testing LEDs under the action of a direct current, the width of the space charge region decreases, the charged centers are diffused towards the metallurgical boundary of the *p-n* junction, while the pronounced inhomogeneities on the CV-profile near the boundary of the space charge region are formed: for the LED A (Figure 7a) three additional maxima were formed, and for LED B (Figure 7b) an increase in the concentration of charge carriers occurred in the regions of inhomogeneities from 100 nm to 200 nm. According to the degradation model presented in [1], the movement of additional charged centers to the active region of the LED leads to nonradiative recombination probability increase and to optical power decrease.

6. Conclusions

1) LEDs that have been subjected to 4-hour tests at the high-density pulsed current have a lower rate of optical power decrease than the LEDs that did not pass these tests. Current training in the forced pulse mode of green InGaN-based LEDs can be used as a technological operation to stabilize their characteristics and to identify potentially unreliable products.

2) The LEDs that have the values of non-ideality factor *m* close to 2 in the voltage range from 1 to 2 V are characterized by low values of the threshold current and have a lower rate of the optical power degradation than the LEDs with *m*, the values of which reach 10.

3) The external quantum efficiency of the short-wave components of the EL spectrum of the LEDs during the testing decrease more rapidly than the EQE of the long-wave components.

4) When testing under the forward constant current, the width of the space-charge region of LEDs decreases, the diffusion of charged centers toward the *p-n* junction and the formation of inhomogeneities near the heterointerface occurs.

Acknowledgements

The reported study was funded by RFBR, according to the research project No. 16-32-60051 mol_a_dk.

References

[1] Meneghesso G, Meneghini M and Zanoni E 2010 *J. Phys. D: Appl. Phys.* **43** 354007
[2] Huang J, Golubovic D S, Koh S, Yang D, Li X, Fan X, Zhang G Q 2015 *Microelectronics Reliability* **55** 2654
[3] Wang F K, Chu T P 2012 *Microelectronics Reliability* **52** 1332
[4] Wu Y-R, Shivaraman R, Wang K-C, and Speck J S 2012 *Appl. Phys. Lett.* **101** 083505
[5] Sushkov V, Nikiforov S 2011 *Poluprovodnikovaya svetotekhnika* **3** 11
[6] Radaev O A, Sergeev V A, Frolov I V 2016 *Journal of Physics: Conference Series* **741** 012087
[7] Sergeev V A, Frolov I V, Radaev O A 2017 *Technical Physics Letters* **43** 224
[8] Sergeev V A, Frolov I V, Shirokov A A 2014 *Instruments and Experimental Techniques* **1** 137

The impact of 3D compositional fluctuations on the external quantum efficiency and defect generation in nitride alloys with a low In or Al molar fraction

E I Shabunina[1], N M Shmidt[1], A E Chernyakov[2], N A Talnishnikh[2], A Ya Polyakov[3], N B Smirnov[3], I-H Lee[4]
[1]Ioffe Physical Technical Institute, St. Petersburg 194021, Russia
[2]Submicron Heterostructures for Microelectronics Research and Engineering Center RAS, St. Petersburg 194021, Russia
[3]National University of Science and Technology MISIS, Moscow 119049, Russia
[4]School of Materials Science and Engineering, Korean University, Seoul 02841, Korea

Abstract. Peculiarities of the low-frequency noise, the evolution of the full width at half maximum (FWHM) in electroluminescence spectra with the injection current change were studied in InGanN/GaN and AlGaN/GaN MQW containing less than 10% of In and Al components. It was shown that one of the reasons behind the low values of the external quantum efficiency in the LEDs based on these structures is 3D fluctuations in the alloy composition with the local regions of the irregular compound. The results of the investigation of the I-V characteristics and spectral noise density dependences in the LEDs after different aging stages are presented. The defect generation under the injection current during the degradation is shown to occur locally in the extended defects and in the regions of the irregular alloy composition.

1. Introduction

MQW InGaN/GaN and AlGaN/GaN structures with the low In or Al molar fraction (~ 5%) are the foundation for the LEDs of near ultra violet (NUV) spectral range (365-400 nm). These materials are exceptionally promising in a great variety of areas including biological agent detection, document monitoring, and air and water purification. The past decade has witnessed a successful the development of such LEDs due to a significant decrease in current values at < 2 V, enabling one to enhance the external quantum efficiency. However, the quantum efficiency of NUV LEDs is still much lower than that of typical InGaN based LEDs of the visible spectral range. One of the reasons might be a disorder of alloy composition, which is similar to AlGaN alloys containing a high molar fraction of aluminium. However, the experimental studies of fluctuations in alloy composition and phase separation in the compounds with a low molar fraction of In and Al are scarce. The comparative study of electrical and optical properties of LEDs based on InGaN (with 5% of In) and AlGaN (with 6% of Al) and those based on InGaN with 25 % of In was carried out to clarify the impact of 3D compositional fluctuations on the quantum efficiency and defect generation in the nitride based solid solutions a low molar In or Al fraction. All the LEDs under the study had relatively low excess current values ~ 10^{-10} A at < 2V.

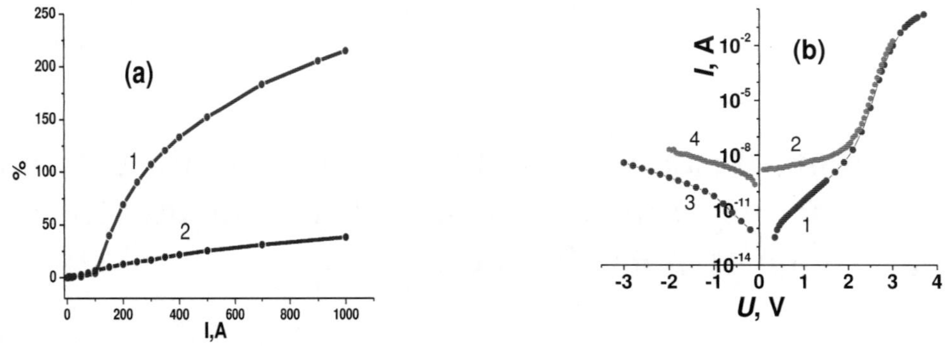

Figure 1(a, b). (a) The relative change in FWHM in the EL spectra for the NUV (curve 1) and blue (curve 2) LEDs; **(b)** Direct (curves 1,2) and inverse (curves 3,4) I-V characteristics of the NUV LEDs before and after aging.

2. Experimental

The study of external quantum efficiency dependences on current density shows the lack of radiative recombination of localized carries in the LEDs containing a low In or Al molar fraction unlike in those based on $In_{0.25}Ga_{0.75}N/GaN$, with the maximum quantum efficiency values reaching 45-50% in $In_{0.25}Ga_{0.75}N/GaN$ LEDs, 24-27% in the NUV LEDs based on InGaN/GaN with 5% of In, and 19-20% in the AlGaN/GaN LEDs with 6% of Al. It should be noted that quantum efficiency in the NUV LEDs peaks at noticeably higher current density values than that in blue InGaN LEDs. These data are in an agreement with the FWHM in electroluminescence (EL) spectra (at < 100 mA) of NUV LEDs being twice as low as in blue LEDs, showing there are few nano-sized regions with fluctuations in the alloy composition in the structures. At the same time, these results as well as the shift into the long wave-length region in the EL spectra and a drastic increase in the relative change in FWHM in the EL spectra at the injection current > 100 mA (figure 1, a) correlates with results obtained for NUV LEDs with high Al concentration. The observed peculiarities differ significantly from those in blue InGaN/GaN LEDs. They were explained [1] by the presence of large regions (up to 1 μm) of fluctuations in the alloy composition which are the source of non-radiative recombination. Moreover, the low-frequency noise investigation shows that in NUV LEDs the non-radiative recombination rate at current values > 100 mA is 10-15 times higher and the crowding effect is more profound than in blue LEDs. The investigation of I-V characteristics, spectral noise density dependences on current at the different aging stages shows a uniform evolution these characteristics for all LEDs under the study. Figure 1 (b) presents the evolution of I-V characteristics in the NUV LEDs with 5% of In during aging. Just like in blue LEDs, there is an increase in the conductivity of shunts related to the extended defect system, which manifests itself in a current growth at voltage values < 2V and an increasing distortion of I-V characteristics at direct current, showing a formation of local regions with a reduced barrier height. For blue LEDs, it was shown earlier that the observed phenomena are caused by the defect generation in local regions with the irregular alloy composition and in extended defects.

3. Conclusion

The obtained results allow us to assume that optimisation of the NUV LED growth conditions intended to reduce 3D compositional fluctuations would lead to an increase in the external quantum efficiency values.

References

[1] Iwata Y, Oto T and Kawakami Y 2015 *J. Appl. Physics* **117** 115702

Optical characterization of mid-infrared range quantum-cascade laser structures grown by MBE

G A Gusev[1], A N Sofronov[1], D A Firsov[1], L E Vorobjev[1], A V Babichev[2,3], A A Usikova[3], N D Il'inskaya[3], Yu M Zadiranov[3], V N Nevedomsky[3], G S Sokolovskii[3], V M Ustinov[3], A G Gladyshev[2], L Ya Karachinsky[2,3], I I Novikov[2,3], A Yu Egorov[2,4]

[1]Peter the Great St. Petersburg Polytechnic University, St. Petersburg 195251, Russia
[2]Connector Optics LLC, St. Petersburg 194292, Russia
[3]Ioffe Institute, St. Petersburg 194021, Russia
[4]ITMO University, St. Petersburg 197101, Russia

Abstract. Optical characteristics of quantum-cascade laser structures were studied in this work. Current-voltage characteristics and dependencies of photoresponse signal on current through the structure were measured together with the spontaneous and stimulated emission spectra. The stimulated emission with wavelength close to 9.56 μm with optical power of tens of milliwatt was observed in a pulsed mode.

1. Introduction

Quantum-cascade lasers (QCL) are extensively developed optoelectronic devices. However, the growth of the multi-stage structures with highly controlled parameters, as well as precise post-processing are still the challenges for nowadays technology.

Basic technologies to create QCL heterostructures are metal-organic vapour-phase epitaxy (MOVPE) and molecular beam epitaxy (MBE). MOVPE allows one to create heterostructures with a greater growth speed. MBE allows implementation of lasers with higher quantum efficiency [1]. Additional improvement of quantum efficiency of QCL is possible to achieve by implementing multi-stage QCL heterostructures containing up to 100 stages [2]. Manufacturing of such multi-stage QCL heterostructures requires maintaining the high identity of chemical composition and thickness of layers throughout the whole long process of epitaxy [3]. The utilization of industrial MBE system satisfies these requirements, as demonstrated by a number of research groups [4-7]. The use of state of the art multi-substrate industrial MBE systems such as Riber 49-7000, Veeco Gen 200-2000 and Oxford VG Semicon V80H-V150 substantially increases the overall yield.

In this work, we present the results of detailed optical characterization of the multi-stage quantum-cascade lasers with emission wavelength close to 9.5 μm at liquid nitrogen temperature. We studied spontaneous emission spectra for below-threshold currents, and time-resolved stimulated emission spectra in pulsed operation mode with different current pulse durations.

2. Samples and experimental setup

The design with the active region based on three-phonon resonance scheme [8] was used. QCL heterostructures were grown by molecular-beam epitaxy on InP (100) substrates using the Riber 49 MBE machine. It contains the 40 stages active region based on $In_{0.53}Ga_{0.47}As/Al_{0.48}In_{0.52}As$ heteropair. The bottom cladding is formed by silicon-doped InGaAs layer (thickness ~ 2000 nm). The top cladding is based on InGaAs and InAlAs solid alloys with thickness of 1830 and 4000 nm, respectively. Contact n++ InGaAs layer is fabricated on heterostructure surface. Grown QCL heterostructures were studied by XRD and TEM. Obtained results indicate high structural quality and precision of the growth technique. Fluctuations of layer composition and thickness were less than 1% for studied structures. To study electroluminescent characteristics, the QCL heterostructures were processed into 22-μm-stripe-width lasers. Cavity length of ~ 1.5 mm was chosen. Lasers were mounted on copper heatsinks.

The experimental setup for optical studies is schematically shown in Figure 1. The samples were measured in a cryostat with ZnSe optical window with 70% optical transmission in the desired spectral range. The bias was applied to the structure in the form of 0.1 – 2.0 μs pulses at a frequency of 0.1 – 10 kHz with gate circuit based on the power MOSFET. Current through the sample was controlled as the voltage on the series-connected 1 Ohm resistor. Waveforms of voltage and current through the sample were recorded with digital oscilloscope.

The spectral measurements were performed with the Bruker Vertex 80v Fourier spectrometer operating in a step-scan mode with a liquid nitrogen cooled HgCdTe photodetector. Sample in the cryostat was placed in the focus of the spectrometer input port. The photodetector signal was measured with a Lock-In amplifier in the case of spontaneous emission studies, and directly recorded with the spectrometer ADC in the case of stimulated emission. Spectral resolution was 16 cm^{-1} and 0.4 cm^{-1} for spontaneous and stimulated emission measurements, respectively. The time resolution of the experimental setup was limited by the photodetector bandwidth (approximately 100 ns). The total dynamic range of the experimental setup is more than 4 orders of magnitude. The average output optical power was estimated with a calibrated mid-infrared power-meter and recalculated to the pulse power taking into account the pumping duty cycle.

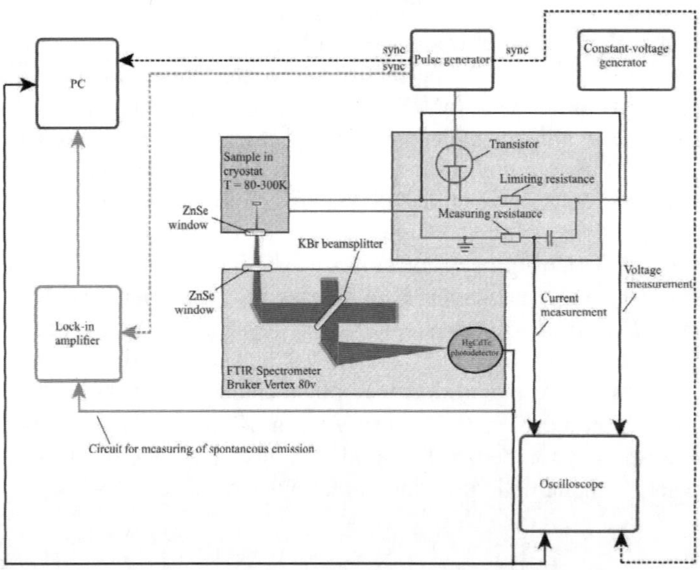

Figure 1. Experimental setup schematics.

3. Experimental Results

The typical current-voltage and current-intensity characteristics are presented in Figure 2(a). The typical threshold current values were found to be in the range of 700 – 800 mA. The spectra of spontaneous emission (Figure 2 (b)) demonstrate a single emission peak with a width at half maximum of about 50 cm^{-1} with a center at 1030-1040 cm^{-1} with slight blue shift with the increase of the current. A typical lasing spectrum measured for the same sample under 200 ns current pumping is plotted in the inset of Figure 2(b). Stimulated emission demonstrates a single line at the frequency of 1043 cm^{-1} (wavelength 9.588 μm).

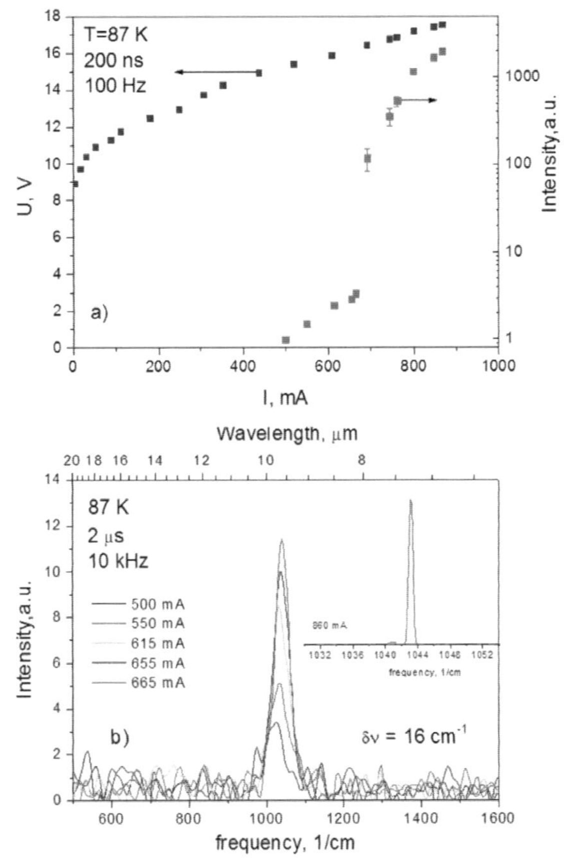

Figure 2. (a) The typical current-voltage and current-intensity characteristics. (b) Typical spontaneous and stimulated (inset) emission spectra at different currents.

The time-resolved measurements of the stimulated emission spectra with relatively long current pulses allowed us to study the temporal evolution of the lasing. The typical lasing transients are plotted in Figure 3 under 800 ns current pulses at different current values.

The characteristic feature of the plots in Figure 3 is a slight tilt of the generation line in the 'time-frequency' domain. It shows the slight red shift of the lasing frequency that is related to the heating of the structure with the current. At higher currents lasing demonstrates jumps of the generation frequency due to mode competition in the Fabry-Perot cavity.

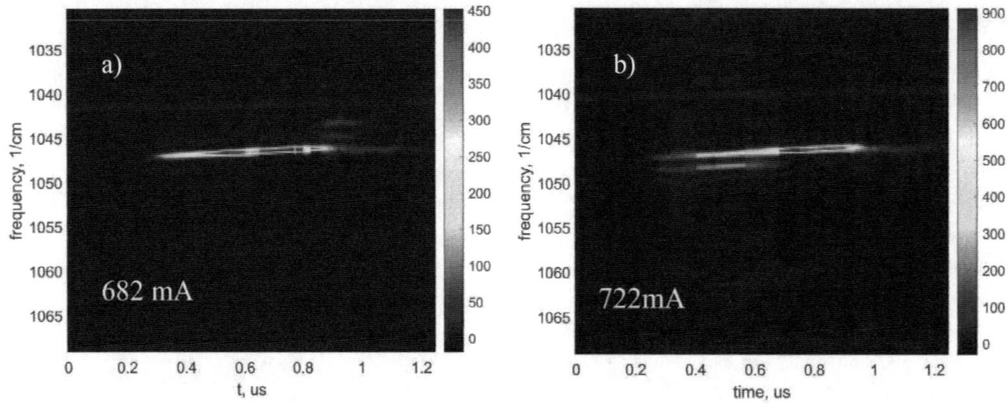

Figure 3. Typical time-resolved lasing spectra (lasing intensity is in arb. units) at pump current values of 682 mA (a) and 722 mA (b).

The typical output optical power at the one facet of the laser structure is estimated as tens of milliwatts.

Acknowledgments
This study was supported by the Ministry of Education and Science of the Russian Federation within the Federal target program "Research and Development in Priority Areas of the Science and Technology Complex of Russia for 2014–2020", code no. 2017-14-588-0004, unique identifier RFMEFI61617X0074.

References
[1] Troccoli M 2015 *IEEE J. Sel. Top. Quantum Electron.* **21** 61-7
[2] Gmachl C, Capasso F, Tredicucci A, Sivxo D L, Kohler R, Hutchinson A L and Cho A Y 1999 *IEEE J. Sel. Top. Quantum Electron.* **5** 808-16
[3] Fedosenko O, Chashnikova M, Machulik S, Kischkat J, Klinkmüller M, Aleksandrova A, Monastyrskyi G, Semtsiv M P and Masselink W T 2011 *J. Cryst. Growth* **323** 484-7
[4] Babichev A V *et al* 2016 *Semiconductors* **50** 1299-03
[5] Garcia M, Vermersch F J, Marcadet X, Bansropun S, Carras M, Wilk A, Chaix C and Sirtori C 2006 *Proc. SPIE* (San Jose, CA, USA) p 613304
[6] Han Y J, Freeman J R, Isac N, Colombelli R, Dean P, Davies A G and Linfield E H 2016 *Opt. Express* **24** 28583-93
[7] Zhu H *et al* 2016 *Appl. Phys. Lett.* **109** 231105
[8] Wang Q J, Pflügl C, Diehl L, Capasso F, Edamura T, Furuta S and Kan H 2009 *Appl. Phys. Lett.* **94** 011103

Fiber-optic transmission system information for the testing of active phased antenna arrays in an anechoic chamber.

I K Saveleiv[1], N V Sharova[2], M Yu Tarasenko[1], T R Yalunina[3], V V Davydov[1, 3, 4], V Yu Rud'[4]

[1]Peter the Great Saint Petersburg Polytechnic University, St Petersburg 195251, Russia
[2] NTC JSC "Zaslon", Saint Petersburg 196084, Russia
[3]The Bonch-Bruevich Saint - Petersburg State University of Telecommunications, Saint Petersburg 193232, Russia
[4]Department of Ecology, All-Russian Research Institute of Phytopathology, 143050, Moscow Region, Odintsovo district, B.Vyazyomy, Russia

Abstract. The results of the research of the developed fiber-optic transmission systems for analog high frequency signal are represented. On its basis, a new method to identify various structural defects in the active phased antenna arrays is elaborated.

1. Introduction.

Now there is a continuous improvement exploited and development of new fiber-optic systems transmission (FOST) of information, both trunk, and local assignment. But in recent years, analog FOST for transmission of a very high frequency of signals (radiophotonics) became an important component of development of local telecommunication area networks. Analog FOST were given the greatest application in radar-tracking systems of different function. Because in locations of radar-tracking systems the large number of various equipment and power stations are concentrated. In addition, the use of FOST data can reduce the weight and dimensions of the radar systems themselves, in cases of their use on aircraft, etc.

At the stage of designing and tuning the active phased arrays antenna (APAA) for radar stations arises many problems. One of them is the distortion of the signal in the transmission from the receiver to the high frequency analyzer at different APAA modes of operation during testing in an anechoic chamber (AC). These distortions are connected with the position of coaxial lines for transmitting signals to nearby high-voltage cables entering the APAA, and also location of the amplifier (without it the signal fades) in the area of direct radiation pattern of high power electromagnetic radiation(since the distance between the transmitter and receiver is smaller than 3 m). In addition, the placement of any signal transform microwave devices (e.g. network analyzer) in the AC violates its anechoic, which makes use of the camera is not expedient .

One solution of this difficult problem is the development of fiber-optic communication line , in which the analog signal from the receiving device placed in AC, transmits the received electromagnetic radiation of APAA in the area of instrumentation location through the region with a complex electromagnetic environment.

2. Fiber-optic transmission system and measurement technique

Since the signals from the microwave generators are also transmitted to APAA via cables placed near the lines with high voltage for a variety of its devices (in AC there is only one entrance - it's also the output for all switching elements), so for this purpose it is also desirable to use FOST. This will improve noise immunity in addition to significantly reduce the size of the connections, and unload the platform on which is placed APAA. Additionally, higher accessibility to the equipment in case of repair will be provided and to testing of APAA and the load of the mechanical elements providing movement of a platform when scanning the APAA pattern will be.

In Figure 1 a block diagram of the test stand for testing APAA in AC is represented. For the implementation developed by us the fiber-optic communication line the high-frequency laser module 1 was selected ("Dilaz" company) – it is the transmitter with direct modulation with wavelength 1310 nm. The receiver is an optical module 2 ("Dilaz" company). Choosing a transmitter with direct modulation is caused by a small length of optical line (350 meters long) and more stability of its parameters to temperature changes. As the APAA testing with use of FOST developed by us must be held in case of ambient temperatures from 233 to 323 K (real operating conditions of radar station). A part of temperature conditions is provided when testing APAA in AC. Remaining temperature conditions are checked when testing APAA in the conditions of a polygon in the presence of opportunities.

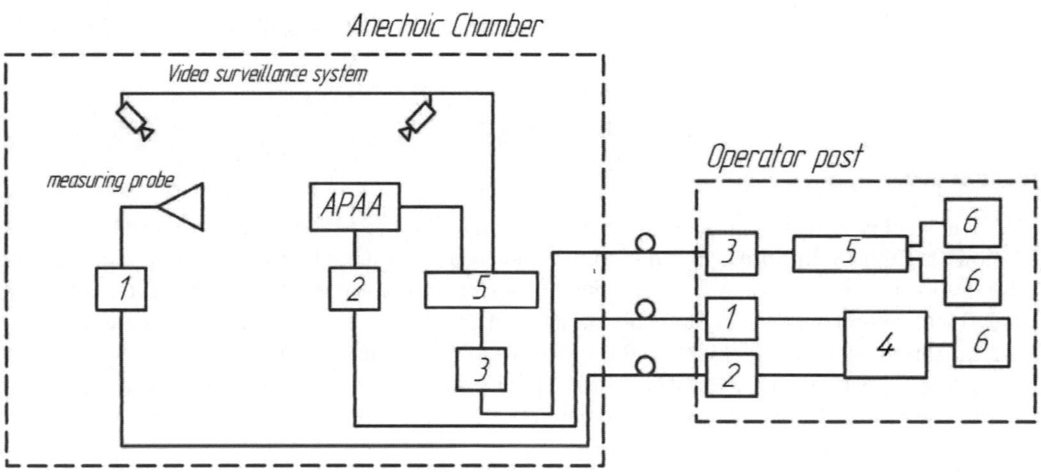

Figure 1. Block diagram of the test stand for testing APAA in AC.; 1- laser module; 2 –receiving module.

The data obtained from measuring probe are coming over FOCL on vector network analyzer 4 with the subsequent processing on the personal computer 6.

3. Results and their discussion.

As an example, in figure 2 AFC in case of transmission of analog signals of a very high frequency on FOST developed by us for different temperature conditions of its operation are provided.

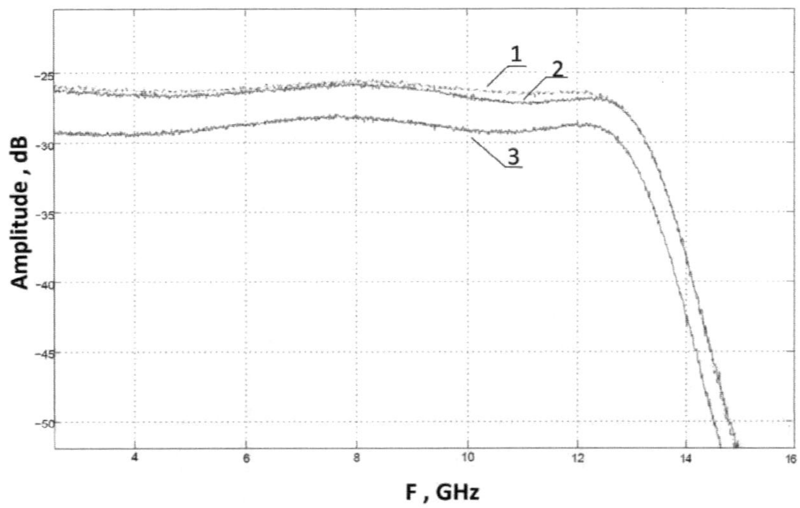

Figure 2. AFC, charts 1, 2 and 3 corresponds to T in K: 288.15; 298.15; 328.15.

The analysis of the received experimental result shows that losses of power during transmission of a very high frequency of a signal on FOST developed by us in the frequency range of operation of APAA in the range of temperatures of its operation – aren't essential. It allows using FOST developed by us for testing of APAA, as in the conditions of AC, and also at a polygon and the place of operation.

As the experiments showed the use of the developed FOST allowed, to research besides range of radiation of all APAA in different operation modes, also the ranges of radiation of its single active elements. These researches weren't possible earlier because of the existence of different noises, in particular the noise in the transferring path.

As an example, Figure 3 shows the emission spectra of a single active element APAA, transmitted to the control sector from the registration device of high frequency signal over coaxial cable, and developed by authors FOST.

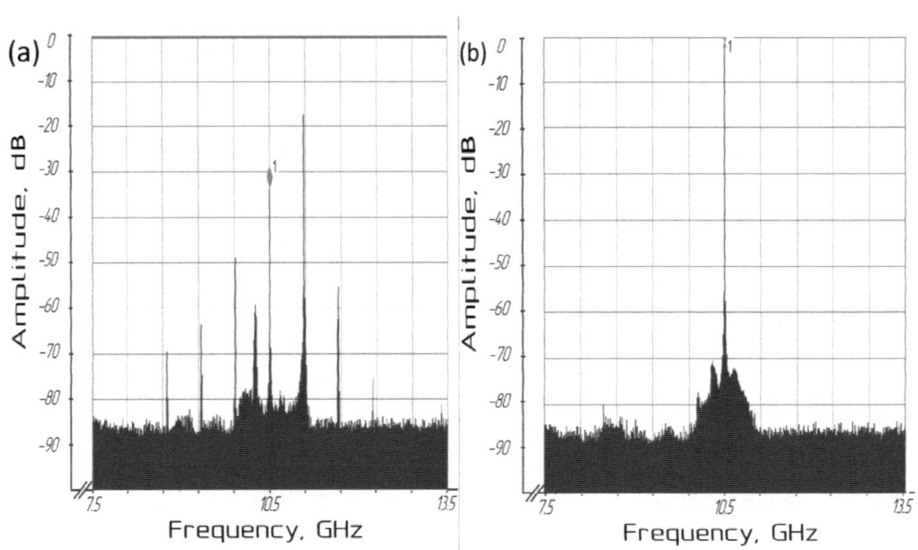

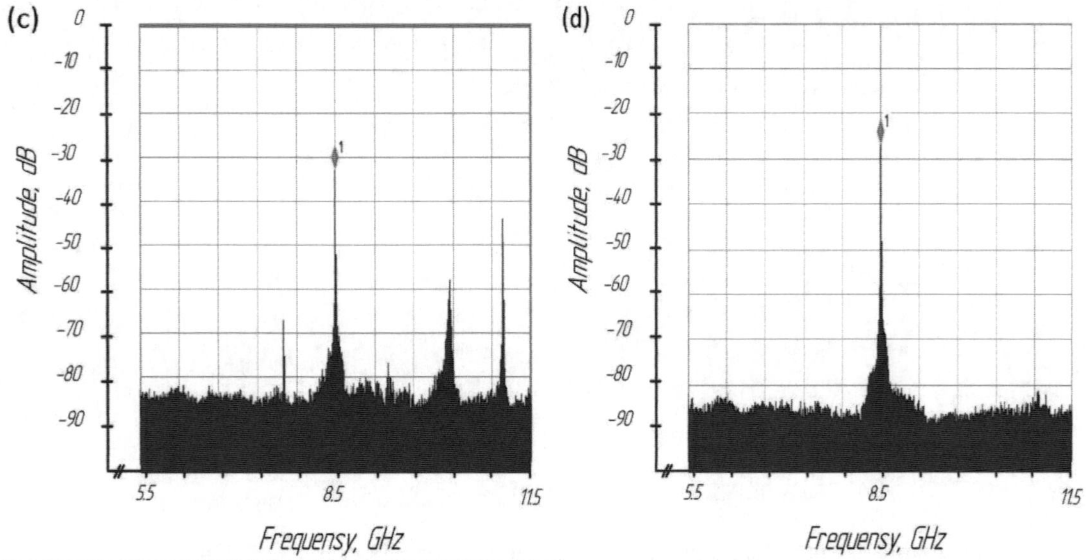

Figure 3. (a, b, c, d). (a, c) The emission spectra of a single active element on 10.5 and 8.5 GHz frequency transmitted through a coaxial cable; **(b, d)** – FOCL.

Analysis of the results, shows that the use of the developed by us FOST allowed to completely eliminate transmission by it unpredictable RF disturbances and different induced signals, taking account of which and their further compensation (for example, subtraction) requires numerous additional measurements. Now changes in the registered pattern of the APAA single elements are connected only with the change of parameter of the very high frequency signal radiated by the antenna or defects of montage, and in the emitter itself. At the presence of the defect of the installation constriction or in the emitter itself, the change of the pattern form takes place (for example reduction of amplitude of a maximum, etc.). Comparing of the constructed patterns of different emitters allows determining these defects.

4. Conclusions

The carried out researches showed that the use of FOST developed by us allows to solve the problems considered earlier arising during testing APAA in AC with use the coaxial cable for transmission of a very high frequency signal. Besides that, the use of FOST allowed to develop a new technique of determination, both defects of assembling of APAA, and defects in the construction of an individual emitter. We can test each emitter of APAA as a part of all operating construction of the APAA. The received results of researches confirmed feasibility of the FOST use for testing APAA and validity of a new developed technique of its different defects determination.

References

[1] Ermolaev A N, Krishpents G P, Davydov V V, Vysoczkiy M G 2016 *Journal of Physics: Conference Series* **741(1)** 012171
[2] Davydov R V, Antonov V I, Kalinin N S 2015 *Journal of Physics: Conference Series* **643(1)** 012107
[3] Davydov V V, Ermak S V, Karseev AYu, Nepomnyashchay E K, Petrov A A, Velichko E N 2014 *Lecture Notes in Computer Science* **8638** 694

[4] Marpaung D, Roeloffzen C, Heideman R, Leinse A, Sales S, Capmany J 2013 *Integrated Microwave photonics* (Laser & Photonics Reviews) **4** 506
[5] Davydov V V, Dudkin V I, Karseev A Yu 2014 *Optical Memory & Neural Networks* **23** 170
[6] Davydov V V, Sharova N V, Fedorova E V, Vologdin V A, Karseev A Yu 2015 *Lecture Notes in Computer Science* **9247** 712
[7] Davydov V V, Dudkin V I, Karseev A Yu 2014 *Optical Memory & Neural Networks* **23** 259

Phosphorus-free mode-locked semiconductor laser with emission wavelength 1550 nm

E S Kolodeznyi[1], I I Novikov[1,2], A V Babichev[1,2], A S Kurochkin[1,2], A G Gladyshev[1,2], L Ya Karachinsky[1,2], I M Gadzhiev[1,3], M S Buyalo[1,3], A A Usikova[3], N D Ilynskaya[3], V E Bougrov[1] and A Yu Egorov[1]

[1] ITMO University, Kronverkskii pr. 49, St. Petersburg, 197101, Russia
[2] Connector Optics LLC, Domostroitelnaya 16, St. Petersburg, 194292, Russia
[3] Ioffe Physical Technical Institute, Polytekhnicheskaya 26, St. Petersburg, 194021, Russia

E-mail: e.kolodeznyy@gmail.com

Abstract. We have fabricated passive mode-locked laser diodes based on strained InGaAlAs/InGaAs/InP heterostructures with crystal lattice mismatch parameter of +1.0 % between quantum well and barrier. The laser with temperature stabilization at 18 °C was demonstrated 10.027 GHz optical pulse repetition rate with 6 ps pulse duration time. Timing jitter of optical pulses in mode-locked regime was 0.145 ps.

1. Introduction

Development of high-speed internet, entry of optical interconnects in supercomputers and data centers, ubiquitousness of cloud technologies accelerates substitution of electronic components by photonic in communication technologies [1]. Optoelectronic semiconductor devices based on A_3B_5 compounds are widely used in the field of communication due to the demonstrated reliability, low cost, energy efficiency and integration capabilities [2]. Development of optical analog-to-digital (ADC) converters can be distinguished among the basic trends of progress in the photonic components. Conversion of analog signals with a frequency greater than 10 GHz and sampling time less than 10 fs cannot be done without optical components [3]. Role of clock-pulse generator in the optical ADC can play an edge-emitting semiconductor laser with passive mode-locking [4–6]. Active region of the laser emitting at telecommunication wavelength of 1550 nm can be made on the basis of elastically strained heterostructure InGaAs/InAlGaAs on InP substrate. It is known that the using the strained InGaAs quantum wells (QWs) reduce threshold current density and increase differential gain [7–9].

2. Experiment

The laser heterostructures were grown on the InP substrate (100) by molecular-beam epitaxy using Riber MBE49 system. Our first attempts to fabricate laser heterostructure with misfit stresses between QW and barriers started from determination of necessary lattice mismatch parameter. In the works [10,11] we experimentally determined the critical thickness of QW and lattice mismatch parameter showing best interface morphology and lasing characteristics, i.e. 31 Å and 1%, and the results agreed with the theory [9,12]. In the work [13] we studied threshold and gain characteristics of laser heterostructures with different number of QWs in active region, the investigated lasers showed a high modal gain of

11 cm^{-1} and a low transparency current density of 46 A/cm^2 per QW, the maximum threshold gain in the laser with 8 QWs of 175 cm^{-1} was obtained. To improve threshold characteristics and to reduce optical losses we developed second generation of laser heterostructures with optimized number and thickness of QWs, doping level and thickness of emitter and doping level of waveguide. Active region of the second generation contained 3 $In_{0.67}Ga_{0.33}As$ QWs separated by $In_{0.53}Al_{0.20}Ga_{0.27}As$ barriers with crystal lattice mismatch parameter of +1.0% between QW and barrier. The $In_{0.53}Ga_{0.27}Al_{0.2}As$ waveguide layer had 0.3 um width without doping. The $In_{0.53}Al_{0.47}As$ emitters were lattice-matched with the lattice parameter of InP substrate and thickness of p-emitter was 1.5 um while thickness of n-emitter was 1 um. Detailed information about first and second generations of laser heterostructures is shown on the figure 1. Comparison of threshold current and optical losses of first and second generation is represented on the figure 2 and 3. One can see that threshold current density J_{th} was reduced twofold and optical losses α_i were reduced from 103 cm^{-1} to 33 cm^{-1}.

We utilized the second generation laser heterostructures to evaluate a passive mode-locking regime in two sections laser configuration [14]. The fabricated edge-emitting mode-locked laser had a cavity length of 4 mm, stripe width of 5 um and etching depth of 250 nm. Passive mode-locking regime was obtained by utilizing of saturable absorber i.e. negatively biased part of gain section with 120 um length electrically separated from the gain section. Schematics of the fabricated laser is represented on the figure 1.

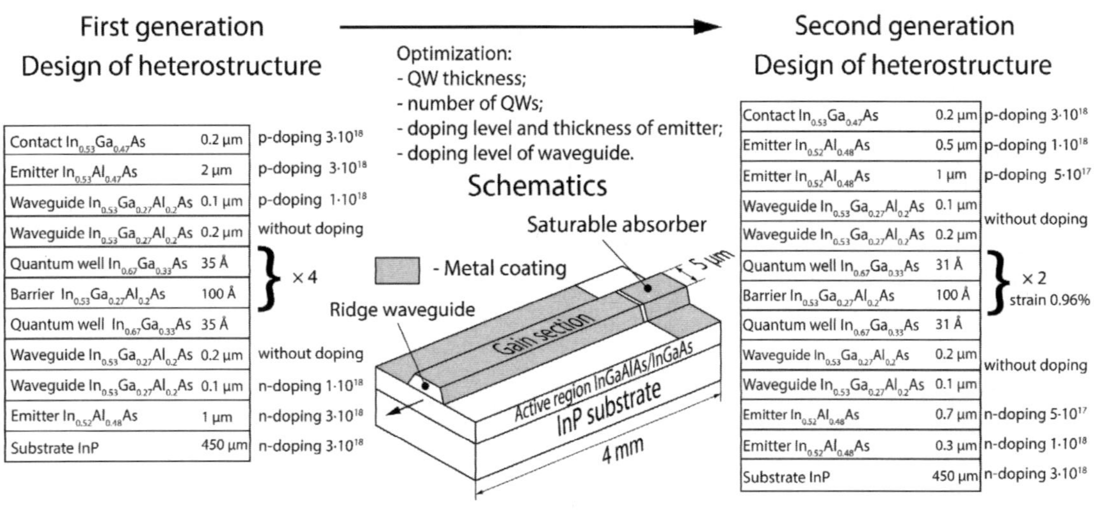

Figure 1. Optimization of the laser heterostructure and schematics of the mode-locked semiconductor laser with two-section configuration.

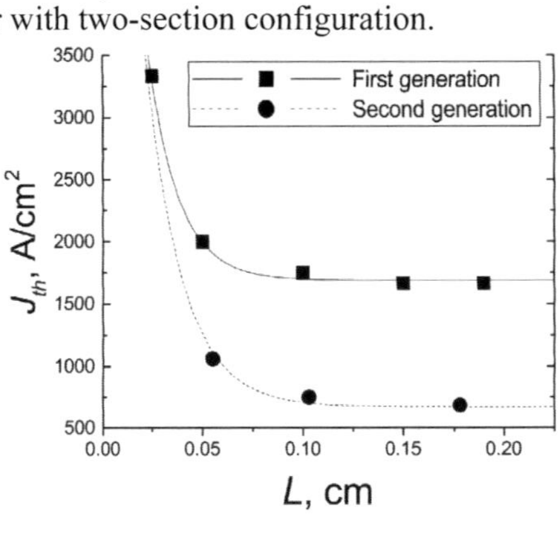

Figure 2. Threshold characteristics of the fabricated laser generations.

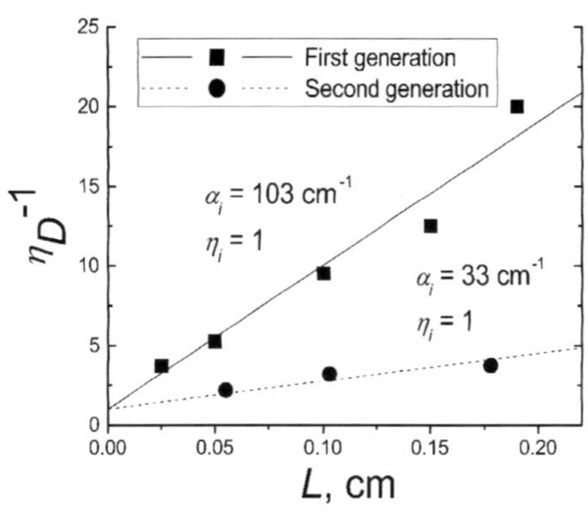

Figure 3. Reciprocal differential losses of the fabricated laser generations.

3. Experimental results and discussion

The fabricated mode-locked lasers were mounted on copper heatsink. Measurements carried out at stabilized temperature 18 °C. The lasers were pumped by direct current I = 340 mA, and saturable absorber was biased at V_{SA} = 2.8 V, the laser had optical power of 6 mW at that bias point. Optical spectrum had the centre at 1539 nm.

The pulse duration was measured using a photodetector whose response to a δ-pulse is 18 ps and an oscilloscope with a bandwidth of 50 GHz at -3 dB. The laser radiation was carried into a multimode 50/125 optical fiber, part of the radiation was removed by a fiber splitter to the synchronization channel of the oscilloscope and a monochromator. In multimode fiber the optical pulse is tightened by 2 ps due to the fiber length of 20 m as well as the spectral width and wavelength of radiation, the bending radius of the fiber and the eccentricity of the optical connectors.

Using equation (1) the reconstructed pulse duration at half-height τ is equal to 6 ps:

$$\tau = \sqrt{\tau_{pulse}^2 - \tau_{ph}^2 - \tau_{osc}^2} - 2 , \quad (1)$$

where τ_{ph} = 18 ps is the response time of the photodetector (18.5-picosecond detector 1454-50 from NewFocus), τ_{osc} = 7 ps is the response time of the oscilloscope and τ_{pulse} = 20.9 ps is the half-amplitude duration of optical pulse from the fabricated laser measured by oscilloscope (insert to the figure 4). Losses on the connectors were not considered. Optical pulse repetition rate was equal to 10.027 GHz and given from radio-frequency spectrum of optical pulse train (see figure 4) from the fabricated mode-locked laser.

An indirect method can be used to calculate the jitter of the fabricated mode-locked laser. One way is based on the fact that with large delay relative to synchronization pulse, the jitter is proportional to the following equation:

$$\tau_{ocs} = \sqrt{\tau_0^2 + N \cdot \tau_{ML}^2} , \quad (2)$$

where τ_{ocs} is the measured oscilloscope jitter, τ_0 is the internal oscilloscope jitter, N is the number of pulses, τ_{ML} is the jitter of mode-locked laser. The estimation gave the value of τ_{ML} about 0.145 ps. Figure 5 is shown dependence of measured oscilloscope jitter on number of pulses and approximation by formula (2).

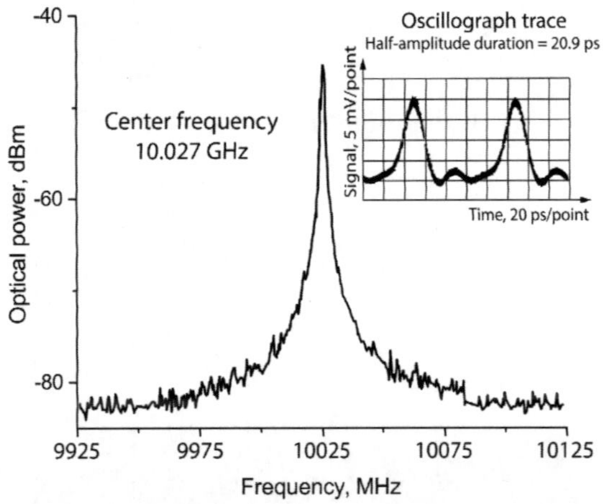

Figure 4. Radio frequency spectrum of optical pulse train. Oscillograph trace of the generated pulses is on the insert to the figure.

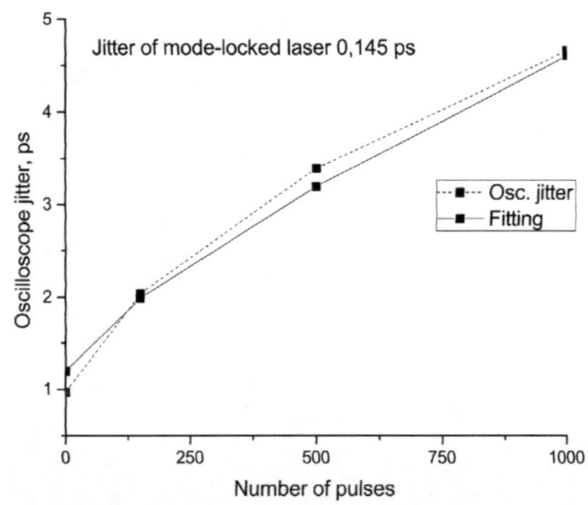

Figure 5. Oscilloscope jitter as a function of number of optical pulses and fitting curve determinates jitter of the mode-locked laser.

4. Conclusions

Threshold current density of tested mode-locked lasers based on strained InGaAlAs/InGaAs/InP heterostructures with crystal lattice mismatch parameter of +1.0% between quantum well and barrier was lower 600 A/cm^2, optical losses were ≈ 33 cm^{-1}. At stabilized temperature 18 °C in passive mode-locking regime the laser with the cavity length of 4 mm and the saturable absorber length of 120 um demonstrated generation of optical pulses with pulse duration time 6 ps and repetition rate 10.027 GHz. Estimation gave the value of timing jitter 0.145 ps.

Acknowledgments

This work was supported by the Ministry of Education and Science of the Russian Federation in the framework of the Federal Target Program "Research and development on priority directions of scientific-technological complex of Russia for 2014-2020", the code 2015-14-582-0038, agreement № 14.581.21.0013 of August 4, 2015, unique ID RFMEFI58115X0013.

References

[1] Iezekiel S 2009 *Microwave photonics: devices and applications* (Chichester: John Wiley & Sons, Ltd)
[2] Yvind K and Jørn M H 2003 *Semiconductor mode-locked lasers for optical communication systems* (Technical University of Denmark)
[3] Khilo A, Spector S J, Grein M E, Nejadmalayeri A H, Holzwarth C W, Sander M Y, Dahlem M S, Peng M Y, Geis M W, DiLello N A, Yoon J U, Motamedi A, Orcutt J S, Wang J P, Sorace-Agaskar C M, Popović M A, Sun J, Zhou G-R, Byun H, Chen J, Hoyt J L, Smith H I, Ram R, Perrott M, Lyszczarz T M, Ippen E P and Kärtner F X 2012 Photonic ADC: overcoming the bottleneck of electronic jitter *Opt. Express* **20** 4454–69
[4] McMaster S 2010 *Monolithically integrated mode-locked ring lasers and Mach-Zehnder interferometers in AlGaInAs* (University of Glasgow)
[5] Avrutin E A, Marsh J H and Portnoi E L 2000 Monolithic and multi-gigahertz mode-locked semiconductor lasers: constructions, experiment, models and applications *IEE Proc. Optoelectron.* **147** 251–78
[6] Hasegawa M, Satoh T, Nagashima T, Mendez M and Konishi T 2015 Below 100-fs timing jitter seamless operations in 10-GSample/s 3-bit photonic analog-to-digital conversion *IEEE Photonics J.* **7** 7201007
[7] Yablonovitch E and Kane E O 1986 Reduction of Lasing Threshold Current Density by the Lowering of Valence Band Effective Mass *J. Light. Technol.* **4** 504–6
[8] Thijs P J A, Dongen L F, Tiemeijer P I, Kuindersma J J, Binsma M and Van Dongen T 1991 High-performance 1.5 μm wavelength InGaAs-InGaAsP strained quantum well lasers and amplifiers *IEEE J. Quantum Electron.* **27** 1426–39
[9] Mitsuhara M and Oishi M 2004 *Long-wavelength infrared semiconductor lasers* (New Jersey: Wiley)
[10] Babichev A V, Kurochkin A S, Kolodeznyi E S, Gladyshev A G, Novikov I I, Karachinsky L Y and Egorov A Y 2015 Molecular Beam Epitaxy Grown Strained Heterostructures for Active Region of Laser Diode With Emission Wavelength 1520-1580 nm *Mater. Phys. Mech.* **24** 284–8
[11] Novikov I I, Babichev A V, Kolodeznyi E S, Kurochkin A S, Gladyshev A G, Nevedomsky V N, Blokhin S A, Blokhin A A and Nadtochiy A M 2016 The structural properties of elastically strained InGaAlAs/InGaAs/InP heterostructures grown by molecular beam epitaxy *Mater. Phys. Mech.* **29** 76–81
[12] Loehr J P and Singh J 1993 Effect of strain on CHSH Auger recombination in strained In(0.53+x)Ga(0.47-x)As on InP *IEEE J. Quantum Electron.* **29** 2583–8
[13] Novikov I I, Karachinsky L Y, Kolodeznyi E S, Bougrov V E, Kurochkin A S, Gladyshev A G, Babichev A V, Gadzhiev I M, Buyalo M S, Zadiranov Y M, Usikova A A, Shernyakov Y M,

IOP Conf. Series: Journal of Physics: Conf. Series **917** (2017) 052021 doi:10.1088/1742-6596/917/5/052021

Savelyev A V, Nyapshaev I A and Egorov A Y 2016 On the gain properties of “thin” elastically strained InGaAs/InGaAlAs quantum wells emitting in the near-infrared spectral region near 1550 nm *Semiconductors* **50** 1412–5
[14] Otto C 2014 Mode-Locked Laser *Dynamics of Quantum Dot Lasers* (Springer International Publishing) pp 191–262

Solar water splitting with III-N nanocolumn structures

O S Medvedev[1], M V Puzyk[2,3], A S Usikov[2,4], H Helava[4], Yu N Makarov[4]

[1]St. Petersburg State University, Research Park, IRC for Nanotechnology, St. Petersburg, Universitetskaya nab. 7-9, St. Petersburg 199034, Russia
[2]University ITMO, Kronverkskiy pr. 49, St.Petersburg 197101, Russia
[3]Herzen University, Nab.r.Moyki 48, St.Petersburg 194186, Russia
[4]Nitride Crystals Inc., 9702 Gayton Road, Suite 320, Richmond, VA 23238, USA

Abstract. HVPE – grown GaN/AlGaN p-n structures with nanocolumns fabricated on the top (1.0 μm of height and 100-130 nm of diameter) were used in a photoelectrochemical process to investigate direct water splitting. The H_2 production rate of 0.56 ml/cm^2 × h was measured for the GaN/AlGaN structures in KOH electrolyte under the Xe-lamp illumination. H_2 evolution was observed at the Pt electrode. However an etching and a small amount of generated gas was observed at the working electrode. Possible mechanisms of corrosion in GaN/AlGAN nanocolumn structures are discussed. A sacrificial agent (like HBr) or the surface passivation by nitrogen atoms is needed to prevent oxidation and etching of the III-N materials.

1. Introduction

Direct water photoelectrolysis using III-N materials is a way for the clean technology development for hydrogen production. In this process, III-N semiconductor, which is immersed in an aqueous solution of electrolyte, allows decomposition of water into H_2 and O_2 gases by solar irradiation of its surface. Required energy to split water molecule is generated by sunlight absorption in the semiconductor. The method is simple in comparison with other methods of hydrogen production. However, the technology is in the development stage.

The main challenges to enable broad application of this method include: research and development of the materials used as photoelectrodes for water electrolysis; slowing down the corrosion of electrodes for a long-term work; increase of the solar-to-hydrogen conversion; research and development of new materials and approaches for effective H_2 and O_2 storage.

Two major approaches of the direct water splitting under solar radiation are currently under investigation at nowadays. The first one is to separate electron-hole (e-h) pairs created by the solar irradiation inside the III-N materials in order to get oxygen evolution reaction and hydrogen evolution reaction at different electrodes in an electrochemical cell. It that case, electrons and holes participate in proper reactions separately at different electrodes, H_2 and O_2 gases generate separately too and can be collected in different reservoirs [1]. The second one is to transport electron-hole pairs created by the solar irradiation to the III-N material surface in order to get oxygen evolution reaction and hydrogen evolution reaction simultaneously on the same surface. It that case, H_2 and O_2 gases need to be separated in different reservoirs [2].

When a p-n structure is immersed into an electrolyte under illumination, an electric field created by space charge at the p-n junction in the volume of III-N structure should improve the efficiency of the e-h pairs separation to compare with the band bending at the semiconductor-electrolyte interface.

However, the same carriers separated by the p-n junction and the band bending have opposite flow direction in the semiconductor that reduces the total efficiency of the water splitting. When all illumination absorbs at the p-n junction (with p-layer on the tip), the flow of holes separated at the p-n junction toward the interface with electrolyte may be larger the opposite flow of the holes separated at interface by the band-bending. In that case, diminution of the barrier high at the interface with the electrolyte and increasing of the surface chemical resistance are the major tasks to be studied for III-N p-n structures for application for the direct water splitting.

In this work we used as working electrodes GaN/AlGaN p-n structures with nanocolumns on the top to investigate the hydrogen generation rate and the material corrosion during the photoelectrochemical (PEC) process.

2. Experimental details

The initial - thick AlGaN/GaN structures were grown by chloride HVPE on c-plane 2-inch sapphire substrates [3]. First, the growth procedure included in-situ sapphire substrate treatment followed by multilayer structure growth consisting of AlN-based buffer layer, $Al_xGa_{1-x}N/Al_yGa_{1-y}N$ ($x > y$) stress control structure. Then, a p-n structure was grown on the top in the same run. The basic structure included a 50- 100 nm-thick AlGaN active region, which was sandwiched between p- and n- $Al_xGa_{1-x}N$ barriers ($x \sim 0.05$-0.12). A 0.5 µm - thick Mg-doped p-type GaN contact layer covers the structure. A net acceptor concentration N_A-$N_D \sim (2$-$4) \times 10^{18}$ cm^{-3} was determined on the surface by C-V measurements using Hg probe.

The nanocolumns on the structure top structure were processed by reactive ion etching in Cl_2 plasma using self-aligned Ni/SiO_2 masks as described in detail in [4]. A sketch of the structure used in the electrochemical experiments is shown on figure 1.

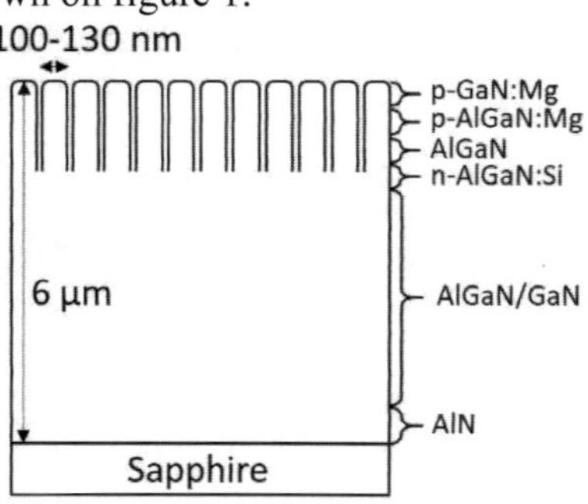

Figure 1. The design of AlGaN/GaN p-n structure with nanocolumns

The samples with nanocolumns used as working electrodes in the photoelectrochemical experiments. The samples were immersed in a KOH aqua electrolyte (0.75M KOH) under external electrical bias from 0 V to +0.5 V and irradiation with a Xe lamp (AM 1.5 standard spectrum, 3500 W/m^2). Pt plate was used as counter electrode. The electrical contact to the sample (working electrode) was made through a metallic In pads. The In contact was protected by a fluorine-based varnish to prevent its etching in the electrolyte. The structures were partly immersed in the electrolyte. An area irradiated by the Xe lamp was about 2.5 cm^2.

A digital multi-meter and an UV-radiometer were used to measure the photo - current and the luminous flux (the optical power of light) of the incident light from the light sources on the GaN-electrode surface. Sample surfaces were characterized by field emission scanning electron microscope (SEM) at an accelerating voltage of 5 kV and with an electron beam current of 300 pA.

3. Results and discussion

Figure 2 shows scanning electron microscope (SEM) images of one of the S3 group GaN/AlGaN structures with nanocolumns (1.0 µm of height and 100-130 nm of diameter) before and after PEC process. The corrosion (etching) process penetrates deep into the structure and results in much thinner

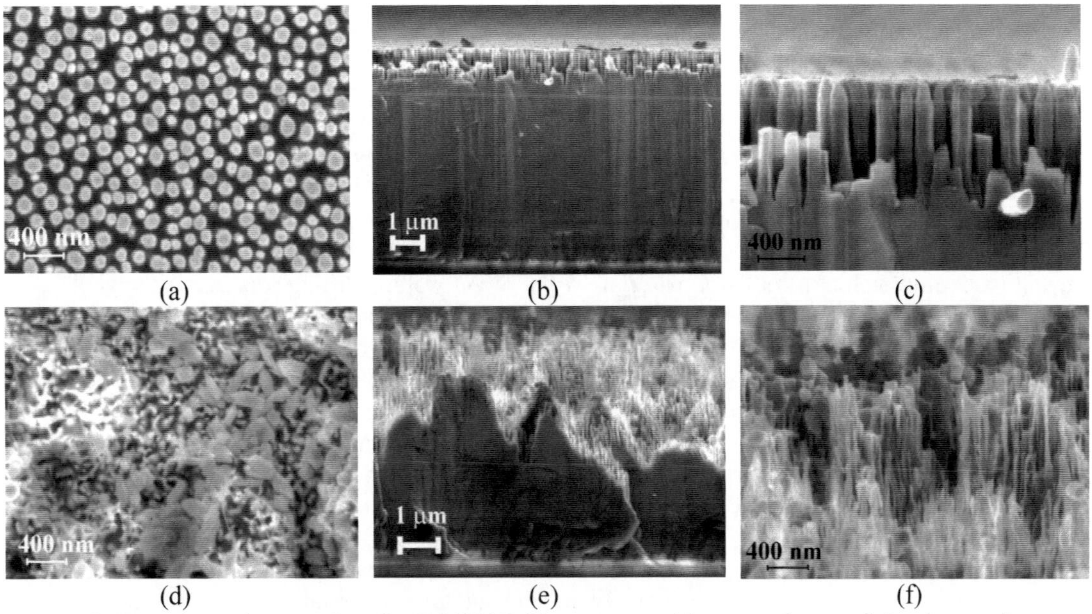

Figure 2. Scanning micrographs of a GaN/AlGaN structure with nanocolumn: plain view and cross-section view before (a-c) and after (d-f) PEC process. The nanocolumn density is near 7×10^8 cm^{-2}. Corrosion of the nanocolumn structures is clearly seen.

nanocolumns. SEM plan-view images (figure 2 (a) and (d)) show that p-type top of nanocolumns was not etched and fell down between the columns. An investigation of cleavage section of the sample shows that the corrosion process was heterogeneous and the sample surface was covered by thinner columns of a 10-20 nm diameter and a 200-400 nm height. An approximately half of the nanocolumns volume was etched away during 1-hour operation of the PEC cell.

Obviously, the observed photoelectrochemical corrosion (etching) follows paths of the photo-current flow. The current flow paths in III-N GaN/AlGaN or GaN/InGaN materials, especially at low current, are non-uniform and can be associated with shunt conductivity in an extended defect system of grain boundaries, threading dislocations, local regions with irregular alloy composition enriched by metallic atoms (Ga or In), and other extended defects [5]. Note that the nanocolumn density is 7×10^8 cm^{-2} that is close to the threading defects density in the initial structure ($3\text{-}5\times10^8$ cm^{-2}) estimated by X-ray diffraction measurements. It means that near 40-60% of nanocolumns may have a threading defect that may act as a leakage path for the photocurrent promotion the structure corrosion.

In addition, dry etching at the nanocolumns processing produces radiation damage defects in the near surface area of nanopillars. These defects consist of donor-like centers, such as nitrogen vacancies, gallium interstitials, and acceptor defects including nitrogen vacancy complexes, interstitial nitrogen, gallium vacancies and their complexes. The net result depends on the dry etching conditions and post-etching annealing. The dry etching defects in nanocolumns create another path for the photocurrent leakage that is favorable for the electrochemical etching but not for controllable water splitting and H_2 generation.

It is know that spontaneous and piezoelectric polarization between AlGaN/GaN layers in the structure results in formation of a net positive charge at their interface [6]. The GaN/AlGaN structures in this work had a stack of doped GaN and AlGaN layers that also influences the net positive charge in the structure. In addition, a thin n-AGaN layers in the structure are not fully relaxed and may have additional positive piezoelectric polarization charge at the n-AlGaN /AlGaN layer interfaces.

An additional positive charge at the layer interfaces can be associated with positively charged ionized donors, which are Si^+ presumably, in a space charge region of the p (AlGaN)-n (AlGaN) junction. This positive charge is located in the n-AlGaN active region too. Note that negatively charged ionized acceptors (Mg^-) in the p-GaN and p-AlGaN layer may screen positive polarization charge at the p-AlGaN barrier layer/n-AlGaN active layer interface.

The net positive charge at the GaN/AlGaN n-n and p-n interfaces causes OH^- ions from the electrolyte to form $Ga(OH)_3$ and $Al(OH)_3$ or their soluble forms [3, 7] at the nitrogen vacancy sites and to continue the etching process along the positive charge sites resulting in narrowing of the nanocolumns.

H_2 evolution was observed at the Pt electrode. However no O_2 was observed at the working electrode. Obviously, a sacrificial agent (like HBr) or passivation of the surface by nitrogen atoms are needed to prevent oxidation (etching) the III-N materials at the direct water splitting process. As a whole, GaN based structures grown on c-plane sapphire have Ga-terminated surfaces, as a result one to be easily oxidized to Ga_2O_3, which may even dissolve in acidic conditions, thereby leading to instability and degradation of the PEC cells. On the other hand the reverse crystal polarization (N-terminated) could enhances an efficient of separate and transport of electron-hole pairs to the photoelectrode/electrolyte interface so improving the direct water splitting process [2]. N-terminated surfaces protect the GaN-based nanowire surfaces against attack by the electrolyte (oxidation and photocorrosion) and thus, the engineered optimum surface band bending can be maintained during the photocatalytic overall solar water splitting reaction.

Figure 3 shows changing of a photocurrent at the electrochemical process under irradiation by Xe lamp. External bias of +0.5 V was applied. Hydrogen evaluation reaction was observed on Pt counter electrode. The H_2 production rate of 0.56 ml/cm^2 × h was measured for GaN/AlGaN structures with nanocolumns fabricated on the top in KOH electrolyte under the Xe-lamp illumination. The nanocolumn GaN/AlGaN p-n structures worked as a sacrificial anode. The nanocolumns size decreased remarkably, as shown SEM images. Accordingly, a decrease of the photo-current was observed too during the experiment. At the end of the experiment (after 1 hour), the photocurrent decreased almost twice. We believe it is due to the electrochemical corrosion of nanocolumns. A small amount of gas was released on the GaN/AlGaN photo-anode. This gas can be O_2 and/or N_2, taking into account the corrosion of GaN/AlGaN structure (as shown in figure 2 (d-f)).

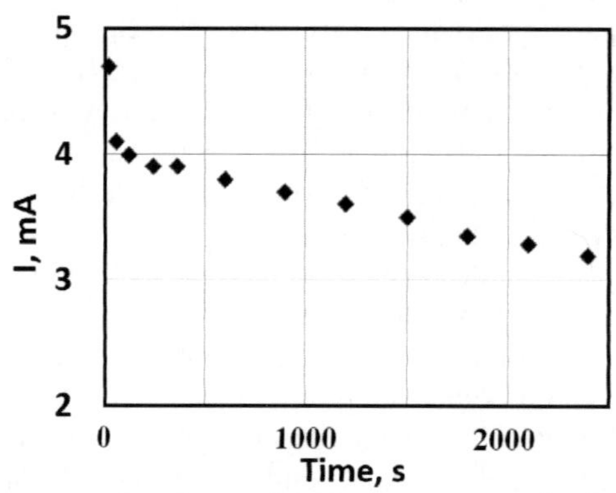

Figure 3. A photocurrent at the electrochemical process under irradiation by Xe lamp

4. Conclusions

Corrosion of the p-n GaN/AlGaN nanocolumn structures at the electrochemical process in a KOH aqua electrolyte under the Xe-lamp illumination penetrates deep into structure and results in much thinner nanocolumns. The top portions of the nanocolumns that had p-type conductivity were not etched and simply fall dawn from the top. H_2 evolution was observed at the Pt electrode at an external bias of +0.5 V. The H_2 production rate of 0.56 ml/cm^2 × h was measured. However, only a few gas bubblers generated on the working electrode. Obviously, a sacrificial agent (like HBr) or passivation of the surface by nitrogen atoms are needed to prevent oxidation (etching) the III-N materials at the direct water splitting process.

Acknowledgments
The authors thank to Dr A. Y Polaykov and Prof. I.H. Lee for help in the nanocolumn fabrication. Work at University ITMO was financially supported by the Ministry of Education and Science of Russian Federation (grant agreement 14.575.21.0054, unique identifier of research activities is RFMEFI57514X0054).

References
[1] Ohkawa K, Ohara W, Uchida D, Deura M 2013 *Jpn. J. Appl. Phys.* **52** 08JH04
[2] Kibria M G, Qiao R, Yang W, Boukahil I, Kong X, Chowdhury F A, Trudeau M L, Ji W, Guo H, Himpsel F J, Vayssieres L, and Mi Z 2016 *Adv. Mater.* **28** 8388–8397
[3] Usikov A, Helava H, Nikiforov A, Puzyk M, Papchenko B and Makarov Y 2016 *American J. of Appl. Sci.* **13** (7) 845-852
[4] Polyakov A Y, Jeon D W, Smirnov N B, Govorkov A V, Kozhukhova E A, Yakimov E B, Lee I H 2012 *J. Appl. Phys.* **112** 073112
[5] Shmidt N, Shabunina E, Usikov A, Chernyakov A, Kurin S, Helava H and Makarov Y 2015 *Phys. Status Solidi C* **12** 1136.
[6] Mishra U K, Parikh P and Wu Y F 2002 *Proc. IEEE* **90** 1022.
[7] Zubenko T K, Puzyk M V, Stozharov V M, Ermakov I A, Kovalev D S, Ivanova S A, Usikov A S, Papchenko B P, Medvedev O S, Kurin S Y, Antipov A A and Chernyakov A E 2016 *J. of Phys.: Conf. Ser.* **741** (1) 012050

IOP Conf. Series: Journal of Physics: Conf. Series **917** (2017) 052023 doi:10.1088/1742-6596/917/5/052023

Layer-by-layer deposition of nanostructured $CsPbBr_3$ perovskite thin films

A A Reshetnikova, L B Matyushkin, A A Andronov, V S Sokolov, O A Aleksandrova, V A Moshnikov
Department of micro- and nanoelectronics, St. Petersburg Electrotechnical University "LETI", St. Petersburg 197376, Russia

Abstract. Layer-by-layer deposition of nanostructured perovskites cesium lead halide thin films is described. The method of deposition is based on alternate immersion of the substrate in the precursor solutions or colloidal solution of nanocrystals and methyl acetate/lead nitrate solution using the device for deposition of films by SILAR and dip-coating techniques. An example of obtaining a photosensitive structure based on nanostructures of ZnO nanowires and layers of $CsBbBr_3$ nanocrystals is also shown.

1. Introduction

The development of solar cells with an active layer of a thin perovskite film became one of the most promising technology for the solar energy application in the past several years [1]. Perovskites (general chemical formula ABX_3) allow increasing the power conversion efficiencies larger than 20% [1, 2]. The technology of perovskite thin film synthesis is relatively simple: a standard technique is the spin-coating of a solution containing A, B, X elements, then desiccation in a drying cabinet. For example, $CH_3NH_3PbI_3$ films usually sensitized from CH_3NH_3I and PbI_2 precursors soluble in the polar fluids, such as dimethylformamide or dimethyl sulfoxide.

However, the thermal and chemical instability of the hybrid (organic-inorganic) perovskites leads to a search for similar structure and efficiency, but more suitable materials. In the perovskite materials group, it can be considered a family of lead cesium halides $CsPbX_3$ (X = Cl, Br, I). The use of cesium Cs instead or in addition to methylammonium group CH_3NH_3 allows obtaining dramatically more stable structures [3].

Materials of this group in the form of nanocrystals (NCs) and nanorods also have an effective photo- and electroluminescence [4]. The halogen type defines the band gap, and the luminescence spectra position: $CsPbCl_3$, $CsPbBr_3$ and $CsPbI_3$ NCs emit in the blue, green and red regions respectively. So, in the case of $CsPbX_3$ materials, it is possible to achieve fine tuning of absorption edge and luminescence maximum in the whole visible spectral range as a result of the selection of $CsPbX_3$ solid solution composition or nanocrystals size [5]. It is noteworthy that $CsPbX_3$ perovskite NCs have the cube shape, which facilitates better carrier transport conditions compared with commercial CdSe/ZnS QDs having a spherical shape.

Thus, using $CsPbX_3$ system one can create light emitting devices and solar cells contemporaneously. In the case of creating solar cells, it is possible to use FTO or ITO as transparent conductive oxide, TiO_2 or ZnO as electron transport layer; $CsPbX_3$ as perovskite layer ($CsPbI_3$ in the case of solar cells), CuO or NiO as hole transport layer, and Au or Ag as a back electrode.

The above-described spin-coating method's hard to scale and leads to material losses during the deposition process. In this work, we proposed solution-based perovskite deposition technique by immersion of substrates in the precursor solutions or colloidal solutions of NCs using the device used for SILAR technique deposition of films [6, 7].

2. Description of the method

The device for the deposition of thin films is represented in Figure 1. It has horizontal and vertical actuators transferring the substrate holder respect from one thermostable beaker with a solution to another by predetermined synthesis parameters. The solutions are heated to temperatures controlled by detectors in chemically inert shells of stainless steel. The whole device is operated by Arduino controller connected to a computer with a program designed in NI LabVIEW environment using LabVIEW Interface for Arduino. An operator sets in the program the technological parameters of thin film deposition: the number of cycles, volume, and temperature of precursors, the dimensions of the substrate, the delay times between operations, etc. [7].

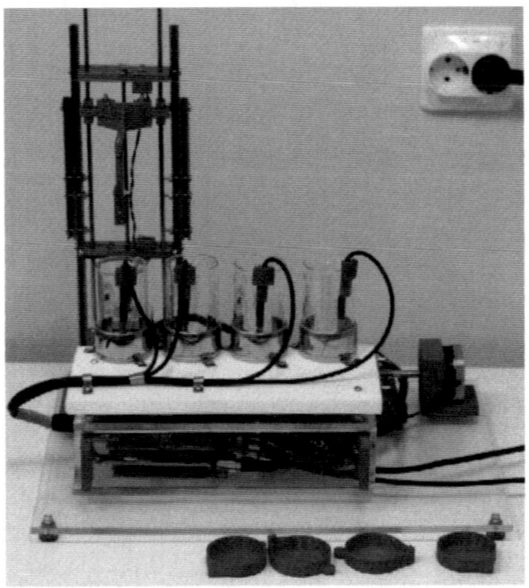

Figure 1. The device for the deposition of thin films

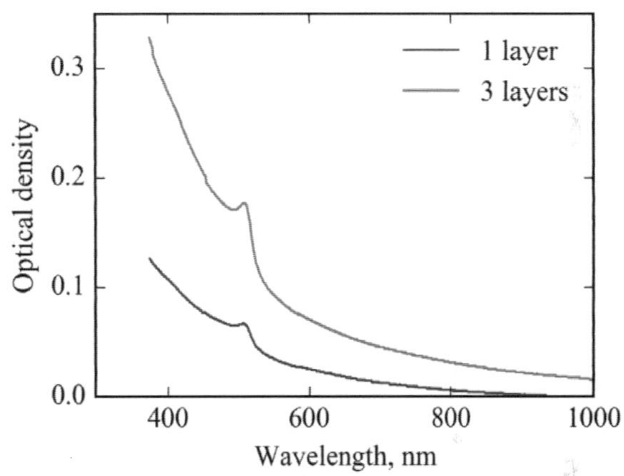

Figure 2. Optical density spectra of thin films after 1 and three cycles of layering of $CsPbBr_3$ nanocrystals

Colloidal crystals of $CsPbBr_3$ were synthesized by the method of work [8], purified and dispersed in toluene. The surface of nanocrystals during the synthesis was covered with molecules of oleic acid and oleylamine. Although in solution these ligands provide such valuable properties as sedimentation resistance and passivation of surface states, the presence of these ligands is critical for precipitation. The general problem with the deposition of such colloidal NCs from solutions by dip-coating is that the following dipping results in partial or full dissolution of the previous layer due to colloidal nature of nanocrystals. To remove ligands, we used methyl acetate as antisolvent that removes excess of ligands [9] and results in deposition of material. To passivate the surface states due to the lead vacancies remaining when some of the ligands are removed, a small amount (<1 wt. %) of lead nitrate was added to the methyl acetate solution.

Thus, the procedure for depositing nanocrystals consists of successively immersing the substrate in a solution of nanocrystals and a solution of lead nitrate in methyl acetate. The number of repetitions determines the thickness of the resulting layer. To assess the layer deposition procedure, a cover slip was used as a test sample. As can be seen from Figure 2, a three-fold repetition of the procedure leads to a proportional increase in the optical density, which is commensurate with the thickness of the film. The spectrum feature near 530 nm corresponds to the first exciton peak of nanocrystals.

When the films were placed under an ultraviolet lamp, it was found that the deposition procedure leads to a significant decrease in the luminescence intensity in comparison with nanocrystals in solution or films obtained without ligands etching. However, the material still represented an interest in light absorbing.

3. Results and discussion

One of the well-known problems for hybrid perovskites is rapid structural degradation of the perovskite/zinc oxide contact. While in the case of titanium oxide such problem is not observed, zinc oxide deprotonates the organic component of hybrid perovskite such as $CH_3NH_3PbI_3$ [11]. Obviously, in the case of the $CsPbX_3$ system, deprotonation is excluded. Therefore, to test the proposed procedure, we deposited a perovskite $CsPbBr_3$ layer of nanocrystals on ZnO nanorods obtained by the hydrothermal method [12] on the surface of a compact ZnO layer deposited by ultrasonic spray pyrolysis. Contacts to the structure represent a transparent electrode of tin oxide doped with fluorine (FTO) and pinch gold contacts, used for the measurements. The corresponding structure is shown in Figure 3.

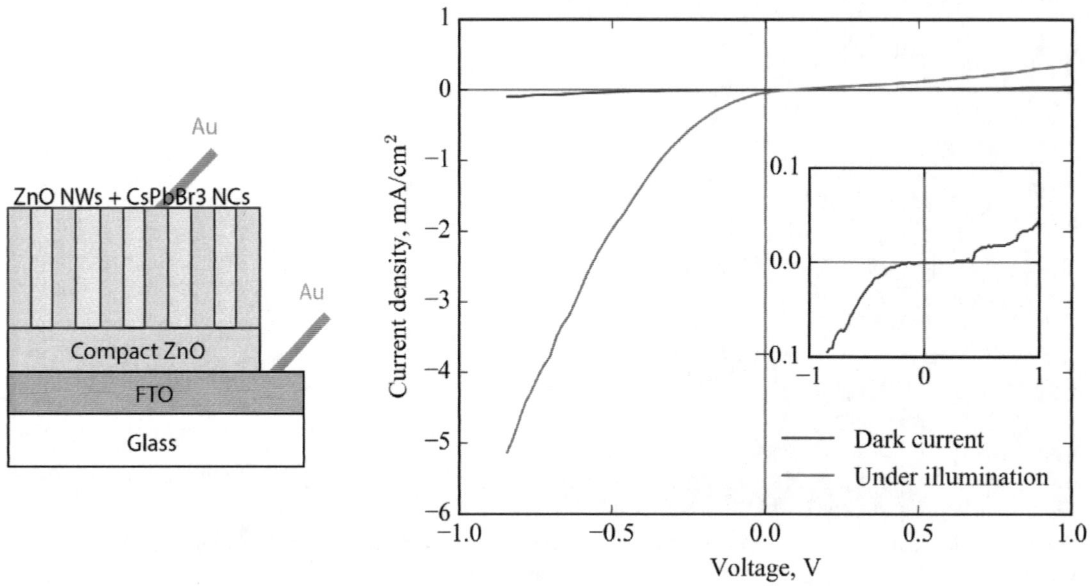

Figure 3. FTO/ZnO/$CsPbBr_3$/Au structure and corresponding current–voltage characteristic

The nanowires obtained by this method are fixed perpendicular to the substrate and have an extended structure (~um). In fact, when measuring such structures, we investigated the effect of perovskite nanocrystals deposited on the surface of charge carriers inside zinc oxide rods. The corresponding current-voltage characteristics in the dark and under illumination are shown in Figure 3. It can be seen that when illuminated, the current through the structure increases approximately 50 times. Since the remaining materials of the structure are wide-gap and do not exhibit a strong photoresponse, it can be concluded that such a significant current change is due to charge injection from the perovskite layer to ZnO NWs. The effect is reversible, and no degradation of the sample was observed during the measurements.

4. Conclusion

Thus, a method for creating perovskite layers from colloidal solutions has been proposed, and the possibility of its application in the composition of a photosensitive structure based on a perovskite/zinc oxide contact is demonstrated., The same device is applicable for the creation of electron and hole transport layers [12]. Thus, this technique can be used for the synthesis of different layers of the final structure.

IOP Conf. Series: Journal of Physics: Conf. Series **917** (2017) 052023 doi:10.1088/1742-6596/917/5/052023

Acknowledgments
The authors thank Andrei Ryabko for providing a sample of zinc oxide for creating the described structure.

References
[1] Protesescu L et al. 2015 *Nano Lett.* **15** 3692
[2] Park, N G 2013 *J. Phys. Chem. Lett.* **4** 2423
[3] Kulbak M, Cahen D, Hodes G 2015 *J. Phys. Chem. Lett*. **6** 2452
[4] Nedelcu G, et al. 2015 *Nano Lett.* **15** 5635
[5] Akkerman Q et al. 2015 *J. Am. Chem. Soc.* **137** 10276
[6] Aleksandrova O, Matyushkin L, Hondryukov D 2015 Patent Rus, no. 2015114951
[7] Andronov A O, Matyushkin L B, Hondryukov D V 2016 *Young Researchers in Electrical and Electronic Engineering Conference (EIConRusNW), 2016 IEEE NW Russia* pp 19–22
[8] Protesescu, L et al. 2015 *Nano Letters* **15**(6) 3692
[9] Swarnkar A et al. 2016 *Science* **354(**6308) 92
[10] Yang J et al. 2015 *Chem. Mat.* **27**(12) 4229
[11] Lashkova N A et al. 2016 *Semiconductors* **50**(9) 1254
[12] Matyushkin L, Reshetnikova A et al. 2017 *Semiconductors* **5** 622

IOP Conf. Series: Journal of Physics: Conf. Series **917** (2017) 052024 doi:10.1088/1742-6596/917/5/052024

Photoelectric diagnostics of InGaN-based LEDs in static and dynamic modes

O A Radaev[1,2], I V Frolov[1], V A Sergeev[1,2]
[1] Ulyanovsk Branch of Kotel'nikov Institute of Radio-Engineering and Electronics of Russian Academy of Sciences, Ulyanovsk, 432071, Russia
[2] Ulyanovsk State Technical University, Ulyanovsk, Russia

Abstract. The method and measuring installation for diagnostics of light-emitting heterostructures (HS) with the quantum well (QW) by registration of the photocurrent arising in case of radiation of HS the narrow-band optical radiation of different wavelength in the static and dynamic modes are described. On the example of commercial InGaN-based LEDs it is shown that offset of a maximum of photocurrent spectrum concerning a maximum of electroluminescence spectrum is defined by QW depth. The steepness of photocurrent spectrum on the section corresponding to absorption in QW is defined by structural perfection of heteroboundaries. In case of harmonic modulation of a luminous flux on all probed wavelengths amplitude of a photocurrent of blue LEDs monotonically falls down with growth of modulation frequency with the steepness 0,2 dB for a decade, and at green LEDs to the frequency of 200 kHz the plateau and then recession with the steepness 8 dB for a decade is watched.

1. Introduction

For the non-destructive diagnostic of the InGaN-based LEDs with quantum well (QW) developed various methods, including photoelectric, based on the measurement of parameters and characteristics of the LED during irradiation with monochromatic radiation [1, 2]. The purpose of this operation to show communication of amplitude and frequency responses of the photocurrent arising in an external circuit of a LED in case of radiation by its narrow-band optical radiation with parameters and quality of LED HS.

If energy of the falling photons is more than width E_g^{QW} of the forbidden band of material QW, but it is less than width E_g^{GaN} of the forbidden band of GaN-barriers, then light absorption and generation of excess charge carriers happens only in QW (figure 1a). At the same time the charge carriers generated by light by means of thermal burst, a tunnel effect and photoexcitation [3] can pass from QW into barrier areas and move freely in the direction, perpendicular to the interface (figure 1b), as leads to appearance of a photocurrent in case of switching on of the LED in a closed circuit. The excess charge carriers arising when lighting create a photocurrent if it is generated in the space-charge region (SCR) or at distance of no more diffusion length from its boundaries. Level of a photocurrent is defined by the speed of activation of charge carriers, that is speed of their burst from QW: where η – share of the activated charge carriers forming photocurrent. Speed of activation is defined by different mechanisms and can be found (is estimated) from an equilibrium condition of processes of generation and disappearance of charge carriers at QW:

$$\frac{dn}{dt} = G - R - A, \quad (1)$$

where $G = kW\beta\Phi$ – the speed of photogeneration of charge carriers in QW, k – a radiation absorption coefficient in material QW, W- thickness QW, β – quantum yield, F – density of the luminous flux density falling on QW, R – the full speed of a recombination representing the amount of speeds of a radiant and nonradiative recombination.

Precisely it isn't possible to describe all processes happening in HS in case of its radiation by light, but in the stationary mode ($dn/dt = 0$) from (1) follows that $A = G - R$, and for qualitative analysis it is possible to use expression:

$$I_{\Phi} = \eta(G - R). \quad (2)$$

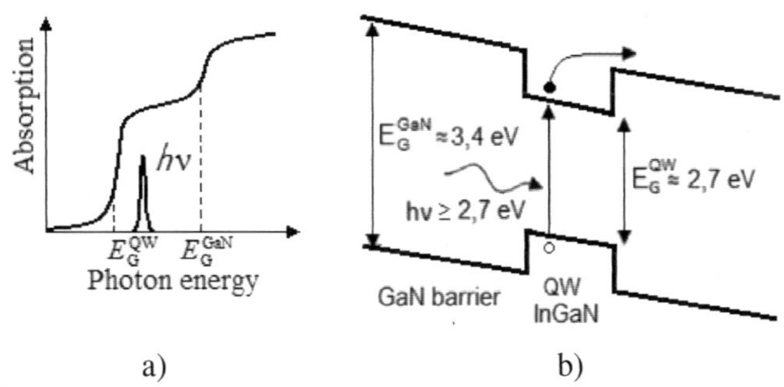

Figure 1. Generation of a photocurrent in heterostructure with a quantum hole (schematically).

When training structure radiation with the energy of photons exceeding width of the forbidden band of a barrier layer, optical radiation is absorbed both in QW, and in barrier layers (figure 1a), and the equilibrium condition will have more difficult appearance, than (1). With a further growth of energy of photons the main part of radiation will be absorbed near a surface of structure, at the distance considerably exceeding the diffusion length of charge carriers, and the photocurrent will decrease.

In the dynamic mode in case of switching on and switching off of a flare of HS LED the photocurrent will rise and fall down not instantaneous, and with growth of modulation frequency of a luminous flux the amplitude of a photocurrent will decrease. The steepness of this recession is defined by effective times of a recombination and activation of charge carriers in QW HS. Qualitative analysis shows that constants of time of rise and recession of a photocurrent shall differ, and on a waveform of a relaxation it is possible to judge the speed of the specified processes.

2. Experimental setup for the study of photocurrent spectra

Measurement of a photocurrent and photovoltage of LEDs was carried out in case of radiation them by narrow-band radiation, the close to monochromatic, created on an output of the spectrophotometer SF-46 with a width of output slot of 6,5 nm, in the range of lengths of waves 350…530 nm and the range of flux density of radiation 0,03…5000 uW/cm^2 (figure 2). Photocurrent spectrum of LEDs was calculated by normalizing the measured spectrum to the spectrum of the radiation source. The central wavelength and width of narrow-band radiation was in addition controlled by OceanOptics USB2000 spectrometer. Emissive power was measured by the integrating sphere of TKA-KK1. As a source of radiation the 100 W glow lamp was used. A photocurrent and photovoltage depending on the diagram of switching on of the researched LED the registered the universal B7-21A voltmeter. The electroluminescence spectrum of the LEDs was measured by the Ocean Optics USB2000 spectrometer.

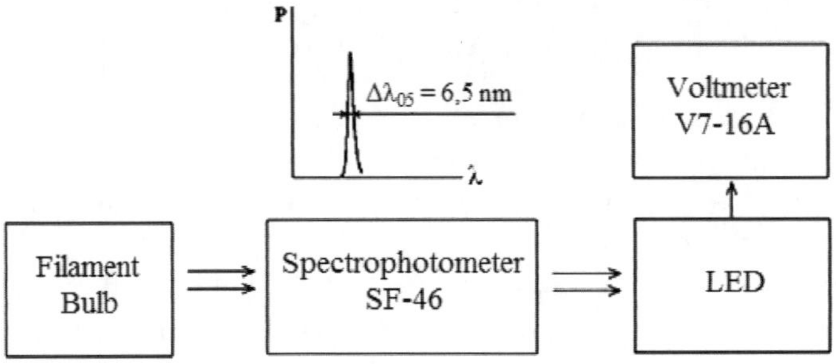

Figure 2. Structural diagram of the apparatus for studying the photocurrent spectra.

3. Measurement and analysis of a photocurrent spectrum of LEDs

We investigated commercial InGaN-based LEDs of two types: blue LEDS C503B-BAN-GY0461 produced by Cree (central wavelength of the EL spectrum 470 nm) and green LEDs ARL-5213 PGC produced by Arlight (central wavelength of the EL spectrum 525 nm).

Figure 3 shows the PC spectrum and EL spectrum of blue LEDs measured at a current of 1 mA. The maximum of photocurrent spectra corresponds to energy of the falling radiation 2,92 eV, and a LED electroluminescence spectrum maximum – 2,64 eV. Stokes's shift characterizing losses in structure makes 0,28 eV. From figure 3 it is visible that the photocurrent spectra plotted in half-logarithmic scale has several line (exponential) sections of different steepness, and, the more steep the spectral characteristic of a photocurrent, the is higher homogeneity of HS [4]. The most steep section is watched in the range from 2,6 eV to 2,68 eV that testifies to homogeneity of the interface of heteroboundaries of QW. Fractures on a photocurrent spectra demonstrate change of a contribution of different energy levels in QW in the general photocurrent. In case of the radiant energy corresponding to an electroluminescence spectra maximum, the photocurrent decreases more than much, and in case of energy less than 2,58 eV photogeneration of charge carriers almost completely stops.

Figure 4 shows the PC spectrum and EL spectrum of green LEDs. Stokes's shift makes 0,56 eV. The left wing of a spectral characteristic in the range of energies 2,35…2,5 eV confirms losses on band-gap absorption of radiation in structure. The steepness of recession of the right wing of a spectral characteristic characterizes parameters of "tails" of frequency curve of statuses in the forbidden band and is evaluated by E_0 parameter value. The low level of orderliness of structure is confirmed by rather great values of the E_0 parameter.

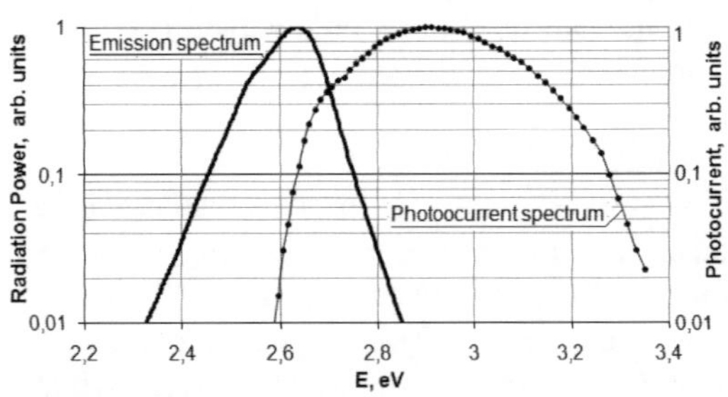

Figure 3. Normalized photocurrent and electroluminescence spectrums of blue LEDs C503B-BAN-GY0461.

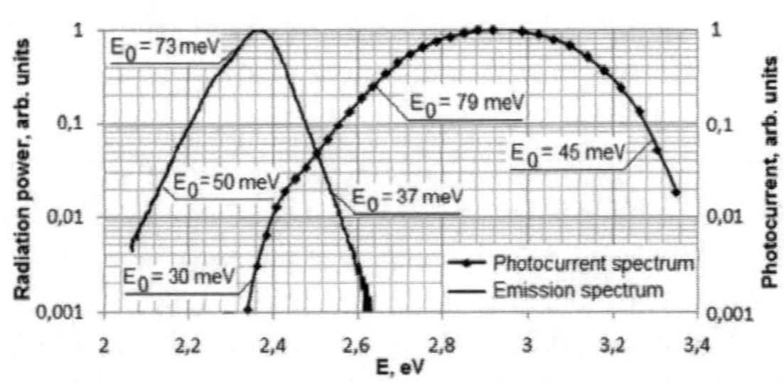

Figure 4. Normalized photocurrent and electroluminescence spectrums of green LEDs ARL-5213 PGC.

As a result of measurement of a photocurrent and photovoltage of the LED in case of different flux density of the falling radiation it is set that value of a photocurrent of the LED linearly depends on flux density of the falling radiation within the probed range of intensity (figure 5). The photovoltage has the linear dependence from the level of the falling radiation in the range of small flux densities of radiation and experiences saturation in the field of big flux densities in case of 1,9...2,1 Century. The same tension in case of live broadcast of a LED corresponds to threshold voltage of the beginning of a glow of a LED.

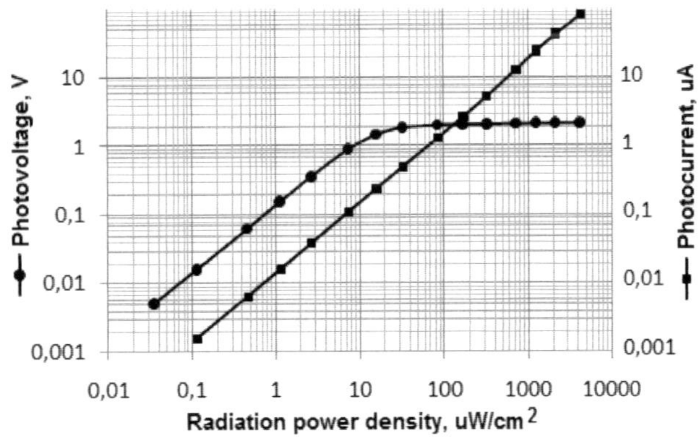

Figure 5. Photocurrent and photovoltage vs radiation power density curves measured on green LED when irradiated with light at λ = 470 nm.

4. Measurement and analysis of dynamic characteristics of a photocurrent

Figure 6 shows the photocurrent frequency dependences of the blue LED when irradiated with alternating luminous flux in the frequency range from 170 Hz to 7 MHz. Two LEDs with central wavelengths of 410 nm (UV) and 470 nm (blue) used as the source of variable optical signal. Register photocurrent was carried out on the collector resistor of 25 Ω. The alternating LED photocurrent was determined by measurement of alternating voltage on collector resistor by selective nanovoltmeter. The amplitude of the alternating optical signal is maintained constant over the entire frequency range. The measuring unit bandwidth is 100 MHz. The level of the photocurrent in the case of illumination of the blue LED is about 6 times less than when illuminated with UV LED, at the same time, in both cases the form of the frequency dependence of the photocurrent is almost identical.

For comparing the frequency dependences of a photocurrent of green and blue LEDs of the specified type are given in figure 7. In case of harmonic modulation of a luminous flux on all probed lengths of waves amplitude of a photocurrent of blue LEDs monotonically falls down with growth of modulation frequency with the steepness 0,2 dB for a decade, and at green LEDs to the frequency of 200 kHz the plateau and then sharp recession with the steepness 8 dB for a decade is watched.

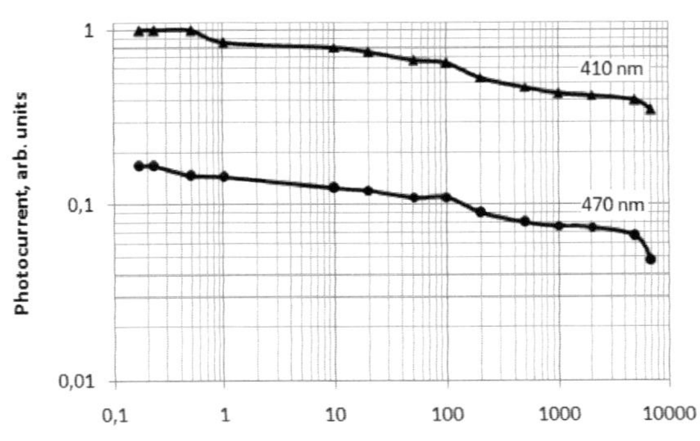

Figure 6. Dependences of the photocurrent blue LED on the frequency of the incident radiation.

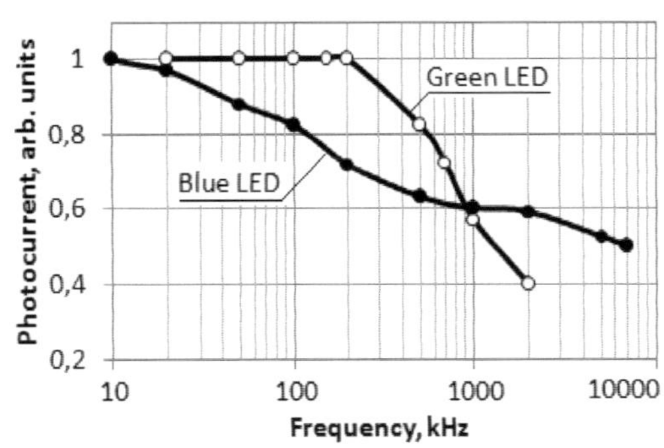

Figure 7. Dependences of the photocurrent of green and blue LEDs on the frequency of the incident radiation with λ = 470 nm.

5. Conclusions

1) Photocurrent level in case of photoexcitation of light-emitting HS low-energy photons decides on QW by the sizes and zonal structure of QW.

2) Value of offset of a maximum of photocurrent spectra in relation to electroluminescence spectra of InGaN/GaN of LEDs is defined by QW depth.

3) Character of photocurrent spectra and in particular the steepness of dependence of value of a photocurrent on energy of photons on different sections of spectrum characterizes homogeneity of HS. The steepness of recession of a photocurrent of blue InGaN/GaN of the LED in the range 460…475 nm is defined by homogeneity of the interface of heteroboundary of QW.

4) In case of harmonic modulation of the falling radiant flux the level of a photocurrent of blue InGaN/GaN of the LED monotonically falls down with the steepness 0,2 dB for a decade of frequency, and at green InGaN/GaN of the LED to the frequency of 200 kHz the plateau, and at frequencies more than 200 kHz recession with the steepness 8 dB for a decade is watched.

Acknowledgments

The reported study was funded by RFBR and Government of Ulyanovsk Region according to research project № 16-47-732159.

References

[1] Baranovskiy M V, Glinskii G F, Mironova M S 2013 *Semiconductors* **47** 58.
[2] Monti D, Meneghini M, De Santi C et. al. 2017 *Proc. Of SPIE* **10124** 10124OT-1
[3] De Santi C, Meneghini M, la Grassa M. et. al. 2016 *Journal of Applied Physics* **119.** 094501
[4] Sergeev V A, Frolov I V, Shirokov A A, Radaev O A 2017 *Nitrides of gallium, indium and aluminum: structures and devices* (Moscow State University) p. 92

Investigation of electrical and optical properties of low temperature titanium nitride grown by rf-magnetron sputtering

D Sosnin[1], D Kudryashov[1], A Mozharov[1]

[1]renewable energy sources Lab, St. Petersburg Academic University, St. Petersburg 194021, Russia

Abstract. Titanium nitride is a promising material due to its low resistivity, high hardness and chemical inertness. Titanium nitride (TiN) can be applied as an ohmic contact for n-GaN and rectifying contact for p-GaN and also as a part of perovskite solar cell. A technology of TiN low temperature reactive rf-magnetron sputtering has been developed. Electrical and optical properties of titanium nitride were studied as a function of the rf-power and gas mixture composition. Reflectance and transmittance spectra were measured. Cross-section and surface SEM image were obtained. 250 nm thin films of TiN with a resistivity of 23.6 µOm cm were obtained by rf-magnetron sputtering at low temperature.

1. Introduction

Titanium nitride is a promising material due to its low resistivity, high hardness and chemical inertness. Titanium nitride is used in optical filters, thin film resistors, protective and decorative coatings. Titanium nitride can be applied as an ohmic contact for n-GaN and rectifying contact for p-GaN and also as a part of perovskite solar cell.

Most scientific articles reports dc-magnetron sputtering at 300-400 ^{0}C [3-5] or rf-reactive magnetron sputtering in 100% nitrogen atmosphere [6]. Heating up to 600 ^{0}C is used for obtain good stoichiometry of TiN. In this work we have reported low temperature rf-magnetron sputtering of titanium nitride.

2. Experimental details

TiN thin layers are generally deposited by rf-magnetron sputtering on glass and silicon substrates. Before the deposition – vacuum chamber was pumped off to a pressure of $6\cdot10^{-6}$ mBar. Operating temperature in the chamber was 40-50 ^{0}C. N_2 and Ar were used as working gases. Optical reflectance and transmittance spectra were measured by spectrometer based on a SOLAR LS M266 monochromator. The morphology of the films was studied by SEM.

3. Results and Discussions

Reactive sputtering of titanium nitride means the use of metal target as the supplier of titanium and the mixture of reactive and working gases. TiN growth conditions require N_2/Ar gas flow ratio of less than 1. The nitrogen flow should not be too small, otherwise the titanium will settle on substrate as nitrogen will not be sufficient for a reaction with titanium. If nitrogen flow is too much, it also becomes a working gas and only titanium growth begins on substrate.

The growth rate of TiN reactive sputtered film linearly increases with increase of rf-magnetron power. At the same time, the relationship between sputtering rate and N_2/Ar ratio (figure 1) or chamber pressure (figure 2) nonlinear. Thus a high concentration of gas molecules prevents sputtered particles of titanium, nitrogen and titanium nitride from settling to the surface of the sample. The growth rate of TiN on Si may reach maximum saturation at low N_2 concentration [4]. In this study, it was not observed.

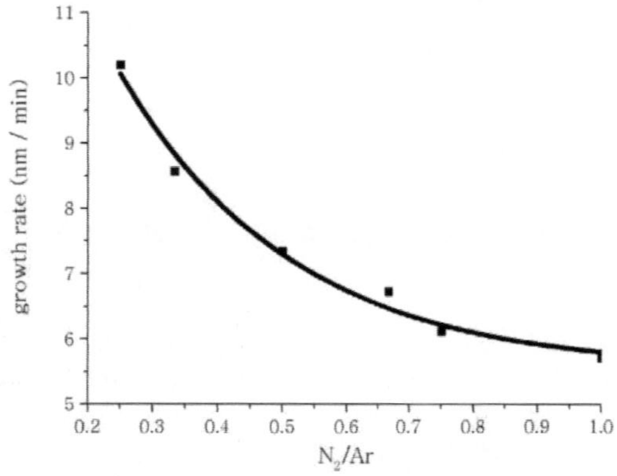

Figure 1. The relationship between TiN deposition rate on Si and N_2/Ar ratio, rf-power of 400 W and $2 \cdot 10^{-3}$ mBar gas pressure.

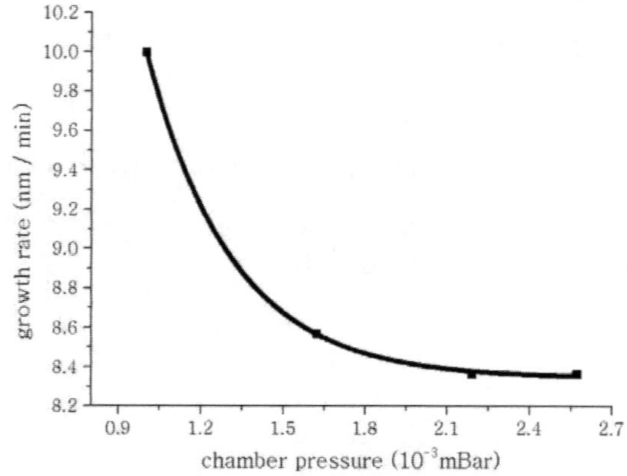

Figure 2. The relationship between TiN deposition rate on Si and chamber pressure at rf-power of 400 W and N_2/Ar = 1/3 ratio.

SEM images of 300 nm TiN film sputtered on Si substrate show nano-crystal structure. All produced films with different thickness demonstrate smooth surface morphology (figure 3 and figure 4). The resulting films have flat surface with a metallic luster. TiN stoichiometry depends on N_2 concentration [4], therefore there is a difference in the structure of the two samples. Morphological studies showed the presence in films of nanoscale structures with a particle size of 16-38 nm.

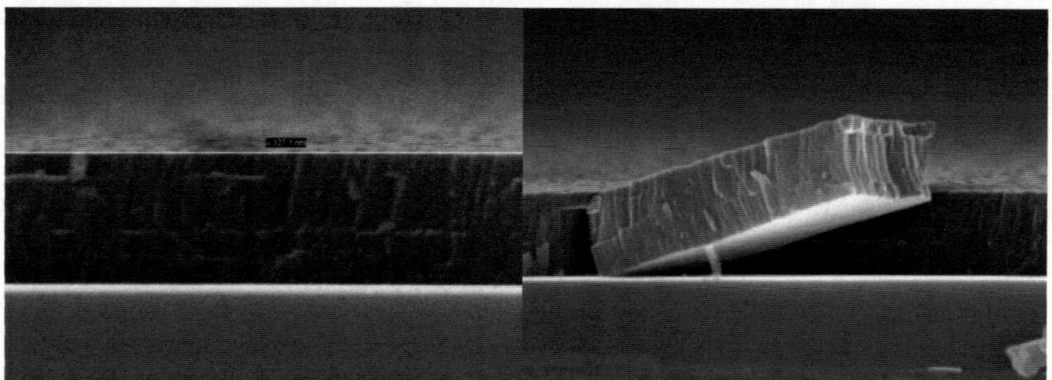

Figure 3. SEM images of 300 nm TiN film on Si sputtered at rf-power of 400 W, 1/3 N_2/Ar gas ratio.

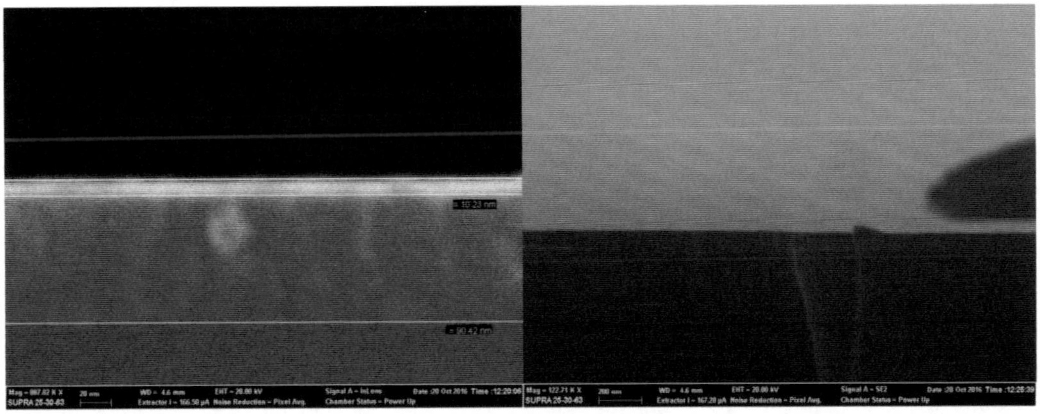

Figure 4. SEM images of 90 nm TiN film on Si sputtered at rf-power of 300 W, 3/4 N_2/Ar gas ratio.

Transmittance and reflectance spectra of produced films are shown on figure 5. These data well correlate with the other references [1,2]. Films exhibit a low reflection band in 350–600 nm. At the same time, no transmittance in visibe range were observed. TiN films demonstrate golden colored mirror surface.

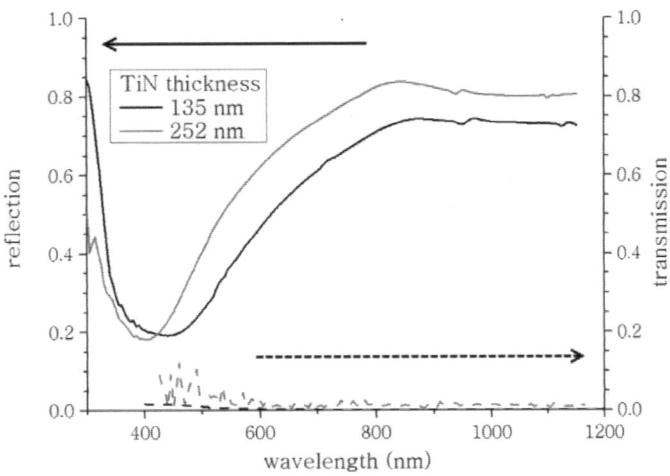

Figure 5. Transmittance and reflectance spectra of TiN with different films thickness

Resistivity of TiN films depends on the growth conditions. It's shown in table. Better resistivity was obtained with rf-power of 400 W N_2/Ar gas flow ratio 0.33. 252 nm thick film has a resistivity 23.6 μOm cm that corresponds to the thicker TiN coatings [3]. First of all the resistivity depends on stoichiometry, which differ for various N_2/Ar ratio value [4], and on film thickness [5]. Actually, resistivity of 100-120 nm thin film 4 times more than 250-300 nm TiN film. As for the stoichiometric coefficient, a better resistance is achieved for N:Ti ratio of 0.98.

Table. Comparison of resistivity of TiN films prepared on Si substrate under different sputtering condition.

Rf-power, W	Thickness, nm	N_2/Ar	Growth rate, nm/min	Resistivity, μOm·cm
300	95	0.75	3.2	108.54
300	153	0.5	3.8	136.12
300	230	0.25	5.1	33.01
400	252	0.33	8.4	23.61

4. Conclusion

In this work we have reported low temperature rf-magnetron sputtering of titanium nitride. As a result 250 nm thin films of TiN with a resistivity of 23.6 μOm·cm and smooth surface morphology were obtained.

References

[1] Guler U, Kildishev A, Boltassevaab A and Shalaev V 2015 *Faraday Discuss.* **178**, 71-86
[2] Newport A, Carmalt C, Parkin I and O'Neill S 2002 *J. Mater. Chem.,* **12**, 1906–1909
[3] Gagnon G, Currie J, Beique C, Brebner J, Gujrathi S, Onllet L. 1994 *J. Appl. Phys*., **75**, 1565
[4] Ellwanger R and Towner J 1988 *Thin Solid Films*, **161** 289-304
[5] Jeyachandran Y, Narayandass S, Mangalaraj D, Areva S, Mielczarski J 2007 *Materials Science and Engineering A* **445–446** 223–236
[6] Vasua K, Ghanashyam M, Padmanabhan K 2011 *Applied Surface Science* **257** 3069–3074

About significance of absolute photocurrent values determination during the solar cell external quantum efficiency measurements

E D Filimonov[1], S A Kozhukhovskaia[1], S A Bogomolova[2] and M Z Shvarts[1]

[1]Photovoltaics Laboratory, Ioffe Institute, St. Petersburg 194021, Russia
[2]Analytical Control Department, The National University of Science and Technology MISiS, Moscow, 119049, Russia
e-mail: efilimonov@mail.ioffe.ru

Abstract. In the work, the link of registered relative and absolute current signals of different amplitude and shape in recording spectral characteristics of the sunlight PV converter external quantum efficiency is investigated, and the need to use absolute current in spectral response studies of multijunction solar cells with clearly pronounced luminescent coupling is established.

1. Introduction

In experimental practice, in studying spectral dependencies of the solar cell (SC) external quantum efficiency (EQE), the comparative approach is used, which is based on direct comparing photocurrent signals from a reference calibrated photo-receiver and from a tested SC. Mainly, conditions for measuring the SC EQE are determined by low levels of monochromatic radiation illumination, what leads to weak values of photocurrent generated by a SC. To register small current signals, specialized devices ensuring enhancement of such signals, their frequency filtration and accentuation are applied. Such devices and instrumentation systems, due to using in them band-pass filters and converting-amplifying signal electronic cascades, can influence the registered values decreasing them compared to the absolute output value. It is obvious that, in using the matching method for comparing two electrical values (in our case, photocurrents of a reference photo-receiver J_{ref} and tested SC one J_{SC} with applying the formula $EQE_{SC} = \frac{J_{SC}}{J_{ref} \cdot EQE_{ref}}$, where EQE_{ref} and EQE_{SC} - quantum efficiency values for the reference photo-receiver and tested SC, respectively), any substantial effect of electronic devices on the result of estimation of the EQE_{SC} should not be expected. The situation changes in investigating complicated cells such as multijunction (MJ) SCs with pronounced optical (luminescent) coupling phenomenon. Such a coupling is known to affect significantly the spectral characteristics and the EQE_{SC} values being determined underestimating the measurement results. To eliminate such a negative effect, a multistage procedure was proposed, which is based only on using the absolute values of photocurrents in determining the EQE values [1]. For this reason, to spread the comparative method advantages for solving the task of determining MJ SC spectral dependences and using it satisfactorily in determining the QE_{SC}values, an unambiguous association between relative signals from a measurement equipment and absolute photocurrents generated by MJ SCs at corresponding conditions of irradiating by monochromatic light should be established.

IOP Conf. Series: Journal of Physics: Conf. Series **917** (2017) 052026 doi:10.1088/1742-6596/917/5/052026

2. Experimental details

In a standard experimental procedure, in determining the MJ SC EQE values, the principle of optical frequency modulation of monochromatic radiation with following amplification-conversion of a current signal from a SC and synchronous detection for weak useful signal isolation is used [2]. A low-noise current preamplifier (Stanford SR570) converts a current signal into voltage with 3-9 orders of magnitude amplification. Together with a useful signal, noise is also amplified. Isolation of a useful carrier-frequency signal is ensured by a Stanford SR830 lock-in amplifier [2]. It, according to the Fourier transformation, presents any input signal in a form of a sum of sinusoidal waves, what leads to some proportional variation of an output value being measured. To determine the proportionality coefficient, the following calibration procedure is applied in the work (the measurement circuit is presented in Fig.1).

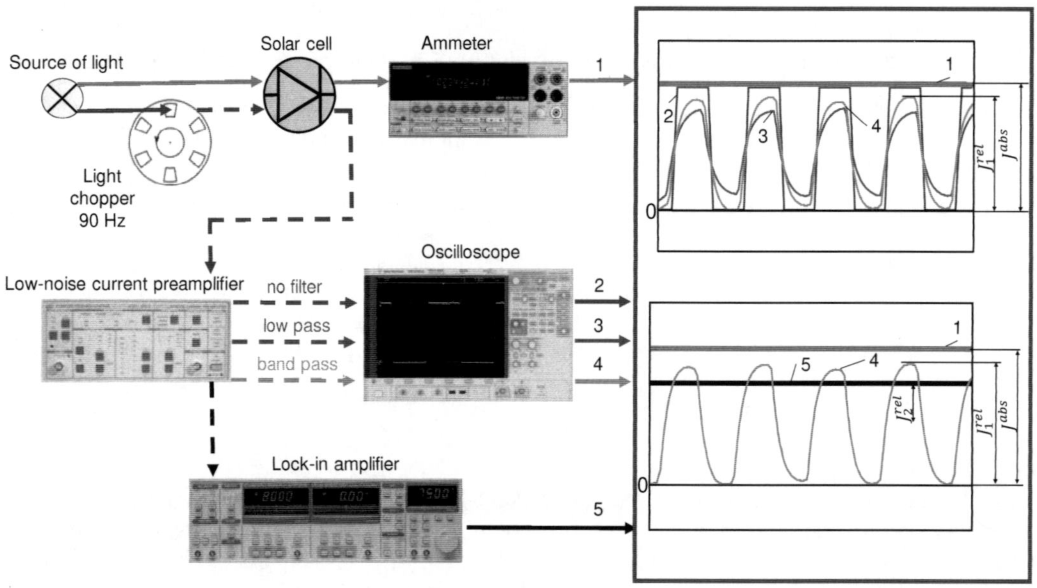

Figure 1. Schematic representation electronic layout and detected signals of the experiment: line 1 – reference signal (J^{abs}); line 2 – amplified signal without application of filters (is equivalent to the reference one); lines 3, 4 – distorted signal with application of filters (J_1^{rel}); line 5 – useful signal obtained at a definite frequency (J_2^{rel}) after lock-in conversion.

In a SC light biasing condition, the absolute current signal J^{abs} being reference one in an experiment is registered. Then the monochromatic radiation flux is modulated by an optical chopper with a preset chopping frequency (in the present work. a frequency of 90Hz was used), and an alternating current signal passes through a low-noise current preamplifier (all filters are switched off) to a registering device (oscilloscope or lock-in amplifier). At inactive frequency filters the current preamplifier does not change the current signal amplitude. Therefore, it does not affect directly the absolute signal value: the output signal oscillation amplitude equal to that of the reference signal is recorded. Application of built-in band-pass filters (30-100Hz) of the preamplifier SR570 ensures more correct signal accentuation on the operating frequency of 90Hz at simultaneous distortion of a signal waveform and amplitude (see the oscillograms in Fig.1). The following relation is valid for the signal being obtained:

$$J_1^{rel} = k_1 \cdot J^{abs}; \quad (1)$$

where J_1^{rel}– amplitude of the output signal measured in arbitrary units; k_1 – coefficient depending on the input signal waveform and on the filter frequency range, J^{abs} - absolute input current signal.

The synchronous detector presents, according to the Fourier theorem, any input signal in a form of a sum of sinusoidal waves, what leads to variation of output values being registered. Thus, the output values under consideration become not proportional to the input photocurrents and last ones are different by the value from the absolute ones. Hence, having amplitude values of the reference (Fig.1, line 1) and modified (Fig.1, lines 3, 4) signals, one can find the coefficient of the reverse conversion from relative to absolute units with obtaining the sought-for photocurrent value for a SC. According to the special feature of action of the algorithm for treating the signal of the synchronous detector, the following is valid:

$$J_2^{rel} = k_2 \cdot J_1^{rel}; \qquad (2)$$

where k_2 – coefficient depending on a waveform of the signal from the preamplifier.
As a result, in allowing for (1) and (2), the following is valid:

$$J_2^{rel} = k_1 \cdot k_2 \cdot J^{abs}; \qquad (3)$$

To determine the conversion coefficient $n_1 = k_1 \cdot k_2$ of a signal of an arbitrary amplitude J_1^{abs} and waveform, it is enough to know at least one reference signal with a known coefficient n_2 and with a corresponding amplitude J_2^{abs}:

$$n_1 = \frac{n_2 \cdot J_2^{rel}}{J_1^{rel}}; \qquad \text{при } J_1^{abs} = J_2^{abs} \qquad (4)$$

This relation has been verified experimentally in studying signals of different configurations, for which the parameter n_i is known in advance:

$$\text{meander} \rightarrow J_1^{abs} = \frac{\pi\sqrt{2}}{2} \cdot J_1^{rel} \qquad (5.1)$$

$$\text{triangular signal} \rightarrow J_2^{abs} = \frac{\pi^2\sqrt{2}}{4} \cdot J_2^{rel} \qquad (5.2)$$

$$\text{sine} \rightarrow J_3^{abs} = \frac{6\pi\sqrt{2}}{3+2\pi} \cdot J_3^{rel} \qquad (5.3)$$

The results of comparison of a signal before and after passing through a synchronous detector with rated data (5.1-5.3) are presented in Fig.2:

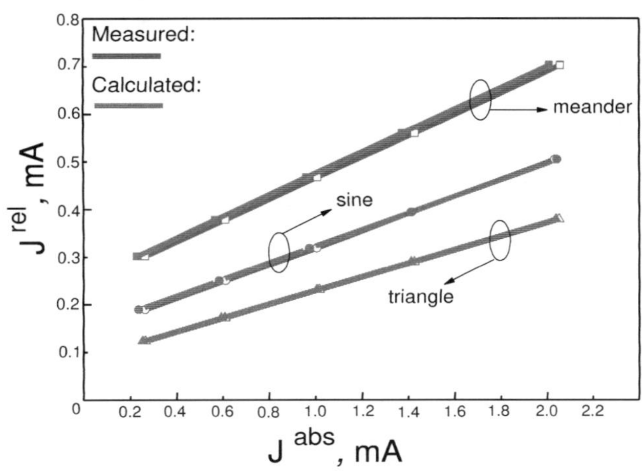

Figure 2. The dependencies (both measured and rated) of relative values of current amplitudes on absolute ones for different input signal waveforms: meander, triangle and sine.

Also, the dependence of the conversion coefficient n on the alternating signal frequency for different levels of current generation (from 10^{-2} to 10^{2} μA) corresponding to different SC illuminations was studied. The investigation has shown that, even for similar amplitudes of the input signal, the conversion coefficient depends strongly on frequency (Fig.3), what, in the whole, introduces the additional uncertainty in searching for its value. It should be noted that, at optimum measurement conditions (optical chopper rotation frequency is 90Hz, frequency range of low-noise preamplifier filters is 30-100Hz, signal range is from several μA to mA), the coefficient of conversion from relative units to absolute values of a tested signal tends to $\frac{\pi\sqrt{2}}{2}$.

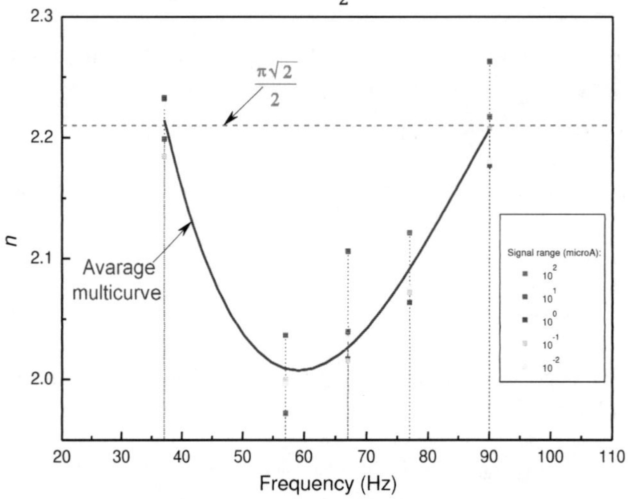

Figure 3. Dependence of the coefficient (n) of converting a relative value into absolute one at active filters on an optical chopper rotation frequency.

3. Conclusions

In the work, the degree of the effect of filtering and detecting elements on the absolute SC photocurrent values has been studied in realizing the procedure for determining the EQE spectral dependence. An association between relative (being registered) signals and absolute (being searched-for) photocurrents, which is determined by the coefficient n depending on the input signal waveform, has been established. The investigation results have shown that, in dependence on the modulated signal waveform, the absolute values, in using a synchronous detector, should be determined according to the expressions (3, 4) и (5.1-5.3) at corresponding selection of a pair of parameters – filters-frequency.

The results obtained in performance of the work are applied in realizing the multistage procedure for determining real values of the quantum efficiency values of MJ SCs with a strong luminescent coupling between p-n junctions [3, 4].

Acknowledgments

This work supported by the Russian Foundation on Basic Research (Grant 17-08-01056).

References

[1] Kozhukhovskaia S A, Filimonov E D, Mintairov M A and Shvarts M Z 2015 Elimination of light coupling negative effect on the accuracy of external quantum yield determination in a MJ SC Journal of Physics: Conference Series 643, p 012064

[2] Armen G B 2008 *Phase sensitive detection: the lock-in amplifier* (Knoxville: The University of Tennessee)

[3] Shvarts M Z, Emelyanov V M, Evstropov V V, Mintairov M A, Filimonov E D, Kozhukhovskaia S A Overcoming the Luminescent Coupling Effect in Experimental Search for the Actual Quantum Efficiency Values in Multi-Junction Solar Cells AIP Conf. Proceedings, v.1766, 2016, pp. 06005, ISSN: 0094-243X, E-ISSN: 1551-7616.

[4] Kozhukhovskaia S A, Filimonov E D and Shvarts M Z Optical leakage versus luminescent coupling in a multijunction solar cell: how to recognize a source of additional photocurrent in subcells Journal of Physics: Conference Series 816, (2017) 012040

IOP Conf. Series: Journal of Physics: Conf. Series **917** (2017) 052027 doi:10.1088/1742-6596/917/5/052027

Capacitance characterization of GaP/n-Si structures grown by PE-ALD

A I Baranov[1,2,a], A S Gudovskikh[1,3], A Darga[2], S Le Gall[2], J-P Kleider[2]

[1] St Petersburg Academic University of RAS, 194021 St Petersburg, Russia.

[2] GeePs (Group of electrical engineering – Paris), UMR CNRS 8507, CentraleSupélec, Univ. Paris-Sud, Université Paris-Saclay, Sorbonne Universités, UPMC Univ Paris 06, 91192 Gif-sur-Yvette CEDEX, France.

[3] St Petersburg Electrotechnical University "LETI", 197376 St Petersburg, Russia.

[a]baranov_art@spbau.ru

Abstract. Thin layers of µc-GaP and a-GaP grown on n- type silicon wafers by plasma-enhanced atomic layer deposition at 380 °C are characterized by space charge capacitance techniques, C-V profiling and deep level transient spectroscopy (DLTS). Two defect levels with activation energies of 0.30 eV and 0.80 eV were detected by DLTS in the µc-GaP/n-Si structure. Measurements performed on Schottky barriers formed on n-Si after selective etching of the GaP layer did not reveal any defect level meaning that the observed defects in the µc-GaP/n-Si structure are related to µc-GaP layer.

1. Introduction

Nowadays, growth of III-V compounds on silicon wafers is a challenge for many scientists in the world. The development of such technology is essential for new silicon photonic applications: optoelectronic integrated circuits, silicon based multijunction solar cells (MJ SC) etc. GaP is one of the most interesting materials for the fabrication of photovoltaic devices. First of all, the GaP/Si heterojunction is a perspective for bottom subcells in MJ SC, because GaP has only 0.4% of lattice- mismatch with silicon and it can be grown on Si by modern epitaxial methods. Furthermore, dilute nitrides such as (In)GaP(NAs) can be grown on GaP layers for fabricating top subcells in view of high efficiency MJ SC. Nowadays, there are many efforts to grow GaP(NAs) layers on silicon wafers [1-3], but the quality of the obtained structures is not sufficient for photovoltaic applications. It is known that molecular beam epitaxy (MBE) and vapor phase epitaxy (VPE) require high temperatures (800-900 °C) during process growth. It leads to the deterioration of the Si wafer quality [4, 5] and of the GaP/Si interface due to different coefficients of thermal expansion and inter-diffusion of atoms at the III-V/Si interface. Recently, a promising technology of GaP growth was presented in [6]. It is based on plasma-enhanced-atomic-layer deposition (PE-ALD) with alternation of P and Ga sources at temperatures below 400 °C. This work presents initial research of GaP/n-Si structures grown on silicon wafers by the PE-ALD method. The main aim of the study is the characterization of the GaP/n- Si structure by the capacitance-voltage (C-V) method and deep-level spectroscopy (DLTS).

2. Experiments and results

Two different types of structures were studied. Initially, thin GaP layers were grown by PE-ALD using an Oxford PlasmaLab System 100 at 380 °C with thickness of 50-75 nm on n-type (100) silicon substrates (2-7 Ω·cm). Phosphine (PH_3) and trimethylgallium (TMG) were used as sources of P and Ga, respectively. An additional flow of silane is considered as a source of Si to obtain donor-doped GaP. As a result, 50 nm of amorphous and 75 nm of microcrystalline GaP layers were grown (a-GaP/n-Si and μc-GaP/n-Si). Mesa-structures were fabricated for capacitance measurements. In a first step, 50 nm of n- type amorphous silicon was deposited on the bottom of the silicon wafer by PECVD. Then silver was evaporated onto a-Si:H to form the bottom contact. This bottom contact became ohmic after thermal annealing during 20 min at 170 °C in the air atmosphere. Gold contact dots with diameter of 0.5 and 1 mm were evaporated on the top of the structures to form Schottky barrier (SB) diodes to GaP. Finally, mesa structures were formed by dry etching of the GaP layer and 2 μm of silicon. The schematic view of the fabricated structure with GaP/n-Si heterojunction is shown in Figure 1a. Furthermore, a second type of structures was fabricated to study the influence of the growth process onto the silicon wafer properties. Firstly, the GaP layer was selectively removed by wet etching. Then the same silver ohmic contact and gold SB were formed on the bottom and top sides of the silicon wafer, respectively. The mesa structures were fabricated in the same way by dry etching of 2 μm of Si. The schematic view of such Au/n-Si SB structures is shown in Figure 1b.

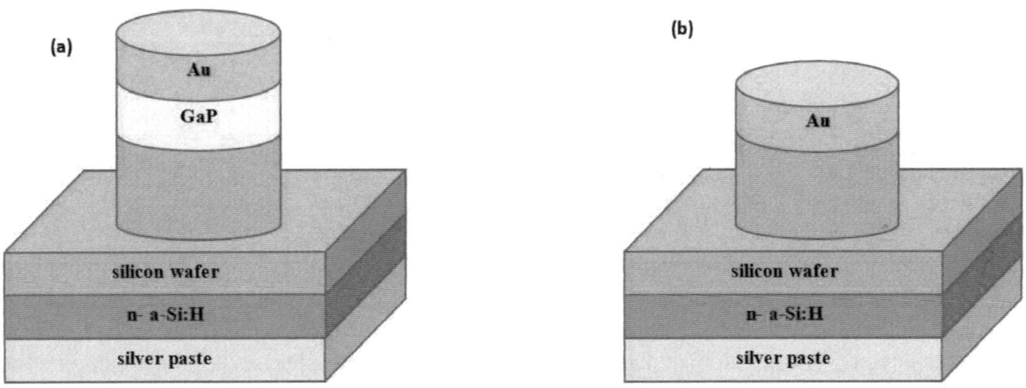

Figure 1 (a, b). Schematic view of GaP/n-Si (a) and Au/n-Si (b) structures.

Measurements of C-V and DLTS were performed using a liquid nitrogen cryostat in the temperature range of 80-400 K. Using C-V characteristics, profiles of effective concentration of free carriers (N_{CV} - W) were estimated in the depletion layer approximation by equations:

$$N_{CV} = \frac{2}{q\varepsilon\varepsilon_0 S^2}\left(\frac{dC^{-2}}{dV}\right)^{-1} \quad W = \frac{\varepsilon\varepsilon_0 S}{C} \qquad (1)$$

where W is the width of the space-charge region, q is elementary charge, ε_0 is the permittivity of vacuum, ε is the relative permittivity, and S is contact area. An automated installation based on a Boonton- 7200B capacitance bridge was used for measurements by the classical DLTS method [7].

3. Results

Capacitance-voltage characteristics of the GaP/n-Si samples at 80 K and Au/n-Si at 300 K are shown in Figure 2a. The capacitance decreases with increasing amplitude of reverse bias voltage for both samples. Consequently, the extension of the space charge region (SCR) inside the structures is observed: this is the conventional behaviour of C-V curves. N_{CV} - V and N_{CV} - W profiles were plotted for the three samples in Figure 2b and Figure 2c, respectively.

The concentration profile for Au/n-Si is almost constant with a value of 1×10^{15} cm^{-3}. It corresponds to the doping concentration of the silicon wafer (2-7 Ω.cm). The experimental profile for a-GaP/n-Si shows the constant value of 7×10^{14} cm^{-3}, which also corresponds to the doping concentration of the silicon wafer. However, for the μc-GaP/n-Si structure a clear peak of concentration is observed at low bias voltage, which corresponds to the lowest depth of SCR. With the increase of reverse bias voltage (increase of SCR depth) the concentration becomes constant with the same value as in the Si wafer. Also, the capacitance is much higher for low bias voltage in μc-GaP/n-Si so the SCR width is lower than in a-GaP/n-Si and Au/n-Si structures.

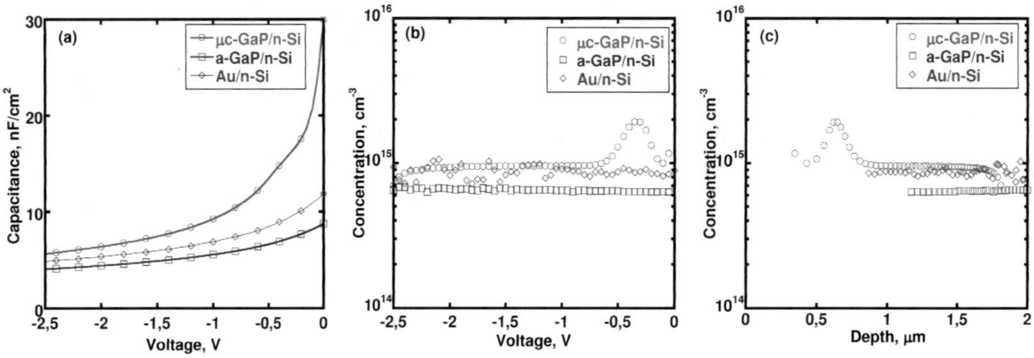

Figure 2 (a, b, c). Analysis of the bias dependence of the measured capacitance of the three types of structures: *C-V* (a), N_{CV}-*V* (b) and N_{CV}-*W* (c).

Two differences are observed from the comparison of profiles for a-GaP/n-Si and μc-GaP/n-Si: smaller width of SCR for μc-GaP/n-Si and existence of a peak on its profile at low reverse bias voltage. One of the possible explanations is the different n-type doping level in GaP. We suggest μc-GaP is more doped compared to a-GaP so the concentration peak appears when SCR extends from GaP to Si through the n-GaP/n-Si interface [8]. Figure 3 shows the band diagram for GaP (n= 1×10^{17} cm^{-3})/Si (n= 1×10^{15} cm^{-3}) heterojunction at zero bias voltage and the concentration profile of electrons at 80 K simulated using AFORS-HET 2.5 software [9]. The position of the peak in the concentration profile corresponds to the GaP/Si interface: for the experimental μc-GaP/n-Si sample similar phenomenon is observed (Figure 2c). However, values of SCR width are higher in the experiment than in the simulation. Disparity of values can come from imperfect bottom contact to silicon wafer. Consequently, estimation of the absolute value of interface position from the experiment can be incorrect and should be considered more critically. On the opposite, the described peak does not appear on the profile for a-GaP. It means that the SCR extends inside the silicon wafer even at zero bias voltage. Consequently, a-GaP should be fully depleted at any reverse bias voltage. It explains why the capacitance of μc-GaP/n-Si is much higher compared to that of a-GaP/n-Si. It is a very important issue because it means that the border of SCR for the a-GaP/n-Si sample is further from the heterojunction compared to the μc-GaP/n-Si one. Therefore in case of the a-GaP/n-Si sample the DLTS experiment is expected to rather detect the response of defects far from the a-GaP/n-Si interface in the bulk silicon.

IOP Conf. Series: Journal of Physics: Conf. Series **917** (2017) 052027 doi:10.1088/1742-6596/917/5/052027

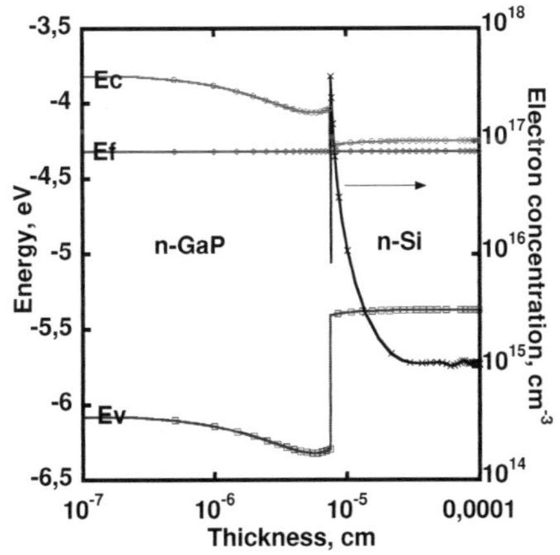

Figure 3. Band diagram of n-GaP/n-Si and profile of electron concentration at 80 K.

Results of DLTS measurements are shown in Figure 4. DLTS spectra of μc-GaP/n-Si (Figure 4a) have two series of peaks, which are characteristic for the response from two defect levels. Their properties are estimated from the Arrhenius plot (Figure 5) by:

$$e = AT^2e^{-E_a/k_BT} \quad (2)$$

where e is emission rate, A is a pre-exponential factor, E_a is the activation energy of the defect level, k_B is Boltzmann's constant. Activation energies for low- and high-temperature levels are equal to 0.3 eV and 0.8 eV, respectively. Both peaks are broadened so non-exponential behavior of the capacitance relaxation due to extended defects can be suggested in this structure [10]. Unfortunately, low thicknesses of GaP layer and width of SCR estimated from *C-V* measurements do not allow us to define exactly the region in the μc-GaP/n-Si structure where the response takes place. The response can be from the GaP/Si interface, from the GaP layer or from the bulk Si. *S*(*T*) spectra for a-GaP/n-Si do not allow one to extract any defect responses within our measurement accuracy (Figure 4b), because the signal is much lower compared to that of the μc-GaP/n- Si sample. Nevertheless, defects described above were not detected in a-GaP/n-Si. However, according to the above explanation the a-GaP layer is fully depleted and the SCR extends deeper in the silicon bulk as opposed to the μc-GaP/n-Si structure, so we can not detect defects in a-GaP. Also, we did not find any peaks on the *S*(*T*) spectra of the Au/n-Si structures in the used temperature range (Figure 4c). It means that the silicon bulk was not deteriorated by the deposition of the GaP layer in our low-temperature deposition method. Consequently, the detected defects in the μc-GaP/n-Si structure are attributed to the μc-GaP layer. Since the detected peaks are broad, the defects can tentatively attributed to threading dislocations in μc-GaP appearing during the growth. Analysis of their nature and value of pre- exponential factors is a complicated issue and it will be explored in the future experiments in details.

Nevertheless, *C-V* and DLTS characteristics show high differences between samples with and without GaP layer. Therefore, these methods allow one to study the structures with thin GaP layers grown by PE-ALD on silicon wafers. The future research will be focused on the interpretation of the obtained results for GaP/Si structures grown in different conditions.

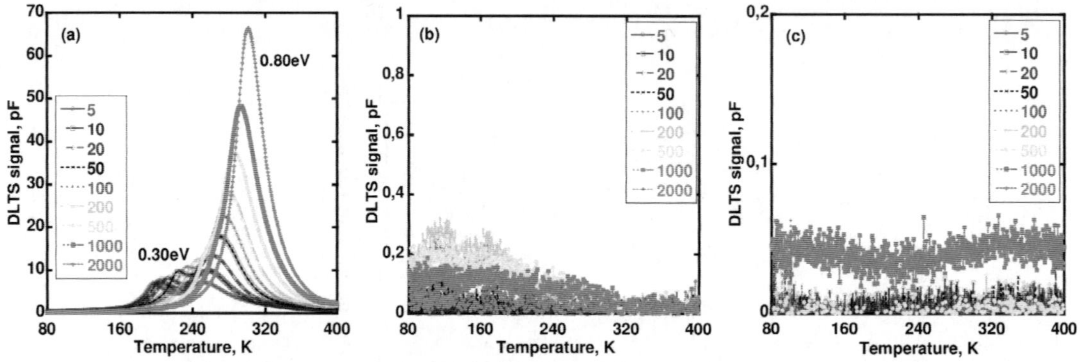

Figure 4 (a, b, c). DLTS spectra *S*(*T*) of μc-GaP/n-Si (a), a-GaP/n-Si (b), and Au/n-Si (c) for different emission rates (in s^{-1}) obtained by changing the DLTS gate window.

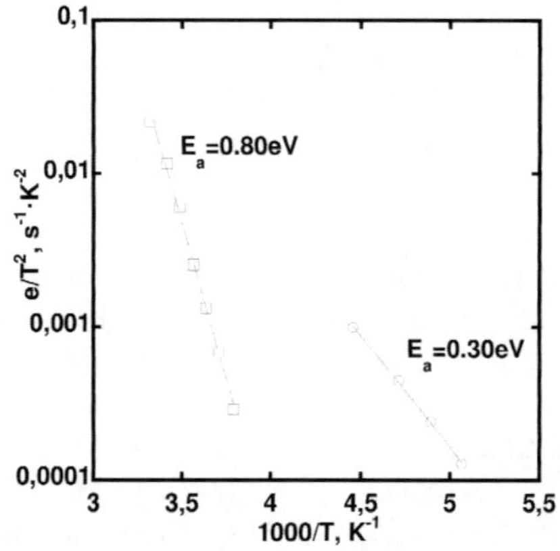

Figure 5. Arrhenius plots of e/T^2 for the two DLTS peaks detected in the μc-GaP/n-Si structure.

4. Conclusion

Thin layers of μc-GaP and a-GaP were grown on n-type silicon wafers by the new low temperature PE–ALD technique. Structures were characterized by capacitance-voltage and DLTS techniques. The a-GaP layer is fully depleted and no defects could be observed in the c-Si wafer. We have observed two peaks in the DLTS spectra of the μc-GaP/n-Si structure associated with defect levels having activation energies of 0.30 and 0.80 eV. The lack of response in Schottky barrier structures formed on silicon after etching of μc-GaP demonstrates that the detected defects originate from the μc-GaP layer and that no deterioration of c-Si is induced by the low temperature PE-ALD GaP process.

Acknowledgements

This work was carried out in the framework of the joint Russian-French project LAMSOL supported by Ministry of Science and Education of Russian Federation (RFMEFI61615X0040) and by PHC Kolmogorov Program (35522TL).

References

[1] Geisz J F et al. 2005 *Proc. of the 31st IEEE Photovoltaic Specialists Conference PVSC2005* (Orlando, USA) p. 695
[2] Sobolev M S et al. 2015 Semiconductors **49** (4) 559
[3] Ping Wang Y et al. 2015 Appl. Phys. Lett. **107** 191603
[4] Ding L et al. 2016 Energy Procedia **92** 617

[5] Varache R et al. 2015 Energy Procedia **77** 493
[6] Morozov I A et al. 2016 Journal of Physics: Conference Series **741** 012088
[7] Lang D V 1974 J. Appl. Phys. **45** 3023
[8] Kroemer et al. 1980 Appl. Phys. Lett. **36** (4) 295
[9] Varache R et al. 2015 Solar Energy Materials and Solar Cells **141** 14
[10] Schröter W et al. 1995 Phys. Rev. B **52** 13726

IOP Conf. Series: Journal of Physics: Conf. Series **917** (2017) 052028 doi:10.1088/1742-6596/917/5/052028

Quantum analyzer of force lines structure at magnetic fields

S E Logunov[1], V V Davydov[1, 2, 3], M G Vysoczky[1], A Yu Koshkin[1], V Yu Rud'[3]

[1]Peter the Great Saint-Petersburg Polytechnic University, Saint Petersburg 195251, Russia
[2]The Bonch-Bruevich Saint - Petersburg State University of Telecommunications, Saint Petersburg 193232, Russia
[3]Department of Ecology, All-Russian Research Institute of Phytopathology, 143050, Moscow Region, Odintsovo district, B.Vyazyomy, Russia

Abstract. A new method for studying the structure of the field lines of the magnetic systems of various configurations. The optimum parameters of ferrofluid cells and laser light for different research variants of force lines structure in the area between the poles of the magnetic system are determined. The results of experimental studies, which allow to define the change ΔB inhomogeneity of the magnetic field over the cross section of the magnetic system and to carry out its adjustment to the minimum ΔB in real time are represented.

1. Introduction.

One of the important problems of fundamental physics is the study of the lines of force of the magnetic field structure [1-4]. The greatest difficulties arise in the case of power lines investigation between the poles of the magnetic structure of the system, especially in fields with high induction B > 1 T [3, 5-7]. Most imaging methods of single magnet lines of force (sawdust, arrow, etc.) can not be used in the interpolar space of the magnetic system or allow to display only the average change in the magnetic field inhomogeneity ΔB (e.g., films, etc.), but not the structure itself of the power lines [2, 3, 5, 7, 8].

The use of different measurement sensors (such as Hall or nuclear magnetic resonance (NMR), etc.) give in most case the information about the induction value B in the measurement area. [7-11]. When using sensors based on NMR are used the line width permits to determine the inhomogeneties ΔB in the volume from which the NMR signal is recorded. But any movement of considered sensors for measuring inhomogeneities in the interpolar space will be associated with a high mistakes of determining the coordinates of their location between the poles. The placement of a grid of these sensors (except NMR) allows to determine the nature of the change in the average heterogeneity along one direction, but it is impossible to draw a picture of the lines of force on the basis of these measurements [2, 3-5, 8, 9]. In [3] we have presented experimental results showing the ability to study the structure of the magnetic field lines with ferrofluid cells and laser radiation. The results obtained in [3, 6], and further conducted experiments allowed to choose the optimal parameters of the ferrofluid cell (type of material from which it is made, the thickness of the magnetic fluid layer in the direction of laser radiation d_k, composition of magnetic field, etc), and the laser radiation wave length λ.

2. Experimental setup and method for studying the structure of magnetic field lines.

Based on our experiments on different ferrofluid cells with different magnetic fluids it has been found that for the study of the structure of the force lines of the magnetic field is most advisable to use

IOP Conf. Series: Journal of Physics: Conf. Series **917** (2017) 052028 doi:10.1088/1742-6596/917/5/052028

ferrofluid cell made of quartz glass with two transparent parallel faces or with one face for various cases of studies. This is due to the peculiarities of different configurations of magnetic systems, in the interpolar space of which the structure of the lines of force is investigated. For the first variant of the cell d_k = 1 mm for the second one - $d_k \approx 3$ mm. As the magnetic fluid in two variants of cell structure the most appropriate is to apply an aqueous solution of single-domain nanoparticles of hematite with a bulk density of 0.054 with a surface active agent (surfactant) tetrametilamoniya hydroxide [6].

Taking into account the various restrictions on placement of ferrofluid cells, as well as the possible impact on her laser radiation in the interpolar space, it was found that in order to explore the magnetic field inhomogeneity nature change along three directions and to build on these results the volume picture of force lines structure taking into account all peculiarities of the magnetic system, it is necessary to use ferrofluid cells of two considered variants. In one case, a diffraction pattern is recorded from the laser radiation transmitted through it with λ = 613 nm - a cell with two transparent faces is used. In the other two - from reflected laser radiation on a structure of nanoparticles of a magnetic fluid, which is a similarity of the diffraction grating [3]. For these two cases, a cell with one transparent face is applied. To record these diffraction patterns, experimental setups have been developed and assembled, and relationships have been obtained that allow the results of measurements to determine the period of the diffraction grating (the distance between the lines of force in the plane of incidence of the laser radiation on the ferrofluidic cell).

In Fig. 1 is a block diagram of an experimental setup for recording a diffraction pattern from a laser radiation transmitted through a ferrofluidic cell.

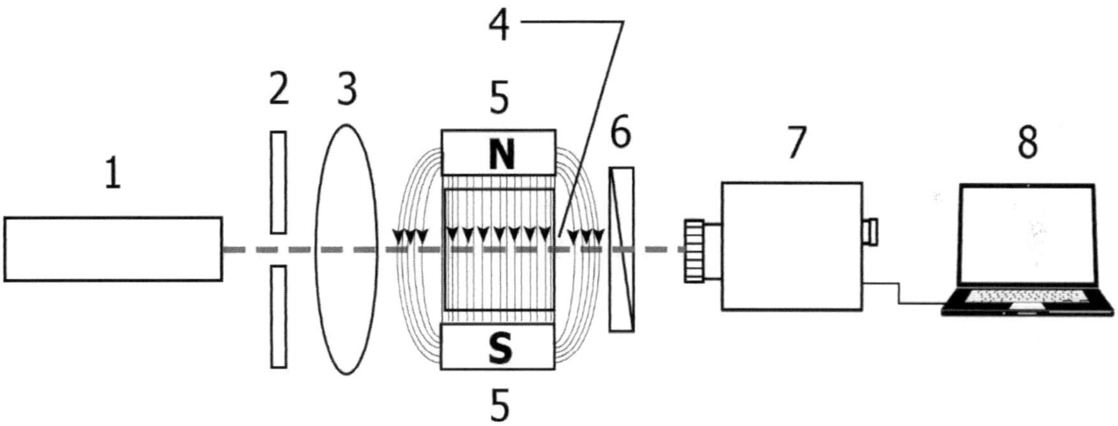

Figure 1. Structural scheme of experimental setup 1 – semiconductor laser; 2 – aperture; 3 – lens; 4 – ferrofluid cell; 5 – magnetic system; 6 – polarizer; 7 – camera; 8 – computer.

The procedure for measuring the period of the diffraction grating d_r, which corresponds to the distance between the lines of force of the magnetic field, in this case partially corresponds to the classical scheme for studying the diffraction pattern from a multi-slit lattice. The difference in the case under consideration is the presence of a transparent layer (glass walls of a ferrofluidic cell) with a thickness d_1 and a layer thickness of a magnetic fluid d_2. In Fig. 2 is a represented diagram of the propagation of laser radiation beams in a ferrofluidic cell placed in a magnetic field.

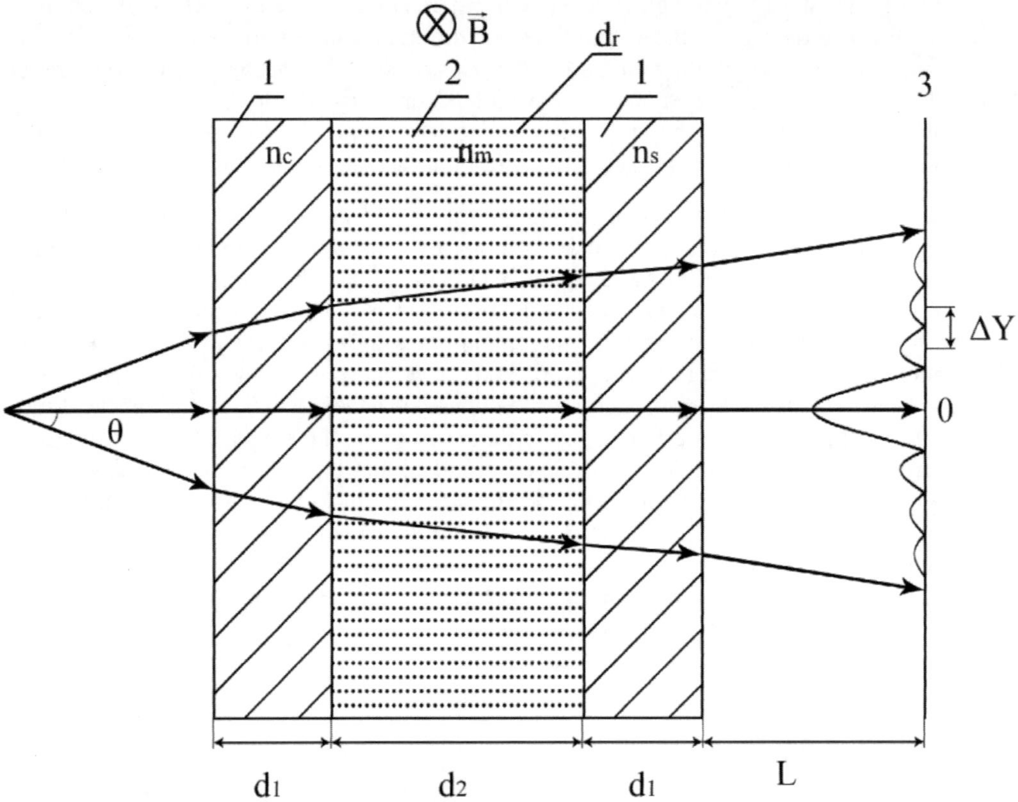

Figure 2. Scheme of propagation of laser radiation rays in a ferrofluidic cell when a diffraction pattern is recorded in transmitted light: 1 - walls of a ferrofluidic cell; 2 - magnetic fluid layer in the direction perpendicular to the magnetic field with the thickness equal to d_r; 3 - the screen.

It was shown in [3, 6] that nanoparticles of a magnetic fluid in a magnetic field are placed on its lines of force, forming transparent and opaque zones in the forrefluidic cell for laser radiation with a period of d_r. If, at a distance L from the center of the ferrofluid cell, the screen 3 (camera) is placed, then it will form a diffraction image in the laser radiation transmitted through the cell. The position of each maximum on the screen relative to its center (point O) will be determined by the diffraction order k, d_r and λ. And also by the distance to the screen L and the divergence angle of the laser radiation θ. For two adjacent maxima their position on the OY axis already was determined. As well as also the value of the additional displacement Δy_k along the axis OY, which is formed due to the refraction of radiation at the boundaries of the magnetic fluid - glass and glass - air. It was found that at distances of L = 20 cm, which were used in the experiment, the value of Δy_k is infinitesimal compared to the distance between two adjacent maxima ΔY. To determine the given distance ΔY, using the conditions for obtaining the maxima of the diffraction pattern, the following relation was obtained:

$$\Delta Y = \lambda L(k_2\sqrt{\frac{L}{d_r^2-k_2^2\lambda^2}} - k_1\sqrt{\frac{L}{d_r^2-k_1^2\lambda^2}}) \tag{1}$$

Where L – distance between the ferrofluid cell and camera, k_1 and k_2 - an integer representing the propagation-mode of interest.

By measuring ΔY from the diffraction image recorded by camera 7 (Fig. 1), for each diffraction order in accordance with (1), the value d_r is determined, which makes it possible to determine establish the distance between the lines of force of the magnetic field. Constructing the dependence of the change d_r

on the distance between the poles of the magnetic system, we can determine the degree of spatial inhomogeneity of the magnetic field.

To study the spatial structure of magnetic field lines using a diffraction pattern in the reflected laser radiation, an experimental setup was developed and assembled, that is shown in Fig. 3.

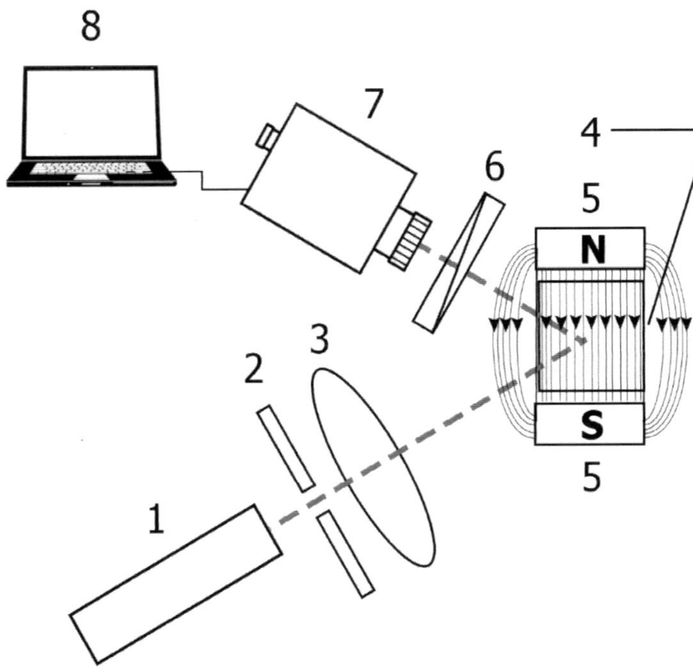

Figure 3. Structural scheme of experimental setup 1 – semiconductor laser; 2 – aperture; 3 – lens; 4 – ferrofluid cell; 5 – magnetic system; 6 – polarizer; 7 – camera; 8 – computer.

In the studies carried out, the ferrofluidic cell 4 with one transparent face is used. The wavelength of the laser radiation is λ = 515 nm. In the Fig. 4 is represented a diagram of the propagation of laser radiation beams in a ferrofluidic cell placed by a magnetic field when conducting studies in reflected light. For the recorded diffraction pattern in the reflected light, the positions of the diffraction maxima on the OY axis were determined and was also determined the magnitude of their displacement Δy_k were also determined due to the presence of refraction of laser radiation at the boundaries of four media: air - glass, glass - magnetic fluid, magnetic fluid - glass, glass - Air (Figure 4). During the calculations, the screen 3 (camera) supposed to be was located at the distance L = 15 cm from the ferrofluid cell. The laser radiation was fed to the ferrofluidic cell at point A at an angle $\alpha_1 = 45^0$. Then, after refraction, on the cell walls the radiation was incident on a diffraction grating formed by nanoparticles with the period dr. The structure of the diffraction grating begins at a very small distance d_k from the inner wall of the ferrofluid cell. The performed calculations showed that the deflection (α_1 - α_2) of laser radiation emitted from the ferrofluidic cell at point B (after all 4 refractions) from the direction of the beam at point C (the classical diffraction case in reflected radiation) is less than $0^0 30$ ', and the displacement Δy_k is less than 10^{-3} mm. Therefore, when determining the position of the diffraction maxima along the OY axis in the reflected laser radiation, these values do not have a significant effect on the measurement result

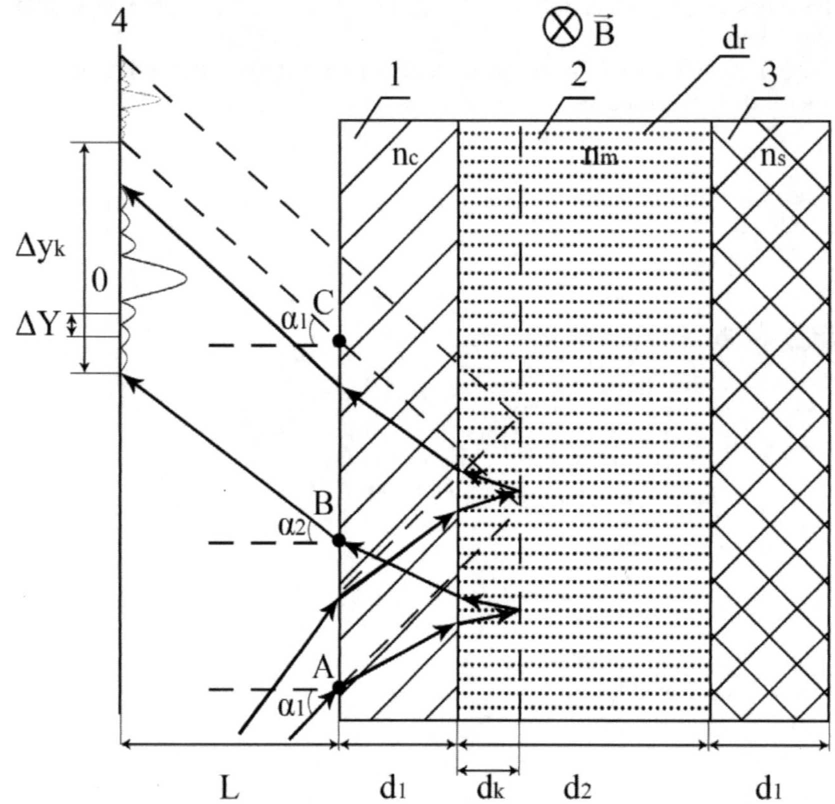

Figure 4. Scheme of propagation of laser radiation rays in a ferrofluidic cell when a diffraction pattern is recorded in reflected light: 1,3 - walls of the ferrofluidic cell; 2 – thr layer of the magnetic fluid with equal to d_r the thickness the direction perpendicular to the magnetic field; 4 - the screen.

The obtained diffraction pattern allows, with an error of not more than 1%, to determine the distance between neighboring maxima ΔY (Fig. 4). The value of ΔY can also be determined using the relation obtained by us:

$$\Delta Y = \frac{d_k}{d_r}\cdot\left[\frac{t_1}{\sqrt{1-\frac{t_1^2}{d_r^2}}}-\frac{t_2}{\sqrt{1-\frac{t_2^2}{d_r^2}}}\right] + d_1\cdot n_m\cdot\left[\frac{\sqrt{1-\frac{t_1^2}{d_r^2}}}{\sqrt{n_c^2-n_m^2\left(1-\frac{t_1^2}{d_r^2}\right)}}-\frac{\sqrt{1-\frac{t_2^2}{d_r^2}}}{\sqrt{n_c^2-n_m^2\left(1-\frac{t_2^2}{d_r^2}\right)}}\right] +$$

$$+L\cdot\left[\frac{\sqrt{n_c^2-n_m^2\left(1-\frac{t_1^2}{d_r^2}\right)}}{\sqrt{1-\left(n_c^2-n_m^2\left(1-\frac{t_1^2}{d_r^2}\right)\right)}}-\frac{\sqrt{n_c^2-n_m^2\left(1-\frac{t_2^2}{d_r^2}\right)}}{\sqrt{1-\left(n_c^2-n_m^2\left(1-\frac{t_2^2}{d_r^2}\right)\right)}}\right] \quad (2)$$

$$t_1 = (2k_1 + 1)\cdot\lambda/2;\ t_2 = (2k_2 + 1)\cdot\lambda/2 \quad (3)$$

where n_m – the refractive index of a magnetic fluid, n_c – refractive index of the transparent glass wall of the cuvette, d_k – distance from the cell wall to the first layer of nanoparticles in a magnetic fluid that forms a diffraction grating, k_1 и k_2 – orders of diffraction maxima ($k_2 > k_1$).

The experiments carried out showed that in the case when the angles of incidence of the laser radiation are significantly different from 45^0, disregarding the displacements Δy_k, connected by the refraction of laser radiation at four media boundaries, can not be ignored. This greatly complicates the calculation, as well as the carrying out of the experiment itself. Measuring the distance between the maxima of the diffraction pattern ΔY, which is recorded with the help of a special chamber 7 (Fig.3), it is possible to calculate with the help of (2) the value of d_r the diffraction grating in the zone of its location in the interpolar space of the magnetic system. It allows to determine the structure of its lines of force for the two planes of the magnetic system.

3. Results and its discussion.

In Figure 5, as an example, a diffraction pattern (after computer processing) of the laser radiation transmitted through the cell in the direction perpendicular to the induction of the magnetic field B is presented.

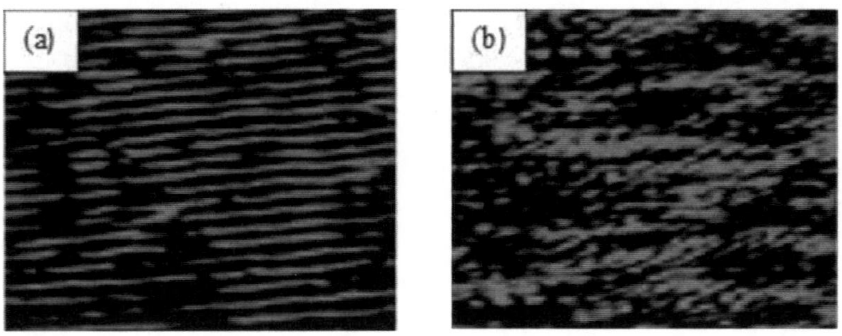

Figure 5 (a, b). The diffraction pattern of the laser radiation in the case of the magnetic fluid placing: (a) in a uniform magnetic field; (b) in an inhomogeneous magnetic field.

The resulting diffraction image allows to determine visually the degree of inhomogeneity of the magnetic field in terms of the structure and relative location of the diffraction maxima. By selecting the line mode on the laptop screen, it is possible to determine with a high accuracy the distance between the neighboring maxima of the diffraction pattern, as well as the position of the central peak in magnitude of the amplitude of the peaks in the diffraction image.

Conclusion

Realizing measurements of the d_r values in three planes of the magnetic system using with two types of ferrofluidic lattices, it is possible to construct a 3D (volumetric) picture of the magnetic field lines in the inter-polar space of the magnetic system. Such an image of the configuration of the field lines can not be obtained even when calculating the magnetic field fields. In addition, the obtained result makes it possible to adjust the magnetic system to the minimum value of the magnetic field inhomogeneity in real time by the changing of the character of the recorded diffraction pattern.

Reference

[1] Suharnikov K V, Rychkov M M 2015 *Radiophysics and Quantum Electronics* **58** 132

[2] Cherepov C V, Moroz O H, Derecha D A 2011 *Radiophysics and Quantum Electronics* **54** 137
[3] Logunov S E, Koskin A Yu, Davydov V V, Petrov A A 2016 *Journal of Physics: Conference Series* **741(1)** 012092
[4] Zaitsev N I, Ilyakov E V, Kulagin I S, Shevchenko A S 2006 *Radiophysics and Quantum Electronics* **49** 120
[5] Davydov V V, Dudkin V I, Karseev A Yu 2015 *Technical Physics Journal* **60** 456
[6] Davydov V V, Dudkin V I 2016 *Technical Physics Journal* **61** 1115
[7] Davydov V V, Dudkin V I, Karseev A Yu 2015 *Instruments and Experimental Techniques* **58** 787
[8] Kudasov Yu B, Makarov I V, Maslov D A, Platonov V V, Popov E Ya, Surdin O M, Voronov S L, Malyshev A Yu, Korotkov S V, Vodovozov V M 2015 *Instruments and Experimental Techniques* **58** 781
[9] Stolypko A L 2013 *Instruments and Experimental Techniques* **56** 649
[10] Cherepov S V, Moroz O Kh, Derecha D A, Hesse O P 2011 *Instruments and Experimental Techniques* **54** 273
[11] Davydov V V, Velichko E N, Dudkin V I, Karseev A Yu 2015 *Measurement Techniques* **58** 556

IOP Conf. Series: Journal of Physics: Conf. Series **917** (2017) 052029 doi:10.1088/1742-6596/917/5/052029

High-speed 1.3 -1.55 um InGaAs/InP PIN photodetector for microwave photonics

O A Kozyreva[1], Y V Solov'ev[2], I S Polukhin[2], A K Mikhailov[1,4], G A Mikhailovskiy[2], M A Odnoblyudov[2], E Z Gareev[1], E S Kolodeznyi[1], I I Novikov[3], L Ya Karachinsky[3], A Yu Egorov[1,3] and V E Bougrov[1]

[1]ITMO University, Kronverkskii pr. 49, Saint Petersburg 197101, Russia
[2]Peter the Great St. Petersburg Polytechnic University, Polytechnicheskaya 29, Saint Petersburg 195251, Russia
[3]Connector Optics LCC, Domostroitelnaya 16, Saint Petersburg 194292, Russia
[4]Electrotechnical University ("LETI"), ul. Professora Popova 5, St. Petersburg 197376, Russia

Abstract. We have fabricated the 1.3-1.55 um PIN photodetector based on InGaAs/InP heterostructure. Measurement results of optical and electrical characteristics of PIN photodetector chip were the following: photoconductivity at 1550 nm was 0.65 A/W and internal capacitance was 0.025 pF. Microwave model of photodetector was developed and verified by measurements of scattering matrix. The implementation of broadband (up to 20 GHz) hybrid integrated matching and biasing circuit for high-speed photodetector is presented.

1. Introduction

Microwave (MW) photonics is a rapidly developing area for application in high-speed communication systems with 1.3–1.55 um spectral band corresponding to II and III transmission windows [1]. One of the key components of photonic links is a high-speed PIN photodetector (PD). In communication systems PD is destined to convert MW-modulated light into high-frequency electrical current, hence low PD capacitance *C* is required for high-speed operation [2-4]. Optimization of the PD structure (the use of PIN structure) and photoconductive operating mode (reverse-biased operating point) are commonly used to reduce the p-n junction capacitance *C* which is a function of active region area *A* and width of depletion layer *W* [5]. Another important characteristic of PD for communication systems is responsivity S (A/W) that also is a function of W, therefore, maximum operating frequency is compromised by setting up parameters A and W to achieve low C and sufficient S simultaneously.

Accurate microwave characterization of PD chip is necessary for the proper design of the external matching circuit. Impedance matching (to system's 50 Ohm) is required for effective data transmission as reverse-biased PD represents a high resistance (tens kOhm) for MW signal[6]. In this work we discuss fabrication of the PD heterostructure, modeling and measuring of optical and electrical characteristics of the 440×370 um PIN PD chip with the PD active area of 18 um diameter. The design of matching and biasing circuit for photodetector operating up to 20 GHz is presented.

IOP Conf. Series: Journal of Physics: Conf. Series **917** (2017) 052029 doi:10.1088/1742-6596/917/5/052029

2. Experiment

Heterostructure for high-speed PIN PD for λ=1550 nm was grown by Molecular beam epitaxy (MBE) on a Riber49 ("Connector Optics", St. Petersburg). Growth of epitaxial heterostructure was implemented on semi-insulating indium phosphide substrate of 650 um thickness. The PIN photodiode heterostructure is shown on figure 1. The main part of light is absorbed by $In_{0,53}Ga_{0,47}As$ layer of 0.8 um thickness. Dimension of absorption layer provides small tunneling time of generated charge carriers resulting in high-speed operation. At the same time high absorption coefficient of $In_{0,53}Ga_{0,47}As$ provides sufficient photosensivity. Several gradient layers are designed to match the lattice constant of contact layer, window layer ($In_{0,52}Al_{0,48}As$) and absorption layer.

Contact layer $In_{0.53}Ga_{0.47}As$	0.1 μm	p-doping $1{\cdot}10^{19}$
Gradient layer $In_{0.53}Ga_{0.47}As$ >> $In_{0.52}Al_{0.48}As$	288 Å	p-doping $1{\cdot}10^{19}$
Window $In_{0.52}Al_{0.48}As$	0.1 μm	p-doping $3{\cdot}10^{18}$
Gradient layer $In_{0.52}Al_{0.48}As$ >> $In_{0.53}Ga_{0.47}As$	288 Å	p-doping $1{\cdot}10^{18}$
Absorption layer $In_{0.53}Ga_{0.47}As$	0.1 μm	p-doping $1{\cdot}10^{18}$
Absorption layer $In_{0.53}Ga_{0.47}As$	0.8 μm	i - without doping
Absorption layer $In_{0.53}Ga_{0.47}As$	0.1 μm	n-doping $1{\cdot}10^{18}$
Gradient layer $In_{0.53}Ga_{0.47}As$ >> $In_{0.52}Al_{0.48}As$	288 Å	n-doping $1{\cdot}10^{18}$
Buffer layer $In_{0.52}Al_{0.48}As$	0.1 μm	n-doping $5{\cdot}10^{18}$
Contact layer $In_{0.53}Ga_{0.47}As$	0.5 μm	n-doping $5{\cdot}10^{18}$
Substrate InP	650 μm	n-doping $3{\cdot}10^{18}$

Figure 1. Heterostructure for PIN PD

Fabrication of PIN PD crystals included the following technological operations. Firstly, we formed ohmic contacts Ti / Au and AuGe/Ni/Au to epitaxial layers of InGaAs of p- and n-type conductivity, respectively, with subsequent burning at a temperature of 360^0C. Then mesa-structures were formed by plasma-chemical etching BCl3/Ar. Mesa configuration was passivated using plasma-chemical deposition of silicon nitride. Contact pads were formed by sputtering of Ti/Au and enhanced by electroplating. The final step of fabrication of the PIN PD chip was the thinning of the substrate to 120 um and cutting of the operating plate to single PD chips. SEM image of PD active area is shown on figure 2 (a).

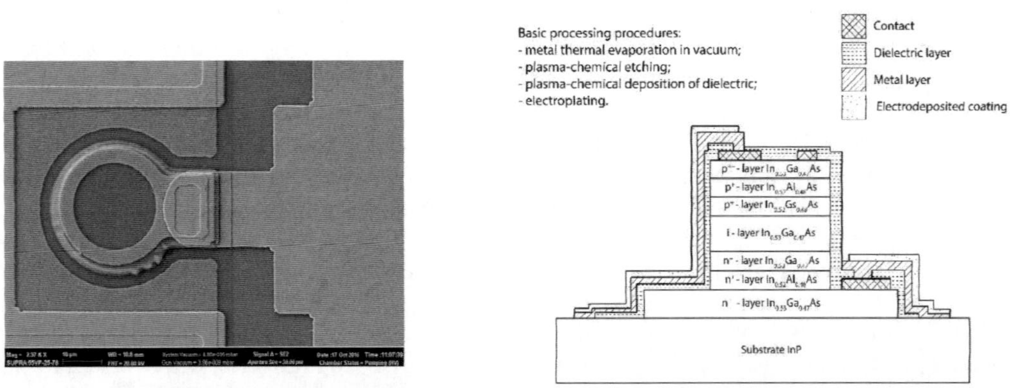

(a) (b)

Figure 2. (a) - SEM image of the PD active area; (b) – cross section of the PD

The coplanar waveguide structure of the PD contacts allowed direct probe measurements for MW characterization of crystal. We used vector network analyzer Keysight PNA N5242A together with MPI

TS150 probe station to measure reflection coefficient form PD output and obtain the PD microwave parameters. Calibration on lumped coplanar waveguide standards (reference impedance, short line and open end) on AlN substrate helped to exclude measurements uncertainty connected with influence of crystal's holder. Modeling of microwave part was conducted using equations based on distributed line theory together with 3D electromagnetic simulation based on method of moments.

3. Results and discussion

Measured PD chips were photosensitive in wide spectral range. Long-wave limit of the PDs was near 1,7 um , corresponding to $In_{0,53}Ga_{0,47}As$ content of active layer. Maximum level of photosensivity was 0,65 A/W and photosensivity above 0,6 A/W was observed in 1400-1600 spectral range (see figure 3). Current-voltage measurements showed 100 nA dark current under 2 V reverse bias typically.

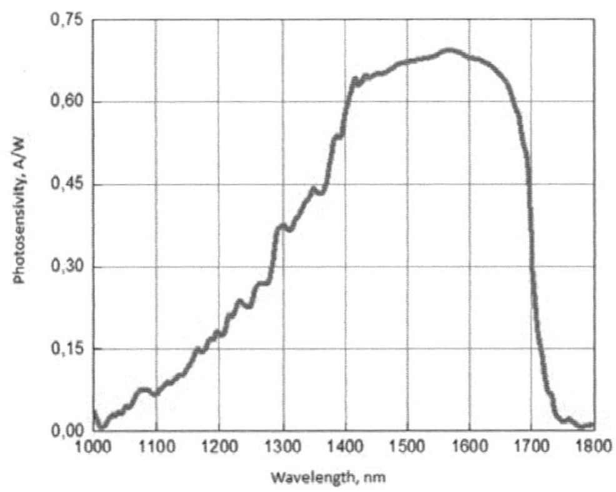

Figure 3. Photosensitivity of the PD

Figure 4 (a) shows reflection coefficient from PD crystals depicted on Impedance Smith Chart, imaginary part of PD impedance and parameters of small-signal equivalent circuit derived from the measurement. Commonly used equivalent small-signal electrical circuit representing parallel connection of intrinsic area capacitance C_j and resistance under reversed bias R_d placed in series with parasitic contact resistance R_s was examined to describe MW properties of PD crystal [1]. Value of capacitance Cj=25 fF, which was previously calculated using equation (1) with modeled resistance Rs=3 Ohm and Rd=10 MOhm obtained from IV-curves fitted the measurement data, depicted on figure 4.

$$C_j = \frac{S(\varepsilon\varepsilon_0 e\, N_d)}{2(\varphi+V)}, \quad (1)$$

where S -p-n junction area, ε – relative permittivity , e - electron charge , , N_d – donor concentration in i-region, φ - contact potential difference, V – reverse bias voltage.
Maximum operating frequency of the PD chips can be evaluated using equation (2):

$$f_c = \frac{1}{2\pi(R_s+R_L)C_j}, \quad (2)$$

where R_L – load resistance, R_L = 50 Ohm. Obtained crystals with values of capacitance C_j = 25 fF in combination with low resistance R_s can provide up to 100 GHz bandwidth.

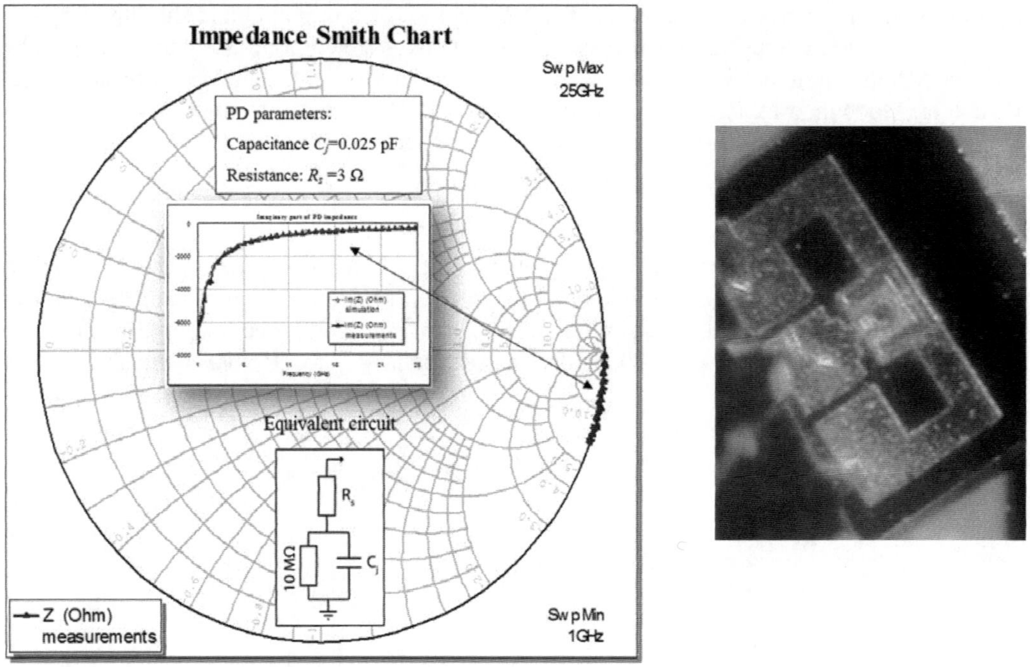

(a) (b)

Figure 4. (a) – Reflection coefficient from PD crystal shown on Impedance Smith Chart, imaginary part of PD impedance and parameters of PD equivalent small signal circuit. (b) – photo of the sample under probe station.

To realize the PD operation at MW frequencies we developed two circuits in microstrip realization. The two printed circuit board (PCB) layouts are designed for operation in circuits without (figure5a) and with (figure 5b) requirements for decoupling MW and biasing channels. We used 0.25 mm Al_2O_3 ceramic substrate to design PCB operating up to 20 GHz. Layout of hybrid monolithically integrated circuits and their equivalent circuits are shown on figure 5.

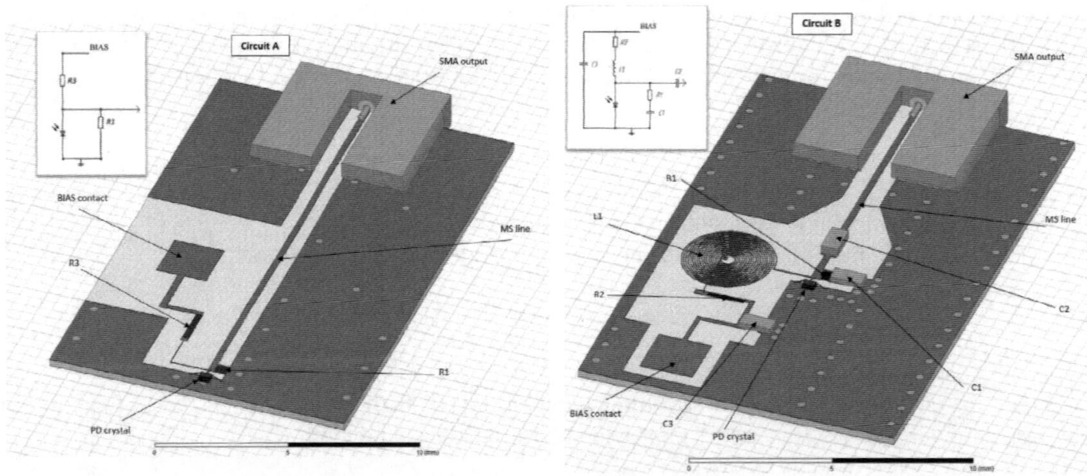

(a) (b)

Figure 5. Layouts of PCB for MW matching and biasing PD crystal: (a) – without DC decoupling, (b) –with decoupling

We considered the contribution of wire bonding, microstrip discontinuities and microstrip-coaxial junction for accurate modelling of reflection coefficient of the matched and biased PD chips. The layout of the PCBs was optimized to minimize reflection coefficient from the PD output. The critical parameters for high-frequency operation were wire-bonding length and distance between PD crystal and resistor R_l representing match load seen at output and providing broadband MW-matching. The use of thin-film resistor R_l in combination with elements on microstrip line sections and broadband surface-mount components helped to achieve 20 GHz bandwidth. The results of modelling reflection coefficient and decoupling are shown on figure 6.

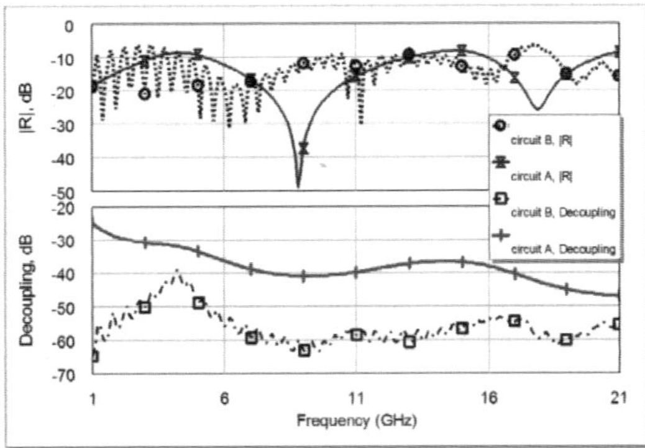

Figure 6. Reflection coefficient R from MW output of PD and decoupling between DC and MW channel

4. Conclusion

Fabricated PD chips has maximum photosensivity 0.65 A/W at wavelength 1550 nm. The good agreement between simulated and measured parameters of PD for high-speed operation is shown. Hybrid monolithically integrated circuits of impedance matching and biasing for PD operating up to 20 GHz are developed.

Acknowledgments

This work was supported by the Ministry of Education and Science of the Russian Federation in the framework of the Federal Target Program "Research and development on priority directions of scientific-technological complex of Russia for 2014-2020", the code 2015-14-582-0038, agreement № 14.581.21.0013 of August 4, 2015, unique ID RFMEFI58115X0013.

References

[1] V.J. Urick Jr., J.D. McKinney, K.J. Williams, Wiley, New Jersey, (2015), p. 473.
[2] A. Beling, J.C. Campbell, J. of lightwave technology, **27**, (2009), p. 343.
[3] Y.Geng, S. Feng, A. Poon, IEEE J. of Selected Topics in Quant. El., **20**, (2014), p. 3801807.
[4] V. Magnin, L. Giraudet, J. Harari, et. al., J. of Lightwave Technology, **20**, (2002), p. 477.
[5] P.S. Menon, A.A. Ehsan, S. Shaari, InTech, Shanghai, (2011), p. 466.
[6] S. Loquai, C. Bunge, O. Ziemann, et. al., J. of Lightwave Technology, **28**, (2009), p. 2646.

IOP Conf. Series: Journal of Physics: Conf. Series **917** (2017) 052030 doi:10.1088/1742-6596/917/5/052030

Influence of dry etching condition to geometry of vertically aligned silicon nanostructures.

I A Morozov [1] A S Gudovskikh[1, 2], D A Kudryashov[1], K P Kotlyar[1]

[1]Saint-Petersburg Academic University RAS, 194021, St.-Petersburg, Russia

[2]Saint-Petersburg Electrotechnical University "LETI", 197376, St.-Petersburg, Russia

Abstract. Recently, a new approach to the creation of solar cells based on a combination of silicon technology and III-V technology is gaining popularity. One of the interesting pair to silicon is GaP. Unfortunately, GaP has a low absorption coefficient and using it as a second junction it is required to grow it sufficiently thick that affects its electrical characteristics. To reduce the thickness of GaP, proposed to use vertically aligned silicon nanostructures, which allow increasing the absorption of light by increasing the absorption length and quantum-size effects with decreasing layer thickness. In this paper we consider the influence of dry etching condition to the geometry of vertically aligned silicon nanostructures.

1. Introduction

The theoretical calculations showed that the use of advanced surface topography could increase by factor of 1.5 the value of the short circuit current of the top junction of multijunction solar cells due to enhanced absorption [1]. This is a way to reach a currents match between the top and bottom junctions. However, technology of vertically aligned Si nanostructures formation with required geometry is a key issue for fabrication of such devices [2, 3]. Polystyrene spheres lithography followed by dry plasma etching are proposed as cost effective mass scale production technology. The aim of this study is to explore the influence of dry etching conditions to geometry of multijunction solar cells based on vertically aligned silicon nanostructures. Influence of chlorine and fluorine based chemistry as well as ICP and RIE [4] etching modes to Si nanostructure geometry was studied in this paper.

2. Experimental details.

Masks were prepared either using conventional photolithography process or using polystyrene spheres lithography with 2 micron size spheres. Etching was performed using Oxford Plasma Plasma Lab ICP 380 with two different etching mode ICP and RIE with different chemistry: Cl_2/BCl_3,SF_6/O_2. Etching temperature was -10 °C and the pressure was varied in the range of 2-7 mTorr. The RF and ICP power was varied in the range of 20-150 W and 0-2000 W, respectively. After the dry etching process the vertically aligned silicon nanostructures were cleaned in hot chloroform. Structures were studied by scanning electron microscopy (SEM).

2.1. *Fluorine silicon etching*

Fluorine etching is the main dry etching method for silicon. In most cases, it is performed using passivation gas. We did not use the passivation gas to reduce amount of process gasses. Etching parameters are shown in table 1.

Table 1. Etching parameters of fluorine process.

SF_6	O_2	P	*RF*	ICP
36 sccm	2-6 sccm	5 mTorr	20-200 W	0-1500 W

It was unable to obtain vertical walls using ICP mode with fluorine chemistry. ICP mode with fluorine chemistry without passivation is more isotropic compare to RIE mode as shown in Fig.1(a,b) for SF_6= 36 sccm, O_2=4sccm, RF power = 200 W, P = 5 mTorr. Thus, RIE mode is preferred to reach the desired geometry. Variation of oxygen flow shown that increasing of flow to 6 sccm leads to more anisotropic etch of silicon but also leads to dramatical increase of etching rate of polystyrene spheres. O_2 flow of 4 sccm was chosen as an optimum between wall passivation and polystyrene etching in oxygen plasma. Etching rate for RIE mode four times lower compared to ICP mode however, vertical walls could be achieved. Variation of RF power shows linear dependence of etch rate on power but power should not be so high to melt spheres if the cooling conditions are not enough. Etching was perfomed with 1μm and 2μm polystyrene mask. An example of SEM images of silicon nanostructures formed using fluorine based chemistry is presented in Fig.1(b,c).

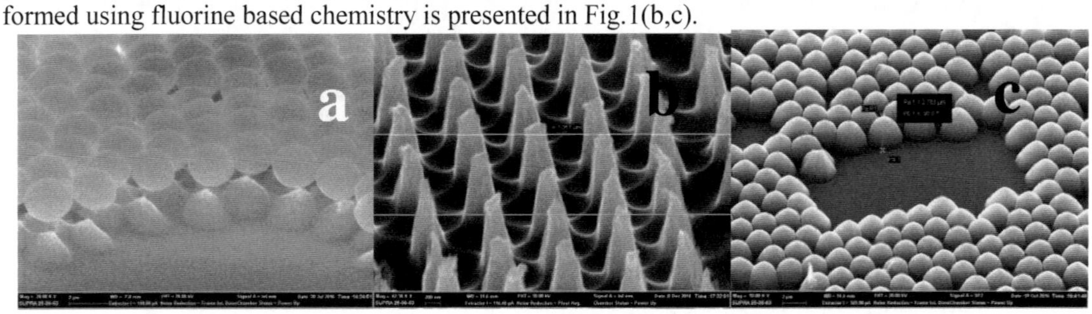

Figure 1. SEM images of fluorine etching in ICP mode without passivation (a) and of periodic vertically aligned silicon nanostructures obtained by fluorine RIE performed with 1μm (b), 2μm(c) polystyrene mask.

2.2. *Chlorine silicon etching*

To test the chlorine etching the samples with a chromium mask with strips of various thicknesses were prepared. This was done to reduce the effect of etching of polystyrene spheres in the chlorine chemistry.

Table 2. Etching parameters of chlorine process.

Cl_2	BCl_3	P	*RF*	ICP
10-30 sccm	2-6 sccm	2-7 mTorr	20-200 W	0-1500 W

Chlorine etching was performed in ICP mode because silicon etching rate in RIE mode 10 times lower. Variation of Cl_2/BCl_3 ratio from 7/1 to 11/1 shows that increase of amount of BCl_3 leads to isotropic silicon etching but etching without BCl_3show rough bottom surface. Increasing ICP power leads to increase of etching rate and increase of isotropy at the same time. Rise of RF power increases etching rate and anisotropy. First we start from Cl_2/BCl_3=7/1 and pressure of 25 mTorr (figure 2a). After some optimization steps, the most sutable etching conditions was found: P=5 mTorr, T=-28 °C, Cl_2=22 sccm, BCl_3=2 sccm, ICP=1500 W, RF=40 W. Etching profile shown in figure 2.b. A sidewalls angle close to 80° and uniform bottom of etching were achieved, but a small trenching effect was observed.

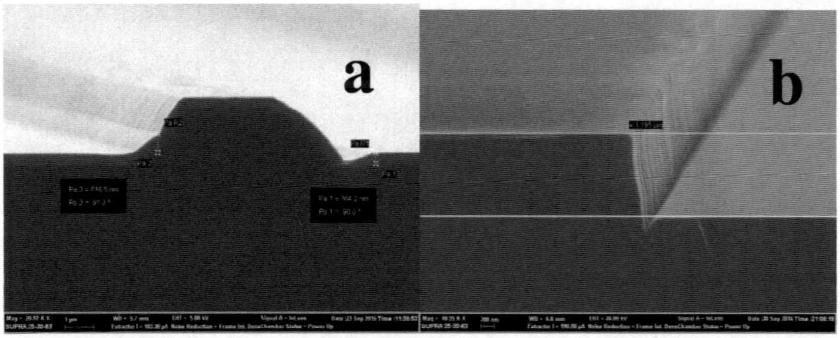

Figure 2. SEM images for structures obtained by chlorine ICP etching with Cl_2/BCl_3=7/1 (a) and Cl_2/BCl_3=11/1 (b).

Geometry of vertically aligned silicon nanostructures obtained with polystyrene mask and chlorine ICP etching is shown in figure 3. On the right a trenching can be seen that affect to electrical characteristics of silicon heterojunction solar cells. Trenching causes leakage and short circuits in the structures. Thus, chlorine shows better sidewalls and etched profile compared to fluorine etching, but its usage for real structure is difficult because of the current leakage.

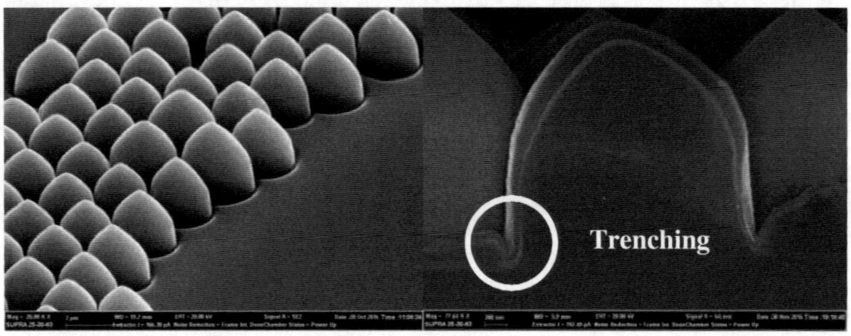

Figure 3. SEM image of vertically aligned silicon nanostructuresstructures obtained by chlorine ICP.

3. Conclusions

Vertically aligned silicon nanostructures using different etching conditions were prepared. Comparative analysis of different etching conditions was performed. Chlorine based chemistry shows better angles of side walls but have trenching effect. Trenching in vertically aligned silicon nanostructures leads to leakage current in multijunction solar cells. Fluorine based chemistry does not exhibit this trenching effects but side walls angles is worse compared to chlorine one. It is related to etching of polystyrene spheres mechanism. For chlorine based process the nanospheres are etched anisotropic, while for fluorine based process rather the isotropic etching in oxygen is observed.

Acknowlegements
This work was carried out in the framework of the joint Russian-French project LAMSOL supported by Ministry of Science and Education of Russian Federation (RFMEFI61615X0040) and by PHC Kolmogorov Program (35522TL).

References

[1] Dmitrii A. Kudryashov, A.S. Gudovskikh. Photocurrent Enhancement in GaPNAs-based Solar Cells With Si Nanowire Array Substrate // Proceedings of 14th International Conference on Numerical Simulation of Semiconductor Optoelectronic Devices. 1-4 September 2014. Palma de Mallorca, Spain. P.171-172

[2] Nanosphere lithography: A materials general fabrication process for periodic particle array surfaces John C. Hulteen and Richard P. Van Duyne Journal of Vacuum Science & Technology A: Vacuum, Surfaces, and Films 13, 1553 (1995); doi: http://dx.doi.org/10.1116/1.

[3] Nanosphere lithography: a powerful method for the controlled manufacturing of nanomaterials P Colson, C Henrist, R Cloots - Journal of Nanomaterials, 2013 - dl.acm.org

[4] Critical tasks in high aspect ratio silicon dry etching for microelectromechanical systemsIvo W. RangelowJournal of Vacuum Science & Technology A: Vacuum, Surfaces, and Films 21, 1550 (2003); doi: http://dx.doi.org/10.1116/1.1580488

IOP Conf. Series: Journal of Physics: Conf. Series **917** (2017) 052031 doi:10.1088/1742-6596/917/5/052031

Correction factor in temperature measurements by optoelectronic systems

N Bikberdina, R Yunusov, M Boronenko, P Gulyaev
Department of Physics and technical disciplines, Ugra State University, Khanty-Mansiysk 628012, Russia

Abstract. It is often necessary to investigate high temperature fast moving microobjects. If you want to measure their temperature, use optoelectronic measuring systems. Optoelectronic systems are always calibrated over a stationary absolutely black body. One of the problems of pyrometry is that you can not use this calibration to measure the temperature of moving objects. Two solutions are proposed in [1]. This article outlines the first results of validation [2]. An experimentally justified coefficient that allows one to take into account the influence of its motion on the decrease in the video signal of the photosensor in the regime of charge accumulation. The study was partially supported by RFBR in the framework of a research project № 15-42-00106

1. Introduction

Currently, the study of fast high-temperature processes is impossible without the use of optoelectronic systems (OES). With the help of OES successfully measure the temperature and speed of fast processes. The accuracy of the results is influenced by the noise and distortion. Global shutter in such systems significantly increases the measurement accuracy. However, apart from noise on the accuracy of measurement is affected by methodological error. All of the ECO calibrated at a fixed standard blackbody [2]. The present work raises the problem of imperfection of the methods of measuring the temperature of moving self-luminous objects with the aid of ECO. One of the most promising non-contact methods for measuring temperature is undoubtedly the method of spectral pyrometry. It allows to determine the temperature from macro to nano-objects with unknown emissivity. However, spectral pyrometry is used to determine the average temperature of a certain surface, in the diagnosis of mainly stationary processes. Brightness pyrometry is good for nonstationary processes, but emissivity must be known. The combination of spectral and brightness pyrometry is complementary [3-4]. However, here again it is necessary to take into account the superposition of the radiation of the detected particles in their spatial displacement.

2. The proposed hypothesis

Let it be necessary to measure the temperature of a moving hot microparticle. The energy of radiation of a particle during the accumulation time of a photodetector:

$$\Delta E = \int_0^t \Omega(t)dt.$$

The input optical signal of the photodetector is converted into an electrical signal described by *I(x, y)* is a bivariate continuous function discrete spatial arguments *x, y*.

Conversion of the input signal optical system and the image formation on the sensing site of the photodetector can be represented as follows:

$$I_{\Omega} = \int_{0}^{l} g(x, y) \cdot I(x, y) dxdy,$$

Where *g(x, y)* is the weight function. The image on the monitor is in the form of an area, the brightness of the pixels that are above the background. After calibrating the ECO over an absolutely black body, certain values of the brightness of the image pixels correspond to the brightness temperature.

$$\bar{I} = \frac{1}{n \cdot m} \cdot \sum_{i,j}^{n,m} I_{i,j},$$

The entire image is represented as a temperature map. If the radiating object (particle) is stationary, the charge accumulation in the photosensitive cell occurs all the time of exposure, illuminating the same cells. During the accumulation of charge in the photosensitive region of the sensor, the flying luminous particles move by a distance depending on their velocities. The track of a moving particle with velocities of 100 - 1140 m/s is less bright. Calibration on a fixed black body can not be used. To overcome the problem, two approaches are possible. The following assumptions are accepted:

- ✓ radiation of a particle uniformly from its entire surface. The brightness of the radiating particle surface is assumed to be equal to the average value of the amplitude of the signal of the photosensitive cells of the illuminated region (the brightness of the pixels of the particle image).
- ✓ during the exposure time, the radiation intensity of the particle remains unchanged. This means that the amplitude of the signal of the photosensitive cells of the illuminated region remains constant.
- ✓ the particle speed does not change during the exposure time.
- ✓ "image" of a spherical fixed particle of radius r in the image is a square with side $n = 2r$, a moving particle illuminates a region of size $n \times m$.

The first approach takes into account the decrease in charge accumulation time by photosensitive sensors due to the decrease in charge accumulation time. This means that the temperature measurement is carried out only by those photosensitive cells that would be illuminated by the same light source if its velocity were zero. So you can apply calibration on a motionless black body. This approach can be used at relatively low speeds of a moving light source. For example, registration of a radiating object (T> 2000C, υ of the order of 100 m/s) can be performed at microsecond exposures. In this case, the brightness of the study is sufficient to form an image. If the speeds are very high, of the order of 1000 m/s, then the track is a long, but very small brightness. In this case, you can not switch to a smaller exposure. The emitter can not be considered stationary. Calibration on a fixed black body can not be used. In this case, it is necessary to apply the second approach.

The second approach is based on the principle of superposition of the charge accumulated during the exposure time. At the length of track *m/n* such particles fit. Therefore, taking into account the superposition principle, the integral brightness, according to which one should judge the temperature of a moving particle in a discrete representation:

$$I' = \frac{m}{n} \cdot \bar{I} = k \cdot \bar{I},$$

Where$\bar{I}$ - is the average luminance value in the track, k≥1. Thus, despite the seeming similarity, the correct use of approaches makes it possible to significantly expand the measurement limits for optoelectronic systems.

3. Technique and procedure of the experiment

It consists of: a mini-oven «JC Small Melting», a standard optical diaphragm (1 mm); Electric drill, video camera SOOCO S70. Carrying out experiments with a high-speed camera Video Sprint (1000 f / s, exposure 20 ns-20 μs) is difficult due to the lack of a test object with the necessary characteristics.

The tube furnace is heated to one of the temperatures of 600-800C, maintained during video shooting. In the clamp of an electric drill instead of a drill, an optical diaphragm was attached. The electric drill was rigidly mounted so that when the diaphragm rotated through the hole, the bottom of the tubular furnace could be seen. The diaphragm is parallel to the bottom of the tubular furnace and to the lens. The distance from the aperture to the video camera l >> 1 mm. The video was taken at a fixed temperature of the tube furnace (120 frames per second, the accumulation time is constant).

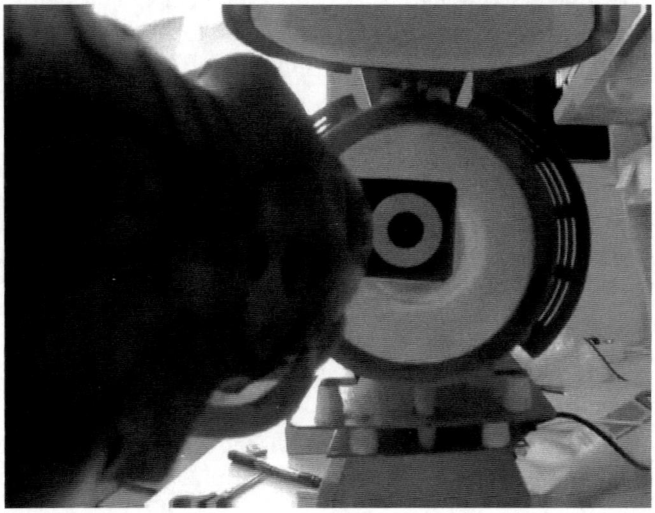

Figure 1. Experimental setup.

This was an example of the figure. The moment of interest corresponds to the same spatial position of the diaphragm between the lens and the tubular stove. Through the opening of the diaphragm, the thermal radiation of the heated bottom of the tube furnace was observed. The movement of the hole simulated the motion of a self-luminous particle. The purpose of the experiments is to compare the brightness of the image of the aperture of the fixed diaphragm and the brightness of the track "of the moving particle and to reveal the connection of the brightness of the track from its area. The resulting video files were analyzed in ImageJ. An example of the received frames can be seen in Figure 2 (in conventional colors). In the program ImageJ, we outlined the tracks, measuring their area and brightness.

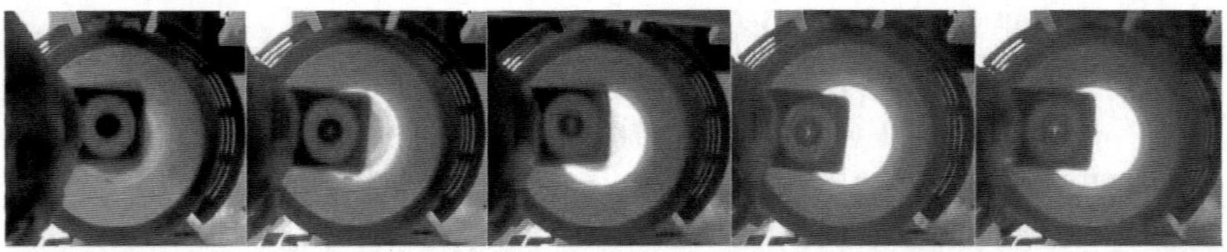

Figure 2. The sequence of frames at different temperatures of the mini-oven (in conventional colors).

The area of tracks strongly depends on the quality of contouring. At low temperatures, the contrast is small, the dispersion of areas is large, if the contours are searched automatically. The values of the track area, normalized by the area of the fixed hole, were plotted along the abscissa axis.

The scatter of the areas and brightness of the tracks with the automatic selection of the contours can be seen in Figure 3. Therefore, before the binarization to the duplicate video file, the technological methods of image processing were applied: "unsharp masking" and "subtract background", etc.

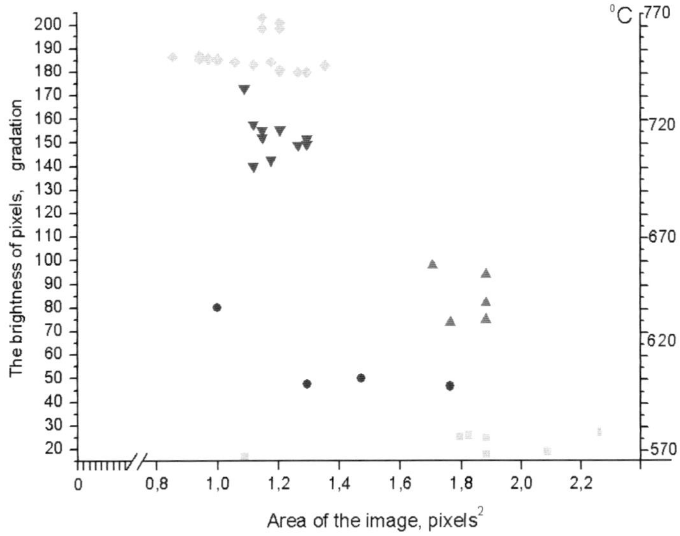

Figure 3. The automatic selection of tracks on the images T=600C, T=650C, T=700C, T=750C.

The average values of the areas of the images were taken as true ("fixed"). For each temperature of the furnace, a hole/track of "fixed" values was found. To find a mathematical model describing the change in the brightness of the image of the hole as it moves, the experimental data were approximated by a linear function. Figure 4 shows the results for some temperatures.

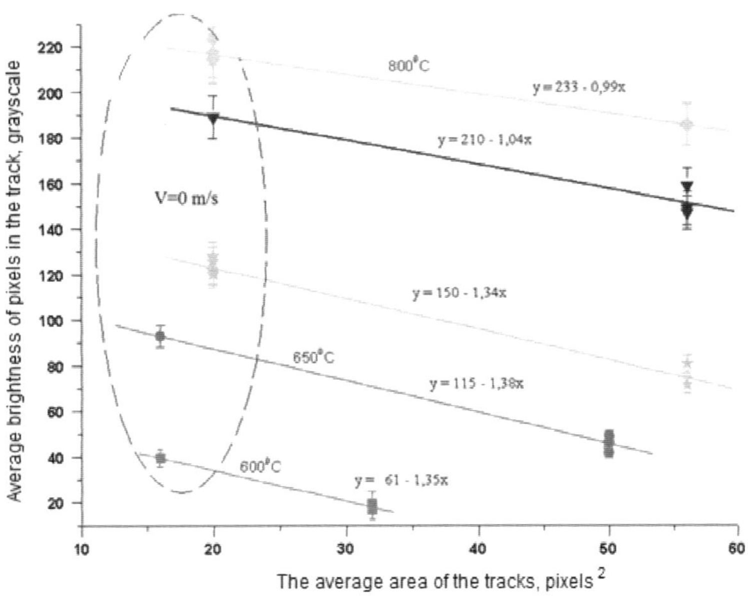

Figure 4. Linear approximation of the data.

Despite the imperfections of the installation used, all straight lines have practically the same angle of inclination. This confirms the correctness of the results, because The accumulation time is constant. For any two points belonging to the same straight line in the coordinates (*S; I*):

$$\frac{\bar{I} - I}{\bar{S}} = -tg\,\alpha = const.$$

To keep points on a straight line, it is necessary that the enlargement of the image area of the hole when moving the diaphragm is the same as reducing its brightness. With sufficient accuracy this requirement is satisfied by the experiment:

at 750C the ratio $\bar{S_0}/S = 0{,}74, \quad \bar{I}/I_0 = 0{,}84;$

at 800C the ratio $\bar{S_0}/S = 0{,}74, \quad \bar{I}/I_0 = 0{,}87;$

Thus, for temperatures T > 650C, possible to obtain contrast images experimentally shown to increase image area by the same factor, how much the brightness is reduced. Therefore, the second approach, based on the principle of superposition of the accumulated charge during the exposure, is correct.

4. Main results and conclusions

The purpose of the experiment was to test the validity of introducing a correction factor in the method of measuring the temperature of a moving object. The introduction of the correction factor k allows one to take into account the influence of the speed of motion on the reduction of the video signal of the photosensor (in the charge accumulation mode). The proposed technique allows to use optoelectronic measuring systems as a micro-pyrometer of static objects and an anemometer micro-pyrometer for two-phase flows with particles dmin = 33μm moving at a speed of 100 m/s and having a temperature of not less than 1500 K.

Acknowledgments

The study was partially supported by RFBR in the framework of a research project № 15-42-00106

References

[1] Boronenko M P, Gulyaev P Yu, Seregin A E, Poluhina K G 2015 Journal of Physics Conference Series **643** 012028

[2] Boronenko M P, Gulyaev P Yu, Seregin A E, Bebiya A G 2015 IOP Conference Series: Materials Science and Engineering **93** 012021

[3] Dolmatov A V, Gulyaev I P, Jordan V I, IOP Conference Series: Materials Science and Engineering 2015 **81** 012041

[4] Gulyaev P Yu, Kotvanova M K, Pavlova S S, Sobol E N, Omelchenko A I 2012 Nanotechnologies in Russia **7** 127

Er:Yb phosphate glass laser with nonlinear absorber for phase-sensitive optical time domain reflectometry

A A Zhirnov[1], A B Pnev[1], C Svelto[2], M Norgia[2], A Pesatori[2], G Galzerano[3], P Laporta[4], D A Shelestov[1], V E Karasik[1]

[1] Bauman Moscow State Technical University, 2-nd Baumanskaya st. 5, 105005 Moscow, Russia
[2] DEIB, Politecnico di Milano, Via Ponzio 34/5, 20133 Milano, Italy
[3] IFN-CNR, Piazza Leonardo da Vinci 32, 20133 Milano, Italy
[4] DFIS, Politecnico di Milano, Piazza Leonardo da Vinci 32, 20133 Milano, Italy

Abstract. A novel laser for phase-sensitive optical time-domain reflectometry (Φ-OTDR) is presented. The advantages of a compact solid-state laser are listed, current problems are shown. Experiments with a microchip single-optical-element laser, from setup construction to usage in Φ-OTDR system, are presented. New laser scheme with two-photon intracavity absorber is suggested and its advantages are described.

1. Introduction

Narrow linewidth lasers at 1.5 μm are very promising devices in different areas, e.g. LIght Detection And Ranging(LIDAR), Φ-OTDR, and as a local oscillator (LO) for optical frequency comb metrology. Nowadays most popular lasers of this kind are: stabilized fiber lasers (FL), semiconductor external-cavity laser diodes (ECLD) with feedback via Bragg grating, and whispering gallery mode (WGM) lasers. High price, defined by complicated production technology, is their common disadvantage. The main goal of this work is to develop optical systems, based on narrow linewidth laser, more widespread by creating less expensive light source.

This paper is organized as follows. In the first part we describe application of narrow linewidth laser in a Φ-OTDR system. In the second part we describe our experiments with single-optical-element microchip laser. Thirdly we present a new scheme and its components. In the fourth part we show our future steps in narrow-linewidth laser development.

2. Φ-OTDR system

Φ-OTDR is nowadays a popular distributed fiber optic sensor [1-3]. Its working principle is presented in figure 1. Light from a laser source (1) is amplified in EDFA-booster (2) to obtain the required power level [4, 5]. Then it is formed into pulses via acousto-optical modulator (3). Through circulator (4) the optical pulses go into sensing fiber (5) with a time period longer than round-trip-time needed to pass fiber (5) forward and backward. From each local section of the sensing fiber a small part of pulse power is backscattered by inhomogeneities – Rayleigh scattering centers. All scattered light waves from these inhomogeneities in the range of one optical pulse are summed in amplitude also taking into account their phases due to the laser source narrow linewidth and the corresponding large coherence length. Since optical phases are randomly distributed, we obtain one-dimensional back-reflected speckle signal or interferogram which attenuates to the far end of the sensor. This returning signal is collected by the circulator and amplified by a second EDFA (6). Amplified spontaneous emission noise generated by the second EDFA is cut off by a narrow-band optical filter (7). Finally we obtain signal by detection on photoreceiver (8), transform it to digital via analog-to-digital converter (9) and then process it on personal computer (10). Received signals are called reflectograms. Their shape remains quite stable while there are no acoustical or vibration sources in proximity of the sensor. Any acousto-mechanic perturbation of the fiber sensor will change the position of backscattering centers and hence the optical phases of reflected waves and the result of their summation, thus altering the reflectogram shape. Analysis of reflectograms fluctuations can give us information about events happening in proximity of the sensor: events "localization", as distance from the modulated laser source, as well as events "strength" can be continuously monitored in this way.

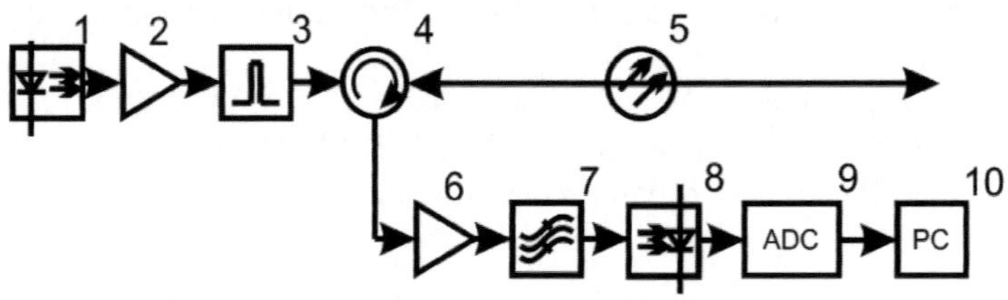

Figure 1. Φ-OTDR scheme: 1 – laser source, 2 – EDFA-booster, 3 – optical modulator, 4 – optical circulator, 5 – sensing fiber, 6 – EDFA-preamplifier, 7 – optical filter, 8 – photoreceiver, 9 – ADC, 10 – personal computer.

In this scheme the quality of the laser source plays a crucial role. It must have sufficiently large coherence length to obtain coherent summation of waves by amplitudes instead of typical summation by power happening in commercial telecommunication OTDRs [6] using low-coherence sources. Also, for reflectogram stability, the laser source must be stable in terms of its optical frequency (or wavelength). Since obtained signal is interference between backscattered light waves, the fluctuations of laser wavelength will strongly influence on summation result. Existing research shows that requirements for the source frequency stability are: laser linewidth less than 1 MHz over 1 ms [7] and frequency fluctuations no more than 100 MHz over 1 min.

Design and construction of such a low-noise 1.54 μm laser source, satisfying the above requirements and quite simple/inexpensive in production, is the goal of this work.

3. Single-element microchip laser

Our first step in the source development was to test a microchip laser assembled according to the scheme in figure 2. Pump at 976 nm from fiber-coupled LD goes through a telescopic focusing system to enlarge spot size diameter from the original 6 μm at fiber output up to 42 μm. The active medium is an athermal phosphate glass doped with Er^{3+} and Yb^{3+} ions (Yb^{3+}:Er^{3+} ratio is 20:1). The microchip thickness is 230 μm as measured from longitudinal modes separation. Coatings parameters at the different laser and pump wavelengths are the followings. For the first surface $R1_{976nm}$=0.2 and $R1_{1540nm}$=0.999, for the second surface $R2_{976nm}$=0.9 and $R2_{1540nm}$=0.98. More detailed data about this laser can be found in [8].

Unexpected feature of this monolithic source was the significantly high amplitude noise due to relaxation oscillations at a Fourier frequency of ~1 MHz. This large intensity instability, up to 20 % of the mean value, is shown in the oscilloscope trace of figure 3. In Φ-OTDR system these fluctuations impact on reflectogram stability resulting in very poor Signal-to-Noise detection. Also this intensity noise makes linewidth measurements [9] a very difficult task. An opto-electronic feedback loop was used to decrease the oscillation peak by 27 dB but has a negative consequence. Laser wavelength of this microchip laser depends significantly on pump power [10], so feedback control by pump power also changes the laser wavelength. Influence of active feedback on frequency stability, at different loop gains, is presented in figure 4 in terms of the laser fractional frequency ($y=\Delta\nu/\nu$) Allan deviation [11] and in figure 5 in terms of frequency deviations plots as a function of time. This frequency instability, as seen in the end of Section 2, is undesirable in such a device.

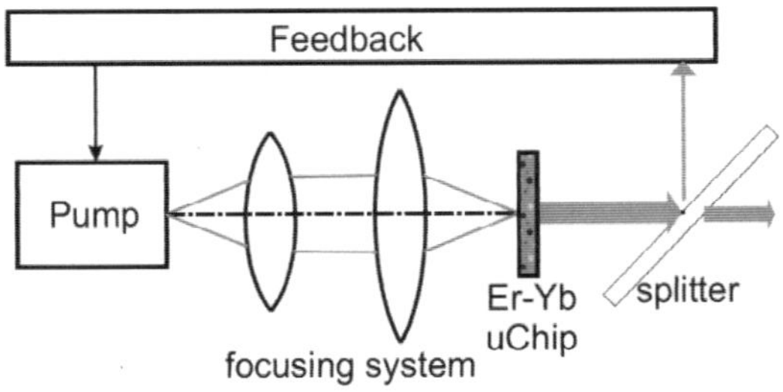

Figure 2. Scheme of single-element microchip laser with optoelectronic feedback loop.

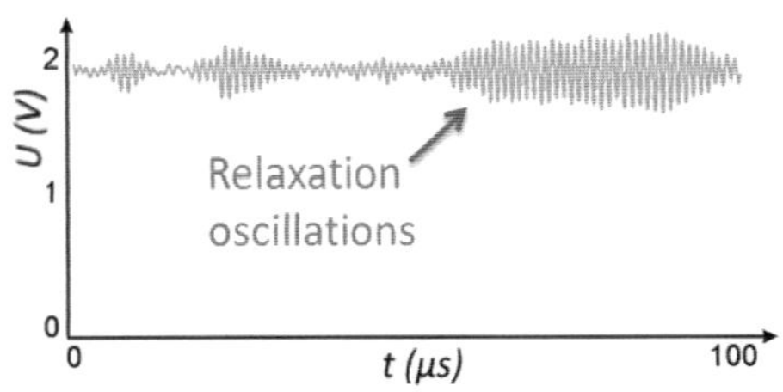

Figure 3. Graph of power intensity instability due to relaxation oscillations.

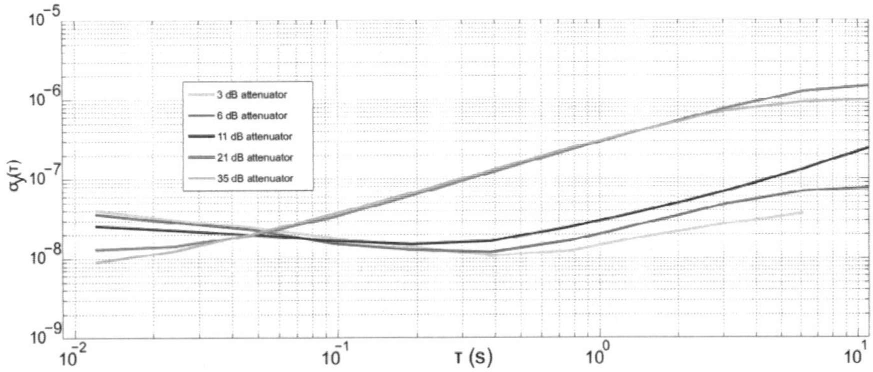

Figure 4. Allan deviation of laser frequency with different gain feedback.

To conclude of this step of laser source design, we tested the microchip laser in the Φ-OTDR scheme shown in figure 6. This scheme is different from the scheme in figure 1. Firstly, we need more gain to amplify radiation entered into fiber because of non-optimal input coupling (can be fixed in future). So instead of one EDFA-booster we have two EDFAs (2, 4) with optical filter 3 to suppress spontaneous emission from EDFA 2. Secondly, microchip laser operates at ~1534 nm, which is erbium emission spectrum peak, instead of 1550 nm of the laser wavelength in figure 1. Because of operation at the erbium emission peak we have higher noise level from EDFA-preamplifier (6 in figure 1) which is even higher than reflectogram level. This is the reason for using detecting part of the system (8, 9 and 10) without preamplifier but with more sensitive and slower photoreceiver. We used LCA-S-400K (FEMTO Messtechnik GmbH) with 400 kHz bandwidth and NEP 75 fW/√Hz instead of JDSU EPM605 with custom electrical amplifier scheme used in system from figure 1, having 5 MHz bandwidth and NEP 5 pW/√Hz. The obtained reflectogram signals are presented in figure 7.

IOP Conf. Series: Journal of Physics: Conf. Series **917** (2017) 052032 doi:10.1088/1742-6596/917/5/052032

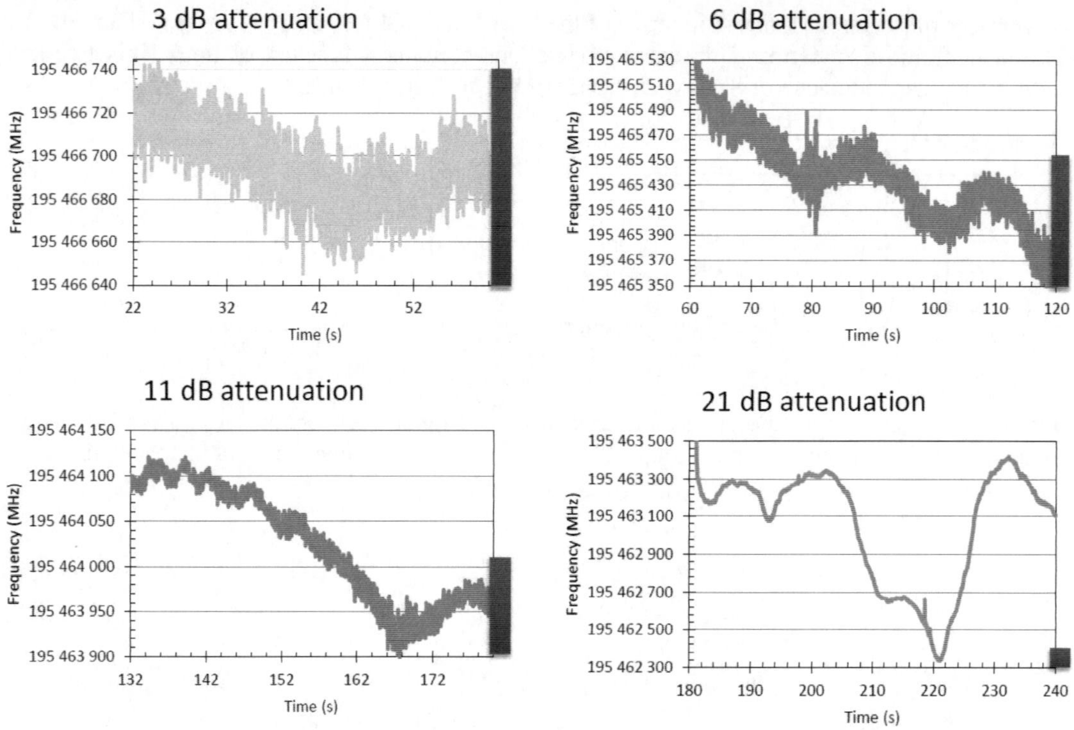

Figure 5. Laser frequency fluctuations during different observation times using different gain in the feedback loop. Violet bar on right side shows 100 MHz fluctuation scale.

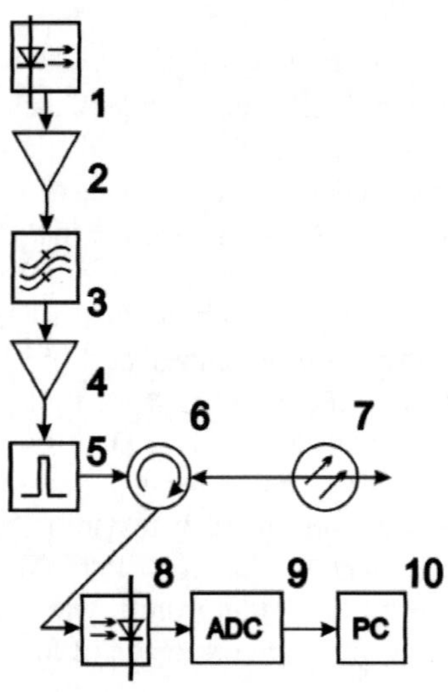

Figure 6. Φ-OTDR scheme for microchip laser testing.

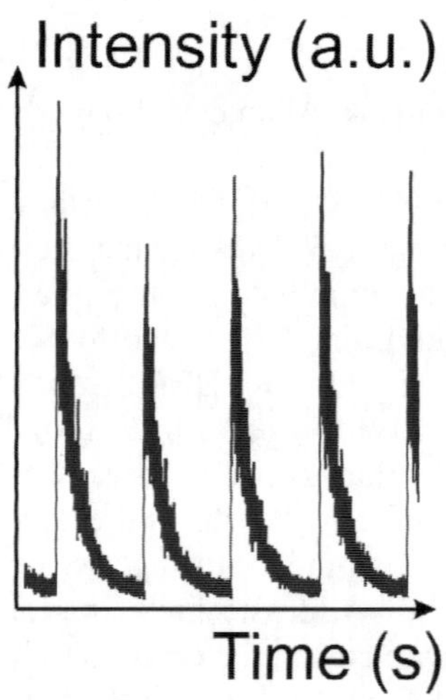

Figure 7. Graph of reflectograms time sequence.

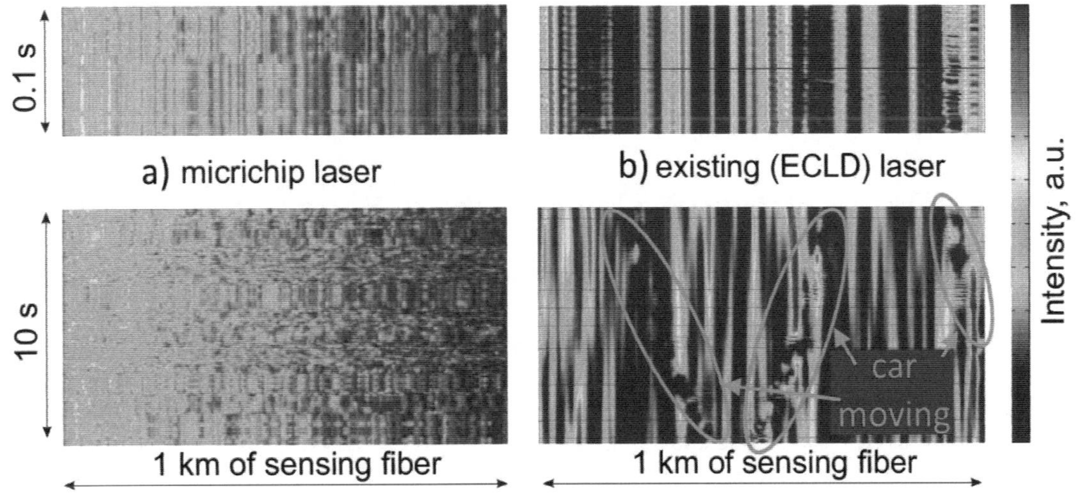

Figure 8. Signals from Φ-OTDR system with microchip laser (a) and ECLD (b).

To enable signal processing of this signal with unstable pulse intensity we used reflectogram normalization. Random nature of signal doesn't allow choosing one special point of reflectogram to compare its power with the others. Moreover, optical and electrical noise has quite stable intensity in the signal from the far sensor end. Accordingly, we calculated normalizing coefficient as follows:

$$NK_i = \frac{\sum_{t=20}^{500} I_{1,t}}{\sum_{t=20}^{500} I_{i,t}} \quad (1)$$

where NK_i is normalizing coefficient for the *i-th* reflectogram, $I_{i,t}$ is intensity from *i-th* reflectogram from *t-th* point of sensor. This equation allows avoiding the influence of large fluctuations in the beginning of sensor and noise from the far end. After normalizing each reflectogram

$$I_{i,t}^{N} = I_{i,t} NK_i \quad (2)$$

we obtained the signal presented in figure 8a. For comparison we also show signal from Φ-OTDR with nowadays used ECLD source in figure 8b. Analysis of figures 8a and 8b allows us to draw the following conclusions. Firstly, light from microchip laser has coherence length large enough to be used use it in Φ-OTDR systems, since we see random intensity modulation along the sensor. However, its contrast is less than 1 (intensity interference minima don't reach values about zero). Secondly, on small periods of time laser frequency is stable enough to analyze fluctuations of reflectograms, *e.g.* in figure 8a (picture with 0.1 s duration) positions of intensity peaks and valleys are stable. But signal from ECLD is much better also due to the lack of imprecise normalization procedure. Thirdly, higher EDFA noise at 1535 nm working wavelength instead of 1550 nm reduces the maximum sensing length to a few kilometers for the microchip laser as compared to the 40 km attainable length with the ECLD. Effect is that with ECLD we easily can see the car movements along the cable and with microchip laser we can not detect such events.

Results of the preliminary work with this laser showed that we need to reduce the laser intensity noise in particular at the relaxation oscillations frequency, make the optical frequency more stable and shift the working wavelength away from the erbium emission peak.

4. Scheme with two-photon absorber

On the basis of previous Section conclusions, we propose a new scheme, presented in figure 9, for the solid-state laser to be used in Φ-OTDR systems.

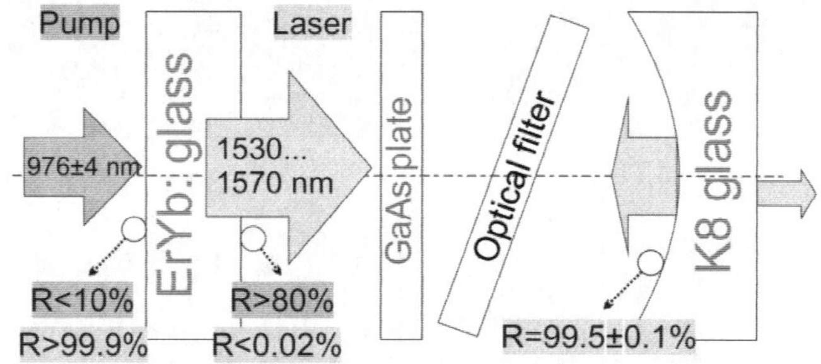

Figure 9. Scheme of Er-Yb:glass laser with two-photon absorber.

To fulfill the amplitude and frequency noise requirements, as well as wavelength tunability, we have to make a cavity with few elements. Laser cavity is formed by two mirrors: one on the first surface of active medium chip and the second one on the concave surface of K8-glass lens. This discrete-cavity design allows inserting in the laser resonator a few needed optical elements. The first element is a GaAs plate used to reduce relaxation oscillations due to the semiconductor nonlinear absorption at 1.54 μm [12]. In such a way we will be able to stabilize output intensity without influence on laser wavelength, that was a negative side of opto-electronic feedback from previous scheme in figure 2. The second intracavity element is an optical filter which allows choosing the laser operating wavelength and shift it away from the erbium emission peak to reduce noise from EDFA-preamplifier (6 in figure 1).

Suggested scheme is more sensitive to optical elements vibrations, so last step can be done to integrate a monolithic multi-element cavity, *e.g.* like in [13]. Before this final step, the optimal design of intracavity elements must be found also from experimental results.

Acknowledgments

Authors are thankful to A Borisova, K Koshelev, E Nesterov, K Stepanov and A Chernutsky for their help with experiments. This work was economically supported under the Grant agreement No 14.577.21.0224 (Unique identifier RFMEFI57716X0224) with RF Education and Science Ministry.

References

[1] Juarez J C and Taylor H F 2007 *Appl. Opt.* **46(11)** 1968-71
[2] Alekseev A E, Vdovenko V S, Gorshkov B G, Potapov V T and Simikin D E 2014 *Laser Phys.* **24(11)** 115106
[3] Fedorov A K, Anufriev M N, Zhirnov A A, Stepanov K V, Nesterov E T, Namiot D E, Karasik V E and Pnev A B 2016 *Rev. Sci. Instrum.* **87(3)** 036107
[4] Nesterov E T, Zhirnov A A, Stepanov, K V, Pnev A B, Karasik V E, Tezadov Y A and Ushakov A B 2015 *J. Phys.: Conf. Ser.* **584** 012028
[5] Martins H F, Martin-Lopez S, Corredera P, Salgado P, Frazão O and González-Herráez M 2013 *Opt. Lett.* **38(6)** 872-4
[6] Gold M P 1985 *J. Lightwave Technol.* **3** 39
[7] Choi K N, Juarez J C and Taylor H F 2003 *Proc. SPIE 5090* **134**
[8] Pniov A *et al* 2016 *ACTA IMEKO* **5(4)** 24-8
[9] Derickson D 1998 *Fiber optic test and measurement* (Upper Saddle River, NJ: Prentice Hall)
[10] Zayhowski J J and Keszenheimer J 1992 *J. Quantum Electron* **28(4)** 1118-22
[11] Allan D W 1966 *Proc. IEEE* **54(2)** 221-31
[12] Watkins L S, van Leeuwen R, Xu B, Wang Q and Ghosh C 2009 *AVFOP* **09** 7-8
[13] Shoji T D, Xie W, Silverman K L, Feldman A, Harvey T, Mirin R P and Schibli T R 2016 *Optica* **3(9)** 995-8

Admittance spectroscopy of InGaAsN based solar cells

A I Baranov[1,2,a], A S Gudovskikh[1,3], D A Kudryashov[1], A M Mozharov[1], K S Zelentsov[1], J-P Kleider[2]

[1] St Petersburg Academic University of RAS, 194021 St Petersburg, Russia.

[2] GeePs (Group of electrical engineering – Paris), UMR CNRS 8507, CentraleSupélec, Univ. Paris-Sud, Université Paris-Saclay, Sorbonne Universités, UPMC Univ Paris 06, 91192 Gif-sur-Yvette CEDEX, France.

[3] St Petersburg Electrotechnical University "LETI", 197376 St Petersburg, Russia.

[a]baranov_art@spbau.ru

Abstract. Single-junction InGaNAs solar cells were grown by MBE with active layers based on a GaAsN/InAs superlattice. Dependence of defect formation on thickness of InGaNAs was explored in the study. Thickness increase from 900 nm to 1200 nm leads to defect formation with two activation energies of 0.20 eV and 0.50 eV but the value of quantum efficiency stays almost the same. Further thickness increase up to 1600 nm leads to the increase of defect concentration in the InGaNAs active layer. These defects are non-radiative recombination centres because a significant decrease of solar cell quantum efficiency was observed. The existence of a critical thickness for defect-free growth of InGaNAs based on GaAsN/InAs superlattice is proposed.

1. Introduction

In 2015 multi-junction solar cells (MJ SC) based on III-V compounds with a record efficiency of 46% were fabricated [1]. Triple-junction SC based on the GaInP (1.85 eV) / GaAs (1.42 eV) / Ge (0.7 eV) system are used in industry for space applications where the key factors are efficiency and resistance to radiation. But its concentrator efficiency has almost reached the theoretical limit (46%) and now its value is more than 40% [2]. Fortunately, according to theoretical estimations an additional subcell with energy bandgap of 1 eV could increase the SC efficiency up to 52% [3]. However, such subcell should be lattice-matched to Ge and GaAs wafers. $Ga_{1-x}In_xN_yAs_{1-y}$ with a small content of nitrogen is one of the most promising alloys to reach this goal. Such III-V-N alloys (GaPNAs, InGaNAs etc.) with nitrogen content less than 5% are called dilute nitrides. It has been shown that the small addition of nitrogen leads to large bowing parameters for the bandgap in quaternary $Ga_{1-x}In_xN_yAs_{1-y}$ alloys: already few percent of nitrogen reduces the bandgap by hundreds of meV [4] and a value of 1 eV can be achieved. Furthermore, these layers can be epitaxially grown on Ge and GaAs wafers when y=0.35x. Various groups [3, 5-8] studied InGaNAs layers grown by epitaxy, but the quality is not sufficient for fabrication of high-efficiency solar cells. The low lifetime of charge carriers is the most crucial issue in dilute nitrides. The main reason is the high concentration of non- radiative recombination centers in active layers. In MOCVD (metal- organic chemical-vapor deposition) it comes from the background doping of carbon and hydrogen during the process. On the other hand, MBE (molecular-beam epitaxy) allows one to avoid these problems and enables to control the flow of nitrogen more precisely. However, a lower process temperature is required for the growth of GaAsN compared with that of GaAs in MBE: it leads to defect formation reducing dramatically the lifetime in InGaNAs. Some researchers try to avoid the defect formation by addition of antimony in active layers of InGaNAsSb [6]. Unfortunately, Sb can deposit on the wall of the chamber which leads to negative background doping in subsequent layers. Consequently it is necessary to have two chambers for the growth of MJ SC, but this renders the technology much more complicated.

Recently, a new method was explored for the fabrication of InGaNAs based compounds without antimony [8]. It is based on the growth of a GaAsN/InAs superlattice by MBE. In this paper properties of i- GaAsN/InAs layers in single-junction SC by admittance spectroscopy will be explored.

2. Experiments and samples

Three p-i-n structures with different thickness (900 nm, 1200 nm, 1600 nm) of active InGaNAs i-layer based on GaAsN/InAs superlattice were grown on n-type GaAs wafers by MBE using a Veeco GEN III setup with a plasma nitrogen source. A p-type GaAs layer was grown on the top as a wide-bandgap window (concentration of acceptors is 10^{19} cm^{-3}). Silicon and beryllium were used for n- and p- type doping in GaAs layers, respectively. The bandgap energy of the InGaNAs layer is equal to 1 eV as confirmed by PL measurements. Metallic Au/Ge and Au/Zn alloys were evaporated on the bottom and top side of structure, respectively, for the fabrication of contacts. Ohmic behaviour was obtained after rapid thermal annealing at 400 °C. Au/Zn was evaporated as circular contacts with diameters of 0.25, 0.5, 1 and 2 mm for capacitance measurements. Then samples were etched in a solution of H_2SO_4:H_2O_2:H_2O=5:1:1 for the fabrication of mesa-structures. The sample structure is schematically presented in Figure 1.

The admittance spectroscopy technique allows one to estimate the activation energy (E_a) and the capture cross section (σ_a) of the defect level from the analysis of the temperature and frequency dependence of the capacitance, C(T,f), and of the conductance, G(T,f) [9]. The measurements were performed using a liquid nitrogen cryostat in the temperature range from 80 K to 360 K and at frequencies from 20 Hz to 1 MHz. Three different value of bias voltage were used (-0.4 V, -0.2 V, 0.0 V) and the test signal amplitude was set to 50 mV.

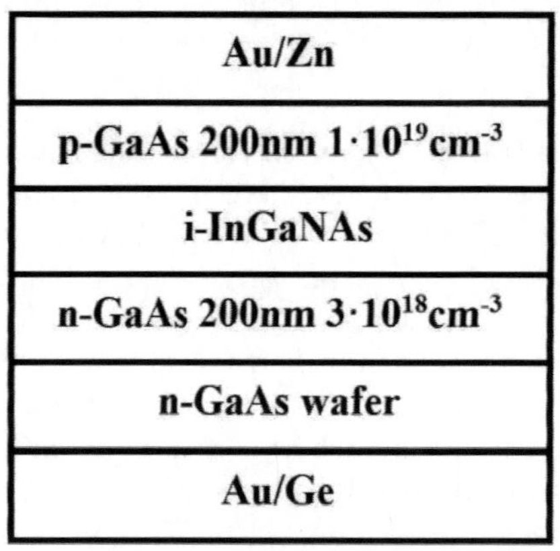

Figure 1. Schematic view of the p-i-n structure with i- InGaNAs layer.

3. Results

The measured temperature and frequency dependences of the capacitance are shown in Figure 2 for zero bias voltage. The capacitance of the sample with the 900 nm thick InGaNAs layer decreases monotonously with the increase of frequency for all temperatures. No step is observed in C(f) curves meaning no response from defect level in this temperature range. However, the shape of C(f) curves is dramatically changed when the InGaNAs layer thickness is increased to 1200 nm. Detailed analysis of C(f) curves allows one to observe two small steps: a first step at low temperatures 100..220 K and a second step at high temperatures 320..360 K. These steps can be caused either by bulk defect levels in the undoped layer or by interface states at InGaNAs/GaAs heterojunctions. The step position in C(f, T) curves does not depend on applied bias voltage (measurements are not shown here) indicating that the response originates from bulk defects rather than from the InGaNAs/GaAs interface. Further increase of thickness to 1600 nm leads to a drastic increase of the amplitude of the first step while the second step in C(f)curves remains the same (Figure 2c).

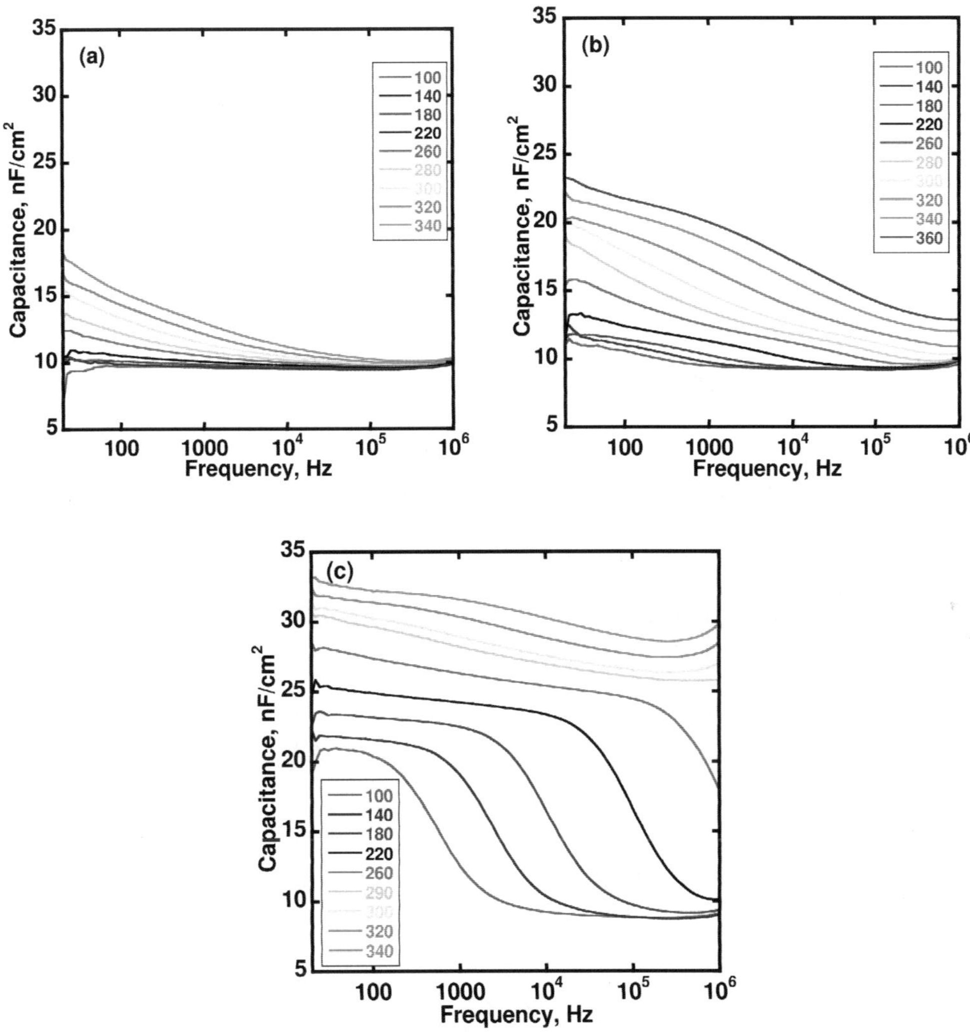

Figure 2 (a, b, c). C(T,f) curves for SC with InGaNAs active layers with thickness of: 900 nm (a), 1200 nm (b), and 1600 nm (c).

Arrhenius plots of f/T^2 (f and T being the turn-on frequency and temperature, respectively, of the capacitance step, or of the maximum of the related conductance peak) are used to derive the activation energy and capture cross section (from the extrapolation to infinite temperature) of the defect responsible for the capacitance step. The Arrhenius plots are shown in Figure 3 for both structures with i-InGaNAs thickness of 1200 nm and 1600 nm. For the sample with i-layer thickness of 1200 nm the low-temperature defect level has an activation energy (E_a) of 0.20 eV and a capture cross section (σ) of 7.2×10^{-17} cm^2 while the high-temperature defect has parameters E_a= 0.46 eV and $\sigma = 6.3\times10^{-16}$ cm^2, respectively. For the sample with 1600 nm thick i-layer the low-temperature defect level has parameters $E_a = 0.18$ eV and $\sigma = 6.4\times10^{-17}cm^2$ and for the high-temperature $E_a = 0.54$ eV and $\sigma = 1.5\times10^{-14}cm^2$. Taking into account the uncertainty in the experimental measurements we can conclude that the parameters of the both high-temperature and low-temperature levels are similar in both samples, meaning that it is a characteristic feature of the i-InGaNAs layer based on the GaAsN/InAs superlattice.

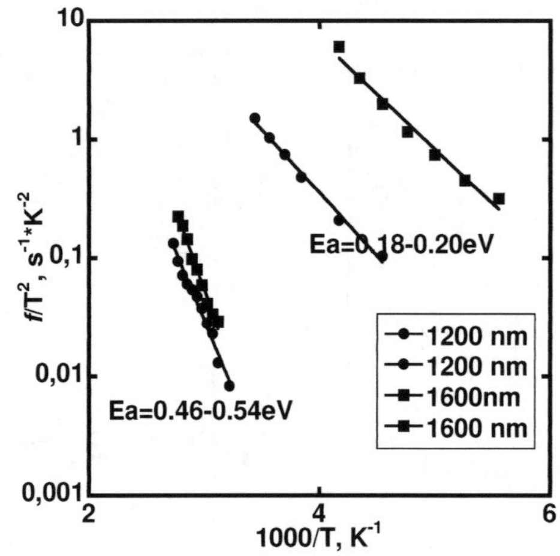

Figure 3. Arrhenius plot of f/T^2 for the two steps detected in the 1200 nm and 1600 nm thick i-InGaNAs samples.

Parameters of the high-temperature defect detected in our i-InGaNAs layers correspond to previously described defects with an activation energy of 0.50 eV in literature [7, 10]. Such defects located in the middle of the InGaNAs bandgap (1 eV) can be considered as centers of strong non-radiative recombination leading to low charge carriers lifetime in these active layers. In our previous study [8] we showed the dependence of external quantum efficiency on thickness for these samples (Figure 4). From experimental results of admittance spectroscopy and quantum efficiency we can conclude that, when the thickness increases from 900 nm to 1200 nm conditions become more favorable for the formation of defects in InGaNAs. But their concentration is too low to affect the quantum efficiency of SC. Quantum efficiency decreases slightly in the short wavelength range due to recombination losses while it slightly increases in long wavelength range because of enhanced absorption with the thickness increase. Further thickness increase to 1600 nm leads to significant increase of defect concentration in active layers and extremal decrease of quantum efficiency of SC. Obviously, the detected defects can be centers of non-radiative recombination which lead to degradation of photoelectrical properties of InGaNAs. Unfortunately, admittance spectroscopy does not allow one to estimate the absolute value of defect concentration directly, but future experiments will be based on deep-level transient spectroscopy (DLTS) technique for a more precise determination of detected defects parameters.

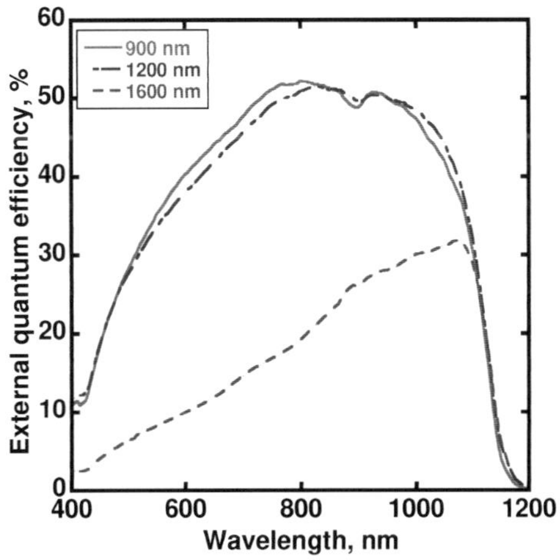

Figure 4. External quantum efficiency of p-i-n structures with i- InGaNAs layers.

4. Conclusion

The dependence of defect formation on the thickness of the i- InGaNAs layer based on GaAsN/InAs superlattices was explored in this study. Increase of thickness from 900 nm to 1200 nm leads to the start of defect formation with two activation energies of 0.20 eV and 0.50 eV but the quantum efficiency does not change significantly. Further thickness increase up to 1600 nm leads to high defect concentrations in the InGaNAs active layers and to a significant decrease of quantum efficiency of solar cells. The existence of a critical thickness for defect-free growth of InGaNAs based on GaAsN/InAs superlattices can thus be emphasized.

Acknowledgements

The reported study was funded by RFBR in the framework of the research project No. 16-38-00791.

References

[1] Green M A, Emery K, Hishikawa Y et al. 2015 Prog. Photovolt. Res. Appl. **23** 1
[2] King R R, Law D C, Edmondson K M et al. 2007 Appl. Phys. Lett. **90** 183 516
[3] Yamaguchi M, Nishimura K I et al. 2008 Solar Energy **82** (2), 173
[4] Weyers M, Sato M, Ando H 1992 Jpn. J. Appl. Phys. **31** (2) 7A, L853
[5] Tzung-Han Wu et al. 2011 Jpn. J. Appl. Phys. **50** 01AD07
[6] Jackrel D B , Bank S R et al. 2007 J. Appl. Phys. **101** 114916
[7] Polojärvi V, Aho A et al. 2016 Sol. Ener. Mat. & Sol. Cells **149** 213
[8] Nikitina E V, Gudovskikh A S et al. 2016 Semiconductors **50** (5) 663
[9] Losee D L 1975 J. Appl. Phys. **46(**5) 2204
[10] Johnston S, Ahrenkiel R, Ptak A et al. 2003 No. NREL/CP-520- 33557, National Renewable Energy Laboratory (NREL), Golden, Colorado, USA

AUTHOR INDEX

AUTHOR INDEX

AUTHOR INDEX

AUTHOR INDEX

AUTHOR INDEX

AUTHOR INDEX

AUTHOR INDEX

AUTHOR INDEX

Institute of Physics
Dirac House, Temple Back
Bristol BS1 6BE UK

ISSN: 1742-6588
ISBN 978-1-5108-5340-9

9 781510 853409

CURRAN ASSOCIATES INC.
proceedings
.com

4th International School and Conference on Optoelectronics, Photonics, Engineering and Nanostructures (Saint Petersburg OPEN 2017)

Journal of Physics: Conference Series Volume 917

Saint-Petersburg, Russia
3 - 6 April 2017

Part 2 of 2

IOP
Institute of Physics

4th International School and Conference on Optoelectronics, Photonics, Engineering and Nanostructures (Saint Petersburg OPEN 2017)

Journal of Physics: Conference Series Volume 917

Saint-Petersburg, Russia
3 - 6 April 2017

Part 2 of 2

Editors:

Alexey Toropov
Leonid Karachinsky
Sergei Alexandrov
Sergey Morozov
Alexander Marmaluk
Nika Akopian
Anatoly Evstrapov
Seppo Honkanen

ISBN: 978-1-5108-5340-9
ISSN: 1742-6588

Printed from e-media with permission by:

Curran Associates, Inc.
57 Morehouse Lane
Red Hook, NY 12571

Some format issues inherent in the e-media version may also appear in this print version.

Printed by Curran Associates, Inc. (2018)

For permission requests, please contact the Institute of Physics
at the address below.

Institute of Physics
Dirac House, Temple Back
Bristol BS1 6BE UK

Phone: 44 1 17 929 7481
Fax: 44 1 17 920 0979

techtracking@iop.org

Additional copies of this publication are available from:

Curran Associates, Inc.
57 Morehouse Lane
Red Hook, NY 12571 USA
Phone: 845-758-0400
Fax: 845-758-2633
Email: curran@proceedings.com
Web: www.proceedings.com

TABLE OF CONTENTS

PART 1

INVITED

CRYSTAL GROWTH AND STRUCTURAL PROPERTIES OF SEMICONDUCTOR MATERIALS AND NANOSTRUCTURES

NANOBIOTECHNOLOGY, BIOPHYSICS AND BIOPHOTONICS

LASERS, SOLAR CELLS AND OTHER OPTOELECTRONIC DEVICES

PART 2

NANOPHOTONICS, SPECTROSCOPY, MICROCAVITIES, OPTICS, PLASMONICS

SPINTRONICS, ELECTRO- AND MAGNETOOPTICS

ELECTRIC, MAGNETIC AND MICROWAVE DEVICES

Manifestation of counteracting photovoltaic effect on IV characteristics in multi-junction solar cells

M A Mintairov[1], V V Evstropov[1], S A Mintairov[1], M Z Shvarts[1], S A Kozhukhovskaia[1] & N A Kalyuzhnyy[1]

[1] Photovoltaics Lab, Ioffe Institute, Saint Petersburg 194021, Russia

Abstract. The existence within monolithic double- and triple-junction solar cells of a photoelectric source, which counteracts the basic photovoltaic p-n junctions, is proved. The paper presents a detailed analysis of the shape of the light IV-characteristics, as well as the dependence V_{oc}-J_{sc} (open circuit voltage - short-circuit current). It is established that the counteracting source is tunnel p^+-n^+ junction. The photoelectric characteristics of samples with different tunnel diode peak current values were investigated, including the case of a zero value. When the tunnel p^+-n^+ junction is photoactive, the V_{oc}-J_{sc} dependence has a dropping part, including a sharp jump. This undesirable effect decreases with increasing peak current.

1. Introduction

Multi-junction solar cells (MJ SC) - a promising direction for the development of semiconductor photovoltaics. Due to its sophisticated design the specific effects are possible. One of such effect is the producing of a photo-electromotive force (photo-emf), which counteracts the operation of basic photovoltaic p-n junctions (sub-cells). This is revealed in the deformation of the normal shape of the V_{oc}-J_{sc} characteristic (open-circuit voltage - short-circuit current), which can have an abnormal portion - the drop of V_{oc} with J_{sc} growth. Such deformations were observed in a number of experiments [1,2]. As shown in [3] the same V_{oc}-J_{sc} deformation may be connected with the effect of incident radiation heating. The heating effect can be eliminated by using the extrapolation method, described in details in 3]. The experiments showed that the heating effect should be taken into account (with an error of <1%) only if the concentration of solar radiation is greater than 2000 suns. Meanwhile, in the process of search and investigation work on the optimization of triple-junction GaInP/GaAs/Ge and double-junction GaInP/GaAs SCs, a number of samples were obtained with the V_{oc}-J_{sc} anomaly observed at concentration less than 2000 suns.

The present work deals with the origin of this anomaly which is the appearance of photo-emf source, counteracting the basic photovoltaic p-n junctions. This source is located in the tunnel region between InGaP and GaAs sub-cells (p^+-$Al_{0.4}Ga_{0.6}As$/n^+-GaAs). The counteracting photo-emf supresses the growth of V_{oc} with the J_{sc} and leads to the appearance of a V_{oc} drop portion. It was found that, this effect is also accompanied by a sharp V_{oc} jump (Figure 1, sample C) when the photocurrent flowing through the tunnel region is equal to tunnel peak current. If there is no jump (Figure 1, sample B), then it means that the peak on the dark IV-characteristics of the tunnel pn junction is practically absent – tunnel peak current equals zero. The counteracting photovoltaic effect was observed on both triple- and double-junction SCs. In all cases, the tunnel region between GaInP and GaAs p-n junctions was grown identically. This effect is most convincingly shown on the family of experimental light IV characteristics for sample C (Figure 2c).

2. Object of study

Three samples of double-junction GaInP/GaAs SCs were studied. All samples were grown by metal-organic vapour phase epitaxy (MOVPE). In all samples the tunnel junctions were grown in the same way - 30 nm p$^+$: $Al_{0.4}Ga_{0.6}As$ doped to a concentration of (1-5)·10^{19} cm^{-3} and 15 nm n$^+$: GaAs doped to (5-7)·10^{18} cm^{-3}). The design of GaAs and GaInP sub-cells differs slightly. Sample C has important structural difference compare to other samples.. Its structure doesn't have a wide-gap barrier between the GaInP sub-cell and the tunnel diode. That makes possible the injection of photogenerated carriers from GaInP sub-cell into the tunnel diode.

Measurements of IV-curves were carried out on a pulsed solar radiation simulator. The simulator is equipped with devices for monitoring the level of illumination and temperature of the SC under examination, as well as a four-probe system for measuring the IV characteristics. The peak current of the tunnel diodes was registered with the voltage decreasing scanning direction. Maximum concentration of solar radiation reaches about 10,000 suns.

3. Light IV characteristics of the tunnel part

For all studied samples, a set of light IV-curves recorded at different illumination was obtained. On the basis of these measurements, the V_{oc}-J_{sc} dependencies (Figure 1) were constructed. Similar to [4] procedure for connecting part IV-curve extraction was performed. The result of the extraction is shown in Figure 2. It can be seen from Figure 1 that all the dependences contain a portion of the linear growth (in the logarithmic current scale) of V_{oc}. In the case of sample A, this region extends up to a current of 25 A/cm^2. This current corresponds to the illumination of the order about 1500 suns. Perhaps the stop of further linear growth of V_{oc} may be due to the aforementioned heating of the sample [4]. The V_{oc} drop on sample C occurs at lower currents (about 3 A/cm^2), and there is also a jump in the region of 20 A/cm^2. This is due to the photoactivity of the tunnel diode, which can be clearly seen in Figure 2c. The presence of a jump is related to the fact that the peak vertex of the tunnel diode becomes below the zero axis, this is discussed in detail in Section 4. The V_{oc} drop for sample B occurs even earlier (about 0.1 A/cm^2), which is also related to the counter photo-emf in the tunnel part. In this case, the value of the counter photo-emf is much larger than for other samples. This is due to the fact that the peak current is zero, so there is no pre-jump portion, which is well observed for sample C. This is demonstrated in detail and discussed in Section 4.

From the experimental light IV-curves (circles of Figure 2), the light IV-curves of the connecting part were extracted. For this purpose, the light IV-curves of the basic photovoltaic (generator) part of the solar cell were reproduced. The basic (generator) part, as in [4], consist of GaInP and GaAs p-n junctions with identical photogenerated currents. Reproduced IV-curves are described by the formulas:

$$J = J_g - J_{dark} \tag{1a}$$

$$J_{dark} = J_{02}\left(\frac{V_{dark}}{2kT} - 1\right) + J_{03}\left(\frac{V_{dark}}{3kT} - 1\right) \tag{1b}$$

where k – Boltzmann constant, T – absolute temperature. The pre-exponents J_{02} and J_{03} necessary for reproduction are taken from the fitting of the linear parts of the experimental V_{oc}-J_{sc} characteristics. The fitting result is shown of Figure 1 (dashed black line). This characteristic functionally coincides with the dark IV-curve ($J_{dark} \rightarrow J_{sc}$, $V_{dark} \rightarrow V_{oc}$) if the photo-generated sub-cell currents are equal [4]. The reproduced IV-characteristics are shown by solid lines in Figures 2a, 2b, 2c. Then a voltaic subtraction of the experimental and reproduced IV characteristics was made. The obtained (extracted) IV-curves are the IV-curves of a SC connecting part (tunnel-junction and resistive elements of SC). The IV-curves are shown in the left part of Figure 2a, 2b,. 2c by triangles symbols. It is seen that connection IV-curves are moving "down", i.e. contrary to the "up" movement of the basic IV-characteristics.

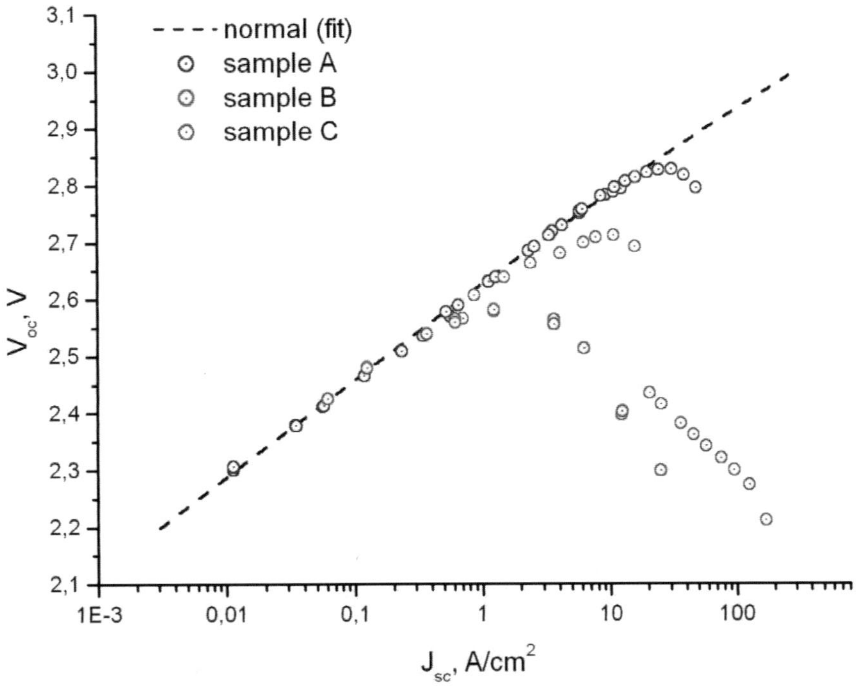

Figure 1. Experimental V_{oc}-J_{sc} dependencies (circles) for GaInP/GaAs SC, deformed by the action of the counter photo-emf. All experimental dependences deviate from a normal dependence (dashed curve). Normal curve obtained by fitting linear parts of the experimental V_{oc}-J_{sc} characteristics with formula 1b. The obtained fitting parameters are J_{02}=1.5·10^{-24} A/cm^2, J_{03}=5.5·10^{-16} A/cm^2.

On Figure 2a and 2b (especially 2b) it is clearly seen that the vertex of the tunnel junction peak moves «down» (in the direction of decrease), while the short-circuit current of SC goes «up». These opposite motions prove that the photoelectric effect is connected with the tunnel AlGaAs/GaAs hetero- p^+-n^+ junction, which counteracts the photoelectric effect of the basic GaInP and GaAs p-n junctions. Note that the sample B (Figure 2b) does not have a peak current, but the motion of the extracted IV characteristics, which is opposite to the base motion, is still observed. The investigated samples differed in the initial coordinate of the peak vertex (25 A/cm^2, 0 A/cm^2, 8 A/cm^2 for A,B,C sample respectively). These differences are reflected in the nature of the fall of V_{oc} with the growth of J_{sc} (Figure 1). The sample C shows the V_{oc} sharp jump (Figure 1, green circles) when the peak vertex of the tunnel junction passes through the zero value (Figure 2c green IV-curves). At the sample B there are no jumps because the sample does not have a peak current. The sample A has a high peak current value, and the voltage drop effect starts at higher currents.

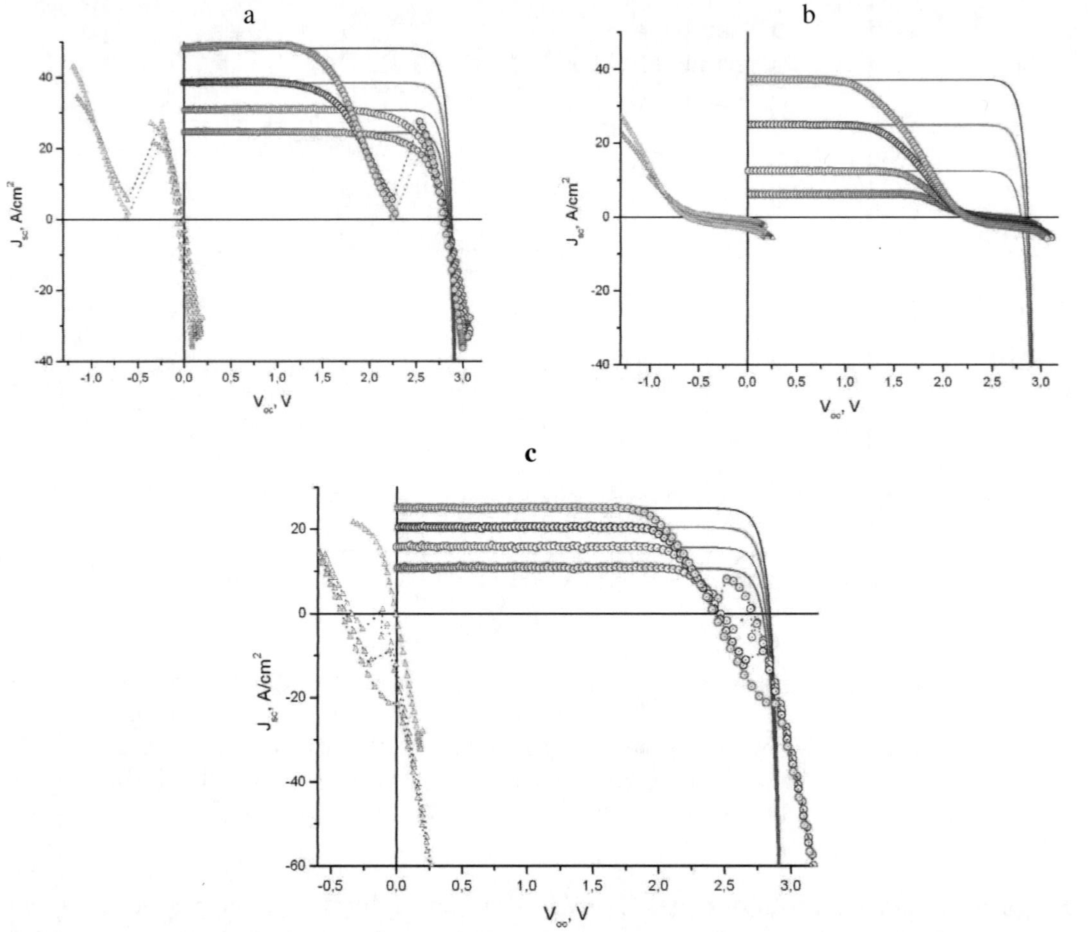

Figure 2. Light IV-curves (a – for sample A, b – for smaple B and c – for sample C) at different illumination: circles - experimental IV-curves; solid lines - IV-curves of the generator part (constructed with the help of formula 1, with parameters taken from the analysis of the V_{oc}-J_{sc} characteristics Figure 1); triangles - IV curves of connecting (tunnel plus resistive) parts, obtained by volt subtraction of IV curves of the experimental and generator parts.

4. The principles of the anomalous of Voc-Jsc dependence formation

The IV-curves of double-junctions SC was calculated to demonstrate the interpretation of the obtained results. The calculation of each IV-curve was carried out as follows. According to eq. 1, the generator IV-curve was calculated, the parameters for calculation were taken as those used to calculate the generator IV-curves in Section 3. To calculate the IV curves of the tunnel part, we used the expression like [5].

$$J = -J_{gt} + \left[J_p \left(\frac{-V}{E_p} \right) \cdot exp\left(1 - \frac{-V}{E_p} \right) + J_{0t} \cdot exp\left(\frac{-V}{E_t} - 1 \right) \right] \quad (2)$$

The expression in square bracket consists of two parts, iso-energetic (given by parameters J_p – peak current, E_p – voltage peak parameter), and tunnel-trap (given by parameters J_{0t} – preexponent, E_t – diode voltage coefficient). The resulting full IV-curve was obtained by voltaic summation of both generator and tunnel parts IV-curves. This calculation was carried out for two cases: in one case the tunnel peak current J_p was equal to 20 A/cm^2, in the other 0. The remaining parameters for calculating the tunnel part were the same in both cases: E_p=0.05 V, J_{0t} = 2.5 A/cm$_2$, E_t=0.2 V.

Figure 3 shows the result of these estimates. On the right there are IV curves of the tunnel parts with a peak current of 20 A/cm^2 (at the upper left, Figure 3) and 0 (lower left). The magnitude of the counter photo-emf, is marked with lines with numbers. On the right there is a calculation of the V_{oc}-J_{sc}. The black dashed line - the dependence in the absence of the counteracting photo-emf, the green and blue dotted lines - the counter photo-emf is generated by the tunnel part with a peak current of 20 and 0 A/cm^2, respectively. Numbers and symbols mark the points corresponding to the points on the graph on the left. As one can see (Figure 3, upper left), the tunnel part with J_p = 20 A/cm^2 generates small photo-emf (points 1, 2, 3 "). This occurs as long as the peak current is above 0 and the generation of photo-emf is going in the iso-energetic regime. Further, the generation of photo-emf occurs on the tunnel-trap regime (points 3 ", 4.5.6). The transition from one regime to the other (between the points 3 "and 3 ') is manifested in a jump of the V_{oc}-J_{sc} dependence. For the case when the tunnel part has a zero peak current (Figure 3, left, bottom), the iso-energetic term of the tunnel junction IV-curve is absent, so a large counter photo-emf is immediately generated.

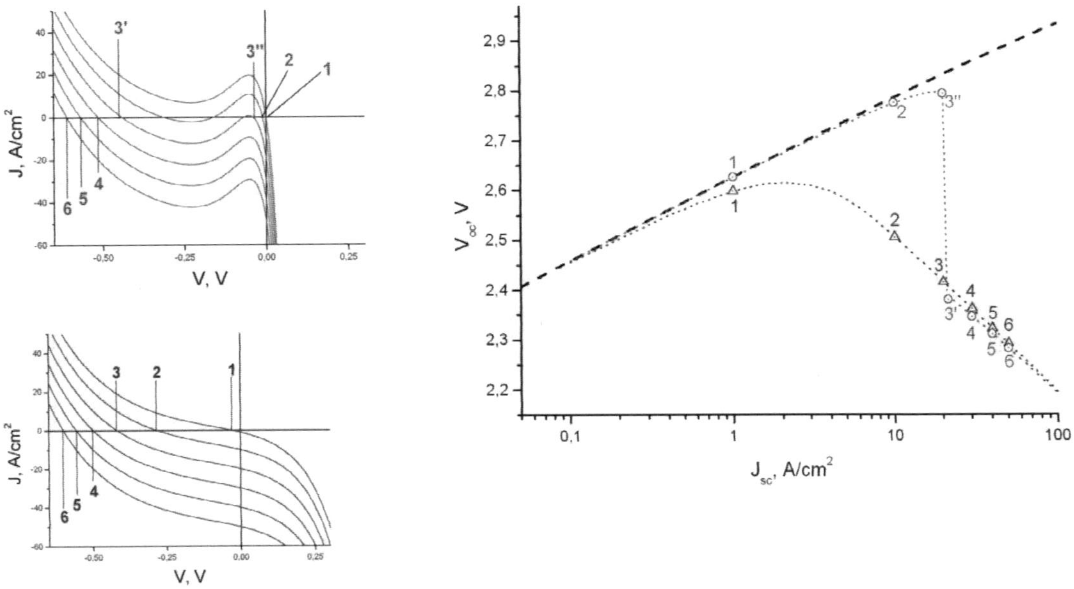

Figure 3. The principle of the anomalous V_{oc}-J_{sc} dependence formation (right) in MJ SC. The principle is illustrated by simulating a double-junction GaInP/GaAs SC with two different types of tunnel part: with the presence of a peak current (left, top) and without it (bottom left). The relationship between the presence of a peak current and the formation of a "jump" on the anomalous V_{oc}-J_{sc} dependence is shown.

The similar behavior as shown in Figure 3 can be seen in the presented experimental V_{oc}-J_{sc} dependencies (Figure 1) and extracted IV curves (triangles in Figure 2). For sample A, the peak vertex of the tunnel part is always above 0, so the "jump" of the V_{oc}-J_{sc} characteristic does not come, and we can expect a delayed "jump". The sample B tunnel part IV-curves does not have peaks (Figure 2b), and the behavior of the V_{oc}-J_{sc} dependence is the same as for 0-peak current simulation on Figure 3. The sample C tunnel part IV-curves have a peak. The peak cross 0-axis at J_{sc} about 15 A/cm^2 (Figure 2c) and its V_{oc}-J_{sc} dependence has a sharp Voc jump. So the behavior of these characteristics is the same as for simulated ones (green on Figure 3).

IOP Conf. Series: Journal of Physics: Conf. Series **917** (2017) 052034 doi:10.1088/1742-6596/917/5/052034

5. Conclusion

The photovoltaic effect in the connecting part of the MJ SC between GaInP and GaAs basic p-n junctions is found and considered. This connecting part includes p^+-n^+ heterojunction – p^+ $Al_{0.4}Ga_{0.6}As/n^+$ GaAs. The photoelectric effect in the connecting part is directed against the action of the basic p-n junctions. This counteraction is manifested as a deviation of the V_{oc}-J_{sc} dependence from the normal shape (linear in a semilogarithmic scale).

Especially convincingly, the effect can be seen on the set of light IV-curves for the sample C (double-junction GaInP/GaAs SC). These IV-curves show the peak current of the tunnel diode. The current vertex of the peak decreases (goes "down"), when the short-circuit current SC increases (goes "up"), i.e. movement is opposite.

IV-curves of the connecting part were extracted from experimental light IV characteristics. Their shape practically reflects the shape of tunnel junction IV-curves, with taking into account the influence of resistivity. It was shown that the shape of V_{oc}-J_{sc} dependence is connected with tunnel junction IV-curve shape. The principles of V_{oc}-J_{sc} dependence formation is considered and it is shown that the V_{oc} jump is appears when the tunnel junction peak current vertex cross zero-axis. After this happen the value of voltage drop becomes dramatically high. The V_{oc} jump can be used to diagnose the counter emf source existence.

Acknowledgments

This work has been supported by the Presidium of the Russian Academy of Sciences.

References

[1] M Stevens, C Downs, D Emerson, J Adler, S Maclachlan and T E Vandervelde, 2015 Proc. of 31 PVSEC, Hamburg (Germany) pp. 1474 – 1477

[2] B Chhabra, S Jacobs, C B Honsberg, 2006 IEEE PV Spec. Conf. pp 791 - 794,

[3] M A Mintairov, V V Evstropov, M Z Shvarts, S A Kozhukhovskaia, S A Mintairov and N A Kalyuzhnyy, 2016, AIP Conf. Proc. Freiburg (Germany) v 1766 p. 050005

[4] M A Mintairov, V V Evstropov, N A Kalyuzhnyy, S A Mintairov, N Kh Timoshina, M Z Shvarts, 2015 AIP Conf. Proc. Aix-les-Bains (France) v 1679 p. 050007

[5] S M Sze, 1981 *Physics of Semiconductor Devices* J. Willey and Sons, New York.

IOP Conf. Series: Journal of Physics: Conf. Series **917** (2017) 052035 doi:10.1088/1742-6596/917/5/052035

Lateral mode control in edge-emitting lasers with modified mirrors

A Payusov[1,2,3], A Serin[1,3], I Mukhin[2,4], Y Shernyakov[1,2] Y Zadiranov[1], M Maximov[2,1,3], N Gordeev[1,2]

[1]Ioffe Institute, 26 Polytechnicheskaya, St Petersburg 194021, Russia
[2]St. Petersburg Academic University, 8/3 Khlopina, St Petersburg 194021, Russia
[3]Peter the Great St. Petersburg Polytechnic University, 29 Polytechnicheskaya, St Petersburg 195251, Russia
[4]ITMO University, 49 Kronverksky, St Petersburg 197101, Russia

e-mail: plusov@mail.ioffe.ru

Abstract. We present a study on lateral mode control in edge-emitting lasers with profiled mirror reflectivity. The object was to eliminate high-order lateral modes in conventional ridge-waveguide InAs/InGaAs QD (quantum dot) lasers with the stripe width of 10 μm. We have used a FIB (focused ion beam) technique to selectively etch windows in the AR (anti-reflection) facet coatings in order to introduce extra mirror losses for the high order modes. This approach allowed us to eliminate the first-order mode lasing without deterioration of the laser parameters. We suppose that further optimisation of the laser heterostructure and window designs may lead to a pure lateral single-mode lasing in the broadened ridge waveguides.

1. Introduction

Today high-power diode lasers are essential for various applications due to their energy efficiency, reliability, compact size, and low cost. However, increasing optical power in conventional ridge-waveguide laser diodes inevitably requires the waveguide expanded in the lateral direction resulting in multi-mode lasing. A number of approaches has been developed to ensure oscillation on a fundamental mode in broadened ridge-waveguides. Among them are tapered waveguides, external cavities, phase-locked arrays, and profiled mirror reflectivity. The reflectivity profile in semiconductor edge-emitting lasers is formed with partial high-reflection (HR) or AR coating of the laser output mirrors. Lateral modes are differ in their spatial distribution, thus, partial coating may be used to create additional loss for high-order modes and to change the lasing conditions in favor of the fundamental mode. Previously several techniques have been used to create modulated reflectivity mirrors in broad-area lasers, and single lobe far-field patterns where demonstrated [1,2]. Nevertheless, obtaining pure single-mode lasing in broadened ridge-waveguides is still a challenging task. In this work, we present a study on the lateral mode control in edge-emitting InAs/InGaAs quantum dot lasers with modulated reflectivity mirrors.

2. Experiment and results

The laser wafer was grown by molecular beam epitaxy. Ten layers of InAs quantum dots capped with InGaAs and separated by 35 nm GaAs where sandwiched between 1.5 μm $Al_{0.35}Ga_{0.65}As$ claddings. The wafer was processed into 10 μm wide shallow mesa ridge-waveguide lasers using standard photo-lithography. All studied devices where mounted on copper heat sinks using indium solder in order to minimize overheating in continuous wave (cw) regime. The 2 mm long devices showed the threshold current density of 100 A/cm^2 and the lasing wavelength of 1.26 μm in cw regime, which corresponds to the lasing via the QD ground state. Plotting reciprocal differential quantum efficiency versus cavity length yielded the internal quantum efficiency of 77% and internal loss of 2.8 cm^{-1}.

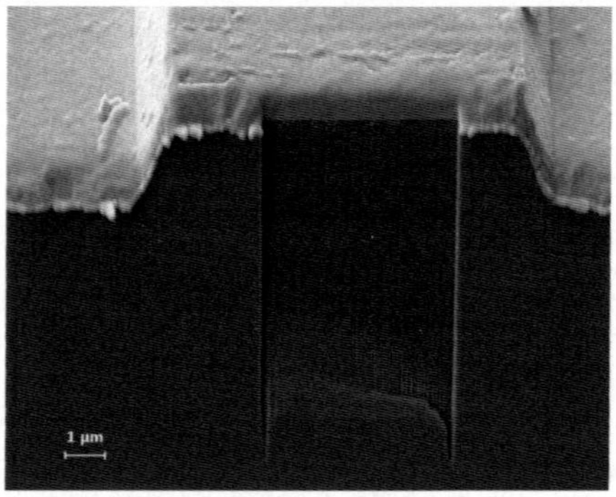

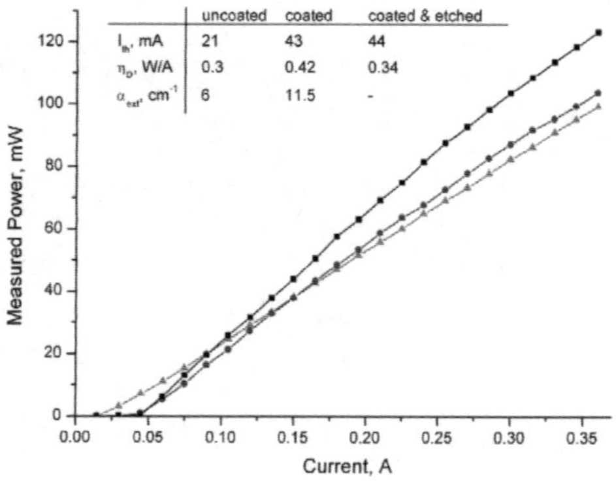

Figure 1. SEM (scanning electron microscope) photograph of the laser with etched mirror.

Figure 2. Light output vs. injected cw current of the initial (red triangles), coated (black squares), and etched (blue dots) devices. The inset shows the laser parameters and estimated output losses.

Coatings where deposited on the both facets of the 2 mm long devices. The HR coating consisted of 5 pairs of SiO_2/Ta_2O_3 with the thicknesses of 217 nm and 157 nm, respectively. Another facet was coated with single 157 nm Ta_2O_3 AR layer which was partially removed (figure 1) using FIB. Opening width was designed to eliminate lasing of the first-order lateral mode.

To ensure that the devices did not degraded after the coating and FIB etching, we measured cw light-current characteristics at each step of fabrication of the modified mirrors (figure 2). The inset in figure 2 shows the laser parameters and estimated output loss. One can see that the threshold current increased twofold after the coating and the differential efficiency increased accordingly due to increased output loss. After the FIB etching threshold current remained unchanged but differential efficiency decreased. We attribute the latter to the lower output loss for the fundamental mode resulted from partial removing of the AR coating.

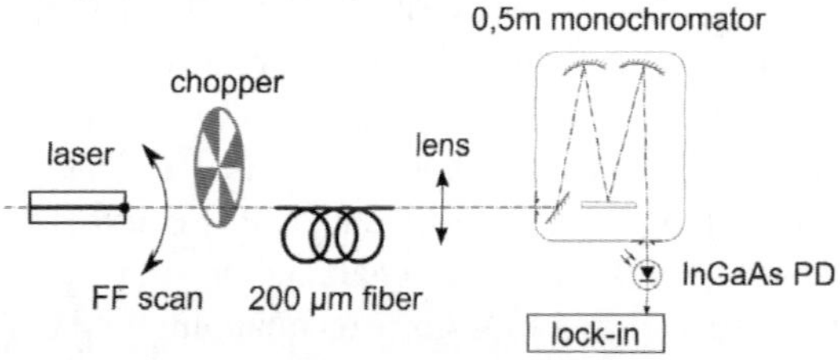

Figure 3. Experimental setup used for the measurement of spectrally resolved lateral far-fields.

To confirm elimination of the first-order mode lasing in the devices with modified mirrors, we studied spectrally resolved far-fields. Figure 3 shows a layout of the experimental setup used for these measurements. A 200 μm fiber was used to collect chopped far-field emission. A lens was used to focus the fiber output to the 0.5 m monochromator. The output emission from the monochromator was detected with an InGaAs photodiode connected to a lock-in amplifier. Spatial and spectral resolutions were 0.06° and 0.4 nm respectively.

Figure 4 shows typical spectra and spectrally resolved lateral far-fields of the uncoated laser (figure 4a) and laser with modulated reflectivity mirror (figure 4b). Fundamental mode lasing took place around 1265 nm in the both types of the devices. The far-field patterns taken at 1265.2 nm (red lines in figure 4 a,b) prove it. We attribute the small fringes around ±8 deg. in the far-field pattern of the device with modified mirror to the diffraction at the AR opening but not to the presence of the first-order mode.

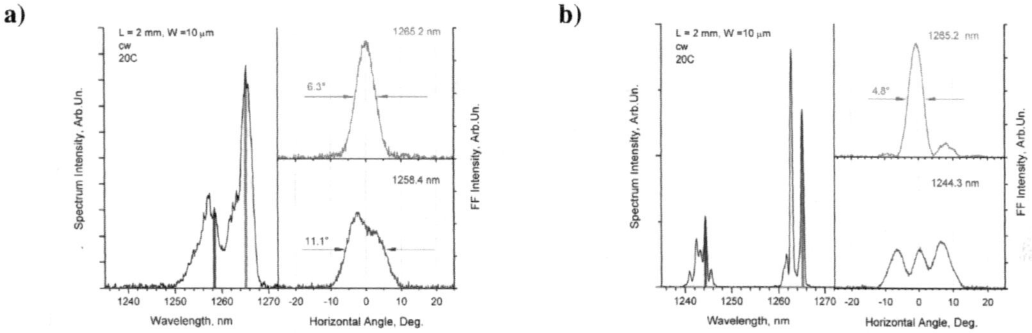

Figure 4. Lasing spectra (left hand side) and spectrally resolved lateral far-fields (right hand side) of the uncoated laser (a) and laser with modulated reflectivity mirror (b) measured in cw regime at 300 mA pump current.

First-order mode lasing appeared around 1258 nm in the uncoated device (blue line in figure 4a). In contrast, the device with modified mirrors showed no lasing at this wavelength. However, an additional group of lines emerged in the lasing spectra around 1244 nm (left hand side of figure 4b). We attribute these lines to the second-order mode. This speculation is in agreement with the far-field pattern taken at 1244.3 nm (blue line in figure 4b). We assume that at the high currents the waveguide effectively expands in the lateral direction due to the current spreading. Therefore, the second-order mode could be eliminated with deeper ridge etching or with optimization of the AR opening width.

3. Conclusion

In conclusion, having used profiled reflectivity mirror we were able to selectively eliminate the first-order mode lasing in the conventional InAs/InGaAs QD laser with 10 μm ridge waveguide without noticeable deterioration of the device performance.

Acknowledgments

The authors acknowledge the support from the Russian Foundation for Basic Research (project № 16-29-03279) and the Ministry of Higher Education and Science of the Russian Federation (project № 3.9787.2017/8.9).

References

[1] Shigihara K *et al* 1994 *IEEE J. Quantum Electron.* **30** 1683–1690
[2] Stryckman D *et al* 1996 *Applied Optics* **35** 5955

Investigation of whispering gallery modes in microlasers by scanning near-field optical microscopy

Yu S Polubavkina[1], N V Kryzhanovskaya[1], A M Nadtochiy[1,2], A M Mintairov[3], A A Lipovsky[4], S A Scherbak[4], M M Kulagina[2], M V Maximov[1,2], A E Zhukov[1]

[1] St Petersburg Academic University, Khlopina 8(3), St Petersburg 194021, Russia
[2] Ioffe Institute, St Petersburg, Russia
[3] University of Notre Dame, Notre Dame, Indiana 46556, USA
[4] Peter the Great St. Petersburg Polytechnic University, St Petersburg, Russia

Abstract. Near-field scanning optical microscopy (NSOM) with a spatial resolution below the light diffraction limit was used to study intensity distributions of the whispering gallery modes (WGMs) in quantum dot-based microdisk and microring lasers on GaAs with different outer diameters. Room temperature microphotoluminescence study (μPL) reveal lasing in microlasers of both geometries.

1. Introduction

Optical whispering-gallery mode (WGM) microcavities [1] having ultrahigh quality (Q) factor and ultralow threshold were proposed for large variety of applications including all-optical switching and signal processing. It is of great importance to know the electromagnetic field distribution inside the cavity for proper device design. The resonant frequencies of the microlasers can be determined by μPL study whereas spatial distribution of the modes requires NSOM technique [2]. NSOM allows direct measurements of a number of WGM intensity maxima that is important for different nanophotonic applications [3, 4]. Near-field images for microdisks with different diameters had been investigating [5], but there is a lack of information about near-field patterns of WGMs in microrings. In this work we investigate μPL and NSOM spectra of microdisk (MD) and microring (MR) lasers deposited on GaAs substrates.

2. Experiment details

The epitaxial structures used for fabrication of the MD and MR lasers were grown by molecular beam epitaxy on a n-doped GaAs substrate. An active region represents 5 layers of InAs/InGaAs quantum dots inserted into a 0.35-μm-thick GaAs waveguiding layer cladded with 400-nm-thick $Al_{0.98}Ga_{0.02}As$ layer from the substrate side. Spectral position of quantum dot ground-state transition was located around 1.28 μm at room temperature. Microresonators were fabricated using photolithography and Ar^+ ion beam etching. Outer diameter of the microlasers was varied in the 4-6 μm range. SEM images of MD and MR are shown in the Figure 1 (a) and 1 (b), respectively. The $Al_{0.98}Ga_{0.02}As$ bottom cladding layer was selectively oxidized to be transformed into an AlGaO oxide. All measurements were done at room temperature. μPL spectra were excited by second harmonic of CW-operating YAG:Nd laser, λ = 532nm and measured by Olympus LMPlan IR objective x100 with NA 0.8. Optical emission was detected with a monochromator Horiba FHR1000 and InGaAs CCD array Horiba

Symphony (spectral resolution 30 pm). NSOM spectra of MDs with diameters 4, 5 μm and MR were excited by using the 488 nm line from an Ar ion laser and measured by using a 270 mm focal length spectrometer and liquid nitrogen-cooled multichannel InGaAs- or single-channel Ge detector. The spectral resolution was 0.3 nm. NSOM spectra and images were measured by using a NANONICS Multi-View 1000 head and scanning probe microscopy controller. Uncoated fiber probes (tips) were used with an apex of 300 nm and a taper angle of 40° –50°; these provided a spatial resolution as high as 300 nm in the near-infrared spectral range. NSOM spectra were excited and collected through the same fiber probe. NSOM images of MD with diameter 6 μm were measured by hollow pyramid cantilever probe with 100 nm aperture. The emission was collected by 100x Mitutoyo Objective with NA=0.5. The microdisk was pumped by YAG:Nd laser, λ = 532nm on the side.

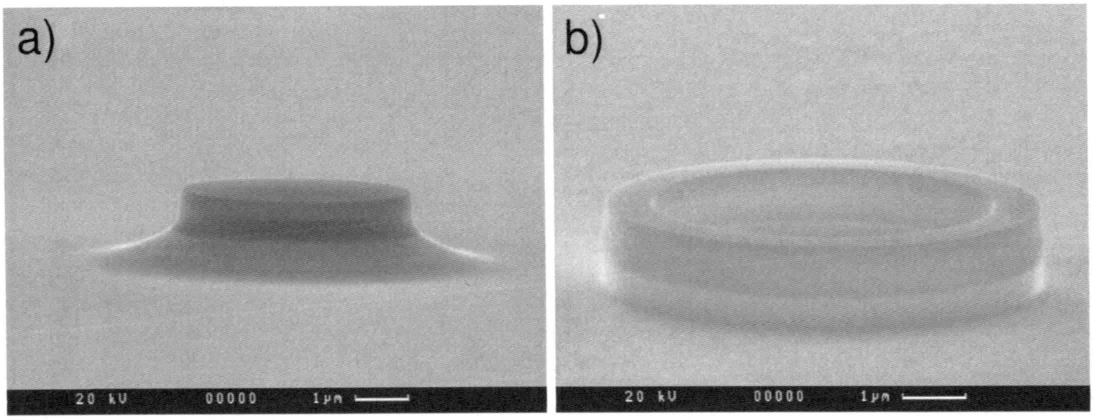

Figure 1 (a, b). (a) SEM image of microdisk; **(b)** SEM image of microring.

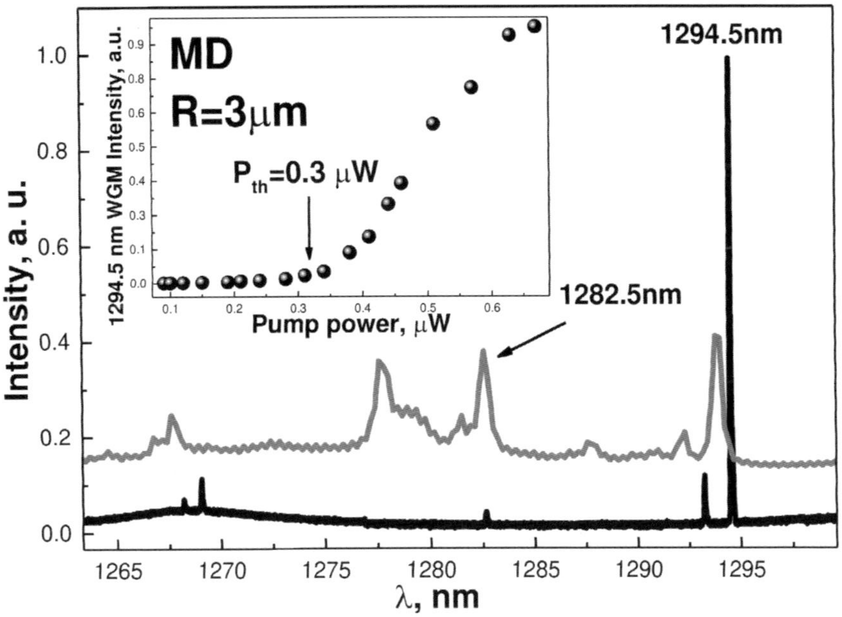

Figure 2. The room temperature μPL (black line) and NSOM (red line) spectra of the MD with diameter 6 μm. The spectra are vertically shifted for clarity. Inset: The dependence of the 1294.5 nm WGM line intensity on the pump power.

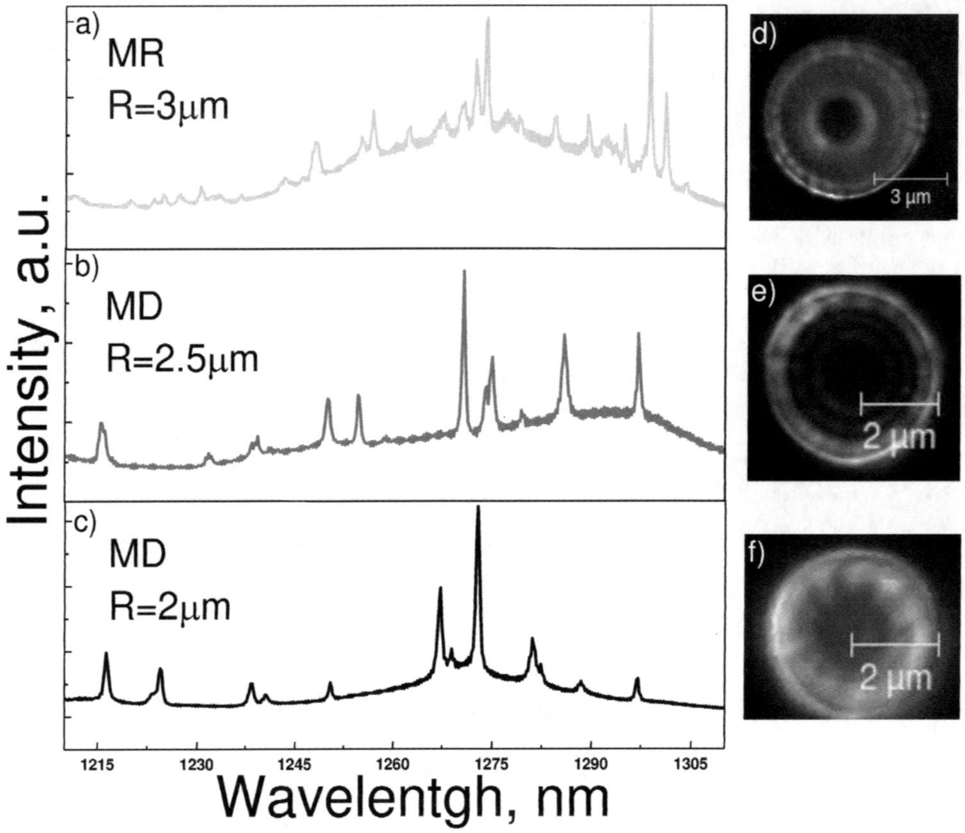

Figure 3. (a) NSOM spectrum for the microring; **(b, c)** NSOM spectra for microdisks; **(d)** NSOM image of WGM line λ=1299nm; **(e)** NSOM image of WGM line λ=1271nm; **(f)** NSOM images of WGM line λ=1281nm.

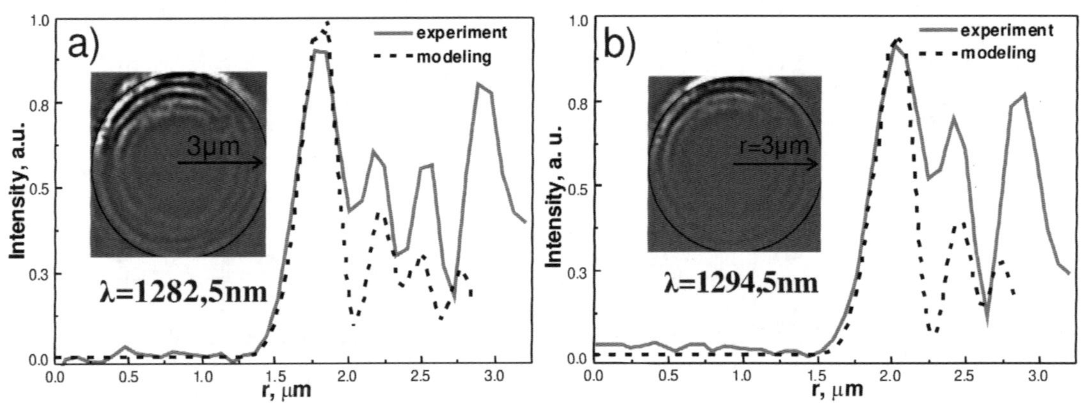

Figure 4 (a, b). (a) NSOM (red line) and calculated (black dotted line) intensity distribution profiles for 1282.5nm WGM line; **(b)** and for 1294.5nm WGM line.

Results

First we investigated μPL spectra of microlasers with different outer diameters. All the microlasers demonstrate room temperature lasing at one of the WGM line (Figure 2). The threshold power for MD and MR was estimated from kink on the dominant mode intensity vs optical pump power dependence to be 200-300μW (insert of Figure 2). Microcavities quality factor (Q~30 000) was estimated from the lasing line full-width at half maximum (about 30pm). NSOM spectra for MDs and MR and corresponding intensity distributions of one of the WGM lines are shown in Figure 3. Bright maxima azimuthally aligned along the disk and ring circumferences are clearly distinguished. One can define that azimuthal numbers m are 13, 23 and 36 for MDs with diameter 4μm, 5μm and MR with outer diameter 6μm, respectively. Increase of the MD diameter (5 μm) results in appearance of WGM with high radial numbers, 3 radial maxima are observed at NSOM image of λ=1271nm resonance (Figure 1(e)). The NSOM intensity distributions of MD with diameter 6 μm have more complicated patterns. NSOM images of 1282.5nm WGM line and 1294.5nm WGM line are shown in Figure 4. Profiles of NSOM intensity distributions for both WGMs were built from NSOM images. Also distributions for these WGMs were calculated and shown with experimental profiles (Figure 4). 1282.5nm and 1294.5nm WGM lines have forth and third radial order, respectively, and are in good agreement with calculated distributions, but one can see that maxima of electromagnetic field located near the MD edge are higher than calculated ones. The reason of this discrepancy is adding emission from side walls to the emission under investigation when the probe tip achieves the MD edge. The black circumferences on the NSOM images of Figure 4 show the MD edge and one can see extra maximum beyond them. NSOM data demonstrate that high-order radial modes are suppressed in MR laser as compared to its MD counterpart. As a result, emission spectrum becomes less complicated thereby facilitating single-frequency lasing.

Acknowledgments

This work was supported by the RFBR (15-02-03624, 16-29-03111) and the Ministry of Higher Education and Science of the Russian Federation (project No 3.9787.2017/8.9).

References

[1] McCall S L, Levi A F J, Slusher R E, Pearton S J, Logan R A 1992 *Appl. Phys. Lett.* **60** 289
[2] Pohl D W, Denk W and Lanz M 1984 *Appl. Phys. Lett* **44** 651
[3] Moreau E, Robert I, Gerard J M, Abram I, Manin L and Thierry-Mieg V 2001 *Appl. Phys. Lett.* **79** 2865
[4] Imamoglu A, Awschalom D D, Burkard G, DiVincenzo D P, Loss D, Sherwin M, Small A 1999 *Phys. Rev. Lett.* **83** 4204
[5] Mintairov A M, Chu Y, He Y, Blokhin S, Nadtochy A, Maximov M, Tokranov V, Oktyabrsky S, Merz J L 2008 *Phys. Rev. B* **77** 195322

IOP Conf. Series: Journal of Physics: Conf. Series **917** (2017) 062001 doi:10.1088/1742-6596/917/6/062001

Optical properties and energy parameters of Gd_2O_3 and Gd_2O_3:Er nanoparticles

Yu A Kuznetsova, A F Zatsepin,
Institute of Physics and Technology, Ural Federal University, Yekaterinburg, Russia

Abstract. The reflection spectra of Gd_2O_3 and Gd_2O_3:Er nanoparticles have been investigated. Several types of absorption centers associated with Er^{3+} activator ions and intrinsic lattice defects have been revealed. It was found that interband absorption is formed by indirect optical transitions with participation of phonons. The values of energy gap and phonons energy were determined.

1. Introduction

Low-dimensional modifications of gadolinium oxide are of interest as functional materials for devices of nanophotonics, optoelectronics, alternative energy and systems of radiation conversion [1,2]. The important fundamental parameter that determines optical properties of material is the band gap. The most common way for obtaining this parameter is to use the data of absorption spectrum. However, in some cases a direct measurement of optical absorption is difficult due to the features of samples structure (nontransparent powders, compacts, ceramics). The main purpose of the present work is to realize an alternative method for determining the band gap value of Gd_2O_3 compacted nanopowders on the basis of reflection spectra analysis.

It is known that development of rare-earth based materials for devices of radiation conversion requires the introduction of single or paired activator ions into the host lattice, for example Eu^{3+}, Tb^{3+}, Yb^{3+}/Er^{3+}, Yb^{3+}/Ho^{3+} [3,4]. In this regard, it is necessary to understand the impact of dopants on the optical characteristics of matrix. The second task of this work is devoted to establishing the influence of Er^{3+} activator ions on formation of energy structure of Gd_2O_3 host lattice.

2. Experimental

The Gd_2O_3 and Gd_2O_3:Er (2 mol%) compacted nanopowders were obtained from aqueous solutions of $Gd(NO_3)_2$ and $Er(NO_3)_2$ precursors by chemical precipitation method followed by pressing and annealing at 1000°C. The diffraction peaks in XRD patterns of both samples correspond to the cubic single-phase structure of Gd_2O_3 (Ia-3 space group). The average crystalline grains sizes were estimated to be of 30-40 nm. Reflection spectra were recorded on a Lambda 35 spectrophotometer equipped by integrating sphere.

3. Results and discussion

3.1 Reflection spectra

Reflection spectrum of Gd_2O_3 (figure 1, blue) contains a number of absorption bands in the near UV region. The most intense band with a maximum at 220 nm corresponds to interband transitions of Gd_2O_3 matrix [5]. The lines at 276 and 315 nm can be attributed to the 4f-4f optical transitions of Gd^{3+}

ion from the ground $^8S_{7/2}$ to the excited ^6I_J and ^6P_J states correspondingly. Selective absorption peaks of the host lattice cations indicate that energy structure of a part of Gd^{3+} ions in Gd_2O_3 is distorted, in other words, such ions represent intrinsic defective optical centers [6].

As reflection spectrum for Gd_2O_3:Er shows (figure 1, red), introduction of Er^{3+} ions into the gadolinium oxide host lattice results in the appearance of additional absorption peaks in visible and near-infrared spectral regions. Along with the bands of interband transitions and absorption in Gd^{3+} ions the lines at 380, 523, 655, 802 and 977 nm are clearly observed. These bands correspond to the optical transitions in Er^{3+} ions from the ground $^4I_{15/2}$ to the excited $^4G_{11/2}$, $^2H_{11/2}$,$^4F_{9/2}$, $^4I_{9/2}$ and $^4I_{11/2}$ energy levels [7]. Based on the primary experimental data obtained from the reflection spectra, one can conclude that there are at least two types of absorption centers in Gd_2O_3:Er nanoparticles: intrinsic defective Gd^{3+} ions and Er^{3+} activator ions.

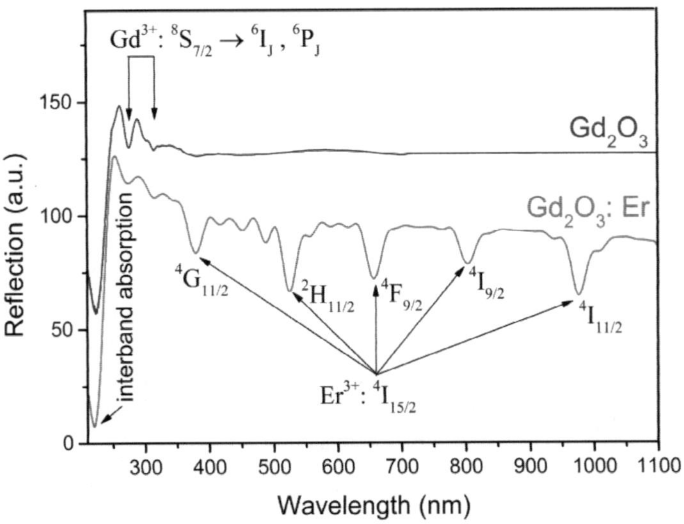

Figure 1. Reflection spectra of Gd_2O_3 (blue) and Gd_2O_3:Er (red) nanoparticles. The optical transitions corresponding to absorption in Gd^{3+} and Er^{3+} ions are shown by arrows.

3.2 Absorption spectra and parameters of energy structure

Spectral dependences of absorption coefficient $\alpha(h\nu)$ can be obtained by the analysis of reflectance experimental data $R(h\nu)$ in accordance with the models proposed by Kumar [8] and Kubelka-Munk [9] by using following analytical equations:

$$\alpha(h\nu) = \ln[(R_{max} - R_{min})/(R(h\nu) - R_{min})]/2d \quad (1)$$

$$F(h\nu) = (1 - R(h\nu))^2 / 2R(h\nu) \quad (2)$$

where R_{max} and R_{min} are maximum and minimum values of reflectance, d is the thickness of the absorption layer, $F(h\nu)$ is the Kubelka-Munk function that is directly proportional to the absorption coefficient. At the same time, the dependence of the absorption coefficient on the energy of the incident photons can be described by the power-law expression [10]:

$$\alpha(h\nu)\cdot h\nu = A(h\nu - E_g)^n \quad (3)$$

where A is a constant, E_g is the energy gap, n is the exponent that determines the type of interband transitions (1/2; 3/2; 2 and 3 for direct and indirect allowed and forbidden transitions respectively).

Since the information about the energy structure and nature of the interband transitions in Gd_2O_3 is limited, we performed the approximation of absorption data obtained in accordance with both Kumar's and Kubelka-Munk's models by equation (3) setting the parameter n as a variable. The best fitting (the correlation coefficient was 0.996) has been achieved with the n values of 2.3 and 2.4 for Gd_2O_3 and

Gd_2O_3:Er samples correspondingly. It indicates the realization of indirect interband transitions as for the undoped Gd_2O_3 matrix as for the sample activated by the Er^{3+} ions.

In order to obtain the energy gap values we carried out the analysis of spectral dependences of absorption determined by using the Kubelka-Munk's and Kumar's approaches in the coordinates of indirect optical transitions (figure 2 and figure 3).

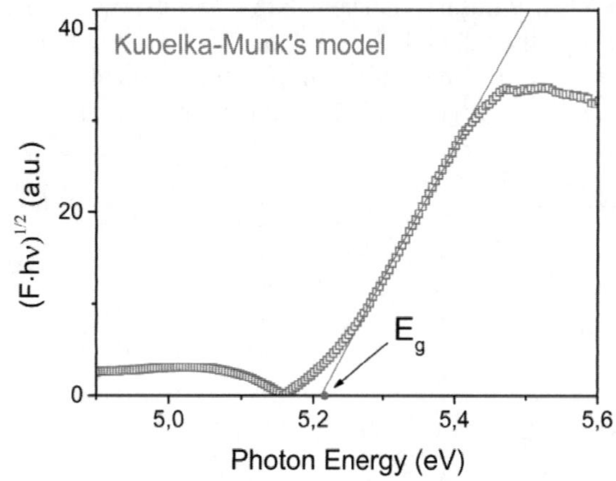

Figure 2. Absorption spectra of Gd_2O_3:Er nanoparticles for indirect transitions in accordance with the Kubelka-Munk's model [9]. Arrow indicates the value of energy gap E_g.

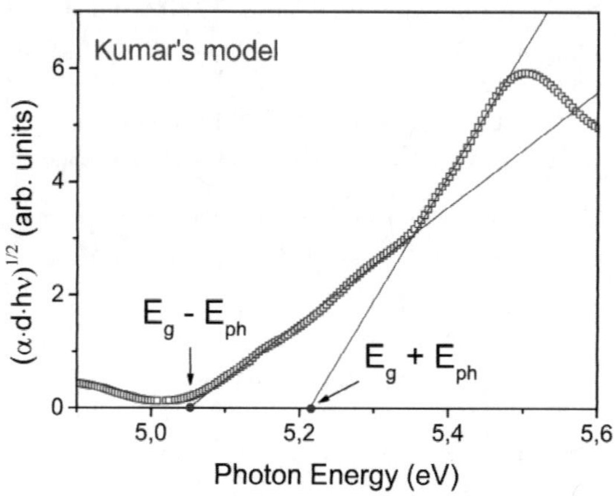

Figure 3. Absorption spectra of Gd_2O_3:Er nanoparticles for indirect transitions in accordance with the Kumar's model [8].The energy gap values (E_g) for transitions with absorption and emission of phonons (E_{ph}) are marked by arrows.

The approximation of linear range in the spectral dependence of Kubelka-Munk's function by straight line up to the crossing with the abscissa axis allows determining the energy gap for indirect transitions that is found to be of 5.09 and 5.21 eV for Gd_2O_3 and Gd_2O_3:Er nanoparticles respectively. In the case of using the Kumar's model a more complex features are observed (figure 3). The absorption spectrum is characterized by the two linear regions. Coordinates of crossing points with the energy axis correspond to the values E_g+E_{ph} and E_g-E_{ph} where E_g is the energy gap between maximum of valence band and minimum of conduction band, E_{ph} is the energy of phonons that participate in the indirect optical transition [11]. In other words, there are two types of indirect transitions: with absorption and with the emission of phonons. In figure 4 the schematic representation of these transitions is presented. The obtained values of energy gap and phonon energy for Gd_2O_3 and Gd_2O_3:Er nanoparticles are summarized in table 1.

Table 1. The values of energy gap and phonons energy obtained under fitting the reflectance experimental data in accordance with the Kumar's and Kubelka-Munk's models for Gd_2O_3 and Gd_2O_3:Er nanoparticles.

Sample	*Kumar's model (Equation (1))*		*Kubelka-Munk's model (Equation(2))*
	E_g, *eV*	E_{ph}, *meV*	E_g, *eV*
Gd_2O_3	4.97	78	5.09
Gd_2O_3:Er	5.13	80	5.21

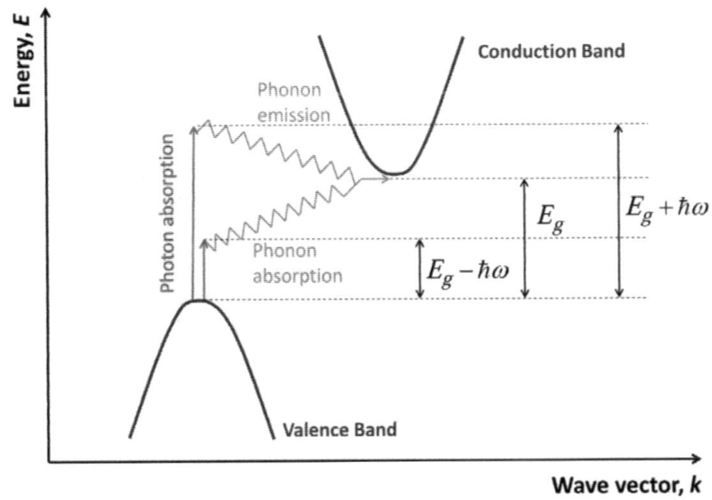

Figure 4. Schematic representation of indirect interband transitions with absorption and emission of phonons in Gd_2O_3 and Gd_2O_3:Er nanoparticles.

4. Conclusion

In present work we have demonstrated an approach to the determination of the spectral characteristics of absorption centers, the type of interband transitions, and the energy gap value for low-dimensional modification of gadolinium oxide based on the data of reflection spectra. Several types of optically active absorption centers have been revealed in Gd_2O_3 and Gd_2O_3:Er nanoparticles: intrinsic defective Gd^{3+} ions and Er^{3+} activator ions. It was found that the absorption edge of Gd_2O_3 at room temperature is formed by the phonon-assisted indirect optical transitions. The effect of increasing of Gd_2O_3 band gap under the activation by Er^{3+} ions has been detected.

Acknowledgments

The work has been funded by the Ministry of Education and Science of the Russian Federation (Government task №3.1485.2017/PP).

References

[1] Tamrakar R K, Bisen D P, Brahme N 2014 *Res. Chem. Intermediat.* **40** 1771-1779
[2] Anishur Rahman A T M, Vasilev K, Majewski P 2011 *J. Colloid. Interf. Sci.* **354** 592-596
[3] You L, Wang S, Ren B et all 2016 *J. Nanosci. Nanotechno.* **16** 9731-9737
[4] Li D, Qin W, Aidilibike T et all 2016 *J. Alloy. Compd.* **675** 31-36
[5] Ferrara M C, Altamura D, Schioppa M et all 2008 *J. Phys. D Appl. Phys.* **41** 225408
[6] Zatsepin A F, Kuznetsova Yu A, Spallino L et all 2016 *Energy Proc.* **102** 144-151
[7] Jia D, Lu L, Yen W M 2002 *Opt. Commun.* **212** 97-100
[8] Kumar V, Sharma T P, Singh V 1999 *Opt. Mater* **12** 115–119
[9] Kubelka P, Munk F 1931 *J. Tech. Phys.* **12** 593
[10] Tauc J 1974 *Amorphous and Liquid Semiconductors* (Plenum, New York)
[11] Wakaki M, Shibuya T, Kudo K 2007 *Physical Properties and Data of Optical Materials* (CRC Press Taylor & Francis Group)

Laser post-processing of halide perovskites for enhanced photoluminescence and absorbance

E Y Tiguntseva[1], I N Saraeva[2], S I Kudryashov[2], E V Ushakova[1], F E Komissarenko[1], A R Ishteev[3], A N Tsypkin[1], R Haroldson[4], V A Milichko[1], D A Zuev[1], S V Makarov[1], A A Zakhidov[1,3,4]

[1]ITMO University, 49 Kronverksky Pr., St. Petersburg, 197101, Russia
[2]Lebedev Physical Institute of Russian Academy of Sciences, 53 Leninsky Pr., Moscow, 119991, Russia
[3]National University of Science and Technology, NUST "MISIS", 2 Leninsky Pr., Moscow, 119049, Russia
[4]The University of Texas at Dallas, 800 W Campbell Rd, Richardson, TX 75080, USA

Abstract. Hybrid halide perovskites have emerged as one of the most promising type of materials for thin-film photovoltaic and light-emitting devices. Further boosting their performance is critically important for commercialization. Here we use femtosecond laser for post-processing of organo-metalic perovskite ($MAPbI_3$) films. The high throughput laser approaches include both ablative silicon nanoparticles integration and laser-induced annealing. By using these techniques, we achieve strong enhancement of photoluminescence as well as useful light absorption. As a result, we observed experimentally 10-fold enhancement of absorbance in a perovskite layer with the silicon nanoparticles. Direct laser annealing allows for increasing of photoluminescence over 130%, and increase absorbance over 300% in near-IR range. We believe that the developed approaches pave the way to novel scalable and highly effective designs of perovskite based devices.

1. Introduction

Organic-inorganic perovskites of methylammonium lead trihalides ($MAPbI_3$) family have attracted considerable attention when their conversion efficiency of photovoltaic (PV) devices raised rapidly from 6.5 to 19% in 2012–2015 [1]. Solution-based perovskite solar cells demonstrated efficiencies of more than 16–18% [2]. These impressive performance allowed perovskites to compete with the leading solar materials of the third generation. Advances in film formation and the optimized perovskite PV architectures have led to further conversion efficiency increase up to 22.1% [3]. Moreover, new applications of this hybrid materials have been investigated, including light emitting diodes, optical amplifiers and lasers [4]. However, these high efficiencies were demonstrated on small samples rather than on large-scale devices, required by modern industry. This fact encourages researchers to develop novel approaches for increase of the efficiencies of the realistic perovskite based devices. One of the most simple method is to introduce resonant nanoparticles fabricated by low-cost femtosecond laser ablation in liquids. Also, direct laser gentle heating of a perovskite film can reduce defects concentration and enlarge size of crystallites.

Enhancement of light harvesting in perovskite based solar cells and photodetectors due to metallic nanoparticles (NPs) has been extensively employed during last few years [5]. The perovskite solar cells with the incorporated plasmonic NPs demonstrate improved efficiencies up to 16.3% [6] as low-cost designs. However, parasitic absorption refers to the loss of photocarriers excited and generated photons in the metal particles themselves that decay via non-radiative channels that produce heat [7]. Therefore, increasing of photoluminescence (PL) from hybrid perovskite films by incorporating resonant NPs is still a challenge. In contrast to plasmonic NPs, all-dielectric nanophotonics based on resonant high-index dielectric NPs is a novel paradigm for light enhancement and manipulation at the nanoscale [8,9], as well as for nanothermometry [10]. Indeed, excitation of Mie-type resonances in such NPs results in strong light scattering and near-field enhancement [11], along with low Ohmic losses. The resonant frequency can be controlled by changing the size and shape of the NPs. Silicon NPs are a building block of all-dielectric nanophotonics, being high-index ($n>3.5$) material in visible and IR ranges, whereas its optical losses are almost negligible as compared with small plasmonic NPs [12]. Therefore, silicon based nanoantennas were successfully used for PL enhancement from dyes [13]. Even organometalic perovskites can be considered as a material with high refractive index, which was recently used for fabrication of perovskite-based resonant nanostructures by high-throughput nanoimprint lithography [14], which demonstrate enhanced PL properties.

In this work, we apply femtosecond laser for improving hybrid perovskite optical properties. In particular, we show that silicon nanoparticles (Si NPs) fabricated by laser ablation and incorporated in a perovskite thin film significantly improve light absorption in near-IR range. Further, we directly apply femtosecond laser pulses for heating and improving the perovskite characteristics. Since we use cheap and high throughput methods, we believe that our findings pave the way for low-cost and highly efficient perovskite based optoelectronic devices.

2. Results and discussion

2.1. Laser fabricated silicon nanoparticles within perovskite film

First of all perovskite films were prepared by solvent engineering method. Films thickness was 300-500 nm. Our study perovskite composition contains organic (MA) and inorganic (PbI_3) components. Laser ablation in liquids. In order to create pure colloids of resonant Si NPs, we apply method of laser ablation in liquids. We use femtosecond laser (Avesta Project) providing laser pulses at λ=800 nm with energy up to 2 mJ and pulse duration of 35 fs at the repetition rate of 1 kHz. Laser pulses are focused by lens with 10 cm focal length. Sample is placed before than the focal distance around 10 cm.

For nanoparticles incorporation into the perovskite layer, several steps have to be done. Firstly, laser ablation of single crystalline Si slab in liquid is provided, generating colloids of spherical crystalline Si NPs with a broad size distribution. Toluene is used as the liquid since it is used during our perovskite film formation and doesn't dissolve perovskite. Second step is the deposition of the colloids on a liquid layer of a spin-coated perovskite. After the annealing, the perovskite layer with Si NPs becomes dark and ready for further utilization. As shown in Figure 1b, Si NPs are distributed randomly inside the perovskite. Average size of the Si NPs is about 140 nm. In order to characterize the $MAPbI_3$ perovskite layer with embedded Si NPs, we study its optical properties. According to our reflection/transmission measurements, absorbance increases from near 0 up to 0.3 at 800 nm (Figure 1b). However, PL intensity under 633-nm cw laser photoexcitation is weaker as compared to the similar perovskite without Si NPs (Figure 1a).

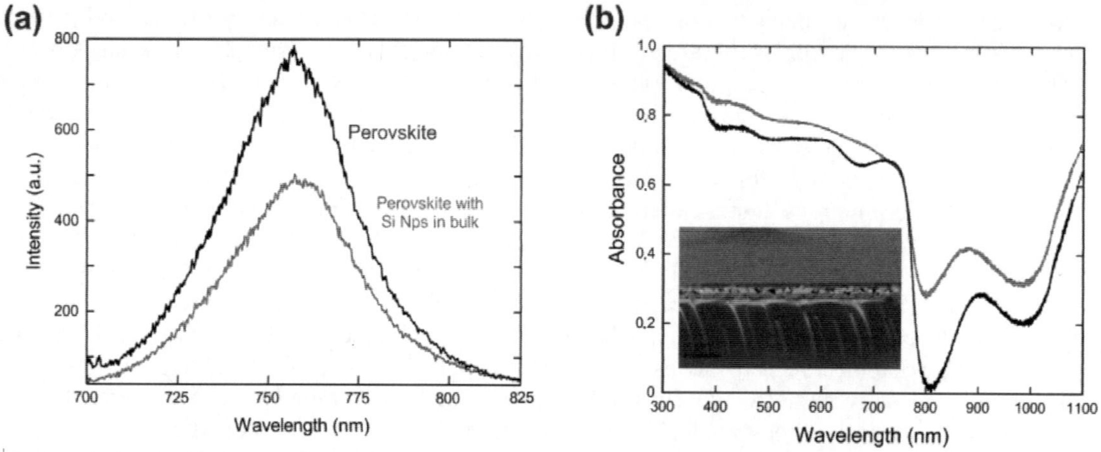

Figure 1 (a, b). (a) Comparison of PL spectra of perovskite film with embedded Si NPs in the bulk (red curve) and without it (black curve). (b) Absorbance spectra for the film with Si NPs (red) and without (black). Inset shows SEM image of perovskite film with Si NPs in the bulk.

2.2. Laser processing on perovskite surface for material modification

Second goal of our research is studying possibility to enhance PL and absorbance properties after the perovskites laser processing. We use Yb^{+3}-doped fiber femtosecond laser with wavelength 1050 nm and pulse width around 300 fs. Film surface was processing at laser fixed average power 35 mW at repetition rate 100 kHz. Different number of laser pulses were applied for the laser processing, resulting in various surface morphologies, as sown in SEM image in Figure 2a.

After laser post-processing of the film, we studied its PL and absorbance spectra. Measurements show different behavior for different post-processing regimes. The regime of irradiation by 1000 pulses has stronger PL spectra than initial plain area (Figure 2b). In absorbance measurements 100 pulses regime shows slight increase from plane area (Figure 2c). The origin of such optical properties modification is in ultrafast and gentle laser annealing of the perovskite film under multipulse femtosecond laser irradiation [15].

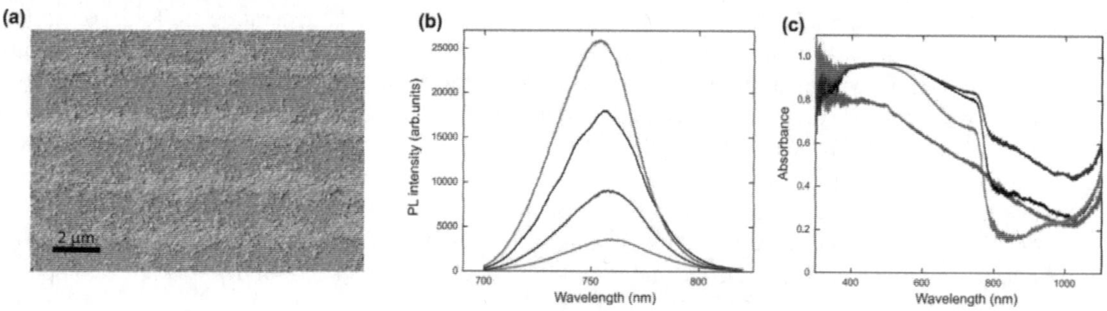

Figure 2 (a, b, c). (a) SEM image of surface after laser-processing (100 pulses). (b) PL spectra for 1 pulse (black), 10 pulses (red), 100 pulses (blue) and 1000 pulses (pink). (c) Absorbance spectra for 1 pulse area (black), 10 pulses (red), 100 pulses (blue) and 1000 pulses (pink).

3. Conclusion

To conclude, we have applied femtosecond laser for postprocessing of organo-metalic perovskite ($MAPbI_3$) films. We have demonstrated high throughput laser approaches, which include both ablative silicon nanoparticles integration into perovskite layer and its laser-induced annealing. By using these techniques, we have achieved 10-fold enhancement of absorbance in a perovskite layer with the silicon nanoparticles. Moreover, direct laser annealing allows for increasing of photoluminescence over 130%, and increase absorbance over 300% in near-IR range. We believe that the developed approaches pave the way to novel scalable and highly effective designs of perovskite based devices.

Acknowledgments
This work was supported by Ministry of Education and Science of Russian Federation (grant № 14.Y26.31.0010).

References

[1] Green M, Ho-Baillie A and Snaith H 2014 *Nature Photonics* **8** 506–514
[2] Jeon N, Noh J, Kim Y, Yang W, Ryu S and Seok S 2014 *Nature Materials* **13** 897–903
[3] Saliba M, Matsui T, Seo J, Domanski K, Correa-Baena J, Nazeeruddin M, Zakeeruddin S, Tress W, Abate A, Hagfeldt A et al. 2016 *Energy & Environmental Science* **9** 1989–1997
[4] Sutherland B and Sargent E 2016 *Nature Photonics* **10** 295–302
[5] Zhang W, Saliba M, Stranks S, Sun Y, Shi X, Wiesnerand U and Snaith J 2013 *Nano Letters* **13** 4505–4510
[6] Saliba M, Zhang W, Burlakov V, Stranks S, Sun Y, Ball J, Johnston M, Goriely A, Wiesner U and Snaith H 2015 *Advanced Functional Materials* **25** 5038–5046
[7] Atwater H and Polman A 2010 *Nature Materials* **9** 205–213
[8] Kuznetsov A, Miroshnichenko A, Brongersma M, Kivshar Y and Luk'yanchuk B 2016 *Science* **354** ag2472
[9] Jahani S and Jacob Z 2016 *Nature Nanotechnology* **11** 23–36
[10] Zograf G, Petrov M, Zuev D, Dmitriev P, Milichko V, Makarov S and Belov P 2017 *Nano Letters* (doi: 10.1021/acs.nanolett.7b00183)
[11] Bakker R, Permyakov D, Yu Y, Markovich D, Paniagua-Domínguez R, Gonzaga L, Samusev A, Kivshar Y, Lukýanchuk B and Kuznetsov A 2015 *Nano Letters* **15** 2137–2142
[12] Krasnok A, Makarov S, Petrov M, Savelev R, Belov P, and Kivshar Y 2015 *SPIE Optics+ Optoelectronics* 950203-950203
[13] Caldarola M, Albella P, Cortés E, Rahmani M, Roschuk T, Grinblat G, Oulton R, Bragas A and Maier S 2015 *Nature Communications* **6** 7915
[14] Makarov S, Milichko V, Ushakova E, Omelyanovich M, Cerdan Pasaran A, Haroldson R, Balachandran B, Wang H, Hu W, Kivshar Y and Zakhidov A 2017 *ACS Photonics* **4** 728-735
[15] Kim S. J. et al. 2016 *ACS nano* **12** 10851-10857

Laser treatment of Se/Bi heterostructure: Bi_2Se_3 nanofilm formation

K G Mikheev, V Ya Kogai, G M Mikheev

Institute of Mechanics, Ural Branch of RAS, Izhevsk 426067, Russia

Abstract. Thin films of Se/Bi heterostructure were obtained by the by thermal vaporization of Se and Bi powders in a vacuum camera. A focused He-Ne laser beam at 632.8 nm is shown to cause local darkening of Se/Bi heterostructure at incident power densities above 3.5 kW/cm^2. The possibility of laser recording of diffraction grating by this technique is demonstrated. It is shown that the resistance between the electrical contacts on the heterostructure reduces as the lines between these contacts are recorded. Data obtained from Raman spectra before and after laser treatment indicate that the darkening is accompanied by Bi_2Se_3 formation.

1. Introduction

Bismuth selenide is a thermoelectric material, possessing features of topological insulator [1,2]. Due to the unique properties Bi_2Se_3 is considered as a promising material for applying in different thermoelectric devices, photosensors, as well as in spintronics, optoelectronics, magnetoelectronics and quantum computation [1,3]. Bi_2Se_3 films could be produced by chemical deposition [4], chemical reaction of sequentially deposited ionic layers [5], electrochemical deposition [6], molecular-beam epitaxy [7], reactive evaporation [8], as well as by organometallic chemical deposition [9]. It was also shown that bi-layer nanofilms, such as Se/Cu, Se/In, Se/Ag and Cu(Ag)/Se, formed on a glass substrate at room temperature could spontaneously and explosively transform into crystalline CuSe, InSe and Ag_2Se nanofilms [10–13]. By analogy with that it was expected that spontaneous crystallization would be observed in Se/Bi heterostructure too. However it does not occur, therefore it is interest to study the crystallization process in Se/Bi heterostructure during local laser heating. Besides that, this study is relevant in terms of development of laser image recording techniques, i.e. for obtaining different diffractive optical elements, all kinds of scales and gratings of optical devices, limbs, matrixes [14–19]. Thus, this work is aimed to study the formation of Bi_2Se_3 nanofilm by laser treatment of Se/Bi heterostructure.

2. Experimental details

Se/Bi heterostructure was formed both on an interelectrode gap of contact pads on a glass substrate (see Figure 1) and on a smooth glass substrate by consecutive thermal vaporization of initial Se and Bi powders in vacuum camera, with the substrate temperature being of 300 K. The contact pads with the interelectrode gap of 0.3×0.5 mm^2 were obtained by mask technology and represent itself bi-layer metal films Cr/Au. Evaporation of samples of initial components of Se and Bi was carried out from tantalum and molybdenum evaporators respectively. The samples were weighed with an accuracy of 0.01 mg. The thickness of Se and Bi layers was measured to be 141 and 45 nm respectively by ellipsometric method.

Thermal treatment of the Se/Bi sample obtained on a smooth glass substrate was implemented by vacuum camera at 10^{-3} Pa and temperature of 493 K. Laser modification of Se(141 nm)/Bi(45 nm) heterostructure was carried out by focused He-Ne laser radiation at 632.8 nm. The Raman spectra of the films were measured by Raman spectrometer (Horiba HR800) at excitation of 632.8 nm before and after laser modification. To avoid thermal effect while Raman studying, the intensity of laser beam focused by 100× objective did not exceed 0.1 kW/cm^2 [18].

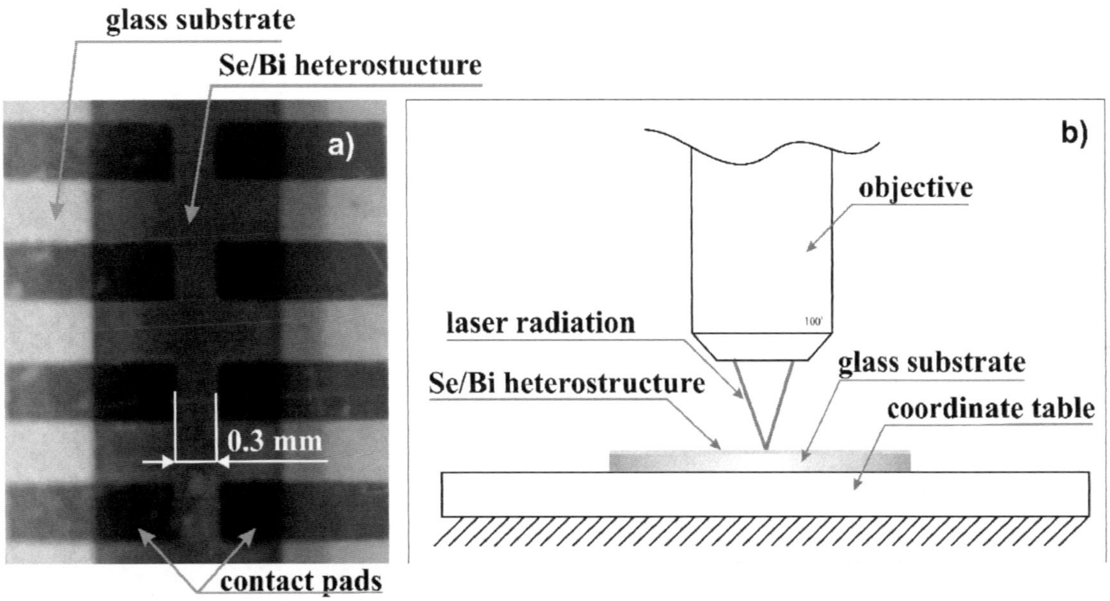

Figure 1. (a) Se/Bi heterostructure formed on an interelectrode gap of contact pads on a glass substrate, (b) sketch of the experiment of laser image recording on Se/Bi heterostructure.

3. Results and discussion

Research into the laser treatment of Se(141 nm)/Bi(45 nm) heterostructure showed that the irradiation of the heterostructure by >3.5 kW/cm^2 laser beam focused by 100× objective led to dark spot formation on the sample surface. With the mechanical laser shutter open, displacing the film using horizontal coordinate table a continuous line can be drawn. The scheme of this experiment is shown in Figure 1b and a set of such lines is presented in Figure 2.

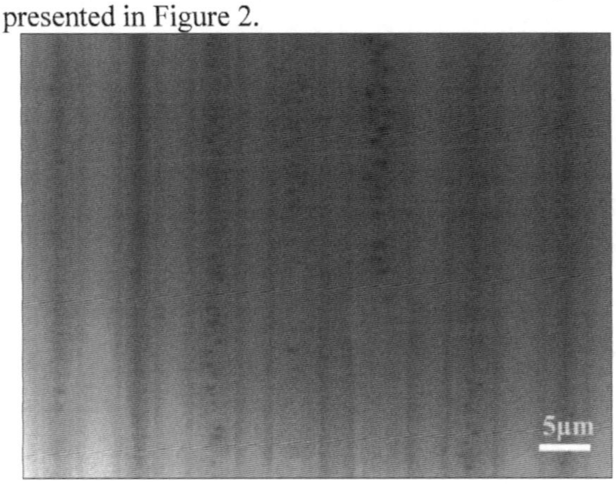

Figure 2. A set of lines drawn on Se(141 nm)/Bi(45 nm) heterostructure by 632.8 nm laser beam focused by 100× objective.

IOP Conf. Series: Journal of Physics: Conf. Series **917** (2017) 062003 doi:10.1088/1742-6596/917/6/062003

The Raman spectra of the film area before and after laser treatment are presented in Figure 3. The Raman spectrum before laser treatment shows the peaks at 72 and 97 cm^{-1}, assigned to the E_g and A_{1g} Raman modes of bismuth, respectively [20]. These peaks are expected since the upper layer of Se/Bi heterostructure is bismuth. After laser treatment the Raman spectrum shows the peaks at 71, 131 and 173 cm^{-1}, assigned to the A_{1g}^1, E_g^2, and A_{1g}^2 vibrational modes of bismuth selenide, respectively [21].

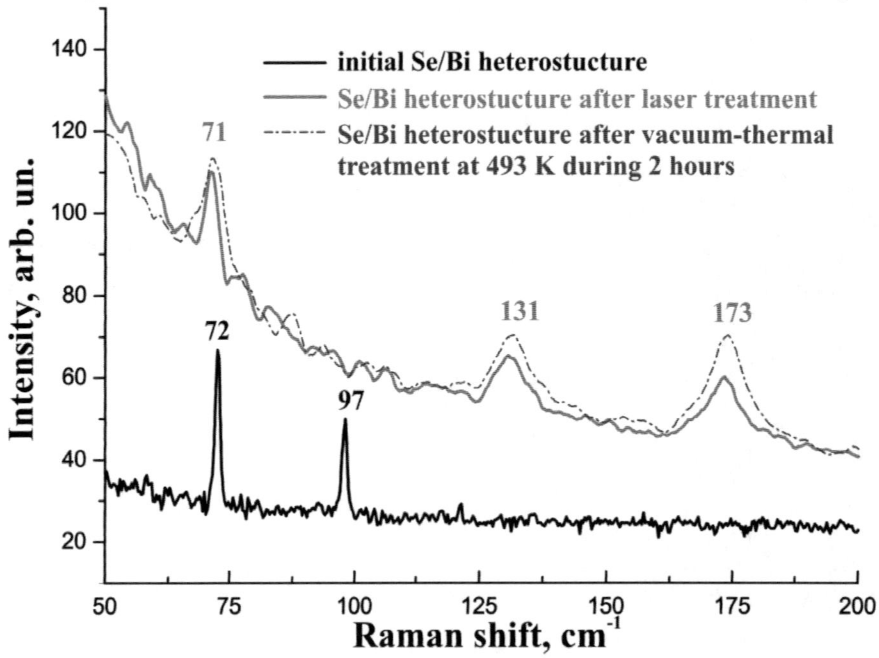

Figure 3. Raman spectra of an area of Se(141 nm)/Bi(45 nm) heterostructure (solid black line) before, (solid red line) after laser treatment, excitation wavelength of 632.8 nm, and (dashed blue line) after vacuum thermal annealing at 493 K during 2 hours.

The obtained results are explained by local laser heating of the heterostructure leading to Bi_2Se_3 film formation in the laser exposure area. Indeed, the Raman spectrum of the sample after annealing at $T = 493$ K during 2 hours consists only of above mentioned vibrational modes assigned to bismuth selenide (see Figure 3, dashed blue line). This indicates on the Se/Bi heterostructure conversion into Bi_2Se_3 during the annealing process and proves that local heating of the Se/Bi heterostructure leads to Bi_2Se_3 film formation.

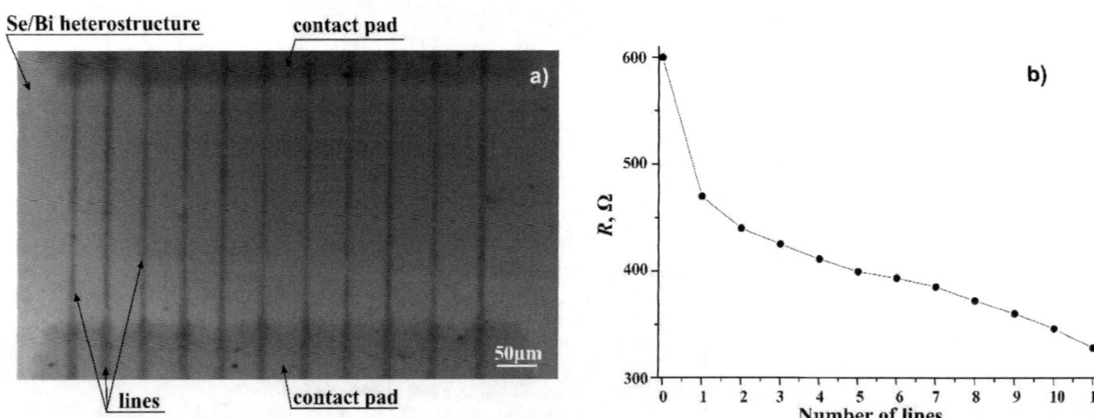

Figure 4. (a) A set of lines drawn on Se(141 nm)/Bi(45 nm) heterostructure formed between contact pads by 632.8 nm laser beam focused by 10x objective, (b) electrical resistance as a function of

number of lines drawn on Se(141 nm)/Bi(45 nm) heterostructure by 632.8 nm laser beam focused by 10× objective.

In Figure 4a a set of lines drawn on Se(141 nm)/Bi(45 nm) heterostructure formed between contact pads by 632.8 nm laser beam focused by 10× objective is presented. The electrical resistance R between contact pads as a function of lines number is shown if Figure 4b. One can see that the resistance R drops down as the first line is drawn. Then it monotonically reduces with increasing the lines number. Thus, increase of Bi_2Se_3 content between contact pads in Se/Bi heterostructure promotes the electrical resistance reduce.

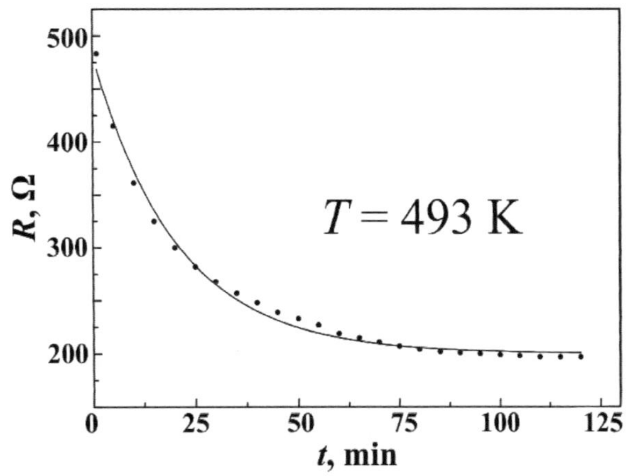

Figure 5. Electrical resistance as a function of annealing time of Se/Bi heterostructure at T = 493 K (dots – experiment, line – approximation).

As it was mentioned above, annealing the Se/Bi heterostructure at T = 493 K during 2 hours leads to formation of Bi_2Se_3 film as well. The electrical resistance as a function of annealing time of the heterostructure shown in Figure 5 suggests that this process proceeds exponentially following to the equation: $R = R_{in}\exp(-t/\tau)+R_{fin}$, where R_{in} = 281.5 Ω, R_{fin} = 200 Ω, and the characteristic time of equilibrium state establishing τ = 20.5 min (see Figure 5).

Figure 2 and Figure 4a demonstrate that the Se/Bi heterostructure transformation into Bi_2Se_3 under laser radiation is accompanied by increase of the film optical density. Figure 4b and Figure 5 show that the mentioned phase transformation leads to the surface film resistance reducing, i.e. to increase of the film conductivity. These results are in agreement with each other since an increase of a medium optical density is accompanied with an increase of its conductivity and vice versa: a reduce of a medium optical density leads to reduce of its conductivity [22]. It is evident that increase of Se/Bi heterostructure optical density under focused laser radiation can be used for formation of optical images, e.g. for obtaining diffractive gratings. It is well demonstrated by the images of diffractive gratings presented in Figure 2 and Figure 4a. To increase the contrast of the diffractive gratings obtained it is necessary to apply chemical etching techniques, the study of which goes beyond the scope of this work.

4. Conclusions

Thus, we have studied laser influence on Se(141 nm)/Bi(45 nm) heterostructure. Laser treatment leads to local laser heating which causes Bi_2Se_3 film formation in the laser exposure area. Increase of Bi_2Se_3 content in the Se/Bi heterostructure promotes the electrical resistance reduce and leads to increase of the film optical density. The Se(141 nm)/Bi(45 nm) heterostructure conversion into Bi_2Se_3 during the annealing process proceeds exponentially, with the characterstric time of equilibrium state establishing being of 20.5 min. The results obtained show the possibility of laser image recording on thin films of Se/Bi heterostructure.

References

IOP Conf. Series: Journal of Physics: Conf. Series **917** (2017) 062003 doi:10.1088/1742-6596/917/6/062003

[1] Mishra S K, Satpathy S and Jepsen O 1997 Electronic structure and thermoelectric properties of bismuth telluride and bismuth selenide *J. Phys. Condens. Matter* **9** 461–70

[2] Xia Y, Qian D, Hsieh D, Wray L, Pal A, Lin H, Bansil A, Grauer D, Hor Y S, Cava R J and Hasan M Z 2009 Observation of a large-gap topological-insulator class with a single Dirac cone on the surface *Nat. Phys.* **5** 398-402

[3] Zhang H, Liu C-X, Qi X-L, Dai X, Fang Z and Zhang S-C 2009 Topological insulators in Bi_2Se_3, Bi_2Te_3 and Sb_2Te_3 with a single Dirac cone on the surface *Nat. Phys.* **5** 438–42

[4] Pramanik P, Bhattacharya R N and Mondal A 1980 A Chemical Method for the Deposition of Thin Films of Bi_2Se_3 *J. Electrochem. Soc* **2** 1857–8

[5] Pejova B and Grozdanov I 2002 Chemical deposition and characterization of glassy bismuth(III) selenide thin films *Thin Solid Films* **408** 6–10

[6] Sankapal B R, Mane R S and Lokhande C D 2000 Preparation and characterixation of Bi_2Se_3 thin films deposited by successive ionic layer adsorption and reaction (SILAR) method *Mater. Chem. Phys.* **63** 230–4

[7] He L, Xiu F, Wang Y, Fedorov A V, Huang G, Kou X, Lang M, Beyermann W P, Zou J and Wang K L 2011 Epitaxial growth of Bi_2Se_3 topological insulator thin films on Si (111) *J. Appl. Phys.* **109**103702(1-6)

[8] John K J, Pradeep B and Mathai E 1993 Electrical properties of Bismuth selenide (Bi_2Se_3) thin films prepared by reactive evaporation *Solid State Commun.* **85** 879–81

[9] John K J, Pradeep B and Mathai E 1992 Polycrystalline bismuth selenide (Bi_2Se_3) thin films prepared by reactive evaporation *Solid State Commun.* **83** 501–3

[10] Kogai V Y, Vakhrushev A V and Fedotov A Y 2012 Spontaneous explosive crystallization and phase transformations in a selenium/copper bilayer nanofilm *JETP Lett.* **95** 454–6

[11] Kogai V Y and Vakhrouchev A V 2013 Spontaneous explosive crystallization and phase formation in a nanosized selenide/indium heterostructure *Tech. Phys. Lett.* **39** 1044–6

[12] Kogai V Y 2014 Explosive crystallization in the course of formation of Se/Ag nanosize film structure *Tech. Phys. Lett.* **40** 656–9

[13] Kogai V Y 2016 Reaction-diffusion-induced explosive crystallization in a metal–selenium nanometer film structure *Tech. Phys.* **61** 461–3

[14] Kononenko V V, Komlenok M S, Pimenov S M and Konov V I 2007 Photoinduced laser etching of a diamond surface *Quantum Electron.* **37** 1043–6

[15] Veiko V P, Korol'kov V I, Poleshchuk A G, Sametov A R, Shakhno E A and Yarchuk M V. 2011 Study of the spatial resolution of laser thermochemical technology for recording diffraction microstructures *Quantum Electron.* **41** 631–6

[16] Volkov A V, Moiseev O Y and Poletaev S D 2013 Precision laser recording on a molybdenum films for diffractive microrelief formation *Komput. Opt.* **37** 220–5

[17] Mikheev G M, Mikheev K G, Mogileva T N, Puzyr A P and Bondar V S 2014 Laser image recording on detonation nanodiamond films *Quantum Electron.* **44** 1–3

[18] Mikheev G M, Mikheev K G, Anoshkin I V and Nasibulin A G 2015 Laser images recording on aerosol-synthesized single-walled carbon nanotube films *Tech. Phys. Lett.* **41** 887–90

[19] Mikheev K G, Nasibulin A G, Gilmutdinov F Z and Mikheev G M 2016 On the mechanism of laser bleaching of aerosol synthesized single walled carbon nanotubes on the polymer substrate *Khim. Fiz. Mezoskopiya* **18** 130–41

[20] Trentelman K 2009 A note on the characterization of bismuth black by Raman microspectroscopy *J. Raman Spectrosc.* **40** 585–9

[21] Zhang J, Peng Z, Soni A, Zhao Y, Xiong Y, Peng B, Wang J, Dresselhaus M S and Xiong Q 2011 Raman spectroscopy of few-quintuple layer topological insulator Bi_2Se_3 nanoplatelets. *Nano Lett.* **11** 2407–14

[22] Mikheev G M, Kuznetsov V L, Mikheev K G, Mogileva T N and Moseenkov S I 2011 Laser-induced diamagnetism in suspension of onion-like carbon particles *Tech. Phys. Lett.* **37** 831–4

IOP Conf. Series: Journal of Physics: Conf. Series **917** (2017) 062004 doi:10.1088/1742-6596/917/6/062004

Modeling of a narrow band pass filter for Bathymetry light detection and ranging (LIDAR) system

M A Butt[1], S A Fomchenkov[1, 2], S N Khonina[1, 2]
[1]Samara National Research University, 34 Moskovskoe Shosse, Samara.
[2]Image Processing Systems Institute of RAS, Branch of the FSRC "Crystallography and Photonics" RAS, Samara, Russia

Abstract. In this work, a narrow band pass Fabry-Perot filter is designed which can be used in an airborne light detection (ALB) and ranging bathymetry. LIDAR is done by reflecting a pulse laser beam from a target and detecting the round-trip propagation time between the source and the target. ALB systems consist of Nd: YAG laser that emits the pulses at two different wavelengths such as 1064 nm and 532 nm. Infrared pulses at 1064 nm are reflected from the water surface and the green pulses at 532 nm which penetrates the water surface and are reflected from the ground. Filters are desirable to suppress the ambient light that is reflected by the surface of the water or an atmosphere which always enter the detector as a noise. The designed filter shows a high quality with an average transmission of more than 95 % at 532 nm which is considered as practically ideal for water penetration in typical coastal waters.

1. Introduction

Airborne lidar bathymetry (ALB) [1] is widely used to measure the topography as well as the bathymetry of near shore coastal waters/lakes from a low elevation aircraft with the help of scanning pulsed laser beam. The ALB system consists of Nd: YAG laser that emits the pulses at two different wavelengths such as 1064 nm and 532 nm [2, 3]. Infrared pulses at 1064 nm are reflected from the water surface, the green pulses at 532 nm which penetrates the water surface and are reflected from the ground. These wavelengths are principally suitable specified the optical properties of water. Laser pulses transmitted at 532 nm wavelength are adjacent to the minimum absorption spectrum of the water and that wavelength is considered as practically ideal for the water penetration in typical coastal waters.

Thin film optics is a well-developed technology and various devices such as pass band filters, stop band filters, polarizers and reflectors are realized with the help of multilayer dielectric thin films [4, 5, 6, 7, 8]. These devices contain the alternating layers of high and low refractive index materials with particular thicknesses with good knowledge of their refractive index and absorption coefficients. These devices are based on the principle of multiple reflections between high and low refractive indexes materials interface. Distributed Bragg Reflectors (DBRs) are quarter wave thickness of the centre wavelength. The high reflection region of a DBR is known as the DBR stop band and can be obtained by the refractive index contrast between the constituent layers. By inserting a half wave spacer between two highly reflecting mirrors governs an output at the designed wavelength [5]. A high Finesse design transmits a narrow pass band with high transmission. This kind of device is known as Fabry-Perot (FP) filter as shown in figure 1. DBRs are the most commonly used mirrors in FP filters, due to their high reflectivity. FP has numerous applications in laser line filtering, astronomical observations, fluorescence microscope imaging and spectroscopic instrumentation.

IOP Conf. Series: Journal of Physics: Conf. Series **917** (2017) 062004 doi:10.1088/1742-6596/917/6/062004

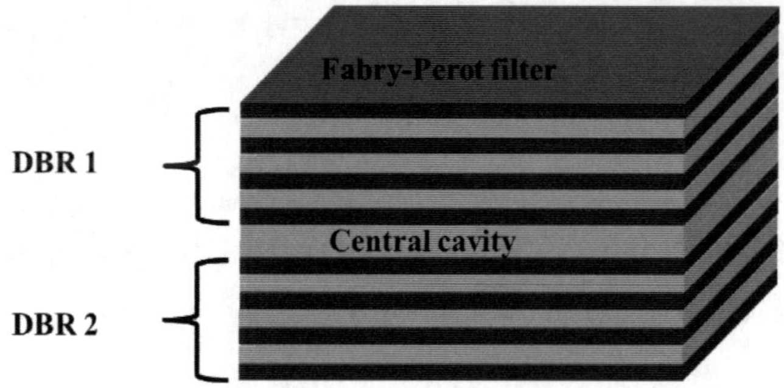

Figure 1. Fabry Perot filter's design.

2. Design and discussion

In this work, the design of a FP filter based on TiO_2/SiO_2 is proposed at a central wavelength of 532 nm with a FWHM as narrow as 1.6 nm. TiO_2 and SiO_2 dielectric materials are selected as high and low refractive index materials, respectively. The choice of these materials is made on the basis of their low absorption and high index contrast in the wavelengths of interest [9, 10, 11]. Open-source software, Open Filters, is used in this work to design and optimize the required filter. Transmission and reflection properties of interference filters are dependent on materials refractive index and layer thickness of the materials. Open filter calculates optical properties of the filters. It uses transfer matrix method to calculate the transmission and reflection properties of filters based on the absorption and refractive indices of the materials [12]. Optimization techniques are available in this software like needle synthesis (Adding an extra layer to give targeted transmission). The FP filter is designed for visible spectrum ranges from 350-700 nm. The optimized thicknesses of the layers are shown in Table I. The total thickness of the filter is 3066 nm.

Table I. Layer thickness of TiO_2/SiO_2 based Fabry Perot filter with a central wavelength at 532 nm.

No. of layers	Dielectric materials	Layer thickness (nm)	No. of layers	Dielectric materials	Layer thickness (nm)
1	TiO_2	41	21	TiO_2	50
2	SiO_2	66	22	SiO_2	97
3	TiO_2	109	23	TiO_2	80
4	SiO_2	168	24	SiO_2	88
5	TiO_2	49	25	TiO_2	49
6	SiO_2	90	26	SiO_2	83
7	TiO_2	55	27	TiO_2	47
8	SiO_2	137	28	SiO_2	81
9	TiO_2	58	29	TiO_2	46
10	SiO_2	75	30	SiO_2	75
11	TiO_2	23	31	TiO_2	113
12	SiO_2	67	32	SiO_2	87
13	TiO_2	47	33	TiO_2	66
14	SiO_2	90	34	SiO_2	83
15	TiO_2	69	35	TiO_2	45
16	SiO_2	105	36	SiO_2	75
17	TiO_2	59	37	TiO_2	41
18	SiO_2	94	38	SiO_2	67
19	TiO_2	51	39	TiO_2	33
20	SiO_2	175	40	SiO_2	132

3. Transmission spectra of TiO_2/SiO_2 FP filter at various angles of incidence

The transmission spectra in the visible range of the filter show a characteristic peak with a maximum of 96 % at a reference wavelength of 532 nm. The width (FWHM) of the transmission spectra is observed at 1.6 nm. The filter is designed with a less possible number of layers with high transmission in pass band region and high reflection is obtained in the stop band. The transmission spectrum of FP filter is shown in figure 2.

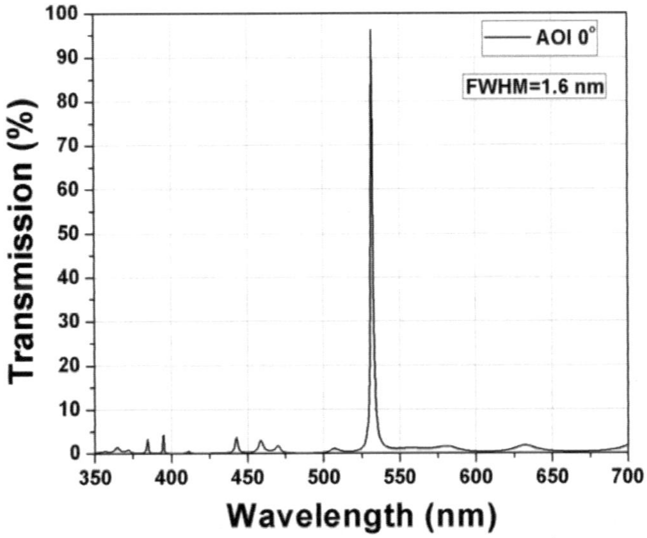

Figure 2. Transmission spectra of TiO_2/SiO_2 based FP filter at zero degrees of angle of incidence.

For all dielectric stack filters, the transmission depends on the angle of incidence. The central wavelength of the FP filter shifts toward the smaller wavelengths as the angle of incidence is increased as shown in figure 3.

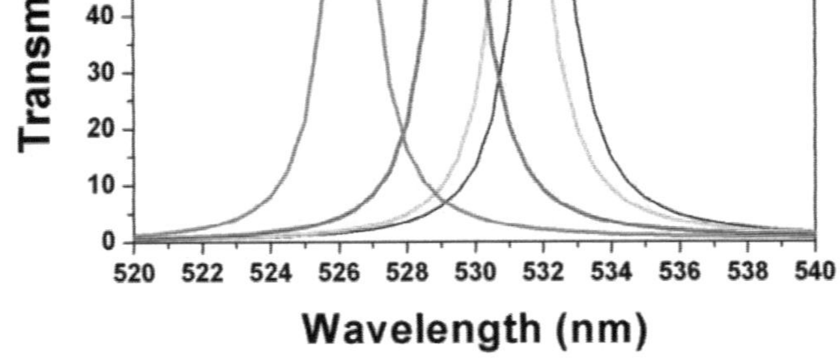

Figure 3. The transmission spectrum of the FP filter at various angles of incidence.

4. Conclusion

In this work, a multilayer narrow band pass dielectric filter is designed which can be used in numerous applications such as laser line filtering, astronomical observations and spectroscopic instrumentations. One of its major applications is in LIDAR bathymetry systems where ambient light is blocked from entering the detector as a noise. The designed filter shows a high quality with a maximum transmission of 96 % at 532 nm which penetrates the water surface and is reflected from the ground and received by the detector. One of the basic limitations of the filters with small FWHM is their dependence on the angle of incidence of the reflected light; therefore these filters should always be adjusted to receive the reflected light at zero degrees of angle of incidence.

Acknowledgements

This work was supported by the Ministry of Education and Science of the Russian Federation and the Russian Foundation for Basic Research (grants no. 16-47-630546, 16-29-11744).

References

[1] Chen Qi 2007 Photogramm. Eng. & Rem. Sens. **73** 109.
[2] Jerlov N G *Marine Optics* 1976 (Elsevier Scientific Publication).
[3] Measures R M *Laser Remote Sensing: Fundamentals and Applications* 1992 (Kreiger Publishing company).
[4] Markus A F, Gary E C, Sheetal C, Sarah A L, Robert L J 2015 *Proc. of SPIE*. 9612 96120K-1
[5] Macleod A H *Thin film Optical Coatings* 2001 (Thin film Center Inc.).
[6] Kazanskiy N L, Kharitonov S I, Khonina S N, Volotovskiy S G, Strelkov Yu S 2014 Computer Optics **38** 256.
[7] Kazanskiy N L, Kharitonov S I, Khonina S N, Volotovskiy S G 2015 *Proc. of SPIE*. 9533 95330L-1.
[8] Doskolovich L L, Golovastikov N V, Bykov D A, Kharitonov S I 2015 Computer Optics **39** 311.
[9] Butt M A, Fomchenkov S A, Ullah A, Habib M, Ali R Z 2016 Computer Optics **40** 674.
[10] Butt M A, Fomchenkov S A 2017 J. Korean Phy. Soc. **70** 169.
[11] Butt M A, Fomchenkov S A, Ullah A, Verma P, Khonina S N 2016 *J. Phy. Conf. Ser*. 741 012136.
[12] Larouche S, Martinu L, 2008 Appl. Opt **47**, C219.

Subwavelength population density gratings in resonant medium created by few-cycle pulses

R M Arkhipov[1,2], **M V Arkhipov**[1], **A V Pakhomov**[3,4], **I Babushkin**[5,6], **A Demircan**[5,7], **U Morgner**[5,7], **N N Rosanov**[2,8,9]

[1] St. Petersburg State University, St. Petersburg, 199034, Russia
[2]ITMO University, 197101 St. Petersburg, Russia
[3]Department of Physics, Samara University, 443086 Samara, Russia
[4]Department of Theoretical Physics, Lebedev Physical Institute, 443011 Samara, Russia
[5]Institute of Quantum Optics, Leibniz University Hannover, 30167 Hannover, Germany
[6]Max Born Institute, 12489 Berlin, Germany
[7]Hannover Centre for Optical Technologies, 30167 Hannover, Germany
[8]Vavilov State Optical Institute, 199053 St. Petersburg, Russia
[9]Ioffe Physical Technical Institute, St. Petersburg 194021, Russia

E-mail: arkhipovrostislav@gmail.com

Abstract. We consider theoretically recently proposed a new possibility of creation, erasing and ultrafast control of population density grating. Such grating can be created in resonant medium when ultrashort pulses with duration smaller than relaxation times in the resonant medium (coherent light matter interactions) propagate without overlapping in this medium. Possible applications in the ultrafast optics such as optical switcher and laser beam deflector are discussed.

1. Introduction

Periodic spatial population density gratings are typically created in resonant medium via pattern produced by the interference of nearly long monochromatic light beams [1]. In this case, two beams overlaps in space and the spatial period of the grating is limited by the light wavelength. These gratings found a lot of applications in linear and nonlinear spectroscopy, Bragg diffraction, and nonlinear optics etc [2, 3].

To date, ultrashort pulses with few-cycle or even subcycle duration became available [4] and found huge amount of applications in ultrafast optics. It allows to observe and control previously inaccessible ultrafast processes in matter such as wave packet dynamics [5]. Furthermore, ultrashort pulses can coherently interact with a resonant medium when their duration is smaller than the medium relaxation times T_1 and T_2 [6].

In this paper, we revise recently proposed [7-11] novel method of creation and ultrafast control of spatial periodic gratings of polarization and inversion by few-cycle pulses coherently propagating in a resonant medium. This method allows achieving the subwavelength gratings by the pulses non-overlapping in space, which is contrast to well-known method, where overlapping of the pulses is needed [1-3]. Such grating are formed due to the interaction of incoming pulses with polarization wave created by the previous pulse. Possible applications of these gratings in ultrafast optics and attosecond science are discussed.

2. Theoretical model and subwavelength gratings created by unipolar sybcycle and bipolar few-cycle pulses

Our theoretical analysis is based on analytical and numerical solution of Maxwell-Bloch equations for the density matrix elements and electric field without use of slowly varying envelope and rotating wave approximations. Resonant medium is described as a two-level system. This system of equations is given by:

$$\frac{d\rho_{12}(z,t)}{dt} = -\frac{\rho_{12}(z,t)}{T_2} + i\omega_0\rho_{12}(z,t) - \frac{i}{\hbar}d_{12}E(z,t)n(z,t)\,, \tag{1}$$

$$\frac{dn(z,t)}{dt} = -\frac{n(z,t)-n_0(z)}{T_1} + \frac{4i}{\hbar}d_{12}E(z,t)\,\mathrm{Im}\,\rho_{12}(z,t)\,, \tag{2}$$

$$\frac{\partial^2 E(z,t)}{\partial z^2} - \frac{1}{c^2}\frac{\partial^2 E(z,t)}{\partial t^2} = \frac{4\pi}{c^2}\frac{\partial^2 P(z,t)}{\partial t^2}\,, \tag{3}$$

$$P(z,t) = 2N_0 d_{12}\,\mathrm{Re}(\rho_{12})\,. \tag{4}$$

Equations (1)-(2) describe the evolution of nondiagonal element of density matrix ρ_{12} and population difference n (inversion) between ground and excited states. Propagation of the electric field E is governed by the wave equation (3). Medium polarization is given by Eq. (4). Other parameters: ω_0 – resonance transition frequency ($\lambda_0 = 2\pi c/\omega_0$ - corresponding wave length), d_{12}– transition dipole moment, n_0 – inversion in the absence of the electric field ($n_0 = 1$ for absorbing medium), N_0 – concentration of two-level atoms, c – speed of light, $\hbar$ – Planck constant.

We investigated the grating dynamics created by the train of bipolar pump pulses (see Refs. [7,8,11]) given by

$$E_p(t) = \sum_{i=1}^{N} E_0 \exp\left(-\frac{[t-\tau_i]^2}{\tau_p^2}\right)\sin\left(\omega_0\left[(t-\tau_i)\right]\right), \tag{5}$$

and the train of unipolar subcycle pulses without carrier frequency given by (see Refs [9,10])

$$E_p(t) = \sum_{i=1}^{N} E_0 \exp\left(-\frac{[t-\tau_i]^2}{\tau_p^2}\right). \tag{6}$$

Here, τ_p is the pulse duration, τ_i is delay between the 1st pulse and the ith pulse, N is the number of pulses. Using of unipolar subcycle pulses (6) seems to have more advantages with respect of bipolar ones (6) seems to have more advantages due to their smaller duration and unipolar character.

Numerical simulations revealed that with proper selection of the delay between the incident bipolar few-cycle bipolar pulses (5) of subcycle pulses (6) and their propagation directions it is possible to create, erase and even multiplicate the spatial period of polarization and inversion gratings, see figure 1. For more details see Refs. [7-10]. Figure 1 illustrates the typical dynamics of inversion n under the action of single-cycle bipolar pulse train (5). In figure 2 the dynamics of polarization is plotted. These results were obtain from the numerical solutions of Maxwell-Bloch equations (1)-(4). The pulses were launched from left to right and from right to left and are not overlapped in the medium.

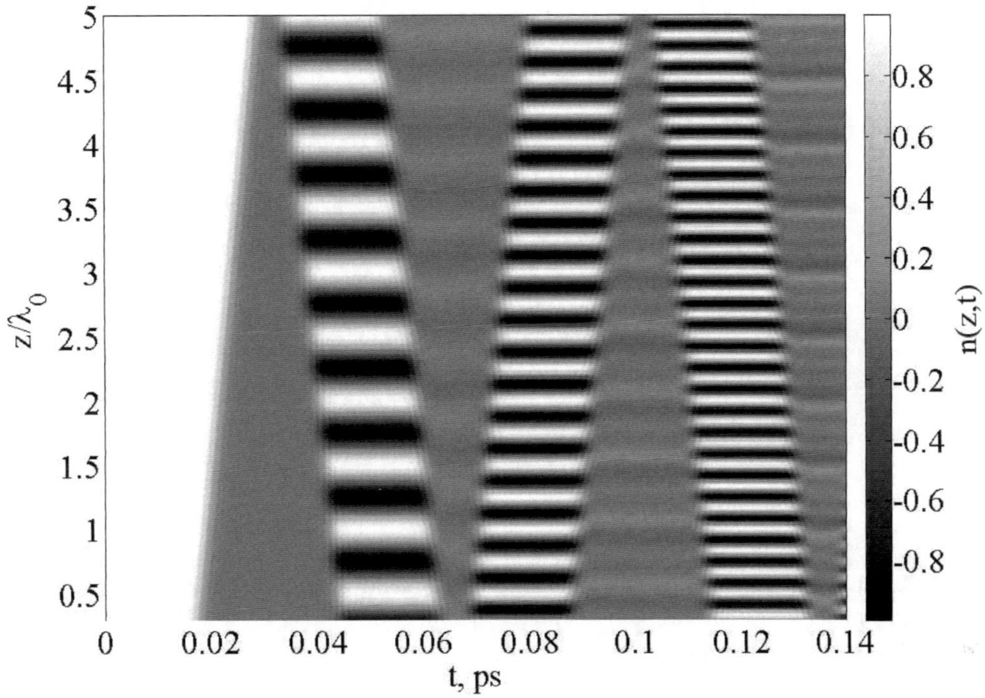

Figure 1. Typical behaviour of spatial and temporal dependence of the population inversion n in a resonant medium demonstrating creation, erasing of periodic gratings and multiplication of its wave vector. Train of few-cycle pulses is launched in the system. Parameters: $\lambda_0 = 700$ nm, $d_{12} = 5$ D, $T_1 = 1$ ns, $T_2 = 2$ ps, $\tau_p = 2$ fs, $N_0 = 10^{17}$ cm^{-3}.

Pulse amplitude was set in a way that pulse acted as $\pi/2$ pulse. From figure 1 and figure 2 it is easy to see the possibility of grating creation, erasing and their spatial frequency multiplication. More detailed description of these grating dynamics can be found in Refs. [7-11]. We remark that this method allows to create the inversion gratings with the spatial frequencies $2mk_0$ ($k_0 = 2\pi/\lambda_0$ is the wave number), $m = 1,2,3...$ and polarization gratings with the frequencies $(2m+1)k_0$, $m = 0,1,2,3...$, see Ref. [8] and figures 1, 2. This method is valid for the pulses with durations smaller than medium relaxation times and can be realized even when the pulse duration approaches subcycle duration [9,10].

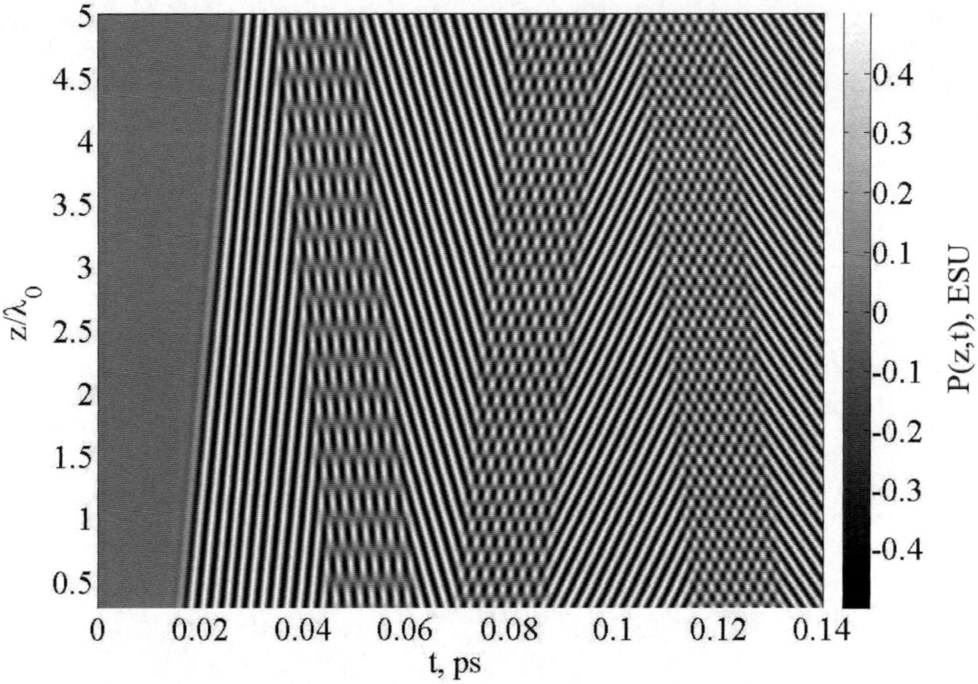

Figure 2. Typical behaviour of spatial and temporal dependence of the polarization P for the parameters of figure 1 in resonant medium demonstrating creation, erasing of periodic gratings and multiplication of its wave vector.

3. Conclusions

To conclude, we briefly described recently proposed mechanism of the ultrafast control of the polarization and inversion gratings in resonant medium by train of few-cycle and subcycle pulses coherently interacting with resonant medium. It allows grating creation, erasing and spatial frequency multiplication. Proposed method is extremely fast, which opens novel opportunities in ultrafast optics. It can be realised in various media where coherent light-matter interactions were observed experimentally – gases, semiconductor bulks, quantum dots. Investigated phenomena, for instance, can find their applications, for laser beam deflectors, ultrafast control of medium properties, attosecond science, spectroscopy, and optical signal processing. A detailed analysis of the grating dynamics can be found in Refs. [7-12].

Acknowledgments

This work was supported by Russian Science Foundation (17-19-01097).

References

[1] Eichler H J, Günter E, Pohl D W, *Laser-Induced Dynamic Gratings* (Springer-Verlag, Berlin, Heidelberg, 1986)

[2] Dao L V, Lowe M, Hannaford P, Makino H, Takai T, Yao T 2002 *Appl. Phys. Lett.* **81** 1806

[3] Scholes G D, Kim J, Wong C Y 2006 *Phys. Rev. B* **73** 195325

[4] Krausz F, Ivanov M, 2009 *Rev. Mod. Phys.* **81** 163
[5] Gallmann L, Cirelli C, Keller U 2012 *Ann. Rev. Phys. Chem.* **63** 447
[6] Allen L, Eberly J H, *Optical resonance and two-level atoms* (Wiley, New York, 1975)
[7] Arkhipov R M, Arkhipov M V, Babushkin I V, Rosanov N N 2016 *Opt. and Spectr.* **121** 758
[8] Arkhipov R M, Arkhipov M V, Babushkin I, Demircan A, Morgner U, Rosanov N N 2016 *Opt. Lett.* **41** 4983
[9] Arkhipov R M, Arkhipov M V, Babushkin I, Pakhomov A V, Rosanov N N, 2017 *Quantum Electronics* **47** 589
[10] Arkhipov R M, Arkhipov M V, Babushkin I, Pakhomov A V, Rosanov N N, 2017 *Laser Physics Letters* , in press
[11] Arkhipov R M, Pakhomov A V, Arkhipov M V, Babushkin I, Demircan A, Morgner U, Rosanov N N, in preparation
[12] Bagaev S N, Egorov V S, Nikolaev V G, Chekhonin I A, Chekhonin M A 2015 *Russian Journal of Physical Chemistry B* **9(4)** 582

Analysis of the Casimir force measurements in systems with Si gratings

Valery N Marachevsky and Alexandra D Nelson

Department of Theoretical Physics, Saint Petersburg State University, Ulianovskaya 1, Petrodvorets, 198504 St. Petersburg, Russia.

E-mail: alexsandranelson94@gmail.com , maraval@mail.ru

Abstract. A comparison of the Casimir force theory and experiments with Si rectangular gratings is performed. We use Rayleigh decompositions to evaluate the Casimir force in Si grating – Au sphere systems for two Si grating samples. Ratios of exact Casimir force to proximity force approximation (PFA) and pairwise additive approximation (PAA) results for the Casimir force are found. The difference in the use of theoretical PFA results and PAA results extracted from experimental data is emphasized.

1. Introduction

The Casimir effect is a quantum fluctuation effect in the presence of boundaries. It was theoretically predicted by H Casimir [1] who derived the formula for the attractive force between two perfectly conducting plates separated by a vacuum slit.

The theory for gratings with 1d spatial periodicity separated by a vacuum slit was developed in [2,3], the Casimir free energy was expressed in terms of grating reflection matrices and Rayleigh coefficients [4]. The theory was found to be in agreement with experiments on the lateral Casimir force between two sinusoidal Au 1d gratings [5, 6]. The difference up to 66% in lateral Casimir force results based on Rayleigh decompositions and the proximity force approximation was predicted and observed [5,6]. Details of a theoretical approach to the Casimir effect based on Rayleigh decompositions are given in [7, 8]. Proximity force approximation is based on Lifshitz theory [9] for two dielectrics with flat surfaces separated by a vacuum slit. The modal approach to the Casimir forces was developed in [10]. Explicit expressions for the Casimir energy between a sphere and 1d grating in terms of the sphere and grating reflection matrices were derived in [11].

Measurements of the Casimir-Polder potential of Rb atom above a grating (made of parallel Au stripes on the sapphire substrate) in the presence of an additional external repulsive potential created by the laser field were done during diffraction of Bose-Einstein condensates consisting of cold Rb atoms [12], theoretical results based on Rayleigh decompositions and experiment were found to be in a good agreement. General analytic properties of Rayleigh coefficients were established in [13]. Existence of repulsive Casimir-Polder potential for an anisotropic atom in the presence of 20 nm thin 1d Au grating was predicted in [13].

Two systems with gratings from [14] were already studied in [2] where scattering theory was applied to grating-grating systems. However, due to a misprint in notations in [14] the calculations in [2] were applied to grating systems different from the ones used in experiments [14] as one can easily check by reading an Erratum to [14]. Therefore we decided to reanalyze

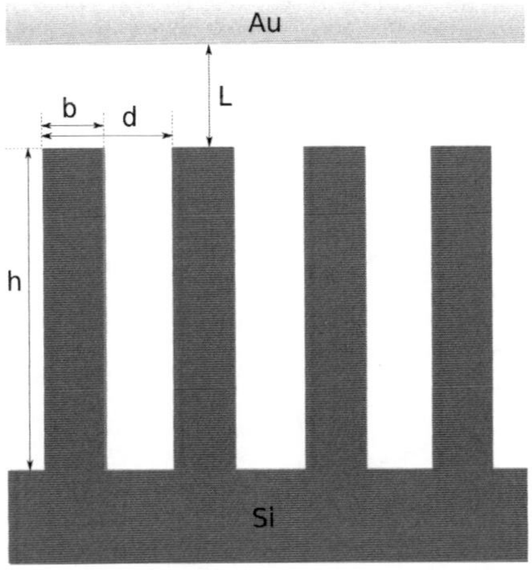

Figure 1. Si grating and Au half-space separated by a vacuum slit.

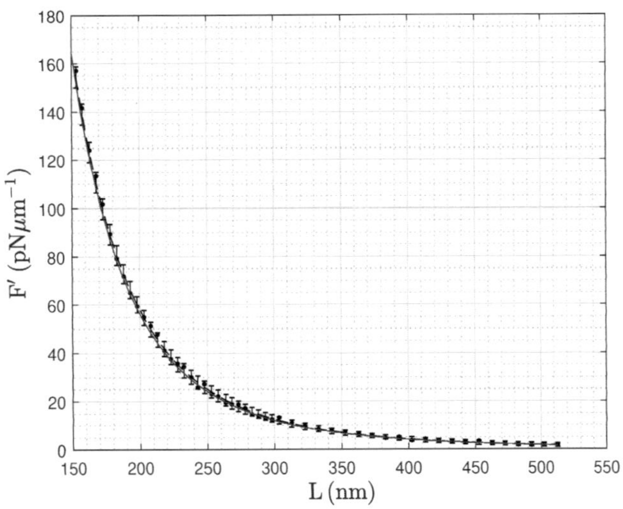

Figure 2. Casimir force gradient between Au sphere and Si sample A grating.

theoretical results related to experiments of [14]. In the current paper we present results of theoretical calculations of the Casimir normal force for two systems with Si gratings from [14] at temperature $T = 300\,\mathrm{K}$. Also we perform analysis of experimental data from [14, 15].

2. Comparison of the theory and experiments

In experiments [14] the force gradient between an Au sphere of radius R and Si rectangular grating separated by a vacuum slit was measured. In proximity force approximation (PFA) one can write

$$F'_{\text{sphere}-\text{g}}(L) = 2\pi R F_{\text{flat}-\text{g}}(L), \tag{1}$$

where $F'_{\text{sphere}-\text{g}}(L)$ is a force gradient in sphere – grating geometry, $F_{\text{flat}-\text{g}}(L)$ is a force in half-space – grating geometry, L is a minimum distance between a grating and respective geometry.

IOP Conf. Series: Journal of Physics: Conf. Series **917** (2017) 062006 doi:10.1088/1742-6596/917/6/062006

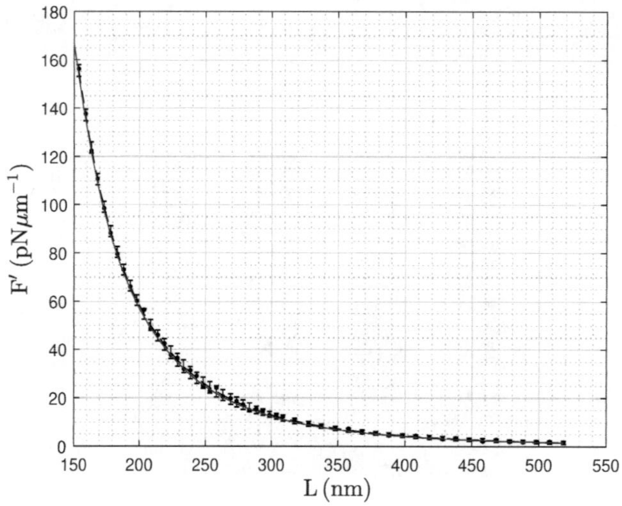

Figure 3. Casimir force gradient between Au sphere and Si sample B grating.

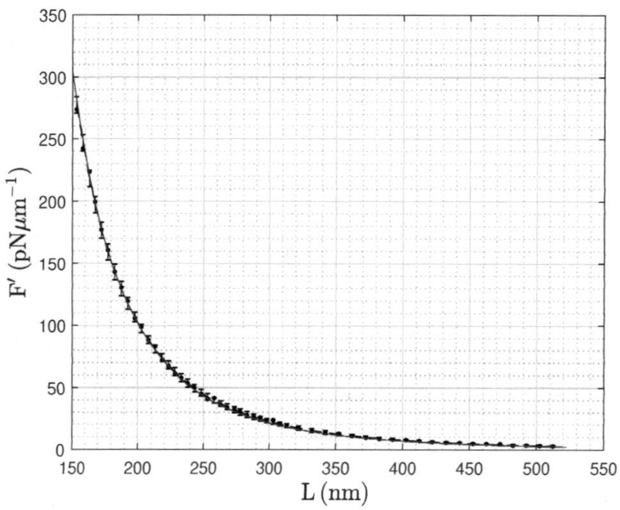

Figure 4. Casimir force gradient between Au sphere and Si half-space.

For $L \lesssim 0.5\mu$m and $R = 50\mu$m from [14] the condition $L \ll R$ is satisfied and PFA approximation is accurate. Therefore one can obtain gradient of the force $F'(L)$ in Au sphere – Si grating geometry by making an exact calculation of the force $F(L)$ in Au half-space - Si grating geometry (Figure 1). We follow this route for comparison of the theory and experiments.

The Casimir force between the grating and the half-space is evaluated by making use of scattering theory [2, 7] at temperature $T = 300\,$K. We use $2N + 1$ terms in every Rayleigh decomposition with $N = 40$ to determine Rayleigh coefficients. A generalized six-oscillator plasma model of Au [16] and one-oscillator model of intrinsic Si [17] are used in calculations. Theoretical results for the Casimir force in Si grating – Au half-space geometry are evaluated with an error 0.1%.

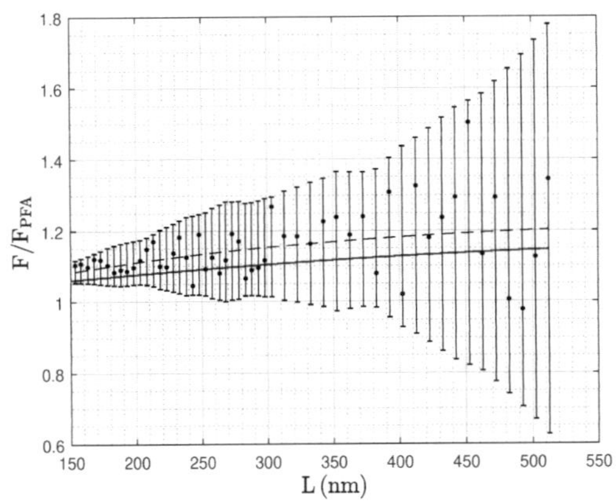

Figure 5. Ratio of the Casimir force F to the theoretical force F_{PFA} for Au half-space – Si sample A grating.

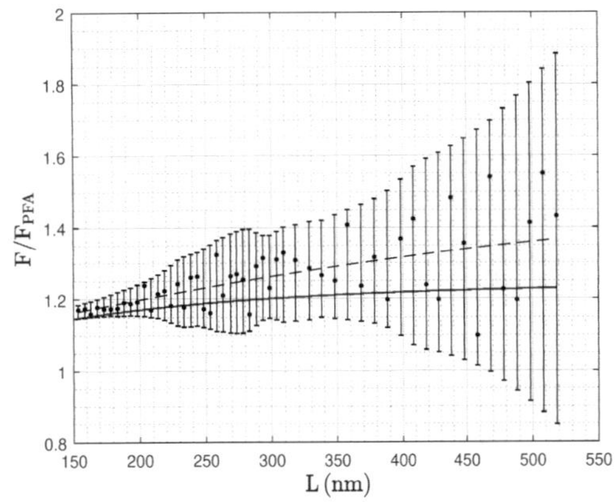

Figure 6. Ratio of the Casimir force F to the theoretical force F_{PFA} for Au half-space – Si sample B grating.

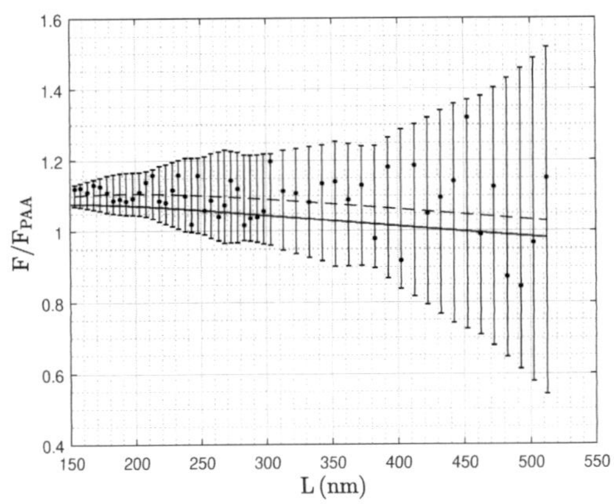

Figure 7. Ratio of the Casimir force F to F_{PAA} for Au half-space – Si sample A grating.

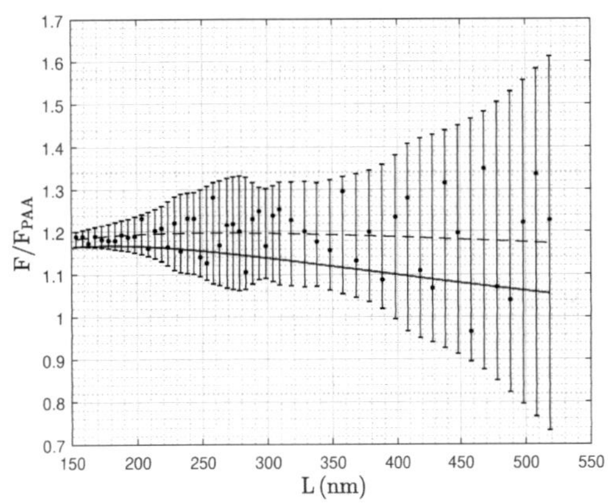

Figure 8. Ratio of the Casimir force F to F_{PAA} for Au half-space – Si sample B grating.

We calculate the Casimir force at an interval of distances $L = [150, 520]$ nm for two sets of parameters for a Si grating – Au half-space system shown on Figure 1: sample A with $b = 510$ nm, $d = 1000$ nm , $h = 1070$ nm (experimental data for the Casimir force derivative $F'(L)$ are plotted on Figure 3b in [14] and on Figure 2 in this paper) and sample B with $b = 191.2$ nm , $d = 400$ nm, $h = 980$ nm (experimental data for the Casimir force derivative $F'(L)$ are plotted on Figure 3c in [14] and on Figure 3 in this paper). Experimental data points are plotted by black dots at distances of experimental measurements on all figures. On Figure 4 we plot theoretical results and experimental data from Figure 3a in [14] for the Casimir force gradient for flat Si surface (half-space) in front of Au sphere. Theoretical results for $F'(L)$ are plotted on Figure 2, Figure 3, Figure 4 by blue solid lines.

We also plot ratios of the Casimir force $F(L)$ to the theoretical force $F_{PFA}(L)$ evaluated in PFA approximation for samples A and B. For geometry of Figure 1 one has

$$F_{PFA}(L) = (b/d)F_{flat}(L) + (1 - b/d)F_{flat}(L + h), \tag{2}$$

where $F_{flat}(L)$ is a theoretical Lifshitz force [9] for two half-spaces separated by a distance L.

Theoretical ratios for samples A and B are plotted by blue solid lines on Figure 5 and Figure 6 respectively. Ratios of experimental Casimir force values to F_{PFA} are shown by black dots at distances of experimental measurements on Figure 5 and Figure 6.

In [14,15] ratios of the measured Casimir force to the Casimir force defined in pairwise additive approximation $F_{PAA}(L) = b/d\, F^{e}_{flat}(L)$ were considered (Figure 3d in [14]) where $F^{e}_{flat}(L)$ is the force in Si half-space – Au half-space system extracted from the experimental force gradient in Si half-space – Au sphere system by the use of (1). To make a connection with ratios F/F_{PAA} considered in [14,15] we extract average values of $F^{e}_{flat}(L)$ from $F'(L)$ on Figure 4 by the use of (1). Ratios of the theoretical Casimir force to F_{PAA} obtained this way for samples A and B are plotted by blue solid lines on Figure 7 and Figure 8 respectively. Ratios of experimental Casimir force values to F_{PAA} are shown by black dots at distances of experimental measurements on Figure 7 and Figure 8.

Next we perform analysis of experimental data and evaluate random, theoretical and absolute errors of experimental data. Random errors are taken into account by making use of a Student's distribution [18]. An uncertainty in the distance $\delta L = 0.2$ nm mentioned in [15] contributes to the theoretical error. An uncertainty in theoretical results due to optical data is set 0.5% at all separations as in [19]. Absolute errors of the difference between theory and experiment are plotted by black error bars at distances of experimental measurements with confidence intervals shown at 95%, a single series of measurements for each sample from [14] is considered to find confidence intervals. Mean experimental values are plotted by dashed black lines, they are extracted from the same series of measurements.

It is instructive to compare Figure 5 with Figure 7 and Figure 6 with Figure 8. Theoreticians typically find ratios to the theoretical force F_{PFA} (shown on Figure 5 and Figure 6) while in [14] ratios to the force F_{PAA} obtained from experimental measurements were considered (they are shown by solid squares and hollow circles on Figure 3d in [14] and on Figure 7 and Figure 8 in the current paper). These ratios do not coincide due to the difference between theoretical values (solid blue line) and average experimental values (dashed black line) of the force gradient on Figure 4.

3. Conclusion

We make a comparison of the Casimir effect theory for Si rectangular grating – Au half-space system at temperature $T = 300\,\mathrm{K}$ with experiments from [14]. The proximity force approximation (1) is used to express Casimir force gradient in Au sphere – Si grating system via the force in Au half-space – Si grating system. We use a generalized six-oscillator plasma model of Au and one-oscillator model of intrinsic Si in Casimir force computations. Ratios of the Casimir force values F to the proximity force approximation theoretical values F_{PFA} and pairwise additive approximation values F_{PAA} obtained from experiment are evaluated. The difference in the use of theoretical PFA results and PAA results extracted from experimental data is emphasized. Error analysis of experimental data from [14] is performed, confidence intervals for absolute errors of the difference between theory and experiment are calculated at 95%. It turns out that for both grating samples A and B the theory is consistent with experiment at all separation distances.

Acknowledgments

We thank Ho Bun Chan for providing us experimental data from [14]. This research was supported in part by Saint Petersburg State University grant 11.38.237.2015. Research was carried out using computational resources provided by Resource Center "Computer Center of SPbU" (http://cc.spbu.ru/en).

References

[1] Casimir H B G 1948 *Proc. Kon. Ned. Akad. Wetensch.* **51** 793-5
[2] Lambrecht A and Marachevsky V N 2008 *Phys. Rev. Lett.* **101** 160403
[3] Lambrecht A and Marachevsky V N 2009 *Int. J. Mod. Phys.* A **24** 1789-95
[4] Rayleigh O M 1907 *Proc. Roy. Soc.* A **79** 399-416
[5] Chiu H C, Klimchitskaya G L, Marachevsky V N, Mostepanenko V M and Mohideen U 2009 *Phys.Rev.* B **80** 121402(R)
[6] Chiu H C, Klimchitskaya G L, Marachevsky V N, Mostepanenko V M and Mohideen U 2010 *Phys.Rev.* B **81** 115417
[7] Marachevsky V N 2012 *J.Phys.* A*: Math.Theor.* **45** 374021
[8] Marachevsky V N 2015 *Theor.Math.Phys.* **185** 1492-501
[9] Lifshitz E M 1956 *Soviet Phys. JETP* **2** 73
[10] Davids P S, Intravaia F, Rosa F S S and Dalvit D A R 2010 *Phys.Rev.* A **82** 062111
[11] Messina R, Maia Neto P A, Guizal B and Antezza M 2015 *Phys.Rev.* A **92** 062504
[12] Bender H, Stehle C, Zimmermann C, Slama S, Fiedler J, Scheel S, Buhmann S Y and Marachevsky V N 2014 *Phys.Rev.* X **4** 011029
[13] Buhmann S Y, Marachevsky V N and Scheel S 2016 *Int.J.Mod.Phys.* A **31** 1641029
[14] Chan H B, Bao Y, Zou J, Cirelli R A, Klemens F, Mansfield W M and Pai C S 2008 *Phys.Rev.Lett.* **101** 030401. Erratum: Chan H B, Bao Y, Zou J, Cirelli R A, Klemens F, Mansfield W M and Pai C S 2011 *Phys. Rev. Lett.* **107** 019901
[15] Chan H B, Bao Y, Zou J, Cirelli R A, Klemens F, Mansfield W M and Pai C S 2010 *Int.J.Mod.Phys.* A **25** 2212-22
[16] Klimchitskaya G L, Mohideen U and Mostepanenko V M 2007 *J.Phys.* A*: Math.Theor.* **40** F339-46
[17] Lambrecht A, Pirozhenko I, Duraffourg L and Andreucci Ph 2007 *EuroPhys.Lett.* **77** 44006
[18] Rabinovich S G 2000 *Measurement Errors and Uncertainties: Theory and Practice* (New York: Springer)
[19] Klimchitskaya G L, Chen F, Decca R S, Fischbach E, Krause D E, López D, Mohideen U and Mostepanenko V M 2006 *J.Phys.* A*: Math.Gen.* **39** 6485-93

Dielectric-Metal-Dielectric (D-M-D) infrared (IR) heat reflectors

M A Butt[1], S A Fomchenkov[1, 2], S N Khonina[1, 2]
[1]Samara National Research University, Samara, Russia
[2]Image Processing Systems Institute of RAS, Branch of the FSRC "Crystallography and Photonics" RAS, Samara, Russia

Abstract. Sunlight is the primary source of light in the buildings and automobiles. However, infrared (IR) radiations in the sunlight result in heat generation. In this work, dielectric-metal-dielectric (D-M-D) heat reflectors are demonstrated which is transparent for visible light 400-700 nm while reflecting IR radiations above 700 nm. These reflectors are based on a TiO_2-Ag-TiO_2 sandwich structure with 22 nm of the thickness of each layer shows the maximum transmission of 83 % theoretically and 81 % experimentally in the visible wavelengths region observed by ellipsometer; a maximum reflection of IR wavelengths was also observed. The study was also extended to experimental verification of heat transmission by testing these reflectors both under IR lamp and sun in the month of June 2016. Experimental results confirm less heat transfer through these filters as compared to the commercially available tinting paper. As a practical application, these reflectors can be used in automobiles and houses during summer to reflect back the excessive heat from the sun.

1. Introduction

Smart buildings and automobiles with minimum consumption of electrical energy are designed so that sunlight can be used as a source of light during daytime in buildings using large windows [1]. However, exposure to sunlight results in heating and eventually consumption of extra electrical energy as air-conditioning to remove trapped heat in buildings [2] [3] and automobiles especially when they are parked without any shadow [4] [5]. Conventional windows pass both visible and Infrared (IR) wavelengths, eventually IR radiations are mainly responsible for heating [6] and also harmful to human skin [7]. Thin film design techniques have been improved over the years for efficient design of optical filters with selective transmission and reflection [8]. Dielectric-Metal-Dielectric (D-M-D) reflectors based on thin film coatings are widely used to transmit visible light and reflect IR wavelengths [9]. These reflectors can be effective to reduce heating in buildings and automobiles. Easily available solutions for heat reflectors include aluminium foil and tinting paper to blind the windows of the car. Tinting paper can reduce the transmission of sunlight resulting in a low transparency. New cars have adopted techniques like vacuum glazing to reduce heating from sunlight [10], however, they are expensive and not beneficial for people with old cars. In this work, we have fabricated D-M-D reflector design which has high transparency with a maximum transmission of visible light and maximum reflection of IR wavelengths based on thin film coatings that can be used in glass windows of buildings and automobiles.

These D-M-D reflectors are basically a filter that transmits visible light and reflects heat causing IR wavelengths. Filters are of two types: absorption filter that absorbs the wavelengths that need to be stopped and interference filter that reflect the undesired wavelengths [11, 12, 13]. Transmission and reflection properties of interference filters are dependent on materials refractive index and layer thickness of

IOP Conf. Series: Journal of Physics: Conf. Series **917** (2017) 062007 doi:10.1088/1742-6596/917/6/062007

materials. To model the IRC, open source software Open Filter is used. Open filter calculates optical properties of filters. It uses transfer matrix method to calculate the transmission and reflection properties of filters/IRC based on the absorption and materials refractive indices [14]. Optimization techniques are available in this software like needle synthesis (Adding an extra layer to give targeted transmission). However, in this work, the designs are optimized manually.

2. D-M-D design and characterization

Based on our modelling results [15], we carefully selected an optimized design which presents a promising visible wavelengths transmission and highly IR reflections. To fabricate these reflectors, thin films of metal and dielectric material were deposited on the glass substrate with the help of Magnetron sputtering AJA International Inc. ATC Orion-8- HV. At first, TiO_2 layer was deposited on glass substrate by using RF power of 150 W, 20 s.c.c.m of Ar flow at 3 mTorr pressure for 30 min, followed by metal (Ag) layer deposition: DC power of 150 W, 20 s.c.c.m at 3 mTorr pressure for less than 1 min and then again TiO_2 layer was deposited on top. Good quality TiO_2 dielectric layers can also be obtained by annealing Ti metal layer [16].

2.1. Measurement of heat transfer through D-M-D reflector by using IR lamp source

To scrutinize the heat transmission through D-M-D reflector, we designed a hollow cylinder made of plastic and a thermocouple was placed inside the cylinder connected to multi-meter in order to record the temperature caused by an IR lamp placed directly over the cylinder as shown in the inset of Figure 1.

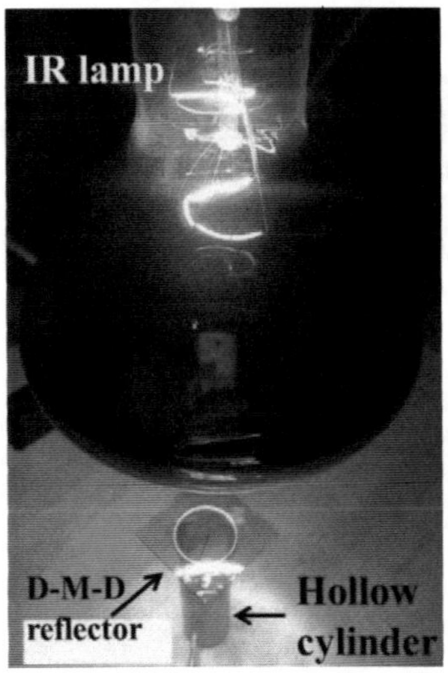

Figure 1. The set-up used for the measurement of temperature in the hollow cylinder. The thermocouple is placed inside the cylinder externally connected to a multimeter.IR lamp used a source.

In that very case, we are only considering the heat transmitted through the open end of the cylinder. Therefore, we are not taking into account the heat transmission through the sidewalls of the box. This phenomenon was considered while measuring the temperature under the sun, as the light falls from all possible directions. We noticed the temperature inside the hollow cylinder rises to 86 °C in 15 minutes when the regular optical glass was placed. On the other hand, when a D-M-D reflector (TiO_2-Ag-TiO_2=25-25-25 nm) was placed over the cylinder, the temperature elevated gradually and arrived at 38 °C in 30 min and stabilised as shown in figure 2. Furthermore, we found that our D-M-D reflector has shown best results as compared to commercially available tinting paper. Temperature is marginally above the room temperature by using D-M-D reflector; this temperature can easily be controlled in buildings and automobiles by using simple fans for cooling which results in an economical and energy efficient solution.

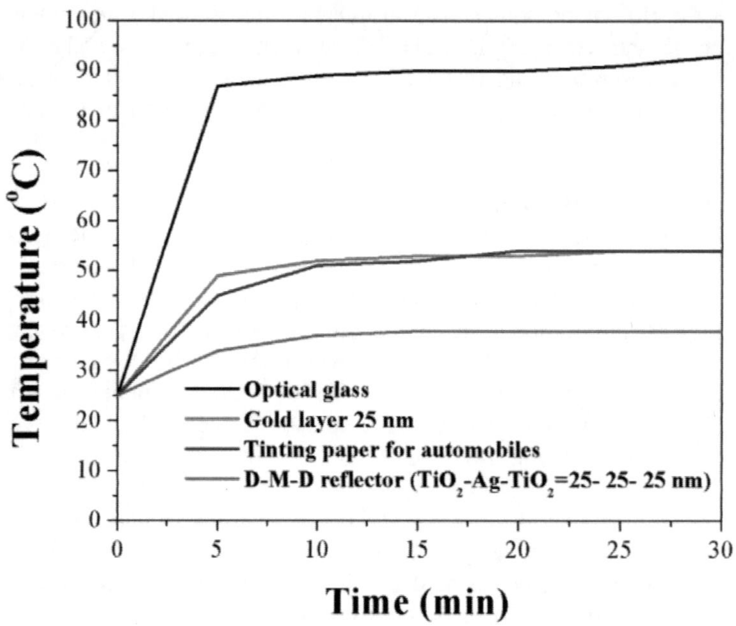

Figure 2. Temperature versus time graph. Heat transfer through the optical glass, a gold layer, tinting paper and D-M-D reflector.

2.2. Measurement of heat transfer through D-M-D reflector under Sunlight
This experiment was conducted under sunlight in the hot month of June in Samara, Russia. The room temperature on that day was 32 °C and weather was partially cloudy. We wanted to observe the temperature through the filter, therefore, we make a special box covered with aluminium foil to protect the heat transfer from the sidewalls of the box. But the light was falling on the box from all directions so we noticed some increase in temperature which was caused by the heat transfer through all four sides of the box as shown in Figure. 3.

Figure 3. Temperature measurement through the D-M-D reflector placed on the open end of the box under sunlight. Whereas the sidewalls of the box are covered with aluminium foil.

The temperature measurement results are shown in Table I. In the experiment, the temperature in the cylindrical box wasn't stabilized for optical glass. It was increasing with time so we can expect that if this experiment is performed on the total sunny day for a long time then the temperature could rise to 50 °C. Contrary to this, the temperature was stabilized when designed D-M-D reflector was used in the experiment.

Table I: Temperature measurement in a cylindrical box under different heat sources

Heat source	Type of filter	Initial temperature (°C)	Final temperature (°C)	Time (min)
IR lamp	Optical glass	25	86	15
	D-M-D reflector	25	38	30
Sunlight	Optical glass	32	42	30
	D-M-D reflector	32	38	30

2.3. Measurement of optical transmission of D-M-D reflector, gold layer and tinting paper

The optical transmission of the gold layer (25 nm), commercially available tinting paper and D-M-D reflector was measured with the help of ellipsometer as shown in figure 4. It can be seen that D-M-D reflector provides the best result as compared to gold layer and tinting paper. It is well noting that the D-M-D reflector provides ~81 % of transmission in visible wavelength region whereas tinting paper and the gold layer can offer even less than 40 % of transmission. So our designed reflector provides two benefits: Enhances visibility at night and heat protective during day time.

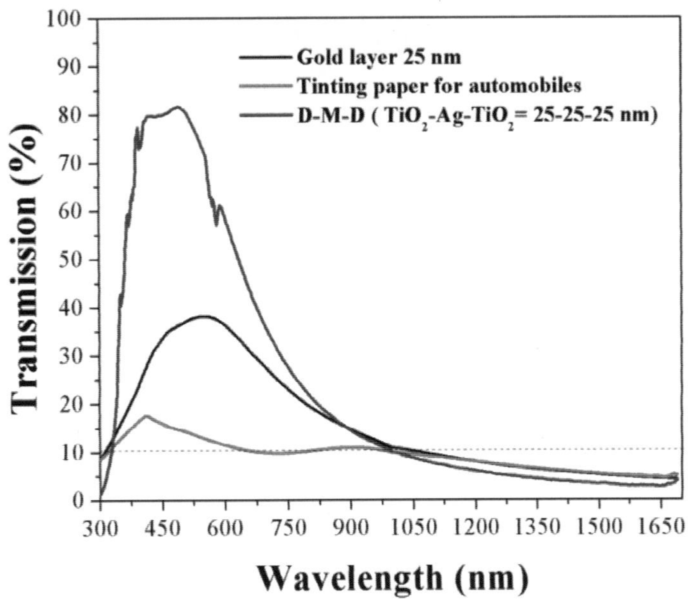

Figure 4. Optical transmission of light of gold layer (25nm), tinting paper and D-M-D reflector in visible and IR region.

3. Conclusions

In this work, we demonstrated a D-M-D reflector based on TiO_2-Ag-TiO_2 thin films which show a significantly high transmission in the visible spectrum and extraordinary reflection in IR wavelengths. These reflectors can be used to control the heat accumulation in the automobiles and houses during the hot weather. Commercially available tinting papers have a high reflection of IR wavelengths but very low transmission of visible light; reduces their transparency which can affect the visibility at night while driving. However, our designed reflectors have high transmission of ~81 % in the visible region and less than 10 % of transmission of IR wavelengths.

Acknowledgements
This work was supported by the Ministry of Education and Science of the Russian Federation and the Russian Foundation for Basic Research (grant no. 16-29-11744).

References

[1] Qu J, Song J, Qin J, Song Z, Zhang W, Shi Y, Zhang H, Zhang R, He Z, Xue X 2014 *Ener. Building* **77** 1.
[2] Perez-Lombard L, Ortiz J, Pout C 2008 *Ener. Building* **40** 394.
[3] Suarez R, Fernandez-Aguera J 2015 *In Build. Simulation* **8** 593.
[4] Aljubury IM, Farhan A A, Mussa M A 2015 *J. Eng.* **21** 1.
[5] Abd- Fadeel W A, Hassanein S A 2013 *Res. Development(IJAuERD)* **3** 75.
[6] Zamanian A, Hardiman C 2005 *H. freq. Elec.* **4** 16.
[7] Cho S, Shin M H, Kim Y K, Seo J E, Lee Y M, Park C H, Chung J H 2009 *In J. Inves. Dermatology Symp. Proceedings* 14 15.
[8] Macloed A H *Thin Film Optical Coatings* 2001 (Thin film Center Inc.)
[9] Joudi A, Svedung H, Cehlin M, Ronnelid M 2013 *Appl. Energy* **103** 562.
[10] Cuce E, Cuce PM 2016 *Renew. Sust. Energy Rev.* **54** 1345.
[11] Asghar M H, Khan M B, Naseem S 2003 *J. Phys.* **53** 1209.
[12] Butt M A, Fomchenkov S A, Ullah A, Habib M, Ali R Z 2016 Computer Optics **40** 674
[13] Butt M A, Fomchenkov S A, Ullah A, Verma P, Khonina S N 2016 *J. Phy. Conf. Ser.* 741 012136.
[14] Larouche S, Martinu L 2008 *Appl. Optics* **47** C219.
[15] Butt M A, Fomchenkov S A, Kazanskiy N L, Ullah A, Ali R Z, Habib M 2017 *Proc. of SPIE* 10342 103420O-1
[16] Butt M A, Fomchenkov S A 2017 J. Korean Phy. Soc. **70** 169.

Quantum random number generator based on quantum nature of vacuum fluctuations

A E Ivanova, S A Chivilikhin and A V Gleim

ITMO University, Kronverkskiy pr. 49, St.Petersburg 197101, Russia

Abstract. Quantum random number generator (QRNG) allows obtaining true random bit sequences. In QRNG based on quantum nature of vacuum, optical beam splitter with two inputs and two outputs is normally used. We compare mathematical descriptions of spatial beam splitter and fiber Y-splitter in the quantum model for QRNG, based on homodyne detection. These descriptions were identical, that allows to use fiber Y-splitters in practical QRNG schemes, simplifying the setup. Also we receive relations between the input radiation and the resulting differential current in homodyne detector. We experimentally demonstrate possibility of true random bits generation by using QRNG based on homodyne detection with Y-splitter.

1. Introduction

Quantum random number generators are based on the physically random processes, therefore, the random numbers generated by these systems are unpredictable, truly random and can be used in applications, requiring high degree of randomness, such as cryptography [1]. There are many different ways to implement quantum random number generation: using beam separation [2], entangled photon states [3], processes of photon emission and detection [4] and quantum noise of lasers [5]. Another method of generating is using of electromagnetic field vacuum fluctuations [6].

This work is focused on investigation of applying fiber Y-splitters in QRNG, based on the on quantum nature of vacuum fluctuations (figure 1). Any practical random number generator operates a bit differently compared to its idealistic description. Hence, it is important to model the effects of any possible disruptions and study their impact on the output statistical features.

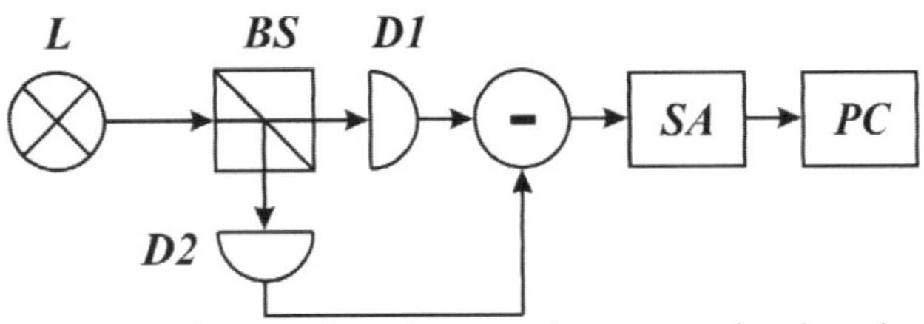

Figure 1. Scheme of random number generation based on homodyne detection principles: L – laser, BS – beam splitter, D1, D2 – detectors, SA – spectrum analyzer, PC – computer.

IOP Conf. Series: Journal of Physics: Conf. Series **917** (2017) 062008 doi:10.1088/1742-6596/917/6/062008

2. Quantum random number generation based on the principles of homodyne detection. Beam splitter

Optical beam splitter (figure 2(a)) is essential for quantum random number generation systems based on the fluctuations of the vacuum [6]. In these schemes to one of beam splitter's input ports strong laser radiation is sent, and second port receive only vacuum fluctuations. Vacuum and high-intensity laser signals are mixed on the beam splitter and splitted into two signals arriving at the detector. After detecting one of these signals is subtracted from other and thus only vacuum noise remains in final signal. It is completely random and can be used to random numbers generation.

We obtain in operator form mathematical description of beam splitter's work in QRNG using fluctuations of vacuum, when a strong laser signal, which can be described by Poisson distribution, is sent to one of its inputs, and vacuum state is sent to other input.

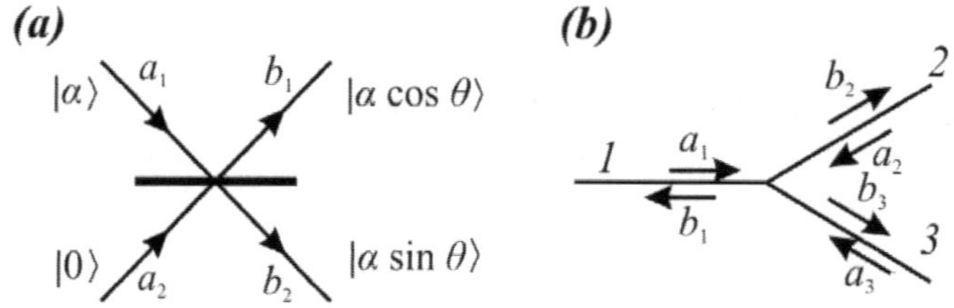

Figure 2. (a) Scheme of a beam splitter with angle θ, where coherent state is sent to the first splitter input, and vacuum state – to the other. (b) Optical Y-splitter. a_1, a_2, a_3 – input signals of the 1st, 2nd and 3rd ports, respectively, b_1, b_2, b_3 – output signals from the splitter.

If input radiation a_1 can be characterized by Poisson distribution with parameter α and radiation passes through a beam splitter with angle θ (figure 2 (a)), then signal at first of beam splitter outputs b_1 can be characterized by Poisson distribution with parameter $\alpha\cos\theta$, and signal at the second output b_2 can be characterized by Poisson distribution with parameter $\alpha\sin\theta$.

In case of symmetric beam splitter we obtain expression, describing signals at both outputs

$$b_1 = b_2 = \frac{1}{\sqrt{2}} a_1 \tag{1}$$

For the case of an asymmetric beam splitter and different quantum efficiencies of detectors we calculate mean differential current, dispersion and amplitude of difference current deviation.

Differential current after detecting can be defined as follows

$$\Delta i = i_2 - i_1 = \gamma_2 b_2^+ b_2 - \gamma_1 b_1^+ b_1, \tag{2}$$

where i_1, i_2 are photocurrents at first and second detectors, γ_1, γ_2 are quantum efficiencies of detectors.

For detectors with equal quantum efficiencies and symmetric beam splitter, mean value of differential current is determined to be zero and amplitude of differential current deviation is directly proportional to intensity of input radiation. In case of using detectors with different quantum efficiencies and asymmetric beam splitter, mean value of differential current is characterized by following equation

$$\langle \Delta i \rangle = \alpha^2 \left(\gamma_2 \sin^2\theta - \gamma_1 \cos^2\theta\right), \tag{3}$$

and in this case amplitude of differential current deviation can be estimated by formula (4)

$$\delta i = \alpha \sqrt{\gamma_2 \sin^2\theta + \gamma_1 \cos^2\theta}. \tag{4}$$

3. Y-splitter. Experimental results.

The fiber splitter with one input and two output ports (figure 2(b)) has been presented as a system with three inputs and three outputs, because signal in each of the ports may be distributed in two opposite

IOP Conf. Series: Journal of Physics: Conf. Series **917** (2017) 062008 doi:10.1088/1742-6596/917/6/062008

directions. Relations between input and output signals of the optical fiber splitter were derived by using a matrix description allowing show the interaction between each pair of signals:

$$\begin{pmatrix} b_1 \\ b_2 \\ b_3 \end{pmatrix} = \begin{pmatrix} -\sqrt{1-2\lambda^2} & \beta & \beta \\ \lambda & -\gamma & \sqrt{1-\beta^2-\gamma^2} \\ \lambda & \sqrt{1-\beta^2-\gamma^2} & -\gamma \end{pmatrix} \begin{pmatrix} a_1 \\ a_2 \\ a_3 \end{pmatrix}, \quad (5)$$

where λ is proportionality factor, connecting input signal at 1st port and output signals at 2nd and 3rd ports; β connects input signals at 2nd or 3rd ports and output signal at 1st port; γ connects input and output signals of 2nd port or input and output signals of 3rd port.

These coefficients are selected in accordance with requirements of unitary property of matrix. Next expressions are also followed from unitarity conditions

$$\begin{cases} -\sqrt{1-2\lambda^2}\,\beta - \lambda\gamma + \lambda\sqrt{1-\beta^2-\gamma^2} = 0, \\ -\sqrt{1-2\lambda^2}\,\beta + \lambda\sqrt{1-\beta^2-\gamma^2} - \lambda\gamma = 0, \\ \beta^2 - 2\gamma\sqrt{1-\beta^2-\gamma^2} = 0. \end{cases} \quad (6)$$

After selection of matrix proportionality factors we can simplify matrix form, using fact, that parameters λ and β can be expressed from system (6) through the γ:

$$\lambda = \beta = \sqrt{2\gamma(1-\gamma)} \quad (7)$$

We consider special case when signal from 1st input port is distributed only between ports 2 and 3 without reflection on the 1st port. In this case $\sqrt{1-2\lambda^2} = 0$ and $\lambda = \frac{1}{\sqrt{2}}$, then by using expressions were obtained above, we can receive values $\beta = \frac{1}{\sqrt{2}}$ and λ=0.5.

We reviewed elements of the matrix describing the relationship between signal is sent to first port of splitter and signals are emanating from second and third output ports.

$$b_1 = b_2 = \frac{1}{\sqrt{2}} a_1 \quad (8)$$

The resulting mathematical expressions were equivalent to expressions obtained earlier for the case of using a beam splitter, when strong laser signal is sent to first input, and vacuum – to second input.

Also in [7] we research case with complex parameters of expression (5). Obtained result, with the exception of phase shift, coincides with results that were obtained for symmetric beam splitter.

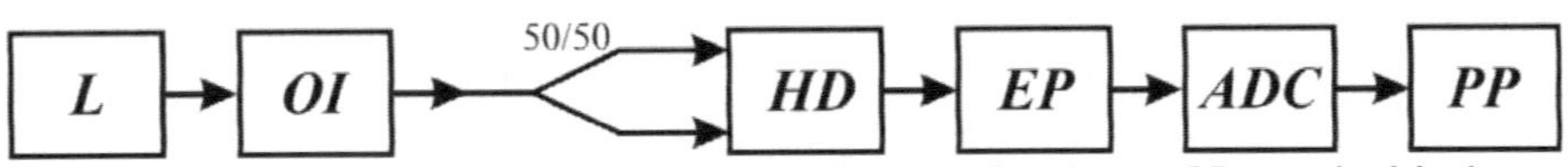

Figure 3. Block diagram of the experimental setup. L – laser, OI – optical isolator, HD – homodyne detector, EP – electronic processing system, ADC – analog-to-digital converter, PP – postprocessing system.

Thus, as descriptions for beam splitter and Y-splitter are equal, we can use results for beam splitter, obtained earlier, to evaluate work of quantum random generation systems, based on vacuum fluctuations, using the Y-splitter. Scheme of our experimental setup is represented in figure 3. We used a Teraxion-NLL laser (L) with wavelength 1550 nm on power 10 mW. After passing optical isolator (OI) radiation goes to fiber Y-splitter with dividing coefficient 50/50. The power of signal on the homodyne detector (HD) is 6 mW. To detection we used p-i-n photodiodes (fid13z81pz) with a quantum efficiency of 60% at 1550 nm, sensitivity S = 0.75 A/W, and a 1 GHz bandwidth. The currents from two photodiodes are subtracted and then resulting signal goes to electronic processing system (EP), where signal is amplified. This circuit contains amplifier in a transimpedance configuration with a gain of 4 kΩ. All components are located on a specially designed board. The frequency band of the circuit is 100 MHz. The bandwidth of the oscilloscope Tektronix dpo70604c (ADC) used to digitizing the received noise is 6 GHz. After digitizing we can choose and use most optimal post processing system (PP).

During the research a linear relationship between the laser power and the noise level was observed, that confirms that noise has quantum nature. Quantum noise, obtained from our system, had next characteristics: mean value of fluctuations $\mu=7\cdot 10^{-6}$, standard deviation $\sigma=0.2$, asymmetry coefficient S =0.01, kurtosis (a measure of sharpness of the random variable maximum) $K=-4.5\cdot 10^{-3}$, probability of the most likely outcome $\max(P_i) = 3\cdot 10^{-3}$ (where P_i – probability of the i-th realization of random discrete variable), min-entropy $H_{min}=-\log_2\{\max(P_i)\}= 8.29$.

In our research we used four different ways to convert samples to sequences of bits: 1) if noise level in count is above 0, then we write "1", else – "0"; 2) we apply XOR to sequence, received by first method; 3) we generate three bits from one sample [6] (convert initial Gaussian distribution to uniform distribution, applying Gaussian error function); 4) we discard most significant bits after analog-to-digital conversion.

If we know probability properties of true random sequence, then we can check how much the generated sequence is similar to random. To do this, we select the appropriate statistics for each of the five tests [8] (monobit test, twobit test, "poker" test, runs test and autocorrelation test) and then compare its value for the ideal sequence and the generated sequence. These tests show [9] that optimal postprocessing technique for our scheme is discarding two or three most significant bits.

4. Conclusions

In this research we derived quantum mathematical description of Y-splitter and beam splitter, which are used in quantum random number generation, based on vacuum fluctuations. Also we obtain statistical parameters of differential current on outputs. In experimental implementation of QRNG systems based on quantum fluctuations of vacuum we use Y-splitter. In our research we consider four postprocessing methods to convert experimental samples to bits and after testing we conclude that optimal postprocessing technique for our system is discarding two or three most significant bits after analog-to-digital convertion.

Acknowledgments

This work was financially supported by Government of Russian Federation, Grant 074-U01 and by the Ministry of Education and Science of Russian Federation (project № 14.578.21.0112 № 02.G25.31.0229).

References

[1] Scarani V, Bechmann–Pasquinucci H, et al 2009 *Rev. Mod. Phys.* **81**. 1301
[2] Jennewein T, Achleitner U, et al 2000 *Rev. Sci. Instrum.* **71**, 1675
[3] Kwon O, Cho Y-W and Kim Y-H 2009 *Appl. Opt.* **48**. 1774
[4] Stipcevic M and Rogina M B 2007 *Rev. Sci. Instrum.* **78** 045104
[5] Qi B, Chi Y-M, Lo H-K and Qian L 2010 *Optics Letters* **35** 312
[6] Symul T, Assad S M and Lam P K 2011 *Appl. Phys. Lett.* **98** 231103
[7] Ivanova A E, Chivilikhin S A, Gleim A V 2016 *Nanosyst.: Phys. Chem. Math.*, **7**(2), 378
[8] Menezes A, van Oorschot P, Vanstone S 1996 *Handbook of Applied Cryptography* (Boca Raton: CRC Press) p 816
[9] Ivanova A E, Chivilikhin S A, Gleim A V 2017 *Nanosyst.: Phys. Chem. Math.*, **8** (2), 239

IOP Conf. Series: Journal of Physics: Conf. Series **917** (2017) 062009 doi:10.1088/1742-6596/917/6/062009

Laser optical information recording in photosensitive silver-containing glasses

V V Gorbyak[1], **A I Sidorov**[2,3]
[1]Department of Computer Photonics and Digital Video Processing, ITMO University, Saint Petersburg 190000, Russia
[2]Department of Optical Information Technologies and Materials, ITMO University, Saint Petersburg 199034, Russia
[3]Department of Quantum and Optical Electronics, St. Petersburg State Electrotechnical University «LETI», Saint Petersburg 197022, Russia

Abstract. It is shown experimentally that laser action on silver-containing photosensitive glasses can be used for optical information recording in octal code. The change of irradiation dose results in the change of luminescence or absorption intensity in irradiated zones.

1. Introduction

Optical data record and storage is really important in the many spheres of human activities. As a rule, this record is done in binary code, that effect negatively on the record density. The use of high order codes allows increasing of data record density significantly. Also, serious part of the storage data is material and lifetime.

The goal of this work is the investigation of the opportunity of optical multilevel laser data recording in photosensitive silver-containing glasses.

For the research the photosensitive silver-containing silicate glass plate with silver were used. Photosensitive silicate glasses based on the Na_2O–ZnO–Al_2O_3–SiO_2–NaF–$NaCl$ system and doped with Ag_2O, the photosensitizer CeO_2 and the reducer Sb_2O_3 were synthesized in ITMO University (St. Petersburg, Russia). The samples were locally irradiated by the third harmonic (λ = 355 nm) of nanosecond YAG:Nd laser. The irradiation dose was varied by the change of laser pulses number. After irradiation in the irradiated areas the visible luminescence appears (figure 1), which is caused by the transformation of charged silver molecular clusters to neutral state [1]. An arrow indicates the sequence with the increase of irradiation dose from 1 up to 7 laser pulses. The intensity of luminescence increases with the increase of irradiation dose (figure 2). The luminescence was excited by semiconductor laser with λ = 405 nm. In this regard, modulation of luminescence intensity is enough to multilevel data record by octal code.

In a subsequent heat treatment of glasses with temperature above the glass transition temperature in the irradiation areas silver nanoparticles are formed which have plasmon resonance [2] and plasmon absorption band in spectral range 405-450 nm. This absorption band changes the color of irradiated zone to weak-yellow, yellow or brown, depending on irradiation dose (figure 3). Optical density spectra of irradiated zones in photosensitive glass after laser irradiation and thermal treatment for different irradiation doses are shown in figure 4. Curve 1 – before irradiation, 2-6 – after irradiation and thermal treatment. Curve 2 – 1 laser pulse, 3 – 5, 4 – 6, 5 – 7, appropriately. Dashed line is wavelength of read-out laser (λ = 405). One can see that plasmon resonance band of silver nanoparticles occupies 370-600 nm spectral region. Maximum of absorption is near λ = 450 nm. The amplitude of the absorption band also depends on laser irradiation dose (figure 5). Excitation wavelength is 405 nm. It also gives the opportunity to optical data recording in the high order codes.

For information read-out in photosensitive glasses with silver nanoparticles semiconductor laser or LED with λ = 405 nm and silicon photodiode can be used. The chosen wavelength is located near the maximum of plasmon resonance band, but this wavelength is at the edge of spectral sensitivity of silicon photodiode. It decreases the contrast during read-out of optical information. The contrast can be increased by the use of spectral conversion of 405 nm radiation to the region of high spectral sensitivity of photodiode. To do this, photosensitive glass, containing cerium ions, after silver nanoparticles formation must be evenly irradiated by UV mercury lamp. The areas of photosensitive glass plate without silver nanoparticles became luminescent with luminescence spectrum like irradiated areas, shown in figure 1. The regions with silver nanoparticles will remain non-luminescent, because they do not contain silver MCs. The maximum of luminescence spectrum of neutral silver molecular clusters after irradiation by UV mercury lamp is near 600 nm. This band matches the spectrum of silicon photodiode sensitivity. The plasmon resonance band is located near the short-wavelength part of the spectrum of silicon photodiode sensitivity. This helps the read-out of information. Photo of PTR glass luminescence after laser irradiation, thermal treatment and UV irradiation by mercury lamp is shown in figure 6. Excitation wavelength is 405 nm. An arrow indicates the sequence with the increase of irradiation dose from 1 up to 7 laser pulses. This method makes possible to increase the contrast during read-out by 2.5-3 times when laser or LED with λ = 405 nm is used .

2. Figures

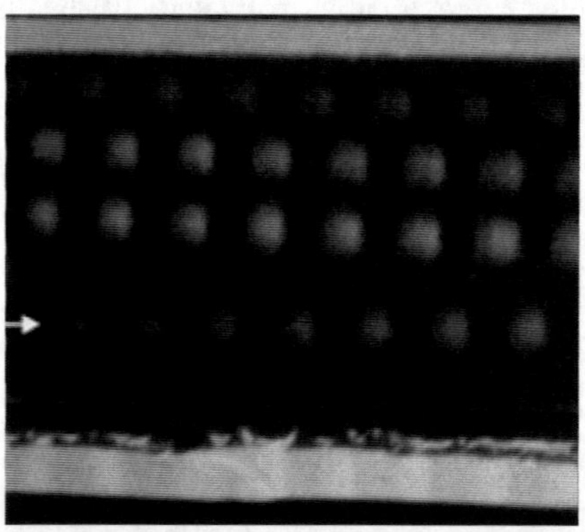

Figure 1. Luminescence of the irradiated areas.

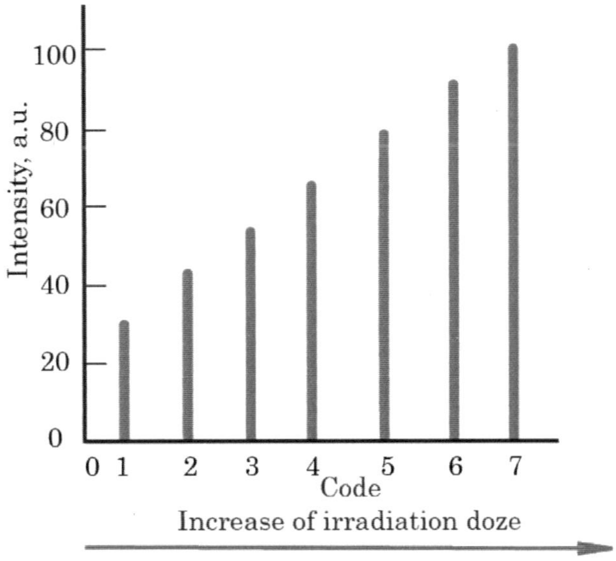

Figure 2. Demonstration of the opportunity of optical multilevel laser data recording in photosensitive silver-containing glass.

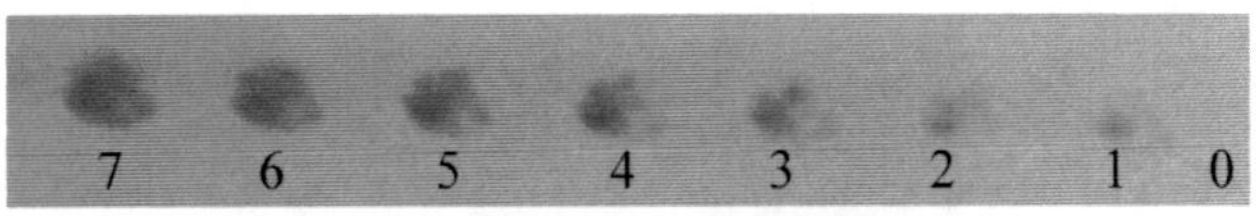

Figure 3. Absorption in the irradiated areas. Numbers from 0 to 7 are code.

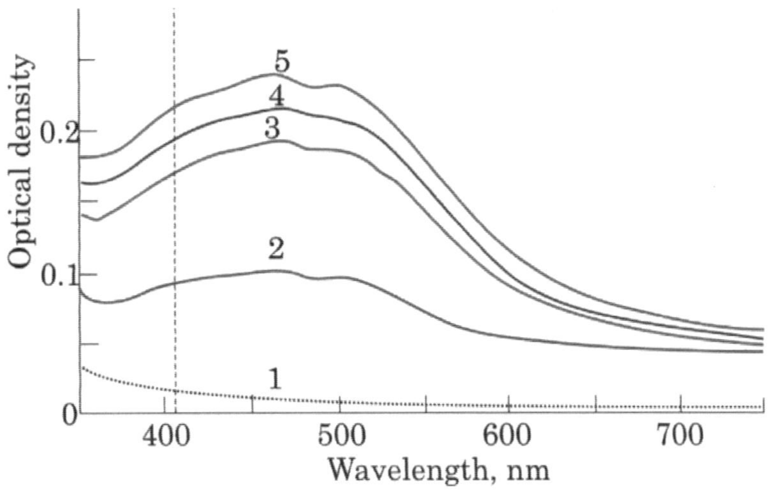

Figure 4. Optical density spectra of irradiated zones in photosensitive glass.

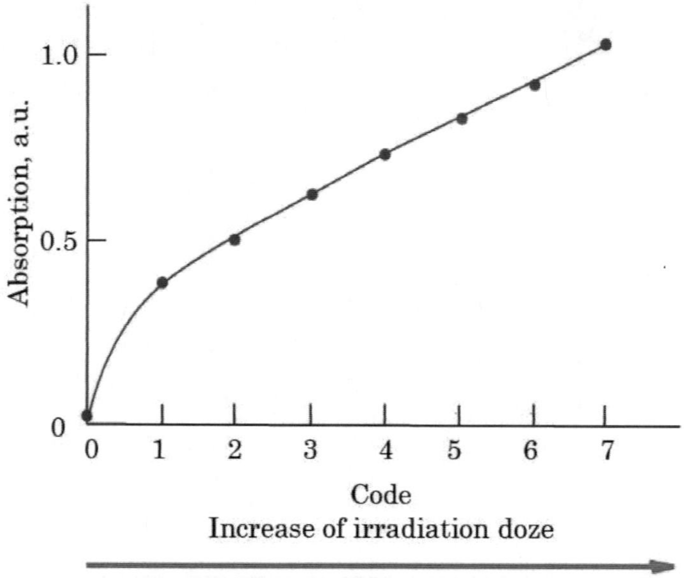

Figure 5. Modulation of absorption in the sequence of irradiated zones in photosensitive glass, shown in Fig. 3.

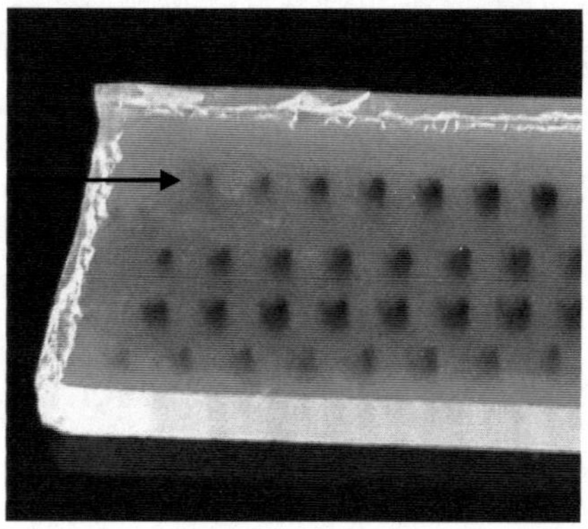

Figure 6. Photo of photosensitive glass luminescence after laser irradiation, thermal treatment and UV irradiation by mercury lamp.

References

[1] V. D. Dubrovin, A. I. Ignatiev, N. V. Nikonorov, A. I. Sidorov, T. A. Shakhverdov, D. S. Agafonova 2014 *Optical Materials* **36** 753
[2] V. V. Klimov 2014 *Nanoplasmonics* (Pan Stanford: Singapore)

IOP Conf. Series: Journal of Physics: Conf. Series **917** (2017) 062010 doi:10.1088/1742-6596/917/6/062010

Generation of orbital-angular-momentum entangled biphotons in twisted nonlinear waveguides

D N Vavulin[1], A A Sukhorukov[2]

[1]ITMO University, 49 Kronverksky Ave., St.Petersburg, 197101, Russia
[2]Nonlinear Physics Centre, Research School of Physics and Engineering, Australian National University, Canberra, ACT 2601, Australia

Abstract. We describe the generation of photon pairs entangled in optical angular momentum (OAM) through spontaneous four-wave mixing (SFWM) in circular arrays of nonlinear waveguides. We show that the incorporation of the array twist along the waveguides can enable robust generation even in presence of defects and inhomogeneities. We present analytical results and numerical simulations based on coupled-mode equations for the classical pump field and a discrete Shrodinger equation for the biphoton wave function.

1. Introduction

Quantum biphoton states with orbital-angular-momentum (OAM) entanglement is one of the most interesting phenomena for application in quantum communications, because they can be used to extend the alphabet of quantum cryptography, which can increase the transmission rate [1]. Also of interest are quantum walks [2], where interference of several walkers can be used to realize various simulations, including quantum teleportation [3,4] and quantum cryptography [5-7], enabling absolutely secure communications.

Quantum walks and generation of biphoton states with OAM in nonlinear straight arrays of waveguides through the effect of the SRS has been discussed in detail and analyzed in paper of Markin et al.[8].

Quantum walks and generation of photon pairs with OAM can be implemented in arrays of coupled waveguides with closed-loop boundary conditions [9-11]. However, in this paper the same OAM of pump photons can produce two different OAM states of biphotons, which may lead to an increase of errors number in quantum cryptography, based on such system. Recently, in a photonic-crystal fiber were demonstrated a new type of coupled waveguides with a twisted geometry [12], which are able to effectively provide the transfer of OAM states. The behavior and generation of biphotons in twisted waveguides can be described by the Shrodinger equation. Also, evolution of the pump profile in such waveguides can be described by the classical coupled-mode equations.

In this paper, we investigate the generation and distribution of photon pairs in nonlinear twisted waveguide arrays, as shown schematically in Fig. 1. In this case signal and idler photons will be strictly defined OAM according to the OAM of pump photons. Also, twisted waveguides allow to control the correlation functions of biphotons at different distances of propagation, depending on the twist angle of waveguides, that could have important applications in quantum cryptography and teleportation.

We consider a linear polarization of light and small twist of waveguides, because in this case the structure of the pump and biphoton modes is essentially the same as in the straight array.

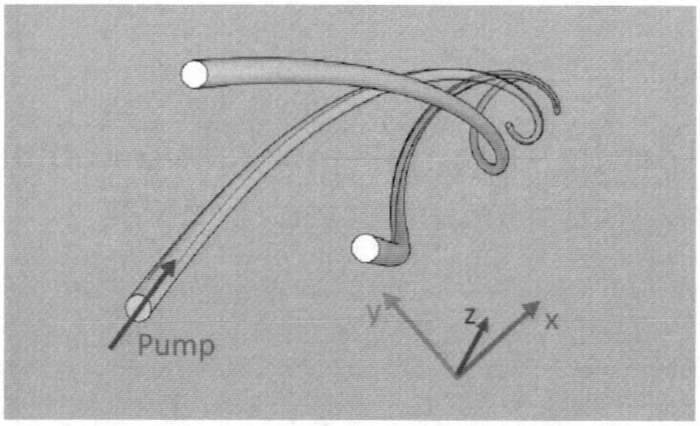

Figure 1. Scheme of twisted circular array of coupled waveguides. Pairs of signal and idler photons are generated through spontaneous four-wave mixing from the pump.

2. Distribution and generation of photon pairs in twisted nonlinear waveguide arrays

We consider an array of closely spaced optical waveguides, which are twisted around a central axis along the propagation direction. Such structure composed of three waveguides is schematically shown in Fig. 1(a).

The normalized pump field profile evolution along the propagation distance z is defined through the classical coupled-mode equations [8]:

$$i\frac{dE_n}{dz}+C(\omega)(\exp[-i\Delta\varphi]E_{n+1}+\exp[i\Delta\varphi]E_{n-1})+\rho_{n_p}E_n=0, \tag{1}$$

where E_n is the complex field amplitude in the nth waveguide, and E_N maps to E_0, and E_{-1} maps to E_{N-1} due to closed-loop boundary conditions [8], n is the waveguide number, N is the total number of waveguides, $\Delta\varphi = \omega d\chi/dz$ is the additional phase, ω - frequency, $d\chi/dz$ - is the periodic waveguide bending profile, $\chi(z) = \chi(z + L)$, L is the modulation period, coefficient $C(\omega)$ defines a coupling strength between the neighboring waveguides (it characterizes diffraction strength in a straight waveguide array with $C_\chi = 0$ [9], and ρ_{n_p} is the mismatch between the waveguides. In our case $d\chi/dz$ = const because waveguide bending profile is the same to all values of z and mismatch is equal to 0.

Generation of photon pairs in cubic nonlinear WGAs through SFWM in the absence of multiple photon pairs can be characterized by the evolution of a bi-photon wave function $\Psi_{n_s,n_i}(z)$ in a Schrodinger-type equation. The equation is obtained from the Hamiltonian, and it has a form similar to that of quadratic media [8]:

$$\begin{gathered}i\frac{d\Psi_{n_s,n_i}}{dz}=-C(\exp[i\Delta\varphi]\Psi_{n_s-1,n_i}+\exp[i\Delta\varphi]\Psi_{n_s,n_i-1}+\exp[-i\Delta\varphi]\Psi_{n_s+1,n_i}+\\+\exp[-i\Delta\varphi]\Psi_{n_s,n_i+1})+i\gamma E^{(p)}_{n_s}(z)E^{(p)}_{n_s}(z)\delta_{n_s,n_i}\exp[i\Delta\beta^{(0)}z]+(\rho^{(s)}_{n_s}+\rho^{(i)}_{n_i})\Psi_{n_s,n_i}\end{gathered}, \tag{2}$$

where n_s and n_i are the waveguide numbers describing the positions of the signal, and the idler photons, and $E_{ns}^{(p)}(z)$ is the pump amplitude in waveguide number ns. Δβ - the linear four-wave mixing phase-mismatch in a single waveguide, γ - is a nonlinear coefficient, δ - Kronecker delta, and $\rho_{n_s}^{(s)}, \rho_{n_i}^{(i)}$ is the mismatch between the waveguides [7]. In our case mismatch is equal to 0.

3. Analysis of photon propagation modes depending on the input state of the pump

A pump beam with a particular OAM will generate only pairs of signal and idler photons which satisfy the angular momentum conservation expressed through the Kronecker delta function. We illustrate this relation by summarizing in Table I the possible states of generated biphotons for different OAM in a three-waveguide structure. In Table I, the first column represents the OAM of a pump beam coupled to the array, and the second column shows the corresponding pump propagation constants. The third column lists the possible OAM states of the generated biphotons, and the fourth column presents the corresponding biphoton propagation constants. For clarity, we explicitly write down the equivalence of pump states due to the discrete periodic symmetry. It then becomes easy to see how the selection rules work based on the OAM conservation for the generated biphotons: $m_s + m_i = 2m_p$ up to multiples of N (number of waveguides). We see that two different momentum states of biphotons can be generated for each pump OAM. Further selectivity in excitation of particular photon states can be achieved by adjusting the phase mismatch.

Table 1. Summary of biphoton states which can be generated by pump beams with different OAM in a three-waveguide structure.

Pump state $\lvert m_p\rangle$	**Propagation constant** β_{m_p}	**Biphoton state** $\lvert m_s, m_i\rangle$	**Propagation constant** β_{m_s,m_i}
$\lvert 0\rangle$	$2C\cos\varphi_0$	$\lvert 0,0\rangle$	$4C\cos\varphi_0$
		$\lvert -1,1\rangle + \lvert 1,-1\rangle$	$-2C\cos\varphi_0$
$\lvert -2\rangle \equiv \lvert 1\rangle$	$-C\left[\cos(\varphi_0) - \sqrt{3}\sin(\varphi_0)\right]$	$\lvert 0,-1\rangle + \lvert -1,0\rangle$	$C\left[\cos(\varphi_0) - \sqrt{3}\sin(\varphi_0)\right]$
		$\lvert 1,1\rangle$	$-2C\left[\cos(\varphi_0) - \sqrt{3}\sin(\varphi_0)\right]$
$\lvert 2\rangle \equiv \lvert -1\rangle$	$-C\left[\cos(\varphi_0) + \sqrt{3}\sin(\varphi_0)\right]$	$\lvert 0,1\rangle + \lvert 1,0\rangle$	$C\left[\cos(\varphi_0) + \sqrt{3}\sin(\varphi_0)\right]$
		$\lvert -1,-1\rangle$	$-2C\left[\cos(\varphi_0) + \sqrt{3}\sin(\varphi_0)\right]$

At the Fig. 2 we present the biphoton correlations $\left|\Psi_{n_s,n_i}\right|^2$, in case of photon generation by SFWM from a pump inside the straight array (1st and 2nd columns, $d\chi/dz = 0$) and twisted array (3rd and 4th columns, $d\chi/dz = 1$) for different distances: at 1st string z = 1.05; at 2nd string z = 2.1; at 3rd string z = 3.15. 1st and 3rd columns corresponds the case without mismatch ($\rho = 0$). 2nd and 4th columns corresponds the case with the mismatch ($\rho = 0.5$). At z = 0 there are no photons. $\lvert m_p\rangle = 2$.

There are strong anti-bunching behavior of photons in the Fig. 2 (intersection of 1st column and 2nd string), which corresponds the violating of Bell-inequality. But in the Fig. 2 (intersection of 2nd column and 2nd string) anti-bunching behavior of photons is vanished because of mismatch. Similarly, at the Fig. 2 (3rd and 4th columns) the behavior of photons is conserved in all the distances, in despite of the mismatch. But strength of correlations may be smaller in the case of twisted waveguides.

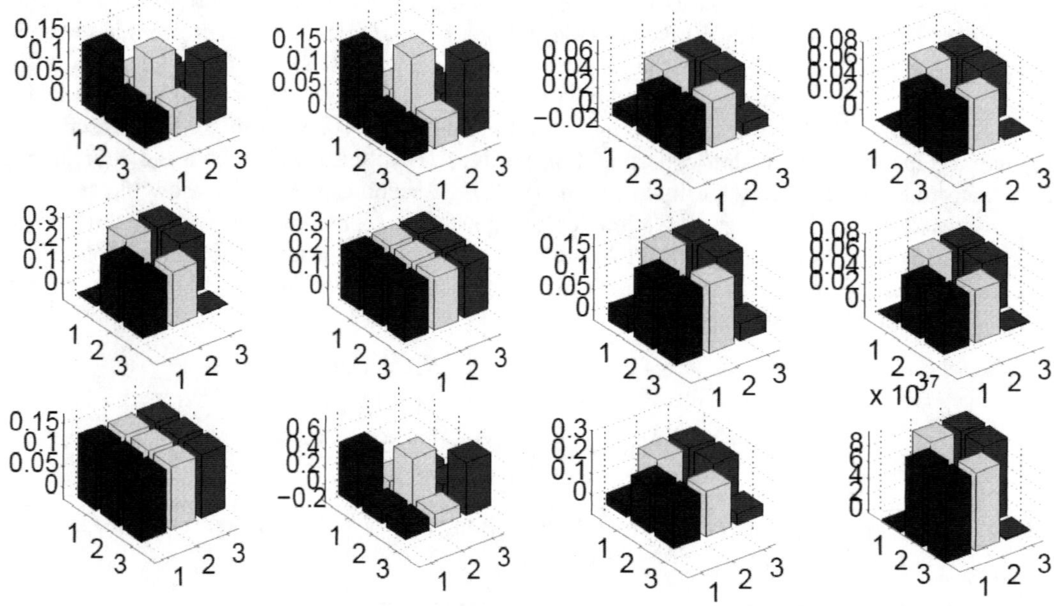

Figure 2. Biphoton correlations $\left|\Psi_{n_s,n_i}\right|^2$, in case of photon generation by SFWM from a pump inside the straight array (1st and 2nd columns, dχ/dz = 0) and twisted array (3rd and 4th columns, dχ/dz = 1) for different distances: at 1st string z = 1.05; at 2nd string z = 2.1; at 3rd string z = 3.15. 1st and 3rd columns corresponds the case without mismatch ($\rho = 0$). 2nd and 4th columns corresponds the case with the mismatch ($\rho = 0.5$). At z = 0 there are no photons. $\left|m_p\right\rangle = 2$.

Thus, the twist of waveguides can effectively provide the generation of photons with certain OAM and its propagation even through the array with the mismatch.

4. Conclusion

Building on a recent experimental demonstration of classical OAM state transmission through twisted multi-core photonic-crystal fibers, we predict that the correlation properties of entangled photon pairs propagating in the regime of quantum walks in closed-loop waveguide arrays can be controlled by the introduction of twist. We also analyze integrated photon generation through spontaneous four-wave mixing and predict that the input pump profile also can control the features of correlation properties of entangled photons. We also show that in both cases photons can demonstrate bunching behavior at some distances and anti-bunching behavior at other distances. These features can be controlled by the amount of twist and the input pump profile.

Thus, our results suggest a potential for developing quantum communications for cryptography and other applications by combining robust photon-pair generation in twisted structures and their subsequent transmission through twisted optical fibers.

5. Acknowledgements
This work was financially supported by the Government of Russian Federation, Grant 074-U01.

References

[1] Hamilton C S, Kruse R, Sansoni L, Silberhorn C, and Jex I 2014 Driven Quantum Walks. *Phys. Rev. Lett.* **113**, 083602-5

[2] Owens J O et al. 2011 Two-photon quantum walks in an elliptical direct-write waveguide array. *New J. Phys.* **13**, 075003-13

[3] Garanovich I L, Longhi S, Sukhorukov A A, and Kivshar Yu S 2012 Light propagation and localization in modulated photonic lattices and waveguides. *Phys. Rep.* **518**, 1-79

[4] Ekert A K, Rarity J G, Tapster P R, and Palma G M 1992 Practical quantum cryptography based on two-photon interferometry. *Phys. Rev. Lett.* **69**, 1293

[5] D N Vavulin, A A Sukhorukov 2015 Quantum walks of photon pairs in twisted waveguide arrays *Journal of Physics: Conference Series* **643**, 012050

[6] Egorov V I, Vavulin D N, Latypov I Z, Gleim A V, Rupasov A V 2013 Analysis of a sidebands based quantum cryptography system with different detector types *NANOSYSTEMS: PHYSICS, CHEMISTRY, MATHEMATICS*, **4 (2)**, pp. 190–195

[7] D N Vavulin, V I Egorov, A V Gleim, S A Chivilikhin 2014 Determining influence of four-wave mixing effect on quantum key distribution *Journal of Physics: Conference Series* **541**, 012066

[8] Markin D M, Solntsev A S, and Sukhorukov A A 2013 Generation of orbital-angular-momentum entangled biphotons in triangular quadratic waveguide arrays. *Phys. Rev. A* **87**, 063814-5

[9] Christodoulides D N, Lederer F, and Silberberg Y 2003 Discretizing light behaviour in linear and nonlinear waveguide lattices *Nature* **424**, 817–823

[10] Solntsev A S, Sukhorukov A A, Neshev D N, and Kivshar Yu S 2012 Photon-pair generation in arrays of cubic nonlinear waveguides. *OPTICS EXPRESS* **20**, No. 24, 27441

[11] Jonathan C F, M and M G Thompson 2012 Quantum optics: An entangled walk of photons. *Nature* **484**, 47–48

[12] Xi X M, Wong G K L, Frosz M H, Babic F, Ahmed G, Jiang X, Euser T G, and Russell P St.J 2014 Orbital-angular-momentum-preserving helical Bloch modes in twisted photonic crystal fiber. *Optica* **1**, 165-169

IOP Conf. Series: Journal of Physics: Conf. Series **917** (2017) 062011 doi:10.1088/1742-6596/917/6/062011

The influence of dipole-dipole interaction on entanglement of two superconducting qubits in the framework of double Jaynes-Cummins model

M M Evseev and E K Bashkirov
Physics Department, Samara National Research University, Samara, 443086, Russia

Abstract. In this paper we investigated the entanglement dynamics between two superconducting qubits interacting with two microwave modes of independent coplanar cavities taking into account the direct dipole-dipole interaction. The model with different qubit-field couplings and detunings is under consideration. Using the dressed-states representation technique we derived the exact solution for considered model with qubits prepared initially in entangled states and vacuum field modes. We have carried out the dependence of the atom-atom entanglement on the strength of the dipole-dipole interaction and other parameters of the considered system such as different coupling constants and detunings. The results showed that the presence of a sufficiently large dipole-dipole interaction leads to stabilization of entanglement.

1. Introduction

Coupling distant qubits is an important goal for quantum information and for its potential applications. This kind of coupling needs the study of interaction between qubits and photons, which has been widely studied in a cavity quantum electrodynamics (QED) [1]. The Jaynes–Cummings model (JCM) is the simplest possible physical model in cavity QED that describes the interaction of a natural or artificial two-level atom (qubit) with a single-mode cavity, and has been used to understand a wide variety of phenomena in quantum optics and condensed matter systems, such as trapped ions, quantum dots, spins, superconducting circuits, optical and microwave cavity QED [1]. In order to explore a wider range of phenomena caused by the interaction of the qubits with the quantum fields in resonators the numerous generalizations of the JCM have been investigated in recent years. Eberly and co-authors [2] have proposed the so-called double JCM (DJCM), consisting of two two-level atoms and two resonator modes, provided that each atom interacts only with one field of the resonator, and investigated the entanglement dynamics of this model. Recently, the DJCM have been extensively investigated. Hu et al. [3, 4] have studied the effect of the Stark shift and detuning and different couplings on entanglement in the framework of DJCM. Du et al. [5] have investigated the sudden birth of entanglement between two initially separate atoms interacting with two entangled photons in a DJCM, and have discussed the influences of different atomic initial states on entanglement among atoms. Xie and Fang [6] have considered the entanglement dynamics of intensity-dependent coupling DJCM with two different initial light fields, which one is in the squeezed vacuum state and another is in the coherent state. Liao et al. [7] have investigated the dynamics of entanglement between two atoms in a two-photon DJCM model when the cavity field is previously in a maximally entangled state. Xie and Fang [8] have examined the entanglement dynamics of the atoms in the DJCM with the

Kerr medium. The entanglement dynamics of a two-photon DJCM with Kerr-like medium have also been studied by Quyang et al. [9]. Vieira et al. [10] have reported on the geometric character of the entanglement dynamics of two pairs of qubits evolving according to the DJCM. Baghshahi et al. [11] have studied the interaction between two two-level atoms and two coupled modes of a quantized radiation field in the form of parametric frequency converter. Zhou and Fang [12] have investigated the analytical solution and entanglement swapping of a DJCM in non-Markovian environments. Zhu et al. [13] have studied the dynamics of quantum entanglement of DJCM model analytically in the framework of corrections to the rotating-wave approximations.

In all previous studies, the direct dipole-dipole interaction between the qubits doesn't takes into account. But for superconducting Josephson qubits the effective dipole-dipole interaction constant may exceed the coupling constant between the qubit and cavity field (see references in [14]). In this paper, we consider a non-resonant DJCM taking into account the direct dipole-dipole interaction between qubits. We investigate the entanglement between qubits for Bell's initial qubits states and vacuum resonators states, and discuss dependence of the entanglement on the parameters of the considered system, such as different intensity of dipole interaction, coupling constants and the detunings between the atomic transition frequency and the cavity field frequencies.

2. The model and its time-dependent dynamics

We consider two identical superconducting qubits labelled *A* and *B*, and two cavity modes of coplanar or LC circuits labelled "a" and "b". Qubit *A* not-resonantly interacts with a single-mode cavity field "a", and qubit *B* not-resonantly interacts with a single-mode cavity field "b". Due to the randomness of the qubits positions in the cavity, it is very difficult to control the couplings between different atom-cavity systems to be the same. Therefore the coupling constants between the atoms and cavities are assumed to be unequal. For superconducting quits interacting with microwave coplanar resonators or LC superconducting circuits the intensity of effective dipole-dipole interaction can be compared with the atom-cavity coupling constant. In this case the dipole-dipole interaction should be included in the model Hamiltonian. Therefore, Hamiltonian for the system under the rotating wave approximation can be written as

$$H = H_0 + H_1,$$

$$H_0 = \hbar\omega_0(\sigma_A^z/2 + \sigma_B^z/2 + a^+a + b^+b),$$

$$H_1 = \hbar\Delta_a a^+a + \hbar\Delta_b b^+b + \hbar\gamma_a(\sigma_1^+a + a^+\sigma_1^-) + \hbar\gamma_b(\sigma_2^+b + b^+\sigma_2^-) + \hbar J(\sigma_1^+\sigma_2^- + \sigma_1^-\sigma_2^+), \quad (1)$$

where $(1/2)\sigma_i^z$ is the inversion operator for the i th qubit ($i = A, B$), $\sigma_i^+ = |+\rangle_{ii}\langle -|$, and $\sigma_i^- = |-\rangle_{ii}\langle +|$ are the transition operators between the excited $|+\rangle_i$ and the ground $|-\rangle_i$ states in the i th qubit, a^+ and a are the creation and the annihilation operators of a cavity mode photon (or plasmon for LC-resonator) of the cavity mode "a", b^+ and b are the creation and the annihilation operators of photons of the cavity mode b, $\hbar\omega_0$ is the superconducting (SC) gap energy, $\gamma_a \equiv \gamma$ is the coupling constant between qubit *A* and the cavity field "a" and γ_b is the coupling constant between qubit *A* and the cavity field "a", Δ_a and Δ_b are the detunigs for mode "a" and "b" and J is the coupling constant of the dipole interaction between the qubits *A* and *B*.

Firstly we take two qubits initially in the Bell-like pure state of the following form $|\Psi(0)\rangle_A = \cos\theta|+,-\rangle + \sin\theta|-,+\rangle$, where $0 \le \theta \le \pi$, and the cavity fields are in a vacuum states $|0,0\rangle$. Then the full initial state for considered model is

$$|\Psi(0)\rangle = (\cos\theta|+,-\rangle + \sin\theta|-,+\rangle) \otimes |0,0\rangle. \quad (2)$$

To obtain the time-dependent wave function of considered model we used the so-called dressed states or eigenvectors of the Hamiltonian H_1 as in [15, 16]

$$|\Phi_i\rangle = \xi_i (X_{i1}|-,-,0,1\rangle + X_{i2}|-,-,1,0\rangle + X_{i3}|-,+,0,0\rangle + X_{i4}|+,-,0,0\rangle) \quad (i=1,2,3,4),$$

where

$$\xi_i = 1/\sqrt{|X_{i1}|^2+|X_{i2}|^2+|X_{i3}|^2+|X_{i4}|^2},$$

$$X_{11} = \frac{1}{\alpha}\left(-\alpha^2+\frac{g^2\lambda_1}{\delta_b-\lambda_1}+\lambda_1^2\right),\quad X_{12}=-\frac{g}{\delta_b-\lambda_1},\quad X_{13}=\frac{1}{\alpha}\left(\frac{g^2}{\delta_b-\lambda_1}+\lambda_1\right),\quad X_{14}=1,$$

$$X_{21} = \frac{1}{\alpha}\left(-\alpha^2+\frac{g^2\lambda_2}{\delta_b-\lambda_2}+\lambda_2^2\right),\quad X_{22}=-\frac{g}{\delta_b-\lambda_2},\quad X_{23}=\frac{1}{\alpha}\left(\frac{g^2}{\delta_b-\lambda_2}+\lambda_2\right),\quad X_{24}=1,$$

$$X_{31} = \frac{1}{\alpha}\left(-\alpha^2+\frac{g^2\lambda_3}{\delta_b-\lambda_3}+\lambda_3^2\right),\quad X_{32}=-\frac{g}{\delta_2-\lambda_3},\quad X_{33}=\frac{1}{\alpha}\left(\frac{g^2}{\delta_b-\lambda_3}+\lambda_3\right),\quad X_{34}=1,$$

$$X_{41} = \frac{1}{\alpha}\left(-\alpha^2+\frac{g^2\lambda_4}{\delta_b-\lambda_4}+\lambda_4^2\right),\quad X_{42}=-\frac{g}{\delta_2-\lambda_4},\quad X_{43}=\frac{1}{\alpha}\left(\frac{g^2}{\delta_b-\lambda_4}+\lambda_4\right),\quad X_{44}=1.$$

The corresponding eigenvalues are

$$E_1=\hbar\lambda_1=\frac{1}{12}\left(6\Delta_+-\sqrt{6}T_1-\sqrt{6(T_2+T_3)}\right),\ E_2=\hbar\lambda_2=\frac{1}{12}\left(6\Delta_+-\sqrt{6}T_1+\sqrt{6(T_2+T_3)}\right),$$

$$E_3=\hbar\lambda_3=\frac{1}{12}\left(6\Delta_++\sqrt{6}T_1-\sqrt{6(T_2-T_3)}\right),\ E_4=\hbar\lambda_4=\frac{1}{12}\left(6\Delta_++\sqrt{6}T_1+\sqrt{6(T_2-T_3)}\right).$$

Here we use the following notations

$$T_1=\sqrt{4C+6\Delta_+^2+2^{2/3}S_2+6S_3},\ T_2=8C+12\Delta_+^2-2^{2/3}S_2-6S_3,\ T_3=12\sqrt{6}\left(B-\Delta_+^3-\Delta_+C\right)/S_1,$$

$$S_1=27B^2-18BC\Delta_+-2C^3+108D\Delta_+^2+72C,\ S_2=\sqrt[3]{S_1+\sqrt{S_1^2-4\chi^3}},\ S_3=\sqrt[3]{2}\chi/3S_2,$$

$$A=\sqrt{\alpha^2+g^2+1},\quad B=2A^2\Delta_+-\delta_1-g^2\delta_2,\quad C=A^2-\Delta_*^2,\quad D=g^2-\alpha^2\Delta_*^2,$$

$$\chi=C^2+6B\Delta_++12D,\quad \Delta_+=(\delta_1+\delta_2)/2,\quad \Delta_*=\sqrt{\delta_1\delta_2},$$

where $\delta_a=\Delta_a/\gamma_a$, $\delta_b=\Delta_b/\gamma_a$, $g=\gamma_b/\gamma_a$, $\alpha=J/\gamma_a$.

To obtain the time dependent wave function one can use the formula $\Phi(t)\rangle = e^{-iHt/\hbar}|\Phi(0)\rangle$. Using the eigenvalues and eigenvectors of Hamiltonian (1) and the initial state (2) we can derive

$$|\Psi(t)\rangle = C_1^{(1)}(t)|-,-,0,1\rangle + C_2^{(1)}(t)|-,-,1,0\rangle + C_3^{(1)}(t)|-,+,0,0\rangle + C_4^{(1)}(t)|+,-,0,0\rangle,$$

where

$$C_1^{(1)}=\cos\theta Z_{11}+\sin\theta Z_{12},\quad C_2^{(1)}=\cos\theta Z_{21}+\sin\theta Z_{22},$$

$$C_3^{(1)}=\cos\theta Z_{31}+\sin\theta Z_{32},\quad C_4^{(1)}=\cos\theta Z_{41}+\sin\theta Z_{42},$$

$$Z_{11}=e^{-iE_1t/\hbar}\xi_1Y_{31}X_{11}+e^{-iE_2t/\hbar}\xi_2Y_{32}X_{21}+e^{-iE_3t/\hbar}\xi_3Y_{33}X_{31}+e^{-iE_4t/\hbar}\xi_4Y_{34}X_{41},$$

$$Z_{12}=e^{-iE_1t/\hbar}\xi_1Y_{41}X_{11}+e^{-iE_2t/\hbar}\xi_2Y_{42}X_{21}+e^{-iE_3t/\hbar}\xi_3Y_{43}X_{31}+e^{-iE_4t/\hbar}\xi_4Y_{44}X_{41},$$

$$Z_{22}=e^{-iE_1t/\hbar}\xi_1Y_{41}X_{12}+e^{-iE_2t/\hbar}\xi_2Y_{42}X_{22}+e^{-iE_3t/\hbar}\xi_3Y_{43}X_{32}+e^{-iE_4t/\hbar}\xi_4Y_{44}X_{42},$$

$$Z_{31}=e^{-iE_1t/\hbar}\xi_1Y_{31}X_{13}+e^{-iE_2t/\hbar}\xi_2Y_{32}X_{23}+e^{-iE_3t/\hbar}\xi_3Y_{33}X_{33}+e^{-iE_4t/\hbar}\xi_4Y_{34}X_{43},$$

$$Z_{32}=e^{-iE_1t/\hbar}\xi_1Y_{41}X_{13}+e^{-iE_2t/\hbar}\xi_2Y_{42}X_{23}+e^{-iE_3t/\hbar}\xi_3Y_{43}X_{33}+e^{-iE_4t/\hbar}\xi_4Y_{44}X_{43},$$

$$Z_{41}=e^{-iE_1t/\hbar}\xi_1Y_{31}X_{14}+e^{-iE_2t/\hbar}\xi_2Y_{32}X_{24}+e^{-iE_3t/\hbar}\xi_3Y_{33}X_{34}+e^{-iE_4t/\hbar}\xi_4Y_{34}X_{44},$$

$$Z_{42}=e^{-iE_1t/\hbar}\xi_1Y_{41}X_{14}+e^{-iE_2t/\hbar}\xi_2Y_{42}X_{24}+e^{-iE_3t/\hbar}\xi_3Y_{43}X_{34}+e^{-iE_4t/\hbar}\xi_4Y_{44}X_{44}.$$

Here $Y_{ij}=\xi_jX^*_{ji}$.

Now we will consider an another type of Bell-like pure initial state of two qubits

$$|\Psi(0)\rangle_A=\cos\theta|+,+\rangle+\sin\theta|-,-\rangle.$$

Then for vacuum cavity field full initial state of the system is

$$\Psi(0)\rangle=(\cos\theta|+,+\rangle+\sin\theta|-,-\rangle)\otimes|0,0\rangle. \tag{3}$$

For initial state (3) the time-dependent wave function can be written in the form

$$|\Psi(t)\rangle=C_1^{(2)}(t)|+,+,0,0\rangle+C_2^{(2)}(t)|+,-,0,1\rangle+C_3^{(2)}(t)|-,+,1,0\rangle+C_4^{(2)}(t)|+,-,1,0\rangle+$$
$$+C_5^{(2)}(t)|-,+,0,1\rangle+C_6^{(2)}(t)|-,-,2,0\rangle+C_7^{(2)}(t)|-,-,0,2\rangle+C_8^{(2)}(t)|-,-,1,1\rangle+C_9^{(2)}(t)|-,-,0,0\rangle.$$

For considered initial state the analytical solution of dynamical equations is a very difficult problem, so for we confine ourselves to obtain numerical solution.

3. Entanglement calculations

For two-qubit system described by the reduced density operator $\rho_A(t)$, a measure of entanglement or negativity can be defined in terms of the negative eigenvalues μ_i^- of partial transpose of the reduced atomic density matrix ($\rho_A^{T_1}$) [17, 18]: $\varepsilon=-2\sum\mu_i^-$. The partial transpose of the reduced atomic density matrix for considered initial state (2) is

$$\rho_A^{T_1}(t)=\begin{pmatrix}0&0&0&H(t)^*\\0&V(t)&0&0\\0&0&W(t)&0\\H(t)&0&0&R(t)\end{pmatrix}. \tag{4}$$

The elements of matrix (4) are

$$V(t)=|C_4^{(1)}(t)|^2,\quad W(t)=|C_3^{(1)}(t)|^2,\quad R(t)=|C_1^{(1)}(t)|^2+|C_2^{(1)}(t)|^2,\quad H(t)=C_4^{(1)}(t)C_3(t)^*.$$

Matrix (4) has only one eigenvalue, which may take a negative value. As a result, the negativity can be written as

$$\varepsilon(t)=\sqrt{R(t)^2+4|H(t)|^2}-R(t).$$

The partial transpose of the reduced atomic density matrix $\rho_A^{T_1}$ for initial state (3) is

$$\rho_A(t)^{T_1} = \begin{pmatrix} U_1(t) & 0 & 0 & H_2(t)^* \\ 0 & V_1(t) & H_1(t)^* & 0 \\ 0 & H_1(t) & W_1(t) & 0 \\ H_2(t) & 0 & 0 & R_1(t) \end{pmatrix}, \quad (5)$$

where $U_1(t) = |C_1^{(2)}|^2$, $H_1(t) = C_1^{(2)}C_9^{(2)*}$, $H_2(t) = C_2^{(2)}C_5^{(2)*} + C_4^{(2)}C_3^{(2)*}$, $V_1(t) = |C_2^{(2)}|^2 + |C_4^{(2)}|^2$,

$$W_1(t) = |C_3^{(2)}|^2 + |C_5^{(2)}|^2, \quad R_1(t) = |C_6^{(2)}|^2 + |C_7^{(2)}|^2 + |C_8^{(2)}|^2 + |C_9^{(2)}|^2 .$$

Matrix (5) has two eigenvalues, which may take a negative value. Then, the negativity can be written as a superposition of two terms. At the same time, each term contributes to the total amount, as long as it takes a positive value. As a result the negativity is

$$\varepsilon(t) = \sqrt{(U_1(t) - R_1(t))^2 + 4|H_2(t)|^2} - U_1(t) - R_1(t) + \sqrt{(V_1(t) - W_1(t))^2 + 4|H_1(t)|^2} - V_1(t) - W_1(t).$$

The results of entanglement parameter calculations for Bell's type initial atomic state $|\psi\rangle_1 = 1/\sqrt{2}(|+,-\rangle + |-,+\rangle)$ are shown in Figures 1 and 2 and these for initial atomic state $|\psi\rangle_2 = 1/\sqrt{2}(|+,+\rangle + |-,-\rangle)$ are displayed in Figures 3 and 4. Figure 1 shows that in the case of exact resonance, the dependence of the negativity evolves periodically between 0 and 1. In this case the inclusion of the dipole-dipole interaction leads to a stabilization of entanglement behaviour. Figure 2 show the effect of dipole-dipole interaction on negativity for non-resonant interaction. If qubits *A* and *B* interact with a single-mode cavity fields via not-zero detuning the presence of dipole-dipole interaction with intermediate strength leads to increasing of the amplitudes of the negativity oscillations. But for large values of dipole-dipole interaction strength one can see the stabilization of entanglement oscillations as in the case of exact resonance for both equal and unequal qubit-field couplings. Figure 3 shows the time dependence of negativity for initial qubits state (3) and different strength of dipole-dipole interaction in the case of exact resonance. This is different from the results obtained for the previous case. The dipole-dipole interaction in the present case does not lead to stabilization of the entanglement, but has only an effect on the periods and amplitudes of the oscillations of entanglement. However, for non resonant interaction between qubits and fields the influence of dipole-dipole interaction on the entanglement is similar to the previous case as one can see from Figure 4. For considered state we show the entanglement behaviour for different detunings and couplings. For sufficiently large values of the dipole-dipole interaction strength we have to deal with the stabilization of entanglement.

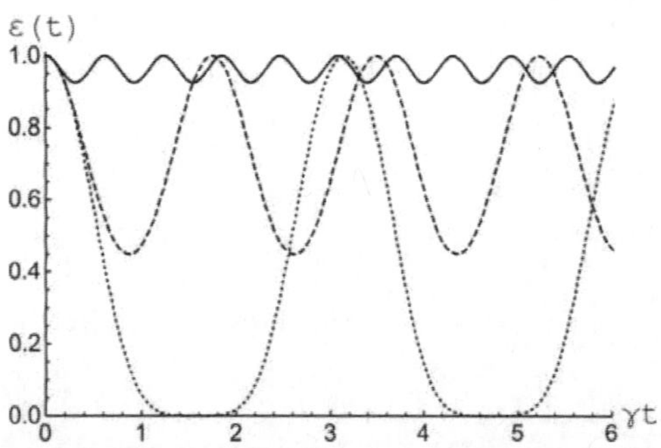

Figure 1. Negativity vs. γt for initial state (2) with $\theta = \pi/4$ and $\delta_a = \delta_b = 0$, $\gamma_b = \gamma_a$. Parameter $\alpha = 0$ (dotted), $\alpha = 3$ (dashed) and $\alpha = 10$ (solid).

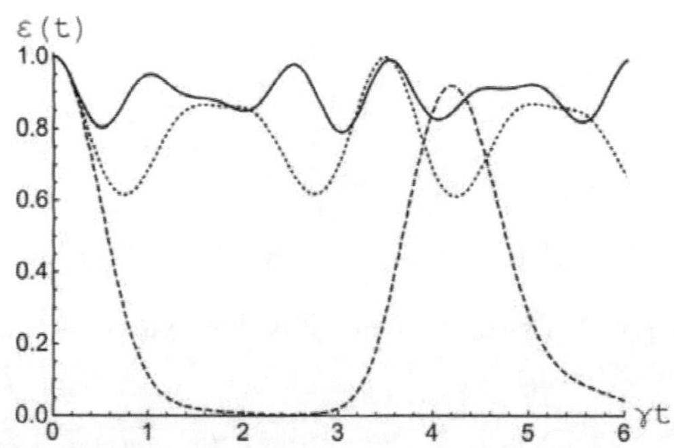

Figure 2. Negativity vs. γt for initial state (2) with $\theta = \pi/4$ and $\delta_b = 5, \delta_a = 3$, $\gamma_b = \gamma_a$. Parameter $\alpha = 0$ (dotted), $\alpha = 3$ (dashed) and $\alpha = 10$ (solid).

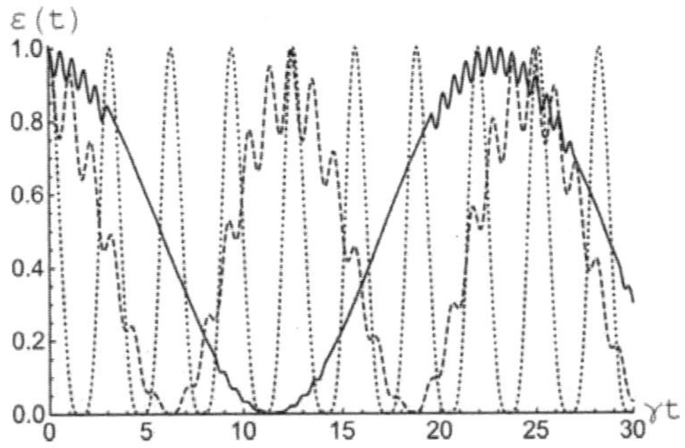

Figure 3. Negativity vs. γt for initial state (3) with $\theta = \pi/4$ and $\delta_a = \delta_b = 0$, $\gamma_b = \gamma_a$. Parameter $\alpha = 0$ (dotted), $\alpha = 3$ (dashed) and $\alpha = 10$ (solid).

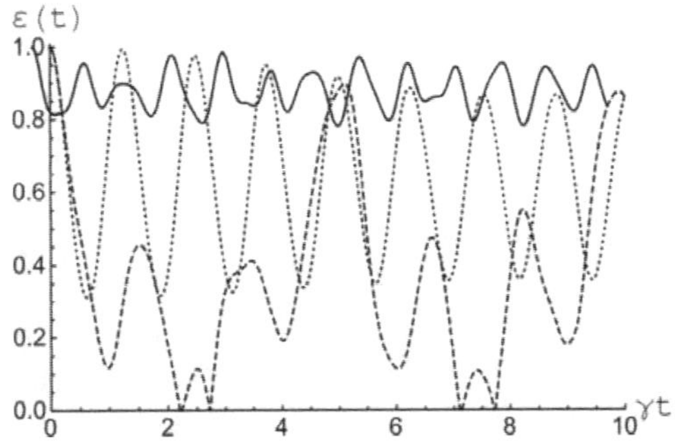

Figure 4. Negativity vs. γt for initial state (3) with $\theta = \pi/4$ and $\delta_b = 5, \delta_a = 3$, $\gamma_b = \gamma_a$. Parameter $\alpha = 0$ (dotted), $\alpha = 3$ (dashed) and $\alpha = 10$ (solid).

4. Conclusions

In this paper, we investigate the entanglement between two superconducting qubits in the framework of a DJCM with different coupling constants and detunings between the atomic transition frequencies and the cavity fields frequencies taking into account the direct dipole-dipole interaction, and discuss dependence of the atom-atom entanglement on strength of the dipole-dipole interaction and other parameters of the considered system such as the different coupling constants and the detunings. The results show that these parameters have great impact on the amplitude and the period of the atom-atom entanglement evolution. In addition, the presence of sufficiently large dipole-dipole interaction leads to stabilization of entanglement for all Bell-types initial qubits states and different couplings and detunings.

References

[1] Buluta I, Ashhab S, Nori F 2011 *Rep. Prog. Phys.* **74** 104401
[2] Yonac M Y, Yu T, Eberly J H 2006 *J. Phys. B: At. Mol. Opt. Phys.* **39** S621
[3] Hu Y H et al. 2008 *J. Mod. Opt.* **55** 3551
[4] Hu Y H., Fang M F, Cai J W, Jiang C L 2008 *Int. J. Theor. Phys.* **47** 2554
[5] Du M, Fang M F, Liu X 2009 *Chin. Opt. Lett.* **7** 443
[6] Xie Q, Fang M F 2012 *Int. J. Theor. Phys.* **51** 778
[7] Liao Q, Nie W, Zhou N, Liu Y, Ahmad M A 2013 *Chin. J. Phys.* **51** 404
[8] Ouyang X C et al. 2010 *Chin. Phys.* **B19** 030309
[9] Vieira A R, de Oliveira Jr. J G G, Peixoto de Faria J G, Nemes M C 2014 *Braz. J. Phys.* **44** 19
[10] Xie Q, Fang M F 2010 *Commun. Theor. Phys.* **54** 840
[11] Baghshahi H R, Tavassoly M Z, Faghihi M J 2015 *Int. J. Theor. Phys.* **54** 2839
[12] Zou H M, Fang M F 2015 *Quan. Inf. Process.* **14** 2673
[13] Zhu W T, Ren Q B, Duan L W, Chen Q H 2016 *Chin. Phys. Lett.* **33** 050302
[14] Bashkirov E K, Mastyugin M S 2014 *Optics and Spectroscopy* **116** 630
[15] Bashkirov E K, Mastuygin M S 2015 *EPJ Web of Conferences* **103** 03002
[16] Bashkirov E K, Mastuygin M S 2016 *J. Phys.: Conf. Ser.* **735** 012025
[17] Peres A 1996 *Phys. Rev. Lett.* **77** 1413
[18] Horodecki R, Horodecki M, Horodecki P 1996 *Phys.Lett.* **A223** 333

Near-field mapping of three-particle plasmonic structures

E S Babich[1], Yu S Polubavkina[2], N V Kryzhanovskaya[1,2] and A A Lipovskii[1,2]

[1] Peter the Great St. Petersburg Polytechnic University, St. Petersburg 195251, Russia
[2] St. Petersburg Academic University, St. Petersburg 194021, Russia

Abstract. We probe the resonance properties of complex plasmonic structures by Scanning Near-field Optical Microscopy (SNOM). Groups of three differently arranged silver nanoparticles covered with Surface-Enhanced Raman Scattering (SERS) analyte Rhodamine 6G were chosen to visualize electric field distribution nearby the nanostructures. The particles were grown on the surface of a silver-ion-exchanged glass using thermal poling and annealing of the processed glass substrate in hydrogen atmosphere. The morphology of the nanoparticles and their arrangement in the groups were characterized with Atomic Force Microscopy (AFM). The near-field SERS maps superimposed on the AFM scans of the studied groups of nanoparticles visualize the Raman hotspots with 100 nm resolution.

1. Introduction

In recent years, silver and gold nanoparticles (NPs) are of great interest for engineers and scientists because of physicochemical and plasmonic properties of such NPs [1, 2]. These Ag and Au NPs are especially promising for biological and chemical sensing [3], photochemistry [4] and nanomedicine [5]. A widely used sensing strategy relies on the Surface-Enhanced Raman Scattering (SERS) mechanism [6]. The phenomenon occurs because of high local electric field (E-field) enhancement provided for both incident and scattered lightwaves by plasmons excited in the NPs. Because of good confinements of the plasmons, the regions of the highest enhancement (so called hotspots) are placed nearby the NPs. Thus, SERS signal from a molecule depends on its positioning relatively to a hotspot [7]. In this regard, spatial distribution of hotspots in ensembles of metal NPs, which strongly depends on shapes and arrangement of the NPs, is critical. In a simple case of one- or two-particle system E-field distribution nearby the nanoparticles of a known shape can be easily computed [8, 9], while a complex plasmonic structures could become an issue. In this case, visualizing hotspots using a near-field (NF) mapping is of interest. Here, we present results of NF characterization of silver nanoisland triplets grown on a glass substrate.

2. Experimental

2.1. Fabrication of plasmonic structures

To fabricate the plasmonic structures we used the electric-field induced patterning technique [10], which allows growing 2D array of groups consisted of two or more silver nanoislands. According to this technique, we immersed a soda-lime glass slides [11] in $AgNO_3/NaNO_3$ melt containing 5 wt.% of silver nitrate and heated up to 325 °C. After 20 minutes of the processing silver ions were supposed to replace sodium ions in the subsurface region of the glass [12], and one of the ion-exchanged slides was subjected to 1 min thermal poling at 300 °C under DC voltage of 500 V. In the poling, we used a profiled anodic electrode patterned with the square net of 300x300 nm^2 hollows with the periodicity of 0.5 μm. We did not pole another slide, which was aimed at the formation of silver nanoisland film.

IOP Conf. Series: Journal of Physics: Conf. Series **917** (2017) 062012 doi:10.1088/1742-6596/917/6/062012

Then both prepared samples passed the last step that was annealing in hydrogen (300 °C, 10 min). In the annealing, silver ions deepened under the poled regions of the glass surface could not reach the surface of the glass contrary to the unpoled regions (correspond to the hollows in the electrode) of this slide and the unpoled glass sample. Thus, the poling prevented silver reduction by hydrogen. This resulted in the formation of a random silver nanoisland film on the surface of the unpoled glass and groups of silver nanoislands only in the unpoled regions of the glass slide, which was poled with the patterned electrode. In this way, we fabricated 0.5 μm in period square array of silver nanoisland triplets.

2.2. Near-field mapping. Micro-Raman measurements

The nanoislands were covered with a dye Rhodamine 6G (R6G) to use SERS for E-field distribution visualization. The drop of 10 μl aqueous solution of R6G in the concentration of 10^{-3} M was dried on the sample surface in the ambient condition with ~ 5 mm diameter of the spot remaining on the substrate.

The SERS near-field mapping was performed via the cantilever-based SNOM setup NTEGRA Spectra (NT-MDT) equipped with linear polarized 532 nm Nd:YAG laser. The aperture of the hollow-pyramid cantilever (SNOM_NC) was equal to 100 nm. The tapping mode of the measurements was used to keep the distance between the cantilever and the sample surface (~10 nm) and to collect SERS spectra from each point of 500 nm x 500 nm scan area with 100 nm step (see figure 1). The incident and the scattered light beams passed through the cantilever aperture were focused/collected by 100x/0.5 Mitutoyo objective. Acquisition time varied from 20 to 60 s/point depending on the laser intensity which was kept in the range 2–4 kW/cm^2 to avoid R6G degradation. Data analysis was performed using the image analysis Nova software (NT-MDT).

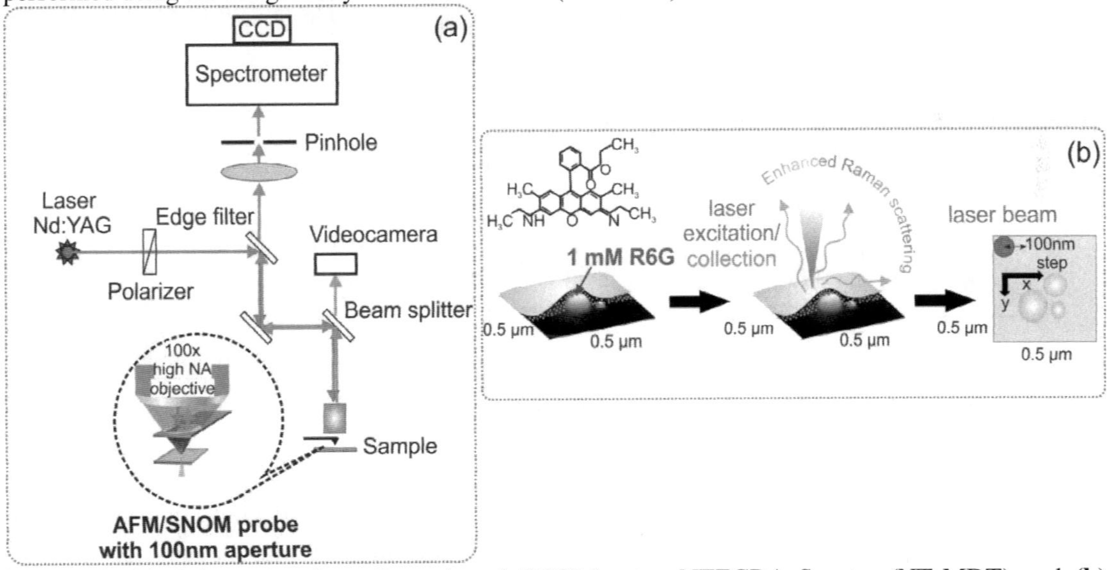

Figure 1. (a) Schematic of cantilever-based SNOM setup NTEGRA Spectra (NT-MDT) and **(b)** methodology of near-field SERS mapping, from left to right: deposition of 10 μl aqueous solution of R6G in the concentration of 10^{-3}M on the substrate; excitation of silver nanoislands with 532 nm laser and collection of Raman signal through 100 nm aperture SNOM cantilever; moving the cantilever with 100 nm step over the 500 nm x 500 nm scan area.

Additionally to the NF cantilever, the setup supported conventional atomic force microscopy (AFM) probe. This allowed us characterizing morphology and arrangement of the prepared nanostructures using AFM with a fine tip (R < 10 nm).

The cantilever could also be excluded from the optical path. In this case, the scattered light was directly collected by the objective and the confocal micro-Raman setup was realized. Corresponding

IOP Conf. Series: Journal of Physics: Conf. Series **917** (2017) 062012 doi:10.1088/1742-6596/917/6/062012

far-field (FF) SERS spectra were obtained at the same acquisition time and laser power density as NF ones. In calculating the laser power density, we accounted for the laser output intensity and the excitation area that was the beam waist in FF geometry (~0.8 μm^2) and aperture size (~0.008 μm^2) in NF geometry.

3. Results and discussion

The analysis of AFM images reveals that the lateral size and height of the nanoislands in triplets were within 60–200 nm and 20–80 nm range, respectively. The interparticle gap varied from 10 to 100 nm. The diameter of grainy silver islands of the nanoisland film falls in the range of tens of nanometers while their height was 10–40 nm.

We probed the nanoisland film and the array of silver nanoisland triplets by micro-Raman spectroscopy to verify their capability to enhance Raman scattering. In FF geometry, we were able to register SERS spectrum of R6G dye from the nanoisland film only. The reason is averaging of signal from an individual group over the large signal collection area. In micro-Raman geometry the excitation/collection area was about 1.0 μm in diameter while the enhanced signal was generated nearby individual triplets (~300 nm in size) separated by 0.5 μm. Thus, at best only 4 groups from the whole array would fall into the collection area whereas hundreds of nanoislands at the nanoisland film. However, we obtained NF SERS spectrum from the three-particle system by positioning the SNOM cantilever in the proper region. Figure 2 shows measured FF and NF SERS spectra of R6G molecules.

In figure 2(left) one can see that the NF SERS spectrum shows a lower signal-to-noise ratio compared to the FF one. We smoothed both spectra by Savitzky–Golay filter (15 point smoothing window radius and the second polynomial order [13]) and superimposed FF and NF spectra for the comparison. Overall the NF R6G spectrum is similar to the FF R6G spectrum with respect to band positions, however, the doublets asterisked in figure 2(right) are better pronounced in the NF spectrum.

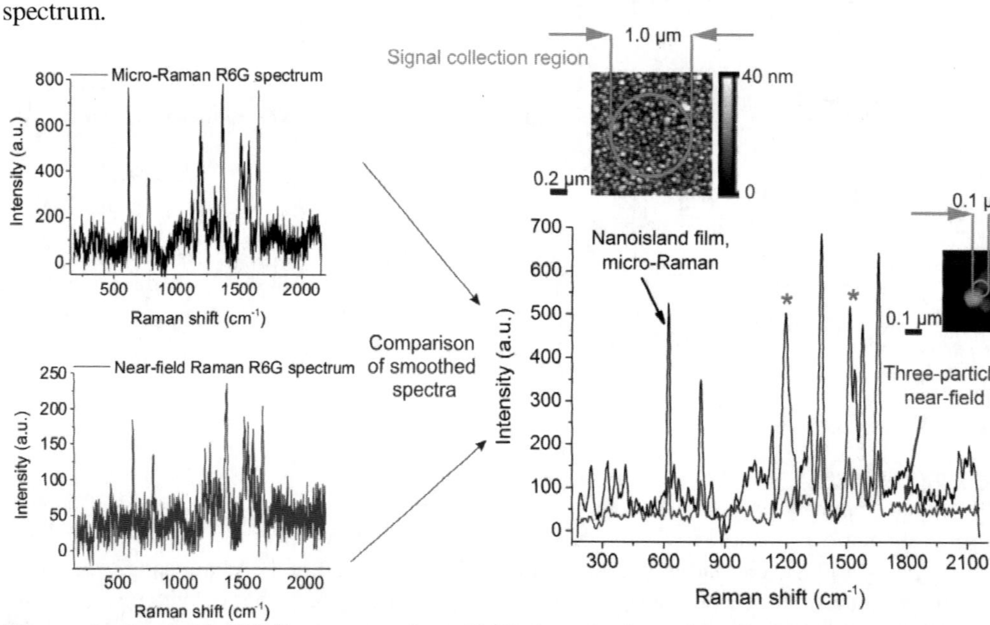

Figure 2. Far-field SERS spectrum from R6G deposited on the silver nanoisland film (upper left) acquired at laser power density of 4 kW/cm^2, 20 s exposition, 1.0 μm beam waist diameter. Near-field SERS spectrum from R6G deposited on the silver nanoisland triplets (lower left) acquired at laser power density of 4 kW/cm^2, 20 s exposition, 0.1 μm aperture. Superimposed FF and NF SERS spectra (right), differences in the spectra are asterisked. Inset: AFM images of the nanoisland film (up) and the three-particle system (right) with tagged signal excitation/collection region.

The relation of the Raman enhancement with the nanoislands morphology and arrangement in the nanoisland triplets is illustrated in figure 3a-d. The NF SERS maps were superimposed on the AFM images of the studied nanostructures. The color indicates detected intensity of the R6G Raman scattering intensity at 1360 cm^{-1}: dark blue is the minimal intensity and bright red is a maximal one. In figure 3a one can see that the maximum enhancement of R6G Raman scattering corresponds to the gap between two bigger nanoparticles. Slight rotation of the group relative to the direction of external E-field corresponds to the rotation of the SERS intensity map (maximal signal area). The absence of a hotspot in the gap between the nanoparticles in the map of similarly arranged group (black arrow in figure 3b), is, most probably, because of the cantilever resolution and the scan step. In the cases of far-separated particles (figure 3c) and particles aligned across the incident lightwave polarization vector (figure 3d) the NF SERS maps are, as expected, mainly uniform and do not show hotspots.

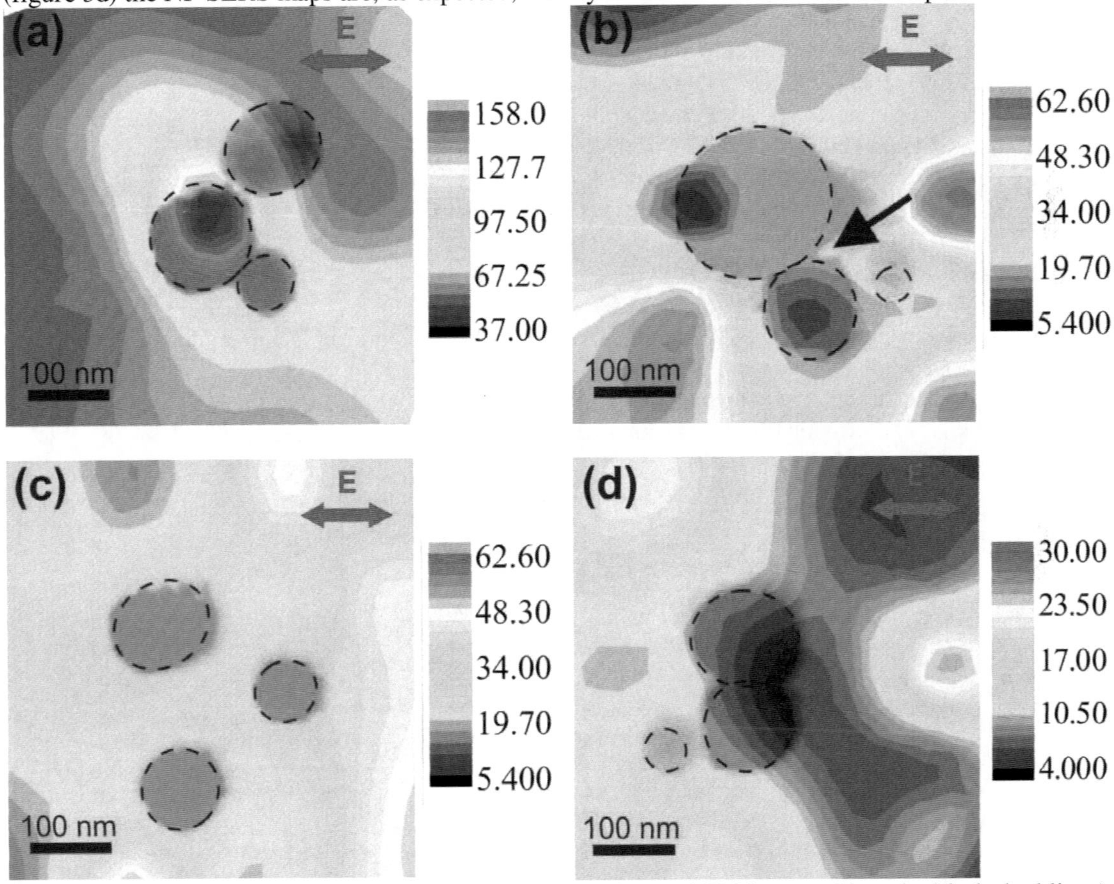

Figure 3(a-d). The near-field SERS maps superimposed onto AFM images (ringed with dashed lines) of the nanoisland triplets with differently arranged particles. **E** arrow denotes the polarization of the incident light. SERS spectra from R6G deposited on the nanoisland triplets acquired at **(a)** laser power of 4 kW/cm^2 , 20 s exposition; **(b)** 2.5 kW/cm^2, 60 s; **(c)** 2.5 kW/cm^2, 60 s; **(d)** 2 kW/cm^2, 60 s. Color bars indicate Raman scattering intensity at 1360 cm^{-1}.

4. Conclusions

We applied near-field optical microscopy to study resonance properties of the triplets with differently arranged silver nanoparticles. The nanoislands were grown via self-assembly of out-diffused silver atoms on a glass substrate patterned with thermal poling, and covered with SERS analyte R6G to visualize electric field distribution nearby the nanostructures. The near-field SERS spectrum included the main R6G bands at 620 cm^{-1}, 1360 cm^{-1} and 1650 cm^{-1}, and was similar to the far-field one collected from the silver nanoisland film. We have obtained near-field SERS maps with 100 nm resolution and demonstrated that spatial variations of Raman signal are in a strong dependence on nanoparticles morphology and arrangement.

Acknowledgements
E. S. B. and A. A. L. appreciate the support by the Ministry of Education and Science of the Russian Federation (project#3.2869.2017). E. S. B. is also grateful to the Foundation for Assistance to Small Innovative Enterprises.

References
[1] Zhang J Z and Noguez C 2008 *Plasmonics* **3** 127–50
[2] Fan Z, Huang X, Tan C and Zhang H 2015 *Chem. Sci.* **6** 95–111
[3] Zeng S, Baillargeat D, Hod H-P and Yong K-T 2014 *Chem. Soc. Rev.* **43** 3426–52
[4] Brus L 2008 *Acc. Chem. Res.* **41** 1742–49
[5] Chen Y and Cheng W 2012 *WIREs Nanomed. Nanobi.* **4** 587–604
[6] Mayer K M and Hafner J H 2011 *Chem. Rev.* **111** 3828–57
[7] Radziuk D and Moehwald H 2015 *Phys. Chem. Chem. Phys.* **17** 21072–93
[8] Heisler F, Babich E, Scherbak S, Chervinsky S, Hasan M, Samusev A and Lipovskii A 2015 *J. Phys. Chem.* C **119** 26692–97
[9] Babich E S, Scherbak S A, Redkov A V, Kamenskii A N, Reduto I V and Lipovskii A A 2017 *J. Nanophoton.* **11** 032503
[10] Reduto I V, Chervinskii S D, Kamenskii A N, Karpov D V and Lipovskii A A 2016 *Tech. Phys. Lett.* **42** 93–5
[11] "Microscope slides," www.menzel.de (1 April 2017).
[12] Redkov A V, Zhurikhina V V and Lipovskii A A 2013 *J. Non-Cryst. Solid.* **376** 152–7
[13] www.origin.com (10 April 2017)

Localized topological states in Bragg multihelicoidal fibers with a twist defect in the presence of a spacer

C N Alexeyev, M C Alexeyeva, B P Lapin, D V Vikulin and M A Yavorsky
V.I. Vernadsky Crimean Federal University, Simferopol 295007, Russia

Abstract. We have studied the influence of a spacer in a multihelicoidal Bragg fiber with a twist defect on the emerging of localized topological states. We have shown that if such a fiber is excited with a Gaussian beam this leads to the appearance of a defect-localized mode, whose topological charge coincides with the order of rotational symmetry of the fiber's refractive index distribution. The influence of the spacer on this mode is studied.

1. Introduction

As has been established periodic dielectric structures can nestle the localized states near areas of their imperfections [1]. Inducing imperfections in photonic band-gap structures enables a number of useful applications [2-7]. For band-gap photonic structures the presence of a single defect results in the emergence of a localized defect mode, the spectral line of which is positioned within the forbidden spectral band [8].

The way of embedding a single defect into a periodic dielectric structure depends on its symmetry. If the structure possesses a rotational degree of freedom then one can insert the defect into it by twisting the first part of the structure with respect to the second. It was demonstrated that localized states exist near twist defects in cholesteric polymer films in the optical range. However, it was shown only for the waves with no wavefront dislocations [9,10]. To the best of our knowledge, the first study concerning the case of the waves with wavefront dislocations or optical vortices (OVs) [11] was carried out by Lobanov et al. [12]. It has been shown that evolution of the incident OV with unit topological charge within disordered array is accompanied by flipping of the topological charge between the values ± 1. As opposed to this study, the influence of a single defect on the evolution of the input beam was considered in the multihelicoidal fibers (MHF), which present one-dimensional photonic band-gap structures, by Alexeyev et al. It has been demonstrated that introducing the twist defect into the Bragg MHF results in the localization of OVs, whose topological charges coincide with the order of the rotational symmetry of the MHF [13]. The intensity of the localized fields in the vicinity of the twist defect turns out to be much greater than the one of the incident beam. Subsequently, it has been shown that combining the twist defect with the pitch jump allows one to control the intensity of the localized states in a wide range [14].

In this paper we make a further step by considering the MHF with the twist defect in the presence of a spacer. We show that exciting such defected fiber with the circularly polarized (CP) Gaussian beam (GB) results in the appearance of defect-nested localized states, which can be controlled by adjusting the thickness of the spacer.

2. The model of the multihelicoidal fiber

The refractive index distribution of a MHF with a twist defect and the spacer is given by [13-19]:

$$n^2(r,\varphi) \approx \begin{cases} n_{co}^2\left(1-2\Delta f(r)\right)-2n_{co}^2\Delta\delta r f_r'\cos l(\varphi-qz), & -d_1/2 \le z \le 0 \\ n_{co}^2[1-2\Delta f(r)], & 0<z<\delta d \\ n_{co}^2\left(1-2\Delta f(r)\right)-2n_{co}^2\Delta\delta r f_r'\cos l(\varphi-qz-\theta), & \delta d < z \le \delta d + d_1/2 \end{cases} \quad (1)$$

where Δ is the height of the profile f, $\delta << 1$ is the dimensionless parameter of the cross-section's deformation, n_{co} is the core's refractive index, $q = 2\pi / H$ is the modulus of reciprocal-lattice vector of the helical grating, H is the pitch of the grating, l is the order of the rotational symmetry of the MHF parts. (r,φ,z) are the cylindrical-polar coordinates. The twist defect is introduced as shown in Figure 1. The spacer between the MHF parts is formed by an ideal fiber.

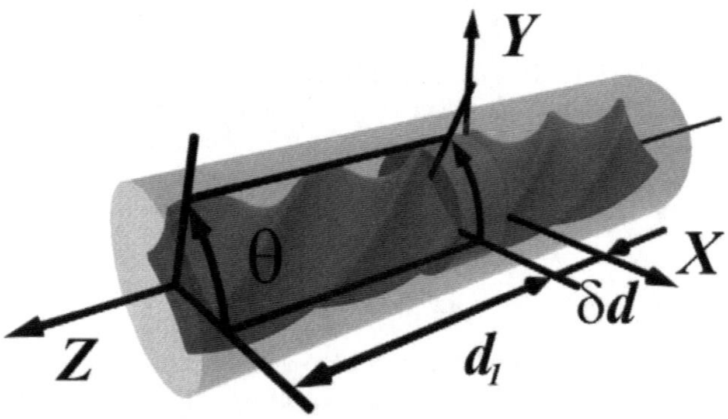

Figure 1. Geometry of the model of a MHF with the twist and a spacer. A MHF of length $2d_1$ (here with a quadrangular type of symmetry $\ell = 4$) is cut in the middle across the XY-plane. Then its marked half is rotated through an angle θ with respect to the other part and moved along z on δd. There is an ideal fiber of length δd inserted between the cut parts.

3. Localized state in the defected multihelicoidal fiber

To study the evolution of the input field one has to decompose it in the modes of a Bragg MHF [14] and ideal fiber modes in the corresponding parts of the system with an appropriate matching the fields and their derivatives with respect to z on the boundaries. This allows one to obtain the system in unknown decomposition coefficients, which can be used while restoring the field's expressions within the defected MHF. Further we will focus our attention on the case where the fiber is excited with the CP GB, which may be approximated near the input end by the fundamental mode $|1,0\rangle$. In the following we set in numerical simulations $l = 4$ and $\theta = \pi / 4$.

As numerical simulations show, there are three localized states within the defected MHF: $|1,0\rangle$ and $|1,\pm l\rangle$, where $|1,l\rangle = F_l(r)\exp(il\varphi)\cdot \mathrm{col}(1,i)$ and $F_l(r)$ satisfies the standard equation. Figure 2a shows relative averaged energy density distribution within the system in study at the wavelength $\lambda = 632.8$ nm. The averaging is carried out over the fast spatial oscillations caused by interference effects. The intensity of each of the modes in the MHF parts falls down exponentially with the distance from the spacer. At the same time, the averaged energy density distribution within the spacer remains constant. As is seen from Figure 2b, the energy of the localized fields stored within the MFH essentially depends on the length of the spacer and is very sensitive to its variations even on the wavelength scale. As the length of the MHF part grows the effect of the field's localization decreases (Figure 3a). The energy of the localized modes also decreases (Figure 3b). Such behavior of the system is connected with the fact that the influence of the single defect on the system's ability to localize fields lessens as the total length of the MHF part grows.

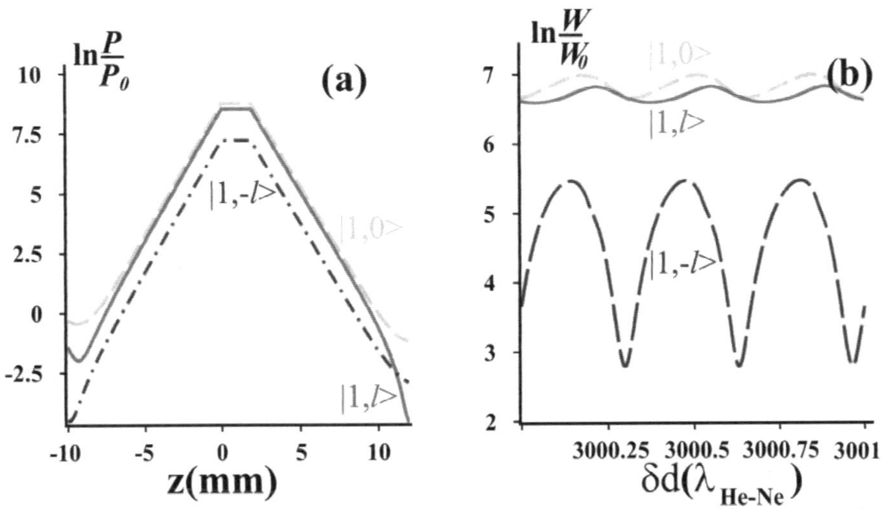

Figure 2. a) Logarithm of the relative energy density distribution within the fiber averaged over fast spatial oscillations vs position z within the fiber, the spacer thickness $\delta d = 3000\lambda_{He-Ne}$; b) logarithm of the relative stored in fiber energy of the localized fields vs the spacer's thickness δd. The type of the field is indicated near the curve, incident beam is $|1,0\rangle$ at the wavelength $\lambda = 632.8$ nm. Fiber parameters: $q = 7.436 \cdot 10^6\ m^{-1}$, $n_{co} = 1.5$, $\Delta = 5 \cdot 10^{-3}$, $\delta = 0.05$, $d_1 = 1\ cm$. Here and throughout energy density P and spectral energy density W are normalized to energy density P_0 and spectral energy density W_0 of the incident GB, respectively.

The stored in the fiber energy also depends on the spacer's thickness and tends to zero as it grows (Figure 4). As is seen from Figure 4 the energy reaches the maximum value at the certain spacer's thickness.

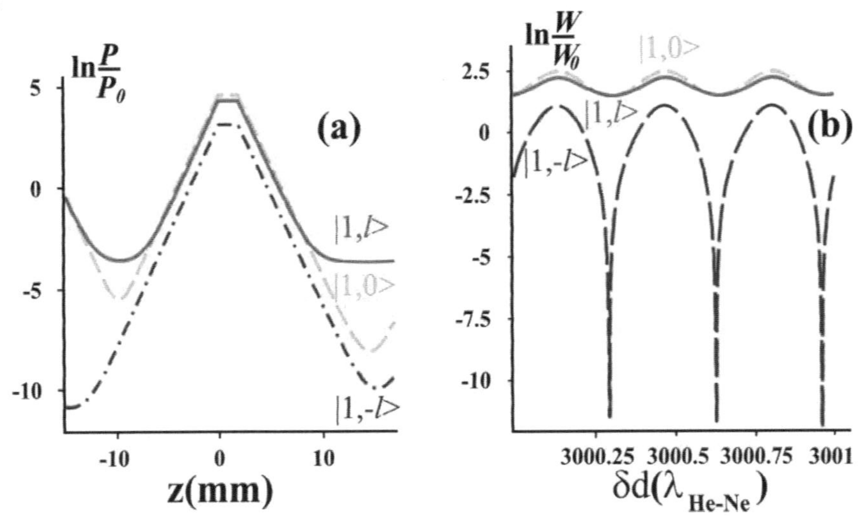

Figure 3. a) Logarithm of the relative energy density distribution within the fiber averaged over fast spatial oscillations vs position z within the fiber, the spacer thickness $\delta d = 3000\lambda_{He-Ne}$; b) stored in the fiber energy of the localized fields vs the spacer's thickness δd. The type of the field is indicated near the curve, incident beam is $|1,0\rangle$ at the wavelength $\lambda = 632.8$ nm. Fiber parameters are the same as in Figure 2 except $d_1 = 1.5\ cm$.

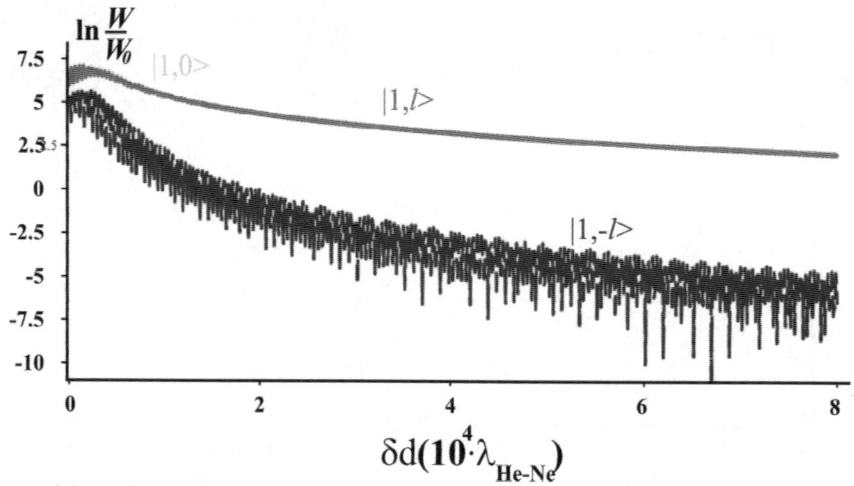

Figure 4. Logarithm of the stored in the fiber energy of the localized fields vs spacer's thickness δd. The type of the field is indicated near the curve, incident beam is $|1,0\rangle$. Fiber parameters are the same as for Figure 2.

4. Conclusion
In conclusion, we have studied influence of the spacer on the emerging of the localized mode. We have shown that the presence of the spacer results in the extension of the localization's zone, but as its thickness grows the effect of localization vanishes.

Acknowledgement
The authors acknowledge the support of the Russian Foundation for Basic Research (Grant No. 16-07-00759 A).

References
[1] Yablonovitch E, 1987 *Phys. Rev. Lett.* **58** 2059.
[2] Amet J, Baida F, Burr G, and Bernal M, 2008, *Photon. and Nanostr. – Fund. and Appl.* **6** 47-59.
[3] Xu Z, Wang J, He Q, Cao L, Su P, and Jin G, 2005 *Opt. Expr.* **13** 5608-13.
[4] Inoue K, Sugimoto Y, Ikeda N, Tanaka Y, Asakawa K, Sasaki H, and Ishida K, 2004 *Jpn. J. Appl. Phys.* **43** L446-48.
[5] Anderson S P, Fauchet P M, 2010 *Proc. of SPIE* **7606** 76060R.
[6] Li H, Phillips D, Wang X, Ho D., et al., 2015 *Optica* **2** 547-552.
[7] Song M, Ha N, Amemiya K, Park B, et al., 2006 *Adv. mat.* **18** 193–7.
[8] Yablonovitch E, 1993 *J. Opt. Soc. Am. B* **10** 283.
[9] Schmidtke J, Stille W, and Finkelmann H, 2003 *Phys. Rev. Lett.* **90** 083902.
[10] Schmidtke J, and Stille W, 2003 *Eur. Phys. J. E* **12** 553–564.
[11] Nye J F, and Berry M V, 1974 *Proc. Roy. Soc. (London) A* **336** 165.
[12] Lobanov V, Kartashov Y, Vysloukh V, and Torner L., 2013 *Phys. Rev. A* **88**, 053829.
[13] Alexeyev C, Lapin B, Milione G, and Yavorsky M, 2016 *Phys. Rev. A* **93** 063829.
[14] Alexeyev C, Lapin B, and Yavorsky M, 2017 *J. Opt.* **19** 045604.
[15] Alexeyev C, Lapin B, and Yavorsky M, 2016 *Opt. Let.* **41** 962.
[16] Alexeyev C, Lapin B, Milione G, and Yavorsky M, 2015 *Phys. Rev. A* **92** 033809.
[17] Alexeyev C, Alexeyev A, Lapin B, Milione G, and Yavorsky M, 2013 *Phys. Rev. A* **88** 063814.
[18] Alexeyev C, Fridman Yu, Lapin B, and Yavorsky M, 2013 *J. Opt.* **15** 125401.
[19] Alexeyev C, Alexeyev A, Fadeyeva T, Lapin B, and Yavorsky M, 2011, *J. Opt.* **13** 095701.

Security conditions for sub-carrier wave quantum key distribution protocol in errorless channel

A Gaidash, A Kozubov, V Chistyakov, G Miroshnichenko, V Egorov, A Gleim

ITMO University, Saint-Petersburg 197101, Russia

Abstract. Article deals with security criterion for sub-carrier wave quantum key distribution (SCW QKD) protocol in errorless channel. SCW QKD is promising type of QKD system in a field of telecommunications due to its compatibility and very efficient using of channel spectral bandwidth. Corresponding expression of security condition was derived for the first time in this work.

1. Introduction

Quantum key distribution (QKD) system allows share secure symmetric key between two users by transmitting single photons encoded in non-orthogonal basis sets [1]. According to the laws of quantum physics, it is impossible for a third party to measure these states without disturbing them and introducing errors, so that legitimate users of such systems cannot share key due to the reduction of its length to zero when error rates reaches critical value during post-processing [2].

One of the practical approaches to the implementation of QKD, with particular potential for optical network implementation, is based on subcarrier wave generation (SCW) [3]. Its main feature lies in a method of single photon generation in which the signal photons are not emitted directly by a source but are generated on subcarrier frequencies, or sidebands, as a result of phase modulation of a classical field on the central frequency, or carrier wave. Advantages of this type of QKD system include a simplification of phase shift matching in the transmitter (Alice) and receiver (Bob) modules with absence of complicated distributed interferometry schemes. Perhaps one of the most valuable features of SCW QKD systems is the exceptionally efficient use of the quantum channel bandwidth (up to 40%, which is one order higher than in other QKD systems) and the ready capability of signal multiplexing of added subcarriers on the same carrier source.

To the best of our knowledge there still no strict mathematical security proof for such a promising protocol. Here we present the first attempt to do so proposing security estimations in errorless channel with losses.

2. Mathematical model

Let us consider firstly the principal scheme of the most danger collective zero-error quantum memory assisted beam-splitting attack as you can see in figure 1. Alice connected to the Bob through optical fiber while illegitimate user (Eve) replaces their channel with ideal one and uses adjustable beam splitter to fake initial loss in the ideal channel. Thus Eve collects the entire "lost" signal. Then she stores all the quantum states in the quantum memory carrying out measurements after reconciliation.

In order to derive expression of the secure key fraction we should consider transmission capacity for Alice-Eve (AE) channel. Therefore, we should consider initial state u prepared by Alice at first modulator as following, however without central mode since it does not carry any phase-coded information:

$$|u(\varphi_A)\rangle = \prod_{n\neq 0} \left|\alpha \cdot \exp(-i\varphi_A n) \cdot d^S_{0,n}(\beta)\right\rangle \quad (1)$$

where φ_A is Alice's phase choice, α is coherent state, $d^S_{n,k}(\beta)$ is d-Wigner matrix, S is number of interacting modes in modulator, β is coefficient related to modulation index.

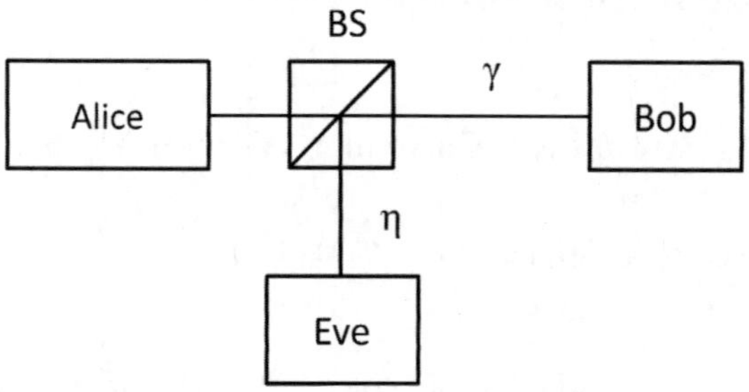

Figure 1. Principal scheme describes Eve's strategy of replacing initial channel with ideal one in a way that she collects the entire "lost" signal while Bob receive expected part of signal. Here η denotes "losses" modeled by Eve, γ denotes "transmissivity", BS denotes adjustable beam-splitter.

Here we use quantum theory of mode-bounded EOM [4] due to useful properties of d-Wigner matrix in later calculation. Next useful expression was derived for overlapping of states with different phases (0 and π) as following for AE channels:

$$\psi = \langle u(0)|u(\pi)\rangle = \langle u(\pi)|u(0)\rangle = \exp\left(-|\alpha|^2 \cdot \eta \cdot \left[1 - d^S_{0,0}(2\beta)\right]\right). \quad (2)$$

We should note that whether there is two or four state protocol Eve only needs to distinguish states within one basis (with relative phases 0 and π) since she operates with them after reconciliation. Hence one can derive Holevo bound for two state SCW QKD protocol in arbitrary errorless channel as follows:

$$X = 1 - \frac{1+\psi}{2} log_2(1+\psi) - \frac{1-\psi}{2} log_2(1-\psi) \quad (3)$$

Secure key can be generated in case when secret key fraction r is positive as follows:

$$r = 1 - X = \frac{1+\psi}{2} log_2(1+\psi) + \frac{1-\psi}{2} log_2(1-\psi) > 0 \quad (4)$$

3. Conclusion

This work dedicated to strict mathematical proof of SCW QKD two states protocol security. The main security condition proposed here for the first time taking into consideration errorless channel and quantum theory of mode-bounded EOM. However, in order to complete security proof one should for instance consider origins of error appearance together with its correction process in the Alice-Bob channel; it will be covered in later papers.

Acknowledgments

This work was financially supported by Government of Russian Federation, Grant 074-U01 and by the Ministry of Education and Science of Russian Federation (projects № 14.578.21.0112 and № 02.G25.31.0229).

References

[1] Bennett C, Brassard G 1984 *Proceedings of IEEE International Conference on Computers, Systems and Signal Processing* 175
[2] Wootters W, Zurek W 1982 *Nature* **299** 802
[3] Gleim A, et al. 2016 *Optics express* **24** 2619
[4] Miroshnichenko G, Gleim A 2015 *arXiv preprint* 1504.01632.

The use of electrochemical deposition of metals at the surface microstructuring by laser ablation in liquids

V S Kazakevich[1], **P V Kazakevich**[1], **P S Yaresko**[1] **and D A Kamynina**[1,2]

[1] Samara branch of P.N. Lebedev Physical Institute of the Russian Academy of Sciences, Samara, Novo-sadovaya 221, 443011, Russia
[2] Samara University, Moskovskoye shosse 34, Samara, 443011, Russia

Abstract. Micron, submicron, surface-periodic and nanostructures were obtained by laser ablation in a liquid method on the titanium plate surface and the nickel foil surface. Electrochemical deposition of the nickel in the modified region of the titanium leads to the formation of conical structures 6 μm in height and filamentary nanostructures on the separated film. During deposition on the laser-modified nickel target, the separable film retains the period of the structures from which it is removed. The component titanium target coated with a nickel film was also irradiated. After separation of the nickel film, perforations with an average diameter of 1 μm and holes up to 50 μm in diameter were detected.

1. Introduction

Thin metal films are widely used in the creation of solar cells, thermocouples, photo resistors, reflective surfaces and many other devices [1]. The most commonly used methods for their production are: chemical or plasmochemical deposition from the gas phase and vacuum spraying. Modification of the thin film surface allows to expand the properties of the material and, as a result, the range of its use. Plasma or laser technologies are usually used to change the surface structure.

The method of laser ablation in a liquid allows, in the presence of a single laser source, to obtain a wide range of structures. So, for example, by varying the energy density of radiation and the number of pulses arriving at the target surface, it is possible to control the dimensionality and growth rate of the structures [2]. And depending on the target material and the choice of the liquid medium in which the irradiation takes place, it is possible to create chemically pure structures or, conversely, initiate the course of chemical reactions [3].

Using laser structuring in common with the technology of thin metal films growth on the massive metal samples surface with their subsequent separation will allow to optimize the process of the creating micron structured metal films. Therefore, the aim of this work was to obtain submicron and nanostructures on the nickel film surface using a combination of laser ablation and electrochemical deposition methods.

2. Experimental technique

The creation of micron and submicron structures on the surface of metal targets was carried out by laser ablation in liquids at cryogenic and room temperature [3]. The thickness of the liquid layer above the target surface was 5 mm. The Nd: YAG laser was used as a source of laser radiation: $\lambda = 1064$ nm, $\tau = 250$ ps, $\nu = 20$ Hz, $Q = 0.3$ mJ, $Q_S = 0.18\text{-}0.6$ J / cm^2, 20000 pulses. The cell with the target remained fixed relatively to the focus spot. Laser treatment was provided on: a titanium plate 1 mm thick, a titanium plate coated with a layer of nickel 6-7 μm thick, and nickel foil 0.3 mm thick. The titanium plate and nickel foil were also coated with a layer of nickel after laser treatment.

The application of a nickel coating was carried out by an electrochemical method. Electrolyte composition was:

Table 1. Electrolyte composition.

Name	Chemical formula	Concentration, g/l
Nickel sulfate 7 aq.	$NiSO_4$+ $7H_2O$	140-200
Nickel 2-chloride 6 aq.	$NiCl_2$+$6H_2O$	30-40
Boric acid	H_3BO_3	25-40
Sodium sulfate non aq.	Na_2SO_4	60-80

Electrolysis current: 1 (± 0.1) A / dm^2; Electrolyte temperature: 45 (± 5) °C; The electrolysis time was 30 minutes.

The experiment of the surfaces structuring was divided on several parts:

- Laser ablation of a titanium plate in ethanol containing 5% water, and coating the synthesized structures with a layer of nickel 6 μm thick;
- Application of a nickel coating on the titanium plate surface and laser ablation of the resulting sample in ethanol containing 5% water;
- Laser ablation of nickel foil 0.3 mm thick in liquid argon, followed by coating of synthesized structures with a six-micron layer of nickel.

To separate the modified nickel coating with the sample to which it was applied, the outer surface of the nickel layer was covered with a glue composition and pressed to a flat metal substrate. After solidification of the glue composition, the metal substrate with the nickel film was separated from the bulk sample surface.

The analysis of the surfaces of the titanium plate, nickel foil and nickel coating at various stages of processing was carried out using a scanning electron microscope Carl Zeiss Evo 50 equipped with a nitrogen-free energy-dispersive detector X-Max 80 (EDX). To determine the period of synthesized structures, a fast Fourier transform (FFT) of SEM images was used.

3. Results and discussion

As a result of the of 20000 laser pulses impact on the surface of a titanium plate in an ethanol medium containing 5% water, conical structures with a characteristic diameter of 6 μm in the central region and 200 nm in the periphery were obtained (figure 1).

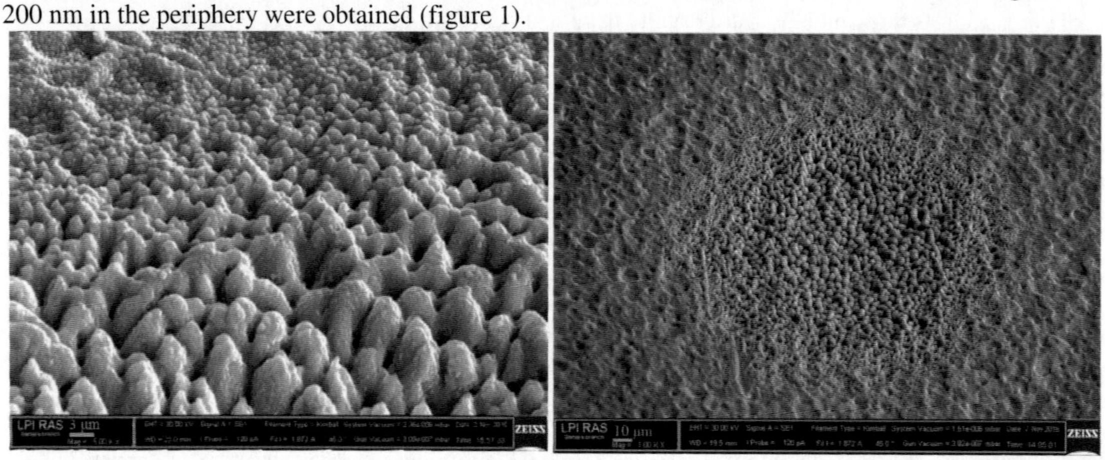

Figure 1. SEM images of structures on the titanium plate surface, obtained by laser ablation in ethanol, containing 5% H_2O, method. 20000 pulses, 0.18 J / cm^2.

The change in the size of the structures along the radius of the focusing spot is associated with the Gaussian distribution of the laser beam energy density. The formation of such structures under a multi-pulse exposure is due to uneven melting of the target surface, removal of the melt into the medium under the influence of a number of instabilities (Kelvin-Helmholtz or Rayleigh-Taylor) and further "dagger penetration" [4]. Uneven melting can be associated with various factors, including the presence of both the original oxide layer on the irradiated sample and the layer formed under laser heating conditions of the metal target in the oxidizing medium, which change the absorbing capacity of the target. Also, upon absorption of the laser radiation by the target, its thermal expansion leads to cracking of the surface oxide layer. This cracking during the irradiation plays the mask role, since the melting point of the oxide is higher than the melting point of pure metal [3]. Since the present work carried out a multi-pulse laser action with an energy density close to the ablation threshold, the removal of matter from the surface and the oxidation of the surface were rival processes. As a result, formation of an oxide layer was observed on the laser-induced structures surface (figure 1).

The next stage consisted in electrochemical deposition of nickel on the structures described above. The growth of a nickel film on the titanium sample surface occurred in the "island" mode [5]. Analysis of the film after separation from the substrate showed that its total thickness was 7 μm (figure 2a).

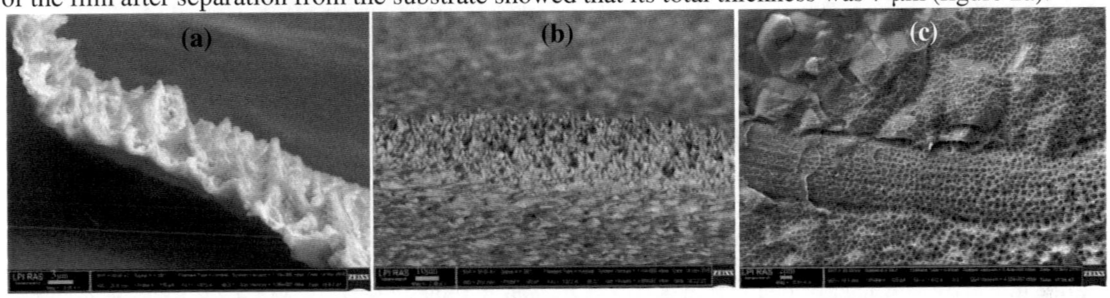

Figure 2. SEM images of a nickel film separated from a laser-structured titanium plate: **(a)** side view. Scale bar is 3 microns; **(b)** general view at an angle of 45°. Scale bar is 10 microns; **(c)** the region of nickel deposition at the periphery of the focusing spot. Scale bar is 2 microns.

As a result of the electrochemical deposition of nickel in the central region of the modified titanium, the formation of nickel microstructures with a height of 6 μm, located on the basis of a thickness of 1 μm, was occurred (figure 2b). During precipitation of the nickel on the modified titanium periphery, a nickel film 7 μm thick was obtained, which repeated the relief of the original irradiated titanium surface and covered by filamentary nickel nanostructures (figure 2c). These structures are formed by the deposition of nickel in the cracks of the oxide layer on the surface of the irradiated titanium sample. Submicron nickel cells with nano inclusions were also found at the periphery (figure 3). Energy dispersive analysis of inclusions showed that they represent titanium dioxide.

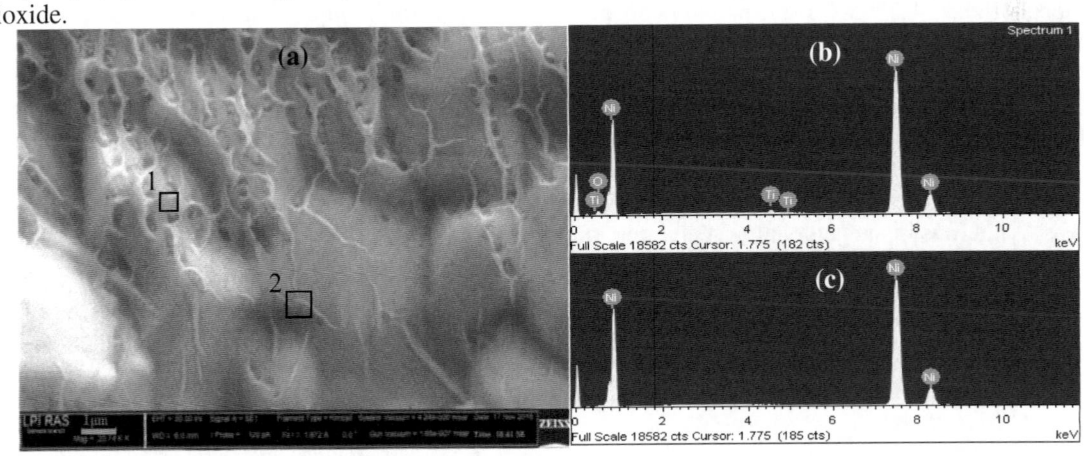

Figure 3. (a) SEM-image of the nickel film surface, formed at the periphery of irradiated titanium target after separation from titanium; **(b)** energy dispersive analysis of the nickel film surface in region 1; **(c)** energy dispersive analysis of the nickel film surface in region 2.

It's can be explained by the fact that during the separation of the nickel layer from the laser-induced structures, covered by an oxide, a part of the oxide layer remained on the film surface.

In order to exclude submission of target material, on which the nickel layer is deposited, into the structure of the synthesized nickel films, a 0.3 mm thick nickel foil was chosen as the mask. To minimize chemical reactions the process of creating laser-induced structures on the target surface was carried out in a liquid argon medium. After exposing 20000 laser pulses on the nickel foil surface in a liquid argon medium, elongated micron structures modulated by surface periodic structures (SPS) were obtained (figure 4a).

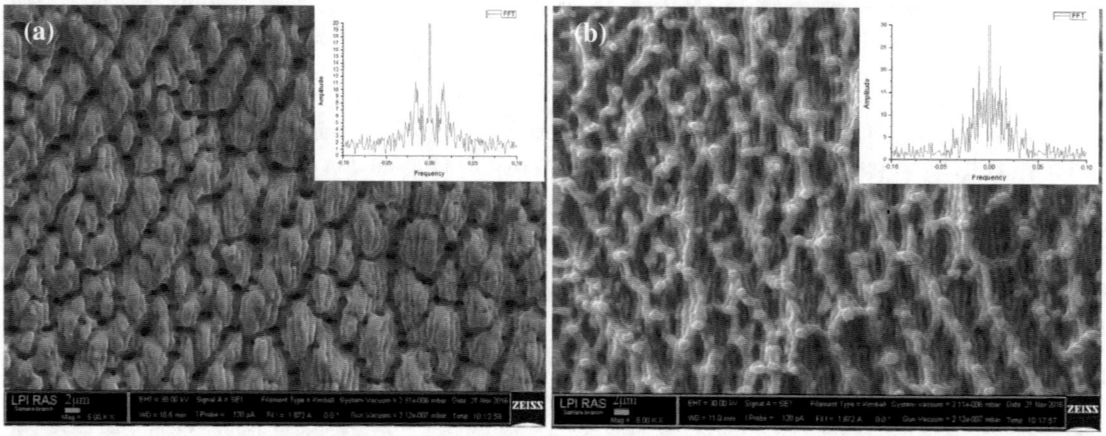

Figure 4. **(a)** SEM-image of characteristic structures on the nickel surface, obtained by laser ablation in liquid argon. 0.2 J / cm^2, 20000 pulses; **(b)** SEM-image of the nickel film surface, formed on an irradiated nickel target, after separation. The inserts show fast Fourier transforms of SEM images taken along a horizontal straight line.

The formation of such structures may be due to the fact that the real surfaces of the materials are not absolutely smooth and the presence of even a slight relief and micro roughness can radically change the character of interaction and absorption of laser radiation by the matter. When an electromagnetic wave falls on a rough surface due to diffraction, surface electromagnetic waves (SEW) arise that propagate along the interface of two media and exist simultaneously in both of them. Interference of the SEW with the falling, reflected and refracted waves determines the nature of the electromagnetic field at the surface and its dissipation (absorption) [6]. The nickel film deposited on these structures and separated from them completely repeats the period of structures on the massive nickel (figure 4b). Two main periods can be distinguished: the period of surface structures, which amounted to 2.7 μm, and the period of modulating SPS, equal to 0.6 μm.

A separate part of the work is irradiation of a thin layer of nickel deposited on a massive metal substrate, in order to its effectively perforation without destroying the film. Irradiation of a titanium target coated with a nickel film leads to the formation of nickel structures, the general view of which is shown in figure 5a. When the average energy density in the laser beam reaches 0.26 J / cm^2, the penetration channels in the nickel layer reach the surface of the titanium and forming a hole with a few tens of microns in diameter in the center and micron perforations at the periphery (figure 5b).

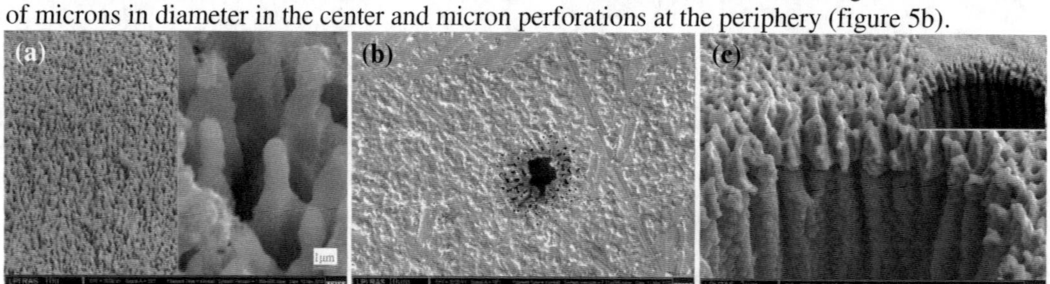

Figure 5. SEM-image of **(a)** nickel structures obtained by laser ablation of a titanium target, coated with a nickel layer, in an ethanol medium; **(b)** nickel film with perforations after separation from the titanium plate. 0.26 J / cm^2, 20000 pulses; **(c)** a hole obtained by laser ablation of a composite sample of titanium, coated with nickel, in ethanol medium. 0.6 J / cm^2, 20000 pulses. The insert show the elemental map.

An increase the energy density up to 0.6 J / cm^2 leads to the formation of a channel in a nickel film and a titanium plate (figure 5c). In addition, on the surface of titanium, in contrast to nickel, structures that are a cracked oxide layer are observed. The distribution map by the elements, presented in the insert, shown that when the composite target is ablated in the liquid, titanium is not deposited on the nickel layer. This can be due to the high liquid and the target vapor pressure in the channel.

4. Conclusion

The combination of methods of the laser ablation of metals in liquids and electrochemical deposition of metals leads to the production of structured nickel films 6-7 microns in thickness. In the case of using a laser-modified titanium plate as a mask on a nickel film, micron cone-shaped nickel structures with an average height of 6 μm in the center and nanodimensional filamentary nickel structures at the periphery were obtained. Also at the periphery, submicron nickel cells with nanoinclusions of titanium dioxide were detected. Nano-layers of titanium dioxide are interest for research, because on them the first memristive element was implemented [7].

Using a structured nickel mask on a nickel film, the formation of micron surface structures modulated by submicron PPS was observed. Metallic masks obtained by the laser ablation method can be repeatedly used to create structures of a given shape and size on thin metal films, which expands the area of this technique use.

By laser ablation of a component target consisting of a 1 mm thick titanium plate and a 6-7 μm thick nickel coating micron titanium and nickel structures in an ethyl alcohol medium were produced. At an average energy density of 0.26 J / cm^2 on a nickel film, a hole a few tens of microns in diameter at the center and micron perforations at the periphery were formed. By modulating the laser field, it is possible to achieve uniform perforation of the nickel film throughout the focusing spot.

References

[1] Konozenko I D 1954 *J. Physics-Uspekhi* **102** 561-602
[2] Barmina E V, Stratakis E, Fotakis C and Shafeev G A 2010 *J. Quantum Electronics* **40** 1012-1020
[3] Kazakevich V S, Kazakevich P V, Yaresko P S and Kamynina D A 2016 *J. IOP conference series* **741** 012188
[4] Bunkin F V and Tribelsky M I 1980 *J. Physics-Uspekhi* **130** 193-239
[5] Bales G S and Chrzan D C 1994 *J. Phys. Rev.* **50** 6057
[6] Ahmanov S A, Emelyanov V I, Koroteev N I and Seminogov V N 1985 *Physics-Uspekhi* **147** 675-745
[7] Strukov D B, Snider G S, Stewart D R and Williams R S 2008 *Nature* **453** 80-83

IOP Conf. Series: Journal of Physics: Conf. Series **917** (2017) 062016 doi:10.1088/1742-6596/917/6/062016

High-brightness single-crystal approach in quantum imaging with undetected photons

D S Tabakaev[1]
[1]Department of Photonics and optoinformatics, ITMO University, St. Petersburg 197101, Russia

Abstract. Quantum imaging with undetected photons is recently appeared promising brunch of quantum optics. Due to properties of photon pairs (signal and idler), generated via spontaneous down-conversion (SPDC) process and phenomenon of induced coherence, one can obtain phase or intensity image of object without detecting of the photon, interacting with it. The theoretical treatment and limitations of such a system with an only crystal and increased brightness are given.

1. Introduction
Quantum imaging with undetected photons [1] arise from quantum interference experiment by Zou, Wang and Mandel [2,3] and their treatment of induced coherence without induced emission as an appearance of sine fringes in coincidence and single counting rates. Lemos et.al. proposed that image of the object could be subtracted from these fringes due to entanglement of SPDC-photon pairs (phase image) and the very idea of quantum interference – indistinguishability as its precondition [4]. This type of systems well could be implemented in spectroscopy under Abbe limit, biophysics and sensing. To improve this system further, higher signal-to-noise ratio (SNR) shorter image-acquisition time are demanded that explicitly means demand of sources with higher brightness [5].

2. Principle of proposed experiment
To provide higher brightness and, consequently, higher SNR and also make whole future system more cost-efficient, nonlinear crystal is placed in cavity, with mirrors M2-M3, aligned to reflect only idler mode (Fig.1). Idler photon propagates through the object and gains some phase information about it, then it reflects back to crystal and be means of the mirror M1, reflecting pumping photons, and the phenomenon of induced coherence and phase memory [2, 3], the new idler photon's phase is "locked" by inducing field. Due to this fact and alignment of M4, reflecting signal mode, one can obtain sine fringes in the count rate of detector D [3]. This experimental arrangement is topologically similar to an experiment [6], and here its quantum imaging application feasibility was investigated, hence, mathematical treatment has common points.
Substantial increase of brightness is inextricably linked with presence of the mirrors M2-M3, that not only multiply the number of idler photons, inducing coherence, but also provide selection of vacuum states, which play essential role in probability of down-conversion and act like a "bridge" between photons in pair through of entanglement.

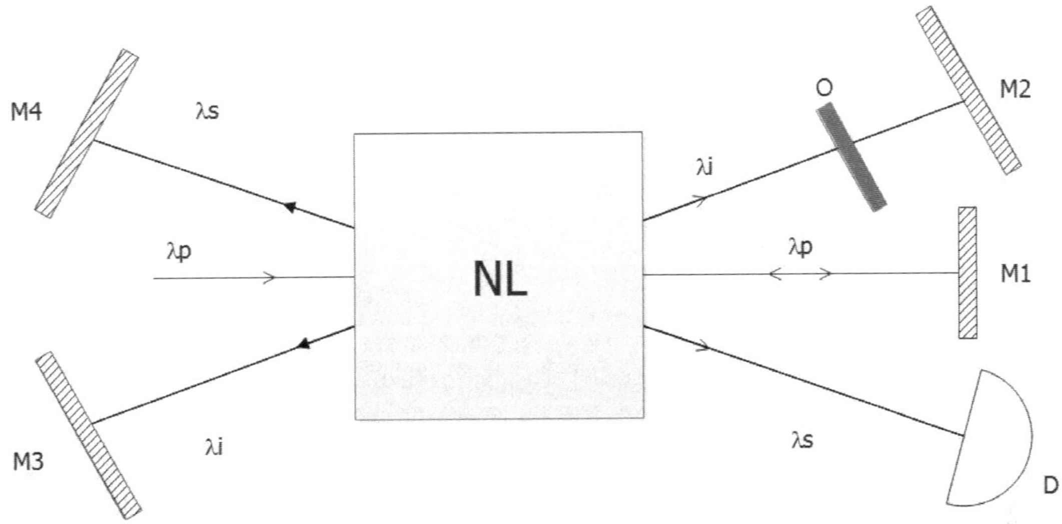

Figure 1. *Schematic of proposed experiment: NL – nonlinear crystal, M1-M4 – mirrors, O -object, D – single-photon detector, $\lambda_{p,i,s}$ – pump, idler, signal photons.*

3. SPDC state and imaging

The process of SPDC and imaging was treated akin developed by Hong and Mandel one [7] and Lahiri et.al [4]: nonlinearity of non-centrosymmetric crystal leads to conversion of pump photon into correlated pair of photons with sum energy equal to the energy of pump photon (phasematching condition

$$k_p = k_s + k_i \qquad (1)$$
$$\omega_p = \omega_s + \omega_i$$

where $k_{p,s,i}$ and $\omega_{p,s,i}$ are pump, signal and idler wave vectors and frequencies respectively. Using approximation of strong enough classical pump field, SPDC-interaction Hamiltonian is

$$\mathrm{H}_{in}(t) = \int_L d^3r\, \chi_{lmq} \mathrm{E}_{pl} \mathrm{E}_{sm}^{(-)} \mathrm{E}_{iq}^{(-)} + H.c., \qquad (2),$$

where χ_{lmq} is nonlinear susceptibility tensor, L – crystal volume, E_p – pump field, $\mathrm{E}_{s,i}$ - negative-frequency parts of quantized signal and idler electric fields, respectively, lmq - mutually orthogonal directions in space. Positive-frequency part of the field, obtained via second quantization formalism, is in general has the form

$$\mathrm{E}_{i,s}^{(+)}(r,t) = \left(\mathrm{E}_{i,s}^{(-)}\right)^\dagger = \sum_{k,\sigma} \mathrm{A}(k,\sigma) \exp\left(\mathrm{i}\left(k_{i,s} r - \omega_{i,s} t\right)\right) a_{i,s}(k_{i,s},\sigma) \qquad (3),$$

where i,s subscripts stand for idler signal photons and A - for all constants "merged" together, σ=1,2 is possible polarization direction, ω – frequency, k – wave vector, $a(k, \sigma)$ – annihilation operator. As far as pump field is classical, then

$$\mathrm{E}_p = \sum_{k,\sigma} \mathrm{V}_p(k,\sigma) \exp\left(\mathrm{i}\left(k_p r - \omega_p t\right)\right) \qquad (4),$$

where V_p stands for complex amplitude of pumping wave. The quantum state of the SPDC-field is described by well-known equation

$$|\psi(t')\rangle = \exp\left(-\frac{i}{\hbar}\left(\int_0^t dt' \mathrm{H}_{in}(t')\right)\right)|0\rangle_i |0\rangle_s \qquad (5).$$

Using (2)-(5), expanding exponential and neglecting second- and higher-order terms due to small amplitude of SPDC and considering spatially well-defined single mode case and propagation directions, we arrive to

$$|\psi\rangle = |0\rangle_{s1}|0\rangle_{i1} + \alpha|1\rangle_{s1}|1\rangle_{i1} \qquad (6),$$

where α stands for SPDC amplitude ($|\alpha| \ll 1$). As far as down-converted modes have strict angle intensity dependence, only mentioned modes will be considered further.

The key point in proposed experiment is to align idler modes $i_{1,2}$ where subscripts stand for idlers generated due to down-conversion of pump field, propagating from left to right and retroflected back (Fig. 1), respectively. Meanwhile, the object (Fig. 1) can well be treated using quantum optical representation of beam splitter with transmission and reflection coefficients T and R (obviously, $|T|^2+|R|^2=1$), respectively. So, annihilation operator for idler photon, generated by down-conversion of retroflected photon after alignment is

$$a_{i2} = (\mathrm{T}a_{i1} + \mathrm{R}a_0)\exp(i\theta_i) \qquad (7),$$

where a_{i1} is annihilation operator for idler photon, generated by down-conversion of pump photon, propagating from left to right, θ_{i1} is the phase that i_1 photon obtained while propagating through the object to mirror and back to crystal and a_0 stands for vacuum field on unused port of object-beam splitter.

Using (5)-(7), again, neglecting terms of order higher than one can find that final SPDC state is

$$|\psi(t)\rangle = |0\rangle_{s_1,i_1,s_2,0} + \alpha_1 V_p|1\rangle_{s_1}|1\rangle_i|0\rangle_{s_2} + \alpha_2 V_p e^{-i\theta_i}(\mathrm{T}^*|0\rangle_{s_1}|1\rangle_i|1\rangle_{s_2}|0\rangle_0 + \mathrm{R}^*|0\rangle_{s_1}|0\rangle_i|1\rangle_{s_2}|1\rangle_0)$$

(8).

As far as crystal is the same, $\alpha_1 = \alpha_2$. Count rate of detector D (i.e. intensity of one mode using perfect-detector approximation) is

$$\mathrm{R}_s = \left\langle \psi_s \middle| \mathrm{E}_s^{(-)}\mathrm{E}_s^{(+)} \middle| \psi_s \right\rangle \propto 2|\alpha_1 V_p|^2|\mathrm{T}|(1 + \cos(\theta_i + \theta_{s1} + \theta_{s2} - \theta_p)) \qquad (9),$$

where $\theta_{s1,2}$ is phase shift of both signal photons obtained during propagation to mirror M4 and to detector D, θ_p stands for phase accumulated by pump photon while propagating to mirror M1 and back. Analysing (9) one can find that counting rate exhibits interference with visibility proportional to T. All together it makes possible to obtain phase and intensity picture of the object from interference fringes without detecting of the photons, interacted with the object. Obviously, interference disappear if the object is opaque.

4. Brightness enhancement

The main advantage of quantum imaging with undetected photons counterpart - classical imaging with undetected photons, proposed by Shapiro et. al. [5] (nevertheless disadvantages, that are intrinsic for this kind of systems, such as absence of phase image acquisition possibility in principle), is higher SNR ratio and lower acquisition time. Both these features are the consequence of high brightness light sources, that are used in classical imaging systems and [5] gives clear analysis of these dependencies.

To explain increase of brightness of down-converted light source, let us analyse (9). Actually, it reflects the bosonic nature of photons and dependence of down-conversion probability on the mode population. Considering perfect alignment of mirrors M2 and M3, idler modes i_1 and i_2, could be treated as one mode with annihilation operator, described by (7), let us write the population of mode after down-conversion as

$$\mathrm{N}_s = \mathrm{F}(\mathrm{N}_i + 1) \qquad (10),$$

where $\mathrm{N}_{s,i}$ – population of signal and idler modes, $\mathrm{F} = \frac{4\pi^2}{c^2}\omega_i\omega_s|\mathrm{V}_p|^2 l^2 = \beta^2 l^2$ – efficiency of down-conversion [8], l – crystal length. It is clear from (10), that population of signal mode rises with population of idler mode. However, if one consider the case of signal and idler mode interaction while propagating in nonlinear crystal (detailed calculation could be found in [8]), then

$$\mathrm{F} = \beta^2 \frac{\mathrm{sh}^2 l\sqrt{\beta^2 - \frac{\Delta^2}{4}}}{\beta^2 - \frac{\Delta^2}{4}} \qquad (11),$$

where $\Delta = k_s + k_i - k_p$. Analysing (11) one can find the pump intensity threshold of SPDC to induced down-conversion, or superluminescence, transition that is of pivotal importance because it let us to estimate the threshold population of N'_i. Moreover, after this transition, photon pairs are no more entangled by vacuum, that makes impossible phase image acquisition. It is clear from (11) that if βl≫1, then

$$N_s = \mathrm{sh}^2\beta l \approx \frac{e^{2\beta l}}{4} \quad (12).$$

To find the threshold population of N'_i, one need to equate (10) to (12) simply

$$N_i^t = \frac{e^{2\beta l}}{4F} - 1 \quad (13).$$

Taking in a count, that N_i rises constantly due to presence of mirrors M2 and M3, the care should be taken about their proper reflection coefficient, avoiding mentioned transition to parametric oscillator regime.

Acknowledgments

This work was financially supported by Government of Russian Federation, Grant 074-U01 and by the Ministry of Education and Science of Russian Federation (project ID 14.578.21.0112 No02.G25.31.0229).

References

[1] Lemos G B, Borish C, Cole G D, Ramelow S, Lapkiewicz R, Zeilinger A 2014 *Nature* **512** 409
[2] Zou X Y, Wang L J, Mandel L 1991 *Phys. Rev. Lett.* **67** 318
[3] Wang L J, Zou X Y, Mandel L 1991 *Phys. Rev. A* **44** 4614
[4] Lahiri M, Lapkiewicz R, Lemos G B, Zeilinger A 2015 *Phys. Rev. A.* **92** 013832
[5] Shapiro J, Venkatraman D, Wong N C 2015 *Sci. Rep.* **5** 10329
[6] Herzog T J, Rarity J G, Weinfurter H, Zeilinger A 1994 *Phys. Rev. Lett.* **72** 629
[7] Hong C K, Mandel L 1985 *Phys. Rev. A* **31** 2409
[8] Klyshko D N 1980 *Photons and nonlinear optics* (Moscow: Nauka)

IOP Conf. Series: Journal of Physics: Conf. Series **917** (2017) 062017 doi:10.1088/1742-6596/917/6/062017

Fabrication of spherical GeSbTe nanoparticles by laser printing technique

M Tajik[1], D A Zuev[1], V A Milichko[1], E V Ubyivovk[2], A B Pevtsov[3], S A Yakovlev[3], M V Rybin[1,3] and S V Makarov[1]

[1]ITMO University, St. Petersburg 197101, Russia
[2]St. Petersburg State University, St. Petersburg 199034, Russia
[3]Ioffe Institute, St. Petersburg 194021, Russia

Abstract. By using the laser printing technique, for the first time we fabricate spherical nanoparticles of germanium-antimony-telluride alloy possessing a high refractive index and amorphous-to-crystalline phase transition. The nanoparticles are examined by means of optical dark field micro-spectroscopy, which shows scatterers with different colours (with different optical resonances). The direct measurements of their diameters by using the transmission electron microscopy reveal that the nanoparticles' colours are related to different sizes of the nanoparticles. While most of nanoparticles are homogeneous, our study shows that some fabricated particles have a core-shell structure.

1. Introduction

Metamaterials are artificial media having extraordinary properties that do not exist in nature. With this concept, metamaterials can control over the parameters of propagating waves in almost arbitrary way. The basic metamaterial functionality is due to resonant behaviour of its structural elements called by meta-atoms [1]. Two-dimensional counterpart of metamaterials, referred to as metasurface, which is formed by arranging meta-atoms in a plane was shown to modify polarization, phase and amplitude of light in this planar geometry [2,3]. Going from microwaves down to optical spectral range, the conventional metal-based metamaterials obtain a number of restrictions including inevitable Ohmic losses and the absent of magnetic response in visible and near infrared. All-dielectric metamaterials with a high refractive index and low losses, for example silicon at near infrared, have attracted a lot of attention in recent years [4]. Manipulation of light due to Mie resonances for spherical nanoparticles, both magnetic and electric modes, make it possible to scale the structure sizes in a wide range for operating from microwaves to optical frequencies [5]. It can open a door for using high dielectric index nanoparticles in the field of colour engineering [6,7], as well as for nonlinear harmonic generation [8,9] and nanothermometry [10].

Higher levels of metasurface functionality require the tuning or switching of their dielectric properties, especially for applications. By exploiting nonlinearities one cannot archive strong modulation of the response. Recently the phase transition between crystalline and amorphous silicon was demonstrated for switching the Mie resonance in silicon nanospheres, however, silicon have relatively weak dielectric contrast between amorphous and crystalline phases and it requires a high temperature for the transition [11]. Among others phase changing optical materials, Ge-Sb-Te (GST) alloy attracts a lot of attention in photonics because its giant dielectric contrast between amorphous and crystalline phases [12]. As this compound has a similar structure between amorphous and crystalline phases, the structural transition can be reversibly triggered very fast with extreme change in high refractive index range from n=4 to 5.5 with low losses in near infrared [13]. According to these properties, GST was used for tuneable nanoantennas and metasurfaces fabricated by lithographic techniques [14]. However the lithography is unsuitable to produce nanoantennas of spherical shape that can be described analytically by the rigorous Mie theory. Recently a laser printing technique [15,16] was shown to synthesize spherical nanoparticles of silicon with magnetic and electric Mie resonances in optical range. In this work we report the fabrication of spherical GST nanoparticles by using the laser printing technique.

2. Results and discussion

The GST nanoparticles are fabricated by the laser-induced forward transfer method [16]: a receiver glass substrate is placed on a donor glass substrate with a deposited thin film of GST (Fig. 1a). The distance between substrates is about 50 μm. The radiation of an ytterbium-doped femtosecond solid-state laser (TeMa-150, Avesta Poject, 1050 nm, repetition rate 80 MHz, pulse duration 150 fs) in a single pulse mode provided by a pulse picker system (Avesta Poject) is focused by an infinity corrected objective (10×NA=0.28 Mitutoyo Plan Apo) on a thin film-glass interface into a spot with a diameter of 4.6 μm. By exploiting laser printing we fabricate about 1000 GST nanoparticles which are studied by means of dark filed microscopy. In the dark-field images, isolated nanoparticles manifest themselves as coloured spots mostly yellow, but there are also blue, purple, red and green nanoparticles (Fig. 1b). Similar different colours in silicon nanoparticles are attributed to a wide distribution of their diameters governing the excitation of Mie resonances at certain wavelengths [15,16].

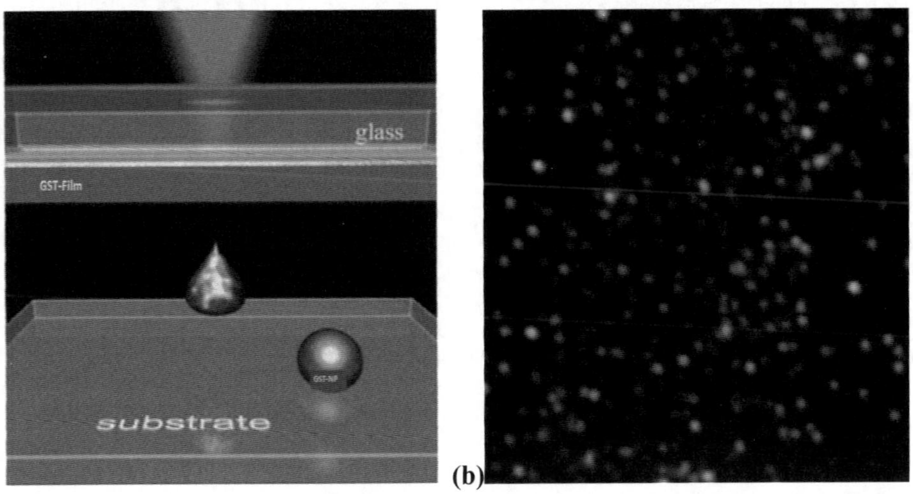

(a) (b)

Figure 1. (a) Schematic Illustration of fabricated GST nanoparticles by laser printing. (b) Dark-field image of GST nanoparticles.

To analyze the presence of different colours in the dark-filed images (Fig. 1b) we simulate scattering cross section of GST nanoparticles by exploiting CST Microwave Studio (commercially available software). The simulation yield the nanoparticle diameters have to lay at least in the interval between 60 and 200 nm. The wide diameter distribution we associate with the relatively high energy per pulse in our fabrication set-up. Indeed, the applied laser fluence is higher than 100 mJ/cm^2, which is related the regime of phase explosion induced by ultrafast overheating of the irradiated material [17, 18].

To measure diameters of GST nanoparticles directly, we use transmission electron microscopy (TEM). The similar scheme is implemented for fabrication of nanoparticles on a special TEM grid which is applied instead of the receiver glass substrate shown in Fig. 1a. Our results show that nanoparticles have nearly spherical shapes and the diameters are in the range of D=40-300 nm with the peak in its distribution at 150 nm. 15% of particles have diameter in the range 140-160 nm which is corresponding to the yellow colours being in a good agreement with the dark field images (Fig. 1b). According to the previous works, this size distribution allows for existing Mie resonances in the visible and infrared ranges [4,5,15]. Furthermore, it is possible for us to study the scattering properties of nanoparticles by using Mie theory.

Figure 2a shows a relatively homogenous phase in amorphous state. However, the TEM images for larger particles are different and they demonstrate a core-shell structure (Fig. 2b). Since the laser energy applies with a gradient profile, one can assume that the ablated particle is not always amorphous completely. The core-shell nanoparticles are defined by two diameters. Figure 2b depicts a core-shell particle having diameters: 105 nm for the core and the external diameter 293 nm for the whole particle, whereas homogeneous GST sphere in Fig. 2a has a diameter of 104 nm. One can speculate that the large particles are cooled down longer than the small particles [9]. This is the main reason why we observe the growth of crystalline grain forming the core of the larger nanoparticles. We also anticipate that after the heating by additional continuous wave irradiation the structure of core-shell particle becomes homogeneous in the crystalline phase owning to temperature-induced amorphous-to-crystalline transition.

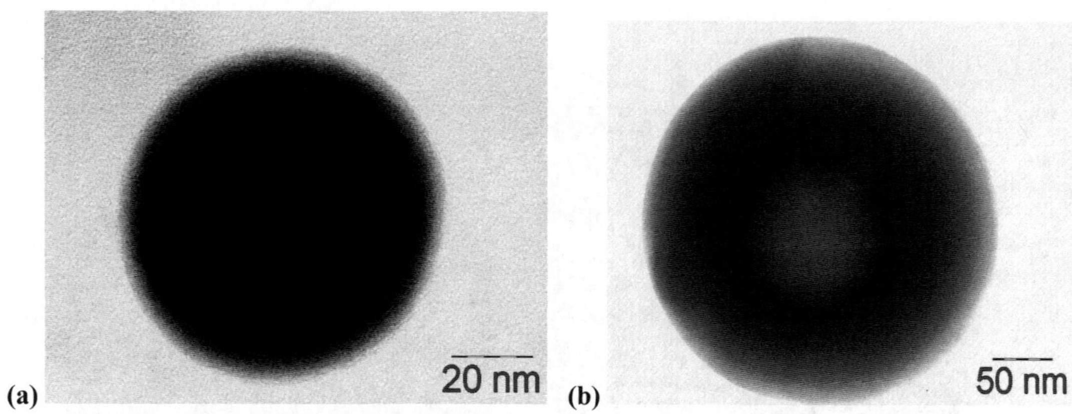

Figure 2. TEM images of homogenous GST nanoparticle of small (a) and large (b) structure of sizes.

3. Conclusion

In this study we have demonstrated novel method of fabrication of GST spherical nanoparticles by the femtosecond laser printing technique. The existence of different colours in dark field image was due to nanoparticle size distribution in our experiment. The investigation of GST nanoparticles by TEM has further confirmed the non-monodisperse size. The TEM images have revealed core-shell structure of the larger particles. However the smaller nanoparticles have exhibited homogeneous internal structure. We believe the developed laser printing technique opens a way for design of tuneable all-dielectric nanostructures and metasurfaces composed from the high refractive index GST nanoparticles. Combining the fascinating properties of GST the laser printing allows GST nanoparticles to compete with the conventional metallic nanoparticles in colour engineering, which has become an important field of research and applications in the last years.

Acknowledgments
This work was supported by the Russian Foundation for Basic Research (16-29-05317 ofi-m, 17-03-00621 a, 17-02-00538 a, 16-02-00684 a and 16-02-00461 a) and by the Ministry of Education and Science of the Russian Federation (3.1500.2017/4.6).

References

[1] Cui T J, Smith D R, Liu R 2010 *Metamaterials: Theory, Design, and Applications* (New York: Springer)
[2] Kildishev A V, Boltasseva A and Shalaev V M 2013 *Science* **339** 1232009
[3] Yu N and Capasso F 2014 *Nature Materials* **13** 139-150
[4] Jahani S and Kacob Z 2016 *Nature Nanotechnology* **11** 23
[5] Kuznetsov A I, Miroshnichenko A E, Brongersma M L, Kivshar Y S and Lukyanchuk B 2016 *Science* **354** aag2472
[6] Proust J, Bedu F, Gallas B, Ozerov I and Bonod N 2016 *ACS Nano* **10** 7761
[7] Zuev D A, Makarov S V, Mukhin I S, Milichko V A, Starikov S V, Morozov I A, Shishkin I I, Krasnok A E and Belov P A 2016 *Adv. Mater.* **28** 3087
[8] Makarov S V, Tsypkin A N, Voytova T A, Milichko V A, Mukhin I S, Yulin A V, Putilin S E, Baranov M A, Krasnok A E, Morozov I A and Belov P A 2016 *Nanoscale* **8** 17809-17814
[9] Makarov S V, Petrov M I, Zywietz U, Milichko V A, Zuev D A, Lopanitsyna N, Kuksin A, Mukhin I, Zograf G, Ubyivovk E, Smirnova D, Starikov S, Chichkov B N and Kivshar Y S 2017 *Nano Ltters* (doi: 10.1021/acs.nanolett.7b00392)
[10] Zograf G P, Petrov M I, Zuev D A, Dmitriev P A, Milichko V A, Makarov S V and Belov P A 2017 *Nano letters* (doi: 10.1021/acs.nanolett.7b00183)
[11] Birkholz M, Selle B and Fuhs W 2001 *Phys. Rev. B* **64** 085402
[12] Rios C, Stegmaier M, Hosseini P, Wang D, Schere T, Wright C D, Bhaskaran H and Pernic W H P 2015 *Nature Photonics* **9** 725-732
[13] Wang. Q, Rogres E T F, Gholipour B, Wang Ch, Yuan G, Teng J and Zheludev N I 2015 *Nature Photonics* **10** 60-65
[14] Michel A U, Zalden P, Chigrin D N, Wuttig M, Lindenberg A M and Taubner T 2014 *ACS Photon.* **1** 833
[15] Kuznetsov A I, Miroshnichenko A, Fu Y H, Zhang J and Lukyanchuk B 2012 *Sci. Rep.* **2** 492
[16] Dmitriev P A, Makarov S V, Milichko V A, Mukhin I S, Gudovskikh A S, Sitnikova A A, Samusev A K, Krasnok A E and Belov P A 2016 *Nanoscale* **8** 5043
[17] Ionin A A, Kudryashov S I, Makarov S V, Seleznev L. V and Sinitsyn D V 2014 *Applied Physics A* **117** 1757-1763
[18] Zayarny D A, Ionin A A, Kudryashov S I, Makarov S V, Rudenko A A, Bezhanov S G, Uryupin S A, Kanavin A P, Emelyanov V I, Alferov S N, Khonina S N, Karpeev S V, Kuchmizhak A A, Vitrik O B and Kulchin Yu N 2015 *JETP Letters* **101** 394-397

IOP Conf. Series: Journal of Physics: Conf. Series **917** (2017) 062018 doi:10.1088/1742-6596/917/6/062018

Formation and transformation of the radiation-induced near-surface color centers in sodium and lithium fluorides nanocrystals

A N Novikov[1], V S Kalinov[1], A V Radkevich[2], L P Runets[1], A P Stupak[1], A P Voitovich[1]

[1]B I Stepanov Institute of Physics, National Academy of Sciences of Belarus, 68 Nezavisimosti ave., 220072 Minsk, Belarus
[2]Joint Institute for Power and Nuclear Research - Sosny, National Academy of Sciences of Belarus, 99 Krasina str., 220109 Minsk, Belarus

Abstract. Near-surface color centers in sodium fluoride nanocrystals have been formed. At pre-irradiation annealing of sodium and lithium fluorides samples at temperatures of 623 K and above, the near-surface color centers in them have not been found after γ-irradiation. Annealing lithium fluoride nanocrystals with the near-surface defects leads to their transformation into bulk ones of the same composition.

1. Introduction

Lithium fluoride (LiF) due to its physical and optical properties is one of the often used crystals in dosimetry of ionizing radiation. Processes of radiation-induced color centers (CCs) formation in the LiF bulk have been investigated for many years and are well studied [1–4]. Recently new radiation CCs in near-surface layers of crystalline plates and in nanocrystals (NCs) of LiF have been found [5, 6]. Their absorption and luminescence characteristics are significantly different from similar characteristics of CCs of the same composition formed in the LiF bulk. Hereinafter we will designate them as near-surface color centers (SCCs). It is shown [7] that presence of nanoscale clusters in the samples is a necessary condition for the formation of SCCs.

The possibility of creating radiation SCCs in other crystals is shown in this work. Sodium fluoride (NaF) crystal was chosen for experiments. It has a cubic lattice, as well as LiF crystal. The influence of pre-irradiation annealing on radiation defects formation in LiF is presented. The effect of after-irradiation annealing on radiation defects transformation in LiF is also investigated.

2. Samples and methods

The experiments were carried out with samples consisting of NCs with dimensions less than 1 μm (at least in one dimension). NCs were prepared by mechanical grinding of nominally pure NaF or LiF crystals. For convenience NCs were pressed into tablets. Radiation defects in tablets were created by γ-rays from a ^{60}Co source. Samples were irradiated at liquid nitrogen temperature $T_{irr} = 77$ K $< T_{\upsilon}$, where T_{υ} is the temperature of anion vacancies mobility. Estimated radiation dose was about $5 \cdot 10^4$ Gy. All samples were kept at room temperature after irradiation until the aggregation processes of created defects was completed.

IOP Conf. Series: Journal of Physics: Conf. Series **917** (2017) 062018 doi:10.1088/1742-6596/917/6/062018

Tablets were annealed at different temperatures before irradiation. Annealing temperature T_{ann} is a constant temperature at which the sample is held in an oven for 1 hour. Unannealed tablets were also investigated.

Two tablets of LiF were used for other experiment. One was annealed before the γ-irradiation at T_{ann} = 623 K for 1 h, the other was not annealed. These samples were annealed at different temperatures after irradiation and aggregation of bulk CCs and SCCs. Annealing was carried out as follows. Sample was annealed 1 minute at a constant temperature. Then the sample was quickly cooled to room temperature and the photoluminescence (PL) and photoluminescence excitation (PLE) spectra were measured. After that the sample was annealed, but at a different, higher temperature. The photoluminescence measurements were repeated. This measurement procedure was repeated several times until it has reached the annealing temperature 553 K.

PL and PLE spectra were recorded by SM-2203 spectrofluorometer (SOLAR, Belarus). In all cases, the PL intensities were proportional to the concentrations of the corresponding color centers. All spectral measurements were performed at room temperature.

3. Results and discussion

3.1. Effect of pre-irradiation annealing on SCCs formation in LiF nanocrystals

The PLE spectra were detected at registration wavelengths λ_{reg} = 540 and 670 nm for the γ-irradiated LiF tablets and for a crystal. Before the irradiation one tablet was annealed at temperature 623 K for one hour. The spectra obtained at λ_{reg} = 670 nm are presented in figure 1, a. The spectrum 1 in figure 1, a demonstrates intense bands belonging to bulk F_2 (the band with the maximum at λ_m = 444 nm) and near-surface F_{S3}^+ (the bands with the maxima at λ_m = 393, 493 and 564 nm) centers [1, 2, 4, 6]. This means that both surface and bulk CCs appear in this tablet after irradiation. SCCs are not formed in a tablet pre-annealed at temperature 623 K and then γ-irradiated (figure 1, a, spectrum 2). Furthermore, the PLE spectrum of this sample (figure 1, a, spectrum 2) is very similar to the spectrum of a LiF crystalline plate (figure 1, a, spectrum 3) which was not subjected to a preliminary heat treatment. There is only one band belonging to F_2 centers in spectra 2 and 3.

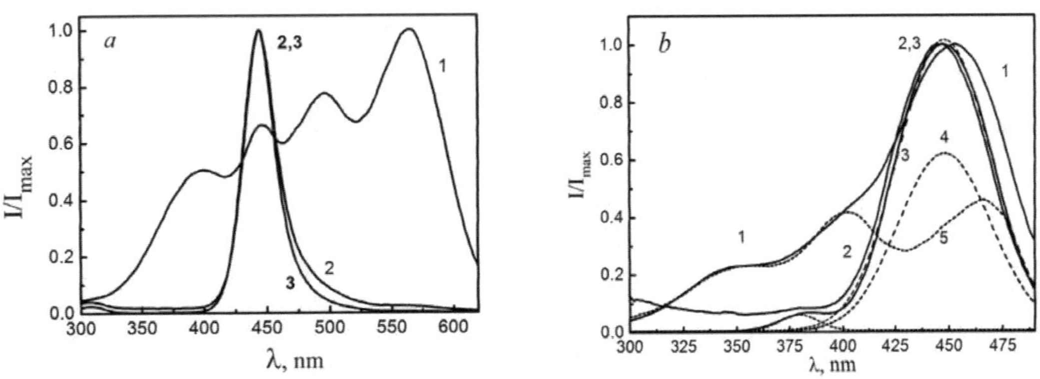

Figure 1. Normalized to their maximum values PLE spectra measured at λ_{reg} = 670 (a) and 540 nm (b) for the following γ-irradiated LiF samples: unannealed (1) and annealed before the γ-irradiation for 1 h at T_{ann} = 623 K (2) tablets, unannealed crystalline plate (3); all spectra were registered after the termination of post-radiation CCs aggregation processes. Explanations are given in text.

The PLE spectra measured at λ_{reg} = 540 nm are presented in figure 1, b. Spectrum 1 belongs to a tablet which was unannealed before γ-irradiation. This complex spectrum was decomposed into separate components, which are represented by dashed lines in the figure 1, b. The PLE bands with the maxima at λ_m = 355, 400 and 463 nm (spectrum 5) belong to F_{S3} SCCs [6], the band with the maximum at λ_m = 448 nm (spectrum 4) belong to F_3^+ CCs [1,4]. Therefore, both surface and bulk defects observed in this sample. PLE spectrum of crystalline plate (spectrum 3 in figure 1, b) was also decomposed into individual components, which are represented by dashed lines. It contains spectral bands with the maxima at λ_m = 380 and 448 nm, which belong to F_3 (R_2) and F_3^+ CCs, respectively [1,4]. The first band practically coincides with the measured spectrum in the range 400-490 nm. PLE spectrum of a tablet (spectrum 2 in figure 1, b) which is pre-annealed at temperature T_{ann} = 623 K before γ-irradiation, is similar to the spectrum of a crystalline plate. Therefore, after the pre-annealing of the samples at T_{ann} = 623 K, the subsequent γ-irradiation does not lead to the formation of the F_{S3} centers. The other measurements show that F_{S2} and F_{S2}^+ SCCs are also not formed in a tablet pre-annealed at temperature 623 K for 1 hour and then γ-irradiated [8].

A similar method was used to detect SCCs in γ-irradiated NaF samples.

3.2. Observation of SCCs in NaF nanocrystals

The PLE spectra were measured at λ_{reg} = 590 nm for the NaF crystalline plate and NCs irradiated at T_{irr} = 77 K (figure 2, a). Before the irradiation, one pellet was annealed at temperature 623 K for one hour. Another pellet was not annealed. The wavelength λ_{reg} of the PL detection was chosen because it is located in the luminescence bands of the F_2 and F_3^+ CCs [9,10].

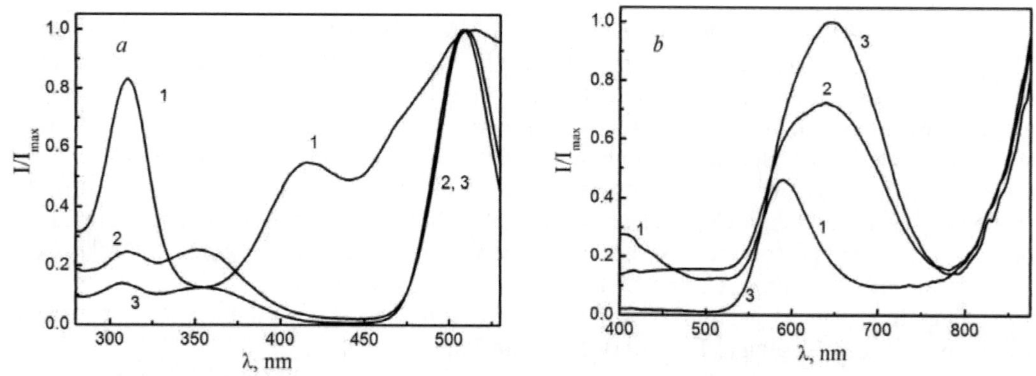

Figure 2. Normalized to their maximum values PLE spectra measured for λ_{reg} = 590 nm (a) and PL spectra at λ_{exc} = 310 nm (b) for the following γ-irradiated NaF samples: unannealed (1) and annealed before the γ-irradiation for 1 h at T_{ann} = 623 K (2) pellets, unannealed crystalline plate (3); all spectra were registered after the termination of post-radiation CCs aggregation processes.

In the PLE spectrum of crystalline plate (figure 2, a, spectrum 3), we observe only spectral bands belonging to bulk F_2, F_3^+ CCs. The most intensive bands are located in the region of 500–520 nm [10]. Also, there are the less intense second absorption bands of F_2 and F_3^+ CCs with maxima at 344 and 310 nm, respectively [10]. The spectrum of the pellet (spectrum 1 in figure 2, a) differs from the spectrum of the crystalline plate. Decomposition of this spectrum into components (not presented here) shows that except the above-mentioned bands, it contains other spectral bands. We consider that these spectral bands belong to near-surface CCs in NaF. The PLE spectrum of pre-annealed before irradiation pellet (spectrum 2 in figure 2, a) contains the same bands as the spectrum of the crystal. The distinction between the intensities of the bands in spectra 2 and 3 can be explained by distortions due to the difference of the scattering in different spectral ranges. We have a strongly scattering sample of nanocrystals (2) and practically no scattering crystalline plate (3).

The PL spectra were measured at λ_{exc} = 310 nm for γ-irradiated at T_{irr} = 77 K NaF crystal and NCs pellets (figure 2, b). We can see that PL spectrum of the annealed before the γ-irradiation pellet differs from that has not been annealed. There is one band with a maximum 590 nm in the PL spectrum of the unannealed pellet (spectrum 1 in figure 2, b). On the other hand PL spectrum of annealed pellet contains two overlapping bands (spectrum 2 in figure 2, b). The decomposition of this spectrum into components indicates the presence of two bands with maxima at 590 and 650 nm, which are PL bands of bulk F_2 and F_3^+ CCs, respectively [10]. Spectrum of the crystalline plate (spectrum 3 in figure 2, b) is similar to the one of the annealed pellet.

Thus, SCCs are present in the NaF pellet. Additional studies are required to establish the composition of these SCCs.

3.3. Effect of after-irradiation annealing on SCCs transformation in LiF nanocrystals

The changes in the concentrations of the F_{S3}^+, F_2 and F_3^+ centers were studied in pellets of LiF NCs annealed at different temperatures T_{ann} after the γ-irradiation. Two pellets were used in each of measurements series. One of them was not annealed before the γ-irradiation. The second pellet was annealed at T_{ann} = 623 K for 1 h and there were no SCCs in this pellet after the γ-irradiation. The results of the PL intensity changes due to the post-radiation annealing are shown in figure 3. They were obtained from the PLE spectra measured for λ_{reg} = 670 nm for the F_{S3}^+ and F_2 centers and from both PL spectra at λ_{exc} = 445 nm and PLE at λ_{reg} = 535 nm for the F_3^+ centers. If necessary, the measured spectra were decomposed into components.

In the pellet which was not annealed before the γ-irradiation an increase in the temperature of the post-radiation annealing above 393 K leads to a rapid drop in concentrations of the F_{S3}^+ SCCs (figure 3 a, data 3). By comparing the data 1, 2 and 3 in figure 3, a, we can see the following: a decrease in the F_{S3}^+ SCCs concentration is accompanied by an increase in the F_2 and F_3^+ CCs concentrations. After the destruction of F_{S3}^+ SCCs, the increase of the F_3^+ CCs concentration stops, while it is still ongoing for the concentration of the F_2 CCs. A similar situation is observed for other samples subjected to the longer duration of the post-radiation annealing [7].

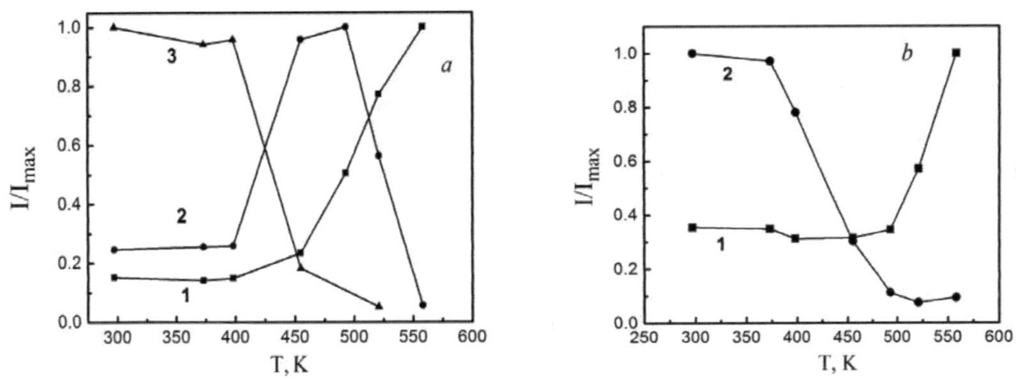

Figure 3. PL intensity changes for the F_2 (1), F_3^+ (2), F_{S3}^+ (3) centers in γ-irradiated non-annealed (a) and annealed at T_{ann} = 623 K for 1 h before the irradiation (b) LiF pellets as a result of the post-radiation heat-treatment for 1 min at each temperature. Points are experimental results. Lines are drawn for the convenience of interpretation.

In the pellet which was annealed before the γ-irradiation, there is a completely different situation. We observe only a monotonic decrease in the concentration of F_3^+ centers with temperature growth (data 2 in figure 3, b). The growth stage of these centers is not observed in contrast to the previous case. An increase in the concentration of F_2 centers is fixed in the temperature range from 493 to 553 K (data 1 in figure 3, b).

Based on these results, we can conclude that post-radiation annealing of the pellets which were not exposed to pre-radiation annealing leads to a transformation of some of the F_{S3}^+ SCCs into the F_3^+ CCs.

4. Conclusion

New spectral bands in PLE spectra of NaF nanocrystals, not subjected to pre-radiation annealing, were found. These bands belong to near-surface color centers in NaF. Near-surface color centers in NaF and LiF nanocrystals not annealed before irradiation are formed as a result of γ-irradiation. For NaF as well as for LiF, pre-annealing at temperatures 623 K and more makes it impossible to create the SCCs as a result of γ-rays exposure.

It was shown [7] that an unannealed pellet of NCs contains clusters with size ranged from 5 to 17 nm. The similar clusters are absent in annealed samples. Mechanical fragmentation of the crystal leads to the creation of clusters. The formation of surface color centers is impossible without these clusters. Pre-irradiation annealing leads to the destruction of clusters and eliminate the conditions necessary for the formation of the surface centers.

Post-radiation annealing of the pellets which were not exposed to pre-radiation annealing leads to a transformation of some of the F_{S3}^+ SCCs into the F_3^+ CCs. It means that SCCs are transformed into CCs of the same composition.

References

[1] Nahum J and Wiegand D A 1967 *Phys. Rev.* **154 (3)** 817
[2] Davidson A T, Kozakiewicz A G, Wilkinson D J and Comins J D 1999 *J. Appl. Phys.* **86** 1410
[3] Salah N, Sahare P D and Rupasov A A 2007 *J. Lumin.* **124** 357
[4] Montereali R M 2002 *Handbook of Thin Film Materials* Vol.3: Ferroelectric and Dielectric Thin Films Nalwa H S ed (Academic Press) Ch.7 399R.M. Montereali, 2002, in Handbook of Thin Film Materials (New York: Academic Press) p. 399
[5] Voitovich A P, Kalinov V S, Korzhik M V, Martynovich E F, Runets L P and Stupak A P 2013 *Radiat. Eff. Def. Solids* **168 (2)** 130
[6] Voitovich A P, Kalinov V S, Stupak A P, Novikov A N and Runets L P 2015 *J. Lumin.* **157** 28
[7] Voitovich A P, Loiko P A, Mateos X, Runets L P, Serres J M and Stupak A P 2017 *J. Lumin.* **188** 75
[8] Voitovich A P, Kalinov V S, Novikov A N, Radkevich A V, Runets L P, Stupak A P and Tarasenko N V 2017 *J. Appl. Spectr.* **83** 970
[9] Kristianpoller N N, Blieden G S and Comins J D 1992 *Nucl. Instr. and Meth. in Phys. Res.* **B65** 484
[10] Chandra A and Holcomb D F 1969 *J. Chem. Phys.* **51** 1509

High sensitive InP emitter for InP/InGaAs heterostructures

K J Smirnov[1,2], V I Medzakovskiy[3], V V Davydov[1,3,4], M G Vysoczky[1], S F Glagolev[3]

[1] Peter the Great Saint – Petersburg Polytechnic University, Saint Petersburg 195251, Russia

[2] OJSC "NRI "Electron", Saint Petersburg, Russia

[3] Bonch-Bruevich Saint – Petersburg State University of Telecommunications, Saint Petersburg 193232, Russia

[4]All-Russian Research Institute of Phytopathology, Moscow Region 143050, Russia

Abstract. Results of indium phosphide structures research, showing the possibility of using in the near infrared range (NIR) photocathodes of InP / InGaAs, are represented. An optimal method of obtaining the atomically clean indium phosphide surface was suggested. The spectral characteristics of indium phosphide and InP/InGaAs heterostructure were given. Researches of the photoemission dependence of the structure with surface grid electrode upon different supply voltages were carried out.

1. Introduction

The creation of a photocathode, which operates in the NIR range, also known as the transferred electron photocathode, was proposed by Bell [1]. Such photocathodes can move the long-wavelength threshold of photoemission by increasing the internal energy of the photoelectrons (with the help of the metal electrode which is applied to the effective surface of the photocathode). This structure has been realized on the basis of heteropair indium phosphide – solid solution of indium – gallium arsenide. The InGaAs forbidden band is narrow enough to detect near–infrared range. This narrow–band material ("absorber") cannot provide by itself the negative electron affinity (NEA) on the surface, which is a determinant factor for obtaining a high quantum efficiency for such structures. However, InGaAs can be perfectly matched to the InP crystal lattice, which, being a conductor of electrons (emitter) generated in the absorber, allows to get on its surface the effective state of the NEA. Semiconductor photocathodes activated to negative electron affinity (NEA) serve as electron beam sources for many applications. InP photocathode is an emitter of InP/InGaAs heterostructure. These heterostructures as compared to other semiconductor materials allows getting greater quantum efficiency in case of detection reflective laser beam in near infrared range [2]. The purposes of the work were to develop the optimal method of obtaining atomic clean indium phosphide surface, to obtain high quantum efficiency of the InP samples, to research the increase of the photoemission with the help of titanium grid electrode and to apply the result of the researches for obtaining high sensitive InP/InGaAs photocathode.

2. Chemical cleaning of InP surface

The structures based on the indium phosphide are characterized by low thermal resistance and heating over with the temperature of 593 K lead to the degradation of these structures, so that the typical cleaning methods for metals and thermally stable materials cannot be used. Sulfur passivation leads to stable surface termination, but the chemisorbed sulfur atoms cannot be removed completely by thermal annealing at temperatures below the decomposition temperature of InP. Process of obtaining a clean InP surface is rather difficult, especially pre–vacuum chemical cleaning step. The treatment with HCl–C_3H_8O and subsequent vacuum annealing is useful method for obtaining the atomically clean GaAs surfaces, another semiconductor of the A_3B_5 group [3]. However, etching with these chemical compounds is ineffective for InP surfaces.

Therefore, we used the efficient three-step method of chemical etching combined with a thermal cleaning in vacuum to obtain the atomic clean surface of InP [4, 5]. Samples of InP (111): Zn, with concentration $p=1\cdot10^{18}cm^{-3}$ were used for studies. Sample of InP structure with surface grid is shown in figure 1.

Figure 1. InP structure with surface grid.

Etching time was increased, as compared to [4], and the mode was adjusted in the direction of the removal rate increase of a thick natural oxide by the use of ultrasound in the etching process. Besides that the intermediate etching step was added: a weak solution of sulfuric acid was used. Strong acid solutions with HCl or H_2SO_4, which are used in third step of chemical cleaning, are able to remove the surface oxide and leave the InP surface passivated with elemental P which is, in turn, terminated with H. This yields a hydrophobic surface and allows lower temperatures to be used during annealing. All steps of chemical cleaning are shown in table 1:

Table 1. Optimized three-step method of chemical cleaning

	Solution	Concentration
1	H_2SO_4:H_2O_2:H_2O	4:1:100
2	H_2SO_4:H_2O	1:10
3	H_2SO_4:H_2O	1:1

Protective layer formed on InP surface are also allows to carry etched structures from chemical laboratory bench to vacuum installation in atmosphere. This is an important achievement in terms of surface oxidation and uncontrolled contamination of surfaces. We studied the effect of etching on the titanium grid and the possibility of etching agents use without degradation of the grid structure. As a result, we found that etching is equally effective for two different acid solutions HCl and H_2SO_4, however, in case of HCl titanium grid is exposed to chemical degradation. Indium phosphide is emitter of the InP/InGaAs heterostructures, accordingly chemical cleaning of this structures identical to the InP.

3. Activation of InP and InP/InGaAs structures

The ultra-high vacuum installation was used to conduct the research in a vacuum, which could reach the level of $1 \cdot 10^{-9}$ Pa. Structural diagram of installation is shown in figure 2.

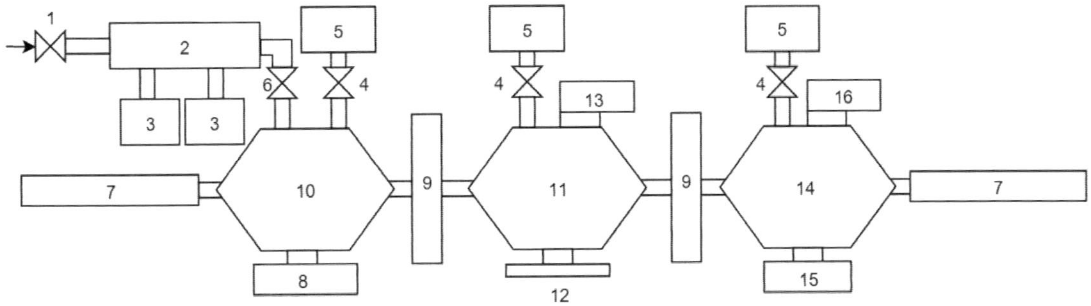

Figure 2. Structural diagram of the ultra-high vacuum installation: 1 – nitrogen valve, 2 – trunk-line, 3 – molecular sieve foreline trap, 4 –ion pump's valve, 5 –ion pump, 6 –trunk-line valve, 7 – manipulator, 8 – entry gate, 9 – gate valve, 10 – loading chamber, 11 – press and housing preparation chamber, 12 – special vacuum press, 13 – viewport, 14 – activation chamber, 15 – sources of Cs and O_2, 16 – sapphire viewport.

As a result of the etching process the hydrophobic surface InP structure was obtained [4]. After heating in a vacuum chamber at a temperature of about 573 K, the net InP atomic surface was received, that's ready for the activation with Cs and O_2, which allows to form effective state of NEA on InP photocathode surface.

The method of the activation structure called «yo-yo» [6], based on a continuous simultaneous supply of cesium and oxygen was used. The process for obtaining the maximum photoemission current was about 90 minutes. The progress of activation is presented in figure 3.

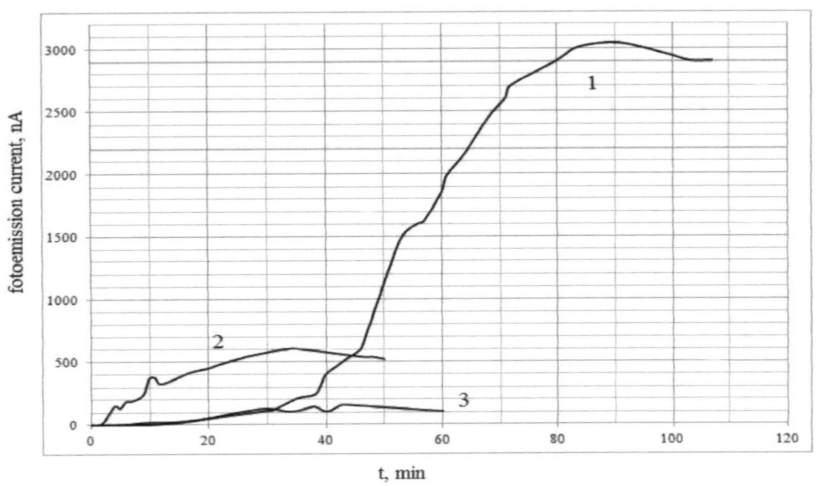

Figure 3. Activation of structure etched by 1. H_2SO_4. 2. C_3H_8O+Br. 3. C_3H_8O+HCl.

Research was conducted on InP structures which etched by solutions of C_3H_8O+Br, C_3H_8O+HCl and our new three stage method of chemical cleaning with solutions of H_2SO_4. After graphical analysis, we found that the results of photoemission current using the new method of chemical cleaning increase 6 times as compared to previous experiments. In addition to pure samples were also studied samples with the grid electrode, used to create the photocathode with the transferred field. The dependence of the photoemission current on the applied bias voltage is shown in figure 4.

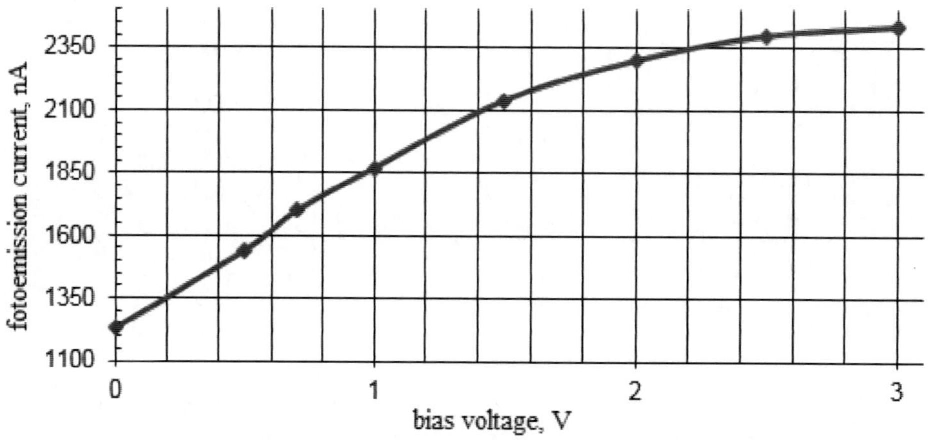

Figure 4. The dependence of photoemission current on the applied bias voltage.

Measurements were carried out in range of 500 to 950 nm. Based on figure 3, quantum efficiency (QE) is calculated using the ratio:

$$QE = \frac{I \cdot h \cdot c}{P \cdot e \cdot \lambda} \cdot 100\% \tag{1}$$

Where I — the photoemission current, P — the monochromator radiation power at wavelength λ, h — the Planck's constant, c — the speed of light e — the elementary charge.

Calculation of the dependence between QE and λ using (1) are represented in figure 5.

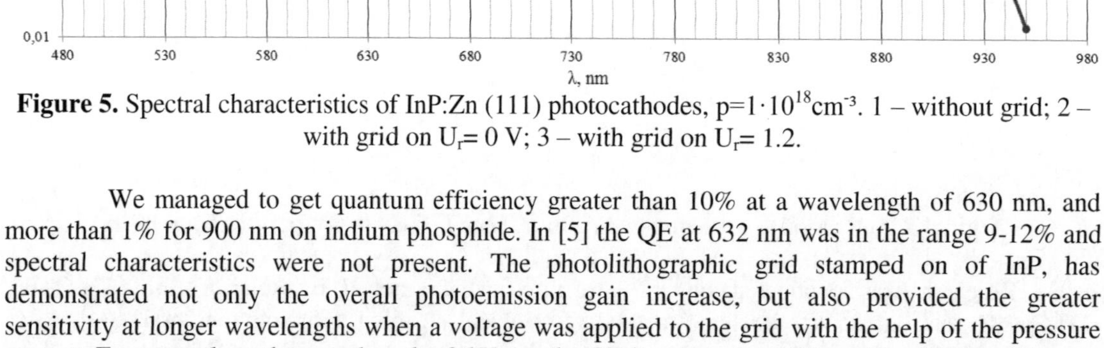

Figure 5. Spectral characteristics of InP:Zn (111) photocathodes, $p=1\cdot10^{18}cm^{-3}$. 1 – without grid; 2 – with grid on U_r= 0 V; 3 – with grid on U_r= 1.2.

We managed to get quantum efficiency greater than 10% at a wavelength of 630 nm, and more than 1% for 900 nm on indium phosphide. In [5] the QE at 632 nm was in the range 9-12% and spectral characteristics were not present. The photolithographic grid stamped on of InP, has demonstrated not only the overall photoemission gain increase, but also provided the greater sensitivity at longer wavelengths when a voltage was applied to the grid with the help of the pressure contact. For example at the wavelength of 650 nm the QE increases by 46%, and by 177% at 900 nm.

The resulting dark current was small (less than 0.05 nA at U_r=3V), but at the applied voltage exceeding 2V the photoemission current has saturation, which makes the voltage increase weakly effective. However, operation mode of the surface grid electrode significantly depends on technical features of it creation. We are going to create titanium electrode with optimal operating level on 8-10 supply voltage.

Results of the investigations of indium phosphide were applied on InP/InGaAS heterostructures, which were grown by MOCVD methods. Heteropair indium phosphide – solid solution of indium – gallium arsenide was realized on substrate InP with thickness 350 mkm. Etching and vacuum annealing were conducted identical to InP structures Spectral characteristics of the InP/InGaAs heterostructure with developed surface grid titanium electrode are shown in the Figure 6. Analyzing obtained dependencies (fig. 6), we found that quantum efficiency of the proposed heterostructure, which operates in reflection mod, reaches 4% at the wavelengths from 900 to 1600 nm. This result is more than 2 times greater than results of previous researches. Quantum efficiency level of structures, which work in transmission mode, was worse than reflection mode. This is important problem for creation real device with those structures. It is connected with non-optimal parametres of the substrate. Theoretical calculations were shown that with the optimal substrare dependence between QE in radiation and transmission mode below 15%.

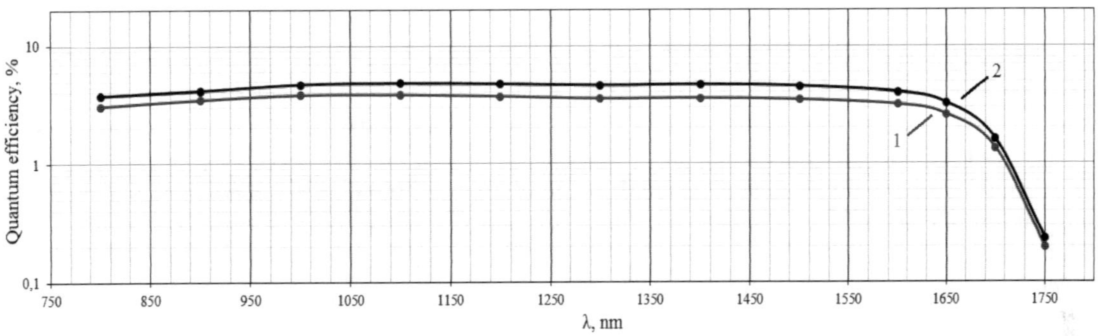

Figure 6. Spectral characteristics of InP/InGaAs, **1** – with grid at U_s= 2 V; **2** – with grid at U_s= 3V

4. Conclusion

The heterostructures of InP/InGaAs work in near infrared range and could be used in devices for special purposes, where high processing speed of detection reflected radiation of special laser systems,(for example quantum-dot lasers) from the object is needed [7]. At present time, solid-state analogs of these devices cannot provide the same processing speed. The received results showed that the developed structure parameters are not optimal. The further increase of QE can be achieved by the optimization of heterostructures InP/InGaAs thickness and the basic carrier concentration.

References

[1] Bell R L 1974 US Patent N23958143
[2] Ermolaev A N, Krishpents G P, Davydov V V, Vysoczkiy M G *2016 J. of Phys: Conference Series* **741** 012171.
[3] Tereshchenko O E, Chikichev S I, Terekhov A S 1999 *Appl. Surface Science* 142 75–80
[4] Sun Y, Liu Z, Machuca F, Pianetta P, Spicer W 2005 *SLAC-PUB*-11018
[5] Sun Y, Liu Z, Pianetta P 2007 *SLAC-PUB*-12710
[6] Chanlek N, Herbert J, Jones R, Jones L, Middleman K, Militsyn B 2014 *J. Phys. D: Appl. Phys.* **47** 055110
[7] Kovsh A, Zhukov A, Livshits D, Krestnikov I 2009 US Patent N7561607

Quantum state engineering with single atom laser

V P Stefanov

Center of Quantum Optics and Quantum Information, Stepanov Institute of Physics of NASB, 68, pr.Nezavisimosti, Minsk-220072, Republic of Belarus

E-mail: v.p.stefanov@mail.ru

Abstract. On the basis of quantum stochastic trajectories approach it is shown that a single atom laser with coherent pumping can generate not only coherent states, but squeezed and Fock states, when different schemes of detection are followed by coherent feedback pulses or feedforward actions.

1. Introduction

A single atom laser (SAL) is the limiting case of lasers. This miniaturized model naturally appears considering the resonant interaction of an atom with the quantum field of high-Q cavity both an optical [1], and microwave [2, 3] ranges. To describe it, over fifty years ago Jaynes and Cummings introduced the model [4], and a numerous theoretical and experimental researches of their model led to the discovery of many quantum effects, including Rabi oscillation [5], collapse-revival effect [6], phase bistability [7], effect of compression (sub-Poisson distribution of the number of photons) [8] and others.

Of particular interest is the consideration of the Jaynes-Cummings model supplemented taking into account the relaxation processes of atomic and field dissipation and pumping. In this case, all radiated photons in the resonator mode are indistinguishable. Thus the conclusions based on the difference between spontaneous and stimulated photons emission, become invalid. In particular, the proposition that the two level system cannot provide the generating radiation. As shown in [9, 10] steady-state distribution of photons emitted by the single atom laser can generate Poisson light.

In this paper by using the method of quantum stochastic trajectories [11, 12] we show that the single atom laser also can be used for engineering another quantum states, for example the Fock states or squeezed states.

2. Model

The model of SAL is a two-level atom (with ground $|g\rangle$ and excited $|e\rangle$ states) in a high-Q cavity interacting with a single field mode via the Jaynes-Cummings Hamiltonian $H_{JC} = \hbar g(a^\dagger\sigma_- + a\sigma_+)$, where g is the interaction constant, $\sigma_+ = |e\rangle\langle g|$, $\sigma_- = |g\rangle\langle e|$ are atomic transition operators and $a^\dagger$, a are field creation and annihilation operators. The field and atomic decay rates $2k$ and Γ are taken into account through the corresponding Lindblad superoperators $2L(x)\rho = 2x\rho x^\dagger - x^\dagger x\rho - \rho x^\dagger x$ in master equation for the density matrix ρ. Also we consider the possibility of switching on and off of the coherent pumping in specific times. For simplicity the amplitude of external field is supposed constant. The coherent pump is described by the

IOP Conf. Series: Journal of Physics: Conf. Series **917** (2017) 062020 doi:10.1088/1742-6596/917/6/062020

interaction operator

$$V = \frac{E_0}{2}(\mathcal{D}_{ge}\sigma_- + \mathcal{D}_{eg}\sigma_+), \tag{1}$$

where $\mathcal{D}_{ge} = \mathcal{D}_{eg} = \mathcal{D}$ is the dipole transition moment of atom and E_0 is the amplitude of external field. Resulting master equation is the following

$$\dot{\rho} = -\frac{i}{\hbar}[H_{JC} + V, \rho] + 2kL(a)\rho + \Gamma L(\sigma_-)\rho. \tag{2}$$

3. Coherent feedback excitation with atomic and field decay detection

It can be shown for SAL with incoherent pumping that possibility of atomic decay and pumping and field decay detection allows to determine the state into cavity in time between clicks of detector as

$$|\Psi\rangle = c(t)|g\rangle \otimes |n\rangle + f(t)|e\rangle \otimes |n-1\rangle. \tag{3}$$

It is also performed for SAL without pumping (click of any detector means decreasing n by one). Because at some time coefficient $f(t)$ in (3) will be zero, we can act on system by short coherent $\pi/2$-pulse [13]. This action increases the amount of excitation by one. Thus, we can represent evolution of the system as quantum stochastic trajectory, which consists of two interchangeable stages. The first is the stage of preparing (we wait time when atom will be in ground state). The second is the stage of pumping (we transfer atom from ground to excited state). After the registration of atomic decay in any time of evolution we pump system by $\pi/2$-pulse at once. After the field decay registration we must wait a specific time for $\pi/2$-pulse again.

For modeling of quantum stochastic trajectory rewrite (2):

$$\dot{\rho} = -iH^{\kappa}_{eff}\rho + i\rho H^{\kappa\dagger}_{eff} + J\rho, \tag{4}$$

where H^{g}_{eff} and H^{d}_{eff} are effective non-hermit Hamiltonians for stages of preparing and pumping respectively:

$$H^{d}_{eff} = (H_{JC} + V)/\hbar - iR, \tag{5}$$

$$H^{g}_{eff} = H_{JC}/\hbar - iR, \tag{6}$$

$$2R = \Gamma\sigma_+\sigma_- + 2ka^\dagger a, \tag{7}$$

and superoperator of reduction J is consisted of two parts:

$$J = \sum_{\varsigma=B,F} J_\varsigma, \tag{8}$$

that act as:

$$J_B x = \Gamma\sigma^- x\sigma^+ \equiv \Gamma\langle b|x|b\rangle|a\rangle\langle a|, \tag{9}$$

$$J_F x = 2kaxa^\dagger \equiv 2k\textstyle\sum_{m,n}\sqrt{mn}\langle m|x|n\rangle|m-1\rangle\langle n-1|. \tag{10}$$

It means that J_B operator by acting on x creates new operator with $R_{ba}\langle b|x|b\rangle|a\rangle\langle a|$. Physical meaning of $J_B\rho(t)$ is committing atomic decay in time t (commits B-type jump). Analogically, J_F operator transforms field components of x ($\langle m|x|n\rangle|m\rangle\langle n|$) into $2k\sqrt{mn}\langle m|x|n\rangle|m-1\rangle\langle n-1|$, and $J_F\rho(t)$ is committing field decay in time t (commits F-type jump).

After each stage of evolution we redefine density matrix: $\rho_{i-1}^{(t_{i-1})} = \rho_i^{(0)}$. We also need probabilities of committing jumps:

$$P_B^{(i)} = \int_0^{t_i} J_B S_\kappa^{(\tau,0)} \rho_i^{(0)} d\tau, \tag{11}$$

$$P_F^{(i)} = \int_0^{t_i} J_F S_\kappa^{(\tau,0)} \rho_i^{(0)} d\tau, \tag{12}$$

$S_\kappa^{(t_i,t_{i-1})}$ is non-unitary evolution operator:

$$S^{(t_i,t_{i-1})}\rho(0) = e^{-iH_{eff}(t_i-t_{i-1})}\rho(0)e^{iH_{eff}^{\dagger}(t_i-t_{i-1})}. \tag{13}$$

Because of the additivity of the probabilities, each stage contributes to the general probabilities of committing jump:

$$P_k = \sum_{i=1}^{k}(P_B^{(i)} + P_F^{(i)}), \tag{14}$$

so, if general probability P_{k-1} on $k-1$ stages is more than generating random number χ $(0 \le \chi \le 1)$, jump happens on stage k (if $P_{k-1} < \chi < P_{k-1} + P_B^{(k)}$ then B-type take place, if $P_{k-1} + P_B^{(k)} < \chi < P_{k-1} + P_B^{(k)} + P_F^{(k)}$ then F-type follows).

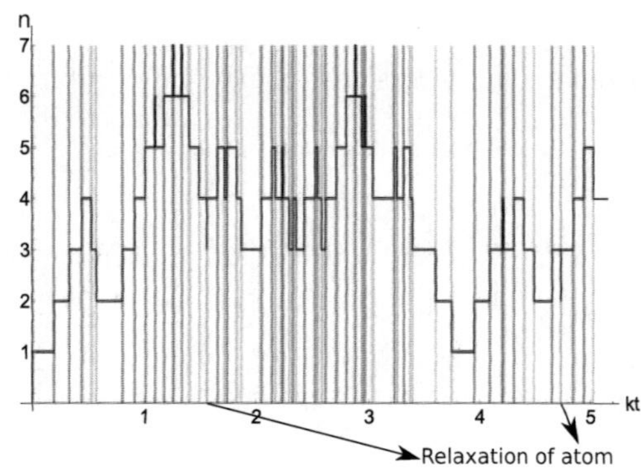

Figure 1. Quantum trajectory with the detection of decay clicks. Black line is the evolution of excitation number m (photon number plus unity if atom occupiers excited state); red lines indicate $\pi/2$-pulses; green lines represent the times of field decay detecting; orange lines represent the times of atom decay detecting and immediately after their detection runs $\pi/2$-pulse.

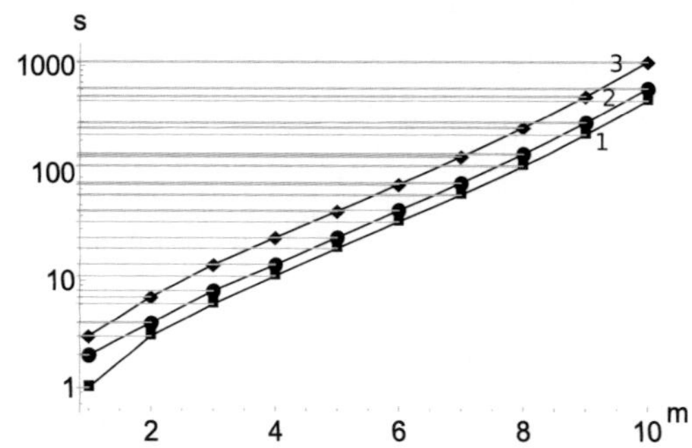

Figure 2. The quantity of relaxation acts for obtaining the m excitation level with probability 1-90%, 2-95%, 3-99.5%.

The second random number β allows to find the time of committing jump t:

$$\beta = \frac{\int_0^t J_\xi S_\kappa^{(\tau,0)} \rho_k^{(0)} d\tau}{\int_0^{t_k} J_\xi S_\kappa^{(\tau,0)} \rho_k^{(0)} d\tau}, \tag{15}$$

where $\xi = B$ or F respectively and t_k is the waiting time counted from time of atomic transition.

We will use parameters for numerical calculation from paper [1]: $2g = 2\pi * 68MHz$, $\Gamma = 2\pi * 2.6MHz$, $k = 2\pi * 4.1MHz$. Normalize parameters are $\eta = g^2/k^2 = 68.8$, $\nu = (\Gamma - 2k)/4k = -0.34$. The value $d = E_0\mathcal{D}/(2\hbar) = 20000$ is taken from consideration that the duration of the pumping stage $(\pi/(2dk) = 19ps)$ was several orders of magnitude less than the duration of the preparing.

On Figure 1 a quantum trajectory is modeled by the approach described above.

The proposed pump method can be used to generate Fock states, but in this case, the obtaining of any state is a statistical problem, and it is necessary to estimate the time at which the probability of generation of a state will be given (for example, at which time will state $|10\rangle$ appear with a probability of $\mathcal{P} = 99.5\%$?).

It is necessary to solve recurrent equation:

$$W_i^k = \sum_{l=i}^{m} \left(P_F^{(l)}(1 - P_F^{(l-1)} - P_B^{(l-1)}) W_{l-1}^{k-1} + P_B^{(l)} W_l^{k-1} \right), \tag{16}$$

$$W_0^0 = W_i^0 = \prod_{l=0}^{m} (1 - P_B^{(l)} - P_F^{(l)}), \tag{17}$$

$$\mathcal{P} = \sum_{j=0}^{s} W_0^j. \tag{18}$$

Then we will know amount of click of detectors s that needed for generating Fock state m with probability $\mathcal{P}$ (see Figure 2).

AS example, for time $51\mu s$ the ten-photons Fock state will be generated with probability 99.5% (see Figure 2, line 3, amount of jumps is $s = 1083$).

4. Photon clicks detection: feedforward strategy

In the absence of detection of acts of atomic decay, the state inside the resonator will no longer be pure. Therefore, the main reason for using the pumping by the short pulses is irrelevant. For this case we consider continuous coherent pumping.

Similar to the previous section we can modulate the quantum stochastic trajectory of single atom laser. Rewrite (2):

$$\dot{\rho} = -i\tilde{H}_{eff}\rho + i\rho\tilde{H}_{eff}^\dagger + J_F\rho, \tag{19}$$

here J_F is jump operator (10) and $\tilde{H}_{eff}$ is effective Hamiltonian from which the non-unitary operator is built:

$$-i\tilde{H}_{eff}\rho + i\rho\tilde{H}_{eff}^\dagger = \frac{i}{-\hbar}[H_{JC} + V, \rho] + \Gamma L_{\sigma_-}\rho - ka^\dagger a\rho - k\rho a^\dagger a, \tag{20}$$

$$\tilde{S}^{(t_i - t_{i-1})}\rho(0) = e^{-i\tilde{H}_{eff}(t_i - t_{i-1})}\rho(0)e^{i\tilde{H}^\dagger{}_{eff}(t_i - t_{i-1})}. \tag{21}$$

We need only one random number to find the time of committing F-type jump:

$$\chi = \int_0^{t_i} \tilde{J}\tilde{S}^{(\tau,0)}\rho_i^{(0)} d\tau, \tag{22}$$

since probability of committing jump always is equal to one.

Figure 3 shows a simulated quantum trajectory. The effect of the decay of the field on the state of the system leads to the fact that short intervals between jumps transform the state inside the cavity into a state with almost exact Poisson statistics (inset a). And even a strong change in the state of the field due to the long absence of decay (inset b) is leveled by a further evolution (inset c).

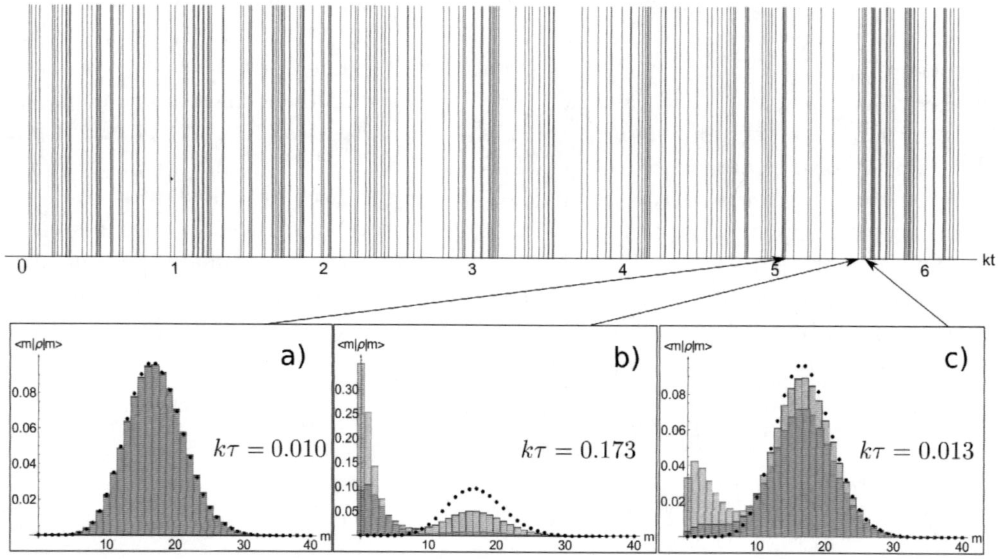

Figure 3. Quantum trajectory with the detection of field decay clicks only. The blue vertical lines indicate the times at which the detector that detects the decay of the photon from the cavity has triggered. Insets *a*), *b*), *c*) show the distribution of the system from photon numbers at the time before (red histogram) and after (blue histogram) of the jump ($k\tau$ is time waiting of jump); black points marked coherent state with $|\alpha^2| = 17.04$.

Thus, it is possible to generate states whose statistics coincide with the statistics of coherent states with great accuracy.

5. No clicks strategy

The last considered case is a single atom laser without detection of atomic and field decays.

It is of interest to analyze the state of the system under the action of alternating stages of switching on / off coherent pumping. We represent the solution of the master equation (2) in the form:

$$\rho(t) = \mathcal{S}_{\kappa}^{(t,0)}\rho, \tag{23}$$

where $\mathcal{S}_{\kappa}^{(t,0)}$ with $\kappa = d$ is evolution operator with coherent pumping, and with $\kappa = g$ without pumping.

Then the state of the system through n alternating stages can be written as:

$$\rho(t_n, t_{n-1}, ..., t_1) = \mathcal{S}_{g}^{(t_n,t_{n-1})}\mathcal{S}_{d}^{(t_{n-1},t_{n-2})}...\mathcal{S}_{g}^{(t_2,t_1)}\mathcal{S}_{d}^{(t_1,0)}\rho(0). \tag{24}$$

The obtained state can be investigated from different points of view, for example, find the values $t_1, ...t_n$ for which the state inside the system will be the most squeezed in the coordinate quadrature:

$$(\Delta X_1)^2 = \langle \left(\frac{a + a^\dagger}{2} \right)^2 \rho(t_n, t_{n-1}, ..., t_1) \rangle - \langle \left(\frac{a + a^\dagger}{2} \right) \rho(t_n, t_{n-1}, ..., t_1) \rangle^2. \quad (25)$$

Numerical calculations establish the values $t_j, j = \overline{1,6}$, at which the minimal value of the coordinate quadrature is equal to $\langle \Delta X_1^2 \rangle = 0.12559$ (see Figure 4). It is achieved by three pumping pulses, and further alternating pumping / absence of pumping will not reduce it. The distribution of excitations of the system at time of maximum of squeezing is shown on the Figure 5.

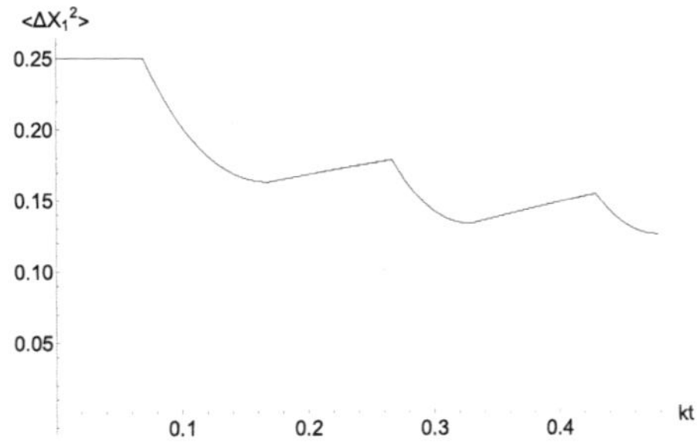

Figure 4. The value of the coordinate quadrature of the state of the system as a function of time; blue segments are pumping stages, red are stages of absence of pumping. The minimum value of $\langle \Delta X_1^2 \rangle = 0.12559$ is reached.

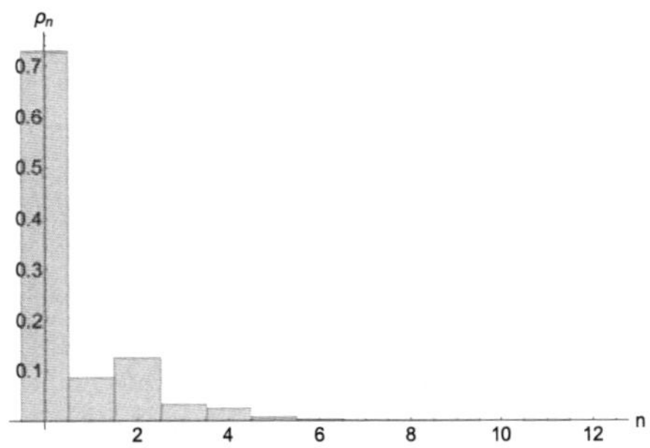

Figure 5. The distribution of excitations of the system at time of maximum of squeezing.

6. Conclusion

In this paper it is shown that a single atom laser can be of interest as a practical device generating a wide class of quantum states. The class type of states depends on the observation of the occurring relaxation processes in the system, which was demonstrated using the parameters realized in a particular experiment [1].

By registering all decays of the system into both the cavity and non-cavity modes, the state of the system becomes localized. The use of pulsed coherent pumping allows us to change the values of the number of excitations inside the resonator, and if we can provide sufficient time observations of the system, we obtain the desired Fock state.

By registering only decays of the system into the cavity mode, with continuous coherent pumping, we can generate coherent states.

Without detecting of the relaxation of the system, the generated states do not belong to any nominal class of states, but we can influence their statistical properties by switching on and off coherent pumping, and as a result obtain states with the necessary properties. For example, it has been shown that a squeezed state can be obtained.

References

[1] Boca A, Miller R, Birnbaum K M, Boozer A D, McKeever J and Kimble H J 2004 *Phys. Rev. Lett.* **93** 233603
[2] Brune M, Raimond J M, Goy P, Davidovich L, and Haroche S 1987 *Phys. Rev. Lett.* **59** 1899
[3] Meschede D, Walther H, and Müller G 1985 *Phys. Rev. Lett.* **54** 551
[4] Jaynes E T and Cummings F W 1963 *Proc. IEEE* **51** 89
[5] Thompson R J, Rempe G, and Kimble H J. 1992 *Phys. Rev. Lett.* **68** 1132

[6] Rempe G, Walther H, and Klein N 1987 *Phys. Rev. Lett.* **58** 353
[7] Walls D F and Zoller P 1980 *Opt. Com.* **34** 260
[8] Mu Y and Savage C M 1992 *Phys. Rev. A* **46** 5944
[9] Kilin S Ya and Karlovich T B 2002 *JETP* **95** 805
[10] Kilin S Ya and Mikhalychev A B 2012 *Phys. Rev. A* **85** 063817
[11] Kilin S Ya 1995 *Quantum Optics: Fields and Their Detection* (London: Inst of Physics Pub Inc)
[12] Carmichael H J 1993 *Phys. Rev. Lett.* **70** 2273
[13] Allen L and Eberly J H 1975 *Optical Resonance and Two-level Atoms* (New York: John Wiley & Sons Inc)

Experimental study of THz electro-optical sampling crystals ZnSe, ZnTe and GaP

M Zhukova[1], E Makarov[1], S Putilin[1], A Tsypkin[1], V Chegnov[2], O Chegnova[2], V Bespalov[1]

[1]Department of Photonics and Optical Information Technology, ITMO University, Saint Petersburg, 197101, Russia
[2]Research Institute of Materials Science and Technology, Moscow, 124460, Russia

Abstract. The application of optoelectronic techniques to the generation and detection of THz radiation is now well established. Wide gap semiconductor crystals of groups II-VI, III-V and III-VI are abundantly used. However, some limitations are occurred while using powerful laser systems. In this paper we introduce experimental results of two-photon absorption (2PA) in ZnSe, ZnTe and GaP studied with femtosecond pump-probe supercontinuum spectroscopy. Using of supercontinuum helps us to measure 2PA absorption dynamics and nonlinear index of refraction in wide frequency ranges. Besides influence of Fe concentration in ZnSe:Fe crystals on transmitted THz radiation is described.

1. Introduction

Semiconducting crystals of groups II-VI, III-V and III-VI are widely used as sources and detectors of terahertz radiation. In time domain spectroscopy systems, that are very popular in different fields of applied research [1], electro-optical method for the detection of THz radiation is often used. The main advantage of this method is a wide dynamic range [2]. The process is based on measuring the change in polarization of probe optical pulse under the influence of terahertz field in electro-optic media [3].

There are several ways to increase the efficiency of THz spectroscopy. The first one is to use more powerful laser sources in the optical and near-IR range to generate and detect the THz radiation. However, it leads to a non-linear response of electro-optic semiconductor crystals. Second way, is to increase the emission of THz radiation by raising the number and mobility of the electron-hole pairs that can be implemented with crystal impurities.

Consequently, we presented two main ideas: to investigate the dependence of crystal response and nonlinear processes arising from the source power improvement [4], as well as to study the efficiency of THz radiation by changing the crystal structure, for example, by densify impurities [5].

2. Investigation of two-photon absorption and refractive index dispersion

One of the main processes influencing on response of semiconductor crystals is a two-photon absorption (2PA). 2PA reduces the intensity of light propagated through the crystal and the ultra-fast response [6] of semiconductor materials. Here we introduce experimental results of 2PA studied with femtosecond pump-probe supercontinuum spectroscopy. By usage of supercontinuum, 2PA absorption dynamics and nonlinear index of refraction variation with wavelength could be detected.

The experiments were carried out on pump-probe supercontinuum spectrometer (figure 1). It was based on a Ti:Sapphire regenerative amplifier Regulas 35F1K, which produces 30 fs pulses with energy up to 2.3 mJ and repetition rate of 1 kHz. Beam was split into two parts.

One was sent through the delay line and used as a probe beam, while the second part of the beam was used for supercontinuum (SC) generation. To produce high-intensity ultra-wide SC in the IR and UV spectral regions the water jet was used [7]. The range of frequencies generated in SC was 350-1400 nm. By using pump-probe technique, we recorded pump-probe scans, pumping at 800 nm and probing at different frequencies and power levels of beams. Variation of probe beam characteristics results on change of the width and the position of 2PA peak. Shift of band peak dependence on probe wavelength can be used to quantify refractive index dispersion. Band width at half-maximum wave dependence helps to determinate the relaxation of induced excitation in material.

2.1. Femtosecond pump-probe supercontinuum spectroscopy experiment results

Using wavelength evolutions of 2PA absorption bands at half-maximum for ZnSe, ZnTe [8] and GaP crystals, extracted from various power dependences, and with knowledge of sample thickness, we obtained the dispersion of refractive index (figure 2). Here we present results only ZnTe crystal. It follows that the laser intensity on the order of 10 GW/cm^2 doesn't lead to non-linear response in ZnTe, ZnSe and GaP crystals in considered frequency range. However, radiation with the intensity on the order of more than 20 GW/cm^2 may cause nonlinear response of detection systems and make experimental findings unreliable.

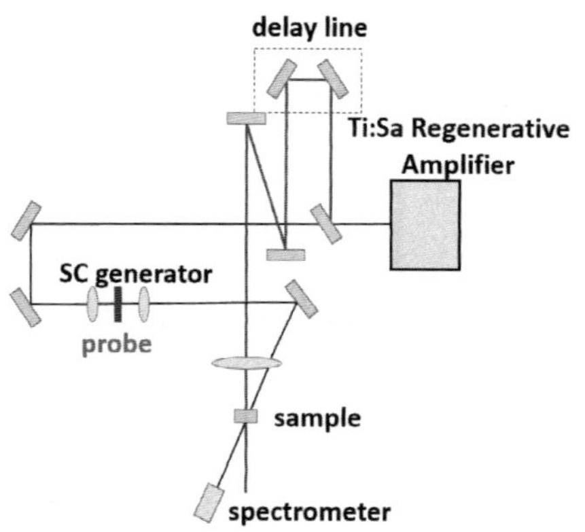

Figure 1. Scheme of pump-probe supercontinuum spectrometer.

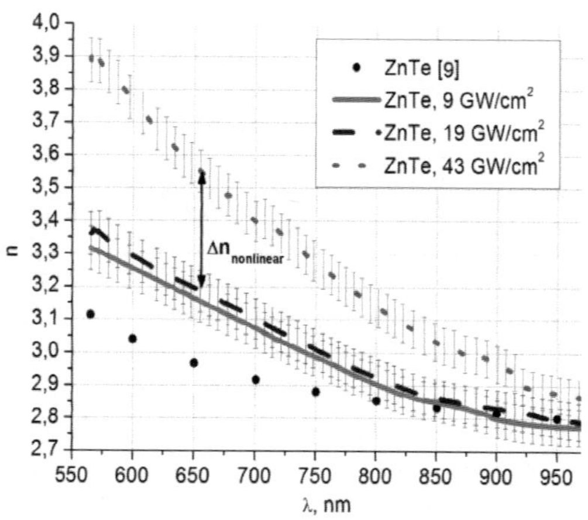

Figure 2. Dispersion of refractive index for different pump power levels for ZnTe (black dot values given in [9]).

3. Influence of crystal impurities on transmitted THz radiation

The next studies will help us to understand the way how to increase the THz emission from doped crystals. As it was shown in [5], increase in the concentration of impurities can both enhance and reduce THz transmission through the sample.

In our experiment we investigate two samples: pure ZnSe and ZnSe:Fe* Fe=[0,23%] wafers. The orientation of samples was the same (111), as well as thickness – 610 μm. To find out the influence of crystal impurities on transmitted THz radiation typical THz spectrograph was use. The experimental setup can be seen in figure 3.

IOP Conf. Series: Journal of Physics: Conf. Series **917** (2017) 062021 doi:10.1088/1742-6596/917/6/062021

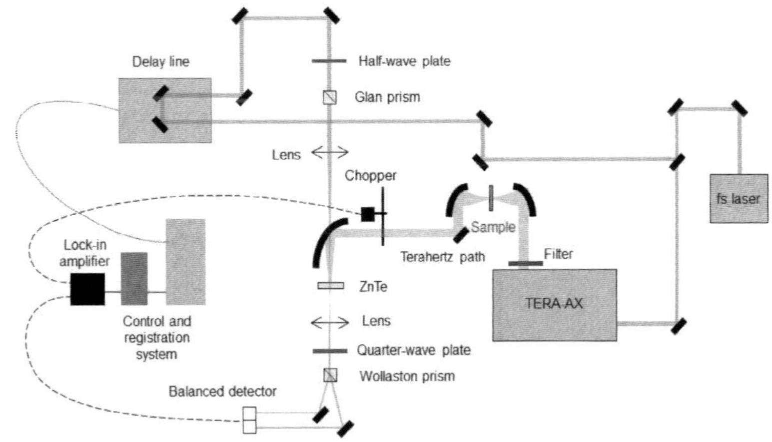

Figure 3. Experimental setup of THz time-domain spectrometer.

The light source is the same as in the pump-probe supercontinuum spectrometer. THz generator (Tera-AX Avesta) has the central frequency of 1 THz with the spectrum width of 1.5 THz, pulse duration <1 ps and THz pulse energy up to 1 uJ. Detection of THz radiation is performed by an electro-optical method using a ZnTe crystal and a balanced photodetector. As a detector Nirvana Auto-Balanced Photoreceiver is used with an operating wavelength range of 400–1070 nm and rise time <3 μs.

3.1. Terahertz time-domain spectroscopy experiment results

With the help of terahertz time-domain spectrometer we were able to get transmitted THz signals through two different samples and to compare their waveforms (figure 4). Fe impurities presence of course increase absorption of the sample, but on some time regions incensement of transmittance can be seen. It is our belief that because of low concentration of imputes this incensement is negligible.

Then the forward Fourier transformation was used to obtain spectral information. Figure 5 shows the amplitude of the terahertz spectrum.

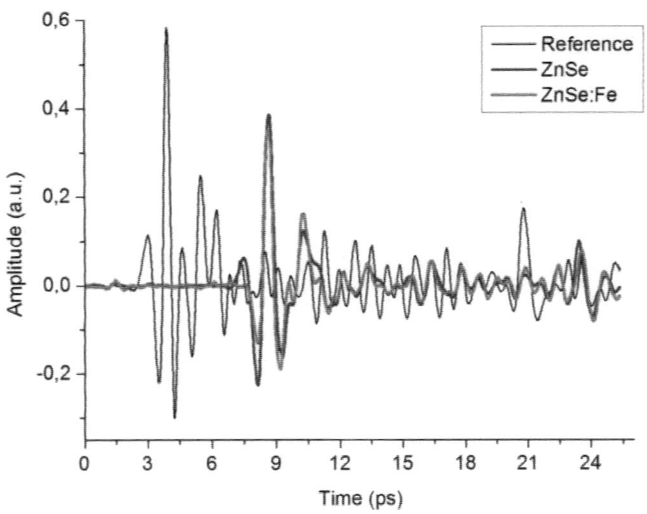

Figure 4. Temporal shapes of THz pulses for reference, transmitted through ZnSe and ZnSe:Fe signals.

If we compare the THz spectra of reference and transmitted through ZnSe and ZnSe:Fe signals, it is seen that at certain frequencies the transmission of ZnSe:Fe increases, as was shown in [5]. The presence of impurities in the sample makes the medium more transparent in the THz range. As it was mentioned before, because of low concentration of imputes this incensement is negligible. This effect can increase the generation of THz radiation. This can be explained in such way: when the ZnSe:Fe is doped to a certain concentration, the polarizability of the molecule increases. For more accurate descriptions and conclusions, further research is needed.

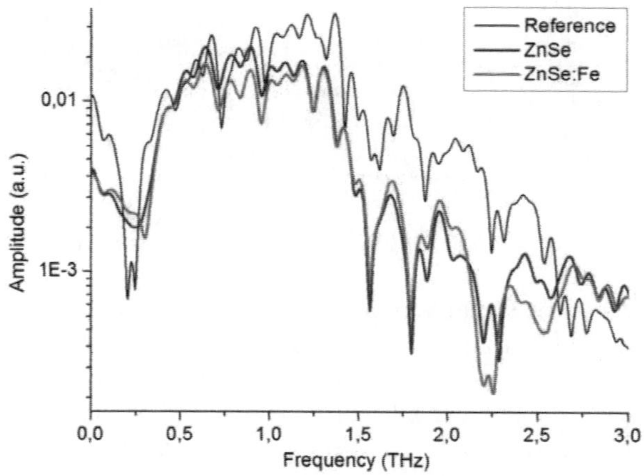

Figure 5. THz spectra of reference and transmitted through ZnSe and ZnSe:Fe signals.

4. Conclusion

With the help of femtosecond supercontinuum pump probe spectroscopy we were able to study nonlinear response of electro-optical semiconductor crystals at several excitation intensities ranging from 5 to 50 GW/cm^2 and in frequency range of 500-1000 nm. This enables to measure refractive index dispersion and to determine the optimum intensity at which no nonlinearity arises – it's around the order of 20 GW/cm^2. We also carried out experiments to study the transmission of THz radiation through ZnSe and ZnSe:Fe crystals using THz time-domain spectrometer. The presence of impurities in the sample makes the medium more transparent on some frequencies of THz range. Dependencies obtained from the experimental data matches well with one from previous works.

References

[1] Withayachumnankul W, Naftaly M 2014 J. *Infrared Millim. TE* **35** 610
[2] Grachev Y, Osipova M, Bespalov V 2014 *J. Quantum Electronics* **44** 1170
[3] Zhang X C, Xu J 2010 *Introduction to THz wave photonics* (New York: Springer)
[4] Osipova M, Makarov E, Bespalov V 2016 Proc. Int. Conf. of LO (St. Petersburg) vol 1 (IEEE) p. R8-45
[5] Red'kin R, Bereznaya S, Korotchenko Z, Sarkisov S 2015 Proc. Int. Conf. SIBCON vol 1 (IEEE) p.1
[6] Hase M 2009 *J. Applied physics letters* **94** 112111
[7] Tcypkin A, Putilin S, Melnik M, Makarov E, Bespalov V, Kozlov S 2016 *J. Applied optics* **55** 8390
[8] Avetissov I, Mozhevitina E, Khomyakov A, Avetisov R, Davydov A, Chegnov V, Chegnova O, Zhavoronkov N 2015 *J. CrystEngComm* **17** 561
[9] Li H 1984 *J.of phys. and chem. ref. data* **13** 103

Light propagation through conventional and extreme-2D van-der-Waals resonant photonic crystals

D R Kazanov[1], **A V Poshakinskiy**[1], **S V Sorokin**[1] **and T V Shubina**[1]

[1]Ioffe Institute, 26 Politekhnicheskaya, St Petersburg 194021, Russia

Abstract. We report on theoretical studies of short optical pulse propagation through resonant photonic crystals (RPCs) based on conventional II-VI quantum wells (QWs) and van-der-Waals (vdW) Bragg heterostructures composed of extreme-2D monolayers. The parameters needed for modelling were determined by fitting of the experimental reflectance spectra. The delay of the transmitted picosecond optical pulses is predicted to be larger in vdW structures than in the structures with QWs due to the higher radiative width of the exciton resonance. Moreover, in the vdW RPCs, the pulse shape undergoes smaller modification owing to the almost linear dispersion of the "slow" mode. High-quality extreme-2D monolayers will allow for even stronger delays without distortion and small intensity decrease, needed for nanophotonics applications.

1. Introduction

Quantum information processing has been playing an important role in the world since quantum computers came into sight. Therefore, the study of the light pulse propagation through structures becomes crucial. The optical pulse can change its shape and slow down while travelling through crystals. Slowing down of light frequently exploits a sharp change in the refractive index near a resonance, which results in the decrease of a group velocity [1]. The refractive index corresponding to "slow" modes in photonic crystals can be increased by dozens of times. Yet this approach often ties with a strong attenuation of the pulse. One of the promising methods is to use optical modes in RPCs, where exciton resonances coincide in frequency with optical modes [2, 3]. In the RPCs, light dispersion is strongly modified near the Bragg frequencies that influences the slow mode performance. Previous studies experimentally demonstrated that RPC with a simple unit cell can provide the delay up to half of the incoming pulse width with attenuation of the intensity up to 1% [4]. We have shown that varying the parameters of the QWs and the design of the unit cell one can control the attenuation of light, its shape change and the group velocity [5, 6]. In such structures, slight detuning of exciton resonance from Bragg condition can provide a smooth control of the dispersion law of the "slow" modes. Hence, we can achieve light velocities as low as $v = c/50$ with only 3-5 times attenuation.

In this paper, we investigate a novel type of RPCs – van-der-Waals (vdW) Bragg structures. They consist of extreme two-dimensional (2D) vdW monolayers, e.g., transition metal dichalcogenides WSe_2, MoS_2, etc. Such extreme-2D materials have begun to draw attention since the appearance of graphene [7]. Heterostructures based on the extreme-2D monolayers offer a unique opportunity to design nanodevices with desired optical properties. We present a comparative study of light propagation through the conventional RPCs based on QWs and through vdW heterostructures comprising stacks of 2D monolayers, whose position in a dielectric matrix (SiO_2) corresponds to the Bragg condition.

2. Structures and parameters

To model light propagating through RPCs we need the material parameters such as the exciton energy ($\hbar\omega_0$), the radiative width (Γ_0) and the non-radiative width (Γ) of the exciton resonance. These parameters can be obtained by fitting the experimental reflectance spectra with theoretical equations obtained using the transfer matrix technique. In this approach, each layer in a structure, i.e., a quantum well or a barrier, is described by a transfer matrix relating the amplitudes of the electric field at the left and right edges of the layer. The transfer matrix through several layers is the product of the transfer matrices of these layers. As a result, the dispersion law, the reflection and transmission coefficients can be calculated from the matrix equation. The reflectance from the structure comprising a 2D layer or a QW, a substrate and a coating layer reads:

$$R = |r_c + \frac{\Gamma_0}{\omega - \omega_0 + i(\Gamma_0 + \Gamma)}|^2 \quad (1)$$

The term r_c in the right-hand-side of the Eq. (1) describes Fresnel reflection of the incoming light from the substrate and the coating layer. It depends on the width of layers and dielectric contrast between them. The second term in Eq. (1) shows the reflection associated with exciton resonance in the 2D layer or QW. Multiple light reflections are neglected in Eq. (1), while taken into account natively in the transfer matrix simulation.

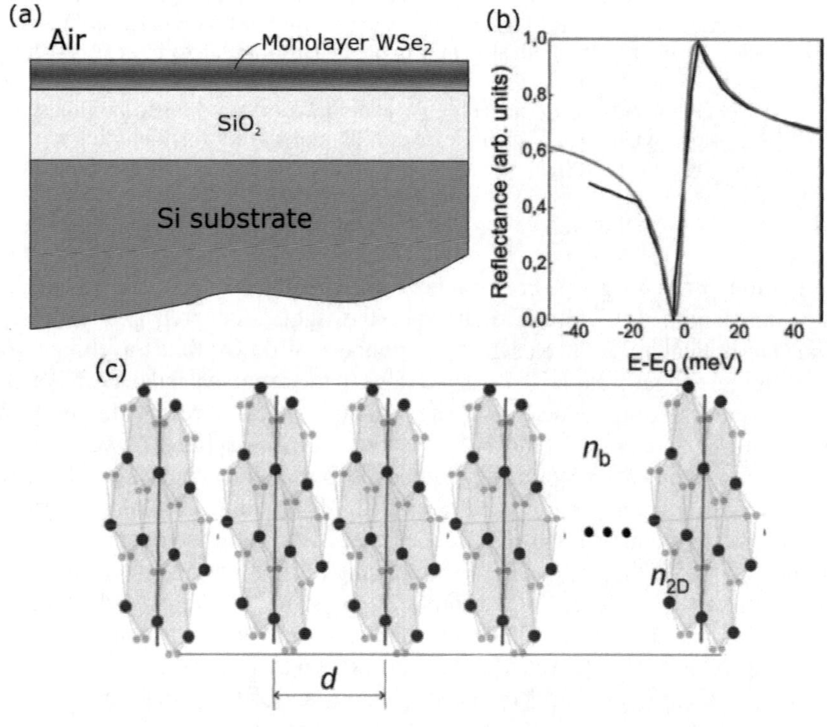

Figure 1. (a) Structure of the extreme 2D WSe_2 crystal on a substrate. (b) Experimental reflectance spectra of WSe_2 (black line, from Ref. [9]) and its fit by Eq. (1) (red line). (c) A sketch of vdW RPC.

As a model system for the conventional RPCs we investigated the spectra of the ZnSe/ZnMgSSe structure with a single quantum well grown on GaAs epitaxial buffer layer pseudomorphically to GaAs (100) substrate by molecular beam epitaxy (MBE) [8]. Exciton properties turned out to be $\hbar\omega_0$=2.83 eV, $\hbar\Gamma_0$~0.17 meV, and $\hbar\Gamma$~1.9 meV. For vdW heterostructures, we have taken the reflection spectra measured at the low temperature (4 K) in a monolayer of the transition metal dichalcogenide WSe_2, deposited on top of SiO_2 layer on Si substrate (Figure 1a), reported in [9]. Our fit of the experimental spectrum is shown in Figure 1b. The determined parameters of the exciton resonance are $\hbar\omega_0$=1.75 eV, $\hbar\Gamma_0$~0.3 meV and $\hbar\Gamma$~3.0 meV. The large value of the non-radiative exciton width is due to inhomogeneous broadening of the exciton resonance. The obtained parameters correspond to the typical extreme 2D structures of moderate quality. We should notice that radiative and non-radiative parameters of the exciton resonances for extreme-2D vdW monolayers can be obtained from the other reflectance experiments [10, 11]. The variation of parameters indicates that 2D crystals are still not mature in terms of fabrication.

Knowledge of the parameters of the exciton resonances can allow us to model pulse propagation through RPC Bragg structures. Our conventional RPCs consist of ZnSe quantum wells located between ZnMgSSe barriers of a finite width. In vdW RPCs, monolayers of WSe_2 are separated by SiO_2 spacers, see Figure 1c. Note that background reflection of the monolayer cannot be neglected. While the thickness of the monolayer is small, dielectric contrast with spacer is strong. The period of both conventional and vdW RPCs is chosen in such way that Bragg condition is fulfilled at the frequency close to exciton resonance.

3. Results and discussions

Using the transfer matrix technique, described in detail in [6], we have calculated the light dispersion inside the vdW and conventional RPCs and modelled pulse transmittance through the structures with 70 resonant layers, see Figure 2. We considered two values of the non-radiative exciton resonance width, $\hbar\Gamma$=0.3 meV (red line) and $\hbar\Gamma$=3.0 meV (black lines), while keeping the parameter $\hbar\Gamma_0$=0.3 meV fixed for both types of RPCs. Incident pulses had the carrier frequency slightly detuned from the exciton resonance ensuring the balance between pulse delay and attenuation.

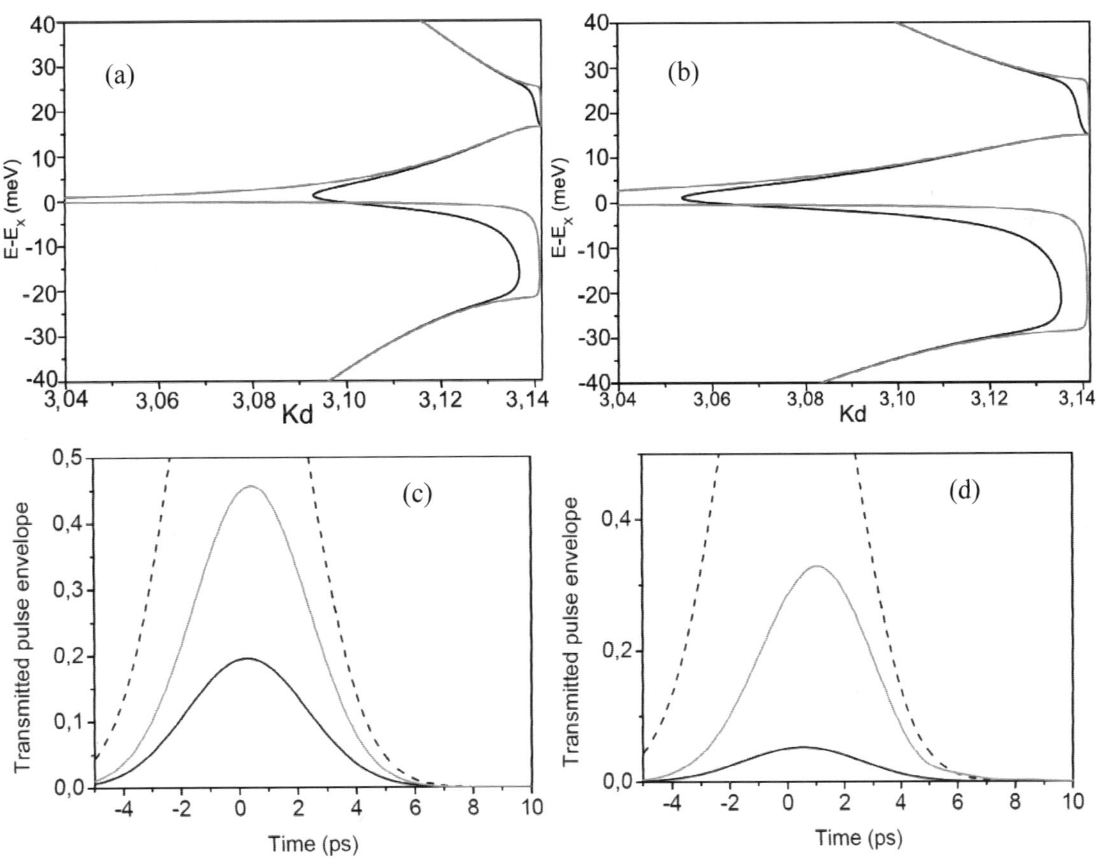

Figure 2. The light dispersion and transmitted pulse envelope for the conventional RPC (a, c) and the extreme-2D vdW RPC (b, d) consisting of 70 layers. For both RPCs, the non-radiative exciton width is

chosen ħΓ=0.3 meV (red line) or ħΓ=3.0 meV (black line), $\hbar\Gamma_0$=0.3 meV. The initial pulse envelope is a Gaussian with 2 ps half width, see the dotted lines.

Comparing of the results for the conventional and the vdW RPCs we conclude that vdW 2D Bragg structure has an advantage in the light delay over the conventional II-VI RPCs. The pulse delay is about 2 ps for the vdW RPC as compared to 1 ps for the conventional RPC. Such delay is comparable with the initial pulse half width (2 ps). On the other hand, transmitted pulse intensity in the conventional PRC is 1.5 times larger than in vdW structures. We note that the increase of structure quality (i.e., decrease of the non-radiative exciton broadening) will lead to the improvement in the pulse delay and attenuation, see red lines in Figure. 2d. Interestingly, light dispersion of the vdW RPC reveals the almost linear behavior of the "slow" mode throughout the half width of the stopband. This ensures minimal distortion of the pulse. Contrary, in the conventional RPCs, the linear behavior is not traceable.

Finally, the Bragg structure consisting of 20 periods ZnSe (9.7nm)/ZnMgSSe (64.4nm) has been grown by MBE on GaAs (001) substrate via 0.2 μm-GaAs buffer layer using double-chambers MBE setup. The total thickness of the structure was 1.51 μm. Modelling of the reflection spectra shows that peculiarities related to interaction of exciton and Bragg resonances are very small for a such thin RPC Bragg structure. However, the exciton resonance was clearly observed in the reflection experiments. The MBE growth of the 70-100 periods ZnSe/ZnMgSSe RPC Bragg structure with the total thickness of 5-7 μm looks as a rather complicated task because of the strong requirements for both growth temperature and Se flux stability during the growth run. In case of MBE growth of S-containing alloys the time drift of both parameters results in growth rate (and, in turn, in thickness) and sulfur content variations throughout the RPC structure [12]. We hope that better quality and longer length of the RPCs will find someday various applications in nanophotonics.

Acknowledgments
The studies of vdW and II-VI RPCs structures were supported, respectively, by the Government of the Russian Federation (Project # 14.W03.31.0011) and RFBR (Project № 16-02-00204).

References:
[1] Shubina T V *et al.*, *Phys. Rev. Lett.* **100**, 087402 (2008)
[2] Ivchenko E L, Nesvizhskii A I, and Jorda S, *Phys. Sol. State*, **36**, 1156 (1994)
[3] Ivchenko E L, Voronov M M, Erementchouk M V, Deych L I, and Lisyansky A A, *Phys. Rev.* B **70**, 195106 (2004)
[4] Prineas J P, Johnston W J, Yildirim M, Zhao J, and Smirl A L, *Appl. Phys. Lett.* **89**, 241106 (2006)
[5] Poshakinskiy A V, Poddubny A N, Pilozzi L, and Ivchenko E L, *Phys. Rev. Lett.* **112**, 107403 (2014)
[6] Kazanov D R, Poshakinskiy A V, and Shubina T V, *JETP Lett.* **105**, 8-12 (2017)
[7] Geim A K and Grigorieva I V, *Nature* **499**, 419 (2013)
[8] Kotova L V, Platonov A V, Kats V N, Kochereshko V P, Sorokin S V, Ivanov S V, and Golub L E, *Phys. Rev* B **94**, 165309 (2016)
[9] Wang G, Bouet L, Lagarde D, Vidal M, Balocchi A, Amand T, Marie X, and Urbaszek B, *Phys. Rev.* B **90**, 075413 (2014)
[10] Cadiz F *et al.*, *Phys. Rev.* X **7**, 021026 (2017)
[11] Manca M *et al.*, *Nat. Commun.* **8,** 14927 (2017)
[12] Ivanov S V, Sorokin S V, Kop'ev P S, Kim J R, Jung H D, Park H S, *J. Cryst. Growth* **159**, 16 (1996)

Emission properties of individual InAs/$Al_{0.44}Ga_{0.56}$As quantum dots

M V Rakhlin[1], K G Belyaev[1], G V Klimko[1], I S Mukhin[2,3], S V Ivanov[1], and A A Toropov[1]

[1]Ioffe Institute, Saint Petersburg 194021, Russia

[2]St. Petersburg Academic University, St. Petersburg 194021, Russia

[3]ITMO University, St. Petersburg 197101, Russia

maximrakhlin@mail.ru

Abstract. We report on micro-photoluminescence studies of InAs/AlGaAs single self-assembled quantum dots grown by molecular beam epitaxy. A variation of the exciton fine structure splitting induced by the anisotropic part of the electron-hole exchange interaction is systematically studied depending on the exciton emission energy and the insertion of a GaAs interlayer prior formation of the InAs QDs.

1. Introduction

In recent years, an electron-hole exchange interaction in semiconductor quantum dots (QDs) has been the subject of careful studies [1] because detailed understanding of the relevant exciton fine structure is of crucial importance for applications of the individual QDs as single photon emitters possessing non-classical photon statistics, and especially as sources of the entangled pairs of single photons. In principle, the polarization-entangled photons can be generated by recombination of a biexciton cascade in a single QD. However, in the QD structures with low enough confinement symmetry (C_{2v} or lower), the anisotropic part of the electron-hole exchange interaction lifts twofold degeneracy of the exciton bright states, resulting in the exciton fine structure splitting (FSS). If the FSS is larger than the exciton homogeneous linewidth, it inhibits the emission of two entangled photons [2]. The issues of controlling both the FSS and the exciton linewidth have been relatively well understood for conventional InAs/GaAs QDs [1,2], whereas for InAs/AlGaAs single QDs which are promising non-classical light sources of visible light, this has been yet poorly investigated [3,4].

In this work, we apply micro-photoluminescence (μ-PL) spectroscopy to elucidate the emission properties of individual InAs/$Al_{0.44}Ga_{0.56}$As QDs, grown by molecular beam epitaxy (MBE). In particular, we investigate the influence on the FSS of the presence of a thin GaAs interlayer deposited on a AlGaAs bottom barrier prior to the InAs QDs formation.

2. Experiment

We have studied emission properties of individual InAs/$Al_{0.44}Ga_{0.56}$As QDs either with (sample A) or without (sample B) a 2-monolayer (ML)-thick GaAs interlayer inserted between the bottom AlGaAs

IOP Conf. Series: Journal of Physics: Conf. Series **917** (2017) 062023 doi:10.1088/1742-6596/917/6/062023

barrier and the InAs QDs. The QDs were formed by MBE on GaAs (001) substrates, using the Stranski-Krastanow growth mode. Special measures were taken to reduce the density of QDs down to around 10^{10} cm^{-2}. The emission of a limited number of individual QDs was registered by the confocal μ-PL set-up in cylindrical mesa-structures of 200-800 nm in diameter, which were fabricated by electron beam lithography and Ar^+-ion etching. Scanning electron microscopy studies have confirmed formation of perfect cylindrical mesas with vertical sidewalls. The sample is mounted in a He-flow cryostat with an Attocube XYZ piezo-driver inside, which allows us to optimize and precisely maintain the positioning of the chosen mesa with respect to a laser spot during a long time (few hours). The μ-PL measurements were carried out under optical excitation by a *cw* laser line (405 nm). The laser power density was 4 W/cm^2 before a cryostat window. Photon correlation measurements were performed in a Hanbury Brown-Twiss detection scheme exploiting two single-photon avalanche photo diodes possessing the photon timing resolution of about 40 ps.

3. Results and discussion

Typical *cw* PL spectra (integrated over the area of ~200 μm^2) of samples A and B, measured at 77 K, are shown in Figure 1a. The emission spectrum in sample A with a GaAs interlayer reveals two peaks attributed to the emission of QDs and a wetting layer (WL). The QD PL line is centered between 900 and 1000 nm, that is similar to the emission spectrum of most common InAs/GaAs QDs. However, the blue wing of the PL line spreads over the significant spectral range right up to the narrow PL peak from the WL, located near 1.8 eV (~690 nm). Thus, most part of the InAs/AlGaAs QDs in sample A emit in the same spectral range as InAs/GaAs QDs, possessing, however, much larger activation energy for the confined charge carriers, that is promising for higher temperature operation. Only one strongly asymmetric peak attributed to the QDs emission can be seen in the PL spectrum of sample B. The peak is markedly blue-shifted relative to the QD PL line in sample A, that is a signature of a smaller QD size in sample B, confirmed also by atomic force microscopy (not shown here), and/or a different proportion of Ga and Al atoms incorporation into the QDs due to inter-diffusion. Thus, the absence of the GaAs interlayer in sample B causes the shift of the PL spectrum towards higher energy and the lack of the WL emission. This sample reveals a wide PL spectrum extending over the energy range between ~1.4 and ~2 eV (890 – 620 nm), which allows one to tune the individual QDs emission wavelength over the extremely wide wavelength range from visible red to near infrared.

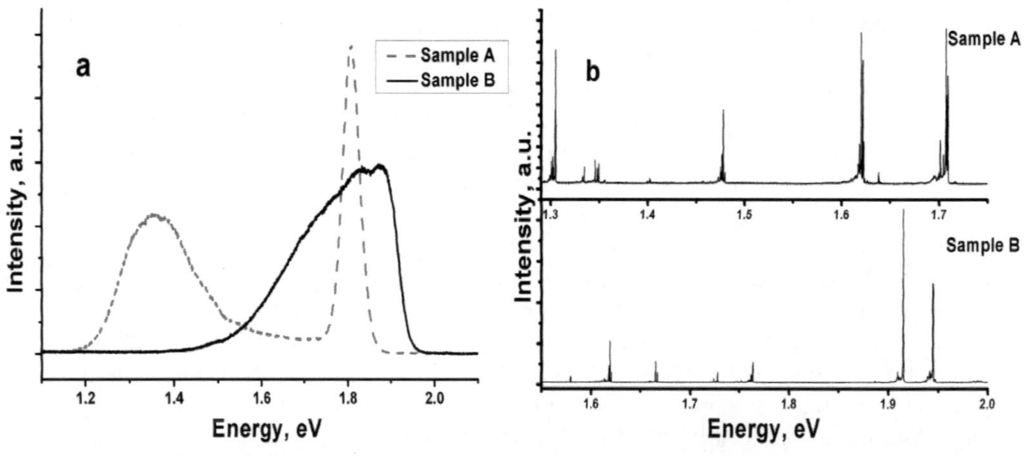

Figure1. (a) Normalized low temperature (77 K) spatially integrated PL spectra of the structures grown either with a GaAs interlayer (sample A) or without it (sample B). (b) μ-PL spectra measured at 8 K in InAs/AlGaAs QDs, using 500-nm etched mesas.

The μ-PL spectra measured at 8 K reveal a number of relatively narrow lines assigned to the emission of excitons and other electron-hole complexes – biexcitons, trions, and charged excitons [5] – in individual QDs (Figure 1b). In both samples, the ballpark number of narrow excitonic lines in a μ-PL spectrum is estimated as about several tens in the mesas with 500-nm diameters, that corresponds to the QDs density as small as $1.7\times10^{10}cm^{-2}$ (sample A) and $1.3\times10^{10}cm^{-2}$ (sample B). The obtained QDs densities are low enough and suit perfectly for creation of single-photon sources based on the emission of single QDs.

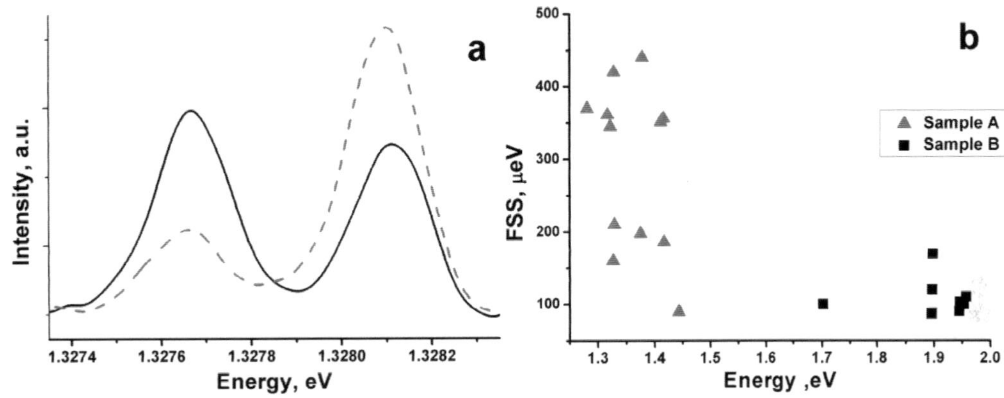

Figure 2. (a) Typical single-exciton FSS as large as 440 μeV, measured in sample A at 8 K. Solid and dashed lines represent the spectra for orthogonal linear polarizations corresponding to [110] and [1-10] crystal directions. (b) Measured FSS as a function of the exciton recombination energy for samples A and B.

It was found that different QDs in sample A, located in the same mesa-structure, display various FSS ranging from the values less than 70 μeV (set-up spectral resolution) up to extremely large FSS of the order of 500 μeV. A typical μ-PL spectrum with FSS equal to 440μeV is shown in Figure 2a. Solid and dashed curves are the spectra for orthogonal linear polarizations corresponding to [110] and [1-10] crystal directions of the QD sample. Such a large FSS is consistent with the results of Finley *et al.* who reported on the FSS values up to 1 meV [3]. In contrast, different QDs in sample B demonstrate approximately the same FSS equals to 100-150 μeV. This value is quite comparable with the typical exciton linewidth (~130 μeV) detected in this sample. Figure 2b displays the measured exciton FSS as a function of the exciton recombination energy for the samples with and without the GaAs interlayer. We found out that there is no pronounced dependence of FSS on the exciton recombination energy in both samples. The presence of the GaAs interlayer in sample A results in enhanced scattering of the FSS magnitude among different QDs. Nevertheless, most of the excitonic lines are not split in both samples, at least within the set up spectral resolution. On average, there are 11 unpolarized exciton lines per one line with FSS.

Figure 3 shows the autocorrelation function $g^{(2)}(\tau)$ of a single QD, obtained at T= 8 K in sample B. The experimental data on photon coincidences is fitted with the equation

$$f(\tau) = a - b\exp\left(-\frac{|\tau|}{c}\right) \quad (1)$$

and sequentially normalized by count rates of the detectors, width of time bins, and the integration time [6]. The autocorrelation function exhibits an antibunching dip at zero time delay with the average $g^{(2)}(0)$ as low as 0.24 that is a clear signature of the single-photon nature of the emission. The obtained value c=650 ps gives estimation of the intrinsic width of the antibunching dip, which is essentially due to the lifetime of the involved QD exciton.

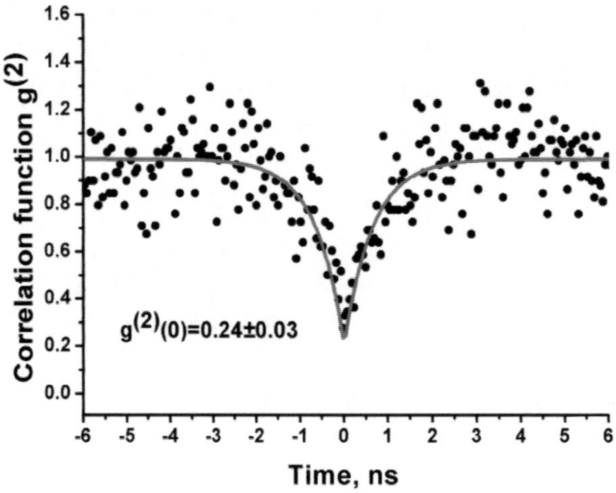

Figure 3. Normalized second-order correlation function ($g^{(2)}$) of single-photon emission, measured at T= 8 K in a 500-nm mesa in sample B.

4. Conclusions

We have represented comparative studies of self-assembled $InAs/Al_{0.44}Ga_{0.56}As$ QDs grown either with or without a GaAs interlayer, whose emission spectrum overlaps the extremely wide wavelength range from visible red to near infrared (1.1-2 eV). The performed measurements of exciton fine structure splitting reveal that the predeposition of the 2 ML-thick GaAs interlayer leads to enhanced scattering of the FSS values ranging between 70 μeV (the set up spectral resolution) and ~ 500 μeV. The measured second-order correlation function at zero delay $g^{(2)}(0)$ as low as 0.24 is a promising value for applications in effective systems of quantum cryptography.

Acknowledgments

The authors gratefully acknowledge the financial support of the Russian Science Foundation (project #14-22-00107).

References

[1] Seguin R, Shliwa A, Rodt S, Pötchke K, Pohl U W, Bimberg D, 2005 *Phys. Rev. Let.* **95** 257402
[2] Santori C, Fattal D, Pelton M, Solomon G S, Yamamoto Y, 2002 *Phys. Rev. B* **66** 045308
[3] Finley J J, Mowbray D J, Skolnick M S, Ashmore A D, Baker C, Monte A F G, Hopkinson M, 2002 *Phys. Rev. B* **66** 153316
[4] Grijseels S C M, van Bree J, Koenraad P M, Toropov A A, Klimko G V, Ivanov S V, Pryor C E, Silov A Yu, 2016 *J. of Lum.* **176** 95
[5] Karlsson K F, Oberli D Y, Dupertuis M A, Troncale V, Byszewski M, Pelucchi E, Rudra A, Holtz P O, Kapon E, 2015 *New J. of Phys.* **17** 103017
[6] Brouri R, Beveratos A, Poizat J-P, Grangier P, 2000 *Opt. Lett.* **25** 1294

Transparent glass-ceramics with Yb^{3+},Ho^{3+}:$YNbO_4$ nanocrystals for green phosphors

A Volokitina[1], P Loiko[1,*], O Dymshits[2], M Tsenter[2], S Zapalova[2], K Bogdanov[1], A Baranov[1] and A Zhilin[2]

[1]ITMO University, Kronverkskiy pr., 49, 197101 St. Petersburg, Russia
[2]NITIOM S.I. Vavilov State Optical Institute, 36 Babushkina St., 192171 St. Petersburg, Russia

Abstract. We report on synthesis, structure, Raman and optical spectroscopy of transparent glass-ceramics (GCs) containing nanocrystals of rare-earth orthoniobates, Yb^{3+},Ho^{3+}:$YNbO_4$ and β-quartz solid solution. Under the near-IR excitation, the GCs exhibit intense upconversion luminescence which color properties can be tuned by the heat-treatment regime due to the structural changes in the orthoniobate nanocrystals. The developed GCs are promising as thermal shock resistant green phosphors.

1. Introduction

Materials containing couples of Yb^{3+} ion and other RE^{3+} ion are attractive for the development of upconversion phosphors [1]. In the particular case of $RE^{3+} = Er^{3+}$, Tm^{3+} or Ho^{3+}, this is supported by direct and efficient energy-transfer (ET) process $Yb^{3+} \rightarrow RE^{3+}$ facilitated by the resonance in energy of the excited states of the considered ions. Transparent glass-ceramics (GCs) containing nanosized crystals and an amorphous residual phase are promising matrices for (Yb^{3+}, RE^{3+}) codoping [2-4]. With a proper selection of the composition of the initial glass and the regime of the secondary heat-treatment leading to crystallization, one can ensure predominant entering of the active ions in the nanocrystalline phase [5]. This opens the possibility of tuning of emission color according to structural transformations of these nanocrystals [6].

Rare-earth orthoniobates, $RENbO_4$ (where RE = La, Y, Gd, and Lu are the host-forming "passive" ions), doped with the active ions as specified above, are promising luminescence materials [7-10]. A clear example of such a material is fergusonite, $YNbO_4$, belonging to the class of ABO_4-type oxides, and exhibiting two main modifications, namely the tetragonal (scheelite-like) and monoclinic ones [11,12]. Rare-earth orthoniobates possess low phonon frequencies (for oxide matrices), $h\nu_{ph}$~800 cm^{-1}, they offer high available doping concentrations with a weak quenching of luminescence, long emission lifetimes, strong interionic interactions, high refractive index (n ~2.1) and relatively broad bandgap of ~4.3 eV.

GCs containing the nanocrystals of rare-earth orthoniobates, $RENbO_4$, have been recognized as efficient luminescent materials [13,14]. To date, GCs containing Yb^{3+}, Er^{3+}:$YNbO_4$ and Yb^{3+}, Eu^{3+}: $YNbO_4$ nanocrystals have been prepared showing intense green and red upconversion luminescence (UCL) under the near-IR excitation by InGaAs laser diodes, respectively. In the present work, we have developed and studied GCs based on the nanocrystals of Yb^{3+},Ho^{3+}:$YNbO_4$ and β-quartz solid solution, for the first time, to the best of our knowledge.

2. Synthesis and structure

The initial glass with the composition of 18 Li_2O – 27 Al_2O_3 – 55 SiO_2 (mol%) [13] doped with the rare-earth oxides, 2.2 Y_2O_3, 0.5 Yb_2O_3, 0.5 Ho_2O_3, and niobium oxide, 3.2 Nb_2O_5 (mol%) was prepared by the melt-quenching technique. Batch to produce 300 g of glass was melted in a quartz ceramic crucible in a laboratory electric furnace at 1580 °C for 4 h with stirring, then the glass melt was bubbled by oxygen for 0.5 h to remove the OH-groups from the melt. The melt was poured onto a metal plate and the as-cast glass was annealed at 620 °C. The initial glass was transparent and had a yellow coloration. The calculated concentration of the RE^{3+} active ions in the initial glass was $N_{Yb} = N_{Ho} = 1.97\times10^{20}$ cm^{-3} (measured glass density, $\rho = 2.76$ g/cm^3). To produce transparent GCs, the initial glass was heat-treated at temperatures ranging from $T = 720$ to 1000 °C for 6–24 h in air.

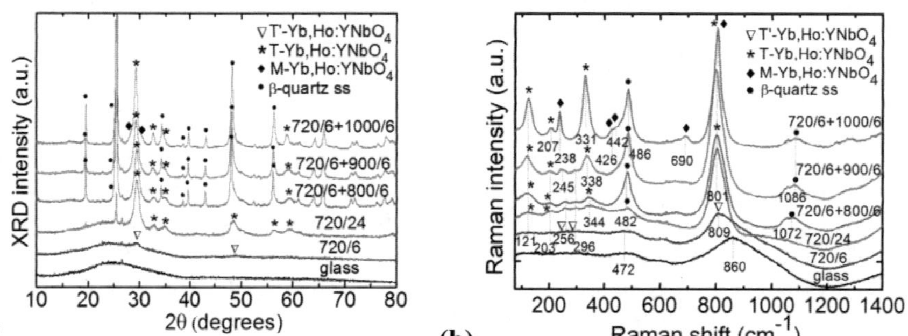

(a) (b)

Figure 1(a, b). (a) XRD patterns and **(b)** Raman spectra [combined data for λ_{exc} =514 nm (<600 cm^{-1}) and 488 nm (>600 cm^{-1})] of the initial glass and GCs containing Yb^{3+},Ho^{3+}:$YNbO_4$ nanocrystals. The numbers denote the heat-treatment regime, °C / hours.

The structure of GCs was studied with X-ray diffraction (XRD) and Raman spectroscopy, figure 1. The initial glass is X-ray amorphous. Crystals of $RENbO_4$ with a disordered fluorite-type cubic structure (T') (sp. gr. $Fm\overline{3}m$) evolve in the glass heat-treated at 720 °C for 6 h, see figure 1(a). The prolongation of this heat-treatment for 24 h leads to crystallization of tetragonal phase with orthorhombic distortion (T) $RENbO_4$ (sp. gr. *Ima*2); traces of β-quartz solid solution (ss) appear as well. For T-phase, the RE^{3+} cations occupy the S_4 symmetry site with an VIII-fold O^{2-}-coordination. After heat-treatments at 800–1000 °C, the GCs contain two crystalline phases, $RENbO_4$, and β-quartz ss. High thermal shock resistance of GCs is ensured by crystallization of the latter phase. Monoclinic (M) $RENbO_4$ crystals (fergusonite-type structure, sp. gr. *I*2/*a*) appear additionally to the distorted tetragonal ones after heat-treatment at 1000 °C. For M-phase, the RE^{3+} cations occupy the C_2 symmetry site also with an VIII-fold O^{2-}-coordination. The large RE^{3+}-RE^{3+} distances for the $RENbO_4$ crystals and the closeness of ionic radii of Y^{3+} (1.019 Å), Yb^{3+} (0.985 Å) and Ho^{3+} (1.015 Å) [15] determine the possibility for the active ions to enter easily into the $YNbO_4$ lattice where they are expected to replace the Y^{3+} "passive" ones.

Raman spectra of the initial glass and GCs are shown in figure 1(b). Raman spectrum of the initial glass contains two broad bands centered at ~472 cm^{-1} and 860 cm^{-1}. Heat-treatment at 720 °C / 6 h induces the modification of the spectrum due to precipitation of $RENbO_4$ nanocrystals with a disordered fluorite-type cubic structure. After the heat-treatment at 720 °C / 24 h, the Raman spectra show enhanced 801 cm^{-1} band and some other signs of the T-$RENbO_4$ crystals, namely several low-frequency bands at 121-344 cm^{-1} [16]. The increase of the heat-treatment temperature to 800 °C leads to the appearance of two bands at 482 cm^{-1} and at 1086 cm^{-1} indicative for β-quartz ss [17]. Finally,

IOP Conf. Series: Journal of Physics: Conf. Series **917** (2017) 062024 doi:10.1088/1742-6596/917/6/062024

the heat-treatment at 1000 °C results in the appearance of several bands at 238, 426, 442, 690 and 802 cm^{-1} representing precipitation of M-RENbO$_4$ crystals [18].

3. Optical spectroscopy

The absorption spectra of the initial glass and GCs containing absorption bands of the Yb^{3+} and Ho^{3+} ions are shown in figure 2. The spectra vary accordingly to the structural transformations of the GCs. This proves that the active ions, Ho^{3+} and Yb^{3+}, are embedded in the $YNbO_4$ nanocrystals.

As for Yb^{3+} ions, the broad absorption band spanning from 850 nm to 1050 nm is due to the $^2F_{7/2} \rightarrow {}^2F_{5/2}$ transition, and it can be used for their excitation with the InGaAs laser diodes. For the initial glass, the maximum absorption cross-section of Yb^{3+} ions is $\sigma_{abs} = \alpha_{abs}/N_{Yb} = 2.1\times10^{-20}$ cm^2 at 977.1 nm (zero-phonon line, ZPL). For GCs obtained by heat-treatment at 1000 °C, the position of the ZPL shifts to 974.5 nm, accompanied by the rise of absorption at shorter wavelengths, so that the maximum $\sigma_{abs} = 1.29\times10^{-20}$ cm^2 at 953.2 nm. For Ho^{3+} ions, the maximum $\sigma_{abs} = 0.43\times10^{-20}$ cm^2 at 1947 nm (for the initial glass) and it increases to 0.60×10^{-20} cm^2 at 1936 nm (for GCs obtained by heat-treatment at 1000 °C).

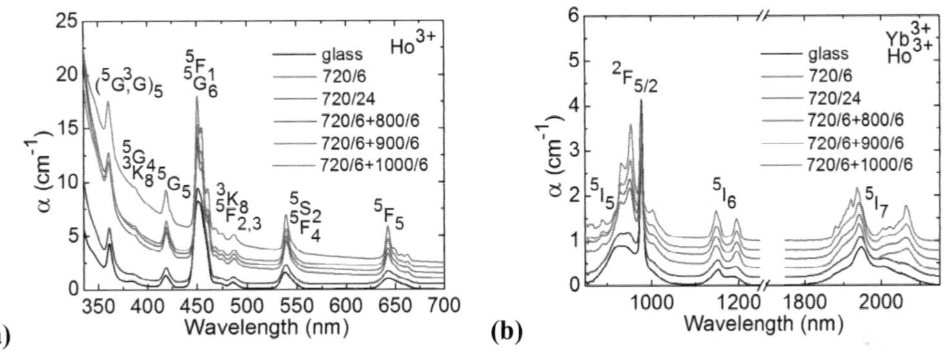

(a) (b)

Figure 2(a, b). Absorption spectra of the initial glass and GCs containing Yb^{3+},Ho^{3+}:$YNbO_4$ nanocrystals **(a)** in the visible and **(b)** in the near-IR. The numbers denote the heat-treatment regime, °C / hours. Each spectrum is shifted on 0.5 cm^{-1} (a) or 0.2 cm^{-1} (b) for the convenience of observation.

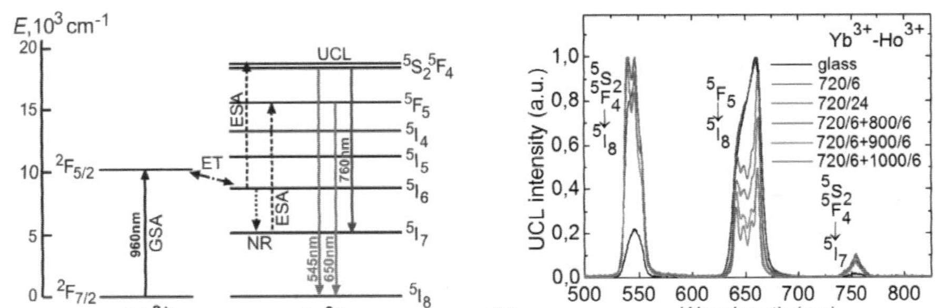

(a) (b)

Figure 3(a, b). (a) Scheme of the energy-levels of Yb^{3+} and Ho^{3+} ions with the relevant processes: GSA and ESA – ground- and excited-state absorption, respectively, NR – non-radiative relaxation, UCL – upconversion luminescence, ET – energy-transfer. **(b)** Upconversion luminescence spectra for the initial glass and GCs containing Yb^{3+},Ho^{3+}:$YNbO_4$ nanocrystals, λ_{exc} = 960 nm (the spectra are normalized to unity).

Under the excitation by a commercial InGaAs laser diode emitting at ~962 nm (to the $^2F_{5/2}$ state of the Yb^{3+} ion), intense visible emission is detected from the initial glass and GCs, see figure 3(b). It is related to the upconversion in the (Yb^{3+}, Ho^{3+}) ion couple, see figure 3(a). This process includes ground-state absorption (GSA) by Yb^{3+} ions, energy-transfer (ET) $^2F_{5/2}(Yb^{3+}) \rightarrow {}^5I_6(Ho^{3+})$, excited-state absorption (ESA) from the $^5I_6$ and $^5I_7$ intermediate states, and emission from the higher-lying $^5F_5$ and $^5S_2$+$^5F_4$ excited-states of Ho^{3+}. A remarkable feature of the GCs is the redistribution of intensity between the green (535-555 nm, the $^5S_2+{}^5F_4 \rightarrow {}^5I_8$ transition) and red (635-670 nm, the $^5F_5 \rightarrow {}^5I_8$ transition) upconversion luminescence (UCL) with the heat-treatment thus leading to the tunability of color properties.

The color coordinates (*Commission internationale de l'eclairage*, CIE 1931) for UCL of the initial glass are x = 0.534, y = 0.441 (dominant wavelength, λ_d = 584 nm, color purity, p = 93%, orange-yellow color) and for the GCs prepared by the heat-treatment at 1000 °C, they are x = 0.317, y = 0.648 (λ_d = 553 nm, p = 99%, green color).

4. Conclusion

Transparent thermal shock resistant glass-ceramics containing (Yb^{3+}, Ho^{3+}) ion couple entering the fluorite-type, distorted tetragonal and monoclinic yttrium orthoniobate nanocrystals, $YNbO_4$, are synthesized for the first time, to the best of our knowledge. The possibility to excite intense color-tunable (orange-green) upconversion luminescence is demonstrated with the synthesized GCs. Such materials are of interest for green phosphors. Further work will focus on the determination of the efficiency of the $Yb^{3+} \rightarrow Ho^{3+}$ ET.

Acknowledgments

This work was partly supported by the RFBR (Grant 16-03-01130). P.L. acknowledges financial support from the Government of the Russian Federation (Grant 074-U01) through ITMO Post-Doctoral Fellowship scheme.

References

[1] Auzel F 2004 *Chem. Rev.* **104** 139
[2] Wang Y and Ohwaki J 1993 *Appl. Phys. Lett.* **63** 3268
[3] Biswas A, Maciel G S, Friend C S and Prasad P N 2003 *J. Non-Cryst. Solids* **316** 393
[4] Dymshits O S, Loiko P A, Skoptsov N A, Malyarevich A M, Yumashev K V, Zhilin A A, Alekseeva I P, Tsenter M Ya and Bogdanov K 2015 *J. Non-Cryst. Solids* **409** 54
[5] de Pablos-Martin A, Patzig C, Höche T, Durán A and Pascual M J 2013 *Cryst. Eng. Comm.* **15** 6979
[6] Liu F, Ma E, Chen D, Yu Y and Wang Y 2006 *J. Phys. Chem. B* **110** 20843
[7] Nazarov M, Kim Y J, Lee E Y, Min K I, Jeong M S, Lee S W and Noh D Y 2010 *J. Appl. Phys.* **107** 103104.
[8] Liu X, Lü Y, Chen C, Luo S, Zeng Y, Zhang X, Shang M, Li C and Lin J 2014 *J. Phys. Chem. C* **118** 27516
[9] Li K, Zhang Y, Li X, Shang M, Lian H and Lin J 2015 *Phys. Chem. Chem. Phys.* **17** 4283
[10] Lu Y, Tang X, Yan L, Li K, Liu X, Shang M, Li C and Lin J 2013 *J. Phys. Chem. C* **117** 21972
[11] Blasse G and Bril A 1970 *J. Lumin.* **3** 109
[12] Mather S A and Davies P K 1995 *J. Amer. Ceram. Soc.* **78** 2737
[13] Dymshits O S, Alekseeva I P, Zhilin A A, Tsenter M Ya, Loiko P A, Skoptsov N A, Malyarevich A M, Yumashev K V, Mateos X and Baranov A V 2015 *J. Lumin.* **160** 337
[14] Loiko P A, Dymshits O S, Alekseeva I P, Zhilin A A, Tsenter M Ya, Vilejshikova E V, Bogdanov K V, Mateos X and Yumashev K V 2016 *J. Lumin.* **179** 64
[15] Shannon R D 1976 *Acta Cryst.* **A32** 751
[16] Blasse G 1972 *J. Solid State Chem.* **7** 169
[17] Chuvaeva T I, Dymshits O S, Petrov V I, Tsenter M Ya, Shaskin A V, Zhilin A A and Golybkov V V 2001 *J. Non-Cryst. Solids* **282** 306
[18] Yashima M, Lee J-H, Kakihana M and Yoshimura M 1997 *J. Phys. Chem. Solids* **58** 1593

IOP Conf. Series: Journal of Physics: Conf. Series **917** (2017) 062025 doi:10.1088/1742-6596/917/6/062025

FTIR photoreflectance of narrow-gap heterostructures based on $Al_xIn_{1-x}Sb$ alloys

D D Firsov*[1], O S Komkov[1], V A Solov'ev[2], A N Semenov[2], and S V Ivanov[2]**

[1]Micro- and Nanoelectronics department, St. Petersburg Electrotechnical University "LETI", Professora Popova 5, St. Petersburg 197376, Russia
[2] Ioffe Institute, Politekhnicheskaya 26, St. Petersburg 194021, Russia

Abstract. Photoreflectance (PR) spectra of $Al_xIn_{1-x}Sb$-based heterostructures have been obtained by using a novel photomodulation Fourier-transform infrared (FTIR) spectroscopy method. The studied samples - bulk $Al_xIn_{1-x}Sb$ epilayers and $InSb/Al_xIn_{1-x}Sb$ heterostructures - were grown by molecular beam epitaxy on semi-insulating GaAs substrates via AlSb buffer layers. The critical-point energies E_0 and $E_0+\Delta_0$ of $Al_xIn_{1-x}Sb$ alloys of various direct-gap compositions were defined from the PR spectra features. It was found that the E_0 values are greater than the observed $Al_xIn_{1-x}Sb$ photoluminescence peak energy, with the difference increasing with x. For the $InSb/Al_xIn_{1-x}Sb$ heterostructures, PR signals corresponding to Franz-Keldysh oscillations in the alloy barrier have been observed. Analysis of their period has allowed one to determine the intensity of the internal electric field in $Al_xIn_{1-x}Sb$ layers. This result enables evaluation of the surface Fermi level pinning, and elucidation of the effect of doping in such heterostructures.

1. Introduction

Narrow-gap $Al_xIn_{1-x}Sb$ alloys are the optimal barrier material for quantum well (QW) heterostructures based on InSb – a III-V semiconductor with the highest electron mobility and the largest g-factor – since their lattice parameters are close to that of InSb, and the band offsets increase rapidly with x. The $InSb/Al_xIn_{1-x}Sb$ QW heterostructures are promising for numerous applications, such as super high-frequency, low-voltage high-electron-mobility transistors (HEMT) with cutoff frequency above 300 GHz [1, 2], mid-infrared (IR) light-emitting diodes of the 4-5 μm range [3, 4], and even detectors of terahertz radiation [5]. The use of $Al_xIn_{1-x}Sb$ instead of InSb for IR focal plane arrays shifts their spectral range towards shorter wavelengths, and enables an increase of operating temperature from 80 K to 110 K [6, 7]. In addition, Sb-based III-V compounds are considered to be promising for silicon-based optoelectronic devices grown directly on Si substrates [8].

However, the fundamental properties of the $Al_xIn_{1-x}Sb$ alloys, as dependent on x, have only been briefly investigated, and the available data are quite scarce [9] and based on studies of polycrystalline samples with a high x gradient. Moreover, there exist only few publications dedicated to studying the optical properties of $InSb/Al_xIn_{1-x}Sb$ heterostructures (see e.g. [10-12]).

In this work, a novel Fourier-transform infrared (FTIR) photoreflectance (PR) spectroscopy method [13] is applied to study the optical properties and band structure of epitaxial $Al_xIn_{1-x}Sb$ alloys, as well as $InSb/Al_xIn_{1-x}Sb$ heterostructures. To the best of our knowledge, the photoreflectance of $Al_xIn_{1-x}Sb$ is observed experimentally for the first time. The PR spectral features are analyzed in order to determine the interband transition energies, and the intensity of the electric field in the $Al_xIn_{1-x}Sb$ barriers.

2. Samples and experimental setup

The studied $Al_xIn_{1-x}Sb$ epitaxial layers and $InSb/Al_xIn_{1-x}Sb$ heterostructures were grown by molecular beam epitaxy (MBE) on lattice-mismatched semi-insulating (001) GaAs substrates via AlSb buffer layers. A RIBER 32P MBE setup was employed, and x was determined by using electron probe microanalysis and X-ray diffraction measurements [14]. Due to the large thickness of the $Al_xIn_{1-x}Sb$ epilayers (~1 μm), as determined using both an optical technique described in [15] and scanning electron microscopy measurements, the elastic strain is assumed to be fully relaxed [16].

In the $InSb/Al_xIn_{1-x}Sb$ heterostructures, a 30 nm-thick InSb QW was grown at 50 nm beneath the surface of thick (1.6 – 3.3 μm) $Al_xIn_{1-x}Sb$ epitaxial layers delta-doped with Te above the QW. A $\{3nm\text{-}InSb/3nm\text{-}Al_xIn_{1-x}Sb\}\times 12$ superlattice was incorporated in the $Al_xIn_{1-x}Sb$ buffer to suppress propagation of defects to the surface layer.

The optical setup is based upon a VERTEX 80 FTIR spectrometer equipped for measurements in the 0.9 – 16 μm range. During PR measurements, the samples were placed into a liquid nitrogen cryostat and excited by a 809 nm laser diode (P_{las} up to 300 mW), which was electrically modulated at 2.5 kHz. The PR component of the signal was amplified by using a SR-830 lock-in amplifier tuned to the modulation frequency, digitized by internal ADC of the spectrometer, and processed by using an algorithm described in Ref. [17]. Photoluminescence (PL) spectra of the studied samples were measured by using the same spectrometer and excitation laser, and a CCS-150 closed-cycle helium cryostat.

3. Results and discussion

The photoreflectance spectra $\Delta R/R$ (T = 100 K) of $Al_xIn_{1-x}Sb$ epitaxial layers with different x, which are supposed to be the only available in literature to date, are presented in Figure 1, along with their low-temperature (T = 10 K) PL. The PL spectra at T = 100 K are not presented here since most of the samples did not show PL signals from $Al_xIn_{1-x}Sb$ layers at that temperature.

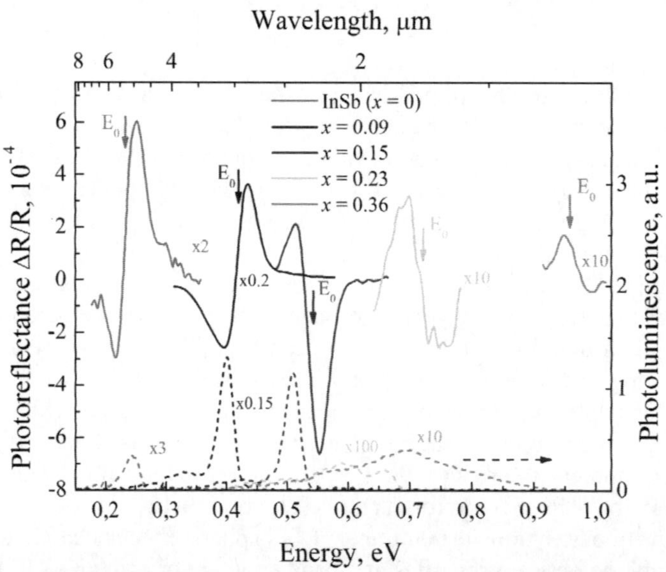

Figure 1. Photoreflectance (solid lines) and photoluminescence (dashed lines) spectra of $Al_xIn_{1-x}Sb$ alloy layers of various composition x.

IOP Conf. Series: Journal of Physics: Conf. Series **917** (2017) 062025 doi:10.1088/1742-6596/917/6/062025

The PR spectra exhibit characteristic derivative-like features, attributed to the Γ_6 - Γ_8 interband critical point E_0 of the band structure (corresponding to the direct band gap of $Al_xIn_{1-x}Sb$). In modulation spectroscopy, such kind of lineshape is typical for so-called low-field regime [18] when a characteristic electro-optic energy $\hbar\Omega$ is less than or equal to a broadening parameter. In this case $\hbar\Omega \sim F^{2/3}$, where F is a surface electric field, and the broadening parameter is defined by structural quality of the samples. The critical point energies E_0 (marked by arrows on Figure 1) were determined from the PR spectra using a technique employing a Kramers-Kronig transformation, described in Ref. [19]. The obtained E_0 values match reasonably well to the PL peak position in case of the InSb film, and are noticeably higher than the observed PL peak energies E_{max}^{PL} in the case of the $Al_xIn_{1-x}Sb$ layers, as illustrated in Figure 2a. This difference tends to increase with x, and might be related to the disordering of the $Al_xIn_{1-x}Sb$ alloy or radiative transitions involving impurity or defect levels within the band gap.

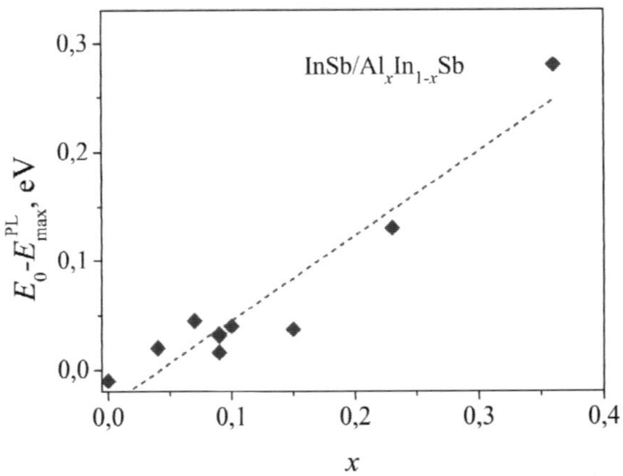

Figure 2a. Difference between E_0 from PR and PL peak energy of $Al_xIn_{1-x}Sb$ as a function of x. A linear trend line is added as a guide for eye.

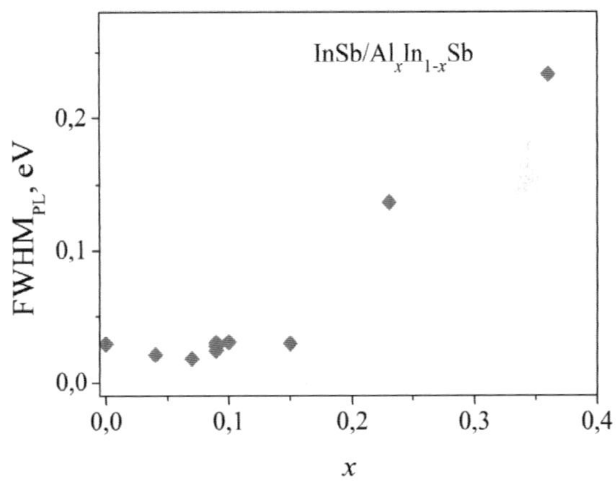

Figure 2b. Full width at half maximum (FWHM) of the $Al_xIn_{1-x}Sb$ alloy PL peak.

The correlation between the observed energy difference and the $Al_xIn_{1-x}Sb$ PL linewidth (Figure 2b) implies that the PL peak energy might be decreased relatively to the interband optical transition energy E_0 due to the contribution of a band tail in $Al_xIn_{1-x}Sb$ to the PL lineshape.

The PR spectra of $InSb/Al_xIn_{1-x}Sb$ heterostructures have exhibited a variety of spectral features, comprising of intensive signals from $Al_xIn_{1-x}Sb$ barriers not only at E_0 critical energy, but also at $E_0+\Delta_0$ transition from the spin-orbit split valence band (shown in Figure 3).

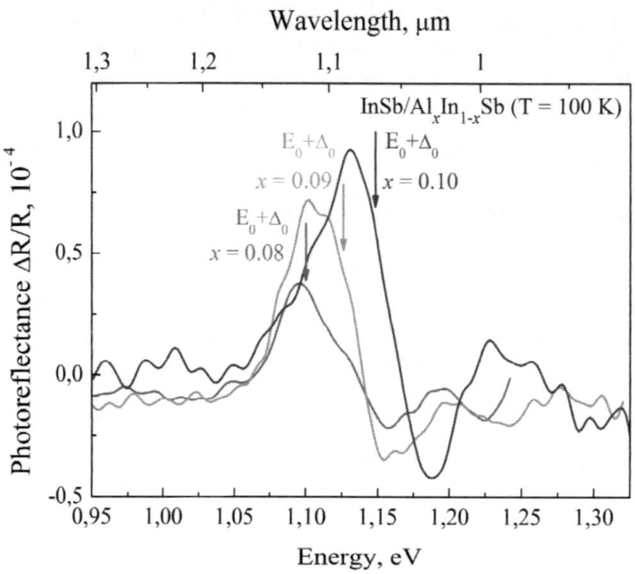

Figure 3. PR of $Al_xIn_{1-x}Sb$ alloy heterostructures near the spin-orbit split band transition.

The energies of the $E_0+\Delta_0$ transitions were also determined from the PR spectra by using the above-mentioned Kramers-Kronig transformation technique [19]. The resulting E_0 and $E_0+\Delta_0$ values for the $Al_xIn_{1-x}Sb$ layers are plotted as a function of alloy composition in Figure 4.

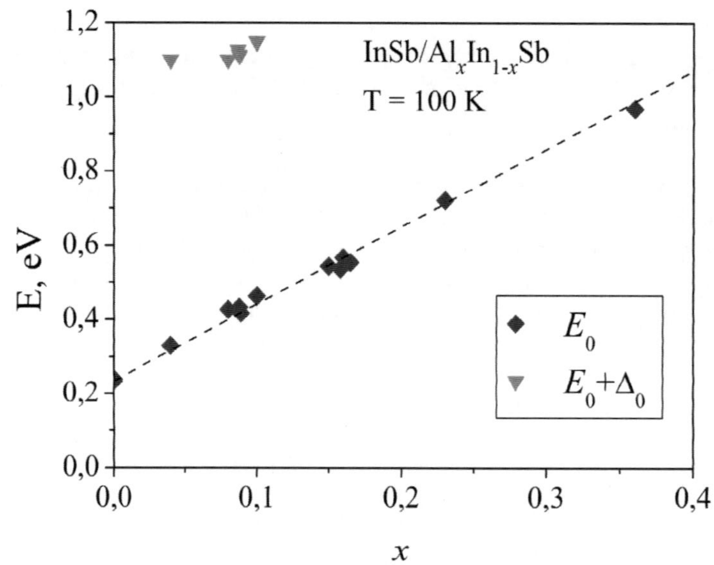

Figure 4. Dependence of the interband critical point energies of the $Al_xIn_{1-x}Sb$ alloy epilayers on the composition *x*.

The $E_0(x)$ dependence can be described by a conventional quadratic relation with a rather small bowing parameter, in agreement with our previous work [20]. The spin-orbit splitting energy values (obtained as $\Delta_0 = (E_0+\Delta_0) - E_0$) are in the 0.68 - 0.78 eV range, which correlates well with the available data for InSb and AlSb [9].

The samples with InSb/$Al_xIn_{1-x}Sb$ heterostructures have exhibited a periodic oscillating structure in the PR spectra above the E_0 energy, caused by Franz-Keldysh oscillations (FKOs) in $Al_xIn_{1-x}Sb$. Example of such PR spectrum is shown in Figure 5. This lineshape differs from the derivative-like features observed above, and corresponds to intermediate-field regime [18] when $\hbar\Omega$ is greater than or approximately equal to the broadening parameter. Electric field F (and hence $\hbar\Omega$) in the InSb/$Al_xIn_{1-x}Sb$ heterostructure samples is expected to be greater than in the $Al_xIn_{1-x}Sb$ alloys (because of additional Te doping). At the same time the broadening should be smaller, since the density of defects is reduced owing to the presence of the superlattice.

Compared to the PL of similar structures [11], PR measurements provide additional information about the barrier layers. In case of heterostructures with QWs, the PR spectra also exhibit signals from exciton transitions in the wells, which will be reported elsewhere.

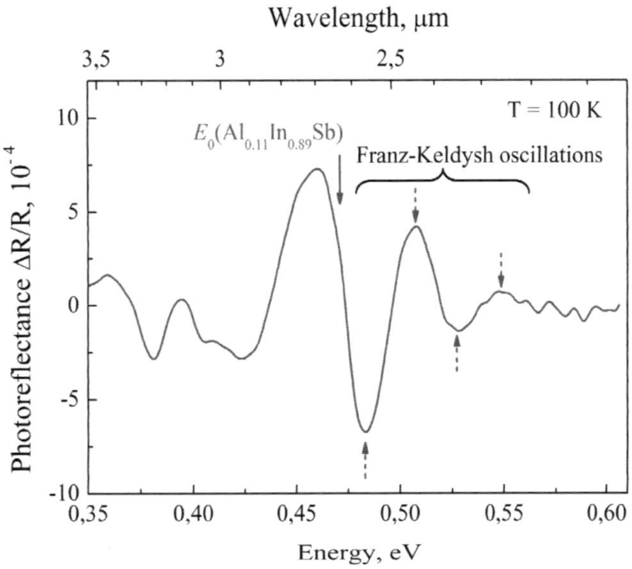

Figure 5. Franz-Keldysh oscillations in PR spectrum of a typical InSb/$Al_xIn_{1-x}Sb$ epitaxial heterostructure. The surface field intensity determined from FKO equals to 24kV/cm.

The lineshape of the FKOs can be described with a known relation (see e.g. [21]). The period of the oscillations is increased with the intensity of internal electric field F. The result of determining the field intensity F in $Al_xIn_{1-x}Sb$ from the PR spectrum using the approach described in Ref. [22] is presented in Figure 5 caption. The intensity of the surface electric field is governed by the concentration of free carriers in the semiconductor layer (and hence, doping), as well as by the surface states (which define the Fermi level pinning).

Observation of Franz-Keldysh oscillations enables one to extract the value of surface Fermi level pinning in $Al_xIn_{1-x}Sb$, provided that the carrier concentration in the studied samples would be known from independent measurements (e.g. Hall effect). And vice versa, the known surface pinning value E_F would enable one to obtain the free carrier concentration from the filed intensity derived from the FKO period in the PR spectra. Therefore, a contactless technique of determining the free carrier concentration in $Al_xIn_{1-x}Sb$ epilayers can be developed, which is the subject of future research.

In conclusion, it was demonstrated that the PR spectra of $Al_xIn_{1-x}Sb$ measured with FTIR modulation spectroscopy method provide information on the energies of critical-points corresponding to both the direct band gap (E_0) and the spin-orbit split valence band transition (E_0+Δ_0). Compared to the observed PL peak energies E_{max}^{PL} of the $Al_xIn_{1-x}Sb$ alloy, the obtained E_0 values are systematically higher. Observation of Franz-Keldysh oscillations in the PR spectra of InSb/$Al_xIn_{1-x}Sb$ heterostructures has allowed us to determine the intensity of the internal electric field in $Al_xIn_{1-x}Sb$ layers.

Acknowledgments
D.D. Firsov acknowledges the support of RFBR, project № 16-32-60076 mol_a_dk. O.S. Komkov acknowledges the support of Ministry of Education and Science of Russia, research project 16.1750.2017/PCh.

References
[1] Ashley T and Buckle L 2007 *Electronics Letters* **43** 14
[2] Yi W, Kiselev A A, Thorp J, Noah R, Nguyen B M, Bui S, Rajavel R D, Hussain T, Gyure M F, Kratz P, Qian Q, Manfra M J, Pribiag V S, Kouwenhoven L P, Marcus Ch M and Sokolich M 2015 *Applied Physics Letters* **106** 142103
[3] Nash G R and Mirza B I 2013 *Applied Physics Letters* **102** 011127
[4] Meriggi L, Steer M J, Ding Y, Thayne I G, MacGregor C, Ironside Ch N and Sorel M 2015 *Journal of Applied Physics* **117** 063101
[5] Gouider F, Vasilyev Yu B, Bugár M, Könemann J, Buckle P D and Nachtwei G 2010 *Phys. Rev. B* **81** 155304
[6] Glozman A, Harush E, Jacobsohn E, Klin O, Klipstein P, Markovitz T, Nahum V, Saguy E, Oiknine-Schlesinger J, Shtrichman I, Yassen M, Yofis B and Weiss E 2006 *Proc. of SPIE* **6206** 62060M
[7] Lyu Y, Si J, Cao X, Zhang L, Peng Z, Ding J, Yao G, Zhang X and Reobrazhenskiy V 2016 *Proc. of SPIE* **9819** 98191K
[8] Rodriguez J B, Madiomanana K, Cerutti L, Castellano A, Tournié E 2016 *Journal of Crystal Growth* **439** 33
[9] Vurgaftman I, Meyer J R and Ram-Mohan L R 2001 *Journal of Applied Physics* **89** 5815
[10] Litvinenko K L, Murdin B N, Allam J, Pidgeon C R, Bird M, Morris K, Branford W, Clowes S K, Cohen L F, Ashley T and Buckle L 2006 *New Journal of Physics* **8** 49
[11] Mironova M S, Komkov O S, Firsov D D and Glinskii G F 2014 *Journal of Physics: Conference Series* **541** 012085
[12] Chen G, Sun W and Lv Y 2017 *Infrared Physics & Technology* **81** 262
[13] Komkov O S, Firsov D D, Lvova T V, Sedova I V, Semenov A N, Solov'ev V A, Ivanov S V 2016 *Physics of the Solid State* **58** 2394
[14] Semenov A N, Meltser B Y, Solov'ev V A, Komissarova T A, Sitnikova A A, Kirylenko D A, Nadtochyi A M, Popova T V, Kop'ev P S and Ivanov S V 2011 *Semiconductors* **45** 1327
[15] Komkov O S, Firsov D D, Semenov A N, Meltser B Y, Troshkov S I, Pikhtin A N and Ivanov S V 2013 *Semiconductors* **47** 292
[16] Komkov O S, Semenov A N, Firsov D D, Meltser B Ya, Solov'ev V A, Popova T V, Pikhtin A N and Ivanov S V 2011 *Semiconductors* **45** 1425
[17] Firsov D D and Komkov O S 2013 *Technical Physics Letters* **39** 1071
[18] Aspnes D E 1973 Surface Science **37** 418
[19] Hosea T J C 1995 *Physica Status Solidi B* **189** 531
[20] Komkov O S, Firsov D D, Pikhtin A N, Semenov A N, Meltser B Ya, Solov'ev V A and Ivanov S V 2011 *AIP Conference Proceedings* **1416** 184
[21] Hughes P J, Weiss B L and Hosea T J C 1995 *Journal of Applied Physics* **77** 6472
[22] Pikhtin A N, Komkov O S and Bazarov K V 2006 *Semiconductors* **40** 592

IOP Conf. Series: Journal of Physics: Conf. Series **917** (2017) 062026 doi:10.1088/1742-6596/917/6/062026

Fabrication of amplitude-phase type diffractive optical elements in aluminium films

S A Fomchenkov[1, 2], **M A Butt**[1]
[1]Samara National Research University, Samara, Russian Federation
[2]Image Processing Systems Institute of the RAS – Branch of the FSRC "Crystallography and Photonics" RAS, Samara, Russian Federation

Abstract. In the course of studies have been conducted a method of forming the phase diffractive optical elements (DOEs) by direct laser writing in thin films of aluminum. The quality of the aluminum films were investigated depending on the parameters of magnetron sputtering process. Moreover, the parameters of the laser writing process in thin films of aluminum were optimized. The structure of phase diffractive optical elements was obtained by the proposed method.

1. Introduction

One of the traditional methods of forming the phase diffraction optical elements is a method based on the formation of micro-relief structure in the substrate itself [1]. Typically, this process begins with the deposition of thin metal film such as chromium on the quartz substrate, which acts as a hard mask [2]. Moreover, the patterns are formed on the hard mask with either by photolithography or by laser writing system [3]. Further, the desired patterns are obtained by etching the substrate with the help of plasma-chemical etching [4]. For example, a quartz substrate is physically etched in a gas medium (SF_6) through a chromium oxide (Cr_2O_3) mask. After the mask is removed by chemical etching. The main idea of our proposed method is to be eliminate the plasma etching process which is responsible for the surface roughness of the substrate and consequently reduces its efficiency [5, 6]. Furthermore, it will allow us to reduce the number of fabrication steps and consequently reduce the time and costs of production. Therefore, we propose to make a direct laser writing process in the aluminum films with the help of circular laser writing system "CLWS-200S".

2. Fabrication process

Initially quartz substrates were cleaned in a special solution to improve the adhesion of deposited films on the substrate. Aluminum films were deposited by magnetron sputtering system "Caroline D12A" at a pressure of $5x10^{-4}$ Pa, with a gas flow rate of 1 *l*/h at power of 700 W [7, 8]. Total view of this system shown on figure 1.

IOP Conf. Series: Journal of Physics: Conf. Series **917** (2017) 062026 doi:10.1088/1742-6596/917/6/062026

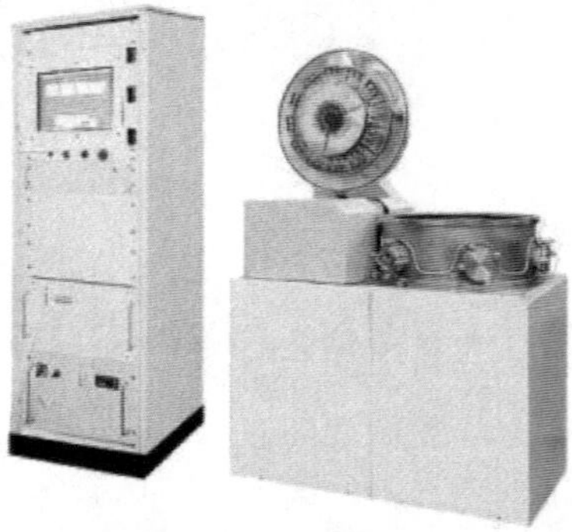

Figure 1. Automated magnetron sputtering system "Caroline D12A".

Aluminum thin films of 50 nm were obtained in 4 min. The film thicknesses were measured by using the profiler "KLA Tencor P-16 +" (shown on figure 2).

Figure 2. Stylus profiler "KLA Tencor P-16 +".

The surface roughness of 1 nm rms value was measured by using the interferometer "Zygo NewView 7300" (shown on figure 3).

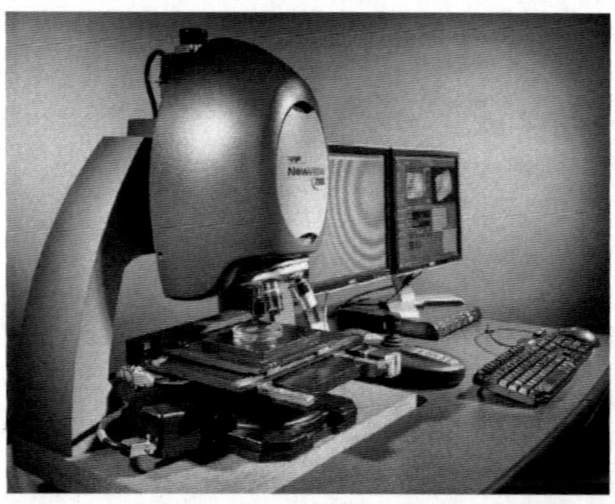

Figure 3. Optical 3D profilometer "Zygo NewView 7300".

Circular diffraction gratings (axicon) with a period of 4 μm and ½ duty cycle were fabricated for the desired wavelength of 532 nm. The overall diameter of the structure was 15 mm. The fabrication was done on a circular laser writing station "CLWS-200S" at 10 rpm. Total view of this system shown on figure 4.

Figure 4. Circular laser writing station "CLWS-200S".

When aluminum thin film was exposed by focused laser radiation, it oxidized into crystalline Al_2O_3. Unexposed aluminum metal was removed with the help of hot NaOH solution. Laser writing parameters were chosen such that the effect was similar to the annealing process at temperature 600-700 °C. Consequently, the obtained oxidized films were expanded up to 1.2-1.5 times. The obtained Al_2O_3 film was optically transparent and allows 85-90% of visible radiations. In the visible wavelength spectrum, Al_2O_3 has a refractive index of 1.8 which is higher than the refractive index of quartz 1.5 that provides a significant refractive index contrast. As a result, this fact reduces the requirement of high aspect ratio of the diffractive element structures in the Al_2O_3 film as compared to the structures in quartz by 20%. The general scheme of the technological manufacturing process is shown in figure 5. The fabricated sample is immediately available to be used in phase diffractive optical element.

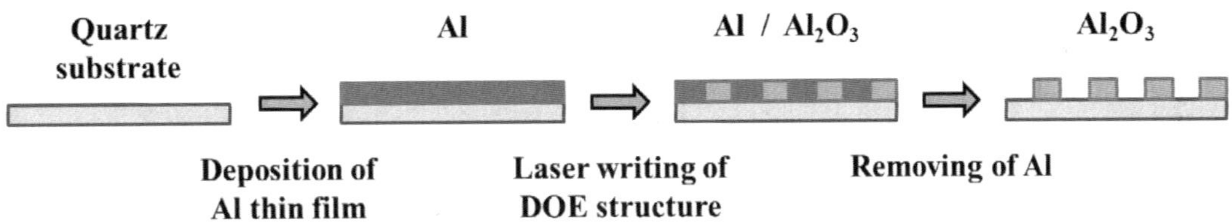

Figure 5. The general scheme of the technological process of the manufacturing of DOE.

3. Conclusion

As a result, it can be said that this method of forming a phase diffraction optical elements substantially reduces the time and significantly improves the quality of the elements. This allows to obtain greater efficiency in comparison with similar structures, made on quartz, using plasma-chemical etching.

Acknowledgments

This work was partially funded by the Ministry of Education and Science of the Russian Federation and Russian Federation Presidential grants for support of young candidates of sciences (MK-2390.2017.2), and Russian Foundation for Basic Research grant No. 17-42-630008, 16-47-630546.

References

[1] Fomchenkov S A, Butt M A, Podlipnov V V, Poletaev S D, Skidanov R V and Kazanskiy N L, 2016 J. of Physics: Conf. Ser. **741**, 012150.
[2] Butt M A, Fomchenkov S A, Ullah A, Habib M, Ali R Z, 2016 Computer Optics **40**, 674.
[3] Kazanskiy N L, Kharitonov S I, Khonina S N, Volotovskiy S G, Strelkov Yu S 2014 Computer Optics **38**, 256.

[4] Kazanskiy N L, Kharitonov S I, Khonina S N, Volotovskiy S G 2015 Proc. of SPIE 9533 95330L-1.
[5] Butt M A, Fomchenkov S A 2017 J. Korean Phy. Soc. **70** 169.
[6] Butt M A, Fomchenkov S A, Ullah A, Verma P, Khonina S N 2016 J. Phy. Conf. Ser. 741 012136.
[7] Butt M A, Fomchenkov S A, Kazanskiy N L, Ullah A, Ali R Z, Habib M 2017 Proc. Of SPIE 10342 103420O-1.
[8] Korolkov V., Shimansky R., Cherkashin V., Denk D., 2003 Computer Optics **25**, 79.

IOP Conf. Series: Journal of Physics: Conf. Series **917** (2017) 062027 doi:10.1088/1742-6596/917/6/062027

Luminescence and carrier concentration in Sb-containing narrow bandgap quantum wells under optical excitation

A V Selivanov[1], **M Ya Vinnichenko**[1], **I S Makhov**[1], **D A Firsov**[1], **L E Vorobjev**[1], **L Shterengas**[2] **and G Belenky**[2]
[1] Peter the Great St. Petersburg Polytechnic University, 29 Polytechnicheskaya str., St. Petersburg 195251, Russia
[2] Department of Electrical and Computer Engineering, State University of New York at Stony Brook, New York, 11794 USA

Abstract. We have experimentally studied the photoluminescence from InGaAsSb/AlGaAsSb quantum wells of different width under different levels of interband optical pumping. The dependences of photoluminescence intensity at spectral maxima on carrier concentration were theoretically calculated. This dependence was received using the rate equation taking into account the role of radiative and nonradiative recombination. Comparison of the theoretical and experimental results allowed to determine the nonequilibrium carrier concentration in quantum wells at certain experimental conditions.

1. Introduction

The InGaAsSb/AlGaAsSb quantum well (QW) nanostructures can serve as a basis of sufficiently powerful semiconductor injection lasers emitting in the wavelength range of 2–4 μm operating in continuous-wave mode. Lasers emitting in the mid-infrared range can be widely applied in the fields of spectroscopy of various substances, transmission of information via wireless communication lines, in security and fire control systems, in medical, military and other industries. The nonequilibrium carrier concentration and dependence of concentration upon pumping intensity are the important factors determining the operation of the laser. In this work, we present the method that helps to determine the nonequilibrium carrier concentration using the functional dependences of photoluminescence intensity on pumping intensity taken from experiment and theoretical calculations.

2. Samples and experimental setup

The InGaAsSb/AlGaAsSb QW structures with different quantum well width (4, 5, 7 and 9 nm) were studied. Growth details and band structure calculations are given in Ref. [1]. The interband photoluminescence spectra were studied using a vacuum Fourier transform infrared spectrometer Bruker Vertex 80v. The samples were mounted into an optical cryostat cooled with liquid nitrogen ($T = 77$ K). Electron-hole pairs were excited directly in QWs by radiation of pulsed semiconductor laser with a photon energy of 1.17 eV. The duration of optical excitation pulses was $\Delta t = 100$ ns, the repetition frequency was 800 Hz, and the maximum average power was 2.4 mW. An InSb photodetector was used as a detector of radiation.

3. Results and discussions

The concentration of non-equilibrium charge carriers involved in radiative recombination can be obtained from analyzing the photoluminescence spectra at different levels of optical pumping. We measured photoluminescence spectra at different pumping intensity J_{PUMP} and experimentally obtained dependences of photoluminescence intensity at spectral maximum J_{PL} on pumping intensity. Obviously, the spectral maximum corresponds to the effective bandgap. Dependence $J_{PL}(J_{PUMP})$ for the 9 nm wide QWs is presented in figure 1 by dots. Using the method described in Ref. [2], we received the theoretical dependences of photoluminescence intensity at spectral maxima on carrier concentration N (see figure 2). For this purpose, we wrote down the number of photons emitted per unit volume per unit time, which is proportional to the derivative of Kane matrix element, overlap integral and Fermi-Dirac distribution function with N as a parameter. It should be noted that non-equilibrium charge carriers in our experiments were excited directly in the quantum wells, i.e., the pump photon energy (1.17 eV) was less than the barrier band gap (1.72 eV). Under this type of excitation, the distance between the quantum confinement levels at which electrons and holes are produced in a quantum well is less than the pump photon energy value.

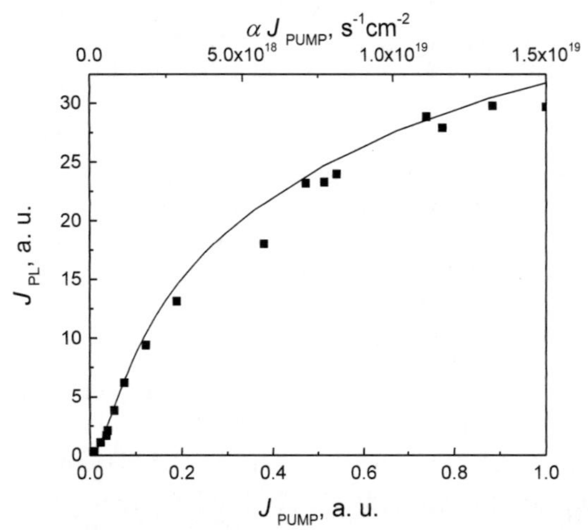

Figure 1. Experimental (dots) and theoretical (line) dependences of photoluminescence intensity at spectral maximum on pumping intensity for 9 nm wide QWs.

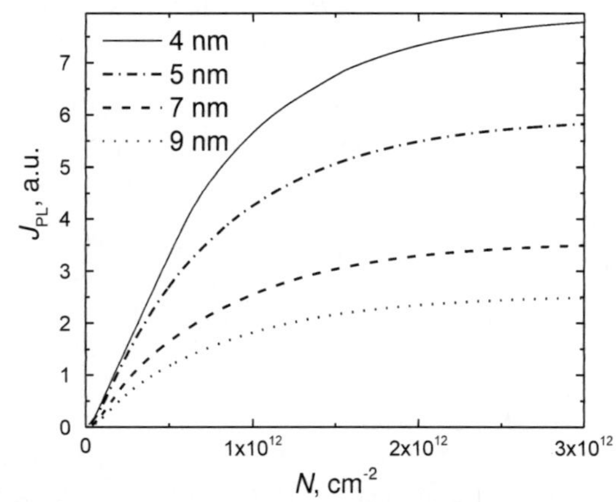

Figure 2. Theoretical dependence of photoluminescence intensity at the energy near effective bandgap versus carrier concentration for all structures at $T = 77$ K.

In order to find relation between carrier concentration and pumping intensity we used the rate equation:

$$\alpha \cdot L \cdot J_{PUMP} = AN + BN^2 + CN^3, \quad (1)$$

where α – absorption coefficient, L – QWs width, $A = 0{,}5\ \text{ns}^{-1}$ [1] – non-radiative Shockley-Read-Hall recombination rate, $B = 10^{-3}\ \text{cm}^2/\text{s}$ [2] – radiative recombination rate, $C = 10^{15}\ \text{cm}^4/\text{s}$ [1] – non-radiative Auger-recombination rate for 5 nm wide QW. The coefficient C for other QWs was taken in two order of magnitude less. All this approximations are correct because samples have similar design and band parameters except the QW width. Quantum wells width affects only the energy spectrum and the condition of Auger recombination (see below).

The calculated dependence of $J_{PL}(\alpha \cdot J_{PUMP})$ for 9 nm wide QW is presented in figure 1 by line. By scaling the abscissa we have found the relation between experimental and theoretical value of pumping intensity. Using similar dependences $J_{PL}(\alpha \cdot J_{PUMP})$ for 4, 5, 7 and 9 nm wide QWs and the rate equation (1), we determined the dependences of carrier concentration N on the pumping intensity. They are presented in figure 3.

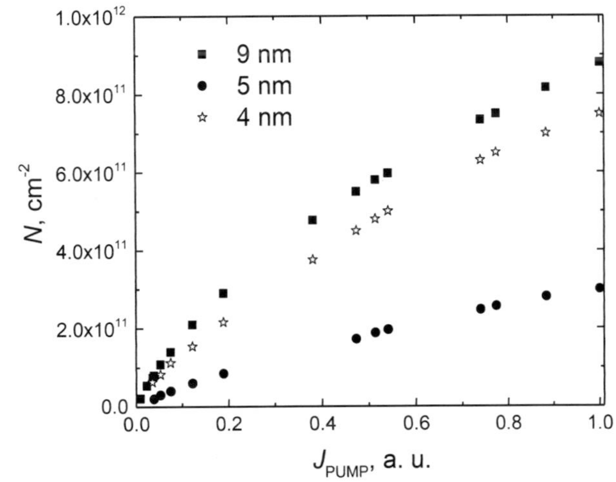

Figure 3. Dependences of carrier concentrations on pumping intensity for structures with different QW widths.

The carrier concentration in 5 nm wide QW is lower because of more probable nonradiative processes. This approach is correct because samples differ only by QW width which causes the difference in the band structure. This does not influence the non-radiative Shockley-Read-Hall recombination rate, but could significantly influence the non-radiative Auger recombination [1, 3]. Band structure and energy levels were calculated using the Schrödinger equation and Kane's model taking into account the nonparabolicity of the subbands. This approximation must be used, since the electron energy in subbands is of the order of the band gap value. The calculation shown that in 5 nm wide QWs the condition of resonant non-radiative Auger recombination is satisfied because the energy separation between the first heavy hole level and the spin-orbit split-off level is equal to the effective bandgap (see inset in figure 4):

$$E(hh1) - E(so1) \cong E(e1) - E(hh1). \quad (2)$$

So, in structure with 5 nm wide QW the concentration of carriers is less than in other samples due to satisfaction of resonant Auger condition. It should be noted that the resonant Auger recombination was early observed in the structure with 5 nm wide QWs when analyzing the photoluminescence dynamics measured by the up-conversion method [1]. Here we got the same results using another experimental method.

It is well-known that shortwave tail of photoluminescence spectra is determined with the distribution function of carriers and can give information about carrier heating. Using the method described in [4], we determined the carrier temperature for all structures at different pumping intensities. The carrier temperature increase (ΔT_C) is approximately 10 K for 5 nm wide QWs for the maximum pump level. For other structures, $\Delta T_C \approx 20$ K (see figure 4). It is associated with more frequent electron-electron collisions and greater carrier concentration in structures without resonant nonradiative Auger recombination.

IOP Conf. Series: Journal of Physics: Conf. Series **917** (2017) 062027 doi:10.1088/1742-6596/917/6/062027

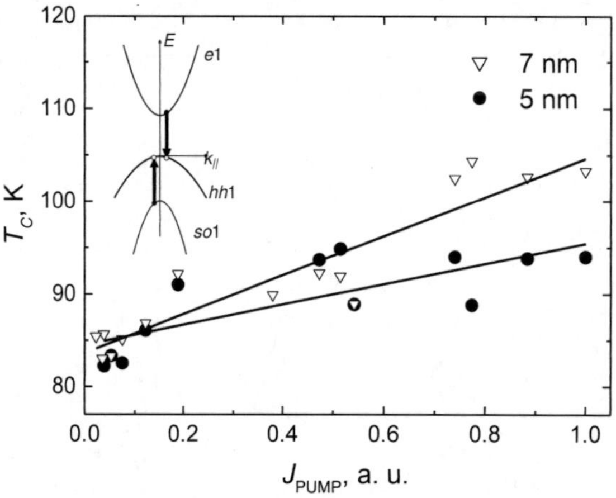

Figure 4. Dependences of carrier temperature on pumping intensity for structures with different QW widths. The inset shows the condition of resonant Auger recombination.

4. Conclusion

Finally, we presented the method allowing the determination of nonequilibrium carrier concentration dependence on optical pumping intensity in narrow-band nanostructures. Using this method, we found the carrier concentrations in InGaAsSb/AlGaAsSb nanostructures with different QW widths under certain experimental conditions.

Acknowledgments
This work was supported by the Russian Foundation for Basic Research (grants 16-32-60085, 16-02-00863), Russian Federation President grant for young scientists No MK-4616.2016.2 and the Ministry of Education and Science of the Russian Federation (state assignment).

References

[1] Vinnichenko M Ya, Firsov D A, Vorobjev L E, Mashko M O, Shterengas L and Belenky G 2013 *Semiconductors* **47** 146
[2] Vorob'ev L E, Zerova V L, Borshchev K S, Sokolova Z N, Tarasov I S and Belenky G 2008 *Semiconductors* **42** 737
[3] Danilov L V and Zegrya G G 2008 *Semiconductors* **42** 550
[4] Vorobjev L E, Vinnichenko M Ya, Firsov D A, Zerova V L, Panevin V Yu, Sofronov A N, Thumrongsilapa P, Ustinov V M, Zhukov A E, Vasiljev A P, Shterengas L, Kipshidze G, Hosoda T and Belenky G 2010 *Semiconductors* **44** 1402

IOP Conf. Series: Journal of Physics: Conf. Series **917** (2017) 062028 doi:10.1088/1742-6596/917/6/062028

Polariton condensation, superradiance and difference combination parametric resonance in mode-locked laser

S N Bagayev[1], R M Arkhipov[2,3], M V Arkhipov[2], V S Egorov[2], I A Chekhonin[2], M A Chekhonin[2]

[1]Institute of Laser Physics, Novosibirsk, Russia
[2]St. Petersburg State University, St. Petersburg, Russia
[3]ITMO University, St. Petersburg, Russia

E-mail: chekhonin@mail.ru

Abstract. The generation of the ring mode-locked laser containing resonant absorption medium in the cavity was investigated. It is shown that near the strong resonant absorption lines a condensation of polaritons arises. Intensive radiation looks like as superradiance in a medium without population inversion. We studied theoretically the microscopic mechanism of these phenomena. It was shown that in this system in absorbing medium a strong self-induced difference combination parametric resonance exists. Superradiance on polaritonic modes in the absorbing medium are due to the emergence of light-induced resonant polarization as a result of fast periodic nonadiabatic quantum jumps in the absorber.

1. Introduction

Exciton-polaritons systems in semiconductors placed in microcavity are called polariton laser. These systems pumped by single ultra-short pulse are actively studied so far. For pumping of atomic polariton laser it was proposed for the first time in [1, 2] and further in [3, 4] to use laser radiation with periodical amplitude modulation. It was shown that in the system "field + matter" in the strong light-matter coupling regime when modulation frequency $\Omega_m = 2\omega_c$ there is a difference combination parametric resonance. Here Ω_m - frequency of amplitude modulation of the pump field, ω_c is a cooperative (collective) the frequency of the resonant medium (1)

$$\omega_c \equiv \sqrt{2\pi\omega_{12} \frac{d_{12}^2}{\hbar} N_o} . \quad (1)$$

Here N_o - atomic density, ω_{12} - transition frequency of the two-level particles, d_{12} - transition dipole moment.

In this paper, we present a new efficient way of parametric excitation of the polariton laser and superradiance excitation using coherent pump with *periodic phase (frequency) modulation*. The method is based on the following idea. When system is pumped by this FM field the population difference $N(t)$ can be rapidly changed. Namely, there are periodic surges nonadiabatic quantum energy states between quantum energy states of medium. Furthermore, in this method, energy of the all modes of the pump laser can be used, regardless of their position on the frequency axis. Generation of the single mode polariton laser was investigated by the numerical solution of Maxwell-Bloch equations.

IOP Conf. Series: Journal of Physics: Conf. Series **917** (2017) 062028 doi:10.1088/1742-6596/917/6/062028

Under the condition of strong light-matter coupling self-consistent motion of the Bloch vector is equivalent to the motion of a physical pendulum, oscillating near the equilibrium point with the cooperative frequency ω_c. Nonadiabatic quantum jumps of atoms lead to a periodic change in the population difference $N(t)$ and the parametric excitation of oscillations of a pendulum. Simultaneously with the parametric build up the amplitude of the pendulum increases. The amplitude of the Bloch pendulum oscillations corresponds to radiation of two polaritonic modes - upper polariton and lower polariton.

The condition of the difference combination parametric resonance for the Bloch pendulum is follows: $\Omega_{mod} = 2\omega_c/m$, m = 1, 2, 3... Here Ω_{mod} – frequency of the phase modulation of the pump field. In the numerical simulations, periodic modulation of the phase (frequency) of the pump field was used (FM mode-locking), $\Delta\omega(t) = \Delta\omega_{12} + d\varphi(t)/dt$, $\varphi(t) = \varphi_0\cdot\cos(\Omega_{mod}\, t)$. Here $\Delta\omega(t)$ is the frequency detuning between the carrier frequency of the FM pump field and atomic transition frequency.

It is shown that when FM pumping field is applied the efficiency of excitation of atomic polariton laser can be increased by factor $10^6 \div 10^8$ with respect to the case of single pulse pumping (when the energy of FM and non FM pulses are equal). The investigated phenomena opens opportunity to use the laser diodes of small power for the pumping of polariton laser.

2. Polariton condensation and superradiance in a ring mode-locked laser

The effect of polariton condensation occurs on the resonant transition of absorption atomic cell located in the cavity of mode-locked laser was studied in [5-7]. In general case, the spectrum of laser radiation has the form of spectral doublets which contains upper polariton and lower polariton branches. However, discussing the nature of the microscopic mechanism of this phenomenon is still ongoing. In our studies, it was shown that in a multimode laser without mode locking in such case appears the mode-locking regime. The laser spectrum is close to the resonant frequency of the absorber ω_{12}, and it generates 0π-pulses [8, 9].

A superradiance on the resonant lines D_1 and D_2 of rubidium was observed in [10] in experiments when rubidium vapor was pumped by non-resonant femtosecond pulses of a Ti-sapphire laser. By increasing the density of the resonant particles N_0 the strong light-matter coupling was realized. Superradiance on polaritons modes arose and spectrum takes the form of a spectral doublet. As it was for the first time shown in [1, 2] in a single mode cavity model, this effect can be greatly enhanced by using periodically modulated pump laser. This effect increases due to the occurrence of fast periodic nonadiabatic quantum jumps between the energy states of the resonant medium.

In this work, we investigate the mechanism of occurrence of the gain on the polaritonic modes near the absorber resonance ω_{12} in the case when laser operates in the mode-locking regime. It turns out that a significant increase of this effect is possible by using of non-resonant laser pulses with self-consistent periodic phase (frequency) modulation: $\Delta\omega(t) = \Delta\omega_{12} + d\varphi(t)/dt$. The variable parameter in this parametric resonance is $\Delta\omega(t)$.

Generation of the ring mode-locked laser containing intracavity resonant absorption cell was investigated in the frame of the numerical solution of Maxwell-Bloch equations.

This model takes into account a wide class of phenomena: multimode regimes of generation, mode competition and mode-locking effects, chaotic regimes of generation, soliton formation, and many others. This model is based on the system of Maxwell-Bloch equations, which has the following form [11]:

$$\frac{d}{dt}Pc(t) = -\frac{Pc}{T_2} - \Delta\omega\cdot Ps(t) - \frac{d_{12}}{2\hbar}\cdot\Delta\rho(t)\cdot B(t), \tag{2}$$

$$\frac{d}{dt}Ps(t) = -\frac{Ps}{T_2} + \Delta\omega\cdot Pc(t) + \frac{d_{12}}{2\hbar}\cdot\Delta\rho(t)\cdot A(t), \tag{3}$$

$$\frac{d}{dt}\Delta\rho = -\frac{(\Delta\rho - \Delta\rho_o)}{T_1} - \frac{2d_{12}}{\hbar}\cdot(A(t)\cdot Ps(t) - B(t)\cdot Pc(t)), \tag{4}$$

$$\frac{\partial A(z,t)}{\partial t} + c \cdot \frac{\partial A(z,t)}{\partial z} = -4 \cdot \pi \cdot \omega \cdot d_{12} \cdot No \cdot Ps(z,t), \tag{5}$$

$$\frac{\partial B(z,t)}{\partial t} + c \cdot \frac{\partial B(z,t)}{\partial z} = 4 \cdot \pi \cdot \omega \cdot d_{12} \cdot No \cdot Pc(z,t) \cdot \tag{6}$$

Here T_1 is population difference relaxation time, T_2 is polarization relaxation time, c is velocity of light.

Bloch equations (2)-(4) describe the behavior of the slow envelopes of inphase and quadrature components of polarization $Pc(z,t)$ и $Ps(z,t)$ and population difference Δρ between ground and excited states:

$$P(z,t) = Pc(z,t)\cos(\omega t - kz) + Ps(z,t)\sin(\omega t - kz). \tag{7}$$

Similar Bloch equations are solved for the amplifyimg medium of laser. Maxwell equations (5)-(6) describe the behavior of slow envelopes of the electric field $A(z,t)$ and $B(z,t)$ in gain and absorbing laser media in the cavity:

$$E(z,t) = A(z,t)\cos(\omega t - kz) + B(z,t)\sin(\omega t - kz). \tag{8}$$

For numerical simulations the parametrs which are realized in the experiment [1, 2, 8] were used. Numerical simulations showed that superradiance and polariton condensation on the polaritonic modes started from the arising of phase modulation of laser pulses from mode-locked pump laser. When the nonresonant pump pulse propagates the frequency deviation of the field Δω(t) decreases.

At the initial stage of the lasing process, $t = 0$, $d\varphi(t)/dt = 0$ and $\Delta\omega(0) = \Delta\omega_{12}$. Figure 1 shows the $d\varphi(t)/dt$ of the laser pulses near the threshold of the parametric instability. Resonantly absorbing medium develops a self-consistent field with the phase modulation. As shown in figure 1, the phase modulation has not simple harmonic form. When pump laser frequency $\omega(t) \approx \omega_{12}$ nonadiabatic quantum jumps occur, leading to the creation of superposition states in a resonant absorption medium and the generation short pulses of superradiance arises.

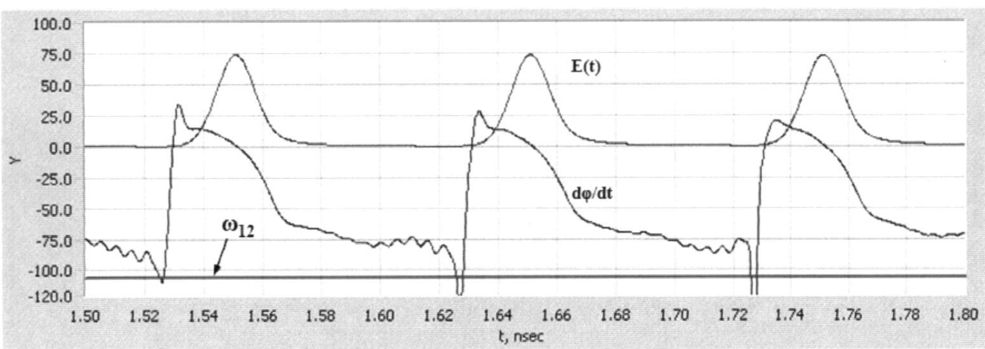

Figure 1. The upper curve - pumping pulses E (t), a.u. The lower curve - $d\varphi(t)/dt$ in GHz. $\omega_c = 6.7 \cdot 10^{10}$ rad/sec, $T_1 = T_2 = 10$ nsec, the laser cavity length - 3 cm. $\Delta\omega_{12}/2\pi = -115$ GHz.

Figure 2 has the spectrum of the laser after the build up of parametric resonance. The spectrum of the mode-locked laser becomes asymmetrical and expands to the direction of the resonant absorption.

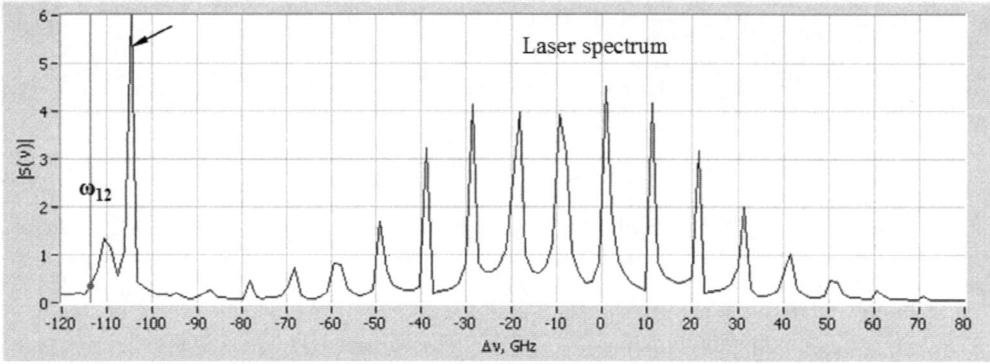

Figure 2. The laser spectrum at t = 6 nsec. Axial cavity modes are visible in the laser spectrum. The arrow shows the spectrum of polariton condensation at the absorption line. $\Delta\omega_{12}/2\pi$ = -115 GHz.

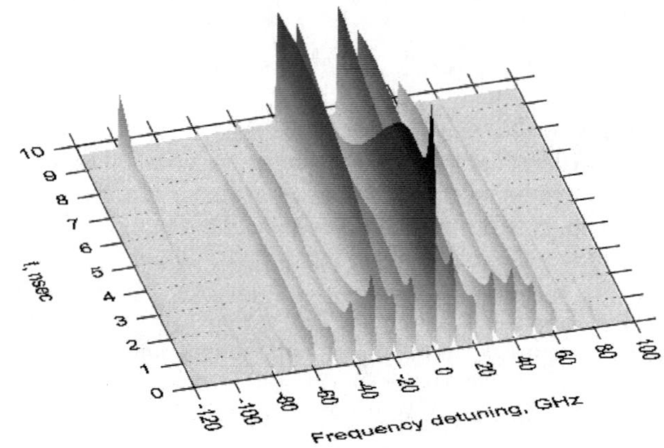

Figure 3. Laser spectrum evolution and arising of polariton condensation near the atomic absorption line ($\Delta\omega_{12}/2\pi$ = -100 GHz). $N_o = 10^{12}$ cm^{-3}, λ_{12} = 700 nm, $d_{12} = 5D$, $T_1 = T_2$ = 10 nsec, the laser cavity length - 3 cm.

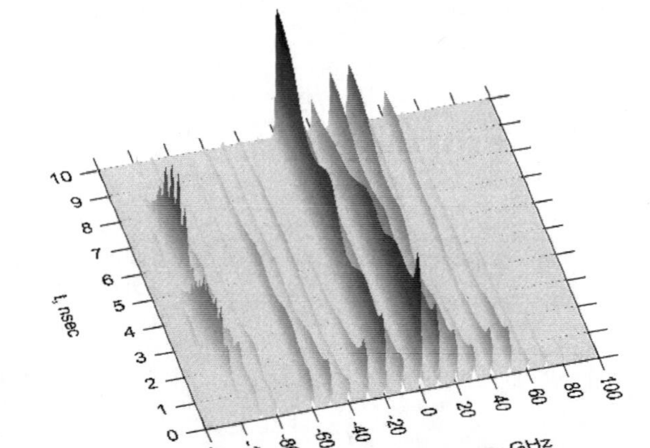

Figure 4. Laser spectrum evolution and arising of polariton condensation near the atomic absorption line ($\Delta\omega_{12}/2\pi$ = -100 GHz). $N_o = 10^{13}$ cm^{-3}. Other parameters are same as in figure 3.

Figure 3 and figure 4 illustrates the influence of the atomic density N_0 on the duration of the process of the formation of phase modulation of pulses in mode-locked laser. Increase of atomic density from $N_0 = 10^{12}$ cm^{-3} (figure 3) to $N_0 = 10^{13}$ cm^{-3} (figure 4) leads to the descrease of its duration from 6 ns to 2 ns. This result is in agreement with the experimental observations [1, 2, 12].

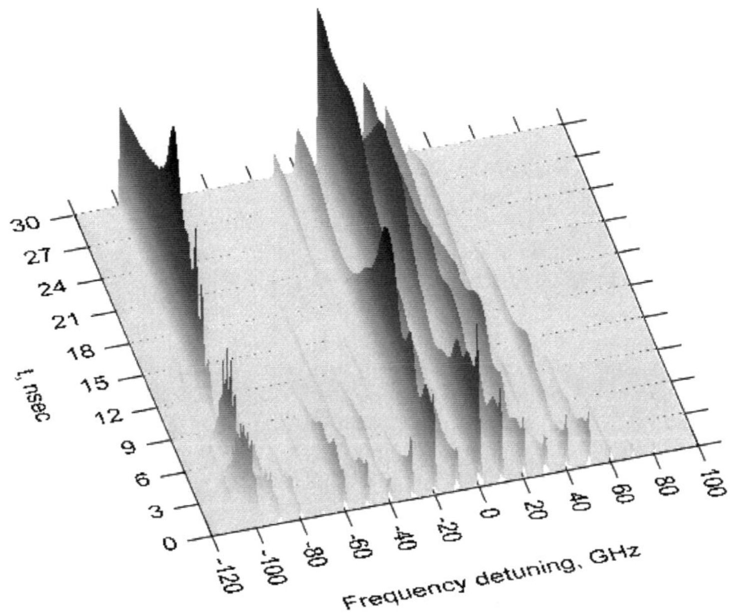

Figure 5. The build up of the steady state of polariton condensate near the atomic transition ($\Delta\omega_{12}/2\pi$ = - 100 GHz) during the time of laser generation, $N_o = 10^{13}$ cm^{-3}. Other parameters are same as in figure 3.

Figure 5 shows the stage of the build up of the steady state of amplitude of the polariton condensate near absorber transition frequency ω_{12} for a long time. The increase of intensity of polaritons radiation near the absorption lines is limited by the saturation effect.

3. Conclusion

In conclusion, the new microscopic mechanism of the formation of condensation of polaritons near a strong absorption line of a medium in the ring optical cavity in the mode-locking regime was considered theoretically.

It was shown that this process has two stages. In the first stage, a self-consistent phase modulation of the laser pump pulses occurs. The duration of this process depends on the density of the absorbing atoms N_o. The interaction of phase modulated pulses with absorber atoms leads to the appearance of fast nonadiabatic quantum jumps in the atoms in the absorbing cell and to the formation of coherent superposition states of atoms.

In the second stage, the induced polarization of the medium emittes the radiation in the form of short pulses of superradiance at the frequencies of the polaritonic modes. The fulfillment of the condition of difference combination parametric resonance leads to a rapid increase in the amplitude of the polaritonic modes. The increment of the parametric instability depends on the frequency detuning $\Delta\omega_{12}$, amplitude and duration of the laser pulses.

The dynamics of the field $E(z,t)$, polarization $P(z,t)$, and population difference $N(z,t)$ in the absorbing and amplifying cells of a ring laser were analyzed via numerical solution of the Maxwell-

Bloch equations. Numerical calculations performed for realistic conditions realized in the experiment showed a qualitative agreement between the results of simulations and experiments.

Acknowledgements
This work was supported by Russian Science Foundation (project №17-19-01097).

References

[1] Vasilev V V, Egorov V S, Chekhonin I A 1985 Opt. Spectr. **58** (4) 944-946
[2] Vasilev V V, Egorov V S, Fedorov A N, Chekhonin I A 1994 Opt. Spectr. **76** (1) 134–146
[3] Svidzinsky A A, Yuan L, Scully M O 2013 Phys. Rev. X **3** (11) 041001-1 - 041001-25
[4] Chen G, Tian J, Bin-Mohsin B, Nessler R, Svidzinsky A, Scully M O 2016 Phys. Scr. **91** (6) 073004-1 - 073004-15
[5] Runge P K 1971 Opt. Commun. **4** (3) 195-198
[6] Rubinov A N, Belokon' M V, Adamushko A V 1979 Sov. J. Quant. Electron. **9** (4) 433–436
[7] Baev V M, Eschner J, Weiler A 1989 Appl. Phys. B: Lasers and Optics **49** (4) 315-322
[8] Egorov V S, Chekhonin I A 1982 Opt. Spectr. **52** (4) 355-356
[9] Arkhipov M V, Arkhipov R M, Shimko A A, Babushkin I 2015 JETP Letters **101** (4) 232–235
[10] Bagaev S N, Egorov V S, Pastor A A, Preobrazhenskii D Yu, Preobrazhenskaya A A, Serdobintsev P Yu, Chekhonin I A, Chekhonin M A 2016 Opt. Spectr. **121** (3) 391-394
[11] Allen L, Eberly J H, *Optical resonance and two-level atoms* (Wiley, New York, 1975)
[12] Bertseva E, Campargue A 2004 Opt. Commun. **232** (1-6) 251–261

Two-photon absorption cross section for Coumarins 102, 153 and 307

A S Melnikov[1], P Yu Serdobintsev[1], A D Vedyaykin and M A Khodorkovskii[1]
[1] Peter the Great St.Petersburg Polytechnic University, Saint-Petersburg, 195251, Russia

Abstract. Two-photon absorption cross sections of Coumarins 102, 153, 307 solutions in methanol in the wavelength range of 730-870 nm were measured. The maximum cross section for two-photon absorption in this wavelength range were 26±7, 36±9, 28±7 GM for Coumarins 102, 153, 307, respectively. The absolute value of the cross sections was determined by comparison with a fluorescein (43 GM at a wavelength of 790 nm). The dependence of the two-photon absorption cross section on the excitation wavelength coincides with the shape of the single-photon absorption cross section.

1. Introduction

Many novel research methods based on two-photon absorption (TPA), in particular multiphoton fluorescence microscopy, photodynamic therapy with TPA are currently being intensively developed. These methods use fluorescent probes, which are usually organic phosphors. Coumarins as a class of compounds are among the most versatile and widely used fluorescent probes. In biological preparations, coumarins show a high degree of sensitivity to the environment [1]. The photophysical properties of coumarins are well understood, however data on the TPA of coumarins are poor. The aim of this work was to study TPA of Coumarins 102, 153 and 307 in methanol solutions.

It was found in paper [2] that the shape of the TPA cross section coincides with the one-photon absorption for the Coumarin 102 solution in methanol. This effect is possible in the absence of a central symmetry in the molecule and was observed earlier for certain organic phosphors [3]. The use of this type of phosphors as standards makes it possible to simplify the relative measurements of the TPA cross section. The Coumarins 102, 153 and 307 with a single-photon absorption maximum in the 375-435 nm region were chosen as the object of measurements, since their two-photon absorption corresponds to the widely used 730-870 nm range.

2. Experimental setup

A Ti:Sa laser was used for dyes excitation. Parameters of laser radiation were: pulse width - 50 fs, repetition rate – 80 MHz, power - 0.45 W, wavelength tuning range - 730-870 nm. To adjust the intensity of the laser radiation, a $\lambda/2$ plate and a Glan-Taylor prism were used. The photodiode was used to monitor the laser intensity.

The laser beam was focused into a quartz 10x10 mm cuvette. The cuvette was placed in such a way that the distance between beam caustic and the cuvette surface did not exceed 0.5 mm to lower reabsorption. Diameter of the caustic (FWHM) was approximately 20 microns. Liquid in the cuvette was stirred vigorously to eliminate any thermal effects. Concentration of the solutions did not exceed 100μmol/L. Emission spectra were recorded using miniature spectrometer (Ocean Optics HR2000).

IOP Conf. Series: Journal of Physics: Conf. Series **917** (2017) 062029 doi:10.1088/1742-6596/917/6/062029

Halogen lamp LS-1-CAL (Ocean Optics) and deuterium lamp DH2000 (Ocean Optics) were used to calibrate the absolute spectral response of the spectral system in the 200-1100 nm range.

The following phosphors were used to prepare the solutions: Coumarin 307 (hereinafter referred to C307, $C_{13}H_{12}NO_2F_3$, MW = 271.24, manufactured by Exciton), Coumarin 153 (C153, $C_{16}H_{14}NO_2F_3$, MW = 309.29, manufactured by LambdaChrom), Coumarin 102 (C102, $C_{16}H_{17}NO_2$, MW = 255.32, manufactured by LambdaChrom), Fluorescein (Fl, $C_{20}H_{12}O_5$, MW = 332.31, manufactured by Sigma-Aldrich), Rhodamine B (RhB, $C_{28}H_{31}N_2O_3Cl$, MW = 479.02, manufactured by Exciton).

To measure the TPA cross section the two-photon-induced fluorescence method was used. The number of photons (R) emitted after the laser pulse from the volume $dldxdy$ can be written in the form [4] :

$$R = \frac{1}{2}\frac{1}{(h\nu)^2} N\sigma_{TPE} I^2 \tau \, dl dx dy \quad (1),$$

where $\boldsymbol{I}$ is the laser radiation intensity, τ is the pulse length, $h\nu$ is the quantum energy, N is the concentration of molecules, dl is the length of the absorbing layer, and σ_{TPE} is the cross-section of two-photon luminescence. The last is related to cross section of two-photon absorption (σ_{TPA}) as:

$$\sigma_{TPE} = q_2 \sigma_{TPA} \quad (2),$$

where q_2 is the quantum yield of two-photon luminescence. As a rule, the quantum yield of two-photon luminescence will coincide with the quantum yield of single-photon luminescence q [4]. In the present paper we assume that $q_2 = q$. The absolute emission quantum yield was determined by comparison with the solution of Rhodamine 6G in methanol as a standard.

3. Results and discussion

Figure 1 shows the dependence of the luminescence intensity on the pump power for the investigated phosphors (the pump power square is plotted along the abscissa axis). The luminescence intensity for two-photon excitation depends quadratically on the pump power and in this scale should be expressed in direct proportionality. As a rule, deviations from the quadratic dependence are observed at a high radiation density and clearly indicate the presence of other nonlinear processes. It can be seen that deviations from the quadratic dependence are absent taking into account the experimental error.

As mentioned above, it was found in paper [2] that the shape of the TPA cross section coincides with the one-photon absorption shape for the C102 solution in methanol. We measured the shape of the TPA cross sections of solutions C307 and C153 in methanol using C102 as a reference. It was found that the shape of the TPA cross sections for C153 and C307 in the wavelength range 730-870 nm coincides with single-photon absorption cross sections within the error of the experiment. Thus, C102, C153 and C307 can be used as a reference for relative measurements of the TPA cross section in the range of 730-870 nm.

It is more convenient to carry out relative measurements of the TPA cross sections using C307 solution in methanol as a standard, since at $\lambda > 800$ nm its TPA cross section decreases more slowly upon increasing wavelength than that of C102. Figure 2 shows the relative TPA cross sections for C102 and C153 in methanol, while C307 was used as a reference. It supports that good agreement between the shape of one- and two- photon absorption cross sections takes place.

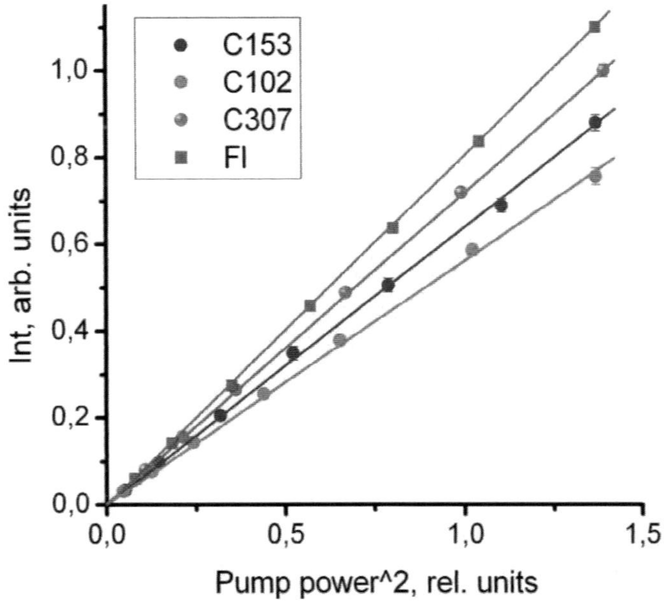

Figure 1. Dependencies of luminescence intensity on pump power. C102, C153, C307 – solution in methanol, Fl – solution in water (pH = 11), excitation wavelength 790 nm.

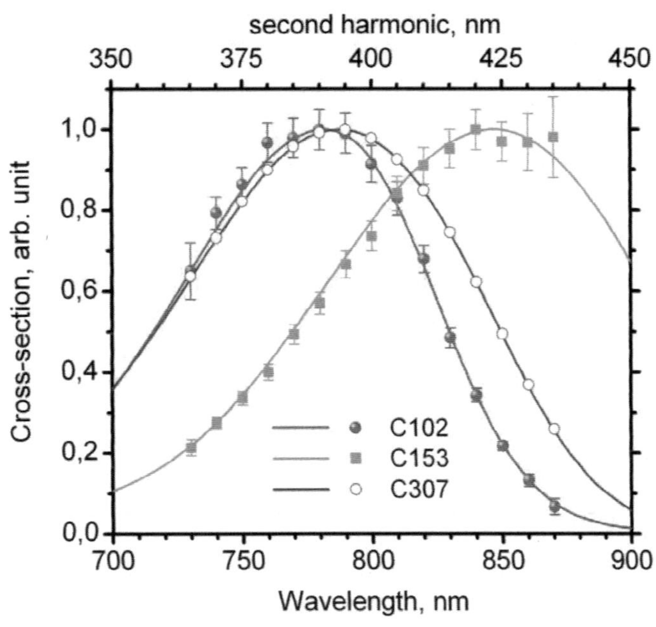

Figure 2. Dependencies of the single-photon absorption cross-section (line) and the TPA cross-section (symbols) on the excitation wavelength for C102, C153 solutions in methanol. Solution C307 in methanol was used as a reference.

Fluorescein and Rhodamine B were used for absolute calibration as the most studied luminescent substances. We carried out relative measurements of the TPA cross sections of an aqueous solution of fluorescein (pH = 11), as a reference we used a solution of C307 in methanol. Figure 3 presents the results of this work, as well as the results of [5, 6]. It can be seen that the data of [5, 6] is consistent well with our data in the regions 740-750 nm and 800-840 nm. Outside this wavelength range , the differences between the values of [5] and [6] reach 1.5 times. For absolute normalization of the data of this work, the value $\sigma_{TPA} = 43$ GM was chosen at the maximum of the cross section (wavelength 790 nm). With this normalization in the spectral regions 740-750 and 820-840 nm, the results of this work are in good agreement with the results of [5, 6]. The maximum difference with the results of [5] is observed in the wavelength region 760-800 nm and does not exceed 16%, which agrees with the magnitude of the error in the work [5]. In the range 750-840 nm, the differences between the values obtained in this work and in [6] do not exceed 15%, in the 850-870 nm region our data differ from the [6] by not more than 50%. Results of earlier works can be found in [5].

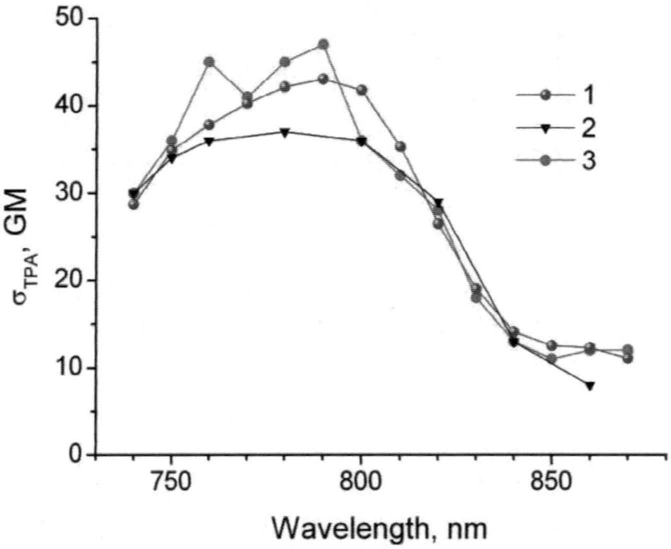

Figure 3. Dependencies of the TPA cross section on the wavelength in water solution of Fl (pH = 11), curves 1 - this work, 2 – [6], 3 – [5];

Using the above-described normalization of the fluorescein cross-section (σ_{TPA}(790 nm) = 43 GM), the C307 cross-section at the maximum at λ = 790 nm is 28±7GM. This normalization of the TPA cross section for the C307 solution in methanol was used for all subsequent measurements.

Figure 4 shows the results of measurements of the TPA cross-section in a solution of Rhodamine B in methanol obtained in this paper and in papers [5, 7]. Note that in the work [5] there is a discrepancy between the data shown in the graph and in the table, so here we give both results. It can be seen that the results of this paper are in good agreement with the data given in the graph in [5]. The maximum discrepancy is observed in the region of 770-780 nm and reaches 20%. In the region of 730-840 nm, the data of this paper are in good agreement with the work of [7]. In the 850-870 nm area, the values of work [7] are below the values of this paper and paper [5].

Thus, the TPA cross-sections of the Fluorescin and Rhodamine B measured using a solution of C307 in methanol as a standard with an absolute value 28 ± 7 GM at maximum and the shape of TPA cross-section coinciding with single-photon absorption show good agreement with the most reliable results obtained in the works [5, 7]. It should be emphasized that using the proposed calibration, the TPA cross-sections for C102, C153, RhB and Fl are smooth and do not contain local peaks of width ~ 10 nm on the wavelength dependencies. Such peaks appear, for example, on the dependences of the TPA cross sections for C102 and C153 when using the data from [5] for fluorescein as reference.

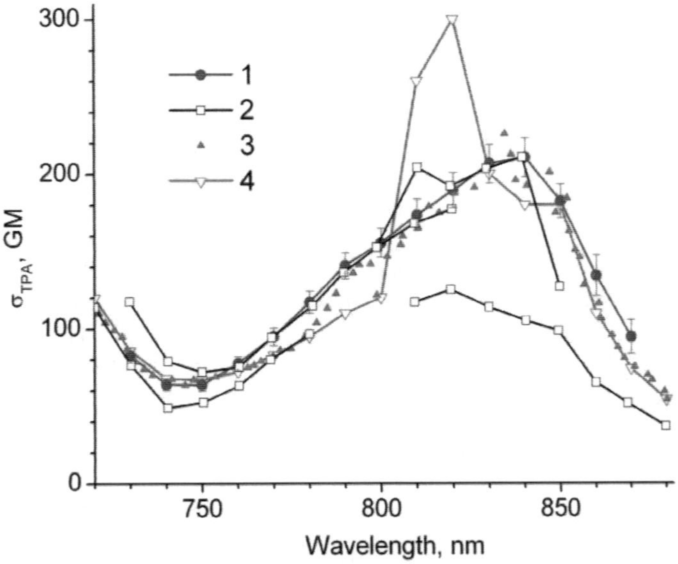

Figure 4. Dependencies of the TPA cross section on the wavelength in Rhodamine B methanol solution, curves 1 - this work, 2 – paper [7], 3 - [5] (Fig.4), 4 - [5] (Table 4).

The TPA cross sections at the maximum for solutions of C102, C153, C307, RhB in methanol and Fl in water are shown in the Table 1. The values of the quantum yields measured in the present work were used to calculate the TPA cross-section, see Table 1. The value q = 0.925 obtained in the work [8] was used for fluorescein.

Table 1. Two-photon absorption cross-sections at the maximum value in range 730-870 nm.

Substance	Solvent	q	λ_{max}, nm	σ_{TPA} (λ_{max}), GM
C102	MeOH	0.88±0.06	780	26±7
C153	MeOH	0.47±0.05	850	36±9
C307	MeOH	0.81±0.06	790	28±7
RhB	MeOH	0.64±0.05	840	210±30

Next, we compare the results of this paper with the previous publications for coumarins. As far as we know, the absolute value of TPA cross-section for C102 in methanol was evaluated only in one work [2]. In work [2], a comparison with bis-MSB at a wavelength of 725 nm was used to estimate the absolute value of the C102 cross-section. The TPA cross-section for bis-MSB at 725 nm was calculated by extrapolation from data from paper [9], resulting in a $\sigma_{TPE}(\lambda = 725\text{ nm}) = 4.8$ GM. To calculate the cross section for two-photon absorption, C102, the quantum yield of both molecules was assumed to be 1, and the spectral sensitivity of the recording system was not taken into account. As a result, for the TPA cross-section, 160 GM was obtained at the maximum ($\lambda = 781$ nm). This value is more than five times higher than value obtained in the present study. However, in paper [7] a direct measurement of the TPA cross section for bis-MSB was carried out. It was shown that $\sigma_{TPA} = 1.0$ GM at the wavelength of 725 nm. If we recalculate the results of [2] with the results of [7],

IOP Conf. Series: Journal of Physics: Conf. Series **917** (2017) 062029 doi:10.1088/1742-6596/917/6/062029

we get σ_{TPA} = 33 GM at λ = 780 nm. This value is consistent with that obtained in this paper. In paper [10], as a result of calculation, the maximum value of 21.7 GM at 725 nm was obtained, which agrees well with the data of the present work.

The two-photon emission cross sections for C307 in methanol reported in [7] is σ_{TPE} = 19.5 GM at λ = 776. Taking into account the quantum yield of 0.81, the TPA cross-section is σ_{TPA} = 24 GM, which coincides with the results of this paper within the experimental error. Note that in the [7], the data from the table are contradictory to the figure from the same paper. The maximum cross-section in Figure 2 in [7] is 16 GM. In the paper [10], as a result of the calculation, a value of 25.1 GM was obtained, a maximum position of 782 nm, which agrees well with our results.

We could not find the results of the measurement of TPA cross-section C153 in methanol. In the work [5], TPA cross-section was measured in a solution of C153 in CCl_4. The maximum cross section of the C153 solution in CCl_4 is 45 GM at a wavelength of 800 nm. Since there is significant change of the spectra of single- and two-photon absorption of C153 in methanol and CCl_4, it can be assumed that there is good agreement between the results of this work and the results of [5].

References

[1] Wagner B D 2009 Molecules **14** 210
[2] Fisher W G, Watcher E A, Lytle F E, Armas M, Seaton C 1998 Appl. Spectroscopy **52** 536
[3] Bredikhin V I, Galanin M D, Genkin V N 1973 Phys. Usp. **16** 299
[4] Rumi M, Perry J. W. 2010 Advances in Optics and Photonics **2** 451
[5] Makarov N S, Drobizhev M, Rebane A 2008 Opt. Express **16** 4029
[6] Albota M A, Xu C, Webb W W 1998 Applied Optics **37**, 7352
[7] Xu C, Webb W W 1996 Opt. Soc. Am. B **13** 481
[8] Magde D, Wong R, Seybold P G 2002 Photochem. Photobiol. **75** 327
[9] Kennedy S M, Lytle M F 1986 Anal. Chem. **58**, 2643
[10] Nguyen K A, Day P N, Pachte R 2007 J. Chem. Phys. **126**, 094303

IOP Conf. Series: Journal of Physics: Conf. Series **917** (2017) 062030 doi:10.1088/1742-6596/917/6/062030

Formation of 2D-PhCs with missing holes based on Si-layers by EBL

D E Utkin[1], A A Shklyev[1,2], A V Tsarev[1] and A V Latyshev[1,2]
[1]Rzhanov Institute of Semiconductor Physics SB RAS, Novosibirsk 630090, Russia
[2]Novosibirsk State University, Novosibirsk 630090, Russia

Abstract. The fabrication of the periodic structures, that is two-dimensional photonic crystals (2D PhCs) based on Si-materials by electron beam lithography (EBL) technique has been studied. We have investigated basic lithography processes such as designing, exposition, development, etching and others. The developed top-down approach allows close-packed arrays of elements and holes to be formed in nanometre range. This can be used to produce 2D PhCs with emitting micro-cavities (missing holes) with lateral size parameters with an accuracy of about 2% in the Si (100) substrate and in silicon-on-insulator structures. Such accuracy is expected to be sufficient for obtaining the cavities-coupling radiation interference from large areas of 2D PhCs.

1. Introduction

Fabrication and study of samples with artificially created variable in space directions (refraction index) permittivity and optical resonators based on the photonic crystal are perspective in photonics and optoelectronics [1]. The 2D photonic crystal with emitting micro cavities (missing holes) is used to create effective Si-based light emitters. Light emission spectrum with multiple peaks is inherent feature of these structures, whereas single-peak emission is usually required for commercial applications. To create the Si-based PhC structure we have investigated lithography processes such as designing, exposition and development.

2. Method and material

To investigate the properties of the emission spectra we created Si-based two-dimensional photonic crystals (2D PhCs) with differently ordered and partially disordered single hole cavity arrays. Their design is based on results of three-dimensional finite-difference time-domain method (3D FDTD). Also, in future it is planned to measure light emission spectra of these structures.

We used lithography with a strongly focused electron beam (systems of RAITH 150 and RAITH Pioneer) to form the mask in resist PMMA-950K on Si substrates. The pattern design of 2D-PhC (shown in figure 1) is determined by reviewing the results of the 3D FDTD method [2].

The substrate surface is preliminary cleaned by dimethylformamide and deionized water, then the drying process is used at 170 °C for 10 min which is followed by resist deposition with spin-coating technique.

To obtain the required elements dimensions in the photonic crystal mask several factors should be considered. Firstly, exposure parameters should be determined. So we used test doses with values ranging from 100 to 200 μC/cm2 at 20 kV. Secondly, the development mode is required to be evaluated. In our case we used a standard procedure of development with parameters of 30 s and the substance of MIBK:IPA (1:3). After the exposure process, we applied method of plasma chemical (BCI3: Argon) etching in order to transfer this mask in Si substrate.

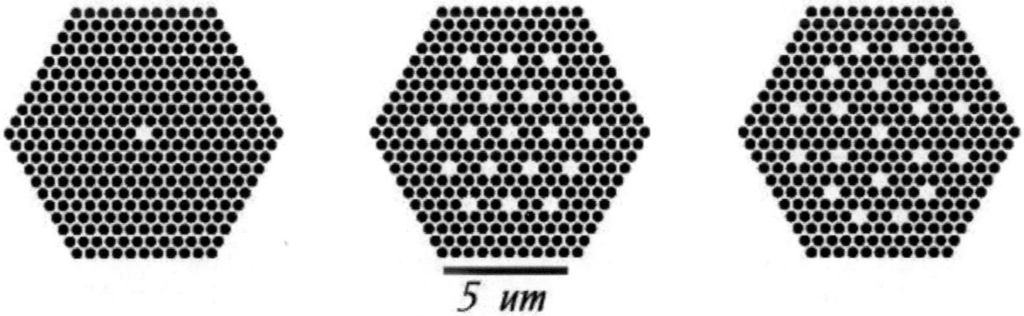

Figure 1. Pattern design of 2D PhCs with missing holes.

3. Results and conclusion

The obtained structures of three types design (H1, H19 and H19d) with emitting micro cavities (missing holes) in Si shown in figure 2.

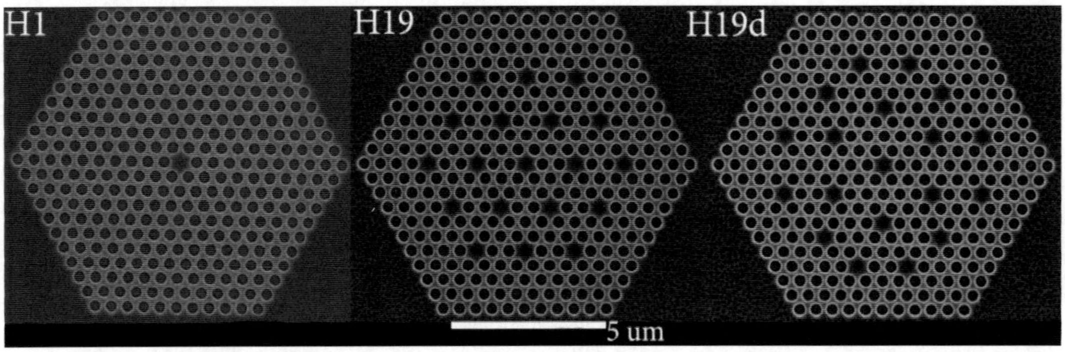

Figure 2. SEM-images of 3 types of 2D PhCs with missing holes in Si.

The holes diameter in Si is 384 nm and their period is equal to 550 nm that corresponds to the specified CAD (shown in figure 3).

Figure 3. SEM-images of areas 2D PhC with missing holes in Si.

The etch H19 2D PhC in Si depth of 87 nm was measured by AFM technique, the results are shown in figure 4.

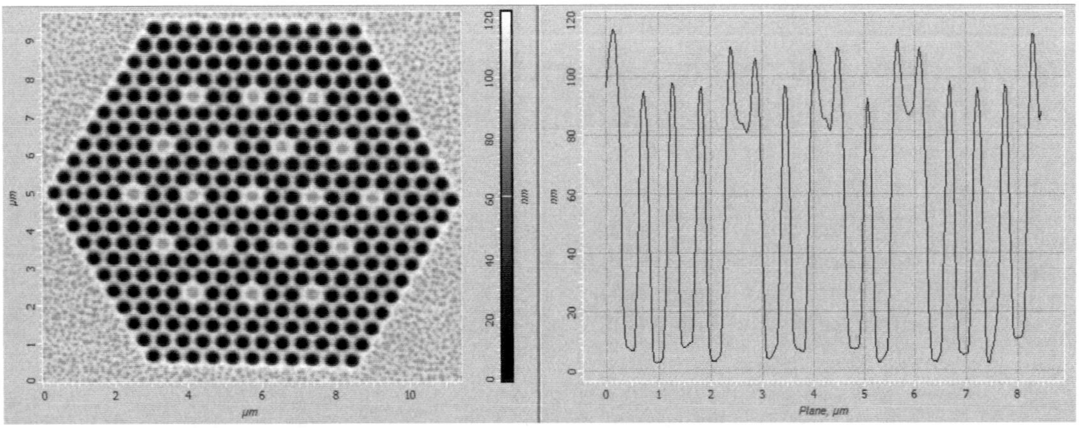

Figure 4. SEM-images of areas 2D PhC with missing holes in Si.

This technique can be used to produce hexagonal 2D PhCs with controlled parameters in the Si (100) substrate and prospective on a SOI structures. These structures have been optimized for the laser irradiation for the photoluminescence measurement. The three types design (H1, H19, H19d) of photonic crystal region with emitting micro cavities (missing holes) in Si are shown in figure 1. In figure 3 closed PhC in Si region is shown.

It is found that the accuracy of the 2D PhC formation is high, giving the considering that the expected spectral and spatial photoluminescence distributions, obtained for such structures, should be of sufficient quality for the adequate comparison with the results of 3D FDTD method calculations.

Acknowledgments
This work was supported by the Russian Science Foundation (Grant No. 14-22-00143). The authors are members of scientific school No. 10211.2016.8.

References
[1] B Jalali, S. Fathpour, J. Lightwave Tech. 2006, V.24, P. 4600-4615.
[2] A Tsarev, A Shklyaev, J. Lightwave Tech, 2014, V. 32, P. 4277

Optical absorption edge parameters of zirconium dioxide nanotubular structures

A V Kozhevina[1], A S Vokhmintsev[1], R V Kamalov[1], N A Martemyanov[1], A V Chukin[1] and I A Weinstein[1,2]
[1]Ural Federal University, NANOTECH Center, Ekaterinburg,620002, Russia
[2] Institute of Solid State Chemistry, Russian Academy of Sciences, Yekaterinburg 620990, Russia

Abstract. Nanotubular layer of zirconia with an thickness of 10 ± 1 μm and an outer/inner diameters of nanotubes of 80/60 ± 5 nm was synthesized by anodization in the electrolyte based on ethylene glycol with small amount of water and ammonia fluoride. It is shown that the obtained samples before and after annealing in air at 400 °C consist of 90 % and 79 % tetragonal, 10 % and 21 % monoclinic phases, respectively. Analysis of diffuse reflectance spectra showed that the thermal treatment leads to a decrease in concentration of surface Zr^{3+}-centers. Under the assumption of direct allowed transitions, the bandgap width was estimated to be E_g = 5.41 ± 0.01 eV and 5.66 ± 0.01 eV for as-grown and annealed samples, respectively.

1. Introduction

Self-organized nanotubular structures of metal oxides obtained by anodizing are promising matrices for gas sensors, solar cells, photochemical and photoelectrochemical cells [1, 2]. It is known [3] that ZrO_2 with the energy gap of E_g = 5 – 5.8 eV [1, 4, 5] has one the largest photocatalytic activity under UV exposure (254 nm) in the series of transition metal oxides. However, the photocatalytic efficiency of ZrO_2 under sunlight drops considerably in comparison, for example, to TiO_2 (E_g = 3.1 to 3.2 eV [6]) that limits its practical use.

Electrochemical oxidation of Zr is mainly carried out in the potentiostatic mode at a voltage of ≤ 120 V in both inorganic aqueous media [7] - sulfates [8, 9], phosphates [1] and their mixtures [10, 11], and organic electrolytes based on ethylene glycol [9, 12], glycerol [13], mixtures of glycerol and ethylene glycol [14], glycerol and formamide [14 – 16]. However, it remains unclear how the anodizing conditions have an effect on the defectiveness of the synthesized ZrO_2 nanotubular layers and, consequently, on photocatalytic properties of zirconia. At the same time, it is known that self-organized highly ordered nanotubular structures of TiO_2, obtained in organic electrolytes based on ethylene glycol with fluoride ions and subjected to thermal treatment, have more efficient electron transport and, as a result, exhibit higher photocatalytic activity [17].

During the excitation of the samples by photons with energy of $h\nu < E_g$ the main role in electron-hole pairs generation is played by impurities and/or intrinsic defects forming a system of allowed energy levels within the forbidden gap and distort the optical absorption edge [18, 19 – 21]. The surface states with uncompensated chemical bonds of metal atoms (for example, Zr^{3+}-ions) and oxygen vacancies in a different charge state (F-type centers) are the predominant defects for ZrO_2 nanostructures. For this reason, a thermal treatment of samples under oxidizing conditions can reduce the concentration of intrinsic defects influencing the optical absorption edge.

In despite of the considerable interest in zirconium dioxide in various structural and phase states [18, 12, 22 – 27], a knowledge about the absorption characteristics of anodized ZrO_2 in the region of the optical absorption edge [1, 10] is insufficient. As an example, the energies $E_g = 4.8$ [1] and 4.13 eV [10] corresponding to ZrO_2 nanotubes fabricated by anodization process are significantly different from the energies corresponding to nanopowders and thin films obtained by other methods [1, 4, 5].
Thus, the aim of this work was to study structural and optical features of the zirconia layers synthesized by anodization in an electrolyte based on ethylene glycol with fluorine ions using scanning electron microscopy, X-ray diffraction and diffuse reflection techniques.

2. Samples and experimental techniques

2.1 Synthesis of ZrO_2 nanotubular layers
Before the anodizing a 300 μm thick zirconium foil (99.6%) was treated with acetone, etched in a solution containing $HF/HNO_3/H_2O$ (1:6:20), washed with distilled water and dried in air. The electrochemical oxidation was carried out in a two-electrode cell with a thermostatic control. The cathode was a stainless steel plate. The anodization was conducted at a constant voltage of 50 V and a temperature of 20 °C for 9 hours in the solution of ethylene glycol containing 5 wt. % H_2O and 0.5 wt. % NH_4F. After anodization the samples was washed and dried in air. The thermal treatment of synthesized structures was performed in air at 400 °C for 1 h. Thus, as-grown (S1) and annealed (S2) samples of anodized zirconia were obtained.

2.2 Characterization
The surface morphology of the anodized ZrO_2 was studied using scanning electron microscopy (SEM) by CarlZeiss SigmaVP. Structural characterization of as-grown and annealed samples was performed on PANalytical X'Pert PRO MPD diffractometer with a copper anode using the Rietveld method. The scanning step was fixed at 0.05°.

2.3 Diffuse reflection spectroscopy
An investigation of the absorption properties at room temperature was carried out using a Shimadzu UV-2450 two-beam spectrometer and an ISR-2200 integrating sphere. Scanning was conducted in the range of 190-850 nm with a spectral slit width of 2 nm at a rate of 90 nm/min.

3. Results

3.1 SEM characterization
The SEM images of the surface and cleavage of sample S1 is shown in Figure 1. It can be note that the synthesized oxide layer consists of highly ordered nanotubes with an external 80 ± 5 nm and an internal 60 ± 5 nm diameters. Some tubes (Figure 1a) are closed, presumably by a barrier layer of metal oxide [6] or remnants of a loose precipitated zirconium hydroxide [8]. The layer thickness of nanotubular ZrO_2 is 10 ± 1 μm.

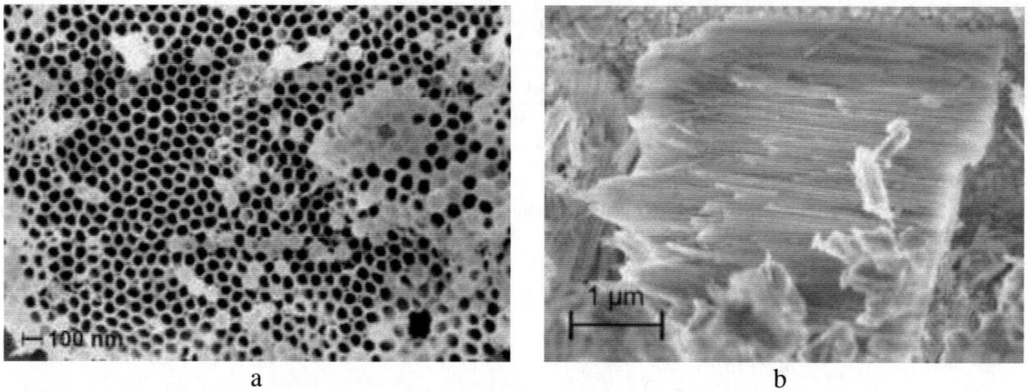

Figure 1. SEM images of surface (a) and lateral view (b) of anodized ZrO_2 layer.

3.2 XRD analysis

Figure 2 shows the XDR patterns of as-grown and annealed ZrO_2 samples. It can be seen that synthesized oxide layers have a crystalline structure. These spectra show characteristic peaks of both tetragonal (t) (scattering angles $2\theta = 30.2, 50.2, 60.2, 74.5, 85.2°$), and monoclinic (m) (at $2\theta = 24.1, 28.2, 41.4, 55.9, 74.7, 83.1°$) phases. The figure also presents the Zr peaks caused by the metal substrate.

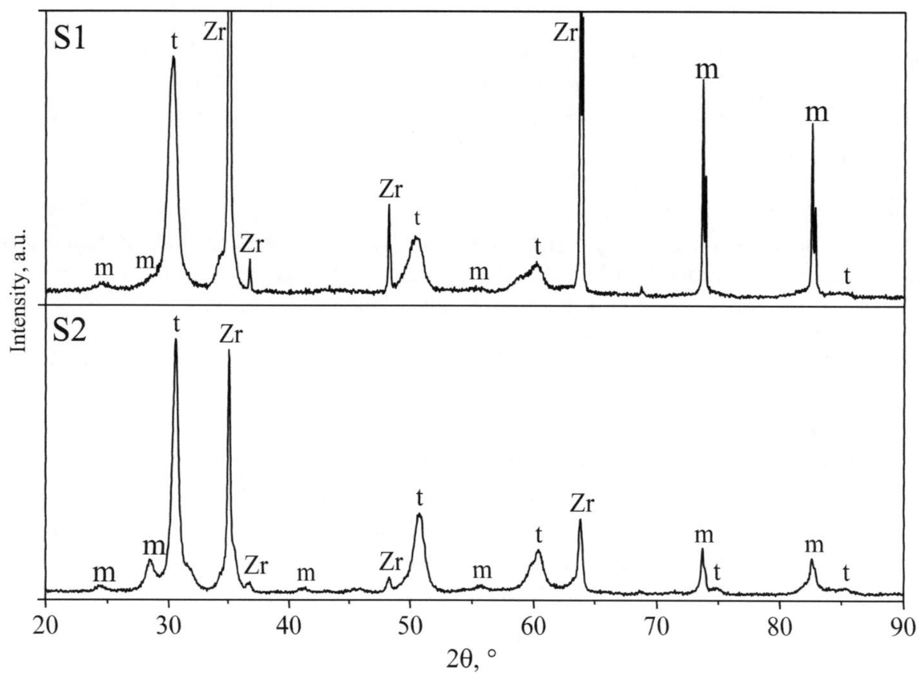

Figure 2. XRD patterns for as-grown (S1) and annealed (S2) samples.

3.3 Diffuse reflectance spectra

Figure 3 demonstrates the diffuse reflectance spectra $R(\lambda)$. It is seen that spectra have a sharp decline at $\lambda \leq 250$ nm due to the optical absorption edge. In the $\lambda = 250 - 850$ nm region, the R coefficient for the S2 sample is 5 to 15 % higher than for S1. This fact indicates the concentration decrease of the optically active color centers.

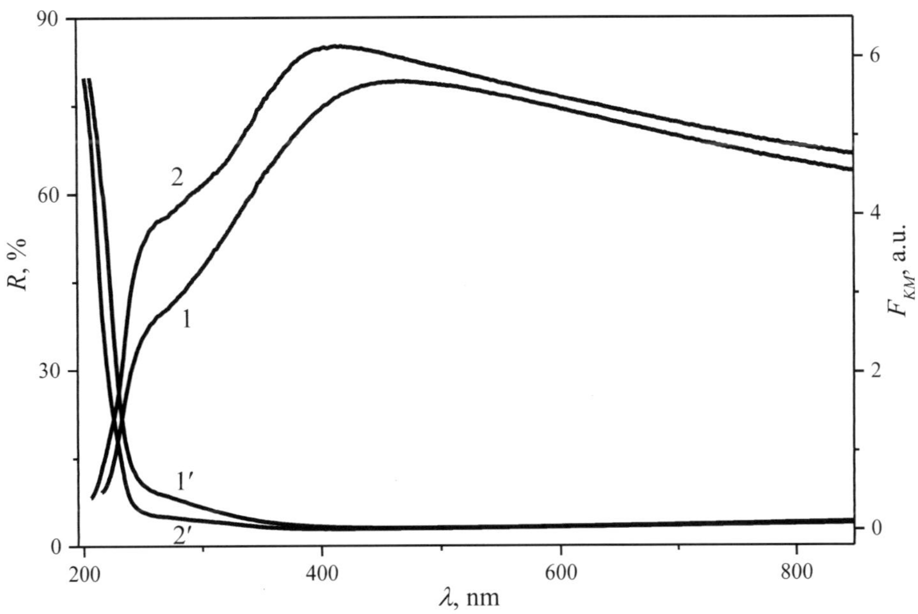

Figure 3. Diffuse reflectance spectra (1, 2 curves) and the Kubelka-Munk function (1′, 2′ curves) of S1 (1, 1′ curves) and S2 (2, 2′ curves) samples.

4. Discussion

A quantitative evaluation of the phase composition of the two-component system was carried out by a direct method. The experimental intensities of the main peaks for t- (30.2°) and m-ZrO2 (28.2°) were compared with the literature data. The 90 and 79 % of t-ZrO_2, 10 and 21 % of m-ZrO_2 were detected in samples S1 and S2, respectively. The obtained experimental data are in satisfactory agreement with studies of phase transformations in the low-temperature sintering process (200 – 350 °C) during the crystallization of an amorphous ZrO_2 nanopowder performed by decomposition of zirconium carbonate [28]. The authors of this work report that the crystallization proceeds with the formation of predominantly t- with the addition of m-ZrO_2 (15 – 20 %) with an average grain size of 36 ± 6 nm. It is also known that for macrosamples of ZrO_2 the phase transition of the monoclinic structure to the tetragonal one takes place at a temperature of 1170 °C [29]. In our case, the dominance of the tetragonal symmetry even in the as-grown samples and after annealing at 400 °C of ZrO_2 may exhibit the presence of structural features < 30 nm. This is consistent with the work data since the thickness of the zirconia nanotubes walls, according to SEM study, is near 20 nm, see 3.1.

For the analysis of the diffuse-reflectance spectra the Kubelka-Munk function was used, which is proportional to the optical absorption coefficient of the material:

$$F_{KM} = \frac{(1-R)^2}{2R} \qquad (1)$$

where R is the diffuse reflection of the material measured with respect to the absolutely white body.

The calculated dependences of $F_{KM}(\lambda)$ are shown in figure 3, curves 1′ and 2′. Within the $\lambda = 250 - 350$ nm a shoulder is observed for the samples with clearer presence for the S1 structure. According to [18], the absorption band with a maximum at λ = 280 nm is due to Zr^{3+} centers. In this connection, it can be assumed that annealing in the air atmosphere leads to the oxidation of surface defects and, as a consequence, to bonds compensation and decrease in the concentration of Zr^{3+} centers.

The Tauc plot was used to determine the bandgap width [30] using:

$$\left(h\nu \cdot F_{KM}\right)^{1/n} = A\left(h\nu - E_g\right), \quad (2)$$

where n is a constant depending on the type of electronic transition in the material; A is a proportionality coefficient.

Analysis of the optical absorption spectra using expression (2) under the assumption of direct allowed transitions (n = 1/2) is shown in figure 4. Estimations of E_g for S1 and S2 samples were 5.41 ± 0.01 and 5.66 ± 0.01 eV, respectively. Obtained results are in satisfactory agreement with independent study of crystalline (E_g = 5.65 – 5.74 eV) and amorphous (E_g = 5.65 eV) zirconia films on Si-substrate obtained by magnetron sputtering [4] and atomic-layer deposition [5], respectively. However, for crystalline films of anodized ZrO_2, lower values of E_g = 4.8 [1] and 4.13 eV [10] are reported. This fact indicates a significant influence of the samples synthesis conditions on the characteristics of nanostructural ZrO_2 optical absorption edge.

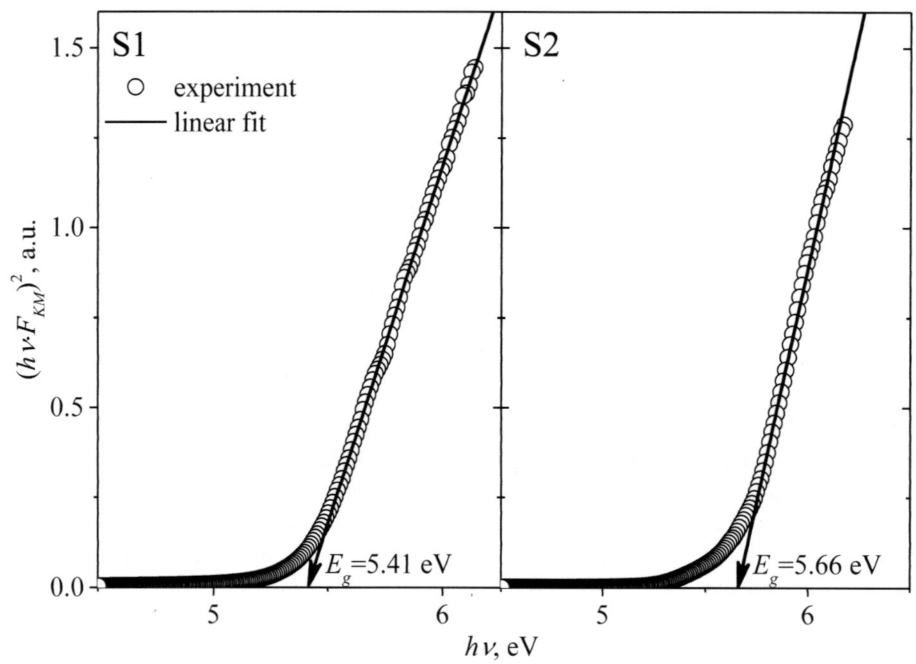

Figure 4. Tauc plot of optical absorption edge spectra for as-grown (S1) and annealed (S2) samples.

5. Conclusion

In the present research of nanotubular ZrO_2 samples synthesized by the anodic oxidation method was carried out. The morphology of the anodized oxide was studied by scanning electron microscopy. The thickness of the nanotubular layer was 10 ± 1 μm, the outer diameter of the nanotubes was 80 ± 5 nm, the inner diameter was 60 ± 5 nm. XRD analysis showed that as-grown samples and after 400 °C annealing consist of a mixture of tetragonal and monoclinic phases. The analysis of diffuse reflection spectra was performed. The bandgap width of E_g = 5.41 ± 0.01 and 5.66 ± 0.01 eV for nanotubular ZrO_2 samples before and after thermal treatment was determined under the assumption of direct allowed transitions using the Tauc plot. It was concluded that annealing of the investigated samples in the air atmosphere leads to compensation of break bonds of surface defects and a decrease in the concentration of Zr^{3+} centers.

Acknowledgments
The work was supported in part by Act 211 Government of the Russian Federation, contract № 02.A03.21.0006.

References
[1] Trivinho-Strixino F, Guimarães F E G and Pereira E C 2008 *Chem. Phys. Lett* **461** 82
[2] Kamalov R, Vokhmintsev A, Dorosheva I, Kravets N, Weinstein I 2016 *Adv. Sci. Lett* **22** 688
[3] Karunakaran C and Senthilvelan S 2005 *J. Mol. Catal. A-Chem* **233** 1
[4] Zhao S, Ma F, Xu K W and Liang H F 2008 *J. Alloy. Compd* **453** 453
[5] Puthenkovilakam R and Chang J P 2004 *Appl. Phys. Lett* **84** 1353
[6] Roy P, Berger S and Schmuki P 2011 *Angew. Chem.-Int. Edit* **50** 2904
[7] Lee W-J and Smyrl W H 2008 *Curr. Appl. Phys* **8** 818
[8] Tsuchiya H, Macak J M, Sieber I and Schmuki P 2005 *Small* **1** 722
[9] Tsuchiya H, Akaki T, Nakata J, Terada D, Tsuji N, Koizumi Y, Minamino Y, Schmuki P and Fujimoto S 2009 *Electrochim. Acta* **54** 5155
[10] Jiang W, He J, Zhong J, Lu J, Yuan S and Liang B 2014 *Appl. Surf. Sci* **307** 407
[11] Hahn R, Berger S and Schmuki P 2010 *J Solid State Electrochem* **14** 285
[12] Shin Y and Lee S 2009 *Nanotechnology* **20** 105301
[13] Muratore F, Baron-Wiecheć A, Hashimoto T, Skeldon P and Thompson G E 2010 *Electrochem. Commun* **12** 1727
[14] Berger S, Faltenbacher J, Bauer S and Schmuki P 2008 *Phys. Status Solidi-Rapid Res. Lett* **2** 102
[15] Fang D, Luo Z, Liu S, Zeng T, Liu L, Xu J, Bai Z and Xu W 2013 *Opt. Mater* **35** 1461
[16] Zhao J, Wang X, Xu R, Meng F, Guo L and Li Y 2008 *Mater. Lett* **62** 4428
[17] Kahnt A, Oelsner C, Werner F, Guldi D M, Albu S P, Kirchgeorg R, Lee K and Schmuki P 2013 *Appl. Phys. Lett* **102** 233109
[18] Emeline A, Kataeva G V, Litke A S, Rudakova A V, Ryabchuk V K and Serpone N 1998 *Langmuir* **14** 5011
[19] Ilin D O, Vokhmintsev A S and Weinstein I A. 2016 *AIP Conf. Proc* **1767** 020028
[20] Savchenko S S, Vokhmintsev A S and Weinstein I A 2017 *Opt. Mater. Express* **7** 354
[21] Savchenko S S, Vokhmintsev A S and Weinstein I A 2017 *Tech. Phys. Lett* **43** 297
[22] Liang J, Deng Z, Jiang X, Li F and Li Y 2002 *Inorg. Chem. Commun* **41** 3602
[23] Singh A K and Nakate U T 2014 *The Scientific World Journal* **2014** 349457
[24] Lucovsky G 2007 *J. Mol. Struct* **838** 187
[25] Rivera T, Olvera L, Azorín J, Sosa R, Barrera M, Soto A M and Furetta C 2006 *Radiat. Eff. Defects Solids* **161** 91
[26] Clament Sagaya Selvam N, Manikandan A, John Kennedy L and Judith Vijaya J 2013 *J. Colloid Interface Sci* **389** 91
[27] Nikiforov S V, Kortov V S, Savushkin D L, Vokhmintsev A S, Weinstein I A 2017 *Rad. Meas* in press DOI: 10.1016/j.radmeas.2017.03.020
[28] Gabelkov S V, Tarasov R V, Poltavcev N S, Logvinkov D S i Mironova A G 2004 *Voprosy atomnoj nauki i tekhniki* **3** 116 (in russian)
[29] Ackermann R J, Garg S P and Rauh E G 1977 *J. Am. Ceram. Soc* **60** 341
[30] Tauc J, Grigorovici R and Vancu A 1966, *Phys. Status Solidi B-Basic Solid State Phys.* **15** 627

IOP Conf. Series: Journal of Physics: Conf. Series **917** (2017) 062032 doi:10.1088/1742-6596/917/6/062032

Waveguide integrated superconducting single-photon detector for on-chip quantum and spectral photonic application

V Kovalyuk[1,2,3], S Ferrari[2], O Kahl[2], A Semenov[1], Yu Lobanov[1,4], M Shcherbatenko[1,4], A Korneev[1,4,5], W Pernice[2], G Goltsman[1,3,5]

[1]Department of Physics, Moscow State Pedagogical University, 119992, Russia
[2]Institute of Physics, University of Münster, 48149, Germany
[3]Zavoisky Physical-Technical Institute of the Russian Academy of Sciences, 420029, Russia
[4]Moscow Institute of Physics and Technology (State University), 141700, Russia
[5]National Research University Higher School of Economics, Moscow 101000, Russia

Abstract. With use of the travelling-wave geometry approach, integrated superconductor-nanophotonic devices based on silicon nitride nanophotonic waveguide with a superconducting NbN-nanowire suited on top of the waveguide were fabricated. NbN-nanowire was operated as a single-photon counting detector with up to 92 % on-chip detection efficiency in the coherent mode, serving as a highly sensitive IR heterodyne mixer with spectral resolution (*f/df*) greater than 10^6 in C-band at 1550 nm wavelength.

1. Introduction

Hybrid superconductor-nanophotonic devices (SND) have been proven to be one of the most promising approaches for development of complex quantum photonic integrated circuits (QPICs) [1]. One of the main components of SND is a supercounducting nanowire single-photon detector (SSPD or SNSPD), which combines high detection efficiency, low dark count rate, high temporal resolution and a compact design [2]. Since SSPD is a threshold detector, it detects the presence or absence of photons, but misses information about its energy or frequency. However, for some applications knowledge about the wavelength of photons is desirable, and optical heterodyne technique was recently implemented to overcome this imperfection of SSPDs [3]. Here we adopt this approach for QPICs and demonstrate both single-photon counting as well as coherent detection with high resolution. Unlike previous studies, here we investigated the coherent detection in the vicinity of the superconducting transition, heating up the electron subsystem with a high power optical local oscillator (LO).

2. Device design and fabrication

Fabricated device includes two focusing grating couplers (FGCs) for 1550-nm light coupling, 50:50 Y-splitter for calibration of optical power as well as U-shaped superconducting NbN nanowire (4 nm thickness and 80 nm width) as a single-photon detector (Figure 1a,b). For fabrication of devices we used three steps of electron beam (e-beam) lithography. In the first step, Au-contact pads and alignment marks are formed using PMMA resist and standard lift-off technique. In the second step, U-shaped NbN nanowires are made with use of HSQ resist by reactive ion etching (RIE) in CF_4. Finally, negative electron resist maN-2403 and etching in atmosphere of CHF_3 are employed for nanophotonic waveguides formation.

3. Experimental results

3.1. Measurement of critical temperature and critical current density

After fabrication process, the critical temperature (T_c) and the critical current density (j_c) of the NbN nanowires were measured. Our devices were installed on to a cold plate of a closed-cycle refrigerator operating with the minimum reachable temperature of T_b=1.78 K. Measurements of the T_c were carried out at a current of less than 100 nA, so as to exclude the Joule heating. As the width of the nanowire

IOP Conf. Series: Journal of Physics: Conf. Series **917** (2017) 062032 doi:10.1088/1742-6596/917/6/062032

decreased, the critical temperature of the film decreased from 9.5 K for wire 120 nm wide and 8.5 K for wire width 60 nm, which we associate with the effect of the proximity effect at the edges of the nanowire (Figure 2a). For the same the nanowires we measured the critical current also. In Figure 2b shown measured critical current (black dots) and their linear approximation (red line) is made. Using a nominal value of 4 nm as the nanowire thickness, we found the critical current density j_c = 1.7 MA/cm^2, which is comparable with the values obtained on other substrates.

3.2. Single-photon counting

Tunable fibre-coupled laser (LS) (New Focus TLB-6600) was used as a light source, which was routed to the device through attenuator (AT_1), polarization controller (PC_1) and beam splitter (BS), with real-time measurement of power (P_{in}) by the first channel of optical power meter (PM_1). The light, coupled by the first FGC into the optical waveguide, was divided by the on-chip Y-splitter into two parts. One half was directed through the calibration arm (1) for the measurement of output power (P_{out}) by the second channel of the optical power meter (PM_2), the other half routed to the nanowire (2). We determined the coupling efficiency (C_{eff}) from the input and output powers as: $C_{eff} = \sqrt{P_{out}/(P_{in} \times S \times WT_1)}$, where WT_1 refers to the waveguide transmission for the calibration arm and S is the splitting ratio of the on-chip Y-splitter (S = 0.5). The photon flux reaching the detector can be estimated as $\Phi_{ph} = (P_{in}/hf) \times S \times C_{eff} \times WT_2$. Absorption of single photons by the NbN nanowire leads to electrical pulses, which are amplified (A) and registered by a counter (C) (Figure 1c). The on-chip detection efficiency of the integrated SSPD ($\eta_{oc}^{counter}$) can be determined as ratio of the count rate of the detector (CR), with the exception of dark count rate (DCR), to the photon flux reaching the SSPD:

$$\eta_{oc}^{counter} \equiv \frac{CR-DCR}{\Phi_{ph}}. \tag{1}$$

The dependence of the on-chip detection efficiency on the normalized bias current, measured at a telecommunication wavelength λ = 1.55 μm is shown in the Figure 2e. We found saturated efficiency at the level of 92 % in a linear scale close to the critical current as well as DCR not exceeded of 10^4 (s^{-1}).

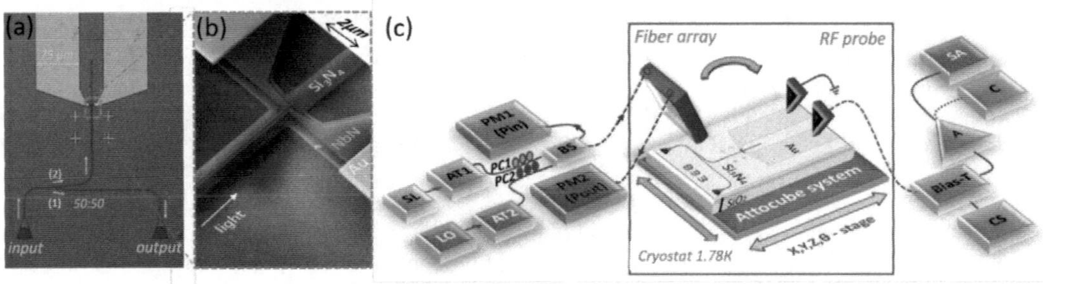

Figure 1 (a-c). (a) Optical image of a nanophotonic circuit in false colors; **(b)** SEM image of NbN nanowire atop the nanophotonic waveguide in false colors; **(c)** Schematic view of the experimental setup for measurement both single photon counting and coherent mixing in travelling wave geometry approach. Equipment marked blue and gray colors correspond to single photon counting regime, blue and red equipment correspond to heterodyne mixing regime. Abbreviation of the equipment: SL – signal laser; LO – local oscillator; AT_1, AT_2 – tunable attenuators; PM_1, PM_2 – optical power meter for input/output power; CS – current source; A – amplifiers; C – counter; SA – spectrum analyzer.

3.3. Coherent detection

In order to demonstrate on-chip coherent detection, we have introduced an additional tunable and stable laser, which is referred to as the LO, and whose radiation of almost same wavelength also passes a power attenuator and polarization controller, and a beam-splitter is used to combine power of both the lasers into a single-mode optical fiber (Figure 1c). We used rather large LO power in order to heat the electron temperature of the nanowire from the bath temperature (T_B = 1.78 K) to the superconducting transition temperature. In this case, the detector begins to operate not in a single-photon mode, but in a bolometric mode.

For the dependence of the total electric field $E(t)$ of the heterodyne (E_{LO}) operated at the frequency f_{LO} and the signal laser (E_S) at frequency f_S, one can write:

$$E(t) = E_{LO}\, cos(2\pi f_{LO} t) + E_S\, cos(2\pi f_S t). \quad (2)$$

The total field power is proportional to the $E^2(t)$ and defined as

$$P(t) = E^2(t) = E_{LO}^2\, cos^2(2\pi f_{LO} t) + E_S^2\, cos^2(2\pi f_S t) + 2E_{LO}E_S\, cos(2\pi f_{LO} t)\, cos(2\pi f_s). \quad (3)$$

The bolometer is not fast enough to follow the RF, so the dissipated power at this frequencies is the time averaged values (P_{LO}, P_S) However, if the intermediate frequency (IF) $f_{IF} = |f_S - f_{LO}|$ is low enough, the bolometer can follow this variation, so the time dependent term presents in this frequency.

$$P(t) = P_{LO} + P_S + 2\sqrt{P_{LO}P_S}\, cos(2\pi f_{IF} t). \quad (4)$$

Since near the superconductor transition, the bolometer has a strong resistance-temperature dependence, in general, it can be shown that the absorbed power leads to a change in temperature, a change in temperature to a change in the resistance, which in turn leads to a voltage dependence at IF on the bolometer on the radiation power:

$$V_{IF} = 2\, S\sqrt{P_{LO}P_S}, \quad (5)$$

where S is the voltage responsivity:

$$S = \frac{I(dR/dT)}{G\sqrt{1+(2\pi f_{IF})^2\tau_{th}^2}}, \quad (6)$$

where I is the bias current, G is the thermal conductance to the bath temperature, dR/dT is the derivative of film resistance vs temperature and τ_{th} is the thermal response time. For low enough $(2\pi f_{IF})^2\tau_{th}^2 < 1$, the bolometer can follow the IF power modulation, but for $(2\pi f_{IF})^2\tau_{th}^2 > 1$, the IF voltage will decrease as well as conversion efficiency

$$\eta_{oc}^{bolometer} \equiv \frac{P_{IF}}{P_{RF}} \quad (7)$$

will decrease.

With use of an RF spectrum analyzer (SA) placed instead of the counter, we have observed a beat signal which is generated due to superposition of the two waves of the signal (f_S) and local oscillator laser sources (f_{LO}) at the difference (“intermediate”) frequency $f_{IF} = |f_S - f_{LO}|$. In Figure 2d is shown the electrical signal generated by the nanowire and measured by the (SA). The signal full width half maximum (FWHM) equal to 4 MHz, which is limited by the frequency stability of the laser sources. Nevertheless, even with such stability, the spectral resolution of our setup is $193.5\times10^{12} / 4\times10^6 \approx 5\times10^7$, which can be used for high-resolution spectroscopy on a chip.

To choose the operating point with maximum on-chip conversion efficiency, we measured the IV characteristics at various P_{LO}. Three of them (dash lines), as well as P_{IF} for the 80 nm nanowire width and 140 μm total length are shown in Figure 3a.

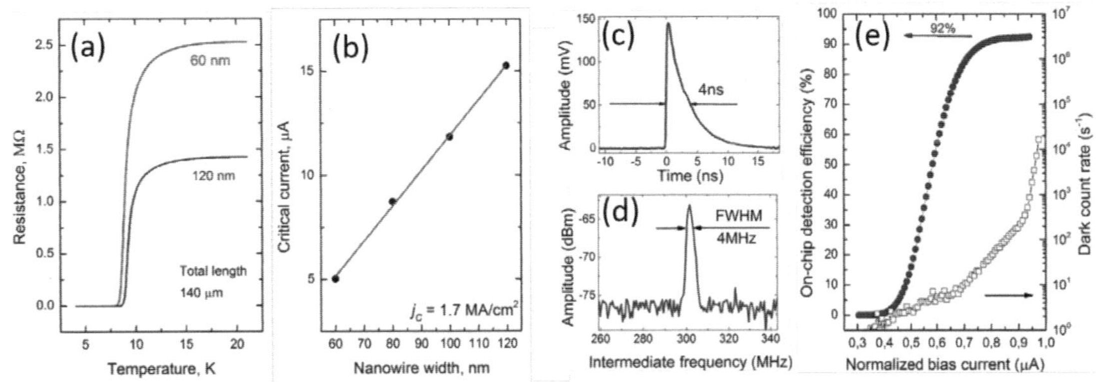

Figure 2 (a-e) (a) The measured dependence of the resistance on temperature for nanowires 60 nm and 120 nm wide; **(b)** The measured dependence of the critical current for nanowires with the different widths; **(c)** Electrical output of NbN nanowire after amplification in a single-photon regime; **(d)** Electrical output of NbN nanowire in a coherent regime; **(e)** On-chip detection efficiency (blue dots) and dark count rate (black squares) measured for the fabricated integrated detector.

The f_{IF} were fixed at 200 MHz, far from cut-off frequency. The maximum value of signal occurs at the first IV curve without critical current. Above that curve the negative differential resistance causes unstable operation of the mixer. At fixed P_{LO}, corresponding the negative differential maximum P_{IF}, we also measured the dependence of the bandwidth of the mixer. In Figure 3b the dependence of P_{IF} versus IF is shown (color points). The experimental data are fitted by formula $P_{IF} = -10\log_{10}(1 + (f_{IF}/f_{3dB})^2)$. The dotted line indicates the cut-off frequencies f_{3dB} corresponding to 3 dB decrease of the P_{IF}. The bandwidth drastically changes with applying dc power and rises up to 3.5 GHz at 2.5 µA, but at the cost of significantly reduced signal power.

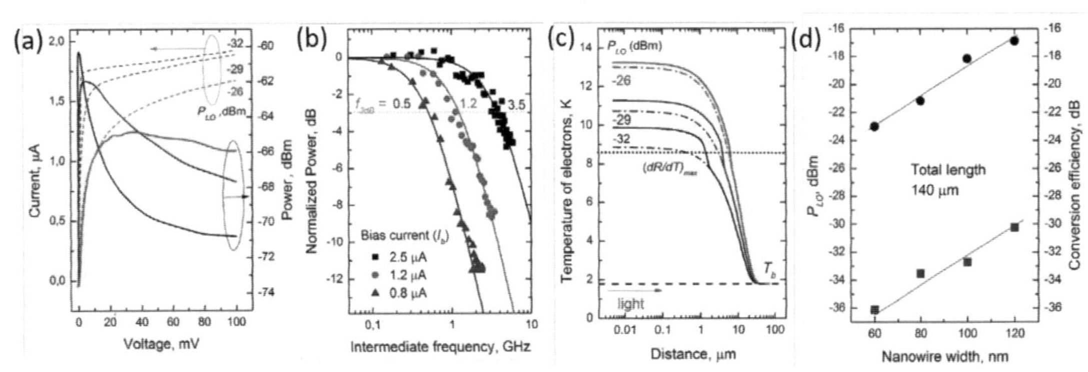

Figure 3 (a-d). (a) The measured IV curves (dash line) and the power at the intermediate frequency (P_{IF}) versus voltage on the nanowire at different local oscillator power (P_{LO}) (solid line); **(b)** Normalized P_{IF} versus intermediate frequency (f_{IF}) for different bias currents; **(c)** Numerically calculated temperature profile along U-shaped NbN nanowire. The arrow indicates the light propagation in the waveguide; **(d)** Measured dependence of conversion efficiency versus nanowire width at the same length (black circles) and the local oscillator power (blue squares).

4. Discussion

For further analysis we used a numerical calculation of the temperature profile. One of the main features of travelling wave geometry approach is that the absorption along the nanowire decreases exponentially with the nanowire length. For this reason the front parts of the nanowire have higher electron temperature (T_e) than subsequent. To model the T_e distribution along the nanowire, we have adapted the heat balance equation by introducing the dependence on coordinates.

The simulated electron temperature profiles along NbN nanowire for IV curves from Figure 3a are shown in Figure 3c. Black line corresponds to the maximum observed P_{IF}. Such temperature profile of 70 μm nanowire (140 μm the total length) can be divided into three distinct parts. The first part of the nanowire, starting from the device input point, is equal about 1 μm long and is heated close to the middle of the superconducting transition $(dR/dT)_{max}$ without bias current (dash-dot line). DC bias (solid line) corresponding to the maximum signal, leads to a substantial increase in the electron temperature in this nanowire section, as well as to the increase in its length. The second part of the nanowire in the range from 2 to 30 μm has lower temperature. The temperature in this section is substantially below the transition temperature and the current does not affect the T_e. The last and the longest nanowire part with 40 μm length is not heated with radiation and remains at the bath temperature equal to 1.78 K.

The same picture is held higher levels of P_{LO}. Regarding the superconducting transition temperature, in all these cases, there are "overheated" parts in the beginning of the nanowires, "unheated" parts in the middle and "cold side" at the back end. Due to differences in the electron temperature in different parts of a nanowire, each part contributes to the measured conversion efficiency and the $\eta_{oc}^{bolometer}$ can be generally found only numerically. For simplicity we did not put into our numerical model several effects like electro-thermal feedback, which play important role in gain bandwidth. However even with such rough estimation we can qualitatively identify the factors influencing the $\eta_{oc}^{bolometer}$.

One can expect higher contribution to $\eta_{oc}^{bolometer}$ in those parts of a nanowire, where the electron temperature is close to the middle of the superconducting transition and dR/dT has the maximum value. But in the case of U-shaped nanowire its normal resistance vary from about 0.5 MOhm for 120 nm to 1 MOhm for 60 nm nanowire width at the same total length of 140 μm. This leads to the fact that the maximum signal is not observed when all nanowire parts are uniformly heated to the middle of the superconducting transition, but rather a small portion at the beginning of nanowire, so that the resistance of the nanowire is close to the 50 Ω impedance of the amplifier.

Figure 3d shows the measured dependence of $\eta_{oc}^{bolometer}$ (black circles) and required P_{LO} (blue squares) versus nanowire width. As with conventional NbN HEB can be noted in the ultra-low P_{LO} at operating point. The figure also shows that an increase in the width of the nanowire increases both $\eta_{oc}^{bolometer}$ and optimal P_{LO}. Qualitatively, such dependence can be explained by the following main reasons using the formula for

$$\eta(\omega) = 2\beta S^2(\omega) P_{LO} P_L^{-1}, \tag{8}$$

where β is the coupling factor of radiation to the nanowire and R_L is the load resistance (50 Ω).

Firstly, by increasing the width of the nanowire the area of overlap between evanescent waveguide modes and a nanowire also increases, which leads to an increase in absorption. When the width of the nanowire changes from 60 nm to 120 nm, absorbance per micron length ranges from 0.058 dB/μm to 0.266 dB/μm. This leads to the increasing the length of nanowire with temperature close to critical temperature. And a greater part of nanowire starts to work in the mixing mode, contributing to heterodyne signal conversion.

Secondly, when the width of a nanowire increases, its critical current also increases, and consequently increases the required P_{LO} (Figure 3d) to heat to the first IV curve without negative differential resistance slope.

The third reason is connected to the decrease in resistance of the nanowire when width increases. This allows the point with maximum dR/dT going closer to point with 50 Ω, leading reducing mismatch with the amplifier.

These three reasons contribute to the increase the conversion efficiency. But on the other hand an increase of the nanowire width leads to a decrease in $S(\omega)$ due to increase volume that reduces $\eta_{oc}^{bolometer}$.

5. Conclusion

Dividing the power of the signal at the intermediate frequency by the incident optical power, we found on-chip conversion efficiency in the best operating point to be equal to −17 dB. We found a conversion bandwidth of up to 3.5 GHz by tuning frequency of the LO laser. Direct comparison with the data presented in [3] shows significant increase of the conversion bandwidth under hard illumination of the LO power, but at the expense of the sensitivity. Therefore choice of the detection regime depends on both available optical power, and desired sensitivity.

Acknowledgments

A. Korneev and V. Kovalyuk acknowledge support by the Russian Science Foundation (project 16-19-10633; design and NbN film deposition). W. Pernice acknowledges support by the DFG grants PE 1832/1-1 & PE 1832/1-2 and the Helmholtz society through grant HIRG-0005 (fabrication of nanophotonic circuits). Yu Lobanov, M. Shcherbatenko. We also acknowledge M. Finkel for help with numerical modelling and R. Ozhegov for helpful discussions about hot electron bolometer operation principle.

References

[1] Bogdanov S, Shalaginov M Y, Boltaseeva A, and Shalaev V M 2016 *Opt. Mater. Express* **7** 1
[2] Pernice W H P, Schuck C, Minaeva O, Li M, Goltsman G N, Sergienko A V, and Tang H X 2012 *Nat. Commun.* **3** 1325
[3] Shcherbatenko M, Lobanov Y, Semenov A, Kovalyuk V, Korneev A, Ozhegov R, Kazakov A, Voronov B M, and Goltsman G N 2016 *Opt. Express* **24** 26

Single photon emission from charged excitons in CdTe/ZnTe quantum dots

K G Belyaev[1], M V Rakhlin[1], S V Sorokin[1], G V Klimko[1], S V Gronin[1], I V Sedova[1], I S Mukhin[2,3], S V Ivanov[1], and A A Toropov[1]

[1]Ioffe Institute, Saint Petersburg 194021, Russia
[2]St. Petersburg Academic University, St. Petersburg 194021, Russia
[3]ITMO University, St. Petersburg 197101, Russia

Abstract. We report on micro-photoluminescence studies of individual self-organized CdTe/ZnTe quantum dots intended for single-photon-source applications in a visible spectral range. The quantum dots surface density below 10^{10} per cm^2 was achieved by using a thermally activated regime of molecular beam epitaxy that allowed fabrication of etched mesa-structures containing only a few emitting quantum dots. The single photon emission with the autocorrelation function $g^{(2)}(0)<0.2$ was detected and identified as recombination of charged excitons in the individual quantum dot.

1. Introduction

Quantum dot (QD) heterostuctures based on wide bandgap II-VI semiconductor compounds are promising for fabrication of single-photon sources of visible light, required for quantum cryptography. CdTe/ZnTe self-organized QDs are especially suitable for such applications since they can be grown by molecular beam epitaxy (MBE) with a reasonably low surface density [1]. Further decreasing of the QD density below $\sim 10^{10}$ cm^{-2} makes it possible to use routine lithography techniques and micro-photoluminescence (μ-PL) spectroscopy for even simpler extraction of the single-QD emission. The CdTe/ZnTe QD emission spectrum typically represents a set of sharp discrete lines, which correspond to multiexciton cascade recombination, i.e. recombination of excitons, biexcitons, charged excitonic states (trions), etc. [2,3]. In this work, we investigate the emission properties of single CdTe/ZnTe QDs in heterostructures grown on InAs(100) substrates by MBE, using a previously developed version of a thermal activation technique [4], which allows fabrication of the QDs arrays with the surface density noticeably lower than $\sim 10^{10}$ cm^{-2}.

2. Samples and experimental techniques

The studied CdTe/ZnTe QDs were formed by MBE within Zn(Mg)Te structures pseudomorphically grown on InAs(100) substrates, by using a so-called "thermal activation technique" which comprises deposition of a strained CdTe 2D layer, covering it with amorphous Te, and fast thermal desorption of the Te layer [1,4]. The CdTe/ZnTe QD structure includes a 5 monolayer (ML)-thick ZnTe buffer layer deposited on the (2x4)As surface of an InAs buffer layer, a 10 nm-thick $Zn_{0.9}Mg_{0.1}Te$ carrier confining layer, a 30 nm-thick ZnTe barrier layer followed by a single CdTe QD sheet with a nominal thickness of ~3 MLs, and a 20 nm-thick ZnTe top barrier. The Zn(Mg)Te-based layers were grown at ~300°C.

The individual QDs were spatially localized by forming cylindrical mesa-structures of several diameters ranging between 1 μm and 200 nm on the same QD structure. The mesas were fabricated by dry Ar^+-ion etching through the masks formed by an electron-beam lithography. The QDs emission was extracted and spectrally resolved in a confocal μ-PL optical scheme with a 50x micro-objective and a monochromator equipped with a liquid nitrogen cooled CCD camera and diffraction grating with 1800 grooves per mm (spectral resolution is estimated as 60 μeV). A *cw* laser with the emission wavelength 404 nm was used for the PL excitation. A set of calibrated neutral-density filters allowed variation of the excitation power. Photon correlation measurements were performed in a Hanbury Brown – Twiss intensity interferometer exploiting two single-photon avalanched Si photodiodes providing a time resolution better than 40 ps. The measurements were performed in the 8 - 80 K temperature range in a continuous flow He cryostat with a nanopositioning piezo-driver.

3. Results and discussion

The PL spectrum of the CdTe/ZnTe QDs consists of a wide emission band spreading between 530 and 630 nm, which is transformed into a set of descrete narrow lines with a characteristic spectral width of ~ 0.3 meV at 8 K by improving the PL spatial resolution either down to 1.5-3 μm using a standart μ-PL technique with a 50 μm pinhole located in a microscope image plane or down to 200 nm provided by formation of the mesa structures (figure 1).

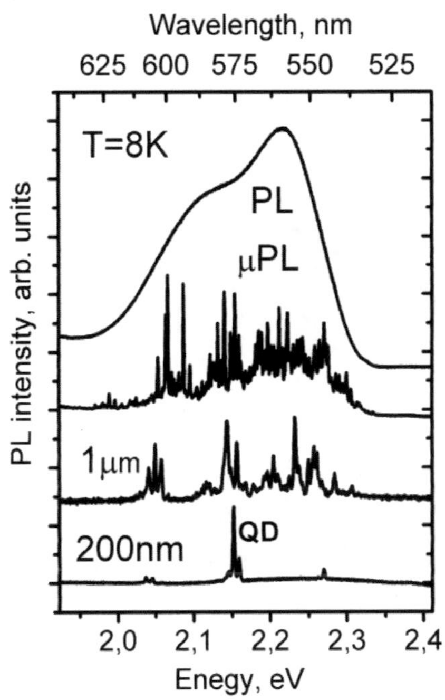

Figure 1. PL and μ-PL spectra of the CdTe/ZnTe QDs. Lines marked as PL, μ-PL, 1 μm and 200 nm represent, correspondingly, a typical PL spectrum (spatial resolution 200-500 μm), a μ-PL spectrum measured within a 1.5-3 μm area, and μ-PL spectra taken from the mesas with either 1 μm or 200 nm in diameter.

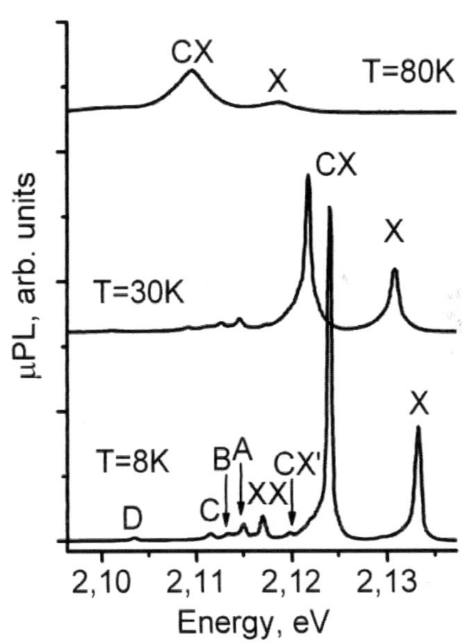

Figure 2. Typical spectra of the same single CdTe/ZnTe QD, measured at 8, 30 and 80 K. X, CX, CX' and XX labels mark exciton, charged excitons (positive and negative), and biexciton peaks. The peaks from other transitions originating from an exciton cascade recombination are marked as A, B, C, and D.

At 8 K, most of the mesas with the diameter 200 nm display the spectra represented by a set of lines with fixed spectral distances bitween them (like shown in figure 2). Moreover, a noticeable part of the mesas show absolutly identical shape of spectra with equal intensity relations between the lines measured at the same excitation power density. Such sets include up to 8 spectrally resolved lines (the lines number depends on the excitation power density), which correspond to excitons (X), positivly and negativly charged excitons (CX and CX'), biexcitons (XX) and other charged exciton and biexciton complexes (marked as A, B, C, D lines in figure 2). Excitation power dependencies allowed us to identify the line with the highest emission energy as an exciton state, because only it emerges in the spectrum at the lowest excitation power (figure 3). CX and CX' lines showing smaller intesity rising with excitation power as compared to XX and A-D lines are identified as charged exciton states (trions). In terms of a multiexciton model, probability of formation for biexcitons and other complexes including more than three carriers (electons and holes) should be lower at low excitation power density, comparing to the formation probability for excitons and trions. As one can see in figure 3, XX and A-D lines are visible only at the highest excitation powers. In adition to the identification by excitation power dependencies, the QDs spectra represent linear polarization of X and XX emission lines with the fine structure splitting (FSS) up to 240 μeV, caused by anisotropic exchange splitting. FSS of the X and XX lines exhibit oposite signs of the spectral shift for two linear polarizations in each QD. Superposition of these factors allowed us to conclude that the XX line corresponds to a biexciton state.

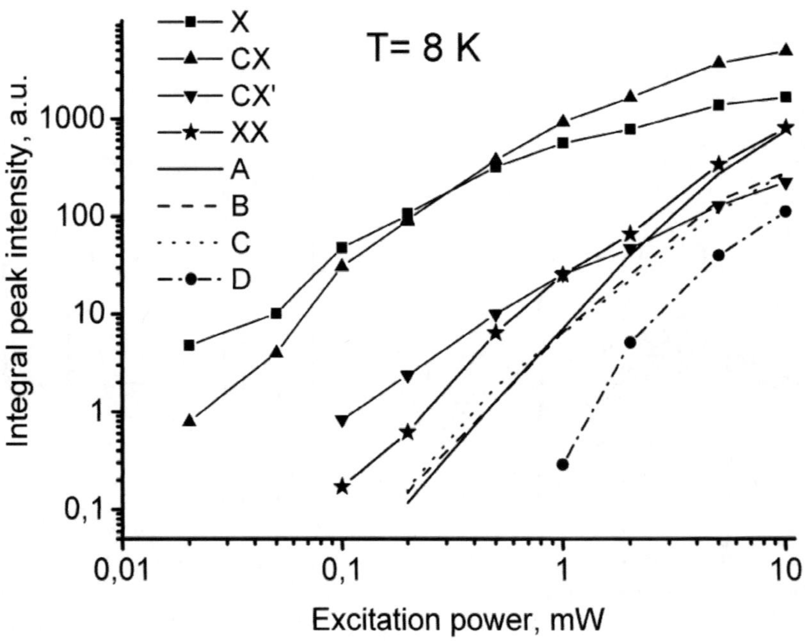

Figure 3. Excitation power dependencies for the peaks of the single-QD emission. Peaks labels correspond to labels in figure 2. The excitation laser emission was focused into a 20-50 μm^2 area.

We assume that in the studied CdTe/ZnTe structures the formation probability for the positivly charged exciton is higher than that for the negatively charged one. This is in complete agreement with the fact that ZnTe grown by MBE has usually p-type conductivity [5]. In such conditions, a typical CdTe/ZnTe QD spectrum exhibits the CX line with noticeably higher intensity as compared to all other lines. This line is spectrally separated from the X line by 9 meV on one side and by 7 meV from the XX line on other side. The CX' line intensity is lower than that of the CX one by two orders of magnitude. As a result, only the X, CX, and XX lines are well resolved up to the liquid nitrogen tempretature (figure 2).

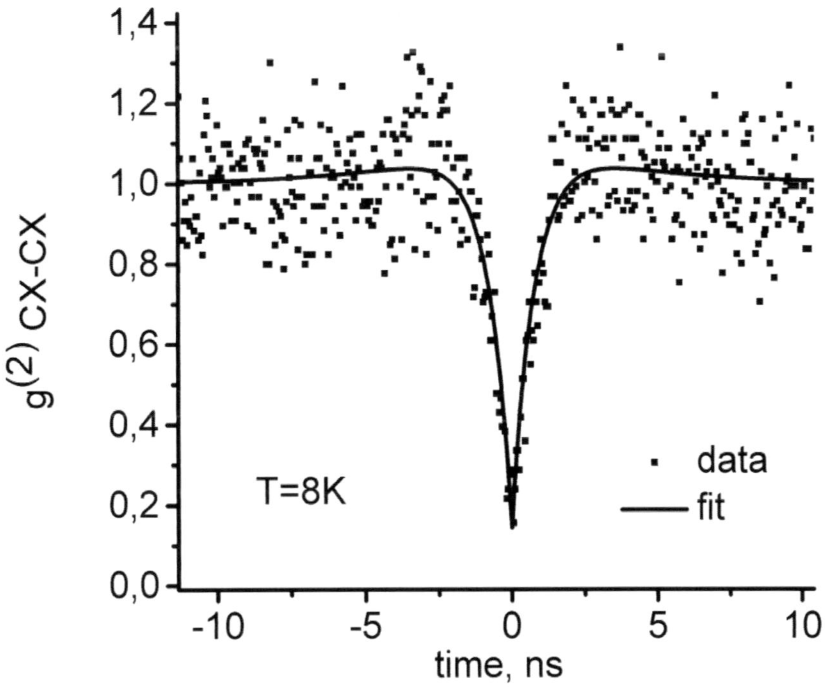

Figure 4. The second-order correlation function for the CX charged-exciton peak. The points represent experimental data while the solid line shows fitting by equation (1).

The measured second-order correlation function of the CX state emission demonstrates an antibunching dip with a minimal value $g^{(2)}(0)$ as low as ~0.2 that is a clear signature of the non-classical single-photon nature of the emission (figure 4). To obtain this function, the measured histogram of coincidences was normalized by count rates of the detectors, width of the time bins, and the integration time [6]. The solid line represents fitting by the expression

$$g^{(2)}(t) = 1 - A \cdot \mathrm{e}^{-t/t_a} + B \cdot \mathrm{e}^{-t/t_b}, \quad (1)$$

where t is the delay time. The estimated values for t_a and t_b are 0.8±0.1 ns and 2-5 ns, correspondingly.

4. Conclusions

We have demonstrated that CdTe/ZnTe QDs with extremely low surface density grown by MBE with thermal activation are suitable for single-photon applications in a visible spectral range. Etched mesa-structures with a 200 nm lateral size allow easy separation of the emission of the single QDs with the value of $g^{(2)}(0)$ below 0.2.

Acknowledgments
The authors acknowledge the financial support of the Russian Science Foundation (project #14-22-00107).

References

[1] Tinjod F, Gilles B, Moehl S, Kheng K and Mariette H 2003 *Appl. Phys. Lett.* **82** 4340
[2] Suffczyński J, Kazimierczuk T, Goryca M, Piechal B, Trajnerowicz A, Kowalik K, Kossacki P, Golnik A, Korona K P, Nawrocki M, Gaj J A and Karczewski G 2006 *Phys. Rev.* B **74** 085319
[3] Lee H S, Rastelli A, Benyoucef M, Ding F, Kim T W, Park H L and Schmidt O G 2009 *Nanotechnology* **20** 075705
[4] Sorokin S V, Sedova I V, Gronin S V, Klimko G V, Belyaev K G, Rakhlin M V, Mukhin I S, Toropov A A, and Ivanov S V 2017 *J. Cryst. Growth* (to be published)
[5] Yao T, Amano S, Makita Y, and Maekawa S 1976 *Jap. J. Appl. Phys.* **15**(6) 1001
[6] Brouri R, Beveratos A, Poizat J P, and Grangier P 2000 *Opt. Lett.* **25** 1294

Dependence of *InGaAs/InP* avalanche photodiode based single photon detector's noise characteristics on the photodiode's active area

V Zavodilenko, A Losev, A Miller, V Kurochkin, Y Kurochkin

Russian Quantum Center (RQC), Business Center «Ural», 100, Novaya st., Skolkovo, Moscow reg., 143025, Russia

Abstract. Nowadays the best single photon detectors from a practical view are those based on InGaAs/InP avalanche photodiodes, operating at a wavelength of 1.55 μm. In this work the characteristics of such a detector of our development are studied, while operating in Geiger mode with a synchronization frequency of 312.5 MHz. The dependence of quantum efficiency and noise levels on the active area are studied.

1. Introduction

Single photon detectors (SPDs) are used in various fields of science and technology. They must have a high quantum efficiency, low noise, and sufficiently high count rate. Detectors based on semiconductor avalanche photodiodes (APDs) use to be the most convenient for practical use [1]. Quantum efficiency is a probability to detect a single photon and includes both a probability to get a secondary electron appeared after a photon was absorbed in the active area of a diode and a probability to get then an avalanche. The first probability depends only on the semiconductor composition and structure, whereas the second one may be influenced by the applied voltage. The applied bias voltage increase causes the increase in the probability of getting an avalanche from one charge carrier. Dark count probability is a probability that a detector clicks in the absence of a photon. This kind of noise is of a temperature effect, and the dark count probability decreases with a temperature. The applied bias voltage increase causes both growth of the probability of getting an avalanche and increase in the dark count probability. There can be a non-zero probability that a SPD registers a signal and then clicks after a while again even when there is no a photon. Such a probability is called an afterpulsing probability, and it is more typical for *InGaAs/InP* avalanche photodiodes. This phenomenon is caused by defects which are in the crystal lattice of a semicoductor. The defects cause the appearance of potential "hollows" in the crystal which leads to charge carriers become caught after an avalanche. Further, such caught charge carriers can escape and cause a new avalanche which is no due to a photon detection.

Nowadays, at the wavelengths range of 400-900 nm, *Si* based avalanche photodiodes are the best APDs to develop single photon detectors. Present quantum communication systems use to operate at a wavelength of 1550 nm which corresponds to the lowest attenuation and dispersion in single mode optical fibers [2]. To detect single photons at the wavelength of 1300-1500 nm, *InGaAs/InP* based avalanche photodiodes are being used [1,3]. To be able to detect single photons, an APD must operate in Geiger mode when one photon is sufficient to cause an avalanche [4-6]. For this purpose, the reverse bias voltage applied to an APD has to be more than the breakdown voltage.

IOP Conf. Series: Journal of Physics: Conf. Series **917** (2017) 062034 doi:10.1088/1742-6596/917/6/062034

One the one hand, the more the bias voltage is, the higher the probability to detect a photon is. On the other hand, the increase in the applied bias voltage leads to the increase in dark count probability as well as the afterpulsing probability. To suppress these negative effects a number of techniques, like the APD cooling, are used. Usually, the temperature of an APD is reduced to -40 – -70 °C using microcoolers based on Peltier elements, which allows to improve the signal-to-noise ratio significantly. To reduce the afterpulsing probability, active quenching circuits for reducing avalanche charge are used [2]. Also, there can be a pulsed bias voltage technique where the bias voltage is being kept under the breakdown voltage most part of time, whereas it is increased to a level higher than the breakdown voltage only for a short time, ~1 ns, when a photon is expected to fall [4,6].

In this work, we present a newly developed single photon detector operating at the wavelength of 1.55 μm with a sinusoidal gating scheme. A schematic diagram of the detector is shown in Figure 1. In our detector we gated the bias voltage with a sinewave signal at a frequency of 312.5 MHz. Inhibition of the gating voltage response in the output signal was done via a compensation method at the main frequency and a filtration method at the closest harmonics. The avalanche response is amplified with a transimpedance amplifier and two sequential cascade amplifiers to a level of ~1 V. Wherein the ratio of the avalanche response and the gating voltage response with its harmonics in the output signal is provided at the level of ~20 dB. Avalanche registration is accomplished with the comparator of a cymometer, tuned to a level no less than 6 dB below the level of the gating voltage response. The diodes are cooled with three-stage Peltier elements, the temperature is measured with a platinum thermoresistor, located near the diode, and is controlled with the thermoregulator RMT DX5000.

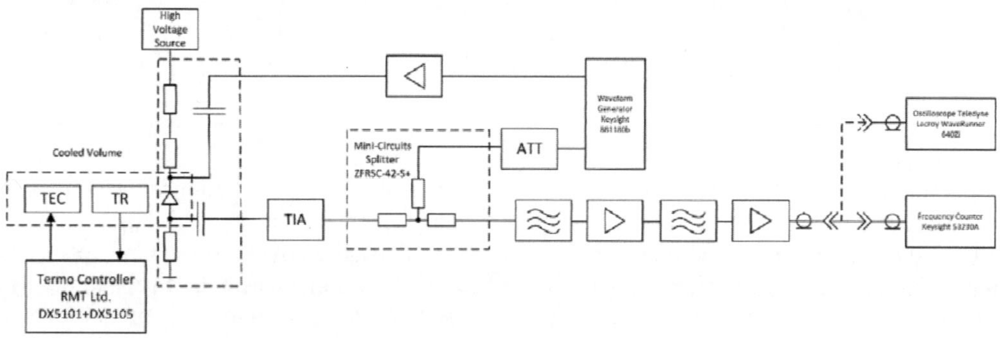

Figure 1. The schematic diagram of a single photon detector with the bias voltage gated with a sinewave signal at a frequency of 312.5 MHz.

2. Results and discussion

We tested the detector with various APDs, which have different semiconductor structure and active area. The following three avalanche photodiodes were under test. It was PGA-308u and PGA- 025u made by Princeton Ligthwave company with 16 and 25 μm active area diameters, correspondingly, and APDI-30 with 30 μm active area diameter. One can see the corresponding dependencies of the dark count probability on quantum efficiency in Fig. 2. The best signal-to-noise ratio was obtained when testing PGA-025u cooled to the temperature of 203 K (□). Since PGA-308u diode has its own cooling Peltier element, a temperature that could be achieved when testing PGA- 308u was limited by the ultimate power for the Peltier cooler, and was no more than 228 K. Nevertheless, the noise level for PGA-308u taken at a temperature of 228 K (Δ) does not exceed that taken at the same temperature for PGA-025u (○). One can note, that APDI-30 demonstrates the dark count probability 3 orders of magnitude more than that obtained for PGA-025u. Thus, for APDI-30 it is equal to 4×10^{-5} at a temperature of 200 K and a quantum efficiency of 10 %, whereas for PGA-025u it does not exceed 4×10^{-8} under the same conditions.

The dark count probability becomes a "cornerstone" when using single photon detectors in quantum communications. Indeed, the dark count results in the appearance of errors in a quantum key distributed between a sender and a receiver. The higher a dark count probability is, the more errors the sender and receiver have. If they get too many errors, the errors that might be caused by a possible eavesdropper may go unnoticed in the background of the dark counts. The error level of 11 % is known to be the maximum level of errors caused by natural factors when a secret quantum communication is still ensured [7]. The signal-to-noise ratio depends on how much the optical signal is attenuated while it follows through a quantum communication line to the receiver. Present single mode fibers used in quantum communication lines attenuate an optical signal at 1550 nm by about 0.18-0.20 dB per a km. So, taking into account this value, it becomes possible to estimate the maximum length of a quantum communication line that could be used to a secure communication. Our calculations show that it is possible to implement a quantum key distribution over the distances of up to 130 km using PGA-025u at a temperature of ~200 K and a quantum efficiency of at least 10 %.

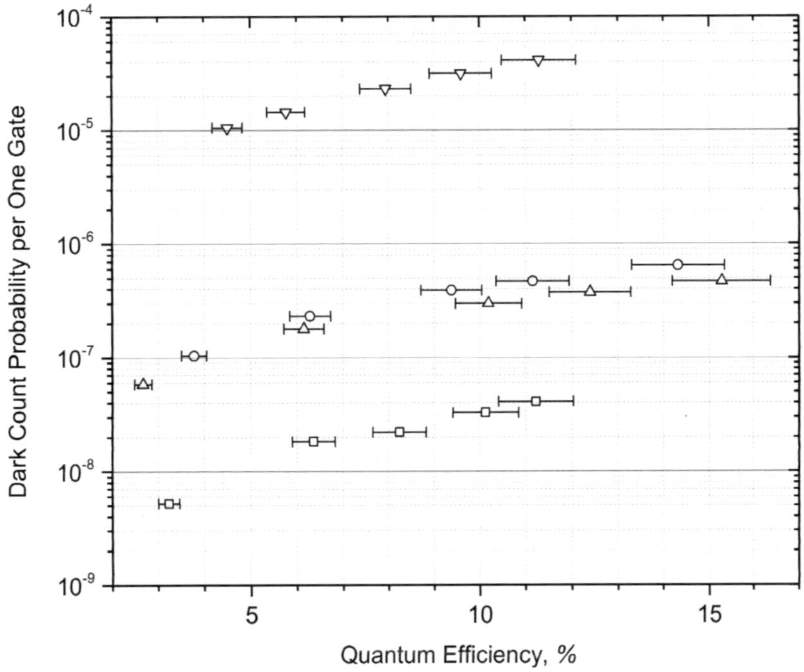

Figure 2. The dependencies of dark count probability on quantum efficiency obtained for APDI-30 at a temperature of 200 K (∇), PGA-025u at temperatures of 203 (□) and 228 K (○), PGA-308U at a temperature of 228 K (Δ).

Such a characteristic as an afterpulsing probability is no of less importance when we consider an SPD for use in quantum communications. The results obtained on the afterpulsing for three types of APDs under test are presented in Figure 3. The greatest afterpulsing probability was found for a PGA-025u diode at a temperature of 203 K. The afterpulsing probability was found to decrease with increasing a temperature, what is an agreement with the data obtained earlier. A comparison of different types of APDs at the same temperatures shows that PGA-025u demonstrates much more higher afterpulsing probabilities than PGA-308u, while testing the detector with APDI-30 we found no afterpulsing events even at a temperature of 200 K.

The appearance of afterpulses leads to the increase in the number of false detector's counts. As one can see looking at Figures 2 and 3 the probability of such counts can be much more higher than dark count probability in the absence of an afterpulsing effect. It can result in too high level of errors in the data transmitted through a quantum communication line. To avoid this, a mechanism of a "dead time", that is a time when a SPD cannot receive an optical signal and does not outputs any counts, must be applied. Such a mechanism might suppress the largest part of afterpulses if the value of the "dead time" is sufficiently high to let charge carriers escape.

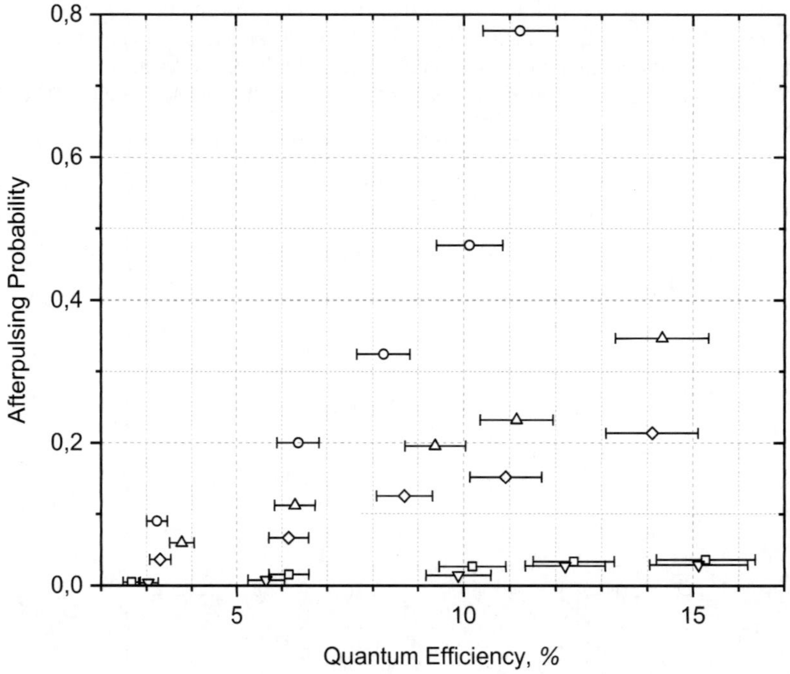

Figure 3. The dependencies of afterpulsing probability on quantum efficiency for PGA-025u at the temperatures of 203 (○), 228 (Δ), and 253 K (◊), and for PGA-308u at the temperatures of 228 (□) and 253 K (∇).

Thus, a newly single photon detector with a sinusoidal gating scheme operating at a frequency of 312.5 MHz and a wavelength of 1.55 μm has been developed. Now it operates in a real device for quantum key distribution over a quantum channel in the laboratory. The quantum channel is a coil with single mode optical fiber with the length of 25.5 km and total losses of 4.8 dB which is only in the laboratory. Now it is hard to implement a quantum key distribution over significantly greater distances because we use APDI-30 showing a relatively high noise level. Thus, even at a distance of 25.5 km the corresponding contribution to the quantum bit error rate is equal to 2.4 %. However, now we are working on a mechanism of "dead time" which might allow to use PGA-025u avalanche photodiodes in the detector. As it was found during our studies, these APDs have the smallest dark count probability, which can allow to use them in quantum communication systems to distribute a quantum key over the distances of up to 130 km.

Acknowledgments
The support from Ministry of Education and Science of the Russian Federation in the framework of the Federal Program (Agreement 14.582.21.0009, ID RFMEFI58215X0009) is greatly acknowledged.

IOP Conf. Series: Journal of Physics: Conf. Series **917** (2017) 062034 doi:10.1088/1742-6596/917/6/062034

References

[1] G S Buller and R J Collins 2010 *Meas. Sci. Technol.* **21** 012002
[2] R H Hadfield 2009 *Nature Photonics* **3** 696
[3] J Jan, A Migdall, and S V Polyakov 2011 *Review of Scientific Instruments* **82** 071101
[4] Ribordy G, Gautier J, Zbinden H and Gisin N 1998 *Appl. Opt.* **37** 2272
[5] Rarity J G, Wall T E, Ridley K D, Owens P C M and Tapster P R 2000 *Appl. Opt.* **39** 6746
[6] Cova S, Ghioni M, Lotito A, Rech I and Zappa F 2004 *J. Mod. Opt.* **51** 1267
[7] N Gisin, G Ribordy, W Tittel and H Zbinden 2002 *Rev. Mod. Phys.* **74**, 145

IOP Conf. Series: Journal of Physics: Conf. Series **917** (2017) 062035 doi:10.1088/1742-6596/917/6/062035

Investigation of vortex laser beam injection into an optical fiber

D A Savelyev[1,2], **S N Khonina**[1,2]

[1]Department of Engineering Cybernetics, Samara National Research University, Samara 443086, Russia
[2] Laboratory of Laser Measurements, Image Processing Systems Institute - Branch of the Federal Scientific Research Centre "Crystallography and Photonics" of Russian Academy of Sciences, Samara 443001, Russia

Abstract. We investigate Laguerre-Gaussian vortex laser beam injection into an optical fiber. Modelling of radiation entering an optical fiber with plane (cylinder) and axicon (cone with diffrent apex angle) micro-relief is numerically investigated by the finite difference time domain (FDTD) method.

1. Introduction

The state-of-the art technology of optical communication systems based on the conventional time/frequency division multiplexing is on the brink of reaching the capacity limits [1]. One way to achieve a manifold increase in the communication line capacity is through the use of mode division multiplexing (MDM), which was proposed for optical fibers relatively long ago [2]. Using the MDM, different transverse modes of the same optical fiber are used as a data carrier of a communication system. The data transmitted can be contained either in the mode content or in a proportion of energy carried by a particular laser mode.

A particular advantage for the enhancement of the information channel capacity is delivered by vortex optical beams carrying orbital angular momentum and having an infinite number of possible quantum states [3]. A substantial success of employing such method of channel diversity has been already demonstrated in optical fibers [4] and in free space [5].

Generation and propagation of laser vortices in free-space are studied well enough [6, 7], however, a single vortex excitation or vortex superposition generation in an optical fiber is a big problem [8-12]. As distinct from the classical LP-modes (modes in optical fibers), the vortex modes are scale invariant when entering into the fiber and when analyzed at the fiber output using diffractive microstructures [11-14]. This gives more freedom in choosing parameters of an optical scheme.

However, a particular problem is the introduction of the laser radiation into the optical fiber. And vortex beams with a screw phase singularity and zero intensity at the center, may lead to additional complications. In [15], various types of polarizations of laser radiation and their influence on the focusing of a laser beam were considered in detail. We choose circular polarization for this research.

In this paper, we consider the possibility of vortex laser beam injection into an optical fiber. To the numerical simulation of diffraction considered laser beams used the FDTD-method with high-performance computing [16]. Calculations were made on the computational cluster with power of 775 GFlops.

IOP Conf. Series: Journal of Physics: Conf. Series **917** (2017) 062035 doi:10.1088/1742-6596/917/6/062035

2. Investigation of vortex laser beam injection

We considered input into an optical fiber (with conical axicon in the first case and with cylinder in the second case at the end of fiber) of two kind of laser beam: Gaussian beam and Laguerre-Gaussian vortex with the circular polarization. Simulation parameters: the wavelength $\lambda = 0.532$ microns, the size of the computational domain x, y, z $\in [-5.83\lambda; 5.83\lambda]$. The thickness of the absorbing layer PML $\sim 1.13\lambda$ (0.6 micron), the sampling step of space – $\lambda/31$, the sampling step of time – $\lambda/(62c)$, where c is the velocity of light, α - the apex angle of the conical axicon. The radius of the cone base is 2.5 microns. The considered laser beams are shown in figure 1.

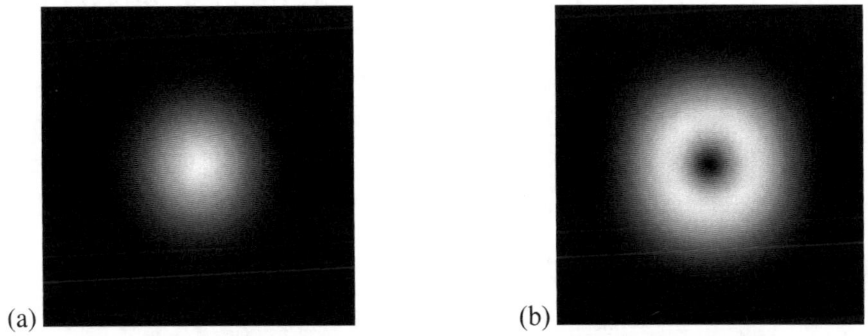

Figure 1. The considered laser beams: (a) Gaussian beam; **(b)** Laguerre-Gaussian vortex.

The numerical results of investigations are given on figure 2 in the case with conical axicon at the end of fiber. The total intensity in the yz plane is shown for different angle α.

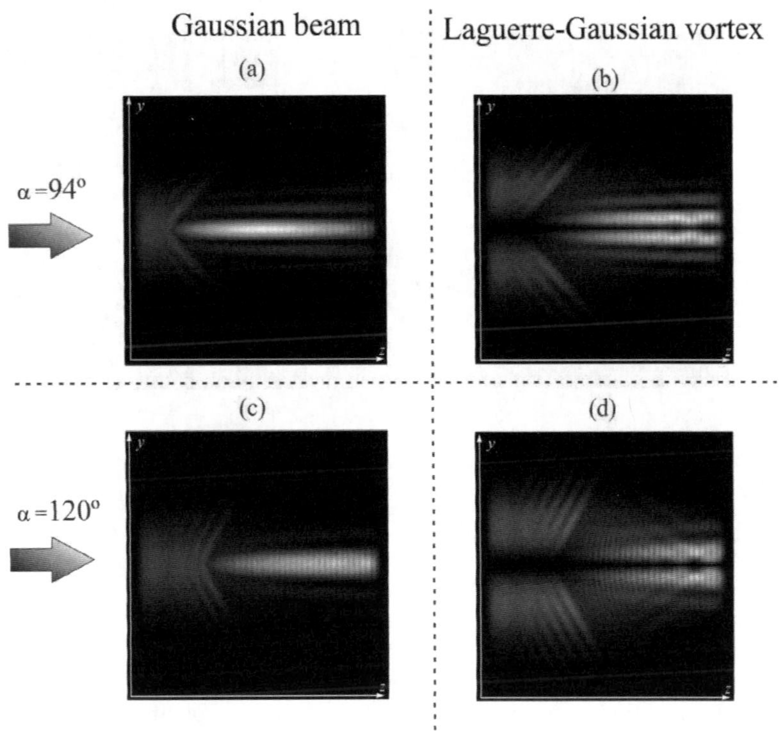

Figure 2. Injection into an optical fiber (cone) with diameter 5 microns: (a) Gaussian beam, $\alpha = 94^o$; **(b)** Laguerre-Gaussian vortex, $\alpha = 94^o$; **(c)** Gaussian beam, $\alpha = 120^o$; **(d)** Laguerre-Gaussian vortex, $\alpha = 120^o$.

It should be noted that the decrease in angle gives better focusing for the considered laser beams. This applies both to the case of a Gaussian beam and to the case of the Laguerre-Gaussian: for Gauss, a narrower light needle is obtained, for a vortex – a ring.

Let us show, for comparison, the Injection of the laser beams into an ordinary cylinder with a radius of 2.5 microns. Figure 3 shows the diffraction of laser beams with injection into an optical fiber. The top line is the total intensity, the bottom line is the z-component. It should be noted a narrow light needle, consisting of the z-component, obtained for the case of Laguerre-Gaussian vortex.

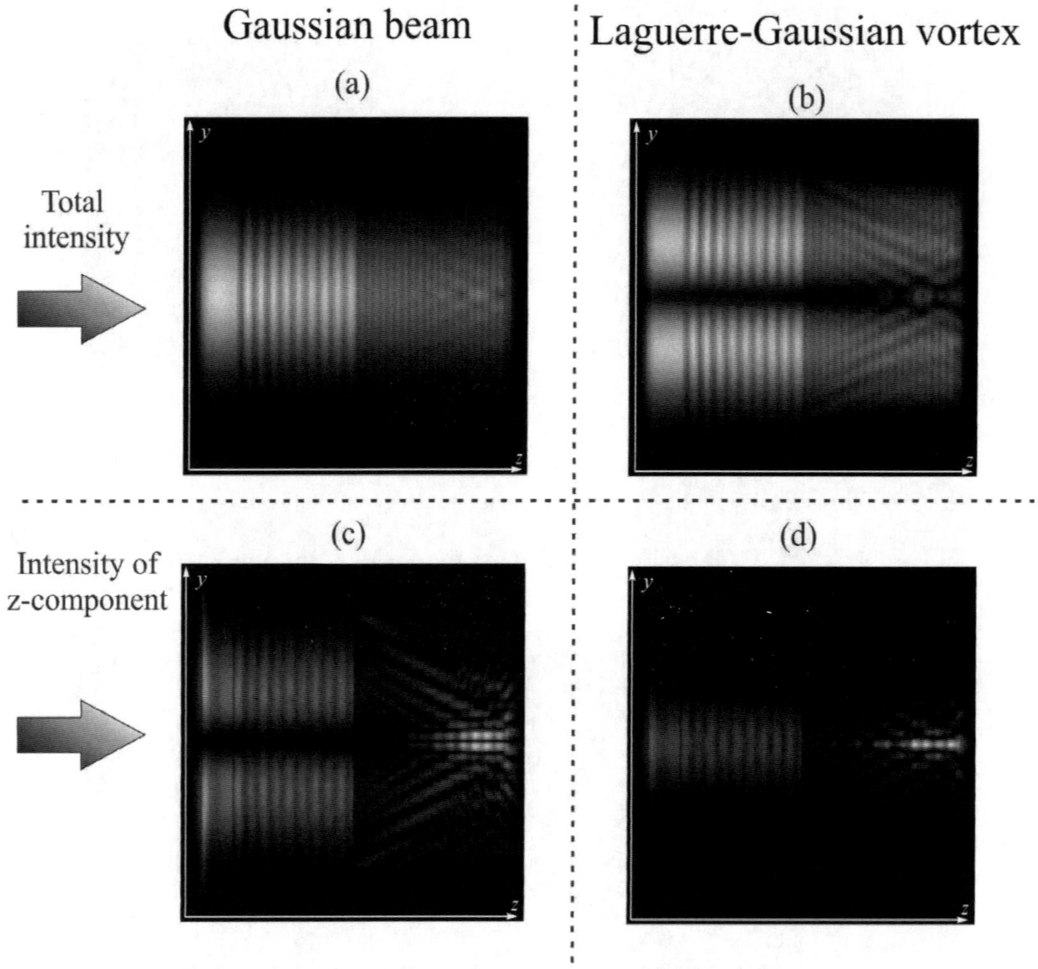

Figure 3. Injection into an optical fiber (cylinder) with diameter 5 microns: (a) Gaussian beam, total intensity; **(b)** Laguerre-Gaussian vortex, total intensity; **(c)** Gaussian beam, intensity of z-component; **(d)** Laguerre-Gaussian vortex, intensity of z-component.

3. Conclusion

In this paper shows the effect on the diffraction of considered laser beams when changing the angle of the conical axicon (at the end of optical fiber) with using the FDTD method. It is shown that the presence of the relief in the form of a conical axicon allows getting a narrower light needle compared to entering a conventional cylinder for Gaussian beams and vortex Laguerre-Gaussian beam.

Acknowledgments

This work was supported by the Russian Foundation for Basic Research (grants 16-07-00825, 16-29-11698) and by the Ministry of Education and Science of Russian Federation.

References

[1] Essiambre R, Kramer G, Winzer P J, Foschini G J, Goebel B 2010 *J. Lightw. Technol.* **28** 662
[2] Berdague S, Facq P 1982 *Appl. Optics* **21** 1950
[3] Wolf E 2001 *Progress in Optics* (Amsterdam, North Holland: Elsevier Science)
[4] Bozinovic N, Yue Y, Ren Y, Tur M, Kristensen P, Huang H, Willner A E, Ramachandran S 2013 *Science* **340** 1545
[5] Yan Y, Xie G, Lavery M P J, Huang H, Ahmed N, Bao C, Ren Y, Cao Y, Li L, Zhao Z, Molisch A F, Tur M, Padgett M J, Willner A E 2014 *Nature Communications* **5876** 4876
[6] Gbur G, Tyson R K 2008 *J Opt Soc Am A* **25** 225
[7] Soifer V A, Korotkova O, Khonina S N, Shchepakina E A 2016 *Computer Optics* **40** 605
[8] Dubois F, Emplit Ph, Hugon O 1994 *Opt. Lett.* **19** 433
[9] Thornburg W Q, Corrado B J, Zhu X D 1994 *Opt. Lett.* **19** 454
[10] Bolshtyansky M A, Savchenko A Yu Zel'dovich B Ya 1999 *Opt. Lett.* **24** 433
[11] Karpeev S V Khonina S N 2007 *Opt. Mem. & Neur. Netw. (Inf. Opt.)* **16** 295
[12] Yasin M, Harun S W, Arof H. 2012 *Recent Progress in Optical Fiber Research*, (Rijeka: InTech)
[13] Lyubopytov V S, Tlyavlin A Z, Sultanov A K, Bagmanov V K, Khonina S N, Karpeev S V, Kazanskiy N L 2013 *Computer Optics* **37** 352
[14] Savelyev D A, Khonina S N 2014 *Proc. In Optical Technologies for Telecommunications 2013* (International Society for Optics and Photonics) p. 915609
[15] Khonina S N, Savelyev D A, Kazanskiy N L 2015 *Optics Express* **23**(14) p. 17845
[16] Savelyev D A, Khonina S N 2014 *Journal of Physics: Conference Series* **490** p. 012213

IOP Conf. Series: Journal of Physics: Conf. Series **917** (2017) 062036 doi:10.1088/1742-6596/917/6/062036

Plasmonic absorption of THz radiation in graphene structure with a metal grating

V S Melnikova[1,2], O V Polischuk[1], V V Popov[1,2,3]
[1]Kotelnikov Institute of Radio Engineering and Electronics (Saratov Branch), Russian Academy of Sciences, Saratov 410019, Russia
[2]Saratov State University, Saratov 410012, Russia
[3]Saratov Scientific Center of the Russian Academy of Sciences, Saratov 410028, Russia

Abstract. The plasmonic absorption spectrum of THz radiation in the graphene with periodic metal grating is theoretically studied. It is shown that graphene exhibits strong plasmon absorption resonances at the frequencies of plasma oscillations in graphene for narrow-slit metal grating and a thin barrier layer.

1. Introduction

In recent years, studies of graphene, a two-dimensional graphite monolayer, have been of great interest on account of the unique electronic properties of this material following from the linear (Dirac) gapless energy spectrum of charge carriers [1]. Graphene possesses a strong plasmonic response at terahertz (THz) frequencies due to both a high density and low collective effective mass of free charge carriers [2, 3]. Using the plasma oscillations of charge carriers in graphene is attractive because it allows for concentrating the electromagnetic field near the graphene layer, which considerably enhances the efficiency of the interaction between THz radiation and graphene.

Nanoperiodic graphene-based plasmon structures are physically interesting objects. Since the wavelength of the plasma wave excited in such a structure is comparable with the structure period, this structure forms a planar plasmonic crystal. On the other hand, since the spatial period of the metal grating is shorter than the wavelength of the incident THz radiation by two or three orders of magnitude, this structure may be considered as a planar resonant THz metasurface strongly coupled with THz radiation. In this case, the entire structure can be characterized by the effective surface impedance that experiences resonances at the excitation frequencies of plasma waves in graphene [4]. This allows one to find the conditions for perfect matching of the structure with the incident THz radiation which ensures the most effective excitation of plasmons in graphene [5].

2. Theoretical model

In this paper, we consider the absorption of a THz wave by the resonant plasmons in a homogeneous graphene sheet located in x-z plane on the surface of a dielectric substrate (see figure 1). A one-dimensional planar metal grating is isolated from the graphene sheet by a thin dielectric barrier slab. Each ideally conducting strip of the planar metal grating is infinite in the z direction and has zero thickness in the y-direction. External THz wave is incident normally to the plane structure along y-direction with the polarization of the electric field E_0 across the metal grating contacts.

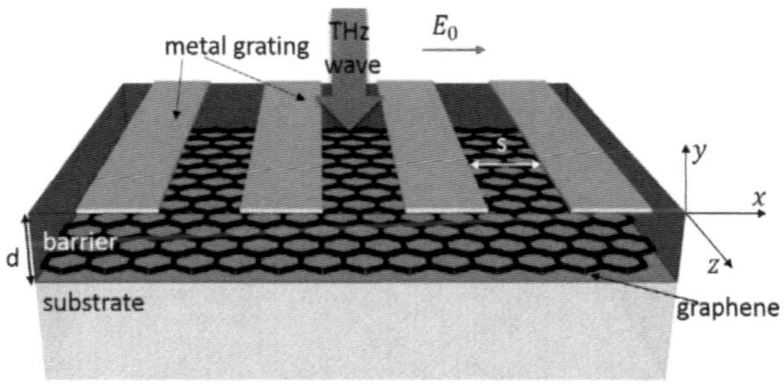

Figure 1. Schematic view of the graphene sheet with the metal grating and the coordinate system. The dielectric constant for the substrate and the barrier are the same, $\varepsilon_s = \varepsilon_b = 11.7$.

The optical problem of interaction between THz wave and the graphene structure with metal grating has been solved by using a self-consistent electromagnetic approach based on a plane wave decomposition of the full system of the Maxwell equations, similar to that described in [6].

The response of graphene is described by the complex dynamic surface conductivity [7]:

$$\frac{\sigma_{Gr}(\omega)}{\sigma_0} = \frac{8ik_BT}{\pi\hbar(\omega+i\gamma)}\ln\left[2\cosh\left(\frac{E_F}{2k_BT}\right)\right] + G\left(\frac{\hbar\omega}{2}\right) + \frac{4i\hbar\omega}{\pi}\int_0^{\infty}\frac{G(\varsigma)-G(\hbar\omega/2)}{(\hbar\omega)^2-4\varsigma^2}d\varsigma, \quad (1)$$

where

$$G(\varsigma) = \frac{\sinh(\varsigma / k_BT)}{\cosh(E_F / k_BT)+\cosh(\varsigma / k_BT)},$$

$\sigma_0 = e^2/4\hbar$, E_F is the Fermi energy, the temperature T is set to 300K, e is the electron charge, k_B is the Boltzmann constant, and $\hbar$ is the reduced Planck constant. The first term describes a Drude-like response due to the intraband processes involving the phenomenological intraband electron scattering rate γ. The rest terms describe the interband transitions in graphene. The real part of the conductivity equation (1), which is responsible for energy dissipation and takes into account both (intraband and interband) scattering mechanisms, is presented in figure 2 (a). The main contribution to the conductivity originates from the Drude losses, which increases with increasing the Fermi energy and decreases with increasing the frequency (as seen in figure 2 (a)). Losses due to the interband transitions contribute to the conductivity only for small values of the Fermi energy (figure 2 (b)).

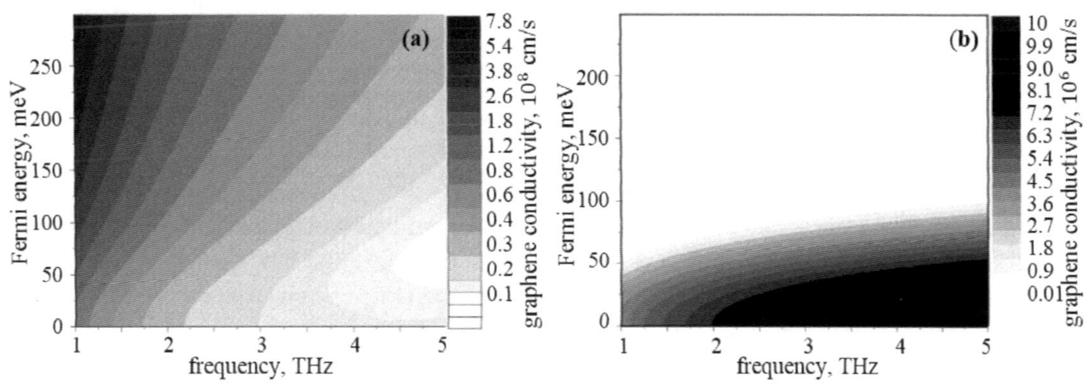

Figure 2 (a, b). **(a)** the real part of graphene conductivity as a function of the frequency and Fermi energy for $\gamma = 10^{12}\ \mathrm{s}^{-1}$ and $T = 300$ K; **(b)** the real part of graphene conductivity for $\gamma = 0$.

3. Results and discussion

Figure 3 shows the absorption spectrum calculated for realistic parameters of the graphene structure as a function of the Fermi energy and frequency at room temperature. As can be seen in figure 3, graphene exhibits strong plasmon response at the frequencies of the fundamental and higher plasmon resonances in graphene.

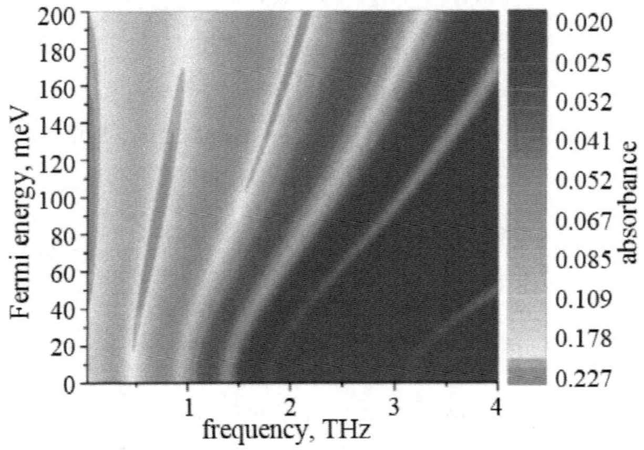

Figure 3. Absorption spectrum of the graphene structure with metal grating with period 2 μm and metal strip width 1.5 μm, and the barrier layer is 20 nm thick. The phenomenological intraband electron scattering rate is $\gamma = 10^{12}\,\mathrm{s}^{-1}$. for temperature 300K. The red colour shows the area in which the absorption is maximal.

We define the absorption coefficient as

$$A = P_{\mathrm{abs}} / P_{\mathrm{in}},$$

where

$$P_{\mathrm{abs}} = \frac{1}{2L}\mathrm{Re}\left[\sigma_{\mathrm{Gr}}(\omega)\right]\int_{-L/2}^{L/2} |E_x(x,0)|^2 dx,$$

is the absorbed power of the THz wave per unit area of the graphene structure and P_{in} is the power density of the incident THz wave. The maximal possible absorption is given as [8]

$$A^{\max} = 0.5\left(1-\sqrt{R_0}\right), \text{ where } R_0 = \left(\sqrt{\varepsilon_s}-1\right)^2\left(\sqrt{\varepsilon_s}+1\right)^{-2}$$

is the reflectance of the structure without metal grating and graphene. In our case $A^{\max} = 0.227$ (for the substrate dielectric constant $\varepsilon_s = 11.7$ and dielectric constant of the ambient medium equals 1) which is consistent with the calculated results (red color lobe in Figure 3).

The total width of the absorption line is composed of the Drude dissipative damping (intraband processes involving the phenomenological intraband electron scattering rate γ), damping due to interband transition, and radiative damping. The radiative damping increases proportionally to the concentration of the charge carriers [8], which is defined by the value of the Fermi energy in graphene. On the other hand, the Drude dissipative damping of the plasmons is $\gamma/2$ and does not depend on the concentration of charge carriers in graphene [10]. The absorption of THz radiation in the structure becomes maximal when the sum of the Drude dissipative and interband transition damping is equal to radiative damping.

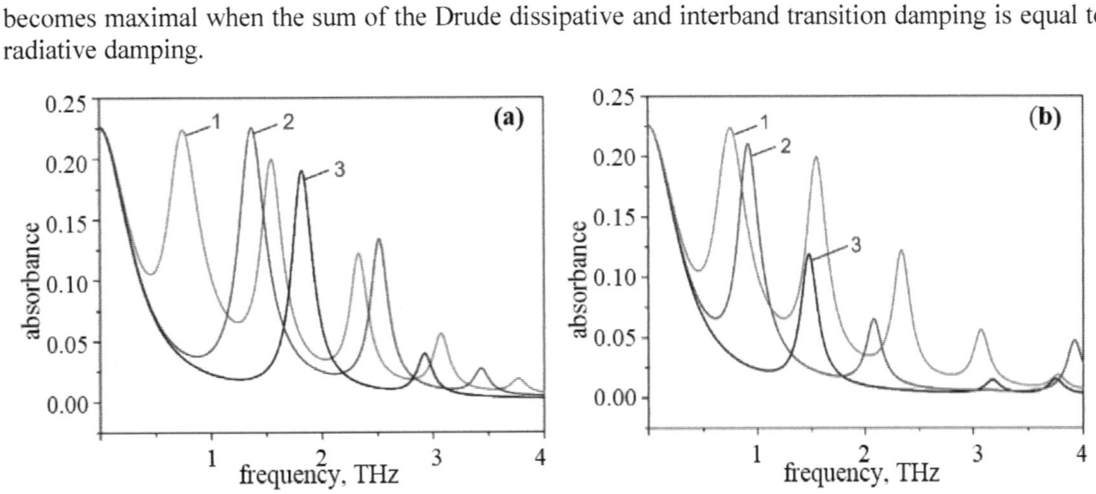

Figure 4 (a, b). (a) Absorption spectrum for different widths of the gaps between the metal strips of the metal grating: 1 – 0.5 μm; 2 – 1 μm; 3 – 1.5 μm. The metal grating period is 2 μm, Fermi energy is 100 meV, and the thickness of the barrier layer is 20 nm. **(b)** Absorption spectrum for different barrier thicknesses: 1 – 20 nm; 2 – 80 nm; 3 – 200 nm. The Fermi energy is 100 meV, the metal grating period is 2 μm, and the metal strip width is 1.5 μm.

As can be seen in figure 4, graphene exhibits strong plasmon response for a narrow-slit metal grating and small thickness of the barrier layer. Reducing the thickness of the barrier layer or the width of the gaps between the metal strips result in increased screening of graphene. Since the screening lowers the frequency of plasmons [9], plasmon resonances of the same orders are excited at lower frequencies (for the same values of the Fermi energy in graphene).

Because the positions of the plasmon absorption resonances can be tuned electrically by applying the electrostatic potential to the metal grating strips, the results of this paper can be used for creating tunable THz resonant absorbers.

Acknowledgments
Financial support was provided by the Russian Foundation for Basic Research (Project No. 16-02-00814).

References

[1] Novoselov K S, Geim A K, Morozov S V, Jiang D, Katsnelson M I, Grigorieva I V, Dubonos S V and Firsov A A 2005 *Nature* **438** 197
[2] Chen J, Badioli M, Alonso-González P, Thongrattanasiri S, Huth F, Hillenbrand R and Koppens F 2012 *Nature* **77** 487
[3] Fei Z *et al.* 2012 *Nature* **82** 487
[4] Polischuk O V, Melnikova V and Popov V V 2016 *Semiconductors* **50** (11) 1566
[5] Zhang L, Tang L, Wei W, Cheng X, Wang W and Zhang H 2016 *Optics Express* **24** (18) 20008
[6]Popov V V, Polischuk O V, Nikitov S A, Ryzhii V, Otsuji T and Shur M S 2013 *J. Optics* **15** 114009
[7] Falkovsky L A and Varlamov A A 2007 *Eur. Phys. J.* **56** 281
[8] Popov V V, Polischuk O V, Teperik T V, Peralta X G, Allen S J, Horing N J M and Wanke M C 2003 *J. of Applied Phys.* **94** (5) 3557
[9] Popov V V 2011 *J. Infrared Millim. Terahertz Waves* **32** 1178-91
[10] Popov V V, Bagaeva T Yu, Otsuj T and Ryzhii V 2010 *Phys. Rev. B* **81** 073404

Development of high-effective superconducting single-photon detectors aimed for mid-IR spectrum range

P I Zolotov[1, 2], A V Divochiy[2], Yu B Vakhtomin[1, 2], P V Morozov[2], V A Seleznev[1, 2] and K V Smirnov[1, 2, 3]

[1]Moscow State Pedagogical University, Moscow 119435, Russia
[2]JSC «Superconducting Nanotechnology» (SCONTEL), Moscow 119021, Russia
[3]National Research University Higher School of Economics, Moscow 101000, Russia

E-mail: zolotov@scontel.ru

Abstract. We report on development of superconducting single-photon detectors (SSPD) with high intrinsic quantum efficiency in the wavelength range 1.31 – 3.3 μm. By optimization of the NbN film thickness and its compound, we managed to improve detection efficiency of the detectors in the range up to 3.3 μm. Optimized devices showed intrinsic quantum efficiencies as high as 10% at mid-IR range.

1. Introduction

Recent years have seen ultimate growth of interest in sphere of quantum communication and quantum computation systems. Such experiments usually require single-photon-counting solutions with ultimate timing and efficiency characteristics. Perfect example of such devices are SSPDs which took the leadership over other photon-counting technologies in near-IR spectrum range due to its excellent values of the timing resolution, counting rate and detection efficiency [1, 2, 3, 4]. Fiber-optic-based setups, which used in a wide range of experiments [5, 6], significantly reduce possible field of implementation of SSPDs, which are expected to have wide spectrum band up to mid-IR range. Due to the fact, that absorbance and quantum output (the number of emerged from absorbed photon quasiparticles) are the only restrictive parameters for SSPD's wideband operation, the optical cavity usually employed to increase poor absorbance of the structured film [7]. However, improvement of the intrinsic quantum efficiency of the device stays not a trivial task. As it has been shown [8, 9], decreasing the strip width below 50 nm results in extending of the sensitivity of SSPD at mid-IR range. Because decreasing the strip width usually leads to the appearance of constrictions of the nanowire and therefore of significant drop of its critical current density (J_c), it is important to develop a new technology of fabrication of the device, which will allow its ultimate spectrum capabilities and high quantum efficiency simultaneously.

This study involves optimization of the films deposition process. Aiming to increase the quantum output of the photons with energies in the range 0.9-0.4 eV (1.3-3.3 μm) we decided to reduce films' critical temperature (T_c) by means of decreasing its thickness and by small variation of its compound, which led to the growth of the intrinsic quantum efficiency of the devices. Our goal was to make first fast and sensitive single-photon-counting solution for spectrum range above 1.7 μm, which could significantly accelerate development of single- and multi-photon sources in this spectrum range, for example quantum dots and semiconductor lasers, required in a broad band of applications, such as free-space optical communication and photoemission microscopy [10, 11, 12].

IOP Conf. Series: Journal of Physics: Conf. Series **917** (2017) 062037 doi:10.1088/1742-6596/917/6/062037

2. Implementation methods

The single-photon detection efficiency (η) is defined as the ratio of the measured number of counts to the number of photons incident on the device active area. η can be written as the product of the device absorbance α to the nanowire intrinsic single-photon detection efficiency ($\eta_{in.}$), i.e. the probability that absorbed photon triggers the output voltage pulse. Nowadays, problem of poor absorbance of ultra-thin structured films can be fixed using integrated cavity designed to concentrate electric field in the nanowires, which increases the absorption up to the values of 95% [13, 14, 15]. However, achievement of high detection efficiency is impossible in case, when $\eta_{in.}<1$. Being associated with the quantum output, intrinsic quantum efficiency is dependent on photon energy. Thus, increasing the wavelength of incident radiation should be compensated by increasing of the current density or decreasing of the nanowire cross-section. As it was mentioned earlier, decreasing the width of the stripe comes with technological difficulties, however, decreasing film thickness could be done without major drawbacks. It is known that reduction of the film thickness leads to reduction of its Tc. Owed to the proportional decreasing of Δ and T_c parameters, quantum output of the photon is increased in this case, which results in growth of intrinsic quantum efficiency of SSPD.

To optimize sheet resistance (R_s), Tc and Residual-Resistance Ratio (RRR) values we made a series of film deposition processes. Films for our study were deposited using reactive magnetron sputtering technique with AJA INTERNATIONAL Inc. Orion-8 setup, which allows background pressure of $6 \cdot 10^{-8}$ Torr. By changing N_2 to Ar ratio in gas mixture for each deposition process, we managed to obtain films with various values of the main parameters. Variation of RRR allowed us to obtain films with high values of R_s at close-to-transition temperatures and to optimize proximity of our detectors to saturation of intrinsic quantum efficiency at biasing currents (I_b) close to the critical value (I_c). The authors elsewhere will publish detailed research on relation between film parameters and SSPD performance soon. Exact values of T_c, R_s and RRR for selected film in given research were 9.4 K, 630 Ohm/□ and 0.74 respectively. Fabrication of SSPD continued with formation of the meander pattern with Jeol GSM6380 scanning electron microscope self-altered to e-beam lithography setup with approx. 3.5 nm resolution. Meanders had 100 nm wide strips with filling factor of 0.5 and cover-area of 15x15 μm^2. Ti-Au metallization layer was deposited by e-beam evaporation and then formed to the contact metallization by means of *lift-off* process. Unwanted NbN areas were plasma-etched in SF_6 using Corial 200R. At final step wafer was scribed into separate chips.

3. Experimental results and discussion

In order to sort-out defective devices, we electrically and optically characterized them. Devices with close-to-calculated values of resistance we tested in liquid helium in order to determine its T_c and I_c which allowed us to choose detectors with maximal uniformity of the strip (which corresponds to maximal density of critical current J_c) and lack of overetched regions (maximal T_c). Figure 1 represents histograms of resistance, critical current and quantum efficiency distributions for devices within the same batch.

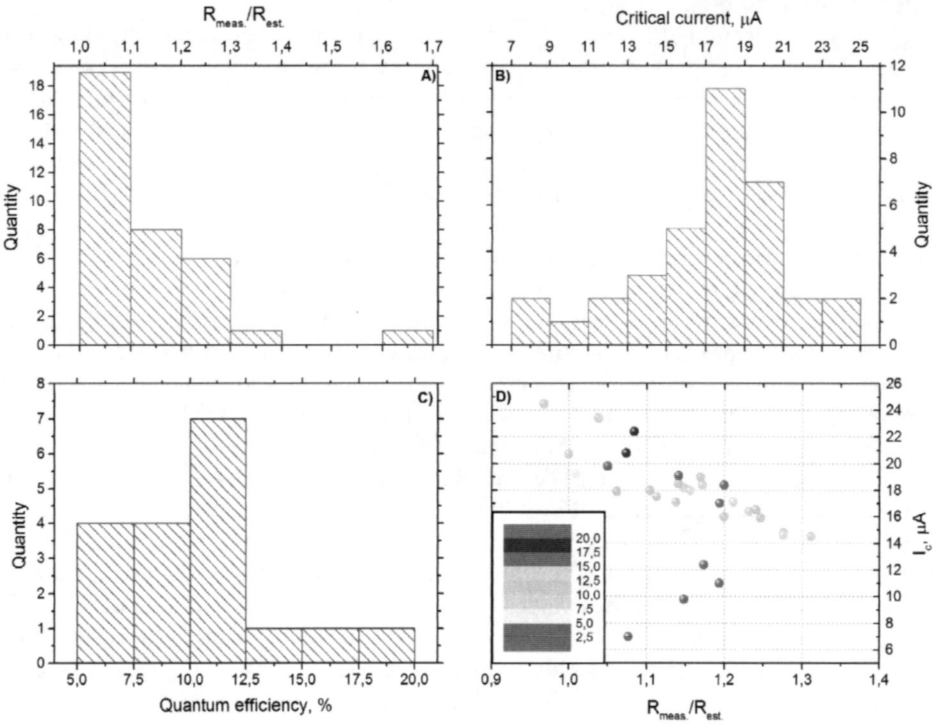

Figure 1. Statistics of the main parameters of fabricated devices within the same batch. A): Distribution of resistance of the devices measured at room temperature; B): histogram of critical currents of the devices at 4.2 K; C): distribution of quantum efficiencies (QE) of the devices at 1.55 μm at 4.2 K; D): critical current and resistance dependence of QE (marked in color).

Histogram **A)** of the resistance, i.e. relation of measured resistance of devices $R_{meas.}$ to thereof estimated values $R_{est.}$, distribution at room temperature show that 77% of SSPDs had ~20% deviation from calculated value of resistance. Histogram **B)** presents that majority of the devices from the batch (65%) had critical currents ranging from 15 to 21 μA. Taking into account film thickness (6 nm) and strip width, we can conclude that calculated J_c value of $2.5\text{-}3.3\cdot10^6$ A/cm^2 fits to the highest standards for superconducting NbN films [16].Quantum efficiency histogram at 1.55 μm **C)** has pronounced maximum which corresponds to 10-12.5% of measured quantum efficiency using standard technique [17] with free-space configuration which only allows initial characterization of the devices because of poor coupling of the detector with radiation, which estimated on the level of 30%. Presented histograms demonstrate that only 10% of the devices from the batch had low QE, with respect to the majority. Graph **D)** shows that for the greater number of devices from the batch J_c linearly drops with the growth of the nanowire resistance. For such detectors its QE corresponds to the maximal value from the histogram. This explains hypothesis of the presence of narrowed regions of the stripe for some devices with lower than optimal critical currents. Thus, it could be summarized that based on demonstrated histograms our SSPD batch had high from-chip-to-chip uniformity with almost 90% of the devices having similar parameters.

Among tested specimens, we selected those, which had close parameters. Each of these devices was placed in holder made of oxygen-free copper and fixed on cold plate of Gifford-McMahon (GM) cryocooler. As a light source we used grating monochromator with wavelength selectivity of 0.1 nm. Optical input of the cryostat with Si window allowed transmittance of radiation in selected wavelength range. Si window also acted as a high-effective filter for short-wave half-wavelength harmonics of desired wavelength region, which usually presents on monochromator output. For measurements at wavelengths, starting from 2.2 μm as filter for half-wavelength harmonics we additionally used 3-mm thick Ge plate. To electrically bond SSPD with input/output channel, we used tpt HB12 wire bonder. Because our goal did not require any measurements of the detector input optical power, we did not create any precision coupling setup. Photon flux to the cryostat input was maintained at power level corresponding to 0.01-1 MHz counts of the device, minimizing saturation of the count rate and influence of dark counts on measured dependencies. In our experiment, we studied proximity of quantum efficiency of our devices to saturation level, i.e. proximity of intrinsic quantum efficiency to 1. Our study was performed in spectrum range 1.31-3.3 μm at temperature of 2.5 K. Results of the experiment are presented on Figure 2.

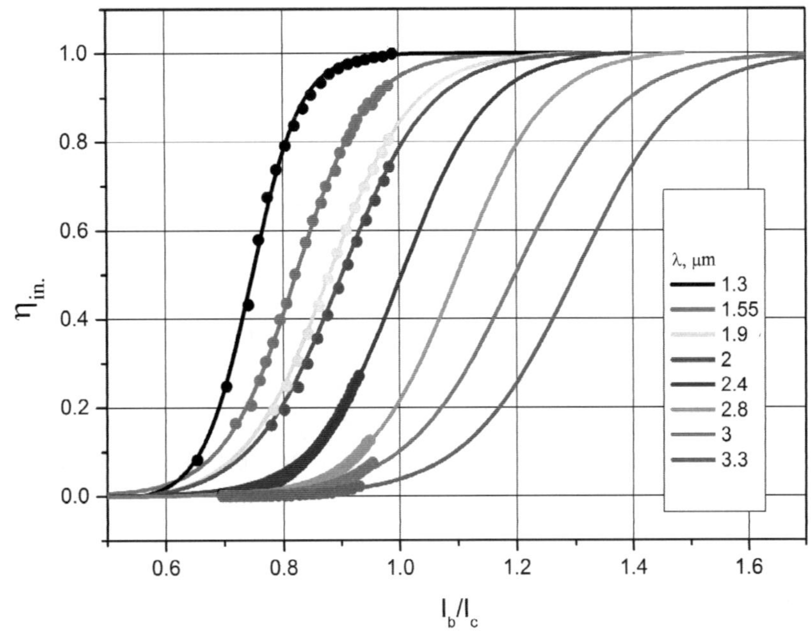

Figure 2. Results of bias current dependence of intrinsic quantum efficiency of SSPD at different wavelengths. Lines represent fitting curves and scatter – experimental results.

Presented graph show experimental curves fitted with sigmoidal function of obtained results allowing calculating exact level of $\eta_{in.}$ at selected wavelength. Values of intrinsic quantum efficiency varied from 1 to 0.04 corresponding to the spectrum range 1.31-3.3 μm, accordingly to the drop of photon energy. The shifting of the dependency to the region of higher currents with wavelength growth, as well as slope scaling, could be noticed. For example, curves corresponding to the wavelengths range of 2.8-3.3 μm matched values of internal efficiency ≤ 0.2 at I_c and reached values close to unity at currents as high as $1.5I_c$. Obtained results allow us to calculate possible detection efficiencies of the devices in cases of presence of optical resonator structure allowing to achieve high absorption at a given wavelength and in case of efficient coupling of the radiation to the sensitive area of the detector. Both of these goals have already been successfully implemented for visible ÷ near-IR range [13, 14, 18, 19], and therefore creating of the similar solutions for mid-IR region occur as a technical task.

Values of the fitting curves corresponding to the critical current of the detector at 2.5 K are plotted as a function of the wavelength on Figure 3. On the basis of our simulations, quantum efficiency of our detectors could be estimated as 40% and 10% at 2.5 μm and 3 μm respectively.

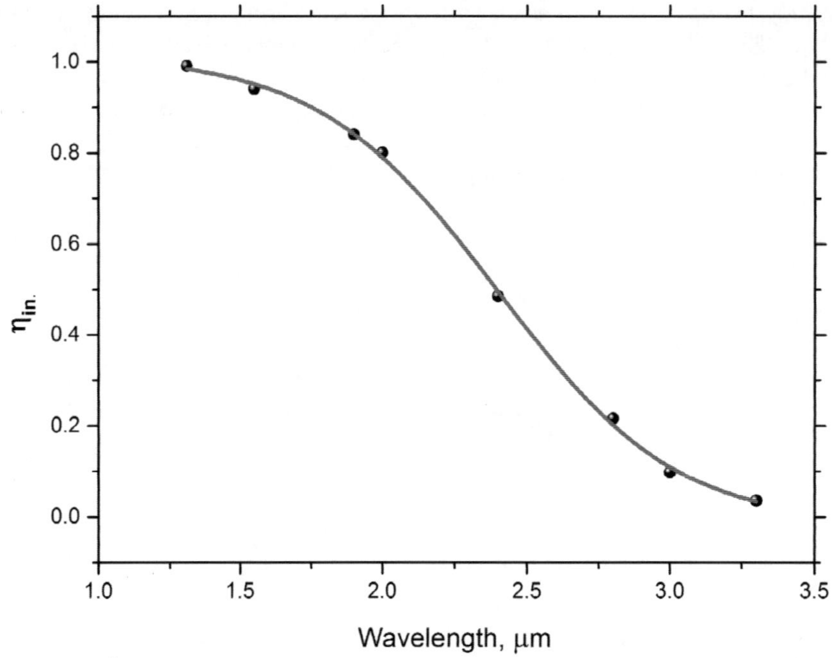

Figure 3. Spectral dependence of intrinsic quantum efficiency (dots) and fitting of the obtained results (line).

4. Conclusion

We have demonstrated capabilities of the SSPD fabricated from film with optimized parameters at mid-IR range. Obtained values of intrinsic quantum efficiency verify our improvements of the device qualities and enable the development of first fast and sensitive single-photon-counting systems for mid-IR spectrum range. By selection of different coupling method, this detector system could efficiently cover wavelength region from 1.3 to 3.3 μm.

References

[1] Zadeh, I.E., Los, J.W., Gourgues, R., Steinmetz, V., Dobrovolskiy, S.M., Zwiller, V. and Dorenbos, S.N., 2016. Single-photon detectors combining near unity efficiency, ultra-high detection-rates, and ultra-high time resolution. arXiv preprint arXiv:1611.02726.

[2] V. Shcheslavskiy, P. Morozov, A. Divochiy, Yu. Vakhtomin, K. Smirnov and W. Becker "Ultrafast time measurements by time-correlated single photon counting coupled with superconducting single photon detector" Rev. Sci. Instrum. 87, 053117 (2016).

[3] Junjie Wu, Lixing You, Sijing Chen, Hao Li, Yuhao He, Chaolin Lv, Zhen Wang, and Xiaoming Xie, "Improving the timing jitter of a superconducting nanowire single-photon detection system," Appl. Opt. 56, 2195-2200 (2017)

[4] K. Smirnov, Yu. Vachtomin, A. Divochiy, A. Antipov and G. Goltsman "Dependence of dark count rates in superconducting single photon detectors on the filtering effect of standard single mode optical fibers" 2015. Appl. Phys. Express 8, 022501.

[5] N.R. Gemmell, A. McCarthy, B. Liu, M.G. Tanner, S.D. Dorenbos, V. Zwiller, M.S. Patterson, G.S. Buller, B.C. Wilson, and R.H. Hadfield, "Singlet oxygen luminescence detection with a fiber-coupled superconducting nanowire single-photon detector," Opt. Exp., vol.21, no.4, pp.5005–5013, 2013.

[6] "Superconducting Devices in Quantum Optics," Eds. R.H. Hadfieldand G. Johanson, Springer 2016.

[7] Anant, V., Kerman, A.J., Dauler, E.A., Yang, J.K., Rosfjord, K.M. and Berggren, K.K., 2008. Optical properties of superconducting nanowire single-photon detectors. Optics Express, 16(14), pp.10750-10761.

[8] Marsili, F., Najafi, F., Dauler, E., Bellei, F., Hu, X., Csete, M., ... & Berggren, K. K. (2011). Single-photon detectors based on ultranarrow superconducting nanowires. Nano letters, 11(5), 2048.

[9] Marsili, F., Bellei, F., Najafi, F., Dane, A.E., Dauler, E.A., Molnar, R.J. and Berggren, K.K., 2012. Efficient single photon detection from 500 nm to 5 μm wavelength. Nano letters, 12(9), pp.4799-4804.

[10] Vadim V. Vorobyov, Alexander Yu. Kazakov, Vladimir V. Soshenko, Alexander A. Korneev, Mikhail Y. Shalaginov, Stepan V. Bolshedvorskii, Vadim N. Sorokin, Alexander V. Divochiy, Yury B. Vakhtomin, Konstantin V. Smirnov, Boris M. Voronov, Vladimir M. Shalaev, Alexey V. Akimov, and Gregory N. Goltsman, "Superconducting detector for visible and near-infrared quantum emitters [Invited]," Opt. Mater. Express 7, 513-526 (2017)

[11] Yamashita, T., Liu, D., Miki, S., Yamamoto, J., Haraguchi, T., Kinjo, M., Hiraoka, Y., Wang, Z. and Terai, H., 2014. Fluorescence correlation spectroscopy with visible-wavelength superconducting nanowire single-photon detector. Optics express, 22(23), pp.28783-28789.

[12] Grein, M., Dauler, E., Kerman, A., Willis, M., Romkey, B., Robinson, B., Murphy, D. and Boroson, D., 2015. A superconducting photon-counting receiver for optical communication from the Moon. SPIE Newsroom, July.

[13] F. Marsili, V.B. Verma, J.A. Stern, S. Harrington, A.E. Lita, T. Gerrits, I. Vayshenker, B. Baek, M.D. Shaw, R.P. Mirin, and S.W. Nam, "Detecting single infrared photons with 93% system effi- ciency," Nat. Photon. vol.7, no.3, pp.210–214, 2013.

[14] V.B. Verma, B. Korzh, F. Bussieres, R.D. Horansky, S.D. Dyer, A.E. ` Lita, I. Vayshenker, F. Marsili, M.D. Shaw, H. Zbinden, R.P. Mirin, and S.W. Nam, "High-efficiency superconducting nanowire singlephoton detectors fabricated from MoSi thin-films," Opt. Exp., vol.23, no.26, pp.33792–33801, 2015

[15] Yamashita, T., Waki, K., Miki, S., Kirkwood, R.A., Hadfield, R.H. and Terai, H., 2016. Superconducting nanowire single-photon detectors with non-periodic dielectric multilayers. Scientific reports, 6.

[16] Semenov, A., Günther, B., Böttger, U., Hübers, H.W., Bartolf, H., Engel, A., Schilling, A., Ilin, K., Siegel, M., Schneider, R. and Gerthsen, D., 2009. Optical and transport properties of ultrathin NbN films and nanostructures. Physical Review B, 80(5), p.054510.

[17] Sidorova, M.V., Divochiy, A.V., Vakhtomin, Y.B. and Smirnov, K.V., 2015. Ultrafast superconducting single-photon detector with a reduced active area coupled to a tapered lensed single-mode fiber. Journal of Nanophotonics, 9(1), pp.093051-093051.

[18] W. Slysz, M. Wegrzecki, J. Bar, M. Gorska, V. Zwiller, C. Latta, P. Bohi, I. Milostnaya, O. Minaeva, A. Antipov, O. Okunev, A. Korneev, K, Smirnov, B. Voronov, N. Kaurova, G. Gol'tsman, A. Pearlman, A. Cross, I. Komissarov, A. Verevkin, R. Sobolewski, "Fiber-coupled single-photon detectors based on NbN superconducting nanostructures for practical quantum cryptography and photon-correlation studies", Appl. Phys. Lett. 88, 261113 (2006)

[19] Korneev, A., Korneeva, Y., Manova, N., Larionov, P., Divochiy, A., Semenov, A., Chulkova, G., Vachtomin, Y., Smirnov, K. and Goltsman, G., 2013. Recent nanowire superconducting single-photon detector optimization for practical applications. Ieee Transactions on Applied Superconductivity, 23(3), pp.2201204-2201204.

Surface plasmon-phonon polaritons in GaAs

G A Melentev, V A Shalygin, D A Firsov, M Ya Vinnichenko and L E Vorobjev

Department of Physics of Semiconductors and Nanoelectronics, Peter the Great St. Petersburg Polytechnic University, 29 Polytechnicheskaya Street, St. Petersburg 195251, Russia

E-mail: gamelen@spbstu.ru

Abstract. We investigate surface plasmon-phonon polaritons in *n*-doped GaAs with the aim to obtain terahertz radiation emission. Dispersion of the surface plasmon-phonon polaritons is theoretically simulated taking into account contributions of the lattice and electron subsystems to the dielectric permittivity of *n*-GaAs. It is shown that regular grating at outer surface of GaAs provides a possibility to convert surface plasmon-phonon polaritons into terahertz photons and vice versa.

1. Introduction

Currently bulk and surface polaritons are employed to obtain terahertz (THz) and mid infrared (MIR) radiation emission. In particular, THz radiation associated with excitation of surface plasmon polaritons in electric field was achieved in GaN [1]. Emission of MIR radiation due to surface phonon polaritons was investigated in SiC [2]. Incoherent THz radiation related to bulk plasmon-phonon polaritons was observed in GaAs [3]. In the present work we investigate *surface* plasmon-phonon polaritons (SPPPs) in doped GaAs with the aim to obtain THz radiation emission.

2. Results and discussion

Surface plasmon-phonon polaritons are electromagnetic waves excited at the interface of two media with dielectric permittivities of different sign. They have the following dispersion law [4]:

$$k_{\mathrm{SPPP}}^2 = \left(\frac{\omega}{c}\right)^2 \frac{\varepsilon_1(\omega)\varepsilon_2}{\varepsilon_1(\omega)+\varepsilon_2}, \tag{1}$$

where k_{SPPP} is surface plasmon-phonon polariton wave vector, ω is angular frequency, c is speed of light, $\varepsilon_1(\omega)$ is dielectric permittivity of the first medium (*n*-doped GaAs in our case), ε_2 is dielectric permittivity of the second medium (vacuum in our case). It should be emphasized that the wave vector of SPPP is larger than the photon wave vector at the same frequency. Thus, SPPPs are interface-localized non-radiative modes. Nevertheless, SPPPs can be converted into photons and vice versa by means of a regular surface grating which provides the opportunity to fulfill the phase-matching condition for SPPP and photon (see [1] and references therein):

$$k_{\mathrm{SPPP}} = k_{\mathrm{Ph}} \sin\theta + m k_{\mathrm{G}},\ m = \pm1, \pm2, \pm3, \ldots, \tag{2}$$

where $k_{Ph} = \frac{\omega}{c}$ is the wave vector of photon in vacuum, θ is the incidence angle, $k_G = \frac{2\pi}{a}$ is the reciprocal vector of the grating, a is the grating period.

The necessary conditions for existence of surface plasmon-phonon polaritons at the GaAs/vacuum interface are the following: $\mathrm{Re}(\varepsilon_1) < 0$, $|\mathrm{Re}(\varepsilon_1)| > 1$. In this way let us consider the dielectric permittivity of n-doped GaAs (ε_1). We assume that $\varepsilon_1(\omega)$ consists of two contributions, namely the lattice term and electron component [5]:

$$\varepsilon_1(\omega) = \varepsilon_\infty + \frac{\varepsilon_0 - \varepsilon_\infty}{1 - \left(\frac{\omega}{\omega_{TO}}\right)^2 - \frac{i\omega\gamma}{\omega_{TO}^2}} - \frac{\varepsilon_\infty \omega_P^2}{\omega^2 + i\omega\tau^{-1}}, \tag{3}$$

where ε_∞ is high frequency permittivity, ε_0 is low frequency permittivity, ω_{TO} is transverse optical phonon frequency, γ is damping constant for transverse optical phonon, $\omega_P = \sqrt{\frac{4\pi e^2 N_e}{\varepsilon_\infty m_e}}$ is plasma frequency, N_e is free electron concentration, e is the elementary charge, m_e is electron effective mass, τ is electron relaxation time determined by the mobility μ: $\tau = \mu m_e / e$. To improve the model under consideration, we took into account the dependences of μ and γ on doping level [6, 7]. The electron-concentration dependence of m_e due to nonparabolicity of GaAs conduction band [7] was accounted as well. We have taken free electron concentration N_e of $2 \cdot 10^{18}$ cm^{-3} and mobility μ of 2500 cm^2/V·s in our calculations. Figure 1 demonstrates calculated dielectric permittivity of the n-GaAs crystal. In accordance with simulation, the real part of the dielectric permittivity $\varepsilon_1(\omega)$ is less than –1 in the spectral ranges (0 ... 31.7 meV) and (33.2 ... 56.9 meV). Thus, surface plasmon-phonon polaritons can exist just in these spectral ranges.

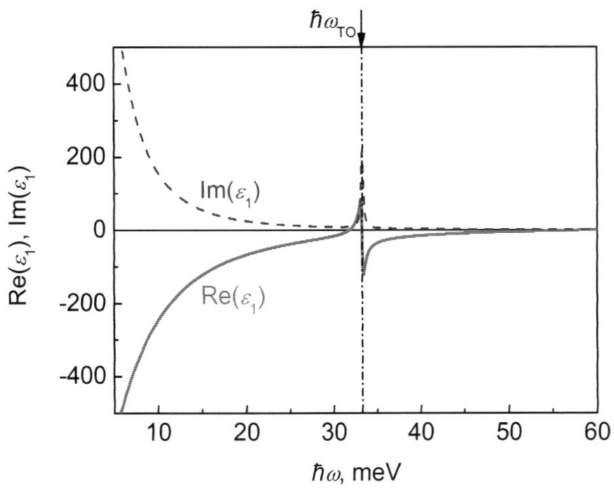

Figure 1. Frequency dispersion of the real (solid line) and imaginary (dash line) parts of the n-GaAs dielectric permittivity.

Using expressions (1) and (3) we have calculated dispersion dependences of SPPPs at the GaAs/vacuum interface. The calculation results are presented in figures 2(a) and 3(a). One can see that SPPP dispersion dependence consists of two branches separated by a narrow gap. The origin of this gap is clearly seen from figure 1: there is a narrow spectral range in the vicinity of optical phonon energy where the real part of GaAs dielectric permittivity exceeds –1 and surface plasmon-phonon polaritons can not exist.

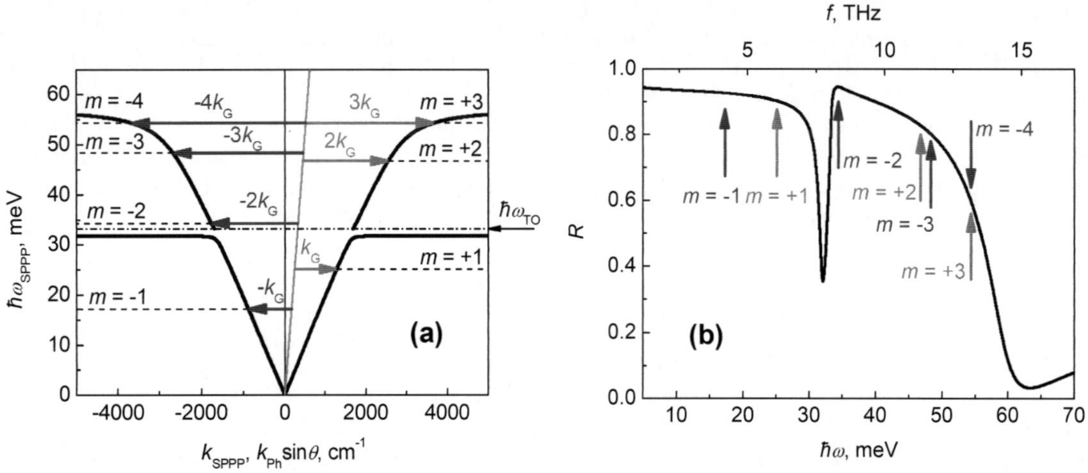

Figure 2. (a) Dispersion dependencies of the surface plasmon-phonon polaritons at GaAs/vacuum interface (solid thick lines) and photons in vacuum at $\theta = 11°$ (solid thin line). Horizontal dashed lines show spectral positions of the SPPPs resonances of different orders *m*. (b) Reflectivity spectrum of semi-infinite GaAs with a plane surface.

We have simulated the reflectivity spectrum of semi-infinite GaAs with a plane surface (see, e.g., [5]). We have considered the case of oblique incidence of radiation (angle of incidence $\theta = 11°$). The calculation results are displayed in figure 2(b). The spectrum demonstrates two distinct dips associated with the bulk plasmon-phonon polariton excitations [8]. Fabrication of a regular grating at the semiconductor surface (for instance, by means of photolithography technique) will result in a significant change of the reflectivity spectrum. It can provide conversion of photons into surface plasmon-phonon polaritons and vice versa (see above). The result of such conversion is the emergence of the resonant dips in the reflectivity spectrum. We have calculated the spectral positions of these dips using SPPP dispersion law (1) and the phase-matching condition (2). We have taken the surface grating period *a* of 60 μm. The spectral positions of peculiarities associated with SPPP excitation are shown in figure 2 by arrows.

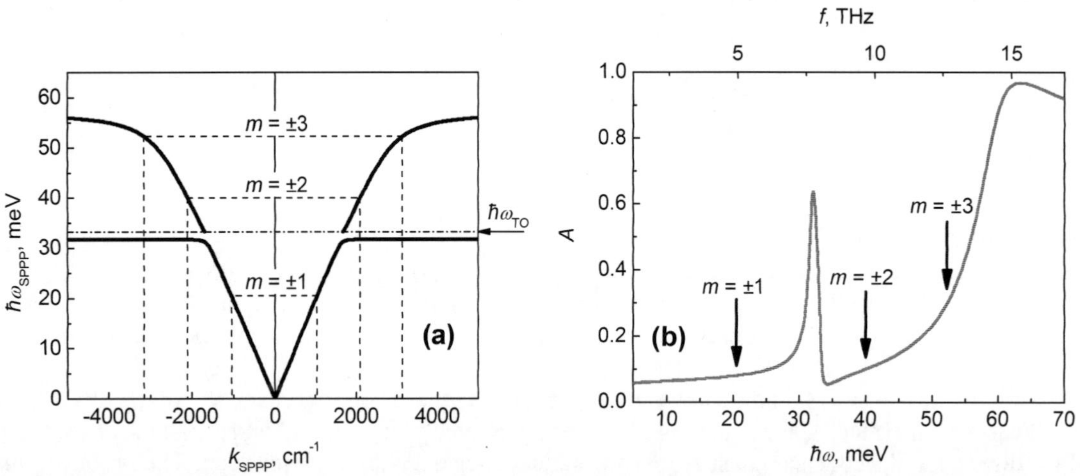

Figure 3. (a) Dispersion law of surface plasmon-phonon polaritons at GaAs/vacuum interface. Horizontal dashed lines show spectral positions of the SPPPs resonances of different orders *m* for the case of normal incidence of radiation ($\theta = 0°$). (b) Absorptivity spectrum of semi-infinite GaAs with a plane surface.

We have simulated the absorptivity spectrum of semi-infinite GaAs with a plane surface. The case of normal incidence of radiation was considered. As known, the absorptivity A and reflectivity R for any semi-infinite medium are connected by the following expression: $A = 1 - R$. The simulated dependence $A(\hbar\omega)$ is displayed in figure 3(b). Surface grating should change the absorptivity spectrum significantly. Non-equilibrium SPPPs can scatter on surface grating and transform into photons. The result of such transformation is the emergence of the resonant peaks in the absorptivity spectrum. We have calculated the spectral positions of such peaks according SPPP dispersion law (1) and the phase-matching condition (2). The surface grating period a was 60 μm. The results are shown in figure 3(b) by arrows. Then from Kirchhoff's law follows that one can expect an effective emission of THz and MIR radiation at the resonant frequencies under condition of surface plasmon-phonon polariton excitation in n-GaAs.

3. Conclusions

Surface plasmon-phonon polaritons can be applied to produce THz and MIR radiation emission. It can be realized by means of a regular grating at outer surface of GaAs. When phase-matching condition for SPPP and photon is fulfilled, surface modes can be converted into photons and vice versa. We have calculated corresponding spectral positions of resonant dips/peaks in the reflectivity/absorptivity spectra associated with SPPPs. Performed studies can be utilized for the development of solid state THz emitters based on surface plasmon-phonon polaritons in n-GaAs.

Acknowledgments

This work was supported by the Ministry of Education and Science of the Russian Federation (state assignment) and RFBR (grant 16-32-60085).

References

[1] Melentev G A *et al* 2016 *J. Appl. Phys.* **119** 093104
[2] Greffet J-J, Carminati R, Joulain K, Mulet J-P, Mainguy S and Chen Y 2002 *Nature* **416** 61
[3] Širmulis E, Šilėnas A, Požela K, Požela J and Jucienė V 2014 *Appl. Phys.* A **115** 199
[4] Maier S A 2007 *Plasmonics: Fundamentals and Applications* (New York: Springer)
[5] Yu P Y and Cardona M 2010 *Fundamentals of Semiconductors* (Berlin–Heidelberg: Springer)
[6] Holm R T, Gibson J W and Palik E D 1977 *J. Appl. Phys.* **48** 212
[7] Chandrasekhar H R and Ramdas A K 1980 *Phys. Rev.* B **21** 1511
[8] Orlov E Yu, Melentev G A, Shalygin V A and Suihkonen S 2017 *J. Phys.: Conf. Ser.* **816** 012004

IOP Conf. Series: Journal of Physics: Conf. Series **917** (2017) 062039 doi:10.1088/1742-6596/917/6/062039

Anti-corrosion treatment of metal surfaces based on photonics methods

J S Ruzankina[1], O S Vasiliev[1,2], V A Parfenov[1,3]
[1]Department of Laser technologies and systems, ITMO University, Saint Petersburg 197101, Russia
[2]Laser center, Saint Petersburg 195176, Russia
[3]Department of Quantum Electronics and Opto-Electronic Devices, St.Petersburg State Electrotechnical University, St. Petersburg Academic University, St. Petersburg 197376, Russia

Abstract. Metal corrosion protection is the main problem of all metal structures. Estimated to NASA, the loss of metal around the world is about of 2.5 trillion dollars, equivalent to 3.4% of global GDP. In our work we used a CW fiber laser with the wavelength of 1064 nm and a power up to 18, 4 W for laser irradiation of metal surfaces. We report on the optimal treatment of the metal corrosion with laser power density in the range from $7*10^4$ to $9*10^4$ W/cm^2. After the process of laser treatment of steel surface we observe decreased roughness of steel and a small change in its chemical composition.

1. Introduction

Metal corrosion is the main problem of all metal constructions and buildings. The main problem is not the deterioration of the metal itself, but the huge cost of metal products which can be destroyed by corrosion. Annual losses resulting from corrosion in industrialized countries are estimated in the range from 2% to 4 % of gross national product. Furthermore, the loss of metal, incorporating many of the failed metal structures, products, equipment, is in the range from 10 % to 20 % per year of steel production. Therefore, to achieve the economic effect (reducing significantly the cost of repairs or full replacement of equipment), it is necessary to protect the metal structures from corrosion.

For the past ten years, in industry there has been a tendency to find new ways to change surface properties of metals to increase corrosion resistance. One of the breakthrough methods for protecting the materials against corrosion is laser treatment. But despite many years research activity in this field there still is a number of significant challenges to be addressed, for example, investigation of physico-chemical processes occurring on the surface of the material treated by laser radiation. So, goal of the work - study of influence of laser treatment on the workpiece surface.

2. Experiment and results

In our work we used a CW fiber laser with the wavelength of 1064 nm and a power up to 18, 4 W. The power density of laser radiation was varied in the range $q=7*10^4$ W/cm^2 and $q=9*10^4$ W/cm^2. Experimental samples (steel plates) were irradiated for 20-35 seconds by laser. Surface treatment of metal was held at a room temperature of 23 ℃ and a relative humidity of air of 55%.

Analysis of obtained results does allow to conclude that laser irradiation of steel surface resulted in forming the thin film protecting steel from corrosion processes. Figure 1 presents the microphotography of the steel surface before and after laser treatment.

(a) 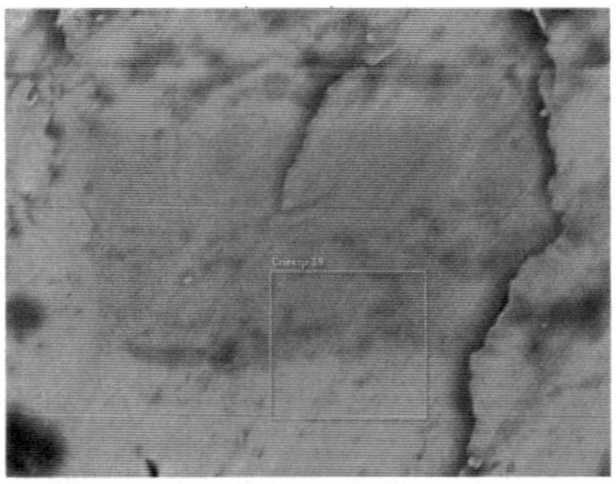 **(b)** 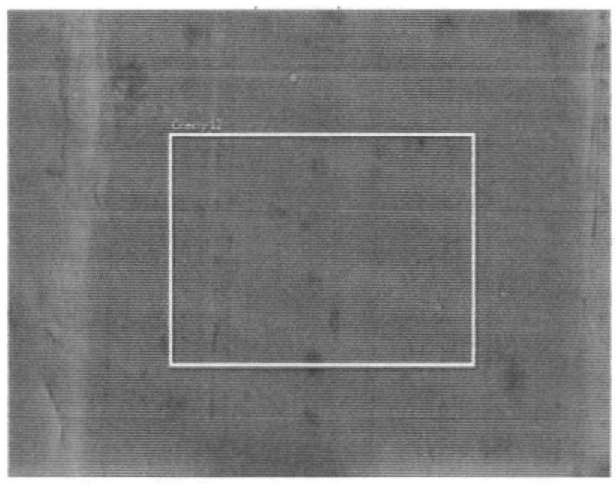

Figure 1(a, b). (a) Microphotography of the steel surface before laser treatment; **(b)** Microphotography of the steel surface after laser treatment.

Table 1 presents steel characteristics. The impact of laser radiation on the surface of the steel has contributed to a small change in its chemical composition. We did find the presence of such chemical elements as Al, F and O. It is worth noting that the presence of F there was originally in the studied steel grade, but the impact of laser radiation significantly increased its concentration in the metal is from 0.01 atom. % to 5.24 atom. %. We assume that this is due to diffusion. Studies are still being conducted in this direction.

Table 1. Steel characteristics before and after laser treatment.

	C	**O**	**F**	**Al**	**Si**	**Cr**	**Mn**	**Fe**	**Ni**	**Cu**	**As**
Before atom.%	13,5	1,4	0	0	0,7	0,05	0,5	83,4	0,2	0,3	0,05
After atom.%	12,4	2,7	0	0	0,7	0,04	0,4	78.1	0,19	0,3	0,02

3. Conclusion

The obtained results allow us to find the optimal parameters of laser treatment of the steel surface. Protective coatings with best resistance to corrosion were obtained with laser power density in the range $q=7*10^4$ W/cm^2 and $q=9*10^4$ W/cm^2. It should be noted that presented results are preliminary ones and authors plan to carry out more detailed analysis of steel surfaces treating by radiation of CW fiber laser.

References

[1] Ruzankina J S and Vasiliev O S 2016 Journal of Physics: Conference Series 735
[2] Ruzankina J S, Kascheev S V, Vasiliev O S and Parfenov V A 2016 Laser Optics (LO), 2016 International Conference
[3] Ruzankina J S, Iakovlev A A, Kascheev S V, Vasiliev O S and Parfenov V A 2017 Press. SPIE

IOP Conf. Series: Journal of Physics: Conf. Series **917** (2017) 062040 doi:10.1088/1742-6596/917/6/062040

Numerical simulation of eigenmodes of ring and race-track optical microresonators

A V Raskhodchikov[1], D V Raskhodchikov[1], S A Scherbak[1,2] and A A Lipovskii[1,2]

[1] Department of Physics and Technology of Nanoheterostructures, St. Petersburg Academic University RAS, St. Petersburg 194021, Russia
[2] Institute of Physics, Nanotechnology and Telecommunications, Peter the Great St. Petersburg Polytechnic University, St. Petersburg 195251, Russia

Abstract. We have performed a numerical study of whispering gallery modes of ring and race-track optical microresonators. Mode excitation was considered and their spectra and electromagnetic field distributions were calculated via numerical solution of the Helmholtz equation. We pay additional attention to features of eigenmodes in race-tracks in contrast with ring resonators. Particularly, we demonstrate that modes in race-tracks are not "classic" WGM in terms of total internal reflection from a single boundary, and an inner boundary is essential for their formation. The dependence of effective refractive index of race-tracks modes on the resonator width is shown.

1. Introduction

Optical resonators with disc-like shapes are of particular interest in many areas of applied optics and photonics, e.g. lasing [1, 2, 3], sensing [4, 5], nonlinear frequency mixing [6], frequency comb generation [7], etc. The distinctive feature of disc, ring or alike shaped resonators is the existence of so-called optical whispering gallery modes (WGM) [8, 9], which are strongly localized in a resonator volume and have an extra high quality-factor [10] and therefore very narrow spectral peaks. However, there is a lack of fundamental knowledge about mode structure of resonators of any shape except disc and sphere, where the well-known sets of eigenfunctions [8] are applicable. That is why various numerical methods [11, 12, 13] are being successfully applied for simulation of WGM in resonators of other shapes. In this study we present a numerical analysis of modes of ring and race-track resonators via the finite elements method (FEM).

2. Approach

To analyze spectral and spatial structure of WGM in ring and race-track resonators we used numerical simulation in COMSOL Multiphysics environment. According to COMSOL Application Gallery [14], in the frames of FEM there are two ways to find eigenfrequencies and corresponding field distributions. The first way is to solve an eigenvalue problem directly:

$$\nabla^2 \mathbf{E}_\mathrm{n} + k_\mathrm{n}{}^2 \mathbf{E}_\mathrm{n} = 0 . \qquad (1)$$

Despite this is the most obvious and direct choice, this approach has several disadvantages which were mentioned in [14]. First, the results can contain nonphysical modes, e.g. modes of free space. To identify them one has to examine manually the modes obtained. Second, an eigenvalue problem is solved by a memory-intensive direct solver, which can become an issue for complicated 3D problems.

We consider an alternative way [14] which is to excite a structure at different frequencies and to analyze the system response. In accordance with this approach we solve the frequency-dependent Helmholtz equation with specified external sources for each desired frequency:

$$\nabla^2\mathbf{E} + k^2\mathbf{E} = \mathbf{F}(\mathbf{r},\omega), \qquad (2)$$

and calculate integral of electromagnetic energy density over the resonator volume. In the state with excitation frequency close to an eigenfrequency the energy in the resonator increases greatly. This allows constructing a mode spectrum where only real, physical modes are presented. In our previous paper [15] we simulated experimental results that required solving 3D problem. Here we pay attention mostly to methodology of the problem and consider 2D model to evade calculations overloading.

First, we simulated mode spectra of a microring resonator with inner/outer radii 2.5/3.5 μm using different numbers of dipole antennas (five, four, three, two and one), randomly oriented and randomly scattered throughout the resonator volume. Essentially, we varied the form of source-function $\mathbf{F}(\mathbf{r},\omega)$ in Eq. (2). The decrease in the number of dipole sources did not lead to considerable degradation of the mode spectra. Then we considered 5 different positions of a single dipole source (see the inset in figure 1): two outside the resonator and three inside (two near the sidewalls and one near the centre). Spectra presented in figure 1 show that placement the antenna outside the resonator (positions 1 and 5) results in significant spectra degradation, whereas internal positioning gives much better outcome. However, the spectrum corresponding position 2 (near the inner sidewall) is worse. Therefore, we conclude that a single dipole antenna is enough for effective excitation of most modes, and its preferable positioning is somewhere between the centre and the outer sidewall of the resonator. Note that we consider two-dimensional problem and the dipoles are also two-dimensional. Therefore, all the modes demonstrated here and below are of TM kind where **H**-field is out-of-plane and **E**-field is in-plane. Consideration of TE modes does not bring brand new results in the frames of the present study.

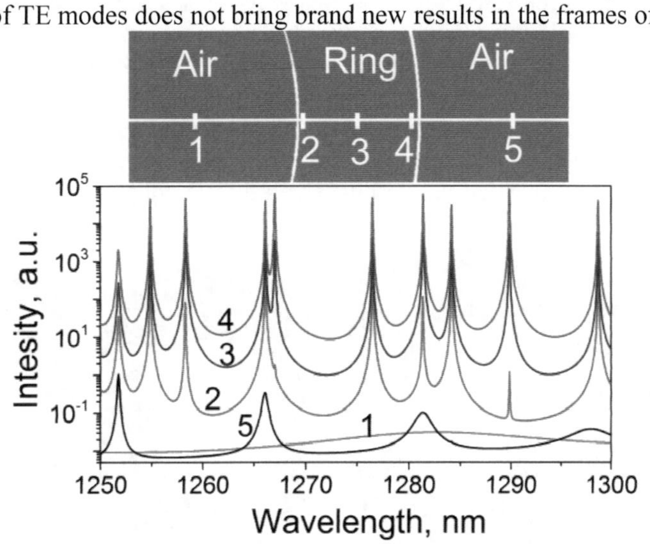

Figure 1. Spectra of a microring resonator calculated for different positions (marked by numbers 1-5 in the inset) of a dipole antenna. Spectra 2, 3 and 4 are shifted along y-axis for convenience.

3. Eigenmodes field distribution

In this section we demostrate results of the simulation of eigenmodes in ring, race-track and stadium-shaped microresonators. The attention is mostly paid to features of WGM and their dependence on geometrical parameters of the resonators.

In figure 2 we present spatial distribution of |H|-field of WGM in the ring resonator with inner/outer radii 3/3.5 μm. These modes are exemplary for further discussion. Refractive index of the resonator material here and below is 3.5, and all the modes considered are in near-infrared range, around 1250-1300 nm. We denote modes as TM(x,y) where x and y are radial and azimuthal order of a mode, respectively.

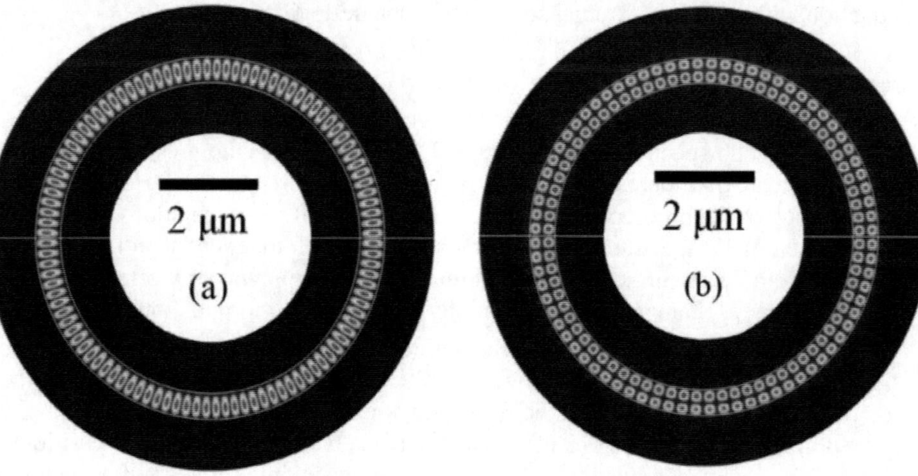

Figure 2. Spatial distribution of |H|-field of a first radial mode TM(1,52) (a) and a second radial mode TM(2,38) (b) in the ring resonator with inner/outer radii 3/3.5 μm.

Indeed, the modes in figure 2 are very typical whispering gallery modes. The field is distributed evenly along the resonator sidewalls. However, this high-symmetrical distribution degrades if a straight section is added to the resonator (so-called race-track shape). We simulated eigenmodes of race-track resonators with different widths w (the distance between inner and outer side walls) and straight section lengths L. The resonator width was varied from 0.15 to 0.7 μm and straight section length from 1.0 to 5.0 μm. In contrast to a ring resonator, electromagnetic field of the first radial modes in race-tracks is not concentrated near a border, but tends to oscillate between inner and outer sidewalls (so-called zigzagging) as was observed in the experiments [15]. Calculated distributions of |H|-field of first radial modes in race-track resonators with different parameters are shown in figure 3.

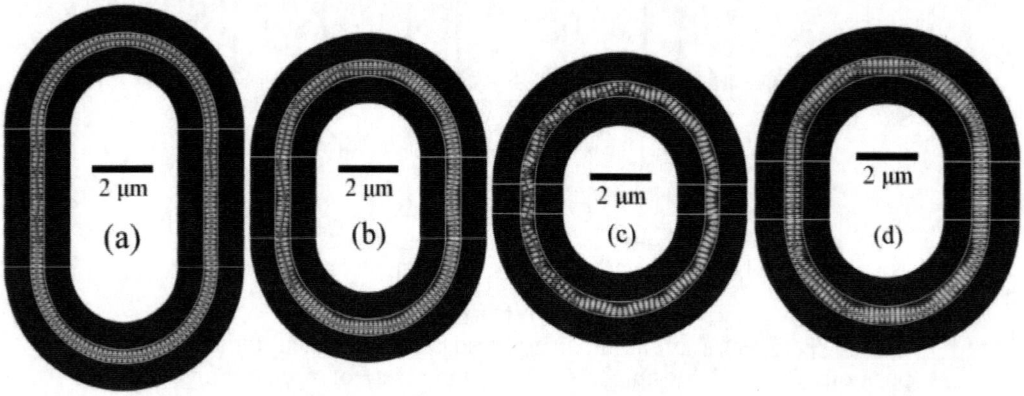

Figure 3. Field distribution of first radial modes in race-track resonators with inner radius 3 μm. (a) L=5 μm, w=0.5 μm; TM(1,80) mode; the field is distributed almost evenly in a bent section of the resonator and oscillates only in a straight section; (b) L=3 μm, w=0.6 μm; TM(1,71) mode; the wave refracts mainly from the outer sidewall of the resonator in the bent section and from both walls in the straight one; (c) L=1 μm, w=0.5 μm; TM(1,58) mode; the wave refracts from both sidewalls along all the parts of the resonator; (d) L=2 μm, w=0.64 μm; TM(1,67) mode; zigzagging is observed in the bent section only, whereas in the straight section the field is distributed almost without deviations in its direction. Arrows schematically show trajectory of the wavefront of the eigenmodes (guide for eyes only) and white lines indicate straight and bent parts of the resonators.

Here we emphasize that the zigzagging behaviour essentially depends on the resonator parameters, most typical behaviours being presented in figure 3. Knowledge about these distributions can help in optimizing systems using these resonators, like resonator-waveguide directional coupling, etc. Field distributions of second radial modes in two race-tracks differing in parameters are shown in figure 4.

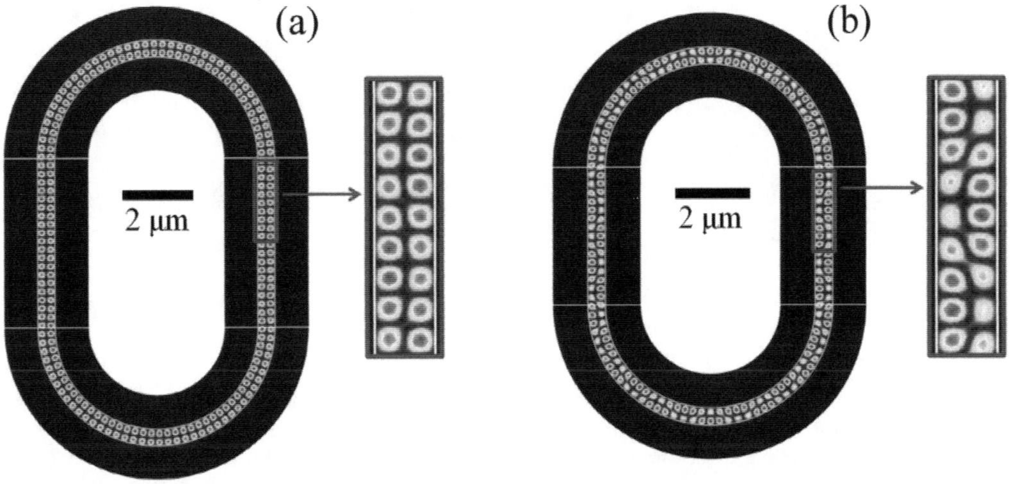

Figure 4. Two types of field distribution in second radial modes in race-tracks with inner radius 3 μm: (a) L=5 μm, w=0.5 μm; TM(2,61) mode; intensity of energy redistribution between radial maxima is barely noticeable; (b) L=4 μm, w=0.5 μm; TM(2,56) mode; the redistribution is well-pronounced. White lines indicate straight and bent parts of the resonators.

A feature of the second radial order modes in the race-track resonators is that they can exhibit energy redistribution between first and second radial maxima. This was also observed in SNOM experiments [15]. However, the power of this redistribution can differ depending on resonator parameters, e.g. it is barely noticeable in figure 4a and well pronounced in figure 4b. Moreover, in wider resonators these modes also tend to the zigzagging. Besides, in the modeling an interesting correlation was also revealed: first radial modes do not exhibit zigzagging behavior if a resonator is narrow enough to forbid second radial modes existence.

Both presented in figures 3 and 4 phenomena are periodic, i.e. these features of eigenmodes have the other characteristic period besides the wavelength. It is worth to note that the results obtained in the 2D case are qualitatively the same as in the 3D case. In this study we demonstrate all the phenomena, which were revealed in [15] where 3D modeling and related experiments were presented. However, in this study we have performed more detailed consideration which became possible because of less computational time and powers required for the 2D modeling.

The features of WGM in race-tracks shown above make us assume that these are not "classic" WGM, which distributed along a single border due to total internal reflection phenomenon. To support this item we show the transformation of mode-field distribution under the elongation of initially disk-shaped (without space in the center) resonator in figure 5. Indeed, the images in figure 5 demonstrate that a "classic" WGM in a disc-shaped resonator (figure 5a) suffers noticeable degradation with inserting straight section and increasing its length (figures 5b,c). For longer sections WGM are almost gone. Therefore, we assume that an inner sidewall of race-track resonators is essential for the eigenmodes formation, in contrast to ring and disc resonators where WGM are distributed along the outer border only.

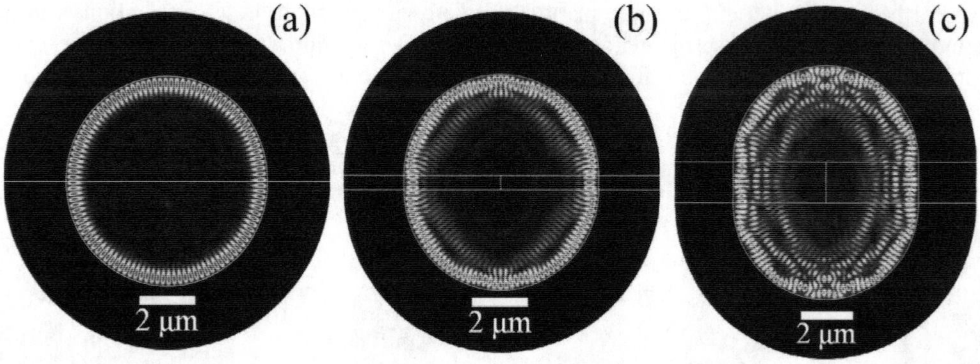

Figure 5. Field distribution of an eigenmode in: (a) disc resonator; (b) stadium-shaped microresonators with straight section lengths of 0.5 μm; (c) stadium-shaped microresonators with straight section lengths of 1.5 μm. Radius of the resonators is 3.5 μm.

4. Race-track resonator: effective mode index

In this section we present a dependence of an effective mode index on a resonator width (figure 6). This kind of dependence is convenient to demonstrate what kinds of radial modes are supported by a resonator of a certain width and how well the modes are localized in the resonator volume. An effective index is defined by a proportion of electromagnetic energy in a resonator volume and an outer medium and described by equation:

$$n_{eff} = \frac{1}{\int\limits_{V_{out}+V_{res}} W_{em} dv} \left[n_{res} \int\limits_{V_{res}} W_{em} dv + n_{out} \int\limits_{V_{out}} W_{em} dv \right], \tag{3}$$

where W_{em} - electromagnetic energy density; n_{res} and n_{out} - refractive indices of the resonator material and the outer medium; V_{res} and V_{out} - volumes of the resonator and the outer medium, respectively.

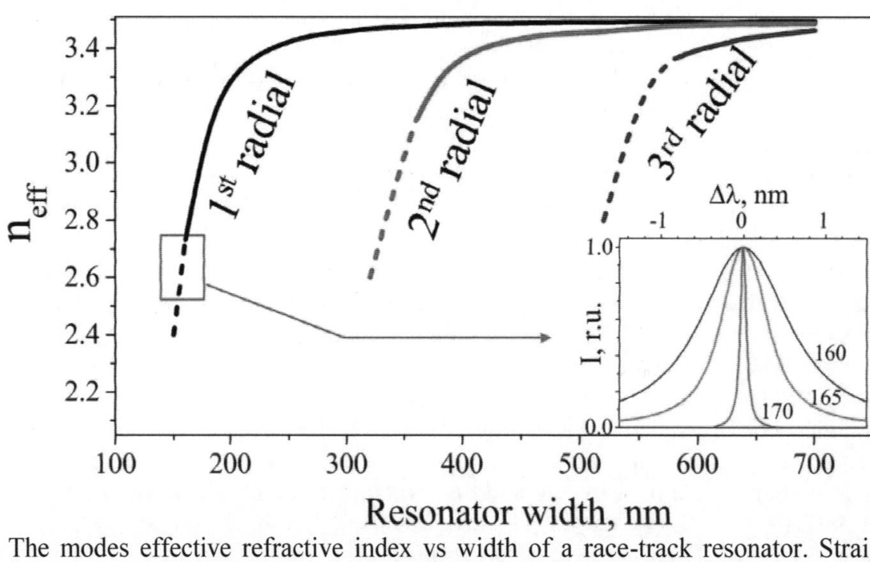

Figure 6. The modes effective refractive index vs width of a race-track resonator. Straight section length is 2 μm, inner radius is 3 μm. The dashed lines correspond to the case when a mode is barely seen in a calculated spectrum. The inset: spectral shape of the first radial modes corresponding to the dashed line; resonator widths in nm are labeled near the curves; the spectra are normalized and centered relatively their resonant wavelengths.

In a resonator narrower than 150 nm almost no modes are presented, second radial modes can be excited in a resonator wider than 350 nm and third radial – wider than 550 nm. In figure 6 areas where corresponding modes are barely seen are marked as dashed lines. At these widths the modes suffer abrupt decrease in their quality factor – see the inset in figure 6. A spectral shape of the mode at width w=170 nm qualitatively does not differ from ones of wider resonators. Whereas even slightly decrease in the width to w=165 nm considerably broadens the resonant curve. Further narrowing of the resonator results in mode disappearance – this approximately corresponds to the area where the dashed lines break. We believe that this relates to radiative losses caused by the bent section [16].

Generally, the relation of the effective index and material index describes the part of energy propagating in the resonator material. One can see that above 80% of the first radial modes energy are localized in the resonator volume if the resonator width exceeds 170 nm. The curves in the figure 6 saturate quickly, and the saturation level is close to 3.5, which is the resonator material refractive index. This means that for wider resonators the modes are strongly localized in the resonator volume.

5. Conclusion

We numerically simulated mode spectra and corresponding electromagnetic field distributions of microring and race-track resonators and effective mode indices of a race-track using 2D model. So-called zigzagging phenomenon in the race-track resonators, when the optical field is not concentrated near one boundary but tends to oscillate between sidewalls, was considered. It was shown that that the appearance, behaviour and power of the zigzagging strongly depend on the resonator geometry. It was demonstrated that second radial modes exhibit energy redistribution between first and second radial maxima. We have shown that WGM in stadium-shaped resonators drastically degrades with increasing straight section length. This makes us presume that WGM in race-tracks are not "classic" WGM and an inner sidewall of a resonator is essential for their formation.

Acknowledgements

This work was supported by Russian Foundation for Basic Research, project #16-29-03111.

References

[1] Ide T, Baba T, Tatebayashi J, Iwamoto S, Nakaoka T and Arakawa Y 2005 *Opt. Express* **13 (5)** 1615
[2] Nadtochiy A et al. 2013 *Tech. Phys. Lett.* **35** 830
[3] McCall S, Levi A, Slusher R, Pearton S and Logan R 1992 *Appl. Phys. Lett.* **60** 289–91
[4] Boyd R and Heebne J 2001 *Appl. Opt.* **40 (31)** 5742
[5] Foreman M, Swaim J and Vollmer F 2015 *Adv. Opt. Photon.* **7** 168-240
[6] Sinha R, Karabiyik M, Ahmadivand A, Al-Amin C, Vabbina P, Shur M and Pala N 2016 *J Infrared Milli Terahz Waves* **37** 230
[7] Del'Haye P, Schliesser A, Arcizet O, Wilken T, Holzwarth R and Kippenberg T 2007 *Nature* **450** 1214-17
[8] Rayleigh L 1910 *Philos. Mag* **20** 1001–04
[9] Mie G 1908 *Annalen der Physik* **25** 377
[10] Grudinin I, Ilchenko V and Maleki L 2006 *Phys. Rev.* A **74** 063806
[11] Boriskina S, Benson T, Sewell P and Nosich A 2003 *Opt. and Quant. Electron.* **35** 545–59
[12] Ctyroky J, Prkna L and Rigorous M 2004 *Proc. 6th Int. Conf. on Transparent Optical Networks* (*Wroclaw, Poland*) vol 2 (IEEE) pp 281–286
[13] Mintairov A, Chu Y, He Y, Blokhin S, Nadtochy A, Maximov M, Tokranov V, Oktyabrsky S and Merz J 2008 *Phys. Rev.* B **77** 195322
[14] https://www.comsol.com/model/download/344161/models.woptics.fabry_perot.pdf
[15] Kryzhanovskaya N et al. 2017 *J. Appl. Phys.* **121** 043104
[16] Unger H G 1977 *Planar Optical Waveguides and Fibres* (Oxford: Oxford University Press)

IOP Conf. Series: Journal of Physics: Conf. Series **917** (2017) 062041 doi:10.1088/1742-6596/917/6/062041

Synthesis and postsynthetic anion exchange of $CsPbX_3$ (X = Cl, Br, I) quantum dots

A Percova, L B Matyushkin
Department of micro- and nanoelectronics, St. Petersburg Electrotechnical University LETI, St. Petersburg 197376, Russia

Abstract. $CsPbX_3$ (X = Cl, Br, I) quantum dots were synthesized using hot-injection method at various temperatures (50–200°C), resulting in quantum dots size change and corresponding photoluminescence shift. During anion exchange, continuous formation of intermediate solid solutions was observed resulting in fine adjustment of photoluminescence peak position. Dynamics of photoluminescence spectra under continuous laser irradiation for anion exchange was studied. It is necessary for anion exchange that halide should be in ionic, not molecular, form.

1. Introduction

Nanomaterials never cease to amaze. Even when we think we have seen all the terrific properties and thought about all possible applications, there appears something spectacular. High quantum yields of up to 90%, possibility to tune photoluminescence over the entire visible spectra, the absence of the blinking effect and narrow emission spectral widths – all these interesting properties are present in novel optoelectronic materials, $CsPbX_3$ (X = Cl, Br, I) quantum dots (QDs) [1]. Fully inorganic perovskites attracted emphasis since organic/inorganic hybrid perovskites, such as $CH_3NH_3PbI_3$, showed staggering power conversion efficiencies above 20%, so it became the only one logical step to synthesize all-inorganic perovskite QDs because of a higher stability they show compared to hybrid ones [2].

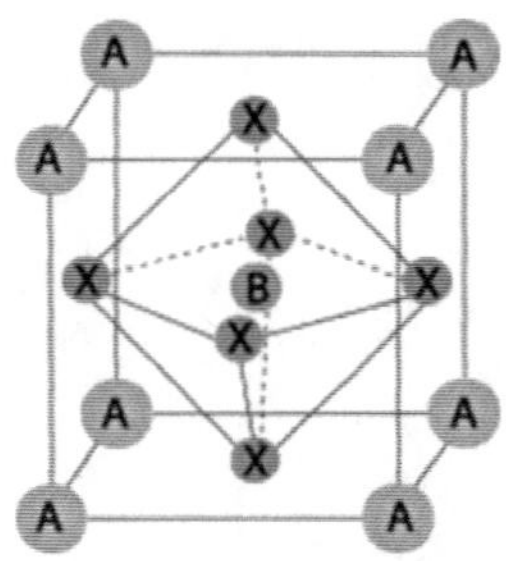

Figure 1. Cubic ABX_3 perovskite lattice (A = Cs, B = Pb, X = Cl, Br, I).

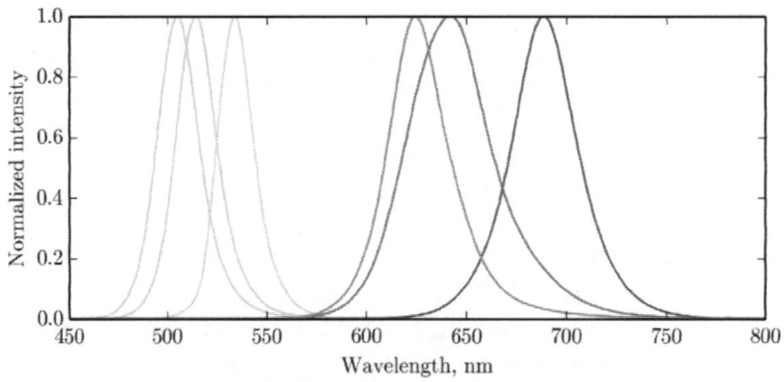

Figure 2. Photoluminescence spectra of $CsPbBr_3$ (left group of graphs) and $CsPbI_3$ (right group of graphs) QDs synthesized at different temperatures (50–200°C).

2. Synthesis of $CsPbX_3$ QDs

All $CsPbX_3$ QDs crystallize in cubic phase of the perovskite lattice (Figure 1), which is the high-temperature state for bulk compounds. QDs were synthesized via hot injection method using Cs-oleate and PbX_2 as precursors and dodecane as a medium. Since the growth of QDs happens exceedingly fast (within 1–3 s), the average size of QDs is rather controlled by reaction temperature. Photoluminescence spectra of $CsPbBr_3$ and $CsPbI_3$ QDs synthesized at various temperatures (50–200°C) are shown in Figure 2.

3. Postsynthetic anion exchange

The enchanting special thing about $CsPbX_3$ (X = Cl, Br, I) QDs is that we can significantly tune photoluminescence spectra using postsynthetic anion exchange [3]. Photoluminescence peak position depends on which halogen atom (Cl, Br or I) is in the QD structure. Halogen atoms ability to easy substitute each other can be used to obtain intermediate energy gap values. Due to the significant difference between Cl^- and I^- ionic radii there is no possibility to obtain $CsPb(Cl/I)_3$, yet $CsPb(Cl/Br)_3$ and $CsPb(Br/I)_3$ systems are feasible. The best option, in this case, is to synthesize $CsPbBr_3$ QDs with subsequent addition of lead chloride and lead iodide solutions to carry out anion exchange. Corresponding photoluminescence spectra are shown in Figure 3. As it can be seen fast anion exchange provides the possibility to the fine tuning of photoluminescence spectra over the entire visible spectra. In Figure 4 are shown dynamics of photoluminescence spectra under continuous laser irradiation for anion exchange from $CsPbBr_3$ to $CsPbI_3$. The following shift of photoluminescence spectra and broad peaks between initial and final states are connected with the continuous formation of $CsPbBr_xI_{3-x}$ solid solutions. Note that photoluminescence intensity of $CsPbI_3$ stays comparable to the $CsPbBr_3$. However, change in solid solution's composition leads to absorption spectra shift, which causes absorption coefficient change on excitation wavelength (405 nm). Especially clearly it is seen in case of $CsPb(Cl/Br)_3$ QDs: chlorine concentration increase in these solutions leads to absorption coefficient decrease on excitation wavelength and applicable luminescence intensity reduction (Figure 5). Talking about $CsPb(Br/I)_3$ QDs, the same effect causes absorption coefficient increase and leads to the fact that charge carrier excess energy converts to the thermal energy, accelerating diffusion balance. Therefore, the process time for $CsPb(Cl/Br)_3$ QDs is three times slower.

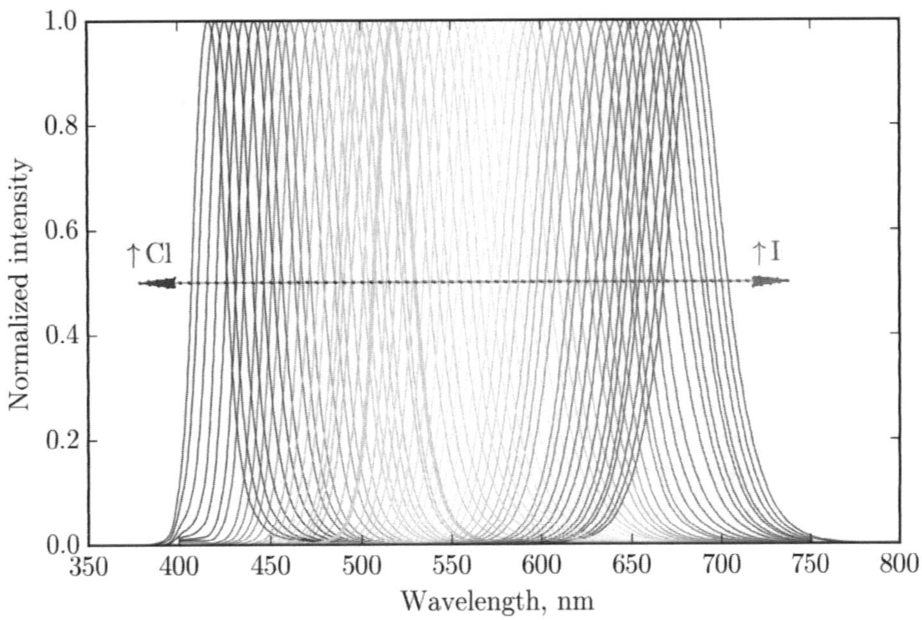

Figure 3. Photoluminescence spectra of $CsPbX_3$ (X = Cl, Br, I) synthesized using anion exchange.

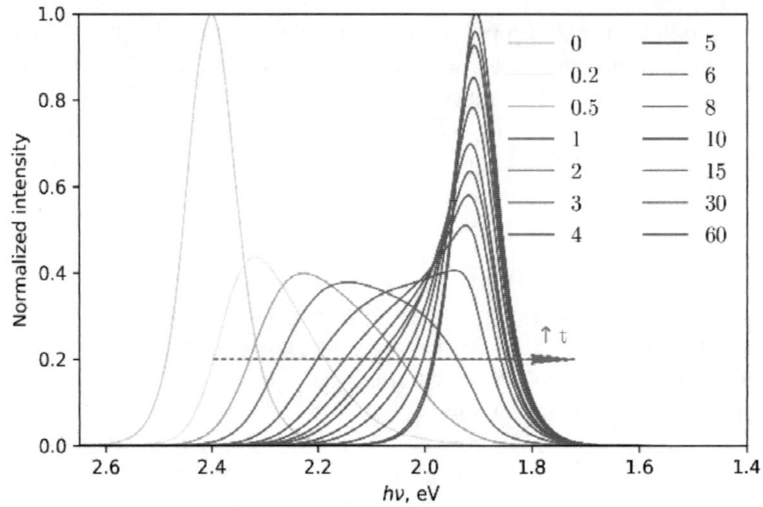

Figure 4. Dynamics of PL spectra under continuous laser irradiation for anion exchange $CsPbBr_3$ to $CsPb(Br/I)_3$, the legend corresponds to time (in minutes) since injection of PbI_2 precursor.

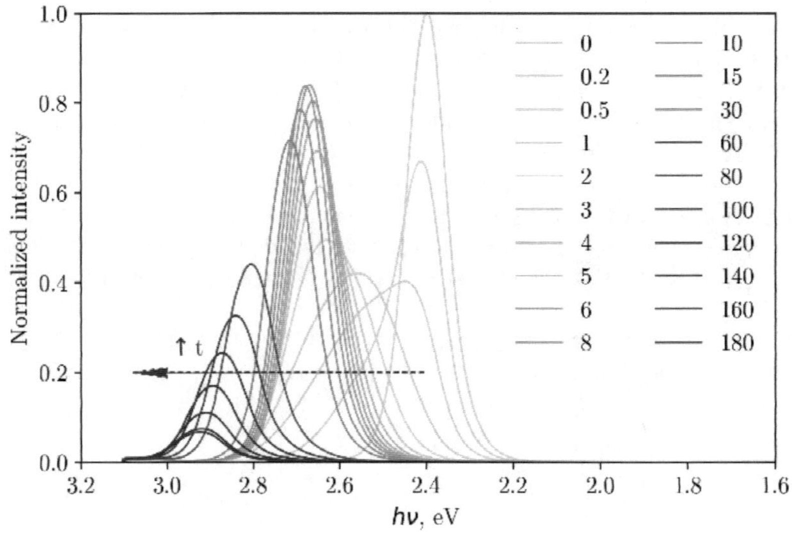

Figure 5. Dynamics of PL spectra under continuous laser irradiation for anion exchange $CsPbBr_3$ to $CsPb(Cl/Br)_3$, the legend correspond to time (in minutes) since injection of $PbCl_2$ precursor.

Note that it is necessary for anion exchange that halide should be in ionic, not molecular, configuration. When adding molecular iodine to the colloidal solution of $CsPbBr_3$ QDs instead of anion exchange etching of surface ligands occurs leading to luminescence intensity decrease (Figure 6). Surface etching is also confirmed by the fact that precipitation of more large-scale $CsPbBr_3$ QDs happens as the result of interaction with molecular iodine. Thus, $CsPbBr_3$ can be used as sensors in iodine electrolytes – shift of peak position indicates dissociated iodine presence, intensity modification – concentration of molecular iodine.

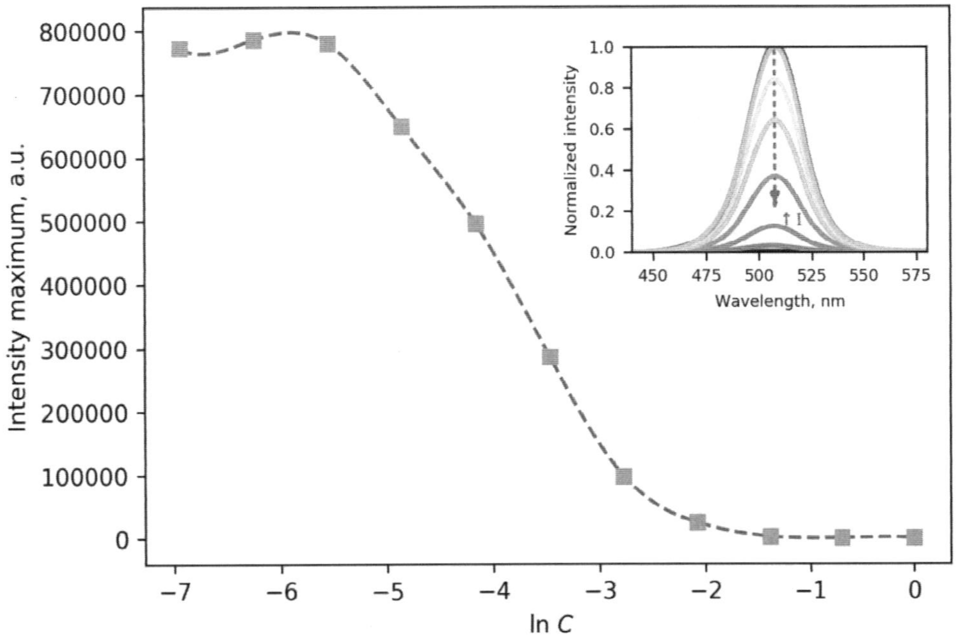

Figure 6. Dependence of photoluminescence intensity maximum versus iodine concentration C (the inset shows photoluminescence spectra evolution increasing iodine concentration).

4. Conclusion

As it was shown in this work, postsynthetic anion exchange allows to use one initial halogen source solution to obtain different solid solution systems, which means capability to achieve wanted luminescence wavelength. Possibility of anion exchange using molecular iodine was studied and it can be seen that anion exchange happens only if halogen is in ionic form. Adding molecular iodine leads to luminescence intensity decrease.

Taking into account all the captivating properties of these materials, it is easy to see possible applications. For LEDs and lasers, we can use the ability to adjust photoluminescence spectra, high quantum yield, and narrow emission linewidths, for solar cells – significant light absorption by these structures.

References

[1] Protesescu L, Yakunin S, Bodnarchuk M I, Krieg F, Caputo R, Hendon C H, Yang R Xi, Walsh A and Kovalenko M V 2015 *J. Nano Lett* **15** 3692
[2] Kim Y, Yassitepe E, Voznyy O, Comin R, Walters G, Gong X, Kanjanaboos P, Nogueira A F and Sargent E H 2015 *J. Appl. Mater. Interfaces* **7** 25007
[3] Akkerman Q A, D'Innocenzo V, Accornero S, Scarpellini A, Petrozza A, Prato M and Manna L 2015 *J. Am Chem Soc.* **137** 10276

Integrated Bragg waveguides as an efficient optical notch filter on silicon nitride platform

E Zubkova[1,2], P An[1,2], V Kovalyuk[1,2], A Korneev[1,2,3], S Ferrari[4], W Pernice[4], G Goltsman[1,5]
[1]Department of Physics, Moscow State Pedagogical University, 119992, Russia
[2]Zavoisky Physical-Technical Institute of the Russian Academy of Sciences, 420029, Russia
[3]Moscow Institute of Physics and Technology (State University), 141700, Russia
[4]Institute of Physics, University of Münster, 48149, Germany
[5]National Research University Higher School of Economics, 101000, Russia

Abstract. We modeled and fabricated integrated optical Bragg waveguides on a silicon nitride (Si_3N_4) platform. These waveguides would serve as efficient notch-filters with the desired characteristics. Transmission spectra of the fabricated integrated notch filters have been measured and attenuation at the desired wavelength of 1550 nm down to -43 dB was observed. Performance of the filters has been studied depending on different parameters, such as pitch, filling factor, and height of teeth of the Bragg grating.

1. Introduction

Continuous development of micro- and optoelectronics is inevitably connected with a substantial decrease of the planar structure dimensions. State-of-the-art nanotechnology allows creation of complex nanophotonic circuits with a high performance and a compact design on chip, which currently are replacing free-space optical schemes build up on the optical tables. Integrated on chip circuits have a number of advantages, such as low optical losses, no need for optical alignment, insensitivity to vibrations, small size, weight, and power consumption, but the most important is their scalability [1]. This allows development of quantum-optics integrated circuits (QPICs), which would be beneficial for quantum cryptography in the quantum key distribution systems, quantum simulations of complicated molecules, and quantum computing development increasing the data processing and computation rate gradually. One of the key prerequisites for the QPICs development is the on-chip implementation of a highly efficient single-photon source. Silicon nitride (Si_3N_4) is a promising material for nanophotonic circuits, which combine good mechanical properties, low optical absorption in the infrared (IR) and visible wavelength ranges as well as possibility for creation of single photon sources integrated into the nanophotonic circuit. One possible realization of the single-photon source is the four-wave mixing [2]. One of the main problem for such sources is an efficient pump power rejection down to -120 dB. Here, we have numerically simulated and fabricated nanophotonic waveguides with the Bragg grating of different geometrical configurations on silicon nitride platform. Such waveguides can be effectively used as notch-filters integrated into the nanophotonic circuit for the pump power suppression. For the fabricated structures, we have studied their transmission properties and the power suppression efficiency at the desired wavelength depending on the Bragg grating parameters. Data analysis demonstrates reasonable agreement between the measured and simulated characteristics.

2. Device design and fabrication

Proposed devices have a complicated structure both for the longitudinal and transverse directions, which suggests that a proper numerical simulation is required in order to obtain parameters (the width and thickness of the waveguide, the Bragg grating parameters) that would ensure the desired characteristics. We have estimated an effective refraction index of the Bragg waveguide as $n_{eff} = 1.66$. The grating is defined by a few parameters: pitch of the Bragg grating, $\Lambda = \lambda/(2\times n_{eff})$, filling factor, $ff = a/\Lambda$, height of teeth, h, with a being the width of a tooth (Figure 1(c)). By choosing proper parameters, one could efficiently suppress a spectral component in the transmitted spectra, or even tune up the reflectivity for that component [3]. With increase of Λ or ff, wavelength of the spectral component is also increased, and increase of h affects the efficiency and controls the width of the notch. According to the numerical simulation, a telecommunication-wavelength (1550 nm) notch filter should have Bragg's grating pitch of 466 nm, filling factor – 0.5 and the waveguide width W = 1 μm. Basing on the provided above relations, we have developed and fabricated a nanophotonic structure, which included a pair of focusing grating couplers (FGCs) and the Bragg grating in between them, as shown in Figure 1.

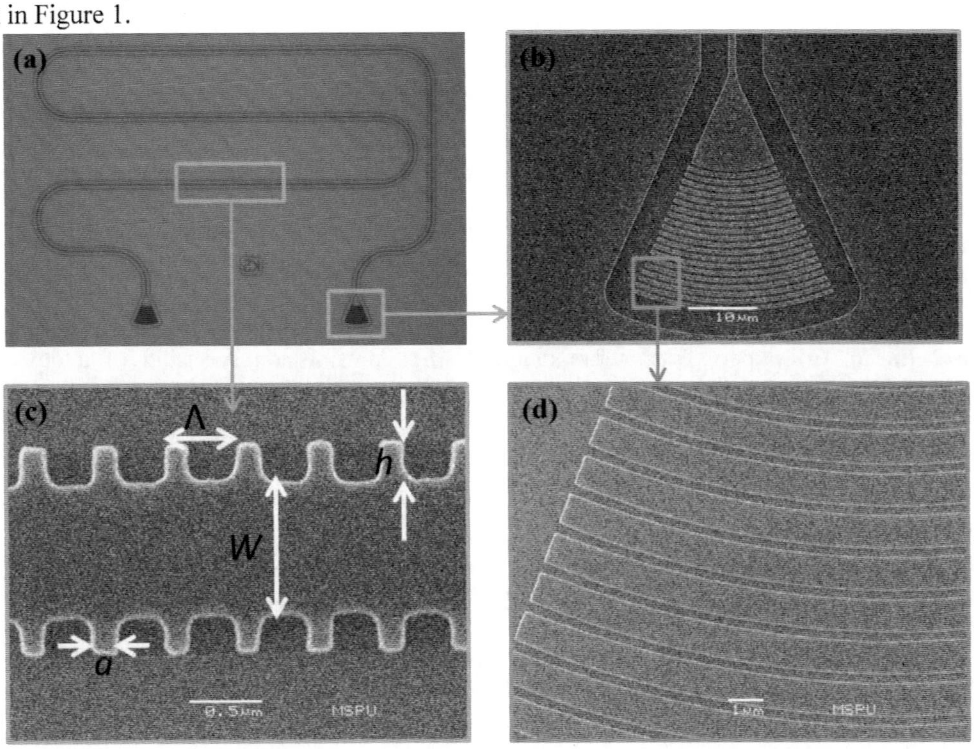

Figure 1 (a, b, c, d). (a) Optical photograph of a fabricated structure; **(b)** SEM image of the FGC; **(c)** Zoom-in in the Bragg grating (SEM) (Λ – pitch, W – width of the waveguide, h – height of the teeth, a – width of a tooth); **(d)** Zoom-in in the FGC (SEM).

On the horizontal parts of the structure, the Bragg grating was formed, as depicted in Figure 1(c). Also shown the waveguide and Bragg grating parameters. Coupling of the incident power to the waveguide structure was done with use of a FGC depicted in Figure 1(b) and shown in greater details in Figure 1(d). For the devices, multi-layered substrates were used, with thicknesses of silicon (Si), silicon oxide (SiO_2), and silicon nitride (Si_3N_4) of 450 μm, 2.6 μm, and 450 nm, respectively. The rib waveguide was formed from Si_3N_4 layer by means of e-beam lithography with use of a positive resist ZEP 520A providing good contrast during the lithography. NanoMaker software [4] was used for proximity effect correction and precise adjustment of the e-beam exposure time. Removing of the exposed material was done by reactive-ion-etching (RIE) in CHF_3-Ar mixture. The resist residuals were cleaned out with help of the oxygen plasma. In order to evaluate performance of the notch-filter depending on the Bragg structure parameters, we have intentionally varied the Bragg grating pitch, filling factor, and height of the teeth during fabrication of the devices.

3. Experimental setup and results

In order to characterize the devices, we have experimentally analyzed their transmission at room temperature, and precisely measured attenuation at the desired wavelength (around the telecommunication wavelength, λ =1.55 μm). Experimental setup incorporated wide-band sources of light, whose power were fed to the device through a polarization controller and a fiber array which includes a set of SM-28e fibers fixed with a precise spacing of 250 μm and tilted by 8° to the device is shown in Figure 2(a). The device itself was mounted on to a table equipped with micrometer-step 3D positioner needed for the device – to – fiber array alignment.

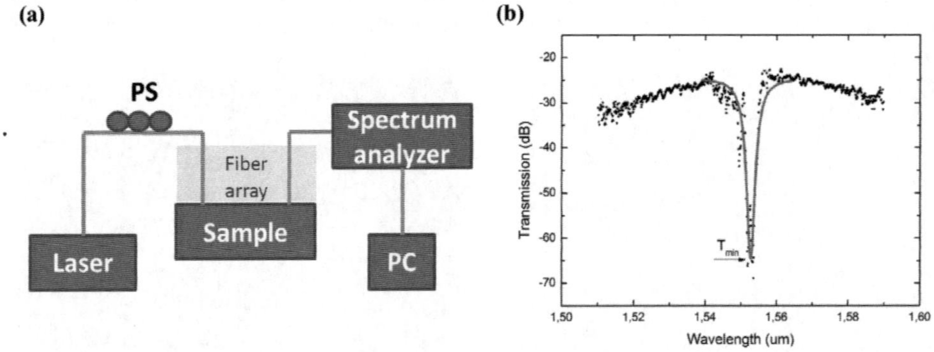

Figure 2 (a, b). **(a)** Experimental setup diagram; **(b)** Transmission spectrum of a device with Λ = 466 nm, *ff* = 0.5, *h* = 100 nm, demonstrating desired attenuation at wavelength of ~1.55 μm. 25 dB attenuation as well as the power decreasing at higher and lower wavelengths is due to introduced by the two FGC in the nanophotonic circuit.

Coupling to the device was realized through the FGCs, one of them was used to irradiate the device, and the other one – to measure the power transmitted through the Bragg grating. Transmitted power was analyzed by an optical spectrum analyzer (OSA) controlled by a PC. A measured curve is depicted in Figure 2(b). Experimentally evaluated dependence of the notch wavelength on the grating pitch is represented in Figure 3(b): with increase of the pitch the notch-wavelength is also increased. The same effect is observed by increasing the filling factor which is in line with the simulations and allows one to fabricate a notch-filter at a desired wavelength. Dependence of the transmitted power versus the grating periodicity is given in Figure 3(c) showing increase of the notch-filter efficiency from -5 dB to -43 dB with increase of the grating periodicity from 150 to 2350 at the wavelength of 1.55 μm and the other Bragg's grating parameters fixed (Λ = 466 nm, *ff* = 0.5, *h* = 100 nm). Grating periodicity of 2350 takes a length of ~ 1 mm. With linear approximation, we are able to produce notch-filters with the desired attenuation at the selected wavelength. Also, changing the teeth height one is able also to tune the notch-filter efficiency and its width as shown in Figure 3(a).

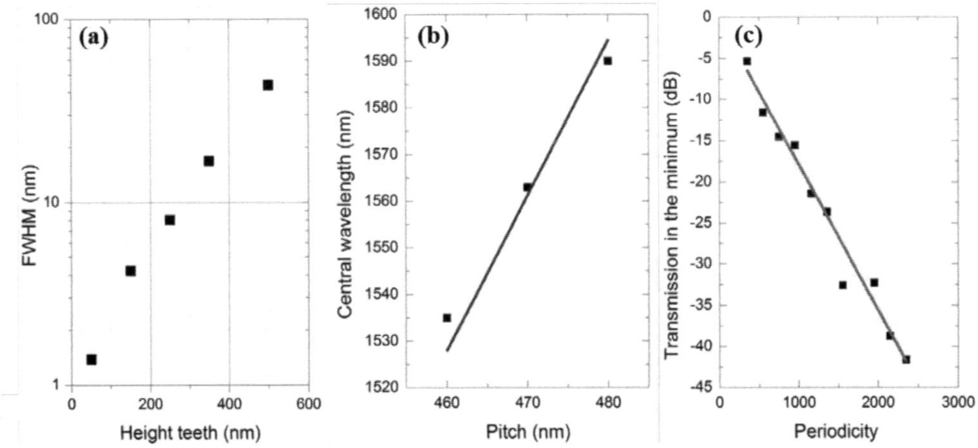

Figure 3 (a, b, c). (a) Dependence of the notch-filter spectral width (defined as FWHM) on the teth height of the Bragg grating; **(b)** Centre-wavelength dependence on the grating pitch, experimental data points are linearly fitted using n_{eff} = 1.66; **(c)** Transmission dependence at λ = 1550 nm on the grating periodicity (i.e., number of periods), experimental data points are fitted using linear interpolation.

With decrease of the teeth height, the FWHM is also decreased. The data were obtained for devices with fixed Bragg's grating pitch and filling factor (Λ = 466 nm, ff = 0.5). FWHM is decreasing from 43 nm at h = 500 nm down to 1.5 nm at h = 50 nm. This result ensures development of notch-filters with desired selectivity.

4. Conclusion

To conclude, we have shown that Bragg's waveguides made from silicon nitride could be efficiently used as filters operated in complex quantum-optics integrated circuits, such as for filtering out pumping power driving single-photon sources. By varying the Bragg grating pitch, filling factor and the height of the teeth, a notch-filter can be designed and fabricated with the desired characteristics, such as notchwavelength, spectral selectivity (i.e., FWHM of the notch-line), and efficiency. By extrapolation, one may predict that a 3.2-mm long Bragg's waveguide would be needed for unwanted power suppression down to -120 dB. Further work will be devoted to maturating of the technological steps, which is crucial for technological repeatability and improvement of the spectral properties of the devices. As a next step, on-chip integration of the filter with a multi-element circuit including a single-photon source based on four-wave mixing should be considered.

Acknowledgments

E Zubkova, P An, G Goltsman acknowledge support of the Russian Science Foundation (project 16-12-00045, waveguide fabrication and transmission measurements). W. Pernice acknowledges support by the DFG grants PE 1832/1-1 & PE 1832/1-2 and the Helmholtz society through grant HIRG-0005 (design and modeling). We also are thankful to Yu Lobanov (MSPU) for helpful discussion and proofreading of the English version of the text.

References

[1] Silverstone J W, Bonneau D, O'Brien J and Thompson M G 2016 Silicon Quantum Photonics IEEE J. Sel. Top. Quantum Electron. 22 1–13

[2] Ramelow S, Farsi A, Clemmen S, Orquiza D, Luke K, Lipson M, and Gaeta A L 2015 *ArXiv:1508.04358v1*

[3] Shi W, Yun H, Lin C, Greenberg M, Wang X, Wang Y, Fard S T, Flueckiger J, Jaeger N A F, and Chrostowski L 2013 *Opt. Express* **21** 6

[4] http://www.nanomaker.com

IOP Conf. Series: Journal of Physics: Conf. Series **917** (2017) 062043 doi:10.1088/1742-6596/917/6/062043

Band-pass filters based on photonic crystal

S A Khodenkov[1], **I A Yushkov**[2]

[1]Department of Physics, Reshetnev Siberian State Aerospace University, Krasnoyarsk 660037, Russia
[2]Laboratory of electrodynamics and microwave electronics, Kirensky Institute of Physics, Krasnoyarsk 660036, Russia

Abstract. Multilayer photonic crystal structures with bleaching layers are being investigated. In order to calculate the characteristics of ultra-wideband filters on their basis, T-lines lossless model was used. Amplitude-frequency characteristics for the synthesized filters of 5th, 11th and 17th orders are given. It is proved that by a significant increase in filter *N* order, the difference between the connection coefficients of central resonators' layers' becomes negligible. This makes it possible to develop 27-order filter, in which almost half of the layers are realized by periodic interchange of only two identical high-contrast materials. The investigated band-pass filters, including the ones on a glass substrate, have high frequency-selective properties at a relative bandwidth of 80%.

1. Introduction

It is well known [1, 2] that unidimensional (1D) photonic crystals are dielectric structures composed of alternating layers of different materials and having radiolucent «windows» for electromagnetic waves in certain frequency range, between which there are bandgaps - photonic stop band.

Such multilayer dielectric structures are the base for the production of mirrors for laser techniques [3], which have less losses and higher reflectance coefficient in comparison with metal mirrors, besides they are the base for the development of band-pass filters and high-*Q* resonators [4].

The given paper presents the results of investigation of multilayer dielectric structures with not more than 27 layers that are ultra-wideband ones with high frequency-selective properties.

2. Modelling results

The amplitude-frequency characteristics (AFCs) of all 1D structures were calculated by means of *T*-line without loss model. It was assumed that electromagnetic waves from free space having wave resistance Z_0 = 377 Ω incident orthogonally on thin films which are under investigation. Having passed through the structures at certain frequencies electromagnetic waves propagate further in free space. Falling (P_{fal}), reflecting (P_{ref}) and passing (P_{pas}) power of electromagnetic waves are connected by the law of energy conservation:

$$P_{fal} = P_{ref} + P_{pas} \tag{1}$$

The tune of band-pass filters is done by «manual» parametric synthesis, during which the central frequency of passband f_0=300 GHz as well as the relative bandwidth $\Delta f/f_0$ = 80% were recorded. Meanwhile, the passband of single-dimensional structure from each dielectric layer is formed by one resonance. To improve frequency-selective filters properties, the first and the last selected layers were chosen as radiolucent ones. They are quarter-wave resonators (λ/4) with electrical length (optical film thickness) $\theta_1 = \pi/2$. The remaining layers in the structures are half-wave ($\lambda/2$) with electric length $\theta_2 = \pi$. The scheme of such photonic crystal structure with 17 dielectric layers is shown in figure 1.

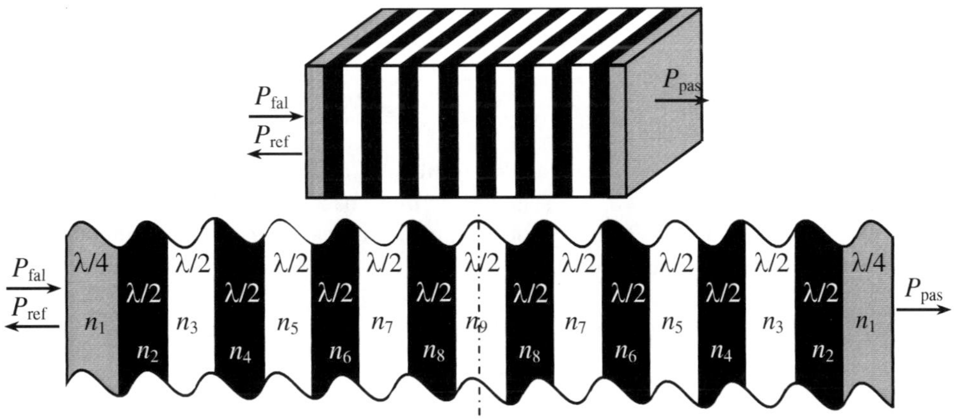

Figure 1. Schematic representation of 17-layers structure.

The filters of 3th, 5th, 7th, 9th, 11th, 13th, 15th and 17th orders were synthesized with the same maximum level of return loss ($L_R = -30$ dB) in the passband. The comparison of amplitude-frequency characteristics of these band-pass filters of 5th, 11th and 17th orders is shown in figure 2.

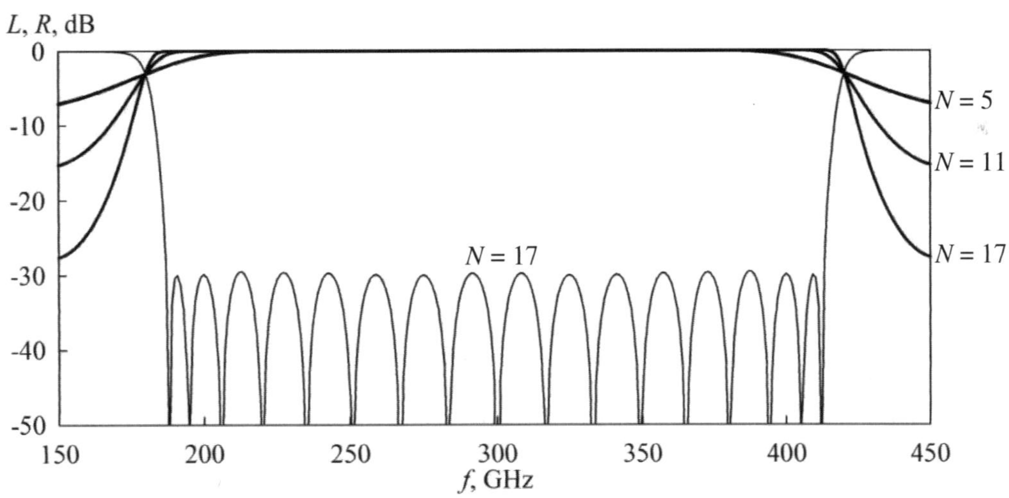

Figure 2. AFCs of the 5th, 11th and 17th orders filters.

Here for $N = 17$ is the value of relative refractive indices of dielectric layers' diffraction: $n_1 = 1.249$, $n_2 = 1.763$, $n_3 = 1.257$, $n_4 = 1.885$, $n_5 = 1.175$, $n_6 = 1.990$, $n_7 = 1.129$, $n_8 = 2.040$, $n_9 = 1.116$.

As it is seen from the above given figure, with an increase of N order of such ultra-wideband filters, the increase of power suppression in the stop bands occurs slowly. Therefore, the improvement of their frequency-selective properties requires the use of larger number of layers-resonators in the structures and, correspondingly, an increase in the number of tuning parameters. However, this significantly exaggerates «manual» parametric synthesis of filters. Therefore, another approach to develop such band-pass filters with strong suppression of parasitic power was proposed.

In this connection, for all 8 synthesized filters, the coupling coefficients between pairs of adjacent layers-resonators were calculated (figure 3). For this purpose, the formula that had been obtained by the author of the work [5] for a filter of series-connected half-wave resonators formed by line-segments with «low» Z_- and «high» Z_+ wave resistance was used:

$$|k_{+-}| = \frac{2\pi \operatorname{arctg}\sqrt{Z_-/Z_+}}{\pi^2 + \operatorname{arctg}^2\sqrt{Z_-/Z_+}}. \qquad (2)$$

It is obvious that connection value between the filter and the input and the output will be determined by the difference in the wave resistance in free space (Z_0 = 377 Ω) and the extreme structure layers that the first and last resonators form. This also allows to estimate coupling coefficient k_{01} of the first resonator with the «space», using the above mentioned formula.

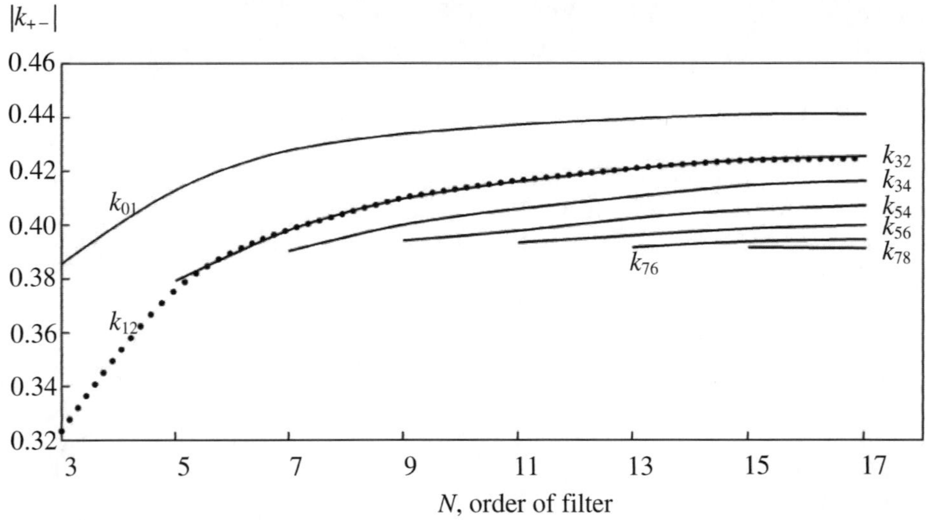

Figure 3. The dependence of coupling coefficients between the adjacent layers on the basis of filter order.

It is seen that during significant increase in filter N order, the difference between coupling coefficients of central resonator layers becomes negligible. Therefore, thirteen inner layers can be realized in the form of periodic interchange of only two high-contrast materials with the refractive index n_8 and n_9 (figure 4).

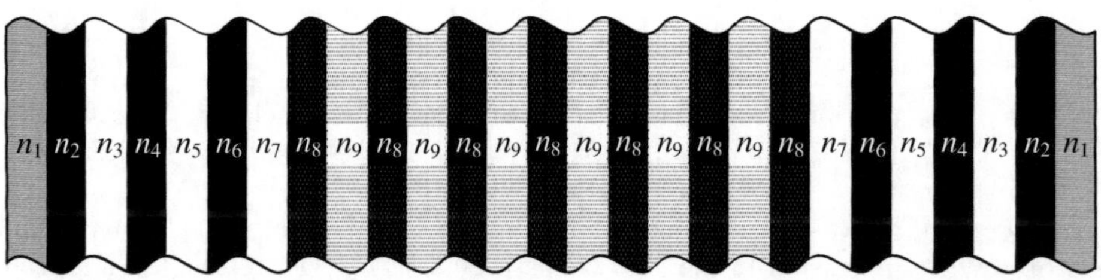

Figure 4. Schematic representation of 27-layers structures.

This improves structure manufacturability and reduces its cost in mass production, but, meanwhile, the filter with a relative bandwidth of 80% realised on its base has high frequency-selective properties (figure 5).

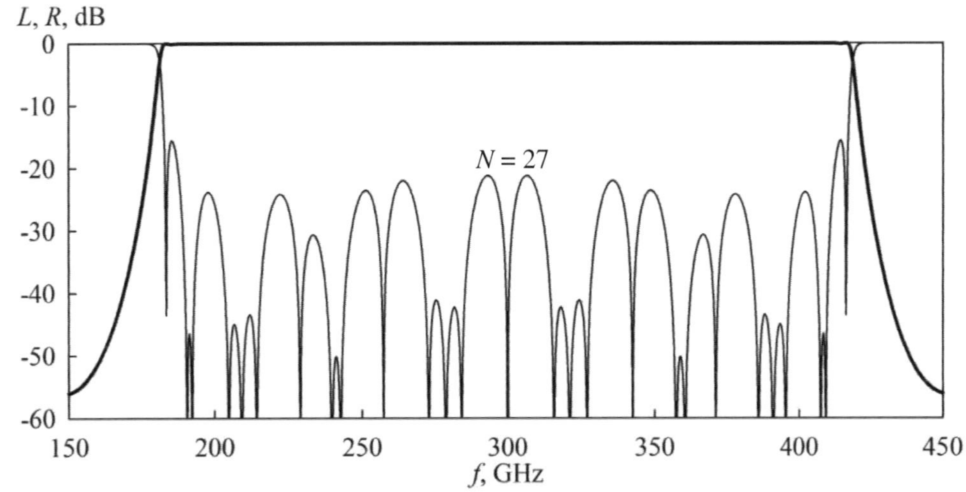

Figure 5. AFCs of 27-layers 1D structure.

Here for $N = 27$ is the value of relative refractive indices of dielectric layers' diffraction: $n_1 = 1.255$, $n_2 = 1.760$, $n_3 = 1.252$, $n_4 = 1.888$, $n_5 = 1.170$, $n_6 = 1.982$, $n_7 = 1.105$, $n_8 = 1.964$, $n_9 = 1.031$. It is worth noting, the initial filters investigation can be performed notwithstanding the substrate in the models. Further the substrate of infinite thickness with a certain refractive index, for example $n_s = 1.300$ (glass is the material) is added to model of T-lines (figure 6).

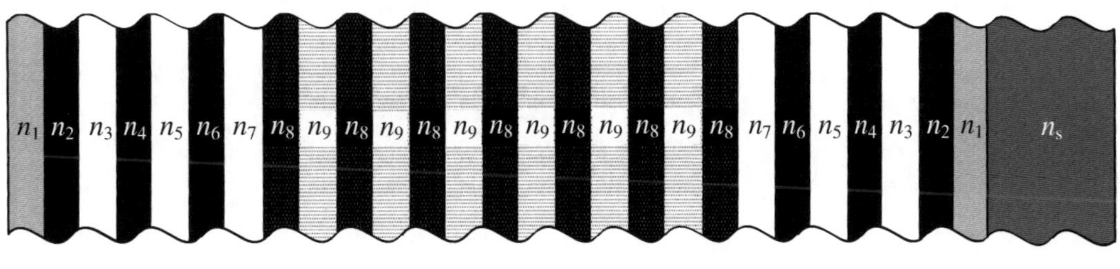

Figure 6. Schematic representation of 27-layers structures with the use of glass substrate.

Frequency-selective properties of the ultra-wideband filter of the twenty seventh order based on one-dimensional dielectric photonic crystal change slightly (figure 7).

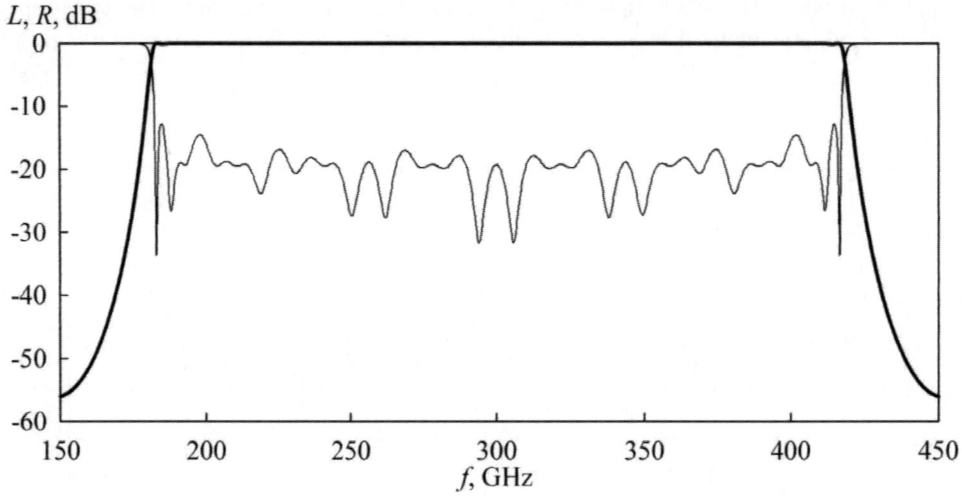

Figure 7. AFCs of 27-layers 1D structure with the use of glass substrate.

3. Conclusions

Therefore, the proposed ultra-wideband filters based on one-dimensional photonic crystal are manufacturable and have high frequency-selective properties including a strong suppression of power at frequencies of the stop bands.

Acknowledgments

This study was supported by the Ministry of Education and Science of the Russian Federation, grant MK-9119.2016.8.

References

[1] Belyaev B A, Voloshin A S and Shabanov V F 2004 *Dokl. Phys.* **49** 213
[2] Escorcia-Garcıa J and Mora-Ramos M E 2009 *PIERS Online* **5** 36
[3] Tikhonravov A V and Trubetskov M K 2007 *Journal of Optical Technology* **74** 845
[4] Belyaev B A, Voloshin A S and Shabanov V F 2005 *Dokl. Phys.* **50** 337
[5] Tyurnev V V 2007 *Int. Conf. Electronic means and control systems. Experience of innovative development* vol 1 (Tomsk: V-Specter) p 88

IOP Conf. Series: Journal of Physics: Conf. Series **917** (2017) 062044 doi:10.1088/1742-6596/917/6/062044

Nonlinear optical properties of hybridized CdS/ZnS-PVP sols

A S Kulagina[1,2], **S K Evstropiev**[3], **A I Khrebtov**[1]
[1]Laboratory of Epitaxial Nanotechnologies, St. Petersburg Academic University, St. Petersburg 194021, Russia
[2]Laboratory of Physics of Epitaxial Nanostructures, ITMO University, Saint Petersburg 197101, Russia
[3]Department of Design and Safety of Computer Systems, ITMO University, Saint Petersburg 197101, Russia

Abstract. Hybrid composites of CdS-core ZnS-shell nanoparticles embedded in polyvinylpyrrolidone (PVP) matrixes have been prepared and characterized. Cadmium sulfide (CdS) nanocrystals were grown in water-propanol-2 solutions containing high-molecular (Ms=1300000) polyvinylpyrrolidone (PVP) at room temperature using cadmium nitrate and sodium sulfide as the cadmium and sulfur sources, respectively. The CdS/ZnS-PVP suspensions have promising optical properties for nanocomposite films based on. Nonlinear optical properties of diluted CdS/ZnS sols were studied at 532 nm and 5 ns laser pulses by using the Z-scan technique. Dependence of the nonlinear-optical coefficients on the CdS weight has been obtained.

1. Introduction

The quantum-sized CdS nanocrystals has an exceptional third-order nonlinear optical properties useful for optical switches or limiting devices [1,2]. The ZnS shell stabilizes the size of quantum dots (QDs) and improves the quantum yield of luminescence [3]. To realize optical devices, the nanoparticles should be integrated into a solid-state matrix, e.g. glass or polymer [4]. The nanocrystals must be homogeneously distributed in the matrix and save stable properties. Polymers are considered a good choice as host materials due to their relatively architectural flexibility, simple fabrication process and low cost [5]. Among all the useful polymeric materials, polyvinylpyrrolidone (PVP) belongs to a neutral charged polymeric dispersant and has outstanding colloidal and complexing properties [6,7]. PVP has been widely used as the protecting agent or stabilizer for synthesizing nanoparticles [8,9] and the matrix for manufacturing the optical devices [10,11]. In this paper we studied nonlinear optical properties of hybridized CdS/ZnS-PVP sols (*in situ*) by exciting of laser radiation at λ = 532 nm and 5 ns pulse with higher intensity. Nonlinear coefficients of refraction γ (cm^2/W) and two photon absorption (TPA) β (cm/W) were extracted from the analysis of the z-scan technique data [12]. The excitation was carried out in the transparency region of QDs and an effective two-photon absorption process was observed. Investigated sols have promising nonlinear properties for switching devices and for creating films based on.

2. Methods

Cadmium sulfide (CdS) nanocrystals were grown in water-propanol-2 solutions containing high-molecular (Ms=1300000) polyvinylpyrrolidone (PVP) at room temperature using cadmium nitrate and sodium sulfide as the cadmium and sulfur sources, respectively. CdS nanocrystals were perfectly stabilized by PVP molecules without the formation of precipitates, the concentration of CdS was about

10^{-5} M. The band gap values of semiconductor nanoparticles in liquid solutions and composite coatings were obtained by the analysis of optical spectra by using Tauc's equation [13]. From these data was calculated the diameter of quantum dots. A field-emission scanning electron microscope was employed to examine the size and distribution of the resultant CdS nanocrystals in the PVP matrix.
Z-scan is an effective technique to investigate nonlinear absorption and nonlinear refraction. The experiment was performed by method, described in [10]. Nanoparticles were excited at the wavelength of the second harmonic of YAG:Nd^{3+} laser (λ_{ex} = 532 nm), τ_{pulse} = 5 ns, E = 100 μJ. Gaussian beam focused by a lens with F = 15 cm. The cell with the solution (d = 2 mm) was moved automatically along the main optical axis passing successively to positions before the focal plane (- z) and after it (+ z). The output energy of the cell was measured by a photometer (OPHIR) in two regimes "closed aperture" and "open aperture", that allowed us to separate the nonlinear refraction from the nonlinear scattering and absorption.

3. Results and discussion

The absorption spectra measurements of sols were carried out on a spectrophotometer Lambda 1050 Perkin Elmer (Figure 1).

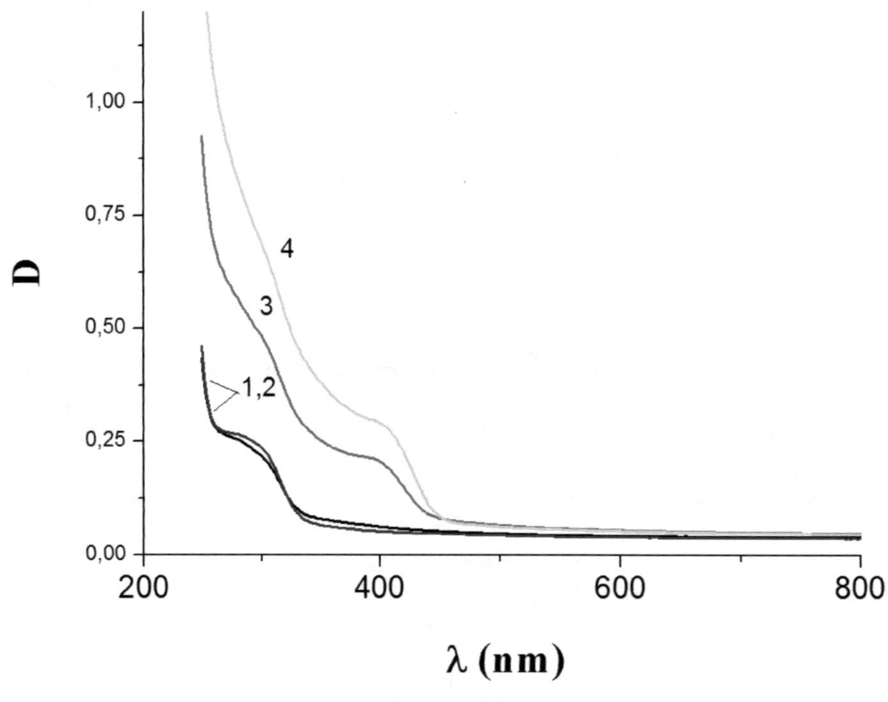

Figure 1. The absorption spectra of CdS/ZnS-PVP sols with different CdS mass concentration: 1 – 0.8 mM, 2 – 1.4 m M, 3 – 2.8 mM, 4 – 4.2·mM.

In Figure 1, the main exciton absorption peak of quantum dots near 400-420 nm is clearly visible, which corresponds to the $1S_{1/2}(e)$-$1S_{3/2}(h)$ transition. The energy of the band gap and the diameters of the quantum dots were calculated in accordance with the Tauc`s equation [13]. The diameters of the synthesized quantum dots were about 2-4 nm. On the QDs spectra 3 and 4 (Figure 1) the bands corresponding to the $1P(e)-1P_{3/2}(h)$ transition with an energy of about 4.1 eV are also resolved. The synthesized sols showed high stability over time.
The position of the laser wavelength λ_{ex} = 532 nm corresponds to the transparency region of the investigated sols, but at high intensities we have the increased transitions probability with a doubled

photon energy is 4.6-4.7 eV. Nonlinear absorption, self-focusing and defocusing, generation of the third harmonic, etc. are described by the cubic nonlinear susceptibility tensor $\chi^{\langle 3 \rangle}$, it is a complex quantity $\chi^{\langle 3 \rangle} = Re\chi^{\langle 3 \rangle} + iIm\chi^{\langle 3 \rangle}$. The $Im\chi^{\langle 3 \rangle}$ is responsible for nonlinear absorption, and the real part describes a nonlinear change in the refractive index γ (γ is nonlinear refraction index that is due to the bound electrons and is related to the usual nonlinear index n2 through n2=cn0 γ /40 π [14]). If we confine ourselves to considering only third-order nonlinearity, then the following relations are satisfied for the coefficients γ (cm^2/W) and β (cm/W):

$$\chi_{Im}^{(3)} = \frac{n_0^2 \varepsilon_0 c^2}{\omega} \beta \quad (1),$$

$$\chi_{Re}^{(3)} = 2 n_0^2 \varepsilon_0 c^2 \gamma \quad (2)[12].$$

Figure 2 shows the experimental dependences of the transmission of laser radiation as a function of the sample z position with respect to the focal plane of the CdS/ZnS-PVP sol with a maximum concentration. The closed aperture-measurements for the sample near the focus plane demonstrates the valley-peak behavior, it characterizes the positive refraction (self-focusing of the radiation).The removal of the aperture showed the contribution of two-photon absorption and scattering.

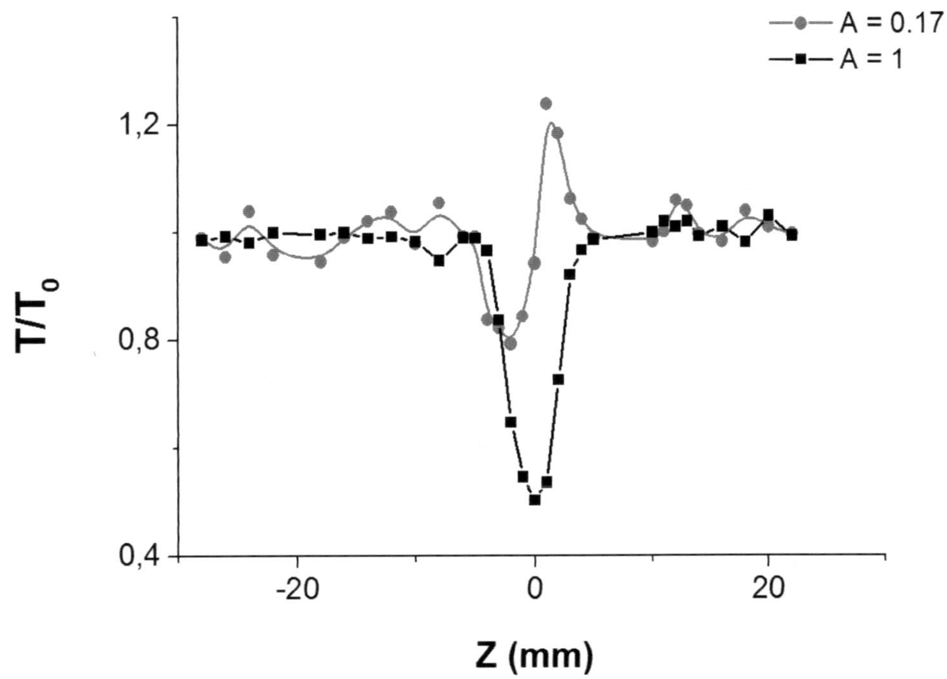

Figure 2. The Z-scan results of CdS/ZnS-HMW-PVP sol (C=8.4·10^{-3} M): squares represent normalized transmittance for the open aperture; dots are data by the closed aperture A≈0.2 (with removed open data). The lines are drown for clarity. Parameters of pulse λ_{ex} = 532 nm, t=5 ns, I = 0.4 GW/cm^2.

The calculated coefficients of two-photon absorption and nonlinear refraction for all available CdS mass concentrations are presented in Table (below). In the case of an open aperture, the relative error was 4.8%, in the case of a closed aperture, 10%. The main contribution to the discrepancy was determined by the spatial instability of the laser beam. Each point in Fig. 4 is obtained by averaging over 5 mono pulses.

Table. TPA and nonlinear refraction coefficients of CdS/ZnS QDs, stabilized by PVP (I = 0.4 GW/cm^2)

Concentration CdS *(in situ)*, M	T_0, %	β, cm/GW	γ, cm^2/W
$1.7\cdot10^{-3}$	88.2	1.9	-
$2.8\cdot10^{-3}$	86.7	2.9	$3\cdot10^{-14}$
$5.6\cdot10^{-3}$	83.0	4.2	-
$8.4\cdot10^{-3}$	82.8	5.3	$7\cdot10^{-14}$

Taking into account the low excitation energy and the predominance of bound electrons refraction, the positive refraction sign means, that excited states of QDs are connected by two-photon transition with energy close to Eg value. Similar mechanism was confirmed in studies of CdS and other nanoparticles [14,15]. In our opinion, the influence of the molecular environment cannot be ruled out. The positive refraction is described also in [16,17] under the femtosecond laser radiation at wavelengths 750-800 nm, while the γ values deviate both towards higher and lower values depending on the environment matrix (see table in [18]).

The two-photon absorption coefficients β show a well-marked linear dependence on the CdS QDs concentration (Figure 5) (with accuracy of 98%). It is obvious, because of increasing in absorbing centers per unit irradiated area.

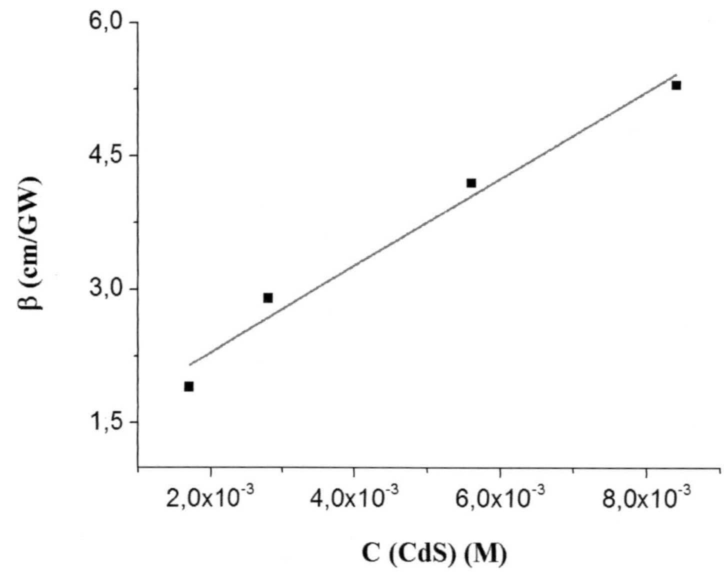

Figure 3. The linear dependence of experimental TPA coefficients β (cm/GW) on CdS mass concentration in high-molecular PVP-sol.

We see, that at the highest CdS concentration, TPA coefficient β of nanoparticles is comparable with that for bulk CdS 5.4 cm/GW [19]. However, the molar concentration Cμ of the bulk material is 10^4 times higher (for bulk CdS Cμ = ρ/M = 33.4 M), which indicates an improvement in the nonlinear properties of nanoparticles compared to a bulk semiconductor. Performed experiments with third detector showed the presence of nonlinear scattering. The quantitative contribution to the transmittance decrease of it in relation to TPA in investigated sols remains discussion.

4. Conclusion

Synthesized highly stable hybridized CdS/ZnS-PVP sols have non-linear absorbing properties exceed considerably the values for the bulk material under the same conditions. Two photon absorption coefficients depend on CdS concentration in direct ratio. The nonlinear refraction mechanism for about 2 nm diameter synthesized quantum dots excited by the 532 nm laser wavelength appeared to be a positive lens. The hybrid CdS/ZnS-PVP sols can be used to create polymer coatings in photoelectronic devises.

Acknowledgments

The work was supported by the project of the Education and Science Ministry of Russian Federation №16.2483.2017/PCH.

References

[1] Venkatram N, Kumar R S S, Rao D N2006 *J. Appl. Phys***100** 074309
[2] Jing C, Xu X, Zhang X, Liu Z, Chu J 2009 *J. Phys. D: Appl. Phys* **42** 075402
[3] Dabbousi B O, Rodriguez-Viejo J, Mikulec F V, Heine J R, Mattoussi H, Ober R, Jensen K F, Bawendi M G, 1997 *J. Phys. Chem. B***101** 9463-9475
[4] Shik A, Ruda H and Sargent E H 2000 *J. Appl. Phys.* **88** 3448
[5] Beecroft L L,Ober C K 1997 *Chem. Mater* **9** 1302
[6] Yu Chen Y, Wei W C J 2001 *J. Eur. Ceram. Soc* **21** 2535
[7] Du T and Ilegbusi O J 2004 *J. Mater. Sci.* **39** 6105
[8] Pattabi M, Saraswathi Amma B, Manzoor K 2007 *Mater. Res. Bull* **42** 828
[9] Evstrop'ev KS, Gatchin Yu A, Evstrop'ev S K, Dukels'skii K V, Kislyakov I M, Pegasova N A, Bagrov I V, 2016 *Opt. Spectrosc.* **120** 434
[10]Sanchez C, Juli´an B, Belleville P, Popall M 2005 *J. Mater. Chem* **15** 3559
[11] Yoshida M, Prasad P N 1996 *Appl. Opt* **35** 1500
[12] Sheik-Bahae M, Said A A, Wei T-H, Hagan D J, Van Stryland E W 1990 *IEEE J. Quantum Electron* **26** 760
[13] Tauc J 1968 *Materials Research Bull* **3** 37
[14] Said A A, Sheik-Bahae M, Hagan D J, Wei T H, Wang J, Young J, Stryland V E W 1992 *J. Opt. Soc. Am. B.* **9** 405
[15] Ganeev R A, Ryasnyansky A I, Tugushev R I, Usmanov T. 2003 *J. Opt. A: Pure Appl. Opt.* **5** 409
[16] He J, Ji W, Ma G H, Tang S H, Kong E S W, Chow S Y, Zhang X H, Hua Z L, Shi J L*J. Phys. Chem. B* 2005 **109** 4373
[17] Gong H M, Wang X H, Du Y M, Wang Q Q*J. Chem. Phys.* 2006 **125** 024707
[18] Szeremeta J, Nyk M, Wawrzynczyk D, Samoc M 2013 *Nanoscale* **5** 2388
[19] Li H P, Kam C H, Lam Y L, Ji W 2001 *Opt. Commun.* **190** 351

IOP Conf. Series: Journal of Physics: Conf. Series **917** (2017) 062045 doi:10.1088/1742-6596/917/6/062045

Graphene plasmonic terahertz detector with high responsivity

K V Mashinsky [1,2], **D V Fateev** [2], **V V Popov** [1,2,3]
[1] National Research Saratov State University, 410012 Saratov, Russia
[2] Kotelnikov Institute of Radio Engineering and Electronics of the Russian Academy of Sciences (Saratov Branch), 410019 Saratov, Russia
[3] Saratov Scientific Center of the Russian Academy of Sciences, 410028 Saratov, Russia

Abstract. Rectification of terahertz radiation in a spatially periodic graphene structure with dual grating gate is theoretically studied. The calculated current responsivity is an order of magnitude greater than the responsivity of conventional plasmonic detector based on the field-effect-transistor array.

1. Introduction

Unique properties of graphene [1] such as high mobility and drift velocity of charge carriers allows for utilizing graphene plasmons for detecting terahertz (THz) radiation in graphene micro- and nanostructures [2]. Due to nonlinear properties of plasma waves in graphene [3] there are two different physical mechanisms of THz plasmonic rectification which are caused by the effects of differential plasmonic drag and electron-hole plasmonic ratchet [4].

2. Results and discussion

Let us consider a graphene monolayer screened by an interdigitated metal dual-grating gate (figure 1). Graphene is located on SiO_2 substrate and separated from the dual-grating gate (DGG) by Al_2O_3 barrier.

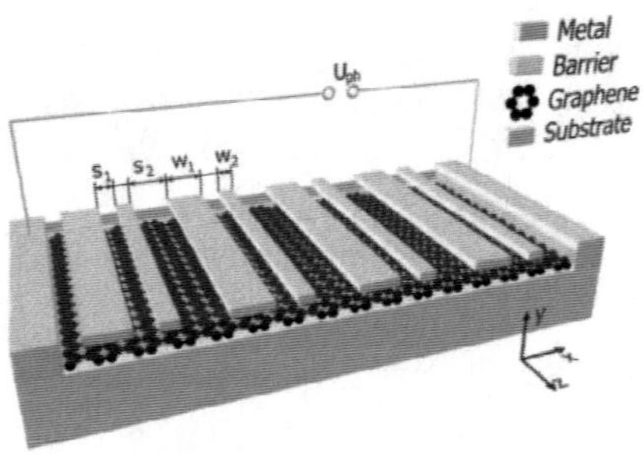

Figure 1. Schematic view of the DGG graphene structure.

Two sub-gratings of the DGG are laterally shifted in respect to each other in order to introduce asymmetry into the unit cell of the periodic DGG-graphene structure. Applying the constant voltage between DGG and graphene allows for controlling the equilibrium carrier density in gated areas of graphene. Type of carriers in gated areas of graphene depends on the gate voltage sign.

Nonlinear dynamics of the charge carriers in graphene is considered in the hydrodynamic approximation [5]. We solved the hydrodynamic equations in the perturbation approach [6] by expanding carrier velocity, Fermi energy and current density in powers of the electric field amplitude and keeping only linear and quadratic terms in this expansion. We used approximations of zero temperature and small carrier hydrodynamic velocity. Plasmon field induced by normally incident THz wave in graphene was calculated by using the self-consistent electromagnetic approach [7].

For the harmonic electric field $E(x,t)=E_{\omega}(x)\exp(-i\omega t)/2+\text{c.c.}$, with ω being the frequency of the incident wave, the total rectified DC current density in spatially periodic graphene can be written as [4]

$$j_0=-\frac{|e|^3 v_{\rm F}^2}{2\hbar^2\pi\omega\gamma\left(\omega^2+\gamma^2\right)}\left\langle \operatorname{sgn}(e)\operatorname{Re}\left(\left(\omega-i\frac{\gamma}{2}\right)E_{\omega}\frac{\partial\left(E_{\omega}\right)^*}{\partial x}\right)\right\rangle, \tag{1}$$

where $|e|$ is the elemental charge ($e<0$ for electrons and $e>0$ for holes), γ is the carrier relaxation rate, $v_{\rm F}$ is the Fermi velocity of charge carriers in graphene, and angular brackets denote the space averaging.

We assume that, in the graphene regions under the metal strips of width w_2, there is a conductivity of one (for example, hole) type while the conductivity of another (electronic) type is in the rest of the unit cell. Such a spatial distribution of carriers in graphene is described by the distribution of the equilibrium Fermi energy $\varepsilon_{{\rm F},w_1}^{(0)}=\varepsilon_{{\rm F},s_1}^{(0)}=\varepsilon_{{\rm F},s_2}^{(0)}=\varepsilon_{\rm F1}$ and $\varepsilon_{{\rm F},w_2}^{(0)}=\varepsilon_{\rm F2}$. For a periodic structure with a given distribution of charge carriers, equation (1) transforms into the form $j_0=j_0^{dr}+j_0^{rt}$, where

$$j_0^{dr}=\frac{|e|^3 v_{\rm F}^2}{\hbar^2\pi\omega\left(\omega^2+\gamma^2\right)}\left(\operatorname{sgn}(\varepsilon_{\rm F1})\left(1-\frac{w_2}{l}\right)+\operatorname{sgn}(\varepsilon_{\rm F2})\frac{w_2}{l}\right)\sum_p q_p\left|E_{x,\omega,p}\right|^2, \tag{2}$$

$$\begin{aligned} j_0^{rt}=&\frac{|e|^3 v_{\rm F}^2}{\hbar^2\pi\omega\left(\omega^2+\gamma^2\right)}\frac{4}{\gamma l}\left(\operatorname{sgn}(\varepsilon_{\rm F2})-\operatorname{sgn}(\varepsilon_{\rm F1})\right)\\ &\times\operatorname{Re}\left[\left(i\omega+\frac{\gamma}{2}\right)\sum_p\sum_{p'\neq p}E_{x,\omega,p}E_{x,\omega,p'}^*\frac{p'}{p-p'}\sin\left(\pi(p-p')\frac{w_2}{l}\right)\exp\left(i\pi(p-p')\frac{s_1-s_2}{l}\right)\right], \end{aligned} \tag{3}$$

where the expansion of the electric field in the Fourier space series $E_{x,\omega}(x)=\sum_p E_{x,\omega,p}\exp\left(iq_p x\right)$ is used, $q_p=2\pi p/l$, and p,p' are integer. Rectification of THz radiation in a periodic graphene structure can occur due to two different physical mechanisms [4]: the effect of differential plasmon drag of charge carriers in graphene described by equation (2) and the effect of the plasmon electron-hole ratchet described by equation (3).

The efficiency of rectification of THz radiation due to the effect of the nonlinear plasmonic effects is affected by both the geometric asymmetry of the graphene structure and by the interaction of the spatial harmonics of the electric field of different orders. Effective interaction of spatial harmonics of the electric field of different orders becomes stronger when “hybrid” plasmonic resonances are excited in the structure due to interaction of plasmon modes in different plasmon resonators in the unit cell of the graphene structure [7, 8]. In addition to the "radiative" plasmon modes in spatially symmetrical graphene structure, the "nonradiative" plasmon modes (having zero net dipole moment in spatially symmetrical structures) can be also excited [8] in a spatially asymmetric structure.

The current responsivity of THz rectification is calculated by the expression $R=j_0/Pl$, where $P=1$ W/cm^2 is the incident THz wave power density and $l=1$ mm is the width of graphene structure (along the z-axis).

Plasmon rectification of THz radiation has been investigated for a graphene structure with a double grating gate (Figure 1) with parameters $w_1 = 1\ \mu\text{m}$, $w_2 = 0.25\ \mu\text{m}$ and $\gamma = 0.2 \cdot 10^{-12}\ \text{s}^{-1}$ for the distribution of the Fermi energy $\varepsilon_{\text{F},w_1} = \varepsilon_{\text{F},s_1} = \varepsilon_{\text{F},s_2} = -\varepsilon_{\text{F},w_2} = 39\ \text{meV}$. The widths of the gaps between the gate electrodes s_1 and s_2 vary, while their sum remains the same, $s_1 + s_2 = 0.375\mu\text{m}$. The electric fields of plasmons excited by the incident THz wave and the absorbance of the graphene structure are calculated in the framework of a self-consistent electromagnetic approach based on the integral equation method [7]. The absorption spectra and current responsivities are calculated for different coefficients of the structure asymmetry $K = 1 - \frac{s_1}{s_2}$, $0 < K < 1$.

Figure 2 demonstrates the calculated absorption and current responsivities of DGG graphene structure with the Fermi energy distribution of charge carriers in graphene unit cell as. The resonant

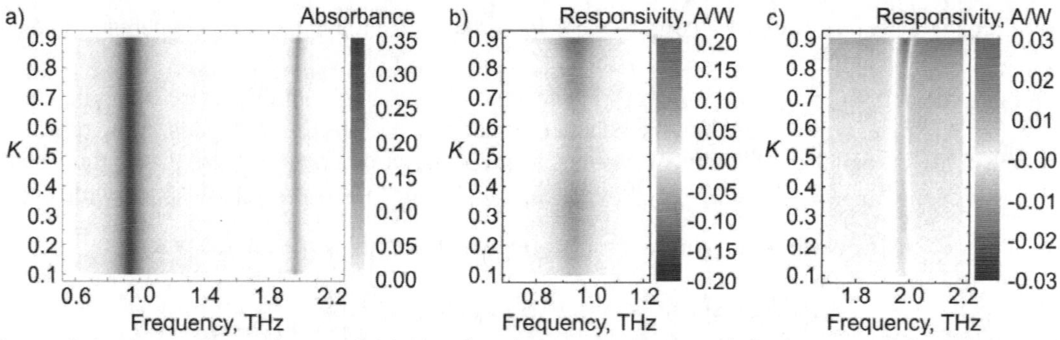

Figure 2 (a, b, c). Dependence of **(a)** the absorbance spectrum and **(b,c)** responsivity of the DGG graphene structure on the asymmetry coefficient *K* for the fundamental and second-order plasmon resonances.

peaks correspond to the excitation of gated fundamental and second-order plasmon modes under wider gate electrodes. Each of these plasmon modes weakly interact with the other plasmon modes. In this case, the plasmon fields is almost symmetric [8], and therefore the plasmon rectification effect is determined mainly by the spatial geometric asymmetry of the graphene structure.

The effect of the asymmetry of the electric field on the plasmon rectification is seen in figure 3 for the third and fourth plasmon resonances at 2.4 THz and 2.7 THz. Resonances at these frequencies are the "nonradiative" plasmon mode (not excited in a symmetric structure with $K = 0$) and the "radiative" plasmon mode at higher frequency. As the geometric asymmetry of the structure increases, the influence of the "nonradiative" plasmon mode increases (figure 3), which leads to the increase in the asymmetry of the electric field at the hybrid resonance and to a significant increase in the rectified current (figure 3(b)).

At the hybrid resonances in the frequency range 3.1-3.5 THz (figure 4), the maximum rectified current arises for the asymmetry coefficient 0.4 (figure 4(b)). The "nonradiative" mode (with a higher frequency) in the weakly asymmetric structure has a small dipole moment and therefore weakly interacts with the other modes and weakly excited by the incident THz wave (figure 4). As the asymmetry of the structure increases, the dipole moment of the "nonradiative" mode increases. As a result, this mode can be efficiently excited by the incident THz wave. In this case, the "nonradiative" mode becomes optically active and the effect of anti-crossing of the "radiative" and "nonradiative" plasmonic modes occurs (figure 4). For a strong asymmetry of the structure, the interaction of these modes decreases, as a result of which the asymmetry of the spatial distribution of the field decreases, which leads to decreasing rectified current (figure 4(b)). The greatest rectification effect is achieved in

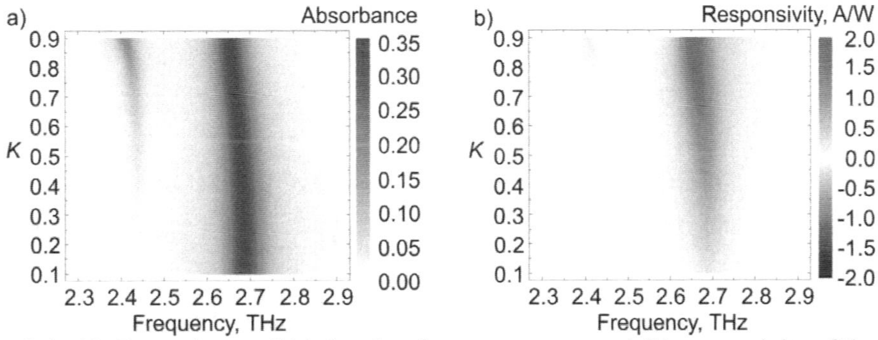

Figure 3 (a, b). Dependence of **(a)** the absorbance spectrum and **(b)** responsivity of the DGG graphene structure on the asymmetry coefficient *K* for the third- and fourth-order plasmon resonances.

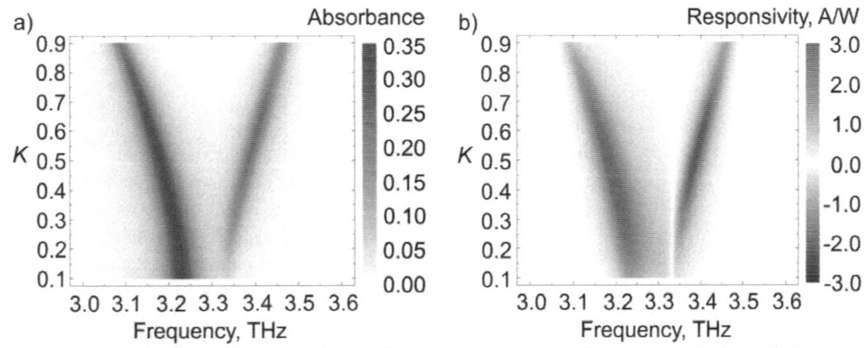

Figure 4 (a, b). Dependence of **(a)** the absorption spectrum and **(b)** responsivity of the DGG graphene structure on asymmetry coefficient *K* for the fifth- and sixth-order plasmon resonances.

the case when the "radiative" and "nonradiative" modes effectively interact (figure 4), leading to the formation of a hybrid mode with a Fano-type resonance.

In conclusion, in this work we have investigated the asymmetry dependence of the current responsivity of plasmonic DGG graphene structure. It is shown that the responsivity of THz rectification of hybrid plasmon modes is an order of magnitude higher than for weakly interacting plasmonic modes. Due to high current responsivity, the DGG graphene structure can be used as an effective THz resonant detector.

Acknowledgements
This work was supported by the Russian Foundation for Basic Research (Grant Nos. 15-02-02989 and 17-52-53063) and the Grant of the President of the Russian Federation for Young Scientist (Grant No. MK-5447.2016.2).

References

[1] Geim A K and Novoselov K S 2007 *Nature Mater.* **6** 183–91
[2] Olbrich P *et al.* 2016 *Phys. Rev. B* **93** 075422
[3] Tomadin A, Tredicucci A, Pellegrini V, Vitiello M S and Polini M 2013 *Appl. Phys. Lett.* **103** 211120
[4] Fateev D V, Mashinsky K V and Popov V V 2017 *Appl. Phys. Lett.* **110** 061106
[5] Rudin S 2011 *Int. J. High Speed Electron. and Systems* **20** 567–81
[6] Aizin G R, Fateev D V, Tsymbalov G M and Popov V V 2007 *Appl. Phys. Lett.* **91** 163507
[7] Fateev D V, Popov V V and Shur M S 2010 *Semiconductors* **44** 1406–13
[8] Popov V V, Fateev D V, Ivchenko E L and Ganichev S D 2015 *Phys. Rev. B* **91** 235436

IOP Conf. Series: Journal of Physics: Conf. Series **917** (2017) 062046 doi:10.1088/1742-6596/917/6/062046

Modulation spectra of volume holograms recorded in additively colored fluorite crystals

V N Borisov[1], A E Angervaks[2], A I Ryskin[2], and A V Veniaminov[1]
[1]Department of Photonics and Optical Information Technologies, ITMO University, Saint Petersburg 197101, Russia
[2]Department of Optical Information Technologies and Materials, ITMO University, Saint Petersburg 197101, Russia
E-mail: `borisov.itmo@gmail.com`

Abstract. Additively colored fluorite crystals are of considerable interest as a holographic recording medium. Due to very high stability as well as potentially large refractive index modulation, its possible to use holograms recorded in these crystals as high quality diffractive elements and photonic-crystalline structures. Therefore, it is important to be able to correctly and accurately analyze such structures. Spectral dependencies of absorption coefficient and refractive index profiles are the most precious characteristics in this regard. Unfortunately, they are also the most elusive ones. In this study the required parameters were obtained by a method based on the analysis of selectivity hypercontours (combined spectral and angular dependencies of diffraction efficiency) using the Kogelnik's coupled wave theory in wide optical spectral range. Hologram profile was reconstructed from first three spatial harmonics. The results were crosschecked with monochromatic analysis to prove the validity of proposed technique.

1. Introduction

In the modern world holography has serious relevance in terms of its applications. Volume holograms can be used as narrowband spectral selectors, holographic prisms, beam combiners, holographic memory elements, and, if provided refractive index modulation is sufficient, as photonic-crystalline structures. A holographic medium for such applications must possess certain qualities. First of all, it should be stable enough to remain for a long time and persist a strong energy impact. Also, it must have the variability on the modulation of the refractive index along the spectrum. It should be also possible to make a sample thick enough for low-frequency applications. Finally, in the case of large physical thickness that ensures really high selectivity, moderation of refractive index modulation may be sufficient (about 10^{-5} for narrowband spectral needs and from 10^{-3} for photonic crystals production).

Additively colored crystals of calcium fluoride (CaF_2), in which holograms are recorded through phototransformation and photoinduced displacement of color centers, matches those requirements because of very large thickness and extreme stability. The ability of tuning refractive index and absorbance modulation by means of nondestructive post-exposure conversion of the centers makes CaF_2 one of the most promising photosensitive media. Moreover, last studies indicate the possibility of forming metal holographic structure in CaF_2 if concentration of color centers is sufficient, thus pointing out the chance to achieve a significant modulation of refraction [1].

IOP Conf. Series: Journal of Physics: Conf. Series **917** (2017) 062046 doi:10.1088/1742-6596/917/6/062046

Certainly, careful and accurate analysis of hologram properties is necessary for such high-precision applications. From this standpoint, the hologram profiles - spatial distributions of its optical parameters across the grating are of significant value. The profiles can be obtained by means of optical microscopy or reconstructed basing on coupled-wave analysis of angular selectivity contours [2, 3]. Despite obtaining hologram profiles for a single wavelength is not tough, same task for wide spectral range, which is so necessary for the completeness of the study, has not yet acquired an optimal realization [4, 5]. This work describes the technique capable of obtaining rather complete angular-spectral-spatial dependencies of volume holograms parameters.

2. Methods

2.1. Coupled Wave Theory Application

Since the aim of the work is to obtain information on the profiles of holograms, it is necessary to choose the correct method for their analysis. Coupled wave theory approximation approach has proved its accuracy and reliability in volume hologram analysis as it provides inartificial access to both amplitude and phase modulation components as well as phase shift between them [6, 7]. Coupled wave theory calculates angular and spectral diffraction response from hologram properties. Therefore, it can be applied to determine hologram properties from spectral and angular selectivity contours by means of an approximation procedure.

The first mathematical expressions of coupled wave theory were presented by Hervig Kogelnik in 1969 [8], and they were developed rigorously since then [9, 10]. The amplitude-phase nature of the hologram is revealed in the Borrmann effect [11] (zeroth order diffraction enlargement at the Bragg angle for pure amplitude holograms, and shifted with respect to the Bragg angle - for mixed ones [12, 13]) which can be best seen on model angular selectivity contours (figure 1).

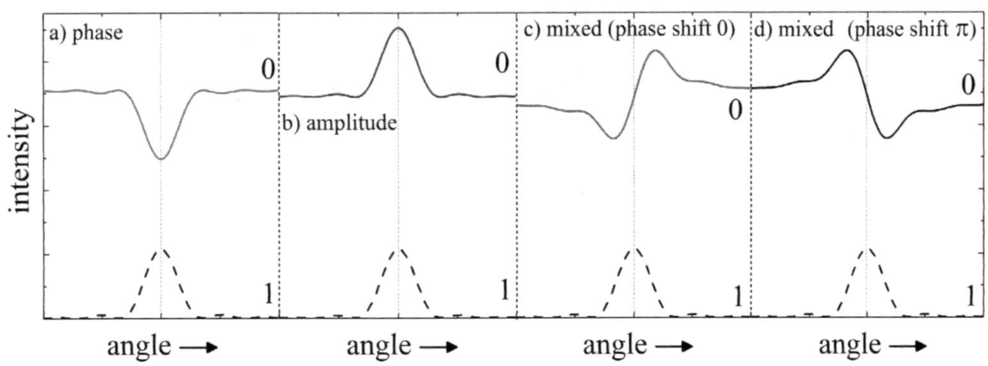

Figure 1. The shape of all simulated 1st order (dashed) selectivity contours looks the same; information about amplitude-phase nature of hologram can be extracted from 0th order (solid).

2.2. Contours Measurement Technique

The main experimental technique broadly used for recording angular selectivity of volume holograms is generally based on their rotation with respect to a collimated light beam [6, 12, 13, 14, 15] and broadly used for last decades. Sufficiently low beam divergence of about 0.1 mrad can be achieved by means of a collimator and a coherent light source. To study spectral selectivity, either wavelength-tunable or broadband source with a monochromator or spectrum analyzer become necessary. Tunable or broadband lasers (such as supercontinuum generators) are not always affordable, while proper collimation of non-laser white-light source may be challenging.

Reading a grating with a beam whose divergence exceeds the grating's angular selectivity requires complicated deconvolution procedure at the stage of data analysis, which makes the approach fully impracticable. Therefore, despite dramatic loss of light energy, proper collimation of incoherent source seems to be the most digestible option. Since key information on hologram properties can be extracted from zero order diffraction, it is sufficient to measure only the transmitted light intensity.

2.3. Experimental Realization

Hologram in question was recorded with CW DPSS Nd:YAG laser (532 nm) in an additively colored fluorite (CaF_2) crystal at 190^oC with spatial period of 4.55 μm. For measuring both spectral and angular selectivity contours a bright white-light LED (CREE XM-L T6) was employed as a light source. Its light was collimated with two sequential telescopic beam expanders to bring the divergence down to about 0.1 mrad. The beam passed through the hologram under incidence angle varied with a rotation stage. Transmittance spectra were recorded at each angular step with a matrix spectrometer (Ocean Optics USB4000-UV-VIS) and, divided by LED emission spectrum, finally were combined into 3D surfaces of zeros order diffraction efficiency plotted vs. angle and wavelength - spectral-angular (spectrangular) selectivity contour (hypercontour) (figure 2).

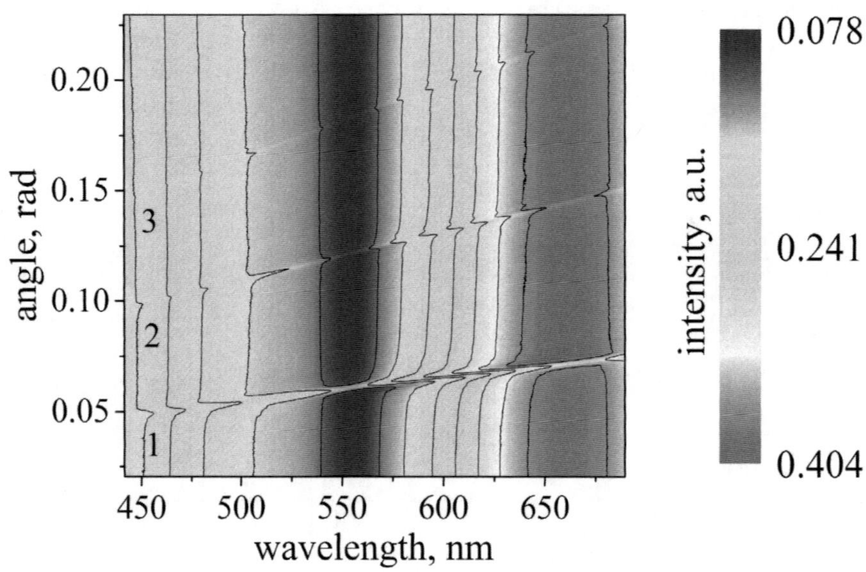

Figure 2. Three "valleys" with lifted edges at hologram zero order diffraction hypercontour manifest nonlinear phase-amplitude recording.

The hypercontour was approximated by coupled wave theory expressions as a series of angular selectivity contours along the spectrum. In this work we use the formulæ previously discussed in [16]. Angular selectivity contours shape at different wavelengths (figure 3) shows how the hologram appears a mixed one with absorptive and refractive contributions in counterphase (a) and in-phase (c), purely amplitude (b), and almost phase one (d), depending on reading wavelength. The depicted contours almost coincide with the simulated model ones in figure 1.

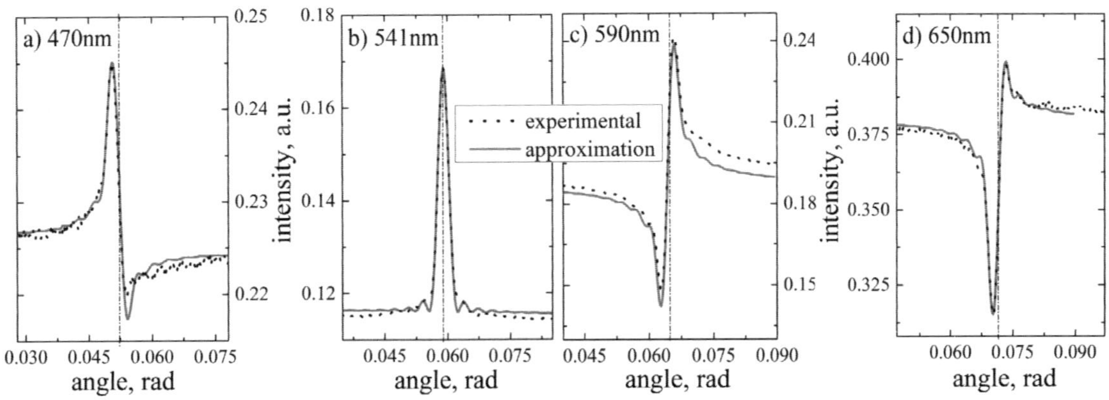

Figure 3. Approximation of angular selectivity contours for the hologram allowed to determine the values of absorption and refraction modulation amplitudes (figure 4).

3. Results

3.1. Hypercontour Analysis

Fitting with the expressions of the coupled wave theory was consistently applied to each angular selectivity contour. Therefore, refractive index and absorption coefficient modulation amplitudes were determined for the 1st spatial harmonic of the hologram along the studied spectrum with 1 nm-step (figure 4).

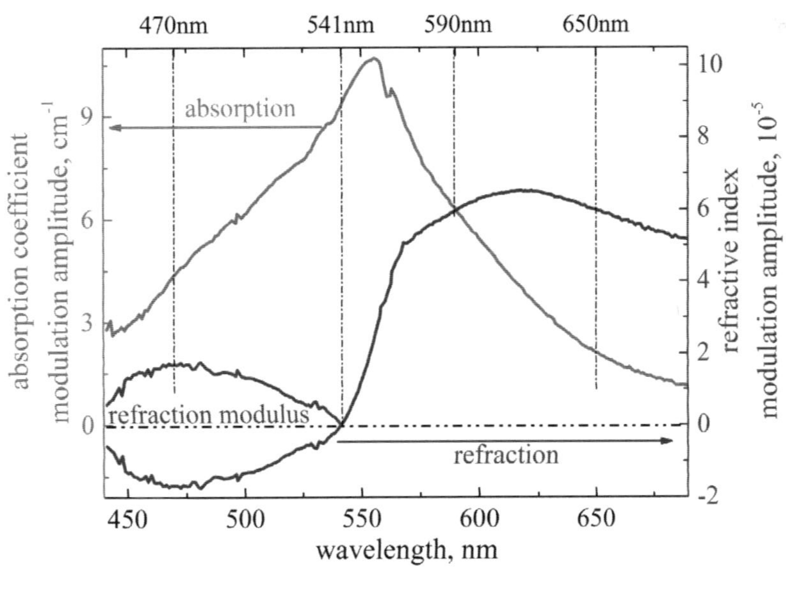

Figure 4. Spectral dependencies of 1st harmonics of absorbance and refraction modulation determined as selectivity hypercontour fit-parameters confirm previous observations in figure 3.

3.2. Profile Reconstruction

Since each diffraction order corresponds to a certain spatial harmonic of a volume hologram, the refractive index and absorption coefficient amplitudes of corresponding harmonics can be obtained by approximation of zero order diffraction contour in the vicinity of the given diffraction order. Therefore, spectra of amplitudes were obtained for the first three spatial harmonics whereupon hologram profile at different wavelengths was reconstructed by partial Fourier synthesis (figure 5).

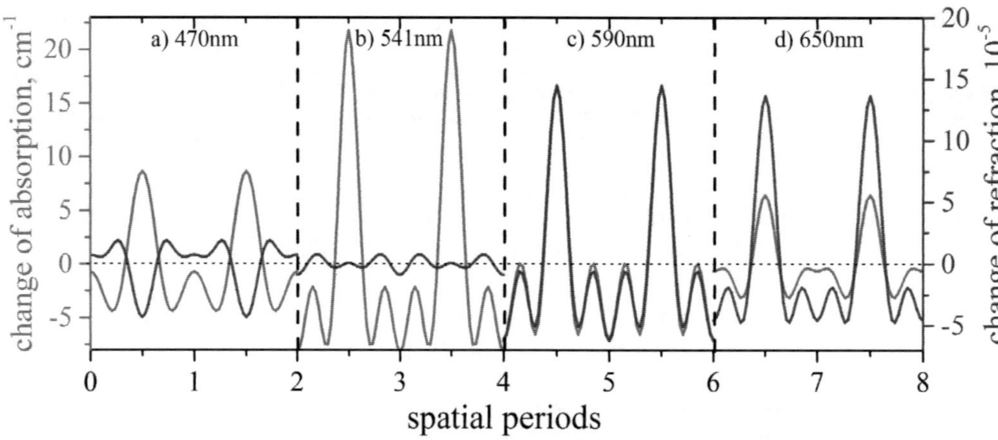

Figure 5. Spatial profiles of hologram absorbance and refraction reconstructed as a sum of three first harmonics demonstrate how the phase and amplitude contributions change with wavelength.

3.3. Results Crosscheck

To confirm validity of the proposed technique, the particular angular selectivity contours were re-measured using laser radiation. For this crosscheck two different laser sources were used: CW DPSS Nd:YAG laser at 532 nm where hologram is almost amplitude and CW He-Ne laser at 633 nm where hologram looks an in-phase mixed one with predominant phase contribution.

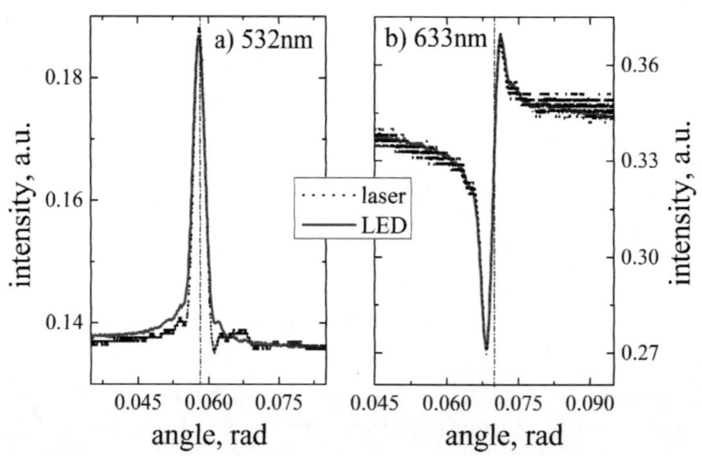

Figure 6. The coincidence of selectivity contours measured with monochromatic (laser) light and obtained with white light (LED) source confirms the validity of proposed technique.

4. Conclusions

In the course the study, experimental setup and approximation algorithm for automated acquisition and analysis of volume hologram selectivity hypercontours were designed. An exemplary transmission hologram in CaF_2 was analyzed by means of this algorithm, so that the spectral distributions of the parameters were obtained. The hologram demonstrated change of phase-amplitude nature along the spectrum as well as evidence of nonlinear recording which was revealed in the appearance of higher diffraction orders. Series of hologram profiles with wavelength scan were reconstructed from first three spatial harmonics of the hologram.

As an approval of the method, crosscheck measurements of monochromatic angular selectivity contours were made by means of laser sources at two different wavelengths. Therefore, the proposed spectral-angular (spectrangular) technique allows fast and accurate determination of spectral characteristics of materials and holographic structures recorded in them. The method is a promising tool for analysis of elementary holograms, and it deserves further development to be applied to more complex holographic structures, such as holographic photonic crystals.

Acknowledgments

This work was partially financially supported by the Government of the Russian Federation (grant 074-U01).

References

[1] Ryskin A I, Angervaks A E and Veniaminov A V 2017 *Holographic Materials and Optical Systems* (InTech) chap 18, pp 405–33
[2] Sabel T, Orlic S, Pfeiffer K, Ostrzinski U and Grützner G 2013 *Optical Materials Express* **3** 329–38
[3] Shcheulin A S, Angervaks A E, Zakharov V V, Veniaminov A V and Ryskin A I 2014 *Journal of the Optical Society of America B* **31** 248–54
[4] Voit K M and Imlau M 2013 *Materials* **6** 334–58
[5] Vojtísek P, Kveton M and Richter I 2016 *Journal of the European Optical Society* **11** 16009(11)
[6] Fally M, Ellabban M and Drevenšek-Olenik I 2008 *Optics Express* **16** 6528–36
[7] Fally M, Ellabban M and Drevenšek-Olenik I 2009 *Optics Express* **17** 23350
[8] Kogelnik H 1969 *Bell System Technical Journal* **48** 2909–47
[9] Moharam M and Gaylord T 1983 *Journal of the Optical Society of America* **73** 451–5
[10] Neipp C, Francés J, Gallego S, Bleda S, Martínez F J, Pascual I and Beléndez A 2014 *The Scientific World Journal* **2014** 513734(7)
[11] Russell P S J and Solymar L 1980 *Applied Physics* **22** 335–53
[12] Carretero L, Madrigal R, Fimia A, Blaya S and Beléndez A 2001 *Optics Letters* **26** 786–8
[13] Neipp C, Pascual I and Beléndez A 2002 *Optics Express* **10** 1374–83
[14] Alekseev-Popov A, Knyazkov A and Saikin A 1983 *Soviet Technical Physics Letters* **9** 475–7
[15] Borisov V and Veniaminov A 2015 *Proc. SPIE* vol 9545 p 954513(7)
[16] Borisov V, Klepinina M, Veniaminov A, Angervaks A, Shcheulin A and Ryskin A 2016 *Proc. SPIE* vol 9885 p 98851(9)

Phase shifting profilometry with optical vortices

B Sokolenko[1,2], D Poletaev[2], S Halilov[2]

[1]Department of Science and Research, V.I. Vernadsky Crimean Federal University. Simferopol, 295007, Russia
[2]Institute of Physics and Technology, V.I. Vernadsky Crimean Federal University. Simferopol, 295007, Russia

Abstract. In this work we review principles and applications of a method of phase shifting profilometry with using of optical vortices imbedded into the probe beam. High spatial resolution caused by vortex phase sensitivity is analysable to retrieve the 2D and 3D shape of optically transparent and reflecting surfaces with exceeding of optical diffraction limit. This method applicable for non-destructive testing of thin films, live cells and biological tissues in real-time regime. Automatic processing of vortex interferograms with vortex phase shift analysis allow to achieve a vertical resolution down to 1,75 nm.

1. Introduction

The field of singular optics and its applications is a relatively new branch of development and investigation, but it already is very active and significant research effort has been invested in its physical implications and possible applications, especially in microscopy and digital imaging.

The diffraction limit, discovered in 1873 by Ernst Abbe, is the smallest possible size of a light spot that can be obtained by focusing electromagnetic radiation of a given wavelength in a medium with a refractive index n. For more than a century this limit of spatial resolution was an insurmountable obstacle to the improvement of the technique of optical imaging in the far field. Overcoming the diffraction limit is the topic of many fundamental and applied studies of modern optics. High interest is due to the fact that the field of application of super resolution is not limited to the improvement of image quality, but is widely used to recording on optical media, in lithography and nano-structuring, optical manipulation [1-4].

In the present paper we focus our attention on the interferometric analysis of phase evolution of structured light and its application to profilometry. Interferometry is used for control of micro-relief and defects of scattering surfaces, roughness of solid bodies like microchip subtract and thin films [5]. This technique is able to perform measurements of thickness and surface topology without direct contact with an object in non-destructive way with high accuracy. The phase shifting interferometry is widely used and actively developing method significantly increased prevalence of interferometric measurements in research and production applications due to its high noise sustainability, especially with using of low-coherence light sources where measurements are based on analysis of oscillating light intensity envelope during the interference process which is described by function of time and special coherence [6]. The field of special interest is using of low-coherence differential interferometry for diagnostics of stepped structures by 3D-profilometry, optical tomography of semi-transparent biological objects and other applications because it allows to measure high difference (thickness) of the sample which is exceeding $\lambda/2$ [7].

Using of interferometric methods in coupling with computer processing leads to a new opportunities in interferometry of 3D reconstruction, measurement and visualisation of structural and morphological features of studying complex surfaces at micro- and nanoscale [8].

Last researches in field of nanotechnology and nano engineering gave rise to the corresponding control and testing equipment which is used in fundamental research and optical device production [9]. Intensive developing of this brunches need a new technique for fast and low-error measurements, therefore optical methods and technical apparatus requires further improvement and attraction of novel physical mechanisms [10-11].

2. Phase shifting profilometry with optical vortices

Methods of obtaining and decoding of interferograms have been widely used in interferometry systems construction in recent years due to the efficiency of phase-sampling. Phase shifting interferometry for fast and non-destructive analysis is extensively adopted in optical sensing and metrology [5]. This method is based on registration of several interference patterns while the reference beam phase was changed on certain values. In this technique the main task in phase recovering procedure remains a definition of phase difference between interfering wave fronts by the values of registered intensity. For this purposes we need no less than a three interferograms for each point of imaging object. Numerous papers considered methods with using of interferograms sequence to extract the phase and surface relief [2-4]. First known algorithms of phase reconstruction used formulas based on three and four phase steps. Increasing of computing performance gave a possibility to use algorithms with a high number of phase shifts: from 15 up to 101 and the objects must remain stationary during the capture of these fringe sequence images. Therefore, any motion of the object could introduce phase error and thus measurement error. In contrast, profilometry with Fourier transform of interference pattern has the advantage of only requiring one single image for phase retrieval, but steel need high quality of interference fringes [9]

Practical application of optical vortex in the vortex scanning optical imaging allows to study, for example, the surface geometry and optical density of the sample by analysis of phase singularity's distortion [10]. It was shown that vortex phase analysis carrier information about sample topology, and depends of the features of incident beam and different aperture systems. This study demonstrates the response of the optical vortex imbedded in focused Gaussian beam to a produce a small phase shift after spreading through the probe and may be used for high-resolution profilometry [11].

In this paper we consider theoretically and experimentally monochromatic linearly polarized along x-axis singular beam with the wavelength λ bearing the single-charged optical vortex (a so-called vortex beam) and transmitting through the isotropic glass plate with complex surface relief, which provokes additional phase transformations. Let us consider first the propagation of the paraxial beam along the z-axis. The transverse E_x component have a wavenumber $k_x = nk_o$, where k_o is a wavenumber in a free space and $n=\sqrt{\varepsilon}$ – refractivity index of a test sample. In the paraxial approximation, we can treat the linearly polarised component as $E = \tilde{E}_x(x,y,z)\exp(-ik_x z)$. Then, the paraxial equations for the complex amplitude $\tilde{E}_x$ is represented in the form:

$$d_x^2\tilde{E}_x + d_y^2\tilde{E}_x - 2ik_x d_z\tilde{E}_x = 0. \quad (1)$$

A particular solution to the paraxial wave equation (1) for the vortex beam can be written in the reference frame x, y, z of the sample:

$$\tilde{E}_x = \left(\frac{x - i\xi y}{w_o\sigma_o}\right) \times \exp\left[-\left(x^2+y^2\right)/w_o^2\sigma_o\right]/\sigma_o, \quad (2)$$

where: $\sigma_o = \frac{1-iz}{z_o}$, $z_o = \frac{k_x w_o^2}{2}$, w_o – is the radius of the beam waist at the plane ($z=0$), $\xi = \pm 1$ – is

IOP Conf. Series: Journal of Physics: Conf. Series **917** (2017) 062047 doi:10.1088/1742-6596/917/6/062047

the vortex topological charge. The vortex position is described by the equation $\mathrm{Re}\,\tilde{E}_x(x,y,z) = \mathrm{Im}\,\tilde{E}_x(x,y,z) = 0$

Numerical calculation of intensity distribution for Gaussian envelope (according to the second term in equation (2) and intensity of whole vortex beam with typical axial minima are shown on figure 1 (a, b). The phase pattern on figure 1 (c) depicts vortex phase as a helix.

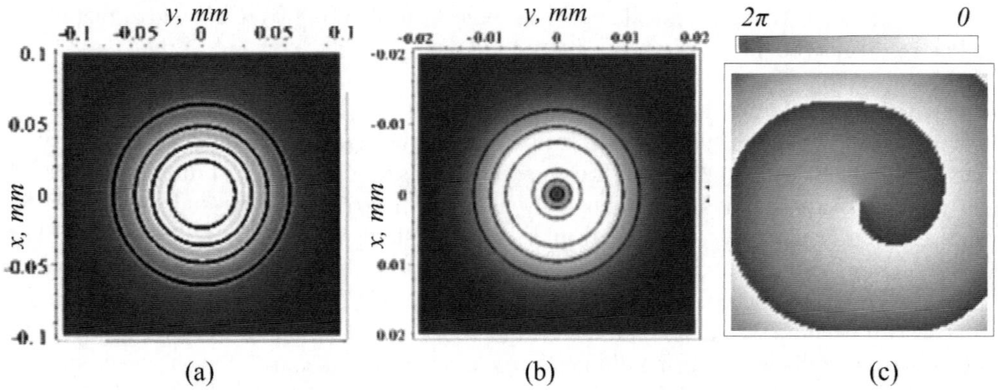

Figure 1. Intensity distribution of Gaussian beam (a) and vortex beam (b) with topological charge $\xi = +1$ and its phase profile (c). Other beam parameters are: $\omega_o = 30\ mcm$, $n_x = 1.54$, $z = 20\,mm$, $\lambda = 632.8\ nm$.

Singular beam phase is extremely sensitive to changes of surface and material parameters, like thickness and refractivity index, therefore this phenomenon can be exploited for metrology with high sensitivity which allow to study different physical processes at the nanoscale [12]. Let us consider a case when singular beam propagates through the isotropic plate with a stepped profile so that the reference (zero-order) plane coincides with the lower edge of the surface as shown in figure 2 (a), whereas due to the optical path difference from lower to height surfaces we can observe a phase shift [13]. The singular beam with helical wavefront has a unique feature – the spiral interference pattern and its rotation (illustrated in figure 2 (b, c) can be easy recovered with computer processing in real-time regime.

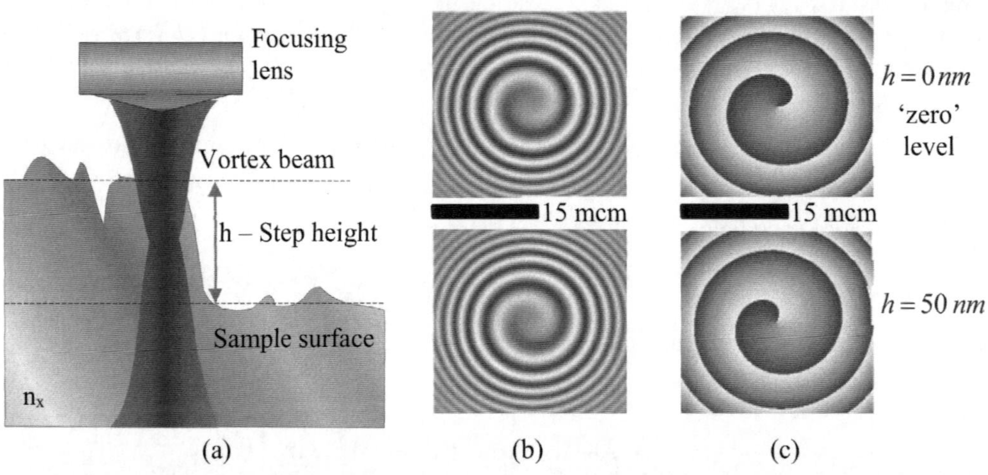

Figure 2. Qualitative sketch of singular beam with topological charge $\xi = +1$ spreading through the quartz plate with step profile (a) of height h. Intensity distribution (b) of interference between Gaussian and singular beams after a step with $h = 50\,nm$ in comparison to the zero level $h = 0\,nm$ and its phase profiles respectively (c); $n_x = 1.54$, $z = 20\,mm$, $\lambda = 632.8\ nm$.

IOP Conf. Series: Journal of Physics: Conf. Series **917** (2017) 062047 doi:10.1088/1742-6596/917/6/062047

3. Experimental results and its discussion

We have experimentally considered the propagation of a paraxial monochromatic vortex beam through the isotropic plate with complex surface relief. En experimental set-up with specially designed interferometer is illustrated in figure 3. Linearly polarized Gaussian beam with a waist radius $w_0 \approx 0.02\ mm$ at the wavelength $\lambda = 0.6328$ μm acquires an optical vortex with the topological charges $\xi = \pm 1$ after passing through the optical wedge. Then the vortex beam is launched into the sample. The 60× micro-objective projected the central beam region at the output face of the plate onto the input pupil of the CCD camera.

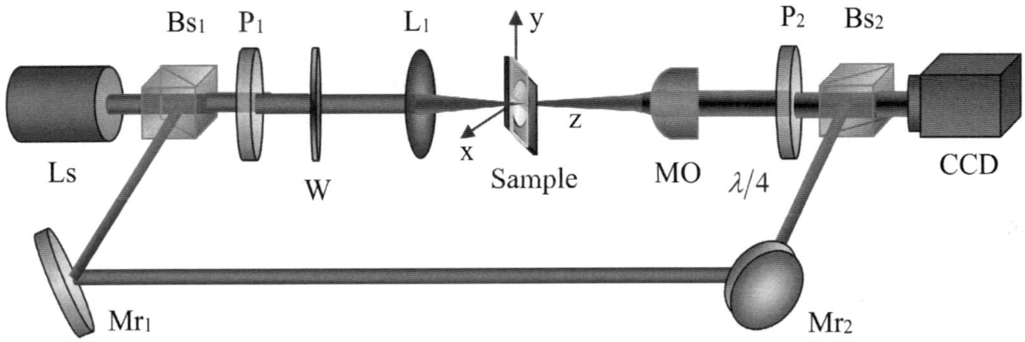

Figure 3. Scheme of experimental set-up: Ls – He-Ne laser source, Bs_1, Bs_2 – beam splitters, P_1, P_2 – polarisers, L_1 – focusing lens with focal length 2 *cm*, MO- microobjective 60×, Mr_1, Mr_2 – mirrors, W – optical wedge and CCD – recording camera.

Due to the high sensitivity of singular beam phase to the small distortions of the wavefront, the interference of the reference beam and the vortex beam (Laguerre-Gaussian beam) transmitted through the isotropic thin plate coated with a wedge-shaped layer of $0 \div 500\ nm$ thickness makes possible to analyse rotation of the spiral phase, depending on the thickness of applied layer, as shown on figure 4.

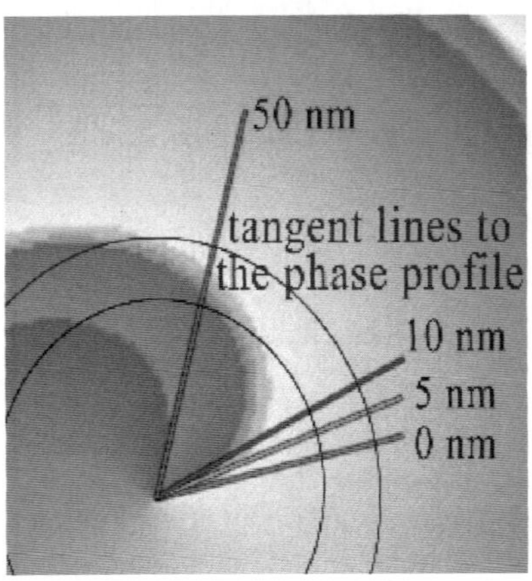

Figure 4. Overlay of four vortex phase profiles received during the passing through an isotropic plate with complex profile. Tangent lines from central region to the periphery of phase profile shows a rotational angle for different optical paths within observing sample: the maximal distinguished experimental resolution for current set-up configuration is 5 nm.

To enable rapid implementation and universal phase recovery method of vortex beam with images from the CCD camera, we used the method proposed in the works [14, 15]. For the comparison of object beam with the reference one the Mach-Zehnder interferometer was used (Figure 3). The sharp thick curves with a characteristic "spiral", corresponding to optical vortices were imaged by the camera and assayed. Total phase shift which is observable due to angular rotation of interference spiral can be calculated from simple equations: $2\pi \sim \lambda, \Delta\varphi \sim n(\mathrm{d}_1 - d_2) = \mathrm{nh}$, where $\mathrm{d}_1 - d_2$ is a difference between observable and neighboring levels of sample surface, n – is refractivity index of sample material.

4. Conclusion

Overcoming the diffraction limit is the topic of many fundamental and applied studies of modern optics. High interest is due to the fact that the field of application of nanoscale resolution is not limited to the improvement of image quality, but is widely used to recording on optical media, in lithography and nano-structuring and optical manipulation. The field of special interest is using of interferometry for diagnostics of stepped structures by 3D-profilometry, optical tomography of semi-transparent biological objects and other applications because it allows to measure thickness of the sample which is exceeding diffraction limits.

We have theoretically and experimentally considered evaluation of phase sensitivity and have shown that the distinguishable spiral phase rotation occurs at the isotropic plate thicknesses equals to $\lambda/130$, where λ – is a wavelength. Proposed technique may be applied to optically transparent and reflecting surfaces exceed optical diffraction limit. Experimental measurement accuracy was reached up to 5 nm and may be improved by using of optical system with short focal distances and blue-light laser sources.

Moreover, this method applicable for non-destructive testing of live cells and biological tissues in real-time regime. Automatic processing of vortex spiral interferograms allow to achieve a vertical resolution down to 1,75 nm for visible light with $\lambda = 310\,nm$ and longitudinal resolution down to 7 nm.

Acknowledgments

The authors thank to Agnieszka Popiołek-Masajada, Prof. Jan Masajada and research group of Faculty of Fundamental Problems of Technology, Wroclaw University of Science and Technology for formulation of the problem. The authors gratefully acknowledge financial support according to V.I. Vernadsky Crimean Federal University developing program in framework of GSU AMMUR 2016 project: academic mobility of young scientists of Russian Federation. Basic experimental research was performed at Laboratory of Photonics ITMO University, Saint-Petersburg, Russia. The authors are grateful to Nikolay V. Petrov for his supervision, useful discussions and for help with the experiments.

References

[1] Zhenrong Z, Jing Z, Peifu G 2010 Optica Applicata **40** 4, 811–818
[2] Nevliudov I Sh, Vtorov E P, Tokarev V V 1998 Messenger of the Kharkov economic university
[3] Nevliudov I Sh, Filipenko A I 2004 News of Engineering Sciences Academy of Ukraine: Mechanical engineering and progressive technologies **4** 24, 81–87
[4] Lukin E S, Anufrieva E V, Popova N A, Morozov B A 2009 Functional Ceramics, 35–40.
[5] Tian C, Liu S 2016 Optics Express Vol. **24**, Issue 16 18695
[6] Xing S, Guo H 2017 Applied Optics Vol. **56**, Issue 6 1591
[7] Pham Q D, Hayasaki Y 2015 Applied Optics Vol. **54**, Issue 1 A39
[8] Tychinsky V P 1989 Opt. Commun **74** 41
[9] Masajada J, Leniec M, Jankowska E, Thienpont H, Ottevaere H, Gomez V 2008 Opt. Express **16**
[10] Augustyniak I, Popiołek-Masajada A, Masajada J, Drobczynski S 2012 Appl. Opt. **51**, C117–C124
[11] Popiolek-Masajada A, Sokolenko B, Augustyniak I, Masajada J, Khoroshun A and Bacia M 2014 Optics and Lasers in Engineering **55** 105
[12] Plociniczak Ł, Popiołek-Masajada A, Szatkowski M, Wojnowski D 2016 Opt. Las. Technol. **81**, 127–136.
[13] Plociniczak Ł, Popiołek-Masajada A, Masajada J, Szatkowski M 2016 Appl. Opt. **55**, B20–B27
[14] Takeda T M, Ina H, Kobayashi S 1982 J. Opt. Soc. Am. **72** 156
[15] Takeda M 1990 *Spatial-carrier Fringe Pattern Analysis and Its Applications to Precision Interferometry and Profilometry: An Overview* (Industrial Metrology) p 79

IOP Conf. Series: Journal of Physics: Conf. Series **917** (2017) 062048 doi:10.1088/1742-6596/917/6/062048

Suppression of slow decaying emission in II-VI quantum dots with Förster resonance energy transfer

T N Mikhailov[1,2*], E A Evropeytsev[1], K G Belyaev[1], V H Kaibyshev[1], A A Toropov[1], A V Rodina[1], S V Ivanov[1], G Pozina[3], and T V Shubina[1]
[1]Ioffe Institute, 26 Politekhnicheskaya, St Petersburg 194021, Russia
[2]St Petersburg Academic University of RAS, Russia
[3]Linköping University, S-581 83 Linköping, Sweden

Abstract. We report on time-resolved photoluminescence studies of Förster resonance energy transfer (FRET) in structures with two arrays of epitaxial Cd(Zn)Se quantum dots (QDs) of different sizes separated by the ZnSe barrier of a variable width. The acceleration of recombination rate of both fast and slowly decaying components of emission from the energy-donating small QDs with the decrease of the barrier width is well consistent with the FRET mechanism. The found Förster radii turn out to be different for the fast and slow components. The rate acceleration is accompanied by the strong suppression of the slow emission component related, presumably, to the dark excitons. These findings open a way to control the characteristic of QD-based devices.

1. Introduction

Epitaxial Cd(Zn)Se/ZnSe quantum dots (QDs) are very promising as an active region of nano-emitters for the visible range. Recently, it has been demonstrated that the Förster resonance energy transfer (FRET) can take place in arrays of the QDs, possessing high lateral densities [1,2]. It is suggested that the FRET occurs from the ground levels of small QDs (so-called “donors”) towards the excited levels of large QDs (“acceptors”) with fast excitation relaxation to their ground levels. This mechanism can enhance the emission of the large QDs by harvesting the energy from a huge number of the small ones. The efficiency of FRET is strongly dependent on the distance between the donor and acceptor. Therefore, in this study we focus on the heterostructures with double insertions of QDs of markedly different sizes. By variation of the barrier width between these insertions we change the average distance between the donors and acceptors. Using time-resolved photoluminescence (TR PL) measurements we investigate how the decrease of the distance effects the characteristic decay times of exciton transitions. We demonstrate that the FRET apparently controls both the fast and slowly decaying components of excitonic emission related, presumably, to bright and dark excitonic states.

2. Basic consideration

The studied systems with two insertions comprise the donors giving energy to neighboring QDs, acceptors taking this energy, and separated QDs which do not participate in the energy transfer. Details on the peculiarities of the energy transfer, which is controlled by the average donor-acceptor distance and exciton radiative time can be found, e.g., in Ref. [3]. The excitation transfer can occur via FRET if the ground level of the donor-QD coincides in energy with the excited level of the acceptor-QD and if the condition of close spatial position is satisfied. The acceleration of recombination rate of the donor-QD in the presence of the acceptor-QD located at the distance R_{da} is approximately expressed as

$$t^0/_t = 1 + \left(R_0/_{R_{da}}\right)^6, \quad (1)$$

where t^0 is the donor-QD lifetime in the absence of the acceptor-QD, R_0 is the Förster radius defined as the distance where the transfer efficiency is 0.5.

Previously, it has been well established that the slowly decaying emission component is indicative of emission involving the dark exciton eigenstates in QDs [4,5]. It has been demonstrated also that this emission is promising for qubit application [6]. Because our Cd(Zn)Se/ZnSe QDs have predominantly spheroidal shapes [7], the lowest dark exciton state is characterized by the momentum projection ±2 on the growth direction and is spin-forbidden. Its radiative recombination, however, can be activated due to the small admixture of the bright exciton states caused by the in-plane asymmetry [8] or by the interaction with acoustic or optical phonons [9]. Such admixtures of the bright exciton dipole moment allows the dark exciton states to be involved into the FRET process as shown by figure 1.

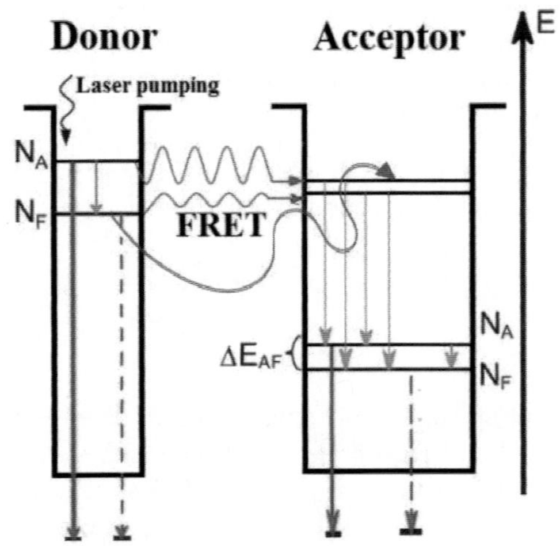

Figure 1. Schematic of the energy transitions in a donor-acceptor system. The relaxation and recombination processes are shown by blue and red lines, respectively.

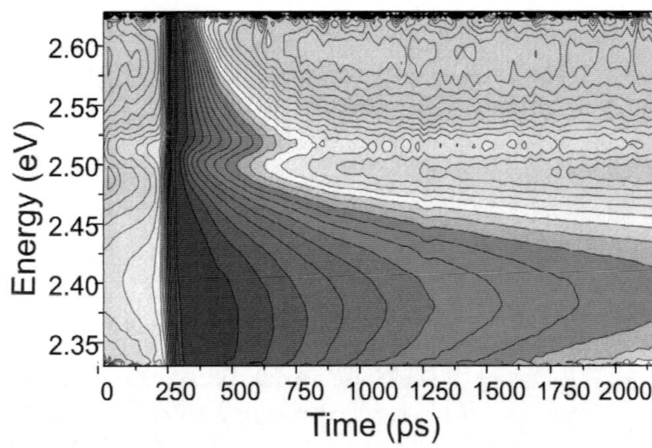

Figure 2. Typical TR PL image (normalized in intensity) showing the dependence of emission intensity spectra on time in a structure with two insertions of QD arrays.

The total PL intensity I can be characterized by the expression $I \sim N_A\Gamma_A + N_F\Gamma_F$, where N_A (N_F) and Γ_A (Γ_F) are the exciton population and radiative decay rate, respectively, for the allowed (forbidden) excitonic states. The slow decaying component at low temperatures can be observed only in the case $\Gamma_F \neq 0$. With the temperature increase, the long living dark exciton states populates the bright exciton states and the two exponential decay of the PL can be observed even for $\Gamma_F = 0$. If the relaxation rate between the bright and dark states is faster than Γ_A, with the rise of temperature T, the populations N_A and N_F become related by the thermodynamic equilibrium condition $N_A/_{N_F} = \exp\left(-\Delta E_{AF}/_{kT}\right)$, where ΔE_{AF} is the energy difference between the forbidden and allowed states. In this case, one can observe at room temperature the PL decay with some average rate which is approximately equal to

$$\Gamma = 1/_2\,(\Gamma_A + \Gamma_F). \quad (2)$$

IOP Conf. Series: Journal of Physics: Conf. Series **917** (2017) 062048 doi:10.1088/1742-6596/917/6/062048

3. Experimental

The Cd(Zn)Se/ZnSe structures were grown by MBE on GaAs substrates. Each of them has two insertions of QDs with a nominal CdSe thickness of ~1.8 and 2.7 monolayers (MLs). The barrier width, w, between them is varied from ~8 nm to 18 nm. Similar structures have been studied previously; however, the arrangement of double insertions was different [10]. TR PL measurements were performed at low temperature (5 K) using a Hamamatsu streak-camera with temporal resolution of ~15 ps. The second harmonics of a tunable Ti: sapphire femtosecond pulsed laser was used for excitation. The laser beam with 2-4 mW averaged power before a cryostat window was focused into a spot of 0.5-1 mm in diameter. At room temperature, the measurements were done using a fast photomultiplier with time-correlated photon counting system.

4. Results and discussion

The emission spectra measured in these structures consist of two wide bands related to the donor QDs (higher-energy) and acceptor QDs (lower energy). The typical image of the normalised emission intensity measured at low temperature in the structures with the narrowest w show the strong suppression of the slow-decaying emission in the donor QD band (figure **2**). The remnant part is mostly pronounced at the energies near 2.52 eV, where QDs should be out of the resonance condition between the excited and ground excitonic levels in the donor and acceptor QDs [2]. Generally, the maximum amplitude of the slow-decaying component decreases, respectively, by about 20 and 4 times for the higher-energy and lower-energy PL band, as compared with the maximal amplitude of the corresponding fast component. The decrease is even more pronounced at elevated temperatures.

The PL decay curves derived from the low-temperature TR PL images are modeled using the following expression

$$I(t) = -I_{rise}e^{-t/t_{rise}} + I_1e^{-t/t_1} + I_2e^{-t/t_2} \quad (3)$$

Here, t_{rise} is the rise time, t_1 and t_2 are the decay times of the fast and slow decaying components, respectively, related presumably to the bright and dark excitonic states; I_i denotes the corresponding intensity amplitude. Both the fast and slow decay times decrease with diminishing the barrier width.

We use equation (1) to analyse our results focusing on the characteristic decay times measured at the maxima of the PL bands. With data processing, we assume R_{da} being equal to the width w; t^0 is taken equal to either fast or slow decay time value measured in the structure with the maximal w, which is assumed to be not influenced by FRET. The results of the modeling are well consistent with the linear dependence anticipated for FRET (figure 3). The Förster radius R_0 derived for the bright and dark exciton states turn out to be different: 10.5 nm and 8 nm, respectively. This difference is suggestive of the independent relaxation channels at low temperature for the bright and dark excitonic states. It implies also that the energy transfer channels for these states are different (see Fig.1). Indeed, the theoretical consideration predicted the strong dependence of the Förster radius on the acceptor radiative and nonradiative life times as well as on the detuning between donor and acceptor energy levels [3].

At room temperature, the decay curves are modified by the strong influence of non-radiative centers and thermally-induced carrier escape. It especially concerns the fastest time which does not reflect more the intrinsic decay. Moreover, the fitting of the room-temperature decay curves needs three decaying exponents instead of two. In such situation, it is difficult to ascribe which component with either “middle” or “slowest” decay times is related to the mixed (bright-dark) excitonic state. Nevertheless, in the structure with w =18 nm, which is almost not-influenced by FRET, both respective decay rates are close to the half-sum of the maximal and minimal rates at low temperature in accordance with equation (2). The deviation from the expected half-sum values is much higher in the structure with w=8 nm that indicates likely the stronger modification of decay rates by the FRET mechanism.

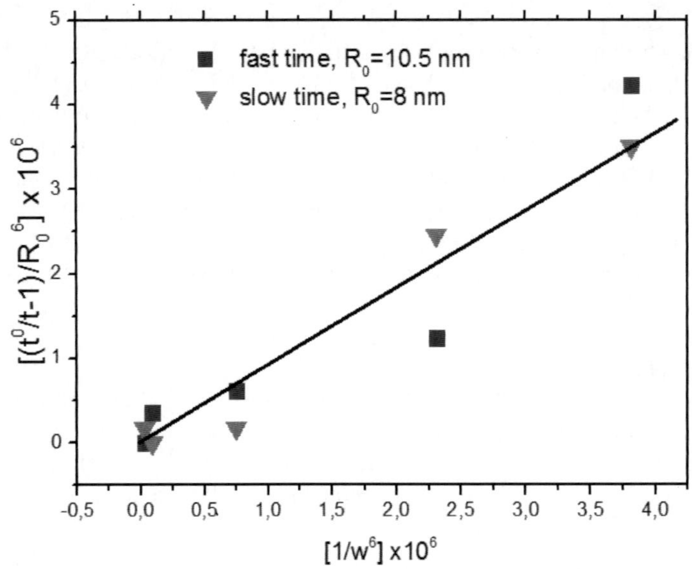

Figure 3. Acceleration of fast (squares) and slow (triangles) PL decay times with the barrier width decrease at low temperature. Fitting (dashed line) is done with R_0 = 10.5 nm and t^O = 550 ps for the fast component and with R_0 = 8 nm and t^O = 1100 ps for the slow component.

Summarizing, our studies of recombination dynamics in structures with two QD insertions separated by a barrier of a variable width confirms the existence of FRET between dots of different sizes. This mechanism influences both fast and slowly decaying emission components ascribed, presumably, to the bright and dark excitonic states. This effect is accompanied by the strong attenuation of the slowly decaying emission component in the donor-QDs. The Förster radius R_0 is found to be different for the bright and dark excitonic states (10.5 nm and 8 nm, respectively). Room-temperature measurements have exhibited the tendency to averaging the decay rate in the mixed (dark-bright) exciton states. We highlight that the results need more thorough analysis which should takes into account the trapping of carriers by non-radiative centres and thermally-induced carrier escape. However, the demonstrated possibility to control the intensity of slow component by FRET may be useful for set of nanophotonic applications.

Acknowledgement
This work has been supported by Russian Science Foundation, grant № 14-22-00107.

References

[1] Shubina T V *et al. Phys. Stat. Sol. B* **253**, 1485 (2016).
[2] Shubina T V *et al*. J. Electronic Materials (2016). DOI: 10.1007/s11664-016-5007-2.
[3] Poddubny A N, Rodina A V, JETP. 122, 531 (2016).
[4] Liu F et al., Phys. Rev. B **92**, 125403 (2015).
[5] Nirmal M *et al*., Phys. Rev. Lett. 75, 3728 (1995).
[6] Poem E. *et al*., Nature Physics 6, 993–997 (2010)
[7] Golovatenko A A, Semina M A, Rodina A V and Shubina T V, APP A **129**, 107 (2016).
[8] Don Y, Zielinski M, and Gershon D, ArXiv: 1601.05530 (2016)
[9] A.Rodina and Al.L. Efros, Phys. Rev. B **93**, 155427 (2016).
[10] Mikhailov T N *et al*., Journal of Physics: Conference Series **741**, 012153-4 (2016).

IOP Conf. Series: Journal of Physics: Conf. Series **917** (2017) 062049 doi:10.1088/1742-6596/917/6/062049

Experimental sol-gel fabrication and theoretical simulation of one-dimensional photonic crystals with a defect state

V A Romanova, P A Somov, L B Matyushkin
Department of Micro- and nanoelectronics, St. Petersburg State Electrotechnical University “LETI”, St. Petersburg, 197376, Russia

Abstract. One-dimensional multilayer structures of the periodically alternating low refractive index (silica) and high refractive index (titania) materials have been deposited by sol–gel spin-coating. The possibility of creating an optical filter by introducing a structural defect in the form of a layer of double thickness is shown theoretically and experimentally. All experimental spectra are in agreement with theoretical spectra calculated by transfer matrix method.

1. Introduction
Photonic band gap structures (PBGs), also known as photonic crystals (PCs), are structures whose refractive index is periodic on a length scale of the order of optical wavelengths preventing light propagating through the structure due to Bragg reflection. Depending on the dimensionality of this periodicity, PBGs can be one, two or three dimensions (1D, 2D or 3D). In the case of sol-gel derived materials, the primary emphasis has been on 1D- and 3D-structures [1]. Distributed Bragg reflectors (DBR), also called Bragg mirrors, are examples of 1D-PBGs.
Sol-gel technology is one of the simplest and cheapest technologies for the synthesis of photonic crystals. Sol-gel DBRs can be made, for example, by dip- [2] or spin-coating [3] of colloidal nanoparticles. Silica (SiO_2) and titania (TiO_2) are usually chosen because of the high optical contrast of their refractive indices (~1.46 and ~2.4, respectively).
In optics, filters often need narrowband transmission of photons in a narrow range of wavelengths in units of nanometers wide. To obtain such a narrow-band filter structure PCs some defects can be introduced: 1) the layers with double (triple, quadruple, etc.) or some aliquant thicknesses, 2) the layers of different material.

2. Experimental section
An experimental technique for the synthesis of multilayer structure was spin-coating of silica and titania sols on a glass substrate, followed by annealing of layers in a muffle furnace. The technique in details has been described previously [4].
The transfer matrix method was used to calculate multilayer structure transmission spectra [5]. Matrix A describes light propagation through the layers boundary, and matrix B describes light propagation through current material layer:

IOP Conf. Series: Journal of Physics: Conf. Series **917** (2017) 062049 doi:10.1088/1742-6596/917/6/062049

$$A = \frac{1}{2n_2}\begin{pmatrix} n_1 + n_2 & n_2 - n_1 \\ n_2 - n_1 & n_1 + n_2 \end{pmatrix},$$

$$B = \begin{pmatrix} \exp\left(2\pi i \frac{nL}{\lambda}\right) & 0 \\ 0 & \exp\left(-2\pi i \frac{nL}{\lambda}\right) \end{pmatrix},$$

where n_1 and n_2 are refractive indices of alternating materials, n and L – refractive index and thickness of the current layer respectively, λ – the light wavelength.

The dispersion curves of the refractive indices $n_1(\lambda)$ and $n_2(\lambda)$ used in the calculations respectively for SiO_2 [6] and TiO_2 [7] are following:

$$n_1 = \sqrt{1 + \frac{0.6962\lambda^2}{\lambda^2 - 0.0684^2} + \frac{0.4079\lambda^2}{\lambda^2 - 0.1162^2} + \frac{0.8975\lambda^2}{\lambda^2 - 9.9961^2}},$$

$$n_2 = \sqrt{5.913 + \frac{0.2441}{\lambda^2 - 0.0803}}.$$

The described matrices are multiplied for all layers of the structure and all layers boundaries. The result of the multiplication is the transfer matrix M. Elements of alternate matrix multiplication M ($M = A_1B_1A_2B_2 \ldots A_nB_n$) allows to describe the optical properties of the multilayer structure. This common method of spectrum simulation can be used not only for photonic crystals but also for any multilayer structures including aperiodic ones.

To investigate producing of photonic crystals filters by sol-gel technology it has been researched 1D-PC structures with different numbers and thicknesses of layers. The structural defect was achieved by doubling the layer of titanium dioxide in the center of the structure.

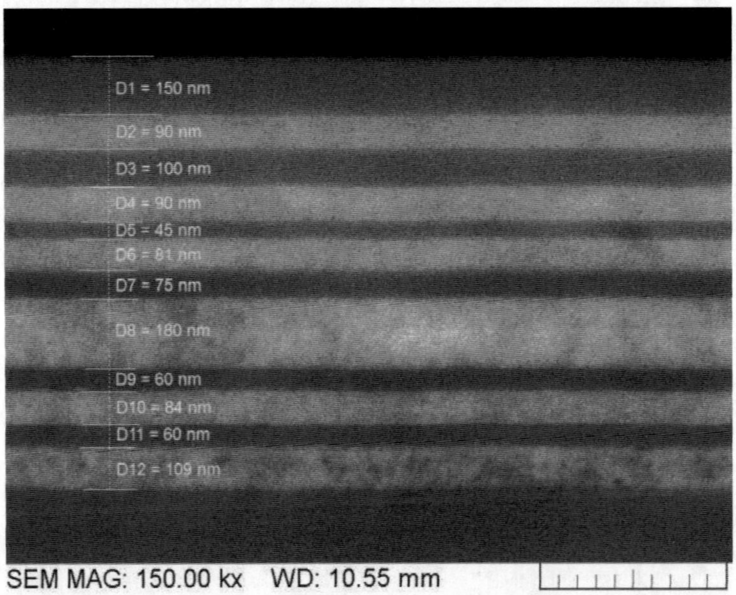

Figure 1. Scanning electron microscopy (secondary electrons) of photonic crystals with thicknesses of alternating layers (bar is 500 nm)

3. Results and discussion

The PC structure with defect has been characterized by scanning electron microscopy (Fig. 1). The light bands correspond to titanium dioxide, and dark ones are for silica. Therefore, thinner silicon dioxide stripes interspersed with thicker layers of silicon oxide. The average thicknesses are 70 and 100 nm respectively.

Theoretical end experimental transmission spectra of the photonic crystal with double titania defect layer are compared in Fig. 2a. The position of the defect energy level inside the stop band and the quality factor of the photonic crystal is different in theory and experiment. The quality factor difference is explained due to ideal wall approximation in the transfer matrix model while there are non-zero porosity and roughness of layers in the experiment. When the porosity is taken into account, the maximum associated with a defect state shifts to shorter wavelengths because of the lower refractive index of titania and silica films in Fig.2a.

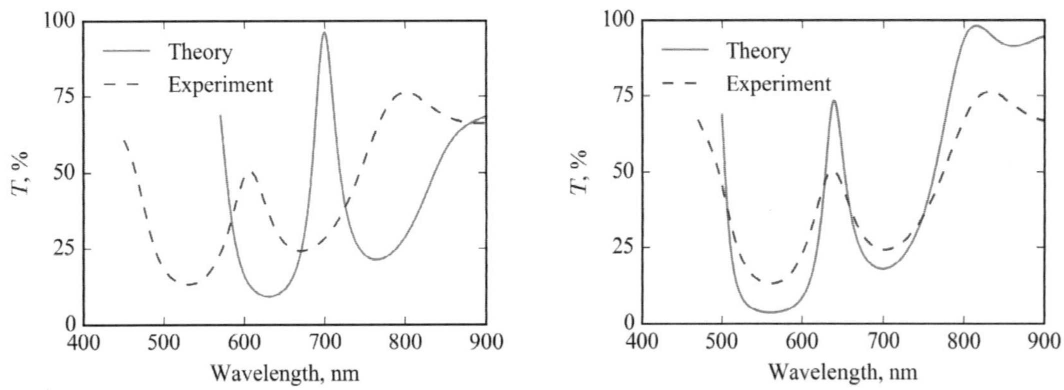

Figure 2. Theoretical end experimental transmission spectra of the photonic crystal with defect double titania layer in the center of the structure (a) in the case of continuous material dispersion curve, (b) in the case of refractive indices taking into account porosity some porosity of films.

The porous structure of the thin films composing the photonic crystal can be taken into account by modifying the expressions for the refractive indices $n_1(\lambda)$ and $n_2(\lambda)$.

$$n_1 = \sqrt{0.5 + \frac{0.6962\lambda^2}{\lambda^2 - 0.0684^2} + \frac{0.4079\lambda^2}{\lambda^2 - 0.1162^2} + \frac{0.8975\lambda^2}{\lambda^2 - 9.9961^2}},$$

$$n_2 = \sqrt{4.913 + \frac{0.2441}{\lambda^2 - 0.0803}}.$$

The proposed correction allows one to obtain the wavelength of the defect state corresponding to the experiment (Fig. 2b). However, this approach does not take into account the roughness of the layers, which leads to an experimentally observed decrease in the quality factor. Also, it is worth noting that the difference in layers is also due to the technique of structure synthesis – different times of the layers in the muffle furnace lead to small differences in the dispersion curves of the layers for the same material.

4. Conclusions

Although sol-gel technology is one of the cheapest methods for the production of one-dimensional photonic crystals, using this technique and simulating photonic crystals, one must take into account the porosity of the layers. Since annealing time affects variate refractive index values, instead of ellipsometric measurements the simulation of the inverse problem can be used for reconstructing refractive indices from transmission spectra with known layer thicknesses.

Acknowledgments

This study was supported by the Ministry of Education and Science of the Russian Federation within the framework of the project part of the state assignment for the St. Petersburg Electrotechnical University "LETI" (project 16.2112.2014/K, "Preparation and study of porous systems, functionalized nanomaterials, applications in photonics, sensor technology and medicine").

References

[1] Almeida R M, Sabine P2003 *Curr. Opin. Solid State Mater. Sci.***7**(2)151
[2] Bellessa J, Rabaste S, Plenet J C, Dumas J, Mugnier J, Marty O 2001 *Appl. Phys. Lett.***79** 2142
[3] Xin M, Orignac X, Almeida R M 1995 *J. Am. Ceram. Soc.* **78**(8) 2254
[4] Ilinykh V A, Matyushkin L B *Young Researchers in Electrical and Electronic Engineering Conference (EIConRusNW), 2016 IEEE NW Russia p. 47-50*
[5] Ilinykh V A, Matyushkin L B 2016 JPCS 741(1) 012008
[6] Malitson I H 1965 J. Opt. Soc. Am. 55 1205
[7] Devore J R 1951 J. Opt. Soc. Am. 41 416

IOP Conf. Series: Journal of Physics: Conf. Series **917** (2017) 062050 doi:10.1088/1742-6596/917/6/062050

Coexistence of type-I and type-II band line-ups in 1-2 monolayer thick GaN/AlN single quantum wells

E A Evropeytsev[1], V N Jmerik[1], D V Nechaev[1], S Rouvimov[2], T V Shubina[1], V Kh Kaibyshev[1], G Pozina[3], S V Ivanov[1] and A A Toropov[1]

[1]Ioffe Institute, Saint Petersburg 194021, Russia
[2]University of Notre Dame, Notre Dame, Indiana 46556, USA
[3]Department of Physics, Chemistry and Biology, Linköping University, Linköping S-581 83, Sweden

Abstract. GaN/AlN quantum wells (QWs) with varied nominal thickness of 0.5-4 monolayers have been studied by time-resolved photoluminescence (PL) spectroscopy. The structures demonstrate an emission peak with the thickness-dependent wavelength in the range 225-320 nm. The observed temporal behavior of PL between 225 and 280 nm can be described as a superposition of fast and slow decaying components with characteristic decay time constants of the order of 0.1-0.7 ns and 7-30 ns, respectively. The fast PL component with the decay time smaller than 1 ns dominates in the thicker GaN insertions and tends to vanish in the thinnest ones, where the slow PL component becomes progressively longer. These observations imply formation in the GaN/AlN monolayer-thick layers of an inhomogeneous excitonic system involving both direct and indirect in space excitons.

1. Introduction

GaN/AlN (or GaN/AlGaN) quantum wells (QWs) have been recently considered as a promising active region of light-emitting devices operating in mid- and deep-ultraviolet regions [1, 2]. Wavelength of emission from such QWs can be varied from ~320 to ~220 nm due to strong quantum confinement by varying the QW thickness from 4 to 1 monolayer (ML) and thinner. There are several advantages of using such ultrathin GaN/AlN QWs comparing with a conventional design. First, more conventional AlGaN-based QWs with the thickness of several nanometers, grown on *c*-sapphire, usually suffer from intrinsic electrostatic fields, which reduce an electron-hole overlapping as a result of the quantum-confined Stark effect (QCSE). Decreasing the QW thickness results in the reduction of the QCSE that was observed and theoretically described for several ML thick GaN/AlN QWs [3-5]. Another advantage is efficient light extraction from the *c*-plane of the GaN/AlN QWs, which resulted from TE-polarization of the emitted light (electric field is perpendicular to the *c*-axis) [6]; that is because the lowest quantized hole energy level in such structures originates from a heavy hole band [4, 5]. On the contrary, light extraction from the c-plane AlGaN with high Al content is hampered due to inherent TM-polarization of the emitted light [7, 8].

In this work, we represent growth of GaN/AlN QWs with nominal thickness of 0.5-4 ML and study of their electronic properties by means of time-resolved photoluminescence (TR PL) spectroscopy.

2. Results and discussion

The samples were grown on *c*-sapphire by plasma-assisted molecular beam epitaxy (PA MBE) using a Compact 21T Riber setup. Metal-rich conditions and a maximum substrate temperature T_S=780°C were used for the growth of a 65 nm thick AlN nucleation layer by migration enhanced epitaxy and a 1.5-µm-thick AlN buffer layer by metal-modulated epitaxy. The buffer layer comprised four strained 3.5 nm thick GaN insertions, whose incorporation aimed at decreasing density of threading dislocations in the QW region [9, 10]. Then, 3 ML of AlN, 0.5-4 ML of the GaN QW, and 5 nm of an upper AlN barrier were formed at T_S=700°C. Finally, a 35 nm thick AlN cap layer was grown by standard PA MBE at T_S=770-780°C. The samples with 1 ML, 2 ML, and 4ML thick QWs were rotated during the growth. Also, similar samples with the GaN QW nominal thickness of 0.5 and 1 ML were fabricated without rotation during the growth of the QW that resulted in a gradient of the QW thickness along a certain direction on the sample surface. Further we will refer to the samples grown with or without rotation as R or N, respectively, appending the value of the QW nominal thickness in MLs.

For PL excitation, we used a fourth harmonic of a femtosecond Ti-sapphire laser with the emission wavelength 210 nm and pulse frequency 76 MHz. The PL decay curves were measured using a fast photomultiplier operating in a time-correlated single photon counting mode with temporal resolution ~145 ps. Additionally, samples R2 and R4 were studied with the excitation by a third harmonic of the Ti-sapphire laser (260 nm) and TR PL images were obtained with overall resolution ~25 ps using a Hamamatsu streak camera. The high angle annular dark field (HAADF) scanning transmission electron microscopy (STEM) analysis has been carried out on FEI Titan 80-300 TEM with spatial resolution of 1.3 Å.

The PL spectra of 4 ML and 2 ML thick GaN/AlN QWs at 77 K demonstrate a multiexponential decay on the time scale ~1.5 ns and a single exponential one with the decay time constant of 0.15 ns, respectively (figure 1). The relation of the PL intensity at 77 K to those at 300 K equals 3 for the 2-ML-QW and 10 for the 4-ML-QW. Both shortening the PL lifetime and improving the internal quantum yield apparently indicates the above mentioned reduction of the QCSE in the thinner QW [3].

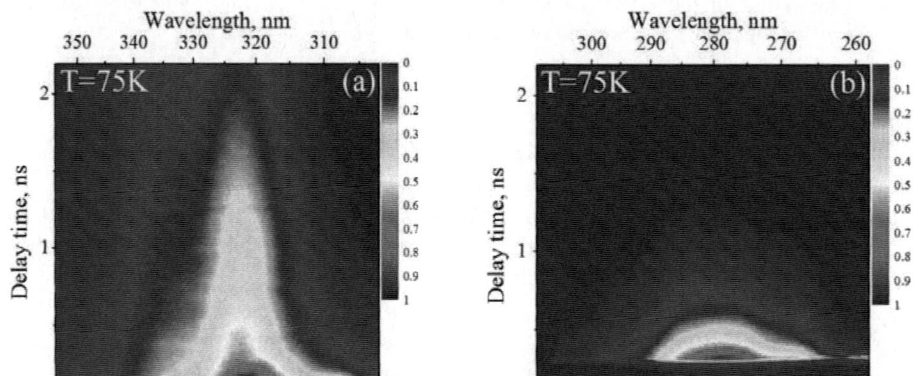

Figure 1. TR PL images taken from the 4 ML thick GaN QW in sample R4 (a) and form the 2 ML thick QW in sample R2 (b).

Figure 2 shows HAADF STEM images, measured in two different points of sample N1. Although the STEM images in figure 2 look quite similar, the corresponding PL bands are spectrally separated by~0.7 eV. Integrated STEM profiles of the QWs are presented in figure 3. As can be seen, the QW thicknesses are close to the nominal values. Smaller height of the STEM profiles in the 1-2 ML thick QW in comparison with the 4 ML thick one is probably related to formation of an AlGaN/AlN QW instead of the intended binary GaN/AlN one.

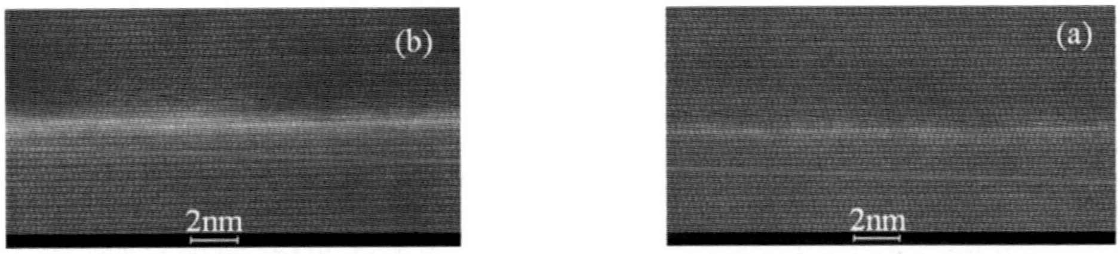

Figure 2. HAADF STEM images of sample N1 measured in two points: (a) the point with the corresponding PL emission at ~280 nm (point 3); (b) the point with the PL emission at ~240 nm (point 2).

Temporal behavior of the PL emission from the QWs with the nominal thickness ≤ 2 ML can be described by two decay time constants, τ_1~0.1-0.7 ns and τ_2~10-30 ns. Spectral positions of the QW PL bands and respective extracted decay time constants are shown in figure 4. A prominent trend of an increase of the time constant τ_2 and a decrease of τ_1, while thinning the QW, is observed for all studied samples. We suppose that the longer PL component results from the formation of type-II band line-ups, when the electron and hole are separated in space.

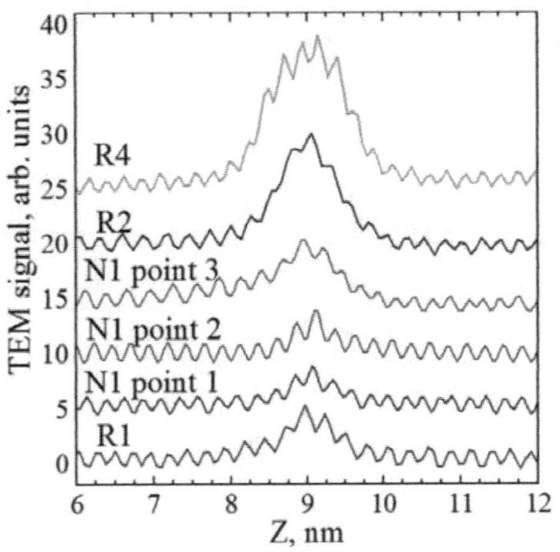

Figure 3. HAADF STEM profiles. The curves are vertically shifted for clarity. Central peaks correspond to the GaN QW and backgrounds – to the AlN matrix.

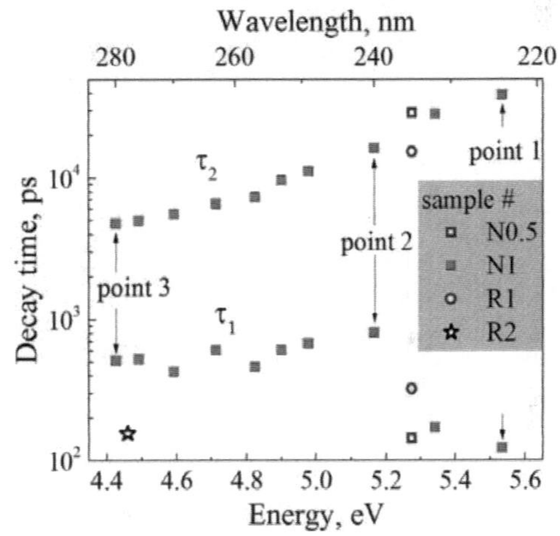

Figure 4. Dependence of the fitted QW PL decay time constants τ_1 and τ_2 on the PL peak spectral position at 77 K.

According to Ref. [5], a type-II band alignment is predicted for sufficiently thin (less or equal to 4 ML) GaN/AlN QWs and corresponds to the situation, when a recombining electron is localized in the GaN insertion whereas a hole from the crystal-field split valence band is located in the surrounding AlN. On the other hand, a type-I band alignment (recombining heavy hole states are localized in the QW) is confidently predicted for thicker QWs. In this way, at a certain average QW thickness both types of transitions may have approximately the same energy, owing to unavoidable QW thickness fluctuations. In terms of this model, the observed increase of τ_2 and vanishing of the fast PL component (see figure 5) can be explained by progressive sinking of the heavy and light hole energy levels in the GaN insertion below the level of the crystal-field split hole level in the AlN barriers with decreasing the average GaN thickness.

An additional confirmation of this model follows from temporal behavior of PL registered in the GaN QWs with thickness ~1.5-2 ML, emitting in the wavelength range 260-280 nm (figure 5 (a) and (c)). We found that the fast and intense PL component detected in these sample is spectrally shifted toward lower energies with respect to the slow one. This appears in the TR PL image as a shift of the

PL peak toward higher energies during first nanoseconds after the excitation pulse (figure 5 (c)). This rather unusual behavior cannot be explained neither by hopping of localized excitons nor by the QCSE screening. We propose that the observed rapidly decaying bright emission corresponds to type-I transitions, which originate from thicker areas of the QW. On the other hand, type-II transitions are most probable for thinner QW areas, and, therefore, they possess higher energy due to stronger quantum confinement of electrons in the thinner part of the QW comparing with the thicker one. That additionally confirms coexistence of type-I and type-II band line-ups in the studied GaN/AlN QW.

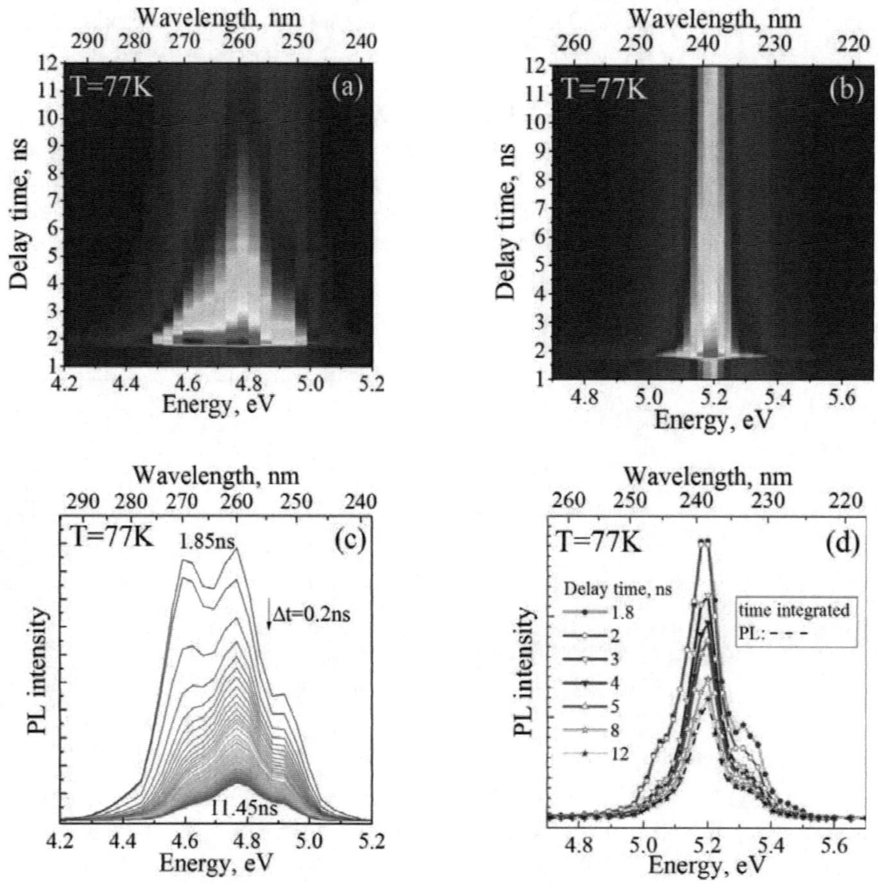

Figure 5. TR PL images detected from sample N1 in point 3 (a) and in point 2 (b). Dependence of the PL spectrum on delay time for point 3 (c) and point 2 (d). Respective STEM images are shown in figure 2.

Acknowledgments

This work was supported by the Government of the Russian Federation (Project # 14.W03.31.0011).

References

[1] Rong X, Wang X, Ivanov S V *et al* 2016 *Adv. Mater.* **28** 7978

[2] Islam S M, Lee K, Verma J, Protasenko V, Rouvimov S, Bharadwaj S, Huili (Grace) Xing and Jena D 2017 *Appl. Phys. Lett.* **110** 041108

[3] Langer R, Simon J, Ortiz V, Pelekanos N T, Barski A, André R and Godlewski M 1999 *Appl. Phys. Lett.* **74** 3827

IOP Conf. Series: Journal of Physics: Conf. Series **917** (2017) 062050 doi:10.1088/1742-6596/917/6/062050

[4] Kamiya K, Ebihara Y, Shiraishi K and Kasu M 2011 *Appl. Phys. Lett.* **99** 151108 (2011)
[5] Strak P, Kempisty P, Ptasinska M and Krukowski S 2013 *Journal of Applied Physics* **113** 193706
[6] Taniyasu Y and Kasu M 2011 *Appl. Phys. Lett.* **99** 251112
[7] Nam K B, Li J, Nakarmi M L, Lin J Y and Jiang H X 2004 *Appl. Phys. Lett.* **84** 5264
[8] Kawanishi H, Senuma M, Yamamoto M, Niikura E and Nukui T 2006 *Appl. Phys. Lett.* **89** 081121
[9] Jmerik V N, Lutsenko E V and Ivanov S V 2013 *Phys. Status Solidi* A **210** 439
[10] Ivanov S V, Nechaev D V, Sitnikova A A, Ratnikov V V, Yagovkina M A, Rzheutskii N V, Lutsenko E V, Jmerik V N 2014 *Semicond. Sci. Technol.* **29** 084008

IOP Conf. Series: Journal of Physics: Conf. Series **917** (2017) 062051 doi:10.1088/1742-6596/917/6/062051

Size-dependent blinking of molecular aggregate total emission

V V Savchenko[1,2], A A Zimin[1,2], N Yu Ignatova[1,2], V Kimberg[1,2], F Gel'mukhanov[1,2], S P Polyutov[1]

[1]Institute of Nanotechnology, Spectroscopy and Quantum Chemistry, Siberian Federal University, Krasnoyarsk, 660041, Russia
[2]Department of Theoretical Chemistry and Biology, KTH Royal Institute of Technology, Stockholm, Sweden

E-mail: vvsavchenko@sfu-kras.ru

Abstract. Molecular aggregates are well known for their customizable optical properties. Vibronic coupling in monomers forming such aggregates offers rich opportunities for property tuning. We study generic molecular aggregate models of growing complexity (from a dimer up to a decamer) and report how vibronic coupling affects aggregate fluorescence intensity. The total aggregate fluorescence intensity is a measure sensitive to both vibronic coupling and Coulomb coupling between monomer transition densities. Using an exact diagonalization approach in the two-particle basis set, we show how the interplay between Coulomb and vibronic coupling affects aggregate fluorescence. Moreover, for H-aggregates we predict a periodic variation of the fluorescence intensity with aggregate size and show that vibronic interaction decreases the effect.

1. Introduction

The investigation of molecular aggregates (MAs) opens the possibility of creating new promising materials for optical technologies [1, 2, 3]. Photophysical properties of vibronic excitation in MAs are of crucial importance for systems ranging from organic dye solutions to artificial photosynthetic devices. One of widely used methods to employ exciton-vibrational coupling in dynamics and spectroscopy of MAs is to use the theory of open systems by putting low-frequency vibrational degrees of freedom in a thermal reservoir characterized by its spectral function [2], whereas only the important degrees of freedom (typically only electronic) are included in a system Hamiltonian. However, recent research [4, 5, 6, 1, 7] show that this way is not always appropriate for calculations of optical response and study of exciton dynamics of such aggregates. These recent contributions have spurred interest to the exciton-vibrational coupling problem in molecular aggregates and resulted in a revision of theories of energy transfer and coherence in MAs, in particular, by explicit inclusion of a quantized vibrational mode in the system Hamiltonian.

In recent years methods of direct diagonalization of Hamiltonian with respect to the vibronic problem have attracted much attention [1, 8, 5, 9, 10, 11, 7, 6] especially in the context of the role of vibronic coherence in the light-harvesting machinery [12, 13, 14, 15, 4]. Another application of this method is a modelling of MAs structure with specified properties and revealing of MAs structure and the dynamics of aggregation [16]. One of the most interesting of MAs is a perylene bisimide (PBI) aggregate. An experiment with this dye shows unusual temperature-dependent emission spectra behaviour and motivates to study this aggregate in details [17, 16]. Here we study the fluorescence intensity of vibronic H-aggregates on a generic model of growing complexity using the exact diagonalization approach in the two-particle basis set [18].

2. Theory

The Hamiltonian of the molecular aggregate is represented as:

$$H = \sum_{A} |g^A\rangle H_g^A \langle g^A| + |e^A\rangle H_e^A \langle e^A| + \sum_{A>B} \left(|g^A e^B\rangle J^{AB} \langle e^A g^B| + c.c. \right), \qquad (1)$$

where $|g^A\rangle$ and $|e^A\rangle$ are electronic eigenstates for monomer A in ground and excited states, respectively, indices A and B run over all monomers in the aggregate, H^A is the appropriate Hamiltonian for isolated monomer A, J^{AB} represents the Coulomb coupling term between electronic transition densities on monomers A and B, c.c. stands for complex conjugate terms (see [1] for theoretical details). Strength of the vibronic coupling is quantified by the Huang-Rhys parameter S. The population of energy states is described by Bose-Einstein distribution $N_{BE} = (exp[(E - E_F)/k_B T] - 1)^{-1}$, where k_B=3.16681 Eh/K with temperature T, E is state energy, E_F is the chemical potential (Fermi level). Monomer ground and excited potential energy surfaces (PES) are taken to be harmonic, with excited state minimum being displacement by $(2S)^{1/2}$ relative to the ground state. Emission intensity is computed as: $F(\omega) = \Sigma_{i,f} |\langle f|d_{el}|i\rangle|^2 \delta(\omega - \omega_{if})$, where the summation goes over all initial and final states, d_{el} is the electronic transition tor, and ω_{if} is the transition energy. Integrated emission is then obtained as $I_{ems} = \Sigma_\omega F(\omega)$. Note that we simplify our definition of emission by dropping out all factors excepts for transition amplitude.

The aggregate Hamiltonian is then expressed in the two-particle basis set of the form (here shown for tetramer) $|e^A, \nu^A\rangle \otimes |g^B, \nu^B\rangle \otimes |g^C, 0\rangle \otimes |g^D, 0\rangle$, where capital letters are used to index monomers and ν denotes a variable number of vibrational quanta in a particular mode for ground (g) or excited (e) PES. Only single-exciton manifold is required for our purposes. "Two-particle" means that no more than two vibrationally excited monomers enter the basis function at the same time, one of them being in electronically excited state. Physically, two-particle basis set is required to take excitonic polaron formation into account, resulting from deformation (vibrational excitation) of monomers surrounding a vibronically excited monomer. Strictly speaking, a general N-particle basis is required for an aggregate of size N, quickly making exact diagonalization approach unfeasible for $N > 3$. Two-particle basis set is sufficient for description of low-energy aggregate eigenstates when Coulomb coupling is comparable to vibrational mode energy ω_{vib}. Details of our theoretical approach can be found in [1]. Our methodology is formulated in the eigenstate representation and is not limited to the case of harmonic potential energy surfaces – vibronic coupling is hidden in the Franck-Condon overlap integrals. In particular, application to energy transfer problems with explicit inclusion of a quantized vibrational mode in condensed systems such as liquid water solutions or doped semiconductors are possible, given the availability of the monomer potentials.

3. Results and discussion

We begin by comparing energy levels of excitons in purely electronic aggregates with vibronic aggregates, limiting ourselves to dimer, trimer and tetramer (Figure 1 (a), (b) and (c), respectively). Aggregates are modeled as a linear chain of monomers with their transition dipole moments aligned either in a head-to-tail fashion (J-aggregates and negative Coulomb coupling), or in a parallel arrangement orthogonal to the aggregate axis (sandwiched H-aggregates). Excitonic states are labeled $|\alpha_1\rangle$ through $|\alpha_4\rangle$. Square of the transition dipole moment ($d_i = \langle 0|\mathbf{d}_{el}|\alpha_i\rangle$) from each of these states to the (vibrationless) ground state is shown in the figures in the middle, whereas excitonic energy levels for the corresponding vibronic aggregates are located at the bottom. Here E is the eigenstate energy, E_{00} is the electronic vertical transition energy. Vibrational mode frequency is 1500 cm^{-1}.

Electronic dimer behaves symmetrically with respect to the sign change of Coulomb coupling (Figure 1(a)). Transitions to the ground state form either $|\alpha_1\rangle$ or $|\alpha_2\rangle$ are allowed for J-dimer and H-dimer, respectively. Energy levels are separated by twice the Coulomb interaction strength $2J$. Situation becomes more complicated when the vibronic coupling is included. When there is no Coulomb coupling between monomers, we obtain a set of degenerate monomer vibronic energy levels located at integer multiples of vibrational mode frequency. These "bands" lose degeneracy as Coulomb coupling strength is increased.

Energy levels for electronic trimer, as well as transition dipole moments to the ground state and vibronic trimer energy levels are shown in Figure 1(b). Note the antisymmetry of $|\alpha_2\rangle$ excitonic state energy with respect to change of Coulomb coupling sign, whose energy behaves analytically as $E_{00} - J_{13}$, where J_{13} is the strength of Coulomb coupling between edge monomers. This example illustrates that by taking only nearest-neighbor Coulomb coupling into account for vibronic trimer we get an error in eigenstate energies on the order of J_{13}. In contrast to H-dimer, H-trimer does not have full compensation of the transition amplitude form the lowest-energy eigenstate $|\alpha_1\rangle$. This antisymmetry further manifests itself in the energies of excitonic states for vibronic trimer on the bottom of the Figure 1(b). Electronic and vibronic tetramers follow the same tendency, as shown in Figure 1(c).

The total absorption intensity is independent of the temperature and interaction strength in accordance with the total dipole moment square conservation law in the described systems, but this is not the case for emission. The vibrational energy redistribution of exciton states in accordance with the temperature can lead to the fact that a significant transition portion from populated states to the ground vibronic global state can be forbidden by dipole moment. Thus, the total emission intensity is not retained. In Figures 2 (a, b) and 3 (a,b) we can observe features that forbidden transitions in H-aggregate become allowed due to odd number of monomers in molecular aggregates. The odd number of monomers line up in a molecular aggregate so that it leads to intermediate molecular configuration. This effect appear not in all cases.

Next, we study the integrated aggregate fluorescence intensity as a function of three parameters: number of monomers (N), vibronic coupling strength (S) and Coulomb coupling strength between monomers (J). Figure 2 shows a (S,N) plane for two Coulomb couplings $J = \omega_{vib}/3$ and $J = \omega_{vib}$, whereas (J,N) plane is shown in Figure 3 for two Huang-Rhys factors $S = 0.05$ and $S = 0.4$. Units are chosen relative to the monomer integrated fluorescence intensity.

The role of Coulomb coupling for vibronic H-aggregates is twofold: weak coupling ($\omega_{vib}/3$, Figure 2(a)) regime allows to gain about 20-30% of monomer fluorescence intensity for all N, given sufficiently large vibronic coupling strength. Increasing aggregate size requires increase in S to obtain such gains. However, when Coulomb coupling strength is increased, we see a rapid drop to only about 14% of monomer intensity at best for dimer at $S = 1$. Aggregates of larger size show heavily quenched fluorescence (Figure 2(b)). Overall, increasing vibronic coupling strength opens additonal emission channels and lifts electronic selection rules.

Fluorescence intensity in the (J,N) plane for two Huang-Rhys factors, $S = 0.05$ and $S = 0.4$, clearly shows the drastic role of vibronic coupling. For $S = 0.05$ vibronic aggregates behave very similar to the electronic aggregates, as one indeed might expect. Note the oscillatory behavior of the fluorescence intensity with the aggregate parity. On the contrary, up to about 15% of monomer fluorescence intensity can be regained when $S = 0.4$.

Let us note that a dimer has forbidden lowest exciton state (Figure 1), while for a trimer lowest exciton state is allowed and, consequently, a dimer has smaller emission intensity than a trimer.

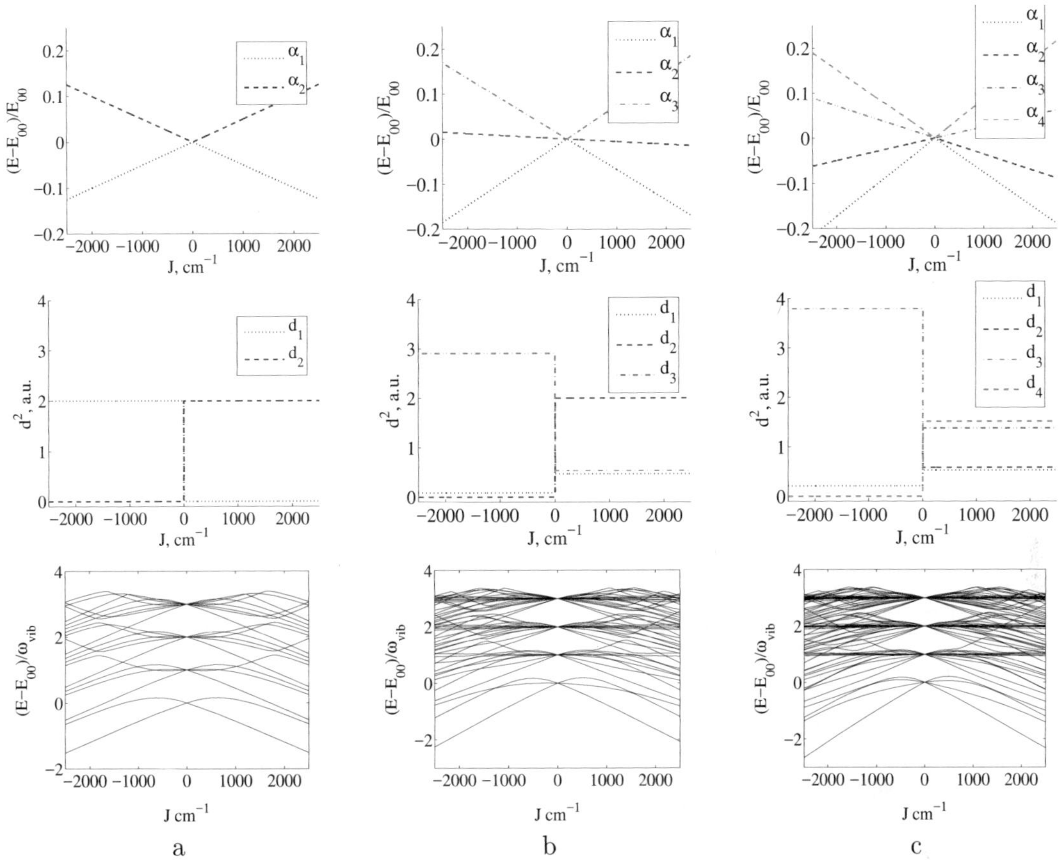

Figure 1. Energy levels of dimer (a), trimer (b) and tetramer (c) for the electronic model (upper panel) and vibronic model (lower panel), respectively. For vibronic case S=0.4. The squared transition dipole moments to the vibrationless aggregate ground state are presented on the middle panel.

The decrease in total emission intensity with increasing J for all aggregate sizes appears due to several factors. First, exciton-vibrational coupling results (a) in exciton states splitting and (b) in transition dipole moments redistribution making some of transitions forbidden. Second, the splitting between exciton states results in different thermalization of exciton states dependent on the coupling. The competition of these factors finally results in observed behaviour.

It is important to note that role of finite-size effects becomes apparent from Figures 2 and 3. For smaller aggregates (up to 5 monomers) fluorescence intensity changes drastically both in (S,N) and in (J,N) planes as number of monomers is increased. As one might expect, for larger aggregates finite-size effects become less pronounced.

Finally, we point out that interplay between Coulomb and vibronic coupling in molecular aggregates is not so straightforward to unravel. Vibronic coupling could enhance aggregate fluorescence intensity for a right choice of parameters – relatively weak Coulomb coupling and strong vibronic coupling. Finite-size effects as well as Coulomb coupling between non-nearest neighbors are important for understanding of the underlying photophysical aspects of molecular aggregates. Detailed studies and simplified theoretical models are required to further explore possibilities for rational design of molecular aggregates with required optical properties.

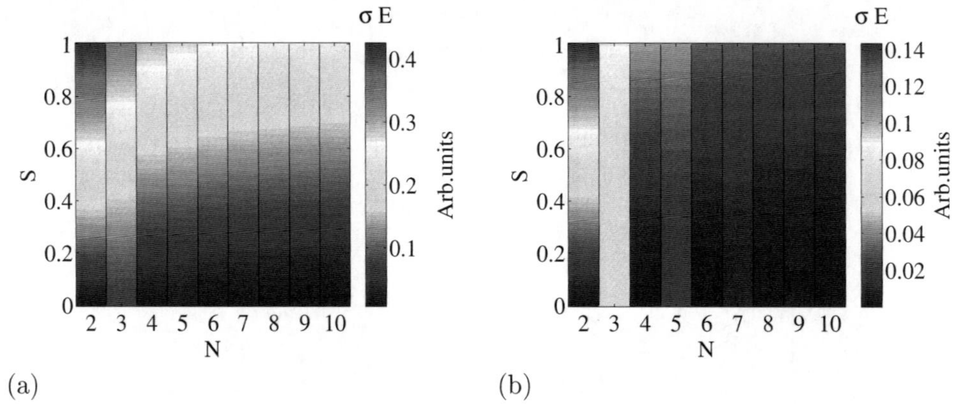

Figure 2. Total emission for J=500 cm^{-1} (panel (a)) and J=1500 cm^{-1} (panel (b)) for different number of monomers N.

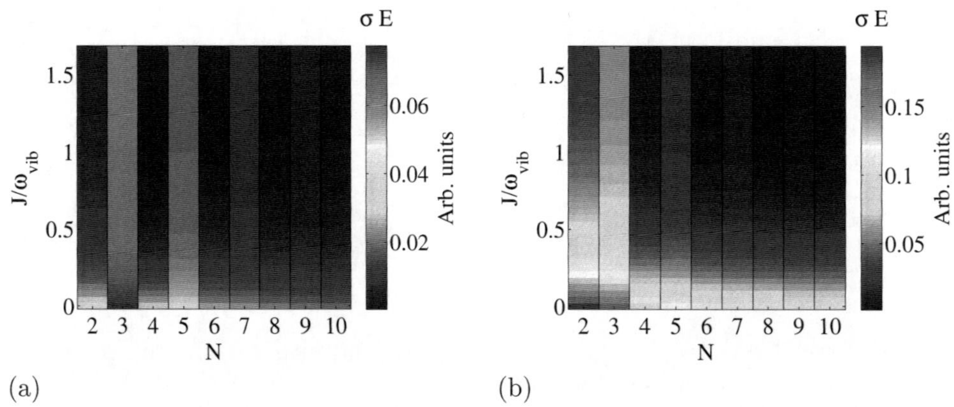

Figure 3. Total emission for Huag-Rhys factors S=0.05 (panel (a)) and S=0.4 (panel (b)) for different number of monomers N.

4. Acknowledgments
This work was supported by the RF Ministry of Education and Science (project 3.2662.2017/ПЧ).

References
[1] Schröter M, Ivanov S, Schulze J, Polyutov S, Yan Y, Pullerits T and Kühn O 2015 *Physics Reports* **567** 1–78
[2] May V and Kühn O 2008 *Charge and energy transfer dynamics in molecular systems* (John Wiley & Sons)
[3] Scholes G D, Fleming G R, Chen L X, Aspuru-Guzik A, Buchleitner A, Coker D F, Engel G S, van Grondelle R, Ishizaki A, Jonas D M *et al.* 2017 *Nature* **543** 647–656
[4] Tiwari V, Peters W K and Jonas D M 2013 *Proceedings of the National Academy of Sciences* **110** 1203–1208
[5] Womick J M and Moran A M 2011 *The Journal of Physical Chemistry B* **115** 1347–1356
[6] Polyutov S, Kühn O and Pullerits T 2012 *Chemical Physics* **394** 21–28
[7] Bašinskaitė E, Butkus V, Abramavicius D and Valkunas L 2014 *Photosynthesis research* **121** 95–106
[8] Yang M 2005 *The Journal of chemical physics* **123** 124705
[9] Guthmuller J, Zutterman F and Champagne B 2009 *The Journal of chemical physics* **131** 154302
[10] Guthmuller J, Zutterman F and Champagne B 2008 *Journal of chemical theory and computation* **4** 2094–2100
[11] Myers Kelley A 2003 *The Journal of chemical physics* **119** 3320–3331
[12] Kobayashi T 1996 *J-aggregates* vol 1 (University of Tokyo, Japan: World Scientific)

[13] Spano F C and Mukamel S 1991 *Physical review letters* **66** 1197

[14] Burghardt I, May V, Micha D A and Bittner E R 2009 *Energy Transfer Dynamics in Biomaterial Systems* (Springer)

[15] Christensson N, Kauffmann H F, Pullerits T and Mancal T 2012 *The Journal of Physical Chemistry Part B* **116** 7449–7454

[16] Plötz P A, Polyutov S, Ivanov S, Fennel F, Wolter S, Niehaus T, Xie Z, Lochbrunner S, Würthner F and Kühn O 2016 *Physical Chemistry Chemical Physics* **18** 25110–25119

[17] Fennel F, Wolter S, Xie Z, Plötz P A, Kühn O, Würthner F and Lochbrunner S 2013 *Journal of the American Chemical Society* **135** 18722–18725

[18] Philpott M R 1971 *The Journal of Chemical Physics* **55** 2039–2054

IOP Conf. Series: Journal of Physics: Conf. Series **917** (2017) 062052 doi:10.1088/1742-6596/917/6/062052

Surface enhanced Raman spectroscopy of fullerene C_{60} drop-deposited on the silvered porous silicon

N Khinevich[1], K Girel[1], H Bandarenka[1], V Salo[2], A Mosunov[2]
[1]Belarusian State University of Informatics and Radioelectronics, Minsk, 220013, Republic of Belarus
[2]Sevastopol State University, Sevastopol, 299053, Russian Federation

Abstract. Surface enhanced Raman spectroscopy (SERS) of fullerene C_{60} drop-deposited from the $1.4 \cdot 10^{-4}$ M aqueous solutions on the silvered porous silicon (Ag/PS) is reported for the first time. The used concentration is found to be not detected by the ordinary Raman spectroscopy. It is shown that SERS-spectrum of the fullerene deposited from the air-aged solution are characterized by less intensity than that of the fullerene solution kept out of the air. This indicates degradation of the fullerene solution due to oxidation. The results are prospective for the fast qualitative and quantitative analysis of the fullerene-based materials.

Introduction

Closed cage structures of carbon atoms such as C_{60}, C_{70}, C_{76}, C_{84} molecules are called fullerenes. In these molecules, atoms of carbon are located at the vertices of regular hexagons or pentagons that cover the surface of sphere or spheroid. C_{60} occupies special position among all the fullerenes. This molecule demonstrates the highest symmetry and, as a consequence, the greatest stability. Fullerenes are characterized by a number of unique properties including high sorption capacity, photoconductivity, mechanical strength, low surface energy, biocompatibility, etc. Thorough study of these features can lead to the breakthrough in the areas of materials' strengthening, energy storage, catalytic systems, micro- and nanoelectronics and biomedicine [1]. Accurate qualitative and quantitative analysis of the fullerenes is an urgent task to realize their successful application. Considering biomedicine, this problem has to be overcome at very low concentrations to control fullerene transfer through biological membranes, accumulation in proteins or cells and interaction with other submicron bioorganic objects. Scanning tunneling microscopy, nuclear magnetic resonance spectroscopy, different methods of vibrational spectroscopy have been used to study the fullerenes [2]. Since the beginning of fullerene discovery its Raman frequencies have been calculated [3] and measured [2]. Theoretical analysis [4] showed that 46 fundamental modes can be distinguished from the 174 possible types of C_{60} molecular vibrations. Two of these modes are characterized by A_g symmetry, one – A_u, three – T_{1g}, four – T_{1u}, five – T_{2u}, six – G_g, six – G_u, eight – H_g and seven – H_u. Vibrational modes of the T_{1u} symmetry are active during registration of the optical absorption spectra, those of the H_g, A_g symmetries are active in Raman spectroscopy, while the other types of oscillations can't be detected by optical methods. Remarkable advantage of Raman spectroscopy comparing to other techniques is in possibility to realize SERS effect. This means enormous enhancement of the Raman intensity by study of analyte adsorbed on the nanostructured metallic surfaces that are usually called SERS substrates. SERS spectroscopy allows to detect and investigate trace amounts of substances and thus is very attractive for the study of fullerenes. Several approaches of registration of C60 SERS spectrum in nonaqueous Ag and Au colloidal systems have been already developed [5, 6]. These methods provide good sensitivity, however, the solvent media contributes to the resulting Raman spectrum. That is why it requires to make additional manipulations to see only the Raman spectra of fullerenes. In this work, we propose to solve such a problem by using SERS substrate based on porous silicon (PS). PS covered with metallic nanostructures can be used as an effective SERS substrate [7]. In this work, we were able to register intensified Raman scattering of the C_{60} fullerene thanks to the SERS substrate composed of silver nanoparticles on PS (Ag/PS). According to our knowledge, such results have not been previously published.

Experimental

Fabrication of SERS substrates based on PS consists of two major steps: formation of porous template and deposition of metallic nanostructures on its surface. PS was fabricated by an electrochemical anodization of n^+-type silicon wafer in electrolyte of HF, deionized H_2O and C_3H_7OH mixed in the volume ratio 1:3:1. Prior to anodization process silicon wafer was cleaned in a solution NH_4OH, H_2O_2 and H_2O mixed in a volume ratio of 1:1:4. To remove native oxide from silicon wafer 4,5 % HF aqueous solution was used. Thickness of the PS layers is 5 μm and an average pore diameter is 60 nm. Silver nanostructures were formed on PS by a chemical deposition from aqueous solution of $AgNO_3$. Deposition process was carry out at room temperature. Before applying of analyte SERS-substrates were rinsed in HCl aqueous solution to remove contaminants which adsorbed on the surface of the substrates from environment. Fullerenes C_{60} were drop-deposited on the Ag/PS substrate from the air-aged and kept out of the air aqueous solutions and then dried at 21℃. Concentration of the fullerene solutions was $1.4 \cdot 10^{-4}$ M. SERS-spectra of fullerene were recorded with the 3D scanning confocal microscope Confotec NR500 using 633 nm laser and signal accumulation time 1 s. The laser spot varied in the 500 nm range while power of laser beam coming from the microscope objective was about 0.68 mW.

Results and discussion

Combinations of PS morphologies and regimes of metal deposition give rise to create the metallic nanostructures with fascinating SERS characteristics. In this work we used regimes of Ag/PS samples formation favourable for the most effective SERS according to [7]. SERS substrates presented silver nanostructures on the external surface of PS with an average size of silver nanoparticles in the range varying from 50 to 300 nm. Figure 1 shows SEM images of the virgin Ag/PS samples and those after drop deposition of the fullerene C_{60} at different magnifications. It is well-seen that the analyte is non-uniformly distributed on the Ag/PS surface (Fig. 1, b). Figure 1, f presents enlarged top view of the fullerene-coated Ag/PS. Following this image fullerene molecules formed the fullerite crystals with dimensions of 150 – 360 nm that are coalesced in aggregates of sizes varied in the range from 50 to 100 μm. Figure 2 shows the results of the SERS measurements of fullerene C_{60} drop-deposited on the SERS substrates. Measurements of the ordinary Raman spectra of fullerene deposited on the glass plate, samples of the silicon wafer and Ag-free PS were impossible due to low concentration of the carbon nanostructures. Both spectra in Figure 2 have three prominent bands of fullerene C_{60} such as $H_g(1)$, $A_g(1)$ and $A_g(2)$ [8]. The last one has the highest intensity and corresponds to the "pentagonal pinch" mode. It is known that the shift of this mode is caused by the symmetry breaking of the fullerene molecule. For instance, this can be observed at the molecule polymerization. Linear polymerization is accompanied by the $A_g(2)$ shift from 1470 to 1459 cm^{-1} [9]. The peak position at 1464 cm^{-1} is usually connected with formation of C_{60} dimer or $C_{60}O_2$ [10]. The $A_g(2)$ band of the kept of air fullerene is most intensive near the 1466 cm^{-1} while the maximum of the same band of the air-aged sample is shifted to the 1464 cm^{-1} position indicating the stronger oxidation of the last one. Both samples have a left shoulder in the $A_g(2)$ band extending to 1459–1460 cm^{-1} which shows a presence of a small amount of the polymerized fullerene molecules. This occurs due to C_{60} photopolymerization at the natural light. The bands $H_g(1)$ and $A_g(1)$ of the air-aged sample are slightly red-shifted. The kept out of the air sample is also characterized by weak bands $H_g(2)$ – $H_g(8)$ typical for fullerene C_{60} [8]. The marked bands in the SERS spectra correspond to ten vibrational modes of fullerene C_{60} (A_g, H_g) which are

IOP Conf. Series: Journal of Physics: Conf. Series **917** (2017) 062052 doi:10.1088/1742-6596/917/6/062052

Raman-active [4]. SERS intensity of the air-aged sample is about four orders of magnitude lower than that of the other sample. These results are caused by the changes in the fullerene structure due to oxidation of the air-aged sample and its destruction.

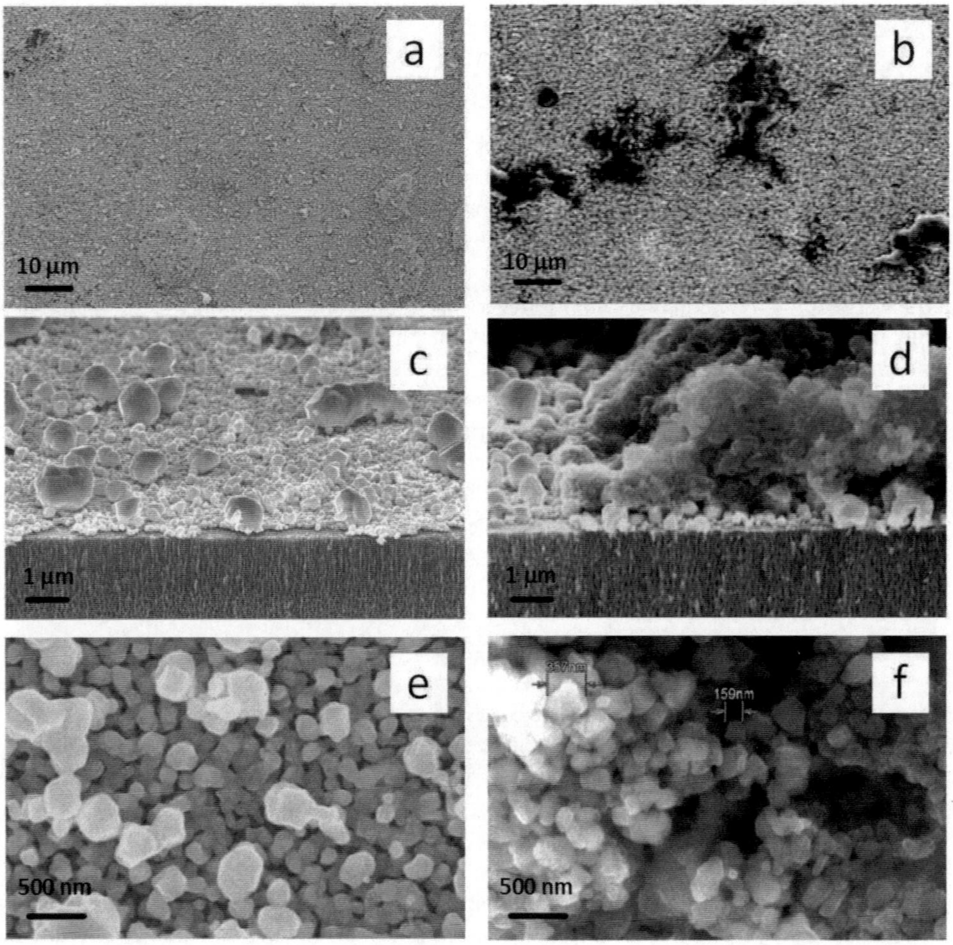

Figure 1. SEM images of the Ag/PS substrates (a, c, e) before and (b, d, f) after C_{60} deposition.

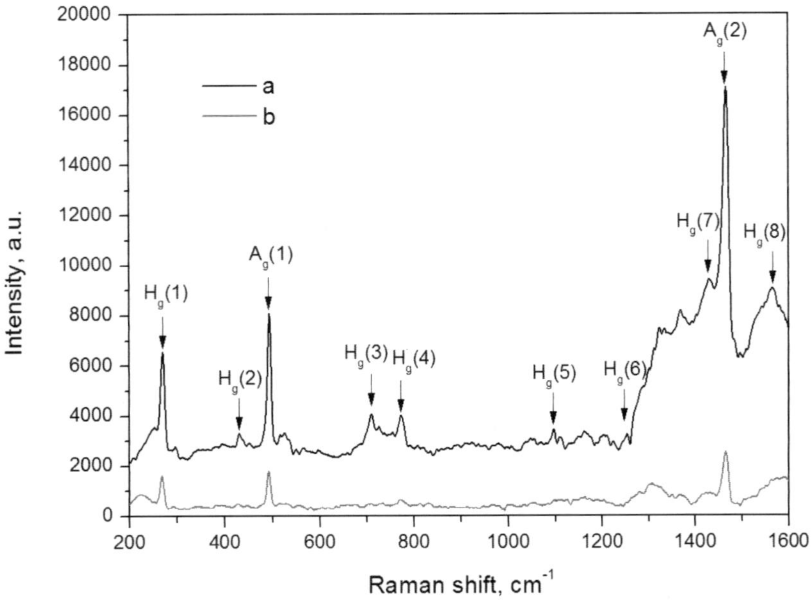

Figure 2. SERS-spectra of fullerene C_{60} deposited on Ag/PS from (a) kept out of air and (b) air-aged solutions.

Conclusion

We showed that Ag/PS can be used as effective SERS substrates for the detection and study of the structure of the submolar concentration of the fullerene C_{60} aqueous solutions. Structural changes of the air-aged fullerenes in comparison with the kept out of the air samples were revealed. The both fullerene samples were photopolymerized. At the same time, the air-aged sample was much more oxidized in comparison with the kept of the air fullerene solution. The obtained results open new opportunities for the precise control of the fullerene properties in different materials.

References

[1] Dresselhaus M S, Dresselhaus G, Eklund P C 1996 *Science of Fullerenes and Carbon Nanotubes, Academic Press,* ch. **10** p 292.
[2] Meijer G, Bethune D S, Tang W C, Rosen H J, Johnson R D, Wilson R J, Chambliss D D, Golden W G, Seki H, De Vries M S, Brown C A, Salem J R, Hunziker H E, Wendt H R 1991, *Mat. Res. Soc. Symp. Proc.* **206** 619.
[3] Weeks D E, Harter W G 1988, *Chemical physics letters* **144** 4.
[4] Negri F, Orlandi G, Zerbetto F 1988, *Chemical physics letters* **144** 31.
[5] Luo Z, Fang Y 2006, *Journal of Colloid and Interface Science* **301** 184.
[6] Baibarac M, Mihut L, Preda N, Baltog I, Mevellec J Y, Lefrant S 2005, *Carbon* **43** 1.
[7] Bandarenka H V, Girel K V, Bondarenko V P, Khodasevich I A, Panarin A Yu, Terekhov S N 2016, *Nanoscale Research Letters* **11** 262.
[8] Pospisil L, Gal M, Hromadova M, Bulickova J, Kolovoska V, Cvacka J, Novakova K, Kavan L, Zukalova M, Dunsch L 2010, *Phys. Chem. Phys.* **12** 14095.
[9] Talyzin A V, Dzwilewski A, Wagberg T 2006, *Sol. St. Commun.* **140** (3) 178.
[10] Kompanets V O, Chekalin S V, Melnik N N, Hess B C 1998, *JETP Lett.*, **68** (4) 344.

Magneto-optical microcavity with Au plasmonic layer

T V Mikhailova, S D Lyashko, S V Tomilin, A V Karavainikov, A R Prokopov, A N Shaposhnikov and V N Berzhansky
Institute of Physics and Technology, V.I. Vernadsky Crimean Federal University, Simferopol 295007, Russia

Abstract. Optical and Faraday rotation spectra of magneto-optical microcavity coated with Au plasmonic layer of gradient thickness were investigated theoretically and experimentally. It was shown that the Tamm plasmon-polaritons mode forms near the long-wavelength edge of photonic band gap. The presence of Au coating of thickness of 90.4 nm increase the Faraday rotation at Tamm plasmon-polaritons and cavity resonances in 1.3 and 7 times, respectively. By transfer matrix method it were found that the incorporation of SiO_2 buffer layer with a thickness in the range from 155 to 180 nm between microcavity and Au coating leads to the strong coupling between cavity mode and Tamm plasmon-polaritons. In this case, one or two resonances arise in the vicinity of the cavity mode depending on the thickness of plasmonic layer. The Faraday rotation for coupled mode in twice less than the value of rotation for single cavity mode.

1. Introduction

Microcavity (MC) is the most common element of modern optoelectronic devices. Cavity structure allows to enhance the different optical effects exhibited of embedded materials. One of the first proposed structures in magnetophotonics was the structure of MC based on nonmagnetic dielectric Bragg mirrors and a layer of bismuth-substituted iron garnet [1, 2]. In such structure the increasing of Faraday and Kerr rotation angles of the polarization plane is achieved most effectively. Later, the photonic and magnetophotonic crystals with localized Tamm surface states were investigated [3-11]. The excitation of optical Tamm state also leads to an amplification of magneto-optical (MO) effects. One of the effective methods for exciting optical Tamm states is to use the structure of a photonic crystal with a plasmonic coating or layer [4, 6, 8-11]. In this case, the optical Tamm state or Tamm plasmon-polaritons (TPP) can effectively be combined with resonances of another nature. For example, in the work [10] the hybrid state of Tamm and surface plasmon-polaritons were observed. The coupling of TPP and cavity modes occurs in MC with metal layers [8, 11]. In this paper, we propose to consider the influence of TPP excitation on properties of MO MC in Faraday effect geometry.

2. Experimental technique and modelling

The investigated MO MC consists of symmetrically dielectric Bragg mirrors of titanium and silicon oxides, between which the double garnet layer is placed. Two following configurations, schematic shown in Figure 1, are considered

$GGG / [TiO_2/SiO_2]^4 / Bi_{1.0}Y_{0.5}Gd_{1.5}Fe_{4.2}Al_{0.8}O_{12} / Bi_{2.8}Y_{0.2}Fe_5O_{12} / [SiO_2/TiO_2]^4$ / Au (No. 1) and

$GGG / [TiO_2/SiO_2]^4 / Bi_{1.0}Y_{0.5}Gd_{1.5}Fe_{4.2}Al_{0.8}O_{12} / Bi_{2.8}Y_{0.2}Fe_5O_{12} / [SiO_2/TiO_2]^4 / SiO_2$ / Au (No. 2).

There GGG is substrate of gadolinium gallium garnet with crystallographic orientation of (111). The MC No. 1 were synthesized and investigated experimentally and theoretically. Synthesis of MO MC performed according to the technology described in [12, 13]. Iron garnet layers were fabricated by reactive ion beam sputtering of corresponding ceramic targets in argon-oxygen mixture and crystallized in the annealing process at the air. TiO_2 and SiO_2 layers were synthesized by electron beam evaporation. Plasmonic Au coating with a gradient of thickness along the chosen direction on the sample surface, was deposited by thermal evaporation in vacuum [14, 15]. Using a unique technique of coating synthesis allowed in the experiment to choose different surface areas along a thickness gradient in order to reveal dimensional effects on the properties of structures. As a result, it became possible to compare the theoretical and experimental data obtained for different parameters of the layers of the structure. The MO MC No. 2 were considered only theoretically to investigate the capabilities of modes control.

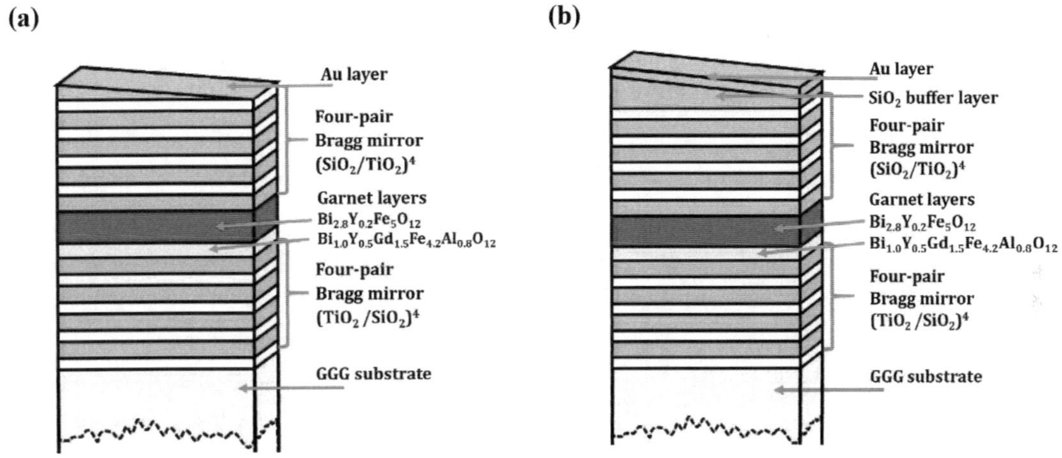

Figure 1 (a, b). Schematic diagram of investigated MO MC with configurations No. 1 **(a)** and No. 2 **(b)**.

The layers thicknesses of the synthesized structure determined by fitting the transmission and Faraday rotation (FR) spectra were $h_{M1} = 66$ nm for garnet layer of composition $Bi_{1.0}Y_{0.5}Gd_{1.5}Fe_{4.2}Al_{0.8}O_{12}$, $h_{M2} = 161$ nm for garnet layer of composition $Bi_{2.8}Y_{0.2}Fe_5O_{12}$, $h_{SiO2} = 117$ nm and $h_{TiO2} = 79$ nm for SiO_2 and TiO_2 layers of Bragg mirrors, respectively. The 4×4 transfer matrix method [16] was used for numerical solution of Maxwell's equations in modelling of light propagation. The components of layers permittivity tensors, taken into account, are given in the publication [13].

To measure the thickness of the deposited Au coating, a microinterferometer MII-4 with a digital data processing complex were used. The thickness of the wedge of Au coating from one edge of the sample to other varied from 0 to 100 nm. Investigation of transmittance was carried out in the wavelength range from 400 to 980 nm using an automated spectrophotometer KFK-3. Measurements of MO FR in the wavelength range from 400 to 750 nm was performed using hand-made computer-control spectropolarimeter by compensation method. The range of variation of the constant magnetic field was from 0 to 5 kOe. The sample was placed in a saturated field for Faraday effect measuring.

Buffer SiO_2 layer of thickness from 110 to 330 nm was incorporated in model configuration No. 2 between MO MC with the same layer parameters as in the structure No. 1 and Au coating of a certain thickness (only 10 and 40 nm are considered).

The comparison of calculated and experimental optical and FR spectra for structure No. 1 are given in the subsection 3.1. The results of modeling of the properties of structure No. 2 are given in the subsection 3.2.

3. Results and discussion

3.1. Microcavity with Au plasmonic layer of gradient thickness (No. 1)

Measured and calculated transmittance and FR spectra of MO MC with Au thickness h_{Au} from 6.7 to 90.4 nm are shown in Figure 2. The experiment is in a good agreement with the theoretical model. The center of photonic band gap (PBG) of the structure is placed at 690 nm. Wavelength of cavity mode is 653.5 nm. While the thickness of Au layer increases, the TPP mode forms near the long-wavelength edge of PBG in the vicinity of the wavelengths from 770 to 790 nm. TPP resonance is not observed at thicknesses h_{Au} less than 25 nm and has a maximum optical figure of merit at $h_{Au} = 65.8$ nm in experimental spectra. Theoretical spectra show that TPP mode is clearly separated from PBG edge at the thickness of $h_{Au} = 65.8$ nm.

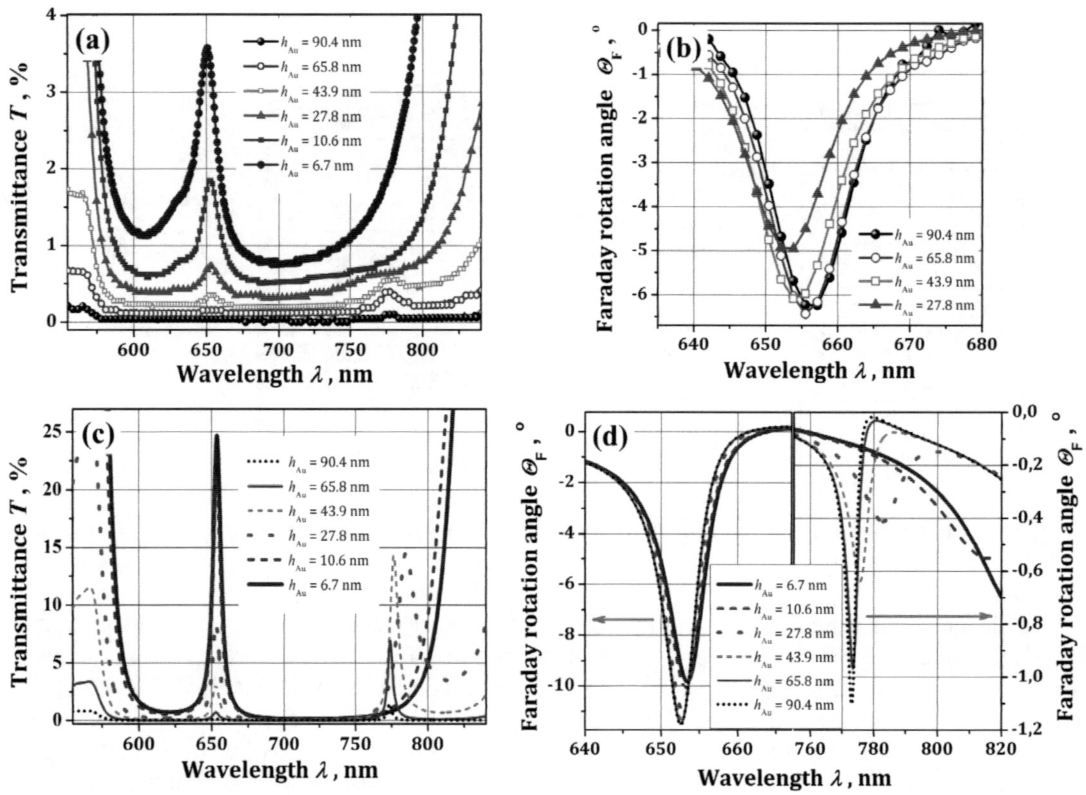

Figure 2 (a, b, c, d). Measured **(a, b)** and calculated **(c, d)** transmittance and FR spectra of synthesized MO MC No. 1 as a function of thickness of Au coating h_{Au}.

TPP resonance is more clearly expressed in theoretical transmittance spectra, since a higher absorption of Au coating that was used in calculation was realized in the experiment. The amplitudes of cavity and TPP resonances are compared in Figure 3 (a). TPP peak has amplitude higher than the amplitude of cavity peak for thickness of Au layer $h_{Au} > 35$ nm. Calculations shown that enhancement of FR at TPP mode for MO MC with $h_{Au} = 90.4$ nm is 7 times in comparison with the value for MO MC without Au layer. For cavity mode, the resonant value of transmittance decreases sharply at increasing of h_{Au} to 90.4 nm. Simultaneously, FR of cavity mode increases only by 1° in theory and 1.5° in experiment (Figure 2). These values correspond to enhancement factors of 1.15 and 1.3 times, respectively. According to the calculated spectra, blue shift of resonant modes takes place with increasing thickness h_{Au}. Shifting of TPP is more pronounced. Nevertheless, the red shift of resonant modes is observed in the experimental spectra. A possible reason for the difference in modes shift in the experimental and theoretical spectra is the structural changes in Au coating with increase of its thickness, which leads to a change of optical properties of coating. So, the corrections of the components of the permittivity tensor of Au layer for each thickness are needed. In the calculation presented in Figure 2, the same values of the components for different thicknesses h_{Au} were used [13].

Figure 3 (b, c, d) shows calculated spatial distributions of electric field intensity in MO MC No. 1 with Au coating of thickness $h_{Au} = 43.9$ nm and without Au coating for resonance modes. The distribution presented in Figure 3 (b) is classical for TPP mode. The maximum intensity is observed in the layer adjacent to the plasmonic coating, and an exponential decay of the electromagnetic field is formed inside the structure. Minor localization in the vicinity of the MO layer is present. Distribution for cavity mode is characterized by asymmetric localization in the vicinity of the boundaries of the MO layer (Figure 3, c, d). The intensity values are smaller for a configuration with an absorbing Au coating (Figure 3, c).

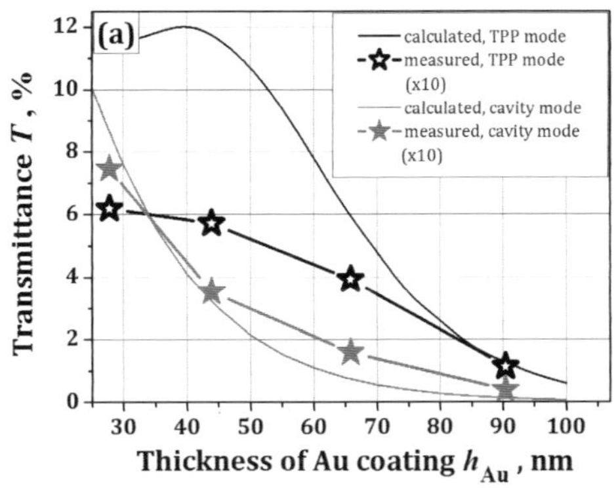

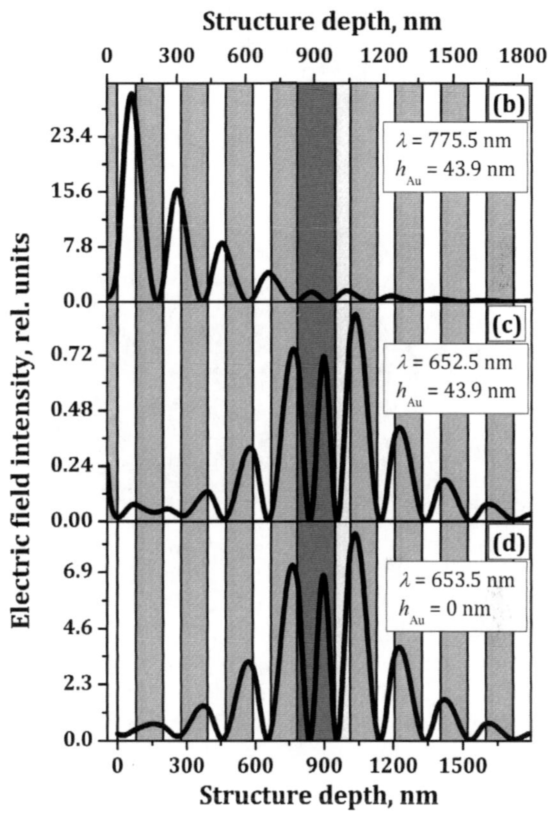

Figure 3 (a, b, c, d). Calculated and measured amplitudes of TPP and cavity resonances in transmittance spectra of MO MC No. 1 **(a)**. Calculated spatial distributions of electric field intensity in MO MC No. 1 with Au coating of thickness $h_{Au} = 43.9$ nm for TPP **(b)** and cavity **(c)** modes and without Au coating $h_{Au} = 0$ nm for cavity mode **(d)**; the incident electric field has unit amplitude, coming from the left.

3.2. Microcavity with buffer SiO_2 layer of gradient thickness and Au coating (No. 2)

As next step we investigated the configuration No. 2, in which the strong coupling between cavity mode and Tamm plasmon-polaritons occurs. Figure 4 shows image plots of the calculated transmittance and FR spectra for MO MC No. 2 with Au coating thicknesses $h_{Au} = 10$ nm and $h_{Au} = 40$ nm. The shift of TPP mode within the PBG is caused by varying thickness of buffer SiO_2 layer h_{bSiO2} from 110 to 330 nm. The shift of cavity mode is caused by the coupling to the TPP mode. Positions of modes and region of coupled mode are illustrated in the Figure 4 (a). The strong coupling between modes occurs at thicknesses $h_{bSiO2} = 155$ nm for $h_{Au} = 10$ nm and $h_{bSiO2} = 180$ nm for $h_{Au} = 40$ nm. The thickness of the Au coating strongly affects the characteristics of the coupled mode. Evidence of this coupling is formation of two nearly symmetric resonances with an increase of the thickness h_{Au} above 20 nm. The splitting of resonances is 17 nm for $h_{Au} = 40$ nm and $h_{bSiO2} = 180$ nm. The coupled mode has higher values of transmittance and almost in two times smaller values of the Faraday rotation angle.

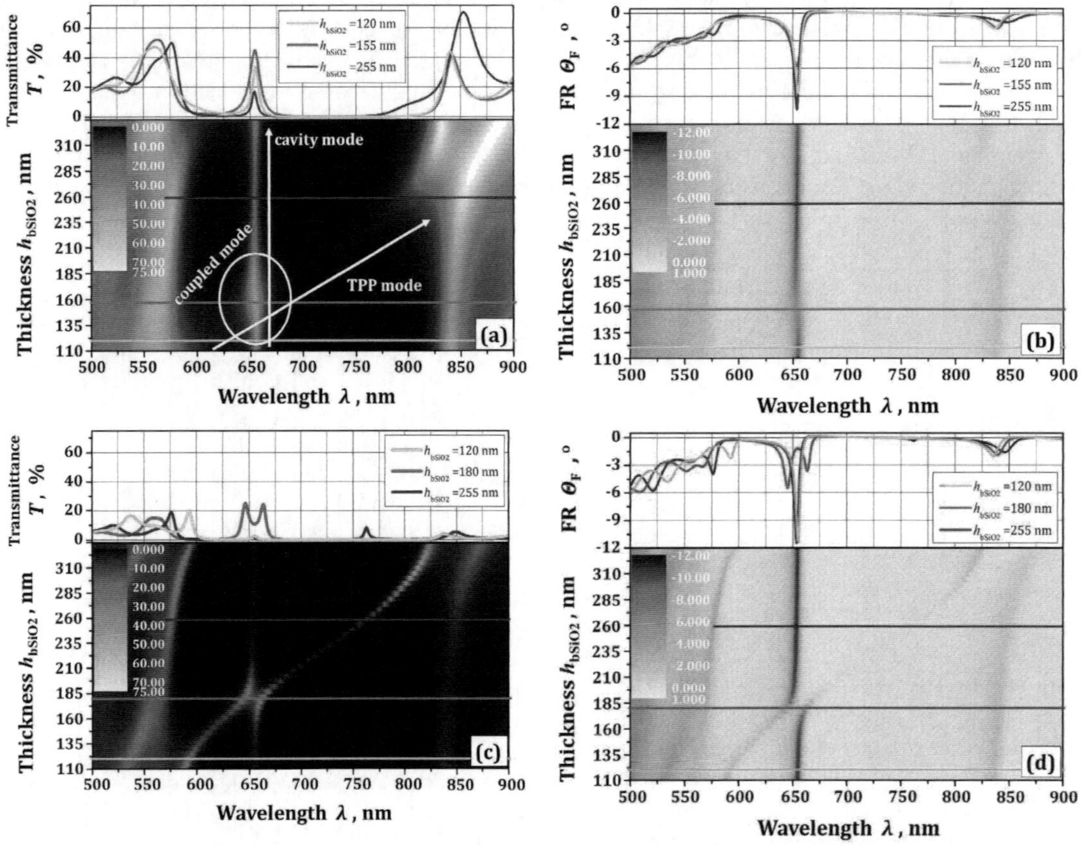

Figure 4 (a, b, c, d). Image plots of the calculated transmittance and Faraday rotation (FR) spectra for MO MC No. 2 with Au coating thickness h_{Au} = 10 nm **(a and b, respectively)** and h_{Au} = 40 nm **(c and d, respectively)** as a function of thickness of buffer SiO_2 layer h_{bSiO2}. The upper graphs show the spectra corresponding to the cross sections in the plots.

4. Conclusions

Optical and Faraday rotation spectra of magneto-optical microcavity with structure of $[TiO_2/SiO_2]^4$ / $Bi_{1.0}Y_{0.5}Gd_{1.5}Fe_{4.2}Al_{0.8}O_{12}$ / $Bi_{2.8}Y_{0.2}Fe_5O_{12}$ / $[SiO_2/TiO_2]^4$ on GGG substrate, coated with Au plasmonic layer of gradient thickness, were investigated theoretically and experimentally. It was shown that the Tamm plasmon-polaritons mode forms near the long-wavelength edge of photonic band gap. The presence of Au coating of thickness of 90.4 nm increase the Faraday rotation at Tamm plasmon-polaritons and cavity resonances in 1.2 and 7 times, respectively. The configuration of structure with SiO_2 buffer layer between microcavity and Au coating, in which the strong coupling between cavity mode and Tamm plasmon-polaritons arises, are investigated theoretically. In this case, one or two resonances arise in the vicinity of the cavity mode depending on the thickness of plasmonic layer. The Faraday rotation for coupled mode in twice less than the value of rotation for single cavity mode.

Acknowledgments

The authors acknowledge support by the RF Ministry of Education and Science in the framework of the state task (project no. 3.7126.2017/8.9).

References

[1] Lyubchanskii I L, Dadoenkova N N, Lyubchanskii M I, Shapovalov E A, Rasing Th 2003 *Journal of Physics D: Applied Physics* **36** R277

[2] Inoue M, Baryshev A V, Goto T, Baek S M, Mito S, Takagi H, and Lim P B 2013 Magnetophotonic Crystals: Experimental Realization and Applications *Magnetophotonics* eds. M. Inoue et al. (Springer-Verlag Berlin Heidelberg) chapter 7 pp 163–190

[3] Vinogradov A P, Dorofeenko A V, Erokhin S G, Inoue M, Lisyansky A A, Merzlikin A M and Granovsky A B 2006 *Phys. Rev. B* **74** 045128

[4] Kaliteevski M, Iorsh I, Brand S, Abram R A, Chamberlain J M, Kavokin A V, and Shelykh I A 2007 *Phys. Rev. B* **76** 165415

[5] Goto T, Dorofeenko A V, Merzlikin A M, Baryshev A V, Vinogradov A P, Inoue M, Lisyansky A A and Granovsky A B 2008 *Physical Review Letters* **101** 113902

[6] Kaliteevski M, Brand S, Abram R A, Iorsh I, Kavokin A V and Shelykh I A 2009 *Applied Physics Letters* **95** 251108

[7] Vinogradov A P, Dorofeenko A V, Merzlikin A M 2010 *Uspekhi Fizicheskikh Nauk* **180** 249

[8] Brückner R, Sudzius M, Hintschich S I, Fröb H, Lyssenko V G and Leo K 2011 *Phys. Rev. B* **83** 033405

[9] Zhou Hai-Chun, Yang Guang, Wang Kai, Long Hua, Lu Pei-Xiang 2012 *Chin. Phys. Lett.* **29** 067101

[10] Afinogenov B I, Bessonov V O, Nikulin A A and Fedyanin A A 2013 *Applied Physics Letters* **103** 061112

[11] Rahman SK S-U, Klein T, Klembt S, Gutowski J, Hommel D and Sebald K 2016 *Sci Rep* **6** 34392

[12] Berzhansky V, Shaposhnikov A, Prokopov A, Karavainikov A, Mikhailova T, Lukienko I, Kharchenko Yu, Golub V, Salyuk O, Belotelov V 2016 *JETP* **123** 744

[13] Mikhailova T V, Berzhansky V N, Karavainikov A V, Shaposhnikov A N, Prokopov A R and Lyashko S D 2016 *J. Phys.: Conf. Ser.* **741** 012121

[14] Tomilin S V, Yanovsky A S 2013 *J. of Nano- and Electronic Physics* **5** 03014

[15] Tomilin S V, Berzhansky V N, Shaposhnikov A N, Prokopov A R, Milyukova E T, Karavaynikov A V and Tomilina O A 2016 *J. Phys.: Conf. Ser.* **741** 012113

[16] Yin C P, Wang T B and Wang H Z 2012 *Eur. Phys. J. B* **85** 104

IOP Conf. Series: Journal of Physics: Conf. Series **917** (2017) 062054 doi:10.1088/1742-6596/917/6/062054

Dielectric Yagi-Uda nanoantennas driven by electron-hole plasma photoexcitation

S Li[1], S Lepeshov[1], R Savelev[1], D Baranov[2,3], P Belov[1], and A Krasnok[4]

[1]ITMO University, Saint Petersburg 197101, Russia
[2]Moscow Institute of Physics and Technology, Dolgoprudny 141700, Russia
[3]Department of Physics, Chalmers University of Technology, 412 96 Gothenburg, Sweden
[4]Departement of Electrical and Computer Engineering, The University of Texas at Austin, Texas 78712, USA

Abstract. All-dielectric nanophotonics based on high-index dielectric nanoparticles became a powerful platform for modern light science, providing many fascinating applications, including high-efficient nanoantennas and metamaterials. High-index dielectric nanostructures are of a special interest for nonlinear nanophotonics, where they demonstrate special types of optical nonlinearity, such as electron-hole plasma photoexcitation, which are not inherent to plasmonic nanostructures. Here, we propose a novel type of highly tunable all-dielectric Yagi-Uda nanoantennas, consisting of a chain of Si nanoparticles exciting by an electric dipole source, which allow tuning of their radiating properties via electron-hole plasma photoexcitation. We theoretically and numerically demonstrate the tuning of radiation power patterns and the Purcell effect by additional pumping of several boundary nanoparticles with relatively low peak intensities of fs-laser.

1. Introduction

In the last several years high-index dielectric nanoparticles and nanostructures [1] proved to be a promising platform for various nanophotonic applications, in particular for the design of functional nanoantennas [2], enhanced spontaneous emission [3], photovoltaics [4], frequency conversion [5], Raman scattering [6], and sensing [7]. The great interest in such nanostructures is caused mainly by their ability to control the electric and magnetic components of light at the nanoscale [1], while exhibiting low dissipative losses inherent to the materials with a negligible concentration of free charges [7]. In particular, it has been demonstrated that nanoantennas composed of high-index nanoparticles have the ability to realize directional scattering of the incident light and to effectively transform the near field of feeding quantum sources into propagating electromagnetic waves [2].

Modification of the spontaneous emission rate of a quantum emitter induced by its environment, known as the Purcell effect [8], is not so pronounced in all-dielectric structures [9], as in microcavities [10] or plasmonic nanoantennas [11]. This is due to the fact that optical resonances of high-index nanoparticles are characterized by relatively low-quality factors and large mode volumes, which results in low efficiency of light-matter interaction. However, it was recently shown that this disadvantage can be overcome by relying on the *Van Hove singularity* of a chain of high-index nanoparticles [12]. Such approach allows to substantially enhance the local density of optical states (LDOS) at the location of a quantum source and thus achieve high values of the Purcell factor with relatively small dielectric nanostructures while leaving unaltered all their other advantages.

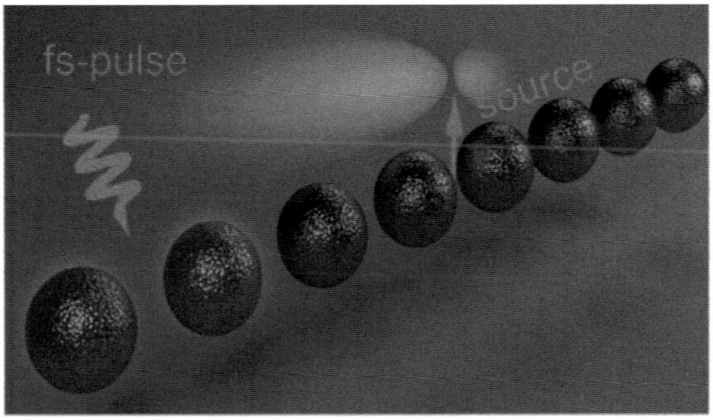

Figure 1. General view of the considered all-dielectric nanoantenna driven by the dipole source (orange arrow).

High-index dielectric nanostructures are also of a special interest for nonlinear nanophotonics, because they exhibit strong nonlinear responses. It was recently predicted and experimentally demonstrated that photoexcitation of dense electron-hole plasma (EHP) in silicon (Si) nanoparticles [13] and nanodimers [14] by femtosecond laser (fs-laser) pulses is accompanied by a dramatic modification of the radiation properties, whereas generation of EHP in germanium nanoantennas can even turn them into plasmonic ones in the mid-IR region [15]. Here, we propose a highly tunable all-dielectric nanoantenna, consisting of a chain of Si nanoparticles excited by an electric dipole source, which allows for tuning its radiation properties via electron-hole plasma photoexcitation. We theoretically and numerically demonstrate the tuning of radiation power patterns and Purcell factor by pumping several boundary nanoparticles in the chain with relatively low peak intensities of fs-laser pulses.

2. Result and Discussion

It is known that in periodic one-dimensional it is possible to observe Van Hove singularity, which is occurring at zero group velocity leading to infinite factor Purcell according to Ref. [16]:

$$F \cong \frac{1}{\pi}\left(\frac{\lambda}{2}\right)^2 \frac{c}{A_{\text{eff}} V_{\text{gr}}}, \tag{1}$$

with A_{eff} being the effective area of the resonant guided mode, λ the free space wavelength, V_{gr} the group velocity of the mode, and c is the speed of light. Here, we consider a chain of N spherical dielectric nanoparticles excited by an electric dipole (orange arrow), placed in the center of the chain and perpendicularly oriented to the chain axis, Figure 1. Crystalline silicon (c-Si) can be used as a material of the nanoantenna, with the real part of the dielectric constant being 16 in the considered frequency range [17]. All nanoparticles have the same radius r and center-to-center distance between neighboring particles is a.

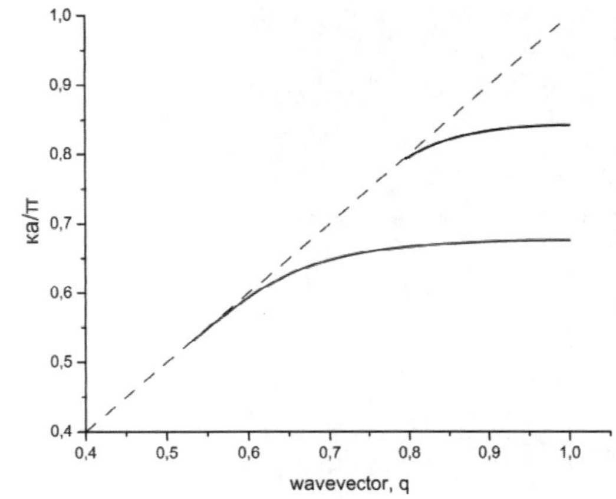

Figure 2. Dispersion curves of the eigenmodes in an infinite dielectric chain.

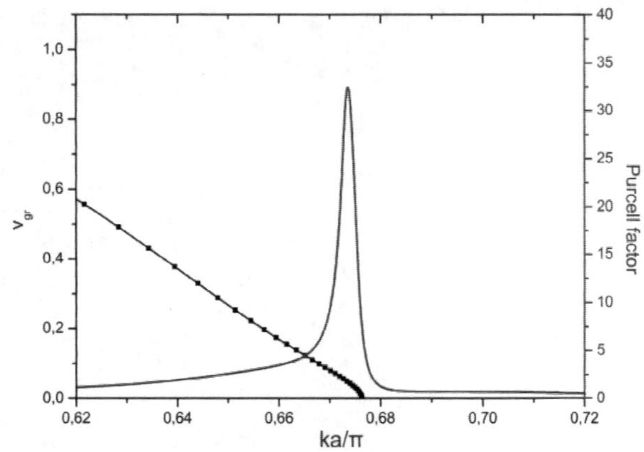

Figure 3. Spectral dependence of the Purcell factor (blue curve) and group velocity (black curve) for 8 dielectric nanoparticles.

To show existence of Van Hove singularity, we start by calculating dispersion of waveguide eigenmodes $\omega(k)$ in infinite chain of nanoparticles. For this we use well-known coupled-dipole model, where each particle is modeled as a combination of magnetic $\boldsymbol{m}$ and electric $\boldsymbol{p}$ dipoles oscillating with frequency ω. In the CGS system we can get the linear system of equations:

$$\mathbf{p}_i = \alpha_{ei} \sum_{j \neq i} (\hat{C}_{ij}\boldsymbol{p}_j - \hat{G}_{ij}\boldsymbol{m}_j), \tag{2}$$

$$\boldsymbol{m}_i = \alpha_{mi} \sum_{j \neq i} (\hat{C}_{ij}\boldsymbol{m}_j - \hat{G}_{ij}\boldsymbol{p}_j),$$

where, $\hat{C}_{ij} = A_{ij}\hat{I} + B_{ij}(\hat{r}_{ij} \otimes \hat{r}_{ij})$, $\hat{G}_{ij} = -D_{ij}\hat{r}_{ij} \times \hat{I}$, $\otimes$ is the dyadic product, $\hat{I}$ is the unit 3x3 tensor, $\hat{r}_{ij}$ is the unit vector in the direction from i-th to j-th sphere, and

$$A_{ij} = \frac{\exp(ik_h r_{ij})}{r_{ij}} \left(k_h^2 - \frac{1}{r_{ij}^2} + \frac{ik_h}{r_{ij}}\right), \tag{3}$$

$$B_{ij} = \frac{\exp(ik_h r_{ij})}{r_{ij}} \left(-k_h^2 + \frac{3}{r_{ij}^2} - \frac{3ik_h}{r_{ij}}\right),$$

$$D_{ij} = \frac{\exp(ik_h r_{ij})}{r_{ij}} \left(k_h^2 + \frac{ik_h}{r_{ij}}\right),$$

where, r_{ij} is the distance between the centers of i-th and j-th spheres, ε_h is the permittivity of the host medium, $k_h = \sqrt{\varepsilon_h}\, \omega/c$ is the host wavenumber, $\omega = 2\pi\nu$, ν is the frequency. The quantities α_e and α_m are the magnetic and electric polarizabilities of a spherical particle [18]:

$$\alpha_e = i\frac{3\varepsilon_h a_1}{2k_h^3}, \qquad \alpha_m = i\frac{3b_1}{2k_h^3}, \tag{4}$$

where a_1 and b_1 are electric and magnetic Mie coefficients.

Dispersion of infinite chain with r = 70 nm and a = 200 nm in free space is presented on Figure 2. Here, we use the dimensionless wavenumber $q = {}^{\beta a}/_{\pi}$, where β is the Bloch propagation constant. One can observe two modes in this frequency range. To characterize this mode, we calculate electric and magnetic moments and field distribution inside of nanoparticles. It turns out that blue and black curves correspond to TM-mode and TE-mode, respectively. In case of transverse magnetic mode, magnetic moments of each nanoparticles oscillating with π phase difference, which matches to magnetic field of dipole, orientated in the z direction. For this reason, we may expect significant enhancement of the Purcell factor for a finite system around the frequency, where group velocity of TM vanishes.

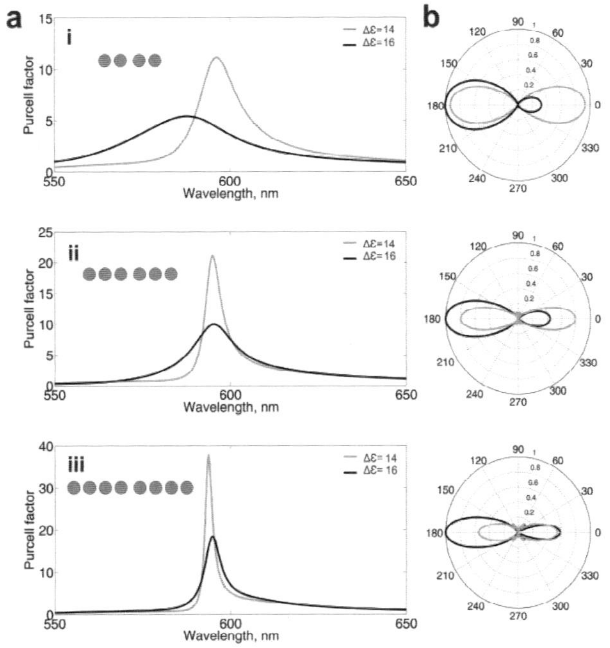

Figure 4. (a) Spectral dependence of the Purcell factor for the chains of dielectric nanoparticles for different number of nanoparticles N: (i) N = 4; (ii) N = 6; (iii) N = 8. (b) Radiation power patterns (E-plane) of the electric dipole source at the radiation wavelength of 600 nm. Green curves correspond to the unaffected chains Δε=0, whereas the black ones correspond to the chains with photo-excited boundary particles (Δε=-2).

To confirm this expectation, we calculate the Purcell factor using the Green's tensor approach [19]:

$$F = \frac{3}{2k_h^3}\mathbf{z}\cdot \mathrm{Im}[\boldsymbol{G}(0,0,\omega)]\cdot \mathbf{z}, \tag{5}$$

with $G(0;\ 0;\ \omega)$ being Green's tensor of an electric dipole in the center of chain (point of the dipole source localization) and z being the unit vector pointing in the z direction (Figure 1). Figure 3 shows the group velocity (black line) of TM-mode and calculated Purcell factor (blue line) as a function of wavelength for finite chain consisting of 8 nanoparticles with r = 70 nm and a = 200. We can observe a maximum of factor Purcell around the Van Hove singularity.

On the next step we have shown that Van Hove singularity is very sensitive to electron-hole plasma (EHP) photoexcitation in boundary particles. Due to this excitation, one can control optical properties of the material and consequently the radiation properties (intensities of emission and power patterns) of the nanoantenna. To describe EHP-induced tuning of the nanoantenna, we employ the analytical approach developed in Ref. [14]. The dynamics of volume-averaged EHP density ρ_{eh} is modeled via the rate equation:

$$\frac{d\rho_{\mathrm{eh}}}{dt} = -\Gamma\rho_{\mathrm{eh}} + \frac{W_1}{\hbar\omega} + \frac{W_2}{2\hbar\omega}, \tag{6}$$

where, $W_{1,2}$ are the volume-averaged absorption rates due to one- and two-photon processes, and Γ is the EHP recombination rate which depends on EHP density [20]. The absorption rates are written in the usual form as $W_1 = \frac{\omega}{8\pi}\langle|\tilde{E}_{\mathrm{in}}|^2\rangle \mathrm{Im}(\varepsilon)$ and $W_2 = \frac{\omega}{8\pi}\langle|\tilde{E}_{\mathrm{in}}|^4\rangle \mathrm{Im}\chi^{(3)}$ where angle brackets denote averaging over the nanoparticle volume, and $\mathrm{Im}(\chi^{(3)}) = \frac{\varepsilon c^2}{8}\beta$, with β being two photon absorption coefficient. The relaxation rate of EHP in c-Si is dominated by Auger recombination $\Gamma = \Gamma_{\mathrm{ag}}\rho_{\mathrm{eh}}^2$, $\Gamma_{\mathrm{ag}} = 4\cdot10^{-31}\mathrm{s}^{-1}\mathrm{cm}^6$. Now the permittivity of photoexcited Si should be related to time dependent EHP density:

$$\varepsilon(\omega,\rho_{\mathrm{eh}}) = \varepsilon_0 + \Delta\varepsilon_{\mathrm{bgr}} + \Delta\varepsilon_{\mathrm{bf}} + \Delta\varepsilon_D, \tag{7}$$

where ε_0 is the permittivity of non-excited material, while $\Delta\varepsilon_{\mathrm{bgr}}$, $\Delta\varepsilon_{\mathrm{bf}}$, $\Delta\varepsilon_{\mathrm{D}}$ are the contributions from bandgap renormalization, band filling, and Drude term, respectively. The detailed expressions for all contributions in Eq. (7) can be found in Ref. [14].

We calculated Purcell factor for different number of particles N before and after EHP photoexcitation (Figure 4). The calculations are performed for the change of the real part of Si permittivity $\Delta\varepsilon=-2$, where $\Delta\varepsilon = \varepsilon(\omega,\rho_{\mathrm{eh}}) - \varepsilon_0$, which is achieved at the wavelength of 600 nm upon excitation of EHP with density $\rho_{\mathrm{eh}} = 1.5\cdot10^{21}\mathrm{cm}^{-3}$. The decrease of the Purcell factor approximately by a factor of 2 in all cases (i–iii), along with the spectral broadening, are caused by symmetry breaking of the chain and corresponding decrease of the quality factor of the Van Hove singularity mode. The EHP photoexcitation also modifies the radiation pattern of the nanoantenna, Figure 4(b). Before the plasma excitation ($\Delta\varepsilon = 0$) the radiation pattern has two symmetric lobes directed along the chain axis in forward and backward directions (green curves). In this case, the maximal value of directivity grows with increasing of *N*. After the plasma excitation ($\Delta\varepsilon = 2$), nanoantenna radiates mostly in the direction of the affected particles. We note that the degree of modification of the radiation pattern grows with increasing number of particles. For example, in the case of *N=8* [Figure 4(b)iii], the directivity in the left direction is two times larger than in the right one. Thus, the EHP photoexcitation can be applied for all-optical switching of radiation patterns. Such dramatic tuning of the radiation pattern is caused by the Van Hove singularity regime of the initially unaffected nanoantenna.

3. Conclusion

We have proposed highly tunable all-dielectric nanoantennas, consisting of a chain of Si nanoparticles excited by an electric dipole source, that allow tuning their radiation properties via electron-hole plasma photoexcitation. We have theoretically and numerically demonstrated the tuning of radiation power patterns and the Purcell effect by additional pumping several boundary nanoparticles with relatively low peak intensities.

Acknowledgments

This work was financially supported by Russian Science Foundation (Grant 15-19-30023) and by Russian Foundation for Basic Research (Project 16-37-60076). This work was also partially supported by the Air Force Office of Scientific Research.

References

[1] Kuznetsov A I, Miroshnichenko A E, Brongersma M L, Kivshar Y S, and Luk'yanchuk B 2016 Science vol 354 p 2472
[2] Rolly B, Geffrin J-M, Abdeddaim R, Stout B, and Bonod N 2013 Scientific Reports

[3] Regmi R, Berthelot J, Winkler P M, Mivelle M, Proust J, Bedu F, Ozerov I, Begou T, Lumeau J, Rigneault H, Garc´ia-Paraj´o M F, Bidault S, Wenger J, and Bonod N 2016 Nano Letters vol 16 p 5143
[4] Brongersma M L, Cui Y, and Fan S 2014 Nature Materials vol 13 p 451
[5] Carletti L, Locatelli A, Stepanenko O, Leo G, and De Angelis C 2015 Optics Express vol 23 p 26544
[6] Dmitriev P A, Baranov D G, Milichko V A, Makarov S V, Mukhin I S, Samusev A K, Krasnok A E, Belov P A, and Kivshar Y S 2016 Nanoscale 8 p 9721
[7] Caldarola M, Albella P, Cort´es E, Rahmani M, Roschuk T, Grinblat G, Oulton R F, Bragas A V, and Maier S A 2015 Nature Communications vol 6 p 7915
[8] Purcell E M 1946 Phys. Rev vol 69 p 681
[9] Bozhevolnyi S I and Khurgin J B 2016 Optica vol 3 p 1418
[10] Vahala K J 2003 Nature vol 424 p 839
[11] Russell K J, Liu T-L, Cui S, and Hu E L 2012 NaturePhoton. vol 6 p 459
[12] Krasnok A, Glybovski S, Petrov M, Makarov S, Savelev R, Belov P, Simovski C, and Kivshar Y 2016 Applied Physics Letters vol 108 p 211105
[13] Makarov S, Kudryashov S, Mukhin I, Mozharov A, Milichko V, Krasnok A, and Belov P 2015 Nano Letters vol 15 p 6187
[14] Baranov D G, Makarov S V, Krasnok A E, Belov P A, and Al`u A 2016 Laser & Photonics Reviews vol 10 p 1009
[15] Fischer M P, Schmidt C, Sakat E, Stock J, Samarelli A, Frigerio J, Ortolani M, Paul D J, Isella G, Leitenstorfer A, Biagioni P, and Brida D 2016 Phys. Rev. Lett. vol 117 p 047401
[16] Manga Rao V S C and Hughes S 2007 Physical Review B vol 75 p 205437
[17] Aspnes D E and Studna A A 1983 Physical Review B vol 27 p 985
[18] Bohren C F and Huffman D R 1998 Absorption and scattering of light by small particles New York: Wiley
[19] Novotny L and Hecht B 2006 Principles of Nano-Optics Cambridge University Press
[20] Yu P and Cardona M 2005 Fundamentals of semiconductors Springer

IOP Conf. Series: Journal of Physics: Conf. Series **917** (2017) 062055 doi:10.1088/1742-6596/917/6/062055

Purcell effect in complex photonic structures based on optical Fibonacci lattices

K M Morozov[1,2]*, A R Gubaydullin[1,2], K A Ivanov[1,2], M A Kaliteevski[1,2,3]
[1]*St. Petersburg Academic University, St. Petersburg, 194021 Russia*
[2]*ITMO University, St. Petersburg, 197101 Russia*
[3]*Ioffe Institute, St. Petersburg, 194021 Russia*

*E-mail: morozovk@spbau.ru

Abstract. The modification of probability of spontaneous emission for a dipole emitter that placed in Fibonacci lattice-based palindrome structures was examined. Localization properties of eigenstates, that appears in photonic band gaps of palindrome structure was investigated. It was shown, that for a dipole placed in centre of palindrome structure an increase in the spontaneous emission rate will be observed and, in addition, calculated value of Purcell factor is significantly greater than for critical localized states of usual Fibonacci lattices, and close to microcavity eigenmode case.

1. Introduction

Photonic quasicrystals (PQC) have some interesting properties like fractal density of states and connected with that property light localization [1,2]. It is well known, that environment (media with spatially inhomogeneous dielectric constant) modify spontaneous emission rate of a point dipole [3]. Environment of Fibonacci-type PQC gives increasing of spontaneous emission rate if frequency and direction of a dipole emission correspond to critical localized eigenstate of structure, but effect is less expressed than in planar microcavities [4].

Photonic structures with complex geometrical properties that formed by stacking the various types of PQC can bring new aspects in light behavior in dielectric media. In the review [5] was reported about results of analysis of optical properties of complex structures, based on PQC structures (Fibonacci) and some aperiodical types structures (Thue–Morse, Rudin–Shapiro, period-doubling sequences). Some eigenstates, that appears in photonic band gap (PBG) of hybrid photonic multilayer structures may correspond to perfect transmission resonances. That fact may be useful in design of non-reciprocal optical devices [6,7].

The localization behavior of these, symmetry-induced eigenstates can be different from critical localized states, and has nature close to microcavity eigenmode. The aim of present paper is to investigate, how environment of Fibonacci palindrome structures make change spontaneous emission rate for a dipole, and, particularly contribution of PBG states to effect by using S-quantization formalism.

2. Results and discussion

Optical Fibonacci lattice is a sequence of layers of two types: a large one with thickness L, and a small one with thickness S, and with refractive indexes n_L and n_S respectively. Layers parameters are satisfied to relation (τ – the golden ratio)

$$\frac{L}{S} = \frac{n_S}{n_L} = \tau \approx 1.618. \quad (1)$$

Fibonacci structure of order F_i is constructed using the recurrent rule $F_{i+1} = \{F_{i-1}, F_i\}$ with initial values $F_0 = L$ and $F_1 = S$. Spatial scale of structure (L and S) sets characteristic frequency

$$\omega_0 = \pi c/(n_s S + L). \quad (2)$$

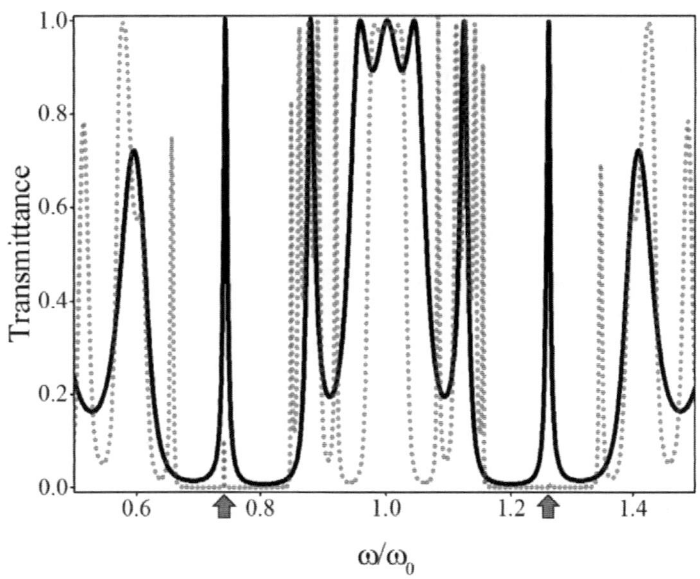

Figure 1. (color online) Transmission spectrum of Fp_6 palindrome structure (solid curve) and Fp_8 structure (dotted curve). Arrows indicate two eigenmodes in PBG with frequencies ω_1=0.743ω_0 and ω_2=1.261ω_0.

Now consider the Fp_6 palindrome structure based on F_6 lattice. The structure was constructed by stacking of two mirror-mapped ($\overleftarrow{F_6}$ and $\overrightarrow{F_6}$) optical lattices as shown in figure 2 (a). As you can see, in the center of Fp_6 palindrome structure appears $\lambda/2$ defect, which gives enhancing of EM field localization (like in microcavities) in structure. The transmission spectrum of Fp_6 and Fp_8 structures, obtained by transfer matrix method shown in figure 1. Spectrum looks like the case of regular Fibonacci structure, except of presence of two peaks in PBG's, that corresponds to states with frequencies ω_1=0.743ω_0 and ω_2=1.261ω_0. These peaks are having place in Fp_8 spectrum too, but expressed less. This fact indicates the presence of that states in structures of higher order.

IOP Conf. Series: Journal of Physics: Conf. Series **917** (2017) 062055 doi:10.1088/1742-6596/917/6/062055

The states in PBG are appears only when parts of structure are composing in mirror-mapping style (for example, if compose the $\overrightarrow{F_6}$ lattice with itself, as a result will be only modified spectrum of states of the higher order lattice). In case of stacking lattices of odd order, due to edge layers of lattice has n_L refractive index there is no way to get $\lambda/2\ n_S$ cavity in the centre, so there are no states in PBG.

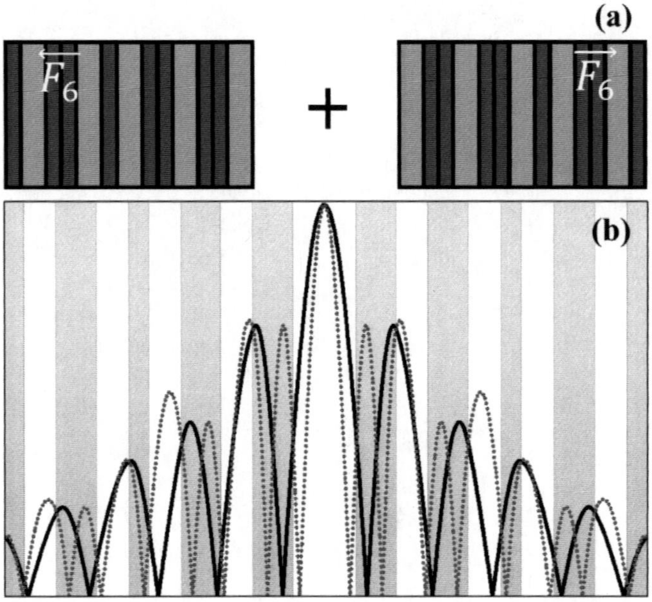

Figure 2. (color online) (a) Schematic drawing of the Fp_6 palindrome structure forming (b) Symmetric modes $\tilde{\boldsymbol{\epsilon}}^{(1)}$ profile of electric field amplitude of Fp_6 structure with corresponding frequencies ω_1 (solid curve) and ω_2 (dotted curve). On background – refractive index profile (n_S - grey color, n_L-white color).

S-quantization formalism - rigorous self-consistent procedure, that equating eigenvalues of scattering matrix of the system to unity to calculate the directional dependence of spontaneous emission rate [8,9]. Procedure of *S*-quantization method is based on analysis of changing of spatial eigenmode profile in layered structure $\tilde{\boldsymbol{\epsilon}}^{(1,2)}$ in relation to uniform media $\epsilon^{(1,2)}$. For TE polarization eigenmodes of structure $\tilde{\boldsymbol{\epsilon}}^{(1)}$ (symmetric) and $\tilde{\boldsymbol{\epsilon}}^{(2)}$ (antisymmetric) relate to modes in uniform media by coefficients $Y^{(1,2)}$

$$\tilde{\boldsymbol{\epsilon}}^{(1,2)} = \begin{pmatrix} 0 & \tilde{\epsilon}_y^{(1,2)} & 0 \end{pmatrix} = \begin{pmatrix} 0 & Y^{(1,2)}\epsilon_y^{(1,2)} & 0 \end{pmatrix}. \quad (3)$$

In the case of TE polarization, for the dipole oriented along y-axis (emitter is co-directed with modes electric field), directional Purcell factor (Purcell factor) is

$$F_\theta^{(TE)} = \sum_{i=1,2}\left|Y^{(i)}\right|^2. \quad (4)$$

Figure 2 (b) shows distribution of electric field amplitude in Fp_6 corresponds to symmetric modes $\tilde{\boldsymbol{\epsilon}}^{(1)}$ with frequencies $\omega_1 = 0.743\omega_0$ and $\omega_2 = 1.261\omega_0$. As it might be seen, amplitude of field grows exponential to the centre of structure. That type of field distribution usually inherent in microcavities. Field is localized in center cavity of media, which means for a dipole placed in center layer should be observed the enhancing of spontaneous emission rate.

Figure 3 shows dependency of the directional Purcell factor $F_\theta^{(TE)}$ on the emission angle and frequency for Fp_6 palindrome structure, when emitter is positioned at the centre of the structure. It can be seen that when frequency and angle corresponds to the considering modes, there is maximum value of Purcell factor (396). Contribution of PBG modes in picture of effect is higher than contribution of critical localized states, due to higher localization of electric field.

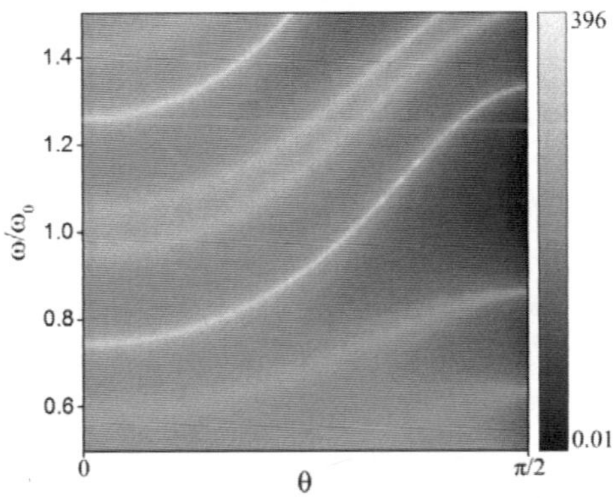

Figure 3. (color online) Dependency of the directional Purcell factor $F_\theta^{(TE)}$ on the emission angle θ and frequency for Fp_6 palindrome structure; emitter is positioned at the centre of the structure.

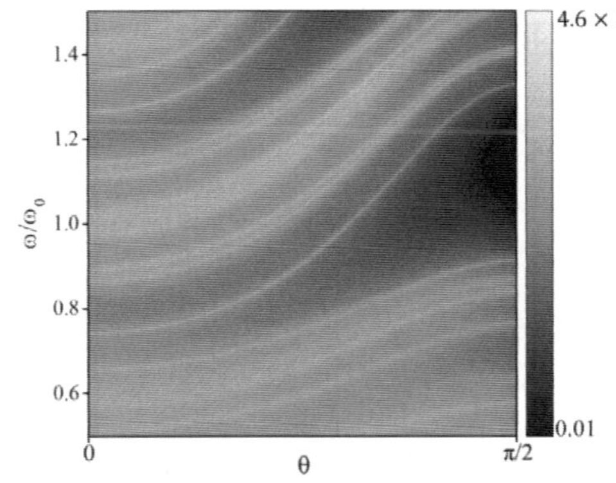

Figure 4. (color online) Dependency of the directional Purcell factor $F_\theta^{(TE)}$ on the emission angle θ and frequency for Fp_8 palindrome structure; emitter is positioned at the centre of the structure.

When the order of palindrome structure is increasing, in spectrum of eigenmodes appears more and more critical localized states (figure 1), but frequency of considered states almost the same. These facts correlate with results of calculation of directional Purcell factor $F_\theta^{(TE)}$ frequency for Fp_8 palindrome structure, when emitter is positioned at the centre of the structure (figure 4). As you can see, dispersion of considered PBG states is similar to case of Fp_6 palindrome structure, and maximum value of Purcell factor again falls on frequencies and angles, that corresponds to PBG states and reaches to 4.6×10^5. This value is significantly greater than values that can be achieved in usual Fibonacci lattices. For clarity, maximum of Purcell factor value $F_{max}^{(TE)}$ was built as a function of length for a set of planar photonic structures eigenstates (figure 5). Due to strong (exponential) localization of PBG modes, behavior of dependency $F_{max}^{(TE)}(D/S)$ in palindrome structures is closer to planar microcavities eigenmode scenario, and approximately like exponent function of normalized length. When the case of usual Fibonacci lattices dependency is closer to behavior of Bragg reflectors edge state (as power law of normalized length), due to less marked field localization. The considered symmetry-induced states, that appears in PBG's of palindrome structures are characterized by high localization, and their maximum of electric field are isolated from field other critical localized states, that would be useful in design of some new optical devices.

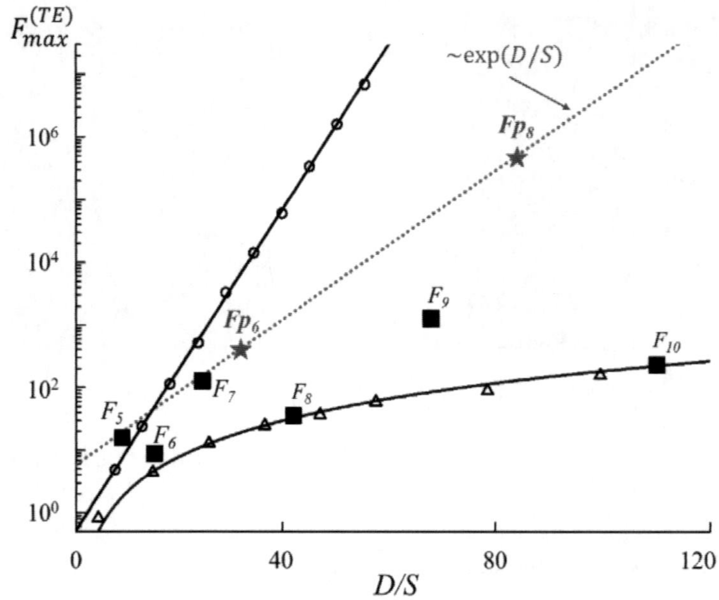

Figure 5. (color online) Dependencies $F_{max}^{(TE)}$ value as functions of normalized length for set of photonic structures eigenstates: microcavity eigenmode (circles), Bragg reflector edge state (triangles), Fibonacci lattice critical state (squares), Fibonacci lattice-based palindrome structures state (stars).

3. Conclusions

Calculation of spontaneous emission rate for emitter placed in Fibonacci lattice-based palindrome structures using *S*-quantization formalism was carried out. Was studied how directional Purcell factor for TE polarization depend on palindrome structures order when the dipole is co-directed with electric field of eigenmode. It was shown, that couple of states, that appears in palindrome structures, behavior is close to microcavity eigenstate, namely described by strong localization and gives great contribution in Purcell effect in photonic structure.

Acknowledgments

This work has been supported by Russian Science Foundation (Project no. 16-12-10503)

References

[1] W. Gellermann, M. Kohmoto, B. Sutherland, P.C. Taylor, Phys. Rev. Lett. **72** 633 (1994).
[2] M.A. Kaliteevski, R.A. Abram, S. Brand, V.V. Nikolaev, Opt. Spect. **91** 109-118 (2001).
[3] E. M. Purcell, Phys. Rev. **69**, 681 (1946).
[4] K.M. Morozov, K.A. Ivanov, A.R. Gubaidullin, M.A. Kaliteevski, Opt. Spect. **122**, *No. 2, 235–242* (2017).
[5] Enrique Maciá, *Rep. Prog. Phys.* **75** 036502 (2012).
[6] S V Zhukovsky, *Phys. Rev. A,* **81**, 053808 (2010).
[7] R. W. Peng, X. Q. Huang, F. Qiu, Mu Wang, A. Hu, S. S. Jiang & M. Mazzer, *Appl. Phys. Lett.,* **80**, 3063-3065 (2002).
[8] M.A. Kaliteevski, V.A. Mazlin, K.A. Ivanov, A.R. Gubaidullin, Opt. Spect. **119**, 832-837 (2015).
[9] M.A. Kaliteevski, A.R. Gubaidullin, K.A. Ivanov, V.A. Mazlin, Opt. Spect. **121**, 71-81 (2016).

The motion of nanoparticles under the non-conservative forces mediated by surface plasmon polaritons

I D Toftul[1], A A Bogdanov[1,2], M I Petrov[1,2]
[1]Department of Theoretical Physics, St Petersburg National Research Academic University of the Russian Academy of Sciences, St. Petersburg 194021, Russia
[2]The Metamaterials Laboratory, University ITMO, Saint Petersburg 199034, Russia

Abstract. We have developed the theoretical and numerical modelling of nanoparticle dynamics near planar metallic interface under light radiation. Using our model, we employed the Green's function formalism to simulate the dynamics of nanoparticles under the action of surface plasmon polariton mediated optical forces. By varying the illumination conditions, we determined the different regimes of motion, such as surface attraction and repulsion, and optical pulling regime. We showed that the topology of the trajectories dramatically change, when the surface plasmon polaritons are excited.

1. Introduction
Nowadays optical tweezing and light manipulation of nanoobjects became one of the common tools in modern optical instrumentation [1]. The unique opportunities for non-invasive control of objects at the nanoscale made it highly prospective for biological and biochemical applications [2]. Currently, the active studies in the field of optical forces are devoted to the forces generated via momentum exchange with surface waves [3-5]. This approach gives new prospectives for developing methods of nanoparticles transport and optical sorting. While almost all the studies are devoted to stationary computations of optical forces, the exact dynamics of the nanoparticles motion in the presence of the substrate is not studied yet. In our work, we suggest the dynamical model of nanoparticle motion near the interface between two media. We apply the model for simulating nanoparticle trajectories near metallic interfaces, which support surface plasmon polariton (SPP) waves. Using the developed model, we show the different regimes of the nanoparticle dynamics due to SPP modes interaction, which strongly depends on the illumination conditions.

2. Forces acting onto a point dipole near interface
We start by examining the problem in the dipole approximation when the nanoparticle size is smaller than the wavelength and typical scales of electric field variations. We consider a spherical dielectric nanoparticle with permittivity ε_p and radius R located inside a medium with permittivity ε_m. The particle is placed at the height z_0 above a substrate with permittivity ε_s, as shown in Figure 1. The incident wave with wavelength $\lambda = 2\pi/k_1$ impinges onto the surface at the angle ϑ.

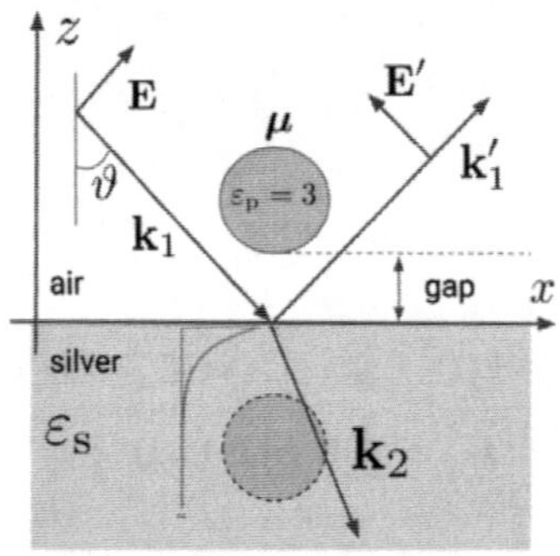

Figure 1. (a)The inset depicts the geometry of the problem: a dielectric particle is located near silver planar interface. Incident electromagnetic wave is propagating from the upper half-space.

Assuming that the nanoparticle is small in comparison to the wavelength, $R \ll \lambda$, one can consider it as a point electric dipole with polarizability $\alpha_0 = i6\pi\varepsilon_0 a_1/\left(k_1\sqrt{\varepsilon_m}\right)^3$, which is given by the Mie theory, where $a_1(\varepsilon_p, \varepsilon_m, k_1, R)$ is the scattering coefficient [8]. The averaged optical force which is acting onto a particle located in $\mathbf{r}_0$ in an arbitrary electric field $\mathbf{E}^{\text{loc}}$ can be written as [6]

$$F_j = \frac{1}{2}\text{Re}\left[\mu_i^* \nabla_j E_i^{\text{loc}}\right], \tag{1}$$

where Einstein summation notation was used and $\boldsymbol{\mu} = \alpha_0 \mathbf{E}^{\text{loc}}$ is the induced dipole moment. The local electric field acting on a particle may be written as $\mathbf{E}^{\text{loc}} = \mathbf{E}^{\text{inc}} + \mathbf{E}^{\text{ref}} + \mathbf{E}^{\text{D}}_{\text{scat}}$, where $\mathbf{E}^{\text{D}}_{\text{scat}}$ is scattered dipole field which includes the field induced by local fields generated in the substrate. The last component may can be found to be

$$\mathbf{E}^{\text{D}}_{\text{scat}}(\mathbf{r}) = \omega^2 \mu_0\, \hat{G}^R(\mathbf{r}, \mathbf{r}_0)\boldsymbol{\mu}, \tag{2}$$

where ω is the frequency of the incident wave, μ_0 is the vacuum permeability and $\hat{G}^R(\mathbf{r}, \mathbf{r}_0)$ is the dyadic Green's functions for the reflected dipole fields near planar interface [6]. Self-action of the dipole on itself may be taken into account by introducing the effective polarizability

$$\hat{\alpha} = \alpha_0\left[\hat{I} - \alpha_0\omega_0\mu_0\hat{G}^R(\mathbf{r}_0, \mathbf{r}_0)\right]^{-1}, \tag{3}$$

where $\hat{I}$ is the unity tensor. In this particular case $\hat{G}^R(\mathbf{r}_0, \mathbf{r}_0)$ has only diagonal terms. Due to the symmetry of the problem $\hat{\alpha}_{xx} = \hat{\alpha}_{yy}$. After that the dipole moment can be calculated directly as $\boldsymbol{\mu} = \hat{\alpha}(\mathbf{E}^{\text{inc}} + \mathbf{E}^{\text{ref}})$.

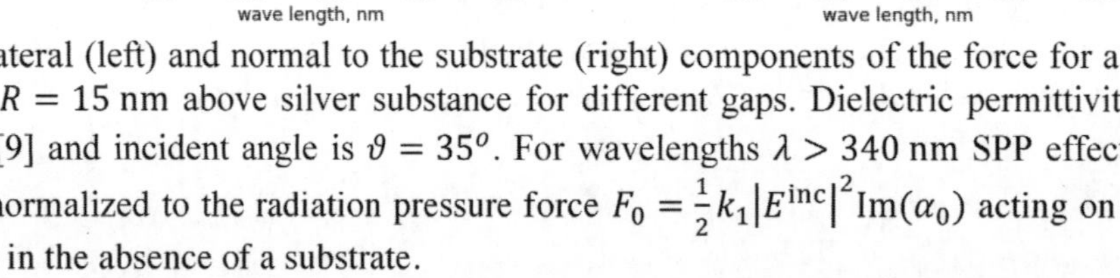

Figure 2. Lateral (left) and normal to the substrate (right) components of the force for a particle with $\varepsilon_p = 3$ and $R = 15$ nm above silver substance for different gaps. Dielectric permittivity of silver is taken from [9] and incident angle is $\vartheta = 35^o$. For wavelengths $\lambda > 340$ nm SPP effects take place. Values are normalized to the radiation pressure force $F_0 = \frac{1}{2}k_1\left|E^{\text{inc}}\right|^2\text{Im}(\alpha_0)$ acting on the spherical nanoparticle in the absence of a substrate.

Analysis of the optical force (1) shows that a so-called recoil force regime is possible [4]. It is an effect of emergence of a pulling force towards the light source, in this particular case it means $F_x < 0$.

The physical background of this effect is in the directional radiation of surface plasmon waves. The important conditions are i) excitation of SPP wave, and ii) small gap between particle and the metallic surface. Moreover, the less particle size is, the more significant is the SPP contribution. The most efficient angle to observe the largest negative pulling force is $\vartheta = 35^o$ [4].

3. Dynamic simulation results and discussions

In this section we find the specific parameters corresponding to the overall pulling force. For further simulations particle radius was chosen $R = 15$ nm. To indicate the parameters of potentially different regimes of motions, the optical force was calculated within the expression (1). In Figure 2 it is shown that for $\lambda \gtrsim 340$ nm SPP effects are important, and for significantly small gap F_x gets negative. For further simulation of nanoparticle dynamics, two wavelength were chosen: $\lambda_1 = 320$ nm for a non-SSP regime and $\lambda_2 = 340$ nm for a strong SSP regime. The equation of motion for the particle is

$$m\ddot{\mathbf{r}} = \frac{1}{2}\mathrm{Re}\left[\mu_i^* \nabla E_i^{\mathrm{loc}}(\mathbf{r})\right] - \gamma\dot{\mathbf{r}}, \qquad (4)$$

where m is the mass of the particle, and γ is a phenomenological viscosity damping parameter. Differential equation (4) was solved numerically for λ_1 and λ_2 with different initial particle positions z_0. The laser power was chosen to be 100 mW and with the beam cross section of 10 mm^2, which defines the amplitude E^{inc} of the incident wave.

The solution of the dynamical equation (4) is presented in Figure 3. One can see that the topology of the trajectories are dramatically changed when the wavelength of the incident light is changed from λ_1 to λ_2. In the non-plasmonic regimes the nanoparticle moves along the direction of light incidence, finally getting into the position of equilibrium with respect to vertical coordinate at approximately 30 nm height. At the same time in the regime of SPP excitation (see Figure 3, right panel) one can see that nanoparticle is first pulled, being displaced along the negative horizontal direction. Then it also ends at the z-equilibrium position around 45 nm above surface.

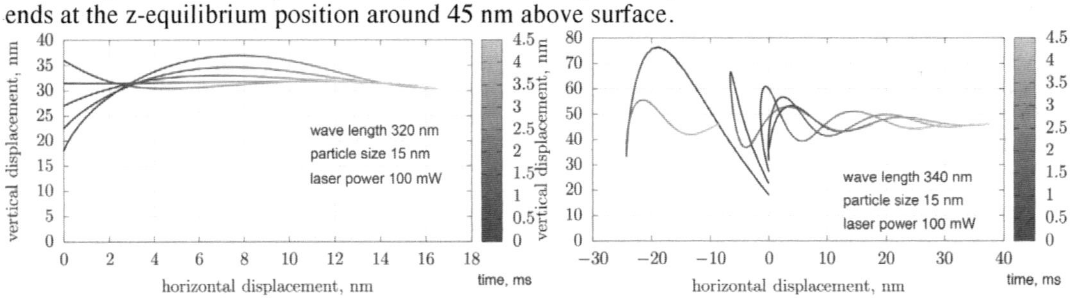

Figure 3. The nanoparticle trajectories in the non-plasmonic (left) and plasmonic (right) regimes for different initial height of nanoparticle above the interface. The color denotes the time during the nanoparticle motion.

4. Conclusions

In this work we show that nanoparticle dynamics placed close to the metallic interface is significantly affected by the excitation of surface plasmon polariton modes. Depending on the condition of the exciting plane wave incidence one can achieve different conditions of SPP generation, and, in particularly, one can see the pulling action and displacement of nanoparticle in the direction opposite to the direction of light incidence.

Acknowledgement
This work was supported by Russian Foundation of Basic Research 16-32-60167, 16-37-60064.

References

[1] Maragò O M, Jones P H, Gucciardi P G, Volpe G, Ferrari A C 2013 *Nat. Nanotech* **8(11)** p 807

[2] Neuman K C, Nagy A 2008 *Nat. Meth.* **5**(6) pp 491–505
[3] Wang S B, Chan C T 2014 *Nat. comm.* **5** 3307
[4] Petrov M I, Sukhov S V, Bogdanov A A, Shalin A S, Dogariu A 2016 *Las. and Phot. Rev.* **10**, 116
[5] Rodríguez-Fortuño F J, Engheta N, Martínez A, Zayats A V 2015 *Nat. Comm.* **6** 8799
[6] Novotny L, Hecht B 2012 *Principles of nano-optics* (New York: Cambridge University Press)
[7] Han X, Jones P H 2015 *Optics Letters* **40**(17) pp 4042–45
[8] Doyle W T 1989 *Physical Review* B ***39***(14)
[9] Johnson P B, Christy R W 1972 Optical Constants of the Noble Metals *Physical Review* B ***6***(12) pp 4370–4379

X-ray studies of doped bismuth titanates

N N Shomysov[1], M S Koroleva[2], A E Mingaleva[1], S V Nekipelov[1,3], O V Petrova[1], I V Piir[2], E N Shustova[3], V N Sivkov[1]

[1]Komi Science Center Ural Division, RAS, 167982, Syktyvkar, Russia
[2]Institute of Chemistry Komi Science Center Ural, RAS, 167982, Syktyvkar, Russia
[3]Syktyvkar State University, 167000, Syktyvkar, Russia

Abstract. X-ray diffraction and NEXAS- and XPS-spectroscopy have been investigated thermostable solid solutions Mn, Cu- doped bismuth titanate with a pyrochlore-type structure. It was shown that manganese and copper in said solid solution of bismuth titanate is present mainly in the oxidation state +2, and titanium - in +4.

1. Introduction and experimental

Doped bismuth titanates, depending on the nature of the dopant and the type of crystal structure, are characterized by wide variability of electrical properties and practical applications as dielectrics, mixed and ionic conductors for solid oxide fuel cells, oxygen permeable membranes and gas sensors, catalysts for oxidative processes.

The pyrochlore $Bi_2Ti_2O_7$ is thermodynamically unstable at high temperature and decompose at t>650°C [1]. In recent years, there has been a significant structural disordering in the metal-containing bismuth titanates (pyrochlore) and the displacive disorder. In a recent studies of manganese and copper –bismuth titanate pyrochlore we have synthesized a broad range of solid solutions by solid state reaction method at the temperatures 900–1150 °C [2-3].

A feature of the pyrochlore structure with the general formula unit $A_2B_2O_6O'$ is independent of the coexistence of two lattices: A_2O' и BO_6. A-positions of large cations, for example bismuth, are surrounded by oxygen atoms O` and both sublattices (cationic and anionic) can be defective. Cations, which are located in the B-positions (Ti) are in octahedral coordination of oxygen atoms (second anionic sublattice), and these cation and anion positions are completely filled by atoms of metals and oxygen. The M-metal atoms doped in the structure of the pyrochlore-type can be distributed on the two cationic sublattices (positions) [4]. In the present work, we consider the results of an investigation of the antistructural disorder in cation sublattices and the defect structure in the oxygen sublattice of manganese and copper-containing bismuth titanates with pyrochlore structure. For the obtained samples was performed x-ray phase analysis. Also were determined the lattice parameters of the pyrochlore and measured pycnometric density. For individual samples performed a full-profile analysis of diffraction patterns obtained at room temperature using the Fullprof package to specify variable parameters, including the filling of atomic positions. To determine the charge state of atoms of doped metals, all samples were investigated by NEXAS and XPS spectroscopy using the BESSY-II synchrotron source (Berlin) [5]. All of the NEXAS spectra were recorded in total electron yield (TEY) mode.

2. Crystal structure

IOP Conf. Series: Journal of Physics: Conf. Series **917** (2017) 062057 doi:10.1088/1742-6596/917/6/062057

Diffraction patterns of series of solid solutions are presented in Figure 1. We undertook an attempt to describe the structure of the $Bi_{1,6}Mn_{0,8}Ti_{1,6}O_{6.4}$ sample within the limits of reported models for the pyrochlore $Bi_2Ti_2O_7$. The ideal high symmetry sites occupied by pyrochlore $A_2B_2O_6O`$ (Fd3m) are the A cation at the 16c site (0, 0, 0), the B cation at the 16d site (1/2, 1/2, 1/2), the O anion at the 48f site (x, 1/8, 1/8) and the O` anion at the 8a site (1/8, 1/8, 1/8). We fixed the fractional occupation of the atom sites for the studied sample according to the formulas. The formula of the sample can be presented in two ways: manganese occupies both A (Bi) and B (Ti) sites equally with full occupation of the A position—$(Bi_{1.6}Mn_{0.4})(Mn_{0.4}Ti_{1.6})O_{6.4}$ (1)—and taking into account the partial deficiency in the bismuth sublattice, the sample formula is $(Bi_{1.54}\,Mn_{0.31}\Delta_{0.15})(Mn_{0.46}Ti_{1.54})O_{6.16}$ (2) (Δ - vacancy). The similar results were obtained for the sample $Bi_{1.93}Cu_{0.58}Ti_2O_{7-\delta}$. It shows distribution of copper atoms over the both A- and B-sites: $(Bi_{1.67}Cu_{0.25}\,\Delta_{0.08})(Cu_{0.25}Ti_{1.75})O_{6.53}$.

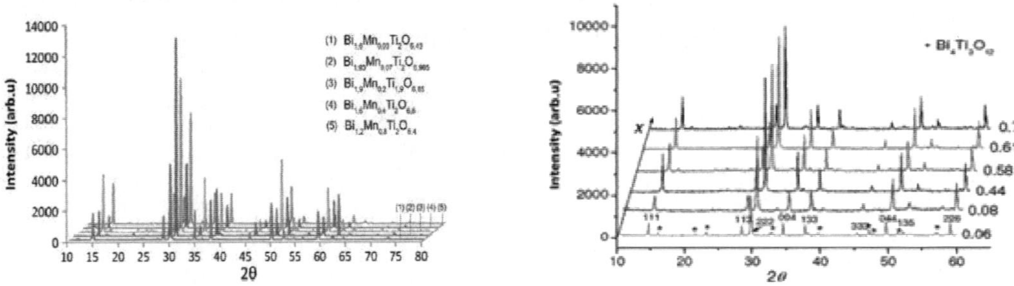

Figure 1. Diffraction pattern of solid solutions of $Bi_yMn_xTi_2O_{7-\delta}$ (left) and $Bi_{1,6}Cu_xTi_2O_{7-\delta}$ (right).

3. XPS-spectroscopy

We have also investigated the XPS-spectra of pyrochlore (Fig.2). All the spectra obtained were decomposed into peaks approximated by Gaussian curves, and the background was approximated by the Sherley-approximation. As can be seen from the XPS spectra of titanium all the atoms in the bismuth titanates are in the same charge positions. This follows from the absence of splitting in the corresponding peaks. At the same time, in the XPS spectrum of oxygen in $Bi_{1,9}Mn_{0,2}Ti_{1,9}O_{6,9}$ shows the splitting of the 1s-state of oxygen - strips A and B (Fig.2, left). The strip C is most likely associated with satellite transitions, so in the spectra of various compounds its position varies greatly/ The low-energy strip A corresponds to the more electronegative state of oxygen. Apparently, this is due to atoms from the octahedral environment of titanium (sublattice BO_6.) with less electronegativity. The higher-energy strip B is associated with non-equivalent oxygen atoms from the bismuth environment (sublattice A_2O'). The correctness of this assumption is also indicated by the fact that the ratio of the integral intensities of the peaks A and B is 5.5, which is comparable to the ratio of the number of oxygen atoms O/O', which is 6 in the pyrochlore structure.

4. NEXAFS spectroscopy

The absorption spectra studied are shown in Fig. 3-4. The NEXAFS Mn2p-spectra analysis points to the following: (i) the spectra of the studied bismuth titanate pyrochlores with various amounts of manganese are similar both in the fine structure, in the shape of the absorption lines and in their energy positions, and (ii) the structure is identical to the spectra of MnO (Fig. 3 left). Based on the above information, we believe that the oxidation states of the manganese atoms in these solid solutions are Mn^{2+}. A similar analysis of the absorption spectra of titanium in these compounds indicates their similarity (Fig. 3 right) both to each other and to the structure of the spectra TiO_2 in which the titanium atom is in an octahedral cluster of oxygen atoms $[TiO_6]^{8-}$. It should be noted that there is a broadening of the B-strip associated with the transition of 2p electrons to the e_g-orbitals of titanium in the octahedral field of the oxygen environment. In the spectrum of amorphous titanium oxide (a-TiO_2) is also a noticeable widening of the strip and in the spectrum of anatase and rutile observed splitting of this strip. Apparently, this indicates some distortion in the octahedral environment of titanium. At the same time, we believe that the oxidation states of the studied atoms are Ti^{4+}. Similar analysis of the absorption spectra of copper and titanium in Cu- doped bismuth titanate (Fig. 4) led us to the conclusion that their degree of oxidation of Cu is +2 and Ti – +4.

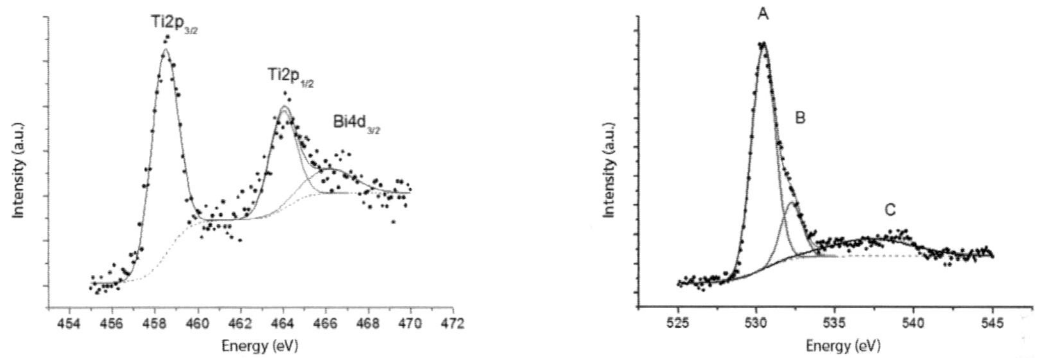

Figure 2. XPS-spectra of titanium (left) and oxygen (right)in $Bi_{1,9}Mn_{0,2}Ti_{1,9}O_{6,9}$. The points are the experimental data, the thin black line is the sum of the approximation curves.

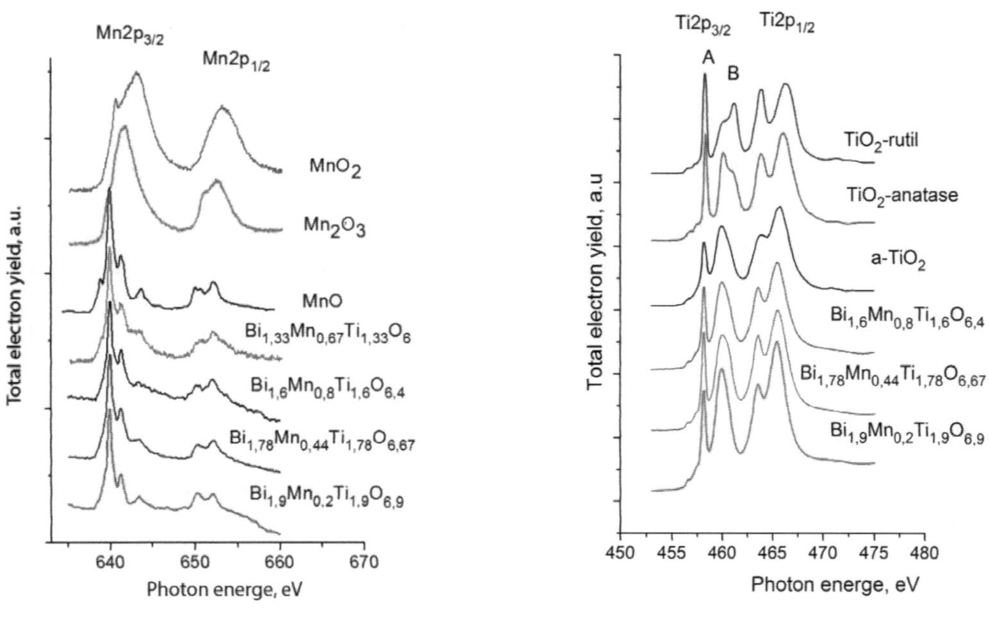

Figure 3. NEXAFS Mn2p- (left) and Ti2p- (right) spectra of manganese doped bismuth titanates.

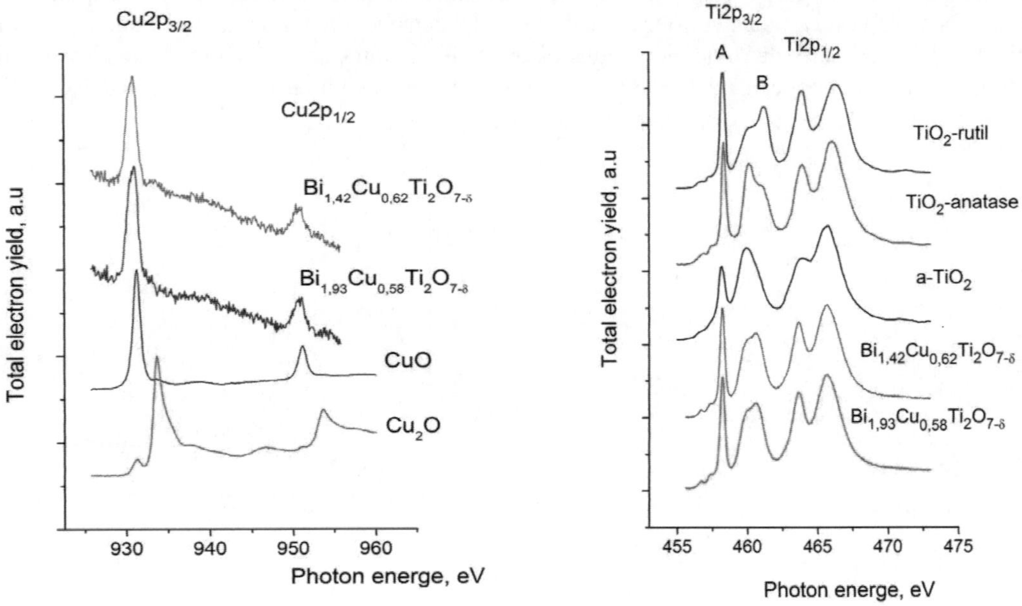

Figure 4. NEXAFS Cu2p- (left) and Ti2p-(right) spectra of copper doped bismuth titanates.

5. Results

The thermostable bismuth Mn, Cu titanate solid solutions with the pyrochlore structure in a wide range of compositions obtained by the ceramic technique.

The crystal structure of prepared samples and the chemical formula normalized to pyrochlore type ones are simulated and obtained.

The ability of the manganese atoms to be distributed over two cationic crystallographic positions increases the stability of the pyrochore crystal structure of the bismuth titanates, reduces displacive disorder in the bismuth sublattice and causes the formation of solid solutions in a fairly wide range of compositions.

Manganese and cupper in solid solutions of bismuth titanate is present mainly in the oxidation state +2, and titanium +4.

Acknowledgments

This work was supported by grants RFBR and Komi Republic 1216-32-00441 mol_a, 15-03-09173 a, 16-42-110610 r-a, 16-43-110350 r-a and by the Bilateral Program of the Russian-German Laboratory at BESSY II.

References

[1] A L Hector, S B Wiggin, J Solid. State. Chem.177 (2004) 139 –145.

[2] Piir I V, N A Sekushin, V E Grass, Y I Ryabkov, N V Chezhina, S V Nekipelov, V N Sivkov, D V Vyalikh. Solid State Ionics. 2012. V. 225. P. 464–470.

[3] I V Piir, M S Koroleva, Yu I Ryabkov, E Yu Pikalova, S V Nekipelov, V N Sivkov, D V Vyalikh. Solid State Ionics. 2014. V.262.P. 630–635.

[4] J Roberto Esquivel-Elizondo, Beverly Brooks Hinojosa, Juan C. Nino, Chem. Mater. 23 (2011) 4965–4974.

[5] J Stöhr, NEXAFS Spectroscopy, Springer, Berlin, 1992, p. 403.

QKD using polarization encoding with active measurement basis selection

A Duplinskiy[1,2], V Ustimchik[1], A Kanapin[1,3], Y Kurochkin[1]

[1]Russian Quantum Center (RQC), Business Center «Ural», 100, Novaya st., Skolkovo, Moscow reg., 143025, Russia

[2]Moscow Institute of Physics and Technology, 9 Institutskiy per., Dolgoprudny, Moscow Region, 141700, Russian Federation

[3]Lomonosov Moscow State University, GSP-1, Leninskie Gory, Moscow, 119991, Russian Federation

Abstract. We report a proof-of-principle quantum key distribution experiment using a one-way optical scheme with polarization encoding implementing the BB84 protocol. $LiNbO_3$ phase modulators are used for generating polarization states for Alice and active basis selection for Bob. This allows the former to use a single laser source, while the latter needs only two single-photon detectors. The presented optical scheme is simple and consists of standard fiber components. Calibration algorithm for three polarization controllers used in the scheme has been developed. The experiment was carried with 10 MHz repetition frequency laser pulses over a distance of 50 km of standard telecom optical fiber.

1. Introduction

Quantum cryptography is a part of quantum information technology which has been successfully practically implemented within the laboratories all over the world. Quantum key distribution (QKD) allows secret key sharing between two distant users – the transmitter (Alice) and the receiver (Bob), while guaranteeing unconditional security based on the fundamental laws of nature. In this paper we present a new optical scheme for the practical implementation of the BB84 protocol and present out preliminary experimental results.

Nowadays QKD distribution is performed at GHz frequencies, nevertheless polarization tracking at such speed is a challenging task. The simplest scheme uses four independent laser sources, one for each state of polarization, required for BB84 protocol implementation [1]. However it has been shown that proving the indistinguishability of light emitted from different sources is non-trivial, while it is vital for key secrecy. Another approach has been proposed, using fiber phase modulators' Pockels cells for state of polarization changing [2-5]. Electro-optical modulators based on $LiNbO_3$ crystals may be used as free space Pockels cells, while operating at GHz frequencies and requiring low driving voltages ($V\pi<5V$). Alice produces two pairs of orhogonal states – two circular and two diagonal, depending on the phase shift applied two one one of the components with the modulator and sends pulses to Bob, who selects which basis – diagonal or circular is used. Once Alice's and Bob's bases match – they get the same result with high probability, according to the scheme extinction ratio. If the bases are different the probability of both detectors click is equal. These bits are cancelled during the sifting procedure.

The main drawback of this solution is polarisation mode dispersion which occurs within the lithium niobate crystal due to its birefringence. It causes pulse polarization distorsion, resulting in a high quantum bit error rate (QBER). Different methods of compensating have been proposed including HiBi

fiber [2] and Faraday mirrors [5]. In the presented scheme polarization controller is adjusted so that two identical modulators of Alice and Bob compensate each others optical path difference along different axes.
The BB84 protocol was chosen due to its proven unconditional security [1]. Polarization encoding provides opportunity to use the scheme for both fiber and free space quantum channels, while the simplicity and compactness makes it perspective for satellite implementations. One-way optical scheme based on the lithium niobate modulators has a potential to work at GHz laser pulse repetition frequencies. Calibration algorithms for polarization controllers allow the system to compensate the drifts of polarization state both for quantum channel and for Alice's and Bob's devices. This procedure applied if needed and maintains the (QBER) at a level suitable for key transmission.

2. Optical scheme

We present a polarization encoding QKD system where both Alice and Bob use fast electro-optical $LiNbO_3$ phase modulators. The whole scheme consists of standard fiber elements (Fig.1). Alice creates one of four polarization states – two diagonal and two circular, appling voltages to her modulator, corresponding to 0, $\pi/2$, π and $3\pi/2$ phase shifts. Modulator exploitation for generating these states allows Alice to use a single laser source, solving the problem of pulse indistinguishability. Bob uses his modulator for active basis selection, appling 0 and $\pi/2$ phase shifts. This allows Bob to use only two detectors in contrast to four in the passive basis selection scheme. The polarization mode dispersion issue with $LiNbO_3$ crystals described in previous works [2,5] has been solved with a proper calibration of the polarization controller between the modulators of Alice and Bob (PC 2), allowing two crystals to compensate each other's' birefringence by rotating polarization vector by 90°.

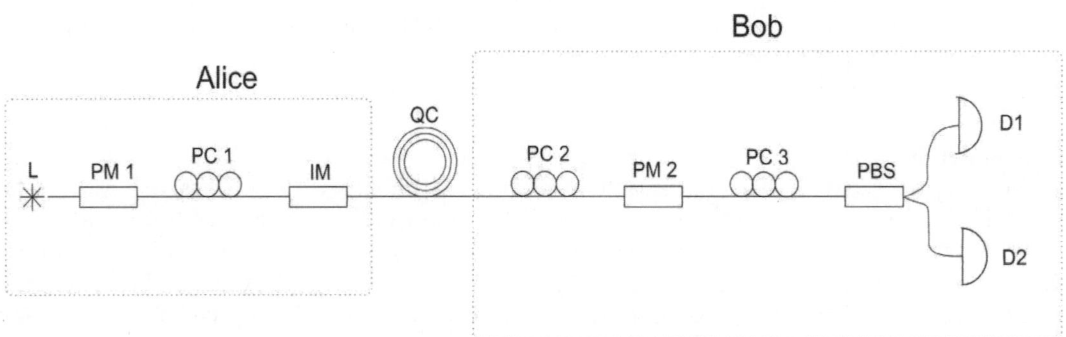

Figure 1. Optical scheme. Laser (L) generates optical pulses, which pass through the polarization controller (PC 1), that guides the light at an angle of 45° into the phase modulator (PM 1) to generate two pairs of orthogonal states. Variable optical attenuator (IM) has two modes of operation – calibration and key distribution. PC 2 compensates the polarization drifts within the quantum channel (QC) and rotates the polarization state by 90° to compensate lithium niobate birefringence. Bob's phase modulators (PM 2) is used for basis selection and PC 3 transforms the polarization state so that polarization beam splitter (PBS) apportion the states to single-photon detectors (D1, D2).

3. Calibration procedure

Three polarization controllers shown on the scheme have to be adjusted properly in order to make the QKD procedure possible. A calibration algorithm has been developed in order to carry out this procedure automatically. The system uses bright laser pulses in order to increase information collecting speed from the single-photon detectors. The calibration program processes the data and adjusts the voltages of the polarization controller channels. The system monitors the QBER and launches the calibration system automatically when the error value increases above a certain level. The intensity of the optical pulses is switched with an optical attenuator.

The controllers are adjusted one by one, to compensate polarization drifts within the devices of Alice and Bob (PC 1 and PC 3, respectively) and the quantum channel (PC 2).

4. QKD experiment

The proof-of-principle experiment has been conducted with laser pulses repetition frequency of 10 MHz. The key distribution has been carried out over a distance of 50 km of standard single-mode optical fiber. The system automatically performs the calibration procedure and maintains QBER below 5% for hours, applying recalibration if needed.

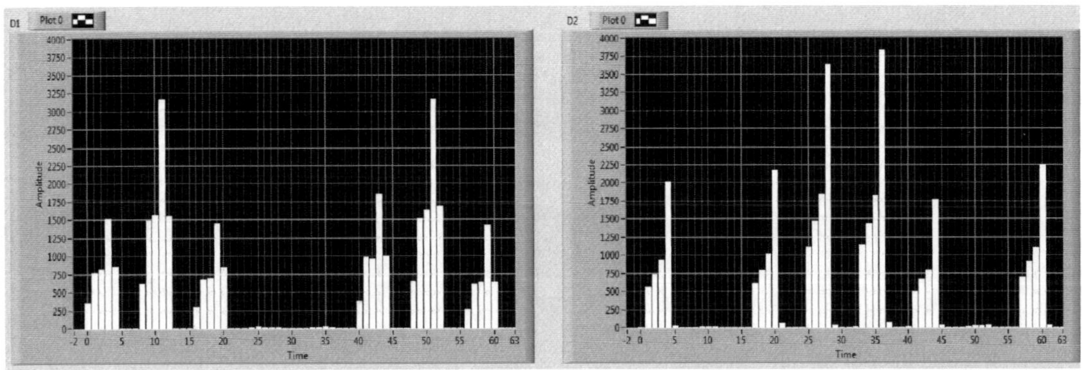

Figure 2. Histogram, illustrating all eight possible combinations of states, generated by Alice and Bob, using their modulators. The picture show statistics for both detectors. Bins with logical 1 on the first detector correspond to logical 0 on the second detector and vice versa. Each pulse consists of five columns in order to monitor the polarization mode dispersion effects.

The approximate rate of calibration regime is about 20% of total time, which means that 80% of time system generates the key.

To conduct the experiment modular QKD setup, developed within the laboratory has been used. Electronics is driven by National Instruments FPGA, that provides control of electro-optical modulators, laser source and polarization controllers, while also caring out data acquisition, and active feedback calibration loop.

Histograms for all possible states measured by Bob is shown on the Fig.2.

5. Conclusion

Novel polarization encoding scheme based on fiber lithium niobate modulators proposed in the paper has opportunity to reach GHz frequencies with proper driving electronics. Current experiment proove it's robustness with the developed calibration techniques, making the scheme promising for future commerial applications.

Acknowledgments

The support from Ministry of Education and Science of the Russian Federation in the framework of the Federal Program (Agreement 14.582.21.0009, ID RFMEFI58215X0009) is acknowledged.

References

[1] Gisin N, Ribordy G, Tittel W, & Zbinden H 2002 Reviews of modern physics 74(1) p. 145

[2] Jofre M, Gardelein A, Anzolin G, Molina-Terriza G, Torres J P, Mitchell M W, & Pruneri V 2010 Journal of Lightwave Technology 28(17) p 2572

[3] I Lucio-Martinez, P Chan, X Mo, S Hosier and W Tittel, "Proof-of-concept of real-world quantum key distribution with quantum frames", New J. Phys. 11, 095001 (2009).
[4] Zhizhong Yan, Evan Meyer-Scott, Jean-Philippe Bourgoin, Brendon L. Higgins, Nikolay Gigov, Allison MacDonald, Hannes Hübel, and Thomas Jennewein, "Novel High-Speed Polarization Source for Decoy-State BB84 Quantum Key Distribution Over Free Space and Satellite Links," J. Lightwave Technol. 31, 1399-1408 (2013).
[5] Jindong Wang, Xiaojuan Qin, Yinzhu Jiang, Xiaojing Wang, Liwei Chen, Feng Zhao, Zhengjun Wei, and Zhiming Zhang, "Experimental demonstration of polarization encoding quantum key distribution system based on intrinsically stable polarization-modulated units," Opt. Express 24, 8302-8309 (2016).

IOP Conf. Series: Journal of Physics: Conf. Series **917** (2017) 062059 doi:10.1088/1742-6596/917/6/062059

Methods of non-destructive testing in studies of self-organization processes in protein films

M A Baranov[1], **E N Velichko**[1], **E T Aksenov**[1]

[1]Institute of Physics, Nanotechnology and Telecommunications, Peter the Great Saint Petersburg Polytechnic University, Saint Petersburg 195251, Russia

Abstract. A non-destructive optical control of dynamics of protein aqueous solution drying under different conditions is considered. The processes of organization of different structures in protein films are discussed.

1. Introduction

Today there are many methods of non-destructive control of a variety of physical, chemical and biological processes [1]. Such a control is especially important for creation and maintenance of essential products, components and structures. To identify various defects, such as corrosion, rusting, and cracking, various methods of non-destructive control are used. There are several types of non-destructive testing: magnetic, electric, radio wave, eddy-current, thermal, optical [2], radiation, acoustic and penetrant. The optical visual observation method is one of the main non-destructive methods used for testing of biological objects, as well as for the study of self-organization processes.

2. Methods of investigation

In the analysis of self-organizing systems the main task is to explain the physics of structure organization. To solve this problem, researchers put forward certain theories and hypotheses. In addition, mathematical modelling and computational approaches provide support for such hypotheses and theories. For example, the "reaction-diffusion" theory provides a well-developed conceptual basis for understanding the patterning and morphogenesis in developing organisms [3]. For the practical study of the processes of self-organization, natural science uses two main methods. These are chemistry and optics, or, to be more exact, the study of the interaction of matter with light in the entire wavelength range: from x-rays to radio waves. Chemistry deciphers the primary structure of biopolymers, the structure of the functional centres of protein globules, etc. However, chemistry cannot establish the spatial structure of a protein or nucleic acid. [4]

To accomplish this task, a large number of physical methods for studying the processes of self-organization are used:

1. X-ray analysis of structure. Radiography gives direct information about the arrangement of atoms in molecules and crystals. With the help of this method, the structure of some vitamins have been revealed, as well as the structure of the first protein, myoglobin. Denaturation of the molecules of this protein was established (globule-coil transition). Predominantly, protein structure formation is studied at the molecular level by this method. It is able to establish a molecular lattice with strong covalent chemical bonds. Owing to this method, a great progress has been made in visualizing protein structures at level A. This method still dominates in the rapidly developing field of biological science called proteomics [5].

2. Some methods of nuclear physics. Synchrotron radiation is a very promising method for studying biological processes associated with conformational and other structural transformations at the molecular and supramolecular levels.

3. Voltammetry. The electric field affects a wide variety of processes, for example, phase transitions of substances. The electric field reduces the heat of water evaporation and increases the rate of heat transfer in liquids. Under the influence of electric field a conformational transformation occurs in the solution of a spiral single-stranded polynucleotide into a glomerulate one. The degree of transition is a linear function of the field strength [6].

4. Non-invasive optical methods. These methods include almost all types of optical and electron microscopy. They allow one to observe biological processes in real time with little or no effect on them [2].

The most accessible and simple method for studying the processes of self-organization is the method of visual registration of the protein condensation dynamics and its phase transitions under nonequilibrium conditions: in an open protein-water system, far from the thermodynamic equilibrium.

This method consists in evaporation of water from the colloidal protein-water system (dehydration) with further dynamic visualization of the process of condensation and self-organization of the protein in equilibrium and in nonequilibrium conditions in vitro. The colloidal protein-water system of various natures (albumin, thrombin) is placed on a solid substrate (glass) and dried in an open system at room temperature and atmospheric pressure. The dynamics of the process is visualized with an optical microscope and a sensitive CCD camera.

Advantages of this method:

1. This method does not affect the sample properties and the kinetics of the processes proceeding inside it;

2. The method allows one to evaluate the entire dynamics of the process;

3. The method allows one to change process conditions during the experiment (temperature, composition, other external factors);

4. The method does not require special sophisticated equipment.

Thus, it can be concluded that optical microscopy is the most accessible and quite informative method for studying the self-organization of protein films.

3. Self-organization

Self-organization is seen as a vital concept of intracellular architecture. From simple examples in physics and chemistry, one can note that self-organization can spontaneously appear in systems with a large number of interacting components.

Analysis of the investigations on the protein film preparation by the dehydration of their aqueous solutions reveals several general principles.

In most of the studies of drying out of an aqueous solution, the latter was placed on a solid base made of various wettable materials - glass, plastic, etc. Then the process of dehydration, the formation of structures in protein films was observed.

4. Samples and Experimental Results

The interaction of albumin in water solution with a magnetic fluid (magnetite) and kinetics of protein film structure formation were studied by drying the water from the solution and visual control. In our experiments the aqueous solution of albumin and Fe_3O_4 ferrofluid solution (at a concentration of 1 ml of pure substance to 1 ml of distilled water) were used. The components were mixed in special cuvettes. Experimental data were recorded in the form of photos on the PC through a microscope, with a connected CCD camera. Several series of experiments aimed at studies of the kinetics of protein structure formation of dry films under the influence of a constant magnetic field and pH changes were conducted. Experiments were performed at room temperature (27 °C) and at atmospheric pressure. Fig. 1 shows various pictures of spiral cores in the albumin protein films formed under different initial conditions.

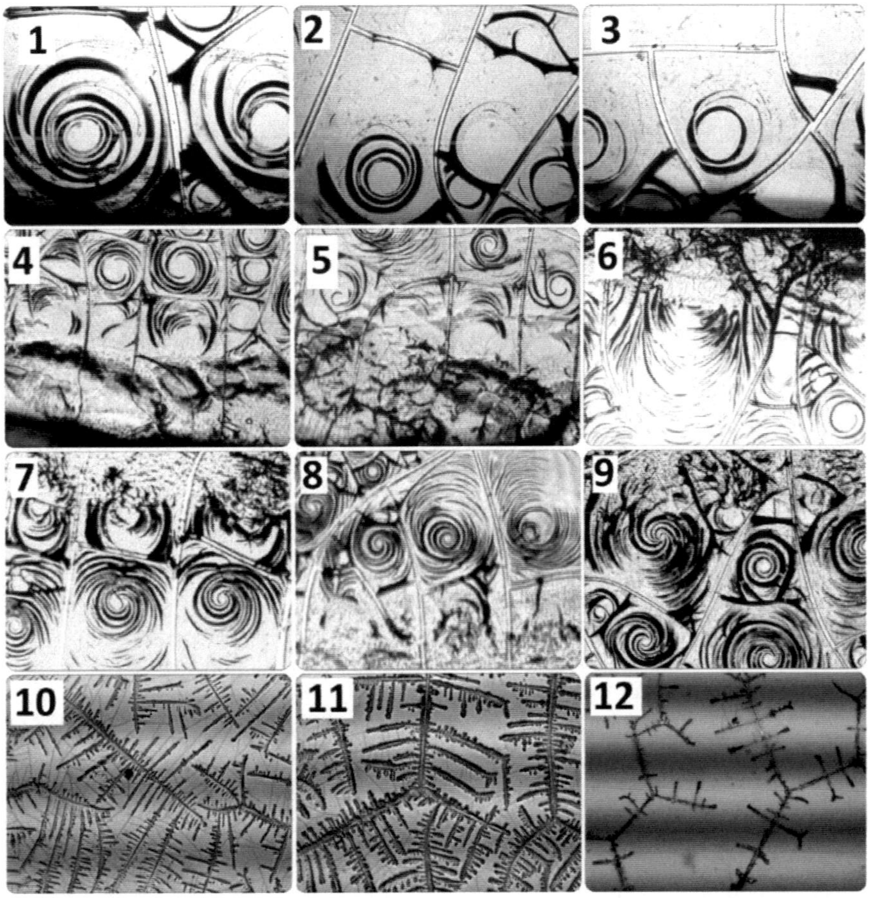

Figure1. Some results of self-organization investigation of dehydrated protein films, different forms of spiral cores in a film of the solution under different conditions: 1 – aqueous solution of albumin with magnetite without any influence, 2 – aqueous solution of albumin with magnetite in magnetic field H = 10 Oe, 3 – aqueous solution of albumin with magnetite in magnetic field H = 20 Oe, 4 – aqueous solution of albumin with addition of 0.5 μl of acetic acid, 5 – aqueous solution of albumin with addition of 1 μl of acetic acid, 6 – aqueous solution of albumin with the addition of 2 μl of acetic acid, 7 – aqueous solution of albumin with magnetite with the addition of 2 μl of acetic acid, 8 – aqueous solution of albumin with magnetite with the addition of 2 μl of acetic acid in magnetic field H = 10 Oe, 9 – aqueous solution of albumin with magnetite with the addition of 2 μl of acetic acid in magnetic field H = 20 Oe, 10, 11, 12 – solution of albumin in NaCl

Experiments have shown that a constant magnetic field affects the structure formation of spiral nuclei: the higher the magnetic field, the smaller the characteristic rings around the central nuclei, and the smaller the size of the "cells".

Experimental data revealed that the increased acidity of the initial solution has a significant effect on the formation of protein structures, violating their “helicity”. As it is seen from the figures, the increased acidity of solutions in an applied external magnetic field does not have a significant effect on the structure formation of spiral nuclei. However, it is clear that the number of turns has increased.

Solution of sodium chloride has an entirely different effect on the formation of protein films, in contrast to distilled water.

To sum up, we revealed that magnetic field is a key factor in formation and appearance of spiral vortices in the film structures.

5. Conclusion

Experimental data show that described optical method of research is the basis for the study of self-organization processes. It allows one to visualize the emergence of a cooperative magnetic phenomenon and turbulent (chaotic) properties of the protein in the transition phase and opens a new and promising approach to the scrutiny of the mechanism of self-organization.

References

[1] Tuchin V.V. 2007 *SPIE Tutorial Texts in Optical Engineering TT38, Bellingham, WA, USA* 882 p.
[2] Nepomnyashchaya E., Savchenko E., Velichko E., Aksenov E. 2017 *Proc. SPIE* **10336** 103360H
[3] Rajagopal K. and Schneider J.P. 2004 *Current Opinion in Structural Biology* **14** 480–486
[4] Volkenstein M.V. 1998 *Biophysics.* 130 p.
[5] Rapis E. 2004 *Journal of Technical Physics*, 2004, **74** 4
[6] Clauss C., Schweigart H. J. 1976 *Electrochem. Soc.* **123** 951-959

IOP Conf. Series: Journal of Physics: Conf. Series **917** (2017) 062060 doi:10.1088/1742-6596/917/6/062060

Optimization of Nanoantenna-Enhanced Terahertz Emission from Photoconductive Antennas

S I Lepeshov[1], A A Gorodetsky[1,2,3], N A Toropov[1], T A Vartanyan[1], E U Rafailov[3], A E Krasnok[4], and P A Belov[1]
[1] ITMO University, St. Petersburg, Russia
[2] Lancaster University, Lancaster, UK
[3] Aston Institute of Photonic Technologies, Aston University, Birmingham, UK
[4] Department of Electrical and Computer Engineering, The University of Texas at Austin, Austin, USA

Abstract. We present the results of hybrid photoconductive antenna THz emission enhanced by silver nanoantenna arrays. By varying the size of nanoantennas and the distance between them, we obtain the greatest value of optical-to-THz conversion efficiency reached so far. The results of experimental investigations are in a good agreement with numerical simulations. The conversion efficiency reveals over 5-fold improvement at certain frequencies, if compared with similar photoconductive antenna without silver nanoparticles, while previous results for this type of antenna barely exceeded 2-fold conversion efficiency gain.

1. Introduction

The terahertz spectral band of electromagnetic waves has a wide range of perspective applications for spectroscopy [1], biological sensing [2], security imaging [3], detection of dangerous or illicit substances [4,5] and potentially for ultrafast data transmission [6]. However, the widespread use of THz technologies is hampered by the absence of effective THz sources operating at room temperature. The most common source of coherent pulsed terahertz radiation is so-called THz photoconductive antennas (PCAs) [7]. The principles of THz PCAs operation are based on the effect of ultrafast surface photoconductivity of semiconductor substrate under femtosecond laser irradiation. Since such antennas are broadband, they have been applied for spectroscopic investigations of organic molecules [8] and material science [9] as well as wireless THz transmitters and receivers [10]. However, THz PCAs still have low efficiency that prevents data transmission over long distances and do not provide a permissible signal-to-noise ratio. The low efficiency is mainly related to the photocarrier screening effect [11] and the low absorption coefficient of the surface layer of the semiconductor substrate. To overcome these limitations, optical nanoantennas [12] have been proposed to be placed in the gap of THz PCA [13]. Optical nanoantennas are resonant nanostructures capable of transforming incident optical waves in a strong near-field. Nowadays, nanoantennas are used to enhance the absorption coefficient of a semiconductor substrate of PCAs. Nanoantenna-based PCAs have been called hybrid terahertz-optical PCAs (Figure 1). It has been shown that this solution provides high fs-laser pump absorption, shorter photocarrier lifetime and excellent thermal efficiency [13]. Despite the fact that many nanoantenna designs have been studied for enhancement of THz generation from PCA, an optimized geometry has not yet been proposed. Here, the results of optimization of nanoantenna-enhanced THz emission from hybrid PCA are presented. We show that an increase of more than 5-fold in the high-frequency spectral range can be achieved for an optimal design of nanoantenna array.

IOP Conf. Series: Journal of Physics: Conf. Series **917** (2017) 062060 doi:10.1088/1742-6596/917/6/062060

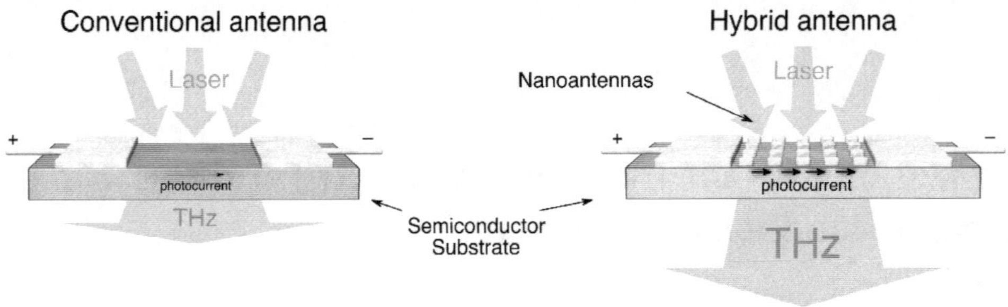

Figure 1. Illustration of the conventional (left) and hybrid (right) photoconductive antenna operation.

2. Numerical optimization of nanoantenna array

First, we carried out numerical calculations of silver (Ag) spheroid nanoantenna array on the high-index Gallium Arsenide (GaAs) semiconductor substrate. It is well known that metallic (particularly Ag) nanoparticles provide localized plasmonic resonances in the optical range [12]. This type of resonances is characterized by a strong localization of the electric field near a metal nanoparticle due to conversion of far-field of the incident wave to near-field of the oscillating electron plasma on the metal surface, which can be considered as an oscillating electric dipole if the size of the nanoparticle is sufficiently small. While nanoparticles are on the surface of the high-index substrate (GaAs), the electric field is mostly concentrated in a thin layer of the substrate surface that increases the absorption of fs-laser radiation near nanoantennas resulting in enhancement of THz emission.

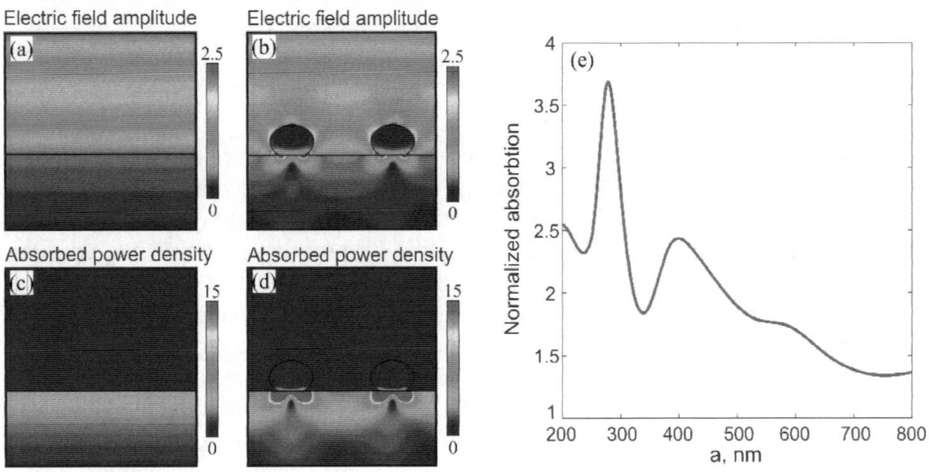

Figure 2. (a) Electric field distributions in the pure GaAs substrate and (b) in the substrate containing Ag nanoantennas at the wavelength of 800 nm. (c,d) The same for absorbed power density. (e) The dependence of the normalized absorbed power on the period of the nanoantenna array *a*.

The results of the numerical calculations of the electric field distribution at the air-GaAs interface with Ag spheroid nanoantenna and without nanoantenna are presented in Figure 2(a,b). For the simulation, we used CST Microwave Studio software package. The structure is excited by the normally incident plane electromagnetic wave at the wavelength of 800 nm. The geometrical parameters of nanoparticle correspond to the most optimal for the excitation of plasmonic resonance in this structure. The large semi-axis of the spheroid is D = 168 nm, the minor semi-axis of the spheroid is d = 106 nm. It can be seen that the magnitude of the electric field under the spheroid (Figure 2(b)) is larger by a factor of 3 than the magnitude of the field at the interface between air and GaAs without nanoparticle (Figure 2(a)). The absorbed power P in a volume V can expressed as

$$P = \frac{1}{2}\int \sigma |\mathbf{E}|^2 \, dV,$$

where $\mathbf{E}$ is a vector of electric intensity in the medium, σ is a medium conductivity. The absorbed power is directly proportional to the value of $|\mathbf{E}|^2$, and thus, the strong localization of the electric field in the semiconductor near nanoparticle causes increase of the absorbed power, which results in photoexcitation of free carriers. Figure 2(c,d) shows a comparison of absorbed power densities in the semiconductor with and without nanoparticles. One has to note that the power absorbed in GaAs with silver nanoantenna is significantly higher than in pure GaAs near the surface.

Then, we have studied the dependence of the absorption of the light in the substrate surface layer on the distance between nanoparticles. For this purpose, we calculated quadratic periodic array of spheroid nanoparticles of the previously obtained optimal size. By varying the period of the array a, different values of the absorbed power density were received and the maximum was found at the period of 270 nm.

3. Results and discussions

To prove our theoretical findings, we performed experimental investigations of hybrid PCA with Ag nanoantennas and compared the obtained results with previous studies [14]. We fabricated the conventional log-periodic PCAs as a basis for next stages. Such antenna construction provides a broad radiation spectrum that allows studying the enhancement in the high-frequency range. The log-periodic antenna electrodes were deposited onto the quantum-dot based InAs:GaAs semiconductor substrate [15,16] by optical lithography. The antennas have a smallest gap of 8 μm, and overall diameter of 1.8 mm. Then, a 20-nm silver film has been arranged onto the substrate surface of one of the fabricated antennas. Upon thermal dewetting process caused by heating, the silver film has been dewetted into a disordered array of spheroid nanoparticles. The size of resulting nanoparticles depends on the thickness of the silver film [17]. The size of the fabricated nanoparticles corresponds to calculated optimal ones and the average distance between nanoparticles is 280 nm.

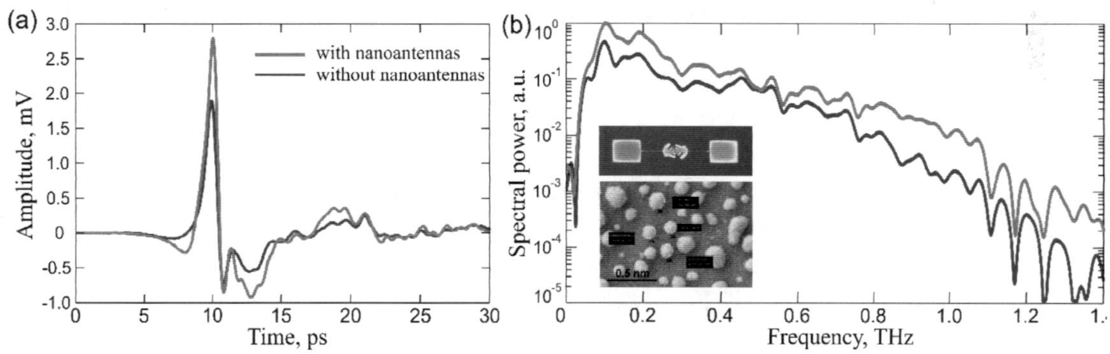

Figure 3. (a) The terahertz signals from the log-periodic photoconductive antenna with silver nanoantennas (red curve) and without nanoantennas (blue curve); (b) the spectra of terahertz signals. Insert: image of log-periodic photoconductive antenna on quantum-dot based InAs:GaAs semiconductor substrate and SEM-image of the silver nanoparticles in the gap of the antenna.

For experimental verification of the proposed hybrid antennas, a standard THz time-domain spectroscopic (TDS) system has been used. THz-TDS is pumped with Sprite-XT (M Squared Ltd.) femtosecond Ti:sapphire laser that delivers pulses of 120 fs duration at 80 MHz repetition rate with central wavelength of 800 nm. As THz detector, LT-GaAs photoconductive antenna made by Teravil Ltd. was used, and the beam was guided between transmitter and detector by two off-axis parabolic mirrors. To estimate the effect of silver nanoantennas, signals from standard and nanoantenna-enhanced transmitters were measured. Figure 3 summarize the obtained experimental results.

Figure 3(a) shows the terahertz signals from the log-periodic photoconductive antenna with silver nanoantennas (red curve) and without nanoantennas (blue curve). Figure 3(b) demonstrates that antenna reveals uneven enhancement across the spectrum, reaching its maximum around 1 THz. Negligible effect from nanoantennas at 0.5 THz can be associated with the change in antenna impedance induced by highly conductive silver in the gap. Thus, our results demonstrate significant amplification of the emitted THz signal spectral power. The results for nanoantennas demonstrate over 5-fold increase in comparison with the case of nanoantenna absent that in 3 times higher than previously reported [14].

4. Conclusion

We have demonstrated, both theoretically and experimentally, an unprecedented enhancement of photoconductive antenna operation by silver plasmonic nanoantennas. The resulting hybrid PCA demonstrates over 5-fold increase in generated THz signal around 1 THz, and over 2-fold increase in overall generated THz power, which perfectly coincides with theoretical estimations. We belive that our results may be useful for many vital applications, including spectroscopy, biological sensing, security imaging, and ultrafast data transmission

Acknowledgments

This work was supported by Russian Foundation for Basic Research (Projects 16-07-01166 a, and 16-37-60076).

References

[1] Jepsen P, Cooke D and Koch M 2011 *Laser&Photonics Reviews* **5** 124-166

[2] Yang X, Zhao X, Yang K, Liu Y, Liu Y, Fu W and Luo Y 2016 *Trends in Biotechnology* **34** 810-824

[3] Federici J, Schulkin B, Huang F, Gary D, Barat R, Oliveira F and Zimdars D 2005 *Semiconductor Science and Technology* **20** 266-280

[4] Massaouti M, Daskalaki C, Gorodetsky A, Koulouklidis A and Tzortzakis S 2013 *Applied Spectroscopy* **67** 1264-1269

[5] Zhang L, Zhong H, Deng C, Zhang C and Zhao Y 2008 *Applied Physics Letters* **92** 91117

[6] Nagatsuma T, Hisatake S, Fujita M, Pham H, Tsuruda K, Kuwano S and Terada J 2016 *IEEE Journal of Quantum Electronics* **52** 1-12

[7] Auston D, Cheung K and Smith P 1984 *Applied Physics Letters* **45** 284286.

[8] Fischer B, Walther M and Jepsen P 2002 *Physics in medicine and biology* **47** 3807.

[9] Grischkowsky D, Keiding S, Van Exter M and Fattinger C 1990 *JOSA B* **7** 2006-2015.

[10] Akyildiz I, Jornet J and Han C 2014 *Physical Communication* **12** 16-32

[11] Loata G, Thomson M, Löffler T and Roskos H 2007 *Applied Physics Letters* **91** 232506.

[12] Krasnok A, Maksymov I, Denisyuk A, Belov P, Miroshnichenko A, Simovskii C and Kivshar Yu 2013 *Phys. Usp.* **56** 539-564

[13] Lepeshov S, Gorodetsky A, Krasnok A, Rafailov E and Belov P 2016 *Laser&Photonics Reviews* **11** 1600199

[14] Park S, Choi Y, Oh Y and Jeong K 2012 *Optics Express* **20** 25530-25535.

[15] Leyman R, Gorodetsky A, Bazieva N, Molis G, Krotkus A, Clarke E and Rafailov E 2016 *Laser&Photonics Reviews* **10** 772-779

[16] Fedorova K, Gorodetsky A and Rafailov E 2017 *IEEE Journal of Selected Topics in Quantum Electronics* **23** 1-5

[17] Toropov N, Gladskikh I, Parfenov P and Vartanyan T 2017 *Optical and Quantum Electronics* **49** 154

Blue photoluminescence of sponge-like highly porous alumina synthesized in hydrofluoric acid based electrolytes

D O Ilin[1], **A S Vokhmintsev**[1], **N A Martemyanov**[1], **I A Weinstein**[1, 2]
[1] NANOTECH Center, Ural Federal University, Yekaterinburg 620002, Russia
[2] Institute of Solid State Chemistry, Russian Academy of Sciences, Yekaterinburg 620990, Russia

Abstract. For the first time nanostructures of anodized aluminum oxide (AAO) were synthesized using hydrofluoric acid based electrolytes under potentiostatic mode with varied oxidation conditions. As-grown oxide layers were amorphous and had sponge-like disordered structure with ramified pores system of 50 to 300 nm diameters. All samples under daylight demonstrated blue emission with power up to 10 μW that was seen by naked eye. It was shown that integral intensity of photoluminescence (PL) emission band in 350 – 650 range increases up to 7 – 60 times depending on synthesis details. Observed PL spectra were approximated by superposition of two components with E_{max} = 2.74 and 2.44 eV, FWHM = 0.63 and 0.53 eV. It was substantiated that studied emission have intrinsic origin and can be attributed to F_2^{2+}- and F_2-centers.

1. Introduction

Porous structures of AAO are widely used as templates for fabrication of ordered periodic arrays of 0D and 1D nanomaterials [1]. Unique physical and chemical properties allow to utilize it as functional medium for forming of composite structures [2], producing of filters [3], thermoconducting coatings [4], active elements of optoelectronics [5] etc.

Numerous studies considering luminescent properties of anodic aluminum oxide synthesized in electrolytes of various type and composition have been conducted up to date [6-22]. The most intensive emission with maximum in 430 - 500 nm range was demonstrated by amorphous as-grown nanostructures anodized in oxalic acid based solutions [6-9]. A major contribution in mechanisms of observed luminescence were made by defects attributed to oxygen vacancies in different charge state [7, 10-18] and/or oxalate-ions impurities [8, 9, 19-22]. Therefore, it is of current interest to search for optimal anodizing modes and electrolytes composition in order to obtain nanostructured oxide layers with high concentration of optically active defects formed due to developed surface with big specific area. The aim of this work was to study photoluminescent characteristics of highly porous AAO structures synthesized using hydrofluoric acid (HF) based electrolytes under varied conditions of electrochemical oxidation.

2. Experimental

A foil of A5 (99.5 %) aluminum with thickness of 500 μm was used as original material. A surface was cleaned preliminarily using ultrasonic bath for 10 min. One-step anodic oxidation was conducted under potentiostatic mode at T_A = 1 °C constant temperature. CV Digma two-electrode electrochemical cell was used with AKIP-1125 power supply and UT-2.6 thermocontroller. A solution of hydrofluoric acid in ethylene glycol was used as electrolyte. Two groups of samples using different conditions were obtained. A constant C_{HF} = 0.9 M electrolyte concentration and varied voltage in U_A = 75 - 400 V range was applied for the first one. A constant U_A = 150 V and altered concentration in C_{HF} = 0.9 – 10 M range was used for the second group. All samples were cleaned after the process. Anodized region was a circle with 2.5 cm diameter. For more details regarding synthesis process see [23].

Surface images of AAO were obtained by using Carl Zeiss SIGMA VP scanning electron microscope with InLens detector in high vacuum. Phase composition was analyzed by means of PANalytical X'Pert Pro diffractometer with Cu K_α (0.1541 nm) line.

Emission spectra of photoluminescence within 350 - 650 nm range were registered at room temperature in phosphorescence mode using Perkin Elmer LS 55 spectrometer. All samples were excited with photons in 277 ± 10 nm band [23]. Slit widths of monochromators in excitation and registration channels were adjusted to 10 and 5 nm, respectively. Speed of spectra scanning was 120 nm/min. To evaluate photometric characteristics Vega OPHIR meter with PD 300 standard photodiode based detectors was used.

3. Results

Synthesized alumina was in amorphous state according to XRD analysis within 2θ = 10 - 90 °range. Figure 1 shows SEM images of grown structures surfaces. It is seen that anodized layer is characterized by sponge-like structure with ramified system of pores with various diameters of 50 to 300 nm. Oxide layer of sample with C_{HF} = 2.5 M and U_A = 150 V was ≈ 20 μm thick.

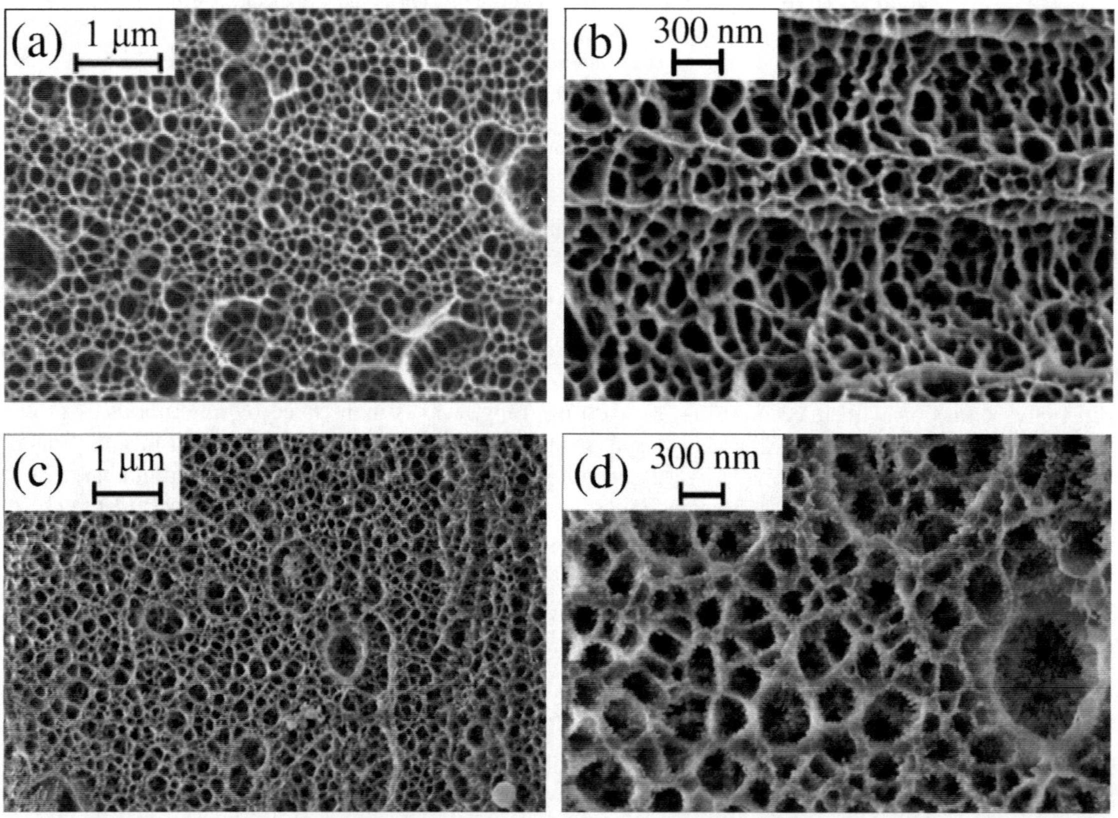

Figure 1. SEM surface images of AAO: 0.9 M, 150 V (a, b) and 5.1 M, 150 V (c, d).

It was found that under daylight highly porous oxide has blue emission with power up to 10 μW that was seen by naked eye. Figure 2 and 3 shows measured spectra for varied U_A and C_{HF} anodizing parameters. The integral intensity of PL emission in studied range increases up to 60 times due to voltage changes from 75 to 400 V, see inset in Figure 2. Relative integral intensity grows by 7 times within concentration change range, see inset in Figure 3. Samples with the most intensive emission were the ones synthesized in electrolytes with C_{HF} = 5.1 and 6.2 M. Concentration and anodizing voltage increase can lead to growth of oxide layer thickness, specific effective surface area and, therefore, cause a rise for luminescence response of samples. Further studies with use of independent methods are necessary to reliably assess indicated correlation.

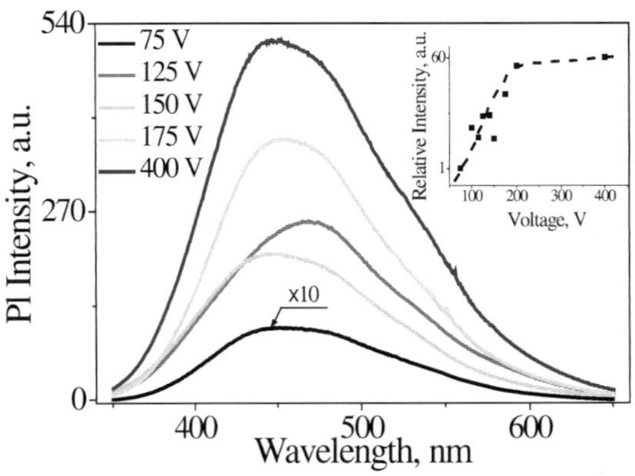

Figure 2. PL emission spectra for different U_A. Inset: relative changes of integral PL intensity.

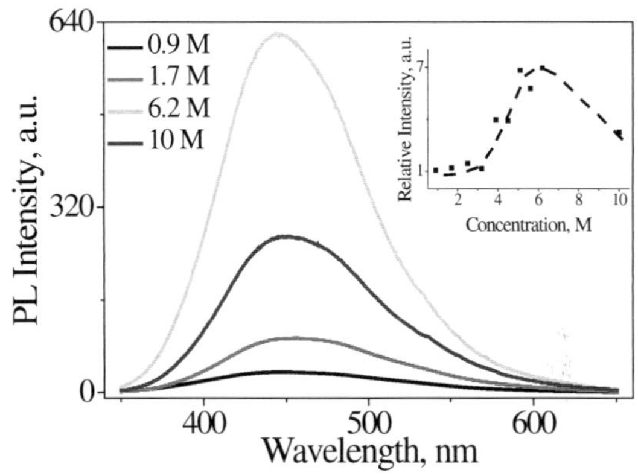

Figure 3. PL emission spectra for different C_{HF}. Inset: relative changes of integral PL intensity.

4. Discussion

Normalized experimental PL spectra of samples are shown in Figure 4. It can be concluded that registered peak is non-elementary since λ_{max} maximum position changes in 440 to 480 range and differences in shape are present for various synthesis modes. It is in agreement with known data regarding values of λ_{max} for AAO synthesized in oxalic acid – 420 nm [16], 430 nm [12], 450 nm [11], 480 nm [15] etc. PL spectra were approximated by superposition of two Gaussian bands, see examples in Figure 5. Precision of quantitative description of all experimental data for resulting curves was $R^2 > 0.999$ (coefficient of determination). Calculated values of E_{max} energy and FWHM halfwidth for each Gaussian component after averaging are listed in Table 1. It is to be noted that obtained values are in satisfactory agreement with each other in shown ± 0.05 eV scatter.

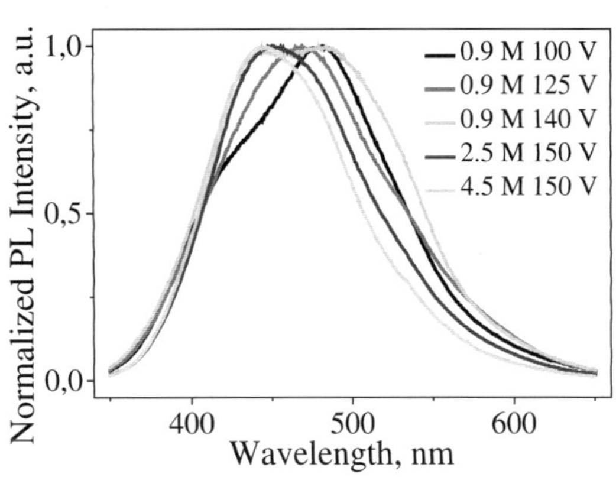

Figure 4. Normalized PL spectra of AAO.

It is known that luminescence centers of AAO synthesized in $(COOH)_2$ are complexes based on oxygen vacancies and/or oxalate-ions impurities formed during oxidation process [7-22]. Electrolyte composition in current work excluded presence of oxalate-ions. As far as data about spectral characteristics of vacancy based centers with different charge states in single crystalline α-Al_2O_3 is concerned it can be supposed that observed emission has intrinsic origin and can be attributed to F_2^{2+}-centers (2.66 eV [24-26]) for G1 peak and to F_2-centers (2.4, 2.46 eV [24-26], 2.38 eV [27]) for G2 one. Also, various relative concentrations of optically active centers can cause scatter of spectral and kinetic characteristics of photo- and thermally stimulated processes in aluminum oxide [28]. Thus, observed geometrical parameters variations of studied nanostructures spectra could be due to different concentration ratios of corresponding defects because of anodizing modes alteration.

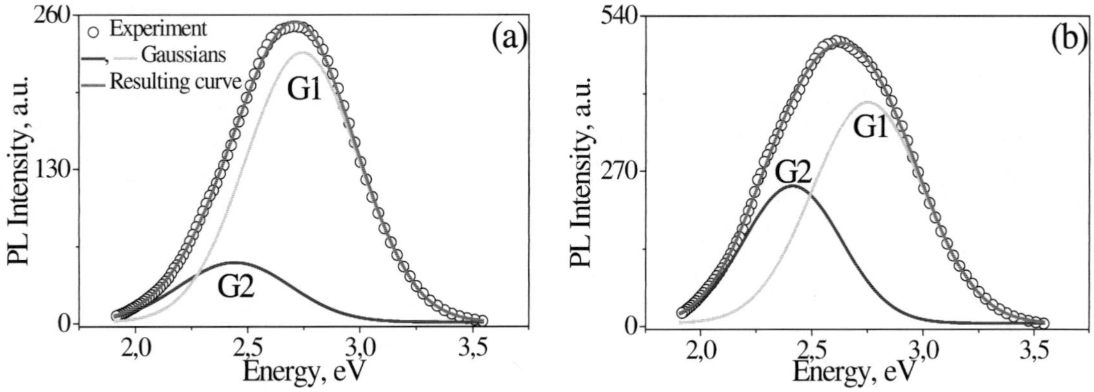

Figure 5. Deconvolution of PL spectra: 5.6 M HF, 150 V (a) and 0.9 M HF, 200 V (b).

Table 1. Calculated parameters of spectral components.

	E_{max} (±0.05 eV)	FWHM (±0.05 eV)
G1	2.74	0.63
G2	2.44	0.53

5. Conclusions

For the first time structures of sponge-like highly porous AAO were fabricated by anodic oxidation under potentiostatic modeusing hydrofluoric acid based electrolytes. Voltage was varied within U_A = 75 - 400 V range and concentration were altered in C_{HF} = 0.9 - 10 M diapason during nanostructures synthesis. It was shown that obtained oxide layers are characterized by unordered sponge-like morphology with ramified system of poreswith 50 to 300 nm diameters. Under daylight grown oxide structures had blue emission that was seen by naked eye. It was found that integral intensity of PL emission in 350 - 650 nm range grows by 7 - 60 times due to variation of anodizing parameters. Experimental spectra were approximated by superposition of two Gaussian bands with high degree of precision. Calculated parameters of spectral components E_{max} = 2.74 and 2.44 eV, FWHM = 0.63 and 0.53 eV were analyzed in comparison with independent literature data. It was concluded that studied PL has intrinsic origin and can be attributed to F_2^{2+}- and F_2-centers. Concentration ratios of discussed vacancy based defects are varied depending on anodizing conditions and cause observed changes of maximum position and luminescence band shape in synthesized samples.

Acknowledgements

The work was supported by Act 211 Government of the Russian Federation, contract № 02.A03.21.0006.

References

[1] Kohli P, Wharton J E, Braide O and Martin C R 2003 *J. Nanosci. Nanotech.* **4** 605–10
[2] Savchenko S S, Vokhmintsev A S and Weinstein I A 2016 *J. Phys.: Conf. Ser.* **741** 0120151
[3] Vlassiouk I, Krasnoslobodtsev A, Smirnov S and Germann M 2004 *Langmuir* **20** 9913–5

[4] Zhang B J, Kim K J and Yoon H 2012 *Int. J. Heat Mass Transf.* **55** 7487–98
[5] Dudem B, Leem J W, Choi M and Yu J S 2015 *Appl. Phys. B-Lasers Opt.* **118** 439–47
[6] Ilin D O, Vokhmintsev A S and Weinstein I A 2016 *AIP Conf. Proc.* **1767** 020028
[7] Gopal Khan G, Singh A K and Mandal K 2013 *J. Lumines.* **134** 772–7
[8] Mukhurov N I, Zhvavyi S P, Terekhov S N, Panarin A Yu, Kotova I F, Pershukevich P P, Khodasevich I A, Gasenkova I V and Orlovich V A 2008 *J. Appl. Spectrosc.***75** 214–8
[9] Li Z, Huang K 2007 *J. Phys.: Condens. Matter.* **19** 216203
[10] Li G H, Zhang Y, Wu Y C and Zhang D 2003 *J. Phys.: Condens. Matter.* **15** 8663–71
[11] Ning X, Kai-di M and Qian W 2013 *Proc. of SPIE - The International Society for Optical Engineering* **9068** 906811
[12] Vrublevsky I, Chernyakova K, Ispas A, Bund A, Gaponik N and Dubavik A 2011 *J. Lumines.* **131** 938-42
[13] Huang G S, Wu X L, Mei Y F, Shao X F and Siu G G 2003 *J. Appl. Phys.* **93** 582–5
[14] Mukhurov N I, Zhvavyi S P, Gasenkova I V, Terekhov S N, Pershukevich P P and Orlovich V A 2010 *J. Appl. Spectrosc.* **77** 549–55
[15] Li Y, Li G H, Meng G W, Zhang L D and Phillipp F 2001 *J. Phys.: Condens. Matter.* **13** 2691–9
[16] Li Z J and Huang K L 2007 *J. Lumines.* **127** 435–40
[17] Reddy P R, Ajith K M and Udayashankar N K 2016 *J. Mater. Sci.-Mater. Electron.* **27** 5331–44
[18] Komarov F F, Mudryi A V,Vlasukova L A, Mukhurov N I and Ivanyukovich A V 2008 *Opt. Spectrosc.* **104** 235–7
[19] Li Y B, Zheng M J and Ma L 2007 *Appl. Phys. Lett.* **91** 073109
[20] Fang D *et al.* 2014 *Mater. Sci. Eng. B-Solid State Mater.* **179** 71–6
[21] Vrublevsky I, Jagminas A, Hemeltjen S and Goedel W A 2008 *Appl. Surf. Sci.* **254** 7326-30
[22] Gao T, Meng G and Zhang L 2003 *J. Phys.: Condens. Matter.* **15** 2071–9
[23] Ilin D O, Vokhmintsev A S and Weinstein I A, 3rd April 2017, *Application for RU Patent №2017111101*
[24] Jheeta S, Jain D C, Kumar R, Singh F and Garg K B 2006 *J. Nucl. Mater.* **353** 190–192
[25] Evans B D 1995 *J. Nucl. Mater.***219** 202–223
[26] Li B, Oliveira S L, Rand S C, Azurdia J A, Hinklin T R, Marchal J C and Laine R M 2007 *J. Appl. Phys.* **101** 053534
[27] Vokhmintsev A S, Weinstein I A and Kortov V S 2013 *Radiat. Meas.* **56** 228–31
[28] Weinstein I A and Popko E A 2007 *J. Lumines.* **122–123** 377–80

IOP Conf. Series: Journal of Physics: Conf. Series **917** (2017) 062062 doi:10.1088/1742-6596/917/6/062062

High resolution photolithography using arrays of polystyrene and SiO_2 micro- and nano-sized spherical lenses

L N Dvoretckaia[1], A M Mozharov[1,2] and I S Mukhin[1,2]

[1] St. Petersburg Academic University, Khlopina 8/3, 194021 St. Petersburg, Russia
[2] ITMO University, 197101 St. Petersburg, Russia

E-mail: liliyabutler@gmail.com

Abstract. Photolithography mask made of close-packed array of micro- and nano-sized spherical lenses allows to obtain the ordered structures and provides highest "optical resolution/cost" ratio between all existing photolithography and laser direct writing methods. In this letter, we present results of modeling the propagation of a plane wave falling on the array of quartz (SiO_2) microspherical lenses and focusing in the image reverse photoresist layer. We present here experimental results on fabrication of ordered arrays of submicron wells and columns and substrate preparation for growth of monocrystalline nanowires on metal surface using photolithography with mask of SiO_2 microspheres. Such ordered nano-sized arrays of wells and columns can be used in fabrication of further growth of monocrystalline nanowires, quantum dots and production of plasmon structures.

1. Intoduction

Ordered arrays of micro- and nanostructures are objects of growing interest because of their unique properties and wide range of applications in nanoelectronics and plasmonics. Production of plasmonic and dielectric metasurfaces involves fabrication of large area periodic arrays of metallic and dielectric submicron structures. The latter fact impose extra requirements on photolithographic systems resolution leading to rise of equipment cost, lithography process cost and as a result to end-product price increase.

Industrial scanner-steppers are capable of forming the structures with 14 nm resolution. However, those systems are very expensive and are not suitable for scientific research. Research institutes and universities are commonly equipped with projection or laser lithography systems, which typical resolution is under the diffraction limit and optical scheme with corresponding resolution limit in 0.5-1 μm range.

Electron-beam lithography allows nanoscale resolution structure fabrication. However, it is very area-limited as a direct writing method. Projection lithography method through array of micro- and nanosized spherical lenses is based on focusing of incident plane wave under the lens. When the spheres are closed-packed an ordered array of micro- or nanostructures can be produced.

The purpose of this work was to develop fabrication method of ordered arrays of nanoscale columns or holes by means of photolithography through arrays of microspheres. These structures can

IOP Conf. Series: Journal of Physics: Conf. Series **917** (2017) 062062 doi:10.1088/1742-6596/917/6/062062

be used to create an antireflection coating or plasmon structures as well as a mask for selective area growth of monocrystalline NWs or quantum dots [1].

2. Experiment

In this work we developed technology of fabricating a mask for synthesis of NWs or quantum dots via deposition of close-packed polystyrene (PS) and SiO2 microspherical lenses one monolayer thick layer with further photolithography.

We carried out photolithography on glass substrate covered with layer of aluminum and layer of diluted AZ1505 optical photoresist over it. SiO2 microspheres 1.5 µm in diameter were deposited on the surface of the photoresist. Due to polarity of SiO2 microspheres and hydrophobicity of the photoresist , before the deposition of microspheres on wafer the sample was processed in the TMAH based developer. Treatment of the structure with the developer makes its surface hydrophilic due to reduction of hydrophobic groups and deposited microspheres uniformly distribute along the surface of the photoresist.

Deposition of microlenses was carried out via spin-coating method [2]. In order to obtain uniform distribution of the spheres we optimized the deposition parameters and found optimal balance between deposition time and rotation speed. In our experimental series we varied both parameters: the speed of the first centrifugation step was 300 rpm and dwell time time was 40 s while the last step was 2000 rpm for 10 s. From SEM images shown in Figure 1 (a, b) it is seen that during the deposition of microspheres with the spin-coating method homogeneous large area arrays of self-organized SiO_2 microspheres with a diameter of 1.5 µm were obtained.

After that the photoresist covered with ordered array of microspheres was subjected to projection photolithography with 405 nm wavelength. The photoresist was then developed and the aluminum layer was removed through the opened windows in the photoresist with etching in acid solution. As a result, openings in aluminum with diameter of about 550 nm were formed on the substrate as shown in Figure 1 (c). This ordered array of holes can be used as a mask for synthesis of NWs or quantum dots.

In the next part of our experimental work we fabricated nanoscale holes array in SiO_2 layerdeposited on Si wafer partially covered with different metal layers. Formation of such a mask is desirable for synthesis of NWs or quantum dots. Initial approach to solving this problem was photolithography through the mask of SiO_2 microspheres with the following etching of SiO2 through the prepared resist mask. In the course of our work it was found that size of the holes depends on several factors: size of the microspheres, type of photoresist, photoresist exposure duration. It is well known, that SiO_2 etching provides broadened hole edges, so in order to obtain holes in SiO_2 without etching involvement we used the reverse photoresist approach. The optimum photoresist thickness was determined via simulation described as followsin Section 3.

To obtain mask with desired morphology we deposited three metals (Cr, Ti, Mo) on different regions of Si (111) wafer via thermal evaporation in a vacuum. The reverse photoresist and SiO_2 microspheres were then deposited with spin-coating on the wafer and exposed to radiation . As a result of the photoresist development ordered array of photoresist columns on the silicon surface and various metals was obtained. As one can see from Figure 1 (d) images obtained with SEM diameter of the columns is in the 220-300 nm range. Further, silicon oxide was deposited on the nanostructured substrate and after lift-off the resist was removed with SiO_2 and ordered array of holes was obtained as shown in the AFM image shown in Figure 1 (e). This structure is suitable for growth of NWs or quantum dots.

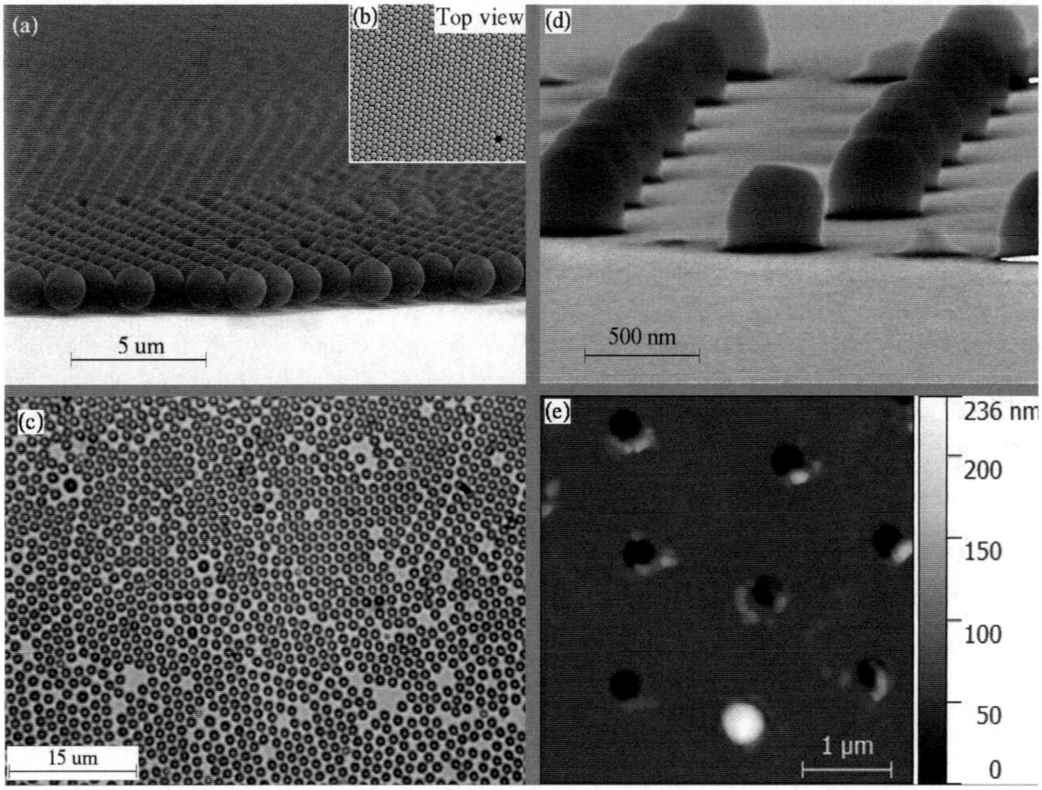

Figure 1 (a, b, c, d, e). SEM images of ordered SiO_2 microsphere lens array on metal-covered glass substrate (a), top view of the SEM image of SiO_2 microsphere lens array (b), optical image of nano-structure array on metal-covered glass substrate after sphere-enhanced photolithography and metal etching (c) an ordered array of columns of the reverse photoresist on silicon and the diameter of the columns is in the range 220-300 nm (d), and AFM image of an ordered array of holes obtained by lift-off the removed resist with SiO_2 (c)

3. The modeling

In the course of the work 3D modeling of the propagating a plane incident electromagnetic wave through SiO2 microspherical lenses into a photoresist layer with different photoresist thicknesses from 2 µm to 0.2 µm was carried out to determine the optimal radiation absorption by an optical reverse photoresist. For modeling, the length of the incident plane wave was chosen equal 405 nm in that the projection photolithograph, with which were carry out an experiments described above, has an i-line mercury lamp with a wavelength of 405 nm. On the figures 1 (a-f) presented different resist thicknesses (2 µm, 0.45 µm, 0.2 µm, 0.3 µm, 0.4 µm, 0.5 µm) for comparison. Comparing the simulation results shown in figure 1 (a - f) we conclude that for the wavelength equal to 405 nm and for diameter SiO2 microspheres equal 1.5 µm the optimum photoresist thickness is 0.45 µm. These obtained data of modeling were used in the experiment to obtain ordered columns by photolithography using microspheres as shown in figure 1 (g). The profile of the absorbed light in the photoresist layer in figure 2 (b) has two expressed maxima which coincides with the column profile obtained experimentally as shown in figure 2 (g) This result makes it possible to use a reverse photoresist to obtain an array of ordered columns with nanometer sizes.

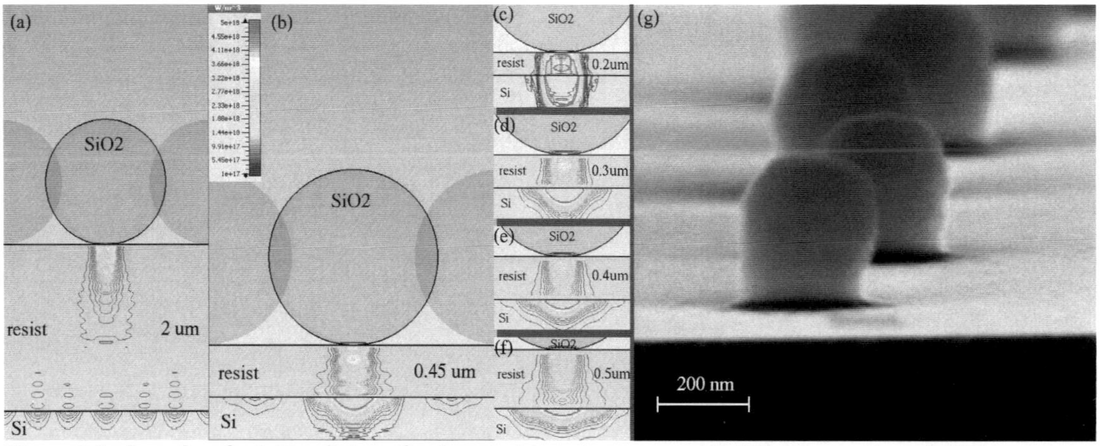

Figure 2 (a, b, c, d, e, f, g). Simulation of a plane incident electromagnetic 405 nm wave through SiO_2 microspherical lenses 1.5 μm propagation into the reverse photoresist layer of thickness 2 μm (a), 0.45 μm (b), 0.2 μm (c), 0.3 μm (d), 0.4 μm (e) and 0.5 μm (f), SEM image of ordered column array experimentally obtained on silicon substrate (g).

4. Conclusion

Results of the numerical simulation to determine the optimum thickness of the optical image reverse photoresist (0.45 μm) which allows to make a high quality photolithography when the photoresist is exposed through 1.5 μm SiO_2 microlens array . Here we present ordered layers of microspherical SiO_2 and PS lenses deposition technology on substrates with photoresist. Using the developed technology ordered openings array with a diameter about 550 nm as well as columns with a diameter of the order of 220-300 nm are obtained. The substrate preparation technology for growth of NWs using photolithography with SiO_2 microspheres mask on silicon and different metals was tested. With use of this method nanostructured substrates have been fabricated that can be used for further realization of plasmon structures as well as growth of monocrystalline NWs and quantum dots.

Acknowledgements

This work was carried out with the support of President grant (14.Z56.17.3632-MK), the Russian Foundation for Basic Research (15-02-06839 A, 16-32-60094 mol_a_dk and 16-32-00560 mol_a), the leading universities of the Russian Federation (grant 074-U01), grant of government of the Russian Federation (16.2593.2017/8.9, 16.2593.2017/4.6)

References

[1] J Zhu, X Zhu et al. 2012 *Applied Physics Letters* **100** 143109
[2] Z Zhang, C Geng, Z Hao, T Wei and Q Yan 2016 *Advances in Colloid and Interface Science* **228** 105

Giant Magnetoresistance in NiO/Co/Cu/Co/Ti spin valve fabricated by EBPVD

F Matroodi[1], M G Shoar[2]

[1] Department of Physics, Shahid Chamran University of Ahvaz, Golestan BLV, Ahvaz, Iran, 61357-83151

[2] Laser and Plasma Research Institute, Shahid Beheshti University, G. C., 1983963113, Evin, Tehran, Iran

Abstract. In this paper, glass/NiO/Co/Cu/Co/Ti magnetic multilayers were deposited by electron beam evaporation system. A magnetic field of a few tens Oe was applied during the deposition of some of the layers in order to pin the magnetic moments of the ferromagnetic layer to the adjacent antiferromagnetic layer. The thickness of NiO layer is 140 nm and Co/Cu/Co/Ti layers thickness is 15 nm. The dc 4-point probe technique was used to measure the magnetoresistance (MR) of the prepared samples while the magnetic field was applied in the film plane and in the current direction. The giant magnetoresistance values up to 12% and 8% achieved respectively in samples deposited in presence and absence of magnetic field.

1. Introduction

Fret et al and Grunberg et al separately discovered giant magnetoresistance (GMR) in 1988 respectively in Fe/Cr multilayer and Fe/Cr/Fe three layer [1,2]. For some thicknesses of Cr layer around 1 nm, Fe layers may show antiferromagnetic coupling and have antiparallel magnetic moments in the absence of external magnetic field. By applying magnetic field, electrical resistance of the structure reduces as their magnetic moments start to align in a same direction toward the applied magnetic field. Because this resistance reduction is very large in comparison to other phenomena they call it Giant. The ratio of the resistance change to the minimum resistance, $\Delta R/R$, is called GMR. In 1991 Dieny et al introduce a more practical GMR structure, spin-valve. This new structure consists of a soft ferromagnetic layer called free layer, a nonmagnetic layer and a second ferromagnetic layer called pinned layer which has exchange coupling with an antiferromagnetic layer adjacent to it. The interaction between these two last layers is the origin of an asymmetric exchange that pins the magnetization of the second ferromagnetic layer in a definite direction [3].

In the present work, we exhibit a cost-effective spin-valve structure prepared by electron beam evaporation on a glass microscope slide and functions at room temperature still shows a high GMR effect.

2. Materials and methods

Figure 1. Evaporation chamber with a coils for applying magnetic field of 50 Oe.

Figure 2. Cobalt pallet & Titanium granules in graphite boats.

2.1 Sample fabrication

Figure 1 show the deposition chamber used in this study. For raw material in powder form some pellets have been made by press the powder with 25 ton weight to the use in the deposition chamber. An image of a cobalt pellet and titanium granules in graphite boats are seen in figure 2. We used glass microscope slide as a substrate. Cobalt, nickel oxide and titanium have been deposited using electron beam evaporation. For copper deposition we used thermal evaporation and a tungsten boat with an electrical current order of 100 A. deposition parameters are indicated in table 1. A pair of coils with magnet cores inserted inside the deposition chamber to create in plane magnetic field in the order of 50 Ga during cobalt layer deposition. To avoid electron beam deflection caused by this magnetic field, aluminium foil used as a shield around the mentioned coils as could be seen in figure 1.

Table 1. Deposition parameters.

Material	Chamber pressure ($\times 10^{-5}$ mbar)	Substrate Temprature ℃)	Layer thickness (nm)
NiO	4	27-42	140
Co/Cu/Co	3	29-44/-/45-51	10
Ti	3.5	27-42	-

2.2 Magnetoresistance measurement

The prepared sample is shown in figure 3. To measure the change in sample resistance by applying magnetic field, four-point probe technique (figure3 right) was used. In this technique, the two outside contacts used for conducting the current through the sample and the inside ones for voltage measurement. Keithley instrument 225 direct current power supply applied the current in the range of 50-150 nA through the sample and Keithley 160 B digital multimeter used to control the current. The sample voltage in the range of 20-65 μV was measured by 163 Digital Voltmeter Keithley. Magnetic field was applied in plane of the multilayers by 20 mT steps and up to 250 mT.

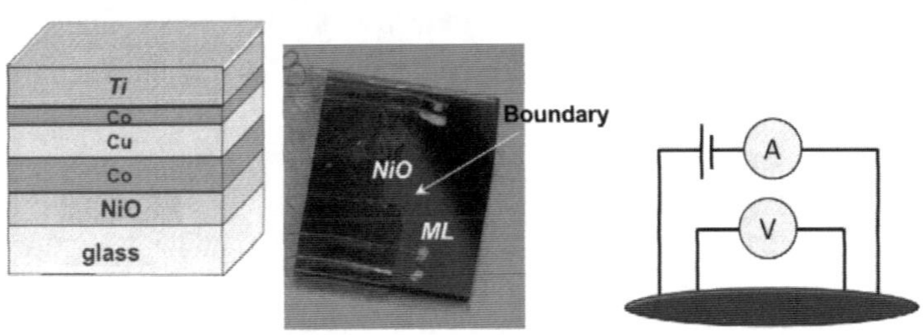

Figure 3. left: schematic sketch of the spin-valve sample, center: spin-valve sample with electrical contacts, right:schematic sketch of a four-point probe technique .

IOP Conf. Series: Journal of Physics: Conf. Series **917** (2017) 072001 doi:10.1088/1742-6596/917/7/072001

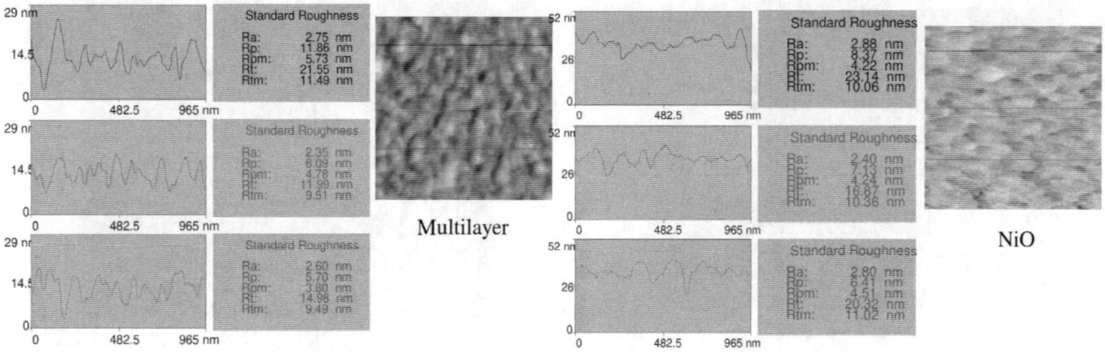

Figure 4. AFM image of NiO/Co/Cu/Co/Ti multilayer and NiO layer.

3. Results

Roughness and inter layers diffusion have a great effect on GMR. Some studies show that in Fe/Cr layers magneticresistance increases with roughness [4,5]. On the other hand it has been shown that as a result of the large resistance in mixed NiFe/Cu layers, interface roughness of NiFe/ Cu/NiFe/FeMn spin valve leads to reduction in magneticresistance [6,7]. In Co/Cu layers, this inter layer diffusion effect may cause different GMR results [8-10]. AFM images from the NiO and the multilayer (figure 4) show that surface roughness in both sides is around 2.8 nm. Magnetoresistance measurement result is shown in figure 5. Magnetization of the sample measured by magneto optical Kerr effect (MOKE) technique and the arising hysteresis loop is shown in figure 6. Magnetoresistance curve and hysteresis loop indicate an extreme consistency. Both increase in the same manner. This show that in-plane easy axis was developed as a result of NiO layer and its adjacent Co layer coupling. Applying the magnetic field causes magnetic moment of the soft cobalt layer to rotate toward magnetic field direction and increases the angle between the two Co layers.

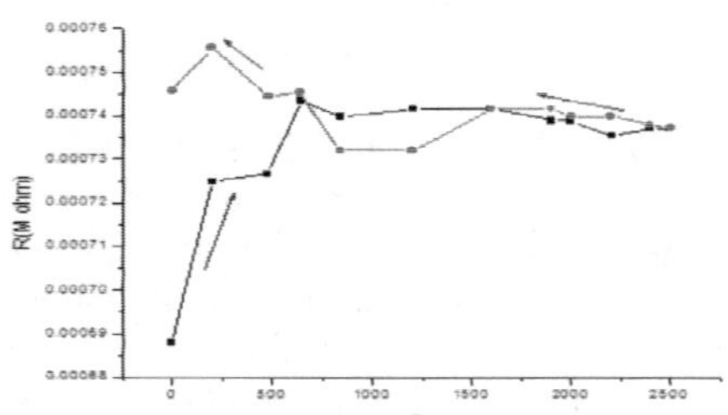

Figure 5. Resistacne versus applied magnetic field. The arrow indicates increasing or decreasing magnetic field

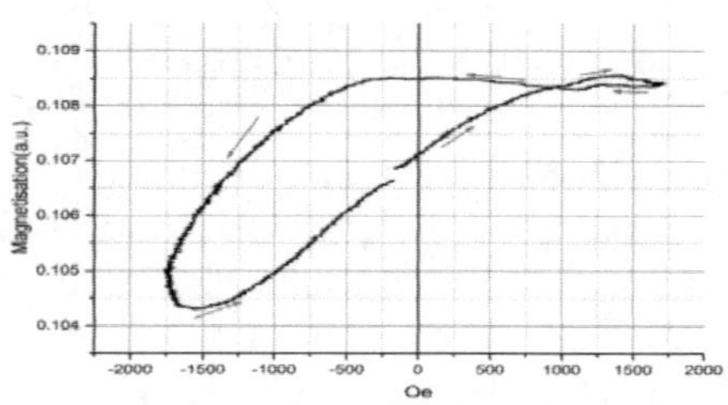

Figure 6. Hysteresis loop of the sample by MOKE.

4. Coclusion

Although we used a cost-effective technique and substrate, our structure shows spin-valve effect at high value. Magnetoresistance curve and hysteresis loop show the same behavior and support the interpretation of GMR as caused by spin-valve effect..

References

[1] Baibich MN, Broto JM, Fert A, Van Dau FN, Petroff F, Etienne P, Creuzet G, Friederich A, Chazelas J. Giant magnetoresistance of (001) Fe/(001) Cr magnetic superlattices. Physical review letters. 1988 Nov 21;61(21):2472.

[2] Binasch G, Grünberg P, Saurenbach F, Zinn W. Enhanced magnetoresistance in layered magnetic structures with antiferromagnetic interlayer exchange. Physical review B. 1989 Mar 1;39(7):4828.

[3] Dieny B, Speriosu VS, Parkin SS, Gurney BA, Wilhoit DR, Mauri D. Giant magnetoresistive in soft ferromagnetic multilayers. Physical Review B. 1991 Jan 1;43(1):1297.

[4] Petroff F, Barthelemy A, Hamzić A, Fert A, Etienne P, Lequien S, Creuzet G. Magnetoresistance of Fe/Cr superlattices. Journal of Magnetism and Magnetic Materials. 1991 Feb 28;93:95-100.

[5] Fullerton EE, Kelly DM, Guimpel J, Schuller IK, Bruynseraede Y. Roughness and giant magnetoresistance in Fe/Cr superlattices. Physical review letters. 1992 Feb 10;68(6):859.

[6] Huang TC, Nozieres JP, Speriosu VS, Lefakis H, Gurney BA. X-ray reflectivity analysis of giant-magnetoresistance spin-valve layered structures. Applied physics letters. 1992 Mar 30;60(13):1573-5.

[7] Speriosu VS, Nozieres JP, Gurney BA, Dieny B, Huang TC, Lefakis H. Role of interfacial mixing in giant magnetoresistance. Physical Review B. 1993 May 1;47(17):11579.

[8] Suzuki M, Taga Y. Giant magnetoresistance in Co/Cu superlattices with mixed interfaces. Journal of applied physics. 1993 Oct 1;74(7):4660-3.

[9] Willekens MM, Rijks TG, Swagten HJ, De Jonge WJ. Interface intermixing and magnetoresistance in Co/Cu spin valves with uncoupled Co layers. Journal of applied physics. 1995 Dec 15;78(12):7202-9.

[10] Bensmina F, Humbert P, Dinia A, Muller D, Speriosu VS, Gurney BA. Annealing effect on the magnetic properties of Ta 50Å/Cu 50Å/Co 75Å/Cu 50Å/Ta 50Å sandwiches. Journal of magnetism and magnetic materials. 1999 Jun 1;198:338-40.

IOP Conf. Series: Journal of Physics: Conf. Series **917** (2017) 072002 doi:10.1088/1742-6596/917/7/072002

The study of response of electro-optic phase modulator based on $LiNbO_3$ with the aim of improving the accuracy of fiber-optic gyroscope

D A Pogorelaya, M A Smolovik, S M Aksarin, V E Strigalev, V A Shulepov, A B Muhtubaev
Department of Light-Guided Photonics, ITMO University, St. Petersburg 197101, Russia

Abstract. The work offers the method for a study of parasitic refractive index drift (RID) of electro-optical modulator (EOM) based on the lithium niobate (NL) crystal under the action of applied electric voltage. As a result there is proposed a method of compensation EOM`s inertial behavior. Described method can be used for improving the accuracy of fiber-optic interferometric sensors and of fiber-optic gyroscope in particular.

1. Introduction
EOMs based on NL crystal ($LiNbO_3$) are widely used in electro-optical devices due to unique properties of NL, such as a large electro-optical coefficient, wide transparency range, mechanical and chemical stability, as well as the possibility of formation of channel waveguides in NL crystal [1, 2].

Despite many advantages of modulators based on NL, they have several disadvantages. Because of the dielectric nature of the NL crystal parasitic refractive index drift (RID) is observed under action of the applied electrical voltage [3, 4, 5]. This effect has an unwanted influence on the optical wave propagating through the modulator and thus it makes an additional phase shift, which leads to increased measurement error [6]. That is why the investigation of this parasitic effect in EOMs based on NL crystal is currently an urgent task to improve the accuracy of fiber-optic interferometric sensors in general and of fiber-optic gyroscope in particular.

2. The study of parasitic RID of phase EOM based on NL crystal
In this work we have proposed and have implemented the method of study of parasitic RID of phase EOM based on NL crystal. Parasitic RID can be measured only in case of using the phase EOM as a part of the interferometer, where parasitic RID leads to the interferometer`s phase shift and to the according to its intensity change. In order to measure RID time and applied voltage dependences transient processes were investigated. The experimental setup (figure 1) represents the Mach-Zehnder fiber interferometer with a phase EOM in one of its arm. We used the phase EOM from multifunctional integrated optical scheme (MIOS), fabricated on X-cut NL. MIOS includes a polarizer, X-type coupler and two phase EOMs. Optical waveguides axes are oriented along Y axis of NL. An external electrical field was applied along Z axis of NL, so EO coefficient r_{33} was used.

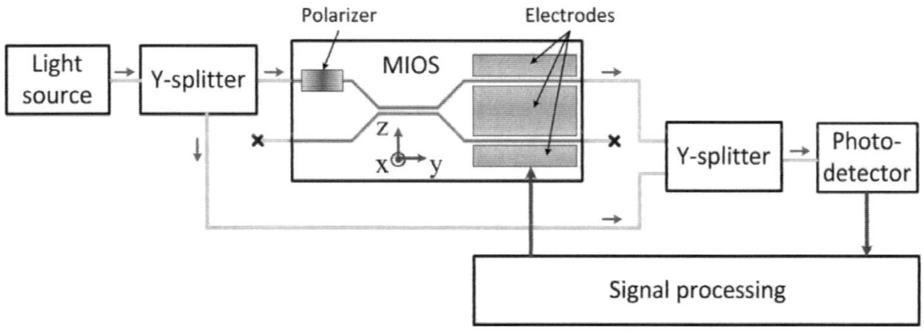

Figure 1. The experimental setup for studying of the parasitic RID of the phase EOM based on NL crystal.

Since a process of parasitic RID is most clearly observed in case of a jump of the applied voltage, in other words, in case of a sharp change in the magnitude and/or direction of the external electric field, the shape of the modulating signal for measuring transient processes of the modulator should be a step function of the applied voltage.

The amplitude of the step was chosen to be according to the phase shift equal $\pi/2$, and there were analyzed only those interferometer`s responses which were received in quadrature operating point. Such a principle was applied in order to measure transient process in the linear part of the interferometric response, where normalized intensity changes are equal to the phase changes. Figure 2 shows the EOM`s step response which was recalculated from interferometer`s response in operating point $-\pi/2$ in case of applying a jump of voltage to electrodes of the phase EOM. The jump of voltage was counted in such a way to shift interferometer's operating point from $-\pi$ to $-\pi/2$.

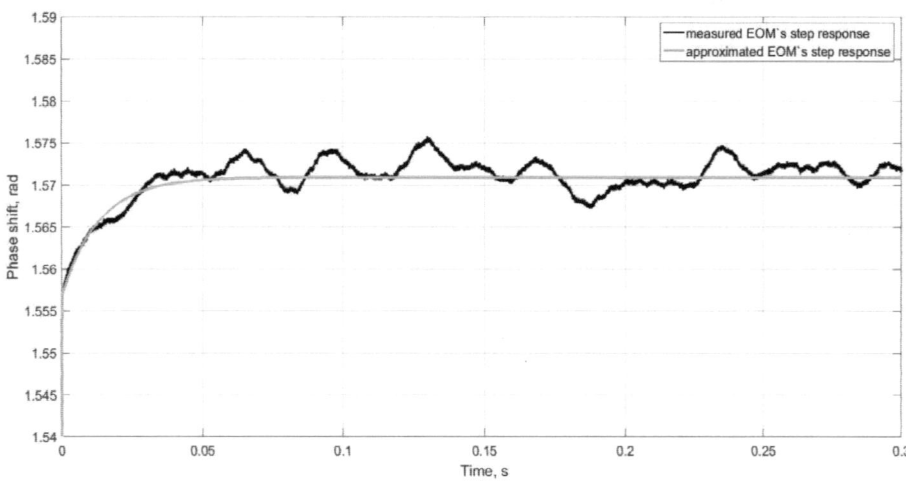

Figure 2. EOM`s step response in case of applying a jump of voltage to its electrodes.

Revealed parasitic RID of phase EOM in case of changing the modulating voltage may be caused by mobile charges formed as a result of penetration of the metal electrodes in the thickness of the substrate and by defects in the surface layers of the crystal [6]. They move under the influence of an electric field generated by electrodes, and also create its own field which can increase or, conversely, weaken external field. Such charges have limited mobility, so they exhibit inertia in relation to changes in the external field. Thus, the process of relaxation of charges caused by the field generated by electrodes creates a drift of the magnitude of the total electric field and, as a result, it creates parasitic RID of phase EOM, which makes an additional phase difference, which leads to increased measurement error.

3. The method of compensation of parasitic RID of phase EOM based on NL crystal

Transient response of the EOM in time domain pictured at figure 2 can be performed in frequency domain. The calculated transfer function in z-domain is expressed as:

$$Y(z) = \frac{0.9821\, z^{-1} - 0.9815\, z^{-2}}{1 - 1.006\, z^{-1} + 0.004902\, z^{-2} + 0.002045\, z^{-3}}.$$

The knowledge of the transfer function gives us an opportunity to calculate the inverse filter which can be used for pre-emphasis of the modulating voltage signal. Frequency responses of the EOM and of the inverse filter are shown at figure 3.

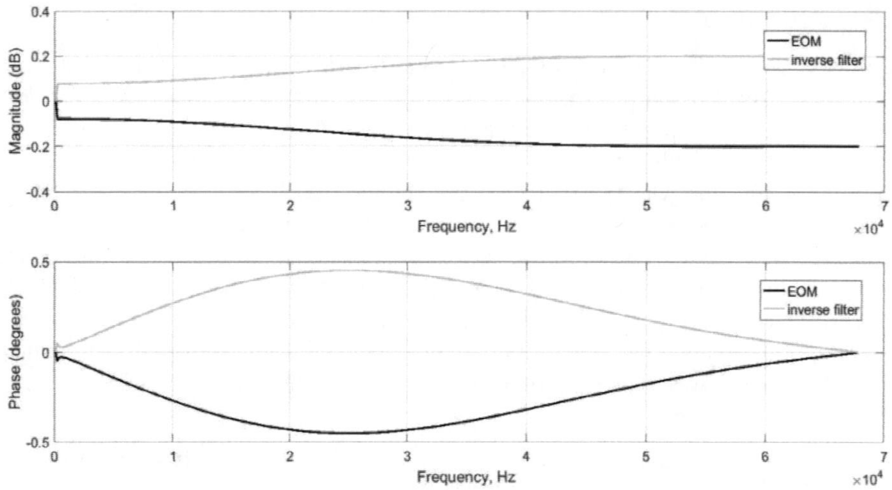

Figure 3. Frequency response of the EOM and of the inverse filter.

Pre-emphasis of the modulating voltage allows to compensate EOM`s parasitic phase drift. Comparing of transient response of the EOM and of the system consisting of EOM and inverse filter is shown at figure 4.

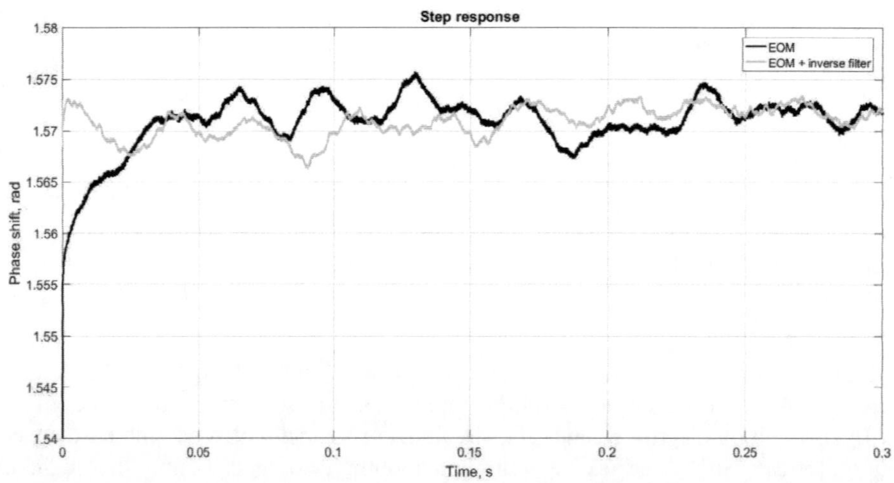

Figure 4. EOM`s step response and compensated by inverse filter step response.

Figure 4 shows that application of inverse filter for pre-emphasis of the modulating voltage signal for compensation the parasitic phase shift. Such compensation method can improve the accuracy of fiber-optic interferometric sensors in general and especially of fiber-optic gyroscope.

4. Conclusion

Revealed parasitic RID of phase EOM in case of changing the modulating voltage may be caused by mobile charges in the surface layers of the NL crystal. They move under the influence of an electric field generated by electrodes, and also create its own field which can increase or, conversely, weaken external field. Such charges have limited mobility, so they exhibit inertia in relation to changes in the external electric field. In order to compensate the parasitic phase shift caused by the process of relaxation of charges there is proposed to describe the inertia process of EOM by filter and to pre-emphasis the modulating voltage signal by the inverse filter. Described method can be used for improving the accuracy of fiber-optic interferometric sensors.

Acknowledgments

This work has been done at the ITMO University and supported by the Ministry of Education and Science of the Russian Federation (unique identifier of the project: RFMEFI57816X0202, contract № 14.578.21.0202).

References

[1] Porte H, Le Roux P 2015 *Photonics* **49** 90–99
[2] Wong K.K. 2002 *Properties of Lithium Niobate* (London: IET)
[3] Chen A, Murphy E J 2011 *Broadband optical modulators: science, technology, and applications* (Boca Raton: CRC Press)
[4] Nagata H, Yagang Li, Bosenberg W R, Reiff G L 2004 *IEEE Photonics Technology Letters* **16** 2233–5
[5] Salvestrini J P, Guilbert L, Fontana M, Abarkan M, Gille S 2011 *J. Lightwave Technol.* **29** 1522–34
[6] Wang W, Wang J 2010 *Opt. Eng.* **49** 114401

IOP Conf. Series: Journal of Physics: Conf. Series **917** (2017) 072003 doi:10.1088/1742-6596/917/7/072003

Magnetic phase transitions and linear magnetic dichroism in manganese-doped copper metaborate (Cu,Mn)B_2O_4

A D Molchanova[1], K N Boldyrev[1], A S Erofeev[2], E M Moshkina[3], L N Bezmaternykh[3]

[1]Laboratory of Fourier-spectroscopy, Institute of Spectroscopy RAS, Troitsk, Moscow108840, Russia
[2]Moscow Institute of Physics and Technology, Dolgoprudny, Moscow Region, 141701, Russia
[3]Kirensky Institute of Physics, Federal Research Center KSC SB RAS, Krasnoyarsk, 660036 Russia

Abstract. The work presents a study of manganese-doped copper metaborate (Cu,Mn)B_2O_4 using optical spectroscopy. The temperature of the antiferromagnetic phase transition $T_N = 19$ K has been defined according to the absorption spectra. Polarization studies (Cu,Mn)B_2O_4 in isotropic *ab*-plane show the presence of linear antiferromagnetic dichroism in the magnetically ordered state previously observed in pure copper metaborate CuB_2O_4.This measurement allows to find the magnetic phase transition into an elliptical structure at the temperature $T^* = 7.0$ K.

1. Introduction

Magnetic materials with complicated crystal and magnetic structures attract great attention because of their interesting properties and features. One of the brightest representatives of such compounds is a copper metaborate CuB_2O_4. This compound demonstrates a unique combination of magnetic, magnetoelectric, linear and nonlinear optical properties. CuB_2O_4 has a complex crystal structure (S.G. *I42d*, Z = 12) in which the magnetic ions Cu^{2+} (S = 1/2) occupy two different crystallographic positions, *4b* and *8d* [1], forming two magnetic subsystems. The “strong” magnetic Cu (*4b*) subsystem orders at the Néel temperature T_N= 21 K into an antiferromagnetic commensurate (C) structure with the spins lying in the easy *ab*-plane and then at the temperature $T^* \sim 8$ K into an incommensurate helical (IC) structure while the "weak" quasi-one-dimensional subsystem Cu (*8d*) not completely ordered and fluctuate even at the lowest temperatures [2-6]. Multiple frustrated and nonfrustrated antisymmetric exchange interactions within and between the *4b* and *8d* magnetic sublattices result in a rich complex magnetic phase diagram. The study of electronic absorption spectra of CuB_2O_4 revealed narrow zero-phonon (ZP) lines for all transitions between the crystal-field-split *3d*-states of Cu^{2+} ions [7]. Recently, sublattice-sensitive optical linear dichroism in the crystallographically isotropic *ab*-plane attributed to a magnetic Davydov splitting has been detected below the temperature of an antiferromagnetic ordering [8]. It should be noted that LD was observed only on ZP exciton lines associated with the magnetic subsystem Cu(*4b*). Also, using the LD method a splitting of the phase transition at T^* into two transitions (at $T_1^* = 8.5$ K and $T_2^* = 7.9$ K) not previously registered by other methods has been established.

In a study of the magnetic properties of copper metaborate, a special interest is the investigation of CuB_2O_4 compounds in which the copper Cu^{2+} ion is partially replaced by another magnetic *3d*-ion. For example, a possibility of the weak ferromagnetic moment in Ni-doped CuB_2O_4 rotation up to ±30°by applying an electric field was demonstrated [9]. A recent paper dedicated to the investigation of magnetoelectric properties has also revealed the possibility of induction and control of electric polarization in (Cu, Ni)B_2O_4 under a magnetic field [10]. The authors explain observed phenomenon in the framework of spin-dependent metal-ligand hybridization. These results indicate a strong correlation between magnetic and electric orders and shed light on the mechanism of the magnetoelectric effect in (Cu, Ni)B_2O_4. So it was expected that a doping with another magnetic *3d*-ion could lead to new interesting magnetic and magnetoelectric effects. However, LD studies have not been performed for doped metaborates up to now. Thus, our work is aimed at studying magnetic properties of hybrid metaborates, namely manganese-doped copper metaborate (Cu,Mn)B_2O_4. In this paper, we report results of high-resolution optical spectroscopy investigation of (Cu,Mn)B_2O_4 with linearly polarized light, in the same geometry that was used in the experiments [8].

2. Experimental details

Single crystalline samples of (Cu,Mn)B_2O_4 were grown by the flux method in the system $Bi_2Mo_3O_{12}$ - B_2O_3 - Na_2O - Mn_2O_3 – CuO at spontaneous crystallization. The crystals have dark-blue color and a maximum size of 3x4x5 mm^3. Mn^{2+} concentration was estimated to be 2.0±0.2% by X-ray fluorescence.

Optical absorption measurements were performed with a resolution 1.2 cm^{-1} using a Bruker IFS 125HR Fourier spectrometer and a closed-cycle Cryomech ST403 cryostat at the temperatures between 3.2 and 300 K. The polarization of an incident light was controlled by a Glan-Taylor prism.

3. Results and discussion

Figure 1(b) shows the LD spectra shifted from each other along the *y*-axis in the region of the first Cu (*4b*) ZP line. Figure 1(a) shows the same spectra as an intensity color map with the frequency-temperature axes.

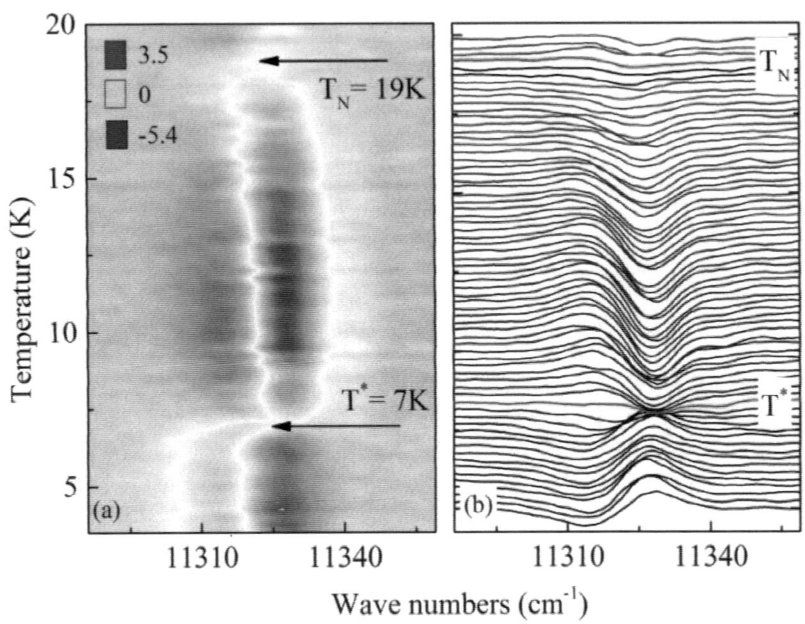

Figure 1.LD spectra in the region of the first *4b* ZP line of (Cu,Mn)B_2O_4 at temperatures below T_N, (for the light polarized along and perpendicular to the [110] direction; **k**||z), presented as (a) intensity color map, (b) shifted spectra.

The emergence of the LD below the temperature $T_N = 19.0$ K indicates a magnetic phase transition into an antiferromagnetic commensurate (C) structure as well as in CuB_2O_4 studies [8]. The reverse of LD has also been observed at the temperature $T^* = 7.0$ K.

As mentioned above, the features of the CuB_2O_4 magnetic structure at $T < T^*$ were considered earlier in [8]. The LD method made possible to clarify the previously proposed model of a simple circular helical magnetic structure. It has been shown that elliptical magnetic structures are realized below T^* with the large axis of the ellipse reorientation by $\pi/2$ at T_1^* and T_2^* sequentially. Thus, it can be assumed that $T^*$is the temperature of the (Cu,Mn)B_2O_4 magnetic phase transition into a simple helical structure, below which in the temperature interval $T < T^*$ an elliptical helical magnetic structure takes place.

Figure 2 shows a comparison of the pure and manganese-doped copper metaborates LD signals in the temperature region of 3.5 – 21 K.

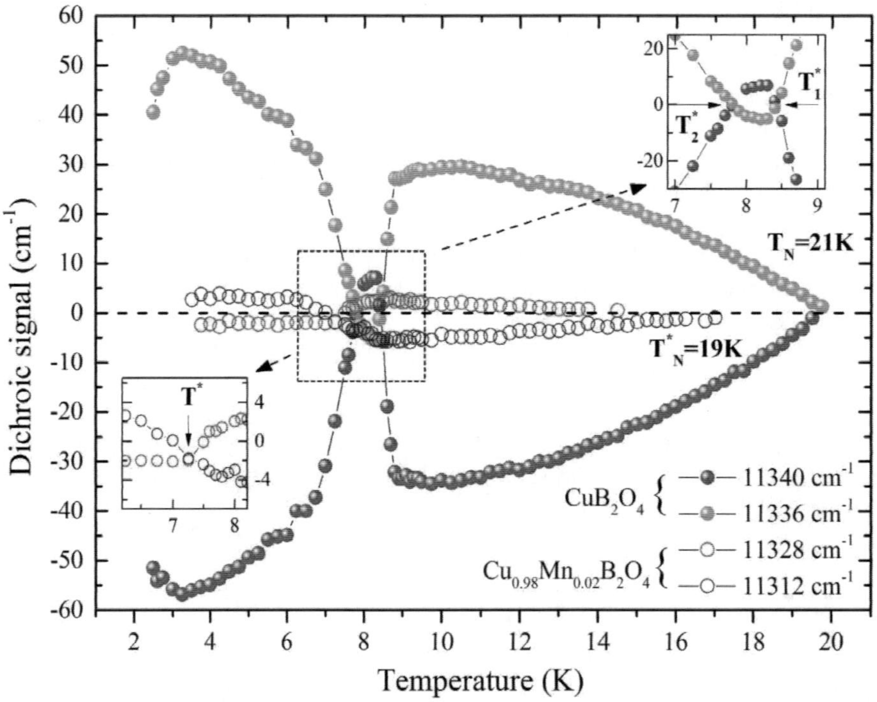

Figure 2. Temperature dependences of (Cu,Mn)B_2O_4 and CuB_2O_4 LD signals (for the light polarized along and perpendicular to the [110] direction). The inserts show the close-lying areas of the magnetic phase transitions of (Cu,Mn)B_2O_4 (T^*) and CuB_2O_4 (T_1^*, T_2^*) separately.

We can see that the LD signal of (Cu,Mn)B_2O_4 is about 10 times less than that in undoped metaborate. Doping with manganese results in a decrease of the CuB_2O_4 magnetic phase transitions temperatures ($T_N = 21$ K and $T^* \sim 8.0$ K for pure metaborate). The LD of (Cu,Mn)B_2O_4 undergoes a sign reverse at $T^*$and then does not disappear or change sign at least up to the lowest measured temperature of 3.5 K. In contrast to the pure metaborate, no splitting of T^* transitions are observed.

4. Conclusions

The absorption spectra of manganese-doped copper metaborate (Cu,Mn)B_2O_4 has been studied in the region of the first *4b* ZP line at temperatures below T_N. Linear dichroism associated with a magnetic

Davydov splitting has been observed in the magnetically ordered state of (Cu,Mn)B_2O_4, the same as in undoped metaborate CuB_2O_4. According to the LD spectra the temperatures of magnetic phase transitions into an antiferromagnetic commensurate (C) and an incommensurate helical (IC) structures has been established (T_N = 19.0 and T^* = 7.0 K respectively). The presented results reveal interesting opportunities of studying complicated magnetic structures by sublattice-sensitive optical linear dichroism method.

Acknowledgments
This work was supported by the Russian Foundation for Basic Research (Grant No 15-02-07451a) and the President of the Russian Federation (Grant MK-3577.2017.2).

References

[1] Martinez-Ripoll M, Martinez-Carrera S and Garcia-Blanco S 1971 *Acta Crystallogr. Sect. B* **27** 677
[2] Petrakovskii G, Velikanov D, Vorotinov A, Balaev A, Sablina K, Amato A, Roessli B, Schefer J and Staub U 1999 *J. Magn. Magn. Mater.* **205** 105
[3] Boehm M, Roessli B, Schefer J, Ouladdiaf B, Amato A, Baines C, Staub U and Petrakovskii G 2002 *Physica (Amsterdam)* **318B** 277
[4] Roessli B, Schefer J, Petrakovskii G, Ouladdiaf B, Boehm M, Staub U, Vorotinov A and Bezmaternikh L 2001 *Phys. Rev. Lett.* **86** 1885
[5] Boehm M, Roessli B, Schefer J, Wills A, Ouladdiaf B, Lelievre-Berna E, Staub U and Petrakovskii G A 2003 *Phys. Rev. B* **68** 024405
[6] S. Martynov, G. Petrakovskii, and B. Roessli, J. Magn. Magn. Mater. 269, 106 (2004).
[7] Pisarev R V, Kalashnikova A M, Schöps O, Bezmaternykh L N 2011 *Phys. Rev. B* **84** 075160
[8] Boldyrev K N, Pisarev R V, Bezmaternykh L N, Popova M N 2015 *Phys. Rev. Lett.* **114** 247210
[9] Saito M, Ishikawa K, Konno S, Taniguchi K and Arima T 2009 *Nat. Mater.* **8** 634
[10] Khanh N D, Abe N, Kubo K, Akaki M, Tokunaga M, Sasaki T and Arima T 2013 *Phys Rev. B* **87** 184416

Scanning probe microscopy of high-coercive iron garnet films

Y V Danishevskaya [1,2], A S Krikun [1], A S Nedviga [1], T V Mikhailova [1] and V N Berzhansky [1]

[1] Institute of Physics and Technology, V.I. Vernadsky Crimean Federal University, Simferopol 295007, Russia
[2] Crimean University of Culture, Art and Tourism, Simferopol 295017, Russia

Abstract. Mono-crystalline uniaxial anisotropy high-coercive garnet films are used for thermomagnetic recording by laser beams or contact printing. High-coercive properties of garnet films are due to a significant mismatch between the lattice parameters of the film and the substrate. The mismatch leads to appearance of facet crystal structure in these films. In the work, the facet crystal and domain structures of high-coercive garnet films were established by methods of scanning probe microscopy.

1. Introduction

High-coercive garnet films are transparent for infrared light and absorb blue-green beams that makes them promising for applications for atom cooling and manipulating and magnetic patterns thermo-recording. Real magneto-optic image of atom traps is shown in Figure 1 (a) [1-3].

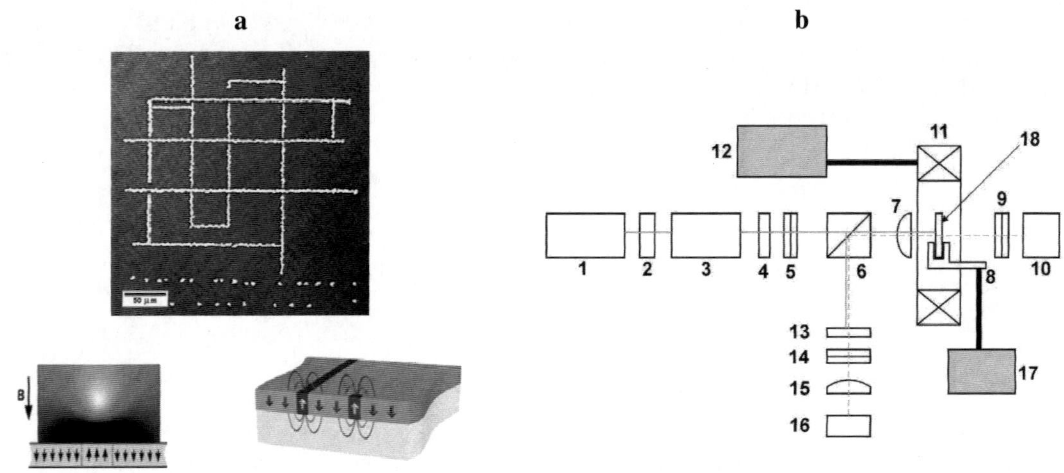

Figure 1 (a, b). (a) Real magneto-optic image of atom traps and formation of magnetic trap in uniform external field [1]. **(b)** The scheme of experimental installation for thermomagnetic recording (1 – laser (532 nm), 2 – λ/2 plate, 3 – electro-optical modulator, 4 – λ/4 plate, 5 and 9 – polarizers, 6 – beam splitter, 7 – lens, 8 – x-y table with control unit 17, 10 – LED blue spectrum, 11 – coil to create a magnetic field with a DC controller 12, 13 – a light filter, 14 – an analyzer, 15 – a projection lens, 16 – a CCD camera).

In the presented work authors investigated the garnet films mostly of $(Yb, Bi, Sm, Lu, Ca)_3(Fe, Al)_5O_{12}$ and $(Bi, Sm, Lu)_3(Fe, Al, Ga)_5O_{12}$ compositions on $Gd_3Ga_5O_{12}$ substrates with a lattice parameter 12.383 Å by magnetic force microscopy. The films were synthesized by liquid phase epitaxy, using isothermal dipping technique. The investigated samples had only positive film-substrate lattice constants mismatch in a range of Δa = 0.04-0.12 Å, coercivity H_c = 50-350 Oe, Faraday rotation θ = 0.9-2.3 °/μm (on wavelength λ = 532 nm), thickness h = 4-5 μm and Curie temperature T_C = 50-210 °C. Uniaxial anisotropy field was H_k = 1-1.7 kOe.

2. Properties of high-coercive garnet films

The measurements were performed on a scanning probe microscope NTEGRA with optical head by two-pass magnetic force microscopy. For a start, we have tried to investigate the domain structure by a polarization optical scanning microscopy, but the developed surface prevents obtaining a clear image. Therefore, a method of magnetic force microscopy was proposed for research. The initial morphology of films with a great Δa is shown in Figure 2 (a, b).

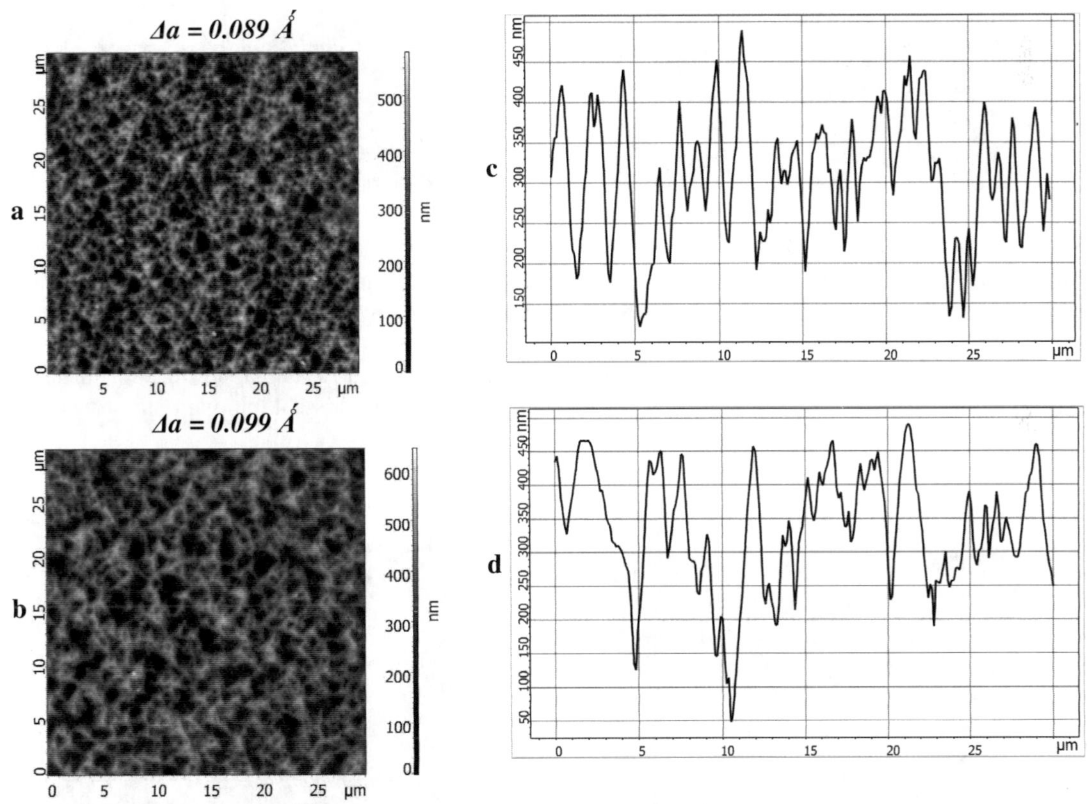

Figure 2 (a, b, c, d). Topography of facet surfaces for films with film-substrate lattice constants mismatch 0.089 Å **(a)** and 0.099 Å **(b)**. Figures **(c)** and **(d)** present the cross sections of correspondently surfaces.

The "faceting" after the liquid phase epitaxy is formed by intersection of dislocations with garnet films surface. All of the samples are characterized by block structure. It forms a mosaic. The size of its elements is about 1-8 microns. Block sizes reveal dependence on mismatch of film-substrate crystal constants Δa. The boundaries of the blocks consist of dislocation clusters. For films surfaces in Figure 3 (a) and (b) peak-to-peak values are 579 and 641 nm and the RMS roughness are 74.25 and 75.34 nm, respectively.

IOP Conf. Series: Journal of Physics: Conf. Series **917** (2017) 072004 doi:10.1088/1742-6596/917/7/072004

Domain structure of the films is characterized by thermal hysteresis. Figures 3 (a) and 4 (a) or Figures 3 (b) and 4 (b) show the magnetic force microscopy images of two types of domain structures, which are observed in the same films. Domain structure in Figure 3 (a, b) obtained after demagnetization of the film by an alternating magnetic field. This domain structure is stable. It retains its form during remagnetization.

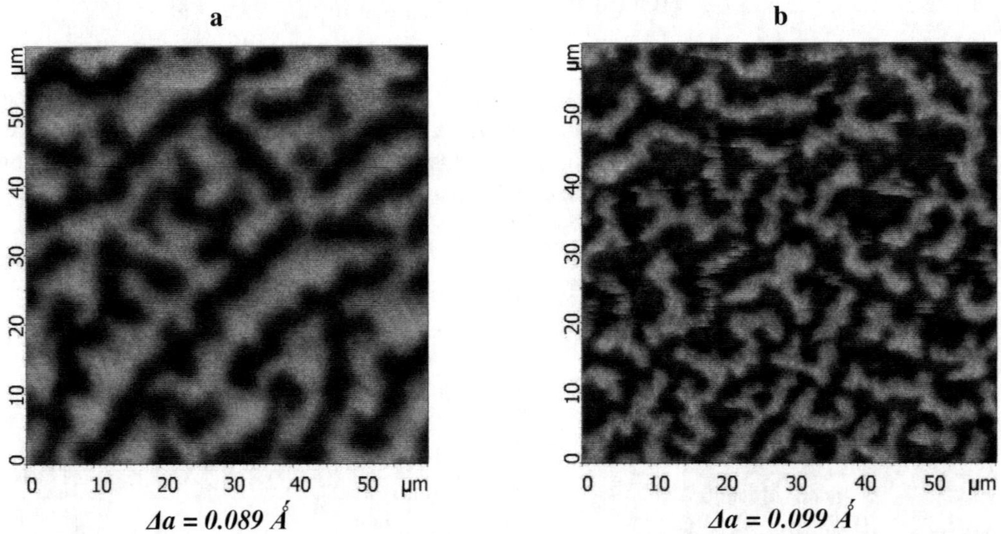

Figure 3 (a, b). Magnetic force microscopy image of domain structure of films with film-substrate lattice mismatch 0.089 Å **(a)** and 0.099 Å **(b)** after demagnetizing by variable magnetic field.

Domain structure in Figure 4 (a, b) obtained by method of thermo-demagnetization, during the heating above T_C. Such domain structure is metastable, It turns into a stable domain structure during remagnetization.

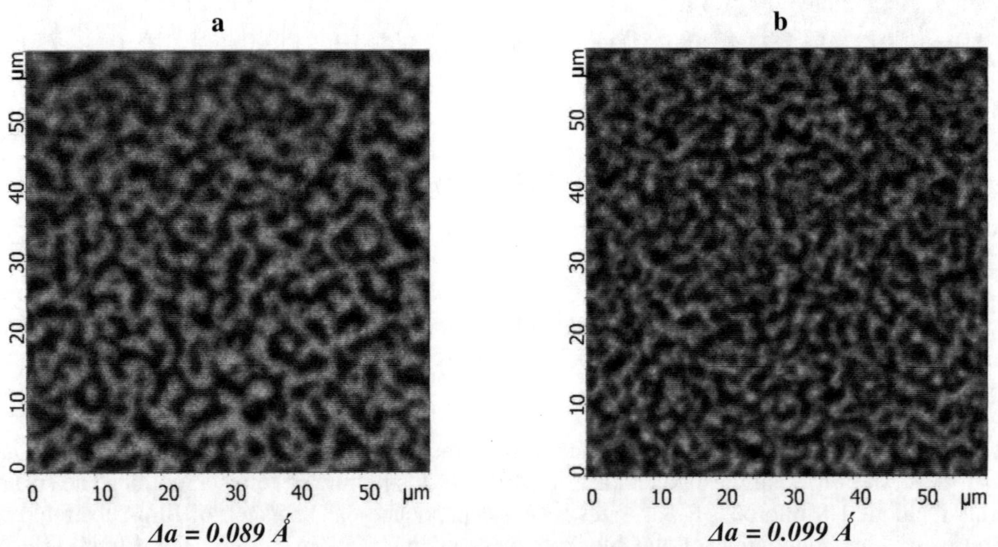

Figure 4 (a, b). Domain structure of high-coercive films with film-substrate lattice mismatch 0.089 Å **(a)** and 0.099 Å **(b)** after demagnetizing by heating. Magnetic force microscopy image.

Thus, the investigations of iron garnet films with different film-substrate lattice constants mismatch 0.089 Å and 0.099 Å showed that the periods of domain structures were respectively 11.2 and 8.74 μm after demagnetizing by variable magnetic field and 5.7 and 4.14 μm after demagnetizing by heating. Also, using magnetic force microscopy it was found that the period of the metastable structure is less than the period of the stable domain structure. At the same time, a film with a smaller film-substrate lattice constants mismatch has smaller elements of the facet structure and a longer period of the domain structure.

3. Conclusions

Structural characteristics of surface and formed magnetic domains of high-coercive garnet films were revealed by magnetic force microscopy. The facet structure of mono-crystals increases the coercivity, which allows using the film for thermo-magnetic recording by laser beams or contacting printing.

Acknowledgments

The authors acknowledge support by the RF Ministry of Education and Science in the framework of the state task (project no. 3.7126.2017/8.9). Measurements were carried out in the Center of collective use "Electron and Probe Microscopy" of V.I. Vernadsky Crimean Federal University.

References

[1] Berzhansky V N, Vishnevskii V G, Milyukova H T, Nedviga A S, Nesteruk A G, Danishevskaya H V 2010 *ACTA PHYSICA POLONICA A* **118** 864
[2] Shin K-H, Mizoguchi M, Inoue M 2007 *Journ. of Magnetics* **12** 129
[3] Berzhansky V, Nedviga A, Vishnevskii V, Prokopov A 2009 *Solid St. Phenomena* **152-153** 11

Resistive contact MEMS switch in a "hot" operation mode

I V Uvarov, V V Naumov, A N Kupriyanov, O M Koroleva, E I Vaganova and I I Amirov

Yaroslavl Branch of the Institute of Physics and Technology RAS, 150007, Universitetskaya Street 21, Yaroslavl, Russia

Abstract. Electrostatically actuated MEMS switch with the active contact breaking mechanism is tested in a "hot" regime. In this mode the lifetime of the switch is sufficiently restricted by the electrical failure mechanisms occurring at the contacts. Input DC voltage is applied to the source electrode during the actuation, output voltage at the drain electrode is registered. Measurements of the working characteristics of the switch including actuation voltage, contact resistance and lifecycle are performed. Experimental data is compared with the results of finite element simulation.

1. Introduction

Electrostatically actuated MEMS switches are widely used in radio frequency and microwave systems [1]. In comparison with switches based on pin-diodes and field-effect transistors, MEMS switches have advantages of low power consumption, small insertion loss in the closed state, high isolation in the opened state, low harmonic distortion and high radiation resistance. One of the main disadvantages of electrostatically actuated MEMS switches is the high actuation voltage, which is typically around a few tens of volts [2]. The most effective way to make the actuation voltage lower is to reduce the stiffness of the movable electrode. However, this approach increases the probability of stiction of the electrodes. Therefore, design of the switch must contain mechanisms allowing, if necessary, to overcome stiction and to detach the electrodes (active contact breaking).

MEMS switch with the active contact breaking mechanism was designed and fabricated by our group. Preliminary testing in the "cold" mode confirmed the significant mechanical reliability of the structure [3]. An important parameter of the switch is the ability to work in the "hot" mode, i.e. to be actuated repeatedly with applying input power during actuations. In this regime the additional failure mechanisms like temperature, current density and material transfer take place at the contact area [4]. They cause rapid degradation of contacts and increase the probability of stiction. Here we present the first results of testing the switch with the active contact breaking mechanism in the "hot" mode.

2. Design and operation principle of the switch

Design of the MEMS switch is schematically shown on figure 1. The movable electrode (source) is a metallic beam, suspended on torsion hinges. Gate and drain electrodes are fixed and located under each arm of the beam symmetrically with respect to the clamping point. Thus, the switch has two output channels. Each gate occupies the area from the clamping point of the beam almost to its free end, thereby providing a large area of an electric field. The beam has the contact bumps on its bottom side. These bumps localize the area of contact of the beam with the drain and prevent the short circuit with the gate when triggered. The beam also has the longitudinal ribs on the upper side. They allow to increase the stiffness of the beam without sufficient growth of its mass and to reduce the bending of the beam under the influence of residual stress [5]. To speed up the removal of the sacrificial layer from under the beam and to partially take off the residual stress in the material the perforation holes are included in the design. They also reduce the air damping [6] and the mass of the beam, thereby increasing the switching speed. The beam has the length of 102 μm, the width from 8 to 32 μm depending on the design, and the thickness of 2 μm in the rib region. The gap between the beam and the gate is 1.5 μm. Detailed dimensions of the structure are given in the earlier work [3]. SEM image of the switch with the beam width of 32 μm is presented on the figure 2.

The switch operates as follows. Initially the beam has the horizontal position, as shown on figure 1. When the certain voltage is applied to one of the gates, the beam tilts towards the electrode under the electrostatic force and comes in contact with the drain. This voltage is usually called pull-in voltage V_{PI}. Elastic force is generated in the torsion hinges, which tries to return the beam to its initial position. If the elastic force is insufficient to overcome the adhesion forces and to detach the beam from the drain, the switch retains its state when the gate voltage is removed. To break the contact the gate located under the raised arm of the beam is used. Applying the voltage to it, the beam tilts to the opposite direction and comes in contact with another drain. This voltage is called recovery voltage V_{REC}. Thus, two states of the switch are realized during operation, and the horizontal state of the beam is not reached anymore. This design provides a mechanism of active contact breaking, which is realized by the presence of the gate under the each arm of the beam. It is important that the beam should be sufficiently stiff and should not bend when trying to detach it from the drain.

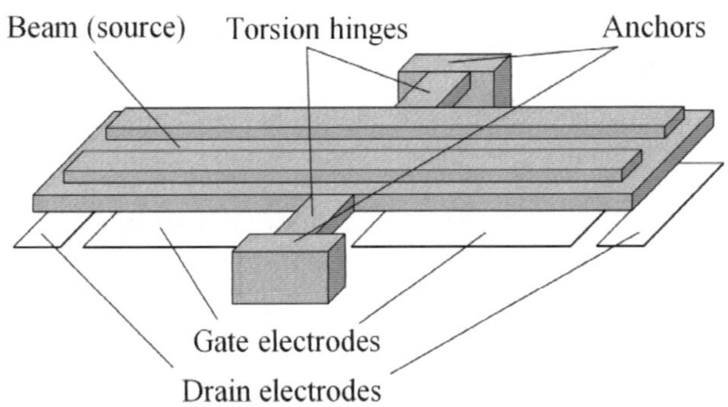

Figure 1. Design of the switch.

Figure 2. SEM image of the switch.

3. Fabrication procedure

The switch was fabricated by surface micromachining. The main steps of the process are shown on figure 3. At first the dielectric layer was formed on the silicon wafer. The wafer was thermally oxidized in wet oxygen to a SiO_2 layer thickness of 0.9 μm. Next, driving and signal electrodes of the switch were performed on the SiO_2 layer (figure 3a). For this purpose, adhesive Ti layer and Pt layer with the thicknesses of 10 and 100 nm, respectively, were deposited on the wafer by magnetron sputtering followed by lift-off.

In the next step the wafer was coated by a sacrificial layer of amorphous silicon (a-Si, figure 3b). In order to obtain the contact bumps of the beam, the dimples with the depth of nearly 500 nm were formed on the surface of the sacrificial layer (figure 3c). To do this, the round windows of 2 μm in diameter were formed in the photoresist layer. Through these windows the a-Si layer was etched in SF_6 plasma. Then 100 nm thick Pt layer was deposited and the lift-off was performed (figure 3d).

Further, anchors, torsion hinges and the beam were formed on the sacrificial layer. For this purpose the first aluminum layer with 1 μm thickness was deposited, followed by lithography and wet etching. Ribs of the beam were formed from this layer (figure 3e). Next, the second Al layer with 1 μm thickness was deposited over the ribs. The anchors, torsion hinges and the beam were made from it (figure 3f). The final fabrication stage was the removal of the a-Si sacrificial layer from under the beam by SF_6 plasma etching (figure 3g).

In the previous work [3] the mechanical reliability of the switch design was investigated, and the contact materials (Cr and Al) were used for reasons of the fabrication simplicity. In the present work the contact region played a significant role, and Pt-to-Pt contact was realized in the switch. Platinum has been selected as the contact material due to its chemical inertness and relatively high values of hardness (5.1 GPa) and melting point (2045 K) [7].

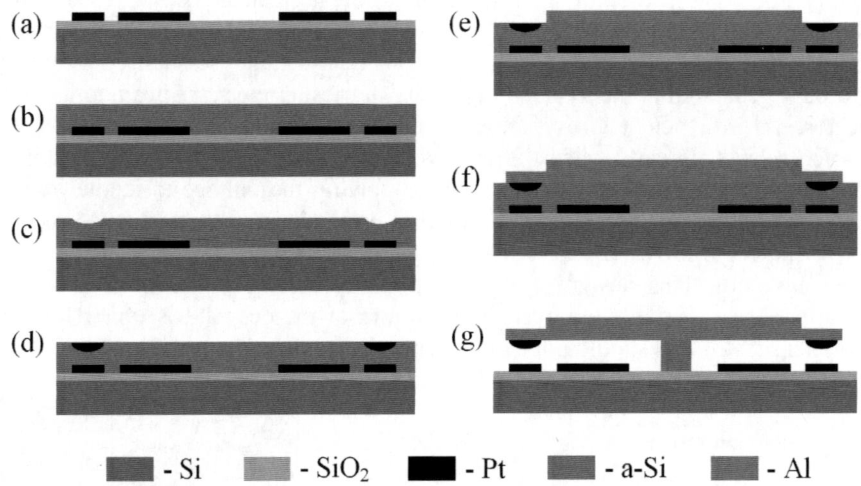

Figure 3. Procedure of the MEMS switch fabrication.

4. Experimental setup

Testing of the switches was performed in ambient air under normal conditions. Wiring diagram is shown on figure 4. Driving DC voltage (gate voltage, V_G) was applied from the power supply Agilent E3647A in order to actuate the switch. Input DC signal (source voltage, V_S) was applied from the power supply Mastech HY3005D. Output signal (drain voltage, V_D) was observed using the oscilloscope PicoScope 5000 and measured by the multimeter Keysight 34461A. The equipment was controlled by LabView software via GPIB interface using National Instruments PCI-GPIB controller. All the measurements were done in the low-current mode, the source-drain current was no higher than 0.1 mA. In this regime no stiction was observed, and breaking of the contact was carried out in a passive way.

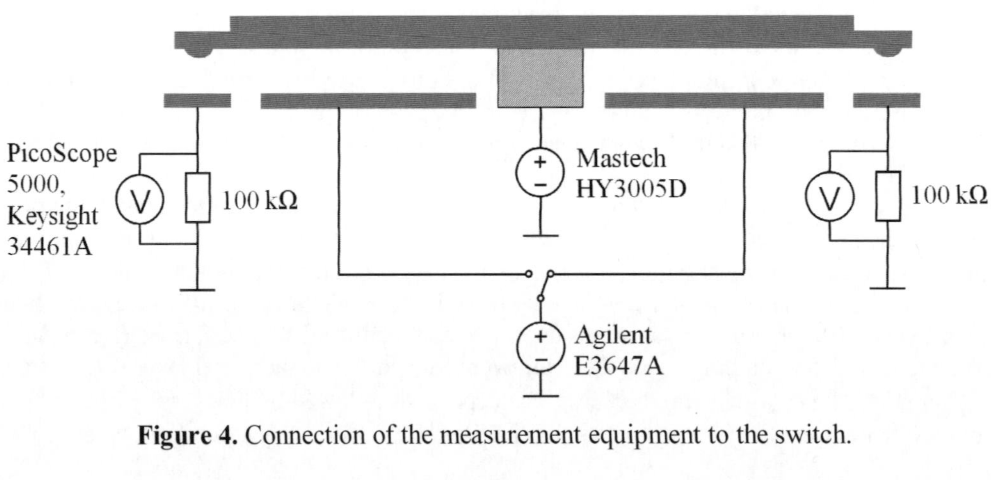

Figure 4. Connection of the measurement equipment to the switch.

5. Results and discussion

To determine the pull-in and pull-out voltage (V_{PI} and V_{PO}), the gate voltage was gradually increased from 0 to 60 V and then decreased as shown on figure 5. The moments of closing and opening the switch are clearly visible by the sharp change of the drain voltage. Experimental results are given in table 1. Pull-in voltage was in the range of 29÷46 V and depended on the beam width. Finite element method (FEM) simulation was performed using the validated software, details can be found in the previous work [3]. Simulation results age also given table 1. Experimental values of V_{PI} were 25% lower than it was predicted by the modeling. Probable reason of discrepancy was the reduction of the air gap between the beam and the fixed electrodes caused by the initial deformation of the beam under the residual stress. Pull-out voltage was in the range of 21÷33 V which is about 30% lower than the pull-in voltage.

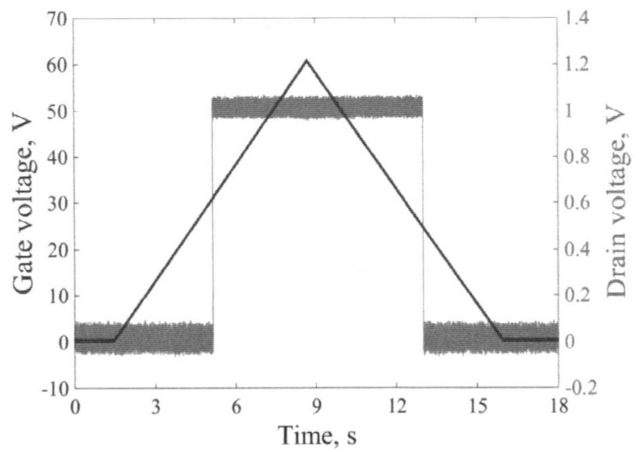

Figure 5. Gate and drain voltages during the pull-in and pull-out measurements. $V_S = 1$ V.

Table 1. Simulated and experimentally obtained values of the pull-in and pull-out voltage.

	FEM[b]	Experiment	
Switch type[a]	V_{PI} (V)	V_{PI} (V)	V_{PO} (V)
1	68	45.7 ± 6.1	32.8 ± 8.7
2	53	42.7 ± 3.7	30.6 ± 5.0
3	45	31.6 ± 4.5	22.6 ± 3.3
4	40	29.3 ± 2.2	20.7 ± 2.5

[a]The types of the switch differ from each other by the width of the beam [3].
[b]Details of the FEM simulation are given in [3].

To measure the lifecycle of the switch, rectangular voltage pulses were alternately applied to the each gate and the output voltage at both drains was recorded. The frequency of the driving pulses was 1 Hz. V_G and V_D at the same channel of the switch are shown on figure 6. It is worth noting that the drain signal was not rectangular. This indicates that the contact was not completely stable. Typical dependence of the drain voltage on the number of actuation cycles is shown on figure 7. During the first few dozen cycles the output voltage was $V_D = 0.99\ V_S$, and after approximately 1500 cycles it decreased to $V_D = 0.5\ V_S$. Contact region degraded completely after 9000 cycles, which corresponded to $V_D < 0.1\ V_S$.

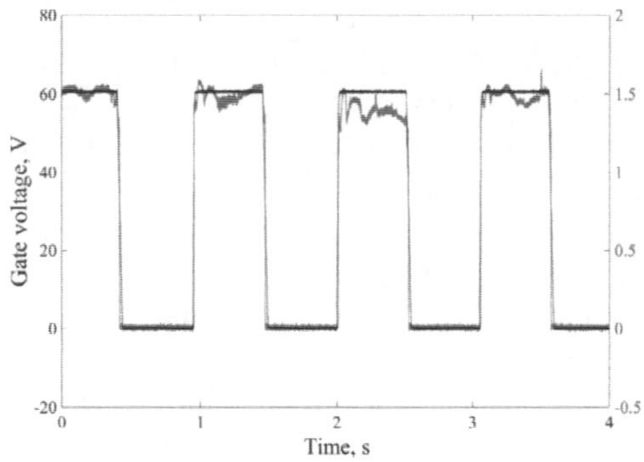

Figure 6. Gate and drain voltages at the one channel of the switch during the lifecycle measurement. $V_S = 1.5$ V.

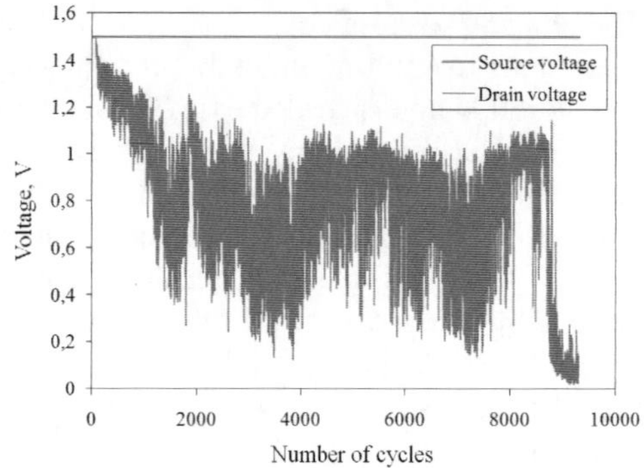

Figure 7. Dependence of the drain voltage on the number of actuation cycles. $V_S = 1.5$ V.

Contact resistance R_C of the switch was calculated using the following equation:

$$R_C = R_0\left(\frac{V_S}{V_D} - 1\right) - R_S - R_D,$$

where $R_0 = 100$ kΩ is the resistance limiting the source-drain current (figure 4), $R_S \approx 1$ Ω and $R_D \approx 50$ Ω are the resistances of the lines connecting the source and the drain with the corresponding contact pads of the sample. The initial value of R_C was about 1 kΩ, which is high for the ohmic MEMS switches. Contact resistance exceeding 5 Ω is considered as a resistive failure [1]. After several hundred cycles R_C increased to 100 kΩ. It is clear that the lifecycle of the switch in the "hot" mode has to be improved. We plan to investigate the degradation of the contact in detail and to choose the materials allowing to reduce the contact resistance and to increase the reliability.

6. Conclusions

MEMS switch with the active contact breaking mechanism was tested in a "hot" regime for the first time. All the measurements were performed in the low-current mode, so that no stiction was observed and the contact breaking was carried out passively. Pull-in voltage was 25% lower than it was predicted by FEM simulation probably because of the initial deformation of the beam. Lifecycle of the switch was about several hundred cycles, and it was restricted mainly by degradation of the contact area. Further we plan to investigate the resistive failure mechanisms in order to increase the lifecycle.

Acknowledgments

This work was supported by RFBR research project No. 16-37-60065 mol_a_dk and performed using the equipment of Facilities Sharing Centre "Diagnostics of Micro- and Nanostructures". Ilia Uvarov thanks Denis Pukhov and Leonid Mazaletskiy for the SEM measurements.

References

[1] Rebeiz G M 2003 *RF MEMS: Theory, Design, and Technology* (Hoboken, New Jersey: John Wiley & Sons, Inc.)
[2] Sharma A K, Gautam A K, Farinelli P, Dutta A and Singh S G 2015 *J. Micromech. Microeng.* **25** 035014
[3] Uvarov I V, Naumov V V, Koroleva O M, Vaganova E I and Amirov I I 2016 *Proc. SPIE* **10224** 102241A
[4] Toler B F, Coutu R A and McBride J W 2013 *J. Micromech. Microeng.* **23** 103001
[5] Gupta A, Barron L, Brainin M and Lee J-B 2014 *J. Micromech. Microeng.* **24** 065023
[6] De Pasquale G, Veijola T and Soma A 2010 *J. Micromech. Microeng.* **20** 015010
[7] Ekkels P, Rottenberg X, Puers R and Tilmans H A C 2009 *J. Micromech. Microeng.* **19** 065010

A study of 4H-SiC diode avalanche shaper

A V Afanasyev[1], B V Ivanov[1], V A Ilyin[1], A A Smirnov[1], A F Kardo-Sysoev[2], S A Shevchenko[1]

[1] Saint-Petersburg Electrotechnical University "LETI", Saint-Petersburg 197376,
[2] Ioffe Institute RAS, Saint-Petersburg 194021, Russia

Abstract. The results of the simulation of an ultrashort pulse generator with a 4H-SiC diode avalanche shaper implemented as a superfast closing switch are presented. The possibility of shaping a voltage pulse with an amplitude of 3300 V and 19 picosecond rise time is demonstrated.

1. Introduction

A study of the impact ionization phenomenon in reverse-biased silicon junction diodes resulted in a discovery of the superfast switching effect [1], which occurs at extremely high voltage rise velocities at *p-n* junctions (dU/dt > 10^{12} V/s). The voltage drop across the diode may reach values exceeding its static breakdown voltage up to two times, and, after that value is reached, it decreases in a very short time (less than 100 ps). This process is accompanied by a significant growth of the current through the diode and links with a superfast transition to a conductive state. In fact, it has been shown in [2] that such a transition may be possible as a result of the formation and propagation of a delayed shockwave front. Based on this unique idea the silicon avalanche shapers capable of switching powers up to hundreds of kilowatts at short times (70…100 picoseconds) were designed [3].

A detailed analysis of this superfast switching effect is undoubtedly of both practical and fundamental interest as is the problem of the design of a superfast closing switch based on new and perspective materials, such as 4H-SiC. For instance, the results of the design, simulation and experimental studies of the nano- and subnanosecond opening switches (the drift step recovery diodes, or DSRDs) based on 4H-SiC that are reported in [4…6] approve the possibility of a qualitative improvement of the 4H-SiC devices over silicon devices.

The following paper states for the first time the prospects of shaping voltage pulses with 20…30 picosecond rise times and magnitudes of kilovolt orders in an ultrashort pulse generator with the 4H-SiC diode avalanche shaper implemented as a superfast closing switch.

2. Simulation

The examined diode avalanche shaper has a p^+–p–n^+ structure with the thicknesses of the epilayers, respectively: 10 um, 10 um and 100 um. The concentrations of acceptors (Al) and donors (N) in *p*-type and *n*-type layers are, respectively: 10^{19} cm^{-3}, $8 \cdot 10^{14}$ cm^{-3} and 10^{19} cm^{-3}. The area of the top contact is 0.1 mm^2, likewise in [7]. The doping profile of the structure is presented in Figure 1. The static breakdown voltage of the following structure is 1500 V. The operation of the avalanche shaper had been studied in the ultrashort pulse generator circuit with inductive energy accumulation shown in Figure 2.

The simulation has been carried out using Synopsys Sentaurus TCAD software in the drift-diffusion approach with the following effects accounted: bandgap narrowing at high doping levels, avalanche generation in high fields, incomplete dopants ionization, Shockley-Read-Hall recombination, Auger recombination and various mobility models including mobility lowering in highly doped regions and carrier velocity saturation at high fields.

The simulation had been executed in the mixed mode in order to combine the actual physical simulation of the diodes D1…D3 and D4 and the SPICE simulation of the ultrashort pulse generator circuit with the avalanche shaper D4 implemented as a superfast closing switch.

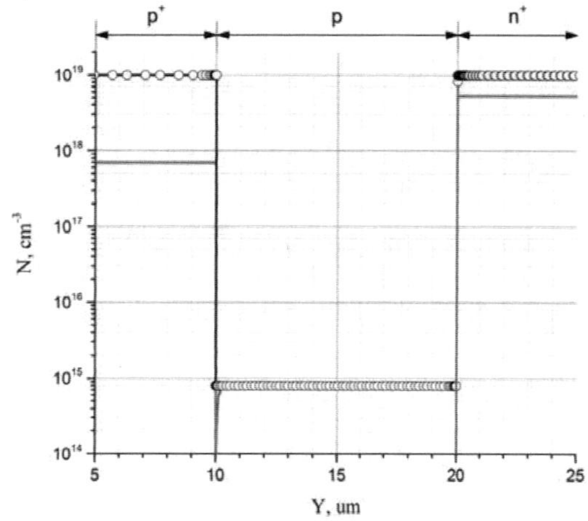

Figure 1. Doping profile of the avalanche shaper (marked line) and ionized doping distribution (dashed line)

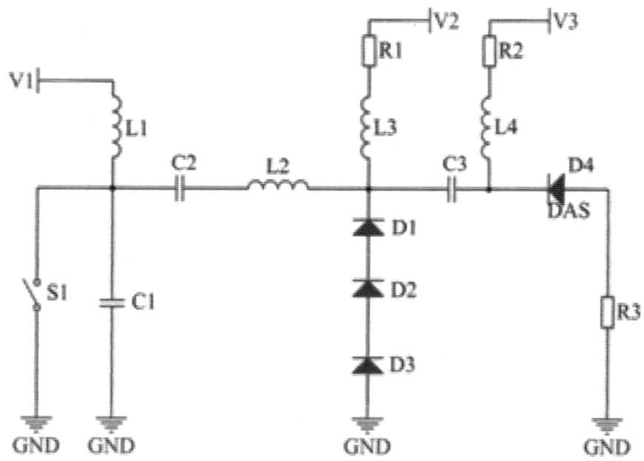

Figure 2. The ultrashort pulse generator circuit

In the Figure 2 the D1…D3 diodes stand for a high-voltage diode stack consisting of three single-crystal drift step recovery diodes (DSRDs) – the ultrafast opening switches. It was shown in [4] that it is possible to implement single-crystal 4H-SiC DSRDs in ultrashort pulse generators. The magnitudes of the generated pulses may reach the values varying from 1500 V to 2000 V with rise times less than 500 picoseconds. The article [4] also deals with information on the possibility of assembling single-crystal DSRDs into the high-voltage diode stacks in order to raise the switching voltages keeping the same subnanosecond rise times.

The implemented DSRDs D1…D3 have p^+-p-n^+ structures with the thicknesses of the epilayers respectively: 5 um, 8 um and 100 um. The concentration of acceptors and donors in *p*-type and *n*-type layers are respectively: 10^{19} cm^{-3}, 10^{16} cm^{-3} and 10^{19} cm^{-3}. The areas of the top contacts are 0.9 mm^2. The static breakdown voltage of a single DSRD in a stack is estimated 1000 V, so the whole stack's static breakdown occurs at 3000 V.

The operation principle of the generator circuit is described below. At the initial moment of time the C2 capacitor is charged to the potential of voltage difference between the V1 and V2 sources, so when the switch S1 is closed, the diode stack D1…D3 is biased forward with a current pulse duration of T_+ (50 ns). The period T_+ is defined by the values of L2 inductance and C2 capacitance. Throughout the T_+ period the injection of the non-equilibrium carriers into the base regions of D1…D3 diodes occurs, so the whole diode stack is in a conductive state.

By the end of the forward current pulse the switch S1 opens and the period of oscillations in the circuit is now defined by the values of L2 inductance and C1, C2 capacitances. The L2 coil accumulates energy, which is proportional to the reverse current flowing through the diode stack D1…D3. The voltage at the stack changes its polarity, so that the extraction of the non-equilibrium carriers could begin, so for the T_- period the voltage drop at the stack is low.

By the end of non-equilibrium carriers extraction the diodes' D1…D3 recovery process begins: majority carriers are being synchronously extracted from the structures' base regions with saturated velocities in a short time t_F, which is followed by a significant rise of the whole stack's resistance and accompanied by shaping an impulse with a very steep front at the load resistor R3. The amplitude of the impulse greatly exceeds the power supply voltage V1.

The switching process of the avalanche shaper D4 is presented in Figure 3. The voltage pulse shaped at the D1…D3 stack acts as a driving pulse for the avalanche shaper D4. By the end of the slow part of the driving pulse the base region of the shaper D4 is fully depleted with majority carriers, so at the stage of the steep voltage rise at the diode stack the current at the shaper is defined by only the displacement current and by the time t = 64.93 ns the voltage at the structure is $U_{D4\ MAX}$ = 2620 V, which is 1.75 times greater than the static breakdown voltage. After the voltage at D4 has reached its peak value the propagation of the shockwave in the base region of the structure is initiated and during the following 19 picoseconds the device is switching to a conductive state, the latter process being accompanied by shaping the voltage pulse at the R3 load resistor with the same rise time as that of the D4 switching.

The V3 source is implemented with the purpose of biasing the avalanche shaper D4 in the reverse direction in order to speed up its recovery into a non-conductive state after the switching.

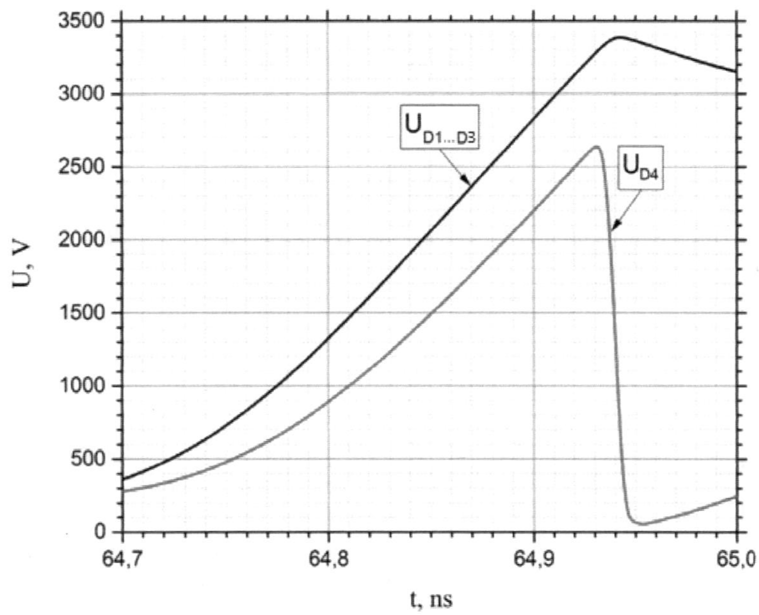

Figure 3. The process of the D4 shaper switching (red line) with the driving pulse at the DSRD stack D1…D3 (black line)

3. Results and conclusion

In the following study we have successfully demonstrated the 4H-SiC diode avalanche shaper's operation in an ultrashort pulse generator circuit where the device has been implemented as a superfast closing switch. The simulation results are presented in Figures 4 and 5.

The voltage pulse shaped at the DSRD stack D1…D3 (Figure 4, black line) is of 3400 V amplitude and 160 ps rise time measured at $(0.2 – 0.9)U_{STACK\ MAX}$.

A mentioned earlier, the process of the shaper's transition into a conductive state goes along with a very steep voltage rise at the R3 load resistor (Figure 4). The amplitude of the voltage pulse at R3 made up to 3300 V. The rise time at $(0.2 - 0.9)U_{LOAD\ MAX}$ is estimated as 19 picoseconds.

While the pointwise derivation of the impulse (Figure 5, dashed line) demonstrated that the shaper's switching velocity is time dependent, the mean voltage rise velocity at the efficient part of its switching process is of 121 V/ps.

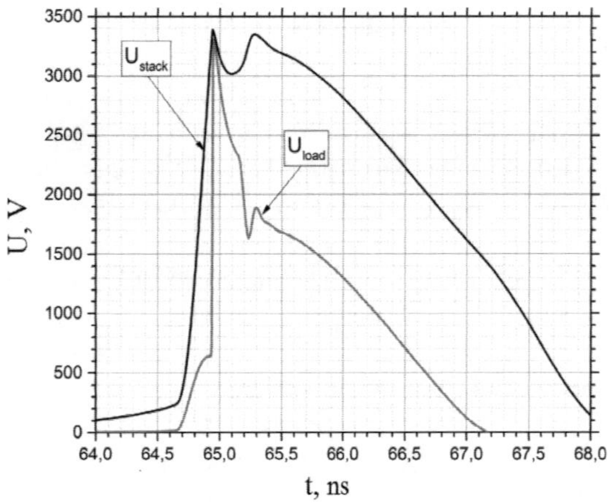

Figure 4. Voltage pulses shaped at the DSRD stack (black line) and the R3 load (red line)

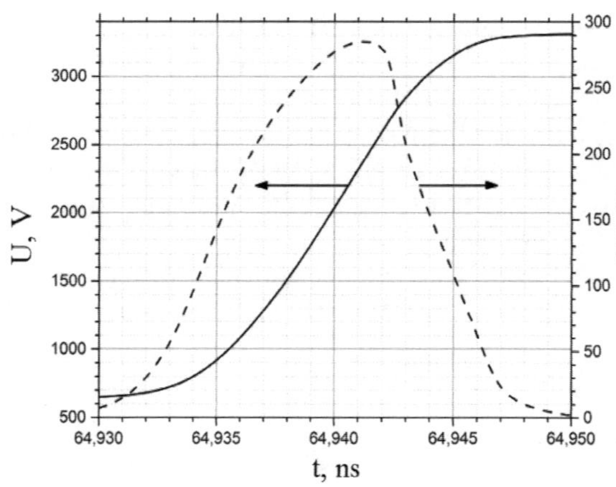

Figure 5. The voltage pulse obtained at the R3 load resistor (solid line) and its time derivative (dashed line)

References

[1] Grehov I V, Kardo-Sysoev A F. 1979 Technical Physics. 5 v 15 p 950.
[2] Grehov I V, Kardo-Sysoev A F. 1981 Technical Physics Letters. 3 v 3 p 1709.
[3] Grehov I V, Efanov V M, Kardo-Sysoev A F, Shenderei S V. 1986 Appl. Phys. 1 p 93.
[4] Afanasyev A V, Ivanov B V, Ilyin V A, Kardo-Sysoev A F, Luchinin V V, Reshanov S A, Schöner A, Sergushichev K A, Smirnov A A. High-voltage ultra-fast pulse diode stack based on 4H-SiC. Materials Science Forum Volume 858, 2016, 16th International Conference on Silicon Carbide and Related Materials, ICSCRM 2015; Sicily; Italy, pp. 786 – 879.
[5] Afanasyev A V, Demin Y A, Ivanov B V, Ilyin V A, Kardo-Sysoev A F, Luchinin V V, Smirnov A A. 2015 «Proceedings of the Russian Universities: Radioelectronics» **3** p 21.
[6] Afanasyev A V, Ivanov B V, Ilyin V A, Kardo-Sysoev A F, Luchinin V V, Smirnov A A, Shevchenko S A. 2015 Proceedings of IVth All-Russian Scientific and Technical Conference «Microwave Electronics and Microelectronics» 1 p 43.
[7] Zhang F, Li C. 2004 Chin. Phys. Lett. Vol. 21, No. 11. pp. 2305 – 2307.
[8] Afanasyev A V, Ivanov B V, Ilyin V A, Kardo-Sysoev A F, Kuznetsova M A, Luchinin V V. 2013 Materials Science Forum 740-742, 1010 – 1013.

Numerical simulation of induction heating considering the surface oxidation of cp-Ti samples

A Fomin[1], M Fomina[1], A Voyko[1], A Zorkin[1], V Koshuro[1], V Perinskii[1], I Rodionov[1]
[1]Yuri Gagarin State Technical University of Saratov, Saratov 410054, Russia

Abstract. Induction heating and oxidation of cp-Ti samples surface using high frequency currents are studied. The dependency of heating up to the temperature from 800 to 1500 °C is determined depending on the current values from 1.5 to 3 kA on the inductor at 90 kHz. Experimental results are compared to the data of numerical simulation of heat transfer in titanium samples.

1. Introduction

Induction heating is widely used in the treatment of metals, e.g. in heat treatment of ferromagnetic (steel) and some paramagnetic materials (aluminium, titanium) [1-3]. There is experimental data on hardening and surface modification with induction heating on titanium and its medical alloys [4,5]. Simulation of the physical processes is frequently applied when it is necessary to assess the distributed parameters in the elements of the studied products (sensors, medical products) [6,7]. In the description of fast processes of electromagnetic emission the development of models is an important task [8-10].

Control of technological parameters, temperature in particular, is important in the course of treatment with high-frequency currents (HFC) of metal products. In the accelerated mode of heating of metal products, the application of non-contact temperature control methods, infrared (IR) pyrometry and colorimetry (by colors of burning) in particular, is considered to be the most appropriate [4]. During the treatment of small-sized products using HFC the temperature measurement is significantly complicated by a number of factors:

– the characteristic size of the products is equal or smaller than the area where the temperature data are recorded using IR pyrometry (which is a constructive feature of pyrometers);

– the heated product is located in the treatment area (heating and heat treatment chamber), the access to which is limited;

– when the heating temperature exceeds 1200 °C, the application of colorimetry is not effective as the heated product is characterized by a bright light yellow or white color of burning.

According to the factors mentioned above, it is necessary to develop a complex calculation procedure for the reliable analysis of temperature and heating rate based on experimental data on heating kinetics and analysis of the processes of changing the structure of metals occurring during the high temperature treatment. The last statement is related to the fact that many metals when heated up to more than 600 °C, are characterized by intensive formation of an oxide coating [4].

Medical products from titanium alloys are machined, e.g. they are subjected to turning or abrasive blasting (Figure 1a). According to the obtained experimental results on the analysis of surface morphology of commercially pure titanium (Grade VT1-00) porous crystalline coatings with nanograins and nanopores of various shapes are formed on its surface due to the high temperature treatment with HFC [4] (Figure 1b).

Figure 1(a, b). Macrostructure of the surface after grinding (left) and abrasive blasting (right) **(a)**; nanostructure of the oxide coating treated with HFC at the temperature of 800 °C (×200,000) **(b)**.

Thus, at a high temperature of treatment the actual surface area *S* of the article, where heat exchange with the environment occurs, is several orders of magnitude higher than its ideal area determined by macrogeometry, e.g. a flat disc surface (Figure 1a).

As the choice of treatment parameters for the samples appears to be difficult numerical simulation of distribution of AC magnetic field and heat transfer can be used to make it easier. Thus, the aim of this study is to establish the dependency of heating $T(I,t)$ of commercially pure titanium samples by induction heating on the current values from 1.5 to 3 kA of the inductor *I* and to compare the experimental results of heating to the temperature *T* values from 800 to 1500 °C with numerical simulation data.

2. Methodology

Numerical simulation was performed using software "Elcut 6". In this program the geometry of the system "inductor – sample" was drawn, further a finite element mesh was generated and AC magnetic field distribution and non-stationary heat transfer were modeled depending on the peculiarities of heat source and heat loss, due to convection, radiation, and oxidation. Main elements of the studied system comprised an inductor (1), water (4), air (5), a quartz camera (3), a titanium sample (2) with its central part [0; 0] and periphery [0; Y] (Figure 2a). The model of the system "inductor – sample" also contained boundary elements, on which heat exchange with the environment (7, 8) took place, and the boundary of the considered system (6). The boundary of the sample (8) was characterized by several processes – convective and radiant heat exchange, as well as the formation of a morphologically heterogeneous oxide coating.

A thin oxide layer at the initial stage of treatment with HFC (up to 500–600 °C) of titanium samples practically did not affect the heating kinetics $T(t)$. However, in the course of further treatment with constant current on the inductor *I* the heating efficiency decreased sharply (Figure 2b). This was reflected in a significant (about 650–700 °C) difference in temperatures of high temperature exposure during the treatment with HFC (above 800 °C). Thus, intensive high temperature oxidation and the formation of the TiO_2 coating were associated with the appearance of an additional heat sink caused by a change in the actual area of the sample and an intensification of the convection and radiation processes. The mentioned heat sink can be associated with volumetric heat losses qv. The volume *V* of this sink was determined by the increased (by several orders) surface area S and the depth h that can equal the thickness (from 0.5–2 to 5–50 μm) of the oxide coating as well as the energy cost Q depending on the inductor current I (Figure 3a). This circumstance is particularly important for small-sized products such as titanium implants (dental implants and bone screws), the characteristic size D (wall thickness or diameter) of which varies from 2 to 6 mm.

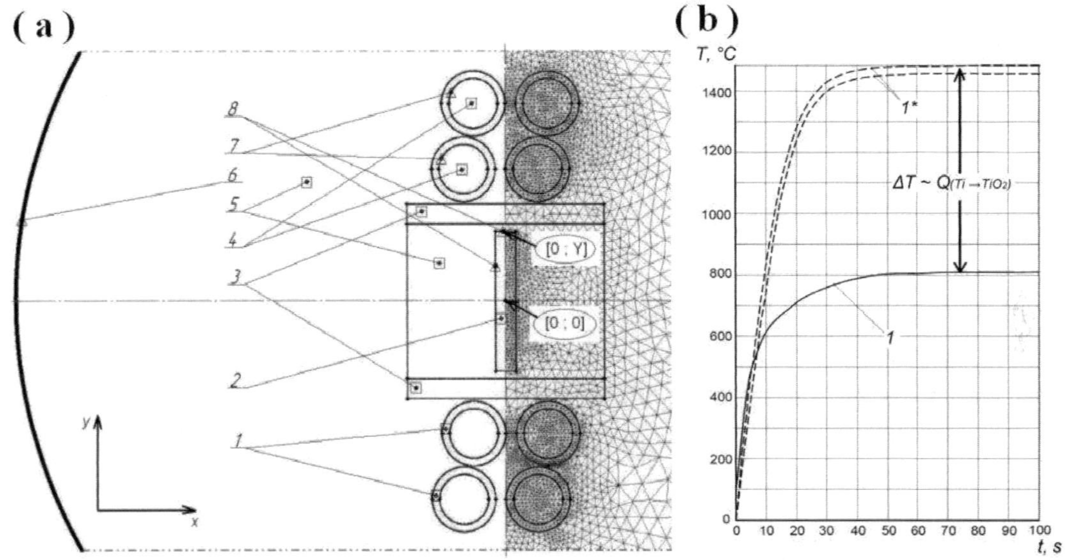

Figure 2(a, b). Model of the system "inductor – sample" (**a**); temperature differences between the experimental 1 and the calculated 1* data (**b**).

The laboratory apparatus for induction heat treatment included power supplies with rectified high (up to +300 V) and stable low (+15 V) voltage, a generator unit, and an induction heating unit. It comprised a filter, AC-DC and DC-AC converters, and load – the system "inductor – sample". The generator unit included an oscillator with adjustable frequency in the range of 89.5±0.2 kHz [11].

The temperature of titanium samples with the diameter of 14 mm and length of 2 mm was measured using a non-contact method. For this purpose infrared pyrometer "MS6550B" with the upper limit of 1650 °C was used. To study high temperature area of the samples characterized by the radiance from light yellow (around 1000–1050 °C) to dazzling white (1200–1400 °C) an optical camera with a filter was applied, which enabled to extend the study area [4]. Hence, the exact blackness coefficient, which mainly depends on the surface state in case of commercially pure titanium and other metals, was not necessary.

3. Results

When the heat transfer problem was solved the curves of heating of a titanium sample in its central part and periphery – the hottest area of the skin-layer (Figure 3b) were built. The calculation results *1'-3'* were compared to the experimental data *1-3* at current values on the inductor 1.5, 2, and 3 kA, respectively.

In order to solve the heat transfer problem the influence of the oxidation process was considered. The power of the source q_V was negative and it depended on the current in the inductor and the corresponding temperature of the heated titanium sample.

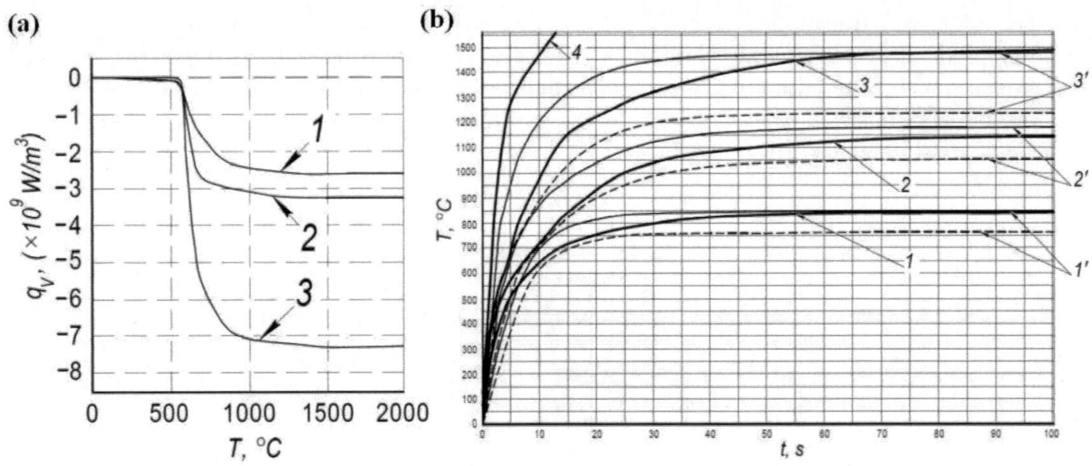

Figure 3(a, b). Dependencies of losses on the surface oxidation (**a**); heating curves (**b**).

4. Conclusions

Thus, the experimental data on heating $T(I,t)$ at 1.5–2 kA correspond to the simulation results. At the maximum current of I = 3kA there was a significant difference between the mean temperature of a titanium sample and the calculated temperature in the central part of the sample. The presented simulation results prove that oxidation can be considered within the temperature values around T = 1150–1200 °C.

Acknowledgments

The research was supported by the Ministry of Education and Science of the Russian Federation in the framework of the Program of Scientific Research in Universities (project No. 11.1943.2017/PP). A Fomin expresses his gratitude to the company "TOR" for the test access to the software "Elcut 6".

References

[1] Demidovich V B, Chmilenko F V and Rastvorova I I 2015 *Acta Tech. CSAV.* **60** 107
[2] Kuvaldin A B, Lepeshkin S A and Lepeshkin A R 2014 *Acta Tech. CSAV.* **59** 279
[3] Demidovich V B 2012 *Therm. Eng.* **59** 1023
[4] Fomin A, Dorozhkin S, Fomina M, Koshuro V, Rodionov I, Zakharevich A, Petrova N and Skaptsov A 2016 *Ceram. int.* **42** 10838
[5] Fomin A, Fomina M, Koshuro V, Voyko A and Rodionov I 2016 *J. Phys.: Conf. Series.* **741** 012170
[6] Schmidt M-P, Oseev A, Engel C, Brose A, Aman A and Hirsch S 2014 *Procedia Eng.* **87** 88
[7] Aman A, Majcherek S, Schmidt M-P and Hirsch S 2014 *Procedia Eng.* **87** 124
[8] Aman A, Majcherek S, Hirsch S and Schmidt B 2015 *J. Appl. Phys.* **118** 164105
[9] Majcherek S, Aman A and Fochtmann J 2016 *J. Micromech. Microeng.* **26** 025013
[10] Aman S, Aman A and Morgner W 2013 *Compos. Sci. Technol.* **84** 58
[11] Fomin A A, Fomina M A, Steinhauer A B, Petrova N V, Poshivalova E Yu and Rodionov I V 2014 *Proc. 55th Int. Sci. Conf. on Power and Electrical Engineering of Riga Technical University (Riga)* (Riga: IEEE) p 111

IOP Conf. Series: Journal of Physics: Conf. Series **917** (2017) 082004 doi:10.1088/1742-6596/917/8/082004

The influence of power consumption and current in the inductor on the temperature of heated titanium samples

A Fomin[1], M Fomina[1], A Voyko[1], S Kalganova[1], I Rodionov[1]
[1]Yuri Gagarin State Technical University of Saratov, Saratov 410054, Russia

Abstract. The measurement results for power consumption and current in the inductor and heating temperature of titanium disc-shaped and cylindrical samples when treated with high frequency currents within 90 to 120 kHz have been presented. The influence of the current in the inductor in the range of 0.3 to 8.5 kA on the heating kinetics of disc-shaped samples within 800 °C to 1500 °C in the air has been identified. A regression model describing the temperature change depending on the power consumed by the heat treatment device as well as its duration has been built.

1. Introduction

Determination of the current-voltage and watt-ampere characteristics of various electronic devices and electrotechnological processes, e.g. the ones accompanied by the occurrence of gaseous electrical discharges (electric arc welding, magnetron sputtering, electric spark alloying, plasma spraying) or high-frequency treatment by electromagnetic waves is important for the development of technological recommendations and optimal modes [1-3]. In order to investigate various systems such as sensor modules the characteristics of the transformation of the input-output signal should be studied as well [4]. When the system "inductor-product" is studied the transfer characteristics of the device for heating with high frequency currents (HFC) are determined together with those of the heat treatment process. The current in the inductor is necessary to develop the technological process with the use of HFC heating devices having the parameters different from the ones described in this work [2]. However, in this case a technological parameter should be used – the consumed power, which is determined by the product of voltage on the laboratory transformer output by consumed effective current. The aim of this work is to establish the dependence between the power consumed by the HFC heating device and the current in the inductor as well as to determine the heating kinetics at different power values.

2. Methodology

To heat the titanium samples (mass 1.1–1.3 g) of the disc-shaped (diameter 14 mm, length 2 mm) and cylindrical (diameter 3.5–4 mm, length 10–13 mm) products special designs of inductors of a cylindrical spiral shape were used [2].

Clamp meter "UT 205" (Uni-Trend Group Ltd.) was used to measure the current in the inductor *I*. This clamp enables the measurement of the effective current *I* from 0 and 4 kA in the high frequency range (within 150 kHz) (Figure 1a). This meter enables the increase of the measurement range from 4 to 10 kA in the short-term use. Current clamp *2* was located close to the inductor so that the water-cooled copper conductor line *1* passed through the measuring part (clamp) of the device (Figure 1b).

IOP Conf. Series: Journal of Physics: Conf. Series **917** (2017) 082004 doi:10.1088/1742-6596/917/8/082004

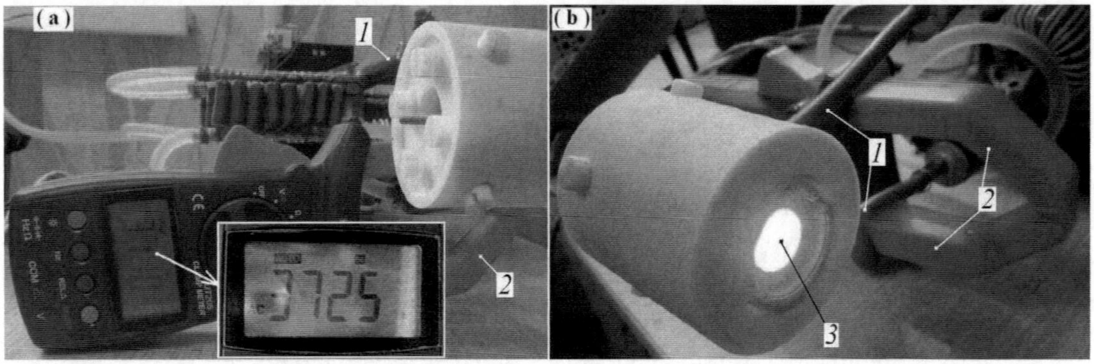

Figure 1(a, b). Measurement of the inductor current (the actual value on the indicator is about 3.7 kA) **(a)**; position of the measuring part of the current clamp from the side of the heated disc sample **(b)**.

The temperature *T* of the heated samples in the range from the room +20 °C to +1100 °C (+1650 °C) was controlled by a non-contact method with the use of infrared (IR) pyrometers "DT-8828" (CEM Instruments) and a high-temperature one – "MS6550B" (Precision Mastech) [2] (Figure 2).

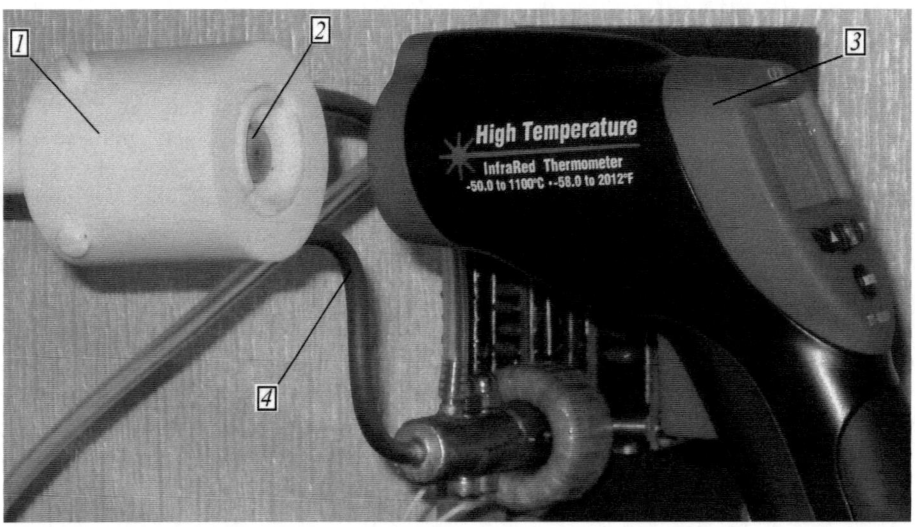

Figure 2. Non-contact measurement of the sample temperature.

The disc sample *2* located in the inductor *1* with an insulating cover and water-cooled copper current-carrying parts *4* were located at a sufficient distance from the pyrometer *3* in order to exclude its damage (Figure 2).

3. Results

The applied circuit design of the HFC device allowed regulation of the amount of the consumed electrical power *P* with the help of voltage *Uc* (Figure 3a). At the input of the device the value of the consumed current *Ic* was also controlled.

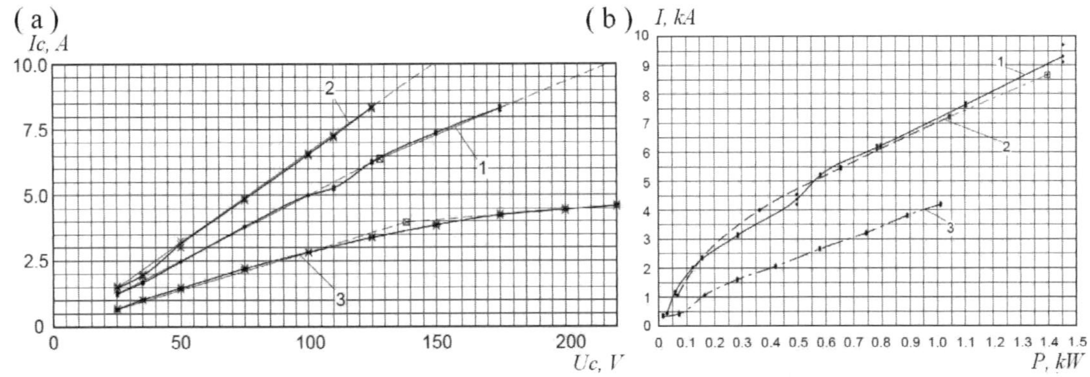

Figure 3(a, b). Volt-ampere characteristics of the device for treatment with HFC with various systems "inductor–product" **(a)**; dependency of the current in the inductor on the power **(b)**.

When the technological modes of the treatment with HFC are determined it is necessary to know the current in the inductor *I* (Figure 3b). The dependency of the current *I* on the consumed power *P* was received for different systems "inductor-product": 1 – for titanium disc-shaped samples (mass 1.1–1.3 g); 2 – for cylindrical products with a smaller (less than 4 mm) diameter (mass 1.0–1.2 g); 3 – for a product with a mass 80 times bigger compared to the disc-shaped sample.

Approximation of the experimental data on the temperature dependence *T(P,t)* on the consumed electric power *P* and the treatment duration *t* for the considered temperature range was presented in the polynomial form (1).

$$T(P, t) = -1700{,}15 + 315{,}67\times\ln(P) + 64{,}85\times t - 2{,}27\times t^2 + \\ + (0{,}04\times10^{-2})\times t^3 - (3{,}4\times10^{-4})\times t^4 + (1{,}1\times10^{-6})\times t^5 \,, \quad (1)$$

This type of model made it possible to establish a direct-proportional temperature dependence on the natural logarithm of the consumed electric power *P* (coefficient of multiple determination $R^2 \approx$ 0.91). In the graphical form the regression model *T(P,t)* was characterized by the presence of heating and exposure areas (Figure 4).

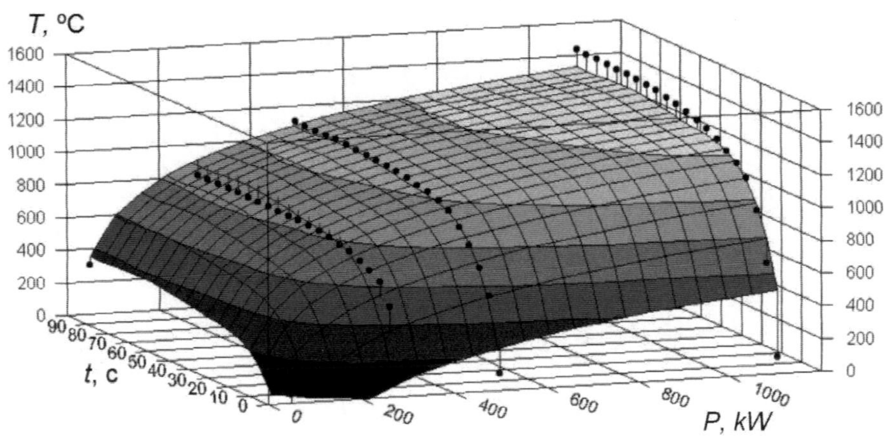

Figure 4. Heating kinetics at different power values.

Knowing the watt-ampere characteristics of the device for the heating with HFC the technological modes for these types of the treated disc-shaped samples and cylindrical products can be determined. When the modes of treatment of titanium products using HFC are determined the kinetic curves of heating at the chosen consumed power P should be received if the developed device for the treatment with HFC is applied.

4. Conclusions

The summarized data on the technological parameters (power P and current in the inductor I) and kinetic temperature characteristics $T(P,t)$ allow the development of the technological recommendations for the treatment of small-sized titanium products with the use of HFC.

According to the previously obtained data on the composition, structure, and physico-mechanical properties of oxide coatings produced on commercially pure titanium, the application of temperature range from 800 to 1000 °C appears to be most reasonable [5]. The samples should be heated to the given temperatures at the current in the inductor around 7.3±0.2 kA, which corresponds to the device power 1.25–1.3 kW. The exposure at temperature T = 800±10 °C should be performed at the power values of 0.28-0.3 kW and current on the inductor of 3.2±0.1 kA whereas for T = 1000±20 °C it is necessary to use a mode with parameters P = 0.36–0.4 kW and I = 3.9±0.1 kA.

Acknowledgments

The research was supported by the Ministry of Education and Science of the Russian Federation in the framework of the Program of Scientific Research in Universities (project No. 11.1943.2017/PP).

References

[1] Fomin A A, Fomina M A, Koshuro V A, Rodionov I V, Voiko A V, Zakharevich A M, Aman A, Oseev A, Hirsch S and Majcherek S 2016 *Tech. Phys. Lett.* **42** 932
[2] Fomin A A, Fomina M A, Steinhauer A B, Petrova N V, Poshivalova E Yu and Rodionov I V 2014 *Proc. 55th Int. Sci. Conf. on Power and Electrical Engineering of Riga Technical University (Riga)* (Riga: IEEE) p 111
[3] Morozova M Yu and Kalganova S G 2004 *Fibre Chemistry*. **36** 186
[4] Aman A, Majcherek S, Hirsch S and Schmidt B 2015 *J. Appl. Phys.* **118** 164105
[5] Fomin A, Dorozhkin S, Fomina M, Koshuro V, Rodionov I, Zakharevich A, Petrova N and Skaptsov A 2016 *Ceram. int.* **42** 10838

Features of transmission at analog intermediate frequency signals on fiber – optical communication lines in radar station.

D L Filatov[1], A A Galichina[2], M G Vysoczky[1], T R Yalunina[3], V V Davydov[1, 3, 4], V Yu Rud'[4]

[1]Peter the Great Saint Petersburg Polytechnic University, St Petersburg 195251, Russia
[2]AO "NII Vector", Saint Petersburg 196084, Russia
[3]The Bonch-Bruevich Saint - Petersburg State University of Telecommunications, Saint Petersburg 193232, Russia
[4]Department of Ecology, All-Russian Research Institute of Phytopathology, 143050, Moscow Region, Odintsovo district, B.Vyazyomy, Russia

Abstract. Features of transmission at analog intermediate frequency signals on fiber – optical communication lines are considered. In view of these features the technique for check of developed optical transmission systems at analog intermediate frequency signals is offered. The results of experimental investigations are presented.

1. Introduction.

On of the principle problems that are solved with the help of radiophotonics is the reliability ensuring of radar (RLS) work at different conditions of operation [1-4]. Besides that in many cases it is necessary the weight loading on the scanning antenna platform, especially in case when it is placed on flying device or at the top part of the ship mats. And one needs ensure a free access to equipment of its repair at unfavorable conditions [2, 4]. It is necessary to note also, that in case of radar placing the non – stable object the electromagnetic disturbances level in the area of the communication network location is several times greater than that observed at the ground station. The modern methods of the coaxial connections screening almost exhausted their resources from the point of view of creating new methods of protecting from electromagnetic disturbances. The additional protection is realized in the main by the increase both the number screening cover and its thickness. It leads to increase of the cable weight and volume, also to decrease its flexibility that creates difficult problems during radar operation [2, 4-7].

In the given situation the effective protection of signal, coming to the antenna complex especially the intermediate frequency signal (IF) that is the most sensitive to different disturbances appears to be rather difficult problem. As the IF f_i in the given case is taken frequency f in which the signal frequency transformed at the intermediate period of its processing in the radioelectron setup (superheterodyne receiver). Depending upon the radar exploitation condition f_i represent the sum or different between the received signal frequency and that of heterodyne. As a result all signal processing is carried out at the fixed frequency that is much easier than to readjust the whole receive path in the order to find the necessary signal.

In transmission IF signal tract occurs the frequency f_i offset value on random law, if there is interference in it. The automatic frequency control system (AFC) in most cases does not have time that to follow up bias f_i, that leads to the appearance of various errors (for example, In determining the distance to the target, the size and speed of the target, etc.).

One of its solution this task can be constructed by us fiber – optical transmission system (FOTS) of analog IF signal from the master generator to heterodyne located at the antenna past.

2. Peculiarities of analog intermediate frequency signals transmission through fiber – optical communication system.

But during the FOTS application for the IF signal transmission a number of peculiar properties appear that must be taken into account during its construction. The investigations carried out showed that they can be divided into three groups. The first group is connected with energy and frequency characteristics: amplitude frequency characteristic (AFC), amplitude characteristic, the loss coefficient at the frequency of the transmitted analog signal, etc. It is connected with the fact that the heterodyne is placed at the antenna past in the area of great number of different disturbances. That is why the signal from the optical receiver must be immediately sent to the heterodyne input. In case of a low power of the optical receiver output signal the application of the low – noise amplifier, as was shown by the experiments carried out, must be eliminated. The measurements error, for example, of the distance to the target during the received SHF signal, reflected from it, can be increased by the order and greater.

The second group is connected with the fact, that a part of FOTS is located in the area where the temperature change from 213 to 332 K is possible. The performed experiments showed that in transmitting and receiving optical modula (Dilaz company) the supply driver with the thermostabilizer doesn't provide the necessary stability for the IF signal transmission. Therefore is necessary to search for new solutions of this problem.

The third group - is connected with high requirements which is applied to the IF signal itself. During transmission it to FOTS should not happen pulse shape distortion (even minor) compared to other analogue signals which are also transmitted in RLS by the FOTS. Especially in case of impulses with high pulsing ratio. On figure 1 the structural scheme for research of FOTS developed by us is presented.

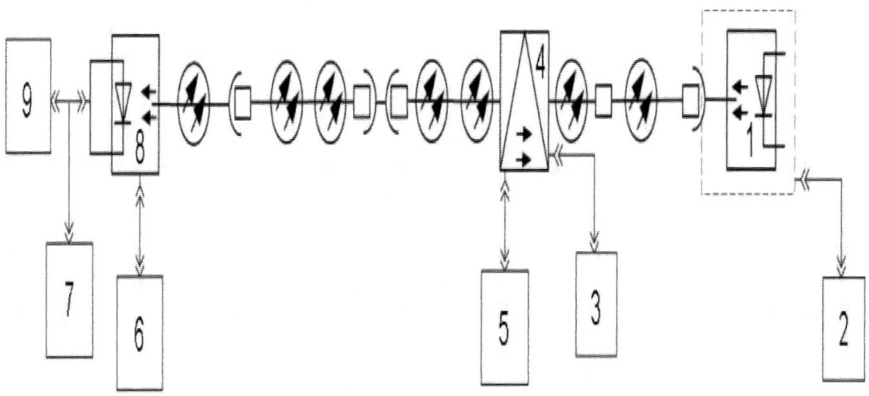

Figure 1. The structural scheme of IF signal transmission: 1 – laser (λ = 1550 nm); 2 - power supply; 3 – generator IF; 4 - Mach -Tsendera modulator (Oplilab company); 5 - source DC bias; 6 - bias source (+12 V); 7 – heterodyne; 8 - optical detector (Miteq); 9 – oscilloscope.

The division of the transmitting laser module allowed to make for 1 and 4 more reliable stabilization circuits comparing to the previous single optical transmitting module, and also to increase the optical laser radiation power up to 12 dBm. In addition, experiments have shown that the use of the displacement mode in the optical modulator (in case of temperature shift) enabled to adjust the point of minimum attenuation of signals IF for different f_i at the level of the output signal from the optical sensor up to a certain limit.

In the range of operating temperatures of the radar (RLS) from 223 to 333 K the location of the operating point in the linear portion of the modulation transfer characteristic will provide quite difficult. Therefore, to solve this task, a new scheme of operating point control was developed. On figure 2 the his structural scheme is presented. Through the optical divider (points 3 and 4) part of input and output radiation arrives at the control photodiodes. The output voltages from the photodiodes through the amplifier 6 and low frequency filter 7 are transmitted to the adder inputs 8. At its output the error signal is formed (the operating point shift from the given value). The error signal is fed to the proportional regulator 9. Integral part of the regulator allows to take into account the history of changes output value.

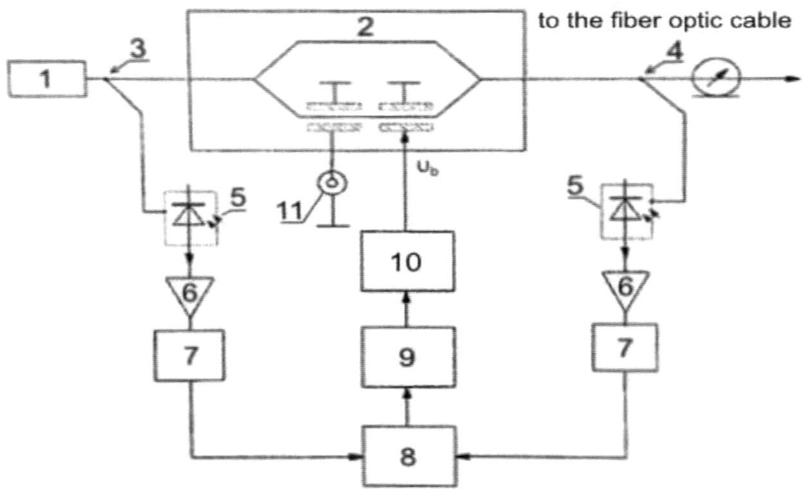

Figure 2. The structural scheme of modulator operating point control: 1 – transmitting laser module; 2 – electro-optical modulator; 3, 4 – optical divider; 5 – photodetector; 6 – operational amplifier; 7 – low frequency filter; 8 – adder; 9 – proportional regulator; 10 – zone switch; 11 – Input of high frequency signal.

In the absence of temperature fluctuations the constant bias at the output of the switch 10 is stabilized. The integral part of the regulator 9 provides a constant bias voltage U_b.

3. Results and its discussion.

In order to determine the functional possibilities of the application in RLS the constructed FOTS were investigated following characteristics: AFC; dynamic characteristic; the loss coefficient at the frequency of the transmitted analog signal on the environment temperature T change. As an example the AFR of the output signal for the different T are represented in the figure 3.

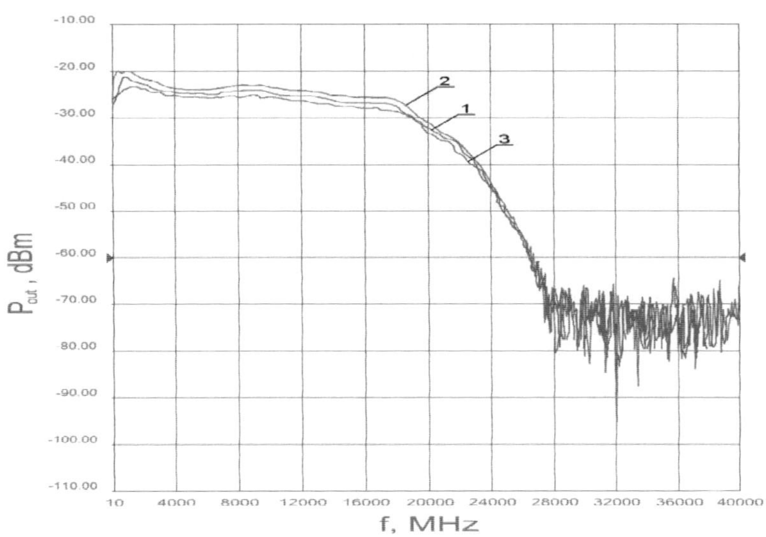

Figure 3. AFC, charts 1, 2 and 3 corresponds to T in K: 231.2; 293.1; 333.4.

The analyses of the received results shows that the developed by us device of electro-optical modulator 2 operating point and the scheme thermal stabilization of the transmitting laser module 1 allows to make insignificant the influence of temperature T change (over the whole range of the RLS operation) upon the value of optical signal power loss at it travels along the FOTS. It gives a possibility to transmit IF signals along it without loss and distortions.

As an example in the figure 4 are represented the IF signals in the from rectangular impulses transmitted to the heterodyne input from the master generator by its developed FOTS and coaxial cable.

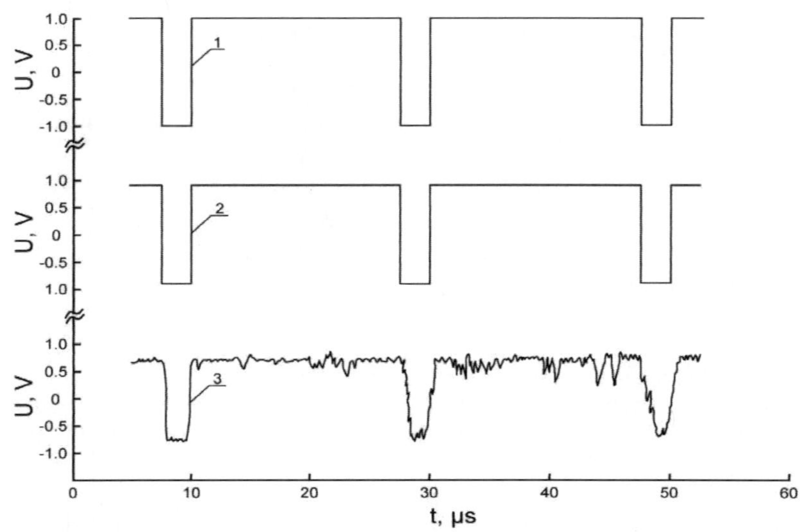

Figure 4. The form of the pulses of the output signal: 1 – generator 3; 2 - at the exit FOTS (input oscilloscope 9); 3 - at the exit a coaxial cable.

The analyses of the received results shows that there are not distortions of the impulses from in the IF signal supplied to the heterodyne 7 input from the optical detector 9 output. It allows to use this signal for the processing of the received RLS signal, reflected from different targets. In the coaxial cable of distortions the substantially changes the IF signal impulses form, that does allow to use them in this form without additional processing in the heterodyne.

Besides it was found in results of performed experiments that for ensuring reliable work of radar (RLS) with using the FOTS in a given part of it's the following has to be carried out:

1. Unevenness of amplitude frequency characteristic FOTS in the range of the transferred frequencies has to be less than ±3 dB.
2. Change of coefficient of losses in FOTS from the frequency of the transmitted microwave signal shouldn't exceed ± 2 dB
3. Phase deviation of modulation at distribution of a signal on FOTS from temperature in the range from 238 to 323 K shouldn't exceed 3 degrees.

4. Conclusion

The obtained results show that technical solutions offered by us allowed to make insignificant the influence of considered negative factors connected with features to transmission of signal IF on heterodyne work results. The developed FOTS can be used for various types of RLS in the entire frequency range of IF signals.

References

[1] Marpaung D, Roeloffzen C, Heideman R, Leinse A, Sales S, Capmany J 2013 *Integrated Microwave photonics* (Laser & Photonics Reviews) **4** 506
[2] Davydov V V, Velichko E N, Dudkin V.I, Karseev A Yu 2015 *Journal of Optical Technology* **82** 132
[3] Petrov A A, Davydov V V 2017 *Journal of Communications Technology and Electronics* **62** 289

[4] Davydov V V, Sharova N V, Fedorova E V, Vologdin V A, Karseev A Yu 2015 *Lecture Notes in Computer Science* **9247** 712
[5] Friman R K 2012 *Fiber-optic communication systems* 496
[6] Davydov V V, Dudkin V I, Karseev A Yu 2014 *Optical Memory & Neural Networks* **23** 259
[7] Ermolaev A N, Krishpents G P, Davydov V V, Vysoczkiy M G 2016 *Journal of Physics: Conference Series* **741(1)** 012171

Fast electrochemical membrane actuator: Design, fabrication and preliminary testing

I V Uvarov[1], A V Postnikov[1], P S Shlepakov[1], V V Naumov[1], O M Koroleva[1], M O Izyumov[1] and V B Svetovoy[1,2]

[1]Yaroslavl Branch of the Institute of Physics and Technology RAS, 150007, Universitetskaya Street 21, Yaroslavl, Russia
[2]Zernike Institute for Advanced Materials, University of Groningen - Nijenborgh 4, 9747 AG Groningen, The Netherlands

E-mail: i.v.uvarov@bk.ru

Abstract. An actuator based on water electrolysis with a fast change of voltage polarity is presented. It demonstrates a new actuation principle allowing significant increase the operation frequency of the device due to fast termination of the produced gas. The actuator consists of a working chamber with metallic electrodes and supplying channels filled with an electrolyte. The chamber is formed in a layer of SU-8 and covered by a flexible polydimethylsiloxane membrane, which deforms as the pressure in the chamber increases. Design, fabrication procedure, and first tests of the actuator are described.

1. Introduction

Membrane actuator is a wide spread component of microfluidic systems. It is a key part of pumps [1], valves [2], lenses, mirrors [3] and other devices. Several mechanisms driving the membrane motion are known: pneumatic, piezoelectric, electromagnetic and others [4]. Typically the actuator is integrated with other microfluidic components, so its structure should be as simple as possible. Electrochemical actuation is very attractive because it allows a simple design and provides a strong response. It is based on the electrochemical gas production with the following termination of the gas. The main disadvantage of the electrochemical actuators is a long response time due to slow recombination of the gas inside of the working chamber [1]. Recently a short-time electrolysis using alternating polarity voltage pulses was demonstrated, which allows fast generation and termination of the gas in the chamber [5]. Here we describe design, fabrication, and first results of testing of the electrochemical membrane actuator working on the new actuation principle.

2. Design and operation principle of the actuator

Design of the actuator is schematically shown in figure 1. Actuator has a chamber with metallic electrodes inside of it, channels for filling the system with the electrolyte, and holes for connection with microfluidic tubes. The diameter of the chamber is 500 μm and the height is 8 μm. An oxidized silicon substrate is used as the bottom wall, the side walls are made of cured SU-8 resist, and the upper wall is the flexible polydimethylsiloxane (PDMS) membrane with a thickness of 30 μm. The actuator has a planar configuration, the chamber and the channels have the same height. The electrolyte is a molar solution of Na_2SO_4 in distilled water. Photograph of the fabricated actuator with close-up view of the working chamber is shown in figure 2. The electrodes have interdigitated structure to increase the efficiency of the electrolytic bubble generation [6]. The diameter of the overall electrode area is 450 μm, which is close to the diameter of the working chamber. Interdigitated structure of electrodes and their size are chosen to saturate the entire volume of the chamber evenly with the gas bubbles.

Operation principle of the actuator is schematically shown in figure 3. A series of short (~10 μs) voltage pulses of alternating polarity is applied to the working electrode (the other one is grounded). These pulses produce hydrogen and oxygen due to water electrolysis. The alternating polarity regime differs in that sense that all the gas is collected in nanobubbles, which are invisible because they do not scatter light [7]. The pressure increases in the chamber and pushes the membrane up. When the pulses are switched off, the gas in nanobubbles recombines into water in 1 ms or so releasing the pressure and the membrane returns to its initial state. Unlike the conventional operation mode when a single polarity voltage pulse is applied to the electrodes, in the alternating polarity mode hydrogen and oxygen molecules are not spatially separated and form not only H_2 and O_2 bubbles, but also bubbles containing mixtures of H_2 and O_2 [5]. The latter disappear in the reaction, which is ignited in nanobubbles spontaneously. The mechanism of the reaction is still not completely clear but the reaction happens with the assistance of the gas-liquid interface [5]. When the pulses are switched off the pressure is reduced in about 1 ms. Short recombination time makes it possible to reduce significantly the duration of the actuation cycle.

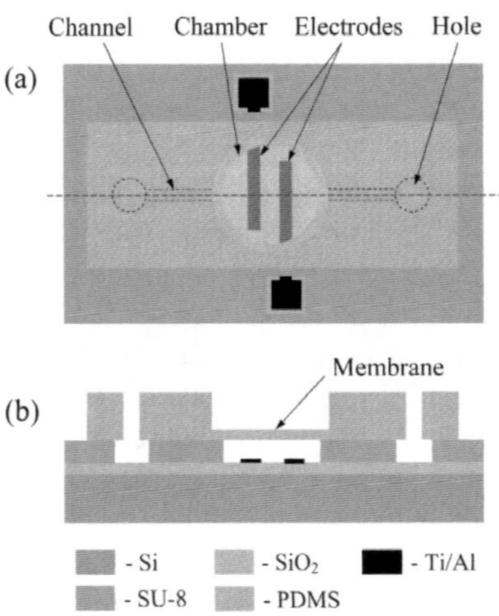

Figure 1. Design of the actuator: top view (a) and the cross-section (b).

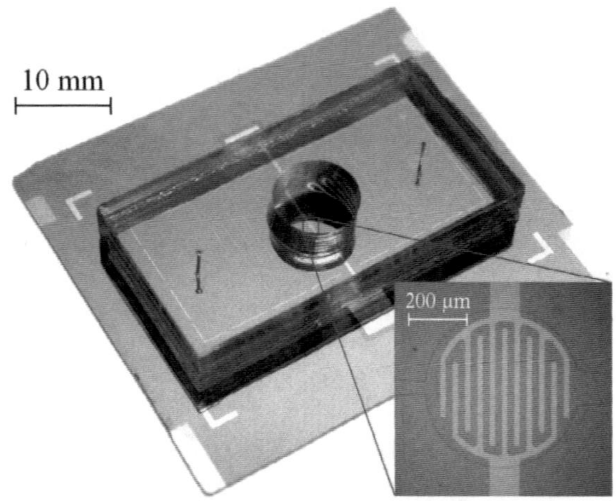

Figure 2. Ready to use actuator. The hole in PDMS layer is much larger than the working chamber covered with the membrane. The inset shows a zoomed view of the chamber.

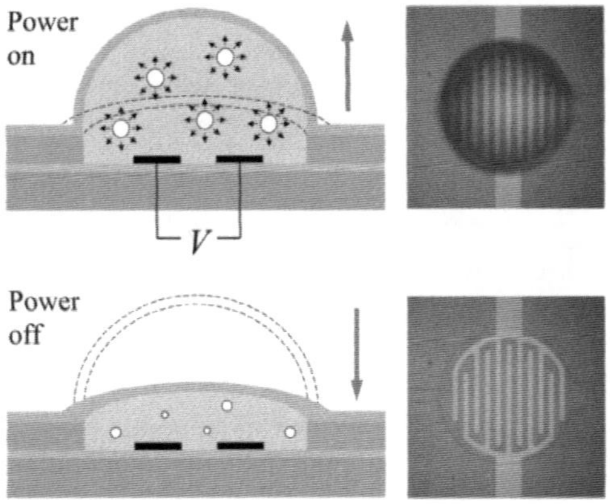

Figure 3. Operation principle of the actuator. Gas produced in the chamber electrochemically pushes the membrane up. When the voltage is switched off the gas in nanobubbles is terminated fast and the membrane is going down.

3. Fabrication

The main steps of the fabrication process are shown in figure 4. The process includes three stages. At the first stage a silicon substrate containing metallic electrodes and a layer of SU-8 containing the chamber and channels is prepared. The wafer is thermally oxidized in wet oxygen so that a dielectric layer of a thickness of 0.9 μm is formed on the wafer (figure 4a). Next, the electrodes of the actuator are fabricated (figure 4b). The electrodes contain 10 nm thick adhesive Ti layer, 500 nm thick Al layer, and 100 nm thick Ti layer deposited by magnetron sputtering. The substrate is covered by the photoresist Shipley S1813 and the UV contact photolithography is performed followed by the wet etching of Ti/Al/Ti layers through the photoresistive mask. Further, the walls of the chamber and channels are formed in 8 μm thick layer of the negative photoresist SU-8 3005 (figure 4c). The post-exposure baking of SU-8 is done at a temperature no higher than 95 °C in order to preserve the residual epoxy groups on the SU-8 surface [8].

At the second stage a PDMS structure holding a membrane is fabricated. The silicon wafer is treated in hydrofluoric acid solution for 10 s to make its surface hydrophobic. It is necessary to reduce the adhesion of PDMS to silicon. A PDMS layer with a thickness of 30 μm is applied to the wafer by spin-coating of liquid compound (Sylgard 184, a mixture of a base and a curing agent at 10:1 ratio) and curing at the temperature of 100 °C during 35 min (figure 4d). Then 4 mm thick PDMS structure with a through hole of 8 mm in diameter is formed separately and bonded to the thin PDMS layer (figure 4e). The bonding is performed using oxygen plasma treatment of PDMS at the pressure of 0.5 mbar and 300 W power during 8 s. Next, the PDMS structure is detached from the wafer with the through hole being closed by a membrane. After that the inlet and outlet holes with a diameter of 0.5 mm are made in the structure by a biopsy puncher (figure 4f).

At the final stage the PDMS structure containing the membrane and the silicon substrate containing the electrodes, chamber, and channels are assembled together using irreversible bonding of PDMS to SU-8. The bonding is based on the chemical reaction between PDMS and SU-8 by generating amino groups on PDMS surface with nitrogen plasma treatment and allowing the amino groups to react with the residual epoxy groups on SU-8 surface at elevated temperatures [8]. The PDMS structure is exposed to N_2 plasma with the pressure of 0.5 mbar and 700 W power during 1 s. Exposed surface is brought into contact with the SU-8 surface aligning the structure in such a way that the chamber is covered by the membrane and the channels are sealed with a thick layer of PDMS (figure 4g). The assembled parts are treated at 100 °C for 30 min to drive the chemical reaction. After cooling the samples are ready for filling with the electrolyte and testing.

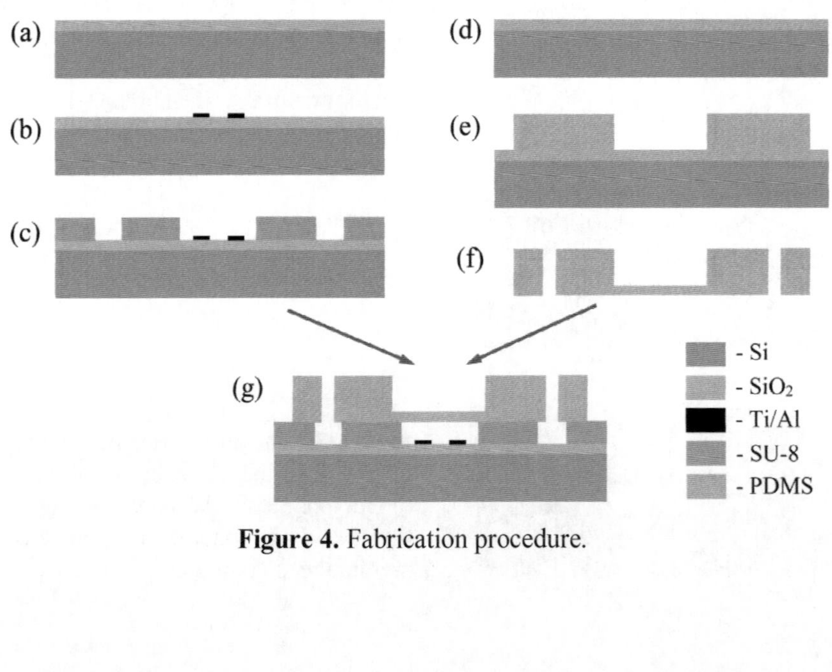

Figure 4. Fabrication procedure.

4. Preliminary testing of the actuator

Performance of the actuator is measured using a homemade Fabry-Perot interferometer. The PDMS membrane has poor reflectivity even after deposition of a semitransparent metallic layer. Laser beam passing through and reflecting from the bottom wall of the chamber creates parasitic interference. To avoid this, a thin rectangular piece of silicon acting as a working mirror is mounted on the membrane in such a way that the membrane deflection tilts the piece as shown in figure 5. The mirror has lateral dimensions of 2×1.5 mm^2, a thickness of 30 µm, and a mass of 0.2 mg. Photograph illustrating the position of the working mirror on the actuator is shown in figure 6. The edge of the mirror is located as close as possible to the center of the membrane to increase the accuracy of the measurement.

The interference pattern is a system of stripes of equal thickness created by interfering laser beams (wavelength $\lambda = 630$ nm) reflected from the reference and working mirrors. The pattern is observed by a microscope. An example of stripes without applying the driving voltage is shown in figure 7a. A series of pulses applied to the electrodes deforms the membrane and changes the angle between the mirrors increasing the number of stripes (figure 7b). During the test the pattern is recorded with a framerate of 25 fps by the video camera ToupCam U3CMOS03100KPA mounted on the eyepiece of the microscope. The deflection of the membrane from the initial state d at each frame can be calculated as follows:

$$d = \frac{L}{l}\frac{\lambda}{2}\left(N_2 - N_1\right),$$

where $L = 2$ mm is the length of the working mirror, $l = 0.8$ mm is the part of the mirror captured by the frame, N_1 and N_2 are numbers of stripes at the initial state and at a given frame, respectively.

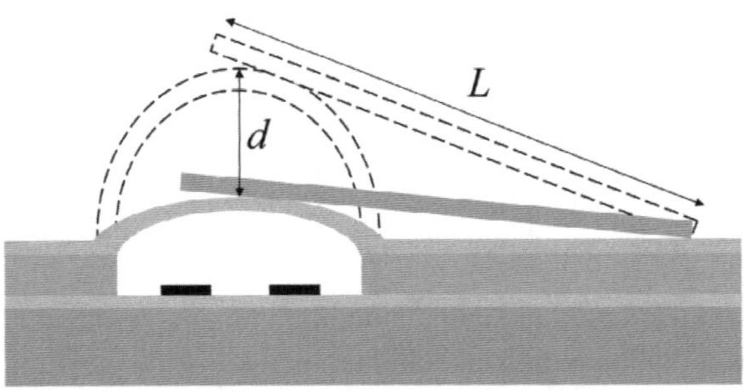

Figure 5. Schematic side view of the working mirror mounted on the actuator. Dashed lines indicate the position of the membrane and the mirror when a series of pulses is applied to the device.

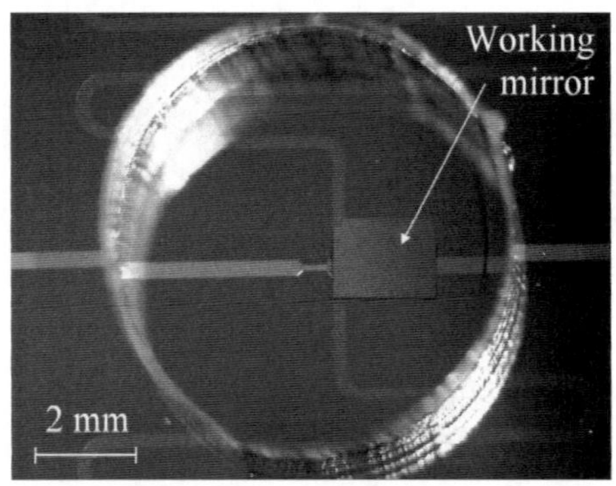

Figure 6. Photograph of the actuator with the working mirror installed (top view).

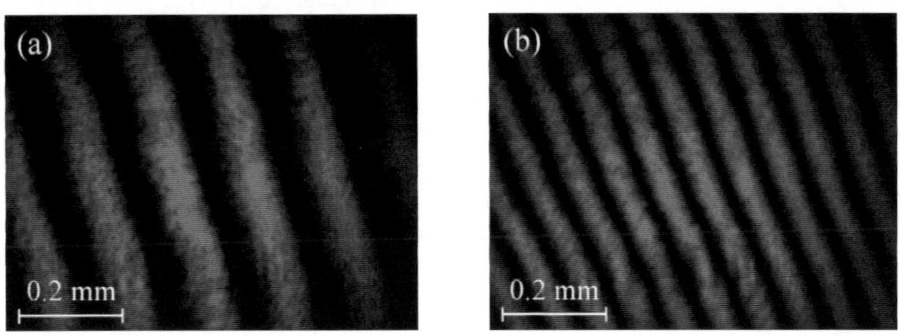

Figure 7. The interference pattern: initial state (a) and the state with the deformed membrane (b).

To drive the actuator a series of $N = 200000$ rectangular voltage pulses of alternating polarity with a duration of $t_0 = 1.25$ μs (frequency of pulses $f = 400$ kHz) is applied to the electrodes. Duration of the series is $t_s = Nt_0 = 250$ ms. Maximal deflection of the membrane as a function of the pulses amplitude U is shown in figure 8. The highest deflection of 7.6 μm is obtained at $U = 13.5$ V. This deflection corresponds to an increase in the chamber volume by approximately 1.5 times in comparison with the nominal value (1.6 nl). Further increase in the amplitude does not lead to larger membrane deflection probably because of the formation of pinned microbubbles in the working chamber. Unfortunately, the working mirror does not allow to observe the chamber. Relatively high amplitude of pulses is needed to drive the actuator that is explained by the formation of TiO_2 on the electrodes [9].

Deflection of the membrane during several actuation cycles at $U = 13.5$ V is shown in figure 9. The time interval between series of pulses is 3-5 s. The membrane reaches the maximal deflection for less than 200 ms and returns to the initial state first very fast but then the process slows down. The total relaxation time is approximately 5 s so that the actuator can operate at frequency of about 0.15 Hz. The cycle duration for existing electrochemical actuators is several minutes [1, 6, 10], which corresponds to a frequency of 0.01 Hz. Thus, a short-time electrolysis produced by the alternating polarity voltage pulses demonstrates the increase of the operating frequency by more than one order of magnitude. It is worth noting that during the first three series of pulses the membrane does not reach the initial position. It returns to the initial state more slowly than expected. From the previous experiments with the fast electrochemical micropump it is known that the gas recombination time must be shorter than 1 s [9]. Slow return of the membrane can be partly connected with the formation of the pinned bubbles in the chamber, which dissolve slowly. The phenomena described above need further investigation.

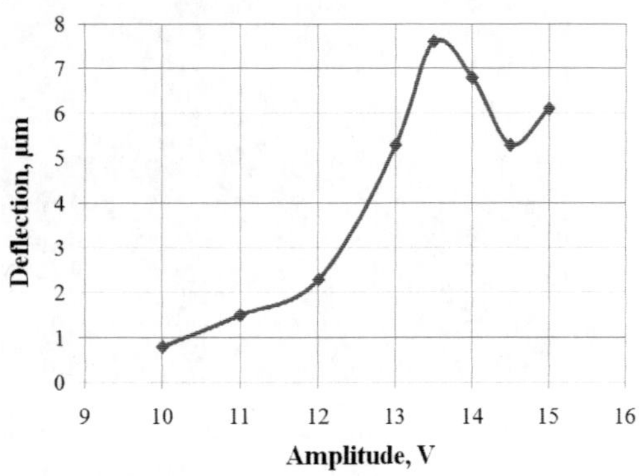

Figure 8. Dependence of the maximal membrane deflection on the amplitude of driving pulses.

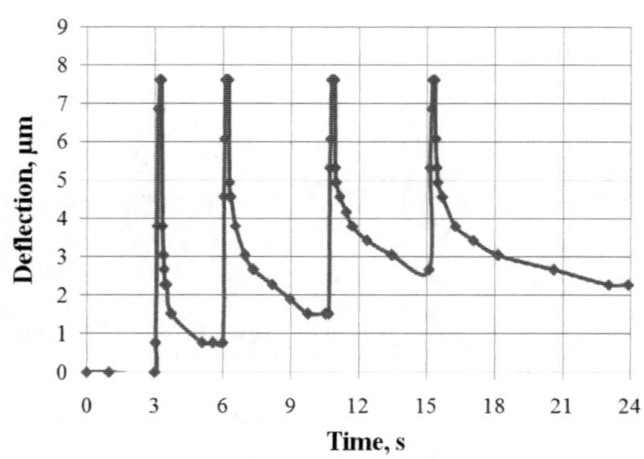

Figure 9. Deflection of the membrane with time at $U = 13.5$ V. Peaks correspond to the application of driving voltage.

Gradual wear of the electrodes is observed during the test. The working chamber after several hundred cycles is shown in figure 10. The fingers are strongly damaged. To all appearance the damage is caused by a significant energy release in the close proximity to the electrode surface as the result of the combustion reaction between hydrogen and oxygen in nanobubbles [11]. Damage is accompanied by the decrease in the Faraday current and the decrease in the membrane deflection. Such a strong wear of Ti/Al electrodes was not observed previously [9]. Probably the electrodes become less stable due to the application of SU-8 and subsequent chemical treatment. We plan to try circular electrodes similar to those used in [12]. It is expected that they will be more stable since they have less edges.

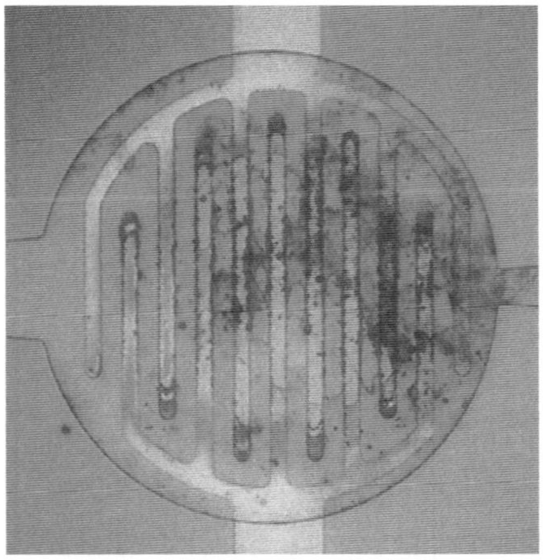

Figure 10. Working chamber after several hundred cycles.

5. Conclusions

An electrochemical membrane actuator was presented. It was designed to demonstrate a new actuation principle based on water electrolysis with a fast change of voltage polarity. The device consisted of the oxidized silicon substrate with metallic electrodes, working chamber and channels for liquid and the PDMS structure supporting the membrane. Preliminary tests of the actuator were performed, deflection of the membrane was detected using a homemade interferometer. Recently proposed alternating polarity mode allowed to achieve sufficiently short actuation cycle due to very short gas termination time in comparison with conventional electrochemical actuators. However, the cycle was longer than expected. A probable reason is the formation of pinned bubbles in the chamber, which dissolve for a relatively long time. The other problems are fast degradation and oxidation of the electrodes. An additional work has to be done to choose an appropriate material and shape for the electrodes.

Acknowledgments

This work is supported by the Russian Science Foundation (Grant No. 15-19-20003).

References

[1] Yi Y, Buttner U, Carreno A A A, Conchouso D and Foulds I G 2015 *J. Micromech. Microeng.* **25** 105011
[2] Lee D E, Soper S and Wang W 2008 *Microsyst. Technol.* **14** 1751–56
[3] Werber A and Zappe H 2008 *J. Microelectromech. Syst.* **17** 1218–27
[4] Au A K, Lai H, Utela B R and Folch A 2011 *Micromachines* **2** 179–220
[5] Svetovoy V, Postnikov A, Uvarov I, Sanders R and Krijnen G 2016 *Energies* **9** 94
[6] Li P-Y, Sheybani R, Gutierrez C A, Kuo J T W and Meng E 2010 *J. Microelectromech. Syst.* **19** 215–28
[7] Svetovoy V B, Sanders R G P, Ma K and Elwenspoek M C 2014 *Sci. Rep.* **4** 4296
[8] Zhang Z, Zhao P, Xiao G, Watts B R and Xu C 2011 *Biomicrofluidics* **5** 046503
[9] Uvarov I V, Lemekhov S S, Melenev A E and Svetovoy V B 2016 *J. Phys. Conf. Ser.* **757** 012008
[10] Gensler H, Sheybani R, Li P-Y, Mann R L and Meng E 2012 *Biomed Microdevices* **14** 483–96
[11] Svetovoy V B, Sanders R G P, Lammerink T S J and Elwenspoek M C 2011 *Phys. Rev.* E **84** 035302(R)
[12] Postnikov A V, Uvarov I V, Lokhanin M V and Svetovoy V B 2016 *Sci. Rep.* **6** 39381

Design and technology optimization of SiC-based RF MEMS switch

A V Lagosh, A V Korlyakov

Department of Micro- and Nanoelectronics, Saint Petersburg Electrotechnical University "LETI", St. Petersburg 197376, Russia

Abstract. This paper presents design and technology optimization of RF MEMS switch based on the silicon carbide film. The universal optimization criterion for electromechanical parameters was calculated. The experimental results of determination of the technological critical factors are discussed. Recommendations for the subsequent technology improvement are presented.

1. Introduction

Silicon, its compounds and metals are conventional materials to use as movable elements in RF MEMS switches [1]. At the same time, silicon carbide (SiC) is one of most challenging materials for various RF MEMS devices. The use of SiC film as a material for the movable MEMS element provides the following advantages:

- The higher vibration stability due to the great value of Young's modulus.
- The lesser material density in comparison with metals ensures higher resonance frequency and speed of operation.
- The extended temperature operational range due to the low values of temperature coefficients of thermal expansion and Young's modulus.

On the other hand, the technology of SiC film deposition by magnetron plant is not optimized to produce movable MEMS elements. One more problem is to determine the best design of a MEMS switch within the specified technology.

2. Design optimization

Main electromechanical parameters of RF MEMS switch are the driving voltage (U_c) and the resonance frequency (f_0), that determines the speed of device operation. A large quantity of researches are held at the present time. Some of them are directed to get lower driving voltage and others to heighten a device speed [2]. So, one could find a big amount of various switch designs.

Thus, the optimization criterion was derived to compare different switch designs. This criterion allows one to select the design with the most optimal ratio of the electromechanical parameters. The optimization criterion has the following form:

$$\frac{U_c^2}{f_0^2} = \frac{4\pi^2}{3} C \frac{y_0^3 \rho h}{\varepsilon_0}$$

where y_0, ρ, h are the gap between driving electrodes, material density, film thickness. These quantities are defined by selected technology. The C parameter is defined by the type of switch design. It was found during the calculations that the C parameter is a constant with the value of 8/9 for designs where bending deformation dominates (C_k). However, if one uses movable elements with torsion suspensions (figure 1), the C parameter begins to depend on the length of the movable mass (C_α) (figure 2):

$$C_\alpha = \left[\frac{1}{1-\left(\frac{l_2}{l_3}\right)} - \frac{1}{1-\left(\frac{l_1}{l_3}\right)} + \ln\left(\frac{1-\left(\frac{l_2}{l_3}\right)}{1-\left(\frac{l_1}{l_3}\right)} \right) \right]^{-1}.$$

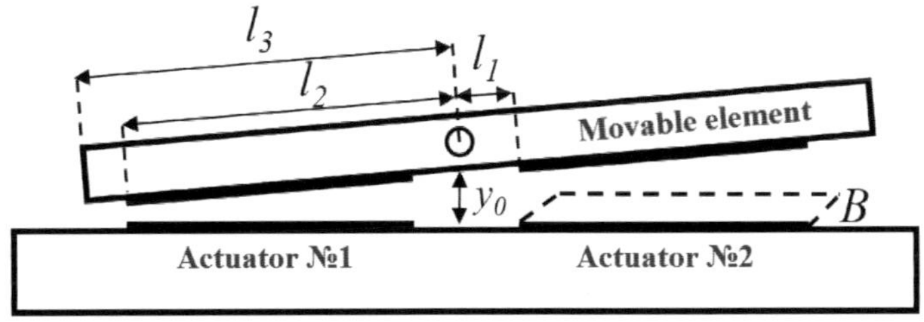

Figure 1. Sketch of actuator with movable element based on torsions.

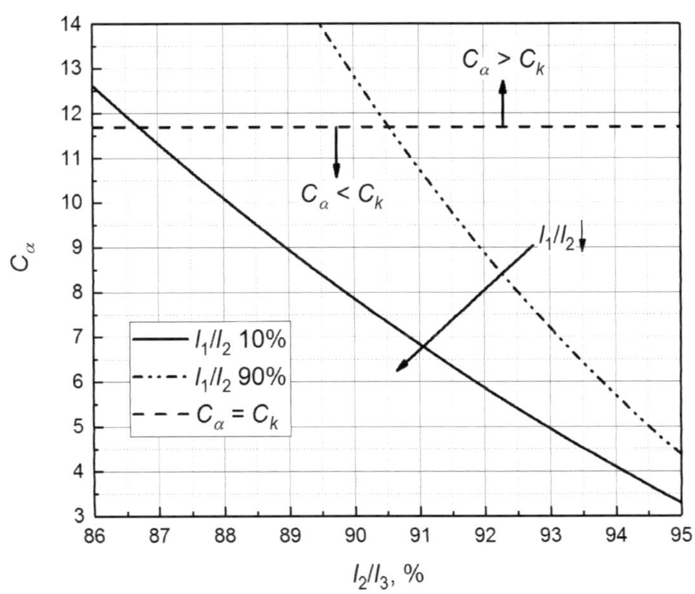

Figure 2. Parameter C_α vs geometrical dimensions.

Thus, the use of elements with torsions allow one to obtain the gain (region below dashed line on figure 2) according to the optimization criterion within the same technological process.

The use of a SiC film as a material for a movable element constrains the structure design. It is known that SiC film deposited by magnetron plant has compressive mechanical stresses [3]. So, the structure with interior mount of torsion suspensions was designed to compensate negative stresses (figure 3).

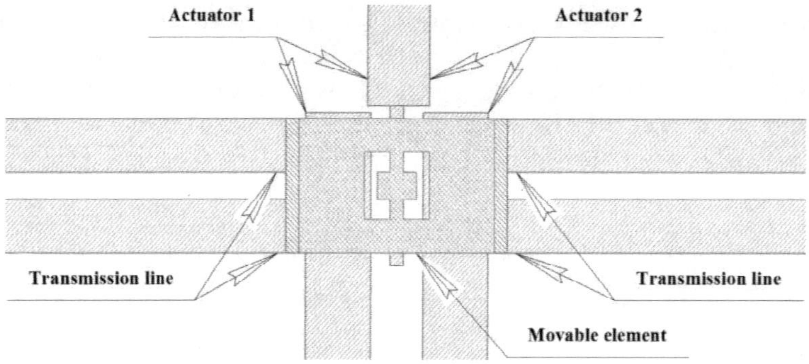

Figure 3. Sketch of actuator with interior mount of torsion suspensions.

The parameter C for this design will be as follows:

$$C=\left[\frac{1}{1-\left(\frac{l_2}{l_3}\right)}-\frac{1}{1-\left(\frac{l_1}{l_3}\right)}+\ln\left(\frac{1-\left(\frac{l_2}{l_3}\right)}{1-\left(\frac{l_1}{l_3}\right)}\right)\right]^{-1}\left[1-\left(\frac{l_1}{l_3}\right)^3\right]$$

Thus, moving the anchor inside the movable element allows one to obtain an additional gain of several percent order due to the term $1-\left(\frac{l_1}{l_3}\right)^3$.

3. Technology optimization

To implement designed structures, the basic technology was developed for manufacturing the RF MEMS switch based on a silicon carbide film.

The main features of the basic technology:

- isolating substrate with two-sided polishing;
- 4 photolithography operations;
- one two-level sacrificial layer;
- movable element based on SiC layer which is deposited by magnetron sputtering;
- dry etching of sacrificial layer;
- Au-Au contact area;
- vacuum sealing.

The main disadvantage that limits usage of SiC films in MEMS devices is the appearance of residual mechanical stresses. These stresses are non-uniformly distributed along the thickness of SiC film. As a result, undesirable bend of movable element occurs (figure 4). This leads to the increase of driving voltage, deterioration of operation speed and device reliability.

Figure 4. SEM image of RF MEMS switch with undesirable bend of movable element based on torsions.

To eliminate these defects, a number of technological experiments were held to determine the critical factor which has an influence on the magnitude of mechanical stresses in a silicon carbide film. The following critical factors of the technological process were picked out:

- substrate heating after shutter opening;
- difference between residual atmosphere pressures before and after the deposition process;
- surface charging;
- impurity diffusion into growing layer.

The experiments have shown that if one reduces the difference between residual atmosphere pressures before and after the deposition process by a factor of N times, the maximum gap also decreases N times and the radius of surface curvature increases $N^{0.5}$ times. The experiments with other factors showed their minimum influence on the nonuniformity of mechanical stress distribution along the thickness of SiC film.

4. Conclusion

The universal optimization criterion for electromechanical parameters of RF MEMS switch was worked out. The use of this criterion allows one to choose optimal switch design within the same technological process.

It has been found experimentally that the difference between residual atmosphere pressures before and after the SiC film deposition process is the main critical factor. The reduction of this difference allows one to produce devices with a minimal deviation from the design parameters.

References

[1] Rebeiz G M 2003 *RF MEMS: Theory, Design, and Technology* (Hoboken: Wiley)
[2] Iannacci J 2013 *Practical Guide to RF-MEMS* (Weinheim: Wiley)
[3] Astashenkova O N, Korlyakov A V, Luchinin V V 2013 *Materials Science Forum* **740-742** 998

IOP Conf. Series: Journal of Physics: Conf. Series **917** (2017) 082008 doi:10.1088/1742-6596/917/8/082008

Carbon nanotubes based vacuum gauge

N N Rudyk, O I Il'in, M V Il'ina, A A Fedotov, V S Klimin and O A Ageev

Institute of Nanotechnologies, Electronics, and Equipment Engineering, Southern Federal University, 2 Schevchenko Street, Taganrog, Russia

Abstract. We have created an ionization type Vacuum gauge with sensor element based on an array of vertically aligned carbon nanotubes. Obtained asymmetrical current-voltage characteristics at different voltage polarity on the electrode with the CNTs. It was found that when applying a negative potential on an electrode with the CNTs, the current in the gap is higher than at a positive potential. In the pressure range of 1 ÷ 10^3 Torr vacuum gauge sensitivity was 6 mV/Torr (at a current of $4.5 \cdot 10^{-5}$ A) and in the range of 10^{-5} ÷ 1 Torr was 10 mV/Torr (at a current of $1.3 \cdot 10^{-5}$ A). It is shown that the energy efficiency of vacuum gauge can be increased in the case where electrode with CNT operates as an emitter of electrons.

1. Introduction

The formation of carbon nanotubes (CNTs) with given geometric parameters and the study of their properties are considered to be a really important task, both in terms of basic research and their future application in electronic devices [1, 2]. CNTs possess a number of unique features, such as stable emission characteristics, electrical conductivity, high aspect ratio and chemical stability, which allows us to use them to create sensitive elements for micro- and nanoelectronic devices, gas sensors and vacuum gauges [3-8].

To determine the degree of vacuum we can use capacitive, thermocouple, thermistor and ionization gauges [9-12]. Thermistor and thermocouple ones are often used in microelectronic process equipment where they operate in the pressure ranges of 10^{-3} – 760 and 10^{-12} - 10^{-1} Torr, respectively [13-14]. Ionization vacuum gauges, whose operating principle is based on the ionization of gas molecules in the discharge gap and the following measurement of the ion current flowing between the electrodes, are also considered promising [15]. The use of an array of vertically aligned CNTs to ionize gas molecules can significantly reduce the operating voltage and power consumption of the ionization vacuum gauges [16].

In order to create the arrays of vertically aligned CNTs on contact pads and thus make such structures suitable for ionization gauge sensor elements, plasma-enhanced chemical vapor deposition (PECVD) can be used [17].

Our aim is to create and study the parameters of a vacuum gauge prototype with a sensor element based on an array of vertically aligned CNTs.

2. Experiments and methods

To create experimental samples, we used Si (100) wafers with resistivity 0.001 - 0.005 Ω·cm (Universitywafer, USA). RCA chemical cleaning was carried out by means of Amerimade (Amerimade Technology Inc., USA). The deposition of the contact layer (Ti, 250 nm), the barrier layer (Cr, 20 nm) and the catalytic layer (Ni, 10 nm) was carried out using the AUTO 500 magnetron sputtering set (BOC Edwards, UK). The topology of the lower electrode with CNTs was created by means of photolithography on the Ti film using the SUSS MJB4 set (Suss MicroTek, Germany). Cr and Ni films were deposited in a single cycle through a mask. The resulting Ni/Cr/Ti/Si structure was later placed into the chamber of the PECVD module of the nanotechnology complex NANOFAB NTK-9 where we grew the array of vertically aligned CNTs with a height of about 3 μm. The description of process condition can be found in [18].

To manufacture the upper electrode, we used a Si (100) wafer with resistivity 0.001 - 0.005 Ω·cm where, similarly to the first step, we formed a Ti film with a thickness of 250 nm. Then we deposited a Si3N4 layer with a thickness of ~6 μm using the plasma chemical deposition set STE ICPd81 (SemiTEq, Russia). After that, we formed dielectric supports by means of photolithography. At the final stage the upper electrode was connected to the lower one. The design of the prototype provides a ~3 μm gap between the upper electrode and the CNTs, which is shown in Figure 1.

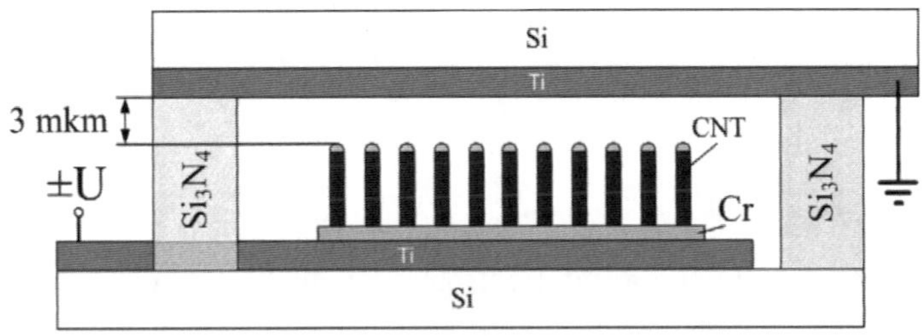

Figure 1. The design of the prototype.

We studied the work of the gauge using a measuring stand constructed on the basis of a chamber with the ultimate vacuum of $8 \cdot 10^{-6}$ Torr. The measurement was carried out using the VAC picoammeter Keithley 6487 (Keithley Instruments Inc., USA) with the upper electrode grounded and a voltage of ±20 V applied to the lower electrode with CNTs at a temperature of 300 K. The pressure in the range of 760 – 10^{-5} Torr was controlled using a calibrated gauge MKS Series 999 Quattro. To carry out the comparative analysis and qualitative assessment of the influence of the electrode with CNTs on the gas gap brakedown, we created a prototype of the sensor with a similar structure (Figure 1) but with no CNTs on the lower electrode. To ensure that the measurement conditions were identical, the gap between flat electrodes was decreased to 3 μm.

3. Results and discussion

Having analyzed the SEM images of the lower electrode with CNTs (Figure 2), we found that the CNTs we had grown had a diameter of 80±9 nm and a height of 2.9±0.13 μm, which, taking into account the thickness of the contacts, confirms the fact that the value of the gap between the CNTs and the upper electrode was ~3 μm.

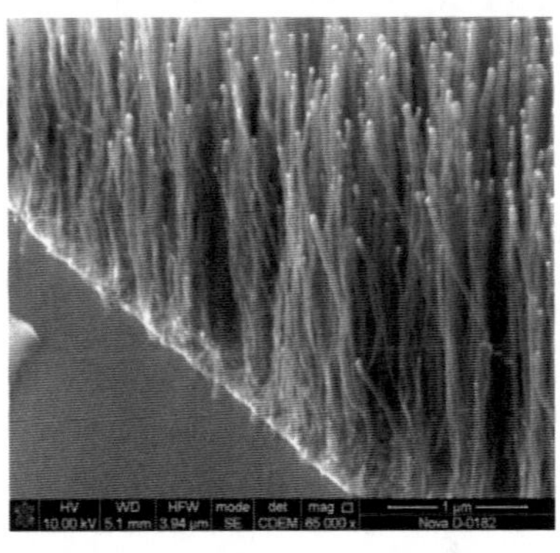

Figure 2. The CNT array grown on the lower electrode.

The analysis of the current-voltage characteristics of the prototype (Figure 3) shows the effect of the pressure on the flowing current. Notably, the forward and reverse branches of the CVC graph are asymmetrical.

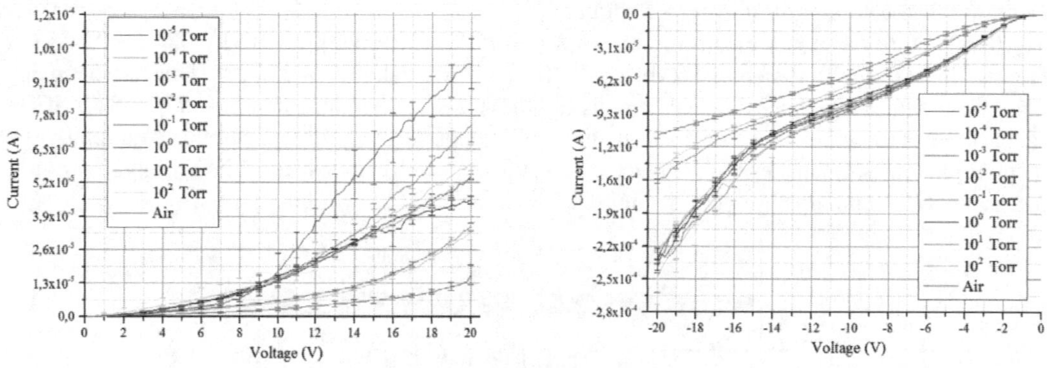

a) b)

Figure 3. The CVC of the prototype at different pressures with a) positive and b) negative voltage polarity.

Having analyzed the current-voltage characteristics of the prototype (Figure 3), we got the opportunity to plot the current against the pressure at a range of voltages (±20 V) applied to the lower electrode with the CNTs (Figure 4).

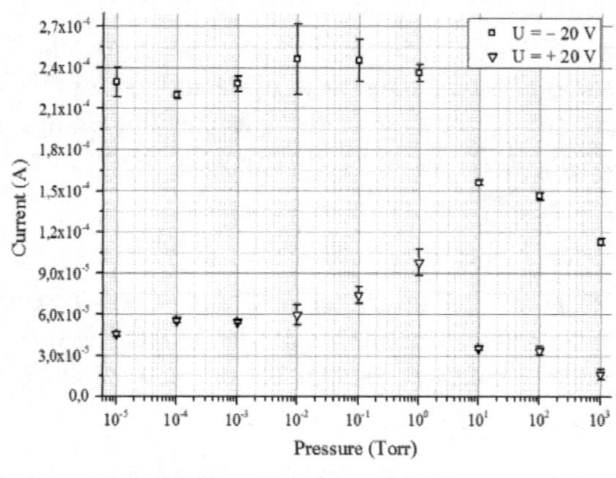

Figure 4. The dependence of the current on the pressure at different voltages applied to the electrode with CNTs.

Figure 4 shows that, with equal absolute values of the voltage, the current has greater values when a negative voltage is applied to the electrode with the CNTs compared to the case when we applied a positive voltage. This effect is associated with an additional electron emission taking place as a result of the concentration of the electric field at the tops of CNTs due to their small diameter and high aspect ratio [19]. Notably, the prototype of the sensor with planar electrodes, i.e. with no CNTs on the lower electrode, did not show the presence of any current even when the range of the voltage applied was 0 ÷ ±50 V.

Figure 5 shows the dependence of the voltage on the pressure at constant currents and a positive voltage applied to the electrode with CNTs. Its shape can be accounted for by the fact that at a fixed value of the gap between the electrodes and the pressure values increased from 10^{-1} to 10^{3} Torr, the number of collisions between electrons and neutral atoms and molecules increases accordingly, which results in the reduction of energy of electrons accumulated along the mean free path. We should also note that it is necessary to increase the voltage applied in order to maintain the current.

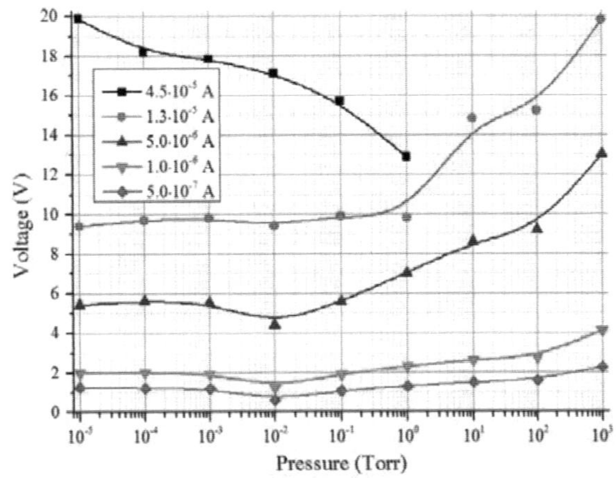

Figure 5. The dependence of the voltage on the chamber pressure at various currents.

According to [20], at pressures of less than 10^{-1} Torr the mean free path increases and so does the accumulated energy of the electrons whereas the number of collisions drops, which reduces the probability of impact ionization. Therefore, to maintain a constant current it is necessary to increase the electron energy on the mean free path by increasing the voltage. However, our experiments showed that within the pressure range of 10^{-5} – 10^{-1} Torr and at currents of $5\cdot10^{-7}\div1.3\cdot10^{-5}$ A the voltage remains practically unchanged, which does not show consistency with the results obtained in [20]. The result we achieved may be explained by the fact that the distance between the electrodes (~3 μm) in our prototype is much smaller than the mean free path of electrons, which is about 500 μm at low vacuum [21]. Thus, there are only emitted electrons which contribute to the total current, and they cover the interelectrode distance almost unhindered in vacuum. In order to maintain a predetermined current when the CNTs act as electron emitters (there is a negative voltage applied to the CNT electrode), it is necessary to apply a voltage 4 times lower than in the case where electrons are emitted from a planar electrode (Figure 6).

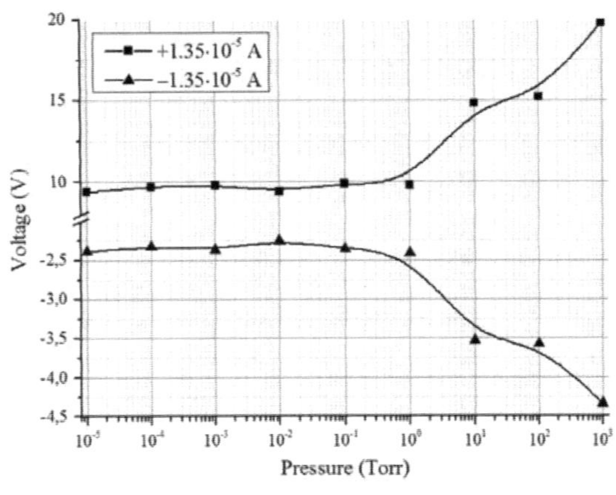

Figure 6. The dependence of the voltage on the pressure.

The analysis of the results presented in Figure 5 shows that for the prototype discussed we can talk about two sensitivity ranges: 10^{-5} – 1 Torr and 1 – 10^{3} Torr, for which the sensitivities are 6 mV/Torr (at an operating current of $4.5\cdot10^{-5}$ A) and 10 mV/Torr (at an operating current of $1.3\cdot10^{-5}$ A), respectively.

4. Conclusion

We created and studied a prototype of the vacuum gauge with a sensor element based on an array of vertically aligned CNTs. Within the pressure ranges of 1 – 10^{3} Torr and 10^{-5} – 1 Torr the prototype showed a sensitivity of 6 mV/Torr at an operating current of $4.5\cdot10^{-5}$ A, and 10 mV/Torr at an operating current of $1.3\text{x}10^{-5}$ A, respectively. We discovered that the energy efficiency of the sensor can be increased when the CNT electrode acts as electron emitter. In such a mode it is necessary to apply a lower voltage (4 times lower for our prototype) in order to maintain a predetermined current value, compared to the case where we used a planar electrode without CNTs to act as electron emitter. The power consumption of our prototype was less than 7 mW, which provides an opportunity to use it in mobile devices.

Acknowledgments

The study was carried out using the equipment of the Centre of Collective Usage and the Research and Education Center "Nanotechnologies" of Southern Federal University. The financial support was provided by the Russian Foundation for Basic Research [grant number 16-29-14023 ofi_m] and by SFedU grant № VnGr-07/2017-26.

References

[1] V.N. Popov, Mater. Sci. Eng. R Reports **43**, 61–102 (2004).
[2] O. A. Ageev et al, "Development of New Metamaterials for Advanced Element Base of Micro- and Nanoelectronics, and Microsystem Devices" in *Advanced Materials. Manufacturing, Physics, Mechanics and Applications*, edited by Parinov, Chang, Topolov (2016), pp. 563-580.
[3] M. Kosaka, Electronic Properties of Carbon Nanotubes Probed by Magnetic Measurements, (1999) 76-88.
[4] O.A. Ageev et al., Phys. Solid State. **57**, 825–831 (2015).
[5] J.T.W. Yeow and Y. Wang, J. Sensors (2009).
[6] Y. Yang, L. Liu, K. Jiang, S. Fan, Vacuum **86**, 885–888 (2012).
[7] X. Hao, K. Maenaka, H. Takao, K. Higuchi, Sensors Actuators, A Phys. **205**, 92–102 (2014).
[8] S. Wilfert, C. Edelmann, Vacuum **86**, 556–571 (2012).
[9] E. Gavignet, F. Lanzetta, Vacuum **100**, 18–21 (2014).
[10] A. Górecka-Drzazga, Vacuum **83**, 1419–1426 (2009).
[11] Z. Hou, D. Xu, B. Cai, Appl. Phys. Lett. **89**, 21–24 (2006).
[12] A. Modi, N. Koratkar, E. Lass, B. Wei, P.M. Ajayan, Nature **424**, 171–174 (2003).
[13] W. Jitschin, S. Ludwig, Vacuum **75**, 169–176 (2004).
[14] X. Sun, D. Xu, B. Xiong, G. Wu, Y. Wang, Sensors Actuators, A Phys. **201**, 428–433 (2013).
[15] S. Lakkis, R. Younes, Y. Alayli, M. Sawan, Sens. Rev. **34**, 24–35 (2014).
[16] R.B. Sadeghian, M. Kahrizi, Sensors Actuators, A Phys. **137**, 248–255 (2007).
[17] I. Ibrahim, A. Bachmatiuk, J.H. Warner, B. Büchner, G. Cuniberti, M.H. Rümmeli, Small. **8**, 1973–1992 (2012).
[18] A.A. Fedotov, "Study of Growth Processes of PECVD Carbon Nanotubes, Used in Micro- and Nanosystem Technologies" in *Piezoelectric and Nanomaterials*. (2015) pp. 62–85.
[19] N. Doostani, S. Darbari, S. Mohajerzadeh, M.K. Moravvej-Farshi, Sensors Actuators, A Phys. **201**, 310–315 (2013).
[20] M. Alimpijević, Z. Rajović, D. Brajović, M. Vujisić, K. Stanković, Vacuum **99**, 89–94 (2014).
[21] S. Jennings, J. Aerosol Sci. **19**, 159–166 (1988).

IOP Conf. Series: Journal of Physics: Conf. Series **917** (2017) 082009 doi:10.1088/1742-6596/917/8/082009

Electrical properties of polycrystalline materials from the system Cu-As-Ge-Se under high pressure condition

V E Zaikova, N V Melnikova, A V Tebenkov, A A Mirzorakhimov, O P Shchetnikov, A N Babushkin and G V Sukhanova

Institute of Natural Science and Mathematics, Ural Federal University, Ekaterinburg 620000, Russia

Abstract. The paper deals with electrical properties of polycrystalline materials $(GeSe)_x(CuAsSe_2)_{1-x}$ (x= 0.5 and 0.7) under high pressure (up to 45 GPa) conditions. The phenomenon of negative magnetoresistance was observed for studied materials.

1. Introduction

Multicomponent copper chalcogenides are actively studied materials due to their interesting from the fundamental and practical point of view electrical, optical and thermoelectrical properties [1-5]. The materials in the system Cu-As-Ge-Se demonstrate the electrical switching effects, which can be associated with the transition from an amorphous state to a crystalline state. In particular at small value of x, $(GeSe)_x(CuAsSe_2)_{1-x}$ can be obtained in glassy and crystalline state, wherein glassy and crystalline materials with same composition have significant difference in electroresistance. Materials from the system $(GeSe)_x(CuAsSe_2)_{1-x}$ are thermoelectric materials some of which have high value of thermoemf at high pressures up to 50 GPa [3].

This paper is devoted to study of high pressure influence (up to 45 GPa) on electrical properties of polycrystalline materials $(GeSe)_x(CuAsSe_2)_{1-x}$, x= 0.5 and 0.7, in constant electric field and in a transverse magnetic field (up to 1 T). The negative magnetoresistance phenomenon is investigated.

2. Materials and experiment details

Material $(GeSe)_x(CuAsSe_2)_{1-x}$, x= 0.7, further Sample 1, crystallizes in cubic syngony, material $(GeSe)_x(CuAsSe_2)_{1-x}$, x= 0.5, further Sample 2, crystallizes in tetragonal syngony [2]. More detail information about structure is presented in the table 1. The Sample 2 structure were detected by homology theory taking in account transformations of cubic lattice with Ge content changing in the system $(GeSe)_x(CuAsSe_2)_{1-x}$.

Table 1. Crystal structure and the lattice parameters for Sample 1 and Sample 2.

	$(GeSe)_x(CuAsSe_2)_{1-x}$	Syngony	Lattice parameters
Sample 1	x = 0.7	Cubic	a = 0.554 nm
Sample 2	x = 0.5	Tetragonal	a = 0.394 nm, c = 0.548 nm

High-pressure conditions were achieved with high-pressure cell with rounded cone-plane-type anvils made of carbonado-type artificial diamonds, which are good conductors with resistance less than

10 Ohm [6]. Detailed description of the pressure estimation method is presented in works [6, 7]. The powdered materials were used for high pressure measurements, the samples with a diameter of ~ 0.2 mm and thickness from 10 to 30 μm were obtained by the compression of powdered materials in the high-pressure cell. For magnetoresistance measurements the high-pressure cell was placed in the transverse magnetic field. The pressure and magnetic field were measured and controlled directly during the experiment. Magnetoresistance was calculated from the data obtained in experiment by formula (1)

$$MR = \frac{R_P(B) - R_P(0)}{R_P(0)} \quad (1),$$

where *MR* – magnetoresistance, $R_P(B)$ – electroresistance in fixed magnetic field and at fixed pressure, $R_P(0)$ – electroresistance at the corresponding pressure without magnetic field.

The Seebeck coefficient *S* was calculated by measuring the thermoelectromotive force (thermoemf) ΔU. For creation of temperature gradient ΔT one of the side of the sample was warmed. The accuracy of temperature measurements was about 0.05 K. Thermoemf measurements into interval 300-400 K at residual pressure of 0.1 Pa were carried out using ProboStat cell and the Solartron ModuLab Materials Test System, and the Seebeck coefficient was calculated by formula (2)

$$S = \frac{\Delta U}{\Delta T} \quad (2).$$

The Seebeck coefficient at high pressure conditions was calculated by measuring thermoelectromotive force induced in the sample via the copper branches of the thermocouples (used simultaneously for measurement the differences in the temperature between cold and hot faces of a sample) by formula (3)

$$S = S_{cu} + \frac{\Delta U}{\Delta T} \quad (3),$$

where S_{cu} – Seebeck coefficient of copper (1.8 μVK^{-1}).

3. Results and Discussion

Analysis of experimental results revealed significant (on 2 orders) electroresistance decreasing with pressure (*P*) increasing from 16 GPa to 42 GPa for both Sample 1 and Sample 2. The negative magnetoresistance (*MR*) phenomenon was observed for both materials and there are some extrema on curves *MR*(*P*). Absolute value of negative *MR* increase with magnetic field increasing. Typical field dependences of *MR* at two fixed pressures (these pressures belong to the baric interval where *MR*<*0*) for the Sample 1 are presented on figure 1.

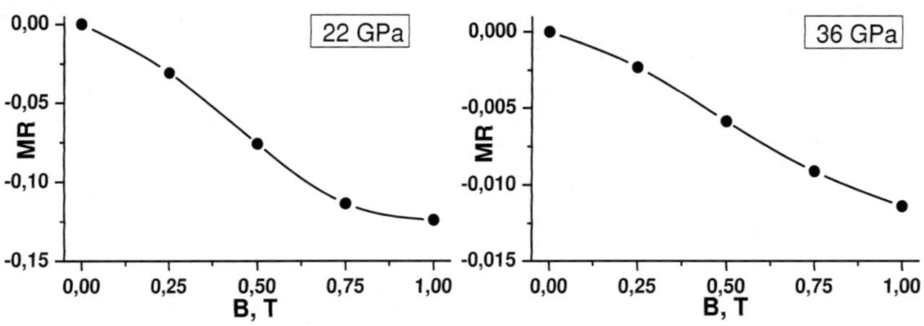

Figure 1. Field dependences of magnetoresistance for Sample 1 at pressures 22GPa and 36 GPa.

The electroresistance relaxation from applying the static pressure was studied for additional characterization of high pressure influence. It was found that electroresistance of studied materials decreases before achieving the constant (corresponding to applied pressure) level. The elecroresistance relaxation curves can be approximated by exponential low in two ways. The first is approximation by standard decreasing exponential function (formula 4) and the second one, which is more accurate, is approximation by sum of two decreasing exponents (formula 5). When double exponential function is used for approximation, there are two relaxation times: smaller (t_1) and bigger (t_2), and using one exponent function for approximation, we get average relaxation time (t_a).

$$R = R_0 + a \cdot e^{-\frac{t_a}{t}} \tag{4}$$

$$R = R_0 + a \cdot e^{-\frac{t_1}{t}} + b \cdot e^{-\frac{t_2}{t}} \tag{5}$$

We can suggest that smaller relaxation time (t_1) corresponds to electroresistance relaxation process caused by carriers' concentration and band gap changes, and bigger relaxation time (t_2) corresponds to the relaxation process in lattice. The average relaxation time t_a at different pressures can be closer to either t_1 or t_2. It means that at some pressures the relaxation of the electronic subsystem makes the main contribution to the whole relaxation process, at the other pressures – the relaxation of the lattice is prevailing. Curves $t_a(P)$ also have some extrema, herewith the baric intervals of $t_a(P)$ curves extrema correspond to the *MR*(*P*) curves extrema (figure 2 and figure 3). Times of relaxation process corresponded to the changing of structure characteristics of crystal lattice (t_2) take theirs maximum values ($5 \cdot 10^3$ s for Sample 1 and $2 \cdot 10^3$ s for Sample 2) at pressure values corresponded to the *MR*(*P*) extrema. Thus, we can assume that significant changes in crystal lattice cause changes in electronic structure, increase of carriers' concentration which in turn entails electroresistance decrease and manifests itself in appearance of *MR*(*P*) extrema.

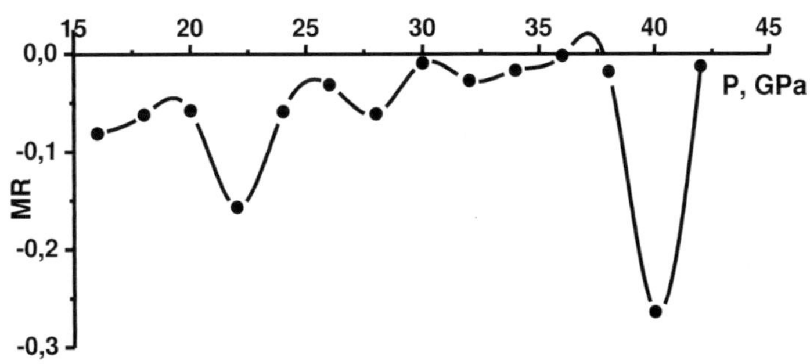

Figure 2. Baric dependence of magnetoresistance for Sample 2 at magnetic field 1 T.

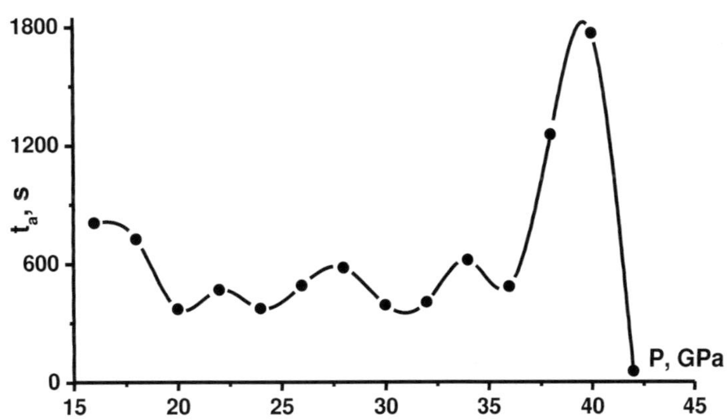

Figure 3. Baric dependence of average relaxation time (t_a) for Sample 2.

For studied materials, the extrema on curves *MR*(*P*) correspond not only to the extrema on curve of relaxation time baric dependence, but also to the baric intervals of the other electrical properties significant changes. The *MR* baric dependence for Sample 1 has main extremum at pressure range 20-22 GPa (figure 4) and there are substantial changes in the electroresistance and in dielectric loss tangent behavior near 20-22 GPa (figure 5). The decreasing of electroresistance of Sample 1 almost ceases and the significant increase of dielectric loss tangent begin near 22 GPa.

Similar behavior of *MR* is known for materials $CuInAsSe_3$ and $(GeS)_x(CuAsS_2)_{1-x}$, x =0.7 [8 – 10]. These materials can be obtained by replacement of germanium atoms with indium atoms in studied Sample 2 and by isovalent replacement of selenium atoms to sulfur atoms in Sample 1. As we can see in table 2, for both pair of materials the atoms replacing leads shift of *MR(P)* extrema, wherein replacement of atoms by others with bigger radii shifts extrema on *MR(P)* curve to the lower pressure ranges.

Table 2. Comparison of the pressure intervals in which the features *MR* behavior in the form of extrema were observed for studied materials $CuGeAsSe_3$ and $(GeSe)_x(CuAsSe_2)_{1-x}$, x =0.7, and $CuInAsSe_3$ and $(GeS)_x(CuAsS_2)_{1-x}$, x =0.7.

	$CuAAsSe_3$		$(GeA)_x(CuAsA_2)_{1-x}$, x =0.7	
	A = In	A = Ge	A = S	A = Se
MR(P) extrema	20 GPa, 24 GPa, 36 GPa [8]	22 GPa, 26 GPa, 40 GPa	22-24 GPa [10]	20-22 GPa
A atomic radii	80 pm	73 pm	182 pm	193 pm

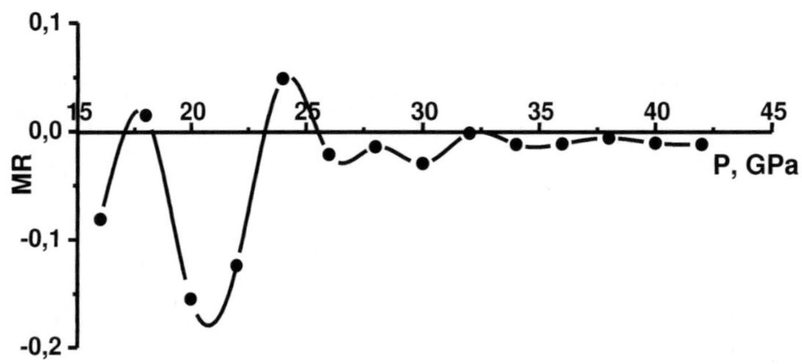

Figure 4. Baric dependence of the magnetoresistance for Sample 1 at magnetic field 1 T.

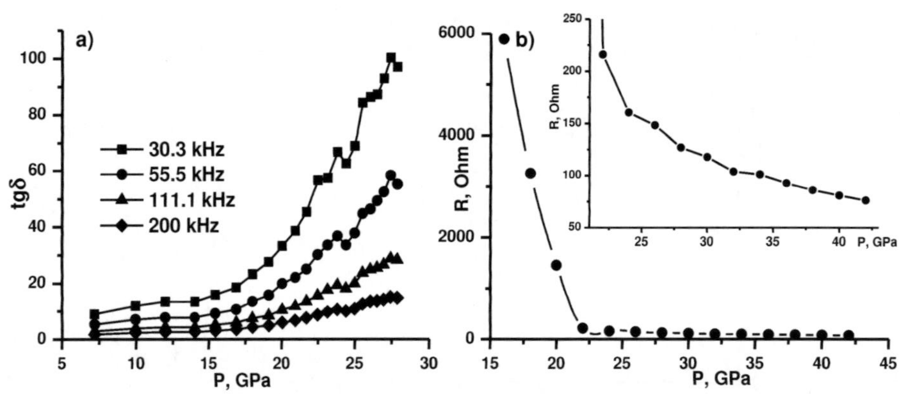

Figure 5. Baric dependences of a) dielectric loss tangent and b) the electroresistance for Sample 1.

Appearance of negative magnetoresitance for the material $CuInAsSe_3$ was explained by loose crystal structure, small number and mobility of carriers [8, 11]. The *MR(P)* extrema for material $(GeS)_x(CuAsS_2)_{1-x}$, x =0.7, can be related with significant lattice parameters changing (the results of X-ray studies under high pressure condition are presented in [10]). Because of the similarity of the crystal structures and the baric behavior of the studied properties (including pressure *MR* behavior) of the research materials to the corresponding parameters of $CuInAsSe_3$ and $(GeS)_x(CuAsS_2)_{1-x}$, x =0.7, we can associate the negative magnitoresistance with the structure change and features of carriers behaviors under high pressure. With pressure increasing the band gap can decrease, the carriers' concentration and mobility can change, as well as the effects determined by significant increase of numbers of defects and inhomogeneitis, which are induced by deformation, can play significant role in *MR* behavior.

Exploration of carriers type and theirs behavior under pressure can help verify this assumption. By this reason, the study of thermoemf at atmospheric and high pressures and study of conductivity temperature dependence at some pressures were carried out for Sample 1.

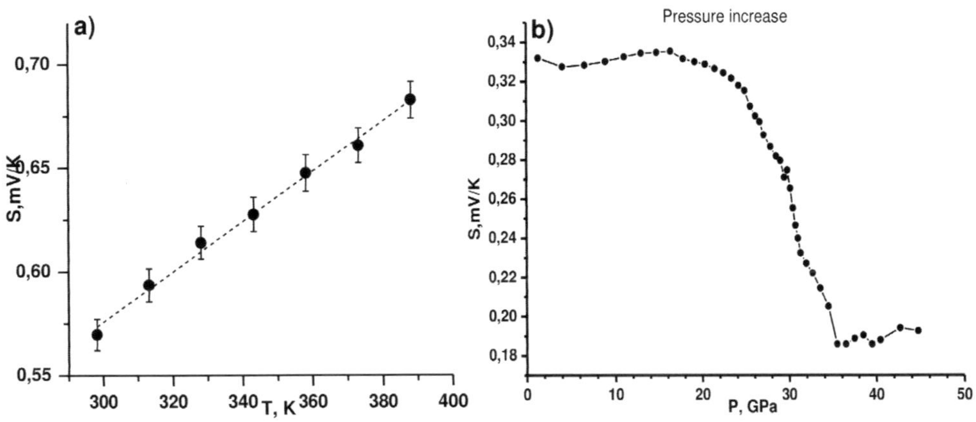

Figure 6. Dependences of Seebeck coefficient vs temperature at atmospheric pressure (a) and vs pressure at 298 K (b) for Sample 1.

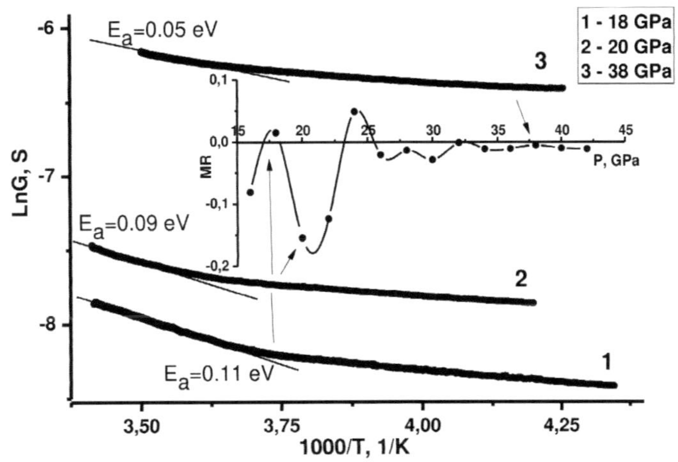

Figure 7. Temperature dependences of Sample 1 conductivity at pressures 18 GPa, 20 GPa and 38 GPa. On the insert the *MR*(*P*) graph for Sample 1 is present.

The results of thermoemf study for Sample 1 at atmospheric pressure are presented at figure 6a. It is seen that studied material has high Seebeck coefficient value 0.57-0.68 mV/K into the temperature interval 298-388 K. Positive sign of thermoemf indicates that studied materials have p-type conductivity. Study of the temperature dependences of Sample 1 conductivity was carried out at pressures 18 GPa, 20 GPa and 38 GPa. These pressures correspond to the small positive value of *MR*, the main *MR*(*P*) extremum, and the closeness of *MR* to zero (at P>32 GPa) respectively (figure 7). At all investigated pressures, material is characterized by activation type of conductivity, the activation energy values at room temperature are presented in figure 7. It is seen that the activation energy decreases with increasing pressure, we can assume that band gap becomes narrower and at pressures bigger 32 GPa *MR* weakly depends on pressure.

The results of pressure dependence of thermoemf for Sample 1 are presented on figure 6b. Material has high value of thermoemf at high pressure. The thermoemf decreases with pressure increasing but saves its sign. We can see correlation between significant changes in *MR* (figure 4) and thermoemf behaviors. Baric interval 20-22 GPa corresponds to the main extremum on *MR*(*P*) curve and beginning of thermoemf decreasing. After 32-34 GPa *MR* becomes close to zero and thermoemf achieves its minimum and almost stops changing as well as *MR*.

Described results of thermoemf studies and conductivity temperature studies let us make the following conclusions and confirm assumption made above. With increasing of pressure and deformation of the crystal lattice the defect number increases, the activation energy decreases, the band gap decreases, which manifest themselves in high negative magnetoresistance. Summarizing all described above, the phenomenon of negative magnetoresistance and presence of *MR*(*P*) extrema with high negative value of *MR* in the studied materials are caused by significant changes in the crystal and the electronic structures with pressure increasing.

4. Conclusions

The experimental study of electrical properties of $(GeSe)_x(CuAsSe_2)_{1-x}$, x= 0.5 and 0.7, at high pressure shows that electroresistance of studied materials decreases with the pressure increasing and both materials exhibit the negative magnetoresistance in studied pressure range. Magnetoresistance pressure dependences have some extrema witch correspond to the baric intervals of features in the other electrical properties behaviors (electroresistance, dielectric loss tangent, electroresistance relaxation time, thermoemf). The suggestion that phenomenon of negative magnetoresistance and magnetoresistance extrema presence in studied materials are due to significant changes in the crystal and the electronic structures with pressure was made.

Acknowledgments

The study was supported by the Russian Foundation for Basic Research, projects № 16-02-00857.

References

[1] Melnikova N V et al 2015 *Physics of the Solid State* **57** №10 2025
[2] Mel'nikova N V et al 1995 *Technical Physics Letters* **21** №1 3
[3] Melnikova N V et al 2015 *Azerbaijan Journal of Physics* **XXI** №2 3
[4] J-l Adam and X. Zhang 2014 *Chalcogenide glasses* (Cambridge: Woodhead Publishing Limited)
[5] Chetty R et al 2016 *Intermetallics* **72** 17
[6] Yakovlev E N et al 1980 *Rev. Phys. Chem. Jpn.* **50** 243
[7] Babushkin A N 1992 *High Pressure Res.* **6** 349
[8] Melnikova N V et al 2015 *Journal of Physics: Conference Series* **653** 012098
[9] Melnikova N V et al 2016 *Book of abstracts of 54th International conference EHPRG* 28
[10] Zaikova V et al 2016 *Journal of Physics: Conference Series* **741** 012200
[11] Seeger K 1982 *Semiconductor Physics* (Berlin: Springer)

Silicon carbide of Ni/6H-SiC and Ti/4H-SiC type Schottky diode current-voltage characteristics modelling

P V Panchenko, S B Rybalka, A A Malakhanov, A A Demidov, E Yu Krayushkina, O A Shishkina
Bryansk State Technical University, Bryansk, 50 let Oktyabrya 7, Russia
E-mail: sbrybalka@yandex.ru

Abstract. On the base of the physical analytical models based on Poisson's equation, drift–diffusion and continuity equations the forward current–voltage characteristics of 6H-SiC and 4H-SiC type Schottky diode with Ni and Ti Schottky contact have been simulated. It is shown on the base of analysis of current–voltage characteristics in terms of classical thermionic emission theory it is shown that the proposed simulation model of Schottky diode corresponds to the almost "ideal" diode with ideality factor n equals 1.1. Because of this it is determined that the effective Schottky barrier height φ_B equals 1.57 eV and 1.17 eV for Ni/6H and Ti/4H silicon carbide Schottky diode type, respectively.

1. Introduction

Silicon carbide belongs to a class of materials commonly referred to as wide-bandgap semiconductors. At first studying of silicon carbide (SiC) as material for semiconductor electronics has been made in Leningrad (St. Petersburg) by the O.V. Losev's at the A.F. Ioffe Institute at the beginning of the 1930s [1]. Now silicon carbide represents an excellent candidate for high-temperature power electronic device applications because of its high breakdown voltage, low series resistance, and stability under harsh chemical and high temperature conditions [1].

SiC Schottky diodes are of special interest since these unipolar devices avoid reverse recovery effects of bipolar devices, thereby offering higher frequency operation In particular, SiC Schottky diodes for power electronics in future must be produced by domestic company the ZAO «GRUPPA KREMNY EL» (Bryansk). It is obviously that for development of component base on the base of SiC studying and optimisation of such important device as Schottky diode it is necessary. Therefore with used of TCAD program in present work simulation of current-voltage (*I-V*) characteristics in Schottky diode on base of 6H-SiC and 4H-SiC type silicon carbide with Ni and Ti Schottky contact has been carried out.

2. Materials and methods

In Figure 1 is presented the schematic silicone carbide Schottky diode structure for modelling. In particularly, for simulation were chosen the following the materials parameters: the concentration of donors (nitrogen) in the substrate equals $N_D^+ \approx 10^{18}$ cm^{-3}, in the epitaxial layer equals 8×10^{15}, anode material is Ni (nickel) and Ti (titanium), thickness of the epitaxial layer (4H-SiC and 6H-SiC) equals z=15 μm, radius of the structure equals r=200 μm. For simulation model of current-voltage characteristics has been solved electrostatic Poisson's equation in cylindrical coordinates together with continuity equations for electrons and holes. Because of this for calculation current-voltage (*I-V*) characteristics in Schottky diode was applied of the classical thermionic emission theory [1].

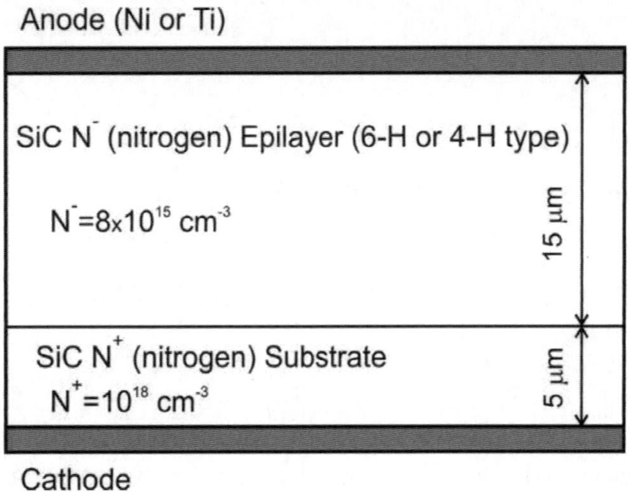

Figure 1. Schematic silicone carbide Schottky diode structure for modelling.

3. Results and discussion

For simulation model of current-voltage (*I-V*) characteristics has been solved electrostatic Poisson's equation in cylindrical coordinates together with drift–diffusion and continuity equations. In particular, we assumed that in within the framework of Fermi-Dirac statistics can be approximated by the Maxwell-Boltzmann distribution and mobility and diffusion coefficients can be described by Einstein's equations. Finally the system of the basic equations for simulation model of Schottky diode must be given by expressions of the form [1]:

$$\frac{1}{r}\frac{\partial}{\partial r}\left(\varepsilon_r r\frac{\partial \varphi}{\partial r}\right)+\frac{\partial}{\partial z}\left(\varepsilon\frac{\partial \varphi}{\partial z}\right)=-q\left(p-n+N_D^+-N_A^-\right) \quad (1)$$

$$\frac{dn}{dt}=-\frac{1}{q}div(\mathrm{J}_n)+(G-R), \quad (2)$$

$$\frac{dp}{dt}=-\frac{1}{q}div(\mathrm{J}_p)+(G-R), \quad (3)$$

$$\mathrm{J}_n=q\mu_n n\mathrm{E}_n+qD_n\nabla n, \quad (4)$$

$$\mathrm{J}_p=q\mu_p p\mathrm{E}_p+qD_p\nabla p, \quad (5)$$

where r – the radius of diode, z – the height of diode, q – the elementary charge, n and p – the concentration of electron's and holes, φ – the electrostatic potential, ε_r – the dielectric relative permeability of SiC epitaxial layer, N_D^+ – the donor impurity concentration, N_A^- – the ionised acceptors concentration, J_n and J_p – the electron's and holes electric current density, G and R – the generation and recombination rates for both electrons and holes, μ_n and μ_p – the electron and hole mobilities, $D_n=\frac{k_BT}{q}\mu_n$ and $D_p=\frac{k_BT}{q}\mu_p$ – the electron and hole diffusivities according to Einstein's relationship, k_B – the Boltzmann constant, T – the temperature, E_n and E_p – the effective driving electrical field to electrons and holes, which related to with local band diagram. In the presented model also were included Shockley-Read-Hall model recombination, Auger recombination and anisotropic impact ionization [2,3]. Boundary conditions between Ni and Ti metal contacts for 4H-SiC and 6H-SiC layer on a lateral surface of the cylinder were used as standard for Schottky diode [1,3]. All materials parameters data [2,3] used for simulation are shown in Table 1.

Table 1. 4H-SiC and 6H-SiC type materials parameters for simulation.

Parameter	4H-SiC	6H-SiC
Band-gap energy, $E_g\,(eV)$	3.23	3.0
Dielectric relative permeability, ε_r	9.7	9.66
Donor impurity concentration, N_D^+ (cm^{-3})	10^{18}	10^{16}
Density of state in conduction band, N_C(cm^{-3})	1.7×10^{19}	8.9×10^{19}
Density of state in valence band, N_V(cm^{-3})	2.5×10^{19}	2.5×10^{19}
Electron affinity, χ(V)	3.7	3.3
Electron mobility, μ_n(cm^2/(V$\cdot s$))	800	400
Mobility of holes, μ_p(cm^2/(V$\cdot s$))	115	95

Further, for simulation forward current-voltage characteristics were chosen the following temperature: 300, 400, 500 and 600 K. Simulation results of forward current-voltage characteristics for Ni/6H-SiC and Ti/4H-SiC [4,5] Schottky diodes in TCAD program are presented in Figure 2 and Figure 3.

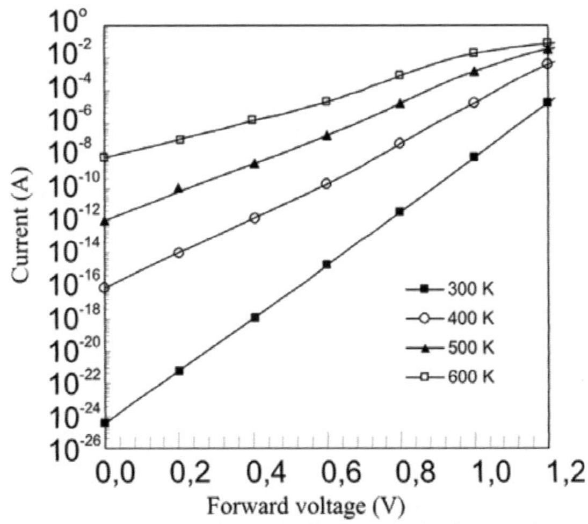

Figure 2. Ni/6H-SiC Schottky diode forward *I-V* characteristics.

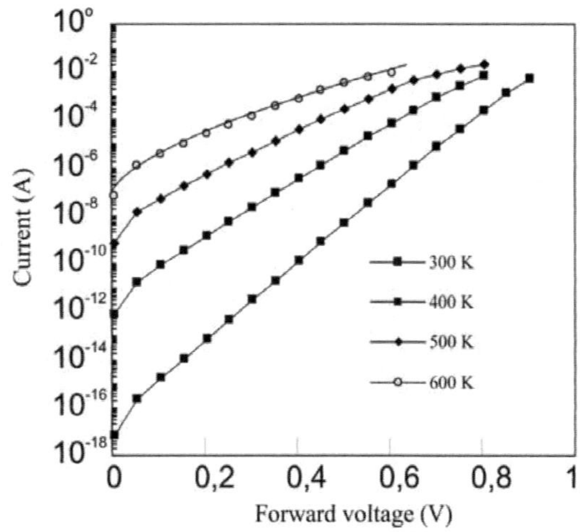

Figure 3. Ti/4H-SiC Schottky diode forward *I-V* characteristics.

Afterward the results of forward current-voltage characteristics for Ni/4H-SiC and Ti/4H-SiC Schottky diodes were analyzed in framework of classical diode theory. According to the thermionic emission classical diode theory [1,3] current-voltage characteristics the dependence of forward current I on voltage V for semiconductors can be described by the following equation:

$$I = I_o \exp\left(\frac{qV}{nk_BT}\right)\left(1-\exp\left(-\frac{qV}{k_BT}\right)\right) \quad (6)$$

where I_o – the saturation current [A]; T – the absolute temperature, [K]; V – the voltage, [V]; q – the elementary charge, [C]; k_B –Boltzmann constant, [J/K]; n – the Schottky diode ideality factor. It is possible to obtain [6] the saturation current I_o and the ideality coefficient n if plots dependence of ln$I$

on voltage *V*. Data for ideality coefficient *n* and saturation current I_0 for different temperatures ant type of diode are presented in Table 2.

Table 2. The ideality coefficient and saturation current value.

Schottky diode type	Ideality coefficient, *n*	Saturation current, I_0 (A)	Temperature (K)
Ni/6H-SiC	1.035	3.63×10^{-25}	300
Ti/4H-SiC	1.078	4×10^{-17}	300
Ni/6H-SiC	1.145	7.91×10^{-17}	400
Ti/4H-SiC	1.058	4×10^{-12}	400
Ni/6H-SiC	1.131	1.07×10^{-12}	500
Ti/4H-SiC	1.139	6×10^{-9}	500
Ni/6H-SiC	1.368	6.6×10^{-9}	600
Ti/4H-SiC	1.08	4×10^{-7}	600

As can be seen from Table 2 the average ideality coefficient *n* equals 1.16 for Ni/6H-SiC Schottky diode and 1.09 for Ti/4H-SiC Schottky diode. From the value of the coefficient *n*, it is used to estimate the Schottky diode quality which for good Schottky diode is equal ~1.0÷1.1. In our case the obtained ideality coefficients *n* are close to good Schottky diode.

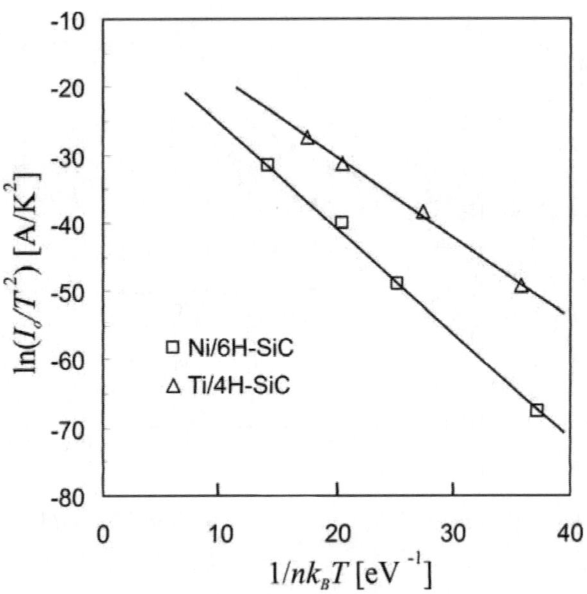

Figure 4. $\ln(I_o/T^2)$ dependence on $1/nk_BT$ for Ni/6H-SiC and Ti/4H-SiC Schottky diode.

Thus, the forward *I–V* characteristics in the entire temperature range proved to be reason ably close to the ideal ones. On the other hand the forward current can be rewritten in general case by following equation:

$$I = I_0\left[\exp\left(\frac{qV}{nk_BT}\right)-1\right], \tag{7}$$

where I_o is the saturation current and *n* the ideality coefficient [6,7]. In this case if plots the dependence of $\ln(I_o/T^2)$ on $1/(nk_BT)$ (the Richardson plot) it is possible to obtain the effective barrier height for Schottky diode [6,7]. In Figure 4 from the slope of line were determined the effective barrier height φ_B for Ni/6H-SiC Schottky diode – 1.57 eV and consequently for Ti/4H-SiC Schottky diode – 1.17 eV. Above-mentioned parameters are very good agreements with experimental data for such type of silicon carbide Schottky diode [8-12].

4. Conclusions

Finally, forward current-voltage characteristics in Ni/6H-SiC and Ti/4H-SiC Schottky diode with Ni and Ti Schottky contact have been obtained on the base of simulation model in TCAD in framework of the physical analytical models based on Poisson's equation, drift–diffusion and continuity equations. It is shown that forward current–voltage characteristics in terms of the Schottky diode simulation model corresponds to the "ideal" Schottky diode describes in framework of the classical thermionic emission theory with the ideality factor of Schottky diodes is close to good Schottky diode.

It is established that effective Schottky barrier height φ_B=1.57 eV for Ni/6H-SiC and φ_B=1.17 eV for Ti/4H-SiC Schottky diode. Thus, it is suggested that matches of model results with published experimental results indicate that the simulation model is correct and can be used for calculation of such type SiC Schottky diode current-voltage characteristics determination.

Acknowledgements

This work at Bryansk State Technical University was supported by the Russian Ministry of Education (Grant No. 02.G25.31.0201). Authors also would like to thank Dr. Surin B.P. for help in carrying out of simulation.

References

[1] Shur M 1990 *Physics of semiconductor devises* (New Jersey: Prentice–Hall Int.)
[2] Kimoto T, Cooper J A 2014 *Fundamentals of Silicon Carbide Technology. Growth, Characteriztion, Devices, and Applications* (New York: Wiley–IEEE Press.)
[3] Zhao J H, Sheng K and Lebron-Velilla R C 2006 *SiC materials and devises* vol 1, ed. M Shur, S Rumyantsev and M Levinshtein (Singapore: World Scientific) p. 117
[4] Panchenko P V, Rybalka S B, Malakhanov A A, Krayushkina E Yu, Rad'kov A V 2016 *Proc. of the International Conference "Micro- and Nanoelectronics-2016" ICMNE-2016* (Moscow: Maks Press) p 210
[5] Panchenko P V, Rybalka S B, Malakhanov A A, Krayushkina E Yu, Rad'kov A V 2016 *Proc. SPIE "International Conference on Micro- and Nano-Electronics" 2016* vol. 10224. (Washington: SPIE) p 102240Y-1
[6] Rhoderick E H 1978 *Metal–Semiconductor Contacts* (Oxford: Clarendon Press)
[7] Ivanov P A, Grekhov I V, Kon'kov O I, Potapov A S, Samsonova T P, Semenov T V 2011 *Semiconductors* **45** 1374
[8] Itoh A, Matsunami H 1997 *Physica Status Solidi (a)* **162** 389
[9] Potapov A S, Ivanov P A, Samsonova T P 2009 *Semiconductors* **43** 612
[10] Itoh A, Kimoto T, Matsunami H 1995 *IEEE Electoron Device Lettters* **16** 280
[11] Benmaza H, Akkal B, Abid H, Bluet J M, Anani M, Bensaad Z 2008 *Microelectronics Journal* **39** 80
[12] Perrone D, Naretto M, Ferrero S, Scaltrito L, Pirri C F 2009 *Materials Science Forum* **615-617** 647

Temperature Effects on the Magnetization and Magneto-impedance in Ferromagnetic Glass-Covered microwires

A Uddin[1], S A Evstigneeva[1], A Dzhumazoda[1], M M Salem[1,3], M G Nematov[1], A M Adam [1,4], LV Panina[1,2] and A T Morchenko[1]

[1] National University of Science and Technology, MISiS, Moscow 119991, Russia
[2] Institute for Design Problems in Microelectronics RAS, Moscow 124681, Russia
[3] Physics Department, Faculty of Science, Tanta University, Tanta 31527, Egypt
[4] Physics Department, Faculty of Science, Sohag University, Sohag 82524, Egypt

Abstract. The effect of temperature on static and dynamic magnetization in Co-based amorphous microwires was investigated with the aim of potential applications in miniature temperature sensors. The wires of two compositions with different magnetostriction and Curie temperature in glass-cover and after removing the glass layer demonstrated very different temperature behaviour of the magnetization loops and magnetoimpedance. The mechanisms of the temperature effects are related to the residual stress distribution due to fast solidification, the difference in thermal expansion coefficient of metal and glass and the proximity to the Curie temperature. The interplay of these factors may result in a very strong temperature dependence of magnetoimpedance in a moderate temperature range (room temperature –373K). Such elements may be incorporated in various composite materials for a local temperature monitoring.

1. Introduction

The development of magnetic sensing materials with improved performance, microsize and wireless operation are increasingly critical for different applications. The present work tackles this problem through the investigation and design of temperature properties of micron-sized amorphous magnetic wires as embedded sensing elements. Ferromagnetic microwires have a number of specific magnetic properties such as a magnetic bi-stability [1-3] and very large magnetoimpedance (MI) [4-6]. All these effects can find applications in various sensors. For commercial applications, the thermal stability of sensing elements is mandatory; however, the magnetic properties of amorphous alloys may show the temperature dependence even for temperatures which are much lower than the Curie temperature. This is related with the internal stress distribution which is changed by heating. The other source of internal stresses is related with the presence of interfaces like in glass-coated microwires where there is a difference in thermal expansion coefficients of glass and metal. Finally, approaching the Curie temperature which can be tuned by the alloy composition will impose very strong temperature variations of all magnetic parameters. Therefore, understanding the influence of various sources of internal stresses on the temperature dependence of magnetic structure in amorphous elements is important for sensor design.

The magnetic anisotropy in amorphous alloys is mainly determined by magnetoelastic interactions [7], which explains why the internal stresses coupled with the magnetostrictive parameter strongly affect the magnetic structure. The saturation magnetostriction which depends on the alloy composition [8] may also depend on the internal stresses which further enhance the stress-dependence of the magnetization process. Tuning the magnetic structure with external stimuli will result in a corresponding dependence of magnetoimpedance [9].

Since the presence of glass coating adds an additional source of magnetoelastic anisotropy it would be interesting to compare the magnetic properties in as-prepared state and after glass removing [10, 11]. The removal of glass showed a remarkable impact on the magnetisation process in low magnetostrictive wires owing to the interface stress release. Consequently, the microwires without glass may possess higher temperature stability of magnetic properties and more sensitive magnetoimpedance [12]. Along with this, the temperature stability will be also improved by appropriate heat treatments [13].

In this paper, we are interested in the temperature effects on the magnetic anisotropy and the related static and dynamic magnetic properties including magnetization loops and magnetoimpedance in glass-covered amorphous microwires of Co-rich compositions having a small saturation magnetostriction. We have studied the effect of temperature on magnetic hysteresis loop and the magnetoimpedance for as-prepared microwires and after glass removing. We have used the wires made of alloy compositions with a relatively high and low curie temperatures.

2. Materials and Experimental details

We investigated the effect of temperature (from room temperature up to 373 K) on the hysteresis loops and magnetoimpedance in glass coated microwires before and after glass removing. Glass-coated microwires were fabricated by a modified Taylor–Ulitovsky method [14, 15]. Microwires of two compositions $Co_{66.94}Fe_{3.83}Ni_{1.44}B_{11.51}Si_{14.59}Mo_{1.69}$ (sample No.1,$T_c = 573\ K$) and $Co_{23.67}Fe_{7.14}Ni_{43.08}B_{13.85}Si_{12.26}$ (sample No.2, Tc=335K) were used. In as-prepared stay, the wires had a different magnetostriction coefficients: small and negative for sample No. 1 and positive for sample No. 2. The glass cover was removed by chemical method with hydrofluoric acid solution. The geometric parameters of microwires were examined by an optical microscope.

The Curie temperature and crystallization temperature were estimated from DSC curves using standard IT application. The experimental magnetization loops were measured by an inductive method having two detection differential coils having 3mm of the inner diameter and 5mmof length. The magnetization coils were excited by a current of frequency up to 500Hz producing a magnetization field with an amplitude of 1000A/m. To obtain the hysteresis curves, the generated voltage signal was digitally integrated wires were inserted into a very small detection coil

The high-frequency impedance was deduced from S21-parameters measured using a vector network analyser (Hewlett Packard 8753E) in a special cell designed for temperature measurement at frequencies of 1-100 MHz.

3. Result and discussion

The magnetization curves for wires in as-prepared state and after glass removal are shown in figure 1. For sample No.1 after removing the glass the hysteresis changes dramaticly: from an inclined loop to a loop of a bi-stable type. This indicates a change in the easy anisotropy direction: from nearly circumferential to nearly axial. In the case of sample No.2 (positive magnetostrioction) the type of hysteresis after glass removing does not change but the coercivity reduces from 51 down to 20 A/m. this indicates that indeed the metal-glass interface imposes an additional stress. In the case of a negative magnetostriction in as-prepared state, the glass removal and associated stress release may cause a change in the sign of the magnetostriction- it becomes positive which results in the abrupt change of hysteresis. In the case of a positive magnetostriction in as-prepared state, the glass removal and the associated stress release decreases the magnetostrictive anisotropy and coercivity.

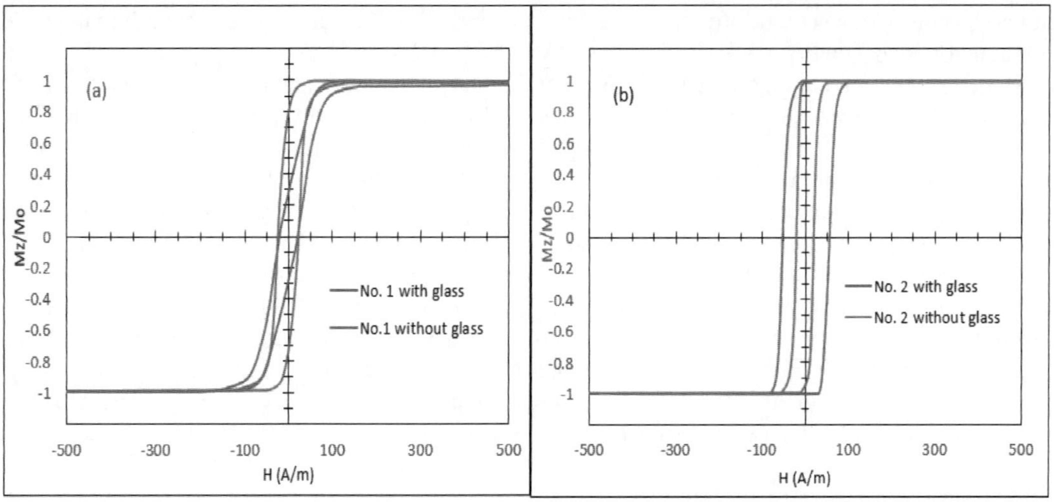

Figure 1. Hysteresis loops of microwires with composition $Co_{66.94}Fe_{3.83}Ni_{1.44}B_{11.51}Si_{14.59}Mo_{1.69}$ (sample No.1, Tc=573K) and $Co_{23.67}Fe_{7.14}Ni_{43.08}B_{13.85}Si_{12.26}$ (sample No.2, Tc=335K) in as prepared state (with glass,depicted by red line) and after glass removal (depicted by blue line).

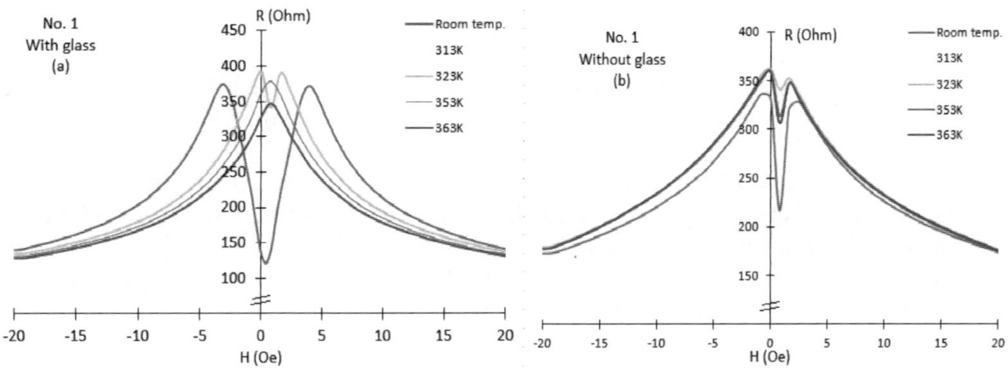

Figure 2. Real part of the impedance of glass-coated microwires of composition $Co_{66.94}Fe_{3.83}Ni_{1.44}B_{11.51}Si_{14.59}Mo_{1.69}$ (sample No.1) before (a) and after (b) glass removing for different temperatures. Applied frequency is 80 MHz

MI of glass-coated wires of alloy No. 1 exhibits strong variation with temperature: two symmetrical peaks gradually merge and a single peak at a zero field is seen at a temperature of 335K (which is still much lower than the Curie temperature) as seen in figure 2 a. However, after glass removal, the impedance temperature dependence considerably weakens which indicates the major role of glass-metal interface in forming the magnetoelastic anisotropy and associated temperature effects. For sample No.2 which has a positive magnetostriction, an axial anisotropy and a relatively low Curie temperature, in both as-prepared and after glass removal states a considerable temperature dependence of MI is observed. In this case the major temperature effect is associated with approaching the Curie temperature.

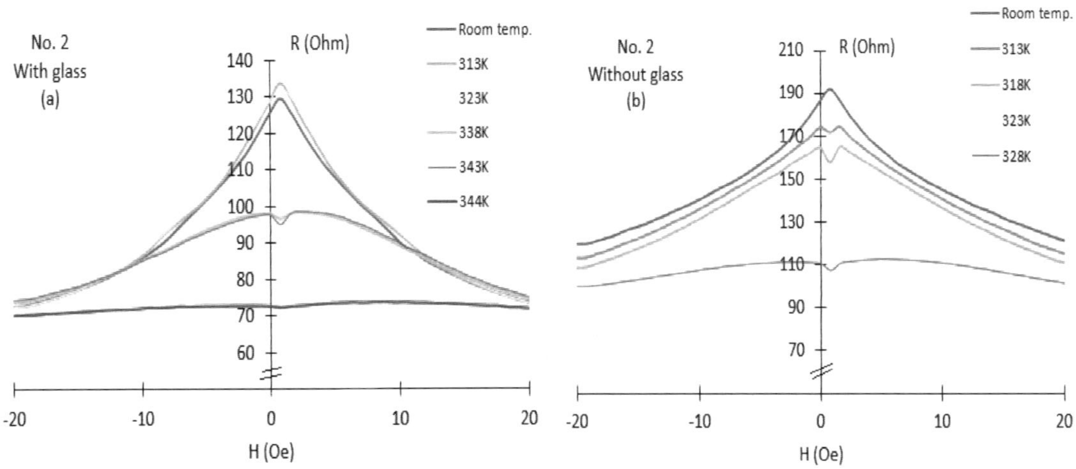

Figure 3: Real part of the impedance for of glass-coated microwire of the composition $Co_{23.67}Fe_{7.14}Ni_{43.08}B_{13.85}Si_{12.26}$ with Tc=335K (sample No.2) before and after glass removing for different temperatures at 80 MHz

In this case, the major factor determining the temperature sensitivity is related with the interplay of a low Curie temperature and reduced anisotropy. Controlling the value of T_c the impedance temperature dependence can be realized in the desired temperature range.

4. Conclusion

We have determined that the magnetization and magnetoimpedance in glass-coated microwires may show remarkable temperature dependence. This is related with the temperature dependent internal stress distribution and with the variation of magnetic parameters when approaching the Curie temperature. In the case of high Curie-temperature alloys, the effect of internal stresses on the temperature stability of magnetoimpedance in intermediate temperature range may be reduced by glass removal. This is important for commercial applications of MI wires. However, for alloys with a relatively high Curie temperature the temperature influence on MI may be even stronger after glass removal owing to reduced anisotropy? Which can be interesting for designing temperature sensitive elements.

Acknowledgment

This work is acknowledged by L. V. Panina under the Russian Federation State Contract for organizing a scientific work.

The work was carried out in the framework of Increase Competitiveness Program of NUST «MISiS» (№ K4-2016-60), with financial support from the Ministry of Education and Science of the Russian Federation implemented by a governmental decree dated 16th of March 2013, N 211.

The authors are also very thankful to Dr. V. Larin, MFTI Ltd (http://www.microwires.com) for donating the wire samples.

References

[1] Chiriac H and Ovari T A 1996 *Applications Progr. Mater.* Sci. **40** 333
[2] Zhukov A and Zhukova V 2009 *Nova Science Publishers, New York* ISBN: 978-1-60741-770-5
[3] Vazquez M, Chiriac H, Zhukov A et.al 2011 *Phys. Status Solidi A* **208** 493
[4] Panina L and Mohri K 1994 *Appl. Phys. Lett.* **65** 1189
[5] Knobel M and Pirota K R 2002 *J. Magn. Magn. Mater.* **242/245** 33
[6] Tannous C 2004 *Materials in Electronics* **15** 125
[7] Zhukov A, Ipatov M, Garcia C, Gonzalez J, Panina L, Blanco J M and Zhukova V 2008 *PIERS*

Proceedings, Hangzhou, China
[8] Zhukov A 2006 *Adv Func Mat* **16** 675
[9] M, Roy R K, Panda A K, Greve D W, Ohodnicki Jr.P R, McHenry M E 2014 *J. Appl. Phys.* **105** 222407
[10] Zhukova V, Ipatov M and Zhukov A 2009 *Sensors* **9** 9216
[11] Shevyrtalov S, Chichay K, Ershov P, Khovaylo V, Zhukov A, Zhukova V and Rodionova V 2015 *Acta Physica Polonica A* **127** 603
[12] Panina L V and Mohri K 1994 *Applied Physics Letters* **65** 1189
[13] García C, Zhukov A, Gonzalez J, Zhukova V and Blanco J M 2006 *Journal of Optoelectronics and Advanced Materials* **8** 1706
[14] Taylor G F 1924 *Physical Review* **23** 655
[15] Va´zquez M, Advanced magnetic microwires, in: H.Kronmuller, S.Parkin 2007 (Eds.), *Hand book of Magnetism and Advanced Magnetic Materials*, Wiley, Chichester, UK 2193

Effect of current limitation on the adaptive behavior of memristive nanostructures

D S Korolev, A N Mikhaylov, A I Belov, E V Okulich, I N Antonov, O N Gorshkov, and D I Tetelbaum

Lobachevsky University, 23/3 Gagarin prospect, Nizhny Novgorod, 603950, Russia

Abstract. Adaptive behavior of resistive state in response to electrical stimulation has been studied for the silicon oxide based memristive nanostructures subjected to electroforming in the conditions of current compliance. The limitation of current and temperature during electroforming affects the parameters of growing conductive filaments and reduction-oxidation reactions resulting in a higher dynamic range of gradual resistance change important for neuromorphic applications.

1. Introduction

The rapid development of information technologies and electronics is associated with innovations in the information storage devices and neuromorphic (biosimilar) systems based on memristive nanomaterials [1]. Memristive effect is often implemented in a thin-film nanostructure of the metal-oxide-metal type as a reproducible resistance change between the low-resistance and high-resistance states under application of electric stress [2, 3]. The resistance change is analogous with the change in transmissive efficacy of synaptic junction and considered as one of the main conditions for the application of memristive devices in neuromorphic systems.

The ability of memristive nanostructures to continuously change the resistance in dependence on input electrical signal is equivalent to the adaptive behavior (plasticity) of synapse as a key mechanism underlying the phenomena of learning and memory in neural networks. The realization of adaptive behavior of metal-oxide memristive nanostructures [4] is based on a continuous spectrum of resistive states related to the different configurations of conductive silicon filaments grown in oxide film during electroforming (EF) and on a gradual character of switching between the different resistive states. The range and sensitivity of a resistive state to electrical stimulation should depend on the kinetics of microscopic reduction-oxidation phenomena driven by electric field and temperature. Effect of current compliance (CC) applied during EF on the character and parameters of resistive switching (RS) was investigated in several works for different compositions of materials in a memristive nanostructure [5, 6]. For the HfO_2-based memristive nanostructures [7], it was established that the character of RS (gradual or abrupt) in the processes of increase (SET) or decrease (RESET) of conductivity also depends on the use of CC. Obviously, such peculiarities of a memristive device must depend on specific materials used in its thin-film structure. To the best of our knowledge, for the CMOS-compatible SiO_2-based nanostructures [8], the detailed study of CC effect on the adaptive behaviour of resistive state has not been reported yet.

In the present work, the adaptive change of resistance in a wide range is studied for the silicon oxide based memristive nanostructures in dependence on the application of CC during EF and RS.

2. Experimental

The capacitor-like metal-oxide-metal nanostructure was deposited on the CMOS-compatible TiN (25 nm) / Ti (25 nm) / SiO_2 (500 nm) / Si substrate by magnetron sputtering using the Torr International® MagSputt™ 3G2 and 2G1-1G2-EB4-TH1 vacuum thin-film deposition systems. The SiO_x film (40 nm) was deposited by RF-magnetron sputtering of fused silica in the argon-oxygen gas mixture (30% oxygen content) at the substrate temperature of 300 ºC. The top Au electrodes (40 nm) with area of $1.2 \cdot 10^{-3}$ cm^{-2} were deposited through the mask by the method of DC-magnetron sputtering in an argon atmosphere and substrate temperature of 300°C. The electrical characteristics in continuous and pulse regimes were measured by the Agilent B1500A semiconductor device analyzer. The sign of bias on the device corresponds to the potential of top electrode (Au) relative to the grounded bottom electrode (TiN). The EF was performed during the first current-voltage sweep at the negative bias with or without application of CC.

3. Results and discussion

Typical curves of EF realized in the range of 2-6 V of negative bias and subsequent RS hysteresis of the valence-change type [8] are shown in Figure 1 for the fabricated SiO_2-based memristive nanostructure.

If no CC is used during EF, the "strong" silicon filaments are grown in oxide film, and the structure can be switched between the low-resistance state (LRS) and high-resistance state (HRS) in the so-called high-current mode. The corresponding RS hysteresis is shown by the red current-voltage curves in Figure 1 and characterized by the gradual or abrupt stepwise current changes in the SET (HRS to LRS) and RESET (LRS to HRS) transitions, which are poorly reproducible from cycle to cycle.

If the CC at 100 μA is applied during EF, the SET transition becomes abrupt but is followed by a monotonous RESET process to the "deeper" HRS state with the lower threshold voltage. This gradual RESET transition is reproducible and is much more appropriate for the demonstration of adaptive behavior compared to the SET transition in the high-current RS mode studied in our previous work [9]. It should be noted that the observed deep HRS state cannot be achieved in the high-current RS mode, even if CC is applied during the preceding SET transition (e.g. CC at 1 mA shown in Figure 1).

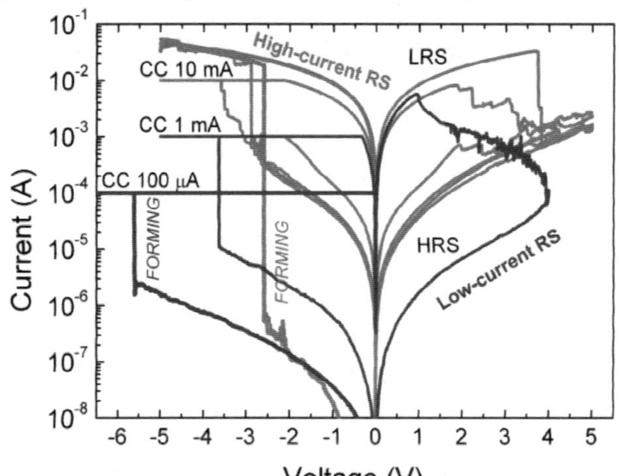

Figure 1. The current-voltage characteristics of one and the same memristive nanostructure measured in cyclic sweeping mode with different conditions of EF (with or without CC).

The averaged dependence of resistance on the amplitude of positive (RESET) programming pulse (5 ms) is shown in Figure 2a for the memristive nanostructure subjected to EF in the conditions of CC at 100 μA. The resistance was measured at a reading voltage of +0.5 V after each RESET pulse, which was also followed by the negative pulse (–6 V) used to set the structure back to LRS.

In contrast to adaptive behavior measured in the SET region (Figure 2b – see detailed discussion in [9]), the adaptive behavior related to the gradual and deeper RESET transition is characterized by the almost linear dependence on the voltage amplitude in the range of 0-5 V and the much higher dynamic range of resistance change from 10^2 to 10^5 Ω important for the application as a programmable weight in artificial neural networks.

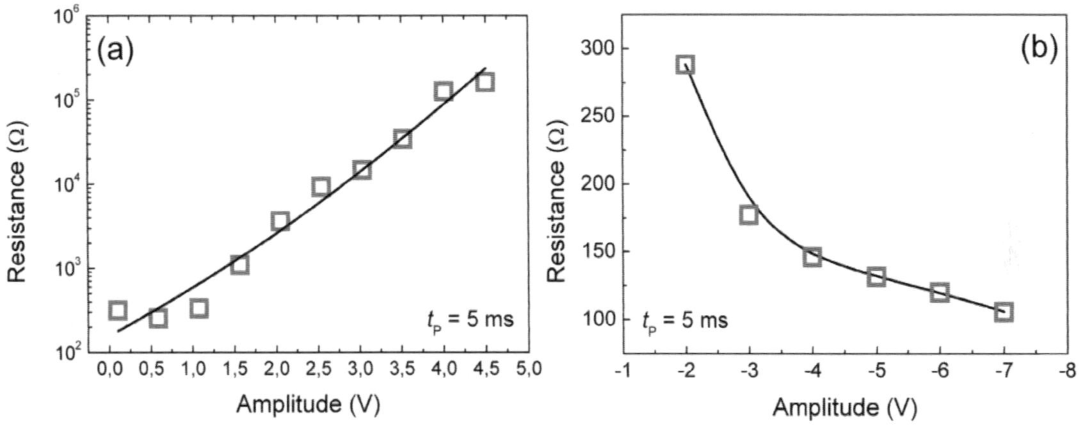

Figure 2. The resistance of memristive nanostructure vs. the RESET pulse amplitude (a) compared to the analogous dependence of resistance on the SET pulse amplitude [9] (b).

The observed effect of CC applied during EF can be interpreted from the viewpoint of previously proposed RS model [3, 8] based on the processes of formation and partial oxidation of a silicon filament accompanied by the migration of interstitial O^{2-} ions produced during the breaking of Si-O-Si bonds. Without CC, the high current value through the forming filament provides the strong Joule heating in a local area adjacent to filament. As a result, the thermal diffusion moves oxygen far away from the filament. When current is limited by the CC value, the temperature around the filament at the stage of EF is smaller, and the distance of oxygen removal from the filament is lower too. In this case, the RESET process associated with the filament oxidation near the electrode begins at lower voltages and leads to a more expressed oxidation, which cause the decrease in RESET voltage and increase in HRS resistance. The RESET process becomes gradual because the oxidation takes place in a broad range of voltages, whereas the "high-current" RESET transition is realized only in a narrow voltage range and has an abrupt character.

The obtained dependence of resistance of memristive device on the amplitude of voltage pulse (Figure 3) provides the following technique of programming the resistance (weight) of a memristor for the application in neural networks. The desired value of resistance is set by the application of programming pulse with the amplitude in the range of 1-5 V. To set the next value (in any direction of resistance change), the initializing pulse of opposite polarity with the amplitude of 5-8 V (which turns the device into initial resistance state) is applied before the programming pulse. The amplitude of a voltage pulse, which does not change the resistance, should be lower than ±1 V.

The histograms of resistive states obtained by multiple (100 times) programming of each value (programming pulses: 1 V; 2 V; 3 V; 4 V) in the RESET mode for a memristive device subjected to EF with CC are shown at Figure 3. By using Gaussian fitting of the resistance distributions for different programming voltages, it is obtained that the variation (standard deviation) of resistance increases with increasing the resistance of memristive device, but in relative units (relative to the mean value) it remains approximately constant and does not exceed 30%. It is evident that the observed variation of memristive states is caused by the stochastic nature of RS process [3] and should be taken into account in the design of artificial neural networks based on memristive devices.

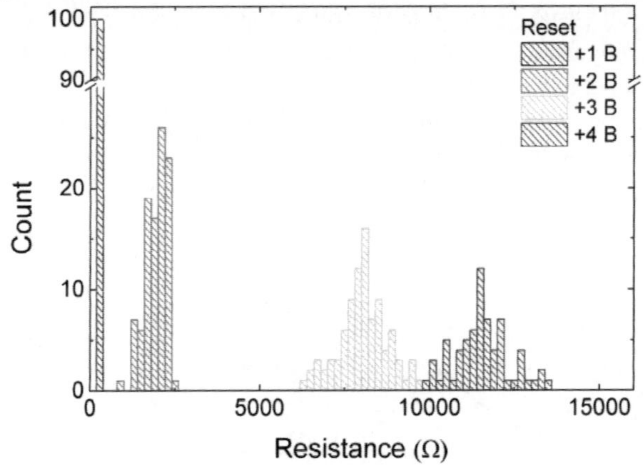

Figure 3. The histograms of resistive states of memristive device programmed by the application of different RESET voltage pulses.

4. Conclusions

In summary, the limitation of current during electroforming process allows controlling the parameters of adaptive behavior of SiO_2-based memristive nanostructures. Programming the resistive state in a wide range with the 30% accuracy is demonstrated in the region of gradual transition from low-resistance to high-resistance state of memristive device. The developed memristive devices on the basis of CMOS-integrable materials and technology can be used in the design and hardware implementation of artificial neural networks and neuroprocessors.

Acknowledgments

The study is supported by the Russian Science Foundation (project № 16-19-00144) and the Russian Foundation for Basic Research (project № 16-37-00360).

References

[1] Adamatzky A and Chua L 2014 *Memristor Networks* (Switzerland: Springer International Publishing)

[2] Ielmini D 2016 *Semicond. Science and Technology* **31** 063002

[3] Mikhaylov A N, Gryaznov E G, Belov A I, Korolev D S, Sharapov A N, Guseinov D V, Tetelbaum D I, Tikhov S V, Malekhonova N V, Bobrov A I, Pavlov D A, Gerasimova S A, Kazantsev V B, Agudov N V, Dubkov A A, Rosário C M M, Sobolev N A and Spagnolo B 2016 *Phys. Stat. Sol. C* **13** 870

[4] Ha S D and Ramanathan S 2011 *J. Appl. Phys.* **110** 071101

[5] Kim S and Park B 2016 *Appl. Phys. Lett.* **108** 212103

[6] Sowinska M, Bertaud T, Walczyk D, Thiess S, Calka P, Alff L, Walczyk Ch and Schroeder T 2014 *J. Appl. Phys.* **115** 204509

[7] Brivio S, Covi E, Serb A, Prodromakis T, Fanciulli M and Spiga S 2016 *Appl. Phys. Lett.* **109** 133504

[8] Mikhaylov A N, Belov A I, Guseinov D V, Korolev D S, Antonov I N, Efimovykh D V, Tikhov S V, Kasatkin A P, Gorshkov O N, Tetelbaum D I, Gryaznov E G and Yatmanov A P 2015 *Mat. Sci. Eng. B* **194** 48

[9] Korolev D S, Mikhaylov A N, Belov A I, Sergeev V A, Antonov I N, Gorshkov O N and Tetelbaum D I 2016 *J. Phys.: Conf. Series* **741** 012161

IOP Conf. Series: Journal of Physics: Conf. Series **917** (2017) 082013 doi:10.1088/1742-6596/917/8/082013

Bistable arch-like beams with modulated profile as perspective supporting structures of a microelectromenchanical actuator

Y B Enns[1], E N Pyatishev[1], A Glukhovskoy[2]

[1]Peter the Great Saint-Petersburg Polytechnic University, St. Petersburg, Russia
[2]Institut für Mikroproduktionstechnik, Leibniz Universität Hannover, Garbsen, Germany

Abstract. The static mechanical properties of a bistable electromechanical drive with arcuate suspension is investigated. The behaviour of an arcuate suspension and the influence of the beam thickness modulation on the stability of both stationary states is studied experimentally and theoretically. The influence of the beam shape modulation on its potential energy was determined with finite element analysis. The simulation results were verified experimentally. In the study, the optimal modulation configuration is determined. With help of this modification the sustainability of both mechanically stable states is significantly improved.

1. Introduction

The major element of micro-electro-mechanical systems (MEMS) is an elastic microstructure in the form of spring, a beam or a membrane. These elements are usually the linear elastic systems, wherein its linearity enables mechanical – electrical energy conversion both in sensors and in actuators. However, in the last decade, the use of nonlinear supporting beams becomes more popular. The active actuation of the nonlinear suspensions enables the formation of multistable systems. The work is particularly focused on arch-shaped suspensions, which profile is determined by the mode of buckling. The nonlinearity of such suspensions causes a sharp increase in the axial load of the suspension under an applied lateral load of the suspension [1]. This structure has two minimum of the potential energy i.e. the mechanical structure has two stable states. The transition between the potential energy minima of such mechanical system causes loss of stability during the transition between states. However, in such systems a serious asymmetry of bifurcation points is noted, i.e. the difference of the potential energy minima is large, which indicates reduced stability of one of the states [2]. The method of localized stiffness enlargement of the arcuate suspension beam augment at the sustainability of the "weaker" state. Here we consider the influence of the arch thickness modulation of and optimization of the initial profile of an arcuate suspension.

2. Method description

The profile (w(x)) of an arch beam of thickness t is determined by the first form of buckling (Figure 1a). The arch ends are clamped at x=0 and x=l. The stability loss of the arched beam which occurs under the influence of an axial restriction leads to a sudden change of the beam shape. The second stable state is a result of the beam shape transformation (known as snap-through) under the lateral load (F)governed by the axial one (p). The potential barrier between the stable states can be enlarged introducing two stiff segments into the arch structure (Figure 1b).

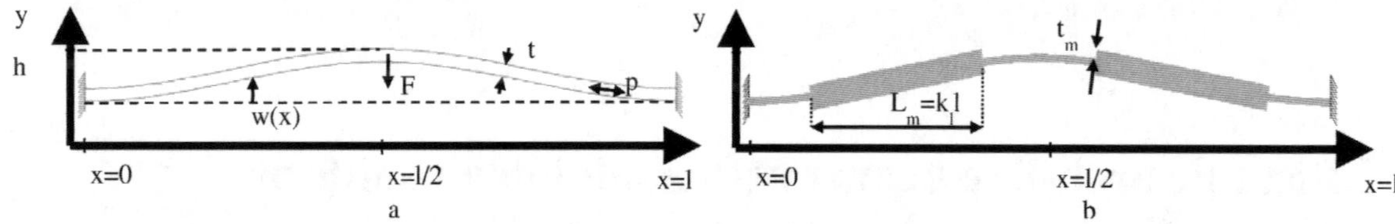

Figure 1. a) A uniform beam arch-like beams profile; b) An arch-like beam with modulated profile.

The stability of the snapped-through state of a "thin-thick-thin-thick-thin" shaped arch beam is significantly better compared to a uniform arch beam [3]. The mechanical stability of an arcuate beam is determined with the length of the ridged beam section L_m.

3. Design and fabrication

Bistable microdrive is an electromechanical system comprising an array of comb electrodes, where the motile group of electrodes is supported with the elastic arcuate beam system. The prototype of the bistable microactuator is fabricated using bulk technology via DRIE of silicon in a 100 μm thin structural layer of a silicon-on-glass wafer substrate (Figure 2). The initial shape of the actuated beam are directly controlled through the lithography and DRIE.

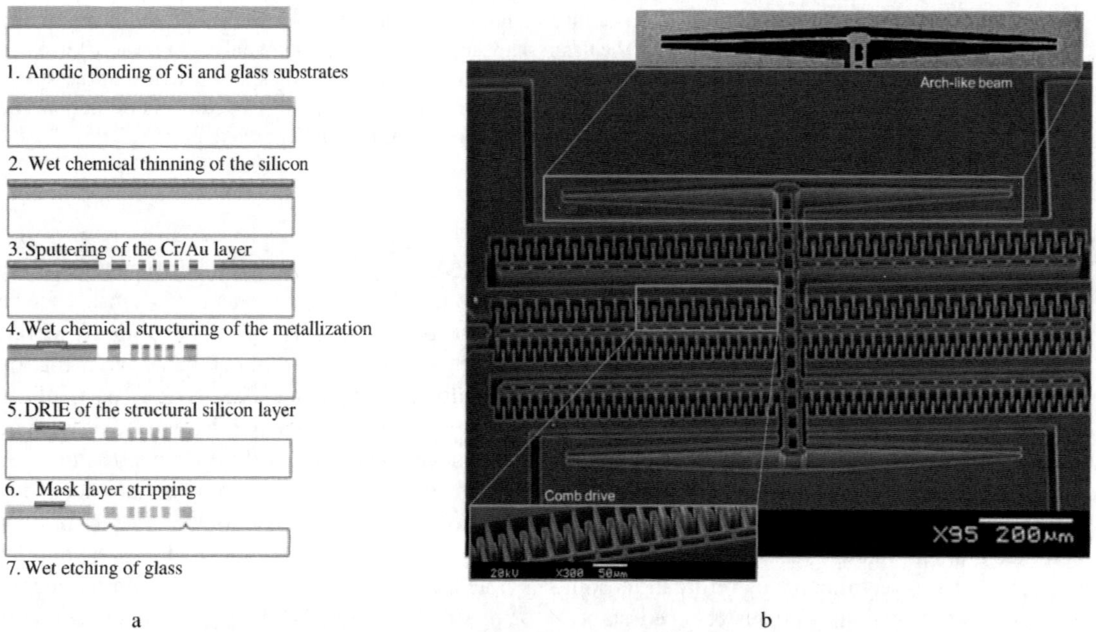

Figure 2. a) Fabrication process; b) Bistable actuator (REM image).

4. Numerical and experimental study of a uniform beam behavior

The reaction of the arcuate structure to the central lateral load is estimated analytically using the method of J Qiu et al. [2] and the finite element analysis (FEA) (Figure 3). The forces at snap-through transition are determined experimentally via the potential at which up and down transitions occur. These critical forces are denoted with horizontal lines (Figure 3). The resulting theory/experiment mismatch is mostly justified by the geometrical imperfection of the resulting etched beam.

IOP Conf. Series: Journal of Physics: Conf. Series **917** (2017) 082013 doi:10.1088/1742-6596/917/8/082013

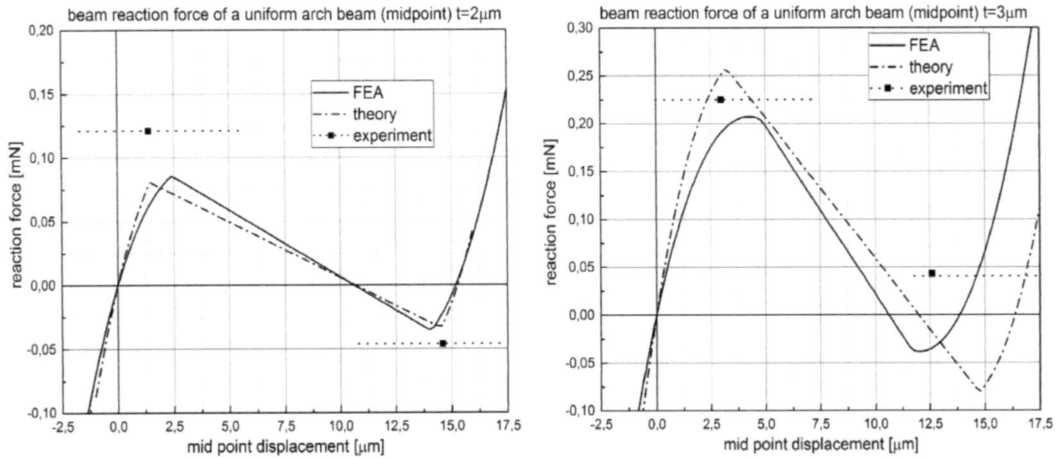

a b

Figure 3. The uniform arch beam reaction force vs displacement. Analytical estimation, FEA simulation, and experiment: a) beam thickness 2um; b) beam thickness 3um.

The results achieved by means of the numeric simulation match well to the analytical solution of the problem with a large ratio of the height of the deflection to the thickness of the beam (Figure 3a). However the beam thickness enlargement brigs a deviation between the analytical and FEA models. The experiment demonstrates deprivation of second stable state as shown on figure 3b.

The relation of axial and lateral reaction forces within the beam determine the stability of this secondary state. This relation can be expressed as the bow of the arch to thickness ratio. If the bow is fixed, the depth of the potential energy minimum of the arch decreases with its thickening, as it is illustrated below (Figure 4a, b).

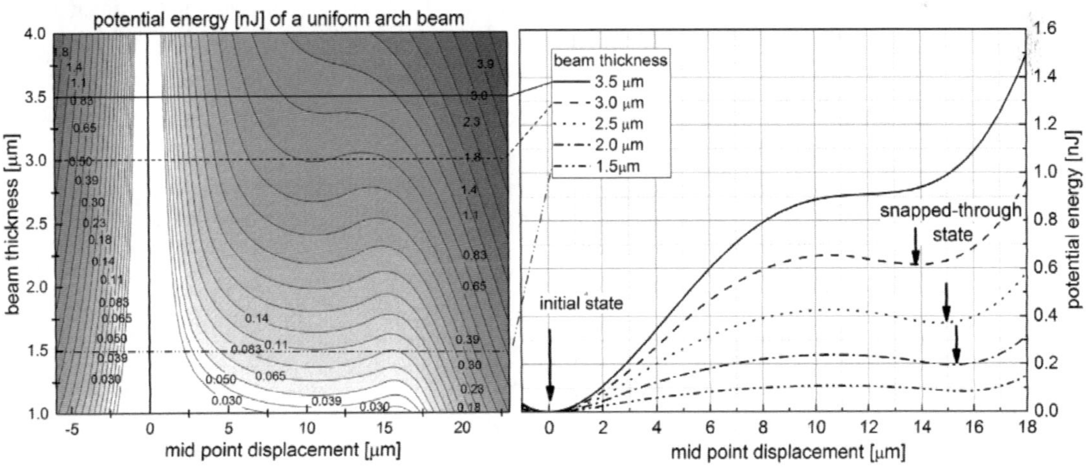

a b

Figure 4. Estimated potential energy of the uniform arch beam with a bow of 8 μm; a) potential energy topography in space arch thickness/displacement b) 2D slices of the elastic potential energy of selected thickness.

The relative mechanical stability R of the snap-through state can be expressed as the ratio of the top potential energy minimum depth to the bottom one. One can notice that at a beam height of the arch rise of 9 μm and thickness of 3.5 μm and above the stable state of the structure can no longer be reached (Figure 4). On the other hand thinning of the arch will reduce the absolute depth of the second minimum.

5. FEM Stability Analysis

5.1 The arch beam with rectangular modulation

The potential barrier between the stable states can be enlarged by introducing two stiff segments into the arch structure (Figure 1b). Figure 5-7 represents the numerically simulated potential energy of such arch structures of modulated thickness and stiffness. The most important optimization parameters of such thickness modulation are the length of the modulation segment Lm and its thickness, tm. The proper set of this parameters help to improve the stability of the snap-through stable state. The ratio of the potential energy barrier height to the minimum depth R is at Figure 5a. The optimal length of the modulation segment km is achieved at k_l=0,25. The real value of the potential energy well depth is highest at this point as well Figure 5b.

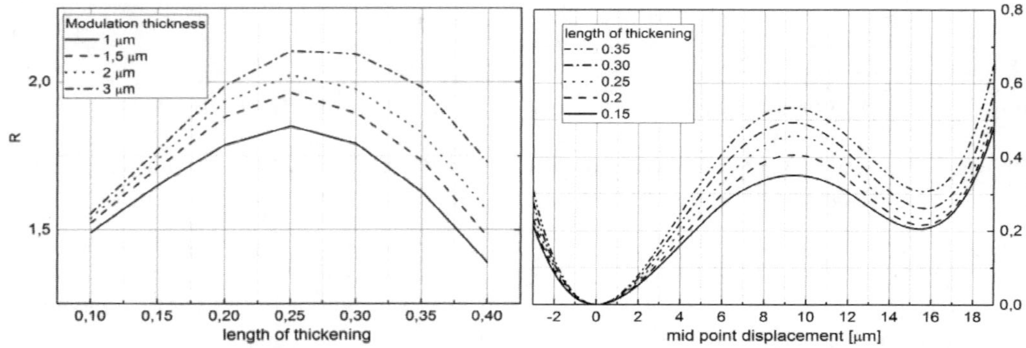

a b

Figure 5. Estimated potential energy of an arch beam with rectangular modulation.
a) 2D slices of the relative depth of the elastic potential well vs length of thickening (k_l); b) 2D sections of the elastic potential energy beam at t_m=1.5 μm at selected length/ displacement values.

The stiffness of the "shoulder" beams of the arch plays an important role as well. However the major contribution of the thickening occurs mostly at its lower end 0.5 μm to 2.5 μm at the initial thickness of beam 2 μm (Figure 6a.).

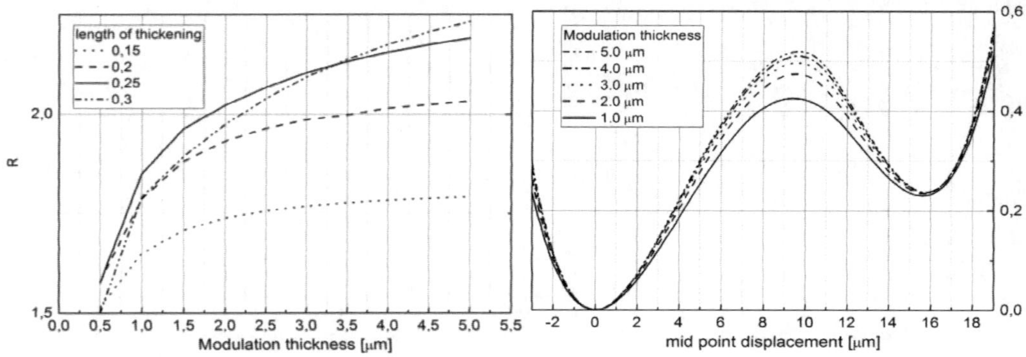

a b

Figure 6. Estimated potential energy of aт arch beam with rectangular modulation.
a) 2D sections of the relative depth of the elastic potential wellvs thickness (t_m); b) 2D cross section of the elastic potential energy beam of k_l=0.25 regarding the selected thickness/ displacement.

On the basis of results represented at Figure 5 and Figure 6 the optimal shape of the arcuate suspension beam is chosen. The optimization criterion is in maximizing the relative depth of the potential well.

5.2 The arch beam with sinusoidal modulation

Another beam rigidity modification of the arch thickness is accomplished using a smooth function as a cosine. The beam thickness modulation can be written as

$$t_m = \frac{t_b}{2}\left(1 - \frac{3}{5}\cos\left(4\pi\frac{x}{l}\right)\right), \tag{1}$$

where t_b is modulation amplitude. As it is shown in Figure 7, the arcuate beam with modulation of the cosine shape demonstrates higher specific energy well depth. The R factor decreases with modulation amplitude. At the same the energy well depth increases (Figure 7).

The figure 8 represents the comparison of the elastic reaction force of the beams with various modulation patterns (uniform, rectangular and cosine). Reaction force of the beams modulated with rectangular insets and the cosine one are similar, whereas the energy, required to switch the beam to the upper energy state is higher for the cosine modulated arch.

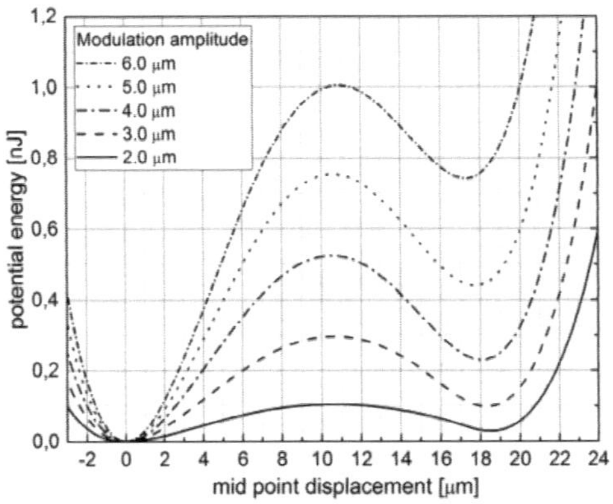

Figure 7. Estimated potential energy of an arch beam with cosine modulation of selected modulation amplitude vs. beam displacement.

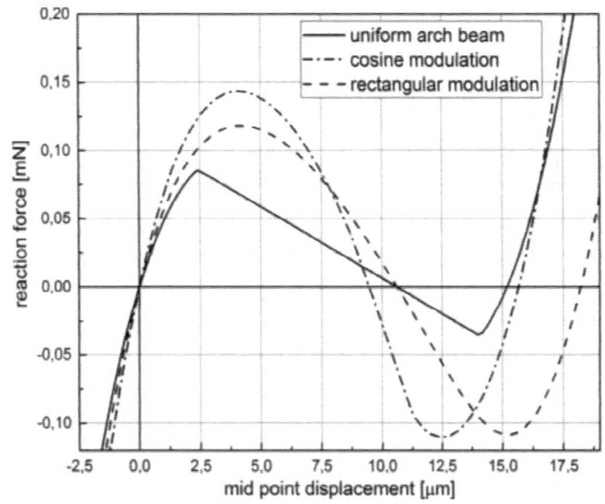

Figure 8. FEA simulation of the beam reaction force: uniform arch beam (t=2um); arch beam with rectangular modulation (k_l=0,25μm; t_m=1,5μm); arch beam with cosine modulation (t_b=3,5μm).

The stability of the snapped-through state of a "thin-thick-thin-thick-thin" shaped arch beam is significantly better compared to a uniform arch beam. This stability can be achieved without compromising the force, required for this transition.

6. Conclusion

This work illustrates that the initial profile of the arched beams and external or internal axial loads are the key factors that determine behavior and nonlinear properties of the arched beams. The modulation of the arch beam profile thickness recesses the depth of the potential energy minimum of the second stable state of the pre-shaped arch beam. Thus increasing the sustainability of the second stable state.

Acknowledgments

This research was carried out within the Federal Program "Research and development in priority areas for the scientific and technological complex of Russia for 2014-2020" (RFMEFI57815X0123) 14.578.21.0123

References

[1] Zhang Y, Wang Y, Li Z, Huang Y, Li D 2007 *J. Microelectromech. Syst.* **16** 684
[2] Qiu J, Lang J H, Slocum A H 2004 *J. Microelectromech. Syst.* **13** 353
[3] Brenner M, Lang J H, Li J, Qiu J, and Slocum A 2002 Proc. 5th International Conf. on Modelling and Simulation of Microsystems, USA.

Effect of thermal stabilization on surface traps in the HEMT-transistors based on AlGaN / SiC

A S Evseenkov[1], S A Tarasov[1], V G Tikhomirov[1], V E Zemlyakov [2]
[1]Saint-Petersburg Electrotechnical University "LETI", Prof. Popova 5, St. Petersburg 197376, Russia
[2]National Research University of Electronic Technology – MIET, Shokina area, 1. Moscow, 124498, Russia

Abstract. HEMT transistors based on the structures of the AlGaN/SiC were created. All the basic characteristics of the transistors were studied and their operating parameters were determined, including power, leakage current, current saturation, etc. The influence of surface traps on the characteristics of the transistors by optical activation was investigated. The method for the passivation of traps based on thermal stabilization was developed.

1. Introduction

Currently heteroepitaxial structures of group III nitrides are actively replacing traditional $A^{III}B^{V}$ materials on the basis of gallium arsenide in the optical [1-3] and microwave [4] electronics. Large value of the electron concentration in the channel in combination with high breakdown fields make it possible to provide a microwave power density in GaN-based field effect transistors 5-10 times larger than in GaAs-based devices. There are several problems in the operation of HEMT-nitride based transistors in microwave mode. One of the major problems is the presence in transistor structures of various defects, significantly reducing both the performance and power of the device. The critical impact on the operation of HEMT-transistors have surface traps that arise at the Si_3N_4-AlGaN interface. Actively search is being made for methods of reducing the effects of such traps on the parameters of microwave devices. The paper proposed a method of surface passivation of traps based on the use of thermal stabilization. Thus, HEMT-transistors based on AlGaN/SiC structures were created, the influence of traps on their performance was studied and ways of dealing with traps, in particular, by thermal stabilization of HEMT-transistors, were developed.

2. The samples

For the experiment, several AlGaN / SiC plates were made, on which HEMT transistors with a gate width of 300 μm and a length of 0.5 μm are placed. SiC was used as a substrate, on which the epitaxial transistor structures [5], shown in Fig. 1, were grown. The main functional layers are: barrier layer $Al_{0.23}Ga_{0.77}N$, sub-buff layer of AlN, GaN channel layer and the top layer $Al_{0.35}Ga_{0.65}N$. The plate was divided into separate transistors from which for further experiments control samples were selected to the following criteria: the saturation current was from 150 to 170 mA and a cut-off voltage was not less than 5 V, the leakage current at 60 V was no more than 300 microamps.

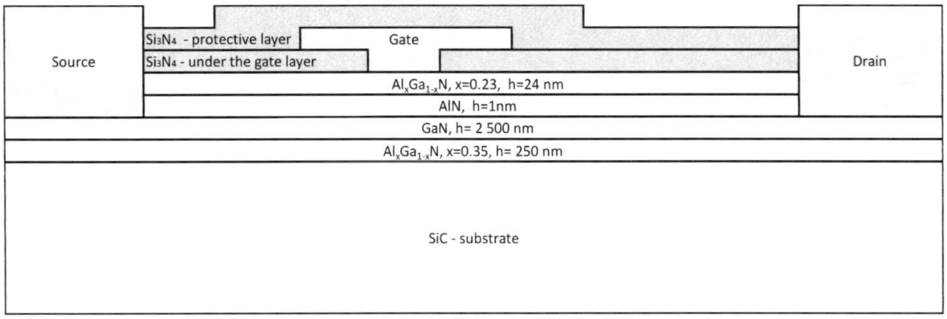

Figure 1. Structures of created HEMT-transistor

In the work, special attention was paid to the creation of dielectric layers in the structure and investigation of the influence of their structure on the parameters of the created HEMT-transistors. The dielectric was Si_3N_4. In the structure, two dielectric layers were created - an under the gate layer and a protective layer (Fig. 1). The layers were grown by the method of plasma-chemical deposition from the gas phase by the decomposition of silane and nitrogen. This method is relatively simple and low-cost, that determines its widespread use in the technology of micro- and nanoelectronics. At the same time, the composition and structure of the silicon nitride layers grown using this method vary greatly with the change in process parameters. A frequent problem is the deviation of the composition of such films from stoichiometry. In addition, in such dielectric layers in most cases, a significant amount of hydrogen and oxygen atoms remains. This leads to the appearance in the structure of traps for charge carriers and a significant decrease in the transistor speed. In this paper, the properties of such traps were considered and a method for effectively reducing their influence on device parameters was proposed.

3. Experiment

Created HEMT-transistors underwent comprehensive testing, including a study of the basic characteristics and determination of operating parameters. In particular, their power parameters, delay times, saturation currents, etc. were determined. Special attention was paid to detecting and investigating the properties of traps in nanostructures, as well as determining their influence on the performance characteristics of devices. The detection and diagnostic technique was based on the activation of traps by exposing them to radiation with a photon energy of 1 to 4.4 eV. During the diagnostic process, the HEMT-transistors operated in a pulsed mode at a frequency of 3.3 GHz. The characteristic photon energies corresponding to the appearance of the trap activation effect for structures of various types were determined, and the relative power increase (up to three times) and the relaxation times of surface traps, in some cases exceeding 20 s, were calculated. As a method of passivation of traps, thermal stabilization was used at a temperature of 300 degrees Celsius.

In addition to the traps, thermal effects on HEMT transistors can also have an effect on the dielectric, in this case Si_3N_4. Properties of the created dielectric layers of silicon nitride before and after the procedure of thermal stabilization were studied by optical spectroscopy, transmission, refractometric and other methods of analysis. A modification of the IR absorption spectra [6] is observed with a change in the temperature of the processes and the duration of the thermal impacts. The change in the refractive index from 1.875 to 1.84 and a small decrease in the thickness of the dielectric, on the order of 1.5 nm, was also established. The analysis of the obtained array of experimental data made it possible to single out the main changes in the structure of the Si_3N_4 layers, which occur when they are modified by the thermal stabilization. The changes in the ratio of the chemical bonds during stabilization (Figure 2) are investigated. An increase in the process temperature leads to the removal of a significant amount of hydrogen from the N-H bonds in the dielectric layer.

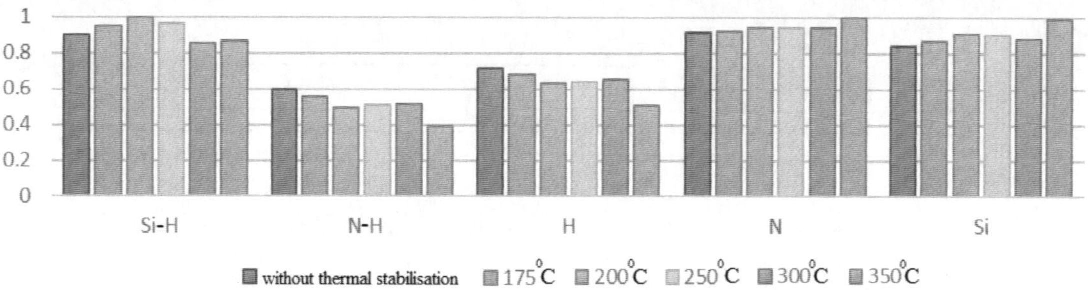

Figure 2. Distribution of IR responses from bonds at different process temperatures

The impact of thermal stabilization on surface traps were investigated in detail. For this purpose, the microwave power of the transistors was measured when they were irradiated with photons with different energies. The photon flux for all energies was kept constant at 0.2 mW.

On the basis of the obtained data, special UV LEDs [7] and a measuring bench attachment on their basis were created [8-10], which makes it possible to measure the effect of external radiation on the power characteristics of a HEMT transistor. Spectral characteristics of UV LED are shown in Fig. 3.

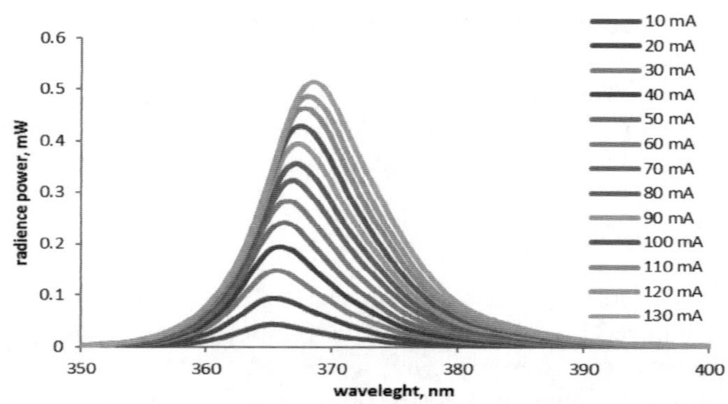

Figure 3. Spectral characteristics of the irradiator based on the created UV LEDs at different values of the transmitted current.

Transistors were exposed before and after thermal stabilization. Before thermal stabilization, the microwave power of the transistors was about 400 mW. After stabilization, the power of the transistors increased significantly and could reach 1200 mW. Figure 4 shows the dependence of the additional power on the energy of the incident photons. It can be seen that when irradiated, the power of the nonstabilized transistor increases 2.5 – 3 times, while in the case of stabilized transistors the power does not change much.

It is important that this effect is observed in a wide range of the energies of the incident photons (Figure 4), including for energies smaller than the band gap of GaN, the most narrow-gap semiconductor layer. This confirms the assumption that the reduction in microwave power in nonstabilized transistors is primarily due not to the properties of semiconductor layers, but to the effect of surface traps.

Another important fact is that when illuminated, the power in nonstabilized transistors increased exactly to the values reached in the microwave mode for stabilized structures. This suggests that the effect of the traps in the structure was almost completely eliminated as a result of stabilization.

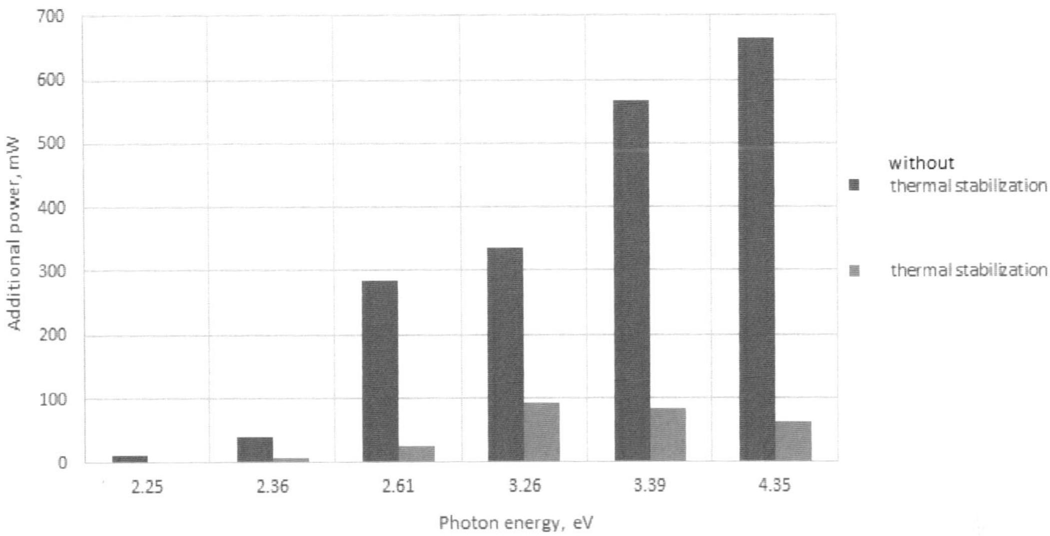

Figure 4. The dependence of the additional power of the irradiated HEMT transistor before and after thermal stabilization on the energy of the incident photons

Thus, a method of thermal stabilization was developed, which by eliminating the effect of surface traps [11] has made it possible to significantly improve the most important parameters of the HEMT-transistor, including their microwave power and speed. The method is based on a single or multi-stage impact on structures of elevated temperatures (from 200 to 350 C^0). The total duration of impact ranged from a few minutes to 5 hours. Detailed studies of the effect of changes in the time and temperature parameters of stabilization on the characteristics of HEMT transistors have been carried out, and optimal treatment modes for structures have been determined, which make it possible to achieve the greatest improvement in their operating parameters.

One of the most important results achieved is the stabilization of the operating parameters of structures located in different parts of the plate. It can be seen that after thermal stabilization, the spread of the characteristics of different transistors was minimized. (Figure 5). This effect was observed not only for microwave power, but also for saturation currents and other electrical characteristics of transistors.

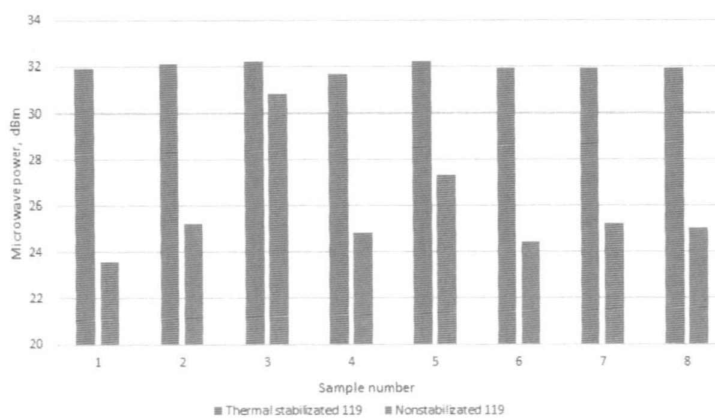

Figure 5. Transistors power before and after thermal stabilization

4. Conclusion

Thus, during the work, HEMT-transistors based on AlGaN/SiC structures were created. The gate width was 300 μm, the length was 0.5 μm. All the main characteristics of transistors were investigated and their operating parameters were determined, including power, cut-off voltage, leakage current, saturation currents, etc. The effect of surface traps on the characteristics of transistors by the optical activation method was investigated, a laboratory attachment for its implementation was created and their characteristic parameters were determined. The technique of passivation of traps based on the use of thermal stabilization has been developed. As a result, a significant improvement in the parameters of the structures was achieved, in particular, the power of transistors increased on average from 26 to 32 dBm, stabilization of the saturation currents was achieved. The stabilization of the operating parameters of structures located in different parts of the plate was achieved.

Acknowledgments

The work was supported by the Ministry of Education and Science of the Russian Federation in the framework of the project of the state task in the field of scientific activity, project number 16.1750.2017/PCh.

References

[1] Evseenkov A S, Lamkin I A, Tarasov S A, Solomonov A V 2015 Journal of Physics: Conference Series 643 1 012033
[2] Evseenkov A S, Tarasov S A, Lamkin I A, Solomonov AV, Kurin S Y 2015 Proceedings Of The 2015 IEEE North West Russia Section Young Researchers In Electrical And Electronic Engineering Conference, Elconrusnw 27-29
[3] Zubkov V I, Evseenkov A S, Orlova T A, Zubkova A V 2015 Russian Physics Journal 2015 58 8 1172-1180
[4] Faraclas E, Anwar A AlGaN/GaN 2006 Solid-StateElectron 2006 50(6):1051–6
[5] Tikhomirov V, Zemlyakov V, Volkov V, ParnesYa, Vyuginov V, Lundin W, Sakharov A, Zavarin E, Tsatsulnikov A, Cherkashin N, Mizerov M, and Ustinov V 2016 Semiconductors 50 2 244–248
[6] Ki Seok Kim, Nishant Sirse et al, 2016 Journal of Physics D: Applied Physics 49 39
[7] Kurin S, Antipov A, Barash I, Roenkov A, Makarov Y, Usikov A, Helava H, Solomonov A, Tarasov S, Evseenkov A, Lamkin I 2015 Physica Status Solidi (C) Current Topics in Solid State Physics 2015 12 4-5 369-371
[8] Kurin S Y, Makarov Y N, Usikov A S, Papchenko B P, Helava H, Evseenkov A S, Tarasov S A, Solomonov A V 2016 Journal of Physics: Conference Series 741 1 012107
[9] Evseenkov A S, Tarasov S A, Solomonov A V, Kurin S Yu, Makarov Yu N, Usikov A S, Papchenko B P, Helava H 2015 Journal of Physics: Conference Series 661 1 012038
[10] Evseenkov A S, Tarasov S A, Solomonov AV, Altimime S M, Obukhova A S 2016 Journal of Physics: Conference Series 741 1 012209
[11] Tirado J M, Sánchez-Rojas J L, and Izpura J I 2007 IEEE Transactions on Electron Devices 54 3

IOP Conf. Series: Journal of Physics: Conf. Series **917** (2017) 082015 doi:10.1088/1742-6596/917/8/082015

Suppression of the self-heating effect in GaN HEMT by few-layer graphene heat spreading elements

V S Volcheck and V R Stempitsky

Department of Micro- and Nanoelectronics, Belarusian State University of Informatics and Radioelectronics, 6 Brovki, Minsk 220013, Belarus

Abstract. Self-heating has an adverse effect on characteristics of gallium nitride (GaN) high electron mobility transistors (HEMTs). Various solutions to the problem have been proposed, however, a temperature rise due to dissipated electrical power still hinders the production of high power and high speed GaN devices. In this paper, thermal management of GaN HEMT via few-layer graphene (FLG) heat spreading elements is investigated. It is shown that integration of the FLG elements on top of the device structure considerably reduces the maximum temperature and improves the DC and small signal AC performance.

1. Introduction

Gallium nitride (GaN) high electron mobility transistors (HEMTs) have been contemplated to be very promising devices for applications in power electronics, microwave systems, and sensing technologies. High electron saturation velocity and breakdown voltage of GaN allow HEMTs to sustain large current densities, leading to extensive and non-uniformly distributed Joule heating, resulting in deterioration of the drain current, radio frequency gain and output power, as well as an increase in the gate leakage current. Numerous approaches to decrease the device temperature have been attempted, including atomic bonding or direct growing of the transistor structure on a diamond substrate, a conventional flip-chip technique to attach a heat sink, integration of heat spreading elements formed out of materials with high thermal conductivity, such as diamond and graphene/graphite combination. Nevertheless, in spite of these efforts, the self-heating effect still persists.

In this communication, we demonstrate that local thermal management of GaN HEMT can be significantly improved by means of additional top-surface heat spreading elements composed of few-layer graphene (FLG). The thermal conductivity of suspended exfoliated FLG is estimated in the range of 13 to 28 W/(cm·K) [1] and order-of-magnitude higher than that of GaN. Heat conduction of FLG is dominated by phonons and its quantitative properties are preserved even if layers of graphene stacked up have a total thickness of only several nanometers. Contrary to FLG, heat conduction in metallic films is limited by electrons and falls sharply with the decreasing film thickness. Advance in single- and few-layer graphene synthesis and patterning on arbitrary surfaces suggests that the proposed heat spreading elements can become technologically and economically expedient in the near future.

In comparison with single-layer graphene, the thermal conductivity of FLG is less prone to degradation caused by extrinsic effects, such as defects and disorder at the junctions. In addition, heat conduction of FLG is characterized by strong anisotropy and, as a result, the FLG elements take the heat away rather than dissipate it at the interface. FLG can also sustain large heat flow through its cross-section while retaining its mechanical flexibility.

2. Simulation details

To estimate the attainable improvement in thermal management of GaN HEMT, we performed a numerical simulation in Wachutka's thermodynamically rigorous model of lattice heating [2] with the use of Silvaco software.

2.1. Structure

A two-dimensional representation of the initial (without any heat spreading elements) GaN HEMT structure is schematically illustrated in figure 1.

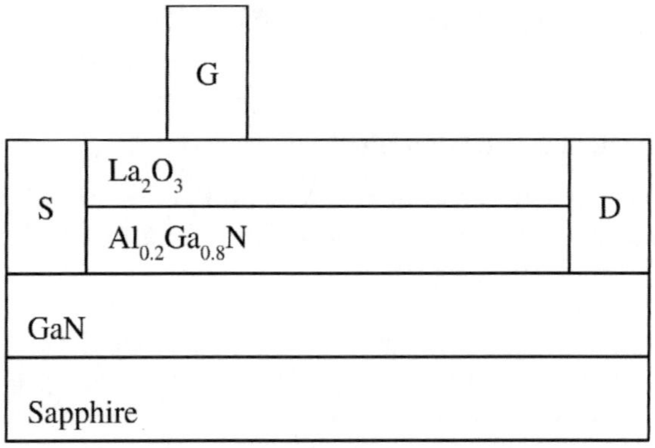

Figure 1. Initial GaN HEMT structure.

The device structure consists of a 1 μm thick GaN buffer layer, a 20 nm thick aluminum gallium nitride ($Al_{0.2}Ga_{0.8}N$) barrier layer, and a 6 nm thick lanthanum oxide (La_2O_3) as the gate dielectric. The thickness of the sapphire substrate simulation domain is 18 μm. The source-to-gate and gate-to-drain distances equal 1 and 4 μm, respectively, and the gate length is 1 μm.

The spontaneous and piezoelectric components of polarization in $Al_{0.2}Ga_{0.8}N$ are taken into account. The relatively thick GaN layer is assumed to be unstrained, and so polarization in it has only the spontaneous component. The two-dimensional electron gas density is adjusted to $6.4 \cdot 10^{12}$ cm^{-2}.

A La_2O_3/AlGaN interface is reported to accumulate a negative fixed charge, with its quantity depending heavily on annealing parameters [3]. To imitate this effect, a negative fixed charge with a density of $3 \cdot 10^{12}$ cm^{-2} is placed on the gate dielectric/barrier layer interface. Since the position of the negative fixed charge is 20 nm away from the two-dimensional electron gas, the contribution of remote Coulomb scattering is neglected.

2.2. Wachutka's model of lattice heating

Simulation of the self-heating effect in Wachutka's model implies addition of a lattice heat flow equation to the set of basic semiconductor device equations, namely Poisson's equation, the continuity and transport equations. The lattice heat flow equation has a form

$$C \frac{\partial T}{\partial t} = \nabla(\kappa \nabla T) + H, \tag{1}$$

where C is the volumetric heat capacity (J/(cm^3·K)), T is the temperature (K), κ is the thermal conductivity (W/(cm·K)), and H is the heat generation (W/cm^3).

Since thermal conductivity is generally temperature dependent, it is calculated in accordance with a model

$$\kappa = \kappa_{300} \left(\frac{T}{300} \right)^{\alpha}, \tag{2}$$

where κ_{300} is the thermal conductivity at the temperature of 300 K and α is the temperature dependence coefficient.

The κ_{300} and α parameters for the materials used in the simulation are given in table 1.

Table 1. Thermal conductivity model parameters.

Parameter	Material				
	La_2O_3	Sapphire	GaN	AlN	FLG
κ (W/(cm·K))	0.058	0.446	1.3	2.85	20
α	-0.892	-1.092	-0.28	-1.64	0

The values for La_2O_3 are derived from non-linear (power) regression equations obtained on a basis of experimental data given in [4]. The thermal conductivity of $Al_{0.2}Ga_{0.8}N$ is interpolated as a linear function of composition fraction.

The thermal boundary condition is specified at the bottom surface of the substrate as a perfect heat sink to the temperature of 300 K. The adiabatic boundary conditions are set up at the top surface and the sidewalls of the structure.

2.3. Mobility models

The drift-diffusion transport model is used in the calculations with the Albrecht low field and the Farahmand high field mobility models for electrons in the device channel.

The Albrecht model [5], which treats the mobility in a low field as a function of doping and temperature, is described for GaN as follows:

$$\frac{1}{\mu_A}=\frac{2.61\cdot10^{-4}N}{10^{17}}\left(\frac{T}{300}\right)^{-3/2}\ln\left[1\div3\left(\frac{T}{300}\right)^{2}\left(\frac{N}{10^{17}}\right)^{-2/3}\right]+2.9\cdot10^{-4}\left(\frac{T}{300}\right)^{3/2}+\frac{0.017}{\exp(1065/T)-1}, \qquad (3)$$

where μ_A is the electron low field mobility (cm²/(V·s)) and N is the doping concentration (cm^{-3}).

The Farahmand high field mobility model [6] is based on a fit to Monte Carlo data and described for GaN as follows:

$$\mu_F=\frac{\mu_A+1.9\cdot10^{7}\dfrac{E^{6.2}}{7.6\cdot10^{16}}}{1+6.2\left(\dfrac{E}{220.9}\right)^{0.8}+\left(\dfrac{E}{220.9}\right)^{7.2}}, \qquad (4)$$

where μ_F is the electron high field mobility and E is the electric field intensity (V/cm).

3. Results

When biased at high drain voltages, GaN HEMT is remarkable to have a hot spot formed at the drain-side gate edge. This region of the structure reveals decreased electron mobility, thus degrading device characteristics.

For the purpose of suppression of the self-heating effect, two FLG heat spreading elements are integrated on top of the initial GaN HEMT structure on both sides of the gate to specifically target the hot spot. An accurate placement of FLG in predetermined locations can be realized, for example, by a polymethyl methacrylate-assisted method. The thickness of the heat spreaders is chosen to be 4 nm, corresponding roughly to four-layer graphene. Since the heat spreading elements are supposed to connect directly with heat sinks outside of the device, two additional thermal boundary conditions are specified at their extreme sides.

A two-dimensional representation of the modified (with the FLG heat spreading elements) GaN HEMT structure is schematically illustrated in figure 2.

IOP Conf. Series: Journal of Physics: Conf. Series **917** (2017) 082015 doi:10.1088/1742-6596/917/8/082015

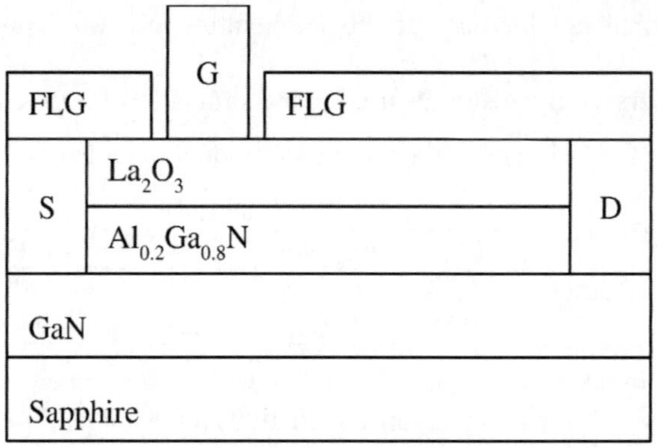

Figure 2. Modified GaN HEMT structure.

In figure 3, the temperature distribution profiles along the $Al_{0.2}GaN_{0.8}$/GaN interface of the HEMT without and with the heat spreaders at different gate voltages V_g are shown. L is the distance from the source contact. As follows from figure 3, the peak temperature of the basic device structure rises from 439 K to 517 K when the gate voltage is raised from 0 V to 2 V at the drain voltage V_d of 20 V. The FLG heat spreaders allow to reduce the peak temperature to 372 K and 402 K at V_g = 0 V and 2 V, respectively.

In figure 4, the corresponding mobility distribution profiles are shown. It is clear that the higher the temperature a point of the structure has, the lower the mobility of the respective electrons. At the hot spot, the mobility drops to a value as low as 21 cm^2/(V·s) at V_g = 0 V and 40 cm^2/(V·s) at V_g = 2 V. When the FLG heat spreaders are integrated, the hot spot electron mobility equals 24 cm^2/(V·s) (an increase of 14.29 %) at V_g = 0 V and 55 cm^2/(V·s) at V_g = 2 V (an increase of 37.5 %).

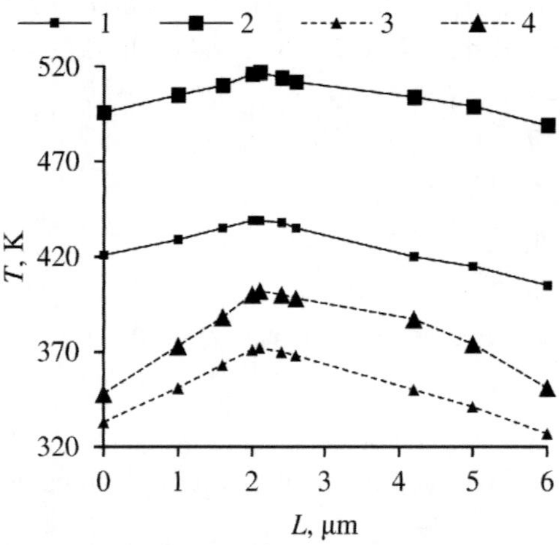

Figure 3. Temperature distribution profiles along the $Al_{0.2}Ga_{0.8}N$/GaN interface without (solid lines) and with (dashed lines) the FGL heat spreaders at V_g = 0 V (lines 1 and 3) and 2 V (lines 2 and 4), V_d = 20 V.

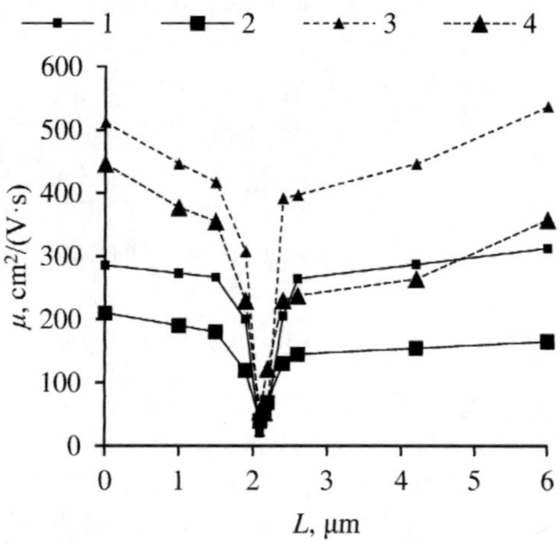

Figure 4. Mobility distribution profiles along the $Al_{0.2}Ga_{0.8}N$/GaN interface without (solid lines) and with (dashed lines) the FGL heat spreaders at V_g = 0 V (lines 1 and 3) and 2 V (lines 2 and 4), V_d = 20 V.

The transfer and transconductance characteristics of the simulated GaN HEMT without and with the FLG heat spreading elements are presented in figures 5 and 6, respectively. The initial transistor exhibits the maximum transconductance of 0.147 S/mm at V_d = 5 V, while the modified one has the maximum value of this parameter equal to 0.158 S/mm (an improvement of 7.48 %).

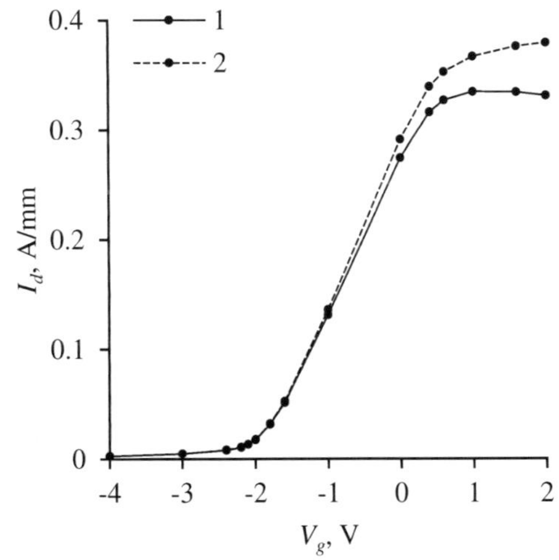

Figure 5. Transfer characteristics of the GaN HEMT without (line 1) and with (line 2) the FLG heat spreaders at $V_d = 5$ V.

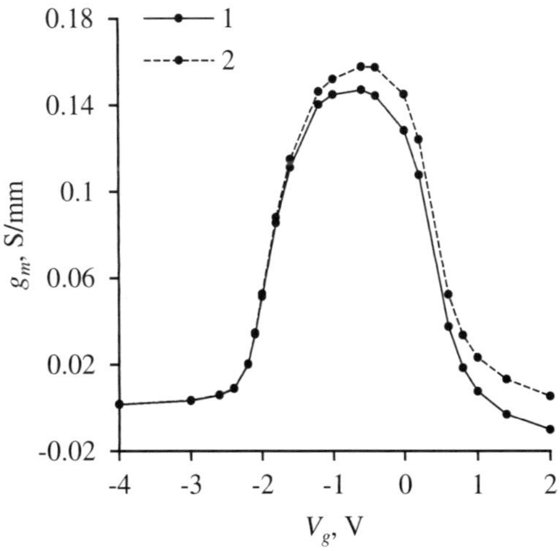

Figure 6. Transconductance characteristics of the GaN HEMT without (line 1) and with (line 2) the FLG heat spreaders at $V_d = 5$ V.

In figure 7, the output current-voltage characteristics for both variants of the GaN HEMT are shown.

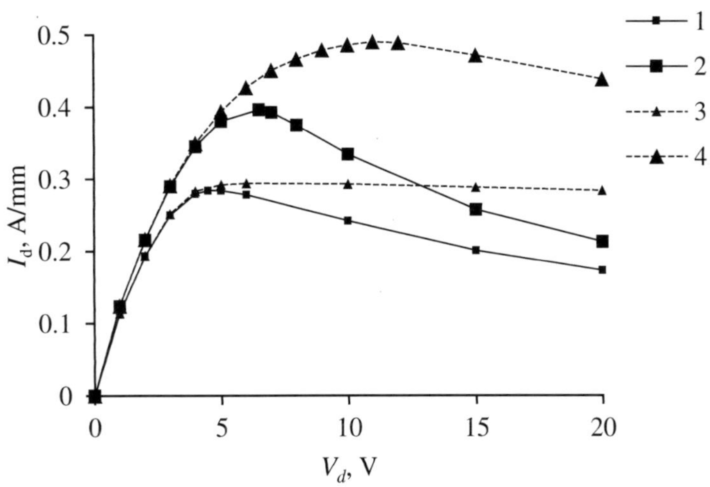

Figure 7. The output characteristics of the GaN HEMT without (solid lines) and with (dashed lines) the FLG heat spreaders at V_g = 0 V (lines 1 and 3) and 2 V (lines 2 and 4).

As follows from figure 7, the self-heating effect brings about negative differential conductance through the decreased mobility. When the basic device is biased at V_g = 0 V and V_d = 20 V, the drain current equals 0.173 A/mm, a degradation of 39.08 % from its peak value of 0.284 A/mm. At V_g = 2 V and V_d = 20 V, the drain current is 0.213 A/mm, a decrease of 46.21 % from the maximum value of 0.396 mA/mm. The GaN HEMT with the FLG heat spreading elements demonstrates much improved output characteristics. When it is biased at V_g = 0 V and V_d = 20 V, the negative differential conductance is almost eliminated as the drain current reduction equals to only 3.40 %. At V_g = 2 V and V_d = 20 V, the drain current is 0.438 A/mm, a decrease of 10.61 % from its peak value of 0.49 A/mm.

After solving for the DC condition, a small signal sinusoidal AC analysis is performed. The GaN HEMT is biased at V_g = 0 V and V_d = 20 V. The AC performance is evaluated via the cut-off and maximum oscillation frequencies. The frequency dependencies of the AC current gain (*h21*) and unilateral power gain (*U*), derived from calculated S-parameters, are presented in figures 8 and 9. The cut-off frequency of the transistor equals 6.4 GHz and is increased by 51.56 % to the level of 9.7 GHz in case of the presence of the FLG heat spreading elements. The maximum oscillation frequency equals 16.1 GHz and is increased by 56.52% to the value of 25.2 GHz.

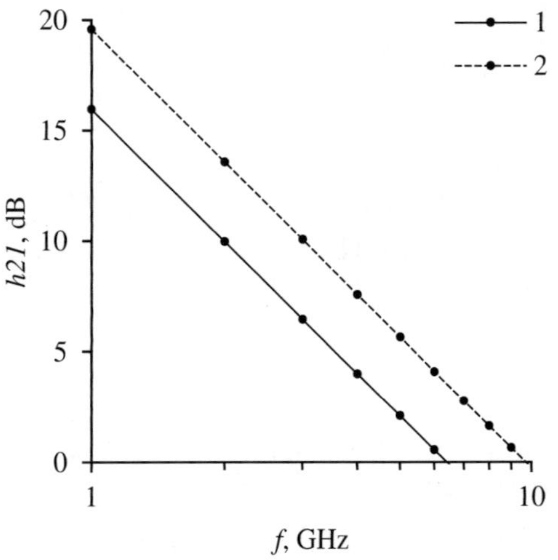

Figure 8. Frequency dependence for the AC current gain of the GaN HEMT without (line 1) and with (line 2) the FLG heat spreaders.

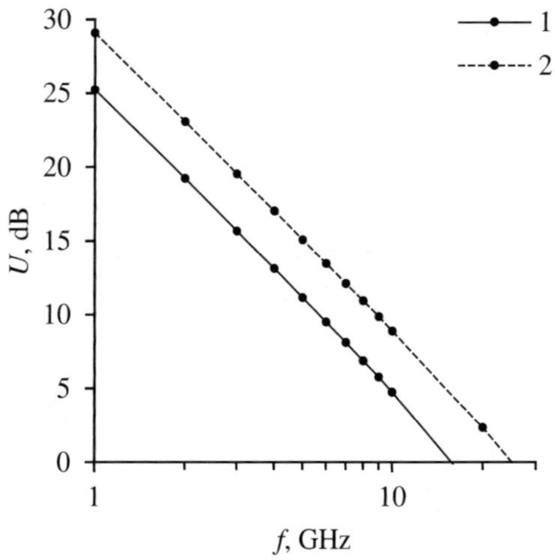

Figure 9. Frequency dependence for the AC unilateral power gain of the GaN HEMT without (line 1) and with (line 2) the FLG heat spreaders.

4. Conclusions

FLG heat spreading elements allow to suppress the self-heating effect in GaN HEMT, thus improving device DC and small signal AC performance. Progress in single- and few-layer graphene synthesis and patterning on arbitrary surfaces suggests that the proposed in this paper heat spreaders can become technologically and economically expedient in the near future.

Acknowledgments

This work was supported by the grants 1.8.05 of Belarusian National Scientific Research Program "Informatics, Space, Safety" and 3.2.01 of Belarusian National Scientific Research Program "Photonics, Opto- and Microelectronics".

References

[1] Balandin A A 2011 Thermal properties of graphene and nanostructured carbon materials *Nat. Mater.* **10** 569–81

[2] Wachutka G K 1990 Rigorous thermodynamic treatment of heat generation in semiconductor device modeling *IEEE Trans. Comput.-Aided Des. Integr. Circuits Syst.* **9** 1141–9

[3] Chen J, Kawanago T, Wakabayashi H, Tsutsui K, Iwai H, Nohata D, Nohira H and Kakushima K 2016 La_2O_3 gate dielectrics for AlGaN/GaN HEMT *Microelectronics Reliability* **60** 16–9

[4] Fornarini L, Conde J C, Alvani C, Olevano D and Chiussi S 2008 Experimental determination of La_2O_3 thermal conductivity and its application to the thermal analysis of a-Ge/La_2O_3/c-Si laser annealing *Thin Solid Films* **516** 7400–5

[5] Albrecht J D, Wang R P, Ruden P P, Farahmand M and Brennan K F 1998 Electron transport characteristics of GaN for high temperature device modeling *J. Appl. Phys.* **83** 4777–81

[6] Farahmand M, Garetto C, Bellotti E, Brennan K F, Goano M, Ghillino E, Ghione G, Albrecht J D, Ruden P P 2001 Monte Carlo simulation of electron transport in the III-nitride wurtzite phase materials system: binaries and ternaries *IEEE Trans. Electron Dev.* **48** 535–42

IOP Conf. Series: Journal of Physics: Conf. Series **917** (2017) 082016 doi:10.1088/1742-6596/917/8/082016

AlGaAs/GaAs HBTs with C-doped base and undoped emitter-base spacer layer

M A Bobrov[1,2], N A Maleev[1], A G Kuzmenkov[2], S A Blokhin[1], A P Vasil'ev[2], V I Egorkin[3], V E Zemlyakov[3], V M Ustinov[2]

[1]PSH Laboratory, Ioffe Institute, St. Petersburg, 194021 Russia
[2]Submicron Heterostructures for Microelectronics Research and Engineering Center of the RAS, St. Petersburg, 194021 Russia
[3]Department of Electronics and Computer Technologies, National Research University of Electronic Technology, Moscow, 124498 Russia

Abstract Effect of undoped GaAs spacer layer between AlGaAs emitter layer and C-doped GaAs base layer on DC characteristics of AlGaAs/GaAs HBTs with ledge structure was investigated. Devices with spacer layer thickness (~3-5 nm) and optimal ledge geometry demonstrate stable high current gain (β_{max} = 120-140) combined with collector-emitter breakdown voltage of 23-25V.

1. Introduction

Heterojunction bipolar transistors (HBTs) are widely used as microwave power devices due to their intrinsic high-power density, linearity and efficiency [1]. Initially beryllium (Be) was used as an p-dopant for GaAs base layers in AlGaAs/GaAs HBT epitaxial structures. However it is difficult to provide high p-doping level due to Be surface segregation and diffusion in GaAs and AlGaAs layers grown by molecular beam epitaxy (MBE) [2]. To prevent negative effect of uncontrollable Be redistributing the thin undoped GaAs spacer layer may be introduced between n-doped AlGaAs emitter layer and p-doped GaAs base layer but it is not guarantee long-term reliability at high current density and high temperatures [3]. Therefore, currently carbon (C) is widely used in MBE for p-doping due to its low diffusion coefficient and high solubility in GaAs layers [4].

It is well known that high surface recombination velocity for highly doped p-type GaAs base layers leads to degradation of the HBT current gain. The insertion of a thin depleted layer with a wide band gap and a low surface recombination velocity over the extrinsic base region around the emitter periphery (so called ledge) is an effective approach for recombination current reduction. Using a depleted ledge over the extrinsic base surface improve gain stability [5], device reliability [6] and significantly reduce l/f noise of AlGaAs/GaAs HBT's [7].

Proper ledge design is critical since thick or heavily doped ledge as well as too thin or lightly doped one may result in degradation of initial DC characteristics and poor device reliability. In C-doped GaAs HBTs n-doped emitter and p-doped base usually are in direct contact without any spacer layer between them [8]. Although the insertion of undoped GaAs layer between emitter and base can improve the ledge passivation effectiveness, it may also reduce emitter efficiency simultaneously. In this paper, we investigate effect of undoped GaAs spacer thickness on DC characteristics of C-doped AlGaAs/GaAs HBTs.

2. Results and discussion

The layer sequences of HBT epitaxial structures under investigation are given in table 1. Thin undoped GaAs spacer layer separates n-doped AlGaAs emitter layer with graded interfaces from C-doped p-GaAs base layer. Cross-section of HBT emitter-base region is schematically shown in Figure 1. Ledge structure includes bottom undoped GaAs part and top n-doped AlGaAs part. The ledge length can be roughly estimated as distance between emitter mesa edge and base contact.

Table 1. Epitaxial layers structures parameters of investigated HBTs.

Layer	Material	Thickness (nm)	Doping (cm^{-3}), dopant
Cap	n+ InGaAs/GaAs	10	>$5x10^{18}$, Si
	gradient n AlGaAs	30	gradient, Si
Emitter	n $Al_{0.25}Ga_{0.75}As$	80	$5x10^{17}$, Si
	gradient n AlGaAs	20	$5x10^{17}$, Si
Spacer	u.d. GaAs	0-10	–
Base	p GaAs	80	$2x10^{19}$, C
Collector	n GaAs	700	$2x10^{16}$, Si
Sub-collector	n+ GaAs	600	$3x10^{18}$, Si
Buffer		200	–

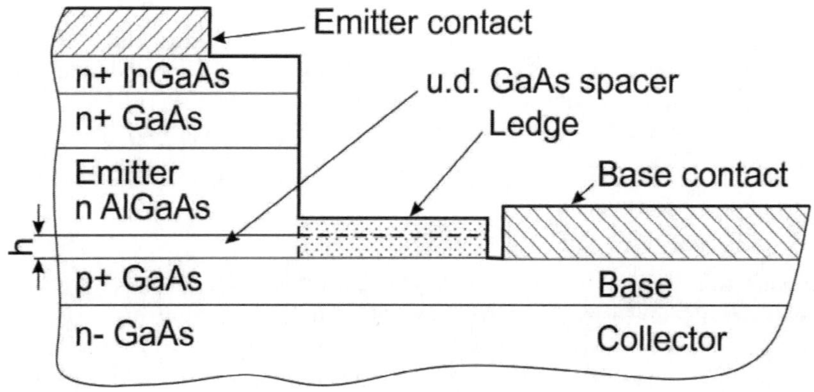

Figure 1. Schematic cross-section of HBT around the emitter ledge.

The numerical modeling of HBTs with the emitter mesa width of 4 μm and the ledge length of 3 μm was carried out. Figure 2 depicts the calculated dependences of the current gain β on the total ledge thickness for different thickness of undoped GaAs spacer layer. On the one side, for the fixed thickness of undoped GaAs spacer layer, the increase in the total ledge thickness leads to the reduction of the surface recombination at the ledge (due to higher Al composition at the ledge surface) and, as a result, the rise of the current gain. When the total ledge thickness becomes greater than the depth of the surface depletion region, the electron concentration in the ledge increases (see Figure 3), which leads to the increase in the effective emitter width and increasing of the current gain. However, increased effective emitter width leads to significant growth of the emitter-base capacity and RF characteristics degradation. Therefore the total ledge thickness (including bottom undoped GaAs and top n-doped AlGaAs parts) is optimal when the ledge area is nearly full depleted and the surface recombination is damped. In our case optimal total ledge thickness is roughly about 5-10 nm (depending on the undoped GaAs spacer thickness).

On the other side, for the fixed ledge thickness, the increase in the thickness of undoped GaAs spacer layer leads to the rise of the surface recombination at the ledge (due to lower Al composition at the ledge surface) and, as a result, the reduction of the current gain (see Figure 2). However the high-power HBT's undergoes significant heating and thermal cycling during operation. Therefore the decrease in the chemical activity of the surface layer due to the reduction of the Al composition (<15%) should lead to the improvement of the device reliability and life-time.

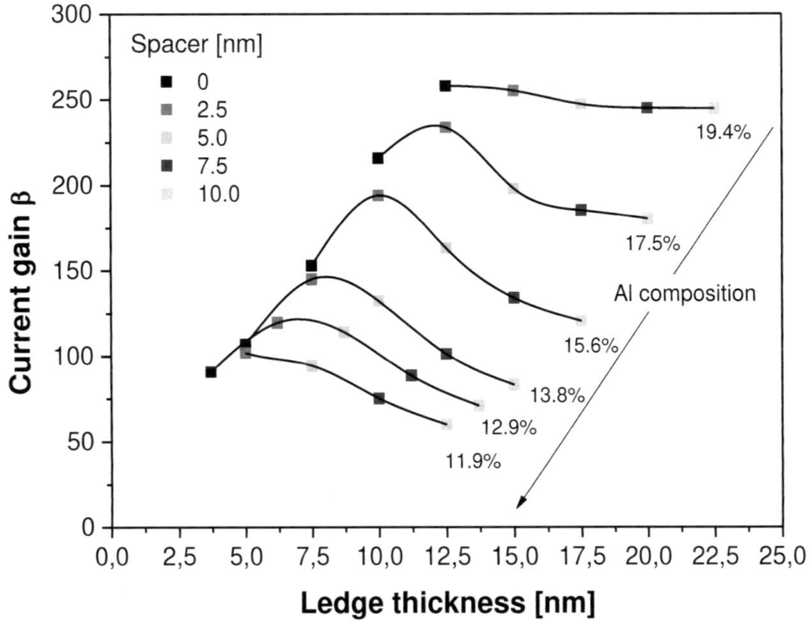

Figure 2. Calculated current gain as a function of total ledge thickness for different spacer thickness. The solid lines correspond to the fixed Al composition on the ledge surface.

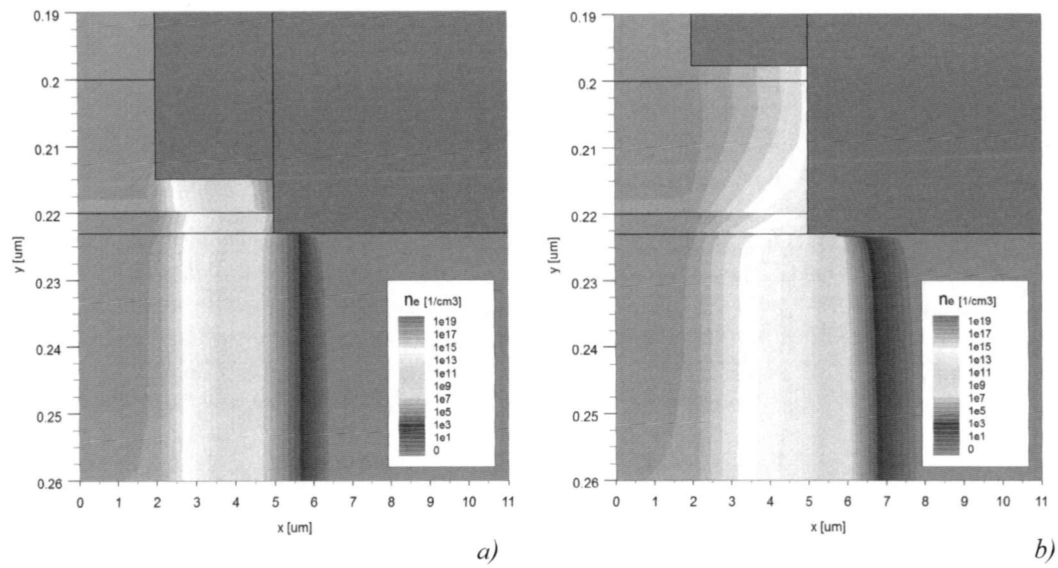

Figure 3. Calculated 2D distributions of electron concentration at $I_C = 3.6$ mA/μm for HBTs based on the epitaxial structure with 3-nm-thick spacer layer and 8 nm-thick (a) and 25 nm-thick (b) ledge thickness.

According with modeling results the maximum current gain $\beta_{max} \approx 150$ as well as better surface passivation and stability can be achieved when the thickness of undoped GaAs spacer layer is about 3-5 nm. Experimental device fabrication included the following basic steps: emitter contact deposition, emitter mesa etching down to the ledge surface, ledge mesa etching down to the base layer and base contact deposition, base mesa etching down to the sub-collector layer and collector contact deposition, isolation mesa etching down to substrate, dielectric passivation, contact pad deposition, emitter air bridge formation. To obtain the optimal ledge thickness the etching depth of the emitter mesa was controlled by measurement of leakage current between neighboring emitter contacts.

The effect of the surface recombination in the ledge can be neglected in HBTs with large emitter size. Test HBTs with emitter size ~ 50x100 µm based on the epitaxial structures with undoped GaAs spacer thickness of 10, 5 and 0 nm demonstrate the maximum current gain β_{max} of about 200, 240 and 315, respectively. It is clearly indicated that the thicker undoped GaAs spacer layer results in the emitter efficiency degradation.

In case of HBTs based on the epitaxial structure without any undoped GaAs spacer layer, the devices with the emitter mesa width of 4 µm and the ledge length of 3 µm demonstrate the current gain β_{max} in the range of 90-150. According to the numerical modeling results, the obtained high dispersion of DC characteristics can be attributed to the strong impact of the ledge thickness on the current gain. Moreover, the lower experimental value of the maximum current gain as compared to the calculated data can be attributed with the partial oxidation of the ledge surface (typical oxidation depth is 2-4 nm depending on the Al composition) which reduces the effective ledge thickness. At the same time, HBTs based on the epitaxial structure with 5 nm-thick undoped GaAs spacer layer and the same device geometry demonstrate the more uniform distribution of the maximum current gain from chip to chip (β_{max} = 120-140). The collector-emitter breakdown voltage of 23-25V was obtained for all investigated HBTs. Such combination of high current gain and reasonable breakdown voltage is suitable for most RF applications.

Acknowledgments
This work was supported by the Ministry of education and science of the Russian Federation (grant agreement № 14.578.21.0212 from September 29, 2016, unique ID RFMEFI57816X0212).

References

[1] Reisch M 2003 *High-Frequency Bipolar Transistors*, (Berlin: Springer-Verlag)

[2] Ilegems M 1977 Beryllium doping and diffusion in molecular-beam epitaxy of GaAs and AlxGa1−xAs *J. Appl.Phys.* **48** 1278

[3] Borgarino M, Menozzi R, Tasselli J, Fontini F 1997 A study of the DC and RF degradation of Be-doped AlGaAs/GaAs HBTs under constant current stress *Proc. GaAs Reliability Workshop* p 2

[4] Huanga Q F, Yoona S F, Tana K H, Suna Z Z, Zhanga R, Jiangb J, Leeb L H 2003 Carbon doping in GaAs using carbon tetrabromide in solid source molecular beam epitaxy *J. Cryst. Growth* **252** 37

[5] Camnitz L H 1991 Improved current gain stability in C-doped HBT's with an AlGaAs surface passivation layers *Proc. GaAs Reliability Workshop*

[6] Henderson T S, Ikalainen P K 1995 Reliability of self-aligned, ledge passivated 7.5 GHz GaAs/AlGaAs HBTs *Proc. 17th Annual GaAs IC Symposium IEEE* p 151

[7] Costa D, Harris J S Jr. 1992 Low-frequency noise properties of n-p-n AlGaAs/GaAs heterojunction bipolar transistors *IEEE Trans. Electron Devices* **39** 2283

[8] Pelzel R 2013 A Comparison of MOVPE and MBE Growth Technologies for III-V Epitaxial Structures *Proc. CS MANTECH Conf. (*New Orleans, Louisiana, USA, *13–16 May)* p 105

IOP Conf. Series: Journal of Physics: Conf. Series **917** (2017) 082017 doi:10.1088/1742-6596/917/8/082017

Modeling the semiconductor devices with negative differential resistance based on nitride nanowires

A M Mozharov[1,2], F E Komissarenko[2], A D Bolshakov[1], V V Fedorov[1], G A Sapunov[1], G E Cirlin[1,3,4,5] and I S Mukhin[1,2]

[1] St.Petersburg Academic University, Khlopina 8/3, 194021 St. Petersburg, Russia
[2] ITMO University, 197101 St. Petersburg, Russia
[3] Institute for Analytical Instrumentation RAS, 190103 St. Petersburg, Russia
[4] St. Petersburg State University, 199034 St. Petersburg, Russia
[5] St. Petersburg State Polytechnical University, 195251 St. Petersburg, Russia

E-mail: mozharov@spbau.ru

Abstract. We proposed a numerical model of the Gunn diode and resonant diode based on a single nitride nanowire. Important model parameters needed for development of the considered devices, namely layers thickness, composition and doping level were evaluated. It has been shown theoretically that the single GaN nanowire based Gunn diode is capable to operate in over 1 THz frequency regime.

1. Introduction

Semiconductor nitride compounds are very promising for developing up-to-date high efficiency electronic devices. The most commonly used in device production nitride compounds are InN, GaN, AlN and their solutions (known as III-N). Due to high electron mobility and thermal conductivity nitride materials are good candidates for use in radio frequency (RF) power devices such as RF FET, Gunn diode and other.

The Gunn diode is an RF current generator which operation is based on non-linear dependence of electron velocity on electric field in the material. This dependence leads to negative differential resistance regime when specific voltage is applied. The most successful devices of this type are based on GaAs and can emit waves with frequency up to 100 GHz. Theoretically, use of nitride semiconductors allows to extend this range over 1 THz due to fast interband transitions and large critical electric field.

There are several technological problems in production of III-N based planar devices, probably main of them is absence of lattice-matched substrate. Another important problem is piezoelectric field along c-direction caused by the wurtzite crystall structure. The first point limits crystal quality and therefore mobility, efficiency and maximum operation frequency of the devices while the second point makes inapplicable many common successful designs. Use of nitride nanowires allows, firstly, to improve the device crystal quality and, secondly, to select nonpolar crystal direction. One important feature about the III-N nanowires is a variety of applicable growth techniques such as metal-organic chemical vapour deposition and molecular beam epitaxy, as well as variety of growth substrates including sapphire, silicon, polished metal wafers and some other [1].

2. Modeling

In this work we considered model of Gunn diode based on a single GaN nanowire with modulated doped layers in polar directions. The calculation was done in Silvaco TCAD program. The majority of bulk semiconductor parameter values were taken from [2], while the mobility value of MOCVD GaN nanowires was taken from [3].

It is known that Gunn effect can be observed in materials that have negative differential mobility in high electric fields. In such case, high electric field mobility depends on ionized impurity concentration as well as temperature. Experimental measurement of this dependence is a difficult task due to need of many structures fabrication differing with doping level as well as necessity to carry out time resolve volt-ampere measurements to register the Gunn effect.

Instead of experimental measurements, we have calculated electron velocity dependence on electric field along the Monte Carlo method which is often used for this purpose. Carrier differential mobility can be found as derivative of electron velocity on electric field. In our electron velocity calculation we took into account ionized impurity scattering and intervalley phonons. In figure 1 one can see the calculated dependence of electron velocity on electric field for different NW doping levels under 300 Kelvins. The calculations show us that electron velocity decrease takes place for doping level under 10^{19} cm^{-3}, but not for 10^{20} cm^{-3}. This might be an identification of threshold doping concentration existence around 10^{19} for Gunn effect in case of GaN NW.

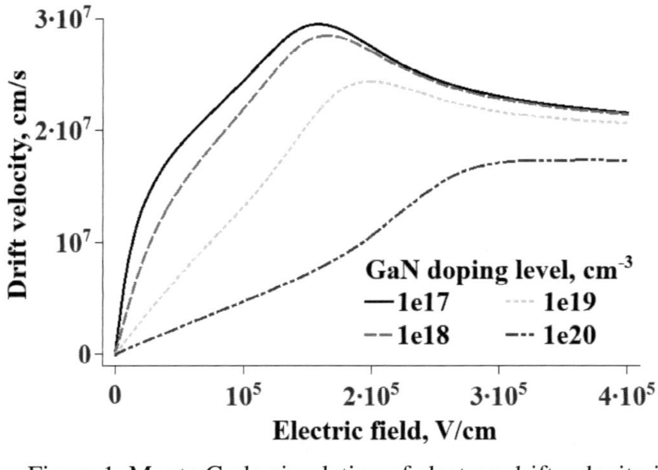

Figure 1. Monte Carlo simulation of electron drift velocity in GaN for different doping levels.

Results of the Monte Carlo simulation were adapted to TCAD negative differential mobility model to study numerically Gunn effect in GaN NWs [4]. We should mention here, that method above is applicable to calculate high electric field behavior of GaN NW conductivity but not for low field one.

In the calculations we considered single GaN NW with typical for MBE parameters: diameter (D) of NW was 200 nm, active region thickness (H_{AO}) was from 200 to 2000 nm and the doping level of active region was from 10^{17} to 10^{19} cm^{-3}. Existence of the surface states (N_{SS}) in GaN was also taken into account. Energy level of the surface states in GaN was considered as 0.6 eV and their density about (1-2)10^{11}eV^{-1}cm^{-2}[5]. Figure 2(a) shows schematic image of the NW model. Emitters in the model were assumed GaN with 10^{19} cm^{-3} doping level and 50 cm^2/(V·s) fixed mobility. Anode and cathode were assumed metallic with total resistance of 25 Ohm each.

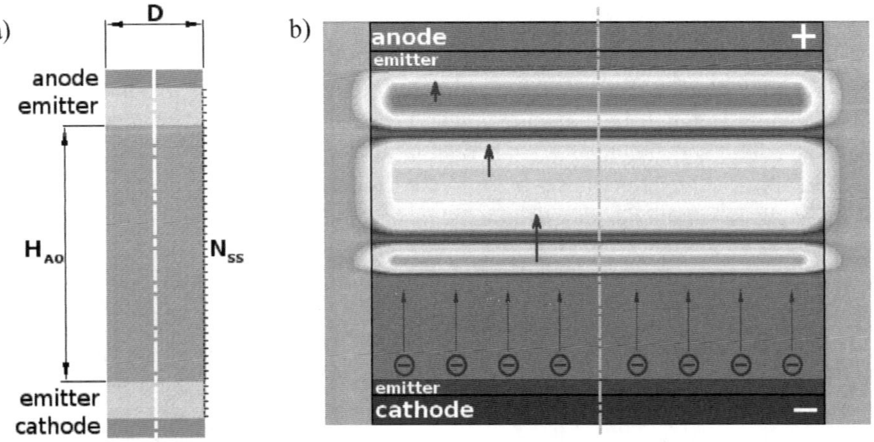

Figure 2(a, b). Schematic image of GaN NW model (a) and electron density waves in NW due to Gunn effect (b).

In our Gunn effect numerical simulation we calculated time resolve volt-ampere characteristic for each set of model parameters. Simulated voltage was applied with linear ramp from zero to $V_{max} = E_{max} / H_{AO}$, where E_{max} is maximum electric field value for active region of specified thickness. The ramp time and stabilization time were 100 ps each. Stabilization time is needed to steady the current oscillations induced by Gunn effect. After stabilization of the simulated oscillations the program calculated them for 50 modeled ps to obtain percentage of RF signal among all the current through NW and to make Fourier transform in order to obtain frequency dependence instead of time dependence. To obtain RF current percentage, standard deviation of oscillating current was calculated and divided by mean current value.

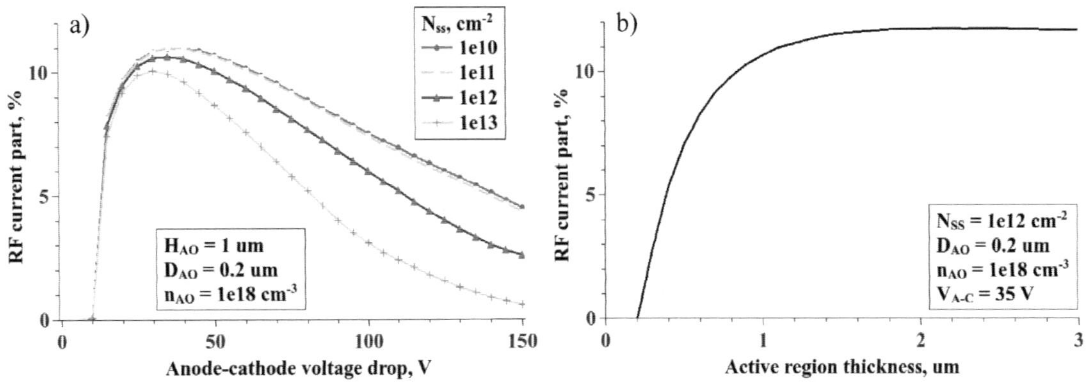

Figure 3(a, b). RF current part dependence on: (a) anode-cathode voltage drop, (b) one of active region thickness.

Figure 3(a) shows RF current part dependence on applied voltage between anode and cathode for different surface states density. According to the calculation results surface states lead to energy bands bending what reduces the electron concentration near NW edge. This phenomenon works as an effective decrease of the NW diameter . Readers can see electron domains as well as reduction of the electron concentration inside the domain edge in figure 2(b). Electron domains consist of heavy electrons from non-gamma valleys. The thickness of electron domains correlates with electron density.

If the active region thickness is smaller than the domain thickness then the Gunn diode efficiency will be reduced similar to fall of the RF current part with decrease of active region thickness in figure 3(b).

Fourier transform of the current oscillations data allows to distinguish higher order modes from the first mode. According to the calculations oscillation frequency decreases with active region thickness increase as shown in figure 4(a). This phenomenon is intuitively understandable as current oscillation period is a time between two consequitive acts of domain formation. One domain drifts to anode with saturation velocity of heavy electrons which is fixed for specific doping level. So when the thickness decreases the oscillation time decreases too.

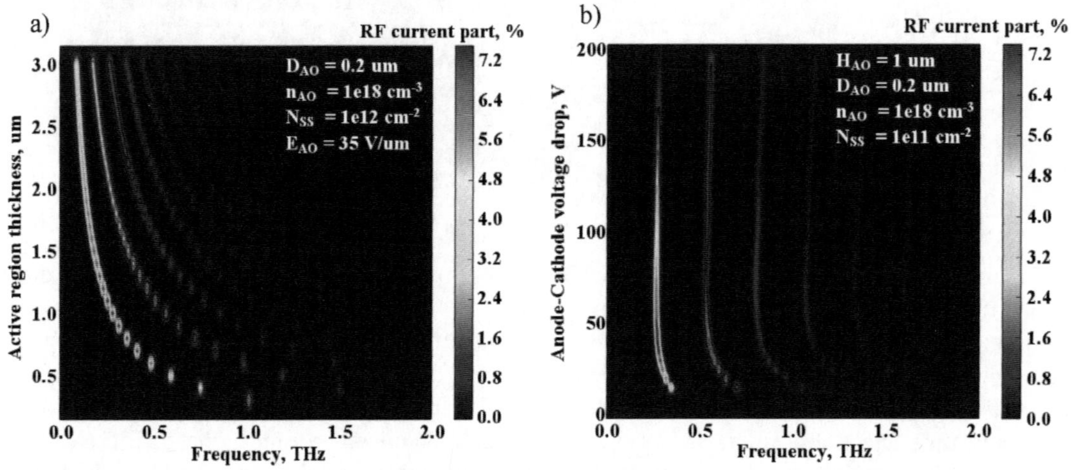

Figure 4(a, b). Current oscillation spectra in Gunn diode for different applied voltages and active region thickness.

As it shown in figure 2(b) there is a distance between cathode and the first electron domain. In this region electrons accelerate until they obtain enough energy to become heavy electrons which form the domains. The lower applied voltage, the lower electric field in the structure, so the thickness of the acceleration region is higher. For this reason, the domain thickness should be not more than active region thickness minus thickness of this region which leads to increase of the oscillation frequency with decrease of the voltage as it shown in figure 4(b).

In spite of possibility to increase the oscillation frequency by decreasing the active region thickness and increasing the doping level there is a critical frequency value for the material. The first reason for that is existence of the critical electron density in domain that depends on critical electric field (point of maximum velocity)[6]. The critical electric field related to energy difference between gamma-valley and a second valley. So, for GaAs this difference is about 0.29 eV, but for GaN it is more than 1.1 eV. The second reason is existence of the impulse relaxation time for the material. This is time needed to form the electron domain. Thus, the way to increase the oscillation frequency is use of material with lower impulse relaxation time and higher critical electric field.

3. Summary

We shown theoretically that specially designed nitride NW can act as an RF device despite existence of the surface state level. We demonstrate that Gunn diode based on single GaN NW is able to operate in a frequency range over 1 THz.

Acknowledgment

This work was carried out with the support of President grant (14.Z56.17.3632-MK), the Russian Foundation for Basic Research (15-02-06839 A, 16-32-60094 mol_a_dk and 16-32-00560 mol_a), the leading universities of the Russian Federation (grant 074-U01), grant of government of the Russian Federation (16.2593.2017/8.9, 16.2593.2017/4.6)

References

[1] J May Brelon, A T M Golam Sarwar and Roberto C Myers 2016 APL **108** 141103
[2] V Bougrov et al. 2001 Properties of Advanced Semiconductor Materials GaN, AlN, InN, BN, SiC, SiGe (Wiley & Sons)
[3] A A Vasiliev et al. Study of electrical properties of single GaN nanowires grown by MOCVD with a Ti mask 2016 J. Phys.: Conf. Ser. **741** 012007
[4] J J Barnes et al IEEE Trans. Electron Devices **23** 1042-1048
[5] M Sawada et al. Proc. Second Int. Conf. Nitride Sem. p.706
[6] E Alekseev et al. GaN-based NDR Devices for THz Generation 2000 Proc. of the Eleventh International Symposium on Space Terahertz Technology p.162

IOP Conf. Series: Journal of Physics: Conf. Series **917** (2017) 092001 doi:10.1088/1742-6596/917/9/092001

XPS Depth Profiling of Air-Oxidized Nanofilms of NbN on GaN Buffer-Layers

A.V. Lubenchenko[1], A.A. Batrakov[1], S. Krause[2], A.B. Pavolotsky[2], I.V. Shurkaeva[1], D.A. Ivanov[1], O.I. Lubenchenko[1]
[1]National Research University "Moscow Power Engineering Institute", Moscow, 111250, Russia
[2]Chalmers University of Technology, Göteborg, 41296, Sweden

Abstract. XPS depth chemical and phase profiling of an air-oxidized niobium nitride thin film on a buffer-layer GaN is performed. It is found that an intermediate layer of Nb_5N_6 and $NbON_x$ under the layer of niobium oxide is generated.

1. Introduction

The hot electron bolometer mixer (HEB) based on a NbN nanofilm is probably the most sensitive detectors for spectroscopy in the terahertz frequency range above ca. 1 THz [1]. Performance of HEB's depends on uniformity of the 3-6 nm thin NbN layer [2]. The best results have been obtained for monocrystalline NbN films epitaxially grown on GaN buffer-layers. The critical temperature of films on the GaN buffer-layer reached 13.2 K when grown on the hot substrate [3], and 10.4 K when grown over the substrate at the ambient temperature [4].

As a result of oxidation films of niobium and its compounds (initially homogeneous), multilayer and multiphase films are generated. Superconductive functional properties on nano-sized films of niobium compounds worsen by increase of thickness of niobium oxide layers on film surfaces. A sufficient number of works is dedicated to study of niobium oxide films (e. g. [5]). However, there are no papers reporting on phase layer profiling of oxidized nano-sized films of niobium compounds.

At present, depth profiling of thin films is performed using both destructive and non-destructive methods. A review of different profiling methods can be found in [5]. If layers more than tens of nanometers thick are explored, ion sputtering together with some non-destructive method of surface analysis is most frequently used. Ion-surface interaction results not only in sputtering but, due to selective sputtering and ion mixing, in modification of the rest of solid layers up to depths about the projective ion path (few nanometers).

One of the non-destructive profiling methods for nano-sized films is X-ray photoelectron spectroscopy (XPS). The surface area is not only multilayer but layer-multicomponent and multiphase. Layer profiling based on interpretation of XPS spectra of such targets is a complex inverse problem with many parameters that are unknown a priori. To solve this problem correctly, in this work we suggest: 1) a background subtraction method considering difference of energy losses on surface and in volume; 2) using of constant parameters for background calculation and line profiling in all the range of photoelectron spectra; 3) using parameters of line profiles from the Handbook of X-Ray Photoelectron Spectroscopy for pure homogeneous targets; 4) simultaneous interpretation of different lines of the same element using the same model.

IOP Conf. Series: Journal of Physics: Conf. Series **917** (2017) 092001 doi:10.1088/1742-6596/917/9/092001

2. Experimental Details

In this work, a 5 nm film of NbN on a GaN buffer-layer was studied. Niobium nitride was sputtered onto the substrate by the magnetron method in the experimental setting AJA Orion-5-U-D. The film thickness was controlled by the known sputtering rate confirmed by TEM. After the target was unloaded from the vacuum chamber, it was oxidized in room air at the ambient temperature, and consequently, the film structure changed. This structure change was of interest. Depth profiling of such targets was performed by means of XPS. The XPS spectra were recorded with the help of an electron and ion spectroscopy module based on the platform Nanofab 25 (NT-MDT) by a semi-spherical energy analyzer SPECS Phoibos 225. The energy analyzer was calibrated by samples of Cu, Ag and Au. The energy resolution of the spectrometer at the line Ag 3d5/2 was 0.78 eV for non-monochromatic X-rays of Mg *K*α. All spectra were recorded using a Mg anode, while the energy analyzer worked in the FAT (Fixed Analyzer Transmission) mode. For survey spectra, the deceleration energy setting of the energy analyzer lens was $E_{pass} = 80$ eV, for detailed spectra it was $E_{pass} = 20$ eV.

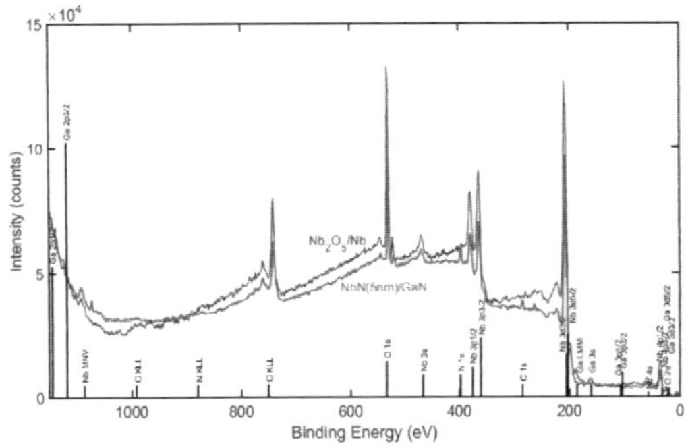

Figure 1. Survey XPS spectrum

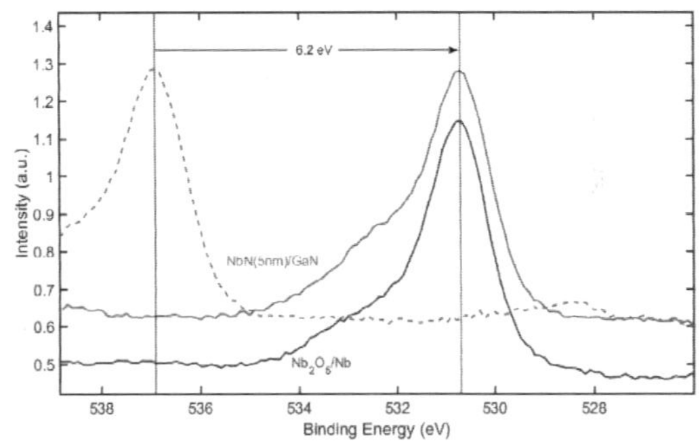

Figure 2. XPS spectrum of line O1s. Dashed line: uncorrected spectrum

Targets with a NbN film on GaN experienced electrical charging during recording of the spectra. The spectra were corrected by the position of the O1s line. For that, spectra of a Nb 100 nm film on a silicon substrate were preliminarily recorded. That film was exposed to natural oxidation; as analysis showed, a layer of Nb_2O_5 9 nm thick was formed over it. Fig. 1 shows survey XPS spectra, Fig. 2 shows a correction procedure of a NbN (5 nm)/GaN. For a target NbN (5 nm)/GaN, relative concentrations calculated by the standard XPS methods: C – 23%, O – 41%, N – 10%, Nb – 24%, Ga – 2%. However, as far as these data were obtained in assumption of a depth-uniform target, these are only qualitative.

3. Theory

Theoretical explanation of layer profiling methods by means of XPS may be found in papers [7-9]. These methods are based on the standard model of photoelectron scattering in solid [10]. One of the factors of influence on layer profiling calculation accuracy is accuracy of separation of peaks generated by elastically and inelastically scattered electrons. As scattering angles approach sliding angles, surface effects have more and more influence onto the background formed by inelastically scattered electrons. The widely used methods of background subtraction (Shirley [11] and Tougaard [12]) don't consider the difference of inelastic electron scattering in volume and on surface. So the calculation accuracy of layer profiling will be indefinite.

To find the background within an energy range from E_{min} up to E_{max}, considering inelastic scattering on surface as in volume, let us use an approach described in [7]. The Tougaard formula [12] is derived by assumption of depth homogeneity of a target. However, for inelastic scattering a target is always inhomogeneous. Energy losses in surface layers and deep in the target obey different laws. In the first approximation, in a semi-infinite target, let us select a very thin near-surface plane-parallel layer d_S, in which energy losses are determined by excitation of a surface plasmon. In this layer, photoionization probability is negligible. Out of this layer, energy losses are losses for excitation of a volume plasmon, ionization losses. Using such model and approach [7] we get a formula for background calculation:

$$\text{Background}(E) = A \int_{E}^{E_{\max}} j(E') x_{\text{SB}}(E - E') dE', \tag{1}$$

where $j(E)$ is an experimentally measured flux of photoelectrons of an energy E, the value A is calculated after finding the background by the energy $E_{\min}$, $x_{\text{SB}}(\Delta)$ is the differential cross-section of inelastic scattering, orthonormalized per unit of length, named inelastic indicatrix, Δ is the energy loss. The function $x_{\text{SB}}(\Delta)$ depends on inelastic indicatrices of electron scattering in volume $x_{\text{B}}(\Delta)$ and on surface $x_{\text{S}}(\Delta)$ and on $SEP = d_S / \lambda \cos\theta$ (SEP – surface excitation parameter), λ is the inelastic mean free path (IMFP), θ is the angle between the direction to the energy analyzer and the surface normal:

$$x_{\text{SB}}(\Delta) = x_{\text{B}}(\Delta) - \int_{0}^{\Delta} L_{\text{S}}(\Delta - \varepsilon) x_{\text{B}}(\varepsilon) d\varepsilon + L_{\text{S}}(\Delta), \tag{2}$$

where the function $L_{\text{S}}(\Delta)$ is calculated by the formula

$$L_S(\Delta) \approx \sum_{n=1}^{N} (-1)^{n+1} \frac{SEP^n}{n!} y_n(\Delta), \tag{3}$$

N is the maximal scattering order considered, $y_n(\Delta)$ are the multiple inelastic indicatrices, $y_1(\Delta) = x_S(\Delta)$, $y_n(\Delta) = \int_0^{\Delta} y_{n-1}(\Delta - \varepsilon) y_1(\varepsilon) d\varepsilon$. If in formula (2) surface energy losses are not considered ($SEP = 0$), the Tougaard formula [14,15] for background is obtained. The calculation formulae for *SEP* are given in [19,20]. The *SEP* depends on photoelectron energy, compound and state of the surface.

To calculate the inelastic scattering indicatrix, we suggest to use a formula that is proved to be good in calculation of characteristic electron energy losses [13]:

$$x(\Delta) = \text{norm} \frac{\Delta^{\alpha}}{\left(\Delta^2 - \varepsilon_{pl}^2\right)^2 + (\gamma \Delta)^{4-\beta}} \tag{4}$$

where ε_{pl} is the energy of plasma oscillations (plasmon energy); α и β are parameters determining dependence of the function on energy loss; γ is the parameter determining the peak width; the value of norm is found from the normalization condition $\int_0^{\infty} x(\varepsilon) d\varepsilon = 1$.

Fig. 3 shows the results of background subtraction using different methods, for a target of Nb_2O_5/Nb, line Nb 3*d*. The dash-and-dotted line displays the background according to Shirley, the dotted line is for the Tougaard method using the "universal" inelastic scattering function, the solid line shows the result of the method (1) – (4). The numbers denote the extreme points of various ranges.

The reconstruction method of layer-by-layer profile is based on a target model that results from the history of its production and life. The principle statements of the target model used in this paper:

1) The target consists of few plane layers on a substrate. Each layer is homogeneous and may be multicomponent. Such statement is reasonable because an XPS signal is recorded from a surface area which dimensions are by orders greater than the probing depth. This area depends on focusing of the energy analyzed and geometry of the experiment and it is about 0.1 mm^2. The probing depth is about 10 nm. So the signal will be measurement area-averaged.

2) Inhomogeneities (islets, interlayer asperities, inclusions, etc.) are layer-averaged. Inhomogeneity rate will define relative concentrations of element phases.

3) After unload from the chamber, the surface is oxidized and oxide and suboxide layers are generated. As far as oxidation progresses from the surface, the oxidation level decreases with depth. Besides that, at the top a hydrocarbon layer will be precipitated.

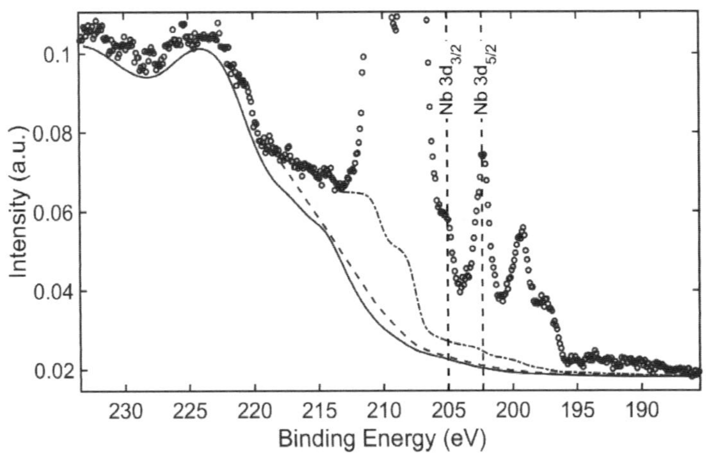

Figure 3. Background calculated by different methods for target of Nb_2O_5/Nb, line Nb 3d. Dash and dots: background calculated using the Shirley method, dots: the Tougaard method, solid line: the method (1) – (4).

The shape of a spectral line of photoelectrons will be determined by a convolution of functions describing the natural line shape and instrumental broadening. The natural line shape will be described the Doniach-Šunjić expression [14] and the instrumental broadening function by the Gauss function. We suggest to take the binding energy and spin-orbital interaction energy for chemically pure elements from the experimental data of the Handbook of X-ray Photoelectron Spectroscopy [15]. The chemical shift energy is almost linearly proportional to the oxidation level, so it is enough to find the chemical shift energy for the most oxidized element. For example, we used a value of 5.31 eV for niobium oxide [16].

Layer thicknesses will be calculated by the formula [17]:

$$d_i = \lambda_i \cos\theta \ln\left(\frac{I_i / \left(n_i \omega(\gamma) \lambda_i\right)}{\sum_{j=0}^{i-1} I_j / \left(n_j \omega(\gamma) \lambda_j\right)} + 1\right) \tag{5}$$

where d_i is the thickness of the i-th layer, n is the atomic concentration, $\omega(\gamma)$ is the differential cross-section of photoelectron production [18], γ is the angle between the directions of incident radiation and to the energy analyzer, λ is the IMFP (IMFP is calculated by the TPP2M formula [19]), θ is the angle between the direction to the energy analyzer and the surface normal, I_i is the i-th peak intensity. The layers will be numerated upwards from the substrate. Number 0 means the substrate.

Let us find the depth sensitivity and probing depth of this method. For that a homogeneous layer of niobium oxide Nb_2O_5 on a niobium substrate will be regarded. As far as we determine partial intensities by our method with an accuracy of 1%, let us assume $I_{Nb_2O_5} / I_{Nb} = 1/100$. The we get from (9) $d_{min} = 0.1$ nm. Probing depth can be found if assumed $I_{Nb_2O_5} / I_{Nb} = 100/1$. Under formula (9) we calculate $d_{max} = 13.9$ nm.

To decrease the uncertainty of the number of layers, we suggest to introduce the minimal thickness of a homogeneous layer equal to 0.2 nm. Let us begin calculation from the greatest number of layers possible assuming that every layer is of the same oxidation level. If the calculation shows that a layer is less than the minimal thickness thick, this layer is added to the nearest one with a greater oxidation level of the main element. The newly constructed layer will be multicomponent. This layer may be regarded as inhomogeneous averaged to a homogeneous layer with an effective thickness. The inhomogeneity degree can be evaluated with the help of relative concentration of the main element of the layer. After that, the layer thicknesses are to be re-calculated. Such approach enables to decrease the number of calculation variants significantly. Then the variants are regarded as profiles calculated with different depth detailing.

4. Results and Discussion

Fig. 4 shows XPS spectra of the line Nb3d, Fig. 5 shows spectra of N1s; the circles show experimental data, the solid line is for theoretical interpretation of the spectrum, the dashed line shows partial theoretical spectra. The line N1s displays different phases of niobium nitride: NbN and NbN_x. Partial intensities on the lines N1s и Nb3d enabled to found a stoichiometric coefficient $x \approx 1.2$ that corresponds to Nb_5N_6.

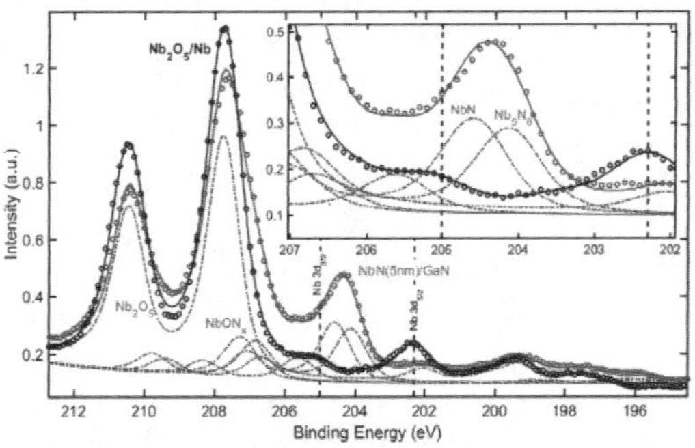

Figure 4. XPS spectra of line Nb3d

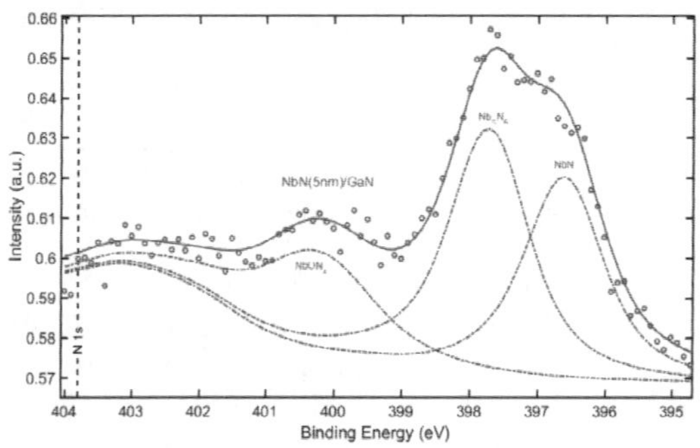

Figure 5. XPS spectra of line N1s

Table 1. Chemical and Phase Depth Profiling of a NbN(5nm)/GaN Target.

	d, nm	Formula
Σ	6.22	
6	0.78	hydrocarbons
5	1.45	Nb_2O_5
4	0.23	$0.07\ NbO_2 + 0.93\ Nb_2O_3$
3	0.78	$NbON_x$
2	1.30	NbN_x ($x \approx 1.2$)
1	1.68	NbN
Substrate		GaN

Table 1 shows computing results for thicknesses following the formula (5), which accounts for partial intensities obtained by decomposition of the lines N1s, C1s and Nb3d. From the data above, it is found that NbN thickness decreases and various additional phases of niobium nitride appear by oxidizing of the film. Such result is obtained for the first time. Influence of these layers onto HEB's performance needs further research.

5. Conclusions

In this work, a method of chemical and phase layer profiling of multicomponent multilayer films is suggested. The method includes a new subtraction method of background of multiply inelastically scattered electrons considering inhomogeneity of inelastic scattering by depth; a new method of photoelectron line decomposition into constituent peaks considering physical nature of different decomposition parameters; simultaneous solution of the problems of background subtraction and decomposition of a photoelectron line; determination of layer thicknesses of a multilayer target using a simple formula. On the base of the suggested method, depth chemical and phase profiling of an air-oxidized niobium nitride thin film on a buffer-layer GaN was performed.

References

[1] Meledin D. et al. 2009 *IEEE Transactions on Microwave Theory and Techniques*. **57.1**. 89.
[2] Sergeev A., Reizer M. 1996 *Int. J. Mod. Phys.*, **B 10**. 635-667
[3] Krause S. et al. 2014 *Supercond. Sci. Technol*. **27**. 065009
[4] Krause S. et al. 2016 *IEEE Transactions on Applied Superconductivity*. **26.3** 1-5
[5] Darlinski A., Halbritter J. 1987 *Surface and interface analysis*. **10**. 223.
[6] Galindo R. E. et al. 2010 *Analytical and bioanalytical chemistry*. **396.8.** 2725-2740.
[7] Cumpson P. J. 1995 *Journal of Electron Spectroscopy and Related Phenomena*. **73.1.** 25-52.
[8] Palacio C. et al. 2003 *Journal of Electroanalytical Chemistry*. **545**. 53-58.
[9] Oswald S. et al. 2005 Applied Surface Science. **252**. 3–10.
[10] Macak K. 2011 *Surface and Interface Analysis*. **43.3**. 1581
[11] Shirley D. A. 1972 Physical Review. **B 5.12** 4709
[12] Tougaard S. 1989 Surface Science Letters. **216.3**. A330.
[13] Afanas'ev V. et al. 2004 *The European Physical Journal B-Condensed Matter and Complex Systems*. **37.1**. 117.
[14] Doniach S., Sunjic M. 1970 *Journal of Physics C: Solid State Physics*. **3.2**. 285.
[15] Chastain Jill, King Roger C., Moulder J.F. 1995 Handbook of X-ray photoelectron spectroscopy: a reference book of standard spectra for identification and interpretation of XPS data.
[16] Naumkin A. V, Kraut-Vass A., Powell C.J. 2008 NIST X-ray photoelectron spectroscopy database.
[17] Trifonov A.S. at al. 2015 *Journal of Applied Physics*. **117.12.** 125704.
[18] Yeh J.J., Lindau I. 1985 *Atomic data and nuclear data tables*. **32.1.** 1.
[19] Tanuma S., Powell C.J., Penn D.R. 1994 *Surface and Interface Analysis*. **21.3**. 165.

The Formation of Quantum Dots - Liquid Crystal monolayers by Langmuir-Blodgett method

A Chumakov[1], A J Al-Alwani[1,2], I Gorbachev[1], A Ermakov[3], O Shinkarenko[1], N Begletsova[1], A Kolesnikova[1], E Glukhovskoy[1]

[1]National Research Saratov State University named for N.G. Chernyshevsky, Astrakhanskaya st. 83, 410012, Saratov, Russia
[2]Babylon University, Babylon, Iraq
[3]Institute of Materials Research and Engineering, A*STAR, 2 Fusionopolis Way, Singapore 138634

Abstract. The article describes the features of the formation of CdSe/CdS/ZnS quantum dots - liquid crystal Langmuir monolayers on a water subphase surface by different temperature value. Monolayers structure was investigated by analyzing the compression isotherms and surface potential value. Films on the solid substrate (glass plates) were investigated by atomic force microscopy method for morphology, thickness and roughness. Film structure was studied as a function of the pressure at which the transfer was carried. It is shown that the structure of films obtained by a higher temperature of subphase is more perfect. The film is monolayered and uniform in thickness.

1. Introduction

Nowadays one of the fastest growing areas in nanotechnology is provide materials for biological applications and purposes of electronics. Colloidal quantum dots (QDs) of the semiconductor materials are one of the most interesting materials for research and applications projects. They are used extensively for the modification of biological objects [1], as well as photovoltaic devices and as a promising material for solar cells [2]. And one of the critical issues for practical using of QDs is obtaining ordered structures with them.

One method, which allows obtaining the ordered layers of nanoobjects, is a Langmuir-Blodgett method. It allows to obtaining the ordered layers of amphiphilic substances or objects using self-assembly principle. This method is not expansive and not required specific conditions in process of organization of layers.

In our work we obtained Langmuir monolayered films consisting of CdSe/CdS/ZnS QDs and 4′-octyl-4-biphenylcarbonitrile (8CB) liquid crystal (LC). We choose these materials by few reasons. In one hand semiconductor QDs is photoactive material perspective for solar cells and photovoltaic devices in other hand 8CB is material which can change properties (by phase shifting) in temperature near to normal room. And we can control structure of monolayers (and films) by changing temperature in Langmuir trough. Molecules of 8CB have large dipole moment [3] and we can get information about orientation molecules on the surface by measuring surface potential value.

We study formation of QDs-8CB monolayers by changing few control parameters: temperature of subphase, concentration of QDs in monolayer and surface pressure of deposition monolayers to solid substrate.

2. Materials and methods

In all experiments we use deionized water (R about 18 Mom×cm) as a subphase. QDs obtained by chemical reduction method and have three-layered structure with CdSe/CdS/ZnS composition. Diameter of QD is about 7-8 nm and they stabilized by oleic acid [4]. Solution of 8CB (was purchased from Sigma-Aldrich) prepared in chloroform (Vecton, Russia) with concentration 10^{-3} M. We used different ratios QDs:8CB (4:1, 2:1, 1:1 1:2, 1:4) as material of monolayers. For formation, deposition and measure surface potential of monolayers we used KSV Nima LB trough KN 1003 (KSV Nima, Finland). Temperature in the trough was controlled by LOIP thermostat (18 and 35°C). Deposition of monolayers was realized by different value of surface pressure: 3, 6, 9, 12 mN/m. Glass plates (previously cleaned) was used for substrates. Atomic force microscopy (AFM) picture taken in semicontanct mode by Nanoeducator II AFM microscope.

3. Results and discussion

In few series of experiments (with different temperatures of subphase and ration of QDs:8CB) we received a series of monolayers compression isotherms. Most interesting result we obtain by ratio QDs:8CB 4:1 and 35°C temperature of subphase. We choose these parameters of LB process for deposition because the compression isotherm not shown strongly breaking point (figure 1) which is typical for 8CB monolayers and corresponds formation of three-layers complex LC structure [5].

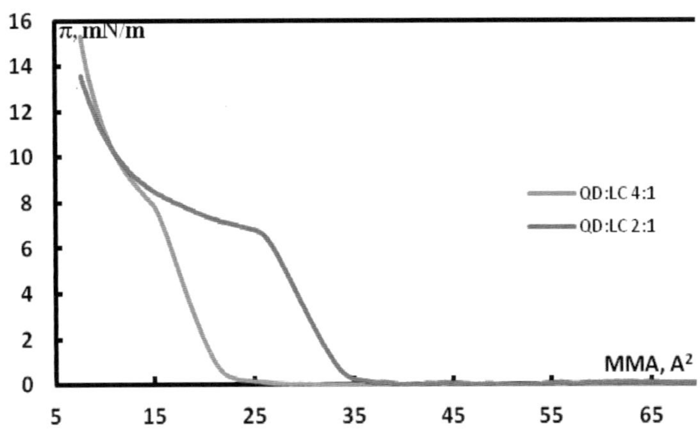

Figure 1. Compression isotherms for QD:8CB 4:1 and 2:1 ratios

And at 35°C, 8CB molecules change their orientation on the water surface. We can assert this for the reason, because at higher temperatures the sign of the surface potential was reversed (figure 2).

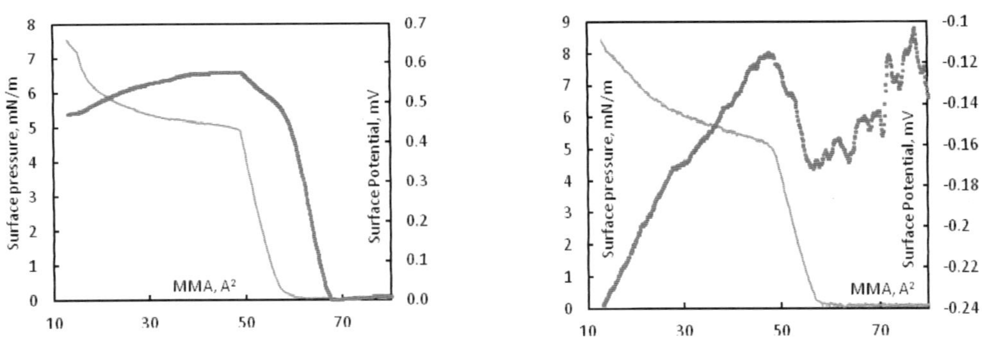

Figure 2. Surface pressure isotherms (blue curves) and surface potetntial of monolayer graph (red) for 8CB by 18°C (left) and 35°C (right)

Moreover, after passing through the phase transition pressure at 35°C (about 5 mN/m) the absolute value of the surface potential increases. This indicates the compaction of molecules in the monolayer in the same orientation. On the contrary, at 18°C after the phase transition point the absolute value of the surface potential short time is constant and decreases after. This can be due to a decrease in the surface density of the molecules (which is impossible - with the decrease in the working area surface of the Langmuir trough, the surface density of the molecules increase, if we have surfactant molecules) or due to changes in the orientation of molecules. Thus, as the temperature rises, the molecules of a liquid crystal behave like typical surfactant molecules (the same behavior of 8CB molecules we can see at 18°C in first part of isotherm before breaking point) and we can obtain monolayered structure with them. Thus in our experiment we obtain complex organic-inorganic films by using high temperature and specific ration of QDs:8CB.

For further study monolayers were deposited on glass substrates and studied by AFM method by semicontact mode. The transfer was carried out at different values of surface pressure: 3, 6, 9, 12 mN/m by Langmuir- Schaefer method and dried at room temperature.

For each sample we study morphology of surface and structure, height profiles were measured. On figure 3 and figure 4 shown AFM picture for few samples. Film obtained by different value of surface pressure at deposition.

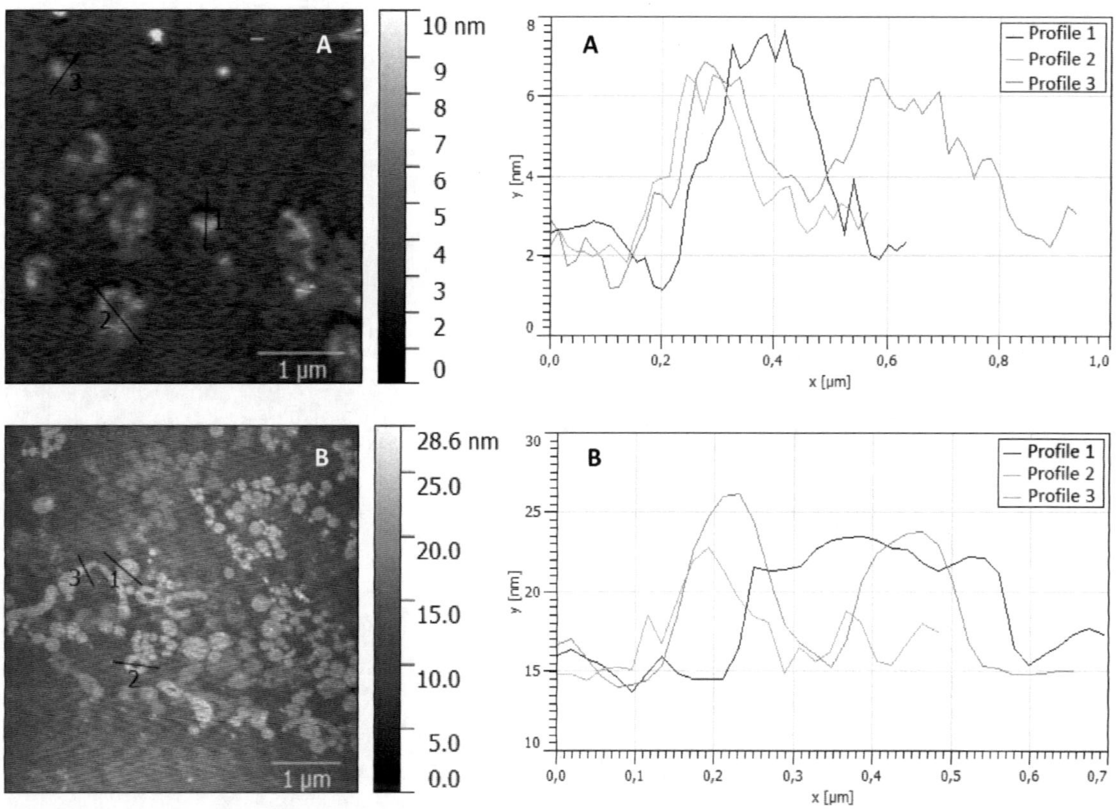

Figure 3. AFM pictures and hight profile for films deposited at 3 mN/m (A) and 6 mN/m (B)

These pictures shown that in the case of transfer a monolayer at 3, 6 or 9 mN/m surface pressure value, the resulting films has an island structures. Moreover, as the value of the pressure of transport increases, the boundaries of the islands became clearer and the size of the islands is larger. However,

IOP Conf. Series: Journal of Physics: Conf. Series **917** (2017) 092002 doi:10.1088/1742-6596/917/9/092002

in each case the thickness of the film in the island is of the order of the diameter of the quantum dot used in the monolayer. This fact allows us to say that the resulting films are monolayers. When the film is transferred at 12 mN/m, it covers almost the entire sample and has a smaller roughness. However, it has some heterogeneity in the morphology of the surface, in contrast to films with a island structure.

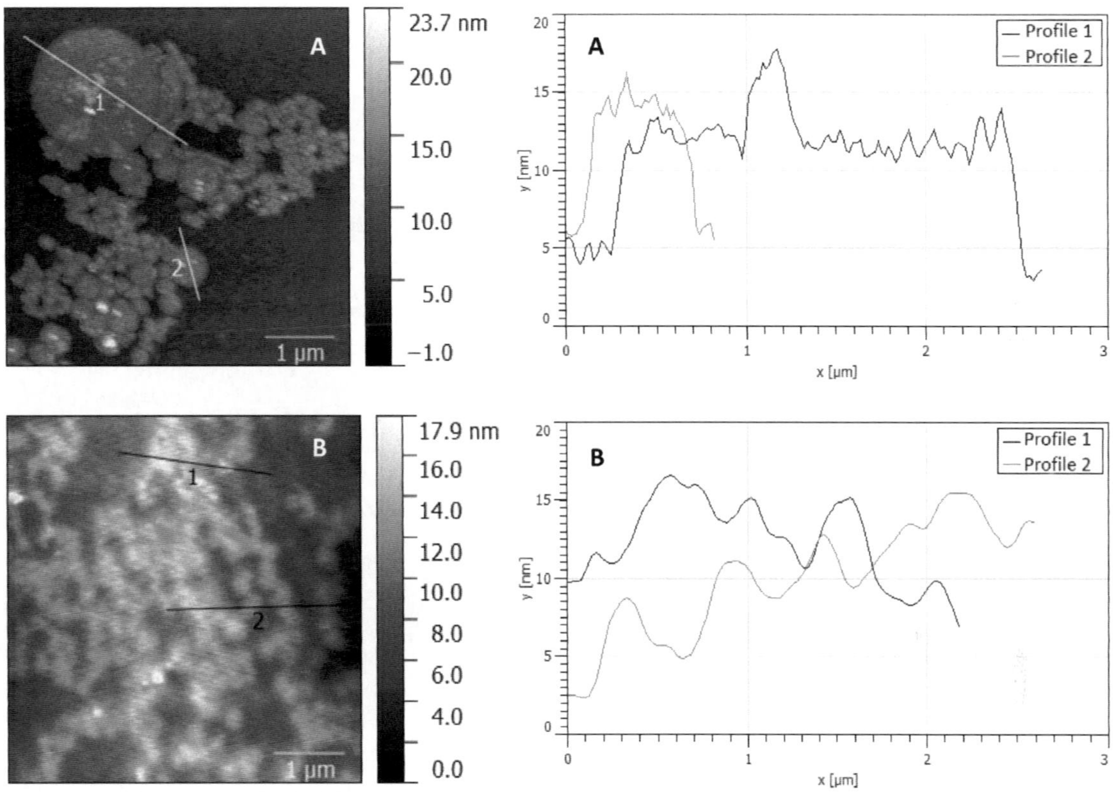

Figure 4. AFM pictures and hight profile for films deposited at 9 mN/m (A) and 12 mN/m (B)

4. Conclusion

Research results allow us to say that an increase in temperature in Langmuir trough and changing ratio of QDs:8CB allows us obtaining a film with monolayer structure. It was confirmed by analysis of compression isotherms and using AFM method. Structure of the film dependent on the surface pressure by deposition, and if we transfer monolayer by surface pressure more 10 mN/m we will obtain monolayer structure.

Acknowledgments

This work was supported by grants from the Russian Foundation for Basic Research project № 16-07-00093/16 and № 16-07-00185/16.

References

[1] Barroso M. 2011 *J. of Histochem.and Cytochem.* **59(3)** 237

[2] Kimberly A. Sablon, J. W. Little, Mitin V., Sergeev A., Vagidov N., Reinhardt K. 2011 *Nano Letters* **11 (6)** 231

[3] Weiss K, Wöll C, Johannsmann D 2000 *Journal of Chemical Physics* **113** 11297

[4] Gorbachev I, Goryacheva I Y, Glukhovskoy E G 2016 *BioNanoScience* 1

[5] Collins J, Funfschilling D, Dennin M 2006 *Thin Sol. Films* **496** **(2)** 60

Diffusion theory and optimization of ohmic contacts to n-layer of bipolar nanoheterostructures

A V Nezhentsev, V E Zemlyakov, V I Egorkin and V I Garmash

National Research University of Electronic Technology, Moscow, Russia, 124498

Abstract. Ohmic contacts to n-layers of gallium arsenide-based heterobipolar nanoheterostructures obtained by layer electron-beam evaporation Ge/Au/Ni/Au are studied. Time and temperature dependencies of diffusion profiles of doping Ge distribution are calculated. The interface of metal-semiconductor is analyzed with SEM, then an RTA installation design and methodology of RTA are suggested based on the results of this study. This allows to obtain ohmic contacts with low resistance and minimum transition layer.

1. Introduction

Ohmic contact to the n-layers nanoheterostructures based on gallium arsenide are widely used in microwave, digital and optical devices. Despite the enormous amount of work on metallization systems based on thin layers of Ag, Ni, Pt, Pd, Ti and other materials, [1-3] Au/Ge based metallization systems are most widely used at the moment [4,5]. To obtain ohmic contact an electron beam vacuum deposition and a special heat treatment - RTA, in which diffusion processes take place, in order to create a heavily doped region with low resistance between the semiconductors and metallization, - are used. This mechanism is still not fully comprehensible. In practice, it turns out that the doping mechanism that occurs at the interface metal/GaAs section is very complicated, and the resistivity of the Ni-Au-Ge/GaAs contact is very sensitive to heat exposure processing procedure. Developed metallization schemes allow to obtain the minimum of contact resistance at the level of 0.1-0.2 Ohms*mm, but literature data indicate that the extent of the transition layer in this case is a few tenths of a micron [6]. We assumed that the thickness of the material forming eutectic alloy determines transition layer thickness. We modeled distribution of impurities in the transition layer. On the basis of the calculation of diffusion profiles, we have optimized the time and temperature of ohmic contacts annealing.

2. Theoretical background

To calculate the diffusion of thin layer of the germanium to gallium arsenide was examined diffusion from a limited supply in to the semi-infinite body at a relatively short time. Since the modeling problem consists in calculating the impurity distribution profile in depth, we consider the one-dimensional case. The calculation of diffusion processes reduces to solving a differential equation with given initial and boundary conditions. For a description of diffusion process, we use the Fick's second law:

$$\frac{\partial N}{\partial t} = D\frac{\partial N^2}{\partial x^2} \tag{1}$$

where *N*– the concentration of diffused particles, *D* is the diffusion coefficient, *t* is the diffusion time and *x* is the doping depth. This equation is used to describe diffusion processes. The maximum surface concentration of the diffusing impurity corresponds to the limit of its solubility N_0 for a given body. During this process, we are creating a highly doped layer in the surface region of the semiconductor. Border conditions:

$$N = N_0, \text{ at } x = 0, t > 0. \tag{2}$$

The concentration tends to zero in the volume of the body. Initial conditions:

$$N = 0, \text{ at } x \gg 0, t > 0. \tag{3}$$

The solution of the diffusion equation has the form:

$$N(x,t) = N_0 \exp\left(-\frac{x^2}{4Dt}\right). \tag{4}$$

The diffusion coefficient of the impurity depends on the temperature according o the Arrhenius law:

$$D = D_0 \exp\left(\frac{E_a}{kT}\right) \tag{5}$$

where E_a is the diffusion activation energy, D_0 is the diffusion coefficient value at $E_a = 0$, T is the absolute temperature, and k is the Boltzmann constant.

We used the package MATLAB applications. The program builds a dopant distribution profiles in the gallium arsenide. Profile defined by the expression (5). The purpose of the calculations was to form a transitional layer of minimum depth. This is necessary to avoid puncturing the emitter layer of the heterobipolar structure. The greater depth of the transition layer will lead to the penetration of the contact layer, which is less to the contact with a rather high level of resistance. We must learn to obtain a transitional layer of a certain depth with high accuracy. Germanium is an electrically active impurity for gallium arsenide, in order to obtain an ohmic contact, the concentration of the dopant in the near-contact region should be at least 10^{19} sm^{-3}. The solubility limit of Ge in GaAs is determined by the maximum concentration of Ga vacancies in GaAs. These vacancies are formed during diffusion and are replaced by germanium atoms. In the case of diffusion of pure Ge, the vacancy concentration and, accordingly, the solubility limit are 10^{17} sm^{-3} [7]. To increase the degree of doping, the diffusion of germanium is carried out in the presence of gold, since in this case gallium leaves the semiconductor more actively, forming AuGa compounds, the concentration of vacancies reaches $5*10^{19}$ sm^{-3}. In only one of five cases, germanium occupying a vacancy is an electrically active impurity, in the remaining cases the impurity is neutral [7,8]. The donor impurity concentration was calculated, since it determines the parameters of the contact. Accordingly, the solubility limit in this case will be 10^{19} sm^{-3}. We have received doping profiles (figure 1).

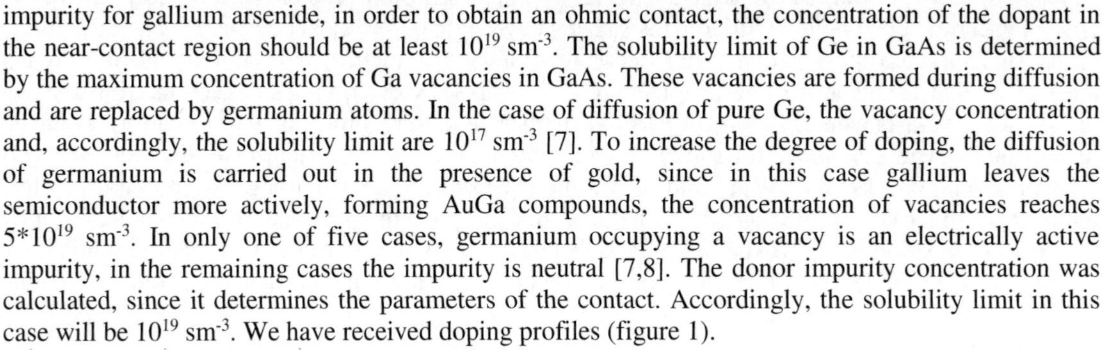

Figure 1. The germanium diffusion profiles in gallium arsenide. A) RTA time 30 seconds, B) RTA time of 60 seconds, C) RTA time 120 seconds.

The calculation shows a significant change in the depth of the impurity from the RTA time and the RTA temperature. There is a strong diffusion of impurities at higher temperatures and times. This allows for smooth transition layers, if you can create a rapid heating and cooling mode.

3. Experimental data

An experiment was conducted to verify the calculations. During the experiment contacts were made to the n-layers of bipolar nanoheterostructures based on gallium arsenide. The wafer was cut into pieces and subjected to heat treatment. A feature of this type of structure is that the depth of the transition layer should not exceed the thickness of the contact layer (not more than 50 nm). After the formation of metallization and mesa, the plate was cut into pieces measuring 1*1 sm^2 and annealed, the time and temperature of the process were set in accordance with the results of the calculations. In our furnace the cooling rate is comparable to the heating rate [9], which allows us to fix or "freeze" the diffusion front defined by time and heating temperature. Measurement of the contact resistance was made by the method of the transmission line measurement (TLM) with a width of the rectangular metallized regions 100 um [10].

4. Results

Figure 2 shows the dependence of the contact resistance on the temperature and time of RTA. It is a typical shaping of contact resistance process to gallium arsenide [11]. There is a minimum contact resistance. Also, the contact resistance increases with the level of RTA time. This dependence is reduced at high temperatures. Figure 2 shows that at an incineration temperature of 398 °C, the contact resistance is practically independent of the RTA time. This makes it possible to fabricate ohmic contacts with a minimum transition layer. Most likely, the rapid heating and cooling allows you to fix or “freeze" the diffusion front, set the time and temperature of heating.

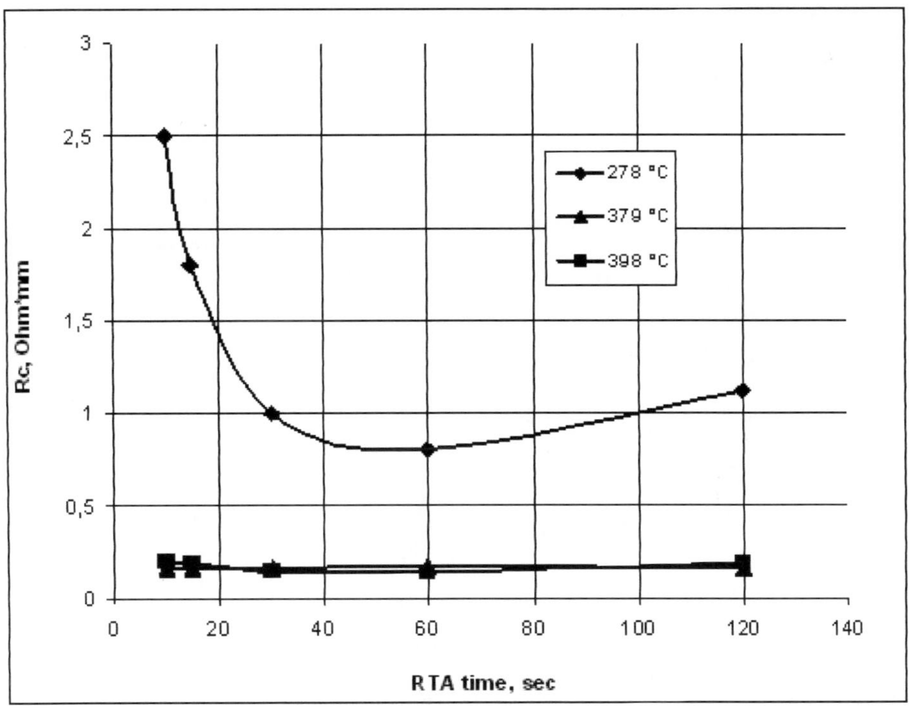

Figure 2. The dependence of contact resistance on RTA time and RTA temperature this case simply justify the caption so that it is as the same width as the graphic.

IOP Conf. Series: Journal of Physics: Conf. Series **917** (2017) 092003 doi:10.1088/1742-6596/917/9/092003

To determine the depth of doping, a SEM microscopy was used. Figure 3 shows a SEM image of the metal-semiconductor boundaries for different parameters of heat treatment. The cross-sections show the correspondence to the calculated profiles. The transition layer of the contact obtained by annealing for 60 seconds at 398 ° C has a depth of about 50 nm (figure. 3A), which agrees well with the theory. It can be seen from figure 3B that an increase in the RTA time leads to a blurring of the transition layer, pulling it down into droplets, and also to an increase in its depth. The latter is very critical for heterobipolar structures.

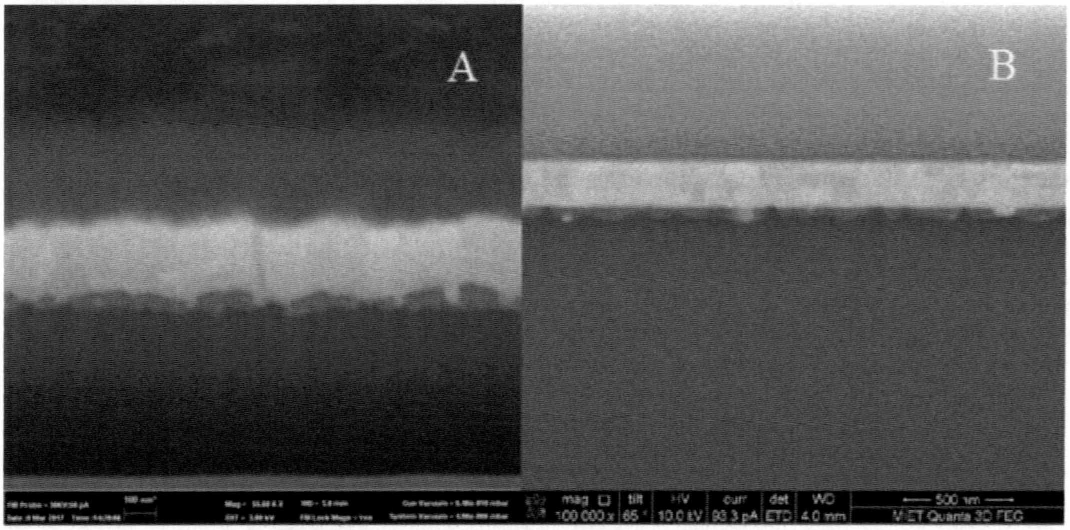

Figure 3. Cross-section SEM image of test sample interfaces at different heat treatment parameters: A) RTA time 60 seconds, temperature 398 ^{0}C, transition layer 50 nm. B) RTA time 480 seconds, temperature 398 ^{0}C, transition layer 80 nm.

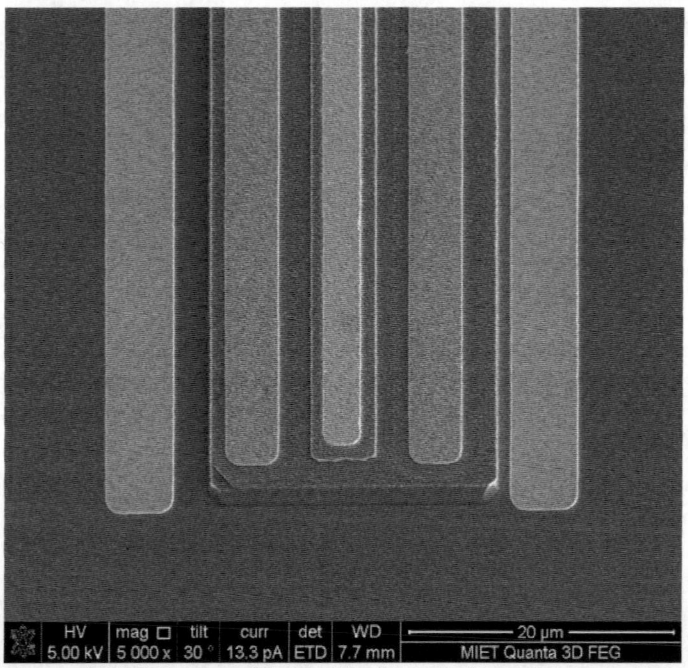

Figure 4. SEM image of heterobipolar transistor.

Figure 4 shows an image of the surface of a heterostructural bipolar transistor (HBT). Ignition time was 60 seconds, RTA temperature 398 °C, contact resistance 0.1 Ω*mm. As a result, ohmic contacts have a smooth surface morphology and smooth edges.

5. Conclusion

We spent RTA process optimization of ohmic contacts to n-layers GaAs-based bipolar nanoheterostructures, based on results of our calculations. Optimization of the heat treatment of ohmic contacts made it possible to reduce the transition layer thickness of up to several hundreds of angstroms, while maintaining the level of contact resistance 0.1 Ohm*mm to n-layers of GaAs. The research on the impact of changes in the contact resistance to nanoheterostructures under the brazing temperature and time revealed the existence of the optimal parameters of the process in which the level of the contact resistance is minimal and depends weakly on variations of the parameters.

Acknowledgments

This work was supported by the Ministry of education and science of the Russian Federation (grant agreement № 14.578.21.0212 from September 29, 2016). Unique identificator RFMEFI57816X0212.

References

[1] Iliadis A and Singer K E 1984 Metallurgical Behaviour of Ni/Au-Ge Ohmic Contacts to GaAs *Solid State Communications* **49** 99-101
[2] Wang L C, Lau S S, Hsieh E K and Velebir J R 1989 Low-Resistance Nonspiking Ohmic Contact for AlGaAs/GaAs High Electron Mobility Transistors Using the Ge/Pd Scheme *Applied Physics Letters* **54** 2677-79
[3] Jones K A, Linfield and E H Frost J E 1996 Contact Resistances of NiGeAu, PdGeTiPt, and TiPd Ohmic Contacts to GaAs and their Temperature Dependence from 4.2 to 350 K *Applied Physics Letters* **69** 4197-99
[4] Kim T J 1997 Ohmic contacts to GaAs epitaxial *layers Critical Rewiews in Solid State and Materials Sciences* **22** 239-273
[5] Lin H C, Senanayake S and Cheng K-Y 2003 Optimization of AuGe–Ni–Au Ohmic Contacts for GaAs MOSFETs *IEEE TRANSACTIONS ON ELECTRON DEVICES* **50** 880-885
[6] Koop E J 2013 On the annealing mechanism of AuGe/Ni/Au ohmic contacts to a two-dimensional electron gas in GaAs/AlxGa1− xAs heterostructures *Semiconductor Science and Technology* **28** 1-9
[7] Nebauer E and Trapp M 1984 Ge and Au profiles in GaAs produced from AuGe contacts and studied by SIMS *Physica status solidi (a)* **84** 39-42
[8] Gupta R P and Khokle W S 1985 Gallium-vacancy-dependent diffusion model of ohmic contacts to GaAs *Solid-state electronics* **28** 823-830
[9] Egorkin V, Zemlyakov V, Nezhentsev A and Garmash V 2017 Optimization of ohmic contact to GaAs-based n-layers of heterobipolar nanoheterostructures *Russian_Microelectronics* **44** 1-5
[10] Shur M S 1987 *GaAs devices and circuits* (Minneapolis: University of Minnesota)
[11] Christou A 1979 Solid phase formation in Au: Ge/Ni, Ag/In/Ge, In/Au: Ge GaAs ohmic contact systems *Solid-State Electronics* **22** 141-149

IOP Conf. Series: Journal of Physics: Conf. Series **917** (2017) 092004 doi:10.1088/1742-6596/917/9/092004

Degradation study of AlAs/GaAs resonant tunneling diode IV curves under influence of high temperatures

M O Makeev, S A Meshkov and V Yu Sinyakin
Bauman Moscow State Technical University, Moscow, 105005, Russia

Abstract. In the present work the thermal degradation of IV curves of AlAs/GaAs resonant tunneling diodes using artificial aging method was investigated. The dependency of AuGeNi specific ohmic contact resistance on time and temperature was determined.

1. Introduction
One of the ways to improve quality of radio electronic systems is the use of semiconductor devices operating on the base of quantum-size charge-transport effects. Resonant-tunneling diodes (RTDs) based on AlAs/GaAs nanoscale heterostructures with transverse charge transport are among such devices. The prospects of RTD for modern radio electronics is caused by the following circumstances:

- The limit frequency for RTD spreads up to units of THz [1-4], which makes the RTD a promising device for SHF, EHF, and THF electronics.
- RTD operates in the range of temperatures and other external factors necessary for technical applications.
- Technologies and the equipment for RTD production exist at Russian enterprises producing heterostructural electronics devices.
- It is possible to control the shape of a current-voltage (IV) curve and create a diode with the optimum IV curve shape for the specific type of nonlinear conversion by changing the parameters of the heterostructure layers (thickness, chemical composition). The set of possible nonlinear transformations using RTD is very wide: generation of radio signals, frequency modulation, mixing of radio signals, amplitude detection, rectification, frequency labels grid generation and etc.

The extensive bibliography is devoted to the application of RTD as an active element of various wireless devices, in particular the benefits of using RTD for the micropower SHF signal mixers and rectifiers is shown in [5-8]. However, there are not many works devoted to the resistance of RTD to the destabilizing factor influence. This work is aimed at filling this gap, the main task of it is to research the thermal degradation of the IV curves of RTD.

2. Methods
Experimental research was carried out for the line of AlAs/GaAs RTD of 30 pieces. The diodes structure is shown in Figure 1. RTD consist of resonant tunneling structure (RTS) (active region, AlAs/GaAs layers), near contact regions (doped GaAs layers) and ohmic contacts.

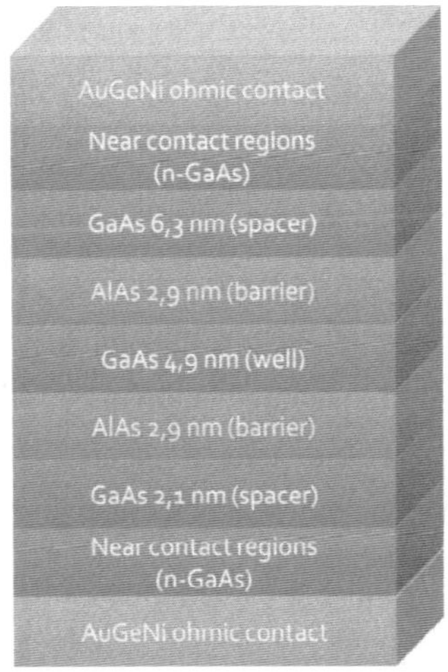

Figure 1. RTD structure.

The equivalent RTD circuit, whereby the IV curve kinetics analysis was carried out, is shown in Figure 2. R_{OC} is the upper and lower ohmic contacts resistance; R_{NCR} is near contact regions resistance; R_{RTS} is the RTS resistance, which is nonlinear and is determined by the RTS quantum mechanical properties.

The total ohmic contacts resistance R_{OC} is defined as

$$R_{OC} = \rho_C \cdot \frac{(S_{OC\_U} + S_{OC\_L})}{S_{OC\_U} \cdot S_{OC\_L}} \tag{1}$$

where ρ_C – ohmic contact resistivity, Ohm·cm^2; S_{OC_U} and S_{OC_L} – the upper and lower contacts areas, cm^2.

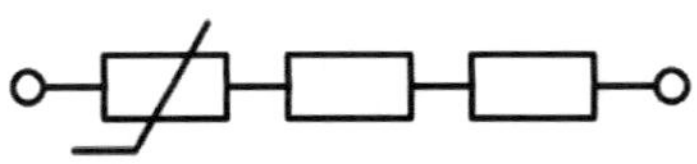

Figure 2. The equivalent RTD circuit.

For experimental research, the technique of accelerated aging by means of the thermal annealing on the sample was used. The temperature of annealing was 300 °C. The authors assumed that at this temperature, the failure mechanisms are the same as on normal operating conditions [9-10].

RTDs annealing was carried out on laboratory air thermostat, which provides to vary the temperature in the range from 40 to 300 °C with accuracy ± 5 °C. Total time of thermal annealing was 8 hours.

IV curve measurements of RTDs before and after thermal annealing were performed using microprobe bench, which consists of a microprobe device, power supply Agilent E3641A and personal computer. This bench provided IV curve measuring of the diodes in voltage range from 0 to 35 V (accuracy ΔU = ± 1 mV) and current range from 0 to 0,8 A (accuracy ΔI = ± 10 μA).

3. Results and discussion

RTD IV curves before and after the thermal annealing were measured, and the statistical analysis was performed. Dispersion of IV curves lies within the measurement error range. In Figure 3 only forward-bias region of IV curves, averaged over whole line, are shown.

As the result of degradation IV-curves shift on voltage axis was took place. IV curve changes could be caused by degradation processes in the resonant tunneling structure and contact areas as well as ohmic contact degradation.

In [11-12] it was indicated, that diffusion blur in the resonant tunneling structure affects RTD IV curve shape. Diffusion processes modelling the for the selected parameters of thermal impact [13] indicates that the diffusion blur of the researched resonant tunneling structure is negligibly small and does not change RTD IV curve shape significantly. This means that R_{PTC} can be considered to be constant under the conditions of this experiment.

The diffusion blur of Si in the near contact region is more observable, but the analysis showed that it also does not significantly affect the series resistance and IV curve of RTD. R_{NCR} is considered to be a constant under the conditions of this experiment.

Therefore, it is expected that RTDs IV curve shift under the experiment conditions is causes by an increasing of AuGeNi ohmic contact resistance as the result of its thermal degradation. The dependence of AuGeNi ohmic contact resistivity ρ_C on time t and temperature T can be described by the following relationship:

$$\rho_C = \rho_{C0} + \chi \cdot \exp\left(-\frac{E_a}{2kT}\right) \cdot \sqrt{t} \tag{2}$$

where ρ_{C0} – initial ohmic contact resistivity (after the production), Ohm·cm^2; χ – coefficient, depending on the design and technology of ohmic contacts, (Ohm·cm^2)/s$^{1/2}$; E_a – activation energy, eV. Activation energy E_a of degradation phenomena of AuGeNi ohmic contacts equals 1 eV [14]; T – temperature, K; k – Boltzmann constant (k = 8,617 10^{-5} eV·K^{-1}); t – time, s.

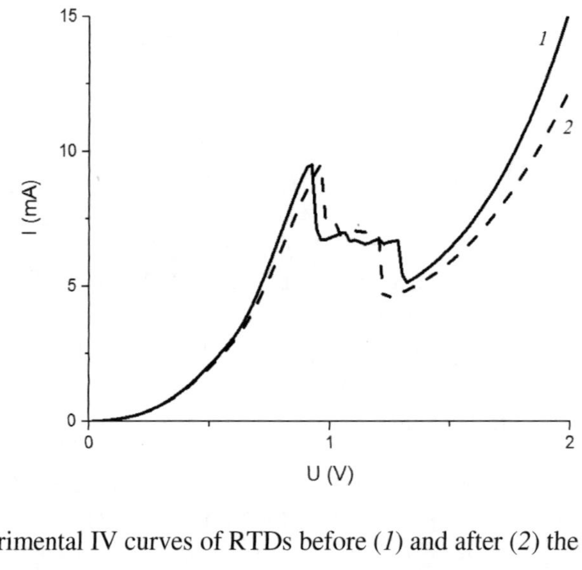

Figure 3. Experimental IV curves of RTDs before (*1*) and after (*2*) the thermal annealing.

The technique was developed to determine the dependence of ohmic contacts resistivity ρ_C on time and temperature and to estimate RTD's time to failure at specified conditions. It is structurally composed of the following basic steps:

- RTDs IV curves measuring before and after the selected number of cycles of thermal annealing;
- RTS diffusion blur mathematic simulation;
- RTS IV curve mathematic simulation based on quantum-mechanical description;
- determination of the ohmic contacts series resistance increasing $\Delta\overline{R}_{OC}$ after each cycle of thermal annealing;

$$\Delta \overline{R}_{OC} = \frac{\sum_{i=1}^{N} \frac{\Delta \overline{U}_i}{I_i}}{N} \tag{3}$$

where $\Delta \overline{U}_i$ – difference in the voltage values of the averaged IV curves at a fixed current value I_i before and after of thermal annealing, N – IV curve points number;

- determination of χ coefficient corresponding to a specific contact resistance change in (2).

The simulation of the RTS diffusive blurring and IV curve is necessary to confirm the fact that the change in the diode IV curve caused by the RTS degradation is much less than its changes caused by the contact degradation. Otherwise, it is necessary to separate the change in the RTD IV curve caused by the RTS degradation from changes caused by contact degradation. The RTS degradation of the investigated diodes is small under the experiment conditions, therefore, it is not considered in this work.

As a result of the processing of experimental data the coefficients χ was determined, which shows the degradation velocity of ohmic contacts. It was equal $(2{,}95 \pm 0{,}01)\cdot 10^{-3}$ $(\text{Ohm}\cdot\text{cm}^2)/\text{s}^{1/2}$ for the investigated line of diodes.

Under the assumption of a high-quality manufacturing of resonant tunneling structure, the developed technique allows to control manufacturing quality of ohmic contacts and to forecast RTD reliability (for the gradual degradation) and devices based on it. In addition, this technique will also be useful in developing new ohmic contacts.

4. Conclusion

RTDs IV curve shift under the experiment conditions is basically causes the ohmic contacts degradation. Obtained dependence of ohmic contact resistivity on time and temperature can be used for determination of RTD IV curve kinetics and to forecast the reliability of resonant tunneling diodes and devices based on it and for the low degradation ohmic contacts engineering.

The resonant tunneling structure degradation of the researched RTD is small under the experiment conditions. Nevertheless, Al and Si diffusion rate will be higher if the semiconductor heterostructure imperfection is higher. That can cause more severe and rapid changes of RTD IV curve shape and indices of devices based on RTD. In this case it is necessary to take into account not only ohmic contacts degradation, but also RTS degradation, for prediction the reliability of RTDs and devices based on it.

Acknowledgments

The research work was supported by Ministry of Education and Science of the Russian Federation under state task № 16.1663.2017/4.6.

References

[1] Mizuta H and Tanoue T *High-speed and functional applications of resonant tunnelling diodes. In The Physics and Applications of Resonant Tunnelling Diodes* 2006 (New York: Cambridge university press) p 133
[2] Suzuki S, Asada M, Teranishi A, Sugiyama H and Yokoyama H 2010 *Appl. Phys. Lett.* **97** 242102
[3] Kanaya H, Shibayama H, Sogabe R, Suzuki S and Asada M 2012 *Applied Physics Express* **5** 124101
[4] Maekawa T, Kanaya H, Suzuki S and Asada M 2016 *Applied Physics Express* **9** 024101
[5] Ivanov Y A, Meshkov S A, Fedorenko I A, Fedorkova N V and Shashurin V D 2010 *J. Commun. Technol. Electron.* **55** 921
[6] Ivanov Y A, Gudkov A G, Agasieva S V, Meshkov S A, Sinyakin V Y and Makeev M O 2014 *Microwave & Telecommunication Technology (CriMiCo), 2014 24th Int. Crimean Conf.* (Sevastopol) pp. 1063-1064
[7] Gudkov A G, Ivanov Yu A, Meshkov S A, Agasieva S V, Petrov V I, Sinyakin V Yu and Shchukin S I 2015 *Biomedical Engineering* **49** 98

[8] Sinjakin V Y, Makeev M O and Meshkov S A 2016 *J. Phys.: Conf. Ser.* **741** 012160
[9] Gorlov M I, Emelyanov V A and Strogonov A V 2004 *Gerontology of silicon integrated circuits* (Moscow: Nauka) 241 p
[10] Swingler J 2014 *Reliability Characterisation of Electrical and Electronic Systems* (Cambridge: Elsevier) 256 p
[11] Makeev M O, Ivanov Y A and Meshkov S A 2015 *J. Phys.: Conf. Ser.* **584** 012014
[12] Makeev M O, Ivanov Y A and Meshkov S A 2016 *Semiconductors* **50** 83
[13] Makeev M O, Meshkov S A and Ivanov Y A 2017 *Key Engineering Materials* **724** 48
[14] Vashchenko V A and Sinkevitch V F 2008 *Physical limitations of semiconductor devices* (Berlin: Springer) 330 p

The influence of the chemical and physical component of the plasma etching of the surface of gallium arsenide on the etching rate in the chloride plasma of the combined discharge

V S Klimin, R V Tominov, A V Eskov, S Y Krasnoborodko and O A Ageev

Department of Nanotechnologies and Microsystems, Institute of Nanotechnologies, Electronics and Electronic Equipment Engineering, Southern Federal University, 2 Shevchenko Street, Taganrog, 347928, Russia

Abstract. In this paper, experimental studies were carried out on the formation of a microrelief on the surface of gallium arsenide substrates. The surface was modified by the method of plasma etching. Etching was carried out in a medium of chloride gas. In the work, the etching rates of the surface of gallium arsenide depend on the power of the inductively coupled plasma source. The etching rate of gallium arsenide was 28.2 nm / min, 24.9 nm / min, 27.8 nm / min, at inductively coupled plasma power values of 200 W, 400 W and 600 W. In the same work, the patterns of etching rate of gallium arsenide surface Taking into account the capacity of the capacitive plasma and the voltage of the shift. The etching rate of gallium arsenide was 516.7 nm / min, 559.5 nm / min, 603.3 nm / min, with capacitive plasma power values of 10 W, 35 W and 70 W, respectively. These results can be used to form a modified surface for the subsequent epitaxial growth of quantum dots based on gallium arsenide.

1. Introduction

The purpose of this work was to study, by atomic force microscopy, the dependence of the rate of etching of the surface of gallium arsenide on the chemical and physical components of plasma etching in an atmosphere of BCl_3. In modern micro- and nanoelectronics semiconductor compounds of the AIIIBV group are widely used in various areas from discrete diodes to devices on heterojunctions [1-7]. High resolution in the formation of structures based on gallium arsenide can be achieved through the use of plasma processing methods. With the reduction of topological standards of modern products of micro- and nanoelectronics, the requirements to surface quality after plasma treatment are very high. A study of the chemical composition of the surface of gallium arsenide substrates by means of Auger spectrometry after plasma-chemical treatment showed that the surface contains up to 5 atomic percent of chlorine. At a depth of about 2.5 nm, the chlorine content is practically nonexistent [8-14]. For plasma BCl_3, much lower concentrations of chlorine atoms are characteristic, but precipitation of solid products of plasma reactions on the walls of the reactor and the surface of the material being treated is often observed. However, the widespread use of plasma etching for gallium arsenide-based structures is not fully realized due to the poor knowledge of the mechanisms of interaction of active plasma particles with the surface [15-18]. The lack of information on the types of active particles providing interaction, the limiting stages of processes, their kinetic characteristics (probabilities, rate constants) does not allow to effectively manage the result of processing, to optimize the process regimes, and also to conduct an adequate comparison of the output parameters of the process with more studied systems [19- 22].

2. Description of the experiment and research methods

In this work we used the method of plasma-chemical etching in a combined chlorine plasma of a capacitive and inductive discharge. This method makes it possible to combine two modes of plasma-chemical etching: reactive-ion etching and etching in an inductively coupled plasma. As experimental samples, plates of own gallium arsenide having a surface after standard liquid polishing were used. The preparation of the gallium arsenide substrate consisted of applying a protective mask of a plasma-resistant photoresist on the surface. Surface control after plasma treatment was carried out using NTegra probe laboratory using a "semi-contact" mode. For samples processed in the plasma in the first stage, AFM images were obtained. The depth of the etched layer was determined from the images obtained by scanning the AFM. Also, with the help of a specialized package «Image Analysis», the roughness of the etched surface was estimated. Before conducting measurements on a microscope, the photoresist mask was removed. In this paper, experimental studies were carried out to determine the etching rate and the roughness of the etched region from the power of various plasma sources, as well as the anisotropy of the plasma etching in a combined plasma. The first batch of samples was processed in a plasma at a pressure of 1-2 Pa, the power of the capacitive plasma source W_{RIT} was 35 W, the flow rates of the process gases were N_{Ar} = 100 cc/m and N_{BCl3} = 10 cc/m. The values of the power of the inductively-coupled plasma source W_{ICP} varied 200, 400, 600 W. The second batch of samples was etched at a pressure of 1-2 Pa, the power of the source of the inductively coupled plasma W_{ICP} - 300 W, the process gas flow rate was N_{Ar} - 100 cc/m, and N_{BCl3} 15 cc/m. The values of the power of the source of the capacitive plasma W_{RIT} varied 10, 35, 70 W.

3. Results and discussion

Based on the results of the experimental studies, the dependencies of the thickness of the etched layer on the etching time for various values of the inductively coupled plasma source power shown in Figure 1 were constructed.

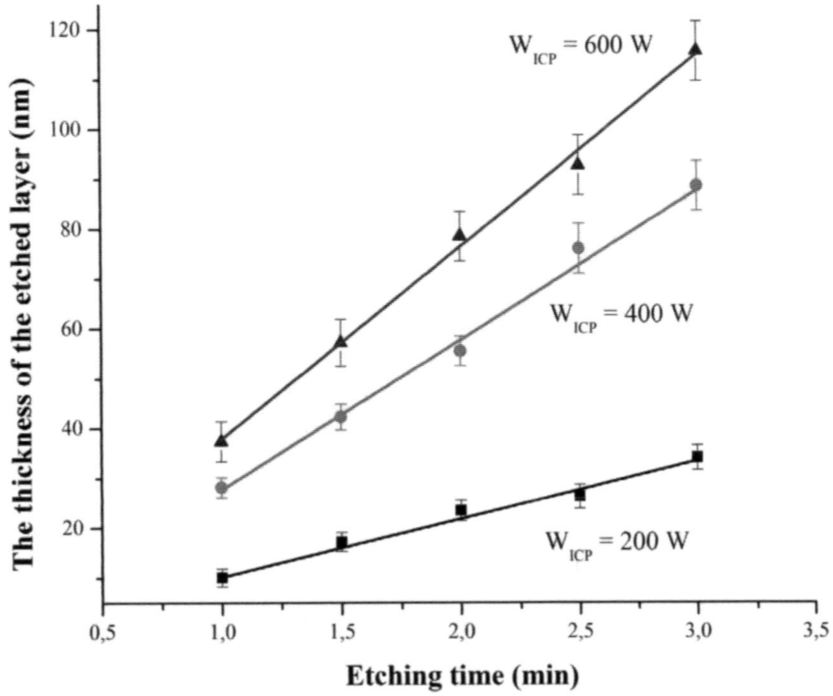

Figure 1. The thickness of the substrate layer is etched GaAs at different values of the power source of the inductively coupled plasma.

Thus, based on the results of the conducted studies, it is possible to estimate the etching rate from the dependencies presented in Fig. 1. The etching rate of gallium arsenide was 28.2 nm/min, 24.9 nm/min, 27.8 nm /min, at inductively coupled plasma power values of 200 W, 400 W and 600 W, respectively. At the second stage, the thickness of the etched layer from the power source of the capacitive plasma was also estimated. Figure 2 shows an AFM image obtained by scanning a sample of a capacitive plasma V_{RIE} = 70 W etched at a power of a capacitive plasma for t = 1 minute. The thickness of the etched layer was h = 480.4 nm, the mean square root roughness of the etched region was S = 3.196 nm. The bias voltage for this mode was U_{bias} = 154 V.

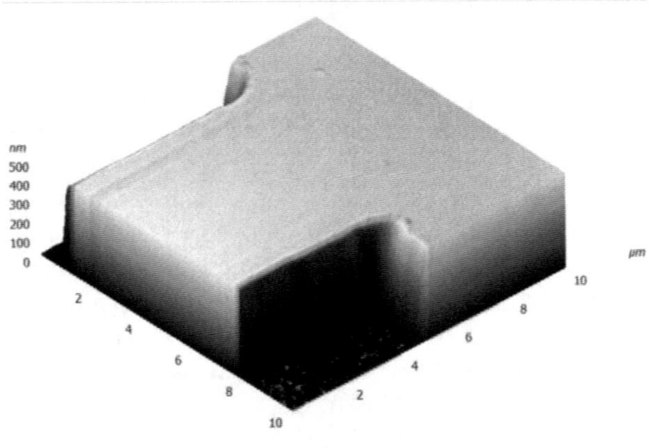

Figure 2. AFM image of etched layer the results of experimental studies of etching GaAs in the combined plasma, on the effect of capacitive plasma source power.

Based on the results of the experimental studies, the dependence of the thickness of the etched layer on the etching time was constructed for different values of the capacitive plasma source power providing the basis of reactive ion etching, shown in Figure 3.

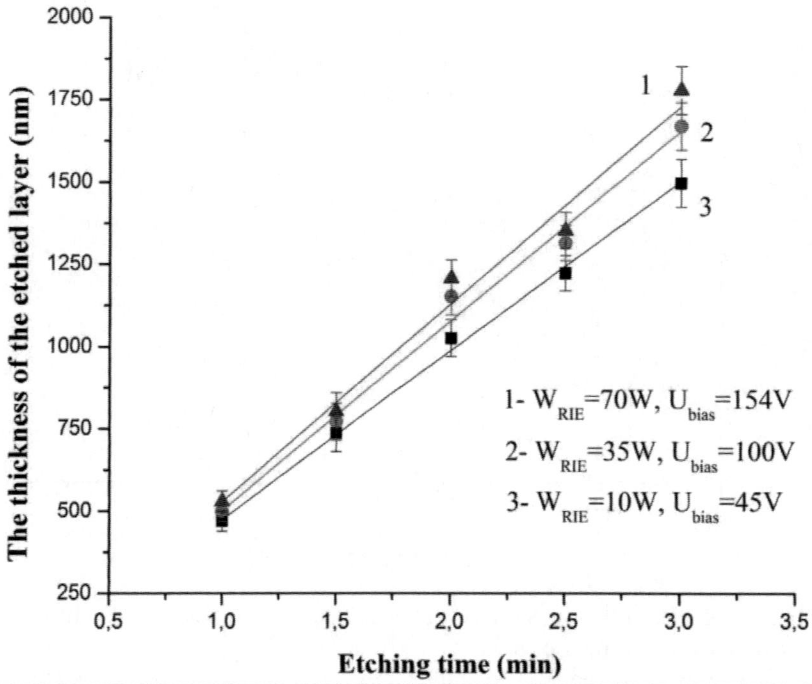

Figure 3. The thickness of the substrate layer is etched GaAs for various values of the power source of capacitive plasma.

It can be seen from the dependences that when the power of a source of capacitive plasma increases, the etching rate increases, but the etching rates at thicknesses not exceeding 500 nm differ insignificantly. The etching rate of gallium arsenide was 516.7 nm/min, 559.5 nm/min, 603.3 nm/min, with capacitive plasma power values of 10 W, 35 W and 70 W, respectively.

4. Conclusion

A study was made of the effect of the power of various plasma sources on the thickness of the etched layer, when etching the surface of gallium arsenide by the plasma method in a BC13 gas medium. It is shown that for large values of the power of an inductively coupled plasma source, the thickness of the etched layer is deeper, which is associated with a large number of particles responsible for the formation of volatile compounds with products of reactions of reactive ion etching. It has also been shown that the rate of etching of gallium arsenide over time with increasing power, and, accordingly, the bias voltage increases. It was found that when the power of the inductively coupled plasma source increases, the roughness remains relatively unchanged, and when the power of the capacitive plasma source changes, the average square roughness of the etched surface increases. When combining the modes, it is possible to select the optimal etching mode, which will allow, most accurately control the thickness of the etched layer and the roughness of the etched surface.

Acknowledgments
This work was supported by the Russian Science Foundation Grant No. 15-19-10006. The results were obtained using the equipment of Common Use Center and Education and Research Center "Nano-technologies" of Southern Federal University.

References

[1] LaBella V P, Bullock D W, Ding Z, Emery C, Harter W G and Thibado P M 2000 *J. Vac. Sci. Tech.* A **18** 1526.

[2] Ageev, O.A., Klimin, V.S., Solodovnik, M.S., Eskov, A.V., Krasnoborodko, S.Y. / The study of influence of the gas flow rate to etched layer thickness, and roughness of the anisotropy field of gallium arsenide is etched in the plasma chemical etching process/ Journal of Physics: Conference Series. 2016. T. 741. № 1. C. 012178.

[3] Ledentsov N N *et al.* 1996 *Sol. St. Electron.* **40** 785.

[4] Riel B J, Hinzer K, Moisa S, Fraser J, Finnie P, Piercy P, Fafard S and Wasilewski Z R 2002 *J. Cryst. Growth* **236** 145.

[5] Viktor S. Klimin, Maxim S. Solodovnik, Vladimir A. Smirnov, Andrey V. Eskov, Roman V. Tominov, Oleg A. Ageev/ A study of the vertical walls and the surface roughness GaAs after the operation in the combined plasma etching / Proc. SPIE 10224, International Conference on Micro- and Nano-Electronics 2016, 102241Z (December 30, 2016).

[6] Marina V. Ilina ; Yuriy F. Blinov ; Oleg I. Ilin ; Viktor S. Klimin ; Oleg A. Ageev /Resistive switching of vertically aligned carbon nanotube by a compressive strain / Proc. SPIE 10224, International Conference on Micro- and Nano-Electronics 2016, 102240U (December 30, 2016).

[7] Morgan C G, Kratzer P and Scheffler M 1999 *Phys. Rev. Lett.* **82** 4886.

[8] Murdick D A, Wadley H N G and Zhou X W 2007 *Phys. Rev.* B **75** 125318.

[9] Shiraishi K and Ito T 1998 *Phys. Rev.* B **57** 6301.

[10] Amrani A, Djafari Rouhani M and Mraoufel A 2011 *Appl. Nanosci.* **1** 59.

[11] Kangawa Y, Ito T, Taguchi A, Shiraishi K, Irisawa T and Ohachi T 2002 *Appl. Surf. Sci.* **190** 517

[12] Ageev O.A., Solodovnik M.S., Balakirev S.V., Eremenko M.M. Effect of interaction in the Ga–As–O system on the morphology of a GaAs surface during molecular-beam epitaxy // Physics of the Solid State – 2016 – V.58, №5 – pp. 1045–1052.

[13] Foxon C T and Joyce B A 1977 *Surf. Sci.* **64** 293

[14] Tok E S, Neave J H, Zhang J, Joyce B A and Jones T S 1997 *Surf. Sci.* **374** 397.

[15] Kley A, Ruggerone P and Scheffler M 1997 *Phys. Rev. Lett.* **79** 5278.

[16] Avery A R, Dobbs H T, Holmes D M, Joyce B A and Vvedensky D D 1997 *Phys. Rev. Lett.* **79** 3938.

[17] Ageev O.A., Smirnov V.A., Solodovnik M.S., Rukomoikin A.V., Avilov V.I. A study of the formation modes of nanosized oxide structures of gallium arsenide by local anodic oxidation // Semiconductors – 2012 – V.46, №.13 – pp. 1616–1621.

[18] Nishinaga T 2004 *Prog. Cryst. Growth Charact. Mater.* **48-49** 104.

[19] Higuchi Y, Uemura M, Masui Y, Kitada T, Shimomura S, Hiyamizu S 2003 *J. Cryst. Growth* **251** 80.

[20] Schmidt M, Johari Z, Ismail R, Mizuta H, Chong H 2012 Microelectron. Eng. 98 313.

[21] Tseng A 2004 J. Micromech. Microeng. 14 15.

[22] Han J, Lee H, Min B, Lee S 2010 Microelectron. Eng. 87 1.

Graphene modification via cobalt and silicon intercalation

G S Grebenyuk[1], M V Gomounova[1], E Yu Lobanova[1], D A Smirnov[2,3], I I Pronin[1,4]
[1] Ioffe Institute, 26 Politekhnicheskaya, 194021 St Petersburg, Russian Federation
[2] St. Petersburg State University, Lieutenant Shmidt Emb. 11, 198504 St. Petersburg, Russian Federation
[3] Institute of Solid State Physics, Dresden University of Technology, D-01069 Dresden, Germany
[4] ITMO University, 197101 St. Petersburg, Russia

E-mail: georgijmail@gmail.com

Abstract. Cobalt and silicon intercalation under graphene formed on the Co(111) surface has been investigated. The thickness of deposited films was varied within the range of up to 2 nm and the temperature of their annealing was 400-450°C. The evolution of atomic and electronic structure of the surface due to intercalation was studied by low-energy electron diffraction and high-resolution photoelectron spectroscopy with synchrotron radiation. It is shown that annealing of deposited Si layer leads to formation on the sample surface of both intercalated and nonintercalated areas. On the intercalated areas, the Co-Si solid solution is formed covered with segregated Si. The presence of a silicon phase strongly affects the electronic structure of graphene and substantially weakens its interaction with the substrate. Subsequent intercallation of Co leads to the restoration of strong coupling between graphene and the substrate.

1. Introduction
The problem of the controlled modification of the graphene properties is a key issue in the study of this material since its discovery [1,2]. One of the possible ways to change these properties is intercalation of graphene with atoms or molecules of other substances, which means the introduction of these particles into the interlayer space between the substrate and graphene. This process has been used successfully for the modification of the electronic and magnetic properties of graphene/substrate interface. For example, the intercalation of graphene with noble metals [3, 4] has been used to decrease the interaction between the graphene and the substrate and to recover the unique electronic properties inherent in free-standing graphene. At the same time, the intercalation of graphene with alkali metal atoms turned out to be an effective method for doping of graphene with electrons [5], and the intercalation with magnetic metals makes it possible to change the magnetic moment of the carbon atoms [6]. On the other hand the graphene-capped metal films are inertness that was demonstrated by the exposure of the system to oxygen, which had no impact either on the composition of the intercalated layers or their electronic structure [7]. In this work, we have analyzed the initial stages of cobalt and silicon intercalation under a graphene layer formed on the surface of the epitaxial cobalt film.

2. Experimental details

The experiments were carried out at Helmholtz-Zentrum Berlin (BESSY II) at the Russian-German beamline. All experiments were done under ultra-high vacuum (UHV) conditions. At first, the W(110) surface was cleaned by the standard technique of annealing in oxygen at a pressure of 1×10^{-7} mbar and a temperature of 1100 °C, followed by flash heating to temperatures of ~2000 °C in an UHV. Then, single-crystal Ni(111) film with the thickness of 10 nm was grown on a clean W(110) substrate to enhance the quality of 3 nm Co film which was grown on Ni. Graphene was grown on Co(111) by chemical vapor deposition. This method consists in exposing the surface to an atmosphere of 2×10^{-6} mbar of C_3H_6 for 10 min at a temperature of 600 °C as previously described in the literature [8]. The high quality of synthesized graphene was confirmed by low-energy electron diffraction (LEED) (Fig. 1).

For preparation of intercalation systems, high-purity Co and Si layers were deposited on top of graphene. The amount of deposited elements was controlled by means of a quartz microbalance and photoelectron spectroscopy (PES). Intercalation of Co and Si underneath graphene was carried out by thermal annealing of the system at temperatures of 400-450°C for 10 min. All stages of sample preparation were controlled in situ by LEED and PES.

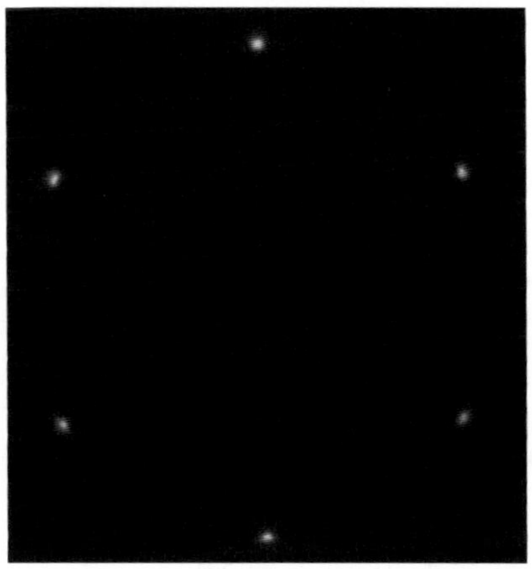

Figure 1. Low-energy electron diffraction pattern of graphene/Co(111) system.

3. Results and discussion.

Information about the structure of graphene and its coupling with a substrate can be obtained from analysis of the C 1s line-shape. The evolution of the C 1s spectra with increasing amount of intercalated materials is shown in Fig. 2. The spectrum of native Gr/Co(111) at the bottom of Fig.2 looks very similar to those previously reported for this system [7, 9]. The main component of this line has binding energy of 284.8 eV, which is ~0.5 eV more than that of graphite due to a sufficiently strong interaction of graphene with a metallic substrate.

The intercalation of the system with Co evidently has a slight influence on the C 1s spectrum, while intercalation of even small amount of silicon results in its strong modification. For example, after intercalation of 0.1 nm Si, the C 1s line splits into two components. The first one has the same binding energy as that prior to the deposition of silicon. Evidently, it corresponds to a contact of graphene with cobalt. The second component is shifted to lower binding energies (284.3 eV) and refers to those parts of graphene that are directly in contact with intercalated silicon. Thus, the sample surface consists of both intercalated and non-intercalated areas. In places without Si atoms, graphene is contacted with cobalt, while in the areas subjected to intercalation, graphene contacts with the silicon atoms, but scarcely interacts with the substrate. Subsequent intercalation of 1nm Co results in increase of the left component which indicates the restoration of the strong coupling between graphene and the substrate.

Such behaviour of the system is similar to that of the graphene-silicon-iron system which was observed previously [10].

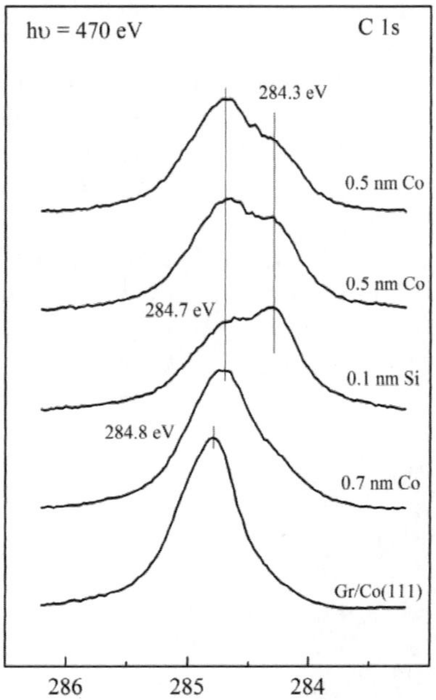

Figure 2. C 1s spectra for graphene on Co(111), and graphene intercalated with Co and Si.

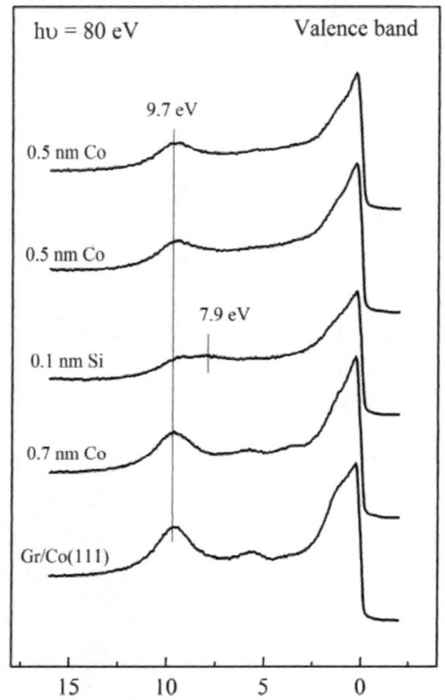

Figure 3. Valence band spectra for graphene on Co(111), and with increasing doses of intercalated Co and Si.

The data illustrating the dynamics of changes in the spectra of valence electrons during the intercalation process are shown in Fig. 3. In the initial valence band spectra for graphene on cobalt there is a well-notable peak at 9.7 eV which corresponds to the π-state of graphene [9]. After intercalation of 0.1 nm of silicon intensity of this peak considerably weakens. This attenuation is compensated by the appearance of a second component with a lower binding energy of 7.9 eV. In accordance with the results of the paper [10], this component can be interpreted as π-state of graphene in contact with intercalated silicon. The 2 eV shift of this component in compare with the component of graphene on cobalt indicates that for the graphene-silicon-cobalt system the Dirac cone of π-states of graphene shifts to the Fermi level by 2 eV, which also confirms the weakening of the interaction of graphene with the substrate due to the silicon intercalation. This conclusion is also supported by the fact that subsequent cobalt intercalation leads to the restoration of the peak at 9.7 eV.

More detailed information about the chemical composition of the intercalated areas of the sample surface can be obtained from the analysis of the Si 2p photoelectron spectra measured at different stages of the intercalation process (Fig. 4). The component with the highest binding energy of 99.3 eV (denoted as A) corresponds to the solid solution of silicon in cobalt according to [11]. The most intensive component with a binding energy of 98.9 eV (denoted as C), which was also observed in the previous study [11], is attributed to silicon on the metal surface but under the graphene layer. The remaining two components with energies of 98.7 and 99.1 eV are explained by the presence of silicon with different cobalt content in the nearest coordination sphere.

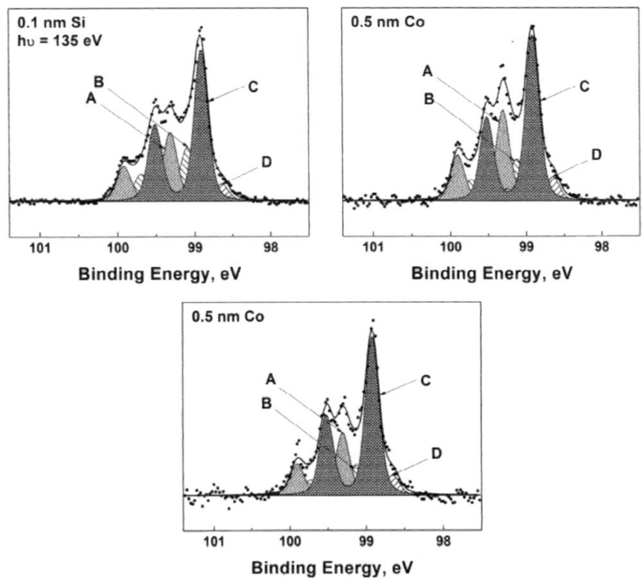

Figure 4. Si 2p spectra measured after the intercalation with silicon and cobalt under graphene formed on the cobalt film.

Thus, it is shown that, at a temperature of 400°C, silicon intercalates under graphene with the formation of two-dimensional islands of silicon and solid solution Co-Si. The capping graphene layer is weakly electronically coupled to silicon. The subsequent intercalation of cobalt leads to the restoration of strong coupling between graphene and substrate.

Acknowledgments
We thank HZB for the allocation of synchrotron radiation beamtime. G.S.G., E.Yu.L. and I.I.P. are grateful to the Russian Foundation for Basic Research for financial support (grant № 16-02-00387).

References
[1] K. Novoselov, A. Geim, S. Morozov, D. Jiang, M. Katsnelson, I. Grigorieva et al., Nature 438, 197 (2005).
[2] A. Varykhalov, J. Sanchez-Barriga, D. Marchenko, P. Hlawenka, P.S. Mandal & O. Rader, Nature Comm. 6, 7610 (2015).
[3] A. Varykhalov, J. Sanchez-Barriga, D. Marchenko, P. Hlawenka, P. S. Mandal, and O. Rader, Nat. Commun. 6, 7610 (2015).
[4] A. Y. Tontegode, Prog. Surf. Sci. 38, 201 (1991).
[5] Yu. S. Dedkov, A. M. Shikin, V. K. Adamchuk, S. L. Molodtsov, C. Laubschat, A. Bauer, and G. Kaindl, Phys. Rev. B: Condens. Matter 64, 035405 (2001).
[6] C. Riedl, C. Coletti, T. Iwasaki, A. A. Zakharov, and U. Starke, Phys. Rev. Lett. 103, 246804 (2009).
[7] M. Weser, E.N. Voloshina, K. Horn, Y.S. Dedkov, Phys. Chem. Chem. Phys. 13, 7534 (2011).
[8] A. Nagashima, N. Tejima, C. Oshima, Phys. Rev. B, 50, 17487 (1994).
[9] G. S. Grebenyuk, M. V. Gomoyunova, O. Yu. Vilkov, B. V. Sen'kovskii, I. I. Pronin, Physics of the Solid State 58, 2135 (2016).
[10] G.S. Grebenyuk, O.Yu. Vilkov, A.G. Rybkin, M.V. Gomoyunova, B.V. Senkovskiy D.Yu. Usachov, D.V. Vyalikh, S.L. Molodtsov, I.I. Pronin, Applied Surface Science 392, 715 (2017).
[11] O. Vilkov, A. Fedorov, D. Usachov, L. V. Yashina, A. V. Generalov, K. Borygina, N. I. Verbitskiy, A. Grü neis, and D. V. Vyalikh, Sci. Rep. 3, 2168 (2013).

Surface changes of poly-L-lactic acid due to annealing

P Juřík[1], I Michaljaničová[1], P Slepička[1], Z Kolská[2] and V Švorčík[1]
[1]Department of Solid State Engineering, University of Chemistry and Technology, Prague, Czech Republic
[2]Faculty of Science, University of J. E. Purkyně, Ústí nad Labem, Czech Republic

Abstract. Surface modifications are very important part of both current cutting-edge research and modern manufacturing. Our research is focused on poly-L-lactic acid, which is biocompatible and biodegradable polymer that offers applications in modern medicine. We observed morphological changes of the surface of metalized samples due to annealing and studied effect of modifications on total surface area and pore surface and volume. We observed that annealing of non-metalized samples had most pronounced effect up to the 70°C, after which all observed parameters dropped significantly. Metallization has changed behaviour of the samples significantly and resulted in generally lower surface area and porosity when compared to non-metalized samples.

1. Introduction
Surface of material plays very important part in it functional properties and final use and therefore modifications of various substrates are thoroughly researched. Many methods ranging from lithography through micro-printing to plasma treatment are implemented in both cutting-edge research and modern manufacturing processes [1-4]. This paper will focus on application of self-organization as this process is currently very attractive for research for several reasons. First, emergence of order from disorder is fascinating subject on its own. Second, practically every biological structure was created due to self-organization and therefore understanding mechanisms behind these processes can significantly help us understand existence of life [5-8]. Finally, self-organization has potential to revolutionize every aspect of technology and manufacturing as it offers way to improve properties of materials on macro-scale due to carefully organized structure on nano-scale, which is practically impossible to achieve by any other process [9-11].
In this paper we would like to present our work on unveiling mechanism behind self-organization due to metallization and subsequent thermal annealing of poly-L-lactic acid, a biocompatible and biodegradable polymer with promise of applications in medicine. Combination of these two methods is used for altering morphology of this polymer and formation ripple-like structures [12-14]. Modification of surface includes chemical and physical changes which both have very important role in bioapplications as they are crucial for cell growth [15]. Careful examination of the changes in total surface area, pore volume and area [16] gives us valuable data for better understanding what occurs during this modification.

2. Experimental
Biopolymer poly-L-lactic acid (PLLA, density 1.25 g•cm-3, glass transition temperature Tg = 60°C, crystallinity 60-70 %, thickness 50 µm, Goodfellow, Ltd., UK) was used in the presented experiments. Gold layers were deposited from Au target (99.999 %) by diode sputtering technique (BAL-TEC SCD 050 equipment, Switzerland). Typical sputtering conditions were: room temperature, sputtering times

10–300 s, argon (purity 99.997 %) pressure of about 5 Pa, electrode distance of 50 mm, electric current of 20 mA. Part of the samples were annealed at 60°C for 60 min (BINDER thermostat, Germany) and then cooled to room temperature in air.
Surface morphology and roughness of the samples were examined by means of atomic force microscopy (AFM). The AFM images were taken on VEECO CP II setup (Veeco, USA) in a tapping mode. Si probe RTESPA-CP with the spring constant 20-80 N•m-1 was used. Roughness (Ra) represents the arithmetic average of the deviations from the centre plane of the sample [17].
Properties of porous CNPs (total surface area, total pore volume and pore size distribution) of were determined using a Colter SA 3100 (Beckman Colter) analyzer. The results were obtained by evaluation of the shape of BET isotherms of nitrogen absorption and desorption [16].

3. Results and discussion

Our samples preparation is shown in Figure 1. Pristine PLLA was sputtered with gold to form metal nanolayer on the surface. Part of these samples was studied as-sputtered, the rest was subsequently annealed. For AFM scans samples were treated only from one side, for surface area and porosity measurements both sides of the material were treated to achieve uniformity of the surface for this method.

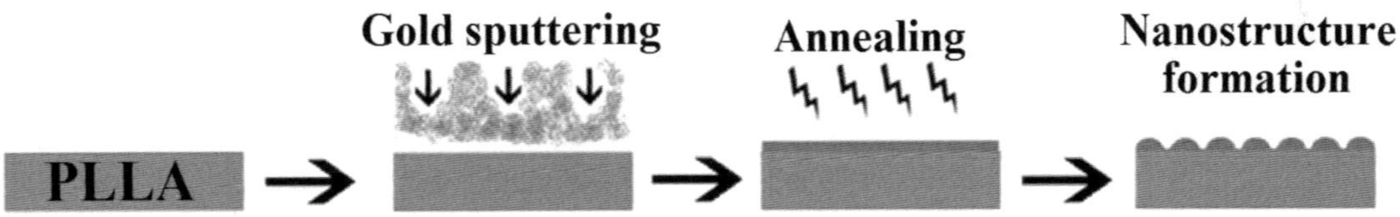

Figure 1 Scheme of poly-L-lactic acid (PLLA) samples preparation. Pristine PLLA foil is sputtered with gold and subsequently annealed, leading to formation of ripple-like structure.

Figure 2 shows AFM scans of PLLA samples. As-sputtered sample have smooth surface with few protruding "spikes", which are more prominent on the sample with thinner metal layer. These structures are either impurities on the polymer surface or crystals that grew through the surface. This morphology changes drastically after annealing. Smooth surface is replaced by wrinkles formed over the surface. Size of the wrinkles directly depend on amount of sputtered gold. We can also observe preferential orientation of the wrinkles, which is particularly prominent on samples with 10 nm thick gold layer.

Figure 3 shows surface area, pore volume and surface as obtained by BET isotherms of nitrogen absorption and desorption. We observed that pristine PLLA has small total surface area and porosity. These properties significantly increase due to annealing above glass transition temperature peaking at 70°C. These changes are probably due to uneven expansion of polymer and formation of pores inside the material. At this temperature, longer polymer chains are still capable of holding the structure during annealing therefore we can observe this large change in both total surface area and porosity. On the other hand, on the samples annealed at higher temperatures (80 and 90°C) we observed sharp drop of both surface area and porosity when compared to samples treated at lower temperatures. We assume that after this temperature material viscosity is either low enough to compensate for thermal caused expansion or longer polymer chains are not capable of holding pores open. Both of these effects, or their combination, lead to one observed effect: pores are not present in material after cooling back to room temperature and measurement.

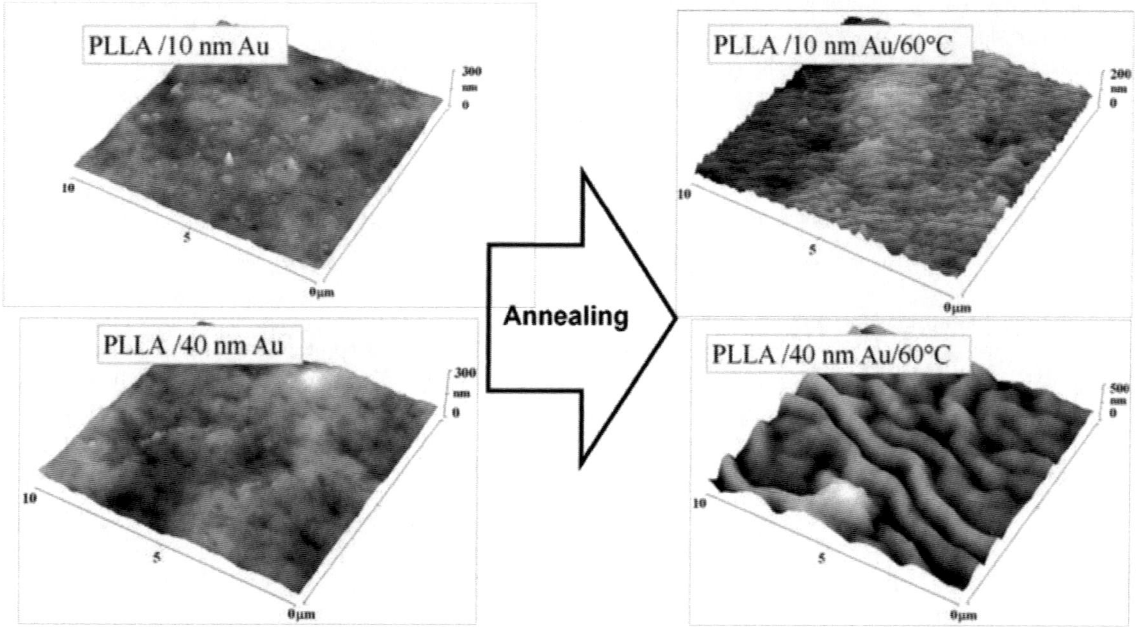

Figure 2 Afm scans of as-sputtered (left) and annealed (right) samples with 10 and 40 nm thick gold layer.

To compare, we also tested metalized samples. They showed in general lower both surface area and porosity when compared to their non-metalized counterparts. We assume that changes in material due to annealing are transformed into wrinkling on the surface thanks to the metal layer. We also observe sharp increase in total surface area for the sample with thickest metal layer, while porosity remains very similar to samples with thinner metal layers. In this case surface structure is responsible for this additional surface, while porosity changes only slightly.

This change of behaviour led us to the idea of principle behind wrinkle formation. We assume that there is some residual stress in the polymer, probably from manufacturing process, which is responsible for wrinkle formation. To support this assumption we prepared four samples with carefully marking their original orientation on the polymer foil. These samples were subsequently sputtered with gold layer and annealed with various combinations of deviations from original orientation to rule out that orientation of the samples during either sputtering or annealing have any effect on the final wrinkle orientation. Afterwards samples were arranged so that they had same orientation as in the original polymer foil and observed by means of AFM. Observed orientation of wrinkles differed by less than 10° from each other and therefore orientation during sample preparation has no significant effect on final wrinkle orientation, therefore reason behind ripple-like structure formation must be intrinsic property of polymer foil: most probably internal stress from manufacturing process as we suggested. This corresponds well with our findings. Material directly from manufacturing has low porosity in comparison with annealed samples. Annealing allows release of the internal stress by rearrangement of polymer and formation of pores as shown in Figure 3. Higher temperatures (more than 80°C) may allow release of stress by creep or flow of polymer chains in addition to pore formation (and their subsequent collapse), therefore observed increase is significantly smaller. In comparison to this, metal allows another mechanism of stress release: by wrinkling. Due to different thermal expansion coefficients of PLLA and gold local instabilities appear during annealing that force polymer rearrangement, resulting in release of internal stress. This corresponds well with observed almost non-existent porosity of metalized samples.

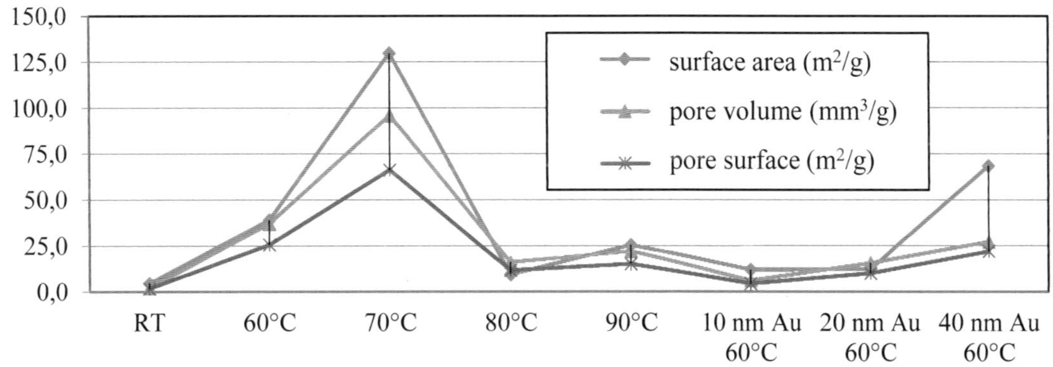

Figure 3 Graph showing total surface area, pore volume and surface obtained by evaluation of BET isotherms

4. Conclusion

Annealing of samples causes significant changes to surface morphology and porosity of PLLA. Annealing of pristine PLLA resulted in significant changes to both surface area and porosity of the samples, while most pronounced change was on samples treated at 70°C. Higher temperatures caused collapse of polymer structure and porosity sharply dropped. On the other hand metalized samples had only small changes in porosity after annealing but we observed increase of surface area due to wrinkling, especially on samples with thicker gold layer. Based on the obtained data, we suggest that both increase in porosity and wrinkling are mechanisms of stress release that is inherent to original polymer foil and originates probably from manufacturing process. These data offers us insight into this mechanisms and base for other experiments with PLLA.

5. Acknowledgments

This work was supported by the Czech Science Foundation (GA CR) under the project 108/12/G108.

6. References

[1] Malloy M and Litt L C 2011 *J Micro Nanolithogr MEMS MOEMS.* **10**
[2] Colson P, Henrist C and Cloots R 2013 *J. Nanomater.*
[3] Vandrovcova M, Vacik J, Svorcik V, Slepicka P, Kasalkova N, Vorlicek V, Lavrentiev V, Vosecek V, Grausova L, Lisa V and Bacakova L 2008 *Phys. Status Solidi A-Appl. Mat.* **205** 2252
[4] Slepicka P, Trostova S, Kasalkova N S, Kolska Z, Malinsky P, Mackova A, Bacakova L and Svorcik V 2012 *Polym. Degrad. Stab.* **97** 1075
[5] Philp D and Stoddart J F 1996 *Angew. Chem.-Int. Edit.* **35** 1155
[6] Dullemeijer P 1999 *Science.* **284** 1627
[7] Ball P 2012 *Archit. Des.* **82** 22
[8] Pascal R 2015 *Isr. J. Chem.* **55** 865
[9] Lash M H, Fedorchak M V, McCarthy J J and Little S R 2015 *Soft Matter.* **11** 5597
[10] Gentili D, Valle F, Albonetti C, Liscio F and Cavallini M 2014 *Acc. Chem. Res.* **47** 2692
[11] Athanasiou K A, Eswaramoorthy R, Hadidi P and Hu J C 2013 *Self-Organization and the Self-Assembling Process in Tissue Engineering* (Palo Alto: Annual Reviews)
[12] Jurik P, Slepicka P, Mistrik J, Janicek P, Rimpelova S, Kolska Z and Svorcik V 2014 *Appl. Surf. Sci.* **321** 503
[13] Jurik P, Slepicka P, Kolska Z, Rimpleova S, Svorcik V and Ltd T 2015 *Nanocon 2015: 7th International Conference on Nanomaterials - Research & Application.* 666
[14] Juřík P, Slepička P, Nagyová M and Švorčík V 2017 *Surf. Coat. Technol.* **311** 344

[15] Sridhar S, Venugopal J R, Sridhar R and Ramakrishna S 2015 *Colloid Surf. B-Biointerfaces.* **134** 346
[16] Brunauer S, Emmett P H and Teller E 1938 *J. Am. Chem. Soc.* **60** 309
[17] Jurik P, Slepicka P, Kolska Z and Svorcik V 2016 *Mater. Lett.* **165** 33

Resistive switching in ZnO/ZnO:In nanocomposite

D A Khakhulin[1], Z E Vakulov[1], V A Smirnov[1], R V Tominov[1], Jong-Gul Yoon[2] and O A Ageev[1]

[1] Research and Education and Centre "Nanotechnology", Southern Federal University, Taganrog, 347922, Russian Federation
[2]Department of Physics, Natural Sciences College, University of Suwon, Wau-ri, 445-743, Republic of Korea

Abstract: A lot of effort nowadays is put into development of new approaches to processing and storage of information in integrated circuits due to limitations in miniaturisation. Our research is dedicated to one of actively developed concepts – oxide based resistive memory devices. A material that draws interest due to its promising technological properties is ZnO but pure ZnO lacks in performance in comparison with some other transition metal oxides. Thus our work is focused on improvement of resistive switching parameters in ZnO films by creation of complex nanocomposites. In this work we report characterisation of a nanocomposite based on PLD grown ZnO films with inclusions of In. Such solution allows us to achieve improvements of main parameters that are critical for ReRAM device: R_{HRS}/R_{LRS} ratio, endurance and retention.

1. Introduction

Current crisis in microelectronics, caused by fundamental limitations that occur during further miniaturisation could be solved in a number of different ways [1], including implementation of $A^{III}B^{V}$ materials [2], carbon nanotubes [3], local oxidation technology [4] and some other concepts [5]. In that regard, one of the most widely discussed topics is non-volatile memory devices based on materials with resistive switching behaviour. Among a vast range of such substances certain attention is drawn by transition metal oxides [6], and a material of particular interest is ZnO, due its technological accessibility [7]. However, resistive memory devices (ReRAM), that implement pure ZnO, considered to be less effective, in comparison with devices implementing HfO_x, SiO_x and TiO_x [8]. Thus we experiment with ZnO/ZnO:In nanocomposite to improve both R_{HRS}/R_{LRS} and endurance, as well as retention characteristics for an implementation in a memory device.

2. Experiment

To obtain ZnO/ZnO:In nanocomposite we used pulsed laser deposition (PLD) equipment Pioneer 180 (Neocera, USA). ZnO and ZnO:In targets were consequently ablated by excimer KrF laser (λ=248nm). Energy density on target surface was maintained at 2J/cm^2. The quantity and frequency of laser pulses were 50.000 and 10Hz respectively. Target-substrate distance was 80mm. Nanocomposite films were deposited on silicon substrates. To characterise structural and electrical properties of these films we used following equipment. Structural studies were conducted using atomic-force microscopy (AFM) system Solver P47-Pro (NT-MDT, Russia), dual-beam scanning electron microscopy (SEM) system with focused ion beam module Nova Nanolab 600 (FEI, Netherlands) and Rigaku Miniflex 600 (Rigaku, Japan). For electrical studies electrometer Keithley 617 and semiconductor characterisation system Keithley 4200-SCS (Keithley Instruments, USA) were used. We implemented top-bottom measurements layout: one of the electrodes was placed on an electrical contact to the bottom of the film and the second one to the film itself.

3. Results

3.1. Structural properties of ZnO/ZnO:In nanocomposite

To characterise morphology and thickness of the ZnO/ZnO:In films we conducted SEM and AFM studies. They showed, that we obtained nanocrystalline film with thickness around 80 nm and uniform roughness with average meaning of about 19 nm. Both SEM and AFM images of the film are given in figure 1 in part (a) and (b), respectively. Optimal thickness and roughness that allow to achieve optimal film properties for implementation as a ReRAM structure were determined during our previous research, which was dedicated to the influence of PLD process parameters on resistive switching in ZnO films.

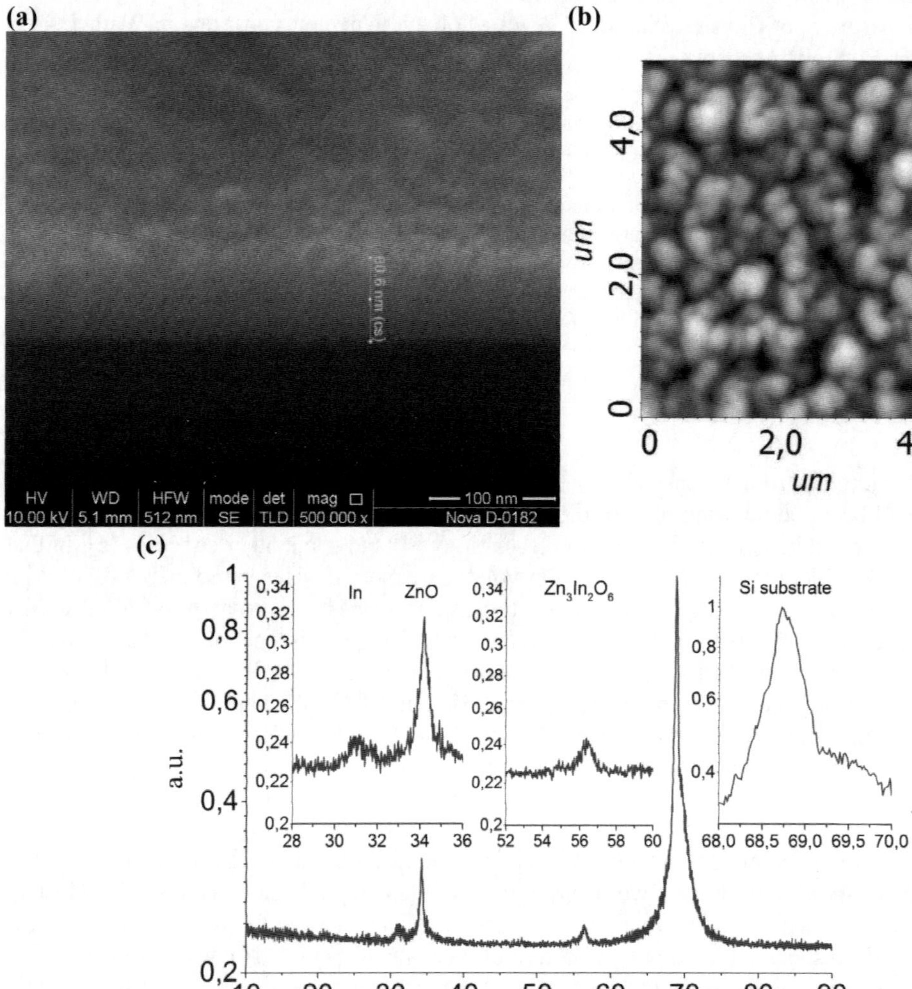

Figure 1(a,b,c). (a) SEM image of the films with a cross-section cut, demonstrating thickness and nanocrystalline structure of the film, and **(b)** AFM scan, showing surface morphology. **(c)** XRD pattern of ZnO/ZnO:In film on silicon substrate

We analysed composition of the film by XRD to exclude possibility of major influence of indium based compounds on resistive switching properties of ZnO/ZnO:In films. Structure-wise, films of ZnO/ZnO:In nanocomposite that were grown at target-substrate of 80mm show minor inclusions of $Zn_3In_2O_6$, lack of In_2O_3 and $Zn_3In_2O_6$, unlike samples, that were grown with another target-substrate distances. Main peaks on XRD patterns correspond to ZnO, In and substrate. XRD pattern of ZnO/ZnO:In film on silicon substrate is shown in figure 1(c).

3.2. Electrical properties of ZnO/ZnO:In nanocomposite

ZnO/ZnO:In nanocomposite shows high resistance state to low resistance state switch ratio (R_{HRS}/R_{LRS}) of about 2 at U_{read}=0.25V and up to about 500 at U_{read}=2.25V. Resistive switching occurs at U_{write}=2.5V and rewrite process at $U_{rewrite}$=-3.5V. Current-voltage characteristics (CVC) of the film, that allow to estimate U_{write} and $U_{rewrite}$ and R_{HRS}/R_{LRS} at different voltages is provided in figure 2(a). In terms of endurance characteristics film keeps switching stably from LRS to HRS for 25 cycles at U_{read}=0.25V then switching properties degrade: R_{HRS}/R_{LRS} falls down due to decrease of R_{HRS}, yet, resisting switching continues to occur for up to 100 cycles even with a lower resistances ratio. Endurance characteristic of the ZnO/ZnO:In nanocomposite is depicted in figure 2(b). Despite better R_{HRS}/R_{LRS} higher U_{read} is not applied, since it leads to a faster degradation of endurance characteristics. Retention tests showed, that film could remain in LRS for about 10^3sec.

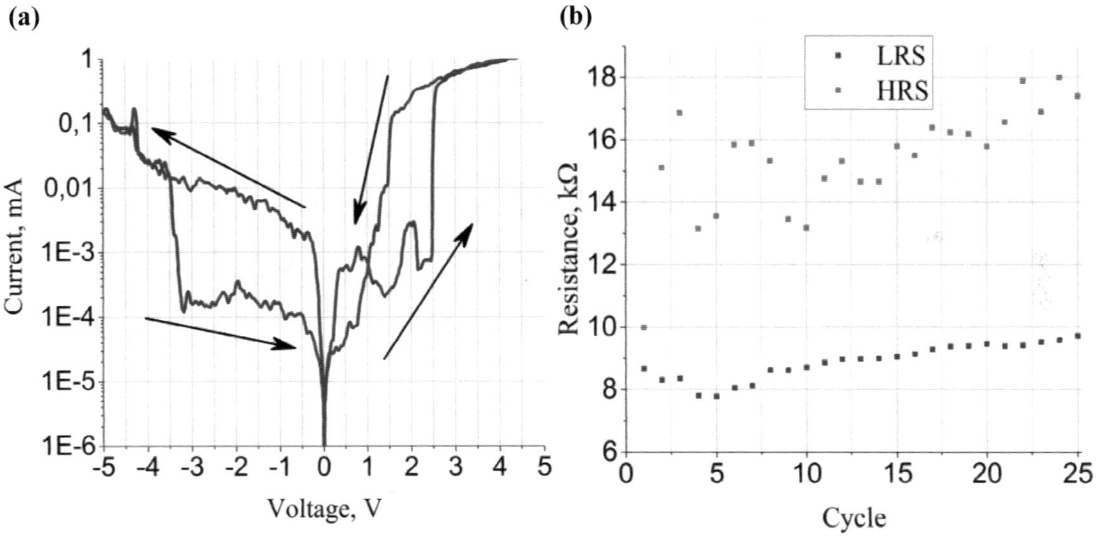

Figure 2(a,b). (a) CVC characteristics of the ZnO/ZnO:In, and **(b)** edurance test, obtained at U_{read}=0.1V.

4. Conclusion

We studied resistive switching in PLD grown films of ZnO/ZnO:In nanocomposite. Our structural studies showed that switching properties that were obtained are linked to the processes in ZnO and ZnO:In rather than other compounds, that are inevitably formed in presence of oxygen. Electrical properties measurements showed that nanocomposite films show both good R_{HRS}/R_{LRS} and retention characteristics, while endurance characteristics are a matter of further improvement.

Thus, there are two main conclusions to be made. First of all, PLD allows us to form films of nanocomposites with controlled structural and electrical properties to be used as a part of a ReRAM device. Secondly, we may conclude, that inclusion of additional components to the film has a positive impact on its resistive switching properties. Further experiments with a ZnO:In as a nanocomposite component and introduction of other materials into ZnO could possibly lead to a major improvement of endurance characteristics, as well as R_{HRS}/R_{LRS}

Acknowledgements
This work was financially supported by Southern Federal University by grant VnGr-07/2017-02 and within the program of stimulation of international mobility. The work was executed on the equipment of Research and Education Centre "Nanotechnology" and Collective Use Centre "Nanotechnology", Southern Federal University and Laboratory of Physics of Natural Sciences College, University of Suwon.

References

[1] Carballo J A, Chan W T J, Gargini P A, Kahng A B and Nath S 2014 ITRS 2.0: Toward a re-framing of the Semiconductor Technology Roadmap *32nd IEEE Int. Conf. Comput. Des. ICCD 2014* 139–46
[2] Ageev O A, Solodovnik M S, Balakirev S V and Eremenko M M 2016 Effect of GaAs native oxide on the Ga droplets formation during GaAs MBE growth *J. Phys. Conf. Ser.* **741** 1–2
[3] Ageev O A, Blinov Y F, Il'in O I, Kolomiitsev A S, Konoplev B G, Rubashkina M V., Smirnov V A and Fedotov A A 2013 Memristor effect on bundles of vertically aligned carbon nanotubes tested by scanning tunnel microscopy *Tech. Phys.* **58** 1831–6
[4] Avilov V I, Ageev O A, Kolomiitsev A S, Konoplev B G, Smirnov V A and Tsukanova O G 2014 Formation of a memristor matrix based on titanium oxide and investigation by probe-nanotechnology methods *Semiconductors* **48** 1757–62
[5] Cavin R K, Lugli P and Zhirnov V V. 2012 Science and Engineering Beyond Moore's Law *Proc. IEEE* **100** 1720–49
[6] Pan F, Gao S, Chen C, Song C and Zeng F 2014 Recent progress in resistive random access memories: Materials, switching mechanisms, and performance *Mater. Sci. Eng. R Reports* **83** 1–59
[7] Ageev O A, Dostanko A P, Zamburg E G, Konoplev B G, Polyakov V V. and Cherednichenko D I 2015 Effect of the processes in the laser ablation plume on the resistivity and morphology of nanocrystalline ZnO films *Phys. Solid State* **57** 2093–8
[8] Ha S D and Ramanathan S 2011 Adaptive oxide electronics: A review *J. Appl. Phys.* **110**

Nonlinear electron emission in ultrashort laser pulses and rescattering suppression

E A Mikhin[1], M A Preobrazhenskii[1] and P A Golovinski [1,2]

[1]Physics Research Laboratory, Voronezh State Technical University, Voronezh, 20-Letiya Oktyabrya st., 84, 394006, Russia
[2]Moscow Physical-Technical Institute (State University), Moscow region, Dolgoprudny, Institutskii per., 9, 141700, Russia

Abstract. The process of nonlinear electron emission from a metal surface under the action of femtosecond laser pulse with moderate intensity ~10^{11} W/cm^2 is considered. One-dimensional model is formulated, taking into account the advantage of the *p*-polarized light in the nonlinear emission. The time dependent Schrödinger equation with fixed equilibrium boundary conditions is solved in the half-space using the Laplace transform technique. The energy spectrum of emitted electrons is presented, including analysis of dependence on the laser pulse parameters. In the framework of classical mechanics, the impact of additional dc electric field on the rescattering of emitted electron on the metal surface is investigated. The model reproduces the fiches of electron energy distribution and can be used at arbitrary Keldysh parameters.

1. Introduction

The nonlinear cold emission of electrons from a metallic nano-needle under the action of a short laser pulse has a number of unique features that are very attractive for possible practical applications. For its experimental realization, it is sufficient to have moderate intensities of the laser field, since in the nano-needle, surface plasmons are generated at the metal-vacuum interface, which leads to a significant enhancement of the laser field near the tip [1]. Arising surface plasmon resonance is a collective oscillation of conduction band electrons that typically occurs at optical frequencies in noble metals. The characteristic size of the amplification region is several tens of nanometers, which makes it possible to of obtain a practical point source of electrons.

A strong nonlinear dependence of the photoemission on the laser field strength leads to the formation of free electrons mainly at small intervals of time near the amplitude values of the field. As a result of nonlinear photoemission, electron bunches are formed with duration less than the period of oscillations of the laser pulse. Nonlinearity is also manifested in multiphoton above-threshold photon absorption, which leads to the ejection of electrons with energies much greater than the energy of the light quanta.

The emission of electrons from the nano-needle caused by the field of an ultrashort laser pulse can be used in a number of laser applications related to monitoring, control and diagnostics of various nano-sized objects [2, 3]. Generated in this way ultrashort coherent electron bunches are crucial to the free electron lasers, laser acceleration of relativistic electrons, picosecond cathodoluminescence, and femtosecond electron diffraction. They would enable exciting technological developments such as four-dimensional time-resolved electron microscopy, spectroscopy, and holography.

2. Theoretical model

The experimental data [4] show a dependence on the laser polarization, indicating that the emission yield is about one order of magnitude larger for the *p*-polarized light electric field normal to the surface then for the *s*-polarization. This result could apparently suggest, according to the model proposed by Broudy [5], that the extraction mechanisms are directly related to the component of the electric field normal to the surface. Appropriate analytical one-dimensional model of electron emission from a metal surface, under the influence of both a dc electric field and laser field illumination with exact solution was derived to electron emission for arbitrary values of laser field, laser frequency, dc electric field, and metal work function [6]. However, the considered electron emission from a metal-vacuum interface driven by a combination of a dc electric field and a light field was limited to a single carrier frequency.

We improve the model, describing the motion of electrons in a half-space under the influence of arbitrary pulsed laser field. The boundary conditions will ensure the asymptotic behavior of the metal electronic states outside its surface. We write down the one-dimensional Schrödinger equation for an electron in the laser pulse field, described by the vector potential $A(t)$ for the half-space outside the metal

$$i\frac{\partial\psi(x,t)}{\partial t}=\frac{1}{2}\left(\hat{p}+\frac{1}{c}A(t)\right)^2\psi(x,t),\quad x>0. \tag{1}$$

Here $\hat{p}$ is the momentum operator, $\psi(x,t)$ is a wave function. We use the atomic system of units in which $m=\hbar=e=1$.

Consider laser field with a strength $F(t)<<1$, and assume that external electromagnetic field weakly distorts electron states in the metal. The initial states belonging to the solid target band structure are taken as exponentially decreasing into the vacuum. Under our assumptions, the wave function $\psi(x,t)$ satisfies the boundary conditions

$$\begin{aligned}&\psi(0,t)=B\exp(-iE_0t),\\&\psi'_x(0,t)=-\kappa B\exp(-iE_0t).\end{aligned} \tag{2}$$

Here B is the normalization constant, $\kappa=\sqrt{2|E_0|}$, E_0 is the energy of an electron inside the metal, the prime in the second line of Eq. (2) denotes the derivative with respect to the space variable.

The solution of Eq. (1) with the boundary conditions (2) in the positive half-space ($x>0$) can be found using the Laplace transform $f(s,t)=\int_0^\infty\exp(-sx)\psi(x,t)\,dx$, where $f(s,t)$ is transform of the function $\psi(x,t)$, s is a complex variable. In the transform space Eq. (1) takes the form

$$i\frac{\partial f(s,t)}{\partial t}=\frac{1}{2}(-is+v(t))^2 f(s,t)+\left\{s\frac{\alpha(t)}{2}+\frac{\beta(t)}{2}+iv(t)\alpha(t)\right\}. \tag{3}$$

For more evident written we introduce notations: $\alpha(t)=\psi(0,t)$, $\beta(t)=\psi'_x(0,t)$ and $v(t)=A(t)/c$. When enter additional parameters for a classical electron displacement under the action of the laser field and the integral from the kinetic energy:

$$a(t,t_1)=\int_{t_1}^{t}v(t_2)dt_2 \text{ and } S(t,t_1)=\int_{t_1}^{t}\frac{v^2(t_2)}{2}dt_2 \tag{4}$$

respectively, the solution of Eq. (3) can be written as

$$f(s,t) = -i\int_0^t \left(s\frac{\alpha(t_1)}{2} + \frac{\beta(t_1)}{2} + iv(t_1)\alpha(t_1) \right) \exp\left\{ i\left(\frac{s^2(t-t_1)}{2} + isa(t,t_1) - S(t,t_1) \right) \right\} dt_1 . \qquad (5)$$

Substitution $s = ip$ into Eq. (5) gives the solution of the original Eq. (1) in the momentum representation, and the calculation of spectral distribution for the emitted electrons as a function of the pulse parameters is reduced to the integral

$$f(p,T) = -iB\int_0^T \left(\frac{ip-\kappa}{2} + iv(t_1) \right) \exp(-iE_0 t_1) \exp\left\{ i\left(-\frac{p^2(T-t_1)}{2} - pa(T,t_1) - S(T,t_1) \right) \right\} dt_1 . \qquad (6)$$

Here T is duration of the pulse. With the aid of the obtained function $f(p,T)$, the electron energy spectrum can be written as

$$\frac{dN}{dE} = \frac{\left| f(\sqrt{2E},T) \right|^2_{T=\infty}}{\sqrt{2E}} . \qquad (7)$$

Eq. (7) describes a spectrum of emitted electrons having definite initial energy. However, we need take into account that electrons in the metal are distributed over the initial energies. According to the Sömmerfeld theory, the number of electron states in a unit volume of a metal per unit energy interval is determined as

$$\frac{dn}{dE_0} = \frac{1}{\pi^2} \frac{\sqrt{2(E_0 - U_0)}}{1 + \exp\left(\frac{E_0 - E_f}{k\theta} \right)} , \qquad (8)$$

Here θ is the metal temperature, E_f is the Fermi energy, U_0 is the energy corresponding to the bottom of the conduction band, k is the Boltzmann constant.

After integration over the initial electronic states energy and taking into account Eq. (8), we obtain the final equation that gives the energy spectrum of the emitted electrons

$$\frac{dN}{dE} = \frac{1}{\pi^2} \int_{U_0}^{E_f} \sqrt{\frac{E_0 - U_0}{E}} \left(1 + \exp\left(\frac{E_0 - E_f}{k\theta} \right) \right)^{-1} \left| f_{E_0}(\sqrt{2E},T) \right|^2_{T=\infty} dE_0 . \qquad (9)$$

The integration in Eq. (9) is carried out from the bottom of the metal conduction band U_0 to the energy E_f . At the room temperature, the population of electronic states above the Fermi level can be neglected.

3. Electron spectrum

To consider the impact of the laser field on the spectrum of emitted electrons, let us determine the shape of the pulse. In the few-cycle regime, the electric field $F(t)$ can be written as $F(t) = f(t)\cos(\omega t + \varphi)$, where $f(t)$ is the pulse envelope, ω is the laser carrier frequency. In this parameterization the electric field depends on the carrier absolute phase φ, although the envelope is the same for all the pulses. Let us assume for further examination the Gaussian envelope, so

$$F(t) = F_0 \exp\left(-\frac{t^2}{\tau^2} \right) \cos(\omega t + \varphi) . \qquad (10)$$

Here F_0 is the amplitude of the laser field strength, τ determines the laser pulse duration.

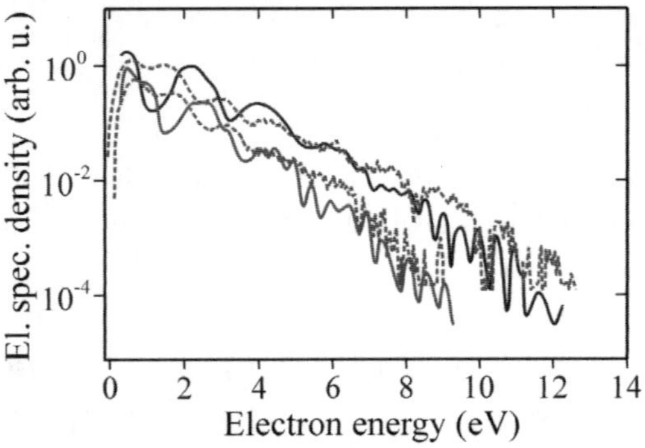

Figure 1. Theoretical and experimental [8] photoelectron spectrums for tungsten target averaged over the carrier-envelope phase.

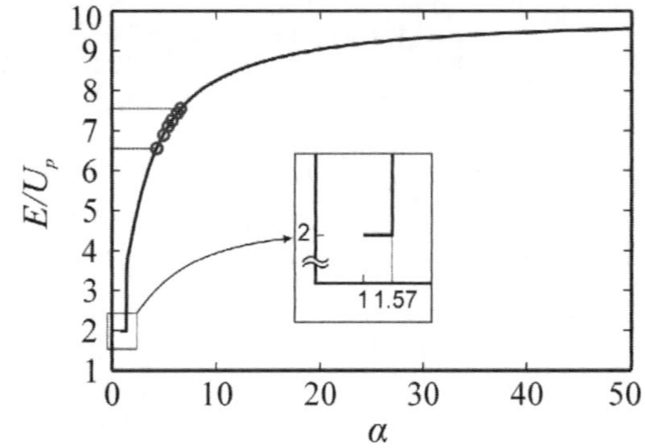

Figure 2. The maximum electron energy as a function of the ratio the ac field strength to the dc electric field.

The magnitude of the laser pulse field influences efficiently on the electron emission dynamics. Increasing the field strength leads to the greater yield of electrons according to higher photoemission probability and stretching the electron energy spectrum as the manifestation more effective electron acceleration by the laser field. Emitted electron spectrum is shown in figure 1 for different strengths laser pulse. Solid curves represent the results of calculations within the framework of our model. Two amplitude values of the electric strength of the laser field of 6.9 V/nm and 8.7 V/nm correspond to these curves in the order from below to the top. Other parameters taken to calculate the electron spectrum were $\lambda = 800\,\text{nm}$, $\varphi = 0$, $\tau = 6.5\,\text{fs}$ и $E_f = -4\,\text{eV}$. The Keldysh parameter was 1.8 and 2.3, and the nonlinear emission was intermediate between the multiphoton emission and the tunneling, more similar to the first one. This is confirmed also by existence characteristic peaks in the electron spectrum, which of them demonstrates the result of fixed number absorbed photons of the laser field. The minimum number of photons required for above-threshold electron emission is $K = \langle |E_f| / \omega + 1 \rangle$, where $\langle ... \rangle$ is the integer part of a number. In the present case $K = 3$, and the position of the first peak in figure 1 is located according to this number. Subsequent peaks show a lower yield of electrons and are separated by almost equidistant gaps $\sim \omega$. The photon spectrum of incident laser light is broad enough when one has a very short pulse, and in addition some contribution to the photoemission is produced by electrons having initial energy below the Fermi level. As a result, the energy spectrum of the emitted electrons is broadened, and peaks in the spectrum are not strictly equidistant.

Experimental realization of electron nonlinear cold emission from a metal surface implies usage relatively large laser fields. However, increasing laser intensity leads to heating of target and growth of thermoemission instead of the nonlinear field process. To overcome this difficulty, cozily to operate with ultrashort laser pulses of moderate intensity and magnify the field, using effect of plasmon superfocusing on the tip of a nanoneedle. The gain coefficient for the electric field strength by this mechanism can reach ~10 [7], and its magnitude strongly depends of a tip radius, thus specially prepared targets with nanostructured surfaces are exploited. Nevertheless, even with special treatment of the target surface and excitation of plasmons in it, the electron yield per laser pulse is small, and for a set of representative statistics, a large number series of laser pulses are used. Along with this, an additional dc electric field is applied, which sometimes is comparable in magnitude to the laser field strength. The field gain effect for electromagnetic wave we incorporate in the model as additional factor in Eq. (10), at the same time, the impact of ad substantial electric field remains out of frame present consideration.

IOP Conf. Series: Journal of Physics: Conf. Series **917** (2017) 092009 doi:10.1088/1742-6596/917/9/092009

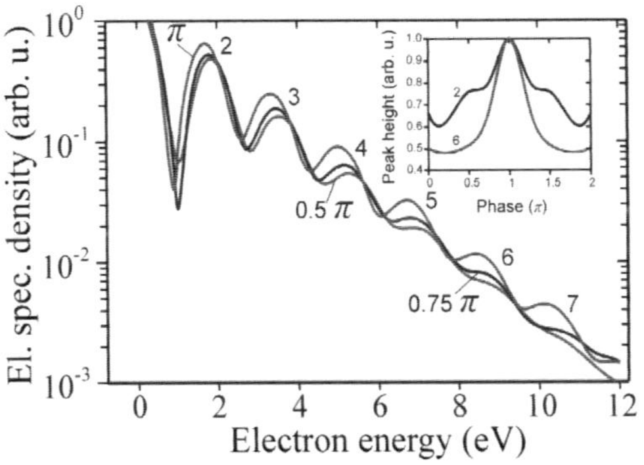

Figure 3. Photoelectron spectrums for different carrier-envelope phases: $F_0 = 8.7$ V/nm , λ = 800 nm, τ = 6.5 fs. The phases for different curves are indicated in the figure.

To demonstrate the typical shape of the experimental electron spectrum, we plotted two of them in the figure 1, following the data [8] for cold nonlinear electron emission from the tungsten nanoneedle with radius of tip ~50 nm stimulated by laser radiation. Under these conditions, plasmons generation gives a considerable magnification of the field near the needle apex. The experimental ac electric field strengths of 4.32 V/nm and 4.98 V/nm correspond to these curves in the order from below to the top. The dc electric field strength in experiment was 0.8 V/nm. Other laser field and metal parameters are exactly the same as in the numerical simulations. Experimental electron spectrums have the local peaks separated with the gaps nearly equal the laser frequency as it is seen also in theoretical dependencies that indicates domination the laser field over the dc field and shows multiphoton character of emission. Because experimental conditions essentially differ from the theoretical model parameters, it would be very odd to expect exact coincidence of the results, and we plotted these curves together only to show evident correlation of the dependencies, but not to explain directly the experimental data.

The above calculations do not take into account the reflection of electrons on the metal surface, whereas it is known that this effect substantially increases the maximum final energy in the analogous process of nonlinear ionization of atoms [9]. A complete model involving the rescattering process should be quantum mechanical, but the upper limit of the electron spectrum can be estimated using the classical equations of motion, taking into account the vibration motion of an electron in the laser field. After the emission from the metal, when the direction of the field is changed, the electron can return and be reflected from the surface of the metal. This process increases the final drift velocity of the emitted electrons and influences on the observed electron spectrum. Thus, rescattering leads to the appearance of a high-energy spectral plateau, beginning near the energy $2U_p$ and abruptly terminating at the energy $\sim 10\ U_p$ [10].

An additional accelerating dc electric field in the nonlinear emission experiments impedes return of electron back to the metal surface and at some critical ratio α of the laser field amplitude to the dc field this effect completely disappears. The result of the classical simulation of the maximum electron energy as a function of α is shown in figure 2. When $\alpha < 1.57$, the dc electric field completely suppress rescattering, and the maximum energy of the electron is $2U_p$. With a decrease in the relative strength of the dc electric field, part of the emitted electrons returns to the metal surface and is rescattered on it. However, the dc electric field, in this case, also inhibits the returning electrons, making the collision conditions with the metal surface far from optimal to obtaining the maximum possible energy. With the decrease of the electric field (at $\alpha > 1.57$), the maximum energy of the emitted electron increases to $\sim 10\ U_p$.

In figure 2 the small circles indicate the estimation of the maximum electron energy for experimental field ratios [8]. These energies have evident correlation with experimental cutoff. Significantly to note that due to the nonlinearity of emission, a large part of the free electrons is formed near the maximum of the laser field strength. For this group of electrons, the rescattering disappears at a relatively weak dc electric field, since before changing the direction of the laser field they already gain some velocity in the direction from the metal surface. Thus, the ones shown in figure 2 the estimates refer only to the limiting values of the electron energies and do not characterize the very form of the spectrum.

Strong nonlinear dependence the probability of the electron photoemission on the laser field strength leads to a sharp change in the electron currents during a period of field oscillation. When the laser pulse duration greatly exceeds the period of oscillation, series of sub-femtosecond electron pulses are formed, which follow one after the other with the laser field period. For the few-cycle laser pulses, the process looks different since a variation a carrier wave phase with respect to the envelope strongly modulates the peaks of the laser field strength impact the spectrum of the emitted electrons. This effect does not influence the main parameters of photoemission and the overall shape of the electron spectrum, but changes the positions and heights of the local energy maxima. Figure 3 shows the results of calculation the electron spectrum for different carrier phases, where the peaks modulation is clearly-visible on all three curves.

Increasing the electron energy the phase effects are manifested to a greater extent. To demonstrate this, the inset in figure 3 shows the normalized to the unity maximum magnitude of the second and sixth peak as a function of phase. These dependences are qualitatively consistent with experimental data [11].

4. Conclusions

We have formulated the one-dimensional analytical model for the nonlinear field electron emission from a metal surface under the ultrashort laser pulse action. In this framework we investigated the influence the laser pulse parameters on the energy spectrum of the emitted electrons. The model is predicted the general shape of the electron energy distribution and it gives a reasonable qualitative description the dependences on the laser pulse parameters. Our model was developed for ac electric field, but experimental measurements were produced sometimes in the presence of essential additional dc electric field. Accelerating electrons by additional dc electric field reduces the rescattering effect, and decreases their maximum energy. Thus, dc electric field can be used to control the form of the energy spectrum. For the case of complete suppression of the rescattering, our model allows us to make a qualitative comparison with the experimental electron spectrum and give its interpretation. In the next paper we shall present a more complete quantum-mechanical description of the nonlinear electron emission taking into account wave packet rescattering as the next step in the development of the theory.

Acknowledgments

This work was supported by RFBR (grant No 16-32-00255) and Ministry of Science and Education RF (project No. 3.6369.2017).

References

[1] Manuilovich E S, Astapenko V A and Golovinskii P A 2016 *Quantum Electronics* **46** 50
[2] Paarmann A, Gulde M, Muller M, Schafer S, Schweda S, Maiti M, Xu C, Hohage T, Schenk F, Ropers C and Ernstorfer R 2012 *J. Appl. Phys.* **112** 113109
[3] Germann M, Latychevskaia T, Escher C and Fink H W 2010 *Phys. Rev. Lett.* **104** 095501
[4] Kupersztych J and Raynaud M 2005 *Phys. Rev. Lett.* **95** 147401
[5] Broudy R M 1971 *Phys. Rev.* B **3** 3641
[6] Zhang P and Lau Y Y 2016 *Sci. Rep.* **6** 19894
[7] Golovinski PA, Astapenko V A and Manuylovich E S 2016 *J. Nanophoton.* **10** 033511
[8] Schenk M, Krüger M and Hommelhoff P 2010 *Phys. Rev. Lett.* **105** 257601
[9] Becker W, Grasbon F, Kopold R, Milošević D B, Paulus G G and Walther H 2002 *Adv. At. Mol. & Opt. Phys.* **48** 35
[10] Smirnov M B and Krainov V P 1998 *J. Phys. B: At. Mol. Opt. Phys.* **31** L519
[11] Kling M F, Rauschenberger J, Verhoef A J, Hasovic E, Uphues T, Milosevic D B, Muller H G and Vrakking M J J 2008 *New Journal of Physics* **10** 025024

IOP Conf. Series: Journal of Physics: Conf. Series **917** (2017) 092010 doi:10.1088/1742-6596/917/9/092010

Infra-red mediated synthesis of C-Cu-Zn nanocomposites using metal oxides and metal nitrates precursors

K B Rogushin[1], E N Sidorova,[1], M N Efimov[2]

[1]Department of Functional Nanosystems and High-Temperature Materials, National University of Science and Technology "MISiS", Moscow 119049, Russia
[2]Chemistry of Polyconjuncted Systems Laboratory, A.V. Topchiev Institute of Petrochemical Synthesis RAS, Moscow 119991, Russia

Abstract. The objective of this study is to synthesis C-Cu-Zn nanocomposites with infrared radiation. Metal-carbon nanopowders structural characteristics were analyzed using X-ray diffraction (XRD), transmission electron microscopy (TEM). The results reveals spherical metal particles with the mean size of 3 – 74 nm.

1. Introduction

Currently there are lots of issues in modern society due to the depletion of conventional oil resources development, in requirements for environmental performance are getting stricter every year, in this context there is a growing interest in the production of synthetic liquid hydrocarbons.

Methanol is one of the alternative energy sources, which can easily replace fossil fuels. Methyl alcohol provides an opportunity to use it as an alternative source as well [1, 2].

Industrial methanol production is detained with mixing hydrogen and carbon monoxide. One of the stages includes Fischer-Tropsch synthesis.

Fischer-Tropsch (FT) synthesis is a catalytic process that converts syngas (CO and H_2) into components of liquid fuels [4]. There is a wide range of metals, that applied as commercial catalysts. Nowadays such companies as Shell (Malaysia), Sasol (South Africa) and Qatar Petroleum (Qatar) are making researches in obtaining the most efficient and cost-effective catalyst [5]. Synthesis of metal-carbon nanosystem with metal nanoparticles which already recovered, and there is no stage of recovering to pure metal allows to reduce expenses. The FT synthesis requires catalyst, which is the major contributor to the final product of reaction and has an impact on the stoichiometry of hydrocarbon.

Recently nanoscale catalysts are widely used, hence it leads to the aim of this work to synthesize metal-carbon system with nanoparticles of Cu-Zn, further studying its properties due to its original materials and annealing temperature [6].

2. Materials and methods

As the result of infra-red mediated synthesis metal-carbon systems were synthesized. Precursors were based on solutions of polyacrylonitrile alcohol (PAN) as well as copper and zinc hydronitrates ($Cu(NO_3)_2 \cdot 3H_2O$ and $Zn(NO_3)_2 \cdot 6H_2O$), and similarly metal oxides precursors were taken as CuO, ZnO. Samples were annealed at 5 different temperatures (300, 400, 500, 600, 700 °C). The molar ratio of metal components was 7 : 3. The total metal concentrations in polymer was 20 % wt.

IR-chamber of pulse photon annealing was the device, where pyrolysis was conducted [7]. IR-annealing was held in two stages: at first stage for 15 minutes at 150 – 200°C preliminary annealing in air during which there was removal of the solvent and initial structuring of PAN and then the main annealing in rarefied atmosphere at a specified temperature. The main annealing time was 2 minutes.

IOP Conf. Series: Journal of Physics: Conf. Series **917** (2017) 092010 doi:10.1088/1742-6596/917/9/092010

The morphology and size of the metal nanoparticles in system C-Cu-Zn were established by TEM (JEM-1011) analysis.
X-ray diffraction analysis was conducted at room temperature on a DIFREY-401 diffractometer using Cr - Kα radiation source with the Bragg-Brentano focusing geometry. Spherical crystallite diameter distribution was estimated from X-ray diffraction data.

3. Results and discussion
Figures 1 and 2 show TEM images for samples synthesized at different temperatures, using nitrates and oxides as precursors. It is visually evident that the carbon matrix in which the metallic particles are located differs in morphology for samples synthesized from oxides and nitrates. Thus, more dense carbon flakes are observed in samples synthesized from nitrates, while carbon spheres enveloping metal-containing particles are found for samples of oxides. For samples from nitrates, precipitation of particles from the carbon matrix is characteristic. At an annealing temperature of 700, the morphology of the samples is elongated thin-walled carbon fibers. It is established that the metal-containing nanoparticles of Cu-Zn are spherical and located in a carbon matrix.

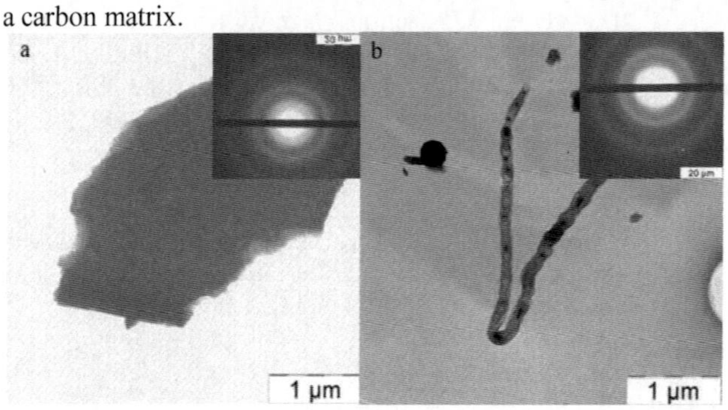

Figure 1. TEM images of C-Cu-Zn nanocomposites using nitrates (a) 600°C, (b) 700°C

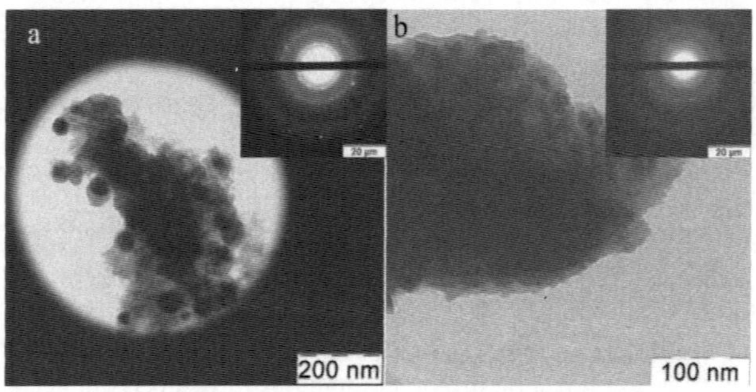

Figure 2. TEM images of C-Cu-Zn nanocomposites using oxides (a) 400°C, (b) 600°C

The histograms of the particle size distribution (Fig. 3, 4) were obtained as a result of processing data from images with a TEM. In Figure 3, the metallic nanosized particles located in the carbon matrix of the sample synthesized at an annealing temperature of 300 ° C have an average size of 74 nm, while the sample annealed at 600 ° C has metallic particles with an average size of 39 nm. The reasons for such a relationship are currently being established. Metal nanoparticles in a sample synthesized from oxides (Figure 4) at an annealing temperature of 300 ° C have an average size of 3 nm, and in a sample annealed at 700 ° C, the average size is 15 nm. This dependence of the average particle size can be attributed to the fact that at a higher annealing temperature the role of coagulation processes increases, which leads to a similar dependence of the average size of metallic particles.

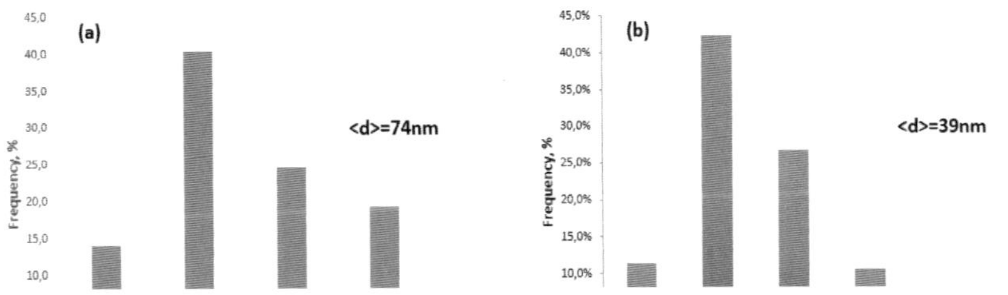

Figure 3. Particle size distribution histograms of Cu-Zn nanoparticles obtained with nitrates at: (**a**) 300 °C; (**b**) 600 °C.

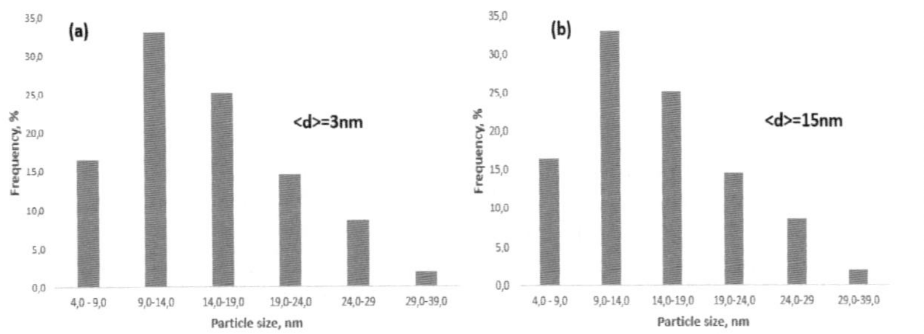

Figure 4. Particle size distribution histograms of Cu-Zn nanoparticles obtained with oxides at: (**a**) 300 °C; (**b**) 600 °C.

Figures 5 and 6 show the results of X-ray diffractometry. Figure 5 shows a phase analysis of samples obtained from nitrates at annealing temperatures of 300, 400, 500, 600, and 700 ° C. During the research, the presence of such phase as: graphite-like structure and copper was established. Presumably, zinc was not detected during the phase analysis, since a solid solution of copper-zinc formed. All reflection peaks in the range from 300 to 700 ° C have the same shape and are not offset from each other.

Figure 6 shows the phases contained in the samples synthesized from oxides. In these samples, phases are established: copper, graphite-like structure and zinc oxide. The phase of zinc oxide increases with the annealing temperature, which is due to the reduction processes of the oxide, a solid solution of copper-zinc is formed.

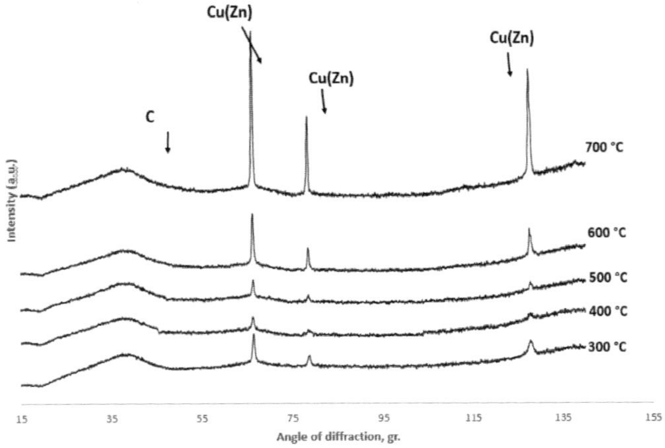

Figure 5. X-ray diffraction patterns of C-Cu-Zn nanocomposites from nitrates

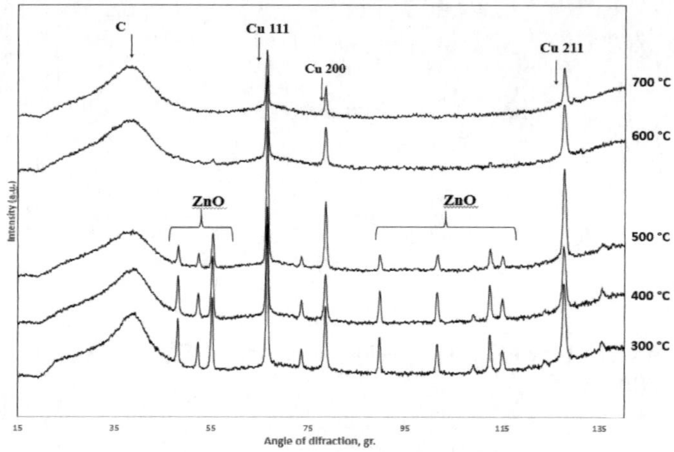

Figure 6. X-ray diffraction patterns of C-Cu-Zn nanocomposites from oxides

Figure 7 shows the coherent scattering region (CSR) size distribution by the method of Selivanov-Smyslov [8] for samples where the starting materials were nitrates. From the data obtained, it can be seen that the overall temperature change affects the distribution of CSR. As the annealing temperature increases, the CSR distribution expands, which is well reflected in the distribution at 700 ° C. In the opposite case, at a minimum annealing temperature, the size distribution can be narrowed. Such a picture may be due to the fact that when the annealing temperature in the sample increases, the intensity of the reaction increases. Also worth noting is the possible increase in the cases of coagulation of particles, which is clearly visible on the graph (the tails of the lines).
Figure 8 shows the CSR distribution by size for a series of samples synthesized from oxides. The CSR ZnO shifts to a larger area when the annealing temperature rises, and the CSR region expands. For Cu, the CSR does not have any explicit dependence and ranges from 10 to 60 nm over the entire temperature range.

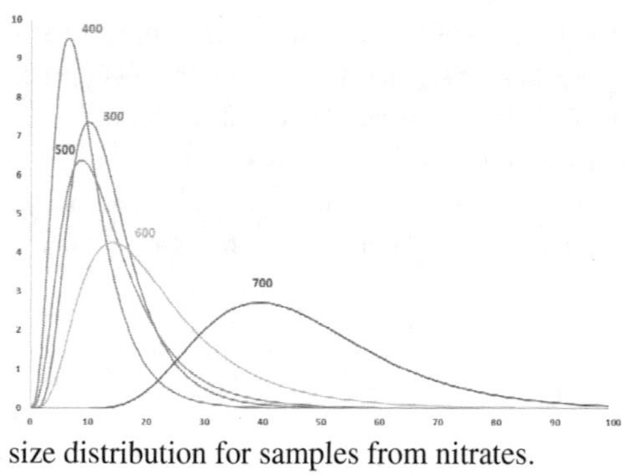

Figure 7. Crystallite size distribution for samples from nitrates.

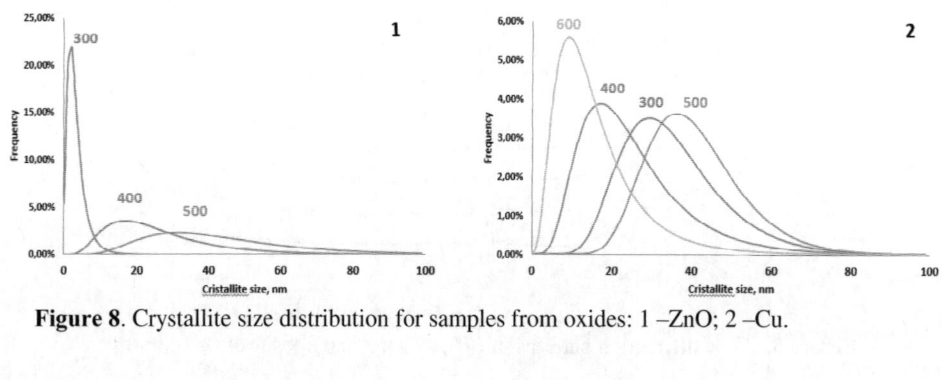

Figure 8. Crystallite size distribution for samples from oxides: 1 –ZnO; 2 –Cu.

Table 1 shows the calculation results of the structural characteristics of CuZn phase according to X-ray analysis. The lattice constant *a* in the material is smaller than in a bulk analogue due to specific features of nanoscale metal condition.
Within the framework of the work, the composition of the solid solution was evaluated according to the procedure [9]. Figure 8 shows the calculated dependence of the lattice constant as a function of the Zn content of the CuZn solution. The starting point was obtained as a result of the study of a sample of the C-Cu system of a nanoscale material. It is assumed that the plot of the lattice period of the substance is analogous to the massive CuZn system. The obtained data for the samples fit adequately into the deduced dependence (Table 1) [10].

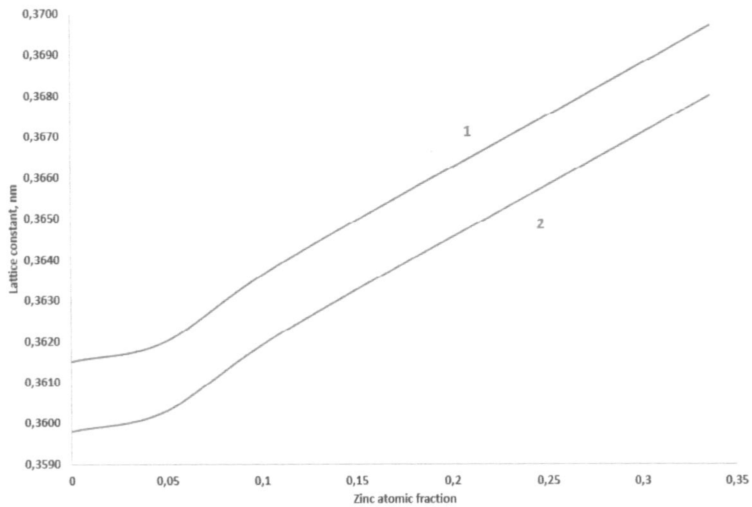

Figure 9. The curves of CuZn lattice constant dependence on zinc content; 1 – bulk material, 2 – nanosize material.

Table 1. Dimensional and structural characteristics of CuZn phase

T, °C	*a*, nm	Cu : Zn in mass percent	Zn percentage from the graph
300	0.3608	70 : 30	0.33
400	0.3604	70 : 30	0.31
500	0.3601	70 : 30	0.30
600	0.3607	70 : 30	0.32
700	0.3609	70 : 30	0.33
700	0.3626	50 : 50	0.44
700	0.3600	80 : 20	0.28

4. Conclusion

In the course of the studies it was established that in the samples synthesized from nitrates and oxides the morphology of the carbon matrix in which the particles are located differs. In samples from oxides, as a result of the phase analysis, it was established that along with the graphite-like structure and zinc oxide, a metallic phase of copper is present. For the samples synthesized at 600 and 700 ° C, there is no ZnO phase, which is associated with the reduction of zinc and the formation of a solid solution of Cu-Zn. For samples synthesized from nitrates, a graphite-like phase is established, as well as a solid solution of copper-zinc. The composition of the solid solution CuZn is estimated. As a result of the transmission electron microscopy, metal nanosized particles with a spherical shape were found. The average particle size for a sample synthesized from oxides and annealed at 300 ° C is 3.3 nm, and for a sample annealed at 600 ° C it is 15 nm, which is due to coagulation processes with increasing temperature. In a sample obtained from nitrates upon annealing at 300 ° C, the average particle size was 74 nm, while for the sample annealed at 600 ° C, the average size was 39 nm. For samples synthesized from nitrates, CSR distributions by size were calculated and it was established that an increase in the annealing temperature entails an expansion of the CSR region and its shift toward larger dimensions.

References

[1] Markhotskiy Y L. Fundamentals of ecology and energy conservation. –Litres. 2016
[2] Kuznetsov A M. Industry GTL. Status and prospects. NK “Rosneft”. – 2012. Pp. 44 – 49.
[3] Kashkarov A P. Modern bio-, benzo-, diesel-generators and other useful structures. Moscow. DMK Press. 2013. 136 p.
[4] Krylova A U, Kulikova M V. 2014. *Solid Fuel Chemistry* **48** 230
[5] Sineva L V, Mordkovich V Z, Ermolaev V S. 2012 *Catalysis in Industry* 13
[6] The Active Site of Methanol Synthesis over Cu/ZnO/Al2O3 Industrial Catalysts Malte Behrens, Felix Studt, Igor Kasatkin1, Stefanie Kühl1, Michael Hävecker, Frank Abild-Pedersen, Stefan Zander1, Frank Girgsdies, Patrick Kurr, Benjamin-Louis Kniep4, Michael Tovar, Richard W. Fischer4, Jens K. Nørskov, Robert Schlögl. Science 18 May 2012: Vol. 336, Issue 6083, pp. 893-897
[7] Zemtsov L M, Karpacheva G P, Efimov M N 2006 *J. Polymer Science* **48** 977
[8] Selivanov V N, Smislov E F 1993 *Crystallography Reports* **38** 174
[9] Pat. 2597935 C1 Russian Federation, IPC G01N 23/20 B & 2B 1/00/ Method for determination of the composition of a solid solution / E.L. Dzidziguri, M.N. Efimov, LM Zemtsov, etc .; The applicant and the patent owner of INKS RAS. - No. 2015123269/28; Claimed. 06/17/2015; Publ. 09/20/2016, Byul. № 26 - 29 p.
[10] Barabash O M, Koval Y N 1986 *Crystalline Structure of Metals and Alloys* (Kiev: Naukova Dumka) p 598

IOP Conf. Series: Journal of Physics: Conf. Series **917** (2017) 092011 doi:10.1088/1742-6596/917/9/092011

Formation of multicomponent matrix metal oxide films in anodic alumina matrixes by chemical deposition

G G Gorokh[1], A I Zakhlebayeva[1], A I Metla[1], V V Zhilinskiy[2], A N Murashkevich[2], N V Bogomazova[2]

[1]Nanotechnology Research Laboratory, Belarusian State University of Informatics and Radioelectronics, Minsk 220013, Belarus

[2]Department of Chemistry, Electrochemical Production Technology and Materials for Electronic Equipment, Belarusian State Technological University, Minsk 220006, Belarus

Abstract. The metal oxide films of $Sn_xZn_yO_z$ and $Sn_xMo_yO_z$ systems deposited onto anodic alumina matrixes by chemical and ion layering from an aqueous solutions were characterized by scanning electron microscopy, Raman spectroscopy, electron probe X-ray microanalysis and IR spectroscopy. The obtained matrix films had reproducible composition and structure and possessed certain morphological characteristics and properties.

1. Introduction

The semiconductor metal oxide films such as SnO_2, ZnO, WO_3, In_2O_3, MoO_3, and others due to their high chemosensitive properties are among the most promising materials for gas sensors [1-2]. The structuring of metal oxides by applying of high ordered dielectric anodic alumina matrixes (AAM) allows to form the systems possessing a large active surface area and a heightened sensitivity to various gases [3]. The formation of mixed metal oxide films with various morphology and an aspect ratio of the surface area of formed structures to their thickness can improve the selectivity of these materials [4]. Therefore, the structuring of the metal oxides in combination with the selection of the composition allows to create a matrix system possessing certain morphological characteristics and properties.

This paper presents the results of matrix-forming of metal oxide film by chemical sedimentation of transition metals into the porous anodic alumina as well as investigations of their morphology and microstructure.

2. Experimental

2.1. Formation of nanoporous anodic alumina matrixes

The initial specimens were pure (99.99%) aluminum films, 1.2 μm thick, sputtering-deposited onto silicon substrates. Nanoporous AAM were formed by anodizing of the aluminum films in 0.4 mol/dm^3 malonic acid electrolyte with current density of 6 mA/cm^2 at anodic potential of 85 V. To improve pore regularity and enlarge the surface of the uppermost pore layer in the films, a multi-step anodizing process was developed and applied here, following generally the procedure described in Ref. [3]. The first anodizing step was performed in the above electrolyte, followed by selective dissolution of the porous alumina formed. This is due to the specific swelling and widening of the distances between the concaves during the compact oxide formation, which confines effectively the surface available for pore nucleation and, ultimately, makes the pores grow straight and vertical just from the film surface.

The second anodizing step was done to the concaved aluminum surfaces pre-textured by anodic oxidation at anodic potential of 90 V, to cover the concaved aluminum surface with a layer of compact alumina of 1 μm thickness. To increase the surface-to-volume ratio, the films were subjected to pore modification in mixture of phosphoric and chromic acids at 323K during 5 minutes. The pore diameters were ~ 120 nm.

The schematic diagram showing the main steps for forming of AAM is presented on Figure 1.

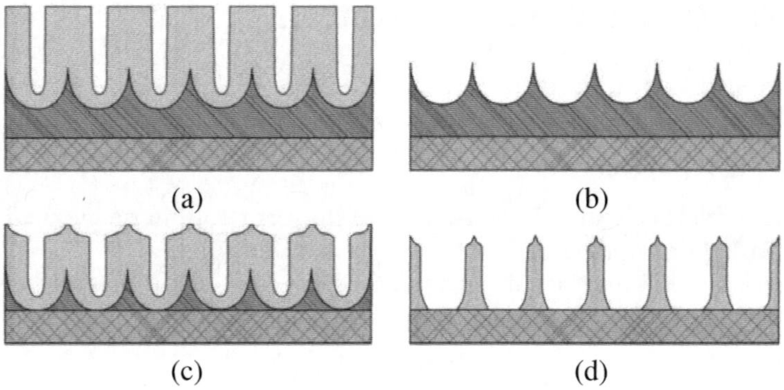

Figure 1. Schematic diagram showing the main steps for forming of anodic alumina matrixes: first stage of anodizing the Al layer (a), selective removal of the anodic alumina (b), second anodizing step (c), pore widening treatment of anodic alumina (d).

2.2. Formation of metal oxides matrix systems

Matrix $Sn_xZn_yO_z$ systems were formed by chemical deposition metal oxide films from aquatic solutions on the AAM [5]. At the first step, the samples were subjected to chemical precipitation $Sn(OH)_2$ in 1% $SnSO_4$ solution with subsequent annealing to form SnO_2 films. At the second step, the $Zn(OH)_2$ layers were prepared by the chemical precipitation method using aqueous solution of 0,01 M $ZnSO_4$ and high-temperature annealing of formed structures at T = 750°C for 30 minutes to form $Sn_xZn_yO_z$ system.

Matrix $Sn_xMo_yO_z$ systems were formed by ion layering interleaved films of molybdenum and tin hydroxides [6]. The films $Sn(OH)_2$ were prepared by electrophoretic deposition on AAM from an aqueous solution of 0.01 M $K_2[Sn(OH)_4]$ at pH = 8. To make the transition of Sn^{2+} to Sn^{4+} the resulting films were annealed at T = 750°C. Further, the molybdate ions were deposited on the resulting film from 0.01 M solution of $(NH_4)_2MoO_3$. Generated, thereby a film of the insoluble tin polymolybdates were annealed at T = 750°C for forming the $Sn_xMo_yO_z$ compound.

Such sequences of operations for the formation of $Sn_xZn_yO_z$ and $Sn_xMo_yO_z$ layers were repeated from 10 to 30 times for a complete filling of the alumina templates pores.

The schematic diagram showing the filling of AAM of metal oxide films is presented on Figure 2.

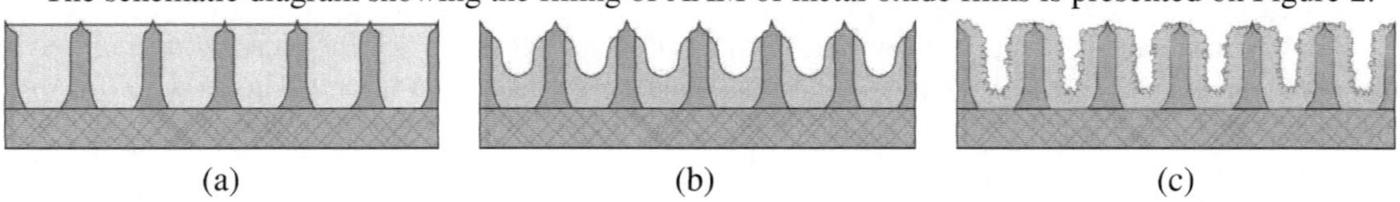

Figure 2. Schematic diagram showing the filling of AAM of metal oxide films: deposition metal hydroxides into the matrixes from aqueous solutions (a), low-temperature drying of filled matrixes (b) high-temperature annealing of formed structures (c).

3. Results and discussion

3.1. Investigation of morphology and structure of formed systems

The surface morphology and cross-sections of the AAM with metal oxides films were examined in a Hitachi S-806 scanning electron microscope (SEM) operated at 15 kV of accelerated voltage. Figure 3 shows SEM images of the surface morphology (a-c) and cross-section (d-f) of the initial AAM (a,d) AAM with $Sn_xZn_yO_z$ (b,e) and AAM with $Sn_xMo_yO_z$ (c,f) films.

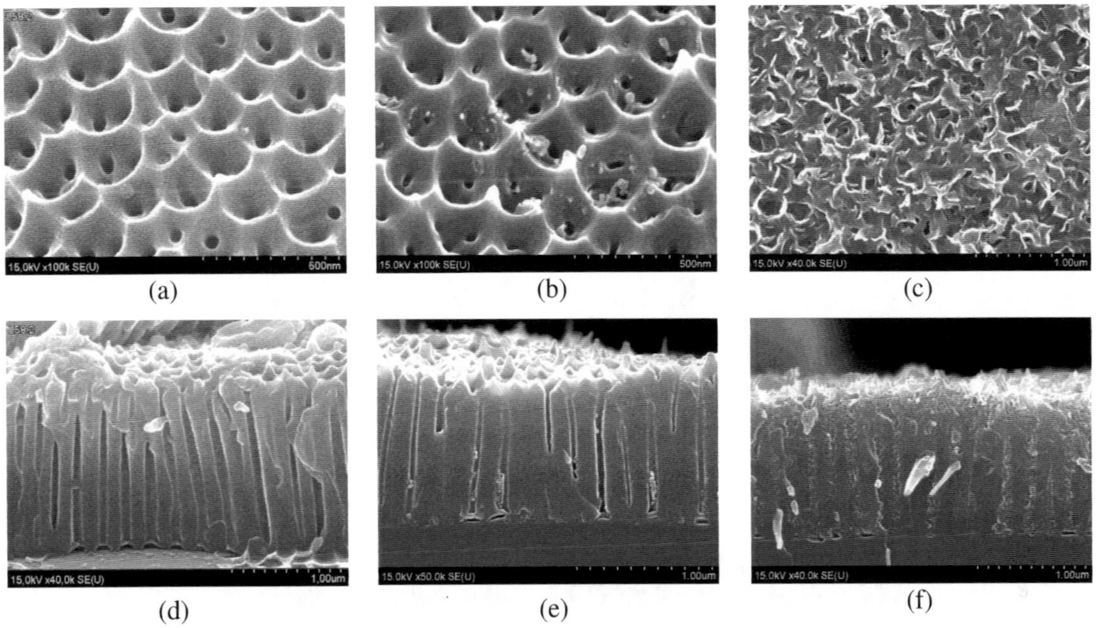

Figure 3. SEM microphotographs of the surface morphology (a-c) and cross-section (d-f) of the initial AAM (a,d), AAM with $Sn_xZn_yO_z$ (b,e) and AAM with $Sn_xMo_yO_z$ (c,f) films.

As we can see from these pictures, in the matrices filled by the method of chemical deposition, $Sn_xZn_yO_z$ grains of various sizes are formed, wherein the pores of the anodic oxide are filled non-uniformly. The most uniform filling of matrixes is observed during the formation of structures by ion layering.

3.2. Study of composition of mixed metal oxides

The composition of the AAM with $Sn_xZn_yO_z$ and $Sn_xMo_yO_z$ films were clarified by Raman spectroscopy (Figure 4) and the electron probe X-ray microanalysis (Figure 5).

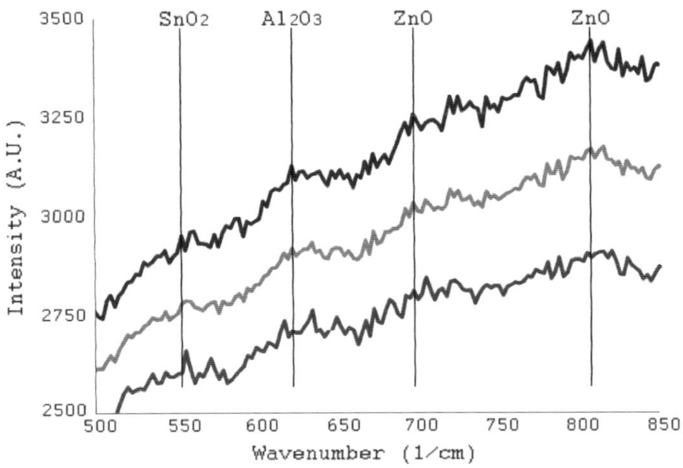

Figure 4. The Raman spectra of AAM/$Sn_xZn_yO_z$ structures.

The Raman spectra of AAM/$Sn_xZn_yO_z$ structures show the presence of crystalline phases with SnO_2 (550 cm^{-1}), Al_2O_3 (625 cm^{-1}), and ZnO (705 cm^{-1}, 820 cm^{-1}), which corresponded to the composition of metal oxide films deposited on the AAM.

IOP Conf. Series: Journal of Physics: Conf. Series **917** (2017) 092011 doi:10.1088/1742-6596/917/9/092011

The electron probe X-ray microanalysis of $Sn_xMo_yO_z$ layers deposited onto the AAM demonstrated the presence of the following elements: Al (1.432 keV), O (0.56 keV), Si (1.77 keV), Mo (2.31 keV), and Sn (3.44 keV), and their quantitative ratio in the composition of the films (Figure 5).

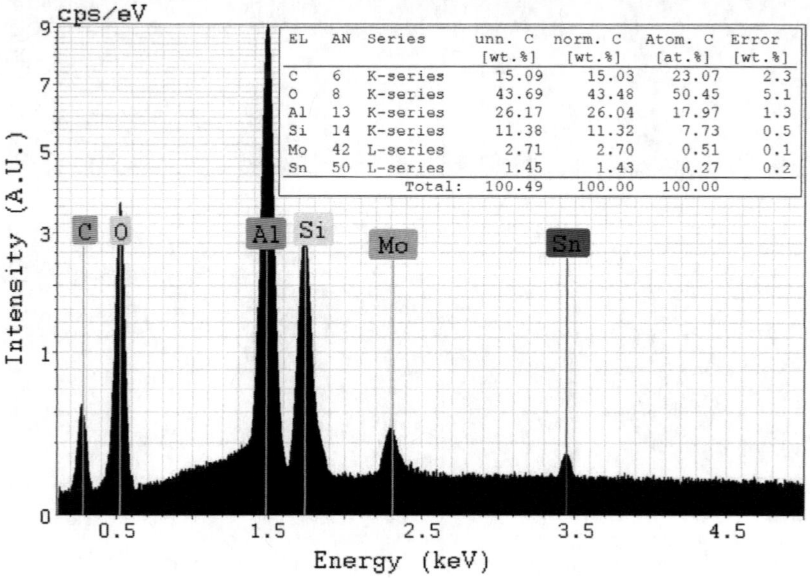

EL	AN	Series	unn. C [wt.%]	norm. C [wt.%]	Atom. C [at.%]	Error [wt.%]
C	6	K-series	15.09	15.03	23.07	2.3
O	8	K-series	43.69	43.48	50.45	5.1
Al	13	K-series	26.17	26.04	17.97	1.3
Si	14	K-series	11.38	11.32	7.73	0.5
Mo	42	L-series	2.71	2.70	0.51	0.1
Sn	50	L-series	1.45	1.43	0.27	0.2
		Total:	100.49	100.00	100.00	

Figure 5. The energy dispersive X-ray spectrum of $Sn_xMo_yO_z$ layers deposited onto the AAM.

The identification of chemical bonds present in the formed $Sn_xMo_yO_z$ layers was carried out using IR spectroscopy (Figure 6).

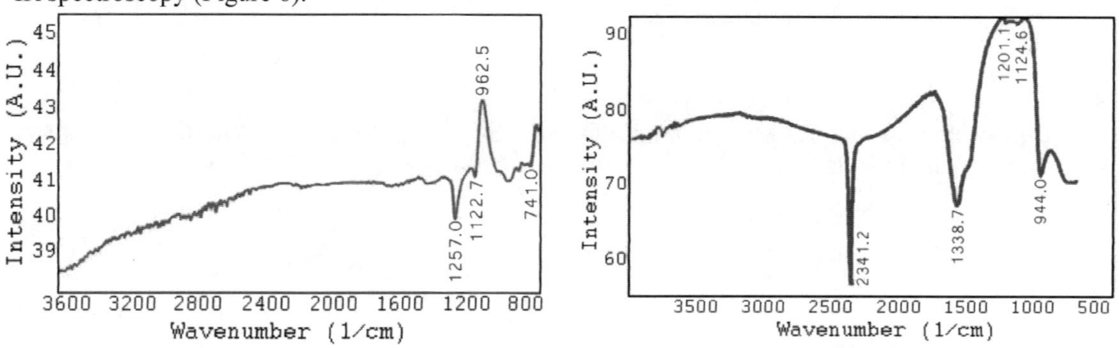

(a) (b)

Figure 6. The Infrared spectra of initial AAM (a) and $Sn_xMo_yO_z$ films (b).

The initial surface of the AAM is characterized by peaks which corresponding to the bonds of hydrated alumina [7]. The surface of $Sn_xMo_yO_z$ layers, as we can see from IR spectra, contains molecular water (3300 cm^{-1}) and hydroxyl forms (1440 cm^{-1}), which bonds Sn-OH and Mo-OH forms with surface metal ions.The bands at 960, 876, and 710 cm^{-1} correspond to the Mo-O groups in Mo=O and Mo-O-Mo, the band at 540 cm^{-1} correspond to the Sn-O-Sn bonds [8]. The absence of 1640 cm^{-1} band is associated with dehydration of the layers during their thermal annealing.The presence of 1260 cm^{-1} peak is typical to formation of polycoordination bonds in nanostructured films of complex composition [9].

4. Conclusion

So, the variation of the qualitative and quantitative composition of the solutions, the heat treatment regimes, the configuration and microgeometry of the anodic alumina matrixes allows for a directed change in the phase composition of synthesized metal oxide structures. The developed technique based

IOP Conf. Series: Journal of Physics: Conf. Series **917** (2017) 092011 doi:10.1088/1742-6596/917/9/092011

on the using of ordered AAM allows the functional films forming of different complex composition compounds with reproducible structure and properties. The formed multicomponent films can be used as chemosensitive layers of promising gas sensors and sensory microsystems.

References

[1] Rumjanceva M, Kovalenko V, Gas'kov A and Pan'e T 2007 *Rus. Chem J.* **LI(6)** 61.
[2] Waitz T, Becker B, Wagner T, Sauerwald T, Kohl C-D, Tiemann M 2010 *Sens. Actuat. B* **150** 788.
[3] Khatko V, Mozalev A, Gorokh G, Solovei D, Guirado F, Llobet E, Correig X 2008 *J. Electrochem. Soc.* **155** K116.
[4] Gorokh G, Mozalev A, Solovei D, Khatko V, Llobet E, Correig X *Electrochim. Acta* 2006 **52** 1771.
[5] Bogomazova N, Pyanko A, Murashkevich A, Gorokh G 2016 *Proc. Int. Sci.Conf. "Modern electrochemical technologies and equipment" (Minsk)* (Minsk: BSTU) p 241.
[6] Gorokh G, Zakhlebayeva A, Zhilinskiy V,Pyanko A 2016 *Proc. of 26th Int. Crim. Conf. "Microwave & Telecommunication Technology" (Sevastopol)* (Sevastopol: Veber) p 1478.
[7] Jurchenko Je, Kustova G, Bacanov S 1981 *Vibrational spectra of inorganic compounds* (in russian) (Novosibirsk: Science).
[8] Gulina L, Tolstoy V 2003 *Thin Sol. Films* **440** 74.
[9] Jurchenko Je 1986 *Methods of molecular spectroscopy in the chemistry of coordination compounds and catalysts* (in russian) (Novosibirsk: Science).

Electronic properties of graphene-based heterostructures

V A Skachkova, M S Baranava, D C Hvazdouski, V R Stempitsky

Lab "CAD in Micro- and Nanoelectronics", Micro- & Nanoelectronics Department, Belarusian State University of Informatics and Nanoelectronics, Minsk 220015, Belarus

Abstract. Heterostructures based on graphene were investigated by using density function theory (DFT). It is found that energy gap appear in ZnS-graphene heterostructure. Heterostructures graphene-ZnO and graphene-phosphorene have almost zero-band gap. Binding energy between heterostructure layers was determined.

1. Introduction

It is known that graphene has zero-band gap that makes difficult using of this perspective material for micro- and nanoelectronic applications [1]. There are numerous different approaches to increase graphene band gap [2]. One of these approaches lies in formation of heterostructures based on graphene and graphene-like materials. This paper presents DFT-investigation of graphene-based heterostructures with zinc oxide (ZnO), zinc sulfide (ZnS) and phosphorene monolayers. The main criterion for choosing these materials is presence non-zero band gap in it and suitable lattice parameters.

ZnO and ZnS have semiconducting properties. Zinc refers to class of transition metals with completely filled 3d-shell. Heterostructures consisting of graphene and zinc-containing materials open new possibilities for their applications, because these structures are observed to have energetic gap and high charge carrier mobility via electronic interactions [3, 4]. Phosphorene, black-phosphorus monolayer, is also semiconductor with direct band gap of 2.0 eV. Phosphorene has high carrier mobility of (10^3 $cm^2V^{-1}s^{-1}$)[5, 6] and it may be prepared by scotching method [5, 7] like graphene.

The feature of layered materials is presence van-der-Waals interactions between layers. So it becomes very important to take into account this interactions when simulate properties of layered heterostructures to get an adequate results. To define which of the functional would be more suitable for layered heterostructure simulations, the investigation of interlayer distance in black phosphorus was carried out.

Thus, theoretical investigations of the graphene-based heterostructures have a great importance to learning and understanding their new properties and capabilities for application in micro- and nanoelectronic devices.

2. Description of method

Simulation of heterostructure electronic properties was performed using density function theory (DFT) which is realized in program VASP [8]. Calculations were carried out using different approaches to description of electronic density functionals, such as local density approximation (LDA) and generalized gradient approximation (GGA), using Perdew-Burke-Ernzernhof functional (PBE),

DFT-D2 method of Grimme and vdW-DF functional of Dion. The last two take into account the dispersion interactions (van-der-Waals forces), which play important role in layered systems.

The dependence of system full energy vs. interlayer distance for different functional is presented on figure 1. LDA functional shows interlayer distance to be of 2.92 Å. PBE, DFT-D2 and vdW-DF show interlayer distance of 3.54, 3.1 and 3.57 Å, respectively. The better conformity to the experimental value, which is 3,11 Å [9], gives DFT-D2 method. Therefore DFT-D2 functional was chosen for further simulations.

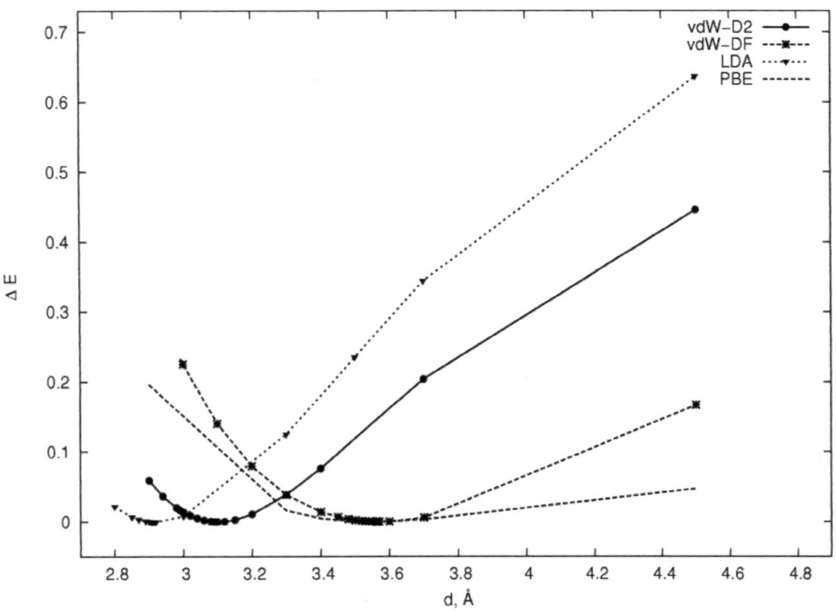

Figure 1. Dependence system full energy vs. interlayer distance in black phosphorus for different functional.

Interaction between atomic core and valence electrons was described by augmented plane wave method (PAW). Structure optimization converged when difference of full energy between the last two steps was equal to $1 \cdot 10^{-6}$ eV. Since VASP translates specified cell in all directions, vacuum layer was set to exclude energetic influence of two-layered structures on each other. Vacuum layer was added along the normal to layer surface. K-point mesh of reciprocal space was chosen 2×4×1, 3×3×1 and 8×8×1 for graphene-phosphorene (G-P), graphene-ZnO (G-ZnO) and graphene-ZnS (G-ZnS) heterostructures respectively. Supercells of these heterostructures are shown on figures 2, (a), (b) and (c), respectively. Cutoff energy was 400 eV for G-P and 500 eV for G-ZnO and G-ZnS heterostructures.

Results

2.1. Graphene-phosphorene heterostructure electronic properties

Model of G-P heterostructure was created from 3×1 phosphorene unit cells and 4×1 graphene unit cells. Lattice mismatch is equal to 1.35 and 5.86% along to "zigzag" and "armchair" directions respectively. Vacuum layer is of 15 Å. Because phosphorene electronic properties highly depend on strain, it was preserved unchanged while graphene lattice was stretched up to phosphorene lattice constants. Investigation of G-P interaction using PBE and DFT-D2 functionals is shown that PBE functional slightly take into account this interaction (full energy minimum is observed when interlayer distance is 4.3 Å). DFT-D2 gives interlayer distance equal to 3.4 Å, that is corresponding to results showed in [10]. Investigation of band structure revealed that energy gap in G-P system is equal to 0.01 eV that is too small for micro- and nanoelectronic applications (figure 3 (a)). Binding energy between graphene and phosphorene layers is -0.57 eV.

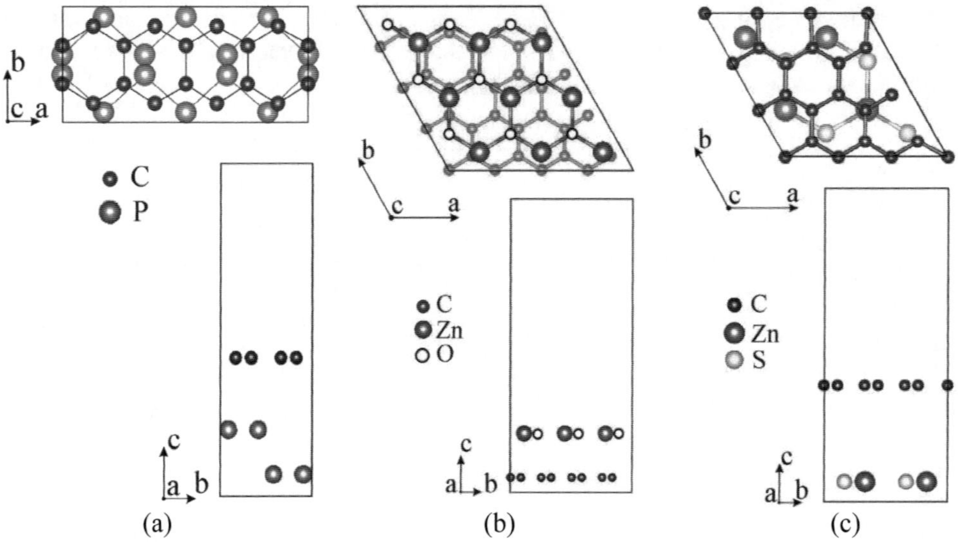

Figure 2. Supercells of (a) G-P, (b) G-ZnO and (c) G-ZnS heterostructures.

2.2. Graphene-ZnO heterostructure electronic properties

Model of G-ZnO heterostructure was created from 4×4 graphene unit cell and 3×3 ZnO unit cells. Lattice mismatch was less than 2%. Vacuum layer is of 20 Å. After atomic structure optimization was performed, ZnO lattice constant decreased on 1.85% and graphene lattice constant stayed unchanged. DFT-D2 functional shows energy minimum when distance between layers equals to 3.16 Å, while LDA functional doesn't show adhesion of graphene on ZnO. Band structure investigation reveals that energy gap in G-ZnO heterostructure becomes equal to $1 \cdot 10^{-4}$ eV, that is also too small to be used for micro- and nanoelectronic devices (figure 3(b)). Binding energy between graphene and ZnO layers is -0.5 eV.

2.3. Graphene-ZnS heterostructure electronic properties

Model of G-ZnS heterostructure was created from 3×3 graphene unit cell and 2×2 ZnS unit cells. Lattice mismatch is 1.72%. Vacuum layer is of 20 Å. Using DFT-D2 functional it was found that equilibrium interlayer distance is equal to 3.45 Å. Band structure investigation reveals that energy gap of G-ZnS structure is equal to 0.35 eV (figure 3 (c)). Zones have parabolic character, but curvature near the Dirac point is preserved that tells about slight losses of charge carrier mobility. This may be explained by energetic influence of wafer (ZnS). Binding energy between graphene and ZnS layers is 0.45 eV.

3. Conclusion

Electronic properties of heterostructures based on graphene (G-P, G-ZnO, G-ZnS) were investigated. It was shown that the best description of van-der-Waals forces which hold heterostructure layers together gives DFT-D2 functional. Also it was found that only ZnS opens considerable energy gap in graphene (0.35 eV), while in cases of G-ZnO and G-P energy gap equals to $1 \cdot 10^{-4}$ and 0.01 eV respectively, that is too small for micro- and nanoelectronic applications.

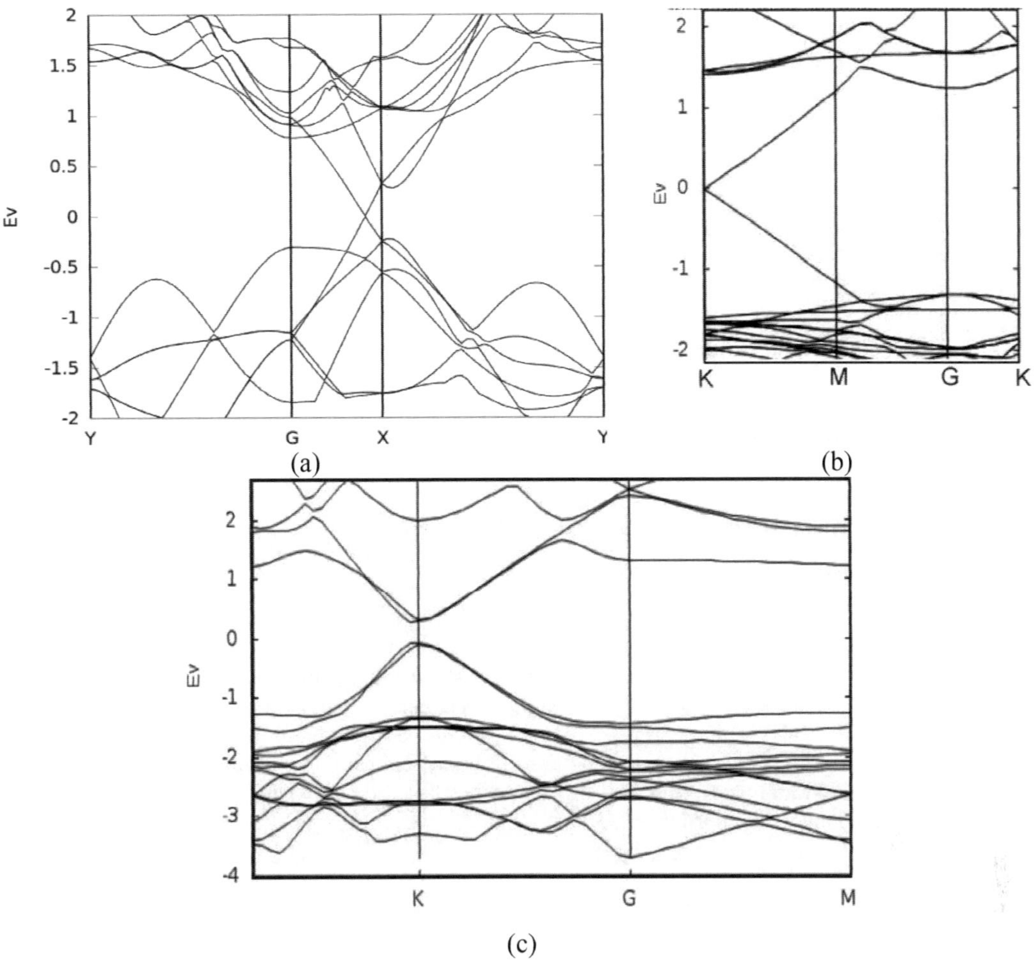

Figure 3. Band structure of (a) G-P, (b) G-ZnO and (c) G-ZnS heterostructures.

Acknowledgement

This work was supported by the grants 2.53 of Belarusian National Scientific Research Program "Physical Materials Science, Novel Materials and Technologies" and 3.02 Belarusian National Scientific Research Program "Convergence 2020".

References

[1] Cooper D R, D'Anjou B, Ghattamaneni N, Harack B, Hilke M, Horth A, Majlis N, Massicotte, Vandsburger L, Whiteway E, Yu V 2012 *ISRN Cond. Mat. Phys.* **2012** 56

[2] Liua L, Shen Z 2009 *Appl. Phys. Lett.* **95** 252104

[3] Xu T, Zhang L, Cheng H, Zhu Y 2011 *Appl. Catal., B* **101** 382

[4] Pan S, Liu X 2012 *J. Solid State Chem.* **191** 51

[5] Li L,Yu Y, Ye G J, Ge Q, Ou X, Wu H, Feng D, Chen X H, Zhang Y 2014 *Nat. Nanotech.* **9** 372

[6] Xia F, Wang H, Jia Y 2014 *Nat. Comm.* **5** 4458

[7] Liu H, Neal A T, Zhu Z, Luo Z, Xu X, Tománek D, Ye P D 2014 *ACS Nano* **8** 4033

[8] Kresse G, Hafner J 1994 *Phys. Rev. B.* **49** 14251

[9] Brown A, Rundqvist S 1965 *Acta Crystallogr.* **19** 684

[10] Padilha J E, Fazzio A, da Silva A J R 2015 *Phys. Rev. Lett.* **114** 066803

IOP Conf. Series: Journal of Physics: Conf. Series **917** (2017) 092013 doi:10.1088/1742-6596/917/9/092013

Investigation of the influence of polymerization temperature on the agglomeration process of the CNT in the composite matrix

A AGoshev[1], M K Eseev[1,2]

[1]Northern Arctic Federal University named after M.V. Lomonosov, Severnaya Dvina Emb. 17, 163002, Arkhangelsk, Russia
[2]Federal Center for Integrated Arctic Research, the Russian Academy of Science, Severnaya Dvina Emb. 23, 163000, Arkhangelsk, Russia

Abstract
The goal of this work is experimental study of electrophysical properties of polymer nanocomposites reinforced with multiwalled carbon nanotubes (MWCNTs) in alternating electric field in low frequency band of 0.01 Hz – 10 MHz. In this paper we present investigation concentration properties, different conduction mechanism, agglomeration effects and activation carrier process in composites based on epoxy resin with CNTs as addition of temperature region from 150÷300 K.

1. Introduction
At present, intensive research is being carried out in the field of nanocomposite materials, that is, one of the phases of which is nanoscale. It should be noted that the properties of such composites are not additive quantities of their phases and can differ very much from the properties of each component separately. Moreover, this concept allows to create a material with the required mechanical, electrophysical and other properties. A special place for such a filler is given carbon nanotubes (CNTs). The structure of CNT –tubular in the form of folded graphene sheets, it features a high aspect ratio ($\alpha \sim 100$– 1000) and a large surface area. Also, CNTs have good conductivity. This is related to the characteristic electronic band structure of grapheme π–electrons and the type of its convolution (chirality) [1]. CNTs can have semimetallic or semiconductor conductivity, high individual specific strength, high adsorption capacity and other special properties. Recently, attention has been paid to composites with ultra-low concentration of CNTs. In this aspect, study the interphase layer in polymer composites with the addition of small concentrations of CNTs presented in [2] is of interest. The task of reducing the propensity of CNTs to agglomeration and the increase connection with the polymer matrix itself is also of interest [3], [4], [5].

2. Method
For the preparation of polymer composites we used epoxy ED–20 with solidifier of cold solidification PEPA as a composite matrix. MWCNTs Taunit–MD were used as a filler. For uniform distribution in the polymer matrix in the sample tubes were dispersed by ultrasonic disperser (MEF 91.1). The

solidifier was added at fixed temperature after this. We have investigated the frequency dependence of our nanocomposite electrical characteristics by dielectric relaxation spectroscopy method [4, 6], according to which the sample is placed between the plates of the capacitor. The sample was exposed to an alternating electric field with a frequency varying in the band of 0.01 Hz – 10 MHz. The measured values were real and imaginary part of permittivity (ε', ε") and conductivity σ. To identify the activation properties of carriers in composites, we also used the dielectric relaxation spectroscopy method in the range of temperatures T from 150 to 400 K.

3. Concentration dependence and agglomeration effect

To identify percolation processes, we prepared and investigated samples of different concentrations of CNTs. Below we show the dependence of the conductivity from the frequency of the electromagnetic field for the sample of different concentrations (Figure 1a) and dependence of the conductivity from the concentration at different frequency (Figure 1b).

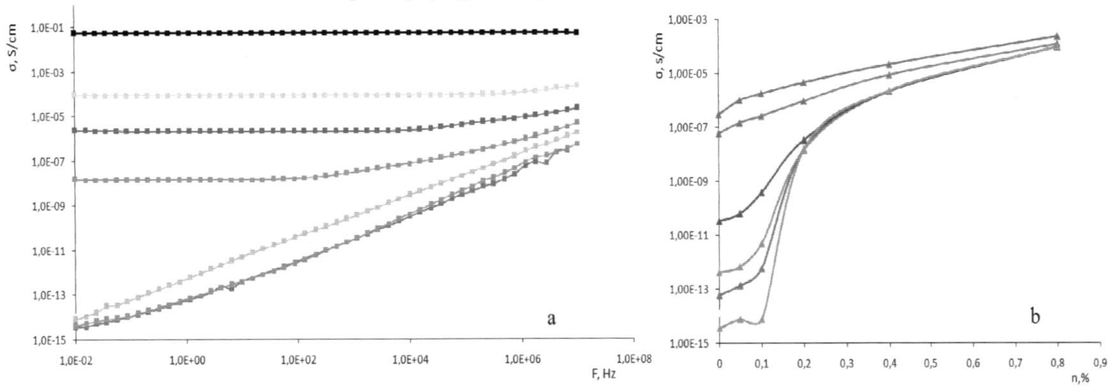

Figure 1.a)- Dependence of the conductivity from the frequency of the alternating field for polymer composites of different concentrations of CNTs ■–0%, ■- 0,025%, - 0,05% , ■- 0,1%, ■-0,2%, ■–0,4%, ■- 0,8%, ■- 100%,forT=22°C. **b)**-Dependences of the conductivity of the polymer composite from the concentration for different frequency:▲-0.01Hz, - 0.013Hz, ▲-1.15Hz, ▲- 9.98Hz, ▲-1.15 10^3Hz,▲-1.15 10^6Hz, ▲- 10^7Hz.

Analysis of the data shows that the addition of CNT less than 0.1% practically does not contribute to the conductive properties of the composite (figure 1a). The transition to through conductivity occurs quite sharply, which is typical for such systems. The criterion in this case is the inflection point of the graph of the function σ (n) (figure 1b). In the low–frequency region the percolation threshold is in the concentration range 0.12÷0.14% and disappears at high frequencies (figure 1b). Conditionally, figure 1a can be divided into three intervals for CNT concentrations: the first from 0% to 0.1%, a slight increase in σ due to the hopping conductivity, the second from 0.1% to 0.15% formation of the through conduction channel (percolation threshold), the third interval roughly from 0.15% to 0.8% is a smooth increase in σ, associated with the increase in the quantity channels of through conductivity. As mentioned earlier, CNTs tend to form agglomerates in a polymer matrix. To detect this phenomenon, a series of samples of equal concentration but different temperature input of the curing agent was prepared and measured. The more the temperature of the mixture where the hardener is injected, the faster the polymerization process takes place. Therefore, CNTs have less time to form agglomerates.
As can be seen from figure 2, the conductivity of composites of equal concentration of CNT varies over a wide range of 10^3, depending on the temperature input of the hardener entering the mixture (polymerization time). This in turn is a criterion for the degree of agglomeration of CNTs.

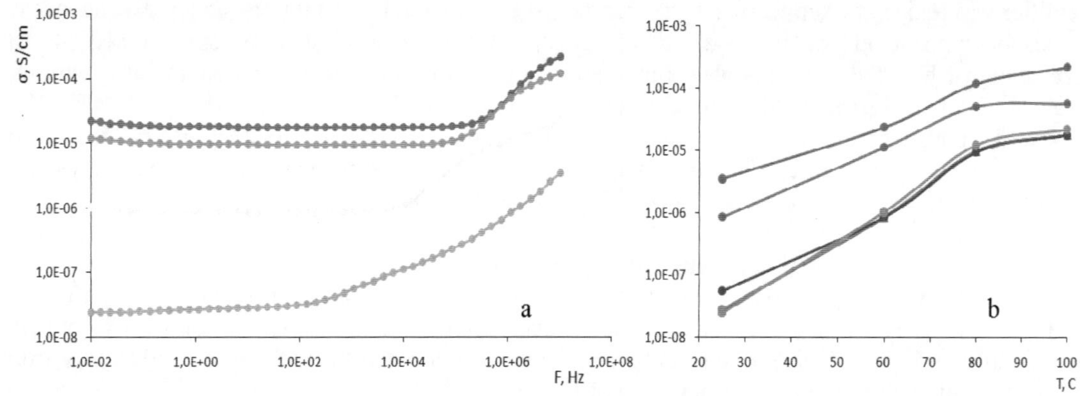

Figure 2 (a,b). a- The dependence of the conductivity of polymer composites equal to the concentration of CNTs (0.2%) from the frequency of the alternating field, for samples of different temperatures hardener input •–100°C, •- T=80°C, - 60°C,•- 25°C. **b-** Dependences of the conductivity of from the temperature of the hardener input at frequencies: •-0.01Hz, -0.013Hz, •-1.15Hz, •-9.98Hz, •-1.15 10^3Hz, •-1.15 10^6Hz, •-10^7Hz.

As can be seen from the above results (figure 1, 2), the σ(F) dependences have a similar character. It is known that the frequency (f) dependence of conductivity for composite is well described following expression [6]:

$$\sigma(f) = \sigma_{dc} + Bf^t \quad (1)$$

Where σ_{dc} – is direct conductivity, B, t – empirical parameters. The coefficients B, t were chosen by fitting the experimental curves with a good degree of correlation. The results are shown in the tables below. The presented in tables results make it possible distinguish two groups of the samples: were conductivity is linearly or non–linearly dependents from *f*. As can be seen linearity with respect to *f* is observed in a sample with good conductivity (n = 0.8%), or on the contrary in dielectric sample (n = 0%, n = 0.1%).

Table 1. Fitting of Curves of concentration by the relation (1).

n%, T=Const	σ_{dc}S/cm	B S/(cm·Hz^t)	*t*
0.8	$9.2\cdot10^{-5}$	$2\cdot10^{-11}$	1
0.4	$2.1\cdot10^{-6}$	$6\cdot10^{-10}$	0.67
0.2	$1.5\cdot10^{-8}$	$1\cdot10^{-10}$	0.67
0.1	$3\cdot10^{-15}$	$4\cdot10^{-13}$	1
0	$3\cdot10^{-15}$	$4\cdot10^{-14}$	1

Table 2. Fitting of Curves of temperature by the relation (1).

T°C,n=0.2%	σ_{dc}S/cm	B S/(cm·Hz^t)	*t*
100	$1.8\cdot10^{-6}$	$3\cdot10^{-11}$	1
80	$0.95\cdot10^{-5}$	$1.8\cdot10^{-11}$	1
60	$0.8\cdot10^{-6}$	$7\cdot10^{-10}$	0.67
25	$2.5\cdot10^{-8}$	$8\cdot10^{-11}$	0.67

As suggested by the authors the reason for non–linearity with respect to w is associated with an equal contribution of two types of conductivity in the transition region: residual hopping and through conductivity, which is evident from figure 1a.

4. Activation processes in polymer composites

To study the activation processes in the samples with the same concentration but different time of polymerization, we researched temperature dependenc these composit at the temperature range from 150 to 400 K .The following are the dependencies of conductivity σ, real and imaginary part of permittivity E_1, E_2 from the frequency of the external field F, at different temperatures, show in figure 3 (a), (b), (c) respectively.

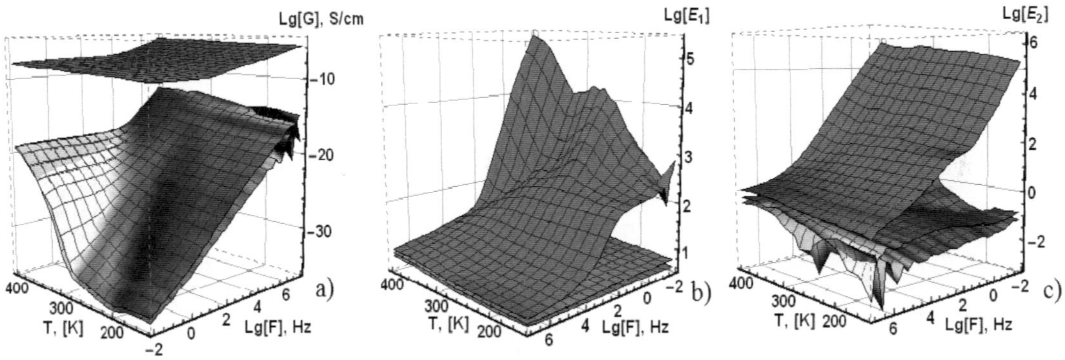

Figure 3. The dependence of conductivity (a), real part of permittivity (b),and the imaginary part of permittivity (c), from the frequency of the external field Lg [F], at different temperatures. Green plot ■ – clean epoxy resin, yellow – composite with CNT concentration n = 0.1%, polymerized at T_{start}=40 C°, red ■– composite with CNT concentration n = 0.1%, polymerized at T_{start}=100 C°.

From the present results (figure 3), the agglomeration effect is also evident, which affects both quantitatively and qualitatively on the electrophysical properties of the composite. It is note that at a hardener input temperature T_{input}= 40 C°, the composite in its electrophysical properties is almost similar to pure resin, in contrast to a composite of the same concentration but with T_{input} = 100 C°. This is explained by the large dielectric layer between the conductive structure of the CNT in the matrix, and consequently the inhomogeneity of the distribution of individual CNTs in the composite (high degree of agglomeration). For example, in the low-frequency region at room temperatures, the conductivity of the sample with T_{input}= 100 C° exceeds the sample of the same concentration but with T_{input}= 40 C° at 10^6 times, and in the frequency range 10^7 Hz by almost 2 orders of magnitude (Figure 3a).Calculation of the activation energy of carriers for certain regions can be carried out within the framework of the follow relation:

$$\sigma_{dc} = A \cdot \exp\left[\frac{-E_a}{kT}\right] \quad (2)$$

To calculate the activation energy of carriers, it is convenient to represent the results in Arrhenius coordinates (see figure 4). As seen (from figure 4) the temperature dependence of the composite (n = 0.1%, T_{input} = 100 °C), is linear in the Arrhenius coordinates, in contrast to both the pure polymer and the composite (with n = 0.1% and T_{input} = 40 °C). In the latter samples, two regions are clearly pronounced. The region of low temperatures (140–280 K), where the conductivity weakly depends on temperature. And the activation region (280–400 K) with activation energy Ea = 1.22 eV (calculated from formula (2)). For a composite with n = 0.1%,T_{input} = 100 °C, the activation energy of carriers Ea = 0.041 eV. This in turn also indicates a significant contribution to the conductivity of the tubes, and a reduction in the potential barrier between them. For pure resin, as well as for a composite (n = 0.1%, T_{input} = 100 °C), there is a phase transition point T_g, in which character the temperature dependence of the conductivity varies. In the temperature region below the phase transition T_g, a conduction mechanism is realized that is associated with tunneling induced by thermal fluctuations. According to this mechanism, the electrical conductivity is due to the tunneling of electrons through a potential barrier of variable height, which is determined by local temperature fluctuations [6].

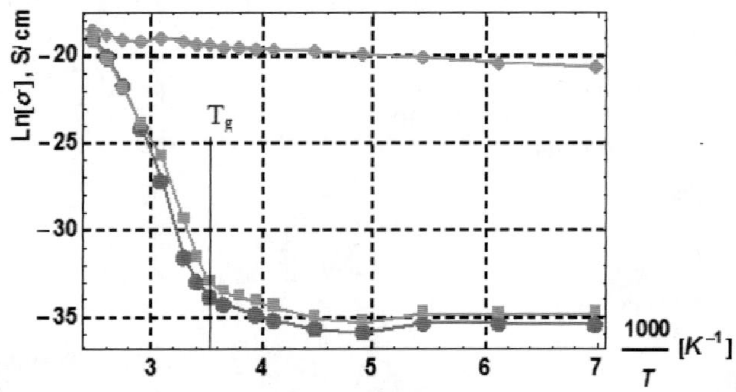

Figure 4. Dependences of the conductivity of polymer composites from temperature in Arrhenius coordinates: ● – clean epoxy resin, ■ – composite with CNT concentration n = 0.1%, polymerized at T_{input}=40 C°, ♦– composite with CNT concentration n = 0.1%, polymerized at T_{input}=100 C°. T_g– glass transition temperature.

5. Conclusions

We formulate main conclusions and results.

- A group of composites of various concentrations was studied by relaxation spectroscopy method. In the low-frequency region the percolation threshold is in the concentration range 0.12÷0.14% and disappears at high frequencies . Concentration dependence are presented, where find characteristic conductivity mechanisms in specific concentration regions. Such as area with hopping conductivity, area where through conduction channel is form (percolation threshold area) and area where directly increase in the quantity channels of through conductivity.
- A group of composites of equal concentration but different in the temperature hardener input was investigated. It is shown that the frequency dependences arising in this case are similar to the concentration dependences. In the framework of the well-known model (1) the fitting of the curves was carried out. A region deviating from linearity in w is revealed. The reason for this behavior is associated with an equal contribution of two types of conductivity in the transition region: residual hopping and through conductivity.
- The processes of agglomeration of CNTs in a polymer composite are examined, it is shown that they strongly depend on the initial polymerization temperature of the sample. A method for minimizing the concentration of CNTs and their uniform distribution over a polymer matrix is proposed.
- Activation processes in a composite at various degrees of agglomeration are investigated. Qualitative and quantitative differences in activation processes of composites polymerized at different temperatures were revealed. Characteristic activation energies of carriers in the system are found.

IOP Conf. Series: Journal of Physics: Conf. Series **917** (2017) 092013 doi:10.1088/1742-6596/917/9/092013

References

[1] MeunierV, et al2016*Rev. Mod. Phys.* **88** (2) 025005
[2] Mikitaev A K, Kozlov G V 2016 Tech. Phys.**61** (10) 1541
[3] Dyachkova T P, et al 2013 *Nanosystems: Phys., Chem., Math.***4** (5) 605
[4] Goshev A A, Eseev M K,et al 2016*J. Phys.: Conf. Series***741** 012191
[5] Eseev M K, Vinnik L N,et al 2016 AIP Conf. Proc. **1767** 020026
[6] Eletskii A V,et al 2015 *Phys. Usp.***58** (3) 209

IOP Conf. Series: Journal of Physics: Conf. Series **917** (2017) 092014 doi:10.1088/1742-6596/917/9/092014

Copper nanoparticles obtained by chemical reduction stabilized by micelles of various surfactants

N N Begletsova[1,2], O A Shinkarenko[1,2], A S Chumakov[1,2], A J K Al-Alwani[3], A A Selifonov[4], E I Selifonova[2], M V Pozharov[5], A M Zakharevich[2], R K Chernova[2,5], A S Kolesnikova[2] and E G Glukhovskoy[1,2]

[1]Department of Nano- and Biomedical Technologies, Saratov State University, Saratov 410012, Russia
[2]Education and Research Institute of Nanostructures and Biosystems, Saratov State University, Saratov 410012, Russia
[3]Department of Physics, Babylon University, Babylon, Iraq
[4]Department of Physics, Saratov State University, Saratov 410012, Russia
[5]Institute of Chemistry, Saratov State University, Saratov 410012, Russia

Abstract. We carried out the synthesis of copper nanoparticles via chemical reduction of micelles stabilized by various surfactants of different nature, such as cetylpyridinium chloride (CPC) and cetyltrimethylammonium bromide (CTAB). We investigated the sustainability of the products of synthesis, i.e. copper nanoparticles stabilized by micelles of various surfactants, after the period of 50 days. Surface morphology and elemental composition of obtained copper nanoparticles were characterized by scanning electron microscopy (SEM).

1. Introduction

The advances in nanoscale entities (particularly, nanoparticles dots) synthesis have gained much prominence in chemistry, physics, medicine, instrument engineering and other fields of science and industry. People planning to use nanoparticles in their field of work usually expect these entities to have properties that are significantly different from large-scale materials. The lower is size of said particles, the greater share of atoms form their surface, therefore, surface energy contribution to total energy increases which may lead to high reactivity, unique optical properties, lower melting temperature etc. Therefore, development of effective methods of synthesis and further study of nanoscale entities (such as nanoparticles) is very important.

Some of the most interesting materials of this type are copper nanoparticles. As a material, copper functional characteristics (such as heat and electric conductivity) are not worse than those of other widely used metals (such as silver or gold), while the price of copper is much lower.

Moreover, slight changes in copper chemical composition, namely its oxidation, turn it from a highly efficient conductor to semiconductor. This feature is unique for copper and finds many practical applications in nanoelectronics, photonics, medicine, instrument engineering, agriculture etc. [1]. Currently, copper nanoparticles gained much prominence as catalysts used in methane conversion and ethanol oxidation. They are an important component of composite materials [2] and synthetic dyes [3, 4]. They are used to produce medical equipment and antibacterial agents and films [5] as well as sensor devices. In electronics copper nanoparticles are used in production of integrated circuits [6], contact layers and electrodes [7], active layers of solar cells, transistors, storage batteries, sensors, capacitors etc.

There are various methods of production of active conductive layers with embedded nanoparticles [8-10]. Some of them are rather widespread, i.e. spin coating method [11, 12], which is often used for serial production. Another method – Langmuir-Blodgett technique [13-15] is usually considered a laboratory technique, so it is less widespread, however, this methods allows formation of monomolecular layers of particles with organic surfactant matrix. Another specific feature of this method is its ability to produce multilayered structures with unique and controllable composition of each layer.

Preparation of working nanoparticles solutions for each of these methods can be highly specific and may consists of multiple stages. Development of synthesis methods and further production, modification and study of nanoscale particles remain a very important and difficult task.

Out of all numerous methods of copper nanoparticles production, the most simple and cheap one is nanoparticles synthesis via reduction of copper (II) ions using various stabilizers [16, 17], as this method does not require complex technical performance and allows controlling the size and morphology of obtain products. A specific modification of this method is micelle synthesis using surfactants as stabilizers. Micelles used as «nanoreactors» lead to formation of colloid metal particles and the formed coating serves as a deterrent for metal particles growth which leads to creation of a stabile system inside the micelle and prevents oxidation of metal particles [18].

The purpose of this work is to examine a stable of copper nanoparticles for some period of time using a various surfactants.

2. Material and methods

2.1 Chemicals

The following chemicals were used in the process copper nanoparticles synthesis: copper (II) chloride [$CuCl_2{\cdot}2H_2O$] (Mr ≈ 171 g/mol), cetylpyridinium chloride (CPC) [$C_{21}H_{40}ClNO$] (Mr ≈ 357.5 g/mol) and cetyltrimethylammonium bromide (CTAB) (Fluka) as a cationogenic surfactants [$C_{19}H_{42}BrN$] (Mr ≈ 364.46 g/mol), hydrazine hydrate (HH) [$N_2H_4{\cdot}H_2O$] (Mr ≈ 50 g/mol) (Vekton), distilled water [H_2O] and aqueous ammonia [NH_4OH] UHP 25-5 (Vekton).

2.2 Instruments

Solution mixing during the process of copper nanoparticles synthesis and heating the substrate with copper nanoparticles on silicon substrate was performed using ICA RCT basic magnetic stir bar. Centrifugation of copper nanoparticles (in order to separate them from surfactant) was performed using Elmi CM-70M centrifuge. The morphology of a surface and the chemical composition were determined by means of the scanning electronic microscope (SEM) (Tescan Mira II LMU) in the mode of detecting of secondary electrons (SE) (at 20-30 kV). UV-visible absorption spectra were registered on the spectrophotometer (SHIMADZU UV-2550) in the wavelength range of 500-750 nm. All measurements were made at normal conditions.

2.3 Cetylpyridinium chloride (CPC)

50 ml of aqueous solution of CPC (0.01 M) was divided evenly between two conical flasks. 0.0428 g of copper (II) chloride $CuCl_2{\cdot}2H_2O$ was added to the first flask resulting in light blue solution. Then 3 ml of aqueous ammonia NH_4OH was added to the flask until pH 11. The color of solution slightly changed. 25 ml of HH was added to the second flask. Dissolution of HH in CPC solution led to heat generation; the colour of resulting solution was light yellow. The solution in the first flask was then poured to the solution in the second flask. The resulting solution was mixed by magnetic stir bar at room temperature (23 °C) with rate of revolution being 500 rpm. The colour of solution changed from light yellow to beetroot red. The solution was mixed for 2 hours.

2.4 Hexadsyltrimethylammonium bromide (CTAB)
50 ml of aqueous solution of CTAB (0.01 M) was divided evenly between two conical flasks. 0.0428 g of copper (II) chloride $CuCl_2 \cdot 2H_2O$. Then 2 ml of aqueous ammonia NH_4OH was added to the flask until pH 11. The color of solution slightly changed. 25 ml of HH was added to the second flask. Dissolution of HH in CPC solution led to heat generation; the resulting solution was transparent. The solution in the first flask was then poured to the solution in the second flask. The resulting solution was mixed by magnetic stir bar at room temperature (23 °C) with rate of revolution being 500 rpm. The colour of solution changed from light yellow to beetroot red. The solution was mixed for 2 hours.

2.5 Preparation of the sample solution for application to the Si substrate
In order to improve visualization of copper nanoparticles on the silicon substrate, the sample was specifically prepared. 1.5 ml of copper nanoparticles solutions obtained in accordance with sections 2.3-2.4 was put into an eppendorf microtube. These solutions were then centrifuged for 5 minutes (without adding water). After that 0.5 ml of centrifuged solution were taken from the surface area and diluted with 0.5 ml of water. The resulting solution was centrifuged afterwards for 5 minutes. This operation was performed three times. The prepared solution containing copper nanoparticles was applied to the Si substrates which were then heated at 100 °C for 5 minutes.

3. Results
3.1 Cetylpyridinium chloride (CPC)
The resulting solution was stirred for 2 hours using thermal magnetic stirrer at 30 °C and 500 rpm. Every 20 minutes the record of the aqueous solution color was made. This color changed from light yellow to brown, which presumably indicates the formation of copper nanoparticles whose presence has been proven spectrophotometrically.

Identification of copper nanoparticles in the product of synthesis was performed via UV-visible absorption spectrum of the solution (Figure 1(a), curves 1 and 2). UV-visible spectrum has a pronounced absorption peak at 580 nm that is characteristic for copper nanoparticles. In this work we also studied stability of obtained nanoparticles for 50 days after synthesis (Figure 1(a), curve 2). Figure 1(a) shows that over time the absorption peak position has not changed, but it became less pronounced which may be connected with oxidation and chemical changes of nanoparticles in time.

Figure 1(b) shows SEM image of copper nanoparticles obtained at the day of the experiment on the silicon substrate for solution centrifuged in accordance with section 2.5 of this article and for the conventional solution, respectively. Figure 1(b) shows that copper nanoparticles are very visible. Nanoparticles have a spherical round shape with size range varying from 60 to 200 nm.

(a)

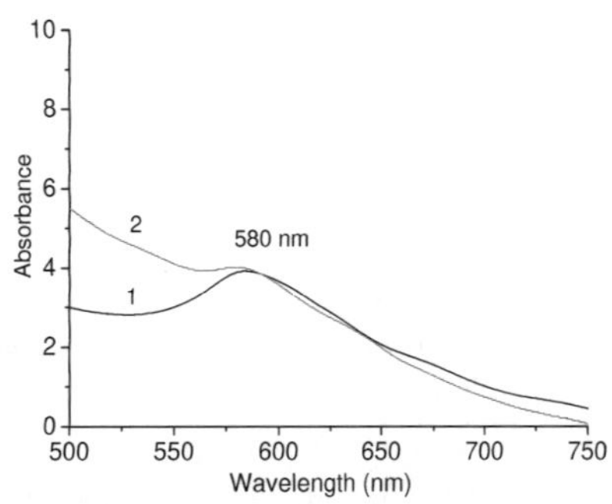

(b)

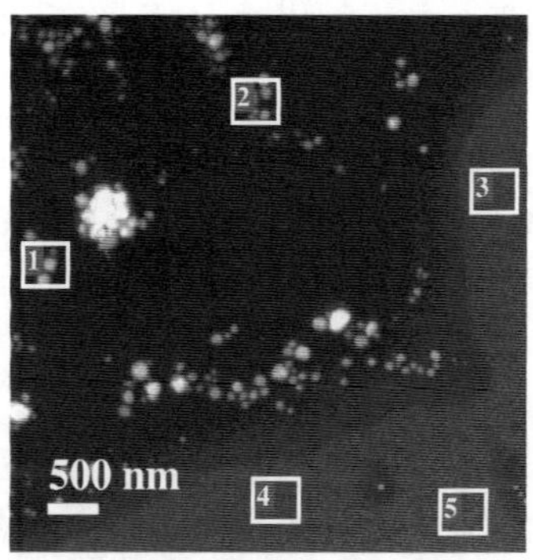

Figure 1(a, b). (a) UV-visible spectrum: 1 – spectrum obtained at the day of the experiment; 2 – spectrum obtained after 50 days; **(b)** SEM image of copper nanoparticles obtained at the day of the experiment.

The chemical (elemental) composition of sample of copper nanoparticles on the substrate was studied by means of SEM. The samples for this study were taken from various places of substrate surface. The results of this study are provided in Table 1.

Table 1. Chemical composition of products of copper nanoparticles synthesis on the Si substrate (all results in weight %).

Point	C	Cu	Result
1	83.43	16.57	100.0
2	87.18	12.82	100.0
3	100.0	–	100.0
4	100.0	–	100.0
5	100.0	–	100.0

3.2 Hexadsyltrimethylammonium bromide (CTAB)

The resulting solution was stirred for 2 hours using thermal magnetic stirrer at 30 °C and 500 rpm. Every 20 minutes the record of the aqueous solution color was made. This color changed from light yellow to brown, which presumably indicates the formation of copper nanoparticles, whose presence has been proven spectrophotometrically.

Identification of copper nanoparticles in the product of synthesis was performed via UV-visible absorption spectrum of the solution (Figure 2(a), curves 1 and 2). UV-visible spectrum has a pronounced absorption peak at 588 nm that is characteristic for copper nanoparticles. In this work we also studied stability of obtained nanoparticles for 50 days after synthesis (Figure 2(a), curve 2).

Figure 2(a) shows that over time the shape of absorption spectrum of copper nanoparticles in CTAB solution remained unchanged which indicated that the particles are present in the solution after 50 days. At the same time there is a strong decrease in intensity of absorption bands, which indicates of a decrease in the concentration of nanoparticles in solution. This may be explained by their sedimentation and oxidation. Nanoparticles have various irregular geometric shapes with dimensions from 60 to 200 nm.

Figure 2(b) shows SEM image of copper nanoparticles obtained at the day of the experiment on the silicon substrate for solution centrifuged in accordance with section 2.5 of this article. Figure 2(b) shows that copper nanoparticles are very visible.

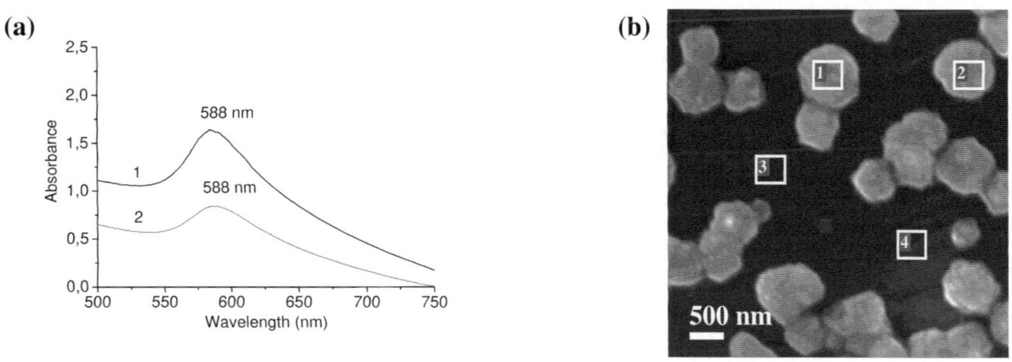

Figure 2(a, b). **(a)** UV-visible spectrum: 1 – spectrum obtained at the day of the experiment; 2 – spectrum obtained after 50 days; **(b)** SEM image of copper nanoparticles obtained at the day of the experiment.

The chemical (elemental) composition of sample of copper nanoparticles on substrate was studied by means of SEM. The samples for this study were taken from various places of substrate surface. The results of this study are provided in Table 2.

Table 2. Chemical composition of products of copper nanoparticles synthesis on the Si substrate (all results in weight %).

Point	C	N	O	Al	Cl	Cu	Br	Result
1	48.06	18.23	–	–	7.47	9.91	16.32	100.0
2	62.78	–	4.04	–	5.35	9.17	18.66	100.0
3	85.91	–	–	11.55	2.54	–	–	100.0
4	100.00	–	–	–	–	–	–	100.0

4. Discussion

Metal nanoparticles can be produced from micellar surfactant solutions where micelles can be regarded as a sort of nanoreactors allowing to regulate the shape and size of nanoparticles. These systems can form in both ionic and non-ionic surfactant solutions where the surfactant has a diphilic structure: hydrocarbon radicals (C_nH_{2n+1}, C_nH_{2n-1}, C_nH_{2n+1}, C_6H_9 etc.) containing 8-18 carbon atoms form the hydrophobic part while polar groups of various nature form the hydrophilic part. Dissolution of surfactants in water lead to water molecules cross-bonding around the non-polar hydrocarbon radicals which is turn leads to decrease of system entropy. Minimal energy corresponds to concentration specific for each surfactant that is called the critical concentration for micelle formation (CCM). At the same time, ions (molecules) of surfactant begin their spontaneous aggregation to micelles which leads to increase in system entropy. The value of CCM depends on the length and degree of branching of hydrocarbon radical, the presence of electrolytes and organic compounds, solution acidity, hydrophilic-lypophilic balance of surfactant and other parameters.

If the concentration is close to CCM, micelles have spherical forms where polar groups are oriented towards the water medium and hydrophobic radicals are located within the sphere forming a non-polar core. These ionic surfactant micelles are charged due to dissociation of polar groups and attract up to 80 % of counterions which are then connected to micelle surface and can be regarded as a single entity from the kinetic point of view.

It is known that the hydrophobic micelle cell can include up to 33-46 volume % of water (as is the case for sodium dodecyl sulfate). Micelles of ionic surfactants at concentration close to CMC form stable aggregates (average radius is 24-30 Å and more, n ~ 20-100).

Micelles are dynamic systems: the lifetime of a surfactant ion (molecule) within a micelle ranges from 10^{-8} to 10^{-5}. Micelles migrate, collide with each other, collapse and form again.

The most important property of micelles which becomes evident after they reached the CCM is their solubilization – spontaneous co-solvation of substances in micelles that have low solubility or do not dissolve in water. Ways of inclusion of substances in ionic micelles vary and depend on the nature of both solubilized compounds and surfactant.

Copper ammonium complexes and hydrazine hydrate are very soluble water. The process of copper ion reduction to nanoparticles within the CTAB and CPC micellar nanoreactors is possible in the presence of water microphases with dissolved copper ammonium complex and hydrazine within hydrophobic micelle core. In this case copper nanoparticles are formed within the hydrophobic micelle core. Their volume is then limited by the volume of water microphases and they are protected from oxidation by external (surface) and internal (core) micelle layers.

5. Conclusions

We have synthesized and studied copper nanoparticles stabilized by various micelles. Cationogenic surfactants (CPC and CTAB) allow to produce particles whose size ranges from 60 to 200 nm despite the differences in the length of hydrocarbon chain in said surfactants (C_{21} and C_{19}), respectively. The synthesized nanoparticles have only slight differences in size which can serve as an evidence of close radii of hydrophobic micelle cells produced by a similar type of cationogenic surfactants. Copper nanoparticles produced in CPC solution remain stable even after 50 days have passed since the date of synthesis. The absorption spectrum of such nanoparticles shows that absorption peak retains its intensity and position but starts to broaden. This can be an indicator of formation of other products of copper nanoparticles oxidation. Analysis of absorption spectra of copper nanoparticles procured in CTAB solution also revealed that the spectra retained their form. This shows that the nanoparticles are present in the solution after 50 days. However, the intensity of spectral band significantly decreases which indicates of a decrease in nanoparticles concentration in said solution. This can be explained by nanoparticles sedimentation and oxidation.

Acknowledgements
This work was supported by grants from the Russian Foundation for Basic Research project № 16-07-00093/16 and № 16-07-00185/16.

References
[1] Gates B C, Guezi L, Knosinger H 1986 *Elsevier* 234
[2] Woo K, Kim D, Kim J S, Lim S, Moon J 2009 *Langmuir* **25** 429
[3] Park B K, Kim D, Jeong S, Moon J, Kim J S 2007 *Thin Solid Films* **515** 7706
[4] Jang S, Seo Y, Choi J, Kim T, Cho J, Kim S 2010 *Scr Mater* **62** 258
[5] Arunachalam D K, Kannappan G 2013 *J. Appl. Pharm. Sci.* 16
[6] Eastman A, Choi S U S, Li S, Yu W, Thompson L 2001 J *Appl. Phys.* 718
[7] Brege J J, Hamilton C E, Crouse C A, Barron A R 2009 *Nano Lett.* **9** 2239
[8] Yang G, Hu L, Keiper T D, Xiong P, Hallinan D 2016 *Langmur* 36
[9] Rao C N R, Kulkarni G U, Agrawal V V, Gautam U K, Ghosh M, Tumkurkar U 2005 *Elsevier* 305
[10] Gorbachev I A, Goryacheva I Yu, Glukhovskoy E G 2016 *Bio Nano Sci.* 4
[11] Eita M, Wagberg L, Muhammed M 2012 *ACS Appl. Mater. Interfaces* **4** 2920
[12] Leck K S, Divayana Y, Zhao D, Yang X, Abiyasa A P, Mutlugun E, Gao Y, Liu S, Tan S T, Sun X W, Demir H D 2013 *ACS Appl. Mater. Interfaces* **5** 6535
[13] Yolanda J, Moreels I, Lambert K, Hens Z 2010 *Nanotechnology* **21** 295606
[14] Lambert K, Capek R K, Bodnarchuk M I, Kovalenko M V, Thourhout V D, Heiss W, Hens Z 2010 *Langmuir* **26** 7732
[15] Lambert K, Wittebrood L, Moreels I, Deresmesb D, Grandidierb B, Hens Z 2006 *J. Colloid Interface Sci.* **300** 597
[16] Zhang X, Yin H, Cheng X, Hu H, Yua Q, Wang A 2006 *Materials Research Bulletin* **41** 2041
[17] Begletsova N N, Shinkarenko O A, Selifonova E I, Tsvetkova O Yu, Zakharevich A M, Chernova R K, Kletsov A A, Glukhovskoy E G 2017 *Adv. Mater. Lett.* **8** DOI: 10.5185/amlett.2016.1423
[18] Cason J P, Roberts C B 2000 *J. Phys. Chem. B* **104** 1217

Photosensitive structures on the basis of magnetic semiconductors $FeIn_2S_4$

V B Fadeenko[1], V V Davydov[1], V Yu Rud'[2, 3], A P Glinushkin[2], Yu V Rud'[3], V Ch Shpunt[3].

[1]Peter the Great Saint Petersburg Polytechnic University, Saint Petersburg 195251, Russia

[2]Department of Ecology, Russian Research Institute for Phytopathology, 143050, Moscow Region, Odintsovo district, B.Vyazyomy, Russia

[3]Department of Solid State Physics, Ioffe Physico-Technical Institute, Russian Academy of Sciences, St Petersburg 194021, Russia

Abstract. In this work, which is in line with the new promising areas of semiconductor electronics, we present the results of a study of the photovoltaic properties of photosensitive structures based on Fe and In_2S_4 to define the application prospects of this new magnetic semiconductor for use in photonics.

1. Introduction

The study of magnetic semiconductors $A^{II}B_2^{III}C_4^{IV}$ (A^{II}-Mn, Fe, Co, Ni, B^{III}-Al, Ga, In, C^{IV}- S, Se, Te) has received increased attention in recent years [1]. We can say that these materials are new and unexplored. These materials are promising for photonics. This work is dedicated to the creation and study of photosensitive structures on the basis of $FeIn_2S_4$ single crystals one of the most promising materials for magnetic optical nanodevices for recording, storing and transmitting information.

The problems with growing single crystals of these substances, development of electronic

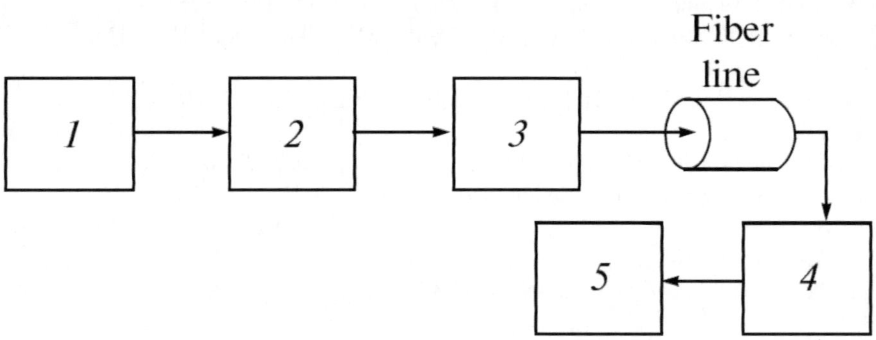

Figure. 1. The block diagram of the optical system with photodetector: (*1*) emitter; (*2*) modulator; (*3*) polarizer; (*4*) photodetector; (*5*) amplifier.

structures based of them, and comprehensive studies of their physical properties remain unresolved so far for the majority of the new multicomponent substances. Such ternary compound $FeIn_2S_4$ is one of these substances, which, in addition, possesses magnetism. In the literature, there is only information on some crystallochemical characteristics of these compounds [2, 3]. In this work, which lies in the tideway of a new promising trend of semiconductor electronics, we show the results of first studies on the growth of $FeIn_2S_4$ single crystals and creation of the first photosensitive structures on their basis. Such photosensitive structures made it possible to begin the first studies of their photoelectric properties and to determine the applied prospects of this new magnetic semiconductor. Optical system is an optoelectronic product, which includes photodetector that is designed to convert the input optical signals to electronic equipment in electrical. A typical block diagram (Fig. 1) of such a system includes an optical radiation source, a modulator, a polarizer, a fiber optical line, photodetector, an amplifier. As such, the device may convert optical signals from the optical linear path, electric signal conversion device, for example, the electronic equipment of the receiving station. The detector uses different photosensitive structures on the basis of isotropic and anisotropic semiconductors. These photo detectors can become more interesting for use as photodetectors in the emerging areas spintronics and Photonics, due to the presence in the composition of Fe.

2. The method

The single crystals of $FeIn_2S_4$ were grown by the directional crystallization method of an almost stoichiometric melt of this compound. The synthesis of such crystals was carried out by the two temperature method from the parent elements of purity no less than 99.999 wt %. The ingot synthesized was ground and placed into a double quartz ampoule; the internal ampoule had a cylindrical capillary at its end, which ensured the formation of the single crystal seeding. After the evacuation of the ampoule to a residual pressure of $\sim 10^{-3}$ Pa, it was placed into a vertical one zone furnace. The temperature in the furnace was increased at a rate of ~100 K/h to 1400 K and, for the homogenization of the melt, was maintained at this temperature for 2 h, after which the directional crystallization of the melt was conducted by reducing the temperature at a rate of ~2 K/s until complete solidification of the melt. For the homogenization of the ingots obtained, they were annealed at ~1020 K for 150 h. This regime made it possible to reproducibly grow single crystal ingots with a diameter of ~14 mm and a length of ~40 mm. The elemental composition of the single crystals was determined by electron microprobe analysis on a Cameca SX 100 device; an error of determining the element concentrations did not exceed ±5 wt %. The structure of the crystals obtained was determined by X-ray diffraction. The X-ray diffraction patterns were recorded on an automated computer aided DRON-3M diffractometer in Cu Cu*Ka* radiation with a graphite monochromator.

The results of the analysis of composition of the $FeIn_2S_4$ ingots showed that the content of elements in the single crystals (Fe: In : S = 13.94 : 28.34 : 57.72 at %).

The schematic diagram for the study of the photosensitivity based on the method of measurement of the photocurrent short-circuit depending on the length of the incident light beam. Light is incident on the receiving plane of the structure along the normal to the receiving plane. For the production of such studies, we used a source of monochromatic radiation with the degree of linear polarization of approximately 100 %. To do this, use lamp RV-24 and a monochromator (SPM-2). The circuit includes a modulator for monochromatic radiation (frequency 40 Hz) and the adjustment device (STF-1) with a semiconductor pattern and ohmic contacts, as well as the scheme of synchronous detection system with computer control of the experiment and display the results on screen.

3. Results and discussion

As was established in the course of the first studies of contact phenomena, the photosensitive struc tures can be reproducibly created using thermal vacuum deposition of thin layers of indium and aluminum onto mechanically and chemically polished substrates, as well as on the cleavage surfaces of $FeIn_2S_4$. An ohmic contact was produced by the chemical deposition of copper from an aqueous solution of Cu_2SO_4 or by applying a silver paste. According to the carried out measurements of the stationary current–voltage characteristics of the In(Al)/$FeIn_2S_4$ structure, they indicate a rectification with a coefficient K ≈ 5 at the bias voltages U ≈ 0.5 V at 300 K. The forward direction corresponds to the positive polarity of the external bias on the $FeIn_2S_4$ crystal. Upon illumination of surface barrier

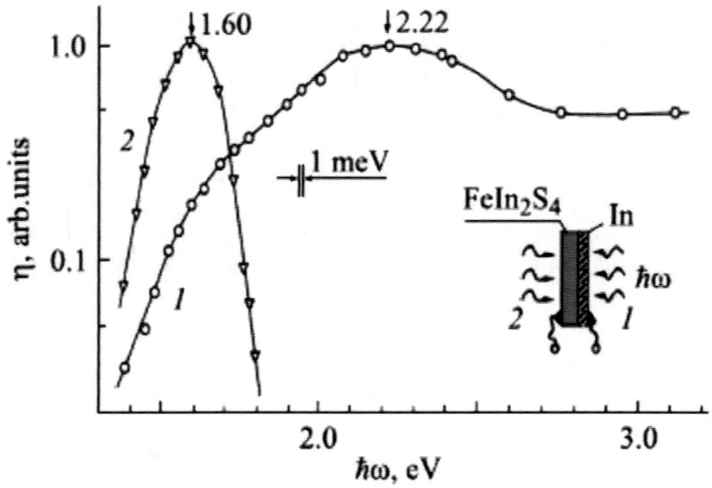

Figure 2. Relation quantum efficiency spectra of In/FeIn2S4, when illumination from the side of semiconductor (1), and when illumination from the side of surface (2).

structures, there reproducibly appears a photovoltaic effect. This photovoltaic effect have sign agrees with the direction of rectification and does not depend on the localization of the light probe (of the diameter of ~3 mm) on the surface of the structure, as well as on the intensity of illumination and photon energy. These specific features serve as a ground to connect them with the energy barrier created on the contact of the metal with the semiconductor. The voltage photosensitivity always predominates upon the illumination of the structures from the side of the barrier film, and in the best structures it reaches ~50 V/W at T = 300 K. Figure 2 displays the first spectra of the relative quantum efficiency of the photoconversion η(ω) of the In/$FeIn_2S_4$ structures calculated as the ratio of the short circuit photocurrent to the number of incident photons depending on the geometry of photodetection. Upon the illumination of the structures from the side of the semiconductor, the η(ω) spectra have a clearly pronounced shortwavelength edge (Fig. 2, curve 2), whereas upon illumination from the side of the barrier there occurs an increase in η instead of a drop. As a result, the δ of the η(ω) spectra grows from 0.2 to 1.2 eV, and the maximum of the photosensitivity ω^m is displaced into the shortwave length region from 1.6 to 2.2 eV. The short wavelength drop in η at ω > 1.6 eV (Fig. 2, curve 2) is displaced with decreasing thickness of the $FeIn_2S_4$ plate into the short wavelength region because of a reduction in the radiation absorption in the substrate of this structure and, therefore, this can show its connection with the beginning of the fundamental absorption in $FeIn_2S_4$. The high value of δ (1.2 eV) upon the illumination of the In(Al)/$FeIn_2S_4$ barriers from the side of the barrier layers demonstrates the high efficiency of the barriers obtained, which makes it possible to strongly suppress the effect of surface recombination upon the photoconversion in geometry 1 (Fig. 2). Consequently, as can be seen from Fig. 2, upon the illumination of the In(Al)/$FeIn_2S_4$ structures from the barrier side the photosensitivity spectra are wideband and ensure photodetection in the range from 1.3 to 3.2 eV at 300 K. An analysis of the spectra of photoactive absorption of the In(Al)/$FeIn_2S_4$ structures from the positions of the theory of fundamental absorption [4,5] in the coordinates $(\eta\omega)^{1/2}$ versus ω, and $(\eta\omega)^2$ versus ω (Fig. 3) suggests that the edge of the interband absorption in $FeIn_2S_4$ is formed by direct (Fig. 3, curve 1) and indirect (Fig. 3, curve 2) transitions, respectively, and the corresponding values of the energy gaps for direct ( ) and indirect ( ) interband transitions in $FeIn_2S_4$ could be estimated from the extrapolation of the appropriate dependences $((\eta\omega)^2$ and $(\eta\omega)^{1/2}$ to zero; then, from the cutoffs on the energy axis we obtain = 1.68 eV and = 1.38 eV for the undoped single crystals of $FeIn_2S_4$ at 300 K.

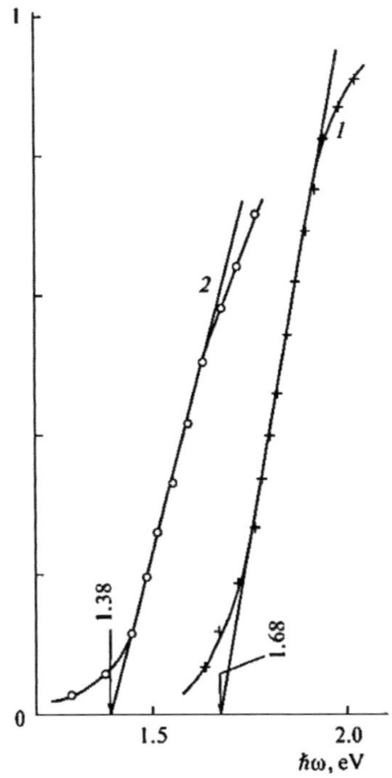

Figure 3. Dependences $(\eta h\omega)^2$ = f(hω) (curve *1*) and $(\eta h\omega)^{1/2}$ = f(hω) (curve 2) for the In/ $FeIn_2S_4$ structure at *T* = 300 K.

4. Conclusion

Thus, using the method of directional crystallization of the melt, we have possibility to grow single crystals of $FeIn_2S_4$ with a conductivity of the p type and obtained for the first time photosensitive In(Al)/ $FeIn_2S_4$ Schottky barriers. We established that upon the illumination of the new structures from the barrier side, a wideband photovoltaic effect is reproducibly observed. It has been shown that, depending on the geometry of photodetection, a wideband or selective photosensitivity can be obtained. If we illuminate this structure from the side of barrier the photosensitivity spectra are wideband and demonstrates photodetection in the range 1.3-3.2 eV at room temperatures. Based on the results of the studies performed, it has been established that the edge absorption in $FeIn_2S_4$ is formed by direct and indirect interband transitions; the width of the energy gap for these transitions has been estimated. A conclusion is made on the possibility of application of the new magnetic semiconductor $FeIn_2S_4$ as wide band photodetectors for photonics and spintronics.

References

[1] Bairamov B Ch, Rud' V Yu, Rud' Yu V 1998 *MRS Bull* 23 91
[2] Bodnar I V, Pauliukavets S A , Trukhanov S V, Arushanov E K 2011 *Proceedings of European Congress and Exhibition on Advanced Materials and Processes «EUROMAT 2011»* vol A21 P-2 (Montpellier) 40.
[3] Shpunt V Ch, Rud' Yu V, Tsendin K D 1993 *Journal Technical Physics* **19** 41
[4] Kesamanly F P, Rud' V Yu, Rud' Yu V 1999 *Semiconductors* **5** 483
[5] Rud`V Yu, Rud` Yu V, Shpunt V Ch, Terukov E I 2016 *Optical Memory & Neural Networks (Information Optics)* **25** 40

Nanoscale matrices to transport high-energy beams

E N Muratova[1], A A Shemukhin[2]
[1]Saint Petersburg Electrotechnical University "LETI", ul. Prof. Popova, 5, Saint Petersburg, 197376, Russia
[2] Skobeltsyn Institute of Nuclear Physics, Moscow State University, 1(2), Leninskie gory, Moscow, 119991, Russia

Abstract. It has been studied the process of charged particle transmission through nanoporous alumina membranes with a pore diameter of about 20 and 200 nm. Experiments with irradiation were carried out on the AN_2500 accelerator complex. It was shown that membranes based on nanoporous alumina provide transporting functions to the high-flow helium ions with energies of 1.5-2 MeV. It has been shown than it is possible to realize spatial localization of ion impacts on a substrate with a nanoscale resolution.

1. Introduction

Several recent papers [1-3] have considered the possibility of transporting the accelerated beams of charged particles using a dielectric channel without loss of energy and without losing the initial charge state. It has recently studied actively the use of nanoscale dielectric matrices for the Rutherford backscattering method (RBS). These matrices according to their geometrical parameters can empower the RBS. Namely, they can serve as the basis (template) for ion lithography, focusing the system for a charged beam or may be part of the construction for the study of complex objects to a vacuum (e.g., biological), promoting the exit of the ion beam into the air.

Nanoporous alumina is more and more popular in this area of research due to the technological, low time and resource-intensive, it is also possible to obtain a wide range of different geometrical parameters of porous layers [3–5].

2. Experiment

In order to obtain a porous anodic alumina membrane the electrochemical anodization of aluminum foil (40 μm) was applied in a potentiostatic mode in electrolytes based on aqueous solutions of phosphoric (H_3PO_4) and sulfuric (H_2SO_4) acids for 10–60 min at temperatures from 0 to 15°C. When using electrolytes with different acidities, the speeds of an anions introduction into the porous layer are different. This effect determines the voltage of the pore formation process and consequently the pore diameter. The anodizing voltage was within the range of 15–30 V (for H_2SO_4) or 130–160 V (for H_3PO_4) depending on the electrolyte type. The pore diameter is 18–25 nm and 90–200 nm, respectively.

Experiments with irradiation were carried out on the AN_2500 accelerator complex at Moscow State University [6, 7]. For carrying out the experiment, a special experimental chamber (Figure 1) was created, which includes a collimating aperture, a system of backscattered ion detection and a goniometer system. The intensity of the analysing ion beam was monitored by using a rotary aluminum propeller, its blades, which periodically block the beam, were spattered with a layer of gold a few hundred angstroms thick. Particles, back-scattered on the spattered layer of the monitor target positioned to block the beam, are recorded by the silicon surface-barrier detector. On the basis of the registered back scattered particles, the intensity of the beam falling on the target is calculated. Ions having passed the monitoring system (while the position of the beam transmission) fall into the experimental chamber.

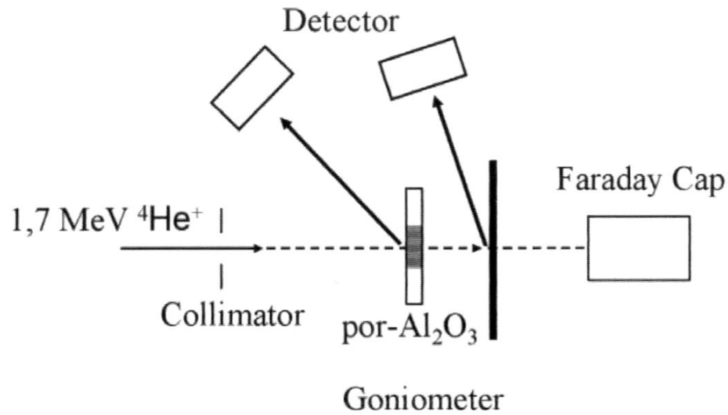

Figure 1. Schematic representation of the experimental process.

The sample under study is installed in the chamber using a goniometer system. A semiconductor detector identifies the signal passed through the membrane of He^+ ions and back scattered ions. The angular position of the detector relative to the target can be changed in the scattering plane on a circle which center coincides with the center of rotation of the goniometer.

Alumina membranes with small (20 nm) and large (200 nm) pores were chosen as researched samples. The foil thickness (10 micrometers) was chosen so that it exceeded the projective range of analyzing particles. According to the program TRIM 2011 calculations, the projective path of He^+ ions in amorphous alumina is 3.37 micrometers. The membranes were fixed in a Teflon frame to improve strength and for the ease of use [3]. In this paper the RBS spectra for the porous alumina membrane were measured at the detecting scattering angle 165° и 105°, the energy of the analyzing He^+ beam was 1,7 MeV, the ion beam diameter after passing the diaphragm system was 1 mm. The residual pressure in the chamber did not exceed $5 \cdot 10^{-4}$ Pa, the current density on the target was maintained constant and amounted to 4 nA.

3. Results and discussion

It is known that after anodizing, the oxide on the back side of the film remains "a barrier layer". As passed through the membrane, the ions will lose energy and scatter at large angles, this layer will negatively affect the transport of ions through the dielectric matrix. Usually for the estimation of value either destructive techniques are involved (SEM, TEM AFM, etc.), or integral methods giving information on the depth (energy dispersive analysis, RSMA, etc.). The present work for a more detailed study used the Rutherford backscattering method. For this purpose the following experiments were conducted (Figures 2, 3): the membrane was located in the chamber so that the ion beam could fall on the front side of the membrane (up), then the sample was taken from the camera, it was turned upside down, so that the beam fell on the opposite side of the membrane (down). Thus, we could assess the elemental composition of the membrane on both sides.

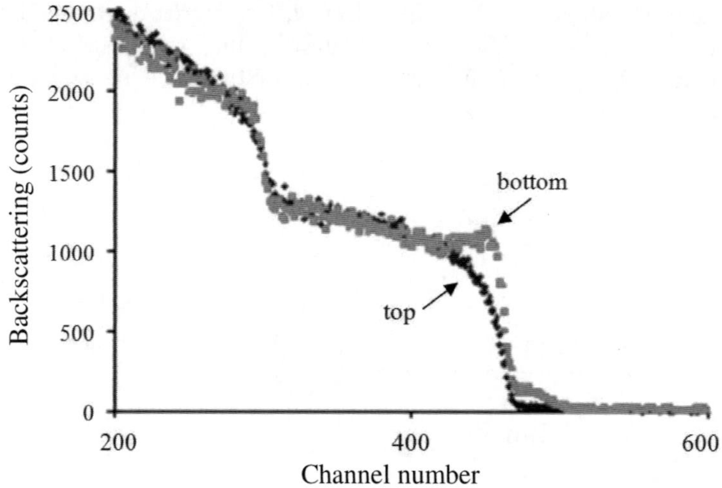

Figure 2. The energy spectrum of backscattered He^{+} ions with energies of 1.7 MeV for the scattering 120 °angle through the membrane 200 nm pore diameter.

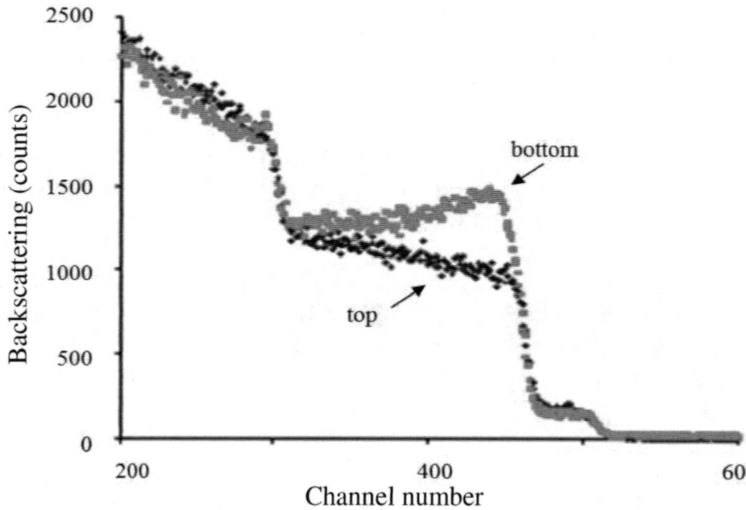

Figure 3. The energy spectrum of backscattered He^{+} ions with energies of 1.7 MeV for the scattering 120 °angle ° through the membrane 20 nm pore diameter.

The obtained results (figures 2 and 3) show that the first (up to channel 300) and the third steps repeat exactly each other (close to channel 500), but the second step (channel 300 – 470), corresponding to aluminum, has a "blockage". This suggests that oxygen and phosphorus (sulfur) are completely uniformly distributed throughout the thickness, but for aluminum a certain concentration gradient is observed, by which the full or partial depletion of the aluminum substrate after the anodization can be judged. Based on these results, it is possible to evaluate the analyzed samples on elemental composition and its distribution in depth. To study the passage of the ions through the sample some samples with the most uniform concentration of the elements in depth were used.

To evaluate the transmission coefficient and the potential energy change of the ion beam transmitted through the nanoporous membrane is possible by measuring spectra of backscattered ions from the membrane surface and the target surface located outside the membrane. It should be noted that in this case we register only those ions that pass through the membrane.

During the first experiments the target was selected solid aluminium alloy (duralumin) sample. Figure 4 shows the spectra of backscattered ions from both the membrane and that have passed through the membrane and scattered from the target. The transmission coefficient is 0.28. The shift of the backscattered ions from aluminum spectrum from channel 420 (blue curve) to 450 channel (red curve) is due to the peculiarity of the experiment (the detector during the experiments is located at different angles). In addition, the spectrum for the membrane (red curve) shows that the edge of the step corresponding to the scattering of ions from aluminum is flatter. This indicates a decrease of the method resolution in energy. Estimative the resolution decreases at 40 keV.

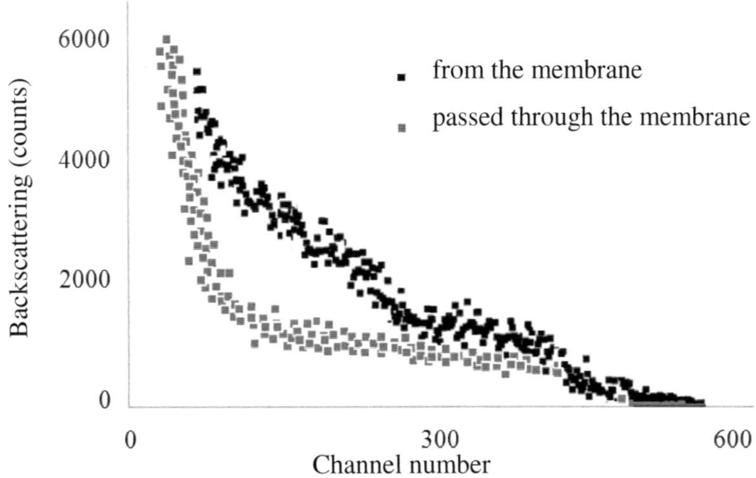

Figure 4. The energy spectrum of backscattered He^+ ions from the membrane and passed through the membrane.

It is known that beam energy resolution determines the RBS depth resolution, which depends on the mass of the element. The heavier the element is, the greater is the depth resolution. Therefore, for detecting the energy spectrum experiments on backscattering of ions passing through the nanoporous membrane were carried out with "heavy" elements. In further experiments as the "registering" environment - target a sample containing a heavy element, in particular hafnium (Hf)) - HfO_2 (24nm) / Si (100) was located behind the membrane. The given RBS spectrum (Figure 1) strongly pronounces the peak of HfO_2 located behind the membrane. This indicates that in a direction to the normal to the surface the He^+ beam passes through the membrane without losing energy. At that method the depth resolution worsened by only 24 keV. Probably part of the transmitted beam has deviated from the original beam axis, so that the resolution of analysis method deteriorated. The spectrum of a signal from the backscattered oxygen ions from the HfO_2 films did not show. Probably this is due to the fact that some of the ions in the vicinity of the channel wall were scattered while moving through the nanoporous membrane. Thus analyzing ions approaches the target at different angles, which adversely affects the RBS resolution.

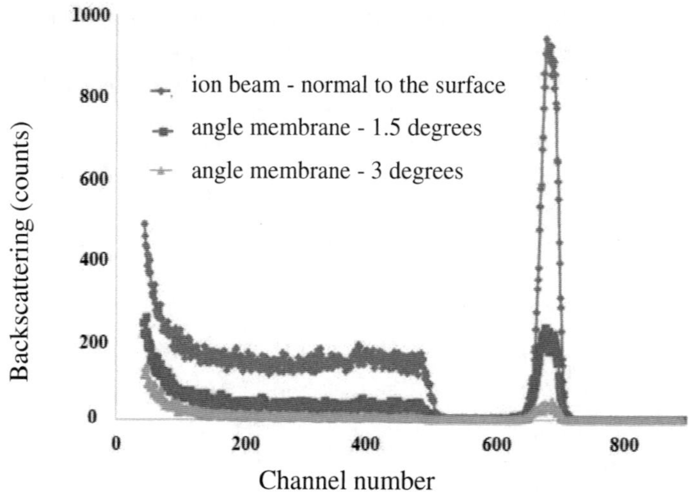

Figure 5. The energy spectrum of backscattered He^+ ions with energies of 1.7 MeV for the scattering angle 120°.

It is found that even a slight deviation of the membrane away from the normal (1.5–3 degrees) with respect to the propagation direction of the ion beam results in a reduction of backscattered ions exit HfO_2 target and hence to reduce the detected current. This allows to realize the spatial localization of the ion impacts on a substrate with nanoscale resolution.

4. Conclusion

It has been shown that the Rutherford backscattering method can help define the barrier layer thickness and define its composition with high spatial resolution without destroying the sample.
It has been shown that nanoporous membranes based on alumina containing nanoscale uniaxially-oriented system of capillaries provide the functions of transporting the He+ ion stream with energy of 1.5–2 MeV.

It has been found that even a slight deviation of the membrane from the normal (1.5 to 3 degrees) relative to the direction of propagation of the ion beam reduces the yield of backscattered ions from the HfO_2 target and, hence, a reduction in apparent current. This enables the spatial localization of ion impact on a substrate with nanoscale resolution.

Acknowledgments

This work was supported by RFBR research project №16-38-60110 mol_a_dk and partly 15-07-99656 A.

References

[1] Komarov F F, Kamyshin A S and Grishin P A 2013 *J. Nano-Elektron. Fiz.* **5** **(1)** 01015
[2] Vokhmyanina K A, Zhukova P N, Irribarra E F, et.al. 2013 *J. Surf. Investig.: XRay, Synchrotron, Neutron Tech.* **7** **(2)** 271
[3] Shemukhin A A and Muratova E N 2014 *JETP Letters* **40** 219
[4] Muratova E N, Luchinin V V, Moshnikov V A, et.al. 2017 *Glass Physics and Chemistry* **43** **(2)** 163
[5] Matyushkin L B, Muratova E N, Panov M F 2017 *Micro & Nano Letters* **12(2)** 100
[6] Shemukhin A A, Nazarov A V, Balakshin Yu V, Chernysh V S 2015 *Nuclear Instruments and Methods in Physics Research, Section B: Beam Interactions with Materials and Atoms* **354** 274
[7] Korolev D, Mikhaylov A, Belov A I et.al. 2016 *Semiconductors* **50(2)** **27**

Features of electrophoretic deposition process of nanostructured electrode materials for planar Li-ion batteries

N A Melkozyorova[1], K G Zinkevich[1], E A Lebedev[1], A V Alekseyev[1], D G Gromov[1], E P Kitsyuk[2], R M Ryazanov[2], A V Sysa[1]

[1]National Research University of Electronic Technology, 124498, Russia, Moscow, Zelenograd, Bld. 5, Pas. 4806
[2]Scientific Manufacturing Complex "Technological Centre", 124498, Russia, Moscow, Zelenograd, Bld. 5, Pas. 4806

Abstract. The features of electrophoretic deposition process of composite $LiCoO_2$-based cathode and Si-based anode materials were researched. The influence of the deposition process parameters on the structure and composition of the deposit was revealed. The possibility of a local deposition of composites on a planar lithium-ion battery structure was demonstrated.

1. Introduction

In recent years researchers focused their interest on the development of miniature electronic devices, such as MEMS, sensors and implantable systems. All of these devices require an energy supply. However, traditional batteries are difficult to scale. Therefore, it is necessary to develop fundamentally new electrochemical power sources with planar structure, in which both cathode and anode located in the same plane. Such problem can't be resolved with traditional methods, that currently in use. So it is demanded to work out a novel technique of the electrode materials creation.

From the overview of different possibilities electrophoretic deposition (EPD) seems to be incredibly perspective for the applying. Electrophoresis is the electrokinetic phenomenon of the particles movement in the liquid or gaseous medium under the applying of the external electric field [1]. Main advantages of this method are simplicity, low operating temperature and cost, possibility of a local and multicomponent deposition. The schematic depiction of the EPD cell is presented in Figure 1. The most simple one includes two electrodes, descended in suspension, and a potential source.

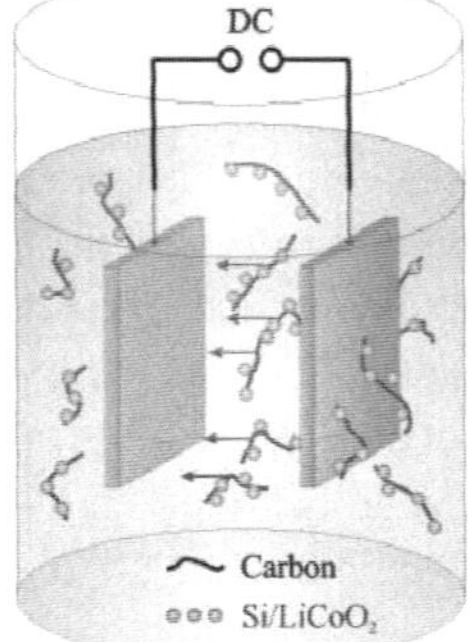

Figure 1. Schematic representation of the electrophoretic deposition.

The necessity of the composite materials creation is determined by charge storage mechanism characters in lithium-ion accumulators. In Table 1 essential characteristics of mostly common anode materials are presented [2, 3, 4].

Table 1. Characteristics of Li-ion anode materials

	Si	C	Sn	Li
Lithiated phases	$Li_{4,4}Si$	LiC_6	$Li_{4,4}Sn$	Li
Theoretical specific capacity (mAh g^{-1})	4200	372	994	386
Volume change (%)	320	12	260	100
Potential vs. Li (V)	0,4	0,05	0,6	0

It is clearly seen, that silicon based compound many times overtops other anode materials because of the high specific capacity, that comes out due to possibility of the compound formation, in which 1 Si atom bonds about 4 Li atoms. But in the same time triple volume increasing of the material takes place when charging. This problem can be solved by the creation of the carbon nanotubes-based composite, wherein CNTs will constitute the connected network, meanwhile Si particles will be smoothly distributed in it. So silicon can provide high capacity with CNTs procuring high conductivity and mechanical sustainability.
By analogy, characteristics of traditional cathode materials [5, 6], presented in Table 2, can be improved.

Table 2. Characteristics of Li-ion cathode materials

Cathode material	Avg.Potential Li/Li+	Th.capacity (mAh g^{-1})	Pract.capacity (mAh g^{-1})
$LiCoO_2$	3.9	274	120-130
$LiNiO_2$	3.8	275	120-150
$LiMn_2O_4$	4.0	148	120
$LiFePO_4$	3.3	170	120-140

Advantageous position is occupied by lithium cobalt oxide ($LiCoO_2$), being able to provide accumulators working in lower temperatures, saving prolonged life time, having relatively high theoretical capacity, characteristics stability and synthesis simplicity [7]. In traditional batteries $LiCoO_2$ is employed in composition of multicomponent material, which also includes graphite as conducting agent and polyvinylidene difluoride (PVDF) as binder. The development of the composite with CNTs will allow not only refuse from graphite, but also partially from PVDF, because the CNTs-formed connected network will provide mechanical strength. In the actual paper an approach in composite materials, based on Si-CNT and $LiCoO_2$-CNT, creation with electrophoretic method for usage in Li-ion accumulators is studied. Investigation of process, materials' structure and consistency features is carried out.

2. Experimental

Composites of silicon and graphite or CNT has been aimed as an anode and composites of $LiCoO_2$ spinel with PVDF and graphite or CNT has been aimed as a cathode materials. The deposition was carried out from suspension made of solvent, listed powder components and citric acid as a charger. Formulations of suspensions is represented in Table 3. Coatings were shaped by EPD on Ti foils. As the counter electrode in the electrophoresis cell gold foil was settled in parallel 1 cm apart. EPD was performed in potentiostatic regime, using a high voltage power source. The voltage applied in all cases was 5-140 V during times from 30 to 300 s. Process was periodic and consisted of deposition and drying stages with different durations. Number of cycles was ranged from 1 to 10. All of these processes were realized at room temperature. Morphology and composition of received layers were controlled by scanning electron microscopy and energy dispersive X-ray analysis JEOL model JSM-6010 PLUS/LA.

Table 3. Formulations of suspensions for cathode and anode materials deposition

Material	Active component	Carbon component	Binder	Charger	Solvent
Anode	Si, 12.5mg	CNT, 5mg	-	Citric acid, 12.5 mg	Acetone, 50ml
		Graphite, 5mg			Isopropyl alcohol, 50ml
Cathode	$LiCoO_2$, 250mg	CNT, 10mg	PVDF 10-40mg	Citric acid 10 mg	Acetone, 50ml
					Ethanol, 50ml
		Graphite, 10mg			n-methyl-pyrrolidone, 50ml
					Isopropyl alcohol, 50ml

Planar Li-ion battery is based on the structure which consists of two nested into each other metal combs on the surface of the substrate. This structure allows to place two electrodes in the same plane and dispose of separator. All of suchlike experimental samples were prepared using standard integral technology and equipment. The schematic depicture of key manufacturing stages is presented on Figure 2.

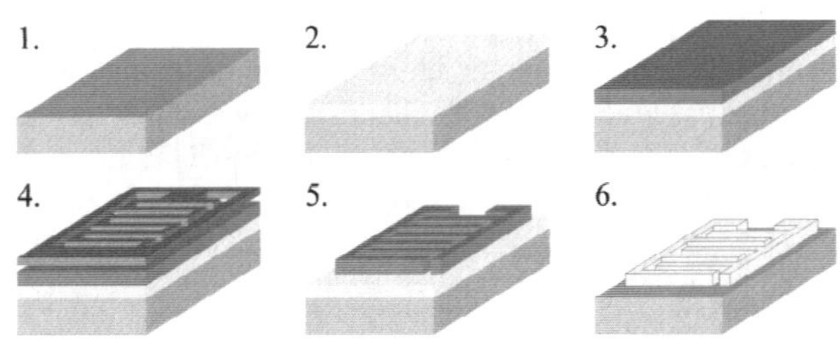

Figure 2. Technological route of interdigital structure fabrication process.

The fabrication process was began with wet cleaning and thermal oxidation of a bare silicon wafer (Figure 2, 1) in order to isolate the substrate from the electrodes. After that, Ti was evaporated onto the substrate till the thickness of 300 nm (Figure 2, 2). Ti acts as a current collector. Photoresist was spin-coated on substrate (Figure 2, 3) and then photolithography was carried out to pattern two comb-like electrodes and contact pads (Figure 2, 4). A photoresist lift-off (Figure 2, 5) followed to remove the metal on unwanted areas using wet etching processes (Figure 2, 6).

3. Results

Multiple test processes of the anode composite materials deposition on titan foil was carried out. When as a solvent isopropyl alcohol was used, very porous layers of the deposit were obtained, which tended to detach from substrate. The maximum precipitate thickness was 80 μm.

In the Figure 3a morphology of the Si-graphite composite, obtained at 60 V in EPD, surface is represented. Figure 3b represents Si-CNT surface morphology. It can be observed, that CNTs-including material has more porous structure. This has positive effect on the anode properties, as it makes easier for lithium ions to traverse during the cycling process. Also, the CNTs-constituded matrix, as we consider, helps to get down mechanical strain of silicon during the lithium ions intercalation in the accumulator charging process.

Figure 3(a, b). **(a)** SEM image of the composite anode material based on Si-graphite; **(b)** SEM image of the composite anode material based on Si-CNT.

In the Figures 4a and 4b a Si-graphite and Si-CNT composite materials layers thickness relations on applied voltage after 10 EPD cycles of 30 seconds each are represented. It is clearly seen, that in both cases results laying in linear dependency, and the deviations can be explained by the thickness unevenness and the measurement peculiarity. With increasing of the potential under 120 V the deposit structure becomes more friable and begins to crumble, that makes it difficult to determine the thickness correctly. With potential less than 5 V the deposition wasn't implemented because of the too low rate. Significant difference in the deposits thicknesses for Si-CNT and Si-graphite composites can be explained by the distinguish inequality in CNT and graphite particles sizes.

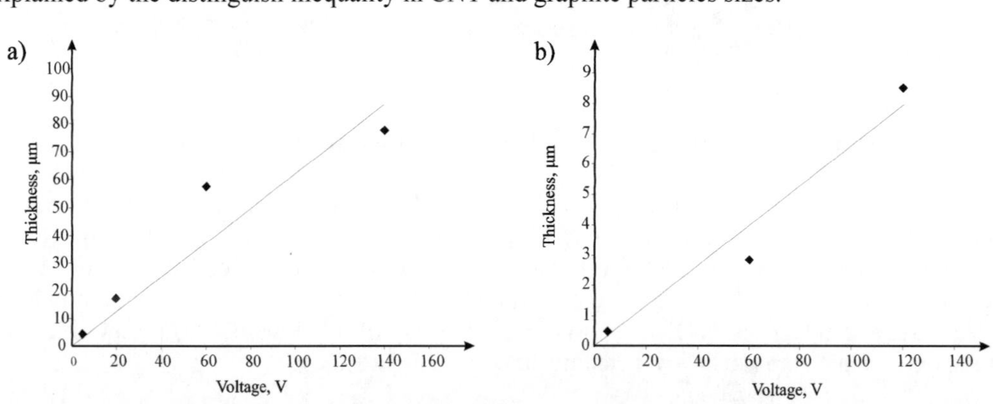

Figure 4(a, b). **(a)** Dependence of the thickness on the applied voltage for Si-graphite; **(b)** Dependence of the thickness on the applied voltage for Si-CNT.

IOP Conf. Series: Journal of Physics: Conf. Series **917** (2017) 092017 doi:10.1088/1742-6596/917/9/092017

Also, series of $LiCoO_2$-graphite and $LiCoO_2$-CNT cathode composite materials deposition experiments has been carried out. The most successful one happened to be with ethanol as the solvent. Good adhesion, high deposition rate and continuity of the deposit was observed. The maximum stated precipitate thickness was 100 µm.

In Figure 5a SEM image of the obtained $LiCoO_2$-graphite composite material is depicted. As can be seen a porous layer was received. In the Figure 5b SEM image of the $LiCoO_2$-CNT composite material, where $LiCoO_2$ particles covered with CNTs, which form connected network, can be observed.

Figure 5(a, b). (a) SEM image of the composite cathode material based on $LiCoO_2$-graphite; **(b)** SEM image of the composite cathode material based on $LiCoO_2$-CNT.

In the Figure 6a the plot of the 10-cycles deposit specific weight dependence from the applied potential is represented. The deviation from a linear dependency can be associated with the deposition area measurement error. With increasing of the potential under 120 V the deposit structure changes and becomes cellular, and after the complete drying begins to crumble, that makes it difficult to determine the weight correctly.

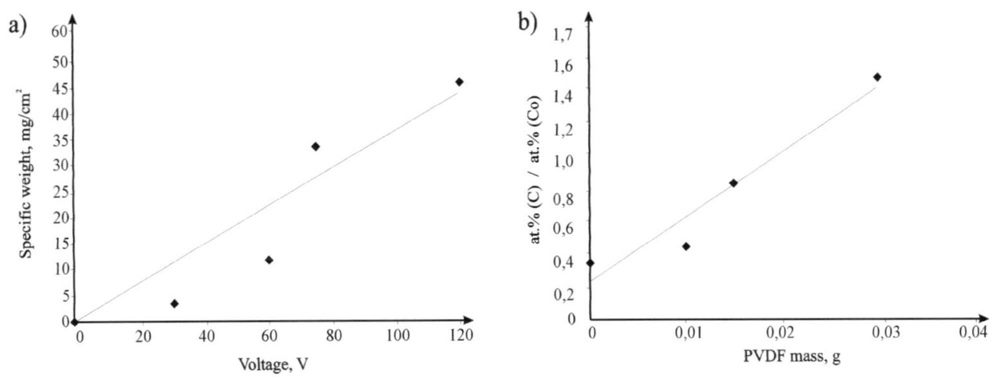

Figure 6(a, b). (a) Dependence of the specific weight on the applied voltage for $LiCoO_2$-CNT; **(b)** Dependence of the carbon to cobalt ratio from PVDF presence.

Also, series of $LiCoO_2$-CNT with PVDF as binder cathode composite materials deposition experiments has been carried out. So an interesting dependency has been discovered, presented in Figure 6b. It shows the increment of the carbon to cobalt atomic percent content ratio in the deposit with the increment of PVDF substance. This can be associated with co-deposition of PVDF and CNTs because of their possible binding in the suspension.

Furthermore, test experiments on the local deposition at the planar accumulator structure of the Si-CNT composite cathode material has been carried out. The results are presented in Figure 7.

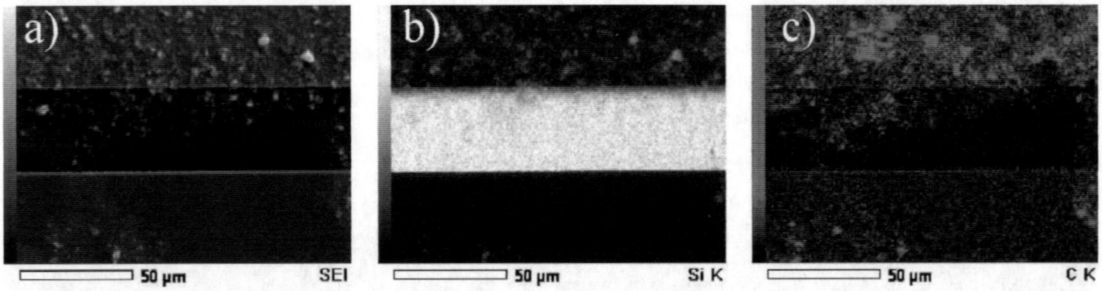

Figure 7(a, b, c). (a) SEM image of the planar structure after the local EPD of the Si-CNT composite; **(b)** Silicon EDX mapping; **(c)** Carbon EDX mapping.

It can be clearly seen (Figure 7a), that the deposition occurred mainly on the top current collector, which was connected to the potential source. Elemental analysis confirmed the presence of silicon (Figure 7b) and carbon (Figure 7c) in the deposit.

4. Conclusions

In this paper an approach for composite Li-ion accumulator electrode materials with different content creation through the EPD technique is represented. Influence of suspension substance and technical parameters on deposits characteristics is researched. Planar Li-ion accumulator structure is created, and the local deposition of the Si-CNT composite electrode material is carried out on its surface.

Acknowledgments

The Russian Science Foundation supported this work (project №16-19-10625).

References

[1] Zeiner J, Clasen R 2006 *Key Engineering Materials* 57–62
[2] Zhang W J 2011 *Journal of Power Sources* **196** 13
[3] Wu H, Zheng G, Liu N, Carney T J, Yang Y, Cui Y 2012 *Nano Letters* **12** 904
[4] Szczech J R, Jin S 2011 *Energy & Environmental Science* **4** 56
[5] Ellis B L, Lee K T, Nazar L F 2010 *Chemistry of Materials* **22** 691–714
[6] Whittingham M S 2004 *Chemical Reviews* **104** 10 4271–4302
[7] Winter M, Brodd R J 2004 *Chemical Reviews* **104** 4245–4270

IOP Conf. Series: Journal of Physics: Conf. Series **917** (2017) 092018 doi:10.1088/1742-6596/917/9/092018

Molecular dynamics simulations of GaAs-crystal surface modifications during nanoindentation with AFM tip.

N D Prasolov[1,2], **P N Brunkov**[1,2] **and A A Gutkin**[2]
[1]ITMO University, Saint Petersburg, 197101, Russia
[2]Ioffe Institute, Saint Petersburg, 194021, Russia

Abstract. The nanoindentation model of atomically flat surface of GaAs with the AFM tip was developed on the base of Molecular Dynamics. It was found that as a temperature rises above 100 K the nanoindention results in increase of number of atoms with higher number of neighbours, i.e. point defect appears in the topmost atomic layers of GaAs. The observed results can be explained with the kinetic concept of the mechanism of fracture of solid state where the generation of native point defects caused by the fluctuation of thermal energy and the external stress results in enhancement of the defect generation rate.

1. Introduction

Tribological phenomena occur by mechanical interaction of sliding surfaces during their relative motion. Important role in this process is played by the individual multiple point contacts between the rubbing surfaces, whose total area is much smaller than would follow from the geometrical sizes of contacting parts. At the initial stage of the development of tribology, the main tasks were primarily of an engineering nature, since they were associated with the study of the properties of macroscopic contacts of details of various mechanisms. Many specific tribological problems, cannot be solved analytically, therefore, numerical simulation methods in the mechanics of contact interaction and friction physics are needed. Tribological study have reached a new level with the invention of atomic force microscopy (AFM) technique, when there was an opportunity to explore (with high spatial resolution) physical properties of a single point nanocontact, formed between an apex of AFM tip and the sample surface. It was found that the physical processes in nanotribology are much broader and include many new phenomena, such as phase transitions caused by shear ordering of thin films in tip-surface contact, chemical effects associated with moisture, superconductivity, electrochemical and triboelectric effects, *etc* [1].

One of the main instruments of nanotribology is molecular dynamics (MD) simulations, which makes it possible to move from mechanics of continuous media to atomistic modeling [2, 3]. MD is based on classical dynamics, where atoms are represented as points of mass interacting with each other according to interatomic potentials. The trajectories of atoms and molecules are determined by numerical solution of Newton's equations of motion for a system of interacting particles, where the forces between particles and the potential energy are determined by interatomic potentials.

Recently, the process of triboelectrization of atomically flat epitaxial n-GaAs layer after scanning by AFM tip in contact mode was studied [4, 5]. The surface of GaAs (001) epitaxial layers grown by MOCVD technology was selected as the test object with a high crystalline quality of the both surface and bulk. Changes in local surface potential of these samples after treatment with the AFM tip was explained by nanotribological generation of native point defects in a few atomic layers below the GaAs surface. They produce additional energy states within the GaAs bandgap. Free electrons can be captured on the surface states, which results in additional negative charge on the surface and decreasing its potential [4, 5].

This work presents results of the MD simulations of nanoindentation process of atomically flat GaAs surface with the apex of AFM tip, as a function of temperature.

2. Details of simulation

The MD calculations were carried out using the LAMMPS Molecular Dynamics Simulator [6]. To describe the atomic interaction in GaAs sample with a perfect crystalline structure we used the bond-order potential (BOP), based on quantum mechanical theory incorporating both σ and π bonds [7]. The GaAs slab has the following parameters: height is 12 nm, depth and width are 56 nm each. The simulated GaAs slab consists of 1 700 000 atoms. The bottom monomolecular Ga-As layer is fixed. The orientation of the top surface of GaAs is (001) plane. The MD simulation was carried out with the periodic boundary conditions for vertical planes at the edges of the GaAs slab. The atomic system was described as canonical (NVT) ensemble, which means constant number of atoms, volume and temperature. Discretization time in MD simulations is 10 fs.

The apex of the AFM tip is modelled as absolutely rigid nanoindentor based on semisphere with radius of 10 nm. The normal velocity of nanoindentor is 1 m/s. During MD simulations the indenter moves with a constant velocity into the top surface of the GaAs slab and stops at the depth of 1 nm, where it stays until the end of the simulation process. The modeling of the nanoindentation process was carried out without lateral movement of the indenter. Also the action of capillary and adhesion forces was neglected.

3. Results and Discussions

Figure 1 represents 2D distribution of the ε_{zz} component of Green-Lagrangian strain tensor [8] calculated at temperature 10 K with nanoindentation depth of 0.5 nm (Fig. 1a) and 1.0 nm (Fig. 1b). At a depth of indentation of 1 nm the most strained region is located at the depth of 1.5 nm under the indenter (Fig. 1b). This indicates the most probable region of point defect generation. Even at maximum indentation depth of 1 nm there are no any signs of surface destruction under the indenter at temperature 10 K (Fig. 1b). The normal force applied in this case can be calculated on the base of Hertz approach which describes the contact between the inelastic ball with the elastic surface [9]. For the nanoindentation process it gives the normal force of 90 nN at the depth of 0.5 nm and 280 nN at the depth of 1.0 nm.

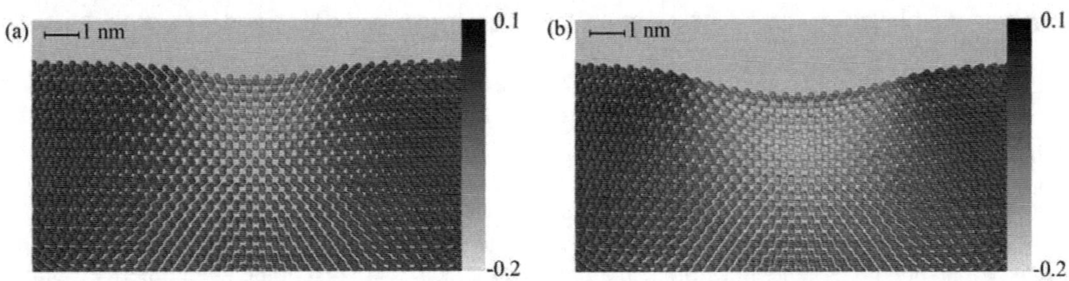

Figure 1 (a,b). 2D distribution of the ε_{zz} component of Green-Lagrangian strain tensor (indicated by colour) calculated at temperature 10 K with nanoindentation depth of 0.5 nm (a) and 1.0 nm (b).

We also performed the MD simulation for different temperatures in the range from 10 K to 900 K (Fig. 2a, b). The strained region still exist at the depth of 1.5 nm below the surface, but it becomes blurred with an increase of temperature. But nevertheless the appearance of the strained area is still noticeable. It is worth to note the MD simulations shows that at temperatures above 100 K the nanoindentation to the depth of 1.0 nm results in formation of point defects in the 2-3 topmost atomic layers (Fig. 2a, b). And these defects do not disappear after release of the nanoindentor. To estimate the point defect concentration we calculated the number of atoms (N_S) located under the indenter within the cylinder with the diameter of 6.4 nm and the depth of 10 monolayers as a function of number of nearest neighbors of one atom within a cutoff radius of 3 Å (Fig. 3). For a perfect GaAs crystal the typical number of neighbors is 4 for the atoms in the bulk and 2 or 3 for atoms located at the surface. The MD simulation at the indentation depth of 1.0 nm carried out at temperature 10 K shows no any defects in topmost atomic layers under indenter (Fig. 1a, Fig. 3). As a temperature rises the nanoindention results in increase of number of atoms N_S with higher number of neighbours, i.e. point defect appears in the topmost atomic layers of GaAs (Fig. 3).

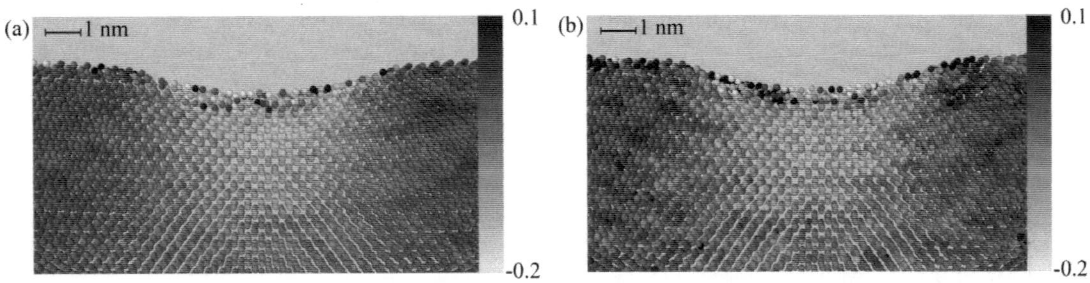

Figure 2 (a,b). 2D distribution of the ε_{zz} component of Green-Lagrangian strain tensor (indicated by colour) calculated with nanoindentation depth of 1.0 nm at temperature 300 K (a) and 900 K (b).

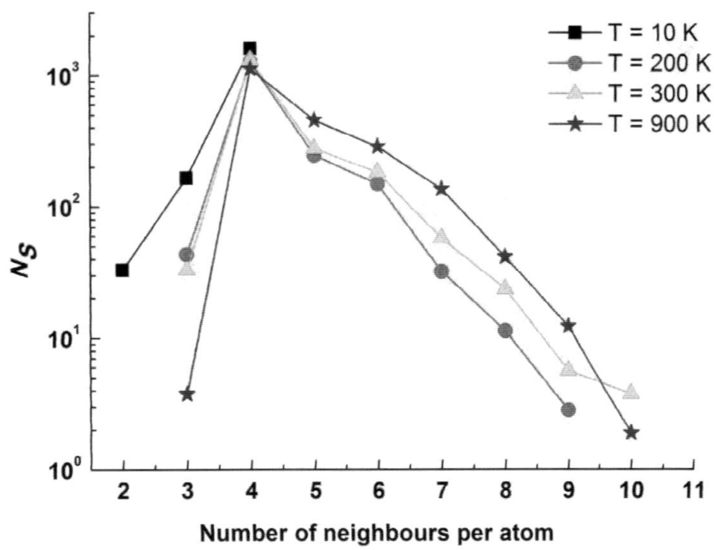

Figure 3. The number of atoms N_S located under the indenter as a function of number of nearest neighbors at different temperatures. The depth of indentation is 1 nm.

The results could be explained using the kinetic concept of the mechanism of fracture of solid state [10]. It is based on two statements that the driving force of the fracture of solids is the fluctuation of thermal energy and the external stress induces enhancement of the rate of thermal production of point defects. The external force does not directly breaks the interatomic bonds, but only prevents the restoration of broken bonds.

4. Conclusions

The nanoindentation model of atomically flat surface of GaAs with the AFM tip was developed on the base of Molecular Dynamics. The distribution of the ε_{zz} component of Green-Lagrangian strain tensor was calculated at different temperatures. It was found that at temperature 10 K indentation up to the depth of 1.0 nm shows no any defects in topmost atomic layers under indenter. As a temperature rises the nanoindention results in increase of number of atoms with higher number of neighbours, i.e. point defect appears in the topmost atomic layers of GaAs. The observed results can be explained with the kinetic concept of the mechanism of fracture of solid state where the generation of native point defects caused by the fluctuation of thermal energy and the external stress results in enhancement of the generation rate. The findings allow to explain the observed earlier the experimental results on modification of surface potential of GaAs after the scanning with the AFM tip.

Acknowledgments

This study was supported in part by the Russian Foundation for Basic Research, project no. 15_01_05903.

References

[1] ed. Bhushan B 2011 *Nanotribology and Nanomechanics: Measurement Techniques and Nanomechanics* (Berlin-Heidelberg, Springer-Verlag) Vol. 1
[2] Frenkel D, Smit B 2002 *Understanding Molecular Simulation: From Algorithms to Applications* (Academic Press)
[3] Pokluda J, Sandera P 2010 *Micromechanisms of Fracture and Fatigue: In a Multi-scale Context* (Springer-Verlag London)
[4] Brunkov P N, Goncharov V V, Rudinsky M E, Gutkin A A, Gordeev N Yu, Lantratov V M, Kalyuzhnyy N A, Mintairov S A, Sokolov R V and Konnikov S G 2013 *Semiconductors* **47** 1170
[5] Baklanov A V, Gutkin A A, Kalyuzhnyy N A and Brunkov P N 2015 *Semiconductors* **49** 1057
[6] Plimpton S 1995 J. Comp. Phys. **117** 1.
[7] Murdick D A, Zhou X W, Wadley H N G, Nguyen-Manh D, Drautz R and Pettifor D G 2006 Phys. Rev. B 73 045206
[8] Falk M L and Langer J S 1998 Phys. Rev. E **57** 7192
[9] Popov V L 2010 *Contact Mechanics and Friction. Physical Principles and Applications* (Springer-Verlag Berlin Heidelberg)
[10] Zhurkov S N 1984 *Int. J. Fract.* **26** 295

Mechanisms of charge transport and resistive switching in composite films of semiconducting polymers with nanoparticles of graphene and graphene oxide

A S Berestennikov[1]*, **A N Aleshin**[1]
[1] *Ioffe Institute RAS, Saint Petersburg 194021, Russia*

*E-mail: a.berestennikov@metalab.ifmo.ru

Abstract. We have investigated the effect of the resistive switching in the composite films based on polyfunctional polymers – PVK, PFD and PVC mixed with particles of Gr and GO with the concentration of ~ 1 - 3 wt.%. We have developed the solution processed hybrid memory structures based on PVK and GO particles composite films. The effect of the resistive switching in Al/PVK(PFD; PVC):Gr(GO)/ITO/PET structures manifests itself as a sharp change of the electrical resistance from a low-conducting state to a relatively high-conducting state when applying a bias to Al–ITO electrodes of ~ 0.2–0.4 V. It has been established that a sharp conductivity jump characterized by S-shaped current-voltage curves and the presence of their hysteresis occurs upon applying a voltage pulse to the Au/PVK(PFD; PVC):Gr(GO)/ITO/PET structures, with the switching time in the range from 1 to 30 μs. The mechanism of resistive switching associated with the processes of capture and accumulation of charge carriers by Gr(GO) particles introduced into the matrixes of the PVK polymer due to the reduction/oxidation processes. The possible mechanisms of energy transfer between organic and inorganic components in PVK(PFD; PVC):GO(Gr) films causes increase mobility are discussed. Incorporating of Gr (GO) particles into the polymer matrix is a promising route to enhance the performance of hybrid memory structures, as well as it is an effective medium for memory cells.

1. Introduction

Hybrid (organic–inorganic) nanocomposite materials are currently of great interest due to the prospects of their use as nonvolatile memory devices because of their high operation speed, high-density data storage, low power consumption, and low cost. Therefore they exhibit significant advantages over inorganic memory materials in the fabrication of volatile and nonvolatile memory devices [1-3]. Such polymeric and hybrid materials exhibit easy processability, good flexibility, high mechanical strength, and good scalability. They require very simple solution processes such as spin coating, dip coating and inkjet printing, which can be carried out at low cost and further are applied for high density memory devices.

In recent years, of particular interest is the use of new hybrid materials based on graphene (Gr) as OFETs and nonvolatile memory devices, in particular, composites based on Gr particles and particles of graphene oxide (GO) enclosed in a polymer matrix. Owing to the unique properties of Gr, such as high charge carrier mobility, high thermal conductivity, and the possibility of the effective transport of charge carriers, including the ballistic transport at room temperature [4,5], it is a promising material for optoelectronics. Graphene flakes, being an effective medium for the accumulation of charge carriers, are considered as a promising material for hybrid (polymer–Gr particles) non-volatile WORM memory cells based on the effects of resistive switching. A similar role is played by GO particles with the best solubility in polymer matrices as compared to Gr particles [3].

In this paper we have investigated the effect of the resistive switching in the composite films based on polyfunctional polymers – PVK, PFD and PVC mixed with particles of Gr and GO with the concentration of ~ 1 - 3 wt. %. The effect of resistive switching in Al/PVK(PFD; PVC):Gr(GO)/ITO/PET structures at applying biases to Al–ITO electrodes ~ 0.2–0.4 V is investigated. The observed conductivity jump is accompanied by the appearance of S-shaped current–voltage–characteristics and the presence of hysteresis. The switching time, t, in such structures is in the range from 1 to 30 μs. The mechanism of resistive switching associated with the processes of capture and accumulation of charge carriers by Gr(GO) particles introduced into the matrixes of the PVK polymer due to the reduction/oxidation processes.

2. Experimental

We investigated the electrical and optical properties of composite films based on soluble in organic solvents multifunctional polymers: poly(9-vinylcarbazole) – PVK - $(C_{14}H_{11}N)_n$, average molecular weight M_w ~ 1.1 × 10^6, band gap E_g ~ 3.6 eV; poly(9,9-di-n-dodecylfluorenyl-2,7-diyl) – PFD, $(C_{31}H_{44}B_2O_5)_n$, M_ω is about 5.18 × 10^2, and E_g is about 3.36 eV; polyvinyl chloride – PVC $(C_2H_3Cl)_n$, M_ω is in the range from 4 × 10^4 to 1.45 × 10^5, and E_g is about 4.86 eV; and their composites with Gr particles from 200 to 300 nm in diameter (high temperature reduced graphene - HTRG) and GO particles from 400 to 500 nm in diameter. The concentration of Gr and GO particles in the polymer matrix of PVK was varied in the range of 1–3 wt. %. The PVK, PFD polymers and ITO/PET substrates used in our study were purchased from Sigma-Aldrich, PVC was purchased from FGUP VNIISV (Tver) and then used without further treatment. The Gr and GO flakes were purchased from AkKoLab (Moscow, Russia) and were used without further modification.

Polymers were dissolved in chloroform, which was also used for the preparation of a colloidal solution of Gr and GO particles. The solutions were mixed and sonicated (Bandelin Sonopuls HD 2070, f ~ 20 kHz) for 3 - 5 min. The obtained 20% solution of polymers with Gr (or GO) particles in chloroform was applied to flexible PET substrates with ITO (In_2O_3/SnO_2) conducting layer, ITO/PET (the surface resistivity of the ITO layer was ~ 60 Ω/cm^2) by spin coating at 2500–3000 rpm, followed by drying at ~ 80°C for 20 min under nitrogen atmosphere. The current–voltage (I-Vs) characteristics of the composite films for memory structures were measured under direct current at 300 K in a sandwich geometry after thermal evaporation of Al electrode on top of the PVK(PFD, PVC):Gr(GO)/ITO/PET structure using an automated measurement setup based on a Keithley 6487 picoammeter in the voltage range from –5 to +5 V. The typical structure of a Al/PVK(PFD, PVC):Gr(GO)/ITO/PET samples for I-V measurements is shown in figure 2(a). The I-Vs of PVK(PFD, PVC):GO-based films were measured in vacuum ($3\cdot10^{-3}$ torr) in the dark at 300 K using a holder of liquid N_2 cryostat and a dc electronic computer controlled measuring system with a Keithley 6487 picoammeter/voltage source and AKIP-1124 programmable voltage source. The voltages were varied between – 15 V and + 15 V in variable steps.

Current–voltage characteristics of the composite films in the pulse mode were measured on a unit with a PCG10A current generator at 300 K. A voltage pulse from the generator was sent to the sample and serially connected load impedance. The ITO layer served as a lower electrode and the gold clamping contact with the area, S ~ 10^{-4} cm^2, as an upper electrode. A triangular pulse with duration of 2 ms was used; the maximum current value was 100 μA, and a clock-frequency period of 4 μs was chosen. The voltage was measured using a PS500 dual channel oscilloscope. The electric current

IOP Conf. Series: Journal of Physics: Conf. Series **917** (2017) 092019 doi:10.1088/1742-6596/917/9/092019

strength was determined from the voltage drop at the load and the voltage on the sample from a difference between the input voltage, U_0, and the voltage at the load.

3. Results and discussion

The above described PVK:Gr(GO) composite films were used in our study of the resistive switching effect which can be utilized for designing of nonvolatile WORM memory. The structure of the studied samples is shown in figure 1.

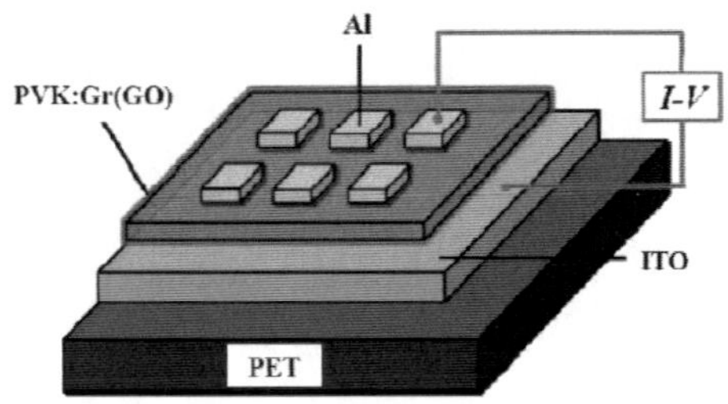

Figure 1. The structure of the Al/PVK:Gr(GO)/ITO/PET samples for I-Vs measurements.

The typical experimental I-Vs of these structures are presented in figure 2 and inset (a) in figure 2. As it evident from figure 2, the memory effect for the Al/PVK(PFD, PVC):Gr(GO)/ITO/PET films consists of in switching the conductivity of the composite structure from the low conducting to high conducting state under applying a bias to Al - ITO electrodes of ~ 0.1 - 0.4 V, which corresponds to the electric field strength $E \sim 3 - 4\times10^4$ V/cm. As can be seen from inset b) in figure 2, the I-Vs of the samples follow a linear law in both the initial and final states and the ratio of the final electric current density to the initial current through the sample (j_2/j_1) varies in the range of ~ 2 – 450. We have obtained similar results for several other samples of the PVK:Gr(GO) composite films with concentrations of Gr and GO particles in the range of 1–3 wt. % as well as for some other polymer matrixes mixed with Gr or GO particles [6]. It is important to note that the switching voltage in the Al/PVK:Gr(GO)/ITO/PET structures proved to be considerably less than the threshold voltage obtained for the polymer–nanoparticle and polymer - Gr composites by other authors [7]. For all the studied samples, the switching effect occurs with a sharp (by one to two orders of magnitude) increase in the current through the structure, which indicates a transition from the "off" state to the "on" state (process of "writing"). The sample remains in the "on" state in the reverse scanning of the voltage, as well as in multiple scanning both at a positive bias and at a negative bias applied to the electrodes. In some cases, we observed the effect of reverse switching of the Al/PVK:Gr(GO)/ITO/PET thin films to a relatively low conducting state. The analysis of I-Vs of the Al/PVK:Gr(GO)/ITO/PET structures, in a log–log scale (see inset (b) in figure 2) reveal the fact that the slopes of the I-Vs are close to unity in both the low conducting and high conducting states when Gr and GO particles are embedded in the PVK matrices. This indicates that the composite films have the ohmic conductivity following the law $I(V) = \alpha V$ in both states. The absence of I-Vs of the shape $I(V) \sim \beta V^2$, which is typical of the mechanism of space charge limited currents, can indicate that this mechanism is not responsible for the observed resistive switching effect at a relatively low (less than ~ 1 V) biases.

Current-voltage characteristics of the PVK:Gr, PFD:Gr, and PVC:GO composite films by using a pulsed bias are given in figures 3(a) and 3(b). As is seen from figure 3, the I–V curves for all the investigated samples have S-shapes after increasing the voltage. The transition from the LC to HC state occurs at the threshold voltages, UT, which are different in magnitude for different types of polymer films; moreover, a sharp voltage drop on the samples without significant changes in the electric current flowing through them is observed in all the cases, which can be explained by the current crowding effect. As follows from the given results, the threshold switching voltages, UT, for the PVK:Gr, PFD:Gr, and PVC:GO composite films are about 3 to 3.5, 7 to 8, and 15 to 16 V, respectively. The resistivity of the films after the switching is changed by an order of magnitude, i.e., the resistivity state is in the range from 200 to 500 kΩ and in the HC state from 20 to 40 kΩ. As a rule, the samples do not return to the LC state after reducing the voltage, but remain in the HC state.

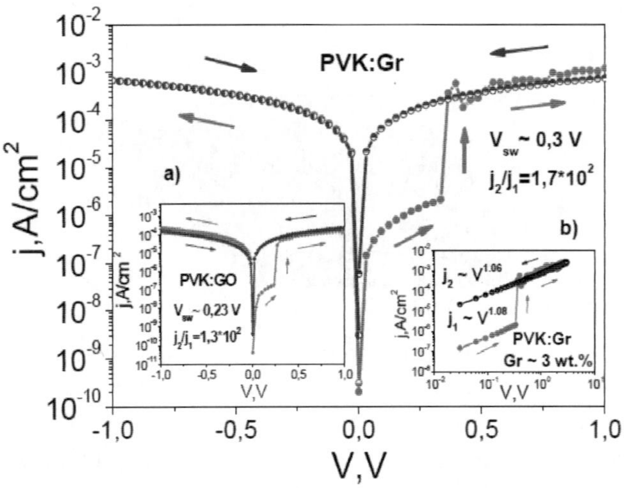

Figure 2. I-Vs of the composite film Al/PVK:Gr/ITO/PET demonstrating the effect of current switching at low voltages; insets: a) I-Vs of Al/PVK:GO/ITO/PET film; b) I-Vs of Al/PVK:Gr/ITO/PET film on a log-log scale.

As is seen from figure 3, the current-voltage characteristic of the PVK:Gr film in the HC state shows a superlinear dependence unlike the I-V curves of the PFD:Gr and PVC:GO films, which are linear. The electric current changes upon the switching are about 0.2 μA for the PFD:Gr sample and about 5 μA for the PVC:GO sample, whereas the electric current change in the PVK:Gr film is zero (within the measurement accuracy). The observed differences perhaps are possibly connected with different switching times of the samples that are about 20, 30, and 4 μs for the PFD:Gr, PVC:GO, and PVK:Gr films, respectively. In particular, the switching time of the PVK:Gr film for the selected clock frequency period equal to 1 μs was estimated at about 1 μs (perhaps less than a few hundreds of nanoseconds). Repeated pulses of the same or opposite polarity sent to the samples do not lead to switching back to the LC state.

Instabilities with S-shaped I–V curves observed in the PVK:Gr and other films resemble the switching effects in chalcogenide glass semiconductors (CGSs) by their characteristics. These kinds of switching effects have been also observed in a series of composite films on the basis of polymers containing inorganic Si [8] and ZnO [9] particles. According to the model described in [9], the electric current instability effect in the case of CGS is caused by the electrothermal phenomena occurring in films under strong electric fields. For the generation of an instability in the form of the S-shaped I–V curve, the positive feedback should be established between the processes of heating and increasing the electric current strength in the material with an activation-type temperature dependence of the conductivity, σ, expressed by the following formula:

$$\sigma(T) = \sigma_0 \exp(-\Delta E/kT), \quad (1)$$

where ΔE is the activation energy of the conductivity, and k is the Boltzmann constant. Since the electric current flow leads to heating of the sample, an increase in the temperature in the presence of the semiconducting dependence, σ(T), in the form of Eq. (1) causes an increase in the conductivity, which in turn leads to a further increase in the current strength. The positive feedback effect and temperature rise are finished upon reaching the temperature, T_n, at which $(\Delta E/kT_n) = 1$, determined by the phase transition temperature in semiconductor–metal structures. These kinds of processes are typical for some types of CGSs, in which the glass-crystal phase transition occurs near the area of current crowding upon reaching the crystallization temperature; furthermore, the conductivity in the indicated area is switched by several orders of magnitude, i.e., from the LC (glass) state to the HC state (crystal) [9].

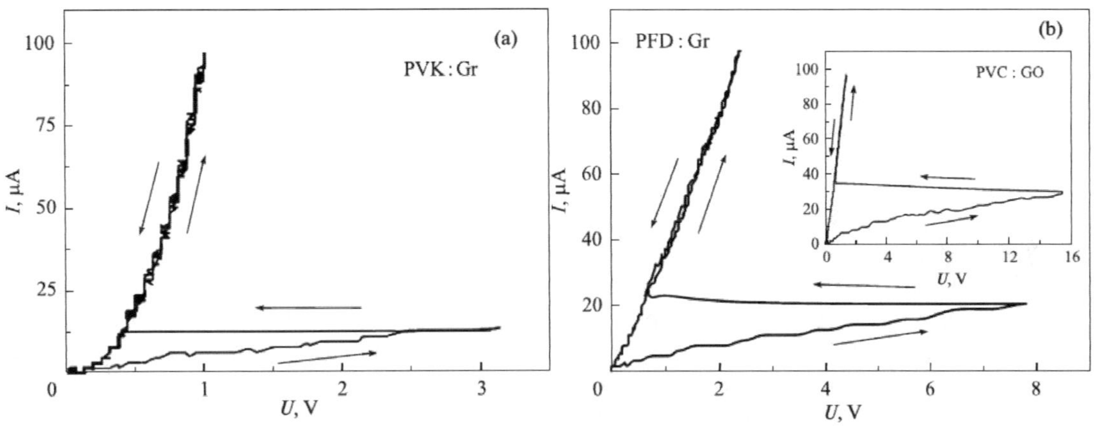

Figure 3. The I–V curves of the (a) Au/PVK:Gr/ITO/PET and (b) Au/PFD:Gr/ITO/PET composite films upon applying a triangular pulse to the ITO–Au electrodes. Current–voltage characteristics of the Au/PVC:GO/ITO/PET film upon applying a triangular pulse to the ITO–Au electrodes are shown in the insertion.

It should be noted that the mechanisms of current crowding in the other inorganic semiconductors, giving rise to S-shaped I–V curves, were fully described in [10]. To some extent, the electrothermal switching mechanism may be responsible for the S-shaped form of I–V curves in the studied films on the basis of polymer composites containing Gr and GO particles. However, one should take into account the peculiarities of thermomechanical and electrical properties of polymers in this case, as well as the occurrence of redox processes on the Gr and GO particles embedded into the polymer matrix. Mechanical features of polymers include their ability to stay in the vitreous, superelastic, or viscous-flow states. An increase in the conductivity with an increase in the temperature, described by Eq. (1), is typical for a glassy (amorphous) polymer phase. The transition of a polymer into a superelastic state, in which the electrical conductivity does not dependent on the temperature and remains constant, occurs upon reaching the glass transition temperature, T_g. At the melting temperature, T_f, a polymer experiences a phase transition to the viscous-flow state, in which an increase in the conductivity with an increase in the temperature is also observed [8]. For the starting PVK and PVC polymers, the glass transition and melting temperatures are the following: T_g is about 200 and 80°C, and T_f is about 290 and 270°C, respectively; at the same time, the T_g and T_f values can vary by 20 to 30°C depending on the film thickness and the chemical composition of the polymers. As was shown for CGSs in [9], the S-shaped I–V curves and switching effect at temperatures above 300 K might be caused by the current crowding effect over the sample. One can assume that a polymer in the area of current crowding experiences a phase transition to the viscous-flow state at the temperatures $T > T_f$. At the same time, there is no degradation of the polymer in the indicated area, since its supramolecular structure changes at high temperatures without breaking the molecular structure due to reducing the viscosity coefficient by several orders of magnitude. After cooling the area of current crowding, the electrical resistivity of the sample may decrease by an order of magnitude. If the temperature of the area of current crowding was in the range $T_g < T < T_f$, i.e., within the superelastic phase, in which the electrical conductivity does not depend on the temperature, then the polymer would return to its initial LC state after cooling, and the HC state would not persist after removing the electric displacement field from the sample.

The considered effect of resistive switching in the Al/PVK:Gr(GO)/ITO/PET composite films makes it possible to create solution processible, i.e. compatible with the technology of flexible organic printing electronics, WORM memory cells based on such composites with electrical recording of information.

4. Conclusions

The effect of the resistive switching in the composite films based on PVK mixed with particles of Gr and GO with the concentration of ~ 1 - 3 wt.% is investigated. The solution processed hybrid memory structures based on PVK and GO particles composite films are developed. It was shown that the effect of the resistive switching in Al/PVK(PFD; PVC):Gr(GO)/ITO/PET structures manifests itself as a sharp change of the electrical resistance from a low-conducting state to a relatively high-conducting state when applying a bias to Al–ITO electrodes of ~ 0.2–0.4 V. The reversible switching of the source-drain current, I_{SD}, is also found at positive and negative V_{SD}. We suggest that the mechanism of resistive switching associated with the processes of capture and accumulation of charge carriers by Gr(GO) particles introduced into the matrix of the PVK polymer due to the reduction/oxidation processes. The mechanism of energy transfer between organic and inorganic components in PVK(PFD; PVC):GO(Gr) films may causes the increase mobility. One may conclude that incorporating of Gr (GO) particles into the polymer matrix is a promising route to enhance the performance of hybrid memory cells.

Acknowledgments

The authors are grateful to P.S. Krylov, I.N. Trapeznikova, S.A. Fefelov and A.S. Komolov for cooperation. A.N. Aleshin and A.S.Berestennikov were partially supported by the Russian Foundation for Basic Research project No. 15-02-01897 and by the Fundamental Program of Presidium of the Russian Academy of Sciences, No 1.8P; subprogram: "The fundamental principles of the creation and investigation of new materials and materials for molecular electronics and spintronics"

References

[1] Y. Yang, J. Ouyang, L. Ma, R.J. Tseng, C.W. Chu, Adv. Funct. Mater. 16 (2006) 1001–1014.
[2] J.C. Scott, L.D. Bozano, Adv. Mater. 19 (2007) 1452–1463.
[3] W.P. Lin, S.J. Liu, T. Gong, Q. Zhao, W. Huang, Adv. Mater. 26 (2014) 570–606.
[4] A.K. Geim, K.S. Novoselov, Nat. Mater. 6 (2007) 183–191.
[5] K.S. Novoselov, A.K. Geim, S.V. Morozov, D. Jiang, Y. Zhang, S.V. Grigorieva, I.V. Dubonos, A.A. Firsov, Science 306 (2004) 666–669.
[6] Krylov P. S., Berestennikov A. S., Aleshin A. N., Komolov A. S., Shcherbakov I. P., Petrov V. N., & Trapeznikova I. N., Phys. Sol. State *57* (2015) 1669-1675.
[7] Lin W.P., Liu S.J., Gong T., Zhao Q., & Huang W., Adv. Mater. 26 (2014) 570-606.
[8] E. A. Lebedev, E. L. Alexandrova, and A. N. Aleshin, Phys. Solid State 51 (1) 208 (2009).
[9] K. D. Tsendin, E. A. Lebedev, and A. B. Shmel'kin, Phys. Solid State 47 (3), 439 (2005).
[10] A. F. Volkov and Sh. M. Kogan, Sov. Phys. Usp. 11 (1968) 881.

IOP Conf. Series: Journal of Physics: Conf. Series **917** (2017) 092020 doi:10.1088/1742-6596/917/9/092020

Study of aerosol jet printing with dry nanoparticles synthesized by spark discharge

A A Efimov, P V Arsenov, I A Volkov, M N Urazov, V V Ivanov
Department of Physical and Quantum Electronics, Moscow Institute of Physics and Technology, Dolgoprudny 141700, Russia

Abstract. A new method of aerosol jet printing utilizing dry (solvent-free) airborne nanoparticles generated by spark discharge is proposed. This method was applied to fabricate thin conducting lines (60–160 μm) composed of silver nanoparticles on the surface of glass substrates. It has been demonstrated that the line width is determined by a sheath flow rate, while its thickness and cross-sectional area can be scaled up by a number of printing runs. The resistivity of printed lines after the annealing was found to be five times higher than that of bulk silver that is attributed to the porosity and the interparticle contact resistance. The proposed method holds promise for the application in technologies of printed electronics.

1. Introduction

In recent years there is an active development of new methods for the production of cheap electronic devices by printing equipment [1]. Aerosol jet printing is one such method. The conventional approach of aerosol jet printing is based on the selective deposition of focused beams of microdroplets on a substrate with nanoparticles inside. The use of nano-ink causes some significant problems, such as environmental pollution due to the use of organic solvents and necessity of removing and recycling solvent and surfactant after deposition. In this regard, researchers are developing and investigating the new environmentally friendly and versatile source of nanoparticles for aerosol jet printing, which does not require the use of inks with solvents. For example, the spark discharge generator [2] is a promising source of dry nanoparticles for aerosol jet printing. The spark discharge generator does not require the use of solvents and is a great alternative for pneumatic and ultrasonic atomizers. However, the use of the spark discharge generator for aerosol jet printing is insufficiently investigated due to the complexity of the design of the spark discharge generator and the novelty of the approach. In this work we used the spark discharge generator and investigated the characteristics of the lines formed in different modes of aerosol jet printing.

2. Experimental

The experimental setup consisted of a multi-spark discharge (m-SDG) generator and a commercial system for the particle deposition AJ 15XE with coaxial nozzle 100 μm in outlet diameter (figure 1a). Aerosol nanoparticles were synthesized by the m-SDG as a result of electrical erosion of silver electrodes in air [2]. The aerosol flow Q_a entered the coaxial nozzle through the inner cylindrical channel, while the sheath flow Q_{sh} was inserted through the outer conically converging axisymmetrical channel. The distance between the nozzle and substrate were fixed in the experiment as 0.5 mm. Then the focused aerosol beam was directed to the glass substrate for deposition. In the process of moving the substrate relatively to the focused aerosol beam, we formed the line of deposited silver nanoparticles on the substrate.

Characteristics of the lines were studied depending on changes in sheath flow rate (from 35 to 100 sccm), number of printing runs (from 9 to 36) and speed motion of the substrate (from 3 to 40 mm/min). Printed lines were annealed in a muffle furnace at 450°C for 1 hour at air atmosphere. Aerosol nanoparticles were characterized by the aerosol spectrometer SMPS 3936 before their entering the nozzle. Figure 1b demonstrates the particle size distribution of aerosol nanoparticles measured before their entering the nozzle. It has been found that while entering the nozzle the size of agglomerates is in the range from 15 to 100 nm, and their mean size is about 44±6 nm. These particles have high inertia due to the high density and high velocity values at the nozzle exit more than 150 m/s.

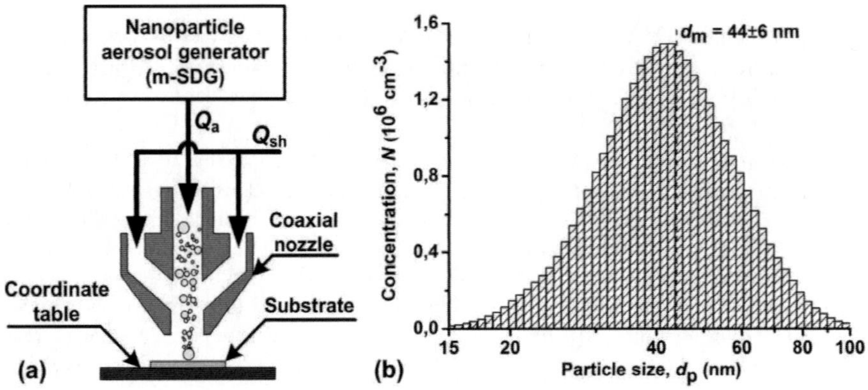

Figure 1(a, b). (a) The scheme of the aerosol jet printing with dry nanoparticles; **(b)** Particle size distribution of aerosol nanoparticles measured before their entering the nozzle by the aerosol spectrometer.

The width of formed lines was measured by using the optical microscope KH-7700. The microstructure, cross-sectional profile and thickness of the line were studied using scanning electron microscope (SEM) JSM-7001F and profilometer DektakXT. Electrical resistivity of the printed and sintered lines was measured by four-probe method.

3. Results and discussions

It is found that an increase of the sheath flow rate Q_{sh} from 35 to 100 sccm at a constant flow rate of aerosol Q_a leads to a decrease in the diameter of the aerosol beam and as a consequence a decrease of the line width W from 160 to 85 μm (figure 2a).

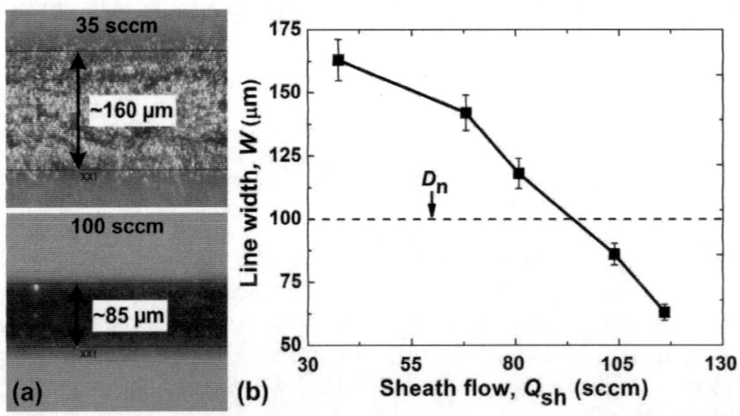

Figure 2(a, b). (a) The image of the printed lines in the optical microscope related to the sheath flow rate Q_{sh}; **(b)** The dependence of the width of the line W from the sheath flow rate Q_{sh}.

It was found that the dependence of the line width W on the flow rate is described by the following expression:

$$W \sim k \cdot D_n \sqrt{\frac{Q_a}{Q_a + Q_{sh}}} \tag{1}$$

where k – is the function of increasing the line width due to the diffusion displacement of the nanoparticles [3] and the spreading of the material on the surface during the deposition process [4]; D_n is nozzle outlet diameter.

Equation (1) shows that an increase of the sheath flow rate Q_{sh} at a constant flow rate of aerosol Q_a leads to a decrease in the line width W. Experimental confirmation of this dependence is presented in figure 2b. The minimum line width is reduced up to 60 μm at high values of sheath flow rate.
The cross-section profile of the printed line has a bell-shaped according SEM image of the cross-section of the line and data surface profilometry see figure 3a and b, respectively. We found that the average thickness t and cross-sectional area A_s of the lines can be raised more than 2 times by increasing the number of printing runs N_p from 9 to 36 at Q_{sh} = 90 sccm, V_s = 10 mm/min (figure 3b). A similar dependence was observed by decreasing the speed of motion of the substrate V_s at Q_{sh} = 90 sccm, N_p = 6 (figure 4a). These both results are explained by the growth amount of material which is deposited per unit time onto the substrate.

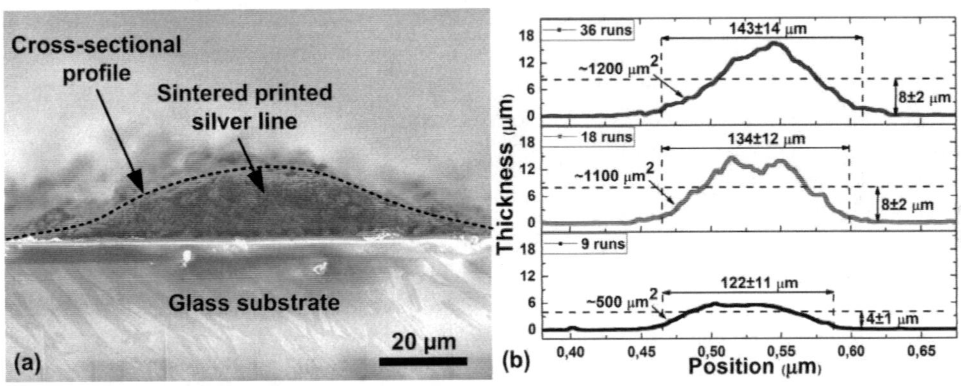

Figure 3 (a, b). (a) Cross-sectional SEM image of a sintered line formed by aerosol jet printing with dry nanoparticles; **(b)** Effect of number of printing runs on cross-sectional profile, average thickness and line width measured by surface profilometer at Q_{sh} = 90 sccm, V_s = 10 mm/min.

It is known that the amount of material exiting from nozzle and deposited on a substrate is determined by the following equation taken from [4]:

$$Wt \approx A_s = \left(\frac{\rho_e}{\rho_s}\right)\frac{V_e A_e}{V_s} \tag{2}$$

where A_s – cross sectional area of the printed line, w and t – width and thickness of the as-printed line, respectively, ρ_e and ρ_s – densities of the aerosol beam and the coalesced as-printed material, respectively, V_e and V_s – exit velocity of the aerosol beam and stage speed, respectively.

Thus, decreasing the stage speed V_s at a fixed aerosol Q_a and sheath flow rate Q_{sh} or exit velocity V_e leads to an increase in the line width W and thickness t, as illustrated in figure 4a. Moreover, figure 4b shows that the width of lines W changes significantly at low stage speed of 3-10 mm/min and become almost constant at high stage speed. Thus, it has been found that at low stage speed V_s, the effect of spreading the line in the lateral direction is observed. The spreading effect of the line disappears at high speed of substrate of more than 20 mm/min and the line width W depends mainly on the diameter of the aerosol beam (figure 4b).

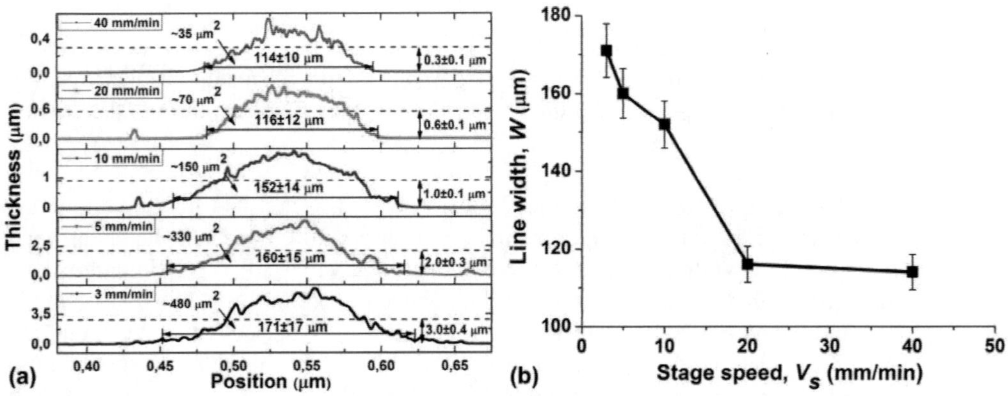

Figure 4. (a, b). Effect of speed substrate on cross-sectional profile, average thickness, line width and cross-sectional area measured by surface profilometer at Q_{sh} = 90 sccm and N_p = 6.

Using the results of measurements of cross-sectional area, the length and line resistance, the electrical resistivity of the formed lines, which amounted to about 7.5 μΩ·cm was determined. Thus, electrical resistivity of sintered lines of silver nanoparticles is 5 times higher than that of bulk silver is determined by the results of electrical measurements. This result is due to the presence of pores in lines and the high contact resistance between the nanoparticles. The obtained results indicate the need for new optimization research of the sintering lines.

4. Conclusion

This work presents a study on the characteristics of silver lines printed at various modes of new solvent-free method of aerosol jet printing. It has been established that when controlling parameters of aerosol jet printing one can form functional lines in a wide range of width values from 60 to 160 μm. The same printed lines can be used in printed electronics devices.

Acknowledgments

This work was supported by a grant from the Russian Science Foundation (project № 15-19-00190).

References

[1] Hoey J M, Lutfurakhmanov A, Schulz D L and Akhatov I S 2012 A Review on Aerosol-Based Direct-Write and Its Applications for Microelectronics *J. Nanotechnol.* **2012** e324380
[2] Efimov A, Lizunova A, Sukharev V and Ivanov V 2016 Synthesis and Characterization of TiO2, Cu2O and Al2O3 Aerosol Nanoparticles Produced by the Multi-Spark Discharge Generator *Korean J. Mater. Res.* **26** 123–9
[3] Efimov A A, Potapov G N, Nisan A V and Ivanov V V 2017 Controlled focusing of silver nanoparticles beam to form the microstructures on substrates *Results Phys.* **7** 440–3
[4] Mahajan A, Frisbie C D and Francis L F 2013 Optimization of Aerosol Jet Printing for High-Resolution, High-Aspect Ratio Silver Lines *ACS Appl. Mater. Interfaces* **5** 4856–64

Automation of the micro-arc oxidation process

P E Golubkov, E A Pecherskaya, O V Karpanin, Y V Shepeleva, T O Zinchenko, D V Artamonov

Polytechnic Institute, Penza State University, Penza 440026, Russia

Abstract. At present the significantly increased interest in micro-arc oxidation (MAO) encourages scientists to look for the solution of the problem of this technological process controllability. To solve this problem an automated technological installation MAO was developed, its structure and control principles are presented in this article. This device will allow to provide the controlled synthesis of MAO coatings and to identify MAO process patterns which contributes to commercialization of this technology.

1. Introduction

Micro-arc oxidation (MAO) is a perspective technological process, demanded in many sectors of industry: automobile, aircraft and aerospace manufacturing, electronics, medicine, photonics and nanotechnology. Its essence lies in the surface treatment of parts made of valve metals and alloys, in particular aluminum, to impart them unique properties: high micro-hardness, wear resistance and corrosion resistance (increase the reliability and durability of machine parts); high thermal conductivity (allows us to produce smaller radiators with the same power dissipation); good biocompatibility (used in prosthetics); ordered pores structure of the top layer of the MAO coating (allows to grow arrays of nanowires in a matrix of aluminum oxide, for example, to obtain meta-materials).

However, the complexity and multifactorial of MAO process prevents the commercial implementation of this technology, because the mode of equipment operation for each application has its own characteristics. To resolve this problem, it is necessary to install the relationship between MAO technological process parameters and properties of obtained coatings.

Works of this kind have already been held in Russia [1] and abroad [2] – [5], but they could not provide enough information for a full and comprehensive analysis of the MAO process (the effect of only one or a group of parameters of MAO process on one or some more coating properties was studied and other factors were not considered). In addition, in these works the investigation of the MAO coatings properties was carried out only after receiving them. It was not possible to change the conditions for obtaining or control of coatings properties during the manufacturing process.

2. Problem definition

The purpose of the study is to establish the laws of micro-arc oxidation process and to develop the controlled synthesis of MAO coatings technology. For this an automated process installation of micro-arc oxidation with the ability to select the optimum mode of operation for each type and scope MAO coating was developed. It allows you to automate the process and contributes to its commercialization.

3. Structure of automated installation MAO

The structure of the automated installation of MAO is shown in Figure 1.

IOP Conf. Series: Journal of Physics: Conf. Series **917** (2017) 092021 doi:10.1088/1742-6596/917/9/092021

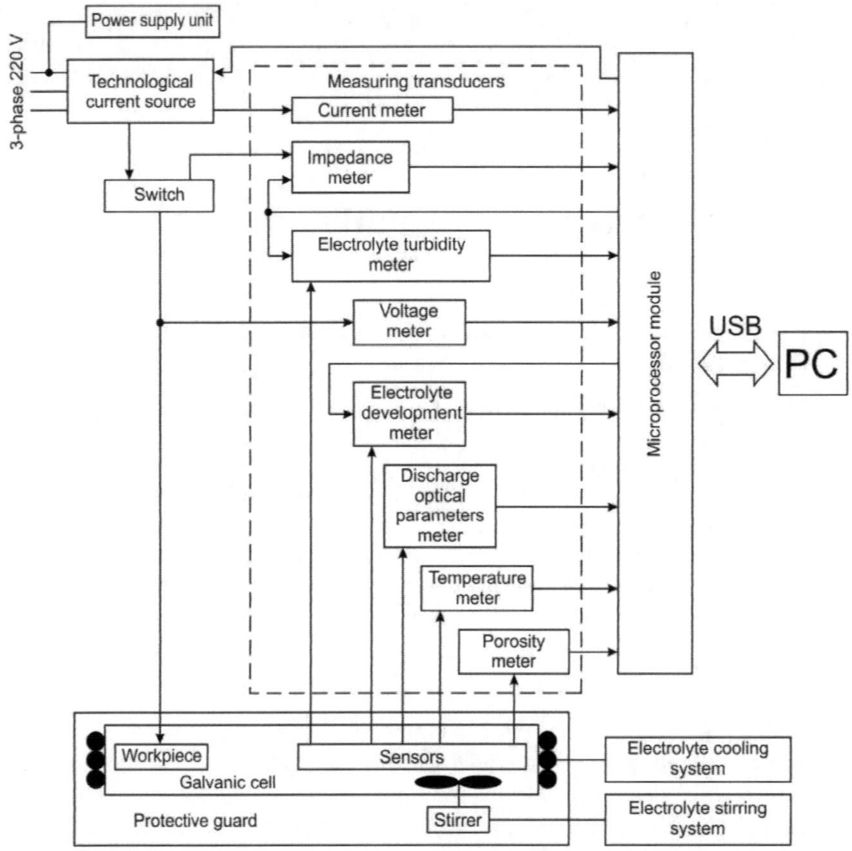

Figure 1. The structure of the automated installation of MAO.

The device consists of a technological current source (TCS), a power supply unit, a supply switch, a protective guard, measuring transducers, a microprocessor module and a galvanic cell.

The technological current source is built on the principle of width pulse modulation (WPM) and is designed to generate current pulses of oxidation of particular shape in a galvanic cell, which provide the MAO process flow .The managing WPM signal goes to the TCS from the microcontroller, which is a part of the microprocessor module that allows you programmatically to set any current pulse form, including arbitrary.

Galvanic cell (GC) is a bath with electrolyte, in which two electrodes are immersed – an anode and a cathode, where the anode is a workpiece from a valve metal or alloy. To GC are connected sensors (temperature, pressure, turbidity, production of electrolyte, etc.) generating a signal for the respective transmitters. The electrolyte is typically a mixture of NaOH or KOH and technological waterglass Na_2SiO_3, but in order to study the influence of the electrolyte parameters on the MAO process other similar electrolytes can be applied. GC is equipped with an electrolyte stirring system which can act as a magnetic stirrer and a cooling system (it is better to use a water one) as the electrolyte in a bath may reach a high temperature.

Microprocessor module (MPM) is used to process signals from the measuring transducers and to control MAO process. The structure of the microprocessor module is shown in Figure 2. The microprocessor module includes a microcontroller, having in his part an analog-to-digital (ADC) and a digital-to-analog converter (DAC), a multiplexer connecting the different measuring channels to the ADC, a digital signal synthesizer, a digital potentiometer, a galvanic isolation assembly and a USB-controller. By means of a galvanically isolated USB-port the communication with a PC is arranged. It meets the requirements of safety. Assign remaining elements of MPM will be discussed further in the description of the measuring transducers.

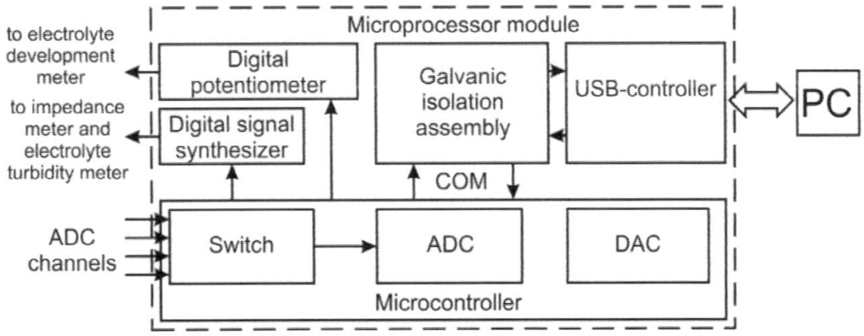

Figure 2. Structure of the microprocessor module.

Measuring transducers are used to convert signals from sensors and to transmit them to the MPM for the subsequent processing. The number of transducers include voltage meter in GC, current meter, impedance meter, discharge optical parameters meter, electrolyte turbidity meter, electrolyte temperature meter, electrolyte development meter, MAO coating porosity meter.

Voltage and current meters are used to construct the mold curves and dynamic current-voltage characteristics of the MAO process. The signal from the voltage meter is a voltage drop across the sample in GC. The signal from the current meter is proportional to the current of TCS flowing through the sample.

Impedance meter is a capacitor voltage divider where one capacitor is a sample in GC and the other is a reference and is used to set the reference voltage. The capacity of the reference capacitor can be changed by switching limits of the impedance measurement. Impedance measurement requires a test signal of a certain frequency coming from MPM signal synthesizer. This signal is applied to the sample, and the voltage is measured on the sample. In this case the current flowing through the sample using a current-voltage converter is converted into a voltage. The impedance can be calculated from Ohm's law.

The switch is designed to switch GC between TCS and the impedance meter.

Discharge optical parameters meter is designed to measure the radiation intensity and the general level of illumination. As you know, during the MAO micro-discharges on the sample surface change its color depending on the capacity. Thus, measuring the radiation intensity you can judge the MAO process stage. Taking into account the fact that MAO is usually conducted with natural lighting, to correct the first photo-sensor readings we need to know the general level of illumination, for that the second photo-sensor can be used outside of GC. The signal from these sensors is supplied to the current-voltage converter and then transmitted to the ADC of the microcontroller.

Measuring the electrolyte turbidity the later is illuminated by LED, and the transmitted or reflected light is recorded by a photo-detector. This signal is supplied to the LED with the frequency 1 kHz from the synthesizer. This transmitter can be operated in two modes: "transmission" and "reflection" depending on the location of the sensors (in front of and perpendicular to each other, respectively). To eliminate the influence of external factors on the turbidity measurement Fourier software transform is used.

Electrolyte temperature meter is presented in the form of a temperature sensor and an amplifier, which output signal is supplied to the ADC of the microcontroller. Metrological analysis of a measuring channel may be conducted in a similar manner as described in article [6].

For MAO process such parameter is important as electrolyte development. It is related to the fact that during the MAO processing a part of electrolyte anions is built into the coating, and the electrolyte, respectively, is depleted. Over time this leads to the fact that the electrolyte becomes unable to provide the desired amount of ions inflow to the coating and must be replaced, or the coating quality is severely degraded and the item is rejected. To make timely replacement of the electrolyte, an electrolyte output meter has been developed, which is based on measurement of electrical conductivity. A variant of this converter design is based on a Wheatstone bridge. At the same time as an unknown resistance an electrolyte in GC acts, a comparator is connected to the bridge diagonal and one of the resistors is a digital potentiometer which is controlled by the microcontroller. By varying the resistance of the digital potentiometer the balancing bridge is achieved which is shown by the signal from the comparator output and then measure the resistance. The unknown resistance in this case is calculated by the formula:

$$R_x = \frac{R_2 R_3}{R_4}, \tag{1}$$

where R_x - electrolyte resistance, R_2, R_3, R_4 are resistances included in the arms of a Wheatstone bridge. For the measurement of electrolyte production the placement of two additional electrodes in GC may be required.

The device is powered by a three-phase 220 V (basically it is necessary for TCS, which forms the technological current pulses with a voltage of 600 V). Also an additional power supply that provides power to control electronics and measuring transducers is available.

In the construction of this installation a protective guard must be provided to ensure the safety requirements working with high voltage. It can be organized as a limit switch on the lid of GC, which disconnects the unit when the cover is opened during the MAO process.

4. MAO process automation

MAO process automation includes the development of the overall structure of the software (SW) and MAO installation governance. The SW of this automated system is based on the "client-server" technology and consists of microcontroller software, client and server software. The general information exchange scheme between the constituent parts is shown in Figure 3.

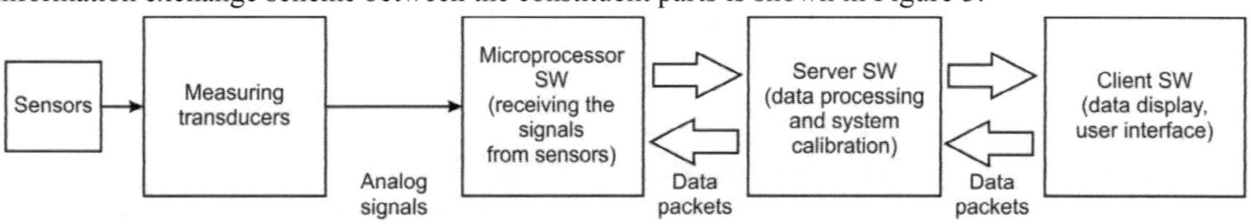

Figure 3. Information exchange between the components of the installation software.

As it was shown, the microprocessor receives an analog signal from the measuring transducers, converts it into digital form and transmits it to the server in the form of a data packet. Server software carries out the processing of incoming information from the microcontroller, displays it on the screen in the form of schedules and manages MAO modes by the user signal or automatically. Client software is used to create a user-friendly interface.

In general, the operation algorithm of the installation is shown in Figure 4. At first MAO mode is set by a user: current value and forms (both anode and cathode). At the same time there is a choice of both standard forms of current pulses (sinusoidal, rectangular, saw-tooth, etc.) and an arbitrary one when the user defines the form of the current curve (so-called "technological current designer") himself on the computer screen. Next the device readiness is checked. If the GC cover is opened, the message about the lack of readiness appears on the screen. It should be noted that the protective guard can interrupt the MAO process at any time, as it operates independently of the program.

Before the MAO process, turbidity and electrolyte development monitoring are carried out and if the electrolyte does not meet the technical standard the message about its replacement is displayed. The electrolyte production measurement can also be used to obtain information about its composition during the MAO process as it is possible to determine the concentration of the electrolyte components by the results of conductivity measurements.

It should be noted that the measurement channels of turbidity, electrolyte development, temperature and porosity are independent on the MAO process, while it is advisable to carry the measurement of

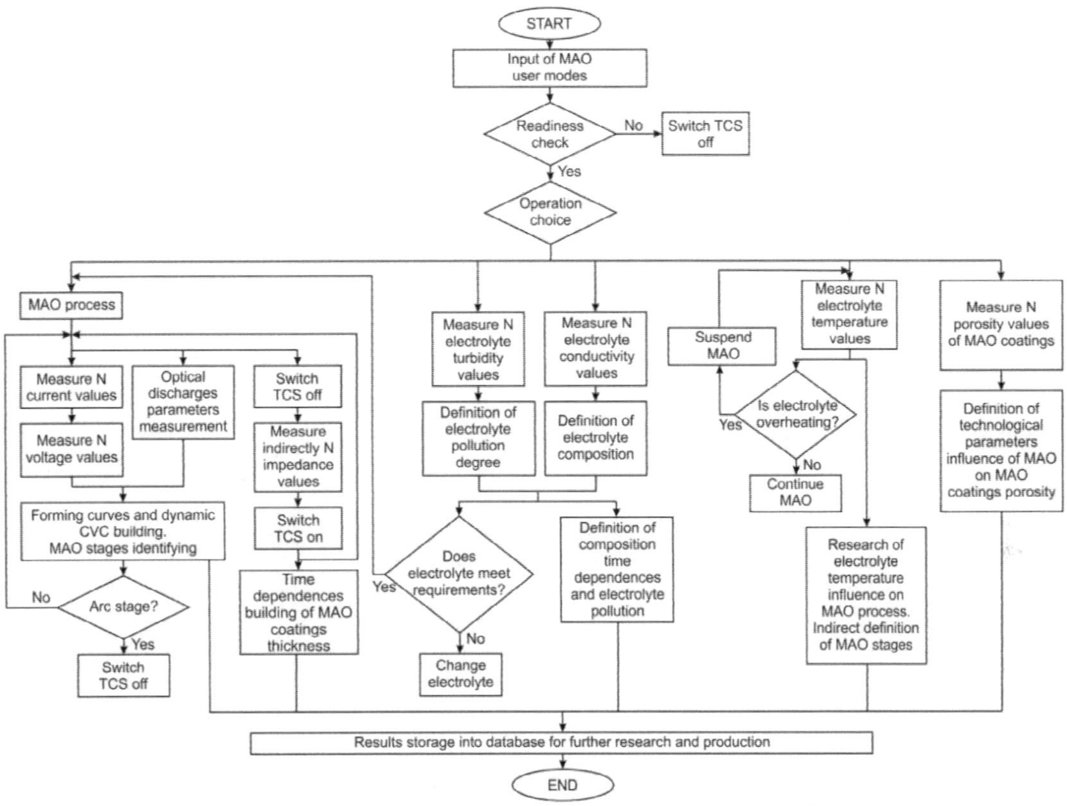

Figure 4. Operation algorithm of the installation.

current, voltage, optical parameters of discharges during the MAO. As the anode and the cathode in GC are used by the TCS and the impedance meter, measurement processes and detail processing ones simultaneously can be not implemented. At the same time it is necessary to carry out the impedance measurement during the detail processing to control this process better. Therefore, it was decided to alternate between the MAO-processing and impedance measurement using the switch. Thus the impedance meter is virtually independent on the MAO process, but it is used during this process.

After all the preparatory measures (readiness check and electrolyte production) you can proceed directly to the MAO process. This microcontroller includes TCS and passes the control WPM signal, the current form of which corresponds to the MAO mode selected by the user. TCS amplifies and averages this signal and generates a technological impact applied to the sample.

At this time the measurement of several values of the current flowing through the sample, the voltage drop across it as well as the intensity of micro-discharges radiation are performed. Further, the switch commutes and indirect impedance measurement take place, and then the MAO process goes again. If it is necessary you measure the turbidity, conductivity and the electrolyte temperature. All these measurements occur cyclically to provide current information about the properties of the coating directly during its formation.

Data obtained from measurements are transmitted to the server and then are appropriately treated. With the help of voltage and current through a sample forming curves and dynamic current-voltage characteristics (CVC) of the system metal-oxide-electrolyte are constructed. Forming curves as well as the dependence of the intensity of micro-discharges radiation from the time allow to determine the MAO process stage at any given time. These relationships are of great practical importance since the arc stage characterized by high power micro-discharges damages the coating and makes it unsuitable for practical use. Therefore, the installation provides the automatic completion of MAO process (TCS shutdown) at the transition to the arc stage.

From the values of the impedance of the test sample obtained by measuring the MAO coating thickness and its dependence on time are calculated. This makes possible to interrupt the MAO process when the desired thickness of the coating is reached, which is critical for some applications (such as nanotechnology and photonics).

From the measured values of the electrolyte conductivity you can judge its structure, as with a known conductivity you can determine the concentration of the electrolyte components and the turbidity measurement gives the information about its contamination (in MAO process as a result of micro-discharges fine particles can enter the electrolyte coating). Thus, the time dependences of conductivity and electrolyte turbidity constructed from the measured data enable us to determine the degree of electrolyte development.

Electrolyte temperature measurement is necessary for the investigation of its effect on the MAO process and also to prevent overheating. In addition, you can indirectly determine the MAO process stage as the micro-discharges capacity on the sample surface influence on the electrolyte temperature. According to the porosity measurements the porosity dependences on the MAO process time are built. It allows not only to determine the effect of process parameters on the coating porosity but also to create a nanoporous matrix by stopping the MAO process in the early stages.

After the MAO completion it is possible to store information about technological modes and the results of measurements into a database for the further study or use in the industrial production.

5. Conclusion

Thus, the application of the proposed installation will allow not only to identify patterns of the micro-arc oxidation process and carry out the controlled synthesis of MAO coatings with different properties but also to increase efficiency of this process on technical-and-economic indexes due to its automation.

References

[1] Kazantsev I A, Rosen A E, Krivenkov A O and Chugunov S N 2007 *News from higher educational institutions. Volga region* **3** 138–42

[2] Dehnavi V, Luan B L, Shoesmith D W, Liu X Y and Rohani S 2013 *Surf. & Coat. Technol.* **226** 100–07

[3] Dehnavi V, Liu X Y, Luan B L, Shoesmith D W and Rohani S 2014 *Surf. & Coat. Technol.* **251** 106–14

[4] Dehnavi V, Luan B L, Liu X Y, Shoesmith D W and Rohani S 2015 *Surf. & Coat. Technol.* **269** 91–9

[5] Dehnavi V, Shoesmith D W, Luan B L, Yari M, Liu X Y and Rohani S 2015 *Materials Chemistry and Physics* **161** 49–58

[6] Ryabov D V, Pecherskaya E A, Shepeleva J V and Pecherskaya R M 2014 *J. of Phys.: Conf. Series*. **541** 012012.

Evaluation of numerical characteristics of the current load distribution on the surface of multi-tip field emitters

S V Filippov[1], E O Popov[1], A G Kolosko[1] and R N Vinnichek[1]

[1]A.F. Ioffe Physico-Technical Institute of Russian Academy of Sciences, ul. Polytekhnicheskaya 26, St.-Petersburg, 194021, Russian Federation

Abstract. The uniformity of the emission sites distribution over the surface of the multi-tip field emitter is one of the main factors affecting its electrical characteristics. Increase of the uniformity leads to increase of the thermal stability and allowable current limit. The paper describes an experimental setup and a method for monitoring uniformity of the emission sites based on analysis of a field emission site images registered on the phosphor screen of a computerized field-emission projector. The analysis includes estimation of the statistical distributions of the sites by radius, by angle and by intensity, as well as the distribution of the current load over the emitter surface. The technique was tested with the perspective nanocomposite cathode MWCNT / polystyrene.

1. Introduction

Field emitters based on nanostructured materials are very promising for the development of vacuum nanoelectronics [1]. Unlike thermal cathodes, they do not require heating and cooling devices, they are based on the inertia-free mechanism of the electron tunneling and allow the formation of a large area cathodes. The introduction of a field emission electronics in modern nanotechnology has a number of fundamental challenges. The main drawback of these quantum structures is the thermal instability of an individual emission sites on their surface. These sites withstand an excessive current load in the course of the field emission experiment and explode in unpredictable way with appearance a vacuum discharge in the interelectrode gap. These processes don't allow to obtain high emission currents and require ultrahigh vacuum conditions, purity and homogeneity of the emitter surface. Therefore, search of the optimal distribution of the current load on the emission sites is one of the main technological problems in creating the nanostructured field cathodes [1-3].

One of the main tools for analyzing the uniformity and purity of the surface of field emitters is the registration of a "glow pattern" with emission site images [4, 5]. For this purpose, a parallel cathode-anode system is used, where the anode is glass with a transparent conductive coating (usually ITO) and a phosphor layer (usually phosphorus-based). The glowing spots observed on the surface of the phosphor layer show the electron beams from the corresponding emission centers located on the surface of the cathode and give information about the dimensions, emissivity and mutual arrangement of these sites.

Using this technique, technological optimization of the process of creating a field cathode is often carried out. For example, in [6] the concentration of the nanotube solution in the PMMA polymer was optimized, as well as the number of layers of the solution applied to the plastic substrate using the spraying method (PMMA solution to the MWCNT-dispersed solution).

It was shown in [7] that nanocomposites based on vertically aligned multiwalled carbon nanotubes with a polymer (PMMA) have enhanced emission properties, including increased brightness and uniformity of the "glow patterns". The effect was due to a decrease in the mutual screening of emission centers. The coating of nanotubes with strontium titanate molecules (emitter PMMA-STO-CNT) also improved the efficiency of field emission.

In [8] improved emission and more uniform "glow patterns" were observed when nickel atoms were added to the nanotube array. To create emitter, the electrophoresis method (EPD) was applied.

In [9] the "glow pattern" was used to characterize the emissivity of a transparent coating of nanotubes. A coating thickness of ~ 100 nm was obtained by the rapid evaporation of a solution placed on a glass substrate or flexible plastic. "Glow patterns" were obtained by shooting through the sample.

Curiously enough, a quantitative analysis of the emission site images on the "glow pattern" is performed infrequently.

In [10] an attempt was made to estimate the emission area of individual emission sites on the surface of emitter from an array of carbon nanotubes using digitizing "glow patterns" and computer simulation (MWCNT film was synthesized by microwave plasma enhanced CVD system).

In [11] a nanocomposite with multiwalled carbon nanotubes and organic binder was applied to the ITO coating by screen printing in the form of an array of round points. Analysis of the "glow patterns" at different voltage levels gives a histogram of the nanocenter distribution in terms of the threshold voltage level. After sample annealing, the distribution was substantially narrowed, so that emission sites with low threshold voltages were disappeared.

In [12] the emission properties of nanotubes grown by CVD (using ferrocene as single source precursor) on a porous aluminum membrane were studied. Emission sites were divided into three groups: stable, flickering and dying out. This allowed estimating the degree of stability of the cathode emitting surface.

2. **Experimental**

To register the emission characteristics, we used a multichannel computerized setup [13]. It consists of a plane-parallel diode system with field emission projector equipped with a long-focus USB microscope. The developed method for estimating the distribution of the current load is based on synchronous recording of the applied to the sample voltage level, of the emission current and the "glow pattern" of the field emission projector. Software processing of registered data was performed using the LabView 2016 directly during the experiment.

In this work, we considered a model emitter based on the promising nanocomposite MWCNT / polystyrene (MWCNT by Samsung, according to the SEM length of tubes is ~ 10 μm and diameter is ~ 19 nm) and numerically estimated the quality of its emission surface.

3. Results and discussion

Figure 1a, b shows the SEM image of the model emitter and its I-V characteristic and the time dependence of the current in the stability regime (~ 400 μA).

Figure 2 shows stages of the process of recording the maxima of the standard "glow pattern" obtained at the emission current 100 μA. To separate the brightness peaks that correspond to the emission centers, from the digital noise of the video camera, we used a smoothing of the glow pattern (the linear filter IMAQ Convolute VI). Search for the maximum brightness on the filtered picture was made using a sliding window (3x3 pixels). The resulting coordinates were displayed in a black and white "maximum / not maximum" picture, which was then subjected to automatic analysis (IMAQ Particle Analysis VI tool) with finding the geometric centers of white areas and rectangular zones circumscribing them.

Note that our processing cannot be carried out for high emission current levels because of phosphor screen illumination at which emission centers cannot be distinguished. Moreover, at high currents the vacuum breakdown is expected (in our case the threshold current is 1 mA).

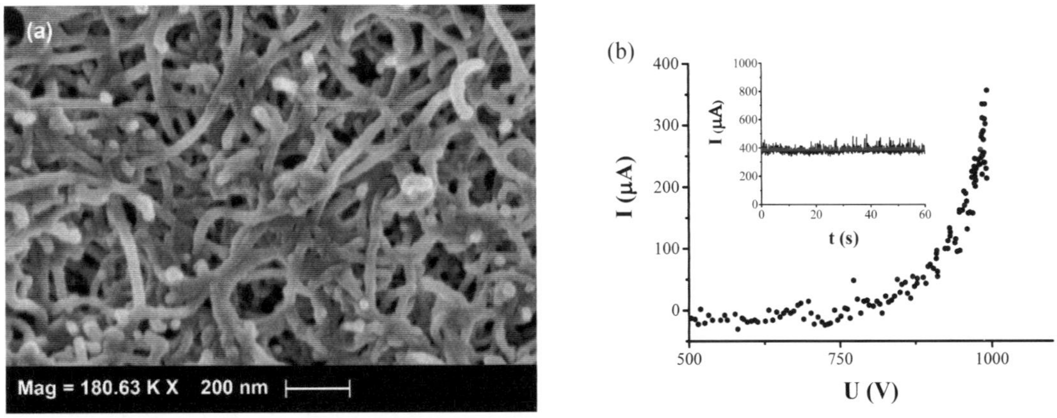

Figure 1. SEM image of the model nanocomposite emitter MWCNT / polystyrene (a) and I-V characteristic of the model emitter (b). The inset shows the current versus time in the mode of stable emitter operation.

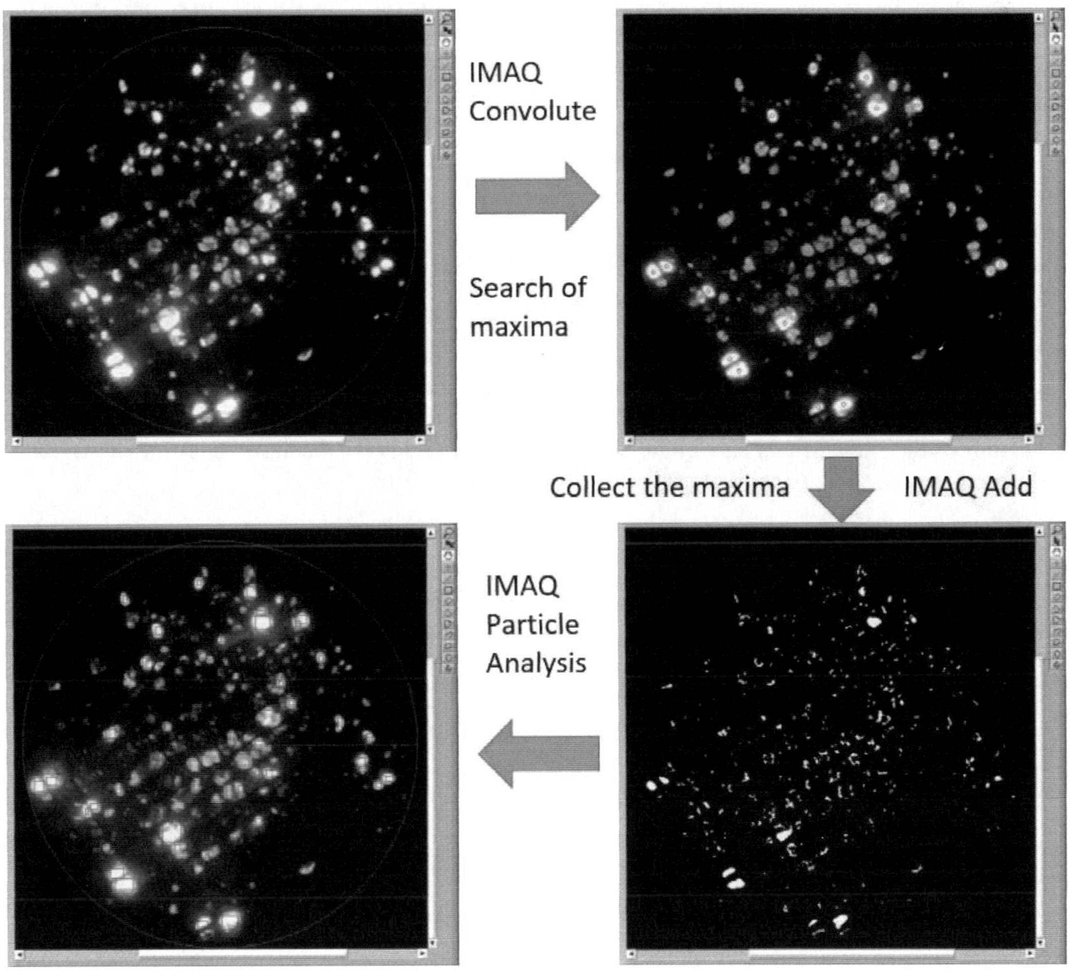

Figure 2. The "glow patterns" processing with emission site images at stable current level 400 μA.

IOP Conf. Series: Journal of Physics: Conf. Series **917** (2017) 092022 doi:10.1088/1742-6596/917/9/092022

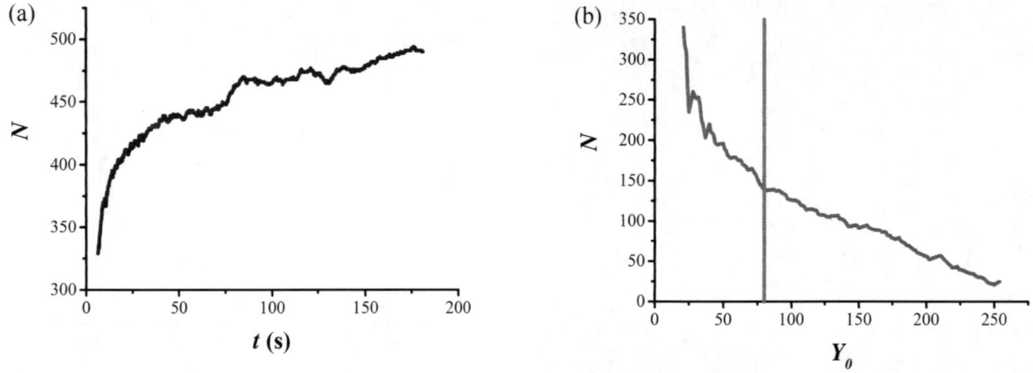

Figure 3. Time dependence of number of the distinct emission sites on the total black and white pattern collected over investigation time (a) and dependence of number of the distinct emission sites on the black and white "glow pattern" on the brightness level Y_0 (b).

Since the adsorption-desorption processes cause the change of "glow pattern" over time, the addition of black-and-white fluctuation patterns (the IMAQ Add VI tool) in stable emission current mode was performed before finding geometric centers. Accumulation was carried out until reaching the horizontal level of the time dependence of the total number of registered sites (figure 3a).

The brightness level Y_0 below which the maxima were not searched, was determined from the brightness level Y of the number of isolated spots in the black-and-white "glow pattern" (obtained from the instantaneous "glow pattern" by the IMAQ Threshold VI tool). Beginning with a certain value of $Y_0 = 80$, the number of spots with brightness $Y > Y_0$ begins to grow rapidly due to the appearance of noise fluctuations in brightness on the picture (see figure 3b), furthermore below 80 in the halo area non-existent sites appear.

The brightness of each of the registered emission sites was considered to be the maximum brightness in the region of the corresponding rectangular zone (inside this district the maximum position and brightness fluctuation). The current load on the emission sites was calculated from the current level I with the condition that the brightness of the sites on the "glow pattern" is proportional to the local current (total sum of the site currents equal total emission current I). Figure 4a shows the time dependence of the current load of one emission site found using this technique.

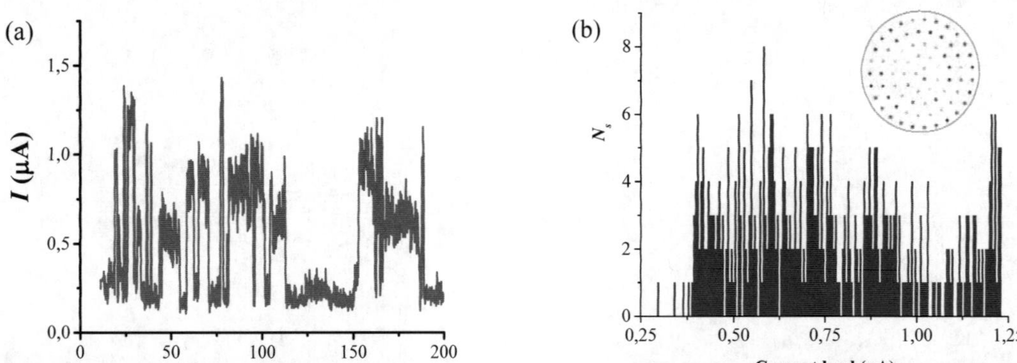

Figure 4. Time dependence of the current level of a certain rectangle (one emission site) at total current level I = 400 μA (a) and distribution of the emission sites on current loading levels (b). In the inset: flat diagram of the current distribution over the surface.

The maximum current load at each of the emission site was calculated in the same way as the current one, but instead of the current level of brightness, the maximum level was taken over a certain period of time (several minutes of operation of the emitter in stable mode), and instead of the current level, the average current was taken (~ 400 μA).

The calculated parameters of each emission site in the "glow pattern" were subjected to statistical processing. Figures 5a, b and 6a show the distribution of emission sites over brightness, the distance to the emitter site, and the angle relative to the center, and comparison of them with the ideal distributions. Ideal distribution corresponds to some ideal emitter which have the same number of the emission sites distributed on the emitter surface uniformly and the same average brightness. It is expected that ideal emitter have an optimum current load distribution and optimum heat dissipation. So numerical comparison of distributions presents the goodness of the investigated emitter.

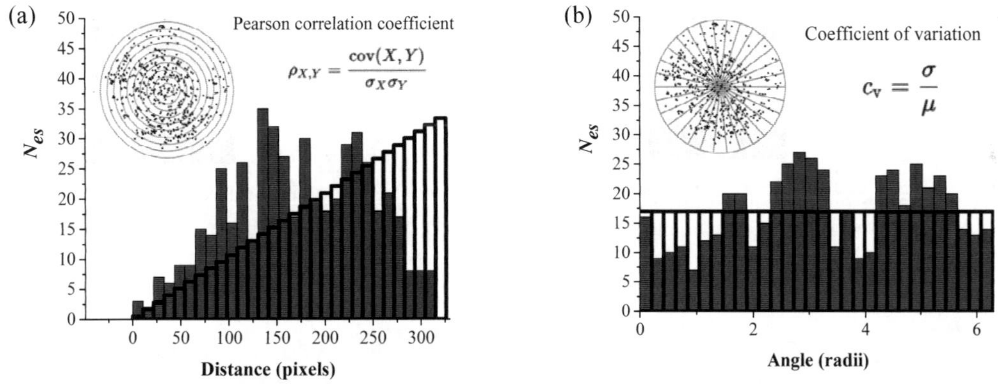

Figure 5. The histogram of the distance of the centers to the emitter center (a) and the histogram of the spread along the angle (b). Triangle represent an ideal distribution. Horizontal rectangle represent an ideal distribution.

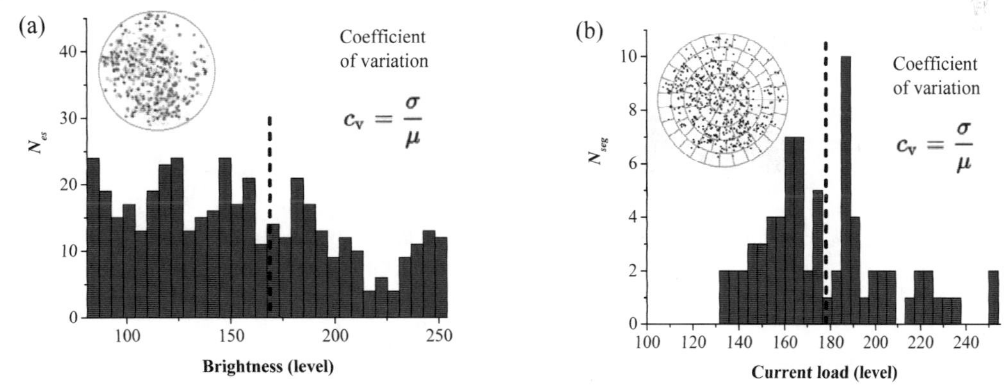

Figure 6. Histogram of the spread in brightness level (a) and an "ideal" emitter – emitter with uniform distribution of emission sites (b). The mean value on brightness distribution corresponding to ideal emitter.

For the distance distribution to the emitter center, a Pearson correlation coefficient was calculated: $\rho_{XY} = 3$ % (maximal correlation between experimental and ideal histograms is 100%).

For the distribution of sites along the rotation angle around the center of the emitter, the coefficient of variation was used: $c_{vA} = 36\%$. It describes the deviation of the distribution from the mean angle 2π/N (maximal correlation between experimental and ideal histograms is 0%).

For the brightness distribution, the coefficient of variation was also taken, but the deviation was considered from the level of average brightness Y_{cp}: $c_{vY} = 33\%$.

All three estimates of the distribution of emission sites can be applied to the technological optimization of the manufacturing process of a multitip emitter. Moreover, each of the estimates has its own priority in different methods a field cathode manufacturing.

For example, the distance distribution from the emitter center is important for methods using radially symmetric systems as a spin coating and spraying techniques.

The distribution of the sites over rotation angles plays an important role in methods where the slope of the substrate or the uneven distribution of temperature over it can create angular unevenness of the emission surface as CVD or arc discharge methods.

The distribution of brightness is important for estimating the uniformity of sites over height which can be determined by the unevenness of the profile of the original substrate or by the presence of protrusions on the emitter surface.

To estimate the thermal stability of the sample, we calculated the current load distribution on its surface by virtual separation of the emitter surface into segments of equal area and plot corresponding histogram (figure 6b). The corresponding coefficient of variation was calculated: $c_{vD} = 72$ % (deviation was considered from the level of average current loading).

The total distribution of the emission sites on current loading levels is presented in figure 4b with the flat diagram of the current load distribution over the emitter surface in the insert. Red spots on the diagram point to the regions where the emitter overheating with vacuum breakdown at high emission currents is expected.

4. Conclusion

We developed the method for the numerical evaluation of the emission current distribution on the surface of multi-tip field emitters. The model sample based on a nanocomposite "carbon nanotubes / polystyrene" was investigated.

The numerical evaluation of the spatial uniformity of the emission sites distribution is produced. We plot histograms of the site distributions by angle, distance, brightness and density of the current load and derive the coefficients ρ_{XY}, c_{vA}, c_{vY} and c_{vD} which present the conformity of the experimental emission site distribution to the ideal one. The flat diagram of the current load distribution over the emitter surface revealed most unstable regions of the cathode.

The method allows to carry out a numerical comparison of the field emitters quality and optimize their technological processing.

References

[1] Milne W I, Teo K B K and Amaratunga G A J 2004 *J. Mater. Chem.* **14** 933-943
[2] Parmee R J, Collins C M, Milne W I and Cole M T 2015 *Nano Convergence* **2:1** 1-27
[3] Zhu W, Baumann P K and Bower C A 2001 *Vacuum Microelectronics* (New York: John Wiley & Sons) pp 247-288
[4] De Heer W A, Bonard J-M, Fauth K and Ugarte D 1997 *Adv. Materials* **9:**1 87-89
[5] Lin P-H et al. 2015 *Nanoscale Research Letters* **10**:297 1-6
[6] Jung H, An S Y, Jang D M, Kim J M, Park J Y and Kim D 2012 *Carbon* **50** 987-993
[7] Pandey A et al. 2013 *ACS Nano* **7:**1 117–125
[8] Xiao X, Ye Y, Zheng L and Guo T 2012 *J. Semicond.* **33**:5 053004-1-6
[9] Kuznetzov A A et al. 2010 *Carbon* **48** 41–46.
[10] Patra R, Singh A, Vankar V D and Ghosh S 2016 *Adv. Mater. Lett.* **7**:10 771-776
[11] Liu W, Zeng F, Xin L, Zhu C and He Y 2008 *J. Vac. Sci. Technol.* B **26**:1 32-35
[12] Lysenkov D et al. 2005 *J. Vac. Sci. Technol.* B **23**:2 809 - 813
[13] Popov E O, Kolosko A G, Filippov S V, Romanov P A and Fedichkin I L 2016 *Nanomaterials and Nanostructures - XXI* **7**:1 14-26

IOP Conf. Series: Journal of Physics: Conf. Series **917** (2017) 092023 doi:10.1088/1742-6596/917/9/092023

Research of influence of the underlayer material on the growth rate of carbon nanotube arrays for manufacturing non-volatile memory elements with high speed

V S Klimin, M V Il'ina, O I Il'in, N N Rudyk and O A Ageev
Department of Nanotechnologies and Microsystems, Institute of Nanotechnologies, Electronics and Electronic Equipment Engineering, Southern Federal University, 2 Shevchenko Street, Taganrog, 347928, Russia

Abstract. This experimental work is devoted to the regimes of obtaining arrays of carbon nanotubes. Arrays of perpendicular nanotubes perpendicular to the surface were obtained by the method of Plasma-enhanced chemical vapor deposition. In this paper, geometric and electronic parameters of carbon nanotubes were investigated depending on the material of the sublayer. The rates of growth of carbon nanotubes on various structures were also determined. In the experiments for growth, structures such as Ni-Al-Si, Ni-V-Si, Ni-Ti-Si, Ni-Cr-Si were used. The growth rates for the intensive section were for the Ni-Cr-Si structure, the growth rate is about 1 μm / min, for the Ni-V-Si structure it is 0.55 μm / min. The growth rates for the saturation region for the Ni-Cr-Si structure, the growth rate is about 0.2 μm / min, for the Ni-V-Si structure 0.16 μm / min. The results obtained in this paper can be used to optimize the growth regimes perpendicularly oriented to the substrate carbon nanotubes, which are used as various elements in modern nanoelectronics.

1. Introduction

The use of carbon nanomaterials in modern instrument engineering makes it possible to obtain a device with smaller geometric parameters and to reduce the power consumption [1- 3]. Of all modern carbon materials, carbon nanotubes stand apart, which have unique electronic and geometric properties [4- 7]. However, the processes of obtaining tubes with given geometric and electrical parameters have not been fully studied. In this paper, topical problems of forming nonvolatile memory elements with high speed on the basis of arrays of carbon nanotubes are considered. In the developed instrument, arrays of carbon nanotubes oriented perpendicular to the surface were applied. To form an array of carbon nanotubes with given parameters and a formed lower contact, it is necessary to carefully select the material of the sublayer [8- 14]. Despite the fact that the catalytic material for the formation of carbon nanotubes is the same, the effect of the catalytic material is obvious. This effect is associated with a different coefficient of thermal expansion and chemical interaction of the catalytic material with the sublayer material [15- 17]. It is also very important that the material of the sublayer interacts with the gas atmosphere at which the growth of carbon nanotubes passes. Four different sublayer materials were investigated in this paper [18- 23].

2. Description of the experiment and research methods

To obtain carbon nanotubes, the method of plasma enhanced chemical vapour deposition. This method allows obtaining carbon nanotubes formed perpendicular to the surface. To select the materials of the sublayer, in our previous works we carried out a theoretical thermodynamic analysis of various structures for the growth of carbon nanotubes [24]. The data of the conducted thermodynamic analysis made it possible to identify the most promising structures. Perspective was defined as the possibility of growth, so the quality of grown carbon nanotubes. In the experiments for growth, structures such as Ni-Al-Si, Ni-V-Si, Ni-Ti-Si, Ni-Cr-Si were used. Carbon nanotubes were formed in an ammonia atmosphere with a feed rate of N_{Ar} = 150 cc / min, and a carbonaceous gas of acetylene N_{C2H2} = 70 cc / min. The reactor pressure was 4.5 Torr, the growth time was 20 minutes, and the reactor temperature was 750 ° C. The parameters of the formed catalytic centers and the electrical parameters of the grown carbon nanotubes were monitored using the «Ntegra» nanotechnology laboratory using a "semi-contact" regime. Also, the height of the formed arrays of carbon nanotubes was estimated using a scanning electron microscope «NovaNanoLab 600».

3. Results and discussion

Figure 1 shows a SEM image of an array of carbon nanotubes obtained in the course of experimental work on a Ni–Ti–Si structure.

Figure 1. SEM image of an array of carbon nanotubes obtained on a Ni-Ti-Si structure.

Figure 2 shows a SEM image of an array of carbon nanotubes obtained in the course of experimental work on a Ni–Cr–Si structure.

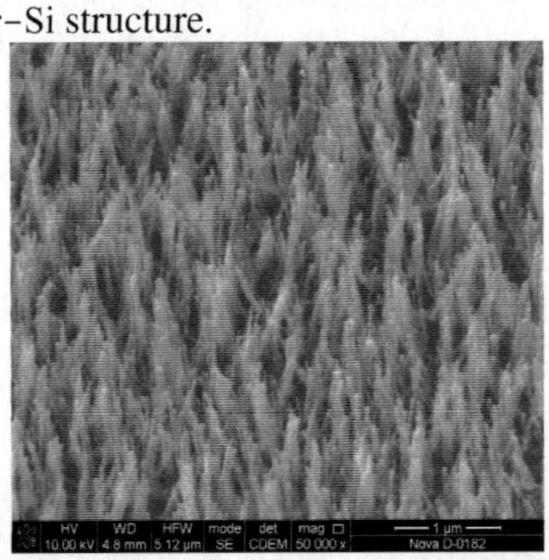

Figure 2. SEM image of an array of carbon nanotubes obtained on a Ni-Cr-Si structure.

Figure 3 shows a SEM image of an array of carbon nanotubes obtained in the course of experimental work on a Ni–V–Si structure.

Figure 3. SEM image of an array of carbon nanotubes obtained on a Ni-V-Si structure.

Figure 4 shows a SEM image of an array of carbon nanotubes obtained in the course of experimental work on a Ni–Al–Si structure.

Figure 4. SEM image of an array of carbon nanotubes obtained on a Ni-V-Si structure.

From the obtained SEM images it is possible to determine the geometric parameters of the carbon nanotubes entering the array. For the Ni-Ti-Si structure (Figure 1), the height of the array of carbon nanotubes was about 1 μm, and the diameter of 90-100 nm, the distance between the tubes in the array is the largest among the scans presented, and the tubes in this array have the smallest dispersion of geometric parameters. For the Ni-Cr-Si structure (Figure 2), the height of the array of carbon nanotubes was about 3 μm, the diameter of the nanotubes was 20-50 nm. Pipes in this array are characterized by a large aspect ratio, but they are assembled into bundles of 3-7 tubes. For the Ni-V-Si structure (Figure 3), the height of the array of carbon nanotubes was about 6.5 μm, and the diameter was 30-70 nm. This array is characterized by the maximum length of carbon nanotubes, but the tubes in it have the largest dispersion in diameter. For the Ni-Al-Si structure (Figure 4), the height of the array of carbon nanotubes was about 5.5 μm, and the diameter was 20-70 nm. This array is characterized by a high aspect ratio, the tubes in the array are assembled into bundles of 3-7 tubes. Using the obtained results, experimental studies of the influence of the growth time on the power of the 2 W plasma source on the geometric parameters of the carbon pipes in the array were carried out. For this experiment, Ni-Cr-Si and Ni-V-Si structures were used, on which the arrays with the longest carbon nanotubes were obtained. After measuring the geometric parameters of the grown arrays, the dependencies of the lengths of CNTs in the array were plotted against the time of its formation (Figure 5).

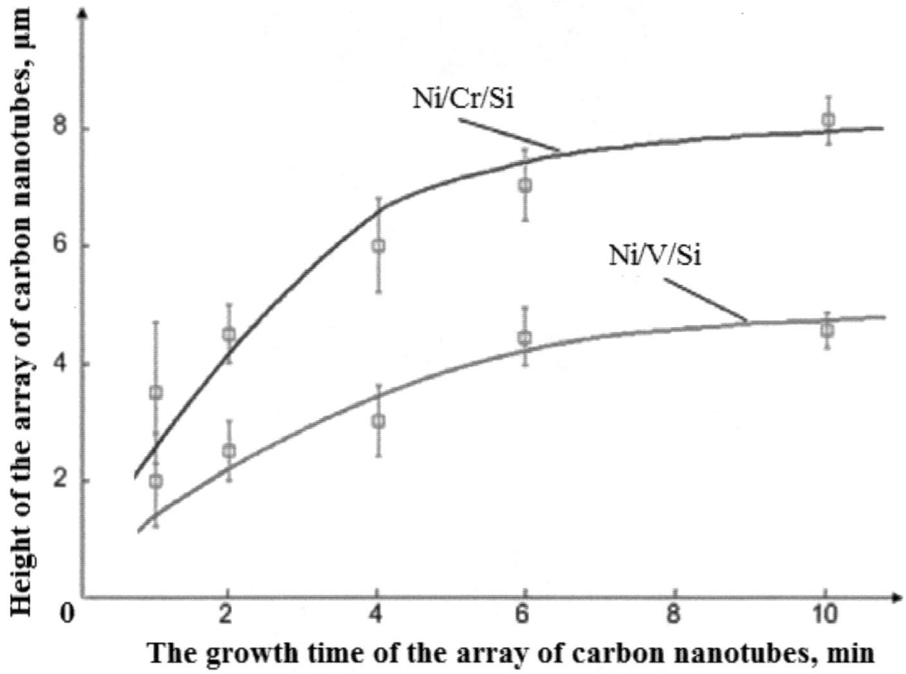

Figure 5. Dependence of the height of the array of carbon nanotubes on the growth time for various sublayer materials.

The obtained experimental dependences characterize the growth rates of the carbon nanotubes array on Ni-Cr-Si and Ni-V-Si structures. The obtained dependence has two growth areas. The first section is characterized by intensive growth and lasts about 5 minutes, for the structure of Ni-Cr-Si the growth rate is of the order of 1 μm / min, for the Ni-V-Si structure it is 0.55 μm / min. In the second section, the curve begins to saturate and the growth slows down significantly for the Ni-Cr-Si structure, the growth rate is about 0.2 μm / min, for the Ni-V-Si structure 0.16 μm / min.

4. Conclusion

In this paper, the regimes of formation of arrays of carbon nanotubes on various structures were studied. It was shown that using different material of the sublayer, different results of geometric parameters and electronic properties of arrays of carbon nanotubes were obtained. The substrate material and the catalytic centers remained the same. In this paper, comparative experiments were conducted to determine the rate of growth of arrays of carbon nanotubes on various sublayer materials. It was shown that with the growth of an array of carbon nanotubes, two stages are distinguished from the gas phase by plasma enhanced chemical vapour deposition. Using experimental results, we can find optimal regimes for growing carbon nanotube arrays for manufacturing nonvolatile memory elements with high switching speeds.

Acknowledgments

This work was supported by the Russian Foundation for Basic Research Project № 16-29-14023 ofi_m. The results were obtained using the equipment of Common Use Center and Education and Research Center "Nano-technologies" of Southern Federal University.

References

[1] Ageev, O.A., Klimin, V.S., Solodovnik, M.S., Eskov, A.V., Krasnoborodko, S.Y. / The study of influence of the gas flow rate to etched layer thickness, and roughness of the anisotropy field

of gallium arsenide is etched in the plasma chemical etching process/ Journal of Physics: Conference Series. 2016. T. 741. № 1. C. 012178.

[2] Viktor S. Klimin, Maxim S. Solodovnik, Vladimir A. Smirnov, Andrey V. Eskov, Roman V. Tominov, Oleg A. Ageev/ A study of the vertical walls and the surface roughness GaAs after the operation in the combined plasma etching / Proc. SPIE 10224, International Conference on Micro- and Nano-Electronics 2016, 102241Z (December 30, 2016).

[3] Orloff J 2009 CRC Press New York.

[4] Morgan C G, Kratzer P and Scheffler M 1999 *Phys. Rev. Lett.* **82** 4886.

[5] Murdick D A, Wadley H N G and Zhou X W 2007 *Phys. Rev.* B **75** 125318.

[6] Shiraishi K and Ito T 1998 *Phys. Rev.* B **57** 6301.

[7] Amrani A, Djafari Rouhani M and Mraoufel A 2011 *Appl. Nanosci.* **1** 59.

[8] Kangawa Y, Ito T, Taguchi A, Shiraishi K, Irisawa T and Ohachi T 2002 *Appl. Surf. Sci.* **190** 517.

[9] Daweritz L and Ploog K 1994 *Semicond. Sci. Tech.* **9** 123.

[10] Foxon C T and Joyce B A 1977 *Surf. Sci.* **64** 293.

[11] Tok E S, Neave J H, Zhang J, Joyce B A and Jones T S 1997 *Surf. Sci.* **374** 397.

[12] Kley A, Ruggerone P and Scheffler M 1997 *Phys. Rev. Lett.* **79** 5278.

[13] O. A. Ageev, O.I. Il'in, M.V. Rubashkina, V.A. Smirnov, O.G. Tsukanova, A.A. Fedotov Determination of the Electrical Resistivity of Vertically Aligned Carbon Nanotubes by Scanning Probe Microscopy // Technical Physics, 2015, Vol. 60, No. 7, pp. 1044–1050.

[14] Nishinaga T and Shen X Q 1994 *Appl. Surf. Sci.* **82-83** 141.

[15] Nishinaga T 2004 *Prog. Cryst. Growth Charact. Mater.* **48-49** 104.

[16] Higuchi Y, Uemura M, Masui Y, Kitada T, Shimomura S, Hiyamizu S 2003 *J. Cryst. Growth* **251** 80.

[17] Schmidt M, Johari Z, Ismail R, Mizuta H, Chong H 2012 Microelectron. Eng. 98 313.

[18] Tseng A 2004 J. Micromech. Microeng. 14 15.

[19] Han J, Lee H, Min B, Lee S 2010 Microelectron. Eng. 87 1.

[20] O. A. Ageev, B.G. Konoplev, M.V. Rubashkina, V.A. Smirnov, M.S. Solodovnik, A.V. Rukomoikin Studying the Effect of Geometric Parameters of Oriented GaAs Nanowhiskers on Young's Modulus Using Atomic Force Microscopy // Nanotechnologies in Russia, Vol. 8, No. 1–2, 2013, pp. 23–28.

[21] Orloff J, Swanson L W, Utlaut M 2003 Springer New York.

[22] Nagase M, Nakamatsu K, Matsui S, Namatsu H 2005 Japanese J. of Appl. Phys. 44(7) 5409.

[23] Orloff J 2009 CRC Press New York.

[24] Marina V. Ilina ; Yuriy F. Blinov ; Oleg I. Ilin ; Viktor S. Klimin ; Oleg A. Ageev /Resistive switching of vertically aligned carbon nanotube by a compressive strain / Proc. SPIE 10224, International Conference on Micro- and Nano-Electronics 2016, 102240U (December 30, 2016).

Iron ions distribution profile obtained by irradiating the silicon single crystal

A A Shemukhin[1], A V Kozhemiako[2], Y V Balakshin[1], V S Chernysh[2]

[1]Skobeltsyn Institute of Nuclear Physics, 119991, Moscow, Russia.
[2]Lomonosov Moscow State University, Department of Physics, 119991, Moscow, Russia.

Abstract. Iron ions with energies of 90 and 250 keV and the irradiation dose of 10^{16} ions/cm^2 and xenon ions with energies of 100 keV and dose 7,7×10^{14} см$^{-2}$, 200 keV and dose 2,6×10^{14} см$^{-2}$ were implanted in the single crystal of silicon (110). The distribution profiles of implanted impurity, as well as the distribution profiles of the radiation defects in the crystal lattice, were studied by the Rutherford backscattering method in combination with channeling. The experimental results were compared with the results of simulations of binary collisions of the Monte Carlo method in the TRIM program. It is shown that the difference between the experimental data and the calculation program TRIM is more than 35% in all cases.

1. Introduction

Today, tens of nanometers semiconductor structures are the electronic components' foundation. In such structures, the active element is realized in several and even single atom. Prototypes of monoatomic single-electron devices were demonstrated: a quantum bit [1-3], a quantum logic gate [4], devices for quantum metrology [5, 6], ultrasensitive charge sensors for biological applications [7]. The accuracy of simulation implemented impurities in depth largely determines the performance of the microcircuits elements. Implantation energies in the range of tens to hundreds keV are used to produce such structures.

There is a well-developed theory of interaction for accelerated charged particles with a solid body and there is an extensive experimental database for higher energies. Despite this, the programs that simulate penetration of impurities in the target, often exhibit the result characterized by tens of percent of that observed in the experiment [8-10].

The effect of irradiation on the target structure, especially in the case of single crystals, is an extremely important factor in the application of semiconductors. Creation of thin films of high crystallinity level is required for high-frequency semiconductor technology with a high degree of integration. In some cases, the creation of thin-film structures requires the removal of crystal lattice defects obtained during the growth of layers of one material on a crystalline substrate of another material. The ion implantation method in combination with high-temperature recrystallization makes it possible to obtain multilayer structures of a high crystallinity level when the parameters of the film and substrate lattices vary within 10%. In particular, the works [11-13]shows the effect of defect formation under the effect of irradiation on the quality of the recrystallized silicon film. The defects formation in the crystal structure of silicon at different irradiation parameters is considered in work [14]. According to the results of works, the direction of recrystallization and the quality of the resulting film depend of destruction level of the defective layer, as well as the location and thickness of the undamaged layer serving as a seed for recrystallization. Since the defect formation process under ion irradiation has not been sufficiently studied. It is necessary to select ion irradiation parameters for obtaining films of the best quality, which is a resource-consuming task.

This paper present the experimental results on the study defect distribution profiles formed after the implantation of iron ions with energies of 90 and 250 keV into a silicon single crystal. The experimental data were compared with calculations obtained with the TRIM program. The choice of iron ions for modifying the silicon conditioned by that this material has a promising magnetic and optical properties [11,15,16]. Iron can form silicides. It is necessary to determine the impact of this effect on the defect distribution profile. For this, implantation of xenon ions with energies of 100 and 200 keV was carried out.

2. Experimental technique

Implantation was performed based on HVEE-500 heavy ion accelerator complex [17]. Experiments of implantation iron ions Fe^+ with the ion beam energies of 90 and 250 keV and xenon ions with energies of 100 and 200 keV in silicon Si(110) were carried out. The implantation energy is chosen to explore the expected difference between the experimental distribution profiles implanted particles with theoretical calculations at an energy in the range from 1 to 4 keV/nucleon. To minimize the effect of ordering the target structure on the dopant distribution profile, the dose of 10^{16} ions/cm^2 was selected. At xenon implantation, the doses were 7.7×10^{14} cm^{-2} at 100 keV and 2.6×10^{14} cm^{-2} at 200 keV.

Upon irradiation, the sample was scanned by a beam in two mutually perpendicular directions to achieve uniformity of the lateral distribution of the impurity. The samples were studied using the Rutherford backscattering method by 1800 keV He^+ ions with a backscattering angle 165° [18].

The calculation of the implemented particles distribution profiles was performed using the TRIM program [19,20]. Program is based on the Monte Carlo method with the following assumptions: collisions are considered paired, target material - completely amorphous, not taken into account the recombination of interstitial atoms with vacancies, changes in the structure and composition of the target at the time of implantation are absent, that is the ions are always implanted into the net and a static target.

3. Results and discussion

Table 1 shows the parameters of the experiments and the corresponding value of average projected range. Doses are given in units of dpa - the number of displacements per atom. This method is universal. This makes it possible to compare the profiles of defects obtained by irradiation with different elements at different energies, doses [ion/cm^2], and other parameters of irradiation. Dose recalculation from units [ion / cm2] to dpa units was carried out using TRIM.

Table 1.

Ion type	Energy		Dose		Average projective range, nm	Straggling, nm
	keV/ nucleon	keV	DPA	ion/cm^2		
Fe^+	1,6	90	48,6	10^{16}	50	80
	4,6	250	41,2	10^{16}	220	170
Xe^+	0,8	100	6,0	$7{,}7\times10^{14}$ см$^{-2}$	65	40
	1,5	200	2,1	$2{,}6\times10^{14}$ см$^{-2}$	120	50

A comparison of the defect profiles, determined experimentally, with the defect profiles calculated by the TRIM program was performed. The selected dose in dpa units for 90 keV energy was 48.6 dpa, and for energy 250 keV - 41.2 dpa. A comparison of the defect distribution determined experimentally (curve A) with a defect profile calculated by TRIM (curve B) for a beam energy of 90 keV is shown in Figure 1. The RBS amorphisation parameter shows the ratio of the signals to the spectra taken in the channeling direction and in the direction not containing the open channels. In comparison with the theoretical calculations, the authors considered the silicon single crystal to be completely disordered when the number of displacements is equal to the number of atoms in the layer. That is, the completely disordered region on the experimental defect distribution corresponds to 4.9×10^{22} vacancies/cm^3 on the graph of the theoretical distribution (shown by a solid red line). According to TRIM calculations, such a number of displacements is observed at a depth of 155 nm. However, according to experimental data, amorphization is observed only to 110 nm, which is 30% less than the theoretical value. Further, the disordering level falls to 35% at 180 nm from the surface. The absence of a fall in the yield of the backscattered particles to the level of the initial structure in the channeling regime is explained by the dechanneling of the helium ion beam as it passes deep into the crystal through the region destroyed by the ion irradiation.

Figure 2 compares the defect distribution profiles determined experimentally (curve A) with the defect profile calculated by TRIM (curve B) for the beam energy of implanted ions of 250 keV. According to the experimental curve, amorphization occurred only at 54 nm. And up to a depth of 360 nm, the average RBS disorder was 95%. According to calculations, amorphization occurs at a depth of up to 365 nm. In addition, for these experiment parameters, the formation of defects deeper than 450 nm does not occur. However, according to the experimental data at 450 nm, the RBS amorphisation parameter is 60%. In this case, in contrast to figure 1, there are more decanalized particles, since the ion irradiation modified layer is thicker.

A similar difference in the experimental data with the TRIM calculation is also observed in the analysis of the impurity distribution profiles [21].

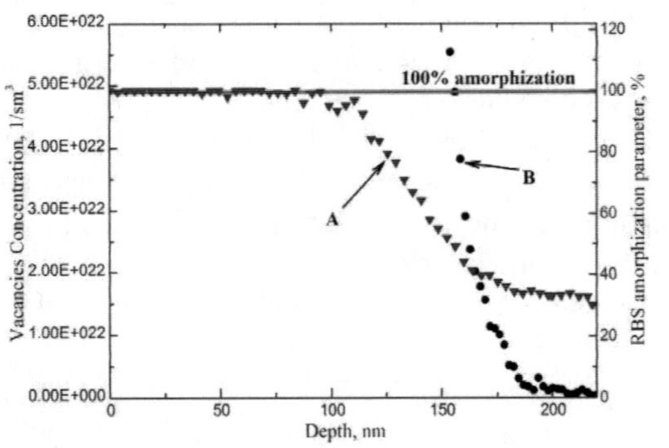

Figure 1. Defects distribution profile formed after 90 keV Fe^+ ions irradiation. A – experimental data; B – TRIM simulation.

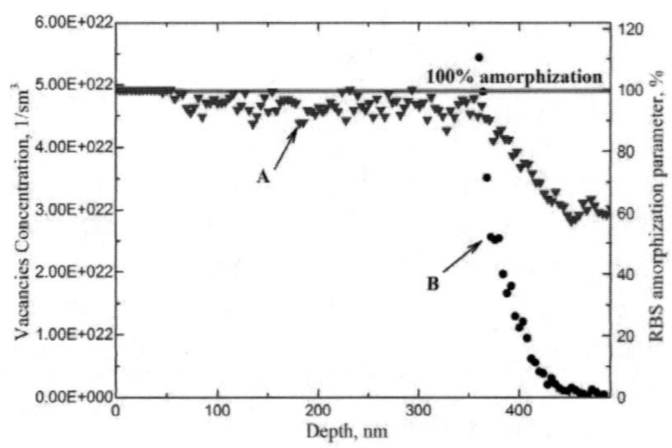

Figure 2. Defects distribution profile formed after 250 keV Fe^+ ions irradiation. A – experimental data; B – TRIM simulation.

The observed difference between the results can be caused by the fact that TRIM does not take into account the effect of the radiation dose on the character of defects accumulation. Therefore, smaller doses were chosen for the following experiments. That the peak from the impurity was as well visible on the RBS spectra, xenon was chosen as the impurity. It is an element with a larger atomic mass and, consequently, with a large scattering cross section. Similarly, since xenon is a noble gas, unlike iron, it does not form silicides, which could also affect the defect profile.

Consider the profile of the defects formed during the 200 keV xenon implantation of xenon (Figure 3). What is 1.5 keV/nucleon. This energy is close to the case of 90 keV (1.6 keV/nucleon) iron implantation. The dose was 2.1 dpa, which is an order of magnitude less than with the corresponding of iron implantation. The figure shows that according to TRIM, the number of displacements corresponding to complete amorphization is observed at a depth of 105 nm, and according to the experimental data – 65 nm. The discrepancy between the theoretical and experimental data was 38%. A similar discrepancy was observed in the 90 keV iron implantation.

Now consider the defects profile produced by implantation with a larger dose, but less energy: 7.7×10^{14} cm^{-2}; 100 keV (0.8 keV/nucleon). The corresponding profile is shown in Figure 4. According to the experimental curve, amorphization occurred only to a depth of up to 35 nm. According to calculations, amorphization occurs up to 77 nm. In this case, the discrepancy between the theoretical and calculated data is 55% and It is even greater than in the previous case.

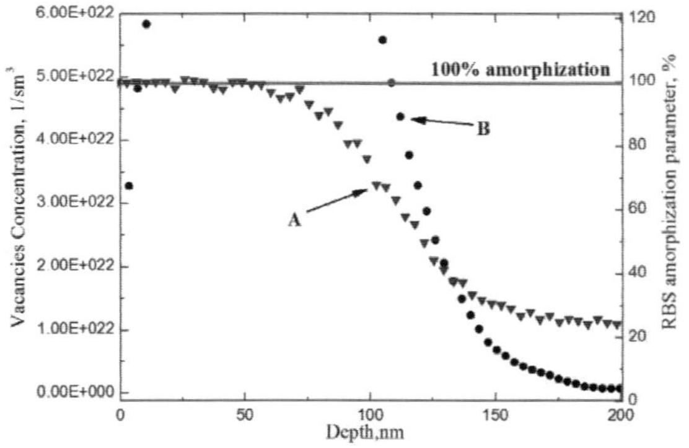

Figure 3. Defects distribution profile formed after 200 keV Xe^+ ions irradiation. A –experimental data; B – TRIM simulation.

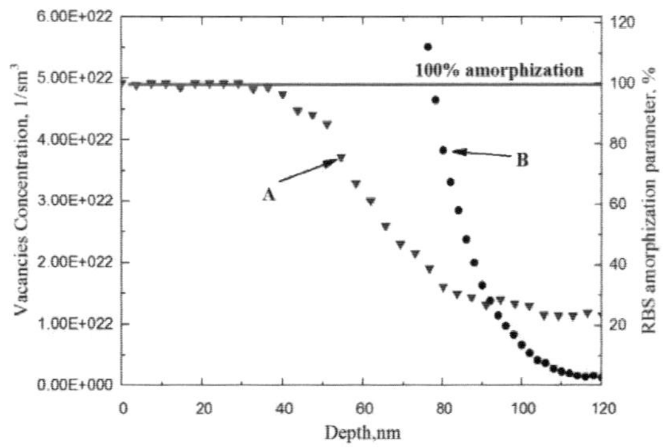

Figure 4. Defects distribution profile formed after 200 keV Xe^+ ions irradiation. A – experimental data; B – TRIM simulation.

4. Conclusion

An analysis of the defects formed at 1.6 keV/nucleon and a dose of 48.6 dpa shows that TRIM gives an overestimate of 30%. According to experimental data, with an energy increase to 4.6 keV/nucleon, there is no complete disordering of the crystal structure of silicon by the penetration depth of iron ions. However, the calculation for these implantation parameters in the TRIM program shows the formation of a significantly larger number of vacancies. The use of the dpa dimension in the study of defect formation is not clear and convenient.

In the study of defects formed after 1.5 keV/nucleon Xe^+ implantation with dose of 2,1 dpa, also there is a significant discrepancy (38%) of the experimental data with the calculation results. With a further decrease in energy to 0.8 keV/nucleon, the divergence increases to 55%.

Acknowledgements
The study was supported by a grant from the Russian Science Foundation (project No. 16-12-00072).

References

[1] Kane B E 1998 *Nature* **393**, 133
[2] Pla J J, Tan K Y, Dehollain J P, Lim W H, Morton J J, Jamieson D N, Dzurak A S and Morello A 2012 *Nature* **489** 541
[3] Fuechsle M, Miwa J. A, Mahapatra S, Ryu H, Lee S, Warschkow O, Hollenberg L C, Klimeck G and Simmons M Y 2012 *Nat. Nanotechnol* **7** 242
[4] Veldhorst M, Yang C H, Hwang J C C, Huang W, Dehollain J P, Muhonen J T, Simmons S, Laucht A, Hudson F E, Itoh K M, Morello A, Dzurak A S, et al. 2015 *Nature* **526** 410
[5] Yamahata G, Nishiguchi K and Fujiwara A 2014 *Nat. Commun.* **5** 1
[6] Tettamanzi G C, Wacquez R and Rogge S 2014 *New J. Phys.* **16** 63036
[7] Presnov D E, Shorokhov V V, Amitonov S V, Trifonov A S and Krupenin V A 2013 *Proc. of the 8th General Meeting of ACCMS-VO* (Tohoku Univ., Japan), PS-3
[8] Grande P L, Zawislak F C, Fink D and Behar M 1991 *Nucl. Instrum. Methods Phys. Res.* **61** 282-90
[9] Naas A, Ntsoenzok E, De Sousa-Meneses D, Hakim B and Beya-Wakata A 2014 *Nucl. Instrum. Methods Phys. Res.* B **339** 46-52
[10] Lakshantha W J, Kummari V C, Reinert T, McDaniel F D and Rout B 2014 *Nucl. Instrum. Methods Phys. Res.* B **332** 33-6
[11] Shemukhin A A, Nazarov A V, Balakshin Yu V and Chernysh V S 2015 *Nucl. Instrum. Methods Phys. Res.* B **354** 274
[12] Shemukhin A A,BalakshinYu V, Chernysh V S, Golubkov S A,Egorov N N and Sidorov A I

2014 *Semiconductors* **48** 517–20
[13] Shemukhin A A, Balakshin Yu V, Chernysh V S, Patrakeev A S, Golubkov S A, Egorov N N, SidorovA I, Malyukov B A, Statsenko V N and Chumak D V 2012 *Semiconductors* **38** 907–9
[14] Shemukhin A A, Nazarov A V, Balakshin Yu V and Chernysh V S 2014 *Surf. Invest.: X-ray, Synchrotron Neutron Tech.* **8** 251
[15] Lakshantha W J, Dhoubhadel M S, Reinert T, McDaniel F D and Rout B 2015 *Nucl. Instrum. Methods Phys. Res*. B **365** 114-9
[16] Ishimary M 2007 *Nucl. Instrum. Methods Phys. Res*. B **258** 490-2
[17] Shemukhin A A, Chernykh P N, Chernysh V S, Balakshin Y V and Nazarov A V 2013 *Applied Physics* **5** 59-62
[18] Nastasi M, Mayer J W and Wang Y 2015 *Ion beam analysis, fundamentals and applications* (Boca Raton, FL, CRC Press)
[19] Ziegler J F, Ziegler M D and Biersack J P 2010 *Nucl. Instrum. Methods Phys. Res.* **268** 1818-23
[20] Ziegler J F, Biersack J P and Littmark U 1985 *The Stopping and Range of Ions in Matter* (Pergamon, New York)
[21] Kozhemyako A V, Balakshin Yu V, Shemukhin A A and Chernysh V S 2017 *Semiconductors* **51** 745–50

IOP Conf. Series: Journal of Physics: Conf. Series **917** (2017) 092025 doi:10.1088/1742-6596/917/9/092025

Empirical evaluation of the field enhancement factor as a function from electrode spacing for LAFE and single emitter

M A Chumak[1], A M Chikova[2], S V Filippov[3], A G Kolosko[3*], E O Popov[3]

[1]Peter the Great St.-Petersburg Polytechnical University, ul. Polytechnitscheskaya 29, St.-Petersburg, 195251, Russia
[2]The Bonch-Bruevich SPb State University of Telecommunications, pr. Bolshevikov 22, St.-Petersburg, 193232, Russia
[3]Ioffe Institute, ul. Polytechnitscheskaya 26, St.-Petersburg, 194021, Russia
E-mail: agkolosko@mail.ru

Abstract. We have done a study of the electric field distribution on the surface of macroscopic and nanoscale field emitters. The electric field distribution was plotted by using simulation software COMSOL Multyphysics 5.2a. A numerical evaluation of the field enhancement factor dependence on the interelectrode distance was produced. We calculated parameter λ which describes this relationship for two different emission systems: elongated mono-tip metallic emitter with macro and micro scales. We conducted an experiments with model field cathodes: macroscopic tungsten tip and large area field emitter based on nanocomposite "carbon nanotubes / polystyrene".

1. Introduction

Vacuum nanoelectronics - one of the priority directions of modern technologies development. Electron sources based on field emission effect has a number of advantages over other types of electron sources: energy efficiency, inertialess, compactness [1].

Operation of the field emitters is strongly dependent on the distribution of electric field in the interelectrode space. To determine the electric field distribution in the interelectrode space a sufficiently complicated analytical calculations are required. The complexity of these calculations does not allow obtaining a field distribution for multiparameter systems, such as multi-point emitters based on nanocomposites, and even trivial electrode shapes require a series of mathematical approximations. So researchers often use computer simulation techniques.

Usually one of two emission systems is studied: a single tungsten tip or a flat emitter consisting of an array of carbon nanotubes. The crucial parameter of these systems is the distance between the flat anode and the cathode vertex. This distance affects the distribution of electric field and causes emission characteristics, such as field enhancement factor (FEF).

In [2] an analytical calculation of the FEF dependence on ratio of the geometric parameters of the mono-tip emission system "hemisphere on a post" (extended cylinder with a hemisphere at the end) was made.

The electric field distribution for metallic emitter "hemisphere on a post" was simulated using Ansys-Maxwell 2015 software [3]. The FEF dependence on the aspect ratio was constructed.

In [4] the mutual influence of the emission centers was analytically studied with a simpler point charge model and their optimal arrangement was calculated.

In [5] a system of several macro emitters was simulated using software ICEPIC. The FEF dependences on their mutual arrangement and geometric parameters were obtained taking into account the screening effect.

The current voltage characteristics (IVC) of the CNT uniformly dispersed on a Ta-metal substrate coated with a thin Ti-film were registered experimentally [6]. The shift of IVC with rise of the anode-cathode distance (up to 1 mm) was obtained. Authors explained this effect by the presence of the macroscopic projections on the nanocomposite surface, which create the effect of mutual screening of the emission centers.

In [7] the scale invariance of the electron tunnel transition with respect to the change in the distance from the tip of the emitter to the anode plane (from several nm to several mm) was shown. A displacement of the IVC of tungsten wire (emitter with a few mm of length and 250 μm diameter) with change in the interelectrode distance was observed.

The studies [8] showed that the magnitude of the electric field on the surface of the one emitting tip can be dependent on the distance between electrodes by the typical power law:

$$\Phi(z,d) \propto \Phi_d \cdot (z/d)^{\lambda} \tag{1}$$

where $\Phi(z, d)$ - electrostatic potential for point with coordinate z in cylindrical symmetric system, Φ_d - potential on the anode plane, d - anode-cathode distance.

Using COMSOL Multyphysics 5.2a software package, as well as the experimental setup with system of multichannel registration of the field emission characteristics, we conducted a study of the electric field distribution on a surfaces of macroscopic and nanoscale field emitters.

2. Modeling

Figure 1 presents the model system "tip-cathode / plate-anode" for a macroscopic emitter "ellipsoid on a cylindrical post" with a mesh. To improve the quality / time ratio of computer calculations the vertex of the tip and the opposite part of the anode were placed in a separate virtual cylinder with mesh of high resolution.

System parameters: plate diameter D_a = 5 mm, tip height H = 5 mm, tip diameter D_o = 10 μm, vertex of the tip - ellipsoid with small radius r_1 = 5 μm and large radius r_2 = 15 μm, distance between tip vertex and plate of the anode D_{ao} was varied from 5 to 500 μm.

Figure 2 shows the electric field distribution near the tip vertex for a distance D_{ao} = 100 μm and applied voltage U = 10 kV.

The calculation of the field E in the tip vertex at a fixed distance D_{ao} makes it possible to calculate the form factor β of the given emission system:

$$\beta = \frac{E}{U} D_{ao} \tag{2}$$

where U is the voltage applied to the system.

The obtained form factor in turn makes it possible to calculate the voltages U_0, which needed to obtain a given value of electric field E_o in the tip vertex (at a fixed value of D_{ao}):

$$U_o = \frac{E_o}{\beta} D_{ao} \tag{3}$$

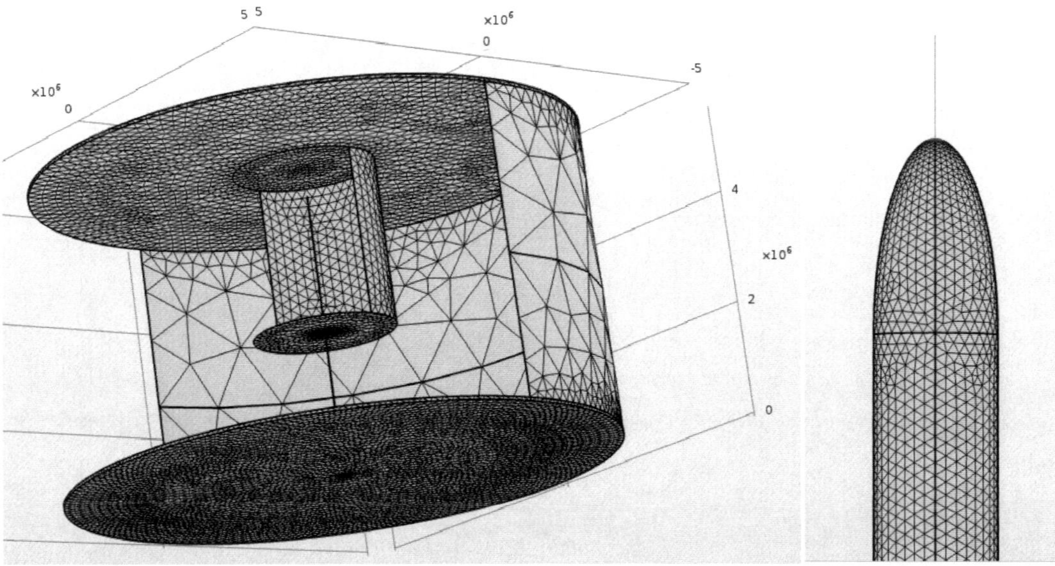

Figure 1. Computing mesh for electric field evaluation in the system with monotip emitter.

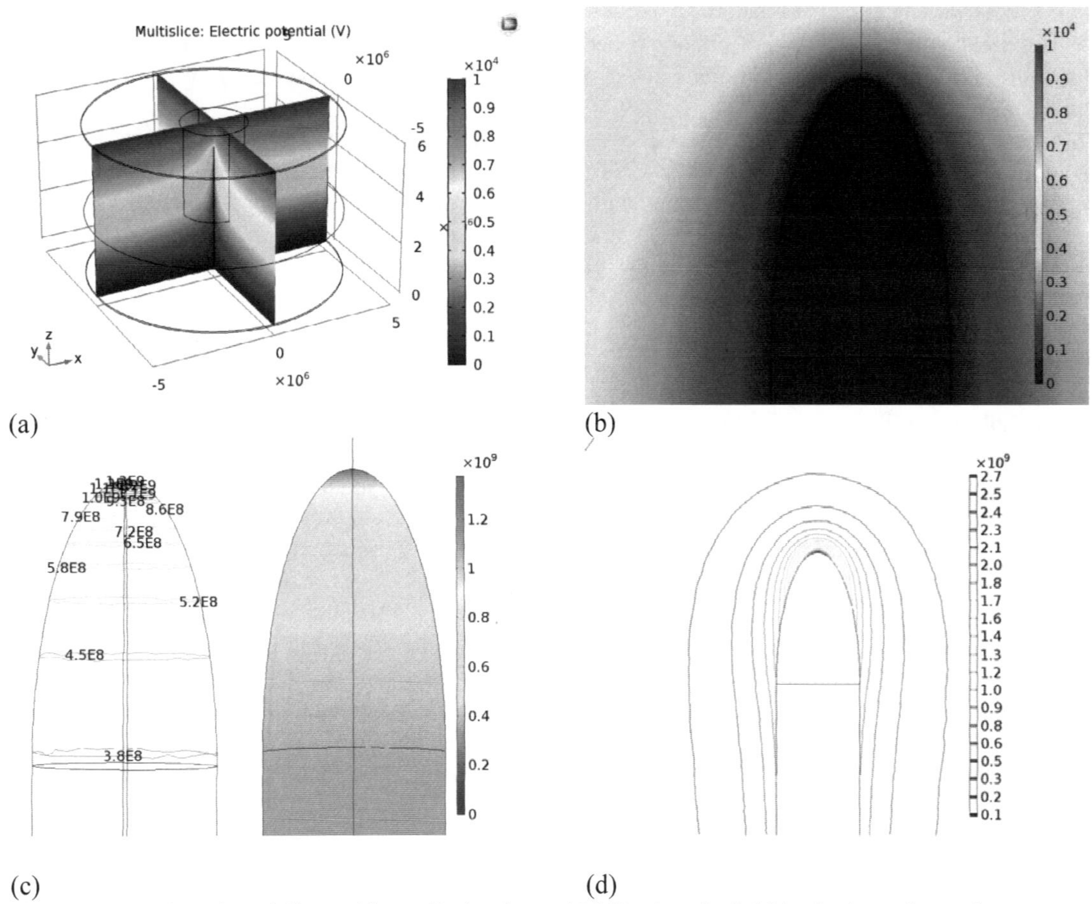

Figure 2. Results of modeling with applied voltage 10 kV: electric field in the interelectrode gap (a), electric potential near the tip vertex (b), field at the tip surface (c), field near the tip surface (d).

Figure 3a shows the dependence of the electric field E_o in the vertex of the tip on the distance D_{ao} at applied voltage $U = 10$ kV (red line) and also dependence of the voltage U, which needed to obtain the field value $E_L = 10^{10}$ V/m (it is sufficiently for the electron tunneling occurs) on the distance D_{ao} (blue line). Data calculated with the COMSOL software. It is worth noting that a change in the given value E_L leads to a corresponding linear change in the voltage U.

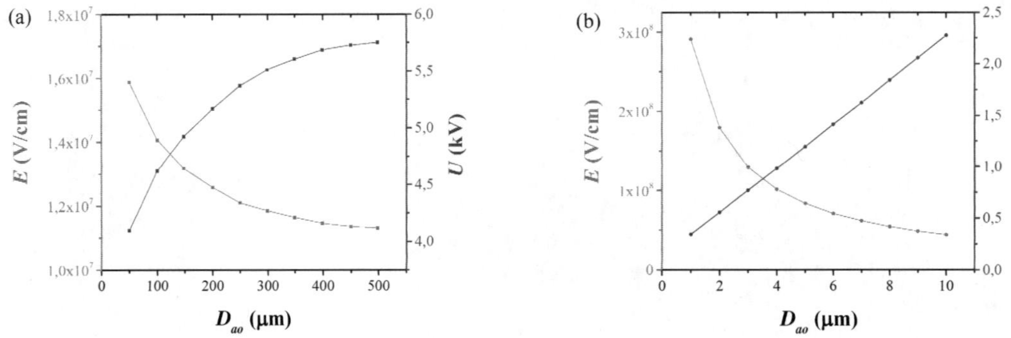

Figure 3. Dependence of the electric field in the tip vertex and voltage U, which needed to obtain given field E_L in the tip vertex, on the anode-cathode distance. (a) Macroscopic emitter "ellipsoid on a cylindrical post", (b) microscopic emitter "hemisphere on a cylindrical post".

The obtained dependences are in good agreement with the theoretical concepts of the electric field distribution in similar emission systems, which predict a power-law dependence of the electric field on the distance (see (1)).

Figure 3b shows analogous dependences obtained for a nanoscale emitter "hemisphere on a cylindrical post", which simulating a MWCNT. Parameters of the system: anode plate diameter $D_a = 4$ μm, cathode plate diameter $D_c = 4$ μm, tip height $H = 500$ nm, tip vertex - hemisphere with radius $r = 4$ nm, distance between tip vertex and anode plane (D_{ao}) ranged from 1 to 10 μm. The graphs indicate the absence of a power-law dependence of the electric field on the distance D_{ao} for the nanotubes at distances of micron scale.

3. Experimental

For an experimental study of the influence of the interelectrode distance on the field distribution, we used a multichannel computerized setup which measures field emission characteristics [9].

In the experiment two model emitters were studied: the tungsten tip cathode and multi-tip emitter (10 mm in diameter) based on nanocomposite multiwall carbon nanotubes (Taunit M produced in NanoTechCenter, Tambov) in polystyrene.

The experimental setup and photographs of the samples (SEM images) are shown in Figure 4 and Figure 5. The distance between the plates of anode and cathode was manually varied by means of micrometer screw and additionally controlled with a long-focus USB microscope.

The performed experiments showed that the IVC shape depends on the interelectrode distance for both studied emitters (see Figure 6a, b). The received data was processed by a program written in the LabView. For all IVCs it calculates a voltages U_i corresponding to a given total emission current (90 μA for a tungsten needle and 1.4 mA for a nanocomposite with MWCNTs).

The equality of the currents (given total emission current) at different interelectrode distances assumes the equality of the corresponding electric fields E on the emitter surface, so the known power-law dependence of the field on the interelectrode distance can be used to calculate the value of the degree λ from any two measured IVCs at distances d_1 and d_2:

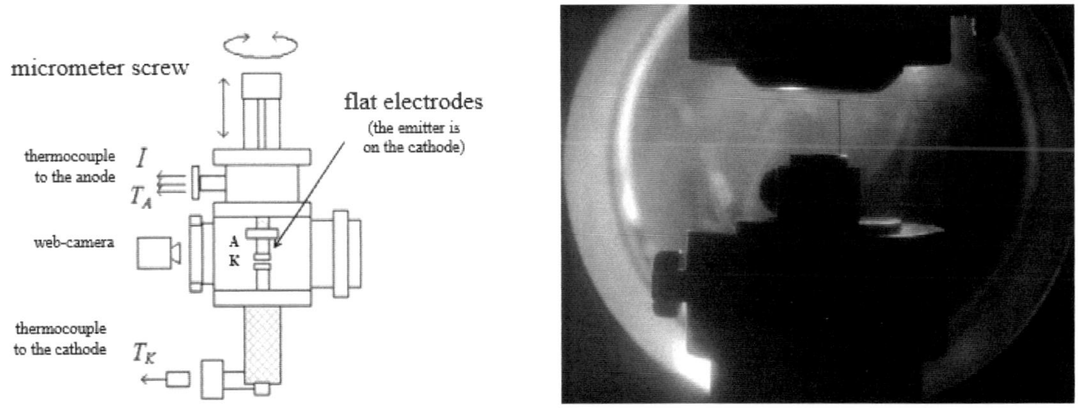

(a) (b)
Figure 4. The experimental setup: scheme (a) and photo of the measuring chamber (b).

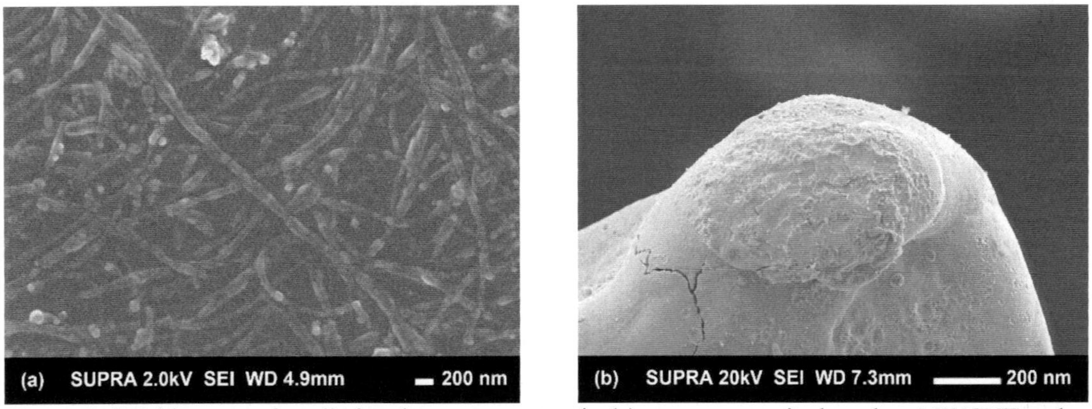

Figure 5. SEM images of studied emitters: tungsten tip (a), nanocomposite based on MWCNT and polystyrene (b).

$$E = \frac{U_1}{d_1^{\lambda}} = \frac{U_2}{d_2^{\lambda}} \quad (4)$$

To increase the accuracy of the calculation, we used a series of experimental IVCs. The values of U_1 and d_1 were fixed (they corresponded to the IVC with a minimum threshold voltage, i.e., the minimum interelectrode distance D_{ao}) and values of U_2 and d_2 changed. The constructed dependences of U_2 on d_2 was approximated by a power-law dependence with obtaining the desired values λ for both studied emitters (Figure 6c, d).

The obtained experimental dependences are consistent with the results of the above simulation. However, there is no complete numerical coincidence because the shape of the real tungsten tip is still different from the elliptical shape of the virtual emitter.

4. Conclusion

In this paper we present the results of experimental data processing and simulations performed by software LabView and COMSOL Multyphysics 5.2a. Calculation of the λ was made on the basis of series of experimental data which were obtained by using a multi-channel research setup for two types of the samples: tungsten tip cathode and multi-tip emitter based on nanocomposite "carbon nanotubes / polystyrene". The resulting empirical λ was ~0.3 for the

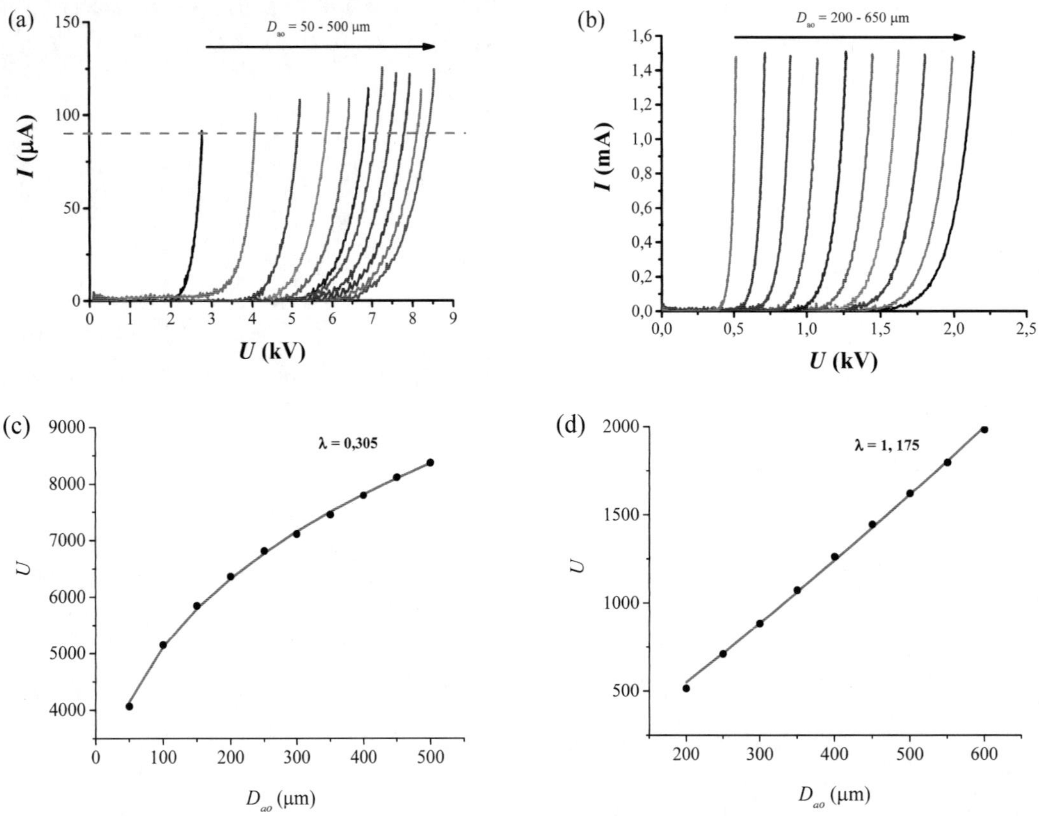

Figure 6. IVC variation with rise of the anode-cathode distance for tungsten tip emitter (a) and nanocomposite based on MWCNT (b). Corresponding dependences of voltage at I = 90 μA on the anode-cathode distance for tungsten tip (a) and nanocomposite based on MWCNT (b). The red lines is trend lines: $y = y_o \cdot x^{\lambda}$.

tungsten tip and ~1.1 for the nanocomposite emitter. Last one is near the 1 that indicates the proximity of nanocomposites to the planar capacitor system, but difference can be connected with some protrusions on the sample surface.

References

[1] G.N. Fursey, Nanotechnology Perceptions 9(3), (2013), pp. 167-185.
[2] C.J. Edgcombe, U. Valdre, Solid State Electronics. 45(6), (2001), pp. 857-863.
[3] D.S. Roveri et al., Ultramicroscopy, 160 (2016), pp. 247–251.
[4] J.R. Harris, K.L. Jensen, D.A. Shiffler, J. Vac. Sci. Technol. B 34 (2016), pp. 041215(10).
[5] D.A. Shiffler, K.L. Jensen, K. Golby, M. LaCour, J.J. Petillo, J.R. Harris, J. Appl. Phys. 118 (2015), pp. 083302(6).
[6] H. Liu, S. Kato, Y. Saito, Nanotechnology 20 (2009), pp. 275206(6).
[7] C.T. Thomas, Th. Michaels, H. Cabrera, D.A. Zanin, L. De Pietro, U. Ramsperger, A. Vindigni, D. Pescia, A. Kyritsakis, J.P. Xanthakis, Fuxiang Li, Ar. Abanov, Phys. Rev. B 87, (2013), pp. 115436(6).
[8] C.T. Thomas, Th. Michaels, H. Cabrera, D.A. Zanin, L. De Pietro, U. Ramsperger, A. Vindigni, D. Pescia, Proc. Math. Phys. Eng. Sci. 470(2167), (2014), pp. 20140014(26).
[9] E.O. Popov, A.G. Kolosko, S.V. Filippov, P.A. Romanov, I.L. Fedichkin. Nanomaterials and Nanostructures - XXI, 1(14), (2016), pp. 14-26.

Simulation of spontaneous emission in hyperbolic media

A R Gubaydullin[1,2,3], K A Ivanov[2], M A Kaliteevski[1,2,3]

[1] St. Petersburg Academic University, Khlopina 8/3, 194021, St. Petersburg, Russia
[2] ITMO University, Kronverkskiy pr. 49, 197101, St. Petersburg, Russia
[3] School of Engineering and Computing Sciences, Durham University, Durham DH13LE, UK

e-mail: gubaydullin.azat@gmail.com

Abstract. We study the probability of spontaneous emission enhancement of the dipole emitter imbedded in the stratified metal-dielectric metamaterial, characterized by the hyperbolic isofrequency surface. We analyze the performance of finite-difference time-domain (FDTD) method implementations for a 3D simulation of the spontaneous emission of the dipole source embedded in the metal dielectric stratified media. We discuss boundary conditions applicability.

1. Introduction

Metamaterials have attracted significant scientific interest because they offer a way to engineer many fascinating applications, including negative refractive index materials [1], metamaterial antennas [2], and enhancement of spontaneous emission [3,4]. Spontaneous emission is in particular interest since it plays an essential role in many phenomena of light–matter interaction. Experimentally the enhancement of spontaneous emission and superradiance are demonstrated in periodic dielectric nanostructures [5].

Commonly studied model of a hyperbolic metamaterial is a stratified structure which consists of infinite number of metal-dielectric periods. This system is expected [4] to demonstrate special optical properties emanating primarily from hyperbolic dispersion of isofrequency surface at certain frequency ranges. According to the conventional effective-medium approach a metal–dielectric nanostructure for certain frequency ranges can be analyzed as a uniaxial anisotropic medium, which effective permittivity tensor According to the conventional effective-medium approach a metal–dielectric nanostructure for certain frequency ranges can be analysed as a uniaxial anisotropic medium, which effective permittivity tensor ε_{eff} of the diagonal form:

$$\varepsilon_{eff} = \begin{pmatrix} \varepsilon_\perp & 0 & 0 \\ 0 & \varepsilon_\perp & 0 \\ 0 & 0 & \varepsilon_{||} \end{pmatrix}, \quad (1)$$

where components $\varepsilon_\perp$ and $\varepsilon_{||}$, depend on the thickness of layers d_m (metal) and d_d (dielectric) and corresponding permittivity, are given by:

$$\varepsilon_\perp = \frac{d_m\varepsilon_m + d_d\varepsilon_d}{d_d + d_m}, \quad \varepsilon_{||} = \frac{(d_m + d_d)\varepsilon_d\varepsilon_m}{d_m\varepsilon_d + d_d\varepsilon_m}. \quad (2)$$

And dispersion relation of a layered metal–dielectric nanostructure is given by:

$$\frac{k_x^2+k_y^2}{\varepsilon_{||}}+\frac{k_z^2}{\varepsilon_{\perp}}=\left(\frac{\omega}{c}\right)^2 \qquad (3)$$

For a certain frequency ranges and a certain ratio of the thickness of layers can demonstrate a hyperbolic shape of isofrequency surface.

Nowadays, numerical modeling is being widely used in research activities as well as in engineering. The finite-difference time-domain (FDTD) method was first introduced by Kane Yee in 1966 [6]. And nowadays, when highly-efficient computer facilities have been developed FDTD method attracted significant scientific interest for electromagnetic phenomena simulations. There are a number of applications based on the FDTD method applied to solve parabolic equations which provide a numerical solution of wave propagation in anisotropic, nonlinear, and dispersive media. To perform calculations of spontaneous emission we use the commercial software simulation package Lumerical FDTD Solutions.

2. Theory and model

The concept of the radiative decay enhancement in a cavity first proposed by Purcell in 1946 [7] within the framework of nuclear magnetic resonance and is known as the Purcell effect. This effect is described by the Purcell factor, which in the broad sense is defined by the ratio:

$$F_p = \tau_{bulk}/\tau_{cav} \qquad (4)$$

between lifetimes of spontaneous emission of a point light source in the infinite homogeneous medium to that inserted into a resonant cavity. According to the Fermi golden rule, the radiative decay rate $1/\tau$ in a homogeneous lossless medium is defined by:

$$\frac{1}{\tau}=\frac{2\pi}{h}\sum_{\boldsymbol{k},\sigma}|\langle f\,|H_{int}(\boldsymbol{k},\sigma)|i\rangle|^2\,\delta(h\,\omega_{\boldsymbol{k},\sigma}-h\omega) \qquad (5)$$

where $\langle f\,|H_{int}(\boldsymbol{k},\sigma)|i\rangle$ is the matrix element between the initial state $|i\rangle$ and the final state $|f\rangle$ $h\omega$ is the transition energy with the emission of the photon $(\boldsymbol{k},\sigma)$, characterized by the wave vector $\boldsymbol{k}$ and polarization σ. The emitter is commonly assumed as an atom and is initially assumed to be in an excited state, and can be characterized by the dipole moment $\boldsymbol{d}$, henceforth simply referred to as either dipole or emitter. For the quantum transition characterized by dipole moment $\boldsymbol{d}$ the Fermi golden rule can be written:

$$W=\frac{2\pi}{\hbar}|\langle f|\boldsymbol{Ed}|i\rangle|^2\rho \qquad (6)$$

Where in the matrix element $\langle f|\boldsymbol{Ed}|i\rangle$ for the transition between states $|i\rangle$ and $\langle f|$, $\boldsymbol{E}$ describe spatial distribution of the electric field of the mode, and ρ the density of states. The density of photonic states in hyperbolic medium diverges providing an enhancement of the spontaneous emission [4], which allows to tune the emission properties of the light source, potentially useful for a variety of applications. In essence, the problem is leading to calculation of the distribution of the electric field of the mode in the structure and the density of states.

The FDTD method solves Maxwell's equations on a discrete spatial and temporal grid applying specific boundary conditions which should be accurately chosen depending on the particular model. The FDTD solver supports a range of boundary conditions (BC), such as perfectly matched layer (PML) [8,9], periodic [10], and Bloch BC (useful for oblique incidence in the periodic systems). PML BC absorb electromagnetic waves incident upon them. Since we assume our structure infinite in x and y directions we perform FDTD calculations using PML BC on these boundaries.

Periodic BC's are useful when studying periodic systems, they afford calculating the response of the whole system by only simulating one unit cell. Periodic BC's copy the electromagnetic fields which occur at one boundary of the simulation area and inject them at the opposite boundary. Periodic BC's provide fast simulation, however, their applicability is limited only for completely periodic systems, when both the physical structure and the electromagnetic fields are periodic.

In this paper we consider a periodic configuration of alternating silver and dielectric layers with permittivity $\varepsilon_{m,d}$ and layer thicknesses $d_{m,d}$ respectively, with a dipole emitter imbedded in the central dielectric layer. The system is shown schematically on figure 1. The permittivity of silver is defined by the Drude model: $\varepsilon_m = \varepsilon_\infty - \frac{\omega_p^2}{\omega(\omega+i\gamma)}$, which was fitted to a particular frequency range of the Johnson and Christy data [11] with the following parameters: $\varepsilon_\infty \approx 4.96$, the plasma frequency $\omega_p \approx 8.98$ eV, and γ=0.018 eV. In complex metamaterials, the effective plasma frequency can be significantly lower than that which belongs to bulk metal [12]. The dielectric constant ε_d=1, the structure period $D = d_m + d_d$=30 nm, and the ratio of the thickness of silver and dielectric layers is varied $\eta = \frac{d_m}{d_d}$. It is convenient to study two dipole orientations parallel and perpendicular to layers of the structure.

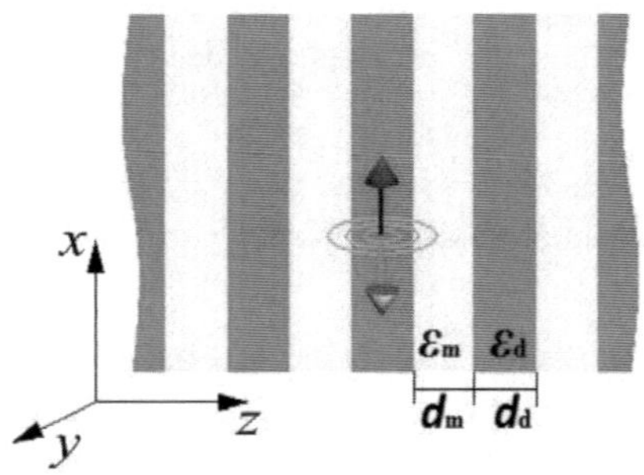

Figure 1. The dipole emitter centered in the stratified metal-dielectric structure.

3. Results and discussion

The Purcell factor or the decay rate enhancement is commonly defined as a ratio between the power emitted by a dipole source imbedded into a photonic structure to the power emitted by the dipole in a homogeneous environment. To calculate the Purcell factor Lumerical proposed several options. The first is an estimation of the Purcell factor by using the "dipolepower" command, which returns the power actually radiated in the environment.

We study hyperbolic media as a stratified structure which consists of infinite number of metal-dielectric periods. Therefore we start our modelling with periodic BC's in the z direction, results calculated by using the dipolepower command are illustrated on figure 2 (a,b). An enhancement of the spontaneous emission and a sharp dip near the Surface Plasmon Polariton (SPP) [13,14] frequency could be seen, as it was presented in [4]. Interesting to note that our results are in good agreement with results presented in [4], although authors used approach proposed by F.de Martini [15]. The original approach [15] is based on the integration over the incidence angle inside the light cone, a direct application of this approach to our metal-dielectric system was illustrated in [16].

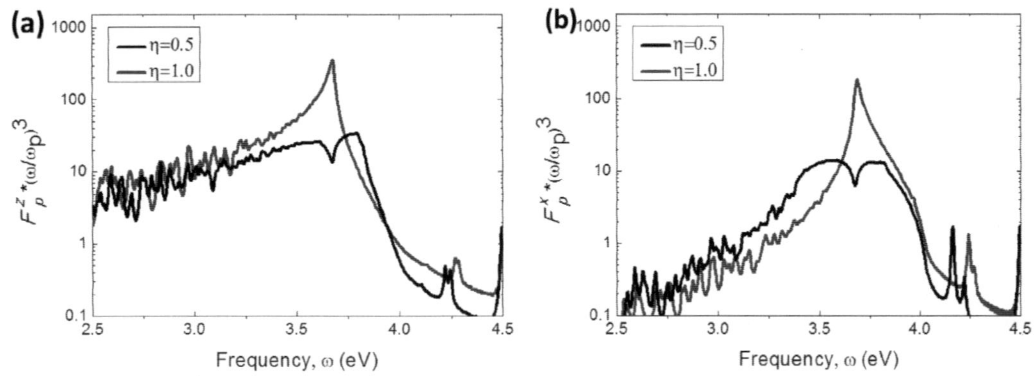

Figure 2. The Purcell Factor of the stratified structure calculated with "dipolepower" command in the FDTD Solutions package of Lumerical. The period of the structure is fixed D=30 nm, while the ratio of the thickness of layers $\eta = \frac{d_m}{d_d}$ is varied. Two orientations of the dipole emitter: (a) perpendicular and (b) parallel with respect to the interface.

We should note that periodic BC's are applicable only for infinite models, in the case of finite structures (as in experiment) application of periodic BC's may lead to inaccuracies in the analysis of systems of finite size, as was demonstrated in [17]. Moreover, since we are interested in simulating a system with one single dipole source the system can not be simulated as periodic, since it does not have one dipole per each period, therefore hereinafter we will focus on the structure which is constructed from 6 periods with a dipole source embedded in the central dielectric layer, which results obtained with "dipolepower" command are presented on figure 3.

The second method is based on the Green's function formulation [18,19] which relies on the fact that the emission rate is proportional to the local density of optical states. Lumerical has implemented Green's function approach to calculate the spontaneous decay rate for a two-level quantum system at r_0 and ω by:

$$\gamma = \frac{2\omega}{3\hbar\varepsilon_0}|\mu|^2\rho_z(r_0,\omega), \quad (7)$$

where the partial local density of states along Z axes can be obtained from the imaginary part of Green's function:

$$\rho_z(r_0,\omega) = \frac{6\omega}{\pi c^2}[Im\{G_{zz}(r_0,r_0;\omega)\}] \quad (8)$$

Results of FDTD simulation obtained by using the Green's function technique, implemented in the FDTD Solutions package of Lumerical, are illustrated on figure 3 by dashed green curve. The frequency dependence of the Purcell factor calculated by the "dipolepower" method is illustrated on figure 3 by red solid curve.

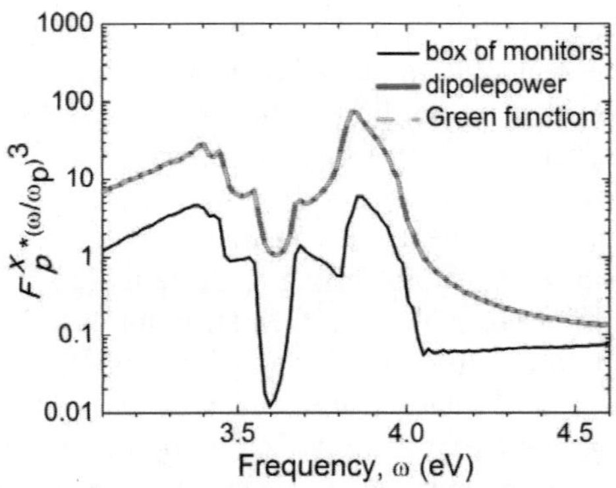

Figure 3. The frequency dependence of the Purcell factor of the stratified structure calculated by the "dipolepower" method is illustrated by the red solid curve. The Purcell Factor calculated by using the Green's function technique - dashed green curve, and by using the "box of monitors" technique - black solid curve. The period of the structure is fixed D=30 nm, the ratio of the thickness of layers η =1. The dipole orientation is assumed to be parallel to layers.

However, the dipolepower technique and the Green's function formulation only work if the dipole is in a lossless dielectric medium. If the dipole is imbedded in a lossy media it is recommended by Lumerical to measure the power flow out of a small box of monitors surrounding the source in the environment. We also perform calculation using a transmission box of monitors enclosing our model. The dependence of the Purcell factor on frequency obtained by using the box of monitors is illustrated by black solid curve on figure 3.

The comparison of results obtained by all three methods are illustrated on figure 3. As one can note the method based on the dipolepower function and the method based on the Green's function illustrate similar estimations of the Purcell factor, however, as was noted by Lumerical those two methods are not applicable for a lossy media. The method based on measurement of the power flow through a small box of monitors illustrates values of the Purcell factor, which for certain frequencies are by several orders of magnitude smaller than results predicted by the Green's function method. We should note that results depend on many simulation parameters and require a lot of computational effort for convergence tests, and variation of simulation parameters leads to strong changes of calculated results, that confirms the need to develop a stable approach for calculating the probability of spontaneous emission in arbitrary layered structures.

In our recent work we have developed a rigorous self-consistent procedure of quantization of electromagnetic field for inhomogeneous media (S-quantization) [17], which provide accurate calculation of the probability of spontaneous emission for arbitrary, asymmetric layered metal/semiconductor structures [20]. S-quantization also allows analysis of the Purcell effect in the waveguide mode regime [21].

4. Conclusions

We performed simulation of the hyperbolic metamaterial constructed from one-dimensional metal-dielectric nanostructure. We study the probability of spontaneous emission enhancement of the dipole emitter imbedded in the stratified metal-dielectric metamaterial by FDTD simulation methods. We have shown that the probability of spontaneous emission depends on the simulation parameters and applied formalism. Therefore it is of particular interest to develop reliable approach to calculate the Purcell factor for arbitrary structures.

Acknowledgments

The work was supported by Russian Science Foundation grant № 16-12- 10503.

References

[1] Veselago V G 2003 *Physics–Uspekhi* **46** (7) 764–768
[2] Enoch S, Tayeb G, Sabouroux P, Guerin N and Vincent P 2002 *Phys. Rev. Lett.* **89** (21) 213902
[3] Kuttge M, Javier García de Abajo F and Polman A 2010 *Nano Lett.* **10** (5) 1537–1541
[4] Iorsh I, Poddubny A, Orlov A, Belov P and Kivshar Y S 2012 *Physics Letters A* **376** 185187
[5] Pozina G, Kaliteevski M A, Nikitina E V, Denisov D V, Polyakov N K, Pirogov E V, Gorai L I, Gubaydullin A R, Ivanov K A, Kaliteevskaya N A, Egorov A Yu and Clark S J 2015 *Sci. Rep.* **5** 14911
[6] Kane Y 1966 *IEEE Transactions on Antennas and Propagation* **14** 302-307
[7] Purcell E M 1946 *Phys. Rev.* **69** 681 Abstract B10
[8] Berenger J P 2007 *Perfectly Matched Layer for Computational Electromagnetics* Morgan & Claypool Publishers
[9] Gedney S D and Zhao B 2010 *IEEE Transactions on Antennas and Propagation* **58** 3
[10] Yariv A and Yeh P 2002 *Optical Waves in Crystals: Propagation and Control of Laser Radiation* Wiley Classics Library
[11] Johnson P B and Christy R W 1972 *Phys. Rev.* B **6** 4370
[12] Brand S, Abram R A and Kaliteevski M A 2007 *Phys. Rev. B* **75**, 035102
[13] Jun Y C, Kekatpure R D, White J S and Brongersma M L 2008 *Phys. Rev. B* **78** 153111
[14] Maier S A 2007 Plasmonics: Fundamentals and Applications Springer (New York: Springer)
[15] Martini F, Marrocco M, Mataloni P, Crescentini L and Loudon R 1991 *Phys. Rev.* A **43** 2480
[16] Gubaydullin A R, Mazlin V A, Ivanov K A, Kaliteevski M A and Balocco C 2016 *Applied Physics A: Materials Science and Processing* **122** (4) 425
[17] Kaliteevski M A, Gubaydullin A R, Ivanov K A and Mazlin V A 2016 *Optics and spectroscopy* **121** (3) 410–419
[18] Tomas M S 1995 *Phys. Rev.* A **51** 2545
[19] Novotny L, Hecht B 2006 *Principles of Nano-Optics* Cambridge
[20] Gubaydullin A R, Symonds C, Bellessa J, Ivanov K A, Kolykhalova E D, Sasin M E, Lemaitre A, Senellart P, Pozina G and Kaliteevski M A *Scientific Reports* in press
[21] Ivanov K A, Gubaidullin A R, Morozov K M, Sasin M E and Kaliteevski M A, *Optics and Spectroscopy* **122** (5) 835-842

AUTHOR INDEX

AUTHOR INDEX

AUTHOR INDEX

AUTHOR INDEX

AUTHOR INDEX

AUTHOR INDEX

AUTHOR INDEX

AUTHOR INDEX

Institute of Physics
Dirac House, Temple Back
Bristol BS1 6BE UK

ISSN: 1742-6588
ISBN 978-1-5108-5340-9

CURRAN ASSOCIATES INC.
proceedings.com